MANUAL
OF
PHOTOGRAMMETRY

Fourth Edition

Manual
of
Photogrammetry

Fourth Edition

Editor-in-Chief
Chester C Slama

Associate Editors
Charles Theurer
Soren W. Henriksen

**American Society of
Photogrammetry**

Fourth Edition

Library of Congress Cataloging in Publication Data

American Society of Photogrammetry.
Manual of Photogrammetry.

Includes index.
1. Photographic surveying—Handbooks, manuals, etc.
2. Photogrammetry—Handbooks, manuals, etc. I. Slama, Chester C
II. Theurer, Charles. III. Henriksen, Soren W. IV. Title.
TA593.25.A48 1980   526.9′82   80-21514
ISBN 0-937294-01-2

PUBLISHED BY
AMERICAN SOCIETY OF PHOTOGRAMMETRY
105 N. Virginia Ave.
Falls Church, Va. 22046

Foreword

THE PRIME FUNCTION of the American Society of Photogrammetry is that of advancing the science of photogrammetry and remote sensing. Over the years, our Society has discharged this function in large measure by the publication of its world-renown MANUALS. The list is quite imposing. It includes:

MANUAL OF PHOTOGRAMMETRY, FIRST EDITION, 1944

MANUAL OF PHOTOGRAMMETRY, SECOND EDITION, 1952

MANUAL OF PHOTOGRAPHIC INTERPRETATION, 1960

MANUAL OF PHOTOGRAMMETRY, THIRD EDITION, 1966

MANUAL OF COLOR AERIAL PHOTOGRAPHY, 1968

MANUAL OF REMOTE SENSING, FIRST EDITION, 1975

HANDBOOK OF NON TOPOGRAPHIC PHOTOGRAMMETRY, 1979

MANUAL OF PHOTOGRAMMETRY, FOURTH EDITION, 1980

This newest FOURTH EDITION, as with all of the previous publications, is the result of hard work, dedication, and generosity on the part of many professionals in the field who have written sections and chapters of the MANUAL. These authors have positioned themselves in the forefront of the advances which have been made in our science since the THIRD EDITION was published in 1966. Consequently, we are assured of coverage of the latest instruments, techniques, and photogrammetric processes which have been developed during this period.

Because the MANUALS of the American Society of Photogrammetry enjoy world-wide distribution, the publication of this latest edition will help to update the teaching and the practice of photogrammetry in the schools, and governmental and private agencies not only in the United States, but also in countries all over the world. It will also help to set the standards of practice world-wide, and will encourage a greater exchange of information on a uniform basis.

The preparation of this MANUAL represents a tremendous amount of effort on the part of all of those involved. Writing a section or a chapter as a contributing author is a labor of love. The reward is the satisfaction in having made a most significant contribution to the advancement of the science of photogrammetry and to the dissemination of information pertaining to that science. The members of the American Society of Photogrammetry are most grateful to the authors and can take pride in knowing that such dedicated men and women are numbered in our ranks.

The task of the Author-Editors of each of the chapters is not an easy one. This task of eliciting manuscript in suitable form from all of the authors must be one of gentle persuasion. They are to be commended for their patience and understanding throughout the period of development of the text.

And finally, the Society is deeply indebted to the Editor-in-Chief and his Associates for the many hours they devoted to handling the details of such a large undertaking as the production of this MANUAL. It has been a continual effort on their part from the moment that the go-ahead for the publication of this FOURTH EDITION was approved by the Board of Direction of the American Society of Photogrammetry.

Presidents of the American Society of Photogrammetry during development of the Fourth Edition:

Joe E. Steakley, 1974
John H. Wickham, Jr., 1975
Hugh B. Loving, 1976
Vern W. Cartwright, 1977
Clifford J. Crandall, 1978
Francis H. Moffitt, 1979
Rex R. McHail, 1980

Editor's Preface To The Fourth Edition

OVER THE YEARS, Photogrammetry has been a dynamic science as evidenced by the frequent publication of new editions to the MANUAL OF PHOTOGRAMMETRY since the first in 1944. The justification for every new edition has been the need to "update" the procedures, instrumentation and theory as presented in the previous issue. The reason for these continual changes is because the practice and understanding of photogrammetry itself incompasses a wide variety of other disciplines. For instance, to trace the individual steps of a very basic application of photogrammetry, one must study the physics of light, the atmosphere and optics, the chemistry of photography, the mechanisms of mensuration and the mathematics of adjustment theory. Each of these fields is experiencing continual change, especially today in the area of electronics and computers. Therefore, a manual on photogrammetry must be periodically updated to maintain a ready reference to which one can refer which represents the latest in the description of the science.

Realizing this, in 1974 President Steakley submitted a recommendation by the Publications Committee (chaired by Captain L. W. Swanson) to the Board of Directors of the American Society of Photogrammetry. In essence, he recommended that a new edition of the MANUAL OF PHOTOGRAMMETRY be prepared under the direction of Charles Theurer, who would act as Editor-in-Chief. The Board approved both recommendations and, as a result, we proudly present the 4th Edition of the MANUAL OF PHOTOGRAMMETRY. Immediately thereafter, an Advisory Committee was established whose function was to define the chapter titles and content, prepare time schedules and select contributing chapter author-editors. The committee was composed of twelve of the most eminent and knowledgeable photogrammetrists in the ASP which included Fred Doyle, Morris M. Thompson, Capt. L. W. Swanson, Dominic Bucci, Marshall S. Wright, Jr., Kenneth E. Reynolds, G. C. Tewinkel, H. M. Karara, A. O. Quinn, Clifford J. Crandall, Brig. Gen. Lawrence P. Jacobs, and Joe E. Steakley. Using the 3rd Edition as a guide, the committee decided that the new edition should delete some of the material that described procedures which are no longer used, combine some of the chapters that tended to overlap, and present the material in nineteen chapters to be bound in a single volume. From an exhaustive list of prospective authors, an author-editor was selected for each chapter for which he was free to select additional contributors.

The contributing authors for this manual represent a wide range of professionals from the various fields within photogrammetry. They practice this science in a variety of commercial firms, educational institutions, and governmental organizations. Because of an occasional change in affiliation and the embarrassing possibility of an omission, the individual organizations will not be listed here. These authors spent many of their free hours in the preparation of the original text and their organizations, in every case, gave them complete support. To each of these organizations, and to the nearly 100 individual contributors to this manual, the Society owes a great debt of gratitude.

In 1976 Charles Theurer elected to retire from government service. As a retiree, he felt he could not do full justice to the task of Editor-in-Chief and submitted his resignation to the Society. I was asked to step in as a replacement and I reluctantly agreed after coercing him into agreeing to remain active as an Associate Editor. The remainder of the editorial staff consisted mainly of colleagues from the National Ocean Survey and included Soren Henriksen, Alison Barrett, and Frank Wright. Associate Editor Theurer served in an advisory capacity and helped with the many decisions that could only be resolved at the Editor-in-Chief level. Associate Editor Henriksen devoted his principal efforts to technical review of the manuscripts; Editor Barrett reviewed the manuscripts for syntax; and Production Manager Wright handled the scheduling and liaison with the printer. Many others in the organization were called upon to perform tasks leading to the completion of this manual and gave freely of their time. To all, the Society is indeed grateful.

The subject matter for this edition of the manual has been divided into nineteen chapters. An attempt was made to assemble the material in such a way that its order of presentation would follow, as closely as possible, the sequence one would encounter in the application of photogrammetry. That is, starting with the sensor and proceeding through reduction of the data. As an introduction, this material is preceeded by background information in the form of history, mathematics and optics. To stir the imagination, the description of basic processes is followed by an expanded discussion of non-traditional applications of photogrammetry culminating with applications brought about through man's recent adventures into space. To complete the manual, the last chapters are devoted to additional background information in photogrammetry such as education, definitions and an index. Each author approached and presented material for his own area of specialization with enthusiasm to the extent that the major effort of the editorial staff was concentrated on maintaining balance and limiting the overall size of the volume without gross omission.

One major problem in the planning of a text that uses more than one author in its preparation, is that of overlap and duplication. A certain amount of overlap could be eliminated in the initial stages of production by comparing the material contained in outlines submitted by each author. This did not entirely eliminate duplication because a certain amount had to remain in order to allow the authors the freedom of continuity in his writing. Another problem is that of common notation in both the written descriptions of processes and in the symbolization contained in mathematical forms. In most cases, the author's text was modified to agree with those descriptive terms and symbolizations adopted by the Nomenclature Committee of the ASP. Some obvious differences in notation were retained and are identified in the text through use of footnotes.

The publication of a text of this nature is by no means a small task. Many hours have been spent in the review of manuscripts, galleys, page proofs and final blue-line copy. No matter how carefully each review is made, the editorial staff is aware of the fact that there will be errors in the final product. These errors can appear in many forms such as spelling, misstatement of fact, labeling of illustrations or errors in some of the many complex mathematical expressions. If the reader detects any such errors, he is asked to notify the American Society of Photogrammetry so that a list of errata can be prepared.

Chester C Slama
Editor-in-Chief

Rockville, Maryland, October 1980

Contents

CHAPTER I

CHAPTER II

CHAPTER III

CHAPTER IV

CHAPTER V

CHAPTER VI

CHAPTER VII

PLANNING AND EXECUTING THE PHOTOGRAMMETRIC PROJECT . . . .
AUTHOR-EDITOR: John E. Combs
CONTRIBUTING AUTHORS: James W. Crabtree, Jindo Kim, Maurice E. Lafferty

CHAPTER VIII

FIELD SURVEYS FOR PHOTOGRAMMETRY . . . .
AUTHOR-EDITOR: S. W. Henriksen
CONTRIBUTING AUTHORS: Stanley H. Schroeder, Ronald K. Brewer

CHAPTER IX

CHAPTER X

CHAPTER XI

CHAPTER XII

CHAPTER XIII

CHAPTER XIV

CHAPTER XV

CHAPTER XVI

NON-TOPOGRAPHIC PHOTOGRAMMETRY 785
AUTHOR-EDITOR: H. M. Karara
CONTRIBUTING AUTHORS: M. Carbonnell, W. Faig, S. K. Ghosh, R. E. Herron,
   V. Kratky, E. M. Mikhail, F. H. Moffitt, H. Takasaki, S. A. Veress

CHAPTER XVII

SATELLITE PHOTOGRAMMETRY 883
AUTHOR-EDITOR: Donald L. Light
CONTRIBUTING AUTHORS: Duane Brown, A. P. Colvocoresses, Frederick J. Doyle,
   Merton Davies, Atef Ellasal, John L. Junkins, J. R. Manent, Austin
   McKenney, Ronald Undrejka, George Wood

CHAPTER XVIII

CHAPTER XIX

INDEX

CHAPTER I

Foundations of Photogrammetry[1]

Author-Editor: MORRIS M. THOMPSON
Contributing Author: HEINZ GRUNER

1.1 What is Photogrammetry?

1.1.1 DEFINITIONS[2]

THE WORD "photogrammetry" came into general usage in the United States about 1934 when the American Society of Photogrammetry was founded, although the term already had been widely used in Europe for several decades. It is derived from three Greek words, *photos* meaning "light," *gramma* meaning "something drawn or written," and *metron* meaning "to measure." The root words, therefore, originally signified measuring graphically by means of light.

In the previous editions of the MANUAL OF PHOTOGRAMMETRY, the definition of photogrammetry was given as "the science or art of obtaining reliable measurements by means of photographs." In the Third Edition, the MANUAL included a statement that this definition might well be amplified to include "interpretation of photographs" as a function of nearly equal importance, for the ability to recognize and identify an object by its photographic image is often as important as the ability to derive its position from the photographs.

In the decade following the publication in 1966 of the Third Edition of the MANUAL OF PHOTOGRAMMETRY, the science of photogrammetry has been broadened by the application of a new scientific tool, *remote sensing*. In applications of remote sensing, imagery may be acquired, not only through the use of a conventional camera (which in itself is a remote-sensing system), but also by recording the scene through one or more special sensors. These special sensors usually operate by electronic scanning,

using radiations outside the normal visual range of the film and camera—microwave, radar, thermal infrared, ultraviolet, as well as multispectral. Special techniques are applied in order to process and interpret remote-sensing imagery for the purpose of producing conventional maps, thematic maps, resources surveys, etc., in the fields of agriculture, archaeology, forestry, geography, geology, and others.

So important is the remote-sensing application, that the American Society of Photogrammetry has published a separate MANUAL OF REMOTE SENSING (1976). Moreover, the name of the official journal of the American Society of Photogrammetry was changed in 1975 from *Photogrammetric Engineering* to *Photogrammetric Engineering and Remote Sensing*. In view of these developments, the definition of photogrammetry, as of 1979, may be stated as follows:

> Photogrammetry is the art, science, and technology of obtaining reliable information about physical objects and the environment through processes of recording, measuring, and interpreting photographic images and patterns of electromagnetic radiant energy and other phenomena.

In this Fourth Edition of the MANUAL OF PHOTOGRAMMETRY, the concept of measurement will be regarded as implicit in the term "photogrammetry."

The best known application of photogrammetry is the compilation of topographic maps and surveys, complete with contour lines, based on measurements and information obtained from aerial and space photographs; the compilation is usually performed by means of optical analog instruments and/or analytic computations. Similar topographic principles of precise measurement are applied in close-range photogrammetry to map (measure) objects which are difficult to study in other ways, such as the shape of an astronomic radio-reflector subject to environmental deformations; similarly, they are used for synoptically recording measurable deformations in engineering models, and for the medical study (*in situ*) of live specimens, and so on.

In mapping, measurements on photographs

---

[1] The historical summary (section 1.2) in this chapter is the product of painstaking research and authorship by Heinz Gruner. The remainder of the chapter is adapted from chapter I of the Third Edition of the MANUAL OF PHOTOGRAMMETRY. That chapter, titled "Introduction to Photogrammetry," was prepared by George D. Whitmore, author-editor, and Morris M. Thompson, contributing author.

[2] For a comprehensive glossary, see chapter XIX, Definitions of Terms and Symbols Used in Photogrammetry.

replace field surveys, in whole or in part; consequently, the use of photographs and photogrammetry in mapping is often referred to by such terms as "aerial survey" or "photogrammetric survey."

1.1.2 CATEGORIES OF PHOTOGRAMMETRY

Photogrammetry is frequently divided into specialties or categories according to the types of photographs or sensing systems used, or the manner of their use. For instance, the type of photogrammetry used when the photographs are taken from points on the ground surface is called *terrestrial photogrammetry* or *ground photogrammetry*. Photographs taken on the ground with the optical axis of the camera horizontal are called *horizontal photographs*. *Aerial photogrammetry* denotes the use of photographs or other sensor data which have been obtained from an airborne vehicle, and these may be either vertical aerial photographs or sensor data, or oblique aerial photographs or sensor data. If the sensor is radar, the technology is sometimes referred to as *radargrammetry*. Other special technologies include *X-ray photogrammetry* (using an X-ray sensing system), *cinephotogrammetry* (using motion pictures), *hologrammetry* (using holographs), and *monoscopic photogrammetry* (using single photographs, with the stereoscopic effect, if any, produced by reflected mirror images). If the camera or sensing system is borne in a space vehicle, the process may be called *space photogrammetry, satellite photogrammetry,* or *extraterrestrial photogrammetry*.

In *stereophotogrammetry*, overlapping pairs of photographs are observed and measured, or interpreted, in a stereoscopic viewing device, which gives a three-dimensional view and creates the illusion that the observer is viewing a relief model of the terrain. In *analytical photogrammetry*, problems are solved by mathematical computation, using measurements made on the photographs or sensor images, or digital data, as input.

Vertical aerial photographs are best known and most used at the present time, although much of the early development of the basic theory of photogrammetry was evolved from horizontal and oblique photographs. *Vertical* photographs are those taken with the optical axis of the lens pointing vertically downward at the time of exposure. *Oblique* photographs are those taken with the optical axis intentionally deviated from the vertical. A *low oblique* has a relatively small or low angle of deviation from vertical, and does not include the apparent horizon (the visible junction of earth and sky as seen from the camera station). A *high oblique* has a relatively large or high angle of deviation from the vertical and includes the apparent horizon. A low oblique has the optical axis nearer to vertical, and a high oblique has the optical axis nearer to horizontal. Vertical aerial photographs are actually only nearly vertical, because at present there is no practical means available of holding the optical axis in an exactly vertical position at the instant of exposure. Each so-called vertical photograph is therefore tilted, in some degree, from the true vertical. For many practical purposes, however, good near-vertical aerial photographs are so nearly vertical that they may be used as such without corrections or rectifications. In precise photogrammetric instruments, means are provided for determining conveniently the amount of tilt in each photograph so that rectification to the equivalent of the true vertical can be accomplished by simple adjusting devices. Therefore, the lack of true verticality in the pictures should not affect the accuracy of the final results.

1.2 Historical Summary

1.2.1 INTRODUCTION

Photogrammetry and its inseparable partner, remote sensing, comprise an art, a science, and a technology. The steady growth of photogrammetry comes from roots that reach down to ancient cultures and philosophies and are anchored in the fertile strata of periods of great discoveries, arts, and inventions. These roots have furnished a solid foundation on which we safely stand and continue to build photogrammetry's impact on human civilization.

Photogrammetry, in its simplest definition as the art of scribing and measuring by light, was practiced long before the invention of the photographic process, as was remote sensing in its most direct and nature-given form of human vision. Most of the rudiments are well documented and preserved in institutions, museums, and libraries of the world and constitute the heritage that fuels the progress which we call our "accomplishments."

1.2.2 PRECURSORS OF PHOTOGRAMMETRY

As the reality of an image of any form or origin is the fulcrum of the photogrammetric axiom, one may associate the beginning of the history of photogrammetry with the era of a great Italian genius, who was a scientist, an engineer, an inventor and a master of creative imagery: *Leonardo da Vinci*, (fig. 1-1). His mind explored the sciences of geometry, optics, mechanics,

FIGURE 1-1. Leonardo da Vinci (1452–1519).

and geophysics, and his intuition graphically demonstrated in 1492 the principles of aerodynamics and of optical projection (Mac-Leish 1977). He also designed grinding and polishing mechanisms for lenses.

Leonardo da Vinci's German contemporary was *Albrecht Dürer* (1471–1528), painter and master of the graphic arts. Dürer produced an outline of the laws of perspective and in 1525 he constructed samples of mechanical devices with which he made true perspective drawings of nature and studio scenes (fig. 1-2). His devices included an apparatus for producing stereoscopic drawings.

About 1600 the German astronomer *Johannes Kepler* (1571–1630) gave a precise definition of "stereoscopy," and a Florentine painter *Jacopo Chimenti* produced what was probably the first hand-drawn stereo-picture pair (fig. 1-3), which has been preserved in the Wicar Museum of the city of Lille, France. *Aughtread* of England constructed the first slide rule in 1574, and shortly thereafter *John Napier* (1550–1617) published tables of logarithms and *Blaisé Pascal* (1623–1662) gave the world a desk calculator. The graphic arts were endowed in 1630 with the invention of the "Stork's beak," the first pantograph. An outstanding pair of mathematicians, *Isaac Newton* (1642–1727) and *Gottfried von Leibnitz* (1646–1716), presented the early scholars with the opportunity to master differential and integral calculus.

The concept of stereoscopic drawing was used in its first practical survey application by a Swiss physician, *F. Kapeller,* in 1726. He constructed a topographic map from drawings of Mt. Pilatus on Lake Lucerne. French Admiral *Beautemps-Beaupré* constructed topographic maps from pairs of perspective sketches utilizing the instructions of the mathematician *Henry Lambert* (1728–1777), who in 1759 wrote a classical treatise on "The Free Perspective." Lambert's treatise dealt with the concept of inverse central perspective and space resection of conjugate image rays; it contained the geometric fundamentals of the process which 100 years later was named "photogrammetry."

In the same year, 1759, *Schultze,* a Nuremberg alchemist, observed that silver nitrate blackens when exposed to sunlight. He started the long, frustrating search by chemists in Germany, France, and England to find the answer to the problem of permanently retaining the elusive image that an optical system so generously produces. Optician *J. Dollond* had already created achromatic lens doublets, and by the end of the century, multi-element objective lenses had been put together by painstaking probing methods.

With the invention of several forms of optical prisms, the pinhole camera was supplanted by the camera lucida, and several types of stereoscopes (Wheatstone, 1838, Brewster, 1844) came into use. With the invention of the zonal lens in 1800 by *A. J. Fresnel* (fig. 1-4), the variable diaphragm by *Edward Bausch* in 1810, and the fundamental design of a 4-element portrait lens in 1844 by Viennese Professor *J. Petzval* (1807–1891), all requisites for the photographic camera were at hand except the image-retaining photographic emulsion. With prophetic insight, *Guido Schreiber* had rendered a treatise in 1829 on "The process and formulae for air topographic equations and determination of the camera station," envisioning the time when the earth's image would be produced from the birdseye view.

While scientists in French, English, and German laboratories were secretly struggling to extract the latent image from their compounds and coatings on paper, glass, and metal plates, *Fox Talbot* (1800–1877) of England succeeded in producing with his camera obscura some negative paper pictures, and *Arago* and *Niepce* announced a "Heliographic Process." Based on their pioneer work, *Louis Daguerre* (fig. 1-5), French painter and physicist, presented to the French Academy of Arts and Sciences in 1837 the first tangible photographic results, which were positive pictures on metal plates that he named "daguerrotypes." Still quite cumbersome and time-consuming, the process was soon improved and created a great sensation. It stimulated all Europe's productive forces and created enormous new business.

It is not surprising that the new process took to the air from its beginning. Already, in 1783 *Montgolfier* had made successful flights with his hot-air balloon near Paris. Military forces

FIGURE 1-2. Perspective device of Albrecht Dürer (1471–1528).

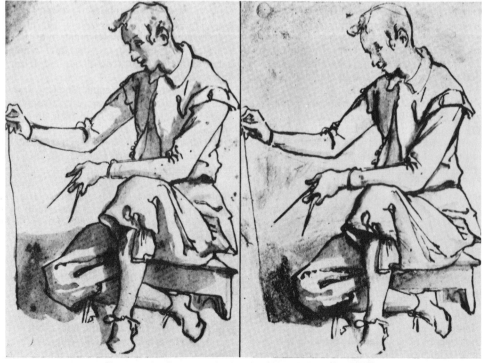

FIGURE 1-3. Stereoscopic drawing by J. Chimenti (c. 1600).

FIGURE 1-4. Augustin J. Fresnel (1788–1827).

quickly realized the value of an observation station in the sky. The French army used a captive balloon in the battle of Fleurus in southern Belgium (1794) as a signal post for their commander. The caricaturist, photographer, and fashionable

FIGURE 1-5. Louis J. M. Daguerre (1789–1851).

sports balloonist *Tournachon* (nicknamed *Nadar*) took his camera and darkroom aloft, became the first aerial photographer, and ardently agitated to use his art for mapping the countryside (1858). Emperor Napoleon III ordered Nadar in 1859 to furnish reconnaissance photography in preparation for the battle of Solferino in northern Italy. Inventors in England, Germany, and Russia concentrated their experiments on kites and rockets to carry single and multiple cameras and even gyro-stabilized cameras high over target areas.

1.2.3 EARLY PHOTOGRAMMETRY IN CENTRAL AND WESTERN EUROPE

The second half of the 19th century signaled the dawn of European industrialization and the birth of several scientists of great vision and industrial leadership, resulting in scientific and technological progress of lasting significance. Some of the pertinent milestones have been recorded by history as follows:

the *founding* of the Zeiss Works (1846), the Schott Glass Works (1884), and other plants capable of manufacturing instruments or components for photogrammetry;

the *advancement* of optical design, including the aplanatic lens by *Steinheil* and the Protar lens of 90° field by *Rudolph;*

new *inventions,* including the floating measuring mark by *Stolze* (1892), stereoviewing with anaglyphic filters by *D'Almeida* (1851), and stereoviewing by polarized light (1891);

new *instrumentation,* including the photographic planetable by *Chevallier* (1858), phototheodolites by *Paganini* (1884), *Koppe* and *Thiele* (1888);

new *techniques,* including stereo-pictures on daguerrotypes (1843), experiments with photosculpture, the first survey of a harbor from shipboard, submarine photography at a depth of 900 meters, and cloud measurements in England;

developments of photogrammetric *literature and archives,* including the coining of the term "photogrammetry" in 1855 by geographer *Kersten* and its introduction by *Meydenbauer* in 1867 to international literature; the first German textbook on photogrammetry by Koppe (1889); the founding of the library for stereograms of monuments and architectures by *Meydenbauer* (1883) in Berlin; and *Laussedat's* classic work on French photogrammetry (1898).

1.2.3.1 FRANCE

Spearheading the application of the photogrammetric technique during this period was French Colonel *Aimé Laussedat,* (fig. 1-6), designer of terrestrial equipment for photographic surveying (fig. 1-7) and of a process which he called "iconometry." Laussedat applied this process to the reconstruction of classical architecture and to a survey of Paris by rooftop photography in 1864. He earned a science award and a gold medal in a contest held in Madrid (1861) on "the errors of maps constructed from

FIGURE 1-6. Aimé Laussedat (1819–1907). (Courtesy Institut Géographique National, Paris.)

perspective photographs.'' His many scientific and technical contributions as well as his success in winning official acceptance of photogrammetry as a surveying tool by the French government and the public earned him the title ''Father of Photogrammetry,'' a distinction which is widely recognized in European literature.

1.2.3.2 AUSTRIA

By the turn of the century the center of photogrammetric progress was Austria, where government survey agencies had begun to organize planetable photogrammetry in the territories of the Alpine, Carpathian, and Balkan mountains. Precise photosurveys for the engineering masterworks of mountain railroads and transalpine tunnels are still admired accomplishments of that period. The driving spirit was *Eduard Doležal* (fig. 1-8), geodesy professor, mathematician, and lecturer at the University of Vienna (Doležal 1932). He founded the Austrian

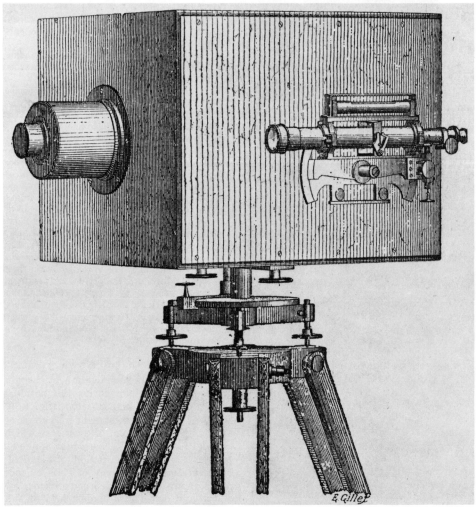

FIGURE 1-7. Laussedat's iconometric equipment (1859).

FIGURE 1-8. Eduard Doležál (1862–1955).

FIGURE 1-9. Theodor Scheimpflug (1865–1911).

Society of Photogrammetry in 1907 and the International Society of Photogrammetry in 1909, and became the first President of both societies. He also created the International Archives of Photogrammetry, which today constitute the most comprehensive depository of professional literature of some 62 national member societies around the world.

In 1897, Captain *Theodor Scheimpflug* (fig. 1-9), one of Doležál's countrymen, published a paper on opto-mechanical coincidence in stereo-orientation of balloon and kite photography, in which he developed the theory of double projection as a possible solution to a direct-projection instrument (Scheimpflug 1946). Scheimpflug did not pursue the realization of this theory due to the limitations of optical equipment then available. He discovered the law governing the sustained focus in projective rectification which became known as the "Scheimpflug Condition." This led to the invention by French and German designers of several forms of mechanical computers, called "inversors" (Schwidefsky 1935), which, when combined with similar linkages for maintaining the projective conditions of the process, are important mechanisms for autofocus rectifiers of German, Swiss, French, and American design (Wiebrecht 1960). As an independent researcher, Scheimpflug made contributions to the related problem of zonal transformation, and to aerophotogrammetry and aerotriangulation. He

also designed an eight-lens camera consisting of seven oblique lenses grouped around a central vertical lens (fig. 1-10). When transformed into a single composite with precise alignment of the components, the eight views gave an extremely wide-angle picture of approximately 140°.

From another Austrian officer, *E. von Orel* (fig. 1-11), stems an invention which marks a first approach to automation of a mapping instrument: he devised a mechanism which solves the equations governing the "normal case" of terrestrial photogrammetry.

FIGURE 1-10. Scheimpflug's 8-lens camera (1900).

FIGURE 1-11. Eduard von Orel (1877–1941).

FIGURE 1-12. Ernst Abbe (1840–1905).

1.2.3.3 GERMANY

In 1871, *Ernst Abbe* (fig. 1-12), cofounder of the Zeiss Works, had started intense studies and laboratory tests to place the design of optical elements and their combinations on the basis of rigorous mathematical analysis. His success superseded the earlier approaches by cumbersome trigonometric schemes that had seldom yielded reliable results with compound systems of more than four refractive surfaces. In the early 1900's, newly developed types of glass with improved homogeneity aided further in the design and production of photographic lens systems having a higher number of refractive surfaces, larger apertures, and better overall performance (Price 1976). At this state of the art, the chemical industry had accomplished the production of orthochromatic and panchromatic emulsions coated on unbreakable nitrate-film bases.

Meanwhile, *F. Stolze* had discovered the principle of the floating mark (1892) and *C. Pulfrich* (fig. 1-13) developed a practicable method of measuring and deriving spatial dimensions from photographic images by the floating mark. In 1901, the optical works of C. Zeiss in Germany supplemented von Orel's first prototype with the Zeiss-Pulfrich stereocomparator and a system of mechanical elements which bears the name "Zeiss Parallelogram." A number of these instruments were soon acquired by newly organized private enterprises in European countries and South America.

Sebastian Finsterwalder (fig. 1-14), professor, geographer, and mathematician, was a researcher and authority on alpine glacier surveys

FIGURE 1-13. Carl Pulfrich (1858–1927).

FIGURE 1-14. Sebastian Finsterwalder (1862–1951).

FIGURE 1-15. Henry Fourcade (1865–1948). (Courtesy Prof. L. P. Adams, University of Cape Town, South Africa.)

undertaken in 1888, in which he used a light-weight mountaineer's phototheodolite of his own design. His contributions to the photogrammetric literature included farsighted analytical studies (1903) covering the rudiments of photogrammetry and its problems when applied to photography from balloons (Finsterwalder 1937). In keynote addresses to the Academy of Sciences of Bavaria, Finsterwalder gave a concise description of the geometric relations which govern the relative orientation of stereophotographs by corresponding bundles of image rays. He predicted the future possibility of nadir point triangulation and the application of aerophotogrammetry to astronomic-geodetic measurements for the development of continental triangulation nets, and formulated the derivation of the laws of error propagation in long triangulation chains.

1.2.3.4 GREAT BRITAIN

In 1901, *Henry G. Fourcade* (fig. 1-15), an imaginative surveyor in the British Forest Service of South Africa, disclosed to the Philosophical Society of Cape Town his invention of a stereocomparator, which was quite similar to Pulfrich's instrument of 1902. In 1904, Fourcade made a topographic map of a test survey using stereophotographs taken with a phototheodolite of his own design. In his retreat at the edge of the African jungle he developed his theory of relative orientation of photographs of unknown spatial relationship. To prove his theory, published in 1926, he constructed a stereo-photo-goniometer with a mapping attachment, and

designed a bench-type rectifying printer and a stereoprojector for rectified photographs. Unfortunately, the project failed to receive financial support and was abandoned in 1940 (Adams 1975).

Similarly ill-fated was the invention in 1908 of a stereoplotter for terrestrial photographs, named "stereoplanigraph," by *F. Vivian Thompson* (fig. 1-16), a Captain in the Royal Engineers. His concept was a rigorous mathematical-mechanical solution employing a stereocomparator (Thompson 1908). The parallax displacement needed for solving the equation of the object distance was obtained automatically from a machined inversor curve. Had its design been carried to completion, the instrument could have permitted continuous plotting of map features; however, the proposal remained filed away and unavailable for publication until 1974.

1.2.3.5 IMPACT OF AVIATION AND WORLD WAR I

The event that signaled the beginning of organized civilian aviation was the introduction in the early 1900's of dirigible airships of considerable carrying capacity and cruising range: the Zeppelins and Parsivals opened new horizons to exploration by the surveying camera. At this

FIGURE 1-16. F. Vivian Thompson (1880–1917). (Courtesy Royal Engineers' Museum, Catham, U.K.)

same time, heavier-than-air flying machines had outgrown their experimental stage and the airplane offered a marvelous platform for the aerial camera; however, the promising outlook for greater progress suffered a serious setback by the outbreak of World War I in 1914.

At the beginning of hostilities the emphasis shifted to equipment useful to the war effort. Aerial cameras on both sides of "no-man's-land" were initially of medium to long focal length; they had focal-plane shutters and were hand operated. A newcomer in the uniform of either side was the "remote sensor" with telescope and sketchbook, hanging in the sky in a basket under a big tethered bag of hydrogen. With the advent of the shooting pursuit plane he offered an exciting sight when floating to earth suspended from a parachute with a ball of blazing hydrogen following him. As the number and quality of cameras and the daily volume of aerial photography increased, interpretation keys were developed (trench installations, camouflage patterns, etc.) for photointerpreters who watched over changes in trench systems, gun emplacement, and logistics behind the lines.

In 1915 an automatic film camera (by *O. Messter,* Berlin) went into service on the German side. This camera produced broad-coverage

strip mosaics along the flight line (fig. 1-17). It is reported that 240 of these cameras covered 7 million square kilometres of French, Belgian, and Russian territories.

The invention of the Stereo-Skiagraph plotter by *A. Hasselwander* of the German Anatomy Institute saved thousands of human lives by locating shrapnel in wounded bodies by measurements from X-ray stereograms.

Max Gasser (fig. 1-18) was an energetic young Bavarian geodesist and teacher of aeronautics and photogrammetry in military schools, who founded the German Society of Photogrammetry in 1909. In 1917, Gasser had succeeded in constructing a mapping apparatus for vertical aerial photography, which became known as the "Double Projector Gasser," as an embodiment of Scheimpflug's idea of 1897. Unfortunately, his instrument was confiscated and his wartime patent classified "secret," thus preventing further development. The merit of this creative invention was not realized until decades later when advanced technologies and reinventors made use of the same principles to produce practicable plotting machines.

The recovery after 1918 of the production capacity of Europe's major industries, particularly that of Germany, was forcefully accelerated by the reparation requirements of the peace treaty. Revised maps were urgently needed at this time of Europe's recovery. It is therefore not surprising that photogrammetric developments re-

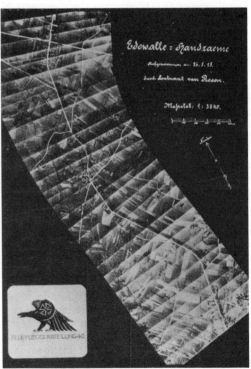

FIGURE 1-17. Oscar Messter's strip-camera flight-line mosaic (1915).

FIGURE 1-18. Max Gasser (1872–1954).

ceived a considerable boost in the ensuing decades.

1.2.4 MODERN PHOTOGRAMMETRY IN CENTRAL AND WESTERN EUROPE

During the period from World War I to the present, the modern era of development in photogrammetry, advances in the science took place in various countries in a complex, interlocking pattern. In this text, developments within each country making a significant contribution are treated separately. This treatment poses some difficulties, for the efforts to solve a given photogrammetric problem often proceeded simultaneously in different countries, and developments in country A certainly affected parallel work in country B. It is therfore impossible to assign each development to a neat cubicle of scientific principle, time, place, and people without affecting the logic of other cubicles. A special factor is the impact of World War II, which had a major effect on developments in all countries. Yet within each country, postwar developments, generally speaking, were a continuation of pre-war efforts, accelerated by the exigencies of the war. The impact of World War II is discussed separately in section 1.2.4.8, although this placement departs somewhat from strict chronological sequence.

1.2.4.1 GERMANY

Reinhard Hugershoff (fig. 1-19), geodesist, explorer, and professor at the Technical Univer-

sity of Dresden, Germany, was the inventor of numerous geodetic and photogrammetric instruments. In 1921, Hugershoff introduced the first universal analog plotter, the Autocartograph, embodying a photogoniometer solution using the Porro-Koppe principle of observation. (The photograph is observed through a lens identical in distortion characteristics to the taking camera lens.) It made plotting of terrestrial as well as of aerial stereograms of any orientation possible. At the Second Congress of the International Society of Photogrammetry (1926) he exhibited the Aerocartograph, a lighter instrument of the universal type, which incorporated capabilities for extension of control and aerotriangulation (Gruner 1929). Among his later creations of major significance were:

a planetable-stereoplotter for continuously delineating a contour in the terrain and simultaneously plotting it on the board (this instrument was successfully used in 1929 in the survey of the trans-Persian railroad);
a "Small Autograph" for terrestrial and short-range applications; and
a pendulous aerial camera for exploration surveys (Hugershoff 1930).

In 1931, Hugershoff joined *Otto von Gruber* (fig. 1-20) on the scientific staff of the Zeiss Works.

In 1923 the Zeiss Works produced a universal plotting instrument, the Zeiss Stereoplanigraph, based on direct optical projection. This invention was credited to *W. Bauersfeld* (fig. 1-21), chief engineer. Under von Gruber's guidance, eight consecutive models of this instrument received many refinements reflecting advancing technology which have contributed to the high rating that this plotter has been given over the course of time since then. In 1965, over 250 of

FIGURE 1-19. Reinhard Hugershoff (1882–1941).

FIGURE 1-20. Otto von Gruber (1884–1942).

FIGURE 1-21. Walter Bauersfeld (1879–1959).

these instruments were said to be in operation throughout the world.

Otto von Gruber was a professor, geodesist, and scientific collaborator of *C. Pulfrich* and the latter's successor as head of the photogrammetric and geodetic instrument department of the Zeiss Works. He became internationally known for his landmark publications, "Single and Double Point Resection in Space (1924)," and "Theory and Practice of Aerotriangulation," and for his "Collected Lectures and Essays" (von Gruber 1932). For many years he conducted the annual "Jena Photogrammetric Weeks" for international audiences. On the practical side, inventive and operational contributions were made by von Gruber and *K. Schwidefsky* (1934) in the development of several models of Autofocus Rectifiers (Schwidefsky 1959).

A simplified descendant of the Stereoplanigraph was the Aeroprojector Multiplex, a small-dimensioned, portable plotter of modular design for mapping of a continuous strip of overlapping vertical photographs. The structural simplicity and the capability for extension of control and aerotriangulation attracted many governmental and military users. Improvements over a period of 25 years added wide-angle and superwide-angle, oblique and convergent projectors. When projectors and instrument components were also manufactured in England and the United States, beginning about 1940, the Multiplex was the most mass-produced piece of mapping equipment in the field of photogrammetry. It was extensively used in all theaters of operation by European and American armies in World War II.

The Photogrammetry GmbH of Munich, a consortium of photogrammetric surveyors, produced a unique instrumentation system, designed by *C. Aschenbrenner* in 1929–34. This complete mapping system included a short-focus 9-lens camera, a transforming printer, an autofocus rectifier, and a simple plotter for rectified composite diapositives tinted with anaglyphic colors for stereoviewing and measurements. The system was extensively used for large-area coverage of colonial territories and for mapping at small scales.

1.2.4.2 SWITZERLAND

Heinrich Wild (fig. 1-22), a Swiss engineer and inventor, surprised the Second International Congress on Photogrammetry (1926) at Berlin with a modification of his plotter prototype of 1920 (Police Autograph), newly equipped with design components for mapping from pairs of vertical or near-vertical aerial photography. He was known to land surveyors for his innovative design of a small geodetic theodolite, which he had developed in the Zeiss Works. He subsequently founded a factory in Switzerland, today known as Wild Heerbrugg Ltd., which developed several models of the well-known and

FIGURE 1-22. Heinrich Wild (1877–1951).

FIGURE 1-23. Umberto Nistri (1895–1962).

widely used opto-mechanical Autographs distinguished by their observation system, based on orthogonal stereoviewing. By 1968, the worldwide distribution of these instruments had passed the 1,000 mark.

The Kern Co. of Aarau, Switzerland, joined the photogrammetric industry in 1930 with a stereoautograph type of plotter for terrestrial work, which was based on the concept of the Spanish Colonel *Ordovas,* named the Ordovas-Kern Photocartograph. In 1948 followed a double-projection type PG-1 stereoplotting instrument and a new version of a mechanical-projection instrument, PG-2, providing a stationary optical system with orthogonal viewing; by the end of 1976, their total production had approached the 800 mark.

1.2.4.3 ITALY

A second concept of double projection, somewhat different from Gasser's, was developed in 1919 by *Umberto Nistri* (fig. 1-23), of Rome, Italy. His first experimental model, Photocartograph I, used alternating projection on rear-view screens. Later models II and III of the direct-projection type were succeeded by a small multi-projector type (1934). In his later designs, Nistri changed to opto-mechanical solutions, the Photostereograph Alpha (1948) and Beta (1952) (Nistri 1952).

Ermeneguildo Santoni (fig. 1-24), member of the Military Geographic Institute of Florence,

Italy, was a gifted inventor and designer of photogrammetric equipment. He constructed single- and multi-lens aerial cameras and created an auxiliary unit (solar periscope) recording the sun's image in flight as a celestial control point for aerotriangulation (1919). He developed a series of Models I (1925) to V (1964) of Stereocartographs based on the principle of opto-mechanical projection. The last model of

FIGURE 1-24. Ermeneguildo Santoni (1896–1970).

this sequence, presented at the Tenth International Congress on Photogrammetry (1964) at Lisbon, incorporated mechanical correctors for residual, systematic, and random errors within the mapping system. In developing a second series of plotters, the Stereosimplex Models I (1934) to IIId (1964), Santoni excelled as a master designer of antiflex guide rods and pivoting components sustained in permanent equilibrium (Santoni 1971).

1.2.4.4 FRANCE

French inventors added two important contributions to the parade of competing stereoplotters. In 1923, *G. Poivilliers* (fig. 1-25), scientist of the Société D'Optique et de Méchanique de Haute Précision (SOM) in Paris constructed a Stereotopograph Model A, later followed by a redesigned Model B (1937). Both instruments used the photogoniometer concept and therefore resembled the Hugershoff Autocartograph of 1921 to a degree. These instruments served the Institut Géographique National (IGN) for many years as the backbone of its intense photogrammetric activity. They were supplemented in 1947 by Poivilliers' Stereotopograph Model D which employs an orthogonal viewing system with space-rod guidance of the rectilinear movements of the photograms.

Another French instrumental and methodical contribution came from *R. Ferber* (fig. 1-26), associated with the firm of Gallus. The Gallus-Ferber device was a direct-projection plotter (1933), initially using alternating image projection, followed by stereoviewing and by a model featuring a scanning attachment for the production of orthophotographic prints.

FIGURE 1-26. R. Ferber's plotter.

1.2.4.5 GREAT BRITAIN

Great Britain's survey institutions had refrained until about 1945 from participation in the European modernization of surveying practices in the belief that the heavy continental plotters could not meet the Empire's requirements for economical small-scale mapping in its colonial territories (Thompson 1964). The pioneering efforts of two outstanding British surveyors, Henry Fourcade and F. Vivian Thompson, had therefore ended in disappointment, as mentioned in section 1.2.3.4. After World War II, the world's atlas was being considerably revised, and British scientists and practitioners became energetic partners in the photogrammetric community.

Edgar H. Thompson (fig. 1-27), a graduate of the Royal Military Academy at Woolwich and of Downing College in Cambridge, was research officer in the War Office Air Survey Committee from 1934 until 1938. He then became assistant to *Martin Hotine* (fig. 1-28), the author of "Stereoscopic Examination of Air Photographs" (1927) and "Calibration of Surveying Cameras" (1929) (Whitten 1973). In Hotine's company, Thompson became thoroughly acquainted with photogrammetry. In 1951 Thompson became professor of Photogrammetry at the University College, London, which was the first academic chair in the country teaching the science.

Among Thompson's many contributions are: a

FIGURE 1-25. George Poivilliers (1892–1967).

FIGURE 1-27. Edgar H. Thompson (1910–1976).

FIGURE 1-28. Martin Hotine (1898–1968).

series of publications on the evaluation of aerial photographs; design of the first Barr & Stroud projection plotter (1935); design of the Cambridge stereocomparator (1938); and design of the Thompson-Watts Plotter (1950), an opto-mechanical solution of the category of first-order continental machines, followed by two later models, which were extensively used by the British Ordnance Survey. His finest contributions were made to analytical (matrix algebra and procedural) processes, reseau techniques, and aerotriangulation by the method of independent models (Thompson 1964). He edited the British journal "The Photogrammetric Record" for 14 years.

1.2.4.6 THE NETHERLANDS

To *Willem Schermerhorn* (fig. 1-29), who became a professor at the Institute of Technology at Delft in 1926, may be accorded the title "Photogrammetrist of the Netherlands" for his educational efforts in spreading knowledge and acceptance of the new technology to many countries of the world. In 1932, Schermerhorn began systematic tests of aerotriangulation techniques, which he applied in 1936 to uncharted land in the Netherlands colonies of the East Indies. In close cooperation with O. von Gruber he contributed to the clarification and understanding of error sources and error propagation in phototriangulation, which stimulated mathematicians of two continents in their search for manageable solutions. Schermerhorn was also the initiator of the tri-lingual journal "Photogrammetria," and the founder (1950) of the International Training Center for Aerial Survey in Delft (now located at Enschede) (Schermerhorn 1964). This institution, well equipped with modern instrumentation, has produced an

international *collegium* of thoroughly trained experts of photogrammetry and remote sensing.

1.2.4.7 FIFTH INTERNATIONAL CONGRESS OF PHOTOGRAMMETRY

The Fifth International Congress of Photogrammetry, held in Rome, Italy, in 1938, offered a comprehensive summary of the European state of the art. The latest developments were impressively displayed in an exhibition, which showed

FIGURE 1-29. Willem Schermerhorn (1894–1977).

applications far beyond its main field of surveying and mapping. Government and private organizations of Belgium, Poland, Hungary,
Greece, Turkey, Germany, Switzerland, Italy,
Spain, France, Sweden, Denmark, Norway, and
the United States showed their accomplishments
in applying photogrammetry to the fields of construction, testing of engineering structures,
medical and X-ray research, exploration of uncharted regions by the German-Russian Arctic
expedition with a Zeppelin airship (1936), and the
Antarctic Expedition (1938/39) using flying boats.

This exhibition also offered a historical review
of the working procedures which had paralleled
or followed the development trend of photogrammetric instrumentation. The opto-
mechanical process of orienting stereo-imagery
to form a virtual model had fully replaced the
semi-computational process that had preceded
the aerial mapping performance in the days of
the Hugershoff Autocartographs of the 1920's.
Otto von Gruber's fundamental analysis of coincidence setting of corresponding image points
had become the common guideline of the
world's stereo-operators and instrument designers. Hugershoff's simple optical switching of
consecutive plate pairs (1927) had been incorporated in several plotters by other optical or
mechanical expedients. In attempts to minimize
troublesome error propagation in the triangulation process, attachments to aerial cameras
gathering flight and orientation data, horizon
pictures, time, temperature and statoscope recordings were used to augment the airborne
equipment. Santoni's system of recording images of the sun synchronized with the aerial
camera exposures appeared very promising. The
end products of all plotting machines were still
graphical in nature, lines drawn on paper or
scribed on emulsions, but developments had
been started to provide for color separation to
simplify the subsequent reproduction process.

However, the promise of continued cooperation, encouraging the scientific and technological advancement of photogrammetry among
member nations of this assembly, was abruptly
destroyed soon after the 1938 Congress as
growing international political tensions created
an uneasy atmosphere in the City of the Seven
Hills.

1.2.4.8 IMPACT OF WORLD WAR II

Soon after the 1938 Congress was held in
Rome, most European military forces were engaged in World War II. They began to exploit
photogrammetry's potential in attack and defense on land, at sea, and in the air. The development of devices for photoreconnaissance
and aerial navigation was vigorously accelerated, and military photointerpretation, almost
forgotten in 20 years of peacetime, rose to the
status of a tactical weapon. The heavy stationary
plotters, wherever they were found to be operable, were pressed into service, and mobile

equipment of the Multiplex type was carried
with field units.

By 1945, destruction of manufacturing
facilities on the continent and in the British Isles
had brought European industrial production to a
near standstill. The German photogrammetric
industry was hardest hit; its largest private mapping organization, Hansa Luftbild, had van-
ished. However, revised military strategy for the
postwar period and political reorientation among
the victorious allies, aided by the undefeated
energies of the surviving populations, prevented
a total European lapse into chaos. The stigma of
photogrammetry as a weapon of war was cast
off; the expropriated Zeiss Works at Jena were
resurrected and a second Zeiss establishment
was constructed in West Germany. The French
and Italian establishments resumed their work
and introduced new optical, mechanical, and
electronic components. Application of the
photogrammetric technique spread to new
fields, from microscopic to macroscopic dimensions, and the methodology began to utilize a
widening fan of the electromagnetic spectrum
for its sensors and image-acquisition devices.
During this phase of evolution the center of
gravity for photogrammetric development
shifted to the North American continent and its
relatively untapped scientific and industrial resources, as we shall see in section 1.2.6.

1.2.5 NORTHERN AND EASTERN EUROPEAN DEVELOPMENTS

1.2.5.1 SCANDINAVIA

The Scandinavian countries, Norway, Sweden, and Finland, began to use the new technique of photogrammetry in its early stages for
land surveying and construction engineering
projects. They contributed significantly to scientific progress in the field and to the development of instrumental components. In 1929, the
Norwegian government launched a successful
aerial expedition to the eastern coastal regions of
Greenland; the German Research Institute for
Aviation cooperated on this project by mapping
large areas covered by the Norwegian oblique
photography.

The Finnish Society of Photogrammetry was
founded in 1931. *Nenonen* and *Vaisälä* developed a method of aerotriangulation aided by
photographic records of the horizon and statoscope readings (1936).

Sweden made extensive use of terrestrial
photosurveying in its rugged mountain ranges
and derived design and construction data for
road and railroad systems from aerial mapping
(1924). The Swedish Technical University
founded its photogrammetric institute in 1930.
One of its early professors, a brilliant orator,
was *A. v.Odencrants,* designer of the Wild-
Odencrants autofocus rectifying instrument in
1934.

Bertil Hallert (fig. 1-30), geodesist and

FIGURE 1-30. Bertil Hallert (1910–1971).

mathematician, became Sweden's leading photogrammetrist by his investigations in instrument calibration, orientation procedures for plotting machines, and by the setting of standards for the testing of cameras, comparators, and stereo instruments. He also established the mathematical term "standard error of unit weight"; in 1968, he was invited to the United States by the American Society of Photogrammetry on a "distinguished lecture tour."

1.2.5.2 RUSSIA

Several early Russian uses of reconnaissance aerial photography are noteworthy. In 1886, Commander *Kovanka* made photographic balloon flights over the cities of Kronstadt and St. Petersburg (now Leningrad). In 1900, the topographers *R. Tiele* and *Uljanin* lifted an assembly of 8 coupled cameras, called a "Panoramagraph," into the sky with seven huge kites. An attempt at photoreconnaissance was made in Manchuria during the Russo-Japanese war. Subsequently, the application of photogrammetric processes in Russia followed developments in Western Europe closely. Lacking specialized industrial installations and skills for domestic production, Russian instrumentation for terrestrial work in pre-aviation days originated largely in Germany. Its design and use was carefully analyzed in several research centers, and equipment of greatest economic efficiency was imported in large quantities.

In 1924, the Institute for Topography was founded as a Division of the Russian military topographic agency. A German specialist, Professor *O. Lacman*, acted as advisor for several years. Air surveys of the city of Moscow started in 1925. A ten-year survey program by terrestrial photogrammetry in the mountains of Pamir, Caucasus, and Ukraine was started in 1926, fol-

lowed by aerial surveys of planned transcontinental rail lines.

The Central Scientific Research Institute for Geodesy, Photogrammetry, and Cartography began its work in 1928 in Moscow. Since that time, it has attracted leading scientists endeavoring to keep pace with western European and American scientific progress in all disciplines. A first all-Union Congress of Aerotopography was held in 1929. The construction of first prototypes of phototheodolites and of a stereoautograph (by *Drobyshev* in 1957) marks the beginning of domestic developments of precision instruments and optical glass types in the Soviet Union. Aerial photographic terrain coverage in Karelia, Central Asia and Transcaucasia was reported to amount to one million square kilometres in 1934; this coverage was doubled by 1938 under government control. A wide-angle camera lens "Liar" designed in 1931 by *M. Russinov*, followed by the "Russar" in 1940, designed to cover a 120° field (Pestrecov 1954), as well as aerial films, were produced and extensively used to complete photocoverage of USSR territories for the preparation of the continental map at 1:100,000.

In the period following World War II, several plotter prototypes were developed by Russian designers which show certain departures from stereocomparators and autographs of German origin. The Stereoplotter SPR by *Romanovsky* (Schoeler 1961) differs in that it uses affine projection of the aerial exposure as a first stage, followed by a corrective stage in which orthogonality of projection is established. This system made use of superwide-angle photography for small-scale plotting. Other notable developments were a Multiplex type instrument by *Wictorow,* a stereocomparator, and the leading textbook "Aerophotogrammetria" by *M. D. Konchine* (Editions NEDRA, Moscow 1967).

An attempt to produce orthophotographs was made about 1960 by Zhukov. A terrain model produced with Multiplex projectors was manually scanned in parallel profiles. The scanning movements were duplicated under a third projector holding one of the stereopictures in identical orientation. A slit traveling over a sensitized surface produced the orthographic print.

The voluminous publications by Russian authors in the bi-monthly journal "Geodesy and Aerotopography" (available in English translation) provide an insight into the current highly diversified Soviet research in all scientific fields related to photogrammetry (American Geophysical Union 1970).

1.2.6 DEVELOPMENTS IN THE WESTERN HEMISPHERE

The development of photogrammetry and its practical application on the North American Continent took a somewhat later and different course from that in the Old World for quite natural reasons. The relatively sophisticated implements that were required to surpass inherited practices called for specialized sources of supplies and training institutions for practitioners, few or none of which existed in nineteenth-century America. Excluding a few exceptional events, the American chapter of photogrammetry had its apparent beginning almost a half century after the barnstorming period in Europe (Landen 1962).

1.2.6.1 UNITED STATES

One of the early American uses of photography for the purpose of gaining some kind of dimensional information was made in 1862 by the Union Army, employing tethered balloons (fig. 1-31) for reconnaissance. Photographs of surprising quality were obtained of enemy-held terrain during the Peninsular Campaign. By the use of grid overlays, target points were transferred

to existing maps. These early applications of photogrammetry in the United States are described in the first American textbook on the subject: "Photography Applied to Surveying," published in 1888 by Lt. Henry A. Reed, a West Point professor.

In 1893, *C. B. Adams* suggested taking overlapping photographs from separated balloon stations (fig. 1-32) and locating target points by graphical intersections, a forerunner of the radial-line method. In 1894, the Canada-Alaska boundary survey made use of terrestrial photography. Sensational aerial photography was obtained by *G. R. Lawrence* with a large camera suspended from seven kites over San Francisco after the city was destroyed by the earthquake in 1906.

Two industrial leaders were destined to become strong supporters of a prosperous development of American photogrammetry. One

FIGURE 1-31. Union soldiers inflating a tethered balloon for observing across Confederate lines (1862).

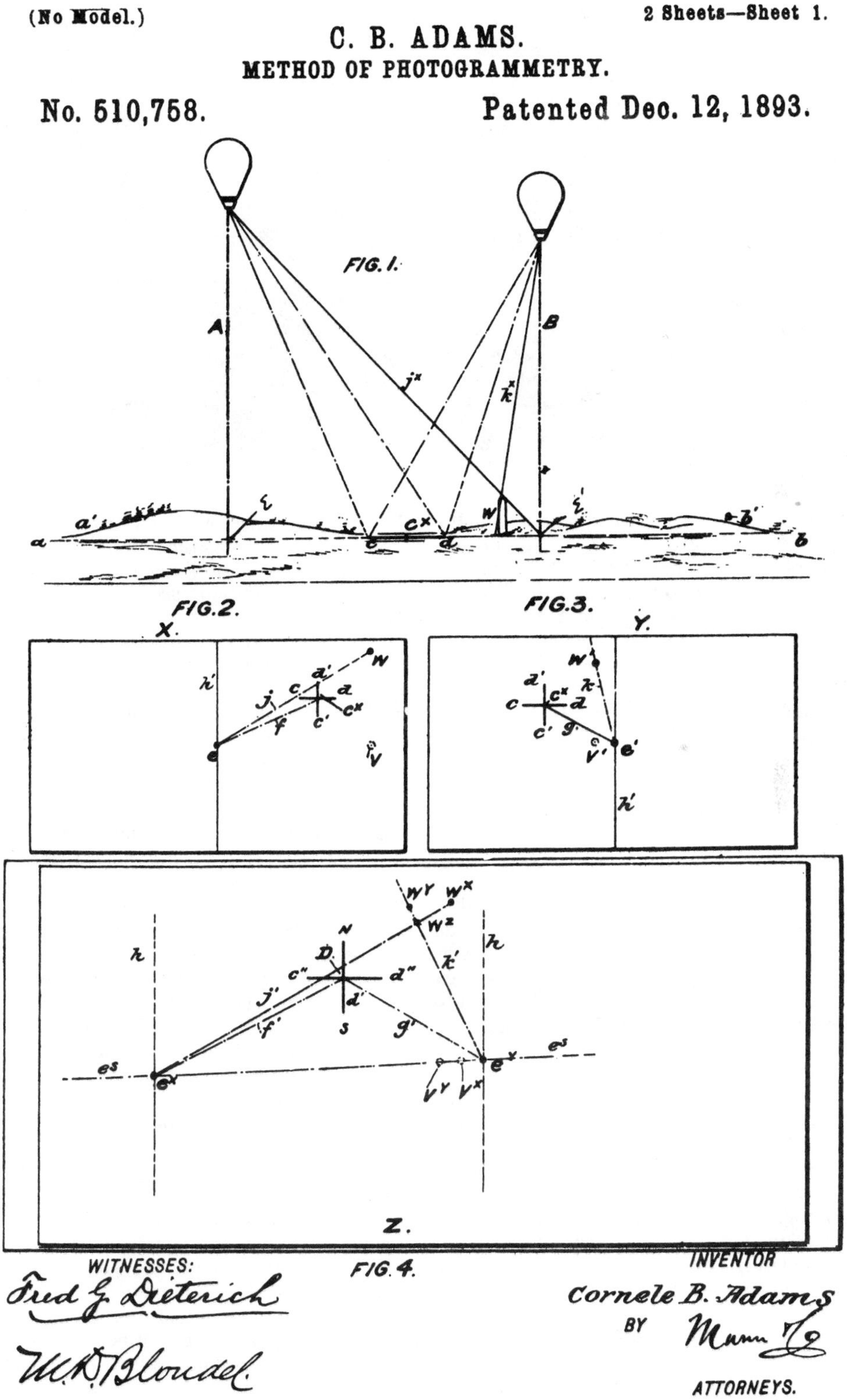

FIGURE 1-32. C. B. Adams' system of mapping from balloon photography.

FIGURE 1-33. George Eastman (1854–1932).

FIGURE 1-34. Sherman Fairchild (1897–1971).

was *George Eastman* (fig. 1-33), who had built a photographic dry-plate factory in Rochester, New York, in 1880, and 10 years later put roll film in his "Kodak Box" that gained world fame. The second was *Sherman Fairchild* (fig. 1-34), who produced the first aerial cameras for American military use in 1917. The Fairchild Camera Corporation developed the initial 2-lens "Bagley Camera" into 3-, 4-, and 5-lens cameras. Fairchild went on to include in the steady growth of his multi-lateral enterprises navigational instrumentation, airplanes, photogrammetric cameras, and accessories serving aerial photography.

1.2.6.1.1 COAST AND GEODETIC SURVEY

The Survey of the Coast was established in 1807 as the first survey agency of the United States. In 1878, the name of the agency was changed to U.S. Coast and Geodetic Survey (USC & GS) when its charter was expanded to embrace mapping of the coasts and intracoastal waterways and harbors, recording tides, making nautical charts, and establishing geodetic surveys and level lines across the continent. When aerial photographs became available in 1918, the Coast and Geodetic Survey revised its procedures to take advantage of this new data source. From 1928 on, all mapping activities were based on photogrammetric processes, including the construction of aeronautical charts and their periodic updating. This organization developed its own photogrammetric system in an effort to expedite the survey of U.S. tidal shorelines, which extend over 88,000 miles.

Captain *O. S. Reading* of the Coast and Geodetic Survey conceived a system, similar to the Aschenbrenner short-focus concept of 1929, comprising a unique 9-lens camera (fig. 1-35) of 8¼-inch focal length covering a field angle of 140°, a photo-transformer for projective conversion of eight peripheral oblique photocomponents to the plane of the central vertical photograph, and a rectifying instrument dimensioned to accept the compound film negative of 35″ × 35″ format. These components, which went into operation about 1938, produced a then unequalled terrain coverage and photographic image quality. Pairs of rectified prints were placed in the "Reading Plotter," a stereocomparator-type instrument with which a complete topographic map was produced. Large portions of the western coastal regions and much of Alaska's uncharted wilderness were mapped with this system.

After World War II, the Coast and Geodetic Survey received aerial mapping equipment of European origin, and participated thenceforth in the development of advanced phases, adding electronic components and high-speed computers to its routines. The photogrammetric scientists of the organization (known since 1970 as the National Ocean Survey) developed analytical approaches to the problems of aerial triangulation and strip and block processing, and assumed a leading role in computer programming.

1.2.6.1.2 GEOLOGICAL SURVEY

The U.S. Geological Survey (USGS) was founded in 1879 with a simple charter which has been interpreted to include nationwide topographic and geologic mapping and inventorying of natural resources. Realizing the monumental task involved, given the continental dimensions, this organization set out to investigate and avail itself of all advances in science that could expedite its task, and to organize its own research, testing, and the development of the inventive ideas of its experienced workforce.

FIGURE 1-35. O. S. Reading's 9-lens camera ready for installation in airplane (1936).

In 1904, USGS photosurveying began with a combination of a panoramic camera and a planetable in Alaska. In the following three decades, USGS workers produced a number of prototypes of in-house photogrammetric inventions. When vertical aerial photography became available, systems of graphic evaluation (McNeil 1953) and extension of control by the radial-line method were utilized. Stereoscopes and measuring attachments manufactured in the USGS workshop or by local machine shops were investigated, including such devices as contour finders (parallax bars), the Stereocomparagraph (designed by B. B. Talley), and the first American plotters offering approximate solutions of the restitution problem by mirror stereoscopes (e.g., the KEK- and the Wernstedt-Mahan plotters). The "Slotted Templet" system, based on the classical radial-line principle, was invented by C. W. Collier in 1935 and manufactured by Fairchild (Kelsh 1940). This was an important and very effective graphical step in two-dimensional strip and block triangulation; a large area deficient in ground control, consisting of a block of 4,400 square miles covered by 2,700 photos, was used for a test of the slotted-templet system by the Department of Agriculture. (This solution of the control extension problem was a classical forerunner to the analytical solution of the 1960's not envisionable at that time.)

A modification of the slotted-templet system using metal strips, pin-hinged together in spider-like fashion (Lazy Daisy), was developed by USGS about 1950. The Lazy Daisy was extensively used for reconnaissance mapping over vast areas of Alaska. A refinement of the templet method, developed by M. B. Scher about 1960, was obtained by the "stereotemplet" derived from the area of stereomodels set up on Kelsh plotters; this system had the inherent accuracy of rectified templets.

The first Chief Topographic Engineer of the USGS, Colonel C. H. Birdseye (fig. 1-36), was a strong believer in the developing role of photogrammetric instrumentation, which offered a rigorous three-dimensional solution to the problem of topographic map production. He seized the opportunity to test automated terrestrial photogrammetry in 1921. Birdseye made two surveys, one in the White Mountains of New Hampshire and one in the vicinity of Washington, D.C., employing an Orel-Zeiss Stereoautograph and a phototheodolite. Contour maps meeting U.S. standards of accuracy were obtained, but the method was judged to be more expensive than the existing survey practice.

In 1927, Professor Hugershoff (Gruner 1971) was invited by USGS to demonstrate his Aerocartograph. A test of its aerotriangulation capability was convincing, although systematic and random errors traceable to sources unrelated to the plotter required considerable cor-

FIGURE 1-36. Claude H. Birdseye (1878–1941).

rections and repetitions of the process. After removal of these error sources and extensive training of operators, the Aerocartograph was successfully employed in routine mapping of rugged mountain regions of national parks. In 1934, the Geological Survey acquired the first Multiplex Aeroprojectors. With the formation of the Tennessee Valley Authority, a large photogrammetric survey of the Tennessee River Watershed began in 1936 under the direction of *T. P. Pendleton* (fig. 1-37). On this project, extensive use was made of aerotriangulation with Multiplex instruments.

Universal plotters (Zeiss Stereoplanigraphs) and related equipment from Germany were as-

signed to the Geological Survey after the termination of World War II. From 1950 on, domestic survey operations of USGS were increasingly based on the photogrammetric process. The research section and model shop headed by *R. K. Bean* (fig. 1-38) provided the necessary adaptations to the foreign equipment and began intensified design and construction of new versions of equipment, including a control extension instrument for convergent low-oblique photography called Twinplex (1954), a new Multiplex type ER-55 (Balplex) plotter (1956), and a new concept of a three-projector instrument combined with a slit scanner generating orthophotographic prints named "Orthophotoscope" (1959). Bean's staff also completed the refinement of the anaglyphic direct-projection plotter, which in its earlier stages had been developed by *H. Kelsh,* and which is, therefore, known as the "Kelsh Plotter"; this plotter is based on the Scheimpflug-Gasser-Ferber concepts. In 1967, the anaglyphic light filters were replaced by revolving shutters (the Stereo-Image-Alternator, designed by *J. W. Knauf*), permitting model observation in white light. These substantial improvements gave the Kelsh plotter a wide popularity in the Western Hemisphere.

1.2.6.1.3 MILITARY PHOTOGRAMMETRY

In the early 1900's, the survey battalions of the U.S. Army Corps of Engineers (CE) became the suppliers of topographic maps of strategic areas for the armed forces. Experiments conducted during and after World War I with aerial photography for military mapping of areas of difficult access produced successful results (Goddard 1969).

FIGURE 1-37. Thomas P. Pendleton (1887–1954).

FIGURE 1-38. Russell K. Bean (1900–1976).

Major *J. W. Bagley,* CE, promoted the use of multi-lens aerial cameras for extension of control by the classical radial-line method, and designed equipment for obtaining the position and elevation of target points from oblique photographs by graphical means.

From 1929 to 1934, Captain *B. C. Hill,* CE (fig. 1-39), headed a Research Detachment at Wright Field with the assignment of exploring triangulation capabilities with photographs from single- and multiple-lens cameras using the Aerocartograph. In 1934, the Research Detachment investigated the Zeiss Multiplex equipment and its adaptability to military photography and supervised training of military personnel in its field use. The Multiplex was made standard equipment of the survey battalions of the U.S. Army Corps of Engineers. Its domestic production, incorporating design changes based on research by the Wright Field detachment and a TVA-USGS team, was initiated about 1942 by the Bausch & Lomb Optical Co.

Successors to the Corps of Engineers' Research Detachment of Wright Field were the U.S. Engineers' Research and Development Laboratories, Fort Belvoir, Va. Their photogrammetric section provided the Army Map Service with test results of foreign photogrammetric innovations and assisted in new developments of Kelsh-type projection plotters. At the middle of the century, they initiated experiments with sophisticated electronic and automated mapping systems. One of their early prototypes was the UNAMACE, a rapid analytical mapping system designed to require a very minimum of operator assistance in a chain of stages ending in a reproducible map manuscript for the use of the armed services.

FIGURE 1-39. Bruce C. Hill (1906–1943).

The Army Map Service (now the Defense Mapping Agency Hydrographic/Topographic Center) was charged during World War II with the monumental task of supplying the military command with up-to-date map information concerning war theaters distributed around the world. It represented—and still represents—an organization of widest diversification and ingenuity in the use of photogrammetric and geodetic techniques. Its instrumental arsenal was substantially augmented after the war by analog plotters, rectifiers, etc., collected from the European scene. The operations of this organization have been broadened in the post-war period in response to extensive requests for training of personnel for aerial photographic missions in connection with defense commitments in Africa and Asia and for extended radar mapping in South America.

Similar organizations created by the U.S. Air Force (Aeronautical Chart and Information Center) and U.S. Navy (Oceanographic Office) utilized and refined the state of the art, and designed a great variety of photogrammetric apparatus commensurate with their specialized assignments and services within the armed forces. Outstanding examples are the shutterless strip camera (by Sonne), night photography, and sea and beach floor mapping techniques. In 1972, the elements in the Department of Defense engaged in the production and distribution of maps and charts were brought together under one command—the Defense Mapping Agency (DMA), operating through two principal elements: DMAH/T (Hydrographic/Topographic Center) and DMAAC (Aerospace Center).

1.2.6.1.4. COMMERICAL PHOTOGRAMMETRY

Private mapping organizations in the United States have made valuable contributions in pioneering new techniques, in educating the public, and in achieving acceptance of their photogrammetric products in the scientific, administrative, and engineering professions, and even in the courts.

Major *Edward H. Cahill* (fig. 1-40) and *A. Brock* of Philadelphia independently designed a mapping system which deviated distinctly from European concepts in being well adapted to the American practice of division of labor. All system components, including a precision hand-operated plate camera, a rectifying optical enlarger, a stereometer (stereocomparator), and a projector for zonal transformation of the central perspective map into an orthogonal end product, were fabricated in the tool works of Arthur and Norman Brock. Cahill had developed the mathematics by which the instrument settings for the sequential steps of production had to be computed. Mapping operations with this system, which became known as the "Brock and Weymouth Method," started in 1923. The sys-

FIGURE 1-40. Edward H. Cahill (1885–1974).

FIGURE 1-41. Leon T. Eliel (1894–1974).

tem produced maps of acclaimed planimetric accuracy and topographic fidelity. The instrumentation was acquired in 1938 by *Virgil Kaufman* (Aero Service Corporation). After some technical and procedural improvements it was placed back in operation and was used to produce very large scale maps for construction work during World War II. The producers of this unique system were memorialized by *H. Tubis* in an address to the ASP Convention of 1976 (Tubis 1976).

Leon T. Eliel (fig. 1-41), an aviator and aerial photographer, was an early photogrammetry pioneer active on the West Coast. He developed an artistic technique for the preparation of mosaics of hilly terrain. By applying zonal rectification with his scaling projector, Eliel corrected displacements caused by elevation differences in urban areas. He added phototopography with improvised stereometer-type equipment. Eliel also developed an airborne "Solar Navigator" and a phototransformer for a 4-couple camera acquired from Zeiss. As manager of Fairchild Aerial Surveys in Los Angeles, Eliel added a Hugershoff Aerocartograph (1932) and a Zeiss Stereoplanigraph (1934) to his office equipment. The latter was equipped for topographic mapping from untransformed photocomponents of the 4-couple camera. A unique project carried out with this equipment was the periodically repeated survey of the shoreline of Lake Mead as it grew to its present configuration after completion of the Hoover Dam on the Colorado River.

The Aerotopograph Corporation of America

was organized in 1929 in Washington, D.C., with Colonel C. H. Birdseye as president. The company's photogrammetric office was equipped for terrestrial and aerial surveys using Hugershoff instrumentation. Among the various mapping projects successfully completed, the survey of the Hoover Dam site in the Black Canyon of the Colorado River serves as a classic example of accomplishment by terrestrial photogrammetry under extremely difficult topographic conditions which would have defeated any attempt to employ conventional ground or air survey methods.

Birdseye's talent as an eloquent missionary to scientific and professional audiences acted as a stimulant to growing demands for photogrammetric surveys throughout the land. His strong advocacy of photogrammetric techniques led to research and to training of photogrammetrists in American universities and encouraged specialized industries; it also resulted in the founding of the American Society of Photogrammetry in 1934, and in the ASP's association with the International Society of Photogrammetry, which offered its membership access to worldwide information and experience (Gruner 1972).

When the Bausch & Lomb Optical Co. undertook the production of Multiplex equipment in 1942, it became the first American industrial enterprise to engage in a comprehensive manufacturing program of photogrammetric optics, photointerpretation equipment, and map-producing devices. This production line was set up just in time for a crucial period that began with World War II, when the United States was in dire need of optical instrumentation for aerial reconnaissance and mapping. Bausch & Lomb became the chief supplier of aerial camera lenses and of completely redesigned and integrated Multiplex plotting equipment.

Much of the optical research and development by Bausch & Lomb led in later years to a production line of rectifying printers, photointerpretation and map-revision instruments, and direct-projection plotters for wide-angle and superwide-angle photography. The company did not, however, undertake to compete with European manufacturers in the field of the more sophisticated stereomapping equipment.

The scientific development at Bausch & Lomb was essentially spearheaded by the chief optical designer, Dr. *Wilbur R. Rayton* (1882–1946), and his successor, Dr. *K. Pestrecov* (fig. 1-42). Among their major contributions to optical science was the design of the Metrogon lens, an aerial reconnaissance lens, and the design of the wide-angle Cartogon lens for topographic mapping. Both lenses were based on a tentative design formula of the Zeiss Topogon prototype.

The first American institute of photogrammetry was established in 1929 at Syracuse University. Professor *E. Church* (fig. 1-43) was its first head. His approach to the photogrammetric process was mathematical. In 20 years of teaching, Church analyzed the fundamental problems of photogrammetry step by step in 19 University Bulletins (Quinn 1975). These publications may be regarded as one of the American stimulants of the analytical treatment of the basic problem of photogrammetry, which in connection with the rapid progress in electronic computer technology paved the way for the birth, in 1959, of an entirely new breed of photogrammetric instrumentation, the analytical plotter. To acquaint students with practical aspects of the science and with the technological advances of instrument design, the institute conducted yearly excursions to the Bausch & Lomb factory in Rochester. This institute was later integrated with the educational system of the State University of New York.

FIGURE 1-43. Earl Church (1890–1956).

1.2.6.2 CANADA

Canadian photogrammetry owes much of its success to *Edouard G. Deville* (fig. 1-44), who became Surveyor General of Dominion Lands in 1885. In the mountain regions of the West he used monoscopic terrestrial photogrammetry and the method was applied in the Canada-Alaska boundary survey of 1893. In 1895 he conceived, and in 1902 he described, a mapping instrument incorporating a modified Wheatstone stereoscope for use with terrestrial stereophotography (fig. 1-45). As a member of the "Air Board" from 1920 on, he turned to mapping from oblique aerial photographs obtained by flying boats and covering large areas of eastern Canada's lake-strewn plains. For the transfer of photographic detail into correct map position, he made use of perspective grids prepared in incremental steps of flight altitude and camera tilt, as developed by *McKay;* this procedure became known as the "Canadian Grid Method." Over 100,000 exposures were taken before 1930 and their information content was transformed into planimetric maps.

The magnitude of these operations and Deville's determination to improve the accuracy of the mapping process led to the installation of a camera testing and calibration laboratory (National Research Council 1952). Experiments began in 1937 with multi-lens cameras; these were soon replaced by single-lens wide-angle types. Extensive mapping operations followed, based on the use of graphical radial-line triangulation. In 1975, Edouard Deville was commemo-

FIGURE 1-42. Konstantin Pestrecov (1903–1967).

FIGURE 1-44. Edouard Deville (1849–1924). (Courtesy Canadian Institute of Surveying, Ottawa).

rated with a monument erected at the entrance to Yoho National Park near the town of Field, B.C. (Andrews 1976). Nearby Mt. Deville had been named after him in 1886.

When the national mapping program advanced into the central and western mountain regions, the Canadian mapmakers still operated for some time without the aid of any ground-control-saving, three-dimensional stereoplotter. In 1934, Canada acquired a "Radial Stereo Plotter," and in 1937 a "Stereograph." The Maps and Surveys Research Committee decided to acquire an instrument designed by Hotine (Whitten 1973) of the British War Office, which embodied the 1929 concept of South African pioneer, H. Fourcade; however, the prototype was totally destroyed in 1939 by war action in Southampton Harbor. Pressing need for advanced control extension equipment led to the importation of Multiplex equipment. After World War II, the Department of Energy, Mines and Resources added Wild Autographs. Oblique photography was revived to meet demands for aeronautical charts and extensive use was made of trimetrogon photography.

The Canadian National Research Council, founded in 1933, made considerable contributions to the advancement of photogrammetric practices by inspecting and calibrating aerial mapping and navigation equipment, and by designing and experimenting with new instruments (e.g., an airborne radar altimeter and a radial stereoplotter). NRC scientists added the

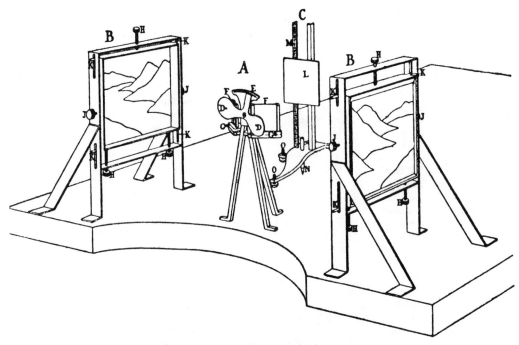

FIGURE 1-45. Deville's mapping instrument.

"Stereomate" to the orthophoto print and created an "Orthophoto-Stereoplotter" developed by *T. Blachut*. By contributing mathematical innovations to analytical photogrammetry and programming, *U. V. Helava* provided the basis for development of the first analytical plotter which he unveiled in 1958 (Helava 1958 and Doyle 1964).

In 1959, *Gilbert Hobrough* of Photographic Survey Corporation, Toronto, exhibited an ingenious system for electronic image correlation superimposed on a Nistri "Photomapper." With this system, Hobrough accomplished the automatic relative orientation of a pair of overlapping aerial photographs. In several years of painstaking research he developed the electronic program for a faceted rectification process, the "Gestalt System," which possesses the capability to produce a contoured orthophoto print and a digitized terrain model.

1.2.7 PHOTOGRAMMETRY IN OTHER PARTS OF THE WORLD

Although the need for maps has been increasingly felt in all developing countries of the world, the absence of specialized industries and professional skills has prevented them from producing photogrammetric devices of inherent complexity on their own soil. Most of these countries organized and equipped their governmental survey agencies with systems of Swiss, Italian, American, or German origin, or they invited private engineering concerns to fill their mapping needs. Many major projects for road and railroad design, hydro and electric installations, oil and mineral exploration in *Asia, Africa,* and *South America* have been carried out by modern photogrammetric methods. Most of them are well documented in the Archives of the International Society of Photogrammetry.

The government of *Japan* acquired the first laboratory prototype of the Hugershoff Autocartograph in 1921. Colonel *Ujihusa Kimoto* accomplished the difficult task of assembling and calibrating this instrument from written instructions with a study group of Japan's first photogrammetrists in the Nippon Department of Land Survey. Mapping experiments began with oblique aerial photographs by computation of the "aerial pyramid" of each exposure station to obtain the exterior orientation of the individual exposures. An earthquake in the test area in 1923 interrupted the operation. A second experiment with vertical aerial photography and the advanced method of stereo-orientation by parallax elimination yielded more encouraging results. After acquisition of an early Zeiss Stereoplanigraph and a 1927-model Hugershoff Aerocartograph, the old Autocartograph was retired. It is still preserved in the Science Museum of Tokyo.

Japanese engineers organized a Photogrammetric Society in 1948, later combined with the Geodetic Society. The present Japan Photogrammetric Society (Nippon Syashin Sokurjo Gakkai) was founded in 1961. Its quarterly technical journal informs its members about the photogrammetric field and the related fields of ground, air and space developments.

China's Central Bureau of Land Survey introduced photogrammetric practices soon after the appearance of the first European plotters. It opened its first training school for stereoplotter operators in 1930 and imported stereoplanigraphs, autographs, and rectifiers. Multiplex instruments were used with wide-angle aerial photography by the General Staff for the national 1:50,000-scale map series. After China's conflict with Japan in 1937, American instruments of the stereoscope and stereometer type were introduced and trimetrogon photography was extensively used to aid photographic missions for cadastral mapping at larger scales. In the search for faster and more economical solutions, studies were under way in 1946 on the feasibility of applying analytical solutions to map-production problems.

Countries of the British Commonwealth have widely replaced the traditional surveying practices in the post-war period by photogrammetric technology. Their inventory of aerial cameras and restitution equipment of Swiss, German, Dutch, Italian, and American origin in government and private offices as well as in educational institutions essentially increased after World War II.

International cooperation entered a new phase in post-war years when the leading industrial countries awakened to the need of inventorying their natural resources, protecting the environment from destruction, conserving the fertility of the soil, and limiting wasteful energy-consuming practices. In 1964, the United Nations convened one of its quadrennial Cartographic Conferences in which the execution of an inventory of natural resources for Asia and the Far East was considered. A continuing review was established of the cartographic technology available to improve living standards, production, and economy in developing nations of Asia and the Far East. The urgency of intensified mapping and the role of photogrammetry were impressively demonstrated (United Nations 1966).

The Pan American Institute on Geography and History (PAIGH), the Organization of American States (OAS), the combined forces of U.S. photogrammetric engineering companies, and the U.S. Inter-American Geodetic Survey (IAGS) have effectively contributed to the dissemination of advances in photogrammetric technologies and to the training of photogrammetrists of Central American and South American countries. They have also aided in disaster relief and rehabilitation in these countries, by rapidly providing maps of devastated areas.

1.2.8 A NEW ERA IN PHOTOGRAMMETRY

During the last three decades of photogrammetry's historical development, a new era has appeared, with new horizons that presage the future. These horizons are calling for new terminologies, are paving new avenues of approach to photogrammetry's most complex mathematical problems, and are demanding ultimate accuracies of measurement, speed and economy of automated processes. Domestic and foreign industries are eagerly following the trend, and equip their stereoplotters, now collectively called "analog instruments," with coordinate readouts and recorders; with electronic equipment for data storage and retrieval, for on-line, off-line, graphic, automatic data presentation; and with opto-mechanical or electronic attachments for model scanning in parallel profiles and for reproduction of the central perspective aerial photography in orthographic map projection.

Aircraft on photographic missions are being guided on long-and-straight or arched-and-parallel flight lines by high-precision positioning systems, e.g., HIRAN. Highly developed instruments, such as the Canadian airborne profile recorders, register terrain undulations along flight paths. Advanced instrumentation uses wave lengths far beyond the visible segment of the electro-magnetic spectrum (e.g., radar-grammetry, multi-waveband exposure of special emulsions, and electronic storage of signals). By these means, scientists probe the earth's crust and its waters' depths, its atmospheric and climatic movements, and its life-sustaining riches above and below the ground.

The classical comparator of Pulfrich's time has made a comeback (e.g., the Mann Precision Comparator) to serve as a key component in the new analytical processes which are striving for pre-eminence over the classic "analog routine." By the miniaturization of electronic components and design of high-speed electronic computers, the mathematical accomplishments presented by H. Schmid, D. Brown, C. Tewinkel (USA), and G. Schut (Canada) have gained economic significance and widespread application in extensive aerial strip and block triangulation schemes.

In solving geodetic problems, the application of precise photogrammetric measurements on photographs obtained with satellite-tracking cameras has resulted in the establishment of a globe-girdling triangulation system (L. Swanson, H. Schmid). By this means, the continents and islands anchored to the various floating tectonic plates of our planet's fragile shell have been coordinated in an orderly geodetic framework.

Helava's analytical plotting instrument, the prototype of which had been assembled in the Canadian National Research Council laboratories in 1958, was soon placed on the market by OMI Corporation, Italy. Subsequently, Bendix Research Corp., Southfield, Michigan, perfected the electronic components so that the analytical plotter provides a means for performing relative and absolute orientation, extension of control, aerotriangulation, compensation of optical and mechanical anomalies, and selection of plotting scale and type of map projection. Special high-speed computers sustain the continuous flow of mapping operations, while the operator guides the floating mark in accordance with the conventional routine of map compilation within the area of the stereo model.

Similar systems have been developed by Raytheon and IBM Corporation. European manufacturers have integrated electronic systems built by Bendix and Itek Corporation into their hybrid plotters, most of which are based on the comparator principle.

Human knowledge and skill have enabled us to learn to utilize the forces that govern our galaxy for the development of satellite photogrammetry; this technique provides the capability of complete surveillance over the entire globe and of probing deep into its crust for sensitive secrets above and below. Great benefits are certain to be derived by all mankind, provided that international restraint and goodwill among nations unite in guiding the future of the technique.

1.2.9 AMERICAN SOCIETY OF PHOTOGRAMMETRY

The American Society of Photogrammetry (ASP), with approximately 8,000 members (in 1979), is the largest affiliate of the International Society of Photogrammetry (ISP), which in 1976 comprised 62 national member organizations. In 1952, ASP was host to the Seventh International Congress of Photogrammetry in Washington, D.C. The circulation of its monthly journal, Photogrammetric Engineering and Remote Sensing, has passed over the 8,000 mark.

Since its founding in 1934, the Society has published the following Manuals, which comprehensively portray the scientific and technologic advance of American photogrammetry and its allied developments:

> Manual of Photogrammetry, First Edition, 1944, one volume, 841 pages;
> Manual of Photogrammetry, Second Edition, 1952, one volume, 876 pages;
> Manual of Photogrammetry, Third Edition, 1966, two volumes, 1199 pages;
> Manual of Photographic Interpretation, 1960, one volume, 868 pages;
> Manual of Color Aerial Photography, 1967, one volume, 550 pages;
> Manual of Remote Sensing, 1975, two volumes, 2047 pages;

Handbook of Non-Topographic Photogrammetry, 1979, one volume, 210 pages.

These publications are the result of voluntary contributions of several hundred authors from all parts of the United States.

The Society presents the following awards at its Annual Meetings:

Sherman Fairchild Photogrammetric Award for outstanding achievements in the science of Photogrammetry;

Talbert Abrams Award for authorship of current and historical engineering and scientific development in Photogrammetry;

Photographic Interpretation Award, by AIL Information Systems, for contributions to advancement of photointerpretation;

Autometric Award, by Raytheon Company for outstanding publication on photointerpretation;

Alan Gordon Memorial Award for achievements in Remote Sensing and Photographic Interpretation;

Luis Struck Award for advancement of Pan American cooperation in photogrammetry;

Bausch & Lomb Photogrammetric Award for the best paper on photogrammetry by a college student;

Wild-Heerbrugg Photogrammetric Fellowship Award for encouragement of graduate photogrammetric studies;

Ford Bartlett Award for stimulation of interest and membership in ASP;

Honorary Membership Award in recognition of outstanding accomplishments in the science of photogrammetry.

ASP is also custodian of the Norman Brock Award, donated to the International Society of Photogrammetry by Virgil Kaufman, Aero Service Corporation. It consists of a gold medal commemorating the pioneering deeds of Norman and Arthur Brock, to be presented for a "Landmark Contribution" to Photogrammetry at the quadrennial Congresses of ISP.

1.3 Major Technical Problems of Photogrammetry

Many problems face the photogrammetrist in his efforts to obtain precise measurements from imagery collected photographically or by other remote sensing techniques. The principal causes of these problems include: conditions are rarely ideal for obtaining and preserving the original imagery; the subject of the photogrammetric survey is never perfectly level, flat and smooth; and complete accuracy in the transference of the imagery and other data to the final product is rarely attained.

To appreciate the variety of difficulties that can arise from these sources, the requirements for one of the most exacting uses of aerial photographs will be considered: the preparation of an accurate topographic map. Some of the problems that will be discussed are not applicable to terrestrial photogrammetry. If the source material is not photographic, there will likely be additional problems. They are treated in the MANUAL OF REMOTE SENSING, and will not be discussed here.

1.3.1. OBTAINING THE BASIC DATA

To illustrate why it is seldom possible to obtain the original photographs (or other basic data) for a mapping project under ideal conditions, several requirements are listed below. All of these requirements would have to be fulfilled in order to obtain perfect original photographs and to prepare the map in the most economical manner.

1. Each exposure should be at exactly the ideal position and altitude, determined in advance.
2. The optical axis of the lens should be exactly vertical at the instant of exposure (for vertical photography), or the angle of deviation from the vertical should be known exactly.
3. The camera should be oriented in azimuth at the instant of exposure, precisely as planned in advance.
4. There should be no forward movement of the aircraft relative to the ground during the time of exposure.
5. The camera lens should be free of distortion and otherwise optically perfect.
6. The camera should have stable and known metrical characteristics, and should be in perfect adjustment.
7. The emulsion-bearing surface of the photographic film or plate should be perfectly flat and perfectly oriented with respect to the lens at the instant of exposure.
8. The emulsion should be of uniform thickness and should give infinite resolution.
9. The film base should be perfectly stable in dimensions.
10. Atmospheric conditions while taking the photographs should be perfect.

It is obvious that even the most expert team of pilot and photographer cannot consistently maneuver the aerial camera into the exact position and elevation desired for each given exposure unless and until perfect flight-control devices are developed. The aircraft tends to drift laterally off course, and may run into head winds or be carried along by tail winds. It alternately gains and loses altitude. The nose of the aircraft dips up and down. The wings oscillate about an axis running the length of the fuselage, and the whole aircraft swings back and forth around a vertical axis.* All of these erratic movements are ac-

---

* Even in the earliest days of balloon photography, attempts were made to render the camera independent of the unavoidable oscillation of its carrier. In 1858, F. Tournachon obtained a French patent on a device to hold the camera vertical by means of a pendulum or

companied by continual vibration from the engines. There is, therefore, no practical possibility of regularly meeting requirements 1, 2, and 3 with the present operating equipment. The best that can be expected now is that the extent of the deviations from these ideal conditions can be ascertained, so that appropriate corrections can be applied. Fortunately, this is practical and has been accomplished. Each of the major photogrammetric systems has provided a method of determining the positions of the original exposure stations and the orientation of each photograph at the instant of exposure by mathematical, optical, or mechanical means.

The term "instant of exposure" is ordinarily taken to mean that the period of time during which the camera shutter remains open is practically zero. In actual practice, however, this period is a measurable quantity of time which is of great importance in photogrammetry. To illustrate, suppose an aerial camera to be set so that the shutter remains fully open for $1/200$th of a second, which is within the range of practical values for exposure time. (With the opening and closing time included, the shutter is actually at least partly open for a somewhat longer period.) This seems like a short space of time, yet during the time of exposure of an aerial photograph taken from a jet aircraft traveling at a speed of 600 miles per hour, the camera has been moved about four feet along the flight course, calculated as follows:

$$D = \frac{600 \times 5{,}280}{60 \times 60 \times 200} = 4.4 \text{ feet.}$$

The result is that every image on the photograph is "smeared" the equivalent of about four feet on the ground because of the relative movement of the camera with respect to the ground. In small-scale photography, this effect may not be appreciable, but in large-scale photography it is one of the limiting factors. The possible remedies for this effect are reduced speed of aircraft, faster emulsions and faster camera lenses, or a camera in which the film moves during exposure at a rate which compensates for the relative movement of camera and ground.

All the information that is recorded on the negative is obtained through the medium of the camera lens. The inaccuracies in the lens are carried forward in the operational procedures of compiling from the photographic data, unless some compensation is made for them. All lenses now in use have measurable distortions and

---

gimbal suspension of the camera with suitable stabilizing devices. Devices of this type, never sufficiently effective to attain the desired stability, have been gradually improved through years of research until they now approach acceptable quality, under ideal conditions.

other optical defects. It is important to strive to eliminate these lens inaccuracies, but it is even more important that the nature and magnitude of the errors in the lenses which are actually used on each project be recognized and known. It is essential, therefore, that all aerial-camera lenses be carefully tested and calibrated, and that the relationship of lens to focal plane be kept in correct adjustment.

Apart from the optical properties of the lens, the camera must be in proper mechanical working order. For example, the advantages of an excellent lens are completely lost when the film is in a warped or bent position at the instant of exposure. Unless the film or plate surface is perfectly flat and precisely perpendicular to the optical axis of the lens, the assumed geometrical relationships do not hold.

The lens and film should be of complementary characteristics and qualities, so far as is feasible. For example, a lens of high resolution is not fully effective unless used with a film emulsion of correspondingly high resolution. The advantages of a nearly distortion-free lens are likewise ineffective if the negative film emulsion "creeps" or is otherwise unstable. Photographic film has been greatly improved in recent years, but it has by no means been perfected, and its defects remain one of the major challenges in photogrammetric research.

Even if aerial photographs could be taken with a camera that was perfectly calibrated and adjusted, correctly oriented, motionless, and loaded with the best film, a perfect photograph would seldom result, because it still would be necessary to contend with atmospheric conditions. Good photographic days do occur occasionally, when the atmosphere is entirely free of clouds, haze, and smoke; but atmospheric refraction is always present, so that the light rays entering the lens are bent in varying amounts, causing the images of the ground to be distorted. The amount of this refraction has been evaluated, and for many practical purposes its effect is negligible in amount, but its existence must certainly be recognized in critical uses of the photographs, especially in oblique and other wide-coverage photography and in fundamental research.

1.3.2 PROCESSING THE DATA

The second general source of difficulties connected with the preparation of accurate maps from aerial photographs stems from the several procedures involved in making measurements on the original photographs, or negatives, and in transferring quantitative information from the photographs to the map-compilation sheet. Some of the operations which can and do frequently entail difficulties at this stage of the photogrammetric process are: developing the original negative films or plates; making positive prints from the negatives; and operating the photogrammetric plotting instruments.

The processing or developing of the original negative films or plates, particularly films, is an extremely critical operation, and requires meticulous care, thorough workmanship, proper equipment, and professional skill. The original efforts of the flight crew to secure the best possible exposures are wasted if the films are later carelessly processed and improperly developed. Not only must the chemical treatments be applied properly to bring out the maximum in image quality and to prevent "creeping of emulsion," but it is equally important that the film base be handled carefully and properly in order to reduce to a minimum the expansion, shrinkage, stretching, warping, curling, and even tearing of this material. Unskilled personnel and improper chemicals and equipment should never be used in this operation.

In most aerial survey procedures, the original negative films or plates are not used in the actual compilation. Instead, for convenience, positive copies of one type or another are used. These positives may range from ordinary prints on photographic papers to the transparent positives on glass plates required in the precise plotting instruments. These glass plates are variously called diapositive plates, lantern slides, or transparencies. They are sometimes made by exposure through a lens in a ratioing printer or reducing camera, but, in other cases, the positive plates are made by being exposed while in direct contact with the negatives. The precautions and requirements which were noted as being necessary for the best results in developing the original negative films are equally necessary in making the positive prints.

Further difficulty and loss of accuracy are possible, actually probable, in the operation of the various instruments used for transferring map details from the photographs to the map sheet. These instruments are exceptionally efficient optical and mechanical devices, but are necessarily complicated and delicately balanced, and hence can be and frequently are the source of many troubles and problems. Not the least of these problems is that of training well-qualified personnel. Although the instruments produce excellent results, it is still true that much of the effectiveness of the compilation work is dependent upon the exercise of good technical judgment by the operator of the instrument, and upon the acuity of his stereoscopic vision.

In the most advanced compilation instruments, image correlation is achieved by automated systems and the acuity of an operator's stereoscopic vision is no longer a factor; however, the advantage of automated compilation is sometimes nullified because it eliminates the opportunity for an operator to exercise his judgment in overriding anomalies in compilation that occur in automated systems.

There are, in addition, various extraneous problems and difficulties which occur during the compilation stage of an aerial survey project. An example of a major problem of this sort is the not uncommon situation in which there is a deficiency of suitable ground control. Deficiency in this case may mean insufficient quantity, poor distribution, low accuracy, or difficulty of proper identification. Suitable ground control is required for all photogrammetric surveys. The problem of acquiring this control is an important consideration affecting the possible volume of map output for each organization. This is an economic as well as a technical problem because, in some instances, the cost of control has rendered the use of certain photogrammetric methods prohibitive. The reduction to a minimum of the required amount of ground control is still a widely-pursued objective of photogrammetric research.

1.3.3 NATURE OF EARTH'S SURFACE

The character of the surface of the earth itself causes difficulties for the photogrammetrist. If it were possible to take a truly vertical photograph of flat and level terrain, and if the effects of atmospheric refraction and lens distortion were compensated, the perspective view shown on the aerial photograph would be identically the same as an orthographic projection. It would thus be a true map, and the photogrammetrist's problems would be greatly reduced. There has been, in fact, a notable and successful effort to produce modified photographs which are nearly orthographic; these orthophotographs* are coming into rapidly increasing use, but they still do not represent the conventional situation in photogrammetry.

The effect of the curvature of the earth's surface enters into many photogrammetric calculations, such as in the case of high-altitude, orbital, oblique, or other wide-coverage photographs, or in photogrammetric control-extension operations. A factor of even greater importance, however, is relief of the terrain. To illustrate how the difference in elevation displaces the position of a photographic image, consider an aerial camera with lens at point O (fig. 1-46) and the tower BC rising above flat ground surface BM. In nature, $MB = NC$, but on the negative these distances are recorded as $M'B'$ and $M'C'$, respectively, the distance $M'C'$ being inconsistent in scale with other distances on the negative because of the elevation of C above the surrounding terrain. It should be noted, however, that the displacement of photographic images because of relief is the basis for determining elevations and contours from the photographs.

The aerial photograph differs essentially from

---

* See chapter XV.

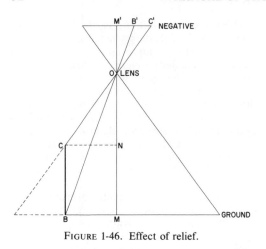

FIGURE 1-46. Effect of relief.

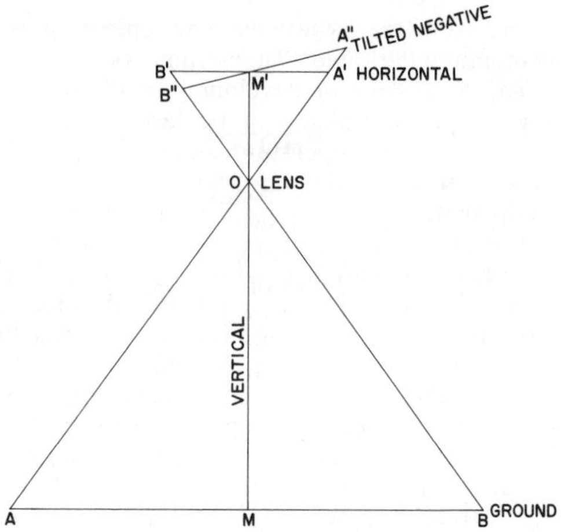

FIGURE 1-47. Effect of tilt.

a map because the negative, at the time of exposure, is seldom parallel to the ground it represents, and because relief of the terrain causes images of points on the ground surface to be displaced from their orthographic, or map, position. Because of these conditions, the photograph may have an infinite number of scales. The effect of relief has already been demonstrated. The effect of tilt is indicated in fig. 1-47. The ground distance MB, which projects on a horizontal negative plane as $M'B'$, becomes $M'B''$ on the tilted negative, while MA which projects on the horizontal negative plane as $M'A'$, becomes $M'A''$ on the tilted negative.

1.3.4 SOLUTIONS FOR THE PROBLEMS

The foregoing problems are some of the major ones which confront the photogrammetrist in his efforts to carry out accurate surveys from aerial photographs. Much of the material in this Manual is devoted to the solution of these problems. The descriptions of these basic problems are not intended to have a discouraging effect nor to paint a pessimistic picture of the future of photogrammetry. Actually, the picture appears quite optimistic, as remarkable results are being achieved in spite of the handicaps of imperfect source materials and of other difficulties which have been described. These results could be better in many cases, however, and should be attained with less effort and at lower costs. It is well for all those connected with or interested in photogrammetry to be aware of these problems and shortcomings in order to avoid disillusionment and to be better prepared to help solve the problems.

1.4 Products of Photogrammetry

Photogrammetric techniques are used to bring forth an almost endless variety of useful products that convey information about given surfaces or objects. These products can be grouped in five general categories: conventional maps, photomaps, numerical data, data from remote sensors, and special products.

1.4.1 CONVENTIONAL MAPS

A very important and perhaps the best known application of photogrammetry is in the field of topographic surveying and mapping. Although photogrammetry is only a part of the total activity of mapping, it forms the heart of the operation because it is the means of performing the actual delineation of the map details, and it is also a means of triangulating for intermediate control. It is still necessary to obtain geodetic control in the field and to inspect, check, and complete the photogrammetric compilations by field surveys; but the total labor for field surveys has been greatly reduced.

The aerial photograph with tilt and relief included replaces the field notebook of the planetable topographer. Very little geometrical data can be obtained reliably from a single photograph; but from two overlapping photographs and certain control points located in the overlap area, a topographic map of the area included within the overlap can be compiled by means of stereoscopic compilation instruments.

The conventional map compiled by photogrammetry represents cartographic detail by means of standard symbols. This has the advantage of including the compiler's interpretation of the data from which the map is compiled. On the other hand, it may sometimes have the disadvantage of omitting detail desired by some map users or emphasizing various details in a proportion not in accord with the needs of map users.

1.4.2 PHOTOMAPS

A photomap differs from a conventional map in that the detail is shown photographically instead of by conventional symbols. A photomap has the basic attributes of a map: a known scale and a known orientation system. It may be cartographically enhanced by the addition of standard map symbols, names, and other data.

Photomaps are generally derived from conventional (perspective) photographs that have been rectified to remove image displacements caused by camera tilt and ground relief. If the image displacements remaining after rectification are negligible, the rectified photograph is called an orthophotograph. The rectification procedure that removes only the effect of tilt is called simple rectification; the product of simple rectification is not an orthophotograph unless the terrain is relatively flat. The rectification procedure that removes the image displacements caused by both tilt and relief is called differential rectification; in this procedure, the perspective photograph is usually scanned in such a manner that tilt and relief corrections are made for each differential element of the photograph and each differential image is exposed at the corrected position. The product of correctly executed differential rectification is an orthophotograph.

Orthophotographs may be printed at a predetermined scale and assembled to form an orthophotomosaic. An orthophotoquad is produced by bringing an orthophotograph or orthophotomosaic to a defined scale, relating it to a geodetic reference system, and providing a map border in quadrangle format. By applying color to the photoimagery, using appropriate colored inks and masking techniques, ground features can often be shown in more recognizable colors than nature provides; the color-enhanced version of the orthophotoquad is called an orthophotomap. The orthophotomap also has cartographic symbols—including contours, elevations, boundaries, and labels—tailored to suit the area and intended use of the map.

The subject of photomaps is treated in detail in chapter XV.

1.4.3 NUMERICAL DATA

Advances in using and processing numerical data have, to a large extent, revolutionized photogrammetric procedures and products in the 1966–79 period. Whereas formerly the end product of photogrammetric processes was usually a map, a chart, a measurement or a photographic image, the modern product is often an array of coordinates, a collection of data, or a numerical model.

The production of numerical data by photogrammetric methods is by no means entirely new, for photogrammetry has long been used to derive the x, y, z coordinates of desired image points by aerotriangulation. Likewise, distances, elevations, areas, volumes, and the like have been determined photogrammetrically for many years. The difference lies in the great shift that has been made from showing terrain data by graphical means (principally maps) to storing the same data digitally for automatic processing as needed. The numerical map thus exists only in a latent state—in computer programs and stored data—until the particular information needed is called for, at which time the automated output system produces it in the form desired, graphically as a map, or digitally as a printed array of numbers.

1.4.4 REMOTE-SENSOR DATA

The kinds of products which are obtained from remote-sensing systems can perhaps best be understood by reviewing the definition of remote sensing as given in the MANUAL OF REMOTE SENSING:

In the broadest sense, the measurement or acquisition of information of some property of an object or phenomenon, by a recording device that is not in physical or intimate contact with the object or phenomenon under study; e.g., the utilization at a distance (as from aircraft, spacecraft, or ship) of any device and its attendant display for gathering information pertinent to the environment, such as measurements of force fields, electromagnetic radiation, or acoustic energy. The technique employs such devices as the camera, lasers, and radio frequency receivers, radar systems, sonar, seismographs, gravimeters, magnetometers, and scintillation counters.

Although remote sensors can produce numerical data, in practice, the bulk of the data is in the form of imagery. This imagery may be color, color infrared, or multiband photography; multispectral scanner imagery; or imagery produced from electro-optical or electronic devices. The imagery can be enhanced or manipulated by the interpreter, using various techniques such as automatic classification or theme extraction to display specific features such as water, snow, vegetation, or works of man. These techniques permit the extraction of information that cannot be derived by conventional photointerpretation techniques. The MANUAL OF REMOTE SENSING has separate chapters explaining how remote sensing can be used to develop information on crops and soils, forestlands, terrain and minerals, water resources, range resources and wildlife, urban studies, regional planning, engineering site selection and route surveys, marine environment and resources, archaeology and demography.

1.5 Various Applications of Photogrammetry

Although the principal use of photogrammetry at present is still in mapmaking, particularly in topographic mapping, other applications are steadily increasing in importance. Whatever the purpose of the photogrammetric work, the general principles remain the same. Equipment designed for one particular purpose is frequently adapted with ease for use in other fields. A few of the well-known applications of photogrammetry, other than mapping, are as follows.

Geology.—Geologists and geophysicists apply photogrammetry to structural-geology studies, investigations of water resources, analysis of thermal patterns on the Earth's surface, studies in geomorphology including investigations of shore features, engineering geology of investigations, stratigraphic studies, general geologic mapping, study of luminescence phenomena, and the recording and analysis of "catastrophic" events, such as earthquakes, floods, and eruptions.

Forestry.—Aerial photographs are used as a basis of timber inventories, "cover maps," acreage studies, and fire control. The U.S. Forest Service has developed instruments and techniques which are especially suited for the particular needs of that organization.

Agriculture.—The U.S. Department of Agriculture has a continuing program of photographic coverage of large areas of American farmland. These photographs are used for such purposes as the study of soil types, soil conservation, crop planting, crop diseases and damage, and crop-acreage determinations.

Design and construction.—Much of the data required in connection with site and route studies, particularly when several alternate schemes are being considered, may be obtained quickly and accurately by photogrammetric means for use in the design and construction of dams, bridges, transmission lines, and other projects of a similar nature.

Planning of cities and highways.—State highway departments make extensive use of aerial photography and photogrammetry in the selection of new highway locations and in detailed design for construction contracts. Slum-clearance authorities, waterfront development agencies, park commissions, and street and sewer departments benefit from photogrammetry in the planning of civic improvements.

Cadastre.—Photogrammetry has a wide application in the solution of cadastral problems. Many counties and municipalities use aerial photographs, orthophotographic maps and large-scale aerial maps as a basis for determining landlines, for assessments, and for real estate tax studies. In Europe aerial photography and photogrammetry are used extensively in preparing large-scale cadastral plats, which are used for the reapportionment of land.

Environmental studies.—To develop an intelligent program of environmental management, planners need a reliable means of land-use analysis. Photogrammetry provides a vital tool for such studies. In recent years, the potential for land-use analysis and other studies related to environmental management has been greatly enhanced by the use of sophisticated technologies such as high-altitude-aircraft photography, satellite photography, remote-sensing techniques, and revolutionary new methods for interpretation of data.

Exploration.—Imagery from aircraft and space vehicles, and photogrammetric analysis of the imagery, are among the most important tools of modern exploration. Wide-coverage types of photography and frequently repeated near-worldwide imagery and other data obtained via spacecraft are helping to unlock the secrets of the last unexplored areas on earth. Moreover, space vehicles have been launched to carry data-collecting systems which provide information for exploring the Moon, Mars, Jupiter, and Venus.

Military intelligence.—The first practical application of photogrammetry was probably in the field of military reconnaissance, and its importance is well recognized as a key element of military science. Photographic reconnaissance and photogrammetric analyses are accepted methods of determining the deployment of forces, obtaining detailed information for planning of maneuvers, assessing the effect of operations, and solving military problems involving topography, terrain conditions, or works.

Medicine and surgery.—Precise stereoscopic measurements on the human body, often by X-ray photography, are utilized in the diagnosis and treatment of certain medical conditions. Typical applications are in the location of foreign matter lodged in the body and in the location and examination of fractures and growths. Medical applications entail the use of close-range photogrammetric systems in which the cameras are placed at short distances from the subject, in contrast to the thousands of feet usually separating an aerial camera from the ground in obtaining photography for an aerial survey. These applications have given rise to the discipline of "Biostereometrics," which was the subject of a specialized ASP symposium in 1974, with proceedings published in a separate volume.

Miscellaneous.—There are many other applications of photogrammetry, such as in the fields of crime detection, traffic studies, oceanography, meteorological observations, architectural and archaeological surveys, and visual education. Among the more unusual applications are a method used by tailors in obtaining customers' measurements for individually tailored suits, and a method of contouring beef cattle for animal husbandry studies. Miscellaneous applications such as these are treated in depth in chapter XVI, Non-Topographic Photogrammetry, which has also been published separately as the HANDBOOK OF NON-TOPOGRAPHIC PHOTOGRAMMETRY.

1.6 Future Role of Photogrammetry

Photogrammetry is a comparatively modern science; many of its problems are still to be solved and many uses for it are yet to be discovered. Yet even as classical photogrammetry approaches a mid-stage of development, the very nature of the science is in a state of radical change. Classical photogrammetry based on conventional photography is yielding some of its preeminence under the impact of sensing systems other than photography. There is growing recognition of the fact that photography is only one of a number of remote-sensing systems. The various systems are subject to handling in a manner analogous to the handling of classical systems, based on conventional photography; but each system entails its own parameters which require specialized treatment, as explicitly documented in the MANUAL OF REMOTE SENSING.

Developments in the science of photogrammetry that may reasonably be anticipated in the near future include:

increased use of high-altitude aircraft and space vehicles for the acquisition of basic imagery and other data;

vastly increased use of sensors other than optical photography;

further improvements in the quality of imagery from the various kinds of sensors;

reduction to practice of inertial positioning and height-finding systems to give the precise position of a sensor at any desired instant;

reduction in the amount of ground control needed for surveys, especially by the use of inertial position and height-finding systems in the field and the automation of analytical aerotriangulation in the office;

reduction to practice of automated photogrammetric systems which produce orthophotographs, contours, and digital terrain models simultaneously;

development of advanced methods for manipulation of imagery from various sensors to accomplish theme extraction, automatic classification, or derivation of specialized information;

increasing applications of automation techniques, particularly in the development of integrated, automated cartographic systems in which the photogrammetric system is a central element of a continuously automated mapping process; and

development of new kinds of products in graphical, digital, or image form.

In reviewing these forecasts, it is evident that the key words and phrases in the future of photogrammetry are: inertial positioning, automation, sensing systems, digital systems, image quality, image manipulation, and space systems. There can be no doubt that advances and improvements in the techniques and instruments of photogrammetry will continue into realms beyond our surmise in 1979. Yet, though the state of the art may advance through a succession of new frontiers, the basic truths presented in this MANUAL will continue to be fundamental to an understanding of the science of photogrammetry.

REFERENCES

Adams, L. P., 1975, Henry Georges Fourcade Memorial: *Photogrammetric Record*, v. 8, no. 45: 287–297.

American Geophysical Union, 1970, *Geodesy and Aerotopography*, alphabetical index of Russian edition. American Geophysical Union, Washington, D.C.

Andrews, G. S., 1976, Edouard Gaston Daniel Deville (memorial lecture): *Canadian Surveyor*, 3(1): 36–41.

Doležal, E., 1932, *Festschrift Österreichischer Verein für Vermessungswesen*, Vienna.

Doyle, F. J., 1964, The historical development of analytical photogrammetry: *Photogrammetric Engineering*, v. 30, no. 2: 259–265.

Finsterwalder, S., 1937, *Festschrift, Deutsche Gesellschaft für Photogrammetrie*: Verlag Herbert Wichman, Karlsruhe.

Goddard, G. W., 1969, Overview—A lifelong adventure in aerial photography: Doubleday, New York.

Gruner, H., 1929, Der Aerokartograph: Konrad Witwer, Stuttgart.

———1971, Reinhard Hugershoff (memorial lecture): *Photogrammetric Engineering*, v. 37, no. 9: 939–947.

———1972, Colonel Claude H. Birdseye (memorial lecture): *Photogrammetric Engineering*, v. 38, no. 9: 869–875.

Helava, U. V., 1958, New principle for photogrammetric plotters: *Photogrammetria*, v. 14, no. 2: 89–96.

Hugershoff, R., 1930, Photogrammetrie und Luftbildwesen: Band VII, Handbuch der wissenschaftlichen und angewandten Photographie, Julius Springer, Vienna.

Kelsh, H. T., 1940, The slotted templet method: U.S. Department of Agriculture Publication No. 404, U.S. Government Printing Office, Washington, D.C.

Landen, D., 1952, History of photogrammetry in the United States: *Photogrammetric Engineering*, v. 18, no. 5: 854–898.

MacLeish, K., 1977, Leonardo da Vinci: A man of all ages: *National Geographic*, v. 153, no. 3: 296–319.

McNeil, G. T., 1954, Photographic measurements: Pitman Publishing Corp., New York.

National Research Council, 1952, History of Photogrammetry in Canada: NRC Publication no. 2809, Ottawa.

Nistri, U., 1952, History of Photogrammetry: Stabilimento Tipografico M. Danesi, Rome.

Pestrecov, K., 1954, Notes on Russian photogrammetric optics: *Photogrammetric Engineering,* v. 20, no. 3: 488–492.

Price, W. H., 1976, The photogrammetric lens: *Scientific American,* v. 235, no. 5: 72–83.

Quinn, A. O., 1975, Professor Earl Church (memorial lecture): *Photogrammetric Engineering,* v. 41, no. 5: 595–602.

Santoni, E., 1971, Selected works, (special edition): Societa Italiana di Fotogrammetria, Florence.

Scheimpflug, T., 1956, *Festschrift, Österreichischer Verein für Vermessungswesen,* (special edition), Vienna.

Schermerhorn, W., 1964, Jubilee volume: International Training Center for Aerial Survey, The Delft, Netherlands.

Schöler, H., 1961, Bemerkungen zum Stereoprojektor Romanowski SPR-2: *VEB Verlag für Bauwesen,* v. 9, no. 10, Berlin (GDR).

Schwidefsky, K., 1935, Das Entzerrungsgerät: Verlag Herbert Wichmann, Karlsruhe.

———1959, An outline of photogrammetry: Sir Isaac Pitman and Son, London.

Thompson, E. H., 1964, An essay on analytical photogrammetry: Contained in Schermerhorn 1964.

Thompson, V. F., 1908, Sterophoto-Surveying, *The Geographical Journal,* v. 31, May 1908: 534–557.

Tubis, H., 1976, The Brock brothers and the Brock process (memorial lecture): *Photogrammetric Engineering,* v. 42, no. 8:1017–1034.

United Nations, 1966, U.N. Regional Cartographic Conference for Asia and the Far East, v. 2.: United Nations, New York.

von Gruber, O., 1932, Photogrammetry, collected lectures and essays: Chapman and Hall, London. Also 1942, American Photographic Publishing Co., Boston.

Weibrecht, O., 1960, Über die Möglichkeiten zur Erfüllung der Perspektivbedingungen an Entzerrungsgeräten: *Jenaer Jahrbuch,* VEB Carl Zeiss, Jena.

Whitten, C. A., 1973, Martin Hotine (memorial lecture): *Photogrammetric Engineering,* v. 39, no. 8: 821–828.

CHAPTER II

Basic Mathematics of Photogrammetry

Author-Editor: K. W. WONG

2.1 Introduction

PHOTOGRAMMETRY, as a discipline of applied science, employs a wide spectrum of mathematical principles and methods which include but are not restricted to the following: analytic and solid geometry, calculus, probability, statistics, vector algebra, linear algebra, and numerical analysis. Methods of analytic and solid geometry and vector algebra are used to describe the fundamental geometric relationships between a photograph and the object field. Applications of the methods of calculus range from simple problems such as coordinate transformation, polynominal modeling and linearization of equations, to highly complex problems dealing with orbital motion of artificial satellites and wave motion in holography. Theory of probability and statistical analysis methods such as regression analysis, analysis of variance, cluster analysis and classification are used extensively in the adjustment of redundant measurements, multispectral analysis and image correlation. In computation problems which involve the use of electronic computers, linear algebra and numerical analysis methods are needed to develop efficient computing algorithms and computer programs.

It is beyond the scope of this chapter to present a thorough treatment of all the mathematical principles and methods that have been applied in photogrammetry. Many textbooks have already been written on each of the subjects mentioned above. Therefore, this chapter is devoted primarily to the basic mathematics that are most commonly used in practice. Including the introduction, this chapter consists of six sections. The fundamental geometry of a photograph will be discussed in section 2.2. One section is then devoted to each of the following topics: matrix algebra, probability and statistics, and methods of least squares. Finally, in section 2.6, a general mathematical solution for the least squares adjustment of photogrammetric blocks will be developed.

The contents of this chapter represent a major expansion from the same chapter in the previous edition (3rd) of the MANUAL OF PHOTOGRAMMETRY. Nevertheless, much of the materials that were included in the previous edition have been preserved here with little or no modification. The section on the geometry of the photograph and on matrix algebra are taken directly from the 3rd edition with only some minor revision and deletion. Due credits should therefore be extended to the authors of chapter II in the 3rd edition; namely: G. C. Tewinkel (author-editor), Dr. Helmut H. Schmid, Dr. Bertil Hallert, and George H. Rosenfield.

2.2 Geometric Principles

The fundamental characteristic of a photograph is that each image point on the photograph corresponds to a unique point in the object scene. A definite geometric relationship exists between the relative spatial positions of the image points in the two-dimensional photograph and their corresponding positions in the three-dimensional object space. This section will describe the fundamental geometry of both vertical and tilted aerial photographs.

2.2.1 THE VERTICAL AERIAL PHOTOGRAPH

An aerial photograph is said to be vertical if the camera axis is plumb or nearly so. If the axis is exactly plumb, the photograph is said to be truly vertical and to have zero tilt. Vertical photographs constitute the predominant mode of photography used for topographic mapping. The layman is often impressed with the striking resemblance of a vertical aerial photograph to a map or plan view of the same area. A "vertical" aerial photograph is invariably tilted accidentally a small amount in practice; nevertheless, the principles expressed here are very useful in dealing with "near" verticals and also serve to demonstrate important photogrammetric characteristics which might be obscured if perfectly general conditions were assumed.

2.2.1.1 SCALE OF A VERTICAL PHOTOGRAPH

The scale S of a vertical aerial photograph is stated by the equation (fig. 2-1),

$$S = \frac{1}{H/f} \quad (\text{or } S = f/H) \qquad (2.1)$$

where f is the focal length of the aerial camera and H is the flight height above mean ground elevation. The values of f and H must be expressed in the same units of measurement. The result is a representative fraction corresponding to map scale.

EXAMPLE 1: Determine the scale of a vertical aerial photograph if the focal length of the camera is 203.2 mm, the flight altitude is 4,250 metres above sea level, and the terrain is about 275 m above sea level.
SOLUTION:

$$S = \frac{1}{\dfrac{4,250 - 275}{0.2032}}$$

$$= \frac{1}{19,560}.$$

Or, the scale is 1:19,560. (Caution: In equation (2.1), H is the flight height. If flight altitude is given, subtract the ground elevation to obtain H).

Although the scale of a photograph (or a map) is customarily referred to in conversation by a whole number, such as 20,000, this number is actually the denominator of an arithmetic fraction whose numerator is 1. Consequently, a large number indicates a small scale. Large scale refers to the value of the entire fraction, which could be obtained for comparison by expressing the scale as a decimal fraction instead of an arithmetic fraction. For example, if the scale fractions 1/11,000 and 1/12,000 are considered, their values in decimal form are 0.000,091 and 0.000,083, respectively, and hence the former is said to be the larger of the two scales. Moreover, an image of a given object is actually larger on the larger scale photograph (or map).

Equation (2.1) indicates that the scale of a vertical aerial photograph depends on the relative values of the factors f and H. For a given flight height H, the longer the focal length, the larger will be the scale of the photograph. Conversely, for a camera of a given focal length, the higher the flight altitude, the smaller will be the scale of the photograph. For example, at a flight height of 3,000 m, a 300-mm focal length camera will yield a photographic scale two times (2×) as large as that provided by a 150-mm focal length camera; and for a 300-mm focal length camera, the photographs taken at a flight height of 3,000 m will have a scale two times as large as those taken at a flight height of 6,000 m.

The scale of a vertical aerial photograph can also be determined by using the following simple formula:

$$S = \frac{1}{AB/ab} \quad (\text{or } S = \frac{ab}{AB}) \qquad (2.2)$$

in which ab (scale-check line) is the measured distance between two image points, a and b, on the photograph, and AB is the corresponding distance on the ground where A and B are of equal elevation or where the difference in elevation is small enough so that it can be safely assumed to have no significant effect. The two distances must be expressed in the same units of measurement.

EXAMPLE 2: The distance between two points on a vertical aerial photograph is 178 mm and the corresponding distance between the two objects is 6 kilometres. What is the scale of the photograph?
SOLUTION:

$$S = \frac{1}{\dfrac{6,000}{0.178}} = \frac{1}{33,708} \quad \text{or } 1:33,708.$$

Equation (2.2) assumes that the line ab on the photograph is parallel to the line AB on the ground; that is, the photograph is perfectly vertical and that points A and B have the same elevation on the ground. These conditions are rarely satisfied in practice. Therefore when equation (2.2) is used to compute the average scale of a near-vertical photograph, the scale should be based on the average of two scale-check lines instead of on one; and it is preferable to have lines that are relatively long, that intersect approximately at right angles, and that are centrally located on the photograph.

The scale of a contact print which is illustrated in figure 2-1 has, by definition, the same scale as the photographic negative. Photographic enlargement or reduction at a ratio M has exactly the same geometric effect on an aerial photograph as taking the photograph with a camera the focal length of which is M times f. But because of topographic relief, the effect is not identical to taking the photograph from a flight altitude H/M.

The flight height, H, is ordinarily controllable in flight to within only 1 percent. Therefore, there usually are small variations in the scales of the photographs obtained in any given photographic mission. Moreover, because of topographic relief and the unavoidable presence of small tilts, the exact scale varies from image point to image point on a single photograph. For making photo mosaics, the method of rectification is used to correct for the effects of tilts so that all the photographs can be corrected to have relatively uniform average scale. The principles and methods of rectification will be discussed in chapter XIV.

The magnitude (d_s) of image displacement caused by scale difference may be expressed by the following relationship:

$$d_S = r_p \left(\frac{S_c}{S_p} - 1\right) = r_c \left(1 - \frac{S_p}{S_c}\right)$$

where r_p is the radial distance from the principal point of the photograph to an image point; S_c is

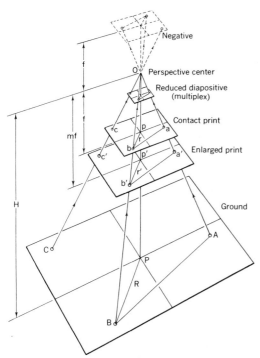

FIGURE 2-1. Relationship of various forms of an aerial photograph.

the desired (or correct) scale of the photograph; S_p is the actual scale of the photograph; and r_c is the corresponding radial distance on a photograph of the desired scale. Letting

$$M = S_c/S_p = r_c/r_p,$$

the above equation may be rewritten as follows:

$$d_S = r_p (M - 1) = r_c (1 - 1/m). \qquad (2.3)$$

EXAMPLE 3: What is the scale displacement of an image 75 mm from the center of a photograph if the photograph scale is 1:41,000 when it should have been 1:40,000?

SOLUTION:

$$M = \frac{1/40,000}{1/41,000} = 1.025$$

$$d_S = 75(1.025\text{-}1) = 1.88 \text{ mm.}$$

It can be seen from equation (2.1) that the flight height is equal to the focal length times the denominator of the scale fraction; *i.e.*

$$H = f(1/S) = fS^{-1}. \qquad (2.4)$$

For example, if the focal length is 152 mm and the scale is 1:20,000, the required flight height above the ground is

$$H = 152 \text{ mm} \times 20,000 = 3,040,000 \text{ mm}$$
$$= 3,040 \text{ m.}$$

2.2.1.2 THE EFFECTS OF TOPOGRAPHIC RELIEF

One of the ways in which a photograph differs from a map is in the effect caused by the ground

being hilly or mountainous instead of flat. Equation (2.1) indicates that the scales of photographic images vary in accordance with the elevations of the corresponding objects on the ground; hence features on the top of a mountain are imaged at a larger scale than features in a valley nearby because the mountain is closer to the camera. On the contrary, all features on a *map* are shown at the same scale, regardless of their relative elevations. The phenomenon is known quite commonly. A familiar trick of amateur photography is to take a close-up photograph of a person in a seated position showing his feet in the foreground and his head in the background so as to give the impression of two huge feet and a relatively small head. As with the mountain top, the image of the feet is larger than that of the head because the former is closer to the camera. This is also what the eyes actually see, but the eyes usually do not focus on both the head and the feet at the same time, and furthermore the mind is familiar with this condition of perspective and makes an appropriate interpretation.

The effect of relief can also be considered as a component of image displacement. Figure 2-2 illustrates image displacement in which AA' is an elevated object on the ground not vertically beneath the camera, and aa' is the image. The image a is of the top A of the object, and a' is the image of the base. Moreover, a' is presumably the correct position of the image, hence the distance from a to a' is the displacement of the image caused by the height h of the object. In

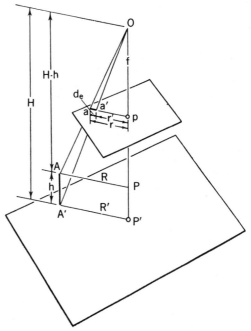

FIGURE 2-2. Displacement of the photograph image due to the elevation of the object.

practice, the object A is more often a station on gently rolling elevated ground, hence the base A' is not visible and its image does not appear on the photograph. Nevertheless, the image a is considered as displaced relative to its correct map position.

The *direction* of relief displacement is radial with respect to the *nadir* point, regardless of intentional or accidental tilt of the aerial camera. This is a fundamental concept of photogrammetry. The nadir point coincides with the principal point p (geometric center) in the case of a truly vertical aerial photograph. Hence, the principal point is frequently considered to be the radial center for relief displacement on near-vertical photographs—those that are intended to be vertical.

The *magnitude* d_e of relief displacement on a truly vertical photograph is expressed by the equations:

$$d_e = rh/H \tag{2.5}$$
$$d_e = r'h/(H - h). \tag{2.5a}$$

The parameters in the equations are indicated in the figure: r is the distance on the photograph from the center to the image of the top of the object, r' is the corresponding distance for the base of the object, h is the ground elevation of the object, and H is the flight altitude of the camera relative to the same datum as h. If equation (2.5) is solved for h,

$$h = d_e H/r. \tag{2.6}$$

Hence it is possible, by this method, to determine elevations of objects from measurements made on photographs, although a different method (section 2.2.1.3) is usually employed in practice.

Considerable latitude exists with respect to the units of measurement employed in equations (2.5), (2.5a), and (2.6). In the first two, h and H must be expressed in the same units, but r (or r') can be in different units, and d_e will result in the same units as r (or r'). In equation (2.6), d_e and r must be in the same units of measurement but H can be in different units, and h will be in the same units as H.

The principal use of equation (2.5a) is to make a templet for use in optically rectifying a tilted photograph. A map or radial plot furnishes the distance r', from which d_e can be computed if h is known. Then the displacement can be applied graphically to the map position to arrive at the correct position of the photographic image. The value of H in this equation is determined from equation (2.1).

EXAMPLE 4: An image of a hill is 90 mm from the center of a photograph. The elevation of the hill is 2,000 metres and the flight altitude is 10,000 metres with respect to the same datum level. How much is the image displaced because of the elevation of the hill?

SOLUTION: Applying equation (2.5),

$$d_e = \frac{90 \times 2,000}{10,000} = 18 \text{ mm.}$$

EXAMPLE 5: A radial plot at a scale of 1:20,000 shows an image 100 millimetres from the position of the center of a photograph and the elevation of the corresponding object is known to be 305 m. If the focal length of the camera is 152.4 mm, what is the relief displacement needed to change the plotted point for use in rectification?

SOLUTION: Using equation (2.1) to find H,

$$H = \frac{f}{S} = \frac{0.1524 \text{ metres}}{\dfrac{1}{20,000}} = 3,048 \text{ metres.}$$

Then, applying equation (2.5a),

$$d_e = \frac{100 \times 305}{3,048 - 305} = 11.12 \text{ mm.}$$

EXAMPLE 6: In planning aerial photography, the flight map indicates that the base position of a 1,500 m mountain will appear 75 mm from a flight line at the anticipated scale of photography. If the flight altitude is to be 6,000 m above the datum, how far from the base position will the image of the mountain top occur, and how near the edge of a 228.6 mm × 228.6 mm photograph will the image be?

SOLUTION: Using equation (2.5a), the image displacement is

$$d_e = \frac{75 \times 1,500}{6,000 - 1,500} = 25 \text{ mm.}$$

The distance from the edge of the photograph to the image is

$$114.3 - 75 - 25 = 14.3 \text{ mm.}$$

EXAMPLE 7: A comparison of a two-diameter enlargement of an aerial photograph with a radial plot at the same scale shows that an image 178 mm from the center is displaced 35 mm due to the elevation of the object. If the aerial camera was actually flown at an altitude of 6000 m above the datum, what was the elevation of the object?

SOLUTION: Applying equation (2.6)

$$h = \frac{35 \times 6,000}{178} = 1,179.8 \text{ m.}$$

Although equation (2.5) is a relatively simple relation, it is sometimes misunderstood. Large relief displacements are objectionable in pure planimetric mapping by graphic methods, whereas large displacements are generally advantageous for contouring with stereoscopic plotting instruments. The most effective way to control relief displacement is to select the proper flight altitude. This is indicated by the presence of H-squared in the differential form of the equation, regarding d_e and H as variables:

$$dd_e = -\frac{rh}{H^2} dH.$$

Thus, the best theoretical way to control d_e is to select an adequate value of H, to use whatever focal length is best with respect to ground-control stations, to control the scale of the resulting photographs by enlargement in the laboratory, and to control r by properly planning the overlaps and flight-line spacings.

2.2.2 THE TILTED AERIAL PHOTOGRAPH

Nearly all so-called vertical photographs obtained for mapping purposes are taken with the camera axis accidentally tilted a small though significant angular amount away from the true vertical. Under good operating conditions, with existing methods, about 50 percent of the vertical photographs taken for domestic mapping are tilted less than 2°, and very few are ordinarily tilted more than 3°. The effect of tilt is most noticeable where precise elevation measurements are to be made from aerial photographs. Hence the design of the better stereoscopic plotting instruments incorporates the very involved and complicated mechanisms needed to correct properly for tilt. Even planimetric maps of flat terrain cannot be traced directly from aerial photographs without making some compensation for the effect of tilt. The lack of a mechanism which can hold the camera axis in the perfectly vertical direction while the aircraft is in flight is the principal causes of tilt. Level vials are not dependable inasmuch as they are affected by inertia forces created by normal air turbulence and by the pilot keeping the aircraft on a straight and level course. Various gyroscopic controls have been proposed whereby it may eventually be possible to confine the amount of tilt to 15 minutes of arc or less.

The definitions and nomenclature of a tilted photograph are shown in figure 2-3. The figure illustrates some commonly used terms for tilted photographs.

A mark, called a *fiducial mark,* usually shows in each of the four borders of an aerial photograph. The positions of the fiducial marks are adjusted in the aerial camera so that the intersection of lines through opposite fiducial marks identifies the intersection of the optical axis of the camera with the image plane; hence, the images or shadows of the marks on a photograph indicate the location of the *principal point.* The points O and O' denote the incident and emergent nodes, respectively, of the aerial camera lens. The dotted portion of the figure pertains to the camera and the negative film. Inasmuch as some form of the printed photograph is nearly always employed in photogrammetry, the lower part of the figure will almost always be referred to, and the dotted part will not be shown in subsequent diagrams. No loss of generality or accuracy is imposed on the mathematical formulas by such a procedure.

The point O is the *perspective center,* or point of perspective, of the photograph. The line Op represents the optical axis, which is perpendicu-

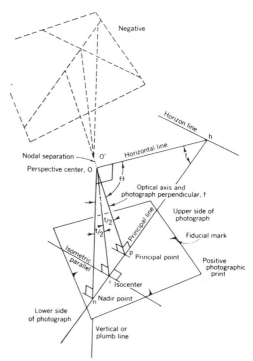

FIGURE 2-3. Nomenclature for a tilted photograph.

lar to the plane of the photograph at the principal point p or geometric center. Thus, every photograph has a perspective center that is geometrically just as much a part of the photograph as the images themselves. The length of the line Op is correctly called the *principal distance* of the photograph. The principal distance of a contact print made from distortionless negative aerial film and on distortionless paper is equal to the focal length of the aerial camera. An enlarged print has a principal distance equal to the focal length of the camera times the factor of enlargement or magnification. The letter f is used here to signify the principal distance of a photograph whether or not the distance is equal to the focal length of the taking camera.

The angle of *tilt,* of an aerial photograph can be defined in two ways: (1) tilt is the angular deviation pOn of the photograph perpendicular from the vertical; (2) tilt is the dihedral angle Ohp between the plane of the photograph and a horizontal plane. The vertical line or plumb line through the perspective center O pierces the plane of the photograph or its extension at the *nadir point n.* The nadir point is always on the depressed or lower side of the printed photograph with respect to the principal point. The *principal line* is in the plane of the photograph and passes through the principal point and nadir point. Incidentally, the principal line is oriented in the direction of the steepest inclination of the photograph: if a marble is released at the principal point of a photograph in its tilted position, the marble will roll down the principal line through the nadir point. The plane *Ohpin* is the *principal plane*. The principal plane is a vertical plane containing the optical axis, and therefore is: (1) perpendicular to the plane of the photograph; (2) determined by three points—the principal point, the perspective center, and the nadir point; and (3) determined by any two of the three lines—the photograph perpendicular, the plumb line, and the principal line. The principal plane also includes the horizon point and the isocenter. Obviously, the principal line can also be defined as the line of intersection or trace of two intersecting planes—the principal plane and the plane of the photograph.

The bisector of the angle *t* or *pOn* at the perspective center pierces the plane of the photograph at the *isocenter i* on the principal line. The *isometric parallel* is in the plane of the photograph and is perpendicular to the principal line at the isocenter. The isometric parallel is a horizontal line inasmuch as all lines perpendicular to the vertical principal plane are horizontal. The isometric parallel is sometimes called the "axis of tilt." In this edition, the term "axis of tilt" is restricted to mean the line which passes through the perspective center and is perpendicular to the principal plane.

An equivalent vertical photograph is the theoretical, truly-vertical photograph taken at the same camera station *O* with a camera whose focal length is equal to that of the corresponding tilted photograph. The isometric parallel is thence also the line of intersection of the plane of the tilted photograph with the plane of the equivalent vertical photograph. The scale of an equivalent vertical photograph is simply $S = f/H$. Inasmuch as the isometric parallel is a line that is common to both the equivalent vertical photograph and the tilted photograph, the scale along the isometric parallel and at the isocenter of a tilted photograph is provided by the same simple expression.

The term *upper side* is used here with respect to a tilted photograph to indicate the raised half (or thereabouts) of the printed photograph (not of the negative). The isometric parallel is usually regarded as the line of division between the upper and lower sides.

The horizontal plane through the perspective center intersects the plane of a tilted photograph or its extension in the *true horizon line*. The *horizon point h* is the intersection of the horizon line and the principal line. The tilt of an oblique photograph is often indicated by the depression angle *pOh* of the optical axis below the horizontal, denoted by θ. Thus, θ and *t* are complementary angles: $\theta + t = 90°$. A few of the more useful trigonometric relations are:

$$pn = f \tan t = f \cot \theta$$
$$pi = f \tan (t/2)$$
$$ph = f \cot t = f \tan \theta$$
$$On = f \sec t = f \csc \theta$$

$$Oh = hi = f \csc t = f \sec \theta$$
$$hn = f(\tan t + \tan \theta) = f(\tan t + \cot t)$$
$$= f(\tan \theta + \cot \theta)$$
$$hn = f \sec t \csc t = f \sec \theta \csc \theta$$
$$= f \sec t \sec \theta = f \csc t \csc \theta.$$

If the tilt of a near-vertical photograph is less than 5°, the values of *pn* and *pi* can be computed from approximate equations with a maximum error of about 1 part in 300:

$$pn \approx f \sin t$$
$$pi \approx \tfrac{1}{2}pn. \qquad (2.8)$$

EXAMPLE 8: Compute the distances from the principal point to the nadir point and to the isocenter using both the exact and approximate relations, if the focal length is 152.6 mm and the resultant tilt is 4°.

SOLUTION: Using the exact relations, the distance from the principal point to the nadir point is

$$pn = 152.6 \tan 4° = 152.6 \times 0.06993 = 10.67 \text{ mm}$$

and the distance from the principal point to the isocenter is

$$pi = 152.6 \tan 2° = 152.6 \times 0.03492 = 5.33 \text{ mm}.$$

Using the approximate equations, the solutions are:

$$pn \approx 152.6 \sin 4° = 152.6 \times 0.06976 = 10.65 \text{ mm}$$
$$pi \approx \tfrac{1}{2} \times 10.65 = 5.32 \text{ mm}.$$

EXAMPLE 9: The depression angle of a high-oblique photograph is 28° and the focal length is 151.1 mm. Compute the distances from the horizon point to the principal point, isocenter, and nadir point.

SOLUTION:

$$hp = f \tan \theta = 151.1 \times 0.53171 = 80.34 \text{ mm}$$
$$hi = f \sec \theta = f/\cos \theta = 151.1/0.88295 = 171.1 \text{ mm}$$
$$hn = f(\tan \theta + \cot \theta)$$
$$= 151.1 (0.53171 + 1.8807) = 364.5 \text{ mm}.$$

The tilt of a photograph may be completely defined by two parameters: the magnitude of the resultant tilt angle (*t*), and the swing angle (*s*) which defines the photographic direction of the tilt. A system of rectangular coordinates, which is illustrated in figure 2-4, is adopted for an aerial photograph to define the position of an image point and to define tilt and swing. The principal point is the origin of the coordinate system, one of the fiducial marks and the principal point define one of the axes, and the other axis also passes through the principal point and is at an angle of 90° from the first axis. The *x* axis is in the general direction of the line of flight, with +*x* being considered to the right, presumably the general direction of flight, but this is not important. The *x* axis of a photograph is then not exactly in the direction of flight because the exact direction is usually not known at the time the coordinate system is adopted for image measurement. (At least two other systems of axes are sometimes used but are not considered just yet: in one the +*x* axis is exactly in the line of flight; in the other the +*y* axis is in the direction of the upper end of the principal line and the nadir point is the origin.)

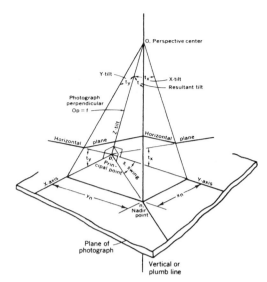

FIGURE 2-4. Components of tilt.

Swing (fig. 2-4) is the angle at the principal point measured from the $+y$ axis clockwise to the nadir point. Tilt, swing, and focal length are then spherical coordinates of the nadir point, and pn and swing are its polar coordinates. The term swing is also used with reference to the multiplex and other stereoscopic plotting instruments, where the term has an entirely different meaning. Instrument swing is a seldom-measured angle or rotational movement in the plane of a photograph that is required to achieve coincidence of the common flight lines of two overlapping photographs, and is not influenced by the amount of tilt nor the direction of the principal line of either photograph.

Resultant tilt can also be considered as composed of two angular components. In this analysis, the components are called x tilt, t_x, and y tilt, t_y. The x tilt is the component that corresponds to a rotation of the photograph about the adopted x axis, is related to the y coordinate y_n of the nadir point, and is the angle that the y axis makes with a horizontal plane. The x tilt corresponds to the wing-up or wing-down orientation of the airplane and hence is sometimes called *list* and also *roll*. The algebraic sign of the direction of x tilt is defined as positive if the y coordinate of the nadir point is positive, which corresponds to the right wing of the airplane being raised. Many multiplex operators still call this component "tilt" which is very confusing if one is accustomed to think of tilt as meaning resultant tilt. However, multiplex list is not exactly analogous to analytic x tilt because "list" is the angular rotation about a fixed instrumental direction which is not necessarily parallel to either the line of flight or the fiducial axes of the photograph (diapositive). In certain other plotting instruments, such as the stereoplanigraph, and in

computational operations, this tilt component is called omega (ω) and is the corresponding angle related to the adopted x axis of the line of flight or other adopted x axis.

The y tilt is the component that corresponds to a rotation of the photograph about the adopted y axis; it is plus if the x coordinate of the nadir point is plus, which corresponds to the nose of the airplane being lowered. This component is sometimes called *tip* in multiplex work, but the two terms are not exactly alike because the tip axis of the instrument is inclined at the list angle relative to the horizontal, and the axis is perpendicular to the adopted x axis of a strip of photographs. In stereoplanigraph and computational work the y tilt component is called phi (ϕ); its reference axis is horizontal and is perpendicular to the adopted x axis, y tilt is also referred to as *pitch*.

The construction of the multiplex is such that the ϕ component logically is defined before the ω, because ϕ is tilt about a primary axis which is approximately parallel to the adopted y axis of the system, whereas ω is about a secondary axis which is inclined to the horizontal at an angle ϕ. The same general description of ϕ and ω applies to the stereoplanigraph; the ϕ axis is primary and is fixed precisely parallel to the y axis of the instrument, whereas the ω axis is secondary inasmuch as it assumes an inclination equal to ϕ. In analytical aerotriangulation the sense is frequently reversed; ω is primary, and ϕ is secondary, as in sections 2.2.3.2.3 and 2.6.

The following equations express the relationship between the spherical coordinates t, s, and f, the rectangular coordinates x_n, y_n, and f, the spherical coordinates t_x, t_y, and f of the nadir point, and the components ϕ and ω defined in section 2.2.3.2.2

$$\overline{pn} = \sqrt{x_n{}^2 + y_n{}^2}$$
$$\tan t = \overline{pn}/f$$
$$\tan s = x_n/y_n = \sin t_y/\sin t_x$$
$$\sin s = x_n/\overline{pn}$$
$$\cos s = y_n/pn$$
$$\sin t = \sqrt{\sin^2 t_x + \sin^2 t_y}$$
$$\sin t_x = \sin t \cos s = (y_n/f) \cos t$$
$$\sin t_y = \sin t \sin s = (x_n/f) \cos t$$
$$\cos t = \cos \phi \cos \omega.$$

If the resultant tilt is less than 5°, which is true for almost all near-vertical photographs, the following approximate equations result in errors less than 1 minute, or about one part in 300. The angles t, t_x, t_y are considered as being expressed in minutes for convenience rather than in radians, or in degrees, minutes, and seconds.

$$\tan s \approx t_y/t_x$$
$$t \approx \sqrt{t_x{}^2 + t_y{}^2}$$
$$t \approx \sqrt{\phi^2 + \omega^2}$$
$$t_x \approx t \cos s$$
$$t_y \approx t \sin s$$
$$t \approx 3{,}438\overline{pn}/f$$

$$t_x \approx 3{,}438 y_n/f$$
$$t_y \approx 3{,}438 x_n/f.$$

EXAMPLE 10. The coordinates of the nadir point are +7.62 and −5.34 mm and the focal length of the camera is 152.61 mm. Determine the tilt, swing, x tilt, y tilt, and the sine and cosine of the swing. Also determine the approximate values of the resultant tilt, x tilt, y tilt, and swing.

SOLUTION: The exact solutions are:

$$\overline{pn} = \sqrt{7.62^2 + 5.34^2} = 9.305 \text{ mm}$$
$$\tan t = 9.305/152.61 = 0.06097$$
$$t = 3°29.3'$$
$$\tan s = +7.62/-5.34 = -1.4270$$
$$s = 125°01.4'$$
$$\sin t_x = \sin t \cos s = (0.06086)\,(-0.57389)$$
$$= -0.03493$$
$$t_x = 2° 00.1'$$
$$\sin t_y = \sin t \sin s = (0.06086)\,(+0.81891)$$
$$= +0.04984$$
$$t_y = +2° 51.4'$$
$$\sin s = +7.62/9.305 = +0.81891$$
$$\cos s = -5.34/9.305 = -0.537389.$$

The approximate solutions are:

$$t_x \approx 3{,}438(-5.34)/152.61 = -120.2 \text{ minutes}$$
$$= 2° 00.2'$$
$$t_y \approx 3{,}438(+7.62)/152.61 = +2° 51.6'$$
$$t \approx 3{,}438(9.305)/152.61 = 209.5 \text{ minutes}$$
$$= 3° 29.5'$$
$$t \approx \sqrt{120.2^2 + 171.6^2} = 209.5 \text{ minutes}$$
$$= 3° 29.5'$$
$$\tan s \approx +171.6/-120.2 = -1.4276$$
$$s \approx 125° 01.6'.$$

2.2.2.1 DISPLACEMENT OF IMAGES DUE TO TILT

As the result of tilt alone, images appear to be displaced radially toward the isocenter on the upper side of the photograph (figure 2-5), and radially outward or away from the isocenter on the lower side. Along the isometric parallel (line through the isocenter perpendicular to the direction of tilt) there is no displacement relative to

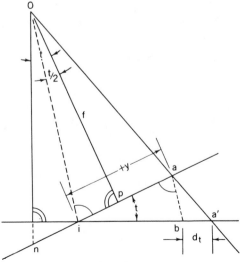

FIGURE 2-5. Image displacement due to tilt.

an equivalent untilted photograph. The amount of displacement d_t is given by the equation

$$d_t = \frac{y^2}{(f/\sin t) - y} \tag{2.10}$$

where y is the component along the principal line of the distance from the image to the isocenter. An approximate form of this equation is

$$d_t \approx (y^2 \sin t)/f \tag{2.10a}$$

inasmuch as the value of y in the denominator of the exact equation is usually small relative to the value of $f/\sin t$. These equations indicate that the amount of displacement varies nearly as the square of the y distance.

Wide-angle photographs, on which the relative values of y with respect to f can be large, are subject to greater tilt displacements.

EXAMPLE 11: Compute the image displacement relative to an untilted photograph if the focal length is 152.4 mm, the tilt is 3°, and the image is 120.3 mm from the isocenter, measured parallel to the direction of tilt. What displacement is obtained using the approximate equation?

SOLUTION:

$$d_t = \frac{120.3^2}{\dfrac{152.4}{0.05234} - 120.3} = 5.18 \text{ mm.}$$

If the approximate equation is used,

$$d_t = \frac{120.3^2 \times 0.05234}{152.4} = 4.97 \text{ mm.}$$

It should be noted that if equation (2.10a) is solved for $\sin t$,

$$\sin t \approx f d_t/y^2 \tag{2.10b}$$

which forms the basis of the displacement method of tilt determination discussed in section 2.2.2.6.

2.2.2.2 INCLINATION OF ANY LINE ON A TILTED PHOTOGRAPH

The principal line is a line of maximum inclination on a tilted photograph. The principal line is inclined at the tilt angle t with respect to a horizontal plane. The inclination of any other line parallel to the principal line is also equal to t. The inclination of any line perpendicular to the principal line is zero; that is, the line is horizontal and its degree of tilt is zero. The isometric parallel is a horizontal line as is also any line that is parallel to it. A line parallel to the isometric parallel is called a line of equal scale inasmuch as distances between images on the line are arranged according to the same scale insofar as tilt is concerned.

Any random line (see fig. 2-6) on a photograph that is neither parallel nor perpendicular to the principal line is inclined at some angle t', which is less than t. If the angle between a random line and the principal line in the plane of the photo-

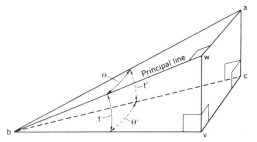

FIGURE 2-6. The inclination of any line.

graph is θ, the inclination of the random line is given by

$$\sin t' = \sin t \cos \theta \qquad (2.11)$$

If θ' is the dihedral angle between vertical planes containing the principal line and the random line (θ' is also the orthogonal projection of θ onto a horizontal plane),

$$\tan t' = \tan t \cos \theta' \qquad (2.12)$$

$$\tan \theta = \tan \theta' \cos t. \qquad (2.13)$$

The random line may intersect the principal line at any point. The angle θ can be measured to the right or left of the principal line, on either the upper or lower side of the point of intersection: angle θ is never more than 90°. Any line parallel to the random line is inclined at an equal angle t' with respect to the horizontal.

Any image on a tilted photograph can be considered as lying on a random line that passes through the isocenter. The position and displacement of an image on its random line through the isocenter is the same as though the image were on the principal line of a photograph that was tilted at an angle t'. Consequently, given t, θ, and y of an image, the displacement due to tilt can be found by (1) determining t' by means of equation (2.11), and (2) using equation (2.10) or (2.10a) with t' in place of t. The idea is used further in rectification where each image is rectified or displaced as though it lay on a principal line of its own but tilted a specified related amount.

EXAMPLE 12: What is the radial image displacement due to a tilt of 3° if the focal length of the photograph is 152 mm, the image is on the lower side of the photograph, 100 mm from the isocenter, and the line from the image to the isocenter makes an angle of 50° with the principal line?

SOLUTION: By equation (2.11),

$$\sin t' = \sin 3° \cos 50° = 0.0523 \times 0.64279 = 0.03364.$$

By equation (2.9),

$$d_t = \frac{100 \times 100}{\dfrac{152}{0.03364} - (-100)} = 2.17 \text{ mm}.$$

The displacement is outward on the lower side.

2.2.2.3 RATE OF CHANGE IN SCALE

The scale of a tilted aerial photograph changes in a regular manner throughout the picture. If the scale near the center is correct, then the scale is too small on the side that is tilted upward and too large on the side that is tilted downward. Tilt is considered here to be the single resultant tilt that is a combination of components frequently referred to as tip and tilt, (ϕ and ω). Although the scale changes across the photograph in the direction of the tilt (along the principal line), the scale does not change along any line (line of equal scale) that is perpendicular to the direction of tilt. The scale of any image on a tilted photograph (fig. 2-5) can be expressed as

$$S = \frac{f - y \sin t}{H - h} \qquad (2.14)$$

where t is the tilt angle, y is the distance of the image from the isocenter measured in the direction of the tilt, positive on the upper side, and where the other terms have already been defined. (The isocenter is a point on the principal line a distance $pi = f \tan t/2$ from the principal point toward the lower edge of the photograph.) Thus, the scale depends on the amount of the tilt and the position of the image on the photograph relative to the direction of the tilt, as well as on the focal length, flight height, and relief. Tilt then compresses the images on the upper side of the photograph and expands those on the lower side.

EXAMPLE 13: An aerial photograph taken with a camera which has a focal length of 152.4 mm from a height of 6,500 m has a tilt of 3° in the direction of one of the corners of the 228.6 mm × 228.6 mm format. What are the scales of the image of a sea-level object if the image appears (a) at the isocenter, (b) 125 mm from the isocenter on the upper side of the inclined photograph, and (c) at the same distance on the lower side?

SOLUTION: (a) At the isocenter $y = 0$ and

$$S = \frac{0.1524 - 0}{6,500 - 0} = \frac{1}{42,650}$$

(b) At the upper point, $y = +125$ mm and

$$S = \frac{0.1524 - 0.125 \sin 3°}{6,500 - 0} = \frac{1}{44,564}$$

(c) At the lower point, $y = -125$ mm and

$$S = \frac{0.1524 - (-0.125) \sin 3°}{6,500 - 0} = \frac{1}{40,895}.$$

The magnitude of the change in scale ΔS across a tilted photograph is directly proportional to the change Δy in the distance y of an image measured in the direction of the tilt. By calculus, the incremental form of equation (2.14) is, regarding S and y as variables:

$$\Delta S = -\frac{\sin t}{H - h} \Delta y, \quad \frac{\Delta S}{\Delta y} = -\frac{\sin t}{H - h} \qquad (2.15)$$

where $\Delta y = y_2 - y_1$, and $\Delta S = S_2 - S_1$. Thus, the change in scale from image to image in the direction of tilt is equal to a constant times the change in the y distance, and the negative sign verifies that the scale decreases as y increases.

EXAMPLE 14: Compute the rate of change of scale for the conditions in the previous example:
SOLUTION:

$$\frac{\Delta S}{\Delta y} = -\frac{\sin 3°}{(6,500 - 0)\,1000}$$

$$= -8.05 \times 10^{-9} \text{ per mm.}$$

Note that the value is the same as the difference in the two scales (of previous problem) at the upper and lower points expressed in decimal form and divided by the distance between the points:

$$1:44,564 = 0.00002244$$

$$1:40,895 = 0.00002445$$

$$\frac{0.00002244 - 0.00002445}{250} = -8.05 \times 10^{-9} \text{ per mm.}$$

2.2.2.4 SCALE POINT

A *scale point* is a point on a photograph where the scale is known. A scale point is considered to be an infinitesimal segment of a line ab, called a *scale check line*, whose scale can be defined

$$S = ab/AB \text{ or } S^{-1} = AB/ab \qquad (2.16)$$

where AB is the length of the corresponding level line on the ground. The scale at a point can also be expressed as (*see* also 2.2.2.3):

$$\left.\begin{aligned}S &= \frac{f \sec t - y \sin t}{H - h} \\[2mm] S^{-1} &= \frac{H - h}{f \sec t - y \sin t}\end{aligned}\right\} \qquad (2.17)$$

where f is the focal length, t is the resultant tilt, y is the coordinate of the point regarding the nadir point as the origin and the upper side of the principal line as the positive y axis, H is flight altitude above sea level or other selected datum, and h is the elevation of the object above the datum. If the elevations of the ends of the ground scale check line are not equal, the horizontal projection of the line must be used.

S is considered as the representative map fraction usually expressed with 1 as the numerator, but in scale-check problems S is more conveniently manipulated in the form of a decimal fraction with the decimal point displaced an arbitrary number of places: $1:20,000 = 0.000,05 = 5,000 \times 10^{-8}$. However, S^{-1} is used more frequently than S in scale check problems. S^{-1} is usually expressed in terms of feet per inch, or metres per millimetre. Thus, S^{-1} is the reciprocal of S.

The principle of scale points is significant because, neglecting relief, the scales at points on any random line on a photograph change at a uniform linear rate with respect to distance along the line: the same change in scale value

per unit distance along the line is effective everywhere on the line. As was shown in the preceding section, the rate of change of scale with respect to distance is a constant. Thus, the scales at equally spaced points along any line on a photograph form an arithmetic progression. (These statements are not quite true for S^{-1}, but for the relatively small tilts usually encountered, the deviation of S^{-1} from an arithmetic progression is ordinarily insignificant.)

Consequently, if there are two points of known scale on a photograph, the scale of any new point on the line through the first two can be determined by simple linear interpolation or extrapolation. Conversely, the location of a point having any desired or given scale value can be determined in the same manner. Furthermore, if the scale is known for three points not on the same straight line, the scale of any point on the photograph can be determined by not more than two linear interpolations or extrapolations.

EXAMPLE 15: The scale value (S^{-1}) at image a is 70 m per mm, at b it is 190, and the two points are 60 mm apart. The rate of change of scale along this line is (refer also to equation 2.15):

$$\frac{\Delta S^{-1}}{\Delta y} = \frac{S_b^{-1} - S_a^{-1}}{ab} = \frac{190 - 70}{60}$$

$$= 2 \text{ m per mm per mm.}$$

This means that the scale values expressed in this manner change in the amount of 2m for each mm along this line, increasing in the direction from a toward b.

EXAMPLE 16: A point c, 40 mm from a and 20 mm from b, has the scale value of a plus the change that takes place between a and c:

$$S_c^{-1} = S_a^{-1} \pm \left(\frac{\Delta S^{-1}}{\Delta y}\right)\overline{ac} = 70 + (2 \times 40)$$

$$= 150 \text{ m per mm.}$$

EXAMPLE 17: A point d that has a scale value of 90 m per mm is required to be located on the line ab. By direct proportion,

$$\overline{ad} = \left(\frac{\overline{ab}}{S_b^{-1} - S_a^{-1}}\right)(S_d^{-1} - S_a^{-1})$$

$$= \frac{60}{190 - 70} \times (90 - 70) = 10 \text{ mm.}$$

EXAMPLE 18: A third point e having a known scale value of 130 is not located on the line ab, and the scale value is desired at a fourth random point f. Lines be and af are constructed, intersecting at g, and all the line segments are measured: $ag = 43$ mm; $gf = 27$ mm; $bg = 28$ mm; $ge = 22$ mm. The scale at g is determined by interpolation between b and e:

$$S_g^{-1} = 130 + \left(\frac{190 - 130}{22 + 28}\right) \times 22$$

$$= 156.4 \text{ metres per mm.}$$

Then the scale value at f is found by extrapolation on the line through a and g either (1) by using a as the initial point:

$$S_f^{-1} = 70 + \left(\frac{156.4 - 70}{43}\right) \times (43 + 27)$$

$$= 210.7 \text{ metres per mm}$$

or (2) by using g as the initial point:

$$S_f^{-1} = 156.4 + \left(\frac{156.4 - 70}{43}\right) \times 27$$

$$= 210.7 \text{ metres per mm.}$$

The preceding discussion deals with the value of the scale at a point. Let us now consider the location of a point on a scale check line where the scale is equal to the average scale of the line. The scale point (equation 2.16) can be located by an empirical procedure of graphic construction, as follows (see fig. 2-7): (1) a perpendicular is dropped from the isocenter to the scale check line (in the absence of the isocenter the principal point is used instead without causing much error); (2) the distance from either end of the scale check line to the scale point is equal to the distance from the opposite end to the foot of the dropped perpendicular, the two distances being measured in an opposite sense of direction. Thus, a point for which the scale value is known can be located with a straightedge, triangle, and dividers. Although the method is empirical, it is sufficiently correct with the usual small tilts to be very useful.

These principles of the scale point are significant inasmuch as they form a basis for determining the amount and direction of tilt of an aerial photograph. Equation (2.15) can be solved for $\sin t$ and expressed in the "delta" notation:

$$\sin t = (H - h) \frac{\Delta S}{\Delta y} \qquad (2.18)$$

where ΔS is the change in scale value between two points that are a distance Δy apart measured parallel to the principal line. The values of H and

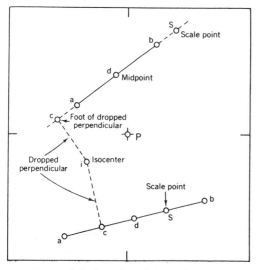

FIGURE 2-7. Location of the scale point.

h are considered known. Two scale check lines (preferably more) are sufficient data to determine ΔS. A third scale check line is sufficient data to find two scale points of equal scale (by linear interpolation) that fix the orientation of a line of equal scale to which the principal line is perpendicular. Hence ΔS and Δy are both determinable, from which t can be computed. The scale point method of tilt determination is a practical application of these principles.

2.2.2.5 OTHER RELATED EFFECTS OF TILT

Because of this changing displacement and scale effect of tilt, the shapes of areas on tilted photographs are not similar to the corresponding shapes on a map or on the ground. An extreme example of the manner in which shapes are affected by tilt is shown by the shapes of the quadrilaterals in the different parts of a perspective grid where each quadrilateral represents the same size square on the ground. Also, any horizontal circle on the ground is shown as an ellipse on a tilted photograph. Consequently, in laying mosaics, tilted photographs are often rectified photographically so that an area near the edge of one photograph will have the same size and shape as on an adjoining photograph. In graphic compilation of map detail from photographs, the photogrammetrist is restricted to tracing images near the isocenters of the photographs where the displacements and scale differences are small because the y values are small.

Angles at a central point (radial center) subtended by photographic images are assumed in radial plotting to be equal to the corresponding angles on the ground. This condition is a fact on truly vertical photographs if the principal point (which is also the nadir point) is used as the radial center. If there is no ground relief, and the isocenter is used as the central point, the condition is also true on tilted photographs. But if both tilt and relief are present, no central point exists at which all the radial directions are correct, because relief displacements are radial from the nadir point and tilt displacements are radial from the isocenter. The two displacements are entirely independent both in direction and magnitude. Relief displacement is related to the first power of the radial distance, to the height of the object, and is normally positive over the entire photograph. Tilt displacement varies with the square of the radial distance, with the cosine of the angle that a line through the image and the isocenter makes with the direction of tilt, and is positive on one side of the photograph and negative on the other side.

Tilt is most objectionable where the photographs are being used to determine elevations of points on the ground, or to draw contours without precision stereoscopic plotting instruments. Equation (2.6) ($h = d_e H/r$) shows that heights are directly related to the relief displacement d_e

measured on the photograph. Whether stereoscopic parallax or a plotting instrument is implied, relief displacement is the fundamental element that makes height determination possible.
The desired elevation differences, together with
the corresponding displacements, are relatively
small quantities. For example, the relief displacement for a 10-metre contour interval where
the flight height is 6,000 metres and where the
image is 30 mm from the center, is only 0.05
mm. The small size of the quantity causes it to
be difficult to measure accurately. If, in addition, the value of the apparent displacement of
an image is combined with a totally unrelated tilt
displacement that can be larger than the relief
displacement, an accurate analysis of the effect
is very complicated. The simplest instrument
made to assist in solving the problem in a comparatively crude, near-accurate manner costs a
few hundred dollars, whereas an instrument that
gives a complete and accurate solution costs a
few tens of thousands of dollars. Graphic methods are not applicable, but analytic computations are coming into common use.

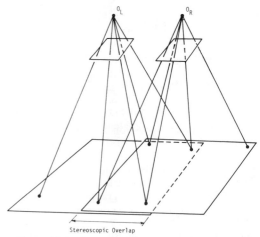

FIGURE 2-8. Geometry of stereoscopic coverage.

2.2.2.6 TILT DETERMINATION

The tilt of an aerial photograph can be
computed if the ground position of three or more
image points in the photograph are known.
The Church Method of space resection, which
will be described in section 2.2.4.2, is suitable
for tilt determination using either desk or electronic computers. This method has sufficient
accuracy for most practical applications relating
to photo-interpretation and quantitative
analysis. If a large number of control points
are available, then the method of least squares
adjustment may be applied. The mathematical
solution for the simultaneous solution of large
photogrammetric blocks, which will be presented in section 2.6, may also be used to determine the tilts of all the photographs which provide complete stereoscopic coverage of an area.

Several other methods of tilt determination
were discussed in chapter II of the Third
Edition of the MANUAL OF PHOTOGRAMMETRY.
These methods included: (1) the image-displace-
ment method; (2) the Anderson scale-point
method; and (3) the Morse Method. These
methods are not in common use and interested
readers are referred to the previous edition of
this MANUAL.

2.2.3 CONCEPTS OF ORIENTATION

The geometric principle of stereophotogrammetry is illustrated in figure 2-8. The area to
be mapped is photographed from two different
camera positions, O_L and O_R. The area of
common coverage by the two photographs is
called stereoscopic overlap. Each photograph
may be considered as a record of the bundle of
light rays which travel from the object space,

pass through the nodal point of the camera lens
system, and register on the photographic film.
In the laboratory, an optical model of the
stereoscopic overlap area can be constructed in
an instrument called stereoplotter. Each bundle
of rays is reconstructed by inserting either a
glass plate diapositive or film negative into a
projector. The process of reconstructing the
internal geometry of the bundle of rays in a
projector is called interior orientation. The two
projectors may each be translated and tilted
until they assume the same relative position and
attitude as that of the camera in its two positions
during the photography. This process is called
relative orientation. At the completion of relative orientation, corresponding light rays in the
two bundles intersect in space and a three-
dimensional optical model is formed. Finally, in a
process called absolute orientation, points of
known ground positions (called control points)
are used to scale the model and to level it
with respect to the reference plane in the instrument. Once absolute orientation is completed,
the position of any point in the stereo model
may be measured at the intersection of the two
corresponding rays from the two projectors.
Further details on the operation and principle
of the stereoplotters will be included in chapters
XI and XII.

In computational photogrammetry, the path
of each ray of light may be described by a
mathematical expression which is a function of
the position of the point in the object space,
position of the image point in the photograph,
position of the exposure center in the ground
reference system, direction of the optical axis
of the camera and the perspective geometry of
the camera. If the perspective geometry of the
camera has been determined by camera calibration and if three or more control points are
imaged on a photograph, the position of the
camera and its attitude with respect to the
ground control reference system can be deter-

mined. Once the orientation of both of the photographs of a stereoscopic pair is known, the position of any object point which is located in the overlap area may be computed as the point of intersection of two rays.

Therefore, in both the instrumental and analytical approach of photogrammetric measurements, determination of the orientation of the camera at the moment of exposure is a necessary step in the measurement process. There are basically four orientation problems: (1) interior orientation; (2) exterior orientation; (3) relative orientation; and (4) absolute orientation.

2.2.3.1 INTERIOR ORIENTATION

The interior orientation of a camera refers to the perspective geometry of the camera and is defined by the following parameters: (1) the calibrated focal length; (2) the position of the principal point in the image plane; and (3) the geometric distortion characteristics of the lens system. The interior orientation of a camera can be determined by either field or laboratory calibration procedure (see chapter IV).

2.2.3.1.1 PURPOSE OF FIDUCIAL MARKS

An aerial mapping camera is equipped with either four or eight fiducial marks which are permanently mounted in the camea housing and located in front of the image plane. Images of the fiducial marks appear on each photograph. The primary purpose of the fiducial marks is to define the location of the principal point of the photograph. The fiducial marks are positioned so that the intersection of the lines joining diametrically opposite fiducial marks coincide with the principal point. In stereoplotting instruments, the plate carriage in each projector are also equipped with at least four fiducial marks which define the position of the principal point in the carriage plate. Thus, by positioning the photographic plate on the carriage so that the fiducial marks on the photographic plate coincides with those in the carriage, the photograph is centered properly with respect to the optical axis of the projector.

2.2.3.1.2 PHOTO-COORDINATE SYSTEM

Figure 2-9 illustrates a three-dimensional photo-coordinate system $(\bar{x}, \bar{y}, \bar{z})$ which is commonly used to define the location of image points with respect to the exposure center, O. In the plane of the photograph, point O' is the point of intersection of the lines joining fiducial marks A and C and B and D. If the four fiducial marks are perfectly aligned, the lines AC and BD should be orthogonal with each other and the point of intersection (O') should be in coincidence with the principal point. This condition is difficult to achieve in practice. For the purpose of generality, the x-axis on the plane of the photograph is defined as the line joining

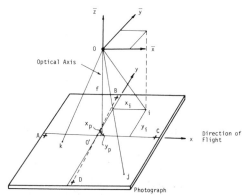

FIGURE 2-9. A photo-coordinate system.

point O' and fiducial mark C. The positive x-axis points in the general direction of flight of the aircraft. The y-axis passes through O' and is perpendicular to the x-axis.

The photo-coordinate system is defined by the axes $\bar{x}$, $\bar{y}$ and $\bar{z}$ with the origin of the system situated at the exposure center (O). The $\bar{z}$-axis coincides with the optical axis of the camera and is positive along the direction towards the image plane of the camera. The $\bar{x}$-axis is parallel to the x-axis on the plane of the photograph and is positive towards the direction of flight. The $\bar{y}$-axis is parallel to the y-axis on the photographic plane. Thus, the $\bar{x}$-$\bar{y}$ plane is parallel to the photographic plane. Assuming that the principal point is not exactly in coincidence with point O', its position on the plane of the photograph is then defined by its coordinates x_p and y_p. The position of an image point i in the same plane can be defined by its coordinates x_i and y_i. The position of the same image point with respect to the exposure center (O) is then defined by its photo-coordinates $\bar{x}_i$, $\bar{y}_i$ and $\bar{z}_i$ as follows:

$$\bar{x}_i = x_i - x_p$$
$$\bar{y}_i = y_i - y_p \qquad (2.19)$$
$$\bar{z}_i = -f \, .$$

2.2.3.1.3 MATHEMATICAL DEFINITION OF INTERIOR ORIENTATION

The interior orientation of a photogrammetric camera is said to be mathematically defined if the following parameters are known:

(1) focal length, f;
(2) coordinates of the principal point, x_p and y_p; and
(3) geometric distortion characteristics of the lens system.

One commonly used model for correcting lens distortion is that developed by D. Brown:

$$\Delta x_j = \bar{x}_j [l_1 r^2 + l_2 r^4 + l_3 r^6]$$
$$+ [p_1(r^2 + 2\bar{x}_j^2) + 2\, p_2\bar{x}_j\bar{y}_j][1 + p_3 r^2]$$

$$(2.20)$$

$$\Delta y_j = \bar{y}_j \left[l_1 r^2 + l_2 r^4 + l_3 r^6 \right]$$
$$+ \left[2p_1 \bar{x}_j y_j + p_2 (r^2 + 2\bar{y}_j^2) \right] \left[1 + p_3 r^2 \right]$$

where Δx_j and Δy_j are corrections for geometric lens distortions present in the coordinates $\bar{x}_j$ and $\bar{y}_j$ of image point j; and $r = (\bar{x}_j^2 + \bar{y}_j^2)^{1/2}$. The coefficients l_1, l_2, l_3, p_1, p_2 and p_3 may be determined as a part of the camera calibration process. The model accounts for both symmetric radial distortion and asymmetric distortions caused by lens decentering. The terms which include the coefficients l_1, l_2 and l_3 represent symmetric radial distortion, and the terms which include p_1, p_2 and p_3 represent asymmetric distortion.

The photo-coordinates are corrected for lens distortion by the following expression:

$$\bar{x}_j' = \bar{x}_j + \Delta x_j$$
$$\bar{y}_j' = \bar{y}_j + \Delta y_j. \tag{2.21}$$

2.2.3.2 EXTERIOR ORIENTATION

The exterior orientation of a camera during the moment of exposure of a particular photograph is defined by the geographic position of the exposure center and the direction of the optical axis. In computational photogrammetry, the geographic position of the exposure center is most conveniently defined by its coordinates in a three-dimensional rectangular coordinate system, and the direction of the optical axis is usually defined by three rotation angles (either ω, ϕ and κ or tilt, swing and azimuth).

2.2.3.2.1 A THREE-DIMENSIONAL RECTANGULAR COORDINATE SYSTEM

Figure 2-10 shows that the locations of points in the object space may be defined by a three-dimensional rectangular coordinate system. The origin and orientation of the coordinate system may be arbitrarily defined. In analytical aerotriangulation, the origin is usually chosen to be located near the center of the area of concern so that the number of digits in each coordinate number can be kept at a minimum. The positive Y-axis is usually directed towards true north. The location of any point j in the object space can therefore be defined by its three coordinates X_j, Y_j and Z_j. The position of the exposure center of a photograph can similarly be defined by its coordinates X_i^c, Y_i^c and Z_i^c.

In geodetic surveying, the positions of survey stations are usually defined in geographic coordinates (longitude, latitude, and elevation above mean sea level), or in a state plane coordinate system (X, Y, and elevation above mean sea level). Neither of these coordinate systems are convenient for computation purposes. When geodetic control points are defined in either one of these coordinate systems, their coordinates are usually first transformed to a locally defined rectangular coordinate system.

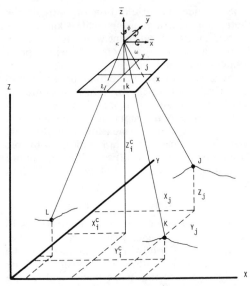

FIGURE 2-10. Exterior orientation.

The definition of various ground reference coordinate systems and transformation equations for converting from one system to the other are given in chapter VII.

2.2.3.2.2 OMEGA (ω), PHI (ϕ) AND KAPPA (κ)

As the spatial position of the exposure center is defined by its coordinates X_i^c, Y_i^c and Z_i^c, the direction of the optical axis may be defined by the rotation angles omega (ω), phi (ϕ) and kappa (κ) about the $\bar{x}$-, $\bar{y}$- and $\bar{z}$-axis respectively of the photo-coordinate system. All the rotations are defined positive in the counter-clockwise direction. When $\omega = \phi = \kappa = 0$, the optical axis is perpendicular to the X-Y plane and the $\bar{x}$-, $\bar{y}$- and $\bar{z}$- axis are parallel to the X-, Y- and Z-axis respectively. However, it should be noted that because of the curvature of the earth's surface, zero rotation about the $\bar{x}$-, $\bar{y}$- and $\bar{z}$-axis does not mean that the photograph is truly vertical.

2.2.3.2.3 PROJECTIVE TRANSFORMATION EQUATIONS

Assuming that light rays travel in straight lines, that all the rays entering a camera lens system pass through a single point and that the lens system is distortionless, then a projective relationship exists between the photographic coordinates of the image points and the ground coordinates of the corresponding object points as illustrated in figure 2-10. It will be shown in this section that this projective relationship can be represented by the following set of projective transformation equations:

$$X_j - X_i^c = \lambda_{ij} \left[m_{11} (x_{ij} - x_p) \right.$$
$$\left. + m_{21} (y_{ij} - y_p) + m_{31} (-f) \right]$$

$$Y_j - Y_i^c = \lambda_{ij} \left[m_{12} (x_{ij} - x_p) \right.$$
$$\left. + m_{22} (y_{ij} - y_p) + m_{32} (-f) \right] \quad (2.22)$$
$$Z_j - Z_i^c = \lambda_{ij} \left[m_{13} (x_{ij} - x_p) \right.$$
$$\left. + m_{23} (y_{ij} - y_p) + m_{33} (-f) \right]$$

in which

$m_{11} = \cos \phi_i \cos \kappa_i$
$m_{12} = \cos \omega_i \sin \kappa_i + \sin \omega_i \sin \phi_i \cos \kappa_i$
$m_{13} = \sin \omega_i \sin \kappa_i - \cos \omega_i \sin \phi_i \cos \kappa_i$
$m_{21} = - \cos \phi_i \sin \kappa_i$
$m_{22} = \cos \omega_i \cos \kappa_i - \sin \omega_i \sin \phi_i \sin \kappa_i \quad (2.23)$
$m_{23} = \sin \omega_i \cos \kappa_i + \cos \omega_i \sin \phi_i \sin \kappa_i$
$m_{31} = \sin \phi_i$
$m_{32} = - \sin \omega_i \cos \phi_i$
$m_{33} = \cos \omega_i \cos \phi_i.$

X_j, Y_j and Z_j are the object space coordinates of object point j; X_i^c, Y_i^c and Z_i^c are the object space coordinates of the exposure center of photo i; x_{ij} and y_{ij} are the image coordinates of object point j on photo i; f is the focal length of the camera; λ_{ij} is the photo scale factor at image point j in the newly rotated photo-coordi- about the $\bar{x}$-, $\bar{y}$- and $\bar{z}$-axis of the photo coordinate system.

The derivation to be presented in the following paragraphs will first assume that the photograph is parallel to the X-Y plane; i.e. $\omega_i = \phi_i = \kappa_i = 0$ as shown in figure 2-11. Then a small ω_i-rotation is introduced. It is followed by a small ϕ_i-rotation and a κ_i-rotation. This sequence of rotation is illustrated in fig. 2-12.

Let $\omega_i = \phi_i = \kappa_i = 0$, and let $\bar{x}_o$, $\bar{y}_o$ and $\bar{z}_o$ be the photo coordinates of image point j. It can be seen from figure 2-11 that the following relationships exist:

$$X_j - X_i^c = \lambda_{ij} \bar{x}_o$$
$$Y_j - Y_i^c = \lambda_{ij} \bar{y}_o \quad (2.24)$$

and

$$Z_j - Z_i^c = \lambda_{ij} \bar{z}_o.$$

Next, assume a small rotation ω_i and $\phi_i = \kappa_i = 0$. Let $\bar{x}_\omega$, $\bar{y}_\omega$ and $\bar{z}_\omega$ denote the coordinates of image point j in the newly rotated photo-coordinate system. It can be easily derived from figure 2-13a that

$$\bar{x}_o = \bar{x}_\omega$$
$$\bar{y}_o = \bar{y}_\omega \cos \omega_i - \bar{z}_\omega \sin \omega_i$$
$$\bar{z}_o = \bar{y}_\omega \sin \omega_i + \bar{z}_\omega \cos \omega_i.$$

Next introduce an additional rotation ϕ_i about the $\bar{y}_\omega$-axis as shown in figure 2-12b, and let $\bar{x}_{\omega\phi}$, $\bar{y}_{\omega\phi}$ and $\bar{z}_{\omega\phi}$ denote the coordinates of the same image point in the newly rotated system. The following equations can be derived from figure 2-13(b):

$$\bar{x}_\omega = \bar{x}_{\omega\phi} \cos \phi_i + \bar{z}_{\omega\phi} \sin \phi_i$$
$$\bar{y}_\omega = \bar{y}_{\omega\phi} \quad (2.26)$$
$$\bar{z}_\omega = -\bar{x}_{\omega\phi} \sin \phi_i + \bar{z}_{\omega\phi} \cos \phi_i.$$

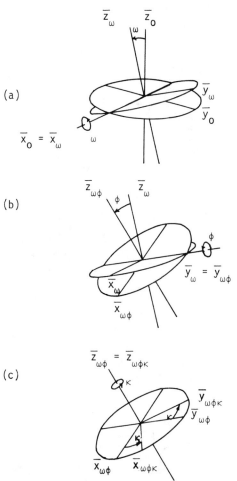

FIGURE 2-12. The rotations defined by the ω-ϕ-κ sequence.

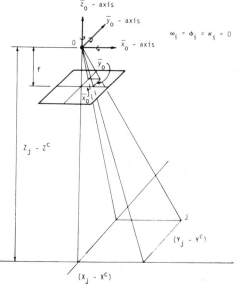

FIGURE 2-11. Zero-rotation case.

Substituting equations (2.26) into equations (2.25) yields

$$\bar{x}_o = \bar{x}_{\omega\phi} \cos \phi_i + \bar{z}_{\omega\phi} \sin \phi_i$$

$$\bar{y}_o = \bar{x}_{\omega\phi} \sin \phi_i \sin \omega_i + \bar{y}_{\omega\phi} \cos \omega_i$$

$$- \bar{z}_{\omega\phi} \cos \phi_i \sin \omega_i \qquad (2.27)$$

and

$$\bar{z}_o = -\bar{x}_{\omega\phi} \sin \phi_i \cos \omega_i + \bar{y}_{\omega\phi} \sin \omega_i$$

$$+ \bar{z}_{\omega\phi} \cos \phi \cos \omega.$$

Finally, introduce a κ-rotation about the $\bar{z}_{\omega\phi}$-axis as shown in figure 2-12c, and let $\bar{x}_{\omega\phi\kappa}$, $\bar{y}_{\omega\phi\kappa}$ and $\bar{z}_{\omega\phi\kappa}$ denote the coordinates of the image point in the newly rotated system. The following equations can be derived from figure 2-13c:

$$\bar{x}_{\omega\phi} = \bar{x}_{\omega\phi\kappa} \cos \kappa_i - \bar{y}_{\omega\phi\kappa} \sin \kappa_i$$

$$\bar{y}_{\omega\phi} = \bar{x}_{\omega\phi\kappa} \sin \kappa_i + \bar{y}_{\omega\phi\kappa} \cos \kappa_i \qquad (2.28)$$

and

$$\bar{z}_{\omega\phi} = \bar{z}_{\omega\phi\kappa}.$$

Substituting these equations into equations (2.27) yeilds

$$\bar{x}_o = \bar{x}_{\omega\phi\kappa} \cos \phi_i \cos \kappa_i - \bar{y}_{\omega\phi\kappa} \sin \kappa_i \cos \phi_i$$

$$+ \bar{z}_{\omega\phi\kappa} \sin \phi_i$$

$$\bar{y}_o = \bar{x}_{\omega\phi\kappa} (\cos \kappa_i \sin \phi_i \sin \omega_i + \sin \kappa_i \cos \omega_i)$$

$$+ \bar{y}_{\omega\phi\kappa} (\cos \kappa_i \cos \omega_i - \sin \kappa_i \sin \phi_i \sin \omega_i)$$

$$- \bar{z}_{\omega\phi\kappa} (\cos \phi_i \sin \omega_i) \qquad (2.29)$$

and

$$\bar{z}_o = \bar{x}_{\omega\phi\kappa} (\sin \kappa_i \sin \omega_i - \cos \kappa_i \sin \phi_i \cos \omega_i)$$

$$+ \bar{y}_{\omega\phi\kappa} (\sin \kappa_i \sin \phi_i \cos \omega_i + \sin \kappa_i \cos \kappa_i)$$

$$+ \bar{z}_{\omega\phi\kappa} (\cos \phi_i \cos \omega_i).$$

In equations (2.29), $\bar{x}_{\omega\phi\kappa}$, $\bar{y}_{\omega\phi\kappa}$ and $\bar{z}_{\omega\phi\kappa}$ denote the photo-coordinates of an image point on a tilted photograph. Since in practice all aerial photographs can be assumed to have some small tilt, the subscripts for these coordinates can be dropped without any los of generality; that is,

$$\bar{x}_j = \bar{x}_{\omega\phi\kappa} = x_j - x_p$$

$$\bar{y}_j = \bar{y}_{\omega\phi\kappa} = y_j - y_p \qquad (2.30)$$

$$\bar{z}_j = \bar{z}_{\omega\phi\kappa} = -f.$$

Substituting equations (2.30) into (2.29) and then substituting the resulting equations into (2.24) will yield the projective transformation equation (2.22) which is stated at the beginning of this section.

It is recalled that the following assumptions were made in the above derivation: (1) the image (x, y, z) and object (X, Y, Z) coordinate systems are both right-handed; (2) the sequence of angular rotations is ω, ϕ, κ, where ω is the first rotation and is about the X axis; (3) the x, y, z coordinates refer to a contact print of a photograph (diapositive with emulsion upward); and (4) one visualizes a vertical or near-vertical photograph as though looking downward from an airplane. But one is reminded that these assumptions do not always apply: (1) in certain European systems the geodetic rectangular coordinate system is different; (2) ω need not be the first of the sequence of rotations; (3) where glass plates are used in a camera, measurements are usually made on the original negative instead of the diapositive, and if aerotriangulation is being performed for the eventual use of a compilation instrument like the Kelsh plotter, the diapositives may be printed through the film base and the emulsions are not in contact; (4) for horizontal (terrestrial) photographs certain terms are already defined by convention and for astronomical work the camera "looks" upward instead of downward; and (5) the measuring comparator may impose arbitrary considerations with regard to the designation of the x and y axes and the plus and minus directions (Rosenfield 1959).

The subject of coordination of the various rotational systems was the basis for a resolution

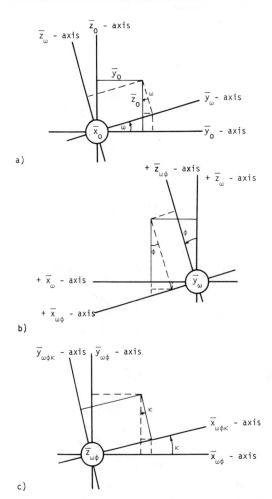

FIGURE 2-13. Plane cross-sections about axis of rotation.

adopted by the International Congress on Photogrammetry which met in Stockholm in 1956. The ideas of the resolution are expressed by Schermerhorn (1955).

The right-handed system adopted has the Z axis downward and the X axis to the right; the position of the Y axis is thus predetermined. The Stockholm system is equivalent to the geodetic coordinate system of many European countries: X north, Y east, and Z downward.

In the United States, the geodetic coordinate system upon which surveying and mapping is based regards the Z axis as upward, the X axis to the East, and Y axis to the North. This is also a right-handed system. Again, in this country, the survey (azimuth) and navigation angles are measured clockwise for the positive direction. Thus, the basis for the recommendation of the Stockholm resolution can be adopted in this country by assuming a photogrammetric system based on our geodetic system.

It is readily seen that the two systems are the same in abstract space and the formula for any rotation about a given axis in either system would therefore be the same.

2.2.3.2.4 TILT, SWING AND AZIMUTH

The orientation of a photograph can also be defined by the three rotation angles tilt (t), swing (S) and azimuth (α), which are illustrated in figure 2-14. The tilt angle (t) is measured in the principle plane from the optical axis of the camera to the plumb line and always has a positive sign. The direction of tilt with respect to the photographic axes is defined by the swing angle (S), which is measured clockwise from the positive y-axis of the photograph to the lower part of the principal line which passes through the nadir point (n). The direction of tilt with respect to the ground reference coordinate system (X, Y, Z) is defined by the azimuth angle (α), which is measured in the X-Y plane clockwise from the positive Y axis to the projection of the principal line on the X-Y plane.

The elements m_{ij}'s in projective transformation equations (2.22) may be expressed in terms of t, S and α as follows:

$$m_{11} = -\cos S \cos \alpha - \sin S \cos t \sin \alpha$$
$$m_{12} = \sin S \cos \alpha - \cos S \cos t \sin \alpha$$
$$m_{13} = -\sin \alpha \sin t$$
$$m_{21} = \cos S \sin \alpha - \sin S \cos t \cos \alpha$$
$$m_{22} = -\sin S \sin \alpha - \cos S \cos t \cos \alpha \quad (2.31)$$
$$m_{23} = -\cos \alpha \sin t$$
$$m_{31} = -\sin S \sin t$$
$$m_{32} = -\cos S \sin t$$
$$m_{33} = \cos t.$$

A major disadvantage of this system of defining orientation is that it breaks down for the ideal aerial photograph which is truly vertical. When there is no tilt, the two planes are either parallel or coincident, and there is no line of intersection (or principal line); thus, the angles of swing and azimuth are undefined.

2.2.3.2.5 TERRESTRIAL PHOTOGRAMMETRY

In terrestrial photogrammetry, the orientation of the photograph can often be more conveniently described by the three rotation angles omega (ω), alpha (α), and kappa (κ) which are illustrated in figure 2-15. Omega is a rotation about the $\bar{x}$-axis of the photo-coordinate system and is measured positive in the counter-clockwise direction. Alpha (α) is a rotation about the $\bar{y}$-axis of the photo-coordinate system and is measured positive in the clockwise direction. Kappa (κ) is a rotation about the $\bar{z}$-axis and is measured positive in the counter-clockwise direction. When $\omega = \alpha = \kappa = 0$, the $\bar{x} - \bar{z}$ plane is parallel to the $X - Y$ plane and the $\bar{z}$-axis is parallel to the Y-axis but points in the opposite direction. The ground coordinate system is commonly defined so that the Z-axis is the local vertical and the Y-axis points in the direction of true north. Then, the alpha (α) angle is the horizontal azimuth of the optical axis measured clockwise from true north; and the omega (ω) angle measures the tilt angle of the optical axis with respect to the horizontal plane.

The elements m_{ij}'s in projective transformation equations (2.22) may be expressed in terms of ω, α and κ as follows:

$$m_{11} = \cos \alpha \cos \kappa$$
$$m_{12} = -\sin \omega \sin \kappa - \cos \omega \sin \alpha \cos \kappa$$

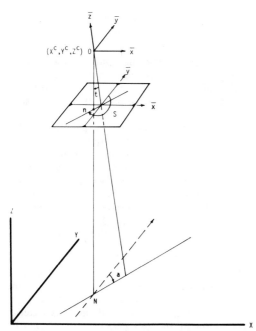

FIGURE 2-14. Rotation angles tilt ($\pm$), swing (S) and azimuth (a).

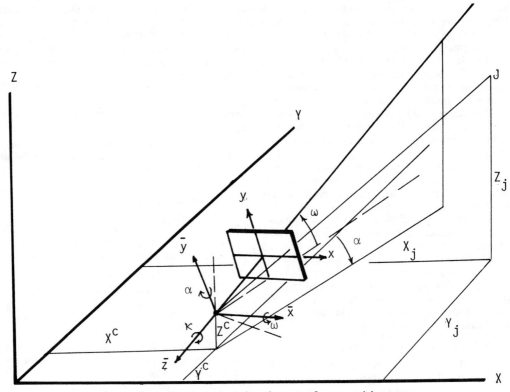

FIGURE 2-15. Exterior orientation elements of a terrestrial camera.

$m_{13} = \cos \omega \sin \kappa - \sin \omega \sin \alpha \cos \kappa$

$m_{21} = -\cos \alpha \sin \kappa$

$m_{22} = -\sin \omega \cos \kappa + \cos \omega \sin \alpha \sin \kappa$ (2.32)

$m_{23} = \cos \omega \cos \kappa + \sin \omega \sin \alpha \sin \kappa$

$m_{31} = -\sin \alpha$

$m_{32} = -\cos \omega \cos \alpha$

$m_{33} = -\sin \omega \cos \alpha.$

The sequence of rotations used in deriving the above expressions is as follows: ω-rotation first, α-rotation second, and κ-rotation last.

2.2.3.3 RELATIVE ORIENTATION

Relative orientation is the determination of the relative position and attitude of the two photographs in a stereoscopic pair with respect to each other. The primary purpose of relative orientation is to orient the two photographs so that each corresponding pair of rays from the two photographs intersect in space. This condition is achieved at the completion of relative operation only if the optical lens system in both the camera and the projector of the stereoplotter are distortionless, if the light rays truly travel in straight lines through the atmosphere, and if no geometric distortion is introduced into the photographic image during the photographic processing. These conditions are rarely satisfied in practice, and usually the corresponding pair of rays from the two photographs cannot be all made to intersect exactly in space. The objective of relative orientation is then to orient the two photographs so that the condition of intersection is as nearly achieved as possible.

2.2.3.3.1 GEOMETRIC CONDITIONS FOR RELATIVE ORIENTATION

Assuming the absence of geometric distortions caused by various sources, the relative orientation of a stereoscopic pair of photographs is accomplished by making five pairs of rays intersect. That is, if five pairs of rays intersect, then every pair of rays in the two bundles will intersect. Figure 2-16 shows the general location of the six pairs of rays that are universally used for performing relative orientation in a stereoplotter. Five of the six pairs are absolutely needed for performing relative orientation, and the sixth pair is used for checking purposes. This geometric condition will become obvious from examination of the coplanarity equation which will be derived in section 2.2.3.3.3.

2.2.3.3.2 MATHEMATICAL DEFINITION

Let X_L, Y_L and Z_L be the object space coordinates of the exposure center and ω_L, ϕ_L and κ_L be the rotation angles of the left photograph. These six parameters then define the exterior orientation of the left photograph. Similarly, let X_R, Y_R, Z_R, ω_R, ϕ_R and κ_R be the exterior orientation parameters of the right photograph.

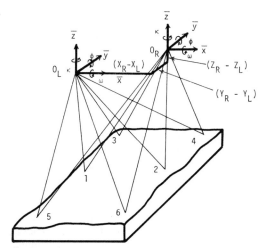

FIGURE 2-16. Geometric conditions for relative orientation.

Then mathematically speaking, a stereoscopic pair of photographs is said to be relatively oriented with respect to each other if the following five parameters are known: $(Y_R - Y_L)$, $(Z_R - Z_L)$, $(\omega_R - \omega_L)$, $(\phi_R - \phi_L)$, $(\kappa_R - \kappa_L)$. These parameters are illustrated in figure 2-16. It is assumed here that the X-axis is approximately along the direction of flight. The separation of the two photographs along the X-direction, $X_R - X_L$, controls the scale of the model and is not a parameter of relative orientation.

2.2.3.3.3 COPLANARITY EQUATION

In figure 2-17, $\bar{A}_i$ denotes the vector which originates from the exposure center (O_i) of the left photo, passes through image point a_i, and

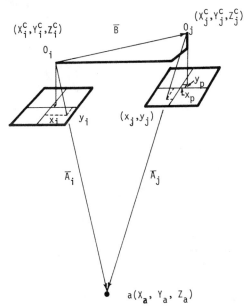

FIGURE 2-17. Condition of coplanarity.

ends at point a in the object space. Similarly, $\bar{A}_j$ denotes the vector from the right exposure center (O_j) to the same point a. The vector $\bar{B}$ extends from O_i to O_j. Then, the conditions that the two vectors, $\bar{A}_i$ and $\bar{A}_j$, intersect in space is expressed by the following equation:

$$(\bar{A}_i \times \bar{A}_j) \cdot \bar{B} = O. \tag{2.33}$$

From figure 2.17, the vector $\bar{A}_i$ may be expressed in terms of its components as follows:

$$\bar{A}_i = (X_a - X_i^c)\bar{i} + (Y_a - Y_i^c)\bar{j} + (Z_a - Z_i^c)\bar{k}. \tag{2.34}$$

From the projective transformation equation (2.22),

$$X_a - X_i^c = \lambda_i[m_{11}(x_i - x_p) + m_{21}(y_i - y_p) + m_{31}(-f)]$$
$$Y_a - Y_i^c = \lambda_i[m_{12}(x_i - x_p) + m_{22}(y_i - y_p) + m_{32}(-f)] \tag{2.35}$$

and

$$Z_a - Z_i^c = \lambda_i[m_{13}(x_i - x_p) + m_{23}(y_i - y_p) + m_{33}(-f)]$$

where the m_{ij}'s are functions of the rotation angles ω_i, ϕ_i and κ_i of photo i; x_p and y_p are photo-coordinates of the principal point; x_i and y_i are photo-coordinates of the image point a_i; and λ_i is a scale factor.

Let

$$u_i = m_{11}(x_i - x_p) + m_{21}(y_i - y_p) + m_{31}(-f);$$
$$v_i = m_{12}(x_i - x_p) + m_{22}(y_i - y_p) + m_{32}(-f); \tag{2.36}$$

and

$$w_i = m_{13}(x_i - x_p) + m_{23}(y_i - y_p) + m_{33}(-f).$$

Then Equation (2.35) may be written as

$$X_a - X_i^c = \lambda_i u_i$$
$$Y_a - Y_i^c = \lambda_i v_i \tag{2.37}$$

and

$$Z_a - Z_i^c = \lambda_i w_i.$$

Substituting Equation (2.37) into (2.34) yields

$$\bar{A}_i = \lambda_i u_i \bar{i} + \lambda_i v_i \bar{j} + \lambda_i w_i \bar{k}. \tag{2.38}$$

A similar equation can be derived for the vector $\bar{A}_j$ from the right photograph:

$$\bar{A}_j = \lambda_j u_j \bar{i} + \lambda_j v_j \bar{j} + \lambda_j w_j \bar{k}. \tag{2.39}$$

Furthermore, the following expression for vector $\bar{B}$ can be derived directly from figure 2.17:

$$\bar{B} = (X_j^c - X_i^c)\bar{i} + (Y_j^c - Y_i^c)\bar{j} + (Z_j^c - Z_i^c)\bar{k}. \tag{2.40}$$

The coplanarity condition stated in equation (2.33) is then satisfied if the determinant

$$\begin{vmatrix} (X_j^c - X_i^c) & (Y_j^c - Y_i^c) & (Z_j^c - Z_i^c) \\ \mu_i & v_i & w_i \\ \mu_j & v_j & w_j \end{vmatrix} = 0 \tag{2.41}$$

That is,

$$(X_j^c - X_i^c)(v_i w_j - v_j w_i) + (Y_j^c - Y_i^c)(u_j w_i - u_i w_j)$$
$$+ (Z_j^c - Z_i^c)(u_i v_j - u_j v_i) = 0 \tag{2.42}$$

which is commonly known as the coplanarity equation. There are twelve parameters in the equation: X_i^c, Y_i^c, Z_i^c, ω_i, ϕ_i, κ_i, X_j^c, Y_j^c, Z_j^c, ω_j, ϕ_j and κ_j. The problem of relative orientation only requires the solution of the five differential parameters $(Y_j^c - Y_i^c)$, $(Z_j^c - Z_i^c)$, $(\omega_j - \omega_i)$, $(\phi_j - \phi_i)$ and $(\kappa_j - \kappa_i)$. Therefore, arbitrary values can be assigned to the parameters X_i^c, Y_i^c, Z_i^c, ω_i, ϕ_i, κ_i and X_j^c, and then the only unknowns to be solved in a relative orientation problem will be Y_j^c, Z_j^c, ω_j, ϕ_j and κ_j. Each pair of rays for which the image coordinates x_i, y_i, x_j and y_j have been measured gives rise to one coplanarity equation. Thus five pairs of rays would yield five equations which are the minimum required to solve for the five unknowns of relative orientation. Of course, equation (2.42) is non-linear and must first be linearized be-

fore it can be used for solution. In addition, when more than five pairs of image coordinates are measured, the method of least squares may be used to determine the most probable solution. The methods of linearization of equations and least squares will be discussed in section 2.5.

2.2.3.4 ABSOLUTE ORIENTATION

After relative orientation is accomplished, the stereo model must be scaled, translated and leveled with respect to a ground reference coordinate system. The process of orienting a stereo-model into an absolute reference system is called absolute orientation. Mathematically, the problem may be defined simply as a problem of coordinate transformation. Referring to figure 2-18, let x_j, y_j and z_j represent the coordinates

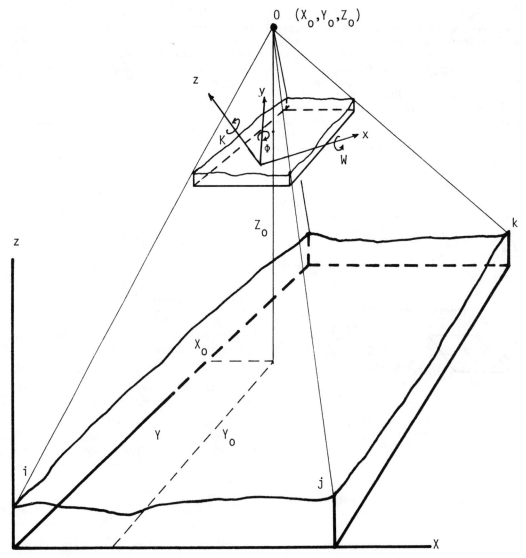

FIGURE 2-18. Absolute orientation.

of point j in the model coordinate system, and let X_j, Y_j and Z_j be the coordinates of the same point in the ground reference system. Then the relationship between the model coordinates (x_j, y_j and z_j) and the ground coordinates (X_j, Y_j and Z_j) may be expressed by the projective transformation equations derived in section 2.2.3.2.3; $i.e.$

$$X_j - X_o = \lambda \left[m_{11} x_j + m_{21} y_j + m_{31} z_j \right]$$
$$Y_j - Y_o = \lambda \left[m_{12} x_j + m_{22} y_j + m_{32} z_j \right] \quad (2.43)$$
$$Z_j - Z_o = \lambda \left[m_{13} x_j + m_{23} y_j + m_{33} z_j \right]$$

in which

$$m_{11} = \cos \Phi \cos K$$
$$m_{12} = \cos W \sin K + \sin W \sin \Phi \cos K$$
$$m_{13} = \sin W \sin K - \cos W \sin \Phi \cos K$$
$$m_{21} = -\cos \Phi \sin K$$
$$m_{22} = \cos W \cos K - \sin W \sin \Phi \sin K \quad (2.44)$$
$$m_{23} = \sin W \cos K + \cos W \sin \Phi \sin K$$
$$m_{31} = \sin \Phi$$
$$m_{32} = -\sin W \cos \Phi$$
$$m_{33} = \cos W \cos \Phi.$$

The transformation parameters include a scale factor (λ), three translation parameters (X_o, Y_o, and Z_o) and three rotation parameters (W, Φ and K). Each control point, for which the ground coordinates are known, gives rise to three equations when its model coordinates are measured. Thus, a minimum of three control points in the model area will be needed to perform the absolute orientation. Again, equation (2.43) must be linearized before it can be used in the solution for the absolute orientation parameters and the method of least squares may also be used.

The procedure for performing absolute orientation in a stereoplotter will be discussed in chapters XI and XII.

2.2.4 GEOMETRICAL SOLUTIONS TO SOME PHOTOGRAMMETRIC PROBLEMS

Photogrammetric methods are commonly used to map the topography of the terrain, size and shape of objects as well as the precise location of discrete points in an object space. The accuracy and precision of the measurements depend on the quality of the hardware and software used in both data collection (such as photography) and data reduction. Photogrammetric cameras of outstanding geometric fidelity have been developed for both aerial and terrestrial mapping projects. Mono- and stereocomparators having a precision in the order of 1 μm are commonly available, and rigorous computer softwares have been developed for analytical aerotriangulation yielding a precision approaching that of first-order geodetic surveys.

Fully automated stereoplotting instruments of high precision have been used in routine production for many years. Detailed dicussions on these hardwares and softwares will be presented in later chapters in this manual. It will suffice to state here that the types of instruments and the data reduction procedures to be used in a given project situation depends largely on the accuracy that is required. This section will present a few computational methods that are derived from simple geometric principles and which have been commonly used in applications where high-order of accuracy of measurement is not required.

2.2.4.1 STEREOSCOPIC PARALLAX AND ELEVATIONS

The stereoscopic parallax of a point A which is imaged in the overlap area of a stereoscopic pair of photographs is defined as the difference in the x-components of distances in the two photographs which are measured from the principal point to the image point A. Because stereoscopic parallax is measured along the x-axis which is in the direction of flight of the aircraft, it is commonly referred to as x-parallax. In a given pair of vertical aerial photographs, the difference in parallax or parallax difference between two image points is directly related to the difference in elevations of the two points on the ground.

In figure 2-19, two truly vertical photographs of equal focal length f are shown a distance $OO' = B$ apart and at an altitude H above a horizontal reference plane. An object A has an elevation h and images of A occur at a on the left photograph and at a' on the right one. An x axis is adopted on each photograph parallel to OO', and n and n' are both the principal points and nadir points of the respective photographs. The ordinates aa_1 and $a'a_1'$ are perpendicular to the x axis. Triangle $OO'A_1$ is in the vertical plane that contains the two perspective centers (camera stations), AA_1 is perpendicular to plane $OO'A_1$, and the elevation of A_1 is also h. The absolute stereoscopic parallax of A is defined as the algebraic difference of the absissas na_1 and $n'a_1'$:

$$p = x - x'.$$

The parameters in the above equation are illustrated graphically in the smaller figure which is composed of triangle Oa_1n of the left photograph and triangle $O'a_1'n'$ of the right one. It can be shown by similar triangles that

$$\frac{p}{f} = \frac{B}{H - h} \quad (2.45)$$

$$h = H - \frac{Bf}{p}. \quad (2.46)$$

It is to be noted that the derivation of these equations is based on (1) utilted photographs,

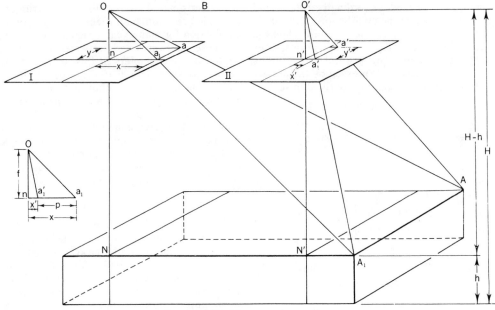

FIGURE 2-19. Stereoscopic parallax.

(2) equal flight altitudes, and (3) equal focal lengths.

The foregoing equations serve principally to define stereoscopic parallax. It is parallax difference that is used to determine elevations. It can be derived from equation (2.46) by differential calculus that

$$\Delta h = \frac{\Delta p (H - h_1)}{p_1 + \Delta p} \qquad (2.47)$$

$$\Delta p = \frac{\Delta h p_1}{(H - h_1) - \Delta h} \qquad (2.48)$$

in which Δh means "change in the value of h," h_1 is the elevation of the lower of two objects, and p_1 is the parallax of the lower point.

In practice, the average height (H') above the terrain is frequently used for the term ($H - h_1$), and the average distance (photographic base b) between principal point and conjugate principal point is used for p_1. (A conjugate principal point is the image of the principal point of a given photograph on an overlapping photograph.) Thus, equations (2.47) and (2.48) are often used in the following simplified forms

$$\Delta h = \frac{H'}{b + \Delta p} \Delta p \qquad (2.49)$$

$$\Delta p = \frac{b}{H' - \Delta h} \Delta h \ . \qquad (2.50)$$

For cases where Δp is relatively small, the following approximations are sometimes useful:

$$\Delta h = \frac{H'}{b} \Delta p \qquad (2.51)$$

$$\Delta p = \frac{b}{H'} \Delta h \ . \qquad (2.52)$$

EXAMPLE 19: Images of two objects on a pair of photographs have a parallax difference of 1.37 millimetres and the average photographic base is 92.3 millimetres. Compute the difference in elevation of the two objects if the flight altitude is 4,000 metres above the average ground level.

SOLUTION: Applying exact equation (2.49),

$$\Delta h = \frac{4,000 \times 1.37}{92.3 + 1.37} = 58.5 \text{ metres} \ .$$

If approximate equations (2.51) is applied,

$$\Delta h = \frac{4,000 \times 1.37}{92.3} = 59.4 \text{ metres} \ .$$

EXAMPLE 20: The average photographic base of a pair of overlapping photographs is 90 mm and the flight height is 7,000 metres above ground. What is the parallax difference for a contour interval of 20 metres?

SOLUTION: Applying approximate equation (2.52),

$$\Delta p = \frac{90 \times 20}{7,000} = 0.26 \text{ mm} \ .$$

In stereoplotting instruments, parallax differences are not measured but are converted mechanically into elevation differences. For approximate measurement of elevation differences, parallax differences can be measured graphically on paper prints of the photographs. The basic procedures are as follows: (see figure 2-20).

(1) Identify and mark on both photographs the principal points m, n', conjugate principal points n_1, n_2, and the images a, a', c, c' for which elevations are desired.

(2) Scale the photographic base lengths b and compute the average value.

(3) Fasten the two photographs to a drawing

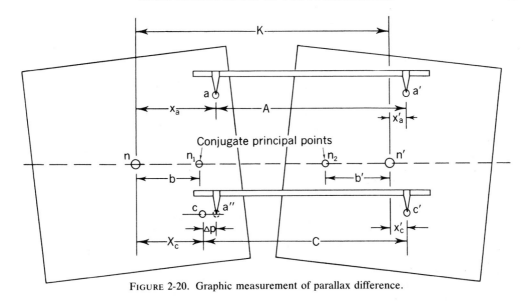

FIGURE 2-20. Graphic measurement of parallax difference.

board so that all four center points (principal points and conjugate centers n, n_1, n_2, n') lie in the same straight line. The separation of the photographs is immaterial as long as one photograph does not lie on top of the other.

(4) Draw a short line (about one-half inch to the right and to the left) through one of the images c parallel to the line of centers.

(5) Set a pair of dividers, or a beam compass, to span the distance between the other two corresponding images aa' (not the one with the line through it of step 4).

(6) Transfer the dividers to the other pair of images, setting one point on the image c' that does not have the line through it.

(7) With the other end of the divider point, prick a small hole a'' on the short line.

(8) The parallax difference is the distance $a''c$ from the prick hole of step 7 to the nearby image, which distance can be measured with a scale.

Instead of using a divider, a parallax bar may be used for more accurate measurement of the parallax. It is basically a precision ruler which has a fixed reticle of glass containing a measuring mark on one end and a mobile reticle of glass also containing a measuring mark at the other end. The movable end is equipped with a micrometer so that turning the micrometer screw changes the distance between the measuring markers by exactly the same amount shown on the micrometer. Some parallax bars give micrometer readings directly to 1/100 mm. The parallax bar is usually used in combination with a mirror stereoscope which, by permitting the user to view the overlap area stereoscopically, increases the accuracy of centering the measuring mark on the same point in each photograph. Figure 2-21 illustrates the basic set-up of a parallax bar and a mirror stereoscope.

The major source of error in the elevations computed from using equation (2.49) and the graphical method of measuring parallax differences is the unavoidable presence of tilt. In stereoplotting instruments, the tilt of the photographs is accounted for by tilting the projectors an appropriate amount. In the graphical method of parallax measurement, the photographs are placed on a flat surface and the tilt is completely ignored. Paper shrinkage caused by temperature and humidity variations introduce additional errors to the measurement. Consequently, if two images are located more than one inch apart on a photograph, errors in the elevation differences computed by the above method can be so large as to render the measurement completely unreliable. However, if several points of known elevations are imaged in the overlap area, the trend of errors can be diagrammed in the form of a parallax correction graph.

Directly analogous to x-parallax, y-parallax is defined as $p_y = y - y'$. Under the ideal con-

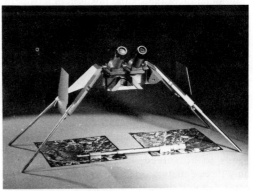

FIGURE 2-21. Parallax bar and mirror stereoscope. (Courtesy Wild Heerbrugg Instruments, Inc.)

ditions of zero tilt, complete absence of geometric distortions, equal H and equal f, the y parallax of any pair of images is zero. However, if any one of these ideal conditions is violated, y parallax becomes a definite value. The presence of y parallax is objectionable because it hinders stereoscopic vision, It indicates error in the system, usually tilt. The elimination of y parallax is the first operation in adjusting a pair of photographs in a stereoscopic plotting instrument. At least one general method of tilt determination is based on the magnitude and direction of y parallax in specified parts of the overlap area.

The relative accuracy in determining elevations and drawing contours is indicated by equations (2.51) and (2.45). The former equation shows that difference in elevation is a factor times the parallax difference. Both terms in the factor are relatively strong: H' and b are usually known within, say, 2 percent. But the value of Δp for a contour interval is weak: in multiplex work, for example, the probable error in the contour lines is about 10 percent of the contour interval because of lack of photographic definition. Consequently, one is concerned with increasing the size of Δp as a means of increasing the relative accuracy. Equation (2.45) shows the parallax is almost directly proportional to f and B, and inversely proportional to H. Thus, to increase p, f and/or B can be increased, and/or H can be decreased. But to increase f necessitates an increase in film size if the value of B is to remain the same; to increase B necessitates a lens of wider angular field of view; to decrease H necessitates a larger number of photographs of a given area which, in turn, requires more field control stations. A practical balance of the values of these elements, together with a consideration of map accuracy requirements, poses one of the most difficult economic problems in photogrammetry.

2.2.4.2 CHURCH METHOD OF SPACE RESECTION

The problem of space resection involves the determination of the spatial position of a camera exposure station. Solution of the problem usually also yields the rotation angles of the optical axis of the camera during the moment of exposure. Thus, space resection is a method of determining the six exterior orientation parameters of an aerial photograph. The method of space resection to be presented here was developed by Professor Earl Church at Syracuse University (1945, 1948). The method is particularly suited for manual calculation because the largest system of simultaneous equations that needs to be solved in this method involves only three equations. The method requires three control points imaged on the photograph, and a unique solution is obtained by assuming that no geometric distortion exists at the three image

points. The formulation can be extended to account for image distortion. Then four or more control points are needed to determine the most probable solution by the method of least squares. The Church method is presented here both because of its applicability for manual calculation and because its formulation is derived from some fundamental geometric properties of aerial photography.

Church's solution is based on the fundamental condition that the phase angle subtended at the exposure station by any two points on the ground is equal to the phase angle subtended at the exposure station by the images of the two points on the aerial photograph. In figure 2-22, this condition states that

$$\text{angle } AOB = \text{angle } aOb;$$
$$\text{angle } BOC = \text{angle } bOc; \qquad (2.53)$$

and

$$\text{angle } AOC = \text{angle } aOc.$$

Let

$$\begin{aligned} K_1 &= \cos A\hat{O}B; & k_1 &= \cos a\hat{O}b; \\ K_2 &= \cos B\hat{O}C; & k_2 &= \cos b\hat{O}c; & (2.54) \\ K_3 &= \cos A\hat{O}C; & k_3 &= \cos c\hat{O}a. \end{aligned}$$

Then, from equation (2.53),

$$\begin{aligned} K_1 &= k_1, \\ K_2 &= k_2, & (2.55) \end{aligned}$$

and

$$K_3 = k_3,$$

which are the condition equations used in Church's formulation.

Three control points will form a space pyramid with the apex angle at the exposure center O. Similarly, the three image points also form a space pyramid at the exposure center O. Since interior orientation of the camera is usually determined from camera calibration, the space pyramid formed by the three image points is geometrically defined. The solution to the problem follows a successive iteration procedure. An initial position of the exposure center O and initial attitude of the photograph are assumed. Spatial discrepancies between the two space pyramids are computed and corrections are applied to the assumed parameters. The procedure is repeated until the two pyramids coincide; that is, equations (2.55) are satisfied.

Figure 2-23 illustrates the computation form developed by Church. It contains all the instructions that are needed to perform the computation. The procedure is briefly described by means of the example included in figure 2-23.

The ground coordinates of the control points can be defined in the State Plane Coordinate System, Universal Transversal Mercator (UTM) projection, or any local rectangular coordinate system. The initial estimated coordinates of the

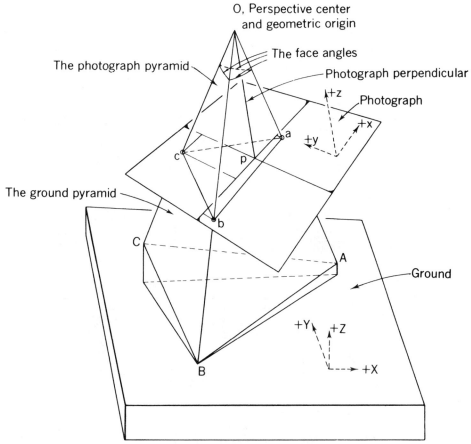

FIGURE 2-22. Geometry of space resection.

camera station are in the same system as those adopted for the ground control and can be obtained from a flight map of the area, or scaled carefully from a good map showing the estimated location of the picture center, or determined by the tracing paper method of graphic resection (radial plotting). Only a brief time period should be spent with this step, because it is not possible to obtain precise results, and errors of ordinary magnitude do not affect the accuracy of the final results.

The values d are the distances Oa, Ob, and Oc from the perspective center to the photograph images (equation 2.67). The reciprocal of d is used as a multiplier instead of repeated divisions by d itself. The quantities l, m and n are the three direction cosines of the lines Oa, Ob, and Oc (equations 2.68) and k is the cosine of the angle between two of the lines: $k_1 = \cos aOb$ (equation 2.70). These values are final and are not recomputed. Since any error in their determination has a direct effect on the solution, they should be checked carefully. These values, incidentally, are independent of the tilt of the photograph and of the coordinate system in which the images are measured.

The first three columns of row II-2 are the coordinates of the control stations in terms of the camera station considered as the origin. They are found merely by subtraction as indicated, which is an analytical translation of the origin from the ground control system to the camera station system. The remaining elements of the row are found in exactly the same manner as in the row above. If the estimated values of the coordinates of the camera station are correct, the values of K will be equal, respectively, to those of k. If they are not equal, the coordinates of the camera station are adjusted systematically by the subsequent steps until the desired equality is obtained.

Row III follows row II-2, and II-3 is returned to later. The adjustment equations are a result of partial differentiation of the expression for K with respect to X_c, Y_c, Z_c, resulting in corrections ΔX, ΔY, ΔZ which caused the discrepancies ΔK or $(k-K)$. I_i and J_i are auxiliary values used in the formation of the coefficients A, B, C of the adjustment equations. (See instruction at bottom of example.) The subscripts i, j, k are convenient brief mathematical expressions that represent the numbers 1, 2, 3 or letters a, b, c, sometimes used. The system enables one to visualize the formation of all of

COMPUTATION FORM

SPACE RESECTION AND ORIENTATION*

		Ground Coordinates (Ft.) X_g	Y_g	Z_g		Camera Station (Ft.) X_0	Y_0	Z_0
I	1	57934	20972	612	Estimated Values	49 200	29 600	19 100
	2	31 378	30 476	107	First Adjusted Values	50 005	29 992	20 049
	3	54 204	40 103	2734	Second Adjusted Values	50 002	30 000	20 002
		///	///	///	Third Adjusted Values	50 002	30 002	20 000
		///	///	///	Fourth Adjusted Values	50 001	30 002	20 000

		Image Coordinates (mm) x	y	z = -f	d	$1/d.10^{-8}$	ℓ	m	n	k
II-1	1	+10.74	+98.28	-152.40	181.66	550 479	+.05912	+.54101	-.83893	+.37587
	2	+75.91	-105.47	"	200.28	499301	+.37902	-.52661	-.76093	+.48468
	3	-101.53	-22.69	"	184.52	541 947	-.55024	-.12297	-.82593	+.59384

		$X_g - X_0$	$Y_g - Y_0$	$Z_g - Z_0$	D	$1/D.10^{-10}$	L	M	N	K
II -2	1	+8734	-8628	-18488	22193	450593	+.39355	-.38877	-.83306	+.32494
	2	-17822	+876	-18993	26060	383730	-.68388	+.03361	-.72882	+.44117
	3	+5004	+10503	-16366	20080	498008	+.24920	+.52306	-.81504	+.57370
II -3	1	+7929	-9020	-19437	22848	437675	+.34703	-.39478	-.85071	+.37774
	2	-18627	+484	-19942	27293	366394	-.68248	+.01773	-.73066	+.48642
	3	+4199	+10111	-17315	20486	488138	+.20497	+.49356	-.84521	+.59531
II-4	1	+7932	-9028	-19390	22812	438366	+.34771	-.39576	-.84999	+.37589
	2	-18624	+476	-19895	27256	366892	-.68330	+.01746	-.72993	+.48476
	3	+4202	+10103	-17268	20442	489189	+.20556	+.49423	-.84473	+.59389
II-5	1	+7932	-9030	-19388	22811	438384	+.34773	-.39586	-.84994	+.37587
	2	-18624	+474	-19893	27254	366919	-.68335	+.01739	-.72991	+.48468
	3	+4202	+10101	-17266	20440	489237	+.20558	+.49418	-.84472	+.59382

		I	J	Adjustment Equations $A(\Delta X_0)$	$+ B(\Delta Y_0)$	$+ C(\Delta Z_0) = \Delta K = k-K$	ΔK	ΔK	ΔK
III	1	-2.14	-3.06	+1.25	+0.73	+4.01	5093	-187	-6	0
	2	-3.12	-1.42	+1.78	-0.85	+3.43	4351	-174	-8	0
	3	-1.55	-2.30	-1.29	+0.08	+3.18	2014	-147	-5	2

		E	F	G	Δ	Adjustment to Camera Stations	ΔX_0	ΔY_0	ΔZ_0
IV-1	1	-2.36	-2.85	+5.91	-14.88	1st Adjustment	805	392	949
	2	-0.95	+10.09	-2.98	///	2nd Adjustment	-2	+8	-47
	3	-1.04	-9.15	-2.00	///	3rd Adjustment	0	+2	-2
		///	///	///		4th Adjustment	-1	0	-2

		e	f	g	Δ	Elements of Orientation u	v	w	tan s =
IV-2	1	-.23619	+.27299	-.85346	.698257	-.61953	+.78357	+.04717	-1.56383	t 2° 51'
	2	-.33637	-.73174	+.34137	///	-.78386	+.62076	+.01630	tan a =	s 302°36'
	3	-.29042	+.51044	+.55000	///	+.04202	-.02687	+.99876	2.89387	a 250°56'

I. First adjusted values values = Est. Values + adjustments of IV-1

II-1. $d = \sqrt{x^2 + y^2 + z^2}$
$\ell = (1/d)x;\ m = (1/d)y;\ n = (1/d)z$
$k_i = \ell_i \ell_j + m_i m_j + n_i n_j$

II-2,3,4,5. Exactly like Corresponding steps of II-1. II-3,4,5 used for subsequent adjustments only.

III. $I_i = k_i(1/D_i) - (1/D_j)$
$J_i = k_i(1/D_j) - (1/D_i)$
$A_i = L_i I_i + L_j J_i$
$B_i = M_i I_i + M_j J_i$
$C_i = N_i I_i + N_j J_i$

IV-1. $E_i = A_i B_j - A_j B_i$
$F_i = A_i C_j - A_j C_i$
$G_i = B_i C_j - B_j C_i$
$\Delta = E_1 C_3 + E_3 C_2 + E_2 C_1$
$\Delta X = (G_2 \Delta K_1 + G_3 \Delta K_2 + G_1 \Delta K_3)/\Delta$
$\Delta Y = (F_2 \Delta K_1 + F_3 \Delta K_2 + F_1 \Delta K_3)/(-\Delta)$
$\Delta Z = (E_2 \Delta K_1 + E_3 \Delta K_2 + E_1 \Delta K_3)/\Delta$

IV-2. Follows last L's, M's, N's of row II-3, 4 or 5
$e_i = \ell_i m_j - \ell_j m_i$
$f_i = \ell_i n_j - \ell_j n_i$
$g_i = m_i n_j - m_j n_i$
$\Delta = e_1 n_3 + e_2 n_1 + e_3 n_2$
$u_1 = (L_1 g_2 + L_2 g_3 + L_3 g_1)/\Delta$
$v_1 = (L_1 f_2 + L_2 f_3 + L_3 f_1)/(-\Delta)$
$w_1 = (L_1 e_2 + L_2 e_3 + L_3 e_1)/\Delta$

(u_2 and u_3 are formed by replacing L with M and N, resp.; similarly for the v's and w's.)
$\cos t = w_3$
$\tan s = (-u_3)/(-v_3)$
$\tan a = (-w_1)/(-w_2)$

(i, j, k represent cyclicly the subscripts 1, 2, 3 which in turn refer to the rows of the computation: if i = 1, then j=2, k=3; if i=2, then j=3, k=1; if i=3, j=1, k=2)

*Method of Earl Church "Revised Geometry of Aerial Photography," Syracuse University Bulletin 15, 1945.

FIGURE 2-23. Computation form for the Church method of space resection and space orientation.

the eight quantities with but two rather simple cyclical expressions. The system merely indicates that if i is any one of the values 1, 2, 3, the j represents the value that follows. It is considered that k follows j, and that i follows k. Each of these auxiliary terms can be computed completely, without writing or reentering a value, by means of a modern calculating machine that is equipped with a positive and negative accumulative multiplication device. Algebraic signs are rigidly observed.

The nine coefficients are formed from the values of I, J, L, M, and N, according to the equations given at the bottom of the form. These values also are usually final and not computed again unless the approximate values of the estimated coordinates are grossly in error. The ΔK values, however, change with each successive adjustment.

The terms E, F, G in row IV-1 are auxiliary values used in the simultaneous solution of the three adjustment equations by means of Cramer's Rule and determinants. The equations can be solved equally as well by any other method, but the indicated method is concrete and direct, which the non-mathematician may prefer. The results of the solution are placed in the three spaces opposite *First Adjustment*, and these three values are added algebraically to the *Estimated Values* at the top of the page to obtain the *First Adjusted Values*.

Row II-3 is then computed just like II-2, except the *First Adjusted Values* are used in place of the *Estimated Values*. The resulting new values of K in II-3 yield new values of ΔK, placed in row III.

This time the solution of the adjustment equations of row IV-1 consists only of the three direct calculations for ΔX, ΔY, ΔZ from the previously determined values of E, F, G, as indicated at the bottom of the form.

Rows II-3, -4, and -5 are repeated until the values of ΔK are small enough to be of no consequence, or until the errors in the K's are relatively comparable to the accidental errors of measurement of the image coordinates and control point identification. The last computed values of L, M, N are considered to be final and correct and are used later in row IV-2.

Row IV-2 is the direct solution of space orientation. Three sets of three simultaneous equations in terms of the nine unknown elements u, v, w are not specifically formed although all the coefficients and constant terms are already tabulated in the l, m, n and final L, M, N values above. The equations are:

$$l_1 u_1 + m_1 v_1 + n_1 w_1 = L_1$$
$$l_2 u_1 + m_2 v_1 + n_2 w_1 = L_2$$
$$l_3 u_1 + m_3 v_1 + n_3 w_1 = L_3$$
$$l_1 u_2 + m_1 v_2 + n_1 w_2 = M_1$$
$$l_2 u_2 + m_2 v_2 + n_2 w_2 = M_2$$
$$l_3 u_2 + m_3 v_2 + n_3 w_2 = M_3$$
$$l_1 u_3 + m_1 v_3 + n_1 w_3 = N_1$$
$$l_2 u_3 + m_2 v_3 + n_2 w_3 = N_2$$
$$l_3 u_3 + m_3 v_3 + n_3 w_3 = N_3.$$

Inasmuch as the coefficients are alike for each of the three sets of equations, and only the constant terms vary, Cramer's rule is again a convenient method of solution. The terms e, f, g, Δ are auxiliary terms computed just like those in row IV-1 bearing corresponding capital letters. The tilt, swing, and azimuth are given specifically, although they may not be required in some computations. Moreover, only the bottom row of the matrix need be computed if only the tilt and swing are required.

Checks against arithmetic errors can be applied in several places. In every line of row II,

$$l^2 + m^2 + n^2 = 1, \text{ and } L^2 + M_2 + N^2 = 1.$$

The values of the k's can be checked only by repetition, which is highly advisable inasmuch as the remainder of the work is based on these values. Mistakes in the K's, although disconcerting, cause no errors in the final resuls. The solution of the adjustment equations can be checked by substituting the residual values of ΔX, ΔY, ΔZ of IV-1 in any one of the equations of III, which should yield the same value of ΔK as was used in the solution.

The adjustments should become successively smaller with each adjustment, but all of them need not be smaller—one value sometimes increases. The rate of decrease in the residual adjustments depends somewhat on how ideally the control points are situated, and on the accuracy of the *estimated values*. In certain peculiar circumstances where the exposure station is near the surface of the cylinder or the sphere containing the three control points, the residuals might fail to converge to zero and the solution fails, or is indeterminate. Ordinarily, the sum of the residuals becomes about 1/10 as large with each successive adjustment.

The values of u, v, w of IV-2 can be checked by substitution in equations III on the form.

The accuracy of the six resulting elements of the computation depends only on the accuracy of the measured image coordinates and on the number of adjustments used. In the example, after the third adjustment, the small size of the residuals $(-2, -1, 0)$ indicates that the last adjusted values have an error of about one part in 10,000, which is comparable to the accuracy of the given image coordinates.

2.2.4.3 COORDINATES OF UNKNOWN POINT

Once the exterior orientation parameters of the two photographs in a stereoscopic pair are known, the spatial position of any point in the overlap area can be computed from the photocoordinates of the image point in the two photographs by using equation (2.22). The geometric principle is basically that of spatial intersection of two rays from two fixed points in space. The exposure centers of the two photographs represent the two fixed points. The direction of the photographic axis (defined by either ω, ϕ and κ or t, S and α rotation angles) together with the photo-coordinates of the image point define the direction of the light rays from one photograph. Thus two such rays uniquely define a point of intersection in space.

Let X_i^c, Y_i^c, Z_i^c, ω_i, ϕ_i and κ_i denote the exterior orientation parameters of the left photograph in the stereoscopic pair, and let X_j^c, Y_j^c, Z_j^c, ω_j, ϕ_j and κ_j denote the exterior orientation parameters of the right photograph as shown in figure 2-17. Let X_A, Y_A and Z_A, be the unknown ground coordinates of point A. From equation (2.22), the ray of light which travels from the exposure center of the left photograph to the ground point A is expressed in the following projective transformation equations

$$X_A - X_i^c = \lambda_i \left[m_{11}^i (x_i - x_p) + m_{21}^i (y_i - y_p) + m_{31}^i (-f) \right]$$
$$Y_A - Y_i^c = \lambda_i \left[m_{12}^i (x_i - x_p) + m_{22}^i (y_i - y_p) + m_{32}^i (-f) \right]$$
$$Z_A - Z_i^c = \lambda_i \left[m_{13}^i (x_i - x_p) + m_{23}^i (y_i - y_p) + m_{33}^i (-f) \right]$$
$$(2.56)$$

in which λ_i is the scale factor at the image of point A in the left photograph; x_i and y_i are the photo-coordinates of the image point; and m_{11}^i, m_{12}^i, ... and m_{33}^i are functions of ω_i, ϕ_i and κ_i as expressed in equation (2.23).

Let

$$u_i = m_{11}^i (x_i - x_p) + m_{21}^i (y_i - y_p) + m_{31}^i (-f);$$
$$v_i = m_{12}^i (x_i - x_p) + m_{22}^i (y_i - y_p) + m_{32}^i (-f);$$

and

$$w_i = m_{13}^i (x_i - x_p) + m_{23}^i (y_i - y_p) + m_{33}^i (-f).$$
$$(2.57)$$

Then

$$X_A - X_i^c = \lambda_i \, u_i$$
$$Y_A - Y_i^c = \lambda_i \, v_i$$
$$(2.58)$$

and

$$Z_A - Z_i^c = \lambda_i \, w_i.$$

Rearranging terms yields

$$X_A = \lambda_i \, u_i + X_i^c;$$
$$Y_A = \lambda_i \, v_i + Y_i^c;$$
$$(2.59)$$

and

$$Z_A = \lambda_i \, w_i + Z_i^c.$$

Similarly by changing the subscript and superscript i to j, the following expressions can be used to describe the ray from the right photograph:

$$u_j = m_{11}^j (x_j - x_p) + m_{21}^j (y_j - y_p) + m_{31}^j (-f)$$
$$v_j = m_{12}^j (x_j - x_p) + m_{22}^j (y_j - y_p) + m_{32}^j (-f) \quad (2.60)$$
$$w_j = m_{13}^j (x_j - x_p) + m_{23}^j (y_j - y_p) + m_{33}^j (-f)$$

$$X_A = \lambda_j \, u_j + X_j^c$$
$$Y_A = \lambda_j \, v_j + Y_j^c \qquad (2.61)$$

and

$$Z_A = \lambda_j \, w_j + Z_j^c.$$

Equating equation (2.59) to (2.61) yields

$$\lambda_i \, u_i + X_i^c = \lambda_j \, u_j + X_j^c \qquad (2.62)$$
$$\lambda_i \, v_i + Y_i^c = \lambda_j \, v_j + Y_j^c \qquad (2.63)$$
$$\lambda_i \, w_i + Z_i^c = \lambda_j \, w_j + Z_j^c. \qquad (2.64)$$

Solving equations (2.62) and (2.63) for λ_j yields

$$\lambda_j = \frac{(X_i^c - X_i^c) v_i - (Y_i^c - Y_i^c) u_i}{v_j \, u_i - u_j \, v_i}. \qquad (2.65)$$

Thus, equation (2.65) can be used to compute λ_j which is then substituted into equation (2.61) to yield the coordinates (X_A, Y_A and Z_A) of the unknown point A. The computation procedures may be summarized as follows

1. Compute m_{11}^i, m_{21}^i, m_{31}^i, m_{12}^i, m_{22}^i, m_{32}^i, m_{13}^i, m_{23}^i and m_{33}^i from equation (2.23).
2. Compute u_i, v_i, and w_i from equation (2.57).
3. Compute m_{11}^j, m_{21}^j, m_{31}^j, m_{12}^j, m_{22}^j, m_{32}^j, m_{13}^j, m_{23}^j and m_{33}^j from equation (2.23).
4. Compute u_j, v_j and w_j from equation (2.57).
5. Compute λ_j from equation (2.65).
6. Compute X_A, Y_A and Z_A from equation (2.61).
7. As a computation check, λ_i can be computed by the following expression

$$\lambda_i = \frac{(X_j^c - X_i^c) \, v_j - (Y_j^c - Y_i^c) \, u_j}{(v_j \, u_i - u_j \, v_i)}. \qquad (2.66)$$

Then by substituting λ_i into equation (2.61), X_A, Y_A and Z_A can again be computed.

Although the above computation can be tedious if performed by hand, it can be easily performed in a programmable desk calculator or an electronic computer.

2.2.5 SOLID ANALYTIC GEOMETRY

A few of the most common principles of solid analytic geometry are included here. Considerations are limited to perpendicular, right-handed coordinate axes.

The coordinate axes of a three-dimensional rectangular coordinate system consist of three mutually perpendicular lines which intersect at a common origin O (fig. 2-18). The axes are denoted in a specific ordered sequence, such as X, Y, Z, or x_1, x_2, x_3, or first, second, third. One end of each axis is considered to have a positive, or plus, direction (OX, OY, OZ) and distances along each of the axes are measured in equal units of length. The system of axes is considered to be *right handed* in the sense that if the positive end of the first axis is rotated into the positive axis of the second, then the positive end of the third axis is in the direction that a right-hand screw advances if so rotated. It is significant that the right-hand rule continues to apply to a set of axes if the numbering of the axes is changed in a cyclic manner. Thus, the positive direction of angular measurement in analytic geometry usually appears to be counterclockwise on diagrams and illustrations.

The position of an object in space may be designated by its three ordered coordinate numbers: $P(x,y,z)$; $Q(2,1,9)$.

The distance $P_1 P_2$ between two points in space $P_1 (x_1, y_1, z_1)$ and $P_2(x_2, y_2, z_2)$ is given by

$$\overline{P_1P_2} = [(x_2 - x_1)^2 + (y_2 - y_1)^2 + (z_2 - z_1)^2]^{1/2}. \tag{2.67}$$

The orientation of a line P_1P_2 is specified by the three *direction angles* α, β, γ (each less than 180°) the line makes with the positive directions of the three axes. The cosines of these angles, called *direction cosines*, are given by

$$\cos \alpha = (x_2 - x_1) / \overline{P_1P_2}$$
$$\cos \beta = (y_2 - y_1) / \overline{P_1P_2} \tag{2.68}$$
$$\cos \gamma = (z_2 - z_1) / \overline{P_1P_2}.$$

It is significant that

$$\cos^2 \alpha + \cos^2 \beta + \cos^2 \gamma = 1.$$

Any three numbers a_1, a_2, a_3 which are proportional to (a common factor of) the direction cosines of a line are called the *direction numbers* of the line:

$$a_1: a_2: a_3 = \cos \alpha: \cos \beta: \cos \gamma. \tag{2.69}$$

The angle θ between two intersecting directed lines is given by

$$\cos \theta = \cos \alpha_1 \cos \alpha_2 + \cos \beta_1 \cos \beta_2 + \cos \gamma_1 \cos \gamma_2. \tag{2.70}$$

Note that $\cos \theta = 1$ for two parallel lines and $\cos \theta = 0$ for two perpendicular lines.

The *general* linear equation

$$Ax + By + Cz + D = 0 \tag{2.71}$$

represents a plane surface. The intercept form of the equation is

$$x/a + y/b + z/c - 1 = 0 \tag{2.72}$$

where a, b, c represent the points where the plane intersects the three axes. The *normal* form is

$$x \cos \alpha + y \cos \beta + z \cos \gamma - p = 0 \tag{2.73}$$

where the angles are the direction angles of a line perpendicular to the plane and p is the perpendicular distance from the origin to the plane.

The equation of a plane that passes through the point x_1, y_1, z_1 and has the direction numbers A, B, C is

$$A(x - x_1) + B(y - y_1) + C(z - z_1) = 0. \tag{2.74}$$

Note that A, B, C may also be direction cosines. The equation of a plane passing through three points is stated by means of the determinant

$$\begin{vmatrix} x & y & z & 1 \\ x_1 & y_1 & z_1 & 1 \\ x_2 & y_2 & z_2 & 1 \\ x_3 & y_3 & z_3 & 1 \end{vmatrix} = 0. \tag{2.75}$$

The equation of a straight line in space is stated by means of the equations of two planes that intersect in the line. For example, the *general* form is

$$A_1x + B_1y + C_1z + D_1 = 0$$
$$A_2x + B_2y + C_2z + D_2 = 0. \tag{2.76}$$

The equation of a line passing through two defined points is:

$$\frac{x - x_1}{x_2 - x_1} = \frac{y - y_1}{y_2 - y_1} = \frac{z - z_1}{z_2 - z_1}. \tag{2.77}$$

Although three equations can be derived from equation (2.77), only two are significant and the third is redundant. The point-direction form of the equation of a line passing through x_1, y_1, z_1 and having the direction numbers a, b, c is

$$\frac{x - x_1}{a} = \frac{y - y_1}{b} = \frac{z - z_1}{c} \tag{2.78}$$

in which a, b, c may also be the direction cosines of the line.

Curved surfaces may be stated by equations of higher degree. Conic surfaces are specified by equations of second degree. As in plane analytic geometry, the relations of the coefficients in the equation specify the type of conic surface and its orientation. Equations of curved lines are stated in the form of two (parametric) equations of surfaces.

If the origin of the coordinate system is translated to a new location (a, b, c) so that the coordinates x, y and z of a point are transformed into x', y' and z', the relationship can be stated in terms of the three coordinates a, b and c as follows:

$$x' = x - a$$
$$y' = y - b \tag{2.79}$$
$$z' = z - c.$$

If the coordinate axes are rotated so that the coordinates x, y and z of a point are transformed into x', y' and z', the relationship of the two sets of coordinates can be stated in terms of nine direction cosines m_{ij}:

$$x' = m_{11} x + m_{12} y + m_{13} z;$$
$$y' = m_{21} x + m_{22} y + m_{23} z; \tag{2.80}$$

and

$$z' = m_{31} x + m_{32} y + m_{33} z;$$

and inversely,

$$x = m_{11} x' + m_{21} y' + m_{31} z';$$
$$y = m_{12} x' + m_{22} y' + m_{32} z'; \tag{2.81}$$

and

$$z = m_{13} x' + m_{23} y' + m_{33} z';$$

in which m_{11} is the cosine of the angle between the x- and x'-axes; m_{12} is the cosine of the angle between the x- and y'-axes, *etc*. The following relationships hold true for the m_{ij}'s:

(1) The determinant

$$\begin{vmatrix} m_{11} & m_{12} & m_{13} \\ m_{21} & m_{22} & m_{23} \\ m_{31} & m_{32} & m_{33} \end{vmatrix} = 1;$$

(2) the sum of the squares of the three terms in any row or column in the above determinant is equal to 1;

(3) the sum of the products of corresponding terms in any two rows (or columns) is zero, for example,

$$m_{11} m_{21} + m_{12} m_{22} + m_{13} m_{23} = 0$$
$$m_{12} m_{13} + m_{22} m_{23} + m_{32} m_{33} = 0;$$

and

(4) each m_{ij} is equal to its cofactor, *i.e.*

$$m_{11} = m_{22} m_{33} - m_{32} m_{23},$$
$$m_{12} = m_{31} m_{23} - m_{21} m_{33}, etc.$$

The m_{ij}'s can be expressed as functions of three rotation angles such as *omega* (ω), *phi* (ϕ) and *kappa* (κ) (section 2.2.3.2.2), and tilt (t), swing (S) and azimuth (α) (section 2.2.3.2.4).

2.3 Matrix Algebra

Matrix algebra is used extensively in computational photogrammetry for the representation of linear equations. It is particularly useful in the development of solution algorithms to complex least squares adjustment problems and in the development of efficient algorithms for processing in electronic computers.

A matrix is a rectangular arrangement of m rows and n columns of numbers called *elements*. Such a matrix is said to have dimension $m \times n$ or be of order $m \times n$. Matrices may be systematically added, subtracted, and multiplied in a way analogous to ordinary (or scalar) algebra.

An upper case letter, such as A, is used to represent a matrix consisting of the elements a_{ij}, where i denotes the number of the *row* and j the number of the *column* in which the number a is located. A matrix is shown as bounded by either brackets or parentheses:

$$A = \begin{bmatrix} a_{11} & a_{12} & \cdots & a_{1n} \\ a_{21} & a_{22} & \cdots & a_{2n} \\ & & \cdots & \\ a_{m1} & a_{m2} & \cdots & a_{mn} \end{bmatrix} \qquad (2.82)$$

The value of i has the range 1, 2 $\cdots$ m, and j, 1, 2 $\cdots$ n.

A matrix may consist of a single column or a single row. A column matrix is frequently called a *vector* or *column vector*. A single row matrix may likewise be referred to as a *row vector* and is frequently considered to be a special form of a column vector. A matrix consisting of a single row and a single column (a single element) is called a *scalar*, which corresponds to a numerical multiplication factor in ordinary algebra.

With reference to a given matrix A, the matrix A^T which consists of the elements of A interchanged column for row is said to be the *transpose* of A

$$A^T = \begin{bmatrix} a_{11} & a_{21} & \cdots & a_{m1} \\ a_{12} & a_{22} & \cdots & a_{m2} \\ & & \cdots & \\ a_{1n} & a_{2n} & \cdots & a_{mn} \end{bmatrix}. \qquad (2.83)$$

(Some authors use A' instead of A^T to indicate the transpose.) Thus,

$$X^T = (x_1, x_2 \cdots x_m)$$

indicates that X is a *column vector* consisting of m elements. (Some authors also use the term *row vector*.)

Two matrices of the same order $m \times n$ can be added (or subtracted). The sum of two matrices of the same order is another matrix of like order whose elements are the sums of the corresponding elements of the two original matrices:

$$A + B = \begin{bmatrix} a_{11} & a_{12} & \cdots & a_{1n} \\ a_{21} & a_{22} & \cdots & a_{2n} \\ & & \cdots & \\ a_{m1} & a_{m2} & \cdots & a_{mn} \end{bmatrix}$$

$$+ \begin{bmatrix} b_{11} & b_{12} & \cdots & b_{1n} \\ b_{21} & b_{22} & \cdots & b_{2n} \\ & & \cdots & \\ b_{m1} & b_{m2} & \cdots & b_{mn} \end{bmatrix}$$

$$= \begin{bmatrix} a_{11} + b_{11} & a_{12} + b_{12} & \cdots & a_{1n} + b_{1n} \\ a_{21} + b_{21} & a_{22} + b_{22} & \cdots & a_{2n} + b_{2n} \\ & & \cdots & \\ a_{m1} + b_{m1} & a_{m2} + b_{m2} & \cdots & a_{mn} + b_{mn} \end{bmatrix}. \qquad (2.84)$$

Example 21: Given the following matrices:

$$A = \begin{bmatrix} 1.1 & 2.0 & 3.4 \\ 5.0 & 7.3 & 4.8 \\ 1.3 & 2.4 & 3.0 \end{bmatrix} ; B = \begin{bmatrix} 1 & 2 & 5 \\ 7 & 6 & 3 \\ 4 & 8 & 2 \end{bmatrix} ;$$

$$C = \begin{bmatrix} 1.9 & 21.3 & 4.0 \end{bmatrix} ; D = \begin{bmatrix} 4 \\ 3 \\ 1 \end{bmatrix}$$

$$A^T = \begin{bmatrix} 1.1 & 5.0 & 1.3 \\ 2.0 & 7.3 & 2.4 \\ 3.4 & 4.8 & 3.0 \end{bmatrix} ; C^T = \begin{bmatrix} 1.9 \\ 21.3 \\ 4.0 \end{bmatrix} ;$$

$$A + B = \begin{bmatrix} 2.1 & 4.0 & 8.4 \\ 12.0 & 13.3 & 7.8 \\ 5.3 & 10.4 & 5.0 \end{bmatrix}$$

$$C^T + D = \begin{bmatrix} 5.9 \\ 24.3 \\ 5.0 \end{bmatrix}.$$

The scalar multiplication of a scalar and a matrix implies that each element of the matrix is multiplied by the scalar value:

$$s\,A = \begin{bmatrix} sa_{11} & sa_{12} & \cdots & sa_{1n} \\ sa_{21} & sa_{22} & \cdots & sa_{2n} \\ \vdots & \vdots & & \vdots \\ sa_{m1} & sa_{m2} & \cdots & sa_{mn} \end{bmatrix}. \qquad (2.85)$$

Matrices are said to be *conformable* with respect to addition and multiplication if the orders of the matrices are suitable for the operations. Two matrices are conformable for addition if they are of the same order; they are conformable for multiplication if the number of columns in the first is equal to the number of rows in the second.

The product of an $m \times n$ rectangular matrix and an $n \times 1$ column vector whose elements consist of the sum of the products of the elements of the ith row of the matrix multiplied by the corresponding elements of the vector:

$$\mathbf{Ax} = \begin{bmatrix} a_{11} & a_{12} & \cdots & a_{1n} \\ a_{21} & a_{22} & \cdots & a_{2n} \\ \vdots & \vdots & & \vdots \\ a_{m1} & a_{m2} & \cdots & a_{mn} \end{bmatrix} \begin{bmatrix} x_1 \\ x_2 \\ \vdots \\ x_n \end{bmatrix}$$

$$= \begin{bmatrix} a_{11}\,x_1 + a_{12}\,x_2 + \cdots + a_{1n}x_n \\ a_{21}\,x_1 + a_{22}\,x_2 + \cdots + a_{2n}x_n \\ \vdots \\ a_{m1}\,x_1 + a_{m2}\,x_2 + \cdots + a_{mn}\,x_n \end{bmatrix}. \qquad (2.86)$$

The reverse order of combination x**A** is not defined and has no meaning because the matrices are not conformable. The product of a $1 \times m$ row matrix and an $m \times n$ rectangular matrix is similarly defined as a row matrix:

$$\mathbf{y^T A} = (y_1\,y_2 \cdots y_m) \begin{bmatrix} a_{11} & a_{12} & \cdots & a_{1n} \\ a_{21} & a_{22} & \cdots & a_{2n} \\ \vdots & \vdots & & \vdots \\ a_{m1} & a_{m2} & \cdots & a_{mn} \end{bmatrix}$$

$$= (y_1 a_{11} + y_2 a_{21} + \cdots + y_m a_{m1} \\ y_1 a_{12} + y_2 a_{22} + \cdots + y_m a_{m2} \\ y_1 a_{1n} + y_2 a_{2n} + \cdots + y_m a_{mn}). \qquad (2.87)$$

The product of an $m \times k$ rectangular matrix and a $k \times n$ rectangular matrix is an $m \times n$ matrix where an element in the ith row and j^{th} column of the product matrix is the sum of the products of the corresponding elements of the ith row of the first matrix and the jth column of the second matrix, (as though the second matrix consisted of j column vectors). Two matrices can be multiplied only if the number of columns k in the first matrix is equal to the number of rows of the second. In general, if two matrices are multiplied in reverse order, the product (if conformable) is a different matrix.

EXAMPLE 22: For the matrices defined in Example 21,

$$\mathbf{BD} = \begin{bmatrix} 1 & 2 & 5 \\ 7 & 6 & 3 \\ 4 & 8 & 2 \end{bmatrix} \begin{bmatrix} 4 \\ 3 \\ 1 \end{bmatrix} = \begin{bmatrix} 4 + 6 + 5 \\ 28 + 18 + 3 \\ 16 + 24 + 2 \end{bmatrix}$$

$$= \begin{bmatrix} 15 \\ 49 \\ 42 \end{bmatrix}$$

$$\mathbf{CD} = \begin{bmatrix} 1.9 & 21.3 & 4.0 \end{bmatrix} \begin{bmatrix} 4 \\ 3 \\ 1 \end{bmatrix} = \begin{bmatrix} 7.6 + 63.9 + 4.0 \end{bmatrix}$$

$$= \begin{bmatrix} 75.5 \end{bmatrix}$$

$$\mathbf{AB} = \begin{bmatrix} 1.1 & 2.0 & 3.4 \\ 5.0 & 7.3 & 4.8 \\ 1.3 & 2.4 & 3.0 \end{bmatrix} \begin{bmatrix} 1 & 2 & 5 \\ 7 & 6 & 3 \\ 4 & 8 & 2 \end{bmatrix}$$

$$= \begin{bmatrix} 28.7 & 41.4 & 18.3 \\ 75.3 & 92.2 & 56.5 \\ 30.1 & 41.0 & 19.7 \end{bmatrix}$$

$$\mathbf{BA} = \begin{bmatrix} 1 & 2 & 5 \\ 7 & 6 & 3 \\ 4 & 8 & 2 \end{bmatrix} \begin{bmatrix} 1.1 & 2.0 & 3.4 \\ 5.0 & 7.3 & 4.8 \\ 1.3 & 2.4 & 3.0 \end{bmatrix}$$

$$= \begin{bmatrix} 17.6 & 28.6 & 28.0 \\ 41.6 & 65.0 & 61.6 \\ 47.0 & 71.2 & 58.0 \end{bmatrix}$$

$$2\mathbf{B} = 2 \begin{bmatrix} 1 & 2 & 5 \\ 7 & 6 & 3 \\ 4 & 8 & 2 \end{bmatrix} = \begin{bmatrix} 2 & 4 & 10 \\ 14 & 12 & 6 \\ 8 & 16 & 4 \end{bmatrix}$$

A *unit* matrix has *ones* for the values of the elements on the principal diagonal and zeros elsewhere:

$$\mathbf{I} = \begin{bmatrix} 1 & 0 & \cdots & 0 \\ 0 & 1 & \cdots & 0 \\ \vdots & & \ddots & \vdots \\ 0 & 0 & \cdots & 1 \end{bmatrix}. \qquad (2.88)$$

The product of a given matrix and a unit matrix is the given matrix itself unchanged and is analogous to multiplication by 1 in ordinary algebra. If the product of two matrices is a unit matrix, then either of the matrices is called the *inverse* of the other:

$$\mathbf{AA^{-1}} = \mathbf{A^{-1}A} = \mathbf{I}.$$

The inverse of a given matrix can ordinarily be found through a computational routine which is quite lengthy if the matrix is large.

A *reversal* rule applies both to the *transpose* and *inverse* operations. The transpose (inverse) of the product of two matrices is equal to the product of the transposes (inverses) of the separate matrices taken in *reverse* order:

$$(\mathbf{A\,B})^T = \mathbf{B^T A^T}$$

$$(\mathbf{A\,B})^{-1} = \mathbf{B^{-1}A^{-1}}. \qquad (2.89)$$

The *determinant* of a *square* matrix may be expressed in the same manner as in ordinary algebra. The determinant of a rectangular matrix is undefined and nonexistent.

As an example of the use of matrix algebra, a system of n simultaneous linear equations in n unknowns can be expressed as

$$\mathbf{A}x = l \qquad (2.90)$$

where $\mathbf{A}$ is the matrix of the coefficients, x is a vector consisting of the list of unknowns, and l is the vector consisting of the list of constant terms. To represent the solution of the system, both sides are multiplied on the left by the inverse of $\mathbf{A}$:

$$\mathbf{A}^{-1}\mathbf{A}x = \mathbf{A}^{-1}l$$

and since

$$\mathbf{A}^{-1}\mathbf{A} = \mathbf{I}, \text{ and } \mathbf{I}x = x,$$

$$x = \mathbf{A}^{-1}l \qquad (2.91)$$

which states that the list of unknowns can be found by the matrix multiplication of the inverse and the constant terms. Further, if in $\mathbf{A}x = l$ there are more equations than unknowns, a least squares solution requires that $\mathbf{A}$ first be "normalized." It can be demonstrated that the square matrix of the coefficients of the normal equations can be computed by premultiplying $\mathbf{A}$ by its transpose:

$$\mathbf{A}^T\mathbf{A}x = \mathbf{A}^Tl.$$

Then the solution is, as before,

$$(\mathbf{A}^T\mathbf{A})^{-1}(\mathbf{A}^T\mathbf{A})x = (\mathbf{A}^T\mathbf{A})^{-1}\mathbf{A}^Tl$$

$$x = (\mathbf{A}^T\mathbf{A})^{-1}\mathbf{A}^Tl \qquad (2.92)$$

$$x = [(\mathbf{A}^T\mathbf{A})^{-1}\mathbf{A}^T]l.$$

EXAMPLE 23: Consider the three simultaneous linear equations

$$3x - 4y + 5z = 10$$
$$2x - y + z = 3$$
$$x - 3y + 2s = 1.$$

These can be represented in matrix notation as

$$\begin{bmatrix} 3 & -4 & 5 \\ 2 & -1 & 1 \\ 1 & -3 & 2 \end{bmatrix} \begin{bmatrix} x \\ y \\ z \end{bmatrix} = \begin{bmatrix} 10 \\ 3 \\ 1 \end{bmatrix} \text{ or } \mathbf{A}x = l$$

The solution similarly is

$$\begin{bmatrix} x \\ y \\ z \end{bmatrix} = \begin{bmatrix} 3 & -4 & 5 \\ 2 & -1 & 1 \\ 1 & -3 & 2 \end{bmatrix}^{-1} \begin{bmatrix} 10 \\ 3 \\ 1 \end{bmatrix} \text{ or } x = \mathbf{A}^{-1}l$$

The inverse of $\mathbf{A}$ is shown as follows (the routine for its determination is beyond the scope of this treatment):

$$\begin{bmatrix} x \\ y \\ z \end{bmatrix} = \begin{bmatrix} -1/10 & +7/10 & -1/10 \\ +3/10 & -1/10 & -7/10 \\ +1/2 & -1/2 & -1/2 \end{bmatrix} \begin{bmatrix} 10 \\ 3 \\ 1 \end{bmatrix}.$$

Then

$$x = 10(-1/10) + 3(7/10) + 1(-1/10)$$
$$= +1$$
$$y = 10(3/10) + 3(-1/10) + 1(-7/10)$$
$$= +2$$
$$z = 10(+1/2) + 3(-1/2) + 1(-1/2)$$
$$= +3$$

which is the solution to the three equations.

The projective transformation equation (2.22) may be written in matrix form as follows:

$$\begin{bmatrix} X_j - X^c_i \\ Y_j - Y^c_i \\ Z_j - Z^c_j \end{bmatrix} = \lambda_{ij} \begin{bmatrix} m_{11} & m_{21} & m_{31} \\ m_{12} & m_{22} & m_{32} \\ m_{13} & m_{23} & m_{33} \end{bmatrix} \begin{bmatrix} x_{ij} - x_p \\ y_{ij} - y_p \\ -f \end{bmatrix}. \qquad (2.93)$$

By manipulation of the matrices,

$$\begin{bmatrix} x_{ij} - x_p \\ y_{ij} - y_p \\ -f \end{bmatrix} = \frac{1}{\lambda_{ij}} \begin{bmatrix} m_{11} & m_{21} & m_{31} \\ m_{12} & m_{22} & m_{32} \\ m_{13} & m_{23} & m_{33} \end{bmatrix}^{-1} \begin{bmatrix} X_j - X^c_i \\ Y_j - Y^c_i \\ Z_j - Z^c_j \end{bmatrix} \qquad (2.94)$$

in which the m_{ij}'s are functions of three rotation angles as expressed in equations (2.23). Let

$$\mathbf{M} = \begin{bmatrix} m_{11} & m_{21} & m_{31} \\ m_{12} & m_{22} & m_{32} \\ m_{13} & m_{23} & m_{33} \end{bmatrix} \qquad (2.95)$$

which is commonly referred to as the orientation matrix. For a given set of rotation angles, it is always true that

$$M^{-1} = M^T. \qquad (2.96)$$

That is, equation (2.94) may be simply written as follows:

$$\begin{bmatrix} x_{ij} - x_p \\ y_{ij} - y_p \\ -f \end{bmatrix} = \frac{1}{\lambda_{ij}} \begin{bmatrix} m_{11} & m_{12} & m_{13} \\ m_{21} & m_{22} & m_{23} \\ m_{31} & m_{32} & m_{33} \end{bmatrix} \begin{bmatrix} X_j - X^c_i \\ Y_j - Y^c_i \\ Z_j - Z^c_i \end{bmatrix} \qquad (2.97)$$

A square matrix $\mathbf{A}$ is said to be an orthogonal matrix if $\mathbf{A}^{-1} = \mathbf{A}^T$. An orthogonal matrix is characterized by the following properties:

(1) its determinant is equal to $+1$;
(2) the sum of the squares of the terms in any row or column is equal to 1;
(3) the sum of the products of corresponding terms in any two rows (or columns) is zero; and
(4) each element in the matrix is equal to its cofactor; for example: $m_{12} = m_{31}m_{23} - m_{21}m_{33}$.

2.4 Probability and Statistics

Photogrammetric measurements can be broadly subdivided into two categories.

(1) Direct measurements—the unknown parameter is measured directly. For example, the coordinates of image points on an aerial photo can be measured directly using a comparator with rectangular coordinate axes.

(2) Indirect measurements—the unknown parameter is computed from the measured values of one or more other physical parameters. For example, the ground coordinates (X_j, Y_j, Z_j) of an object point is computed from the measured photo-coordinates of an image of the same point on two or more photographs.

In order to minimize the influence of unavoidable errors in the measurement, repeated measurements are usually made under a wide range of conditions. In the case of indirect measurements, redundant measurements are usually made to provide more than one solution of the desired parameters. Consequently, one of the major problems in photogrammetric measurements is to determine the most probable solution from a set of repeated and/or redundant measurements and to estimate the accuracy of the solution. The method of least squares adjustment, which is the most commonly used method of adjustment in photogrammetry, will be discussed in section 2.5. This section will present some fundamental principles of probability and statistics which form the basis to all methods of adjustment.

2.4.1 TYPES OF MEASUREMENT ERRORS

The errors in quantitative measurements can be broadly classified into the following four types:

(1) blunders,
(2) constant errors,
(3) systematic errors, and
(4) random errors.

Blunders are simply mistakes caused by human carelessness. A blunder can be of any sign and magnitude and its occurrence is unpredictable. Blunders are often detected by repeated measurements of the same quantity. Constant errors are errors which always have the same sign and same magnitude. The most common source for constant error is the measuring instrument. For example a 100-foot tape may in fact only measure 99.9 feet. Then, every tape length with this tape would contain a constant error of 0.1 foot. Constant errors of this type can be detected and corrected for by accurate calibration of the measuring instrument. Personal bias of the observers may also be considered as constant errors. However, such errors are more difficult to calibrate.

Systematic errors occur according to some definite pattern which may or may not be known. When the law of occurrence of the systematic errors is known, it can sometimes be modeled by a mathematical expression and the measurements can be corrected accordingly. Lens distortions in aerial cameras is a good example of systematic error. The symmetric radial distortion of the lens system can be modeled by an odd order polynomial in terms of the radial distance from the principal point.

Random errors are caused by the inherent incapability of instruments and human observers to make exact measurements and by uncontrollable variations in the operating conditions during the measurements. Random errors are generally very small in magnitude, but they can be of any sign. To minimize the effect of random errors, measurements should be made under as wide a range of operating conditions as possible and with the greatest care.

2.4.2 TRUE ERROR AND RESIDUAL ERROR

The *true error* (ϵ_i) of an observation (l_i) is defined as its deviation from the true value (L) of the measured parameter; *i.e.*

$$\epsilon_i = l_i - L. \tag{2.98}$$

Since the true value is rarely known, the term true error is only of theoretical significance.

The *residual error* (v_i) of an observation (l_i) is defined as its deviation from the most probable value $(\bar{L})$; *i.e.*

$$v_i = l_i - \bar{L}. \tag{2.99}$$

2.4.3 STATISTICAL EVALUATION OF DIRECT MEASUREMENTS

Let $l_1, l_2, l_3, \ldots l_n$ be n repeated measurements of a parameter which has a true value L. These n values form the sample space of the measurement. The *sample mean* is the most probable estimate of the true value (L) based on this set of observations. It is computed by the following formula:

$$\bar{L} = \frac{\sum\limits_{i=1}^{n} l_i}{n}. \tag{2.100}$$

The *sample variance* (m^2) of this set of observations is defined as follows:

$$m^2 = \frac{\sum\limits_{i=1}^{n} (l_i - \bar{L})^2}{n - 1}. \tag{2.101}$$

The standard deviation (m) or root-mean-square (RMS) error of the individual observations is defined as follows:

$$m = \pm \sqrt{\frac{\sum\limits_{i=1}^{n} (l_i - \bar{L})^2}{n - 1}}. \tag{2.102}$$

Furthermore, the *standard error of the sample mean* is computed according to the following formula:

$$m_{\bar{L}} = \pm \sqrt{\frac{\sum_{i=1}^{n}(l_i - \bar{L})^2}{n(n-1)}} = \pm \frac{m}{\sqrt{n}}. \quad (2.103)$$

The statistical significance of these terms will be explained later in this section. In the meantime, it will suffice to state that the standard deviation provides a measure of the precision of the individual measurements. The standard error of the mean gives a measure of the precision of the mean value. Precision and accuracy do not have the same meaning in adjustment. Precision is the degree of repeatability of a measurement. A measurement is said to be of high precision if its value can be repeated within close tolerance at repeated measurements. Accuracy is a measure of the nearness of a measured value to the true value. A measurement of high accuracy means that its value is very close to the truth. In the absence of systematic error, precision and accuracy then have the same meaning.

The maximum error of a set of observations is the limit beyond which random errors seldom occur and is the limit normally set for rejection of observations. An error greater than the maximum error is considered as a blunder. A commonly adopted value for the maximum error is about 3 or 4 times the standard deviation; *i.e.*

$$\text{Maximum error} = 3m \text{ or } 4m. \quad (2.104)$$

The statistical significance of this factor will also be explained later in this section.

    EXAMPLE 24: Given that the six repeated measurements of the x-coordinate of an image point are: 10:45 mm, 10.40, 10.35, 10.42, 10.43 and 10.00 mm. Compute the sample mean, sample variance, standard deviation, and the standard error of the sample mean.

	$V_i = l_i - \bar{L}$	V_i^2
$l_1 = 10.45$ mm	0.108	0.011664
$l_2 = 10.40$	0.058	0.003364
$l_3 = 10.35$	0.008	0.000064
$l_4 = 10.42$	0.078	0.006084
$l_5 = 10.43$	0.088	0.007744
$l_6 = 10.00$	−0.342	0.116964

$$\sum l_i = 62.05$$

Sample mean $(\bar{L}) = \dfrac{\sum l_i}{n} = 10.342$ mm

Sample variance $= m^2 = \dfrac{\sum V_i^2}{n-1} = 0.0291768$

Standard deviation $= m = \pm 0.2$ mm

Standard error of the sample mean $= m_{\bar{L}} = \dfrac{m}{\sqrt{n}} = \pm 0.08$ mm

Maximum error $= 3m = \pm 0.6$ mm

Thus $\bar{L} = 10.34 \pm 0.08$ mm.

Any observation which has an error (v_i) the absolute value of which is greater than 0.6 mm can be considered as blunder and rejected from the sample. Although l_6 has a large residual error and "appears" to be a blunder, the magnitude of the error is still within the range of maximum error and should be included in the computation for the mean.

2.4.4 PROBABILITY FUNCTION OF DISCRETE VARIABLES

Probability provides a measure of the likelihood that an event will occur or that a random variable will take a certain value. It always refers to an experiment. For example, the tossing of a coin represents an experiment. There are two possible outcomes (or events) which can occur: head (H), or tail (T). The probability of the event H occurring, represented by the symbol $P(H)$, then indicates the likelihood that the coin comes up head. As a second example, the tossing of a die has six possible outcomes (1, 2, 3, 4, 5, or 6) which form the sample space of the experiment.

Let there be n mutually exclusive, exhaustive and equally likely cases. If m of these are favorable to an event A, then the probability of A occurring is m/n. Mutually exclusive means that no two cases can occur at the same time and exhaustive means that all possible cases are considered in the n possibilities.

    EXAMPLE 25: In one toss of a coin,

$$P(H) = P(T) = \frac{1}{2}$$

    EXAMPLE 26: In one roll of a die,

$$P(1) = P(2) = P(3) = P(4) = P(5) = P(6) = \frac{1}{6}$$

    EXAMPLE 27: For two tosses of a coin, the possible outcomes are:

$$(H, H), (H, T), (T, T), (T, H)$$

Thus,

$$P(\text{1st one } H, \text{ 2nd one } T) = \frac{1}{4}$$

and

P(one H and one T regardless of order)

$$= \frac{1}{4} + \frac{1}{4} = \frac{1}{2}.$$

The probability function always has a value range between 0 and 1. A probability of 0 means the event will surely not occur, and a probability of 1 will mean sure occurance. Moreover, the summation of the probabilities of all the possible events in an experiment is equal to 1. That is,

$$1 \geq P(X = x_i) \geq 0 \quad (2.105)$$

and

$$\sum_i P(X = x_i) = 1. \quad (2.106)$$

Let X and Y be two discrete random variables. The probability that $X = x_i$ and $Y = y_j$ at the same time is called the joint probability of X and Y and is denoted by $P(X = x_i, Y = y_j)$ or $P_{X,Y}(x_i, y_j)$. The joint probability function of two variables must also satisfy the conditions of a probability function; *i.e.*

$$1 \geq P_{XY}(x_i, y_j) \geq 0, \qquad (2.107)$$

and

$$\sum_i \sum_j P_{XY}(x_i, y_j) = 1. \qquad (2.108)$$

The marginal probability of $X = x_i$ is defined as follows:

$$P_X(x_i) = \sum_j P_{XY}(x_i, y_j) \qquad (2.109)$$

and the marginal probability of $Y = y_j$ is defined as

$$P_Y(y_j) = \sum_i P_{XY}(x_i, y_j) . \qquad (2.110)$$

If X and Y are two random variables which are mutually independent of each other (that is, the outcome of one does not depend on the outcome of the other), then the joint probability of the two variables is equal to the product of the two marginal probabilities. That is,

$$P_{XY}(x_i, y_j) = P_X(x_i) \cdot P_Y(y_j). \qquad (2.111)$$

EXAMPLE 28: X is a discrete random variable which can take on the values 0, 1 or 2; and Y is another discrete variable which can take on the values of 0, 1, 2, or 3. Suppose that the joint probabilities of X and Y are those given in the following table. It can be easily proved that X and Y are mutually independent.

Y \ X	0	1	2	$P_Y(y_j)$
0	1/30	1/30	4/30	1/5
1	3/30	3/30	12/30	18/30
2	0	0	0	0
3	1/30	1/30	4/30	1/5
$P_X(x_i)$	1/6	1/6	2/3	1

2.4.5 PROBABILITY FUNCTION OF CONTINUOUS RANDOM VARIABLES

In photogrammetry, random variables (such as random errors in coordinate measurement) usually can take on any value within a certain range. Such a random variable is said to be continuous. The above definitions on probability can be easily extended to this type of variables.

The density function $f(X)$ of a continuous random variable X is a normalized frequency distribution function such that $f(X) \geq 0$ for all possible value of X and

$$\int_{-\infty}^{\infty} f(X) \, dX = 1 . \qquad (2.112)$$

The probability that a continuous random variable X will take on a value between b and c is defined as

$$P(b \leq X \leq c) = \int_b^c f(X) \, dX . \qquad (2.113)$$

The significance of these two definitions is illustrated in figure 2-24. It should be noted that $P(X = b) = 0$ for a continuous variable.

Given a set of continuous random variables: $X_1, X_2, X_3, \ldots$ and X_n. They are mutually independent if, and only if, their joint density function is equal to the product of their individual density functions; *i.e.*

$$f(X_1, X_2, X_3 \ldots X_n) = f(X_1) \cdot f(X_2) \cdot f(X_3) \ldots f(X_n) \qquad (2.114)$$

for all values within the ranges of these random variables for which $f(X_1, X_2, X_3, \ldots X_n)$ is defined.

2.4.6 MATHEMATICAL EXPECTATION

Let X be a discrete random variable from a population consisting of $x_1, x_2, x_3, \ldots x_m$; that is, these are all the possible values that X can take. Let $P(x_i)$ denote the probability that $X = x_i$, then the mathematical expectation of X is defined as

$$E(X) = \sum_{i=1}^{n} x_i P(x_i) = \mu . \qquad (2.115)$$

$E(X)$ is also called the population mean of X, or simply the mean of X. It is not to be confused with sample mean $\bar{x}$ which is the mean of a set of measurements obtained from actual experiments.

Mathematical expectation has the following operational properties: (1) if a and b are two constants, then

$$E(aX + b) = aE(X) + b; \qquad (2.116)$$

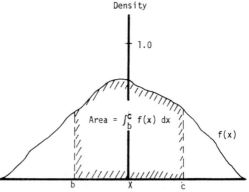

FIGURE 2-24. Density function of a continuous random variable.

and (2)

$$E(X + Y) = E(X) + E(Y). \quad (2.117)$$

The second moment of the variable X about its population mean μ is defined as

$$E(X - \mu)^2 = \sum_{i=1}^{n} (x_i - \mu)^2 P(x_i). \quad (2.118)$$

It is more commonly known as the population variance of the random variable X and is denoted as σ^2. It is useful as a measure of the variability or spread of the population distribution. The population variance (σ^2) is not to be confused with sample variance (m^2). The former refers to the population distribution of a random variable X, whereas the latter refers to the sample variance of a set of measurements of the parameter X. The relationships between σ^2 and m^2 and between μ and $\bar{x}$ will be explained in section 2.4.8.

For computational purposes, the following equation for the population variance (σ^2) is sometimes preferred from equation (2.118):

$$\sigma^2 = E(X - \mu)^2 = E(X^2) - \mu^2. \quad (2.119)$$

Let X and Y be two discrete, random variables. Then the covariance between X and Y is defined as

$$E[(X - \mu_X)(Y - \mu_Y)] =$$
$$\sum_i \sum_j (x_i - \mu_x)(y_j - \mu_y) P_{X,Y}(x_i, y_j) \quad (2.120)$$

where μ_X and μ_Y are the population means of X and Y respectively. The covariance measures the amount of correlation between the two variables. If there is a high probability that large values of X implies large values of Y and small values of X implies small values of Y, the covariance is positive. If there is a high probability that large values of X implies small vales of Y and *vice versa*, then the covariance is usually denoted by σ_{XY}. It can be easily shown that

$$\sigma_{XY} = E(XY) - \mu_X \mu_Y. \quad (2.121)$$

If X and Y are mutually independent, then their covariance $(\sigma_{XY}) = 0$.

The above definitions can be easily extended to describe the distributions of continuous random variables. Let X be a continuous random variable with the density distribution $f(X)$. Its population mean and population variance are then defined as follows:

$$\mu = E(X) = \int_{-\infty}^{\infty} x f(x) \, dx \quad (2.122)$$

and

$$\sigma^2 = E(X - \mu)^2 = \int_{-\infty}^{\infty} (x - \mu)^2 f(x) \, dx. \quad (2.123)$$

Similarly, the covariance between two continuous random variables X and Y with a joint distribution $F_{X,Y}(x, y)$ is defined as

$$\sigma_{XY} = E[(X - \mu_X)(Y - \mu_Y)] =$$
$$\int_{-\infty}^{\infty} \int_{-\infty}^{\infty} (x - \mu_X)(y - \mu_Y) f_{X,Y}(x, y) dx dy. \quad (2.124)$$

2.4.7 NORMAL DISTRIBUTION

A continuous random variable X is said to have a normal distribution with mean μ and variance σ^2 if it has the following frequency distribution function:

$$f(x) = \frac{1}{\sqrt{2\pi}\sigma} e^{-\frac{1}{2}\left(\frac{x-\mu}{\sigma}\right)^2}, \quad (2.125)$$

which is also often referred to as the Gaussian distribution. It describes a bell-shaped curve which is symmetrical about the mean μ (figure 2-25). The normal distribution of X is denoted as $N(\mu, \sigma^2)$.

The probability that X takes on a value between a and b is then computed as

$$P(a \leq X \leq b) = \int_a^b \frac{1}{\sqrt{2\pi}\sigma} e^{-\frac{1}{2}\left(\frac{x-\mu}{\sigma}\right)^2} dx . \quad (2.126)$$

In the special case where X is normally distributed with $\mu = 0$ and $\sigma^2 = 1$, X is said to have a *standard normal distribution* which is denoted as $N(0, 1)$. The standard normal distribution function is thus defined as

$$f(X) = \frac{1}{\sqrt{2\pi}} e^{-\frac{x^2}{2}} ; \quad (2.127)$$

and

$$P\{a \leq X \leq b\} = \int_a^b \frac{1}{\sqrt{2\pi}} e^{-\frac{x^2}{2}} dx. \quad (2.128)$$

Probability tables for the standard normal distribution are readily available. The following probabilities are of special interest in engineering measurements:

$$P\{-\sigma \leq (X - \mu) \leq \sigma\} = 0.6827,$$

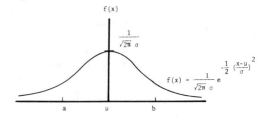

$$f(x) = \frac{1}{\sqrt{2\pi}\,\sigma} e^{-\frac{1}{2}\left(\frac{x-\mu}{\sigma}\right)^2}$$

FIGURE 2-25. Normal distribution.

$$P\{-2\sigma \le (X - \mu) \le 2\sigma\} = 0.9545,$$
$$P\{-3\sigma \le (X - \mu) \le 3\sigma\} = 0.9973,$$

and

$$P\{-4\sigma \le (X - \mu) \le 4\sigma\} = 1.00.$$

The last two probabilities show that 99.7% and 100% of the measurements on X should fall within 3σ and 4σ of the mean respectively. It is for this reason that $3m$ or $4m$ are usually taken as the tolerance limit for random error. Errors greater than the chosen limit may be considered as blunders.

Figure 2-26 illustrates the significance of σ. A normal distribution having a large σ means that the possible values of the random variable are spread over a wide range on both sides of the mean. A small σ means that the possible values of the random variable are concentrated about the mean.

The following two theorems concerning the normal distribution are of particular importance and are presented here without any proof:

THEOREM 2.1. If X is a random variable having a normal distribution $N(\mu, \sigma^2)$, then $Z = X - \mu/\sigma$ has a standard normal distribution; i.e. $\mu_Z = 0$ and $\sigma_Z = 1$.

THEOREM 2.2. Let $l_1, l_2, l_3, \ldots l_n$ be n mutually independent random variables and each is normally distributed with mean μ and variance σ^2. Then their mean

$$L = 1/n \sum_{i=1}^{n} l_i$$

is also normally distributed with mean μ and variance σ^2/n; i.e. $N(\mu, (\sigma^2/n))$.

The normal distribution function was derived from the phenomena of random errors in physical measurements. Although random errors appear to occur in an irregular nature; when a large number of observations are made to measure a parameter, it can be found that the random errors in the observations conform to the normal law of error. The law states that purely random errors must have the following properties:

(1) positive and negative errors of similar magnitude occur with similar frequency, i.e.
$$\Sigma \epsilon_i = 0;$$

(2) small errors occur more frequently than large errors;
(3) zero errors occur most frequently; and
(4) there is a practical limit beyond which random errors seldom occur.

The above conditions are adequately fulfilled by the following normal distribution function.

$$f(\epsilon) = \frac{1}{\sqrt{2\pi}\sigma} e^{-\frac{1}{2}\frac{\epsilon^2}{\sigma^2}}. \qquad (2.129)$$

2.4.8 POPULATION VS. SAMPLE MEANS AND VARIANCE

Let $x_1, x_2, x_3 \ldots$ and x_n be n independent observations all of which were made with equal care and accuracy. The sample mean ($\bar{x}$) and sample variance (m^2) of the set of observations can be computed using equations (2.100) and (2.101) respectively; i.e.

$$\bar{x} = \frac{\sum_{i=1}^{n} x_i}{n}$$

and

$$m^2 = \frac{\sum_{i=1}^{n} (x_i - \bar{x})^2}{n - 1}.$$

It can be shown that $\bar{x}$ and m^2 are unbiased estimates of the population mean (μ) and population variance (σ^2), i.e.

$$E(\bar{x}) = \mu, \qquad (2.130)$$

and

$$E(m^2) = \sigma^2. \qquad (2.131)$$

If n is very large and approaches infinity, then the sample mean approaches the population mean and the sample variance approaches the population variance.

2.4.9 LAW OF PROPAGATION OF ERRORS

The law of propagation of errors states that if Y is a function of the variables $X_1, X_2, X_3, \ldots X_n$; that is,

$$Y = f(X_1, X_2, X_3, \ldots X_n);$$

then the sample variance of Y is related to the sample variance of the X_i's according to the following expression:

$$m_Y^2 = \left(\frac{\partial f}{\partial X_1}\right)^2 m_{X_1}^2 + \left(\frac{\partial f}{\partial X_2}\right)^2 m_{X_2}^2 + \ldots$$

$$+ \left(\frac{\partial f}{\partial X_n}\right)^2 m_{X_n}^2. \qquad (2.132)$$

EXAMPLE 29: In distance measurement using a subtense bar, the following equation is used to

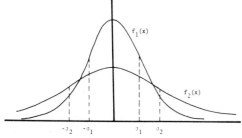

FIGURE 2-26. Significance of σ.

compute the distance d from the measured angle γ:

$$d = \frac{s}{2}\cot\frac{\gamma}{2} = f(s,\gamma)$$

According to the law of propagation of error,

$$m_d^2 = \left(\frac{\partial f}{\partial s}\right)^2 m_s^2 + \left(\frac{\partial f}{\partial \gamma}\right)^2 m_\gamma^2$$

i.e.

$$m_d^2 = \frac{1}{4}\cot^2\frac{\gamma}{2}m_s^2$$
$$+ \frac{s^2}{16}\csc^4\frac{\gamma}{2}m_\gamma^2.$$

Let

$$\gamma = 3°,\ m_\gamma = \pm 1'' = 0.4848 \times 10^{-5}\ \text{radians}$$
$$s = 2\ \text{metres, and }m_s = \pm 0.00005\ \text{m.}$$

Then

$$m_d^2 = \frac{1}{4}(38.188)^2\,(5\times10^{-5})^2$$
$$+ \frac{4}{16}\left(\frac{1}{(0.02618)}\right)^4(0.48\times10^{-5})^2$$
$$= 1318 \times 10^{-8}\ \text{m}^2$$

i.e.

$$m_d = \pm 0.004\ \text{m.}$$

It was stated in section 2.4.2 that the standard error of the sample mean is computed according to the formulae

$$m_{\bar{l}} = \pm\sqrt{\frac{\sum\limits_{i=1}^{n}(l_i-\bar{l})^2}{n\,(n-1)}}.$$

This formulae can be easily derived from the sample variance of the observations l_i using the law of propagation of error. Since

$$\bar{l} = \frac{1}{n}(l_1+l_2+l_3\ldots+l_n)\ ;$$

by the law of propogation of error,

$$m_{\bar{l}}^2 = \left(\frac{1}{n}\right)^2 m_{l_1}^2 + \left(\frac{1}{n}\right)^2 m_{l_2}^2 + \ldots + \left(\frac{1}{n}\right)^2 m_{l_n}^2$$

If all the observations are made with equal accuracy, then

$$m_{l_1}^2 = m_{l_2}^2 = \ldots = m_{l_n}^2 = m^2.$$

Thus,

$$m_{\bar{l}}^2 = \frac{m^2}{n}.$$

Since

$$m^2 = \frac{\sum\limits_{i=1}^{n}(l_i-\bar{l})^2}{(n-1)}$$

$$m_{\bar{l}}^2 = \frac{\sum\limits_{i=1}^{n}(l_i-\bar{l})^2}{n(n-1)}.$$

Therefore,

$$m_{\bar{l}} = \pm\sqrt{\frac{\sum\limits_{i=1}^{n}(l_i-\bar{l})^2}{n(n-1)}}.$$

2.4.10 MULTIVARIATE NORMAL DISTRIBUTION

Let $l_1, l_2, l_3 \ldots$ and l_n be n mutually independent observations of a parameter which has a true value μ. Suppose that these observations were made with equal care and accuracy and that all systematic errors have been eliminated. Then each observation l_i is said to belong to a normal distribution with a population mean μ_i and a variance σ_i^2. Since the observations are on the same parameter, we have $\mu_1 = \mu_2 = \ldots = \mu_n = \mu$; and since all observations are of equal accuracy, we have $\sigma_1^2 = \sigma_2^2 = \ldots = \sigma_n^2 = \sigma^2$. The probability density function of one observation l_i is described by the following expression:

$$f_i(l_i) = \frac{1}{\sqrt{2\pi}\sigma}e^{-\frac{1}{2}\frac{(l_i-\mu_i)^2}{\sigma^2}}. \qquad (2.133)$$

Since the observations are all mutually independent, the joint probability function of the n observations is the product of the individual distribution functions; i.e.

$$f(l_1,l_2,l_3,\ldots l_n) = f_1(l_1)\cdot f_2(l_2)\cdot f_3(l_3)\ldots f_n(l_n) =$$

$$\left(\frac{1}{\sqrt{2\pi}\sigma}\right)^n e^{-\frac{1}{2}\sum\limits_{i=1}^{n}\left(\frac{l_i-\mu}{\sigma}\right)^2}. \qquad (2.134)$$

Thus,

$$P(\mu - a_1 \leq l_1 \leq \mu + b_1, \ldots\ldots, \text{and } \mu - a_n \leq l_n \leq \mu + b_n)$$

$$= \int_{a_1}^{b_1}\int_{a_2}^{b_2}\ldots\int_{a_n}^{b_n}\left(\frac{1}{\sqrt{2\pi}\sigma}\right)^n e^{-\frac{1}{2}\sum\limits_{i=1}^{n}\frac{(l_i-\mu)^2}{\sigma^2}}dl_1dl_2dl_3\ldots dl_n. \qquad (2.135)$$

In photogrammetric adjustments, many different and correlated variables are often encountered in a given problem. For example, the ground coordinates (X_j, Y_j, Z_j) of a point j are three variables that have strong correlations.

Another example is the exterior orientation parameters of an aerial photograph: ω, ϕ, κ, X^c, Y^c, and Z^c. In general, let $X_1, X_2, X_3 \ldots X_n$ be n random variables that are not necessarily mutually independent and each has a $N(\mu_i, \sigma_i^2)$ distribution. Let the variance-covariance matrix of these variables be defined as follows:

$$\boldsymbol{\sigma}_X = \begin{bmatrix} \sigma_1^2 & \sigma_{X_1 X_2} & \sigma_{X_1 X_3} & \cdot & \cdot & \sigma_{X_1 X_n} \\ \sigma_{X_2 X_1} & \sigma_2^2 & \sigma_{X_2 X_3} & \cdot & \cdot & \sigma_{X_2 X_n} \\ \cdot & & \cdot & \cdot & \cdot & \cdot \\ \cdot & & & \cdot & \cdot & \cdot \\ \cdot & & & & \cdot & \cdot \\ \sigma_{X_n X_1} & \cdot & & \cdot & \cdot & \sigma_n^2 \end{bmatrix} \quad (2.136)$$

where $\sigma_{X_i X_j}$ denotes the covariance between X_i and X_j. It is to be noted that $\sigma_{X_i X_j}$ is not equal to $\sigma_{X_i} \sigma_{X_j}$. The joint distribution function of these n variables is defined as follows:

$$f(X_1, X_2, \ldots X_n) = \frac{1}{(2\pi)^{n/2} (\det \boldsymbol{\sigma}_X)^n} \exp - \frac{1}{2} \left[(\mathbf{X} - \boldsymbol{\mu})^{\mathrm{T}} \boldsymbol{\sigma}_X^{-1} (\mathbf{X} - \boldsymbol{\mu}) \right]$$

$$(2.137)$$

where

$$\mathbf{X} = \begin{bmatrix} X_1 \\ X_2 \\ \cdot \\ \cdot \\ X_n \end{bmatrix} \quad \text{and} \quad \boldsymbol{\mu} = \begin{bmatrix} \mu_1 \\ \mu_2 \\ \cdot \\ \cdot \\ \mu_n \end{bmatrix}$$

2.4.11 CHI-SQUARE (χ^2) DISTRIBUTION

A random variable X is said to have a chi-square distribution with n degrees of freedom (χ_n^2) if it has the following distribution function:

$$f(x) = \frac{1}{(2^{n/2}) \gamma(n/2)} x^{\frac{n-2}{2}} e^{-x/2} \text{ for } x > 0$$

$$= 0 \quad \text{elsewhere}$$

$$(2.138)$$

where

$$\gamma \frac{n}{2} = \int_0^x e^{-x} x^{\frac{n}{2} - 1} dx .$$

The shape of the χ^2-distribution is not symmetrical and is illustrated in figure 2-27.

The probability that the random variable X takes on a value equal to or greater then $\chi_{\alpha n}^2$ is defined as follows:

$$P(X \ge \chi_{\alpha n}^2) = \int_{\chi_{\alpha n}^2}^{\infty} f(x) dx = \alpha . \quad (2.139)$$

The value of $\chi_{\alpha n}^2$ for various combinations of α and n can be obtained directly from a χ^2 - table.

THEOREM 2.3 If $X_1, X_2, X_3, \ldots$ and X_n are independent random variables having standard normal distributions, then the new variable Y defined by

$$Y = \sum_{i=1}^{n} X_i^2$$

has the χ^2 - distribution with n degrees of freedom.

THEOREM 2.4. Let $x_1, x_2, x_3 \ldots$ and x_n be n independent observations of a parameter. Assuming that the errors in these measurements are truly random, each observation x_i may be considered as a random variable having a normal distribution with mean μ and variance σ^2. Furthermore, let m^2 denote the sample variance of these n observations. Then, $(n - 1) m^2/\sigma^2$ has a χ^2 - distribution with $(n - 1)$ degrees of freedom. It is beyond the scope of this Chapter to present the proof for these two theorems. However, the proofs can be found in any standard textbook in statistics. Theorem 2.4 is particularly useful for testing hypothesis on the population variance σ^2 or for computing the confidence limits for σ^2.

EXAMPLE 30: The 20 repeated measurements in an adjustment problem have a sample variance of $m^2 = 4$. What is the probability that the population variance $\sigma^2 \le 5$?

$$P(\sigma^2 \le 5) = P \left(\frac{\sigma^2}{(n-1)m^2} \le \frac{5}{(n-1)m^2} \right)$$

$$= P \left(\frac{\sigma^2}{(n-1)m^2} \le \frac{5}{(19)(4)} \right)$$

$$= P \left(\frac{(n-1)m^2}{\sigma^2} \le 15.2 \right) = 0.71.$$

Since $(\sigma^2)/(n - 1)m^2$ has a χ_{n-1}^2 - distribution, it can be found from a χ^2 - table that the probability is 0.71.

EXAMPLE 31: In the problem cited above, what

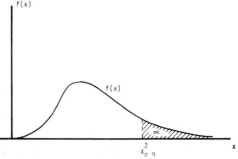

FIGURE 2-27. Chi-square distribution function.

is the 90% confidence limit for the sample variance m^2?

That is, what is c such that

$$P(m^2 \leq c) = 0.90?$$

SOLUTION:

$$P\left(\frac{\sigma^2}{(n-1)m^2} \leq \frac{c}{(n-1)m^2}\right) = 0.90$$

$$P\left(\frac{(n-1)m^2}{\sigma^2} \geq \frac{(n-1)m^2}{c}\right) = 0.90$$

i.e.

$$P\left(\frac{(n-1)m^2}{\sigma^2} \geq \frac{(19)(4)}{c}\right) = 0.90$$

From χ^2 − table, it can be found that

$$\frac{(19)(4)}{c} = 11.651 \ ,$$

i.e.

$$c = \frac{76}{11.651} = 6.52 \ .$$

Thus, there is 90% confidence that the sample variance is less than or equal to 6.52.

2.4.12 STUDENT-t DISTRIBUTION

Let Z and Y be two random variables such that Z has a standard normal distribution ($N(0,1)$) *and* Y has a χ_n^2 − distribution. Then the variable defined as

$$X = \frac{Z}{\sqrt{Y/n}} \tag{2.140}$$

has a student-t distribution with n degrees of freedom. Symbolically, it can be written as

$$t_n : \frac{N(0,1)}{\sqrt{\dfrac{\chi_n^2}{n}}} \tag{2.141}$$

The t-distribution curve is simular in shape to that of the normal distribution, but it has a much flatter curvature about the mean. As n becomes greater than 30, the t-distribution can be approximated by the normal distribution.

THEOREM 2.5. Let x_1, x_2, x_3.. and x_n be n observations having the same $N(\mu, \sigma^2)$ distribution. Let

$$\bar{x} = \frac{\sum\limits_{i=1}^{n} x_i}{n} \text{ and } m^2 = \frac{\sum\limits_{i=1}^{n}(x_i - \bar{x})^2}{n-1} \ .$$

Then

$$\frac{\bar{x} - \mu}{m/\sqrt{n}} : t_{n-1} \ .$$

In measurement problems, the student-t distribution is frequently used to compute the confidence interval of the population mean. Suppose that a set of n observations has a sample mean $\bar{x}$ and a sample variance m^2, and it is desired to find the lower and upper limits (b and c respectively) such that

$$P(b \leq \mu \leq c) = \alpha$$

where α is to be specified. We have

$$P(b \leq \mu \leq c) = P(-c \leq \mu \leq -b)$$

$$= P\left(\frac{\bar{x} - c}{m/\sqrt{n}} \leq \frac{\bar{x} - \mu}{m/\sqrt{n}} \leq \frac{\bar{x} - b}{m/\sqrt{n}}\right)$$

Let

$$c' = \frac{\bar{x} - c}{m/\sqrt{n}}$$

and

$$b' = \frac{\bar{x} - b}{m/\sqrt{n}} \ .$$

Then,

$$P(b \leq \mu \leq c) = P\left(c' \leq \frac{\bar{x} - \mu}{m/\sqrt{n}} \leq b'\right) = \alpha \ .$$

Since

$$\frac{\bar{x} - \mu}{m/\sqrt{n}}$$

is t_{n-1}, the values of c' and b' for a given specific value of α can be found from a t-distribution table. The meanings of c' and b' is illustrated in figure 2-28. Having found c' and b', the lower limit b and the upper limit c for the population mean can be found from the following expressions:

$$b = \bar{x} - \frac{m}{\sqrt{n}} b' \ ,$$

and

$$c = \bar{x} - \frac{m}{\sqrt{n}} c' \ .$$

EXAMPLE 32: Given $\bar{x} = 2.5$, $m = 0.5$, and $n = 20$, what is the 90% confidence interval of the population mean μ?

SOLUTION:

$$P(b \leq \mu \leq c) = P\left(\frac{\bar{x} - c}{m/\sqrt{n}} \leq \frac{\bar{x} - \mu}{m/\sqrt{n}} \leq \frac{\bar{x} - b}{m/\sqrt{n}}\right)$$

$$= P\left(c' \leq \frac{\bar{x} - \mu}{m/\sqrt{n}} \leq b'\right) = 0.90 \ .$$

From t-distribution table,

$$c' = -1.729 \text{ and } b' = +1.729$$

Thus,

$$c = \bar{x} - \frac{m}{\sqrt{n}} c' = 2.5 - \frac{0.5}{\sqrt{20}}(-1.729) = 2.693 \ ,$$

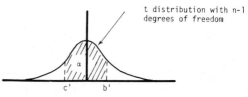

FIGURE 2-28. Confidence interval using student-t distribution.

and

$$b = \bar{x} - \frac{m}{\sqrt{n}} b' = 2.5 - \frac{0.5}{\sqrt{20}} (+1.729) = 2.307 .$$

Therefore,

$$P(2.307 \leqslant \mu \leqslant 2.693) = 0.90.$$

2.5 Methods of Least Squares Adjustment

2.5.1 BASIC PRINCIPLES

The method of least squares adjustment was developed over 150 years ago independently by Legrandre and Guass, both of whom were attempting to predict the orbits of celestial bodies from limited and imperfect observations. It has been used extensively in both geodetic surveying and photogrammetric engineering as a method of handling redundant measurements.

The method of least squares is based on the principle of maximum likelihood estimators. It has two fundamental assumptions on the nature of the observations; which are:

(1) the observations contain only random errors which follow normal distributions; and
(2) the observations are mutually independent of each other.

The principle is best illustrated by a simple measurement problem. Let $l_1, l_2, l_3 \ldots$ and l_n be n independent observations of a parameter μ, and let $v_1, v_2, v_3 \ldots$ and v_n be the corresponding errors in these n observations; i.e.

$$\begin{aligned} v_1 &= l_1 - \mu \\ v_2 &= l_2 - \mu \\ v_3 &= l_3 - \mu \\ & \cdot \\ & \cdot \\ v_n &= l_n - \mu. \end{aligned} \qquad (2.142)$$

The problem is to determine the most probable value of μ based on the given set of observations $(l_1, l_2, \ldots l_n)$, which is equivalent to finding the most probable set of errors $(v_1, v_2, \ldots v_n)$ in the given set of observations. The v_i's are random errors, each having its own normal distribution described by the following function:

$$f(v_i) = \frac{1}{\sqrt{2\pi}\sigma_i} e^{-\frac{1}{2}\left(\frac{v_i}{\sigma_i}\right)^2} .$$

Moreover, since these observations are mutually independent, their joint distribution function is the product of their individual distribution function; i.e.

$$f(v_1, v_2, v_3 \ldots v_n) = \frac{1}{\sqrt{2\pi}\sigma_1} e^{-\frac{1}{2}\left(\frac{v_1}{\sigma_1}\right)^2} \cdot \frac{1}{\sqrt{2\pi}\sigma_2} e^{-\frac{1}{2}\left(\frac{v_2}{\sigma_2}\right)^2} e \cdots$$

$$\cdot \frac{1}{\sqrt{2\pi}\sigma_n} e^{-\frac{1}{2}\left(\frac{v_n}{\sigma_n}\right)^2}$$

$$= \left(\frac{1}{\sqrt{2\pi}}\right)^n \left(\frac{1}{\sigma_1} \cdot \frac{1}{\sigma_2} \cdots \frac{1}{\sigma_n}\right) e^{-\frac{1}{2} \sum_{i=1}^{n}\left(\frac{v_i}{\sigma_i}\right)^2} .$$

Then by the definition of distribution function,

$$P(v_1 - \frac{\epsilon_1}{2} \leqslant v_1 \leqslant v_1 + \frac{\epsilon_1}{2}, v_2 - \frac{\epsilon_2}{2} \leqslant v_2$$

$$\leqslant v_2 + \frac{\epsilon_2}{2}, \ldots, v_n - \frac{\epsilon_n}{2} \leqslant v_n \leqslant v_n + \frac{\epsilon_n}{2})$$

$$= \left(\frac{1}{\sqrt{2\pi}}\right)^n \left(\frac{1}{\sigma_1} \cdot \frac{1}{\sigma_2} \cdots \frac{1}{\sigma_n}\right) e^{-\frac{1}{2} \sum_{i=1}^{n}\left(\frac{v_i}{\sigma_i}\right)^2} d\epsilon_1 \, d\epsilon_2 \, d\epsilon_3 \cdots d\epsilon_n$$

where the ϵ_i's are some infinitely small values. The most probable set of errors $(v_1, v_2, v_3 \ldots v_n)$ will be that which maximize the above probability function. However, to maximize the probability function, we must minimize the term

$$\sum_{i=1}^{n} \left(\frac{v_i}{\sigma_i}\right)^2 .$$

That is, the most probable value of μ based on a given set of observations $(l_1, l_2 \ldots l_n)$ is that value of μ which makes

$$\sum_{i=1}^{n} \frac{v_i^2}{\sigma_i^2} = \text{minimum}.$$

Since such a solution of μ minimizes the sum of the squares of the residuals v_i's, it is called a least squares solution. Let σ_o be some arbitrary constant, it does not alter the least squares solution by specifying that the solution must satisfy the condition that

$$\sum_{i=1}^{n} \frac{\sigma_o^2 v_i^2}{\sigma_i^2} = \text{minimum}.$$

The constant σ_o^2 is used primarily as a scaling factor. Furthermore, let

$$P_i = \frac{\sigma_o^2}{\sigma_i^2}, \qquad (2.143)$$

then the least squares condition is

$$\sum_{i=1}^{n} P_i v_i^2 = \text{minimum}. \qquad (2.144)$$

The parameter P_i is commonly referred to as the weight of observation l_i. Thus, the weight of an observation is inversely proportional to the variance of that observation. Weights are completely relative, and their common scaling factor is σ_o^2. Since an observation having a variance equal to σ_o^2 will have a weight of 1, σ_o^2 is referred to as the variance of unit weight.

Returning to the problem of finding the most probable value of μ from a given set of observations $(l_1, l_2, \ldots l_n)$, the most probable solution is that which minimizes

$$\sum_{i=1}^{n} P_i v_i^2 .$$

From equation (2.142)

$$P_1 v_1^2 = P_1 l_1^2 + P_1 \mu^2 - 2P_1 l_1 \mu$$

$$P_2 v_2^2 = P_2 l_2^2 + P_2 \mu^2 - 2P_2 l_2 \mu$$

.

.

.

$$P_n v_n^2 = P_n l_n^2 + P_n \mu^2 - 2P_n l_n \mu .$$

Therefore,

$$\sum_{i=1}^{n} P_i v_i^2 = \sum_{i=1}^{n} P_i l_i^2 + \mu^2 \sum_{i=1}^{n} P_i - 2\mu \sum_{i=1}^{n} P_i l_i .$$

The term

$$\sum_{i=1}^{n} P_i v_i^2$$

is minimized if

$$\frac{\partial}{\partial \mu} \left(\sum_{i=1}^{n} P_i v_i^2 \right) = 0;$$

i.e.

$$\frac{\partial}{\partial \mu} \left(\sum_{i=1}^{n} P_i v_i^2 \right) = 2\mu \sum_{i=1}^{n} P_i - 2 \sum_{i=1}^{n} P_i l_i = 0.$$

Therefore,

$$\mu = \frac{\sum_{i=1}^{n} P_i l_i}{\sum_{i=1}^{n} P_i} \qquad (2.145)$$

which is the well-known formula for computing the weighted mean of n observations. Thus, the weighted mean is also the least squares solution (i.e. the most probable solution).

2.5.2 CONCEPTS OF CONDITION EQUATIONS AND OBSERVATION EQUATIONS

The first problem in least squares adjustment is the formulation of a mathematical model of the problem. Condition equations and observation equations are often used for this purpose. A condition equation is a mathematical expression which expresses the physical or geometrical relationship that must be satisfied by a set of measured parameters. Characteristically, a condition equation involves two more measured parameters, but it may or may not include any unknown parameters other than the unknown residual errors in the measurements. An obser-

vation equation is a special type of condition equation which includes only one measured parameter in each equation. It is used to express the mathematical relationship between a measured parameter and some unknown parameters. Simple problems can be modeled by using either condition equations or observation equations. But for more complicated problems, the two types of equations are often used in combination.

The basic concept of these two types of equations can best be illustrated by considering a simple example. Figure 2-29 shows a level net consisting of five elevation bench marks, among which B.M. 1 is known to have an elevation of 100 feet above a vertical datum. The l_i's denote the most probable elevation differences between bench marks. Since there are more observations than is required for a unique solution, the problem is to find the most probable set of elevations for the other four bench marks.

2.5.2.1 PROBLEM MODELING BY CONDITION EQUATIONS

The geometrical condition that must be satisfied within the level net is that the sum of the elevation differences along any closed loop must be equal to zero. That is,

$$l_1 + l_2 + l_3 + l_4 + l_5 \quad\quad = 0$$

$$l_1 \quad\quad\quad + l_4 + l_5 + l_6 = 0$$

Let l_i^{00} be the measured value of l_i and v_i be the corresponding residual in l_i^{00}, that is,

$$l_i = l_i^{00} + v_i.$$

The two condition equations can be written as

$$(l_1^{00} + v_1) + (l_2^{00} + v_2) + (l_3^{00} + v_3) + (l_4^{00} + v_4)$$

$$+ (l_5^{00} + v_5) = 0$$

$$(l_1^{00} + v_1) + (l_4^{00} + v_4) + (l_5^{00} + v_5) + (l_6^{00} + v_6) = 0.$$

Rearranging terms and putting them into matrix form, we have

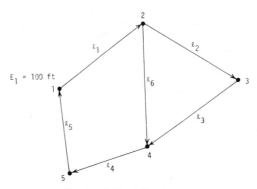

FIGURE 2-29. A level net.

$$\begin{bmatrix} 1 & 1 & 1 & 1 & 1 & 0 \\ 1 & 0 & 0 & 1 & 1 & 1 \end{bmatrix} \begin{bmatrix} v_1 \\ v_2 \\ v_3 \\ v_4 \\ v_5 \\ v_6 \end{bmatrix} = \begin{bmatrix} -l_1^{00} & -l_2^{00} & -l_3^{00} & -l_4^{00} & -l_5^{00} \\ -l_1^{00} & & & -l_4^{00} & -l_5^{00} & -l_6^{00} \end{bmatrix}$$

i.e. $\qquad\qquad$ **AV = L.** $\qquad\qquad$ (2.146)

The problem in the adjustment is to find the most probable set of v_i's which satisfy these two condition equations. After having found such a set of residuals to the measured elevation differences, the elevations of the unknown bench marks can be easily computed. Equation 2.146 represents a typical form of the problem model when condition equations are used. The characteristic feature of this type of problem model is the coefficient matrix **A** in front of the residual matrix **V**.

2.5.2.2 PROBLEM MODELING BY OBSERVATION EQUATIONS

The level net in fig. 2-29 may also be modeled using observation equations. Let E_1 and E_2 denote the elevation of bench marks 1 and 2 respectively, then

$$l_1^{00} + v_1 = E_2 - E_1,$$

i.e.

$$v_1 - E_2 = (-100 - l_1^{00}).$$

This is an observation equation because it includes only one measured parameter (l_1^{00}). A similar equation can be written for each of the six measured elevation differences. Thus, the following system of observation equations also adequately constitutes a mathematical model of the level net

$$\begin{aligned}
v_1 - E_2 &= -(100 + l_1^{00}) \\
v_2 + E_2 - E_3 &= -l_2^{00} \\
v_3 + E_3 - E_4 &= -l_3^{00} \\
v_4 + E_4 - E_5 &= -l_4^{00} \\
v_5 + E_5 &= (100 - l_5^{00}) \\
v_6 + E_2 - E_4 &= -l_6^{00}.
\end{aligned}$$

In matrix notation, this set of equations may be written as follows:

$$\begin{bmatrix} v_1 \\ v_2 \\ v_3 \\ v_4 \\ v_5 \\ v_6 \end{bmatrix} + \begin{bmatrix} -1 & 0 & 0 & 0 \\ 1 & -1 & 0 & 0 \\ 0 & 1 & -1 & 0 \\ 0 & 0 & 1 & -1 \\ 0 & 0 & 0 & 1 \\ 1 & 0 & -1 & 0 \end{bmatrix} \begin{bmatrix} E_2 \\ E_3 \\ E_4 \\ E_5 \end{bmatrix} = \begin{bmatrix} -(100 + l_1^{00}) \\ -l_2^{00} \\ -l_3^{00} \\ -l_4^{00} \\ (100 - l_5^{00}) \\ -l_6^{00} \end{bmatrix}$$

i.e. $\qquad$ **V + B E = C.** $\qquad$ (2.147)

In this case, the residual matrix **V** is not premultiplied by a coefficient matrix. This is the characteristic feature of a problem model consisting of observation equations. A least squares solution of equation (2.147) yields directly the most probable set of elevations (E_2, E_3, E_4 and E_5) and the most probable set of residuals (v_1, v_2, v_3, v_4, v_5, and v_6).

2.5.3 LINEARIZATION OF EQUATIONS

In general, a problem model, consisting of m independent equations, involving n unknowns and r measured parameters, may be expressed as follows:

$$\begin{aligned}
f_1 (X_1, X_2, X_3 \ldots X_n, Y_1, Y_2, \ldots Y_r) &= 0 \\
f_2 (X_1, X_2, X_3 \ldots X_n, Y_1, Y_2, \ldots Y_r) &= 0 \\
f_3 (X_1, X_2, X_3 \ldots X_n, Y_1, Y_2, \ldots Y_r) &= 0 \quad (2.148)
\end{aligned}$$

$$\cdot \qquad\qquad \cdot$$

$$f_n (X_1, X_2, X_3 \ldots X_n, Y_1, Y_2, \ldots Y_r) = 0.$$

If these equations are non-linear functions, they must be linearized before they can be used in the adjustment solution. This can be achieved using Newton's first order approximation. Let Y_1^{00}, $Y_2^{00}, \ldots$ and Y_r^{00} be a set of measured values for the set of parameters $Y_1, Y_2, Y_3 \ldots Y_r$ with the corresponding residuals $v_1, v_2, v_3 \ldots v_r$; *i.e.*

$$\begin{aligned}
Y_1 &= Y_1^{00} + v_1, \\
Y_2 &= Y_2^{00} + v_2, \\
&\quad\cdot \qquad (2.149)
\end{aligned}$$

$$Y_r = Y_r^{00} + v_r.$$

Furthermore, let $X_1^{0}, X_2^{0}, \ldots X_n^{0}$ be some approximate values of $X_1, X_2, X_3 \ldots X_n$ respectively such that

$$\begin{aligned}
X_1 &= X_1^{0} + \Delta X_1, \\
X_2 &= X_2^{0} + \Delta X_2, \quad (2.150)
\end{aligned}$$

$$X_n = X_n^{0} + \Delta X_n$$

where ΔX_i's are the corrections needed for the approximations. Assuming that the function f_i $(X_1, X_2, X_3 \ldots X_n, Y_1, Y_2 \ldots Y_r)$ is linear about the point defined by ($X_1^{0}, X_2^{0}, \ldots X_n^{0}, Y_1^{00}, Y_2^{00}, \ldots Y_r^{00}$), this function can be approximated as follows:

$$f_i (X_1, X_2, X_3 \ldots X_n, Y_1, Y_2 \ldots Y_r)$$

$$= f_i (X_1^{0}, X_2^{0}, X_3^{0}, \ldots X_n^{0}, Y_1^{00}, Y_2^{00} \ldots Y_r^{00})$$

$$+ \left(\frac{\partial f_i}{\partial y_i}\right)^0 v_1 + \left(\frac{\partial f_i}{\partial y_2}\right)^0 v_2 + \ldots + \left(\frac{\partial f_i}{\partial y_r}\right)^0 v_r$$

$$+ \left(\frac{\partial f_i}{\partial X_1}\right)^0 \Delta X_1 + \left(\frac{\partial f_i}{\partial X_2}\right)^0 \Delta X_2 + \ldots + \left(\frac{\partial f_i}{\partial X_n}\right)^0 \Delta X_n.$$

In this manner, the set of non-linear equations in (2.148) may be replaced by a set of linear equations as follows:

$$a_{11} v_1 + a_{12} v_2 + \ldots a_{1r} v_r + b_{11} \Delta X_1$$
$$+ b_{12} \Delta X_2 + \ldots b_{1n} \Delta X_n = -f_1^o$$

$$a_{21} v_1 + a_{22} v_2 + \ldots a_{2r} v_r + b_{21} \Delta X_1$$
$$+ b_{22} \Delta X_2 + \ldots + b_{2n} \Delta X_n = -f_2^o$$

$$\text{(2.151)}$$

.
.
.

$$a_{m1} v_1 + a_{m2} v_2 + \ldots a_{mr} v_r + b_{m1} \Delta X_1$$
$$+ b_{m2} \Delta X_2 + \ldots + b_{mn} \Delta X_n = -f_n^o$$

where the a_{ij}'s and b_{ij}'s are the partial derivatives, and

$$f_i^o = f_i (X_1^o, X_2^o, \ldots X_n^o, Y_1^{oo}, Y_2^{oo}, \ldots Y_r^{oo}).$$

In matrix notation,

$$\begin{array}{ccccc} \mathbf{A} & \mathbf{V} & + & \mathbf{B} & \Delta & = & \mathbf{C} . \\ (m,r) & (r,1) & & (m,n) & (n,1) & & (m,1) \end{array} \qquad \text{(2.152)}$$

A least squares solution of this set of equations yields the most probable set of residuals (v_i's) and corrections (ΔX_i's) to the approximations. However, since equations (2.151) are only linear approximations of equations (2.148), several iterations of the solution must be made to eliminate the errors due to the linearization process. That is, the computed corrections are applied to the approximations at the end of each iteration according to equations (2.150). These corrected values are used as new approximations in the next iteration. The iteration process is repeated until the corrections (ΔX_i's) become negligibly small within the desired accuracy standard.

One special type of condition equations is the observation equations which contain only one observed parameter in each equation. It can be easily seen from equations (2.151) that a problem model using only observation equations will have the following form:

$$v_1 + b_{11} \Delta X_1 + b_{12} \Delta X_2 + \ldots + b_{1n} \Delta X_n = -f_1^o$$
$$v_2 + b_{21} \Delta X_1 + b_{22} \Delta X_2 + \ldots + b_{2n} \Delta X_n = -f_2^o$$
$$v_3 + b_{31} \Delta X_1 + b_{32} \Delta X_2 + \ldots + b_{3n} \Delta X_n = -f_3^o$$

$$\text{(2.153)}$$

.
.
.

$$v_m + b_{m1} \Delta X_1 + b_{m2} \Delta X_2 + \ldots + b_{mn} \Delta X_n = -f_m^o$$

which may be expressed in matrix notation as follows:

$$\mathbf{V} + \mathbf{B} \Delta = \mathbf{C}. \qquad \text{(2.154)}$$

Instead of using Newton's first order approximation, linearization can also be performed by using Taylor's series expansion and omitting all second and higher order terms. The formulation is identical to that given in this chapter.

2.5.4 ADJUSTMENT BY OBSERVATION EQUATIONS

2.5.4.1 DEVELOPMENT OF NORMAL EQUATIONS

Let Y be a random variable the expected value of which may be expressed as a linear function of n variables $X_1, X_2, X_3, \ldots$ and X_n; i.e.

$$E(Y) = a_1 X_1 + a_2 X_2 + a_3 X_3 + \ldots + a_n X_n \qquad \text{(2.155)}$$

where the X_i's are the independent variables and Y is the dependent variable. Linear equations of this form are often encountered in geodetic surveying and photogrammetry. The problem generally concerns the determination of the most probable values for the set of coefficients $a_1, a_2, a_3 \ldots$ and a_n. To do so, observations can be made on Y at various values of the parameters ($X_1, X_2, X_3 \ldots$ and X_n). For example, suppose that by setting $X_1 = x_{11}, X_2 = x_{12}, X_3 = x_{13}, \ldots X_n = x_{1n}$, we observe that $Y = y_1$. This set of observations represents one experiment and gives rise to one observation equation as follows:

$$v_1 + y_1 = x_{11} a_1 + x_{12} a_2 + x_{13} a_3 + \ldots + x_{1n} a_n$$

where v_1 is the residual in the observation y_1. By transferring y_1 to the right-hand side, this equation may also be written as

$$v_1 = x_{11}a_1 + x_{12}a_2 + x_{13}a_3 + \ldots$$
$$+ x_{1n} a_n - y_1. \qquad \text{(2.156)}$$

In general, the experiment is repeated m times to obtain m independent measurements of Y for m different combinations of the values of X_i's. Since each experiment gives rise to one observation equation as in (2.156), the mathematical model will consist of m observation equations as follows:

$$v_1 = x_{11} a_1 + x_{12} a_2 + x_{13} a_3 + \ldots + x_{1n} a_n - y_1$$
$$v_2 = x_{21} a_1 + x_{22} a_2 + x_{23} a_3 + \ldots + x_{2n} a_n - y_2$$
$$v_3 = x_{31} a_1 + x_{32} a_2 + x_{33} a_3 + \ldots + x_{3n} a_n - y_3$$

$$\text{(2.157)}$$

.
.
.

$$v_m = x_{m1} a_1 + x_{m2} a_2 + x_{m3} a_3 + \ldots + x_{mn} a_n - y_m.$$

If $m = n$, no least-squares adjustment is possible or necessary since there are as many equations as unknowns. In such a case, it is necessary to assume that no error is present in the measurements (y_i's) by setting $v_i = 0$ for $i = 1$ to m. Equations (2.157) are then reduced to n equations involving n unknowns, and a unique solution can be obtained.

If $m > n$, the least squares method may be used to find the most probable values of the residuals and the coefficients; that is, $v_1, v_2, v_3 \ldots v_m, a_1, a_2, a_3 \ldots$ and a_n. However, the measurements (y_i's) should satisfy the two fundamental assumptions of least squares: (1) the residual (v_i) has a normal distribution with a mean of zero and a variance σ_i, for $i = 1$ to m; and (2) the observations y_1, y_2, ... and y_m are mutually independent. Assuming that these two conditions are satisfied, the joint probability distribution function of the residual v_i's is as follows:

$$f(v_1, v_2, v_3 \ldots v_n) = \frac{1}{(\sqrt{2\pi})^n}\left(\frac{1}{\sigma_1}\frac{1}{\sigma_2} \cdots \frac{1}{\sigma_n}\right)e^{-\frac{1}{2}\sum_{i=1}^{m}\left(\frac{v_i}{\sigma_i}\right)^2}$$

According to the derivation presented in section (2.5.1), the most probable set of v_i's is that which minimize the term

$$\sum_{i=1}^{m}\left(\frac{v_i}{\sigma_i}\right)^2$$

with respect to the unknown parameters a_1, a_2, a_3 ... and a_n. That is, the following conditions must be satisfied by the most probable values of a_1, a_2, a_3 ... a_n:

letting $Q = \sum_{i=1}^{m} p_i v_i^2$ where $p_i = \left(\frac{1}{\sigma_i}\right)^2$;

$$\frac{\partial Q}{\partial a_1} = 0$$

$$\frac{\partial Q}{\partial a_2} = 0 \qquad (2.158)$$

$$\frac{\partial Q}{\partial a_3} = 0$$

$$\vdots$$

$$\frac{\partial Q}{\partial a_n} = 0 \ .$$

From equation (2.156)

$p_1 v_1^2 = p_1 x_{11}^2 a_1^2 + p_1 x_{12}^2 a_2^2 + p_1 x_{13}^2 a_3^2 + \ldots$
$\quad + p_1 x_{1n}^2 a_n^2 + p_1 y_1^2$
$\quad + 2p_1 x_{11} a_1 [x_{12} a_2 + x_{13} a_3 + \ldots + x_{1n} a_n - y_1]$
$\quad + 2p_1 x_{12} a_2 [x_{13} a_3 + x_{14} a_4 + \ldots + x_{1n} a_n - y_1]$
$\quad \vdots$

$\quad - 2p_1 x_{1n} y_1 a_n.$

Thus,

$$Q = \sum_{i=1}^{m} p_i v_i^2 = \left(\sum_{i=1}^{m} p_i x_{i1}^2\right)a_1^2 + \left(\sum_{i=1}^{m} p_i x_{i2}^2\right)a_2^2 + \ldots$$

$$+ \left(\sum_{i=1}^{m} p_i x_{in}^2\right)a_n^2 + \sum_{i=1}^{m} p_i y_i^2$$

$$+ 2a_1 \sum_{i=1}^{m} p_i x_{i1}[a_2 x_{i2} + a_3 x_{i3} + \ldots + a_n x_{in} - y_i]$$

$$+ 2a_2 \sum_{i=1}^{m} p_i x_{i2}[a_3 x_{i3} + a_4 x_{i4} + \ldots + a_n x_{in} - y_i]$$

$$+ \ldots + (2\sum_{i=1}^{m} p_i x_{in} y_i)a_n. \qquad (2.159)$$

Then, from the conditions stated in equations (2.158),

$$\frac{\partial Q}{\partial a_1} = (\sum p_i x_{i1}^2)a_1 + (\sum p_i x_{i1} x_{i2})a_2 + \ldots$$
$$+ (\sum p_i x_{i1} x_{in})a_n - \sum p_i x_{i1} y_i = 0$$

$$\frac{\partial Q}{\partial a_2} = (\sum p_i x_{i1} x_{i2})a_1 + (\sum p_i x_{i2}^2)a_2 + \ldots$$
$$+ (\sum p_i x_{i2} x_{in})a_n - \sum p_i x_{i2} y_i = 0$$

$$\vdots$$

$$\frac{\partial Q}{\partial a_n} = (\sum p_i x_{i1} x_{in})a_1 + (\sum p_i x_{i2} x_{in})a_2 + \ldots$$
$$+ (\sum p_i x_{in}^2)a_n - \sum p_i x_{in} y_i = 0. \qquad (2.160)$$

These are the so-called normal equations. They include as many equations as unknowns. Thus, this set of equations gives a unique solution to the unknown parameters, a_i's. This unique solution is also the most probable solution based on the given set of observations on Y. Having solved for the a_i's in equations (2.160), the most-probable residuals can then be computed by substituting the values of the a_i's into equations (2.157). The normal equations (2.160) may be expressed in matrix notation as follows:

$$\begin{bmatrix} \sum p_i x_{i1}^2 & \sum p_i x_{i1} x_{i2} & \sum p_i x_{i1} x_{i3} & \cdots & \sum p_i x_{i1} x_{in} \\ \sum p_i x_{i2} x_{i1} & \sum p_i x_{i2}^2 & \sum p_i x_{i2} x_{i3} & \cdots & \sum p_i x_{i2} x_{in} \\ \sum p_i x_{i3} x_{i1} & \sum p_i x_{i3} x_{i2} & \sum p_i x_{i3}^2 & \cdots & \sum p_i x_{i3} x_{in} \\ & & & \ddots & \\ \sum p_i x_{in} x_{i1} & \sum p_i x_{in} x_{i2} & \sum p_i x_{in} x_{i3} & \cdots & \sum p_i x_{in}^2 \end{bmatrix} \begin{bmatrix} a_1 \\ a_2 \\ a_3 \\ \vdots \\ a_n \end{bmatrix}$$

$$= \begin{bmatrix} \sum p_i x_{i1} y_i \\ \sum p_i x_{i2} y_i \\ \sum p_i x_{i3} y_i \\ \vdots \\ \sum p_i x_{in} y_i \end{bmatrix}. \qquad (2.161)$$

Note that the coefficient matrix of the normal equations is symmetrical about the diagonal.

2.5.4.2 MATRIX FORMULATION

The development of the normal equations as described in the previous section can be performed more conveniently using matrix notation. The set of observation equations in (2.157) may be written in matrix notation as follows:

$$V = B\delta - C; \qquad (2.162)$$

where

$$
B = \begin{bmatrix}
x_{11} & x_{12} & x_{13} & \cdots & x_{1n} \\
x_{21} & x_{22} & x_{23} & \cdots & x_{2n} \\
\cdot & & & & \cdot \\
\cdot & & & & \cdot \\
\cdot & & & & \cdot \\
x_{m1} & x_{m2} & x_{m3} & \cdots & x_{mn}
\end{bmatrix}
$$
$$(m,n)$$

$$
\delta = \begin{bmatrix}
a_1 \\
a_2 \\
a_3 \\
\cdot \\
\cdot \\
a_n
\end{bmatrix}, \text{ and } C = \begin{bmatrix}
y_1 \\
y_2 \\
y_3 \\
\cdot \\
\cdot \\
y_m
\end{bmatrix}.
$$
$$(n,1) \qquad\qquad (m,1)$$

According to the principle of least squares, the most probable solution for the δ-matrix is that which minimizes

$$Q = \sum_{i=1}^{m} \left(\frac{v_i}{\sigma_i} \right)^2 .$$

It can be shown that

$$Q = \sum_{i=1}^{m} \left(\frac{v_i}{\sigma_i} \right)^2 = V^T W V, \qquad (2.163)$$

where W is the weight matrix and is defined as follows:

$$
W = \sigma_o{}^2 \begin{bmatrix}
\sigma y_1^2 & & & & \\
& \sigma y_2^2 & & & \\
& & \sigma y_3^2 & & \\
& & & \cdot & \\
& & & & \sigma y_m^2
\end{bmatrix}^{-1} . \quad (2.164)
$$

Since the observations (y_i's) are assumed to be uncorrelated, the off-diagonal elements in the W-matrix are equal to zero.

Substituting equation (2.162) into (2.163) yields

$$Q = (B\delta - C)^T W(B\delta - C);$$

i.e.

$$Q = (\delta^T B^T - C^T)(WB\delta - WC).$$

Therefore,

$$Q = \delta^T B^T W B \delta - \delta^T B^T W C - C^T W B \delta + C^T W C. \qquad (2.165)$$

Since the problem is to find the δ-matrix which minimizes Q, the following condition must be satisfied by the solution:

$$\frac{\partial Q}{\partial \delta} = 0$$

From equation (2.165),

$$\frac{\partial Q}{\partial \delta} = 2B^T W B \delta - B^T W C - (C^T W B)^T$$

Hence,

$$2B^T W B \delta - 2B^T W C = 0;$$

or

$$(B^T W B) \delta = B^T W C \qquad (2.166)$$

which is the normal equation for the model stated by equation (2.162). It can be easily seen that equations (2.166) and (2.161) are equivalent. Letting

$$N = B^T W B, \text{ and } K = B^T W C \qquad (2.167)$$

the normal equations can be simply written as

$$
\begin{array}{ccc}
N & \delta & = K . \\
(n,n) & (n,1) & (n,1)
\end{array}
\qquad (2.168)
$$

The most probable solution for δ is then obtained by solving this system of normal equations.

2.5.4.3 VARIANCE-COVARIANCE MATRIX OF THE COMPUTED PARAMETERS

The variance-covariance matrix of the unknown δ-matrix in equation (2.168) may be expressed as follows:

$$
\sigma_\delta = \begin{bmatrix}
\sigma_{a_1}^2 & \sigma_{a_1 a_2} & \sigma_{a_1 a_3} & \cdots & \sigma_{a_1 a_n} \\
\sigma_{a_2 a_1} & \sigma_{a_2}^2 & \sigma_{a_2 a_3} & \cdots & \sigma_{a_2 a_n} \\
\sigma_{a_3 a_1} & \sigma_{a_3 a_2} & \sigma_{a_3}^2 & \cdots & \sigma_{a_3 a_n} \\
\cdot & & & & \cdot \\
\cdot & & & & \cdot \\
\sigma_{a_n a_1} & \sigma_{a_n a_2} & \sigma_{a_n a_3} & \cdots & \sigma_{a_n}^2
\end{bmatrix} \quad (2.169)
$$

in which $\sigma_{a_i}^2$ is the variance of the parameter a_i, and $\sigma_{a_i a_j}$ is the covariance between parameters a_i and a_j.

If the original set of observation equations is exactly linear as in the case in equation (2.157), and if the normal equation is the same as that given in equation (2.168), then it is theoretically valid to compute the variance-covariance matrix (σ_δ) of the δ-matrix by the following expression:

$$\sigma_\delta = \sigma_0^2 N^{-1} \qquad (2.170)$$

where σ_0^2 is the variance of unit weight. Most of the least squares adjustment problems encountered in analytical photogrammetry require the use of an iterative procedure because of the nonlinearity of the observation or condition equations. In this type of iterative least squares solution, equation (2.170) is theoretically valid only if the unknown corrections in the δ-matrix converge to zero in the last iteration.

However, experimental results have shown that equation (2.179) could provide reliable estimates to the root-mean-square (RMS) errors of the computed parameters as long as the correction parameters converge to a value which is less than the computed estimate of the corresponding RMS error. If the corrections in the last iteration exceed the computed RMS errors in the last iteration, then the variance-and-covariance matrix is not a reliable estimator of the adjustment accuracy even though the solution has stabilized. Experimental evidence has shown that the formulation in equation (2.170) could not reflect any rapid accumulation of systematic effects which are caused by the random errors in the measurements.

2.5.4.4 SUMMARY

In general, when observation equations are used to relate the measurements with the unknown parameters, the mathematical model takes on the following form:

$$\mathbf{V} = \mathbf{B} \quad \boldsymbol{\delta} - \mathbf{C};$$
$$(m,1) \quad (m,n)(n,1) \; (m,1)$$

(2.171)

where m = number of measurements, n = number of unknowns, and m > n. The normal equations will then be given by the following expression:

$$(\mathbf{B}^{\mathsf{T}}\mathbf{W}\mathbf{B})\,\boldsymbol{\delta} = \mathbf{B}^{\mathsf{T}}\mathbf{W}\mathbf{C}. \quad (2.172)$$

Letting $\mathbf{N} = \mathbf{B}^{\mathsf{T}}\mathbf{W}\mathbf{B}$, and $\mathbf{K} = \mathbf{B}^{\mathsf{T}}\mathbf{W}\mathbf{C}$; equation (2.172) becomes

$$\mathbf{N}\,\boldsymbol{\delta} = \mathbf{K}. \quad (2.173)$$

The least squares solution for $\boldsymbol{\delta}$ is thus obtained as

$$\boldsymbol{\delta} = \mathbf{N}^{-1}\,\mathbf{K}; \quad (2.174)$$

and the variance-covariance matrix for $\boldsymbol{\delta}$ is given by

$$\boldsymbol{\sigma}_{\delta} = \sigma_o^2\,\mathbf{N}^{-1}. \quad (2.175)$$

2.5.5 ADJUSTMENT BY CONDITION EQUATIONS

2.5.5.1 DEVELOPMENT OF NORMAL EQUATIONS

It was shown in section 2.5.2.1 that a closed circuit within a leveling net may be described by a condition equation as follows:

$$a_{i1}v_1 + a_{i2}v_2 + a_{i3}v_3 + \ldots + a_{in}v_n = \epsilon_i \quad (2.176)$$

where v_i is the residual in the level line (l_i), and the a_{ij}'s are either 0 or 1, and

$$\epsilon_i = \sum_{j=1}^{n} a_{ij}l_j.$$

Let there be m independent closed circuits within the net. Then the complete mathematical model consists of the following equations:

$$a_{11}v_1 + a_{12}v_2 + a_{13}v_3 + \ldots + a_{1n}v_n = \epsilon_1$$
$$a_{21}v_1 + a_{22}v_2 + a_{23}v_3 + \ldots + a_{2n}v_n = \epsilon_2 \quad (2.177)$$
$$\cdot$$
$$\cdot$$
$$a_{m1}v_1 + a_{m2}v_2 + a_{m3}v_3 + \ldots + a_{mn}v_n = \epsilon_m.$$

The problem is to find the most probable set of residuals (v_i's) which satisfy all of the above conditions. Again assuming that the observations are mutually independent and the v_i's satisfy the normal law of errors, then the most probable set of ($v_1, v_2, v_3 \ldots v_n$) is that which minimizes

$$\sum_{i=1}^{n} p_i v_i^2 \quad \text{where} \quad p_i = \frac{\sigma_o^2}{\sigma_i^2}.$$

Let $k_1, k_2, \ldots$ and k_m be m arbitrary variables. According to equation (2.176),

$$a_{i1}v_1 + a_{i2}v_2 + \ldots + a_{in}v_n - \epsilon_i = 0.$$

Therefore,

$$Q = \sum_{i=1}^{n} p_i v_i^2$$

$$= \sum_{i=1}^{n} p_i v_i^2 - 2k_1\,(a_{11}v_1 + a_{12}v_2 + \ldots + a_{1n}v_n - \epsilon_1)$$

$$- 2k_2\,(a_{21}v_1 + a_{22}v_2 + \ldots + a_{2n}v_n - \epsilon_2)$$

$$\cdot$$
$$\cdot \quad (2.178)$$

$$- 2k_m\,(a_{m1}v_1 + a_{m2}v_2 + \ldots + a_{mn}v_n - \epsilon_n).$$

The set of residuals which minimizes Q must satisfy the following n conditions:

$$\left.\begin{array}{l} \dfrac{\partial Q}{\partial v_1} = 0 \\[2mm] \dfrac{\partial Q}{\partial v_2} = 0 \\[2mm] \cdot \\ \cdot \\ \dfrac{\partial Q}{\partial v_n} = 0 \end{array}\right\} \; n \text{ condition equations.} \quad (2.179)$$

From equation (2.178)

$$\frac{\partial Q}{\partial v_i} = 2p_i v_i - 2k_1 a_{1i} - 2k_2 a_{2i} - 2k_3 a_{3i} \ldots - 2k_m a_{mi} = 0$$

i.e.

$$v_i = \frac{a_{1i}}{p_i}k_1 + \frac{a_{2i}}{p_i}k_2 + \ldots + \frac{a_{mi}}{p_i}k_m.$$

Therefore, the n conditions in equation (2.179) give rises to the following n equations:

$$v_1 = \frac{a_{11}}{p_1}k_1 + \frac{a_{21}}{p_1}k_2 + \frac{a_{31}}{p_1}k_3 + \ldots + \frac{a_{m1}}{p_1}k_m$$

$$v_2 = \frac{a_{12}}{p_2}k_1 + \frac{a_{22}}{p_2}k_2 + \frac{a_{32}}{p_2}k_3 + \ldots + \frac{a_{m2}}{p_2}k_m$$

$$\cdot$$
$$\cdot \quad (2.180)$$
$$\cdot$$

$$v_n = \frac{a_{1n}}{p_n}k_1 + \frac{a_{2n}}{p_n}k_n + \frac{a_{3n}}{p_n}k_3 + \ldots + \frac{a_{mn}}{p_n}k_n.$$

These expressions can be substituted back into equations (2.177). For example, after the substitution, the ith equation in (2.177) becomes

$$a_{i1}\left(\frac{a_{11}}{p_1}k_1 + \frac{a_{21}}{p_1}k_2 + \ldots + \frac{a_{m1}}{p_1}k_m\right)$$

$$+ a_{i2}\left(\frac{a_{12}}{p_2}k_1 + \frac{a_{22}}{p_2}k_2 + \ldots\right.$$

$$\left. + \frac{a_{m2}}{p_2}k_m\right) + a_{in}\left(\frac{a_{1n}}{p_n}k_1\right.$$

$$\left. + \frac{a_{2n}}{p_n}k_2 + \ldots + \frac{a_{mn}}{p_n}k_m\right) = \epsilon_i.$$

Collecting terms, we have

$$\sum_{h=1}^{n}\left(\frac{a_{ih}a_{ih}}{p_h}\right)k_1 + \sum_{h=1}^{n}\left(\frac{a_{ih}a_{2h}}{p_h}\right)k_2 + \ldots$$

$$+ \sum_{h=1}^{n}\left(\frac{a_{ih}a_{mh}}{p_h}\right)k_m = \epsilon_i.$$

Thus, equations (2.177) may be transformed to m equations consisting of m unknowns as follows:

$$\sum_{h=1}^{n}\left(\frac{a_{1h}^2}{p_h}\right)k_1 + \sum_{h=1}^{n}\left(\frac{a_{1h}a_{2h}}{p_h}\right)k_2 + \ldots$$

$$+ \sum_{h=1}^{n}\left(\frac{a_{1h}a_{mh}}{p_h}\right)k_m = \epsilon_1$$

$$\sum\left(\frac{a_{2h}a_{1h}}{p_h}\right)k_1 + \sum\left(\frac{a_{2h}^2}{p_h}\right)k_2 + \ldots$$

$$+ \sum\left(\frac{a_{2h}a_{mh}}{p_h}\right)k_m = \epsilon_2$$

$$\cdot \qquad\qquad\qquad\qquad (2.181)$$

$$\sum\left(\frac{a_{mh}a_{1h}}{p_h}\right)k_1 + \sum\left(\frac{a_{mh}a_{2h}}{p_h}\right)k_2 + \ldots$$

$$+ \sum\left(\frac{a_{mh}^2}{p_h}\right)k_m = \epsilon_m.$$

This is a set of normal equations with respect to the arbitrary variables (k_i's). Having solved for the k_i's from equations (2.181), their values can be substituted into equations (2.180) to give the most probable residuals (v_i's).

In geodesy, this method is often referred to as the method of adjustment by correlates, and the variables $k_1, k_2, \ldots$ and k_m are called correlates. The k_i's are also often called the Lagrange multipliers.

2.5.5.2 MATRIX FORMULATION

Equations (2.177) may be expressed in matrix notation as follows:

$$\begin{array}{ccc} \mathbf{A} & \mathbf{V} = \mathbf{C} \\ (m,n) & (n,1) \ (m,1) \end{array} \qquad (2.182)$$

The most probable solution for $\mathbf{V}$ is that which minimizes $Q = \mathbf{V}^T\mathbf{W}\mathbf{V}$. Let $\mathbf{K}$ be a row matrix of m variables; i.e.

$$\begin{array}{c} \mathbf{K} = [k_1\, k_2\, k_3 \ldots k_m]. \\ (m,1) \end{array} \qquad (2.183)$$

Since $\mathbf{AV} - \mathbf{C} = 0$,

$$Q = \mathbf{V}^T\mathbf{W}\mathbf{V} - 2\mathbf{K}\,(\mathbf{AV} - \mathbf{C}).$$

Taking partial derivative of Q with respect to $\mathbf{V}$ and equating the result to zero yields

i.e.
$$\frac{\partial Q}{\partial \mathbf{V}} = 2\mathbf{W}\mathbf{V} - 2\mathbf{A}^T\mathbf{K}^T = 0; \qquad (2.184)$$

$$\mathbf{V} = \mathbf{W}^{-1}\mathbf{A}^T\mathbf{K}^T.$$

Substituting equations (2.184) into equations (2.182) yields

$$\begin{array}{cc} (\mathbf{A}\mathbf{W}^{-1}\mathbf{A}^T) & \mathbf{K}^T = \mathbf{C}; \\ (m,m) & (m,1) \quad (m,1) \end{array} \qquad (2.185)$$

which is a set of normal equations consisting of m unknowns. It corresponds directly with the set of normal equations (2.181). Solving for $\mathbf{K}^T$, we get

$$\mathbf{K}^T = (\mathbf{A}\mathbf{W}^{-1}\mathbf{A}^T)^{-1}\,\mathbf{C}. \qquad (2.186)$$

Finally, substituting equation (2.186) into equation (2.184) yields

$$\mathbf{V} = \mathbf{W}^{-1}\mathbf{A}^T\,(\mathbf{A}\mathbf{W}^{-1}\mathbf{A}^T)^{-1}\,\mathbf{C} \qquad (2.187)$$

which is the expression for the most probable solution of $\mathbf{V}$.

2.5.5.3 VARIANCE-COVARIANCE MATRIX

It can be proven that if $\mathbf{X}$ and $\mathbf{Y}$ are two variable vectors such that

$$\mathbf{X} = \mathbf{AY};$$

then the variance-covariance matrix of $\mathbf{X}$, denoted as σ_X, can be expressed as a function of the variance-covariance matrix of the variable vector $\mathbf{Y}$ as follows:

$$\sigma_X = \mathbf{A}\,\sigma_Y\,\mathbf{A}^T \qquad (2.188)$$

where σ_Y is the variance-covariance matrix of $\mathbf{Y}$. From equation (2.186), since

$$\mathbf{K}^T = (\mathbf{A}\mathbf{W}^{-1}\mathbf{A}^T)\ \mathbf{C},$$

the variance-covariance matrix of the Lagrange multipliers ($\mathbf{K}$-matrix) can be obtained directly from the following expression:

$$\sigma_K^T = \sigma_0^2\,(\mathbf{A}\mathbf{W}^{-1}\mathbf{A}^T)^{-1}.$$

According to equation (2.184),

$$\mathbf{V} = \mathbf{W}^{-1}\mathbf{A}^T\mathbf{K}^T.$$

$$\mathbf{V} = \mathbf{W}^{-1}\mathbf{A}^T\mathbf{K}^T.$$

$$\sigma_V = (\mathbf{W}^{-1}\mathbf{A}^T)\sigma_K^T\,(\mathbf{A}\mathbf{W}^{-1})$$

i.e.

$$\sigma_V = \sigma_0^2\,\mathbf{W}^{-1}\mathbf{A}^T\,(\mathbf{A}\mathbf{W}^{-1}\mathbf{A}^T)^{-1}\,\mathbf{A}\mathbf{W}^{-1}. \qquad (2.189)$$

2.5.5.4 SUMMARY

When only condition equations are used to model a least squares adjustment problem, the set of condition equations may take the following form:

$$\mathbf{AV} = \mathbf{C}.$$

The least squares solution for **V** may be obtained in two steps:

(1) solve for the lagrange multipliers in the $\mathbf{K}^T$ matrix from the following expression:

$$(\mathbf{A}\mathbf{W}^{-1}\mathbf{A}^T)\,\mathbf{K}^T = \mathbf{C};$$

(2) solve for **V** using the following expression:

$$\mathbf{V} = \mathbf{W}^{-1}\mathbf{A}^T\mathbf{K}^T.$$

The variance-covariance matrix of the V-matrix may then be computed from equation (2.189):

$$\boldsymbol{\sigma}_V = \sigma_o^2 \mathbf{W}^{-1}\mathbf{A}^T\,(\mathbf{A}\mathbf{W}^{-1}\mathbf{A}^T)^{-1}\,\mathbf{A}\mathbf{W}^{-1}.$$

2.5.6 A GENERAL FORMULATION FOR THE PROBLEM OF LEAST-SQUARES ADJUSTMENT

In general, a problem model in least squares adjustment may be represented by the following matrix equation:

$$\underset{(m,r)\ (r,1)}{\mathbf{A}\quad \mathbf{V}} + \underset{(m,n)\,(n,1)}{\mathbf{B}\quad \boldsymbol{\delta}} = \underset{(m,1)}{\mathbf{C}}; \qquad (2.190)$$

where **V** is the residual vector, and $\boldsymbol{\delta}$ is a vector of unknown parameters. Without going into the details of the derivation, it can be simply stated here that the normal equations for this model are represented by the following expression:

$$\mathbf{B}^T\,(\mathbf{A}\mathbf{W}^{-1}\mathbf{A}^T)^{-1}\,\mathbf{B}\,\boldsymbol{\delta} = \mathbf{B}^T\,(\mathbf{A}\mathbf{W}^{-1}\mathbf{A}^T)^{-1}\,\mathbf{C}. \qquad (2.191)$$

Furthermore, the variance-covariance matrix for the solution of $\boldsymbol{\delta}$ may be computed from the following expression:

$$\boldsymbol{\sigma}_\delta = \sigma_o^2\,[\mathbf{B}^T\,(\mathbf{A}\mathbf{W}^{-1}\mathbf{A}^T)^{-1}\,\mathbf{B}]^{-1}. \qquad (2.192)$$

2.5.7 SPECIAL SOLUTION MODELS

2.5.7.1 OBSERVATION EQUATIONS WITH CONSTRAINTS

Consider a mathematical model which consists of m observation equations and t constraint equations as follows:

observation equations

$$\underset{(m,1)}{\mathbf{V}} + \underset{(m,n)\,(n,1)}{\mathbf{B}\quad\boldsymbol{\delta}} = \underset{(m,1)}{\mathbf{C}};$$

constraints $\qquad\qquad\qquad\qquad (2.193)$

$$\underset{(t,n)\ \ (n,1)}{\mathbf{A}\quad\boldsymbol{\delta}} = \underset{(t,1)}{\mathbf{D}}.$$

The least squares solution to this model is given by the following expressions:

$$\boldsymbol{\delta} = \bar{\boldsymbol{\delta}} - \boldsymbol{\delta}', \qquad (2.194)$$

$$\boldsymbol{\sigma}_\delta = \boldsymbol{\sigma}_{\bar\delta}\,(\mathbf{I} - \boldsymbol{\sigma}_{\delta'}), \qquad (2.195)$$

$$\bar{\boldsymbol{\delta}} = (\mathbf{B}^T\mathbf{W}\mathbf{B})^{-1}\,\mathbf{B}^T\mathbf{W}\mathbf{C}, \qquad (2.196)$$

$$\boldsymbol{\sigma}_{\bar\delta} = \sigma_o^2\,(\mathbf{B}^T\mathbf{W}\mathbf{B})^{-1}, \qquad (2.197)$$

$$\boldsymbol{\delta}' = (\mathbf{B}^T\mathbf{W}\mathbf{B})^{-1}\,\mathbf{A}^T\,[\mathbf{A}(\mathbf{B}^T\mathbf{W}\mathbf{B})^{-1}\mathbf{A}^T]^{-1}\,[\mathbf{A}\,\bar{\boldsymbol{\delta}} - \mathbf{D}], \qquad (2.198)$$

and

$$\boldsymbol{\sigma}_{\delta'} = \mathbf{A}^T\,[\mathbf{A}(\mathbf{B}^T\mathbf{W}\mathbf{B})^{-1}\,\mathbf{A}^T]^{-1}\,\mathbf{A}\,(\mathbf{B}^T\mathbf{W}\mathbf{B})^{-1}. \qquad (2.199)$$

A comparison of the above equations with equations (2.166) to (2.168) shows that contributions of the set of constraint equations in (2.193) to the solution model are the $\boldsymbol{\delta}'$ and $\boldsymbol{\sigma}_{\delta'}$ matrices. If the mathematical model consisted of only observation equations, *i.e.* $\mathbf{V} + \mathbf{B}\,\boldsymbol{\delta} = \mathbf{C}$; the most probable solution is

$$\boldsymbol{\delta} = \bar{\boldsymbol{\delta}}$$

and

$$\boldsymbol{\sigma}_\delta = \boldsymbol{\sigma}_{\bar\delta}.$$

With the addition of a new set of constraints, $\mathbf{A}\boldsymbol{\delta} = \mathbf{D}$, the solution is simply changed to

$$\boldsymbol{\delta} = \bar{\boldsymbol{\delta}} + \boldsymbol{\delta}', \text{ and}$$

$$\boldsymbol{\sigma}_\delta = \boldsymbol{\sigma}_{\bar\delta}\,(\mathbf{I} - \boldsymbol{\sigma}_{\delta'});$$

where $\boldsymbol{\delta}'$ and $\boldsymbol{\sigma}_{\delta'}$ are computed according to equation (2.198) and (2.199) respectively.

By separating $\boldsymbol{\delta}'$ and $\boldsymbol{\sigma}_{\delta'}$ from $\bar{\boldsymbol{\delta}}$ and $\boldsymbol{\sigma}_{\bar\delta}$, this algorithm provides an easy means of analyzing the influence of the constraint equations in the solution. It also provides a simple means for adding constraint equations to an existing solution. As long as the matrices $(\mathbf{B}^T\mathbf{W}\mathbf{B}^T)^{-1}$ and $(\mathbf{B}^T\mathbf{W}\mathbf{C})$ of an existing solution are kept in computer storage, the effects of new constraints can be easily accounted for using the $\boldsymbol{\delta}'$ and $\boldsymbol{\sigma}_{\delta'}$ matrices.

2.5.7.2 ADDITION OF CONSTRAINT EQUATIONS TO THE GENERAL MODEL

Suppose that a set of constraint equations is added to the general model represented by equation (2.190); *i.e.*

$$\mathbf{A}\mathbf{V} + \mathbf{B}\boldsymbol{\delta} = \mathbf{C} \qquad (2.200)$$

$$\mathbf{D}\boldsymbol{\delta} = \mathbf{C}_c. \qquad (2.201)$$

The new solution can then be computed by

$$\boldsymbol{\delta} = \bar{\boldsymbol{\delta}} - \boldsymbol{\delta}' \qquad (2.202)$$

where

$$\bar{\boldsymbol{\delta}} = \mathbf{N}^{-1}\,\mathbf{E} \qquad (2.203)$$

$$\mathbf{N} = \mathbf{B}^T\,(\mathbf{A}\mathbf{W}^{-1}\mathbf{A}^T)^{-1}\,\mathbf{B}$$

$$\mathbf{E} = \mathbf{B}^T\,(\mathbf{A}\mathbf{W}^{-1}\mathbf{A}^T)^{-1}\,\mathbf{C}$$

and

$$\boldsymbol{\delta}' = \mathbf{N}^{-1}\mathbf{D}^T\,(\mathbf{D}\mathbf{N}^{-1}\mathbf{D}^T)^{-1}\,(\mathbf{D}\mathbf{N}^{-1}\mathbf{E} - \mathbf{C}_c). \quad (2.204)$$

The new variance-covariance matrix becomes

$$\boldsymbol{\sigma}_\delta = \boldsymbol{\sigma}_{\bar\delta}\,(\mathbf{I} - \boldsymbol{\sigma}_{\delta'}); \qquad (2.205)$$

where

$$\boldsymbol{\sigma}_{\delta'} = \mathbf{D}^T\,(\mathbf{D}\mathbf{N}^{-1}\mathbf{D}^T)^{-1}\,\mathbf{D}\mathbf{N}^{-1}, \qquad (2.206)$$

and

$$\boldsymbol{\sigma}_{\bar\delta} = \sigma_o^2\,\mathbf{N}^{-1}. \qquad (2.207)$$

2.5.7.3 ALGORITHM FOR DATA REJECTION

From equation (2.190), the general adjustment model is described by the following expression:

$$\begin{array}{cccc} A & V & + & B & \delta & = & C. \\ (m,r) & (r,1) & & (m,n) & (n,1) & & (m,1) \end{array}$$

The normal equation for such a model is

$$B \ (AW^{-1}A^T)^{-1} \ B \ \delta = B^T \ (AW^{-1}A^T)^{-1} \ C, \quad (2.208)$$

and the least-square solution is

$$\delta = [B^T \ (AW^{-1}A^T)^{-1}B]^{-1} \ B^T \ (AW^{-1}A^T)^{-1} \ C.$$

Letting

$$N = [B^T \ (AW^{-1}A^T)^{-1} \ B]$$

and

$$E = B^T \ (AW^{-1}A^T)^{-1} \ C;$$

then,

$$\delta = N^{-1}E, \qquad (2.209)$$

and

$$\sigma_\delta = \sigma_o^2 \ N^{-1}. \qquad (2.210)$$

Suppose that after having obtained a solution by equations (2.209) and (2.210), it is discovered that there are g observations which have very large residuals. In order to obtain a more reliable solution, it is decided to reject these g observations from the model and to repeat the adjustment. By employing the present algorithm, this means that the entire solution has to be repeated from the very beginning. In the case of a large system of equations, this can involve a considerable amount of computations. It would be desirable to have an algorithm so that the effects of the rejected observations can be subtracted directly from the already computed N^{-1} and E; i.e.,

$$(N^{-1})_{New} = (N^{-1})_{Old} - N_r^{-1} \qquad (2.211)$$

and

$$(E)_{New} = (E)_{Old} - E_r \qquad (2.212)$$

where N_r^{-1} and E_r are functions of the g rejected observations only.

Let the system of condition equations in equation (2.190) be subdivided into two groups as follows:

$$\begin{array}{cccc} A_e & V_e & + & B_e & \delta & = & C_e \\ (s,t) & (t,1) & & (s,n) & (n,1) & & (s,1) \end{array} \qquad (2.213)$$

$$\begin{array}{cccc} A_r & V_r & + & B_r & \delta & = & C_r \\ (k,g) & (g,1) & & (k,n) & (n,1) & & (k,1) \end{array} \qquad (2.214)$$

such that

$$(s + k) = m$$

$$(t + g) = n.$$

Equation (2.214) represents the group of k equations which are to be rejected before the next solution. Then, the matrices N_r^{-1} and E_r in equations (2.211) and (2.212) respectively can be computed from the following expressions:

$$E_r = B_r^T \ (A_r \ W_r^{-1}A_r^T)^{-1} \ C_r \qquad (2.215)$$

and

$$N_r^{-1} = N^{-1} \ B_r^T \ [-A_r W_r^{-1}A_r^T + B_r N^{-1}B_r^T]^{-1} \ B_r N^{-1}. \qquad (2.216)$$

2.5.7.4 ALGORITHM FOR DATA ADDITION

Suppose that after solving the system of equations

$$AV + B\delta = C$$

by

$$\delta = N^{-1}E$$

where

$$N = B^T \ (AW^{-1}A^T)^{-1} \ B$$

and

$$E = B^T \ (AW^{-1}A^T)^{-1} \ C \ ;$$

it is desired to add the following new group of equations to the problem model:

$$A_a V_a + B_a \delta = C_a.$$

It can be shown that the new solution may be computed from the following expressions:

$$(\delta)_{New} = (N^{-1})_{New} \ (E)_{New}$$

$$(E)_{New} = E + B_a^T \ (A_a \ W_a^{-1}A_a^T)^{-1} \ C_a \qquad (2.217)$$

$$(N^{-1})_{New} = N^{-1} - N^{-1} \ B_a^T \ [A_a W_a^{-1}A_a^T + B_a N^{-1}B_a^T]^{-1}B_a N^{-1}. \qquad (2.218)$$

2.5.8 ERROR ELLIPSE AND ELLIPSOID

Error ellipse and ellipsoid are used to evaluate the accuracy of photogrammetric determination of positions in two- and three-dimensional spaces respectively. For example, let the variance-and-covariance matrix (σ_P) of the computed X and Y coordinates of a point be given as follows:

$$\sigma_p = \begin{bmatrix} \sigma_x^2 & \sigma_{xy} \\ \sigma_{xy} & \sigma_y^2 \end{bmatrix} \qquad (2.219)$$

Then, the family of error ellipses about the computed position (X,Y) is given by the following expressions:

$$\frac{X_t^2}{\sigma_{x_t}^2} + \frac{y_t^2}{\sigma_{y_t}^2} = C_1 \qquad (2.220)$$

$$\sigma_{x_t}^2 = \frac{(\sigma_x^2 + \sigma_y^2) + \sqrt{(\sigma_x^2 - \sigma_y^2)^2 + 4\sigma_{xy}^2}}{2} \qquad (2.221)$$

$$\sigma_{y_t}^2 = \frac{(\sigma_x^2 + \sigma_y^2) - \sqrt{(\sigma_x^2 - \sigma_y^2)^2 + 4\sigma_{xy}^2}}{2} \qquad (2.222)$$

and

$$\tan 2\theta = \frac{2\sigma_{xy}}{(\sigma_x^2 - \sigma_y^2)} \qquad (2.223)$$

in which x_t is the major axis of the ellipse; y_t is the minor axis: $\sqrt{C_1}\sigma_{x_t}$ and $\sqrt{C_1}\sigma_{y_t}$ are the semi-major and semi-minor axes respectively; θ is the angle which the x_t-axis makes with the X-axis;

and C_1 is a constant. Figure 2-30 illustrates these parameters.

Statistically, there is 38 per cent chance that the true position of the point will fall within the error ellipse defined by $C_1 = 1$. This ellipse is called the standard ellipse. Similarly, there is 90 per cent and 99 per cent probability that the true position would fall within the ellipse defined by $C_1 = 4.6$ and $C_1 = 9.2$ respectively.

In geodetic surveying, the survey network and measurement procedures are designed so that the semi-major and semi-minor axes of the ellipse about each unknown point are approximately equal and that their magnitudes fall within the accuracy specification of the survey.

The computation of the parameters of an error ellipsoid in three-dimensional space is considerably more involved. Let the variance-and-covariance matrix (σ) of a point j which has the coordinates X_j, Y_j, and Z_j be defined as follows:

$$\sigma = \begin{bmatrix} \sigma_x^2 & \sigma_{xy} & \sigma_{xz} \\ \sigma_{xy} & \sigma_y^2 & \sigma_{yz} \\ \sigma_{xz} & \sigma_{yz} & \sigma_z^2 \end{bmatrix} \qquad (2.224)$$

Let the three orthogonal axes of the ellipsoid be denoted as x', y', and z'. Furthermore, let the orientation of the ellipsoid be defined by the rotations ω, ϕ, and κ about the $x'-$, $y'-$, and $z'-$ axes respectively, and let $\mathbf{R}$ be a rotational matrix such that

$$\begin{bmatrix} \sigma_{x'}^2 & 0 & 0 \\ 0 & \sigma_{y'}^2 & 0 \\ 0 & 0 & \sigma_{z'}^2 \end{bmatrix} = \begin{bmatrix} r_{11} & r_{21} & r_{31} \\ r_{12} & r_{22} & r_{32} \\ r_{13} & r_{23} & r_{33} \end{bmatrix}$$

$$\begin{bmatrix} \sigma_x^2 & \sigma_{xy} & \sigma_{xz} \\ \sigma_{xy} & \sigma_y^2 & \sigma_{yz} \\ \sigma_{xz} & \sigma_{yz} & \sigma_z^2 \end{bmatrix} \begin{bmatrix} r_{11} & r_{12} & r_{13} \\ r_{21} & r_{22} & r_{23} \\ r_{31} & r_{32} & r_{33} \end{bmatrix} \qquad (2.225)$$

i.e.,

$$\sigma' = \mathbf{R}\sigma\mathbf{R}^{\mathrm{T}} \qquad (2.226)$$

in which the elements r_{ij}'s are functions of the three rotation parameters ω, ϕ, and k.

Equation 22 consists of the following six independent equations:

$$\begin{aligned} r_{11}(a_{11}r_{11} + a_{12}r_{12} + a_{13}r_{13}) &+ r_{12}(a_{21}r_{11} + a_{22}r_{12} + a_{23}r_{13}) \\ &+ r_{13}(a_{31}r_{11} + a_{32}r_{12} + a_{33}r_{13}) \\ &= \sigma_{x'}^2, \qquad (2.227) \end{aligned}$$

$$\begin{aligned} r_{21}(a_{11}r_{21} + a_{12}r_{22} + a_{13}r_{23}) &+ r_{22}(a_{21}r_{21} + a_{22}r_{22} + a_{23}r_{23}) \\ &+ r_{23}(a_{31}r_{21} + a_{32}r_{22} + a_{33}r_{23}) \\ &= \sigma_{y'}^2, \qquad (2.228) \end{aligned}$$

$$\begin{aligned} r_{31}(a_{11}r_{31} + a_{12}r_{32} + a_{13}r_{33}) &+ r_{32}(a_{21}r_{31} + a_{22}r_{32} + a_{23}r_{33}) \\ &+ r_{33}(a_{31}r_{31} + a_{32}r_{32} + a_{33}r_{33}) \\ &= \sigma_{z'}^2, \qquad (2.229) \end{aligned}$$

$$\begin{aligned} r_{11}(a_{11}r_{21} + a_{12}r_{22} + a_{13}r_{23}) &+ r_{12}(a_{21}r_{21} + a_{22}r_{22} + a_{23}r_{23}) \\ &+ r_{13}(a_{31}r_{21} + a_{32}r_{22} + a_{33}r_{23}) \\ &= 0 \qquad (2.230) \end{aligned}$$

$$\begin{aligned} r_{11}(a_{11}r_{31} + a_{12}r_{32} + a_{13}r_{33}) &+ r_{12}(a_{21}r_{31} + a_{22}r_{32} + a_{23}r_{33}) \\ &+ r_{13}(a_{31}r_{31} + a_{32}r_{32} + a_{33}r_{33}) \\ &= 0 \qquad (2.231) \end{aligned}$$

$$\begin{aligned} r_{21}(a_{11}r_{31} + a_{12}r_{32} + a_{13}r_{33}) &+ r_{22}(a_{21}r_{31} + a_{22}r_{32} + a_{23}r_{33}) \\ &+ r_{23}(a_{31}r_{31} + a_{32}r_{32} + a_{33}r_{33}) \\ &= 0 \qquad (2.232) \end{aligned}$$

These six equations involve six unknowns: $\sigma_{x'}$, $\sigma_{y'}$, ω, ϕ, and x. The three rotation parameters can be obtained by solving equations (2.230), (2.231) and (2.232). Since these equations are non-linear, they must first be linearized and then solved by an iterative procedure. The computed values of ω, ϕ, and κ can then be substituted into equations (2.227), (2.228) and (2.229) to solve for $\sigma_{x'}$, $\sigma_{y'}$, and $\sigma_{z'}$.

There is only a 20 per cent probability that the true position of point j would fall within the ellipsoid defined by $C = 1$ in the following expression:

$$\frac{x'^2}{\sigma_{x'}^2} + \frac{y'^2}{\sigma_{y'}^2} + \frac{z'^2}{\sigma_{z'}^2} = C. \qquad (2.233)$$

There is 90 per cent and 99 per cent probability that the true position would fall within the ellipsoid defined by $C = 6.25$ and $C = 11.34$ respectively.

Most photogrammetric mapping problems involve the determination of positions in three-dimensional space. However, because of the involved computation, it is neither practical nor necessary to perform analyses using error ellipsoids for all points in the photogrammetric solution. In many applications, the diagonal elements (σ_x^2, σ_y^2, and σ_z^2) in the variance-and-covariance matrix (σ) should provide sufficient insight into the mapping accuracy. For cases in which extremely high accuracy is required, the error ellipsoids for a few carefully selected points may be computed for analysis.

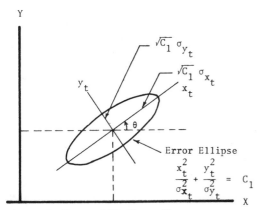

FIGURE 2-30. Error ellipse.

2.6 Least Squares Adjustment of Photogrammetric Blocks

A simultaneous least squares adjustment of all the measurements in a photogrammetric mapping problem can be formulated by the use of condition and observation equations. In the case of analytical aerotriangulation, the basic measurements include: (1) the photo-coordinates of the relevant image points on the photographs; (2) the ground coordinates of at least three control points; and (3) auxiliary data on the exterior orientation of the photographs. The purpose of the least squares adjustment is then to determine the most probable solution for the ground coordinates of all the unknown points and the exterior orientation parameters of all the photographs.

In the case of camera calibration, the basic

$$\begin{bmatrix} x_{ij} - x_p \\ y_{ij} - y_p \\ -f \end{bmatrix} = \frac{1}{\lambda_{ij}} \begin{bmatrix} m_{11} & m_{12} & m_{13} \\ m_{21} & m_{22} & m_{23} \\ m_{31} & m_{32} & m_{33} \end{bmatrix} \begin{bmatrix} X_j - X_i^c \\ Y_j - Y_i^c \\ Z_j - Z_i^c \end{bmatrix} \quad (2.97)$$

in which x_p and y_p are the photo coordinates of the principal point; f is the focal length of the camera; λ_{ij} is a scale factor; the m_{ij}'s are function of three rotation parameters (ω, ϕ, and κ) as expressed in equations (2.23); X_j, Y_j and Z_j are the ground coordinates of point j; and X_i^c, Y_i^c and Z_i^c are the ground coordinates of the exposure center of photo i.

Dividing the first two equations by the third, and rearranging terms, the above set of equations is reduced to the following two equations:

$$x_{ij} - x_p + \frac{f\left[m_{11}(X_j - X_i^c) + m_{12}(Y_j - Y_i^c) + m_{13}(Z_j - Z_i^c)\right]}{m_{31}(X_j - X_i^c) + m_{32}(y_j - Y_i^c) + m_{33}(Z_j - Z_i^c)} = 0$$

$$\hspace{10cm}(2.234)$$

$$y_{ij} - y_p + \frac{f\left[m_{21}(X_j - X_i^c) + m_{22}(Y_j - Y_i^c) + m_{23}(Z_j - Z_i^c)\right]}{m_{31}(X_j - X_i^c) + m_{32}(Y_j - Y_i^c) + m_{33}(Z_j - Z_i^c)} = 0 \ .$$

photogrammetric measurements include the photo-coordinates and ground coordinates of a large number of control points. The purpose of the adjustment is to determine both the interior and exterior orientation parameters of the camera. Instead of absolute positional control, directional control points are also frequently used in camera calibration.

The formulation of a mathematical model for least squares adjustment involves the following three basic steps:

1. develop observation equations for each type of observation;
2. generate all the necessary observation equations to build a mathematical model of the photogrammetric problem; and
3. develop an efficient computation algorithm.

The steps will be illustrated in this section by developing a simultaneous solution model for the problem of analytical aerotriangulation. The same mathematical model can be applied, with little or no modifications, to phototriangulation problems in close-range and terrestrial photogrammetry. The model can also be easily adopted for space resection and camera calibration.

2.6.1 DEVELOPMENT OF OBSERVATION EQUATIONS

2.6.1.1 COLLINEARITY EQUATIONS

Let x_{ij} and y_{ij} denote the photo coordinates of the image of point j in photograph i. From equation (2.97), the following relationship exists:

These two equations are called the collinearity equations. They are the two most fundamental equations in analytical photogrammetry. These two equations together describe the absolute orientation and length of the light ray joining the exposure center i, the location of ground point j, and the position of the image of point j on photograph i. These equations are derived from the projective transformation equation of equation (2.22) which is based on the fundamental assumption that the exposure center, the ground point and its corresponding image point all lie on a straight line.

The collinearity equations can be used to solve a wide variety of photogrammetric problems. Since each light ray can be described by two collinearity equations, a complete mathematical model of the rays forming a photogrammetric model can be easily constructed. However, before equations (2.234) can be used in adjustment, they must be reduced to a linear form and the random errors that are always present in the measurements must also be added to the mathematical model.

Let

$$x_{ij} = x_{ij}^{oo} + V_{x_{ij}},$$

$$\hspace{10cm}(2.235)$$

and

$$y_{ij} = y_{ij}^{oo} + V_{y_{ij}};$$

where x_{ij}^{oo} and y_{ij}^{oo} are the measured image coordinates, and $V_{x_{ij}}$ and $V_{y_{ij}}$ are the corrections needed to account for the random error in the measured coordinates. Similarly, let

$$\omega_i = \omega_i^o + \Delta\omega_i, \ X_i^c = (X_i^c)^o + \Delta X_i^c, \ X_j = X_j^o + \Delta X_j,$$

$$\phi_i = \phi_i^o + \Delta\phi_i, \ Y_i^c = (Y_i^c)^o + \Delta Y_i^c, \ Y_j = Y_j^o + \Delta Y_j,$$

$$\kappa_i = \kappa_i^o + \Delta\kappa_i, \ Z_i^c = (Z_i^c)^o + \Delta Z_i^c, \ Z_j = Z_j^o + \Delta Z_j;$$

$$(2.236)$$

where ω_i^o, ϕ_i^o, κ_i^o, $(X_i^c)^o$, $(Y_i^c)^o$, $(Z_i^c)^o$, X_j^o, Y_j^o and Z_j^o are some approximations, and $\Delta\omega_i$, $\Delta\phi_i$, $\Delta\kappa_i$, ΔX_i^c, ΔY_i^c, ΔZ_i^c, ΔX_j, ΔY_j and ΔZ_j are their corresponding corrections. The super-subscript ^{oo} will be used to denote measurements, and the super-subscript ^{o} will be used to denote approximations. Furthermore, it is assumed that the corrections in equations (2.236) are small and the collinearity equations are linear over the small intervals between the true values of these parameters and their corresponding approximations. Then, equations (2.236) can be linearized by Newton's first order approximation as follows:

ϕ_i^o, . . . and Z_j^o. Similarly, the partial derivatives are also to be computed with these approximations. It is to be remembered that these two equations are only linear approximations of the two exact equations in (2.234). As the approximations ω_i^o, ϕ_i^o, . . . Y_j^o and Z_j^o approach their corresponding true values, the differences between equations (2.234) and (2.238) become negligible. Therefore, when the linearized collinearity equations are used in computation, it always requires an iterative solution. The computed corrections ($\Delta\omega_i$, $\Delta\phi_i$ . . . . ΔY_j and ΔZ_j) from one iteration are applied to their corresponding approximations. The corrected parameters are then used as approximations in the next iteration, and the whole iteration process is repeated until the corrections ($\Delta\omega_i$, $\Delta\phi_i$, $\Delta\kappa_i$ . . . . ΔY_j and ΔZ_j) become negligibly small.

In mathematical formulation, it is more convenient to express equation (2.238) in matrix notation. Let

let

$$F_x = x_{ij} - x_p + \frac{f\left[m_{11}(X_j - X_i^c) + m_{12}(Y_j - Y_i^c) + m_{13}(Z_j - Z_i^c)\right]}{m_{31}(X_j - X_i^c) + m_{32}(Y_j - Y_i^c) + m_{33}(Z_j - Z_i^c)},$$

$$(2.237)$$

and

$$F_y = y_{ij} - y_p + \frac{f\left[m_{21}(X_j - X_i^c) + m_{22}(Y_j - Y_i^c) + m_{23}(Z_j - Z_i^c)\right]}{m_{31}(X_j - X_i^c) + m_{32}(Y_j - Y_i^c) + m_{33}(Z_j - Z_i^c)};$$

then

$$V_{x_{ij}} + \left(\frac{\partial Fx}{\partial\omega_i}\right)^o \Delta\omega_i + \left(\frac{\partial Fx}{\partial\phi_i}\right)^o \Delta\phi_i$$

$$+ \left(\frac{\partial Fx}{\partial\kappa_i}\right)^o \Delta\kappa_i + \left(\frac{\partial Fx}{\partial X_i^c}\right)^o \Delta X_i^c$$

$$+ \left(\frac{\partial Fx}{\partial Y_i^c}\right)^o \Delta Y_i^c + \left(\frac{\partial Fx}{\partial Z_i^c}\right)^o \Delta Z_i^c$$

$$+ \left(\frac{\partial Fx}{\partial X_j}\right)^o \Delta X_j + \left(\frac{\partial Fx}{\partial Y_j}\right)^o \Delta Y_j$$

$$+ \left(\frac{\partial Fx}{\partial Z_j}\right)^o \Delta Z_j + F_x^o = 0 \qquad (2.238)$$

$$V_{y_{ij}} + \left(\frac{\partial Fy}{\partial\omega_i}\right)^o \Delta\omega_i + \left(\frac{\partial Fy}{\partial\phi_i}\right)^o \Delta\phi_i$$

$$+ \left(\frac{\partial Fy}{\partial\kappa_i}\right)^o \Delta\kappa_i + \left(\frac{\partial Fy}{\partial X_i^c}\right)^{o} \Delta X_i^c$$

$$+ \left(\frac{\partial Fy}{\partial Y_i^c}\right)^o \Delta Y_i^c + \left(\frac{\partial Fy}{\partial Z_i^c}\right)^o \Delta Z_i^c$$

$$+ \left(\frac{\partial Fy}{\partial X_j}\right)^o \Delta X_j + \left(\frac{\partial Fy}{\partial Y_j}\right)^o \Delta Y_j$$

$$+ \left(\frac{\partial Fy}{\partial Z_j}\right)^o \Delta Z_j + F_y^o = 0$$

$$b_{11} = \left(\frac{\partial F_x}{\partial\omega_i}\right)^o, \quad b_{21} = \left(\frac{\partial F_y}{\partial\omega_i}\right)^o,$$

$$b_{12} = \left(\frac{\partial F_x}{\partial\phi_i}\right)^o, \quad b_{22} = \left(\frac{\partial F_y}{\partial\phi_i}\right)^o,$$

$$b_{13} = \left(\frac{\partial F_x}{\partial\kappa_i}\right)^o, \quad b_{23} = \left(\frac{\partial F_y}{\partial\kappa_i}\right)^o,$$

$$b_{14} = \left(\frac{\partial F_x}{\partial X_i^c}\right)^o, \quad b_{24} = \left(\frac{\partial F_y}{\partial X_i^c}\right)^o,$$

$$b_{15} = \left(\frac{\partial F_x}{\partial Y_i^c}\right)^o, \quad b_{25} = \left(\frac{\partial F_y}{\partial Y_i^c}\right)^o, \quad (2.239)$$

$$b_{16} = \left(\frac{\partial F_x}{\partial Z_i^c}\right)^o, \quad b_{26} = \left(\frac{\partial F_y}{\partial Z_i^c}\right)^o,$$

$$b_{17} = \left(\frac{\partial F_x}{\partial X_j}\right)^o, \quad b_{27} = \left(\frac{\partial F_y}{\partial X_j}\right)^o,$$

$$b_{18} = \left(\frac{\partial F_x}{\partial Y_j}\right)^o, \quad b_{28} = \left(\frac{\partial F_y}{\partial Y_j}\right)^o,$$

$$b_{19} = \left(\frac{\partial F_x}{\partial Z_j}\right)^o, \quad b_{29} = \left(\frac{\partial F_y}{\partial Z_j}\right)^o.$$

where F_x^o and F_y^o are the functions F_x and F_y computed with the approximate values ω_i^o,

Then, equations (2.238) may be written in matrix notation as follows:

$$\begin{bmatrix} V_{x_{ij}} \\ V_{y_{ij}} \end{bmatrix} + \begin{bmatrix} b_{11} & b_{12} & b_{13} & b_{14} & b_{15} & b_{16} \\ b_{21} & b_{22} & b_{23} & b_{24} & b_{25} & b_{26} \end{bmatrix} \begin{bmatrix} \Delta\omega_i \\ \Delta\phi_i \\ \Delta\kappa_i \\ \Delta X_i^c \\ \Delta Y_i^c \\ \Delta Z_i^c \end{bmatrix}$$

$$+ \begin{bmatrix} b_{17} & b_{18} & b_{19} \\ b_{27} & b_{28} & b_{29} \end{bmatrix} \begin{bmatrix} \Delta X_j \\ \Delta Y_j \\ \Delta Z_j \end{bmatrix} = \begin{bmatrix} -f_x^o \\ -f_y^o \end{bmatrix} \qquad (2.240)$$

or simply

$$\begin{array}{ccccc} V_{ij} & + & \dot{B}_{ij} & \dot{\Delta}_i & + & \ddot{B}_{ij} & \ddot{\Delta}_j & = & \epsilon_{ij} \\ (2,1) & & (2,6) & (6,1) & & (2,3) & (3,1) & & (2,1) \end{array} \qquad (2.241)$$

The superscript "one dot" is used to denote corrections to the exterior orientation parameters, and "two dots" is used to denote corrections to the ground coordinates.

2.6.1.2 GROUND CONTROL EQUATIONS

Let X_j^{oo}, Y_j^{oo}, and Z_j^{oo} be the ground coordinates of a point j determined previously from control surveys. Let V_{X_j}, V_{Y_j} and V_{Z_j} be the unknown residuals associated with X_j^{oo}, Y_j^{oo} and Z_j^{oo} respectively such that the ground coordinates of point j can be expressed as follows:

$$\begin{aligned} X_j &= X_j^{oo} + V_{X_j} \\ Y_j &= Y_j^{oo} + V_{Y_j} \\ Z_j &= Z_j^{oo} + V_{Z_j}. \end{aligned} \qquad (2.242)$$

Furthermore, it was defined in equation (2.236) that

$$\begin{aligned} X_j &= X_j^o + \Delta X_j \\ Y_j &= Y_j^o + \Delta Y_j \\ Z_j &= Z_j^o + \Delta Z_j. \end{aligned}$$

Therefore,

$$\begin{aligned} V_{X_j} + X_j^{oo} &= X_j^o + \Delta X_j \\ V_{Y_j} + Y_j^{oo} &= Y_j^o + \Delta Y_j \\ V_{Z_j} + Z_j^{oo} &= Z_j^o + \Delta Z_j. \end{aligned}$$

Rearranging terms and putting the equations into matrix form yields

$$\begin{bmatrix} V_{X_j} \\ V_{Y_j} \\ V_{Z_j} \end{bmatrix} - \begin{bmatrix} \Delta X_j \\ \Delta Y_j \\ \Delta Z_j \end{bmatrix} = \begin{bmatrix} X_j^o - X_j^{oo} \\ Y_j^o - Y_j^{oo} \\ Z_j^o - Z_j^{oo} \end{bmatrix}. \qquad (2.243)$$

These are then the observation equations for the known (or measured) ground coordinates of point j. They may be simply written as a single matrix equation as follows:

$$\begin{array}{ccccc} \ddot{V}_j & - & \ddot{\Delta}_j & = & \ddot{C}_j . \\ (3,1) & & (3,1) & & (3,1) \end{array} \qquad (2.244)$$

2.6.1.3 EXTERIOR ORIENTATION PARAMETERS

Observation equations for the exterior orientation parameters can be derived in exactly the same manner as for the ground coordinates.

Let ω_i^{oo}, ϕ_i^{oo}, κ_i^{oo}, $(X_i^c)^{oo}$, $(Y_i^c)^{oo}$ and $(Z_i^c)^{oo}$ be the measured values of the exterior orientation of photograph i and let V_{ω_i}, V_{ϕ_i}, V_{κ_i}, $V_{X_i^c}$, $V_{Y_i^c}$, and $V_{Z_i^c}$ be their corresponding unknown residuals. Thus, the true values of these six orientation parameters may be expressed as follows:

$$\begin{aligned} \omega_i &= \omega_i^{oo} + V_{\omega_i}, \\ \phi_i &= \phi_i^{oo} + V_{\phi_i}, \\ \kappa_i &= \kappa_i^{oo} + V_{\kappa_i}, \\ X_i^c &= (X_i^c)^{oo} + V_{X_i^c}, \qquad (2.245) \\ Y_i^c &= (Y_i^c)^{oo} + V_{Y_i^c}, \\ Z_i^c &= (Z_i^c)^{oo} + V_{Z_i^c}. \end{aligned}$$

But according to equation (2.236),

$$\begin{aligned} \omega_i &= \omega_i^o + \Delta\omega_i, \\ \phi_i &= \phi_i^o + \Delta\phi_i, \\ \kappa_i &= \kappa_i^o + \Delta\kappa_i, \\ X_i^c &= (X_i^c)^o + \Delta X_i^c, \\ Y_i^c &= (Y_i^c)^o + \Delta Y_i^c, \\ Z_i^c &= (Z_i^c)^o + \Delta Z_i^c. \end{aligned}$$

Therefore, equating the two sets of equations and expressing the result in matrix notation yields

$$\begin{bmatrix} V_{\omega_i} \\ V_{\phi_i} \\ V_{\kappa_i} \\ V_{X_i^c} \\ V_{Y_i^c} \\ V_{Z_i^c} \end{bmatrix} - \begin{bmatrix} \Delta\omega_i \\ \Delta\phi_i \\ \Delta\kappa_i \\ \Delta X_i^c \\ \Delta Y_i^c \\ \Delta Z_i^c \end{bmatrix} = \begin{bmatrix} \omega_i^o - \omega_i^{oo} \\ \phi_i^o - \phi_i^{oo} \\ \kappa_i^o - \kappa_i^{oo} \\ (X_i^c)^o - (X_i^c)^{oo} \\ (Y_i^c)^o - (Y_i^c)^{oo} \\ (Z_i^c)^o - (Z_i^c)^{oo} \end{bmatrix}. \qquad (2.246)$$

These are the final observation equations for the exterior orientation parameters of photograph i. Although all six parameters are included in the above set of equations, observation equations need only be written for the parameters that actually have measured values. Equation (2.246) may be represented by a single matrix equation as

$$\begin{array}{ccccc} \dot{V}_i & - & \dot{\Delta}_i & = & \dot{C}_i . \\ (6,1) & & (6,1) & & (6,1) \end{array} \qquad (2.247)$$

2.6.2 FORMULATION OF THE MATHEMATICAL MODEL

Consider a block of m photographs covering an area in which there are n points of basic interest. Moreover, let r of these n points be ground control points the positions of which have been determined by other survey techniques. For the sake of generality, it is assumed that each of the n points is imaged on all of the m photographs. In reality, only a small percentage of the n points are imaged on each photograph. However, in order to avoid the problem of having to specify in the model which ground points are imaged on each of the photographs, it is assumed that each point is imaged on all of the photographs. It will be a simple matter to ignore, during the computation phase, the fictitious observations and generate equations only for the parameters for which there are actual measurement data.

Assuming that point j is imaged in all m photographs, the complete set of collinearity equations generated by its image coordinates on the m photographs is as follows:

$$\begin{bmatrix} V_{1j} \\ V_{2j} \\ V_{3j} \\ \cdot \\ \cdot \\ \cdot \\ V_{mj} \end{bmatrix} + \begin{bmatrix} \dot{B}_{1j} \\ \dot{B}_{2j} \\ \dot{B}_{3j} \\ & \cdot \\ & & \cdot \\ & & & \cdot \\ & & & & \dot{B}_{mj} \end{bmatrix} \begin{bmatrix} \dot{\Delta}_1 \\ \dot{\Delta}_2 \\ \dot{\Delta}_3 \\ \cdot \\ \cdot \\ \cdot \\ \dot{\Delta}_m \end{bmatrix}$$

$$\begin{array}{ccc} (2m,1) & (2m,6m) & (6m,1) \end{array}$$

$$+ \begin{bmatrix} \ddot{B}_{1j} \\ \ddot{B}_{2j} \\ \ddot{B}_{3j} \\ \cdot \\ \cdot \\ \cdot \\ \ddot{B}_{mj} \end{bmatrix} \ddot{\Delta}_j = \begin{bmatrix} \epsilon_{1j} \\ \epsilon_{2j} \\ \epsilon_{3j} \\ \cdot \\ \cdot \\ \cdot \\ \epsilon_{mj} \end{bmatrix} \qquad (2.248)$$

$$\begin{array}{ccc} (2m,3) & (3,1) & (2m,1) \end{array}$$

which may also be simply expressed as

$$V_j + \dot{B}_j\,\dot{\Delta} + \ddot{B}_j\,\ddot{\Delta}_j = \epsilon_j. \qquad (2.249)$$

Since it is assumed that each point j is imaged in all m photographs, each point gives rise to a complete set of collinearity equations as above. Thus, the complete collection of colinearity equations for all n points are:

$$\begin{bmatrix} V_1 \\ V_2 \\ V_3 \\ \cdot \\ \cdot \\ \cdot \\ V_n \end{bmatrix} + \begin{bmatrix} B_1 \\ B_2 \\ B_3 \\ \cdot \\ \cdot \\ \cdot \\ B_n \end{bmatrix} \dot{\Delta} + \begin{bmatrix} \ddot{B}_1 \\ \ddot{B}_2 \\ \ddot{B}_3 \\ & \cdot \\ & & \cdot \\ & & & \cdot \\ & & & & \ddot{B}_n \end{bmatrix} \begin{bmatrix} \ddot{\Delta}_1 \\ \ddot{\Delta}_2 \\ \ddot{\Delta}_3 \\ \cdot \\ \cdot \\ \cdot \\ \ddot{\Delta}_n \end{bmatrix} = \begin{bmatrix} \epsilon_1 \\ \epsilon_2 \\ \epsilon_3 \\ \cdot \\ \cdot \\ \cdot \\ \epsilon_n \end{bmatrix}.$$

$$\begin{array}{ccccc} (2mn,1) & (2mn,6m) & (6m,1) & (2mn,3n) & (3n,1) & (2mn,1) \end{array} \qquad (2.250)$$

That is

$$V + \dot{B}\dot{\Delta} + \ddot{B}\ddot{\Delta} = \epsilon. \qquad (2.251)$$

Next, observation equations for exterior orientation parameters and for known ground coordinates are added to the model. Again, to keep the model as general as possible, it is assumed that measured values are available for the exterior orientation parameters of all photographs and for ground coordinates of all points.

The complete set of observation equations for the exterior orientation parameters can be stacked together as follows:

$$\begin{bmatrix} \dot{V}_1 \\ \dot{V}_2 \\ \dot{V}_3 \\ \cdot \\ \cdot \\ \cdot \\ \dot{V}_m \end{bmatrix} - \begin{bmatrix} \dot{\Delta}_1 \\ \dot{\Delta}_2 \\ \dot{\Delta}_3 \\ \cdot \\ \cdot \\ \cdot \\ \dot{\Delta}_m \end{bmatrix} = \begin{bmatrix} \dot{C}_1 \\ \dot{C}_2 \\ \dot{C}_3 \\ \cdot \\ \cdot \\ \cdot \\ \dot{C}_m \end{bmatrix} \qquad (2.252)$$

$$\begin{array}{ccc} (6m,1) & (6m,1) & (6m,1) \end{array}$$

which may be stated as

$$\dot{V} - \dot{\Delta} = \dot{C}. \qquad (2.253)$$

Similarly, the complete set of observation equations for the ground coordinates of all n points is

$$\begin{bmatrix} \ddot{V}_1 \\ \ddot{V}_2 \\ \ddot{V}_3 \\ \cdot \\ \cdot \\ \cdot \\ \ddot{V}_n \end{bmatrix} - \begin{bmatrix} \ddot{\Delta}_1 \\ \ddot{\Delta}_2 \\ \ddot{\Delta}_3 \\ \cdot \\ \cdot \\ \cdot \\ \ddot{\Delta}_n \end{bmatrix} = \begin{bmatrix} \ddot{C}_1 \\ \ddot{C}_2 \\ \ddot{C}_3 \\ \cdot \\ \cdot \\ \cdot \\ \ddot{C}_n \end{bmatrix} \qquad (2.254)$$

$$\begin{array}{ccc} (3n,1) & (3n,1) & (3n,1) \end{array}$$

which may also be stated as

$$\ddot{V} - \ddot{\Delta} = \ddot{C}. \qquad (2.255)$$

Thus, combining equations (2.251), (2.253) and (2.255) gives a complete mathematical model of the photogrammetric problem; i.e.

$$V + \dot{B}\dot{\Delta} + \ddot{B}\ddot{\Delta} = \epsilon$$
$$\dot{V} - \dot{\Delta} \qquad\quad = \dot{C}$$
$$\ddot{V} \qquad\quad - \ddot{\Delta} = \ddot{C}.$$

Expressing this equation as a single matrix equation, we have

$$\begin{bmatrix} V \\ \dot{V} \\ \ddot{V} \end{bmatrix} + \begin{bmatrix} \dot{B} & \ddot{B} \\ -I & 0 \\ 0 & -I \end{bmatrix} \begin{bmatrix} \dot{\Delta} \\ \ddot{\Delta} \end{bmatrix} = \begin{bmatrix} \epsilon \\ \dot{C} \\ \ddot{C} \end{bmatrix} ; \quad (2.256)$$

i.e.

$$\overline{V} + \overline{B}\Delta = \overline{C}. \qquad (2.257)$$

Based on the derivation presented in section 2.5.4, it is obvious that a least squares solution to the model in equation (2.257) results in the following normal equation:

$$(\overline{B}^T \ \overline{W} \ \overline{B}) \ \Delta = \overline{B}^T \ \overline{W} \ \overline{C} \qquad (2.258)$$

where $\overline{W}$ is a weight matrix. Hence, the least-squares solution will be

$$\Delta = (\overline{B}^T \ \overline{W} \ \overline{B})^{-1} \ \overline{B}^T \ \overline{W} \ \overline{C} \cdot \qquad (2.259)$$

2.6.3 STRUCTURE OF THE WEIGHT MATRIX

The $\overline{W}$-matrix in equation (2.258) consists of a series of submatrices which serve as the weight matrices assigned to the observed parameters. They must be stacked and organized in exactly the same manner as the V-matrix. In this respect, it may be beneficial to re-examine the structure of the V-matrix. According to equations (2.248), (2.250), (2.252) and (2.254),

$$\overline{V} = \begin{bmatrix} V \\ \dot{V} \\ \ddot{V} \end{bmatrix} = \begin{bmatrix} V_1 \\ V_2 \\ \cdot \\ \cdot \\ V_{mn} \\ \dot{V}_1 \\ \dot{V}_2 \\ \cdot \\ \cdot \\ \cdot \\ \dot{V}_m \\ \ddot{V}_1 \\ \ddot{V}_2 \\ \cdot \\ \cdot \\ \ddot{V}_n \end{bmatrix} \begin{matrix} \text{image coordinates} \\ \text{residuals} \\ \\ \\ \\ \\ \text{exterior orientation} \\ \text{residuals} \\ \\ \\ \\ \text{ground coordinates} \\ \text{residuals} \end{matrix} \qquad (2.260)$$

where

$$V_j = \begin{bmatrix} V_{1j} \\ V_{2j} \\ V_{3j} \\ \cdot \\ \cdot \\ V_{mj} \end{bmatrix}, \quad \dot{V}_i = \begin{bmatrix} V_{\omega_i} \\ V_{\phi_i} \\ V_{\kappa_i} \\ V_{X_i^c} \\ V_{Y_i^c} \\ V_{Z_i^c} \end{bmatrix}, \quad \text{and} \quad \ddot{V}_j = \begin{bmatrix} V_{X_j} \\ V_{Y_j} \\ V_{Z_j} \end{bmatrix}.$$

It was shown in section 2.5.1 that the weight of an observation is inversely proportional to the population variance of the measurement. Let w_i denote the weight of measurement i, then from equation (2.143),

$$w_i = \frac{\sigma_o^2}{\sigma_i^2}$$

where σ_o^2 is a constant scaling factor and is called the variance of unit weight. A value can be arbitrarily chosen for σ_o^2 in a given problem. However, once defined, the same constant factor σ_o^2 must be used to scale the weights for all other observations in the adjustment. It can be easily seen that an observation having a variance $\sigma_i^2 = \sigma_o^2$ will have a weight of 1.

Let the variance-covariance matrix for the pair of image coordinates (x_{ij}, y_{ij}) be defined as follows:

$$\sigma_{ij} = \begin{bmatrix} \sigma_{x_{ij}}^2 & \sigma_{x_{ij}y_{ij}} \\ \sigma_{x_{ij}y_{ij}} & \sigma_{y_{ij}}^2 \end{bmatrix}^{-1} \qquad (2.261)$$

where $\sigma_{x_{ij}}^2$ and $\sigma_{y_{ij}}^2$ are the variances of the measurements x_{ij} and y_{ij} respectively; and $\sigma_{x_{ij}y_{ij}}$ is the covariance of the two measurements. The corresponding weight matrix is then defined as

$$\underset{(2,2)}{W_{ij}} = \sigma_o^2 \begin{bmatrix} \sigma_{x_{ij}}^2 & \sigma_{x_{ij}y_{ij}} \\ \sigma_{x_{ij}y_{ij}} & \sigma_{y_{ij}}^2 \end{bmatrix}^{-1} = \begin{bmatrix} a_{11} & a_{12} \\ a_{21} & a_{22} \end{bmatrix} \qquad (2.262)$$

Stacking the weight matrices for all the image coordinates on point j yields

$$\underset{(2m,2m)}{W_j} = \begin{bmatrix} W_{1j} & & & \\ & W_{2j} & & \\ & & W_{3j} & \\ & & & \cdot \\ & & & & W_{mj} \end{bmatrix} . \qquad (2.263)$$

Proceeding one step further and stacking the weight matrices for all n points yield

$$\underset{(2mn,2mn)}{W} = \begin{bmatrix} W_1 & & & \\ & W_2 & & \\ & & W_3 & \\ & & & \cdot \\ & & & & W_n \end{bmatrix} \qquad (2.264)$$

Both the W_j and the W matrices are diagonal matrices because it is assumed that there is no correlation among the measurements of the individual image points.

Similarly, let the variance-covariance matrix of the exterior orientation parameters of photograph i be denoted as

$$\underset{(6,6)}{\dot{\sigma}_i} = \begin{bmatrix} \sigma_{\omega_i}^2 & \sigma_{\omega_i\phi_i} & \sigma_{\omega_i\kappa_i} & \sigma_{\omega_iX_i^c} & \sigma_{\omega_iY_i^c} & \sigma_{\omega_iZ_i^c} \\ \sigma_{\phi_i\omega_i} & \sigma_{\phi_i}^2 & \sigma_{\phi_i\kappa_i} & \sigma_{\phi_iX_i^c} & \sigma_{\phi_iY_i^c} & \sigma_{\phi_iZ_i^c} \\ \sigma_{\kappa_i\omega_i} & \sigma_{\kappa_i\phi_i} & \sigma_{\kappa_i}^2 & \sigma_{\kappa_iX_i^c} & \sigma_{\kappa_iY_i^c} & \sigma_{\kappa_iZ_i^c} \\ \cdot & & & & & \\ \cdot & & & & & \\ \cdot & \cdot & \cdot & \cdot & \cdot & \sigma_{Z_i^c}^2 \end{bmatrix} . \qquad (2.265)$$

Then, the weight matrix associated with the exterior orientation parameters of photograph i is defined as follows:

$$\mathbf{W}_i = \sigma_o^2 \, \dot{\boldsymbol{\sigma}}_i^{-1} \qquad (2.266)$$
$$(6,6)$$

Assuming no correlation among the orientation parameters of different photographs, the complete set of weight matrices for all m photos is as follows:

$$\mathbf{W} = \begin{bmatrix} \dot{\mathbf{W}}_1 & & & & \\ & \dot{\mathbf{W}}_2 & & & \\ & & \dot{\mathbf{W}}_3 & & \\ & & & \ddots & \\ & & & & \dot{\mathbf{W}}_m \end{bmatrix} \qquad (2.267)$$
$$(6m,6m)$$

Furthermore, the variance-covariance matrix of the ground coordinates of point j may be expressed as follows:

$$\ddot{\boldsymbol{\sigma}}_j = \begin{bmatrix} \sigma_{X_j}^2 & \sigma_{X_j Y_j} & \sigma_{X_j Z_j} \\ \sigma_{Y_j X_j} & \sigma_{Y_j}^2 & \sigma_{Y_j Z_j} \\ \sigma_{Z_j X_j} & \sigma_{Z_j Y_j} & \sigma_{Z_j}^2 \end{bmatrix} \qquad (2.268)$$
$$(3,3)$$

Then,

$$\ddot{\mathbf{W}}_j = \sigma_o^2 \, \ddot{\boldsymbol{\sigma}}_j^{-1}. \qquad (2.269)$$

Again, assuming no correlation among the measured coordinates of the different points, the weight matrix of the ground coordinates is as follows:

$$\ddot{\mathbf{W}} = \begin{bmatrix} \ddot{\mathbf{W}}_1 & & & & \\ & \ddot{\mathbf{W}}_2 & & & \\ & & \ddot{\mathbf{W}}_3 & & \\ & & & \ddots & \\ & & & & \ddot{\mathbf{W}}_n \end{bmatrix} \qquad (2.270)$$
$$(3n,3n)$$

Stacking equations (2.264), (2.267) and (2.270) together yields the total weight matrix $\overline{\mathbf{W}}$; $i.e.$

$$\overline{\mathbf{W}} = \begin{bmatrix} \mathbf{W} & & \\ & \dot{\mathbf{W}} & \\ & & \ddot{\mathbf{W}} \end{bmatrix}. \qquad (2.271)$$

2.6.4 STRUCTURE OF THE NORMAL EQUATIONS

The set of normal equations (2.258) may be simply expressed as

$$\begin{array}{ccc} \mathbf{N} & \boldsymbol{\Delta} & = & \mathbf{K} \\ (6m + 3n,\ 6m + 3n) & (6m + 3n,1) & (6m + 3n,1) \end{array}$$
$$(2.272)$$

The direct solution of this set of equations will involve solving $(6m + 3n)$ simultaneous equations. Even for a small block of photographs, the size of the normal equation can be so large as to make its solution impractical. For example, for a block of 20 photographs involving 100 points, $(6m + 3n) = 420$. The storage of the N-matrix alone calls for $420 \times 420 = 176{,}400$ locations in the computer for single-precision computation. Thus, there is a need for developing an efficient algorithm for collecting the observation equations together to form the normal equations and for solving for the unknowns. In fact, a detailed study of the structure of the matrices N, $\boldsymbol{\Delta}$ and K in equation (2.272) will reveal some characteristic features which may be used to great advantage in computation. By a process of successive substitution using the equations that have already been developed, it can be shown that the three matrices N, $\boldsymbol{\Delta}$ and K have the following composition:

$$\mathbf{N} = \left[\begin{array}{cccc|cccc} \dot{\mathbf{N}}_1 + \dot{\mathbf{W}}_1 & & & & \overline{\mathbf{N}}_{11} & \overline{\mathbf{N}}_{12} & \cdots & \overline{\mathbf{N}}_{1n} \\ & \dot{\mathbf{N}}_2 + \dot{\mathbf{W}}_2 & & & \overline{\mathbf{N}}_{21} & \overline{\mathbf{N}}_{22} & \cdots & \overline{\mathbf{N}}_{2n} \\ & & \ddots & & & & & \\ & & & \dot{\mathbf{N}}_m + \dot{\mathbf{W}}_m & \overline{\mathbf{N}}_{m1} & \overline{\mathbf{N}}_{m2} & \cdots & \overline{\mathbf{N}}_{mn} \\ \hline \overline{\mathbf{N}}_{11}^{\mathrm{T}} & \overline{\mathbf{N}}_{21}^{\mathrm{T}} & \cdots & \overline{\mathbf{N}}_{m1}^{\mathrm{T}} & \ddot{\mathbf{N}}_1 + \ddot{\mathbf{W}}_1 & & & \\ \overline{\mathbf{N}}_{12}^{\mathrm{T}} & \overline{\mathbf{N}}_{22}^{\mathrm{T}} & \cdots & \overline{\mathbf{N}}_{m2}^{\mathrm{T}} & & \ddot{\mathbf{N}}_2 + \ddot{\mathbf{W}}_2 & & \\ & & \ddots & & & & \ddots & \\ \overline{\mathbf{N}}_{1n}^{\mathrm{T}} & \overline{\mathbf{N}}_{2n}^{\mathrm{T}} & \cdots & \overline{\mathbf{N}}_{mn}^{\mathrm{T}} & & & & \ddot{\mathbf{N}}_n + \ddot{\mathbf{W}}_n \end{array} \right] \qquad \boldsymbol{\Delta} = \left[\begin{array}{c} \dot{\boldsymbol{\Delta}}_1 \\ \dot{\boldsymbol{\Delta}}_2 \\ \vdots \\ \dot{\boldsymbol{\Delta}}_m \\ \hline \ddot{\boldsymbol{\Delta}}_1 \\ \ddot{\boldsymbol{\Delta}}_2 \\ \vdots \\ \ddot{\boldsymbol{\Delta}}_n \end{array} \right] \qquad \mathbf{K} = \left[\begin{array}{c} \dot{\mathbf{K}}_1 - \dot{\mathbf{W}}_1 \dot{\mathbf{C}}_1 \\ \dot{\mathbf{K}}_2 - \dot{\mathbf{W}}_2 \dot{\mathbf{C}}_2 \\ \vdots \\ \dot{\mathbf{K}}_m - \dot{\mathbf{W}}_m \dot{\mathbf{C}}_m \\ \hline \ddot{\mathbf{K}}_1 - \ddot{\mathbf{W}}_1 \ddot{\mathbf{C}}_1 \\ \ddot{\mathbf{K}}_2 - \ddot{\mathbf{W}}_2 \ddot{\mathbf{C}}_2 \\ \vdots \\ \ddot{\mathbf{K}}_n - \ddot{\mathbf{W}}_n \ddot{\mathbf{C}}_n \end{array} \right]$$

$$(2.273)$$

where

$$\dot{\mathbf{N}}_i = \sum_{j=1}^{m} \dot{\mathbf{B}}_{ij}^{\mathrm{T}} \mathbf{W}_{ij} \dot{\mathbf{B}}_{ij}, \quad \overline{\mathbf{N}}_{ij} = \dot{\mathbf{B}}_{ij}^{\mathrm{T}} \mathbf{W}_{ij} \ddot{\mathbf{B}}_{ij},$$

$$\ddot{\mathbf{N}}_j = \sum_{i=1}^{m} \ddot{\mathbf{B}}_{ij}^{\mathrm{T}} \mathbf{W}_{ij} \ddot{\mathbf{B}}_{ij},$$

$$\dot{\mathbf{K}}_i = \sum_{j=1}^{n} \dot{\mathbf{B}}_{ij}^{\mathrm{T}} \mathbf{W}_{ij} \boldsymbol{\epsilon}_{ij},$$

and

$$\ddot{\mathbf{K}}_j = \sum_{i=1}^{m} \ddot{\mathbf{B}}_{ij}^{\mathrm{T}} \mathbf{W}_{ij} \boldsymbol{\epsilon}_{ij}. \qquad (2.274)$$

The following interesting observations can be made directly from equations (2.273):

(1) the N matrix is symmetrical about the principal diagonal;

(2) the upper left-hand corner of the N matrix consists of (6×6) - submatrices along the principal diagonal with each submatrix corresponding to one particular photograph, and all the other elements outside of these submatrices being zeros;

(3) the lower right-hand corner of the N matrix consists of (3×3) - submatrices along the principal diagonal with each submatrix corresponding to one particular photograph, and all the other elements outside of these submatrices being zeros;

(4) the inclusion of observation equations for the exterior orientation parameters contributes the $\dot{W}_i$ submatrices in the N matrix and the $\dot{W}_i\dot{C}_i$ submatrices in the K matrix;

(5) the observation equations for ground coordinates contribute only the $\ddot{W}_j$ submatrices in the N matrix and the $\ddot{W}_j\ddot{C}_j$ to the K matrix; and

(6) the contribution of each set of the collinearity equations can also be identified within the normal equation.

Because of the structure of the normal equations, as each set of observation equations is written, the contributions of these equations can be added directly into the normal equations. For example, suppose that point 2 is imaged in photograph 1. This gives rise to two collinearity equations resulting in the matrices $\dot{B}_{12}$, $\ddot{B}_{12}$, $\bar{C}_{12}$ and $\bar{W}_{12}$. Then, using equations (2.274), the $\bar{N}_{12}$ matrix and the contributions to the $\dot{N}_{11}$, $\ddot{N}_2$, $\dot{K}_{11}$ and $\ddot{K}_2$ matrices can be computed. This pair of collinearity equations will have no contribution to other parts of the normal equations. When this procedure is used, the elements of the normal equations should be initially set equal to zero. If point 9 is not imaged on photograph 1, no contribution need be added to the normal equations and $\bar{N}_{19}$ would remain a null submatrix. Therefore, although it was assumed in the derivation that each point is imaged on all photographs, observation equations need to be written only for the actually measured image points. The same is true for the assumptions that measurements are also available for all the exterior orientation parameters and for the ground coordinates.

2.6.5 AN EFFICIENT COMPUTING ALGORITHM

For equations (2.273), the normal equations may be subdivided into two matrix equations as follows:

$$\dot{N}\dot{\Delta} + \bar{N}\ddot{\Delta} = K \tag{2.275}$$

$$\bar{N}^T\dot{\Delta} + \ddot{N}\ddot{\Delta} = \ddot{K}. \tag{2.276}$$

Solving for $\ddot{\Delta}$ in equation (2.276) yields

$$\ddot{\Delta} = \ddot{N}^{-1}(\ddot{K} - \bar{N}^T\dot{\Delta}). \tag{2.277}$$

Substituting this expression into (2.275) yields

$$\dot{N}\dot{\Delta} + \bar{N}\ddot{N}^{-1}\ddot{K} - \bar{N}\ddot{N}^{-1}\bar{N}^T\dot{\Delta} = \dot{K};$$

i.e.

$$(\dot{N} - \bar{N}\ddot{N}^{-1}\bar{N}^T)\dot{\Delta} = (\dot{K} - \bar{N}\ddot{N}^{-1}\ddot{K}), \tag{2.278}$$

which is often referred to as the reduced normal equation.

Again through a process of direct substitutions, it can be shown that equation (2.278) may be expressed in the following form:

$$\begin{array}{ccc} S & \dot{\Delta} & = & E \\ (6m,6m) & (6m,1) & & (6m,1) \end{array}, \tag{2.279}$$

in which

$$S = \sum_{j=1}^{n}[\dot{N}_j - \bar{N}_j(\ddot{N}_j + \ddot{W}_j)^{-1}\bar{N}_j^T] + \dot{W}, \tag{2.280}$$

and

$$E = \sum_{j=1}^{n}[\dot{K}_j - \bar{N}_j(\ddot{N}_j + \ddot{W}_j)^{-1}(\ddot{K}_j - \ddot{W}_j\ddot{C}_j)] - \dot{W}\dot{C}. \tag{2.281}$$

Similarly, because of the diagonal structure of the $\ddot{N}^{-1}$ matrix, equation (2.277) may be decomposed into independent equations for each ground point; that is,

$$\ddot{\Delta}_j = (\ddot{N}_j + \ddot{W}_j)^{-1}(\ddot{K}_j - \ddot{W}_j\ddot{C}_j - \bar{N}_j^T\dot{\Delta}) \tag{2.282}$$
$$(3,3)$$

for $j = 1, n$.

Thus, by using equations (2.279) and (2.282), the largest set of simultaneous equations to be solved is $6m$. After having solved for Δ from equation (2.279), the corrections to the ground coordinates of the pass points can be computed point-by-point using equation (2.282). The latter operation only involves the solution of a maximum of 3 simultaneous equations.

In phototriangulation, only a few of the total n points are imaged on each photograph. This results in a banded structure for the S-matrix in equation (2.279) which can be further exploited to reduce computation time as well as computer storage. It can be easily derived from equations (2.273) and (2.278) that the reduced normal equations of (2.279) have the following detailed structure:

$$\begin{bmatrix} \dot{N}_{11}+\dot{W}_1+S_{11} & S_{12} & \cdot & \cdot & & S_{1m} \\ S_{12}^T & \dot{N}_{22}+\dot{W}_2+S_{22} & \cdot & \cdot & & S_{2m} \\ \cdot & & \cdot & \cdot & & \cdot \\ \cdot & & & \cdot & & \\ \cdot & & & & \cdot & \\ S_{1m}^T & & \cdot & & \cdot & \dot{N}_{mm}+\dot{W}_m+\dot{S}_{mm} \end{bmatrix} \begin{bmatrix} \dot{\Delta}_1 \\ \dot{\Delta}_2 \\ \cdot \\ \cdot \\ \cdot \\ \dot{\Delta}_m \end{bmatrix} = \begin{bmatrix} E_1 \\ E_2 \\ \cdot \\ \cdot \\ \cdot \\ E_m \end{bmatrix} \tag{2.283}$$

where

$$S_{ik} = -\sum_{j=1}^{n} \overline{N}_{ij} (\ddot{N}_j + \ddot{W}_j)^{-1} \overline{N}_{kj}{}^T,$$

and   (2.284)

$$E_i = \dot{K}_{ii} - \dot{W}_i \dot{C}_i - \sum_{j=1}^{n} \overline{N}_{ij} (\ddot{N}_j + \ddot{W}_j)^{-1} (\ddot{K}_j - \ddot{W}_j \ddot{C}_j).$$

Each submatrix S_{ij} consists of the contributions of the collinearity equations from all the image points common in both photographs i and j. Thus if photographs i and j do not contain any common points, $S_{ij} = 0$. In conventional photo-triangulation, each photograph overlaps with only a few other photographs in the block. Thus, by proper numbering of the photographs and the pass points, the S matrix can be designed to have all its non-zero elements within a narrow band about the principal diagonal.

Figure 2-31 shows a 4 × 7-block with the photographs numbered consecutively down the strip, and figure 2-32 shows the locations of the non-zero elements in the S-matrix. Figure 2-33 shows the same block with the photographs numbered cross-strip, and figure 2-34 shows the structure of the corresponding S-matrix. For a block having 60% overlap between successive photographs along each strip and 20% sidelap between adjacent strips, the band-widths of the S-matrix are:

for down-strip numbering, $P_d = 6(r + 3)$;   (2.285)
for cross-strip numbering, $P_c = 6(2S + 2)$;   (2.286)

where r = number of photographs per strip and S = number of strips in the block.

The variance-covariance matrices of the unknown parameters (exterior orientation and pass-point coordinates) can also be obtained as by-product of the least-squares solution. Let $\sigma_{\dot{\Delta}}^{\cdot}$ denote the variance-covariance matrix of the exterior orientation parameters. From equation (2.279),

$$\sigma_{\dot{\Delta}}^{\cdot} = \sigma_o{}^2 \, S^{-1}$$ (2.287)

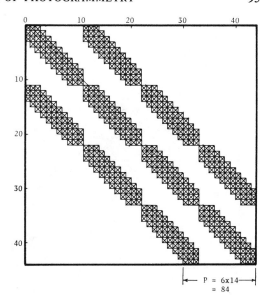

$$\leftarrow \, P = 6 \times 14 \, \rightarrow$$
$$= 84$$

⊠   6x6 – submatrix having some non-zero elements

FIGURE 2-32. Structure of S-matrix for down-strip numbering.

in which $\sigma_o{}^2$ is the variance of unit weight, and is arbitrarily defined in the solution for the purpose of weighting the observations (see equation 2.143). Similarly, let $\sigma_{\ddot{\Delta}_j}^{\cdot}$ denote the variance-covariance matrix of the computed coordinates $(X_j, Y_j$ and $Z_j)$ of ground point j. From equation (2.282),

$$\sigma_{\ddot{\Delta}_j}^{\cdot} = \sigma_o{}^2 (\ddot{N}_j + \ddot{W}_j)^{-1} + Q_j \, \sigma_{\dot{\Delta}}^{\cdot} \, Q_j{}^T$$ (2.288)

where

$$Q_j = (\ddot{N}_j + \ddot{W}_j)^{-1} \overline{N}_j{}^T.$$ (2.289)

Figure 2-35 shows a general flow chart of the computation process. After all the data have been read into the computer, observation equations are computed one by one and their contributions are entered immediately into the S and E matrices of the normal equations. The solu-

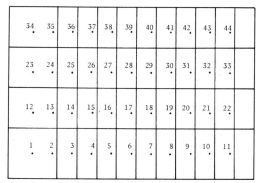

FIGURE 2-31. 4 × 11-photo block with down-strip numbering of photographs.

34	35	36	37	38	39	40	41	42	43	44
23	24	25	26	27	28	29	30	31	32	33
12	13	14	15	16	17	18	19	20	21	22
1	2	3	4	5	6	7	8	9	10	11

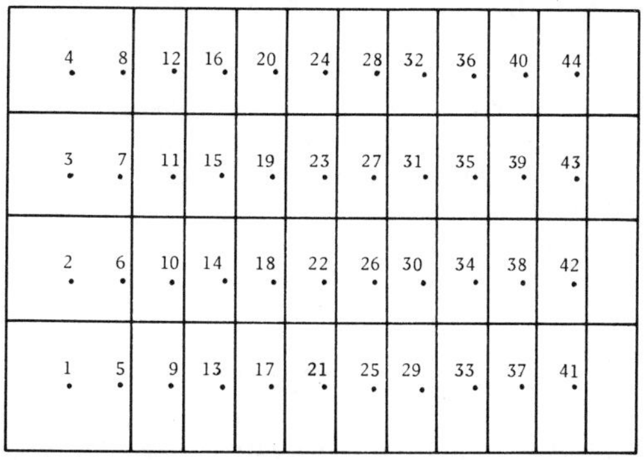

FIGURE 2-33. 4 × 11-photo block with cross-strip numbering of photographs.

4	8	12	16	20	24	28	32	36	40	44
3	7	11	15	19	23	27	31	35	39	43
2	6	10	14	18	22	26	30	34	38	42
1	5	9	13	17	21	25	29	33	37	41

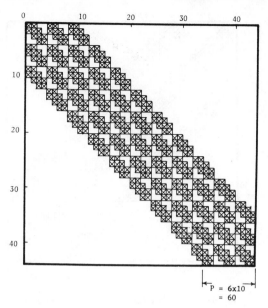

☒   6x6-submatrix having some non-zero elements

FIGURE 2-34. Structure of S-matrix for cross-strip numbering.

tion to this set of reduced normal equations gives the most probable corrections ($\dot{\Delta}$) to the exterior orientation parameters. The corrections to the ground coordinates (ΔX_j, ΔY_j, ΔZ_j) are next computed using equation (2.282). The computed corrections are then applied to the approximate values of the unknown parameters; *i.e.*

$$\omega_i^o = \omega_i^o + \Delta\omega_i,$$
$$\phi_i^o = \phi_i^o + \Delta\phi_i,$$
$$\kappa_i^o = \kappa_i^o + \Delta\kappa_i,$$
$$(X_i^c)^o = (X_i^c)^o + \Delta X_i^c,$$
$$(Y_i^c)^o = (Y_i^c)^o + \Delta Y_i^c,$$
$$(Z_i^c)^o = (Z_i^c)^o + \Delta Z_i^c,$$
$$X_j^o = X_j^o + \Delta X_j,$$
$$Y_j^o = Y_j^o + \Delta Y_j,$$
and $$Z_j^o = Z_j^o + \Delta Z_j.$$

These newly corrected approximations are used as new approximations and the complete solution is repeated in a second iteration. The iterative procedure is repeated until the corrections become negligibly small as compared to the accuracy requirement of the solution. This iterative procedure is necessary because linear approximations are involved in the derivation of the collinearity equations.

In the following paragraphs, three different convergent criteria are discussed.

CRITERION 1: Convergence limits $\Delta\omega_i$, $\Delta\phi_i$ and $\Delta\kappa_i$.

Stop the iterative procedure if $\Delta\omega_i$, $\Delta\phi_i$ and $\Delta\kappa_i$ are less than or equal to some predefined constant (k) for all the photographs in the solu-

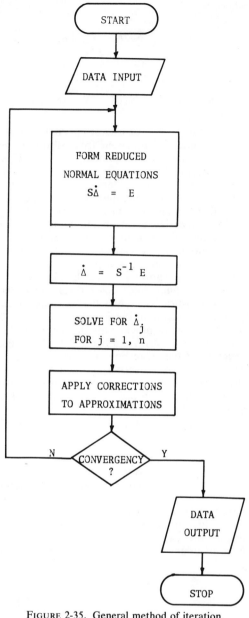

FIGURE 2-35. General method of iteration.

tion. The exact value of k will depend on the desired accuracy of the solution. In conventional aerotriangulation, the value for k is usually defined in arc seconds. This is not a good criterion, because the solution would not stop even if a single correction parameter (say $\Delta\omega_i$ of photograph i) exceeds k. It is quite possible that large measurement errors exist in the image coordinates of one photograph and this results in an inaccurate determination of the exterior orientation parameters of the photograph. Consequently, even though the corrections to the remaining photographs have become less than k, the correction for photograph i may vary within a range larger than k.

CRITERION 2. Difference of m_o between successive iterations.

Stop the iterative process if the difference between the computed standard error of unit weight of two successive iterations is less than some prescribed value k. The standard error of unit weight (m_o) may be computed by the following expressions.

$$m_o = \pm \left(\frac{\bar{V}^T \bar{W} \bar{V}}{n - u} \right)^{1/2} =$$

$$\pm \left(\frac{\sum_j^m \sum_i V_{ij}^T W_{ij} V_{ij} + \sum_{i=1}^{} \dot{V}_i^T \dot{W}_i \dot{V}_i + \sum_{j=1}^{n} \ddot{V}_j^T \ddot{W}_j \ddot{V}_j}{n - u} \right)^{1/2}$$

$$(2.290)$$

where n = number of observation equations, and u = number of unknowns in the solution. The quantity $(n - u)$ is called the number of degrees of freedom in the solution. The residual errors must be computed at the end of each iteration using equations (2.241), (2.244) and (2.247). It will necessitate the recomputation of the coefficient matrices $\dot{B}_{ij}$ and $\ddot{B}_{ij}$. Considerable amount of computation can be saved by recognizing the fact that as the solution converges, the corrections approach zero and $V_{ij} \approx \epsilon_{ij}$, $\dot{V}_i \approx \dot{C}_i$, and $\ddot{V}_j \approx \ddot{C}_j$. Therefore, equation (2.290) may be approximated by the following expression:

$$m_o \approx \pm \left(\frac{\sum_j^m \sum_i \epsilon_{ij}^T W_{ij} \epsilon_{ij} + \sum_{i=1}^{m} \dot{C}_i^T \dot{W}_i \dot{C}_i + \sum_{j=1}^{n} \ddot{C}_j^T \ddot{W}_j \ddot{C}_j}{n - u} \right)^{1/2}. \qquad (2.291)$$

It is then a simple matter of keeping a sum of the squares of the constant terms $(\epsilon_{ij}, \dot{C}_i, \ddot{C}_j)$ as the observation equations are formed during each iteration.

The value of m_o as computed using equation (2.290) or (2.291) is an estimate of the real standard error of unit weight (σ_o). Since a value of σ_o must be defined in the solution in order to define the weights, the convergence limit may be set as $0.1 \sigma_o$ or $0.01 \sigma_o$.

This is a better convergence criterion than criterion 1, because m_o always stabilizes in a solution.

CRITERION 3: Fixed Number of Iterations.

The third criterion is to simply force the solution to go through a fixed number of iterations. It is not a good criterion because the number of iterations required for a particular solution cannot be easily predicted. However, it can be used in combination with criterion 1 or 2 to impose a limit on the maximum allowable number of iterations.

Figure 2-36 is a detailed flow chart of the computational procedure. It is noted that the matrices $\dot{N}_j$, $\bar{N}_j$, $(\bar{N}_j + \ddot{W}_j)$, $\dot{K}_j$, and $(\dot{K}_j - \ddot{W}_j \ddot{C}_j)$ are not kept in permanent storage for each and every point. Immediately after they are computed for a given point j, their contributions $(S_j$ and $E_j)$ to the normal equations are computed and added directly to the S and E matrices. For example, the same memory locations for storing $\dot{N}_1$ are used for $\dot{N}_2$, $\dot{N}_3$, ... and $\dot{N}_n$. Although this algorithm requires that these matrices be recomputed several times during each iteration, it results in tremendous saving in computer memory storage. Without this algorithm, the number of photographs that can be included in the solution will be severely limited. For example, $\dot{N}_j$ is a $(6m \times 6m)$ matrix. If a memory block of $6m \times 6m$ locations is used to store $\dot{N}_j$ for each and every point, it will take $(6\, mn \times 6\, mn)$ locations to store the $\dot{N}_j$ matrices alone. For $m = 10$, and $n = 100$, it will require $6000 \times 6000 = 36$ million locations for $\dot{N}_j$'s.

The following formulas are used in the flow chart in figure 2-36:

(1) $\dot{N}_{ij} = \dot{B}_{ij}^T W_{ij} \dot{B}_{ij}$, $\bar{N}_{ij} = \dot{B}_{ij}^T W_{ij} \ddot{B}_{ij}$,

$\ddot{N}_{ij} = \ddot{B}_{ij}^T W_{ij} \ddot{B}_{ij}$, $\dot{K}_{ij} = \dot{B}_{ij}^T W_{ij} \epsilon_{ij}$, and

$\ddot{K}_{ij} = \ddot{B}_{ij}^T W_{ij} \epsilon_{ij};$ (2.292)

(2)

$$N_j = \begin{bmatrix} \dot{N}_{1j} & & & \\ & \dot{N}_{2j} & & \\ & & \cdot & \\ & & & \dot{N}_{mi} \end{bmatrix}, \quad \bar{N}_j = \begin{bmatrix} \bar{N}_{1j} \\ \bar{N}_{2j} \\ \cdot \\ \bar{N}_{mi} \end{bmatrix}, \quad \dot{K}_j = \begin{bmatrix} \dot{K}_{1j} \\ \dot{K}_{2j} \\ \cdot \\ \dot{K}_{mj} \end{bmatrix}$$

$$\ddot{N}_j = \sum_{i=1}^{m} \ddot{N}_{ij}, \quad \text{and} \quad \ddot{K}_j = \sum_{i=1}^{m} \ddot{K}_{ij} ; \qquad (2.293)$$

(3) $S_j = \ddot{N}_j - \bar{N}_j (\dot{N}_j + \ddot{W}_j)^{-1} \bar{N}_j^T$ (2.294)

$E_j = \ddot{K}_j - \bar{N}_j (\dot{N}_j + \ddot{W}_j)^{-1} (\dot{K}_j - \ddot{W}_j \ddot{C}_j) ;$

(4) $S = \sum_{j=1}^{n} S_j + \dot{W}$ (2.295)

$E = \sum_{j=1}^{n} E_j - \dot{W}\dot{C}$

(5) $S \Delta = E;$

and

(6) $\ddot{\Delta}_j = (\dot{N}_j + \ddot{W}_j)^{-1} (\dot{K}_j - \ddot{W}_j \ddot{C}_j) - Q_j \dot{\Delta}$ (2.296)

for $j = 1, n$

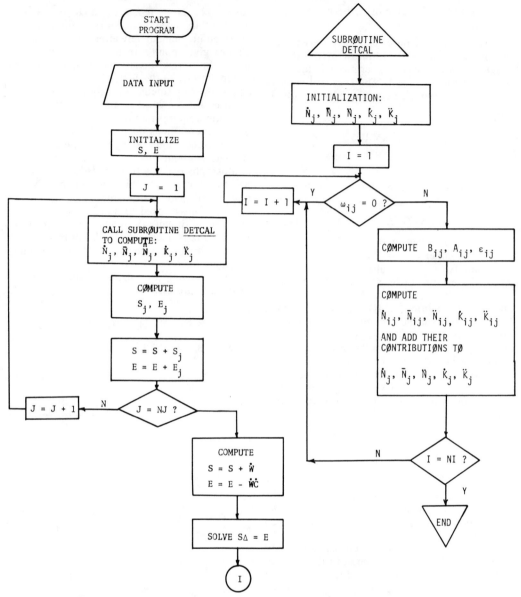

FIGURE 2-36. Flow chart for simultaneous solution of photogrammetric block.

where

$$Q_j = (\check{N}_j + \check{W}_j)^{-1} \bar{N}_j^T.$$

2.6.6 SOLUTION OF NORMAL EQUATIONS

In a least squares adjustment, the coefficient matrix S of the normal equation, as represented in equation (2.279), is always symmetric about the diagonal; *i.e.* $S_{ij} = S_{ji}$. Furthermore, if the photographs are arranged in a regular pattern, which is usually the case in aerial photography, the numbering of the photographs and the ground points can be designed so that all the non-zero elements in the S-matrix fall within a

narrow band along the diagonal of the matrix. Examples of the structure of the S-matrix have been illustrated in figures 2.32 and 2.34. By employing an algorithm which takes full advantage of these properties of the S-matrix in the solution of the normal equations, considerable amount of computing time can be saved.

The method of recursive partitioning has been found to be particularly suited for the solution of normal equations in the adjustment of large photogrammetric blocks. It is a simple variation of the fundamental Gauss elimination method of solving simultaneous equations. Consider a system of n normal equations represented in matrix notation as

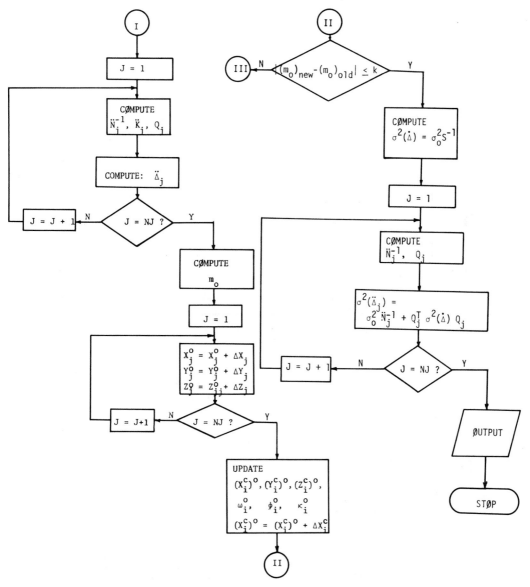

FIGURE 2-36. Flow chart for simultaneous solution of photogrammetric block.

$$\begin{matrix} \mathbf{S} & \mathbf{X} & = & \mathbf{E} \\ (n,n) & (n,1) & & (n,1) \end{matrix} \qquad (2.297)$$

where $\mathbf{S}$ is a symmetric banded matrix with a bandwidth p, $\mathbf{X}$ is a matrix of the unknown parameters, and $\mathbf{E}$ is a matrix of constants.

This system of equations may be partitioned as shown in figure 2.37. The partition is such that $\mathbf{S}_{11}$ is a square matrix with a dimension of $q \times q$. The dimension q must be chosen so that $(n - p)$ is divisible by q. Furthermore, $\mathbf{S}_{22}$ must have a dimension of $p \times p$ and $\mathbf{S}_{13}$ must be a null matrix.

This partition breaks up equation (2.297) into the following three equations:

$$\mathbf{S}_{11}\mathbf{X}_1 + \mathbf{S}_{12}\mathbf{X}_2 + \mathbf{OX}_3 = \mathbf{E}_1 \qquad (2.298)$$

$$\mathbf{S}_{12}^T \mathbf{X}_1 + \mathbf{S}_{22}\mathbf{X}_2 + \mathbf{S}_{23}\mathbf{X}_3 = \mathbf{E}_2 \qquad (2.299)$$

$$\mathbf{OX}_1 + \mathbf{S}_{23}^T \mathbf{X}_2 + \mathbf{S}_{33}\mathbf{X}_3 = \mathbf{E}_3. \qquad (2.300)$$

Solving equation (2.298) for $\mathbf{X}_1$ yields

$$\mathbf{X}_1 = \mathbf{S}_{11}^{-1} (\mathbf{E}_1 - \mathbf{S}_{12}\mathbf{X}_2). \qquad (2.301)$$

Substituting this expression into equations (2.299) and (2.300) yields

$$\bar{\mathbf{S}}_{22}\mathbf{X}_2 + \mathbf{S}_{23}\mathbf{X}_3 = \bar{\mathbf{E}}_2 \qquad (2.302)$$

$$\mathbf{S}_{23}^T \mathbf{X}_2 + \mathbf{S}_{33}\mathbf{X}_3 = \mathbf{E}_3 \qquad (2.303)$$

where

$$\bar{\mathbf{S}}_{22} = (\mathbf{S}_{22} - \mathbf{S}_{12}^T \mathbf{S}_{11}^{-1} \mathbf{S}_{12}) \qquad (2.304)$$

$$\bar{\mathbf{E}}_2 = (\mathbf{E}_2 - \mathbf{S}_{12}^T \mathbf{S}_{11}^{-1} \mathbf{E}_1). \qquad (2.305)$$

Equations (2.302) and (2.303) now consist of $(n-q)$ equations and $(n-q)$ unknowns, and can be conveniently represented as

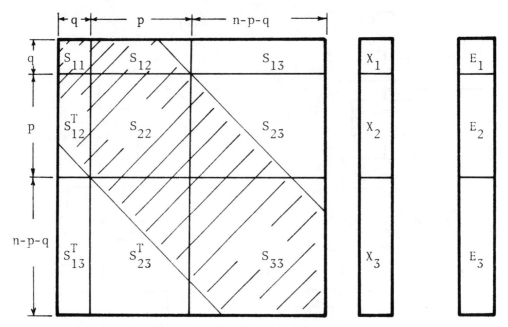

FIGURE 2-37. Partition of normal equations.

$$\begin{matrix} \mathbf{S}^1 & \mathbf{X}^1 & = & \mathbf{E}^1 \\ (n-q,n-q) & (n-q,1) & & (n-q,1) \end{matrix} \quad (2.306)$$

in which the superscript denotes the number of times the original set of equations in equation (2.397) has been partitioned, or the number of sets of q equations that have been eliminated.

At stage 2, equation (2.306) is partitioned and reduced to $(n-2q)$ equations in exactly the same manner. This partitioning process is repeated successively until only p equations remain; *i.e.*

$$\begin{matrix} \mathbf{S}^m & \mathbf{X}^m & = & \mathbf{E}^m \\ (p,p) & (p,1) & & (p,1) \end{matrix} \quad (2.307)$$

This is accomplished after m stages where $m = (n-p)/q$.

Stages 1 to m in the above partitioning process is identically equivalent to the forward solution of the standard elimination procedure. The backward solution is initiated by solving equation (2.307); *i.e.*,

$$\mathbf{X}^m = (\mathbf{S}^m)^{-1}\,\mathbf{E}^m. \quad (2.308)$$

Equation (2.301) can now be used successively in backward substitution to solve for one set of q unknowns at each stage. In the first backward stage, the solution can be represented as follows:

$$\mathbf{X}_1^{m-1} = (\mathbf{S}_{11}^{m-1})^{-1}(\mathbf{E}_{11}^{m-1} - \mathbf{S}_{12}^{m-1}\mathbf{X}^m). \quad (2.309)$$

It will also require m stages to complete the entire solution.

During the backward solution, instead of solving directly for the unknowns, the banded portion of the inverse matrix $\mathbf{S}^{-1}$ can be computed.

Let the inverse matrix of the reduced matrix $\mathbf{S}'$ at the i-th stage be denoted as follows

$$(\mathbf{S}')^{-1} = \begin{bmatrix} \mathbf{D}_{11} & \mathbf{D}_{12} & \mathbf{D}_{13} \\ \mathbf{D}_{12}^T & \mathbf{D}_{22} & \mathbf{D}_{23} \\ \mathbf{D}_{13}^T & \mathbf{D}_{23}^T & \mathbf{D}_{33} \end{bmatrix}. \quad (2.310)$$

The submatrices $\mathbf{D}_{12}$ and $\mathbf{D}_{11}$ can be computed by the recursive equations

$$\mathbf{D}_{12} = -\bar{\mathbf{S}}_{11}^{-1}\,\mathbf{S}_{12}\mathbf{D}_{22}^{-1} \quad (2.311)$$

$$\mathbf{D}_{11} = \mathbf{S}_{11}^{-1}\,(\mathbf{I} - \mathbf{S}_{12}\mathbf{D}_{12}^T). \quad (2.312)$$

These expressions are used recursively during each backward stage, adding q rows to the inverse during each stage.

This algorithm can be easily extended to compute the entire inverse matrix $\mathbf{S}^{-1}$. However, as the inverse is used primarily for error analysis, interest is centered primarily on the elements bordering the diagonal. By not computing the elements outside of the diagonal band, large savings in both storage space and computer time can be achieved without sacrificing any computational accuracy.

SELECTED REFERENCES

American Society of Photogrammetry, *Manual of Photogrammetry,* Third Edition, Volumes I and II, American Society of Photogrammetry, 6269 Leesburg Pike, Falls Church, Virginia 22044, 1966.

Bevington, P. R., *Data Reduction and Error Analysis for the Physical Sciences,* McGraw-Hill Book Co., New York, N.Y., 1969.

Brown, D. C., *A Matrix Treatment of the General Problem of Least Squares Considering Correlated Observations,* Ballistic Research Laboratory Report

No. 960, Aberdeen Proving Ground, Maryland, October 1955.

————, *Research in Mathematical Targeting, the Practical and Rigorous Adjustment of Large Photogrammetric Nets,* RADC-TDR-353, Rome Air Development Center, October 1964.

————, *Decentering Distortion of Lenses,* Photogrammetric Engineering, Vol. 32, No. 3, May 1966, pp. 444–462.

————, *A Unified Lunar Control Network,* Photogrammetric Engineering, 34:12, December 1968, pp. 1272–1292.

Church, Earl, 1945, *Revised Geometry of the Aerial Photograph,* Syracuse University Press, Bulletin 15.

————, 1948, *Theory of Photogrammetry,* Syracuse, N.Y., Syracuse University Bulletin No. 19.

————, 1950, *Illustrative Solutions of Analytic Problems in Aerial Photogrammetry,* Ohio State University Research Foundation, Technical Paper No. 97.

de Jong, S. H., and S. S. Tezcan, *The Electronic Computer in Survey Adjustments at the University of British Columbia,* The Canadian Surveyor, 19:1, March 1965, pp. 87–89.

Freund, John E., *Mathematical Statistics,* Prentice Hall, Inc., Englewood Cliffs, N.J., 1965.

Hallert, B., *Photogrammetry,* McGraw-Hill Book Company, Inc., New York, 1960.

Hald, A., *Statistical Theory with Engineering Applications,* John Wiley & Sons, Inc., N.Y., 1967.

Keller, M., and G. C. Tewinkel, *Aerotriangulation: Image Coordinate Refinement,* U.S.C. & G.S. Technical Bulletin No. 25, March 1965.

————, *Block Analytic Aerotriangulation,* ESSA Technical Report C. & G.S. 35, Environmental Science Services Administration, Rockville, Md., November 1967.

Kreyszig, E., *Advanced Engineering Mathematics,* John Wiley and Sons, Inc., New York, 1962.

Kunz, K. S., *Numerical Analysis,* McGraw-Hill Book Company, Inc., New York, 1957.

Lindgren, B. W., Statistical Theory, The MacMillan Company, New York, 1965.

Linnik, Yu. V., *Method of Least Squares and Principles of the Theory of Observations,* Pergamon Press, New York, 1961.

Madkour, M. F., *Precision of Adjusted Variables by Least Squares,* Journal of the Surveying and Mapping Division, Proceedings of the ASCE, 94:SU2, September 1968, pp. 119–136.

Moffitt, Francis H., *Photogrammetry,* 2nd Edition, International Book Company, Scranton, Pennsylvania, 1967.

Pope, A. J., *Some Pitfalls to be Avoided in the Iterative Adjustment of Nonlinear Problems,* Proceedings of the 38th Annual Meeting, American Society of Photogrammetry, Washington, D.C., March 1972, pp. 449–477.

Rosenfield, George H., 1959, *The Problem of Exterior Orientation in Photogrammetry,* Photogrammetric Engineering, v. 25, no. 4, p. 536.

Schermerhorn, W., 1955, *Proposal for the Introduction of Uniformity in the Signs of Spatial Plotting Instruments,* Photogrammetria, v. 12, no. 1, p. 18–21.

Schmid, Hellmut H., 1953, *An Analytical Treatment of the Orientation of a Photogrammetric Camera,* Aberdeen, Md., Ballistic Research Laboratories, Report No. 880.

————, 1955, *An Analytical Treatment of the Problem of Triangulation by Stereophotogrammetry,* Aberdeen, Md., Ballistic Research Laboratories, Report 961.

————, 1959, *A General Analytical Solution to the Problem of Photogrammetry,* Ballistic Research Laboratories, Report No. 1065.

Schmid, Hellmut H., and Erwin Schmid, *A Generalized Least Squares Solution for Hybrid Measuring Systems,* The Canadian Surveyor, Vol. XIX, No. 1, March, 1965.

CHAPTER III

Elements of Photogrammetric Optics

Author-Editor: JAMES G. BAKER

Contributing Authors: ARTHUR A. MAGILL, WILLIAM P. TAYMAN,
KONSTANTIN PESTRECOV (*Deceased*), FRANCIS E. WASHER

3.1 Fundamentals of Lens Design

THE SCIENCE OF OPTICS is old. It originally covered the phenomena associated with light (the visible radiation) and vision. Then, as new discoveries were made, the field of optics expanded to cover the invisible radiation such as ultraviolet and infrared. Later, some authors included in the field of optics such diverse subjects as X-rays, electron microscopy, and cathode-ray tubes. However, during the past several decades the science of optics has been rapidly expanded to include an even wider variety of subjects, including Fourier transform spectroscopy, laser science, holography, coherence, electronic image tubes, solid state research, thin films, and many more derived from cross-field research and applications. Optics applied to medicine is but one example from such cross-field regions of knowledge.

Because of this ever widening coverage, the term "optics" evades rigorous definition (Ronchi, 1957). Perhaps without trying to be exact, we can define the term as a branch of physical science which deals broadly with electromagnetic radiation and with the interactions of radiation with matter, as brought forth by observation, experiment and theory, and with associated technologies and instrumental devices.

In modern science the boundaries of the electromagnetic spectrum employed in research and technology in optics are very wide indeed and certainly extend from the X-ray region at the short wavelength end of the spectrum to the longest wavelength microwaves that can be used in image-forming devices. If one were to include long baseline radio interferometry, there is practically no longest wavelength limit except that set by size and distance.

The question of what to include in a definition of the science of optics is further complicated by the fact that various optical phenomena either require treatment by different methods or at least may be more conveniently explored by special approaches. As a result of this, we may divide optics into a number of subfields, such as mathematical, physical, geometrical, quantum mechanical, physiological, electro-optical, and engineering. There is, of course, an unavoidable overlapping between these "different optics".

More narrowly, we shall be interested here in that portion of optical science having to do with the formation and characteristics of images. In particular, we shall examine how light waves can be made to perform useful tasks for photogrammetric purposes which in turn involve some of the most precise uses of technology employed to the present time.

The art and science of lens design have been based for a long time on the methods developed by geometrical optics. While the newer subjects are of ever-increasing importance, one must continue to recognize that for purposes of lens design, a secure foundation still rests upon geometry, where one relies on the following principles abundantly confirmed by observation and theory:

(1) *In vacuo* light travels in straight lines along the normals to the moving wavefront. A plane wave propagates as a plane wave. A spherical wave from a central point source continues to expand outwards as a spherical wave. (In the broadest terms one can say that in free space at large, whether "flat" or relativistically curved, light travels between two points along a geodesic path in such a way as to require the least time of transit).

(2) In a transparent medium light also travels between two points along a path of least time of transit. However, when reflective and refractive boundary surfaces are included between or around points in the same or separate media or even *in vacuo*, light may also travel between such points along a path of maximum time. In many situations light also travels between two points in such a way that adjacent paths connecting the two points contain the same number of wavelengths to a high degree of approximation, that is, are in phase. Such a path may be neither a maximum nor a minimum but an inflection point in a curve of time of transit as a

EDITOR'S NOTE: Much of the text and many of the illustrations presented in this chapter are derived from the Third Edition, MANUAL OF PHOTOGRAMMETRY, Chapter III, edited by Heinz Gruner.

function of the integrated lengths of arbitrary paths between the two points. The several possibilities derive from what are called "stationary paths" between the two points.

(3) In a homogeneous isotropic medium light travels in straight lines along normals to the moving wavefront. For purposes of geometrical optics these normals are called the "light rays". Optical glass is nearly such a homogeneous isotropic medium.

(4) Different rays may be treated as proceeding independently without interaction with one another except that in the propagation of a bounded light-wave or in the neighborhood of an image, diffraction and interference phenomena must be taken into account.

(5) In a physical medium, light waves travel more slowly than *in vacuo*. The ratio of the velocity of light *in vacuo* to that in a medium is called the refractive index of that medium. Owing to the nature of physical substances the refractive index is dependent also on the wavelength of the light and also to some extent on the temperature of the medium.

The product of the geometrical pathlength between two points in a medium and its index of refraction is called the "optical path-length" or simply, the "optical path", and is equal to the distance light would have traveled *in vacuo* in the same time interval used by transit between the two points in the medium. The total number of wavelengths of the light-wave remains the same, whether *in vacuo* or in the medium. The path pursued by light in proceeding from one point to another point even in different media or in non-homogeneous media, is also called the optical path. The total optical path between such points in separate media is the sum (or integral) of the optical path segments in the respective media. Optical paths can be virtual as well as real, and can be added and subtracted according to mathematical convention.

(6) When a ray is reflected from a surface, the angles of incidence and reflection are equal, where these angles are measured from the normal at the point of intercept on the surface. The incident and reflected rays and the normal are coplanar.

(7) When a ray passes from one medium to another of different refractive index, it becomes refracted; this means that the direction of the refracted ray is changed. The incident and refracted rays are coplanar with the normal at the point of intercept on the surface, but the angles of incidence and refraction with respect to the normal are not equal. If the angle of incidence is i and the angle of refraction is r, the following equation holds:

$$\sin i/\sin r = n'/n \qquad (3.1)$$

where n and n' are the refractive indices of the first and second media, respectively.

(8) For laboratory purposes the indices of refraction recorded for optical glasses and other optical media are often referred to air under standard conditions of pressure and temperature, instead of to *vacuo*. Correction to *vacuo* can readily be made when necessary, or to ambient air. For photogrammetric equipment flown in aircraft or carried in space vehicles, appropriate corrections are to be applied.

While designers have found it convenient to employ the mathematical means of geometrical optics, nevertheless from very early times other concepts were known and occasionally employed. In the middle of the 17th century Fermat stated his principle of least action, leading later to the concept of stationary paths, described above. Descartes not long afterward investigated the analytic theory of what we now call Cartesian surfaces. These are optical surfaces, both reflective and refractive, separating points in the same or different media and providing a constant optical path between source point and image point for all rays from the source point capable of entering the medium containing the image point. The image may be real or virtual. The reflective surfaces are conics and their geometrical properties have been known from ancient times. The refractive surfaces are called Cartesian ovals and in some cases are also conics. These have applications even today. Further, a necessary and sufficient condition for the formation of an image point by the coalescing of many ray-normals in phase can be stated by the simple requirement that every optical path from source point to the image point along any initial ray-normal accepted by the optical system must have the same total optical path. This amounts to saying also that the family of ray-normals must pass precisely through the desired image point, and indeed this latter principle forms the basis for the theory of geometrical optics. The two concepts are complimentary to one another and in a thoroughly corrected instrument merge into one another. The specialized field of geometrical optics of itself is primarily a mathematical rather than a physical science, but under modern conditions comprises but one aspect of optical science.

Many hundreds of photographic lenses of satisfactory performance have been designed during the past hundred and thirty years and until 1940, at least, almost exclusively by application of the principles of geometrical optics. However, modern developments impose new and much more rigid requirements on performance of optical systems than heretofore. These requirements necessitate the use of some of the methods and criteria of physical optics (wave optics), information or communication theory, and the mathematical theory of images. The new approaches are described at greater length in the appropriate sections of this chapter.

The goal of optical designers has always been to combine a number of optical elements into a system which yields performance adequate for the purposes for which the system is intended, and to accomplish this so reliably that an accurately constructed prototype will perform in agreement with the calculations.

Optical systems may be divided into the following three basic types:

(1) dioptric systems—these contain only refractive elements (lenses, windows, aspheric plates, prisms, filters);
(2) catoptric systems—these contain only reflective elements (mirrors);
(3) catadioptric systems—these contain any necessary mixture of refractive and reflective elements and components.

Most of the older systems were of the dioptric type, and indeed most modern systems also are dioptric, particularly where wide-angle coverage is required. For some newer and specialized purposes, catoptric and catadioptric systems have proved to provide unique advantages. These advantages include, for example, a minimum number of elements or components, freedom from significant chromatic aberrations, and insensitivity to environmental factors such as temperature and air-density fluctuations. The largest optical systems are necessarily catoptric or catadioptric. On the other hand, these systems have certain inherent difficulties such as the need to have the image surface in an accessible location outside of the system, or the problems encountered in the obstruction of the light by one component with respect to another.

In meeting assigned requirements optical designers necessarily must accept some compromises that fall short of perfection in an optical system. Still, they usually have in mind the goal at some later time of designing more nearly perfect systems. From the point of view of geometrical optics, a perfect system would reproduce all of the points in an object plane into corresponding points in an image plane, and the image would conform to a constant scale; the latter condition means also that a straight line in the object plane will be reproduced as a straight line in the image plane, the relationship being known in mathematical treatment as a collineation. An idealized optical system is illustrated in figure 3-1.

Two reservations have to be taken with regard to this definition of an ideal system. The first is that an object point cannot be strictly a mathematical point, particularly for a source at a finite distance. Therefore, an optical designer asked to design a collimator must not expect to find a system that will produce a mathematically perfect parallel beam of light in image space from a point in object space, however desirable for some purposes that might be. While the modern laser beam to some extent approximates such an ideal, small errors of several kinds prevent attaining the mathematically idealized parallel beam, and in any case, the propagated beam is ultimately diluted by diffraction spreading. In observations from artificial satellites, distant stars approximate point sources very closely and yet recent optical researches have succeeded in revealing structure in the stellar discs of some of the nearer stars.

Another reservation is that even if a physical object-point were to exist, its image cannot be a mathematical point because of the mathematical nature of the formation of images (*see* section 3.1.3).

A basic procedure used in lens design is that of ray-tracing. This means that one determines the paths of various rays passing through the elements of a system, and finally in image space examines the pattern of intercepts of any given set of rays in the vicinity of a desired image point. The work is then extended to other image points in the desired field. The designer then by various means proceeds to vary the parameters of the system toward the goal of bringing all the rays from an object point, as accepted by the entrance pupil of the system, as close as possible to a point intersection at a desired location in the image plane. In earlier decades the culmination of this process led to what was considered a completed lens design. Under modern conditions further calculations are required to bring about the best state of compromise for the actual physical formation of the image and for adapting the system optimally to the purposes for which it is designed.

Obviously, at the beginning of a design effort the designer must make a decision as to what type of system he is going to use and how many elements the system should contain. This decision may often depend on economic factors rather than on performance criteria alone, and one must also take environmental factors into account. Generally speaking, one finds that the stricter the requirements set forth, such as lens speed, angle of coverage, and resolution, the more elements may be needed to bring about a successful conclusion. However, exceptions can be found.

Some lens forms yield better performance than others, even though fewer elements may be involved. For example, the 6-inch f/6.3 Metrogon lens (an American version of the Zeiss Topogon, the latter designed in 1939 by Richter) has only four elements. Nevertheless, it covers a total field of 90 degrees and for many years

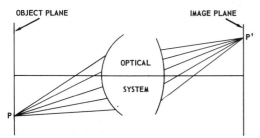

FIGURE 3-1. An ideal optical system. Every point P in the object plane is imaged as a point P' in the image plane, and the image scale is constant.

served very well as a basis for aerial mapping. Later, improved versions appeared by the designations of Planigon and Cartogon; the modification consisted of the addition of a plano-parallel plate following the fourth element. During the past 25 years, however, other lens forms have appeared in abundance following the introduction of the Aviogon design by Berthele which in turn was based on an earlier concept by Rusinov (Pestrecov, 1954). Indeed, some quite recent designs have employed as many as 11 elements to effect desirable improvements in speed, resolution, illumination and mapping precision.

In view of the multiplicity of lens forms already known, the designer must give studied attention to the question of the general starting form for any new design. In most cases he selects a prior design of great promise and tries to improve it to meet new requirements. In earlier times the designer had to proceed fairly cautiously in introducing any additional element, or in making changes in shapes and separations, or in introducing new types of optical glass. However, under modern conditions with electronic computers the designer enjoys a far greater flexibility than heretofore in exercising design judgments and in making probes into the unknown, inasmuch as no very considerable time is lost in the event of failure. The modern designer also benefits greatly by having at his disposal a very large variety of types of fine optical glasses.

Occasionally, the designer will introduce some entirely new idea or concept, such as that referred to above by Rusinov, a concept that tended to set aside all previous mapping lens designs. Rusinov found that by inserting elements with very strong negative menisci in the front and rear of a positive grouping around a central stop, he was able to increase the angle of coverage markedly and yet at the end of the optimized design to arrive at a reduced distortion, and to achieve a more uniform distribution of illumination in the image plane than had theretofore been possible.

Such events seldom happen in optical design today in the mapping lens field because with the multitude of lens forms previously explored, the probability of finding some quite novel form is very small. This is not necessarily true in other forms of optical systems such as zoom systems but for reasons of precision one cannot expect mapping systems to go much beyond the complexities of already known systems.

In all the earlier decades of optical design, designers relied heavily on trial and error in seeking improvements over previously designed systems, and made frequent use of differential corrections to effect small gains. Perhaps even with modern day computers a similar process might still be at work except for the fact that the computational speed is so very great that years of the older work can now be compressed into seconds of time. Convergence to a usable answer is now relatively quick and arrived at by mathematical processes. From a very broad point of view, the modern designer adopts a family of parameters and has an assigned set of conditions to satisfy to various degrees of precision. The computer tackles this problem with little lost motion.

One must look back with sincere appreciation for the intensive efforts of the early designers preceding the introduction of the electronic calculator. Indeed, the earliest designs were carried out by theory and supported by ray-tracings based on calculations with logarithms. Some of this work with log tables was still going on as late as 1945 in some locations. However, the introduction of desk calculators and improved tables greatly aided designers through the 1930's and 1940's until the electronic calculator displaced all previous methods. Now, of course, these very limited earlier procedures are a matter of history, and practically all modern design is carried out on the electronic machines (*see* section 3.1.3.8).

The final specifications for a lens system are called in optical parlance a "lens formula." This is a summary of data which shows the radii of the lens surfaces, lens thicknesses, diameters and clear apertures, the separations, and the optical glasses employed. For maximum convenience this lens formula can be added to a properly scaled layout drawing of the system. Wherever possible this drawing should comply with the convention that light is to travel from left to right, but according to the form of the instrument exceptions necessarily occur.

3.1.1 OPTICAL TERMINOLOGY AND SYMBOLS

The terminology and symbols used in this chapter are based mainly on the following documents: Military Standard "Photographic Lenses," MIL-STD-150A of 12 May 1959, revised to 28 January 1963; and American Standard "Letter Symbols for Physics," ASA Z 10.6-1948. Nevertheless, there are some deviations arising from the fact that there is still no general adherence to these standards. There is a very strong tendency for one to deviate from standards when new terms are introduced to supersede previously well-known terms in general use. For example, "illuminance" was accepted as a standard term to replace the generally used term illumination, and "luminance" to replace the term brightness.

The situation is similar with respect to sign conventions, inasmuch as no lasting agreement has been reached on their standardization. Indeed, the subject is under continuing review by interested societies. In the meantime one should try to employ standard conventions when no compelling reason exists to do otherwise. The sign conventions used here are based on those of analytic geometry. They are explicitly stated in section 3.1.2 (also Sears, 1949).

A comprehensive compilation of definitions of photogrammetric terms is given in chapter XIX of this MANUAL. Yet the needs of optical formulations make it desirable to give here a special listing of symbols and a selection of basic definitions without which the subsequent text may not be readily understandable. The wording given here will differ in some instances from the wording given in chapter XIX.

3.1.1.1 OPTICAL TERMS

Optical system—Any device which operates on light to produce a specific, desired effect. The optical systems most used by photogrammetrists are those intended to produce images and are composed of mirrors, prisms, and/or lenses in various combinations and arrangements. A "lens system" is an optical system consisting either solely of lenses or containing mirrors and/or prisms that do not affect the size or shape of the image.

Member—A member is a group of parts considered as an entity because of the proximity of its parts, or because it has a distinct but not always an entirely separate function.

Component—A component is a subdivision of a member. It may consist of two or more parts cemented together or separated by a small distance.

Element—An element of a compound lens is a single, uncompounded lens; that is, a part constructed of a single piece of glass. An element of an optical system is a single piece of optical material such as glass which is part of the imaging process.

Optical axis—This is the rotational axis of the optical system and passes through the centers of curvature of the surfaces of the system. Conversely, the centers of curvature of the surfaces of the system lie on the optical axis. The optical axis may, however, be "bent" by means of optically flat mirrors or prisms.

Cardinal points—These are the points which are of particular importance to the lens designer, and which are used as reference points for determining object and image distances. The following points are considered cardinal.

(1) *Principal planes and points*—Principal planes are planes perpendicular to the optical axis, so located that the lateral magnification is unity and positive. Principal points are the axial points of these planes.

(2) *Nodal points*—These are axial points whose properties are such that in Gaussian optics any ray passing through the first nodal point emerges from the second nodal point in the direction parallel to that of the original ray. For a single spherical interface between different media the nodal points necessarily lie at the center of curvature. When the media in object and image space of an optical system are the same, the principal and nodal points coincide. This is usually the case in photogrammetric optics. The term "nodal point" is more generally in use by photogrammetrists than the term "principal point." The latter term should be used only to designate the

center of the photograph, unless its usage as an optical term is clearly defined and in proper context.

(3) *Focal points*—These are the axial points at which the images of axial "object points" at infinity are focused. According to the convention accepted here, the first focal point is the image of an infinitely distant axial object point to the right of the optical system. The second focal point is the image of an axial object point infinitely distant to the left of the optical system. Because we assume that the light is coming from the left, we shall usually be dealing with the second focal point. Focal points are occasionally called principal foci.

Focal planes—These are the planes perpendicular to the optical axis and passing through the focal points. An object having an extended angular coverage at infinity is imaged in the focal plane.

Equivalent focal length (or simply focal length)—The numerical value of focal length in object or image space is equal to the distance of the focal point from the corresponding principal point. When the initial and final media are the same, the focal lengths are equal and become a constant f, an important characteristic of the optical system. The quantity f enters into the equations determining image magnification (scale) and the image distance from the reference points such as the nodal points or the focal points. The constant is considered to be positive if the optical system produces a convergent beam (this is the case of a positive lens) from an object at infinity; it is negative if the beam is divergent (a negative lens). The term "equivalent focal length" means that the focal length of a compound optical system behaves the same as for a simple lens having a focal length of the same numerical value.

Wherever distances are involved, at least for some computations algebraic signs have to be considered. Because each lens system has two nodal points and two focal points, one each in object space and image space, and because the nodal points are used for reference, the focal length considered as a distance in front of a positive lens is negative according to the sign conventions stated in section 3.1.2, but positive in image space. This appears to introduce an ambiguity with regard to the sign of focal length used as a constant. To eliminate this ambiguity the sign of the constant used in the optical equations always agrees with the sign of the focal length as a distance in the rear of the lens; this is also in agreement with the sign convention established in section 3.1.2.

Some optical instruments have infinite focal length, or in other words, zero power. They are said to be afocal. In these instruments a parallel beam of light entering the system emerges as a parallel beam of light. Under these conditions such an instrument used alone cannot produce an image on a photographic film or on a screen. However, if the object is at a finite distance, the optical system also will form an image at a finite distance.

Lateral magnification—This is the ratio m of the image size to the object size. In accordance with the sign conventions established in section 3.1.2, lateral magnification is negative if the image and the object are on opposite sides of the optical axis. This is the usual case for imagery produced by a photographic lens.

Angular magnification—This is the ratio of the

tangent of the slope angle (angle made to the optical axis) of a ray in image space to the tangent of the slope angle of the same ray in object space. The frequently used symbol for angular magnification is γ. The relationship between lateral and angular magnification is that $\gamma = 1/m$. Angular magnification has analytical significance to lens designers, but is of little interest to photogrammetrists. Some authors use the same term to denote magnifying power of visual instruments. The latter is the ratio of the tangent of half the angle subtended at the eye by the image formed by an optical instrument to the tangent of half the angle subtended at the eye by the object if it were to be observed without the instrument. Occasionally angles are used instead of their tangents in defining angular magnification in a given instance. This procedure is only approximately correct unless the angles are very small, or unless object and image planes are replaced by spherical surfaces observed from their respective centers of curvature.

3.1.1.2 LIST OF OPTICAL SYMBOLS

C	The diameter of a circle of confusion; also of the circle of least confusion.
d	Distance between the second principal plane of the first lens and the first principal plane of the second lens (this is used in equations for combination of lenses); also used to designate the depth of focus.
D	Effective diameter (free or clear aperture) of a lens; also, radial distortion (linear).
D_r	Radial distortion in percentage.
f	Equivalent focal length (or simply focal length) of a lens.
F	Focal point.
F'	The second focal point for a combination of lenses.
h	Height of a ray on a surface of the lens.
H	The first principal plane (or point) of a lens (object space).
H'	The second principal plane (or point) of a lens (image space).
HH'	Separation of principal planes or points.
i	Angle of incidence (the angle between a ray and the normal to the optical surface).
I	Image plane.
L	Overall distance from the object to the image plane.
m	Lateral magnification.
n,n'	Refractive indices of adjacent media.
p	Power of a lens (the reciprocal of the focal length).
P	Object point.
P_o	Axial object point.
P'	Image point.
P_o'	Axial image point.
r	Angle of refraction (the angle between a refracted ray and the normal to the optical surface).
R	Radius of curvature of an optical surface. Also resolution-usually in lines per millimeter.
R_o	Axial visual resolution with an ideal lens.
R_α	Off-axis visual resolution with an ideal lens.
R_g	Ground detail that can be resolved by an ideal lens.
R_s	Resolution of a system containing a number of components as, for example, a combination of an aerial photograph, a projection lens, and a projection screen.
s	Axial object distance (usually from the first principal point of a lens).
s'	Axial image distance (usually from the second principal point of a lens).
S	Sagittal focus (image).
t	Thickness of a lens.
T	Tangential focus (image).
u	Slope angle of a ray in the object space (the angle between a ray and the optical axis).
u'	Slope angle of a ray in the image space (the angle between a ray and the optical axis).
V	Vertex of a lens.
x	Axial object distance from the first focal point.
x'	Axial image distance from the second focal point.
y	Radial distance of an object point from the optical axis.
y'	Radial distance of an image point from the optical axis.
y_o'	Radial distance of an image point from the optical axis (in absence of distortion).
α	Slope angle of a principal ray in object space (the angle between the ray and the optical axis).
γ	Angular magnification.
λ	Used to designate a wavelength of light.

3.1.2 FORMS OF THE OPTICAL EQUATIONS

If directional quantities are involved, hardly any equation of physics can be used reliably unless one knows the sign conventions, and this is the usual case in geometrical optics.

On the basis of the convention that light travels from left to right, the following sign conventions (the same as those of analytical geometry) are used in this chapter (*see* fig. 3-2).

(1) Distances are considered positive if they are measured in the direction to the right from a reference point. They are negative if measured to the left. The reference point should be indicated in each instance.

(2) Angles are measured from a reference line which, as the case may be, is usually either the optical axis or the radius of curvature (the normal) of the surface. They are positive if the reference line must be rotated counterclockwise through less than 90 degrees to bring it into coincidence with the line of interest, such as the light-ray.

(3) The radius of curvature of a surface is considered positive if the center of curvature is on the right side of the surface; otherwise the radius of curvature is negative.

(4) Distances measured upward are positive; those measured downward are negative.

The algebraic and trigonometric expressions governing the precise path of a chosen initial ray through an optical system are known as ray-tracing equations. After application of these equations one can determine the exact intercept of the ray on the image plane or indeed on any chosen image surface and can compare that

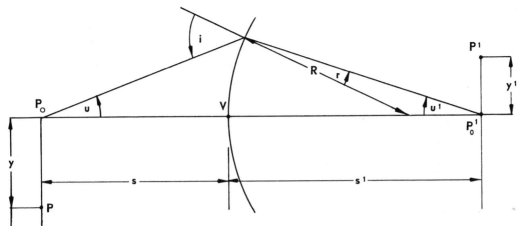

FIGURE 3-2. An illustration of the sign conventions used. The reference point is V. Distance s to the object point P_o is negative. Distance s' to image point P'_o is positive. Slope angle u is positive, u' is negative. Angle i and angle r are positive. Radius R is positive. Distance P_oP is negative, P'_oP' is positive. Lateral magnification, $m = y'/y$ is negative.

intercept with the desired location of the image point. The displacement of one from the other in the image surface can be described as an image error.

Except for a few relatively simple types of ray-tracing situations, one cannot obtain closed analytic expressions for the coordinates of the intercept of a ray on the image surface as a function of the initial conditions and parameters of the system. Instead, one must proceed by making use of power series approximations to control the preliminary form of the optical system and then follow by adequate ray-tracing and differential corrections applied in successive approximations. Under modern conditions these methods take on a high degree of sophistication that are entirely different from the trial-and-error methods used in former times.

Nevertheless, for any given optical system with known constants, whether final or not, one can trace a pattern of rays sufficient to provide a knowledge of the performance of the system. Even though in the final optimization of the optical system one must take into account the physical nature of the images and the relationships of these images to the desired state of performance with respect to resolution and contrast, one can obtain a very large amount of instructive analysis from ray-tracing results alone.

Indeed, for systems not required to approach the critical performance levels of diffraction-limited designs, it may often be sufficient to use only the ray-tracing results plotted in graphical form as "spot" diagrams. These diagrams may be obtained by automated calculations and plotting procedures and show at a glance the condition of image performance over the field and spectrum. Additional rays can be obtained either by amplifying the original ray pattern or by interpolation in image space from the group of

precise ray results already at hand. More will be given on this subject in section 3.1.3.

Much preliminary information can be obtained from adaptations of the ray-tracing equations to first-order expansions. One assumes that the slope angles and angles of incidence and refraction are sufficiently small and replaces the sines by the angles and the cosines by unity. In addition, heights of intercept of rays on the surfaces of a lens system expressed in terms of ratios to equivalent focal length or to some commensurable unit length can also be treated as of first order. These assumptions lead to equations representing the behavior of the optical system in the "paraxial" region surrounding the optical axis.

In the limit a paraxial ray becomes an axial ray different from the optical axis by infinitesimals of the first order. For such a ray all angles can be taken as zero. In consequence, the associated ray-tracing equations become very straightforward. Such axial expressions thus yield the position of the image plane and the value of the focal length. The locations of all of the cardinal points of the system also become readily obtainable and useful for varying applications.

Furthermore, when axial rays are traced for different wavelengths, the designer obtains information with regard to the state of axial color correction or achromatization. Achromatization refers not only to the stability of the position of the image plane with wavelength but also to the stability of image size (or equivalent focal length) as a function of wavelength.

The first-order terms are of considerable value even for advanced systems. Even the most highly corrected optical system has to have an image plane and an equivalent focal length. Such a system is also corrected for chromatic variations of image location and image size. The first-order terms represent these attributes even for the simplest systems and therefore are of basic importance.

Some relationships in the paraxial region were derived by Newton toward the end of the 17th century, but the first complete theory was developed by Gauss in 1844. Hence, the first-order relationships based on his theory are called Gaussian. These early results were very much extended by power-series developments, including the third powers of angles and ray-heights, principally by Seidel in the middle of the 19th century. However, Fraunhofer, Petzval and others had worked out various improved but limited treatments earlier. More will be given on this overall subject in section 3.1.3 and particularly in section 3.1.3.7.

It would be superfluous to present all of such ray-tracing equations here, inasmuch as photogrammetrists are usually not involved in optical design. Those photogrammetrists who may become interested in lens design will find a number of books covering this subject, some of which are referred to at the end of this chapter. Of much greater interest to photogrammetrists are the equations of optics which determine the basic properties of a lens and of image formation. These equations are given in the following sections.

3.1.2.1 Basic, Thin-Lens Equations

Many optical systems are comprised of one or more individual elements separated from one another along the optical axis. Ordinarily, positive lens elements have central thicknesses adequate for reasonable edge thicknesses conforming to the adopted clear apertures or, if negative, have central thicknesses large enough for practicable fabrication. Similarly, lens elements nearly in contact must have axial separations at least large enough to allow for any undesired physical contact or overlapping of the lens elements off the axis.

If one begins to write down the exact equations for even a single element with a practicable central thickness, the algebraic difficulties become immediately evident. In section 3.1.2.3 a few equations for a lens element with a finite central thickness will be given. For the time being, however, the device of the "thin" lens will be adopted. One means thereby that central thicknesses will be set equal to zero. Accordingly, the equations that govern the basic optical behavior of the thin lens can be written in simple form. These provide a relatively quick means for estimating the optical characteristics of one or more such elements on a common optical axis. Similarly, if desired, thin lenses can be in contact. That is, one can assign zero separations between any two or more such elements without regard for physical contact or overlapping off the axis. It will be of value to summarize below the basic Gaussian equations governing the behavior of such an element or group of elements.

The ray geometry of a single thin lens is shown in figure 3-3. In accordance with the conventions adopted above, R_1 is positive and R_2 is negative. Inasmuch as one is now assuming mathematically that the central thickness is negligible, the object and image distances can be taken from the center. In the diagram the object distance s is negative and the image distance s' positive.

The focal length and the power (the reciprocal of the focal length) are determined by the following equation:

$$1/f = p = (n - 1)(1/R_1 - 1/R_2), \qquad (3.2)$$

where n is with respect to the surrounding medium, usually air.

The reciprocals of radii are called "curvatures". Quite often in optical formulas one will find it more convenient to calculate with curvatures rather than with radii, particularly for large radii or for planar surfaces. For the latter the

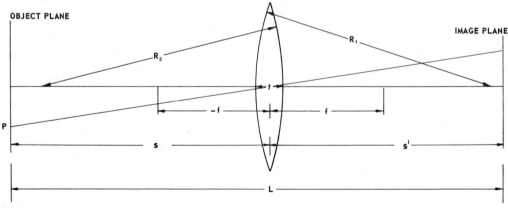

FIGURE 3-3. A thin lens. The radii of curvature of its surfaces are R_1 and R_2, as indicated. Thickness t is thought of as negligible. The focal lengths f on both sides are indicated. L is the overall distance from the object to the image plane. Graphical presentation is also given of the ideal image formation for an object at a finite distance.

curvatures are zero. Curvatures of both wavefronts and lens surfaces, when multiplied by paraxial heights, are equivalent to paraxial angles and are much more likely in practice to be additive, or at least commensurable, than are radii, object and image distances. Nevertheless, one usually retains the latter, particularly toward the close of a computation, as having recognizable geometrical meaning.

The relationship between object and image distance for the thin lens of focal length f or power p is given by:

$$1/s' - 1/s = 1/f. \tag{3.3}$$

This is called the image or focal equation. The lateral magnification m by definition is given by

$$m = s'/s. \tag{3.4}$$

Accordingly, one can find a few useful variants of equations (3.3) and (3.4), namely:

$$s = fs'/(f - s') = s'/m \tag{3.5}$$

$$s' = fs/(f + s) = ms \tag{3.6}$$

$$f = ss'/(s - s') \tag{3.7}$$

$$m = s'/s = f/(f + s) = (f - s')/f. \tag{3.8}$$

Similarly, the overall length L by definition is:

$$L = s' - s \tag{3.9}$$

$$s' = L/2 \pm \sqrt{(L/2)^2 - Lf} \tag{3.10}$$

$$f = s'(L - s')/L = -ss'/L \tag{3.11}$$

$$L = -f(m - 1)^2/m \tag{3.12}$$

$$s' = mL(m - 1) \tag{3.13}$$

$$f = -mL(m - 1). \tag{3.14}$$

Differential forms are often convenient for estimating rates of change of one parameter with respect to others. Two such equations of frequent usefulness are the following:

$$ds' = (f^2ds + s^2df)/(f + s)^2 \tag{3.15}$$

$$dm = (sdf - fds)/(f + s)^2. \tag{3.16}$$

One notes also from equation (3.3) that as the object distance s goes to large negative values,

$$ds' \approx -f^2/s. \tag{3.17}$$

This equation is of particular interest to photogrammetrists inasmuch as it determines quickly and with excellent accuracy the change of focus from infinity focus of even a complicated photographic objective when the object plane lies at a large finite distance instead of at infinity. For example, let the value of f be given as 152.4 mm. The position of the second focal point F' will already have been determined in the laboratory from photographic tests on a collimator. (In practice one usually chooses the plane of best definition.) If the altitude of the airplane is 1500 m, the image will be shifted beyond the calibrated infinity focus by the amount $ds' = + 0.0001 f$, or $+ 0.015$ mm, a generally harm-

less change. Similarly, if the airplane is at an altitude of only 150m above the terrain, the corresponding shift of the image beyond the calibrated infinity focus will be $ds' = + 0.152$ mm. This latter change in focus is so large that one would have to refocus to prevent quite noticeable loss of performance. However, so low an altitude is rarely used for photogrammetric purposes.

3.1.2.2 NEWTONIAN FORM OF EQUATIONS

When given in Newtonian form the equations differ from those of the preceding section. In section 3.1.2.1 the object and image distances are taken with respect to the coincident vertices of the thin lens. In the Newtonian form one uses instead the respective focal points (see figure 3-4.) The resulting expressions apply not only for the case of the thin lens but also for the general coaxial optical system, except for the overall length L. For the moment L will be considered as for thin lenses only.

The most useful Newtonian equations are:

$$xx' = -f^2 \tag{3.18}$$

$$m = f/x = -x'/f \tag{3.19}$$

$$x = -x'/m^2 \tag{3.20}$$

$$x' = -m^2x \tag{3.21}$$

$$L = -(x - f)^2/x = (x' + f)^2/x'. \tag{3.22}$$

The differential expression below will prove to be useful:

$$dx' = f^2dx/x^2 - 2fdf/x = m(mdx - 2df). \tag{3.23}$$

3.1.2.3 THICK-LENS EQUATIONS

In the preceding section it was noted that the power of a thin lens is proportional to the difference of the curvatures. The shape of the lens, whether bi-convex or meniscus, does not appear. In the case of a "thick" lens, that is, a lens element having a finite thickness that must be taken accurately into account, the shape of the lens does appear.

Before the equations for the optical behavior of the thick lens are given, let us consider beforehand the case of refraction at a single spherical face between different refractive

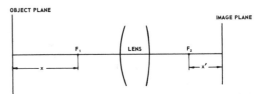

FIGURE 3-4. Object and image distances used in the Newtonian form of equations. The front focal point is F_1; the second focal point is F_2. The distance to the object is x (negative in this case); the distance to the image is x' (positive).

media. The relationship analogous to that given for a thin lens (3.3) becomes:

$$n'/s' - n/s = (n' - n)/R$$
$$= - n/f$$
$$= n'/f'. \tag{3.24}$$

In section 3.24 the rule has been applied that f has the algebraic sign of f', as given earlier, where f' is the focal length in image space. For this single surface the principal planes coincide at the vertex and the nodal planes coincide at the center of curvature. The lateral magnification in this instance becomes:

$$m = ns'/n's. \tag{3.25}$$

If this expression is applied to the nodal planes for which $s = s'$, then in the nodal planes $m = n/n'$. This in turn leads to an example of the usefulness of paraxial expressions. If a hemispherical magnifier is made up and placed with the flat side down on printed matter, the image presented to the eye will appear magnified by $m = n$ irrespective of R, where n is the index of the glass. Inasmuch as the printed matter lies in the nodal plane in its own medium, the other being coincident thereto, the viewed image will remain stationary as seen from any angle. It is also a matter of experience that such a magnifier has very little distortion. This follows from the fact that along every line of sight passing through the center of curvature and therefore through the the nodal points, the eye of the observer can be located approximately on a spherical surface centered on the nodal points which comprises the locus of all the points F, (for normal glass f is about equal to $2R$) on and off axis. For any one such point F the corresponding rays in the medium containing the printed matter will be approximately parallel and effectively free from distortion.

Equation (3.24) can be applied to any number of refracting surfaces in succession and therefore is a basic relationship for all coaxial optical systems. In particular, one can apply the expression twice over to the case of a single thick lens, that is, a lens element with a finite central thickness of any magnitude.

Accordingly, for the single thick lens

$$1/f = p = (n - 1)(1/R_1 - 1/R_2) + (n - 1)^2 \, t/nR_1R_2. \tag{3.26}$$

When $t = 0$, this equation reduces, of course, to the thin lens equation (3.2). For finite t the change in the power depends also on the sign of the product R_1R_2. However, one should note that for a thick lens *in vacuo* or immersed in a single surrounding medium such as air the respective front and rear focal lengths remain equal. This fact is evident in the above expression if one substitutes $- R_2$ for R_1 and $- R_1$ for R_2, as for the same lens turned around.

In the case of a thick lens and indeed for that of a general, coaxial optical system the following

three new quantities appear and become important for optical computations:

Back focal distance.—This is defined as the distance along the optical axis from the vertex of the back surface of the lens to the back focal point.

Front focal distance.—This is the distance along the axis from the vertex of the front surface of the lens to the front focal point.

Separation between principal points or nodal point separation.—This separation is denoted by HH' where H is the first and H' the second principal point. The second usually follows the first in the direction from left to right, for which HH' is positive (*see* figure 3-5). In some systems HH' can be negative. If NN' represents the separation of the nodal points for systems immersed in different media in object and image space, while the nodal points and the principal points do not then coincide, nevertheless, $NN' = HH'$. If the initial medium has the index n and the final medium n', then

$$HN = f'[(n' - n)/n'] \quad \text{and}$$
$$H'N' = f[(n' - n)/n]. \tag{3.27}$$

In those systems where HH' and therefore NN' are negative, the principal points or nodal points are said to be crossed.

For the single thick lens the distance of the first principal point H from the first vertex V_1 is:

$$V_1H = - (n - 1)tf/nR_2$$
$$= tR_1/[n(R_1 - R_2) - (n - 1)t]. \tag{3.28}$$

In accordance with the sign convention the distance V_1H is positive when H is to the right

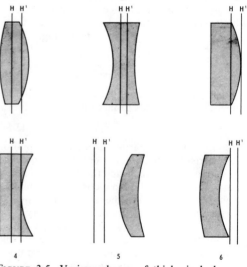

FIGURE 3-5. Various shapes of thick single lenses. Lens 1 is binconvex, 2 is biconcave, 3 is planoconvex, 4 is planoconcave, 5 is a positive meniscus, and 6 is a negative meniscus. The principal planes of the lenses are indicated by H and H'; note that they may fall outside the meniscus lenses as illustrated here.

of V_1. Similarly, the distance of the second principal point H' from the second vertex V_2 is:

$$H'V_2 = (n - 1)tf/nR_1$$
$$= -tR_2/[n(R_1 - R_2) - (n - 1)t]. \quad (3.29)$$

which is positive when H' is to the left of V_2. The principal points are separated by the distance

$$HH' = NN'$$
$$= t - (R_1 - R_2)t/[n(R_1 - R_2) - (n - 1)t] \quad (3.30)$$

which is positive when H is to the left of H'.

The back focal distance is of particular importance in that it determines the space available between the lens and the film plane of the camera. Whenever a new lens is to be used in a camera, the value of the back focal distance of the lens should be ascertained in advance to enable one to avoid a possible mechanical interference or focus problem.

All the symbols and equations of section 3.1.2.1 on the thin lens apply equally to a thick single or compound lens. In the latter case, however, the "lens" ceases to be the reference point. The reference point in object space becomes H (the first principal point), and in image space H' (the second principal point). Similarly, for the Newtonian form of equations the focal points remain the reference points.

Furthermore, the overall distance L from object to image plane remains equal to $(s' - s)$ when used in optical formulas. One should, however, include the separation HH' of the principal points in determining the total physical length of an optical system. If one makes these reservations, then all of the optical equations given above apply to thick lenses.

3.1.2.4 PERSPECTIVE CENTERS AND PUPILS

In some situations optical points and planes additional to those mentioned in section 3.1.1 and section 3.1.2.3 become of interest. Among these are the perspective centers (occasionally called centers of projection) and the associated entrance and exit pupils and planes.

Every optical system has a real stop (aperture stop) which is used to limit the diameter of the image-forming bundles of rays accepted by the system. The stop may be simply a circular hole in a metal plate, or the iris diaphragm itself, or possibly simply a chosen mounting rim of a lens element located in a favorable place within the system. Occasionally the real stop may even lie outside the system.

By definition, the entrance pupil is the image of the stop formed by the part of the optical system preceding the stop. The exit pupil is the image of the stop formed by the part of the optical system following the stop. Thus, the entrance pupil lies in object space and the exit pupil in image space. The two pupils are conjugate to one another and to the real stop.

The rays passing through the axial point of the real stop are called the chief rays (sometimes called principal ray by photogrammetrists). Each chief ray comes initially from some one object point. Within the accuracy of Gaussian optics the chief rays also pass through the axial point of the entrance pupil in object space and through the axial point of the exit pupil in image space. As long as the real stop is circular in form and in the absence of vignetting or obscuration the chief rays are central in the bundles of rays transmitted by the system from the respective object points. The chief rays therefore aid greatly in defining the effective course of the various bundles of rays through the system to the image surface. Their total spread in solid angle is limited by the assigned field of view. The limit is normally set by the size of format whereafter the clear apertures of the surfaces preceding and following the stop can be determined.

In an optical system, perspective centers are conjugate axial points whose characteristics are that the rays passing through the perspective center in object space emerge from the perspective center in image space parallel to their original directions. It is obvious that the perspective centers coincide with the respective nodal points of the optical system. This situation is illustrated in figure 3-6.

Such perspective relationships are clearly obeyed by a pinhole camera. When located between parallel object and image planes, the pinhole serves as the center of projection. Clearly, every chief ray through the pinhole will intersect the image plane without distortion displacements. A square grid on the object plane will be imaged as a square grid on the image plane, whatever the magnification may be. The nodal points are in effect coincident at the

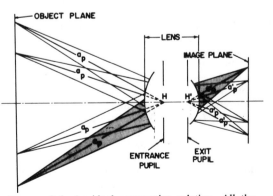

FIGURE 3-6. An ideal perspective relation. All the principal rays a'_p in the image space emerge parallel to the respective principal rays a_p in the object space. H and H' are the nodal points as well as the principal points, and therefore are in coincidence with the respective centers of the entrance and exit pupils.

pinhole and the slope angles of the respective entering and emergent chief rays are identical.

In an actual lens system of finite size the nodal points play the role of the pinhole. Rays in object space incident on N emerge in image space from N' and through paraxial accuracy the slope angles of the respective entering and emergent rays are the same. However, unless the entrance and exit pupils are also located at the nodal points, rays incident in object space on N or emerging from N' are not chief rays and in fact may be quite different.

Without careful design the entrance pupil in general will not be a sharp image of the stop in object space for all rays over aperture, field and spectrum. Even its location on the optical axis may be dependent on the slope angle of the entering chief ray, that is, on the field angle. Similarly, the exit pupil may not be a sharp image of the real stop and may also have a location on the optical axis depending on the slope angle of the respective chief ray. Furthermore, the diameters of the entrance and exit pupils may change from the Gaussian values according to the slope angles of the chief rays.

If the aberrations of the ray bundles also have asymmetric components, then it may be that the intercepts of the chief rays on the image plane in the outer field may not truly represent the effective centroids of the respective bundles of rays focused by the system. It is for such reasons that asymmetric aberrations in the images formed by a mapping lens under design must be eliminated.

Now in the general-purpose optical system it is not required that entrance and exit pupils coincide with either the principal points or nodal points or both, as the case may be. Under these circumstances the conjugate object and image points of the entrance and exit pupils will not in general lie at the nodal points and the entering and emergent angles of the principal rays will differ. Such a circumstance becomes exaggerated in the case of a telephoto lens or in the inverted telephoto lens. For a telephoto lens the rear principal point in an extreme case may actually be located in air well to the left of the entire lens system, whereas the exit pupil usually lies within the lens barrel (Cf. figure 3-5, 5). Accordingly, chief rays in image-space emerging from the exit pupil have slope angles markedly different from those for the respective rays incident on H. Figure 3-7 illustrates a situation of this kind.

If a lens system is designed to be free of distortion for the case where the entrance pupil does not lie near N nor the exit pupil near N', for a given conjugate pair of object and image planes one can design in such a way that the actual intercepts on the image surface fall near the respective distortion-free reference points to some required precision. If only this requirement is met, it is possible that without further design control the locations of entrance and exit pupils on the optical axis will vary markedly from their paraxial values. The correction for distortion arises in a forced way from a functional balance between the variable pupil position and variable slope angle in image space. Such a correction means that for some other widely different conjugate object and image planes, distortion will once again appear. While ray passages within the optical system may be about as before, the new image plane intersects the chief rays in image space at a different location and in a different way. Distortion will result.

A simple example of the absence of distortion for one pair of conjugate planes and its presence at another is shown by the system of two identical plano-convex elements arranged in exact symmetry around a small central stop. The convex surfaces will face outward and the planar surfaces will be adjacent to the stop. For the conjugate pair at 1:1 ratio, that is, at a magnification of -1, there is no distortion at all, even over a very wide angular field. If now the object plane is moved to infinity to the left, the image plane on the right moves to the position of F' on the optical axis. A chief ray traced backward through the first element into object space will have a certain field angle. This same ray traced forward through the rear element will have the same field angle in image space. Indeed, the ray will be just the same as for the 1:1 case. For every such ray there will be an axial point in object space representing the particular location for N for this ray alone, and the same for N'. If, from another direction in object space, the same point N is used as reference, nevertheless in image space N' will have changed again. So also will the slope angle and this will differ from that in object space in spite of the full symmetry of the system. With both N' and the slope angle in error one will find that in the

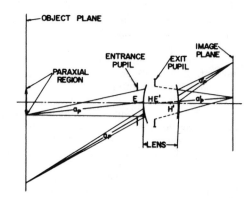

FIGURE 3-7. General perspective relation. The principal rays a_p' in image space do not emerge parallel to the respective principal rays a_p in object space. Only in the paraxial region do the respective principal rays pass through the axial points E and E' of the paraxial entrance and exit pupils. H and H' are the nodal points. The paraxial entrance pupil does not necessarily confine the off-axis cone.

focal plane the intercept of the ray will differ from that required for distortion-free performance.

If instead the optical system is designed in such a way that the locations of entrance and exit pupils over the field remain fixed at their respective paraxial values for all slope angles of the principal rays, a precise correction for distortion requires that the tangents of the initial and final slope angles bear a constant ratio to one another over the field. Such a tangent condition can be met quite exactly by processes of design. However, there is another condition in optical design that will be discussed in further detail in section 3.1.3.2. This condition requires that if the exit pupil is to be a good image of the entrance pupil, the ratios of the sines of the slope angles must be constant. If not, then there is pupil aberration that shows up as a displacement of the principal rays for large field angles. Clearly, the optical system cannot obey both the tangent condition and the sine condition at the same time unless the slope angles in object and image space are the same. In practice, of course, one may be more important than the other according to the requirements but acceptable compromises can often be made.

If the entrance and exit pupils, however, do coincide with the nodal points, as indicated above for the case of the idealized perspective centers, then both tangent and sine conditions can be met simultaneously. That is to say the slope angles in object and image space are the same. Such an optical design calls for specialized attention but when carried out with precision, one will have a system that admits of distortion-free performance for every conjugate pair of object and image planes. Indeed a very excellent state of correction can be achieved.

For example, fully symmetrical lenses that are also corrected for stability of entrance and exit pupil locations (that is, for spherical aberration of the chief rays, as will be discussed in section 3.1.3) have an excellent state of distortion-free performance. Such lenses can be used with somewhat limited apertures not only for mapping purposes but also for projection rectifiers and the like. On the other hand, as happens so often in optical design, one can draw liberally on small departures from perfection for some conditions in order to meet more satisfactorily other more difficult conditions, according to the requirements.

For such reasons the objectives in the Multiplex projectors and other similar instruments are set so that the second nodal point is coincident with the axes of rotation of the projector camera. Thus, when some other conditions are also satisfied, the perspective existing during the taking of the aerial photographs is closely restored in the image space of the projection instruments. In actual practice this presumption has never been strictly realized as has been implied by Pestrecov (1959). Nevertheless, as Pestrecov showed, the fact that the centers of the pupils are not in exact coincidence with the respective nodal points does not introduce significant warpage in the restored model of the terrain. The centers of these objectives happen to be very near the nodal points.

For many optical systems of excellent resolution and correction for distortion the centers of the entrance and exit pupils do not lie very near the nodal points. Therefore, the respective chief rays in object and image space are not parallel to one another. The geometry in image space thus differs from that in object space. Consequently, if such a system is used for forming a terrain model by projection of an overlapping pair of aerial photographs, the model will necessarily be somewhat warped, even if there is no radial distortion of the image in the Gaussian projection plane.

3.1.2.5 EQUATIONS FOR LENSES IN COMBINATION

Usually a photogrammetrist deals with compound lenses which contain a number of elements and components. He cannot perform computations on such a lens unless he knows the lens formula accurately. (*see* section 3.1). Because lens manufacturers consider lens formulas to be proprietary information, they release them only in special cases. Lens formulas can often be found in their patents. However, it is rather difficult to identify the patent applicable to a lens unless the patent number is engraved on the lens, especially because trade names of lenses do not appear in patents.

American manufacturers generally patent the final or nearly final formulas of their lenses. European manufacturers quite often patent the preliminary formulas the use of which may be misleading.

Photogrammetrists who design new instrumentation frequently need such data as the focal lengths of the lenses they plan to use, and the front and back focal distances. Data of this kind are generally available from the manufacturers on request and are all that are needed when a combination of two or more lenses or lens systems are to be used in experimentation.

If in figure 3-8 we denote by subscript 1 the terms pertaining to the first lens (using the convention that this lens is on the left side) and by subscript 2 the terms pertaining to the second lens, we obtain the basic equations listed below. In these equations the terms pertaining to the combination have no subscripts.

$$1/f = 1/f_1 + 1/f_2 - d/f_1 f_2 \qquad (3.31)$$

where d is the distance between the second principal plane H'_1 of the first lens and first principal plane H_2 of the second lens.

The equation acquires a somewhat simpler

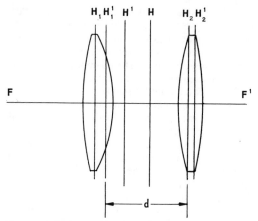

FIGURE 3-8. A combination of lenses. F is the first focal point of the combination. F' is its second focal point.

form if the powers of the lenses (the reciprocals of the focal lengths) are used.

$$p = p_1 + p_2 - dp_1p_2. \tag{3.32}$$

The distance of the first principal plane H of the combination from the first principal plane H_1 is

$$H_1H = d/f_2. \tag{3.33}$$

The distance of the second principal plane H' of the combination from H'_2 is

$$H'_2H' = fd/f_1. \tag{3.34}$$

If a number of lenses is involved, the equations given above may be applied in sequence.

A formula that may be useful, however, is the extension of (3.32) with respect to three or more spaced elements. One finds that if d_1 and d_2 are the respective separations of the adjacent principal points of a trio of elements with focal lengths f_1, f_2, and f_3, then

$$p = p_1 + p_2 + p_3 - d_1p_1p_2 \\ - (d_1 + d_2)p_1p_3 - d_2p_2p_3 + d_1d_2p_1p_2p_3. \tag{3.35}$$

The back focal distance of the combination is given by

$$BF = f[1 - (d_1 + d_2)p_1 - d_2p_2 + d_1d_2p_1p_2]. \tag{3.36}$$

These two formulas can easily be extended to still more lenses or applied in sequence to groups of lenses. The lenses can also be in part thin lenses or thick lenses as long as the separations and powers are properly used.

3.1.2.6 APPLICATIONS OF FIRST ORDER FORMULAS

While one can piece together any given number of single thick elements comprising an optical system of interest by making use of intermediate quantities for individual elements,

nevertheless an appreciable amount of numerical tedium can be avoided simply by applying (3.24) directly and (3.25) indirectly in simple succession.

To avoid numerous additional symbols let us temporarily redefine s and s' as object and image distances from the vertex of the corresponding surface and assume that the individual principal points and focal lengths are not known. Let d be redefined temporarily as the axial separation between successive vertices rather than between the successive and adjacent principal points for individual elements. Accordingly, t as a symbol for central thickness is set aside.

Now let us define a quantity h as the relative height of a Gaussian ray coming from an axial object point. For convenience $h = 1$ at the entrance pupil. For an infinitely distant object h will also be equal to 1. The successive values of h throughout the system provide the designer with a close knowledge of the relative heights of intercept of this special ray from the axial object point.

Let the subscript i refer to the surface number from left to right and let K be the total number of surfaces. The subscript for n, the index of refraction, will refer to the medium following surface i.

In order to calculate the focal length in image space, one proceeds as follows.

Given $[(1/s_1) = 0]$, R_i, d_i, n_i for $i = 1, \ldots, K$, one finds in succession

$$h_1 = 1$$

$$(1/s'_i) = [(1/s_i) - (1/R_i)]n_{i-1}/n_i + (1/R_i) \tag{3.37}$$

$$h_{i+1} = [1 - (1/s'_i)d_i]h_i \tag{3.38}$$

$$(1/s_{i+1}) = (1/s'_i)h_i/h_{i+1}. \tag{3.39}$$

After the computation has been completed through the last surface, $i = K$, one finds the back focal distance s'_K from $(1/s'_K)$. Thus, the position of the focal point F' is known. Moreover,

$$f' = s'_K/h_K \tag{3.40}$$

$$f = n_0f'/n_K. \tag{3.41}$$

The position of the principal point H' then follows from s'_K and f'. A similar calculation through the reversed system yields the position of F, the values of f and f' again, and therefore H. If the initial and final media are the same, f equals f' in both calculations. In the general case the positions of N and N' can now be found from

$$FN = f' \tag{3.42}$$

$$N'F' = f. \tag{3.43}$$

With the positions of F, F', H and H' all determined, one can now return to the standard Gaussian formulas and the standard definitions and usages for s and s'. The location of any image plane conjugate to a desired object plane and the associated magnification can now readily be found.

3.1.3 LENS ABERRATIONS

One can readily prove theoretically and demonstrate experimentally that a single lens element cannot produce a mathematically perfect image point. Even if one disregards diffraction, the image of an object point produced by a simple lens will not be a point but a blur, even in the plane of best focus. Image imperfections are called aberrations and are measured in quantitative terms by the size of the image blur. Qualitatively, image aberrations in monochromatic light are broadly classified as spherical aberration, coma, astigmatism, curvature of field, and distortion. (The last named, however, refers primarily to improper image location and not to image formation). In addition to the five imperfections given above, one finds that there are also chromatic or color aberrations which arise from the inability of a single refractive element to treat all wavelengths of light equally.

All of the aberrations except distortion may be expressed in either longitudinal or lateral (transverse) measure. Longitudinal aberration is measured in the direction parallel to that of the optical axis. Lateral or transverse aberration is measured perpendicularly to the optical axis, usually in the image plane of best focus or in some closely related image plane or surface. A convenient reference plane is the Gaussian (paraxial) image plane determined for a designated mean wavelength. Distortion is the displacement of the image point from its ideal position, radially inward or outward from the optical axis; hence its measure is always lateral. For non-rotational systems the distortion must be defined by lateral displacement errors in two coordinates, such as in x and y, or if small, in terms of radial and tangential displacements in the image plane from the image point.

The two-fold possibility of expressing aberrations either longitudinally or laterally may lead to some confusion in terminology, particularly with respect to chromatic aberrations. Thus, chromatic aberration for an axial beam is usually called longitudinal chromatic aberration, but this aberration can be and often is expressed in lateral measure (image blur) as well as in longitudinal displacements of image points with wavelength. On the other hand, the chromatic aberration of an off-axis beam is usually measured in the lateral or transverse direction, generally along a radial line in the image plane from the center of the field through the image point. This second kind of chromatic aberration is frequently called lateral chromatic aberration, or simply, lateral color. Some authorities prefer to call this aberration oblique, because it is associated with off-axis or oblique pencils of light.

While it is convenient to separate out the two kinds of chromatic aberration, one must note that longitudinal chromatic aberration, measured for convenience for an on-axis image point, persists off-axis as well and may become enhanced or diminished off-axis according to the nature of the optical system. Similarly, the lateral or radial image spread with wavelength caused by the presence of lateral chromatic aberration increases in proportion to the off-axis angle or position, being zero on-axis, and may be modified at large off-axis angles by additional aberrational effects depending on the particular system and state of its optical design. An off-axis image may also be afflicted by chromatic variations of the several aberrations first described, leading to such designations as chromatic spherical aberration, chromatic coma, etc.

The conventional presentation of aberrations is given in figure 3-9. The meaning of the various curves will be explained in detail in the subsequent sections.

The preceding discussion of aberrations and their illustration in figure 3-9 are based on the concepts of geometrical optics, in which aberrations are treated as departures of ray intercepts and image positions from the ideal. Designers often need to improve on such presentations by resorting to the realistic concepts of physical optics in which aberrations are considered as various kinds of deformations of image-forming wave fronts from their ideal forms, spherical or planar, as the case may be.

With the introduction and general use of fully automatic electronic computers, the analytical presentation of aberrations, whether in their geometrical or physical meaning, has become of decreasing value. Instead, one can present spot diagrams which in effect are enlarged image plots of many rays, as discussed in section 3.1.2. A typical spot diagram is given in figure 3-10. However, spot diagrams are insufficient for highly corrected systems and tend to emphasize the geometrical over the physical nature of the image structure. Diffraction images can also be calculated but with a much greater expenditure of computing time.

A powerful modern tool available to designers is the optical transfer function which presents the modulation of an image in terms of the amplitude and phase variations of sine wave frequencies transmitted by the optical system to the image position. The transfer function will be discussed at greater length in section 3.1.3.10 below.

The analytical presentation of aberrations is based intrinsically on the exact formulas needed for ray-tracing. In almost all cases the exact formulas applied to a succession of refractive or reflective surfaces become hopelessly ensnarled with reciprocals and square roots of ever more involved algebraic expressions. Therefore, as in many mathematical situations, one must resort to expanding the various expressions into power

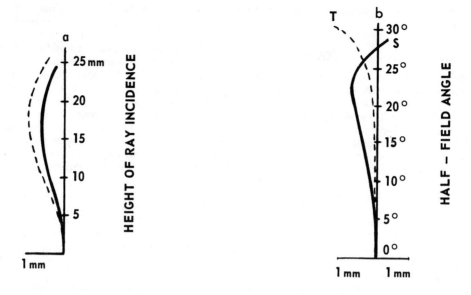

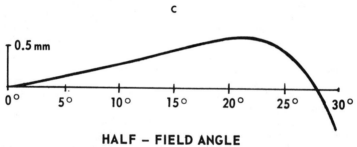

HALF – FIELD ANGLE

FIGURE 3-9. A conventional presentation of monochromatic aberrations. In *a*, the solid curve is the longitudinal spherical aberration, a plot of the position of the image point along the axis against ray height in the pupil; the dashed curve is a plot of the focal length against ray-height; the difference is a measure of coma. Astigmatism is shown in *b*, where the solid curve is the trace of the sagittal foci and the dashed curve of the tangential foci, plotted against off-axis angle. Radial distortion is shown in curve *c*.

series and deal at first with the lower order terms. A very significant expansion of this kind relates directly to the law of refraction (3.1). If *i* is the angle of incidence, one can write down the standard expansion:

$$\sin i = i - i^3/3! + i^5/5! - i^7/7! + \ldots \quad (3.44)$$

The first order term is that ordinarily associated with paraxial optics as discussed in section 3.1.2 above. The second term of the expansion for a purely rotational system leads to the third-order aberrations, having to do with the third power of the radian angle, which in turn form the basis for Seidel theory. The primary aberrations introduced above, namely, spherical aberration, coma, astigmatism, curvature of field and distortion, can all be expressed in Seidel theory as algebraic expressions which

yield quantitative measures of the respective aberrations as functions also of aperture, field-angle and wave-length.

Much more involved are the fifth-order and seventh-order aberrations. These will be discussed under section 3.1.3.7.

A single lens can be corrected for some aberrations only in very special situations, or occasionally by the introduction of one or both aspheric surfaces. Therefore, designers must introduce more elements to correct the respective aberrations to the degree required for a given application. The process of optical design involves the task of reducing the various aberrations to tolerable levels according to the requirements. Accordingly, optical systems may have as few as two elements, as for the objective of the refracting telescope, or perhaps as many as fifty elements for some forms of instrumenta-

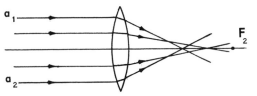

FIGURE 3-11. Spherical aberration. Rays from an object point (in this case at infinity) do not intersect at a single point after passing through the lens. The measure of spherical aberration is the distance from the Gaussian focal point F_2 to the point of intersection of two corresponding rays such as a_1 and a_2.

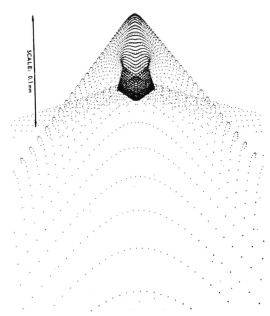

FIGURE 3-10. A typical spot diagram representing a greatly enlarged image blur. Each point represents the intersection of a ray with the image plane. All these rays were assumed to start from a single object point. The effective image point derives from the greatest concentration of energy.

tion. The most elaborate systems, however, usually evolve from simpler forms because of the need for special kinds of manipulation and control. In such cases the task of optical design often becomes one in part of rescuing an acceptable image quality from the overburden of so many surfaces and separations.

3.1.3.1 SPHERICAL ABERRATION

The failure of a lens system to produce a point image in image space of a monochromatic axial point source of light in object space is called spherical aberration. In the presence of this aberration the rays from the object point after refraction through the system fail to meet in the image point. Instead, the rays intersect the optical axis with a longitudinal displacement depending on the magnitude of the spherical aberration and on the height of intersection of the particular ray in the entrance pupil of the system.

Figure 3-11 shows a schematic view of the behavior of the rays in image space after refraction through a simple bi-convex lens. The rim rays, denoted by a_1 and a_2 in the plane of the paper, focus more closely to the rear surface vertex than the intermediate rays and the intermediate rays in turn focus short of the paraxial focus, F_2. The spherical aberration in this instance is said to be under-corrected. In the Seidel theory of third-order aberrations, the progressively shorter intercept along the axis with increasing

ray height in the aperture is represented by a second-degree algebraic expression. In the image plane itself the transverse displacement of the intercept from the paraxial image point is represented in Seidel theory by a third-degree power term.

Figure 3-9a shows a plot of the longitudinal spherical aberration for a typical compound lens system with spherical surfaces. The full curve shows a minimum intercept to the left in the diagram (toward the rear surface vertex) at a ray height in the pupil of about 18 mm. The shape of the curve in the illustration derives from the fact that rays of lesser height show the influence of under-corrected Seidel longitudinal spherical aberration. For rays of increasing height, while the third-order contribution increases as the square, the contributions toward over-correction by fifth and higher order terms (4th and 6th degree longitudinal terms) begin to dominate, and the intercept begins to move to the right. In the internal construction of the corresponding optical system one would expect to find one or more negative refracting surfaces dominating the refractions for the rim rays.

The spherical aberration of any lens system can be eliminated for any given conjugate point pair by the appropriate aspheric figuring of one or more of the surfaces of the system, or by adding a correcting plate with one or both surfaces aspherized. For systems produced in volume such a technique is seldom utilized except possibly for molded parts, usually in plastic. Instead, the task of the designer is to minimize the effects of spherical aberration by drawing upon the interplay of refractive or reflective positive and negative spherical surfaces, while at the same time controlling in the desired way other aberrations described below.

The term spherical aberration is not necessarily associated only with spherical surfaces. Actually, all reflecting or refracting surfaces, whether spherical or aspheric, usually produce spherical aberration except where designed specifically not to do so. Well-known exceptions are the cartesian surfaces, both refractive and reflective, described in section 3.1. The more familiar examples of the cartesian surfaces, such

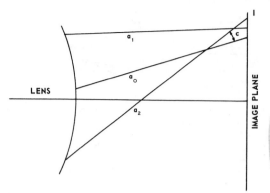

FIGURE 3-12. Coma. The point of intersection of two corresponding rays of equal plus and minus heights in the pupil, such as a_1 and a_2, does not coincide with the point of intersection of the principal ray, a_o, with image plane I. As used by some designers, the measure of coma is the angle c, but with most designers the measure is the radial extent of the comatic fan.

as the conic sections as reflective surfaces, have conjugate focal points free of spherical aberration.

In the presence of spherical aberration, all the rays originating from an object point and otherwise restricted by the pupils of the system are contained within a small circular image in a transverse plane. The diameter of the blur depends on the degree of correction obtained by the designer according to assigned tolerances and may also be minimized by a slight refocusing to an optimum position.

Even in the case of the simple lens, the designer can minimize the diameter of the blur by appropriately "bending" the lens. If the index of refraction is about 1.5, the resulting lens will have a weakly bi-convex shape with the stronger curvature facing the more distant of the conjugate point pair. As the index of refraction is increased, the lens shape passes through the plano-convex form. For still greater indices of refraction, the simple lens becomes of meniscus form, again with the stronger curvature facing the more distant conjugate point.

The presence of spherical aberration in an image will reduce the contrast in the image to a degree depending on the magnitude of the aberration and on the fineness of detail to be recorded. More will be said on this point under section 3.1.3.9.

3.1.3.2 COMA

In the discussion under section 3.1.2.4 the reader has already been introduced to the definition of the chief ray and its importance in perspective relationships. With respect to the aberration known as coma the chief ray has additional importance in defining the location of the apex of the fan of rays shown typically in figure 3-13 as pure coma.

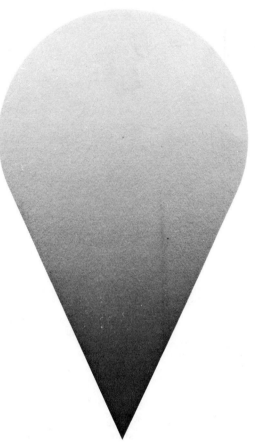

FIGURE 3-13. Pure coma. Note that the pattern is symmetrical only about the meridional plane.

Earlier it has been noted that the chief ray from a given object point by definition is the ray passing through the axial point of the real stop. The plane determined by the chief ray and the optical axis is called the meridional or primary plane of the rotationally symmetrical system. For convenience one ordinarily assigns all object points of design interest to the common meridional plane.

For a rotationally symmetrical system the chief ray from an object point in the meridional plane will always lie in the meridional plane of the refracted rays. This ray will pass through the axial points of both entrance and exit pupils, and in fact will define the locations of these pupils on the optical axis, according to the field angle involved.

The chief ray from a given object point will intersect the image plane at a point away from the axial point of the image plane in accordance with the off-axis location of the object point and with the focal length or magnification of the optical system. Other rays from the same object point to the entrance pupil will strike nearby in the image plane.

In the presence of pure coma, ray intersections in the entrance pupil at equal plus and minus heights will form ray-pairs that intersect the image plane in a common point progressively displaced from the intercept of the principal ray according to increasing zone height in the pupil. In the Seidel third-order theory the displacement from the chief ray intercept will be proportional to the off-axis angle and to the square of the ray-height in the entrance pupil.

If the rim-ray pair of rays intersects the image plane at a point farther from the axial point than the chief ray intercept, the coma is said to be outward. Moreover, the ray-pairing occurs in the plus and minus skew direction in the entrance pupil as well as in the meridional plane and the common point of intercept in the image plane actually lies also in the meridional plane.

Inasmuch as the point of intercept of the chief ray in the image plane depends on the focal length for a given oblique angle, or for finite conjugates, on the lateral magnification m, one can regard the comatic displacement in the image plane of ray-pairs as representing the variation of focal length or of magnification with zone height in the pupil. If coma is zero, all rays will have the same equivalent focal length and in the absence of other aberrations will coalesce at the ray intercept in the image plane.

The comatic fan of rays from a circular pupil shown in figure 3-13 can also be regarded as a miniature map, at the image, of concentric zones in the entrance pupil. Each chosen circle in the entrance pupil, defining a hollow fan of rays from the given object point, maps into the coma pattern as a twice-around circle tangent to envelope lines through the apex at a 60-degree included angle. That is to say, once around in the pupil is twice around the corresponding circle in the coma fan and the latter circle is tangent to the envelope lines. A circle of limited diameter in the entrance pupil lies near the apex of the fan. The circular rim itself maps into the circle farthest out in the fan but remains tangent to the envelope lines.

It is this kind of mapping of the pupil onto the twice-around comatic circles in the image plane that corresponds to the pairing of rays in a common point at the image that are 180 degrees apart in the meridional plane at the entrance pupil. It is clear that the image is asymmetric and that much of the illumination falls in a one-sided way away from the apex of the fan. As a result, comatic images are a serious form of aberration for mapping lenses and designers go to considerable lengths to reduce such asymmetric aberrations to very small residuals.

The magnitude of the coma can be measured in several ways. In the absence of other aberrations one can simply calculate the radial extent of the displacement in the image from the chief ray intercept. One can also recast this meridional length into angular measure in object space by dividing by the focal length, or by dividing the radial extent of the coma by the magnification and then dividing this by the distance along the principal ray in object space.

Coma may also be measured by the variation of focal length with zone height in the pupil, or for finite conjugates, by the variation of the lateral magnification. The governing relationship is known as the Abbé sine condition. For an optical system free of coma, the sine condition requires that

$$\sin u/\sin u' = m. \qquad (3.45)$$

It is evident that this sine condition reduces paraxially to the simple expression already given in equation (3.4). If the object lies at infinity, the relationship reduces to

$$h/\sin u' = -f, \qquad (3.46)$$

where h is the ray-height in the entrance pupil.

The quantity h will be known at the start of a ray-trace and the quantity $\sin u'$ will be found as a result of the calculation. Accordingly, f can be calculated. For the chief ray f will become the limiting value. For finite zone heights in the pupil one can therefore calculate the variation in f and can plot this result on the same diagram as shown in figure 3-9 for spherical aberration. The dashed curve in figure 3-9a therefore indicates the state of correction for coma in this compound optical system. If the error in f is plotted, the curve will contain both the error in location of the principle point with zone height and the longitudinal spherical aberration. If the spherical aberration is zero, the comatic error in focal length in Seidel optics will correspond to the variation in position of the second principal point with zone-height in the pupil. If the position of the second principal point is stable with respect to zone height in the pupil, and the variation in f is due only to spherical aberration, then the two curves in the plot will coincide.

Another plane of special interest is the sagittal or secondary plane. This plane passes through the chief ray but is perpendicular to the meridional plane. This plane in object space intersects the entrance pupil in a line also perpendicular to the meridional plane. This line in the pupil becomes the skew axis. Rays lying out of the meridional plane are in general called skew rays and must be defined in direction and intercepts by four quantities.

The full coma fan therefore involves skew rays but one degree of symmetry remains with respect to the meridional plane. The plus and minus skew rays at the rim intersect the image plane at a point that lies in the meridional plane. To the extent of the Seidel approximation this point lies at a distance from the apex of the fan only one-third that of the full radial extent of the comatic fan. The skew coma in fact is a more basic measure of coma than the full radial extent of the fan.

In applying the sine condition one notes that the slope angles u and u' are defined at the optical axis. Therefore, the sine condition, or more broadly, the sine theorem, relates the size of an image in the immediate vicinity of the optical axis to the slope angle of the ray at the axial point of that image.

This relationship can be derived from the first law of thermodynamics because it relates how energy used in forming an image must be conserved. The longer the focal length for a given pupil size, the smaller is u' and the more the image is spread out in the image plane. The energy density is reduced but the total energy remains the same. Conversely, as the focal length is made ever shorter, the image becomes smaller and the energy density greater, but the total energy remains the same.

Probably the simplest illustration of an optical system free of spherical aberration but afflicted with coma is that of the front surface paraboloidal telescope mirror. An incoming bundle of rays parallel to the optical axis reflects accurately to the focal point of the paraboloid. If a spherical mirror were able to focus a beam of parallel light on F_2 as sharply and if this spherical mirror were centered on F_2, the sine of the slope angle, namely, $\sin u'$, would then equal h/f and there would be no coma.

Instead, the sagittal depth of the paraboloidal surface at zone-height h differs markedly from that of the spherical surface centered on F_2 and coma results. The mathematical expression derived from the properties of the paraboloid shows that u' for any given h is too small and hence the focal length calculated from (3.46) for the zone h is too large. The ray at this zone therefore focuses farther from the axis for a given field angle and the coma is outward. Moreover, the calculation shows that the focal length derived from (3.46) leads to the outward displacement in the meridional plane of the point of intersection of the plus and minus *skew ray pair* at the zone height h in the pupil. The displacement of the meridional ray pair in the pupil, as mentioned earlier, comes out to be three times greater at the image than for the skew ray pair intercept of the same zone-height, a property of the comatic fan shown in figure 3-13. Comatic contributions purely of the fifth order are as simple in principle as for the third order but differ in magnitude and in the included angle of the envelope lines.

3.1.3.3 ASTIGMATISM AND CURVATURE OF FIELD

A first-order portion of a wave-front spreading outward from an object point defines in solid angle a small set of rays called a "pencil" of light. Such a pencil can be thought of as of differential magnitude but in practice can be taken as limited in cross-section only when higher-order corrective terms must be considered also, according to the nature of the optical system.

The reader has already been introduced to the concept of paraxial rays which in effect comprise pencils of rays in the neighborhood of the optical axis. Such paraxial pencils are treated mathematically by use of first-order, Gaussian theory. On the other hand, well-corrected optical systems yield a final image plane with images relatively free of higher-order aberrations. The elementary but important behavior of such corrected systems can also be described by first-order or Gaussian theory, inasmuch as the higher-order aberrations by controlled design have been rendered nearly harmless.

First-power field angles are included in Gaussian theory and lead to the familiar formulas that provide focal length as well as focal position, together with image positions and lateral magnifications between conjugate planes.

One now can ask what happens for pencils of rays that are so inclined to the optical axis that the square and higher powers of the field angle must be taken into account. Even more broadly, one can consider pencils of limited cross-section, centered around respective principal rays, which pass through the various refractive and reflective surfaces of a system at even very large angles. One might consider, for example, the behavior of rays through an $f/32$ real stop over the full field of 95 degrees of a typical mapping lens.

In the general case a pencil of rays from an object point, refracted or reflected at large angles of incidence by an optical surface, refocus into two astigmatic line segment images, displaced somewhat from one another along the principal ray. Each of the two line segments will be perpendicular to the principal ray, and though separated, the two line images will lie at an azimuth of 90 degrees to one another. If the refraction or reflection occurs at the first surface of the system, the stigmatic pencil of rays emanating from the object point becomes astigmatic in the first image space, except for selected conditions, and proceeds thereafter through a compound system, increasing or decreasing in the astigmatic difference along the principal ray in successive image spaces until the final image space is reached.

Ideally, in the final image space the astigmatism, as measured by the separation of the astigmatic line segments, can be designed to vanish and the respective astigmatic foci will lie together on the prescribed image plane throughout the field. Such a system is said to be "anastigmatic" and to have a flat field.

In practice, one finds that there is a "tangential" image surface arising from elementary meridional pencils or fans of rays that focus into line segments perpendicular to a line in the image surface radially directed to the Gaussian

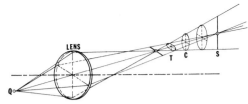

FIGURE 3-14. Astigmatism. Line *T* shows the tangential image of a point. Line *S* shows the sagittal image. *C* is the circle of least confusion and lies approximately half-way between *T* and *S*. The radial line-images *S* lies in the meridional plane.

image point on axis, and also a "radial" image surface arising from the respective foci of sagittal rays of the pencils transmitted by the optical system. One speaks of the tangential image surface as being the locus of foci of fans of rays in primary planes, while the radial image surface is spoken of as comprising the locus of foci of fans of rays in sagittal or secondary planes.

It is customary under modern circumstances to talk of the radial and tangential image-surfaces of an optical system. Thus, while the radial foci arise from skew rays coming to a focus for the respective pencil, the radial line segment so formed has an actual radial extent, plus or minus with respect to the principal ray, arising from the defocusing of the meridional rays at that point and angle. Correspondingly, while the tangential foci arise from the meridional rays of the pencil coming to a focus for the respective pencil, the tangential line segment so formed has an actual tangential extent, plus or minus with respect to the principal ray, arising from the defocusing of the skew rays of the pencil at that point and angle.

The general optical system, therefore, stopped down at the real stop to allow for the transmittal of elementary pencils, has a tangential and a radial image surface. Approximately half-way between the two surfaces along any given principal ray one will find a mean focus and a mean image diameter containing the slightly defocused radial and tangential pencils (outside of focus for one and inside of focus for the other). In strict ray geometry the mean image blur will be circular in the intermediate field but will gradually become elliptical as the field angle increases. The diameter of the circle containing all rays is the diameter of the "circle of least confusion." In practice, the wave nature of light will cause the mean image to show diffraction effects, modified by the presence of any other aberrations and may even produce a cross as the mean image, rather than a circular blur.

A well-designed optical system will have radial and tangential image surfaces that conform on the average to a flat mean-image surface, although the detailed image surfaces may exhibit a variety of shapes not far from planar. If the mean-image surface departs from flatness, the system is said to be afflicted with curvature of field. If the radial and tangential surfaces depart from one another beyond acceptable measure, the system is said to be afflicted with astigmatism.

The aberrations of astigmatism and field curvature comprise two of the five Seidel aberrations. Astigmatism may be measured either by the separation along the principal ray of the astigmatic surfaces as a function of field angle or by the size of the major and minor axes of the astigmatic ellipse in the Gaussian image plane, for a given aperture-ratio, or by the angular equivalent recast into object space by appropriate magnification factors. Similarly, the curvature of field can be measured by the departure of the mean image along the principal ray from the Gaussian image plane, or by the mean value of the major and minor axes of the astigmatic ellipse in the image plane, for a given aperture-ratio, or by the angular equivalent recast into object space.

The user of photographic objectives will often be interested directly in the behavior of the optical system in the Gaussian image plane, refocused slightly as may be necessary to obtain an improved mean performance. More will be written on the testing of photographic objectives in section 3.2.3 below.

Astigmatism can be illustrated by a perspective drawing as shown in fig. 3-14.

Inasmuch as astigmatism can be treated mathematically by use of first-order formulas in the aperture of pencils of rays centered on the respective principal rays, one finds that a detailed ray-trace of principal rays through an optical system at even very great angles with respect to the optical axis contains numerical results that can be used to locate the respective tangential and sagittal images throughout the optical refractions and reflections. These formulas are called the "Coddington" expressions after the originator of the analysis, and in the past, at least, have been of very great importance to the designer. Indeed, plots of the radial and tangential image surfaces obtained from application of the Coddington expressions have long been used as further means for presenting the performance of an optical design. Under modern conditions, with results obtained in other ways from the electronic calculator, the Coddington presentation is of diminished importance but nevertheless remains in the background as a guide to the experienced designer of how a system performs and how it may be improved (*see* also Kingslake, 1939).

3.1.3.4 CHROMATIC ABERRATIONS

The existence of chromatic aberrations in refractive optical systems has already been referred to in section 3.1.3 above. Because of the variation of refractive index with wavelength for all transparent materials, the many quantities associated with Gaussian and Seidel optics having to do with image position, size and quality are all to some degree a function of wavelength.

Photogrammetric lenses usually are intended to produce high quality images over a large portion of the visual spectrum. For the most part the range of wavelengths from blue to deep red is utilized, often modified by the addition of color filters and by use of specially sensitized emulsions. For special purposes photography into the near infra-red had been utilized. Color photography has also been drawn upon for photogrammetric purposes, which in turn requires good optical correction for chromatic errors, even though color balancing is difficult, owing to the variable nature of the atmosphere and to the range of altitudes employed.

Accordingly, it is important in the design of photogrammetric lenses to ensure that both the longitudinal and the lateral color aberrations are reduced to acceptable residuals and that the chromatic variations of the Seidel aberrations are small enough to be tolerated.

A ray of white light in object space refracted at a lens surface becomes a small spectral fan of rays owing to the dependence of the index of refraction on wavelength. In compensation, the designer must so choose the refractions at the following surfaces of the system and the various optical materials involved that the chromatic rays of light come back together again at the required image point. Inasmuch as all object points in the field and all rays through the entrance pupil must be considered, the designer must find a means for correcting the chromatic aberrations of many rays at once.

Fortunately, the designer has at his disposal a large variety of types of optical glass types from a number of manufacturers. The designer is thus free to choose glass types that in combination can correct for chromatic aberration as well as for the monochromatic Seidel aberrations. While, with the modern electronic calculator, methods can be used that minimize the chromatic errors of a pattern of selected rays over the field, aperture and spectrum, one can also obtain a very good first approximation to chromatic correction by combining two widely separated wavelengths to achieve coincidence in the image plane. For longitudinal chromatic aberration one can also adopt a mean zone of the aperture and a mean field angle. For lateral chromatic aberration one can bring the two selected wavelengths to coincidence for the principal ray at a mean position in the field. Such a first approximation is simple and direct. One can improve upon this type of correction by introducing more wavelengths, more rays and more critical requirements but the principles of correction remain the same.

While the ideal goal has always been to eliminate chromatic errors equally over the chosen spectral region, in practice one must deal with chromatic residuals. In simple systems it is usually the case that only two widely separated wavelengths of the spectrum can be brought to a common focus or to zero lateral color. Other wavelengths deviate from focus of the pair of wavelengths in a functional way. With special glass types one can often improve the performance of the system by combining three wavelengths instead of two and indeed in some special systems a long range of the spectrum has been corrected by choosing glass types and four wavelength correction. It is almost inevitable, however, that the improved chromatic corrections are obtained at the cost of an elaborate optical system that is difficult to manufacture. The special materials also seem to have many physical difficulties and greater costs of manufacture.

3.1.3.4.1 Axial (Longitudinal) Chromatic Aberration

As the name implies, this aberration pertains to axial image points but persists off-axis as well. In detailed designwork one must take into account the off-axis variations of longitudinal chromatic aberrations over the field, both in the meridional plane and in sagittal planes. For our present discussion, however, it will be adequate to confine our attention to the chromatic aberration on the optical axis.

A simple positive lens is incapable of focusing a fully color-corrected image of a distant point on the optical axis. Figure 3-15 indicates that the wavelengths on the short wavelength side of the mean are more strongly refracted by the positive element and hence focus closer to the lens than the mean focus. Wavelengths on the long wavelength side of the mean, focus long. One usually plots a curve relating focal position to

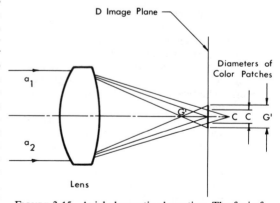

FIGURE 3-15. Axial chromatic aberration. The foci of two corresponding rays a_1 and a_2 are shown for three spectral lines: C(red), D(yellow), and G'(blue-violet). The image patches produced in the D-image plane by the defocused colors are also shown. The sequence of the foci depend on the particular lens design.

wavelength. The simple error curve is called the primary spectrum of longitudinal chromatic aberration.

By combining positive and negative lens elements of properly chosen materials, the designer can bring two selected wavelengths to a common focus on the optical axis. The positive element must be of a material less dispersive than the negative and much lens power is consumed in off-setting the positive action of the one element by the negative action of the other. The resulting curve relating focal position to wavelength now shows an approximately quadratic kind of shape in such a way that pairs of wavelengths focus together.

The residual variation of focal position with wavelength is now called the secondary spectrum of longitudinal chromatic aberration. For almost all standard optical materials so paired, the blue-violet and the deep red focus long, and the mean wavelength, such as yellow, focuses short, with respect to a mean focus. Most mapping lenses do have a residual secondary spectrum owing to limitations on choice of materials when taken in common with the needs of many monochromatic corrections. The designer however can sometimes draw upon non-standard optical glasses to effect a reduction in the amplitude of the residual secondary spectrum but the practical risks are great.

One may assume too quickly that if two elements can bring two wavelengths to a common focus, then a combination of three elements ought to be able to bring three wavelengths together, etc. While this proposition is true in an algebraic sense, the resulting lens curvatures are usually entirely prohibitive. Instead, one must seek combinations of glass types that are more or less matched throughout the spectrum, in which case the residual secondary spectrum becomes smaller in a natural way. However, such glasses usually are so nearly alike in dispersion that again the strong curvatures needed become impracticable in a mapping lens. Nevertheless, in a wide variety of optical systems designers have been able to reduce or even to eliminate secondary spectrum for special purposes.

Images that are inside or outside of the mean focus necessarily cause a loss of contrast in the image and a reduction in resolving power for fine detail in the image. Accordingly, it is important to control not only the longitudinal chromatic aberration, a first-order quantity, but also the chromatic spherical aberration, which is of third-order magnitude. The relatively fast mapping lenses of modern design are therefore highly corrected for the axial chromatic aberrations.

3.1.3.4.2 OBLIQUE (LATERAL) CHROMATIC ABERRATION

This aberration refers to the chromatic dispersion in the lateral position of the principal ray at any given field angle. The ray of white light in object space arrives at the image plane as a spectral spread of colors that left uncorrected would form a radial streak as an image. The designer finds it essential to reduce errors in lateral color to very small residuals.

This oblique or lateral chromatic aberration is occasionally called oblique color, or more formally, the chromatic difference of magnification. That is to say, the variation of focal length with wavelength, or for finite conjugates the variation of lateral magnification with wavelength, leads to the radial spectral image streak indicated above.

In an actual aerial photography, the intensity of the illumination of an object point and its reflectivity may vary with wavelength. Accordingly, its mean position in the image plane will depend on the effective color, a circumstance to the great disadvantage of precision mapping. There may even be a tendency under special circumstances of recording a doubling of an image or three lines instead of two in a target, and the like. More generally, one loses in contrast and resolution.

Fortunately, the reduction or elimination of lateral color is more amenable to design control than is the case for longitudinal chromatic aberration. Even a symmetrical pair of otherwise unachromatized lens elements can be made free of lateral color. The designer looks upon the prismatic refraction of the principal ray as the cause of the dispersion and it is only necessary to offset the first such prismatic refraction by a negative or inverted prismatic refraction elsewhere in the system. Even two positive elements can be so arranged. In the general case, however, the considerations are somewhat involved, and one must allow for higher order chromatic errors as well.

If an optical system is more or less symmetrical and made of quite standard materials, the lateral color when reduced to zero by a wavelength pair commonly leaves very small residuals of a secondary-spectrum nature, even without futher elaboration of the design. However, in unsymmetrical systems, such as tele-

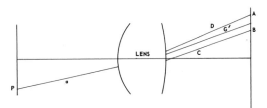

FIGURE 3-16. Lateral chromatic aberration. Ray *a* starting from object point *P* is dispersed into many rays, which are shown for three spectral lines: *C*(red), *D*(yellow), and *G'*(blue-violet). Streak *AB*, produced by these rays in the image plane, is also shown. The sequence of the ray-interactions depends on the lens design.

photo lenses, the residual secondary spectrum in the lateral color can become quite large and its elimination quite difficult.

In general, two widely separated lenses can be corrected for either longitudinal or for lateral chromatic aberration but not for both. Three separated simple elements of properly chosen glass can be corrected for both longitudinal and lateral chromatic aberrations and form the basis for the well-known Cooke triplet.

3.1.3.5 DISTORTION

The various aberrations discussed above have to do with the quality of the optical image. The combined effects of such aberrations are therefore disturbing to the photogrammetrist only when the deterioration of the images prevents him from identifying details of interest or increases the uncertainties of measurement.

Distortion, however, has to do with the position of the image point in the image plane but not with the image quality. The presence of distortion therefore is of great importance to the photogrammetrist and must be taken into account in making plots of locations and in measuring distances between images in the image plane.

Distortion leads also to a variation of the scale of an image as a function of position in the image plane. Clearly its presence is disturbing to the photogrammetrist whose ultimate task is to produce maps of uniform scale from the measured images.

In the simplest case, as for third order or Seidel distortion, an outward displacement of a given image point from its desired location on a mean-image plane is referred to as pincushion distortion. Its magnitude increases as the third power of the field angle from the optical axis of a rotational optical system. An inward displacement, on the other hand, is called barrel distortion and increases in numerical magnitude though negatively also as the third power of the field angle.

If one is photographing a square grid in object space, the recorded image affected by pincushion distortion for a typical centered square in the object is shown typically as *a* in Figure 3-17; correspondingly, barrel distortion is shown as *b* in Figure 3-17. Inasmuch as the scale at any given off-axis point is related to the rate of change of image position in the image plane to that in the object plane, it is evident that pincushion distortion causes the images to spread out in the outer portion of the image and the scale therefore to increase; whereas barrel distortion causes the images to crowd increasingly together in the outer part of the image plane and the scale to decrease.

In more complicated systems the designer may find it possible to balance the inward and outward displacements of the image points to produce zero displacements at one or more nodes in

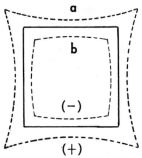

FIGURE 3-17. Distortion. Pin-cushion distortion is shown in *a*; barrel distortion in *b*.

the distortion curve, which in turn is a graph of displacement against position in the image plane. Such balancing usually leads to a minimizing of the intermediate displacements in numerical magnitude, whether outward or inward, and therefore to a best fit of a photographed grid to an ideal grid in the image plane. Typically, the designer can balance residuals of the first order against the third against the fifth and higher orders of displacement, the coefficients of the corresponding terms alternating in algebraic sign.

Note, however, that the scale, which is related to the first derivative of the distortion curve, while having the same number of nodes or cross-overs of the mean scale on a graph of scale against image position, has these nodes in somewhat different positions on the image plane, as compared to those of the distortion curve. Nevertheless, inasmuch as measurements must often be taken between points quite far apart on such an averaged photograph, the designer must minimize the maximum displacements and the photogrammetrist must use a mean scale, as will be discussed more at length below.

The discussion above pertains strictly to radial distortion, which for a purely rotational system causes the actual image point to be displaced radially in the image plane, whatever the azimuth in the field. It is usual to express the distortion as a displacement in tenths of millimeters, or if very well corrected as for modern lenses, in micrometres plotted against position in the field, insofar as photogrammetry is concerned. In other applications one may find it more convenient to express the distortion as a percentage error, plotted also as a function of position in the field, or for very distant objects, as a function of off-axis field angle. The percentage distortion in general will vary as an even ordered power series in the off-axis position. Its lowest order variable term, arising from the third-order distortion, will thus be of the second degree, which means that the distortion in the central field will vary as a quadratic function of the off-axis angle.

It is, of course, a task of the lens designer to keep distortion within specified close limits

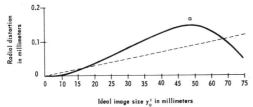

FIGURE 3-18. Distortion calibration. Radial distortion produced by a lens is represented by curve a, computed on the basis of the Gaussian equivalent focal length. The dashed line represents a possible favorable calibration that requires a small change in the focal length that more accurately represents the performance of the lens.

which in turn will allow the lens to be effective for the specific photogrammetric purpose. At the same time, however, the designer will also need to meet specified requirements for image quality. However, distortion is one of the quite stubborn aberrations, partly because of the stringent requirements of photogrammetry. In the past, lens designers were not always able to reduce distortion for wide angle systems to the small maximum displacements required by photogrammetrists. Therefore, some other special means had to be used for compensating measurements made from images having some known form of distortion curve. One of the means is a mechanical compensation incorporated in the plotting instrument (see chapters XIII and XIV).

So far only the residual radial distortion remaining as an aberrational curve in the design of a purely rotationally symmetric optical system of high quality has been discussed. Practically, no manufactured optical system can ever be precisely rotational, although near wonders have been accomplished by skilled optical and mechanical craftsmen. The small errors that remain in the alignment and centering of a precise multi-element optical system may lead not only to further small radial displacements of the principal rays at off-axis positions but this time as a function of azimuth in the field as well. Furthermore, these same errors of fabrication will lead to tangential displacements of image points such that the magnitude will depend on both the azimuth and the off-axis position in the field. Usually one can find a meridian of zero tangential displacement with good symmetry on either side, as shown for example in figure 3-19. In complicated lens systems further asymmetries can be found.

For many years the composite effect has been loosely called tangential distortion. One is to understand thereby that at any given image point in the format, there may be radial and tangential displacements of the effective image point from the desired position, and that the corresponding plus and minus values can be plotted in some

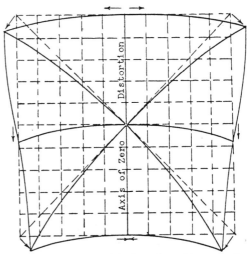

FIGURE 3-19. Tangential distortion. Lines passing through the axial point of the image frame are curved.

convenient way. For example, the displacements can be magnified and therefore exaggerated everywhere in the image field and the distorted grid plotted relative to the true undistorted grid, as in figure 3-19. This kind of graphical presentation, easily programmed into computers, can give the proper impression at a glance of the type of distortion of a given optical system. One must be careful, however, not to impart a false impression to some one else that the exaggerated graphical performance is the true performance.

The presence of tangential distortion was first noted by Pennington (1947). This occurred when exploration of Multiplex models displayed unusual patterns, resulting in inaccuracies not attributable to radial distortion alone. This type of distortion is not to be associated with the usual aberrations of lens design, although designers may often need to evaluate the sensitivity of the particular design to decenterings and misalignments. One of the effects of tangential distortion is that a straight line passing through the center of the field may actually photograph as a weakly curved line. It is obvious also that the local scale on a photograph may vary both radially and tangentially at a given image position from the mean value for the format as a whole.

3.1.3.5.1 RADIAL DISTORTION

The usual procedure for computing the distortion of a plane image formed by a centered lens system is based on two conditions:

(1) the position of the image plane satisfies equation 3.4,
(2) the position of any given image point is determined by the intersection of the corresponding principal ray with the image plane.

Inherent in this second condition is that the bundle of rays converging to the image point is symmetrical around the principal ray or at least, the principal ray is the optimal ray for representing the effective image point of the bundle.

If y is the radial distance of the object point from the optical axis in the object plane, and y_o' the ideal distance of the corresponding image point from the optical axis in the image plane, and if α is the slope angle of the principal ray in object space, then freedom from distortion is defined by the following equations:

$$y_o' = my \text{ (for a finite object distance)}$$
$$y_o' = f \tan \alpha \text{ (for an infinite object distance).} \quad (3.47)$$

One may note that equations 3.47 apply for a simple projection between parallel planes through a perspective center, which after all, is what the lens system, no matter how complex, intends to accomplish. In mathematical terms one speaks of mapping one plane onto another by means of a collineation which optically is represented by the equations of Gaussian theory (see MIL-STD-150A).

In the presence of distortion, the actual position y' of the image point differs from y_o', as may be determined by an exact ray trace. Then the measure of the radial distortion is

$$D = y' - my$$
(for a finite object distance)
$$D = y' - f \tan \alpha$$
(for an infinite object distance). (3.48)

The percentage distortion is given by

$$D_r = 100 \, D/y_o' \quad (3.49)$$

It should be understood that y' is the actual distance in the image plane, of an image point from the optical axis, and that this image location is produced by the optical system itself. On the other hand, in equations 3.48, the paraxial quantities m and f appear. The quantity m is the paraxial lateral magnification and f is the focal length determined from Gaussian theory. While these two quantities are also properties of the optical system and of the object distance, one may regard them as mathematical coefficients that can be altered by the photogrammetrist if, by his doing so, a better match of the image to the ideal can be achieved. The coefficients so adjusted are called, respectively, the calibrated magnification and the calibrated focal length. The resulting distortion is called the calibrated distortion.

The term "calibrated focal length" is firmly entrenched in photogrammetry. The other two terms were introduced by Pestrecov (1959). They have not as yet found general use in photogrammetric literature, but their use is encouraged.

A photogrammetrist can easily determine whether any advantage can be obtained by "calibration." This is carried out as described below (see fig. 3-18).

From computational data supplied by the manufacturer, the radial distortion is plotted against the image height, y_o'. A sloping line of best fit is then drawn through the origin to equalize the plus and minus maximum departures from the mean line. From the slope of the mean line and equation 3.47, one can re-evaluate either m or f as the case may be, and replot the distortion curve around the adjusted horizontal axis. In so doing, the photogrammetrist is carrying out a procedure which the lens designer has also performed in his controlled endeavor to minimize the maximum distortion displacements of the principal rays over the field.

It should be emphasized that "calibration" of distortion in this way does not involve refocusing of the camera or other physical changes. It is purely a mathematical procedure which permits the photogrammetrist to select the most favorable scale in such a manner that the average deviation of the image points from their proper positions will be a minimum (see also section 3.2.3.4).

3.1.3.5.2 TANGENTIAL DISTORTION

Tangential distortion is the displacement of images perpendicular to the radial lines and results from imperfect fabrication of the lens. It is primarily of historic importance and related to difficulties associated with high volume production of the deep meniscus form of the Metrogon and Planigon elements. Where present, it was accompanied by other image displacement anomalies such as an often larger asymmetry in the radial distortion. The more sophisticated lenses required by today's more precise photogrammetry do not show any significant tangential distortion, as indeed was true of the later Planigon lenses. It is customary today to specify all image displacements from the design nominal in terms of a maximum value.

The effect produced by decentering of a lens is, to a certain degree, similar to that produced by a perfectly centered lens in combination with a weak prism. Washer (1957), after exploring this matter, established the term "prism effect" and related it to tangential distortion. The tangential distortion specifically for the Bausch & Lomb Metrogon lens was studied by Pennington (1947), and by Livingston, Cude, and Sewell of the Engineer Research and Development Laboratories. They found that the tangential distortion curves across any diagonal of a photograph are nearly symmetrical, in both magnitude and sign, about the axial point. Furthermore, the curves reach a maximum at a certain distance from the axis.

Sharp (1949), then with Bausch & Lomb, thoroughly investigated the sources of tangential distortion, and introduced new methods of centering lens elements to provide a firm basis for controlling tangential distortion. Gruner (1962), also of Bausch & Lomb, by a mathematical analysis (unpublished) of the adjustment procedures, contributed much to the establishment of practical means for reduction of tangential distortion to the required limits. One of the latest approaches to this problem is described by Deterding (1962).

A typical tangential-distortion pattern is shown in figure 3-19.

3.1.3.6 RESOLUTION

It was established by the astronomer Airy in 1834 that the image of a point source produced by an ideal optical system cannot be a mathematical point but instead consists of a small central disc surrounded by diffraction rings. This central disc of greatest intensity is known as the spurious disc, or often as the Airy disc. If the integrated light energy in the spurious disc is taken as of unit value, the integrated energy of the first bright ring is 8.4%, of the second bright ring, 3.3%, of the third, 1.8% and so on. If, instead, the total light energy in the overall image is taken as of unit value, the spurious disc contains about 84%, the first bright ring about 6%, and so on. In careful laboratory work more than 100 rings of the infinite array have been detected.

Airy also found that the diameter of the spurious disc, taken for convenience as the diameter of the first dark ring, depends on the wavelength of the light and on the aperture-ratio or f-number of the lens. The latter quantity is taken to be the ratio of focal length to diameter of the optical system and is a measure of the angular spread of the light waves converging to the image.

These earliest studies of the resolution of an optical system arose from the need to resolve close double stars in astronomy as observed through a given telescope, and from the question of limits on the details resolvable by a microscope. Further important studies were conducted by Rayleigh (1842–1919) and by Abbé (1840–1905). Rayleigh thought it simplest to adopt a criterion, in the telescopic case, that a pair of equally bright stars can be resolved if the peak of the central disc of the first lies in the center of the trough of the first dark ring of the other. Actually, two such close stars have superimposed diffraction patterns not coherent with one another (*see* Jenkins & White).

The question then arises as to the nature of the detector which must deal with low contrast near the limit of resolution. For the eye certain rules apply, but other kinds of detectors can do better or worse than the eye. Accordingly, one must realize that resolution in the limit depends on the nature of the detector as well as on the quality of image formed by the optical system.

For visual purposes where the eye's performance is aided by an optimum magnification, one may adopt the useful rule that the resolution of a perfectly corrected optical system in white light is given by

$$R_o = 1450/(f\text{-number}) \qquad (3.50)$$

where R_o is the resolution in bright lines per millimetre in the image plane.

Because of the minuteness of the image it is necessary for the observer to view the image of a test target through a microscope of adjustable magnification and focus. Too little magnification will not properly favor the performance of the eye and too much magnification may cause the image to appear diffuse. The actual range of useful power is, however, rather broad and different observers may prefer somewhat different optimum values.

If light of a higher color temperature is used, the value 1450 may be increased. If the light is reddened, as by use of a color filter, or by light of a low color temperature, then the value 1450 will need to be decreased, according to the effective wavelength of the light.

If the eye is replaced by use of a photographic emulsion optimally placed, particularly if a fast emulsion is employed, the numerator given in equation 3.50 will be found to be drastically too large. Indeed, for an optical system used with an emulsion capable of resolving only 100 line-pairs per millimetre, that is, equally spaced bright and dark lines, the effect of diffraction in the live or aerial image is much muted. Clearly, some further treatment is necessary if the combined effects of the aerial image and the photographic emulsion are to be understood.

The formula given in 3.50 is therefore of limited usefulness in photogrammetry. For example, according to the formula, one can anticipate a performance visually of about 230 line-pairs per millimetre on axis from a 152.4-mm $f/6.3$ Metrogon lens, whereas in practice, this same image photographed onto a moderately fast emulsion, will yield no more than 30 line-pairs per millimetre. If, instead, a finer-grain emulsion of greater inherent resolving power is used, one may observe perhaps 40 line-pairs per millimetre in the photographic image, limited now more by the residual spherical aberration and chromatic aberration of the lens than by either diffraction or emulsion alone. Thus, aberrations, diffraction, and nature of the detector are all involved.

Resolution characteristically drops to smaller values according to the position off-axis, although this may not always be true. For some kinds of designs the off-axis resolution may go through a second maximum, usually separately in the tangential or radial direction in the image plane. Often the designer can, however, vary constructional details in such a way as to obtain an optimum, anastigmatic, coma-free image at some selected off-axis point, which then becomes a point of high performance comparing favorably to a well-corrected on-axis image. It is more customary nevertheless for the designer to seek the best average performance over the assigned field. This kind of balanced design may then result in a slowly decreasing resolution in tangential and radial directions as the off-axis distance increases. The present-day manufacturer often supplies both the calculated and the observed performance of a lens at stated off-axis angles in the image field, as affected by filters and by light source.

According to Washer (1945), off-axis resolution with an ideal lens varies in accordance with the following formulas:

$$R = R_o \cos^3\alpha \text{ (for tangential lines)}$$
$$R = R_o \cos \alpha \text{ (for radial lines)} \quad (3.51)$$

where R_o is the axial resolution and α is the angular separation from the axis of the point under consideration.

It is of interest to note that at least some successful lenses yield photographic resolution curves approximating equations 3.42 if R_o is taken to mean the axial lens-film resolution. For example, the tangential resolution curve of the 115-mm Wild Aviogon lens follows rather closely the cosine³ curve up to 45 degrees and its radial resolution curve roughly follows the cosine curve up to 40 degrees.

Attempts have been made to develop formulas from which the resolution of a system containing several components can be determined. For example, in an aerial camera the "components" may be the lens, photographic emulsion, focal setting, filter, aircraft, and the air, all in various kinds of dependence on haze and target contrast. Each of these components may limit resolution in some way. The following empirical expressions have been used for computing the resolution of a system:

or
$$1/R_s = 1/R_1 + 1/R_2 + 1/R_3 + \ldots$$
$$1/R_s{}^2 = 1/R_1{}^2 + 1/R_2{}^2 + 1/R_3{}^2 + \quad (3.52)$$

where R_s is the resolution of a system and R_1, R_2, etc. are the limiting resolutions of the components, assumed known.

Studies at Boston University have shown that the second formula produces somewhat better results, at least in some cases. One might note that in the first formula the reciprocal with proper factor is an angle, and the summation of angles is then taken to be the final angle of resolution. If one had a grainless emulsion insensitive to contrast, perhaps such an addition would be quite exact. In practice, however, there is some interdependence of the effects, leading to the application of the second formula, which more nearly resembles statistical addition by means of an rms of related quantities.

Fortunately a better mathematical treatment is at hand in the form of the transfer function which will be discussed at some length in section 3.1.3.10. For immediate purposes in applying the two simple formulas of 3.52, one must find a means for measuring the resolution curves of the lens by itself, of the emulsion by itself, and so on. Many studies have been made of the individual effects, and manufacturers customarily provide lens-film results for specified emulsions, light-source, filter, and focal setting, all at one or several contrast levels. Specifications are often written around resolution requirements and are given as minimum or threshold values for a production run.

It should be noted that the ability of the eye to evaluate image quality in conjunction with the interpretive powers of the mind is determined not only by resolution but also by density gradients of the edges of detail. Investigations in this direction were started by Cox and expanded by Higgins and Jones (1952). The latter authors introduced the term "acutance" as an expression of the observer's evaluation of the edges of resolved elements and defined this measure quantitatively. Further studies were summarized by Higgins and Wolfe (1955).

The need for photographic systems of ever increasing resolution has long been emphasized by the availability of aircraft capable of flying at much higher altitudes than in earlier decades and by the requirements of instrument-bearing satellites for surveys and mapping of vast regions. In this connection it may be noted that while previous specifications have indicated the resolution required of lens-film combination, actually the end-product is ground detail that can be resolved.

If one assumes that a lens is free of aberrations and that the Rayleigh criterion of resolution applies, the following approximate formula determines the ground detail that can be resolved by the aerial image:

$$R_g = 56 \times 10^{-5} (H/D) \quad (3.53)$$

where R_g is the ground resolution in terms of the center-to-center distance of bright lines in metres, H is the flight altitude in metres, and D the effective clear aperture of the lens in millimetres. The expression 3.53 is derived from observational experience with telescopic images and by such experience leads to noticeably better values than indicated by 3.50 reduced to the same terms. The discrepancy only means that the visual observer equipped with suitable magnification and under the best conditions can see stars resolved whose images are closer together than the radius of the first dark ring.

Of course such ideal results as implied in equation 3.53 can only be approximated. In practice lenses have residual aberrations, errors of fabrication, thermal uncertainties of focus and image quality, and the results are degraded by the presence of image motion, vibration, turbulence and haze. Insofar as the lens-film combination is concerned, modern fine-grain emulsions are capable of resolving some hundreds of line pairs per millimetre in the limit, although to minimize exposure time and thereby image motion, much faster emulsions of more limited resolution are used. Formula 3.53 is therefore to be thought of as a limiting ideal that in practice is much degraded. High quality photogrammetric exposures from an aircraft may be only a tenth as good.

While emulsions of several levels of resolution, graininess and speed are readily available, for a given emulsion the lens designer is often faced with the need to provide a diffraction-limited optical system if only to improve the lens-film results by a few additional line-pairs per millimetre. Although there is thus an imbalance between the capabilities of the fast emulsion and of the lens system, such that slight improvements in the combination can be ever more demanding of near optical perfection for the lens system, modern requirements can be and usually are very stringent. Conversely, the slight gains so made are capable of providing additional information perhaps not available in any other way, particularly in reconnaissance systems. Thus, in modern parlance one hears often of diffraction-limited aerial lenses that approximate perfection within close limits. It is proper that such near perfection be sought after, for constant vigilance is needed to maintain results of the highest quality. The end result is usually worth the additional effort.

3.1.3.6.1 RELATION TO LENS APERTURE

Equation 3.53 contains the quantity D which represents the effective diameter or, as it is also called, the effective free aperture of the lens. Its importance in the equation is derived from the fact that a given clear aperture can be measured in terms of the number of light waves across the diameter, which in turn is closely related to the diffractive resolving power of the aperture. The fact that D is measured in millimetres instead of in wavelengths is only a matter of convenience. Equation 3.53 directly indicates the dependence of the resolution of ground detail on aperture.

Equation 3.53 can be rewritten. If, instead of ground detail, one is interested in detail in the image plane, such that $1/R_o$ is the center to center distance in millimetres of the closest resolved bright lines, then

$$1/R_o = 56 \times 10^{-5} \, (f/D) \qquad (3.54)$$

where f is the focal length in millimetres, and D is once again the clear aperture in millimetres. Then one can replace (f/D) by the f-number and rewrite the equation as:

$$R_o = 1785/(f\text{-number})$$
to be compared to (3.50) above.

As mentioned above, the increase from the Rayleigh value of 1450 to the value of 1785, as obtained from astronomical results, is due to the fact that the peaks of close Airy discs can still be discerned, even when closer together than the radius of the first dark ring of the diffraction pattern.

Diffraction theory developed by Rayleigh shows that the least separation that can be resolved in terms of the Rayleigh adopted criterion such that the peak of one does lie in the center of the first dark ring of the other, in linear terms is given by

$$d = 0.61\lambda \, /\sin u' \qquad (3.55)$$

where λ is the wavelength of the light, and u' is the absolute value of the slope angle of the marginal ray passing through the effective free aperture under consideration.

It can be shown that for an object at infinity,

$$\sin u' = h/f \qquad (3.56)$$

where h is the radius of the effective free aperture of the lens in object space and f is the focal length. This equation is also a measure of the correction of the lens system for coma in that a coma-free system obeys the equation for all values of h from axis to the edge of the aperture.

This equation led many years ago to the concepts of relative aperture and of f-number. The relative aperture is defined as the ratio of the focal length of the lens to the diameter $(2h)$ of its effective aperture. The symbol for relative aperture is $f/$ followed by the numerical value of the ratio just defined. Relative aperture is written as a fraction. For example, $f/2$ signifies that the diameter of the effective aperture is one-half the focal length. Foreign manufacturers prefer the designation in which 1: is used instead of $f/$. Thus, the American $f/2$ is equivalent to the German 1:2.

The f-number is defined as the number in the expression for the relative aperture. Thus, if the relative aperture is $f/2$, the f-number is 2. From these definitions and equations 3.56, it is obvious that the following relationship exists when the object is at infinity:

$$f\text{-number} = 1/(2 \sin u') \qquad (3.57)$$

where the absolute value of u' is used.

Equation 3.45 can be rewritten in terms of the f-number as

$$d = 1.22 \, \lambda \, (f\text{-number}). \qquad (3.58)$$

When the object is at a finite distance, the effective f-number is used instead of the f-number defined above. The effective f-number is determined by the following equation:

$$\text{Effective } f\text{-number} = (1 - m) \, (f\text{-number}) \quad (3.59)$$

where m is the lateral magnification and in accordance with the sign conventions of this chapter is usually negative in photogrammetric instruments (Kingslake, 1951). Accordingly, use of a lens system at finite conjugates increases the f-number and reduces the photographic speed of the lens. Physically, $\sin u'$ is decreased at the image plane at finite conjugates with the same clear aperture of the real stop.

3.1.3.6.2 Relationship to Depth of Focus

The concept of depth of focus is based on the idea, confirmed by observation, that the film or the projection screen does not have to be in the precise theoretical position. Over a range about the theoretical focus the observed diffraction resolution remains virtually constant. It can be shown that within this range the Rayleigh criterion applies, namely, that no optical path length from any given point on the pupil to an image point differ from any other by more than a quarter of the wavelength used. When this condition holds, the diffraction resolving power remains effectively unchanged.

The range through which the image plane may be shifted without loss of resolution is denoted by d. Theoretically, the tolerable shift is the same in both directions.

If one assumes the wavelength of 589 nm, the Rayleigh criterion leads to the following expression:

$$d = 2.4 \ (f\text{-number})^2 \times 10^{-3} \text{ mm}, \qquad (3.60)$$

which is a result derived from diffraction and therefore applies to the live or aerial image. A less specific approach makes use of the concept of circle of confusion. If the actual image plane is shifted from the plane of best imagery by more than $d/2$, the image ceases to be as small as that defined by diffraction theory for an image in focus, particularly if afflicted with aberrations, and instead approximates a circle of a measurable diameter C. If this diameter is below the threshold of resolution of the normal eye from a viewing distance of 250 mm, one cannot therefore distinguish the circle of confusion from a point image.

The depth of focus derived from the threshold of vision is determined by

$$d = 2C \ (f\text{-number}) \qquad (3.61)$$

and is based on simple geometry. On the other hand, if magnification is used or if the viewing distance is decreased, the corresponding depth of focus must be decreased. Even for the diffraction image one can speak of a circle of confusion as one containing some designated percentage of the total light energy in the image and the depth of focus will then depend on the plus or minus focal increment for which this percentage is effectively constant.

A third approach was developed by Washer (1945). According to his results the depth of focus is determined by

$$d = 4 \ (f\text{-number}) \ / R \qquad (3.62)$$

where R is the resolution in line pairs/mm. This expression has the advantage that for an assigned level of resolution the corresponding depth of focus can be stated. Moreover, as long as the particular lens has the resolution R or better, anywhere in the field, the depth of focus

is determined as a by-product of what one needs to know for work at the resolution level R.

A fourth method is dependent on the modulation transfer method which is discussed in section 3.1.3.10 below. This fourth method is also applicable at any given level of resolution R and has the further advantage that the contrast level can also be determined.

Because new photogrammetric applications require near perfection to record small detail from extremely high altitudes, new instruments intended for such critical photogrammetric applications should be focused to provide the best overall image quality, irrespective of the tolerable depth of focus. Nevertheless, some knowledge of the depth of focus at a given level of resolution is desirable as a measure of the sensitivity of the instrument to thermal and chromatic focal changes.

Depth of field is another concept associated with the depth of focus. It is the range in object space corresponding to the focal range in image space. As for the latter, depth of field depends on the adopted circle of confusion, the mean distance for which the lens is focused, the lens focal length, aperture-ratio or f-number and on the state of correction for the various aberrations.

A strong warning is voiced with regard to the significance of the tables for depth of field frequently available from the manufacturer's leaflets describing lenses (see also Hardy & Perrin). The tables generally depend on the adopted circle of confusion which at best is a somewhat arbitrary matter and may not be closely related to the ultimate quality of a lens. Some users of lenses tend, however, to conclude that a smaller circle of confusion used by a manufacturer for a certain lens indicates that this lens is better than a similar lens of some other manufacturer. The only significance of a smaller circle of confusion is that a manufacturer may want to secure for the user very sharp pictures at the far and near ends of the object field. This intent necessarily results in the indication of a depth of field of a lesser extent in comparison to the depth of field that would result if a larger circle of confusion had been adopted.

3.1.3.6.3 Relation to Spectral Region

The spectrum normally associated with aerial photography is restricted. Nevertheless, for modern highly corrected lens systems it is desirable to include an optimal correction for chromatic aberration over a range from the blue-green portion of the spectrum down to the deep red or even infra-red. The diffraction resolution averaged out in equations given above does indeed depend linearly on wave-length. Accordingly, for diffraction-limited systems the ultimate resolution in the blue-green may be as much as 60% greater than in the deep red end of the spectrum, insofar as the aerial image is concerned. Unfortunately, the atmosphere scatters blue light far more than red, particularly under dusty conditions, which in turn through filtering causes the effective wavelength to move toward the yellow and red where the diffraction resolution is lower. Photographically, the emulsion may also have a lowered sensitivity in the green, or frequently also, an enhanced sensitivity in the ordinary red. The effective image is thus likely to be slightly degraded with respect to what the ultimate resolution might be if the same highly corrected lens system were used in the absence of an atmosphere as for space research.

Clearly, it is essential for the lens designer to know the full range of conditions that the lens is intended to meet, whereafter he must strike the best balance possible. Historically, some of the aerial lenses used in the early years of aviation still retained the "photographic" correction for chromatic aberration which had an optimum resolution in the blue-violet, owing to the fact that the designs were largely scaled up versions of lenses used for close range work with emulsions insensitive to the red.

With the advent of World War II, designers introduced a variety of new types of aerial lenses, including wide angle systems for photogrammetry, and optimized these designs for the actual conditions of aerial photography. Under modern circumstances such optimization can be carried out with quantitative knowledge of every aspect of the lens-film-filter combination.

3.1.3.7 HIGHER ORDER ABERRATIONS

The relationships between the x and y intercepts on the image plane, and the x and y intercepts on the object plane and entrance pupil respectively, can be expressed mathematically as functions of several variables. For purely rotational systems as necessarily used in wide angle mapping objectives, the intercept functions in either x or y involve monochromatically three independent rotational variables that can be replaced by an equivalent function of four partially dependent variables. These variables may indeed be the x and y intercepts of the given ray on the entrance pupil, and the x and y coordinates of the object point on the object plane, (or for an infinitely distant object plane, direction numbers or vector components defining the direction of the given point as seen from the center of the entrance pupil.) For both optical and mathematical reasons, however, these simple variables are usually replaced by linear combinations (Seidel variables) that help to refer intermediate quantities to relationships in object space alone.

One finds that because of the rotational symmetry the function in either x or y expressing the intercepts of the ray on the image surface expands into a power series in terms of odd degree.

The first order terms comprise "Gaussian" optics and involve all linear relationships and cardinal points of a lens system. The third order terms comprise the "Seidel" optics and express in sequence spherical aberration, coma, astigmatism, curvature of field, and distortion.

The fifth order terms were first set forth in their full analytical meaning in 1905 by Karl Schwarzschild, the German astronomer, although many fifth and higher order terms had been derived much earlier by Petzval and others. Indeed, in modern theory seventh order terms are not considered intractable for use on computers and expansions of even higher order have been achieved. Nevertheless, methods involving the seventh and higher order terms are of limited value, unless expressed iteratively, and except for special purposes are not as effective as the use of precise ray-tracing analyses combined with multi-parametric differential corrections progressively applied.

There are nine independent, monochromatic, fifth-order aberrations. Schwarzschild noted an axial spherical aberration of the fifth order, a sine theorem type of coma of the fifth order, two forms of oblique spherical aberration, (the first being the reappearance of spherical aberration off-axis, whether or not absent on-axis, and the second being the appearance of a figure-8 kind of aberration), two forms of oblique coma, (the first being the reappearance of third order coma in the outer field, and the second being a new kind of radially extended image or streak), fifth order astigmatism, fifth order curvature of field, and finally fifth order distortion. There are 14 monochromatic terms of the seventh order. In addition, the higher order aberrations are affected by lower order aberrations of the exit pupil.

All of the above mentioned terms are also dependent on the wavelength of light whenever refractive components are employed in the optical system. Even for all-reflective systems the wavelength of light must be considered in the final, fully optimized formulation, inasmuch as the actual physical image is formed by diffraction.

It is possible also to express the aberrational equations from the beginning in terms of the error in optical path at the exit pupil along an arbitrary ray between the actual emerging wavefront and the desired spherical wavefront centered on the indicated image point. Indeed, this procedure in Hamilton's hands evolved into the "characteristic function", and in Bruns' treatment was independently put forth at a much later time as the "eikonal", used by Schwarzschild (1905) as a first step in arriving at his own formulation of the intercept errors. The equations of condition reduced to simplest form are essentially the same for the respective aberrations treated either as coefficients of power series terms in the even-ordered wavefront procedure or the odd-ordered intercept approach. However, for non-zero terms the optimization procedures for balancing lower-order terms against non-zero higher order terms are appreciably different, and indeed in such optimization, at least for systems of diffraction-limited performance, the wavefront method must be used as corresponding most closely to realistic physical processes.

Wide angle mapping lenses are necessarily highly corrected for all asymmetric image errors including lateral chromatic aberration, all coma-like terms, and above all, distortion. In addition, mapping lenses of moderate speed must be well corrected for astigmatism and curvature of field through higher orders. Still faster mapping lenses must further be corrected for longitudinal chromatic aberration, spherical and oblique spherical aberrations, and "fine tuned" against higher order residual errors of all kinds, including chromatic variations and asymmetric displacements of the centroid of the off-axis image affecting the distortion correction.

Mathematical representations of the fifth and higher order terms have been achieved by Kohlschutter, Wachendorf, Herzberger, Buchdahl, Matsui, Bennett, Barakat and others. In addition, other formulations of image errors have been achieved, such as with Zarnike polynomials (Zarnike, 1934), wavefront differential geometry and the like, and extensions of theory have been applied to non-rotational systems, variable index media and holographic equivalents of image structure, to name a few.

3.1.3.8 Lens Design on Electronic Computers

Programs for automatic lens-design on large computers have been in existence for more than 25 years and have received continued attention here and abroad. At the present time there are several excellent computer programs for lens design available on a rental basis. In addition, large optical companies, and many individuals and consulting firms have their own specialized optical design programs of various degrees of flexibility and sophistication.

While there are a few programs that work from first principles of optical design, employing the iterative solution of many equations of condition for the values of the parameters of the lens system, it is clear for reasons of cost, time and viability that some human guidance must be involved. Otherwise, the computer each time must recalculate a multitude of already known relationships and will therefore at considerable cost ultimately reject many unacceptable solutions evaluated in detail before it can arrive at a satisfactory answer to the stated problem.

Generally, the optical designer is free to insert some approximate lens form that he has reason to believe will lead to an acceptable solution for the problem at hand. Such an experienced beginning already sets aside an infinitude of useless automatic computer probings. The computer's task thereafter is to converge as cheaply as possible upon an optimized and hopefully successful solution. If the fully optimized solution still is unsatisfactory, the designer must try to understand why this should be so and introduce the physical changes in the number and general location of refractive or reflective surfaces that may lead to success.

The optimization is often expressed in terms of a "merit function", defined by the designer and generally involving the aberrations expressed in terms of tolerance units or weighting factors. The parameters can be evaluated by some process of least squares minimization of errors or by a path of steepest descent in error space leading to a minimum of the merit function. The performance of the design at this stage can then be analyzed in the fullest detail and in terms of the applicable specifications. The entire process can be included in the programming of a modern computer but good judgment must be used. A few well-selected rays carefully controlled by an experienced designer working with the help of a small calculator can often take the place of a multitude of preliminary operations on a large machine, particularly in the early stages of a design.

It is a function of the optical designer to recognize where his own expertise can be useful and where it is best to allow the computer to do its own kind of work, and then to blend the two capabilities in a favorable way. A version of optical design "chess" ensues. In the final stages of a design the optimization inevitably must be left to the computer, for no one can dependably foresee the results of the interplay of dozens of non-linear equations of condition involving many parameters.

If an additional parameter or set of parameters is needed, or if some other lens form remains to be explored, the designer inserts the requisite starting data or changes and the computer once again is called upon to carry out its role. In complicated cases the designer may find it necessary to introduce one or more aspheric surfaces, even though to do so generally commits the lens system to limited and costly manufacture.

Often one finds that the required performance is more than satisfactorily met by the design completed on the computer. In such instance the lens system may be too elaborate and a simpler solution might well be sought. Frequently, the designer seeks some least costly design which may not always involve a minimum number of lens elements.

Another role of the computer is to analyze existing optical designs, or to analyze a given system for fabrication tolerances, or to perform a much more detailed analysis of diffraction image structure and transfer function performance than can be interpreted by one from the merit figure alone. Some programs employ optimization procedures that are combined with image analysis and transfer function evaluations but such a combination used throughout a design effort may lead to a considerable increase in computing costs without any corresponding effective increase in the final quality of performance.

The most advanced automatic lens design programs allow for continuous and discontinuous changes in parameters, including glass types, exact and least squares handling of the aberrations, weighting factors to express the relative importance of aberrations, soft and hard limits and elastic boundaries in parameter space, physical or shape conditions and parameters, such as clear apertures, edge thicknesses, central obscurations, if any, for non-circular sections and the like and may also allow for desired values of physical quantities expressed as target values in additional equations of condition, in combination with the weighted aberrations.

Automatic lens-design programs are usually stored on disc in the largest computers and are entered through card readers, tape, or keyboard terminals.

In companies and in consulting work the question of overall costs must be kept in mind both in the tasks of design and in the consequences of fabrication of any given design or designs. If production is involved, one must anticipate minimization of production costs quite often in discontinuous ways that cannot be readily programmed. For example, one may often wish to draw upon already existing standard parts selected from a catalog as components of the new design and it may not be possible to know in advance which of perhaps hundreds of available parts may prove most useful. It is for such reasons that advances in design techniques on a theoretical basis are likely to continue to occur in scientific, university and governmental laboratories where computing costs and time expended may not be of primary concern. Conversely, advances in design on a cost basis, that is, in the sense of maximum return or cost effectiveness, are likely to continue to occur in industrial and applied laboratories.

3.1.3.9 TOLERANCING AND ERROR BUDGETS

No matter how mathematically perfect a given optical design may be, the manufacturer is faced with a myriad of practical problems in reducing the design to prototype form and to production. Accordingly, the designer must anticipate as many of these difficulties as possible and plan his design to have minimal sensitivities to errors of fabrication. Indeed, under modern conditions the designer may receive a set of requirements for the optical system to meet that often includes some form of budget for various kinds of errors.

More generally, the designer is included in the planning loop and a final error budget worked out in accordance with practical considerations.

A common difficulty in preparing an error budget is the matter of separating out the allotments to the various kinds of errors, such as tolerances on lens curves, on thicknesses, on indices of refraction, on spacings, on tilt and decenterings, on thermal sensitivity of focus, on weight, position of center of gravity and many others. Manufacturing errors divide into those that are random in nature and those that may have a systematic trend. Random errors are statistical and can be treated mathematically. Systematic errors take different forms and must be given threshold values or limits.

If all errors are treated as if systematic, a set of tolerances spread out over as many as one hundred sources of trouble may become frighteningly small. That is to say, one can investigate what each kind of error can be in the limit before the overall system becomes degraded in some respect to an assigned limit, as for example, resolution, distortion, decentering and so forth. With systematic errors so evaluated, one would then have to divide each error range by the total number of pertinent errors, say, by one hundred, with the result that the range for each error becomes unacceptably small or the overall cost of manufacture unacceptably large.

If, however, the many sources of error are treated as if random, one can divide the individual range by, say, the square root of the number of errors, let us say, by 10 if there are as many as 100 sources of error. The error range for any given parameter thereby becomes far more tractable than before and the cost of manufacture drops accordingly. Such a treatment means that a system assembled from many parts, each having a random error, may as a system have a random spread also in system performance. In a production run a certain percentage of systems will then yield a performance that is degraded beyond the specified system's tolerance in some one respect, say again, resolution, or distortion, or some other matter. These degraded systems may then have to be discarded if the cost of rework is higher than that of the mass-produced satisfactory unit, or else given special attention in a selective recycling and reassembly. Often in the case where a system fails, one can quickly find the few elements that are unusually degraded and once again either rework these or replace them with satisfactory units. A production run so treated and improved will then have fewer rejections of the reworked systems.

If the manufacture of a system is unusually elaborate and spread out among many companies and suppliers, then it becomes necessary to plan an error budget that for the most part is made up of systematic-error ranges. It is for this reason that the overall manufacture of space instrumentation becomes so very expensive. Each supplier must set about meeting critical standards as if there is no possible compensation in some other respect. Conversely, insofar as optical components are concerned, it is therefore important to have as much of a single component manufactured under a single roof as possible.

3.1.3.9.1 INFLUENCE OF ABERRATIONS

If one confines his attention entirely to the subject of optical manufacture, then it is important for the designer to set up his own error budget for the optical aberrations of importance in the problem. Designers of microscopes will need to concentrate therefore on obtaining diffraction-limited performance over a very small field but from a large spectral range at a large numerical aperture ($n \sin u'$). Designers of wide-angle mapping lenses will need to concentrate on meeting critical requirements for distortion correction, while at the same time obtaining an optimum average resolution in the tangential and radial directions at an acceptable level of contrast.

The designer will also need to keep in mind and to incorporate into his calculations suitable criteria for tilt and decentering errors for the purpose of minimizing these during the course of the design effort. An optimum design is therefore likely to be one that can be manufactured with relative ease, although in practice such a design is soon met with tightened requirements, or with need for a faster lens speed, or bigger size, or used with an emulsion of higher resolution in the lens-film combination, and so forth. The designer must also anticipate changing conditions and allow for a trend, if feasible, toward more exacting requirements at a later time. It is important for a designer to realize that he is a part of a larger design loop and that unanticipated requirements down the line may come back to present him with a more difficult design problem with little time remaining for redesign. A successful early design may lead usefully to an improved systems performance that may lead to new possibilities in application that may stimulate others to ask for more critical optical design requirements with even several cycles in the loop.

3.1.3.9.2 COST CONSIDERATIONS

In commercial applications and to some extent in military applications one must always keep in mind any known relationships between design choices and cost or arrive at such relationships during the course of the design effort. For example, a designer of a large optical system may know that use of a certain exotic optical glass of high index would allow him to arrive at a diffraction-limited performance with a minimal number of elements. However, if the cost of a raw blank of optical glass of large size runs into many thousands of dollars, clearly the designer must look for reasonable alternatives. Similarly, many optical systems are associated with high maintenance costs, owing to climatic and environmental conditions. The designer must endeavor to use optical glass that will stand up to prolonged atmospheric attack and even to mishandling, according to the nature of the application.

Optical design is very often a series of successive approximations that include cost factors as well as performance. An initial design with, say, 10 elements may have to be followed by one not so good but with, say, only 7 elements. It may be that a carefully manufactured 7-element system will perform as well as a 10-element system put together with appreciable errors and misalignments. The designer must combine calculations and practical experience to arrive at a favorable result.

3.1.3.10 TRANSFER FUNCTIONS AND LENS ABERRATIONS

One of the most useful tools for the evaluation of the performance of an optical system was developed as early as 1948 by Schade of RCA by analogy with earlier mathematical studies in the field of communication theory. Schade's work was quickly followed by other authors to the extent that the theory and application of the optical transfer function quickly became available to all (*see* Schade, 1948; McDonald, 1961). Before the optical transfer function was introduced, optical systems were tested by viewing a test-target of rather arbitrary form thru the system, and analyzing the characteristics of the image. There was a wide variety of forms for the test targets and an equally wide variety of opinions on how the image should be analyzed and what the results meant. The optical transfer function method of testing replaced other methods because it provided a mathematical basis for the construction of single-type of target applicable to all kinds of systems, it provided a mathematically justifiable method of analyzing the results, and it was very adaptable to use by high-speed computers.

The success of the optical transfer-function method rests on two facts:

(a) any scene, no matter how complicated, can be approximated reasonably well by a Fourier series; and
(b) the Fourier series representing the image of a scene has its constants related to the constants of the series representing the scene in object space by a function (the optical transfer function) that depends on the optical system and not upon the scene or its image. In addition, Fourier series are easily manipulated.

It is therefore convenient to have a test target in the form of a set of sine waves of decreasing wavelength and therefore of increasing frequency. In the target itself, the amplitude of the sine waves remains at unit value, being defined as a contrast function. A relative image-contrast of unity, therefore, means that the high light in the wave is at unit response and the intensity in the trough is zero.

The image formed by an optical system will, to a good approximation, remain a set of sine waves with the same wavelengths but now will be found to be degraded in terms of the contrast. The peaks will no longer be at the high value of an ideal image of the target and the troughs will be partially filled with light. A typical term in a Fourier series has the form $A_n e^{2\pi i (\phi - \phi_n)}$ in which A_n and ϕ_n are constants and n is the "frequency" of the term. A_n gives the amplitude of the term and ϕ_n gives its phase. A full analysis would require determining both the set $\{A_n\}$ and the set $\{\phi_n\}$. It turns out that most of the information, such as resolution, that is photogrammetrically important is contained in the $\{A_n\}$. It is therefore usual to split the optical transfer function into two parts—the modulation transfer function which describes the effect of an optical system on $\{A_n\}$, and a phase-transfer function, which describes the effect of an optical system on the set $\{\phi_n\}$. Only the modulation transfer function is then used in the calculations.

A plot of the modulation transfer function (MTF) can therefore be used to represent the performance of an optical system, whether arrived at purely by calculation or by observation in the laboratory. Different authors may refer to the MTF in other ways, such as the response function or the spatial-frequency function, and the like. These plots may either have the resolution in line-pairs per millimetre, the usual way, or may be presented in linear terms of the decreasing spacing. A sample of such a graph is presented in figure 3-20.

It can be argued as to what the best form of MTF may be for a given application, such as for aerial mapping. One must note that distortion is not a part of such a graph. Insofar as resolution is concerned, the designer may have a system where the MTF reaches quite high resolution values, such as 200 line-pairs per millimetre but at the expense of a lowered contrast even at moderate frequencies. Some other system may have a cut-off at a relatively conservative limit, say at 100 line-pairs per millimetre but may have a better contrast in the lower frequencies, as compared to the first system mentioned. Clearly, if aerial pictures are limited for either system to perhaps a level of 50 line-pairs per millimetre because of haze and vibration or because of the need for a relatively fast emulsion, then the second system may yield better aerial pictures for practical usage than the first.

Some designers have opted for a maximum area under the MTF as a working criterion and others have done the same but to a cut-off frequency determined only by the conditions of use. That is to say, if a resolution of 200 line-pairs per millimetre is not in the least needed, then maximizing the area under the MTF curve from the origin out to, say, 100 line-pairs per millimetre may be the thing to do.

The MTF for a diffraction-limited aperture falls monotonically from unit response at the origin to the cut-off frequency determined by the diffraction considerations discussed earlier. In the presence of lens aberrations the calculated MTF functions may show large dips in the MTF curve at intermediate frequencies, as in figure 3-20. It may be, however, that the resolution will reach a quite high value before the natural cut-off caused either by diffraction or by the onset of overwhelming aberrations. A shadowing of a portion of the aperture, as for example, in a mirror (katoptric) system may also lead to a large dip at intermediate frequencies.

One must consider that the transfer function of the system may be very superior to the transfer functions of the component parts. This is particularly true of transfer functions for the intermediate object and image conjugates throughout a compound system. The individual lens surface gives rise to a very poor MTF, but the task of the designer is to combine many surfaces to give a very favorable final MTF.

On the other hand, transfer functions of unrelated components must be multiplied together which inevitably leads to a degraded final MTF. Thus, a modulation transfer function can be observed for an emulsion which when combined with that from a lens system will lead to an MTF for the lens-film combination that is quite degraded. One can multiply in a transfer function for vibration, image motion, air turbulence and the like.

A simple loss of contrast in the target, such as aerial haze overlying the ground target, can be handled in the final MTF as a direct multiplication, leading to a lowered contrast and lowered cut-off frequency in the final photograph.

A further consideration is that of the so-called

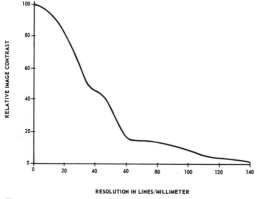

FIGURE 3-20. A transfer function graph. The curve shows that the capability of a lens to resolve smaller detail is necessarily associated with deterioration of contrast.

AIM curve for an emulsion. One can test any given emulsion for the minimum target contrast that will enable an observer to see a given frequency resolved. At low frequency, very little contrast is required and the amplitude of the AIM curve is small. As the line-number per millimetre increases, a given contrast in the test target increases before the observer can see that lines at that frequency are resolved. If the AIM curve is plotted onto the standard MTF graph, it will generally rise slowly to the right until, because of grain and turbidity within the emulsion, at some line frequency no amount of contrast in the target will produce resolution.

The AIM curve can be evaluated using a photograph taken through a microscope objective of a quality far better than that of the aerial lens to be used later, or can be determined also by contact printing of a suitable sine-wave target. The first method is to be preferred but close attention must be given to perfection of technique. Ideally, one should evaluate the AIM curve under perfect imaging conditions for the same f-number, filter and color temperature of the light source as those for the aerial lens to be tested.

If the AIM curve is plotted on the same graph as the MTF of an aerial lens, the intersection point at a given line-frequency of high value will generally correspond to the cut-off frequency of the lens-film combination. In practice, this is a random quantity and may vary from one laboratory to another. The presence of grain which in turn depends on development and type of emulsion may lead to a fluctuation in the AIM curve. The type of light source, filtering and kind of test exposure may also affect the point of intersection. Thus, the results from various laboratories will agree only if sufficient tests are made and if conservative readings are made of the resolution observed.

Some designers have gone to the trouble of incorporating the response-function as a criterion during the course of an optical design. However, this procedure is at best a kind of Monte Carlo procedure in that the response function is an integration of aberrations. Different aberrations may have approximately the same effect on the MTF, or conversely, a particular dip in the MTF does not tell unequivocally what aberration or combination of aberrations may be at fault. The MTF is indeed extremely useful in guiding a designer as to the quality of his results, even along the way, and may demonstrate to the designer that he must work further to obtain a satisfactory MTF curve. The MTF is also even more useful in conveying the results obtained by the designer to another worker downstream. As a design tool, however, it is lacking in analytical power.

One useful aspect of the MTF is its ability to portray the contrast versus focal setting at a given level of resolution. Let us say that one is interested in the focal range at a level of 40 line-pairs per millimetre. If MTF curves are prepared at a sequence of focal settings, both inside of and beyond the optimum focal setting, the contrast can be read from the graphs at the 40-line level, say, and plotted as a function of focal setting. The shape of the curve at different discrete line frequency levels can thus guide the user both as to focal range and to the quality of the design for its intended purpose.

The MTF is useful to the photogrammetrist in providing him with aerial pictures of an optimum level of contrast at a favorable level of resolution. While absence of distortion is of critical importance to the success of the final mapping, along the way easily interpreted images also add to the quality of the final product and to a minimization of errors.

3.2 Lenses for Photogrammetric Applications

3.2.1 DESIGN

3.2.1.1 LENSES FOR IMAGING DISTANT OBJECTS

Photogrammetric lenses designed for imaging objects at great distances, can be found in aerial cameras, ballistic cameras, and in panoramic cameras intended for aerial usage. This class of lenses, among the most difficult to design, is also among the most difficult to manufacture. The optical glass itself must meet rigid standards of homogeneity and minimization of striations, strain, bubbles and seeds. If the production run causes variation of indices and dispersions, then surface curvatures and separations may have to be adjusted selectively to restore the performance above the required threshold.

The techniques employed in the final man-

ufacturing phases of the newer photogrammetric lenses approach the status of an art because the minimizing of tangential distortion and asymmetric radial distortion as well as the correction of other aberrations depend on the precise alignment and spacings of eight or more optical elements.

Utilization of cameras under conditions where vibration or shock may be experienced or where temperature and pressure differ drastically from that of the test laboratory, places added burden on the designer and manufacturer. After the prototype has successfully passed the optical, photographic, performance tests in the laboratory, it is subjected to further performance testing under the extremes of the specified conditions. The results of these tests can reveal problems whose empirical solutions dictate changes in design or mounting. Tied in with environmental problems are those of functional reliability, where the reliability demanded of the lens under a range of conditions (shock, vibration, vacuum, temperature) may be only slightly less than 100 per cent.

Thus, a higher order of quality in the new photogrammetric lenses with respect to both image quality and performance is available to engineers for terrain and space projects. Table 3-1 is a listing of some notable photogrammetric lenses. The data given include only information on photographic performance and a few characteristics significant to engineers or physicists (not actual size, shape, weight, or performance under working environment).

3.2.1.1.1 LENSES FOR MAPPING CAMERAS

Lenses for single-lens or multiple-lens mapping cameras have the following general characteristics.

a) They are designed for an infinite object-distance or under special circumstances may be slightly refocused for a mean altitude, temperature and air pressure in the camera compartment.

b) Three classes are normally distinguished: (1) normal-angle (fields less than 75 degrees); (2) wide-angle (fields between 75 and 100 degrees); (3) super-wide-angle (fields greater than 100 degrees).

c) Maximum relative apertures are as follows: (1) normal-angle lenses - $f/4.0$; (2) wide-angle lenses - $f/4.0$; (3) super-wide-angle lenses - $f/4.0$. In all cases, larger apertures are desirable because they permit the use of fast shutter speeds, thus reducing image motion during exposure. Because of the wide variation in the amount of light passed at large angles by lenses from the Planigon through the Geocon to the Aviogon types, f-numbers should be supplemented by T-numbers if anti-vignetting filters are to be used and if proper comparisons of lens performance are to be made. (*See* table 3-1 and section 3.5.1.5.)

d) A large angular field is preferred if the need for very-high-resolution photography is not paramount. A large field requires less flying, fewer photographs, and less ground control, and gives higher mapping accuracy.

e) Focal lengths are usually short (38 to 305 mm). The standard 23 × 23 cm format furnishes photography which fits the standard projection and stereoplotting instruments.

f) Lenses should be as free of radial and tangential distortions as possible. Lenses with distortion greater than 0.010 mm require compensation or matching with plotting apparatus. (*See* section 3.4.1.5).

g) Lenses should have high resolving power over the entire field when tested with the emulsion to be used. Unfortunately, there usually is a trade-off between large aperture and wide fields

with respect to the maximum distortion, or else the optical system may become large, heavy, expensive, and sensitive to the environment.

h) The focal plane of the lens should be flat so that resolution at the center need not be degraded to increase the resolution at the edge, a procedure that can wipe out the depth-of-focus tolerances on low flying missions.

Table 3-1 includes a number of lenses which meet the above requirements. Much of the history of the development of these lenses, or their forerunners, is available in chapter II of the second edition of the MANUAL OF PHOTOGRAMMETRY.

Several lenses in particular are worthy of note. (1) The Wild Universal Aviogon II and the Super Aviogon II are relatively new lenses that carry the art of wide angle design very close to attainable limits (*see* Schlienger, 1972). The distortion residuals are nearly ideally low, the resolution on a flat focal plane improved over earlier formulas, and the illumination characteristics excellent in view of the large field angles involved. The fall-off in amount of light passed at large angles is far less than that indicated by the $\cos^4$ law, obtained by having the entrance pupil actually larger for inclined rays than for the on-axis bundle. The configuration of the lens is such that considerable positive power lies in the central grouping of elements, which are surrounded front and back by negative elements. This type of construction leads to large lenses but with the optical advantages already indicated. (2) The Zeiss Pleogon and S-Pleogon A lenses are remarkably well-corrected for use with panchromatic, infrared and color films. The older forms were $f/5.6$ and the newer are now $f/4.0$. The resolution and image quality are high and the lens, while not quite as large as those of the Aviogon series, has comparably good illumination over the field (Zeiss Catalog). (3) The Geocon I and IV wide-angle mapping lenses represent a form of lens intermediate between the compactness of the old series, such as the Planigon, and the rather large size of the Aviogon and Pleogon types. While preserving good resolution on a flat focal plane, the Geocons are of moderate weight and diameter and have distortion reduced to minimal values. The illumination fall-off, while much superior to that for the Planigon, cannot be nearly as good as for the Aviogon, owing to the intermediate nature of the lens form. Indeed, the $\cos^4$ law is obeyed rather exactly.

3.2.1.1.2 LONG-FOCAL-LENGTH LENSES

In the earliest practice of aerial photography lenses designed for studio work or for hand-held cameras were scaled up in focal length without appreciable redesign for the new applications. As a result, the impression held for many years that one could not expect to obtain crisp image quality from large lenses. At the same time,

TABLE 3-1. PHOTOGRAMMETRIC CAMERA LENSES

Trade Name	Manufacturer	f and f/no.	Field-coverage	Format of Negative	Max Radial Distortion	Resolution	Remarks
Metrogon	B&L, Fairchild Goerz, AO	152 mm f/6.3	Wide Angle	23 × 23 cm	0.110 mm	Medium	Anti-vignetting filter @ T/2
Planigon (or Cartogon)	B&L, Fairchild Goerz, AO	152 mm f/6.3	Wide Angle	23 × 23 cm	0.025 mm	Medium	Anti-vignetting filter @ T/2
Planigon	B&L	38 mm f/6.3	Wide Angle	57 × 57 mm	Less than 0.010 mm	Medium	Anti-vignetting filter @ T/2
Geocon I	Fairchild, P-E Goerz, A.O.	152 mm f/5.6	Wide Angle	23 × 23 cm	0.010 mm	High	Anti-vignetting filter @ T/1
Geocon IV	Kollsman, Fairchild	152 mm f/5.0	Wide Angle	23 × 23 cm	0.005 mm	High	Anti-vignetting filter @ T/8
S-Pleogon A	Zeiss	85 mm f/4.0	Super-wide angle	23 × 23 cm	0.007 mm	High	Anti-vignetting filter @ T/8
Pleogon A	Zeiss	153 mm f/4.0	Wide Angle	23 × 23 cm	0.002 mm	High	Pan-Infra-Color Anti-vignetting filter @ T/5
Pleogon AR	Zeiss	153 mm f/5.6	Wide Angle	23 × 23 cm	0.005 mm	High	Pan-Infra-Color Anti-vignetting filter @ T/7
Toparon A	Zeiss	210 mm f/5.6	Normal Angle	23 × 23 cm	0.004 mm	High	Pan-Infra-Color Anti-vignetting filter @ T/7
Topar A	Zeiss	305 mm f/5.6	Normal Angle	23 × 23 cm	0.003 mm	High	Pan-Infra-Color
Topar AR	Zeiss	305 mm f/5.6	Normal Angle	23 × 23 cm	0.003 mm	High	Reseau (precision grid plate)
Biogon	Zeiss	75 mm f/4.5	Wide Angle	11 × 11 cm	0.025 to 0.035 mm	High	
Rigel	OMI (Nistri)	152 mm f/6.3	Wide Angle	23 × 23 cm	0.120 mm	High	

Name	Manufacturer	Focal length / Aperture	Angle	Format	Distortion	Resolution	Remarks
Aldebaran	OMI (Nistri)	210 mm f/4.5	Normal Angle	18 × 18 cm			Pan-Infra-Color
Telikon A	Zeiss	610 mm f/6.3	Telelens	23 × 23 cm	0.050 mm	High	Lens on sepcial order
Topar	Zeiss	457 mm f/4.5	Normal Angle	23 × 46 cm	0.005 mm	High	
Universal Aviogon	Wild	152 mm f/5.6	Wide Angle	23 × 23 cm	0.005 mm	High	For pan, infrared and color photography.
Universal Aviogon I	Wild	152 mm f/5.6	Wide Angle	23 × 23 cm	0.004 mm	High	For pan, infrared and color photography.
Universal Aviogon II	Wild	152 mm f/4.0	Wide Angle	23 × 23 cm	0.004 mm	High	For pan, infrared and color photography.
Aviotar I	Wild	304 mm f/5.6	Normal Angle	23 × 23 cm	0.004 mm	High	For pan, infrared and color photography.
Aviotar II	Wild	304 mm f/4.0	Normal Angle	23 × 23 cm	0.004 mm	High	For pan, infrared and color photography.
Astrotar	Wild	304 mm f/2.6	Normal Angle	19 × 22 cm	0.026 mm		Special lens for use in ballistic cameras.
Infragon	Wild	114 mm f/5.6	Wide Angle	18 × 18 cm	0.010 mm	High	Corrected for infrared.
Infratar	Wild	210 mm f/4.0	Normal Angle	19 × 22 cm	0.004 mm	High	Same as Aviotar except for infrared.
Super Aviogon	Wild	89 mm f/5.6	Super-wide angle	23 × 23 cm	0.030 mm	High	
Super Aviogon II	Wild	88 mm f/5.6	Super-wide angle	23 × 23 cm	0.010 mm	High	For pan, infrared and color
Orbigon	Wild	80 mm f/3.5	Wide Angle	11 × 11 cm	0.004 mm	High	For space photography

propeller-driven aircraft were usually afflicted with serious vibrations of both low and high frequency which were imparted to the camera through insufficiently damped camera mounts and their efforts enhanced by inadequate shutters and slow film. Indeed, the existence of rather poor image quality from the available large lenses was tolerated for a long time because of other worse circumstances.

At the beginning of World War II scientists in aerial photography began to insist that improvements be made all along the line. While intensive work at a dozen establishments did succeed in improving lens and camera design, these improvements were slow to reach the field. It was only near the end of the war that large lenses in well designed cameras began to be used in the various theatres of war.

As in other endeavors, one must obtain a precise list of the requirements before undertaking an overall solution of optimum nature. The trouble in the past was that some of the problems were only partly understood and the solutions to others were provisional because of poor data from the field. It even took some time before scientists realized that aerial cameras were often rather roughly handled by relatively inexperienced personnel and that the ideals of the laboratory photographic scientist having to do with carefully aligned and focused cameras often went astray at a later stage. Cameras focused meticuously under stable conditions in the laboratory were "refocused" in the field by users who believed their bad pictures came from improper focusing. In some cases large cameras used under rapidly changing conditions had no reliable mean focal setting.

Cameras and film were often stored in unheated, damp warehouses, for months on end and even for years before being used. The rule seemed to be that only over-age film ever actually reached the final users. Missions were counted as successful if completed, whether or not good pictures were acquired. Tinkerers were often as successful in obtaining results as those who were supposed to be guided by training.

Under modern conditions most of these difficulties have been overcome. The very largest camera installations now are a matter of technical pride and discussions about difficulties are at a fine point far removed from those of World War II.

During the nearly 4 decades that have elapsed since photographic scientists began to delve deeply into the problems of aerial photography, many large lenses of excellent quality have been developed by literally dozens of establishments. The very largest aerial lens has a clear aperture of 762 mm, a focal length of over 6 metres and a diagonal for the format of about 1 metre. In spite of the large size the measured laboratory AWAR on fine grain emulsions is of the order of 80 line-pairs/mm.

Apart from the manufacture of a few highly specialized large lenses, production runs have been made of aerial lenses having as many as 10 elements and clear apertures of the order of 200 mm. Lens barrels have been made of lightweight materials including beryllium and titanium, and much attention has been given to the severe problems of alignment, thermal stability and thermal protection.

Until recently, inadequate attention has been given to the fact that lenses of long focal length need to be focused for the ground distance and air density differences to be encountered. Fixed focus lenses of short focal length have been greatly helped by the accidental fact that a cross-over point exists at a quite useful altitude between curves plotted for change of focus with ground distance and change of focus with air density as a function of altitude, the cross-over point being that corresponding to the laboratory focus setting for infinity under standard conditions. Pressurized compartments such as those in modern aircraft greatly affect the conditions of optimum focus at the same time that the conditions are stabilized. The scientist in charge simply must know what the conditions are to be and in turn the assigned conditions must be maintained. Some lenses mounted in berryllium, for example, have to be held within a narrow temperature range throughout their working life, whether air-borne or in storage.

Unthermostated systems containing large lenses often exhibit surprising changes in focus due to thermal gradients. That is, the focus at any given moment will depend on the rate of change of the temperature within the lens cone. Because of different rates of heat flow, even in thermostated and heated units, thermal changes of focus can be quite appreciable and difficult to predict. Such changes in focus often greatly exceed the depth of focus achieved by the optical design at f-numbers of the order of $f/4$ to $f/8$. In some cases, as will be discussed in section 3.2.1.1.2.2 and section 3.2.1.1.2.3 below, large semi-apochromatic and fully apochromatic aerial lenses contain exotic materials that usually have improved coefficients of thermal expansion and changes of index with temperature (*see* OSRD, Section 16-1).

Thus, if optimal results are to be obtained, the use of large aerial lenses requires a well-trained team of specialists. These specialists must be in direct charge of as many details as possible and must also be kept fully informed of installation problems and conditions of use. Liaison team members must provide feed-back to the laboratory team in a loop that is capable of feeding improvements into the system all down the line.

3.2.1.1.2.1 High Resolution Lenses. One could recite a lengthy history of the design and development of large aerial lenses toward the elusive goal of obtaining diffraction-limited performance when in use. However, it is more important to indicate that at present practically all of the myriad of errors in the error budget can be handled satisfactorily, except possibly the matter of thermal gradients within lenses. Insofar as the optical design is concerned, objectives having as few as 4 elements have been successfully used and some diffraction-limited lenses with enlarged fields of view and focal lengths greater than average have had as many as 12 elements. For example, a fully apochromtic, vacuum-mounted, aerial lens covering a field of view of 18° has been built and combined with a camera mount containing automatic focusing. The lens has 12 elements and operates at $f/4.0$ at a focal length of 600mm, and provides an AWAR (*see* MIL-STD-150A) on high resolution emulsion of over 200 line-pairs/mm. Much larger lenses have been built to comparable standards.

3.2.1.1.2.2 Telephoto Lenses. Telephoto lenses were developed early but first became important for aerial photography in the 1930's. The typical telephoto lens has an overall length (from vertex of front lens to focal plane) that is not greater than the equivalent focal length, and often much less. If the optical axis is a single straight line, the ratio of overall length to effective focal length can be as small as 0.75 before design becomes difficult. Lens systems with ratios less than 0.60 either contain very many elements, large aberrations, or both. Small changes in temperature of a telephoto lens cause comparatively large changes in focus, the changes being greater the smaller the ratio mentioned above.

The considerable attention given in recent years to the design of telephoto lenses has resulted in lenses whose performance approaches that of a diffraction-limited lens system. It is now possible to design a semi-apochromatic telephoto lens having an equivalent focal length of 1 metre, an f-number $f/5$ covering a 23 × 23 cm fine-grained film with an AWAR of more than 100 line-pairs/mm at a contrast-density ratio of 0.2. The depth of focus is that determined by a diffraction-limited system of the same f-number. This means that when such a lens is used, careful attention must be given to keeping the temperature and pressure of the surrounding air constant, using the proper film, emulsion, and focal setting, *etc.*

3.2.1.1.2.3 Apochromatic Lenses. As focal lengths become greater and lens speeds faster, the residual color curve of the typical aerial lens becomes a matter of serious concern. Early aerial lenses when properly adapted from the standard lenses of the studio were designed with more or less normal changes of focus with wavelength, having a minimum focus in the yellow part of the spectrum, and a common mean focus for the blue-green and red. The blue-green to ultraviolet wavelengths were filtered out by use of a minus-blue filter. The very largest lens as mentioned in section 3.2.1.1.2 has a minimum

of its color-focus curve at about 6250 angstroms (in the red) and uses an orange filter.

Both the secondary spectrum and the chromatic changes of aberration, particularly spherical aberration, become serious defects for very large refractive systems. Consequently, attention has been given in design to combinations of special glasses that can result in an improved correction for the several chromatic aberrations. Special glasses were developed more than a century ago by Schott for reduction of secondary spectrum in microscopes and in astronomical systems, and the mineral fluorite was used by Abbé and Zeiss for the same purpose nearly as long ago.

During World War II, synthetic fluorite was developed and boules as large as 100 mm diameter became available. Experimental, fully apochromatic lenses having a focal length of 0.9 metres, a speed of $f/8$ and coverage for the 23 × 23 cm standard format were developed with diffraction-limited performance over the included circle. These lenses consisted of a single central element of synthetic fluorite, cemented between elements of crown glass, the triplet so formed being of net negative power. This triplet in turn lay inside an air-spaced front element of hard crown and an air-spaced rear pair of positive crown elements. Thus, one fluorite and five glass elements were used. This form of system also had negligible distortion and a very satisfactory correction for monochromatic aberrations on a flat image field.

Unfortunately, fluorite has an extraordinarily large change of refractive index with temperature as well as a very large coefficient of thermal expansion. Tests of the large $f/8$ lens in a cold chamber indicated changes in focus as large as 5 mm under temperature conditions normally found in open camera bays of the time (OSRD, Ibid). Aerial tests were made but the change in focus dominated all other considerations. These experimental lenses have since been used for astronomical photography under far better controlled conditions.

The steady improvement of special glasses by the several manufacturers of optical glass for reduction of secondary spectrum has led to the design of many fine semi-apochromatic and fully apochromatic large aerial lenses. Such glass as KzFS-N4, first made available by Schott around 1965, represents a long awaited break-through in the production of a glass combining stability and resistance to corrosion with the properties needed for reduction of secondary spectrum, when used in combination with selected matching types. Other fine glasses are also available for the same purpose, although most are somewhat less resistant to corrosion than KzFS-N4 and require special care by the optician to prevent staining during manufacture.

Introduction of the special types FK-50, FK-51 and FK-52 by Schott has gone far toward making apochromatic design become reasonably standard. These two types in the Schott notation and their equivalents are characterized unfortunately by a very large change of refractive index with temperature (in this instance negative) and by a large coefficient of thermal expansion. Large systems making use of these types must therefore have their temperatures controlled.

The development over the past 4 decades of the rare-earth glasses has also introduced a means for controlling secondary spectrum. Most, but not all of the lanthanum-glass types have a reduced dispersion in the violet. Consequently, when they are used as negative elements, apochromatic correction can be obtained. Regrettably, these same types are needed as positive elements because of their very high indices of refraction. When so used, the secondary spectrum is actually worsened. The designer clearly must pick and choose in his forest of parameters and must combine a sequence of types carefully blended to achieve the best in performance in spite of the conflicting attributes of the optical glasses. At the same time the designer should remain highly appreciative of the variety afforded by nature in combination with the inventiveness of the glass chemist.

While large lenses are rarely used for photogrammetric purposes, at least in terms of the precision expected by photogrammetrists, nevertheless, some of the largest designs are remarkably free of distortion. The elimination of distortion depends not so much on the size of format, which can be very large for long-focal-length lenses, as on moderate field-angles. Thus, the very largest lens that covers a format of about 1 metre (diagonal) has a maximum distortion not exceeding 4 micrometres anywhere in the field. Other lenses not as large have done as well. A so-called Ross 4-element standard lens yields a distortion correction on a 23 × 46 cm format of less than 5 micrometres at any point in the field. This is a type of lens developed during World War I. A 5-element symmetrical system having a focal length of 3.5 metres, a speed of $f/8$ and a diagonal of about 50 cm has also been built and has a distortion over the field of less than 5 micrometres.

3.2.1.1.3 PANORAMIC CAMERA LENSES

There are two basic types of panoramic cameras which use drastically different lenses. The first, and most common is the camera which uses a lens with a relatively narrow field to cover a narrow slit and sweeps the image mechanically over the large (often 180°) angle of view. The second is a frame camera with a concentric lens system which forms an image on a spherical surface which may or may not be flattened into a cylinder by the use of a field-flattening element near the focal surface.

The first type has been made in every conceivable configuration that lends itself to scanning; mechanical and/or electronic, photographic and/or electro-optical, recorded and/or real time display. The forms range from a lens rotating around its nodal point sweeping the image onto a stationary sensor through a coupled slit, to that of stationary lens and slit with moving sensor in synchronism with a scanning element.

A few concentric-lens frame cameras have been built (Baker's Harvard Univ. 1941; OSRD, ibid; Aerojet Delft PACA-70, 1968; Scripto Panatoric, 1970) but have largely been ignored, possibly due to a reluctance to accommodate the curved format. Since the image is everywhere axial, a relatively simple high-speed lens can provide apochromatic, diffraction-limited imagery throughout fields up to about 150°. The relative illumination is sufficiently uniform so as not to require compensation. The image displacements due to aircraft motion during scanning are made negligible by the between-the-lens shutter. The symmetry of the image around the perpendicular chief ray assures that no displacement will occur due to variation in focal distance or exposure.

The mechanically scanning panoramic cameras have specific lens requirements for optimum operation with different forms. Thus the lens for the stationary film camera should be symmetrical around the rear node for dynamic balance, while the scanning reflector, moving-film camera wants a telecentric lens with the pupil forward at the scanning element to minimize the size of the element and the window. In between, the currently popular "split-scan" or "optical bar" configuration permits the use of longer focal lengths with no increase in the film velocity over the scanning reflector. Front and back parallel reflectors at 45° to the lens axis rotate about the axis to provide the scanning motion. The lenses are generally highly telephoto to give the longest focal length within the available space.

All panoramic lenses are inherently distortionless; the concentric lens by its nature, the scanning lens of necessity to avoid image smear during exposure. With suitable timing marks across the frame, the scanning camera can approach the concentric in performance. Both are suitable for most mapping and all charting applications and provide added strength in height from their wide angular coverage. Measurements would be made with goniometers, the data being processed in a computer. Where a mosaic is required, photographic rectifiers have been developed to provide a transformed print.

3.4.1.1.4 LENSES FOR BALLISTIC CAMERAS

Photography has been found to be one of the most reliable tools for the study of ballistics. A missile can be photographed along its trajectory. If the missile carries a flashing lamp, measurements of the negatives, plotted as a function of the absolute time of exposure, establish the curve of the trajectory. The photography is done on clear nights, the light being photographed against the stellar background. The known positions of the stars provide a metric background to which the trajectory can be referred and which can also be used for calibration of the camera. (The camera is usually kept pointed in a fixed direction so that a segment of the trajectory will appear in the photograph. The stars would then appear not as points but as thin, curved streaks. Ballistic cameras used in this way are therefore equipped with rotating shutters which interrupt the streaks at precisely known times (*see* Chapter IV). Two or more cameras photographing the same trajectory from different points provide a complete determination of the trajectory. Some ballistic cameras are not fixed in direction but follow the motion of the moving object, the azimuth and angular elevation of the direction being given by transducers. Such cameras can be used during day or night and do not require a stellar background. Accuracy, however, is not as good as that obtained with the stationary camera because of errors in pointing, errors in the angle-measuring components, etc.

Since weight is not a consideration, ballistic cameras and their lens systems can be more massive and stable than their air-borne versions. Instead of photographic film, ultra-flat glass plates are used in the fixed-position cameras. Concentric lens systems are particularly suitable for ballistic cameras when used with emulsions on spherical segments of glass. Such systems would result in higher-speed lenses nearly diffraction limited and in fields with zero angular distortion up to 150°. The photographs would be measured with a goniometer (modified theodolite) rather than with an x-y comparator.

Lenses in ballistic cameras need large apertures and well-defined images over a narrow field. For photogrammetry, distortion must either be very small or accurately known. However, cameras which track the object have such small angular fields that distortion is not really a serious problem of design. Very good ballistic cameras have been made from long-focal-length aerial lenses re-aligned and re-mounted for the purpose.

3.2.1.2 LENSES FOR IMAGING CLOSE OBJECTS

A lens designed to be used focused at infinity may not produce distortion-free images when focused for finite distances, particularly when object and image distances are nearly equal. Freedom from distortion through the entire range of magnification requires fully symmetrical lenses that are free of spherical aberration of the principal rays over the full field. It is possible for one to design such a lens, at least to a close

approximation. The principal rays in both object and image spaces appear to proceed to and come from fixed perspective centers and to have the same slope angles in both spaces for conjugate object and image points anywhere in the field.

Under these conditions the action of the lens becomes equivalent to that of the simple pinhole which is mathematically free of distortion for all projections. Lens systems that are not fully symmetrical can nevertheless be free of distortion over a wide range of ratios of conjugate distances provided that the ratio of tangents of the slope angles of the respective principal rays in object and image space is a constant over the full field and that the system is corrected both in object and image space for spherical aberration of the pupils.

Conversely, a fully symmetrical design that does have spherical aberration of the principal rays over the field will necessarily be afflicted with distortion when used with the longer conjugate distance set at infinity. An asymmetric design that is corrected for distortion at a given ratio of conjugate distances without regard to correction of the spherical aberration of the principal rays over the field or of the tangent relationship will likely have distortion for other ratios. It is for these reasons that distortion-correcting plates inserted into a fully symmetrical design ought to be placed symmetrically in the front and back spaces to reduce or eliminate the spherical aberration of the principal rays, rather than on one side only. The latter, simpler arrangement is often used to eliminate distortion at a desired ratio but is not optimum for versatile use of the lens system at other ratios.

3.2.1.2.1 PRINTER LENSES

A variety of printers is used in the preparation of diapositives for stereoplotters. Most printers are of the fixed-ratio type with provision for varying the object-to-image ratio within the limits set by focal-length variations of aerial lenses. A few foreign-made printers have provisions for adjustment to several specified object-to-image ratios. Object plane, principal plane, and focal plane are held rigidly parallel so as to preserve the perspective similarity of object and image in the process of transfer.

In the design of a printer lens one must emphasize optical performance up to the highest frequencies so as to minimize loss of information contained in the aerial negative. In special, high-performance printers, the lenses are designed for use within a narrow spectral range. They require monochromatic light and narrow-band filters for the exposure. Radial and tangential distortion compensation and symmetry in the focal plane must be achieved by adherence to the highest standards of homogeneity and annealing of the components, by one's obtaining the best possible quality of spherical and plane surfaces, and by rigorous centering and spacing of the individual elements in the lens mounts. Resolution requirements currently range from 50 to 100 lines per millimetre, referred to the object plane, and tolerances on distortion and asymmetry are at or below the 10 micrometre level in the object plane.

Inasmuch as the object plane containing the aerial negative is flat, one can design the projection printer to operate at a much smaller field angle than that inherent in the negative. Accordingly, the angular coverage and aperture of printer lenses are usually smaller than those of camera lenses. They are chosen to optimize the performance of the printer lens and illumination system.

Table 3-2 lists several types of reduction printers of domestic and foreign makers in use in the United States.

3.2.1.2.2 RECTIFIER LENSES

At first glance the optical problems in the design of rectifier lenses appear greatly intensified by the fact that such lenses are used over a wide range of magnification. At the same time, they are relieved by the fact that the end products of rectification, the rectified prints, are at best a modest substitute for a planimetric map. Being still central projections of the three-dimensional surface of the terrain which they depict, the remaining image displacements of ground features from the orthogonal projection of the map, caused by relief, by far exceed the accuracy standards of photogrammetric mapping.

Therefore it appears justified that the tolerance of radial distortion for a rectifier lens, through the range of magnification at which it is to be used, can be relaxed by a factor of 10 when compared to printer lenses. Greater emphasis can be placed upon resolution and crispness of image projected on the easel throughout the range. Rectifier lenses are designed, therefore, for high resolution and are constructed for freedom from internal reflections. Operation conditions usually demand large and variable aperture

stops and angular field coverage up to 90 degrees.

Formulas of rectifier lenses are frequently derived from those of well-known aerial-camera lenses, such as the Topogon, Metrogon, Reprogon and Dagor. The focal length of these lenses is governed by the focal length of the aerial photography, the angular coverage and the tilt and magnification requirements of the instrument they are to serve. Some of these instruments in use in the United States are listed in table 3-3 (*see* also chapter XVI).

3.2.1.2.3 PROJECTOR LENSES

Projector lenses are used in a variety of direct-projection plotters. The majority of such projectors, such as for Multiplex and Balplex plotters, are constructed with a fixed principal distance to which the aerial negatives are reduced to provide the diapositives.

Another kind of projector, such as the Kelsh and related types, has provisions for varying the principal distance to correspond to the focal length of the aerial cameras whose photography is being used. Lenses for these projectors are designed to produce distortion-free images on a flat screen at nominal magnification; that is, at the nominal projection distance which characterizes the projector type. Some of the projectors currently used in mapping practice are listed in table 3-4.

Above and below the plane of best focus (nominal projection distance) the projected image very gradually deteriorates. By the use of small, fixed apertures, the emerging pencils of image-forming rays are so sharply pointed that detrimental image deterioration is not observed by the unaided eye within a zone of approximately 20 per cent above and below the nominal projection distance.

Since projectors are used in pairs or in larger sets, their optical performance should be as uniform as possible. Manufacturing variations of

TABLE 3-2. REDUCTION PRINTERS

Maker	Max Input Format	Obj./Image Ratio	Focal Length	Aperture Ratio	Max. Output Format	Use
B&L	229 × 229 mm	1/0.222	61 mm	f/16	64 × 64 mm	Multiplex
B&L	229 × 229 mm	1/0.222	68 mm	f/16	54 × 54 mm	Multiplex
B&L	229 × 229 mm	1/0.359	100 mm	f/16	110 × 110 mm	Balplex
Zeiss	229 × 229 mm	1/3.2	200 mm	f/8	229 × 229 mm	Universal
		1/0.15	+ aux. lenses		to 54 × 54 mm	
Wild	229 × 229 mm	1/0.6 to 1/0.5	200 mm		229 × 229 mm	Universal
		1/0.5 to 1/0.3	160 mm		152 × 152 mm	Balplex
		1/0.3 to 1/0.18	60.5 mm		54 × 54 mm	Multiplex
Kelsh	229 × 229 mm	1/1	244 mm		229 × 229 mm	Kelsh Plotter

TABLE 3-3. RECTIFIERS

Maker	Focal Length	Ang. Field	Tilt Range	Maximum Aperture	Input Format	Magnification Range	Model
B&L	138 mm	80°	20°	f/6.3	229 × 229 mm	0.7 to 3.5	Autofocus
B&L	115 mm	80°	30°	f/8	229 × 229 mm	0.8 to 2.5	Manual
B&L	175 mm	80°	30°	f/8	229 × 229 mm	0.9 to 1.5	Manual
B&L	350 mm	80°	76°	f/8	457 × 457 mm	1.0	Two-stage
Zeiss	180 mm	85°	45°	f/6	229 × 229 mm	0.5 to 6	SEG V
Zeiss	180 mm	85°	45°	f/6	229 × 229 mm	0.7 to 5	SEG I
Kargl	150 mm	80°	22°	f/8	229 × 229 mm	0.5 to 5.1	Autofocus
Wild	150 mm	85°	15°	f/5.6	229 × 229 mm	0.8 to 7	Autofocus

the focal length should not exceed 0.25 per cent of the nominal value. Radial and tangential distortion measured in the plane of the diapositive should be compensated to within 10 micrometres; asymmetries should not be in excess of 5 micrometres. Because resolution requirements appear moderate in view of the present trend to extreme demands and the advanced state of the art, finite-projection lenses are available which have resolution on the axis approaching 100 lines/mm.

The most common lens type is the Hypergon, shown in figure 3-21. This simplest of all compound objectives of the wide-angle category has shown surprising results in plotters using anaglyphs for stereoviewing. Although the presence of only four spherical surfaces leaves the optical designer with little choice for correcting the various aberrations, color fringes, which are clearly visible in the projection with white light, vanish when the spectral range of the light source is limited to the selected colors (blue-green and red) (see also Kingslake, p. 39).

Lenses of higher performance are of the quadruplet type. They offer superior correction of aberrations and resolution. The Bausch & Lomb design is used in some of the 90-degree projectors and in all of the 120-degree Balplex projectors. The lenses are specially designed for the spectral range in which they are to be used and are therefore used with the filter transmitting this range. They can be used for photo-

graphic fixation of the projected image and are preferably used in making orthographic prints with the Orthophotoscope. Similar lenses of still higher complexity have been developed by Wild Heerbrugg and the Kelsh Instrument Co.

3.2.2 MANUFACTURE

3.2.2.1 GLASSES

The glasses used in photographic objectives, as in other optical elements, are specially made for that purpose. They differ from ordinary plate glass both in composition and in the manner in which they are made. The composition of the various types of optical glass differs greatly, but the basic ingredient is usually silica in the form of a fine white sand. In some special types, little or no silica is used. For instance, glasses based on the borates of rare elements such as lanthanum have high refractive indices for a given dispersive power.

All the materials that go into a batch of optical glass must be chemically pure and must be weighed and mixed under rigid control. In the classical process still used for small quantities of most glasses, the mixture is fed into a large clay pot (0.9 to 1.5 metres in diameter) which has been pre-heated in a gas furnace. After the batch melts, the resulting molten mass is stirred to remove air bubbles and to insure uniformity of composition. The pot is then removed from the furnace, covered with an insulating material, and

TABLE 3-4. PROJECTORS FOR DIRECT-PROJECTION STEREOPLOTTERS

Name	Prin. Dist.	Nominal Proj. Dist.	Magnification	Type
Multiplex	28.182 mm 30.000 mm	360 mm	12×	90°
Balplex	55.00 ± .01 mm	400 mm	7.3×	90°
Balplex	55.00 ± .01 mm	525 mm	10.5×	90°
Balplex	55.00 ± .01 mm	760 mm	15.0×	90°
Balplex	31.80 ± .01 mm	400 mm	12.5×	120°
Kelsh	150-156 mm	760 ± 75 mm*	4.5-5.5×	90°
Kelsh	150-156 mm	1000-1100 mm	7×	90°
Halcon	152 and 305 mm	1500 and 750 mm	10× and 5×	90° and 60°

* In Kelsh projectors using aspheric cam correctors to vary the principal distance during the plotting operation, the nominal projection distance variation is increased by approximately ±12 mm.

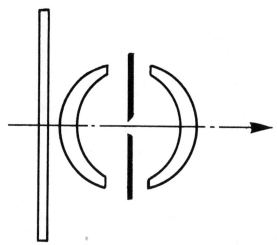

FIGURE 3-21. The Hypergon lens is the simplest wide-angle projection lens. Although it has only four spherical surfaces, thereby limiting the designer's freedom of aberration correction, it performs remarkably well at small aperture (f/16 or smaller) and with filtered light. In one design a thin planar plate is added as a third element for compensation of radial distortion.

allowed to cool slowly. During this cooling period the glass cracks into irregular chunks, which are later inspected for such defects as bubbles, undesirable color, stones, striae, and strain. Alternatively, the pot is cast into large blanks or rolled into sheets.

Glasses required in large volumes are produced in continuous tanks which consist of melting, fining, and homogenizing compartments. The fining and homogenizing compartments are lined with platinum to prevent contamination. In the fining compartment, bubbles are removed at high temperatures. Inhomogeneity created in this chamber is eliminated in the final chamber, in which stirring is conducted by platinum stirrers while the glass flows from the tank.

The continuous process yields either pressed blanks, continuous strips, or gobs, that is, unfinished pieces of sheared and dropped glass of various weights. The higher-index glasses for which no large requirements exist, are now frequently cast from platinum-lined pots. Both processes involving platinum lining are usually carried out using electrical heating, and are characterized by high yields of high quality. Optical glass from all processes is annealed (cooled in a highly controlled manner) with utmost care, since not only survival and strain, but also the homogeneity of refraction as well as the value of refractive index itself, depend upon the schedule of annealing.

3.2.2.2 MOLDING

Glass from the continuous process can be molded in rotary presses right at the exit of the tank. Gobs and pieces cut from strip or rolled sheets can be reheated and pressed. Glass from cooled and cracked pots, the classical process now rarely used except for low-volume, high-quality requirements, requires dropping (that is, sagging irregular chunks in ceramic molds) to obtain blanks suitable for pressing. Large casts are frequently diamond-sawed into suitable blanks.

3.2.2.3 GRINDING AND POLISHING

A blank obtained in one of the ways described above is ready to be ground. Blanks pressed from the molten stage require the least finishing.

3.2.2.3.1 DIAMOND GENERATING

The fastest way to prepare raw glass for optical fining and polishing is to grind the glass to the desired shape to close tolerances on a diamond mill. The diamond particles of a graded size for roughing or fining may be either of natural or synthetic stones. The natural stones tend to grind faster but the synthetic diamonds of a given grade seem to provide a more even grind. The diamond particles are embedded in a sintered and bonded matrix or sometimes are embedded in a copper base.

For diamond surfacing of spherical lenses the most common procedure is to have the diamonds on the rim of a cup-wheel whose diameter is more than the radius of the lens to be ground. The cup-wheel is mounted on a spindle of excellent quality that is heavily loaded to prevent end and side play in the bearings. The cup-wheel ultimately contacts the sphere ground onto the glass blank as a small circle of the sphere, passing over dead center to prevent a central hump or pimple and otherwise tilted at a predetermined angle that goes with the radius desired. The lens blank to be ground usually is mounted to rotate about a vertical axis at a normal speed of rotation, whereas the diamond wheel rotates at very high speed. Good cutting action is obtained if the rim velocity of the wheel is of the order of several thousand feet per minute. It is also necessary to use a special lubricant sprayed onto the work at moderate to high pressure for the purpose of washing away the ground powders that otherwise would tend to clog or glaze over the diamond grit.

The ideal bond of a diamond wheel crumbles away slowly from general wear and tear as the grinding proceeds, allowing the stones to rise slightly above the surface. If too much pressure is used, or if the diamonds are poorly bonded, or if the lubricant is not reaching the cutting line or surface, or if the bond is not crumbling away but is charging the diamonds with a layer of contaminants of one form or another, the surface will glaze over and burn and the cutting action will be slow and ineffective.

If the machine is not too noisy, one with a trained ear can determine whether the mill is working properly. A defective grind may be exceedingly dangerous to a large disc of glass in that local heat generated by poor cutting accompanied by frictional heating may lead to the breakage of the glass.

3.2.2.3.2 SPHERICAL SURFACING

Before grinding, the glass blanks for small lenses (such as those for handheld cameras, etc.) are usually assembled on and fixed by cement or imbedding in a plaster mounting called a block, the block being large enough to hold a number of blanks at one time and shaped so that the blanks are presented to the cutting tool at the correct aspect.

A properly set block of glass blanks can be ground to the desired spherical radius and depth to the precision expected of a fine machine tool with adequate controls. Normally, one must leave a small excess of thickness for fining operations to follow. If pressings are used, the amount of glass to be removed will be small and a comparatively fine grit can be used. If the raw blank or block is far from the desired radius and thickness, the work can be speeded up by use of a coarser grit, followed by a finer one. Experience dictates how rapidly a given kind of glass can be removed. The cutting speed is to some extent dependent on the nature of the glass powders, on the binding energy of the glass molecules of the particular type of optical glass, on the grit size and bonding of the cup-wheel, on the lubricant used and on a safety factor dependent on the value of the work being accomplished. In some cases an individual glass blank may cost thousands of dollars, particularly if of a special type well annealed of the best of homogeneity and selected to be free of noticeable striae, and the optician must be correspondingly careful.

The principle of the diamond cup-wheel is such a good one that the properly ground spherical surface can be prepared to within a few wavelengths of light of the prescribed surface. In special work one sometimes follows use of a coarse or medium diamond wheel with a fine grit grind whereby a near polish can be obtained. The hardest glass grinds the best in terms of the rms quality of the final surfacing and with respect to the punishment that the sub-surface can withstand. The crushing action of the high pressure diamond cut penetrates into the glass and it may be that sub-surface damage will occur which in turn may be glossed over during the following fining and polishing, to the detriment of the final product.

3.2.2.3.3 ASPHERIC TECHNIQUES

Generating surfaces with diamond wheels has been used with excellent results in various optical shops to prepare aspheric shapes from assigned mathematical formulas. The milling action can no longer be that of a cup-wheel unless indeed the wheel is tilted to yield a line contact with the glass. Instead, one can make use of a convex head to the diamond cutter that has a smaller radius than the shortest concave radius that is part of the aspheric shape to be ground. Cutting action can be controlled either by programmed tape on a modern milling machine or by successive approximations, alternating with some means for establishing the shape of the previous cut.

Ideally, both actions are on the same machine, managed by a computer. In this instance the experience of any given pass over the work can be fed into the computer and allowed for differentially on the next pass. So much success has been obtained in this way that if hand operations at a later point in the process cause a departure from the desired shape, a return to the generating machine can provide the following optician a fresh start.

3.2.2.3.4 QUALITY CONTROLS

The proper preparation of glass surfaces requires further fining operations beyond the stage of the diamond wheel. The block of lens elements is ground by introducing fine powders of carborundum or emory between the elements and well-shaped iron or brass tools that already have an adequately smooth surface from prior preparation. For faster fining these otherwise smooth tools may be criss-crossed by etched or graved channels. A good machine operator can hold the radius of the assigned spherical surface to a small variation around the mean according to the settings of his machine and can thus deliver to the polishing room properly fined surfaces that have the correct radius, and that are free of any significant cuts, digs or scratches too deep for removal by polishing.

Good grinding technology is an art of its own and generally skilled workers in this area, at least in a production shop, do not ordinarily carry on their work into polishing. The fining powders are a hazard to the polishing that follows and hence the grinding and polishing rooms ought to be well separated.

Polishing produces smoothness and the desired shape to high optical precision, true to specifications to within a fraction of a wavelength as shown by interference patterns. The interference pattern is obtained by one's placing a test glass of opposite curvature on the lens and viewing the contact surfaces in nearly monochromatic light. The resulting pattern of fringes permits a measuring precision of the order of a tenth of a single fringe which is about one twentieth of a wavelength. A better precision can be obtained in special work.

The polishing action is effected by making use of a pitch-covered tool in contact with the surface of the block and lubricated by a polishing compound of rouge, (iron oxide), or cerium oxide, zirconium oxide, sapphire powders or barnesite. However, barnesite is no longer available because of its thorium content but substitutes now being employed are said to be at least as good. Lack of space prevents going into this interesting subject but polishing action depends on many factors including the type of optical glass being polished. The very finest polishes are produced by allowing the polishing compound to break down into ever finer particles during the wet phase, and near the end by having the lens block run nearly dry but in excellent contact with the conforming pitch. The optician can determine from the sound of the polishing squeal the quality of contact and the quality of polish.

3.2.2.4 CENTERING

The centering operations which follow the polishing of each individual lens element now comprises a number of steps in which each piece of glass now containing two polished surfaces is reduced in diameter to within a small tolerance. Auxiliary bevels and ledges are also added as may be required, by which the lens can be properly related to other components and seated correctly in the lens mount. These auxiliary faces may consist of a circular cylinder whose axis coincides with the line connecting the centers of curvature of the spheres involved (optical axis); or of a true-running beveled edge which is a section of a cone whose axis of rotation coincides with the optical axis; or a true flat, which is an annular flat surface normal to the optical axis; (figure 3-22).

Sometimes all three auxiliary surfaces are needed to hold the lens element snugly in its precisely centered and spaced position with respect to the other elements of the system. In highly corrected photogrammetric lenses for cameras, printers, and projectors, the centering and spacing tolerances are very critical. Residual errors in successive elements cause unsymmetrical distortion and lead to the deterioration of performance, which usually cannot be compensated during the process of assembly.

3.2.2.5 CEMENTING

When two or more lens elements are cemented together to form doublets, triplets, and so forth, their contacting convex-concave surfaces must be ground and polished to identical radii of curvature. According to the type of cement, the cementing operation may require heating of the components. Excellent cements are now available, however, that polymerize at room temperature or that require in addition the irradiation of ultraviolet light. It is important to choose a cement that is clear and that does not set up strain in the lens elements during the process of curing. The contacting surfaces are pressed together to squeeze out air bubbles and

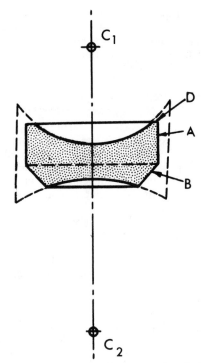

FIGURE 3-22. The over-size, ground and polished element is reduced by centering operations to its final centered shape, including mounting diameter A, bevels B, and flat ledges or seats D. The line connecting the centers C_1 and C_2 becomes the optical axis.

excess material and to avoid non-uniform thickness of the cement layer.

3.2.2.6 TRUING

Before the extremely thin layer of cement between the glass surfaces solidifies, a truing operation is performed by which the centers of curvature of all spherical surfaces involved are made to fall into a straight line, which then forms the optical axis of the compound lens.

3.2.2.7 LOW-REFLECTION COATINGS

The air-glass surfaces of the lens system are to be coated with at least a single layer of magnesium fluoride of a suitable thickness for minimizing by interference the loss of light energy by reflection at the surface. For a single layer, the index of the coating should be the square root of the index of the glass for optimum results. Since magnesium fluoride is the most common material used for low-reflection coating, one actually is saying that most interference occurs when the glass has an index of refraction of about 1.9. This high an index rarely appears in a photogrammetric lens. Thus, the magnesium fluoride coating is not fully efficient and is only partially effective for optical glasses of index as low as 1.5.

Multi-layer coatings can be used to obtain a nearly zero reflection at a desired wavelength and a minimizing of reflected energy throughout a considerable wavelength band. Multi-layer coatings can now be found on many production lenses in the open market and are especially effective in removing the last of glare light from a lens system. Multi-layer coatings are also necessary for the sake of efficiency in complex systems having many air-glass surfaces.

3.2.2.8 Mounting

The assembling of the components of a photogrammetric optical system requires a high degree of mechanical precision of all parts which form the lens mount, great skill of hands, clean assembly rooms, and numerous fixtures and tools. Some lens formulas specify variable spacing of components, by which radial distortion may be compensated or other aberrations may be corrected, so as to obtain best resolution and flatness of field. Lens mounts are, therefore, complex and individually tailored pieces, requiring considerable skill and know-how in their design and construction.

Photographic or projection lens systems in many photogrammetric instruments are approximately symmetrical in design with respect to their aperture stop. The front and rear halves are mounted into cells that are carefully fitted and oriented in a center barrel (lens cone), which in the case of an aerial camera, contains the diaphragm. The barrel is slotted to receive the blades and drive mechanism of the shutter. The shutter can be removed for inspection or repair without disturbing the lens elements. Retaining rings and screws holding critical parts are sealed by the manufacturer in order to prevent loss of calibration, which cannot be restored without the manufacturer's equipment.

3.2.2.9 Baffling and Blackening Techniques

The interior of the mounting of a well-mounted photogrammetric lens should be as thoroughly blackened as possible. Special black lacquers are available that can be baked onto clean metal and glass surfaces and that have little or no gloss on drying. Lacquers have such thickness, however, that they cannot be freely used on precision reference surfaces such as lens seats and separator rings. Special black dyes can be used for blackening the ground edges and ledges of the glass elements but even here, for precision work, one must be careful that the dye dries uniformly. In careful work the optician must go over the dried surface to test for any slight wrinkles of drying or for the possibility that a small piece of lint or dust has become entrapped in the coating. In the best work no irregularities larger than about 0.002 mm can be tolerated in a well-seated lens mount.

The lens cell must also have baffles or have separator rings and retaining rings so designed as to reflect incident light toward blackened areas or forward out of the lens system. The most difficult assignment of this kind arises from relatively thick elements, particularly in large optical systems. Light incident within the glass is reflected from the edge within the critical angle and can be quite bright unless trapped in some studied way. Baffling is therefore an integral part of good cell design.

3.2.2.10 Environmental Considerations

A complex aerial lens must work properly under field conditions as well as in the optical laboratory. Aircraft camera compartments are now usually provided with controlled temperatures and air density, but not always. Sometimes, as for supersonic flight, conditions become extreme. In earlier times aerial lenses were exposed to very low temperatures during flight and often picked up moisture internally on being returned to the ground. It is customary now to specify the range of conditions to be met by mapping lenses both in use and in storage. For the sake of economy one requires in storage only that the lens will not be damaged by prolonged exposure to cold and dampness, and that the lens when thermally stabilized under flight conditions will not have lost any of its performance because of the prolonged storage. Cameras not in use should be placed in storage areas free of atmospheric contaminants, cooking vapors, dust, moisture and insects. Where possible, cameras should be enclosed in properly made storage containers and kept in an upright position.

Aerial lenses must also be able to withstand a certain amount of shock caused by handling, shipping, and by prolonged exposure to vibration. Modern jet aircraft are far freer of vibration than are propeller aircraft. Nevertheless photogrammetric lenses must be able to withstand accelerations of prescribed magnitude, particularly for space applications.

3.2.2.11 Optical Quality Factor

The high-quality, photogrammetric lens system is a work of great precision. However, every process contributing to the final product is capable of introducing some form of degradation. The final product can be assigned an optical quality factor that apart from cosmetic appearance is important in evaluating any loss of contrast or resolution that may further occur under field conditions. Attention will be given later to the presence of residual distortion and asymmetries.

Insofar as the modulation transfer function is concerned, one needs to know just what the MTF curve (section 3.1.3.6) actually is for the lens in the field, including any effects caused by mount and window or windows. The final MTF can be compared to the MTF of the calculated in the laboratory. The degradation can be plotted as the ratio of contrast loss against frequency giving an optical quality curve which in turn can be used to bracket a production run and lead to statistics that control acceptable quality. Usually, the more surfaces one finds in an optical system, the worse the optical quality factor or the greater the expenses will be in ensuring a high standard at each stage of manufacture. Remarkable results have been obtained, however, in most optical shops here and abroad.

3.2.3 TESTING

Lenses that are intended for use in mapping cameras, or in copying work for photogrammetrists, must satisfy certain exacting requirements related to quality of definition in the image and accuracy of reproduction of the scene photographed. The tests for these requirements generally fall into two categories, visual and photographic. The characteristics measured, and the methods used for a particular lens, depend on the intended use of the lens.

Chapter IV of the present MANUAL discusses the testing and calibration of cameras after the cameras have been delivered by the manufacturer to the user or to his testing laboratory. The reader is therefore referred to chapter IV for a discussion of the many details and techniques involved in the precision testing of mapping lenses and in camera calibration. For the moment, however, the discussion given immediately below will be oriented toward the point of view of the designer and manufacturer of optical systems. It is just as important for the manufacturer to have tested and adjusted the system prior to delivery as to have tests performed at a later time when little can be done to improve the system.

The testing of the mapping lens and related photogrammetric equipment in the laboratories of the manufacturer involves teamwork and cooperation by the optical and instrument designers, and by the optical scientists and technicians at various levels. This is all best accomplished under a single roof. The first assembly of a prototype wide-angle lens very often is only temporary. Usually, it becomes necessary to assemble and disassemble the optical system several times for the purpose of troubleshooting and for acquisition of data to guide the team. On the other hand, the state of the art is so far advanced among the major suppliers that even a first assembly, if free of the occasional gross error, leads to performance in the neighborhood of what is desired. Visual and photographic testing, combined with modern types of testing by means of laser interferometry and MTF evaluations, will provide reliable data from which differential corrections can be made.

3.2.3.1 OPTICAL LABORATORY

The optical laboratory of the manufacturer should be equipped either with a nodal-slide optical bench, or with a goniometer, or a multi-collimator type of calibrator (Chapter IV) for the measurement of distortion. A mirror collimator is also used in many laboratories as a reliable source of parallel light for purposes of photographic testing. The laboratory itself should be free of dust and lint and therefore should be maintained with filtered air at a pressure slightly greater than that for adjacent rooms or the out-of-doors. The temperature should be held constant to within a degree or two of 20°C and the humidity should be held between 40 and 50 percent. Technicians and participants should be provided with clean garments and caps.

Insofar as possible, coated lens elements should be protected from contamination in any form, including the myriad of saliva spots that can be caused by nearby conversation. Handling should be only by qualified technicians. Cooking odors and vapors should not be allowed at any time anywhere in the laboratory to prevent the deposit of oil or grease droplets on lens surfaces and on testing equipment. Lens assembly must be carried out under optimum illumination with dust-free equipment and a vacuum means kept at hand for removing electrically charged lint and dust captured on coated surfaces. Dust-free air sprays and air hoses should be available. If compressed air or dry nitrogen is used for cleaning purposes, one should use the quality guaranteed to be free of grease and oil. Finally, the technicians must have at hand all essential gages, shim stock, master optical flats, test plates and anything else that may be needed for evaluating optical and metal parts.

3.2.3.1.1 VISUAL TESTING

It is possible to analyze the optical image on the lens bench and on a precise collimator well enough to determine data to be fed into the same type of computer program that may have been used to design the optical system at the beginning of the activity. A least squares determination by computer can then be used to find what the next steps might be to improve the system. The designer, who necessarily is already familiar with the system, may allow only a few parameters to vary in the expectation that a satisfactory differential correction might be achieved by readjustments in easy-to-reach places. For example, displacements of the first and last elements, guided by calculations, might be sufficient to complete the desired balance for minimizing the maximum distortion over the field. If residual coma is present, perhaps one or more inner air-spaces may have to be shimmed as well. In a difficult case, one or two surfaces may also have to be reworked to slightly altered radii to obtain the necessary effective parametric changes.

The optical team will always need to make a study through the microscope of the visual images formed by the lens under test, on-axis and at selected off-axis angles. The designer will try to interpret what he observes in the various images at a magnification of between say 5× and 100× in relation to the residual aberrations he knows to have been left on completion of the optimum lens design. The visual impression cannot be related with exactitude to the photographic analysis to follow but with extended practice one can anticipate what to expect.

FIGURE 3-23. The Bausch & Lomb 3.7 Metre Collimator. This is a unique $f/7$ reflecting collimator for the testing of infinite conjugate optical systems. The telescope consists of a 3746 mm $f/7$ paraboloidal primary with interchangeable resolution targets and star plates located at its focal point. An electronic flash also provides for photographic tests. Interchangeable, neutral-density filters are used to control the exposure times. Filters can also be interchanged for monochromatic testing at various wavelengths.

The pneumatic anti-vibration mount damps vibration to permit extremely long exposures and high magnification testing.

Lenses have been tested with focal lengths as long as 6000 mm with results recorded on photographic film and aerial images evaluated with a microscope attachment.

If astigmatism is present in the visual image, the designer can make measurements to compare the amount observed with that calculated and will try to determine the nodes, if any, at the off-axis angles as predicted by calculation, where the astigmatism may vanish or where either the radial or tangential image surface crosses the image plane. Spherical aberration, comatic flares, chromatic variations, astigmatism, curvature of field, distortion and higher order errors all can be studied.

Even if a precision optical bench is not at hand, the observer can place a flat glass plate with its front surface in the image plane and use the plate as a reference in measuring radial and tangential field curvatures at various off-axis angles. One can make this kind of measurement either with a microscope provided with a focus scale or by focusing with the target itself, plus or minus with respect to the infinity focus setting of the collimator.

The prime advantage of the visual method of testing is rapidity of measurements with a minimum of effort. The measurements most frequently made are back, front, and equivalent focal lengths, nodal-point separation, radial and tangential distortion, and resolving power.

A typical nodal-slide optical bench is shown in figure 3-24 and is one of the benches currently in use at the U.S. Geological Survey. The optical bench consists of a set of bench ways, upon which are mounted slides carrying a viewing microscope and the nodal-slide assembly. A collimated beam of light is provided by the paraboloidal mirror at one end of the bench ways.

This collimated beam from the mirror is so directed as to be as nearly as possible parallel to the bench ways. The nodal slide holds the lens under test, and is provided with alignment adjustments. Values of lens distortion are now determined with a precision of about 2 micrometres (Washer and Darling, 1954; see also, chapter IV). It is very difficult in the visual type of test to determine the focal length that gives the best average definition over the entire field. For this reason, a photographic type of test is usually required for lenses to be mounted in mapping cameras.

Even without a precision optical bench one can evaluate the distortion residuals visually by means of an auxiliary instrument made use of in the prototype work on the Geocon IV mapping lens (Baker, 1965). At that time, two plane-parallel optical plates were made up to a flatness of an eighth wave on each of the 4 surfaces, and the better surface of each was then coated with a neutral metallic film having a transmission of about 30% and a reflectivity of about 60%. (Ideally, Inconel coatings having a reflectivity of about 80% might have been employed).

The two plates of zero optical power were placed in a special aluminum cell with an air wedge angle between the adjacent coated surfaces of about $5\frac{3}{4}°$, as shown in figure 3-25. The mounted pair of plates then was attached to the front end of the Geocon IV barrel and the combination placed in the parallel beam of light from the collimator mirror. The zero-order beam transmitted directly through the pair of plates was undeviated because of the zero optical power of the respective plates.

FIGURE 3-24. The NBS-USGS visual nodal slide optical bench. The viewing microscope and nodal slide are mounted on the bench ways; the target assembly is fixed to the table beside the ways; and the off-axis paraboloidal mirror collimator is shown at right of the photograph.

The on-axis image had an intensity of about 9% that of the direct beam from the collimator. The first-order beam was reflected back and forth in the wedge space and then was transmitted at an off-axis angle of 11.5 degrees. The second-order beam was transmitted at 23° and so on.

Thus, one obtained a series of images spaced at angular differences of twice the wedge angle, nominally set by careful design and instrument work to 11.5° intervals in the images. Next, a set of one-sided, keenly lapped, stellite slit-jaws was cemented in place on the back side of a glass plate shaped about like a ruler as shown in figure 3-25, placed across a diameter of the field. Clearance holes of 1 cm diameter had already been drilled into the glass plate at the respective 11.5-degree intervals on either side of center to allow the rear surface of the glass slide to lie in the image plane. The respective one-sided slit jaws were then used as knife-edge or Foucault visual testers to enable one to determine the positions of the centroids of the various images.

While the series of images lay only from center outwards along a single radius of the field, the mount in front was designed to rotate about the optical axis, allowing one to work along any desired azimuth of the field. The glass ruler was mounted in ways and driven by a spring mounted micrometer, operated by hand. The exact lateral position of the glass ruler was also determined by a dial gage reading in minimum graduated divisions of 0.0001 inches (about 2.5 micrometres).

The mounting of the respective slit jaws had displacements radially of not more than 0.1 mm from the calculated location based upon the known EFL of the lens and the tangents of the known angles associated with the point images. Thus, the image position measurements required the glass ruler and the reference gage not to read more than about 0.2 mm laterally in total travel. Measurements of the locations of the respective

FIGURE 3-25. A relatively simple means for visual determination of distortion during the process of manufacture.

knife-edges had already been made in the instrument shop, both from the lay-out work and from use of precision reference gages. Nevertheless, further one-time calibrations were obtained both from comparisons with distortion measurements of the same lens at Fairchild and from a calibration carried out at M.I.T. on a precision comparator that provided the exact positions of the various spaced knife-edges.

Visual use of the auxiliary equipment with the Geocon IV gave rather easily determined radial distortion measurements to a precision of about 4 micrometres or better. The equipment made it possible to make many measurements while the lens itself was being worked on and the aspheric figuring completed. As a result, the photographic determinations of distortion carried out at Fairchild were not over-burdened by loss of time and travel and by excessive repetition.

Visual testing is useful in many ways in addition to just an inspection of the image. With respect to aspheric figuring, if required, one finds it possible to use the knife-edge method on and off-axis as a differential guide to the desired shapes of the figured surfaces. Low power examination makes it possible to obtain a better fuel for the distribution of light in the image than if too high a magnification is used. Both, however, contribute to image assessment.

3.2.3.1.2 PHOTOGRAPHIC PERFORMANCE TESTING

The precision lens-testing camera shown in figure 3-26 was developed at the National Bureau of Standards. It is one of the earliest successful devices (Gardner and Case, 1937) developed to measure performance of lenses by photographic means. This apparatus is now at the U.S.G.S.

Lenses intended for use in aerial mapping cameras are usually evaluated in apparatus of this type (Tayman, 1974; *see* also chapter IV for major discussion of photographic testing). The lens-testing camera consists of a bank of ten collimators, spaced at 5-degree intervals, having resolution test charts as reticles (fig. 3-27). The lens under test is mounted at the center of convergence of the multi-collimator bank. The bench ways, upon which are mounted the lens holder, the camera back whereon the photographic recording plate is mounted, and the viewing microscope, can be aimed at either extreme collimators in succession to cover a complete diameter.

A series of exposures is made on two photographic plates, with each row of images being at a different focal setting for the lens under test. The images on the finished negative are examined under a microscope, and a row selected that represents the best overall definition over the entire field. The selected row is measured in a comparator. From the separation of all images in this row, the equivalent focal length and radial distortion can be determined. The resolving power is obtained by examining the row of images under a microscope to determine the finest pattern of the test chart that is distinctly resolved into separate lines. The following is a report issued by the U.S. Geological Survey on a typical wide-angle lens tested with this equipment:

rapidly. It is known that the resolving power of most photographic emulsions goes through a maximum at some density depending on grain and contrast and then falls off again as the exposure and therefore the density is further increased. A very fine-grain, contrasty emulsion may have its maximum resolving power at high target contrast at perhaps a density of 0.6 whereas a fast, coarse-grain emulsion may reach its optimum, though lower, resolution at a

A Bausch & Lomb Optical Company Planigon Lens Number XF6700 having a nominal focal length of 6 inches and a maximum aperture of f/6.3 was tested at maximum aperture with collimated incident light, using incandescent tungsten light source with K-3 filters. The measurements were made using Eastman Kodak spectroscopic emulsion type V-F on ultra flat glass plates, development was in D-19 at 68°F for three minutes with continuous agitation.

I. Focal Lengths

Back Focal Distance	82.19 mm
Equivalent Focal Length	152.62 mm
Calibrated Focal Length	152.59 mm

The values of the focal lengths have been selected to give best average definition across the entire negative. This measurement is considered accurate within 0.02 mm.

II. *Radial Distortion in Micrometers*
   A. *Distortion Referred to the Equivalent Focal Length*

5°	10°	15°	20°	25°	30°	35°	40°	45°
−3	2	−1	−10	−18	−29	−26	−21	−52

   B. Distortion Referred to the Calibrated Focal Length

5°	10°	15°	20°	25°	30°	35°	40°	45°
0	8	7	3	−2	−9	−2	8	−17

The values of the radial distortion are measured in micrometers and indicate the displacement of the image from its distortion-free position. A positive value indicates a displacement from the center of the plate. These measurements are considered accurate within 7 micrometers.

III. *Resolving Power, in Cycles/mm:* Area Weighted Average Resolution 44.2

Field Angle	0°	5°	10°	15°	20°	25°	30°	35°	40°	45°
Tangential	72	64	64	57	51	45	40	36	32	29
Radial	72	64	64	57	51	51	45	40	40	45

The resolving power is obtained by photographing a series of test bars and examining the resulting image with appropriate magnification to find the spatial frequency to the finest pattern in which the bars can be counted with reasonable assurance. The series of patterns has spatial frequencies in a geometric series having a ratio of the sixth root of two. Tangential lines are those perpendicular to the radius from the center of the field. Radial lines are those lying parallel to the radius.

The two surfaces of the yellow filter accompanying this lens are within ten seconds of being parallel. This filter was used for the calibration.

The resolving power obtained by the above method is usually higher than that obtained under practical service conditions, owing to one or more of the following reasons.

   (1) The emulsion of the test plate usually has finer grain than that of the fast panchromatic film used in the aerial camera and may have a different spectral sensitivity.
   (2) The exposure of off-axis points is often increased so as to give uniform illumination to all points on the plate, whereas in the aerial camera, the illumination of off-axis points decreases

density of perhaps 1.2. One may suppose that test plate emulsions may be exposed to an optimum density of perhaps 0.8 to 1.0.

   (3) If a high-contrast target is used, a resolving power is obtained which is much higher than that for the low contrast target, the latter more nearly representing conditions in practice (Washer and Tayman, 1961). Much study has been done (Washer and Tayman, 1955), but a general agreement has not been reached concerning the most suitable type of target for testing lenses used in aerial photogrammetry.
   (4) During the testing procedure, the deleterious

FIGURE 3-26. Precise lens-testing camera. Several of the collimators that provide a fan of collimated beams of light spaced at 5° intervals for values of varying from 0 to 45° appear in the foreground. The lens under test is placed at the point of intersection of the collimated beams and images formed by the lens are registered on a photographic emulsion located in the camera plate holder. The lens holder, camera and viewing microscope are supported on a rotating carriage that can be aimed at any one of 10 collimators.

effects of vibration, image movement, atmospheric haze, and sudden temperature changes are generally absent.

(5) Because of the small amount of light reaching the lens from the collimators, the light scattered in the lens is much less than in practice. However, inasmuch as modern lenses are coated for low reflection (section 3.2.2.7), the problem of scattered light is usually of minor importance.

(6) The aberrations of the collimator lens may affect the test results and therefore one should obtain high quality objectives at the outset.

(7) The light source used in the collimators may

have a different spectral composition than diffusely reflected sunlight.

From the above discussion it can be seen that the comparison of values of resolving power of several different lenses means little unless complete information is available on how the values were obtained. One wishes to use standard testing procedures that correspond as nearly as feasible to those conditions encountered in service.

The apparatus described above is for testing

RESOLVING POWER TEST TARGET

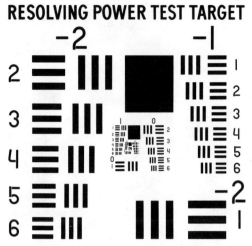

FIGURE 3-27. USAF resolution test chart with resolving power targets spaced as the sixth root of 2. See MIL-STD-150A, Change Notice 2, 28 January 1963, pp. 20–21.

lenses focused at infinity. Certain types of photogrammetric equipment (such as projectors, printers, *etc.*) require lenses for finite distance-ratios ranging from about 1:1 to 5:1. As previously mentioned, lenses must be specially designed for this type of work if optimum optical performance is to be obtained. Also, the distortion and resolving power values usually vary as the ratio is changed.

At present, only a few organizations have precise optical-bench equipment for testing photogrammetric objectives. Lens manufacturers usually conduct tests in their own laboratories (figure 3-28), and the prospective purchaser can often obtain the results of such tests.

At the Bausch & Lomb Company one of the latest developments in lens-testing and evaluation equipment is the 3.5 meter universal optical bench. This bench was conceived and developed at Bausch & Lomb to fulfill the need for a universal, precision bench capable of accepting a wide variety of lens types, operating over a wide

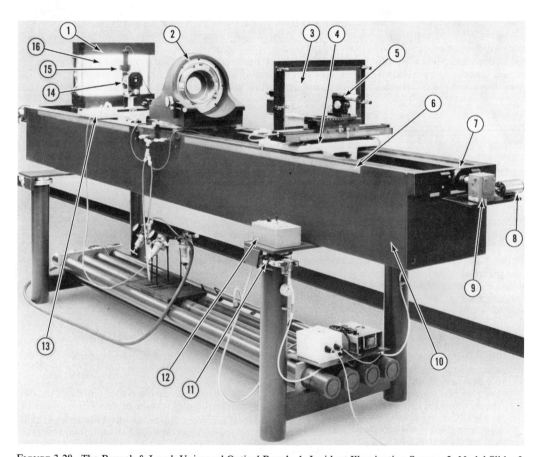

FIGURE 3-28. The Bausch & Lomb Universal Optical Bench. 1. Incident Illumination Source; 2. Nodal Slide; 3. Screen Plane; 4. Output Carriage; 5. Microscope; 6. Scale; 7. Coordinate Measuring Drum; 8. Output Carriage Drive Motor; 9. Gear Box; 10. Granite Rail; 11. Anti-vibration Mount; 12. Drive Control; 13. Input Carriage; 14. Target Holder; 15. Tungsten Source; 16. Transillumination Source.

range from infinity to finite conjugates. Long term stability, use in a normal manufacturing environment and fast operator adaptability were major goals in the development.

The resulting, modular bench has in fact exceeded these goals and has proven invaluable in testing optical systems as simple as an eyepiece to as complex as multi-element diffraction-limited long focal length lenses.

The instrument itself is composed of a 12-foot granite rail supporting a nodal slide and input and output carriages; the longitudinal and lateral motions of the carriages and the rotational motion of the nodal slide are in turn supported and guided by air bearings. The granite rail was chosen for its ground accuracy of ways and inherent, long-term stability. The air bearings were selected because of the load-carrying capacity and reduction of friction and deflection errors in the drive.

The bench consists of five major components. In addition to the granite way, the assemblies include the input carriage, output carriage, nodal slide and target holder.

The input carriage carries a resolution target capable of being illuminated by transmitted or incident light or a star which is illuminated by a microscope-condenser system. The input carriage is translated by a motor-driven lead screw along the granite way, the air bearings supporting and providing tracking accuracy.

The target holder on the input carriage contains provision for inserting interchangeable resolution targets or a star slide. The tungsten light source has a condenser to illuminate the target correctly and interchangeable filters provide for monochromatic lens evaluation.

The entire bench is mounted on air cylinders which reduce the effects of building vibration by 99%.

The output or image evaluation carriage, whose drive is similar to the input carriage, carries a screen and translating microscope for detailed evaluation of the image formed by the optic under test. Removal of the screen permits inspection of the aerial image, the microscope being capable of rotation about a vertical axis and translation to observe the off-axis as well as the on-axis image.

The nodal slide is air-bearing supported and located in the center of the rail. The slide can be rotated about the vertical axis to permit precise location of the nodal point of the test lens. The nodal slide mount will accept a wide variety of lenses up to 150 mm in diameter and 22 kg weight. The precision of the slide is assured by use of a lapped ball and socket arrangement which reduced the vertical axis runout to less than 0.005 mm.

The universal lens bench has been used for a number of lens-system tests, including validating computer-predicted lens performance, confirming centering tolerance and mounting specifications, experimental study and lens analysis, trouble-shooting components and systems, assembly and alignment of prototypes and models and the testing of components and completed optical systems.

In submitting a lens to a testing laboratory, it is necessary to specify at what scale ratios and apertures the tests are to be conducted. For instance, a lens used in a printer or projector should be tested for the scale ratios at which it is to be used, while an aerial camera is tested for infinite object distance.

There are three chief reasons for testing a lens prior to mounting it in an aerial camera: (1) to determine whether its distortion characteristics are near zero or match those of the printing or plotting apparatus with which it is to be used; (2) to determine whether the resolving power is satisfactory; and (3) to determine the correct back focal distance to be set later into the aerial camera.

When a satisfactory lens has been mounted in the camera, the next step is to test the complete unit. The purpose of this test is to determine the distortion and resolving power for the focal length actually set into the camera. The values of the distortion thus obtained are used in any future photogrammetric computations related to photographs taken by the camera.

3.2.3.2 FINITE OBJECT-DISTANCE LENS CALIBRATION

In contrast to aerial cameras, which are used focused at infinity, many photogrammetric lenses are used for projecting a given object (aerial negative or disapositive) onto a screen placed at a short distance from the lens. The calibration of a projection lens, such as a projector of a direct-projection plotter or a projection printer, is therefore based on a given ratio of magnification, that is, a stated ratio of object and image distances.

As the residual distortion of a lens in finite application usually varies with the magnification ratio, the calibration of the lens consists of the measurement and evaluation of its distortion at the given magnification ratio. As freedom distortion is required of the projection instrument in many practical applications, a process of compensation of the measured distortion may accompany the calibration.

A precision optical bench for lens testing at infinite object distance (as shown and described in figure 3-28) may be used with certain reservations. In this case, a two-fold determination of the distortion curve of the lens under test is necessary—at zero and at infinite magnification. If D_0 represents the lens distortion measured at a series of off-axis points in the focal plane while the beam of parallel light from the collimator is incident upon its front element, and D_∞ represents the corresponding distortion quantities while the rear element of the lens is facing the bench collimator, then the distortion D_m of the lens at the operational magnification m can be computed from the equation (Magill, 1955)

$$D_m = D_o - mD_\infty. \qquad (3.63)$$

Inasmuch as very precise measurement is needed for the final compensation, a direct-measuring bench method is preferred. An optical bench specially equipped for the measurement of finite conjugate distances is required.

Another fairly effective method makes use of the projection camera or printer itself with the lens under test finally installed. It is assumed that the positions of both focal points and nodal points have already been determined on an optical bench. A glass scale or preferably a precision grid plate is placed in the object plane and photographed onto a high resolution plate. After suitable measurement one can calculate the radial and tangential distortion in terms of the test scale or grid.

3.2.3.3 CALIBRATED FOCAL LENGTH

3.2.3.3.1 DEFINITION

The focal length determines the scale of the image. For a focal plane it is given by the height of the image divided by the tangent of the half-field angle. If the value varies across the format, it is an evidence of distortion which is normally expressed as the differential in height required to keep the focal length constant. The calibrated focal length is that which makes the positive and negative values of distortion the same (*see* figure 3-29). The equivalent focal length is the limiting value as the field angle goes to zero and is of interest primarily to the lens designer as the focal length equivalent to that of a simple lens. Clearly, if there is no distortion, the calibrated focal length and the equivalent focal length are identical.

For image surfaces that are not plane, the definition of focal length must be clearly stated. A concentric system such as the human eye cannot be considered distorted although it does require that the output be viewed from the center of curvature. For binocular observation this presents a problem unless the output is essentially placed at infinity.

3.2.3.3.2 METHOD OF DETERMINATION

The difference between the equivalent focal length and the calibrated focal length can be expressed as the value of Δf determined from the relation

$$\Delta f = (D_1 + D_2)/(\tan \beta_1 + \tan \beta_2) \qquad (3.64)$$

where D_1 and D_2 are the values of distortion, referred to the equivalent focal length, at angles

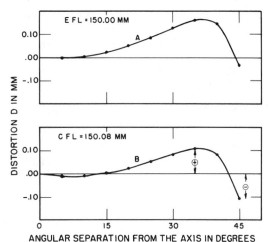

FIGURE 3-29. Determination of calibrated camera focal length from distortion curves. Distortion curve A is based on a camera focal length determined from the image scale within 5 degrees of the central field. The lens distortion in the outer field assumes essentially positive values. Curve B follows from the assumption that image points at 13.5 degrees off-axis are distortion-free. Positive and negative distortion values are balanced throughout the field; image points at 42.5 degrees now are also distortion-free. The focal length related to curve B is therefore accepted as the camera's calibrated focal length, CFL.

β_1 and β_2 for which equal but opposite values of distortion are desired.

Curve B in figure 3-36 shows the distortion referred to the calibrated focal length CFL for $D_{35^\circ} = -D_{45^\circ}$. It is clear that $D_\beta = 0$ at $\beta = 13.5^\circ$ and 42.5° and that the maximum value of plus and minus distortion are equalized. It must be emphasized that no physical shift of the focal plane takes place between the conditions shown as A and B in the figure. The only change is in the choice of a scale factor for use in interpreting measurements of distance separating various image points.

3.2.3.3.3 METHOD OF APPLICATION

The calibrated focal length should be used where the precision of the scale of the image is important such as in a moving-film panoramic camera to determine the velocity of the film. Certain instruments such as the Wild autograph and Kelsh plotter have been designed to compensate the distortion and the equivalent focal length should be used.

Comparator measurements—use the calibrated focal length in computations based on linear measurements on the photograph.

Non-distortion-correcting instruments—use the calibrated focal length. However, its use does not reduce model warpage, but merely gives a very minute improvement in the relationship between the vertical and horizontal scales.

Numerous types of correction curves and graphs (Bean, 1940) have been compiled to take care of this model warpage. From these curves, an elevation or parallax correction is obtained for compensating the distortion error at any given point in the stereoscopic model. However, the presence of relief so complicates the use of these curves that their application becomes very tedious.

Instruments using the Porro-Koppe principle (*see* section 3.4.2)—be consistent by using the equivalent focal length of both aerial camera and projector, or the calibrated focal length of both.

Multiplex reduction printer (*see* section 3.4.3)—the calibrated focal length should be used with Metrogon photography because the reduction ratio in the printer was originally set to correspond to the calibrated distortion values for a nominal Metrogon lens (maximum positive and negative distortion values being balanced).

Aspheric distortion-compensating plates (as in Wild Autograph) (*see* section 3.4.4)—use the equivalent focal length, since the projectors are calibrated without the compensating plate in position.

Mechanical distortion-correcting devices (such as cam in Kelsh plotter) (*see* section 3.4.5)—use the equivalent focal length because that is the one used in the design of the distortion-correction cam. Also, the lens is initially adjusted in the projector (with the aid of a depth gage positioned on the lens axis) so as to recover this equivalent focal length.

3.3 Non-Lens Optics

Optical elements which are not lenses may be divided into three groups. Those in the first group reflect light, those in the second refract light and those in the third both reflect and refract. In this text the third group is covered under the sections dealing with the first two in accordance with the particular primary function.

3.3.1 REFLECTIVE OPTICS

The function of a reflecting device is to redirect the light beam incident on the reflecting surface or surfaces. Plane and curved mirrors, reflecting prisms, beam-splitters and thin-film pellicles fall into this category. The quality required at a given reflection depends on the application and precision desired and may range from low-quality mirrors for irradiance to optical mirrors of extremely high quality.

Mirror reflectance may be obtained from the coating of a metallic film onto a substrate of polished glass, fused silica, or other substance or may be obtained also from diamond point machining of suitably selected metallic solids. Mirror reflectance can also be obtained from the coating of multiple layers of dielectric films onto the substrate such that multiple interference of light energy reflected from the various interfaces add together vectorially to produce a mirror-like response.

Since World War II, grinding techniques have also been carried to an advanced state of the art. Under modern conditions it is possible to obtain a nearly polished surface directly from the grinder, one that requires only a modicum of further polishing prior to coating (section 3.2.2). For best results, however, and to prevent possible aging effects, it is important in the case of some optical glasses and softer materials for one to follow the surface grinding by removal of material at a lower rate by fining powders, prior to polish. In this way sub-surface fracturing can be minimized.

The testing of precision optical surfaces has been carried to a very high technical level and where necessary in the limit can reach residuals from the desired mathematical surface, whether plane or curved, of a very small fraction of a wavelength of light. The introduction of laser into laboratory testing has effected a revolution in the precision testing of quite complex optical surfaces and arrays.

3.3.1.1 PLANE MIRRORS

In particular, the testing of small optical flats is most commonly achieved by use of interference fringes between pairs of flats in surface contact in a set of three, such that each pair under test is allowed to come to thermal equilibrium and the fringes viewed as formed in irradiated laser light or more often in the light from a sodium or mercury vapor lamp. Algebraic reduction of the test residuals over the surface in pairs will then lead unambiguously to a knowledge of the departure from flatness of each of the three mirrors. Such tests are carried out in the optical shop on uncoated surfaces.

Plane mirrors are widely used in optical systems to redirect the incident light waves without the introduction of aberrations in the image-forming characteristics of the waves. Indeed, a plane mirror can be thought of as the only aberration-free optical surface. For practical purposes, resolution, distortion and curvature of field are the same after one of more plane reflections as before.

In practice mirrors may not actually be mathematically plane. Apart from residual errors of manufacture depending on cost, a mirror may not be sufficiently rigid for the application, or sufficiently well mounted. Thermal changes may cause a temporary change of figure of the mirror or mirrors, except for those made of athermal materials. Defects in polish or the acquisition of dust or contamination may affect performance through loss of contrast. In some systems interposition of a mirror may alter the shape or nature of a pupil and affect the ultimate diffraction image.

Some mirror coatings do not reflect all wavelengths of incident radiation equally and therefore may introduce a change in the color temperature of radiation through the system. Other coatings or bare polish may introduce a loss of radiant energy. Indeed, some coatings are best in the violet and others in the red or infrared.

Plane mirrors which do not function in a critical optical array may be rear-surface mirrors, as for a looking-glass. Rear surface coatings are generally less efficient but may be lacquered over on the back and otherwise made permanent against severe environmental exposure.

3.3.1.1.1 PYREX, FUSED SILICA, AND CER-VIT

Mirrors of the highest precision are usually made from glass-like substrates available from suppliers of glass and ceramic products. In earlier decades the very hard borosilicate glasses were generally employed, culminating in low expansion pyrex (*see* Ohara). More recently, in spite of enhanced cost, precision mirrors are generally made of fused silica which has a very low coefficient of thermal expansion. For very critical applications mirror materials having virtually no thermal expansion in a normal range of temperature are employed. These are either of a ceramic composition or are derived from fused silica by admixture of suitable substances such as titanium. Some are clear, some translucent, and some opaque. Very large astronomical mirrors have also been made of such materials, known as ULE (ultra-low expansion), Cer-Vit (ceramic vitreous), and similar materials (*see* Corning, Owens-Illinois, Schott catalogs).

3.3.1.1.2 METAL MIRRORS

Reflecting devices which use only the front surface may be made of materials opaque to light. Among these are the stainless steels, stellite, beryllium, aluminum, copper and other metals. Mirrors may also be made of such materials coated or electro-plated onto some other kind of solid. Single point diamond machining has been carried out to such precision that quite good plane and curved mirrors can be machined directly from the solid ingot or disc.

If necessary, a coating of higher reflectance can be added, either by deposition in a vacuum from a source heated to vaporization by thermal excitation from electrical resistance heating or by electron gun bombardment, or by sputtering at high voltage in a vacuum in an electric field containing the substrate, or by electroforming or deposit of metallic ions from an electrolyte. In general, those materials that can be polished by optical methods to extreme smoothness are to be preferred for the most exacting requirements.

3.3.1.1.3 PELLICLES

A pellicle is a membrane stretched across a rigid mount which for critical imaging systems is generally in the shape of a ring. These are used where the thickness of glass would introduce aberrations which affect the image, or they are used to minimize ghosts. Since the manufacture of pellicles has become very reliable, pellicles coated with various kinds of thin films for reflectance, transmission and for beam-splitting are being used in many kinds of photogrammetric equipment.

3.3.1.2 CONCAVE AND CONVEX MIRRORS

Curved mirrors are defined as those having any reflecting surface whatsoever which departs from a mathematical plane. Generally, they are spherical, ellipsoidal, paraboloidal or hyperboloidal in shape, but they may also have aspheric departures from even the plane or conic reference surface. Light reflected from them will converge or diverge in accordance with the curvature. The dioptric formulas given in the early part of this chapter may be used directly for mirror surfaces, provided that at each reflection, the index n in the backward direction is taken as -1 and distances in the backward direction are considered negative. The guiding rule is that even in a compound system of mirror and lens surfaces, the optical path is always increasing in a positive sense along the actual path followed by the light waves.

Inasmuch as the effective index of refraction at a reflection is taken as -1, mirror surfaces in terms of the value of $(n - 1)$ as used in 3.2 and 3.24 have 4 times the dioptric power of a refractive surface for $n = 1.5$ used as a thin plano-convex lens. Accordingly, mirrors provide a very powerful means for producing convergence or divergence of image-forming waves in an optical system. For any given ray at a reflection the angles of incidence and reflection are equal. This fact in turn tends to reduce aberrations, particularly at high angles, as compared to the case of refraction. Furthermore, the reflected ray is not affected by chromatic aberration in any way and hence mirrors systems can be used over very wide regions of the spectrum.

3.3.1.2.1 SPHERICAL MIRRORS

The simplest case is that of the spherical mirror which is by far the easiest to manufacture. If an object is placed at the center of curvature of a concave spherical mirror, it is imaged back onto itself, though inverted. In projectors the use of a spherical mirror behind the light filament can be used to increase the light through-put of the system and to even out the illumination over the image field.

Spherical mirrors can also be used very effectively in combination with refractive correctors and will be discussed in section 3.3.4. One might note here, however, that spherical concave or convex mirrors do have spherical aberration when used in image forming systems at conjugates different from the 1:1 case at the center of curvature. Because of the dioptric power associated with a reflection the spherical aberration is comparatively small. Quite powerful mirrors can therefore be used in uncritical situations, or conversely, precision performance can be obtained with spherical mirrors if the aperture is trimmed so that the optical path differences over the aperture are less than the quarter-wave Rayleigh criterion. Thus, spherical mirrors at speeds of $f/15$ or slower can be found in precision instrumentation.

3.3.1.2.2 ASPHERIC MIRRORS

Most precision mirrors are ground or figured to mathematical shapes departing from the sphere. In the most common case these mirrors are derived from conic sections and may be ellipsoidal, paraboloidal or hyperboloidal, according to the eccentricity assigned. As is known from the properties of conics, the optical axis of such a mirror is also the major axis of the conic and contains a conjugate pair of focal points. In the limiting case of the spherical mirror where the eccentricity is zero, these focal points coalesce at the center of curvature.

Another useful class of mirror shapes is derived from a rotation of the arc of a conic about the minor axis used as the optical axis. These shapes are termed oblate, insofar as an elliptical arc is concerned. The optical axis contains no mathematically exact focal points, although conjugate pairs on the adopted optical axis can be found through dioptric or gaussian accuracy. In the case of an oblate concave mirror the total depth or sagitta is greater at the edge of the aperture than that for the spherical surface of the same radius of curvature as for the center of the oblate shape. The edge is said to be turned-up.

The most general class of aspheric shapes, however, can be defined only by some functional dependency of x, y and z in three-space. For convenience one can resort to power series to define such shapes or can use power series as a smoothing function of best fit to the desired general shape. Most commonly, the general aspheric surface is still one having an axis of rotation, at least in most optical systems employing mirrors, and hence two of the variables are tied together as a rotational variable, such as $(x^2 + y^2)$.

It is clear that aspheric mirrors have only limited usefulness for photogrammetric applications. For example, paraboloidal mirrors are often used as collimators in that a reflected beam of collimated light from a point-source placed at the finite focal point is altogether free of spherical aberration and chromatic aberration. Thus, such a collimated beam becomes very useful in the testing of refractive systems.

In another such instance, ellipsoidal mirrors can be used where access to either focus is desired. In the Bausch & Lomb projectors the light source is placed at one focus and the aperture of the projection lens at the other. Therefore, all rays reflected from the mirror surface pass through the lens aperture. The specific off-axis ellipsoids direct the light from the light source so that light is distributed evenly over the field of the projector.

3.3.1.3 SEMI-TRANSPARENT MIRRORS

Semi-transparent mirrors are frequently called beam-splitters in accordance with their function in an optical path where an incident beam is to be divided into two beams. If the beam-splitter is neutral in its effect on both subsequent beams, it may be coated in such a way as to have both beams of equal intensity, or by choice, of unequal intensity. Either metallic or multi-layer dielectric coatings can be employed (see OCLI, Liberty Mirror Catalogs).

3.3.1.3.1 BEAM-SPLITTERS

A beam-splitter can be used to provide a visual image as a separate monitor of a photographic or image-detecting device. One can use a visual image, for example, from the normal visual portion of the spectrum at the same time that an infra-red exposure is being made in order to center an object in the field of view or to take advantage of transient phenomena and the like. It may also be that a fairly dense neutral coating can be used to reduce excessive illumination as from the sun, or conversely that a weak reflection can be used for the same purpose. A very familiar application is to have two or more observations going on simultaneously, particularly if the detectors used have separate analytical functions. One can make use of a beam-splitter as a weak wedge to produce image shifts in the field.

While most beam-splitters are made up of thin plates or films coated for beam-splitting purposes on one side and perhaps coated for low reflection on the other, ghost images are altogether avoided if a beam-splitter in the form of a cubical prism is used. One of the hypotenuse faces of the two right-angle prisms making up the cube is thus coated with suitable metallic or dielectric coatings and the two right angle prisms then are cemented together at the hypotenuse faces. The resulting beam-splitting action arises therefore from a single coating or coatings often only a fraction of a wavelength of light thick. The thickness of the prism must be taken into account in the design of the optical system, but otherwise the resulting image is or should be free of ghosts and other aberrations.

A semi-transparent mirror is often useful in the laboratory, and in applications where light efficiency is of small importance, for overcoming the difficulty, in using a mirror system, of gaining access to the reflected rays. That is, any equipment used for recording the energy of the reflected rays ordinarily lies in the way of the on-coming beam. Instead, a full aperture beam-splitter can be used to provide access to at least some of the reflected rays without appreciable loss of contrast or diffraction resolution. A beam-splitter of optical grade is particularly useful in combination with a mirror-type collimator for providing a beam of parallel light of the highest quality without obstruction or shadowing.

3.3.1.3.2 DICHROIC MIRRORS

Dichroic coatings on pellicles or on precision thin optical plates can be used for special purposes to separate desired bands of the spectrum and to direct these to various instruments, optical elements or detectors. Most often the coatings are multi-layer dielectric thin films whose successive reflections from the various interfaces interfere constructively and destructively to produce a reflected band or a transmitted band of high efficiency. As mentioned above, dichroic coatings can be particularly useful in providing a visual monitoring beam at the same time that an ultraviolet or infrared beam is required for some analytical purpose.

Dichroic mirrors and pellicles have received much attention in the fields of color photography and television, and their use extends over a period of several decades. For less precise purposes, dichroic pellicles can be used, for example, as variegated illuminators, heat-rejecting windows, and for decorative effects. In at least one application a dichroic mirror has been used for auto-collimation of an aerial lens in the deep red for automatic monitoring and restoration of focus in real time for ambient photography in the yellow and ordinary red portions of the spectrum.

3.3.1.4 REFLECTING PRISMS

Reflecting prisms can be used to re-direct a light beam or may invert, revert or rotate an image in its own plane. Dispersion of the beam is small or self-compensating according to the circumstances. One can think of the most useful of such prisms as made up optically of a plane reflection and a thick plate of glass with parallel faces. The plane mirror reflection introduces no aberrations. Those introduced by refraction through the thick plate are either self-compensating on re-entering air or are small

enough to be relatively harmless, or the small aberrations associated with the prism can be compensated elsewhere in the instrument.

Reflecting prisms have several advantages over the use of front-surface reflections. First of all, a refraction into the glass by Snell's Law compresses the rays, a factor that can be used to great advantage in obtaining large fields of view or reduced vignetting within an instrument. Secondly, a prism is or should be a nearly permanent optical device, requiring little or no maintenance. In addition, one finds that internal reflection in a prism for rays inclined beyond the critical angle at the hypotenuse face is nearly 100 per cent efficient, as compared to light losses suffered at a reflection from a front-surface mirror or internally from a mirror coating. Specialized prisms can produce optical results not available from ordinary front-surface mirrors, or at least, not compactly. A discussion of a few of the more common types of reflecting prisms (Jacobs 1943; Driscoll and Vaughn 1978) follows.

The simplest form of reflecting prism is the right-angle prism shown in figure 3-30a. The entering ray is normal to one of the sides and in the illustration is shown entering from below. A total internal reflection occurs at the hypotenuse face without loss of energy and without aberrations. The deflected ray is turned 90-degrees with respect to the initial direction. Right angle prisms of this kind are used in many kinds of instruments and are readily available from many manufacturers. For use with wide-angle systems one often requires that the hypotenuse face be silvered and lacquered to include reflections within the critical angle. Reflection from the metallic coating now absorbs a significant percentage of energy.

3.3.1.4.1 AMICI PRISMS

The Amici (roof) angle prism shown in figure 3-30b also produces a deviation of 90 degrees but at the same time inverts the reflected image. The incident beam is actually divided into two beams, each undergoing two reflections internally at the pair of roof faces. The twinned beams cross over and recombine after the reflections. It becomes necessary for a manufacturer to hold such prisms to extremely close tolerances to avoid blurred or doubled images in the following portion of the instrument. The exacting problems of manufacture have long since been solved successfully. Roof prisms accurate to 1″ or better can readily be obtained.

The penta prism shown in figure 3-30c produces a deviation of 90 degrees but no reversion of image as is the case for the simple right-angle prism. Indeed, the deviation of 90 degrees remains unchanged even if the prism is rotated about an axis perpendicular to the plane of the paper. For this reason such prisms are com-

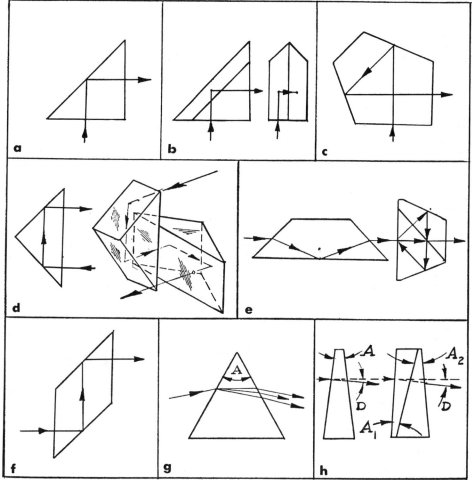

FIGURE 3-30. Basic forms of optical prisms: (a) right-angle prism; (b) Amici roof prism; (c) pentagonal prism; (d) Porro prism and prism system; (e) Dove prism and Pechan prism; (f) rhomboid prism; (g) refracting prism; (h) optical wedge.

monly used in precise instrumentation as for example in range-finders, periscopes and height verniers, where insensitivity to thermal and mechanical rotational errors is required. A penta-prism may also be inserted into the beam just to prevent unwanted reversion of the image. It is also feasible to add a roof to one face of the penta-prism to introduce an additional image inversion.

3.3.1.4.2 PORRO PRISMS

A right-angle prism can be used with a two-fold, internal reflection as shown at the left in figure 3-30d. The returning beam is now deviated 180-degrees from the entering beam and is displaced laterally. Once again, the 180-degree total deviation is unchanged as the prism is rotated about an axis perpendicular to the plane of the paper.

If two such prisms are juxtaposed and even cemented together, as shown at the right in fig-

ure 3-30d, an image may be both reverted and inverted at the same time. That is, one obtains full erection of the image. The ray is undeviated but undergoes a displacement vertically and laterally. Porro prisms of this kind are widely used in prism binoculars to erect the inverted and reverted image formed by a front objective. This also makes prism binoculars quite short and the pairs of prism can be mounted in such a way as to increase the separation of the two optical axes and enhance the stereo effect in the field. Pairs of Porro prisms can also be mounted in another way. One of the prisms is replaced by two smaller right angle prisms that are separated with the other Porro prism in between. The entering light strikes the first of the smaller prisms and is turned through 90 degrees as in a simple right-angle prism. The light then passes through the porro prism and is deviated 180 degrees in the transverse plane. Finally, the light is directed through the second of the smaller prisms that completes its part of the action. The emergent light is parallel to the original direction but is again displaced. The image is properly erected.

3.3.1.4.3 CORNER-CUBE PRISMS

The corner-cube prism, or open, three-mirror cube-corner reflector, is a reflector that returns the light towards its source, provided the light is incident on the diagonal face. It may be thought of as a corner cut from either a hollow or a transparent solid cube by a plane through three diagonally opposite corners. It is useful where other reflectors such as directed diffusing surfaces or beaded screens are not sufficiently efficient.

One of its uses in quantity was in World War II where it was used as a clandestine landing strip delineator, visible only to the pilot of the approaching aircraft with his shielded lamp. Today, it is more useful as a reflector for laser ranging. However, it suffers from diffraction effects and a multiplying of directional errors caused by errors in the reflecting surface. To minimize polarization-dependent phase-shifts at the mirror surfaces, a metallic coating should be applied. Where a plane mirror can be used in collimation, the return signal likely will be several times sharper and brighter than that provided by the corner-cube prism.

3.3.1.4.4 OTHER FORMS OF PRISMS

Another form of prism used to produce an erect image in a telescopic system is shown at right in figure 3-30e. This type of prism is called the Pechan prism, and is becoming increasingly popular for use in binoculars in place of the standard porro prisms. Five internal reflections and a roof face are required, but the emerging optical axis is undeviated and undisplaced.

The rhomboid shown in figure 3-30f is used to displace the optical axis without deviation. Its insertion into a beam is to all effect similar in optical action to insertion of a glass plate of the same glass path. Rhomboids are sometimes constructed at angles other than parallel for the hypotenuse faces in order to obtain a desired small change of direction as well as displacement.

3.3.2 REFRACTING DEVICES

In some forms of prisms, refraction in addition to one or more reflections can be employed usefully to gain some special result. Indeed, as will be discussed below under section 3.3.3, some rather remarkable results can be obtained from refraction that cannot be duplicated by the use of plane mirrors alone.

3.3.2.1 REFRACTING PRISMS

In parallel light the simple right-angle prism can be used in still another fashion, as shown at left in figure 3-30e. This prism is a portion of a right-angle prism and is called a Dove prism. The refraction into the glass at the left face is accompanied by a dispersion of the light. However, after reflection and refraction into air at the right face, the dispersion becomes fully compensated and no chromatic aberration results. There is a side-ways shift of the ray with color which for convenience can be considered as occurring in object-space prior to passage through the prism. That is, the ray after the passage through the prism can be taken as a ray free of color dispersion.

The Dove prism can be used to obtain an inversion or reversion of the image without deviation or displacement of the ray. It is often used to provide for image rotation by the simple expedient of a mounting that rotates about the direction of the entrant ray. In addition, a rotation about an axis perpendicular to the plane of the paper can be used to effect angular excursions of the emergent ray.

The dispersion prism shown in figure 3-30g is most commonly used in the prism spectroscope to produce a desired spectral dispersion. For this reason the glass used in the making of such a prism of itself has a high dispersion. The combination of the dispersive glass and a large prism angle A, as shown in the illustration, can be used to obtain considerable spectral dispersion even from a single prism. One can readily find detailed description and analysis of the dispersion prism in standard optical texts.

The usefulness of a refracting prism in the optical path of an instrument is limited, owing to the dispersion that accompanies the refraction. Even in monochromatic light, refraction may introduce serious asymmetric distortion over the field and may introduce strong astigmatism when placed in the path of converging or diverging rays, except for those pencils of light that traverse the prism at minimum deviation. However, the simple prism and the achromatic prism shown in figure 3-30h may occasionally be of use to the photogrammetrist. If the refracting angle A is less than 2 or 3 degrees, the prism is called a wedge and the angle A is called the wedge angle. For a given application the dispersion of the simple wedge may be tolerable, or if excessive may be reduced or eliminated by a combination of two wedges of different dispersions. The resulting doublet prism is called an achromatic wedge.

The deviation D produced by a single wedge is

$$D = A(n - 1) \qquad (3.65)$$

and the deviation of an achromatic wedge is

$$D = (n_1 - 1)A_1 - (n_2 - 1)A_2 \qquad (3.66)$$

and the condition of achromatism is

$$(n_1 - 1)/v_1 - (n_2 - 1)/v_2 = 0 \qquad (3.67)$$

where A, A_1, A_2 are refracting angles; n, n_1, n_2 are refractive indices, and v_1, v_2 are the Abbé v-values for the respective materials.

Another way of expressing the deviating effect of a wedge prism is in terms of prism diopters. A 1-cm deviation of the light beam from a straight line at a distance of one meter corresponds to one prism diopter.

Refracting prisms are occasionally used in simple lens stereoscopes. Lenses and wedges are ground as one piece. Wedge angles up to 10 degrees are used to permit observation of aerial photographs with an air base larger than the interpupillary distance of the observer. The presence of dispersion can be noticed as color fringes along boundaries of high density-differences.

The simple wedge and the achromatic wedge can be used in other ways. For example, in a camera installation in an aircraft one can make use of a counter-rotating pair of weak wedge-prisms to provide for a fore and aft movement of an image in the line of flight which thereby becomes an external means of image-motion compensation. The achromatic wedge can also be used in place of the rhomboid to produce a small angular deviation of the optical axis in a crowded space. Achromatic wedges are afflicted with small chromatic errors of higher order, related to the secondary spectrum discussed earlier, and therefore are not well suited to work of the highest precision.

3.3.3 SCANNING DEVICES

Many instruments of great analytical power have comparatively small fields of view. It is therefore useful to supplement such instruments with some means of scanning the object and to provide a fixed instrument such as a telescope with a re-directed beam from some object point far off the fixed axis.

3.3.3.1 SCANNING PRISMS

Two Dove prisms (section 3.3.2.1) of the kind shown at left in figure 3-30e can be placed with hypotenuse faces cemented together after each has been coated with silver for use at angles less than the critical angle. The two prisms then form a cube prism that can be used over a full 360° scan in each of two directions to re-direct light from any position of the sphere along the optical axis of a receiving instrument. Commonly, such a cube or double-Dove prism feeds a small telescope that allows an eye in a fixed position to observe the entire sphere.

If one views the entrance pupil of a following telescope through the double-Dove prism backwards from object space, he will find that the pupil has become split, and inverted. The two semi-circular segments are tangent to one another at the curved mid-points. As one scans with the cube prism, the one segment of a circle grows and the other diminishes, on either side as the scan goes one way or the other. At a deviation of 90 degrees only one of the prisms is working and indeed this it true for the complete scan of the inner hemisphere. At a deviation of 180°, the observer is in effect looking at himself and the beam is split by refraction at the roof edge between the side faces.

The splitting of the pupil into two opposed semi-circular segments affects the diffraction resolving power of the telescope. Although each beam traverses the same total glass thickness if the prisms are identical in every respect, in practice small errors of manufacture are present. Accordingly, the use of the double-Dove prism ordinarily produces two images that are made to coalesce in the image plane of the telescope to preassigned tolerances. The diffraction resolution of either pupil will depend on its ambient size and form, and the near coherence of the two images may introduce harmless interference.

3.3.3.2 SCANNING MIRRORS

There are many situations where the designer must resort to one or more scanning mirrors instead of scanning prisms. For example, observations must be made in a spectral region not well suited to use of a scanning prism. It may be that the equipment must be of great size or that it must withstand adverse environmental conditions. The introduction of a scanning mirror, however, involves new problems and new limitations.

The simplest form of scanning mirror is of course the front-surface plane mirror used in a diagonal position. For example, in scanning aerial photography, one can mount the camera with its optical axis horizontal and in the line of flight and rotate the mirror in scan about an horizontal, transverse axis. The line of sight to the ground thus moves in scan, fore and aft, and one can superimpose an image-motion compensation on top of the general scan if a framing camera is used. The scanning mirror may be used to direct the line of sight fore and aft to obtain stereo frames, or can be used for strip photography. In either case the mirror is rotated forward and backwards only within a limited range about a position at 45°.

If the scanning mirror is placed again at the 45° position, one can also cause it to be rotated about an axis in the line of flight and obtain thereby a transverse scan from horizon to horizon through the nadir. This type of scanning produces a rotation of the field that requires considerable engineering skill to accommodate the moving film to the rapidly moving image in the two directions. This type of scan, however, involves no change in the aspect of mirror to camera and has no vignetting over the full sweep of the scan.

In another form of transverse scanning, one can employ a pair of plane mirrors. One of these is fixed at the 45° position in front of the camera as above but tilted about a vertical axis. A scanning mirror on one side is also used at 45° and is rotated about an axis in the line of flight. The line of sight from the ground is thus directed horizontally from the scanning mirror transversely to the fixed mirror and thence into the camera. Scanning is transverse at a high rate without rotation of the field and the film is synchronized to the image motion. Image motion compensation can be accommodated either by a small movement of the camera as a whole or by a shift in azimuth of the scan axis during the scan.

Because of fore-shortening, the range of scan in this kind of mount is limited in angle from the vertical and cannot include the horizon on both sides. In a typical installation within the body of an aircraft, one can arrange for the scan to go through the nadir from a high angle of say 45° from the vertical on the side requiring the higher mirror tilt to perhaps 75° on the side of smaller tilt. In the latter high-angle position the scanning mirror is looking under the fixed diagonal mirror.

In a more elaborate mount of this kind, one can employ a scanning mirror on either side, alternating in action. The central 45° mirror accepts the scan from the first scan mirror from, say, 75° off the vertical to perhaps 10° beyond the nadir. The film motion is synchronized with the image speed and the image motion compensation is introduced by a small non-linear rocking in azimuth of the scan head or of the platen. On completion of the scan through the nadir, the film goes through a loop to give time for the 45° mirror to flip over about an axis in the line of flight to accept continuation of the scan from the other mirror. The film reaches the second slit at the right moment and the second half of the scan is completed out to 75° on the other side of the aircraft. Use of the film loop eliminates need for deceleration and acceleration of the film during the sweep.

There is at least one other way of scanning from horizon to horizon without vignetting and without rotation of the image field. The camera axis is again horizontal in the line of flight and the camera is fitted with a fixed 45° plane mirror tilted about an horizontal transverse axis in such a way as to accept light from the direction of the zenith. Then, two other plane mirrors are fitted into a rigid-roof mirror assembly that rotates in scan about an axis in the line of sight. The two mirrors are at 45° to the optical axis and at 90° to one another with the roof-edge transverse and horizontal at the nadir. If the line of sight comes upwards from the nadir, the light beam reflects from the forward mirror of the roof pair and becomes horizontal in the line of sight directed toward the rear of the aircraft. The beam then strikes the second 45° mirror of the roof pair and reflects vertically downwards. The beam next strikes the fixed 45° mirror which accepts the beam from above and re-directs it into the camera.

The rotation of the rigidly-mounted roof pair about the axis in the line of sight accommodates a scan from horizon to horizon without vignetting, obstruction, change in fore-shortening, or rotation of the field. Indeed, the scan can begin and end above the horizon on either side. This type of mount is somewhat related to the double-Dove prism but there is no splitting of the aperture and no loss of resolution. This specialized device, however, necessarily employs three full-size plane mirrors and therefore is best suited to moderate apertures.

3.3.4 CATADIOPTRIC SYSTEMS

Catadioptric systems have been highly developed but have not been used to any considerable extent in photogrammetry. While combinations of lenses and mirrors have been used for a very long time in military instrumentation, it was only with the introduction of the Schmidt camera around 1929 that optical designers began to appreciate the effectiveness of combining the advantages of each optical form into a single system.

Mirrors of moderate curvature do have substantial dioptric power but also involve difficulties of obstruction. Lenses, however, can be added before and after a single reflection or pair of reflections to obtain additional corrections or to obtain advantages not possible with the mirror system alone.

The Schmidt camera, for example, employs a weakly deformed correcting plate at the center of curvature of a spherical mirror. The working stop is thus at the center of curvature of the mirror and there is in effect no single optical axis. The mirror used in this fashion is afflicted with spherical aberration but is free of coma and astigmatism. The Gaussian focal surface is spherical about the same center of curvature and is therefore concentric with the primary mirror.

The aspheric correcting plate acting as a weak deformed lens is shaped to remove the spherical aberration of the system. The combination thus produces high-quality images on a curved focal surface. One notes that the action of the correcting plate in correcting spherical aberration holds up well, off-axis, purely because the zonal lens power is low and insensitive to the additional refraction of inclined rays or to the fore-shortening of inclined bundles in the meridional plane. Indeed, one can say that the Schmidt camera yields the highest quality performance for the simplest optical combination that can be devised.

3.3.4.1 BALLISTIC OPTICAL SYSTEMS

Reflecting optical systems have been used very widely in the tracking of missiles and rockets. The all-reflecting telescope as a Cassegrainian combination of primary and secondary is well adapted to relatively compact telescopes of long focal length. For many purposes, however, one needs to have a compact system of intermediate focal length but with low f-number and corresponding high lens speed. Various catadioptric combinations have been used and some of these have variable focal lengths, either by some form of zoom action or by use of interchangeable auxiliary optics. Speeds in the neighborhood of $f/2$ to $f/4$ have been used with focal lengths up to several meters with great emphasis placed on precision of tracking and on smoothness and fast slewing of the tracking mount.

3.3.4.2 SATELLITE-TRACKING SYSTEMS

Catadioptric systems have become widely used for precise tracking of artificial satellites. Most of these systems are based on some adaptation of the Schmidt camera, including those that replace the Schmidt correcting plate by a Maksutov, correcting, meniscus lens. In an extreme form, lens speeds to $f/1$ have been used with fields of view up to 30° or more and with apertures up to 0.5 metres. Even larger systems are within the state of the art but are not required currently. Those larger systems that do exist are more or less identical with astronomical telescopes but are mounted to permit tracking at relatively high angular rates in 2 or even in 3 coordinates.

3.3.5 DIFFRACTIVE DEVICES

The wave nature of light permits the use of several highly specialized devices not ordinarily of value to the photogrammetrist. Accordingly, it is sufficient here just to mention that the diffraction grating has been brought to a very high state of perfection. Historically, the diffraction grating began with Joseph Fraunhofer in the early part of the 19th century and became of great value in the physical laboratory through the work of Henry Rowland in the latter part of the same century. Under modern conditions the use of laser interferometry and control systems of the highest quality have made it possible to produce quite large diffraction gratings nearly free of troublesome ghost images in a spectrum of bright emission lines. It has also become possible through the use of lasers and holography to produce very large gratings of high precision by photographic methods.

Similarly, the advanced state of the art of thin film coatings has made it possible to produce interference filters of high quality that will be discussed more at length in section 3.5.

3.3.6 THIN FILM OPTICS

Mention has already been made above from time to time of the use of thin films for reflecting, filtering, or splitting light, etc. While these applications are of very great importance, perhaps the most striking use to date of thin dielectric films lies in reducing or eliminating the stray light reflected otherwise from untreated surfaces. Fortunately, the energy in stray light can be added constructively to the desired refracted or reflected beam as needed.

In the early part of the century optical manufacturers became familiar with the fact that the "browning" of certain kinds of polished optical glasses tended to reduce the intensity of the reflected light. Of course the browning arose from the aging and skeletonizing of the polished substrate in constant contact with atmospheric contaminants but raised the interesting question as to the optical consequences. The uncontrolled drop in mean index of refraction of the skeletonized layer corresponded in a rough kind of way to the use of a thin film.

Suffice it to say that over the decades that have followed, the knowledge and practice of thin films have reached a very advanced state.

By means of interferometry, lasers, control equipment, electron guns, ion beams and the like, optical workers are able to coat as many layers as they wish onto a substrate, knowing at all times the thickness of each layer up to the point of changing the coating to the next layer and so on. Furthermore, the fast electronic computer has made it possible to calculate the thicknesses of the layers to achieve some required result.

One can now obtain optical systems having dozens of surfaces that are so well coated for low reflection as to retain high efficiency in the final image. For example, an aerial lens with as many as 24 surfaces has been coated so well as to have an efficiency as high as 83 percent in its final image. Similarly, one can obtain mirror coatings with more than 97 percent reflectivity over a wide spectral band or can have an optical system with high efficiency in the visual while rejecting the infrared. Manufacturers supply catalogs of various kinds of coatings, hot and cold mirrors, dichroics and beam-splitters. Specialized coatings can be obtained when the requirements are known.

3.3.7 FIBER OPTICS

Fiber optics has played only a minor role to date in photogrammetric systems but nevertheless can be very useful. Fiber optics production has reached quite unanticipated volume as a result of technical advances in the manufacturing processes and because of high demand from the communication industry. For optical purposes, bundles of optical fibers can be used to convey light to hard-to-reach places, to gain compactness, to circumvent obstructions from electronic gear, to transport images from one spot to another, to flatten image fields from image forming systems, and in many other ways.

It would be possible in principle to use a field-flattening element made up of optical fibers to produce a perfectly flat field. The diameter of the individual fiber can be as small as the most critical image obtainable from a camera lens and the light losses can be tolerated as long as the fiber is short. One can visualize a system with distortion corrected by means of optical fibers suitably positioned for the entering and leaving rays.

Optical fibers can be used in range-finders for transporting dual images to a convenient location for the observer around intervening obstructions. They can also be used in illuminators, including those that require flexible bundles for some changing position. They can also be used to produce an enlargement of an exit pupil.

3.3.8 BEAM-EXPANDERS AND LASER SOURCES

Space prevents going into any detailed discussion of the techniques of handling laser beams and associated equipment. However, the monochromatic nature of the laser beam and great intensity of the coherent beam emitted by the laser lead to many useful applications. Particularly in the testing of optical equipment, the laser has proved to be invaluable and the state of the art goes well beyond the fondest dreams of optical workers even of several decades ago.

Because of the small cross-sections of the collimated beams from a laser, one must usually increase the cross-section by adding suitable lenses (or mirrors in afocal systems) to the optical train. For example, one can use an array of refracting prisms quite far from the angle of minimum deviation to cause a laser beam to increase in cross-section without change in its collimation. One can also use afocal arrays of positive and negative lens elements, or concave and convex mirrors, to do the same. Manufacturers can provide such beam-expanders on request.

3.4 Compensation of Lens Aberrations

Plane-parallel glass plates are often used in the optical path of various photogrammetric instruments. These plates may be stage plates (plates which hold the negative flat during exposure). They may also be colored, neutral or gradient-density filters, or the glass plate equivalents or reflecting prisms. Optical elements of this category when used in close proximity to object- or image-planes must be of high optical quality, free from air bubbles, digs and scratches which impair the optical image. Stage plates must meet rigid flatness requirements to avoid small but irregular anomalies.

More generally, the optical designer can often make good use of relatively weak plates or shells in optical systems. Some of these may no longer be plane-parallel or even concentric shells and some may be figured to weak or strong aspheric shapes as required by the application.

3.4.1 OPTICAL-GRADE PLATES USED IN OPTICAL SYSTEMS

Glass plates of all diameters and practicable thicknesses up to quite gigantic sizes have been used at one time or another in optical systems. Glass discs of optical grade have been made in diameters up to several meters and to thicknesses up to 20 cm principally for use in schlieren photography in wind-tunnel work. The largest blanks usually have been made of a hard, borosilicate crown glass. Mirror blanks, of course, have been made in diameters up to 6 metres but while uniformly annealed have no requirements for uniformity of refractive index or transparency.

3.4.1.1 WINDOWS

From time to time it becomes necessary to insert one or more windows into the optical train to effect a boundary between widely different environments, as for example, in supersonic aircraft. Windows of this kind may be exposed to both heat and cold of exaggerated degree, even during a single flight, and may be exposed on the outer side to great heat as from supersonic flow or to great cold as from high altitude flight. The inner side of the window in either kind of flight may be at normal temperatures or may be exposed to heating or cooling air, as the case may be. More will be written about aircraft windows below (*see* section 3.5.4).

Thick glass plates of optical grade have also been used as windows to support the atmospheric pressure for evacuated image-forming optics within. If a full atmospheric pressure of about 1 kg/cm² is involved, the total load on a window of say 30 cm diameter is substantial, namely, about 700 kg. It is known both from calculation and from observation that a window of 30 cm diameter and 2.5 cm thickness will bend to a sagitta of about 0.08 mm under full atmospheric load. Cell support for such a load should not be near the periphery and the window should be uniformly supported around the edge on a firm but slightly flexible seal, resting in turn upon a ledge ground flat to tolerances of the order of 0.005 mm or better. Otherwise, in time such a window may break under load.

Several aerial lenses of long focal length have been enclosed experimentally in a vacuum to provide a stable environment. Quite large windows have been used to enclose collimator systems for testing tunnels, primarily to reduce or to eliminate air turbulence and air thermals that degrade the test beam. In another application a BK-7 window measuring about 0.6 metres diameter by 6 cm thickness is in use on an evacuated solar telescope to support the atmospheric load and to prevent internal heating that would tend to destroy the image quality in a large air-filled tube. In still another instance, both BK-7 and fused silica windows of about 1 metre diameter have been used in connection with a very large vertical installation for solar research. Indeed, the demand for flat windows of high optical quality has persisted over the years beyond all expectations to the point that special polishing machines and polishing methods have been devised. Foremost among these is the art of continuous polishing that has produced large windows of excellent quality.

3.4.1.2 CORRECTING PLATES

If an optical designer needs to incorporate an optical plate into an instrumental design, he may also wish to put the plate to work as a figured correcting plate for the additional correction of aberrations in the system that follows. If the aspheric corrections are relatively weak, as for example, not more than a few wavelengths spread smoothly over the aperture, then centering is not very important and one can make use of a pressure window or window-filter to improve the system that follows. For example, a figured window has been used to reject the infra-red wavefronts from a large refracting telescope to prevent undesirable internal heating.

The most familiar application of the figured window comes from the Schmidt correcting plate widely used in catadioptric systems and discussed in section 3.3.4 above. The Schmidt plate was introduced around 1929 by B. Schmidt in Hamburg, Germany, not as a window, although that purpose is admirably well served, but as a means to correct the spherical aberration of a large spherical primary mirror. In the Schmidt telescope, such as the 122 cm instrument at Palomar in California, a fully enclosed tube is used with the correcting plate-window at the upper end and the large primary mirror at the lower end. No attempt is made to use the plate also as a pressure window but in smaller systems, this triple role can be adopted, that is, correcting plate, window, and pressure seal.

To a good approximation one finds that the figured correcting plate located at the real stop of a system is equivalent, if negative in power, to the use of a concentric glass shell with spherical surfaces, curved around the real stop. Even with intervening optics and many re-imagings, the theorem still holds. The equivalence is actually valid only through the third order of approximation, for the shell has only the two parameters, one radius and one thickness, whereas the correcting plate can make use of several parameters in the form of coefficients of a power series in addition to being allowed a more versatile location within the system. Use of the correcting plate seems to get directly at the heart of the task of correcting aberrations, whereas the concentric shell must be used in special situations that nevertheless by selection can be quite effective.

Thus, one has on the one hand the Schmidt plate and on the other the concentric shell that in turn can be elaborated into the Maksutov negative meniscus of which the concentric shell is a special case. The Maksutov meniscus can be so devised as to be achromatic for rays of importance, or at least nearly so, and can be used in place of a correcting plate to correct for the spherical aberration and coma of a mirror system. The most general situation, however, is for the designer to make use of Schmidt plates or Schmidt-like aspheric corrections here and there in an optical system, or Maksutov menisci, the latter with spherical surfaces or with superimposed aspheric figuring to improve higher order corrections. There is no fixed rule about these applications for they take many forms but suffice it to say that both Schmidt corrections and Maksutov shells play an important role in modern catadioptric systems.

The nearly plane Schmidt plate carries the advantage that little glass is used and that excellent results can be obtained with optical glass not necessarily of the highest grade. Inasmuch as aspheric figuring is being carried out anyway on a rather flexible plate, the optician can overcome by further figuring any minor variations in the index of refraction throughout the disc as long as sudden changes in index do not occur. With the relatively thick Maksutov meniscus, however, at least for large systems homogeneous optical glass of the best grade is usually required.

Schmidt-like corrections can often be inserted into an optical system even without the use of a special glass plate. Quite often a designer can introduce one or more figured surfaces by superposition on the otherwise spherical surfaces of the lens system to obtain improved corrections. As a historical note, one may mention that the optical equations introduced by K. Schwarzschild in 1905 included aspheric terms that might easily have been applied at the time to any optical system, including the as yet uninvented Schmidt system. Indeed, Schwarzschild carried out a very broad investigation of the utility of aspheric terms in an optical system bearing his name that is comprised of two figured concave mirrors. Perhaps it was only an unfortunate choice of algebraic representation or too much emphasis on an all-mirror system that prevented Schwarzschild himself from discovering the Schmidt principle 24 years before Schmidt devised his famous system from experienced optical reasoning in his own right. In any case, the third order aspheric terms are quite general and can readily be applied whenever their use is warranted.

3.4.1.3 FIELD-FLATTENERS

The concept of inserting a field-flattening lens near the image plane was introduced by Piazzi-Smyth as long as as 1866. It is obvious that a lens very close to the image surface has very little influence on the scale of the image unless of exaggerated depth. In particular, a negative plano-concave element placed with its flat side at or just before the image plane can be used to reduce the Petzval sum and thereby to flatten the field, as long as astigmatism is also controlled. In practice the designer thinks of the field-flattening element as an additional element whose aberrations can be compensated elsewhere in the design of the optical system. As a rule the negative field-flattening element introduces distortion but for systems of moderate field angle, quite good results can be obtained by distortion-compensation in the front portion of the optical system during the process of design. For example, a front triplet of crown and flint having of itself strongly under-corrected field curvature and relatively weak surface curvatures can be used with a simple field-flattener of appropriate design to obtain a quite good overall result. In this way a very large triplet of inexpensive glasses can be managed.

Field-flatteners need not be spherical. Indeed, a designer can make use of a relatively thick plate with aspheric surfaces much like the Schmidt plate. To a first approximation, the field-flattening effect depends on the variation in glass thickness rather than on the surface normals and for light crown glass the focal position at any given zone is shifted farther from the lens by about one-third the thickness of the plate. Thus, if the uncorrected field has a zone with sagitta of 2 mm from flatness, the field-flattening aspheric plate in that zone would have a 6 mm thickness plus the base thickness on axis. If the field-flattening effect is strong, however, the surface refractions must also be included in the calculations. All of this is readily available in the standard programming of the electronic computer for optical design.

3.4.1.4 GLASS PRESSURE PLATES AND RESEAUX

Some aerial cameras make use of glass pressure plates to hold the film flat during exposure. Engraved reseaux on glass can also be used to superimpose fine grid lines on the aerial negative at the time of exposure to facilitate precise reductions on negatives, prints or diapositives without regard to later changes of scale. Improvements all along the line in recent years have removed most of the need for precise reseaux except for calibration purposes and for

fiducial marks. These improvements include much-improved, stabilized film bases, more uniform processing, better coating of the negative and copying materials, vacuum platens instead of glass pressure plates, etc. In the case of the Wild Orbigon designed for photogrammetric surveys from space, however, a glass reseau plate is used, engraved with a network of 121 crosses 10 mm part. Similarly, plates with fiducial marks are used in the Wild P31 and P32 Terrestrial Cameras. The Zeiss Pleogon AR and the Topar AR use a plate with a precise grid in front of the image plane.

For either the glass pressure plate or the vacuum platen, the contact between film and platen must be interrupted to permit winding of the film between exposures. Otherwise, the film may become scratched, and ultimately the glass pressure plate also. For this reason the glass pressure plate must be made of a very hard glass such as a boro-silicate glass or else of fused silica and must be easily replaced when scratches do appear.

In other types of optical systems, glass reseaux can be inserted in any conjugate image plane. The designer must choose a location such that engravings can be sharply seen or recorded, which in turn means that the aberrations following the reseau as seen by the eye or sensor must be adequately well corrected at the same time that points in object space are sharply defined.

For work of the highest precision one must engrave the reseau on a substrate of a proper thermal expansion to go along with the rest of the instrument. A reseau on fused silica, for example, while very insensitive to change with temperature, in practice might not be quite as good as a reseau on a type of glass that matches the change in scale with temperature of the instrument as a whole.

3.4.1.5 DISTORTION CORRECTING PLATES

The use of glass plates in the path of image-forming rays without design compensation introduces three types of aberrations: (1) displacement of the image plane; (2) radial distortion accompanied by astigmatism; (3) deterioration of resolving power.

3.4.1.5.1 DISPLACEMENT OF THE FOCAL PLANE

Insertion of a plane-parallel glass plate between a lens and its focal plane displaces the focal plane away from the lens by an amount approximately equal to one-third of the thickness of the plate. In figure 3-31 this displacement is shown by the distance $00'$. It may be computed from the equation

$$00' = T - T/n \qquad (3.68)$$

in which T is the plate thickness and n its index of refraction. For commerical plate glass the

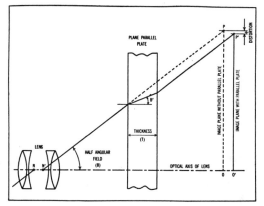

FIGURE 3-31. Radial distortion and image displacement caused by a plane-parallel plate. A plane-parallel plate placed between the lens and its image plane produces (1) displacement of the image plane away from the lens, (2) radial distortion and curvature of field in the displaced image field, and (3) unbalance of lens aberrations.

TABLE 3-5. DISTORTION D INTRODUCED BY A
PLANE-PARALLEL PLATE

B°	(D/T) n = 1.52	B°	(D/T) n = 1.52
5	0.0001	35	0.0532
10	0.0010	40	0.0854
15	0.0035	45	0.1324
20	0.0085	50	0.2006
25	0.0173	55	0.2996
30	0.0315	60	0.4207

given in the table, the following equation may be used (Gardner and Bennett, 1927):

$$D = T\left(\tan B' - \frac{\tan B}{n}\right) \qquad (3.69)$$

where B' may be computed from the relation

$$\sin B/\sin B' = n, \qquad (3.70)$$

and where B and B' are the off-axis angles in air and in the plate. A negative sign for D indicates a shift toward the center of the field. For values of n between 1.48 and 1.56, the tabular values are correct to 3 in the 4th decimal place for angles up to 25 degrees and 6 in the 4th place for angles between 25 and 50 degrees.

If the plate is inserted between object plane and first lens vertex, the distortion displacement in the image plane is outward and can be obtained from table 3-5 if the values for (D/T) are multiplied by the scale ratio, that is, by the ratio of image size to object size, which is the magnification m. Clearly, if the object plane is far away as in aerial photography, a plane-parallel plate on the object side introduces negligible distortion. On the other hand, if copying is being carried out at a low scale-ratio or magnification, the distortion introduced by a plate in the one conjugate can be compensated by the distortion from a plate of suitable thickness inserted in the other conjugate distance, the thicknesses being related by the scale ratio. The compensation is exact in principle, however, for distortion only.

3.4.1.5.3 IMAGE DETERIORATION

When the task of designing an optical system has been accomplished a compromise has been reached whereby the various aberrations have been reduced and balanced in a favorable manner. The additional insertion, removal or thickness change of a glass plate already incorporated into the system may disturb this balance and lead to image deterioration and to undesirable distortion residuals.

3.4.2 THE PORRO-KOPPE PRINCIPLE

Ignazio Porro (1801–1873) and later Carl Koppe (1884–1910) developed a method of distortion correction which is based on the restoration of the geometry existing in the camera at the

value of $n = 1.52$. It is important to note that introduction of the glass plate does not change the Gaussian focal length in any way. When object and image planes are both in air, the front and back focal lengths are equal (3.24). Clearly, for parallel light coming from the right through a plane-parallel plate located in image space, the front focal length will be unaffected. Therefore the rear focal length will also remain unchanged. Thus, introducing the glass plate shifts the focus but does not change the scale. As will be discussed below, distortion however is affected.

Inclined pencils of light coming to a focus through the glass plate far off-axis are affected by a negative astigmatism of small value and by a weak contribution to barrel distortion. For very fast systems one will also encounter a slight over-correction of spherical aberration and an inward coma. There is no change in these small third order aberrations so introduced if the glass plate is shifted forward toward the rear lens vertex unless the glass plate is not optically flat. In some cases introduction of a glass plate can compensate an otherwise inwardly curving tangential focal surface, but one must proceed with caution. The combination is most useful if the designer takes the glass plate into account during the design of the entire optical system.

3.4.1.5.2 IMAGE DISTORTION

The radial distortion introduced by the insertion of a glass plate into an optical system causes a radial displacement of the image point, inward if the plate lies between lens and image, outward if between lens and object plane. The magnitude of this distortion is given in table 3-5 where the distortion D is expressed as a decimal fraction of the plate thickness T. To compute the distortion for specific indices of refraction or for angles not

time of exposure of the negative. It involves the placing of the negative (or diapositive) behind the taking lens at the distance equal to the calibrated focal length of the camera, and centering the picture on the camera's fiducial marks, a process covered by the term "interior orientation". A rotatable sighting telescope focused for infinity is placed in front of the lens. Upon illumination of the aerial photograph, the lens, operating in reverse, emits rays from all image points which have angular relationships identical to those existing when rays from ground objects passed through the lens at the instant of exposure. The rays emerging from the lens are free from distortion, regardless of their path inside of the camera. Specifically, the rays from C and D (figure 3-32) emerge as beams of parallel rays, separated by the angle A. Consequently, the true angle A can be measured by the suitably mounted sighting telescope.

Figure 3-32 which illustrates this principle bears a strong resemblance to figure 3-29, which illustrates the photo-goniometer. (This term is very appropriate since the sighting telescope, equipped with horizontal and vertical circles, functions as a geodetic theodolite). The angle measured between two thus collimated image points on the photograph equals the angle subtended at the taking lens by the same two points on the ground. The principle also extends directly to measurements from oblique photographs and to mathematical changes in perspective applied during reductions.

The Porro-Koppe principle has since been applied to a variety of stereoscopic plotting instruments in which two photo-goniometers, with plateholders for overlapping pairs of aerial photographs, now termed projectors or plotting cameras, are optically and mechanically linked together so as to make stereoscopic measurements possible. They incorporate several design variations, such as rotating telescopes with stationary projectors, rotating projectors with stationary telescopes, or combinations of both solutions.

From the above discussion it may appear as if the Porro-Koppe system is a cure-all for lens distortion problems. Actually this is not the case, for the following reasons: (1) The requirement that the taking and projecting lenses be identical is difficult to attain in practice, because small but troublesome differences in the distortion characteristics always exist. (2) Since a given plotting instrument uses photography taken by several different aerial cameras of the same nominal focal length, it is necessary to provide the projectors with an adjustment to take care of the differences in focal length (usually several millimetres). Thus, the principal distance set in the projector may be shorter or longer than its focal length, and consequently, the emerging rays are no longer parallel but are diverging or converging, respectively. Therefore the point of intersection (Cardan center) of the axes of the sighting telescope cannot be placed arbitrarily within the bundle of emergent rays, but must be located at the front principal (nodal) point of the projector lens. (3) In stereoscopic observation utilizing this principle, horizontal planes appear increasingly inclined upon approaching the edge of the stereomodel, a condition which some observers find disturbing.

3.4.3 COMPENSATING LENSES

Compensating lenses in photogrammetric instrumentation serve to correct the most troublesome lens aberration inherent in aerial photography, namely, radial distortion. The task of eliminating this property from an existing photograph, or of reshaping its distortion curve to adapt it to the requirements of a given plotting machine, is a problem which involves the design of a lens to a predetermined distortion curve. This curve should be the mirror image of the curve expressing the distortion property of the taking lens.

The new era of nearly distortion-free aerial camera lenses, and the perfection of aspheric corrector plates as will be discussed below in section 3.4.4, has practically eliminated the need for compensating lenses in photogrammetric printers. In multiplex printers, they have been replaced by distortion-free, high resolution lenses and aspheric corrector plates. The latter can be tailor-made to correspond to distortion curves, correcting lens distortion and, if required, the influence of atmospheric refraction and earth curvature. They solve the distortion compensation to within ± 5 micrometres.

In the older era Pestrecov attacked the problem of distortion compensation in the early development of Bausch & Lomb multiplex equipment and solved it within the limits of the art in the 1940's. The lens formula developed by him produced a reverted distortion curve which ap-

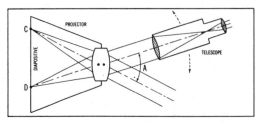

FIGURE 3-32. Schematic of the Porro-Koppe principle. The relationships between lens and photograph are fully restored in a replica of the aerial camera. The bundle of rays from the illuminated photograph, emerging through the lens, is angle-true and unaffected by the distortion of the taking or projection lens. It is received by a telescope universally rotatable about the front principal (nodal) point. Directional observation of image detail is utilized in optical and mechanical solutions of plotting apparatus.

proximated the nominal curve of the Metrogon lens, then the leading camera lens for photogrammetric surveying. With this lens in a projection printer, the projected image from the Metrogon negative was redistorted in such a manner that the sum of the inherent and superimposed distortion values left only small residuals in the entire image field. The match, as shown in figure 3-33, was by no means perfect, but was considered acceptable at the time. The departure of the resulting distortion curve from a straight line was of the order of 30 to 40 micrometres, referred to the plane of the negative. Through continuous research, the initial performance of this lens formula, used for the production of multiplex printer lenses, was materially improved with regard to resolution; but the distortion properties remained nearly unchanged.

Several other methods of converting distortion characteristics are available and have been widely used at times. For the reasons given above, they have lost their initial significance and are, therefore, not discussed further (*see* Sharp, 1949).

3.4.4 COMPENSATING PLATES

Compensating plates for correction of radial distortion are used in several types of plotting machines and projection printers. They are either plane-parallel glass plates or glass plates on which a special aspheric curve has been ground and polished on one surface.

If a curve is selected with ordinates equal in magnitude but opposite in sign to the distortion curve of the lens, it can be used to compensate the lens distortion to zero throughout the field. The uncorrected Hypergon lens, for example, used in multiplex and Balplex projectors, has distortion characteristics almost identical with those of a glass plate of approximately 2 mm thickness (fig. 3-34). By selecting the appropriate thickness for each individual lens, one can compensate all projectors uniformly to nominal

zero distortion, with residuals not exceeding ±3 micrometres. The same method is applied to the compensation of multiplex and Balplex printers to correct the residual distortion of the projection system to within ±5 micrometres.

If the distortion curve to be corrected is of a higher order, containing one or several points of inflection, a plane-parallel plate does not suffice and has to be modified by deformation of one of its surfaces. Sometimes, a slightly spherical surface renders sufficient approximation, but more often the second surface has to given a profile of higher complexity. This calls for a process of curve-fitting in which the several aspheric coefficients of a power series are determined by a least squares procedure from the distortion curve to be compensated. The aspheric shape so determined must then be reproduced in the optical shop to adequate precision.

Aspheric corrector plates are now used in a number of applications. In the Wild Autographs, the distortion is removed from wide-angle photographs by placing a specially ground correction plate between the diapositive and the viewing system (Kasper, 1950). One side of this plate is an aspheric surface of revolution that varies in such a manner as to compensate for the distortion inherent in the photographic transparency placed in contact with the plate. Figure 3-35 is a schematic diagram of the method with the variation in thickness of the correction plate greatly exaggerated. Actually, the thickness variation of one plate inspected that was intended for use with the original Aviogon photography was less than one millimetre. The shape of

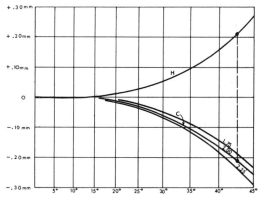

FIGURE 3-34. The compensating plate as an integral part of lens design. Distortion curve *H* depicts the positive distortion of an uncorrected Hypergon lens at the operational magnification ratio 1:10, referred to the object plane. Curve *C* is the negative distortion of a compensating plate placed on the object side of the lens. The algebraic sum of corresponding values along the two curves, if chosen to be zero at an off-axis angle of 42 degrees, departs from zero by only a few micrometers from center to the edge of the field. Aberration corrections in the design of the lens include the compensating plate.

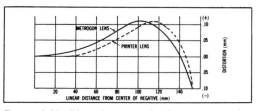

FIGURE 3-33. Distortion correction by a compensating lens. The transformation of distortion caused by a Metrogon lens is accomplished by projective printing of the aerial photograph through a lens of opposite distortion values. The attempted cancellation is not perfect, as shown by the dashed line, drawn as the mirror image of the printer lens curve from the Metrogon curve. A complete reduction to zero of the residual distortion is extremely difficult to obtain by this principle.

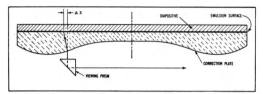

FIGURE 3-35. Aspheric corrector as used in Wild Autographs to go with the older Aviogon forms (cross section, schematic). The aspheric plate is positioned with its flat side in the object plane of the plate holder. The diapositive transparency is in contact with it. The deformation of the lower side of the plate is greatly exaggerated in this illustration. The image point observed in the orthogonal viewing system of the Autograph is displaced from the line of collimation by Δx, which equals the radial displacement of the image point in the photograph, within ±10 micrometres.

the plate can be visualized by rotating the cross-section shown around the central axis.

As can be seen in the figure, the distortion correction Δx varies as the orthographic viewing system scans the diapositive image. The accuracy of compensation by the aspheric surface now extends down to only several micrometres, referred to the plane of the diapositive, for any one particular camera lens. Indeed, the very great advances made in recent years in the design of wide-angle mapping lenses having very much reduced distortion, as for the Universal Aviogons I and II, lead virtually to an elimination of the need for distortion correcting plates, except possibly as a final touch-up to an already very satisfactory situation.

A second application of aspheric corrector plates is found in the projection printers of several manufacturers, which serve to prepare diapositive transparencies for use in plotting instruments. Printers of this type (see table 3-2) are designed for a variety of reduction ratios, input and output formats, and camera focal lengths. They have provisions for use with and without corrector plates and permit simple con-

version from one mode of printing to the other. Wild of Switzerland, and Zeiss of W. Germany produce universal printers with a wide flexibility of application. Bausch & Lomb printers are of the fixed-ratio type, with a small range of variability of their ratios of 1:1 (for Kelsh and Wild plotters), 1:2.78 (for Balplex 90° and 120° projectors), and 1:5 (for multiplex equipment). Because of the increase in path length when the corrector plate is interposed between the film stage and the projection lens, mechanical provisions are made to change the conjugate distance in the object space, in accordance with equation 3.59.

3.4.5 MECHANICAL CORRECTION DEVICES

Contrasting with the optical approaches of correcting radial distortion in mapping systems is a mechanical solution of the problem that may appear to be a better cure than optical expedients. The mechanical solution, which the Italian designer Santoni reduced to practice in 1932 is based on the concept that radial distortion is a result of focal length variation given by the equation

$$\Delta f = Df/r \qquad (3.71)$$

in which Δf is the focal length change that the distortion value D produces at image points of the radial distance r from the center of the picture. Santoni's mechanical device changes the principal distance of the projector automatically as extra-axial image points are sighted or projected in his plotter. A basically identical solution is employed in several direct-projection plotters of the Kelsh type. This mechanical solution is described in sufficient detail in chapters XIII and XIV and needs no further treatment here. The solution, however is not entirely free of optical complications and requires a high degree of accuracy to the mobile linkages in order to be comparable with the purely optical solutions.

3.5 Filters and Windows

While close attention must necessarily be given to the many intricate details involved in the design of a precise, photogrammetric, lens system, one must also keep in mind that the user of such equipment will need to employ both windows and filters in the aerial installation. Many an excellent lens has been degraded in performance by the addition of inadequately made auxiliary optical parts.

Filters of one kind or another have been used in aerial photography from the earliest beginnings. Windows, on the other hand, were generally not required until the combination of high altitude and high speed aircraft appeared at the onset of World War II.

3.5.1 COLOR FILTERS

Filters play an important part in photogrammetric operations. Their purpose generally speaking is to transmit, absorb, or reflect light waves of a certain portion of the spectral colors of a light source, or they may have the purpose of attenuating light energy, or passing light waves of the full spectrum in one specific plane of vibration.

Filters are characterized by their colors when viewed in transmitted light. However, their performance cannot be appraised by their color alone. A graphical plot of transmittance measurements over the full visible range or beyond,

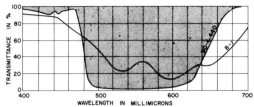

FIGURE 3-36. Spectral transmittance of glass and multilayer filters. Transmittance curves show the spectral region in which light may pass through a filter and the percentage which emerges. B&L glass filter B-1 transmits less than 40 per cent within the visible range of the spectrum and has a minimum at 600 nm. Multilayer filter 90-4-440 is a wide band filter with steep absorption edges at 480 and 640 nm. It practically eliminates all visible light but transmits ultraviolet and infrared very efficiently.

as made in a spectrophotometer, tells the full story. Ordinary glass or gelatin filters show relatively simple transmittance characteristics as compared to multi-layer interference filters (fig. 3-35).

For photographic purposes one finds it convenient to subdivide the spectrum into five regions: ultraviolet, blue, green, red and infrared. Figure 3-37 gives the approximate wavelength limits for each region and also indicates the range of sensitivity of various photographic emulsions. By the use of a filter one can select those portions of the spectrum best suited to each type of work.

The use of a filter in front of a camera lens usually requires an increase in the exposure time above that applicable without a filter. Filter factors by which the exposure time should be multiplied (typically, for light yellow, 1.5; deep yellow, 2; red, 4) depend actually on the given emulsion. Film manufacturers furnish information with their product on selection of filters related to atmospheric conditions and character of the terrain to be photographed. In some camera designs the color filter is an integral part of the lens system and not interchangeable. Indeed, some large camera systems make use of a coated filter on a selected internal surface.

3.5.1.1 DARKROOM FILTERS

These filters are commercially available in a great variety of colors and sizes. They are gelatin filters of the Wratten series or equivalent, mounted between ordinary glass plates. Such filters are selected in accordance with the sensitivity characteristics of the photographic emulsion which they are to protect. Film manufacturers customarily specify the darkroom filter which may safely be used in processing. As a general rule, panchromatic film must be processed in complete darkness.

3.5.1.2 CAMERA FILTERS

These filters as a rule are placed in front of the camera lens on the long conjugate side to prevent change of focus caused by thickness. In some cases, however, where filters of smaller diameter are to be desired, filters are placed in the converging beam immediately behind the lens system or occasionally internally between lens groups by means of a filter slide or Waterhouse type of rotary holder. In both cases the optical designer must take the thickness of the filter into account as to its effect on image aberrations, distortion, and focus. If some uses of the camera lens require no filter, the designer must then require that a clear optical element of the same thickness be used until replaced by the filter. If at all possible, filters and replacement glass plates used internally or behind the lens must be mounted in such a way as to prevent use of the camera lens without either.

Camera filters must have internal and surface qualities equal to or better than those of the lens itself and are therefore to be made of optical materials, free of bubbles and striae, and planeparallel to narrow tolerances measured in seconds of arc. Their spectral transmission must be stated in terms approved by the American Standards Association.

The purpose of filters in aerial photography is either to eliminate or to increase the contrast of the spectral reflection of the terrain features. Atmospheric haze consists of small particles of dust and water vapor floating in the lower layers of the atmosphere, but even a clear atmosphere scatters blue and violet light far more than red. By placing a blue-absorbing filter over the lens (yellow in color), one can eliminate this unwanted scattered light. The exposure is made by means of the remaining colors of light reflected from the ground objects. For maximum penetration of haze, particularly in oblique photography, infra-red emulsions and filters are used.

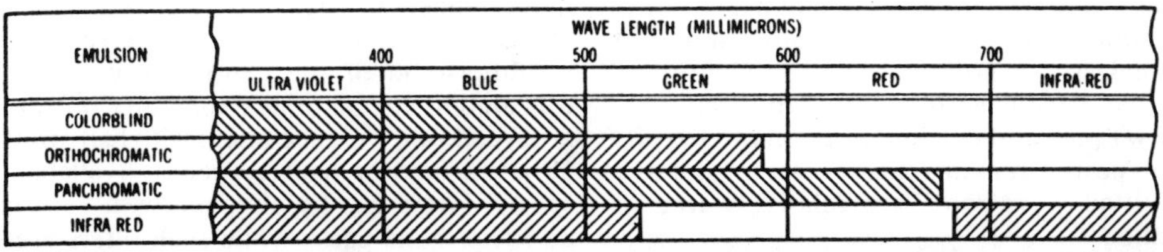

EMULSION	WAVE LENGTH (MILLIMICRONS)					
	ULTRA VIOLET	400 BLUE	500 GREEN	600 RED	700 INFRA-RED	
COLORBLIND						
ORTHOCHROMATIC						
PANCHROMATIC						
INFRA RED						

FIGURE 3-37. Photographic regions of the spectrum. Color sensitivity of commercial photographic emulsions varies with the type and the producer's formula. The above limits for the most widely used types of aerial film are approximate.

Any further loss of contrast still in the transmitted wavefronts can be compensated in part by use of fine-grain materials of enhanced contrast and by suitable processing.

3.5.1.2.1 GELATIN (WRATTEN) FILTERS

Various dyes can be entrapped in gelatin and coated onto supporting thin films of suitable transparent material or by special order onto optical glass substrates. For many decades a wide range of color filters has been available in the Wratten series manufactured by Kodak. To avoid affecting resolution in lens testing by interposition of a Wratten filter in the optical train, one can usually place the gelatin filter over the light source or preceding the test target. If the filter must be used in the optical array, it is usually best to place the filter not far in front of the image surface. For more permanent use in such fashion, a Wratten filter can be cemented between glass plates of all necessary optical quality.

More than 100 types of filters are available in the Wratten series, as described in a technical publication available from Kodak. Color-compensating filters for aerial photography are also listed and much technical data, including filter factors, can be found in this publication.

3.5.1.2.1 GLASS FILTERS

While the Wratten filters have many admirable properties with respect to sharp cut-off of wave-band boundaries at a relatively low cost, they are not easily mounted for precision use in large sizes. If cemented between large glass plates, the differential stresses introduced by the cementing, aggravated by thermal changes, tend to warp the plates and to degrade optical performance. Consequently, such mounted filters are most often used near the image plane.

Manufacturers of optical glass have found it possible to produce melts of glass with the addition of a variety of compounds that have filtering action. Such colored glass filtering is very familiar to everyone in the form of ruby or cobalt glasses for the home or for artwork and there are many kinds. Catalogs of a large variety of filter glasses are readily available from manufacturers in the United States and abroad. Indeed, for the smaller camera lenses one can purchase fully manufactured optical filters of high quality for use in aerial photography. As a rule, these commercial filters are thinner than one might otherwise choose for scientific application, but nevertheless the problems of manufacture of thin filters have been overcome. In spite of residual warping and bending of the thin filter, its optical power remains nearly zero, and the transmitted wavefronts are relatively free of aberration.

Some colored filter glass derives its action by the scattering of the shorter waves of light from the colloidal suspension of particles in the glass.

Such filter glass will usually show much scattered light if illuminated by a bright beam of light and viewed against a dark background. However, a single such filter used in aerial photography will do more good than harm and has been extensively used in the past.

3.5.1.2.3 THIN-FILM FILTERS

Under modern conditions one can obtain precise performance simply from multi-layer coatings on a glass plate or window of optical grade. Such filters are available from companies specializing in thin films and their properties are given in extensive technical catalogs. These coatings are often available as dichroic filters that transmit one band and reflect another. *See* also section 3.3.1.3.2.

3.5.1.2.4 FILTERING BY DOPING OF GLASS OR CEMENT

In a few cases one can obtain filtering action by the doping of a lens cement with a suitable chemical dye. Because of the thinness of the usual layer of optical cement between elements, the filtering action obtained in this way is comparatively weak but may be sufficient, nevertheless, to reduce the ultra-violet transmission of an instrument. Standard optical glasses have in some cases been doped during manufacture to effect ultra-violet absorption and to balance the transmission of a production lens element to compensate excessive ultra-violet sensitivity of a color emulsion.

3.5.1.2.5 FABRY-PEROT FILTERS

Occasionally, one needs to have a filter that transmits a very narrow band for technical purposes. Such filters can indeed become quite complicated, as for example, the crystal-quartz monochromators used in solar research. For more usual purposes one can obtain a very narrow band over almost any portion of the usable spectrum by resorting to the interference filters introduced in the first part of this century by Fabry and Perot.

In the modern version a sandwich is made up of a dielectric layer only a few wavelengths of light thick lying between partially transparent coatings of silver on supporting glass plates. For a chosen wavelength the thickness can be so carefully controlled that a resonance is built up between the interfaces using quite high orders of interference. As a result only constructively interfering wavelengths in the immediate neighborhood of the central wavelength are transmitted and all others are either reflected or absorbed. Interference filters with half-widths of 50 angstroms or so are readily available, but special filters can be obtained with very much narrower band-pass. While such filters are useful in many applications in the laboratory, laser illumination is far more nearly monochromatic. Fabry-Perot filters are therefore most valuable when used in the filtering of natural sources of light, as from the sun and stars.

3.5.1.3 PROJECTION FILTERS

Special color filters are used with direct-projection plotters. These mapping devices produce superimposed images of overlapping photographs on the measuring screen. Filters in the path of the projection rays from an incandescent light source traveling toward the screen, and analyzers in the path of rays reflected from the screen, are needed to separate the two images for separate perception by the eyes of the observer. The simplest and most universally applicable solution is based on dividing the spectrum of visible light into blue and red halves, with the dividing line at about 555 nanometres. One then uses the blue half of the spectrum for projecting one photograph and the red half for projecting the other. Another technique makes use of neighboring peaks in the emission curve of high-pressure mercury vapor lamps.

3.5.1.3.1 COMPLEMENTARY COLOR FILTERS

As the two halves of the full visible spectrum produce white light, so do selected spectral ranges on both sides of the dividing line. For example, cyan (blue-green) and red, or purple and amber. They are called complementary colors. Not all of the possible color combinations are suitable for visual use. The sensitivity of the human eye peaks at 555 nm in the green range of the spectrum. At 600 nm (light red), the sensitivity is about 60 percent. At 525 nm in the blue-green, it is about 80 percent. At 490 nm in the blue, it is only about 20 percent of the peak value. Except for the most brilliant illumination visibility goes practically to zero at 400 nm in the violet and beyond 700 nm in the deep red and infrared.

The ideal combination of complementary colors is that which is perceived by the human eye as of equal brightness or visual effect. Nearest to this ideal case are red and blue-green filters. A residual brightness difference of these colors, occuring after reflection from the screen, is eliminated if one chooses a proper thickness ratio of the red and blue-green filters and analyzer glasses. As a refinement, the blue-green spectacle is usually provided with a $-\frac{3}{8}$ diopter power ground and polished into the filter blank, while the red filter is given zero dioptric power. In this way the focus difference in the human eye between the two colors is compensated. The spectral transmission curves of these filters are shown in figure 3-38.

The advanced technique of evaporating ultra-thin coatings of material on clear glass in vacuum ovens (*see* section 3.5.1.2.5) has made it possible to design and control the transmittance of these layers to a higher degree than has been possible in the production of stained glasses. Multi-layer interference filter transmission

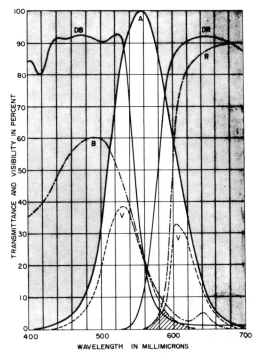

FIGURE 3-38. Visibility curve of the human eye, and transmittance of filter. Curve A shows the ability of the human eye to perceive spectral colors of equal brightness (energy basis). Curves B and R present the transmittance curves of glass filters of the complementary colors cyan (blue-green) and red. Curve V is the resulting visibility curve in the spectral regions of the glass filters B and R. Curves DB and DR are the transmittance curves of dichroic filters of the multilayer type.

curves display steep absorption edges quite close to the 555 nm dividing lines and short tails beyond this line, as in figure 3-38. This accounts for the extinction of satellite images that become disturbing in stereo-vision when the lower ends of the two transmission curves project too far into the region of the opposite color. Filters of this kind are used with some Multiplex projectors.

In many applications complementary color filters are flat discs of stained optical glass. In Balplex projectors, they are sections of spheres with concentric surfaces, enveloping the light bulb. The red spherical filters are equipped with a heat absorbing shell of concentric surfaces.

3.5.1.4 NEUTRAL-COLOR FILTERS

These filters are of the non-reflective type. If truly neutral, the filters do not change the spectral composition of the transmitted light. Neutral filters act solely as attenuators and are well suited for balancing the visual brightness of two images observed in monocular or binocular vision. For such purposes either a linear wedge can be used to provide for a continuous variation of density or else a step wedge can be used with calibration at equal density difference between adjacent steps.

Thin metallic films can be used at appropriate thicknesses to produce neutral filters. Aluminum and chromium are not adequately neutral. The most nearly neutral coating appears to be from evaporated Inconel, an alloy. Indeed, a good Inconel coating leaves little to be desired over the full visual range from 400 nm to 700 nm. Neutral wedges or the equivalent in circular form are now available in density ranges from 0 to 4 in diameters up to 15 cm.

Neutral filters can also be obtained from multi-layer coatings of dielectric films, but these are more nearly of constant transmission or reflection over selected band-widths far less than the visual range. One might find it feasible to have the transmitted beam from one reflect from a second to obtain approximate neutral attenuation over the visual range.

3.5.1.5 ANTI-VIGNETTING FILTERS

Another kind of filter of great interest to photogrammetrists is the neutral filter that has a radial variation of density from axis outwards. Many of the older style mapping lenses, including the Metrogon and Cartogon systems, have a very strong fall-off of illumination with off-axis field angle. Indeed, because of adverse pupil distortion and vignetting, the observed fall-off of image brightness exceeds the already severe losses indicated by the $\cos^4$ law. The modern lenses of the Aviogon form by Wild, Zeiss, Itek and others are in some cases adequately free of fall-off of illuminating power with off-axis angle and are vastly superior to the earlier lens forms, although at the trade-off of physical size and complexity. The Geocon series lies in between.

Anti-vignetting filters often are added in front of the mapping lens to reduce the fall-off as a compromise between uneven illumination and average exposure time required in flight to suppress image motion and vibration. Well made anti-vignetting filters do not affect lens resolution. Such a filter is of course densest at the axis and falls off in density to near zero at the extreme off-axis angle (*see* table 3-1). A filter in front of a wide-angle mapping lens of 153 mm focal length may already be at least 15 cm in diameter when placed nearly tangent to the front lens vertex.

3.5.2 HEAT FILTERS

For the conservation of the metrical properties of survey films, various Government specifications provide an upper temperature limit to which the aerial negative may be exposed in the processes of printing, rectifying, and diapositive making. Others limit temperature gradients in operations where hot-burning lamps are used over sustained periods of time. Protective filters in these applications do not essentially change the color of the transmitted light, but they do stop most of the light energy at the red end of the visible spectrum and beyond. It is also possible to reflect the unwanted infrared heat waves harmlessly away or to a special absorber not actually in the optical train.

3.5.2.1 ABSORPTION FILTERS

If a heat absorbing glass is used, some means must be provided for getting rid of the heat picked up by the glass. The heat absorbing glass generally already has an appreciable visual density that reduces the light efficiency in the instrument. In the region of absorption the glass is effectively black. The radiation received from the nearby tungsten filament may be at a color temperature of 3000°K. The absorbing glass will become warmer and warmer of itself until it radiates away the excess heat at some lower color temperature altogether invisible to the eye. Some help can be provided by mounting the filter in an aluminum frame which in turn picks up heat and readily conducts it away to some larger metallic mass or radiating surface. Further help can be provided for by cooling both filter, mount and light source by circulated cool air.

Heat-absorbing glasses are made in sheets of 5 mm thickness or so by the makers of flat glass and optical glass in this country and abroad.

A water cell is also a good absorber of heat, particularly because of the high specific heat of water in comparison to other substances and because of broad-band infra-red opacity. Indeed, a water cell in the beam provided with slowly moving tap water or with recirculated water through an external heat-exchanger can be used to remove most of the infra-red heat energy from the visually transmitted light.

3.5.2.2 HEAT-REFLECTING FILTERS

If the heat-absorbing filter glass is also coated with a heat rejecting coating, whether of thin gold or of a multi-layer coating calculated for the purpose, a considerable improvement in cooling effect can be obtained. It is important to design the equipment in such a way as to reflect the unwanted infra-red light to some absorber such as a metal mass or a water cell located harmlessly out of the way or separately cooled.

Heat-reflecting multi-layer coatings can be designed to have a very steep boundary between transmission on the visual wavelength side and reflection in the infra-red. Such a reflection is spoken of as a hot mirror. Conversely, a coating that tramsmits the infra-red but reflects the visual waves is called a cold mirror. Cold mirrors are often used in powerful projection equipment to allow escape of the heat waves from the projector.

3.5.3 POLARIZING FILTERS

These filters are nearly neutral with respect to the spectrum. They are physical screens coated onto a plastic film or glass substrate. They pass any wavelength of visual light but limit the passage to a single plane of propagation of the wave motion. Light incident upon such a filter vibrates transversely to the direction of propagation, and if unpolarized, in an infinite number of planes. After passing through the polarizing filter, the light vibrates predominantly in one plane whose orientation is determined by the structure of the polarizing screen. The transmitted light is said to be plane-polarized. If the light is now directed through a second such screen, the transmission is now determined by the azimuth of one with respect to the other. For crossed polarizers the transmission is very low.

There are several limited applications of these filters in photogrammetry. Their use in multi-projection plotters, to present the superposed images to the observer's eyes singly on the tracing screen, is defeated by the prime requirement that the screen surface reflect the incident rays diffusely and adversely to the law of reflection. Polarized light incident on such a screen material becomes severely depolarized upon reflection.

Polarizing filters have been used in aerial photography to prevent sunlight strongly reflected by bodies of water from fogging the exposure. The problem, however, is one of finding the required rotational setting of the polarizing filter in front of the camera lens and may require careful visual monitoring or some form of automatic control.

Analytical instruments often require use of a polarizing filter either to obtain a variable density in an easy way or to provide some partial compensation in observing light already partially polarized. Polarizing films are readily available, both mounted and unmounted, and can be used to obtain a very great range of density. At full density with crossed polarizing films, one may see some violet light leakage that for most purposes is relatively harmless. There may also be some light leakage of this kind in the infrared which makes it unwise to use crossed polarizers for observing any very bright source of radiation having an infrared component. This would be true for the sun, or infrared laser illumination, or even the xenon arc lamp.

The variation in density of the counter-rotating polarizer in a standard mount varies as the $\cos^2$ of the angle of rotation. The maximum transmission occurs at two settings 180-degrees apart with minimum transmission at the 90-degree positions in either sense of rotation.

As in the case of gelatin filters, polarizing films must be mounted between optical plates if resolution is not to be affected. Large films may tend to separate away from optical cements used to establish optical contact between glass and film. For large filters, therefore, one may have to use a semi-fluid optical cement which is actually a very viscous solution of a selected plastic base in xylol. A properly made filter will last indefinitely without separation of film from glass.

3.5.4 WINDOWS

Aircraft photographic windows must be designed and fabricated to a tolerance on surface quality and homogeneity that will allow at most only an insignificant degradation in the performance of the camera system which is to look through them.

It is evident that in many cases, and certainly over some portion of the field of view, aircraft windows in the past have significantly degraded the system performance. In one instance on record a large thick window was made up from fused silica of unknown provenance and used in test flying at supersonic altitudes and speed along with a 12-element precision aerial lens that in the laboratory yielded an average resolution in excess of 150 line-pairs per millimetre. It was found after much flying that the recorded image was afflicted with bad astigmatism and image blur coming from the non-homogeneous fused silica and that the lens system, even though very complex, was innocent of such degradation. The conclusion is that if the window or windows are to be part of the optical train, one must give them as careful attention as any other elements of the train.

The same comment is true also for any given color filter, which too is a kind of window, added to the optics of any instrument. In addition to having the required filtering action, whatever that may be, the color filter must be as well made with respect to the transmitted wave front as any other element of the instrument used at full aperture in the optical train. Indeed, in view of the elementary principles involved, namely, plane-parallel surfaces as boundaries to an homogenous optical medium, one might well require that any given filter or window be made to exacting standards. This is all the more true if multiple windows are needed in an aerial installation, or multiple filters.

During World War II the use of selected polished plate glass was the normal procedure and was generally adequate for the smaller lens apertures of an inch or so in diameter, but serious wavefront deformations no doubt occurred on use with the larger lenses. The permissible wedge for mapping windows as adopted then was probably adequate for a single pane, but with the advent of multi-pane windows stricter parallelism must be maintained.

In any case, the adequacy of polished plate glass is academic since it has largely been replaced by flotation glass. Flotation glass was first developed in England and has admirable vision characteristics for the small pupil of the eye and is perfectly suitable for the construction trade. For the larger apertures of aircraft systems, its surface quality and parallelism are probably not fully adequate except possibly by selection.

Current process windows are being fabricated from glass blanks ground and polished to the required surface tolerances. The cost is, of course, higher than for the previously used selected plate, but there is the assurance that each pane will be within the tolerances demanded by the system performance. Where operating conditions produce excessive thermal gradients, the windows are fabricated from low coefficient materials such as fused quartz.

The nature of the aircraft window installation leads to an unfavorable length to thickness ratio, particularly with regard to flatness. For normal plane-parallel optical flats, this ratio seldom exceeds 10:1, whereas in windows with comparable wavefront deformation limitations, the ratio may be as high as 50:1. It is important, therefore, to recognize that surface flatness in itself is not of primary importance, but rather the smoothness and regularity. A smooth, planar plate can be bent to an operationally absurd extent before there is any practical effect on the transmitted wavefront. Paralellism is also significant in multi-pane windows to avoid imaging splitting and in cartographic cameras to avoid distortion.

The window mount should ensure that the glass is "floating free." The metal mount should be designed to retain, not grab the pane. A layer of flexible material such as silastic, between the metal and the glass will prevent chipping under vibration and shock and absorb the stresses introduced from different thermal expansion coefficients.

In multi-pane windows, it is usually detrimental to use a cement between the edges of adjacent panes. If the bond is sufficiently strong to add significant mechanical stability, it also introduces sufficient stress to produce unacceptable bi-refringence and often fracture. The most successful windows are retained only at the ends with a nominal gap between frames. If a pressure differential across the window is required, a flexible caulking material can be used to seal the seam.

Any non-parallelism or wedge will introduce image-splitting when the entrance beam covers two or more panes. This is caused not only by differences in the wedge angle but by unavoidable differences in the angle of incidence. The permissible image split is a matter of debate, but if it is arbitrarily assumed that the resultant degradation should not result in more than one sixth root of two pattern loss in resolution, then the split should be no greater than about one-fourth the line separation of the maximum lens-film resolution. For a lens-film resolution of 80 lines/mm, this limits the permissible image split to 0.003 mm, or in angular tolerance, 24 seconds of arc divided by the focal length in inches. Thus, for a 153 mm focal length lens, the net image split due to a different wedge vector and incident angle cannot exceed 4 seconds of arc.

In the specification for a specific window requirement, it is well to emphasize transmitted wavefront deformation rather than the details of the glass fabrication. In any case, each window should be treated as an essential component of the system which will look through it, and thus should not be generalized any more, than say, one of the elements of the objective lens.

For normal photogrammetric installations homogeneous optical glass can readily be procured from manufacturers in various countries. Unless unusual conditions are to be encountered, one normally uses a hard boro-silicate crown glass for the window, typically BK-7 in the Schott notation. Such a glass type can withstand normal conditions of wear and from an optical point of view can be made of material homogeneous to small residuals in the sixth decimal place. The transmitted wavefront can be held to within a small fraction of a wavelength of rms variation around the ideal planar wavefront without undue cost.

For unusual conditions, as for supersonic flying, one may replace the BK-7 window by one made of a high-quality, fused silica which also can be obtained from several manufacturers. Fused silica has a very low coefficient of thermal expansion and therefore can withstand sudden changes in temperature. In even more severe cases multiple windows are used and one or more surfaces may be coated with heat-reflecting dichroic films to prevent excessive thermal heating. In some cases a thin, single layer of gold is used which reflects the infra-red and transmits most of the visual light.

Careful studies have been made of window materials in an effort to find a material that will transmit an undisturbed wavefront in spite of changing, localized conditions. The optical path along any given ray is an integral over the product of index of refraction and the physical path length. The index of refraction is a function of temperature that depends on the particular kind of material. The physical thickness too changes with temperature according to the coefficient of thermal expansion, also dependent on the material. For a window a perfect material would have both quantities totally insensitive to temperature.

Fused quartz has an adequately small coefficient of thermal expansion but unfortunately has a rather large variation of index of refraction with temperature. BK-7 has both a significant thermal expansion and variation of index of refraction with temperature. No material has yet been found that is insensitive to temperature in both respects. According to the relationship $(n-1)ds = -s\,dn$, the change ds in the thickness s of the window produces a change in the net optical path that can be off-set at the thickness s by a change dn in the index of refraction with temperature, provided dn is negative. Optical glasses of the FK (fluor-crown) series do have a negative change of index with temperature that for a practicable thickness can off-set the change in optical path caused by thermal expansion. Unfortunately, the FK glasses cannot withstand surface abrasion or strong thermal gradients and are therefore not suited to use as windows under extreme conditions. In practice, therefore, windows of either BK-7 or fused silica or a combination of these are used.

Windows for use with mapping cameras with focal lengths from 86 to 153 mm do not generally have to withstand extreme physical conditions. Therefore, well made windows of BK-7 or fused silica are entirely adequate. It is advisable, however, to employ only windows that have been adequately tested beforehand to exacting standards.

3.6 Summary

This chapter would be incomplete without attention being given in retrospect to the truly enormous advances made in optical science during the past century or so. In a myriad of avenues of optical research, both in theory and technology, progress has been decisive and important.

Credit must also be given to the workers on the foundations of optical science, extending far back into history. While the ancient Greek mathematicians attained a very advanced stage in pure geometry, including a quite thorough knowledge of the properties of conic sections, one must regret that the Greek philosophers who held the stage for so very many centuries took little note of optics or treated optics as illusion. Even the manifold blessings of optometry lay near their grasp but were overlooked through a state of mind that avoided practical matters.

Very likely, however, many artisans even in the earliest periods of the entire ancient world made use of some aspects of optics in their work, possibly as secrets of their trade, silently evident in the exceedingly fine detail wrought into some ancient jewelry. Ancient surveyors must have been very familiar also with visual methods of sighting to the limit of the eye which indeed carries one to accuracies below a minute of arc.

We are all exceedingly fortunate to benefit in all kinds of ways from the advances made by optical scientists and by scientists in closely related disciplines affecting optical endeavors. Only a short sketch has been given throughout this chapter and almost every statement can be usefully amplified. All in all, photogrammetrists in particular benefit by a bringing together of many fine developments. One may hope that the future will bring even greater rewards.

SELECTED REFERENCES AND BIBLIOGRAPHY

REFERENCES

Bean, R. K., 1940, Source and correction of errors affecting Multiplex mapping: Photogrammetric Engineering, Vol. 6, No. 2, pp 63–84.

Berg, A. W., 1961, Reproduction of high-acuity aerial photography: Photographic Science and Engineering, Vol. 5, No. 6.

Brumley, C. H., and Coombs, W. F., 1963, Electronic space rods for large plotters: Photogrammetric Engineering, Vol. 29, No. 5, pp 715–718.

Campbell, Charles E., 1962, The optimization of photographic systems: Photogrammetric Engineering, Vol. 28, No. 3, pp 446–455.

Deterding, Leo G., 1962, High precision, strain-free mounting of large lens elements: Applied Optics, Vol. 1, No. 4, pp 403–406.

Gardner, I. C., 1954, The experimental evaluation of lens performance: International Archives of Photogrammetry XI, p. 224.

Gardner, I. C., and Bennett, A. H., 1927, The compensation of distortion in objectives for airplane photographs: Journal of the Optical Society of America, p. 245.

Gardner, I. C., and Case, F. A., 1937, Precision camera for testing lenses: Journal of Research, National Bureau of Standards, Vol. 18, No. 449, p. 984.

Gruner, H., 1962, The history of the Multiplex: Photogrammetric Engineering, Vol. 28, No. 3, pp 480–484.

Hardy, Arthur C., and Perrin, Fred H., 1932, The Principles of Optics: McGraw-Hill Book Co., Inc., p. 464.

Higgins, G. C., and Jones, L. A., 1952, The nature and evaluation of the sharpness of photographic images: Journal of the Society of Motion Picture Engineers, Vol. 58, No. 4, pp 277–290.

Higgins, G. C., and Wolfe, R. N., 1955, The relationship of definition to sharpness and resolving power in a photographic system: Journal of the Optical Society of America, Vol. 45, No. 2, pp 121–129.

Hotchkiss, R. N., Washer, F. E., and Rosberry, F. W., 1951, Spurious resolution of photographic lenses: Journal of the Optical Society of America, Vol. 41, p. 600.

Jacobs, P. H., 1943, Fundamentals of Optical Engineering: McGraw-Hill, New York, pp 151–165.

Karren, R. J., 1968. Camera Calibration by the Multicollimator Method. Photogrammetric Engineering, Vol. 34, No. 7, pp 706–719.

Kasper, H., 1950, Methods for compensating the distortion of camera objectives in Wild plotting instruments: Photogrammetria, Vol. 2, No. 1, pp 4–5.

Kingslake, Rudolph, 1951, Lenses in Photography: New York, Case-Hoyt Corp.

Magill, A. A., 1955, Variation in distortion with magnification: Journal of Research of the National Bureau of Standards, Vol. 54, No. 3, pp 135–142.

McDonald, Richard K., 1961, Optics and Information Theory: Physics Today, Vol. 14, No. 7, pp 36–41.

National Bureau of Standards, 1949, New precision camera calibrator: Technical News Bulletin, Vol. 33, No. 1, pp 8–10.

National Military Specification for Coatings, JAN-F-675.

Optical Society of America, 1962, Lens Design by advanced automatic computer: Journal of the Optical Society of America, Vol. 52, No. 7, p. 839.

Pennington, John T., 1947, Tangential distortion and its effect on photogrammetric extension of control: Photogrammetric Engineering, Vol. 13, No. 1, pp 135–142.

Pestrecov, K., 1947, Photographic resolution of lenses: Photogrammetric Engineering, Vol. 13, No. 1, pp 64–85.

Pestrecov, K., 1954, Notes on Russian photogrammetric optics: Photogrammetric Engineering, Vol. 20, No. 3, pp 488–492.

Pestrecov, 1959, Radial distortion: its calibration, computation in non-Gaussian image planes, and compensation: Photogrammetric Engineering, Vol. 25, No. 5, pp 702–712.

Ronchi, Vasco, 1957, Optics-the science of vision: New York Univ. Press.

Schade, Otto H., 1948, Electro-optical characteristics of television systems, Part I-Characteristics of vision and visual systems: RCA Review, Vol. 9, No. 1, pp 5–37. Part II-Electro-optical specifications for television systems: RCA Review, Vol. 9, No. 2, pp 245–286. Part III-Electro-optical characteristics of camera systems: RCA Review, Vol. 9, No. 3, pp 490–530. Part IV-Correlation and evaluation of electro-optical characteristics of imaging systems: RCA Review, Vol. 9, No. 4, pp 653–686.

Sears, Francis Weston, 1949, Optics: Addison-Wesley Press.

Sewell, Eldon D., 1948, Field Calibration of aerial mapping cameras: Photogrammetric Engineering, Vol. 14, No. 3, pp 363–398.

Sharp, John V., 1949, Increased accuracy of the multiplex system: Photogrammetric Engineering, Vol. 15, No. 3, pp 430–436.

Sharp, John V., and Hayes, H. H., 1949, Effects on map production of distortions in photogrammetric systems: Photogrammetric Engineering, Vol. 15, No. 1, pp 159–170.

Strong, J., 1938, Procedures in Experimental Physics: Englewood Cliffs, N.J., Prentice-Hall.

Tayman, W. P., 1974. Calibration of Lenses and Cameras at the USGS. Photogrammetric Engineering, Vol. 40, No. 11, pp 1331–1334.

Washer, F. E., 1956, Effect of camera tipping on the location of the principal point: Journal of Research of the National Bureau of Standards, Vol. 57, p. 31.

Washer, F. E., 1956, Source of error in various methods of airplane camera calibration: Photogrammetric Engineering, Vol. 22, No. 4, pp 727–740.

Washer, F. E., E., 1957, A simplified method of locating the point of symmetry: Photogrammetric Engineering, Vol. 23, No. 1, pp 75–88.

Washer, F. E., 1957, The effect of prism on the location of the principal point: Photogrammetric Engineering, Vol. 23, No. 3, pp 520–532.

Washer, F. E., Prism effect, camera tipping, and tangential distortion: Photogrammetric Engineering, Vol. 23, No. 4, pp 721–732.

Washer, F. E., and Case, F. A., 1950, Calibration of precision airplane mapping cameras: Photogrammetric Engineering, Vol. 16, No. 4, pp 502–524: Journal of Research of the National Bureau of Standards, Vol. 45, pp 1–16.

Washer, F. E., and Darling, W. R., 1954, Factors affecting the accuracy of distortion measurements made on the nodal slide optical bench: Journal of the Optical Society of America, Vol. 49, No. 6, pp 517–534.

Washer, F. E., and Darling, W. R., 1959, Evaluation of lens distortion by the inverse nodal slide: Journal of Research of the National Bureau of Standards, Vol. 63C, No. 2.

Washer, F. E., and Darling, W. R., Evaluation of lens distortion by the modified goniometric method: Journal of Research of the National Bureau of Standards, Vol. 63C, No. 2.

Washer, F. E., and Tayman, W. P., 1955, Variation of resolving power and type of test pattern: Journal of Research of the National Bureau of Standards, Vol. 54, No. 3, pp 135–142.

Washer, F. E., and Tayman, W. P., 1960, Location of the plane of best average definition for airplane camera lenses: Photogrammetric Engineering, Vol. 26, No. 3, pp 475–488.

Washer, F. E., and Tayman, W. P., 1961, Location of the plane of best average definition with low contrast resolution patterns: Journal of Research of the National Bureau of Standards, Vol. 65C, No. 3, pp 195–202.

Washer, F. E., and Tayman, W. P.; and Darling, W. R., 1958, Evaluation of lens distortion by visual and photographic methods: Journal of Research of the National Bureau of Standards, Vol. 61, No. 6.

SELECTED BIBLIOGRAPHY

General

Driscoll, W. G. and Vaughn, W. (Editors) Handbook of Optics, McGraw Hill, New York, (1978).

Geometric Optical Theory

Born, Max and Wolf, E., Principles of Optics, Macmillan, New York (1958).

Conrady, A. E., Applied Optics and Optical Design, Dover, New York, Vol. 1 (1957); Volumn II (1960).

Hertzberger, M., Modern Geometrical Optics, Interscience, New York (1958).

Optical Design

Cox, A., A System of Optical Design, Focal Press, New York.

Brixner, B. "Automatic Lens Design for Nonexperts", Applied Optics 2(12): 1281–1286 (1963).

Feder, D. P., "Optical Calculations with Automatic Computing Machinery", J. Opt. Soc. Am. 41: 630–635 (1951).

———— "Automatic Optical Design", Applied Optics 2(12): 1209–1226 (1963).

Kingslake, R., (Editor) Applied Optics and Optical Engineering, Academic Press, New York, 5 volumes (1965–1969).

Shannon, R. R., "Closing the Loop in Optical System Design", IEEE Trans. AES-5 (2): 273–278 (1969).

Transfer Functions

Abbott, Fred, "The Evolution of the Transfer Function: Part I. Concepts and Definitions", Optical Spectra 1970 (March): 54–59 (1970).

Barakat, R., "Computation of the Transfer Function of an Optical System from the Design Data for Rotationally Symmetric Aberrations". J. Opt. Soc. Am. 52(6): 985–997 (1962).

Barakat, R. G., "Numerical Results Concerning the Transfer Functions and Total Illuminance for Optimum Balanced Fifth-order Spherical Aberration", J. Opt. Soc. Am. 54: 38–44 (1964).

Gerasimova, O. A. and Nilov, A. A., "Problems involved in use of Transfer Functions in Optics", Geodesy & Aerophot. 1976 (2): 123–127 (1972).

Noffsinger, E. B., "Image Evaluation: Criteria and

Applications. Paper No. 2: The MTF Criterion'',
S.P.I.E. 9: 95–103 (1971).

Scott, F., Scott, R. M., Shack, R. V., "The Use of
Edge Gradients in Determining Modulation-transfer
Functions'', *Phot. Sci. & Eng. 7:* 345–349 (1963).

Scott, R. M., "Summary and Example", *Photogr.
Sci. & Eng. 9(4):* 261–263 (1966).

Manufacture

Twyman, F., Prism and Lens Making (2nd edition),
Hilger and Watts, London (1952).

Testing

Malacara, Daniel (Editor), Optical Shop Testing, John
Wiley, New York, (1978).

Zoom Lens-systems

Back, F. G. and Lowen, H., "Generalized Theory of
Zoom Systems'', *J. Opt. Soc. Am. 48(3):* 149–153
(1958).

Kienholz, D. F., "The Design of a Zoom Lens with a
Large Computer'', *Applied Optics 9(6):* 1443–1452
(1970).

CHAPTER IV

Aerial Cameras

Author-Editor: ROBERT G. LIVINGSTON

Contributing Authors: CLYDE E. BERNDSEN, RON ONDREJKA, ROBERT M. SPRIGGS, LEON J. KOSOFSKY, DICK VAN STEENBURGH, CLARICE NORTON, DUANE BROWN

4.1 Introduction

AERIAL CAMERAS in the past have been defined as cameras specially designed for use in aircraft. Now that we are in the space age, it is necessary to expand the definition to include cameras used in outer space to photograph the earth, the moon, and other members of our solar system. In addition to aircraft and balloons, high altitude rockets and artificial earth, lunar, and planetary satellites have carried cameras. Pictures from space have been transmitted back to earth by television in some cases, and the photographic film itself has been returned to earth in other cases. Hence, a section of this chapter is devoted to a brief *description* of these special cameras; the *use* of their products is covered elsewhere in the MANUAL.

The primary purpose of this chapter is to supplement the information on aerial cameras contained in the first three editions of the MANUAL OF PHOTOGRAMMETRY. At the same time, an attempt has been made to produce an account which is sufficiently complete to eliminate the necessity for library research.

A considerable amount of information on the frame camera is included in view of the large number of these instruments in use, and because of the great importance of the frame camera to both photogrammetry and photo interpretation. The panoramic camera is covered in detail since its high resolution is particularly desirable for reconnaissance. Strip, multiband, and other special-purpose cameras are described. Special considerations in camera design and development are discussed, accessory equipment such as camera mounts and controls are covered briefly, and camera installations and system configurations are described. Included is a table showing current United States military aerial cameras, commercial domestic and foreign aerial cameras, together with their performance and general physical characteristics. A section on calibration covers the very latest developments and techniques in this highly important procedure which establishes the camera's internal geometry.

Aerial cameras can be classified in a number of ways: by type (frame, panoramic, strip, multiband); by angular field (normal angle—up to 75°, wide angle—75° to 100°, super-wide angle—over 100°); by focal length (short—up to 6 inches, normal—6 to 12 inches, long—above 12 inches); by use (mapping, reconnaissance, special). Developments can be traced through any of these divisions, but consideration is given here only to the broad categories of reconnaissance cameras, mapping (or cartographic) cameras, and special-purpose cameras, particularly with respect to the reasons why different types of cameras and camera features were developed.

The development of *reconnaissance cameras* in the United States received its greatest impetus during World War II. Prior to that time, interest in reconnaissance cameras, or any other kind of aerial camera, was limited. Throughout the war, cameras were developed to meet continually changing requirements. Focal lengths became longer and longer as reconnaissance aircraft were forced higher and higher by antiaircraft fire. Many cameras were built with interchangeable lenses to provide a wider range of uses. For example, the U.S. Air Force Type K-22A, a day reconnaissance frame camera, was made with lenses of 6-, 12-, 24-, and 40-inch focal lengths; and there are three different types of 40-inch lenses.

To meet the needs of the photointerpreter, a wide variety of cameras is required. Cameras are needed for low-altitude, high-speed photography. These must have short-focal-length, wide-angle lenses, fast shutters, image-motion compensating devices, and short cycling times. At a low altitude, even the wide-angle lenses cover only a narrow strip on the ground. Multi-camera installations consisting of vertical and oblique cameras can be used to cover wider paths. Another approach to low-altitude, high-speed photography is the use of vertical and oblique strip cameras. The strip camera makes a continuous exposure on a moving strip of film by passing the film over a stationary slit in the focal plane of the camera at a speed synchronized with the image of the terrain as it moves across the focal plane.

At high altitudes, long-focal-length frame cameras are used to provide a reasonable scale and good ground resolution. However, for long focal lengths, the angular coverage, and hence the ground coverage, becomes very small. Three means have been used to overcome this. First the camera format was expanded from the standard 9 × 9 inches to 9 × 18 inches, with the 18-inch dimension perpendicular to the line of flight. The second was the employment of two or more cameras in a fan arrangement, perpendicular to the line of flight. The third was the development and employment of the aerial panoramic camera. This is a camera which scans the terrain of interest from side to side, normal to the direction of flight. It is capable of wide lateral coverage at very high ground resolution. Its size and weight are much less than those of a fan of long-focal-length frame cameras providing the same lateral coverage and scale.

The continuing struggle for better ground resolution has resulted in the development of improved lenses, shutters, image-motion compensating devices, automatic exposure-control mechanisms, gyroscopically controlled camera mounts, and improved photographic films. The concurrent development of new types of data-acquisition sensors such as infrared cameras, side-looking radar, television, and laser imaging, adds to the total aerial reconnaissance capability for both military and non-military applications. These sensors are covered in the MANUAL OF REMOTE SENSING.

The prime requisite of a *mapping camera* is that certain constant spatial relationships exist between the lens, the focal plane, and the fiducial markers so that once the camera has been calibrated, the calibration data will remain valid for normal use of the camera over a given span of time. This information is used in the data reduction process. In addition to the requirement for high geometric fidelity, mapping photography must be exposed so as to provide a favorable base-to-height ratio for more accurate compilation of terrain relief. These two factors have been dominant in the development of mapping cameras. Improvement also has been made to the photographic resolution capability of the lens-camera-film combination. The wide-angle lens of the mapping camera, which gives broad aerial coverage and contributes to the desirable base-height ratio, inherently produces lower area-weighted average resolutions (AWAR) than does the narrow-angle reconnaissance frame camera.

Early mapping cameras in the United States were of the multi-lens type (*see* Talley, 1938; Bagley, 1941). From the 1920's to the late 1930's, several multi-lens cameras were developed, but they were never produced in large quantities. The first of the single-lens mapping

camera series was the Type T-5, developed in 1938. This camera featured a 6-inch Metrogon lens which had an angular field of 93°, good resolution, and distortion which was correctible in diapositive reduction printers. The format for this, and for most succeeding cameras, was 9 × 9 inches. The T-5 featured an inner cone made of a special aluminum alloy having a low coefficient of expansion. The lens, focal plane, and fiducial markers were rigidly attached to this inner cone. The next military camera in production was the Type T-11, delivered in 1951, which contained a selected 6-inch Metrogon lens and a Rapidyne shutter. Its reliability in operation was much higher than that of the T-5 camera. The 6-inch Planigon lens, which replaced the Metrogon, was installed in the Type KC-1 camera; the KC-1 was delivered to the U.S. Air Force in 1954. It reduced the radial distortion by a factor of 10 and increased photographic resolution substantially (*see* Sewell, 1954). The Type KC-4 camera, which held a 6-inch Geocon I lens developed by Dr. James Baker, was delivered to the U.S. Air Force in 1963. It was capable of producing a resolution about 25 percent higher than that of the Planigon cameras, and could accommodate exposure of infrared and color films in addition to normal panchromatic photography.

The 6-inch f/5.0 Geocon IV lens, a later, more highly color and distortion corrected model, was delivered to the U.S. Air Force in 1966 in the Type KC-6A camera. This camera included forward motion compensation, an automatic exposure control, a platen with a reseau, and a data recording cathode ray tube. In its tie with the inertial unit and mount of the USAF's AN/USQ-28 Mapping System, it was able to stabilize within 4 minutes of arc and to record the angle of tilt to ±30 seconds of arc.

New, low-shrinkage film bases (polyesters) have been developed for precision mapping photography. Mapping lenses are now capable of being used with panchromatic, color, infrared black-and-white, or false color films without the need to change the focal position or introduce significant variance in distortion. The advent of analytical photogrammetry, along with advancements in computer technology, has demanded closer control of the dimensional stability of film. This has led to the incorporation of a camera reseau in some newer cartographic cameras. The reseau (grid) is imaged on the film at the instant of exposure, either by means of a marked glass plate pressed against the emulsion or by projection to the emulsion through tiny holes in the platen.

Cartographic developments in Europe have kept pace with and sometimes exceeded those in this country, and their performance is shown in the tables which are a part of this chapter. The Wild Type RC-10 camera, with a 6-inch Aviogon lens, and the Zeiss Type RMK 15/23 camera, with a 6-inch Pleogon lens, are examples of excellent cameras whose construction and performance are approximately equal to that of the Geocon lens cameras in the United States. In addition, the Europeans have developed and produced two super-wide angle cameras of approximately 3½-inch focal length which expose high quality photography at a 120-degree angular coverage on a 9 × 9-inch format. These are the Zeiss Type RMK 8.5/23 camera, holding an 85-mm f/4.0 S-Pleogon A lens, and the Wild Type RC-9 camera, which holds an 88-mm f/5.6 Super-Aviogon lens.

The one true cartographic camera in NASA's space program is Fairchild's Lunar Mapping Camera, which flew on APOLLO missions 15, 16, and 17. It has a 3-inch f/4.5 lens, exposing on a 4½ × 4½-inch format. It was flown in conjunction with another frame camera which provided time-correlated photography of stars.

Not only must a mapping camera be rigidly constructed so as to maintain certain constant relationships, but these relationships must be accurately known if the camera's product is to be used appropriately in topographic map compilation. The values of these relationships are determined during camera calibration.

In the past several years, a strong effort has been made to apply the high-resolution, wide-coverage capabilities of the reconnaissance-type panoramic cameras to cartographic work. Several approaches have been tried, including calibration of the camera, the use of a reseau, and the superimposition of panoramic image detail over control layouts provided through cartographic photography. All of these means are used in conjunction with rectification of the inherently non-geometric (dynamic) imagery of the panoramic camera. Adding to this the advancement in analytical photogrammetric procedures and the power of computer technology, a certain degree of success has been realized in the data reduction area in approaching equivalent cartographic quality in the final product. This represents a significant achievement.

However, it should be obvious to mappers and other information collectors in general that the data acquisition sensor is a highly important element in the intelligence-production process, whether it is required for finely detailed photo-interpretation or for stringently geometric cartographic use. It is impossible to *improve* the basic resolution of reconnaissance imagery or the geometric accuracy of mapping photography through data reduction procedures. Hence, it is important to select the best sensor for a particular job in order to assure both accuracy and savings in compilation. For example, high quality mapping imagery can be handled generally in existing data reduction equipment without the need for costly modifications or the development of new instruments. If the source material is of inferior quality, then various additional costly manipulations may be necessary during reduction in order to make its product usable.

The aerial camera produces higher resolution in its imagery than any other form of remote sensing device. This is the reason why it is still the prime instrument in the collection of intelligence information.

4.2 Types of Aerial Cameras

4.2.1 FRAME CAMERAS

Most aerial cameras in use in the world can be classified as frame cameras. Of the more common types of cameras, only panoramic and strip cameras are not considered frame cameras. A frame camera is one in which an entire frame or format is exposed through a lens that is fixed relative to the focal plane. Exposures may be controlled by means of a between-the-lens shutter, a focal-plane shutter, a louver shutter, or by illuminating the object for a short interval (night photography). The film is held stationary in the focal plane during exposure, or it is moved to compensate for image motion.

Aerial frame cameras (figure 4-1) are used for both reconnaissance and mapping purposes. Even though the cameras are designed for different purposes, their basic parts are quite similar and will be considered together. Parts peculiar to one type will be indicated.

4.2.1.1 PARTS OF A FRAME CAMERA

The principal parts of an aerial frame camera are: the body (including drive mechanism); the lens (including filter); the shutter assembly (including diaphragm); and the magazine. Mapping cameras generally contain a means for recording auxiliary data on each frame.

4.2.1.1.1 BODY

The camera's body usually houses the drive mechanism, the drive motor, operating handles and levers, electrical connections and switches, and other details which may be necessitated by specific requirements. In reconnaissance cameras, which generally do not contain an inner cone, the camera body, together with the cone and focusing posts, fixes the position of the lens with respect to the focal plane. The *drive mechanism* provides the motion necessary to wind and trip the shutter, to operate the vacuum system and/or pressure-plate system for flattening the film in the focal plane, and to wind the film or change plates between exposures. It is driven either manually by means of a hand crank and tripping level or by a small electric motor which derives its power from the aircraft electrical

FIGURE 4-1. U.S. Air Force Type KC-6A mapping camera (Courtesy of Fairchild Camera and Instrument Corp.).

system. By means of rods and couplings, power is routed to the shutter and to the film magazine. The drive mechanism is located in the body. Rods extending downward from the drive mechanism to the shutter are called wind and trip rods. One or two male couplings extend upward to connect with female couplings in the magazine. Clutches are provided in the drive mechanism which disengage as soon as the shutter or magazine is fully wound. The drive mechanism also serves to transmit the tripping force to the shutter and to the magazine. For electrical operation, a solenoid is contained in the drive. When energized, the solenoid releases a tripping spring in the drive, and an exposure is made. An important characteristic of the drive mechanism is the time required for its complete operating cycle. One complete cycle consists of all the operations which must be performed in order to trip a fully wound camera and restore it to its fully-wound condition again.

In automatic aerial cameras, an electric motor, called the *drive motor,* is mounted in the body near the drive mechanism. Most aerial cameras are designed for a 28-volt aircraft power supply. The power of the motor varies according to the size and design of the aerial camera.

4.2.1.1.2 LENS CONE ASSEMBLY

The *lens cone,* sometimes referred to as "camera cone," serves a very special purpose. It supports the entire lens assembly, including the filter, and prevents any light, except that transmitted through the lens, from striking the film in the focal plane. In some cameras the cone, together with the body, holds the lens at the proper distance from the focal plane, which is then defined by the upper surface of the body.

In some mapping cameras, the body supports an inner frame called the inner cone or spider (*see* figures 4-2 and 4-3). This inner cone holds the lens assembly and also contains the collimation markers (fiducial markers). The inner cone is made of a metal with a relatively low coefficient of thermal expansion so that the lens, the lens axis, the focal plane as defined by the upper surface of the fiducial markers, and the fiducial markers themselves, are all maintained in the same relative positions at operating temperatures. The relative positions of the foregoing components fix the elements of interior orientation of the camera, as calibrated. The inner cone is calibrated as a unit, separately or within the camera body, to determine these orientation elements. Specifications for aerial cameras intended for photogrammetric work are extremely detailed in prescribing the positioning of the lens in the inner cone with respect to the fiducial markers.

The *lens* is the most important single item in the camera. The quality of good photogrammetric work is dependent on the lens, particularly with regard to its distortion characteristics. High quality lenses are essential for reconnaissance and photointerpretation work as well, where ground resolution is of primary concern. Types of lenses are described in chapter III of this MANUAL. In general, the most precise mapping cameras have lenses that are nearly distortionless, or their distortion curves fall within certain narrow envelopes which permit correction by various devices in the photogrammetric reduction process (*see* section 4.8.8.3.9). The function of the lens is to gather a selected bundle of light rays for each of an infinite number of points on the terrain and to bring each bundle to focus as a point on the focal

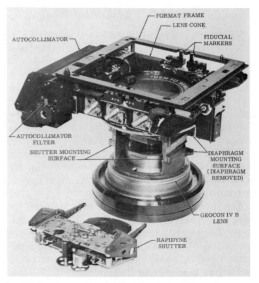

FIGURE 4-2. Inner lens cone assembly for USAF Type KC-6A mapping camera showing parts and shutter (Courtesy of Fairchild Camera and Instrument Corp.).

FIGURE 4-3. Inner cone assembly for KC-6A mapping camera with drawer shutter in place (Courtesy of Fairchild Camera and Instrument Corp.).

plane. The aerial camera is a fixed-focus camera, its focus being set for an object distance of infinity.

The *filter* is not usually a part of the lens assembly, but is involved in the passage of light from the object plane to the focal plane. On cameras with wide-angle lenses, it serves a dual purpose in that it reduces the effect of atmospheric haze on the photographic negative, and also serves as an antivignetting filter to assure the uniform distribution of light on the format. The camera is calibrated with the filter in place, and the camera is not used for mapping photography without it. The orientation of the filter with respect to the lens should never be changed after calibration.

The *focal plane* of a camera is the plane in which, theoretically, all light rays passing through the lens come to a focus. It is usually defined in aerial cameras by the upper surfaces of the fiducial markers, together with the upper surface of the focal-plane frame. Actually, it is located at such distance from the rear nodal point of the lens as to give the best possible image definition throughout the picture. The film may be held in the focal plane by either of two mechanisms—a focal-plane glass plate or a locating back.

The *focal-plane glass plate* is a piece of very high quality selected plate glass at least equal in size to the negative format. The glass is located in the camera with its surfaces perpendicular to the central axis of the lens and at such distance from the lens that its surface away from the lens lies in the exact focal plane. The film is pressed flat against the glass plate when the camera shutter is tripped and the picture taken. Although the focal-plane glass presents a simple

solution to film flatness, it has certain disadvantages. A ray of light passing through the central axis of the lens perpendicular to the glass focal plane passes straight through as if no glass were present. However, any other ray of light which strikes the glass at an angle to the axis is refracted as it passes through the glass in an amount depending on the magnitude of the angle of incidence, the wavelength of the light, and the index of refraction of the glass. Thus, the glass plate must be designed as an integral component of the lens system, or displacements of image points will result. Another disadvantage is its possible contribution to the buildup of static electricity within the camera, resulting from friction between film and glass. If a sufficient amount of static electricity is generated, an electrical discharge will ensue, resulting in the formation of black, branching, tree-like lines on the negative. In serious cases, static renders the photograph useless. This means of holding the film in the focal plane has another inherent disadvantage. When the film is advanced after an exposure, it is pulled across the focal-plane glass. Unless the glass is kept scrupulously clean, scratches will be made on the negatives. Lastly, this system adds one more sizeable piece of glass, always subject to breakage, to the camera.

The *locating back* has replaced the glass plate in all but a very few aerial cameras. The locating back, or platen, is a metal plate in the film magazine positioned just above the focal plane on the side away from the lens. Thus, when the film is stretched across the focal plane opening, and when the locating back presses the film against the focal plane frame, the edges of the film lie in the focal plane. Other portions of the negative are held in close contact with the locating back, and thus in the focal plane, by air pressure (figure 4-4).

Fiducial markers (*see* figure 4-2) of one design or another are included in most aerial cameras. They are usually a minimum of four in number and are located in or near the four corners of the format opening, or in or near the center of each of the four sides. In mapping cameras, the lines joining opposite pairs of fiducial markers should intersect at the principal point of the system. Thus, in the construction of the camera, fiducial markers are located or adjusted to indicate the principal point (the principal point of autocollimation, the point of symmetry, or other defined point) of the system within exceedingly close tolerances. In most other cameras not requiring mapping accuracy, the fiducial markers are located by the manufacturer to indicate only the geometrical center of the format. In most reconnaissance cameras, the fiducial markers are located on the bottom surface of the magazine, and these can be subject to a small shift of location each time the magazine is removed from the camera.

FIGURE 4-4. Type KC-6A camera magazine showing the locating back, or platen (Courtesy of Fairchild Camera and Instrument Corp.).

4.2.1.1.3 SHUTTER ASSEMBLY

The shutter assembly (figure 4-2) of the aerial camera consists of the shutter and the diaphragm. The shutter controls the interval of time during which light is allowed to pass through the lens. The function of the diaphragm is to restrict the size of the bundle of rays which may pass through the lens and strike the emulsion surface of the film in the focal plane. The diaphragm establishes the area of cross-section and the shutter regulates the interval of time in which the light flows through the aperture regulated by the diaphragm.

The two types of *shutters* in general use for aerial cameras made in the United States are the between-the-lens and the focal-plane types. The louver shutter is described briefly, even though its use in aerial cameras is very limited. The speed or total open time of an aerial-camera shutter refers to the duration of the film exposure and is the length of time expressed as a fraction of a second from the instant when the shutter begins to admit light to the film until the instant when the shutter cuts off the light. Shutter speed in an aircraft camera is highly important because the camera is moving while a picture of a stationary object is being taken. For a given aircraft flying speed, improvement of image quality is obtained by reducing the image movement by speeding up the camera shutter while increasing the size of the aperture through which the light flows so that a corresponding quantity of light strikes the film. High shutter speeds also reduce the deleterious effect of aircraft vibrations.

4.2.1.1.3.1 Between-the-lens shutter. This type of shutter is the most widely used in the

United States, and is a necessity in a mapping camera. The ordinary shutter of this type has six major components. (1) The *leaf center* consists, in most cases, of two metal plates between which the shutter leaves and diaphragm are inserted. The shutter leaves are four or five in number and are pivoted at their small end in the leaf center plates. They are connected by links which cause them all to rotate simultaneously when actuated. (2) The *actuating cone assembly* is actuated by the shutter spring and transmits this rotational force to the shutter leaves. (3) The *retard assembly* serves to slow down the operation of the shutter when required. (4) The *spring housing assembly* contains a coil spring which provides the motive power for operating the shutter. The housing also includes means for winding the shutter spring and for preventing the spring from unwinding until released by the trip mechanism. (5) The *trip mechanism* serves to release the power stored in the spring, thereby starting the exposure cycle. (6) The *shutter housing* consists of two parts, the main housing and the cover.

The advantages of this type of shutter are durability, mapping applicability, and absence of obstruction in the lens aperture. Its leaves operate in the relatively small space which separates the front and rear elements of the camera lens. The characteristic of the between-the-lens shutter which enables it to admit light to all parts of the negative simultaneously upon opening and to cut off the light from all parts of the negative simultaneously at the end of the selected exposure time interval, preserves, in the negative, the precise relationship of all object points photographed, and thereby makes it particularly applicable to photogrammetric operations. There are two general types of between-the-lens shutters—the drawer and the nonremovable types.

The *drawer-type shutter* permits insertion of the shutter through a slot in the side of the lens housing in its proper position between the front and rear elements of the lens, and its removal, without disturbing the lens alignment. This capability is a considerable advantage in that the shutter may be removed and repaired in the field without the necessity for recalibration of the camera. An example of this type of shutter is Fairchild Camera and Instrument Corporation's Rapidyne shutter (figure 4-2), only 0.07 inch thick, which is used in the Type KC-6A mapping camera. Instead of having one set of shutter leaves operating in a forward-reverse motion by accelerating to a maximum speed, retarding, and then returning to its original closed position, this device is actually two shutters—one opening and the other closing.

The Zeiss Aerotop shutter is a between-the-lens shutter with *continuously rotating disks.* The operating principle of the shutter is illustrated in figure 4-5. Four disks rotate with a speed corresponding to the shutter time set. The desired exposure moment is selected from the large number of possible exposures by a fifth blade and the capping disk controlled by the intervalometer. The opentime of this shutter is determined by the speed of rotation of the disks, and can be infinitely varied with the aid of a rotary knob. The standard equipment of the shutter permits exposure times from 1/100 to 1/1000 second. The shutter is practically independent of fluctuations of temperature, and has a very favorable degree of light efficiency.

Between-the-lens *nonremovable* shutters differ from the drawer type in that they are mounted firmly in the lens housing assembly and are not considered removable without dismantling the lens structure. Certain reconnaissance cameras, such as the Type KA-1, have this shutter. Such cameras are not used for mapping photography and do not require calibration.

4.2.1.1.3.2 Focal-plane shutter. This shutter is so named because it operates close to the focal plane of the camera. Focal-plane shutters are being used today in some reconnaissance cameras, such as the Type KA-30A, because they permit higher shutter speeds than can be obtained with the between-the-lens shutters. The main component of the focal-plane shutter is the curtain, which is usually composed of a piece of rubberized cloth somewhat wider than the focal plane frame and a little longer than twice the length of the frame. The curtain is attached to rollers, one at each end of the negative area. The curtain has a slit across the narrow dimension of the curtain, which admits light to the film. Before making an exposure, the curtain is wound onto one roller against a spring tension in the opposite roller until the slit is clear of the frame opening. When the shutter is tripped, the slit moves across the frame opening toward the opposite roller, allowing light to pass through the slit and form an image on the film. This shutter also has a capping curtain which forms a light barrier while the primary curtain is being wound back to the starting position. The driving force for the focal-plane shutter is a coil spring which is usually installed in one of the rollers. The fundamental principle of operation of the focal-plane shutter causes the introduction of positional errors in the relationship of photographed images, which makes this shutter unsuitable for photomapping use.

There are a few military reconnaissance cameras that have two shutters—a focal-plane for day use, and a between-the-lens for night photography. When used at night, the terrain is illuminated by a powerful electronic flash or an explosive flash bomb or cartridge. The KS-72A camera used by the U.S. Air Force and Navy is an example of this type. The between-the-lens shutter is controlled by a photoelectric cell that closes the shutter after the flash device has been triggered.

4.2.1.1.3.3 Louver shutter. This shutter acts in a manner similar to a variable, neutral density filter instead of a varying-aperture diaphragm, and may be placed anywhere in the optical system that is convenient. The louver shutter assembly consists of a rectangular frame which supports several louvers (strips of metal about one quarter of an inch wide and sufficiently long to project beyond the cone of light rays which enter, pass through, or leave the lens). Each louver is supported in the shutter frame at its ends by bearings placed on the central longitudinal axis of the louver. The bearing points of the louvers are so spaced that, when the louvers are in a flat (closed) position, each one overlaps the next sufficiently to prevent the passage of light through the lens. All louvers are moved simultaneously by a rack-and-pinion arrangement which is spring actuated. When the louvers are in the wide open position, the light is permitted to pass through the lens to make an exposure. An advantage of louver shutters lies in their high speed for large lens apertures; however, their efficiency and durability are not always as great as that of other shutters. In addition, even in their wide open position, the louvers reduce the light gathering areas of the lens aperture by an appreciable amount.

The *diaphragm* controls the maximum size of the cone of light admitted. It consists of 3 to 20 leaves which have a curvature close to that of the aperture. At each end of each leaf is a stud. One stud is pivoted in one of the leaf center plates while the other pivots in a slot in a rotatable diaphragm-actuating plate. As the plate rotates, the leaves form a nearly circular aperture of variable diameter. It is located approximately midway between the lens elements. The variation in the diameter of the aperture caused by the adjustment of the leaves leads to establishment of the series of lens stops found in shutters. These stops are the f-values which are a measure of the light-gathering power of the aperture. The f-values are the ratio of focal length of the lens to the diameter of the aperture.

4.2.1.1.4 MAGAZINE

The magazine (figure 4-4) serves to hold the exposed and unexposed film (or glass plates), to advance the necessary amount of film (or change plates) between exposures, and to house the

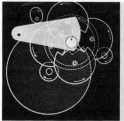 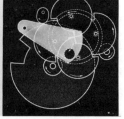

FIGURE 4-5. Zeiss continuously rotating disk shutter. Shutter open on left, closed on right. (Courtesy of Carl Zeiss, Inc.).

film-flattening device. The film magazine contains a driving device which receives power from the drive mechanism and thereby winds the film after each exposure has been made. The device is designed to wind the correct amount of film for each successive exposure. This function is known as *metering*. In addition, the magazine contains a means for holding the film flat in the focal plane while the exposure is being made. The flatness of the film at the moment of exposure is critical, and should receive special consideration in photogrammetric aerial camera design. The magazine fastens into position on top of the camera body by means of several latches. The detachable type of magazine is the most popular for aerial photography. It can be removed from the camera and taken separately to the darkroom for loading and unloading. In the case of magazine failure during a mission, another magazine of similar type can quickly be substituted. For projects of long duration, several loaded magazines can be carried in the aircraft and changed quickly as required.

4.2.1.1.4.1 Film metering. The most elementary type of film metering is obtained when the takeup spool is caused to revolve through a specific number of turns. The number of turns must be enough to pull through sufficient unexposed film for the exposure, regardless of the amount of film on the spool; otherwise, overlap of the negatives will occur. At the same time, the number of turns should be low enough to minimize the space between pictures. As film is wound onto the takeup spool, the diameter becomes greater and each turn of the takeup spool draws more and more film through for each exposure. Special film metering applies a correction for the increasing diameter of the takeup spool. A lever is forced outward by the increasing amount of film on the spool. This lever is connected to a clutch which causes the takeup spool to make fewer turns in proportion to the increase in the spool diameter. Exposures may also be spaced by use of a metering roller, usually located on the takeup-spool side of the magazine. In order to prevent slippage of the film, a pressure roller is used which presses the film against the metering roller at the edges. The metering roller is so designed that, after it has turned a specific amount, a clutch is caused to disengage. This stops the takeup spool from turning. Since the film does not wind up on the metering roller, the diameter remains constant and the same amount of film is advanced for each exposure.

4.2.1.1.4.2 Film flattening. The mechanism in the magazine which holds the film flat in the focal plane at the instant of exposure may do this in one of four ways: by direct tension, by use of a pressure plate against a focal plane glass, by pressure of air introduced on the emulsion side, or by vacuum applied through the platen.

In the case of cameras whose negative does not exceed 4 by 5 inches, where no photogrammetric use for the photography is intended, or where very high resolution is not demanded, the film can be held reasonably flat across the focal plane frame by *direct tension*. During the winding of the film it is stretched across the negative opening. Just before the exposure is to be made, a metal plate clamps the film around the edges of the focal plane frame. The clamping action, plus the slight tension and stiffness of the film, all combine to prevent the film from sagging enough to greatly impair photographic quality.

When a *focal-plane glass* is used, flattening of the film is relatively simple. Just before an exposure is made, a metal pressure plate clamps the film against the focal plane glass over its entire area. After the exposure is made, the pressure plate releases and the film can be advanced. As mentioned earlier, a glass pressure plate in a camera has several disadvantages, such as: the possibility of film scratching by grit particles; the contribution to buildup of static electricity; and the danger of breakage. It must be designed as a part of the optical system if the photography is to be used for photogrammetric purposes.

In the *air-pressure system*, the entire lens cone and camera body must be airtight. The film is pulled across the format space, whereupon the locating back clamps the edges of the film against the focal plane frame. The locating back in the magazine is a metal plate pierced by a large number of small diameter holes. By means of a small hose on the cone, air under pressure is admitted to the cone. The air pressure is obtained by a wind scoop or by a pump. The air pressure which builds up in the cone forces the film flat against the locating back. This system requires an air filter to clean the air before it enters the camera, since any minute particles of dust entering the camera cone may be blown onto the emulsion, causing a harmful scratch.

The method of film flattening used in the greatest number of aerial cameras (including all military mapping cameras) involves a *vacuum system*. In this system, the surface of the locating back which contacts the film is crisscrossed with grooves about 0.5 mm deep and 0.5 mm wide (*see* figure 4-4). Small holes at a number of intersections of the grid grooves lead to a vacuum chamber. As in the air-pressure system, the locating back clamps the film to the shoulders of the focal plane frame. The film in the format space is drawn up against the locating back. The amount of vacuum required to flatten the film in a current mapping camera is equivalent to the force exerted by about 0.07 atmosphere. The grooves must be designed so that air is not trapped between the film and locating back. To prevent distortion of the locating back, a heavy ribbed framework has been incorporated in its design. The vacuum may be obtained by the use of a conventional aircraft venturi, or pitot tube, whose outlet through the skin of the fuselage is located in a shielded position. A vacuum pump or a self-contained bellows system is often used.

4.2.1.1.4.3 Image-motion compensation. Some magazines have been designed to provide film movement during exposure at such a rate as to compensate for the movement of the photographic images. Until the past few years, image-motion compensation (IMC) in aerial mapping operations was not necessary because of the high altitudes, the moderate speeds of propeller aircraft, and the relatively low resolution of the available lens-film combinations. However, IMC mechanisms have been used in reconnaissance cameras since World War II. Since the advent of jet aircraft and the development of slow-speed, fine-grained, high-resolution photographic emulsions, the need for image-motion compensation for photogrammetric work becomes more apparent. For example, for an aircraft at an altitude of 40,000 feet and 500 mph, the image smear for a 6-inch lens at 1/100-second shutter speed is 25 micrometres. The KC-6A camera has an optional IMC capability.

The U.S. Air Force uses two methods for image-motion compensation. One method is to move the film during exposure so that the film and the image move together, thereby preventing image blur. Moving the film to provide compensation helps to produce fast-cycling camera operation. This method, which has been incorporated in several of the later reconnaissance camera systems, is designed so that the mechanism is entirely within the magazine. The second method is to swing the camera in its mount so that it is aimed steadily at the target during the film-exposure interval. Swing mounts can be used only with slow picture-taking rates. They also require larger windows. In a pulse-operated, compensating magazine, the intervalometer pulse causes a variable-speed motor to start driving the film transport system at image-motion-compensation speed. The rate at which the variable-speed motor will run is determined by an adjustment on the camera control system. When the film transport system reaches compensation speed, it produces a pulse of approximately the same value as the standard intervalometer pulse. This shutter pulse operates the camera shutter for a single exposure while the film is moving at IMC speed. At the end of the IMC film movement, the IMC drive ceases and a new motor drive system, called recycle drive, powers the metering rollers. The IMC magazine also differs from conventional magazines in that its drive source is internal instead of coming from the camera body. Although the film is moved during exposure, fiducial markers can be properly recorded by flashing lights at the midpoint of the shutter opening.

4.2.1.1.4.4 Camera reseau. The camera reseau is a precise grid etched on a glass pressure plate, or inscribed on a platen locating back whose surface is in contact with the film as it moves over the focal plane frame. The grid intersections are spaced at approximately equal intervals, usually 1 or 2 centimetres. The principal function of the reseau is to provide a precisely calibrated grid which is superimposed on the negative during exposure. The extent of film shrinkage or expansion may be determined by comparing the grid on the photograph with the known dimensions of the reseau.

A *glass-plate reseau* is used in some of the cartographic cameras of European design. It has the advantage that its grid intersections may be closely established with regard to the fiducial markers and the principal point of the camera; however, it must be designed as a part of the optical system, and its glass surface is subject to scratching.

In the *platen reseau,* the grid is an integral part of the platen. The reseau consists of a series of holes through the platen, arranged in a chosen configuration with some nominal spacing, with lights behind them. Projected dots of light pass through the film base onto the emulsion surface. The size of projected dots usually is between 25 and 50 micrometres in diameter.

4.2.1.1.4.5 Data recording. One of the important features of a cartographic camera is its data-recording system. This places information on the film which aids in correlation of the photography with data produced by positioning equipment, airborne profile recorders, and other airborne instruments. Also, other data useful during map compilation are recorded.

A *conventional data-recording system* consists of such instruments as an altimeter, clock, frame counter, and data tablet, grouped together in an artificially lighted chamber in the camera body. Small lenses project the faces of these instruments onto the film in the area between frames when the chamber lights are flashed at the instant of exposure. In addition to altitude, time, and frame number, data pertaining to the individual camera are also recorded. These include camera type and serial number, lens calibrated focal length and serial number, and a vacuum indicator. The *vacuum indicator* is independent of the data chamber and is located in the magazine. It is a small *V* mark that is exposed on the film when the vacuum is holding the film flat against the platen during the exposure. Absence of this mark indicates the camera vacuum supply has failed. European mapping cameras frequently include a level bubble in the data chamber that is recorded with the other instruments.

Recent military cameras incorporate *automatic data recording* along with the individual data recorded in the conventional manner. The clock and altimeter have been replaced with a data block (*see* figure 4-6). This data block contains much more information in a 1-cm square area on the film than was possible with the con-

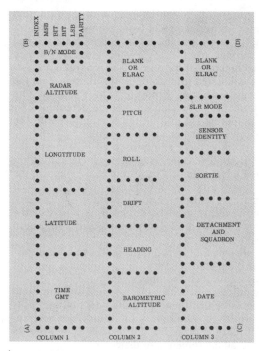

FIGURE 4-6. Camera code matrix data block showing typical recorded data and arrangement (Courtesy of Fairchild Camera and Instrument Corp.).

ventional techniques mentioned previously. The additional information that may be recorded includes: flight line number, date, altitude, time, heading, camera crab angle, geographic coordinates, and the name of the flying organization. Much of this information is obtained from the aircraft navigational system where it is computed and fed into a data-recording-display generator. Two techniques are employed for automatic data recordings. One uses several rows of small lights and data appear on the film coded in binary code. After the film is processed the block is scanned by an automatic reader and the information is then automatically typed on each frame using a goldleaf foil. Another method displays the data on a 1-inch-diameter cathode ray tube. Alphanumeric characters are formed on the tube face and are projected onto the film. The alphanumeric images have the advantage that they can be read by anyone without going through the translation and titling process

4.2.2 PANORAMIC CAMERAS

4.2.2.1 GENERAL

Although panoramic cameras have been used since the 19th century, it was only within the last two decades that intensive efforts began to be made toward their design, development, testing, and routine application to aerial photography.

Panoramic camera designs seek to combine high resolution and wide swath in one camera.

There are three basic means for meeting these otherwise incompatible requirements.

(a) Narrow-angle, fast lens systems are used, and only the portion of the lens system which is on or near the optical axis is employed in photo imaging.
(b) The lens system scans through large angles across the direction of flight.
(c) Film of normal width is advanced parallel to the direction of scanning at rates compatible with the vehicle ground speed in order to obtain continuity of coverage along the flight path.

As a result, panoramic cameras can maintain the inherently high resolution at the center of the field over the total angle scanned, in some cases from horizon to horizon.

Original-negative resolutions exceeding 100 lines per millimetre are not uncommon. The fact that coverage along the flight path is closely related to the cycling rate of the camera suggests that panoramic cameras are best suited for high-altitude missions; however, short-focal-length, moving-film-type panoramic cameras are employed in high speed, low altitude military reconnaissance, and can provide horizon-to-horizon images at a cycle rate over three frames per second. Low-altitude use of panoramic cameras lowers image resolution because of the need for fast (and therefore lower-resolution) films and because of the more erratic flights which create image smear. It is also worth noting that the use of narrow angles to improve resolution and the concomitant high cycling rates to attain continuity of coverage along the flight path make it difficult to obtain adequate stereoscopic coverage with one vertical camera.

Development of panoramic cameras received its impetus primarily from the requirements of aerial reconnaissance. There was little effort toward the development of an airborne panoramic camera until 1949, when a strip camera was modified and tested by Boston University Physical Research Laboratories. The now-famous panoramic photograph of Manhattan was taken by Boston University personnel from an altitude of 10,000 feet (figure 4-7). In this particular version of a panoramic system, the entire camera was rotated about the longitudinal axis of the aircraft and the film was pulled past the slit synchronously with the rotation of the camera body.

In the early 1950's, the Perkin-Elmer panoramic camera was developed for the USAF. This camera had a focal length of 48 inches and used a Dove prism for scanning. The prism rotated synchronously with the translation of the film past a slit, yielding exposures 13 feet long and 18 inches wide (the latter dimension is in the direction of flight). Resolution was limited by the accuracy with which a Dove prism could be made and by the necessity for maintaining synchronization between the rotating prism and the film as the latter moved at high speed through the exposure plane.

FIGURE 4-7. Panoramic photograph of Manhattan from 10,000 feet (Courtesy of Itek Corp.)

Also during the 1950's, the E-2 panoramic camera was developed by Vectron Corporation (Doyle, 1957). This camera consisted of two mirrors and a lens, forming an optical bar which rotated at constant speed about the optical axis. The film was carried on the inside of a cylinder which was concentric with the optical bar, and which rotated at an angular velocity equal to that of the bar but in the opposite direction. Image motion was accomplished by shifting the optical bar longitudinally.

In 1957, a direct-scanning HyAc panoramic camera designed by Itek Corp. consistently produced 100 lines per millimetre of resolution on SO 1213 film. This camera scanned 120° with a 12-inch-focal-length, f/5 lens, the film being stationary on a cylindrical focal surface.

The characteristics of airborne panoramic cameras as evaluated by the Bureau of Naval Weapons in 1962 (Tafel, 1962) are summarized in table 4-1.

The application of panoramic cameras to mapping has been hampered by the facts that the calibration of such cameras is difficult and the photography has comparatively complicated geometry. Both factors are due to the mechanical motions in and of the panoramic camera during exposure. However, with the refinement and increased flexibility of analytical photogrammetric solutions brought about by the advent of high-speed electronic computers and analytical stereoplotters, it has become feasible to use panoramic photography in conjunction with ground control, or simultaneous lower-resolution frame photography, for aerial mapping.

4.2.2.2 THE DESIGN OF PANORAMIC CAMERAS

4.2.2.2.1 BASIC DESIGNS

The numerous mechanical approaches to the design of panoramic cameras may be divided into three general categories: (a) direct-scanning cameras with swinging (or rotating) lenses; (b) cameras that scan by means of rotating mirrors or prisms; and (c) optical-bar-type cameras with folded, rotating optics and moving film.

4.2.2.2.1.1 Direct-Scanning Camera. A typical example of a direct-scanning camera is the HyAc type. As shown in figure 4-8, the design involves rotating the lens about its rear nodal point and accurately positioning the film in the proper focal plane location by means of two small rollers mounted on the end of a scan arm which rotates with the lens. The scan arm also carries the scanning slit. With the lens sweeping at a given rate, the width of this slit (in the direction of scan) determines the amount of exposure given each picture element—the wider the slit, the longer the exposure. In a typical application, this width may be varied from 0.07 to 0.40 inch. The length of the slit across the scan direction must, of course, cover the width of the image strip to be obtained. The maximum scan which can be obtained from a camera of the type shown is necessarily somewhat less than 180°, since at that angle the lens would already be looking back onto the film platen. (The example illustrated here was designed for 140° of angular coverage.)

It should be noted that in the direction scan, only the lens and scan arm move, while the film remains stationary. Furthermore, since the center of the lens rotates as a unit with the scanning slit, the sharpest possible image is always projected onto the film, even if irregularities should develop in the scan rate. At the very

TABLE 4-1. CHARACTERISTICS OF TYPICAL EARLY PANORAMIC CAMERAS

Camera Designation	Camera Type	Focal Length, Inches	Aperture f	Scan Angle, Degrees	Exposure Time, Seconds	Film Width
Fairchild KS-57	Nodal point rotation of objective lens	3.0	2.5	140	1/25– 1/2,000	70 mm
Itek HYAC I	Nodal point rotation of objective lens	12.0	5.0	120		70 mm
Fairchild F-409	Rotary panoramic; entire camera	12.0	3.8	90	1/300– 1/5,000	5 in.
Fairchild F-416 panoramic camera	Rotary prism	48.0	8.0	150	1/250– 1/1,000	9½ in.
Perkin–Elmer Mark II tracking camera and mod. 501-K-2 panoramic camera	Rotary prism	3.0	8.0 6.3	180	1/25– 1/6,400	70 mm

worst, there could be some variation in the exposure because of these irregularities (resulting in corresponding variations in the film density), but the resolution would remain essentially unaffected.

4.2.2.2.1.2 Rotating-Prism Camera. Figure 4-9 is a schematic of a scanning camera employing a rotating prism. A full 180-degree scan, or more, can be achieved with this configuration. A double Dove prism, the diagonal faces of which are aluminized and cemented together, is placed in front of the lens. The prism must rotate 90° for each 180-degree scan. The light path is imaged on the film, which passes around a rotating drum. The rotation of the drum must be accu-

rately coordinated with that of the prism in order to achieve good image quality.

The film passes from a supply spool, around an idler roller and over a pair of loop-forming or dancer rollers to another idler roller, which guides it around the drum. The path to the take-up spool is similar. Because of the speed of the recording drum during exposure, enough space must be allowed so that a full frame can be stored by the dancer rollers.

In the "ready" position, the prism is turned a few degrees beyond the point which allows it to look 90° out to one side of the vehicle. When the mechanism is released, prism and drum rotation are both accelerated up to the desired speed.

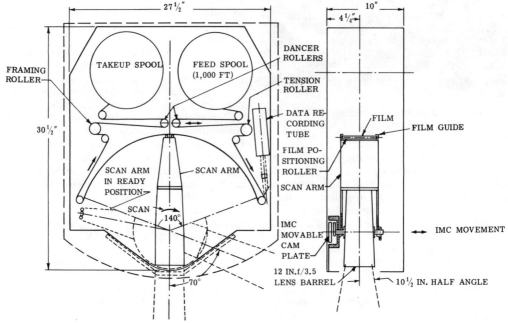

FIGURE 4-8. Direct-scanning panoramic camera design.

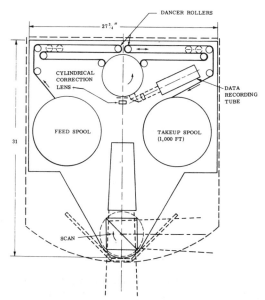

FIGURE 4-9. Scanning panoramic camera design using rotating prism.

The prism turns 90° clockwise at a uniform rate and the drum revolves at a uniform rate during the 180-degree scan. The prism then travels a few degrees beyond for deceleration. When the prism is turned in the counterclockwise direction in preparation for the next photograph, the drum remains stationary. An exception to this stop-start film transport is the KS-69A camera, the design principle of which is illustrated in figure 4-10. Alternate scans are accomplished with two double Dove prisms. To maintain theoretical resolution, the synchronization must be much more precise than the final resolution to be achieved. If resolution on the order of 100 lines per millimetre is desired, relative motion in the image plane cannot exceed approximately 1/200 millimetre. In addition, systems employing Dove prisms cannot be applied successfully where

large apertures are required. The large mass of the glass prism is extremely sensitive to temperature changes, and any distortion of the prism produces image degradation. The fact that the reflection comes from the surface inside the glass adds to the seriousness of this effect. Image-doubling from the two prisms occurs at the slightest provocation. Even under the best laboratory conditions, the theoretical resolution of the lens is cut in half in one direction by the aperture-splitting action of the double Dove prism.

It should be noted that the configurations as shown will result in positive prints that are mirror images of the scenes they represent. If properly oriented photographs are required, an additional mirror must be inserted in the optical path of the camera or enlarger.

4.2.2.2.1.3 Optical-Bar Camera. Figure 4-11 is a schematic of a scanning camera employing a continuously-rotating, folded optical system often referred to as an optical bar. A number of modern panoramic cameras utilize this design principle which permits very compact configurations of long-focal-length, large-scan-angle systems. Since both the entire optical system and the film must move in precise synchronization, the operational realization of the optical bar concept was made possible only through the development of precise electro-optical encoders which direct and synchronize the opto-mechanical functions.

The optical bar concept utilizes a folded optical assembly that constantly rotates about the spin axis. Film is transported in the opposite direction over the roller cage with a precisely maintained image and film velocity. This eliminates high film accelerations and velocities while maintaining the lens in a uniform thermal environment.

Image plane registration is required only over

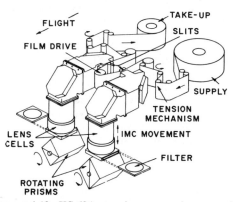

FIGURE 4-10. KS-69A scanning panoramic camera design using two rotating prisms to permit continuous film transport.

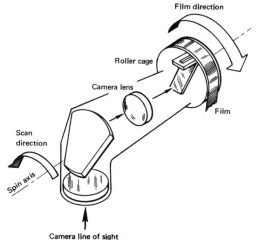

FIGURE 4-11. Scanning panoramic camera design using rotating optical bar.

TABLE 4-2. Characteristics of Selected Modern Panoramic Cameras

Camera Type	Camera Designation	Focal Length mm	Aperture f	Scan Angle Degrees	Film Width
Direct Scanning	Fairchild KA-81	1220	4.0	40	5 inches
	Fairchild KA-82	305	3.8	140	5 inches
Rotary Prism	CAI–145	75	2.8	180	70 mm
	Fairchild KB-29A	75	2.8	41	70 mm
	CAI KS-116A	150–460	2.8–4.0	180–60	5 inches
	Perkin–Elmer KS-69A	460	4.0	180	5 inches
Optical Bar	Itek 5776	150	2.8	180	5 inches
	Fairchild KA-94A	610	4.5	120	5 inches
	CAI KA-90A	610	3.5	105	5 inches
	Perkin-Elmer 2490	610	3.5	90	5 inches
	Itek KA-80A	610	3.5	120	5 inches

the slit area in the optical bar design. Synchronization is accomplished by a positive contact friction gearing system driven directly by the optical bar assembly.

Because the camera lens rotates at a distance from the image surface (the radius of the roller cage) which is not equal to the focal length of the lens, lateral image motions would be introduced. This is the reason why, during the exposure cycle, the film is moved across the roller cage in a direction opposite to the lens rotation in order to compensate for these image motions. The rate of film transport is proportional to the rotation rate of the lens, which can vary as a function of vehicle velocity and altitude.

Because of the compact configuration of optical bar cameras, it is relatively easy to provide convergent stereoscopic photography with a single camera by alternately tipping the entire optical and roller cage assembly forward and aft prior to exposure of each frame. Forward motion compensation (FMC) can be accomplished by rotating the assembly about the pitch axis during the photographic exposure cycle. The rate of pitch is controlled to correct the image smear caused by forward vehicle motion during the panoramic scan.

4.2.2.2.2 Mounting and Stabilization

As with other types of aerial photography, the full potential of panoramic photography can only be realized under conditions of proper stabilization and image motion compensation (IMC), normally only forward motion compensation (FMC).

Because of the wide angular coverage of the panoramic camera, the relative effects of pitch, roll, and yaw vary considerably over the photograph area. For instance, near the horizon the effects of roll and of yaw are maximized while effects of pitch are relatively less. On the other hand, when scanning directly below the vehicle, effects of roll and pitch are much more critical than are effects of yaw. Thus, for maximum information acquisition, a mount which will stabilize the camera in all three axes is important. This would not be necessary in stable high altitude aircraft or spacecraft applications.

Of equal importance is IMC, although this problem is common to all aerial photography. With the use of faster vehicles and requirements for finer ground resolution, the camera will traverse several times the resolution width even during short exposure times (1/400 to 1/2000 second). In order to compensate, the lens must be moved backward, opposite to the direction of flight (that is, at right angles to the direction of scan). In most panoramic cameras this is accomplished by a background mechanical translation of the lens relative to the film; however, some panoramic cameras are now designed to compensate for forward motion by pitching the lens/film assembly in a slight angular nodding motion from fore to aft during scan.

A certain amount of error is unavoidable since the height and ground speed of the vehicles are usually known imperfectly. This error can be compounded further if the vehicle, because of weather conditions, is doing more than a small amount of "crabbing." If it is, the vehicle is not actually traveling in the direction of its heading, and IMC, operating along the vehicle's longitudinal axis, will compensate in the wrong direction. Thus, azimuth correction in the mount is also important for maximum resolution.

A third requirement for all types of photography is exposure control. In a panoramic camera with a fixed scanning speed, this is most readily accomplished by varying the slit width. Required exposure is a factor which usually changes rather slowly and therefore presents a relatively minor problem. Standard techniques are available for measuring available light and setting the required exposure automatically.

4.2.2.3 GEOMETRIC RELATIONSHIPS INHERENT IN PANORAMIC PHOTOGRAPHY

Panoramic photography involves several types of image displacements which are not present in frame photography. The displacements are termed "distortions" because they differ from those of central-perspective geometry. The displacements should not be confused with optical or lens distortion. In the following general discussion, four of these deformations are defined, illustrated, and geometrically analyzed. The nomenclature used in the analysis is as follows:

x and y Coordinates of any image point in the photograph. The $+x$ axis is in the direction of flight

x_p Panoramic distortion component

x_s Scan distortion component

x_{im} IMC distortion component

X and Y Coordinates of any point in the ground datum plane

f Camera focal length

H Flight height above datum

α Camera scan angle

δ Angular velocity of the camera scan arm

t Scan time of the camera

V Velocity of the aircraft

v Velocity of image in the focal plane

ϕ Primary tip angle of convergent camera

4.2.2.3.1 PANORAMIC DISTORTION

Panoramic distortion is the displacement of images of ground points from their expected perspective positions, caused by the cylindrical shape of the negative film surface and the scanning action of the lens. The pertinent geometric relationships are defined in figure 4-12.

$$y_p = f\alpha \qquad (4.1)$$

$$\alpha = \tan^{-1} \frac{Y}{H} \qquad (4.2)$$

$$y_p = f \tan^{-1} \frac{Y}{H} \qquad (4.3)$$

$$x_p = \frac{f}{H} X \cos \alpha. \qquad (4.4)$$

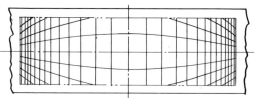

FIGURE 4-13. Panoramic distortion.

Figure 4-13 shows diagrammatically the results of a photograph of a unit grid on the ground with panoramic distortion.

4.2.2.3.2 SCAN POSITIONAL DISTORTION

Scan positional distortion is the displacement of images of ground points from their expected cylindrical positions caused by the forward motion of the vehicle as the lens scans. The pertinent geometric relationships are shown in figure 4-14.

$$x_s = \frac{f}{H} Vt \cos \alpha \qquad (4.5)$$

and since $t = \dfrac{\alpha}{\delta}$,

$$x_s = \frac{Vf \alpha \cos \alpha}{H\delta}. \qquad (4.6)$$

The scan positional distortion is superimposed on the panoramic distortion and modifies the effects of the latter with respect to the image positions of ground points. Figure 4-15 illustrates these effects on the position of the center scan line.

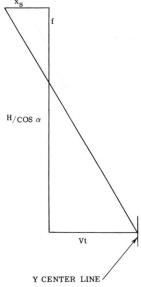

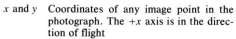

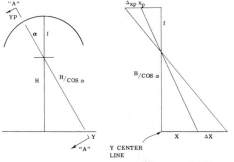

FIGURE 4-12. Geometric relationship of panoramic distortion.

FIGURE 4-14. Geometric relationship of scan positional distortion.

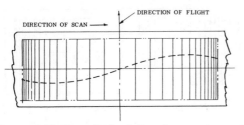

FIGURE 4-15. Scan positional distortion superimposed on panoramic distortion.

4.2.2.3.3 IMAGE MOTION COMPENSATION (IMC) DISTORTION

This refers only to panoramic cameras that employ lens or film translation for image motion compensation. The IMC distortion is the displacement of images of ground points from their expected cylindrical position caused by the translation of the lens or negative surface, a motion used to compensate for image motion during exposure time. The geometry is shown in figure 4-16.

$$v = \frac{dx_{im}}{dt} = \frac{fV \cos \alpha}{H}$$

$$dx_{im} = \frac{fV \cos \alpha \, dt}{H} \qquad (4.7)$$

and since $dt = \dfrac{d\alpha}{\delta}$,

$$x_{im} = -\frac{Vf}{H\delta} \sin \alpha \qquad (4.8)$$

The IMC distortion also affects the two previously discussed distortions. Figure 4-17 shows how the IMC distortion affects the center line scan.

4.2.2.3.4 RESULTANT IMAGE IN THE VERTICAL PANORAMIC PHOTOGRAPHY

The image of a unit grid on the ground obtained with a panoramic camera in flight would show the effects of all the previously described distortions. The residual center line displacement is the algebraic sum of two distorting factors, the scan positional and IMC distortions. The IMC distortion, being always larger, is dominant. The resultant x-displacement is given by:

$$x = x_p + x_s + x_{im}$$

$$x = \frac{f}{H} X \cos \alpha + \frac{Vf}{H\delta} (\alpha \cos \alpha - \sin \alpha). \quad (4.9)$$

The y-component remains the same.

Figure 4-18 shows the combined effects of all distortions on the image.

4.2.2.3.5 TIPPED PANORAMIC DISTORTION

This distortion is the displacement of images of ground points from their expected vertical panoramic positions caused by tipping of the scan axis within the vertical plane of the flight path. This distortion is additive and modifies again the image positions of points already influ-

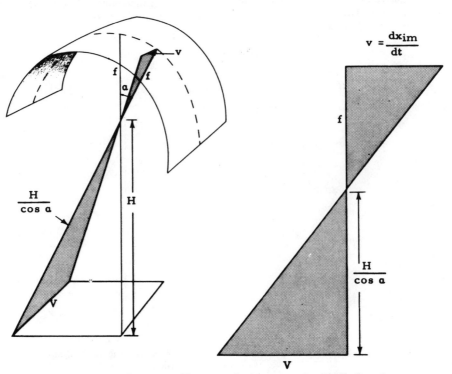

FIGURE 4-16. Geometry of image-motion-compensation (IMC) distortion.

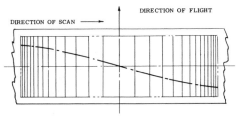

FIGURE 4-17. Effect of IMC distortion of centerline scan.

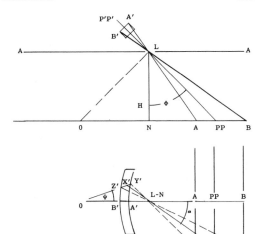

FIGURE 4-19. Geometry of tipped panoramic distortion.

enced by panoramic distortion, scan positional distortion, and IMC distortion.

As was previously mentioned, narrow-angle lens systems seldom provide favorable stereoscopic base-height ratios when the camera is used in a vertical attitude. Therefore, convergent cameras may be used, or a single camera may be rotated fore and aft for stereoscopy and the consideration of the effects of tipped panoramic distortion become important.

It has been shown that ground lines parallel to the line of flight remain parallel in the image, provided the camera is vertical and the photographed ground is flat and horizontal. Furthermore, meridians of the cylindrical focal surface—that is, lines parallel to the cylinder axis—remain parallel when projected down onto the object plane. This is because these lines are produced by central perspective and are not influenced by the scanning. However, when the scanning axis is tipped through an angle in the vertical plane of the flight path, lines parallel to the flight path in the horizontal ground plane are projected onto an oblique cylindrical surface and are no longer parallel in the photography. This is illustrated in figure 4-19 where

α = Scan angle
N = Nadir for "0°" scan
PP = Principal point for "0°" scan
H = Altitude
ϕ = Primary tip angle

Distance $N - PP = H \tan \phi$ (4.10)
Distance $PP - X = H \tan \alpha / \cos \phi$. (4.11)

Ground line $Y - Z$ is parallel to ground line (flight line) $A - B$. In the image, line $Y' - Z'$ forms an image angle ψ with line $A' - B'$.

$$\tan(90° - \psi) = \frac{\tan(90° - \phi)}{\sin \alpha}$$

$$= \cot \psi = \frac{\cot \phi}{\sin \alpha} \quad (4.12)$$

$$\tan \psi = \sin \alpha \tan \phi \quad (4.13)$$

Line $Y' - Z'$ on the film is a trace of the intersection of the oblique scanning plane and the oblique cylinder surface. The developed trace is geometrically a portion of an ellipse. However, as long as it is short, it can be considered a straight line with the determined angle ψ. Figure 4-20 shows diagramatically the image of a unit grid affected by panoramic distortion and by the distortions due to the tip angle ϕ. Compare this with figure 4-13.

For tipped cameras which employ a nodding IMC (rotation about the pitch axis) the angle ψ for each trace will be different even for equal scan angles on opposite sides of the line of flight. This is because the primary tip angle ϕ is modified continuously and in the same direction throughout the scan by changing angle $\Delta\phi$. Angle $\Delta\phi$ will, of course, increase the angle ϕ forward along the line of flight and will decrease ϕ at the end of scan. $\Delta\phi$ is usually a very small

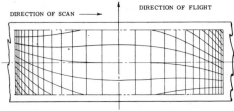

FIGURE 4-18. Resultant image in the vertical panoramic camera.

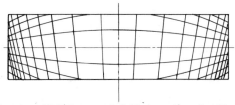

FIGURE 4-20. Diagram of the image of a unit grid affected by panoramic distortion and by the distortion due to tip angle.

angle and the resultant image geometry for the final image will appear similar to figure 4-19. There will be no IMC S-curve as illustrated in figures 4-18 and 4-21.

Figure 4-21 shows diagrammatically a pair of panoramic photographs such as might be obtained for stereoscopic coverage. The images of a unit grid on the ground obtained with a pair of convergent panoramic cameras in flight would include all the previously discussed distortions. For stereoscopic viewing, the photographs must be adjusted so that the epipolar rays are parallel to the eye base.

4.2.2.4 PROBLEMS IN DETERMINING INTERNAL GEOMETRY

The calibration of frame cameras is well understood, and routine procedures have been developed for determining the focal length and the camera principal distance, the location of the principal point and the path of the principal line, and distortions due to optical and/or mechanical causes. Because a panoramic camera exposes elements of imagery during a mechanical scanning cycle, the number of distorting influences is much greater than with frame cameras.

4.2.2.4.1 POSSIBLE SOURCES OF DISTORTION

The sources of distortions which influence the geometric fidelity of panoramic cameras can be grouped into three principal classes: (a) optical, (b) optical-mechanical, and (c) mechanical.

4.2.2.4.1.1 Optical. In this class, the geometry is affected by (a) lens distortions, and (b) errors in the principal distance determination. Lens distortion can be determined according to well-known procedures. The field of view of panoramic cameras is generally quite small in order to maintain the highest possible resolution, usually found in the center of the field. Distortions in this region are normally on the order of only a few micrometres.

In several panoramic cameras (*see* section 4.2.2.2), the film is positioned with respect to the lens by a scanning arm. Due to the floating action of the film over four rollers, the mechanical scan radius is not the principal imaging distance. The true distance from the nodal point of the lens to the film during scanning must be known.

4.2.2.4.1.2 Optical-Mechanical. The following sources of error affect the internal geometry:

(a) failure of lens to rotate about the nodal point or optical bar rotation axis;
(b) irregular motions of the lens scan, scanning arm, scanning prism(s) or mirror(s), slit, or film driver;
(c) wobble of optical and mechanical scanning components across the width of the format.

These sources of error are generally thought to have a negligible effect on the internal geometry. If they were not negligible, these optical-mechanical characteristics would affect the general quality of the photography to an extent incompatible with the high resolution attained with panoramic systems.

4.2.2.4.1.3 Mechanical. The effects of the mechanical sources of error on the internal geometry can be grouped into two main categories:

(a) those associated with the film, such as film motion during scanning when the film is supposed to be stationary, or film out of the focal plane; and
(b) those associated with other important mechanical actions in the camera, such as irregularity of IMC motion, or wrong IMC.

Of the effects listed above, that of film motion during scanning is the most severe. The problem of determining which exposed portion of film belongs with what scan position is at the heart of the more general problem of determining the internal geometry of panoramic cameras.

4.2.2.4.2 CALIBRATION TASKS

The principal task of interior geometry calibration is to make it possible to recover, for any image point, two coordinates which locate this point unambiguously and accurately. These two coordinates are the scan angle α and the cross angle β. The scan angle α is the dihedral angle between the plane through the scan axis and the central principal point, and the plane through the scan axis and the image point in question (it is assumed tacitly that the camera attitude does not change during the scan). The cross angle β is the complement of the angle between the scan axis and a line connecting the image point with the momentary center of projection—that is, the ray of projection. The central principal point is defined as the intersection of the film surface with the intersection of two planes: (a) the plane through the scan axis which bisects the total scan angle and which, hence, is the medial plane of symmetry of the camera and is vertical when

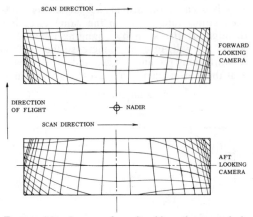

FIGURE 4-21. Images of a unit grid on the ground obtained with a pair of convergent panoramic cameras in flight.

the camera is in its nominal attitude; and (b) the plane normal to the scan axis which contains the rear nodal point of the lens. The central principal point also may be defined as the point of inflection of the IMC trace made by the principal ray on the film (figure 4-17).

4.2.2.4.2.1 Cameras with Film Stationary During Scanning. As was previously mentioned, film motion during scanning when the film is supposed to be stationary is the principal cause of the poor geometric fidelity of panoramic photography. For example, consider a camera as shown in figure 4-8, in which the film is accurately positioned in the focal plane by small rollers mounted at the end of a scan arm. Because the film may slip during one scan, the essential feature of any calibration must be to fix the position, by means of fiducials, of any segment of the film during its exposure with respect to the geometry of the camera. It is not adequate to rely on time markers on the film which indicate the position of the scan arm with respect to the beginning and end of one in terms of time. Time markers do not make it possible to ascertain whether the cause of nonuniform spacing of supposedly equidistant time marks is instability of the film (differential shrinkage), slippage of the film or nonuniform scan rate.

In a camera such as shown in figure 4-8, the film, although positioned exactly at the time of the exposure by the rollers, rests on guide rails. One approach to providing calibration data on such a camera is to drill a series of small holes into the rails, perhaps one every degree. These holes could then be illuminated from behind by small bulbs fastened to the scan arm so that sharp markings on the film would be obtained at the same moment that the adjacent slit element is exposed. During the instant of exposure, the film is lifted from the rails by the rollers. However, the illumination of the fiducial holes can be arranged so that the light passes the holes in a direction normal to the slit plane. Thus, any shift of the fiducial mark (image) due to the lift of the film can be avoided. Because the fiducial holes are a rigid part of the camera, their position in the camera can be expressed in terms of angles α and β. Both of these angles are defined solely with reference to the camera system.

The central problem of calibration thus can be reduced to the problem of measuring the positions of the fiducial holes. This might be accomplished by a combination of mechanical and photogrammetric mensuration procedures and by using the following inherent relationships:

$$\alpha = \frac{x - x_{pp}}{r} \qquad (4.14)$$

$$\beta = \frac{y - y_{pp} - C \sin \alpha}{f} \qquad (4.15)$$

where

x and y are the coordinates of the fiducial mark

x_{pp} and y_{pp} are the coordinates of the principal point

r is the radius of the cylinder to which the film-guide rails are fitted and whose axis is coincident with the scan axis

f is the focal length

$C \sin \alpha$ is the lens shift due to image motion compensation, where C is equal to $fV/H\ \delta$

where v = flight height

H = flight height

f = focal length

δ = rate of scan.

4.2.2.4.2.2 Cameras with Film Moving During Scanning. In a camera such as that shown in figure 4-9, the lens and the slit remain stationary while the scanning prism and the film move synchronously. As in the case of the previously discussed camera design, the key calibration goal remains the unambiguous correlation of any exposed element of film with specific scan angle α and cross angle β.

The generation of fiducial marks which are meaningful with respect to the interior orientation of the camera might be accomplished by several means. One approach might consist of incorporating two fiducial holes with light bulbs behind them into the slit assembly near both film edges. These light bulbs could be actuated or shuttered by a device rigidly connected to the prism drive, but stepped by a factor of 2 (the prism rotates 90° for a 180-degree scan). A fiducial image could be generated for every degree of scan. Mensuration against the fiducial marks would then fix the scan angle which existed for any image detail. Calibration would involve the precise determination of scan angles for the rotating prism, which might be accomplished by photographing an array of collimators with known angular relationships.

In many types of more modern panoramic cameras, whether they are scanning lens or moving film or a combination of these, such as the optical bar type illustrated in figure 4-11, the angular functions of lens scan, IMC nodding, or even film metering are commanded and controlled by precise electro-optical rotary encoders. These encoders can produce a signal at any selected angular interval which in turn can pulse plane light sources which image an angular fiducial reference relative to the exposed scene. Encoders used for control of lens and film motions do not normally have the precision needed for photogrammetric calibration tasks. However, the values are constant and therefore the encoder can be calibrated against a more precise angular reference.

4.2.2.5 REQUIREMENTS FOR USING PANORAMIC PHOTOGRAPHY

The use of panoramic photography in its "raw" form, with all its inherent deformations, poses major inconvenience to the photointerpreter or photogrammetrist. Therefore, one of the first requirements is to provide for the removal of the distortions. One method is to employ the analytical stereoplotter in order to transform sequentially the coordinates on the panoramic photograph to the equivalent of a vertical, fixed-frame photo, using formulation found in section 4.2.2.3, and to reference frame photography on ground control. Special programs must be prepared and the analytical plotter must be modified for differential image rotations and magnification in the viewing system. Another method of preparation of panoramic photography is image rectification.

Four basic approaches have been used to rectify or transform the panoramic image. The first is purely electronic, where the original image is rescanned, converted to an electrical signal, and redisplayed in its corrected or rectified position.

The second approach is a computer-electronic method whereby the image is scanned, digitized as coordinates of grey levels, mathematically operated upon, and reprinted in its rectified position.

A third approach is the computer-operated, optical-electronic method where the image is optically projected from negative to easel, but only a small element at a time. The printing speed and optical functions which determine the final geometry are directed by a computer.

The final basic approach is the purely optical or optical-mechanical projection with a modified form of the conventional optical rectifier.

The optical projection methods appear to offer the greatest number of important advantages, namely, high image resolution, high printing speed, rigid image geometry, and equipment reliability. Two disadvantages might be the limited application and the inability to correct for small variable changes in geometry during the printing procedure. The principles of optical transformation of the panoramic image are explained in the following paragraphs.

4.2.2.5.1 OPTICAL PANORAMIC TRANSFORMING PRINTER

The optical transformation of the distorted negative is accomplished by duplicating in reverse the taking system. A light slit scans across the negative surface and projects the image through a lens onto the easel-mounted printing material. Coincidental with this operation, the easel and platen translate relative to each other to correct for relative ground displacement and IMC during the camera scan.

The following is a brief review of the distortions and the resultant corrections.

(a) Panoramic Distortion: The displacements of images of ground points from their expected perspective positions caused by the cylindrical shape of the negative film platen and the scanning action of the lens.
The optical-mechanical correction for this

distortion is simply a reprojection of the cylindrical negative through a lens onto an easel. The easel or projection plane is the ground surface at rectification scale.

For a more complete solution, the panoramic rectifier easels must be provided with a tilt to correct for ϕ tilt of the scanning plane.

A roll, or ω-tilt, correction is achieved by offsetting the negative within the surface of the input film cylinder.

(b) Scan Positional Distortion: The displacement of the images of ground points from their expected cylindrical positions caused by the forward motion of the aircraft during the scan time of the lens. This distortion is in addition to, and modifies, the displacement of points due to panoramic distortion.
The mechanical correction for this distortion is a displacement or translation of the easel during the reprojection scanning. This motion simulates the apparent ground displacement at rectification scale. The motion is servo-driven and controlled in its magnitude by a V/H (aircraft velocity/aircraft altitude) indicator.

(c) Image Motion Compensation (IMC) Distortion: The displacement of images of ground points from their expected cylindrical positions caused by the translation of the lens or focal plane which is used to compensate for image motion during exposure time. This distortion is in addition to, and modifies, both panoramic distortion and scan positional distortion.
This correction is provided by translating, or displacing, the negative cylinder parallel to the axis of lens scan during the reprojection scanning.

For photo acquisition, the lens of the panoramic camera is translated for the IMC, but, since the reprojection lens must fulfill various complex optical conditions during printing, it is simpler to move the negative platen to achieve the same result.

Although these mechanical motions are complex, the most difficult problem facing the instrument designer is meeting the optical conditions at all times during the printing process.

4.2.2.5.1.1 Focusing Condition. Figure 4-22 shows the geometric relationship prevailing during imaging onto the cylindrical focal "plane" of a panoramic camera. Note that all

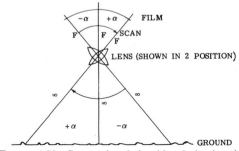

FIGURE 4-22. Geometric relationships during imaging onto the cylindrical focal "plane" of a panoramic camera.

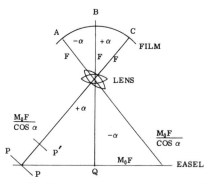

FIGURE 4-23. Geometry of reprojection in printing a panoramic negative.

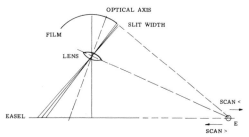

FIGURE 4-25. Geometry of Scheimpflug condition which must be satisfied.

objects are at infinity. The geometry of reprojection in printing is shown in figure 4-23.

If the camera geometry is maintained by keeping the lens at the center of the film radius, the image at point B can be in focus at point Q, but point C will be in focus at plane P' instead of at plane P-P.

In order to overcome this, the field of the rectifier lens must be used to provide the required focusing conjugates. The geometric implications are shown in figure 4-24.

The focusing angle δ can be computed from

$$\cos \delta = \frac{f \text{ rectifier } (F \text{ camera} + d)}{Fd} \quad (4.16)$$

to achieve an approximate value for the selection of a lens with a sufficient angular field.

When the lens is selected, a test target (representing point C in the negative) is placed on an optical bench, conjugates a and b are established, and the lens is rotated about its nodal point until the best image is achieved at plane P-P.

A series of conjugate pairs is tested and the angles determined. From this information, a mechanical cam is produced which rotates the lens axis through the angle γ while the imaging slit is scanned through the scan angle α.

4.2.2.5.1.2 Scheimpflug Condition. Because the negative imaging slit is not infinitely thin,

focus must also be maintained over its width. Figure 4-25 shows the geometry of the Scheimpflug condition which must be satisfied. The optical function performed by the lens rotation in the scanning mechanisms actually satisfies a form of Scheimpflug condition, and thus also satisfies the focusing of the image over the width of the slit, and onto the easel. The easel-lens-negative intersection point E continually changes position during the scan motion.

4.2.2.5.1.3 Other Solutions. In the construction of one panoramic rectifier, where image quality was of utmost importance, the required field angle of the printing lens was minimized by shaping the negative film in a noncylindrical form, translating the lens vertically while scanning (to maintain geometry), and giving the easel a nonflat shape to improve the overall focusing.

If the transforming printer easel can be tilted to correct for ϕ-tilt, the Scheimpflug condition must be fulfilled over the width of the negative (length of the slit). This also varies with the scanning angle, and requires the design of a complex, three-dimensional focusing cam.

4.2.3 STRIP CAMERAS

4.2.3.1 GENERAL

The continuous-strip camera (Kistler, 1946), also known as the "Sonne Camera," was first used in 1932 and further developed prior to World War II. It was used operationally for low-altitude, high-speed reconnaissance. (At that time, frame cameras lacked forward-motion compensation and high cycling rates, because they were intended, primarily, for high-altitude operation.) The image-motion-compensation technique subsequently introduced in frame cameras is an outgrowth of strip-camera technology. Missions requiring object height (or water depth) determinations or route reconnaissance are particularly appropriate for the strip camera. Strip cameras also have been used for night photography with mercury-arc-lamp illumination. Side-looking radar recorders and some infrared recorders are actually strip cameras with the optics looking at a cathode ray tube or other modulated light source, rather than at the ground.

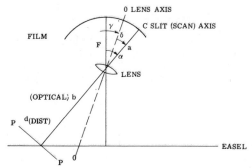

FIGURE 4-24. Geometry of rectifier for providing proper focusing conjugates.

Commercially, strip-camera photography is being successfully used for highway and railroad studies, for selection of rights-of-way for pipelines and power lines, for determination of tree types for forestry applications, and for many other purposes. It also has been used for airport runway inspection, urban traffic studies, and for certain types of charting (Elms, 1962).

4.2.3.2 MODE OF OPERATION

The strip camera exposes a continuous photograph of the terrain by passing the film over a stationary slit in the focal plane of the lens at a speed synchronized with the velocity of the ground image across the focal plane. The rays from any point will be focused as a single point on the film throughout the time of exposure. The duration of the exposure depends upon the film speed and the width of the slit, such that:

$$\text{Exp} = \frac{W}{V_F} \tag{4.17}$$

where

Exp = exposure time;
W = slit width;
V_F = film velocity.

In practice, the slit width usually is quite small (about 0.25 to 0.5 mm) so that at any instant the image of only a narrow "ribbon" of terrain is exposed on the film. As the aircraft moves forward, a long continuous photograph is "painted" onto the film by successive integration of these narrow ribbons.

Two conditions must be met before good continuous strip-photograph can be obtained. Close synchronization between film velocity and image velocity must be maintained to avoid lengthening and shortening of images, and some means for automatically stabilizing the camera in the roll axis must be provided.

Strip cameras may be used either with a single lens system in a nonstereoscopic mode, or with a dual lens system where each lens covers one-half the film width. With a dual system, the stereoscopic base is obtained by displacing the right lens forward and the left lens aft, creating the base in the direction of flight. The lens offset is adjustable on stereo lens systems and is calibrated in "parallax angle." This is the angle between sight lines to the perspective centers of the lenses as seen from the center of the slit, and is related to the base-height ratio by:

$$\text{base-height ratio} = 2 \tan\left(\frac{\text{parallax angle}}{2}\right). \tag{4.18}$$

Larger parallax angles increase vertical exaggeration and the ability to visualize or measure object heights.

The continuous-strip camera has been developed to the point where it can operate automatically and produce good photography under adverse flying and light conditions. Electronic scanners and servo power units are available for automatically measuring and feeding correct image speed information into the camera. Also operational are automatic diaphragm settings for optimum exposure. Stabilized mounts are available for keeping the camera free from harmful vibrations and to prevent motion about the roll axis during operations.

Strip cameras are basically the simplest image-motion-compensating aerial cameras used. They are more trouble-free than other cameras that contain shutters and cyclic changes of mechanisms. The most common trouble with strip cameras is *banding* on the strip photograph. *Banding* is due to cyclic changes of exposure. The fault generally lies in unsteady film velocity, worn gear teeth, and excessive vibration inside the camera.

Strip cameras often are used for side-oblique photography, even though it is not possible to have a side-oblique strip photograph correctly synchronized to allow for image motion on the film throughout the entire picture width. When strip cameras are used for side-oblique photography, the camera slits usually are canted to minimize the adverse effect of obliquity.

In order to provide a fuller understanding of the modern strip camera, a detailed description of the U.S. Air Force KA-18A Strip Camera (figure 4-26) follows.

The KA-18A strip camera is a reconnaissance-type camera engineered to expose roll film in high-speed aircraft at low altitudes. When the camera is used with the Air Force's Simplified Camera Control System (SCCS), it is capable of producing a single strip or stereo strip negative 500 feet long, using

FIGURE 4-26. Type KA-18A strip camera (Courtesy Chicago Aerial Industries).

either automatic or manual operation. The three main components are the cassettes, camera body, and automatic lens cones.

Cassettes for this camera come in two different sizes. Each is a light-tight, heated, roll-film receiver, with film capacities of 250 or 500 feet of Class L-type film. Cassettes are equipped with indicators that show approximate film footage in each cassette. All cassettes are interchangeable units and may be used as supply or takeup receivers.

One of the three safety switches (the cassette interlock switch) is placed so as to prevent operation of the film drive when the cassette is not completely seated. Power to the thermostatically-controlled heaters within the cassette is obtained through connections built into the hinges located on the camera body and attached to the cassette.

The *camera body* for the KA-18A houses, among other parts, the slit mechanism, film-drive motor platen, brake and clutch assemblies, idler roller, drive roller, supply roller, and a very important part of the electronic system, the "conversion adapter."

Slit mechanisms are designed to be used with right-oblique, left-oblique and vertical installations of the camera. The main difference in the mechanisms is in the design of the blades of the slit.

The 28-volt DC film-drive contains a tachometer, the output of which is used to control the exposure and image-motion-compensation film speed.

The platen is of stationary curved-face design and operates without applied vacuum. Film pressure on the platen is maintained through correct adjustment of the torque of the clutch and brake. The clutch operates between the metering roller and the takeup spool. The purpose of the clutch is to adjust the rate of film movement from supply to the takeup spool.

A brake in the camera body operates on the supply mechanism and serves to prevent film from being pulled from the supply spool faster than the image-motion-compensating speed. Film supply also is governed by the idler roller, the drive roller and the supply roller. These units operate to maintain an even supply of film from the supply spool to the takeup spool.

The conversion adapter with its 94 component parts is the heart of the electronic system of the camera. Through the conversion adapter is channeled the feedback from the tachometer, from the iris, and from the slit mechanisms. The output of the conversion adapter is sent to the appropriate component of the Simplified Camera Control System.

Three lens cones are supplied with the KA-18A camera, as follows.

(a) A 6-inch lens cone (LA-79) contains a 6-inch, f/6.3 single lens with an angle-leaf-type iris dia-

phragm, voltage potentiometer, a two-phase motor, and associated wiring.

(b) A 6-inch stereo lens cone (LA-77) contains two lens systems. Both systems are 6-inch, f/2.5. The cone includes a parallax indicator, light baffle, two-phase motor, feedback voltage potentiometer, and necessary wiring.

(c) The 3-inch stereo lens cone (LA-78) contains two 3-inch lens systems, iris diaphragm, feedback potentiometer and two-phase diaphragm drive motor and parallax-angle adjustment with control from 0° to 20°, but normally set at 6°. This setting is necessary to give depth perception.

4.2.4. MULTIBAND CAMERAS

4.2.4.1 GENERAL

Development of cameras designed for multiband photography through the years has followed three paths; (a) multi-camera installations; (b) multi-lens cameras; and (c) beam-splitter cameras.

The multi-camera installation was used in the 1930's to obtain three color separation negatives—blue, green and red—for color aerial photography. Later, a fourth camera was added to record the infrared image.

The multi-lens multiband camera came into use in the early 1960's. This camera has a lens and filter for each spectral band to be photographed. The lenses are contained in an integral camera body.

The beam-splitter multiband camera was developed in the early 1970's. The four images—blue, green, red and infrared—are formed by a single lens and the image from each spectral band is separated by a beam-splitter prism system.

4.2.4.2 MODE OF OPERATION

4.2.4.2.1 MULTI-CAMERA INSTALLATION

Three or more cameras with selected lenses are mounted on a single mounting frame. The cameras are bore-sighted to cover the same field of view. The tripping pulse is sent to each camera and the shutters open and close, in time, as electromechanically possible. The lens of each camera is covered with a filter to pass the desired spectral band. Each camera's shutter speed and aperture stop can be set at the proper exposure for the film-filter combination. Each camera magazine is loaded with the desired film for each spectral band.

4.2.4.2.2 MULTI-LENS CAMERA

The multi-lens camera consists of four lenses, with appropriate filters, mounted on a single camera body. Early camera magazines could be loaded with only one type of film at a time; thus, all four images were recorded on the same film. Later magazines were developed to accommodate two rolls of film. Therefore, the images can

FIGURE 4-27. Spectral Data Model 12 multispectral camera (Courtesy of Spectral Data Corp.).

be recorded on film suitable for the spectral band. This type of camera can be used with image-motion-compensating magazines to improve image quality at low altitudes. The cameras use one of the following types of shutter:

(a) single-slit focal plane, wherein all exposures are not simultaneous,
(b) multi-slit focal plane, wherein each lens format has a separate opening, and all formats are exposed at the same time, and
(c) between-the-lens shutters, which are tripped by the same trip pulse.

Filters are used on each lens to obtain the spectral band transmitted by each lens. Differences in the spectral sensitivity of the films are corrected by adjusting each lens aperture stop for the proper exposure.

An example of a multi-lens camera is shown in figure 4-27. This is the Spectral Data Model 12 Multi-Spectral Camera. It produces four 3½-inch square photographs on a single piece of 9½-inch film.

4.2.4.2.3 BEAM-SPLITTER CAMERA

The single-lens, four-channel multiband camera has a beam-splitter prism assembly between the lens and focal plane of each spectral image. The incoming light passes through the single objective and is spectrally separated by a beam-dispersing prism assembly that is surface-coated with broadband dichroic filters. The image projected by the lens is split at the cyan reflector, which reflects the blue and green images and transmits the red and infrared images. The blue and green images are separated by the blue reflector, and the red and infrared are separated by the red reflector.

Four film magazines are required, and each spectral image is recorded on a separate emulsion. The emulsions can be selected for spectral sensitivity and developed to balance the film speed to the spectral transmission of the lens-prism system for each color band.

Figure 4-28 shows a multiband camera of this type, built by Perkin-Elmer Corp. for the U.S. Army Engineer Topographic Laboratories. It is equipped with 6-inch and 4-inch interchangeable lenses, each of which exposes the four images on 70 mm formats. At ETL, it is being used in support of data base and data bank experiments, and for design and production of experimental image-based terrain intelligence products.

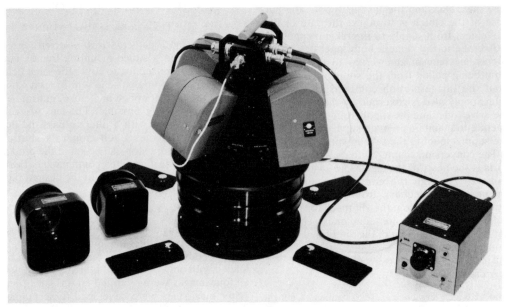

FIGURE 4-28. USAETL multiband camera, built by Perkin-Elmer Corp. (U.S. Army Photograph).

4.3 Special Cameras for Use in Space

Aerial cameras are generally designed for specific circumstances—that is, the operating altitudes, ground speeds, illumination levels, desired ground resolution, and angular coverage. The data are the photographs which are transmitted to the ground by delivery of the film. In the design of space cameras, on the other hand, the data-transmission circumstances assume an equal importance. Some space programs—notably those involving manned space flight—do provide the opportunity to bring back the exposed film. More generally, though, photographs are transmitted from the spacecraft by electromagnetic signals. This always is the case for photographs taken from spacecraft sent to other planets.

Since data transmission involves a time-sequence of signals, a two-dimensional image must somehow be scanned during transmission. Sometimes it is possible to use a camera that scans a scene in step with the transmission. Spacecraft that depend on spin for directional stabilization often carry systems of this kind. However, wherever a scanning camera is not appropriate, a means must be provided for storing the image until it is transmitted.

Storage time can vary from a fraction of a second to several months. Spacecraft television cameras focus an optical image on the photoconductive surface of a Vidicon picture tube, where it is stored as an electrostatic image until is is scanned by an electron beam to provide a signal for transmission. That is all the storage required for a single image. If the circumstances of a particular mission require the acquisition of a series of images before the first one can be transmitted, an additional means of storage is required. A magnetic tape recorder provides appropriate storage for a television camera system, although considerable lengths of tape are required to store moderate quantities of image detail. If fine detail must be stored in large quantities, systems that utilize the vastly greater storage capacity of photographic film must be considered, despite the necessity for processing the film in space and then scanning it for transmission.

The cameras covered in this section are discussed mainly in terms of the individual projects for which they were designed.

4.3.1. CAMERAS DESIGNED FOR RETURN OF FILM

The cameras discussed here were used in NASA's Apollo program.

4.3.1.1 HASSELBLAD DATA CAMERA

This 70-mm film camera was designed for use by astronauts on the lunar surface. The basic camera was the Hasselblad Electric Camera that was normally used in lunar orbit to take photographs through the windows of the spacecraft. Aside from the modifications necessitated by the lunar surface environment and the restrictions on the astronauts' visual field and manual dexterity, the major new feature of the Hasselblad Data Camera was the glass plate bearing a reseau. Preflight calibration of the specially designed 60-mm Zeiss Biogon lens with this reseau satisfied the minimum requirements for a metric camera. The primary use of lunar surface photography for the detailed documentation of the returned lunar samples requires the application of methods of "terrestrial" photogrammetry. The camera also was used with a 500-mm lens for surface photography of selected distant points.

4.3.1.2 APOLLO LUNAR SURFACE CLOSEUP CAMERA

This camera produced stereoscopic pairs, on 35-mm color film, of 3-inch by 3-inch patches of the undisturbed lunar surface. Because it resolved detail as fine as 70 micrometres, it facilitated special studies of the properties and distribution modes of the fine surface powder.

4.3.1.3 APOLLO MAPPING CAMERA SYSTEM

The last three Apollo missions included two photographic systems: the Mapping Camera System and the Apollo Panoramic Camera. The use of these cameras during the period of orbital operations produced a combined coverage, for the three missions, of about one-fourth of the lunar surface.

The Mapping Camera System (*see* figure 4-29) was designed to provide not only metric photography of very high quality, but sufficient auxiliary information to permit the establishment of a network of control points in the absence of any ground survey. The system included a laser altimeter that was aligned with the mapping camera's optical axis, and a stellar camera that photographed the star field at a fixed angle to that axis. Synchronized operation of the three instruments provided an independent determination of the mapping camera's altitude and orientation for each exposure. The inclusion of a data block recording the precise spacecraft time with each exposure permitted the employment of the operational Doppler tracking data in the triangulation scheme.

The mapping camera's 3-inch f/4.5 lens exposed a 4½ by 4½ inch frame on a roll of 5-inch film that was 1500 feet long (enough film for 3600 exposures). During each exposure, the film was pressed against a glass stage plate that bore a reseau. Movement of the stage plate with respect to the optical axis provided image motion

FIGURE 4-29. Lunar mapping camera (Courtesy of Fairchild Camera and Instrument Corp.).

compensation at a rate corresponding to the velocity-to-height (V/H) ratio selected by the astronaut. The interval between exposures was set by the V/H ratio to produce a forward lap of 78%. The rotating-disk shutter was driven by a motor that had seven discrete speeds. An automatic exposure control device selected the correct disk speed, providing exposure durations between 1/15 second and 1/240 second.

The takeup spool for the exposed film was in a drum, outside the camera body, which also held the takeup spool for the 35-mm film exposed by the stellar camera. After the Apollo spacecraft left the moon on the return journey, the astronaut who had accomplished the orbital photography climbed out to the Service Module to retrieve this drum, as well as the drum containing the panoramic photography.

4.3.1.4 APOLLO PANORAMIC CAMERA

A panoramic camera was included with the scientific instruments in the Service Module to provide high-resolution stereoscopic coverage of broad areas of the lunar surface. The camera was a modified version of the Air Force's KA-80A "optical bar" camera of 24-inch focal length (*see* section 4.2.2.).

The optical bar, which included the lens assembly, the film-support rollers, and the exposing slit, rotated continuously during the camera's operation. It exposed film during 108 degrees of each rotation. The exposed frames were 45 inches long (transverse to the flight direction) and 4½ inches wide.

The optical bar's rotational speed was under the continuous control of the camera's V/H sensor, a photoelectric device that measured the apparent ground motion. Another sensor mea-

sured the scene brightness over a field that covered 10° in the flight direction and 30° in the scan direction. This information, combined with the V/H output, provided input to the automatic exposure control system that adjusted the width of the exposing slit.

Image motion compensation was accomplished by rocking the assembly that contained the optical bar in a fore-and-aft direction, at the rate determined by the V/H sensor. The same rocking mechanism provided for stereoscopic coverage by pitching the assembly alternately 12½ degrees forward and 12½ degrees aft of the vertical. The camera cycling rate was controlled so that the ground covered by a given "forward-looking" photograph was covered again five frames later in the sequence by an "aft-looking" photograph.

The camera's capacity of 6500 feet of 5-inch film was sufficient for 1650 frames. The exposed film was taken up in a drum outside the camera body, for transfer to the Command Module.

4.3.2 FILM CAMERAS DESIGNED FOR ON-BOARD PROCESSING, READOUT AND TRANSMISSION

Under circumstances where a large quantity of fine photographic detail must be stored, photographic film may be the preferred storage medium despite the impossibility of getting the film back to earth. While systems of this kind are necessarily complex, the circumstances of some space missions permit a degree of simplification. A relatively simple system will be described briefly before the more elaborate Lunar Orbiter system is discussed.

4.3.2.1 PHOTOGRAPHIC SYSTEM USED ABROAD ZOND-3

When the Soviet ZOND-3 spacecraft flew past the far side of the moon in July, 1965, it carried a photographic system that was designed for photography of the moon at long range. A camera for such long-range photography can operate at a low cycling rate, and need not provide image motion compensation. In this camera the film was continuously pulled past the focal plane and through the processer and dryer at the very low speed required for on-board processing. The interval between shutter operations, which was fixed by the time required to move the exposed film out of the focal plane, was 2 minutes and 14 seconds. The 106 mm, f/8 lens covered a 24 × 24 mm frame. The frames were alternately exposed at 1/100 second and 1/300 second. The operation was not subject to ground control after the turn-on command.

The processer employed viscous solutions having very stable characteristics. Once the entire roll had gone through the dryer, the film could be moved back and forth through the readout apparatus by ground command.

The developed negative was read out by an optical-mechanical scanner that used a rocking mirror to scan a small spot of light across the film gate. Scan in the lengthwise direction was accomplished by driving the film past the gate. The light passing through the negative went through a condenser lens to a photomultiplier tube. The electrical signal from the tube went to the radio transmitter. The diameter of the readout spot was 20 micrometres. Each scan line contained 1100 such picture elements. In the normal readout mode there were 1100 scan lines per frame. It took 34 minutes to read out one frame. There was also a provision for rapid viewing (quick-look), in which only 67 scan lines were read out per frame.

4.3.2.2 LUNAR ORBITER PHOTOGRAPHIC SYSTEM

Because the five Lunar Orbiter spacecraft that operated in 1966 and 1967 were designed from the beginning for high-resolution photography of specific lunar regions, provision was made for highly versatile operation. Exposure, processing, and readout were essentially independent of each other even though they were done with the same roll of 70-mm film. The system is described in considerable detail because Lunar Orbiter photographs continue to provide the definitive coverage for the 75% of the lunar surface that was not photographed by the Apollo mapping and panoramic cameras.

The camera was a dual one, with the two shutters operating simultaneously. The medium-resolution (M) camera employed an 80-mm f/2.8 Schneider Xenotar lens (stopped down to f/5.6) and an intra-lens shutter to expose a 55 × 65-mm frame. The high-resolution (H) camera's 610-mm f/5.6 lens exposed a 55 × 219-mm frame through a focal-plane shutter. A shutter speed of 1/25, 1/50, or 1/100 second was selected by ground command for each picture sequence. Both cameras provided image motion compensation at rates determined automatically by an electro-optical V/H sensor. The V/H sensor also controlled the spacing of exposures in a sequence.

The two camera axes were coincident, so that a single exposure produced a nested ground coverage. The H-camera covered 5° in the direction of flight and 20° perpendicularly. The respective coverage angles for the M-camera were 38° and 44°. As many as 16 exposures could be made in one sequence, at a slow or a fast repetition rate. The fast rate provided a continuous monoscopic strip of H-camera coverage, with 88% forward overlap of the M-frames. At the slow rate, the M-camera overlap was 52%, and the areas covered by the H-frames were separated by evenly spaced gaps.

The layout of the photographic system provided for the temporary storage of up to 20 exposed dual frames between the camera and the processor. In the processor, which employed the Kodak Bimat diffusion transfer process, the film was laminated to a processing web, held in contact for at least 3½ minutes, then separated and dried. A slightly damp gelatin layer of the Bimat processing web held the monobath solution, which developed the film to a negative image and transferred the undeveloped silver ions to a positive image on the Bimat web.

After being dried, the negative film went through the readout scanner to its takeup spool. During readout, the film moved backward through the scanner. A storage loop between the processor and the scanner permitted a "quick-look" readout of selected portions of the film before photography was completed.

Upon completion of photography, the processing web was cut and pulled free of the processor. The negative film then could be read out completely (from the last exposure to the first) and be pulled through the processor and camera onto the supply spool.

The complex design of the readout scanner was dictated by the necessity of preserving, in the transmitted image, all of the fine detail in the film. In order to meet the overall Lunar Orbiter objective of obtaining a ground resolution of one metre from an altitude of 46 kilometres, the negative image on the spacecraft film needed a resolution of 75 line pairs per millimetre. The readout scanner preserved that resolution by reducing the diameter of the scanning spot to 5 micrometres. The light source was a flying-spot scanner (a type of cathode ray tube) that produced a single very bright line. An optical-mechanical scanner moved a greatly reduced image of that line across the width of the film.

The length on the scan line was only 2.68-mm, and the cathode ray scan was repeated 17,000 times during one mechanical scan. When the scanning lens reached one edge of the film, the film was moved 2.54-mm (1/10 inch) in the read-out gate, and the scanner reversed its travel across the film. The resulting sections of scanned film, referred to as framelets, were the basic units of the ground reassembly. The process of scanning a complete dual exposure took 43 minutes.

As the video signal resulting from the readout was received at a tracking station, it was recorded on magnetic tape and simultaneously fed to a ground reconstruction unit. Here the signal was converted into an intensity-modulated line on the face of a cathode-ray tube, and recorded on 35-mm film that was continuously pulled past the image of this line. In order to avoid loss of resolution at this stage, the scan lines were recorded at a magnification of 7 times. The recording film was cut into framelets, which were then reassembled to form a mosaic that was an enlarged replica of the original spacecraft frame.

By segmenting each original frame into framelets, the readout system preserved the image quality at the price of some degradation of the frame's geometric integrity. The reassembly of the frame was facilitated by a repetitive pattern preprinted along one edge of the film, and by sets of measured fiducial marks spaced along the sides of the camera so that at least one mark appeared at each end of every framelet. As a further aid to photogrammetric analysis, a set of reseau marks was preprinted throughout the unexposed film for every mission except the first. Nevertheless, the stereoscopic precision of Lunar Orbiter photography is limited by the readout errors.

4.3.3 SPACECRAFT TELEVISION CAMERAS

Television cameras were employed on lunar missions early in the space era, and they are still widely used in planetary missions. They offer the advantages of mechanical simplicity, well-understood electro-optical design principles, and completely erasable and reusable image storage. Their main limitation is in the number of picture elements that can be recorded on a single frame. If fine ground resolution is desired, the angular coverage must be quite small.

4.3.3.1 Television in Crash-Landing Missions

Television cameras on Ranger missions 7, 8, and 9 provided our first close-up images of the lunar surface. They operated while the spacecraft were hurtling toward the surface at 6000 miles per hour. Since the most important pictures of each mission were to be the ones taken just before the crash, the time required for taking and transmitting these pictures had to be minimized.

Each spacecraft had a package of six television cameras, in two functionally separate sets, called the F (or full-scan) and P (or partial scan) sets. The two cameras of the F-set operated alternately, providing continuouous transmission over one communication channel. A shutter in each camera exposed the face of the Vidicon tube, which then was scanned by an electron beam. The readout scan had 800 active scan lines, with 800 picture elements per line, and required $2^1/_2$ seconds to complete. Thus, the final frame transmitted over the F-channel was exposed between $2^1/_2$ and 5 seconds before the crash. In order to take some pictures closer to the surface, the four cameras of the P-set operated sequentially, with each Vidicon tube scanning only the central portion of its target face. Only 200 lines were scanned, at 200 elements per line, but this entire 40,000-element frame was transmitted in 1/5 second over the P-channel. On the final P-frame of Ranger 9, which was exposed 1/4 second before impact, each picture element covered 25 centimetres of the lunar surface.

4.3.3.2 Television on Soft-Landing Missions

The Surveyor program soft-landed a series of spacecraft on the lunar surface. The camera aboard each landed spacecraft was expected not only to produce panoramas, but also to provide detailed views of selected parts of the scene, and to support other scientific experiments. The versatility of television fulfills these requirements, although other types of camera can also be employed under similar circumstances (for example, see section 4.3.4.1.2 on the Viking Lander cameras). The Surveyor television camera stood about five feet above the footpads and pointed upward into a mirror that could be rotated by ground command to any desired azimuth and elevation angle. The camera used a shutter to expose a Vidicon tube whose target face was scanned in one second. There were 600 active scan lines, each containing 600 picture elements. The picture was transmitted to earth as it was scanned.

In the wide-angle mode, with the camera's zoom lens at a one-inch focal length, it required 200 frames to make a 360° panoramic survey, covering everything from the horizon to the nadir. This could be accomplished with ground commands in 8 minutes. In the narrow-angle mode, at a focal length of four inches, each picture element covered 0.25 milliradian of field angle. A complete panoramic survey in this mode, consisting of 1600 frames, could be accomplished in 75 minutes. On the ground, each frame was reconstructed on the face of a kinescope tube and photographed. The frames then were assembled into mosaics.

4.3.3.3 TELEVISION ON PLANETARY FLYBY MISSIONS

Planetary flyby missions are characterized by a strictly limited time (no more than a few hours) for data acquisition, followed by a much longer time available for transmission. In order to use a television camera on such a mission, a spacecraft must have a tape recorder. The storage capacity required of the recorder depends on the transmission rate—that is, as much photographic detail should be recorded as can be transmitted while the earth is within communication range. Continuing improvements in communications and storage have made it worthwhile to use television cameras capable of increasingly finer ground resolution.

4.3.3.3.1 MARINER 4

This was the first planetary mission to use a television camera and tape recorder. In July, 1965 it flew past Mars and took a total of 22 frames in a three-hour encounter. After each exposure, the image was read off the Vidicon target plate in a 200-line scan, with 200 picture elements per scan. These 40,000 elements were recorded on magnetic tape in digital form. That is, the signal voltages coming from the Vidicon were classified into 64 levels, and each picture element ("pixel") was recorded as being at one of those levels. Since it requires six binary digits to characterize 64 levels, each picture element was recorded as a six-bit word. The total amount of photo data recorded for the 22 pictures was 5 million bits.

After the encounter with the planet, the tape was played back, and the pictures transmitted as pulse-code-modulated digital data. At that stage of space communication technology, the data rate for the 130-million-mile transmission was only $8^1/_3$ bits per second. It took 8 hours to send each frame.

4.3.3.3.2 MARINERS 6 AND 7

By the time Mariner missions 6 and 7 flew past Mars, in July and August of 1969, the rapid development of communication technology (and the completion of a 210-foot-diameter receiving antenna at Goldstone, California) permitted a transmission rate of 16,200 bits per second. In order to take advantage of this 2000-fold improvement, the magnetic tape capacity was increased to about 150 million bits, and each spacecraft had two television cameras capable of recording much finer detail.

Each camera's Vidicon tube had an oblong format, with 704 active scan lines and 945 picture elements per scan. It took four minutes to scan, or to transmit, a single frame. One camera, possessing a 50mm focal length lens and two color filters, was used for wide-angle photogra-

phy. The other camera, with a 508mm focal length, was boresighted so that the high-resolution coverage was at the center of the wide-angle frame. The cameras were mounted on a platform that had two degrees of freedom, permitting them to be pointed during the flyby in order to increase the coverage. The high transmission rate made it possible to transmit several sets of "far encounter" pictures as the spacecraft approached the planet and still have the full tape capacity available for recording the "near encounter" frames during the flyby. In this manner, the two missions obtained a total of 141 far-encounter and 58 near-encounter pictures of Mars. However, the full potential capability of this television system could only be realized in an orbiting mission. Its utilization in such a mission is described in section 4.3.3.4.1.

4.3.3.3.3 MARINER 10 TELEVISION SYSTEM

Mariner 10, which flew by Venus in February, 1974 and photographed the structure and motion of its atmosphere in ultraviolet light, went on to photograph over 40% of the surface of the planet Mercury in three successive flyby encounters in March and September, 1974, and in March, 1975.

The television system had two identically-equipped cameras. The prime lens for each camera was a 1500mm f/8.4 catadioptric telescope that provided a field of view of only 0.36° by 0.48°. The image formed by an auxiliary wide-angle optical system (62mm f/8.5, 11° by 14° field of view) could be placed before the camera by means of a mirror in the filter wheel. The filter wheel also carried untraviolet, blue, orange, polarizing, and clear filters for use with the prime lens. The Vidicon tube had extended response in the ultraviolet (between 300 and 400 nanometres).

Each television frame had 700 scan lines, with 832 picture elements per line. Picture elements were digitally coded into 8-bit words (256 levels of gray). It took 42 seconds to scan one frame, at a rate of 117,600 bits per second. The photographs were recorded on a tape whose total capacity was 1.8×10^8 bits, enough for 36 frames. The tape could be played back for transmission at a rate of 22,050 bits per second. Although this transmission rate (36% higher than that of the Mariner 6, 7, and 9 missions) made it possible to play back all 36 frames in $2^1/_4$ hours, it would not be sufficient for the very ambitious coverage plans during the near encounter with Mercury. A capability therefore was provided for bypassing the tape recorder and transmitting directly at the 117.6 kilobit-per-second scanning rate. Transmission at that rate meant accepting a substantial degradation in picture fidelity, because it increased the error rate from less than one incorrect bit in a thousand to one in forty.

The strategy adopted for the first Mercury encounter made repetitive use of the tape recorder during the approaching and outgoing stages to acquire full-coverage spectral mosaics of the visible surface at each filter setting. During the 8-hour close-encounter period, the cameras operated continuously in the direct-transmission mode, taking a picture through the clear filter every 42 seconds, for a total of 592 frames. Of these, 35 with the highest resolution were recorded on the tape for later transmission at very low bit error rates. Photography during the second and third Mercury encounters was all in the direct-transmission mode because the tape recorder had become inoperable.

4.3.3.3.4 MARINER JUPITER-SATURN MISSIONS

In 1977, NASA launched two Mariner Spacecraft to conduct flyby explorations of the giant planets, Jupiter and Saturn, and their many satellites. The television system designed for these missions resembles the Mariner 10 system, although modified for the special mission circumstances. The system includes a narrow-angle and a wide-angle camera, whose respective focal lengths are 1500mm and 200mm. The Vidicon tube for the narrow-angle camera has some sensitivity in the near ultraviolet. Both cameras are equipped with filter wheels, and both have square formats with 800 scan lines and 800 picture elements per scan. Each picture element is digitally coded into an 8-bit word. The scanned frame either can be transmitted directly (as 115,000 bits per second from Jupiter's distance) or recorded on tape for slower transmission at much lower bit error rates. The tape's capacity of 5×10^8 bits can store about 100 frames.

4.3.3.4 TELEVISION ON ORBITAL MISSIONS

When the early flyby missions to a planet are followed by an orbital mission, an imaging system's capacity for systematic coverage of high quality is multiplied manyfold through repetitive use of the recording tape.

4.3.3.4.1 MARINER 9 MARS ORBITAL MISSION

This television system was essentially the same as that of Mariners 6 and 7. The wide-angle camera had the same lens, but the number of filters in the filter wheel was increased to eight. The narrow-angle optics were unchanged. The scanning format for Mariner 9 was 700 lines, with 832 picture elements per scan. Scanning time was 42 seconds per frame. The tape recorder had a capacity of 32 frames. The maximum transmission rate was again 16,200 bits per second, with lower rates available for use in less favorable circumstances.

Since the use of the full transmission rate depended on the 210-foot antenna at the Goldstone tracking station, the entire orbital mission was designed to maximize the daily data return while Mars was above the Goldstone horizon. The orbit around Mars was adjusted to a period of 11.98 hours that was synchronous with the visibility of Mars from Goldstone. The orbit was elliptical, with a minimum altitude at periapsis of 1650 kilometres. That altitude was chosen so that the acquisition of ten wide-angle frames per revolution would eventually provide complete coverage of the planet with an adequate forward overlap. High-resolution frames, along with the wide-angle frames necessary to pinpoint their locations, utilized the remaining tape capacity.

The routine during the mission's mapping phase was to acquire 32 frames in the vicinity of periapsis on each revolution. After Mars rose above the Goldstone horizon, there were $3^1/_2$ hours available for transmission of the frames recorded during the previous revolution. This was followed by the acquisition and recording of the current revolution's photographs, which were transmitted before the setting of the planet at Goldstone. In this way, 7100 photographs were acquired and transmitted during the first 262 revolutions, Additional photographs, taken later in the mission to fill coverage gaps and to follow up the seasonal disappearance of the north polar ice cap, brought the total number to 7329 frames.

4.3.3.4.2 VIKING ORBITER TELEVISION SYSTEM

At this writing, two Viking spacecraft are on the way to Mars. Each consists of an Orbiter and a Lander. The Lander's imaging system, based on a facsimile (scanning) camera, is discussed in section 4.3.4.1.2.

The Orbiter's television system reflects the differences in orbital imagery objectives between Mariner 9 and Viking. The primary objectives are to verify Viking landing sites and to search for landing sites for future missions. Accomplishing these requires contiguous coverage of broad swaths of terrain at a comparatively high resolution. The system designed for this purpose uses two television cameras with optics of 475mm focal length. The camera axes diverge slightly so that a pair of pictures covers a solid angle 3° by 1.5°.

In order to acquire contiguous coverage from an altitude of 1500 kilometres, each camera must be ready to take a new picture every 4.5 seconds. The Vidicon target is scanned in 1056 lines, with 1182 picture elements per line. This scanning rate is slightly higher than that of the Ranger crash-landing missions to the Moon, but, whereas Ranger could transmit the picture information directly at that rate over a distance of a quarter-million miles, the 200-million-mile distance of Viking makes direct transmission impossible. In fact, the circumstances require a different method of tape recording from those employed in previous space programs.

The recording system digitally encodes 128 levels of gray (seven bits) for each picture element. The bits composing one picture element are recorded simultaneously on the tape, using seven separate tracks of a nine-track tape. In order to keep up with the scanning rate, the picture information is recorded at slightly over two million bits per second. For transmission to earth, the tracks are played back sequentially. Transmission rates of 16,000, 8000, and 4000 bits per second accommodate the differences in distance and antenna-pointing capability as the mission progresses.

4.3.3.4.3 Television in Earth Orbit: LANDSAT Return Beam Vidicon Cameras

The LANDSAT program (probably better known by its earlier name of Earth Resources Technology Satellite—ERTS) is an experimental effort to assess the usefulness of repetitive global multi-spectral imagery of fairly low resolution in studies of the earth's resources. Two image acquisition systems are provided—a Return Beam Vidicon (RBV) camera system, and a Multi-Spectral Scanner (MSS) system. The MSS will be discussed in section 4.3.4.3

A return beam vidicon tube can scan the image on its target plate with much higher resolution than a standard vidicon. The RBV cameras used in LANDSAT scan their images with 4125 active scan lines. The system used in the first two LANDSAT satellites comprises three cameras, each of 126mm focal length, sensing the same scene in three spectral bands.

The optics of the "blue" camera provide a response between 475 and 575 nanometres. The "yellow" camera is sensitive between 580 and 680 nanometres, and the "red" between 698 and 830 nanometres. Simultaneous operation of the camera shutters produces, on the RBV photosensitive surfaces, images of the same 185-by-185-kilometre scene. The shutter operation is repeated at 25-second intervals. The three images are scanned sequentially, with a scanning time of 3.5 seconds for each camera.

The resulting pictures in three spectral bands have essentially the same coverage and resolution as those produced by the Multi-Spectral Scanner in four spectral bands. This functional redundancy has served its experimental purpose. Both systems perform in a very satisfactory manner, but scientific experimenters generally prefer to work with the MSS products. Accordingly, the RBV system has not been operated on a production basis, and the RBV system for the third LANDSAT satellite (launched in March 1978) is designed to function more as a supplementary data source for the MSS.

The later RBV system comprises two cameras with 250mm lenses and panchromatic sensitivity. The camera axes diverge so that the swath covered by a combined pair of frames is 185 kilometres. With a frame repetition rate of 12.5 seconds, a set of four RBV frames will provide the same coverage as a single MSS scene.

An earth satellite has its peculiar data-transmission circumstances. The time available for transmission to any particular tracking station is only a few minutes, but data can be transmitted at extremely high rates if necessary. The requirements for LANDSAT transmission are moderate. The images acquired over the United States (including Alaska but not Hawaii) are transmitted in real time as they are scanned. This transmission to the stations in Greenbelt, MD; Goldstone, CA; and Fairbanks, AK takes an average of 18 minutes per day. The RBV imagery is transmitted in analog form at a bandwidth of 3.2 megahertz. The MSS images are transmitted through the same communication equipment in digital form at a rate of 15,000 bits per second.

Imagery of other parts of the earth's surface is recorded on wideband video tape recorders for delayed transmission to the three receiving stations. The imagery is recorded and played back at the same rate as the real-time scanning. Each of the two wideband recorders can record either the RBV or the MSS data, in analog or digital form respectively. Each recorder tape has a capacity of 30 minutes of data.

The manner in which the LANDSAT missions are operated to acquire the desired ground-coverage pattern will be discussed with the MSS system in section 4.3.4.3.

4.3.4 SCANNING-SPOT CAMERAS

In a scanning-spot camera, a single light-sensitive spot (or small array of such spots) moves across the scene in a systematic scan to produce a sequence of electrical signals that are proportional to the momentary illumination of the spot. The principle is that of the long-familiar facsimile camera that scans news photographs for transmission to a publishing plant. Cameras of this type have been variously designed for use aboard soft-landed spacecraft on a planet's surface, spin-oriented spacecraft as they fly past a planet, and nadir-oriented orbiting spacecraft.

4.3.4.1 Stationary Scanning-Spot Cameras

In order to scan a site from a fixed location, a camera must itself generate motion of its sensitive spot in two dimensions. A simple and a more complex example of this type of camera are described.

4.3.4.1.1 LUNA-9 Camera

The first ground-level views of the lunar surface were supplied by the Soviet LUNA-9 spacecraft, which made its soft landing in February, 1966. The camera was a small vertically-oriented cylinder situated about two feet above the surface. A slow rotation of the camera about its axis effected a scan in azimuth. Within the cylinder was a photosensitive cell whose field of view was restricted to one milliradian. A simple lens brought everything from five feet to infinity to a focus on that spot. A nodding mirror scanned the scene in elevation through a slot in the cylinder.

Each vertical mirror scan covered 29°, or 500 picture elements. A complete 360° panorama, consisting of 6000 scans, required 100 minutes. At that leisurely pace of 500 picture elements per second, transmission of the picture information was easily accomplished in real time.

The camera made three panoramic surveys of the site. A slight stereoscopic separation of the three panoramas was achieved by shifting the attitude of the landed spacecraft between surveys.

4.3.4.1.2 Viking Lander Imaging System

The versatility inherent in the facsimile camera principle is realized to a remarkable degree in the imaging system of the Viking Lander. The imaging system has a variety of assignments: inspecting the geology of the landing sites; observing dust storms and clouds; observing the motions of the Martian satellites; and, most particularly, viewing the operations of the soil sampler and the life-detection experiments. Since the system includes two facsimile cameras mounted 1.3 metres above the Lander footpads and nearly a metre apart, all of the Lander experiments are under stereoscopic observation.

Each camera is essentially a vertical cylinder that scans in azimuth by rotating about its axis. A nodding mirror scans in elevation through a window in the cylinder, and an objective lens images the scene on an array of twelve diode photosensors. All 21 sensors are in the optical field, but during a scan the system is processing the electrical signals from a single sensor. The selection of sensor signals for processing depends on the commanded operating mode.

Four sensors are available for the high-resolution operating mode. They are located at different distances along the optical axis, so that the entire scene depth from 1.7 metres to infinity is in sharp focus on one or another sensor. These sensors have panchromatic sensitivity, and their fields are limited by pinhole apertures to 0.04° (two-thirds of a milliradian). In high-resolution operation, the mirror scans 20° in elevation at a time. Nine center-of-scan elevations can be selected, from 30° above the horizontal to 50° below. Successive scans are 0.04° apart in azimuth.

The visual color operating mode uses three sensors that are fitted with red, green, and blue filters. The size of a picture element is increased to 0.12°. At that resolution, a single focus position is sufficient. The mirror scans 60° in elevation, with five selectable center-of-scan elevations. The signals from the three sensors are processed on successive scans. Since the 0.04° azimuth angle between scans is now only one-

third of a picture element, a full-color picture is transmitted.

An infrared spectral mode uses three sensors whose peak sensitivities are at wavelengths of 0.85, 0.95, and 0.98 micrometres. The field angles and scans in this mode are the same as in the visual color mode. The wavelengths were chosen to permit identification of mineral groups in the surface materials at the site.

The survey operational mode uses the eleventh sensor, which has panchromatic sensitivity and a 0.12° field of view. Since the 60° elevation scans are 0.12° apart in azimuth, this mode produces a panoramic survey with relative rapidity.

The sensitivity of the twelfth sensor is reduced by a filter so that it can image the sun. By providing a record of solar transit times and elevations, the sun mode makes it possible to calculate the site location.

A monochromatic mode permits selection of the signal from any one of the color or infrared sensors for processing, and line rescan mode (in which the azimuth is not changed between scans) records the transits of the two Martian satellites.

Each camera has a stationary post outside the cylinder to protect the window during dust storms. Because of the post, the field is reduced to 342.5° in azimuth. In all modes, the azimuth of the camera is selectable in 2.5° increments.

Each picture element is encoded as a six-bit word. The cameras have two scanning rates, to match the two modes of data transmission to earth. The slow scanning rate, 250 bits per second, is compatible with the direct transmission of data from the Lander to earth. When the Orbiter is overhead, the Lander transmits to it in the relay mode at 16,000 bits per second. That is also the fast scanning rate of the cameras. Real time picture transmission thus is possible in either mode. Data storage capability aboard the Lander can record and play back picture data at either rate for delayed transmission. The Orbiter's tape recorder (*see* section 4.3.3.4.1) records the entire Lander bit stream on one channel.

The Lander's cameras can scan at either bit rate in all operational modes. At 250 bps, it takes 5 minutes and 38 seconds to scan one azimuth degree in the high-resolution and color modes, or a third of that time in the survey mode. At 16,000 bps, the times are reduced to 5.28 and 1.75 seconds, respectively. Thus, a complete panorama can be scanned in the survey mode in just ten minutes when an Orbiter is overhead.

4.3.4.2 A Spin-Scan Camera: Pioneers 10 and 11 Imaging Photopolarimeter

A spacecraft on a long voyage expends a minimum of fuel for attitude control if it is designed to employ spin stabilization. Pioneer 10 (which flew past Jupiter in December, 1973 and then left the solar system) and Pioneer 11 (now on its way to a Saturn encounter after passing Jupiter) carry an Imaging Photopolarimeter (IPP) that functions in its imaging mode as a spin-scan camera, using the spacecraft spin to generate one dimension of scan.

The IPP has two additional modes of data acquisition. As the spacecraft traverses interplanetary space, the IPP functions as a very sensitive photometer, measuring the brightness of the Zodiacal Light and the Gegenschein (counterglow). During planetary encounters the IPP functions as a polarimeter, measuring two polarization components of reflected light in the red and blue wavelength bands over a wide range of phase angles. In its imaging mode, it depolarizes the incoming light, and produces red and blue images of the planets and their satellites.

The main component of the IPP is a Maksutov telescope whose line of sight with respect to the spacecraft's spin axis (the look angle) can be shifted mechanically in small steps. In the imaging mode, the telescope's field of view is restricted to 0.028° (one-half milliradian). The beam is split by a dichroic mirror and spectrally filtered, so that the red and blue light intensities are measured by separate multiplier detectors.

During a planetary encounter the spacecraft's spin, at about 4.8 revolutions per minute, scans the telescope's field across the planet. Detector output is measured every one-quarter milliradian of spin, alternating the detectors. Thus, there is a red and a blue measurement for every picture element. Measurements can be taken for 14° of each 360° revolution.

Scanning in the second dimension is accomplished by stepping the telescope's look angle by one-half milliradian while the revolution is being completed. Measurement starts at the same spin position on successive revolutions, so that a two-dimensional picture is built by repeated stepping.

The intensity measurements are encoded as six-bit words. The spacecraft does not have a tape recorder, but it has a small data-storage capacity in the form of a 6144-bit buffer. That is just sufficient to store the imaging data collected over 14° of each revolution. With a 4.8 rpm spin, twelve seconds are available for transmission of the data from the buffer before the instrument overwrites it with the next scan. An image data rate of 512 bits per second will transmit all of the IPP imagery to earth.

4.3.4.3 A LINE-SCAN CAMERA: LANDSAT MULTI-SPECTRAL SCANNER

If a spacecraft is in a circular orbit and maintains a nadir orientation, a scanning camera on board can take advantage of the uniform forward motion to generate one dimension of scan. Such is the case for the Multi-Spectral Scanner (MSS) aboard the LANDSAT spacecraft.

The MSS scans in the cross-track direction by means of an oscillating mirror. From the spacecraft altitude of about 920 kilometres, the crosstrack swaths are 185 kilometres long. The scene is focused by a reflecting telescope on a 4 × 6 array of glass fibers. Light impinging on each fiber is conducted to an individual detector through a spectral filter. The square ends of the fibers limit the earth-surface field of each detector to 79 metres.

The MSS senses in four spectral bands. Band 1 covers the wavelengths between 0.5 and 0.6 micrometres. Band 2 covers 0.6 and 0.7 micrometres, and Band 3 covers 0.7 to 0.8 micrometres. These detectors are photomultipliers. Band 4 detectors, which are silicon photodiodes, cover 0.8 to 1.1 micrometres.

The arrangement of the glass fibers in the telescope focal plane allows a set of four spectral detectors to cover the same ground swath. The look angles of adjacent sets offset their ground swaths by 79 metres. Thus, each active scan of the mirror generates six contiguous swaths of coverage. The detectors are active on the west-to-east mirror scan. During the east-to-west retrace, a calibrating light source is projected onto the fibers. With an active scan every 73.4 milliseconds, the spacecraft will have moved forward 474 metres, and the six swaths from the second scan will be contiguous with the first six.

The video outputs from each detector in the scanner are sampled and commutated once each 9.95 microseconds. The commutated samples are digitally encoded in six bits, and the 15,000-bit-per-second data stream can be either directly transmitted to a ground station or recorded on tape as described in section 4.3.3.4.3.

The MSS for the third LANDSAT mission will sense a fifth spectral band from 10.5 to 12.5 micrometres. The sensor array will have two additional detectors. These mercury-cadmium-telluride detectors will be radiation-cooled, and each will have a field of view covering 240 metres on the ground. Their combined coverage on each mirror scan thus will equal that of the other six sets of detectors.

The orbit of a LANDSAT mission is carefully designed to produce the desired repetitive global coverage pattern. It is sun-synchronous, so that all the imagery is acquired at about 9:30 a.m. local time on the north-to-south half of each revolution. The orbital period of 103.267 minutes gives slightly less than 14 revolutions per day. The track of the 15th revolution is shifted westward by 1.43° at the equator, which corresponds to 159 kilometres. In the course of an 18-day cycle, the entire surface between the first day's orbital tracks is filled in, and the track of Revolution 252 exactly duplicates that of Revolution 1.

During data processing on the ground, the imagery from the continuously-scanning MSS is divided into frames that have the same centers as the Return Beam Vidicon frames. A limited forward lap is provided by recording the MSS scan lines twice at the frame ends.

4.4 Special Considerations in Camera Design and Development

4.4.1 CAMERAS FOR COLOR PHOTOGRAPHY

For many years aerial reconnaissance has relied on the increased information which is obtainable with infrared and color photography. Such pictorial information has been particularly valuable in the acquisition of military intelligence; as a result, special emulsions (such as color infrared) have been developed to emphasize differences in natural and cultural features. Government agencies and industrial firms now have extended the use of these films into the field of photogrammetry.

The scope of applications of color photography for photogrammetric purposes is limited by the quality of the camera-film combination, the quality of the processing, and the capability of the plotting equipment.

The basic equipment requirements and techniques are as follows.

(a) The camera must be equipped with a lens corrected for color throughout the visual and near-infrared range of the spectrum such that all wavelengths are focussed near the same plane.
(b) Distortion at this focal plane must be substantially the same for each color.
(c) The distribution of light across the format must be uniform.
(d) Exposure must be correct within one-half aperture stop, although information generally will not be lost if fidelity of hue and tone is not maintained.
(e) A well-controlled processing laboratory for color film is required.
(f) Color diapositives are needed.
(g) Plotting equipment must be capable of satisfactorily handling color photographs.

To date, much color and color infrared photography has been obtained with foreign mapping cameras such as those manufactured by Wild and Zeiss. American-made cartographic cameras, such as those equipped with the 6-inch Geocon I or Geocon IV lenses, are capable of producing equivalent high-quality color photography.

4.4.2 CAMERAS FOR NIGHT PHOTOGRAPHY

Night photography does not require a different camera from that used for conventional daylight photography, but additional equipment is needed (USAF Manual 95-3). Basically, the equipment needed for night aerial photography has remained the same since the first successful night photograph was made over sixty years ago. This equipment consists of a camera, an illuminant, and a means for opening the camera shutter at the right time. The cameras today are much improved, with larger lenses, automatically operating shutters and fast cycling rates. The greatest changes in the night-camera systems are in the illuminants. Although most night cameras are of the frame type which make instantaneous exposures, some night strip cameras have been developed. Of course, the latter require continuous illumination. The characteristics of three night-reconnaissance cameras—Types K-37, K-46 and K-47—are shown in section 4-7 of this chapter.

The basic types of illuminants used for night photography are as follows.

(a) Pyrotechnic.
  (1) Photo-flash bomb: for high-altitude photography.
  (2) Photo-flash cartridge (with ejectors): for low-altitude photography.
  (3) Magnesium burner, or "hell roarer": for continuous light source.
(b) Electric-flash, or spark-flash, equipment.
(c) Mercury-vapor-tube illumination: for continuous light source.

4.4.3 CAMERA STABILIZATION

An important means for the preservation of photographic resolution when the camera is moving is the use of a gyroscopically stabilized mount (see section 4.5 of this chapter). Normally, both fixed mounts and gyroscopic mounts in the camera bay rest on vibration isolators which reduce the high-frequency movements which are otherwise transmitted to the camera by the aircraft engines. The *stabilizing* function of the gyroscopic mount reduces the lower-frequency vibrations which are generated by the aircraft in flight.

Some people claim that the value of such a mount is questionable, since the higher-frequency vibrations are the ones which directly affect photo resolution. This may be true if a lower-resolution film, such as aerographic Plus X, is being used. This film, which has a relatively high aerial exposure index (80), is affected less for two reasons. Its lower value of resolution does not change as much because resolution target steps are smaller in that range. In addition, a film of this type requires much less light for exposure, and shutter speeds of 1/400 second or less may be used, thereby reducing the possibility of image blur.

However, if high-definition films having low AEI numbers, such as Eastman 3414 (2.5), are used, then the effects of the lower-frequency vibrations begin to be more critical. In research and development, the value of the gyro mount is unquestioned where testing of high-resolution films is an important part of the program.

Another advantage of the gyroscopic mount is its capability for holding the camera more vertical than the fixed mount can. Many aircraft in flight have what is known as a "Dutch roll," which is an oscillatory movement around the longitudinal axis of the fuselage. Normally, a sidelap of 20% is prescribed for mapping coverages exposed with cameras in gyroscopic mounts. In missions involving certain high-altitude aircraft in which fixed mounts are used, it is necessary to set a 30% sidelap in order to prevent "holidays" (missed areas of coverage) between flight lines.

A larger sidelap necessitates more flight lines for a given coverage, with a resultant increase in flight cost. Furthermore, near-vertical photography is handled with greater facility in analog-type compilation procedures.

4.4.4 CAMERA WINDOWS

An important consideration in photogrammetric operations is the quality of the window mounted in the lower side of the fuselage through which the camera photographs. In addition to the structural considerations which establish the minimum thickness of the glass for a given expanse, the U.S. military specification, "Glass, Window, Aerial Photographic," MIL-G-13660, requires that cartographic-quality Group "M" glass shall have surfaces parallel to within four seconds of arc (2.5 fringes/inch) over a testing length of 2.5 inches. This means a wedge type of aberration, contributing to an increase in tangential distortion effects in the lens-camera-window combination, could be introduced if parallelism of window surfaces deviates beyond the prescribed limits.

4.4.5 THE ENVIRONMENT

The cameras developed for military use in the United States are built to certain specifications and subjected to a number of tests which assure a continued high performance over a wide range of operational environments. These so-called "military environmental qualifications" cover such things as fungus-protection coatings on the lens (for tropical use); resistance to vibration, shock and acceleration; and protection against thermal effects. Two other protective measures which have much to do with flight safety are "explosion proofing" and "electromagnetic interference (EMI) shielding."

Explosion proofing is extremely important because it protects the camera (and the aircraft) from the effects of electrical discharge (or "arcing") which is inherent, to some extent, where electrical motors are concerned. These effects, if uncorrected, increase considerably at the higher altitudes where air pressures decrease, and could cause serious damage, both to the camera and the aircraft, if the camera were placed near a leaking oxygen tank or a flammable object.

The EMI shielding is necessary if the camera is to be used with other electronic equipment in the aircraft. Flight experience has shown that an unshielded camera can make so much "radio noise" in operation that it is impossible to receive instructions from a control tower adjacent to, or within, the photo-coverage area. A U.S. military aircraft may fly an environmentally unqualified camera at high altitude, but it requires a special flight safety waiver.

This is one category in which the American cameras developed for military cartography may possess an advantage over the very high quality European cameras. Generally, European companies do not fly their cameras at extreme altitudes, and the cameras are usually flown as individual items and not as a part of an integrated system. Therefore, it is not necessary to build into the cameras some of the details required for military use which may cost a considerable amount of additional money.

4.5 Accessory Equipment

4.5.1 CAMERA MOUNTINGS AND PLATFORMS

In the strictest analysis, the only function of the aircraft in the aerial photo system is to serve as a means for getting the camera into the desired position. While some types of aircraft are better than others for this purpose, none are to be considered perfect.

Forward motion, engine vibration, the random motions of flight, and the uncertainty of attitude and position, require the use of various compensating devices to achieve the best compromise for camera steadiness and attitude control. The most basic of these devices, the camera mounting, forms the critical intermediary between the aircraft and the camera. Depending on the expected severity of the disturbances and the level of compensation required, mountings can be as simple as a plywood frame, or as complex as a Schuler-tuned, inertially-stabilized platform.

4.5.1.1 FIXED MOUNTS

These are usually metal frames that permit the rigid attachment of the camera to the aircraft in the desired position. In some instances, this kind of installation is preferred, as for example when used to eliminate flexure between the camera and a suitably stable aircraft under precise attitude control. Rigid mounting would be preferred also when a fuselage-mounted gyroscope is used as a reference for camera attitude during exposure. Generally, however, a camera installation incorporates some type of flexible insulation to protect it from landing shock and to attenuate vibration from the aircraft.

4.5.1.2 Adjustable Mounts

Proper camera orientation in flight requires leveling and rotation to compensate for the inclination and crab of the aircraft. For this purpose, adjustable mounts provide freedom around one or more axes (azimuth, roll and pitch). Aircraft crab is compensated by a rotatable azimuth ring, usually marked in degrees. The setting is transferred from a similarly marked viewfinder. Roll and pitch inclinations are compensated by adjustable gimbals or foot screws to a nominally level attitude as indicated by the average position of a level bubble.

Manually-adjusted mounts are useful in compensating for long-term changes in aircraft attitude, but cannot be expected to be effective for keeping a camera level under turbulent flight conditions.

A typical, manually-adjustable camera mount is shown in figure 4-30. This mount (Spectral Data CMA) accommodates trunnion-type cameras specially designed for multi-spectral photography. For interchange of filters in flight, the entire suspension may be tilted.

4.5.1.3 Isolators

For the dual role of protection from shock and isolation from vibration, passive isolators must be rigid enough to avoid excessive sway or bottoming under shock, and soft enough to filter out the expected vibration. Most isolators selected for this purpose have a resonance, under 1G loading, in the range of 5 to 10 hz. A 5 hz isolator, for example, transmits structure vibrations 1:1 between 0 and approximately 4 hz, amplifies to more than 1:1 at resonance and isolates from vibrations above 6 or 7 hz, providing secondary resonance is minimized with suitable damping.

It is fundamentally important (although not always done) that the center of gravity of the camera be located at the geometric center of the pattern formed by the attachment points. This is necessary to avoid introducing rotary motion of the camera, as would result to some extent from an unbalanced mounting.

4.5.1.3.1 Special Anti-Vibration Mountings

An example of a mount incorporating precisely-tuned isolators is the Fairey Surveys mount shown in figure 4-31. The basis of all Fairey isolation systems is either a single or multiple helical spring suspension to support the load, combined with silicone fluid for damping. By this means, both frequency and damping characteristics can be controlled over a wide range of temperature and load conditions. Visco-elastic materials generally are unsuitable for optimum stability because of their inherent changes with time and temperature.

Camera mounts incorporating the Fairey isolators are typically custom-designed for effective performance in the low-frequency range (3 hz to 10 hz). The Type 133 Mount built for the F49 Mk4 Survey Camera permits the camera to be remotely controlled in azimuth and pitch, and provides vibration isolation effective from 4 hz upward.

4.5.1.3.2 Helicopter Iso-Mounts

Another example of a special purpose, isolated-camera suspension is the Robinson Anti-Vibration Helicopter Mount. This device was designed for installation of the Wild RC8 Camera on the outside of a Bell J-2 Helicopter. The mount consists of a rigid frame attached between the fuselage and skid, with the camera installed on isolators selected for the load and expected frequency of disturbance. Isolator selection for helicopter use should take into consideration the beat of the acoustic disturbance generated by the rotor blades as well as the vibrations transmitted through the fuselage.

4.5.1.3.3 Active Vibration Isolators

A unique camera mounting system is the Active Vibration Isolation Stabilizer, developed by

FIGURE 4-30. Adjustable camera mount, Type CMA (Courtesy of Spectral Data Corp.).

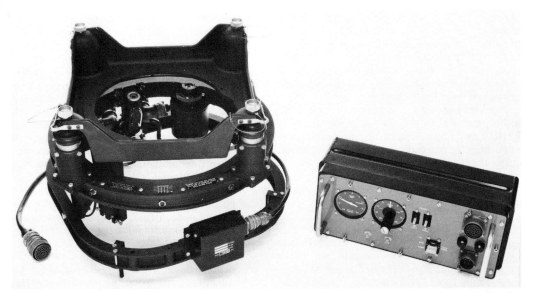

FIGURE 4-31. Remotely controllable camera mount, Type 133, incorporating low-frequency, anti-vibration mountings (Courtesy of Fairey Surveys, Ltd.).

Actron Industries. This special design provides both isolation and limited-range camera attitude control. The wide-band stabilizer systems of Actron were developed for stabilizing USAF tactical reconnaissance cameras. The weight of the camera is supported by springs that remove vibrations above the region of the passive resonance. In parallel with the springs, but supporting no weight, are electromagnetic linear actuators. These are part of a servo loop that incorporates accelerometers and gyros as sensing elements. The passive/active combination is effective over an essentially unlimited frequency range.

4.5.1.4 STABILIZED PLATFORMS

The stabilized platform provides the ultimate in airborne camera steadiness and verticality. The basic arrangement is an azimuth ring on the inside supporting the camera; a roll gimbal supporting the azimuth ring; and a pitch gimbal on the outside supporting the roll gimbal and, in turn, supported by a frame attached to the aircraft by isolators.

The vertical reference generally is a two-axis, pendulum oriented, vertical gyroscope mounted on the same part of the platform as the camera. The gyroscope generates an error signal proportional to its displacement from vertical. Each axis of the vertical reference forms the control element of a closed-loop servo which drives the corresponding gimbal back to null when displaced.

Gear-driven gimbals are suitable for use where rapid angular accelerations are not expected; however, modern stabilized platforms employ electromagnetically driven gimbals which effectively eliminate restrictive mechanical coupling.

High-performance, stabilized mounts are capable of isolating the camera from rotational disturbances over a great range of rates and frequencies.

The precise inertial balance of torque-driven platforms requires that they be shielded from direct air flow. In addition, electrical cables and vacuum lines to the camera must be carefully arranged to avoid imbalance, and a means of automatically adjusting for shifts in the weight of film must be provided.

Stabilized mounts provide steadiness for reduction of image motion and verticality for consistent coverage and freedom from excessively tilted photography. The tolerance for residual angular camera motion varies with focal length. A simple calculation of the allowable angular motion may be expressed as follows:

$$\text{Rotation} = \frac{4060}{\text{Resolution} \times f} \qquad (4.19)$$

where rotation is the angular motion during the exposure period in seconds of arc, resolution is in line pairs/mm, and f is the focal length in inches.

There is little agreement on the permissible residual tilt, as the absolute verticality required for high precision cannot be maintained for 100% of the exposures with available (1976) systems. A verticality within ± 30 minutes and a relative tilt between consecutive exposures within $3-8$ minutes may be expected with present stabilized mounts.

4.5.1.4.1 AEROFLEX T28A CAMERA MOUNT

The T28A is a universal mount designed to be adaptable to a variety of mapping cameras. It is a three-axis, gyro-stabilized, gimbal-type, torque-driven mount, automatically leveled in the roll and pitch axes. Azimuth is remotely controlled by manual setting. The vertical reference is the Type ARG-5 Gyro on the roll gimbal. An automatic weight shifter compensates for imbalance resulting from film transport.

Cameras which may be mounted in the T28A include the Wild RC8, RC9, and RC10; the Zeiss RMK series; and various U.S. Air Force types. Each must be specially adapted for initial balance. The T28A platform maintains the camera axis close to vertical under typical flight conditions where roll and pitch excursions do not exceed ±8 degrees. Ninety percent of all exposures may be expected to be tilted from vertical less than 12 minutes.

The remote-control system includes provision for manual cutoff of gyro erection and weight shifter to minimize the effect of lateral accelerations during turns. This can also be done automatically by a rate switch.

Figure 4-32 shows an RC10 camera installed in the T28A Mount with a Sony television camera on the viewfinder telescope. The viewfinder image is displayed on a monitor at the remote control station.

4.5.2 CAMERA CONTROLS

Aircraft-camera controls are accessories which control the camera's operation according to the variable conditions of flight. They include the display of measured values to guide manual or automatic camera settings.

4.5.2.1 V/H COMPUTERS

A basic accessory is a device for determining the velocity of the image motion at the focal plane caused by the aircraft's forward motion. The image velocity is proportional to the aircraft's ground speed as modified by the aircraft altitude and the focal length. The resultant velocity/height (V/H) ratio for a given focal length is needed for control of the exposure interval for proper stereo coverage and for setting the proper shutter speed to minimize image motion during exposure. On aircraft equipped with an inertial or doppler navigation system and a radar or laser altimeter, these instruments may be used for automatic V/H computation.

An image-velocity detector also may serve this purpose. This device consists of an optical system and photo detectors to produce an electrical output proportional to the speed of the terrain image.

Semi-automatic V/H computers develop the proportional value from dials manually set for aircraft altitude and speed. An instrument of this type is the manual V/H computer, incorporating also switching functions, shown in figure 4-33. The most frequently used accessory for V/H computation and camera interval control is the optical viewfinder.

4.5.2.1.1 INTERVALOMETERS

An intervalometer supplies tripping pulses to the camera at selectable intervals to provide the desired photographic coverage and overlap. The interval required for a given overlap is determined from the focal length, the film format, the aircraft's ground speed, and its mean height above ground.

Intervalometers are basically electrically-

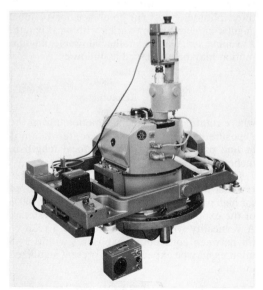

FIGURE 4-32. Vertical stabilized camera mount, Type T28A, with Wild RC-10 camera and Sony television on the viewfinder telescope (Courtesy of Aeroflex Labs.).

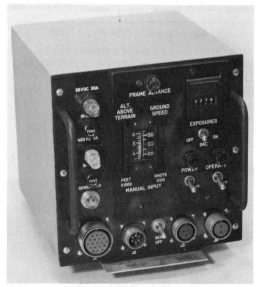

FIGURE 4-33. Image motion compensation control unit, Type CC 110. Altitude and ground speed are entered by the camera operator. (Courtesy of Spectral Data Corp.)

driven timers with a dial for setting the number of seconds between trip pulses. Typical are the Abrams Instrument Company's Type B-5A and Type B-9A Intervalometers, which have a time-interval range of from one to sixty seconds in half-second increments. Where needed, the B-9A may be initially adjusted for use with a scale having an interval range from one to 120 seconds.

4.5.2.2 Viewfinders

The viewfinder provides a clear view of the terrain beneath the aircraft for navigation, selection of photo points, and determination of the camera's field of view and drift alignment. The moving image at the viewing plane provides also a V/H indication for computing the exposure interval. The basic viewfinder permits stopwatch timing of an image point's travel between grid marks to establish the interval to be set manually on an intervalometer. Viewfinders with a manually-controllable, motor-driven moving grid measure the image speed and provide an output proportional to the V/H ratio.

The observer sets the grid speed to coincide with the image speed, and the output may be used directly to control the camera interval. Modern viewfinders are part of the camera control system, and, with the insertion of values for the desired forward overlap and the camera's angular field, they perform the function of automatic overlap regulation.

4.5.2.3 Navigation Sights

The vertical view afforded by camera viewfinders enables the photographer to set the exposure interval and drift and to select targets for pinpoint exposures. The navigation sight provides a forward view for monitoring the flight path of the photo aircraft. Some instruments combine these functions.

For a two-man crew, the photographer/navigator may use a navigation sight to guide the pilot onto the prescribed flight route, and then depress the view to vertical for overlap and drift regulation.

4.5.2.3.1 Jena Navigation Instrument

This is a manually-operated sighting telescope with a vertical field of view extending from 90^g before and 10^g behind the nadir. The operator releases the telescope with a foot switch and rotates it by hand. The range of rotation is 400^g and continuous with a ±50^g scale of drift values.

The objective and folding optics of the instrument extend below the aircraft. Internally, the telescope's length is adaptable to the fuselage thickness and the observer's position. Total magnification is 0.6×.

For ascertaining course deviations, interchangeable graticules with sighting lines, limits of adjacent strips, nadir point, *etc.*, are placed in

the image plane of the 8× eyepiece (*see* figure 4-34).

4.5.2.3.2 Wild NF2 Navigation Sight

This accessory to the RC10 Camera System assumes the function of a vertical viewfinder and overlap control, as well as a flight navigation sight. By turning a knob, the vertical view can be changed to an oblique view of 50° inclination from vertical. At the vertical setting, the NF2 provides a nonreversed field of view ranging from 55° forward to 55° aft of the nadir. The oblique view ranges from 46° ahead of nadir to 4° above the horizon.

The NF2 can be moved 460mm along its axis for installation positioning or landing retraction, can be leveled in its mount through a range of ±5°, and permits drift corrections up to ±30°. The lock of the tube can be released to provide 360° continuous rotation.

The head of the NF2 contains an overlap regulator with a traveling grid and selectable reticles for camera viewfinding and for navigation.

The sight is available in two tube lengths

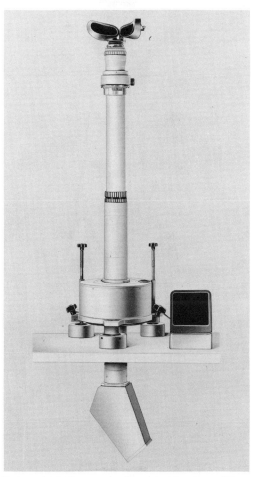

FIGURE 4-34. Jena navigation sight (Courtesy of Zena Company).

(NF2—127cm, and NF2K—94cm), and the head is detachable for mounting in small aircraft. A special version—the NF22—permits simultaneous remote control of two RC10 Cameras. The NF2 Mount incorporates servo sensors to transfer tip, tilt and drift corrections to the RC10 Mount (*see* figure 4-35).

4.5.2.3.3 ZEISS NT1 NAVIGATION TELESCOPE

This instrument for photo flight navigation has a 90° vertical field of view at a fixed 40° forward angle to provide a view ranging from 85° ahead to 5° aft of the nadir. The vertical tube can be rotated through 360°, with click stages each 90°.

Interchangeable navigation graticules provide guide lines for lateral flight strip limits, frame limits, nadir point and axial points of adjacent strips for various lens angular fields.

The NT1 is installed in a manually-leveled mount and provides direct reading of drift-setting information for the aerial camera.

4.5.2.3.4 ZEISS NT2 NAVIGATION TELESCOPE

Similar to the NT1 in viewing and navigation functions, the NT2 also incorporates provision for controlling and monitoring the operation of the Zeiss RMK series camera. V/H values may be determined by synchronizing moving luminous lines with the ground image in the forward-looking position. Camera control functions include push buttons for single picture release and for initiating a series.

When installed with a specially-wired RMK camera, the function monitor displays an indication of shutter speed and aperture, and film advance. The actual overlap is indicated on a digital display for comparison with the overlap ratio set on the separate camera control.

4.5.2.4 EXPOSURE METER/CONTROLS

Correct exposure of aerial films, especially color-emulsion films, requires an accurate de-

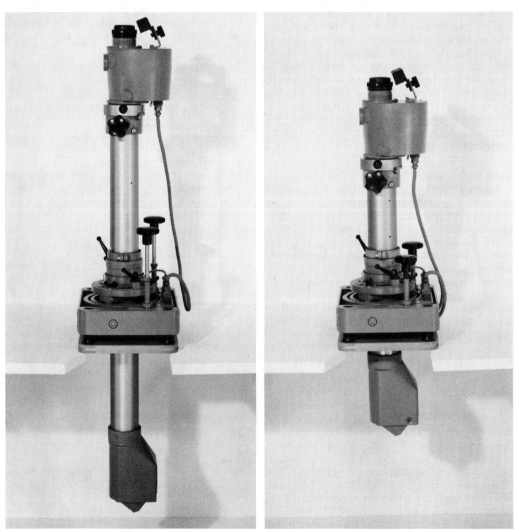

FIGURE 4-35. Wild navigation sight, Type NF2 (left) and Type NF2K (short version, right) with leveling mount for remote camera control (Courtesy of Wild Heerbrugg, Ltd.).

termination of scene brightness. While visual estimation of brightness and haze conditions is used rather generally by photographers who have standardized their procedures, an exposure meter may be employed to give more consistent results. A conventional light meter is useful in determining the brightness value, but the meter readings require careful interpretation. Depending on the spectral sensitivity of the detector and its acceptance angle, the conventional light meter readings may be overly influenced by highlights or non-image-forming light reflected from the intervening atmosphere. The output of light-meter photocells also may be affected by the temperature variation encountered in flight. Scene brightness sensors specially designed for use in aerial photography are now available. These may be either meter types, where the readings are displayed to provide camera setting data, or control types, where either the lens aperture or shutter speed is continually adjusted automatically.

4.5.2.4.1 EXPOSURE METERS, MANUAL

4.5.2.4.1.1 Zeiss EM1-1 Exposure Meter. This integrating instrument consists of a temperature-compensated Cds photoconductor cell which can be mounted on any aerial survey camera, a handheld meter, and an interconnecting cable. The meter serves to determine the proper aperture for a given shutter speed, film speed and multiplying factor. The aperture is then set manually on the camera. Setting ranges are: film speed, 9–24 DIN, equivalent to 6–200 ASA; shutter speed, 1/50–1/1000 sec; apertures from f/4 to f/16.

4.5.2.4.1.2 Jena Aerolux. This instrument consists of the measuring head with a selenium Type D photocell, which is fastened by a flange to the bottom of the aircraft fuselage; the indicating unit; and an interconnecting cable. The indicating unit is attached directly to the camera control unit when used with Jena cameras. The exposure time is read directly on the indicating unit, once the settings of f-stop and film sensitivity have been made. The camera then is adjusted manually. The measuring head circuit contains a thermistor to reduce the effect of temperature on the photocell output. Spectral sensitivity range is from 350nm to 700nm with a peak sensitivity at 580nm. Setting ranges on the indicating unit are: film speed, 9–30 DIN (6 to 800 ASA); shutter speed, 1/25–1/2000 sec; apertures from f/4 to f/16 (*see* figure 4-36).

4.5.2.4.2 EXPOSURE CONTROLS, AUTOMATIC

4.5.2.4.2.1 Wild PEM-1 Exposure Meter/ Control. This instrument was introduced in 1974 as an accessory for the Wild RC10 Camera. It consists of a temperature-stabilized photoelectric element in a streamlined housing for attachment to the underside of the aircraft fuselage, a bulkhead mounting exposure calculator/display,

FIGURE 4-36. Jena aerolux exposure meter mounted on the MRB camera control unit (Courtesy of Zena Company).

and interconnecting cables. The instrument continuously determines the proper exposure time as a function of the selected aperture, effective film speed and multiplying factor. When used with the RC10 Camera in the automatic mode, the calculated shutter speed is transferred automatically to the camera. Used in the manual mode, the displayed exposure for a selected aperture may be transferred manually to any type of camera. The instrument features high sensitivity in the infrared and a negligible effect of atmospheric haze on the measurements. Spectral sensitivity range is from 400nm to 1100 nm with a peak sensitivity at 800nm. The sensor is the integrating type and has a measuring acceptance angle of 60°. Setting ranges on the exposure calculator are: film speed, 25–1600 ASA; aperture from f/4 to f/22; shutter speed, 0 to 1/1000 sec continuous; correction factor, 7 positions from −3/4 stop to +4/4 stop (*see* figure 4-37).

4.5.2.4.2.2 Zeiss EM1-2 Automatic Exposure Control. The control system consists of a pair of temperature-corrected Cds photoconductor cells and an electronic analog computer with integral operating amplifiers incorporated into the bodies of Zeiss RMKA cameras. The camera aperture is automatically controlled by the computer via servo motors as a function of selected shutter speed, film speed and multiplying factor. Manual aperture setting is possible by switching off the computer. The mean acceptance angle of the detector is ±30°, as per DIN 19010. Computer setting ranges are: film speed, 18–30 DIN (equivalent to 50–800 ASA) in 12 steps; multiplying factor, 1×–6×.

4.5.3 AUXILIARY DATA SYSTEMS

In order to assist in the external orientation of aerial photographs, appropriate reference information is recorded on each frame of film, along with the image. Such data as time, altitude, and frame number can be recorded directly from instruments contained within the camera.

Additionally, separate instruments may be

FIGURE 4-37. Wild PEM-1 automatic exposure meter/control for the RC-10 camera system (Courtesy of Wild Heerbrugg, Ltd.).

used to generate auxiliary data useful later in restitution. The output of these instruments may be transmitted electrically or optically to displays in the camera's data-chamber.

Various systems have been developed for recording such useful attitude references as sun and horizon images and the relative attitude of vertical gyroscopes.

Where this recorded information is of sufficient certainty and precision, it may be used to supplement ground control in subsequent plotter orientation.

4.5.3.1 STATOSCOPES

For bridging uncontrolled areas, height differences between photo-stations in a strip may be used for scale control. Absolute pressure altimeters are not sufficiently sensitive for such measurements. Terrain profiling instruments, such as radar altimeters, measure the distance from the airplane to the ground, but this measurement is influenced by the ground profile as well as the flying height. One instrument which is used for Δh recording is the statoscope a particularly sensitive kind of barometer.

Statoscopes measure ambient atmospheric pressure with adequate sensitivity over a limited range. Departures from the initial setting are displayed in the camera data chamber for recording on each frame of film. Usually, an identical reading is displayed also for the pilot's use in maintaining the altitude within the range of the instrument.

In combination with a radar or laser altimeter, the statoscope may be used for continuous terrain profile recording. For proper stability, statoscopes require careful control of the sensor's working temperature.

4.5.3.1.1 ZEISS S-2-C STATOSCOPE

The instrument is a liquid-manometer with coaxially arranged legs of different diameters. Temperature control is by immersion in ice water in a thermos bottle. The measuring range

varies from approximately 40m at sea level, with an accuracy of ±0.40m, to approximately 120m at a 10,000-metre altitude, with an accuracy of ±1.20m.

Electrical dial indicators are supplied for installation in the camera data chamber, on the panel of the S-2-C, and on the cockpit instrument panel.

4.5.3.1.2 JENA REGISCOPE

This recording statoscope is an aneroid barometer. Differences between ambient pressure and pressure in the instrument's measuring compartment result in mechanical changes which are converted into electrical signals. An electrical heater maintains the instrument's temperature at a constant 35°C.

The working range is a constant ±40m at all altitudes. Dial graduation intervals correspond to 2m and can be estimated to ±0.5m.

Displays include dials for the camera recording chamber, cabin, and pilot panel installation (*see* figure 4-38).

4.5.3.1.3 WILD RST2 REGISTERING STATOSCOPE

The Wild Statoscope is an aneroid barometer with an isolated air chamber. Temperature stability is maintained with an electrical heating system. Mechanical meter changes indicate height difference and are transmitted electrically to the camera and cockpit displays.

The measuring range depends on the altitude

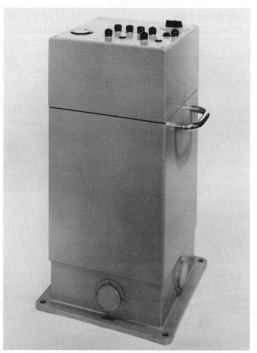

FIGURE 4-38. Jena regiscope registering statoscope (Courtesy of Zena Company).

FIGURE 4-39. Data annotation control/junction box, Type CAU, for automatic film annotation and camera function control (Courtesy of Spectral Data Corp.).

and is approximately ±40m at an altitude of 5000 metres. At this altitude, the height accuracy is ±0.5m within a range of ±20m.

4.5.3.2 ANNOTATION SYSTEMS

Figure 4-39 shows a control unit for manual insertion of data with a seven-digit LED array in the camera.

In aircraft equipped with inertial navigation systems (INS), spatial orientation may be recorded on each frame of film. Digital displays can register time, longitude, latitude, height, heading, drift and camera inclination.

A typical system produced for the Wild RC10 Camera uses a modified Litton LTN-51 Navigator, and a high-range radar altimeter with digital outputs. INS and radar altimeter data are sampled by a digital processing unit which converts from BCD to 7-segment code for display in the camera.

Position information is provided with a reported accuracy of 0.1 minute for latitude and longitude. In order to minimize the effect of long-term drift, the position may be updated by flying over ground marks of known location before and after the photo run. Height accuracy is ±1.5%, which could be improved with statoscope recordings.

4.6 Camera Installations and System Configurations

4.6.1 CAMERA INSTALLATIONS

Each type of aircraft that contains a camera installation has its own specific modification for a particular system. Even though there are many types of aircraft and types of camera, there are only a few basic kinds of camera installation.

Of these, generally only five are in common use in civil and military aircraft for cartographic and reconnaissance photography. They are (a) vertical, (b) oblique, (c) split vertical, (d) convergent, and (e) multi-camera, or fan.

4.6.1.1 VERTICAL CAMERA INSTALLATION

Vertical installation is the kind most widely used for photogrammetric mapping, land-use studies, military reconnaissance and general photography. The camera is installed so that the optical axis of the lens is perpendicular to the surface of the earth in normal aircraft flight. Such a camera may be rigidly attached to the aircraft, mounted in a manual-leveling mount (figure 4-40), or mounted in a gyroscopically stabilized mount (figure 4-41). Normally, this installation consists of only one camera which may be of any focal length or negative size. At times it may be used in combination with other cameras, in which case it becomes a multi- or fan installation. The scale of a photograph from the vertical camera is reasonably uniform and objects will have normal shapes. Thus, the compilation of maps from photographs taken with vertical cameras is relatively straightforward when compared to the reduction of non-vertical photography.

4.6.1.2 OBLIQUE CAMERA INSTALLATION

In the oblique installation, the camera axis is neither vertical nor horizontal, but is at an angle. The size of the angle is determined by the type of photograph required. Some fixed installations are designed to take oblique photographs to the port (left) or starboard (right) sides of the aircraft in flight. In some aircraft, the cameras are installed in mounts which can be pointed manually.

4.6.1.3 CONVERGENT CAMERA INSTALLATION

Oblique camera installations have been in use for a number of years, and many special plotting instruments were designed to accommodate ae-

FIGURE 4-40. Zeiss RMK 15/23 camera in a manual leveling mount attached to floor of aircraft (Courtesy of Carl Zeiss, Inc.).

FIGURE 4-41. USAF Type KC-1 camera installed in a vertical ART-25 stabilized mount with LS-19 vertical reference gyroscope (Courtesy of Aeroflex Laboratories).

rial photography having various degrees of intentional tilt. The most successful oblique system (convergent cameras) for mapping uses two cameras tilted at angles of 20° from the vertical, in opposite directions, along the line of flight. The picture from the forward pointing camera at one exposure station forms a stereo-pair with that of the aft pointing camera at the succeeding exposure station. Several considerations when designing a convergent camera installation for an aircraft are (a) the angular relationship between cameras, once established, should remain fixed, (b) the cameras should be mounted as close together as possible, and (c) the shutters should be synchronized to trip simultaneously.

One major disadvantage in the installation of convergent cameras is the huge window required. The U.S. Air Force RC-130A aircraft were designed for a convergent camera installation which included a window of optical quality glass, 42 inches in diameter and 3½ inches thick. The convergent installation has largely been replaced by ultra-wide-angle vertical cameras which offer nearly the same base/height ratio advantage.

4.6.1.4 THE SPLIT-VERTICAL CAMERA INSTALLATION

When the pair of oblique cameras is placed transverse to the line of flight, the installation becomes a split-vertical configuration. With two cameras oriented perpendicular to the line of flight, the total coverage is increased and flying time is reduced. This is an advantage when using long-focal-length cameras or where clear flying weather is of short duration. The split-vertical installation is used primarily in reconnaissance.

4.6.1.5 THE FAN CAMERA INSTALLATION

The multi-camera, or fan, installation, consists of three or more cameras of the same type and focal length whose shutters are tripped simultaneously. This arrangement produces a fan of photographs perpendicular to the line of flight. Coverage can be increased to include both horizons. One such installation in common use during the 1940-1960 era was called a trimetrogon camera.

The trimetrogon camera consisted of three 6-inch focal length cameras, each equipped with Metrogon wide-angle lenses. The negative was 9 × 9 inches. The lenses had an angular coverage of approximately 74° as measured across the photograph. Thus, three cameras provided more than enough coverage for the required 180° from horizon to horizon.

When installed, one camera was in a vertical position and the other two were mounted with their mating surfaces at an angle of 60° from the horizontal. One camera pointed to the left of the aircraft and the other to the right. Installed in this fashion, the cameras provided for an overlap of approximately 14°. The installation also included all the area from approximately 7° above the true horizon.

The focal-plane frames of all three cameras, being parallel to the longitudinal axis of the aircraft, insured that the optical axis of each oblique camera was approximately perpendicular to the aircraft's longitudinal axis.

In some aircraft, such as fighters, the width of the camera compartment did not permit the installation of the cameras in a single mount. Therefore, the vertical and oblique cameras were installed in two separate bays of the camera compartment. However, the angular relationship between the cameras was preserved.

Military aircraft used fan installations on many reconnaissance missions. In later aircraft, the fan installation generally has been replaced by single panoramic-cameras which combine higher photographic resolution with wide coverage.

4.6.2 PHOTOGRAPHIC SYSTEMS

In the past fifteen years, a considerable amount of progress has been made in the development of systems for obtaining military information from aircraft. As the primary component of such systems, the aerial camera has been integrated successfully into a number of complex systems.

One such system, the U.S. Air Force's AN/USQ-28 Mapping and Surveying System, was fabricated during the 1960's and installed in four RC-135A aircraft. This system, which proved to be highly accurate in its various capabilities, was used operationally until 1972.

The RC-135A/USQ-28 Mapping and Surveying System (figure 4-42) consisted of a highly accurate inertial navigation subsystem, an advanced mapping camera (KC-6A) subsystem, a microwave distance-measuring system, a terrain profiling radar, a digital data-recording system, and various supporting equipment. The entire system was controlled by a computer which also compiled and correlated all the data, and did computations for navigation. Operationally, the RC-135A/USQ-28 system was capable of obtaining high quality cartographic photography and geodetic control data over a 30,000-square mile area in a single day of flying.

4.6.2.1 THE NON-SYSTEM AERIAL CAMERA

Operation of the "non-system" aerial camera is a relatively simple matter. The camera may be in a fixed mount, or in one which permits manual corrective movement with reference to a level attached to the camera body. It also can be installed in a gyroscopically stabilized mount in which corrective action is automatic. The camera's connection to the aircraft system provides electrical current for its operation and for a small heater. The camera may be turned on from a switch in the camera hatch or in the aircraft cockpit. Its intervalometer is preset to a certain rate governed by the height of the aircraft above terrain and a given velocity of flight, or the camera's trip interval may be established automatically by a special V/H sensor. Most of the current aircraft cameras in production lend themselves readily to this kind of operation.

4.6.2.2 THE INTEGRATED SYSTEM CAMERA

The workings of an aerial camera which has been developed as part of a complex system such as the previously mentioned RC-135A/USQ-28 can be very complicated. The 6″ focal length KC-6A camera shared its mount with a highly refined inertial navigation subsystem (figure 4-43). The principal ray of the vertical camera

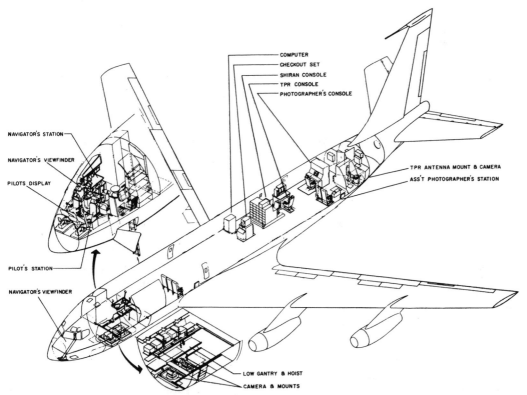

FIGURE 4-42. AN/USQ-28 Mapping and Surveying Subsystem equipment arrangement in the RC-135A Aircraft (Courtesy of Kollsman Instrument Corp.).

FIGURE 4-43. The Hipernas IIB inertial platform in LC-7A gyroscopically stabilized mount, shared with KC-6A mapping camera (Courtesy of Kollsman Instrument Corp.).

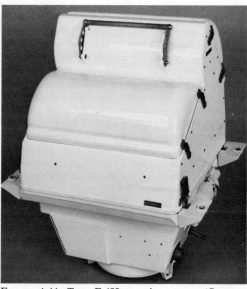

FIGURE 4-44. Type F-489 mapping camera (Courtesy of United States Air Force).

was linked to the axis of the inertial platform by a collimating periscope which recorded directly on each picture frame the angle between the principal ray of the approximately vertical camera and the vertical of the inertial platform. Since the inertial platform also contributed to the stabilization of the gyro mount, it was possible to hold the camera vertical within 3′ or 4′ and to record this amount to an accuracy of 30″.

In the KC-6A camera, a pulse, generated at the mid-point of exposure, froze all flight data for recording on the film. Data, synchronized to one millisecond with the shutter was passed through the system's central computer to the character generator which recorded information in binary-coded decimal form in a small area within the photograph. Additional data were

recorded in the usual manner on the film between frames.

The synchronization of the camera shutter with a very precise continuous wave, line-of-sight distance measuring system permitted trilateration (to within 3 m.) of distances up to about 600 km, and provided a capability for computing aircraft positioning within a 8 m accuracy.

An example of an operational system camera is shown in figure 4-44. This is the 6″ Geoconlens F-489 camera which is flown by the U.S. Air Force.

4.7 Aerial Cameras in Current Use

Table 4-3 lists the types of aerial cameras known to be in use as of 1978, and gives the principal specifications and characteristics of each type.

4.8 Camera Calibration

This section deals with calibration of cameras used to obtain accurate geometric data. It seems advisable, therefore, to establish a firm understanding of the implications of the term "calibration." Eisenhart of the National Bureau of Standards (1962) has investigated the requirements of calibration with sufficient depth to establish solid guidelines. From his paper, "Realistic Evaluation of the Precision and Accuracy of Instrument Calibration Systems," we shall select a few quotations which are most applicable for our purpose.

"Calibration of instruments and standards,"

says Eisenhart, "is basically a refined form of measurement. Measurement is the assignment of numbers to material things to represent the relations existing among them with respect to particular properties. One always measures properties of things, not the things themselves."

"Measurement of some property of a thing is an operation that yields as an end result a number that indicates how much of a property the thing has . . . Viewed thus, it becomes evident that a particular measurement operation cannot be regarded as constituting a measurement process unless statistical stability of the type known as a state of statistical control has been attained. In order to determine whether a particular measurement operation is, or is not, in a state of statistical control, it is necessary to be definite on what variations of procedure, apparatus, environmental conditions, observers, operations, *etc.* are allowable . . . *To be realistic, the 'allowable variations' must be of sufficient scope to bracket the circumstances likely to be met in practice."*

For those contemplating calibration of aerial cameras, Eisenhart's paper is recommended reading.

A condition of "statistical quality control" applied to camera calibration requires that specifications must first be established as to apparatus, operations, sequence, conditions; but prior to all this, the tolerance on the numbers must be set. One talks rather casually about two microns accuracy, but how is this achieved when this is the *sum* of all errors with which we must deal? Some of the questions we must answer are the following.

1. Is the apparatus accurate or are measurements repeatable to a fraction of a second? Thermally stable?
2. Can the human eye accurately select the energy centroid of the image to a fraction of a micrometre, a fraction of an arc second?
3. Is the emulsion surface in the correct plane?
4. Can the correct center be identified if the image is not sharp?
5. Will three operators select the same center?
6. Is the comparator orthogonally true; are the dimensions correct over any distance along the ways?
7. Do the dimensions of the comparator change with prolonged use?
8. How good is the comparator's master grid?
9. What about the photographic surface? Is it square to the entering ray of the zero collimator?
10. How stable is the point of best symmetry?
11. Do the image dimensions change during processing?

All of these questions, and more, suggest the high quality of the control needed; and then, unfortunately, the statistics of such control equate with high costs of labor and maintenance of the equipment, not to mention the quality of data.

This is only the beginning because calibration assumes that the thing being calibrated is stable between calibrations. An aerial camera may be calibrated in a carefully controlled environment, but the majority of surveys are made under drastically different conditions. Even under these conditions, the environments are not stable, but change during use (Norton, 1978).

The answer then is that camera calibration accuracies are limited by equipment, maintenance of equipment, personnel, operations, techniques, and the stability of the camera in varying environments (Carman, 1973; Meier, 1978).

For quality calibration one depends heavily on the designed and "built-in" quality of the camera, and during the past decades, much has been done by manufacturers to produce a reliable product. This is the first step in making calibration meaningful. For a camera manufacturer, the sensitivity of the geometric calibration, showing proved repeatability, is good quality control in the laboratory. The optimum quality control and "true" calibration occurs—remember the long chain of variables on equipment and techniques that add to a *sum* of two micrometres error of calibration—only when laboratory calibration and field calibration agree. At this point we return to Eisenhart's statement, ". . . to be realistic the allowable variations must be of sufficient scope to bracket the circumstances likely to be met in practice," which is a key problem in camera calibration.

4.8.1 GENERAL DISCUSSION

Camera calibration is a process whereby the geometric characteristics of an individual mapping camera are determined. It is performed in order that the photography obtained with the camera can be used to produce accurate maps, to allow measurements whereby ground distances or elevations can be obtained, and to make orthophotographs. It is possible to perform calibration of some order on any camera, but the cameras used to obtain the most accurate geometric data are specifically designed for that purpose. The quality of the lens is most important, where quality includes both well-defined imagery for any color and the near infrared, as well as correct positioning of the image at the film plane. It is desirable that the f/number be as low as possible without introducing excessive distortion. The camera holds the lens fixed (usually) at infinity focus. It has special geometric features—fixed fiducials, for determining a coordinate system and for controlling dimensional variables of film, or perhaps a reseau for the same purpose. The magazine must operate so that the surface of the film is a plane during exposure (there are special exceptions), while at the same time being at the selected (fixed) focus of the lens.

There are a number of methods of calibrating cameras, and all differ in equipment and technique. The physicist involved with calibration has his own preferences, but he must be aware of equipment and technique differences and their effects. There are two basic methods:

(1) To present an array of targets at known angles to a camera which records their images. The targets may be optical (simulating infinite targets) stars, or terrain targets photographed from towers, aircraft, or ground. The sensitized material (on which images are recorded) may be coated on a rigid glass plate or it may be the film in the operating magazine. The recorded images

TABLE 4-3.—AERIAL CAMERAS

Camera Type	Use	Basic Design	Cycling Time	Focal Length	Lens Apert.	Format Size	Shutter Type	Shutter Speed (sec.)	Weight (lbs.)	Film Mag.	Film Size
T-11	Mapping	Frame	2.5 sec	6″	f/6.3	9″ × 9″	Between lens	1/75– 1/500	75	T-11	9½″
KC-1A	Mapping	Frame	2.5 sec	6″	f/6.3	9″ × 9″	Between lens	1/75– 1/500	80	T-11, KC-1	9½″
KC-1B	Mapping	Frame	2.5 sec	6″	f/6.3	9″ × 9″	Between lens	1/75– 1/500	80	KC-1B	9½″
KC-2	Mapping	Frame	4 sec	6″	f/6.3	9″ × 9″	Between lens	1/100– 1/800	86	Integral	9½″
KC-3	Mapping	Frame	2.5 sec	88 mm	f/5.6	9″ × 9″	Between lens	1/100– 1/800	83	Integral	9½″
KC-4A	Mapping	Frame	2.5 sec	6″	f/5.6	9″ × 9″	Between lens	1/25– 1/400	95	Integral	9½″
KC-4B	Mapping	Frame	2.5 sec	6″	f/5.6	9″ × 9″	Between lens	1/50– 1/700	95	KC-1B	9½″
KC-6A	Mapping	Frame	2.5 sec	6″	f/5.0	9″ × 9″	Between lens	1/100– 1/800	100	KC-6 Opt. 1 mc.	9½″
KC-8	Mapping	Frame	2.5 sec	6″	f/6.3	9″ × 9″	Between lens	1/75– 1/500	80	KC-1B	9½″
K&E Aero View 600	Mapping	Frame	3.5 sec	6″	f/6.3	9″ × 9″	Between lens	1/150– 1/500	48	200 400	9½″
K&E Aero View 800	Mapping	Frame	4.5 sec	8¼″	f/6.3	9″ × 9″	Between lens	1/150– 1/250	54	200 400	9½″
K&E Aero View 1200	Mapping	Frame	2.5 sec	12″	f/6.3	9″ × 9″	Between lens	1/150– 1/250	58	200 400	9½″
FS-500	Mapping	Frame		6″	f/6.3	9″ × 9″	Between lens	1/10– 1/500	75	Detachable	9½′
F-489	Mapping	Frame	2.8 sec	6″	f/5.6	9″ × 9″	Between lens	1/25– 1/900	128	Integral	9½″
RC-8	Mapping	Frame	3.5 sec	6″	f/5.6	9″ × 9″	Between lens	1/100– 1/700 continu- ously variable	190	RC-8	9½″
RC-9	Mapping	Frame	3.5 sec	88 mm	f/5.6	9″ × 9″	Between lens	1/150 and 1/300	180	Integ. w/ cassettes	9½″
RC10	Mapping	Frame	1.6 sec	8.8 cm 15 cm 15 cm 21 cm 30 cm	f/5.6 f/4.0 f/5.6 f/4.0 f/4.0	9″ × 9″ 9″ × 9″ 9″ × 9″ 9″ × 9″ 9″ × 9″	Rotary Rotary Rotary Rotary Rotary	1/100– 1/1,000 continu- ously variable	140 kg	Cassettes	9½″
Galileo Santoni Mod. VI	Mapping	Frame	2.5 sec	6″	f/5.6	9″ × 9″	Between lens	1/125 1/200 1/300 1/400	80	Separate	9½″
SÓM Film	Mapping	Frame	4 sec	125 mm	f/6.2	18 × 18 cm	Between lens	1/75 and 1/100 1/125 and 1/150	75	Separate	19 cm
SOM Plate 125 mm	Mapping	Frame	4 sec.	125 mm	f/6.2	18 × 18 cm	Between lens	1/75 and 1/100; 1/125 and 1/150	200	Separate plate	19 cm

IN CURRENT USE (1978)

Film Load	Film Spools		Mount Used	Status	Mfr's Name	General Remarks	Camera Type
	Core Dia.	Flange Dia.					
390'	2⅛"	6⅝"	ART-25; A-28	Ltd Std.	Fairchild	Selected Metrogon lens	T-11
390'	2⅛"	6⅝"	LS-58; A-28; ART-25; ART-21	Ltd. Std.	Fairchild	Planigon lens; 80,000' altimeter	KC-1A
390'	2⅛"	6⅝"	LS-58; A-28; ART-21; ART-25	Std.	Fairchild	Distortions—10 micrometers; Platen flatness—0.0002"	
250' (2)			ART-24	Dev.	Fairchild	Matched twin Planigon lenses	KC-2
390'	2⅛"	6⅝"	ART-25; A-28	Tent. Std.	Aeroflex Labs.	Super Aviogon lens; platen reseau	KC-3
390'	2⅛"	6⅝"	LS-58; A-28; ART-25	Tent. Std.	Fairchild	Geocon I lens; $\sqrt[3]{\quad}$ shutter speeds reseau maga.	KC-4A
390'	2⅛"	6⅝"	LS-58; A-28; ART-25	Exp.	Fairchild	Geocon I lens	KC-4B
600'	2⅛"	7⅝"	LS-7; LS-8	Std.	Fairchild	Reseau platen, IMC avail., AEC	KC-6A
390'	2⅛"	6⅝"	ART-21; ART-25	Sub. Std.	Fairchild	Focused for IR film	KC-8
200' 390'	2⅛"	4" 6⅝"	Aero 2000 w/view finder	Current	Aero serv.	Ericon lens	K&E Aero View 600
200' 390'	2⅛"	4" 6⅝"	Aero 2000	Current	Aero Serv.	Aeroter lens	K&E Aero View 800
200' 390'	2⅛"	4" 6⅝"	Aero 2000	Current	Aero Serv.	Metrogon lens	K&E Aero View 1200
390'	2⅛"	6⅝"	Suspension	Commercial	Fairchild	Improved Planigon lens	FS-500
700' 2.5 mil	2⅛"	6⅝"	Fixed	Std.	Fairchild	Geocon I lens, high altitude	F-489
200'	2⅛"	5³/₁₆"	Integral	Commercial	Wild Heerbrugg	Aviogon lens	RC-8
200'	2⅛"	5³/₁₆"	Integral	Commercial	Wild Heerbrugg	No film spool slot; Super Aviogon lens	RC-9
500' 4 mil	2⅛"	6⅝"	Integral	Commercial	Wild Heerbrugg	Interchangeable lens cones	RC-10
360' 180'	2⅛"	6⅝"	Special	Std.	Officine Galileo, Florence	Orthogon	Galileo Santoni Mod. VI
50 m	4.35 cm	13.5 cm			SOM, Paris	Aquilor	SOM Film
96 Plates	Plate thickness	1.7 mm			SOM, Paris	Aquilor	SOM Plate 125 mm

TABLE 4-3.—AERIAL CAMERAS

Camera Type	Use	Basic Design	Cycling Time	Focal Length	Lens Apert.	Format Size	Shutter Type	Shutter Speed (sec.)	Weight (lbs.)	Film Mag.	Film Size
SOM Plate 210 mm	Mapping	Frame		210 mm	f/5.0	18 × 18 cm	Rotating	1/75 and 1/100; 1/150 and 1/200 and 1/250	220	Separate plate	19 cm
SOM Plate 300 mm	Mapping	Frame		300 mm	f/5.0	18 × 18 cm	Rotating	1/75 and 1/100; 1/150 and 1/200 and 1/250	225	Separate plate	19 cm
RMK A 8.5/23	Mapping	Frame	2 sec	85 mm	f/4.0	9″ × 9″	Rotating	1/50−1/500	60 kg	Separate	9½″
	Mapping	Frame	2 sec	153 mm	f/4.0	9″ × 9″	Rotating	1/100−1/1,000	62 kg	Separate	9½″
RMK A 21/23	Mapping	Frame	2 sec	210 mm	f/5.6	9″ × 9″	Rotating	1/100−1/1,000	45 kg	Separate	9½″
RMK A 30/23	Mapping	Frame	2 sec	305 mm	f/5.6	9″ × 9″	Rotating	1/100−1/1,000	54 kg	Separate	9½″
RMK A 60/23	Mapping	Frame	2 sec	610 mm	f/6.3	9″ × 9″	Rotating	1/100−1/1,000	54 kg	Separate	9½″
RMK 21/18	Mapping	Frame	2.5 sec	210 mm	f/4.0	18 × 18 cm	Rotating	1/100−1/1,000	35 kg	Separate	7½″
RMK 11.5/18	Mapping	Frame	2.5 sec	115 mm	f/5.6	18 × 18 cm	Rotating	1/100−1/1,000	35 kg	Separate	7½″
MRB 21/1818	Mapping	Frame	2 sec	210 mm	f/4.0	7″ × 7″	Rotating	1/100−1/1,000	35 kg	MRB-K 20/120	20 cm
MRB 11.5/1818	Mapping	Frame	2 sec	115 mm	f/4.0	7″ × 7″	Rotating	1/100−1/1,000	45 kg		20 cm
MRB 15/2323	Mapping	Frame	2 sec	150 mm	f/4.5	9″ × 9″	Rotating	1/100−1/1,000	63 kg	MRB-K 24/120	9½″
MRB 9/2323	Mapping	Frame	2 sec	90 mm	f/5.6	9″ × 9″	Rotating	1/50−1/500	44 kg	MRB-K 24/120	9½″
K-17D	Day recon.	Frame	1.25 sec or 3.5 sec	6″ / 12″ / 24″	f/6.3 / f/5.0 / f/6.0	9″ × 9″	Between lens / Between lens	1/25−1/300 / 1/75−1/225 / 1/25−1/150	29½ / 33 / 49½	LA-50 / LA-33 / A-9B	9½″
K-20	Day recon.	Frame	Manual	6⅝″	f/4.5	4″ × 5″	Between lens	1/25 1/250 1/500	11		5¼″
K-22	Day recon.	Frame	2 sec	6″ / 12″ / 24″ / 36″ / 40″ / 40″ / 40″	f/6.3 / f/5.0 / f/6.0 / f/8.0 / f/5.0 / f.8.0 / f.5.6	9″ × 9″	Focal plane	1/150 1/300 and 1/300−1/900	25½ / 28 / 45¼ / 45½ / 107¼ / 45½ / 67	A-18 / A-28 / A-9B	9½″
K-37	Night recon.	Frame	3 sec	12″ / 24″	f/2.5 / f/4.0	9″ × 9″	Between lens		50 / 80	LA-50 LA-31 A-9B	9½″
K-38	Day recon.	Frame	1.6 sec 3 sec	12″ / 24″ / 36″	f/6.3 / f/6.0 / f/8.0	9″ × 18″	Between lens	1/25−1/300 / 1/25−1/150 / 1/25−1/150	37 / 50 / 67½	LA-32 / A-25 / A-8B	9½″
K-46	Night recon.	Frame	0.5 sec	6″ / 7″	f/2.5 / f/2.5	4½″ × 4½″	Focal plane capping	Open flash	33½	A-23	5″

IN CURRENT USE (1978)—Continued

Film Load	Film spools Core Dia.	Flange Dia.	Mount Used	Status	Mfr's Name	General remarks	Camera Type
96 Plates	Plate thickness	1.7 mm			SOM, Paris	Orthor (Orthoscopic)	SOM Plate 210 mm
96 Plates	Plate thickness	1.7 mm			SOM, Paris	Orthor (Orthoscopic)	SOM Plate 300 mm
500'	54 mm	168 mm	AS-5	Std.	Carl Zeiss Oberkochen	Pan, color, and IR films	RMK A 8.5/23
500'	54 mm	168 mm	AS-2; AS-3; AS-5 w/adapt.	Std.	Carl Zeiss Oberkochen	Pan, color, and IR films	RMK A 15/23
500'	54 mm	168 mm	AS-2; AS-3; AS-5 w/adapt.	Std.	Carl Zeiss Oberkochen	Pan, color, and IR films	RMK A 21/23
500'	54 mm	168 mm	AS-2; AS-3; AS-5 w/adapt.	Std.	Carl Zeiss Oberkochen	Pan, color, and IR films	RMK A 30/23
500'	54 mm	168 mm	AS-2; AS-3; AS-5	Std.	Carl Zeiss Oberkochen	Pan, color, and IR films	RMK A 60/23
120 m	55 mm	168 mm	AS-1	Lim. Std.	Carl Zeiss Oberkochen	Pan and color	RMK 21/18
120 m	55 mm	168 mm	AS-1	Lim. Std.	Carl Zeiss Oberkochen		RMK 11.5/18
120 m			MRB-A	Std.	Zena	15 stepped grey wedge recorded on each photograph	MRB 21/1818
120 m			MRB-A	Std.	Zena		MRB 11.5/1818
120 m 150 m			MRB-A	Std.	Zena	15 stepped grey wedge recorded on each photograph	MRB 15/2323
120 m 150 m			MRB-A	Std.	Zena	15 stepped grey wedge recorded on each photograph	MRB 9/2323
390'	2⅛"	6⅝"	A-28	Lim. Std.	Fairchild	General purpose photograph	K-17D
20'	1¼"	2⅛"	Hand-held	Lim. Std.	Fairchild; Graflex	Hand-held	K-20
390'	2⅛"	6⅝"	A-28	Lim. Std.	Fairchild; Chicago Aerial Survey	General purpose	K-22
390'	2⅛"	6⅝"	A-28	Alt. Std.	Fairchild; G.E.	Flash bomb; flash cartridge; electronic flash	K-37
390'	2⅛"	6⅝"	ART-25; A-28	Alt. std.	Fairchild	Medium to high altitude	K-38
250'	2⅛"	5¹⁵/₁₆"	A-28	Alt. Std.	Hycon	Low altitude	K-46

TABLE 4-3.—AERIAL CAMERAS

Camera Type	Use	Basic Design	Cycling Time	Focal Length	Lens Apert.	Format Size	Shutter Type	Shutter Speed (sec.)	Weight (lbs.)	Film Mag.	Film Size
K-47	Night recon.	Frame	0.5 sec	12″ 24″	f/2.5 f/4.0	9″ × 9″	Between lens	1/10– 1/200	50 80	LA-124 A-28	9½″
KA-1	Day recon.	Frame	1.75 sec	12″ 24″ 36″	f/6.3 f/6.0 f/8.0	9″ × 18″	Double action intralens	1/25– 1/300	53 63 80	A-25 LA-23	9½″
KA-2	Day recon.	Frame	0.55 sec	6″ 12″ 24″	f/6.3 f/4.0 f/6.0	9″ × 9″	Double action intralens	1/25– 1/400	53 32 63	A-28 LA-35 LA/50	9½″
KA-3A	Day recon.	Frame	0.5 sec	6″	f/6.3	9″ × 9″	Intralens	1/50– 1/400	24½	A-18 A-28	9½″
KA-18A	Day recon.	Strip	Cont.	6″ 3″ 6″	f/2.5 f/6.3 f/6.3	Stereo Stereo 9″ wide	Slit		75	Cassettes	9½″
KA-20B	Recon.	Frame	0.85 sec	6″ 12″	f/6.3 f/4.0	9″ × 9″	Intralens	1/150– 1/300; 1/50– 1/250	57 60	A-9B	9½″
KA-48	Day recon.	Panoram.	0.5 sec	6″	f/6.3	4½″ × 19″	Stat. slit	1/500– 1/4,000	105	Special	5″
KA-52	Day recon.	Panoram.	0.17 sec	3″	f/4.5	4½″ × 10.8″	Stat. slit	1/500– 1/2,000	85	Special	5″ Perf.
KA-54	Surveil-lance	Panoram.	0.6 sec	3″	f/2.8	2.25″ × 7.38″		1/50– 1/2,000	135	Integral	70 mm
KA-55	Day recon.	Panoram.	1.8 sec	12″	f/5.6	4.5″ × 18.8″	Slit	1/100– 1/3,000	100	Special	5″
KA-56	Recon.	Panoram.	0.17 to 1.0 sec	3″	f/4.5	4.5″ × 9.4″		1/90– 1/5,000	90	Inter-chang.	5″
KA-56A	Day recon.	Panoram.	0.17 sec	3″	f/4.5	4.5″ × 10.8″	Stat. slit	1/100– 1/5,000	90	Special	5″
KA-59	Day recon.	Panoram.	1 sec	12″	f/5.6	4.5″ × 41″	Stat. slit	1/100– 1/5,000	160	Special	5″ Perf
KA-60	Day recon.	Panoram.	0.08 to 1.0 sec	3″	f/2.8	2.25″ × 10″		1/100– 1/10,000	25	Inter-chang.	70 mm
KA-61	Day–night recon.	Frame	0.3 sec	52 mm	f/3.5	2¼″ × 2¼″	Between lens	Bulb– 1/500	6	Integral	70 mm
KB-8a	Day recon.	Frame	0.18 sec	1½″ 3″ 6″	f/4.5 f/2.8 f/2.8	2¼″ × 2¼″	Focal plane	1/500 1/1,000 1/2,000 1/4,000	9	Special	70 mm
KS-67	Day recon.	Frame	0.18 sec	1½″ 3″ 6″ 12″	f/4.5 f/2.8 f/2.8	2¼″ × 2¼″	Focal plane	1/500 1/1,000 1/2,000 1/4,000	15	Self-cont. vacuum	70 mm
KS-72A	Day-night recon.	Frame	0.17 sec	3″ 6″ 12″ 18″	f/4.5 f/2.8 f/4.0 f/5.6	4½″ × 4½″	Intra-lens focal plane	1/25– 1/100 1/100– 1/1,000	44 less lens cone	Special	5″ Perf.

IN CURRENT USE (1978)—Continued

Film Load	Film Spools Core Dia.	Film Spools Flange Dia.	Mount Used	Status	Mfr's Name	General Remarks	Camera Type
390′	2⅛″	6⅝″	A-28; ARA-6	Alt. Std.	Fairchild	Low, medium, and high altitude	K-47
390′	2⅛″	6⅝″	Fixed	Alt. Std.	Fairchild	Medium high altitude	KA-1
390′	2⅛″	6⅝″	Fixed; A-28	Alt. Std.	Fairchild; G.E.	Low, medium, and high altitude	KA-2
390′	2⅛″	6⅝″	A-28	Std.	Fairchild	Medium low altitude	KA-3A
250′	2⅛″	5¹⁵/₁₆″	ART-21	Std.	Chicago Aerial Industries	Low altitude, high speed; IMC = 1″ to 30″ per second	KA-18A
390′	2⅛″	6⅝″	Special		Hycon	Lightweight, compact	KA-20B
1,500′			Fixed		Fairchild	Rotating prism; 180° scan	KA-48
900′			Fixed		Fairchild	Rotating prism; 180° scan; IMC AEC	KA-52
500′			Stabilized		Fairchild	Rotating lens	KA-54
500′ to 1,000′	2⅛″	10½″	LS-58; ART-63		Hycon	Stovepipe; 90° scan; high altitude; AEC	KA-55
250′ 500′ 1,000′			None		Fairchild	Rotating prism	KA-56
250′ to 1,000′	2⅛″	10½″	Fixed or stabilized		Fairchild	Rotating prism; AEC; 180° scan	KA-56A
4,000′	2⅛″	12″	Fixed		Fairchild	Rotating prism; 180° scan	KA-59
250′					Fairchild	Rotating prism	KA-60
50′			Hand-held		Itek	In-flight processing	KA-61
100′			Fixed		J. A. Maurer AEC		KB-8A
85′			Fixed		J. A. Maurer		KS-67
250′ to 500′			ART-63; LS-58		Hycon	AEC; in-flight processing cassette	KS-72A

TABLE 4-3.—AERIAL CAMERAS

Camera Type	Use	Basic Design	Cycling Time	Focal Length	Lens Apert.	Format Size	Shutter Type	Shutter Speed (sec.)	Weight (lbs.)	Film Mag.	Film Size
HR-230	Recon.	Frame	0.33 sec	6″	f/5.6	9″ × 9″	Focal plane	1/200− 1/4,000	67	A-9B	9½″
HR-231	Recon.	Frame	1.5 sec	36″	f/10	9″ × 18″	Intra-lens	1/125− 1/500	122	Special contour platen	9½″
HR-233	Recon.	Frame	1.5 sec	24″	f/8	9″ × 9″	Intra-lens	1/250	100	Mod. A-9B contoured platen	9½″
HR-235	Recon.	Frame	2 sec	12″	f/5.6	4½″ × 4½″	Intra-lens	1/300	21½	Integral	5″
HR-236	Recon.	Frame	2 sec	6″	f/5.6	4½″ × 4½″	Intral-lens	1/250	14½	Integral	5″
HR-320	Recon.	Frame	1.65 sec	40″	f/5.0	9″ × 9″	Focal plane	1/100− 1/1,000	135	Mod. A-9B	9½″
LG-77A	Recon.	Frame	0.33 sec.	48″	f/4.0	4½″ × 4½″	Focal plane	1/100− 1/1,000	415	Integral	5″
KA-94A	Recon.	Panoram.	0.25 to 0.08 CPS	24″	f/4.5	4⅜″ × 50 1/3″	Slit	1/75− 1/10,000	280	Cassettes	5″
KA-97A	Recon.	Panoram.	0.08 to 12 CPS	3″	f/2.8	2.25″ × 9.4″			150	Cassettes	70 mm
KA-65	Recon.	Panoram. strip	6 CPS	3″	f/4.5	4.5″ × 9.4″ 4½″ strip	Slit	1/100− 1/5,000 1/12− 1/1,500	70	Integral	5″
KA-66A	Recon.		1−12 CPS	3″	f/2.8	2 1/4″ × 9 7/16″	Focal plane	1/100− 1/10,000	26.9	Special	70 mm
KR-b 8/24E	Day-night recon.	Frame	0.2 sec	80 mm	f/2.0	71.5 × 71.5 mm 3 ea	Focal plane	1/150− 1/2,000	46 kg	Cassettes	9½″
KR-b 8/24C	Day-night recon.	Frame	0.2 sec	80 mm	f/2.0	71.5 × 71.5 mm 3 ea	Focal plane	1/150− 1/2,000	11.5 kg	Cassettes	9½″
KR-b 6/24	Day-night recon.	Frame	0.14 sec	57 mm	f/2.0	50 × 40 mm 5 ea	Focal plane	1/150− 1/2,000	46 kg	Cassettes	9½″
TR-b 60/24	Recon.	Frame	0.5 sec	610 mm	f/4.0	115 × 230 mm	Between lens	1/150− 1/1,000	75 kg	Cassettes	9½″
KA-80A	Mapping & recon.	Panoram.	3.5 sec	24″	f/3.5	4½″ × 50¼″	Focal plane	0.35−29 msec	255	Integral	5″
KA-80I	Mapping & recon.	Panoram.	1.7 sec	24″	f/3.5	4½″ × 50¼″	Focal plane	0.17−9.5 msec	280	Cassettes	5″
KA-83A	Mapping & recon.	Panoram.	1.74 sec	24″	f/3.5	4½″ × 50¼″	Focal plane	0.18−11.2 msec	230	Integral	5″
F-905	Recon.	Strip	Remarks	6″	f/5.6				80		
Model 10	Multi-spect.	Frame	2 sec	150 mm	f/2.8	2 1/8″ × 4 1/16″	Focal plane 4 slit	1/65−1/150 1/150−1/350 1/350−1/800	75	A-5A A-9B	9½″

IN CURRENT USE (1978)—Continued

Film Load	Film Spools Core Dia.	Film Spools Flange Dia.	Mount Used	Status	Mfr's Name	General Remarks	Camera Type
390'	2⅛"	6⅝"	Special		Hycon		HR-230
1,000'	2⅛"	10½"	Special rocker type		Hycon	Oblique scanning hd. 148° coverage	HR-231
390'	2⅛"	6⅝"	IMC mount		Hycon	Oblique scanning hd. 90° coverage	HR-233
100'	1¼"	3¾"	Special		Hycon	Extreme high-altitude	HR-235
100'	1¼"	3¾"	Special		Hycon	Extreme high-altitude	HR-236
390'	2⅛"	6⅝"	Special		Hycon	High altitude	HR-320
250'	2⅛"	4⅜"	Special		Hycon	High altitude	LG-77A
2,000'			Fixed		Fairchild	Medium high altitude; AEC	KA-94A
3,000'			Fixed		Fairchild	Low altitude; rotating prism; AEC	KA-97A
1,000'			Fixed		Fairchild	Rotating prism; moving film; AEC; FMC	KA-65
250'			Fixed			Rotating prism; low altitude; FMC	KA-66A
75 m	54 mm	151 mm	None	Tent. Std.	Carl Zeiss	144° cos corrected; IMC; elect. operated	KR-6 8/24E
15 m	25 mm	63 mm	None	Std.	Carl Zeiss	144° cos corrected; IMC; elect. operated	KR-6 8/24C
75 m	54 mm	151 mm	None	Tent. Std.	Carl Zeiss	180° cos corrected; IMC	KR-6 6/24
75 m	54 mm	151mm	None	Tent. Std.	Carl Zeiss	Low altitude stand off	TR-6 60/24
6,500'	4"	18"	Vib. isolator	Std.	Itek		KA-80A
2,000'	2⅛"	12"	Vib. isolator	Std.	Itek		KA-80I
2,000'	4"	11"	Vib. isolator	Std.	Itek		KA-83A
			Fixed		Fairchild	0.005 sec/line; real time transmission; mag. tape storage; electron recording	F-905
250'	2⅛"	5¹⁵/₁₆"	RC-8	Std.	Spectral Data Corp.	4 lenses with 4 formats; IMC in mag. C	Model 10

TABLE 4-3.—AERIAL CAMERAS

Camera Type	Use	Basic Design	Cycling Time	Focal Length	Lens Apert.	Format Size	Shutter Type	Shutter Speed (sec.)	Weight (lbs.)	Film Mag.	Film Size
Model 11	Multi-spect.	Frame	2 sec	4″	f/2.8	3½″ × 3½″ 4 ea	Focal plane single slit	1/65–1/150 1/150–1/350 1/350–1/800	75	A-5A A-9B	9½″
Model 12	Multi-spect.	Frame	2 sec	6″	f/2.8	3½″ × 3½″ 4 ea	Focal plane single slit Between lens (optional)	1/65–1/150 1/150–1/350 1/350–1/800	75	A-5A A-9B	9½″
Experimental MB-1 Multiband Camera	Multi-spect.	Frame	1 sec	6″ 4″	f/4.0 f/4.0	2¼″ × 2¼″ 2¼″ × 2¼″	Behind lens	1/25–1/400	131 135½	Model D modified 4 ea	70 mm
KA-89B	Day Recon Panoram		1.6 sec/cy to 3.1 c/s, auto	3″	f/2.8	2.25″ × 9.4″	Focal Plane	1/120–1/6,000	50	P/N 1134R100	70 mm
KA-91B	Day Recon Panoram.		1.33 c/s auto cycle max	18″	f/4.0	4.5″ × 19″	Capping cont var	1/100–1/500	160	Mag LA-452A Cas LA-543A	5″
KA-93	Day Recon Panoram.		0.5 c/s (20°) 0.75 c/s (45°) 1 c/s (70°) 1.25 c/s (95°)	24″	f/5.6	4.5″ × 40″	scan slit cap shut	1/100–1/1,500	170	Mag LA-452A Cas LA-453A	5″
KS-87B	Day and Night Recon	Frame	6 c/s	3″ 6″ 12″ 18″	f/4.5 f/2.8 f/4.0 f/4.0	4.5″ × 4.5″	Focal Plane	1/60–1/3,000 Foc plane 1/25, 1/50 1/100 Bet. the lens	64 61.1 63.75 78.5	Mag LA-325B Cas LA-354B	5″
KS-116A	Day Recon Panoram		1.3 c/s max 1.0 c/s max 1.33 c/s max	6″ 12″ 18″	f/2.8 f/4.0 f/4.0	4.5″ × 18.9″ 4.5″ × 29.5″ 4.5″ × 18.9″	var slit capping	1/100–1/1500	120	Mag LA-452A Cas LA-453A	5″
KS-120A	Day Recon Panoram		1 to 12 c/s	3″	f/2.8	2.25″ × 9.4″		1/100–1/12,000	170		70 mm
KS-121A	Day Recon Frame			1½″ 3″ 6″	f/4.5 f/2.8 f/2.8		Focal Plane	1/250–1/4,000			70 mm
KS-127A	Day Recon Frame		0.5–1.5 c/s	66″	f/8.0	4.5″ × 4.5″		1/30–1/1,500			5″
APOLLO (Stellar)	Mapping	Frame Frame	{ 8.25–33.0 c/s	3″ 3″	f/4.5 f/2.8	4.5″ × 4.5″ 1.25″ dia	Bet lens	1/15–1/250 1.5 sec	225	Integral	5″ 35 mm
MRB 30/2323	Mapping	Frame	1.7–2.6	12′	f/5.6	9″ × 9″	Rotating	1/100–1/1,000			
KA-45A	Day-Night Recon.	Frame	1 to 6 per second	6″	f/2.8	4.5″ × 4.5″	Focal plane	1/60–1/3,000	425	LA-141A Cassettes	5″
KA-57A	Day Recon Panoram.		0.2 sec.	3.15″	f/2.8	2.25″ × 8.247″	Focal plane	1/75–1/600	189	Cassette supply and takeup	70 mm
KA-58A	Day Recon Panoram.		0.87 sec.	18″	f/4.0	4.5″ × 44.1″	Focal plane	1/100–1/3,000	387.2	Cassette	5″

IN CURRENT USE (1978)—Continued

Film Load	Film Spools Core Dia.	Flange Dia.	Mount Used	Status	Mfr's Name	General Remarks	Camera Type
250'	2⅛"	5¹⁵/₁₆"	RC-8	Std.	Spectral Data Corp.	4 lenses with 4 formats; IMC in mag. B	Model 11
250'	2⅛"	5¹⁵/₁₆"	RC-8	Std.	Spectral Data Corp.	4 lenses with 4 formats; IMC in mag. A.	Model 12
100'		3¾"	A-28; LS-58	Experimental	Boller & Chivens, Div. P&E	4-channel beam splitter; 460 mm; 550 mm; 700 mm; 780 mm	Experimental MB-1 Multiband Camera
1800' thin base			Fixed	Std.	Fairchild	Hor to hor for high speed low flying a/c; drone	KA-89B
500'	MS-26565-10		Hard mounted Roll stab gyro	Std.	Chicago Aerial Ind	Large Scale pan; rotating prism; RE-4 a/c	KA-91B
1000'			Hard mounted Gyro roll stab	Std.	Chicago Aerial Ind	Rotating prism pan RE-5 and E-111	KA-93B
500' (Std base) 1000' (thin base)				Std.	Chicago Aerial Ind	Night illuminant synchro; in RF-4 a/c	KS-87B
500' (Std base) 1000' (thin base)			Hard mounted, gyro roll stab	Std.	Chicago Aerial Ind	Rotating prism, moving film	KS-116A
6400' (thin base)			Fixed	Std.	Fairchild	Low altitude; rotary prism; moving film; drone	KS-120A
100' (Std base) 200' (thin base)			Fixed	Std.	Chicago Aerial Ind	RF-5 (Saudi Arabia, Brazil, Jordan)	KS-121A
500' (Std base) 1000' (thin base)			Fixed	Std. Ltd.	Chicago Aerial Ind	Long range oblique photo. (LOROP); RF-4 instal.	KS-127A
1500'(Map) 510' (Stellar)			Special Spacecraft	Comp.	Fairchild	NASA oper. on APOLLOs 15-17 stellar camera control	APOLLO
			MRBA	Std.	Zena		MRB 30/2323
250'	Spec. MS2656-9		Special	Ltd. Std.	Chicago Aerial Ind	Built for USN; 1¾, 3 and 12 in. lenses available	KA-45A
3000'			LA-297 LA-305A	Std.	Perkin– Elmer	Low altitude; USN-RA5C; Twin lens	KA-57A
2500'			LA-301A	Std.	Perkin– Elmer	High alt. hor. to hor.; USN RA-5C	KA-58A

TABLE 4-3.—AERIAL CAMERAS

Camera Type	Use	Basic Design	Cycling Time	Focal Length	Lens Apert.	Format Size	Shutter Type	Shutter Speed (sec.)	Weight (lbs.)	Film Mag.	Film Size
KA-63A	Day vert. Day twiin Night vert	Frame	3 sec per cycle	58 mm 80 mm 6″	f/4.5 f/5.6 f/2.8	2.25″ × 9.45″ 2.25″ × 9.45″ 2.25″ × 4.5″	D-f. plane N-self Capping	1/500− 1/1,000 or 1/2,000 Man. Set	60	LA/364A Cassette	9½″
KA/68A	Day Recon	Panoram.	1 to 6 per second	3″ Biogon	f/4.5	4.5″ × 9.4″	Var. Slit at F. plane	1/92.5− 1/5,000	90	Cassette	5″
KA-74A	Day Recon Night Recon	Frame	1, 2 or 4 f/s 1 or 2 f/s	6″	f/2.8	4.5″ × 4.5″	Focal Plane	1/1,000, 1/2,000	37	LA-384A	5″
KA-76A	Day-Night Recon	Frame	6 per sec	6″ 1.75″ 3″ 12″	f/2.8 f/5.6 f/4.5 f/3.5	4.5″ × 4.5″	Focal Plane	1/60− 1/3,000	52.5 51.75 58.25 68.5	LA-414A Cassette	5″
KA-82A	Day Recon	Panoram	10 sec/cycle to 1.17 c/s, autocy.	12″	f/3.8	4.5″ × 29.3″	Focal Plane	1/30− 1/12,000	216.7	LA/418A Cassette	5″
KA-88A	Day Recon	Frame	2 sec max	24″	f/8.0	9″ × 9″	Focal Plane	1/500, 1/1,000 1/2,000	135	Cassette	9½″

are measured, and the data reduced from measurements provide the elements of interior orientation. Many physical controls are essential.

(2) To clamp a master grid at the focal plane and to measure the observed angles in object space, a visual or photogoniometer technique. The distortion is computed from the focal length and the difference between the image and object angles.

Whichever of the many variations of these two methods is employed, the physicist is concerned with the design, maintenance, and accuracy of sensitive precision equipment and the use of evaluation techniques which are substantially free of error.

The calibrated values and their accuracy are then reported in a camera calibration certificate with tables and graphs.

4.8.1.1 DEFINITIONS

In accordance with recommendations from the International Society of Photogrammetry (ISP), the word "measured" is being used to stop the confusion between image position error, as found in practice, and distortion, the aberration calculated by the lens designer. The latter is always symmetrical and has no tangential component. The ISP recommends use of the word "best" in principal point of best symmetry to emphasize that perfect symmetry is not obtainable, and further that choice of the point depends on the interpretation of "best."

All characteristics that affect the geometry of the photograph are calibrated. These are here termed elements of interior orientation (*see* also chapter II), and, depending on the quality of the camera, may include the following: (*see* figure 4-45).

a. Equivalent Focal length (EFL)

$$\text{EFL} = \frac{d\alpha}{\tan \alpha} \qquad (4.19)$$

with the reasonable assumption that there is no distortion at the small angles. $d\alpha$ is the average distance between the images of the center

RADIAL DISTORTION

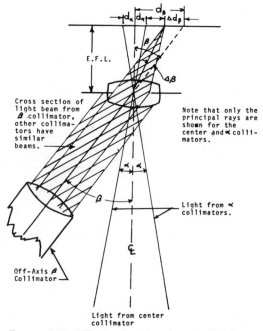

FIGURE 4-45. Effective focal length and radial distortion. This drawing explains the geometry. Using known angles in object space, usually 7.5° or less, and the measured distance between the images, the effective focal length is a computed value. Δd_B is the radial distortion.

IN CURRENT USE (1978)—Continued

Film Load	Film Spools Core Dia.	Flange Dia.	Mount Used	Status	Mfr's Name	General Remarks	Camera Type
180'	75' spool—MS26565-15 200'—spool—MS26565-17		Fixed	Std.	Chicago Aerial Ind	Hor. to hor. day or night vert.; Army	KA-63A
1000' Thin base			Fixed		Fairchild	Low alt hor. to hor.; rotating double dove prism	KA-68A
100' (5.2 mil) 180' (2.5 mil)			LA-366A	Std.	Actron	Vert fwd rt and left obl; part of KS-91A Surv sys	KA-74A
250'	MS-26565-9		LA-408A LA-409A LA-160A	Std.	Chicago Aerial Ind	Part of KS-104A & B photo sys in Army OV-1 ac	KA-76A
2000'			Shock mount	Std.	Fairchild	Med or high alt.; rotery lens	KA-82A
1500' thin base			3 pt vib isol fixed V/H rocking		Actron	Med to high alt.; nodding five position; HC-338A	KA-88A

target and that of a small off-axis image, usually 7.5° or less for a wide-angle mapping lens. α is the angle in object space.

b. Average Radial Measured Distortion: When the incident ray from an off-axis object is deviated while traversing the lens so that the distance d_B between the center and off-axis images, $d_B \neq$ EFL tan B, but

$$d_B = \text{EFL} \tan B \pm \Delta d_B \qquad (4.20)$$

then Δd_B is the radial distortion, being negative when the image distance is less then EFL tan B and positive when greater. When there are several images for similar angular targets, they are averaged. Distortion is plotted in micrometres as a function of image distance in millimetres, or in micrometres as a function of angle (figure 4-46).

c. Asymmetrical Radial Measured Distortion is the variation from the average value.

d. Calibrated Focal Length (CFL)

$$\text{CFL} = \text{EFL} \pm \Delta f \qquad (4.21)$$

When the EFL is adjusted by adding or subtracting Δf to EFL, to provide a preferred balance of the measured distortion curve, usually balancing positive and negative peaks or obtaining a least squares balance, the adjusted value is termed the calibrated focal length and is used in optical projection systems and/or for mathematical corrections.

e. Tangential Measured Distortion is a displacement of the image perpendicular to a straight radial line from the lens axis, and, similarly to radial distortion, is measured in the image plane. It is a manufacturing error which can be minimized by precise centering of lens elements and their selected azimuth rotations.

f. Principal Point of Best Symmetry (see second edition "MANUAL OF PHOTOGRAMMETRY," Camera Calibration, Sewell). A point is selected which reduces the asymmetry of the distortion to a minimum.

g. Fiducial Center or Indicated Principal Point (IPP). If imaginary lines are drawn between opposite fiducials as between A and B and its approximately perpendicular pair C and D, the crossing of the imaginary lines defines the fiducial center, also termed the Indicated Principal Point. A different design features corner fiducials such as E-F and G-H. Fiducials must be small and clearly defined so that the center can be located within a few micrometres. Note that there is no agreement on labeling fiducials (figure 4-47).

h. Angle between the theoretical lines drawn between opposite fiducials: This angle is measured during the calibration tests and is usually required to deviate less than one-minute-of-arc from 90°.

i. Distance between opposite fiducials: In the photographic techniques of camera calibration, the fiducials are recorded on rigid emulsion-coated glass plates when these plates are exposed in the focal plane of the camera. The measured distances (A to B and C to D; or E to F and G to H; or both) are reported in the calibration certificate.

j. Principal Point of Autocollimation (PPA). The PPA is the location at which the image of the zero degree collimator target is recorded when the focal plane of the camera is positioned precisely perpendicular to the direct ray from that target. The location of the PPA is referenced to the IPP.

k. Coordinate System of Fiducials. When the positions of the fiducials are reported with respect to a coordinate system, the geometry provides a better control. In one such case, the fiducial center becomes the origin, the A-B line being coincident with the x-axis. A theoretical line perpendicular to the x-axis and intersecting the IPP is then the y-axis. The remaining fiducials are referenced to this coordinate system. Other arrangements are also used.

l. Vacuum Platen Contour. When the mapping camera is equipped with vacuum to hold the

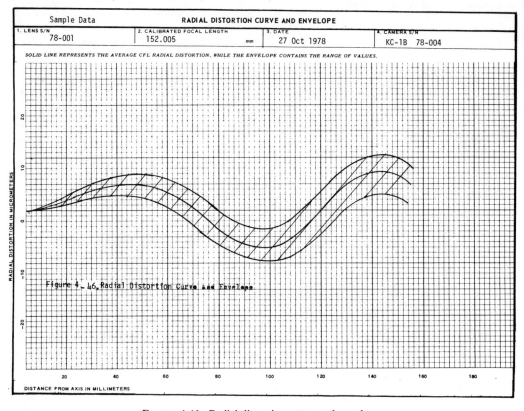

Sample Data		RADIAL DISTORTION CURVE AND ENVELOPE	
1. LENS S/N 78-001	2. CALIBRATED FOCAL LENGTH 152.005 mm	3. DATE 27 Oct 1978	4. CAMERA S/N KC-1B 78-004

SOLID LINE REPRESENTS THE AVERAGE CFL RADIAL DISTORTION, WHILE THE ENVELOPE CONTAINS THE RANGE OF VALUES.

FIGURE 4-46. Radial distortion curve and envelope.

film in contact with the magazine platen during exposure, it is necessary that the contour of the platen follow the intended focal plane of the lens so that distortion is not introduced into the photograph. In most cases, the focal plane of the lens is flat and measurements of the platen are made to show significant deviations. Users can then choose to correct these biases mathematically.

m. Reseau Measurements. There are various patterns of reseau which are recorded as points or crosses on the film during exposure. Their primary purpose is to provide control for film dimensions. When the reseau is fixed with respect to the lens, it may also substitute for the fiducials, certain reseau marks being selected from which to locate the fiducial center, the line of flight, the point of best symmetry, and other control points. Reseau points should be measured directly or from recordings on emulsion-coated plates positioned at the focal plane to assure accuracy.

n. Resolution and Optical Transfer Functions. These are the two best known and most used criteria of image quality. A static test will show the quality of the focus while a dynamic test, camera and magazine operational, will show the quality of focus in conjunction with the film handling characteristics, a measure of operational image quality. Either test may be reported in the camera calibration certificate.

o. Radial Distortion Polynomial. A polynomial of three or four terms may be determined using the measured values of average radial distortion. This is reported in the form:

$$\Delta r = ar + br^3 + cr^5 + dr^7 + \ldots \quad (4.22)$$

Where the r's are average radial image distances in millimetres and Δr is in micrometres.

Note that this equation is anti-symmetrical and gives best results when referenced to the Point of Best Symmetry. If applied with reference to the Principal Point of Autocollimation, the asymmetries will be larger in accordance as the distance between the two points increases.

p. Tangential Distortion Polynomial. A two- or three-term polynomial for tangential distortion may also be derived, but symmetry exists only along each separate diagonal of the format. Along this diagonal the distortion increases with the distance from the center. This is not an aberration peculiar to design, but the result of decentering of the optical elements of which the lens is composed. It is possible for a camera to acquire tangential distortion if a perfect lens is subject to small lateral forces which disturb the colinearity of the elements. Theoretically, there is a diameter of the focal plane along which the tangential distortion is zero, and the diameter perpendicular to it along which the distortion is maximum.

When cameras are calibrated on a multicollimator, the tangential distortion can be measured directly along any diagonal. The

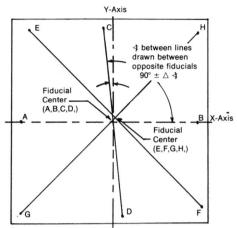

THEORETICAL LINES DRAWN
BETWEEN OPPOSITE FIDUCIALS

Y-Axis

⊰ between lines
drawn between
opposite fiducials
90° ± △ ⊰

Fiducial
Center
(A,B,C,D,)

Fiducial
Center
(E,F,G,H,)

X-Axis

The camera is located in the aircraft such that the
A-B Axis is || to the line of flight.

For fiducials located at the center of the sides this
rectangular coordinate system is recommended. The
fiducial center, crossing of fiducial lines is the origin.
A-B is coincident with the X-Axis.

$$\triangle \; ⊰ \; \leqslant \; \left| \frac{1 \; MIN}{ARC} \right|$$

FIGURE 4-47. Fiducial center or indicated principal
point. A rectangular coordinate system.

maximum radius vector can then be computed
from the data from the two diagonals. Washer
(1957) employed asymmetric radial measurements when he investigated tangential
distortion. Using a thin prism as a model, he
computed the position of a principal point and
the wedge angle of the prism. This was reported
in the Bureau of Standards Camera Calibration
Certificate for many years, during which time
manufacturers of cartographic lenses developed techniques which greatly reduced this
centering error.

Brown also investigated this error, returning
to the early writings of Conrady (1919). The
equations he uses are applicable to all types of
camera calibration and are given in section
4.8.8.2.

4.8.1.2 CALIBRATION OF COMPARATORS

Measurements of images on spectroscopic
plates or film are usually made with a two-
coordinate comparator. A master grid which approximately equals the working area of the two
ways is then employed to calibrate the comparator. The area of the master grid should be
larger than the 23 cm × 23 cm square platens of
the majority of mapping cameras employed. The
standard should also provide means of measuring the orthogonality of the ways. An operator
observes the grid points through the comparator
microscope and selects the measuring value.
The final values are compared with the standard.

Each laboratory has its preferred sequence of
reading the coordinates of the grid, the number
of repeated readings, and the reading pattern.

Data are reduced, usually by iteration using a
least squares program, until a best fit to the grid
is obtained, or until a pattern of errors surfaces.
Corrections can then be made either by adjusting
the comparator or by correcting the measured
numbers.

Temperature control of a comparator is essential in order to obtain reliable measurements.
An example of such control is the constant bathing of the lead screw of a Mann comparator
with cooled oil. The comparator remains accurate during long hours of use because the temperature is kept constant.

There are other types of visual comparators
which do not employ screws but the same principles of calibration and environmental control
apply.

Some automatic comparators use photoelectric settings. The large machine shown in figure
4-48 may be used in either a manual or automatic
mode.

4.8.2 HISTORY

Cameras were first calibrated in the 1930's
using visual techniques. Theodolites made acceptably accurate measurements of narrow
angle lenses while goniometers were used for
wide-angle lenses. When the center of the entrance pupil of the test lens or cone is approximately coincident with the rotation axis of the
goniometer, accurate visual measurements may
be obtained with monochromatic illumination.
The sun, however, is the source of illumination
for aerial surveying, making calibration with
"white" light more meaningful, if not essential.
It is extremely difficult to design and manufacture wide angle mapping lenses covering a 4,000
to 7,000 Angstrom range without some residual

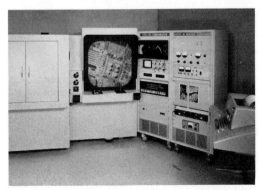

FIGURE 4-48. Mann Type 2405 Automatic Precision
Comparator. This high precision comparator is used
chiefly for measurement of the X and Y coordinate
positions of symmetrical images, as for instance, stars.
Selection of image position is manual or photoelectric
using closed loop feedback serves. The environment is
self-contained, regulating temperature, dust and
humidity within normal levels for accurate measurement.

chromatic aberrations which confuse the observing eye, on which measuring accuracy with visual instruments depends. The visual methods, therefore, were largely replaced by the more realistic photographic techniques during the 1950–1960 time frame. (Carman, Brown 1956). With the present increase of photographic sensitization of films to 9,000 Angstroms (beyond the range of the human eye), calibration on film became essential.

Hotine (1931) in Great Britain; Gardner and Washer, (1937, 1939, 1941, 1944) and Merritt (1948) in the United States; and Field (1949), Howlett (1950) and Carman (1949, 1955, 1956, 1961) in Canada separately investigated equipment and techniques. In the United States, these investigations led to the development of the camera test bench and the precision camera calibrator by Case and Washer at the National Bureau of Standards. The first small calibrator—two perpendicular banks of fixed collimators—became the basic design for modern laboratory test equipment.

At the National Bureau of Standards Gardner and Washer (1937–1956) investigated the basic calibration requirements for mapping cameras and sources of error. The manufacturing error, named tangential distortion, was found by Pennington (1947) to be an uncorrectable error in mapping, and lens manufacturers developed alignment techniques to control its magnitude.

Merritt (1948, 1950, 1951–52) at the U.S. Naval Photographic Interpretation Center developed field and stellar calibration methods which differ from those in use today primarily in the present use of computer programs. Simultaneously, Sewell (1948) at Wright-Patterson Air Force Base developed the widely accepted "point of best symmetry"—about which radial distortion is most symmetric—and a well controlled field calibration method which is detailed in the second edition of the MANUAL OF PHOTOGRAMMETRY.

The first large laboratory calibrator based on the Bureau of Standards table model was established at Fairchild Camera and Instrument (1950) and has been in use for more than three decades. Calibrators of slightly different design were completed during the same period for the Canadian National Research Council and the U.S. Geological Survey. Doyle (Fairchild Camera), Howlett and Carman (Canadian Research Council), Washer (Bureau of Standards), and Bean (Geological Survey), headed this phase of rapidly developing laboratory technology in the United States and Canada.

The publications of Washer, Merritt, Brown, Hallert, Carman, Thompson, Slama, and Schmid (see selected bibliography) cover almost all phases of camera calibration employed today; in some cases these include recent variations of previously used techniques and the extensive use of computer programs. Washer, in particular, has examined every laboratory method of calibration, comparing their accuracies. The steady improvement in lens quality in the United States can be attributed to the professional interest of the Bureau of Standards personnel. Merritt, in turn, has constantly updated the physical methods with mathematical tools. Hallert, in his many evaluations of mapping camera calibration, insisted on a test of the complete camera as necessary for assuring realistic geometric data. Webb used survey tests of each Air Force mapping camera for quality control of camera geometry and selected the plotters which best fitted/corrected the distortion patterns.

Studies of dynamic missile performance emphasized the need for cartographic quality in ballistic cameras. When testing of ballistic cameras became an important factor in missile design, stellar techniques for calibrating cameras were devised by Schmid (1953) at Aberdeen Proving Grounds, and Rosenfield and Brown (1958) at Cape Kennedy. Case detailed the stellar calibration techniques in the third edition of the MANUAL OF PHOTOGRAMMETRY. Cameras at the Coast and Geodetic Survey, now the National Ocean Survey, have been calibrated by stellar techniques for many years (Fritz and Slama, 1976). A comparison of their stellar methods with Bureau of Standards laboratory methods showed close agreement of inner orientation constants in a study conducted by Tayman (then at the Bureau of Standards) and Hull (NOAA).

Techniques for calibrating NASA Surveyor-Lander vidicon cameras were developed by Norton (while at Fairchild Camera), and Ask (Geological Survey, 1963). Further improvements were made by engineers and scientists of the Jet Propulsion Laboratories, particularly in stability of electronics, which improved calibration values. A 25-point reseau etched on the face of the vidicon tube provided the primary geometric control for return beam vidicons. Such techniques have been used in many space cameras where televised imagery must be reassembled in a map format. These were geometrically correct within tolerances which were reasonable for the period of time and design of equipment.

During the last decade, the scientists at J.P.L. added features to improve the reliability and geometry of the observing/recording/transmitting satellite systems. One of the latest papers on calibration by Benesh brings us up to date on the status of the science. The primary objective of the Mariner Mars 1971 mission, he notes (1971), was the observation and mapping of that planet. Based on previous experience with the Surveyor and the Surveyor Lander, the Jet Propulsion Laboratory improved the geometric characteristics of these T.V. cameras and greatly expanded the calibration.

A reseau of 63 and 111 points was measured on the video face plate and the system was calibrated by means of a wide field collimator with a grid of 400 intersections, using conventional techniques. The reseau points were then grouped so that each four or five marks were used as a segment, 48 in all, and a first- or second-order affine transformation was used for a separate adjustment of each segment. Since each point functioned in at least two segments, there was a redundancy such that a least squares adjustment was necessary. The position of the image grid intersections within the reseau segment was, of course, interpolated.

Three calibrations of the two cameras and their orthogonality were made prior to the Mariner Mars 1971 mission. The star calibration used during the mission showed no major changes.

The calibrator built in 1950 for the National Research Council of Canada, similar in basic design and size to that at Fairchild and the Geological Survey, has now been replaced with a single fan of mirror collimators, covering a wider field. A description follows in section 4.8.5.1.2.

A multicollimator calibrator installed at Hill Air Force Base in 1970 has 121 collimators on 12 radial arms. Section 4.8.5.1.1 provides details of equipment and techniques.

In the United States, the Geological Survey Calibration Laboratory added collimators to increase the full angular field to 120° for testing super-wide angle cameras. They have also arranged the collimators so that two exposures through a camera can be overlapped by 60%, forming a neat model, represented by nine well-distributed images. The mathematical solution for adequate geometric accuracy of the camera is contained in the Model Flatness Test (section 4.8.5.4.3) reported by McKenzie of USGS. If the two exposures are projected to form nine stereoscopic images, it is theorized, they should have no significant parallax. The test is done by mathematical intersection rather than by analogue instrument projection. While this is not calibration, it verifies the system geometry, and if exposures are made on film using the operating magazine, it becomes a camera system test.

Brown (1968) uses both stellar calibration techniques and airborne calibration techniques. A description of his methods is contained in section 4.8.8.

Vogenthaller has developed a portable calibrator that employs crossed diffraction gratings as test objects. Images are recorded at the focal plane of the camera, and diffraction theory is used to predict the undistorted positions of the images with which the recorded images are compared.

Close-range photogrammetry has made rapid strides with the establishment of ISP Commission V. Calibration techniques vary from highly precise to simple geometric tests. Calibration of close-range cameras followed the increasing use of non-metric cameras in photographing near objects on which it became desirable to obtain some order of geometric information. The calibration techniques can best be found in the writings of those leading the development of this new science, and section 4.8.5.5, Calibration of Close-Range Cameras (written by Karara, Chairman of the ISP Commission I Working Group), reviews history "in the making."

In an interesting, handwritten letter, October 4, 1966, addressed to Norton, Hallert noted "the new tests and calibration methods being developed for close-up cameras, x-ray instruments, etc . . . calibrations employed three-dimensional tests bodies having a great number of accurately determined points . . . also the method to use a glass grid (negative) of high quality and to image this in three well-defined positions on one and the same photograph. . . ." Hallert foresaw all phases of calibration with the critical eye of the scientific user.

Hallert also took steps toward full camera system testing and verifying laboratory calibration. He believed such proofs were essential, and developed a tall tower at the Royal University of Technology in Stockholm for this purpose.

Similar calibrations on film exposed from aircraft were conducted by various investigators: Salmenpera (1972), Heimes (1972), Anculate and Diacomescu (1976), Merchant (1974). Calibrations made over relatively flat terrain are considered less accurate than those over mixed (altitude) ranges. Also, according to a study being conducted by a Commission I, ISP Working Group (Norton 1978), calibration values are affected by environments, some of which may be extreme and which will also change with the aircraft. Nevertheless, there is a trend toward such calibrations as being more realistic and, in that sense, more accurate. The question of whether adequate testing has been done, to the extent of proving good repeatability of the inner orientation elements so derived, requires more study. Again, if we apply Eisenhart's criteria of calibration, the cost of obtaining such proof may be prohibitive.

Calibration procedures and equipment for underwater cameras were first introduced by McNeil (1965), and later Merritt (1974) suggested the use of equivalent air techniques.

New ideas and modifications of previously employed techniques continue to be advanced, each with the intent of obtaining higher accuracy or more realistic value or both. It cannot be expected at this time that every possible method and its many variations has already been employed. As long as professional men believe that improvements can be made in costs, time, or accuracy, new or modified techniques will continue to be advanced.

A working group under Commission I of the International Society of Photogrammetry, chaired by Ziemann of the National Research Council of Canada, is comparing the result of various methods using two reseau cameras. These methods have been summarized by Jiwalai (1977) and Merchant (1978). The study does show that calibration data seem to correlate with the method employed.

4.8.3 GENERAL GEOMETRIC CHARACTERISTICS OF MAPPING CAMERAS

A mapping camera may be used for many purposes. It may map the terrain beneath an aircraft, or be used for measuring distances between objects, determining shapes, acreage, or volumes. Ground-based, close-range cameras may map body contours for medical studies or any assembly of near objects. Each type of camera has special requirements to which it may be designed and for which it is calibrated.

4.8.3.1 INFINITY-FOCUSED CAMERAS

The aerial mapping camera photographs the ground from a range of altitudes; the scale of the photography (altitude or range divided by camera focal length) is usually greater than 2,000:1. The lens is therefore focused at a rigid focal position as though the targets were at an infinite distance, and the lens is "fixed" (pinned or machined) so that the focus cannot be disturbed or easily changed. This immobilization is necessary in order that the effective and calibrated focal lengths will be constant, the focal position repeatable, and distortions may be quoted with reference to an established center.

4.8.3.1.1 LENS, LENS CONE, CAMERA BODY

The heart of the mapping camera is the high image-quality, low distortion lens. A number of lenses of varying focal lengths and angles have been developed to cover the standard 23 cm square format. The 152 mm lens is therefore a wide-field lens (93° full field) for this format, normally having an aperture of f/4 or (physically) smaller. Modern lenses are corrected for chromatic aberrations over the 4,000Å to 9,000Å range (still maintaining low distortion and good image quality), allowing sensitized films of different spectral ranges and characteristics to be used in the same camera. [The author notes that for wide (93°) and super-wide (125°) angle fields, for larger apertures, and with image quality constantly improving, such lenses are a near miracle made possible by the use of large computers under the direction of experienced cartographic lens designers, along with the use of new critical testing equipment and sensitive alignment techniques.]

American military mapping cameras have separate lens-cones which are individually machined or which have fixed spacers for establishing an unchangeable focus. The cone has rails that define the focal surface, except for the KC-6A which can compensate for forward image motion and requires the platen to rest on roller bearings. The lens-cone is thus a complete unit determining all of the static geometric characteristics other than those of the magazine. It may, therefore, be removed from the camera body, allowing the camera to be repaired, without disturbing the cone geometry. Such a design has resulted in high stability as proved by repeated calibrations.

Other mapping cameras provide focusing and adjustment of the lens system in a camera which has no separate cone. In this case, the fiducials are preset, squared and centered, and the lens which was prefocused is then adjusted laterally by means of visual or photographic criteria until the selected lens center is within the tolerance range for the fiducial center or indicated principal point.

Fiducials are variously located depending on the camera design; they may be at the centers of the sides, at the corners, or at both positions. There are cameras that have supplementary fiducials or marks at other positions around the periphery of the focal plane, all with the purpose of providing a coordinate system for the imagery and correcting for dimensional changes in the film. Determining the relative positions of these marks is part of the calibration.

4.8.3.1.2 MAGAZINE, PLATEN, AND RESEAU

The magazine which holds the film and advances it to the focal plane must correctly position the film at the focal plane, with the surface flat during exposure. Deviations from this requirement introduce distortions. Some cameras are designed with focal surfaces which depart from a plane and these require special measurement techniques. When the surface is highly non-planar, as for example when it is part of a sphere, special instruments may be used for measuring the photograph. If the surface deviates but little from a plane, the platen may be shaped to fit the focal surface and then measured to show its deviation form planarity, these deviations being used to calculate corrections to measurements made as if the focal plane had been planar. Even where the platen is intended to be flat, it may deviate significantly from this state. Its deviations should then be measured. When measurements on photographs are to be accurate better than 5 micrometres, it is good policy to measure the platen for deviations from flatness.

A reseau imprints an array of points on the film for closer dimensional control than would be obtainable with the widely separated fiducials, which are usually outside the format. One type of reseau has an array of etched marks on a glass plate lying at or close to the focal plane, whose shadows record on the film when the shutters are open. There are other reseau patterns, called back-projection reseaus, originating from small projection systems imbedded in the platen, which project their light through the base of the film to record black dots or crosses on the emulsion. Whichever the design, the relative position of each mark of the reseau must be accurately measured and reported in the calibration certificate.

The method of measuring the reseau can be critical. If the reseau surface is also the surface against which the film is in contact with during exposure, it is only necessary to measure the intersections directly. If there is any space between the film surface and the reseau, however, then corrections must be applied for the distance between the focal plane and the reseau. It is also important to correct for the direction of the illuminating beam for each intersection if the distance from the reseau surface to the focal surface is not exactly the same within micrometres. As an alternative, plates of spectroscopic quality may be exposed at the focal position, taking necessary precautions to assure a truly plane surface (Carman, 1955).

Cameras having glass reseaus have the disadvantage of requiring cleaning more frequently than other types, and dust or dirt can introduce random errors. Extra care is therefore needed in handling such cameras in order to retain the reseau's advantage.

4.8.4 GEOMETRY AND MEASUREMENT AS A FUNCTION OF IMAGE QUALITY

Measurement on photographic images is not always a simple, well-defined process. What may be a clean, sharply defined object is not always recorded as a clean, sharply defined image, even on the generally used VF emulsions of spectroscopic plates which are employed in laboratory calibration. When photographs are taken during flight surveys on faster film, the images may be further degraded. The aperture and aberrations of the lens limits the optical quality; the choice of emulsion limits the resolution of the sensor. Vibration or angular motions degrade image quality.

Wight (1968) points out that there are second-order measuring errors due to lateral color aberrations that may be significant, particularly when color films are employed. In addition to the simple lateral aberrations, he notes that the chromatic variations of astigmatism and field curvature contribute further to the displacement of the "center of energy" of off-axis image points. Some residual coma is not unusual in wide-field lenses, and the comatic flare varies as a function of aperture; this will also cause an apparent shift in the centroid. Even with a fixed aperture, coma can still affect measurements, since, for low contrast objects, the energy in the tail may be below the threshold value of the film

(and therefore "lost"), but above the threshold for high contrast objects.

Wight advocates the use of correcting "color" and "magnitude" equations as has been done by astronomers to compensate for the fact that brighter stars will exhibit comatic tails which shift the image centroid. In wide-angle survey cameras, the problem of correcting equations would be more extensive, since coma would vary with the field angle.

The variations in measurement noted by Wight are in no sense gross in modern cartographic lenses, but the few microns can bias the calibration data.

Rosenbruch (1977) points out that there is a correlation between modulation transfer functions (MTF) and field curvature, and that distortion increases with field curvature; thus, tests of field curvature and its symmetry about the optic axis can be used as an order of control for distortion. He also notes that "one can see an additional degradation of the MTF with growing field angles which is caused by increasing amounts of coma," . . . further evidence of measurement problems.

Hempenius (1969) investigated photogrammetric pointing accuracy in a study of mapping techniques. Norton notes that camera calibration, where the centroid of energy is used, needs correlation with respect to uses of edges for pointing when mapping data are extracted and corrected by calibration values.

4.8.5 METHODS OF CAMERA CALIBRATION/EQUIPMENT

Many methods of calibrating mapping cameras have been developed. Highly accurate, expensive, but convenient calibrators exist in national laboratories where the workload is high; manufacturers who design and produce mapping lenses and cameras require special calibrators to support production and quality inspection; stars, because of their high density and known geometry, serve as accurate free targets when the sky is clear; goniometers are used visually and photographically. Calibration has been performed from ground stations, from towers, or from aircraft using a single or mixed range; many of the programs require a large quantity of data, images well-distributed throughout the field (format), and special computer programs.

No matter what the method, the accuracy of calibration depends on the known geometry of the targets which the lens or camera views. This must be known with the highest measurable accuracy; a second of arc is a desirable setting of targets, or a micrometre for linear dimensions of master grids. The camera support during calibration should be rigid and shielded from vibrations; thermal variables should be negligible.

The images must be recorded accurately on rigid, emulsion-coated, plane glass plates, or on

film whose dimensions are corrected by calibrated fiducials or reseau marks, great care
being taken to control film dimensions. If visual
measurement includes chromatic errors, these
errors should be corrected. Measurements on
images must be accurate. And finally, the data
need evaluation by experienced physicists or
mathematicians intimately familiar with calibration equipment and sources of error.

All this is necessary to assure that the unit of
calibration measurement, the micrometre, is
significant. Realizing that the finest spider's
thread, seen only like dust in the sunlight, is
many micrometres thick, one can appreciate the
final accuracy quoted for calibrating a camera.
Only a few micrometres, yet this is the *sum* of
errors of a highly controlled sequence of
factors—equipment, technician, environment,
measurement, computer program, and evaluation. Camera calibration accuracy has always
been level with (if not actually pushing) the state
of the science; as scientists have advanced new
ideas and proposed better controls, as equipment, techniques, and computing programs have
improved, so has the accuracy of calibration.

4.8.5.1 MULTICOLLIMATOR CALIBRATION AND EQUIPMENT

The many methods used differ primarily with
the selection of targets and methods of evaluating data. Large laboratories generally employ
simulated infinity targets (collimators). A col-

limator (*see* figure 4-49) is essentially a highly
corrected lens, free of chromatic error through a
specified spectral range, or a mirror with no
chromatic error, plus a target at the infinity
focus of the lens or mirror. The target is illuminated with white light of "daylight" spectrum,
the energy, from blue at 3800Å to the near infrared at 9000Å, being essentially constant, and
matching the range of the sensitivity of modern
aerial films. The optical objectives of the collimators, the lens or mirror, should be larger in
diameter than the aperture of any lens being
calibrated. The target, in turn, should be small
enough to occupy a narrow angle of the field of
the objective such that the image quality of the
collimator will be near the diffraction limit over
the chromatic range employed.

A number of collimators in a selected rigid
array, pointing to a single center, is termed a
multicollimator calibrator. Such systems
minimize testing time and are therefore
employed where workloads are heaviest.

Other laboratory calibrators have single, or a
few, collimators and may rely on mechanical
settings for angular pointing.

4.8.5.1.1 HILL AIR FORCE BASE CAMERA CALIBRATION FACILITY

The multicollimator facility at Hill Air Force
Base, Utah, was developed to calibrate military
mapping cameras varying from 38 mm in focal
length to 305 mm, on a production basis. The
focal length of the collimators is 610 mm, and the
effective entrance pupils are 60 mm diameter as
set by the diameter of the collimator objectives
(refractors).

The main casting is a hemispherical shell of
cast iron. One hundred and twenty-one collimator tubes, also of cast iron, protrude along 12
radii, providing a full target field of 120° for
super-wide angle cameras. The array is such
that the targets, when photographed through the
mapping lens, produce a photographic field of
images well distributed over the photograph.
The collimator reticules, or targets, are in the
form of crosses at the infinity focus of the
objective lens. They are illuminated with
tungsten lamps at fixed voltage, filtered by a
78Å blue filter to provide a "daylight" spectrum
which covers the range from 3800 Å to 9000 Å
at a kelvin of 5600°. The collimators in each of
the 12 banks have similar angles, except that the
90° azimuth banks have larger angles to accommodate the diagonals of the format of the super-
wide angle cameras. The final precise adjustment of the collimator angles is made by setting
four perpendicular screws bearing on the reticle
adapter, a three-micrometre movement equaling
one second.

The main hemispherical casting is 60 inches in
diameter and rests on three cement posts embedded in three metres of sand. Air mounts be-

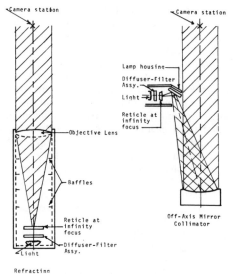

TWO TYPES OF COLLIMATORS
USED IN CAMERA CALIBRATORS

FIGURE 4-49. Two types of collimators used in calibrators. The mirror type has no chromatic aberrations
but is more sensitive with respect to stability.

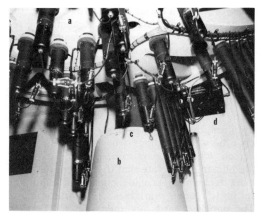

FIGURE 4-50. Hill Air Force Base Calibrator, lower side. Shows main hemispherical casting (a), one support post (b), hydraulic leveler (c), (d) are collimator tubes.

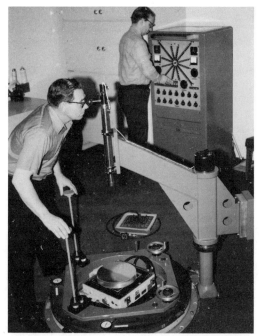

FIGURE 4-51. Hill Air Force Base Calibrator, top level. Autocollimation telescope is mounted directly above the center collimator. The technician is leveling a KC-1B camera, autocollimating from the optical flat which is resting on the focal plane rails.

tween the posts and the main casting keep the casting level and provide further vibration isolation. Isolation above two cycles/second is obtained.

Figure 4-50 shows the lower side of the calibrator as seen through the pit window. The large number of collimators with wires leading to the lamps resembles an octopus. The pit is enclosed to provide a directed movement of temperature controlled air, which enters at the base of the pit, and flows upward and out of the operational area of the calibration room. Temperature and humidity are constantly recorded, temperature control being maintained within 1°F.

Figure 4-51 shows the autocollimating telescope which is used to position the camera or cone being tested in such a way that the focal plane is perpendicular to the center collimator. Under that condition, the recorded image of the center collimator is the principal point of autocollimation.

The console that controls the calibrator lamps is efficient and simple. Each lamp is wired in series with its indicator, which is also the on-off switch. Since the display follows the same geometric pattern as the calibrator, the identification of each angle is immediate. After the lamps are selected, the duration of time for illumination is set; a main switch then operates all lamps (figure 4-51).

Cameras and cones are brought into the calibration area at least 24 hours prior to calibration, having previously been subject to resolution tests that show the infinity setting to meet specifications. The calibration tests are conducted with the cone mounted in the camera body. The first step is to set the autocollimating telescope to be colinear with the center collimator, with an accuracy of $1/2$ to 1 second. The camera is then

mounted on the rotating and leveling ring and an optical flat, whose sides are parallel to one second, is placed on the cone rails when these rails define the focal plane. The leveling ring is then adjusted until the telescope beam and its reflection coincide, thereby making the focal plane perpendicular to the center collimator, a condition required for recording the principal point of autocollimation.

The optical flat is then removed. The lamps illuminating the required collimators are next selected and their timing set. With lights off the unexposed spectroscopic plate is placed on the rails emulsion side down. For the daylight spectrum the VF sensitized emulsion is used. Since the departure of the emulsion surface from a plane will introduce false distortion, plates of micro flat quality are used. In addition, if plate sag is suspected, a specially lapped vacuum plate, flat within two microns over the image area, may be used, with applied vacuum drawing the plate into contact with the platen.

The switch which turns on all the lights for a preset time is then flipped. After exposure, the plate is removed to a light-tight box. With subdued illumination lighting the area, the ring holding the camera is rotated to its 180° position and the stop is inserted. The optical flat is then replaced on the rails, autocollimation rechecked

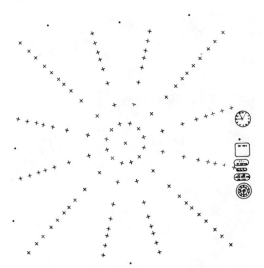

FIGURE 4-52. Exposure of all targets of calibrator with KC-1B camera (schematic). There are 109 positions within a 93° full field.

and corrected if necessary, and a second exposure is made on the same plate, which is slightly offset from the first. Fiducials are exposed at both settings to provide the coordinate data. The plates are then carefully processed, washed, and air dried, and images are examined to assure that exposures are good. Figure 4-52 shows a schematic of one exposure from a KC-1B camera, using all targets.

After a drying period of 24 hours in a glass air-vented case, the plate is aligned on the comparator for reading image positions. The technique used is to make the A and B flight line fiducials of the 0° exposure colinear with the x-axis of the comparator, reading out the posi-

tions of all fiducials in x and y as well as the image of the center collimator. Readings are repeated. The 180° plate is similarly read. Data are compared and averaged.

Each diagonal line of targets is then read, aligning the two 35° targets to the x-axis and reading the images in x and y. The plate is rotated 180° and the same images reread. As noted many years ago by astronomers who read spectral lines, there is sometimes a noticeable difference between readings when the image is reversed, which appears to be due to the bias of the human eye. (Note that these separate alignment techniques provide control of calibrator settings, reading variables, comparator bias.) The punched cards are then sent to the data center.

a. The computer is programmed to reduce the data for inner orientation constants as defined in conjunction with measuring techniques, simultaneously monitoring values which may show error. The inner orientation elements are compiled in a calibration certificate and the error signals provide means of alerting technicians to possible equipment or reader problems.

b. The comparator is a two-coordinate system with the x and y ways accurately perpendicular to two seconds. Since use of the comparator for long periods of time (8 hours without stop) will cause

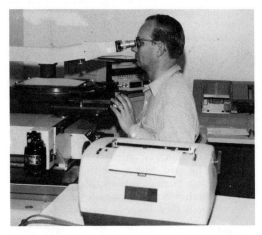

FIGURE 4-53. Comparator. Operator selects reading position and depresses foot pedal. The action causes an IBM card to be punched and a typewriter to print out position which is shown on the data logger display.

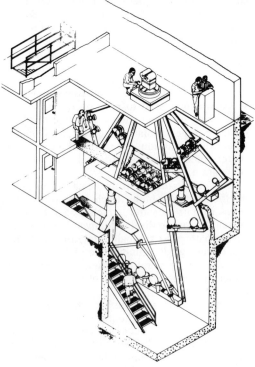

FIGURE 4-54. Schematic of photographic image quality test equipment. All cameras are tested for resolution on this high quality equipment in the Maintenance Directorate of Hill Air Force Base, Utah.

expansion, which gives systematic error, the ways are constantly cooled by oil. Tests using master grids show this stabilization to be maintained.

Figure 4-53 shows the comparator in use. The operator visually selects the center of the image to be measured and depresses a foot pedal. This action causes Hollerith cards to be punched out, with identification, and simultaneously prepared forms are filled in with measured values. The display shows the operator the instantaneous position of the carriage on the x and y axes.

The angles of the calibrator are measured at least once every six months or whenever change is suspected. By setting the collimators symmetrically about the center to an accuracy of one second in elevation, with all collimators in one bank in a plane (*i.e.*, an apparent straight line), and by following symmetric measuring techniques, as described above, it is possible to monitor equipment and operator's readings using the computer program.

Resolution of the operating camera is tested on the Photographic Image Quality Test Equip-

FIGURE 4-55. Photographic image quality test equipment. Technician is cleaning the mirror off-axis parabolas of the 14-foot collimators. The capping covers, operable from the camera station, protect the mirrors when not in use. Both constant and flash xenon illuminations are available.

ment (figure 4-54 and figure 4-55) using the specified film.

4.8.5.1.2 THE CANADIAN NATIONAL RESEARCH COUNCIL (NRC) CALIBRATOR

The NRC calibrator is a photographic instrument with one fan of 43 collimators covering a field of 118.125°. Calibration along various camera azimuths is provided by measured rotation of the camera. A photographic system is used because it provides a direct laboratory simulation of conditions of use and hence gives results which are correct in practical terms. Differences between visual and photographic calibrations were identified and explained by Carman and Brown (1956). The differences can be minimized or allowed for in various ways, but only a photographic calibration with a "daylight" light source, the intended camera filter, and an emulsion with the intended spectral sensitivity provides a straightforward means of obtaining a correct calibration.

The fan is in a vertical plane with its central collimator vertical so that the camera axis is vertical.

The calibrator consists of three main parts: a main casting holding the 43 collimators, a lower bridge holding the camera under test, and an upper bridge supporting three autocollimators used in positioning the camera.

The collimators are reflecting systems with off-axis parabolic mirrors, positioned around the outer arc of the main casting. The reticles are positioned around the inner arc. Mirror collimators have the advantage of no chromatic aberration, so that calibration can be made with equal accuracy at any wavelength, including the infrared. They also have no zonal spherical aberration, so that the position of the entrance pupil of the camera lens is not critical provided all the beams of light from the collimators completely cover the entrance pupil. The mirrors have clear apertures of 63 mm, which is larger than the entrance pupil of the majority of photogrammetric camera lenses.

In each collimator, a reticle provides a clear white cross on a neutral gray background that transmits between 10 and 11.5 percent of the light. The purpose of this gray, rather than black, background is to permit both the cross and its background to be exposed on the straight line portion of the emulsion's characteristic curve, avoiding the possibility of variation of apparent image position with exposure, as can occur with a high contrast target.

The entire calibrator is supported on hydropneumatic anti-vibration mounts. The suspension has a natural frequency of about 0.5 hertz, vertically and horizontally, whereas the supported parts all have natural frequencies of over 100 hertz. Thus the instrument is well isolated from ground vibrations, ensuring that no relative motions occur during a photographic exposure and protecting the calibrator from shocks which might affect its stability.

The angular spacing between each collimator is due to the accurate method selected for positioning the collimators. A ninety-degree cube is used to position opposite 45° collimators. Each angle is then successively bisected until each step is 2°.8125 (90°/32). Positions for the outer collimators are then obtained by projection of the inner angles.

The latest improvement in the system has been to provide azimuth position increments of 90°, and the observer at the microscope has been replaced by closed-circuit TV, reducing errors due to body heat.

A plate flattener (Carman, 1956) bends the emulsion surface to a plane within a few fringes, providing the final control to the very accurate NRC calibration facility.

4.8.5.2 GONIOMETER CALIBRATION AND EQUIPMENT

Calibration at camera manufacturers and some national laboratories is performed with a goniometer. Both Wild Heerbrugg and Zeiss have large, accurate instruments for calibrating their cameras. Hakkarainen reports (1972) that the goniometer developed for Finland is mounted on an optical dividing head in such a way that the projected axis of the dividing head goes through the camera lens. The plane parallel grid plate, whose intersections are known to a high degree of accuracy, is clamped at the predetermined focal plane of the camera. The system is autocollimated using the plane parallel plate as reflector, after which the dividing head is used to read out the angles which the grid subtends. Data may be reduced in the usual manner, since the object angle and image distances of the grid are known. See Hakkarainen (1972).

The Wild AKG Autocollimation Goniometer (visual only) shown in figure 4-56, employs similar techniques. The measuring ring is built into the goniometer base, however, and is not a separate part as in the Finland instrument. With the lens cone, carriage, and mounting ring removed, this becomes a portable piece of equipment.

Bormann notes (1970) that a large amount of control work is done with goniometers at the Wild plant because of their simplicity of use and high precision. In most cases the calibration measurements are made against a "model" during the manufacturing stages while their multicollimator array is available for final testing.

An example of a vertical goniometer and its use in testing and calibrating lenses and cameras is described by Zeiss. The instrument consists of a long-focal-length telescope which observes a scale or grid in the focal plane of the lens or

FIGURE 4-56. Wild AKG Autocollimating Goniometer. This instrument is used in many visual calibration tasks and in quality control during manufacture of cameras. Note the mounting of the fixed table on the main base of the AKG supporting the lead of the lens cone in its mounting ring and carriage. The goniometer base and autocollimator are mounted on a second base which contains the adjustment and alignment of the measuring system.

camera from beneath the camera. Three mirrors bend the observing beam, two of which are rigid with the vertical arm and one of which rotates in azimuth with the ring holding the camera. Elevation angles are obtained by rotation about an axis parallel to the base. An observing telescope for autocollimation from the grid plate is fixed with respect to the sighting telescope.

Zeiss notes the requirement for measuring the symmetry of mapping lenses along four semidiagonals and their tangential distortion. Tolerances are set to assure lens centering is adequate.

Camera calibration provides data for calibrated focal length, fiducial center, principal point of autocollimation, radial distortion, and resolution.

4.8.5.3 STELLAR METHODS OF CAMERA CALIBRATION

Tycho Brahe (1546–1601), the eminent Danish astronomer, was the first physicist to develop equipment for measuring the precise positions of stars. His methodical records set a pattern for those who, over the centuries, added to this fund of information. From the earliest civilized times (perhaps before then), men navigated on land and sea using their empirical knowledge of star positions. It was Brahe's records, made over many years, and those of the devoted astronomers who continued his work, that provided a means of navigation on the unending oceans of Columbus' time and into the unending space of our own age.

Stars also provide free geometric targets for

calibrating ballistic and mapping cameras. With this method, it is necessary to expose film or plates at a camera station whose approximate longitude and latitude are known. The time of exposure within a second is required if the camera is rigidly mounted. The camera is normally pointed toward the zenith and exposures are over an extended period of time so that the motion of the earth, rotating on its axis, produces star trails on the film. Identification of each measured star is necessary and corrections to update stellar positions plus corrections for atmospheric refraction are added to the measurements. Case (section 4.4.3.3, third edition of the MANUAL OF PHOTOGRAMMETRY) explains physical conditions and derives the mathematics for stellar calibration of cameras. These methods are practical when data are reduced on a large computer. Schmid (1953) first calibrated ballistic cameras at the Aberdeen Proving Grounds using a star field when conducting studies of missile characteristics. Later, at the National Ocean Survey, his calibration techniques were refined and expanded and employed as an integral part of the Worldwide Satellite Triangulation Program (Slama, 1972). Each exposure of the satellite trail against a star background served as "calibration" of the internal geometry of the camera. These same procedures were adapted to calibration of the aerial cameras used at the National Ocean Survey (Fritz and Schmid, 1974). Lately, Fritz and Slama have developed field and data reduction techniques that use over 2000 star images each in four orientations of the camera. They believe that measurement errors, atmospheric refraction anomalies, thermal variations, and random departures from plate flatness are then minimized by the data reduction techniques. When reseau cameras are used, the coordinates are calibrated from redundant measurements of multiple photographic flash plates. A Wild RC-8 and a Zeiss RMK 15/23 were calibrated with these methods as part of and ISP Commission I study (Fritz and Slama, 1975).

4.8.5.4 SYSTEM CALIBRATIONS DEFINED

A mapping camera mounted in a vehicle, airborne or in space, becomes a "system." The camera makes exposures from the vehicle, with or without windows, with or without heat, at various altitudes and under various pressures, subject to the vibration of the vehicle. The vehicle may be a balloon, helicopter, airplane, or spacecraft. The real environment, in fact, is usually much different from that of the controlled laboratory. An analysis by Meier (1978) shows that temperature and pressure can change the geometry of a camera lens. They can also change dimensions of the body of a camera, further affecting focal length. Vibration can result in loss of image quality, and rotational motion during exposure can do the same (Carman, 1973). Such circumstances can cause small errors in the val-

ues of the elements of inner orientation as reported by laboratories. As Salmenpera so sufficiently phrased it (1972) " . . . since it is not known with certainty to what extent the results achieved with laboratory measurements hold good in working conditions and what is the accuracy of the image points . . . " Obviously more than a camera is being calibrated when calibration is applied to the "system" as a whole. These opinions are held by many who consider "system" calibration the most realistic of methods. It seems reasonable, however, that the laboratory method should always be used as a basic control, followed perhaps, by a test that closely simulates in environment that actual proposed conditions of use. In "system" calibration the equipment is always subject to varying environments. The ideal situation, of course, would be to calibrate the "system" on the same flight on which the mapping photography was being acquired. Unfortunately, for a majority of users, this would be too time-consuming and too expensive, except perhaps, for research studies.

4.8.5.4.1 SWEDEN'S TALL TOWER AND ICE FIELD METHODS

The Tall Tower method for calibrating aerial mapping cameras was initially developed by Hallert (1957), and has been in use at the Royal University of Technology in Sweden for some years. The method employs concentric grids on the ground beneath the camera station which is at the top of a tall tower. The camera focal plane is adjusted to be parallel to the ground.

The Geographical Survey Office of Sweden keeps a continuing record of the quality of their cameras (Welender and Smedberg, 1972). Yearly measurements are made of the platens, and calibration tests are made using lakes as targets just before the ice melts in the spring. The authors note that the ice surfaces are acceptably flat and rich in detail. The photography is analyzed as a stereo model. A test field consisting of 35 signalled points with elevations determined by leveling is also used.

4.8.5.4.2 AERIAL CAMERA CALIBRATION TECHNIQUES

Salmenpera (1972) gives details of a specially prepared field for calibrating aerial cameras at scales of 1:4000 and 1:9000. The field was prepared by the Institute of Photogrammetry, Helsinki University of Technology, Finland. Called the Seglinge Test Field, the first-order network covers a 2×2 km^2 area. The survey values and accuracies are given in some detail and are used to derive the elements of exterior orientation, flight height, a ground grid, and elevation points. She notes that " . . . the test field had to be three-dimensional if the coordinates of the principal point and the value of the camera constant were to be determined . . . " This requirement for system calibration appears in later papers.

One of the most recent procedures for calibration of the "system," termed the Method of Mixed Ranges (MMR) was invented by Merchant (1974). The method utilizes a three-dimensional control range (mountains), and a conventional flat range, in an adjustment in which the elements of interior orientation are carried as common parameters for all exposures. The use of the three-dimensional range for suppressing high correlations between certain elements of interior and exterior orientation is thought to be unique for the normal aerial case. The author notes that the advantages of the MMR procedure is that no modification to the total "system" is required, other than the choice of the conditional function which includes the parameters of interior orientation.

Merchant's concern is with the calibration of the total "system." With the introduction of reseau cameras, he notes a significant improvement has resulted in film-dimension corrections and the calibration procedures may now be expanded into closer conformance to the concept of "system" calibration.

Anculete and Disconescu (1976), Romania, report a nearly similar method, calling the calibration a "methodology used under normal work conditions." The control points on flat and high ground are again used. Good results in testing and calibration are claimed.

4.8.5.4.3 U.S.G.S. MODEL-FLATNESS TEST

Before the U.S. Geological Survey awards a contract for aerial photography to a successful bidder, the camera to be used must pass a model-flatness test to insure that stereomodel distortion does not exceed 1/5000 the flight height. In the testing, the U.S.G.S. camera calibrator is used to take photographs on film, simulating aerial photographs. Fifteen of the collimators are set so that a stereomodel can be formed by using one photograph as the left plate of a stereopair and the second photograph as the right conjugate plate (figure 4-57). The stereo model so formed contains nine points (images of collimator targets) that would all lie in a plane if errors were not present (figure 4-58). Because errors are present, the nine points appear to lie in different planes. From measured plate coordinates of the images, U.S. Geological Survey first calculates the translations and rotations necessary to make the coordinate system of the right-hand photograph parallel to that of the left-hand coordinate system with its origin the length of the air base from the origin of the left-hand photograph's coordinate system. A plane is then fitted by least-squares to the nine points in the stereoscopic model, and the difference between the measured z-coordinate and the corresponding z-coordinate on the plane calculated. If any of the nine differences exceed 1/5000 of the focal length of the camera, that camera is rejected.

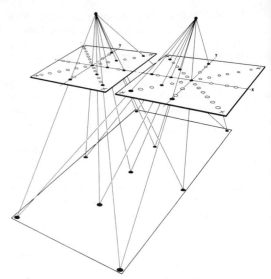

FIGURE 4-57. A stereopair of camera-calibrator plates containing nine conjugate points.

4.8.5.5 CALIBRATION OF CLOSE-RANGE CAMERAS

In contrast to aerial cameras, for which laboratory and field calibration procedures have remained the same over the years and for which standard procedures have been recommended, no standard calibration procedures exist for close-range cameras. Another major difference between these types of cameras is that both metric and nonmetric cameras are used in close-

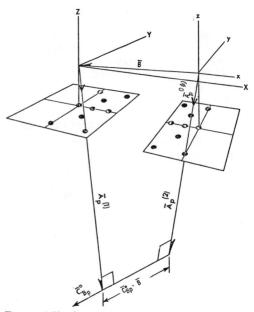

FIGURE 4-58. Geometric relation between the air base of a stereopair, two rays through corresponding images, and a vector perpendicular to the two rays.

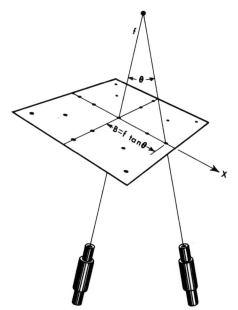

FIGURE 4-59. In order to form a stereomodel at photograph scale, the magnitude of the air base B is computed as the distance at photograph scale from the center target of the left photograph to the outermost target along the flight line. This outermost target corresponds to the center target on the conjugate photograph.

range photogrammetry. In most cases, the optical laboratory methods involving the use of goniometers, collimators, multicollimators, *etc*, are not suitable for close-range camera calibration since such cameras are usually focused (or focusable) at finite distances. As a result, a number of ingenious approaches for calibration of close-range cameras have emerged. Depending on the accuracy requirements in photogrammetric measurements, various degrees of sophistication can be used to define the interior orientation of close-range cameras, ranging from the simple classical approach involving only the reconstruction of the position of the interior perspective center to the highly refined analytical solutions wherein the radial and decentering lens distortions, film deformations, affinity as well as the variation of the lens distortion with

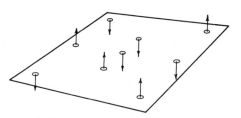

FIGURE 4-60. A regression plane fitted to nine points in a stereomodel. The circles represent the intersection of the error vectors and the regression plane.

object distance within the photographic field are taken into consideration.

Calibration is usually carried out in one of three fashions: in the laboratory, on the job, and by self calibration. Metric cameras are ideally suited for laboratory calibration. Three-dimensional test objects of various kinds have been used, *e.g.*, Abdel-Aziz & Karara (1974), Döhler (1971), Faig (1971), Malhotra & Karara (1975), Torlegård (1967), Wolf & Loomer (1975). The mathematical formulation is generally based on the collinearity equations. Five object-space control points are required to solve for the principal point and principal distance. With the inclusion of additional unknowns, the number of object-space control points has to be increased, *e.g.*, Abdel-Aziz & Karara (1974), Faig (1971), Karara & Faig (1972), Torlegård (1967). The basic resection approach fails for telephoto lenses with cone angles of 2° or less. Merritt (1975) has overcome this problem by applying the Hartman method.

On-the-job calibration utilizes photography taken of the object and of the object-space control simultaneously. At least one full (X, Y, Z) object-space control point should be provided for every two unknown quantities included in the solution. The mathematical formulation is essentially the same as used in laboratory calibration. Several approaches have been reported in the literature on the construction of special control frames which maintain their geometric configuration sufficiently well to be used for control purposes, *e.g.*, Böttinger (1972), Brandow *et al* (1975), Faig (1974), and others.

Although not explicitly a calibration method, the *Direct Linear Transformation* (DLT) method developed at the University of Illinois (*see e.g.* Abdel-Aziz & Karara 1971, 1974) is generally regarded as an on-the-job calibration method. In this method, the solution is principally for the interior orientation of the individual images but the elements themselves are not explicitly obtained from the solution of the equations. The DLT method is particularly suitable for non-metric photography which has no fiducials, since the solution is based on the concept of direct transformation from comparator coordinates into object-space coordinates, thus bypassing the traditional intermediate step of transforming from a comparator system to a photo-coordinate system. The minimum number of unknowns in the solution is eleven (11), but these can be increased to take into account additional unknowns. The recommended mathematical model (Karara & Abdel-Aziz (1974)) involves 12 unknowns, thus requiring a minimum of 6 object-space control point images in the photograph.

The self-calibration approach does not require object-space control as such for the calibration. Multiple (at least 3) convergent photographs are taken of the object. Using the complanarity condition equations and well-defined object points, elements of interior orientation are computed for the camera used, assuming that the elements remain unchanged between photographs. This method is well documented in the literature, *e.g.* Brown (1971, 1972), Kolbl (1972). A self-calibration method which permits the determination of the elements of interior orientation for each photograph has been developed at the University of New Brunswick (Faig, 1976). This method is applicable to metric as well as to non-metric photography.

4.8.5.6 CALIBRATION OF UNDERWATER CAMERAS

Calibration of underwater cameras is done using a multicollimator instrument designed by McNeil (1972), and similar in principle to that originally developed by Washer and Case of the National Bureau of Standards (1949). The distance at which an underwater camera can photograph an object is limited by the turbidity of the water. The focus of each collimator must therefore be easily adjustable to simulate an object distance from six feet to infinity. Like a theodolite, the adjustment must be accomplished without a change in pointing angle. The reticles of the collimators have both fiducial crosses marking the angle and resolution reticles. As the object distances are adjusted in the collimators, the resolution values will change as a function of ratio of collimator to camera focal length.

Both the calibrator and the cameras, with well sealed windows, are under water, preferably water of the same type and clarity as that in which the cameras will be used (*e.g.*, sea water of various clarity). Detailed mathematics and calibration techniques are available in a small book, "Optical Fundamentals of Underwater Photography," McNeil (1972).

Merritt (1974) developed a theoretical basis for calibration of underwater cameras by applying methods for calibration of aerial cameras. The key to the observational procedure is to establish coplanarity between the water surface and the camera focal plane. The refraction angles are then corrected for the air-water interface. Only the camera lens is immersed in a shallow pool of water. Collimators in air, below the pool, are pointed so that their rays, refracted first by the glass window of the pool and then by the water, are incident on the lens in the camera. An auto-collimator, again in air, is mounted vertically above the camera. A simple pool with a plane-parallel plate bottom could easily fit on an infinity-focused calibrator if a 35 or 50 mm underwater camera were to be used.

4.8.5.7 CALIBRATION OF PANORAMIC CAMERAS

Panoramic cameras have "built-in" distortions due to the camera design. If the design is such that during exposure the film is held sta-

tionary against a platen whose radius equals the focal length of the lens, carefully positioned fiducials can be located along the edges of the format to reference the geometry of the format. The lens, scanning the targets of a multicollimator whose angles are known, provides the geometry of the images. Such techniques can be expanded to include the correction for image motion. A second design which requires that the motion of film past a fixed slit move synchronously with the scanning rotation of a prism can also be calibrated. In this case, timing marks are employed. Film transport which might vary from ten to one hundred inches a second might need timing frequencies higher than one thousand cycles per second. The dynamics of panoramic cameras suggests the necessity of calibration with several frames being analyzed at various speeds.

4.8.7 FUTURE CALIBRATION CONSIDERATIONS

The various methods of calibration have been discussed in previous paragraphs, each method having specific advantages. The multicollimator method has high production capability and every phase can be well controlled. The laboratory goniometer techniques are also precise and well controlled when corrections are made for visual chromatic aberrations. Stellar methods have high accuracy by virtue of the dense field of stars of known positions. Camera "system" tests which include the uncontrolled (or partially controlled) environment of the airborne vehicle may appear most realistic but the use of the calibration values so derived needs proof in a reasonable number of applications.

The question as to which method best serves the interest of the map-makers is not yet answered. It begins to be apparent, however, that two additional studies are still needed to evaluate the complete problem. One study should correlate geometry from images on glass plates with geometry from images on film in the operational camera. To explain: there has been general acceptance that the operating camera system is a highly stable instrument geometrically, and unless one is concerned with geometric accuracy to a few micrometres, this assumption is tenable. When, however, isolated tests have shown that good agreement in values of separate calibrations on plates and film is not always attained, then it would seem that the next step in calibration should be a requirement for this test. The operating camera calibration should not be a substitution for calibration of the camera body since this is basic, the role of the national laboratories in quality control of mapping cameras is fundamental, but it should be an additional calibration that requires agreement of geometric values. The stability of the operating camera geometry is then more nearly assured. Such was the path that lead Hallert to his Tall

Tower tests and the aerial survey calibrations which, unfortunately, include unknown variables. The other study is concerned with the effects of the environment on image quality and geometry, a study which is now being conducted by Working Group 3 of the International Society of Photogrammetry. Carman (National Research Council, Canada) is doing further research on vibrations; Meier (Zeiss, West Germany) is continuing the analysis of pressure and temperature on the geometry of selected Zeiss mapping lenses; Worton (Vinton, Gr. Britain) is assisting with aerial surveys; and Norton and Peck (U.S. Air Force) are continuing to compile data on environmental conditions of aerial surveys and will simulate selected environments in laboratory testing. The results of these investigations should provide information on the magnitude of possible geometric changes caused by environmental conditions which differ from those of controlled laboratories.

When the differences in geometry between laboratory calibrations on glass plates and those on film in the operating camera are at an insignificant level, and when environments on the camera system, during survey, can be controlled or monitored; one will have proof of geometric stability. Following such proof, aerial camera calibrations should be performed for testing the laboratory calibration values. Until this path is followed and all of the "more sources of error then you can imagine" corrected (Washer, N.B.C., discussing camera calibration problems with Norton, 1951–2), many reasons for the variables (in the micrometres) can be expected.

At present, the national camera calibration laboratories are issuing calibration certificates showing critical values, and maps of high quality are being made using these values. Just how much more closely the numbers can approach the "true values" depends on our knowledge of truth. One approaches truth asymptotically, sometimes at the cost of great effort; nevertheless, it is necessary to examine the path toward this ultimate goal and select reasonable limits of achievement. Consistent accuracy depends on great stability, and the supersensitive calibration techniques can be used to investigate and determine both.

4.8.8 ANALYTICAL METHODS OF CAMERA CALIBRATION

4.8.8.1 INTRODUCTION

Analytical approaches to calibration have a common starting point, that is, the introduction of specific analytical models for radial and decentering distortion directly into the photogrammetric projective equations. The parameters defining the distortion functions are then recovered simultaneously with the projective parameters $(\alpha, \omega, \kappa, x_p, y_p, c, X^c, Y^c, Z^c)$ in a least squares adjustment leading to the minimization of the quadratic form of the residuals of measured quantities. With the exception of the Analytical Plumb Line Method (to be discussed separately) analytical methods are subject to the following broad classification:

I. SINGLE FRAME METHODS

    A. *Stellar Methods*
    B. *Aerial Methods*

II. MULTI-FRAME METHODS
    A. *Fixed Control*
    B. *Adjustable Control (Self-Calibration)*

The methods of Class II may be regarded as embracing those of Class I as special cases. However, as will be presently seen, derivations of methods of Class II are best approached as direct extensions of derivations of methods of Class I. The development about to be presented is largely extracted from Brown (1956), (1957), (1964), (1966), (1969), (1971a), (1971b), and (1972).

4.8.8.2 MODELS FOR RADIAL AND DECENTERING DISTORTION

The distortion δr of a perfectly centered lens, referred to the Gaussian focal length c, is governed by an expression of the form

$$\delta r = K_1 r^3 + K_2 r^5 + K_3 r^7 + \ldots \quad (4.23)$$

in which

  K_1, K_2, K_3 *are coefficients of radial distortion,*
  r *is the radial distance referred to the principal point.*

More specifically,

$$r = \left[(x - x_p)^2 + (y - y_p)^2 \right]^{1/2}, \quad (4.24)$$

where

  $x, y,$ are the plate coordinates of a photographed point,
  x_p, y_p are the coordinates of the principal point.

Unfortunately, actual lenses (as opposed to their designed counterparts) are subject to various degrees of decentration, that is, the centers of curvature of their optical surfaces are not strictly collinear. This defect introduces what is termed decentering distortion. This distortion has both a radial and tangential component and can be described analytically by the expressions

$$\begin{aligned} \delta_r &= 3(J_1 r^2 + J_2 r^4 + \ldots) \sin (\phi - \phi_0) \\ \delta_t &= (J_1 r^2 + J_2 r^4 + \ldots) \cos (\phi - \phi_0) \end{aligned} \quad (4.25)$$

in which

  δ_r = radial component of decentering distortion,
  δ_t = tangential component of decentering distortion,
  $J_1, J_2, \ldots$ = coefficients of decentering distortion.

The quantity ϕ_0 is the angle between the positive x axis of the photograph and the vector to the coordinates x, y of the photographed point.

Specifically, ϕ_0 is defined by the expressions

$$\phi_0 = \text{arc sin} \frac{(x - x_p)}{r} = \text{arc cos} \frac{(y - y_p)}{r}.$$
(4.26)

The quantity ϕ_0 is the angle between the positive positive x axis and a line of reference termed the *axis of maximum tangential distortion*.

When the elements of interior orientation x_p, y_p, c and the parameters of radial distortion $(K_1, K_2, \ldots)$ and decentering distortion $(\phi_0, J_1, J_2, \ldots)$ are known, the appropriate corrections for distortion can be applied to the photographic coordinates x, y by means of the expressions

$$\Delta x = \bar{x}(K_1 r^2 + K_2 r^4 + \ldots) + \left[P_1(r^2 + 2\bar{x}^2) + 2P_2 \bar{x}\bar{y}\right]\left[1 + P_3 r^2 + \ldots\right],$$
(4.27)

$$\Delta y = \bar{y}(K_1 r^2 + K_2 r^4 + \ldots) + \left[2P_1 \bar{x}\bar{y} + P_2(r^2 + 2\bar{y})\right]\left[1 + P_3 r^2 + \ldots\right],$$

in which

$\Delta x, \Delta y$ = corrections for combined effects of radial and decentering distortion

$\left.\begin{array}{l} \bar{x} = x - x_p \\ \bar{y} = y - y_p \end{array}\right\}$ = photographic coordinates referred to principal point

The newly introduced quantities P_1, P_2, P_3 appearing in the above expressions are the following functions of the decentering parameters $\phi_0, J_1, J_2, \ldots$:

$$\begin{array}{l} P_1 = J_1 \sin \phi_0, \\ P_2 = J_1 \cos \phi_0, \\ P_3 = J_2/J_1. \end{array}$$
(4.28)

The reversal of these relations yields

$$\begin{array}{l} \phi_0 = \text{arc tan } (P_1/P_2), \\ J_1 = (P_1{}^2 + P_2{}^2)^{1/2}, \\ J_2 = (P_1{}^2 + P_2{}^2)^{1/2} P_3. \end{array}$$
(4.29)

In the analytical process of calibration the customary set of elements of interior orientation x_p, y_p, c) is broadened to include the coefficients of radial and decentering distortion K_1, K_2, $K_3 \ldots$; P_1, P_2, $P_3 \ldots$

4.8.8.3 SINGLE FRAME ANALYTICAL CALIBRATION

4.8.8.3.1 ADOPTED FORM OF PROJECTIVE EQUATIONS

The projective equations resulting from an undistorted central projection may be put in the form

$$x - x_p = c \frac{m_{11}\lambda + m_{12}\mu + m_{13}\nu}{m_{31}\lambda + m_{32}\mu + m_{33}\nu}$$
(4.30)

$$y - y_p = c \frac{m_{21}\lambda + m_{22}\mu + m_{23}\nu}{m_{31}\lambda + m_{32}\mu + m_{33}\nu}$$

where the m_{ij} denote elements of the orientation matrix and λ, μ, ν denote the direction cosines of the ray joining corresponding image and object points. Specific expressions for the direction cosines depend on whether the points to be regarded as control are specified in terms of three-dimensional coordinates (X, Y, Z) or in terms of directions (as in the case of stars). With directional control the appropriate expressions are

$$\begin{array}{l} \lambda = \sin \alpha^* \cos \omega^* \\ \mu = \cos \alpha^* \cos \omega^* \\ \nu = \sin \omega^* \end{array}$$
(4.31)

in which the angles α^*, ω^* from the camera to the point are measured in the same sense as the pair of angles α, ω defining the direction of the camera axis. For points in object space specified by Cartesian coordinates the appropriate expressions for the direction cosines are

$$\begin{array}{l} \lambda = (X - X^c)/R \\ \mu = (Y - Y^c)/R \\ \nu = (Z - Z^c)/R \end{array}$$
(4.32)

in which X, Y, Z and X^c, Y^c, Z^c are the coordinates in object space of the point and of the center of projection, respectively, and R is the distance between these points, namely

$$R = \left[(X - X^c)^2 + (Y - Y^c)^2 + (Z - Z^c)^2\right]^{1/2}.$$
(4.33)

The above formulation of the projective equations is advantageous because it facilitates the development of a reduction accommodating either three-dimensional control (X, Y, Z) or directional control (α^*, ω^*) or a combination of both types of control.

4.8.8.3.2 LINEARIZED OBSERVATIONAL EQUATIONS

For present purposes it will suffice to assume that the direction cosines, however computed, have been corrected for the effects of atmospheric refraction by means of standard formulas. It then becomes possible to replace $x - x_p$, $y - y_p$ on the left hand sides of the projective (4.30) by the expressions

$$\begin{array}{l} x - x_p = x^0 + v_x - x_p + \Delta x \\ y - y_p = y^0 + v_y - y_p + \Delta y \end{array}$$
(4.34)

in which

x^0, y^0 = measured values of the image coordinates,

v_x, v_y = residuals corresponding to the measured coordinates.

and Δx, Δy are corrections for total distortion given by (4.27) with the proviso that the unknown, true coordinates x, y in (4.27) are replaced by the measured coordinates x^0, y^0.

When the foregoing substitutions are made, the projective equations explicitly involve parameters of radial and decentering distortion along with the standard elements of interior and exterior orientation. The general term *projective parameters* may be used to denote collectively the nine classical elements of orientation (x_p, y_p, c, α, ω, κ, X^c, Y^c, Z^c) and the coefficients of radial and decentering distortion (K_1, K_2, ... , P_1, P_2, ...). These, in turn may be classified as *exterior projective parameters* consisting of α, ω, κ, X^c, Y^c, Z^c and *interior projective parameters* consisting of x_p, y_p, c, K_1, K_2, ... , P_1, P_2, ... It will serve the present development if the convenient assumption is made that the use of no more than three

$$\dot{u}_i = \dot{u}_i^{00} + \delta\dot{u}_i, \quad i = 1, 2, \ldots, 9,$$
$$\ddot{u}_j = \ddot{u}_j^{00} + \delta\ddot{u}_j, \quad j = 1, 2, \ldots, 6, \tag{4.36}$$

the pair of linearized observational equations may be expressed in matrix form as

$$\underset{(2,1)}{\mathbf{v}_i} + \underset{(2,9)}{\dot{\mathbf{B}}_i} \underset{(9,1)}{\dot{\boldsymbol{\delta}}} + \underset{(2,6)}{\ddot{\mathbf{B}}_i} \underset{(6,1)}{\ddot{\boldsymbol{\delta}}} = \underset{(2,1)}{\boldsymbol{\epsilon}_i} \tag{4.37}$$

in which the subscript i has been introduced to refer to the ith control point. Here, the vector $\boldsymbol{\epsilon}_i$ denotes the projective discrepancies

$$\boldsymbol{\epsilon}_i = \begin{bmatrix} \boldsymbol{\epsilon}_{1_i} \\ \boldsymbol{\epsilon}_{2_i} \end{bmatrix} = \begin{bmatrix} -f_1(x_i^0, y_i^0; \dot{u}_1^{00}, \dot{u}_2^{00}, \ldots, \dot{u}_9^{00}; \ddot{u}_1^{00}, \ddot{u}_2^{00}, \ldots, \ddot{u}_6^{00}) \\ -f_2(x_i^0, y_i^0; \dot{u}_1^{00}, \dot{u}_2^{00}, \ldots, \dot{u}_9^{00}; \ddot{u}_1^{00}, \ddot{u}_2^{00}, \ldots, \ddot{u}_6^{00}) \end{bmatrix}$$

$$\tag{4.38}$$

coefficients of radial distortion K_1, K_2, K_3 and no more than three coefficients of decentering distortion P_1, P_2, P_3 provides a sufficiently accurate model. The extension of the development to include additional coefficients of distortion is a trivial matter. So far, no case has been encountered where more than three coefficients of decentering distortion have been found to be necessary, and two coefficients have been found to suffice in the vast majority of cases. Only in the case of extremely wide angle cameras has it been found that more than four coefficients of radial distortion are needed. As a rough rule, super wide angle cameras generally require four coefficients, wide angle cameras require two or three coefficients and medium- to narrow-angle cameras usually require only one or two coefficients.

It is convenient at this point to introduce the symbol $\dot{u}_i$ to denote the ith interior projective parameter of the 9-element set $\{x_p, y_p, c; K_1, K_2, K_3; P_1, P_2, P_3\}$; and the symbol $\ddot{u}_j$ to denote the jth exterior projective parameter of the 6-element set $\{\alpha, \omega, \kappa, X^c, Y^c, Z^c\}$. It is also convenient to assume at the outset that control that control consists of points with specified X, Y, Z coordinates rather than points specified by angles α^*, ω^* (as with stellar control points). Later, the steps necessary for the admission of directional control will be pointed out.

With the symbolism just adopted the projective equations assume the functional form

$$f_1(x^0 + v_x, y^0 + v_y; \dot{u}_1, \dot{u}_2,$$
$$\ldots, \dot{u}_9; \ddot{u}_1, \ddot{u}_2, \ldots, \ddot{u}_6) = 0,$$
$$\tag{4.35}$$
$$f_2(x^0 + v_x, y^0 + v_y; \dot{u}_1, \dot{u}_2,$$
$$\ldots, \dot{u}_9; \ddot{u}_1, \ddot{u}_2, \ldots, \ddot{u}_6) = 0.$$

These equations can be linearized by the usual process of expansion by Taylor's series about initial approximations. Accordingly, if $\dot{u}_i$, $\ddot{u}_j$ are replaced by approximations plus corresponding corrections in accordance with the expressions

which, under the assumption of a perfect model, would be zero only with the exercise of flawless measurements and perfect approximations to the projective parameters. The vectors $\mathbf{v}_i$, $\dot{\boldsymbol{\delta}}$ and $\ddot{\boldsymbol{\delta}}$ are comprised respectively of residuals of plate coordinates, corrections to approximations to interior projective parameters, and corrections to approximations to exterior projective parameters. Elements of the matrices $\dot{\mathbf{B}}_i$ and $\ddot{\mathbf{B}}_i$ consist of the partial derivatives of the linearized projective equations with respect to interior and exterior projective parameters, respectively. Explicit expressions for the elements of $\dot{\mathbf{B}}_i$ and $\ddot{\mathbf{B}}_i$ are to be found in Brown (1969).

4.8.8.3.3 COVARIANCE MATRIX OF PLATE COORDINATES

Least squares adjustment requires that the precision of the measured quantities be specified in terms of appropriate covariance matrices. The covariance matrix of the measured plate coordinates of the ith point may be represented as

$$\boldsymbol{\Lambda}_i = \begin{bmatrix} \sigma_{x_i}^2 & \sigma_{x_i y_i} \\ \sigma_{x_i y_i} & \sigma_{y_i}^2 \end{bmatrix}. \tag{4.39}$$

The fact that the plate coordinates for a given point are regarded as being possibly correlated (by virtue of the covariance not necessarily being zero in $\boldsymbol{\Lambda}_i$) admits consideration of coordinates established by instruments other than conventional x, y comparators. Although correlation is admitted between the x and y coordinates of a given point, it will be assumed that no correlation exists betwen errors of different points. Accordingly, if a total of n points is measured, the covariance matrix of the coordinates of all n points may be represented as

$$\underset{(2n,2n)}{\boldsymbol{\Lambda}} = \text{diag} \left(\underset{(2,2)}{\boldsymbol{\Lambda}_1} \ \underset{(2,2)}{\boldsymbol{\Lambda}_2} \ \ldots \ \underset{(2,2)}{\boldsymbol{\Lambda}_n} \right). \tag{4.40}$$

4.8.8.3.4 Introduction of *A Priori* Constraints on Projective Parameters

For the sake of fullest practical generality, it will be further assumed that the projective parameters are possibly subject to *a priori* constraints. This implies the existence of supplementary observational equations of the form

$$\dot{u}_i = \dot{u}_i^0 + \dot{v}_i \quad (i = 1, 2, \ldots, 9)$$
$$\ddot{u}_j = \ddot{u}_j^0 + \ddot{v}_j \quad (j = 1, 2, \ldots, 6) \tag{4.41}$$

in which $\dot{v}_i$, $\ddot{v}_j$ are the residuals corresponding to the *a priori* values $\ddot{u}_i^0$, $\ddot{u}_j^0$ of the projective parameters. The covariance matrices of the interior and exterior projective parameters may be denoted by $\dot{\Lambda}$ and $\ddot{\Lambda}$, respectively.

The initial approximations $\dot{u}_i^{00}$, $\ddot{u}_j^{00}$ employed in the process of linearization of the projective equations need not necessarily be taken as equal to the *a priori* values $\dot{u}_i^0$, $\ddot{u}_j^0$. This means that two different expressions exist for $\dot{u}_i$, $\ddot{u}_j$, namely those given by equations (4.36) and (4.41). The elimination of $\dot{u}_i$, $\ddot{u}_j$ from these expressions yields the following alternative set of supplementary observational equations

$$\dot{v}_i - \delta \dot{u}_i = \dot{\epsilon}_i \quad (i = 1, 2, \ldots, 9)$$
$$\ddot{v}_j - \delta \ddot{u}_j = \ddot{\epsilon}_j \quad (j = 1, 2, \ldots, 6) \tag{4.42}$$

in which the supplementary discrepancy terms $\dot{\epsilon}_i$, $\ddot{\epsilon}_j$ represent the differences between the arbitrary approximations to the projective parameters and the corresponding *a priori* values, *i.e.*,

$$\dot{\epsilon}_i = \ddot{u}_i^{00} - \dot{u}_i^0,$$
$$\ddot{\epsilon}_j = \ddot{u}_j^{00} - \dot{u}_j^0. \tag{4.43}$$

The matrix representation of equations (4.51) is given with obvious notation by

$$\dot{v} - \hat{\delta} = \dot{\epsilon},$$
$$\ddot{v} - \ddot{\delta} = \ddot{\epsilon}. \tag{4.44}$$

These equations together with the linearized projective equations (4.37) with $i = 1, 2, \ldots, n$ form the total system of observational equations for the least squares adjustment.

4.8.8.3.5 Normal Equations for Least Squares Adjustment

The formal execution of the least squares adjustment leads to a system of normal equations of the form

$$\begin{bmatrix} \dot{N} + \dot{W} & \bar{N} \\ \bar{N}^T & \ddot{N} + \ddot{W} \end{bmatrix} \begin{bmatrix} \hat{\delta} \\ \ddot{\delta} \end{bmatrix} = \begin{bmatrix} \dot{c} - \dot{W} \dot{\epsilon} \\ \ddot{c} - \ddot{W} \ddot{\epsilon} \end{bmatrix} \tag{4.45}$$

in which

$$\dot{N} = \sum_{i=1}^{n} \dot{B}_i^T W_i \dot{B}_i \qquad \dot{c} = \sum_{i=1}^{n} \dot{B}_i^T W_i \epsilon_i$$

$$\bar{N} = \sum_{i=1}^{n} \dot{B}_i^T W_i \ddot{B}_i \qquad \ddot{c} = \sum_{i=1}^{n} \ddot{B}_i^T W_i \epsilon_i \tag{4.46}$$

$$\ddot{N} = \sum_{i=1}^{n} \ddot{B}_i^T W_i \ddot{B}_i$$

wherein the various weight matrices are defined by

$$W_i = \Lambda_i^{-1}, \quad \dot{W} = \dot{\Lambda}^{-1}, \quad \ddot{W} = \ddot{\Lambda}^{-1}. \tag{4.47}$$

4.8.8.3.6 Analysis of Residuals

The above system of normal equations involves a total of 15 unknowns consisting of corrections to the approximations to the 9 interior projective parameters (the elements of $\hat{\delta}$) and corrections to the approximations to the 6 interior projective parameters (the elements of $\ddot{\delta}$). The corrections resulting from the solution of the normal equations lead to improved approximations which may be employed in an iterative process leading upon convergence to insignificantly small corrections. When $\hat{\delta}$ and $\ddot{\delta}$ have thus been reduced to insignificance, the residuals of the plate coordinates, interior projective parameters and exterior projective parameters, respectively, may be evaluated from

$$v_i = \epsilon_i, \quad \dot{v} = \dot{\epsilon}, \quad \ddot{v} = \ddot{\epsilon} \tag{4.48}$$

in which the discrepancy vectors are those computed from the final set of projective constants. The complete quadratic form q of the residuals is given by

$$q = \dot{v}^T \dot{W} \dot{v} + \ddot{v}^T \ddot{W} \ddot{v} + \sum_{i=1}^{n} v_i^T W_i v_i \tag{4.49}$$

and the statistical degrees of freedom associated with q are equal to $2n\text{-}k$ where k denotes the number of projective parameters for which no effective *a priori* constraints were exercised.

The character and distribution of the residuals of the plate coordinates are primary indicators of the quality of the adjustment. Ideally, of course, they should be perfectly random and free of local systematic tendencies. Failure of this ideal to be realized is a direct indication of some inadequacy of the mathematical model being employed. In some cases difficulties in this regard may be cleared up simply by the introduction of additional parameters for radial and decentering distortion. In other cases, the problem may be more fundamental, being a symptom of what may be termed *anomalous distortion*, a topic discussed more fully in section 4.8.8.4.2.2.

4.8.8.3.7 Practical Application of *A Priori* Constraints

The primary purpose of the covariance matrices $\dot{\Lambda}$ and $\ddot{\Lambda}$ is to provide proper weighting within the adjustment of actual *a priori* values of projective parameters. With regard to any given projective parameter three possible situations arise:

(1) the parameter is indeed known in advance to a worthwhile degree of accuracy;
(2) the parameter is not known in advance to a worthwhile degree of accuracy;
(3) the parameter is either to be given to a prespecified value or is not to be used at all (*e.g.*, as would commonly occur with higher order coefficients of radial or decentering distortion).

To address the first case, one would take pains to be as realistic as possible in assigning specific values to the pertinent elements of the covariance matrix. This would assure that the *a priori* value would be subject to an appropriate degree of adjustment. In the second case, one would simply treat the adopted initial approximation to the parameter as if it were an *a priori* value subject to a very large variance. This would allow the parameter virtually complete freedom to adjust. In the third case, one would treat the value to be enforced as an *a priori* value with an extremely small variance (an appropriate variance would be several orders of magnitude smaller than the variance to be expected for the recovery of the parameter if it were free to adjust). This would render the parameter subject only to an insignificant degree of adjustment.

The foregoing shows that the covariance matrix of the projective parameters can be exploited to exercise general control over the adjustment. This is most advantageous, for it permits the reduction to embrace a wide variety of special options with a minimum special programming. This is perhaps best illustrated by the three examples that follow.

The first concerns the use of the reduction for stellar calibration. This may be accomplished as follows. First, an auxiliary program would be employed to perform the standard astronomical reductions leading to the direction cosines λ μ ν of the stellar control points in accordance with equation (4.31) (details of this reduction are given in the previous edition of the MANUAL OF PHOTOGRAMMETRY). These direction cosines are then used in place of X, Y, Z coordinates of control, and the coordinates X^c, Y^c, Z^c are constrained (by weighting) to zero. The program will then automatically perform the desired stellar calibration.

A second example of the process of exercising *a priori* constraints to effect special control over the adjustment concerns the recovery of parameters for decentering distortion. As can be seen from (4.27), when the higher order coefficient P_3 is considered equal to zero, the expressions in the projective equations accounting for decentering distortion become linear in the leading coefficients P_1 and P_2. This means that

one can adopt values of zero as convenient initial approximations for P_1 and P_2. However, by so doing, one causes the coefficients corresponding to P_3 to be zero in all of the $\dot{\mathbf{B}}_i$ matrices. This in turn renders the normal equations indeterminate. An easy remedy consists of using *a priori* constraints to suppress P_3 to zero in the initial iterations. Once stable, nonzero approximations have been obtained for P_1 and P_2, the *a priori* constraint on P_3 can be relaxed in a final iteration to permit determing this value along with refined values for P_1 and P_2.

The final application considered here of the special application of *a priori* constraints is of particular value in the calibration of cameras having narrow angular fields. Here the projective effects of a small translation δx_p, δy_p of the plate are very nearly equivalent to the projective effects of a rotation of the camera axis. This coupling of translation and rotation can sometimes lead to ill-conditioning of the normal equations sufficient to prevent convergence of the iterative process. The problem is compounded when decentering distortion is being recovered, for decentering coefficients also interact to a moderate degree with x_p, y_p. The remedy consists of subjecting x_p, y_p in the initial reduction to fairly tight *a priori* constraints (*e.g.*, a few hundred micrometres) and of relaxing such constraints by stages in subsequent iterations until they ultimately become inconsequential.

4.8.8.3.8 ERROR PROPAGATION

A decided advantage of the analytical approach to camera calibration is that the error propagation associated with the end results can readily be computed. By contrast, conventional laboratory methods do not lend themselves to . precise statements concerning accuracies of results.

The process of error propagation exploits the fact that the covariance matrix of the adjusted projective parameters is provided directly by the inverse of N, the coefficient matrix of the normal equations. If n^{ij} denotes the element in the ith row and j^{th} column of N^{-1}, it follows from the adopted ordering of the parameters that the variances of the adjusted elements of interior orientation are given directly by

$$\sigma_{x_p}^2 = n^{11}, \ \sigma_{y_p}^2 = n^{22}, \ \sigma_c^2 = n^{33}. \qquad (4.50)$$

The variances of the radial and decentering parameters are provided by elements n^{44} through n^{99}. However, these are of relatively little interest in themselves. Rather, what is of interest is the uncertainty of the calibrated distortion functions throughout the format.

In the case of symmetric radial distortion, the propagation of errors into the distortion function is readily accomplished. Inasmuch as the covariance matrix of the coefficients of radial distortion is

$$\mathbf{\Omega}_K = \begin{bmatrix} n^{44} & n^{45} & n^{46} \\ n^{45} & n^{55} & n^{56} \\ n^{46} & n^{56} & n^{66} \end{bmatrix} \quad (4.51)$$

and the distortion function is defined by

$$\delta r = K_1 r^3 + K_2 r^5 + K_3 r^7$$

it follows from the theory of error propagation that the variance of δr for a specified radial distance r is given by

$$\sigma_{\delta r}^2 = \mathbf{U}_K \mathbf{\Omega}_K \mathbf{U}_K^T \quad (4.52)$$

in which

$$\mathbf{U}_K = \frac{\partial(\delta r)}{\partial(K_1, K_2, K_3)} = (r^3 \ r^5 \ r^7). \quad (4.53)$$

The error propagation associated with decentering distortion is somewhat more involved because the end results are more conveniently expressed in terms of phase angle ϕ_0 and tangential profile

$$P_r = J_1 r^2 + J_2 r^4, \quad (4.54)$$

rather than in terms of the parameters P_1, P_2, P_3 appearing directly in the normal equations. This means that the covariance matrix of J_1, J_2, ϕ_0 must be derived from that of P_1, P_2, P_3 before the error propagation into the tangential profile can be effected. Because the covariance matrix of the adjusted decentering parameters P_1, P_2, P_3 is given by

$$\mathbf{\Omega}_p = \begin{bmatrix} n^{77} & n^{78} & n^{79} \\ n^{78} & n^{88} & n^{89} \\ n^{79} & n^{89} & n^{99} \end{bmatrix} \quad (4.55)$$

it follows that the covariance matrix of the derived parameters J_1, J_2, ϕ_0 is given by

$$\mathbf{\Omega}_0 = \mathbf{U}_p \mathbf{\Omega}_p \mathbf{U}_p^T \quad (4.56)$$

in which by virtue of (4.7)

$$\mathbf{U}_p = \frac{\partial(J_1, J_2, \phi_0)}{\partial(P_1, P_2, P_3)}$$

$$= (P_1^2 + P_2^2)^{-1/2} \begin{bmatrix} P_1 & P_2 & 0 \\ -P_1 P_3 & -P_2 P_3 & -(P_1^2 + P_2^2) \\ \cos\phi_0 & \sin\phi_0 & 0 \end{bmatrix}. \quad (4.57)$$

It then follows that the covariance matrix of the tangential profile and phase angle is given by

$$\begin{bmatrix} \sigma_{P_r}^2 & \sigma_{P_r \phi_0} \\ \sigma_{P_r \phi_0} & \sigma_{\phi_0}^2 \end{bmatrix} = \mathbf{U}_0 \mathbf{\Omega}_0 \mathbf{U}_0^T \quad (4.58)$$

in which

$$\mathbf{U}_0 = \frac{\partial(P_r, \phi_0)}{\partial(J_1, J_2, \phi_0)} = \begin{bmatrix} r^2 & r^4 & 0 \\ 0 & 0 & 1 \end{bmatrix}. \quad (4.59)$$

By virtue of the results expressed by equations (4.50), (4.52) and (4.59), one is able to

establish the degree of confidence to be placed in the results of the calibration. An example of the process of error propagation is given in figure 4-61 which depicts the calibrated radial and decentering distortion functions and associated one sigma error bounds of a KC-6A camera.

4.8.8.3.9 PROJECTIVE TRANSFORMATION OF DISTORTION FUNCTION

The radial distortion function δr generated by the adjustment corresponds to the calibrated principal distance c and is referred to as the Gaussian distortion function. This function, it will be noted, does not involve a linear term. However, a linear term can be introduced through the artifice of altering the calibrated principal distance by an arbitrary increment Δc. This generates a projectively equivalent distortion function Δr that is associated with the specified principal distance $c + \Delta c$. The equation effecting this transformation is given by

$$\Delta r = \frac{\Delta c}{c} r + \left(1 + \frac{\Delta c}{c}\right)\delta r. \quad (4.60)$$

Application of this result to the Gaussian distortion function given by (4.23) leads to the projectively equivalent distortion function

$$\Delta r = K'r + K'_1 r^3 + K'_2 r^5 + K'_3 r^7 + \dots \quad (4.61)$$

in which

$$K'_0 = \frac{\Delta c}{c},$$

$$K'_1 = \left(1 + \frac{\Delta c}{c}\right)K_1, \quad (4.62)$$

$$K'_2 = \left(1 + \frac{\Delta c}{c}\right)K_2, \text{ etc.}$$

The foregoing result is most commonly used to force the transformed distortion function to assume the value of zero at a specified radial distance r_0. Thus if δr_0 denotes the value of δr at $r = r_0$, it follows immediately from (4.60) that the particular choice

$$\Delta c = -\left(\frac{c}{r_0 + \delta r_0}\right)\delta r_0 \quad (4.63)$$

makes Δr equal to zero at $r = r_0$.

Other applications of the transformation involve choice of Δc that serve to

(a) render the mean value of the transformed distortion function equal to zero out to a specified radial distance r_0;
(b) generate a distortion function having the smallest possible mean square value out to a specified radial distance r_0;
(c) produce a function such that the largest posi-

tive and negative values of distortion out to a specified radial distance r_0 are forced to have the same absolute value.

Specific formulae effecting the above transformations are to be found in Brown (1969).

It is to be emphasized that all transformations based on the application of (4.60) yield distortion functions that are projectively equivalent. While the Gaussian distortion function associated with a modern mapping lens may well assume values in excess of, say, 100 μm, an appropriate transformation may yield an equivalent function in which the largest values are suppressed to under, say, 10 μm. For this reason camera manufacturers almost invariably prefer to present transformed distortion functions rather than Gaussian distortion functions. Such cosmetic manipulation is perfectly harmless and undoubtedly conforting to both buyer and seller.

4.8.8.4 MULTI-FRAME METHODS OF CAMERA CALIBRATION

4.8.8.4.1 PRELIMINARY CONSIDERATIONS

The foregoing development was concerned with the reduction of measurements from a single frame for the explicit purpose of determination of interior projective parameters. A frame is defined as consisting of an exposure or series of exposures for which the position and orientation (*i.e.*, the exterior projective elements) are considered to be invariant. In this context, for instance, a stellar plate containing several different exposures may be regarded as consisting of either

(1) a single frame if the orientation of the camera is considered to remain unchanged throughout the series of exposures;
(b) several discrete frames if the orientation is considered to change (however slightly) from one exposure (or group of exposures) to the next.

In the case of aerial photographs each exposure would, of course, ordinarily constitute an independent frame.

When measurements from more than one frame are to be employed for calibration, one can proceed to reduce each frame independently according to the process just outlined. Each frame so reduced would, if determinate, then yield an independent estimation of the elements of interior projection. These could then be averaged in some fashion to yield a final estimation. Such a procedure, though simple, is far from optimal. In the first place, there may not be enough control to produce a worthwhile result from any single frame. Secondly, the strong coupling that exists between interior elements of orientation (x_p, y_p, c) and exterior elements can be expected to result in unacceptably large variances for these particular projective parameters when recovered on a frame-by-

frame basis. Finally, the analytical error propagation associated with the end results no longer emerges as a simple by-product of the reduction but, rather, becomes so cumbersome as to be impractical.

Fortunately, it turns out that a rigorous alternative to the independent frame-by-frame reduction is altogether practical, no matter how many frames may be carried in the reduction. This alternative is embodied in the reduction termed SMAC (Simultaneous Multiframe Analytical Calibration) developed in Brown (1969). The distinguishing characteristic of the SMAC reduction is that it is specifically designed to permit the calibration of the interior projective parameters by means of the simultaneous adjustment of observations from an indefinitely large number of frames having unknown elements of exterior orientation that may vary from frame to frame.

4.8.8.4.2 SIMULTANEOUS MULTI-FRAME ANALYTICAL CALIBRATION (SMAC)

The SMAC reduction is most readily described in terms of the reduction already developed for the single frame. If it is now assumed that a total of m frames is to be subject to a simultaneous reduction, the normal equations (4.45) for the independent reduction of the jth frame may be expressed as

$$\begin{bmatrix} \dot{\mathbf{N}}_j + \dot{\mathbf{W}} & \overline{\mathbf{N}}_j \\ \overline{\mathbf{N}}_j & \ddot{\mathbf{N}}_j + \ddot{\mathbf{W}} \end{bmatrix} \begin{bmatrix} \dot{\delta} \\ \ddot{\delta}_j \end{bmatrix} = \begin{bmatrix} \dot{\mathbf{c}}_j - \dot{\mathbf{W}}\,\dot{\boldsymbol{\epsilon}} \\ \ddot{\mathbf{c}}_j - \ddot{\mathbf{W}}_j\ddot{\boldsymbol{\epsilon}}_j \end{bmatrix}, \quad (4.64)$$
$$j = 1, 2, \ldots, m$$

In anticipation of later results it is instructive to proceed as if each of the above systems were to be solved independently by the method of partitioning, even though the relatively low order of each system would ordinarily warrant an unpartitioned solution. Algorithmically, this entails the following steps. In terms of the matrices appearing directly in the above normal equations, the following four auxiliaries are formed

$$\begin{aligned} \mathbf{Q}_j &= (\ddot{\mathbf{N}}_j + \ddot{\mathbf{W}}_j)^{-1}\,\overline{\mathbf{N}}_j^{\mathrm{T}} \\ \mathbf{R}_j &= \overline{\mathbf{N}}_j\mathbf{Q}_j \\ \mathbf{S}_j &= \dot{\mathbf{N}}_j - \mathbf{R}_j \\ \overline{\mathbf{c}}_j &= \dot{\mathbf{c}}_j - \mathbf{Q}_j^{\mathrm{T}}(\ddot{\mathbf{c}}_j - \ddot{\mathbf{W}}_j\ddot{\boldsymbol{\epsilon}}_j) \end{aligned} \quad (4.65)$$

This process generates the following reduced system of normal equations involving only the interior projective parameters as unknowns

$$(\dot{\mathbf{S}}_j + \dot{\mathbf{W}})\hat{\dot{\delta}} = \overline{\mathbf{c}}_j - \dot{\mathbf{W}}\,\dot{\boldsymbol{\epsilon}}. \quad (4.66)$$

The solution of this reduced system then yields the estimate of the interior projective parameters resulting from the independent reduction of the jth frame. Once the interior projective parameters have been thus determined, the exterior projective parameters for the jth frame can be computed from

$$\ddot{\delta}_j = (\ddot{\mathbf{N}}_j + \ddot{\mathbf{W}}_j)^{-1}(\ddot{\mathbf{c}}_j - \ddot{\mathbf{W}}_j\ddot{\boldsymbol{\epsilon}}_j) - \mathbf{Q}_j\hat{\dot{\delta}}. \quad (4.67)$$

The above steps constitute the essentials of the independent reduction of a single frame. Now, as shown in Brown (1969), the rigorous simultaneous reduction of all m frames involves only the following additional steps. First as S_j and $\bar{c}_j$ in (4.65) are formed for each frame in turn, they are added to the sums of their predecessors to form, ultimately, after all m frames have been processed through this stage, the following sums

$$S = S_1 + S_2 + \ldots + S_m ,$$
$$\bar{c} = \bar{c}_1 + \bar{c}_2 + \ldots + \bar{c}_m . \qquad (4.68)$$

The reduced system of normal equations for the simultaneous reduction is precisely the same as in (4.74) except that S and $\bar{c}$ replace S_j and $\bar{c}_j$, i.e.,

$$(S + \dot{W}) \, \dot{\delta} = \bar{c} - \dot{W} \, \dot{\epsilon} \qquad (4.69)$$

Once $\dot{\delta}$ has been determined from the solution of this system, each $\dot{\delta}_j$ can be recovered from (4.67) as before (no alteration of this equation is needed).

Although the SMAC reduction just outlined does indeed effect the simultaneous least squares adjustment of all observations produced by all m frames (and thus leads to a general system of normal equations of order $9 + 6m$), it paradoxically entails fewer computations, overall, than the independent reduction of all m frames. This is clear when one realizes that the extra steps required by (4.68) actually involve fewer operations than does the evaluation of the m inverses of $S_j + \dot{W}$ required by the independent reductions. Thus not only does the SMAC reduction produce a far more rigorous calibration than the independent reduction of multiple frames, but it also is computationally more efficient. This efficiency of the SMAC reduction is a consequence of the fact that the general normal equations (which were never explicitly formed in the present development) have a bordered, block-diagonal form, which when properly exploited leads to the process just outlined. Full derivational details are to be found in Brown (1969).

The covariance matrix of the interior projective parameters resulting from a SMAC calibration is given by $(S + \dot{W})^{-1}$. Accordingly, the results given in section 4.8.8.3.8 for the error propagation associated with the calibration of a single frame can, with appropriate reinterpretation, be applied directly to the results of a SMAC calibration.

4.8.8.4.2.1 Characteristics of a SMAC Calibration. The interior projective parameters resulting from a SMAC calibration constitute the best possible compromise for the set of frames carried in the reduction. This means that errors, both random and systematic, arising from a variety of sources to be mentioned below are in large measure averaged out insofar as their effects on the resulting interior parameters

are concerned. The nature of the SMAC calibration is such that it becomes pertinent to classify systematic errors into the following two categories:

(a) *transient systematic errors,*
(b) *persistent systematic errors.*

Transient systematic errors are those that are systematic on a given frame but are essentially independent from frame to frame. Persistent systematic errors are those that tend to remain the same from frame to frame.

In a SMAC calibration employing aerial photographs one finds it easy to postulate a large number of possible sources of transient systematic error. A partial listing might include

(a) variations in thickness of film-emulsion combination;
(b) nonuniform, dimensional instability of film;
(c) broad refractive anomalies induced by atmospheric turbulence;
(d) random local failure of film to conform precisely to platen;
(e) dynamic deformations of camera induced by aircraft accelerations, or by corrective accelerations of camera mount; and
(f) deformations attributable to thermal gradients.

As long as one is not pinned down to quantification, such a list could go on and on. The essential point is that while the combined effect of all sources of transient errors may well assume significance on a given frame, the independence of such errors from frame to frame renders them of diminished importance in a SMAC calibration embracing a moderate number of frames. Accordingly, a SMAC reduction of, say, 20 frames having an average of 25 control points per frame is much to be preferred over a reduction of 500 points on a single frame.

Unlike the effects of transient systematic errors, the effects of persistent systematic errors are not diminished by averaging in a SMAC reduction. A partial listing of possible sources of such errors in an aerial calibration might include:

(a) departure of the film platen from a true plane;
(b) persistent, uncompensated component of film deformation;
(c) systematic errors in the adopted survey of ground control;
(d) refractive effects of shock waves of aircraft;
(e) prism effect of optical filter and/or aircraft window;
(f) uncompensated systematic error of comparator used for measurement.

Persistent systematic errors may be partially absorbed by the projective parameters. For instance, a film platen that happens to be essentially spherically concave or convex would introduce an effect that would be absorbed completely by the coefficients of radial distortion. Similarly, possible prism effect of an aircraft window would be absorbed by parameters of decentering distortion. Such compensation may or may not be desirable. It is desirable to the extent that all pertinent elements of the system approximate in routine operations those applying to the photographs used in the SMAC calibration. On the other hand, such compensation would be partially or totally unsound in instances where key contributors to systematic error have characteristics differing significantly from those prevailing in the SMAC calibration. For this reason, attempts should be made to isolate, insofar as practicable, the contributions of those sources of systematic errors that are likely to vary from one operation to another.

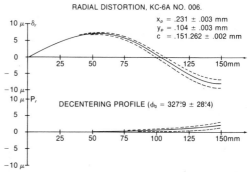

FIGURE 4-61. Illustrating major results from analytical calibration of a specific mapping camera.

4.8.8.4.2.2 Derivation of Anomalous Distortion from SMAC Residuals. That portion of persistent systematic error which is not accommodated by projective parameters would be reflected, along with random error and transient systematic errors, in the final set of least squares residuals resulting from the SMAC calibration. The residual vectors from all frames plotted against x, y plate coordinates on a common graph can be examined for possible uncompensated systematic effects. From numerical and statistical analysis of such residuals one can derive either contour maps or functional representations of residual systematic errors. These can subsequently be used to supplement calibrated interior projective parameters in the correction of operational data.

An example of the foregoing process is taken from Brown (1969). A composite plot was made of some 1200 residuals resulting from an aerial SMAC calibration embracing a total of 40 frames taken by an Air Force KC-6A mapping camera. (The principal results of the calibration of this camera were presented in figure 4-61.) Although the residuals were seen to be essentially random throughout the central two thirds of the format, they displayed definite systematic tendencies about the periphery of the format. It was found that these systematic effects could be largely accounted for through the fit of least squares empirical models defined by general fifth-order polynomials of the form

$$\Delta x = \sum_{i=0}^{s} \sum_{j=0}^{i} \alpha_{ij} x^{i-j} y^j \quad ,$$

$$\Delta y = \sum_{i=0}^{s} \sum_{j=0}^{i} \beta_{ij} x^{i-j} y^j \quad . \qquad (4.70)$$

Contour maps of the fitted error functions for Δx and Δy are shown in figure 4-62. The functions may be regarded as empirically established corrections accounting for those systematic errors that are inadequately modeled in the SMAC calibration. The term *anomalous distortion* has been used to denote this type of "left over" persistent systematic error. As is seen from figure 4-61, anomalous distortion can

assume moderate significance, particularly towards the extremities of the format.

Reasons for the existence of anomalous distortion are not clearly understood at present. Part of the radial component can undoubtedly be attributed to systematic, nonspherical departures of the film platen from a plane. With cameras that do not employ a reseau a sensible degree of anomalous distortion is likely to be induced by the component of film deformation that remains stationary from frame to frame. Still another potential source of anomalous distortion may arise from the lens itself. Modern, highly-corrected mapping lenses generally have rather large front and rear elements intended to reduce the falloff of illumination towards the corners of the field. It is not inconceivable that such factors as mounting strains, gravity loading, inhomogeneities in the glass, and slight, unintended, asymmetric asphericity of optical surfaces could combine to contribute significantly to anomalous distortion. Clearly, the catchall phenomenom termed "anomalous distortion" merits more study than it has been accorded to date.

4.8.8.4.2.3 Derivation of Empirical Weighting Functions from SMAC Residuals. The very large set of residuals obtainable from a comprehensive SMAC calibration of a camera may be employed not only to extract functional representations of possible anomalous distortion but also to generate appropriate empirical functions to represent the dependence on radial distance of the standard deviations of plate coordinates. The primary dependence is best expressed in terms of radial and tangential components of plate coordinates rather than in terms of x, y components. This is partially because the influence on accuracies of variation of photographic resolution is symmetrically dependent on radial distance. In addition, the influence of random departures of the film surface from a perfect plane is also symmetrically dependent on radial distance and, moreover, affects only the radial component of photographic coordinates. The following empirical representations of the variation of the standard

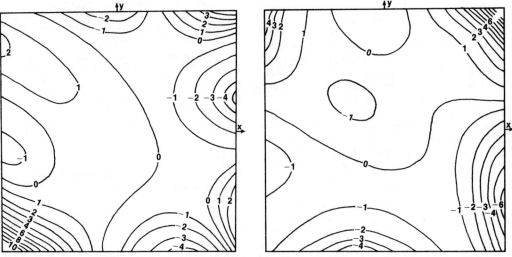

FIGURE 4-62. Illustrating analysis of SMAC residuals for determination of anomalous distortion. Left is a contour map of x component of anomalous distortion and right is a contour map of y component of anomalous distortion.

deviations of radial and tangential components of plate coordinates have been found to be experimentally sound:

$$\sigma_r = \sigma_0 + a_1 r^2 + a_2 r^4$$
$$\sigma_t = \sigma_0 + b_1 r^2 + b_2 r^4. \qquad (4.71)$$

Both expressions contain a common zero order term σ_0 which may be viewed as primarily representing the inherent, isolated contribution of the film measuring process to the error in the coordinates.

Appropriate values for the coefficients in the above expressions may be derived from SMAC residuals by the following process. First, the format is divided into a small number of circular zones of approximately equal area. The rms errors of the radial and tangential components of the residuals are then computed for each zone. The resulting values σ_r, σ_t for a given zone are associated with the mean radius r of the zone, and the coefficients in equation (4.71) are determined by a least squares fit to these values. As presented in Brown (1971b) the application of this process to the particular set of residual vectors leads to the results given in figure 4-63. As can be seen from the figure, two coefficients of the polynomial were found adequate to represent the values of σ_r taken from the accompanying table, whereas all three were needed for σ_t. The rms fit of the polynomials is excellent, being under 0.2 μm overall. As is shown in the table, no significant correlation was found between the radial and tangential components of the residuals.

It will be noted that the empirical functions σ_r, σ_t of the preceding example were derived from residuals generated before the application of corrections for anomalous distortion. These particular functions would therefore be appropriate for the camera in question only if corrections for anomalous distortion were not to be applied in subsequent reductions. Otherwise, the functions for σ_r, σ_t are more properly derived from refined SMAC residuals resulting from the application of corrections for anomalous distortion.

The actual utility of the empirically derived functions σ_r, σ_t lies in their application to realistic weighting of photographic coordinates in the process of analytical aerotriangulation. For this purpose the covariance matrix of the x, y coordinates of a given point may be derived from the functions σ_r, σ_t by means of the relationship

$$\begin{bmatrix} \sigma_x^2 & \sigma_{xy} \\ \sigma_{xy} & \sigma_y^2 \end{bmatrix} = \frac{1}{r^2} \begin{bmatrix} y^2\sigma_r^2 + x^2\sigma_t^2 & xy(\sigma_r^2 - \sigma_t^2) \\ xy(\sigma_r^2 - \sigma_t^2) & x^2\sigma_r^2 + y^2\sigma_t^2 \end{bmatrix}$$

in which σ_r, σ_t are defined by the polynomial expressions (4.71).

4.8.8.4.2.4 Application of SMAC to Multicollimator Calibrations. The SMAC calibration is equally applicable to aerial photographs and stellar photographs. Inasmuch as a multicollimator may be viewed as being nothing more than an artifical star field, it follows that SMAC is also directly applicable to frames made on a multicollimator. In the conventional approach to multicollimator calibration, a time-consuming and painstaking process of autocollimation must be performed in order to orient the camera so that the axis of the central collimator is precisely perpendicular to the focal plane. Such alignment can be totally bypassed when a Stellar SMAC reduction is to be performed. Here, it is sufficient to locate the camera so that its entrance pupil intercepts the converging beams of the collimators. This, in turn, makes it practical to avoid the use of photographic plates (if desired)

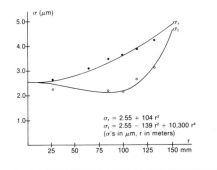

Interval	Mean Radius r	No. Residual Vectors	RMS of Radial Components: σ_r	RMS of Tangential Components: σ_t	*Correlation Coeff: ρ_{rt}
0– 53 mm	26.5mm	269	2.63μm	2.25μm	+.012
53– 75	64.0	251	3.10	2.09	–.035
75– 92	83.5	217	3.48	2.10	–.054
92–106	99.0	208	3.60	2.08	–.117
106–120	113.0	177	3.89	2.66	–.032
≥ 120	131.2	88	4.23	3.13	–.174

*None of the ρ_{rt} differs significantly from zero at the 95% level of confidence.

FIGURE 4-63. Illustrating derivation of radial and tangential weighting functions from residual vectors.

in favor of the use of the film magazine (or magazines) of the camera being calibrated. Alternatively, a special magazine incorporating an ultra-flat platen and a platen reseau (to permit correction of film deformation) could be employed. This latter measure would lead to a calibration that would apply predominantly to the lens itself. In either case, because precise alignment would not be required, a moderate number of frames could be quickly exposed with the camera being rotated about its axis in suitable increments between frames. This procedure would generate a large number of collimator images having an overall distribution far superior to that obtainable from any one frame. Carried in a SMAC reduction, the measurements of all such images would contribute to a common calibration of the interior projective parameters. The large sample of measuring residuals that could thus be generated would provide the material necessary for sound empirical modeling of anomalous distortion and for the derivation of radial and tangential weighting functions appropriate to the camera. Accordingly, laboratory methods of camera calibration are fully compatible with advanced analytical methods.

4.8.8.4.3 CALIBRATION WITH ADJUSTABLE CONTROL

The methods discussed up to this point have in common the fact that available control points are considered to be sensibly free of error and hence not subject to adjustment. In the most advanced level of analytical calibration this assumption is abandoned; calibration and the process of photogrammetric triangulation are simultaneously performed in a single adjustment. Control may range from absolute points to relative points, with all shades in between being admitted through the exercise of appropriate a priori covariance matrices. This

approach has been referred to variously as bundle adjustment with self-calibration, bundle adjustment with additional parameters, and bundle adjustment with block-invariant parameters. The essence of the approach is that the basic bundle adjustment developed in Brown (1958) is extended to embrace parameters accounting for radial and decentering distortion along, possibly, with other parameters intended to account for anomalous distortion (interpreted in the broadest sense to embrace all otherwise unmodeled systematic errors). In this process, calibration of the camera is incidental and not necessarily of primary concern; the primary objective is the performance of the photogrammetric triangulation itself with minimal passthrough of contamination by modelable systematic error.

Two basic formulations of the solution to the bundle adjustment with self-calibration exist. In the first, only a moderate number of frames, usually under ten, are carried. Here, applications are mainly concerned with close-range photogrammetry. As a consequence, high convergence and 100 percent overlap are usually exercised. The general system of normal equations contains not only exterior and interior projective parameters (as in SMAC) but also corrections to approximations to coordinates of photographed points. When the latter are eliminated by the method of partitioning, the resulting reduced system of normal equations generally has a filled coefficient matrix (i.e., one with mainly nonzero elements). Because only a rather limited number of frames are carried in the reduction, the direct solution of this system for the differing exterior projective parameters of the various frames along with the common interior projective parameters of all frames presents no particular difficulties. This straight-forward approach is well suited to a wide range of problems encountered in terrestrial photogrammetry.

The second formulation of the solution to the bundle adjustment with self-calibration is applicable to the typical block of aerial photographs characterized by more or less regular forward and lateral overlap. In the absence of special parameters for calibration, the coefficient matrix of the reduced system of normal equations for such blocks can be made (through proper ordering of unknowns) to assume a banded form in which all nonzero elements are confined to a band about the main diagonal. As shown in Brown (1968), when special parameters for calibration are introduced, the coefficient matrix assumes a bordered-banded form in which the band of the original matrix becomes augmented with a border corresponding to the interior projective parameters common to all frames. This special structure of the normal equations can be exploited in the process referred to as recursive partitioning to effect a practical and efficient reduction, even for blocks containing many hundreds of frames.

Further details of the bundle adjustment with self-calibration are to be found in Brown (1974) and in sections of this manual treating analytical aerotriangulation. Suffice it here to note that the triangulation. Suffice it here to note that the method has proven so effective that it seems increasingly likely to become the preferred method of the future.

4.8.8.4.4 Special Results for Close-Range Applications

Implicit in the foregoing discussions is the assumption that the camera is focused at infinity and that distortion is independent of object distance. In reality, distortion does vary with object distance, though the variation is ordinarily insignificant in mapping and other applications in which the photographed points are removed by a large number of focal lengths from the camera. Here, one can safely employ the infinity calibration of distortion for all photographed points. On the other hand, when object distances are within a few tens of focal lengths of the camera, moderate to severe variation of distortion with object distance may be experienced. Not only does distortion change with

s' = *distance of object plane containing point of interest,*

the process of correcting the radial distortion would proceed as follows.

STEP 1. *Approximate Triangulation.* Using available projective parameters, one employs the measured photographic coordinates to triangulate approximate coordinates X, Y, Z of the photographed point. From these and the coordinates X^c, Y^c, Z^c of the exposure station the approximate distance s' to the object plane containing the photographed point can be evaluated from

$$s' = \left[(X - X^c)^2 + (Y - Y^c)^2 + (Z - Z^c)^2\right]^{1/2} \cos \theta \tag{4.74}$$

where

$$\cos \theta = c_s/(c_s^2 + r^2)^{1/2} \tag{4.75}$$

in which r is the radial distance of the image and c_s is the principal distance of the camera when focused at distance s.

STEP 2. *Computation of Distortion Coefficients for Object Plane at s'.* With the distance s' provisionally determined by approximate triangulation, one evaluates the parameter

$$\alpha_s' = \frac{s_2 - s'}{s_2 - s_1} \frac{s_1 - f}{s - f} \tag{4.76}$$

where s_1 and s_2 correspond to the two object planes for which distortion functions have been calibrated. The coefficients corresponding to the object plane at distance s' are then evaluated from

$$K_{1s'} = \left(1 - \frac{f}{s'}\right)^3 \left\{ \frac{\alpha_s'}{\left(1 - \frac{f}{s_1}\right)^3} K_{1s_1} + \frac{(1 - \alpha_s')}{\left(1 - \frac{f}{s_2}\right)^3} K_{1s_2} \right\},$$

$$\tag{4.77}$$

$$K_{2s'} = \left(1 - \frac{f}{s'}\right)^5 \left\{ \frac{\alpha_s'}{\left(1 - \frac{f}{s_1}\right)^5} K_{2s_1} + \frac{(1 - \alpha_s')}{\left(1 - \frac{f}{s_2}\right)^5} K_{2s_1} \right\}.$$

focal setting, but it also changes throughout the photographic field itself.

It is shown in Brown (1971a) and (1972) that the function defining radial lens distortion for a point in an object plane at distance s' from the camera can be computed from the functions appropriate to two discrete distances s_1 and s_2 (preferably, though not necessarily, $s_1 < s' < s_2$). The Gaussian distortion functions for object-plane distances s_1 and s_2 may be denoted by

$$\begin{aligned} \delta r_{s_1} &= K_{1s_1} r^3 + K_{2s_1} r^5 + \dots \\ \delta r_{s_2} &= K_{1s_2} r^3 + K_{2s_2} r^5 + \dots \end{aligned} \tag{4.73}$$

in which the coefficients are assumed for present purposes to have been pre-established in special calibrations. If it is then further assumed that

f = *Gaussian focal length of lens,*
s = *distance of object plane on which camera is focused,*

Because the camera is not actually focused on the object plane at s', these coefficients require the modification indicated in the next step before they can be employed.

STEP 3. *Evaluation and Application of Distortion Corrections.* From the known distance s of the object plane actually focused on the approximated distance s' of the object plane containing the point of interest one evaluates the parameter

$$\gamma_{s,s'} = \frac{s - f}{s' - f} \frac{s}{s'} \tag{4.78}$$

which is used with the coefficients determined from Step 2 to evaluate $\delta r_{s,s'}$, the distortion function for object-plane s' for a camera focused on object-plane s, in accordance with

$$\delta_{s,s'} = \gamma_{s,s'}^2 K_s' r^3 + \gamma_{s,s'}^4 K_s' r^5 + \dots \tag{4.79}$$

From this and the measured x, y coordinates of the image (referred to principal point as origin), the provisional corrections

$$\delta x = \frac{x}{r} \, \delta r_{s,s'}$$

$$\delta y = \frac{y}{r} \, \delta r_{s,s'}$$
(4.80)

are evaluated.

STEP 4. *Iteration.* The addition of δx, δy to the image coordinates provides an initial correction for distortion. This permits a repetition of analytical triangulation leading to more accurate X, Y, Z coordinates. These, in turn, may be employed in a repetition of Steps 1, 2 and 3 leading to refined corrections for distortion. The iteration of this process to convergence generates the total correction.

The foregoing corrective process may be incorporated into a special close-range version of the bundle adjustment with self-calibration. Here, the distortion function for only one of the two necessary distortion functions need be known in advance (the most convenient would usually correspond to the choice $s_2 = \infty$). The second distortion function would be chosen to correspond to the object-plane s on which the camera is actually focused. The coefficients for this function could then be carried in the bundle adjustment as unknown parameters. As shown in Brown (1972) this process can lead to a sharp recovery of the distortion coefficients.

4.8.8.4.5 ANALYTICAL PLUMB LINE CALIBRATION

Perhaps the most convenient method of calibrating functions of radial and decentering distortion corresponding to close ranges is the analytical plumb line method developed in Brown (1971b). It is based on the projective principle that, in the absence of distortion of any kind, the photographic image of any straight line in object space is itself a straight line. Thus any systematic departure from strict linearity of the image of a straight line can be attributed to distortion. As the name implies, the analytical plumb line method employs exposures of an array of plumb lines in the desired object plane to generate the necessary observational material. The plate coordinates of a moderate number of points are measured on the images of each of several lines well-distributed throughout the format. The set of points measured on the ith line generates a set of observational equations that are functionally of the form

$$f(x_{ij}, y_{ij}; x_p, y_p, K_{1s}, K_{2s}, \ldots; P_{1s}, P_{2s}, \ldots;$$
$$\theta_i, p_i) = 0,$$
$$j = 1, 2, \ldots, n_i$$
(4.81)

in which

x_{ij}, y_{ij} = measured plate coordinates of jth point on i^{th} line;
x_p, y_p = coordinates of principal point;
$K_{1s}, K_{2s}, \ldots; P_{1s}, P_{2s}, \ldots$ = coefficients of

radial and decentering distortion for object-plane s;
θ_i, p_i = constants defining the equation of the undistorted image of the ith plumb line;
r_i = number of points measured on the image of the ith plumb line.

It is noteworthy that the observational equations are independent of all six elements of exterior orientation α, ω, κ, X^c, Y^c, Z^c as well as of the principal distance c_s. This renders the analytical plumb line method unique among analytical methods of calibration. Each measured image of a plumb line introduces two unknowns θ_i, p_i in addition to the set of unknowns common to all plumb lines $(x_p, y_p, K_{1s}, K_{2s}, \ldots; P_{1s}, P_{2s}, \ldots)$ which are here assumed to total q in number. If n then denotes the total number of points measured on the entire set of m plumb lines (i.e., $n = n_1 + n_2 + \ldots + n_m$), the total system of observational equations will consist of n equations in $q + 2m$ unknowns. No matter how large m may be, the system of normal equations generated by the least squares adjustment can be reduced to a $q \times q$ system by applying the same computational algorithms that form the basis for the SMAC calibration. Hence, the analytical plumb line method entails a computational effort that tends to increase only linearly with m rather than with the cube of m as would otherwise be the case. Accordingly, there is no practical limit to the number of images of plumb lines that can be used in the adjustment.

The primary shortcoming of the analytical plumb line method is that it does not yield useful estimates of x_p, y_p, c_s. (Although x_p and y_p, unlike c_s, do appear in the observational equations, they generally cannot be recovered well.) This is not a serious drawback if use of the camera is limited to nonconvergent photography of points lying close to the object plane on which the camera is focused. In this situation, the recovered coordinates of the exposure stations (X^c, Y^c, Z^c) can provide effective projective compensation for errors in enforced values of x_p, y_p, c_s. On the other hand, when moderate to highly convergent photography is employed, it turns out that accurate values of x_p, y_p, c_s can be recovered in the bundle adjustment with self-calibration provided a suitable diversity of swing angles is used (Brown (1972)).

4.8.8.4.6 CONCLUDING CONSIDERATIONS

Conventional laboratory methods of camera calibration are generally limited to:

(a) estimation of sample values of the symmetric radial distortion function;
(b) determination of whether the maximum value of tangential distortion is within acceptable limits (usually $\pm 10 \ \mu m$); and

(c) estimation of elements of interior orientation x_p, y_p, c.

Through the application of analytical methods the scope of calibration can be considerably broadened to include, in addition to the above:

(d) determination of significant coefficients of both radial and decentering distortion;
(e) rigorous propagation of measuring error into all results;
(f) estimation of functions describing anomalous distortion;
(g) estimation of radial and tangential weighting functions.

When fully exploited, the results of an analytical calibration can contribute significantly to improving the results obtainable from photogrammetric triangulation.

BIBLIOGRAPHY

Abdel-Aziz, Yousset I., Asymmetrical Lens Distortion, *Photogrammetric Engineering and Remote Sensing*, v. 41, n. 3, 1975.
———, Lens Distortion at Close Range, *Photogrammetric Engineering*, v. 39, n. 6, 1973.
———, and H. M. Karara, Direct Linear Transformation from Comparator Coordinates into Object-Space Coordinates in Close-Range Photogrammetry, Proceedings of ASP Symposium on Close-Range Photogrammetry, Urbana, Ill., 1974.
Air Force Manual 95-3, Installation and Maintenance of Aerial Photographic Equipment.
Anculete, Eng. Gheorghe, and Eng. Teader Diacomescu, A Methodology Used in Testing and Calibrating Photogrammetric Cameras Under Normal Work Conditions, ISP Congress, Helsinki, Finland, 1976.
Anderson, James M., and Clement Lee, Analytical In-Flight Calibration, *Photogrammetric Engineering*, v. 41, n. 11, pp. 1337–1348, 1975.
Arena, Alfred, and Harry Koper, The KA-92 Camera System, *Photogrammetric Engineering*, v. 40, n. 10, pp. 1225–1235, 1974.
Avera, Harmon Q., The Miniature Camera Calibrator—Its Design, Development, and Use, *Photogrammetric Engineering*, v. 23, n. 3, pp. 601–607, 1957.
Bagley, James W., Aerophotography and Aerosurveying, McGraw-Hill Book Co., New York, N.Y., 1941.
Bean, Russell K., U. S. Geological Survey Camera Calibration, Paper, ASP-ACSM Annual Convention, Washington, D.C., 1962.
Benesh, Milosh, Mariner Mars 1971 Photogrammetric Calibrations, Paper, ISP Congress, Ottawa, Canada, 1972.
Bernath, Hans J., Radiometric Calibration of a Multispectral Camera, *Photogrammetric Engineering*, v. 39, n. 9, 1973.
Berndsen, Clyde E., and Robert G. Livingston, The Aerial Camera, Technical Note 64-1, U.S. Army Engineer GIMRADA Field Office, WPAFB, Ohio, 1964.
———, and Michael W. Onushco, The Business of Looking Down is Looking Up, Proceedings of the Annual ASP-ASCM Convention, Washington, D.C., 1972.
Bormann, G. E., Measurement of Radial Lens Distortion with the Wild Horizontal Goniometer, ISP Commission I, W.G. Study, 1975.
———, The New Wild RC-10 Film Camera, *Photogrammetric Engineering*, v. 35, n. 10, pp. 1033–1038, 1969.
Bottinger, W. V., On Some Aspects of Taking Close-Range Photographs for Photogrammetric Evaluation: Practical Experiences in Photographing the Models of the Cable-Net Roofs for the Olympiad at Munich, International Archives of Photogrammetry, Com. V, ISP, 1972.
Brandow, V. D., et al, Close-Range Photogrammetry for Mapping Geologic Structures in Mines, Proceedings of ASP Symposium on Close-Range Photogrammetric Systems, Champaign, Ill., 1975.
Brown, Duane C., Advanced Methods for the Calibration of Metric Cameras, Proceedings of the Symposium on Computational Photogrammetry, Syracuse University, Syracuse, 1969.
———, An Advanced Reduction and Calibration for Photogrammetric Cameras, Air Force Cambridge Research Laboratories Report 64-40, Cambridge, Mass., 1964.
———, Analytical Aerotriangulation vs. Ground Surveying, paper, Semiannual Meeting of ASP-ACSM, San Francisco, Calif., 1971.
———, Calibration of Close-Range Cameras, paper, XII Congress of ISP, Ottawa, Canada, 1972.
———, Close-Range Camera Calibration, *Photogrammetric Engineering*, August, 1971.
———, Decentering Distortion and the Definitive Calibration of Metric Cameras, *Photogrammetric Engineering*, May, 1966.
———, Evaluation, Application and Potential of the Bundle Method of Photogrammetric Triangulation, paper, ISP Symposium, Stuttgart, Germany, 1974.
———, The Simultaneous Determination of the Orientation and Lens Distortion of a Photogrammetric Camera, RCA-MTP Data Reduction Technical Report No. 33, Patrick Air Force Base, Fla. (AFMTC TR 56-20), 1956.
———, A Treatment of Analytical Photogrammetry with Emphasis on Ballistic Camera Applications, RCA-MTP Data Reduction Tech. Rpt. No. 39 (AFMTC TR 57-22), Patrick AFB, Fla., 1957.
Brucklacher, W. A., Wide-Angle Convergent Photography with Angles of Convergence of 27° or 40°, *Photogrammetric Engineering*, v. 24, n. 5, pp. 786–789, 1958.
Carman, P. D., Camera Calibration Laboratory at N.R.C., *Photogrammetric Engineering*, v. 35, n. 4, pp. 372–376, 1969.
———, Camera Vibration Measurements, *Canadian Surveyor*, v. 27, n. 3, 1973.
———, Control and Interferometric Measurement of Plate Flatness, *Journal, Optical Society of America*, v. 45 p. 1009, 1955.
———, Photogrammetric Errors from Camera Lens Decentering, *Journal, Optical Society of America*, v. 39, p. 951, 1949.
———, and H. Brown, Camera Calibration in Canada, *Canadian Surveyor*, v. 15, n. 8, p. 425, 1961.
———, and H. Brown, Differences Between Visual and Photographic Calibrations of Air Survey Cameras, *Photogrammetric Engineering*, v. 22, n. 4, 1956.
Case, James B., Stellar Methods of Camera Calibration, Manual of Photogrammetry, Third Edition, Chapter IV, p. 180, ASP, Falls Church, Va., 1955.
Conrady, A. E., Decentered Lens Systems, *Monthly

Notices of the Royal Astronomical Society, v. 69, pp. 384–390, 1919.

Corliss, W. R., The Viking Mission to Mars, NASA Special Publication 334, U.S. Govt. Printing Office, Washington, D.C., 1974.

Crouch, L. W., High Performance Mapping Equipment and Material, *Photogrammetric Engineering*, v. 28, n. 3, 1961.

Cutts, J. A., Experiment Design and Picture Data, Mariner Mars 1971 Television Picture Catalog, v. 1, JPL Technical Memo. 33-585, Jet Propulsion Lab., Pasadena, Calif., 1974.

Dohler, M., Nahbildmessung Mit Nicht-Mess Kammern, *Bildmessung und Luftbildwesen*, v. 39, n. 1, pp 67–76, 1971.

Doyle, Frederick J., The Next Decade of Satellite Remote Sensing, *Photogrammetric Engineering and Remote Sensing*, v. 44, n. 2, 1978.

Eisenhart, Churchill, Realistic Evaluation of the Precision and Accuracy of Instrument Calibration Systems, *Journal of Research*, National Bureau of Standards, v. 67c, n. 2, April–June, 1963.

Elms, David G., Mapping with a Strip Camera, *Photogrammetric Engineering*, v. 28, n. 4, pp. 638–653, 1962.

Faig, Prof. Dr. Ing. Wolfgang, Calibration of Close-Range Cameras, ASP Symposium on Close-Range Photogrammetry, Urbana, Ill, 1971.

——, Photogrammetric Potentials of Nonmetric Cameras, *Photogrammetric Engineering and Remote Sensing*, v. 42, n. 1, 1976.

——, Precision Plotting of Nonmetric Photography, Proceedings of Biostereometrics Symposium, Washington, D.C., 1974.

Field, R. H., A Device for Locating the Principal Point Markers of Air Cameras, *Canadian Surveyor*, v. 10, n. 1, pp. 17–21, 1949.

Fimmel, R. O., W. Swindell, and E. Burgess, Pioneer Odyssey: Encounter with a Giant, NASA Special Publication 349, U.S. Govt. Printing Office, Washington, D.C., 1974.

Fritz, Lawrence W., and Hellmut H. Schmid, Stellar Calibration of the Orbigon Lens, *Photogrammetric Engineering*, v. 40, n. 2, pp. 101–115, 1974.

——, and Chester C. Slama, Multi-Plate, Multi-Exposure Stellar Calibration, ISP Commission I WG Study, 1976.

Fulton, Patricia A., and Morris L. McKenzie, Model Flatness Test, Computer Program Documentation, U.S. Geological Survey, Washington, D.C. 1968.

Gardner, I. C., Relation of Camera Error to Photogrammetric Mapping, *Journal of Research*, National Bureau of Standards, v. 22, p. 209, RP 1177, 1939.

——, The Significance of the Calibrated Focal Length, *Photogrammetric Engineering*, v. 10, n. 1, p. 22, 1944.

——, and F. A. Case, Precision Camera for Testing Lenses, *Journal of Research*, National Bureau of Standards, RP 984, 1937.

General Electric Company, ERTS Reference Manual, ERTS-2, General Electric Space Division, Valley Forge, Pa.

Ghosh, S. K., Image Quality vs. Metric Capability, *Photogrammetric Engineering*, v. 39, n. 11, 1973.

Hakkarainen, Juhani, Calibration of Aerial Cameras with a Horizontal Goniometer, *Photogrammetric Journal of Finland*, v. 6, n. 1, 1972.

Hallert, B., The Method of Least Squares Applied to Multicollimator Camera Calibration, ASP-ACSM Semiannual Convention, St. Louis, Mo., 1962.

——, A New Method for the Determination of the Distortion and Inner Orientation of Cameras and Projectors, *Photogrammetria*, v. 11, n. 3, pp. 107–115, 1955.

——, Some Preliminary Results of the Determination of Radial Distortion in Aerial Pictures, *Photogrammetric Engineering*, v. 22, n. 1, pp. 169–173, 1956.

Hartman, W. K., and O. Raper, The New Mars: The Discoveries of Mariner 9, NASA Special Publication 337, U.S. Govt. Printing Office, Washington, D.C., 1974.

Heimes, F. J., In-Flight Calibration of a Survey Aircraft System, ITC, paper, ISP Congress, Ottawa, Canada, 1972.

Hempenius, S. A., The Role of Image Quality in Photogrammetric Pointing Accuracy, European Research Office, U.S. Army, DA-91-591-EUC3721, December, 1969.

Hotine, M., Surveying from Air Photographs, Richard R. Smith, New York, 1931.

Howlett, L. E., Resolution, Distortion and Calibration of Air Survey Equipment, *Photogrammetric Engineering*, v. 16, n. 1, pp. 41–46, 1950.

Itek Laboratories, Panoramic Progress, Part I, *Photogrammetric Engineering*, v. 27, n. 5, pp. 747–766, 1961.

——, Panoramic Progress, Part II, *Photogrammetric Engineering*, v. 28, n. 1, pp. 99–107, 1962.

Journal of Physical Research, The Planet Mercury: Mariner 10 Mission, anonymous, v. 80, n. 17, June 10, 1975.

Karara, H. M., Aortic Heart Valve Geometry, *Photogrammetric Engineering*, v. 40, n. 12, 1974.

——, and W. Faig, Interior Orientation in Close-Range Photogrammetry: An Analysis of Alternative Approaches, International Archives of Photogrammetry, Commission V, ISP, 1972.

Karren, Robert J., Camera Calibration by the Multicollimator Method, *Photogrammetric Engineering*, v. 34, n. 7, pp. 706–719, 1968.

Kistler, Phillip, Continuous Strip Aerial Photography, *Photogrammetric Engineering*, v. 12, n. 2, pp. 219–223, 1946.

Kolbl, O., Selbskalibrierung von Aufgahmekammern, *Bild. u. Luftb.*, v. 40, n. 1, pp. 31–37, 1972.

Kosofsky, L. J., Lunar Stereo Photography, *Photographic Society of America Journal*, December, 1970.

——, Moon Revisited in Stereo, *Photographic Society of America Journal*, April, 1973.

——, and F. El-Baz, The Moon as Viewed by Lunar Orbiter, NASA Special Publication 200, U.S. Govt. Printing Office, Washington, D.C., 1970.

LeResche, John, Analysis of the Panoramic Aerial Photograph, *Photogrammetric Engineering*, v. 24, n. 5, pp. 772–775, 1958.

Levine, Hal, and Seymour Rosin, The Geocon IV Lens, *Photogrammetric Engineering*, v. 36, n. 4, pp. 335–342, 1970.

Lewis, James G., A New Look at Lens Distortion, *Photogrammetric Engineering*, v. 22, n. 4, pp. 666–673, 1956.

Lipskiy, Yu. N., et al, Atlas of the Reverse Side of the Moon, Part II, Nauka Press, Moscow, 1967.

Livingston, Robert G., The Attainment of Quality in the Military Development of Airborne Mapping Systems, *Photogrammetric Engineering*, v. 32, n. 3, pp. 390–407, 1966.

——, A History of Military Mapping Camera Development, *Photogrammetric Engineering,* v. 30, n. 1, pp. 97–110, 1964.

———, A Modern System for Aerial Mapping, *Military Engineer,* v. 59, n. 389, pp. 172–173, 1967.

———, Tangential Distortion in the Metrogon Lens, Technical Report 1219, Engr. R. & D. Labs., Fort Belvoir, Va., 1950.

Magill, A. A., Variation in Distortion with Magnification, *Journal of Research,* National Bureau of Standards, v. 54, pp. 135–142, 1955.

Malhotra, R. C., and H. M. Karara, A Computational Procedure and Software for Establishing a Stable Three-Dimensional Test Area for Close-Range Applications, Proceedings of the ASP Symposium on Close-Range Photogrammetric Systems, Champaign, Ill., 1975.

McNeil, Gomer T., The Normal Angle Calibration, *Photogrammetric Engineering,* v. 28, n. 4, pp. 633–637, 1962.

———, Underwater Camera Calibrator, *SPIE Journal,* v. 4, 1965.

———, Underwater Photography, *Photogrammetric Engineering,* v. 35, n. 11, pp. 1135–1151, 1969.

Meier, H. K., Film Flattening in Aerial Cameras, *Photogrammetric Engineering,* v. 38, n. 4, pp. 367–372, 1972.

———, The Effect of Environmental Conditions on Distortion, Calibrated Focal Length, and Focus of Aerial Survey Cameras, ISP Symposium, Tokyo, Japan, May, 1978.

Merchant, Dean C., An Analysis of Aerial Photogrammetric Camera Calibrations: A Summary Report, The Ohio State University Research Foundation, Geodetic Science Report No. 264, Columbus, Ohio, 1977.

———, Calibration of the Air Photo System, *Photogrammetric Engineering,* v. 40, n. 5, pp. 605–617, 1974.

Merritt, E. L., Application of Air Camera Calibration Procedures to In-Water Cameras, Proceedings of American Society of Photogrammetry, Sept., 1974.

———, Field Camera Calibration, *Photogrammetric Engineering,* v. 14, n. 2, pp. 303–309, 1948.

———, Goniometer Method of Camera Calibration, Report of U.S. Naval Interpretation Center, 1950.

———, Investigation of Nodal Points in Close-Range Camera Systems, Proceedings of American Society of Photogrammetry, Oct., 1973.

———, Methods of Field Camera Calibration, *Photogrammetric Engineering,* v. 17, n. 4, pp. 610–635, 1951.

———, Methods of Field Camera Calibration, Part IV, *Photogrammetric Engineering,* v. 18, n. 4, pp. 665–678, 1952.

———, Procedure for Calibrating Telephoto Lenses, Proceedings of ASP Symposium on Close-Range Photographic Systems, Champaign, Ill., 1975.

———, Terrestrial Exposure Method of Field Calibration, U.S. Naval Photographic Interpretation Center, Report No. 132–150, 1950.

Mutch, T. A., et al, Imaging Experiments: The Viking Lander, Icarus, v. 16, n. 1, February, 1972.

Norton, C. L., Camera Calibration Laboratory, *Photographic Science and Technique,* 17B, 1951.

———, The Fairchild Precision Camera Calibration, *Photogrammetric Engineering,* v. 16, n. 5, pp. 688–695, 1950.

———, Image Properties with Environmental Factors, Interim Report of WG-3, International Society of Photogrammetry, Tokyo, Japan, 1978.

———, Production Control of Factors Affecting the Calibration of a Photogrammetric Camera, *Photogrammetric Engineering,* v. 20, n. 3, pp. 502–506, 1954.

———, Gerald C. Brook, and Roy Welch, Optical and Modulation Transfer Functions, *Photogrammetric Engineering and Remote Sensing,* v. 43, n. 5, 1977.

O'Connor, Desmond C., Factors Affecting the Precision of Measurements on Photographs, USAETL Report, Fort Belvoir, Va., 1968.

Pallme, Ernest H., Photo System Installations in Aircraft, *Photogrammetric Engineering,* v. 21, n. 5, pp. 765–772, 1955.

Pennington, John T., Tangential Distortion and Its Effects on Photogrammetric Extension of Control, *Photogrammetric Engineering,* v. 13, n. 1, p. 135, 1947.

Quick, J. Robert, Aerial Cameras for Mapping, Technical Report 70-3, USAETL Field Office, Wright-Patterson Air Force Base, Ohio, 1970.

Roberts, A., Torquer Stabilized Mount for Convergent Mapping Cameras, *Photogrammetric Engineering,* v. 24, n. 5, pp. 744–750, 1958.

Roelofs, R., Distortion, Principal Point, Point of Symmetry and Calibrated Principal Point, *Photogrammetria,* v. 7, n. 2, pp. 49–66, 1950-51.

Rosenbruch, K. J., Considerations Regarding Image Geometry and Image Quality, *Photogrammetria,* v. 33, pp. 155–169, 1977.

Rosenfield, George H., Fixed Camera Data Reduction Analysis Procedures, RCA Data Reduction Analysis Procedures, 1958.

Salmenpera, Hannu, Camera Calibration Using a Test Field, *Photogrammetric Journal of Finland,* v. 6, n. 1, 1972.

Schmid, Hellmut H., An Analytical Treatment of the Orientation of a Photogrammetric Camera, Report No. 880, Ballistic Research Labs., Aberdeen Proving Ground, Md., 1953.

Sewell, Eldon D., Distortion—Planigon versus Metrogon, *Photogrammetric Engineering,* v. 20, n. 1, pp. 761–764, 1954.

———, Field Calibration of Aerial Mapping Cameras, *Photogrammetric Engineering,* v. 14, n. 3, pp. 363–398, 1948.

———, Field Calibration of Mapping Cameras, Manual of Photogrammetry, Second Edition, Chapter IV, 1952.

———, and Robert G. Livingston, Errors in Ground Positions Caused by Irregularities in the Locating Backs of Aerial Cameras, Technical Report 1137, Engineer R. & D. Labs., Fort Belvoir, Va., 1949.

———, and Robert G. Livingston, Field Resolution Tests with the Metrogon Lens, Technical Report 1176, Engineer R. & D. Labs., Fort Belvoir, Va., 1950.

Slama, Chester C., A Mathematical Model for the Simulation of a Photogrammetric Camera Using Stellar Control, NOAA Technical Report NOS 55, Rockville, Md., December 1972, pp. 138.

Slater, P. N., Multiband Cameras, *Photogrammetric Engineering,* v. 38, n. 6, pp. 543–555, 1972.

Specification, Mann Type 2405 Automatic Precision Comparator, David W. Mann Co., Burlington, Mass., 1972.

Spriggs, Robert M., and Jack Rankin, Air Force Camera Calibration Facility, Technical Report 70-5, USAETL Field Office, Wright-Patterson Air Force Base, Ohio, 1970.

Sweet, Robert M., Lunar Mapping Camera, Optical

Alignment, Calibration and Testing Considerations, Fairchild Space and Defense Systems, Syosset, N.Y.

Tafel, R. W., Comparative Evaluation of Panoramic Camera Reconnaissance Systems, APEL Report No. NADC AP-6204, DDC Document 277908, 1962.

Talley, B. B., Engineering Applications of Aerial and Terrestrial Photogrammetry, Pitman Publishing Co., New York, N.Y., 1938.

Tayman, William P., Analytic Multicollimator Camera Calibration Report, Commission I, Working Group on Image Geometry, ISP, October, 1975.

———, Calibration of Lenses and Cameras at the USGS, *Photogrammetric Engineering*, v. 40, n. 11, pp. 1331–1339, 1974.

Thompson, E. H., The Geometrical Theory of the Camera and Its Application to Photogrammetry, *The Photogrammetric Record*, v. 2, n. 10, p. 241, 1957.

Torlegard, K., On the Determination of Interior Orientation of Close-Up Cameras Under Operational Conditions Using Three-Dimensional Test Objects, thesis, Technical University of Stockholm, 1967.

Trott, Timothy, Development of Aerial Camera Stabilization and Its Effect on Photogrammetry and Photo Interpretation, *Photogrammetric Engineering*, v. 23, n. 1, pp. 122–150, 1957.

Washer, Francis E., Calibration of Airplane Cameras, *Photogrammetric Engineering*, v. 23, n. 5, pp. 890–891, 1957.

———, The Effect of Camera Tipping on the Location of the Principal Point, *Journal of Research*, National Bureau of Standards, RP 2691, v. 57, p. 31, 1956.

———, The Effect of Prism on the Location of the Principal Point, *Photogrammetric Engineering*, v. 28, n. 3, pp. 520–532, 1957.

———, Locating the Principal Point of Precision Airplane Mapping Cameras, RP 1428, National Bureau of Standards, Washington, 1941.

———, The Precise Evaluation of Lens Distortion, *Photogrammetric Engineering*, v. 29, n. 2, pp. 327–332, 1963.

———, Prism Effect, Camera Tipping and Tangential Distortion, *Photogrammetric Engineering*, v. 23, n. 3, pp. 520–532, 1957.

———, A Simplified Method of Locating the Point of Symmetry, *Photogrammetric Engineering*, v. 23, n. 1, pp. 75–88, 1957.

———, Sources of Error in Camera Calibration, *Photogrammetric Engineering*, v. 20, n. 3, pp. 500–501, 1954.

———, Sources of Error in Various Methods of Airplane Camera Calibration, *Photogrammetric Engineering*, v. 22, n. 4, pp. 722–740, 1956.

———, and F. A. Case, Calibration of Precision Airplane Mapping Cameras, *Photogrammetric Engineering*, v. 16, n. 4, pp. 502–524, 1950.

———, and W. R. Darling, Evaluation of Lens Distortion by the Modified Goniometric Method, *Journal of Research*, National Bureau of Standards, v. 63c, n. 2, pp. 113–120, 1959.

———, W. P. Tayman and W. R. Darling, Evaluation of Lens Distortion by Visual and Photographic Methods, *Journal of Research*, National Bureau of Standards, v. 61, n. 6, pp. 509–515, 1958.

Welender, Erik, and Alf Smedberg, The Calibration of Aerial Cameras for Practical Purposes, *Svensk Lantmateritidskrift*, v. 2, 1972.

Wight, Ralph H., Second-Order Distortion Effects in Cartographic Lenses, paper, ASP-ACSM Semiannual Convention, San Antonio, Texas, October, 1968.

Wolf, P. R., and S. A. Loomer, Calibration of Nonmetric Cameras, Proceedings of the ASP Symposium on Close-Range Photogrammetric Systems, Champaign, Ill., 1975.

Wolvin, John H., Precision Automatic Photogrammetric Intervalometer, *Photogrammetric Engineering*, v. 21, n. 5, pp. 773–778, 1955.

CHAPTER V

Aerial Photography

Co-Author-Editors: A. Norman Brew
Herbert M. Neyland

5.1 Introduction

THE AERIAL PHOTOGRAPHY described in this chapter is obtained with conventional optics, using the visible portion of the spectrum and silver bromide emulsions sensitized, when necessary, for specific wave lengths. Its purpose is to provide photographs useful in studying the surface of the earth. The principal uses, discussed in accordance with the objectives of this MANUAL, are for surveying and mapping and other applications involving measurement. The other major use of aerial photography, photointerpretation, is the subject of the MANUAL OF REMOTE SENSING.

Aerial photography is the basic source of data for applying photogrammetry to the making of maps. It is in this field especially that photography becomes a science as well as an art. The aerial photograph is the result of the combined scientific and productive efforts of the optical designer (chapter III), the camera manufacturer (chapter IV), the producer of photographic materials (chapter VI), the airplane manufacturer, the pilot and photographer, and the people who process the various photographic products. The quality of the aerial photograph is greatly dependent upon weather conditions and the position of the sun at the time the photograph is taken.

The net products of any photographic mission are the photographic negatives. The positive reproductions from the negatives are of prime importance, since the positives usually become the medium from which the desired information is derived.

In this chapter, the types and principal uses of aerial photography are described, as well as the factors which influence the planning and execution of the photographic mission and the procurement of aerial photographs in accordance with specifications.

5.2 Types of Aerial Photography

Aerial photography is usually classified according to the following criteria:

orientation of the camera's axis (vertical or oblique);
lens system (single or multiple);
spectral range (infrared, visible, radar); and
mode of scanning (single frame, continuous-strip, panoramic).

5.2.1 ORIENTATION OF CAMERA AXES

5.2.1.1 VERTICAL

A vertical photograph (figure 5-1) is one taken with the axis of the camera lens as nearly vertical as possible. Truly vertical aerial photographs must be considered "fortunate accidents" because of the many factors that make absolute verticality practically impossible, even though many advances have been made in the development of stabilized camera mounts. Gyroscopically controlled mounts and their present-day capabilities are described in chapter IV. Vertical photography is by far the most widely used type of aerial photography for mapping, land-use studies, and acquisition of general information principally because the imagery on the photographs is more easily converted to the orthographic projection of the map. When the axis of the camera deviates from the vertical, a tilted photograph results. The tilt of the photograph must not exceed the ability of the plotting equipment to accommodate it.

The advantages of vertical photography over oblique photography are:

measurements are more easily made since the geometrical relationship is less complicated;
detection and recognition of objects are aided by the more nearly normal shapes of images;
there is less hidden ground in the imaged area because of less obliquity of the rays forming the images.

This chapter is a revision of the chapter "Aerial Photography" by W. E. Harman, published in the Third Edition of the MANUAL OF PHOTOGRAMMETRY.

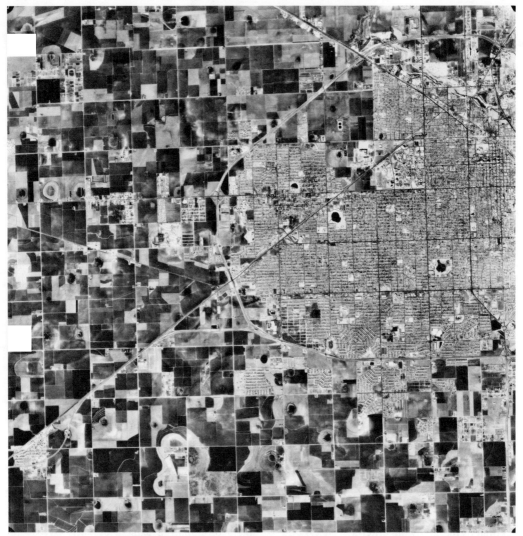

FIGURE 5-1. Vertical photograph taken with a 6-inch focal length camera. The original 9- by 9-inch format showed imagery at an approximate scale of 1:80,000.

5.2.1.2 OBLIQUE

An oblique photograph is one taken with the axis of the camera lens intentionally directed between the horizontal and the vertical. A high-oblique (figure 5-2) is a photograph whose field of view includes a portion of the horizon; the field of view of a low-oblique does not. Oblique photographs are used principally because the total area photographed is increased over that of a vertical picture taken at the same exposure station with the same camera. In a high-oblique, a tremendous area is visible, but detection and recognition of imagery becomes increasingly difficult as the distance from the camera increases. The usefulness and value of an oblique photograph also decrease as the height of the surrounding terrain increases, because some areas may be hidden by high hills or mountains.

High-oblique photographs have been suc-

cessfully used for some kinds of mapping and charting, usually at a relatively small scale. In one method of using high-obliques, developed in Canada, the photographs were usually obtained at an elevation of about 5,000 feet, with the flight lines about 6 miles apart. High-oblique photographs taken at an elevation of over 5,000 feet have been successfully used for reconnaissance maps of unexplored and inaccessible territory.

High-oblique photography was also used extensively during World War II by means of the trimetrogon system. The trimetrogon system was developed in order to obtain as great a coverage as possible for the compilation of maps and charts. In the trimetrogon system, two of the three cameras are mounted obliquely, usually at 60 degrees to the vertical, to cover areas including the horizons on either side of the aircraft and overlapping part of the area covered by the mid-

FIGURE 5-2. High-oblique photograph showing a portion of Washington, D.C. (Photograph courtesy of Air Photographics Inc.)

dle camera which is mounted vertically. The system derived its name from the fact that the three cameras used were equipped with the Metrogon-type lens; however, any type of lens with an appropriate focal length (usually 6 inches) would have served the same purpose. This system has been used by branches of the U.S. Government for aeronautical charts and reconnaissance-type maps of unexplored territory in many parts of the world.

In addition to their use in reconnaissance mapping, high-oblique aerial photographs are often used for pictorial purposes, such as for advertising and for the recording of changes in natural features caused by flooding, hurricanes, etc. or to monitor progress on construction projects.

Low-oblique photographs (figure 5-3) have been successfully used for many mapping and military purposes. Some of the commonly used plotting instruments are designed to use photography having specified degrees of intentional tilt. One of the most successful low-oblique systems was designated "convergent low-oblique." Two cameras were mounted as closely together as possible with their respective axes tilted 20 degrees in opposite directions from the vertical, along the line of flight. The forward-looking camera at one exposure station photographed the same area as was covered by the backward-looking camera at the next station. Using compatible stereoscopic instruments, the lens axes of these two photographs converged to form a stereoscopic model which gave the system its name.

The principal requirements for convergent low-oblique photography, assuming suitable metric and geometric characteristics, were that the angularity between the lens axes, once fixed, should remain the same within a very close tolerance, and that the shutters should be synchronized to trip within 1/200 second of each other.

The principal advantages of this type of photography were that it allowed an increase in

FIGURE 5-3. Low-oblique showing the financial district of New York. (Photograph courtesy of Air Photographics Inc.)

base-height ratio, which resulted in accentuated stereoscopic perception of relief, and that it provided increased area coverage in each stereoscopic overlap. Stereoscopic models from convergent low-oblique photographs covered about 2.2 times the area covered by models from vertical photographs taken at the same flight height with the same type camera. The disadvantages were the loss of resolution and imagery in the transformed prints used for reference (although stereoscopic plotting was done with untransformed diapositives), the relatively small scale in the portion of the photograph representing the area farthest from the camera, and the specialized maintenance and skills required to operate the camera system and the associated photogrammetric equipment.

The twin low-oblique camera installation was occasionally used in the transverse position—that is, with the axes of the two lenses oriented in a plane perpendicular to the line of flight. The total lateral coverage from the outer edge of one photograph to the outer edge of the other was 3.1

times the flight height as compared to 1.6 in line of flight. This type of photography was successfully used by the U.S. Geological Survey for compilation of a large area in northern Alaska in 1957. In that locale, favorable weather is always at a premium and the rough, inaccessible terrain made the location and placement of control points very difficult. Therefore, maximum coverage during each flight mission was a necessity to reduce flight time, and maximum coverage by each photograph was desirable to minimize the amount of control needed. Again, special skills and equipment were required to use this type of photography successfully.

Today, twin low-oblique photography has been supplanted by superwide-angle photography which is taken with a 120 degree lens with a nominal 88 mm focal length. Superwide-angle photography has the advantage of greater area coverage than the 6-inch focal length, convergent system, without the maintenance problems. At the same flight height and with the same, format, a vertical superwide-angle photograph will cover 3 times the area of a vertical photograph taken with a 153 mm focal length camera with a 90 degree wide-angle lens.

The availability of high-altitude jet aircraft for aerial photography has reduced the need for superwide-angle photography in some instances, since the same area can be covered by a 6-inch focal length camera at the higher altitudes with less of the undesirable effects on imagery appearance in the peripheral areas of the photograph caused by the oblique rays of the 120 degree lens.

5.2.2 LENS SYSTEMS

5.2.2.1 SINGLE-LENS SYSTEMS

The single-lens system is by far the most widely used system in aerial photography, whether it be vertical or oblique. The most popular of the single-lens systems is the 153 mm focal length with a 228×228 mm format. Most photogrammetric instruments, even those accepting other focal lengths, are primarily designed for 153 mm focal length photography. The superwide-angle-lens camera is used less frequently and is usually restricted to low relief terrain. The 210 mm and 305 mm focal length cameras give more uniform illumination of the picture format because of the narrower cone of rays accepted by the lens, but heights cannot be measured as accurately from the photographs because the rays in stereoscopic models intersect at more acute angles.

5.2.2.2 MULTIPLE-LENS

Multiple-lens systems may consist of two or more separate cameras mounted or joined together so as to maintain a fixed angle between their respective optical axes, or may consist of two or more lenses mounted in the same camera body, with the respective lens axes systematically arranged at fixed angles in order to obtain at one instant a series of pictures. In the first type, the shutters are usually synchronized to obtain simultaneous exposures. In the second type, the lenses may have separate synchronized shutters or a common shutter for simultaneous exposures. The successful use of multiple-lens photographs is largely dependent upon the accuracy with which the relationships between the lens axes can be calibrated and retained. Suitable plotting equipment capable of the same order of accuracy must be available. The usefulness of the resulting prints is also dependent, in some instances, upon the accuracy with which the component pictures can be rectified and transformed into a single, composite photograph.

In the early days of aerial photography, various types of multiple-lens systems were used for mapping. The number of lenses installed in the camera varied from two to nine. The U.S. Air Force used a camera, designated the T3A, during the early 1920's which contained five lenses, one exposing a vertical negative and the other four exposing oblique negatives. The resulting photographs, one vertical and four oblique, were assembled to make a composite photograph. The obliques were transformed into the equivalent vertical by use of a special transforming printer. Multiple-lens photographs were at one time used extensively by the U.S. Coast and Geodetic Survey (now National Ocean Survey). These photographs were obtained with a 9-lens camera that was developed specifically for the mapping requirements of that organization. The Zeiss Company in Germany also developed and successfully used a 9-lens camera for mapping and charting purposes. Such cameras are seldom used now; modern aircraft and improved vertical photography systems have made them obsolete.

The multiple-lens camera systems in use today are usually termed multispectral because different filters and different films are used to image the same scene simultaneously. This provides the photointerpreter with additional data to analyze ground information. The image format of these cameras is usually about 70 mm square, which keeps the size and weight of the four or more cameras in a cluster, manageable. One system offered incorporates four lenses and filters in a single camera body and provides four separate images on a single frame of 9-inch wide aerial roll film. The MANUAL OF REMOTE SENSING, Volumes I and II, 1975, covers multispectral photography in great detail.

5.2.3 SPECTRAL RANGE

Aerial photography for photogrammetric use is obtained almost exclusively in the optical range (*i.e.*, between 0.4 and 0.8 micrometres) or in the near infrared (from 0.8 to 0.9 micrometres).

Photography further into the infrared requires special emulsions, special conditions and is done for photographic interpretation. It is often referred to as "thermal photography."

There is no "photography" in the radio part of the spectrum. "Radar photography" is the photography of radar information displayed in an oscilloscope.

5.2.3.1 INFRARED PHOTOGRAPHY

There is often some confusion about infrared photography, especially as regards the use of infrared film as a means of detecting thermal changes in the low intensity range. Heat radiation from buildings or thermal variations occurring in water bodies cannot be detected by photographing them with conventional cameras using red filters and infrared film. Unless the object is hot enough so that it begins to emit light at approximately 250° C and is photographed under controlled conditions in a darkened room, it will not register on infrared film. An indirect photographic technique called electronic thermography is employed to detect low-level heat emissions. The infrared waves emitted by the body are collected and directed onto a supercooled photodetector which converts them into electronic signals which can be recorded on magnetic tape or displayed on a cathode-ray tube where it can then be photographed. Such long-wave radiation detection can be carried out in complete darkness. *See* section 5.5 for a discussion of infrared film.

5.2.3.2 RADAR PHOTOGRAPHY

Radar photography is a combination of the photographic process and electronic techniques. Electrical impulses are sent out in predetermined directions and the reflected or returned rays are converted to images on cathode-ray tubes. Photographs are then taken of the information displayed on the tubes. This type of photography and its application to the photointerpretation is described in the MANUAL OF REMOTE SENSING.

5.2.4 MODES OF SCANNING

Most aerial photographs are taken by single-frame cameras. The resulting photographs are the least expensive, simplest to take, and easiest to use. They usually provide the highest accuracy in the transference of data to the map. For some applications, this photography has two characteristics that are disadvantages—the field of view is restricted, and the resolution decreases with distance from the center of the photograph. Two other modes of photography have been developed to remove these deficiencies— panoramic and continuous strip photography. Both modes scan, at each instant, only as much of the ground as can be seen through a very narrow slit.

Panoramic photography (chapter IV, section 4.2.2) is taken by moving a slit perpendicularly to the flight path, followed by the optical axis of the camera lens. Photography is through that part of the lens giving the best resolution and least distortion.

Horizon-to-horizon photography can be obtained by scanning the terrain from most practical altitudes, with high resolution for the entire scan. However, distortions are introduced requiring special and expensive treatment for removal.

Continuous-strip photographs (chapter IV) are taken with the film passing continuously over a narrow slit in the focal plane, at a speed compatible with ground speed of the aircraft. This method dispenses with the conventional shutter. The principal advantage of this type of photography is an extremely sharp picture scanned in one long strip instead of separate frames; it can be taken under conditions that would preclude the use of the ordinary shutter-equipped camera. It is ideal for low-altitude reconnaissance type photography.

5.3 Uses for Aerial Photography

5.3.1 PHOTOINTERPRETATION

Photointerpretation, for both civil and military purposes, is "the art of examining photographs for the purpose of identifying objects and judging their significance." The MANUAL OF REMOTE SENSING, published by the American Society of Photogrammetry, is the standard reference on this subject. Almost everyone who examines a photograph becomes a photointerpreter, since he derives some information from the photograph. His interpretation of the imagery is based on his knowledge and experience. Obviously, a photomicrograph of a specimen of diseased tissue would impart more information to a trained medical technician than it would to a person with slight medical knowledge. The professional photointerpreter, therefore, because of training and experience, is able to detect and recognize objects on photographs which could be overlooked by the amateur.

The amount of information that can be gleaned from an aerial photograph by trained photointerpreters is almost limitless. It is therefore apparent that scientists such as the soil analyst, geologist, forester, geographer and hydrologist should be well versed in photointerpretation.

More recently, land-use and wetland boundaries have become vitally important for observation of economic growth and the conservation of our natural resources. A variety of films and cameras are required in this work. The skills of the photointerpreter are often used by legislators and courts formulating decisions in these matters.

5.3.2 STEREOCOMPILATION PHOTOGRAPHY

Stereocompilation photography is intended for use in stereoscopic instruments (chapter XI) for the production of maps or charts. There are two qualities which each aerial photographic negative must have if it is to be used for mapping purposes. First, it must accurately portray the terrain so that the topographic and planimetric details can be detected, identified, and transferred to the map. Second, the geometrical characteristics of the photograph must be such that the relationship between the images can be measured and their relative positions be established as they existed at the time of the exposure.

5.3.3 ORTHOPHOTOGRAPHY

Aerial photographs intended for orthophotos and photomaps must have all the qualities necessary for stereocompilation photography. In addition, since the image itself becomes the final map product, great care must be used in handling the film. Photomaps vary from the basic image printed with a legend for identification, to a map on which all planimetric and name data is overprinted on the orthophoto image. Photomaps of flat terrain can be made by simple rectification (removal of tilt effect) of the original photograph. In areas of relief, equipment is used which not only removes the image displacement due to camera tilt but also differentially removes the image displacement due to relief (chapter XV).

5.3.4 ANALYTICAL AEROTRIANGULATION

Analytical aerotriangulation (chapter IX) is a mathematical method of extending, with the aid of a computer, horizontal and vertical control (positions and elevations of points on the ground) into uncontrolled areas by means of aerial photographs. The calibration data for the film platen and the lens of the taking camera must be known. Any lens distortions, along with information about the characteristics of the film, data on atmospheric refraction and earth curvature, form part of the computer program applied to obtain the geodetic coordinates of points on the photograph whose coordinates are read on a precision comparator. The adjusted data are then used to establish the photogrammetric control needed for mapping. The aerial photography obtained for aerotriangulation of a specified area is usually required to have unbroken flight lines and to be taken with the same camera, in the shortest time possible.

5.4 Aerial Camera Requirements

The design and construction of the camera and auxiliary equipment, coupled with the efficiency of the lens-shutter combination, have an extremely important influence on the quality of the aerial photographs. A previously calibrated precision-type camera should be required for any photographs to be used in stereocompilation, aerotriangulation, or orthophotography. In all aerial photography, the factors enumerated below are of prime importance and should be the subject of detailed investigation before commencing operations.

5.4.1 THE CAMERA

The camera should be calibrated so that the geometric relationship of the lens, focal plane fiducial marks, etc., are defined and can be taken into account when using the resulting system for photogrammetric work. See chapter IV for camera calibration procedures.

5.4.2 LENS CHARACTERISTICS

The lens should be free of aberrations, it should have high resolving power under operating conditions, and it should be "distortion-free," i.e. have minimal distortion. Information concerning distortion, resolution, focal length, angle of view, and other pertinent characteristics of the lens is obtained from the camera calibration procedures described in chapter IV. The calibration certificate is a record of a series of precise measurements of certain components and their relationships in the camera. If these measured values fall within prescribed limits, it is assumed that the camera lens will produce negatives satisfactory for orthophotos and stereocompilation photographs. The lens is but one of the components of a photogrammetric system and its efficiency as part of the system should be predetermined. An actual performance test should be made, if possible, before assigning the camera to a project; that is, reproductions from negatives taken with the camera should be tested on the equipment to be used in the mapping project.

The test can be made with negatives exposed over an area containing a suitable number of known control points, or with negatives exposed in a camera calibrator. The advantage of testing the camera in the calibrator is that the intrinsic properties of the camera itself can be determined. The anomalies and errors such as vibration, which might be introduced into the test by the aircraft, are eliminated. If, in a later flight or flight test, some deviation is detected in the performance of the camera in the system, the fault can be ascribed to the flight operation.

5.4.3 SHUTTER

The shutter should be sufficiently efficient to assure its proper performance at all times. Shutter speeds as given on the camera should be adjusted based on their efficiency as determined in the camera calibration test. Correct exposure is dependent upon the accuracy of the shutter speed settings.

5.4.4 FLARE

The camera should be so constructed as to admit image-forming light rays only, and to eliminate all flare or extraneous light which might tend to fog the aerial negative. Flare reduces image contrast and image definition.

5.4.5 PLATEN

The platen or plate against which the film is held at the instant of exposure should be flat to the required tolerance. The generally accepted tolerance for a 9″ × 9″ format cartographic camera is that the platen surface shall not depart more than 13 micrometres from a true plane.

5.4.6 METHOD OF HOLDING THE FILM FLAT

The device for holding the film flat at the instant of exposure should perform dependably under all conditions. Film which is not in exact contact with the platen at the instant of exposure will produce warped stereoscopic models. Although there are several methods of ascertaining that the vacuum or pressure is being applied to the film at the instant of exposure, there is no positive method now available to prove definitely that all parts of the film being exposed are in direct contact with the platen. Unless there is enough sag in the film to throw the image out of focus, incomplete contact normally cannot be detected until the models are set up in the plotting instruments. Areas which have not been held flat against the platen will show as depressions in the stereomodel.

5.4.7 MAGAZINE

The magazine should be of such size as to accommodate the amount of film needed for the photographic flight. It should convey the film to the focal plane easily and efficiently and should be easily interchangeable. The film transport mechanism should not put any unusual strain on the film. Regular cleaning of the magazine with compressed air removes particles of dirt or film which can cause scratches on the film.

5.4.8 RECORDING INSTRUMENTS

For many types of surveys, the photogrammetrist finds it advantageous to have recorded on each negative, data such as the altimeter reading, level bubble, time of day, date, camera number, magazine number, and other information which may be helpful.

5.4.9 TYPES OF FILTERS

Filters are of two general kinds.

Color-correction filters—These are filters which absorb certain colors and allow others to pass to the film. The filters most commonly used in aerial photography are some gradation of yellow, in order to filter out the blue color present in the atmosphere, but various other types and combinations are used when conditions warrant. The color of the filter should be such that all pertinent details can be recorded. An excellent reference on filters and their use can be found in the MANUAL OF COLOR AERIAL PHOTOGRAPHY, subchapter 3.5.

Antivignetting filters—Since the ordinary wide-angle lens system allows much more light to pass through the center than passes through the edges, a method has been devised which holds back the amount of light passing through the center and allows the edges to get an exposure almost equal to that of the center. The device used is called the antivignetting filter. Usually the color-correction filter in front of the lens is coated with a vaporized metal in vacuum to the proper density gradation, to give the antivignetting effect. The opposite sides of the filter must be optically flat and should be parallel within a tolerance that will preclude any deformation of the light rays.

5.4.10 MOUNTS

The camera mount and its efficiency have a significant influence on the quality of the resulting aerial photography. An efficient camera mount will, at the instant of exposure, hold the camera free of vibration, in order to prevent deterioration of images due to movement, and vertical within prescribed specifications, to minimize tilt. Most important, it will prevent angular movement of the lens axis during exposure. This angular movement is the most serious cause of blurred photographs. The most desirable camera mount is one in which all three rotational axes pass through the center of gravity.

The camera should be held as vibration-free as possible so as to take advantage of the latest improvements in the resolving power of lenses and the increasing ability of the aerial film manufacturers to produce emulsions having finer grain and thus greater resolution. The isolation of vibration and other movements relative to the frame of the airplane is exceedingly difficult to achieve because of the varying amounts and frequencies of vibrations set up by the engines.

5.5 Film Types

The word "film" as used here means the emulsion and the supporting base. The emulsion should be fine-grained, and have a high sensitivity and a high intrinsic resolving power. The emulsion support should be dimensionally stable.

The film base is a component of the utmost importance in a photogrammetric mapping system. Since the prime objective in any photomapping procedure is to depict the imagery derived from the photograph in its true relationship, any nonuniform deviation in size of the film caused by shrinkage, expansion, and distortion factors is to be avoided. Errors from this source have been largely eliminated with the development of modern dimensionally stable film bases. Although these film-base materials are very tough, great care should be exercised in all phases of handling and processing to preclude any deformation or blemishes.

5.5.1 PANCHROMATIC FILMS

The most popular film emulsion for aerial photography is the panchromatic type. Because its sensitivity extends from the blue through the red portion of the spectrum, it is usually desirable to use a minus blue (yellow) or light red filter to reduce the effect of haze and smog. There is also a greater latitude in exposure and processing of black and white panchromatic films than there is with color films, which assures a greater chance of success in every mission. In addition, all photogrammetric equipment will accept black-and-white imagery, but not all orthophoto equipment and anaglyphic photogrammetric equipment can be used with color films.

5.5.2 INFRARED FILMS

Black-and-white infrared photographic emulsions are sensitized to record mostly in the near infrared portion of the spectrum. But as they also have some sensitivity to blue and green, the camera is usually equipped with either a deep red filter which passes only the red rays, or with a minus blue filter to eliminate blue from haze. One advantage of infrared emulsion is that by utilizing its greater sensitivity to red and the near infrared, it records the longer red light waves which penetrate haze and smoke, and thus it can be used on days that would be unsuitable for ordinary panchromatic films. It is also useful for the delineation of water bodies and for certain types of forestry and land-use studies. Its chief disadvantage is a greatly increased contrast which may tend to cause loss of information. Compare the two photographs in figure 5-4.

5.5.3 COLOR FILM

Color aerial photography entails the taking of aerial photographs in natural color. Both color negative and color positive type films are available. Color photography requires above-average weather conditions, meticulous care in exposure and processing, and color-corrected lenses. High-quality aerial photography in color is possible and practicable whenever the necessary expense and the awaiting of favorable weather conditions are warranted. With color emulsions available on dimensionally stable polyester bases, and color-corrected lenses in aerial cameras and photo mapping equipment, color aerial photography has departed from its former role of a purely pictorial medium and become a mapping tool. (*See* MANUAL OF COLOR AERIAL PHOTOGRAPHY).

5.5.4 COLOR INFRARED

Color infrared has many of the same uses as black-and-white infrared, but in addition, the nuances of color help photoidentification. Because greens are recorded as reds on this emulsion, it is often termed "false-color film." It is used in the detection of diseased plants and trees, identification and differentiation of a variety of fresh and salt water growths for wetland studies, and many water pollution and environmental impact studies. Both black-and-white infrared and color infrared films as well as natural color emulsions require a camera equipped with a color-corrected lens.

5.6 *Factors Affecting Film Quality*

5.6.1 EXPOSURE

The amount of exposure directly controls the density of the negative and the quality of the detail appearing on the negative. The resolution of the emulsion is proportional to the brightness or the contrast ratio of the subject, assuming proper exposure and proper processing. Therefore, the proper exposure will preserve the most detail and will also produce the other qualities necessary for a good negative.

Adjusting the camera for the proper exposure is one phase of the photographic mission which involves many variables, making it difficult to prepare precise instructions. Considerable practical experience is necessary for the best results. This adjustment consists of setting the aperture opening and the shutter speed to be used. The chief factors that must be considered are the following:

    speed of film;
    filters;
    intended method of development;
    atmospheric conditions (temperature at working altitudes, amount of haze, cloud conditions);

color of subject;
contrast ratio or brightness range of subject;
distance between ground and camera;
position of sun (time of year, time of day, locality);
ground speed of the aircraft.

The speed of the particular type of film used is set by the manufacturer and is controlled by the method of development. The first adjustment, therefore, is dependent upon the rated speed of the film and the development method planned. The amount of exposure change due to the filter factor is explained in chapter VI.

The use of automatic exposure controls has been the subject of some debate. Some photographers rely on them completely, while others use them in a semiautomatic way, overriding the controls when the settings do not agree with their judgment. It is obvious that a light-colored subject, like sand or snow, requires less exposure than a dark forest or a burned-over ground area, because of the amount of reflected light. However, exposure determination also involves consideration of the reflectivity of different colors, the color sensitivity and type of film used,

FIGURE 5-4. Portions of two vertical aerial photographs. *Above* - taken with panchromatic film. *Opposite page* - taken with black-and-white infrared film.

and the rapidity of change from one scene to another. As in portrait photography, a close-up requires more exposure, hence flight altitude is another consideration. The position of the sun and the amount of haze also exert a great influence on the amount of actinic light available. The experience and judgment of the photographer are critical in determining proper exposures.

If the shutter speed is kept constant, the light and other above-mentioned conditions prevalent at the time of flight govern the size of the diaphragm opening required. However, the speed

of the aircraft influences the camera shutter setting. At 200-miles-per-hour speed and a photographic scale of 1:20,000, the image moves across the sensitized film at a rate of 0.176 inch per second. Calculations such as this can determine the fastest shutter speed necessary in order to avoid objectionable blurring of images. The effect of the movement causing blurred images can be compensated to some extent by increasing the shutter speed. The shutter speed will be limited by the maximum usable aperture of the lens. Modern aerial cameras, some of which now have a maximum aperture of f/4, are intended to

be used wide open for best overall results. This, coupled with higher shutter speeds, helps to improve image quality by reducing the chances of image motion and the possible effects of vibration.

The best exposure is obtained with a lens aperture setting which gives the highest resolution with the most favorable distortion pattern and with a shutter speed which prevents blurring of the image at the speed of the aircraft.

5.6.2 PROCESSING

Even when the negatives have been properly exposed, correct processing is still absolutely necessary.

A thorough study of chapter VI is recommended before processing of any type of film is undertaken, as the efficiency of any emulsion is dependent upon the type of processing used. Processing methods are meticulously devised by each manufacturer for each product, and very few, if any, improvements are possible in their methods.

Improperly exposed rolls of film have been salvaged in the past by inspection during hand processing (reel type) development; however, many of these rolls have probably been degraded in quality by errors in judgment as to the degree of development required when the film was removed from the developer too soon or too late. Automatic processors and sensitometric equipment are available and these systems should be used whenever possible. They are reliable, and processing can be controlled through the use of pre-exposed step wedges (sensitometric strips) available from film manufacturers. Step wedges can be exposed on the aerial film itself prior to processing by using a sensitometer.

5.6.3 RESOLUTION

Although resolution has been mentioned in connection with both films and lenses, the final resolution or the quality of the small image detail on the final photograph is mainly a function of the lens-film combination and of processing techniques complicated by such other factors as camera vibration, speed of the airplane, temperature, atmospheric pressure, haze, brightness and contrast ratio of the subject area, and the sizes of the images. If the image of a desired object is to appear on the aerial film, it is obvious that the size and shape of the object projected by the lens on the film must be such that it can be recorded and detected. For example, photographs taken for the purpose of making a planimetric map of a city, showing all the houses, must be exposed at a flight height lower than the height at which the combination of film and lens resolution no longer permit positive identification of each structure. Overlooking this fact may result in flight planning that produces photo-

graphs entirely unsuitable for the intended use. Studies have shown that, using the exact aperture for the best resolution of the lens, a suitable distance for the type of object to be photographed, and correct exposure and processing, any subject can be properly photographed to show all desired detail.

5.6.4 TEMPERATURE AND HUMIDITY

Temperature, either excessively high or low, can seriously impair the effectiveness of the lens, the operation of the camera, and the sensitivity of the film. Humidity differences have the greatest effect upon the dimensional stability of the film. For the most favorable end result, provision should be made to control these conditions while photography is in progress.

At excessively low temperatures, contraction of the lens cone has been known to crack the lens elements. Also, the camera operation is seriously impaired by contraction of the operating parts and stickiness of the oil at low temperatures. Special warming devices or the use of pressurized cabins can alleviate many of these problems.

High humidity, in addition to causing changes in film size, causes condensation of moisture in the various parts of the camera if the temperature is lowered abruptly.

5.7 Arrangement and Operation of Equipment

Every photographer has his own preference regarding the way the equipment should be arranged in the aircraft; however, some basic requirement must be met. The relative positions of the camera and viewfinder are of prime importance. The camera must be placed so that it is conveniently accessible to the operator so that leveling, correcting for crab, tripping the shutter, and checking the exposure counter can be readily accomplished. It should also be in such a position that the film magazine can be readily changed. The viewfinder must be so placed that it can be used without the photographer having to move too much. The method most frequently used is for the viewfinder to be mounted between the operator's knees as he sits facing the camera. Some photographers prefer to sit facing in the direction of flight and have the image on the viewfinder move away from them. Others prefer to have the equipment installed so that when they face the rear of the plane and ride backwards, the image in the usual type of viewfinder travels toward them, making it much easier to check map position and compare it with the viewfinder image. Modern cameras are equipped with viewfinder telescopes which can be used to check and adjust forward lap and to control drift.

Many other items of equipment must be arranged for convenience in the aircraft (see figure 5-5). If an intervalometer is used, it must be readily accessible and so mounted that it is easily viewed and there is no parallax between the pointer and graduations of the dial. Spare magazines should be placed within easy reach, and provision should be made so that vibration will not cause them to shift position. If the photographer is to check flight-line position or direct the position of the plane while in flight, a map holder should be provided so that a minimum of effort is required to find the area over which the mission is to be performed. The oxygen equipment must be thoroughly inspected and must be in foolproof condition. Flying above 10,000 feet altitude with no oxygen equipment is quite dangerous (section 5.5.4). Poor oxygen equipment may be worse than none at all. Modern high-altitude jet aircraft have pressurized cabins that protect both the crew and the camera equipment. The communication system between the photographer and the pilot should be thoroughly checked before takeoff.

Any operation that can be performed on the ground lessens the amount of work to be performed in the air, adds that much more to the time available for photography, and gives greater

FIGURE 5-5. Arrangment of photographic equipment in a jet aircraft illustrating a dual camera installation. (Photograph courtesy of Gates Learjet)

assurance of accomplishing the photographic mission satisfactorily. There have been instances in which the crew took off without any film in the magazines. There have also been instances in which the lens cap was not removed from the camera before making exposures.

It is obvious that aerial photographic work requires the utmost in teamwork, skill, and cooperation between photographer and pilot. Even so, much depends upon the element of chance, and some mistakes, defects, and poor execution in aerial survey work are discovered only after the flight has been completed, the aircraft returned to the home base, and the film developed.

5.8 Factors Affecting the Photographic Mission

In addition to the technical requirements of the specifications which determine most of the steps required for a photographic mission, the following factors must be considered, as they greatly affect the conduct and success of the mission: aircraft, weather, position of sun, altitude, and direction of flight.

5.8.1 AIRCRAFT

An airplane, to be suitable for aerial mapping, must have requisite speed, a high rate of climb, good stability while in flight, unobstructed vision in all directions for navigation and identification of landmark features, a range commensurate with the size of the project, and a ceiling higher than the highest altitudes specified. It should be able to remain in the air long enough to take advantage of suitable photographic time, be roomy enough to accommodate all necessary equipment, and be powerful enough to carry its full load to the height required. It should be so designed as to give the pilot and photographer maximum comfort to reduce fatigue and maintain a high level of efficiency. The airplane should be economical in operation, and should be constructed to accommodate the camera or cameras in the position necessary to obtain the type of photography desired.

Much of the aerial photography for mapping purposes is obtained with single- or twin-engine, piston airplanes. These airplanes are standard commercial types modified for aerial photography. The airplane has to be adequately equipped, and capable of performing the mission with very few extras; however, most of these small aircraft are equipped with automatic pilots to aid in flight line navigation.

The great majority of these aircrafts are positioned on the flight line by visual methods. However, inertial guidance systems, now reduced in size by use of minicomputers, are becoming economically feasible, especially for high-altitude jet aircraft. The more efficiently and quickly the mission can be accomplished, the less time is spent in costly waiting for favorable weather conditions.

Some small jet-aircraft, principally Learjets (figure 5-6), have been adapted to aerial photography by the addition of a camera pod which permits the operation of two cameras simultaneously. A simpler arrangement allows the substitution of a camera door in place of the panel holding the aircraft steps. This has made high altitude photography attainable for commercial contractors and much of the United States is now covered by photography taken at altitudes of about 12,500 m (40,000 ft) above mean ground elevation. A program for obtaining high altitude,

quad-centered photographs (chapter XV) initiated by the U.S. Geological Survey in 1969 was primarily responsible for this change in aerial photography procurement. Recent design improvements in small, commercial jet aircraft have raised the possible operating ceiling to about 17,000 m (51,000 ft).

Because of the many personal preferences concerning airplanes and the great variation of requirements in aerial photographic operations, it is not possible to go into detail concerning all of the requisites for low-altitude-mapping airplanes; however, some salient features should be considered.

A high-wing aircraft is more desirable than a low-wing aircraft because the pilot or navigator can see more of the terrain during flight.

The airplane should have good stability in order to eliminate tilt at the time of exposure. This feature is not as necessary when a stabilized mount is used to hold the camera.

The aircraft should have adequate fuel capacity to allow maximum utilization of all suitable photographic weather time.

A single-engine plane can be utilized over the greater portion of the continental United States. Engine failure is not as critical a factor as it would be in territory such as Alaska where emergency

FIGURE 5-6. Photographic configured jet aircraft showing a camera pod modification containing two camera and two viewfinder windows. (Photograph courtesy of Gates Learjet)

landing areas are scarce. A multi-engine airplane which can cruise on one engine is desirable for maximum safety in such areas.

The exhaust manifolds, breathers, and overflows or oil drains should be arranged so that they will not flow or drip in an area that will obscure or cover the camera lens or filter. It is advantageous to have an airplane with the floor close to the outer skin in order to keep the camera hole from being excessively large. Extreme caution should be exercised to insure that the camera does not touch any part of the airplane while in maximum tilt position and that vignetting or cutoff does not occur.

Many pilots using visual navigation have reference marks on various parts of the airplane to help them establish the necessary flight angle or crab and hold it.

The available types of high-altitude, photographic jet aircraft are limited because of the structural change required to accommodate camera mounts and glass camera ports. Few jet aircraft are made specifically for high-altitude photography.

5.8.2 WEATHER

The most uncertain factor in aerial photographic missions is the weather. The ideal photographic day is one in which the air is free of clouds, smoke, and haze; the sun is high enough to shorten objectionable shadows; and the wind velocity and air turbulence at the flight altitude are at a minimum. All of these conditions vary with the time of year and the particular locality. The presence of high clouds such as cirrus sometimes has a beneficial effect on the quality of low-altitude photographs by softening the light and reducing the intensity of the shadows. A highly industrialized area is usually partially obscured by smoke even on cloudless days. It is best to photograph an area of this type the day after a heavy rain which tends to cleanse the air. Sometimes a moving cold front clears away most of the smoke.

Once the flight crew is committed to a project they face a daily decision. If the day starts cloudless, will clouds form, and at what time? Are atmospheric obstructions such as dust, smoke, and haze at a satisfactory minimum? Obviously, the integrity and skill of the flight crew will be an important consideration. They must be guided by the intent of the project specifications, clearly set forth, and not by external pressure for production alone.

The flight crew should establish a base in the project area, or as near it as possible. Their activities can then be governed by direct visual observation of the sky. In addition, they should develop as much skill as possible in the interpretation of current weather reports.

At no time should decisions be based solely on forecasts. If the sky is free of clouds, it should be assumed that it will stay that way and the mission gotten underway. Much photographic

weather is lost because of a forecast of cloudiness that did not materialize. A cloudless day occurs when the relative humidity is low, and the air is stable. There are few localities on earth where such a condition exists for a long period of time.

The quality of the atmosphere known as haze often prevents a cloudless day from being suitable for aerial photography. The exact composition of haze varies with the locality. Weather data from airports include haze conditions but the probability of its occurrence and its extent are based on knowledge of the area.

Haze results from the hydrolyzation of nuclei in the air. For this to occur, both water vapor and nuclei must be present in the air. The most common nuclei in the air contributing to haze formation are those of salt and industrial chemicals. As a result, the most serious haze conditions occur along coastal areas and over and near industrial complexes. True haze tends to diffuse only the blue end of the color spectrum and can be filtered out with a yellow filter. Dust and smoke consist of large particles and not only diffuse the entire color spectrum, but may also absorb a large proportion of the sunlight. Satisfactory photography cannot be accomplished when such pollutants are in the air unless only the long-wave infrared radiation is used for exposure on infrared-sensitive film.

Dust and smoke are often observed in layers; hence, ground visibility measurements do not serve as a suitable guide. Haze, however, is usually uniform from the ground to the dew-point level—the elevation at which the dew-point temperature and the air temperature are equal—hence, ground visibility serves as a suitable guide, and a visibility reading of 15 miles requires a further study of local effects on the area before a final determination is made to conduct photography.

Aerial photographs should not be taken in turbulent air without full recognition that loss in negative quality will result. Turbulence causes image motion on the negative and reduces definition of the photographic image. This motion is not proportional to scale, but at a high altitude where images are small, the loss in interpretability may be serious, even total. As a result, whereas some turbulence may be tolerated at altitudes below 10,000 feet, none can be tolerated above that flight height.

This reference to turbulence is based on the assumption that the camera will move with the airplane. With gyro-controlled camera platforms, a much higher level of turbulence can be tolerated. In addition, there exists a relationship between the characteristics of the airplane and the magnitude of the turbulence that determines the tolerance level. In general, the higher the wing loading of the aircraft and the greater its gross weight, the less response there will be to turbulence. No generalization can be made in

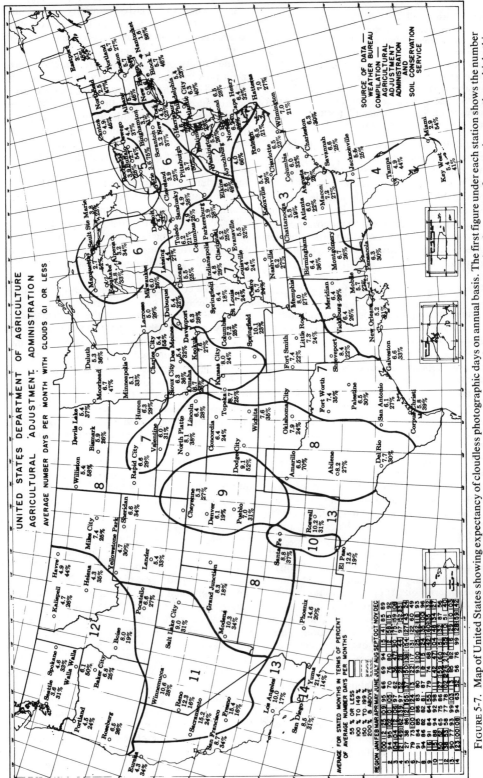

FIGURE 5-7. Map of United States showing expectancy of cloudless photographic days on annual basis. The first figure under each station shows the number of days per month when the ground is 0.1 or less obstructed by clouds, based on a yearly average; to determine the average for any given month, multiply this figure by a factor obtained from the table in the lower left corner. The second figure under each station is the percentage below average which can be expected 1 year in 10 (the worst year). (Diagram courtesy of U.S. Dept. of Agriculture)

this respect regarding aircraft speed, since the response would depend on the character of the turbulence existing at the time.

High winds and air turbulence also vary with the locality and season of the year; forecasting of these phenomena is usually based on past experience.

Photographic crews must be on the alert in studying the reports and forecasts which are normally available from the U.S. Weather Service throughout the United States. Equipment should be ready for use, and in perfect operating condition on likely flying days, so that the utmost advantage can be taken of favorable weather conditions. Judgment as to the suitability of any given day for the acquisition of acceptable photography for a given purpose can be based only on experience.

In 1939, F. J. Sette of the U.S. Department of Agriculture made a detailed study of weather occurrence probabilities in the United States. This study is still the best guide compiled to date for predicting the probably occurrence of good photographic weather.

Figure 5-7 is a map of the United States, with zone lines delineated thereon in such a manner that the number of days in any given month on which the sky can be expected to be not more than 1/10 obscured by clouds can be estimated for a given area. The expectation is based on a study of the weather records for the 37 years prior to 1939. As an example, suppose the area to be photographed lies between Little Rock and Fort Smith, Ark., in region 7 of the map. The average number of days per month with clouds 1/10 or less is 7.3 for Little Rock and 7.4 for Fort Smith. Therefore, the average for the project area is 7.35. This figure is an average for a 12-month yearly period. For some months the figure will be higher, and for some months less. Now, referring to the tabulation in the lower left corner of the map, in the line for region 7 the percentage of the average for each month of the year is given. Thus, for June the figure is 80 percent. The number of days with 1/10 or less clouds in June would be 80 percent of 7.35 or 5.88 days. Referring again to the map, for 1 year in every 10, the number of days at Little Rock can be expected to be 24 percent less, and at Fort Smith 22 percent less. Thus, the worst expectation, as well as the average expectation is given. A day in which the sky is 1/10 or less covered is considered a photographic day.

In using this probability map, an estimate of actual expected photographic hours would be the product of the probable photographic days in any given period multiplied by the number of photographic hours per day (see the next section) for the average latitude of the project.

5.8.3 POSITION OF SUN

Another factor which controls the quality of the photography for most photogrammetric pur-

poses is the position of the sun. The presence of deep shadows tends to obscure detail and may even blank out portions of the photograph. Photography with long shadows is used sometimes for certain types of timber studies, but in most cases it is objectionable for mapping. Small shadows tend to delineate some detail effectively and are generally advantageous in apparently increasing the quality of the photograph. Photographs taken in the middle of the summer at low latitudes when the sun is directly overhead fail to record considerable amounts of detail because shadows are at a minimum. Under such conditions, where house roofs are nearly the same color as the surrounding terrain, or photograph with the same degree of contrast, the buildings blend with the rest of the detail and become indistinguishable.

The illumination by sunlight falling on a unit of horizontal surface varies greatly with the sun's angular altitude, as shown in figure 5-8. The value at an angular altitude of 30° is approximately two-fifths of that at 60°; and when the sun's angular altitude is 25°, it is only one-third of the value at 60°. It is obvious, therefore, that the photogrammetrist planning or requesting aerial photography should carefully consider the effects of solar angular altitude on the specified project.

In determining the time of year during which photographs should be taken, it is important to consider the effect of shadows and the duration of adequate light for satisfactory photography. The National Ocean Survey "Solar Altitude Di-

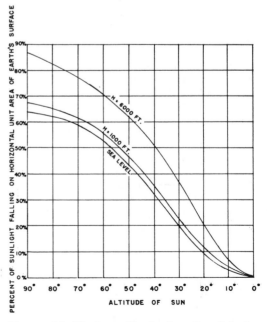

FIGURE 5-8. Maximum illumination of sunlight for photographic purposes as related to the Sun's angular altitude. (Diagram courtesy of the National Ocean Survey)

agram'' (figure 5-9) affords a convenient means of determining the duration of a minimum solar angular altitude of 30° at any time of the year for any part of the United States. The diagram consists of a map of the United States on which has been superimposed a series of curves, each of which shows, for the dates indicated, the variation, according to latitude, of the daily duration of solar angular altitude of 30° or more. The time scale of the curves is horizontal but is not related to longitude. To use the diagram, plot on the map the position of the area to be photographed and draw a horizontal line through this position. Estimate by interpolation the position of a curve for the date of the proposed photography. Drop verticals from the intersections of this estimated curve with the horizontal line already drawn. The intersections of these verticals with the time scale at the bottom of the diagram indicate the times between which the sun is above an angular altitude of 30°.

The times given are local mean time. They must be corrected for difference in longitude between the site of photography and the meridian of the standard time zone. The correction is found by dropping a vertical from the map position of the photography to the correction diagram at the bottom of the map. The point at which this vertical line intersects the diagonal line for the appropriate standard time zone gives the appropriate correction. The proper sign of the correction is indicated. The diagram is constructed for use with standard time. If daylight-saving time is used, the appropriate correction must also be made.

The position of the sun at the time the negative is exposed is an important factor affecting the character of each aerial photograph. Except for special problem areas, sound planning provides for the flying of contiguous strips in sequence. Adjacent photographs should, whenever possible, show similar lighting conditions.

It is good practice to examine the entire project area from the air before beginning photography. This examination permits the flight crew to identify and isolate any terrain features which must be scheduled for photographing at particular times of day. Such terrain features include cliffs and precipitous relief that will cast shadows over pertinent detail.

When the project area includes terrain features which create serious shadow problems and cumulus cloud formation prevents accomplishing midday photography, it is feasible to consider photographing the area under high thin clouds. Cirrus clouds diffuse the sunlight, and at the same time, lower the overall actinic value. As a result, there is increased reflectance from the areas that are in shadow and decreased reflectance from the areas not in shadow.

By increasing exposure time often as much as 100 percent, adequate exposure of shadowed areas is obtained for photogrammetric use. Diffused lighting conditions should, however, never be used over flat, low-contrast terrain. Here, sharp shadows are needed to delineate image detail. When a low sun-angle is desired to accentuate shadows, the project specifications should state the maximum as well as the minimum acceptable solar angular altitude.

The optimum solar angular altitude is not always available to the photographer. In arctic regions, for example, the sun is never very high above the horizon. In the tropics, it is often impossible to obtain mid-day photographs because clouds tend to form early in the day and do not dissipate until late afternoon. Other projects involving low sun-angle include areas which can be photographed only in early spring or late fall, as in the case of a "no-leaf" requirement where deciduous trees lose their leaves in the winter months. There is a popular belief that the high sun-angle photograph is in all ways superior, but research shows that there is an optimum sun-angle for almost all conditions at approximately 45° above the horizon, and any departure from this value in either direction is a necessary compromise.

5.8.4 ALTITUDE

Flight altitude is determined by the desired photograph scale, the contour interval to be used, and the characteristics of the plotting instrument, although it is also controlled by the fact that the quality of photography produced is a function of the scale of the picture. The scale, as we have seen in chapter II, is equal to the focal length of the camera divided by the distance from the camera lens to the ground (both figures should be in the same unit of measure). Flight altitudes for compilation are frequently in a range from 500m (1500 ft) to 7,625m (25,000 ft). Since most piston aircraft used have unpressurized cabins, electrically heated suits and oxygen masks are required for high altitudes. The Air Force chart (figure 5-10) illustrates the problem of low temperatures and the continuous need of oxygen at upper altitudes. Commercial jet aircraft converted to aerial photography work, however, have pressurized cabins, and the photographer and pilot can work comfortably at altitudes up to 16,000m (50,000 ft). This has increased the use of small scale photography for orthophotograpic maps and for map compilation in photogrammetric instruments equipped with earth curvature correction devices.

5.8.5 DIRECTION OF FLIGHT

The determination of direction of flight is based on a careful analysis of the overall mapping project by the photogrammetrist. These considerations are significant:

> the configuration of the area, and the plan that will result in the least number of flight lines and the

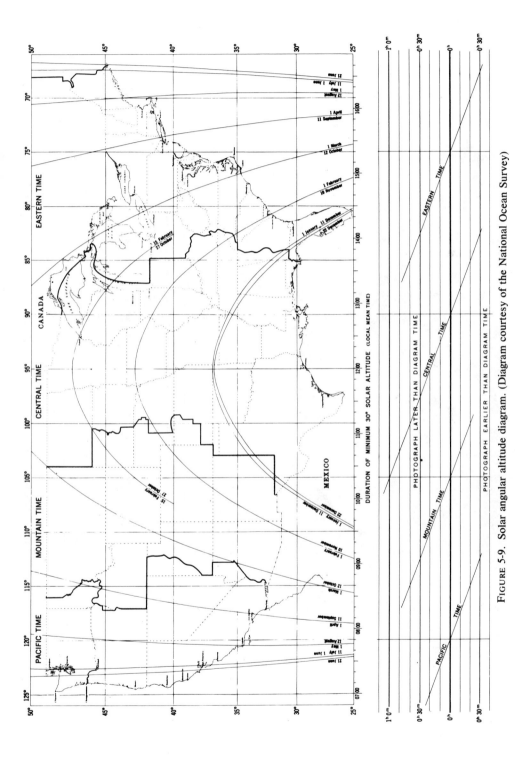

FIGURE 5-9. Solar angular altitude diagram. (Diagram courtesy of the National Ocean Survey)

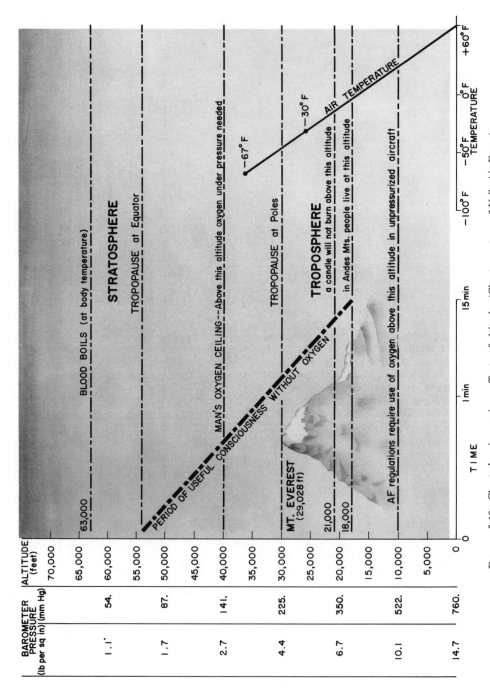

FIGURE 5-10. Chart showing various effects of altitude. (Chart courtesy of U.S. Air Force)

least number of photographs to provide stereo-scopic coverage;

the most efficient production of the photography by the flight crew;

the orientation of the stereomodels relative to the topography and the position of the sun.

The first consideration, configuration of the area, frequently takes precedence since the cost of ground control and compilation is related to the number of individual stereopairs of photographs. If the direction of flight based upon this consideration would result in poor photographic quality or unsatisfactory production of the photography, another direction should be selected to create the best overall plan.

Flying efficiency is of first concern when a limited amount of photographic weather is expected during the time period allocated for the project. Factors affecting production are those which will increase the proportion of actual time in taking photographs, compared to that used in making turns, changing altitude, and deadheading between runs. Flight lines should therefore be as long as possible without breaks for altitude change, and as few in number as possible.

If the area to be photographed is extremely rugged, and it is not necessary to plan the flight lines in the cardinal directions, every effort should be made to orient the flight lines parallel to the major linear topographic features. A minimum number of altitude changes will then be necessary. This orientation provides better utilization of available weather, since clouds usually form along the ridges before they form, if at all, over the valleys. Hence, production can be continued over the valley portions after the mission would otherwise have necessarily been discontinued. Also, the opportunity to fly the high-relief areas under optimum sun-angle is improved, since the shadow areas are isolated. The crew can then start the day's flying over the valley areas, saving the ridges and mountains until the sun is higher.

In addition, the high-relief areas should, when possible, fall on the sides of the model rather than in the center. This can be accomplished by strategic placement of the flight lines over the valleys or the centers of the ridges if the lines are oriented parallel to the relief. It is more important to have such strategic placement than to have uniform side-lap.

It follows that in order to achieve optimum orientation to both the sun and the direction of relief, it is often necessary to specify time limits for production during which the sun and shadow conditions are reasonably compatible with the requirements.

In planning high-altitude photography where maximum relief does not usually present the problems of side and forward lap encountered in low-altitude photography planning, the flight lines are planned in the cardinal directions, preferably north-south, but dependent upon the number of turnarounds required.

5.9 Procurement of Photography

5.9.1 CONTRACTS

A contract is an instrumentality of society developed to establish certain predictable relationships between two or more parties, and it consists of one or more legally enforceable promises. A contract for the procurement of aerial photography usually has two distinct subdivisions, namely, the technical aspects, which specify the quality and quantity of the material to be delivered, and the legal aspects.

The specifications are the essence of this type of contract. The specifications must set forth the relationships by defining the work which the contractor promises to perform and those acts which the purchaser will perform. The proper preparation of the specifications presupposes a concept of the business relationships that will be established and a thorough technical knowledge of the subject, in order to secure the exact product needed. The specifications should be concise, and, to use a famous quotation, "say what they mean and mean what they say." A heterogeneous collection of rules and specifications copied from many divergent sources, placed in no orderly manner and couched in supposedly legal phraseology, do nothing but confuse, and are ineffectual for the purpose. Repetition should be avoided, as well as contradictions.

In the instructions and descriptions that follow, the term "company" shall be understood to mean the contracting party for whom the work is performed; the term "contracting officer" shall mean the company's authorized representative or his duly appointed successor; the term "contractor" shall mean the party performing the work.

5.6.1.1 PREPARATION OF THE DOCUMENTS

Although aerial photography specifications can be prepared as a part of an estimate, or for a proposed project, their preparation is the phase of photogrammetric engineering that usually follows the completion of the steps listed below.

(a) Preliminary investigation has justified the preparation of the map.

(b) The necessary funds will be or have been made available.

(c) The type and accuracy of the map, chart, or mosaic to be produced have been determined.

(d) Technical questions have been decided; that is, type of photography necessary for the method of compilation.

(e) Completion dates have been specified.

Properly prepared, the specifications involve the combined efforts of the administrative, technical, and legal staffs, either of the prospective purchaser or of qualified consultants. Since the Federal government has been the largest purchaser of aerial photography to date, its specifications are the basis for most sets of specifications for aerial photography. The basic differences are chiefly legal, because the government contracting officer is bound by official rules, whereas a private company can make any legal agreement it considers to be of advantage to the organization. The technical requirements must, of necessity, be the same if the accuracy standards for the end product are to be maintained.

Several general considerations are of importance in the preparation of a contract. Utmost care should be taken to preclude ambiguity and errors, either legal or technical. Most contracts have a provision concerning disputes. This provision should be explicit, and should be discussed and understood by both parties before consummation of the contract. It should name a neutral arbiter, if possible, to settle disputes, and his decisions should be mutually agreed to be binding on both parties. In case of a dispute that finally ends in a court of law, each phrase, sentence, and paragraph of the specifications must stand on its own and will be so interpreted. Therefore, although the proper intentions may have existed at the time the specifications were written, if the phraseology is not clear, a dispute could be decided to the disadvantage of the purchaser and force the acceptance of photography which may not be usable.

Photography should be specified for the particular purpose for which it is intended, and, if possible, the specifications should recognize the possibility of other uses. Almost all photography for surveying and mapping purposes is obtained for one of the following two major applications:

(1) for acreage determination, land-use studies, resources studies, the preparation of mosaics, and orthophotos; or
(2) for use in stereoscopic plotting instruments for the preparation of various types of maps.

The essence of all specifications for stereoscopic mapping is complete stereoscopic coverage of the project area within the usable limits of the lens system of the camera/plotting-instrument combinations. Photographs to be used in stereoplotting equipment should be exposed in cameras manufactured and calibrated to a degree of precision equaling or exceeding the precision of the stereoplotting equipment to be used. The lens should be of such focal length and angle of coverage as to insure accuracy for the specified mapping requirements.

5.9.1.2 TYPES OF CONTRACTS

The contract is of paramount importance, for it obligates the contractor to execute the work in accordance with the legal and technical provisions of the specifications, and, in turn, obligates the other parties to the contract to pay the stipulated prices after delivery of the specified items within a specified time. The contract is usually made by one of these methods:

(1) the bid may be signed by the contractor and submitted with all necessary accompanying documents and then signed by the contracting officer to constitute the contract; or
(2) in the absence of bidding, a formal agreement or contract embodying all of the contractual documents can be negotiated by both parties.

5.9.1.3 PERFORMANCE OF THE WORK

It is standard practice today, in most construction and engineering contracts, to require a performance bond to insure completion of the project if the contractor should default. Performance bonds issued by surety or insurance companies guarantee performance of the work by the contractor and relieve the other party to the contract from the responsibility of any claim that may arise from any source in connection with the contract.

At the time the agreement is signed, notice to proceed is usually issued, or a reasonable estimate of the time when notice to proceed will be given is stated. For example, an aerial photographic contract might specify that the photographs should be taken when there are no leaves on the deciduous trees. If a contract of this type is signed in the middle of summer, the contractor may reasonably expect to receive notice to proceed in the fall.

Stipulation should also be made of the circumstances under which notice may be issued that the contract no longer exists, in case the contractor wishes to nullify the bid bond or be enabled to certify in another bid that he is free to undertake the work. The percentage of completion and the equipment involved govern the issuance of this notice.

5.9.1.4 METHODS OF OBTAINING PHOTOGRAPHY

At present, one of the following methods is generally used to obtain aerial photography under contract.

(a) The contractor furnishes all material, labor, equipment, and supervision for the execution of the project, and he assumes all risks.
(b) Same as (a) except that the camera and other specified materials are furnished to the contractor by the purchaser.
(c) Same as (a) except that the purchaser hires the crew and equipment on a standby and flying-hour basis and thus assumes the weather risks. The purchaser may furnish the camera. This means that the contractor furnishes the airplane and camera-crew for a stipulated length of time, during which period he is to be ready to take photographs for a previously determined price per unit. There should be definite provisions as

to when payment should be started and stopped, in addition to a statement of the conditions which preclude payment during the course of the operation. Standby-time payments are usually terminated when a photographic day is missed for reasons other than those defined in the contract. A photographic day is usually defined as "any day, not excluding Sundays and holidays, containing a period of time of not less than two consecutive hours at which the sun is at the proper elevation, during which the day is cloudless over the area to be photographed." There is usually an additional payment of so much per flying hour, and in some cases a specified payment for a stipulated number of acceptable exposures. Supervision in this method of contracting is divided, as there is usually an inspector representing the purchaser who, in addition to his other duties, records flying time and other pertinent data. This system has advantages in that greater flexibility is permitted in assigning priorities, and reflights can be requested for any reason. In ordinary contracts, reflights can be ordered only if the work does not entirely meet the specifications. It is always good policy to re-fly questionable work, since all subsequent mapping procedures are affected by poor photography.

In methods (b) and (c), development of the film and the printing can be accomplished by either party to the contract.

5.9.1.5 BASIS OF BIDDING

The usual basis of bidding and payment for aerial photograph contracts is the number of square miles of acceptable photographic coverage and accompanying materials that are satisfactorily delivered. Another basis for bidding is the theoretical number of flight-line miles to be accomplished, along with the stipulated materials satisfactorily delivered. The final contract is made either for an estimated maximum number of miles or is awarded on a basis of "not to exceed a certain number of miles," while payment is made for the actual number of miles flown. By this method extra flights can be obtained if boundary coverage is indefinite because of poor flight maps, if gaps occur because of relief or other causes that are beyond the control of the contractor, or if flights are needed for other purposes, such as for control extension or for checking.

When the basis for bidding is the theoretical number of miles of flight line, the specifications should be definite as to the location of the beginning and end of each flight line and how the length of the flight lines is to be computed.

5.9.2 FLIGHT MAPS

The previously prepared plans of the photogrammetrist (chapter VII) come to the pilot and photographer in the form of the flight map, which is drawn up in accordance with the calculations such as spacing between flight lines and

between successive exposures necessary to satisfy the technical portion of the specifications. The flight map shows the beginning, end, and location of each flight line, and specifies the flight altitude above mean ground or sea level. The flight map should be accurate, include a title block giving the project designation, and should show all details which the crew will need; also, if specified, the positions of the individual exposures.

The map on which the flight lines is drawn should be the best available map of the area. Topographic maps at scales from 1:24000 to 1:125,000 make good flight-map bases when available. Index maps of previous photography, county maps reduced to appropriate scale, forestry maps, or maps made by local authorities will also serve the purpose. The 1:250,000-scale map series are especially suitable for high-altitude photography using jet aircraft.

The map scale usually preferred is 1:125,000, as this shows sufficient detail for accurate flying, yet does not make the map too bulky to be handled by the photographic crew. However, the map scale is largely a matter of personal choice and availability. When several types of maps are joined or pasted together, the flight lines should be spaced evenly and in the proper position on the individual maps before they are joined, in order to preserve continuity of the flight line. If maps are joined before the flight lines are drawn, paper shrinkage which can cause mismatches of the maps may cause the flight lines to be displaced and may endanger correct flying procedure. It is obvious that a clear, accurate base map, with the lines in their exact position in relation to the ground, is an important factor in securing the best possible execution of the preconceived plan.

5.9.3 INSPECTION

Methods and procedures used in conducting an inspection of the finished aerial photographs vary with many conditions. In any case, the inspection should definitely establish the suitability of the photography for its intended use. To determine the suitability, sample measurements, optical tests, and density readings may be necessary. For those sections of the specifications which stipulate the process necessary to determine its useful life and its ability to resist chemical and physical deterioration, some chemical or physical tests will be necessary.

5.9.4 TITLING

The titling of film consists of marking it with significant symbols, characters, and numbers so that it can be identified with respect to a project or its location on the earth's surface. Most film marking is accomplished with ink, and in some cases printer's ink has proven very successful.

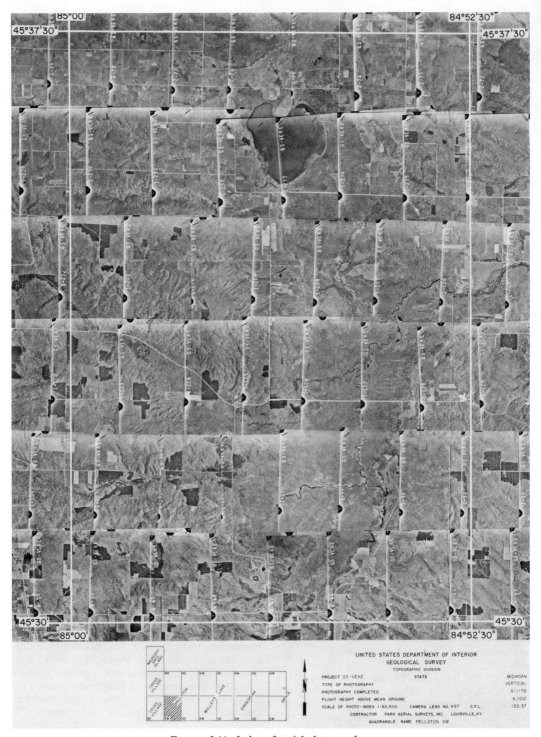

FIGURE 5-11. Index of aerial photography.

Permanence and legibility are the important factors.

The marking may be done by means of an automatic stamping machine (heat transfer or ink), by using lettering guides, or by hand methods. The importance and type of the project for which the film was taken usually determine the care and precision with which the film is marked. The symbols should be placed so that they do not interfere with the useful area of the photograph. If the normal requirements are that the symbols show in a certain area of the film, and if this portion is very dense, the markings should be moved so that they will print on the reproduction. Government agencies and surveying companies usually require the roll number (if more than one roll), the exposure number, the project symbol (if there is one), and date. At the ends of flight lines, additional information such as photoscale or flight height, the camera or lens number, and time of day may also be required. Any symbol or project designation on the film should be of a kind that can be easily identified when referred to the project as a whole.

5.9.5 INDEXING

The purpose of indexing is to show the relation of any one photograph to the others and its relation to major terrain features and to the geographic reference system. Two types of indexes are commonly used—the line index and the photoindex. The line index consists of an overlay to a map of the area, containing lines showing the location of the flights. Sometimes the location of each photograph, with its designating number, is shown as a small square drawn to scale. For most purposes, the numbers of the photographs at the beginning and end of the flight will suffice. If there is a break in flight, the numbers of the photographs at which the break occurs are shown at the appropriate place on the chart.

The photoindex (figure 5-11) is often preferred because of the information which it makes available. It is also useful for other operations, such as reconnaissance, area studies, control planning, and as a flight map. The method of assembling or laying a photoindex should be described in the appropriate sections of the specifications. During the assembly of the index, the accuracy with which the images are matched between the various photographs should be consistent with its intended use. The type of print used (that is, the type of surface, such as glossy, matte, or semimatte) and the method of trimming the prints will also depend on the intended use. The scale should be such that the imagery does not become indistinguishable.

Either type of index should contain all pertinent marginal information, such as the latitude and longitude, strip numbers, relation to the project as a whole, and any other information which would add to its usefulness. Also worthy of mention is the requirement that any photograph beyond the boundary of the index that might possibly be used later should be indicated by appropriate stickup.

REFERENCES

Anonymous, 1971, Photography from Light Planes and Helicopters M-5, Eastman Kodak Company, Rochester, N.Y.

Anonymous, 1970, Kodak Aerial Exposure Computer, No. R-10, Eastman Kodak Company, Rochester, N.Y.

Anonymous, 1976, Thermal Photography, Kodak Bits, V. 1, Eastman Kodak Company, Rochester N.Y.

Anonymous, 1977, Kodak Data for Aerial Photography, No. M-29, Eastman Kodak Company, Rochester, N.Y.

Austin, Alan, 1978, Aerial color and color infrared survey of marine plant resources, *Photogrammetric Engineering and Remote Sensing*, V. 44 (4); 469–480.

Fritz, Norman L., 1977, Filters: An aid in color infrared photography, *Photogrammetric Engineering and Remote Sensing*, V. 43 (1); 61–72.

Gut, D. and Hohle J., 1977, High altitude photography: aspects and results, *Photogrammetric Engineering and Remote Sensing*, V. 43 (10); 1245–57.

Howard, J. S., 1970, Aerial Photo-ecology, American Elsevier Publishing Co., Inc., New York, N.Y.

Keller, Morton, 1975, Aerial photography in the NOS Coastal Division, *Photogrammetric Engineering and Remote Sensing*, V. 41 (8); 1009–1013.

Koopmans, B. N., 1975, Variable flight parameters for SLAR, *Photogrammetric Engineering and Remote sensing*, V. 41 (3); 229–306.

Lockwood, Harold E. and Perry, Lincoln, 1976, Shutter/aperture settings for aerial photography, *Photogrammetric Engineering and Remote Sensing*, V. 42 (2); 239–252.

Smith, John T. (Jr.) and Anson, Abraham (editors), Manual of Color Aerial Photography, 1968, American Society of Photogrammetry, Falls Church, Virginia.

Colwell, Robert N. (ed.), 1960, Manual of Photographic Interpretation, American Society of Photogrammetry, Falls Church, Virginia.

Reeves, R. G. and Anson, A. (eds.) 1975, Manual of Remote Sensing, American Society of Photogrammetry, Falls Church, Virginia.

McEwen, Robert B. et al, 1976, Coastal wetland mapping, *Photogrammetric Engineering and Remote Sensing*, V. 42 (2); 221–232.

McNeil, Gomer T., 1954, Photographic Measurements, Problems and Solutions, Pitman Publishing Corp., New York, N.Y.

Parry, J. T. and Gold, C. M., 1972, Solar altitude nomogram, *Photogrammetric Engineering*, V. 38 (9); 891–899.

Stephens, Peter R., 1976, Comparison of color, color infrared, and panchromatic aerial film, *Photogrammetric Engineering and Remote Sensing*, V. 42 (10); 1273–1278.

Ulliman, Joseph J., 1975, Cost of aerial photography, *Photogrammetric Engineering and Remote Sensing*, V. 41 (4); 491–499.

Walker, Patrick M. and Trexler, Dennis T., 1977, Low sun-angle photography, *Photogrammetric Engineering and Remote Sensing*, V. 43 (4); 493–506.

Welch, R. and Halliday, J., 1975, Image quality controls for aerial photography, *The Photogrammetric Record*, V. VIII (45); 317–325.

CHAPTER VI

Photographic Materials and Processing

Author-Editor: ROBERT G. McKINNEY

Contributing Authors: PETER Z. ADELSTEIN, M. G. ANDERSON, E. C. DOERNER,
ALFRED E. FIELDS, NORMAN L. FRITZ, JOHN GRAVELLE,
JOHN F. HAMILTON, GARY L. ROBISON, JOHN T. SMITH, JR.,
M. RICHARD SPECHT, EDWIN G. TIBBILS, WILLIAM F. VOGLESONG,
ROY A. WELCH

6.1 Introduction

THE PRACTICE OF PHOTOGRAMMETRY is dependent upon several different scientific and technological disciplines. For these several disciplines, the photographic process forms the foundation, since it alone provides the sensitivity and information-storage capacity necessary to obtain, conveniently and economically, the data—in the form of an interpretable representation of the object—upon which all subsequent operations depend.

6.1.1 HISTORICAL BACKGROUND

Photography had its beginning in the early 19th century with the discovery by John Henry Schulze, in 1827, of the sensitivity of silver nitrate to light.

A number of substances were known at the time that would undergo a visible change under the influence of light, but the time required for this change to be produced was impractically long.

A commercial application of the sensitivity of silver to light did not appear until 1839, which brought the announcement of the invention of the daguerreotype. Jacques Mandé Daguerre had accidentally discovered a means of making exposures reasonably short, and the first photographic process was enthusiastically received by the public.

For some time, Daguerre had been experimenting with compounds of silver known to be sensitive to light, and had made plates consisting of an iodized plating of silver on copper. One day, Daguerre had a plate in his camera for only a few minutes when the sky clouded over and, thinking the plate was insufficiently exposed, he placed it in a cabinet to await more favorable conditions. Some time later, he returned and discovered that the scene he had exposed for such a short time was visible on the plate. In the cabinet were a large number of chemicals, including an open dish of mercury. By the process

of elimination, Daguerre discovered that the mercury fumes had formed a visible amalgam with the silver on the plate in proportion to the various intensities of light that had been impressed on the plate during the exposure in the camera (figure 6-1).

Concurrently with Daguerre, William Henry Fox-Talbot was busily conducting similar researches in England, and as soon as he heard of Daguerre's process, he published the results of his own efforts. The sensitivity of Talbot's materials was also very low until he discovered a way of enhancing the affect of the action of light. As finally practiced, Talbot's process started by absorbing potassium iodide into a sheet of paper and then bathing the sheet in a solution of silver nitrate and gallic acid. After exposure, he again bathed the sheet in a silver nitrate/gallic acid solution, which gradually brought out the picture. Sir John Herschel suggested fixing the image by bathing the sheet in a solution of sodium thiosulfate. The process (and modifications of it) was used for years, but it had the disadvantage that the paper fibers destroyed the fine detail that the daguerreotype was able to retain. However, Talbot's process was a negative-to-positive type and is therefore considered to be the true ancestor of modern photographic processes.

The needed phenomenon, the enhancement of the initial action of light, is best described by modern technology as "amplification." Without it, there could be no photography, (certainly no aerial photography as we know it); both men, therefore, can be honored for separately discovering this indispensable phenomenon.

The announcement of a practical method of "painting" a picture by the agency of light naturally excited the imagination of everyone and stimulated many investigators to inquire into the phenomenon. The result was that innovations that increased the versatility of the process were rapidly achieved. In time, the daguerrotype was

305

FIGURE 6-1. A reproduction of a Daguerreotype, the product of the first commercially successful photographic process.

cells containing red-, green-, and blue-colored solutions to serve as filters. The plates available at that time had only the inherent blue sensitivity of the silver halide, so Maxwell had to give very long exposure times through the red and green filters. From these negatives, positives were made on glass plates and the pictures were projected on a screen by means of lantern-slide projectors—each one projecting its picture through the filter solution that had been used in making the negative. The three monochromatic images were superimposed; a colored image was thus obtained of the original object. Maxwell apparently did not pursue this to make a practical photographic system. This was subsequently done by Louis Ducos Duhauron and others (Collins and Giles, 1952).

The discovery by Herman Vogel in 1873 that sensitivity could be extended to longer wavelengths by absorbing certain dyes in the emulsion was very important to the further development of both black-and-white and color photography. Further investigations have provided many sensitizing dyes of varying properties, including those extending sensitivity to the infrared region of the spectrum.

George Eastman produced an emulsion on a base of nitrocellulose in 1889. This base retained the clarity of glass plates, eliminated their fragility, reduced bulk and weight, and provided flexibility. This flexibility permitted the design of smaller, less complex equipment for exposing, processing, and handling the negative materials. The mechanical properties of a clear, flexible film support have proved to be of utmost importance in enlarging the areas of application of photography. However, its importance should not obscure the fact that the versatility of photography has been considerably increased by the coating of sensitive emulsions on a variety of supports. Therefore, the requirements for specific but different mechanical and physical properties demanded by different applications can be satisfied.

Thus, by the time the Wright brothers went aloft in powered flight for the first time in human history (1903), photography had already acquired the various characteristics needed for the art and science of photogrammetry. The organization of the profession awaited only improvements in the airplane.

These characteristics were gained under the impetus of popular interest in the process and came about mostly by invention and innovation, the scientific fraternity having shown little interest in the science of photography as distinguished from its practice. However, the mystery surrounding the change wrought in the silver-halide crystals by the absorption of photons, which initiates the chemical reduction of the whole crystal to silver, could not long escape scientific curiosity. Consequently, toward the end of the 19th century, individuals (and then

displaced by the more convenient negative-to-positive kinds—primarily because it was a single-picture process.

In 1851, Scott Archer prepared a suspension of silver chloride in nitrocellulose and coated it on glass. In his process, he used development with pyrogallic acid to amplify and produce the image. Its advantages over Talbot's process were that it was more sensitive and that it produced better definition than was obtainable from the paper negative. Its disadvantage was that it was necessary to both prepare and process (develop, fix, wash, and dry) the plate immediately. A wet-plate photographer was, therefore, encumbered with considerable bulk, weight, and complexity in the pursuit of his pictures.

Fortunately, the inconvenience of having to process the material immediately after exposure was eliminated by Dr. Richard Leech Maddox. In 1870, he described a gelatin emulsion of silver halide that could be coated and dried before using. Retaining the advantages of Archer's "wet-plate" process, the Maddox gelatin process provided suitable protection for the silver-halide grains so that the invisible image could be developed later—thus eliminating all equipment except the camera and sensitive plates.

Color photography had its start in 1855 when the physicist Clerk Maxwell published a suggestion for a practical color process. Using Thomas Young's three-color theory of vision (1802), Maxwell subsequently prepared three photographs using collodion-coated plates. The plates were separately exposed through plate-glass

organizations) began to inquire into the science of the photographic process. These investigations not only resulted in a clearer understanding of the fundamental phenomenon of the sensitivity of silver halide to light, but they also had a synergistic effect on further improvements in the functional characteristics of the process that are important for practical applications.

6.2 Photographic Properties

6.2.1 THE LATENT IMAGE

The sensitive elements of a photographic material are microscopic crystals of silver halide, having sizes between a few tenths of a micrometre and a few micrometres, and having frequently regular shapes. The largest crystals may contain as many as 10^{10} pairs of silver- and halide-ions.

A normal photographic exposure involves an extremely minute amount of light. Yet there is a change—a *latent image*—produced during this exposure such that when the material is processed for the normal time in a photographic developer, the exposed crystals are completely reduced to elemental silver, whereas most of the unexposed ones are unaffected. In the most sensitive materials, crystals that have absorbed only three or four photons will be reduced. Thus, some crystals contain more than a billion ions of silver. The action of these few photons can be said to have been amplified by a factor of about 3×10^9—one of the largest amplifications by any known means.

The active product of the exposure is thought to be a small aggregate of silver atoms on the surface or interior of the crystal. This silver speck—the latent-image center—acts as a catalyst for the development reaction, increasing its rate manyfold and allowing for discrimination between exposed and unexposed crystals. The silver forms by combination of electrons and silver ions within the crystals. In the dark, the valence electrons are not free to move; but, when a photon of radiation is absorbed, it liberates an electron from a bound state to one in which it is mobile. The electron may become trapped at some point in the crystal where there is an imperfection and a consequent "potential pit." A small fraction of the silver ions in the crystal are mobile even in the dark, as a consequence of thermal excitations. The combination of a mobile silver ion with the trapped electron forms a silver atom and "re-sets" the trap for capture of a second electron, following absorption of another photon. The process continues by repetition of these steps; when a sufficient number of silver atoms have collected at one site, the crystal is developable.

Scientific research into techniques for increasing the efficiency of this process and for extending the absorption through the visible spectrum has resulted in increases of sensitivity of more than four powers of 10, since the inception of photography—some 150 years ago.

6.2.2 SENSITIVITY

The extent or readiness of a photographic material to react to radiant energy is called the *sensitivity* of the material. Sensitivity is different for different wavelengths of radiation; this prop-

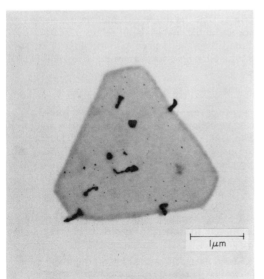
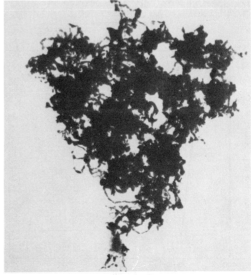

FIGURE 6-2. An exposed silver-halide grain (left) and a developed grain (right). The black specks on the exposed grains are latent-image specks partially developed. The developed grain is not the same grain but illustrates the fundamental principles discussed in the text.

erty is called *spectral sensitivity*. Silver halide, for instance, has an intrinsic sensitivity to radiation in the blue and near-ultraviolet. Some organic dyes can absorb energy at longer wavelengths and transfer this energy to the silver halide. Such dyes are referred to as *optical sensitizers*. In black-and-white photography, the addition of optical sensitizers to emulsions of silver halide extends the sensitivity of the emulsion to longer wavelengths, as shown in figures 6-3 and 6-21. Color emulsion may be made by the superposition of layers of emulsion sensitized individually to red, green, and blue light and containing dye-forming compounds that produce colored material complementary to each sensitizer. The concentration of each colored material is approximately proportional to the amount of light that produced the change. Color film made in this way is called a "tripak." This is described in greater detail in section 6.2.6, "How Color Photography Works."

The sensitivity of an emulsion is one of its most important characteristics, photogrammetrically, but is not measurable directly. It is determined in a series of steps from measurements of the *transparency* (transmittance) of the developed emulsion.

6.2.2.1 DEFINITIONS

The transparency of a material is measured by the ratio—called the *transmittance* (T)—of the amount of light-energy passing all the way through the material per unit time to the amount of light energy per unit time that enters the material. The density (D) of the material is defined as the common logarithm of the reciprocal of the transmittance:

$$D \equiv \log (1/T) = -\log T.$$

The spectral sensitivity S_λ is defined as

$$S_\lambda \equiv I/H_\lambda.$$

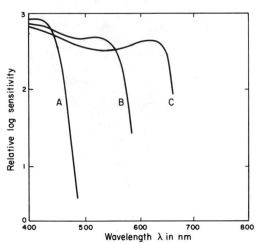

FIGURE 6-3. Spectral Sensitivity of Typical Emulsions. Curve A is for silver halide, B for orthochromatic, and C for panchromatic.

where H_λ is the energy per unit wavelength per square centimetre at wavelength incident on the emulsion that is required to produce some density level selected as reference. The level selected depends on the application intended for the emulsion and on the nature of the emulsion. A density of 1.0 is commonly used. If the material is color emulsion, a density of 1.0 with reference to the processed but unexposed emulsion is standard (called "density of 1.0 above stain"). A photographic transfer-function (also called the "response") (R) can be derived from S_λ and from the spectral distribution Q_λ of energy in the illumination (same units as H_λ):

$$R \equiv \int_{\lambda_1}^{\lambda_2} Q_\lambda S_\lambda d_\lambda .$$

Spectral sensitivities are converted into sensitivity when dealing with silver halide (black-and-white) emulsions:

$$S \equiv \int_{\lambda_1}^{\lambda_2} S_\lambda d_\lambda ,$$

the limits being the wavelengths at which the spectral sensitivity drops to zero and stays there. See section 6.2.3 for a description of instruments for measuring spectral and integrated sensitivity.

6.2.3 SENSITOMETRY

Sensitometry is the science which treats of the response of photographic materials to light. It involves instruments and techniques for precisely exposing, processing, and measuring sensitized samples; and more importantly, it includes the interpretation of data obtained in this way to aid in making correct decisions at each step of the photographic process, from the manufacturing of the film to the making of final prints. The basic sensitometric properties of photographic materials all relate, directly or indirectly, to the degree of darkening or opacity produced by various amounts and qualities of exposure.

6.2.3.1 EXPOSURE

Photographic exposure is the quantity of energy collected by the photosensitive material. The usual definition for exposure[1] (H) is that it is the product of the illuminance (E) (in lux, usually) and the time (t) for which the material is exposed.

$$H = Et$$

Because the illuminance is weighted spectrally to match the energy visible to observers, the

---

[1] Some literature uses E for exposure and I for illuminance. However, the present trend is toward use of the symbolism adopted by the American National Standards Institute and by the ISO, and this symbolism is used in the present Manual (see ISO 1000-1973).

photometric units of exposure will not be an accurate specification for photographic materials. When no spectral weighting is included, the exposure is measured in ergs per square centimetre and the calibration for H is said to be radiometric. Radiometric specifications are not exactly applicable either, since they do not measure that and only that radiant energy to which the emulsion responds. For example, a near-infrared laser cannot expose a blue-sensitive film, although it can flood the film with energy. An "effective" radiometric exposure weights the spectral energy distribution by the emulsion's spectral sensitivity.

$$H_{\text{eff}} = \frac{t \int Q_\lambda S_\lambda d_\lambda}{\int S_\lambda d_\lambda}$$

6.2.3.2 THE SENSITOMETER

The device for metering out precisely controlled amounts of exposure is called a sensitometer. Figure 6-4 shows schematically the elements of a sensitometer. The most common light-source is an incandescent lamp, and modern technology dictates the use of a tungsten-halogen lamp because of its long life and virtually constant output of light. The lamp should be chosen to provide adequate exposure over an interval comparable to that at which the emulsion will be used. The lamp should be placed away from the receiver to provide uniform illumination, and the light should be filtered to provide an approximate match to the energy distribution found in use. The shutter must be designed to open and close rapidly enough that the times required for opening and closing are small compared with the exposure time. Reproducibility of exposure is the key feature of a sensitometer. The exposure modulator is more readily calibrated if it is nonselective spectrally, and it must not change with age.

For calibration and adjustment of a sensitometer, the color-temperature of the source is conveniently controlled by adjustment of the power supplied to the source, while the illuminance is adjusted by changing the lamp-to-film distance. Light sources to be used in sensitometers are specified in ANSI Standard PH2.29-1967, "A Simulated Daylight Source for Photographic Sensitometry" and ANSI Standard PH2.35-1969, "Simulated Incandescent Tungsten Sources for Photographic Sensitometry."

Another type of sensitometer for short-time exposure uses a precisely controlled xenon flash-tube.

6.2.3.3 PROCESSING

If the sensitometer, is used to assess any factor except processing, processing must be well controlled. An exception to this rule may be made when all tests represent side-by-side comparisons and are essentially self-calibrating. Control is established by providing adequate controls, testing, and documentation to ensure that all emulsions of the same kind receive the same kind of treatment.

6.2.3.4 MEASUREMENT OF DENSITY

Sensitivity at the output of the photographic process is measured in terms of density. When the photographic image is spectrally selective, the transmittance is a function T_λ of the wavelength. Density must then be specified in terms of some spectral weighting function. For black-and-white materials, a density-weighting function related to the visual response function $\bar{y}$ is commonly used and this type of density is called visual density. When the photographic image is made up of colorants, the density determined depends on the densitometer's spectral response. Three types of response are commonly used: status A (to evaluate photographic materials normally viewed directly), status M (to evaluate preprint materials) and an arbitrary set of red, green, and blue filters (commonly Kodak Wratten filters Nos. 92, 93, and 94). The detailed specifications of these responses were published by Dawson and Vogelsong (1973).

Because the densitometer measures the transmittance of a material, the instrument's performance must be stated in terms of how the light is transmitted and collected as well as how the instrument responds to different wavelengths. A complete notation for specifying the geometry and spectral distribution of the incoming light as well as the geometry and spectral distribution of the transmitted light was developed by McCamy and is published as ANSI Standard PH2.36-1974, "Terms, Symbols, and Notation for Optical Density."

Errors or variations from instrument to instrument in either geometry or spectral response can be compensated for, on most densitometers, by calibration. This means that compensating errors may produce a false set of readings. There is no substitute for making all functional components of a densitometer meet the specifications of ANSI and ISO standards.

6.2.3.5 THE CHARACTERISTIC CURVE

A photographic material may be characterized by its sensitometric data. The S-shaped curve obtained by plotting the density as a function of

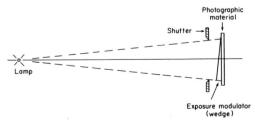

FIGURE 6-4. Sensitometer, schematically represented.

log H from a set of exposed and processed test patches is called the characteristic curve, $D -$ log H curve, or the H and D curve, after the British scientists, Hurter and Driffield, who were the first to introduce it. A typical curve is shown in figure 6-5. Photographic materials whose resultant density increases with amount of exposure are said to be *negative* materials or *negatives* since they record bright images as dark areas. Photographic materials whose resultant density decreases with amount of exposure are said to be *positive* or *reversal* materials since they record bright images as light areas. This terminology carries over into the technology of color photographic materials (section 6.2.3), in which negative materials record colors which are the complement of the colors in the photographed object, and positive or reversal (color-reversal) materials record colors similar to those in the object.

The characteristic curves of all photographic materials have the S-shape with either negative or positive slopes. The exact shape, curvature, slope, and range of density shown will vary with the material or process. The lower part of the curve, A to B, is called the toe region. The top part of the curve, C to D, is known as the shoulder region. Between these regions is a part, B to C, that is more or less straight. The density at A is called minimum density or D-min, while the density at D is called maximum density or D-max.

Curves for color photographic materials are typically drawn by plotting the appropriate red, green, and blue densities on the same graph and require three sets of variables, as shown in figure 6-6 for a color-reversal film.

6.2.3.6 Fog

The minimum density is made up of two components: base density and fog. The base density is determined by the transmittance of the supporting material, while fog is the density of the emulsion and undeveloped material. As devel-

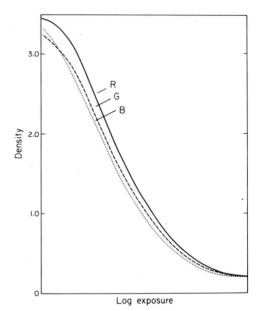

FIGURE 6-6. Characteristic curve of a typical color-reversal film.

opment time increases, it is normal for fog to increase in negative materials. If development is extended beyond normal limits (or if highly active developers are used) the fog may become significant.

Fog tends to increase with the age of unprocessed material and the amount will depend on storage conditions before exposure. It is common to measure densities above fog or D-min since this is essentially a constant for the material.

6.2.3.7 Gamma

The slope of the portion BC of the characteristic curve (fig. 6-5) is known as the *gamma* (γ) of the curve,

$$\gamma = \left(\frac{\Delta D}{\Delta \log H}\right) BC .$$

Gamma is the rate of change of density with log exposure between B and C. The curves with large γ seem to have higher visual contrast under usual viewing conditions. It is common to use γ and "contrast" interchangably, although from a purist's viewpoint, contrast is a subjective or psychological impression and γ is an objective measure.

Gamma may be a characteristic of the material's sensitivity or it may depend on processing. A given material, when developed for varying lengths of time, shows an increase in D-max and in γ as shown in figure 6-7. A similar set of curves might be obtained for four different materials developed for the same length of time.

6.2.3.8 Speed

The most common sensitometric property of photographic materials is their "speed". One

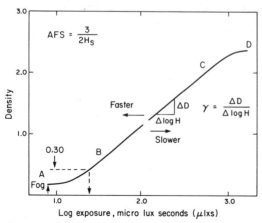

FIGURE 6-5. Characteristic curve and aerial film speed.

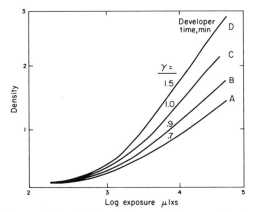

FIGURE 6-7. A family of characteristic curves for the same film and exposure but for development over different lengths of time. Curve *A* is for development over a 2-minute interval; *B*, a 3-minute interval; *C*, a 5-minute interval; *D*, an 8-minute interval.

photographic material is said to have greater "speed" or to be "faster" than another if it acquires a greater density for the same length of exposure than the other. Thus, a pure change in speed would be represented as a translation of the sensitometric curve parallel to the log-*H* axis. Values of speed can be used along with exposure meters to compute the proper time and aperture settings for exposure. The method for determining values of speed of color-reversal films from sensitometric data is given in ANSI Standard PH2.21-1971. Speeds of monochromatic aerial film are specified in ANSI Standard PH2.34-1969.

When values of speed are used to determine exposure, it is highly important that they be consistent with the conditions of use. In aerial photography, differences in illuminance over the image are made smaller by atmospheric haze between the camera and scene. To compensate for such compression, areas with the least illuminance are given exposures that place the resulting densities further up on the toe of the characteristic curve than would be conventional in pictorial photography. When detail is important, some of the needed detail may be just barely resolved. To resolve such detail satisfactorily, minimum density should be fairly up on the toe of the curve. Some American military specifications for aerial films stipulate that the speed should be based on the amount of exposure required to reach the point on the curve where the slope is $\gamma/2$ or the density is 0.3 above fog.

6.2.3.9 DETERMINING FILTER-FACTORS

Filters are frequently used in photography to reduce the effect of light scattered by haze, to reduce glare, and so on. Yellow (minus-blue) filters are most commonly used against the effects of haze, since light is scattered more at short wavelengths than at long. Scattering by air molecules is approximately proportional to the inverse fourth power of the wavelength. Scattering by particles such as smoke, dust, and mist is less spectrally selective as the particle-size increases. Scattered light gives a low, uniform illumination over the entire film, overlying the image of the object or scene. The range of densities produced in the emulsion is correspondingly diminished. The effect is greatest in areas in which the ratio of scattered light to image-forming light is greatest—usually, in the shadowed areas. Haze filters exclude the scattered, short-wavelength radiation and thereby increase contrast of the shadowed areas. As the size of the scattering particles increases, the filter must reject more of the green (and perhaps even yellow) light in order to produce an acceptable, effective contrast.

Light is ordinarily made up of waves lying in many different planes parallel to the direction of motion of the light. When such light is reflected, light in some of these planes is poorly reflected while light in other planes is practically unaffected. Which planes are affected and which are not depends on the geometry and on the nature of the reflector. When reflection is strong and undesired, as when light is reflected from a lake or sea surface and causes glare, a *polarizing filter* can be used to filter out the strongly reflected waves.

All filters exclude some of the light falling on them. The increase in exposure required when a filter is used is frequently expressed as the ratio, called a *filter factor,* of the amounts of exposure needed with and without the filter. Filter factors can be determined in several ways.

1. Empirically, from exposure in a camera. A series of exposures to a single scene is made using the camera both with and without filter. The exposure in each series that produces a just adequate image may be used to compute the filter factor in terms of camera stops or in terms of the exposure ratios producing the best pictures. Since this method will produce a result which varies depending on haze, cloud cover, and solar altitude, the result of a single test may be misleading.
2. Sensitometric determination. A sensitometric scale is exposed both with and without the filter and the log *H* shift of the speed point is used to compute the filter factor.
3. Direct computation. The spectral energy distribution of the source and the spectral sensitivity of the film may be used to compute the filter factor from its measured spectral transmittance T_λ. The effective transmittance (f) of filter (or the filter factor) is given by the ratio of the integrals

$$f \equiv \frac{\int Q_\lambda T_{f\lambda} T_{c\lambda} S_\lambda d\lambda}{\int Q_\lambda T_{c2} S_\lambda d\lambda}$$

where Q_λ is the spectral energy distribution of the source, $T_{f\lambda}$ is the spectral transmittance of the filter, $T_{c\lambda}$ is the spectral transmittance of the camera optics, and S_λ is the spectral sensitivity of the photographic material. The integrals may be evaluated by standard numerical techniques. These data summed at 1-nanometre intervals may be considered an exact solution. The approximations introduced by summing 5-nanometre intervals are still smaller than the experimental errors resulting from sensitometric testing.

6.2.4 PROPERTIES OF IMAGE-STRUCTURE

Image quality is directly related to such properties of the emulsion as granularity and concentration. Granularity, for example, is a measure of the size and distribution of particles of metallic silver or clouds of dye in the developed emulsion and governs the detectability of fine detail; concentration is the amount of light-sensitive material per unit volume. One measure of image quality is the spread function which indicates the degree to which the illuminance in which a point, line, or edge is blurred by the recording medium and is directly related to edge sharpness, resolving power, and modulation transfer function (MTF). Information on granularity, resolving power, and MTF can be easily obtained from manufacturers' data-sheets, and it is now possible for the photogrammetrist to select the best emulsions for his particular tasks. Consequently, in the following paragraphs, consideration is given to those properties governing definition that can be readily analyzed from available data: 1) graininess and granularity; 2) resolving power; and 3) the modulation transfer function (MTF).

6.2.4.1 GRAININESS AND GRANULARITY

Photographic emulsions are often classified according to the impression of grain size obtained by the observer, *e.g.,* fine-, medium- or coarse-grained. This impression is termed graininess and is defined as the subjective sensation of a random pattern apparent to a viewer seeing small local density variations in an area of overall uniform density. Granularity, by contrast, is an objective measurement of these local density variations that is conducted with a microdensitometer and is designed to correlate with graininess (Eastman Kodak Company, 1974; 1975). In practice, granularity values determined by one film manufacturer, Eastman Kodak Company, represent 1000× the standard deviation (σ) of a microdensitometer trace conducted with a 48 μm circular aperture across a film sample with a uniform, diffuse density of 1.00. The granularity values so obtained are calculated with an equation of the general form:

$$\sigma(D) = \left[\frac{n\sum D_i^2 - (\sum D_i)^2}{n(n-1)}\right]$$

D_i = individual density reading
n = number of readings

TABLE 6-1 GRANULARITY AND GRAININESS FOR SONE KODAK AERIAL FILMS

Diffuse RMS Granularity	Graininess	Film
33−42	Coarse	2043
26−30	Moderately Coarse	2405, 2424
16−20	Fine	2402, 3411, 2443
11−15	Very Fine	3410, 2448, 2445
6−10	Extremely Fine	3414/1414

and are referred to as *diffuse rms granularities* (SPSE, 1973). Diffuse rms granularity also indicates the relative sensation of graininess obtained by an observer viewing diffusely illuminated photographic material under 12× monocular magnification. Thus, a photograph recorded on a typical mapping film such as Kodak Plus-X Aerographic film 2402 (Estar base) with a diffuse rms granularity of 19 will appear twice as grainy as one recorded on a high altitude reconnaissance film such as Kodak high-definition aerial film 3414 (Estar thin base) with a value of 9. As magnification is increased, graininess obviously will become more apparent.

Color emulsions also exhibit a grainy structure resulting from the "clouds" of colored dyes which remain in the emulsion's layers after the metallic silver has been removed in the development process. Because the dye clouds are small and semitransparent, the graininess and granularity of color emulsions is usually less than that of black-and-white emulsions.

Table 6-1 provides data relating diffuse rms granularity to graininess for a family of aerial mapping and reconnaissance films, and figure 6-8 indicates the influence of granularity on the detectability of fine detail. Since granularity increases with density, the detectability of small low-contrast details is impaired for objects recorded on the shoulder of the D-log H curve (Brock, *et al.*, 1966).

6.2.4.2 RESOLVING POWER

The resolving power of a photographic emulsion refers to the number of alternating bars and spaces of equal width which can be recorded as visually separate elements in the space of one millimetre, on the emulsion and provides an indication of the ability of the emulsion to provide distinct images of small, closely spaced objects (ANSI, 1969). In photography the combination of a bar and a space is referred to as a *line* or *line pair,* and resolving power is specified in lines/mm (1/mm) or lines pairs/mm (lpr/mm). Thus, a film with a resolving power of 30 1pr/mm will record a pattern consisting of 30 bar-and-space pairs, each having a width of 1/30 mm, and the observer theoretically will be able to distin-

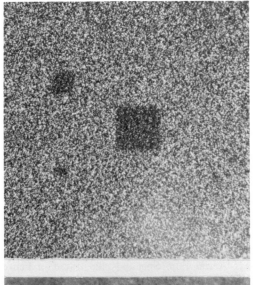

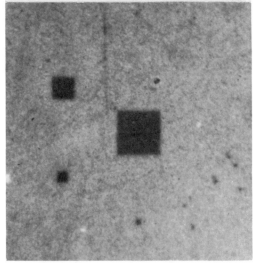

FIGURE 6-8. The detectability of small, low contrast detail is impaired by film granularity. The upper photograph was taken on a typical mapping film, whereas an extremely fine grained reconnaissance film was utilized for the bottom photograph.

guish 60 linear elements (30 bars and 30 spaces) per millimetre when the image is viewed under suitable magnification.[2]

In the United States, resolving power of an emulsion is commonly evaluated under laboratory conditions by photographing standard three-bar resolution targets through a large-aperture, diffraction-limited microscope objective. The image is then visually examined under magnification to determine the number of

---

[2] The resolving power of television and similar instruments is normally defined in terms of the number of individual elements at the resolution limit. Thus, a photographic film with a resolving power of 30 lpr/mm would, in terminology of television, have a resolving power of 60 l/mm.

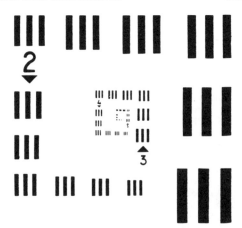

FIGURE 6-9. A typical three-bar resolving power target of high contrast.

lpr/mm resolved (*see* Brock, 1970, pp. 247–249). These targets consist of 3-bar patterns in which the width of the elements decreases in the ratio 1: $\sqrt[6]{2}$ (or 1: $\sqrt[20]{10}$) and the length of the elements is 5× the width (figure 6-9). Since resolving power is highly dependent on target contrast, the luminance differences (or contrasts) of the bars and spaces must be specified. Typically, targets are available with the contrast ratios listed in table 6-2.

Normally, resolving powers of low-contrast emulsions obtained with high-contrast targets range from 30 to 50 lpr/mm for typical mapping films (table 6-3). By using resolving-power values determined for targets of several contrasts, functional relationships termed threshold modulation (TM) or aerial image modulation (AIM) curves can be developed (Lauroesch *et al.*, 1970; SPSE, 1973). When TM or AIM curves are plotted on log-log paper with the ordinate representing target modulation (contrast) as defined in section 6.2.4.3 and the abscissa indicating resolved lines per mm, the resulting functions closely approximate a straight line (figure 6-10). TM or AIM curves indicate resolving power for a range of target contrasts and may be used in conjunction with lens and image-motion MTF's to estimate the resolving power of the camera system under operational conditions (*see* section 6.7.1).

When selecting an aerial film for mapping or when judging its capabilities on the basis of resolving power, values derived from low-contrast targets should be compared, since atmospheric

TABLE 6-2 RESOLVING POWER TARGET
CONTRAST RATIOS

Target Contrast	Contrast Ratio	Log Contrast Ratio
High	1000:1 to 100:1	3.0 to 2.0
Low	1.6:1 or 2:1	0.2 or 0.3

TABLE 6-3. RESOLVING POWER OF SELECTED
KODAK AERIAL FILMS

Film	T.O.C. 1000:1	T.O.C. 1.6:1
PLUS-X AEROGRAPHIC Film 2402 (ESTAR Base)	100	50
TRI-X AEROGRAPHIC Film 2403 (ESTAR Base)	80	25
DOUBLE-X AEROGRAPHIC Film 2405 (ESTAR Base)	100	50
PANATOMIC-X AERECON 2410 Film (ESTAR Thin Base)	250	80
High Definition Aerial Film 3414 (ESTAR Thin Base)	630	250
Infrared AEROGRAPHIC Film 2424 (ESTAR Base)	80	40
AEROCHROME Infrared Film 2443 (ESTAR Base)	63	32
AEROCHROME MS Film 2448 (ESTAR Base)	80	40
AEROCOLOR Negative Film 2445 (ESTAR Base)	80	40

effects normally reduce the contrast ratios of adjacent objects to less than 2:1 at the lens. Another factor to consider is that the resolving power of a typical photogrammetric camera is usually limited by the film (rather than by the lens or image motion) and significant improvements in resolution can sometimes be obtained by switching to a film of better characteristics (Welch and Halliday, 1973).

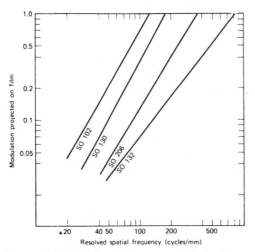

FIGURE 6-10. TM (or AIM) curves indicate the relationship between target contrast and visually determined film resolving power. Contrast is expressed as modulation.

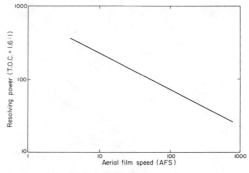

FIGURE 6-11. Relationship between low-contrast resolving power (1.6:1) and film speed for a number of aerial films.

The inverse relationship between film speed and resolving power, which provides a log-log plot of low-contrast (1.6:1) resolving power *versus* Aerial Film Speed, is illustrated in figure 6-11. The data points represent published and unpublished data for aerial camera films manufactured by several companies. The solid curve is a straight-line approximation to the data.

Resolving power is dependent on exposure and development. An exposure that results in all densities being on the toe of the D-log H curve will result in a serious reduction in resolving power. Placing the densities on the shoulder of the D-log H curve generally produces a less severe loss of resolving power, although the increase in density and granularity will impair the detectability of small detail. Maximum resolving power is usually realized at densities of approximately 1.0.

6.2.4.3 MODULATION TRANSFER FUNCTION (MTF)

The modulation transfer function (MTF) of a photographic material indicates the fidelity with which a periodic test pattern that has a sinusoidal intensity distribution of constant amplitude is recorded as a function of spatial frequency (Dainty, 1971; Welch, 1971; Slater, 1974). In practice, a test pattern such as shown in figure 6-12 is recorded on a photographic material and the developed image is traced with a microdensitometer to produce a pattern similar to that given in figure 6-13. At each frequency (v), the values of maximum density (D-max) and minimum density (D-min) are determined and converted to linear relative exposure values by

FIGURE 6-12. Sinusoidal Test Object.

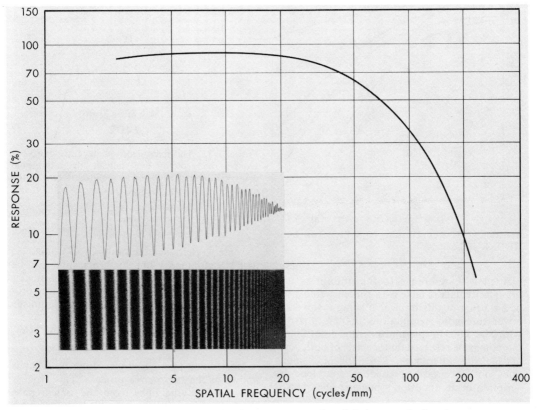

FIGURE 6-13. Sinusoidal image, microdensitometer trace and modulation transfer function curve.

means of the D-log H curve of the emulsion. The contrast, or amplitude, at each frequency is then expressed as image modulation $[M_E(v)]$ as given by the following equation:

$$M_E(v) = \frac{E\max - E\min}{E\max + E\min} \leq 1.0$$

where Emax, Emin = maximum and minimum exposure values converted from image-density values using the D-log H curve of the emulsion.

Using an exactly similar equation, the modulation of the optical-image at the focal plane can be expressed as $M_0(v)$.[3] By definition, the MTF is then equal to:

$$\tau(v) = \frac{M_E(v)}{M_0(v)}.$$

Normally, the MTF is presented as a graph of modulation versus spatial frequency in cycles per millimetre on either linear or log-log graph paper. Since, for practical purposes, cycles/mm may be equated to lpr/mm, the MTF also can be thought of as a graph indicating the reduction in

---

[3] As an example, the modulation of a target with a 2:1 contrast ratio is:

$$M_0(v) = \frac{2-1}{2+1} = 0.33.$$

If $M_E(v) = 0.23$, then $(v) = 0.70$.

contrast which occurs between the elements of a resolving power target when it is recorded on a specific emulsion.

MTF's along with data on granularity and resolving power provide useful information about the imaging capabilities of the emulsion; that is, an emulsion of high resolving power, low granularity and good MTF will faithfully record small detail. In this regard some basic interrelationships between MTF's and other image-structure properties are worthy of consideration. For example, from table 6-3 and figure 6-14, it is evident that MTF's will generally rank black-and-white emulsions in the same order as resolving power does. This is to be expected, as threshold density differences (or modulations) are required to resolve targets recorded on black-and-white emulsions (figure 6-15). If, however, the MTF of a black-and-white film such as Kodak infrared Aerographic film 2424 (Estar base) is compared to that of a color film such as Kodak Aerocolor negative film 2445 (Estar base), the curve for 2424 film indicates significantly higher modulations at frequencies beyond 30 cycles/mm. An inspection of table 6-3, however, reveals that these films deliver identical resolving powers for high- and low-contrast targets (i.e., 80 and 40 lpr/mm respectively). This apparent discrepancy can be ex-

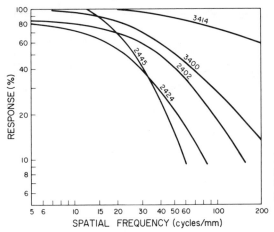

FIGURE 6-14. Representative MTF's for same KODAK aerial films.

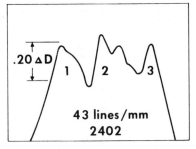

FIGURE 6-15. This microdensitometer trace of a 3-bar group at the resolution limit indicates that a threshold density difference of 0.20 was required to resolve the group.

$$\frac{S/N \text{ Out}}{S/N \text{ In}}.$$

plained with reference to the graininess and granularity data in table 6-1. The very fine grain of 2445 film results in targets being resolved at lower threshold modulations than is possible with 2424 film. Furthermore, with a color film, a target may be resolved on the basis of both color and density differences. Thus, the MTF alone does not necessarily provide a complete description of an emulsion's imaging characteristics.

MTF's are particularly useful when conducting comparisons of different camera systems. For example, given separate MTF's for the camera lens, film, and anticipated image-motion, a *system* MTF can be obtained by multiplying the ordinates of the component curves together frequency-by-frequency. This cascading procedure offers tremendous flexibility since the effects of variable components, such as the film, on the system can be readily ascertained. MTF's may also be derived from microdensitometer traces of aerial photographs of targets. The practical aspects of MTF prediction and analysis are further discussed in section 6.7.1.

6.2.5 RELATIONSHIP BETWEEN SPEED, GRAIN, AND DEFINITION

In the photographic emulsion, the needs for less-objectionable graininess and better picture definition are generally opposed to the need for faster emulsions, or for emulsions that are more sensitive to radiation, because this requirement means larger grain size. The manufacturer of photographic materials, therefore produces different emulsions for different purposes, and the user picks the one for his use based on whatever he needs.

There is, at present, no single, universally used means of characterizing the signal-to-noise ratio of photographic film. DQE or detective quantum efficiency, evaluates the ratio:

Some workers use the ratio γ/σ of gamma (γ) to RMS granularity (σ). Such measures can be helpful in comparing emulsions for a particular task. In aerial photography, the needs for resolution and sensitivity may require compromise.

6.2.6 HOW COLOR PHOTOGRAPHY WORKS

Most color photography is done by what is known as the *subtractive* process. This process works by having light from the object pass through successive layers of emulsion, each layer containing a sensitizer (soupler or dye) that absorbs (subtracts) light from a particular part of the spectrum and passes the light in the rest of the spectrum on to the layers underneath. The usual procedure is (figure 6-16, left part of diagram) to have a blue-sensitive emulsion at the top, since silver halide is most sensitive to blue light. Next comes a thin, non-reactive yellow film to filter out blue light that has passed through the top layer. Then comes a green-sensitive layer, which absorbs light in the green part of the spectrum and passes on the remaining, almost all red, light to the bottom layer where the red light is absorbed. Each layer is sensitive, photographically, to the light it absorbs and is insensitive to the light which it passes on. Consequently, the silver halide in each layer forms a latent image for that, and only that, spectral region to which its layer is sensitive.

The mechanics of the exposure part of the color process and of the following development, and in fact the appearance of the final photo-

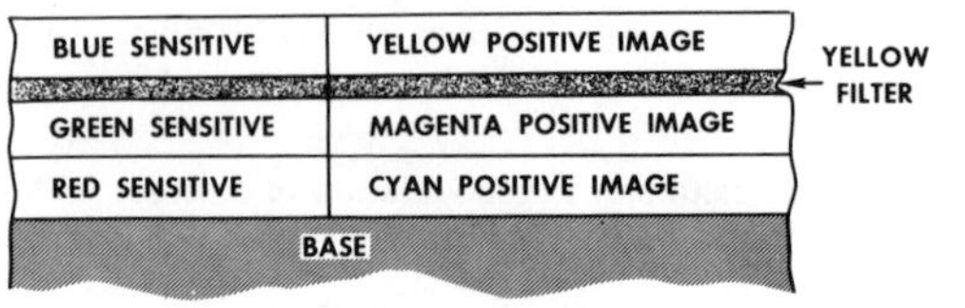

BLUE SENSITIVE	YELLOW POSITIVE IMAGE	YELLOW FILTER
GREEN SENSITIVE	MAGENTA POSITIVE IMAGE	
RED SENSITIVE	CYAN POSITIVE IMAGE	
BASE		

FIGURE 6-16. Schematic representation of the layers of a color film.

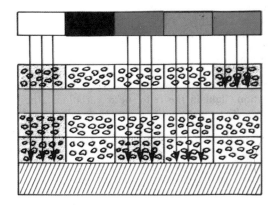

REVERSAL FILM

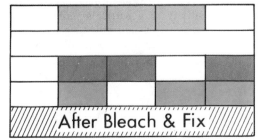

After First Developer

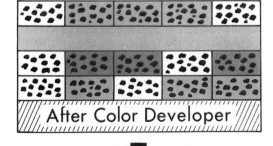

After Color Developer

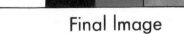

Final Image

NEGATIVE FILM

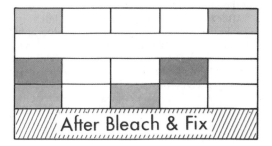

After Color Developer

Final Image

FIGURE 6-17. Reproduction of colors by a reversal color film and a negative color film.

graph, depends on the nature of the sensitizer introduced into the emulsions at the time of manufacture. However, the subtractive process generally involves the presence, either initially or subsequently, of three kinds of dyes: cyan, magenta, and yellow. The cyan dye transmits freely in both the blue and green, but absorbs or controls, by its concentration, light in the red region. In a similar manner, the magenta dye transmits in the blue and red, and absorbs or controls the green; the yellow dye transmits in the red and green regions and controls the blue. A wide gamut of colors can be obtained with these three dyes. For example, if the magenta and yellow dyes are superimposed, the magenta absorbs the green light, the yellow absorbs the blue light, and the combination will appear red. Likewise, combinations of cyan and magenta will appear blue, combinations of yellow and cyan will appear green, and variation in the concentrations of the dyes will give varying shades of the colors. Figure 6-16 (right-hand part) is a schematic cross-section of the usual arrangement of the layers of emulsion with the dyes present. (Other configurations are used, but they will not be described here).

Figure 6-17 will help to explain the basic principles of color photography. At the top of the figure is a representation of a scene that consists of areas that are white, black, red, green, and blue. The white provides radiation in all three primary regions and thus exposes all three layers of the film. There is no radiation from the black; consequently, none of the film layers is exposed. The red light passes through the top layer, the yellow filter, and the middle layer (which are not sensitive to red light) and then exposes the bottom layer. The green radiation passes through the top layer (which is not sensitive to it), through the yellow filter, and exposes the middle layer. The remaining energy passes through the bottom layer, which is not sensitive to green. The blue radiation exposes the top layer and is prevented from reaching the two bottom layers by the yellow filter.

After exposure and formation of the latent images in the layers of emulsion, the photographic material is developed, producing a visible, colored picture. There are basically three different kinds of final picture: color-positive, or positive; color-negative, or negative; and false-color. (The first two are sometimes lumped together under the name *normal-color*). A color-positive picture shows colors which are the complements of those of the object, and a false-color picture contains regions whose colors differ markedly from those in a normal-color picture. There are also, basically three different kinds of development which may be used: the dye-generation (chromogeneric) process, the dye-bleach process, and the image-transfer process. The image-transfer process, first introduced commercially by Polaroid in 1963, is not at present used much in photogrammetry and will not be discussed. The other two processes are widely used, and are discussed in detail in section 6.5.3.

The dye-formation process comes in two variations—the color-reversal process, which produces color-positives, and the direct-color process, which produces color-negatives. In both variations, the coloring dyes are formed during development through the intermediacy of the active silver and a chemical called a *coupler*. In the color-reversal process, images are formed in which the concentrations of dye are *inversely* proportional to the amount of original exposure. If white light is allowed to pass through film developed by the color-reversal process, it can be seen (figure 6-17, left-half, first column) that, since there is no dye, the light transmitted is white. In the second column, the dyes absorb all three primary colors, allowing no radiation to pass; thus, black results. As mentioned previously, the combination of yellow and magenta (column 3) transmits only red radiation, yellow and cyan (column 4) transmit green, and magenta and cyan (column 5) transmit blue. The final reproduction matches closely the colors of the original scene.

In the direct-color process, images are formed in which the concentrations of dye are *directly* proportional to the amount of original exposure. In the right half of figure 6-17, it can be seen that in the first column all three dyes are formed and the resulting color is black (no light transmitted). In the second column, no dyes are formed and white light passes through; the individual cyan, magenta and yellow dyes are formed as a result of exposure by the red, green, and blue light. Thus, each of the colors formed in this way is the complement of the color in the scene. If a photograph or a print is made of this negative on another negative, a reversal to the complementary colors will again be obtained; the final colors will correspond closely to those of the original scene.

It should be noted here that the foregoing descriptions depict color photography as a sort of "go, no-go" system. The emulsion is depicted as having sensitivity to a particular color or as

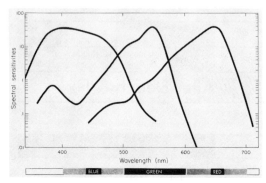

FIGURE 6-18. Spectral sensitivities of the three layers of a typical color film.

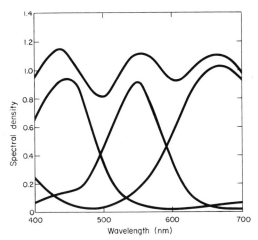

FIGURE 6-19. Spectral densities of a neutral gray and of yellow, magenta and cyan dye components for a typical film developed by the color-reversal process.

having no sensitivity to it. The dyes are said to either transmit or block a color completely. Actually, the emulsion is not as ideal as that. Each of the layers has some sensitivity in regions other than that of its primary sensitivity, and each of the dyes absorbs light of colors other than that which it absorbs primarily. Also, correct exposures to most objects will result in some exposure of all three layers, which provides the transmissions in the three primary regions that will combine to form the gamut of colors possible with color photography.

Figure 6-18 shows the actual spectral sensitivities of the three layers of a typical color film, and figure 6-19 shows the absorptions of the dyes that are formed. The dyes, although primarily absorbing in one spectral region, do have absorption in the other region; they do not absorb uniformly over any spectral region. When these three dyes are combined, as is shown by the upper curve, the neutral gray formed is not a spectral neutral—that is, it does not have the same density at all wavelengths even though it appears neutral to the human eye. Similarly, combinations of the three dyes will form colors that visually match those of an object photographed, but the spectral densities will not be identical. Therefore, it is generally not possible to derive the spectral reflectance of an object from spectral measurements of its photographic image.

6.2.7 SPEED OF AERIAL FILM

Speed is one of the most important sensitometric properties of photographic materials intended for aerial photography. Knowledge of the speed of an aerial film permits the computation of camera settings that will yield optimally exposed images.

Through the years, a number of different sensitometric criteria have been employed to define the speed of films for aerial photography—some based on fractional gradient, others on fixed density. Unfortunately, lack of general agreement within the aerial photographic community resulted in unnecessary confusion and difficulty in relating products of different manufacturers and in comparing results reported by users who employed different criteria. This problem was alleviated with the publication of ANSI PH2.34-1969, "American National Standard Method for Determining the Speeds of Monochrome Photographic Negative Films for Aerial Photography." Speed determined in accordance with this standard is known either as Aerial Film Speed or Effective Aerial Film Speed (the latter being used when speed is determined for films processed by other than the specified standard process).

The standard specifies the exposure and developing conditions to be employed when Aerial Film Speed is to be measured. The reader should refer to the document for full details; but, in brief, the exposure should be made to daylight and the process must include development for eight minutes at a temperature of 20°C in a developer whose formula conforms to the formula for Kodak developer D-19.

The exposure used in determining Aerial Film Speed is the point on the characteristic curve at which the density is 0.3 units above base plus fog (*see* figure 6-5). The exposure (H_s), expressed in micro-lux-seconds, needed to produce this density is used in the following formula to compute the Aerial Film Speed (AFS):

$$AFS = \frac{3}{2H_s} \cdot$$

Speed is generally rounded to the nearest one-third increment in stops, the scale of AFS values being an arithmetic one in which adjacent steps differ by the cube root of two.

When speed is determined for films not developed in the specified manner, the same density (0.3 net density) and formula are used as for AFS, but the resultant speed is referred to as Effective Aerial Film Speed (EAFS) and the method of development should be clearly stated.

As suggested by its title, ANSI Standard PH2.34-1969 pertains to the method of determining speeds for monochromatic aerial films yielding negative images and specifically excludes infrared-sensitive film yielding positive images and color films. Speeds for photographic materials in these classes are generally determined empirically through evaluation of aerial photographs. Although the procedures used may vary, the essential feature of the technique involves a determination of the relative exposures required to produce optimally exposed images on both the film being tested and a black-and-white aerial negative. The AFS (or EAFS) of the latter is determined in accordance with the ANSI Standard, and the difference in exposures required by the two films is applied to this value to yield the speed of the material. The term Effective Aerial Film Speed is then used for the calculated quantity, although this usage extends somewhat the definition given in the ANSI Standard.

Since the technique requires subjective judgment of optimally exposed images on both the standard and test films, a closely spaced series of exposure (*e.g.*, at one-half stop increments) should be made on the two films. The reader will appreciate that the accuracy of the technique de-

scribed depends on the precision and accuracy of the camera(s) used in the test. To avoid errors due to changes in scene illuminance, it is preferable that simultaneous photographs be made on both films. The use of two cameras satisfies this requirement, but accurate calibration of both shutters and apertures is obviously a necessity.

It is appropriate to emphasize that Aerial Film Speed differs from ASA speed, the two being defined and determined in different ways. There is no simple expression of general validity that relates Aerial Film Speed and ASA speed.

6.3 Photographic Materials

6.3.1 BLACK-AND-WHITE FILMS FOR AERIAL PHOTOGRAPHY

Black-and-white films for aerial photography are manufactured in a wide range of photographic speeds and yield images of varying quality. As discussed elsewhere in this chapter, these two factors, speed and quality are approximately inversely proportional to each other—higher-speed films generally having poorer image definition than slower films. When choosing a film for a specific application, the planner should optimize the trade-offs between speed and quality by evaluating the mission requirements and determining existing constraints such as illumination levels, target contrast, maximum camera-lens aperture and shutter speed, image motion, *etc.*

6.3.1.1 SPEED

In terms of Aerial Film Speed (AFS), modern black-and-white aerial films[4] range from fast, high-speed films with AFS values of 400 or greater to slow, high-definition films with AFS values of 12 or less.

The high speed of modern, fast, black-and-white aerial films can be better appreciated when one notes that adequate exposure can be obtained with AFS 500 films and representative mapping cameras at sun-angles of 5° and less. At a latitude of 40°, this capability translates into the availability of a photographic day of at least eight hours in duration throughout the entire year. At latitudes closer to the equator, even longer periods are available during which illumination is adequate for aerial photography with fast films.

The slowest, high-definition aerial films do not have adequate speed for satisfactory use in most mapping cameras, primarily because of limitations in lens aperture and/or lack of image motion compensation. Some recent cameras are, however, able to take advantage of the high quality of slow films. A noteworthy example is the S190B Earth Terrain Camera used in the Skylab missions in 1973 and 1974.

Films designed primarily for general photogrammetric mapping have speeds in the medium-high range (AFS 125 to 400). Faster films are of value under conditions of poor lighting and low sun-angle. Slower, medium-speed films (AFS 40 to 125) can sometimes be used to advantage in small-scale, high-altitude mapping where film graininess may become a limiting factor in detail rendition. Limitations on use of films in this speed class are the relatively small, maximum aperture of mapping cameras and the normal requirement for use of both antivignetting and minus-blue filters with significant filter factors. The recent introduction of mapping cameras with faster (*f*/4) lenses makes the use of medium-speed films a more feasible alternative.

It should be noted that since most low- and medium-low speed aerial films (AFS 40) are designed primarily for reconnaissance, the thickness of the base use for these films is generally less than that commonly found in mapping films (e.g., 2-1/2-mil (64 μm) as compared to 4-mil (102 μm). The differences in dimensional stability and other physical characteristics between films on 2-1/2- and 4-mil polyester base are discussed in detail in section 6.4 of this chapter.

6.3.1.2 CONTRAST

The contrast requirements for an aerial film depend upon the type and contrast of the terrain being photographed and the amount of haze existing between camera and target. Haze increase with higher altitude and lowers the contrast. Black-and-white aerial films are, in general, designed to yield a range of contrasts

---

[4] Up to now, all discussion has applied either to photographic film in general or to such films made to be used in cameras (*i.e.*, for photography). The following discussion will also deal with other kinds of photographic film. However, unless specifically stated otherwise, film for photography (camera film) will be meant.

(*gammas*) through appropriate choice of development. Films manufactured for mapping commonly provide a lower contrast than do films designed for reconnaissance or photointerpretation. A set of characteristic curves for a typical mapping film, when developed at various rates in an automatic developer, is shown in figure 6-20.

Gammas in the range of 1.0 to 2.0 are typically required for general aerial survey and reconnaissance photography at low to medium altitudes. Higher *gammas* are appropriate when photography is obtained at high altitudes or through dense haze to compensate for the contrast-reducing effect of non-image-forming light scattered by the atmosphere.

Specifications for aerial photography for photogrammetric mapping commonly indicate that the *gamma* of the film should be in the range of 0.6 to 1.0. However, since the range of density in the photograph is a function both of the range of effective illuminance in the typical image and the film-process contrast, specifications indicating minimum and maximum image densities are more useful than specifications of film *gamma* alone. For example, the effective log illuminance range of a low-contrast scene on a hazy day could well be as low as 0.5. Restriction of the *gamma* to a maximum of 1.0 would result in an image density scale of 0.5 (assuming the scene was exposed to a given density on the straight-line portion of the characteristic curve), a range unsatisfactory for most purposes. Had the specifications indicated the desired *D*-min and *D*-max values of, let us say, 0.4 and 1.2, respectively, a development yielding appropriately higher contrast (in this case one which

yields a *gamma* of the order of 1.6) would be used to provide the required densities.

6.3.1.3 SPECTRAL SENSITIVITY

Almost without exception, modern black-and-white aerial films are panchromatic (disregarding for the moment infrared-sensitive films). A majority of currently manufactured films have sensitivities that extend out to wavelengths in the neighborhood of 700 nm. Although sensitivity extends well down into the ultraviolet (i.e., below 400 nm), this spectral region is not ordinarily of interest in black-and-white aerial photography since minus-blue filters are commonly used to reduce the effect of haze.

Black-and-white, infrared-sensitive films have sensitivity throughout the visible and into the infrared to about 900 nm. Figure 6-21 shows two spectral-sensitivity curves, one typical of a black-and-white, panchromatic, extended-red sensitivity film and the other of an infrared-sensitive aerial film.

6.3.1.4 IMAGE STRUCTURE

See section 6.2.4 for considerations of image structure on black-and-white aerial films.

6.3.2 COLOR FILMS

As has been described in section 6.2.6, there are two general types of normal-color film: positive-color film, which produces a picture in which the apparent color closely resembles those of the object photographed; and color-negative film, which produces a picture in which the apparent colors are the complements of those of the object. Both types are available for aerial photography.

The sensitivity of a layer and the kind of dye formed or used in that layer do not have to be related. Any combination of sensitivity and dye may be used in each layer, and if different from that used for normal-color films causes the picture to come out in colors not similar to those of the object or its chromatic complement. Photographic film intended to deliberately and systematically falsify an object's color scheme is called "false-color film." The most noted vari-

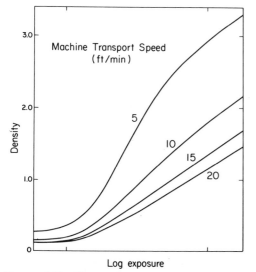

FIGURE 6-20. Characteristic curves of a typical mapping film developed at the indicated rates in an automatic developer.

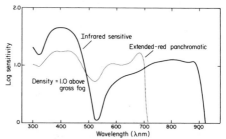

FIGURE 6-21. Spectral sensitivity curves for black-and-white aerial films.

ety of false-color film is infrared-sensitive color film which has one layer sensitive to infrared radiation and the other layers sensitive to light. Such films may be made for either the color-reversal or direct-color process.

Color films may also be categorized in terms of their definition or their speed. Divided this way, they fall into two distinct categories: normal-speed film (which includes a majority of the photographic films available), and the slow, high-definition films for use in high-performance camera-systems at very high altitudes. Section 6.2.4 gives image-structure values for some typical films.

6.3.2.1 COLOR-NEGATIVE FILMS

The great strength of color-negative films lies in their versatility. With them, many copies are easily made; these copies may be either prints or transparencies. In addition, less expensive black-and-white prints may also be made for those applications where color is not needed. Color-negative films are generally used for low-to-medium altitude mapping and reconnaissance.

These films generally have high speed compared to other color films in order to accommodate the illumination provided by low sun-angles or less-than-good atmospheric conditions. Thus, their use will increase the length of the photographic day and efficiently utilize the time available for making the aerial photographs. They tend to have average definition for color films of the normal-speed class mentioned above, having resolution on the order of 40 lines/mm and graininess that depends on the particular photographic material involved. Generally, the contrast values for aerial films are in the medium range. They are higher than those used for amateur photography because of the contrast-reducing effect of haze, but lower than those used for black-and-white or aerial color-reversal films. This characteristic provides considerable latitude in choice of exposure, which makes exposure less critical. Typical characteristics and spectral-sensitivity curves are given in figures 6-22 and 6-23, respectively. For the characteristic curves, the contrasts of the three layers should be as much alike as possible to provide uniform color balance over the exposure range. It is desirable that the speeds of the three layers be similar to maximize exposure latitude, but it is not essential, because color balance can be controlled in printing.

6.3.2.2 COLOR-REVERSAL FILMS

Where direct use will be made of the camera (either for plotting or for interpretation) and many copies will not be needed, a color-reversal film is generally used. Most films used for special applications are also color-reversal, which results in many more color-positive than color-negative films being produced. These films have

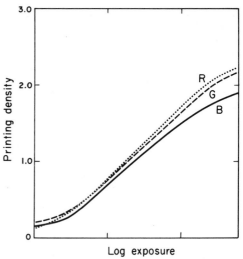

FIGURE 6-22. Characteristic curves of a typical aerial color-negative film.

the advantage that after development they are immediately available for use, and there are no delays caused by printing.

Color-reversal films are available in both the normal-speed range and also as very slow high-definition films. Color-reversal, films of normal speed tend to be somewhat slower than color-negative films but the two kinds have comparable definition as indicated by their similar resolving powers. The high-definition film is about one-tenth as fast as the normal-speed films and has a low resolving power on the order of 100 lines/mm. Its current use is primarily in photo-interpretation, although as camera systems improve, it will find more uses as a very high-altitude mapping film. Typical characteristic and spectral sensitivity curves for the normal-speed films are shown in figures 6-6 and 6-18 (section 6.2.)

Many false-color films have been made for

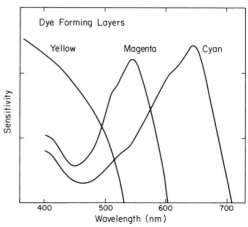

FIGURE 6-23. Spectral sensitivities of the three layers of a typical color-negative film.

special applications, but the most important (and the only ones which will be described here) are the three-layer films in which one layer is sensitive to the infrared. The normal-speed version was originally made for camouflage detection but has since found many applications in remote sensing and earth-resources studies. The high-definition version is made to be used as a counterpart to the high-definition normal-color film in the same camera systems.

Figure 6-24 shows the basic operational characteristics of the normal-speed version of this false-color film compared to that of a normal-color film, to illustrate both the similarities and dissimilarities.

If the spectral region to which photographic films are sensitive is divided into the ultraviolet, blue, green, red, and infrared spectral regions, it will be remembered that normal-color film is sensitive to blue, green, and red light. Associated with these sensitivities are the yellow, magenta, and cyan dyes, respectively, which, after processing, combine to produce the colors blue, green, and red that closely match those of the original scene photographed.

With the infrared-sensitive films, all three layers are sensitive in the blue spectral region; the individual layers are also sensitive to green, red, and infrared radiation, respectively. A yellow filter is always used over the camera lens to prevent blue light from reaching the film. This results in a film-filter combination that has three layers, sensitive to the green, red, and infrared regions, respectively. Again, the same three dyes—yellow, magenta, and cyan—are associated with these sensitivities; and it can be noted that they are used in the same order with respect to increasing wavelength, but their sensitivities have been shifted by one "block" toward longer wavelengths. As with normal-color film, combinations of the three dyes form blue, green, and red colors; but now the blue has resulted from a green exposure, the green from a red exposure, and the red from an infrared exposure. Figure 6-25 shows the actual spectral sensitivities of the three layers of such a film and the transmittance of the yellow filter. The characteristic curves for the infrared-sensitive film are given in figure 6-26. With a normal-color film,

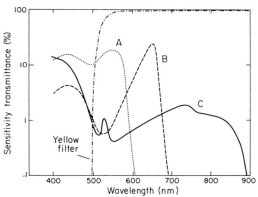

FIGURE 6-25. Spectral sensitivity curves for an infrared-sensitive film and the spectral transmittance curve of the yellow filter through which it is exposed. A is for the yellow-forming layer, B is for the magenta-forming layer, C is for the cyan-forming layer.

the three curves are almost superimposed (figure 6-6), while with this film the cyan-forming layer is about one stop slower than the other two layers.

The reason for the decreased speed of the cyan-forming layer can be explained by considering the high reflectance of most foliage in the infrared. If the cyan layer were as fast as the other two layers and the camera exposure were such that the other two layers were exposed correctly, the infrared exposure would be recorded on the toe of the cyan curve and any variation of infrared exposure would produce negligible change in cyan density. Small differences would not be detected.

Spectral region	Ultraviolet	Blue	Green	Red	Infrared
Normal reversal color film sensitivities		Blue	Green	Red	
Color of dye layers		Yellow	Magenta	Cyan	
Resulting color in photograph		Blue	Green	Red	
Infrared reversal color film sensitivities		Blue	Green	Red	Infrared
Sensitivities with yellow filter			Green	Red	Infrared
Color of dye layers			Yellow	Magenta	Cyan
Resulting color in photograph			Blue	Green	Red

FIGURE 6-24. Principles of operation of normal-color film and of infrared-sensitive color film.

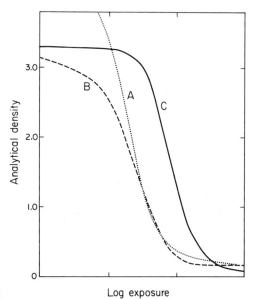

FIGURE 6-26. Characteristic curves of a typical infrared-sensitive film. Curve A is for the yellow layer, B is for the magenta layer, C is for the cyan layer.

6.3.3 PRINT AND DUPLICATING MATERIAL

There are many applications in which several copies of a photograph are needed, or in which the photograph is needed in a different form. The photographic materials used in such applications differ in many ways from those used in cameras. When it is necessary to distinguish them, we will call photographic material used in cameras, "camera material," and refer to the other kinds as print material, duplicating material, *etc.,* according to their application.

Although the terms "print" and "duplicating" are often used interchangeably in photography, strictly speaking, a print material is a material used to make positives from negatives. A duplicating material is one which produces duplicates; that is, it is a material that will produce positives from positives or negatives from negatives.

Print and duplicating films are used to make transparencies, usually positives, from negatives and positives. They are often used when several copies are needed, when it is desired to use the original as little as possible to reduce handling, or when diapositives of different sizes are needed, as for certain plotting devices.

A characteristic necessary for both kinds of film is that they must have considerably higher definition than the original films in order to record as much of the detail from then as is possible. There are always some losses of detail in any printing operation but having as high definition as possible in the print or duplicating film will tend to minimize them—provided that other characteristics of the operation such as lens quality and focus, contrast, *etc.,* are optimized. The fact that higher-definition films tend to be slow is of little consequence because the exposures that can be used for printing are much greater than those available for the original photography.

6.3.3.1 BLACK-AND-WHITE FILMS

Print films currently manufactured cover a fairly broad range of speed and quality. There is no single, universally employed method of determining the speeds of black-and-white print and duplicating films. Some manufacturers report speeds of their products in terms of indices derived from exposures made with tungsten lamps, whereas others publish data obtained using mercury-arc lamps. Because of the significant differences in spectral output of tungsten and mercury arc lamps as well as differences in the formulae used in determining speed, it is usually not possible to compare directly the speed and sensitometric curves of print films published by manufacturers using different illuminants and criteria.

The choice of a print film best suited for a specific application is dependent upon several factors, including the quality of the imagery that is to be printed, the spectral quality and illumination level of the printer, and the required contrast in the print film. Images on fine-grain, high-definition film require higher quality print films than do images on coarser-grained, high-speed films to show detail. In addition, because high-definition images often have a shorter density scale (due to the fact that they were obtained at high altitudes), higher-contrast print films are generally required.

Black-and-white print films, like camera films, can usually be processed to provide a range of *gammas* through appropriate choice of development time (or equivalently, machine transport speed). A *gamma* range of 0.80 to 1.80 is common, with a higher range available with some high-definition films. A set of characteristic curves illustrating the contrast-*versus*-development-time relationship for a representative negative-working print film is shown in figure 6-27. Figure 6-28 provides a similar set of curves for a positive-working duplicating film.

Although there are a few orthochromatic print films available, the spectral sensitivity of print and duplicating films for aerial photograph is typically confined to the UV and blue spectral regions with little sensitivity to green light. Lack of green and/or red sensitivity permits the safe use of appropriate safelights in print rooms.

The availability of some print films with thick, polyester base (*i.e.,* 7 mil) provides an alternative material to glass plates for use in making diapositives as described in section 6.3.5.

6.3.3.2 COLOR FILMS

In addition to having higher definition and slower film speed, color printing and duplicating

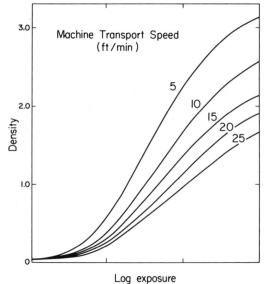

FIGURE 6-27. Characteristic curves of a negative-working print film processed at the indicated throughput rates in an automatic processor.

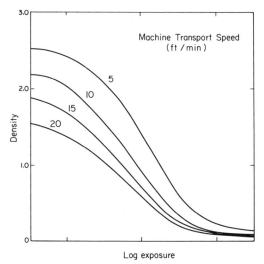

FIGURE 6-28. Characteristic curves of an aerial direct-duplicating film processed at the indicated throughput rates in an automatic processor.

films differ from camera films in their spectral sensitivities. Color camera films are used to photograph the scene with daylight, while color print and duplicating films are used essentially to photograph another photograph (the original) with artificial illumination. Thus, they must have spectral sensitivities that correlate with the spectral densities of the dyes in the camera film rather than with the reflectances of the actual scene itself. The spectral sensitivities of a typical duplicating film, given in figure 6-29, can be compared to those of a camera film in figure 6-18. Finally, these films must provide a fairly faithful representation of either the scene photographed or the photograph being duplicated.

One of the characteristics needed in a color print film for making prints from an aerial photograph is that it have high contrast in order to compensate for the low contrast of the camera film and also for the somewhat low contrast of

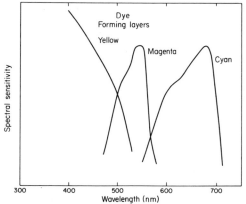

FIGURE 6-29. Spectral sensitivities of a typical color duplicating film.

the scene. The three layers should have matched contrasts to provide uniform color balance over the exposure range. It is not necessary that the three layers have equal speeds because of the latitude in color balance afforded by the use of filters when making the prints.

There are a variety of color-reversal films available for use in reproduction of aerial photography; they are processed by two major processes—dye-formation and dye-bleach. These films generally have low contrast (*gamma* of 1.0 or slightly higher) to provide tones similar to those of the original picture. Also, the low contrast will allow the making of many duplicates without a great contrast buildup. As with print films, it is desirable that contrasts and speeds of the three layers match.

6.3.4 REFLECTION PRINT MATERIALS

Reflection print materials have the emulsion coated on an opaque, white base that is usually paper, but that may also be pigmented acetate or polyester. They are used in making prints from camera negatives or positives, or from duplicated negatives.

The thickness of photographic papers has been designated by the American National Standards Institute under their Specification ANSI PH1.1-1974, as follows.

Single Weight	0.150−0.211 millimetres
Medium Weight	0.211−0.282 millimetres
Double Weight	0.282−0.483 millimetres

6.3.4.1 BLACK-AND-WHITE MATERIALS

Black-and-white papers for reproduction by both contact and production are available; the latter are in more general use for aerial photographs. Except in those cases where special effects are required, prints of aerial photographs are usually made on papers that provide a neutral-black image, a white-tinted base, and a smooth surface.

Photographic papers suitable for aerial prints are available in both glossy and semi-matte (lustre) surfaces. Papers with a glossy surface provide a higher shoulder density and longer density scale (and, as a consequence, denser-appearing blacks and higher contrast shadow detail) than do those with semi-matte surfaces. The same mechanism that produces these characteristics (*i.e.*, the specular nature of the glossy surface) may, under some illumination conditions, prove to be a disadvantage because surface reflections falling within the viewing field tend to mask image detail. These reflections seem to be particularly prevalent and bothersome when two prints are viewed stereoscopically.

The reflectance problem is circumvented through use of papers with semi-matte surfaces from which reflection is largely diffuse in nature.

An additional advantage of semi-matte materials, important in some applications, is the greater surface "tooth," which allows the surface to readily accept penciled annotations.

The speeds of photographic papers are measured and related through use of the ANSI Paper Speeds, ASAP, (ANSI Standard PH2.2-1966), which is defined as:

$$ASAP = \frac{1000}{H_{0.6}}$$

where $H_{0.6}$ is the exposure in micro-lux-seconds required to produce a density of 0.6 above base plus fog density. Speeds of paper used in projectors commonly fall in the range of 100 to 500, but the user should recognize that paper speeds specified by ANSI are derived from sensitometric exposures made with tungsten-filament lamps. Use of other radiation sources may well result in both different absolute and relative speed values.

Paper print contrast, in distinction to that of photographic films, is generally not particularly sensitive to changes in development time. Standard development conditions, as recommended by the manufacturer, should normally be adhered to, although most materials have sufficient latitude in development to provide some compensation for over or underexposure without significant loss in quality of the print.

The log-exposure range of photographic papers is a quantity, related to print contrast, which provides a measure of the difference between the exposures necessary to produce faint highlight and full black tones. The log-exposure range should be approximately equal to the density scale of the negative that is to be printed—a requirement that should be met to avoid "blocking" or loss of shadow or highlight detail.

Contrast control, or more appropriately log-exposure range control, is achieved with all except the panchromatic papers either through choice of the appropriate grade of paper contrast or, in the case of variable-contrast papers, through selection of the correct filter when exposing. Graded printing papers are currently identified either in terms of numbers ranging from 0 or 1 to 4 or 5 or in descriptive terms such as soft, medium, hard, extra hard, and ultra-hard. A set of characteristic curves representing typical characteristics of a graded printing paper is shown in figure 6-30.

Variable contrast printing papers are marketed by several companies as are the filters intended to be used with these materials. Thus, a single paper can be used in place of several of varying grades. The emulsions of these photographic papers contain two components, one of which is sensitive to blue light and the other to blue and green light. One component has a characteristic curve with a long density scale and is used in printing negatives showing a long range

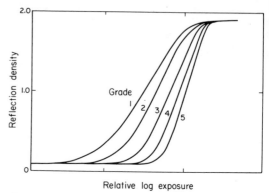

FIGURE 6-30. Characteristic curves for the various contrast grades of a typical photographic paper with a glossy surface.

of densities; the other has a characteristic curve with a short density scale and is for making prints from negatives having a shorter range of densities. Use of the appropriate printing filter limits the exposure to either of the two components or provides the required proportion of blue and green light necessary for correct printing of negatives having an intermediate range of densities.

Except for the variable-contrast materials just described, which have some green sensitivity, photographic papers are usually sensitive only in the ultraviolet and blue spectral regions. An additional exception exists in panchromatic papers that, because they are made to provide correct black-and-white tonal scales from color negatives, have sensitivity to red light as well. It should be recognized that although panchromatic papers are not in general use in aerial photography, paper prints from color-negatives should be made using these materials to ensure that yellow, brown, and red objects are tonally differentiated in the print.

The recent introduction of papers with resin-coated base is of interest. These papers have a water-resistant, plastic resin coated on both sides of the paper, which prevents the porous paper from absorbing too much water (and chemicals) from the developing solutions. Papers incorporating these bases provide shortened washing and drying times. In addition, these materials offer improved dimensional stability, greater resistance to tear, and less tendency to curl than do regular paper base products.

6.3.4.2 COLOR PAPERS

Two different types of color papers are available for making prints from color films; a negative/positive paper for prints from color-negatives, and color-reversal paper for prints from color-reversal transparencies. In contrast to black-and-white papers, each color paper is available with just one contrast and printing speed (enlarging), and with more limited selection of surfaces and support thicknesses. Most papers are now available on resin-coated support for ease of processing and drying. Color-reversal papers are available for both the dye-formation and dye-bleach methods of development. Characteristics of reflecting color print papers are similar to those of the color print and duplicating films in that the color-negative materials have high contrast while the color-positive materials have low, and the speeds of the three layers need not be equal. Similarly, their spectral sensitivities are made to coordinate with the spectral densities of the dyes in the film used in cameras.

6.3.5 FILMS AND PLATES FOR DIAPOSITIVES

Positive transparencies (diapositives) are used in virtually all precise stereomapping methods. In the last few years, emulsions coated on a highly dimensionally stable film have been replacing glass plates for many plotting operations.

Diapositives are usually printed from negatives by either contact or projection, and at various ratios suited to the particular stereoplotting instrument.

In many stereoplotting applications, initial reduction and subsequent enlarged projection of the photographic image is necessary. Full and complete retention of detail is required and fine-grain, positive-type emulsions having high resolving power are used. It is essential that the positive transparency be as stable as possible so that further errors are not introduced. Both flatness and dimensional stability are important, as discussed in section 6.4.

6.3.5.1 CHARACTERISTICS OF PLATES FOR DIAPOSITIVES

Plates to be used as diapositives are ordinarily offered with two different emulsions: a medium contrast, or "medium", emulsion, to accommodate fairly contrasty scenes; and a contrasty, or "contrast", emulsion, to accommodate less contrasty or flatter scenes. Characteristic curves for typical medium and contrast plates are shown in figure 6-31. Direct-viewing plotting equipment can accommodate diapositives with a slightly longer tone scale than can projection plotters.

Diapositive plates are available in three different flatnesses. These are: selected-flat glass, ultra-flat glass, and micro-flat glass. The emulsion characteristics are the same with the three types of plates. The emulsions are blue sensitive and will not record all colors, so they should not be used to make diapositives from color negatives. Information on exposing and developing is available from the manufacturer.

6.3.5.2 CHARACTERISTICS OF FILMS FOR DIAPOSITIVES

In recent years, the technology of film-base manufacturing, as well as that of coating films, have advanced to such a degree that it is now possible to use photographic films for diapositives in many photogrammetric applications. With the introduction of 7-mil (178 μm) thick polyester-film as a support, some of the reasons for plotting from glass plates no longer hold and film is being used as a substitute for the photographic plate. Plotting from color photography is currently being done in the plotter from the original color-reversal films, or from color diapositives printed from the original color negative films. Such diapositives can be printed onto a positive color film such as Kodak Vericolor print film 4111 (Estar thick base). A few words of caution are in order at this point. The laboratory technician should never attempt to expose color materials on conventional black-and-white printers without modifying the printers so that they have a "white" light source. The conventional "blue," or high ultraviolet emitting light sources supplied with most black-and-white printers, are designed to allow short exposures when printing black-and-white materials, and because of their blue color cannot be used to expose full-color materials. Processing of the color diapositives should be done according to the manufacturer's recommendations.

Reference should also be made to the papers by Adelstein (1972) and Young and Ziemann (1972) and the references in these papers, for a discussion of the use of film for diapositives.

6.3.6 UNCONVENTIONALLY PROCESSED AERIAL PHOTOGRAPHIC MATERIALS

6.3.6.1 VESICULAR FILM

Small amounts of vesicular film are currently being used in test programs for the duplication of aerial reconnaissance negatives. Aerial vesicular film consists of a photosensitive layer, 10 to 15 micrometres in thickness, coated on polyester base, in most cases 100 micrometres (3.94 mils) in thickness. The photo-sensitive layer contains a thermoplastic, hydrophobic resin in which is dispersed an aromatic diazonium salt as a sensitizer. The film is sensitive to light in the range of about 350 to 440 nanometres wavelength. Upon exposure, the diazonium salt is decomposed, and yields, as one produce of its photolysis, nitrogen gas. This gas is temporarily trapped within the polymer of the film and pressure-pockets are formed within the layer. These gas molecules constitute the latent image in vesicular photography. While the pressure is insufficient to deform the resin at room temperature and will eventually dissipate as the nitro-

Medium

D-76 Developer

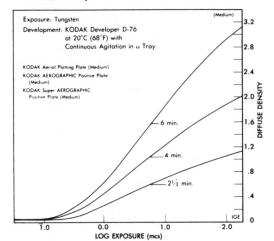

DK-50 Developer

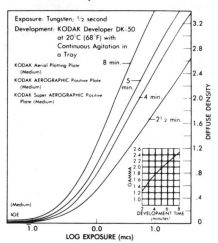

Contrast

DK-50 Developer

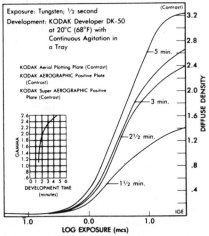

DEKTOL Developer

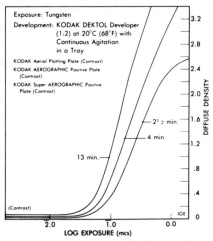

FIGURE 6-31. Characteristic Curves for Medium and Contrast Aerial Plotting Plates.

gen slowly diffuses from the film into the atmosphere, if heat is applied, the combination of softened plastic and increased gas pressure results in the formation of vesicles in the layer, in the size range of 0.5 to 3 micrometres. The developed image, consisting of these vesicles, scatters light rather than absorbing it.

Since the scattering centers are in areas proportional to the amount of incident light, there is a sign change in printing, *i.e.*, negative to positive or positive to negative.

Photographic speed of such film is insufficient for aerial photography. However, it is sufficient so that in conjunction with mercury-vapor lamps, contact prints can be made at a rate of 20 to 50 ft/min in the proper equipment.

The advantages of the vesicular film lie in its dry, rapid processing by light and heat alone; its minimum need for darkroom facilities; the lack of need for chemical-mixing facilities; the fact that there is no problem of disposing of chemical washes; and its use of inexpensive chemicals instead of expensive silver. Moreover, there is no need for the storage and handling of chemicals and no need for water for processing.

A typical vesicular aerial film is Kalvar type 664 film. The characteristic curve illustrated in figure 6-32 shows that 1.58×10^6 ergs/cm^2 are required to obtain maximum density. The density scale is approximately 1.5. The log-exposure scale is 1.5 with a neutral-gray color very similar to that of aerial print films using silver halide.

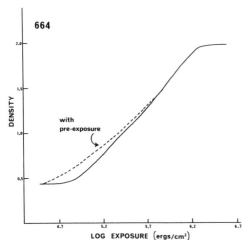

FIGURE 6-32. Characteristic curve of KALVAR type 664 vesicular film.

Extension of the exposure scale to cover higher density scale aerial negatives can be obtained by pre-exposure (dotted line in figure 6-32), which is an integral feature of the #451A printer/processor used in duplication. Resolution is 250 to 300 cycles/mm with a high-contrast target.

A newer Kalvar aerial film, type 684, with characteristic curve shown in figure 6-33, is faster and has a longer exposure scale. It is to be noted that the relatively high D-min seen in both films is related to the requirement that the material have relatively high density when viewed on light tables with diffuse viewing. Since vesicular film scatters rather than absorbs light, the extension of the density scale required for diffuse viewing usually involves changes that raise the minimum density. Side-by-side comparisons with silver halide prints, however, have indicated that this is acceptable to photointerpreters.

The printing capability of these films is illustrated in figures 6-34 and 6-35, in which it can be seen that the printing contrast is such that a fourth copy closely matches the second copy.

The Kalvar model #451A aerial reconnaissance duplicator is an operational printer/processor. It represents the refinements resulting from the testing and evaluation of three previous prototypes. Contact printing and thermal processing are completed in a single pass through the machine. Individual steps in the process include an imaging exposure to ultraviolet light, thermal development at about 250°F (121°C) and an overall ultraviolet exposure to release (fix) the unused diasonium sensitizer.

The printer/processor also includes a provision for pre-exposure when required. There are supply and take-up spools for both the duplicating and master films that accommodate either 5-inch or 70-mm width materials.

Physically, the machine weighs 1980 kg (900 pounds), has a width of 39.4 cm (15½ in), depth of 90.8 cm (35¾ in), and height of 191 cm (75¼ in). The machine requires an operating voltage of 208/220, three-phase, 60 cycle. The operating current is 27 amperes ± 10%. The machine is composed of two units that are bolted together for use. The lower unit contains electrical power supplies and an electronic control system. The upper unit contains a deck assembly, film transport system, processing stations and air blowers. The design of the two separate units is to comply with the U.S. Navy's requirements for convenience in moving the equipment through shipboard hatches.

6.3.6.2 PHOTOTHERMOGRAPHIC MATERIALS

Among the photothermographic materials available for photogrammetric applications is the family of Dry Silver films and print materials manufactured by the 3M Company. These pro-

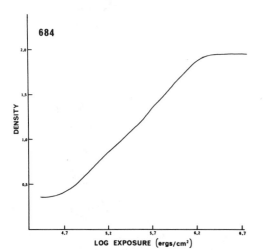

FIGURE 6-33. Characteristic curve of KALVAR type 684 vesicular film.

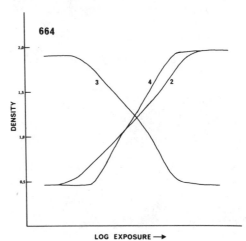

FIGURE 6-34. Printing capability of KALVAR type 664 vesicular film.

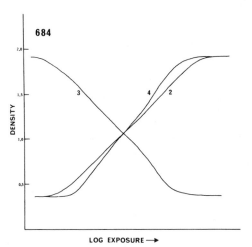

FIGURE 6-35. Printing capability of KALVAR type 684 vesicular film.

vide a simple, photographic means of obtaining black-and-white images on film or paper, line work, or continuous tone with dry chemicals. Originally introduced in the 1960's for electron-beam recording of computer output and for making prints from various forms of microfilm, Dry Silver has evolved into other fields including electro-optical imaging and continuous-tone duplication.

Existing applications of Dry Silver include the following:

laser-imaged film transparencies, line and tonal, on red-sensitive type 7869 film;

contact and projection prints on type 7742, continuous-tone paper; and

continuous-roll prints on type 7842 aerial duplicating film.

In many instances, exposure and processing are found combined into special equipment such as the Associated Press's "Laserphoto" system for transmitting pictures. However, standard darkroom exposing equipment is usable with most Dry Silver papers and films, along with heat processors available from 3M Company and other suppliers. A 25-foot-per-minute continuous roll processor is shown in figure 6-37.

The compound that forms the light-sensitive part of Dry Silver photographic material is silver halide, which responds to light in much the same manner as in wet-processed silver halide materials. The essential elements that make up the material are: 1) a silver halide in catalytic proximity to a light-stable, heat-reducible silver compound, and 2) chemicals that will reduce the silver compounds during heat. Light is used to produce the catalyst and heat is used to produce the silver image. Image density is reversed from the input in each succeeding copy.

When the material is uniformly heated over its entire surface, chemical reaction occurs at an accelerated rate in the exposed areas, analogous to normal photography. Practical developing rates are normally between 3 and 30 seconds and at temperatures between $240-285°F$ ($116-140°C$). In this range, a trade-off of time

FIGURE 6-36. KALVAR model 451A aerial reconnaissance duplicator.

FIGURE 6-37. A compact, continuous heat processor (3M Co. Model 259 BA) which develops Dry Silver films and papers up to 25 feet/minute from rolls up to 9½ inches wide and 1000 feet long.

and temperature can be used to give the same results. *See* figure 6-38. For good reproducibility, temperature is controlled with ±3°F (1.5°C) for most applications. And for very precise results, a variation of less than ±1°F (0.5°C) is achieved in some processors.

Dry Silver is conserving of both silver and electrical energy. One-half to one-third of the silver of similar wet processed imagery is used. This silver is supplied and retained completely in

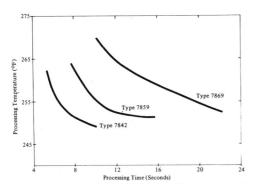

FIGURE 6-38. Processing relationships in the 3M Model 259 BA processor of Dry Silver films, showing time *vs.* temperature trade-off to hold constant image characteristics.

the sheet and can be recovered by ashing and other conventional methods. Likewise, a roll of Dry Silver aerial duplicating film can be printed, processed at 25 feet per minute, and ready to use with consumption of less than one-fifth the electrical energy of a comparable wet system.

Image characteristics of Dry Silver prints are similar to those of their wet-processed counterparts. Maximum density of papers runs 1.6 to 1.8, and in films 2.2 to over 3.0 D-max is standard. Exposure sensitivity is normally from 15 to 500 ergs/cm^2, depending on the particular product and light source used. Resolving power is conservatively estimated from 50 to 300 cycles per millimeter for the various products, usually measured from contact printing with a high contrast target.

Dry silver materials can be stored from six months (paper) to one year (film) at room temperature and for an indefinite period in a deep freeze. The silver image has proven to be quite permanent, and, while background color can change from white or clear to a pink-tan under some environmental extremes, permanence is excellent and measured in years. Shrinkage of the polyester base of the films during heating, even though less than 1%, can at present preclude their use for highly precise photogrammetric mensuration.

6.4. *Physical Properties*

It is essential to photogrammetry that, in addition to having suitable sensitometric properties, the photographic materials have adequate physical properties. Of particular concern to the photogrammetrist are the dimensional stability, flexibility, flatness, and aging characteristics of the various materials used—film, plates, and paper.

6.4.1 PHOTOGRAPHIC FILM

6.4.1.1 PHYSICAL TYPES

Photographic film is used for both the original photography and for duplicating. The physical requirements of these films differ, and depend upon whether they are to be used for photogrammetry or reconnaissance. Film used for photogrammetry must have the ultimate in dimensional stability. Such films must have not only small dimensional changes, but the dimensional changes in all directions of the film should be similar. Such films generally have a base that is 102 micrometres in thickness. Films for reconnaissance, however, have less exacting requirements for dimensional stability and a thinner base can be used. Physical requirements for these films are given in the section on military specifications for aerial films. In this section, emphasis will be on the dimensional stability behavior of films used for photogrammetry.

A listing of the thickness of the bases of various aerial films manufactured in the United States is given in table 6-4.

6.4.1.2 PHYSICAL STRUCTURE

The physical characteristics of photographic materials are influenced by the two main components, the emulsion and the base. The latter carries the main burden of providing the necessary physical properties to the point of resisting certain unfavorable tendencies of the former. Since the physical characteristics of the emulsion are necessarily second in importance to the sensitometric behavior, much effort has been given in searching for plastic bases that will impart improved physical behavior to the whole material.

For many years cellulose nitrate was used as a base until improved cellulose-ester (Calhoun, 1947) formulations had been attained, giving comparable dimensional stability, but with the further advantage of nonflammability and permanence under storage. More recently, polyethylene terephthalate and polycarbonate (Harmon, 1961; Calhoun, 1961) have been introduced. These materials have improved moisture resistance, and polyethylene terephthalate has exceptional strength and stiffness. The vast majority of aerial films is manufactured today using polyethylene terephthalate for the base.

TABLE 6-4. TYPES OF AERIAL FILM MANUFACTURED IN THE UNITED STATES

Film Use	Base	Thickness of Base	
		mils	micrometres
Mapping	Polyethylene terephthalate	4	100
Reconnaissance	" "	1.2	30
	" "	1.5	38
	" "	2.5	64
	" "	4	102
Duplicating	" "	2.5	64
	" "	4	102
	" "	7	178

6.4.1.3 GENERAL PHYSICAL PROPERTIES

6.4.1.3.1 MOISTURE PROPERTIES

One of the most important physical properties of photographic film is its susceptibility to moisture, because of the close relationship between the moisture content and other characteristics. The moisture content of the film as a whole is a constant for a given environmental condition and is almost completely a function of the relative humidity of the atmosphere with which it is in equilibrium. At low relative humidities, shrinkage, curl, and brittleness may be of concern; while at high humidities, friction, ferrotyping, and other effects associated with softening of the gelatin layers may become problems. In manufacture, the films are handled under controlled atmospheric conditions to bring them to final equilibrium in the package at the best condition for all properties concerned. Many aerial films are packaged in equilibrium with air at 50−60% RH, but the individual manufacturer should be consulted about specific information for any particular film.

The components of film (the gelatin layers and the base) differ widely in their moisture capacities. Gelatin absorbs many times as much moisture per unit weight as do the plastic film bases. Among the latter, the polyesters have much lower moisture take-up than cellulose ester. Figure 6-39 shows the water absorption of a typical emulsion layer and of polyethylene terephthalate as a function of relative humidity. Many of the important physical characteristics of film are related to the high moisture absorption of the emulsion and the low moisture absorption of the base.

The exposure of film initially at a given humidity to an atmosphere at a different humidity, immediately results in a change in moisture content in a direction toward equilibrium with the new environment. However, it must be remembered that this is a time-dependent change, and the rate of change is dependent on a number of factors—the air velocity, relative humidity differential, and temperature. The physical form of the film is also a very important factor, a single strip of film changes very much faster than a complete roll. For example, a single strip of film will complete approximately 90% of its total change in ten minutes, while a 180-foot roll of aerial film will require several months.

An important consideration is the amount of water picked up during processing and the consequent drying in the dryer. The bulk of the moisture is in the swollen gelatin layers, since polyethylene terephthalate absorbs very little water. Many of the thinner base (38- and 64-micrometres) films have a backing layer of gelatin. For such films, the film dryer must have provisions for air flow across the film back

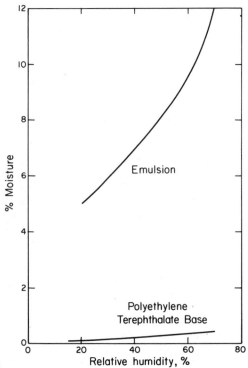

FIGURE 6-39. Moisture capacity of typical components of aerial film.

(Michener, 1963). Drying film to 50% EH equilibrium yields the optimum in physical properties such as curl, brittleness, static, sticking and dimensional stability.

6.4.1.3.2 CURL

The curl of photographic film is caused by the difference in moisture properties between the emulsion layers and the base. At low relative humidities, the emulsion has a greater dimensional contraction than that of the base; although at higher humidities, this contraction is considerably reduced. This contraction causes the emulsion to pull the base into a curved configuration, resulting in the behavior known as film curl. By convention, film curl is called positive curl when the film is bent toward the emulsion side, and negative curl when it is bent away from the emulsion side.

The aerial photographic films using thinner bases have a gelatin backing on the base opposite the emulsion to control curl. The curl for such films is very much less than that for films with gelatin coated on only one side, and is dependent upon the relative pull of the emulsion layer and the backing layer. This is illustrated in figure 6-40. In this graph, curl is expressed according to the American National Standards Institute Method (PH 1.29-1971) in units of $100/R$ inches, where R is the radius of film curvature. Increasing values denote increasing curl.

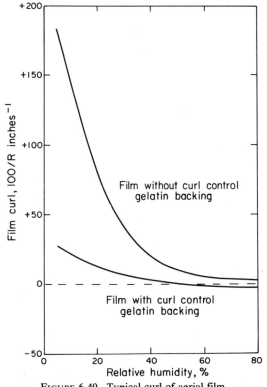

FIGURE 6-40. Typical curl of aerial film.

In addition to the pull of the gelatin layers, film curl can also be influenced by flow of the plastic film support when held in roll form. This phenomenon is known as "core-set," and will result in either positive or negative film curl depending upon whether the film is wound emulsion-in or emulsion-out. It is the practice to wind raw film emulsion-in, and this contributes to a positive curl. Processed film is usually wound emulsion-out and the resulting core-set will decrease the positive curl. The contribution of core-set to film curl is added to or subtracted from the inherent film curl due to the emulsion pull. The effect of core-set is increased with storage time on a roll, with decrease in roll diameter, and with an increase in temperature.

6.4.1.3.3 BRITTLENESS

While film brittleness is ordinarily not of practical importance at moderate environmental conditions, it may sometimes be a problem at very low humidities or low temperatures (Adelstein, 1957). In such cases, brittle failure would occur when the film is bent with the emulsion (or curl control backing, if present) on the outside of the bend. Film that has the coating layer only on one side is very much less brittle when bent gelatin-side in, since brittleness is usually caused by weakness of the gelatin layer when subjected to tensile shock. Brittleness is a rate sensitive property; it is more likely to occur when the film is bent rapidly.

Brittleness can be manifested in two ways. Under severe conditions the film may break into two separate pieces. However, sometimes the emulsion layer may crack, but the crack does not propagate through the base. The film remains in one piece, but the photographic image is damaged. Due to the high strength of polyethylene terephthalate, these films exhibit complete breaks only under very severe conditions.

In practical operation, care should be taken that film is not subjected to excessive tensions or accelerations, nor should it be bent around rollers of small diameter. This is more important at low temperatures or low relative humidities, which may exist under some atmospheric conditions or at high altitude conditions.

6.4.1.4 DIMENSIONAL STABILITY

In many applications of photographic film, dimensional stability is of prime importance. In some instances, it is important that changes in overall size due to processing and storage be kept to a minimum. In others, it is essential that dimensional changes be not only small, but that they also be similar in magnitude in all directions of the film. This latter requirement is very important in films for mapping, otherwise, distortions will occur in the resulting topographic surveys. In this section, both uniform dimensional changes and nonuniform dimensional changes will be discussed.

Although the polyethylene terephthalate bases used for aerial films have excellent dimensional properties, they still show small changes. Some of the changes are caused by the effect of the emulsion on this base. Such dimensional changes are lowered by decreasing the thickness of the emulsion or by increasing the thickness of the base. It is for this reason that the black-and-white films with thinner emulsions have better dimensional stability than the color films with thicker emulsions.

In all considerations of photographic materials, it is important to recognize that there is no such thing as absolute dimensional stability. Even sensitized plate glass shows size-changes under some conditions.

6.4.1.4.1 THEORY

The size changes that occur in photographic film are due to a variety of causes. For this reason, explanations of dimensional changes caused by temperature, humidity, processing, and aging are quite complex. The magnitude of these changes depends on the chemical composition of the film and its mode of manufacture. Before describing the behavior of aerial films, a brief discussion of the most important reasons for film size changes will be given. Changes can be divided into two main categories: temporary or reversible changes, and permanent or irreversible changes.

  (1) Temporary Size Changes
    (a) *Thermal expansion or contraction.* When the temperature increases, film expands; when it decreases, film contracts. This is the same phenomenon observed with the heating and cooling of most other materials. This property is important since exposures are sometimes made in unheated cameras at high altitudes where the temperature may well be below zero. In other types of aircraft, the high outside temperature may result in films being heated during exposure. Under some conditions, the thermal effects may be overshadowed by the humidity effects. When the temperature decreases, the relative humidity usually increases.
    (b) *Humidity expansion or contraction.* Film not only changes size with temperature but also with relative humidity. At higher relative humidities the film is larger, and this is a reversible change. This property is very dependent on the ratio of the gelatin-layer thickness to the thickness of the base. The greater this ratio, the greater is the humidity coefficient of expansion.
  (2) Permanent Size Changes
    All photographic films exhibit some degree of permanent size change—both as a result of photographic processing, and also during subsequent aging of the processed film. There are various causes for these dimensional changes.
    (a) *Plastic flow of the base.* Since the photographic emulsion exerts a compressive force on the film support, this causes, in addition to curl, a flow of the film support with a resulting permanent shrinkage. The magnitude of this change increases with increasing ratio of the gelatin-to-base thickness. Dimensional changes of photographic film due to plastic flow can also occur because of excessive tensions in film handling equipment, especially processing machines. These tensions cause an increase in length with a decrease in width.
    (b) *Release of mechanical strain in the base.* Bases have internal strains that are not completely removed in manufacture. Polyethylene terephthalate base is intentionally stretched during manufacture to give it the desired physical properties. This results in a "frozen-in" strain. Some of this internal strain may be slowly released by the film support upon aging.
    (c) *Mechanical effects of the gelatin (emulsion).* The gelatin layers can exhibit definite differences in behavior because of structural changes as a result of processing or of aging. The effect is to change the compressive force of the gelatin on the base, and consequently this is reflected in the dimensional behavior. These effects are quite involved, and a more complete discussion of these phenomena is given by Calhoun and Leister (1959). It should be noted that their effect on dimensional changes is generally small.

It is evident from the preceding discussion that the dimensional changes that occur in photographic film are the combination of many factors—all of which may occur simultaneously. This makes it difficult to predict exactly, the film size changes in practical applications. However, a general understanding of film behavior can help in estimating effects upon film dimensions and in correcting difficulties.

6.4.1.4.2 OVERALL DIMENSIONAL CHANGES

Approximate values for the characteristic dimensional change of current topographic aerial films are given in table 6-5. Representative behavior is tabulated for both black-and-white and color emulsions on the four common thicknesses of polyethylene terephthalate base. These dimensional change values should only be considered as representative of the materials, and not as indicative of the magnitudes that will always be observed. All measurements were made in the laboratory under carefully controlled conditions, using tray-processed 35 mm × 10-inch film strips and a mechanical-shrinkage gauge (ANSI PH 1.32-1973). It should also be noted that the dimensional measurements represent the average found in the different directions on the film. Differences between directions will be discussed in the next section.

The coefficients of thermal expansion were measured between 21 and 49°C. This property is primarily dependent upon the characteristics of the base. The coefficient for polyethylene

TABLE 6.5 TYPICAL DIMENSIONAL CHANGE CHARACTERISTICS OF AERIAL FILMS ON POLYETHYLENE TEREPHTHALATE BASE

Base Thickness, Mils , Micrometres	1½ 38		2½ 64		4 102		7 178	
Emulsion Type	Black-And-White	Color	Black-And-White	Color	Black-And-White	Color	Black-And-White	Color
Thermal Coefficient of Linear Expansion, (1) % Per C	.002	.002	.002	.002	.002	.002	.002	.002
Humidity Coefficient of Linear Expansion, % Per 1% RH (2)	.0050	.0110	.0035	.0050	.0020	.0030	.0015	.0020
Processing Dimensional Change Range (3) % Shrinkage	−.08	−.10	−.06	−.10	−.03	−.04	−.01	−.01
% Swell	+.04	+.04	+.03	+.03	+.02	+.03	+.02	+.02
Processing and Aging Shrinkage (4) 1 Week at 50 C −20% RH	.09	.12	.08	.10	.03	.06	.02	.02
1 Year at 25 C − 60% RH	.07	.07	.04	.04	.03	.03	.02	.02

The dimensional characteristics were determined in accordance with the appropriate sections of the American National Standards Institute "Methods for Determining the Dimensional Change Characteristics of Photographic Films and Papers", PH1.32-1973. The specific references are:

(1) PH1.32, Section 7 Measurement at 20% RH between 21 and 49C
(2) PH1.32, Section 6 Measurement at 21 C between 15 and 50% RH on unprocessed film
(3) PH1.32, Section 8 Measurement made on tray-processed specimens
(4) PH1.32, Section 9 Measurement made on tray-processed specimens

terephthalate is one of the lowest for plastic materials and is a definite improvement over that for the earlier cellulose ester base. However, it is greater than that of glass (figure 6-41).

The reversible dimensional changes that occur as a result of a change in relative humidity are considerably more complex than those due to temperature changes. They are dependent upon the properties of the gelatin layers, and upon the thickness ratio between these layers. Therefore, the coefficients will differ for different emulsions even when coated on the same base (Calhoun and Leister, 1959; Umberger, 1957; Adelstein 1972). The greater the emulsion thickness, the greater the coefficient, and this is shown in figure 6-42 over the range of base thickness. It should be recognized that film does not necessarily change dimension linearly with relative humidity. This is illustrated by the curves of expansion versus relative humidity in figure 6-43 for several films with polyethylene terephthalate base. For this reason, coefficients given in table 6-5 and figure 6-42 were calculated between 15 and 50% relative humidity. These coefficients also depend upon whether the film is processed or unprocessed, and whether the relative humidity is approached from a point of lower or higher humidity. The latter behavior is known as a hysteresis effect.

The dimensional changes that occur as a result of processing are permanent and are due to the various factors discussed in paragraph 6.4.1.4.1. The size changes reported in table 6-5 were experimentally determined under ideal laboratory conditions. All measurements were made on tray-processed film strips that were not subjected to tension while wet. It is recognized that

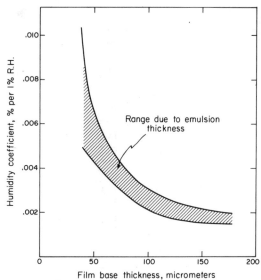

FIGURE 6-42. Effect of base thickness on humidity coefficient of polyethylene terephthalate base aerial films.

these conditions are different from those that occur in usage, but they do permit a comparison between various films.

The dimensional changes that occur in processing also depend upon the relative humidity of both the unprocessed and processed film (which must be the same to eliminate the reversible

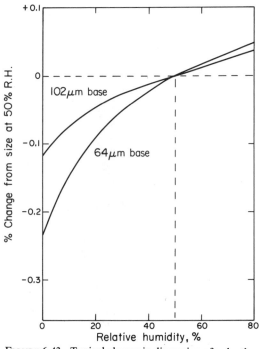

FIGURE 6-43. Typical change in dimension of polyethylene terephthate base aerial films with relative humidity.

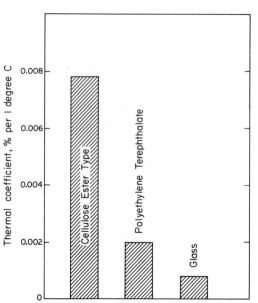

FIGURE 6-41. Typical values of thermal coefficient of linear expansion of photographic materials.

humidity effect) and the previous exposures to moisture. An increase in dimension can sometimes occur as a result of 2 particular history of exposure to moisture.

In addition to the types of dimensional change already discussed, a change in size occurs from aging of the film. The magnitude of the processing and aging shrinkage of aerial films after one year at 25°C − 60% relative humidity is listed in table 6-5 and it is generally small. It does not increase appreciably with longer keeping time. This aging shrinkage is due to plastic flow of the base under the compressive forces of the emulsion. For this reason, it increases with increasing emulsion-to-base ratio.

6.4.1.4.3 NONUNIFORM DIMENSIONAL CHANGES

The previous section dealt with the overall dimensional changes of photographic films. However, it is usually more important in photogrammetry that the dimensional changes be uniform in all directions. Overall uniform changes in dimension can frequently be corrected by a change in magnification, but this is not possible if the dimensional change varies in different directions on the film.

To provide uniformity of dimensional change, mapping films are manufactured as uniaxially as possible—that is, with nearly equal properties in all directions. The maximum dimensional change of polyethylene terephthalate base films may not necessarily be in either the length of width direction, but may be for example in a diagonal direction. To minimize directional differences, aerial films used for mapping are made so that the overall dimensional changes are small.

In addition to the slight directional differences that are inherent in the film, differences can occur as a result of customer usage. In practice, continuous processing machines are frequently used. Some machines place the film under tension while wet, and in some cases while it is drying. Excessive machine tensions could cause a lengthwise stretch and a widthwise contraction. Due to the low water take-up of polyethylene terephthalate base and its relatively high stiffness when wet, the effect of film tension in drying of these films is relatively small.

The behavior discussed in this chapter has so far dealt with average changes in film dimensions in the different directions on the film. Also of concern to the photogrammetrist is any nonuniformity in dimensional change within a given direction. Studies of such random distortions have been made over the years. Most of the early work has consisted of exposing a *reseau* or grid from a glass plate onto the film and subsequently comparing the intersections of this grid with the glass master, using an optical comparator (McNeil, 1951; Brucklacher and Luder,

1956). Subsequently random distortions were studied by registering a halftone tint on film with the original tint, producing a *moiré* pattern (Calhoun *et al*, 1960; Adelstein and Leister, 1963; Adelstein *et al*, 1966). Distortions, if present, are visible as irregularities in this *moiré* pattern.

Illustration of a *moiré* dot pattern for aerial film on 64 μm support is shown in figure 6-44. In this particular example, the *moiré* pattern is very uniform as evident from the overlayed rectangular grid. This type of *moiré* pattern can yield quantitative data on the uniformity of dimensional change. The coordinates of each point at which the *moiré* pattern shows a cancellation are measured and a computer is programmed to calculate the average change in size of the entire film as well as the nonuniform displacement of each *moiré* dot. The latter values can be presented as a vector diagram as shown in figure 6-45 (Adelstein, 1972). In this particular example, the average vector displacement is 2.9 μm and the maximum is 7.7 μm. In general, films with a polyethylene terephthalate base show excellent dimensional uniformity, with linear displacements usually less than 5 micrometres. As will be discussed in a subsequent section, glass plates have even less distortion.

It was also reported (Calhoun et al., 1960) that local distortions as large as 18 micrometres can be caused by water droplets left on the emulsion during drying. This local movement of the image is believed to be caused by distortions of the emulsion caused by local swelling and drying. Excessive heat or tension in processing will also produce distortions in film.

6.4.1.4.4 PRACTICAL CONSIDERATIONS

It should be noted that while each of the different types of film dimensional change has been

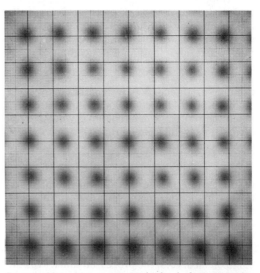

FIGURE 6-44. *Moiré* pattern obtained after processing aerial film on 64 μm polyethylene terephthalate base.

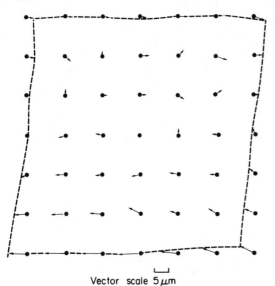

Vector scale 5 μm

FIGURE 6-45. Vector diagram showing non-uniform distortion of aerial film (102 μm polyethylene terephthalate base). Film was exposed at 50% RH processed and re-registered with glass master at 20% RH.

discussed separately, in practical usage these different changes may all occur simultaneously. Moreover, depending on the particular conditions, these changes may be additive or they may partially cancel each other.

Another consideration of practical importance is the change in dimension of film in roll form while inside a camera (Carman, Martin, 1968). The moisture content is not uniform across the film width while the film is being conditioned to a different relative humidity (RH). A roll of film placed in a camera in which the relative humidity of the air is different from that of the packaged film will gradually change in dimension, starting at the edges of the roll. Exposure of a partially conditioned roll of film will obviously result in some image distortion. Similarly, image distortions will result if the film is not completely conditioned to thermal equilibrium.

If the film strip is in the focal plane of the camera, or threaded between the supply and take-up spools, the conditioning times are very much shorter than when the film is wound in roll form. It is well known that film may show greater dimensional changes in the first few exposures at the start of a flight line (Carman, 1946) than in the rest of the roll. This is because the first exposures of the film may remain stationary in the focal plane long enough for moisture conditioning to take place.

Appreciable errors in dimensions also result when processed films are not conditioned to a standard moisture content prior to use. In many commercial dryers, the processed film leaves the dryer with its surfaces sensibly dry to the touch. However, drying film to sensible dryness does

not mean that it is adequately dried with respect to moisture equilibrium. The base may still contain appreciable moisture. Use of film at a different moisture content than that it had in the work area can cause very large errors in photogrammetric work both because of changes in the overall film size and because of distortions due to subsequent nonuniform loss or gain of moisture in the roll (Jaksic, 1972).

6.4.1.5 STORAGE

Of concern to the photogrammetrist is the storage of both unprocessed and processed film.

6.4.1.5.1 UNPROCESSED FILM

Unprocessed photographic film is a perishable item. Its sensitometric properties will deteriorate slowly with time, and this deterioration is accelerated by high relative humidities and high temperatures. Sensitometric degradation is usually reflected in a speed loss, a fog growth, and/or a change in contrast, although the keeping properties vary widely with the film type. Despite these known sensitometric changes in photographic films, practical difficulties are not encountered if film is stored under proper conditions for reasonable periods of time.

If films are packaged in sealed cans that are sufficiently tight to moisture vapor, then control of relative humidity in the storage area is not required. However, high relative humidities (greater than 70% RH) should always be avoided, since damage can occur to labels and cartons from moisture and mold growth, and to the cans from rust. The seal should not be removed from the film can until the film is ready to be loaded into the magazine.

Storage temperature is another important variable in the keeping of unexposed film. The lower the storage temperature, the better the sensitometric characteristics are kept. Moreover, the adverse effect of temperature is cumulative, and favorable keeping conditions following unfavorable keeping conditions will not compensate for any sensitometric deterioration that has taken place. Where possible, the following storage temperature should be maintained for black-and-white and most color films.

For storage periods up to 2 months, 6 months and 12 months, films should be stored at temperatures of 25°C, 15°C and 10°C, respectively. Infrared color films should be stored below −18°C.

Film should not be stored at temperatures greater than 27°C for any length of time. Where it may be necessary to keep film in hot climates and in unconditioned areas, the film should be protected from the direct rays of the sun, and stored for as short a time as possible. Stocks should be rotated so that the oldest film is used first. In applications where it may be necessary to store film for long periods of time, refrigeration at 10°C is recommended. When film is removed from low temperature storage, care must be taken that sufficient warm-up time is allowed before the film is removed from the can. Moisture will condense on the film if the film temperature is below the dew-point of the surrounding air. The necessary warm-up time will vary widely with the film temperature, film size, package, and dew-points of the air. However, a trade-packaged 180-foot roll of aerial film will reach 80% of complete temperature equilibrium in about ten hours.

Once photographic film is exposed, it should be processed as soon as possible. Deterioration of the latent image as well as of the unprocessed film can occur. High temperatures and high humidities should be avoided. Generally, no significant loss of latent image is evident for several days at moderate conditions.

6.4.1.5.2 PROCESSED FILM

For the storage of processed film, controlled relative humidity is usually not required in moderate climates, although the extremes of relative humidity must be avoided. High relative humidities encourage mold growth and cause "ferrotyping" (glossy areas on the emulsion when in close contact with an adjacent lap). Low relative humidities cause a temporary increase in film curl and brittleness.

The most satisfactory range of storage relative humidity is 30–50% RH for black-and-white film (ANSI PH 1.43-1975). The lower end of this range is preferable for color films to reduce dye fading. In applications where dimensional stability is of importance, the film should be stored in an environment with a relative humidity similar to that of the work area. Otherwise, complete moisture conditioning of the film prior to use is required.

The storage of film rolls that are in equilibrium with air at an appreciably different relative humidity than the storage relative humidity may produce edge distortions. Moisture will either escape from, or diffuse into, the edges of the rolls, depending on the humidity difference. This will cause either short or long edges of the rolls and could produce a permanent distortion due to plastic flow. This effect is lowered if the film is stored in covered cans.

High temperatures should be avoided in storage since heat will accelerate film shrinkage, and may cause film distortions. The combination of high temperatures and high relative humidities is particularly bad. Storage temperatures higher than 27°C should be avoided. Low temperature storage is recommended to minimize dye fading (ANSI PH 1.43-1975; Adelstein *et al.*, 1970), but care must be taken that the relative humidity does not go above 60% RH.

Protection of film records from the harmful effects of chemical fumes such as hydrogen sulfide and sulfur dioxide may be necessary in some areas, as these fumes cause slow physical deterioration of photographic film, similar to the behavior of paper records. Ordinarily, air purification would be necessary only if the air is polluted near an archival storage area. The protection of film from the hazards of fire, water, mold, and physical damage is also necessary. This is accomplished by proper design, construction and location of storage rooms.

Another consideration in film storage is that the film have adequate photographic processing. This requires complete fixation and thorough washing of the film. *See* section 6.5.2.4 for information on permanence specifications.

The care and expense that is justified in film storage depends upon the value of the film record and the period of time it is intended to be kept. Reasonable precautions, as already outlined, should provide satisfactory storage for many years. However, if a record is of permanent interest and preservation is desired for possibly hundreds of years, it must be classed as archival. Specific storage areas, air conditioning and purification, analysis of the processed film for residual hypo (ANSI PH 1.43-1973), and periodic inspection should be provided.

6.4.2 PHOTOGRAPHIC PLATES

For photogrammetry, accurate dimensional photographic representation is a fundamental requirement. Glass has yet to be surpassed as a commercially practical base for photographic emulsions where dimensional stability is paramount. In addition, a degree of flatness and rigidity can be provided with glass that is unattainable with any other photographic support commercially available for photogrammetry.

6.4.2.1 GENERAL TYPES

Positive and negative emulsions, with sensitivities ranging from low to high, are available.

6.4.2.1.1 NEGATIVE PLATES

Negative plates (plates with emulsions yielding negative pictures on development) are used in cameras. Automatic cameras such as the Galileo-Santoni model III, the Poivilliers-SOM, and the Wild RC-7a, utilizing plates in magazines, are used for engineering, cadastral, and other types of aerial survey work in Europe and the Western Hemisphere. Cameras such as the Wild BC-4 and the Instrument Corporation of Florida's PC-1000, and a number of types of photo-theodolites also use negative plates.

For automatic cameras, the glass plates should have smoothly beveled edges and slightly rounded corners to facilitate movement through the camera plate-transport system. Plate dimensions must be closely controlled.

6.4.2.1.2 Positive Plates

Positive transparencies on a dimensionally stable base (diapositives) are used in most stereomapping methods. Diapositives are usually printed from negatives by either contact or projection, and at various ratios suited to the particular stereoplotting instrument, as explained in other sections in this manual.

In many applications, initial reduction and subsequent projection of an enlarged image is necessary. Full and complete retention of detail is required and fine-grain, positive-type emulsions having high resolving power are used. During printing of the diapositive, much of the negative photograph is corrected for radial distortion. For high dimensional stability, it is essential that the positive transparency be on a glass plate so that further errors are not introduced. Flatness is also important (Hothmer, 1959; Hallert, 1960).

6.4.2.2 Strength

Glass for photographic plates is intrinsically a material of high strength, and it does not deform plastically before failure. It is always stronger under transient or momentary stresses than under prolonged loading. Because of its high stiffness, it is almost impossible to distort glass plates by external forces other than twisting or bending. To avoid the latter, diapositive plates for stereoplotting should be so mounted that they are neither deformed by clamping nor permitted to sag under their own weight due to poor mounting (Oswal, 1956; Helming, 1960).

6.4.2.3 Optical Properties

The following optical data for photographic glass base represent average values:

Refractive Index: 1.515 ± 0.005
Visible Light Transmission: 90%

In the visible region, from 400 to 700 nanometres, photographic glass shows a fairly flat transmittance curve with a maximum of 91%. It has a useful transmission from about 350 to about 2500 nanometres.

6.4.2.4 Flatness.

Flatness specifications for photographic plates are stated by various users and manufacturers of photographic materials in several ways. In Federal Specifications GG-P-450A, flatness is defined as the departure from a true plane, in decimal parts of an inch per linear inch of surface. However, in many commercial specifications, flatness is defined in terms of "overall limits." This value is the distance between two parallel planes between which lie all points on the emulsion surface.

Four types of glass plates that are available and that differ in their flatness specifications are described in table 6-6. Also included is a formula that can be used to calculate the "overall limit" in inches. Uses of these plate types are as follows.

Selected-flat glass is the standard for plates for, photomicrography, the graphic arts, and photomapping.

Ultra-flat glass is used in photofabrication and photogrammetric stereoplotting equipment. Precision-flat glass is primarily of use in the microelectronics industry.

Micro-flat glass is intended for ballistic and aerotriangulation camera systems and for first-order stereoplotters where the very highest degree of precision is required. The flatness of this material varies with the plate dimensions and is exceeded only by that found on high-grade optical plates.

6.4.2.5 Dimensional Specifications

Glass plates are supplied in thicknesses ranging from 1 to 6 millimetres. However, all plate sizes are not available in each thickness or plate type. An American National Standard (1974) gives the more representative dimensions in general use.

6.4.2.6 Dimensional Stability

6.4.2.6.1 Overall Dimensional Changes

The dimensional stability of photographic plates is unsurpassed by any other common photographic material used in photogrammetry. This will be evident from the data in this section.

The thermal coefficient of expansion of

TABLE 6-6. FLATNESS SPECIFICATIONS FOR PHOTOGRAPHIC GLASS PLATES

Type	Plate Diagonal (inches)	Formula for "Overall Limit"	Overall Limit for $3'' \times 4''$ Plate (inches)
Selected-flat	all	$7.1 \times 10^{-4} \times$ diagonal (*)	3.55×10^{-3}
Ultra-flat	all	$3.5 \times 10^{-4} \times$ diagonal (*)	1.75×10^{-3}
Precision-flat	all	$3.5 \times 10^{-5} \times$ diagonal (*)	8.75×10^{-4}
Micro-flat	<8	$4.0 \times 10^{-5} \times$ diagonal (*)	2.0×10^{-4}
	8–13	$5.0 \times 10^{-5} \times$ diagonal (*)	—
	13–19**	$6.0 \times 10^{-5} \times$ diagonal (*)	—

* diagonal in inches
** applies to center 16-inch diameter

photographic glass is only 0.0008% per 1°C. As seen in figure 6-41, this is lower than that of other photographic supports. Photographic plates frequently are required to transmit heat radiation in various kinds of projection or viewing systems. Most of the energy is transmitted through the glass by radiation but a very small fraction is absorbed. Illumination systems for plates should be designed so that the plate temperature doe not rise unduly and thus introduce dimensional changes. In most partial-field-illuminated projectors and direct-view instruments for stereoplotting, the amount of heat radiated from the lamps is within safe limits. Full-field-illuminated projectors for projection-plotting equipment are usually equipped with forced-air ventilation to prevent thermal gradients from becoming important.

The glass base itself does not absorb water vapor and has no tendency to change dimension with relative humidity. However, as with photographic film, the emulsion does absorb water and exhibits a strong tendency to expand or contract with changes in relative humidity. In the case of photographic plates, the base is inherently very stiff and can resist the expanding and contracting forces of the emulsion layers. In an environment where humidity control is difficult, the absence of dimensional change due to moisture is an outstanding advantage of glass plates.

It is essential that the emulsion be bonded firmly to the glass base to prevent the expanding and contracting of the emulsion. If slipping or reticulation of the emulsion occurs, the inherent dimensional stability of glass plates is, of course, completely useless.

Dimensional change during processing and aging shrinkage are not important factors with glass plates.

6.4.2.6.2 Nonuniform Dimensional Changes

As already discussed in section 6.4.1.4.3, uniformity of dimensional change is very important in photogrammetric work. Measuring the intersections of a grid exposed onto a glass plate with an optical comparator, Brucklacher and Luder (1956) reported nonuniform dimensional changes of ±2.4 micrometres. Subsequent work on photographic plates by Altman and Ball (1961) using a similar experimental technique showed comparable behavior. These authors considered the displacements of the grid intersections to be due to both a systematic (or uniform) and a random (or nonuniform) image displacement. Regression analyses were used to eliminate the systematic displacements in the analysis. The random displacements were calculated to be of the order of 1 micrometre. Generally, the random displacements were slightly greater at the edges and corners of the plate.

Similar studies were made in 1964 by Starbird (1964) on plates used in ballistic-cameras and by Swanson (1964) on color diapositive plates. They showed that errors in a single coordinate, after mathematical corrections for an assignable shift, were less than 2 micrometres. Subsequently, Schmid (1968) determined the random error to be ±1 micrometre. Umbach (1968), Ligterink and Zijlstra (1969), and Burnham and Josephson (1969) also agreed with the earlier measurements.

Nonuniform dimensional changes in photographic plates are consistently less than those found in photographic film. Moreover, color-sensitized photographic plates have the same metric qualities as black-and-white plates (Burnham, Josephson, 1969). However, this high stability of photographic plates is dependent upon proper handling procedures. Local distortions due to residual water spots can be as high as ±6 micrometres. It should be borne in mind that the random image-displacements on photographic plates referred to above are very small and of no interest in most applications. They may be a source of error in fields where positional data or measurements of the highest precision are required, such as analytical photogrammetry or in special technical or scientific applications.

6.4.2.7 Storage

The same general principles that apply to the storage of photographic film also apply to photographic plates. For storage of unexposed plates, temperatures from 20 to 24°C and relative humidities between 40 and 60% are recommended. In general, packages of unprocessed photographic plates are not intended to withstand long periods at high relative humidities. In hot weather, refrigerated storage is recommended for unopened packages. Sufficient warm-up time should then be allowed before opening to avoid moisture condensation on the plates. Opened packages should not be refrigerated because the high humidity will damage the photographic plates.

Photographic plates have excellent stability, but good long time keeping behavior requires satisfactory processing and storage conditions. Processed plates should be stored in areas similar to those recommended for photographic films. Relative humidities above 50% should be avoided (ANSI PH 1.45-1972). Plates can be stored conveniently on edge for greater accessibility.

6.4.3 PHOTOGRAPHIC PAPER

Photographic paper is widely used in printing films with subsequent viewing by reflected light. It is not generally used for measurements since its dimensional stability is much less than that of photographic film and plates.

6.4.3.1 PHYSICAL STRUCTURE

The paper base of photographic paper is manufactured to resist the action of strong alkalies and reducing agents in the developer and acids in the fixing bath, and will remain sufficiently strong to be handled easily, although it imbibes aqueous solutions. At the same time it must not affect the photographic emulsion coated on it. Photographic papers available today for aerial photography consist of paper base that is coated on both sides with a resinous layer. Under recommended processing conditions, these layers essentially prevent moisture absorption during photographic processing, thereby permitting increased processing speeds. However, while these layers significantly improve water resistance and rates of moisture exchange, they do not change the relationship between moisture content and temperature at equilibrium.

6.4.3.2 GENERAL PHYSICAL PROPERTIES

Unprocessed photographic paper is generally reasonably flat between 30 and 60% relative humidity. Curl during photographic processing is not a problem with resin-coated types because the resinous coating prevents wetting of the paper base. Moreover, expansion of the wet emulsion does not exert sufficient lateral force to overcome the stiffness of the base. Although post-process curl on properly processed resin-coated papers is normally at low levels, it can be further reduced by holding the paper flat while drying.

Brittleness in photographic paper can occur for the same reason as in photographic film. The emulsion layers are relatively brittle, particularly compared with the paper base. When photographic paper is flexed with the emulsion layer on the outside of the bend, there is a tensile force on this layer that can cause emulsion cracking. This can be minimized by keeping drying temperatures low after processing, avoiding overdrying, and not using excessive quantities of hardener in the fix path.

6.4.3.3 DIMENSIONAL STABILITY

As with photographic film, photographic paper increases in dimension with increasing humidity. This is due to moisture absorption by both the gelatin and the paper base. The expansion of the latter is greater in the cross direction than in the machine direction of the web. Photographic paper does not change in a linear manner with relative humidity nor does it show the same dimensional change when the humidity increases and decreases. (The latter behavior is due to release of strain in the paper in some cases and to normal hysteresis in others.) For these reasons, values of the coefficient of expansion with humidity can vary widely. Coefficients can range from 0.004 to 0.014% per % RH depending upon the paper type, the paper direction, the

humidity range, and whether the humidity is increasing or decreasing.

The thermal coefficient of expansion for paper is much lower than the humidity coefficient and is of lesser importance.

Photographic paper may change in size from +0.2 to −0.8% owing to processing. This change is very dependent upon the method of drying. For instance, if the paper is dried in a manner that tends to hold it and prevent it from shrinking, such as on a ferrotype tin or a tight belt on a belt dryer, it will tend to retain the dimensions it had when wet. If it is dried too hot, it will tend to have greater contraction than if it is dried at a lower temperature. Because of tension during manufacture, the dimensional changes will generally be greater in the lengthwise than in the widthwise direction of coating.

Processing size changes on resin-coated photographic papers are appreciably less than on regular photographic papers, and the dimensional changes are more alike in the primary directions since some shrinkage occurs in both the length and cross directions. As a function of paper type, paper direction, processing, and drying, size change on resin-coated grades can range from −0.02% to −0.10%.

To obtain the maximum dimensional stability of aerial photographic papers, the following procedure is recommended.

1. Prior to exposure, condition the photographic paper to the same relative humidity that the print will be subjected to in its end use. With resin-coated papers conditioned as individual, free-hanging sheets, this takes approximately seven days.
2. Use the minimum processing times recommended by the paper manufacturer.

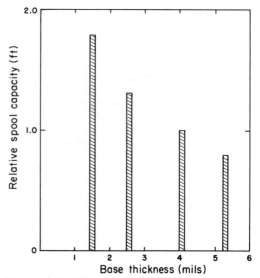

FIGURE 6-46. Effect of base thickness on spool capacity.

3. Air dry the prints at room temperature, or dry on a double belt, heated drum dryer at the minimum temperature.

6.4.4 FILM THICKNESS AND SPOOL CAPACITY

Limitations on available weight and space continue to require attention for photography from high-altitudes and from space. Reduction of format size and reduction of film thickness are the most obvious methods for meeting the limitations. Reduction in size can defeat the purpose where the reduction cannot be accompanied by an increase in definition. So, film-thickness reduction has been the major factor in efforts to increase film payload. Films can be made a little thinner by reducing the emulsion thickness, and this method is most effective in pelloid-backed products where it also permits a proportional reduction in thickness of the curl-balancing gelatin backing. In addition to taking up less space, thinner emulsions have the advantage of sharpness improvement. Reduction in the thickness of the base, however, has the greatest space-saving effect because greater changes are possible. This is illustrated by figure 6-46, which shows that a 25% savings can be affected by a change from 5¼-mil (133 μm) acetate to 4-mil (102 μm) polyester base. A further 30% capacity increase results from a thickness reduction to 2½-mil (64 μm) polyester. Although a few films are available with a 1.5- and 1.2-mil (38- and 30-μm) base, manufacturing problems and difficulty with handling and tracking in cameras and in printing and in processing equipment have limited adoption of this thickness.

6.5 Photographic Processing Principles

6.5.1 GENERAL

The following sections are concerned with black-and-white and color processing principles, process control, and solution conservation and effluent treatment involved with conventional wet-processed (silver halide) photographic materials. For processing of unconventionally processed materials, see section 6.3.6.

At its simplest, processing of the latent image involves the following: a development step in which the exposed silver-halide grains are preferentially reduced to metallic silver, a fixing step to convert the undeveloped silver halides to water-soluble compounds, a washing step to complete the removal of the water-soluble compounds and remaining fixer, and a drying step. In black-and-white reversal processing, that is modified by the inclusion of a bleach bath following the developer to remove the negative image, and a clearing bath to remove residual bleach. Following subsequent re-exposure, a re-developer is used to form a positive image. In processing color films, dye images are formed by development along with the silver of the final image. The silver is subsequently removed with bleach and fix steps.

6.5.2 CONVENTIONAL BLACK-AND-WHITE PROCESSING

6.5.2.1 DEVELOPER

Developers are chemicals that selectively reduce the exposed silver-halide grains to metallic silver at a much greater rate than they reduce unexposed grains. Although the major characteristics of the final image are determined by the composition of the emulsion and are fixed by the emulsion maker, processing (particularly development) has a marked influence on the result.

Thus, the film speed, process contrast, shape of characteristic curve, and micro-structure of the silver image are affected by the conditions of development. In general, when the photographic requirements of a particular job have been determined, it is just as important to establish the correct processing method as it is to select the best photographic material and the proper exposure.

The developer solution contains a number of ingredients to provide the desired activity with the lowest possible fog level and maximum possible useful life. The most important developing agents are organic compounds used singly or in combination. Some of the most commonly used compounds are Metol (p-methylaminophenol), hydroquinone, Phenidone (1-phenyl-3-pyrazolidone), and derivatives of Phenidone such as methyl Phenidone and Dimezone.

Metol is a soft-working developer that produces low contrast. It initiates development quickly and has a long, useful life. Hydroquinone is a slower working developer and, when used alone, provides high contrast with slow emulsion speed. Phenidone (and its derivatives) have properties similar to Metol but are effective at a lower concentration. In some cases when a high-activity developer is needed, Phenidone appears to have unique advantages over Metol and has replaced Metol in many developers.

Other developing agents are occasionally used for special applications, but the agents named cover almost all of the important uses in the development of black-and-white images.

Certain combinations of developing agents show more activity than could be predicted from the activities of the separate agents. This effect, called superadditivity, is of great practical value and is one of the chief reasons for the use of the Metol-hydroquinone and Phenidone-hydroquinone combinations. The versatility of developers using these combinations is so great that almost all the requirements of black-and-white processing can be met by their use.

In addition to the developing agent, the developer solution contains a preservative—usually sodium or potassium sulfite or bisulfite—to minimize oxidation of the developing agent and to transform oxidation products from the reaction into colorless compounds to prevent staining of the emulsion. An alkali with a suitable pH-buffering capacity is present to control the pH (acidity-alkalinity) of the solution, which controls the rate of the development. Commonly used buffers include carbonate, borax, and sodium metaborate. However the use of borax or boron compounds must be restricted when boron concentration in sewer effluent is regulated by legislation.

Since all developers have some tendency to reduce silver halide that has not received an exposure, some type of antifoggant is usually added to limit unselective reduction. The amount of antifoggant that can be added represents a compromise between complete suppression of unselective development and depression of the development rate in exposed areas. Commonly used antifoggants are: potassium bromide, potassium iodide, benzotriazole, and 6-nitrobenzimidazole.

Other constituents may be added for special purposes that may not directly affect the development reaction. Metal-sequestering agents may be added to prevent precipitation of insoluble calcium and magnesium salts in the water supply. A developer-compatible hardening agent is frequently included in formulation of developers designed for use in processing machines that operate at higher-than-ambient temperatures.

In processing, developers are either used once and discarded or are replenished (if appropriate) on either a batchwise or continuous basis to prevent slowdown of the development. Inasmuch as the developer components are not usually all consumed at the same rate, the replenisher contains the components in a different ratio than does the original developer. In addition, certain components from the development that have a profound restraining effect on the development reaction are dissolved into the developer. Different films require different replenishment rates for the above reason and also because of differing amounts of developer constituents that are consumed. Therefore, replenishment rate is always a function of several factors which include: developer composition, film or paper type being processed, percentage exposed area on the material, density resulting from the exposure, and processing conditions. With proper replenishment, a developer solution may be maintained at its proper activity over a sustained period even with different types of material and different

processing times. *See* Section 6.5.4 for information on process control.

6.5.2.2 STOP BATH

A water rinse or an acid stop-bath is frequently used to quickly neutralize alkaline developer remaining in the emulsion, thus stopping development by lowering the pH of the emulsion. This also tends to minimize emulsion swell (if an acid bath is used), remove calcium scums that may have formed in the developer, and preserve the acidity and hardening action of the following fix bath. The acidic stop bath is commonly a dilute solution of acetic acid. The pH of the acidic stop bath should be maintained below a pH of 5.4 by replacement or replenishment.

6.5.2.3 FIXING BATH

The purpose of the fixing bath is to convert the unexposed silver halides to water-soluble compounds that are removable in the fixing bath and the following wash step. The fixing bath must leave the emulsion in the best condition possible for removal of these soluble silver halides as well as for best possible drying—especially in mechanized processors. The fixing baths commonly used contain either sodium or ammonium thiosulfate as the solubilizing agent, sodium sulfite to retard sulfurization of the thiosulfate in acid solution, and a weak acid, usually acetic acid, to control acidity at the proper level. Many fixing baths (especially those designed for use in processing machines) contain a hardening agent (usually an aluminum salt, such as potassium alum) to harden the gelatin and inhibit swelling in the subsequent washing stage. The concentration of this hardener must be balanced so that it is high enough to harden the gelatin sufficiently for drying, yet it should leave the gelatin soft enough for adequate removal of water-soluble silver salts and remaining fixer during washing.

The removal of the unused silver halide is desirable because it leaves the sensitive materials in a much better optical state and also because the silver halide is light-sensitive and will darken gradually to obscure the image. The silver halide must be removed to a very low concentration to prevent the formation of image-degrading silver-sulfide stain by the combination of residual silver salts with sulfur from residual hypo left in the film or by combination with sulfur compounds in the atmosphere. Inasmuch as the effects of incomplete fixing only appear after prolonged storage, its importance is not readily apparent and is frequently overlooked. The concentration of silver compounds necessary to cause image degradation is extremely small, and there is no simple, quantitative method for its determination. A simple test is available, however, to determine the amount of stain that would form with time. The test is as follows.

Prepare the following solution—

| Water | 100 ml |
| Sodium Sulfide, Na$_2$S | 2 grams |

Store in a small stoppered bottle for not more than three months. For use, dilute one part of stock solution with nine parts water. The diluted solution has limited storage life and should be replaced weekly. Place a drop of the working test solution on an unexposed part of the processed negative or print, wait one minute, blot off the surface with a clean blotter. Any yellowing of the test spot, other than a barely visible cream tint, indicates the presence of silver halide. If the test is positive, residual silver halide can be removed by refixing the print or negative in fresh hypo and rewashing for the recommended time.

In fixing films and plates, the old rule of fixing for twice the time required to clear the milky appearance is still valid. However, prolonged fixing should be avoided to prevent degradation of the silver image itself. With paper prints, the point of clearing is not visible, so the fixing recommendations for the product should be followed.

As a fixing bath is used, a number of things take place that lower its efficiency. Thiosulfate is consumed by the formation of the water-soluble silver compounds, so fixing action is slowed down; the concentration of the soluble silver compounds increases, which has a detrimental effect on the complete removal of these compounds in the subsequent wash. The fixer is diluted by carryover from the previous stage of the process (stop bath or developer), which further impairs its effectiveness. Film that has been improperly fixed (either because of insufficient fix time or by fixing in an exhausted fixer) is much more difficult, or even impossible, to adequately wash, which results in impaired image stability.

The use of two fix baths is recommended—especially for paper prints with paper not coated with resin—to assure adequate fixing. Film or paper is fixed for one-half the recommended time in each of the successive baths. This countercurrent flow of fixer is also recommended for multi-tank processing machines. Using the recommended replenishment rate for the product, introduce the replenisher into the last fix tank and overflow from the first fix tank. The use of two or more fix tanks also minimizes carryover of silver compounds into the wash.

6.5.2.4 WASHING

The purpose of washing is to remove the remaining soluble silver compounds formed during the fixing and also to remove fix-bath chemicals that would otherwise slowly decompose and attack the image with the formation of silver-sulfide stain. For image permanence, adequate washing is just as important as adequate fixing.

Washing efficiency is influenced by the nature and condition of the fixing bath, the configuration of the washing container, rate of wash-water flow, agitation, and water temperature. In batch washing, the water flow should be rapid enough to replace the water in the washing container once every five minutes.

Papers not coated with resin require much longer washing time than films because the fibers of the paper base tenaciously hold sufficient hypo to cause image degradation. A washing time of at least one hour is required for these materials. Resin-coated papers, which contain a water-impervious coating on both sides of the paper support, are sufficiently washed within five minutes using the washing conditions described above.

The use of a washing aid such as a salt solution or a commercial preparation such as Kodak hypo-clearing agent hastens removal of chemicals from the emulsion and allows a substantial reduction in washing time or wash-water temperature. Such a treatment is of particular advantage in rewind processing (where access of the wash water to the film is severely limited) and in processing paper prints not resin-coated (where removal of traces of residual hypo is important if long-term storage is contemplated). The washing aid should be preceded and followed by a water wash.

For all practical purposes, films and papers that have been properly fixed and washed (as recommended by the manufacturer) can be expected to keep for many years under normal storage conditions and still maintain acceptable image stability. Where permanence of the processed material is a strong consideration when they are intended to be permanent records, the material must be processed to achieve archival permanence. Specifications for archival permanence are found in American National Standards Institute Standards ANSI: PH 1.28-1973, PH 1.41-1973, and PH 4.32-1974 for cellulose-ester- and polyester-base materials and for black-and-white papers. Standard PH 4.8-1971 describes methods for measuring residual chemicals, and PH 1.42-1969 describes methods for comparing the stability of color films and papers.

6.5.2.5 DRYING

At the end of the washing step, the photographic material carries a fixed amount of water in the emulsion and any gelatin or pelloid backing. In the case of paper not resin-coated it is found in the base as well. Acetate base also carries an appreciable quantity of water, but polyester base contains only a very small amount. The amount of water present in the gelatin is determined by the thickness of the gelatin and the amount of swelling that has occurred during processing. In addition to the minimum quantity of water imbibed by the gelatin, there is always some surface water carried to the dryer. The amount is dependent on the efficiency of the draining or squeegeeing. It is always preferable to remove as much water as possible mechanically, since the energy expended is much less than that required to evaporate the same quantity of water. A frequently used method of removing surface water is treatment in a suitable wetting-agent bath, followed by draining. Some continuous processors use rubber rollers, wiper blades, or air squeegees for removing water.

The actual drying action occurs in essentially two distinct stages—the constant-rate period, and the falling-rate period. The rate during the constant-rate period is primarily a function of the dryer design and is not determined by the nature of the photographic material. The falling-rate period is governed by molecular diffusion. It is strongly affected by the nature of the emulsion, the amount of swelling that has occurred in processing, and other factors. This point is reached when the moisture on the surface has been evaporated to the point that further drying results in dry areas on the surface.

In processing machines, the film ideally should emerge from the dryer with a total moisture content equal to the moisture it would hold if in equilibrium with room temperature air at 50% RH. The temperature in the dryer should be as low as possible for dimensional stability considerations. Set the dryer at 3°C above that required to dry clear film. *See* section 6.4 for the effects of film drying on dimensional stability and other physical properties.

6.5.3 COLOR PROCESSING

There are two processes currently being used to provide an image in the color films or papers used in photogrammetry. These are the dye-formation process and the dye-bleach process. The former is used for both color-negative and color-positive materials, while the latter is used only for color-positives. To help in understanding the mechanism of color processing, it is suggested the reader be familiar with the content of section 6.2.6 "How Color Photography Works."

There are many similarities in the basic ingredients and processing mechanisms of black-and-white and color processing solutions. In the dye-formation process the color developer, in addition to reducing exposed silver-halide grains to metallic silver, forms the image by the combination (coupling) of reacted (oxidized) color developing agent with color couplers. All the color films currently used in aerial photography contain the color couplers incorporated in emulsion layers. Thus a dye image is formed which is proportional to the amount of silver developed by the color developer in that layer. Other color processes exist in which the coupler is contained in the color developing solutions themselves.

6.5.3.1 COLOR-NEGATIVE PROCESSING

The essential ingredients of a color-negative process are as follows.

> Color Developer
> Bleach
> Fixer
> Wash

From the exposed latent image the color developer produces a negative silver image and a negative dye image simultaneously in each color forming layer of the film or paper.

The bleach solution converts the developed metallic silver image to silver salts by the action of a strong oxidizing (bleaching) agent in the presence of halide. The salts are converted to water soluble silver compounds in the following fix bath. These are removed in the fix and final wash leaving only the dye image. Commonly used bleaching agents are potassium (or sodium) ferricyanide, potassium dichromate, and ferric EDTA (chelated Fe^{+++} with ethylenediamine tetraacetic acid). For reasons of economy and legal limitations on waste disposal, the used bleach is often regenerated (oxidized) to its original form for reuse. In some color processes, often limited to materials with lower silver content such as papers and print films, a combination bleach-fix is used. This is a single solution containing both bleaching agent and thiosulfate in which the silver is oxidized and converted to water soluble silver compounds in a single bath.

In addition to the above listed steps of a color negative process, a hardener bath may be present to harden the emulsion to minimize damage during processing—especially in continuous processing machines. In some processes it is the first step of the process, and in others it follows the developer, in which case it also serves to stop action of the developer. A stop bath may be present to stop development and a water wash may follow many of the process steps to avoid contamination as well as to serve as a final wash to remove process chemicals. A stabilizer may be present to stabilize the emulsion to prevent rapid fading of dye images.

6.5.3.2 COLOR-REVERSAL PROCESSING

The essential ingredients of a color-reversal process are as follows.

> First Developer
> Stop Bath or Hardener and/or wash
> Reversal Exposure
> Color Developer
> Bleach
> Fixer
> Wash

The first developer—a negative black-and-white developer—converts the exposed grains of silver halides of the latent image to metallic silver, which forms a negative image. A hardener or stop bath stops the action of the developing agent by neutralizing the alkalinity of the developer. Reversal exposure, often done chemically in the color developer or by exposing the film to white light, is done in order that the silver halide crystals not converted to metallic silver in the first developer will be developable in the color developer. The color developer develops the silver-halide crystals not developed in first developer, producing a positive silver image. At the same time, the oxidized developing agent from this reaction combines with the dye couplers in the emulsion to form a color image in proportion to the amount of silver developed. The rest of the process is similar to a color-negative process.

6.5.3.3 DYE BLEACH PROCESS

In the dye-bleach process, the dyes are initially present in all three layers along with the silver halide. After exposure, the film is first placed in a black-and-white developer, as in the color-reversal dye-forming process, where the exposed silver halide is reduced to form a negative silver image. The film is put in a dye-bleach solution where the dye is selectively reduced in proportion to the amount of silver present. The film is then put through the usual silver bleaching and fixing solutions leaving the positive dye images in which the concentration is inversely proportional to the amount of the original exposure.

6.5.3.4 COLOR-PROCESSING SPECIFICATIONS

In color processing, the classical processing considerations of time, temperature, and agitation are much more critical for consistent quality, so the specifications are usually much tighter. As an example, the temperature variability of first and color developers must often be held within plus and minus 0.15 degrees *Celsius*. Color films have three or more layers which must be balanced for density and contrast. Changing the development conditions affects each of these layers differently.

6.5.4 PHOTOGRAPHIC PROCESS CONTROL

Photographic process control consists of a monitoring procedure to follow and a set of specifications for parameters of the process such as film or paper speed and contrast, time, temperature, amount of agitation, and solution replenishment-rate in continuous processing machines. Proper control assures consistent and reproducible results. It provides data that makes the operator aware of unacceptable deviations as soon as they occur so the process may be corrected to minimize down time and the amount of improperly processed product. It is not the intent of this chapter to go into all the details of sensitometry and process control. This may be found in some of the fine references at the end of the chapter, including SPSE (1973), SMPTE (1960), Kodak Z-99 (1975), and Kodak Z-126 (1972). Chemical methods of control will not be considered either. These involve chemical analysis of the processing solutions and is an exacting, time-consuming, and costly operation requiring trained personnel. Details on such methods may also be found in the references. For black-and-white or color processes involved in photogrammetry, this is not usually required if the manufacturer's recommendations are followed.

The first step in process control involves establishing the specifications for the standard process to produce the desired level of quality. Many photographic manufacturers offer a process-control system for establishing and maintaining good control of their processes. If none is available, specifications from established processes that provide the desired photographic results may be used.

The control strip is made under a laboratory simulation of production conditions, usually in the form of a sensitometric wedge. These strips may be either purchased from the manufacturer or exposed by the user from the same type of film that is most commonly processed by him. It consists of carefully controlled, incremental exposures placed on the film using a sensitometer. Process the control strips under the recommended processing conditions in a process known to be producing the desired results. Measure the control-strip densities on a densitometer and calculate the sensitometric parameters. Once the sensitometric parameters of speed and contrast for this standard process have been established, control-strip data from subsequent processes are compared to it. Process a control strip on start-up and at regular intervals—at least twice daily.

If the results of a subsequent process fall outside the established quality-limits, corrective action is required. In a black-and-white process, it may simply require the addition of replenisher or water to the developer tank to raise or lower process activity. In a color process, because of the greater number of processing steps, corrective action is more complicated. Follow the guidelines offered by manufacturer. Keep a record of each control strip, the time, temperature, etc. and of any corrective action taken.

It is important to keep solution-handling and processing equipment in good, clean, operating condition, to calibrate solution-mixing equipment and replenisher flowmeters, and to mix solutions as recommended. The chemical and mechanical functions of a processor are closely related and interdependent on each other. Some processing difficulties that appear to be chemical may in fact be caused by improperly operated processing equipment. Therefore, in the event of process control problems, it is important to determine whether a malfunction is chemical or mechanical, and to proceed from there to bring the process into control.

6.5.5 WATER AND CHEMICAL CONSERVATION, EFFLUENT TREATMENT

6.5.5.1 WATER CONSERVATION

There are several methods available for reducing wash-water consumption during continuous processing of film. However, the availability of space, energy, capital, analytical facilities and the degree of photograph permanence desired are factors that should be considered when deciding on the method.

The least complex method for reducing wash-water consumption is to reduce the water flow, since most recommended rates of water flow contain a safety factor. Gibson (1972) describes a technique used to reduce wash-water consumption. By using the last fix tank, on a Kodak Versamat film processor, model 11C-M, as an additional wash tank, and by maintaining counter-current water flow, the wash rate was reduced from 15 litres per minute to 100 millilitres per minute. This extreme case was for tactical aerial reconnaissance where film storage was not expected beyond six months. This does show, however, that there is a very large range within which water conservation can be practiced. Where the processing tank configuration permits it, the use of multitank washes with

countercurrent flow will, in general, allow a substantial reduction in the water flow rate.

A commonly more practical way to reduce wash-water consumption with a commonly used aerial film processor (the Kodak Versamat film processor, model 11) involves a change from a spray-wash system to a system that utilizes primarily immersion wash.

The model 11 processor is a roller-transport processor capable of handling films up to 11 inches wide at speeds up to 25 ft/min. The processor contains seven solution tanks; normally two tanks are used for developer, three for fixer, and two for spray wash.

The Kodak Versamat water conservation assembly, model 11C, reduces water consumption from 15 litres per minute to as low as 4 litres per minute. It accomplishes this by converting the spray wash tanks to countercurrent deep-bath wash tanks with a final spray rinse.

As shown in figure 6-47, tempered wash water will fill tank no. 7 (end wash tank) through a spray-wash header on rack 7, to the overflow level where a level-control sensor will energize a relay and automatically turn on the water circulation pump. Water will be pumped from the bottom of tank no. 7, through a regulator valve, and a second flowmeter to tank no. 6.

Tank no. 7 is a combination immersion-wash tank and spray wash tank. Water flow is countercurrent and is pumped into tank no. 6 from tank no. 7; tank no. 6 is an immersion tank only. The combination of immersion and spray washing efficiently utilizes the tempered wash-water and energy, allowing reduced wash rates and a high quality of processed film. This wash system

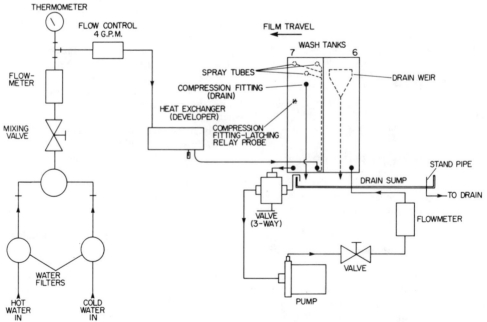

FIGURE 6-47. Schematic of Versamat Water Conservation Assembly to reduce wash water consumption.

is adjustable from 4 to 15 litres per minute; however, most aerial films are adequately washed (but not necessarily to archival quality) at 4 litres per minute.

Ion exchange, evaporation, and reverse osmosis have been evaluated for reusing wash water on the Kodak Versamat film processor. Any one of these techniques is capable of reducing wash-water consumption by at least 95%. At the present time, it appears that the most economical system is reverse osmosis (Gibson, 1972). Reverse osmosis is a process for separating relatively pure water from a less pure solution by letting the solution osmose through a membrane.

The combination of the water-conservation assembly and a recirculating reverse-osmosis system would lower wash-water consumption.

6.5.5.2 CHEMICAL CONSERVATION OF CHEMICALS & TREATMENT OF EFFLUENT

Silver recovery and fixer reuse are the two most important things to be considered in conserving chemicals, reducing operating costs and meeting federal and municipal guidelines on effluent.

Silver occurs in the fixer solution and in the wash as silver compounds. Since many municipal sewer codes prescribe less than 1 ppm (part per million) total silver content, meeting these codes requires treatment of both the fixer solution and wash.

The most popular techniques for recovering silver are:

    (a) sulfide precipitation,
    (b) metallic replacement, and
    (c) electrolytic recovery.

Sulfide Precipitation. Although sulfide precipitation is a classic means of removing silver from solution, development of commercial equipment has been slow due to the possibility of liberating toxic hydrogen sulfide gas during the reaction of silver compounds with sodium sulfide, and because the precipitate is hard to filter.

A paper by LaPerle (1976) describes experimental equipment that utilizes the combination of pH and silver/silver-sulfide probes to control the reaction, preventing the accidental liberation of hydrogen sulfide gas. A two-stage filtration system is also discussed that reduces the amount of silver in the effluent to less than 0.1 ppm.

A small sulfide precipitation unit has the capability of removing all of the silver from the fixer or the effluent from a Kodak Versamat processor, model 11 equipped with a water reduction kit. There is no information on the reuse of fixer after sulfide precipitation. At least a pH adjustment would be required.

Metallic Replacement. Commercially available metallic-replacement cartridges are usually buckets containing steel wool. The buckets are placed between the fixer-solution overflow and the sewer. The silver complex in the fixer solution is passed through the bucket where it is reduced by the iron and the silver precipitates as a sludge. Normally, these cartridges reduce the silver in the fixer-solution overflow to about 100 ppm, but by placing several of these cartridges in series (often two) and maintaining flow rates suggested by the manufacturer, it is possible to reduce the level to less than 0.1 ppm. Fixer is not reused after passing through these cartridges because staining of the photographic film will occur.

Electrolytic Recovery. Electrolytic silver recovery is accomplished by applying a direct current between two electrodes immersed in the fixer solution. Various cell configurations and agitation is used, and at 100% efficiency, approximately 4 grams of silver per ampere-hour are deposited on the cathode.

Commercial electrolytic recovery units are available (Eastman Kodak Co., J-10) that allow for continuous recirculation of the fixer from the processor tank. These units can maintain the silver level in the tank between 0.5 and 1 g/l while making fixer reuse possible. In a processor with more than one fix tank, it is preferable to have countercurrent fix flow with electrolytic silver recovery of the first tank only or of the different tanks independently.

Maximum silver recovery and water conservation can be accomplished by the addition of an electrolytic silver recovery recirculation system to the fix tank, and the addition of a reverse osmosis unit recirculating the wash, followed by passing the combined overflow from the silver recovery unit and concentrate from the reverse osmosis unit through a metallic replacement cartridge.

An additional benefit of continuous electrolytic silver recovery from the fixer is that it may permit some reduction of the fixer replenishment rate.

6.6 *Photographic Processing Facilities*

6.6.1 REQUIREMENTS OF THE PROCESSING LABORATORY

Photographic laboratories incorporate facilities for the specialized operations required in film handling, film and paper processing, photographic printing and related finishing, and handling and storage operations. Photographic laboratory facilities required for photogrammetry are generally similar to those used in conventional photographic practice. In some specific cases, specialized facilities are needed for equipment that is unique in this field.

The need for planning laboratory facilities arises when a new laboratory is being installed or when existing facilities are being modified, enlarged, or moved to another location. In order to insure the most efficient utilization of the space available, it is advisable to consult an architect, a consulting engineer, or a contractor who has architectural assistance available. This will also facilitate compliance with local building codes, sewer codes, safety rules, and other regulations. The rapid growth of color photography has created a general demand for increased processing capacity. With this situation come both the opportunities and the obligation to improve other methods of production and to dispense with improvisations. Certainly, there is no place for makeshift methods in modern color processing.

In planning the layout of a given laboratory, the following factors should be considered: type of work to be done, the anticipated work volume, the number of people who will be working in the laboratory, the major equipment to be installed, and the flow of work throughout the area (Eastman Kodak Company, K-13). Consideration must also be given to long-term needs, growth of the operation, and the addition of new pieces or types of processing and handling equipment.

Since an important feature of a good laboratory is usability, the equipment and methods should be arranged to allow the work to flow efficiently with a minimum of lost motion. Providing for the greatest economy of time and effort, consistent with the production of high-quality results, is the most important factor in laboratory design.

A feature common to all photographic laboratories is the darkroom. Photographic materials are light sensitive, so they must be handled with either properly filtered safelights or in total darkness. Darkrooms require careful design to provide for efficient work in the dark and also to fulfill safety requirements for the worker. They should be separated into wet and dry areas to guard against splashing photographic materials with processing solutions and to minimize the risk of carrying chemicals and dried residues from processing baths to the areas where photographic materials are handled. The wet area includes the chemical-mixing facilities, processing machines, and sinks where processing is carried out. The dry area includes cabinets and work benches where photographic materials are stored and handled, and where printers, enlargers, viewers, and other such devices are located. Ideally, wet and dry areas should be located on opposite sides of the facility with an aisle between them.

The flow of work can be expedited by removing those operations to an adjacent white-light room, that do not require a darkroom. Materials can be transferred between the light and dark areas using a light-tight box or pass-through.

Good design of the processing laboratory assist in proper maintenance and reduces the likelihood that dirt will become a degrading feature in the processing operation. The ventilation system is particularly important as it affects dirt control. If the laboratory is air conditioned, control of dirt may become more difficult because greater movement of air increases the risk of dirt being deposited on the photographic material either before, during, or after processing. The use of proper materials in the construction of processing equipment and processing-solution-handling equipment is important for long life of the processing solutions and the equipment itself.

Filtration of air used in dryers should be as good as, or better than, filtration of air used in the laboratory rooms. Dirt adheres more readily to wet emulsion and pelloid layers and its removal is accomplished only with great difficulty, if at all. In addition to airborne dirt carried into the laboratory in the ventilating air, one must also be concerned with dirt generated within the workrooms by the personnel and the equipment. Not all lint and dirt caused by the personnel comes directly from them to the film. They may dislodge dirt that had already accumulated on the working surface floors and other locations. For this reason, shelves, working surfaces and floors should be kept as clean as possible.

The preceding paragraphs give only rudimentary information about housekeeping procedures for the photographic laboratory. More detailed treatments of these subjects may be found in the references at the end of the chapter.

6.6.2 CONVENTIONAL PROCESSING EQUIPMENT AND TECHNIQUES

The importance of adhering to recommended and tested procedures, and of establishing proper, standardized processing techniques cannot be overemphasized. In processing for photogrammetry, where the investment in personnel and processing equipment may be very high, results of an otherwise well-executed project can be disappointing or lost entirely if the processing is not done properly. Requirements of time, temperature, and agitation must be carefully adhered to. The temperature of all solutions, especially the developer, should be carefully checked before processing. Process time is also critical. In continuous-processing machines, process at the proper transport speed. Proper agitation techniques are given for the various types of equipment and should be carefully followed. For continuous processing, it is also very important to maintain the proper rate of replenishment of solution. The replenishment rate varies with film type, film width, and transport speed. See section 6.5.4 for information on process control in continuous processing.

6.6.2.1 Processing Rolled Aerial Film

Rolled aerial film, can be processed in batch-type apparatus where the entire roll is treated as a unit in each bath in succession, or it can be processed in a continuous processing machine. Black-and-white and color aerial films (excluding those with 2½-mil (64 μm) or thinner, and with 7-mil (178 μm) or thicker base may be processed with rewind processing equipment such as that shown in figure 6-48. Rewind processing equipment is attractive because of the low initial cost, its comparative ease of operation, and its portability. However, the processing quality, especially with color films, is not as good as with a properly operated continuous processing machine since most films have been designed to be processed in continous processors with chemicals particularly appropriate for them. The poorer quality of rewind-processed film is caused by the poorer agitation, differences in treatment time, and poorer ratio of solution volume to film surface, resulting in exhausted solutions. Rewind processing of color films will result in a different color balance than that resulting from continuous processing. However, a quality sufficient for many applications can be obtained with this equipment. It consists essentially of two reels mounted in a frame; during processing the film is wound from one reel to the other by a drive motor while the assembly is immersed in the processing bath. By following instructions carefully, an experienced operator can obtain fairly good quality results with rewind processing—except for a few feet at each end of the roll where the bulge at the area of attachment of the film to the reel produces a heavy cross-banding pattern. Following processing, the film is dried with roll-film dryer. Reel-type processing equipment is also available for processing short lengths of film.

The most efficient and economical method of processing black-and-white and color aerial films, where large volumes of film are handled, is in a continuous processing machine. Figure 6-49 shows a typical continuous roller-transport processor for black-and-white films. Although the purchase and maintenance costs are higher, the processed film quality is much better.

The Kodak Versamat film processor, model 11C-M, is a roller-transport machine that has automatic-threading capability. This machine is designed for the automatic and controlled processing and drying of black-and-white films and papers in long rolls or sheets, up to eleven inches wide. A similar processor (the Kodak Versamat film processor, model 1140), capable of higher transport speeds, is also available.

Processing done in roller-transport type processors offers the advantage of uniform treatment of all portions of the roll, freedom from banding, and lack of significant density variations from ends of the roll to the center. Emulsion speed and contrast characteristics are variable, depending on process conditons.

Aerial films can be processed in a wide variety of chemicals. However, for roller-transport processing, the formulations are somewhat different from those used for tray or rewind processing. The processor can be used with the same processing solutions for both film and paper by changing the bromide content of the developer. Similar processing machines with a larger number of tanks to accommodate the greater number of processing solutions are available to process color aerial films.

6.6.2.2 PROCESSING SHEET FILMS

As discussed in section 6.3.5, in the last few years emulsions coated on a dimensionally highly stable polyester base 7 mils (178 μm) thick have replaced photographic plates for many photogrammetric operations. These materials as well as other sheet films are processed in trays, tanks, rotary tube processors such as the Merz or Colenta rotary-tube processors, or in self-threading roller-transport processors such as one of the Kodak Versamat film processors or Kreonite Krematic processors. With manual tank-and-tray processing, proper agitation is very important to maintain fresh processing solutions at the film surface for optimum processed film quality. Where the volume of tank processing warrants it, gaseous (nitrogen) burst agitation provides high quality agitation. For good, reproducible tank and tray processing the reader is referred to the American National Standards Institute PH4.29-1975 "Methods for Manual Processing of Black and White Photographic Films, Plates and Papers."

6.6.2.3 PROCESSING DIAPOSITIVE PLATES

Diapositive plates are processed in trays or tanks as described in the previous section on sheet film processing. Follow the plate manufacturer's recommendations in selecting the proper plate, developer, and developing time to produce the desired density scale in the processed diapositive. *See* section 6.8.6 for density scale requirements of diapositives.

6.6.2.4 PROCESSING PAPERS

Black-and-white and color photographic paper may be processed in a variety of equipment. Paper sheets may be processed in trays, tanks,

FIGURE 6-48. ZEISS portable rewinding-type FE-120 film developing outfit.

FIGURE 6-49. KODAK VERSAMAT film processor, Model 11.

and rotary-tube processors as described for sheet films, although tray-processing sizes larger than 11 × 14 inches requires considerable skill on the part of the operator to process properly. Sheets may also be processed in a variety of drum processors such as the Kodak rapid color processor; models 11, 16-K, or 30A. Sheet or roll papers may be processed in a host of self-threading roller-transport processors; roll papers are also processable in continuous (leader thread-up required) paper processors such as the Kodak continuous paper processor, model 4D-P (black-and-white) or model 431 (color).

Processing color papers requires the same precision with respect to time, temperature, agitation, and replenishment rate as do color original materials. Deviations from the processing recommendations in these areas will often result in failure to achieve matched contrast in the three or more dye forming layers leading to improper color rendition in the highlights and shadows.

6.7 The Aerial Photographic Image

6.7.1 REQUIREMENTS OF THE PHOTOGRAPHIC IMAGE FOR PHOTOGRAMMETRY

The photogrammetrist must be able to: (1) consistently and uniformly identify on the photographic image the features that are to be mapped, (2) detect and point to the center of targets employed for aerotriangulation, and (3) delineate boundaries of imaged objects so that accurate measurements of size and area may be obtained. His ability to perform these tasks will be influenced by resolving power, MTF (section 6.2.4.3), granularity, and image contrast. Assuming proper exposure and development, the key factors that determine the image quality rendered by a complete photographic system are the camera lens, the film, and image motion. The

TABLE 6-7. PRINCIPAL METHODS OF SYSTEM PERFORMANCE EVALUATION

Prediction	Measurement	Comments
1. Resolution (R) $$\frac{1}{R_{sys}} = \frac{1}{R_{lens}} + \frac{1}{R_{film}} + \cdots$$	1.a. Resolution read from imaged targets in 1pr/mm b. Visual Edge Matching (VEM) c. Edge Spacing	1.a. Resolution values easily and rapidly determined from images to within ±15%. Target contrast is critical. b. Image edge matched to a calibrated edge of similar sharpness and contrast to determine resolution. Expensive microscope and appropriate edge matrices are required. c. Image edge fitted to the space between the elements of a high-contrast resolution target reticle installed in a microscope. Equipment expense relatively low. USAF technique.
2. MTF $(v)_{sys}$ = MET $(v)_{lens}$ × MTF $(v)_{film}$ × $\cdots$	2.a. Microdensitometer measurements of edge or line targets b. Mathematical operation on microdensitometer measurements of same terrain area recorded by small and large scale aircraft and/or satellite images.	2.a. No special man-made targets required. Complex analysis procedures. b. Imagery for the same time period at small and large scales, and digital microdensitometer are required. Involves complex computer routines.
3. System resolution from MTF a. Spatial frequency at the intersection of MTF and film TM curves b. Spatial frequency corresponding to a pre-determined threshold response point on the predicted system MTF.	3.a. Application of % difference in spatial frequency between predicted and measured MTF's at a specified threshold modulation level to resolution values determined from laboratory tests. b. Spatial frequency corresponding to a known threshold modulation.	3. Useful methods of determining system resolution from MTF's.

quality of first-generation aerial photographs and the contributions of the lens, film, and image motion can be determined by the methods listed in table 6-7 and illustrated in figures 6-50 and 6-51 (Hendeberg and Welander, 1963; Scott, Scott, and Shack, 1963; Jones, 1967; DeBelder, *et al.*, 1972; Welch, 1972b,c, 1974a,b, 1976; Corbett, 1974; Schowengerdt, Antos, and Slater, 1974; Welch and Halliday, 1975; Artishevskii and Chalova, 1976; Rosenbruch, 1977; and Tiziani, 1978). When considering the requirements of a photographic image for photogrammetric tasks, the photogrammetrist is interested in three measures of quality, the resolving power, detectability, and measurability of image detail (Welch, 1975, 1977; Norton, Brock, and Welch, 1977; Trinder, 1978; Ziemann, 1978).

6.7.1.1 RESOLVING POWER

The on-axis resolution values obtained with typical photogrammetric camera systems are determined by the film rather than the camera lens (Carman and Brown, 1970; Welch, 1972b). This assertion is easily confirmed by comparing listed resolving power photogrammetric and high-definition reconnaissance film with that obtained on photographs exposed on these films through Wild and Zeiss cameras under laboratory conditions, for example.

Kodak Film	Film RP (1.6:1 TOC)	Wild or Zeiss Camera with listed film (1.6:1 TOC)
2402/2405	50 lpr/mm	40 lpr/mm
3414	250	100

Under operational conditions, the maximum resolving power of photogrammetric camera systems, employed with some Kodak films such as 2405, 2402, 2443, 2445, and 2448 will range from approximately 20 to 30 lpr/mm and 35 to 55 lpr/mm for low- and high-contrast targets, respectively (figure 6-52).

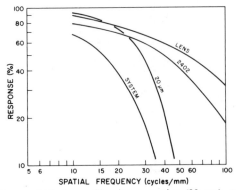

FIGURE 6-50. Lens, film, image motion (20 μm) and total-system MTF curves for a photogrammetric camera system. The system MTF is predicted by cascading the component curves.

The off-axis resolution is determined by the effects of lens aberrations that normally increase with angular distance from the optical axis. For typical, wide-angle camera systems, resolving power remains high to approximately 15° of off-axis; however, maximum resolutions may be limited to about 25 lpr/mm at 30° off the axis and 10 lpr/mm in the extreme corners—regardless of the film employed. For the narrower field angle of reconnaissance cameras, resolution variations due to aberrations are reduced. However, high-resolution systems are very susceptible to degradations because of improper correction of image motion and errors in exposure and processing. Consequently, resolving power is more predictable with photogrammetric cameras (Brock, 1976). The resolving power of modern photogrammetric cameras has been thoroughly investigated by Hakkarainen (1976).

6.7.1.2 DETECTABILITY

The interpretability of a photograph is determined by an observer's ability to detect and to identify small detail in the image (Charman, 1975). Unfortunately, it is extremely difficult to quantify the ability to identify an imaged object because of the many subjective factors involved. In comparison, the detectability of an imaged object at a known location can be evaluated simply (that is, the observer does or does not detect it), and statistics are readily accumulated. Consequently, detectability is a fairly unambiguous measure of image quality.

In considering detectability, two classes of image detail must be distinguished: (1) symmetrical and (2) linear. The detectability of symmetrical images such as squares and circles has been explored by MacDonald (1958), Carman and Charman (1964), and Welch and Halliday (1973). In the latter study, squares of decreasing size with a contrast of 1.6:1 at the camera lens were photographed under laboratory and controlled operational conditions with Wild and Zeiss cameras (figure 6-53). From evaluations of these photographs, it was possible to determine the representative detectability thresholds listed below.

Detectability Thresholds for Photographs Recorded With Wild and Zeiss Cameras

Kodak Film	Detectability Threshold (1.6:1)
3414, SO-242	10 μm
3410, 2402	20 μm
2443, 2424	30 μm

These values will vary with exposure and processing.

From these experiments it is evident that, with the exception of a film such as 3414, the film rather than the lens determines the detectability of small objects imaged near the optical axis. For

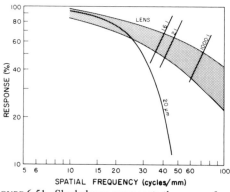
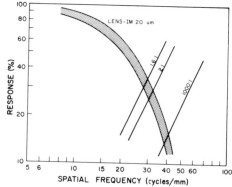

FIGURE 6-51. Shaded area represents the range of on-axis MTF's of Wild and Zeiss camera lenses determined by different investigators. Also shown are TM curves for 2402 film adjusted for target contrasts of 1000:1, 2:1 and 1.6:1 and an MTF for 20 μm of image motion. A static-system resolution value of about 40 lpr/mm is indicated for a low-contrast target by the intersection of the TM curve (1.6:1) with the lens MTF (6-51a) and 30 lpr/mm after the lens MTF's have been cascaded with that for 20 μm image motion (6-51b).

image locations 15° or more off-axis, the detectability threshold for the low-contrast squares is reduced to about 30 μm—regardless of the type of film—because of the effects of lens aberrations. The detectability of high-contrast squares, on the other hand, is considerably less affected by format position and the influence of lens aberrations; that is, the image is diffused but remains detectable. This is a reverse situation from resolution in which the *separation* between imaged lines of high contrast can be severely attenuated by small aberrations.

The detection of linear features (roads, railways, or pipelines) is sometimes cited as a measure of system performance. However, it is well known that the eye can integrate a series of image points that are individually below the threshold of detection (because of granularity of some other factor) to detect a linear feature. Charman (1965) had discussed the relationship between resolving power and the detectability of linear detail. He concludes that the minimum detectable image-line width is always less than that of the resolved distance and that, because of

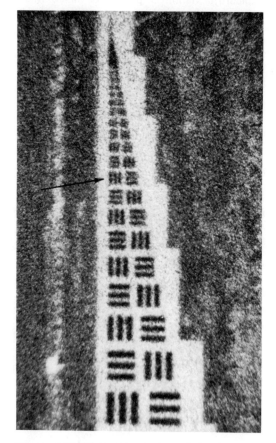

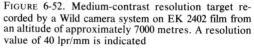

FIGURE 6-52. Medium-contrast resolution target recorded by a Wild camera system on EK 2402 film from an altitude of approximately 7000 metres. A resolution value of 40 lpr/mm is indicated

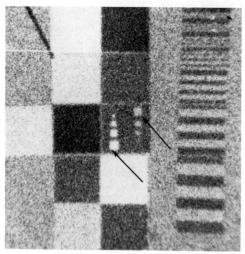

FIGURE 6-53. High- and low-contrast squares recorded on 2402 film at 1:24,000 scale with a Wild RC8 camera.

the interrelated actions of variables such as MTF, granularity, *gamma,* adjacency effects, target contrast, and the observer, a change in the resolving power by a factor of N does not change the minimum detectable linear width by $1/N$. Because of these factors, the detectability of linear objects is difficult to quantify although functions for this purpose have been suggested (Brock, 1970).

Based on both theoretical and empirical studies, it is estimated that low-contrast linear features of 2 to 4 μm width can be detected on photographs recorded on typical air survey films (*e.g.,* 2402) with Wild and Zeiss cameras (Welch, 1972a). The detectability of small, low-contrast, symmetric and linear features can be enhanced by utilizing films of low granularity processed to high *gammas*.

6.7.1.3 MEASURABILITY

The photogrammetrist frequently has two measurement tasks: (1) measurements to the center of the image of a symmetrically shaped object—such as pointing to the images of signalized control points in the aerotriangulation process, or (2) measurements to the imaged edges of an object—such as plotting the edges of a road, measuring the width of a building, or determining the area of a field. In order to assess the measurability of imaged objects, relationships between image quality and measurability must be established.

In aerotriangulation, an instrument operator is frequently required to place a black, circular measuring mark of 20 to 60 μm diameter in the center of the image of a bright circle, square, or cross. The reliability of the measured coordinates so obtained is influenced by several factors, including: (1) size and type of measuring mark, (2) size of the imaged signal, (3) viewing magnification, and (4) image quality (O'Connor, 1967). Contrast between the signal and background is also important; but, since signals are designed to produce high-contrasts, this factor can be considered optimized.

Hempenius (1964, 1969) discussed the factors influencing pointing precision in relation to the system spread function. To simplify his approach, system MTF's and spread functions are assumed to have Gaussian profiles, and his measure of objective image quality is taken as the steepest slope of the spread function. Since the maximum slope of the spread function occurs at the 2σ width of the spread function—at an intensity of 0.61—the 2σ-function width also can be used as a measure of image quality (figure 6-54). The advantage of gaussian-shaped profiles is that they can be convolved—similar to the cascading of MTF's, that is:

$$2\sigma \text{ image} = 2(\sigma^2_{\text{lens}} + \sigma^2_{\text{film}} + \dots)^{1/2}.$$

Large 2σ-spread-function values indicate blurred images. Typical 2σ image spread-function

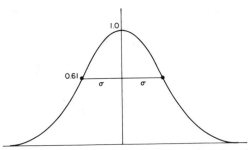

FIGURE 6-54. Width of 2σ of the gaussian-shaped spread function.

widths for photogrammetric camera systems range from 25 to 70 μm, with 40 μm an estimated average value.

In order to obtain reliable pointings, the measuring mark must be smaller than an imaged object leaving a small annulus around images of circular targets. For such targets, Hempenius has derived a "normalized pointing error curve" that can be used to estimate photogrammetric pointing precision (figure 6-55). This curve indicates the relationship between pointing error (standard deviation) in percent of the 2σ-value and an annulus width of some multiple of the 2σ-value. According to the curve, pointing errors within 5 percent of the 2σ-value can be obtained when the annulus width is from 1 to 4 times the 2σ-value. As the 2σ-value for typical photogrammetric systems will be on the order of 40 to 50 μm, pointing errors of 2 to 3 μm are indicated. This work clearly demonstrates that as image blur increases pointing errors also become larger.

Trinder (1971, 1973) extended these concepts by taking a theoretical approach to the establishment of relationships between pointing precision, spread functions, and MTF's. In this work, particular emphasis is given to determining the optimum size for circular signals and the relationship between signal size and the gaussian-shaped spread function of the total system. By taking advantage of previously derived data on pointing precision, the modulation

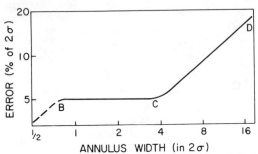

FIGURE 6-55. Normalized pointing error curve plotted as the relationship between pointing error in percent of the 2σ value and the annulus width as a multiple of the 2σ value (Hempenius, 1969).

sensitivity of the eye, and the Fourier transform relationships between spread functions, MTF's, and target frequency spectra, Trinder developed graphical relationships between pointing precision and the σ widths of the Gaussian-shaped spread function. Based on these theoretical studies, Trinder concluded that for a measuring mark of 1 mrad, the optimum ground-target size should be 3 to 4 times the σ width of the spread function—assuming σ is greater than 1 mrad.[5] For σ widths approaching zero, the ground-target size should be only slightly larger than the measuring mark. The precision of pointing was determined to be from 2 to 5 percent of the σ width, or from 0.5 to 1 percent of the ground-target size for a measuring mark of 1.0 mrad. These studies provide the photogrammetrist with an excellent basis for estimating the precision of pointing to signalized targets on imagery for which the spread function or MTF is known.

Measurements to edges are normally undertaken to determine the linear or areal dimensions of small imaged objects. The reliability of such measurements is determined by factors such as: (1) contrast, (2) edge sharpness, and (3) the instrument operator's subjective interpretation of the boundary. Welch (1972a), in a series of experiments involving measurements to the edges of resolution bar-targets recorded by photogrammetric camera systems, noted that for edges with density differences of $0.7-0.8$ units linear dimension errors of $3-7$ μm could be expected. For large spread functions, or when the object was recorded to give densities on the nonlinear portions of the D-log E curve, the linear dimension error increased. The precision (standard deviation) of these edge measurements averaged about 1.5 μm, which correlates with the findings of Thompson (1972) that the subjectivity of edge location by a comparator operator can amount to as much as 3 μm.

In a similar study (Martucci, 1972a,b), attempts were made to relate image quality to measurement errors by using data derived from microdensitometer and monocomparator measurements of resolution bar-target images of high, medium, and low contrast recorded under laboratory conditions with a photogrammetric camera system. Ghosh (1973) suggests that information of this type may prove useful in deriving a weighting function for photogrammetric observations.

Photogrammetrists and photo-interpreters are also faced with the task of deriving reliable area measurements from the images of small objects. Experiments involving thousands of comparator measurements around the perimeters of the images of small squares of high and low contrast, reveal that for a properly exposed target, the linear image dimension must approximate the

width of the system spread-function before a reliable indication of object shape and area can be expected.

6.7.1.4 USEFUL PRINCIPLES

When selecting, comparing, or evaluating a camera system, the photogrammetrist should be guided by the following principles.

1. The lens, film, and image motion are the most important factors governing image quality. Of these, the lens is a constant factor, and image motion will be determined by film speed and the appropriate exposure settings. Consequently, image quality is generally determined by the film characteristics.
2. Image motion of 20 μm or less will not significantly impair the quality of photography obtained with photogrammetric cameras.
3. Photogrammetric camera systems typically deliver image resolutions of 20 to 30 lpr/mm for low-contrast targets.
4. Film *gammas* of approximately 2.0 or higher will improve the detectability of low-contrast detail.
5. Film granularity is the most significant factor influencing the detectability of small, low-contrast detail.
6. Measurability of image detail correlates with image quality.
7. Techniques for assessing resolving power and the detectability and measurability of image detail are available for use by the photogrammetrist.

In summary, for maximum interpretability and measurability of detail, the photogrammetrist should select a film that combines the desired characteristics of good resolving power for low-contrast objects, minimum granularity, and adequate speed and contrast.

6.7.2 INSPECTION AND EDITING

For the inspection of processed aerial films, a table with a rectangular opal-glass top that is illuminated from below is used. It may be convenient to have the top slightly inclined toward the observer. The length of the illuminated area should be at least three times the length of a film frame. Fluorescent lamps are preferable as the light source, and sufficient ventilation should be provided to keep the working surface of the illuminator reasonably cool.

Brackets, attached to the ends of the illuminator and provided with winding cranks, are provided for holding and winding the film.

Any defects found during the inspection should be noted on the record sheet accompanying each roll of negative. In modern aerial cameras, some pertinent information regarding the flight is automatically recorded along the edge of the film frame. Additional information, such as date, project number, and a consecutive number, may be marked on each negative. Negatives at either end of a flight strip may be marked with additional data, such as the ap-

---

[5] At 10x viewing magnification 1 mrad equates to 25 μm.

proximate time of day and the approximate altitude. The information is usually located along the upper edge of the negative.

The editing is best done at a narrow table having a length about twelve times the length of a negative frame. Film-spool brackets with winding cranks are attached to the ends of the table, and a pair of rollers are located just inside the film brackets to keep the film clear of the table surface.

Editing of negatives can be done with either a metal or a rubber stamp, supplied with a special ink that transfers to the film base. The stamping is done on the base of the film so that the lettering will be right-reading on the print. The ink should contain a solvent that slowly attacks the back surface of the film so that the dye will adhere firmly, but not so firmly that it cannot be removed if corrections are required.

In inspection of negatives, it is important to distinguish between defects caused by handling before processing, and those resulting from faulty processing.

6.7.3 IMPERFECTIONS CAUSED PRIOR TO PROCESSING

Adverse atmospheric and ground conditions can cause poor exposure or improper contrast. For good-quality photographs, the following conditions should be avoided if possible: very hazy atmosphere, clouds and cloud shadows, smoky overlays, snow-covered ground, ice-locked harbors, and long shadows due to a low position of the sun in the sky.

Improper camera use can adversely affect the photographic result. The camera design should ensure positive positioning of the film in the focal plane, freedom from static or abrasion, especially from pressure plates, and adequate damping to eliminate vibrational effects. The camera lens should be set at the focal length for best average definition. Filters should be of good quality and in good condition; precautions should be taken to avoid deposition of dust or condensation of moisture or oil on the filter or the lens, and the field of view should be checked to be sure no part of the camera mount or the aircraft is included. Proper exposure requires adequate shutter speed to stop motion, and the correct aperture or diaphragm opening, to pass the correct amount of light while the shutter is open. Underexposed negatives are low in overall contrast and lacking in shadow detail. Overexposed negatives are dense, require long printing times, and may be lacking in highlight detail.

6.7.4 IMPERFECTIONS FROM PROCESSING AND HANDLING

Aerial photographic materials can be expected to produce high-quality results if stored, exposed, handled, and processed as recom-

mended. Assuming correct storage, exposure and handling, poor results can usually be traced to errors in chemical mixings or processing. Processing errors may result from improper process time, temperature, agitation, process control (solution replenishment rate), solution contamination, and washing. As described in section 6.5, the effects of poor washing are often time dependent and may not be obvious with freshly processed film.

6.7.4.1 SENSITOMETRIC IMPERFECTIONS

Space does not allow elaboration of the many kinds of sensitometric processing errors that can occur and their resultant defects. For color materials especially, the reader is urged to consult the processing and process control recommendations offered by the manufacturer. The causes of some processing defects are as follows.

Negatively processed black-and-white or color materials with low density and/or low contrast results from insufficient development time (too fast processor transport speed), or too low developer temperature, or from the process being out of control on the low side (weak or exhausted developer). Film with too high a density or contrast results from excessive temperature or development time, or from the process being out of control on the high side.

Reversally processed film behaves essentially in just the opposite way. Insufficient development in the first developer would result in greater silver available for development in the color developer with resulting higher density. Too low a density in the reversally processed film could result from overdevelopment in the first developer with less than the proper amount of silver available for development in the color developer.

With color films or papers negatively or reversally processed, excessive density can also result from inadequate bleaching that causes retained metallic silver. Inadequate bleaching can be caused by improper bleach pH, insufficient bleach time or temperature or insufficient bleach activity resulting from underreplenishment, improper bleach rejuvenation, or improper mixing.

Irregular fog may be due to light leaks in the film container, the camera, or the processing room. Streaks may be caused by improper agitation in processing. Uniform fog is usually caused by excessive development, abnormally high processing temperatures, or the use of over-age or improperly stored film.

6.7.4.2 PHYSICAL IMPERFECTIONS

A number of processing-related physical defects may be encountered such as the following.

Abrasion streaks—Fine streaks parallel to the edges of the film are usually the result of the emulsion rubbing against a rough surface in the camera.

They may also be produced by the films rubbing in the processing tank, on the workbench, or in other handling operations.

Fingerprints—To prevent fingerprints, the film should be handled before or during processing by the extreme edges only.

Pinholes—These marks are usually caused by dust on the film during exposure, or by chemicals, such as hypo particles, coming in contact with the emulsion before or during development. Tiny air bubbles that form on the film when it is first immersed in the developer and which act locally to prevent the developer from coming in contact with the emulsion also cause marks.

Developer streaks—Such streaks are associated with irregularities in agitation that may result, for example, from stopping the film during processing in rewind apparatus, or from winding a slack roll.

Drying marks—These marks are caused by uneven drying, usually because droplets or streamers of water have been left on one or both sides of the film. They may be avoided to a marked degree by rinsing the film in a wetting agent after washing. Another drying-related defect on some films with a gelatin backing is cross-lines on the gelatin-backed base side caused by the dryer rollers when there is excessive dryer temperature. The dryer temperature must be low enough when processing some gelatin-backed films so this will not happen.

6.8 Printing and Duplicating

6.8.1 CONTACT PRINTING

In contact printing, the original is placed in contact with the print material (film, paper, or plate) to produce a print the same size as the original but reversed in density. When a negative is printed emulsion-to-emulsion with a print material, the resulting print is said to be "right" reading; left-to-right in the negative will be reproduced as left-to-right in the print. When printing from a negative that was exposed emulsion-to-emulsion from a negative, if the print material is printed emulsion-to-emulsion, the resulting contact print will be "wrong" reading. To overcome this situation, the print can be made by exposing the print material with its emulsion in contact with the base of the second negative.

Contact printers designed for aerial negatives usually consist of a "light box" equipped with film-spool holders to accommodate either 250-foot or 500-foot rolls of film. These holders allow the film to be handled in the printing operation without cutting the roll of film into individual negatives.

The following two classes of printers are in common use throughout the industry.

6.8.1.1 THE SINGLE-FRAME CONTACT PRINTER

This is a printer with relatively high-intensity lamps. It may be equipped with as many as 32 separately controlled low-wattage argon lamps for black-and-white contact prints on contact-speed papers. This allows the operator to compensate for any unevenness in the negative caused by vignetting produced by a wide-angle lens. These lamps should be replaced by small, low-wattage tungsten lamps when working with color materials.

A newer type of printer in this class is the electronic dodging printer, manufactured by Log-Etronics, Inc., of Springfield, Virginia. This printer automatically compensates for any unevenness in the negative. The light source is a blue-phosphor cathode-ray-tube that requires an enlarging-speed photographic material for prints.

LogEtronics, Inc. offers a printer for black-and-white photography, as well as a color printer for use when working with color photography. If the original exposures were made on color film, always use a color printer when making black-and-white or color prints from color originals.

6.8.1.2 THE CONTINUOUS-STRIP CONTACT PRINTER

Two types of printers in this class are in general use today. The continuous electronic-dodging type manufactured by LogEtronics, Inc. can expose a 250-foot roll of negatives at rates of 20 feet-per-minute to about 90 feet-per-minute, and will produce automatically dodged prints of the same quality as the single-frame contact printer. Non-dodging printers that were manufactured during World War II are still in use for high-speed contact printing in long rolls. Both of the above types of printers require the use of enlarging-speed papers, and do not generally come equipped for printing color aerial photographs. LogEtronics, Inc. is now marketing a continuous color-contact printer of the electronic dodging type.

6.8.2 PROJECTION PRINTING

When prints other than contact size are required, or when prints are made in a plane not parallel to the plane of the negative, a projection printer is used. These printers are similar in principle to the ordinary enlarger, but they are specially constructed to be very rigid and precise; many handle rolls of negatives. Projection printers frequently incorporate means for tipping and tilting the baseboard that holds the print material, movement and tilting of the lens, and tilting and rotation of the negative holder. There are precise means for varying the distance between the lens and the negative, and between the lens and the baseboard to permit focusing.

Commercially available projection printers provide for a range of enlargement or reduction of approximately 12 to 1.

To obtain enlargements and reductions, the distances from the lens to the negative and from the lens to the copyboard have to be adjusted individually. This double change is usually done simultaneously and automatically by means of a device that connects the plane of the negative, the lens, and the plane of the baseboard (see chapter XIV). With these devices, it can be seen that such a projector is a rather complicated piece of equipment.

Fixed ratio reduction printers are becoming more and more popular in the industry. These printers produce approximately one-half size prints for photo indexes. These photo indexes are then only one-fourth the size of an index made from contact prints, and the photographic material requirements are reduced by 75%.

Most of the projection printers in use today are not equipped with the proper light source for making color prints.

Optical printing of diapositives is covered in section 6.8.6.

6.8.3 REQUIREMENTS FOR CONTACT & PROJECTION PRINTING

Contact papers are required for use with contact printers that are equipped with high-intensity white-light or argon lamps, and enlarging papers are required for use with electronic dodging printers. When printing by contact using a "point" light source, enlarging papers should be used because of the low-light level of the point light source. This is the preferred technique for printing wrong-reading negatives mentioned previously.

As mentioned in section 6.3.4, variable-contrast printing papers are offered with appropriate filters for controlling contrast. This allows prints from different contrast negatives to be printed on one specific type of paper. To take advantage of the variable-contrast capability of the papers, it is necessary to use a "white light" in the printer. While this type of paper can be exposed in an electronic dodging printer equipped with a "blue" light source, the variable contrast properties of the paper are negated by the blue light.

Panchromatic papers are useful for making black-and-white prints from color negatives where all of the detail in the three dye-layers of the color negative will be recorded in the final contact print. By the use of color filters over the light source, detail in any of the three layers can be emphasized or de-emphasized at will. It follows that a white light is required to take advantage of this when using panchromatic papers.

Both the variable-contrast and panchromatic papers are fast, or enlarging-speed papers, and are ideally suited for use on a color printer of the electronic dodging type.

Filtering is no problem, as most electronic dodging printers project only a small dot of light through a lens onto the film/paper combination. Usually large filters are required for use with regular multi-light contact printers.

Color materials are generally intended to be printed with a tungsten or tungsten-halogen lamp, often modified by acetate or dichroic color-correcting filters to balance the color rendition of the final print. Reference should be made to the manufacturer's publications for recommendations for printing a particular product.

6.8.4 COPYING

Generally, copies produced in a photogrammetric laboratory are to be made to a specified scale. This necessitates the use of a precise copying camera similar to the cameras used in the lithographic printing industry. The cameras range in size from small to extremely large. The small cameras will accept films up to about 11" × 14", and the largest will accept films up to 52" wide. The use of large photographic plates for precise copying has largely been abandoned because of the excellent dimensional stability of films with polyester base. The lenses used are specially designed for copying and are known as "process lenses." They are designed to be distortion-free, color corrected, and to give best quality negatives when used for making copies of flat work such as mosaics, photo indexes, or line drawings. Because of their design, they have relatively short depth-of-focus and depth-of-field. The highest resolution with this type of lens is usually obtained when exposures are made at one or two stops less than the largest aperture. Stopping the lens down to its smallest aperture will materially reduce the resolution of the lens.

6.8.5 DUPLICATION OF NEGATIVES

The reasons for duplicating negatives often dictate the method employed for making the duplicates. As described in section 6.3.3, two general types of film are available: printing films of the negative-to-positive-to-negative type, and printing films that provide duplication in one step. Both types are processed with conventional, three-stage processing. These films are available with highly dimensionally stable bases to minimize distortion in the duplicate.

6.8.5.1 NEGATIVE-TO-POSITIVE-TO-NEGATIVE DUPLICATION

A number of films are available for duplicating black-and-white negatives by going from negative to positive to negative. With this method an intermediate positive is made from which the final negative is made. If many negatives are required, or it is desired to use the original as little as possible to minimize the risk of damage to it, the negative-to-positive-to-negative method may be preferred. The emulsions are blue-sensitive and are coated on either 4-mil or 7-mil polyester film base to ensure a high degree of dimensional stability. Corrections can be made if the original negative is over or underexposed, and compensations can also be made for too little or too much contrast in the original negative. The negative-to-positive-to-negative system has the added advantage of better reproduction of tone in shadows and highlights than with the direct method.

6.8.5.2 ONE-STEP DUPLICATION

Duplication in one step is done with films that produce negatives directly from negatives (or positives from a positive) by contact printing in the conventional manner. They are processed in the conventional manner in conventional three-stage processing chemicals without reversal exposure during the process. The emulsion develops to high density without exposure, and with

FIGURE 6-56. Bausch & Lomb Balplex 9 inch × 9 inch reduction printer.

increasing exposure the density decreases. A printer with high-intensity lamps is required.

6.8.6 PRINTING DIAPOSITIVES

A positive picture on a transparent base (plastic or glass) is called a "diapositive," and is generally used in analog-type stereo-plotters. These diapositives are made either by contact, or in projection printers at reduced size, depending on the type of plotter in which they are to be used. The general principles used in the preparation of diapositives are the same regardless of the size or type of plotter in which they are to be used, but there are differences in the sensitometric characteristics required. In all cases, the diapositives must duplicate, as nearly as possible, the detail contained in the original pictures. If the diapositives are to be used in precise mapping, the geometrical, optical, and photographic conditions governing their preparation must be controlled with great precision. Some operators plot from the original stereoscopic pairs (negatives); however, only the production of diapositives will be discussed here.

6.8.6.1 DIAPOSITIVE PRINTERS

For some photogrammetric projectors, diapositives are made in reduction type printers. An example is shown in figure 6-56. The printer is basically a reducing camera of the highest precision that reduces the aerial photograph according to the nominal ratio:

$$\left(\frac{\text{focal length of camera}}{\text{principal distance of projectors}}\right).$$

More precisely, the ratio is based on the principal distance of the negative, rather than on the focal length of the camera.

The principal distance of a negative is the perpendicular distance from the internal perspective center to the plane of that particular finished negative. This distance is equal to the calibrated focal length of the taking camera corrected for film shrinkage or expansion, and maintains the same perspective angles at the internal perspective center to points on the finished negative as existed in the taking camera at the moment of exposure.

If the cone of rays emanating from a projector is to be similar to the original cone that entered the camera at the instant of exposure, the principal distance of the diapositive must be equal to the principal distance of the projectors. Therefore, provision is made for setting the reduction ratio precisely. If the reduction ratio were not accurately set, the projected cone of rays would be either slightly elongated or slightly flattened, with the result that the vertical scale of the model would be slightly in error with respect to the horizontal scale. With the proper reduction ratio, the vertical and horizontal scales are equal.

Another function of the diapositive printer is to compensate for distortion. The optical system of the printer introduces a compensating distortion such that the resultant distortion of camera, printer, and projector combined is as near zero as possible. Figure 6-57 illustrates the functions of distortion compensation and reduction of negative size in recovering the same angle between two projected rays as existed between the corresponding rays in nature.

Diapositives of the same size as the original can be made on contact printers when the distortion characteristics of the camera and plotter are matched. When using a printer equipped with an air bag exerting pressure to keep the photographic materials in good contact, care should be taken that the bag is inflated to the correct pressure. Under- or over-inflation can cause distortion when printing diapositives.

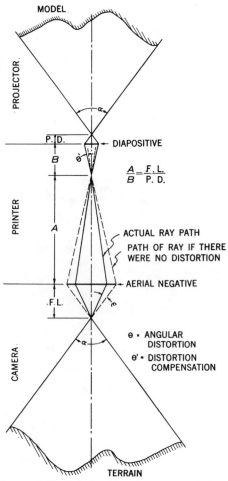

FIGURE 6-57. Relationship of camera, printer, and projector. (U.S. Geological Survey).

6.8.6.2 DIAPOSITIVE CONTRAST

An important factor in the selection of printing exposure and development is the contrast of the positive and its characteristics when processed in the recommended developers.

Diapositive plates are supplied in different contrast grades. The contrast of each grade can be controlled (by modification of development) to give sufficient versatility to the printing operation so that all ranges of densities can be satisfactorily reproduced. *See* section 6.3.5.

The best reproduction of a negative is obtained, in part, by controlling the contrast of the positive plate. A low-contrast negative has a short range of densities and requires a moderately high contrast material to produce the density range necessary to a good diapositive. Conversely, a high-contrast negative has a long range of densities and requires a low-contrast material in order not to exceed the range essential in a good diapositive.

As a general guide, a range of 0.70 is best for projection plotters, and a higher range of approximately 1.20 is more suitable for direct-viewing plotters. The range is calculated by subtracting the minimum density (D-Min) from the maximum density (D-Max). The D-Min on either type of diapositive should be kept at about .30 to get repeatable results in the printing process. Up-to-date recommendations for exposing and processing are available from the manufacturers. 3.0 - 1.0 o

BIBLIOGRAPHY

Adelstein, P. Z., *Dimensional Stability of ESTAR-Base Films*, Photogrammetric Engineering, 38: 55–64 (1972).

—— and D. A. Leister, *Nonuniform Dimensional Changes in Topographic Aerial Film*, Photogrammetric Engineering, 29: 149–161 (1963).

——, P. R. Josephson and D. A. Leister, *Nonuniform Film Deformational Changes*, Photogrammetric Engineering, 32: 1028–1034 (1966).

——, C. L. Graham and L. E. West, *Preservation of Motion-Picture Color Films Having Permanent Value*, Journal of the Society of Motion Picture and Television Engineers, 79: 1011–1018 (1970).

—— *Wedge Brittleness Test for Photographic Film*, Photographic Science and Engineering, 1: 63–68 (1957).

Altman, J. H. and R. C. Ball, *On the Spatial Stability of Photographic Plates*, Photographic Science and Engineering, 5: 278–282 (1961).

American National Standard *Dimensions for Photographic Dry (glass) Plates*, PH1.23-1974.

—— *Method for Determining Curl of Photographic Film*, PH1.29-1971.

—— *Method for Determining the Dimensional Change Characteristics of Photographic Films and Papers*, PH1.32-1973.

—— *Method for Determining the Speeds of Monochrome Photographic Negative Films for Aerial Photography*, PH2.34-1969.

—— *Method for Determining the Resolving Power of Photographic Materials*, PH2.33-1969.

—— *Methods for Manual Processing of Black-and-White Photographic Films, Plates, and Pa* PH4.29-1975.

——*Methylene Blue Method for Measuring Residual Chemicals in Films, Plates, and Papers*, PH4.8-1971.

—— *On Photographic Processing Effluents*, PH4.37-1975.

—— *Practice for Storage of Processed Safety Photographic Film Other than Microfilm*, PH1.43-1971.

—— *Practice for Storage of Processed Photographic Plates*, PH1.45-1972.

—— *Sensitometry of Photographic Papers*, PH2.2-1966, reaffirmed 1972.

—— *Sensitometric Exposure and Evaluation Method for Determining Speed of Color Reversal Films for Still Photography*, PH2.21-1971.

—— *Simulated Incandenscent Tungsten Sources for Photographic Sensitometry*, PH2.35-1969.

—— *Simulated Daylight Source for Photographic Sensitometry*, PH2.29-1967.

American National Standard *Specifications for Photographic Film for Archival Records, Silver-Gelatin Type, on Cellulose ESTAR Base*, PH1.28-1973.

—— *Specifications for Photographic Film for Archival Records, Silver-Gelatin Type, on Polyester Base*, PH1.41-1973.

—— *Specification for Testing the Photographic Inertness of Construction Materials Used in Photographic Processing*, PH4.31-1962, reaffirmed 1970.

—— *Terms, Symbols, and Notation for Optical Transmission and Reflection Measurements* (Optical Density) PH2.36-1974.

Artishevskii, V. I., and V. A. Chalova, *The Determination of the Image Contrast in Optophotographic Systems at their Resolution Limit*, Soviet Journal of Optical Technology, 42: 299–301 (1976).

Aviphot Films, Agfa-Gavaert, Inc., Teterboro, N.J.

The Biological Treatment of Photographic Processing Effluents, Kodak Publication J-46, Eastman Kodak Company, Rochester, New York (1972).

Black-and-White Processing for Permanence, Kodak Publication J-19, Eastman Kodak Company, Rochester, New York (1970).

Brock, G. C., *Image Evaluation for Aerial Photography*, Focal Press Limited, New York, New York (1970).

Brock, G. C., *The Possibilities for Higher Resolution in Air Survey Photography*, Photogrammetric Record, 8: 589–609 (1976).

Brock, G. C., D. J. Harvey, R. J. Kohler, E. P. Myskowski, *Photographic Considerations for Aerospace*, Itek Corporation, Lexington, Massachusetts (1966).

Brucklacher, W. A. and W. Luder, *Untersuchung Uber die Schrumpfung von Messfilmen und Photographischen Platten-material (Investigation Concerning the Shrinkage of Topographic Film and Photographic Plates)*, Deutsche Geodatische Kommission bei der Bayerischen Akademie der Wissenschafen, Munchen, Applied Geodesy, Series B, 31: (1956).

Burnham, J. M. and P. R. Josephson, *Color Plate Metric Stability*, Photogrammetric Engineering, 35: 679–685 (1969).

Calhoun, J. M. and D. A. Leister, *Effect of Gelatin Layers on the Dimensional Stability of Photographic Films*, Photographic Science and Engineering, 3: 8–17 (1959).

—— L. E. Keller and R. F. Newell, Jr., *A Method*

for Studying Possible Local Distortions in Aerial Films, Photogrammetric Engineering, 29: 661–672 (1960).

—— *The Physical Properties and Dimensional Stability of Safety Aerographic Film*, Photogrammetric Engineering, 13: 163–331 (1947).

—— P. Z. Adelstein and J. T. Parker, *Physical Properties of ESTAR Polyester-Base Aerial Films for Topographic Mapping*, Photogrammetric Engineering, 27: 461–470 (1961).

—— Carman, P. D. and J. F. Martin, *Causes of Dimensional Changes in ESTAR-Base Aerial Film Under Simulated Service Conditions*, The Canadian Surveyor, 22: 238–246 (1968).

—— Carman, P. D., *Dimensional Changes in Safety Topographic Aero Film Under Service Conditions*, Canadian Journal of Research, F24: 509–517 (1946).

Carman, P. D. and H. Brown, *Resolution of Four Films in a Survey Camera*, Canadian Surveyor, 24: 550–560 (1970).

—— and W. N. Charman, *Detection, Recognition, and Resolution in Photographic Systems*, Journal of the Optical Society of America, 54: 1129–1130 (1964).

Characteristics of KODAK Aerial Films, Kodak Publication M-57, Eastman Kodak Company, Rochester, New York (1976).

Charman, W. N., *Visual Factors in Photographic Detection, Recognition, and Resolution Tasks, Part I—Resolution*, Photographic Science and Engineering, 19: 228–234 (1975).

—— *Resolving Power and the Detection of Linear Detail*, Canadian Surveyor, 19: 190–205 (1965).

Collins, R. B. and C. H. Giles, *Colour Photography*, Journal of the Society of Dyers and Colourists, 68: 421–457 (1952).

Color as Seen and Photographed, Kodak Publication E-74H, Eastman Kodak Company, Rochester, New York (1972).

Construction Materials for Photographic Processing Equipment, Kodak Publication K-12, Eastman Kodak Company, Rochester, New York (1973).

Control Techniques in Film Processing, Society of Motion-Picture and Television Engineers, New York, New York (1960).

Coote, J. H., Photofinishing Techniques, Focal Press, London (1970).

Corbett, F. J., *Image Evaluation of the Skylab Multispectral Photographic Facility*, SPIE Image Assessment and Specification Seminar Proceedings, 46: 239–246 (1974).

Cox, Arthur A., Photographic Optics, Focal Press, London (1974).

Dagon, T. J., *Photographic Processing Effluent Control*, Journal of Applied Photographic Engineering, 4: 62–71 (1978).

Dainty, J. C., *Methods of Measuring the Modulation Transfer Function of Photographic Emulsions*, Optica Acta, 18: 795–813 (1971).

Dawson, G. H. and W. F. Vogelsong, *Response Functions for Color Densitometry*, Photographic Science and Engineering, 5: 461–468 (1973).

DeBelder, M., R. A. Jones, A. L. Sorem, and E. Welander, *Photographic Modulation Transfer Functions*, Geographical Society Office of Sweden, NR A39 (1972).

DUPONT CRONAR Aerial Films, E. I. duPont de Nemours and Company, Wilmington, Delaware.

Film, Photographic, Aerial, Black-and-White, Military Specification MIL-F-32G (1973).

Fritz, N. L., *Available Color Aerial Photographic Materials*, Photogrammetric Engineering, 40: 1423–1425 (1974).

—— *Basic Principles of Color Photography and Color Aerial Photography*, Proceedings of the Workshop on Aerial Color Photography in the Plant Sciences, pp 1–15, 1969.

—— *New Color Films for Aerial Photography*, Proceedings of the Third Biennial Workshop on Color Aerial Photography in the Plant Sciences, pp 44–68 (1971).

—— Optimum Methods for Using Infrared-*Sensitive Color Films*, Photogrammetric Engineering, 33: 1128–1138 (1967).

—— *The Use of Aerial Photography for Hydrobiological Investigations*, Proceedings of Symposium on Hydrobiology sponsored by the American Water Resources Assn., Urbana, Illinois, pp 255–261 (1970).

GAF Photographic Technical Bulletins, GAF Corporation, New York, New York.

Ghosh, S. K., *Image Quality versus Metric Capability*, Photogrammetric Engineering, 39: 1179–1186 (1973).

Gibson, J. W., *Photographic Water Conservation and Reclamation Processes Study*, Technical Report AFAL-TR-72-273, Air Force Systems Command (1972).

Graininess and Granularity, Tech Bits, 1974—Vol. 3, 1975—Vol. 1, Eastman Kodak Company, Rochester, N.Y.

Hakkarainen, J., *On the Use of the Horizontal Goniometer in the Determination of the Distortion and Image Quality of Aerial Wide-Angle Cameras*, Ph.D. Dissertation, Helsinki University of Technology, Finland (1976).

Hallert, B., *Determination of the Flatness of a Surface in Comparison with a Control Plane*, Photogrammetric Record, 3: 265–274 (1960).

Herman, W. E. Jr., *Recent Developments in Aerial Film*, Photogrammetric Engineering, 27: 151–154 (1961).

Helming, R., *Control of and Improvement On a Phototheodolite*, Photogrammetria, 17: 23–27 (1960).

Hempenius, S. A., *Aspects of Photographic Systems Engineering*, Applied Optics, 3: 45–53 (1964).

—— *The Role of Image Quality in Photogrammetric Pointing Accuracy*, Final Report to European Research Office, U.S. Army, Contract No. DA-91-591-EUC-3721 (1969).

Hendeberg, L. O. and E. Welander, *Experimental Transfer Characteristics of Image Motion and Air Conditions in Aerial Photography*, Applied Optics, 2: 379–386 (1963).

Hothmer, J., *Possibilities and Limitations for Elimination of Distortion in Aerial Photographs*, Photogrammetric Record, 2: 426–445 (1958) and 3: 60–81 (1959).

How Safe is Your Safelight, Kodak Publication K-4, Eastman Kodak Company, Rochester, New York (1975).

Ilford Product Information, Ilford Inc., Paramus, New Jersey.

In Support of Clean Water—Disposing of Effluents from Film Processing, Kodak Publication J-44, Eastman Kodak Company, Rochester, New York (1972).

Introduction to Color-Process Monitoring, Kodak Publication Z-99, Eastman Kodak Company, Rochester, New York (1975).

Jackson, J. E., *Statistics in Processing Control*, Journal of Applied Photographic Engineering, 2: 217–221 (1976).

Jaksic, Z., *Deformation of ESTAR-Base Aerial Films*, Photogrammetric Engineering, 38: 285–296 (1972).

James, T. H. (editor) *The Theory of the Photographic Process*, Fourth Edition, Macmillan Publishing Company (1977).

Jones, R. A., *An Automated Technique for Deriving MTF's from Edge Traces*, Photographic Science and Engineering, 11: 102–106 (1967).

KODAK *B/W Photographic Papers*, Kodak Publication G-1, Eastman Kodak Company, Rochester, New York (1973).

KODAK Data for Aerial Photography, Kodak Publication M-29, Eastman Kodak Company, Rochester, New York (1976).

LaPerle, R. L., *Removal of Metals from Photographic Effluent by Precipitation with Sodium Sulfide*, Journal of the Society of Motion-Picture and Television Engineers, 85: 206–216 (1976).

Ligterink, G. H. and R. Zijlstra, *The Metrical Differences Between Several Contact Prints on Glass, Made in Flow-Production, From One and the Same Film Negative*, Photogrammetria, 24: 23–28 (1969).

Lauroesch, T. J., G. G. Fulmer, J. R. Edinger, G. T. Keene, and T. F. Kerwick, *Threshold Modulation Curves for Photographic Films*, Applied Optics 9: 875–887 (1970).

Macdonald, D. E., *Resolution as a Measure of Interpretability*, Photogrammetric Engineering, 24: 58–62 (1958).

Manual of Color Aerial Photography, John T. Smith, Editor, American Society of Photogrammetry, Falls Church, Virginia (1968).

Martucci, L. M., Image Quality Effects on Image-Geometry of a Mapping Camera, Department of Geodetic Science Report No. 172, The Ohio State University, Columbus, Ohio (1972).

—— *Image Quality and Image Geometry*, Photogrammetric Engineering, 38: 274–276 (1972).

Mason, L. F. A., *Photographic Processing Chemistry*, John Wiley and sons (1975).

Meehan, R. J., *Increasing Productivity in Photogrammetry*, Photogrammetric Engineering and Remote Sensing, 41: 753–759 (1975).

Michener, B. C., *Drying of Processed Aerial Films*, Photogrammetric Engineering, 29: 321–326 (1963).

McNeil, G. T., *Film Distortion*, Photogrammetric Engineering, 67: 605–609 (1951).

Moser, J. S. and N. L. Fritz, *High-Definition Color Films for Terrain Photography*, Photographic Science and Engineering, 19: 243–246 (1975).

Neblette, C. B., *Photography, Its Materials and Processes*, Sixth Edition, D. Van Nostrand Co. Inc. (1977).

Norton, C. L., G. C. Brock, and R. Welch, *Optical and Modulation Transfer Functions*, Photogrammetric Engineering and Remote Sensing, 43: 613–636 (1977).

O'Connor, D. C., *Some Factors Affecting the Precision of Coordinate Measurement on Photographic Plates*, Photogrammetria, 22: 77–97 (1967).

Oswal, H. L., *Flexure of Photographic Plates Under Their Own Weight and Some Modes of Support and Consequent Photogrammetric Errors*, Photogrammetric Record, 2: 130–144 (1956).

Photolab Design, Kodak Publication K-13, Eastman Kodak Company, Rochester, New York (1974).

Plates, Photographic Diapositive, Glass, Black-and-White, Federal Specification GG-P-450A (1970).

Processing Chemicals and Formulas, Kodak Publica-

tion J-1, Eastman Kodak Company, Rochester, New York (1973).

Process Monitoring of KODAK Black-and-White Films, Kodak Publication Z-126, Eastman Kodak Company, Rochester, New York (1972).

Properties of KODAK Materials for Aerial Photographic Systems, Vol. I: *KODAK Aerial Films and Photographic Plates*, Kodak Publication M-61, Eastman Kodak Company, Rochester, New York (1972).

Properties of KODAK Materials for Aerial Photographic Systems, Vol. II: *Physical Properties of KODAK Aerial Films*, Kodak Publication M-62, Eastman Kodak Company, Rochester, New York (1972).

Properties of KODAK Materials for Aerial Photographic Systems, Vol. III: *Physical and Chemical Behavior of KODAK Aerial Films*, Kodak Publication M-63, Eastman Kodak Company, Rochester, New York (1974).

Recovering Silver from Photographic Materials, Kodak Publication J-10, Eastman Kodak Company, Rochester, New York (1972).

Rosenbruch, K. J., *Considerations Regarding Image Geometry and Image Quality*, Photogrammetria, 33: 155–169 (1977).

Schmid, H. H., *Application of Photogrammetry in Three-Dimensional Geodesy*, XI Congress, International Society of Photogrammetry, Lausanne (1968).

Schowengerdt, R. A., R. L. Antos, and P. N. Slater, *Measurement of the Earth Resources Technology Satellite (ERTS-1) Multi-spectral Scanner OTF from Operational Imagery*, SPIE Image Assessment and Specification Seminar Proceedings, 46: 247–257 (1974).

Scott, F., R. M. Scott, and R. V. Shack, *The Use of Edge Gradients in Determining Modulation Transfer Functions*, Photographic Science and Engineering, 7: 345–349 (1963).

SI Units and Recommendations for the Use of Their Multiples and of Certain Other Units, ISO 1000-1973.

Slater, P. N., *Photographic Systems for Remote Sensing*, Manual of Remote Sensing, American Society of Photogrammetry, Falls Church, Virginia (1975).

SPSE Handbook of Photographic Science and Engineering, W. Thomas, Jr., Editor, John Wiley and Sons (1973).

Starbird, H. A., *Study of Emulsion Shift on Ballistic Camera Plates*, Technical Documentary Report No. MTC-64-4, Air Force Missile Test Center, Patrick AFB, Florida (1964).

Swanson, L. W., *Aerial Photography and Photogrammetry in the Coast and Geodetic Survey*, Photogrammetric Engineering, 30: 699–726 (1964).

Tarkington, R. G. and A. L. Sorem, *Color and False Color Films for Aerial Photography*, Photogrammetric Engineering, 29: 88–95 (1963).

Tiziani, H., *Image Quality Criteria for Aerial Survey Lenses*, Proceedings of the ISP Commission I Symposium on Data Acquisition and Improvement of Image Quality and Image Geometry, Tokyo, Japan (1978).

Thompson, L. G., *The Use of Edges of Photographic Images as Specifiers of Image Quality*, Department of Geodetic Science Report No. 179, The Ohio State University, Columbus, Ohio (1972).

Trinder, J. C., *Pointing Accuracies to Blurred Signals*, Photogrammetric Engineering, 37: 193–202 (1971).

—— *Pointing Precision, Spread Functions, and*

MTF, Photogrammetric Engineering, 39: 863–874 (1973).

—— *Measurability and Interpretability of Photogrammetric Details*, Proceedings of the ISP Commission I Symposium on Data Acquisition and Improvement of Image Quality and Image Geometry, Tokyo, Japan (1978).

Umbach, M. J., *Color for Metric Photogrammetry*, Photogrammetric Engineering, 34: 265–272 (1968).

Umberger, J. Q., The *Fundamental Nature of Curl and Shrinkage in Photographic Films*, Photographic Science and Engineering, 1: 69–73 (1957).

Water Conservation in Photographic Processing, Kodak Publication S-39, Eastman Kodak Company, Rochester, New York (1973).

Welch, R., *Quality and Applications of Aerospace Imagery*, Photogrammetric Engineering, 38: 379–398 (1972a).

—— *The Prediction of Resolving Power of Air and Space Photographic Systems*, Image Technology, 14: 25–32 (1972b).

—— *Photomap Image Quality*, Cartographic Journal 9: 87–92 (1972c).

—— *Modulation Transfer Functions*, Photogrammetric Engineering, 38: 247–259 (1972d).

—— *MTF Analysis Techniques Applied to ERTS-1 and Skylab-2 Imagery*, SPIE Image Assessment and Specification Seminar Proceedings, 46: 258–262 (1974a).

——*Skylab-2 Photo Evaluation*, Photogrammetric Engineering, 40: 1221–1224 (1974b).

—— *Photogrammetric Image Evaluation Techniques*, Photogrammetria, 31: 161–190 (1975).

—— *Skylab S-190B ETC Photo Quality*, Photogrammetric Engineering and Remote Sensing, 42: 1057–1060 (1976).

—— *Progress in the Specification and Analysis of Image Quality*, Photogrammetric Engineering and Remote Sensing, 43: 709–719 (1977).

—— and J. Halliday, *Imaging Characteristics of Photogrammetric Camera Systems*, Photogrammetria, 29: 1–43 (1973).

—— and J. Halliday, *Image Quality Controls for Aerial Photography*, Photogrammetric Record, 8: 317–325 (1975).

Young, M. E. and H. Ziemann, *Film Diapositive Deformation*, Photogrammetric Engineering, 38: 65–69 (1972).

Ziemann, H., *Progress Report of the Working Group on Image Geometry*, Proceedings of the ISP Commission I Symposium on Data Acquisition and Improvement of Image Quality and Image Geometry, Tokyo, Japan (1978).

CHAPTER VII

Planning and Executing the Photogrammetric Project

Author-Editor: JOHN E. COMBS

Contributing Authors: JAMES W. CRABTREE, JINDO KIM, MAURICE E. LAFFERTY

IT IS AXIOMATIC that every photogrammetric mapping project is unique. Although the overall method may not vary, the sub-operations or steps that must be accomplished to fulfill a requirement will create certain obstacles and/or situations which will test the ability and experience of the photogrammetrist. This chapter will deal with the basic considerations in conceiving, planning, and executing a successful project, while providing an outline of the basic steps in the photogrammetric process and briefly describing the products derived from that process.

A photogrammetric project may be as basic as the production of a photographic enlargement from an aerial negative, or so complex as to involve a full range of photogrammetric methods and techniques. Most projects fall between these two extremes. Any project, regardless of size or difficulty, must be carefully planned and executed in such a manner as to take advantage of the full potential of photogrammetry for the benefit of the user.

7.1 Basic Elements of the Photogrammetric Mapping Process

Photogrammetric methods have, to a large degree, replaced the planetable method and other ground survey techniques. From the mapping programs performed by governmental agencies to making a map for a bridge site, applied photogrammetry is utilized in a wide variety of mapping projects.

The basic operations in conducting a photogrammetric mapping are:

Photography: obtaining suitable photography for mapping;
Control: obtaining sufficient control through field surveys and/or extension by photogrammetric methods;
Map Compilation: the plotting of planimetric and/or topographic features by photogrammetric methods;
Map Completion: the refinement of the map editing in the office and further, special surveys in the field; and
Final Map Drafting: the completion of the map by drafting/scribing.

7.1.1 BASIC ELEMENTS OF PLANNING FOR A PHOTOGRAMMETRIC MAPPING PROJECT

The first step, as given in the preceding outline and as implied in figure 7-1, is to prepare a complete plan that will make it possible to start the actual mapping operations and that will guide these operations as they are carried out. The plan includes the following *essential* steps.

Conversion of Requirements to Project Specifications. The requirements that are responsible for the project are converted into numbers that specify the region to be mapped, the scale at which it is to be mapped, the accuracy of the final map, the date by which the map should be completed, the approximate cost of the project (or an upper limit on the cost).
Gathering of Materials and People for the Planning. Photographs, maps, survey data, instruments, and personnel are assembled.
Determining Specifications and Conditions for Operations. The materials assembled in the preceding step are analyzed to determine where control should be established, where and how much photography is necessary, what kind of equipment must be used.
Preparing Final Plans. Scheduling; instructions for surveying, photography, compilation, and quality control. This includes preparation of contracts.
Costing and Replanning. This step is actually carried out, in part, along with the preceding steps. If, however, the results of costing are not satisfactory, the plan must be redone until costs are within limits set by requirements.

7.1.2 METHODS AND MATERIALS USED FOR PLANNING

After the first step, that of converting the sometimes unquantified requirements into specifications, the basic information essential to making a plan must be gathered. This information is primarily in the form of (a) aerial photography of the region to be mapped; (b) old maps of the region; and (c) survey data showing the

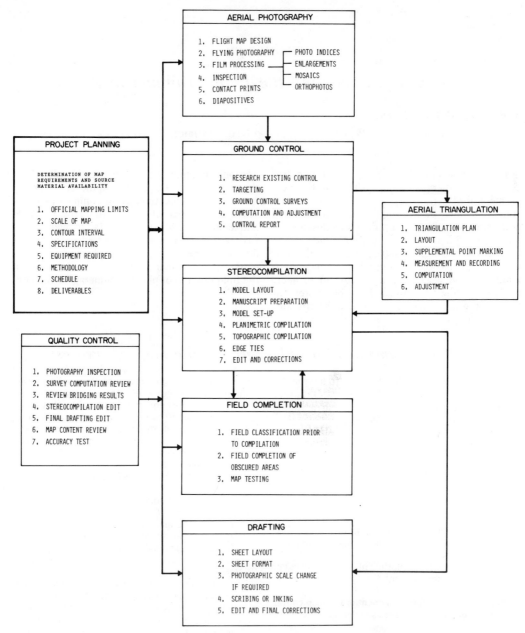

FIGURE 7-1. Basic elements of a photogrammetric mapping project.

locations of horizontal and vertical control in the region (chapter VIII), the accuracy of the control, and its accessibility. Less often, it consists of original survey notes, engineering data on structures in the region, and reports of people familiar with the region. To evaluate this information, experienced photogrammetrists must be available.

7.1.2.1 PHOTOINTERPRETATION

Aerial photography is the best available record of ground surface at the time of exposure. A photointerpreter systematically studies photographs in conjunction with other available materials such as maps and reports and as a result of his study identifies the details appearing in the photographs. Success in photointerpretation largely depends on the training and experience of the interpreter, characteristics of objects to be studied, and quality of photographs being used. Proper selection of emulsion, scale, season of flight, and time of day ensure aerial photographs useful for interpretation.

Stereoscopic viewing, particularly with magnification, is normally superior to monoscopic viewing for interpretation. Many different types of stereoscopes are available which provide very high magnification.

7.1.2.2 PHOTOGRAPHS

Photographs are used, in planning, in several forms: as individual photographs, as stereoscopic pairs, or in assemblages or mosaics as substitutes for or supplements to maps. The photographs in a mosaic may be unrectified, rectified, or differentially rectified (orthoscopic).

7.1.2.2.1 SINGLE AND PAIRED PHOTOGRAPHS

For photointerpretation, photographs are usually used in stereoscopic pairs. Single photographs are used only when stereoscopic pairs are not available, where suitable viewing equipment is not available, or when monoscopic viewing is superior to stereoscopic viewing (a situation that does sometimes occur). In either case, contact prints are usually satisfactory. However, enlargements are more satisfactory for monoscopic viewing without a magnifying device. Enlargements prepared from high-altitude photography are particularly useful, providing a basis for planning routes or for evaluating environmental factors.

Color photographs are generally superior to black-and-white photographs for photointerpretation, particularly when vegetation and land-use are important. For some kinds of terrain, color-infrared may be more suitable than ordinary color, although it is also more costly. It is probably a good idea to start with an inspection of available black-and-white photographs, and decide from this where additional photographs in color or color-infrared are needed.

7.1.2.2.2 PHOTOMOSAICS

A photomosaic is an assemblage of two or more individual overlapping photographs to form a single picture of an area. For flat terrain or for an area which does not require a high standard of accuracy, mosaics provide a great deal of information at minimal cost.

Mosaics are made by assembling photographs to fit ground control details on existing maps, or simply enlarged to a common scale to provide what are called controlled, semicontrolled or uncontrolled mosaics. The individual photographs are generally rectified before assembling.

Although unrectified or rectified mosaics are similar to maps in many respects and often used as substitutes for maps, they suffer the same limitations as contact prints and enlargements, in that it is impossible to eliminate the effects of relief displacement. In cases where a greater degree of accuracy is required, the use of orthophotos is required.

7.1.2.2.3 ORTHOPHOTOGRAPHS

To a casual observer, an aerial photograph with its recording of ground detail is considered equivalent to a map. However, an untreated aerial photograph is geometrically distorted to such an extent that an accurate map cannot be simply made from it. An aerial photograph results by central projection, through the perspective center of a lens onto a plane. A map, on the other hand, is constructed by radial projection onto a sphere and then onto a plane. Orthophotography has been developed to overcome the shortcomings of photographs which show scenes severely distorted by radial displacement even after rectification.

Normally, orthophotographs are prepared by differential rectification using modified versions of photogrammetric mapping instruments. While orthophotographs cost more than enlargements, the cost is generally less than that of planimetric or topographic maps.

7.1.2.3 QUASI-PHOTOGRAPHIC AND NONPHOTOGRAPHIC MAPS

Photomosaics can be assembled by the planner from sets of photographs, and are crude, but often completely adequate, maps of the region to be mapped by the project. More often, greater accuracy is needed than photomosaics, however well controlled and rectified, can provide. Then recourse must be had to maps in which the horizontal and/or vertical distances have their accuracies provided by and guaranteed by extensive ground surveys. When available, such maps should always be consulted. They may be partially photographic, with the planimetric information being provided by orthophotographic photographs and the vertical information provided by contours, or the information may be provided by completely nonphotographic drawings.

7.1.2.3.1 ORTHOPHOTOGRAPHIC MOSAICS WITH TOPOGRAPHIC OVERLAY

This kind of map has topographic information such as contours and spot elevations superimposed on an orthophotograph. Its cost is normally lower than that of a line map, but depends on the kind of terrain depicted. The orthophotographic mosaic is prepared. Contours are compiled on a separate plate and normally enhanced by final drafting or scribing. The orthophotographic negative and the scribecoat are registered one to the other and a composite positive produced photographically in final form.

7.1.2.3.2 PLAN-AND-PROFILE SHEETS

For design of such things as highways, power lines, water lines, gas lines, and so forth, plan-and-profile sheets are useful. The planimetric and/or topographic map is prepared photogrammetrically. In areas of moderate relief, an orthophoto mosaic is sometimes utilized, instead of line mapping, for the plan section of a sheet. The designs centerline is then plotted and prominent features intersecting this line are assigned station numbers. The profile along the centerline is determined by photogrammetric spot elevations and contours, and plotted, usually at an exaggerated vertical scale in the profile section of the sheet.

7.1.2.3.3 PLANIMETRIC MAPS

Planimetric maps are line maps with natural and manmade features shown without elevations. In a broad sense, road maps can be included in this category, along with tax maps, plats and inventory maps. Planimetric maps can be prepared as a by-product of topographic mapping. During the process of compilation for topography, planimetric features are drawn on a separate sheet of material, or after the completion of compilation, planimetric maps can be separated in the process of final drafting. For planning and design a subdued pictures of existing features can be produced by the halftone process.

Planimetric maps maintained by public-utility companies show the outlines of administrative boundaries as well as features such as poles, transformers, manholes and valves. Real estate assessors maintain up-to-date planimetric maps which show boundaries of the properties in their areas of interest. Preparation of planimetric maps has become simplified through the utilization of orthophotos. Orthophotos can be used in place of planimetric maps when the user has some experience in aerial photo-interpretation.

7.1.2.3.4 TOPOGRAPHIC MAPS

A topographic map is defined as follows in various publications: "a map which depicts the configuration and the horizontal and vertical positions of the features represented; distinguished from a planimetric map by the addition of relief in measurable forms."

Scales which are commonly used in many countries of the world for topographic maps are 1:250,000; 1:100,000; 1:50,000; 1:25,000; 1:10,000; 1:5,000; and 1:2,500. However, in the United States and several other countries, English equivalents are still in use, such as 1:63,360 (1″ = 1 mile) and 1:24,000 (1″ = 2000′). In most cases, form follows function in that the contour interval represented on a topographic map is consistent with the map scale and hence the proposed use of that map. Contour interval and corresponding map scales common in the United States are as follows.

Scale	Contour Interval
1:24,000	10 foot or 20 foot
1: 4,800	5 foot or 10 foot
1: 2,400	5 foot
1: 600	1 foot or 2 foot

7.1.2.3.5 SPECIAL PURPOSE MAPS

Any map designed primarily to meet a specific requirement may be called a special purpose map. Usually the map information portrayed on special purpose maps is emphasized by omitting or subordinating non-essential or less important information. The following kinds of special purpose maps are available from various government agencies and private mapping firms.

a. Nautical charts for navigational purposes.
b. Geologic maps.
c. Vegetation maps.
d. Statistical maps.
e. Land use maps.
f. Tax maps.
g. City maps.
h. Transportation maps, showing highways, railroads and airports.
i. Political and historical maps.
j. Cadastral maps.

In Europe, the cadastre is often maintained by photogrammetric methods. Property ownership is thus documented to a high degree of accuracy with map scale and standards determined by the land value. Although in the United States photogrammetrically prepared maps do not generally become legal records of property ownership, photogrammetric surveys for highway and other projects involving acquisition of properties of right-of-way are frequently used to piece together the subject properties and in the preparation of legal descriptions. As in the cadastre, the scale and degree of accuracy are set by the property values.

7.1.3 DETERMINING SPECIFICATIONS AND CONDITIONS FOR OPERATIONS

The selection of photogrammetric procedures and products for a given application requires consideration of the specifications (within the constraints of schedule and cost) derived from the requirements on the project.

7.1.3.1 CHOICE OF SCALE FOR AERIAL PHOTOGRAPHS

Empirically, the choice of scale for the original photographs may be made by determining the size of the smallest object that must be seen on the photographs and dividing that size into the nominal resolving power of the photographic system. For example, if the object is a water valve 12 inches in diameter and the resolving power of the photographic system for such images is 0.004 inch, the scale is determined as follows: 0.004/12 = 1/3,000. Technically then, a scale of 1:3000 (1 inch to 250 feet) is indicated. In practice, however, other factors such as contrast and background influence the effective resolution of the photograph. In the case of the water valve, a scale of 1:3000 is frequently too small because of the lack of contrast between the valve and background. When photographs are to be used for stereoscopic plotting, the selection of scale is influenced by the basic economic consideration that the project be completed at the lowest cost practicable. This is usually accomplished by measuring as few stereomodels as possible, and since the smaller the scale of the photographs the fewer stereomodels required to cover a given area, it is a primary objective to obtain the photographs at the smallest practical scale. Generally, the specifications on the accuracy of the map, the characteristics of the available plotting equipment, and the configuration of the mapping area will fix the scale of the aerial photograph.

7.1.3.2 SCALE AND LEGIBILITY IN PLANIMETRIC MAPPING

The ultimate use of a map should determine its form and method of preparation. For example, large scale may be required in a map of an oil refinery, to permit space for extensive labeling of equipment and for explanatory notations. While a scale of 1:250 is suitable for this pupose, the degree of accuracy for this map, which may be intended to serve as an inventory rather than for new engineering design, is such that some liberty may be taken in determining the ratio of photograph scale to map scale. For a map with standard planimetric accuracy, the enlargement might be restricted to no greater than 8 diameters, while a 10 times enlargement would be adequate in this instance. Conversely, a map prepared at a scale of 1:600 of an urban area, as a base for engineering design might require an upper limit of 5 times enlargement in order to assure the resolution necessary to plot the minute detail required for this kind of mapping.

7.1.3.3 MAP SCALE AND CONTOUR INTERVAL

Although increasingly widespread use of computer-driven plotters permits direct plotting with almost unlimited combinations of scale and contour interval, certain associations are most prevalent. Large-scale urban mapping for a very flat city like Galveston would normally require 30-cm. contours, or a planimetric base with an abundance of spot elevations might suffice. For mapping of rolling Cincinnati, 1:2500, contours at 2-m. intervals should adequately depict the terrain, while for hilly San Francisco, 5-m. contours would be suitable for most of the area, with supplemental 1-m. or 2.5-m. contours where needed. It should be kept in mind, however, that if the interval is dictated by considerations of legibility only, the contours should maintain the same accuracy standard as a 1-m. contour map with respect to spot elevations and the allowable tolerance for contours as well in that an accuracy of ± 80 cm. should be achieved.

For intensive engineering design, precise photogrammetric measurement of stockpiles or of voids in mining and excavation, map scales as large as 1:250 may be required with contour intervals as small as 10 cm. In flat featureless terrain, a contour interval as small as 10 cm. might be required to adequately portray drainage, while a horizontal scale of 1:2500 would suffice due to the lack of planimetric detail. Conversely, measurement of voids in a steep-walled quarry could perhaps best be accomplished with a 7-m. contour interval and a scale of 1:250, as a small horizontal displacement would drastically affect measurement while moderate vertical errors would have little significance in this instance.

Thus it can be surmised from these examples that the proposed use of a map and the type of area being mapped are the determining factors in deciding on the horizontal and/or vertical accuracy required. The examples provided above represent extreme circumstances, and it should be noted that normal relationships of map scale to contour interval will be specified on most projects.

7.1.3.4 BASE MAPPING SUPPLEMENTED BY DETAILED GROUND SURVEYS

When high levels of accuracy are required for only a small portion of an area being mapped, it may prove practical to do a ground survey in this portion only. For example, the flood line of a large reservoir must be determined to an accuracy of ½ metre vertically and horizontally. All other parts of the project require mapping at 1:1,000 with 2-m. contours. Rather than mapping the entire area at 1:500 with 1-m. contours to meet the accuracy requirement for the flood line, it will prove more practical and economical to stay with the smaller scale mapping for all other requirements and determine the flood line by ground survey. Similar situations may be encountered in projects involving drainage such as sewers or irrigation canals.

In preparation of large-scale base maps for engineering, it is normal practice to show all surface utility features visible and identifiable on the photography. This can be accomplished by precompilation field classification of surface features on a set of enlarged prints from the aerial photography. This information is then transferred to the base manuscript during map compilation. The utility features are next verified for location, top and invert elevations determined and underground connections plotted from research record and field investigation.

7.1.3.5 MAPPING FOR PRELIMINARY STUDY AND FOR DESIGN

Mapping for selection of a route for interstate or similar highways is frequently compiled at a scale of one inch to 200 feet with contours at five-foot intervals over a zone approximately one mile in width. In very flat or steep terrain the contour interval may be changed to two feet or ten feet respectively. Based on an analysis of this preliminary map, the engineer will focus his attention on a narrower zone, say

1000 feet, requiring detailed mapping at a scale of one inch to 50 feet with contours at one-foot or two-foot intervals depending upon conditions. Obviously, this approach will provide a great deal of savings compared to preparing large scale mapping of the entire area under consideration at the outset.

7.1.3.6 MAP ACCURACY STANDARDS

The accuracy of a map is determined by comparing the position and/or elevation of a feature on a map with the position and elevation of that feature as determined by field surveys. This comparison presumes that there is no discernible error in the survey.

The Office of Management and Budget has stated:

With the view to the utmost economy and expedition in producing maps which fulfill not only the broad needs for standard or principal maps, but also the reasonable particular needs of individual agencies, standards of accuracy for published maps are defined as follows.

1. *Horizontal accuracy.* For maps on publication scales larger than 1:20,000, not more than 10 percent of the points tested shall be in error by more than 1/30 inch, measured on the publication scale; for maps on publication scales of 1:20,000 or smaller, 1/50 inch. These limits of accuracy shall apply in all cases to positions of well defined points only. "Well defined" points are those that are easily visible or recoverable on the ground, such as the following: monuments or markers, such as bench marks, property boundary monuments; intersections of roads, railroads, *etc.;* corners of large buildings or structures (or center points of small buildings), *etc.* In general what is "well defined" will also be determined by what is plottable on the scale of the map within 1/100 inch. Thus while the intersection of two road or property lines meeting at right angles, would come within a sensible interpretation, identification of the intersection of such lines meeting at an acute angle would obviously not be practicable within 1/100 inch. Similarly, features not identifiable upon the ground within close limits are not to be considered as test points within the limits quoted, even though their positions may be scaled closely upon the map. In this class would come timber lines, soil boundaries, *etc.*

2. *Vertical accuracy,* as applied to contour maps on all publication scales, shall be such that not more than 10 percent of the elevations tested shall be in error more than one-half the contour interval. In checking elevations taken from the map, the apparent vertical error may be decreased by assuming a horizontal displacement within the permissible horizontal error for a map of that scale.

3. The accuracy of any map may be tested by comparing the positions of points whose locations or elevations are shown upon it with corresponding positions as determined by surveys of a higher accuracy. Tests shall be made by the producing agency, which shall also determine which of its maps are to be tested, and the extent of such testing.

4. Published maps meeting these accuracy requirements shall note this fact in their legends, as follows: "This map complies with national map accuracy standards."

5. Published maps whose errors exceed those aforestated shall omit from their legends all mention of standard accuracy.

6. When a published map is a considerable enlargement of a map drawing (*manuscript*) or of a published map, that fact shall be stated in the legend. For example, "This map is an enlargement of a 1:20,000 scale map drawing," or "This map is an enlargement of a 1:24,000 scale published map."

7. To facilitate ready interchange and use of basic information for map construction among all Federal map-making agencies, manuscript maps and published maps, wherever economically feasible and consistent with the uses to which the map is to be put, shall conform to latitude and longitude boundaries, being 15 minutes of latitude and longitude, or 7½ minutes, or 3¾ minutes in size.

The Reference Guide Outline (The Photogrammetry for Highways Committee, 1968)

A. *Contours*—Ninety (90) percent of the elevations determined from the solid-line contours of the topographic maps shall have an accuracy with respect to true elevation of one-half (½) contour interval or better and the remaining ten (10) percent of such elevations shall not be in error by more than one contour interval. This accuracy shall apply only to the contours which are on each map. Thus, in each particular area where the intermediate contours have had to be omitted because of the steepness of the ground slopes and only the index contours are delineated on the maps, the accuracy stipulations apply to contour interval of the index contours. Wherever the intermediate contours are not omitted, of course, the accuracies are applicable to the contour interval specified for the topographic maps. In densely wooded areas where heavy brush or tree cover fully obscures the ground and the contours are shown as dashed lines, they shall be plotted as accurately as possible from the stereoscopic model, while making full use of spot elevations obtained during ground-control surveys and all spot elevations measured photogrammetrically in places where the ground is visible.

B. *Coordinate Grid Lines*—The plotted position of each plane coordinate grid line shall not vary by more than one one-hundredth (1/100) of an inch from true grid value on each map manuscript.

C. *Horizontal Control*—Each horizontal control point shall be plotted on the map manuscript within the coordinate grid in which it should lie to an accuracy of one one-hundredth (1/100) of an inch of its true position as expressed by the plane coordinates computed for the point.

D. *Planimetric Features*—Ninety (90) percent of all planimetric features which are well-defined on the photographs shall be plotted so that their position on the finished maps shall be accurate to within at least one-fortieth (1/40) of an inch of their true coordinate position, as determined by the test surveys, and none of the features tested shall be misplaced on the finished map by more than one-twentieth (1/20) of an inch from their true coordinate position. The true coordinate position shall be determined by making accurate measurements originating and closing on station markers of the project basic control survey, which shall have a closure accuracy conforming with the requirements for the basic control.

E. *Special Requirements*—When stipulated in special provisions that all specified features (planimetry and contours) shall be delineated on the maps, regardless of whether they can or cannot be seen on the aerial photographs and on stereoscopic models formed therefrom, the consultant shall complete compilation of the required maps by field surveys on the ground so as to comply with all accuracy and completeness stipulations.

F. *Spot Elevations*—Ninety (90) percent of all spot elevations placed on the maps shall have an accuracy of at least one-fourth (¼) the contour interval, and the remaining ten (10) percent shall be not in error by more than one-half (½) the contour interval.

Both versions of these accuracy statements are widely used in specifications for mapping projects.

7.2 *Planning and Execution*

The most basic photogrammetric project will prove difficult or costly without proper planning. Larger, more complicated projects can evolve quickly into disjointed efforts which in the end will produce unsatisfactory results both for the photogrammetrist and the client. It is incumbent upon the project planner to obtain all relevant source material; to organize this material into a logical and easily understood system; to decipher critical or ambiguous requirements in the project specifications; to determine the most efficient method for meeting project requirements; and finally to prepare a concise and organized job order, including all of the elements which will govern the conduct of work in each department.

7.2.1 AERIAL PHOTOGRAPHY

The aerial photograph is the base upon which the photogrammetric project is built. The success of the project consequently depends greatly on the availability of suitable photographic coverage.

Suitable coverage for a project depends on many factors of which several are of particular importance:

a. scale of photography;
b. overlap between exposures;
c. optical and mechanical characteristics of the taking camera;
d. film base and emulsion type used; and
e. date of photography.

Existing aerial coverage of the project can be ordered from public or some private mapping agencies when available. Such coverage may not be of optimum quality for photogrammetric use. In such instances, purchase of the coverage would constitute poor planning for execution of the project, although it might prove quite useful for advance project estimating and planning. The user should consider carefully whether the compromises which may be necessary with existing aerial photographs can be offset by an immediate savings in time and costs.

New aerial photography flown especially for the photogrammetric project can be designed to meet exact requirements of the project and has much to recommend it. Projects requiring targeting of ground points would necessarily require new aerial coverage. In instances where the user has decided on acquisition of new aerial photographs, a broad selection of film and cameras is available to meet every need.

7.2.1.1 CHOICE OF EMULSION

Films for color and black-and-white photography are manufactured for aerial use. Both types are also manufactured with limited sensitivity in the near infrared[1] part of the spectrum. Each type has certain inherent advantages and limitations which must be carefully considered before use. Chapter VI covers photographic emulsions in detail. The following is a brief outline to assist the planner.

7.2.1.1.1 PANCHROMATIC EMULSIONS

Panchromatic emulsions are black-and-white types with a color sensitivity similar to that of the human eye. Their sensitivity to blue and ultra-violet usually requires that a yellow anti-haze filter be used for aerial work. These emulsions are produced in a wide range of "film speeds" for different conditions. The slower emulsions generally have high resolution, making them ideal for photo enlargements and for many aerial mapping and measurement applications.

Panchromatic films are inexpensive and have a wide exposure latitude as compared to color. They are easy to process and are widely used throughout the industry. Panchromatic emulsions are considered inferior to color for interpretative uses although they are used for multispectral recording.

---

[1] Not to be confused with infrared thermal detection systems which work on non-photographic principles.

7.2.1.1.2 Infrared Emulsions

Films with extended sensitivities into the red and near-infrared parts of the spectrum are manufactured for special aerial applications. Infrared emulsions, like other aerial emulsions, are exposed by reflected sunlight. Water bodies and wetted areas which absorb most of the infrared from the sun, leave a characteristic "signature" on the infrared emulsion. Black-and-white infrared has been used for this reason to determine riparian and littoral boundaries and to delineate wetted soils.

A color-infrared (color-IR) film is also manufactured for aerial use. An infrared sensitive layer is substituted for one of the color layers, giving a "false" color to the processed film. This film has found wide usage in photogrammetry for agricultural and forestry applications and more recently for wildlife habitat and pollution studies. This film is normally processed directly to a positive transparency from which prints can be made.

While both black-and-white and color-infrared film can be used for topographic or other photogrammetric work, their greatest success has come in the fields of remote sensing and photo-interpretation.

7.2.1.1.3 Color Emulsions

Films designed to record and reproduce normal color tones in aerial photographs are manufactured in two types. The color film commonly used for photogrammetric projects has an emulsion which is processed to a color negative. Positive color prints, enlargements, and transparencies for mapping can be produced from this emulsion.

The second type is usually processed directly to positive transparencies, eliminating intermediate steps. Prints can be made directly from this film. This type can be used for photogrammetric work but is not as widely used for this purpose as the color-negative film.

Color films have gained popularity in recent years for both map making and photointerpretation. Studies indicate that color enhances the ability of the eye to correctly identify images. This capability is felt by many to offset the increased costs of color photography.

7.2.1.2 Film Characteristics

Two types of film base are manufactured. A non-stable plastic base, sometimes referred to as "acetate" base, is used for some aerial reconnaissance films. In this application, dimensional stability of the film base is of little importance. Non-stable film bases should not be used for photogrammetric work requiring precision. A stable "polyester" base is made by several manufacturers to meet the need for an aerial film with high dimensional stability. This base is now used for all mapping films and should be specified for all photogrammetric projects where topographic mapping or precision measurements are to be made.

7.2.2 AERIAL CAMERAS

While any camera capable of being carried aloft could be considered an "aerial" camera, only cameras meeting certain requirements should be considered for photogrammetric projects. Aerial cameras can be separated into general categories of non-precise (non-metric) and precise (metric) cameras. Aerial cameras are fully discussed in chapter IV.

7.2.2.1 Non-Precise Cameras

An aerial camera classified as non-precise can most easily be described as any aerial camera not meeting the requirements described below for a precise mapping camera. Most military surplus aerial cameras fall into this category, as do all but a very few of the commercially available small format cameras. Non-precise aerial cameras have little to recommend their use except for aerial reconnaissance purposes, unless special precautions are taken.

7.2.2.2 Precise Mapping Cameras

Precise cameras are designed principally for photogrammetric work. While a complete discussion on camera design is beyond the scope of this section, some general requirements for such a camera should be noted.

a. High resolution lens with maximum distortion values under .01 mm. Such lenses usually have fixed focus to preserve their precise geometry.
b. Camera body and lens mounting which maintain an exact and consistent relationship between the lens and the film plane.
c. Film platen which can precisely flatten the film and hold it exactly in the focal plane during exposure.

Since the introduction of mapping cameras, many improvements have been made in their design. The earliest models had lenses with distortion considerably in excess of modern day standards. Some compensation for this distortion was usually made during the mapping process.

Modern precise mapping cameras are designed to use roll film with an exposure format of 23 × 23 cm, a format compatible with nearly all photogrammetric plotting equipment. 70-mm., small-format cameras meeting the requirements for precise mapping cameras are also manufactured but have been used mainly in photointerpretation. The smaller format is quite economical when color films are specified. In Europe, mapping cameras using plates have also found application.

Most recently-manufactured mapping cameras are designed to accept interchangeable lenses of different focal lengths. The most commonly used focal length for mapping photography is, nominally, six inches (152 mm).

7.2.3 AIRCRAFT

Normally, the project planner need only determine that an aircraft, to be used as a camera platform, can go the required distance and reach the required altitude. Most aircraft engaged in aerial photography for photogrammetric work fall under government regulations concerning aircraft maintenance, modifications necessary for camera operation and enroute flight control above certain altitudes. There are also requirements concerning qualification of the pilot.

Aircraft used for aerial photography can be grouped into several categories:

  a. fixed-wing, reciprocating engine:
    i. single-engine;
    ii. twin-engine;
  b. fixed-wing turbojet;
  c. rotary wing; and
  d. other.

Most aircraft involved in aerial photography fall into the first category (a).

The turbocharged single-engine aircraft has proven to be both economical and effective for photography at altitudes up to 24,000 feet. Cross-country and climb speeds for this aircraft are relatively low, and this design is better suited to smaller projects near the base of operations. The aircraft is light, and consequently tends to be more affected by turbulence than heavier craft.

Twin-engine aircraft offer a more stable support because of their increased weight. Cross-country speeds are greater as are the climb rates to altitude and service ceilings. For remote or lengthy flying jobs, these aircraft may prove to be more economical than the smaller, less expensive, but limited-capability single-engine craft. Many pilots consider the additional engine to be a desirable safety factor when operating in remote, rugged areas.

Turbojet, fixed-wing aircraft modified for aerial photography offer improved ceilings and speeds over reciprocating-engine aircraft. Extremely high-altitude aerial photography is often beyond the reach of reciprocating-engine aircraft and can be performed only with a turbojet type. All turbojet aircraft have pressurized cabins requiring the installation of an optical "window" through which the camera must view the ground without appreciable photographic distortion. Generally, the increased operating costs of turbojet aircraft prevent them from being used on small projects.

Helicopters are not widely used except for a few very specialized applications. High operating costs and low cross-country speeds prevent them from being economical on most projects. Projects which require very low altitudes, beneath the limit permitted to fixed-wing aircraft, can be done only with a helicopter. Camera mounting and engine vibration have caused some problems in the use of this type of aircraft.

Some experimentation has been done in the use of unpowered aircraft such as lifting balloons and large kites for camera platforms. These techniques are limited to small-format cameras and special needs such as archaeological mapping, and would not normally be considered for most photogrammetric projects.

7.2.4 FLIGHT DESIGN[2]

7.2.4.1 CONSIDERATIONS AND OBJECTIVES

To plan properly for the aerial photography for a new photogrammetric project, a flight designer must carefully evaluate each of the many variables involved. A major consideration in one project may be relatively minor in another; hence, a thorough analysis of each of the elements is essential to the development of a sound flight design. The basic considerations pertinent to the project and the characteristics of the completed photography are discussed elsewhere in this chapter and in chapter V. This section is devoted to the governing factors of flight design for adequate and proper photographic coverage.

When a new photogrammetric project is initiated in a large organization, the general-planning personnel usually determine and furnish the flight designer with basic information covering the project boundaries, details of existing or planned control as applicable to the project, time schedule, and details covering the final product. In addition, before computations can begin, investigations must be completed and decisions must be made regarding such matters as optimum flying season, acceptable ground cover conditions, and requisites of the completed photography.

In his mathematical solution, the flight designer's objective is to establish the plan and standards for the actual photographic mission. More specifically, he determines the optimum conditions for spacing of photographs along the flight line and the number and spacing of flight lines over the area. The plan he evolves must recognize the account for allowable deviations from optimum or ideal conditions and, at the same time, must provide reasonable assurance that complete and acceptable stereoscopic coverage will be achieved over the entire project area. The primary governing condition for acceptability is that complete stereoscopic cover-

---

[2] Sections 7.2.4 and 7.3 have been taken verbatim or with minor revision from chapter VII of the Third Edition of the MANUAL OF PHOTOGRAMMETRY Published in 1965. Chapter VII was authored by Messrs. Eliel, Alster, Halliday, McMillen and Payne.

age be attained within the usable cone of coverage of the camera lens.

7.2.4.2 FLIGHT PATTERNS

In this section the spacing of photographs along a flight line is expressed in terms of the base-height ratio, commonly referred to as B/H, and usually written in decimal form. By definition, B/H is the ratio of the distance between successive exposure stations (air base) to the flight height above local mean ground. In similar fashion, the width-height ratio (W/H) is defined as the distance between adjacent flight lines divided by the flight height above local mean ground.

Another way of expressing the spacing of successive photographs is in terms of percentage of forward overlap, and the spacing of parallel flight lines in terms of percentage of side overlap. While percentage of overlap is perhaps easier to visualize, the ratio approach is more suitable for dealing with nonvertical photography and the expressions base-height ratio and width-height ratio are more convenient to use in some photogrammetric calculations.

The mathematical relationship between the two methods, assuming truly vertical photography, without crab or deviation from constant altitude, over flat terrain, is given below.

Percentage forward

$$\text{overlap} = \frac{G_1 - B}{G_1} \times 100 \qquad (7.1)$$

Percentage side overlap $= \dfrac{G_2 - W}{G_2} \times 100 \qquad (7.2)$

where

G_1 = Ground distance covered by a single exposure, measured *along* the line of flight;
G_2 = Ground distance covered by a single exposure, measured normal to the line of flight;
B = Distance between successive exposures in a flight line;
W = Distance between adjacent flight lines

Further derivation of the desired relationships depends on the metric characteristics of the camera used. Assuming the standard wide-angle mapping camera with 6-inch focal length and 9- × 9-inch format,

$$G_1 = G_2 = 1.5H = \frac{3H}{2} \qquad (7.3)$$

where H = flight height above local mean ground.

Substituting equation (7.3) into equations (7.1) and (7.2),

Percentage forward

$$\text{overlap} = \left[1 - \frac{2}{3}\left(\frac{B}{H}\right) \right] \times 100 \qquad (7.4)$$

Percentage side

$$\text{overlap} = \left[1 - \frac{2}{3}\left(\frac{W}{H}\right) \right] \times 100 \qquad (7.5)$$

These relationships are shown graphically in figure 7–2.

7.2.4.2.1 BASE-HEIGHT RATIO

Several factors influence the choice of the most suitable base-height ratio. These include securing maximum parallax which is favorable to accuracy in reading and setting of the stereoscopic model, reduction of camera lens distortion effect to a minimum, assurance of ample stereoscopic coverage, and maximum area coverage within a stereoscopic neat model. The infuence of various spacings of vertical photographs on model areas and proportions is shown in figures 7-3 and 7-4.

The most accurate model readings and settings result from the greatest parallax, which is obtained by making the air base as long as possible. The amount by which the base can be lengthened is limited, however, by a corresponding loss of stereoscopic coverage.

Increasing the air base also minimizes the effect of camera lens distortion. However, since most plotting systems provide compensation for nominal camera lens distortion, only the slight variations of individual lenses from this nominal value become significant. Thus, lens distortion is a minor factor in flight design.

While shortening the air base increases the gross stereoscopic coverage of a model it also decreases parallax, and consequently accuracy, and is limited by physical and mechanical features of some plotting instruments. With Balplex projectors and vertical photography, a minimum B/H of about 0.45 (70 percent) is possible, while with Kelsh projectors this value is about 0.36 (76 percent). (The figures in parentheses give percentage of overlap.) In general, instruments of the universal type are not particularly limited in this respect.

When a constant altitude above sea level is

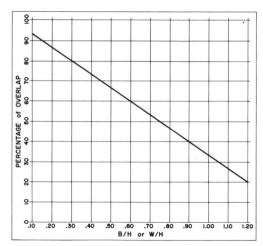

FIGURE 7-2. Overlap conversion chart; standard wide-angle camera, 6-inch focal length, and 9- by 9-inch format.

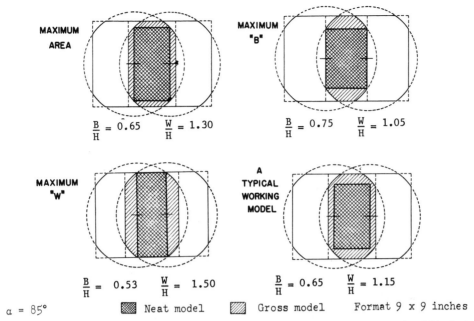

FIGURE 7-3. Effect on model area of spacing of vertical photographs (H is constant; $f = 6.0$ inches).

maintained by the aircraft over an area where the terrain varies in elevation, the value of H increases over lower ground and decreases over high ground. The variations in the air base, B, required by the changing ground elevations are obtained automatically by specifying a constant value of B/H.

7.2.4.2.2 WIDTH-HEIGHT RATIO

The spacing of flight lines, represented by the symbol W, has two slightly different connotations; *i.e.*, the distance between adjacent flight lines, and also the dimension of a neat model measured normal to the flight line. While these may not always be exactly equal in value, they are usually considered so for practical purposes.

7.2.4.2.3 DEVIATIONS FROM IDEAL CONDITIONS

Under ideal conditions of vertical photography with 9- by 9-inch format and a focal length of 6 inches, the most favorable ratios are $B/H = 0.65$ (57 percent) and $W/H = 1.30$ (13 percent) (*see* fig. 7-3). These ratios result in maximum stereoscopic coverage with no gaps at the neat model corners. In practice, however, it is obvious that the designer must make provision for the deviations usually encountered under actual operational conditions. These include accidental tilt or crab in the photography, relief in the terrain, and deviations from the planned flight line or flight height.

Some latitude is generally provided by planning a B/H ratio of 0.63 (58 percent) rather than the theoretical 0.65 (57 percent), and allowing a further variation of ± 0.05 (3 percent) to cover the usual vicissitudes of aerial photography.

This small decrease in the B/H ratio provides leeway in the placement of model pass points and supplemental vertical control. It is also more effective and economical than providing the same safety factor by reducing the W/H ratio.

When an additional safety factor is needed it can be provided by reducing either ratio or both. It is, however, usual practice to maintain the B/H ratio as nearly constant as possible and provide for the more troublesome operational variables by reducing the W/H ratio below its maximum (1.30) value.

7.2.4.2.3.1 Effect of Tilt. In practice, nearly all so-called "vertical" photographs are accidentally tilted slightly at the moment of exposure. Even with the best of camera mounts, tilting is unavoidable because of the inherent instability of the aircraft and the vagaries of gyroscopes. While some of the newer, more sophisticated mounts may control tilts within very small tolerances, flight planning should be based on an average tilt probability of 2 degrees.

In figure 7-5 it can be seen that accidental camera tilt in any direction causes displacement in the ground limits of the photographic coverage. A component of tilt about the flight line (commonly referred to as x tilt) has the effect of displacing these ground limits of coverage in a direction perpendicular to the flight line. Similarly, any component of tilt about a line perpendicular to the flight line (referred to as y tilt) causes displacement of these limits in a direction along the flight line.

Small systematic y tilts cause little trouble because the displacement of coverage is uniform and has little effect upon the B/H relationship. Random y tilts, if held within the 2-degree limit,

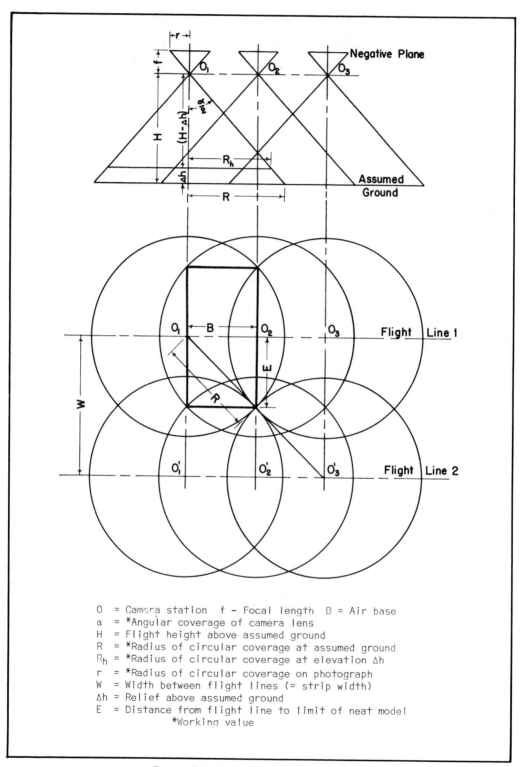

FIGURE 7-4. Spacing of vertical photographs.

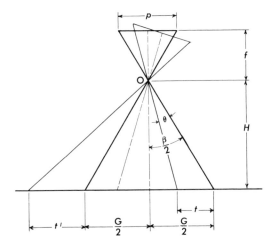

FIGURE 7-5. Displacement of coverage in a single photograph because of tilt.

are absorbed by the ±0.05 tolerance in the B/H ratio. On the other hand, x tilts, which critically affect the stereoscopic coverage at the neat model corners, may be accomodated by using a reduced width-height ratio in the design.

In figure 7-5, t and t' represent displacements of the ground limits of coverage due to an angle of x tilt, θ. Since t' increases the coverage, we are interested only in t, which decreases it. From figure 7-5 it can be shown that, for standard 6-inch vertical photography with θ assumed to be 2 degrees:

$$t = 0.053H = 0.035G \qquad (7.6)$$

Hence, in designing for a proper width-height ratio, the effect of accidental tilt likely to be encountered in standard 6-inch vertical photography can be considered to be about 3½ percent of the strip width.

The flight is designed to absorb this amount of x tilt in the expectation that cumulative tilts in the photographs abreast of each other in adjacent flights will seldom occur in an extent requiring a reflight.

7.2.4.2.3.2 Effect of Relief. In the usual flight planning rocedure for block area coverage, the flight courses are designed to be run in parallel straight lines at fixed elevations above sea level. Under such conditions the design value for the spacing (W) between any two adjacent flights has a fixed value and, if the ground is perfectly flat, the theoretical width-height ratio (W/H) remains constant. But, when there is ground relief, the flight course at a fixed altitude above sea level results in a variable altitude (H) over local mean ground. This causes the width-height ratio to become variable and the theoretical conditions for achieving maximum stereoscopic coverage cannot be maintained.

Ground that rises above the optimum elevation may become critical, for in such areas the altitude (H) over the ground decreases and W/H

increases accordingly. The most unfavorable condition (regarding stereoscopic coverage) occurs when the high ground is at the corner of a neat model. As ground elevations increase in this critical location, the common stereoscopic area within abreast neat models, suitable for the placement of model control, becomes smaller; it approaches zero as W/H approaches its limiting value of 1.30. At that point, complete stereoscopic coverage of the area is barely attained within the usable overlap area. Therefore, the width-height condition over the highest ground point in a project can be considered as a controlling factor in the design for flight-line spacing.

While the effect of relief can be demonstrated by theoretical formulas, the application of these formulas in actual flight design becomes cumbersome and impractical. For this reason, it is considered preferable to adopt empirical W/H limits related to the highest ground point in the project. (Throughout this discussion we continue to refer to standard 6-inch wide-angle mapping photography.)

As a practical matter, two W/H ratios afford an empirical solution to the problems of ground relief. The conditions governing these ratios may be stated as follows.

(a) For projects with low to moderate ground relief (up to about 300 feet), the optimum width-height ratio (the W/H occurring over the optimum ground elevation) should not exceed 1.10, (not less than 27 percent).

(b) For projects involving somewhat higher relief, a maximum width-height ratio (the W/H occurring over the highest ground point) should not exceed 1.15 (24 percent).

In areas with an intermediate amount of relief, where it is not obvious which of these limiting conditions should apply, the design should be computed both ways and based on the one which first tends to be exceeded.

Special consideration of areas of extreme relief is given later in this section.

7.2.4.2.3.3 Effect of Crab. In the values heretofore given for B/H and W/H, it is assumed that the angle of crab will not exceed 10 degrees. Since this angle is seldom even approached by experienced operators, the figure should be recognized as a limit but otherwise does not enter into flight design. Figure 7-6 illustrates how excessive crab may become critical.

7.2.4.2.3.4 Effect of Navigation Errors. Flying requirements for precision aerial photography are difficult to meet. Therefore, some leeway should be provided in tilt, crab, deviation from flight line, and deviation from flight height. In actual aerial photographic operations, the chances are slight that every one of these factors will deviate simultaneously and cumulatively by the full amount of the tolerance. Therefore, a practical flight design need not include a full allowance for every allowable deviation.

A good pilot-navigator can hold very close to a

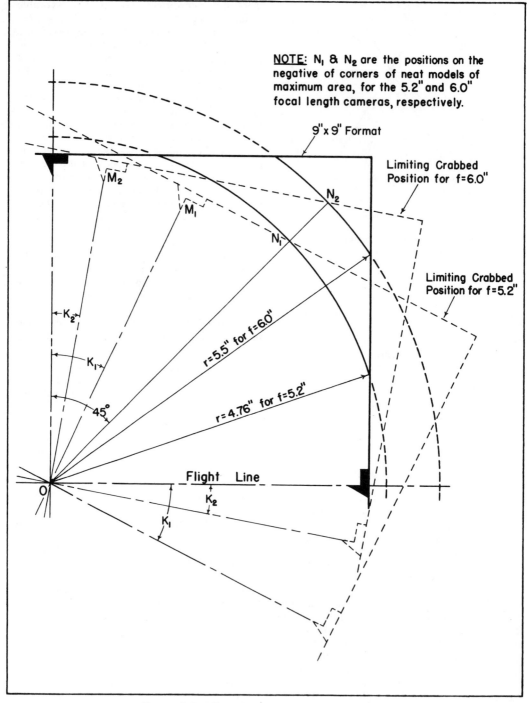

NOTE: N_1 & N_2 are the positions on the negative of corners of neat models of maximum area, for the 5.2" and 6.0" focal length cameras, respectively.

9"x 9" Format

Limiting Crabbed Position for f=6.0"

N_2

N_1

Limiting Crabbed Position for f=5.2"

M_2

M_1

K_2

K_1

45°

r=5.5" for f=6.0"

r=4.76" for f=5.2"

Flight Line

K_2

O

K_1

FIGURE 7-6. Effect of crab in vertical photography.

flight line drawn on a good map. But, if the map is poor or the country devoid of identifiable detail, he may have difficulty. Therefore, in designing the flight, a latitude of $0.10H$ should be provided.

When, because of difficult navigation, the plane drifts somewhat off course, any further re-

duction of the overlap resulting from local high ground may cause a gap. This danger can sometimes be minimized by orienting the flight lines parallel to the "grain" of the topography and centering a flight line on the highest ridge. The flight altitude of adjacent flight lines can be "stepped" down in altitude above sea level so as to fly as closely as possible to the most suitable altitude above ground. The designer should keep in mind that the photographic coverage will be reduced on the high ground side and increased on the low side. Some compensation occurs since the low side of the higher altitude flight overlaps the high side of the adjacent lower "step" flight.

Most flight designs contain a further factor of safety produced by compaction of the flight-line spacing. Since the project width is seldom exactly divisible by the theoretical flight-line separation, the overage may be distributed among the lines, thus bringing them closer than the theoretical spacing.

Consideration should be given to the quality of the best available flight map. Maps of some areas of the world are so poor as to be of little or no help to navigation. Only the most experienced navigators succeed in flying properly spaced lines under such conditions.

7.2.4.2.3.5 Deviations in Flight Height. Specifications for aerial photography generally stipulate allowable deviations below and above the specified flight height. Altimeters are apt to read too low, causing the flight to be high, and since this is good insurance against gaps the operator may actually plan the flight close to the upper limit. It should be remembered that top limits are more likely to be violated than bottom limits.

7.2.4.3 Unusual Flight Conditions

While the above flight planning theory makes provision for deviations from ideal conditions, it is conservative in design because of the many safety factors contained in the assumptions. First, it was assumed that the exposure stations in adjacent flights would always occur opposite each other. This least favorable arrangement for achieving stereoscopic coverage happens only by chance in actual aerial photographic operations. Second, while the theoretical base-height ratio is 0.65, a safer value of 0.63 is generally used in practice. Furthermore, the design presumed that the highest peaks will always fall in the extreme corners of the neat models, where relief has its most detrimental effect. This unfavorable condition can be minimized, however by arranging flights over rugged terrain as previously discussed, so that they will cross directly over the highest peaks. If this is not feasible, safeguard may be added to a standard design by providing extra flights directly over the high peaks.

Under conditions of extreme ground relief, a base-height ratio of 0.63 may be excessive, and a gap in stereoscopic overlap might occur. Furthermore, the flight lines might have to be spaced closer together in order to avoid the possibility of gaps in the side coverage. By halving the base-height ratio to 0.32 (79 percent), however, these dangers are reduced and a more rea-

sonable and economical spacing between flights can be maintained. With such photography the odd-numbered exposures may be used for aerotriangulation and compilation, while the even-numbered pairs are available for completing compilation in occasional gaps. In this way a normal base-height ratio, about 0.64 (57 percent), is maintained for all stereomodels that are used. Costwise, doubling the number of exposures represents only a little extra film, which is a very minor cost.

For some conditions of abrupt relief it may be quite risky to specify the standard base-height ratio of 0.63, yet the half-base procedure just described may not be warranted. In such cases the planner may choose to reduce the base-height ratio (increase the overlap) to a value somewhat less than 0.63 for those particular flights covering the abrupt terrain. This reduced value may be from about 0.50 (67 percent) to about 0.58 (62 percent), depending on the particular circumstances.

The maximum planned width-height ratio, 1.15 (24 percent), specified in section 7.2.4.2.3.2 also represents conservative design. Consider for the moment the highly improbable combination of circumstances wherein the maximum allowable tilt displacement (representing about $0.05H$) and the maximum permissible flight-line deviation ($0.10H$) both occur with cumulative detrimental effect at a model corner containing the highest ground elevation. Even under such conditions the critical width-height ratio for the vertical neat model, 1.30 (13 percent), would not be exceeded; hence, some departure from rigid adherence to the governing width-height conditions stated in section 7.2.4.2.3.2 may be justified at times by circumstances. A certain amount of risk should be taken if, in the judgment of the flight planner, satisfactory coverage at less cost may reasonably be expected to result. For instance, in areas with small concentrations of high relief, the maximum allowable design value for W/H may be extended from 1.15 (24 percent) to 1.20 (20 percent) if this will reduce excessive sidelap conditions over extensive bottomland.

7.2.4.4 Length of Flight Lines

A standard method of calculating the total length of flight lines is generally used by planners. The specific computations to be used depend on the type of photography and, in unusual circumstances, on the shape of the project to be photographed.

7.2.4.4.1 Allowance for Overhang

To assure complete stereoscopic coverage of the project area and/or to reach necessary horizontal or vertical control, planning for vertical photography should provide at least two exposures beyond the project boundary at each end of each flight line. A standard allowance is often useful rather than attempting to measure each actual overhang. A standard allowance equal to 1½ air bases at each end of each flight line represents the average flying requirements for proper overhang.

Since the base-height ratio for vertical photography is usually 0.63, the length of one air base is the flight height above optimum ground elevation multiplied by 0.63. The total theoretical length of one flight line is then equal to the length of the line between project boundaries (as determined from polyconic projection tables where applicable) plus (3 times $0.63H$). This sum multiplied by the number of flight lines provides the total theoretical linear miles of flight line for a square or rectangular project.

In special cases involving control flights, skew flights, flight extensions to control, or flights over irregularly shaped projects such as coastal or island areas, it is usually necessary to measure each flight line on the flight map to determine its length.

7.2.4.4.2 MAXIMUM LENGTH OF LINES

To provide for contingencies that may necessitate additional flying, aerial photography planning often includes an arbitrary limit for maximum length of flight lines. This maximum may be determined by an empirical percentage increase to the theoretical length of flight lines. The following table has been used successfully for this purpose.

Theoretical linear miles	Add
Less than 250	25%
250 to 500	20%
500 to 1,000	15%
Over 1,000	10%

7.2.4.5 STANDARDIZING OF FLIGHT DESIGN

To simplify and standardize the application of the foregoing principles of flight design, and at the same time to have a permanent record of the calculations, the use of a uniform flight-design work sheet is recommended wherever it is applicable. For most projects, the work sheet reduces the flight design procedure to a relatively simple operation. The planner must be alert to situations which are not suitable for the use of a standard work sheet and should always exercise judgment as to the best approach to flight planning.

7.2.4.5.1 SAMPLE DESIGNS AND WORK SHEETS

Two sample designs are presented for vertical photography—one for an area of low relief as controlled by limiting condition "a" described in section 7.2.4.2.3.2 (*see* fig. 7-7); and the other for an area of high relief, governed by limiting condition "b" of the same article (*see* fig. 7-8). Note the safety feature of both designs whereby the final calculated value for the strip width (W)

has been reduced in order to obtain a whole-number value for the number of flights required. It should also be noted that the maximum and minimum values of W/H shown on the work sheets are design values. These would be increased and decreased, respectively, by 0.10 to practical values that allow for the range of tolerable flight-line deviation.

The example given in figure 7-7 is a routine design of no particular difficulty. The optimum ground elevation has the same value as the assumed ground elevation. The range of relief is not sufficient to have an appreciable effect on the design; hence, controlling condition "a" (optimum W/H not to exceed 1.10) prevails. Also, a line-by-line review is not necessary.

The case of figure 7-8 is a more complicated design because of the high relief in the project. In such cases, a preliminary design is computed for the project as a whole to determine tentative values for optimum ground elevation, flight height, and flight spacing. This is followed by a line-by-line review in which appropriate adjustments may be made in the flight height and/or the flight spacing. In the illustration (fig. 7-8), the spacing was held constant and the flight height was adjusted. The flight spacing is within the allowable W/H limit throughout the project and the projection distances are within a favorable range for the plotter to be used.

It should be noted that the maximum W/H computed for this design was 1.05 (30 percent), somewhat closer flight-line spacing than that specified in controlling condition "b" (maximum W/H not to exceed 1.15). This indicates that an alternative solution could be worked out with wider spacing and seven flights. In such a plan, the middle flight would center directly on the 122°48′ meridian. Thus, four long flights would still be needed, and only one short flight would be eliminated.

7.2.4.6 FLIGHT DESIGN NONVERTICAL PHOTOGRAPHY

While the bulk of photogrammetric mapping photography is obtained with precision cameras in vertical arrangment, the designer occasionally encounters unusual situations or special requirements that can be accommodated more satisfactorily by other types of photographic coverage. Generally, these are obtained by twin low-oblique cameras in convergent or transverse arrangement, the high-oblique camera, or the trimetrogon configuration. The convergent and transverse low-oblique types are adaptable to more or less exacting design procedures, while the high-oblique and trimetrogen require a more empirical design approach.

7.2.4.6.1 CONVERGENT LOW-OBLIQUE PHOTOGRAPHY

A low-oblique photograph is one taken with the camera's optical axis intentionally deviated by a fixed amount from the vertical, but not enough for the horizon to appear in the exposure. Twin low-oblique photography is obtained by means of a twin-camera arrangement similar to that shown in figure 7-9. The camera couple (or unit, in some designs) is made up of two wide-angle precision cameras rigidly mounted as close together as possible. While their respective optical axes may be tilted at any arbitrary angle to the plumb-line, the 20-degree value shown in the illustration is assumed throughout this discussion.

When the twin-cameras are arranged so that the plane containing the optical axes is oriented along the flight line, the resulting photography is called convergent low-oblique (see fig. 7-10). In this configuration the forward-looking exposure at one camera station is convergent with the backward-looking exposure at the succeeding camera station, and the overlapping portions of these exposures form the stereoscopic model.

The longer air base that is characteristic of convergent photography enhances the accuracy of spot height reading and contouring in the stereomodel, and makes this type of coverage particularly useful in the mapping of areas with low to moderate amounts of relief.

7.2.4.6.1.1 Convergent B/H Spacing. The base-height ratio B/H for convergent photography is derived from two given conditions: the intentional tilt angle of the cameras, and the convergent photography requirement of 100 percent overlap. An intentional tilt of 20 degrees is preferred by many because it is considered to combine the greatest amount of convergent coverage with a minimum amount of obliquity and a favorable air base. The 100 percent overlap requirement is designed to achieve maximum stereoscopic coverage from the convergent photography.

Assuming that these two conditions shall prevail for standard 6-inch wide-angle cameras with 9- by 9-inch formats, the optimum base-height ratio of 1.23 may be derived for convergent photography (see fig. 7-11). As in the preceding formulations for vertical photography, this derivation presumes ideal conditions (no accidental tilts, no ground relief, and crab held within allowable limits).

Further study of this figure reveals that the maximum model width that can be used in flight planning occurs at either side edge of the neat model, on a *y*-line through either nadir (*N*). Again, assuming all of the conditions previously given, it can be shown that the width-height ratio at these model edges has a value of 1.41.

7.2.4.6.1.2 Deviations from Ideal Conditions. Proper design for convergent low-oblique photography must make adequate provision for accidental camera tilts, ground relief, crab, and deviations from the planned flight lines or planned flight height. As in the preceding design for vertical photography, it is presumed that

the optimum value of the base-height ratio will be maintained as closely as possible, while the deviations from ideal conditions are provided for by varying the width-height ratio.

An allowable accidental tilt angle of 2 degrees (normal to the line of flight) at a single exposure station in convergent photography reduces coverage by an amount equal to about 0.05H (see fig. 7-5). To counter the effects of relief, it is recommended that the two governing conditions set up for optimum and maximum width-height ratios in vertical photography be adopted for convergent coverage. While the limiting value for crab in convergent photography can be shown to be about 14 degrees, a maximum allowable value of 10 degrees should be specified to facilitate later inspection procedures and photogrammetric operations.

To allow for the practical difficulties of maintaining planned flight courses, the recommended adjustments to the maximum and minimum design values for the width-height ration in vertical photography (see section 7.2.4.2.3.2) are equally effective in designing for convergent coverage. Also, since air crews usually fly slightly higher than specified altitudes in order to reduce the possibility of gaps, there is no need to make additional allowance for flight heights that may be too low.

With these design recommendations in mind, it should not be difficult to use the flight-design work sheet form shown in figures 7-7 and 7-8 to calculate the essential data for acceptable convergent low-oblique coverage.

7.2.4.6.2 TRANSVERSE LOW-OBLIQUE PHOTOGRAPHY

To obtain transverse low-oblique coverage, the standard twin-camera arrangement shown in figure 7-9 is oriented so that the plane containing the optical axes is normal to the line of flight (see fig. 7-12). The full angular coverage of 114 degrees normal to the flight line indicates the usefulness of this type of photography in reconnaissance mapping operations, although for some purposes trimetrogon, super-wide-angle, or two 6-inch cameras coupled in a wider angle should be considered.

An optimum base-height ratio of 0.63 and an optimum width-height ratio of 2.31 are generally recommended for design. These two values combine to create a transverse model wherein the long oblique condition of the outer rays tends to weaken stereoscopic viewing and restricts the usable *C*-factor to about 600 or less. However, this reduced *C*-factor is usually offset by the larger contour intervals generally prevalent in reconnaissance mapping, so that the flight height for photography is not adversely affected. Instead, other factors, such as aircraft flight ceilings, range of relief, model scale, and pantograph limitations, are more likely to become the important determinants of flight height.

UNITED STATES
DEPARTMENT OF THE INTERIOR
GEOLOGICAL SURVEY

FLIGHT DESIGN WORK SHEET

PROJECT GS- __VANE__   Name __Lineville__   State __Iowa-Missouri__
(For RT-P only)
Requested by __Central Area__   Designer __C. S.__   Date __5-16-62__

Compilation method __Kelsh__   Compilation scale $(\frac{1}{s})$ __1:15,840__

Contour interval (c.i.) __10'__   Flight direction __E-W__   Focal length __6"__

Season __Fall, spring__   Flight height (H_{opt}) __9000'__

Special conditions __None__

Ground elevations	Flight heights	Projection distances
h_{opt} __1000__ (From best available map)	H_{sl} __10,000__ (See diagram opposite)	d_{opt} __760 mm__ (Kelsh)
h_{max} __1100__	__8,900__	d_{min} __751__
h_{min} __900__ h_{loc} __1000__	H_{min}	
	H_{max} __9,100__ H_{loc} __9,000__	d_{max} __768__

C Factor $=\frac{H}{c.i.}=$ 900	$W_1 = 1.15 H_{min} =$ 10,235	Sq. miles = 452

E-W dimension = __139,060__ (69,530/15 min.) N-S dimension = 91,080
= 26.337 Mi.

$G=1.5H = 13,500$ (1) $L=.25G = 3375$ Use L= 3375 $Q=$ E-W or N-S dimen. whichever applies $-2L = 84,330$

$N=\frac{Q}{W_1}+1 = \frac{84,330}{10,235}+1 = 9.2$ Use N= 10 $W=\frac{Q}{N-1} = \frac{84,330}{9} = 9370$ (2)

$\frac{W}{H}$ (3) Opt 1.04 Min 1.03 Max 1.05 (4) $\frac{B}{H} = .63$ (5)   B= 5670 = 1.073 miles 1.5B= 8500 = 1.609 miles 3.0B= 17,000 = 3.218 miles

Theoretical length of flight lines = $(10 \times 26.337) + (10 \times 3.218) = 296$ miles

Maximum allowable = (6) 296 + 59 = 355 miles   (20 percent allowance)

Remarks

(1) G=1.5H (Vert.) = 1.41H (Conv.)
(2) W should not exceed W_1
(3) $\frac{W}{H}_{opt}$ should not exceed 1.10
(4) $\frac{W}{H}_{max}$ should not exceed 1.15
(5) $\frac{B}{H}$ =0.63 (Vert.) =1.23 (Conv.)
(6) Add to theo.: 0–250=25% 250–500=20% 500–1000=15% Over 1000=10%

REVIEW

Line	h_{min}	h_{max}	h_{opt}	h_{loc}	H_{sl}	H_{loc}	H_{max}	H_{min}	W_1	W (2)
No review required for 200' relief										

FIGURE 7-7. Flight design work sheet (left half), low relief.

When transverse photography is intended for use in map compilation, there must be complete stereoscopic coverage that meets compilation standards. To achieve this, the flight design must provide for the effect of deviations from ideal conditions. While each of these could be analyzed in a manner similar to the treatment in vertical photography, the neat transverse stereoscopic model has more excess stereoscopic coverage available on all sides (see fig. 7-12) than does the vertical arrangement. Hence, the design for transverse photography will be on

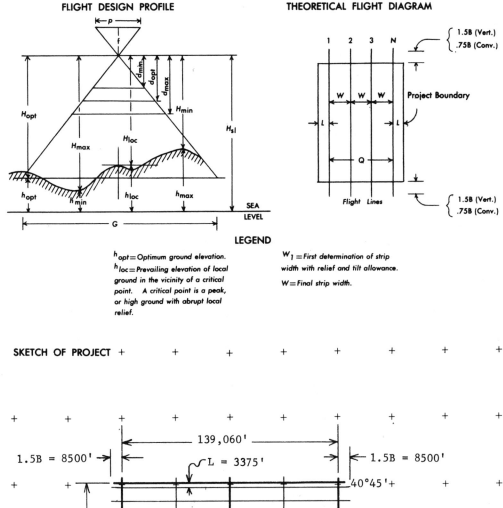

FIGURE 7-7. Flight design work sheet (right half), low relief.

the safe side if it is based on essentially the same limiting conditions for the deviation factors as those specified for vertical coverage.

7.2.4.6.3 HIGH-OBLIQUE PHOTOGRAPHY

A high-oblique photograph is taken with a camera that has its optical axis intentionally tilted high enough to include the apparent hori-zon. This type of photograph may be used for the photogrammetric extension of vertical con-trol, photoalidade mapping, and for reconnais-sance photography.

Under the most favorable general conditions, photoalidade procedures are able to establish elevations from high-oblique photography to an accuracy of 4 to 6 feet; hence, this method is

UNITED STATES
DEPARTMENT OF THE INTERIOR
GEOLOGICAL SURVEY

FLIGHT DESIGN WORK SHEET

PROJECT GS- __GS-VAJM__   Name __Sweet Home__   State __Oregon__
(For RT-P only)
Requested by __Pac. Area__   Designer __J. H.__   Date __8/29/59__

Compilation method __Vert. ER-55__   Compilation scale $(\frac{1}{s})$ __1:24,000__

Contour interval (c.i.) __40/20 Supp.__   Flight direction __N-S__   Focal length __6"__

Season __April 15 - November 1__   Flight height (H_{opt}) __18,700'__

Special conditions __No Snow__

Ground elevations	Flight heights	Projection distances
h_{opt} __1600__ (From best available map)	H_{sl} __20,300__ (See diagram opposite)	d_{opt} __525__ (Vert. ER-55)
h_{max} __4300__	H_{min} __16,000__	d_{min} __455__
h_{min} __300__ h_{loc} __2000__	H_{max} __20,000__ H_{loc} __18,300__	d_{max} __559__ SEE REVIEW
C Factor $=\frac{H}{c.i.}=$ 467 + 935	$W_1 = 1.15 H_{min} = 18,400$	Sq. miles = 1,281

E-W dimension = 65,200/15 min.   N-S dimension = 91,150/15 min.

$G = 1.5H = 28,050$ (1)   L = .25G = 7012 Use L = 7000 $Q = \boxed{\text{E-W or N-S dimen. whichever applies}} - 2L = 51,200$

$N = \frac{Q}{W_1} + 1 = \frac{51,200}{18,400} + 1 = 2.8 + 1 = 3.8$ Use N = 4/15 min. $W = \frac{Q}{N-1} = \frac{51,200}{3} = 17,070$ (2)

$\frac{W}{H}$ SEE REVIEW (3)   Opt .91   Min .86   Max 1.05 (4) $\frac{B}{H} = .63$ (5)   B = 11,780   1.5B = 17,670   3.0B = 35,340

Theoretical length of flight lines $= \dfrac{4\,(6 \times 91,150) + 8\,(35,340)}{5280} = 468$ **mi.**

Maximum allowable = (6)   468 + 20 percent of 468 = 562 mi.

Remarks   Use 8 flight lines, varying flight height as shown in review

(1) G=1.5H (Vert.) = 1.41H (Conv.)

(2) W should not exceed W_1

(3) $\frac{W}{H}_{opt}$ should not exceed 1.10

(4) $\frac{W}{H}_{max}$ should not exceed 1.15

(5) $\frac{B}{H} = 0.63$ (Vert.) = 1.23 (Conv.)

(6) Add to theo.:   0-250 = 25%
250-500 = 20%
500-1000 = 15%
Over 1000 = 10%

REVIEW

Line	h_{min}	h_{max}	h_{opt}	h_{loc}	H_{sl}	H_{loc}	H_{max}	H_{min}	W_1	W (2)
1	300	3000	1300	1500	20,000	18,500	19,700	17,000	19,550	17,070
2	300	2000	1300	1000	20,000	19,000	19,700	18,000	20,700	17,070
3	400	2000	1300	1000	20,000	19,000	19,600	18,000	20,700	17,070
4	400	2400	1300	1300	20,000	18,700	19,600	17,600	20,200	17,070
5	500	3400	1300	1600	20,000	18,400	19,500	16,600	19,100	17,070
6	600	3400	1300	1600	20,000	18,400	19,400	16,600	19,100	17,070
7	600	4300	1800	2300	20,500	18,200	19,900	16,200	18,600	17,070
8	700	4300	1800	2300	20,500	18,200	19,800	16,200	18,600	17,070

FIGURE 7-8. Flight design work sheet (left half), high relief.

feasible only when the contour interval is at least 40 feet. The rough country and difficult access generally associated with the larger contour intervals often makes control extension by photoalidade an attractive, economical substi- tute for costly ground methods. Then too, whenever a short field season aggravates a need to control a large area in a short time, there is added justification for using this method.

7.2.4.6.3.1 Cameras. A standard 8¼-inch-

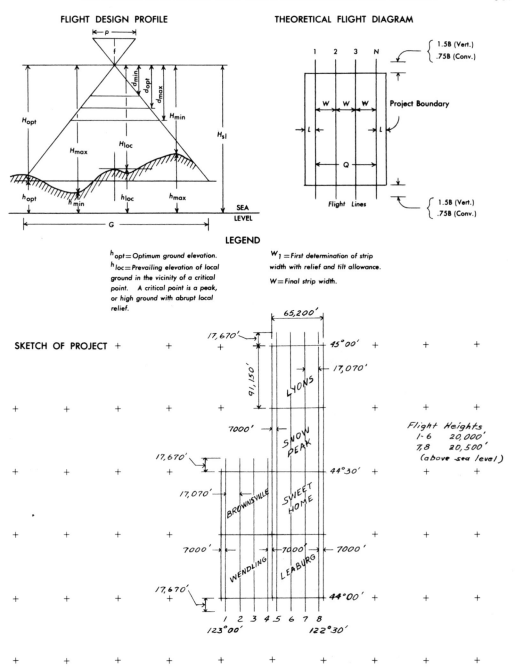

FIGURE 7-8. Flight design work sheet (right half), high relief.

focal-length camera is frequently used in high-oblique work. Its advantage over a standard 6-inch-focal-length camera is that the longer-focal-length camera produces a larger negative scale that facilitates control identification and photoalidade operation. For reconnaissance, the 6-inch camera may be preferable.

7.2.4.6.3.2 *Depression Angles*. When an 8¼-inch-focal-length camera is used, the optical axis should be depressed 20 degrees from the horizontal (*see* fig. 7-13). The depression angle

for a 6-inch-focal-length camera should be 30 degrees. These values insure the imaging of the apparent horizon on the format while allowing for up to 6 degrees of lateral tilt in the aircraft.

7.2.4.6.3.3 *Flight Height*. Since small vertical angles yield more accurate results in photoalidade operations, this type of photography should be obtained at the lowest flight height consistent with terrain conditions. Altitudes of 4,000 to 5,000 feet above mean ground elevation suffice for most projects, but higher altitudes

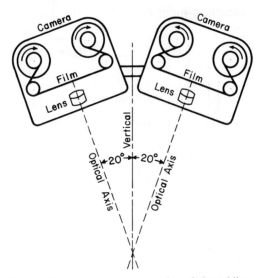

FIGURE 7-9. Camera arrangement for twin low oblique photography.

may be used when the terrain is unusually rugged. Deviations from designed flight heights should be kept to a minimum.

7.2.4.6.3.4 Usable Area of Coverage. At flight heights of 4,000 to 5,000 feet, the usable limits of coverage on an 8¼-inch-focal-length high-oblique photograph extend from about 1 mile to about 20 miles out from the flight line. The area encompassed between these two limits on each exposure is over 200 square miles; however, the portion actually used depends on the location and density of the control.

7.2.4.6.3.5 Base-height Ratio. While the photoalidade operation itself is monoscopic, the photoidentification of the known and supplemental vertical control can be accomplished more accurately from stereoscopic pairs. A proper value for the base-height ratio for side-looking obliques depends on the locations of the existing elevations and the supplemental points to be established. A base-height ratio of 0.64 is a safe design figure, since it guarantees that all

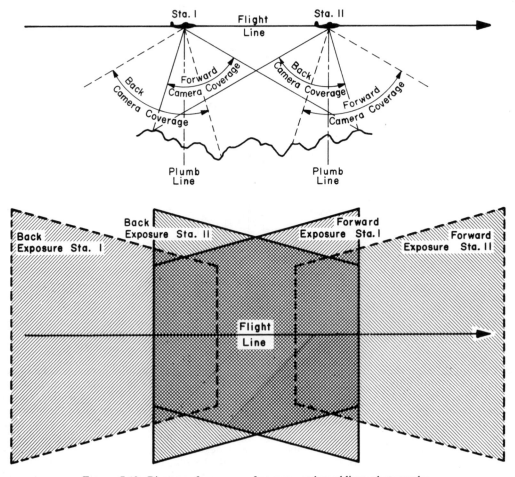

FIGURE 7-10. Diagram of coverage of convergent low-oblique photography.

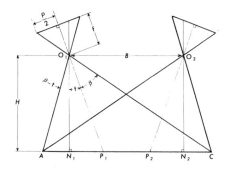

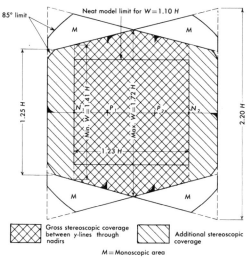

FIGURE 7-11. Geometry of convergent low-oblique photography.

when high-obliques facing south are required in the northern latitudes, they should be taken when the horizontal angle between the sun and the camera axis is as large as possible. For the same reason, eastward-looking exposures are best taken after about 11:00 a.m., while those facing west should be taken before 1:00 p.m. Naturally, any flight requirement to photograph into a 15-minute quadrangle from all four directions requires some compromise with these optimum conditions. In such cases (in the Northern Hemisphere) it is always advisable to fly the mission in the middle of the day and as close to the mid-June season as possible under the given conditions.

7.2.4.6.4 TRIMETROGON PHOTOGRAPHY

Trimetrogon photography is a name commonly applied to almost any three-camera system of aerial photography that furnishes horizon-to-horizon coverage. Although the cameras involved in this configuration were originally equipped with metrogon lenses, any wide-angle lenses of similar quality may be used. This type of photography is generally used for rapid and economical compilation of small-scale maps. Control may be extended by the recto-blique plotter or the photoangulator, while the map compilation is usually accomplished by means of the vertical and oblique sketchmasters or the photoalidade.

7.2.4.6.4.1 Camera Mounts. The three cameras are mounted in a manner shown in figure 7-14, with the middle camera in a vertical position and the two side cameras arranged to photograph obliquely to the left and right of the flight line. The optical axes of the oblique cameras are depressed 30 degrees from the horizontal to enable them to photograph both the horizon and an overlapping portion of the vertical camera coverage. Thus, when the three cameras are exposed simultaneously, a strip of ground extending laterally from horizon to horizon is recorded.

When the cameras are assembled for trimetrogon photography, their respective optical axes and fiducial marks normal to the line of flight should theoretically be in a common plane. While deviations up to about 2 degrees can be tolerated, the installation should be rigid enough to maintain the camera relationship to ±10 minutes of arc.

7.2.4.6.4.2 Flight Height. The flight height for trimetrogon photography is governed mainly by the planned manuscript scale which, in turn, is controlled by the mechanical limitations of the vertical sketchmasters to be used for compilation. These instruments are designed for optimum operation at a 2:1 scale reduction. For 1:125,000 or 1:250,000 publication scales, the manuscript scale is usually about 1:80,000.

ground coverage will be stereoscopic. If none of the control is located in the extreme foreground of the photographs, a somewhat larger base-height ratio can be tolerated.

7.2.4.6.3.6 Flight-line Locations. To extend vertical control for standard quadrangle mapping, the flight lines should be located parallel to and about 3 miles outside of each of the four sides of every 15-minute quadrangle involved. Of course, the camera must be pointed *into* the quadrangle area in a direction perpendicular to the line of flight. To alleviate any problems of hidden ground in rugged terrain, additional interior flights may have to be planned. Whenever unusual circumstances such as coastlines or island areas are encountered, other flight-line locations and directions may have to be adopted.

7.2.4.6.3.7 Time of Photography. To minimize the harmful effects of haze in high-oblique photography, every effort should be made to avoid backlighting conditions. Thus,

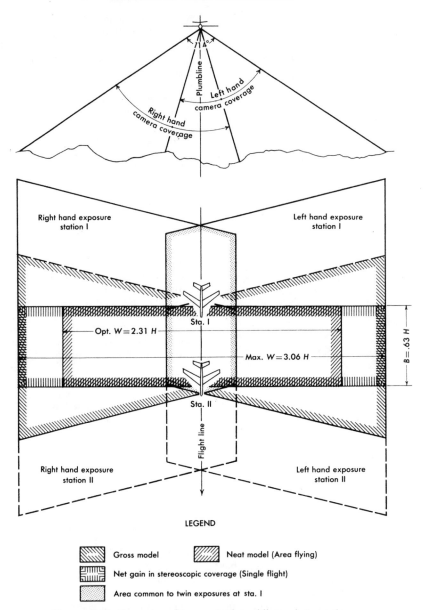

LEGEND

⬚ Gross model   ⬚ Neat model (Area flying)

⬚ Net gain in stereoscopic coverage (Single flight)

⬚ Area common to twin exposures at sta. I

FIGURE 7-12. Geometry of transverse low-oblique photography.

Thus, an aerial negative scale (for the vertical photograph) of about 1:40,000 and, for 6-inch-focal-length cameras, an optimum flight height of 20,000 feet would be required. Deviations from this altitude should not be permitted to exceed 500 feet.

For publication scales smaller than 1:250,000 the altitude above mean ground may be increased to 30,000 feet or higher, depending on the amount of planimetric detail required on the higher altitudes should be maintained within a range of ±750 feet.

7.2.4.6.4.3 Base-height Ratio. To enhance the photoidentification of control, it is desirable that trimetrogon photography be designed for complete stereoscopic coverage. A base-height ratio of about 0.63 will yield such coverage for both the vertical and the oblique exposures.

7.2.4.6.4.4 Flight-line Spacing. The spacing between flights of trimetrogon photography depends on the extent of coverage in the oblique photographs that is considered usable. Such spacing is customarily expressed directly in miles rather than as a function of the flight height. In order to avoid undesirable enlargement of detail when compiling from the oblique photographs, the compilation should not extend beyond the line where the photo scale is approximately equal to the manuscript scale. With a flight height of 20,000 feet and a manuscript

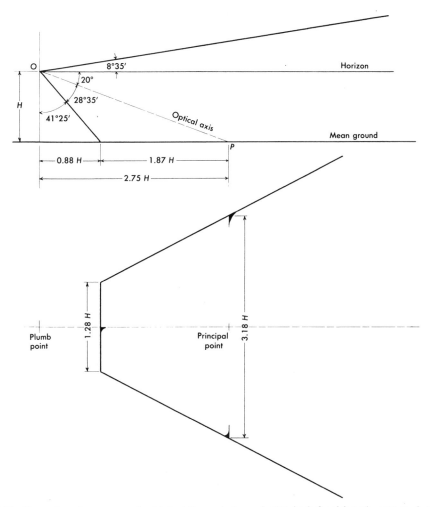

FIGURE 7-13. Geometry of coverage of a high-oblique photograph (8¼ inch focal length camera, 9 × 9 inch format, depressed 20° from horizontal).

scale of 1:80,000, this critical condition occurs about 8 miles out from the flight line. Considering that the facing obliques from the adjacent flights also cover 8 miles, the flight line spacing for these conditions would be about 16 miles.

For the smaller publication scales, where the indicated flight height is 30,000 feet or higher, the spacing between flight lines may be increased to about 24 miles. Both of these values for flight-line spacing should not be exceeded by more than 10 percent, so that the photo scales required for compilation may be retained within reasonable limits.

7.2.4.6.4.5 Flight-line Location. Whenever possible, trimetrogon flights are oriented parallel to the long axis of a project area. However, if high ridges are involved, and compilation is to be by sketchmaster, the flights should be arranged so that the ridge lines appear in profile on the oblique exposures. On the other hand, for compilation by universal plotter, flight lines should parallel the grain of the country and, where pos-

sible, center on high ridges. Whenever simultaneous complete vertical and trimetrogon coverage of an area is desired, this may be accomplished by flying in accordance with the flight design for vertical coverage. The oblique trimetrogon cameras would then operate on only every third or fourth flight, as necessary to obtain proper spacing for trimetrogon coverage.

In areas involving extensive coastline, one flight should be placed parallel to the coast and far enough inland to have the shoreline appear in the foreground of the offshore-looking obliques (about 12,000 feet inshore when the flight height is 20,000 feet). This arrangement facilitates the aerotriangulation and planimetric compilation phases. To aid in the contouring phase, it is advisable to fly an additional coastal flight about 12,000 feet offshore so that the coastline appears in the foreground of the inshore-looking obliques.

7.2.4.6.4.6 Time of Photography. Trimetrogon photography should be limited to those days when local climatic conditions are ideal.

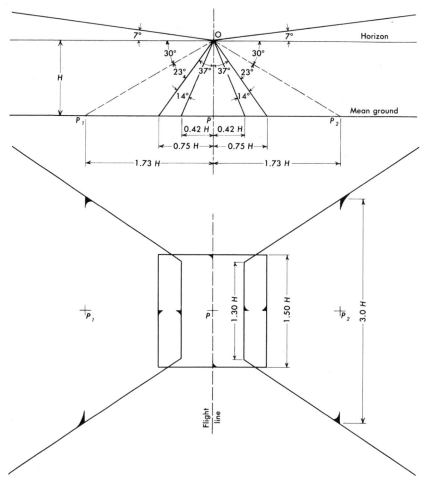

FIGURE 7-14. Geometry of coverage of trimetrogon photography (6 inch focal length, 9 × 9 inch format).

Minimum clouds and haze should extend at least 25 miles on each side of a flight. Also, to minimize heavy shadows and the detrimental effects of any haze that might be present, the flying should be restricted to the season and the part of the day when the sun is as high as possible above the horizon.

7.2.5 GROUND CONTROL

Ground control for mapping consists of a network of photo-identifiable points on the ground, for which values, referred to a horizontal and vertical datum, have been established. Survey requirements will vary with the scope and purpose of the project. Low-order surveys computed from an arbitrary origin with assumed starting coordinates and with a conveniently oriented coordinate system may be adequate for a small, isolated project. For larger projects and for projects in urban or built-up areas it has become increasingly important in recent years that control be of high order, and tied to and computed on a state or national datum (chapter VIII). Mapping at large scale for engineering

studies or design invariably requires that the map user be able to translate his design or locations to the ground. This can be achieved only through the use of the control network established for mapping, and realized by temporary, semi-permanent, or permanent markers set in the course of the original survey. Chapter VIII covers in detail all of the elements related to field surveys for photogrammetry. Thus, in this section, only basic considerations for project planning will be discussed.

7.2.5.1 CONTROL PLANNING FOR MAPPING AND OTHER USES

A control network established for maps at 1:1200 or larger scale should meet at least third-order accuracy standards. It is frequently necessary to establish a second-order baseline within the central region or on the perimeter of a strip-shaped or block-shaped region for later use in detail surveys or in staking out points for construction. This baseline is used for supplemental photo-control surveys and, if timing permits, the baseline can be targeted before photographing.

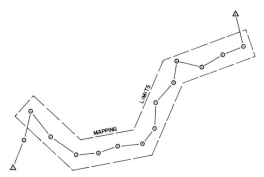

FIGURE 7-15a. Typical control configuration for strip mapping.

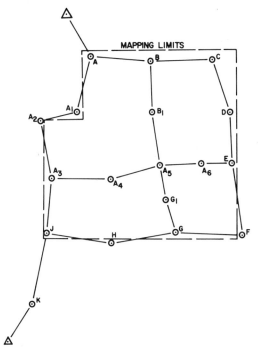

FIGURE 7-15b. Typical control configuration for block mapping.

Figure 7-15 illustrates two typical control-networks, the one in (a) being for a strip-shaped region, the one in (b) for a block-shaped region. The network shown in figure 7-15 (a) is for a map at 1:600 to be used in designing a highway; the one in figure 7-15 (b) for a topographic map at 1:600 with a two-foot contour interval to be used in designing an industrial park. In this second example, the network is arranged to include a boundary survey and stakeouts for construction, as well as control for mapping.

These examples show the careful consideration that must be given to planning control for large-scale mapping.

Terrain will invariably influence the location of a control network and may under certain circumstances dictate the method. Various meth-

ods for control extension, including traverse, triangulation, trilateration, differential and barometric leveling, *etc.*, are discussed in chapter VIII.

7.2.5.2 PHOTO-CONTROL POINTS—LOCATIONS AND CONFIGURATION

Photo-control point locations and patterns differ for photogrammetric projects which are fully-controlled by field surveys and those for which control extension is accomplished by aero-triangulation. An ideal situation requires that each stereoscopic model in a project contain at least three horizontal and four vertical photo-control points. Horizontal and vertical control can be established for the same points in some cases. For example, targeted points normally can serve as horizontal and/or vertical control stations. Errors in control cannot be detected during compilation when fewer than the minimum number of control points are provided. The addition of one horizontal and one vertical control point within a stereoscopic model will permit the isolation of a point in error. When selecting photo-control points, the planner must maintain some adaptability. His first responsibility is to attempt to achieve a strong configuration of control within each model. This consideration must be balanced against the factors of terrain and ground cover. To a large extent, the accuracy of a final map can depend on decisions made in the selection of control. Within economic and practical limits, increasing the density of vertical points in excess of the minimal requirement will enhance the accuracy of map contours and spot elevations.

7.2.5.2.1 FIELD CONTROL FOR MAPPING BY FIELD SURVEYS

When photo-control points are to be targeted prior to the photographic mission, the following factors should be recognized. Control points must be located so that mapping will not be compiled beyond the limits of the control. Figure 7-16 a shows an ideal control scheme for one model and for three. If the air stations shift either way along the line of flight as in figure 7-16 b, insufficient control in one or more models would result. A closer spacing between vertical control points in the central zone would increase costs but also insure that sufficient control exists in each model as shown on figure 7-16 c.

Reducing the interval, along the flight line, between control points to one-half the distance between air stations insures that at least a minimal number of control points appear in each model. Some effort by the aircrew can be made to hold air stations to their planned positions, but some shifting is usually unavoidable. Locations of control points should be planned to allow for this eventuality.

A special situation in mapping strips is illustrated in figure 7-16 d. In this type of project, a

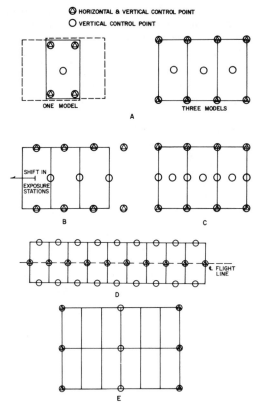

FIGURE 7-16. Control schemes for various configurations of photography.

racy of the ground control, the scale of the aerial photography, the precision of the photogrammetric instruments, and the procedures followed in each phase. A generalized ground-control scheme, for a block of twelve models, suitable for control extension by aerotriangulation, is illustrated in figure 7-16 e. In most cases, an ideal ground-control configuration will not be obtainable within practical limits. In these instances, additional ground-control should be planned for critical and accessable locations throughout the project.

7.2.5.3 HORIZONTAL GROUND CONTROL

Horizontal-control surveys are necessary to establish scale, azimuth and, in most cases, a coordinate system. The accuracies required of these surveys are determined by the uses that will be made of them. Control for mapping only is normally done to third order standards. Traversing with electronic distance-measuring equipment and one-second theodolites is a widely used method for extending horizontal control. Accuracy with this method is normally better than third order when proper observational procedures have been used. Triangulation is best suited to large projects and is generally the most accurate method for horizontal control extension in mountainous terrain. A triangulation net can be strengthened with electronic distance-measurements on additional sides of a figure. Trilateration involves measurement of every side of every triangle within a control net.

7.2.5.4 VERTICAL GROUND CONTROL

Vertical control is required to establish a vertical datum for the project. Vertical coordinates at photo-identifiable locations enable the compiler to orient or "level" a stereoscopic model with respect to that datum. The lowest accuracy required for benchmarks set in the course of a photo-control or baseline survey is normally third order. Vertical coordinates for large-scale mapping are normally required to be accurate to within 1/10th of the specified contour interval. Small-scale mapping, for which elevations are abstracted from various source material and averaged out in orienting the model, is an example of an exception to the rule. Vertical-control for large scale mapping is established by differential levelling. In mountainous terrain and for map-contour intervals of two metres or more, trigonmetric levels can be used. Barometric levelling for small-scale or reconnaissance mapping is economical and, if meteorological conditions are carefully taken into account, reliable.

7.2.5.5 CONTROL POINT IDENTIFICATION

Precision in large scale measurements and mapping by photogrammetric methods cannot be achieved unless an exact correspondence is found between photographic detail and an adequate number of points on the ground for which

line of horizontal control is planned for the central zone of the strip. Vertical-control points are set right and left at the approximate effective model limits. Again, a proper spacing of the control along the line of flight is necessary to insure that at least a minimal number of control points appear in each model.

7.2.5.2.2 CONTROL EXTENSION BY AERIAL TRIANGULATION

Aerial triangulation is a method of increasing the density of photo-control for photogrammetric use. This method is described fully in chapter IX. In general, the advantages of control extension by aerotriangulation are first, a reduction in the amount of field surveying required; second, more versatility in locating control points (inaccessable areas can be accomodated); third, photo-control, derived by this method, can be established at the best locations within the stereo-model; fourth, since ground-control problems are resolved during the computational phase, it can be reasonably assumed that model setup time, for map compilation, will be reduced. Various methods are currently being employed, using disparate types of equipment to accomplish aerotriangulation. It can be surmised that the results will vary and that this variance will be directly related to the density and accu-

coordinates and elevations have been determined precisely. This can be done by setting targets at critical locations throughout a region before photographing, or by careful selection of well defined points after photographing.

7.2.5.5.1 TARGETING

In planning for targeting a project, the following should be considered. Targets should be placed at any existing (and usable) control falling within the proposed limits of photography. If control surveys of a higher order are required by the specifications, targets should be placed in the approximate proposed locations of the baseline stations. Targets should be placed at the approximate neat limits of the model and at a pre-determined interval in the direction of proposed line of flight to serve as wing points for which elevations and, in some cases, coordinates can be later established. Supplemental targets should be placed at regular intervals in the central zone of the proposed flight strip to serve as "extra" horizontal and/or vertical photo ties. In addition, boundary corners can be targeted for precisely relating an actual boundary to a map without resorting to field surveys to corners from the control network. If photogrammetric measurements are to be utilized for computing volume or earthwork quantities for the same area over an extended period of time, a targeting of an adequate number of control points (preferably the same points each time) before every photographic mission will insure correspondence between the various measurements with respect to the original project datum. Targets provide a precise and, unless disturbed prior to photography, reliable image point which can virtually eliminate misidentification problems in orienting the model. Accordingly, targets must be of appropriate size and shape and sufficiently reflective or light-absorbent to create a high contrast against the background. White plastic sheeting works satisfactorily on most dark soils and grasses, but poorly on snow or light reflective surfaces. In these situations, a dark material such as felt building paper would give greater contrast. Highly reflective black or white plastics can create a "hot spot" at critical angles in the photographs, obscuring the image entirely. A matte surface of reinforced plastic sheeting is manufactured with black and white target configurations on opposite sides. Plastic materials do not disintegrate rapidly and can be unsightly if left in the field after use; therefore, clients may require the removal of all targeting material after project completion. Control points can be painted, where legal requirements permit, on hard surfaces such as concrete and asphalt paving. Such surfaces tend to be reflective, however, and require a painted black matte background to improve contrast. The size of targets set is determined by the photographic

scale of the aerial photography. A general rule for minimum size is .01 of the photo scale for the longest dimension. Target sizes larger than this will be more easily visible and may be desirable on some projects. Targets smaller than this may be obscured in the photography. Three- and four-legged crosses are common target designs. Targets with a "T" or "V" shape are occasionally used as well, to differentiate between different types of survey points.

7.2.5.5.2 SELECTION OF PHOTO-IDENTIFIABLE OBJECTS FOR USE AS CONTROL POINTS

When targeting a project is impractical, it becomes necessary to select sharp, well-defined points that can be readily identified in the aerial photography and on the ground, to serve as horizontal and vertical photo-control points. Consideration must also be given, in the selection process, to the amount of surveying required. In selecting horizontal tie-points for mapping at 1:480 or 1:600, a great deal of precision is required. At this scale, a point such as the corner at the end of a paint stripe, a post at the intersection of fence lines, the center of a small manhole cover, or a sharp corner of an asphalt patch or concrete slab makes an excellent horizontal photo-control point. These can also be used as vertical control points, assuming relatively flat ground in the vicinity of the point and, in the case of the fence post, that the elevation of the ground at the base of the post is known, and the inside or outside corner is specified for precise identification. For mapping at 1:1200 or 1:2400, objects such as the center of the base of a transmission pole, building corners unobscured by shadows or overhang, and fence intersections make acceptable horizontal tie-points. At these scales, contours at two and five foot intervals are common. Vertical tie-point identification thus becomes less difficult. Points such as road intersections, the center of a bridge, and the center of a road opposite a transmission pole, make acceptable vertical tie-points.

For mapping at 1:4800 and 1:6000, large physical features such as road intersections, fence intersections, and so on are normally selected as control points. Obviously there is overlap in the type of points that can be used at various map scales. An experienced planner will, of necessity, use a combination of points on any large project. He should select precisely definable objects for control points whenever possible, while recognizing that because of the nature of terrain or other conditions, the next control point may be less than ideal. These less desirable control points can be supplemented by extra points to avoid a scaling or levelling error by a compiler. In any case, where photo-control points are vague, the surveyor should provide a sketch on the back of the control photographs showing the relationship of the points to clearly defined features in the vicinity. Selection of photo-control points should always be made using a stereoscope. Points in the overlap between models should be viewed in both models to insure that shadows, foliage, or building overhang do not obscure the image in either model. Points selected should be pin-pricked using a fine needle which, when correctly used, will pierce the emulsion of the photograph and only slightly pierce the base. This results in a very well-defined and precise hole which can only be seen easily when the print is held up to a lighted background. If photo-control points are selected in the office, it is helpful to do the point-identification and pin-pricking on a light table. This provides a soft, lighted background and a hard surface to keep the pin from piercing the photograph's base too far.

7.2.6 AEROTRIANGULATION

Densification and extension of field control by aerotriangulation has become an accepted practice on most large- and intermediate-sized photogrammetric projects. The decision to use aerotriangulation or full field control on smaller projects is usually based on economic considerations, with a breakeven point around three to four stereoscopic models. Recent improvements in aerotriangulation have brought improved accuracies equalling, in some instances, the results of third-order ground surveys. Fully analytic aerotriangulation also permits the calculation of the orientation elements of each model, shortening stereoplotter set-up time considerably. Aerotriangulation has been applied successfully to cadastral surveys and for establishment of third-order control on construction and similar projects.

7.2.6.1 SELECTION OF METHOD

Two basic types of photogrammetric instruments are used to perform aerotriangulation, and three different methods are used for adjustment of the measurements. A suitably equipped stereoplotter can be used for performing aerotriangulation either by semi-analytic or analogic methods. An optical measuring device called a comparator is used for measurements in the fully analytic method. The following paragraphs outline basic considerations in the selection of a method of aerotriangulation. This subject is covered in depth in chapter IX.

7.2.6.1.1 FULLY ANALYTIC METHODS

Fully analytic methods require more sophisticated computer processing than the other two methods described below. The first step in the procedure is the marking and transfer of common points between models and between adjoining strips. To supplement targeted and/or well-defined image points, non-discrete points are marked by drilling a precise hole in the emulsion of the diapositive. These points are usually located in the strongest possible configuration (classic configuration) of at least six points per model, three each along the approximate extension of fiducial lines defining the effective model. Equipment is now available which combines point-marking and transfer with coordinate measurement. Coordinates are measured on a comparator to 10^{-3} mm (one micrometre). They are measured for both ground-control points and bridging pass points. These coordinates are processed to remove known systematic errors, and are then used to compute the spatial orientation of each photograph in the block. From these computations and a coordinate transformation, the ground coordinates of the unknown supplemental pass points can be calculated, and checks upon the ground control points made. Data computed for the photographs can also be used for stereoplotter orientation. Several programs have been developed for the adjustment of aerotriangulation measurements. Programs which adjust the entire block of photographs simultaneously are more rigorous and provide greater accuracy than programs using approximate or sequential methods. Recent progress has been made in the development of an adjustment program which compensates for factors such as non-uniform film deformation, departures from flatness of camera platen, and changes in lens distortion. This method is known as the bundle adjustment method with self-calibration. Normal horizontal accuracy for photocontrol, established by fully-analytical aerotriangulation, properly conceived and executed, from the photographic mission through final adjustment, is predicted at one part in 10,000 of the photographic flight height. Vertical accuracy results normally are closer to one part in 5,000 of the flight height, although results superior to this have been achieved.

7.2.6.1.2 SEMI-ANALYTIC METHODS

Semi-analytic or independent-model aerotriangulation can be performed with a suitably equipped stereoplotter. The models are relatively oriented in the stereoplotter and measurements of pre-selected pass points are recorded in the model coordinate system. The recording can be done more efficiently if the stereoplotter is connected to a digitizer with output to either a tape or a card punch. Final adjustment and transformation are done by computer, using an adjustment similar to that used in the analytic method. Since the operator performs the relative orientation process, some error may be introduced into the model's coordinates. Correction for systematic camera and film deformations are not usually made in this method. Tests of a simultaneous adjustment for independent models give results that are similar to those achieved with fully analytic methods. Less rigorous, sequential methods of adjustment are often used in semi-analytical aerotriangulation, but may give erratic results when control errors are encountered.

7.2.6.1.3 ANALOGIC METHODS

Some "universal" type stereoplotters are designed to perform aerotriangulation by "holding" the orientation of one of the photographs fixed. A photographic "bridge" is then built by successively connecting stereomodels between widely spaced control points. The individual stereomodels tend to drift and develop an imperfect scale and level as they are connected. Residual errors are measured in the last stereomodel at the control points and adjusted back through the intermediate photographs.

7.2.6.1.4 HYBRID METHODS

Innovations on standard aerotriangulation methods have been developed to meet special needs. Block-adjustment programs have been modified to use distances or directions between points for "control" as well as coordinates of points. This method is designated "bridging with independent horizontal control." With minor changes, the simultaneous adjustment of photographs of different photo scales and focal lengths is feasible. The inherent high accuracy of analytic aerotriangulation has given rise to the "high-low" method of control extension. In this procedure, aerotriangulation is performed on a higher overlapping flight and then transferred stereoscopically to the lower mapping photography. A considerable savings in surveying costs can be realized by using this procedure.

7.2.6.2 CONTROL POINT POINT PATTERNS FOR AEROTRIANGULATION

The problem of planning a control-point network in detail is complex, and while all aspects cannot be covered within the scope of this section, the following is an outline of basic factors. The location and spacing of control points are affected by accessibility, terrain, ground cover, required accuracy of the aerotriangulation, size and shape of area, and by the method of aerotriangulation to be utilized. Networks, consequently, must be planned after careful consideration of each of these factors.

Experience with aerotriangulation has produced some general principles which can be used as a guide to planning.

Horizontal Control Points

a. Strongest configuration results from points around the perimeter of the region.
b. Small survey errors are difficult to detect during an aerotriangulation adjustment when control is widely spaced.
c. Horizontal control points spaced at intervals of 5 models will normally be adequate.
d. Fewer horizontal control points are necessary when using a simultaneous or "bundle" method of aerotriangulation adjustment than with other methods.

Vertical Control Points

a. Should be located throughout the region as well as around the perimeter.
b. Small survey errors are nearly impossible to detect during adjustment.
c. Vertical control points spaced at intervals of 3 to 4 models are normally sufficient.
d. Fewer vertical control points are necessary with simultaneous methods of adjustment.
e. The greater the density of vertical ground control, the better the final results.

Figure 7-16e is a sample control network for fully analytic aerotriangulation of a 12-stereomodel block of photographs. Other networks using more or fewer control points may be desirable depending on project requirements.

7.2.6.3 ECONOMY OF AEROTRIANGULATION

The economy of aerotriangulation is affected by the size of the region and by the amount of surveying needed. The amount of ground control necessary for full control of models on projects of one or two stereomodels is almost the same as would be required for aerotriangulation. Aerotriangulation usually becomes competitive with ground surveying on projects of four or five stereomodels. Projects larger than this show increased savings through the use of aerotriangulation. In ideal situations where the line of sight permits spacing of control at intervals of four or more models, savings of up to 50% in field survey costs can be anticipated. The use of aerotriangulation for projects in areas with extremely rugged terrain and unsuitable ground cover may introduce even greater savings. Modern aerotriangulation methods are usually most economical in terrain where surveying is most difficult. Some savings can also be derived by using the method for detection of errors in ground control and by using the orientation elements derived in the method for setting up the model.

7.2.7 STEREOCOMPILATION

Planning for stereocompilation involves consideration of all prior phases of the photogrammetric process, from the introduction of the requirement through each operational phase. Each phase must be planned in a manner that insures that the stereocompiler and equipment will work within a safe and prudent range of maximum capability.

7.2.7.1 BASIC CONSIDERATIONS

The information to be shown on the map should correlate exactly with the user's needs. Unnecessary or redundant information can obscure important detail and confuse the user. At the other extreme, important cartographic information overlooked during compilation can jeopardize an entire project. A precise understanding of the mapping requirements and the objective of those requirements is essential to a compiler in meeting this responsibility.

Other factors affecting map compilation include:

a. the camera system;
b. scale of the photography;
c. accuracy of the control;
d. type of stereoplotter to be utilized; and
e. nature of the project terrain and ground cover.

7.2.7.1.1 CAMERA FACTORS

The vertical accuracy of topographic mapping is greatly affected by the camera used in the photographic mission. The advantages of precise cameras for mapping projects have been discussed previously in this chapter. Some mapping projects may require the use of aerial photography taken by a non-precise camera. In these instances, calibration of the camera system and subsequent compensation in the stereoplotter is essential. Mapping from aerial photographs taken at different focal lengths is sometimes necessary or desirable. Low distortion, short focal-length photography (90 mm.) can provide improved vertical accuracies in open and flat terrain, but many stereoplotters are not designed to accommodate this type of photography. Low-distortion photography at focal lengths of 152 mm. provide a good compromise between vertical accuracy and the utility of accommodating varied terrain. Most stereoplotters are designed for use with this focal length.

Low-distortion photography at longer focal lengths offers some advantages in mountainous terrain or timbered areas. Some stereoplotters can use this type of photography for mapping, but vertical accuracy is considerably diminished.

7.2.7.1.2 FLYING HEIGHT

Flying height is the distance of the aircraft above the ground at the instant the photograph was taken. Mapping accuracy is inversely proportional to flying height, all other factors being equal. Since photographic scales are proportional to flying heights for cameras of similar focal lengths, it can be surmised that mapping accuracies are also inversely proportional to the photographic scale. This accuracy-height-scale relationship is inherent in the geometry of the aerial photograph and will hold true until other factors such as the stereoplotter type are varied. This relationship has also been used as a method of rating stereoplotters described below.

7.2.7.1.3 GROUND CONTROL

Since ground-control surveys provide the framework within which all photogrammetric work is performed, mapping accuracies are directly affected by ground-control accuracies. While ground-control surveys should always be performed to accepted standards, these standards are usually overdemanding from a photogrammetric viewpoint, and small deviations from the standards will be inconsequential. Generally, horizontal-control errors of less than 1/100 of an inch at the mapping scale, and vertical-control errors less than 1/20 of the contour interval, will not affect mapping accuracies. Control errors greater than these will have an effect on mapping accuracies in direct proportion to the magnitude of the errors. Unless sufficient control points in proper configuration are located in the model, even major survey errors may go undetected. It is of utmost importance that ground-control surveys be performed to specified accuracies, closed and checked to remove the probability of errors before stereocompilation begins.

7.2.7.1.4 C-FACTOR

Most mapping projects must have a pre-determined accuracy, pre-determined in that the map user may specify the accuracy of the map produced. National Map Accuracy Standards (section 7.1.3.6) are the normally accepted guidelines for specifying map accuracy. As a result of experience in using various stereoplotters under different conditions, photogrammetrists have established a rating system which uses a criterion called the "C-factor." The C-factor expresses the ratio of the highest flight height which will achieve a specified contour accuracy to the desired contour interval. The C-factor is related to flying height and the map contour interval by:

$$\text{C-factor} = \frac{\text{Flying Height of Photography}}{\text{Contour Interval of Map}}.$$

The C-factor will differ widely between different combinations of camera, photography, stereoplotter, and plotter operator. A C-factor is normally given for each stereoplotter, although the C-factor is influenced by all of the elements involved in the process. The nature of the terrain, quality of the photography, density and quality of control, instrument calibration and adjustment, and the plotter operator, all combine to influence the C-factor. It can be stated that the C-factor is at best a guide to planning, and should be used in the proper manner with due consideration of the limitations of any photogrammetric system. The normal working range of available equipment should not be stretched to its utmost without giving consideration to the alternative approach of additional control.

Plotter	Normal Working Range (C-factor)
Double-projection Type Optical-Mechanical,	700—1500
Analytical (computerized)	1200—2000+

Under ideal conditions and/or lower map-accuracy requirements, these limits can be and are exceeded. A prudent planner will use these numbers and the experience gained with his own equipment as a guide to designing a proper flight plan.

7.2.7.1.5 Spot-Elevation Factor

Since the C-factor has been derived as a guide to planning for the achievement of a specified contour accuracy, it follows that a corresponding factor can be determined for achieving specified accuracies of spot elevations. This factor can be stated as the ratio of the flight height to the allowable spot-elevation error, or:

$$\text{Spot-elevation Factor} \equiv \frac{\text{Flight Height}}{\text{Allowable Spot Elevation Error}} \equiv \text{S-factor.}$$

For example, at a flight height of 6000′ and an allowable spot-elevation error of 1.25′, the spot elevation factor is 6000/1.25 or 4800. Since plotting contours normally results in an error approximately twice as great as plotting a discrete point, the accuracy of a spot elevation can be assumed to be one-fourth the contour interval. The resultant S-factor can therefore be predicted as four times the C-factor. This increased accuracy results from the fact that the operator has the ability to read the elevation of a specific point with greater precision than to track a contour across terrain of varying characteristics.

7.2.7.2 Map Scale, Contour Interval and the Selection of Flight Height

The maximum allowable flight height is normally governed by three elements: the vertical accuracy requirement of the final map; the horizontal scale of the final map; and the range of C-factors of the plotting equipment to be utilized. The following examples (assuming a nominal focal length of 152 mm.) illustrate the options to be considered in planning.

Example A

Requirement: Map Scale 1:2400; contour interval five feet (5′); rural site moderate to heavy relief; and National Map Accuracy Standards.
1. Available equipment: Double-projection plotter rated at a maximum C-factor of 1200.
   Recommended Approach: C-factor governs—fly at 6000′ CI X C-factor or 5 × 1200), compile at 1:2400, 5′ contour interval.
   Map Scale, contour interval, and available equipment compatible with a flight altitude above mean terrain of 6000′.
2. Available equipment: Optical-mechanical plotter rated at a maximum C-factor of 1800.
   Recommended Approach:
   (a) Based on C-factor, maximum flight altitude is 9000′.
   (b) Based on maximum practical photography-to-map scale ratio of (8X), a flight altitude of 9600′ is indicated. At a flight altitude of 9000′, the plotting ratio is 7½X. C-factor limits altitude to 9000′, or a more conservative approach may be taken using an altitude of 8400′, resulting in an improved C-factor of 1680, a plotting ration of 7X, and a slightly better map.

Example B

Requirement: Map Scale 1:600; contour interval two feet (2′); suburban site moderate to heavy plan;

moderate to minimal relief; and National Map Accuracy Standards.
1. Available Equipment: Double-projection plotter rated at a maximum C-factor of 1500.
   Recommended Approach:
   (a) Based on C-factor, flight altitude is 3000′.
   (b) Based on normal plotting ratios (5X), altitude is 1500′.
   A flight height of 3000′ would take maximum advantage of the instrument's rated C-factor. This would also result in a manuscript scale of 1″ = 100′, which, if enlarged to 1″ = 50′, would introduce horizontal error in the final map as well as an inherent loss of detail caused by the decline in resolution at this higher altitude. Another problem that may be encountered with this approach is the moderate to minimal relief factor. Flat terrain places a severe burden on the plotter operator's ability to make precise measurements of parallax difference (elevation determination) under the best of circumstances. The higher altitude photography in this case would most likely result in vertical accuracy lower than standard.
   A flight height of 1500′ would result in a C-factor of 750. The map could be compiled directly at 1:600. Loss of image resolution would be minimal and the vertical accuracy of the final map would be improved.
2. Available Equipment: Optical-train plotter rated at a maximum C-factor of 1900.
   Recommended Approach:
   (a) Based on C-factor, greatest flight altitude is 3800′;
   (b) Based on greatest practical plotting ratio (8X), flight altitude is 2400′.
   Alternative (b) will result in an effective C-factor of 1200 with acceptable image resolution. Fly at 2400′ above mean terrain.

In some cases, map scale, contour interval, and C-factor are not readily compatible and the flight altitude must be computed to satisfy the most restrictive condition. If mapping at 1:4800 with two-foot (2′) contours is required, a recommended approach is to photograph at 12,000′ above a mean elevation of terrain for planimetric compilation and 1:4800 (or 1:3000 depending on plotter available) for contours. The advantage of compiling planimetric features from the 1:12,000 scale photography is a reduction in the amount of horizontal ground-control required. In some cases, targeting of existing control or use of National Geodetic Survey or Geological Survey intersection points such as water towers, stacks, church spires, etc., will suffice.

7.2.7.3 Sequence of Plotting Operations

The usual order of plotting operations is:

a. planimetry,
b. contours, and
c. spot elevations.

In urban or highly built-up areas, a check list should be prepared of what should be shown on the map. Linear features such as drainage, railroad tracks, and roadways are usually drawn first, followed by buildings and structures. Other miscellaneous planimetric detail such as utilities are then added. Contouring follows the planimetric compilation. Index contours are sometimes drawn first, to identify areas where intermediates are not necessary and where spot elevations may be necessary. Contours are normally started at the edge of a model and followed to the model's limits. Good practice dictates that a compiler draw complete features such as hills and depressions as encountered. Contouring is usually done in one direction, working from a lower elevation upward. When all the contours have been drawn, the plot should be checked for anomalies such as dropped or missing lines around closed topographic features. Spot elevations are then added at hilltops, saddles, and as required to accurately define terrain characteristics. Newer electronic plotting tables have capabilities that permit automatic squaring of buildings and other structures, curve fitting, and point-to-point line plotting. It is possible to produce an inked manuscript requiring little or no final drafting. The stereocompiler becomes effectively an expert draftsman and should plan his work accordingly. With other stereoplotting equipment, the stereocompiler should make notations or symbolize his line work to assist the draftsman in finishing the map correctly. Before the model is removed from the plotter, a complete record of the model, including, readings of all orientation settings, should be prepared. This information insures rapid re-setting of the model and serves as a reference for future questions on map accuracies.

7.2.8 MAP EDIT

Normal quality-control procedures dictate that each compilation manuscript be edited before final drafting or delivery to insure, insofar as is possible, that the manuscript contains no discernable error, is conventional and consistent in expression, and will be able to be interpreted by the user. The editor's function is to produce a map on which the names, labels, and notes are accurate, and to insure that the map conforms to project specifications as to content and format. The editor is responsible for the following operations:

  evaluating and verifying available data in order to place information on the map;
  verifying the placement of names and labels;
  verifying the placement and density of contour values and spot elevations;
  reviewing values for monumented control and grids;
  examining mapped features such as drainage patterns and contours;
  checking ties with adjoining sheets; and
  making an overall final inspection to determine that the map is complete and correct with respect to content.

When a prescribed sequence of editing operations is followed, omissions and errors that will at least prove inconvenient and perhaps also costly to the user can be prevented.

7.2.9 FIELD CLASSIFICATION SURVEYS

When required by the specifications, this operation can be scheduled prior to or after map compilation, and depending on map scale and proposed use, can involve varying amounts of effort. For map scales of 1:2400 or smaller, field classification is normally of a generalized nature. Information such as the identification of cross-county transmission lines, land use, road names, landmark buildings, and cemeteries, might be required. At larger scales such as 1:480, the level of information required, especially in built-up areas, is much greater. Features such as surface utilities and buildings are classified as to size, type, and use; lone, large trees and ornamental plantings are classified as to size and type. These examples do not cover all of the many and varied items that might be requested. In some cases, the information will be supplemented by elevations at drainage structures or at other locations critical to a design. Ties to property corners from the baseline may be required so that this information can be included on the final map. The project specifications are the final determinate in setting forth the level of field classification activity. The project planner should have a firm understanding of the requirements before finishing his plan or job order. Field classification surveys, while feasible after compilation, are most practical and economical when accomplished prior to map compilation. In this way the compiler has the field information at hand and, in the case of larger scale maps, the advantage of being able to locate and identify objects or features which might not be able to be interpreted otherwise. In most instances, this saves reorienting the stereomodel at a later date for addition of detail. The most effective tool in field completion surveys is an enlargement of the aerial photograph. When properly used and interpreted by a thorough and alert team, this item, completely annotated, will aid in the preparation of a concise and useful final map.

7.2.10 MAP TESTING AND FIELD COMPLETION

The accepted method to insure that a map is in compliance with the standard of accuracy is to check it in the field. Contours and spot elevations are checked by the profile method. A profile should be at least five inches long at map scale and should cross at least ten contour lines. The number of profiles required will vary depending on the project. For some mapping, one profile for each final map sheet is normal. Topographic and planimetric features are checked by surveying from the project baseline to well-defined objects and comparing the coordinates of these points as surveyed with values for the same features from the map. Ground control is presumed correct when it satisfies the specified

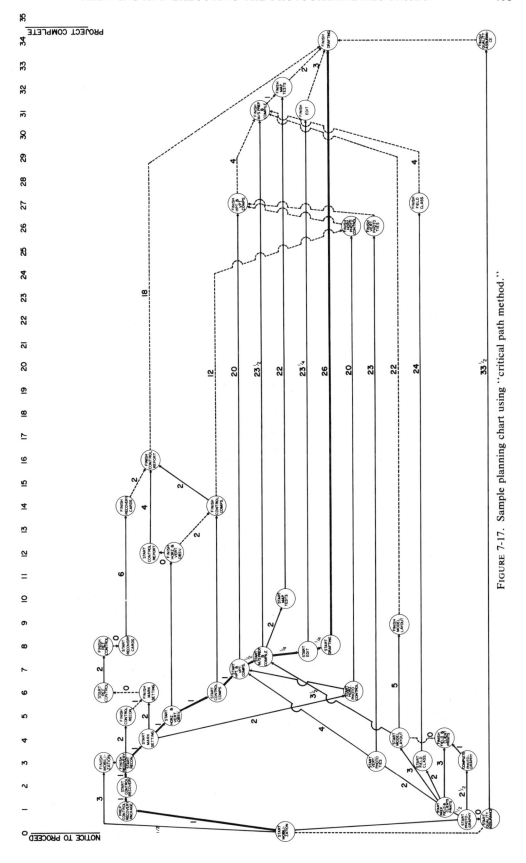

FIGURE 7-17. Sample planning chart using "critical path method."

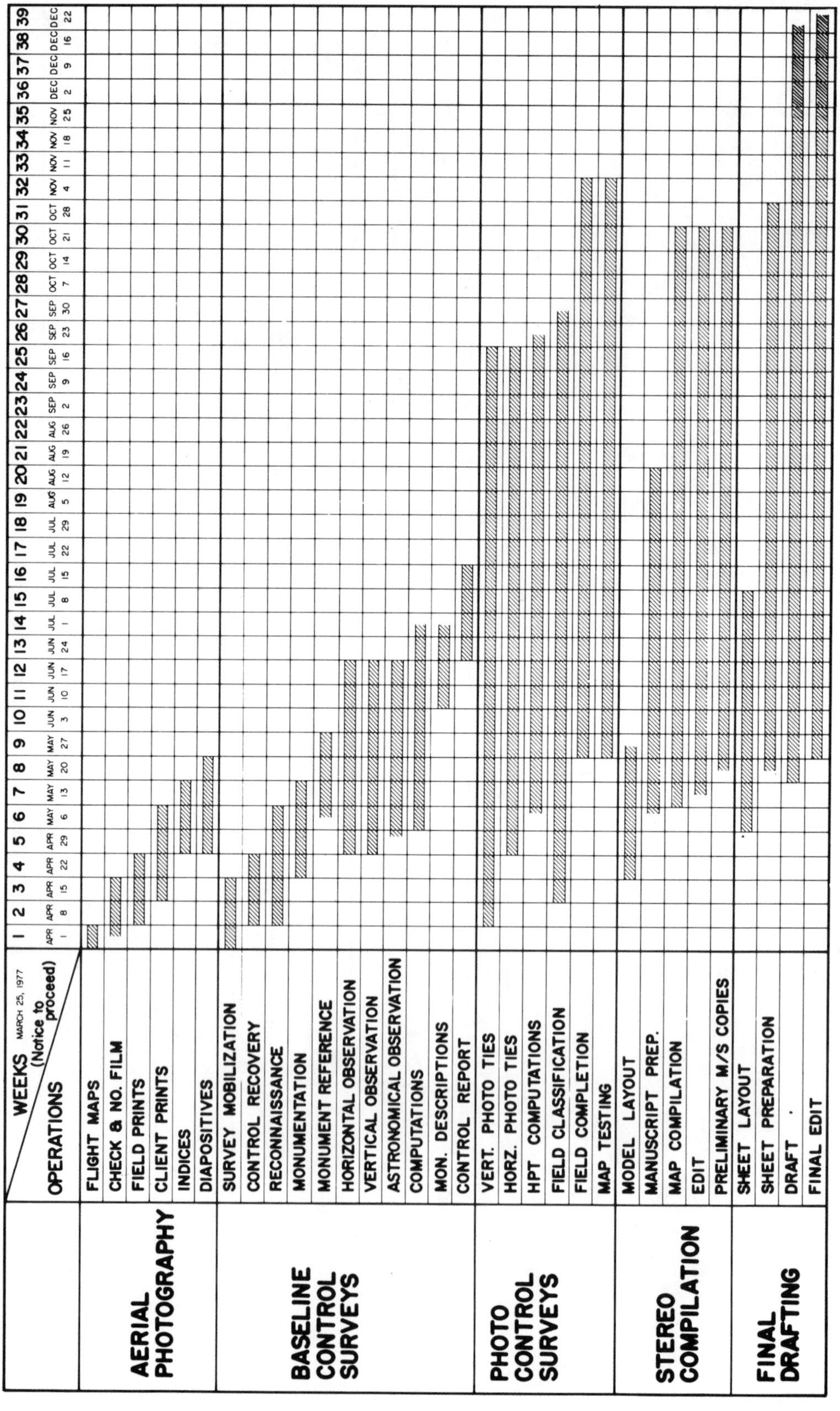

WEEKS MARCH 25, 1977 (Notice to proceed)

1	2	3	4	5	6	7	8	9	10	11	12	13	14	15	16	17	18	19	20	21	22	23	24	25	26	27	28	29	30	31	32	33	34	35	36	37	38	39
APR 1	APR 8	APR 15	APR 22	APR 29	MAY 6	MAY 13	MAY 20	MAY 27	JUN 3	JUN 10	JUN 17	JUN 24	JUL 1	JUL 8	JUL 15	JUL 22	JUL 29	AUG 5	AUG 12	AUG 19	AUG 26	SEP 2	SEP 9	SEP 16	SEP 23	SEP 30	OCT 7	OCT 14	OCT 21	OCT 28	NOV 4	NOV 11	NOV 18	NOV 25	DEC 2	DEC 9	DEC 16	DEC 22

OPERATIONS

AERIAL PHOTOGRAPHY
- FLIGHT MAPS
- CHECK & NO. FILM
- FIELD PRINTS
- CLIENT PRINTS
- INDICES
- DIAPOSITIVES

BASELINE CONTROL SURVEYS
- SURVEY MOBILIZATION
- CONTROL RECOVERY
- RECONNAISSANCE
- MONUMENTATION
- MONUMENT REFERENCE
- HORIZONTAL OBSERVATION
- VERTICAL OBSERVATION
- ASTRONOMICAL OBSERVATION
- COMPUTATIONS
- MON. DESCRIPTIONS
- CONTROL REPORT

PHOTO CONTROL SURVEYS
- VERT. PHOTO TIES
- HORZ. PHOTO TIES
- HPT COMPUTATIONS
- FIELD CLASSIFICATION
- FIELD COMPLETION
- MAP TESTING

STEREO COMPILATION
- MODEL LAYOUT
- MANUSCRIPT PREP.
- MAP COMPILATION
- EDIT
- PRELIMINARY M/S COPIES

FINAL DRAFTING
- SHEET LAYOUT
- SHEET PREPARATION
- DRAFT
- FINAL EDIT

FIGURE 7-18. Sample planning chart using bar chart format.

standards for accuracy. Field-completion surveys are required when areas are obscured from view on the photography.

In these instances the topographic features must be mapped by conventional survey methods. These methods may include planetable or profile-and-cross-section methods, depending on the area and the expertise of the surveyors. In some cases field completion may only consist of the verification of location and classification of certain features suspected of being overlooked in field classification. Some projects require field completion surveys in areas where the ground is only partially obscured and contours would normally be shown as dashed lines. The dashed lines are upgraded to solid lines through field completion surveys and through the correct interpretation of field and photogrammetric information by an experienced editor. Properly planned checks and field-completion surveys can be coordinated and done during the same phase of field activities. It is incumbent upon the editor or quality control manager to fully coordinate the efforts spent on supplemental surveys to obtain the most information in the shortest time.

7.2.11 FINAL DRAFTING

In many instances, in private industry, maps for smaller projects such as subdivision layout are delivered in the form of pencilled manuscripts. When an improved form of the manuscripts is required, the skill and techniques of the cartographic draftsman are required. Final drafting is done by drafting in ink or by manually scribing, or by using an automatic drafting machine.

Prior to final drafting, a sheet layout and format are normally prepared and submitted to the client for additions, deletions, and/or approval. A legend is adopted which identifies the various symbols used to portray detail, and the proposed title block, logo, credit notes, and other marginal data are specified. When scribing is done, a master-sheet format can be produced in negative or positive form and registered with each scribed map sheet during photographic reproduction of the final positive sheet by contact printing in a vacuum frame. Color-separation scribing involves the basic functions of map scribing on a single sheet, except that the various map features are scribed or otherwise compiled on separate plates. As an example, the 7½' quadrangles produced by the Geological Survey are printed in five basic colors: black for most culture, brown for hypsography, blue for hydrography, green for woodland, and red for landlines and road classifications. Additional colors or screened versions of the basic tints are sometimes required for supplemental information. Plates for production of these maps are prepared using a combination of scribed sheets, peel-coat, and stick-up. Each plate will depict only that information or features that are to appear in the specified color. All of the base material for the color-separation plates is pre-punched in registration before starting work, to insure that each completed plate can be oriented precisely relative to all other plates comprising a map sheet.

7.2.12 CHARTING THE PLAN AND PROGRESS OF A PHOTOGRAMMETRIC PROJECT

Charts are an integral part of any large photogrammetric project. It is useful in the early planning stage to develop a chart showing the interrelationship of all operations, the effect of one on another, and the sequence of tasks required to achieve the goal. Another type of chart should be constructed and maintained during the course of a project to record and measure progress in each operation as well as to monitor problems that may affect overall progress. Charts prepared during the proposal stage and included as part of the presentation to a potential client can serve to illustrate an awareness of the project requirements and an understanding of the means being proposed to meet those requirements. Progress charts broken down by operations and sub-operations, and with tasks weighted as to value with respect to the total project fee, can serve as back-up reports for billing during progress of a large project. Operations underway during the billing period are reported at their estimated percentage of completion, and the cumulative value complete for all operations equals the billing amount. Figure 7-17 illustrates a sample diagram based on the "critical path method." Figure 7-18 is a bar chart for a typical, large photogrammetric mapping project.

7.3 Estimating Time and Cost

Preparation of time- and cost-estimates for photogrammetric projects is best done when records of past performance on similar projects are available for review. On similar projects, however, there are variables which may significantly influence the estimate. Beyond the obvious items such as direct-labor rates, material costs, and overhead expense, the estimate must take into account schedule, production efficiency, and degree of difficulty. The schedule often affects project costs not only by requiring the payment of premium rates for labor, but also by dictating the choice of control methods, plotting equipment, and aerial photography scale. Production efficiency is influenced by the suitability of equipment and personnel to the requirements of the project. A production staff equipped and trained to produce maps, at

1:24,000, of a mountainous desert, would likely not be at peak efficiency on a project involving production of 1:2400 scale maps for an urban area. In time, the adjustment would be made, but the estimate should reflect the cost of making this adjustment. Determining a factor for the degree of difficulty involves consideration of technical requirements, climate, and logistics. For example, a mapping project in the rugged mountains of Central America may be directly comparable to one of similar size and specifications in Colorado. Estimates for plotting and drafting might be identical on a unit basis, but figures for field control and aerial photography could show a manifold variation due to unfavorable weather and difficult access. Although estimates in many firms are prepared piecemeal by members of various departments, a better practice is to designate a single competent individual with a broad understanding of the work to prepare all estimates. The estimator can request information on specific operations from the various departments without interfering with production. This arrangement, where feasible, generally results in objective estimates uninfluenced by non-essential factors. The typical estimate (section 7.3.9) shows the figures and operations considered in determining the selling price for a typical photogrammetric project. In practice, the form could provide space to enter actual costs after the project has been completed.

7.3.1 TIME REQUIRED FOR AERIAL PHOTOGRAPHY

Many factors affect the time required to do the aerial photography. Among these are: weather expectancy, distance of project from flying base, length and orientation of flight strips, physical limitations of personnel, flight height, speed of the aircraft in climb and at working altitude, number of aircraft available, and the season of the year.

7.3.1.1 WEATHER EXPECTANCY

Average or normal weather for any place in the United States may be quite reliably anticipated from the Sette Tables of Weather Expectancy shown in chapter V. These tables give averages for the respective reporting stations throughout the country. They are quite reliable if the project extends over several years, but may be far off for a specific effort to be accomplished in one year.

General information concerning weather around the world may be obtained from data in atlases, from the governments of the countries involved, and from consultation with people familiar with the area. Peoples' memory of weather is, however, quite fallible.

It is sometimes specified that photographs be free of snow. Before such a specification is accepted, the area should be studied to make sure that there will likely be a sufficiently long period without snow on the ground to complete the job. Serious trouble is frequently encountered in working to a tight flying schedule when it is discovered that a small amount of reflying is required after the first "permanent" snowfall.

A similar predicament bedevils the photographer who must secure negatives for topographic mapping in areas covered by deciduous vegetation. To achieve the required map accuracy, the photographs must be secured when the ground is not obscured by leaves. This restricts the flying season to two very short periods during which satisfactory photography may be obtained. These are between the time the snow melts and the leaves come out in the spring, and between the time when the leaves have fallen and the onset of snow in the fall. The spring situation is to be preferred, as the winter storms have not only beaten the last leaves from the trees, but they have also beaten them to the ground from the underbrush where they tend to cling. Spring is also better from the standpoint of sun angle, being about two months nearer to the summer solstice.

Care should be exercised in using average weather data recorded on the basis of percentage of cloud cover. Such factors as haze, smoke, and fog are often not recorded. In the Pacific Northwest, the forest-fire smoke sometimes persists for weeks at a time during periods which may be recorded as having many cloud-free days. In the interior valley of California, the same situation prevails in relation to the so-called "tule" fogs. Thus, many regions have their own peculiarities, which can be known only through local acquaintance.

7.3.1.2 DISTANCE FROM FLYING BASE

If the limits of a project can be reached during the period required by the airplane to climb to working altitude, little time is lost in getting started. As the distance to the project from the flying base is increased, the ratio of effective photographing time to travel time getting to and from the area becomes increasingly unfavorable. Ultimately, the entire range of the aircraft might be used to travel to the project, leaving no photographing time. Thus, it is of prime importance to base as close to the area as possible, so that travel time can be held to a minimum and so that the weather can be better evaluated. In this connection, a fast-climbing and fast-flying plane is most valuable when based near a difficult weather area. As soon as a spot of open weather is observed the airplane can get quickly into working position and can obtain the maximum of photography while the weather holds. In tropical areas, it is not unusual to fly 5 hours in attempts for each hour of photography; nor is it unusual under tropical conditions to take off daily and patrol the area in the hope of finding open weather. Weather reporting stations are also sometimes established over the area.

Conversely, when the project is far from the base, the weather at the project may be good but it may be impossible to take off from the base; or the weather may close in while the aircraft is in flight, making it difficult or impossible to return to this far-away base.

It is axiomatic that when trying to photograph a difficult weather area the airplane must get into the air whenever there is even a slight possibility of getting some photography. To illustrate, an expedition may be costing $1,000 a day, while the flying may cost $50 per hour. If, in this case, there is one chance in twenty of getting photography, the effort should be made.

Obviously, the distance between the base and the project is a prime consideration in choosing the airplane for the project. A small, two-seat, low-powered, restricted-range airplane may be ideal to secure photography of terrain close to a suitable base. The same photographic requirements may call for a much larger aircraft, with many hours cruising range, to work at a more distant project.

7.3.1.3 FLIGHT STRIPS

In estimating flying time, the air speed of the aircraft at the particular working altitude should be reduced by the effect of probable winds and the time required to turn around between strips. Thus, an airplane flying 200 mph in a 100 mph crosswind flies only a resultant 173 mph along the desired course. As a rule of thumb, the turn-around between two parallel and even-ended strips takes an average of 6 minutes for most of the types of aircraft in common photographic use when flying at moderate altitudes.

When photographic strips are long and lie somewhat radial to the base, so that the flight is away from the base in one direction and toward it in the other, and they are all about the same length, little can be done to minimize waste. At the conclusion of a certain number of flight lines, there may not be enough fuel left for another. If, on the other hand, the area is irregular, so that there are longer and shorter flight lines, a combination can sometimes be worked out to utilize more of the potential working capacity for that day.

If the base lies within the area, it may be practical to break flight lines to get substantially the maximum production in a given day even though this involves additional costs in the laboratory.

If a flight is being made over or parallel to a river or road with many angles or meandering of such magnitude that the project cannot fall within a single straight strip, it may be necessary to fly the strip in tangents, making a complete turn around at each angle point.

Often there is an economy in orienting the flight strips in one direction rather than another. In the case of a rectangular area the strips should be laid out both ways, to determine the most economical pattern from both the amount of flying and the number of models involved. In fact, it is almost always desirable to consider all the possible flight-line patterns. Each pattern should be carefully measured, and the models counted, before making a final decision. It is sometimes quite misleading to divide the total square miles by the theoretical coverage per model and then add a percentage for "overhang." The very large percentage of total project cost often represented by flying and photography merits the most careful planning of the flight details.

7.3.1.4 ALTITUDE FATIGUE

As discussed in chapter V, oxygen should be taken by the flight crew on all flights above 10,000 feet unless the airplane has a pressurized cabin. In spite of taking oxygen, the crew tires very rapidly in the reduced pressure above 30,000 feet. Without pressure, a crew can work above 35,000 feet only a little over an hour.

7.3.2 AIRCRAFT OPERATING COSTS

Aircraft costs are of two kinds:

(a) Ownership costs, and
(b) Operating costs.

Ownership costs include: interest on investment, depreciation, insurance, housing, and standby maintenance. These costs are a function of time and continue whether the aircraft operates or not.

Operating costs include: gasoline, oil, oxygen, engine maintenance, airframe maintenance, landing fees, guards, and logistical support when basing at a remote spot.

In estimating, it is usual to figure the costs on a hourly basis. Thus, the number of hours the aircraft is flown per unit of time is a very important factor. For example, ownership costs of $10,000 per year reduce to an hourly basis as follows: if the aircraft is only flown one hour per year the ownership cost is $10,000 per hour; if the aircraft is flown 200 hours per year (a reasonable amount for a photographic airplane) the ownership cost becomes $50.00 per hour.

A small aerial photographic operator might rent an airplane, cut a hole in it, and install oxygen and a vacuum system; or he might hire an airplane with a professional photographic pilot, a practice quite common in the trade.

7.3.3 TIME REQUIRED FOR GROUND CONTROL

Weather expectancy of the particular area at the season the work must be done is a major factor in estimating the duration of a field survey. Another factor is the character of the terrain and vegetation. A traverse party might make 4 miles a day under favorable circumstances where the ground is level and clear. The same party might have trouble making 100 yards a day through dense jungle obstructed by a network of fallen trees. A triangulation party might average four stations a day under third-order specifications and favorable conditions of weather and terrain. The same party doing first-order work in difficult terrain and unfavorable weather might consider one station a week very satisfactory. An astronomical party might observe several stars per night in clear desert air, but might have to wait weeks for a single star shot on the west coast of Colombia.

Under average conditions in the United states, a traverse party of six men, working to third-order specifications, might make 2 to 4 miles per day. Over the same terrain, a three-man level party might make 3 to 6 miles a day.

One computer is generally required in the office to support one party in the field.

The wide choice of methods available for the field work, and the infinite variations of terrain and weather conditions, make a careful study of chapter VIII most desirable. The proper planning of any but the simplest survey requires the attention of the professional surveyor with extensive experience in the application of surveying to photogrammetry.

7.3.4 TIME REQUIRED FOR STEREOCOMPILATION

The time required for stereoplotting is generally figured in hours per model. The total man-hours per model usually runs about the same for a given model, whether drawn on the somewhat faster drawing two-man universal instrument or the simpler plotter such as the Kelsh or the Balplex. The time may range between 2 man-hours, or even less, for a model covering a smooth, gently sloping plain with only a few contour lines, little drainage, and no culture, to 30 man-hours or more for a city area like San Francisco, with solid blocks of buildings and considerable areas of closely spaced contours.

For convenience in estimating, the time required for plotting is generally subdivided into four steps: (1) aerotriangulation; (2) model orientation; (3) planimetry; and (4) contours.

Aerotriangulation includes all necessary operations for the extension of horizontal and/or vertical control by means of aerial photographs. It can vary in complexity from the practical situation where horizontal and supplemental vertical control are complete for each model, and where no effort is required, to the project planned with a bare minimum of basic horizontal and vertical control, and where the model control is supplied through analytical aerotriangulation. Exactness in planning the required operations will furnish a clue to the time required for their completion.

Model orientation usually requires 1 to 2 hours for each model.

Planimetry may be drawn quite rapidly in open country where only drainage, roads, and perhaps an occasional ranch house and fence are to be shown. It may take only a matter of minutes. In a city, however, the planimetry includes a large number of streets, buildings, and such features as railroad tracks. The scale of the map is also a governing factor. At 1:24,000, only streets and landmark buildings are generally shown in built-up areas. The built-up area is indicated by a tint. This may require only an hour or so of stereoplotter time per square mile. If the scale of the map is 1:4,800, public buildings may be outlined precisely, but average residences indicated by a conventional square. The stereoplotter operator may indicate these by a dot to show their position, leaving the detail to the draftsman. At scales of 1:2,400 and larger, each building is usually outlined in complete detail. Estimating the time required to plot the map of a city may be approached empirically by knowledge that in the average city there is perhaps one structure for each 2½ inhabitants (considering curbs or property lines around a block as one structure also). The mapping organization may have records indicating that they have been plotting, say, 240 structures per hour. If this average city has 12 blocks per mile there might be 3,460 structures per square mile if solidly built up. Thus, 14.4 man-hours are required to plot the buildings in 1 square mile. If the photography is at a scale of 1:12,000 (for a flight height, H, of 6,000 feet), with $B/H = 0.63$ and $W/H = 1.15$, the net theoretical coverage per model is:

$$\frac{B/H \times W/H \times H^2}{5280^2} = 0.94 \text{ square mile.}$$

Thus, the time required to draw the planimetry of the model is estimated as 14.4 hours × 0.94 = 13.5 man-hours.

Contours. An estimate of the time required to plot the contours can be based on the total length of contours per model. If a contour map of a representative sample in the area is already available, the contours may be counted and the average length estimated. If the map on which the counting is done has 20-foot contours, and the map to be produced is to have 5-foot contours, the factor 4 must be applied. Similarly, if the scales of the two maps are, respectively, 1:24,000 and 1:2,400, a scale factor of 10 must be considered. The product of these two factors, 40, represents the relative total length of contours of the two maps. On steep, uniform slopes, perhaps the compiler may skip contours, leaving them to be added in map finishing. The length of these skipped contours should be subtracted.

More detail will surely be shown on the new larger-scale map, so a factor should be applied to compensate for the meander of contours and increased detail in their shape. The meander represents relatively more inches to plot, and the detail a slowing down of the drawing to get the small expression and changes of direction (fig. 7–19). Knowledge of how much time to add for these factors is rather intuitive, and is only gained by experience.

The rate of plotting per minute may vary between, perhaps, 3 and 15 inches of contour line at the scale of the plotted manuscript. There is, however, the very important consideration of the relationship between the negative scale, manuscript scale, and the final map scale. This relationship affects the rate of drawing on the manuscript, as the following rather extreme example indicates:

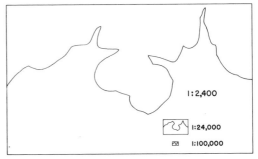

FIGURE 7-19. A contour line drawn at three different scales to illustrate the difficulty in estimation of drafting time.

Average terrain elevation above sea level	5,500 feet
Airplane ceiling above sea level	23,000 feet
Airplane elevation above ground	17,500 feet
Negative scale (6-inch lens)	1:35,000
Manuscript scale:	
Universal plotter	1:24,000
or perhaps	1:16,000
Kelsh	
(without pantograph)	1: 7,000
Final map scale	1:24,000

The Kelsh operator drawing a manuscript at 1:7,000 scale, which must be reduced to the publication scale of 1:24,000, can draw quite fast since any imperfections in his drawing are reduced nearly 4 times in the final map. The universal plotter operator does not have this latitude and must draw more carefully, as his liberties go into the final map at the manuscript scale or only slightly reduced. On the other hand, the Kelsh manuscript will have possibly 16,000/7,000 = 2.3 times as many inches of line to be drawn, so that even if drawn at a faster rate, in inches per minute on the manuscript, the job may take longer. In most cases, of course, negative scale is more suitable for efficient drawing than in this example.

7.3.5 TIME REQUIRED FOR DRAFTING OR SCRIBING

The same factors affecting stereoplotting also govern drafting. The total length of contour lines, their meander and detail, the length of linear planimetric features (railroads and so forth), and the number of structures all contribute to the time required. There is also one additional factor: tightness of detail. If contours are 0.50 inch apart, their proximity has a negligible effect. But, if they are only 0.02 inch apart, the drafting may be slowed fivefold. In such a case, consideration should be given to drafting on an enlarged manuscript.

Color-separation drafting, which is done on several sheets, is slower than showing exactly the same detail on a single sheet for monochrome reproduction. In monochrome drafting, the draftsman sees how close an adjacent line is

to the one he is drawing. If the previous line has drifted slightly toward or away from the one being drawn, he can accommodate slightly. But, in color-separation work, the draftsman drawing drainage on the "blue" sheet must follow the copy with the utmost care; otherwise, the drainage symbol on the "blue" sheet may not fit contour reentrants on the "brown" sheet. There is substantially no drafting tolerance in color-separation drawing, whereas discrepancies of 0.01 or 0.02 inch in monochrome drafting are normal expectancy. Color-separation drafting may take 20 percent to 150 percent longer than monochrome drafting.

The rate of drafting ranges approximately from 20 square inches to 200 square inches per man-day. Estimating by the square-inch-per-day approach is much faster than going through all the steps required in linear estimating. However, it takes years of experience in linear estimating to acquire the ability for square-inch estimating, and even then the chances of error are greater.

A proper balance should be struck between editing and drafting. If a perfect job of editing has been done, the draftsman can work faster than if part of the editing falls upon him. If, for example, the outline of a building has been roughly drawn at the stereoplotting stage, it has to be squared up and perhaps checked against the photography for exact shape and orientation. This may be done by either the editor or the draftsman; in either case it takes time. If the organization is short of editing capacity, the draftsman can do this, or *vice versa*.

7.3.6 TIME REQUIRED FOR EDITING

The faster and rougher the stereoplotting work, the slower the editing; however, the organization which is short on stereoplotting capacity deliberately throws more of this burden on the editors.

A minimum-cost map, made to be used just once on a landscape-engineering project, for example, may require no editing, or at most a quick check to see that contours are correctly numbered and that confusion between contours and planimetric features is minimized.

At the other end of the gamut is color-separation editing, where perhaps every line of the map may have to be checked by the editors. A great deal of time may be spent on checking for completeness, proper spelling and numbering, correct edge matching, and especially the conformance of words and numbers with cartographic detail. For example, a highway to be reproduced as a black casing with a red fill may cross a series of contours, shown on a third sheet, the brown plate. If the contour values are not correctly placed on the brown plate, the road may pass through a contour value when the sheets are combined in the final map (fig. 7–20).

A big task in editing maps of foreign areas is to translate place names to be read and pronounced in the language of the publication. This "transliteration" from Arabic to English, for example, requires time and patience almost beyond estimation.

Editing costs generally fall within the range of 50 percent to 200 percent of drafting costs and, like many of the other facets of photogrammetric estimating, they contain a larger element of intuition than of formula.

By this time the reader must realize that, in the very complex profession of photogrammetry, the amateur is on very dangerous ground. It is hoped that the glimpse of the problems of time and cost estimating and other phases of planning presented in this chapter will lead the reader to the only sound conclusion: if he is not a professional, he should try to get the help of someone who is.

7.3.7 TOTAL COST OF THE PROJECT

The total cost of a project consists of charges for material, labor, and overhead. Estimating the cost of material is quite straightforward after the estimator learns the percentages to allow for waste. Chemicals for the photo-laboratory usually run something less than 10 percent of the other photographic materials. Aerial film has a high percentage of waste because breaks in a flight line involve extra costs throughout almost the entire range of activities. For this reason, short ends of film are seldom used.

Labor estimates are influenced by all of the factors which affect the time required to perform the various operations. A special category of labor cost is that of pilots, which presents an accounting problem. Pilots are notoriously difficult to use effectively during idle periods. On a large project away from the home base, the full salary of the pilot is charged to the project. Projects based at the home base are more difficult to figure. During a slow period, the pilot may fly only one project in a month. His monthly salary may be more than the selling price of the project. Some organizations charge just one day's

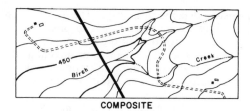

COMPOSITE

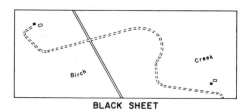

BLACK SHEET

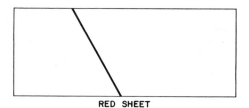

RED SHEET

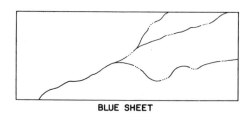

BLUE SHEET

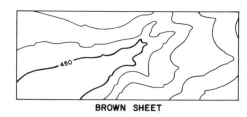

BROWN SHEET

FIGURE 7-20. Editing of color-separation drafting is done from several sheets, one for each color of the published map. Overprint of numbers and symbols is carefully checked at this stage.

time to such a project. The balance of the monthly salary thus becomes overhead.

The estimator ultimately learns from experience how much labor goes into each operation. Perhaps 1,000 or more "check" contact prints can be produced in one man-day by a continuous automatic printer, while only 200 delivery prints can be made manually. One organization with certain operators and equipment may produce 40 ratio prints in a man-day. Another may make 100. One laboratory may make 25 big enlargements in a man-day. Another, with a very efficient enlarging camera, may double this number. An organization doing very meticulous mosaicking to a demanding specification may lay 20 prints per man-day, while just making a so-called "precise" mosaic, without much regard for tone and image match, might proceed at the rate of 100 prints per man-day.

A control layout may be made in a few hours by a trained man, with an accurate and large-enough coordinatograph. The same job, done without the most suitable equipment, can well take days (and even then leave much to be desired).

Copying and reproducing a job with 16- by 20-inch cameras may take four times as long for all steps pertaining to this operation as if 30- by 40-inch cameras were available. "Blending" to give uniformity of tone to a photomap may take many hours per copy negative if done with great skill and care, while under different reqirements no blending at all is done.

Delivery of sheet materials which can be rolled costs much less than hard-backed materials, for which shipping cases must be made. The latter also involve more trucking and shipping cost.

7.3.8 OVERHEAD

Overhead includes all executive, administrative, and supervisory cost and all indirect expense—that is, expense which cannot be readily measured as it pertains to separate projects. Thus water, light, gas, rent, janitorial service, and local telephone service are typical overhead items. Long distance telephone and telegraph charges are marginal items, often included in the overhead because excessive clerical cost would be required to charge these items to their proper projects.

7.3.9 SAMPLE ESTIMATE

Subject: City of Example
Area: 79,300 feet × 71,500 feet; 203± square miles
Products to deliver:

1 set contact prints at 1 inch to 1,000 feet, D.W., glossy, not trimmed.
1 index map at 1 inch to 1 mile
1 set photo-atlas sheets at 1 inch to 400 feet, cloth, standard border, in quads 22¼ × 30½ inches plus borders, 29 × 37 inches overall
1 set topo maps at 1 inch to 200 feet with 5-foot contours, standard specifications; 210 at 24 inches × 30 inches plus border, inked on Cronar

Fly early May (before leaves)
Delivery date: Dec. 31
Plane: Cessna 180 supplied by us. Cost $50/hr. Cruise 150 mph
Base at Example Airport
Cross-country from home base: 200 miles each way
Mean ground elevation: 1,200 feet above sea level
Flight altitude: 7,200 feet above sea level: 6,000 feet above ground
Camera: Precision, low-distortion lens; 9 × 9 inches format.
Film: Estar base; panchromatic

Flying Requirements

Width: 71,500 feet. Overhang = 25%, 1st 2,250 feet inside border, X2 = 4,500 feet. Remaining width 71,500−4,500 = 67,000 feet/6,300 = 10.6 bands, based on 30% (W/H 1.05) sidelap. By compaction W/H = 1.02. W = 6,100 feet. 12 flight strips
Length: 79,300/3,600 = 22 theoretical shots (60%; B/H = 0.60). Heavy lap acct. many tall buildings. Overhang. 3 per strip; safety, 2 per strip. Total photos = 12(22 + 3 + 2) = 324. Diapositives and mosaic 300 shots. Models = 282.
Control: 1st order hor. and vert. available within area for ties. Levels: 6 cross strips at 14 mi. = 84; 2 long. at 15 mi. = 30. Total 114 mi. Traverse: 3 cross strips at 14 = 42; 2 long. at 15 = 30 mi. Total = 72 mi. Bridge on universal 5 & 6 models vert., 9 & 10 models horizontally. Plot map on Kelsh, Wild A8, Jena Stereometrograph.

Flying:	12 strips × 15 miles	= 180 miles		
	at 140 mph on line	= 77 minutes		
	11 turns at 6 min.	= 66 minutes		
	2 climb and descent	= 60 minutes		
	3 attempts at 1 hr.	180 minutes		Materials &
	Cross country, 6 hrs	360 minutes		Other Direct
			Labor	Costs
	Total	743 minutes		$620.
		= 12.4 hours at $50		

	Labor	Materials & Other Direct Costs
Layouts		
Mosaic boards: 7 feet × 8 feet - 8 hrs at $6	48.	30.

Manuscript sheets: 282, 30 inches × 48 inches
  grid and control 0.5 hours at $6, material at $5 per sheet
  (Cronar or equal).
	846.	1,410.

Delivery quad sheets: 24 inches × 30 inches; margins 0.5 hr. at $6. material at $3 per sheet	630.	630.
Printing marginal data on 210 sheets at $3		630.
Mosaic		
Ratio prints: 300 at 5 prints/hr. = 60 hrs × $6	360.	50.
Mosaic 300 prints at 2.5/hr. = 120 hrs at $8	960.	

Copy at 1 inch to 625 feet (1 inch to 600 feet would give
  slightly better image quality but requires several more
  negatives) (15) 30-inch × 40-inch negatives at 1.0 hr × $8
	120.	111.

Reproduce at 1 inch to 400 feet, 56 sheets 24 × 30 inches + margins at 0.25 hr. each = 14 hrs. × $6	84.	44.
Compose marginal data: 8 hrs at $8	64.	20.
Copy marginal data; (1) 30-inch × 40-inch negative	8.	8.
Stereocompilation		
Diapositives: 300 at 4/hr. = 75 hrs. at $6	450.	874.
Bridging: 300 at 1 hr. = 300 hrs. × $8	2,400.	
Computing and adjusting at 0.5 hr. = 150 hrs at $8	1,200.	
Plotting: setup 1.5 hrs., planimetry 9 hrs., contours 4 hrs. = 14.5 hrs/model × 282 × $8	32,712.	
Drafting		
Prepare scribe coats: 210 at 0.5 hr = 105 × $6	630.	3,150.
Population 900,000, structures 360,000/60 per hr. = 6,000 hrs at $6	36,000.	

Contours: Av. density 2.3 per inch, meander 1.1, raggedness 3.0, tightness 1.0 = 8.35 linear-inch factor per square inch

$$\frac{71,500 \times 79,300}{200 \times 200} = 141,500 \text{ in}^2 \times 8.35 = 1,180,000$$

linear inch factors at 180 in./hr.
    = 6,550 hrs at $6

= 6,550 hrs at $6	39,300.	
Print final sheets: 210 at 1 hr = 210 × $6	1,260.	1,260.
Weather expectancy Apr-May: 4 days per month		
Time required to get 2 flights: 0.5 month		
Time awaiting check 0.12 month		
Time preparing, disbanding and cross country 0.13 month		
Duration of operation 0.75 month		
Salary for crew of 2 at $3000/mo. × 0.75	$2,250.	
Room and Board: 2 men: 23 days × $80		1,840.
Landing fees 6 at $10; Storage: 22 nights at $8		236.
Film: 324 shots at 290 shots/roll = 1.25 × $200		250.
Laboratory		
Development 2 shipments of film, clean and number at 3 man hrs. each = 6 hrs × $8.50	51.	
Make 324 check prints = 6 hrs at $8.50 + paper at $.20/sheet	51.	65.
Check flight: 3 hrs at $6	18.	
Index Map		
Staple and number 324 prints = 4 hrs at $6	24.	
Copy index: 1 20 × 24 inch negative at 1 inch to 1 mi. = 1 hr at $8.50	9.	
Reproduce 4 copies; 20 × 24 inches-1 hr at $6	6.	8.

Field Control		Labor	Materials & Other Direct Costs
Research and order materials = 2 hrs at $8.50		17.	20.
Prepare flight map = 1 hr. at $8.50		9.	
Traverse = 72 + 8 (tie run) mi. = 80 mi. (3rd order) 6-man party 30 days	30 days		
Levels—114 + 4 = 118 mi			
Two 3-man parties	20		
Party-base to project and return	2		
Days lost account of weather	6		
Total days at $140/party-day	58 days	8,120.	
Room and board: 6 men at $30/day = $180 × 58 days			10,440.
Vehicle: 9 trips to base and return at 400 mi. each	3,600		
On job 50 days at 30 miles	1,500		
Total car mileage At $.20 mi.	5,100		1,020.
Computing control: 50 man-days at $64		3,200.	
Control and computing supplies			125.

Editing

	Labor	
15% of drafting	11,295.	

Cost Summary

	Labor	Materials & Other Direct Costs
Totals	142,122.	22,841.
Total cost	164,963.	
Overhead 130% on direct labor	184,759.	
Sub-Total	349,722.	
Profit 10% of cost	34,972.	
Selling price	384,694.	
Price per square mile	1,895.04	
Price per acre	2.95	
Price per hectare	7.29	

REFERENCES

Anonymous (1963) "Checking Color-separation Materials" *Topographic Instructions of the United States Geological Survey.* Chapter 4F3. U.S. Geological Survey Center, Reston, Virginia. pages 81–93.

Anonymous (1961) "Color-separation Drafting" *Topographic Instructions of the United States Geological Survey.* Chapter 4E3. U.S. Geological Survey Center, Reston, Virginia. pages 3–8.

Anonymous (1963) "Map Editing" *Topographic Instructions of the United States Geological Survey.* Chapter 4F1. U.S. Geological Survey Center, Reston, Virginia. pages 1–110.

Anonymous (1966) "Map Revision" *Topographic Instructions of the United States Geological Survey.* Chapter 3H6. U.S. Geological Survey Center, Reston, Virginia. pages 3–13.

Anonymous (1979) "Negative Scribing for Color Separation" *Topographic Instructions of the United States Geological Survey.* Chapter 4E2. U.S. Geological Survey Center, Reston, Virginia. pages 3–29.

Photogrammetry for Highways Division, American Society of Photogrammetry (1968) *Reference Guide Outline—Specifications for Aerial Surveys and Mapping by Photogrammetric Methods for Highways.* vii + 109 pages; ill. Superintendent of Documents. Washington, D.C.

BIBLIOGRAPHY—*General*

Aguilar, A. M. (1967) "Cost Analysis for Aerial Surveying" *Photogram. Eng.* 33(1): 81–89.

Aguilar, A. M. (1969) "Management Planning for Aerial Surveying" *Photogram. Eng.* 35(10): 1047–1054.

Anonymous (1978) "Photorevision" *Topographic Instructions of the United States Geological Survey.* Chapter 3H2. U.S. Geological Survey Center, Reston, Virginia. pp. 1–21.

Anonymous (1962) "Planning Aerial Photography" *Topographic Instructions of the United States Geological Survey.* Chapter 3A3. U.S. Geological Survey Center, Reston, Virginia. pp. 1–94.

Anonymous (1978) "Summary of Topographic Mapping Procedures" *Topographic Instructions of the United States Geological Survey.* Chapter 1B2. U.S. Geological Survey, Reston, Virginia. pp. 3–21.

Anonymous (1954) "Supplemental Control Planning and Field Identification" *Topographic Instructions of the United States Geological Survey.* Chapter 2F1. U.S. Geological Survey Center, Reston, Virginia. pp. 3–26.

Bruchlacher, W. (1957) *BEITRAGE ZUR PLANNUNG, VORBEREITUNG, UND DURCHFUHRUNG PHOTOGRAMMETRISCHEN BILDFLUGE* Deutsche Geodatische Kommission. Bayerische Akademie der Wissenschaft. Munchen.

Friedman, S. J. (1961) "American Commercial Practices in Large-scale Topographic Mapping" *Photogram. Eng.* 27(1): 44–47.

Harwood, J. (1957) "Aerial Surveys for Special Purposes" *Photogram. Eng.* 23(4): 746–748.

Lyon, Duane (1957) "Basic Requirements for Charting Photography" *Photogram. Eng.* 23(4): 685–697.

Meier, H.-K. (1966) "Angular Field and Negative Size" *Photogram. Eng.* 30(1): 126–135.

Paterson, G. L. (1971) "Photogrammetric Costing" *Photogram. Eg.* 37(12): 1267–1270.

Prior, W. T. (1959) "Relationship of Topographic Relief, Flight Height, and Maximum and Minimum Overlap" *Photogram. Eng.* 25(4): 572–590.

Soliman, A. H. (1971) "Accuracy and Application" (number of control points needed) *Photogram. Eng.* 37(8): 879–884.

Telford, E. T. (1953) "Photogrammetry Specifications and Practices for the Use of Aerial Surveys" *Photogram. Eng.* 19(4): 569–577.

Wood, G. A. (1972) "Photo and Flight Requirements for Orthophoto Mapping" *Photogram. Eng.* 38(12): 1190–1191.

Woodward, L. A. "Survey Project Planning" *Photogram. Eng.* 36(6): 578–583.

BIBLIOGRAPHY—*Specific Projects*

Case, J. B. (1958) "Mapping of Glaciers in Alaska" *Photogram. Eng.* 24(5): 815–421.

Gamble, S. G. (1958) "Mapping of the Ungava Peninsula" *Photogram. Eng.* 24(3): 410–414.

Gill, E. A. (1971) "Mapping for the Trans-Alaska Pipeline" *Photogram. Eng.* 37(2): 170–176.

Jensen, J. R. and Steele, J. J. (1969) "Production Mapping with Computational Photogrammetry" *Photogram. Eng.* 35(3): 283–296.

Massie, E. S. (1959) "Increasing Productivity through Multiple Use of Basic Data" (mapping procedures of U.S. Forest Service) *Photogram. Eng.* 25(1): 33–41.

Swanson, L. W. (1964) "Aerial Photography and Photogrammetry in the Coast and Geodetic Survey" *Photogram. Eng.* 30(5): 669–726; 8 fold-out colored inserts.

CHAPTER VIII

Field Surveys for Photogrammetry

Author-Editor: S. W. HENRIKSEN

Contributing Authors: STANLEY H. SCHROEDER, RONALD K. BREWER

8.1 Foreword

BEFORE PHOTOGRAPHY was adapted to the purposes of surveying, all maps were the product of the surveyor and his field observations. Office work consisted of little more than the computation of control and the assembly of the field material. The advent of aerial photography, particularly stereoscopic vertical photography, furnished the surveyor with a tool for viewing his subject with thoroughness and consistency from a perspective denied him on the ground, thus greatly relieving him of much arduous field work. The subsequent development of photogrammetry widened the scope of surveying and mapping and made it possible to carry on much of the mapping operation in the office free from the vagaries of weather and the difficulties of transport over inhospitable terrain. Versatile as the applications of photogrammetry are, however, field surveys are still required to supply the basic control (both horizontal and vertical) needed to determine the scale, azimuth, and attachment to datums of the photogrammetric plot, and to secure information which cannot be obtained solely from office examination of photographs. This includes the classification of roads and buildings, the positioning of land corners and civil boundaries, the location of features hidden in trees or shadows, the procurement of place names, and so on. These subjects are discussed in this chapter on field surveys.

In preparing the text of this chapter, the authors have been conscious of a need to supply information for an extremely wide range of mapping over many places on the earth. It has been necessary to consider mapping scales ranging from a ratio of 1:600 with a contour interval of 1 foot to 1:1,000,000 with a contour interval of perhaps 1,000 feet, and the mapping of places where basic horizontal and vertical control systems do not yet exist. In order to attempt such wide coverage in the space of one chapter, it has seemed necessary to make extensive use of references to other publications, listed at the end of this chapter, and to limit the discussion of conventional surveying methods. Most of the space is devoted to supplemental horizontal and vertical control surveys which provide the detailed control for photogrammetry, to new techniques now available for these supplemental control surveys, and to the important question of the identification of control on aerial photographs.

8.2 Basic Geodetic Control

8.2.1 INTRODUCTION

In surveying, as in all measuring, a succession of measurements accumulates the constant, or systematic, error of the measuring device plus or minus the accidental errors of the measuring operations. The amount of error is likely to increase with the distance measured and usually is expressed as a part of it, as 1 part in 25,000. Ordinarily, the more precise methods of measurement require more precautions and more expensive instruments and, therefore, cost more than the less precise. For these reasons, the most efficient surveying practice, when a certain tolerance of error is specified, requires that the longer distances be measured by the more precise methods. Methods of lower precision and less cost may then be used to locate intermediate positions. Thus, if we consider an area 25,000 metres square, methods giving a precision of 1 part in 25,000 might be used to determine the relative positions of points at a spacing of, let us say, about 10,000 metres apart over the area. Then methods having a precision of 1 part in 10,000 could be used to locate positions between the higher-order positions at about 5,000 metres apart. Methods yielding a precision of 1 part in 5,000 could then be used to locate third-order stations between the second-order points at about 1,000 metres apart. In this way, all points in the area would be located with consistent precision at much less cost than if the methods yielding a precision of 1 part in 25,000 were used throughout. Obviously, if methods yielding a precision of 1 part per 1,000 were used throughout, discrepancies and distortions up to 25 metres would arise, which could only be eliminated by more precise methods of survey. It is a

413

cardinal principle of surveying that the more-precise measurements be made first and the less-precise follow, working from the framework to the details. The more-precise positions then control the less-precise surveys.

Thus, a basic framework of high-order geodetic control (both horizontal and vertical) is essential for coordinating surveys and mapping of large areas. Basic geodetic control networks have already been established over the more highly developed areas of the earth and must yet be established in the developing areas. In controlled areas, the control surveys for photogrammetric mapping usually start from stations of the basic network. In undeveloped areas, however, photogrammetric mapping may precede the establishment or completion of a geodetic network, and different procedures used. In the latter case, geodetic control on a local astronomic datum may have to be established for the mapping, or medium- or small-scale mapping might be controlled either by widely spaced astronomic stations or by a combination of astronomic stations and airborne electronic distance-measuring techniques such as Shoran or Hiran.

8.2.2 CONTROL DATA IN CANADA AND THE UNITED STATES

In the United States, the first- and second-order horizontal and vertical control surveys of the U.S. National Ocean Survey (NOS) provide the basic framework of geodetic control. The additional control surveys for mapping fall, generally, into two categories.

(1) The subdivision or extension of the basic network by triangulation, traverse, and leveling. These surveys bring the basic control into the area to be mapped and are usually monumented for future use.
(2) Lower-order triangulation, traverse, and leveling to determine the position and elevation of points visible on the photographs (picture points) that directly control the photogrammetric mapping.

The criteria for horizontal-control and vertical-control surveys of third-order or higher precision in the United States are given in tables 8-1, 8-2, and 8-3.

The U.S. National Ocean Survey (NOS) distributes horizontal and vertical control data for stations of the national basic control network of the United States. A number of other agencies make additional control surveys for mapping as described in (1) above. Control data resulting from these surveys may be obtained from the agency concerned, no one agency being responsible for distribution. However, the National Cartographic Information Center (USGS), Reston, VA. (22092), can provide information about the additional control that may be available in a particular area. Agencies of the United States Government that make control surveys as described in (1) include the Defense Mapping Agency, Washington, D.C.; the Forest Service, Department of Agriculture; the Geological Survey, Department of the Interior; the Tennessee Valley Authority; the Bureau of Reclamation, Department of the Interior (all with offices in Washington, D.C.); and the Mississippi River Commission, Vicksburg, Mississippi. Many of the individual states also make horizontal and vertical control surveys that are connected to stations of the national net and are useful for mapping. Information about this control must be obtained from the individual states.

In Canada, the primary agencies making control surveys are: the Geodetic Survey, the Topographical Survey, and the Hydrographic Service, all in the Department of Energy, Mines and Resources, and the Directorate of Military Survey, Department of National Defense. Various other federal and provincial departments conduct surveys ranging from first-order to supplemental control. The Topographical Survey, with headquarters in Ottawa, can supply or direct inquiries for all information on Canadian control.

8.2.3 BASIC DATUMS FOR CONTROL

8.2.3.1 GEODETIC DATUM

A geodetic datum is, basically, a coordinate system in which the coordinates of control points are given. Control points being located on or close to the earth's surface, the coordinate system must also be related to the earth's surface, and preferably in such a manner as to make coordinates of control points small. Because surveying to establish control points has been carried out by a number of quite different methods and for different purposes, different *kinds* of geodetic datums have been used, as well as a great number of differently located or oriented datums.

There are three different *kinds* of geodetic datums in use today; the general geodetic datum (usually called the geodetic datum); the horizontal-control datum; and the vertical-control datum.

A geodetic datum consists of an ellipsoid of the approximate size and shape of the figure of the earth, and a fixed plane through the smallest axis of the ellipsoid from which longitudes are counted (figure 8-1). Nine constants are needed to completely specify a geodetic datum; three to tell where the center of the ellipsoid is located, three to tell how the ellipsoid is oriented, and three to specify the lengths of the axes of the ellipsoid. At present, ellipsoids with two equal axes are used for geodetic datums; so only eight constants are needed. The location of the center of the ellipsoid may be specified directly as, for instance, by requiring that it be at the earth's center of gravity. The center of gravity is not

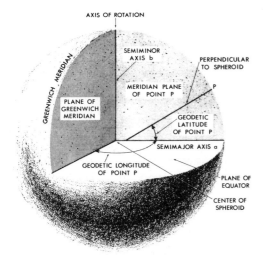

FIGURE 8-1. Spheroid of reference.

accessible, unfortunately, by the methods of surveying used in locating most control; so the center of gravity is used only when data from artificial satellites are available to help in locating it with small error. Most geodetic datums in present use locate the spheroid's center indirectly, by specifying the location of a point (the datum origin), on the surface of the spheroid. When the orientation of the spheroid has been specified, the center of the spheroid falls into place automatically. Two ways of specifying the orientation are in common use. One, also common in horizontal-control datums (*see* below), is to erect a perpendicular at the datum origin and to require that this line make specified angles with the plumb line at the corresponding point on the ground. This provides two angles; the third is obtained by specifying the geodetic azimuth of a point on the ground. The other way is to require that the minor axis of the spheroid be parallel to the average direction of the earth's axis of rotation (two angles), and to specify the longitude of a point on the ground (the third angle).

8.2.3.2 HORIZONTAL-CONTROL DATUM

A horizontal-control datum is used for giving horizontal coordinates of control points. It actually requires as many constants for its specification as does the *geodetic datum*. However, surveying methods and results were sufficiently crude in the 19th and early 20th centuries that many geodesists made simplifying (and often erroneous) assumptions that provided answers with fewer explicitly specified constants than the full number. They reduced the number to five, by assuming that at the datum origin, the direction of the perpendicular, and the direction of the plumb line coincided; thus making it unnecessary to specify two angles. They specified the azimuth, as well as the longitude and latitude,

but did not specify the geodetic height; thus losing another constant.

During the development of the United States of America, Canada, and other large countries, large regions were frequently brought under development before surveys could be made to extend a previously established datum to reach these regions. Therefore, new datums, independent of the older ones, were established. The new datums were usually established in the same way as the old, with their own datum origins, *etc*. Because of the irregular and unpredictable attraction of local masses in each region, the plumb line at any locality usually has a direction that is not geodetically relatable to the direction at another (figure 8-2). Consequently, the various spheroids, each oriented with respect to a plumb line in the region it controlled, were not parallel to each other and their centers did not coincide. Positions and azimuths on one datum (geodetic *or* horizontal-control) cannot be reconciled perfectly with those on another until the networks in the separate datums are expanded until they join or overlap. In North America, transcontinental arcs of triangulation had become sufficiently comprehensive by 1926 to serve as a framework for a unified, continent-wide network. The adjustment of the various individual networks into a single network on a single datum led to the North American Datum of 1927, which is still the governing horizontal-control datum for North America and

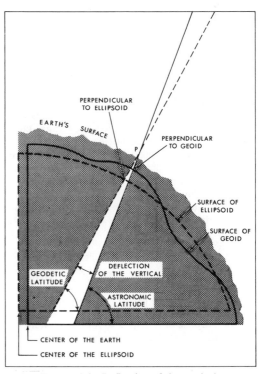

FIGURE 8-2. Deflection of the vertical.

has even been extended through Central America to connect with the datums in South America.

The North American Datum of 1927 (NAD 1927) is specified by the following quantities.

Longitude of Meades Ranch control point, Kansas: 98° 32′ 30″.506 W.
Latitude of Meades Ranch control point, Kansas: 39° 13′ 26″.686 N.
Distance of spheroid below Meades Ranch: unspecified; assumed to be zero.
Deflection of the vertical at Meades Ranch: assumed to be zero.
Azimuth, Meades Ranch to Waldo: 75° 28′ 09″.64.

The spheroid used is the Clarke spheroid of 1866 (figure 8-1), which has the following dimensions.

Length of semi-major axis: 6 378 206.4 m.
Length of semi-minor axis: 6 356 583.8 m.
(Actually, the flattening and not the semi-minor axis length is specified: 1/294.9798).

As a result of the readjustment, now in progress, of the control networks in North America (and extending to Greenland), the North American Datum of 1927 will be superseded, eventually, by a *geodetic* datum rather than by a new horizontal-control datum.

8.2.3.3 VERTICAL-CONTROL DATUM

A vertical-control datum has little relationship to either a geodetic datum or a horizontal-control datum. While the location of a point can be specified completely in a geodetic datum, the height coordinate obtained is unsatisfactory for most practical purposes. This is because "heights" are easily measured as distances along the direction of the plumb line but, except at the datum origin, are nearly impossible to measure along the direction of the perpendicular to the spheroid and usually very difficult to determine by computation. Furthermore, it is convenient to have "heights" along coastlines close to zero, and this would not be the case if geodetic heights were used. So the vertical-control datum was introduced. Distances, called *elevations* (not heights), are measured along the direction of the plumb line from a point down to a reference surface called the *geoid*. The geoid is a level surface which is close (within a couple of metres almost everywhere and within a few tens of centimetres on the average) to mean sea level. In some countries, it is defined to pass through a certain point on a tidal gage. In the United States and Canada, it is approximated by a surface based on mean sea level, and the precise definition of this surface is nearly impossible. For practical work, it is nearly equivalent to a level surface. It is based on mean sea level at 26 tidal gages on the coasts of North America and was for a long time known as "Sea Level Datum of 1929," and so indicated on topographic maps. The name was changed in 1973 to "National Geodetic Vertical Datum of 1929" to avoid the implication that the datum is mean sea level; the datum itself was left unchanged.

Ideally, all ground control should be given in a single, world-wide geodetic datum. Through the use of artificial satellites, this has become theoretically possible and has been done to a limited extent. For practical reasons, geodetic datums of more limited extent, as well as horizontal-control and vertical-control datums, are still in general use. North America, South America, Europe, Australia, and many countries isolated by nature or politics have their own datums to which ground control is referred. Problems can therefore be expected where a photogrammetric project is carried out in a region where two datums meet. The situation is exacerbated by the fact that even within the region controlled by one national datum, there may exist several datums connected to it with varying accuracies; these datums having been established by different organizations for different purposes. In the United States, for instance, the National Geodetic Survey has been responsible for establishing the datums to which the Nation's first-order control is connected. But the U.S. Corps of Engineers, the Tennessee Valley Authority, and other organizations have all, at some time or another, established horizontal-control datums and vertical-control datums independent of the North American Datum of 1927 and the National Geodetic Vertical Datum of 1929, and so have the surveying organizations of local governments. Nor is it uncommon for datums to be set up for specific, aerial-photogrammetric projects. Any photogrammetric mapping project depending on existing control must, therefore, check carefully the datums governing this control and the relations between them.

8.3 Coordinate Systems on Maps

8.3.1 INTRODUCTION

The purpose of a map is to represent symbolically, on a *flat* surface, important facts about the earth's surface and what is on it. The most important facts that maps represent are the locations of points on the surface and distances between them. In preparing a map, the first step is to represent points on the earth's surface by corresponding points on the surface of a spheroid (a slightly flattened sphere) which is approximately the size and shape of the real earth. There are, of course, difficulties in doing this, but they are practical difficulties connected with the technology of surveying and are not mathematical in nature. The second step is to represent points on the spheroid by corresponding points on a flat surface. The pair of mathematical functions which make the transformation from spheroid to flat surface is called a *map projection*. (The term is also applied to the *grid* that results by applying the function to meridians and parallels on the spheroid). Here there is no technological difficulty, but there is an insuperable mathematical difficulty. There is no set of functions which can transform points from a spheroid to a flat surface and still keep the distances between points correct (except for a scale factor). Every map projection is, therefore, a compromise with some distortions being made acceptably small by letting other less important distortions get large. Most map projections in use, at present, adopt one of two compromises.

(a) At each point of the map, the scale of distance is kept the same in all directions. Such a map is said to be *conformal*, because a very small figure on the spheroid is transformed into a figure of the same shape on the map. It probably won't be of the same size, because scale will vary from point to point over most of the map, so a large figure on the spheroid will not have the same shape on the map. But the fact that angles in a small region are correctly given is very important in photogrammetry, so *conformal* map projections are much used.

(b) At each point of the map, a small figure on the spheroid is represented by a figure on the map with the same area, except for scale. This is usually accomplished by compressing the distance scale in one direction to make up for its expansion in another. Such a map is called an *equal-area map*. As one would expect, this simultaneous compression and expansion in two different directions distorts shapes considerably, and the map projection is rarely used in photogrammetry.

The process of going from the spheroid to the flat surface (plane) is often done in two or three steps instead of directly in one. Points on the spheroid are transferred first to the sphere (this step may be omitted); they are then transferred to a developable surface (that is, one that can be unrolled into a flat surface); and the developable surface is then reduced in scale to that wanted for the map and is unrolled, whereby it becomes the flat map. Only three developable surfaces are used in cartography; the cone, the cylinder; and the plane itself (The cylinder can, of course, be thought of as a special case of the cone).

Plane—The simplest form of map projection is onto a plane, tangent to the earth's surface at a point. In such a projection, scale distortion increases rapidly with distance from the point of tangency. By lowering the plane slightly to make it cut the earth instead of being tangent, all the scales are reduced and the effective useful area of the projection may be increased. An example of this general type of projection is the stereographic (section 8.3.2.4).

Cylinder—To extend the capabilities of the planar type of projection in one predominant direction, the plane may be considered as wrapped around the earth to form a cylinder (tangent or secant), and a narrow belt of the earth's surface around the cylinder may be projected onto it. The cylinder may then be unrolled into a plane to form a projection of almost unlimited extent in one direction, but narrow in the other. Different forms of the Mercator projection (sections 8.3.2.2 and 8.3.2.3) are of this type.

Cone—A third common type of projection involves a cone, tangent to the earth (or secant to it) at a parallel of latitude. When the cone is cut along an element and developed, the resulting projection may be extended indefinitely in one direction, but is limited in the other, as with the cylinder. An example of the conic type of projection is the Lambert (section 8.3.2.6).

These geometric explanations of the planar, cylindrical, and conic projections are oversimplified here to illustrate the types of projections in a general way without being absolutely correct. In actual practice, precise development of each type of projection depends on complex mathematical formulas.

8.3.2 MAP PROJECTIONS

Within the limitations of equal-area or conformal map projections, and of projection onto a plane, a cylinder, or a cone, a very large number of map projections have been developed. To a non-cartographer, they appear to be completely different from each other, and new varieties are always appearing. Actually, many of them are merely minor variations of a few ''basic'' types, and of these basic types, only a few are of photogrammetric interest. These are (a) the Mercator map projection, used almost universally for charts; (b) the transverse Mercator map projection, used in one form or another by most countries in the world for military maps and by a majority for civilian maps as well; (c) the oblique Mercator map projection, used by the U.S.A. for coordinate systems in Alaska; (d) the Lambert conformal conic map projection; and (e) the Lambert polyconic map projection, used in modified form for the International Map of the World series. Other map projections are used for special purposes or for regions whose size and shape make a different map projection desirable.

Of the above projections (a) to (e), the projections (b), (c), and (d) are of particular interest for photogrammetry in the U.S.A. For more details on these and for descriptions of the other projections, Deetz and Adams (1944) and J. A. Steers (1965) are recommended.

8.3.2.1 THE MERCATOR MAP PROJECTION

The Mercator map projection takes points on the surface of a spheroid representing the earth's surface and projects them onto the surface of a

cylinder wrapped around the spheroid's equator
(The cylinder sometimes is placed so as to cut
the spheroid on two parallels). However, projection is not from the center of the spheroid but
from points on the equatorial plane removed
from the center, a distance which depends on the
latitude of the point being projected (figure 8-3).
Mercator's map projection introduces considerable distortion into regions at high latitudes. To
avoid this distortion, the transverse and oblique
Mercator map projections were introduced.
These map projections place the cylinder in such
a position that distortion is minimal in the direction of interest, while distortion transverse to
this direction is limited by restricting the width
of the map in the transverse direction.

8.3.2.2 TRANSVERSE MERCATOR MAP PROJECTION

The transverse Mercator map projection may
be produced by wrapping a cylinder around a
meridian of the earth, with its axis lying in the
plane of the equator (*see* figure 8-4). The cylinder is reduced slightly in size, to cut the earth
along two small circles *b* and *c,* parallel to the

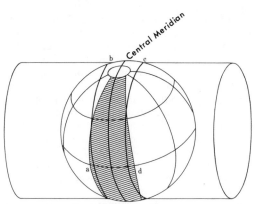

FIGURE 8-4. The Transverse Mercator map projection.

central meridian and equally spaced from it. The
earth (actually a spheroid representing the earth)
is projected mathematically (a simple geometric
projection will not work) onto the cylinder,
which is then cut along an element and developed into a plane. This projection may be extended indefinitely in the north-south direction;
but the east-west dimension is limited to control
the scale error. The scale of the map of a zone
projection is slightly too small between *b* and *c*,
exact along these lines, and too large beyond
them.

8.3.2.3 THE OBLIQUE MERCATOR MAP PROJECTION

Where the principal dimension of an area is
skewed with respect to the meridians and parallels, the oblique Mercator map projection may
be used to depict the area. Like the transverse
Mercator map projection, it is produced by projecting mathematically onto a cylinder that has
been rotated from its original orientation with
the polar axis. However, instead of being rotated
through 90°, the cylinder is rotated until its axis
is perpendicular to the major dimension of the
area. (See Hotine, 1946, 1947.)

8.3.2.4 THE POLAR STEREOGRAPHIC MAP PROJECTION

The polar stereographic map projection is
produced (figure 8-5) by placing a plane tangent
to one of the poles of the spheriod representing
the earth and then projecting the spheroid's
surface onto the plane from the opposite pole.
The projection is used for maps of the polar and
near-polar regions because of the confusion that
would arise there if the transverse Mercator map
projection were to be used; the narrow strips
used for the transverse Mercator overlap completely at the poles and diverge from there.

8.3.2.5 LAMBERT CONFORMAL CONIC MAP PROJECTION

Figure 8-6 illustrates the idea of the Lambert
conformal conic projection. The axis of the cone

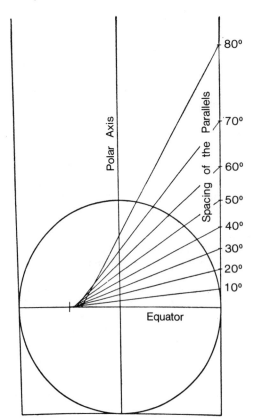

FIGURE 8-3. Principle of the Mercator Map projection. Note that projection is not from the center of the
sphere but from points at varying distances from the
center. From Deetz and Adams, Elements of Map
Projections (1944) page 36.

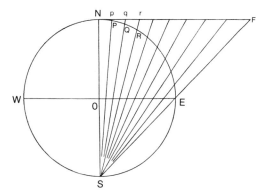

FIGURE 8-5. Principle of the Polar Stereographic map projection. From U.S. C&GS Sp. Publ. 68 (1944).

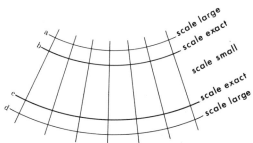

FIGURE 8-7. Lambert conformal conic projection. Scale of grid is too small in area between standard parallels (*b* and *c*) and too large outside these parallels.

is the polar axis of the spheroid, and the cone cuts the spheroid along two parallels of latitude (*b* and *c* in figures 8-6 and 8-7) called "standard parallels." The earth is projected onto the cone, which is then cut along an element and flattened into a plane, carrying with it the parallels and meridians which have been projected onto it (figure 8-7).

8.3.3 GRIDS

Each type of map projection has associated with it a graticule of parallels and meridians to represent the corresponding reference system on

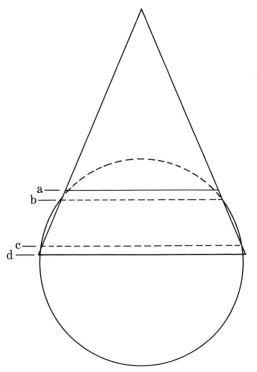

FIGURE 8-6. Lambert conformal conic map projection. Cone cuts spheroid along two parallels of latitude.

the spheroid. Every point on the projection, with its particular latitude and longitude, has one and only one corresponding point on the spheroid. If, now, a rectangular grid be superimposed on the projection in a fixed relation to the geographic graticule, there will be a fixed pair of x, y coordinates corresponding to each pair of geographic coordinates, and interconversion between the two systems can be computed through the known mathematical relation of the systems. Similarly, azimuths and distances between points on one system may be readily converted to the other. These computations are usually accomplished on desk calculators or by logarithms, using prepared computing forms and special tables of factors, but where the volume is large they are often accomplished on electronic computers.

The use of grids based on map projections is very convenient in office plotting and for property surveys and engineering surveys of limited extent. In office plotting, the positions of points (for example, the positions of ground control stations and minor control derived from aerotriangulation) can be plotted on the map sheets much more conveniently by grid coordinates, which permit the use of a coordinatograph, than by latitude and longitude. Property surveys, engineering surveys, and for that matter, any horizontal control survey of limited extent, started and closed on geodetic stations of the basic control network, can be computed on the grid by relatively simple computations employing plane trigonometry; yet, each grid position can be converted to a latitude and longitude position on the spheroid of reference for integration into the basic control network and so is uniquely defined. This is extremely valuable for the replacement of property corners or other stations when monuments are lost. Grids have long been considered essential on military maps for ease in defining map positions. At present, two grids—the Universal Transverse Meractor (UTM) grid and the Universal Polar Stereographic (UPS) grid—are in use for military purposes.

The Universal Transverse Mercator (UTM)

grid was developed after World War II for military use and has been adopted for world-wide use by the NATO countries. It consists of 60 zones, each covering 6° of longitude (about 668 kilometres at the equator) and extending from 80° south latitude to 80° north latitude. The scale factor on the central meridian is 0.9996; near the zone boundaries at the equator it is 1.0010.

The polar regions are most conveniently projected onto a plane perpendicular to the earth's axis and cutting the earth in a circle centered at the pole. In this projection, all meridians radiate from the pole. The rectangular grid is imposed on the projection so that the y axis passes through the pole and coincides with a selected meridian. The Universal Polar Stereographic (UPS) grid was developed to supplement the UTM grid (8.3.2.1) in both polar areas between 80° latitude and the poles.

Most surveying for civilian purposes, including that done for photogrammetric control, is local in character and does not cover the vast regions that must be included in military coordinate systems. In 1933, the Coast and Geodetic Survey (C & GS) introduced the idea of using grids, called "state plane-coordinate systems", for civilian surveying. Because each state needed its own coordinate system and because each state is differently shaped, no one map projection could be made to give satisfactory grids for all states. So three map projections—the transverse Mercator, the Lambert conformal conic, and the oblique Mercator—were found necessary. The grids are known by the names of the map projections with which they are associated.

The transverse Mercator grid is used for those state plane-coordinate systems where the greatest dimension is in the north-south direction. Some of the larger states using this grid are divided into two or more zones. The width of a zone is usually limited to 158 miles (about 254 km) to avoid scale errors greater than 1 part in 10,000.

Where the oblique Mercator grid is used, the rectangular grid is imposed with the y-axis coinciding with a selected meridian near the center of the area. (The only zone with this grid is southeast Alaska).

When a grid is superimposed on a map on the Lambert conformal conic map-projection, the y-axis is placed to coincide with a selected meridian near the center of the area; and the axis of x is placed below the southern limit of the area. The scale of the grid is too small in the area between the standard parallels (b and c of figure 8-7), exact along those parallels, and too large outside of those parallels. The grid may be extended indefinitely east and west. The north-south limit of the grid is restricted to limit the magnitude of the scale error. This grid is also used in the state grid systems in the United States, where the width of the grid is usually limited to 158 miles (about 254 milometres) with a maximum scale error of 1 part in 10,000. The larger states are divided into two or more grid zones.

8.3.3.1 PLANE ANGLES AND AZIMUTHS

Grid azimuths are measured with respect to the y axis of the grid. Because of this, no provision need be made for the convergence of meridians. The difference between grid azimuth and geodetic azimuth therefore increases as the distance from the central meridian or y axis of the grid increases.

All of the projections described in 8.3.2 are conformal; that is, the true shape of any small area of the earth is preserved on the projection. Thus, at any given point the difference between geodetic azimuths and grid azimuths of very short lines is a constant, and angles on the earth represented by such lines are truly represented on the map. For longer lines, the difference varies and the correction to be applied to an observed (geodetic) angle to obtain a corresponding grid angle is the difference of the corrections to the azimuths of the lines, separately derived. Where the observed lines are long, grid and geodetic angles differ significantly because the geodetic line is projected on the grid as a curve which is always concave toward the central line of the projection (central meridian of the transverse Mercator projection or central parallel of the Lambert projection). The geodetic angle is measured between the tangents to these curves, whereas the grid angle is measured between the chords. The small correction angle between the tangent and the chord is a function of the distance of the middle of the line from the central line of construction of the projection (as defined above) and the length and direction of the line. For 2 to 3 km lines, the correction from geodetic to grid angle does not usually exceed about 1".2 on the state plane-coordinate systems. Thus, it may be stated in general that, for all third-order work and for second-order work in practically all areas of the state systems, no significant error is introduced by assuming the observed angles to be grid angles so long as the observed lines are no more than 3 km in length. On the Universal Transverse Mercator grid system, the angular difference between the geodetic and grid azimuths is greater at the outer edges of the zones because the zones are wider than the zones of the state systems. For example, assume an angle of nearly 180° has been measured near the boundary of a Universal Transverse Mercator grid zone at latitude 25°. If the sides of this angle lie in a north-south direction and are 15 km long, a correction of, roughly, 25" must be applied to reduce the observed angle to a grid angle. If the sides are 30 km long, the correction would approximate 50"; if they are 1.5 km long, however, the correction would be only about 2".5.

8.3.3.2 COMPUTATIONS USING PLANE COORDINATES

The computations for extensive horizontal-control surveys (triangulation, trilateration, and traverse) are usually made by rather involved precise geodetic methods. On the other hand, it is simpler, faster, and sufficiently accurate to make the computations for limited horizontal-control surveys in terms of grid coordinates. In this case, the observed angles may be considered as grid angles for short lines that is, for lines of 1 to 3 km between the point of observation and the target. Where the observed lines are long, however, this is not valid and the observed angles must be corrected to grid angles if good positions are to be derived from computations in terms of grid coordinates.

Similarly, distances measured on the ground should be multiplied by the appropriate scale factor listed in the state plane coordinate tables to reduce them to grid distance for these computations. The scale factors may usually be ignored in the state grid systems where lower precision than one part in 10,000 is required and there are adequate datum ties to more precise control. However, such measured distances must be reduced to sea level when this reduction becomes significant. The sea-level corrections, in a proportional sense, vary uniformly with the elevation; for example, at 600 m the sea-level correction is 1:10,000, at 1200 m it is 1:5,000 and so on.

8.3.4 REFERENCES

The basic text in English for map projections is Deetz and Adams (1944) (older editions are satisfactory). Steers's book describes more map projections and does so entirely in terms of simple geometric constructions. Those who are interested in or need the mathematical development and formulations will find the best English texts to be Thomas (1952) on conformal map projections and Adams (1945) on equal-area map projections. The two-volume work by Driencourt and Labord (1932) is encyclopedic in scope and detailed in treatment. For a history of map projections and the philosophy of their use, Eckert's treatise (1921-5) is practically definitive. The essential theory of the transverse Mercator map projections is given in detail in an Army Map Service publication (1967). However, it is also given, together with the theory of the oblique Mercator map projection, and more simply, by Hotine (1946, 1947).

The actual mechanics of using map projections for surveys is tied up, as far as photogrammetric surveys are concerned, in the manuals and tables for construction and use of grids. The construction of Universal Transverse Mercator grid and Universal Polar Stereographic grid are described, with tables for construction and use, in various publications of the U.S. Department of Defense that are available either in the libraries of most geodetic institutions or from the U.S. Defense Mapping Agency, Washington, D.C. Tables for grids between various zones of latitude and on various spheroids have been published. Tables for conversion of geodetic coordinates to coordinates on state plane-coordinate systems and vice versa are obtainable from the U.S. Superintendent of Documents, Washington, D.C. A separate set of tables is issued for each state.

All American textbooks on surveying contain a description of state plane-coordinate systems. However, the photogrammetrist who is going to make extensive use of such systems should read Mitchell and Simmons (1977) and Reynolds (1935).

8.4 Horizontal-Control Surveys

8.4.1 INTRODUCTION

Space does not permit the inclusion in this chapter of instructions for all classes of horizontal-control surveys for photogrammetry and mapping. Reference manuals listed at the end of the chapter contain detailed instructions for horizontal-control surveys and cover a great variety of instruments and procedures for these surveys. A selection can be made from these to fit the economy and size of almost any control-survey problem. Therefore, the intent of this section is to include information not readily found in the reference manuals, to discuss overall procedures, and, insofar as possible, to indicate references that apply to specific types of control surveys.

In this section we are concerned with horizontal control for photogrammetric plotting and mapping rather than with the addition of monumented stations to a basic control network. Nevertheless, since the control for mapping often has to be extended over rather large areas, standards and specifications for first-, second-, and third-order horizontal control surveys are included for reference. They were prepared for horizontal-control surveys in the United States of America. They should provide a reliable general guide even though they may differ somewhat from corresponding standards and specifications used in other countries. Note that the *standards* are the criteria that determine the order and class to which a survey belongs. The *specifications* govern the procedures that must, in the experience of the National Geodetic Survey, be followed for the corresponding standards to be met.

8.4.1.1 STANDARDS AND SPECIFICATIONS FOR HORIZONTAL-CONTROL SURVEYS

Tables 8-1 and 8-2 give the standards and general specifications for first-, second-, and third-order triangulation, trilateration, and traverse, the methods most used for the establishment of ground control.

8.4.1.2 FOURTH-ORDER HORIZONTAL CONTROL

The designation "fourth-order" is applied to horizontal-control surveys of less-than-third-order accuracy. Such surveys are frequently used to locate photo-control points from stations of a higher order of accuracy for photogrammetric plotting. There are no definitive specifications for fourth-order, horizontal-control surveys; the term is general. In practice they must be designed to provide an accuracy of position adequate for the specific purpose. Generally, fourth-order-control surveys, either traverse or triangulation, must be limited in extent.

8.4.1.3 PLANNING HORIZONTAL-CONTROL SURVEYS

In planning horizontal-control surveys for a specific mapping project it is first necessary to determine the positional accuracy required of each of the photo-control points relative to stations of higher order. The accuracy required for the photo-control points will depend upon the scale and accuracy requirements of the mapping, and upon the type of photogrammetric procedure (aerotriangulation) to be used to bridge the ground control and to provide minor control (photogrammetric pass points) for map compilation from the aerial photographs. The accuracy requirement for the photo-control points might be anywhere from ±0.1 m, or less, for the most precise aerotriangulation for projects such as cadastral surveys, to ±0.3 m to ±2 m for general-purpose photogrammetric mapping, or even larger tolerances for medium- and small-scale mapping. For large areas, the control surveys for mapping may include first- or second-order triangulation or traverse; or it may be more economical to use high-order traverse with electromagnetic measurement of lengths to establish the main framework of control.

In planning control surveys, it is well to remember that the accuracy of a section of triangulation or traverse cannot be judged solely by the misclosure in length or position, because there is always the possibility of compensating errors or of systematic errors which will not be disclosed by the misclosure. The accuracy required of a section of triangulation, or of a traverse, is dependent on the precision of the entire survey. The requirements for the precision of the various measurements given in tables 8-1 and 8-2 have been determined from long experience. These requirements, such as the strength of figure and triangle misclosures in triangulation, the quality and frequency of azimuth checks, and the quality of distance measurements in traverse, provide for a certain rigidity in the survey and ensure that the maximum misclosure in position or length will not exceed 1 part in 5,000 for third-order work, 1 part in 10,000 for second-order, and so on. For example, if all of the requirements for precision of the various measurements in the scheme of *second-order* triangulation or traverse are carried out, we can expect the largest error after adjustment to be not greater than 1 part in 10,000 and the *average error in positions established will probably not be greater than 1 part in 30,000.* This latter fact is usually taken into account in planning control surveys for mapping; for example, in determining the permissible length of a section of triangulation or traverse between control stations of a higher order of accuracy, consideration is given to the fact that the maximum error is seldom reached.

8.4.2 TRIANGULATION

Triangulation (figure 8-8) is a method of surveying in which the location of a new point is determined from the mathematical solution of a triangle whose vertices are the new point and two other points whose positions are known together with the length and azimuth of the line between them. The angles of the triangle are determined by field measurements. Using these angle measurements, the lengths and azimuths of the two unknown sides of the triangle and the geographic position of the new vertex point are computed. Successive points are computed through a continuous chain of triangles. For extension of control over large areas, the triangulation is laid out in quadrilaterals, central-point polygons, or other figures, designed so that each new point can be computed through two different triangles as a check. This provides the rigidity and strength of figure required for the particular survey (*see* 8.4.1.1).

Triangulation requires the selection of sites for stations and base lines favorable for use both from topographic and geometric (*i.e.,* strength-of-figure) considerations; it is well adapted to the use of precision instruments and methods in all its operations, and is susceptible of great accuracy in its results. In addition to the actual operations of observing angles and measuring baselines, and the mathematical processing, triangulation also includes the reconnaissance which precedes those operations as well as the astronomic observations which are required in the establishment of a geodetic datum and in the orientation of the triangulation.

The word *triangulation* ordinarily implies geodetic triangulation in which the curvature of the surface of the earth is taken into account. In control surveys for mapping, however, photo-

TABLE 8-1. CLASSIFICATION AND STANDARDS FOR GEODETIC CONTROL AND PRINCIPAL RECOMMENDED USES

Horizontal Control

Classification	First-Order	Second-Order		Third-Order	
	Class I	Class I	Class II	Class I	Class II
Relative accuracy between directly connected adjacent points (at least)	1 part in 100,000	1 part in 50,000	1 part in 20,000	1 part in 10,000	1 part in 5,000
Recommended uses	Primary National Network. Metropolitan Area Surveys. Scientific Studies.	Area control which strengthens the National Network. Subsidiary metropolitan control.	Area control which contributes to, but is supplemental to, the National Network.	General control surveys referenced to the National Network. Local control surveys.	

Vertical Control

Classification	First-Order		Second-Order		Third-Order
	Class I	Class II	Class I	Class II	
Relative accuracy between directly connected points or benchmarks (standard error)	$0.5 \text{ mm } \sqrt{K}$	$0.7 \text{ mm } \sqrt{K}$	$1.0 \text{ mm } \sqrt{K}$	$1.3 \text{ mm } \sqrt{K}$	$2.0 \text{ mm } \sqrt{K}$
			(K is the distance in kilometers between points.)		
Recommended uses	Basic framework of the National Network and metropolitan area control. Regional crustal movement studies. Extensive engineering projects. Support for subsidiary surveys.		Secondary framework of the National Network and metropolitan area control. Local crustal movement studies. Large engineering projects. Tidal boundary reference. Support for lower order surveys.	Densification within the National Network. Rapid subsidence studies. Local engineering projects. Topographic mapping.	Small-scale topographic mapping. Establishing gradients in mountainous areas. Small engineering projects. May or may not be adjusted to the National Network.

TABLE 8-2. GENERAL SPECIFICATIONS FOR HORIZONTAL CONTROL

Classification	First-Order	TRIANGULATION Second-Order		Third-Order	
		Class I	Class II	Class I	Class II
Recommended spacing of principal stations	Network stations seldom less than 15 km. Metropolitan surveys 3 km to 8 km and others as required.	Principal stations seldom less than 10 km. Other surveys 1 km to 3 km or as required.	Principal stations seldom less than 5 km or as required.	As required	As required
Strength of figure					
R_1 between bases					
Desirable limit	20	60	80	100	125
Maximum limit	25	80	120	130	175
Single figure					
Desirable limit					
R_1	5	10	15	25	25
R_2	10	30	70	80	120
Maximum limit					
R_1	10	25	25	40	50
R_2	15	60	100	120	170
Base measurement					
Standard error[1]	1 part in 1,000,000	1 part in 900,000	1 part in 800,000	1 part in 500,000	1 part in 250,000
Horizontal directions[2]					
Instrument	0".2	0".2	0".2 { 1".0	1".0	1".0
Number of positions	16	16	8 or {12	4	2
Rejection limit from mean	4"	4"	5" 5"	5"	5"
Triangle closure					
Average not to exceed	1".0	1".2	2".0	3".0	5".0
Maximum seldom to exceed	3".0	3".0	5".0	5".0	10".0
Side checks					
In side equation test, average correction to direction not to exceed	0".3	0".4	0".6	0".8	2"
Astro azimuths[3]					
Spacing-figures	6-8	6-10	8-10	10-12	12-15
No. of obs./night	16	16	16	8	4
No. of nights	2	2	1	1	1
Standard error	0".45	0".45	0".6	0".8	3".0

TRAVERSE

Classification	First-Order	Second-Order Class I	Second-Order Class II	Third-Order Class I	Third-Order Class II
Recommended spacing of principal stations	Network stations 10-15 km. Other surveys seldom less than 3 km.	Principal stations seldom less than 4 km except in metropolitan area surveys where the limitation is 0.3 km.	Principal stations seldom less than 2 km except in metropolitan area surveys where the limitation is 0.2 km.	Seldom less than 0.1 km in tertiary surveys in metropolitan area surveys. As required for other surveys.	
Horizontal directions or angles[2]					
Instrument	0".2	0".2 or 1".0	0".2 or 1".0	1".0	1".0
Number of observations	16	8 or 12*	6 or 8*	4	2
Rejection limit from mean	4"	4" / 5"	4" / 5"	5"	5"
Length measurements[1]					
Standard error	1 part in 600,000	1 part in 300,000	1 part in 120,000	1 part in 60,000	1 part in 30,000
Astro azimuths					
Number of courses between azimuth checks[5]	5-6	10-12	15-20	20-25	30-40
No. of obs./night	16	16	12	8	4
No. of nights	2	2	1	1	1
Standard error	0".45	0".45	1".5	3".0	8".0
Azimuth closure at azimuth check point not to exceed	1".0 per station or 2"$\sqrt{N}$	1".5 per station or 3"$\sqrt{N}$. Metropolitan area surveys seldom to exceed 2".0 per station or 3"$\sqrt{N}$	2".0 per station or 6"$\sqrt{N}$. Metropolitan area surveys seldom to exceed 4".0 per station or 8"$\sqrt{N}$	3".0 per station or 10"$\sqrt{N}$. Metropolitan area surveys seldom to exceed 6".0 per station or 15"$\sqrt{N}$	8".0 per station 30"$\sqrt{N}$
Position closure[4,6] after azimuth adjustment	0.04m $\sqrt{K}$ or 1:100,000	0.08$\sqrt{K}$ or 1:50,000	0.2m $\sqrt{K}$ or 1:20,000	0.4m $\sqrt{K}$ or 1:10,000	0.8m $\sqrt{K}$ or 1:5,000
Closure in length (also position when applicable) after angle and side conditions have been satisfied, should not exceed	1 part in 100,000	1 part in 50,000	1 part in 20,000	1 part in 10,000	1 part in 5,000

TRILATERATION

	First-Order	Second-Order Class I	Second-Order Class II	Third-Order Class I	Third-Order Class II
Recommended spacing of principal stations	Network stations seldom less than 10 km. Other surveys seldom less than 3 km.	Principal stations seldom less than 10 km. Other surveys seldom less than 1 km.	Principal stations seldom less than 5 km. For some surveys a spacing of 0.5 km between stations may be satisfactory.	Principal stations seldom less than 0.5 km.	Principal stations seldom less than 0.25 km.

(Continued on next page)

TABLE 8-2—Continued

Classification	First-Order	Second-Order Class I	Second-Order Class II	Third-Order Class I	Third-Order Class II
Geometric configuration					
Minimum angle contained within, not less than	25°	25°	20°	20°	15°
Length measurement					
Standard error[1]	1 part in 1,000,000	1 part in 750,000	1 part in 450,000	1 part in 250,000	1 part in 150,000
Astro azimuths[3]					
Spacing-figures	6-8	6-10	8-10	10-12	12-15
No. of obs./night	16	16	16	8	4
No. of nights	2	2	1	1	1
Standard error	0″.45	0″.45	0″.6	0″.8	3″.0
Closure in position[4] after geometric conditions have been satisfied should not exceed	1 part in 100,000	1 part in 50,000	1 part in 20,000	1 part in 10,000	1 part in 5,000

NOTE (1)
The standard error is to be estimated by $\sigma m = \sqrt{\sum v^2 / n \, (n-1)}$ where σm is the standard error of the mean, v is a residual (that is, the difference between a measured length and the mean of all measured lengths of a line), and n is the number of measurements.

NOTE (2)
The figure for "Instrument" describes the theodolite recommended in terms of the smallest reading of the horizontal circle. A position is one measure, with the telescope both direct and reversed, of the horizontal direction from the initial station to each of the other stations.

NOTE (3)
The standard error for astronomic azimuths is computed with all observations considered equal in weight (with 75 percent of the total number of observations required on a single night) after application of a 5-second rejection limit from the mean for first- and second-order observations.

NOTE (4)
Unless the survey is in the form of a loop closing on itself, the position closures would depend largely on the conditions used in the adjustment. The extent of constraints and the actual relationship of the surveys can be obtained through either a review of the computations, or a minimally constrained adjustment of all work involved. The proportional accuracy or closure (*i.e.* 1/100,000) can be obtained by computing the difference between the computed value and the fixed value, and dividing this quantity by the length of the loop connecting the two points.

NOTE (5)
The number of azimuth courses for first-order traverses are between Laplace azimuths. For other survey accuracies, the number of courses may be between Laplace azimuths and/or adjusted azimuths.

NOTE (6)
The expressions for closing errors in traverses are given in two forms. The expression containing the square root is designed for longer lines where higher proportional accuracy is required.

N is the number of stations for carrying azimuth.
K is the distance in kilometers.

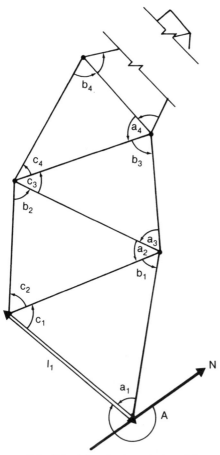

FIGURE 8-8. Principle of triangulation. Measured quantities; azimuth A, length l_1, and angles a_1, b_1, c_1. Calculated quantities; all other lengths, angles and directions.

in the field. Reynolds (1934) will be of interest to those wanting more information about overall theory and about the adjustment of triangulation. Whyte (1969) also contains detailed instructions for triangulation fieldwork and computations. Much computation is now done using electronic computers. Schwarz (1978) describes the program used by the National Geodetic Survey.

8.4.3 TRILATERATION

Trilateration is a method of extending horizontal control whereby the *lengths* of the sides of the triangles are measured instead of the *angles*. Electromagnetic distance-measuring instruments, such as the Electrotape, Geodimeter, or Tellurometer, provide a feasible means of doing trilateration. Trilateration has been extended across wide bodies of water or inaccessible areas (where triangulation or ground-based trilateration would be impossible) by using distance-measuring instruments carried in airplanes (Shoran, Hiran, Shiran). However, where distances greater than those accurately measurable by ground-based instruments are involved, present practice is to use beacons on artificial satellites together with an instrument that measures the Doppler-shift in frequency of the beacon as observed from the ground. In the near future, a considerable amount of ground control will be established using inertial surveying systems and the Global Positioning System. Section 8.6.4 gives a short description of these systems.

8.4.4 TRAVERSE

Traversing (figure 8-9) is a method of surveying whereby the position of a new point is determined by measuring the lengths and directions or azimuths of a series of connecting lines from a point of known position to the point whose position is to be determined. A traverse may also consist of a single line. Occasionally, a traverse is run without connection to points of known position for the purpose of determining only the relative positions of points in the traverse, but for most mapping purposes the traverse starts from and ties to a point of known position; it may form a closed loop by tying back on the starting point or it may close on a second, known point. The angles or directions are measured by means of a transit or theodolite, and the lengths of the lines are usually measured by taping or by electromagnetic distance-measuring instruments. Distances may also be measured by subtense methods or by stadia for surveys of a lower order of accuracy.

Since the advent of electromagnetic distance-measuring instruments (*see* 8.4.6) and the lightweight, optical-reading theodolites, the task of extending horizontal control for mapping has been considerably simplified. Traverse sur-

points close to basic control stations are sometimes located by means of a single triangle. For very short distances, the curvature of the earth may be ignored and the computations made in terms of plane coordinates (*see* also section 8.3.3.).

Instruments for triangulation must be chosen in terms of the accuracy required. Modern optical-reading theodolites are light, fast, and particularly adaptable for triangulation. A one-second optical theodolite is sufficiently precise for second-order triangulation and traverse and is also fast and easy to use for all lower-order surveys in both triangulation and traverse. Night observations on lights are preferable for higher-order triangulation in order to avoid angular errors due to phase and horizontal refraction. On the other hand, angle observations for third-order triangulation are usually made on various types of targets during daylight.

Gossett (1959) provides detailed instructions for field operations, including the abstracting of observations and the computations usually made

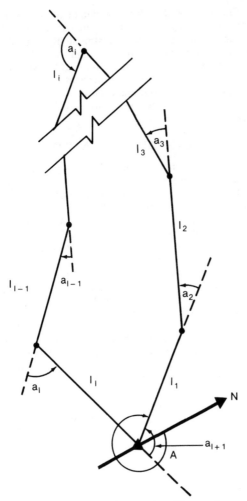

FIGURE 8-9. Principle of traversing. Measured quantities; azimuth A, lengths l_i, and angles a_i. Calculated quantities; all lengths and angles between points.

vides detailed instructions for second- and third-order traverse for mapping and includes a number of procedures for traverse specifically for map control.

Traverses with courses not longer than 1 or 2 miles may be computed in terms of plane coordinates with ordinary methods (Mitchell and Simmons, 1977). (*see* also section 8.3.3.2).

8.4.5 SPECIAL METHODS OF SUPPLEMENTAL HORIZONTAL CONTROL

Control surveys for photogrammetric mapping often involve the location of single points for photo control stations. In this case, the field surveyor is concerned with the location, within the required accuracy, of a point visible on the aerial photographs (photo-point), but is not concerned with setting monumented control stations or extending basic horizontal control. Various methods of locating these photo-points can be used, depending on the ingenuity of the field surveyor. Usually, applications of triangulation or traverse, or a combination of both, are used, based on one or more control stations of known position. These methods include (1) single triangles, preferably with all three angles observed; (2) triangulation intersection from two, but preferably from three, stations of known position; (3) triangulation resection; and (4) short traverses of one or a few courses with the lengths measured by taping, by subtense or short-base methods, by electromagnetic distance-measuring instruments, or by stadia, and with azimuths determined by angle measurements on a horizontal control station of known position, or by observations of Polaris or the sun.

8.4.6 ELECTROMAGNETIC DISTANCE MEASUREMENT

Great advances were made during World War II in the development of methods of detection and ranging to unseen objects using radio or electromagnetic pulses. Generally known as "radar," the basic system depends on an accurate determination of the time it takes a pulse to travel to and be reflected from the intercepted object. Since the rate of wave propagation is known to be equivalent to that of light, determination of distance becomes relatively simple.

From this development has evolved an ever-growing number of radio location and tracking systems, some of which are particularly adaptable to position and height determinations required for surveying and mapping purposes. Originally, ranges were determined from reflected pulses alone: later designs added an artificial reflector which amplified the "echo" permitting longer ranges and more precise measuring characteristics.

veys in which the lengths are measured electromagnetically are rapidly becoming the standard method of extending control for mapping. The electromagnetic distance measurements are not only faster than taping but usually are more accurate. The method is versatile in that distances can be measured wherever unobstructed lines of sight are available, and the many short courses usually associated with a taped traverse can usually be replaced by a few long courses. With ordinary care, the accuracy of electronically measured traverses easily meets second-order requirements and, if the angle and azimuth observations are made with the required precision, first-order accuracies can be approached.

C. V. Hodgson (1955) contains good basic instructions for second- and third-order traverse using invar tapes and spirit leveling for slope corrections. U.S. Geological Survey (n.d.) pro-

8.4.6.1 AIRBORNE EQUIPMENT—SHORAN AND HIRAN

The first important application of this principle to surveying was the development of Shoran equipment, by which distances could be measured up to 300 miles (480 kilometres) with a possible accuracy of 1:60,000. Since the electromagnetic pulses travel lines of sight, direct readings between ground stations are limited in length by the earth's curvature. However, by mounting the equipment in aircraft and using a line-crossing technique, longer distances between ground stations can be measured. After corrections are applied for slope, meteorological conditions, and reduction first to the geoid, and then to the spheroid, the measurements become geodetic distances. Numbers of these lines form a trilateration net which is susceptible to computation in somewhat the same manner as triangulation. The method is of value in areas to which conventional geodetic surveys have not yet been extended.

A modified version of Shoran, known as Hiran, is capable of an accuracy of 1:100,000 over long distances. A large part of northern Canada and the Artic Islands has been controlled by Hiran stations at mean intervals of 200 miles (320 kilometres), providing a reasonably reliable base for modern mapping techniques.

Another application of direct interest to photogrammetrists is Shoran- or Hiran-controlled photography. In this application, position of the aircraft during the photographic mission are determined by ground stations simultaneously with photographic exposures. This results in independent positioning of each photograph and also provides a means of accurate navigation. This form of control is acceptable for medium- and even large-scale mapping in large areas that are otherwise uncontrolled.

8.4.6.2 GROUND-BASED DISTANCE-MEASURING INSTRUMENTS

Since Galileo first tried to measure the speed of light (and decided that it was effectively infinite), the speed of light has been one of the major problems for investigation by scientists. Not until the last few years of the last century, however, was the speed known well enough that serious thought could be given to using the speed of light to determine distance. At about the same time, Maxwell's theory of electromagnetic radiation made it obvious that the speed of all electromagnetic radiation, not just light, was the same constant value, at least in a vacuum. These two developments made it possible to use, in theory, any part of the electromagnetic spectrum for determining distance by measuring travel time and multiplying by the speed of light (properly corrected for the slowing down by the medium).

For technical reasons, radio waves were the first to be used for this purpose, and the development of radio equipment for determining distances of ships and aircraft was accelerated during World War Two. It led, immediately after the close of that war, to application of the equipment and techniques to determining distances between points on the ground and to the development of equipment such as Hiran and Shiran, already mentioned. But all the experience and equipment involved use of radio waves having lengths that made large antennas necessary and required timing and synchronization techniques that were still experimental in nature. The first practical distance-measuring instrument small and accurate enough to be considered of geodetic utility was therefore a return to the concept of using light rather than radio waves for measurement. The instrument, invented by a Swedish geodesist, Dr. Erik Bergstrand, was called the Geodimeter, was the size of a steamer trunk and could be carried by two strong men. The first and early succeeding models therefore were used primarily for measurements of baselines and had to be used at night to get the fullest possible range (50 km). Subsequent refinements in design have produced a variety of lightweight instruments serving various purposes. Instruments for measuring long distances (up to 60 km) use light provided by a laser (the Mekometer ME-3000, made by Wild, uses light from a xenon flash-lamp and has a range of up to 3 km) and a precision of about 1 to 5 mm plus 1 part per million of distance. Instruments for measuring short distances (up to 3 to 10 km) generally use infrared radiation and have precision somewhat less, for most of the instruments, than the precision of light-using instruments. Since most of these instruments use lasers, which give radiation of high intensity, distances can be easily measured in daylight as well as at night, but at only about half the maximum range.

A short time after the Geodimeter went into production and was found satisfactory for geodetic work, the South African Council for Scientific and Industrial Research produced (about 1954) a portable distance-measuring instrument, the Tellurometer, that used radio waves instead of light. By going to very short wavelengths, the inventor, Mr. T. L. Wadley, was able to keep the size of the instrument (including antenna) down to something that allowed the instrument to be mounted on an ordinary tripod and carried by one man. The precision was 1:60,000 over long distances (up to 100 km). Since then, the size of the instrument has been reduced even further and the precision improved. Present instruments (those of the Tellurometer family but also others constructed on the same principle) have a precision of 1 to 3 cm plus 3 parts per million, approximately. The advantage of instruments depending on radio waves is that they operate well under conditions that would prevent measurement with instruments depending on light or infrared radiation, since radio waves are not particularly affected (as far as range is concerned) by haze, fog, rain, light, or darkness.

Tomlinson and Burger (1977) give a detailed description of the appearance, capabilities, and performance of most kinds and models of distance-measuring equipment now (1980) in use.

8.4.6.3 EQUIPMENT USED PRIMARILY FOR NAVIGATION BUT APPLICABLE TO SURVEYING

There are a number of electromagnetic distance-measuring systems such as Shoran, Decca, and LORAN, which were developed and have been used primarily for navigation but which can, if necessary and if conditions are suitable, also be used for establishing photogrammetric control. Their precision, compared to that of systems and instruments just discussed is quite low; 2−5 metre r.m.s. error is considered quite excellent, and r.m.s. errors 10 times as great are the rule rather than the exception. (We do not consider systems such as Hiran or Shiran. These are systems developed from Shoran, are useful primarily for establishing geodetic control between widely separated points and hence not particularly useful for establishing closely spaced photogrammetric control, and have been discussed earlier). These systems are usually referred to as electronic-positioning systems and are classified into two categories: circular-positioning systems, in which the distances of a point from two points of known location are measured; and hyperbolic-positioning systems, in which the differences in distance of a point from three points of known location are measured. (There are also positioning systems in which the *directions* to two or more points are measured, as well as systems which are combinations of the above. Such systems have, however, practically no suitability for determining control.)

8.4.6.3.1 CIRCULAR POSITIONING SYSTEMS

A circular positioning system measures an instrument's distance from a fixed point whose location is known. This is slant range, which must be projected down to the geoid to give horizontal distance. The instrument therefore must lie somewhere on a circle (approximately) which lies on the geoid and has the fixed point as its center. The system actually measures distances from two or more fixed points simultaneously, so the instrument lies at one of the intersections of these several circles. The intersections are usually far enough apart that one of them can be picked immediately as the instrument's location. (If three or more fixed points are used, the method of least squares is applied to give the most probable location.

As can be seen in figure 8-10, and as was discussed in section 8.4.6.1, circular-positioning systems can also be used for determining a

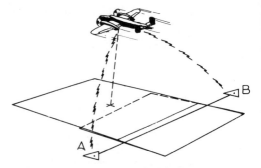

FIGURE 8-10. Circular-positioning system.

single distance between two points. The instrument is moved back and forth between the two fixed points projecting the two measured distances down onto the geoid to get two horizontal distances. It is obvious that when the sum of the two horizontal distances is a minimum, their sum is exactly the same as the total distance between A and B. The technique is, however, of no particular importance photogrammetrically.

Among the circular-positioning systems are Hydrodist (Tellurometer Corporation), Autotape (Cubic Corporation), Sea-Fix (Decca Corporation), and Shoran (Sea-Fix can also be used as a hyperbolic-positioning system). The precision of these systems varies with the configuration of the triangle formed by the mobile instrument and the fixed points; it may be as good as 50 cm. Range is on the order of 70−120 km.

VOR and TACAN are circular-positioning systems used for aerial navigation, and usually in such a way that only one fixed point is available at any one time. They are therefore not particularly useful for establishing control, but they have been used successfully for making circular surveys, in which the aircraft follows a circular path about the fixed point (Wuollet and Skibitzke, 1968; Lafferty, 1968).

8.4.6.3.2 HYPERBOLIC-POSITIONING SYSTEMS

Hyperbolic-positioning systems depend on simultaneous measurements to (or from) three fixed points simultaneously, rather than two as for circular-positioning systems. Furthermore, the distances themselves are not measured, but the differences in distances. Again, slant-ranges are measured, and these must be projected down onto the geoid to get horizontal differences of distance. Just as a distance from one fixed point determines a circle, so a *difference* of distances from two fixed points determines a hyperbola (figure 8-11). So, with differences of distances from *two* pair of fixed points (one point is usually common to the two pair), two sets of intersecting hyperbole are determined on the geoid, and the instrument is located at an intersection (only one of the two intersections of two hyperbolae is anywhere near the instrument's estimated loca-

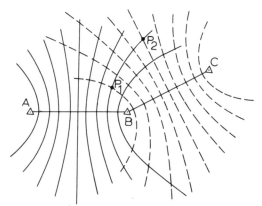

FIGURE 8-11. Hyperbolic-positioning system.

tion, usually). One advantage of the hyperbolic positioning system is that the signal is generated at the fixed points, so that any number of mobile instruments can be in operation at the same time, all making use of the same three fixed points. This king of operation is more difficult with the circular-positioning system.

Among the hyperbolic-positioning systems in common use are LORAN-C, Decca, and Raydist. The highest precision obtainable in average use is on the order of 5–10 m, with greater precision being obtainable using special techniques and under special circumstances, and with favorable configuration of mobile instrument and points. OMEGA is another hyperbolic-positioning system, but its precision is seldom better than 0.5 km and often only several kilometres.

8.4.6.3.3 CORRECTIONS TO MEASURED DISTANCES

Unless the mobile point is on the geoid, measured distances must be reduced to their equivalent horizontal distances. Kroll's method (Kroll, 1946; Carroll, 1963) has been widely used for this. In addition, corrections must be made for the fact that radio waves do not travel with the speed of light in a vacuum but are slowed down by, (a) the air through which they pass, (b) the ground and water over which they travel, and (c) the fact that the path taken by the waves may not be direct but be a set of bounces between ionosphere and land or sea. Corrections for factors (b) and (c) are far from easy, and will not be given here (*see* Dolukhanov, 1963 or Rinner and Benz, 1966). The correction for the slowing down of the waves is simpler.

It is well known that an electromagnetic wave travels in a vacuum at the same speed as light ($c_0 = 299,792.5 \pm 0.1$ km/sec.). However, in the earth's atmosphere it is a function of the refractive index n. The relationship between the refractive index n, the accepted speed in a vacuum c_0, and the speed c of a wave in the atmosphere is given by the following formula: $c = c_0/n$. The

value of n for any given area is computed from either an assumed standard atmosphere or from direct observations of temperature, humidity, and atmospheric pressure, using the following expression:

$$n = 1 + \frac{77.62}{T}P - e\left(\frac{12.92}{T} - \frac{37.19}{(T/100)^2}\right)10^{-6}$$

where:

T is absolute temperature
P is total pressure in millibars
e is vapor pressure in millibars.

The second part of the correction stems from the fact that the electromagnetic wave does not necessarily travel the shortest route between the points being measured. As the ray propagates through the atmosphere (having a varying density) it has a tendency to change direction. Since the density of the atmosphere decreases with increase in altitude, the index of refraction changes, and the ray is bent towards the earth. For most photographic missions where the flight altitude is less than 10,000 m. and the distance from the ground stations is not more than 150 km, the following approximate expression can be used to compute the chord distance S from the measured distance S_0:

$$S = S_0 - \frac{S_0{}^3}{24r^2}$$

where r is the radius of the curved ray path, assumed to be 25,000 km.

8.4.6.3.4 APPLICATION OF SYSTEMS TO PHOTOGRAMMETRIC POSITIONING

Horizontal-positioning devices provide air-station coordinates (X, Y) of each photograph at the instant of exposure which are reduced to nadir positions. The problem, in the application to stereotriangulation, then becomes one of defining and measuring the nadir point in the aerial photograph or spatial model. If the photographs are taken with no tilt, and if the intersection of the camera's fiducial markers falls exactly on the lens axis, the principal point of the photograph will define the nadir. However, since this is not usually the case, the tilt of each photograph must be determined before applying the coordinate positions. In the aeropolygon method of stereotriangulation, the instrument nadir for each photograph (the position of zero x, y coordinate change during a change in instrument z) is found and recorded. These coordinates are thereafter adjusted in horizontal position according to the known slope of the triangulated strip relative to the geoid and as determined by the vertical adjustment. These adjusted instrument x, y coordinates are then adjusted to the electronic positions much the same as when using geodetic ground control.

8.5 *Vertical-Control Surveys*

8.5.1 INTRODUCTION

The basic vertical-control network for any large area consists of lines of first-order and second-order leveling as discussed in section 8.2. Third-order vertical control is established by leveling from stations of the basic network, by the various methods described in this section, as is supplemental (fourth-order or lower) vertical control. The standards for first-, second-, and third-order vertical control are given in table 8-1. Because the precision attainable by leveling is in general much higher than that attainable by aerial photogrammetry, only the standards and specifications for third-order and lower leveling are of general interest for aerial photogrammetry. However, considerable success is being achieved, in special projects, in considerably increasing the precision obtainable by aerotriangulation. Table 8-3 therefore lists some of the specifications important to leveling of second- and third-order.

Fourth-order leveling is a general term for leveling that allows greater misclosures than those allowed for third-order leveling.

The term is applicable to any of the procedures mentioned in subsequent paragraphs, such as fly leveling, trigonometric leveling, and barometric leveling. This class of leveling should not be monumented as part of the basic network. Accuracy requirements of fourth-order leveling depend upon the mapping to be controlled by it.

The usual accuracy requirement for vertical control for contour intervals of 25 feet or less is that the elevations of photo-control points be established within an accuracy of ±0.1 of the contour interval with reference to the higher-order spirit-level lines. For larger intervals the allowable limit is progressively less than 0.1 of the C.I. Thus, the accuracy requirements are determined by the contour interval, which in turn is determined by the steepness of grades in the area to be mapped. Accuracy requirements for supplemental vertical control surveys for topographic mapping in the United States are as follows:

Contour Interval	Maximum Closure Error	
(feet)	(feet)	(metres)
5	0.5	0.2
10	1.0	0.3
20	2.0	0.6
25	2.5	0.8
40	3.0	0.9
50	4.0	1.2
80	5.0	1.5
100	6.0	1.8

8.5.2 SPIRIT LEVELING

Spirit leveling is the determination of elevations of points with respect to each other, or with respect to a common datum, by means of an instrument using a spirit level or plumb bob to establish a horizontal line of sight (figure 8-12).

The term fly leveling is often applied to fourth-order spirit leveling, that is, spirit leveling where greater tolerances are permitted with respect to the balancing of lines of sight, the closures, and so on, to facilitate the rapid establishment of elevations of control points for mapping. The relatively new self-leveling level that depends upon the gravitational pull on a small prism or plumb bob to maintain a horizontal line of sight automatically is particularly adaptable to rapid fly leveling. Fly leveling may also be done with an alidade and planetable, or with a transit having a bubble attached to the telescope. It is practicable in relatively flat terrain.

Some countries, such as Sweden and Germany, have doubled the rate at which spirit-leveling is carried out by mounting the leveling instrument and rods on vehicles. It is their experience, and that of the National Geodetic Survey which is now trying out the method, that the precision by this method is not appreciably lowered and may, in some instances, be raised. Motorized leveling has the disadvantage, of course, that it is practicable only along routes that vehicles can travel easily. Whether it is also economical for putting in photogrammetric control depends on the relative costs of personnel, vehicles and their maintenance, etc.

The practice of the U.S. National Geodetic Survey is covered in (Berry 1981). Leveling of a precision suitable for most photogrammetric work is explained in most standard texts on surveying—*e.g.*, Breed, Bone and Berry (1971).

8.5.3 TRIGONOMETRIC LEVELING

Trigonometric leveling is the determination of differences of elevations by means of observed vertical angles combined with lengths of lines (figure 8-13). Trigonometric procedures are

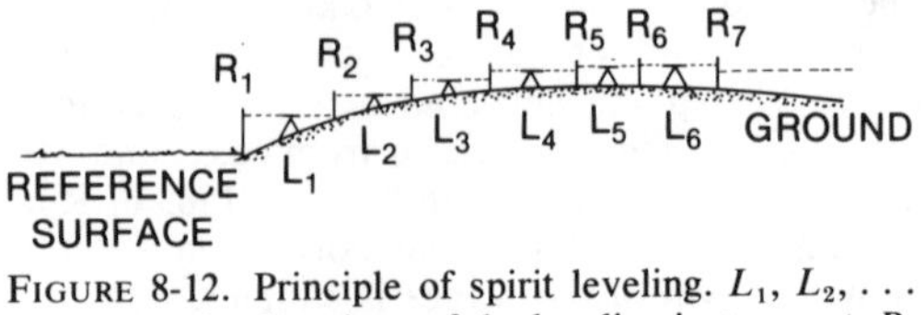

FIGURE 8-12. Principle of spirit leveling. L_1, L_2, . . . are successive locations of the leveling instrument, R_1 and R_2, R_2 and R_3, R_3 and R_4, . . . are the corresponding successive locations of the leveling rods. At each successive location of the leveling instrument, the difference ΔR in readings between rod R_i and R_{i+1} is the difference in elevation between the points on which the rods are placed. The sum of the differences is the difference in elevation between the first point R_1 and the last point R_N.

TABLE 8-3. GENERAL SPECIFICATIONS FOR VERTICAL CONTROL

Classification	Second-Order		Third-Order
	Class I	*Class II*	
Recommended spacing of lines			
National Network	Secondary Net; 20 to 50 km	Area Control; 10 to 25 km	As needed
Metropolitan control; other purposes	0.5 to 1 km	As needed	As needed
	As needed	As needed	As needed
Spacing of marks along lines	1 to 3 km	Not more than 3 km	Not more than 3 km
Instrument standards	Automatic or tilting levels with optical micrometers or three-wire levels; invar scale rods	Geodetic levels and invar scale rods	Geodetic levels and rods
Field procedures			
	Double-run; forward and backward, each section	Double- or single-run	Double- or single-run
Section length	1 to 2 km	1 to 3 km for double-run	1 to 3 km for double-run
Maximum length of sight	60 m	70 m	90 m
Field procedures			
Max. difference in lengths Forward & backward sights			
per set up	5 m	10 m	10 m
per section (cumulative)	10 m	10 m	10 m
Max. length of line between connections	50 km	50 km double-run 25 km single-run	25 km double-run 10 km single-run
Maximum closures			
Section; fwd. and bkwd.	6 mm $\sqrt{K}$	8 mm $\sqrt{K}$	12 mm $\sqrt{K}$
Loop or line	6 mm $\sqrt{K}$	8 mm $\sqrt{K}$	12 mm $\sqrt{K}$

The maximum length of line between connections may be increased to 100 km for double run for Second-Order, Class II, and to 50 km for double run for Third-Order in those areas where the First-Order control has not been fully established.

Check between forward and backward runnings where K is the distance in kilometres.

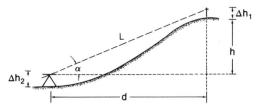

FIGURE 8-13. Principle of trigonometric leveling. Measured quantities; angle α, distance L along slope (dashed line), heights h_1 and h_2. Calculated quantities; height h, horizontal distance d.

widely used for determining control elevations for mapping. Vertical angles are usually measured with a transit, theodolite, or alidade, though they are sometimes determined from horizontal or oblique photographs. The distances, or lengths of lines, are determined in a number of ways: by triangulation, by use of electromagnetic distance-measuring instruments, by taping, by stadia, or by aerotriangulation or scaling from a photogrammetric plot. Trigonometric leveling is often combined with the triangulation or traverse surveys for horizontal control.

8.5.3.1 TRIGONOMETRIC LEVELING BY THE STADIA METHOD

Stadia trigonometric leveling is a very satisfactory and economical method of extending supplemental vertical control for mapping in areas of moderate to low relief. As the name implies, the distances are measured by stadia. This leveling may be done with planetable and alidade or with a transit or theodolite. Usually the instrument is equipped with a telescope bubble; level courses or shots are generally used over relatively level ground and trigonometric leveling is used on the slopes.

It is advisable that all the elevations for control of mapping be included as turning points in the line. Side shots for the determination of these elevations are usually prohibited to avoid blunders; if side shots must be used, each elevation should be checked by a second side shot. Closing errors for lines of stadia trigonometric leveling should be held within the limits listed in 8.5.2, according to the contour interval.

8.5.3.2 TRIGONOMETRIC LEVELING OVER LONG LINES

Trigonometric leveling is also used extensively for the establishment of mapping control in rugged or mountainous areas, particularly where the tops are bare and permit clear lines of sight. In this case, the trigonometric leveling is often combined with the triangulation that provides the lengths of the lines or courses. However, trigonometric leveling is often performed separately from the main-scheme triangulation to extend vertical control points into areas or

places where it is needed for mapping. In such instances, the vertical angles are determined by transit or theodolite and the distances or lengths of lines may be determined by observation with the Tellurometer or similar instruments, or they may be determined by aerotriangulation or scaling from the photogrammetric plot (see also 8.5.4, below, for a special method of trigonometric leveling).

The uncertainty of atmospheric refraction limits the accuracy of trigonometric leveling over longer lines. With nonreciprocal vertical angles, the uncertainty of atmospheric refraction may cause the error in the elevation difference to be as much as 6 cm per km of the length of the line. This uncertainty can be considerably reduced by observing simultaneous reciprocal vertical angles, but that is not always practicable. When simultaneous reciprocal observations cannot be made, reciprocal observations should be made at as nearly the same time of day as practicable. The best time of day for observation of vertical angles is between 12:00 and 16:00. For night observations, the best time is probably between 22:00 and 02:00, when refraction is large but more nearly constant than at sundown. Gossett (1959) provides instructions for trigonometric leveling.

8.5.4 SUPPLEMENTAL VERTICAL CONTROL BY PHOTOGRAMMETRIC METHODS

Trigonometric leveling can also be done by the measurement of vertical angles from horizontal or oblique photographs by means of the photoalidade or photogoniometer. Horizontal photographs are taken with a phototheodolite from ground stations; oblique photographs are taken with a mapping camera from an aircraft following a prescribed flight pattern. In some cases the distances, or lengths of lines, are obtained by a scheme of graphic triangulation utilizing horizontal angles determined from the horizontal or oblique photographs. In other cases, the vertical angles are obtained from the horizontal or oblique photographs but the distances, or lengths of lines, are obtained from a separate photogrammetric plot or aerotriangulation done with vertical photographs. These procedures are described in (Anonymous 1954–1979).

8.5.5 BAROMETRIC LEVELING

In areas of moderate relief where barometric conditions are relatively uniform, supplemental vertical control may be established by barometric techniques where it is impractical to use more precise methods. Modern surveying altimeters transported by helicopter between desired points supply reasonably accurate differences of elevation along lines extending up to 120 km between points of known elevation. Data from self-recording altimeters placed at strategic points of known elevation in the area being surveyed make it possible to adjust approximately for variable atmospheric conditions.

For small-scale photography, a proven method is to secure spot-heights near the corners of each stereomodel, on the lateral overlap between adjoining flights of photography. By this means, each stereoscopic model may be controlled for compilation without having to bridge vertically. For larger-scale photography, a pattern in which every third or fourth model is controlled may be preferred, requiring photogrammetric vertical bridging to control the intervening models.

This barometric method is particularly applicable to extensive lake-covered regions where larger lakes serve to effect ties between lines of barometric leveling. In wooded or dry country, control lines normal to the required lines of barometric leveling assist in arriving at a homogeneous adjustment of the elevations. A "leapfrog" method, using two helicopters, in which one set of altimeters is kept stationary while the other is in flight, has proved very successful in establishing such control lines.

Barometric leveling is not ordinarily used for supplemental control of contouring at intervals smaller than 40 feet. Barometric leveling is covered in (Anonymous 1947).

8.5.6 SPECIAL METHODS OF SURVEYING FOR VERTICAL CONTROL

The preceeding methods are "classical;" they make use of readily available surveying instruments and are described in standard texts on surveying. Two methods using special equipment have been developed for determining elevations when the number of points is huge and the time short. The first, airborne profiling, actually provides much more information on elevations than is needed for vertical control and is supplemental to aerotriangulation rather than complementary. The second, using the elevation meter, can also provide much more data on elevations than is needed for control but can readily be used to give photogrammetrically acceptable vertical control.

8.5.6.1 THE AIRBORNE PROFILE RECORDER (APR)

The APR determines elevations by combining the outputs of two instruments—a very sensitive barometric altimeter (statoscope), which gives the altitude of the airplane above an isobaric reference surface, and a precise radar altimeter, which gives the distance of the terrain below the airplane. The principle is shown in figure 8-14. The pulse is emitted from the radar

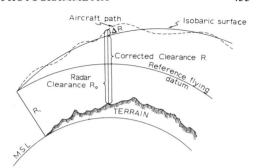

FIGURE 8-14. Principle of the airborne profiling method.

altimeter in a beam about 1° wide. The pulse travels down to the ground, where it is reflected or scattered and some of the energy returns to the radar altimeter, where the length of time it took the pulse to make the round trip is measured, multiplied by the speed of radio waves in the atmosphere, and divided by 2 to give the one-way distance. At an aircraft altitude of 300 m, the beam irradiates a circular area approximately 6 m across, so energy is returned from all parts of this area. If the leading edge of the returned pulse is timed, the distance measured is that of the closest part of the area; if the average position of the pulse is timed, an average distance for the area is determined. In either case, the direction of the beam must be known quite well so that the area or point responsible for the return can be identified.

The barometric altimeter indicates the *altitude* of the airplane above some surface of constant atmospheric pressure (an isobaric surface). It is therefore calibrated, before flying the project, by being brought to a point of known elevation, and the difference between known and indicated elevations used as a correction. For greater accuracy, the instrument can be calibrated at several points whose elevations span the altitudes over which the airplane will be flown. A curve is drawn showing the corrections necessary at different elevations. Because the isobaric surface is probably curved, and not parallel to the geoid, it is well to calibrate the instrument also at several points along the flight path and to distribute the corrections along the flight path between the intermediate points. To get still more accurate results, corrections should be made for those differences in pressure which are caused by winds. If the winds are reasonably steady over the duration of the flight, their velocity can be calculated from the average true airspeed of the airplane and from the amount of drift angle of the airplane. The difference $h\,(a,b)$ of the altitude of the isobaric surface from point a to point b along the flight-path is given by

$$h(a,b) = 2\omega \sin \delta \sin \phi \; V_{av} \; \Delta S/g,$$

where

ω = angular velocity of the earth,
g = average acceleration due to gravity between points a and b,
δ = drift angle,
ϕ = average latitude between points a and b,
V_{av} = average true airspeed of the aircraft between points a and b,
ΔS = distance from a to b.

The elevations must, of course, be tied to horizontal coordinates of some sort. This is done by placing a small camera on the antenna of the radar altimeter. The camera, whose optical axis is alined with the axis of the radar beam, takes pictures of the terrain at frequent intervals during the flight. These photographs are then later correlated with the photography of the mapping camera and with the timing of the radar altimeter to determine the flight path of the airplane and, consequently, the profile.

Results of trials and extensive field operations in Canada and elsewhere (Lyytikainen 1960; Slama 1961) show that elevations can be determined with a mean square error of ±3 metres in flights up to 500 km in length. However, great skill is required in interpretation of the record. In mountainous country and in dense forest, it is not always certainly known that the nadir point, or even the ground itself, is recovered; therefore, a full appreciation of the system's limitation is necessary in order to extract useful data. Experience has shown that an adequate density of control information can readily be obtained by this method for some mapping purposes.

APR should be flown as low as possible, yet high enough that the altimeter is not affected by topography; suitable flying heights range from 2,000 feet (600 metres) to 5,000 feet (1,500 metres), depending on the topography. The flight pattern should give the most effective control for the mapping photography, usually in parallel flights across the photographic flight lines.

APR profiles, keyed to mapping photographs which are obtained simultaneously, can be used to control scale in aerial triangulation; thus, very satisfactory bridging results have been achieved over distances up to 200 miles (320 kilometres). A grid pattern is necessary to reduce deviation in the y-direction, and cross-flight spacing at 60-mile intervals has proved effective for 1:25,000-scale mapping. A project embodying control of this nature requires careful planning and efficient execution.

More detailed description of the airborne profile recorder (also called airborne terrain recorder, etc.) is contained in Blachut and Leask (1952) and in O'Leary (1963).

8.5.6.2 ELEVATION METER

The elevation meter, developed by the Sperry-Sun Well Surveying Co. and used extensively by the U.S. Geological Survey, is particularly adaptable for rapid determination of approximate elevations (±10 to 30 feet) on long, cross-country lines, but can also provide elevations correct to within ±2 feet over closed circuits, provided the lines are carefully planned and the closures are properly adjusted. The instrument is economical on projects which cover relatively large areas and have a suitable network of roads. The instrument operates on the pendulum principle. The equipment is mounted in a four-wheel-drive automotive vehicle provided with four-wheel steering. As the vehicle proceeds along the road, an electromagnetic field acts on a very sensitive pendulum so as to generate an electrical signal whose strength is proportional to the sine of the angle of slope. Another electrical signal is generated by a revolution counter which measures each small increment of distance traveled. An electronic integrator combines these two signals into a continuous record of the difference of elevation from the starting point. Operation of the elevation meter is covered in (Anonymous 1954–1979).

8.6 Surveys Giving Both Horizontal and Vertical Control

8.6.1 AIRBORNE CONTROL (ABC) SYSTEM

A procedure developed by the U.S. Geological Survey for fourth-order, horizontal- and vertical-control surveys, particularly in wooded areas, involves electromagnetic distance measurements and horizontal and vertical angle measurements to a helicopter from at least two, and preferably three, stations of known horizontal position and elevation as the helicopter hovers vertically over the point whose position and elevation are to be determined (figure 8-15). Accurate hovering is accomplished by a sighting

device that enables the pilot to determine when he is directly over the ground point. A special plumbline and drum are used to measure the hovering height. At the instant the pilot signals by radio that he is over the point, theodolite readings are made to a revolving beacon mounted on the helicopter. Distance-measurements are made by using suitable instrumentation of the kind discussed in 8.4.6.2.

8.6.2 THE AIRBORNE PROFILER OF TERRAIN SYSTEM (APTS)

While the airborne profile recorder (section

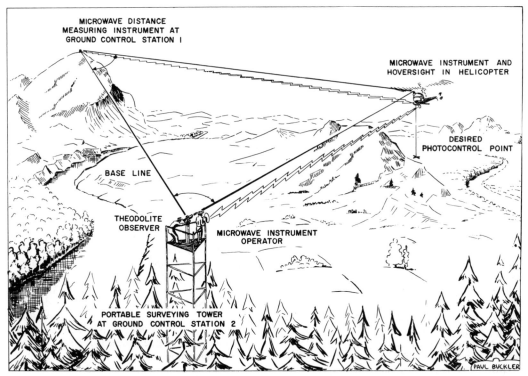

FIGURE 8-15. Airborne control survey system (U.S. Geological Survey).

8.5.6.1) determines only elevations by its own instrumentation, and depends on auxiliary photography (the mapping photography) to provide the horizontal coordinates. APTS will provide all three coordinates of control points, and will do this independently of any photography that may be taken. The key element in this system is (Chapman 1979) a computer-connected, inertial-navigation device (*see* section 8.6.3) which provides, simultaneously and at frequent intervals, the geodetic longitude, latitude, and height of the airplane. The barometric altimeter and camera of the APR are therefore dispensed with; the radar altimeter is replaced by a more accurate laser-type altimeter which measures the travel-time of optical rather than radio pulses. Because an inertial navigation system cannot keep its locating accuracy over the several hours of flight without frequent recalibration, a distance-and-direction measuring device called the tracker is also connected to the computer. Over the area to be surveyed are distributed a number of corner-cube reflectors about 10 km apart, the reflectors having known locations. The tracker measures distance and direction to the reflectors (being pointed at them by the computer) and from this information, corrections to the calculated (inertial navigation system alone) coordinates are made.

8.6.3 PHOTOGRAMMETRIC GEODESY

In section 8.5.4, a method was described by which supplemental vertical control could be established photogrammetrically. The method is over 20 years old and still useful. However, since that method was first developed, considerable work has been done in the improvement of measuring equipment and particularly in the development of programs for solving large systems of photogrammetric equations on electronic computers. At present, it is perfectly feasible to establish second-order or even first-order horizontal control, along with vertical control, with a precision of 5 to 10 cm using standard photogrammetric methods. The requirements are (a) a dense photographic coverage of the area in which control is to be established (each control point appearing on from 4 to 16 photographs), (b) pre-existing control of higher order on the perimeter of the area, and (c) careful control of systematic errors along with access to a large computer having suitable reduction programs. Computational procedures necessary for high precision have been developed. Papers by D. Brown (1976) C. Slama (1978) and J. Lucas (1979) give details on the precision achievable, the economic aspects of this kind of surveying, and the computational procedures being used.

Because of the usefulness of photogrammetric geodesy in establishing horizontal control of geodetic accuracy, a tentative set of standards and specifications for such control has been put together (C. Slama, unpublished paper) and is being considered for adoption by the National Geodetic Survey and other organizations. The standards and specifications are given in table 8-4.

8.6.4 NON-CLASSICAL METHODS

The preceding methods are considered "classical" and are to be used when theodolites, distance-measuring instruments, and/or levels are already available (with personnel) and the conditions in the field are such that surveying with the classical methods is fast and inexpensive. In many projects, however, a very large number of control points may have to be established, or control may have to be established at places costly to reach by classical methods, or the datum of the project may have to be connected to another datum some distance away. Under such conditions, one of the newer methods to be described below may be faster and more economical to use.

For establishing a small number of widely separated, single points, the Doppler-shift method offers advantages. At present, there are in orbit about the earth, at an altitude of about 900 km, four or five satellites, (called TRANSIT or Navy-navigation Satellites) each carrying a radio beacon that broadcasts bursts of CW radio waves at about 150 MHz and 400 MHz. Each satellite broadcasts not only the CW radiation but also, in a readily usable code, its position at frequent intervals. Because the beacon and an observer on the ground are moving with respect to each other, the observer will not receive the signals at their transmitted frequencies but at higher frequencies if the beacon is approaching him and at lower frequencies if it is receding. The difference in transmitted and observed frequencies is called the Doppler-shift, and the number of wavelengths counted over a specific interval of time at this frequency difference is a measure of the distance the beacon has moved toward or away from the observer over that interval. Or, inverting the problem, it is the distance the observer has moved toward or away from the beacon. Given the positions of the beacon and a sufficient number of independent distance-changes, one can solve for the position of the observer in the *coordinate system used for the orbit of the satellite*. Several different makes of receiver are now commercially available for determining position in this way. The most recent versions are easily transportable by one or two persons, and nearly automatic in operation. The accuracy of positions determined in this way depends on whether receivers are used individually or in pairs, how long a position is oc-cupied, and on whether the positions broadcast by the satellite are used or the more accurate but not immediately available positions provided by the U.S. Defense Mapping Agency Topographic/Hydrographic Command are used. A standard deviation of about 1 m in each coordinate is obtainable in one or two days occupancy of a station, this being for the difference in position of two simultaneously occupied stations. In 3−4 days, a standard deviation of less than 0.5 m in each coordinate can be obtained. Since these positions are in the datum used for the satellites' orbits, either all ground control in a photogrammetric project must be in this same datum or coordinates in this datum must be converted to coordinates in the general datum of the project.

The Global Positioning System (GPS) is intended primarily to provide positions for airplanes and ships of the U.S. armed forces. However, it is of a nature such that it can also be used for establishing ground control. At present (1980) it is still in its initial stages of development. If or when it is finished, it will consist of a set of 18 satellites divided into three subsets of 6 satellites each. In each subset, the satellites will be in orbits about 20,000 km above the earth's surface, in a common plane, and equally distant from each other. The three planes will be 120° apart; the resulting geometry is such that at any instant, 3 to 4 satellites will be simultaneously visible from any given point on the earth's surface. Each satellite will carry a beacon giving out coded bursts of radiation at about 1.4 GHz and, like TRANSIT satellites, broadcasting time and position. There is still some question as to how the GPS can best be used for determining positions of ground control, and the accuracy obtainable. Accuracies (standard deviations) of 1−5 m within a few seconds appear readily attainable; there is the possibility of standard deviations below 10 cm being obtainable within an hour or two. Use of GPS, as with use of the Doppler-shift method, gives the same datum-connected problems. Its advantages over the Doppler-shift method would be (a) its greater speed, (b) its possibly lower cost, and (c) its possibly lower standard deviation.

If a large number of ground-control points are to be located to within 0.5 metres within a short time, an inertial-surveying system may be economical or necessary to use. An inertial-surveying system consists of a set of two or three gyroscopes arranged to define the axes of a Cartesian coordinate system, three accelerometers to measure accelerations along these axes, and a computer to turn the output from the gyroscopes and accelerometers into geodetic coordinates. Surveying is done by placing the system on a ground-control point whose coordinates are known and aligning the gyroscopes' axes to a known orientation there. The system is then moved from control point to

TABLE 8-4. CLASSIFICATION, STANDARDS, AND GENERAL SPECIFICATIONS FOR PHOTOGRAMMETRIC GEODESY

Classification	Second-Order Class I	Second-Order Class II	Third-Order Class I	Third-Order Class II
Geometry:				
Forward Overlap	2/3	2/3	2/3	2/3
Side Overlap	2/3	2/3	1/3	1/3
Minimum number of intersecting rays per point[1]	9	9	3	3
Camera:				
Focal length				
Platen flatness[2]	Calibrated to ±1μ			
Dimensional Control[3]	100 or more reseau marks in format	≥4 Fiducials	≥4 Fiducials	≥4 Fiducials
Calibration[4]	Radial Distortion $\sigma \leq 1\mu$ Decent. Distortion $\sigma \leq 1\mu$ Reseau Coordinates $\sigma \leq 1\mu$	Radial Distortion $\sigma \leq 3\mu$ Fiducial Coordinates $\sigma \leq 1\mu$	Radial Distortion $\sigma \leq 5\mu$ Fiducial Coordinates $\sigma \leq 5\mu$	Radial Distortion $\sigma \leq 5\mu$ Fiducial Coordinates $\sigma \leq 5\mu$
Mensuration				
Comparator type[5]	≤1μ	1μ	1μ	1μ
Least count	≥3	≥2	1	1
Pointings per target	≥3	≥2	1	1
Pointing per reseau—fiducial	4			
Minimum number reseau per target				
Calibration	≤1μ over reseau spacing	3μm over fiducial dist.	4μm over fiducial dist.	≤5μ over fiducial dist.
Control:				
Geodetic Accuracy	First-Order	Second-Order Class I	Second-Order Class II	Third-Order Class I
Spacing along periphery	7 air bases	7 air bases	7 air bases	7 air bases
Targets[6]				
Control	Fixed Ground	Fixed Ground	Natural Feature	Natural Feature
Photogrammetric Geometry	Fixed Ground	Natural Feature	Natural Feature	Natural Feature
Reduction:				
Type	Simultaneous (Bundle)	Simultaneous (Bundle)	Independent model	Polynomial strip assembly
RMS residual (plate coordinates)[7]	3μ	5μ	7μ	10μ
Average Propagated σ_{xy}[8]	D/100,000	D/40,000	D/15,000	D/8,000
RMS difference at check distances	D/50,000	D/20,000	D/10,000	D/5,000

(Continued on next page)

TABLE 8-4—*Continued*

NOTE (1)

Each photograph that "sees" a ground point constitutes one intersecting ray for that station. Obviously, the number of such rays intersecting any one point is directly related to the overlap of succeeding photographs along a line and the configuration of the flight lines. Optimum results are obtained when all rays to a point are distributed in a symmetrical pattern.

NOTE (2)

It is assumed that the sensor used in an analytical photogrammetric system will be a well constructed camera of "photogrammetric" quality. That is, in the case of film, a platen will be included onto which the film will be satisfactorily flattened during exposure. Further, the platen will normally be expected to describe a true plane at the focus to better than ±10 micrometers. Deviations of this platen from a true plane shall be calibrated and predictable at the time of exposure with the standard error specified.

NOTE (3)

In the case of image formation on film, some means of dimensional control shall be employed. For reseau, a minimum array (10 × 10) of evenly spaced markers shall appear in the format of the image area. Fiducial markers surrounding the image area may be employed in less accurate methods.

NOTE (4)

Characteristics of the camera's internal geometry shall be determined and applied as corrections to the measured image coordinates. These characteristics (radial symmetric distortion, decentered lens distortion, principal point and point of symmetry coordinates, and reseau coordinates) shall be determined using recognized calibration techniques and presented with statistical evidence to verify their validity.

NOTE (5)

The mensuration device shall indicate measured coordinates to the nearest micrometer. Calibration of the instrument shall be performed using recognized standards and techniques with an output to indicate the statistical significance of the calibrated parameters. Measured image coordinates shall be correctable to the precision indicated.

NOTE (6)

A fixed ground target is defined as a high contrast marker, centered over the monumented ground point, which is visible on all photographs that "see" the ground point and which can be measured by a one-stage comparator. Natural features are those points whose image can be identified on adjacent photographs.

NOTE (7)

"RMS" residual-plate coordinates" is identified as that quantity which results from taking the square root of the sum of the squares of the x and y plate residuals, after the adjustment, divided by two times the number of plate points. This number signifies the internal "agreement" of the observed coordinates with the final block solution.

NOTE (8)

The "average propagated σ_{xy}" is a measure of the average standard error of the determined ground coordinates (X and Y) as computed in the adjusted block. That is, the standard deviation in latitude and longitude for each determined ground station is squared, summed, divided by two times the number of points, and the square root of the result taken. This number should be less than or equal to the distance "seen" by one photograph (D), divided by the factor given. This should represent the average standard error of a distance determination across the areal extent of one photograph. Likewise, determined coordinate differences in distance as compared to independent geodetic measurements of the same points should not differ (as an RMS) by a ratio of D to the factors given.

control point, with a stop every 3 to 10 minutes to recalibrate the accelerometers, until the system is returned either to its starting point or to another control point of known coordinates. The inertial-surveying system acts as an interpolator of position, direction of gravity and magnitude of gravity between the starting and final positions. Carried in a truck, an inertial-surveying system can survey in some 10 to 15 control points per hour, covering a distance of 30−70 km; placed in a helicopter, the same rate of survey can be attained, but covering distances of 100−200 km.

Regardless of the method used, if it does not provide an independent check on all of the field measurements, great care must be taken to avoid blunders. These can easily occur, and could cause an inordinate loss of time later when plotting or computation is undertaken in the office.

8.7 Selection, Targeting, and Identification of Photogrammetric Control Stations Appearing on Aerial Photographs

8.7.1 INTRODUCTION

Control-station identification is defined as the identification, on a photograph (usually an aerial photograph), of the image of a ground point of known horizontal position and/or elevation. The fieldwork for control identification ordinarily includes the recovery of ground stations and either (a) the targeting of those stations prior to the taking of photographs or (b) the selection of natural targets whose images appear on the photographs. A varying amount of surveying is necessary to determine the horizontal position and/or elevation of both artificial and natural targets. Control surveys are covered in earlier sections of the chapter.

Control-station identification is a most important step in photogrammetric mapping. It ties the aerotriangulation and photogrammetric compilation to the ground control. This tie can be no better than the accuracy of the identification. Ground control can be established to a high order of accuracy and extremely accurate positioning can be made on well-defined points in a stereoscopic model; therefore, unless the images of control points are correctly identified on the photograph, the usefulness of the ground survey is lost, and the full capability of the photogrammetric instrument cannot be realized. Misidentification or poor identification of a control station invalidates the usefulness of the field control on which the compilation is based and creates uncertainties that are extremely wasteful of both time and effort in aerotriangulation and in subsequent map compilation.

The accuracy required of the identification in terms of ground distance depends primarily on the scale of the maps being compiled. It may vary from less than 0.3 m to perhaps 3 m. In any case, the identification requirements are extremely rigid because the ground point must often be identified on photographs of relatively small scale. For example, on a photograph at 1:20,000 scale, 0.05 mm corresponds to 1 m on the ground.

8.7.2 TARGETS ON STATIONS PRIOR TO PHOTOGRAPHY

The placement of specially prepared targets on ground control stations prior to photography is the most accurate means of control identification. This practice is required for the most precise photogrammetric surveys such as cadastral surveys, for very large scale mapping, and for special testing purposes. It is always desirable, though not essential, for routine mapping operations.

The field operations for targeting ordinarily include the recovery and targeting of existing control stations as called for by the photogrammetric control plan. If existing control is not adequate, ground surveys are required to determine the positions and/or elevations of new stations. Where possible, control stations are paneled directly on the station mark, but this is often not the case due to the line of sight from the flight line to the target being obstructed by buildings, trees, or ground relief. A nearby substitute station must be paneled when this problem occurs, and its position or elevation must be transferred from the home station by conventional ground methods. Targets must be placed and maintained until the aerial photography has been completed.

Targeting is relatively simple in small, accessible areas. More difficulty is encountered with large areas where many targets have to be maintained until the aerial photography is completed. It is much more difficult and perhaps impracticable to maintain targets in remote areas. Targets are subject to disturbance by weather, animals, and people. This happens more frequently than one might expect, and many targets may be destroyed unless the targeting and the aerial photography can both be completed within a short period of time. In areas where targets might be subject to destruction by nature or vandalism, two photoidentifiable subjects should also be selected. Directions and distances to them from the home station should be observed and recorded with a sketch and description.

8.7.2.1 SHAPE AND SIZE OF TARGETS

Targets should be symmetrical and centered on the station marker. The size and shape will be influenced by such factors as the type of photogrammetric equipment to be used, the altitude at which the photographs will be taken, the nature of the terrain, and the requirements of the specific project. Table 8-5 gives the approximate sizes of panels, of the shapes shown in figure 8-16, that will show as small dots on the aerial photographs at various photographic scales. These sizes are suitable for use in first-order mensuration equipment; for instruments of lesser optical precision, the targets should be larger—perhaps twice as large. Targets of the shapes shown in figure 8-16 have all proven satisfactory. Array number 2 has the advantage that, if one leg is obliterated, the center can still be found. Array number 1, on the other hand, conserves material and is somewhat more distinctive. The size of targets indicated in table 8-5 can be increased without harming the accuracy of identification, so long as the targets are centered and are symmetrical.

8.7.2.2 TARGET COLORS AND MATERIALS

Maximum contrast between the target and its background is a primary consideration. Best results are obtained with a white target on a black background.

Experience shows that white usually is the best color for the target material. Dark-colored earth or rocks, or green turf, provide an excellent background. White targets placed in light scrub or low brush are effective, providing the brush is dense enough to give boundaries of equal density around the target. When targets are placed in small clearings in brush-covered or wooded areas, care must be taken to see that the target is in full sunlight during the hours suitable for the areial photography. If the target must be placed on a light background, as on sand or gravel, a black or dark background material should be used under the target. This might be tar-paper, lamp black, black cloth, or the like.

If a black target is used on a light background, allowance must be made for irradiation (image spread) in the negative, especially if the negative is developed to a high *gamma*. Because of this image spread, the image of a black target on a white background is smaller than that of the same-size white target on a black background. Normally, the linear dimensions of a minimum-size black target on a white background should be two or three times those of a white target on a black background. Table 8-5 is based on white targets with good background contrast.

White, opaque polyethylene film is excellent target material. This film has been used satisfactorily in thicknesses of both 4 and 6 mils (0.1 and 0.15 mm). Unbleached white muslin or bleached cotton sheeting (both 4.6 ounces per square yard or 156 grams per square metre) and white mercerized cotton bunting (flagging material) make good targets. Plywood or masonite, painted flat white, makes excellent target material. They are more expensive than polyethylene or cloth and are more durable, but they present a problem on irregular ground. White cotton cloth tends to lose its reflectivity somewhat with weathering; therefore, targets of this material should be made somewhat larger than those of white polyethylene or white painted plywood or masonite. Black polyethylene should be avoided for either target or background material because its high reflectivity may destroy photographic images.

It is impracticable in this section to include all of the detailed information about targeting that has been accumulated by mapping agencies around the world. Persons having special problems in targeting might request additional information from the following agencies of the United States Government.

Defense Mapping Agency, Washington, D.C.
Federal Highway Administration, Washington, D.C.
U.S. Geological Survey, Reston, Virginia.
National Ocean Survey, NOAA, Rockville, Maryland.
Surveys and Mapping Branch, Department of Energy, Mines and Resources, Ottawa, Canada.

TABLE 8-5. TARGET SIZES

Photography scale	Panel and spacing dimensions (in metres)						
	A	B	C	D	E	F	G
1:10 000	0.5	0.3	1.3	0.2	0.9	0.9	1.5
1:20 000	1.1	0.7	2.6	0.4	1.8	0.9	1.9
1:30 000	1.6	1.0	3.9	0.5	2.7	0.9	2.2
1:40 000	2.2	1.3	5.2	0.7	3.6	0.9	2.5
1:50 000	3.2	2.0	7.8	1.1	5.4	1.8	3.8
1:60 000	3.8	2.3	9.1	1.3	6.3	1.8	4.1
1:70 000	4.4	2.6	10.4	1.4	7.2	1.8	4.4
1:80 000	5.0	3.0	11.7	1.5	8.0	1.8	4.8
1:100 000	6.4	4.0	18.2	2.2	10.8	3.6	7.6

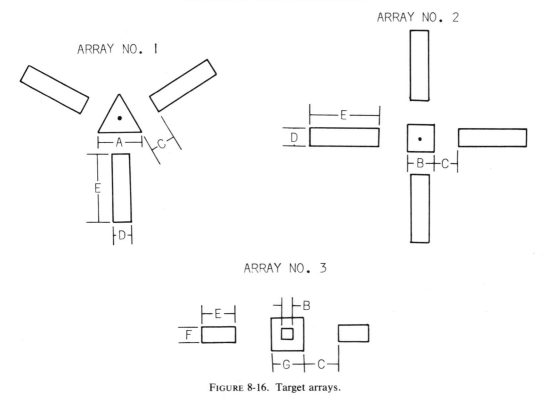

FIGURE 8-16. Target arrays.

Three journals; *Photogrammetric Engineering and Remote Sensing* (formerly *Photogrammetric Engineering*), *The Photogrammetric Record*, and *Photogrammetria*, contain reports on photogrammetric projects that are often the best possible sources of information on targets for use under specific conditions.

8.7.3 NATURAL TARGETS

Not withstanding the advantages of targeting as described in section 8.7.2, much control identification for photogrammetric mapping is still accomplished by selecting existing natural or cultural features in places where the photogrammetric control is needed, determining the horizontal positions and/or elevations of those features as described in sections 8.4 and 8.5, and identifying their images on the photographs. These points are often called substitute points or photocontrol points.

The object so selected, located, and identified must be imaged on the overlapping photographs, both in the line of flight and in adjacent flights, and the point located as the station must be subject to exact identification by the stereoscopic instrument operator, *who may be viewing the model at an 8× or 10× magnification.*

Ideally, the selected natural or cultural feature or object should resemble an artificial target with 2, 3, or 4 legs that form reasonably large angles of intersection and define a positively identifi-

able point, both in nature and on the photographs. Examples are road intersections and fence-line intersections unobscured by vegetation or shadows. Other suitable features include intersections of roads with fence lines, railroad grade crossings, and intersections of water courses. Care must be used in selecting lines of demarcation between white and black features on the ground since such lines may be displayed by "bleeding" tendencies in the photographic emulsion and thus may not truly indicate the station point. Another category of acceptable natural targets consists of isolated lone objects, distinctive both on the ground and on the photographs. These include small isolated bushes and trees, stacks, tanks, towers, corners of buildings, corners of wharves, and so on.

The choice of objects narrows with an increase in the required accuracy of identification; for example, a large object or feature is not good unless the station can be accurately defined on the photograph. Tall objects are often risky in this respect; for example, the instrument operator may not be able to see the exact top of a church spire, or he may not be able to see the exact center or top of a tall conical or spherical tank because of the solar phase effect. These are matters of degree depending upon whether an extreme accuracy of identification within 30 cm or so is required or whether a more liberal tolerance of 2 or 3 m is permissible.

Because the stereoscopic instrument operator

has the advantage of examining the photographs stereoscopically, often under magnification, it follows that stereoscopic examination in the field is essential for the best selection of objects and the best identification of the photocontrol points on the photographs.

Surveyors should use a folding-type pocket stereoscope with a 2× or 4× magnification for viewing the control photos during selection and identification of a natural target. The selected image should be indicated by a small inked circle on the face of the photograph and a larger inked circle with description information on the back of the photo. Only one photograph should be annotated in this manner for each control point.

8.7.4 IDENTIFICATION OF TARGETS

8.7.4.1 CONTROL STATION IDENTIFICATION

A large-scale sketch of the station and surrounding details is essential for the best work. Examples are shown in figures 8-17 and 8-18. These sketches may be made on the back of 2 photographs or on a special form provided for the purpose.

The sketches shown in these figures provide the stereoscopic instrument operator with exact information about the location of the station point with reference to other features imaged on the photographs. The sketch must therefore be made carefully and in considerable detail. The survey measurements made to locate the station point must be recorded on the sketch, along with differences of elevation.

The photocontrol point must also be indicated on the field photographs. Various symbols may be used to indicate the locality of the point, as for example a large circle or triangle. In any case, the symbol must be sufficiently large not to obliterate photographic details in the immediate vicinity of the point. Some organizations require that surveyors mark photocontrol points with a fine needle-hole on the emulsion of the field print. This is satisfactory but should not be overrated. It is usually not possible for the field surveyor to mark the exact photocontrol point as accurately as the stereoscopic operator can point on it. Consequently, the operator depends more on the sketch and written information. The fine hole made on the photograph merely guides him to the place on the photograph.

Measurements should be made for locating photocontrol points from nearby established control to provide a check. Inadequate field measurements are often the sources of error and trouble in office plotting. Because of this, surveyors are usually required to select, locate, and identify two or more separate and distinct photocontrol points wherever one is required to control the photogrammetric plot; thus, one identified station serves as a check on the other.

8.7.4.2 SUPPLEMENTAL PHOTOGRAPHY OF TARGETED CONTROL STATIONS

When it is not practicable to target control stations for the bridging or mapping photographs, targets as described in section 8.7.2 can still be used and recorded on low-altitude, sup-

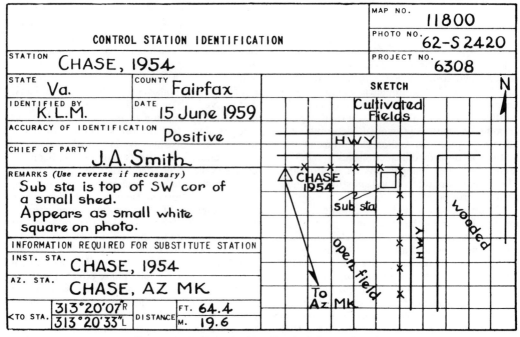

FIGURE 8-17. Form for identification of control station.

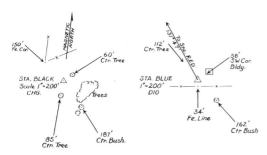

FIGURE 8-18. Identification by reference measurement.

plemental photographs. The identification of control stations is then transferred from the supplemental photographs to the bridging or mapping photographs. With proper precautions, this method provides an accurate means of identification of the control stations on the bridging or mapping photographs. The size of the targets will be determined by the flight height of the supplemental photographs.

The supplemental photographs can be taken from helicopters or small aircraft with hand-held cameras of short focal length (35, 50 or 70 mm). The scale of the supplemental photographs should not be more than 4× the scale of the bridging or mapping photographs. Thus, the flight height for the supplemental photographs may be anywhere from 700 m to 1700 m above ground, depending upon the focal lengths of the cameras used. At least two exposures should be made over each target, preferably in such manner as to provide stereoscopic coverage. While the physical limitations of using hand-held cameras are recognized, every effort should be made to maintain the camera axis as nearly vertical as possible.

The most accurate means of transferring the target identification from the supplemental photographs to the bridging or mapping photographs is by stereoscopic transfer in which one supplemental photograph is fused with a bridging or mapping photograph covering the same area. It is for this reason that it is desirable to take the supplemental photographs at the same scale as the mapping or bridging photographs. This cannot always be done because of flight ceilings of the small aircraft and time limitations. In any case, the scale of the supplemental photographs should be not greater than 4× the scale of the bridging or mapping photographs; reduced-scale prints of the supplemental photographs can then be readily made in the laboratory for the stereoscopic transfer.

8.7.5 IDENTIFICATION OF VERTICAL CONTROL

The preceeding sections relate primarily to the identification of horizontal control, which poses the more critical identification problems, but the methods described may also be used for the identification of vertical control stations.

In most mapping projects, many of the vertical-control stations are separate from the horizontal-control stations and are easier to identify. Images are chosen in selected flat or relatively flat areas. These usually show clearly on the photographs, and the station point can be marked or pricked directly without great difficulty.

The area selected in which a vertical-control station is to be identified on the photographs should be sufficiently level so that a slight horizontal error in setting the floating mark on the identified point does not result in an appreciable error in the elevation reading. Although preferred when such a site is available, it is not essential that the ground be level completely around the vertical control station.

8.8 Field Surveys for Topographic Mapping

8.8.1 INTRODUCTION

Field surveys are required to identify basic control (*see* section 8.7), provide supplementary control, clarify obscured photographic detail, classify cultural features (roads, buildings), locate political boundaries, and check the horizontal and vertical accuracy of the map.

Field surveys are referred to as field inspection or classification surveys (accomplished before compilation) or field edit or completion surveys (after map compilation).

8.8.2 FIELD INSPECTION AND CLASSIFICATION SURVEYS

Field inspection and classification surveys consist of identification of horizontal and vertical control, establishment of supplementary control and providing all data necessary to compile the map. It is sometimes practical and efficient at this time to obtain *all* of the information required and so eliminate the need for a subsequent field completion survey.

All permanently marked horizontal or vertical control points of third-order accuracy or better are shown by proper symbol together with the elevation and designation of each mark.

Spot elevations conforming to special accuracy standards are obtained to provide more accurate elevations for particular points or features than can be interpolated from contours, and to supplement the contour information in flat areas where the contours are widely spaced. These spot elevations should be on important planimetric features or significant topographic points where they are recoverable and will be helpful to the map user.

The entire road net is checked; newly constructed roads are added; roads not mapworthy are deleted; all roads are classified according to specified standards; federal, state, and county routes are labeled; number of lanes, traffic restrictions, and road and median widths, are noted. Buildings intended for human shelter or activities, such as houses, stores, and factories, are shown as class 1. Prominent barns, warehouses, garages, and the like are shown as class 2. Churches and schools are shown with distinctive symbols. Within built-up urban areas on medium-scale maps, only landmark-buildings, such as churches, schools, and public buildings, are symbolized. Buildings are classified on the fieldsheet, and landmark buildings are named or labeled on the information sheet.

Civil boundaries are mapped almost entirely by field methods. Monuments and other ground evidence must be identified and mapped. Legal descriptions are obtained whenever possible. A limited number of section corners are usually located during control operations and are shown on the compiled manuscript. Using a tentative land net as a guide, the compiler may plot recoverable objects near each probable position of a section corner as a guide for the field location. Compiled fence and crop lines, or cut lines in timber, are also guides for locating section corners and boundaries.

Drainage features, including shorelines, are usually compiled with the *intermittent* symbol—dashes separated by three dots; dashed lines are used where the channel is indefinite on the photographs. Fieldmen are responsible for classifying the hydrographic features as perennial or intermittent streams, for determining the normal water-surface level in certain cases, for ascertaining stream widths where the symbol size depends on this, and in general for field-mapping any hydrographic feature not compiled or compiled incorrectly. The objective in field classification of drainage is to represent by symbol those water features that can be expected to contain water. If, under conditions of average flow, the feature contains water all year, it is classified as perennial. Those features that normally contain water only a part of the year are classified as intermittent. In arid and semiarid regions of the West, dry washes or streams which contain water for only a few hours after a rainstorm or brief, heavy snowmelt are classified in a separate category.

Ordinarily, woodland outlines are classified correctly and shown in correct position during stereocompilation, and the field-completion inspection is limited to finding gross omissions or misinterpretations of the photographs. The inspection during completion surveys should cover dark areas on the photographs which may have been plotted as woodland; changes due to new clearings or new growth; consistency in compiling boundaries where woods are bordered by scattered trees or brush; and mapping and classifying orchards and vineyards. Wooded areas that are submerged are also noted.

Fieldmen have full responsibility for collecting name information. This includes obtaining the existing names of all map features; ascertaining the correct form and spelling; determining the location and limits of the features to which the names apply; and, in some cases, specially investigating names that are controversial or conflicting. To carry out this part of the field completion assignment properly, fieldmen should obtain name information whenever opportunities occur during the job. All available sources of information should be consulted—reference works of various kinds, published maps of the area, official records, and especially local residents. The pre-edit information must be carefully checked, and all discrepancies investigated. In the federal mapmaking agencies, a special report is submitted by the fieldman on each discrepancy that cannot be resolved. This report must be filled out completely and well documented, because it is used as evidence for a decision by the Board on Geographic Names. All names to be published are shown on an overlay in the approximate position they will occupy on the published map. If the application of a name is not clear, this should be indicated by encircling in ink the feature in question or connecting it to the name by an arrow.

Areas of special cultural activity, such as industrial, mining, public recreation, and historic areas, require careful treatment during field completion because of the density of detail and the special interest in these map areas. These features are symbolized on the fieldboard and labeled on the oversheet. Linear features, such as powerlines, pipelines, and fencelines, are shown for their landmark character. Large transmission lines assume such landmark importance that individual steel towers are located and shown.

8.8.3 FIELD COMPLETION AFTER MAP COMPILATION

Field completion after map compilation is essentially planetable mapping in which all additions and changes are made concurrently. The planetable worksheet is a copy of the stereocompilation reproduced on metal-mounted paper or on coated plastic at the appropriate field scale. An extra print may also be prepared for recording supplemental information. It is advisable to have a complete set of control notes, the aerial photographs used for supplemental control, copies of township plats (in public-land states), coastal charts, and other maps of the area for guidance in obtaining complete name coverage and other mapworthy information. Comments noted by the compiler are also very valuable in calling attention to weak or doubtful areas of compilation.

A fieldman reviews and evaluates the compilation for completeness of detail and for proper topographic expression of features. He studies intricate drainage areas and special features requiring supplementary contours. He determines the best access routes into the mapped area; possible locations for important spot elevations; boundary lines to be mapped or investigated; and any unusual problems that may arise in connection with drainage, roads, urban areas, or other map features. In public-land states, he prepares a preliminary land-net adjustment as a guide in searching for section corners.

The importance of planning on a day-to-day or week-to-week basis cannot be overemphasized. It is essential to efficient field-completion work since fieldmen must consider such a wide variety of problems. They are concerned with map names, boundaries and sectionizing, contour accuracy, roads and road classification, building classification, landmark features, drainage classification, spot elevations, control marks, and, in fact, every feature that appears on the map, taken separately and collectively. For efficiency, several objectives must be accomplished concurrently; for example, a traverse to map a new road serves also to check and correct the contours, classify buildings and drainage, and possibly to locate section corners or boundaries. Return visits to a local area should be avoided by obtaining all the necessary information required in that locality during the first visit.

Field inspection is a detailed comparison of the map compilation with the ground it portrays. Inspection made at a planetable station may reveal deficiencies in the compilation; if so, measurements are made from the same station to correct them. This is not the only inspection procedure, however. While walking over trails, driving over roads or flying over an area, constant inspection should be a habitual practice because it is very useful in detecting errors or omissions. Fieldmen must visualize the terrain in terms of contours and other map symbols. They should detect errors and repair spots where slight reshaping of contours will more nearly portray the features correctly. Familiarity with all items of map content is necessary for the fieldmen to develop skill as topographers.

The basic operation in field completion is planetable mapping. This means that corrections and additions to and deletions from the manuscript are made while actually viewing the features, using the alidade to determine positions and elevations necessary for the process. Map features are tested in the same manner. Both horizontal and vertical accuracy standards must be met by all planetable surveys.

Plotting or recording the assembled information is a major activity in field completion so that it may be completely and clearly presented in standardized form. This information is recorded on either the fieldsheet or the information sheet. Scribing on coated Mylar sheets provides a dimensionally stable, permanent record. For this purpose, a field scribing kit is needed, containing a fine-line graver, a template for symbols, a french curve, a special blade for scribing roads, a building graver, and a supply of extra scribing needles.

8.8.4 ACCURACY CHECKING

Certification that a map meets National Map Accuracy Standards is based on checking relative accuracy during the field-completion phase and specific tests of absolute accuracy.

8.8.4.1 CHECKING OF HORIZONTAL ACCURACY

Horizontal accuracy is checked on a sample of three percent of the maps produced, by determining third-order positions for at least 20 well-defined map features per test. At least 90 percent of these tested map positions must agree within 1/50 inch of the surveyed position, for scales of 1:20,000 or smaller. For scales larger than 1:20,000, 1/30 inch is allowed.

8.8.4.2 CHECKING OF VERTICAL ACCURACY

Vertical accuracy is checked on each project with the number of test points determined by the size of the project, but there must be at least 20. More testing is done in areas of low relief where contour intervals are smaller. Field elevations are obtained by stadia traverse, trigonometric leveling, or fly levels, and must be accurate with 0.1 of the contour interval. In order to meet National Map Accuracy Standards on the project, at least 90 percent of the elevations tested on a map must agree with the field elevations within half of the contour interval. In determining vertical accuracy, the apparent vertical error may be decreased by assuming a horizontal displacement within the permissible horizontal error for a map at the scale of the one being tested.

8.9 Field Surveys for Coastal Mapping

Over the past 25 years, techniques and procedures have been accumulated and perfected to collectively form the basis for modern coastal mapping. These operations include premarking of ground control, tide-coordinated photography, specialized applications of various photographic emulsions, multi-camera photography, analytical aerotriangulation, and compilation with modern or semi-automated equipment. While they are applied in various combinations to a variety of coastal mapping problems, the largest coastal mapping application is shoreline mapping in support of nautical charting. A second and highly promising application is for photogrammetric bathymetry to supplement or extend inshore hydrographic surveys. Field surveys and operations to support both shoreline mapping and photobathymetry are described in the following sections.

8.9.1 SHORELINE MAPPING

The mean high water line has been charted to represent the shoreline on nautical charts and topographic maps of the coastal zone. Additional importance is placed on the mean high water line since most coastal states use it to define the boundary between state and private property ownership. When the mapping scale will permit, the sounding datum is also mapped. At present, these datums are: mean low water (along the Atlantic coast), Gulf coast low water (in the Gulf of Mexico), and mean lower-low water (along the Pacific coast). This line along the coast of the United States is the baseline from which the territorial sea, the contiguous zone, and the 200 mile limit are measured and represents the boundary between state and federal property. In addition to the shoreline and the sounding datum, requirements for nautical charting call for planimetric mapping of a band of the onshore area adjacent to the shoreline and nearshore features which can be seen on the photography, including fixed aids to navigation.

Field support for shoreline mapping entails two primary operations: (1) identifying selected horizontal-control stations for use in the aerotriangulation process and (2) ground support for tide-coordinated photography. Control identification is accomplished either prior to photography by targeting the required stations or, after aerial photography is available, by photoidentification, as described in section 8.7. Ground support for tide-coordinated photography is discussed in the remainder of this section.

Local tidal datums must already exist or be established prior to tide-coordinated photography. The density of tide stations required over the project area depends upon such factors as the tidal characteristics, the general slope of the foreshore, size of the project area, mapping scale, and accuracy requirements. One tide station may suffice on the open coast in an area where the characteristics of the tide are uncomplicated, *i.e.*, time and range differences from one end of the project to the other are small. On the other hand, an area of the same size displaying a complex tidal system and small foreshore gradients may require 10 or more tide stations.

The tidal datum at each selected tide station must be established by spirit leveling from tidal bench marks to a tide staff installed nearby. Readings on each staff representing the high and low water datums are thus determined for use during photography.

Using tide prediction tables, aerial photography is scheduled when the tide will reach the high and low water datums during daylight hours. An hour prior to the scheduled start of photography, each tide staff is manned by an observer who reports the staff reading to project headquarters via a radio communication system. Tide staffs are read and recorded at 15 minute intervals during the photographic operation. From an analysis of water level observations at all tide staffs, the ground coordinator, who is in radio contact with the aircraft, schedules photography over each flight line when the water is at the appropriate level. The land-water interface is imaged on black-and-white infrared film as near as possible to the high and low water datums for use in compiling these lines on the shoreline map.

A field completion survey (field edit) is performed for each map manuscript compiled. After compilation and review of the manuscript, any discrepancies or inadequacies which cannot be resolved in the office are noted and forwarded to the field for verification. The field editor is also responsible for locating or checking the positions of landmarks and aids to navigation, checking specific geographic names, noting all shoreline features and structures, and indicating where the shoreline is "apparent" (obscured by vegetation) in areas where this condition is not obvious on the photography. In short, he checks the overall completeness and adequacy of the compilation and adds final detail which could not be picked up during office compilation.

8.9.2 PHOTOBATHYMETRY

Until recent years, the responsibility for determination of the low water line was that of the hydrographer. This often presented a formidable task and slowed progress of the survey. The advent of tide-coordinated aerial photography on black-and-white infrared emulsion (section 8.9.1) relieved the hydrographer of this problem to a great extent, but he still had to contend with the frustrating and often hazardous nearshore areas of his survey. A further solution to this problem has been the development of photogrammetric techniques for making water depth measurements; collectively referred to as photobathymetry. Largely through efforts at the National Ocean Survey, an operational photobathymetry system capable of depth measurements, in clear water, of up to 25 metres has been developed.

The operational techniques which comprise a photobathymetric mapping system are primarily those which are combined to provide standard shoreline map products. Reduction of photogrammetric soundings and the need for an extensive vertical-control net are basic additional requirements for photobathymetry.

Horizontal- and vertical-control stations, selected to meet aerotriangulation requirements, are targeted in advance of aerial photography (section 8.7.2). Vertical control must be related to local tidal datums to accommodate the reduction of sounding measurements to the sounding datum. In areas of sparse vertical control, elevations of new stations which are located near the shore may be determined by water level transfer from the nearest tide gage. Vertical-control panels must be carefully constructed to insure they are flat and lie in a horizontal plane.

In some areas, vertical control must be extended into the water; several techniques have been applied successfully. Fixed targets of known elevation can be placed on the bottom where bottom composition permits. Another technique calls for large scale photography (approximately 1:3,000) over each control area located in the water. This film is processed in the field, and selection is made of photidentifiable features which would also appear on the smaller scale aerotriangulation/compilation photography. At a later date, boats are directed over these features by a small aircraft with radio communication to the boats, and leadline soundings are recorded along with the date and time. Once reduced to the sounding datum, these points can be used for vertical control.

Aerial photography for compilation of photobathymetry does not require coordination with a specific tidal datum. The shoreline and the zero sounding curve can be contoured stereoscopically during the compilation phase. However, since the photogrammetric soundings must be reduced to the sounding datum, as in conventional hydrography, the height of water relative to this datum must be known during the time of photography, and tidal observations are required.

There are three possible tidal datum situations that may occur, each of which requires a different level of effort by the field support unit.

1. No tidal data in the project area—In this case, the field unit is required to install and maintain tide gages at specified locations for a period of time sufficient to determine the tidal datums. It is not necessary for the datums to be known during the aerial photography phase since this information is not required until the compilation phase. The operating tide gages will automatically record all water level data needed during the photography.
2. Tide gages already in operation—The only re-

sponsibility of the field unit is to verify that all gages are operating properly at the time of photography.
3. Recoverable tidal benchmarks only—In this case, the field unit must recover the tidal datum by leveling procedures and install a temporary tide staff for use during the project.

All horizontal- and vertical-control panels must be inspected every few days to be sure they are still in place and have not been damaged by the elements or by vandalism. If tide gages are in operation, periodic checks are made to verify that they are running properly. When manual tidal observations must be made during photographic flights, tide staff readings are recorded at 15 minute intervals along with time and date.

After completion of photography, all panels are removed, all remaining ground survey work is completed, and a field report is prepared covering all pertinent operations performed.

REFERENCES AND BIBLIOGRAPHY

8.1 General

Anonymous (1947) *Manual of Instructions for the Survey of the Public Lands of the United States.* viii + 613 pages. Superintendent of Documents. Washington, D.C.

Anonymous (1954–1979) *Manual of Topographic Instruction.* (issued as individual chapters). Technical Information Office. U.S.G.S. National Center. Reston, Virginia.

Bomford, G. (1971) *Geodesy.* 731 pages. Clarendon Press. Oxford.

Bouchard, H. and Moffitt, F. H. (1965) *Surveying* (4th edition). xiv & 644 pages. International Textbook Co. Scranton, Pennsylvania.

Breed, C. B.; Bone, A. J.; and Berry, A. (1971) *Surveying* (3rd edition). xviii + 495 pages. John Wiley Sons. New York.

Henriksen, S. W. (1980; in preparation). *Definitions of Terms Used in Geodesy and Allied Disciplines.* 200 pp. National Geodetic Survey. Rockville, Maryland.

Whyte, W. S. (1969) *Basic Metric Surveying.* x + 312 pages. Butterworths. London.

8.2 Basic geodetic control

Dracup, J. (1978) *National Geodetic Survey Data: Availability, Explanation, and Application.* NOAA Techn. Mem. NOS NGS 5. 39 pages. National Technical Information Center. Springfield, Virginia.

Federal Geodetic Control Committee (1974) *Classification, Standards of Accuracy, and General Specifications of Geodetic Control Surveys.* 12 pages. National Geodetic Survey. Rockville, Maryland.

Federal Geodetic Control Committee (1975) *Specifications to Support Classification, Standards of Accuracy, and General Specifications of Geodetic Control Surveys.* 30 pages. National Geodetic Survey. Rockville, Maryland.

Mather, R. S. (1975) "Mean Sea Level and the Definition of the Geoid" *UNISURV G 23:* 68–79.

Rapp, Richard (1974) "The Geoid—Definition and Determination" *Trans. Am. Geophys. Union (EOS) 55(3):* 118–126.

8.3 Coordinate systems on maps

Adams, Oscar S. (1945) *General Theory of Equivalent Projections.* C. & G.S. Sp. Publ. No. 236. 74 pp. Superintendent of Documents. Washington, D.C.

Anonymous (1951) *The Universal Grid System (Universal Transverse Mercator and Universal Polar Stereographic).* TM 5-241. iv + 324 pp. Superintendent of Documents. Washington, D.C.

Anonymous (1967) *Grids and Grid References.* TM 5-241-1. Government Printing Office. Washington, D.C.

Brown, L. M. and Eldridge, W. H. (1962) *Evidence and Procedures for Boundary Location.* viii + 484 pages. John Wiley. New York.

Deetz, C. H. and Adams, O. S. (1945) *Elements of Map Projection.* C. & G.S. Sp. Publ. No. 68. 226 pages, 15 fold-out plates. Superintendent of Documents. Washington, D.C.

Dracup, J. (1974) *Fundamentals of the State Plane Coordinate Systems.* (preprint). 60 pages. Available on request from National Geodetic Survey (Attn: C13 X4), Rockville, Maryland, 20852.

Dracup, J. (1977) *Understanding the State Plane Coordinate System.* (preprint) 31 pages. Available on request from National Geodetic Survey (Attn: C13 X4), Rockville, Md. 20852.

Driencourt, L. and Laborde, J. (1932) *Traite' des Projections des Cartes Geographiques* (2 vols.) Librarie Scientifique Hermann. Paris.

Eckert, Max (1921–5) *Die Kartenwissenschaft* (2 volumes). Walter de Gruyter. Berlin.

Hotine, Martin (1946–7) "The Orthomorphic Projection of the Spheroid" *Empire Survey Rev. 8:* 281–300; *Empire Survey Rev. 9:* 25–35, 52–70, 112–123, 157–166.

Mitchell, H. C. and Simmons, L. G. (1945) *The State Plane Coordinate System (A Manual for Surveyors)* C. & G.S. Sp. Publ. No. 235. 62 pages. Superintendent of Documents. Washington, D.C.

Raisz, E. (1948) *General Cartography.* xv + 354 pp. McGraw-Hill. New York, N.Y.

Reynolds, W. F. (1935) *Relation between Plane Rectangular Coordinates and Geographic Positions (Local Plane Coordinates).* C. & G.S. Sp. Publ. No. 71. Superintendent of Documents. Washington, D.C.

Steers, J. A. (1965) *An Introduction to the Study of Map Projections.* 292 pages. U. of London Press. London.

Thomas, Paul (1952) *Conformal Projections in Geodesy and Cartography.* C. & G.S. Sp. Publ. No. 251. ix + 242 pages. Superintendent of Documents. Washington, D.C.

8.4 Horizontal-control surveys

Anonymous (1952) *Formulas and Tables for the Computation of Geodetic Positions.* National Ocean Survey Sp. Publ. No. 8. 101 pages. Superintendent of Documents. Washington, D.C.

Anonymous (1967) *Electromagnetic Distance Measurement (Proceedings of the Oxford Symposium 1967).* 449 pages. University of Toronto Press. Toronto.

Bigelow, H. W. (1965) "Electronic Surveying: Accuracy of Electronic Positioning Systems" *Int. Hydrographic Review 6:* 77–112.

Bush, E. and Pappas, M. (1965) "SHIRAN-AN/ USQ29 Microwave Geodetic Survey system". *Int. Archives of Photogrammetry 15 (part 3).*

Carroll, Joel (1963) "On Kroll's Method and Compu-

tation of Shoran Distances" *J. Geophys. Res. 68(19):* 5611–5612.

Dolukhanov, M. P. (1963) *Propagation of Radio Waves* (translation). 480 pages. National Technical Information Center. Springfield, Virginia.

Dracup, J. (1980) *Horizontal Control.* (c. 30 pages). National Geodetic Survey. Rockville, Maryland.

Gardner, C. S. (1976) "Effect of Random Path-fluctuations on the Accuracy of Laser Ranging Systems" *Applied Optics 15(10):* 2539–2545.

Garrett, P. H. "Advances in Low-frequency Radionavigation Methods" *IEEE Trans. AES-11(4):* 562–575.

Gergen, John (ed.) (1978) *Proceedings Second International Symposium on Problems Related to the re-definition of North American Geodetic Networks.* xiii + 645 pages. Superintendent of Documents. Washington, D.C.

Gossett, F. R. (1959) *Manual of Geodetic Triangulation.* xvi + 344 pages. C. & G.S. Sp. Publ. No. 247. Superintendent of Documents. Washington, D.C.

Hodgson, C. V. (1957) *Manual of Second- and Third-order Triangulation and Traverse.* C. & G.S. Sp. Publ. No. 145. Superintendent of Documents. Washington, D.C.

Holmes, A. T. (1961) "The Geodimeter" *Canadian Surveyor 15(8):* 445–451.

Jordan, P. W. "HIRAN Instrumental Developments" *J. Geophys. Res. 65(2):* 462–466.

Kroll, C. W. (1949) "A Rigorous Method for Computing Geodetic Distances from Shoran Observations" *Trans. Am. Geophys. Union 30(1):* 1–4.

Lafferty, R. (1968) "Circular Flight Paths Using DME" *Photogram. Eng. 34(2):* 198–200.

Laurila, Simo (1976) *Electronic Surveying and Navigation.* xiv + 545 pages. John Wiley. New York.

Marussi, A. (ed.) (1960) "International Symposium on Electronic Distance Measuring Techniques" *J. Geophys. Res. 65(2):* 385–527.

McLellan, C. D. (1961) "The Tellurometer in Geodetic Surveying" *Canadian Surveyor 15(7):* 386–397.

McLellan, C. D. and Boal, J. D. (1978) "Readjustment of Canadian Secondary Horizontal-Control Networks" *The Canadian Surveyor 32(4):* 433–442.

Meier-Hirmer, B. (1976) "Untersuchungen zur Lagzeitstabilität des Massstabnormals verscheidener EDM-Geräte" *AVN 85(4):* 121–156.

Munson, R. C. (chairman) (1977) *Positioning Systems.* 33 pages, 4 tables. National Geodetic Survey. Rockville, Maryland.

Nares, J. D. (editor) (1946) "International Meeting on Radio Aids to Marine Navigation" *Hydrographic Rev. 23:* 32–48.

Rice, Donald (1950) "Geodetic Applications of SHORAN" *J. C. & G.S. 1:* 7–12.

Rinner, Karl and Benz, Friedrich (1966) *Die Entfernungsmessung mit Elektro-magnetischen Wellen und ihre geodätische Anwendung. (Handbuch der Vermessungskunde, 10th edition. M. Kneissl, ed. Volume VI).* xvi + 1038 pages. Comprehensive bibliography. J. B. Metzler. Stuttgart.

Rüeger, J. M. (1978) "Misalignment of EDM Reflectors and its Effects" *Australian Surv. 29(1):* 28–36.

Saastamoinen, J. (1962a) "Reduction of Electronic Length Measurements" *The Canadian Surveyor 16(2):* 93–97.

Saastamoinen, J. (1962b) "The Effect of Path Curvature of Light Waves on the Refractive Index Application of electronic Distance Measurement" *The Canadian Surveyor 16(2):* 98–102.

Schwarz, C. R. (1978) *The TRAV-10 Horizontal-network Adjustment Program.* Techn. Mem. NOS NGS-12. 52 pages. National Technical Information Service. Springfield, Virginia.

Tomlinson, R. W. and Burger, T. C. (1977) *Electronic Distance Measuring Instruments.* vi + 79 pages. American Congress on Surveying and Mapping. Falls Church, Virginia.

Vanicek, Petr (ed.) (1974) "Proceedings of the International Symposium on Problems Related to the Redefinition of North American Networks" *The Canadian Surveyor 28(5):* 435–749.

Wermann, G. (1979) "Überprüfung elektro-optischer Entfernungsmessgeräte für Triangulierung 3. und 4. Ordnung" *AVN 1979(7):* 265–283.

Wuollet, G. M. and Skibitzke, H. E. (1968) "Plotting DME Arcs for Aerial Photography" *Photogram. Eng. 34(2):* 189–197.

8.5 Vertical-control surveys

Berry, Ralph M. (1976) "History of Geodetic Leveling in the United States" *Surv. & Map. 36(2):* 137–152.

Berry, Ralph M. (1977) "Observational Techniques for Use with Compensator Leveling Instruments for first-order Levels" *Surv. & Map. 37(1):* 17–23.

Berry, Ralph M. (1981) *Manual of Geodetic Leveling.* c. 200 pages. National Geodetic Survey. Rockville, Maryland.

Blachut, T. J. and Leask, R. C. (1952) *The Radar Profile and Its Application to Photogrammetric Mapping.* Report No. 3006. National Research Council of Canada. Ottawa.

Karren, R. J. (1964) "Recent Studies of Leveling Instrumentation and Procedure" *Surveying and Mapping 24:* 383–397.

Krakiwsky, E. J. and Mueller, I. I. (1965) *Systems of Height.* Report No. 60, Department of Geodetic Science, Ohio State University. xxvi + 157 pages. Ohio State University, Columbus, Ohio.

Lyytikainen, H. E. (1960) *An Analysis of Radar Profiles over Mountainous Terrain.* Report No. 5569. National Research Council of Canada. Ottawa, Canada.

Mather, R. S. (1975) "Mean Sea Level and the Definition of the Geoid" *UNISURV G 23:* 68–79.

O'Leary, W. V. (1963) "A New Development Program for the Airborne Profile Recorder" *Photogram. Eng. 29(5):* 872–880.

Petrov, V. D. (1965) "Use of Mean Sea Level Surfaces in Processing Leveling Results" *Geodesy and Aerophot. 1965(4):* 246–248.

Poetzschke, Heinz (1979) "Motorized Leveling" *Proceedings American Congress on Surveying and Mapping, 39th Annual Meeting March 18–March 24, 1979.* American Congress on Surveying and Mapping, Falls Church, Virginia. pp. 321–329.

Rappeleye, H. S. (1948) *Manual of Leveling Computation and Adjustment.* C. & G.S. Special Publ. No. 240. 178 pages. National Technical Information Service. Springfield, Virginia.

Slama, C. C. (1961) "Evaluation of an APR System for Photogrammetric Triangulation of Long Flights" *Photogram. Eng. 27(4):* 572–578.

Stoughton, H. W. (1977) "Elevations—the Federal Datums" *Surv. and Map. 37(4):* 353–355.

Wemelsfelder, P. J. (1970) "Sea-level Observations as a Fact and as an Illusion" *Report on the Symposium on Coastal Geodesy (R. Sigl, ed.). Technical University, MUNICH.* Institute for Applied Geodesy. Frankfurt, a.M.

Whalen, C. T. (1978) *Control Leveling* NOAA Techn. Report NOS 73 NGS 8. 21 pages. Superintendent of Documents. Washington, D.C.

8.6 Surveys giving both horizontal and vertical control

Anonymous (1976) *Satellite Doppler Positioning.* 902 pages (2 volumes). New Mexico State University, Las Cruces, New Mexico.

Ball, W. E. and Voorhees, G. D. (1978) "Adjustment of Inertial Survey System Errors" *The Canadian Surveyor 32(4):* 453–463.

Brown, D. (1976) "The Bundle Adjustment—Progress and Prospects" Invited paper, Commission III, XIII Congress of ISP, Helsinki, 1976.

Chapman, William H. (1979) "Surveying from the Air Using Inertial Technology" *Proceedings of the American Congress on Surveying and Mapping 39th Annual Meeting.* American Congress on Surveying and Mapping. Falls Church, Virginia. pages 352–365.

Grün, Armin (1979) "Zur Anwendung der modernen Präzisionsphotogrammetrie in der Netzverdicht ung und Katastervermessung" *Zeits. f. Vermess. 104(3):* 85–97.

Lawrence, Myron H. (1967) *Determining Earth Crustal Movements by Precision Analytical Photogrammetry.* v + 55 pages; ill., tables. ESSA Techn. Report RL 61-ESL-2. Superintendent of Documents. Washington, D.C.

Lucas, James R. (1979) "Photogrammetric control densification project" *Proceedings Second International Symposium on Problems Related to the Redefinition of North American Geodetic Networks* (J. Gergen, ed.). Superintendent of Documents. Washington, D.C. pages 507–518.

Mancini, A. and Moore, R. E. (editors) (1977) *Proceedings of the 1st International Symposium on Inertial Technology for Surveying and Geodesy* x + 445 pages. Canadian Institute of Surveying. Ottawa.

Slama, C. C. (1978) "High Precision Analytical Photogrammetry Using a Special Reseau Geodetic Lens Cone", International Symposium, ISP Commission III, Moscow, USSR.

Stansell, T. A. (1978) "The Many Faces of TRANSIT" *Navigation (U.S.A.) 25(1):* 55–70.

8.7 Selecting, targeting, and identification of photogrammetric control stations

Ackerl, F. and Neumaier, K. (1959) "Über die Signalliesierung der Passpunkte für Infrarot-Aufnahme" *Photogrammetria 16(1):* 17–28.

Anonymous (1961) "Field Identification of Horizontal Control" *Topographic Instructions of the United States Geological Survey.* U.S. Geological Survey Center. Reston, Virginia.

Hirn, Albert, et alia (1966) "Zur Signallisierung in Stadtgebieten" *Osterr. Z. Vermess.-wes. 54(5):* 158–163.

Lyon, D. (1957) "Basic Requirements for Charting Photography" *Photogram. Eng. 23(4):* 685–696.

Pastorelli, A. (1959) "Die Signalisierung der Fix- und Grenzpunkte im Gelände als Massanahme der Precisions-photogrammetrie" *Photogrammetria 16(2):* 102–108.

Trinder, J. C. (1971): "Pointing Accuracies to Blurred Signals: A minimum target size should be chosen such that annulus widths are not less than one milliradian subtended at the eye" *Photogram. Eng. 37(2):* 192–202.

8.8 Field Completion Surveys

Anonymous (1962) "Field Mapping and Completion Surveys (draft)" *Topographic Instructions of the United States Geological Survey*. U.S. Geological Survey Center. Reston, Virginia.

Anonymous (1966) "Advance Field Completion (draft)" *Topographic Instructions of the United States Geological Survey*. U.S. Geological Survey Center. Reston, Virginia.

8.9 Surveys for coastal mapping

Brewer, Ronald K., "Project Planning and Field Support for NOS Photobathymetry", *Proceedings of the Coastal Mapping Symposium, American Society of Photogrammetry*, Rockville, Maryland, August 1978, pages 55–66.

Brewer, Ronald K. and Cravat, Harland R., "Baseline Establishment for Positioning Federal State Offshore Boundaries", *Proceedings of the Fourth Annual Offshore Technology Conference*, Houston, Texas, May 1972, pages 377–386.

Brewer, Ronald K. and Heywood, Albert K., "Coastal Boundary Mapping", *Proceedings of the American Society of Photogrammetry, 38th Annual Meeting*, Washington, D.C., March 1972, pages 182–191.

Collins, James, CDR, "Cost Benefits of Photobathymetry", *Proceedings of the Coastal Mapping Symposium, American Society of Photogrammetry*, Rockville, Maryland, August 1978, pages 97–102.

Fritz, Lawrence W. and Slama, Chester C., "Future NOS Analytical Instruments for Photobathymetry", *Proceedings of the Coastal Mapping Symposium, American Society of Photogrammetry*, Rockville, Maryland, August 1978, pages 89–95.

Geary, E. L. (1968) "Coastal Hydrography" *Photogram. Eng. 34(1): 44–50.*

Harris, W. D. and Umbach, M. J. (1972) "Underwater Mapping" *Photogram. Eng. 38(8): 765–772.*

Jones, B. G. "Low-water Photography in Cobstock Bay, Maine" *Photogram. Eng. 23(2): 338–342.*

Keller, Morton, "A Study of Applied Photogrammetric Photobathymetry in the National Oceanic and Atmospheric Administration", *Proceedings of the Coastal Mapping Symposium, American Society of Photogrammetry*, Rockville, Maryland, August 1978, pages 45–54.

Meijer, W. O. J. G. (1964) "Formula for Conversion of Stereoscopically Observed Depth of Water to True Depth, Numerical Examples and Discussion" *Photogram. Eng. 30(6): 1037–1045.*

Moffitt, F. H. (1969) "History of Shore Growth from Aerial Photographs" *Shore and Beach 37(1): 23–27.*

Moore, J. G. (1947) "Determination of the Depths and Extinction Coefficients of Shallow Water by Air Photography Using Color Filters" *Proc. R. Soc. (London) A 240(816): 163–217.*

Smith, John T., "NOS Photographic Operations for Photobathymetry", *Proceedings of the Coastal Mapping Symposium, American Society of Photogrammetry*, Rockville, Maryland, August 1978, pages 67–68.

Stafford, D. B. and Langfelder, J. (1971) "Air Photo Survey of Coastal Erosion" *Photogram. Eng. 37(6): 565–575; 16 refs.*

Tewinkel, G. C. (1963) "Water Depths from Aerial Photographs" *Photogram. Eng. 29(6): 1037–1042.*

Vanderhaven, Gary, "Data Reduction and Mapping for Photobathymetry", *Proceedings of the Coastal Mapping Symposium, American Society of Photogrammetry*, Rockville, Maryland, August 1978, pages 69–88.

Wijk, M. C. Van (1964): "Discussion Paper: 'Water Depths from Aerial Photographs' by G. C. Tewinkel" *Photogram. Eng. 30(4): 647.*

CHAPTER IX

Aerotriangulation

Author-Editor: CHESTER C SLAMA
Contributing Authors: H. EBNER AND LAWRENCE FRITZ

9.1 Introduction

PHOTOTRIANGULATION IS DEFINED by the Nomenclature Committee of the American Society of photogrammetry as "The process for the extension of horizontal and/or vertical control whereby the measurements of angles and/or distances on overlapping photographs are related into a spatial solution using the perspective principles of the photographs. Generally, this process involves using aerial photographs and is called *aerotriangulation* or *aerial* triangulation." The definition then refers the reader to three additional terms, *analytical phototriangulation, radial triangulation,* and *stereotriangulation.*

In the 3rd. Edition of the MANUAL OF PHOTOGRAMMETRY, the subject of phototriangulation was divided into two separate chapters, one called "Mechanical Methods of Phototriangulation" and the other "Analytical Triangulation." The first covered templet methods and photo-triangulation using analog instruments whereas the second covered numerical methods of phototriangulation. In this edition, the two chapters have been combined into one for continuity and the total subject matter has been subdivided into three parts—radial triangulation, analog instrument triangulation, and analytical triangulation. In addition, since most of the procedures that are described refer to current methods used to densify geodetic ground control using aerial photography, the chapter is named aerotriangulation. Since the procedures used for both radial and analog aerotriangulation have not changed to any great extent from those described in the 3rd. Edition, the material contained herein is a condensation of the former chapter IX with some minor modifications. Due credit should therefore be given to Carl Born who served as Author-Editor of that chapter.

9.2 Radial Triangulation

Radial triangulation is a *graphical* method of establishing coordinates of ground objects using the aerial photograph as a basis. The principle of the method is based on the assumption that directions to objects, when reckoned from the center of a vertical aerial photograph, are constant under all conditions of elevation difference and scale change. That is, it is assumed that terrain elevation differences and scale change produce only radial displacements of the images from the photographic perspective center. In the case of a non-vertical photograph, this assumption does not hold; therefore, near verticality is a prerequisite to control extension by radial triangulation. For a complete discussion of the effects of photographic tilt and scale change, refer to chapters II and XIV. As a rule of thumb, however, one can assume that tilts of three degrees or less will not introduce significant errors into the triangulated positions.

Extension of horizontal control by radial triangulation is a very cost- and time-effective substitute for field geodetic methods. Given ground control points in the perimeter of a project area, a network of supplemental control can be established with accuracies which are adequate for many applications in mapping. Radial methods are very well suited for the horizontal control in mapping at large scales such as 1:4,800 to 1:12,000. In some cases, the method has been satisfactorially used to control mapping for scales as large as 1:2,400 for highway planning.

The speed of operation naturally depends a great deal upon the ability and experience of the operator. An experienced operator with good stereoscopic vision can complete all the operations that are necessary to triangulate one photograph in about one-half hour. Compared to field geodetic costs this represents a sizable savings in both time and money.

9.2.1 OPERATIONAL METHODS

9.2.1.1 HAND TEMPLET METHOD

A very simple method of determining control (obtaining secondary or office control) is known as the hand-templet method. It consists of replacing each photograph by a transparent

templet on which lines have been drawn from
the principal point radially through each picture
point. The templets are assembled, and the intersection of all radials common to each identified picture point represents the adjusted coordinate of each point (*see* figure 9-1).

The hand-templet method for the extension of
control by radial triangulation involves the following procedure. The principal point is established on each print as the point of intersection
of fiducial markers. This point will not normally
fall on an identifiable image, so a definite image
as nearly coincident with it as possible is
selected and marked. The point selected, however, should never be more than 3 mm from the
indicated principal point. Under a stereoscope,
this point is then transferred to the succeeding
and preceding photographs resulting in three
azimuth points (*see* figure 9-1) on each photograph (except the first and last in the strip). At
least three additional points, called pass-points
or radial-control points, are picked on each side,
in the overlap area between the flight strips, in
the general positions indicated in the figure.
They should be chosen so that they are readily
identifiable with a stereoscope and then transferred to the photographs in the adjacent flight
strip. This results in at least nine pass-points for
each photograph. It is common practice to select
and mark at least three or four more than the
minimum. The points are identified and marked
by pricking fine holes on the photographs and
encircling these with small inked circles about
5 mm in diameter.

In advance of the office work, the geodetic
ground control has been identified on the photographs (in the field) by pricking the photograph
with a fine needle and encircling with a small
penciled circle. These are usually numbered
and accompanied by a brief sketch and description of each point on the back of the photograph. In the office, these points are then transferred to all adjacent photographs.

After completion of the above steps for all
photographs, transparent plastic sheets (approximately 25 mm larger than the photograph
on each side) are layed on the face of each
photograph. The center point is marked, and radials are drawn or scribed to all ground control
points and pass points. It is preferable to etch
these lines and fill them with an opaque substance, such as graphite. This way, very fine,
distinct lines are obtained. Azimuth lines are
extended from the center of the photograph as
continuous lines, but the radials are merely short
line segments drawn through the pass points,
long enough to accommodate differences of relief and to permit adjustment for slight scale
changes between individual photographs. In the
event that the scale of the map will be larger
or smaller than the scale of the photography, it
will be necessary to use larger or smaller

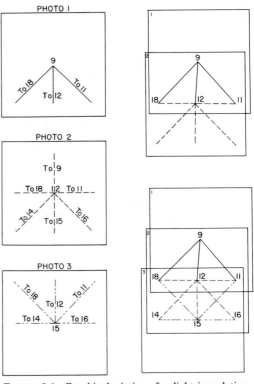

FIGURE 9-1. Graphic depiction of radial triangulation.

templets and extend the radials by a distance
proportional to this difference in scale.

Base sheets for templet assembly are prepared
by constructing grid or graticule lines on a stable
base of masonite, aluminum, or plastic. The
most commonly used projection is the
polyconic. All horizontal control points are precisely plotted and marked on these base sheets
prior to templet assembly.

Assembly of templets is begun with the two
templets whose overlap contains the greatest
number of control points. These templets are
positioned in such a way that the corresponding
azimuth lines on the two coincide, and the radials to the control points intersect at the plotted
positions of the control points on the base sheet.
These templets are then fastened to the base and
to one another with small pieces of removable
adhesive tape. These first two templets establish
directions to additional points and these, in turn,
form the base from which additional templets
can then be positioned, thus carrying the initial
control forward along the flight line. As each
new templet is positioned, it is fixed to the grid
sheet and to the preceding templet by small
pieces of adhesive tape. This procedure is repeated until the strip is tied to all control points.

Regardless of care, a correct assembly will
seldom (even of a single strip) be made the first
time. The discrepancy found between the

templet position and the plotted control point must be proportioned back through the strip by sequentially re-laying the templets to a slightly different scale or orientation. Personal judgment is of great importance, since ordinarily the first problem is how closely to hold the azimuth lines when errors are found in the intersection of the radials, or when the strip does not end on control. When one strip is satisfactorily adjusted, the next strip is laid, whereupon an additional adjustment must be made to bring points common to both strips into coincidence. From this trial and error procedure, an average result is obtained, the time of which depends very largely upon the skill of the operator. As the number of templets increases, slight errors accumulate and the proportioning becomes increasingly difficult, and sometimes impossible beyond a certain point.

After the assembly has been completed, the points of radial intersection are transferred to the base sheet by pricking through the templets. An automatic center punch, similar to those used by a machinist, ground to a long sharp point is a good tool to use.

9.2.1.2 SLOTTED TEMPLET METHOD

The slotted-templet method presents a simple mechanical solution to the templet-assembly method described above. The radial line on the templet is replaced by a radial slot in which a close-fitting round stud is inserted, free to slide along the longitudinal axis of the slot. Since the slot takes the place of the line, the templet material need not be transparent and can therefore be made of stiff material. The rigidity of templet material is one vital factor in the method. It permits all templets with slots intersecting at a common point to be assembled over the stud representing that point. Since the stud and the templets are both of rigid material, motion is possible only along the longitudinal axes of the slots. Thus, the templets will tend to adjust themselves into position.

Preparation of the photographs for slotted templets is basically the same as for hand templets, namely: the selection of the center point (a definite image near the principal point so that no appreciable error is introduced), the selection of pass points in the quantity and location required by the job, and the identification of all horizontal ground-survey control. Excluding control, a minimum of eight slots and a center hole will appear on each templet, two representing the radials to the centers of the preceding and succeeding photographs and six representing the radials (three along each lateral edge) to pass points (*see* figure 9-2).

After all photographs have been prepared, a templet is made of each photograph. The templet material should be three- or four-ply Bristol board or similar cardboard. The photographs are

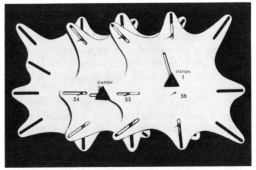

FIGURE 9-2. Slotted templets.

taped to a piece of templet material and the pass points are punched through the photographs onto the templets, then circled with pencil. The photograph number is marked alongside each principal point in orientation with the direction of flight. If the scale of the assembly is to be different from that of the photography, the amount of scale change is proportioned on the radials from the center.

The templets are then punched, using a circular punch for the principal point with a diameter the same as that of the studs. After the center point is punched, the templet is placed with the center hole over a stud on the templet cutter (figure 9-3). This stud slides in a groove in line with the slot-cutting die so that the templet can be rotated on the stud and moved backward or forward until the centering pin of the die is over the marked point. The axis of the slot cut in the templet is in line with the radial from the principal point through the picture point. The length of the slot (4 to 5 cm) is sufficient to accommodate slight differences in scale of the photographs as well as variations in relief. Aluminum, masonite, or plywood sheets, abutted and fastened with transparent tape, have been found to be satisfactory as a base upon which to assemble the

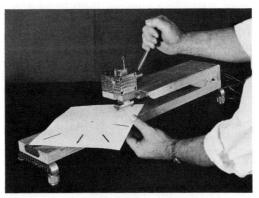

FIGURE 9-3. Slot cutter for preparation of slotted templet.

templets. Sheets of vinyl of sufficient thickness are also acceptable. The largest assembly of slotted templets noted to date has been one representing a ground area of approximately 4,400 square miles assembled at a scale of 1:15,840 which required a base approximately 8 by 11 metres.

After making the templets and constructing the grid base, the next step is to assemble the templets. The assembly can be started from any single control point and with any flight strip, although it is preferable to start with the flight strip having the most ground control points. When the next control point is reached, the scale may be adjusted by squeezing together or stretching the assembly to fit the control.

If the assembly is at approximately the same scale as the photographs, the pass points on 9- by 9-inch photographs are never more than 13 cm away from the center. Consequently, in average terrain a small amount of tilt will cause no appreciable error in the assembly. A templet from a photograph with a large amount of tilt cannot be made to fit unless the accuracy is reduced through the use of slots with a large tolerance between the diameter of the studs and the width of the slots, or unless the material used for templets is of such weak consistency that the studs will indent the sides of the slots. When a templet will not fit into the assembly, it should be checked immediately for errors.

The normal procedure for blocks of photography is to assemble each adjacent flight strip in turn. However, alternate flights may be laid, or the templets may be built up across the lines of flight. Where difficulty in any portion of the area is encountered, the work can be carried around that portion, and eventually the templet or templets causing the trouble will be isolated.

To determine the density and distribution of ground control required for slotted templet methods, a series of tests were conducted by the Soil Conservation Service at Beltsville, Maryland. An area of 155 square miles, roughly rectangular in shape, was chosen for the first of these tests. Two hundred thirty-three (233) photographs, 9 by 9 inches in size, at a scale of 1:12,000, covered the area in twelve strips, with nineteen photographs in each strip. Two hundred seventy-three (273) ground control points of third-order accuracy were used for checking the results.

The first assembly was made using four control points, one in each corner of the area. The distance between control points was 12 miles or 19 photographs in line of flight and 13½ miles or 12 strips across the line of flight. After settling the assembly in what was determined to be a good fit, the average difference in position at the check points was found to be less than 50 feet or, at 1:12,000 scale, 1.3 mm.

In a second test which covered Carter and Murry Counties, Oklahoma, an area of 1,200 square miles of comparatively flat country was used. The photographic coverage averaged 20 photographs per strip in the western part and 40 photographs per strip in the eastern part. The basic control was spaced around the perimeter of the area. Fifty-five check points indicated an average difference in position of 55 feet with a maximum difference of 130 feet.

9.2.1.3 RADIAL ARM TEMPLETS

The radial arm templet is a refinement of the slotted templet where by the slotted templet is replaced by a series of metal arms with slots, bolted at the center. Until tightened, each arm is free to rotate to the desired radial and hence serves as a re-usable templet. Once formed, the process procedes in much the same manner as that just described.

9.2.1.4 STEREOTEMPLETS

Stereotemplet triangulation was developed at the US Geological Survey in 1949. The system evolved from a basic concept that was developed for the purpose of extending the utility of the non-universal analog stereoplotter. That is, an attempt was made to use these stereo-instruments for aerotriangulation. The system employs the concept of templet connection to adjust a series of independently scaled stereomodels to a common two dimensional scale and orientation. The method has a decided advantage over standard templet methods in that the errors resulting from non-verticality of the photograph are supressed by the relative orientation process. Because of this, the method can and has been successfully employed with oblique photography.

9.2.1.4.1 BASIC PRINCIPLES

All points in a relatively oriented stereopair are (within the limits of the photography and instrument) at a common scale and are presented in a three dimensional coordinate system of arbitrary orientation. Unlike the photograph, these points are "rectified" and are not displaced because of relative elevation or tilt of the original photographs. If the stereomodel is near "level," the angles described by any three points are true and the distances between points are at some common ratio of the actual ground distance. The purpose of the templet is, therefore, to provide some means of adjusting the individually scaled models to a common scale and orientation relative to given coordinates of some ground objects. This can be accomplished by preparing two practically identical templets for each stereomodel. Each templet (made from some stable material) contains a plot of at least two well defined features within the model, preferably nearest each corner of the stereo-overlap area. On one of the templets, one point is selected as a "radial point" and punched as a round hole. The other points on the same templet are punched as slots radiating from this hole. On the duplicate templet, the same points are punched, however, a different point is selected as the "radial point" and punched as a hole. Figure 9-4 illustrates both the conventional slotted templet and the stereotemplet for comparison of the two principles.

As in the case of the slotted templet, the stereotemplets are made of stable material which will exhibit a minimum of surface friction. White vinyl-sheets having a thickness of about 0.38 mm are the most commonly used material. Stereotemplet studs have a hollow vertical stem no larger in diameter than 6.35 mm, mounted on a 19 mm flat base. The center hole is about 4.78 mm in diameter through the 12.7 mm long stem of the stud. To maintain precision, the holes and slots in the templet material are punched to be no more than .026 mm larger than the outside diameter of the stem of the stud. The "laydown" process of the stereotemplets proceeds in much

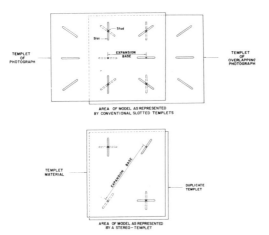

FIGURE 9-4. Comparison of a stereotemplet with conventional slotted templets.

the same manner as the slotted templet described in the previous section.

9.2.1.4.2 EVALUATION OF TEST DATA

Tests of the stereotemplet system of aerotriangulation were performed at the Geological Survey and the results verified by field accuracy checks (see chapter VIII for a description of methods used for field tests). In all cases tested, the map sheets tested met the National Map Accuracy Standards. In addition, simulation studies of stereotemplet techniques carried out at the Geological Survey indicated that: (1) errors in the system are inversely proportional to the density of the perimeter control; (2) errors are inversely proportional to the scale of the assembly; and directly proportional to the stud-slot clearance. The most significant finding from the tests was that the errors increased most rapidly when the ratio of scale between the photography and the templet was greater than 2:3.

The optimum clearance between the outside diameter of the stud and the width of the slot was found to be .001 inch. A smaller clearance caused binding at the time of assembly and a larger clearance, of course, introduced errors in position. Using photography at a scale of 1:24,000 and templets at the same scale, test results showed an RMS difference between the positions of points established by the stereotemplets and geodetic positions to be 12 feet. The density of given control was 30 stereotemplet models per control point. Further testing and operational results have shown that National Standards of Map Accuracy can be achieved using control at each corner of a 7½-minute quadrangle map. For flying heights of less than 5,500 feet, one additional point is needed at the mid-point of the 7½-minute quadrangle corners.

9.3 Analog Instrument Triangulation

Analog instrument triangulation is defined as the process of establishing positions and elevations of points using spatial models oriented in analog instruments. The models are formed in stereoplotters which essentially reverse the process of photographic exposure by re-projection using overlapping photography taken from different camera stations in an ordered sequence. Precise connection between successive models is accomplished in the areas of overlap in such a manner that the model obtained is an accurate scale representation of the object photographed.

9.3.1 TYPES OF STEREOPLOTTING INSTRUMENTS USED FOR TRIANGULATION

In instrument phototriangulation, there are four basic design categories within which most

stereoplotting instruments used for aerotriangulation fall. The design categories relate to method of projection used and are;

> optical projection,
> optical-mechanical projection,
> mechanical projection, and
> analytical projection.

The distinction between particular devices is not always clear, because, to some degree, all stereoplotters incorporate optical-mechanical-analytical mechanisms in their design. Further, stereoplotters are also classified in terms of accuracy, such as first-order or second-order, with the implication that they produce results at different levels of accuracy. This latter distinction is somewhat arbitrary, but has been useful in the past. First-order instruments are more complex, more expensive, and more versatile than second-order devices and, generally, incorporate more overall precision in their design.

As a general rule, plotting devices classed as first-order instruments are designed specifically for use in stereotriangulation and, hence, they provide a direct means of connection between successive models by "base-in" and "base-out" settings. They also have an optical switch which permits continuous relative orientation of a strip of photographs.

9.3.1.1 UNIVERSAL FIRST-ORDER INSTRUMENTS

All of the devices used for aerotriangulation can be used as well for map compilation. Hence, they are commonly referred to as universal devices. The first-order plotters provide for precise recovery of the condition of orientation of the stereomodels. They solve the geometric problem of orientation by using various combinations of optical and mechanical projection or by analytical projection. All first-order instruments are very intricate in their design and represent the optimum in precision optical-mechanical engineering. However, there are practical limits within which they are designed to operate. That is, all optical-mechanical, optical, or mechanical projection devices have a limited range for the motions which represent the interior and exterior parameters of the photograph. These limitations have, in recent years, given rise to new stereoplotters with increased versatility. These newer devices incorporate numerical methods coupled with electronic servo-systems to provide a greater range for both interior and exterior orientation. These newer first-order instruments (analytical stereoplotters) are described fully in chapter XIII.

First-order instruments of the optical-mechanical type are represented by: the Stereotopograph (designed by Poivilliers, produced by Societe d'Optique at de Mechanique, Paris); the Photostereograph (designed by Nistri, produced by Ottico Mechanica Italiana, Rome); and the Thompson-Watts Plotter (designed by Thompson, produced by Hilger and Watts, London). Each uses optical reconstruction for interior orientation and mechanical guides on the outer projection side to obtain the spatial model.

First-order mechanical projection instruments are represented by the Stereocartograph (designed by Santoni, produced by Officine Galile, Florence) and the Autograph A-7 (produced by Wild, Heerbrugg). They use mechanical linkage, in both the interior and exterior cases, which radiates from the projection center. Although optical paths are provided, they serve only as a means of viewing the photographs and do not enter into the geometric reproduction of the perspective conditions.

One first-order stereoplotter uses optical projection for both the inner and outer case. This purely optical projection device is the Stereo-planigraph (designed by Bauersfeld, produced by Carl Zeiss, Oberkochen). In this device, the pencils of light are projected into space to form a real optical model of the area photographed.

All first-order devices provide a complete restitution by compensating for lens distortion, average film shrinkage, and tilt of the exposure. They also provide both a graphical and digital readout of the model coordinates. During the phototriangulation, all first-order devices provide a direct tie between successive models in terms of scale and datum. Although each instrument has only two projectors, a stereomodel from two exposures is established by the projectors and a second, or adjoining, model can be oriented without disturbing the photograph common to the two by means of an optical-mechanical switch. Thus, the orientation of model one is directly connected to the orientation of model two through the undistrubed portion of the photograph common to both models. This "leap-frog" technique is continued so that all models of a strip are referred to a common origin in the instrument.

9.3.1.2 SECOND-ORDER INSTRUMENTS

Other stereoplotting devices, which were fundamentally designed for map compilation, are frequently used for phototriangulation. A second-order instrument, which uses optical projection alone and has been widely used for phototriangulation, is the Muliplex which was first introduced in 1934. Other types of optical plotters of the second-order are the Kelsh and the Balplex. Second-order instruments which use mechanical projection are represented by the Autograph A-8 and the Stereophotograph type D. This list is by no means exhaustive, but does include most of the stereoplotters that have been used for phototriangulation by optical and mechanical projection. A complete description of most stereoplotters available today is given in chapters XI and XII.

9.3.2 PHOTOTRIANGULATION PROCEDURES WITH FIRST-ORDER INSTRUMENTS

There are several different approaches to the problem of strip-orientation for phototriangulation using first-order stereoscopic instruments. In general, however, they can be divided into two classes; the aeropolygon method and, the aeroleveling method. In the ensuing discussion, these two methods will be described. For reasons of clarity this discussion is preceeded, however, with a description of the differences between bridging and cantiliver extension along with model-orientation procedures.

9.3.2.1 BRIDGING AND CANTILEVER EXTENSION

Supplementary control points whose coordinates have been established by phototriangulation are used for the absolute orientation of individual stereomodels for compilation of detail. When this supplementary control is established from given geodetic control which is located at only one end of the strip of photography, the procedure is called *cantilever extension*. On the other hand, when the given geodetic control is located at both ends of the strip (and possibly at intermediate places along the strip) the phototriangulation procedure is referred to as *bridging*. In cantilever extension, only one section of the strip is oriented to geodetic control and this absolute orientation is transferred to subsequent photographs through successive relative orientation of individual models along the strip. Successive relative orientation, however, introduces small systematic errors into each individual orientation which have a tendency to accumulate to the end model. This accumulation increases, of course, with the number of models in the strip and, as such, presents a serious limitation to accuracy of the established supplemental control. In practice, cantilever extensions are generally used only in the cases where subsequent mapping will be at a very small scale or in remote regions where access to geodetic control is limited.

In bridging, the propagation of systematic error is significantly reduced by using geodetic ground control to "correct" for the deformations of the strip. That is, given ground control at either end of a strip of photography, the triangulated data is made to "fit" this control and the error is likewise reduced. In the case of bridging, there normally is insufficient control at any one location to "absolutely" orient a single model. When the "bridge" is completed and adjusted, the relative orientations of intermediate models help provide the necessary supplemental control.

9.3.2.2 MODEL ORIENTATION

There are essentially two methods used to obtain the model orientation in first-order plotters: numerical, and empirical. Both involve the elimination of y-parallaxes between overlapping exposures to form a three-dimensional model. After this relative orientation is completed, the resulting model is absolutely oriented to some given scale and vertical datum (*e.g.*, translation, rotation about three axes and scaled to fit a given ground configuration). This operation is analogous to the minimum conditions of equilibrium as given in engineering statics. That is, for a body to be in equilibrium, the sums of the translations of each of the three axis, the rotations about these three axes, and the scale change of the system must all be zero. In phototriangulation, to satisfy these seven conditions, therefore, one must have a minimum of two ground control points (given in all three coordinates) plus one additional ground control point (not in line with the other two) given in height. This constitutes seven given parameters which *uniquely* satisfy

the seven conditions. Points in excess of the minimum must be introduced into the adjustment through least squares or must be absorbed by additional deformations of the relatively oriented strip.

9.3.2.2.1 NUMERICAL ORIENTATION

Relative orientation between a pair of overlapping photographs can be achieved by making the images of five objects, at selected positions in the overlap area, coincide. In general, the five selected *positions* are those that will produce a maximum in y displacement of the image for the orientation element being adjusted. Five positions are necessary and sufficient for relative orientation and any number greater that will constitute an overdetermination and require "best fit" procedures. The reason for five so-called parallax points becomes clear when one considers the total geometry of the problem. That is, in relative orientation there are twelve unknowns which need to be solved for: those specifying three rotations and three translations of each of the two photographs. As stated above, the conditions of equilibrium require that seven of these unknowns be given or assumed. In the relative orientation, the remaining five unknowns are found by systematic adjustment (or measurement of displacement) of images at five discrete places in the model. That is, each point measured introduces four observations (two coordinates on each photograph), three of which are used to determine the object coordinates, leaving one over-determination which is used to solve for the unknown orientation parameter.

Numerical relative orientation of a model begins with the measurement of y parallaxes (common image displacement in the y direction of a model) at six or more points in the model. A least-squares solution is then used to determine corrections to the instrument orientation elements from dial settings on the projectors. The differential form, in terms of small changes to five elements of orientation of one of the two projectors, is given by:

$$P_y = d\,by + (x - bx)d\kappa_2 - \frac{y}{h}\,dbz_2$$

$$+ \frac{y(x - bx)}{h}d\phi_2 - h(1 + \frac{y^2}{h^2})d\omega_2 \quad (9.1)$$

where

$P_y = y_2 - y_1$ is the measured y parallax at a point where y_2 and y_1 are the measured y model coordinates of the projected image from photographs two and one respectively;

x, y are model coordinates of the point with y being the average of y_2 and y_1;

bx, by, bz are the model coordinate differences in position of the two projectors' perspective centers parallel to the x, y, and z axes of the instrument;

h is the z model height of the first projector and;

κ, ϕ, ω are the three rotations of the projector.

The resulting expressions from all observations of parallax are used to form normal equations whose solution yields corrections to the orientation elements of the relative photograph (in this case, projector two). *See* chapter II for a description of the procedures for forming and solving the normal equation.

Although numerical methods are well developed for relative orientation using measured differences in y parallax, these methods are not applied extensively because of the amount of computations required. With the advent of the small, inexpensive, programmable desk calculator, however, the method may find more application in mapping centers that perform phototriangulation with first-order analog instruments. A decided advantage of the numerical method is that it produces a statistical "best fit" solution for the relative orientation and, in addition, can be extended to provide statistics for the evaluation of the precision of the observations.

9.3.2.2.2 EMPIRICAL ORIENTATION

Since freedom from y parallax at all points in the stereoscopic model is the criterion for good relative orientation, it can be accomplished by simple inspection and systematic elimination. The process is empirical and requires iteration to minimize y parallaxes at all points. When using the empirical method, it is particularly important that the following rules be observed.

The locations of the parallax corrections be symmetrical with respect to the model area;

correction of parallax using that element which produces the maximum y displacement at the location used;

consistent use of specific corrections at specific locations; and

consistent sequence for applying the individual corrections for y parallax.

Like numerical procedures, the empirical method is divided into two techniques. One is usually used for the initial model in an extension, whereas the other is used when one of the projectors must not be disturbed, such as in setting up successive models. In the first case, the relative orientation is accomplished by adjustment of the rotational elements only. In the second case (orientation of a single photograph to one which is to remain fixed), it is necessary to adjust by and bz as well as the rotational elements in the projection of the photograph being oriented. The steps in the procedure for the empirical method are tabulated below.

Steps 1a, 2a, and 3a are for the first case; the others are for the second case. *See* figure 9-5.

1a. Eliminate y parallax at positions 1 and 2 by swing$_2$-swing$_1$.

1b. Eliminate y parallax at position 1 by swinging projector 2 and eliminate y parallax at position 2, by adjusting by_2.

2a. Eliminate parallax at position 4, by adjusting ϕ.

2b. Eliminate parallax at position 4, by adjusting bz_2.

3a. Eliminate parallax at position 3, by adjusting ϕ_2.

3b. Eliminate parallax at position 3, by adjusting ϕ_2.

After completion of steps 1 to 3, points 1, 2, 3, and 4 should be free of y parallax. The remaining parallax at points 5 and 6 is x tilt, which can be removed by adjusting either ω_1 or ω_2. Adjustment of ω_1, however, introduces y parallax at all positions so that to eliminate the parallax at positions 5 and 6 the correction must be such that the overcorrection produces equal parallax at positions 3, 4, 5, and 6 and in the same direction. Although it is possible to estimate the amount of overcorrection, the overcorrection constant (C_ω) is easily computed using,

$$C_\omega = 1/2(1 + \frac{f^2}{y^2}) \qquad (9.2)$$

where

f = projector focal length, and

y = distance from position 1 to 5 or 2 to 6 at photo-scale.

In the general case, for wide-angle photographs, the overcorrection constant is approximately 1.6 and experienced operators usually have little difficulty in making the settings for the correction empirically. Upon completion of the corrections, steps 1 through 3 are repeated and the whole process is iterated until residual y parallaxes are reduced to tolerable limits. Normally, a sixth position, either 5 or 6, is used as a check on the solution and very often a mean is accepted, distributing the residuals between all six positions.

9.3.2.3 AEROPOLYGON PROCEDURE

Both the cantilever extension and the bridge require the orientation of successive stereo-

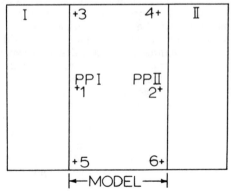

FIGURE 9-5. Positions in stereomodel for observing y parallax.

scopic models; however, the manner in which a model is oriented to its predecessor differs according to the technique being employed.

In the aeropolygon procedure, the projector containing the last exposure oriented is undisturbed while the next exposure is introduced; thus the base length and angular orientation of the model in the strip is preserved. The term "aeropolygon" comes from the analogy between these relationships and the legs of a ground-survey traverse.

9.3.2.3.1 ORIENTATION OF THE STARTING MODEL

The procedure frequently requires absolute orientation of the starting model. The accuracy with which the starting model must be scaled and leveled depends primarily upon the number of models in the extension and the method to be used for the subsequent adjustment (graphical or computational). If the extension is to be adjusted graphically, the starting model should be carefully scaled to the plotted positions of the horizontal ground control. This will minimize the magnitude of horizontal deformation propagated into the stereotriangulated strip. It will also facilitate the graphical adjustment, since the photogrammetric control points must be repositioned as a result of the adjustment. If the horizontal adjustment is to be accomplished by computational methods, absolute orientation of the starting model is not necessary, however, the scale should be known for leveling the model.

9.3.2.3.1.1 Scaling to existing control. Although the model can be brought to the desired scale by trial and error, it is usually more desirable to compute the scale correction when working with first-order stereoplotting instruments. The scale correction can be readily determined after relative orientation by the following procedure. Consider the plotted positions of control points 1 and 3 (*see* figure 9-6). First, the difference in distance x and y between these points at the desired plotting scale and the difference in distance x_i and y_i between the same two points at instrument scale must be determined from which the distances S and S_i can be computed using

$$S = (\Delta x^2 + \Delta y^2)^{1/2}$$
$$S_i = (\Delta x_i^2 + \Delta y_i^2)^{1/2}.$$

From these two distances, a ratio S/S_i can be applied to the instrument base distance (bx_i) to arrive at a corrected base distance (bx) using

$$bx = bx_i \, S/S_i.$$

9.3.2.3.1.2 Method of leveling. The starting model should be leveled, however, it is frequently advantageous to incline the model in the direction of the flight. This will prevent exceeding the physical limits of Δbz in the instrument and reduce the number of times that a translation of bz of the plotting cameras will be required. A common procedure for leveling a model is as follows. Consider the three control stations in figure 9-6, which are assumed to have known elevations and whose horizontal positions are correctly scaled. Axes are drawn through points 1 and 2 respectively, parallel to the instrument x and y axes. Distances S_{1-5} and S_{2-4} are measured, and the errors in elevations ($\Delta h_4, \Delta h_5$) at points 4 and 5 are linearly interpolated between the errors at points 1 and 3, and 2 and 3, respectively. The angular corrections η and ξ can be obtained from the following equations

$$\Delta \eta = \frac{\Delta h_5 - \Delta h_1}{S_{1-5}} \quad \Delta \xi = \frac{\Delta h_4 - \Delta h_2}{S_{2-4}} \quad (9.3)$$

9.3.2.3.1.3 Recording of positions and elevations. Assuming that the starting model has been satisfactorily oriented, it is then necessary to measure the coordinates of all pass points as well as the three carry-over points. The latter are located along the leading edge of the model, with one near the principal point and the remaining two near the lateral edges of the model. If the horizontal and vertical adjustments of the extension are to be computed, it is necessary only to record the three coordinates of each point measured. When graphical adjustment procedures are to be used, the positions of the pass points are plotted with the coordinatograph attached to the stereoplotting instrument. The orientation elements are usually recorded as an aid for resetting a model, but only the bz values have been found to have any practical value when extending with conventional photography. These recordings also aid in determining the source of blunders which occasionally occur.

9.3.2.3.2 ORIENTATION OF SUCCESSIVE MODELS

An important characteristic of all first-order stereoplotting instruments is the capability of

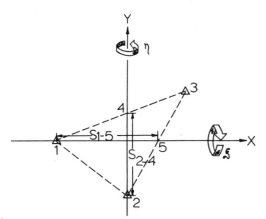

FIGURE 9-6. Starting model orientation in the aeropolygon method.

orienting successive models without disturbing the most recently oriented photographic plate. This is accomplished by means of the principle known as the Zeiss parallelogram which permits the displacement of the plotting cameras in excess of their normal base distance (*see* chapter XII). Thus after the first model is oriented with base-in (first photographic plate on the left side and the second on the right), the setting can be changed to base-out. The first plate in "A" camera is then replaced with the third plate and the optical paths for viewing are interchanged so that the left plate is viewed with the right eye and the right plate viewed with the left eye, thus avoiding a pseudoscopic impression.

The orientation of each successive model will be accomplished using the translational (by, bz) and the rotational motions (κ, ϕ, ω) of the plotting camera which contains the next photographic plate to be joined to the extension. Upon completion of the relative orientation, the newly formed model will be scaled to the previous model by altering the model base. The distance of the model surface from the vertical datum of the instrument varies directly with the change in the x component of the base; therefore, the scale is transferred from the preceding model primarily by changing the base with the bx motion until the elevation of the surface common to both models is the same (*see* figure 9-7). If $by_1 = by_2$ and $bz_1 = bz_2$, then proportionate changes must be made to these base components; that is

$$by_2' = by_2 - \frac{\Delta bx}{bx}(by_1 - by_2)$$

$$bz_2' = bz_2 - \frac{\Delta bx}{bx}(bz_1 - bz_2)$$

where by_2' and bz_2' represent the corrected values of by_2 and bz_2. This transfer of scale is accomplished by making the center carry-over point read the same in both models. The ensuing differences which occur in reading the lateral carry-over points should be recorded. They are normally less than 0.05 mm. Excessive differ-

ences result from a poorly calibrated stereoplotting instrument, distorted photography, or inaccuracies in model orientation. These differences are sometimes taken into account in the vertical adjustment of the photogrammetric extension.

9.3.2.4 AEROLEVELING

In aeroleveling, barometric height measurements of the camera air stations which have been recorded during the photographic mission are used to preset the bz values during the orientation of the successive models on the stereoplotting instrument. Only the differences in flight altitude H are required. These fluctuations are measured with a statoscope. Although the term "aeroleveling" usually refers to the use of statoscopic data in stereotriangulation, it is equally appropriate to consider also in this category, the leveling of models by the presetting of angular elements of orientation. Such orientation data are obtained from the solar periscope, horizon photography, or vertical gyros.

9.3.2.4.1 OPERATIONAL PROCEDURES

The starting model is absolutely oriented as described for the aeropolygon method. The auxiliary data is used in the orientation of the subsequent models. In using statoscopic data, the Δbz is preset for the plotting camera containing the third photographic plate. The model orientation is then effected by using the ϕ motion for both plotting cameras and the by and ω of the third photographic plate. The transfer of scale and the indexing of coordinates on the carry-over points are also accomplished as previously described. When the angular elements of orientation are known, these will be preset instead of the Δbz. Because of the intrinsic inaccuracies of the auxiliary data, coupled with those of the stereoplotting instruments and photography, the want of correspondence (y parallax) cannot be removed completely with by, bz (and κ). The residual y parallax is then removed by orienting the previously oriented plate to the newly oriented plate. This is accomplished using all five elements of orientation, including ϕ and ω. Obviously, the reason for removing the residual y parallax with the previously oriented plate rather than the newly oriented plate is to prevent the transfer of error to all succeeding models in the extension.

9.3.2.4.2 ADVANTAGES AND APPLICATION OF AEROLEVELING

Stereotriangulation by the aeropolygon method has the disadvantage of propagating both systematic and accidental error throughout the photogrammetric extension; conversely, the aeroleveling method eliminates these disadvantages. It is particularly suitable to long strips in which earth curvature would otherwise necessitate the frequent vertical reindexing of the extension and would occasionally result in an ex-

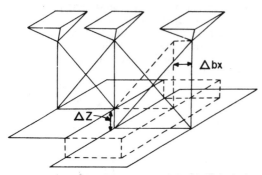

FIGURE 9-7. Effects in a stereomodel with changes in bx (by and bz are assumed to be zero).

ceeding of the bz limitations of the stereoplotting instrument due to excessive drop-off in the longitudinal direction.

9.3.3 STEREOTRIANGULATION WITH SECOND-ORDER INSTRUMENTS

These instruments must be divided into two types when considering their application to stereotriangulation. The first type includes those in which a multiple bank of plotting cameras (commonly referred to as projectors) can be relatively oriented, such as the Multiplex and Balplex. In the second class are those consisting of only two plotting cameras and requiring the transfer of orientation from one plotting camera to the other. Typical of this class of instruments are the Kelsh plotter, Wild Autograph A-8, and the Galileo Stereosimplex III.

9.3.3.1 PROCEDURES WITH MULTIPLEX AND BALPLEX

9.3.3.1.1 INSTRUMENTATION

As a preface to the discussion of stereotriangulation procedures, it may be helpful to visualize a bank of projectors for a strip of relatively oriented models of a stereotriangulated strip. These projectors will assume the relative attitudes and positions of the aerial camera at each exposure station. The series of connected stereoscopic models thus formed represents a segment of the earth's surface. The horizontal reference is a map projection on which control points are plotted, and the vertical reference is the plane of the plotting table, which is obviously not parallel to the geoid or geodetic vertical reference. To minimize the effect of the earth's curvature in the stereotriangulated strip, it is desirable, whenever possible, to scale and level the model in the center of the strip and then successively orient the adjacent models in each direction. For absolute orientation, the positions for the horizontal ground control will normally be plotted on the compilation medium. The ideal distribution of control consists of two points at each end of the stereophotogrammetric bridge and two at or near the center. Care must be exercised in the orientation of the starting model to assure that the direction of the strip will approximately parallel the bar from which the projectors are suspended, so as not to exceed the physical limitations (particularly by) of the instrument. The empirical rather than the computational method of orientation is used with these instruments. The scale is transferred, as previously described, by imposing the same elevation on the center carry-over point in the newly formed model as was read in the previous model. The lateral carry-over points are also read as a check on the relative orientation and for use in the vertical adjustment.

9.3.3.1.2 STRIP ADJUSTMENT

The adjustment of the stereotriangulation data will be discussed in detail in section 9.3.6; however, the horizontal adjustment or scaling of the strip is more easily accomplished with this type instrument by systematically changing the distances between the projectors. If the first model in the strip is absolutely oriented, it will probably be noted that the photogrammetric positions will not agree with the plotted positions for the horizontal control points in the terminal model. The variation in direction (azimuth) can be corrected by pivoting the plotting medium about the centroid of the control in the starting model. The error in length of the strip can be corrected by measuring the longitudinal error in the terminal model and computing the datum elevation adjustment (ΔZ) from the relationship $\Delta Z = (H \times E)/L$, where E is the error in the length of the strip as measured in the terminal model, H is the average projection distance, and L is the length of the strip between the first and last models. It must be realized that this method of adjusting scale will not correct for the nonlinear errors of horizontal bow and differential scale which accumulate in the stereotriangulation process. If greater accuracy is required, these errors must be corrected graphically as described in section 9.3.6.1. In either case the vertical errors must be compensated by a graphical adjustment.

9.3.3.1.3 APPLICATION AND LIMITATION

Although the attainable accuracy with these instruments is considerably less than with the first-order stereoplotting instruments, the economy and simplicity of their operation justify their use whenever the accuracy requirements permit.

The number of models which can be stereotriangulated in a continuous strip is restricted by the physical characteristics of the instruments. On a Multiplex long bar (double bar), the number is approximately seventeen models (depending on the ratio of photograph scale to plotting scale and the percentage of longitudinal overlap between consecutive exposures). The Bausch and Lomb triangulation frame will accommodate eight of the Balplex 525 projectors, six of the Balplex 760 projectors, or eleven Multiplex projectors.

9.3.3.2 INSTRUMENTS WITH TWO PLOTTING CAMERAS

Most second-order stereoplotting instruments consist of two plotting cameras instead of banks of projectors, but differ from most first-order stereoplotting instruments in that no facility is provided for changing from base-in to base-out. Stereotriangulation can, however, be performed with these instruments. The procedure, in general, consists of relatively and absolutely orienting the first model; then the elements of orientation of the right-hand camera, which holds the photograph plate of the second exposure, are transferred to the left-hand plotting camera. The values of ϕ and ω on the right-hand camera are measured with adjustable spirit levels which are designed for mounting and alignment on the plotting cameras. The error in setting the spirit levels is less than the error in ϕ and ω resulting from relative orientation; therefore, this procedure is sufficiently accurate. The element of κ is transferred by dial readings, when such scales exist; otherwise κ is set approximately to zero on the left plotting camera and the plotting medium is swung into coincidence after relative orientation. On the Stereosimplex III of Galileo a special instrument has been designed for the purpose of transferring κ. An alternate procedure for the transfer of the rotational elements is to transfer ϕ and ω as previously described and then to introduce common $\omega(\xi)$ after relative orientation by leveling to the lateral carry-over points. The scale is transferred by altering the base until the distances between the lateral carry-over points coincide with those determined from the preceding model. The scale can also be transferred by height measurements as, for example, is recommended for use with the Wild A-8.

9.3.4 SOURCES OF ERROR

9.3.4.1 FILM

Aerial photographs for stereotriangulation are obtained by film cameras or plate cameras producing photographic images either on film or on glass plates. The dimensional stability of film is generally less than for glass plates. The plate camera, however, has the disadvantage of being heavy, as well as delicate in operation. Also, the number of exposures is much more restricted as compared with a film camera. This is the main reason why film cameras are presently used much more extensively than plate cameras. For this reason, only aerial photography on film is considered in this chapter.

The film, as carrier of the emulsion, is subject to dimensional changes (film distortion) which are a function of the film material, the time (aging of the film), temperature and humidity, and the treatment of the film during processing.

In evaluating film shrinkage in connection with instrumental aerotriangulation, a distinction is made between *systematic film distortion* and *irregular film distortion*.

The *systematic film distortion* is usually defined as displacement of image points toward or away from the center of the photograph (frame). This displacement, $dr_s = \overline{p'p}$, increases or decreases linearly with the radial distance r from the principal point c (*see* figure 9-8). The *systematic film distortion* produces a scale error, dr_s/r = constant for the photograph. Experience has shown that this scale error can reach the amount of 0.2 percent or, in exceptional cases, even

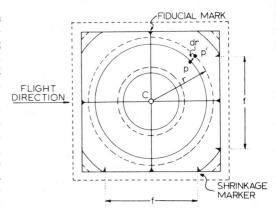

FIGURE 9-8. Systematic film shrinkage.

more. In phototriangulation, the systematic film distortion is eliminated by correct scaling of the initial model or by an appropriate change of the focal length.

The *irregular film distortion* displaces a picture point p' in a more or less arbitrary direction. The amount and the direction of this displacement ds_i is dependent on, among other things, the location of the picture point p' in the photograph. Irregular, or local, film distortion is conditioned by varying elastic properties of the film and can result in picture-point displacements ds_i in the order of 0.01 to 0.02 mm. Due to its complex nature in general, no effort is made in practice to correct for the irregular film distortion in instrumental phototriangulation. These errors follow certain laws of elasticity and do not usually follow a normal distribution. Among a great number of error sources which affect the accuracy of phototriangulation, the film distortion is a significant one.

Another error source directly related to aerial photography on film is a *lack of sharpness* of the photographic image. Such image unsharpness occurs as a result of the limited resolving power of the emulsion (photographic resolution). The photographic resolution depends on the size of the grains of the emulsion and is indicated in lines per millimetre. At present there are emulsions available for aerial photography with a photographic resolution of about 40 to 115 lines per millimetre, or even more in special cases, depending on subject contrast.

In producing aerial photographs on film, it is necessary that the film be absolutely flat during the exposure. Deviations should not exceed ±0.01 to ±0.02 mm. This condition is particularly important when phototriangulation is performed by means of projection-type instruments. Lack of sufficient *flatness of the film* during the exposure is another error source affecting the accuracy of phototriangulation.

9.3.4.2 CAMERA

Photographs obtained by an aerial camera must fulfill the condition of being the result of a central projection. Because of mechanical and optical errors of the camera, this condition is not absolutely fulfilled. In other words, the interior orientation of the camera (relation between photograph and perspective center of the camera) is not quite correct. Since the projectors of the stereoplotting instrument normally do not have the same errors of interior orientation as the aerial camera, the phototriangulation accomplished with the stereoplotting instrument will be affected accordingly.

Basically, for each photograph of a stereoscopic pair, there are six errors of interior orientation, namely: two translation errors (dx and dy) of the photograph with reference to the perspective center O of the camera, a focal length error df, two angular errors $d\omega$ and $d\phi$ resulting from the fact that the image plane is not quite perpendicular to the camera axis, and a small rotation error $d\kappa$ of the photograph around the camera axis. In the case of vertical photographs of relatively flat terrain taken with a camera calibrated to meet official calibration specifications, errors of interior orientation are largely compensated by the procedures of relative orientation, scaling, and absolute orientation of the model in question. In this case, the errors of interior orientation can be treated as errors of exterior orientation. In the cases of convergent photographs and of vertical photographs of mountainous terrain, this compensation is only partially possible, so that the errors of interior orientation are more critical.

Errors of central projection and subsequent image-point displacements are also caused by objective distortion (lens distortion, *see* figure 9-9). Usually a distinction is made between the radial distortion dr and the tangential distortion dt. In general, dt is somewhat smaller than dr. In both cases, a further distinction is made between systematic and irregular distortion. The *systematic radial distortion* dr_s is usually represented by the so-called distortion curve obtained by special calibration procedures; dr_s is constant on a circle with the radius r. The irregular *radial distortion* dr_i can be interpreted as the differential image-point displacement which remains after the systematic radial distortion dr_s is removed. The irregular radial distortion dr_i can be found by determining the true distortion curves in specific directions, such as in the collimating lines and in the image diagonals. The average of these curves yields the standard distortion curve, and the deviations of the individual true distortion curves with respect to the standard distortion curve represent the irregular radial distortion dr_i which is usually expressed as standard error or probable error of the standard distortion curve. The various aerial cameras used for phototriangulation have a maximum systematic radial distortion which varies from about 5 micrometres to about 100 micrometres, while the standard error of the standard distortion curves varies from about 3 micrometres to 20 micrometres. Excessive systematic radial distortion is usually eliminated by computational procedures or by compensation devices, such as compensation plates (special plateholders) in the stereoplotting instrument used for stereotriangulation. Due to the complex nature of the asymmetry of radial distortion, normally no effort is made in practice to correct for this error.

Another error source affecting stereotriangulation is *image unsharpness* caused by the camera optics. The degree of image unsharpness depends on the resolution of the camera optics (*optical resolution*). The optical resolution is indicated in lines per millimetre which are resolved at a certain place on the image plane and is a function of the radial distance r from the principal point c (*see* figure 9-10).

For the best available aerial cameras the maximum optical resolution attains 100 lines per millimetre, or somewhat more, in the center of the image plane, and drops to about 30 lines per millimetre in the corners of the image plane. For other aerial cameras the maximum optical resolution does not exceed 25 to 30 lines per millimetre and decreases 5 to 10 lines per millimetre

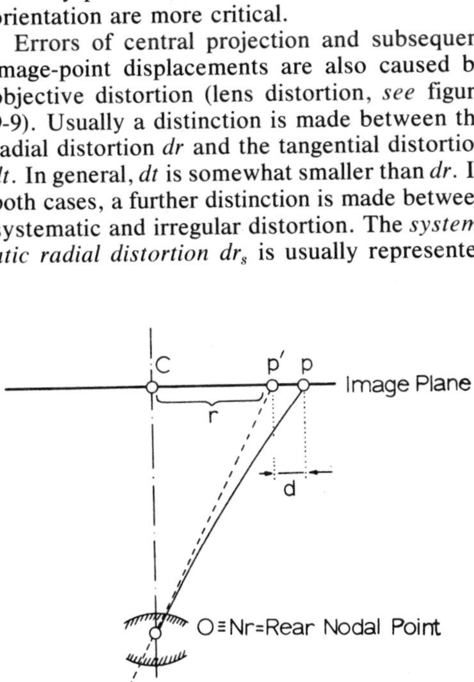

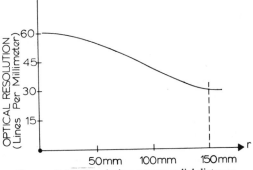

FIGURE 9-9. Image displacement because of lens distortion.

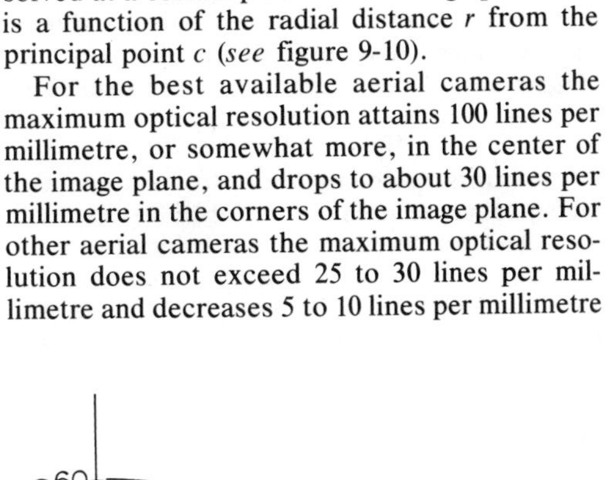

FIGURE 9-10. Resolution versus radial distance.

in the corners of the image plane. The inconsistency of the optical resolution should be taken into account in performing the relative orientation, since, due to this fact, the y parallaxes in the six or more points which are used in performing the relative orientation should not have equal weights. The various procedures of numerical relative orientation which have been published to date assume equal weights for these y parallaxes and are therefore not quite correct.

It is obvious that a high optical resolution contributes to reduction of the errors occurring in phototriangulation. Hence, for precise phototriangulation, aerial cameras with a high optical resolution should be used.

9.3.4.3 STEREOPLOTTING INSTRUMENT

Stereotriangulation has usually been accomplished by means of first- and second-order stereoplotting instruments; at present, analytical stereoplotters are also used. Although it is required that the instruments used for stereotriangulation be well adjusted, small mechanical and optical errors still exist and affect the accuracy of phototriangulation. Such instrumental errors generate *systematic* and *irregular errors* of relative orientation, of scale transfer between adjacent models, and of absolute orientation. Such errors always exist because no instrument is perfect. The magnitude of these errors can only be reduced to the level of the observation errors. Although it is relatively easy to compensate for the systematic errors of stereotriangulation caused by systematic instrument errors, the elimination of irregular errors of stereotriangulation caused by irregular instrument errors is virtually impossible. Such irregular instrument errors are caused by small mechanical and optical deviations and follow a distribution which is, in general, different from a normal distribution.

9.3.4.4 INSTRUMENT OPERATOR

The errors caused by the instrument operator are so-called *personal errors* or *observation errors*. Such errors occur in establishing the interior orientation of the aerial photographs in the plotting cameras of the instrument used in performing stereotriangulation; in establishing the relative orientation, scaling, and absolute orientation of the first model of a strip triangulation; in the co-orientation of the following photographs and the scale transfer between adjacent models; and in measuring machine coordinates x, y and elevations z of points to be recorded or determined by stereotriangulation. Such observation errors include so-called "blunders" caused by lack of care of the operator or by misinterpretation. It is mandatory that "blunders" be kept to a minimum and be detected and eliminated during the instrumental work by independent control measurements.

The remaining observation errors include small interpretation and identification errors by the operator and those caused by the limitations of his stereoscopic perception. Because no point is portrayed on the photographs or in the model with absolute sharpness, the operator sees the images in the instrument as spots and lines with a more or less undefined center, or centerline, instead of as sharp points and lines. He must make arbitrary decisions on where to place the measuring index (measuring mark or "floating mark"). Such *interpretation errors* affect the accuracy of stereotriangulation. They can be reduced to a certain extent by using high-quality aerial photographs with a maximum of resolution.

Even if ground features could be portrayed in the model with absolute sharpness, the operator would commit small *pointing errors*. These errors reflect the accuracy with which the "floating mark" can be located on a sharp model point. Using first-order stereoplotting instruments and high quality aerial photographs, an experienced operator is capable of a standard horizontal and vertical pointing error not greater than about $h/30,000$ (h = flight height above average ground). These pointing or observation errors follow a more or less random distribution (normal distribution). They also show a systematic trend with progressive working time at the plotting instrument due to the eye fatigue of the operator and its effect upon his stereoscopic perception.

9.3.4.5 EARTH CURVATURE

Geodetic positions of points on the earth's surface are based on reference figures such as the geoid, an ellipsoid of revolution, or on a sphere. An aerial photograph results from a central projection of a portion of the earth's surface onto the image plane and a curved surface cannot be projected onto a plane by central perspective transformation without distortion. Hence, an aerial photograph exhibits distortion because of the earth's curvature. The situation is depicted in figure 9-11. In that figure a vertical photograph and a spherical reference surface for the earth are assumed. Given a distance (S) of a ground point (P) from the nadir point (N) of a photograph, the correct position of the image should appear at $\bar{p}$. However, because of the earth's curvature, an image will be displaced an amount ds and appear at p. This displacement is always negative and increase with the third power of S.

In stereotriangulation, the effects of the earth's curvature are not usually detectable in an individual model because the reference plane is usually adjusted relative to known elevations. The situation differs, however, for an entire strip which has been stereotriangulated using conventional (aeropolygon) methods. This is shown in figure 9-12. The stereotriangulation of the strip in the stereoplotting instrument produces heights relative to a plane tangent at the center

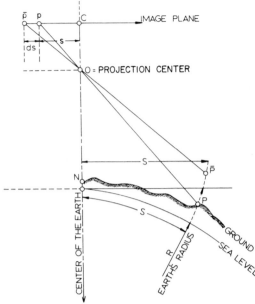

FIGURE 9-11. Displacement because of earth's curvature.

(M) of the first model. That is, the stereo-triangulation results in an elevation z' in place of z. From the figure, it can be seen that as the distance (x) from the point M increases, the measured elevations exhibit a negative trend. This trend (Δz) can be approximated by

$$\Delta z \approx R - (R^2 - x^2)^{1/2} = R(1 - (1 + x^2/R^2)^{1/2})$$

where R is the earth's radius. Thus

$$\Delta z \approx -\frac{x^2}{2R}$$

As a result, the recorded x coordinates correspond to the projection of the distance (S) onto the tangent plane and are, therefore, the same as the geodetic distances. From figure 9-13, it is apparent that with increasing distance from M these values tend to become smaller in comparison with S and the differences can be computed using

$$\Delta x = x - S = R \sin\beta - R\beta$$

$$= R(\beta - \frac{\beta^3}{6} + \ldots - \beta)$$

and, neglecting higher-order terms

$$\Delta x \approx -\frac{R\beta^3}{6}.$$

Also, since

$$\beta = S/R,$$

then

$$\Delta x \approx -\frac{S^3}{6R^2}$$

and, since

$$S \approx x,$$

then

$$\Delta x \approx -\frac{x^3}{6R^2}.$$

For $x = 100$ kilometres, Δx is about 5 metres. From this, even for relatively long strips, Δx can be ignored.

The earth's curvature also requires that the machine coordinates (x and y) determined in the phototriangulation be reduced to sea level datum. That is, the x and y differences between points (transfer points) are either too large or too small when compared with the ground system. In stereotriangulation of a strip the first model is not affected because its scale is determined by coordinates relative to sea level. In all other models of the strip, however, the distance (s) between two points A and B exhibits a small

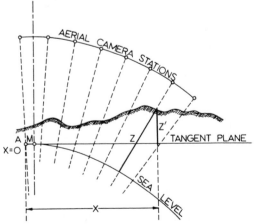

FIGURE 9-12. The Earth's curvature effects on tri-angulated elevations.

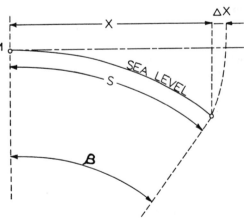

FIGURE 9-13. The Earth's curvature effects on tri-angulated x coordinates.

error ds with respect to the corresponding sea level distance. The magnitude of ds is dependant on the elevations of the points A and B and can be determined using

$$ds = \left(\frac{z_A + z_B}{2} - \bar{z}_i\right)\frac{s}{R}.$$

Here, $\bar{z}_i$ is the mean elevation of all the control points in the first model. This error is relatively small for the y coordinates of a strip triangulation and can be neglected in most cases (see figure 9-14). The situation is different for the x coordinates, where an accumulation of the errors ds takes place (see figure 9-15A). The errors ds for the distances $s_1, s_2, \ldots s_n$ between successive nadir points or successive transfer points $1, 2, 3 \ldots n$ in the direction of the strip are

$$ds_1 = \left(\frac{z_1 + z_2}{2} - \bar{z}_i\right)\frac{s_1}{R} \approx 0,$$

$$ds_2 = \left(\frac{z_2 + z_3}{2} - \bar{z}_i\right)\frac{s_2}{R},$$

and

$$ds_n = \left(\frac{z_n + z_{n+1}}{2} - \bar{z}_i\right)\frac{s_n}{R}.$$

For a transfer point in model N the following closure error ΔS can be computed from

$$\Delta S = ds_1 + ds_2 + \ldots ds_n,$$

which reduces to

$$\Delta S = F/R$$

where R is the earth's radius and F is the algebraic sum of the vertical-plane areas bounded by straight lines connecting points $1, 2, 3, \ldots n$, the normals at points $1, 2, 3, \ldots n$, and the line of equal elevation $\bar{z}_i$ (i.e., F is the shaded area in figure 9-15B). For mountainous areas and long strips, the closure error can be significant. For instance, if F is equal to 100 km², the closure error amounts to 16 m. The functional curve of Δx is purely empirical, depending on ground form, and in each case a decision must be made as to the extent of the correction for Δx.

9.3.4.6 Atmospheric Refraction

The atmosphere causes refraction of light which in turn results in a radial displacement (dr) of image points in a photograph. The geometry is

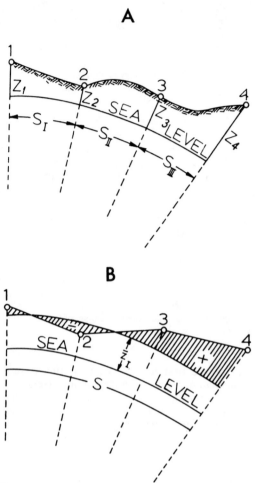

FIGURE 9-15. Reduction of instrument coordinate (x) to sea level.

depicted in figure 9-16. The radial displacement (dr) of an image point p due to atmospheric refraction is discussed in section 9.4.2.4.1. The displacement increases with the third power of r and is similar to that caused by the earth's curvature. Its effect, however, is of opposite sign. In phototriangulation, the effect of refraction is minimized by correct scaling of the initial model or by a corresponding change in the focal length of the projectors.

9.3.4.7 Map Projection

Basically, the measurements in stereotriangulation for the determination of new points are made on the ellipsoid representing the earth. These points then have to be converted to X, Y coordinates of the coordinate system used for the area in question. The ground control points necessary for the adjustment of a stereotriangulation are normally also given in terms of X, Y coordinates. Because these coordinates are usually produced by a conformal map projection, they are affected by the distortions of the

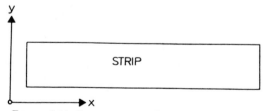

FIGURE 9-14. Instrument coordinate system for strip triangulation.

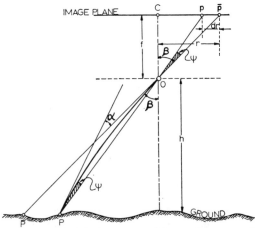

FIGURE 9-16. Displacement of image because of atmospheric refraction.

map projection. This means that distances determined from map coordinates X, Y are not quite identical with the corresponding ground distances. For most conformal map projections, the distortions in X and Y increase with the third power of the distance from the origin of the map coordinate system. This means that correct coordinates X and Y for the points determined by stereotriangulation are obtained only when the formula for the x adjustment of a stereotriangulation includes a term with x^3.

9.3.5 TYPES OF ERRORS

When a phototriangulation (strip phototriangulation; aeropolygon, aeroleveling, or $bz = 0$ method) is performed by means of a stereoplotting instrument, the following operations are affected by errors.

a. Relative orientation, scaling, and absolute orientation of the first model (model I formed by photographs 1 and 2).

b. Co-orientation of the following photographs 3, 4, 5, $\cdots$ forming models II, III, $\cdots$, N.

c. Scale transfer from model I to model II, from model II to model III, etc., by means of the elevations of transfer points.

d. Model connection (connecting model II with model I, model III with model II, etc.)

e. Recording of the machine coordinates x, y and the elevations z of the given ground control points and the transfer points which are the new points to be determined by stereotriangulation.

Phototriangulation is usually performed with vertical photographs having 60 percent overlap. The situation is shown in figure 9-17.

Generally, the machine coordinate system x,y is chosen in such a way that no negative coordinates occur and that the x axis is, as closely as possible, parallel to the strip axis. For the study of the propagation of errors, however, it is preferable to choose the coordinate system x,y with the origin in point 1 and with the x axis coinciding with the strip axis. In general, three transfer points in the triple overlaps are selected. When first-order stereoplotting instruments are used, model I is usually oriented with "base in," model II with "base out," model III with "base in," etc. If the terrain is not too mountainous, the air base b is practically constant.

In studying the error propagation in a stereotriangulation, it is sufficient to consider the effect of the various errors upon the first model I and upon the elements bx, by, bz, ω, ϕ, and κ of each following model (see figure 9-18).

These elements are determined for each model when a phototriangulation is performed on a first-order stereoplotting instrument. In this section, first-order instruments and vertical

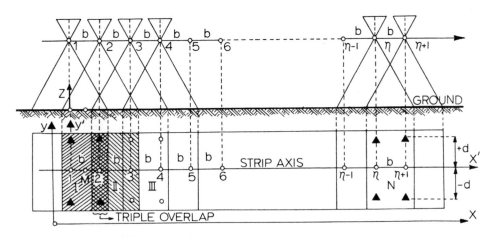

▲ -- Given Ground Control Points
o -- Transfer Points

FIGURE 9-17. Graphic depiction of instrument aerotriangulation.

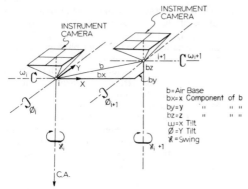

FIGURE 9-18. Exterior orientation elements of a stereo-pair of photographs.

photographs are assumed. Stereotriangulation with convergent photographs and stereotriangulation using second-order plotters show an error propagation which is similar to that for the stereotriangulation of vertical photographs with first-order instruments and therefore are not covered in detail in this section.

9.3.5.1 SYSTEMATIC ERRORS

It is normal to define systematic errors of a stereotriangulation as errors of a constant amount and sign occurring in each model or photograph (systematic errors of the first order). Such systematic errors are the result of image errors, instrument errors, and observation errors. Systematic errors generate a systematic falsification of the stereotriangulation. Such a systematic falsification of the strip's surface is caused by seven initial errors of the first model of the stereotriangulation, by five systematic co-orientation errors in each following model, by a systematic scale transfer error, and by three systematic connection errors between adjacent models. The initial seven errors of the first model are:

$$dS_0 = \text{scale error,}$$
$$dx_0 = x \text{ translation error,}$$
$$dy_0 = y \text{ translation error,}$$
$$dz_0 = \text{datum error,}$$
$$d\Omega_0 = x \text{ tilt error,}$$
$$d\Phi_0 = y \text{ tilt error, and}$$
$$dK_0 = \text{rotation error.}$$

In each succeeding model the following systematic co-orientation errors are to be considered:

$$dby_s = \text{systematic } by \text{ error,}$$
$$dbz_s = \text{systematic } bz \text{ error,}$$
$$d\omega_s = \text{systematic } \omega \text{ error } (x \text{ tilt error}),$$
$$d\phi_s = \text{systematic } \phi \text{ error } (y \text{ tilt error}), \text{ and}$$
$$d\kappa_s = \text{systematic } \kappa \text{ error (swing error).}$$

The systematic scale transfer error between adjacent models is basically a systematic error dbx_s of the x component bx of the air base b. The systematic model connection errors are:

$$dx_s = \text{systematic model connection error in } x,$$
$$dy_s = \text{systematic model connection error in } y, \text{ and}$$
$$dz_s = \text{systematic model connection error in } z.$$

9.3.5.2 ACCIDENTAL (IRREGULAR) ERRORS

If, from the various errors occurring in a stereotriangulation, the systematic errors as defined in section 9.3.5.1 are subtracted, so-called irregular errors remain. They have the property that their magnitude and sign vary from photograph to photograph or from model to model. Such errors are caused by irregular image errors, irregular instrument errors, and observation errors. Irregular errors can follow all kinds of distribution. The mathematical treatment of such errors requires that some mathematical assumptions must be made about the nature of such errors. Usually, the assumption is made that such irregular errors follow a normal distribution (*see* chapter II). This means that small errors are more likely than large ones, and that positive and negative errors are equally probable. Theoretically, even an error of magnitude $\pm\infty$ could occur. The mathematical treatment of such errors on the basis of the mathematical probability yields an error frequency diagram which is a bell-shaped curve and which is expressed by an exponential function. Such errors are abstract in nature and are usually called *accidental errors*. It is obvious that in a stereotriangulation the approximation of irregular errors which are concrete (physical) errors by accidental (mathematically designed) errors is a reasonable one. In general, it can be stated that the quality of this approximation depends on the remaining systematic errors of higher order. If most systematic errors of higher order have been previously eliminated, the remaining errors will more nearly approach a normal distribution. There are, however, limitations with regard to the application of such a procedure, as pointed out in section 9.3.2.1.

The size of accidental errors is usually indicated by the so-called standard error m which is given by the following formulas:

$$m = \pm \sqrt{\frac{\Sigma \epsilon^2}{n}}$$

or

$$m = \pm \sqrt{\frac{\Sigma v^2}{n-1}}$$

where:

$\epsilon_1, \epsilon_2, \epsilon_3, \ldots, \epsilon_n$ = true accidental errors
$v_1, v_2, v_3, \ldots, v_n$ = accidental errors referred to the arithmetic means of n observations (the true errors ϵ being unknown).

It is obvious that, due to the deficiency in approximating irregular errors by accidental errors, the standard error m has to be interpreted cautiously when used as an indication of accuracy.

Irregular errors generate a more or less irregular falsification of the stereotriangulated strip. Such a falsification of the strip's surface is caused by five irregular accidental co-orientation errors in each model, by an irregular (accidental) scale transfer error, and by three irregular (accidental) connection errors between adjacent models. The five co-orientation errors are:

$$dby_i = \text{accidental } by \text{ error,}$$
$$dbz_i = \text{accidental } bz \text{ error,}$$
$$d\omega_i = \text{accidental } \omega \text{ error } (x \text{ tilt)},$$
$$d\phi_i = \text{accidental } \phi \text{ error } (y \text{ tilt), and}$$
$$d\kappa_i = \text{accidental } \kappa \text{ error (swing)}.$$

The accidental scale transfer error between adjacent models is, basically, an accidental dbx_i error of the x component bx of the air base b.

The three accidental model connection errors are:

$$dx_i = \text{accidental model connection error in } x,$$
$$dy_i = \text{accidental model connection error in } y, \text{ and}$$
$$dz_i = \text{accidental model connection error in } z.$$

9.3.6 METHODS OF COMPENSATION FOR ERRORS

The compensation for errors in an aerotriangulation of a strip is done by the so-called strip adjustment. Such a strip adjustment is either performed graphically or analytically (numerically), and its primary goal is to eliminate the systematic-error. By the adjustment procedures described in the following sections, a certain portion of the strip errors, caused by accidental errors, also is eliminated. Graphical adjustment is normally used when no high-speed computer is available. The graphical adjustment has the advantage that the existing blunders are readily visible and detectable. The analytical adjustment has the advantage of being faster, if the computer program is available. However, if blunders in this procedure occur, it is more difficult and time-consuming to eliminate them than with the graphical method. Graphical and analytical strip adjustment require a sufficient number of given ground control points, as is shown in the following sections.

9.3.6.1 GRAPHICAL STRIP ADJUSTMENT

A detailed description of this procedure is given by Brandenberger (1951). It can be readily applied for relatively flat ground where terrain elevation differences $\Delta \bar{z}$ with respect to average ground elevation $\bar{z}$ are less than 10 percent of the average flight height h above ground.

The graphical adjustment is usually based on the error propagation in longitudinal sections of a strip where y is a constant. The following formulas are for the aeropolygon method as well as for aeroleveling:

$$\Delta x = A_0 + A_1 x + A_2 x^2 + A_3 x^3$$
$$\Delta y = B_0 + B_1 x + B_2 x^2 + B_3 x^3$$
$$\Delta z = C_0 + C_1 x + C_2 x^2 + C_3 x^3. \qquad (9.4)$$

For short aeropolygon strips and for most aeroleveling strips, the terms with x^3 can be neglected. These formulas assume relatively flat terrain. The basic idea of the procedure is to approximate the falsified strip surface by three more-or-less equidistant cross-sections and three equidistant longitudinal sections as shown in figure 9-21.

The four cross-sections are necessary to determine the coefficients in formulas (9.4) while the three longitudinal sections permit an approximation of the parabolic strip deformation in the y direction. To construct the four cross-sections, four groups of given ground control points must be available. The instrument coordinates x, y, z of these points are recorded during the triangulation procedure on the stereoplotting instrument. To determine the errors Δx, Δy, and Δz of the given ground control points, the ground coordinates are usually transformed to instrument coordinates using familiar transformation formulas. The coefficients of the transformation formulas are determined by means of the instrument, using the ground coordinates of the given points in the first model and accepting the model scale number as a scale factor. The cross-sections are then determined by constructing the curves of equal errors and by intersecting them with the cross-section planes as shown in figure 9-19.

In this fashion, the errors Δx, Δy, Δz at the points of intersection between cross-sections and longitudinal sections are obtained. By means of these errors, the coefficients in formulas (9.4) can be determined numerically (four equations with four unknowns each), which permits the construction (graphically) of the three longitudinal sections at a convenient scale. Sometimes the longitudinal sections (polyno-

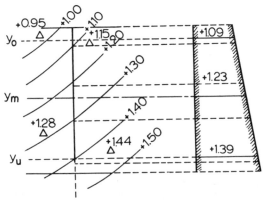

FIGURE 9-19. Cross-section and longitudinal sections of a fictitious strip (second order polynomial).

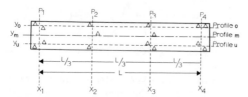

FIGURE 9-20. Construction of a cross-section.

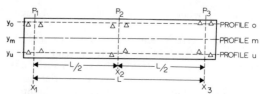

FIGURE 9-22. Cross-sections and longitudinal sections of a strip (parabolic).

mials of third degree) are directly constructed by mechanical means. In this way graphs are obtained for the Δx, Δy, and Δz errors (*see* figure 9-20).

The errors Δx, Δy, Δz of any point P of the strip triangulation can then be obtained from the graph by interpolating a correction from the appropriate curves which is proportional to the lateral distance between the corresponding profiles in figure 9-21. The negative values of Δx, Δy, Δz are the corrections in P and have to be added to the instrument coordinates x, y, z to obtain the adusted instrument coordinates x', y', and the adjusted elevations z'. The adjusted instrument coordinates x', y' are then transformed to the ground coordinate system, yielding the adjusted ground coordinates. For short aeropolygon strips ($N < 20$, $L < 60$ miles) and most aeroleveling strips, the terms with x^3 in formulas (9.4) are omitted. In this case, the following error-propagation formulas for a longitudinal section in which y is a constant, are obtained:

$$\Delta x = A_0 - A_1 x - A_2 x^2$$
$$\Delta y = B_0 - B_1 x - B_2 x^2$$
$$\Delta z = C_0 - C_1 x - C_2 x^2 .$$

This means that, for the determination of the coefficients, only three cross-sections, *i.e.*, three groups of given ground control points, are required (*see* figure 9-22).

For the plotting of parabolic longitudinal sections, simple graphical parabola-construction

methods are available. Reference is made to Brandenberger (1951).

In the case of mountainous terrain, the effect of the y tilt error $d\phi_N$ upon the x coordinates due to the elevation differences Δh and the effect of the scale error dS upon z have to be taken into account. A simple procedure to determine the appropriate corrections is given by Brandenberger (1951). By applying these corrections to the x and z coordinates, instrument coordinates x, y, z are obtained which are corrected for the effects of ground elevation differences Δh. The strip adjustment procedures for flat terrain can then be applied.

The above graphical adjustment procedures require a certain standard distribution of the given ground control points. It is possible, however, to have strip phototriangulations with an irregular distribution of given ground control points. In this case, a strip adjustment using curves of equal corrections might be preferable. Such a procedure is described by Brandenberger (1951).

9.3.6.2 BLOCK ADJUSTMENT PROCEDURES

To provide the necessary control points for stereophotogrammetric mapping of a large area, a so-called block phototriangulation can be performed. Such a block phototriangulation usually consists of three to four principal strips or tie strips (q, in figure 9-23) and a certain number of filler strips L. Very large areas may have to be divided into several blocks.

Such a block phototriangulation is usually performed by first triangulating the principal strips (by aeropolygon or aeroleveling methods) and adjusting them by means of given ground control points as described in the previous sec-

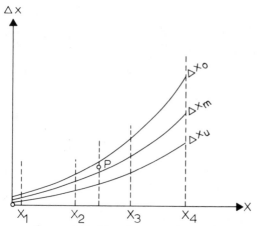

FIGURE 9-21. Plot of x errors along strip.

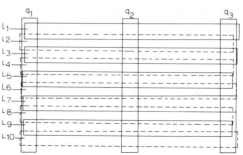

FIGURE 9-23. Graphic of block triangulation.

tions. Then the filler strips are triangulated and adjusted to the adjusted pass points determined in the principal strips. A certain number of pass points in the sidelap between adjacent filler strips are determined in both strips. This double-determination yields two sets of values x, y, and z for each common point and provides an accuracy control on the filler strips. The differences $\Delta x_{r,r+1}$, $\Delta y_{r,r+1}$, $\Delta z_{r,r+1}$ between the corresponding x, y, and z values obtained in filler strips r and $r + 1$ should not exceed certain tolerance values. For these tolerances (T_x, T_y, T_z) the following values can be assumed.

$$T_z = 3\sqrt{2m_x}$$
$$T_y = 3\sqrt{2m_y}$$
$$T_z = 3\sqrt{2m_z}$$

where m_x, m_y, and m_z are the acceptable x, y, and z standard errors of the block phototriangulation. If the differences Δx, Δy, Δz exceed these tolerance values, the filler strips in question should be retriangulated or readjusted. But even if the differences Δx, Δy, Δz are within the tolerance values they might be so large that they cause the map compilation to exceed the allowable accuracy tolerance. It is therefore important that the differences Δx, Δy, Δz are eliminated by an appropriate procedure (block adjustment) which at the same time increases the overall accuracy of the block triangulation. In developing such a procedure, it is well to note that the interior accuracy of a single model of a block triangulation is considerably higher than the position accuracy of the model in the framework of the block. This means that the differences Δx, Δy, Δz are primarily caused by small dx, dy, and dz translation errors, small tilt and orientation errors $(d\theta, d\phi, d\kappa)$, and by small scale errors dS of the models in question. A block adjustment, therefore, should not disturb the shape of the models. This means that the block adjustment has to yield small position and scale corrections $(-dx, -dy, -dz, -d\omega, -d\phi, -d\kappa, -dS)$ for each model which would minimize the differences Δx, Δy, and Δz. This could be done in the most rigorous way by means of a least squares adjustment. The practical performance of such a least squares adjustment, however, represents a tremendous computation task, as in the case of a large block where hundreds of normal equations have to be solved. For this reason, simpler block-adjustment procedures are often used. The simplest procedure, of course, consists in averaging the corresponding coordinates obtained for a sidelap pass point. This might even yield a slight accuracy increase. The procedure, however, has the disadvantage of disturbing the shape of the individual models. Another block adjustment procedure which is more in accordance with the above outlined principles is described by Brandenberger (1951). In this method, x, y, and z translation corrections, as well as a scale correction for each individual model in the block, are determined numerically and graphically, and a discrepancy-free block is obtained.

Other block adjustment procedures use analog computers or templets and are more or less good approximations of the least squares solution.

An essentially different approach to block triangulation and block adjustment was worked out by Nowicki and Born (1960), as well as by Mahoney (1961). Instead of operating with the discrepancies between adjacent adjusted filler strips, Nowicki's and Mahoney's approach consists of tying the unadjusted strips together by means of sidelap pass points. In this way, a distorted block surface is obtained which is then corrected by means of a conformal or nonconformal block adjustment in such a way that the discrepancies in given ground control points are minimized. The system developed by Mahoney allows for insertion of scale and azimuth conditions (measured ground distances and azimuths).

9.3.7 RESTRICTIONS DUE TO ERRORS

The accuracy of phototriangulation is restricted by the limitations of photogrammetric methods. It is interesting to compare this accuracy with that obtained by geodetic methods. Such a comparison can be made by means of so-called directional accuracy, i.e., by the accuracy obtained for the direction pOP where p = image point, O = projection center of camera, P = ground point (see figure 9-24).

For high-quality photographs, an image point p can be located and determined in a well-adjusted first-order stereoplotting instrument

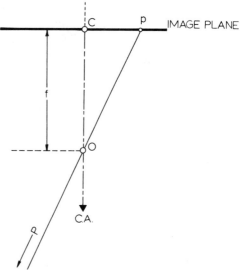

FIGURE 9-24. Directional accuracy for ray $\overline{pOP}$.

with a standard error m_p of about ± 7 micrometres. With a focal length $f = 6$ inches, this yields (at the center of the photograph) the following standard direction error m_d for the direction $p \rightarrow O \rightarrow P$:

$$m_d = \pm \frac{m_p}{f} \approx 30^{cc} \approx 10''.$$

In geodetic triangulation, standard direction errors not exceeding $1''$ of arc are attained. This means that, generally, the accuracy obtained in phototriangulation using stereoinstruments does not approach that obtained with first-order or second-order geodetic triangulation unless the focal length f in the photogrammetric system is considerably enlarged, which would generate operational difficulties due to oversized cameras.

The situation is entirely different, however, when control points for mapping purposes are to be determined. In this case, the control points must only fulfill the graphical map-accuracy requirements, which are considerably lower than the accuracy for geodetic triangulation points. This is explained by the following example: a map of 1:50,000 scale requires a standard horizontal error for a photogrammetric control point on the map to be within 0.1 to 0.2 mm, which represents a tolerable standard horizontal error on the ground of 5 to 10 metres. This accuracy can be easily attained by stereotriangulation using first-order or even second-order stereoplotting instruments, or analytical plotters.

9.3.7.1 DISTRIBUTION OF GROUND CONTROL

In general, it can be stated that the available ground control must be well distributed if stereotriangulation of high accuracy is to be attained. For a strong solution, sufficient control must be available at the corners and along the periphery of the strip as well as of the block to avoid extrapolation (see figure 9-25).

It is also desirable that, in each strip triangulation, a sufficient number of ground control points (four or five) be available in the first model to establish the scale and absolute orientation of the model with as high an accuracy as possible. The maximum errors of such ground control should not exceed 0.1 mm times the map-scale number for planimetry and 1/20,000 times the flight height for elevation. For exam-

ple, when photographs are taken from a flight height of 3000 m for preparation of a 1:24,000-scale map, the horizontal and vertical errors of the control should not exceed 2.4 m and 15 cm, respectively.

In a strip triangulation or in a block triangulation the attainable accuracy is highest when the ground control is evenly distributed so that the distances between adjacent ground control points are nearly equal.

9.3.7.2 DENSITY OF GROUND CONTROL

It is obvious that when more ground control is available, the more accurate the phototriangulation will be if this ground control is used in a sensible way. From a practical viewpoint, the amount of ground control must be limited for economic reasons. The problem that arises is how many ground control points in excess of the minimum should be used for the adjustment of a strip or, eventually, for a block. There are several approaches to the solution of this problem. One consists in performing an analytical adjustment on the basis of the error propagation formulas from a least squares adjustment where the residuals in the superfluous ground control are minimized. This has the advantage that accuracy information in the form of standard residual errors for stereotriangulation is obtained.

From the practical viewpoint, however, such minimized residuals generate local discrepancies in the maps to be compiled and are, therefore, not desirable in case they exceed values of the U.S. National Map Accuracy Standards. This disadvantage can be eliminated by adding more terms of higher order to the error propagation formulas, the number of added terms depending on the number and distribution of superfluous ground control points. Thereby, the residuals in all ground control points are eliminated, and a better approximation of the propagation of accidental (irregular) errors is achieved. This method involves more computation and requires that the available control (determined on the ground and in the stereotriangulation) be highly reliable. Superfluous ground control can also be used to strengthen the graphical adjustment of a stereotriangulated strip. In this case, more cross-sections can be constructed by interpolation (see figure 9-26) permitting a refined construction of the longitudinal sections (usually these longitudinal sections are constructed by mechanical means).

This procedure again yields a better approximation of the propagation of accidental errors and eliminates the residuals in the given ground control.

9.3.7.3 ACCURACY OF PHOTOGRAMMETRIC CONTROL

The accuracy of photogrammetric control determined by phototriangulation depends on a number of factors, such as flight height above ground, number of models per strip or per block,

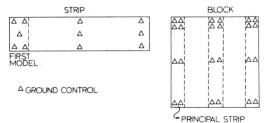

FIGURE 9-25. Distribution of ground control in a strip and a block.

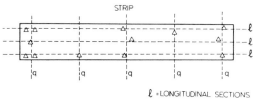

ℓ = LONGITUDINAL SECTIONS
q = CROSS SECTIONS

FIGURE 9-26. Graphic adjustment of a strip with redundant ground control.

and the distribution and density of given ground control. The adjustment of a strip or block eliminates to a greater extent the effect of systematic errors and to a lesser extent the effect of irregular (accidental) errors. The accuracy of a phototriangulation is usually indicated by standard errors m_x, m_y, m_z determined by means of a certain number of test points. These standard errors are obviously affected more by irregular errors which, mathematically, are approximated by accidental errors. It is possible to express the standard errors m_x, m_y, and m_z of an adjusted phototriangulation by means of analytical formulas. If this is done, formulas for m_x, m_y, and m_z are obtained which are square roots of relatively complicated expressions. Such formulas for the standard errors m_x, m_y, and m_z are theoretically interesting. The obtained values, however, very often do not agree with practical results. One of the main reasons for this discrepancy is the fact that the irregular errors in a phototriangulation often do not behave like theoretical accidental errors and show a distribution which is different from a normal distribution.

It is, of course, possible to determine the standard errors m_x, m_y, m_z statistically, by means of a sufficiently great number of tested stereotriangulated strips and assuming a standard distribution for the given ground control. This was done, as described by Bradenberger (1957–1958), for aeropolygon strip triangulations up to 20 models in length. First-order stereoplotting instruments were used with vertical photographs (9- by 9-inch format, $f = 6$ inches) and a standard distribution of ground control as displayed in figure 9-26.

The following formulas for the standard residual errors were obtained:

$$m_p = \pm \sqrt{m_x^2 + m_y^2} = \pm 0.10H\sqrt{N}$$

$$m_n = \pm 0.06H\sqrt{N}$$

$$5 \leqslant N \leqslant 20$$

where

m_p = position standard error in metres
m_h = elevation standard error in metres
H = flight height above ground in kilometres
N = number of models.

In these formulas, the standard errors increase with the 1/2 power of the number of models. Numerous practical examples have shown that the accuracy obtained agrees fairly well with that indicated by the above formulas.

For strips from aeroleveling, a slightly lower accuracy is to be expected due to additional errors resulting from the exposure-station leveling. However, this is not true for high-flown photography ($H > 20,000$ ft) and a sufficiently large number of models. For more information on attainable accuracy with the aeroleveling method, reference is made to Brandenberger (1960).

With regard to the attainable accuracy of block triangulation, little information is available. In general it can be stated that an effective block adjustment yields a more homogenous accuracy for the entire block as compared with a single strip (see Brandenberger and Laurila, 1956). Furthermore, it can be expected that the attainable accuracy in a block triangulation is of the same order, or higher, as compared with a strip triangulation, depending on the block adjustment procedure which is used.

9.3.7.4 INFLUENCE OF ACCIDENTAL ERRORS

The residual errors of an adjusted strip triangulation or block triangulation are primarily residuals from the propagation of irregular errors which are mathematically approximated by the so-called accidental errors. These residuals follow more or less arbitrary rules and in case of a strip phototriangulation fluctuate about the distorted surface of the strip which is primarily generated by the propagation of a longitudinal section of a strip is shown in figure 9-27.

It is obvious that the magnitudes of the residuals increase with the number of models between given ground control. These residuals limit the accuracy of phototriangulation. In practice, usually, little is done to reduce the residuals, due to the complexity of the problem. This does not mean, however, that it would be impossible to reduce the residuals. This can be done, for instance, by taking into account the scale closure error and the by, bz, $b\omega$, ϕ, and κ closure errors which can be determined in the last model of a phototriangulated strip. These closure errors can be used to adjust the scale as well as the by, bz, ω, ϕ, and κ in the individual models, yielding an increased stability of the strip and a reduction of the residuals. Such a procedure, however, requires considerably more work, since new coordinates x, y, z for the various points in the strip have to be computed before the conventional strip adjustment takes place.

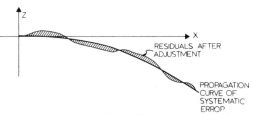

FIGURE 9-27. Residuals after strip adjustment.

9.4 Analytical Triangulation

The subject of analytical photogrammetry was covered separately in chapter X of the Third Edition of the MANUAL OF PHOTOGRAMMETRY. The author, Frederick J. Doyle, divided the subject matter into six distinct parts: introduction, principles, corrections of observed image coordinates, characteristics of systems, instruments and future developments. This chapter leans heavily on data from that publication with a few minor modifications. The section on characteristics of systems is changed to describe three distinct *methods* which have evolved over the years and encompass most of the myriad of *systems* that are employed by various agencies today. These three methods are: sequential adjustment procedures, independant models method, and simultaneous adjustment (sometimes referred to as the bundle adjustment). Additionally, the section on instrumentation has been updated to reflect the latest in mensuration equipment that is available on the market today. The basic mathematics of numerical photogrammetry is given in chapter II and will not be repeated here. The application of analytical methods in the real-time restitution of stereophotographs (analytical plotters) is covered in chapter XIII.

9.4.1 HISTORICAL DEVELOPMENT

The problem of space resection was first attacked by J. H. Lambert (1759), who discussed the geometrical properties of a perspective image and the procedure for finding the point in space from which the picture was made. However, the real foundations of analytical photogrammetry were established by Sebastian Finsterwalder in a series of papers published around the beginning of this century (1899, 1900, 1903a, 1903b, 1932). He applied vectors to investigate the problems of single- and double-point resection in space, topographic mapping from balloon photographs, and the formulation of relative and absolute orientation. Had Finsterwalder possessed the modern capability for extensive computation, the entire development of photogrammetry might have been different.

Another early pioneer was Carl Pulfrich of the Zeiss works in Jena. In 1901 he announced the development of the first stereocomparator designed for the precise measurement of photographic image coordinates. This instrument was originally employed for the measurement of terrestrial photographs, with correspondingly simple computations. Through the course of development, the stereocomparator evolved into Von Orel's Stereoautograph for terrestrial photographs and, with the help of R. Hugershoff, into the Aerocartograph for aerial photographs. In this way, the pattern of instrumental photogrammetry was established and lasted until after World War II. Nevertheless, Pulfrich's stereocomparator is the antecedent of the modern instruments employed in analytical photogrammetry.

Otto von Gruber, lecturing to the Vacation Course in Photogrammetry at Jena, derived the projective equations and their differentials, which are fundamental to analytical photogrammetry. He himself rejected the analytical approach and was directly responsible for the development of the Zeiss Stereoplanigraph. Nevertheless, his lectures, fortunately still available in English reprint (1942), are essential to the development of modern analytical methods.

In 1930, Earl Church, professor of applied mathematics at Syracuse University, obtained a grant from the Guggenheim Fund for Aeronautics to initiate a program in photogrammetry in the civil engineering department. Over the next 20 years he and his students published a series of bulletins developing analytical solutions to space resection, orientation, intersection, rectification, and control extension. In Bulletins 15 and 19 (1945, 1948), Professor Church formalized his work in the direction-cosine notation. While cumbersome by comparison with modern matrix notation, the direction-cosine approach is ideal for organizing machine calculations and outlines explicitly the operations which must be performed in a computer. Professor Church's work was summarized, extended, and applied in an excellent paper by W. O. Byrd (1951).

In 1950 and 1951, Everett Merritt, working at the Naval Photographic Interpretation Center, in Washington, D.C., developed a series of analytical solutions. Although the two men never met, Merritt's work shows the strong influence of Church, both in approach and in notation. It is, however, more complete in the variety of approaches and in the number of problems attacked. Merritt has collected his works in a published book (1958).

Dr. Hellmut Schmid came to the United States from Germany after World War II and became director of the Ballistic Measurements Laboratory of Ballistic Research Laboratories at Aberdeen, Md. In a series of important publications (1956−57, 1959), he developed the principles of modern multistation analytical photogrammetry. Although his investigations were directed initially to ballistic-camera operations, in which several cameras may observe an event simultaneously, the application of these procedures to control extension by strips and blocks of aerial photographs followed immediately. The principal features of Schmid's work are a rigorously correct least-squares solution, the simultaneous solution using any number of photographs, and a complete study of error propagation. He is the first photogrammetrist to plan his solutions in anticipation of the use of high-speed electronic computers.

Although analytical photogrammetry had been applied (unsuccessfully) to a practical mapping program as early as 1920 (Pulfrich, 1919; Hugershoff and Cranz, 1919; and Fischer, 1921), credit for the first operational system of analytical aerotriangulation goes to the British Ordnance Survey (Shewell, 1953). Begun in 1947 as analytical radial triangulation computed by hand, the system evolved into a complete space solution operating efficiently on an electronic computer.

While serving as a consultant to the Mapping and Charting Research Laboratory at Ohio State University, Dr. Paul Herget (1956) developed a system of analytical control extension based upon minimizing the perpendicular distances between pairs of corresponding rays. This system was further developed at Ohio State (Herget and Mahoney, 1957), at Cornell University (McNair, Dodge, and Rutledge, 1958), at the U.S. Geological Survey (Dodge, Handwerker, and Eller, 1959), and at Massachusetts Institute of Technology (1962).

Another formulation for analytical aerotriangulation was developed by G. H. Schut (1955–56), of the National Research Council in Canada. In two papers (1957–58, 1959–60), Schut analyzed the existing methods of analytical triangulation, reduced them to a common notation, and classified them according to three criteria:

    a. triangulation procedures,
    b. type of condition equations,
    c. method of solving condition equations.

For a comprehensive grasp of the various methods of analytical photogrammetry, these two papers are without equal.

Major contributions have been made to the theory and practice of analytical photogrammetry and error analysis by Duane Brown. Developed originally (Brown, 1957) for measurements of ballistic trajectory, his formulation has been adopted worldwide for aerotriangulation. Brown's contributions are numerous. To name a few, his elegant and general treatment of least squares adjustment and error propagation, analytical formulation of internal and external perturbations of imaging systems, algorithms and logic for efficient solution of large systems of equations, close-range applications, mensuration technology, and sensor calibration. His computer programs for the reduction of data from extra-terrestrial missions represent some of the most elegent treatments in operational use today.

The Coast and Geodetic Survey was the first United States mapping organization to have achieved an operational system of analytical aerotriangulation (Harris, Tewinkel, and Whitten, 1962a). The system is based on the formulation of H. Schmid with a preliminary adjustment patterned after that used at the National Research Council of Canada. Since that time many systems have been developed at organizations such as Defense Mapping Agency, Geological Survey and many Universities.

This brief history has emphasized those contributions that were significant to the development of analytical photogrammetry in the United States.

9.4.2 BASIC PRINCIPLES OF ANALYTICAL AEROTRIANGULATION

In general terms, analytical photogrammetry can be considered as the mathematical transformation between an image point in one rectangular coordinate system (image space) and an object point in another rectangular coordinate system (object space). This basic mathematical concept is valid for *all applications* of analytical photogrammetry (*e.g.*, terrestrial, aerial, nontopographic) using *any sensing device* (frame camera, panoramic camera, raster scan, *etc.*) to record directional information to objects in any medium (air, water, *etc.*). Departures from the basic system take the form of mathematical models which describe the internal geometry of the sensor and the external geometry of object space. The two fundamental coordinate systems employed are the image coordinate system and the object space coordinate system.

9.4.2.1 PHOTOGRAPHIC COORDINATE SYSTEM

Normal metric cameras used in photogrammetry contain reference marks (fiducials) in the focal plane around the perimeter of the imaged area. Some cameras carry an array of marks superimposed on the image area (reseau). In either case, these marks are fixed relative to the lens and as such serve to relate the image field to physical properties of the camera. Given three such marks, a coordinate system can be uniquely established in the focal plane (negative) or in the positive plane with its origin at the mathematically defined perspective center O.

The perspective center is the rear nodal point of the camera's lens system. A perpendicular through it to the image plane hits that plane in the principal point and at a distance equal to the camera's principal distance f. When the coordinate system is established in the plane of a negative the z axis is positive toward the image plane. When the coordinate system is established in the plane of a positive, positive z is directed away from the image plane. Any image point then, has the coordinates $x, y,$ and z where $z = f$ in the negative image plane and $-f$ in the positive image plane. This definition assumes that the standard (or zero rotation) position of the camera is with the lens "looking" down as in aerial photography. If one were to assume that the standard position of the camera is with the lens looking up and the coordinate systems remain the same, the signs of z relative to f would be directly opposite, $e.g.$, $z = f$ for a positive and $z = -f$ for a negative. The first definition has more or less been adopted as standard in most analytical systems.

9.4.2.2 CORRECTION OF MEASURED IMAGE COORDINATES

The photographic coordinate system just described is based on the assumption that the camera produces a perfect central projection of the object. That is, the lens can be considered as a point and single rays of light emanating from any object point in the field of view will form an image such that the object, lens, and image fall in a straight line. This is, in fact, the basis for the algorithm that is applied at later stages of adjustment, but in reality the measured coordinates of images hardly ever satisfy this assumption. As a result, a pre-processing routine is normally applied to all raw measurements prior to the adjustment. The details of the pre-processing vary from one organization to another in order to satisfy particular requirements; however all such routines include some or all of the following corrections. The corrections are not necessarily presented in the order in which they should be applied. Also, in order to minimize symbolization, all primed values refer to the true, calibrated or corrected values and the unprimed values refer to the measured or uncorrected (for the particular perturbation under discussion) quantity.

9.4.2.2.1 PRINCIPAL-POINT COORDINATES

In a well adjusted camera, the intersection of lines joining opposite fiducial markers will exactly coincide with the point where a perpendicular from the rear nodal point of the lens meets the focal plane. This point is called the principal point. For a perfect lens, this is the point where a beam of collimated light entering the front of the lens perpendicularly to the focal plane is brought to a focus. In real life, perfect lenses do not exist nor do the intersections coincide exactly with the principal point; therefore, any point measured in the photographic system must undergo a translation of the form

$$\begin{bmatrix} x' \\ y' \\ z' \end{bmatrix} = \begin{bmatrix} x - x_p \\ y - y_p \\ \mp f \end{bmatrix}$$

where x and y are the measured coordinate relative to the fiducial system and x_p and y_p are translations to the principal point. In the normal case, aerotriangulation is performed with photography taken by the same camera. Consequently f, x_p and y_p can be considered as constant for the solution and the effects of the offset are usually applied to the photographic mea-

surements in a pre-processing program prior to the final adjustment. Methods for calibration of the offset and principal distance are given in chapters III and IV and a further discussion on inflight calibration of these and other parameters is covered in chapter IV.

9.4.2.2.2 EMULSION DEFORMATION

The photographic material, film or glass, is subject to deformation between the time it is exposed in the camera and measured in the comparator. The basis for correction of this deformation is dependent upon the construction and calibration of the camera, that is, the method used in the camera to provide a standard to which the deformation may be compared and corrected. The degree to which deformations may be estimated is a function of the amount, distribution, validity and measurability of standards as they existed at the time of exposure. The corrections, thereafter, can take on a variety of forms.

The simplest standard provided by the camera manufacturer is the measured distance between points situated along the edge of the exposure format. A pair of such points, when located along a line parallel to the measurement axes, can be used to compute a linear scale change in that direction. That is, the corrected value for any coordinate x is given by:

$$x' = S_x \cdot x \qquad (9.6)$$

where, S_x = calibrated x distance/measured x distance. If only one such distance is given, the scale change is considered to be common to both the x and y directions and a uniform correction is applied to any second coordinate of the form

$$y' = S_x \cdot y. \qquad (9.7)$$

With two such calibrated distances given, a differential scale change can be computed and the corrected y would be:

$$y' = S_y \cdot y \qquad (9.8)$$

where S_y = calibrated y distance/measured y distance.

In more sophisticated systems, the camera manufacturer may choose to supply an array of reference lighted markers surrounding the image area. In this case, the calibration of these markers is extended beyond mere distances between pairs of points. That is, a glass plate in contact with the focal plane is exposed by the reference marks which in turn are measured in a precision comparator. The result is an x and y coordinate for each image (in some local coordinate system) which can be used to define not only the distance between any pair of markers, but also relative directions to each. These data can be used to infer changes in orthogonality of the film record along with the dimensional change.

The most common expression used to correct

for film deformation when the data is given in coordinate form is the general form of the projective transformation

$$x' = \frac{a_1 x + b_1 y + c_1}{dx + ey + 1}$$

(9.9)

$$y' = \frac{a_2 x + b_2 y + c_2}{dx + ey + 1}.$$

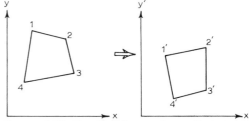

FIGURE 9-28. Graphic of general coordinate transformation.

The coefficients a_1, a_2, b_1, b_2, c_1, c_2, d and e are determined by a least squares fit of coordinates of the markers measured on the film to those given by the calibration. Obviously, to determine all eight parameters, one must have at least four such measured points for a unique solution. A fit using all eight parameters would essentially have the effect of mapping one quadrilateral in the one coordinate system into a quadrilateral in the calibrated system. The transformation is depicted in figure 9-28.

In some cases, the photogrammetrist may wish to restrict the versatility of the transformation by reducing the number of unknowns solved for in the general form. For instance, if the number of parameters is reduced to six (a_1, a_2, b_1, b_2, c_1, and c_2), the transformation will impart a translation, rotation, two scale changes and a nonorthogonality deformation to the measured film coordinates. In this case, the parameters d and e are identically zero, and the other coefficients are carried in the least squares solution as

$$a_1 = S_x \cos \alpha \quad a_2 = S_y \sin \alpha$$

$$b_1 = S_y \sin (\alpha - \beta) \quad b_2 = S_x \cos (\alpha - \beta) \quad (9.10)$$

where S_x and S_y are scalers for the x and y axes respectively, α is a rotation of y toward x and β is the nonorthogonality of the y axis (directed toward the x axis).

Successively reducing the number of parameters, as in the above case for six, will impart the following changes in the transformation.
Five parameters (α, S_x, S_y, c_1, and c_2) where

$$a_1 = S_x \cos \alpha, \ a_2 = -S_y \sin \alpha,$$

$$b_1 = S_x \sin \alpha, \ b_2 = S_y \cos \alpha,$$

(9.11)

will rotate the system, change scales in both x and y and will shift the coordinates by c_1 and c_2.
Four parameters (α, S_x, c_1, and c_2) where:

$$a_1 = S_x \cos \alpha = b_2$$

$$-a_2 = S_x \sin \alpha = b_1$$

(9.12)

will rotate the system, change both axes by a common scale and shift the coordinates by c_1 and c_2.
Three parameters (α, c_1, and c_2) where:

$$a_1 = \cos \alpha = b_2$$

$$-a_2 = \sin \alpha = b_1$$

(9.13)

will rotate the system and shift the coordinates by c_1 and c_2. The three parameter case does not

represent a dimensional change in the data, it merely maintains the same shape and size and transforms it to the calibrated system.

In the cases cited above, the reference marks for correction of film deformation are located outside the image area of the photograph. All corrections to image coordinates are therefore based on an interpolation of the film deformation that takes place on the periphery of the frame. The ideal case, of course, would be to have a reference mark located precisely at each image point in the photograph. In this way, one could eliminate the comparator measurement phase and read the coordinates of each point directly from the film. Since the ideal case is impossible, the next best solution is to provide an array of reference marks inside the format area (normally referred to as a reseau). Such marks, if precisely calibrated for position, not only increase the accuracy of interpolation but provide a reference system whereby the precision of measurement is restricted to the distance between reseau marks only, instead of the full distance across the film format. That is, image coordinates can be referred to calibrated coordinates of one or more reseau marks. Ideally, the array of reseau markers should be located precisely in the focal plane of the camera, e.g., at the precise point where it is recorded by the emulsion. Obviously, since an emulsion is three-dimensional, the image cannot be made to appear (or be calibrated) at the precise position of all grey levels an object can take on. Therefore, the reseau is normally calibrated at some optimum plane as near to the focal plane as possible.

There are two types of systems for generating a reseau array on the exposed film. One is part of the lens system and appears between the lens and film, normally referred to as a projected reseau. The image is formed by opaque marks that block out light from the lens. The second system carries the reseau as part of the pressure platen of the camera and is referred to as a backlighted reseau. Here, an independent light source is generated at the platen and projected onto the emulsion through the film base.

Calibration of either system is a painstaking problem. The optimum is to image the reseau markers onto a superflat glass plate and measure their positions on a precision comparator. In the case of the backlight reseau, one has ignored the effects of refraction through the film base and, in the case of the projected reseau, he has assumed that the glass plate and emulsion form a true plane throughout the format. Additionally, in the projected reseau the image is not necessarily transferred to the focal plane in the form of a linear scale change but does in fact exhibit non-linear characteristics.

The simplest method of applying a reseau for correction of film deformation is to measure the coordinates x_i and y_i of the reseau marker nearest to each measured image point. The corrected image coordinates are then given by

$$x' = x_i' + (x - x_i)$$
$$y' = y_i' + (y - y_i).$$
(9.14)

The values x_i' and y_i' are the calibrated coordinates of the measured reseau marker.

An alternative method of applying the reseau is to measure the coordinates of two or more reseau markers which surround the image point. The adjusted position of the image point may then be found by applying the projective transformation formulas from the preceding section.

Finally, the most intensive use of the reseau can be obtained by measuring all markers within the format area. Correction formulas are then found by using a least-squares method to determine coefficients for a polynomial that expresses the image correction as a function of the image position on the plate.

$$x' = x + a_0 + a_1x + a_2x^2 + a_3y$$
$$+ a_4y^2 + a_5xy + a_6x^2y + \ldots .$$
$$y' = y + b_0 + b_1x + b_2x^2 + b_3y$$
$$+ b_4y^2 + b_5xy + b_6x^2y + \ldots .$$
(9.15)

One disadvantage of this approach is the uncertainty associated with the number of terms to be carried in the polynomial. Another method suggested for the interpolation of the measured reseau into that given by the calibration is through the application of least-squares collocation. This method differs from the ordinary least-squares adjustment by the inclusion of a second random variable, called the signal, in addition to the measuring error or noise. In this case, the signal represents residual distortions that remain after the removal of the main trends by projective transformation.

In the case of the backlighted reseau, one cannot expect the pressure platen to occupy the same position during the exposure of each frame. This means that the relationship between the reseau and some fixed points on the camera (fiducials) must be established anew for each exposure. The measured images of the fiducial markers are treated like all other images within the frame; *e.g.*, they are corrected for film deformation according to the observed coordinates of the reseau. Then, a three-parameter transfor-

mation (rotation α with two translations c_1 and c_2) is applied to all measured image points to bring them into the coordinate system of the calibrated fiducials

$$x' = c_1 + x \cos\alpha - y \sin \alpha$$
$$y' = c_2 + x \sin\alpha + y \cos \alpha.$$
(9.16)

9.4.2.2.3 PLATEN FLATNESS

In the manufacture of a camera pressure platen, one cannot always assume that it is machined perfectly so that the film is assured of lying in a true plane. There will invariably be small undulations that are measurable and usable as a basis for image correction. The effects of unevenness are illustrated in figure 9-29 which shows the cross section of a platen through the principal point. The uneven line represents the surface of the platen and the dashed line represents the mean image plane which comes from the camera calibration for principal distance. Because of the deviation Δh of the surface of the film from the mean image plane (where a positive Δh is directed towards the lens) a correction Δr must be computed from

$$\Delta r = \Delta h \frac{r}{c - \Delta h} .$$
(9.17)

Image coordinate corrections then become

$$\Delta x = \frac{x}{r} \Delta r \text{ and } \Delta y = \frac{y}{r} \Delta r$$
(9.18)

where the corrections Δx and Δy are added to the measured image coordinates.

9.4.2.2.4 LENS DISTORTION

Lens distortion (as described in chapter III) is defined by the lens designer as the failure of a lens to image a straight line in object space as a straight line in image space and to maintain the same metric. Normally, the resultant displacement of the image from that produced by a true projective transformation is referred to as distortion and is characterized by two distinct components—one that is *radially symmetric*

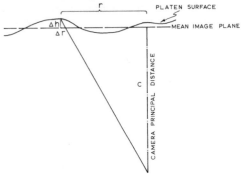

FIGURE 9-29. Image displacement due to uneven platten surface.

about the principal point and one that is *symmetric along a line* directed through the principal point. Terminology adopted by the photogrammetrist to describe these two components is *radial* and *tangential* distortion, respectively.

9.4.2.2.4.1 Radial Distortion. Radial lens distortion can be calibrated and furnished to the user of a camera in a variety of ways. Some calibrations offer the data to the user in the form of radial displacements (Δr) at given intervals of radial distance (r) for one or several directions from the point of symmetry. When one radius is given, it is usually determined from a mean of all the directions of calibration (*see* chapters III and IV). To extend the data for application purposes then, a smooth curve is drawn through the data with Δr plotted as a function of r and intermediate values of Δr are interpolated from the graph. The interval of r for the interpolation is chosen to establish tabular values of Δr such that the increments from one value to the next will be nearly linear or small enough so that they will need no further nonlinear interpolation for accurate use. Application of the data for correction then takes on the simple form

$$x' = x\left(1 - \frac{\Delta r}{r}\right)$$

$$y' = y\left(1 - \frac{\Delta r}{r}\right), \qquad (9.19)$$

where r is given as $r^2 = x^2 + y^2$ and x and y are the measured plate coordinates corrected to the principal point.

In some cases, the radial distortion is given as an odd-powered polynomial such as

$$\Delta r = K_1 r^3 + K_2 r^5 + K_3 r^7 + \ldots \qquad (9.20)$$

This is especially true in the case of a stellar or inflight calibration. Coupled with that type of a calibration is a determination of the camera principal distance c and the resulting distortion curve is sometimes referred to as the camera's *characteristic distortion*. Care must be taken, therefore, to assure that the distortion function chosen is that one which is referred to the focal length being used. For aesthetic reasons, most distortion curves are presented in the form of so-called *calibrated distortion*. The difference between the two methods of presentation is a mathematical transformation of the *characteristic distortion* to provide a distortion pattern over the entire format that will minimize the relative positional errors of the individual images. This transformation takes the form of a linear scale change.

From figure 9-30, it is obvious that a different distortion curve can be presented mathematically for any chosen axis of calibration. For example, if one chooses to "balance" the distortion curve, an oblique axis of calibration can be established that equalizes the areas above and below this axis for some limiting angle of view.

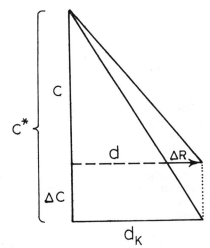

FIGURE 9-30. Diagram of "calibrated" distortion.

This defines the radial distance r_k where the oblique axis intercepts the computed distortion curve. Introducing a linear term $K_0 d$ into expression (9.20) results in the expression for calibrated distortion

$$\Delta r_c = K_0 d + K_1 d^3 + K_2 d^5 + K_3 d^7 + \ldots \qquad (9.21)$$

Since (by definition) the distortion at some desired distance d_K is zero, the coefficient K_0 is obtained from

$$K_0 = -(K_1 d_K^2 + K_2 d_K^4 + K_3 d_K^7 + \ldots). \qquad (9.22)$$

Similarly, the principal distance (c^*) associated with the calibrated distortion curve is, from figure 9-30,

$$c^* = c(1 - K_0). \qquad (9.23)$$

Finally, to resolve the radial distortion (Δr_c) into components x and y of the plate coordinate system, we compute the projections using the angle β (figure 9-33) from

$$\Delta r_x = \Delta r_c \sin \beta = \frac{\Delta r_c}{r}(x_i - x_p)$$

$$\Delta r_y = \Delta r_c \cos \beta = \frac{\Delta r_c}{r}(y_i - y_p). \qquad (9.24)$$

Application of the radial-distortion function to the measured plate coordinates takes on the form (as above)

$$x^1 = x\left[1 - (K_0 + K_1 r^2 + K_2 r^4 + K_3 r^6 + \ldots)\right]$$

$$y^1 = y\left[1 - (K_0 + K_1 r^2 + K_2 r^4 + K_3 r^6 + \ldots)\right] \qquad (9.25)$$

where x and y are understood to be corrected to the principal point.

9.4.2.2.4.2 Tangential Distortion. The distortion described above is considered to be radially symmetric about some point in the format of the photograph. Tangential distortion, on the other hand, is a displacement of the image that is usually attributed to errors made in assembling the lens so that the centers of curvature of the individual elements are not made to fall on a straight line. The result is combined radial and tangential displacement of the image that varies with radial distance and azimuth from some point near the center of the photograph. Washer (1957) showed that these effects (lens decentering) could be simulated by a thin prism in combination with a centered system and coined the phrase "prism effect." No attempt was made to correct for decentering defects in applications of analog instruments; but with the advent of numerical methods, the subject was revived. Naturally, the adopted mathematical simulation first took the form of a prism or wedge; the resulting equations were written as

$$\Delta x_w = P_w \cos \phi$$

and   (9.26)

$$\Delta y_w = P_w \sin \phi$$

where P_w is the distortion profile that defines the amount of displacement as a function of radial distance and ϕ is the angle of the line along which the displacement has no radial component. A graphic example of the displacements from the wedge model for one radial distance is shown in figure 9-31.

Brown (1966) suggested the use of a decentering model (developed by Conrady, 1919) that had gone unnoticed by most photogrammetrists. Using the same methods he used to develop the Seidel sums, Conrady established algebraic expressions for displacements resulting from decentrations of lens elements. In his notation, these expressions are

$$r = 3p_3 V^2 \cos\chi$$

and   (9.27)

$$t = p_3 V^2 \sin\chi$$

where V is the half angle of view of the image, p_3 is a constant peculiar to the lens system, and χ is

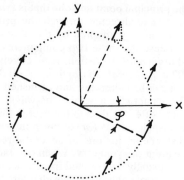

FIGURE 9-31. Wedge distortion.

an angle indicating the orientation of the image relative to the direction of decentration. (In this case, decentration is understood to mean the resultant direction when more than one element is involved.) The component r in the expression is taken as positive away from the origin along the radial line of the field, and the component t is perpendicular to this radial line forming a right-handed system. Evaluating equation (9.27) for any given distance from the center of the field, we see that the displacement of the image would be radial and equal to $3p_3 V^2$ in the direction $\chi = 0°$ and equal to $-3p_3 V^2$ at $\chi = 180°$. All intermediate values of χ result in displacements in both the radial and tangential directions (as indicated in figure 9-32) with the maximum tangential effects occurring at $\chi = 90°$ and $270°$.

To relate this distortion model to our plate coordinate system, we arbitrarily assign one of the directions of maximum tangential distortion ($\chi = 270°$) to our plate positive y axis and introduce a new variable ϕ_τ as the angle of deviation from this standard position. From figure 9-32 then, the angle χ of Conrady takes on the meaning

$$\chi = \phi_\tau + \beta - 90°;$$ (9.28)

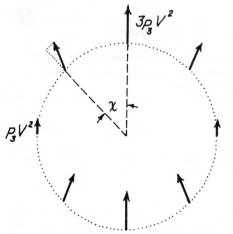
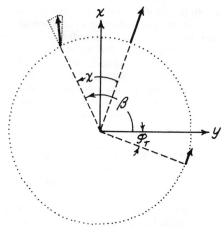

FIGURE 9-32. Conradi's decentering distortion.

and the expressions (9.27) in terms of ϕ_r and β become

$$r = 3p_3V^2 \sin(\phi_r + \beta)$$

(9.29)

and

$$t = p_3V^2 \cos(\phi_r + \beta).$$

Further, to resolve the Conrady displacements into components in our plate system, we premultiply expressions (9.29) by a rotation matrix (rotate the r, t system through an angle β $-90°$, r toward t) and replace the distortion profile p_3V^2 by a polynomial in distance d given by

$$p_3V^2 = P_c = K_4d^2 + K_5d$$

resulting in the final form for decentered lens distortion (ΔT)

$$\begin{bmatrix} \Delta T_x \\ \Delta T_y \end{bmatrix} = \begin{bmatrix} \sin\beta & -\cos\beta \\ \cos\beta & \sin\beta \end{bmatrix} \begin{bmatrix} r \\ t \end{bmatrix} = \begin{bmatrix} P_c(2\sin^2\beta + 1)\cos\phi_r + (2\sin\beta\cos\beta)\sin\phi_r \\ (2\sin\beta\cos\beta)\cos\phi_r + (2\cos^2\beta + 1)\sin\phi_r \end{bmatrix}.$$

(9.30)

Finally, from figure 9-33 and substitution of the relationships

$$\cos\beta = \frac{dy}{d}$$

and

$$\sin\beta = \frac{dx}{d}$$

(9.31)

where

$$dx = (x_i - x_p)$$

and

$$dy = (y_i - y_p),$$

expressions (9.29) reduce to the more convenient form

$$\Delta T_x = (x_i - x_p)(K_4 + K_5d^2)\left(\frac{3dx^2 + dy^2}{dx}\cos\phi_r + 2\,dy\,\sin\phi_r\right) = (x_i - x_p)D_x$$

$$\Delta T_y = (y_i - y_p)(K_4 + K_5d^2)\left(\frac{3dy^2 + dx^2}{dy}\sin\phi_r + 2\,dx\,\cos\phi_r\right) = (y_i - y_p)D_y.$$

(9.32)

Figure 9-34 is a graphic example of the displacements resulting from a combination of both radial and lens decentering distortion evaluated at a distance r from the point of origin of the distortion profiles. As formulated by Conrady, the displacements from each distortion source can be considered as independent vectors in which the resultant is the sum of their individual x and y components. In matrix notation, then, the resultant shifts δ_x and δ_y of any point i can be written as

$$\begin{bmatrix} \delta_x \\ \delta_y \end{bmatrix} = \begin{bmatrix} \dfrac{\Delta R}{r} + D_x & 0 \\ 0 & \dfrac{\Delta R}{r} + D_y \end{bmatrix} \begin{bmatrix} x_i - x_p \\ y_i - y_p \end{bmatrix}$$

(9.33)

where the coordinates of the corrected image are given by

$$x' = x - \delta x$$

(9.34)

and

$$y' = y - \delta y.$$

9.4.2.3 OBJECT-SPACE COORDINATE SYSTEM[1]

In the definition of analytical photogrammetry in paragraph 9.4.2 we stated that it could be considered as the mathematical transformation of a point from a rectangular coordinate system in object space to a rectangular coordinate in image space. The previous discussion has been concerned with the coordinates and their corrections in image space. The following will describe some of the coordinate systems in object space and their corrections. These systems are used in analytical photogrammetry to describe the spacial (X_0, Y_0, Z_0) position of the camera's perspective center and the relative positions of objects on the ground. Additionally, this system provides the reference axes about which the three rotations are given when one is transforming from object to image space.

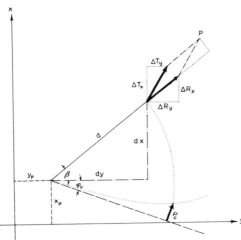

FIGURE 9-33. Combined radial and decentering distortion for single ray.

[1] A description is given for some of these systems in chapters II and VIII. They are repeated here, however, for continuity of discussion.

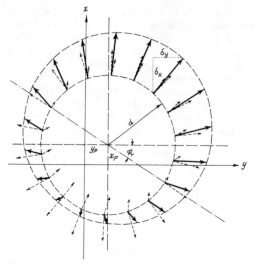

FIGURE 9-34. Combined radial and decentering distortion for one radius.

9.4.2.3.1 GEODETIC COORDINATES

The geodetic coordinate system is the basic reference by which the geodesist describes the relative positions of points on the earth's surface. The system is based on a specified reference spheroid (an ellipsoid of revolution) whose minor (or shorter) axis coincides with the earth's axis of rotation. The reference figure is usually defined by the dimension of the semi-major axis, a, and the semi-minor axis, b. For convenience, the ratio of a and b is given in the form of a flattening, f, where $f = (a - b)/a$. Some reference figures are defined by a and the eccentricity e where the eccentricity is given by

$$e^2 = 2f + f^2. \qquad (9.35)$$

The coordinates of a point in the geodetic system are latitude, ϕ, and longitude, λ. A third coordinate, h, is given which indicates the height of the point above the reference ellipsoid.

9.4.2.3.2 GEOCENTRIC COORDINATES

In most analytical photogrammetric systems the object space is given in a three-dimensional rectangular coordinate system. The most basic of these is the geocentric or earth-centered which is right-handed with the origin at the center of the reference spheroid. The Z axis is parallel with the minor axis and has its positive direction through the north pole. The X axis is directed in a positive sense toward the meridian which passes through Greenwich and the Y axis completes the right-handed system.

9.4.2.3.3 LOCAL VERTICAL COORDINATES

To decrease the numerical size of the coordinates and to make the axes of the system more nearly coincide with known directions of the

photograph, most analytical systems employ the local vertical rectangular coordinate system. This system is merely a rotation and translation of the geocentric coordinates into a system whereby the Z axis is first rotated to pass through some arbitrarily selected point in the area photographed. The origin is translated along the rotated Z axis to this point and the Y axis is made to coincide with the meridian through this point.

In some cases, to make the X or Y axis parallel to flight lines which are not necessarily aligned in a north or east direction, the system is rotated away from the meridian through some angle α. The coordinate system is then commonly referred to as the *local secant rectangular system*.

9.4.2.3.4 UNIVERSAL TRANSVERSE MERCATOR (UTM) COORDINATES

The Universal Transverse Mercator system of coordinates is based on a conformal projection of geodetic coordinates to a cylinder whose axis is perpendicular to the earth's axis of rotation. The result is a set of plane coordinates (easting and northing) along with a height coordinate that is identical to its value in the geodetic coordinate system. In the UTM System, the world is divided into 60 zones of 6-degree increments from the Greenwich meridian and numbered from 1 at 180° west to 174° west longitude eastward to zone 60 at 174° east to 180° east longitude. All zones are limited in latitude by 80 degrees north and south. The central meridian and 0 degrees latitude of each zone is the origin of the transformed system and each zone is identical with the exception of the displacement in longitude. The scale error of projection is balanced such that the scale factor along the central meridian is .9996. UTM coordinates are not recommended as the basic object space system for analytical adjustments. They are, however, commonly used in the output stages of triangulation to present the control point data to the compiler.

9.4.2.3.5 STATE PLANE-COORDINATES

Another commonly used plane coordinate system, which is based on conformal projection, is the state plane-coordinate system. For states with greatest dimension in an east and west direction, the map projection used is the Lambert conformal with two standard parallels. The other states are based on the transverse Mercator map projection. The reason for the two map projections is that in the Lambert map projection the scale varies with the distance from the central parallel; hence it is suitable for states with lesser north-south extent. The transverse Mercator map projection, on the other hand, varies in scale as a distance from the central meridian and is therefore used in states with lesser east-west extent. Florida is the one exception that employs both projections.

9.4.2.3.6 CONVERSION OF COORDINATE SYSTEMS

In the development of an analytical photo-triangulation system, it is customary to establish a series of computer routines and subprograms for the conversion of coordinates from one system into another. In some cases, the conversions serve as pre- or post-processing routines to convert the data to a form compatible with an adjustment or to a set of coordinates most convenient for data extraction and compilation. The following are the most commonly used conversions.

9.4.2.3.6.1 Geodetic-Geocentric. In general, coordinates of ground stations are usually given in a geodetic system. Likewise, the position (ϕ and λ) of the camera station can be approximated from maps along with the known approximate height of the aircraft. The conversion then, from geodetic to geocentric, follows as the most favored for extensive aerotriangulation schemes.

Given the ellipsoidal parameters of the reference figure and the geodetic coordinates (ϕ, λ and h) of any point, the transformation from geodetic to geocentric coordinates is given by

$$N = \frac{a}{(1 - e^2 \sin^2 \phi)^{1/2}}$$
$$X = (N + h) \cos \phi \cos \lambda$$
$$Y = (N + h) \cos \phi \sin \lambda$$
$$Z = \{N(1 - e^2) + h\} \sin \phi. \tag{9.36}$$

The inverse transformation (from geocentric to geodetic) is not direct, but requires an iteration for a correct solution for ϕ and h. That is, given geocentric coordinates of a point (X, Y, and Z) the corresponding longitude (λ) is found from

$$d = (x^2 + y^2)^{1/2}$$

and

$$\lambda = \arcsin \frac{Y}{d} = \arccos \frac{X}{d} = \arctan \frac{Y}{x}. \tag{9.37}$$

For ϕ and h, however, we first compute an initial approximation to tangent ϕ using

$$\tan \phi \approx \frac{Z}{d}. \tag{9.38}$$

With this ϕ we start an iterative loop to compute a better approximation to ϕ using

$$\tan \phi = \frac{Z + Ne^2 \sin \phi^0}{d}$$

where ϕ^0 is the value obtained in the last iteration (initially equal to zero) and N is given by

$$N = \frac{a}{(1 - e^2 \sin^2 \phi^0)^{1/2}}. \tag{9.40}$$

These two computations are continued until there is no change in ϕ. Finally, h is computed from the expression

$$h = \frac{d}{\cos \phi} - N. \tag{9.41}$$

9.4.2.3.6.2 Geocentric-Local Vertical. As pointed out earlier, the local vertical system has an advantage (in the computation phase) over the geocentric system in that the coordinates in question have been converted to a more manageable size. In other words, a local system is no more than the geocentric system translated and rotated to some local position. In general the transformation is given by

$$\begin{bmatrix} X \\ Y \\ Z \end{bmatrix}_L = M \left\{ \begin{bmatrix} X \\ Y \\ Z \end{bmatrix}_G - \begin{bmatrix} X \\ Y \\ Z \end{bmatrix}_O \right\} \tag{9.42}$$

where the vectors L, G and O represent the local system, the geocentric system and the translation. The matrix M is the rotation matrix and is

$$M = \begin{bmatrix} -\sin \lambda_0 & \cos \lambda_0 & 0 \\ -\sin \phi_0 \cos \lambda_0 & -\sin \phi_0 \sin \lambda_0 & \cos \phi_0 \\ \cos \phi_0 \cos \lambda_0 & \cos \phi_0 \sin \lambda_0 & \sin \phi_0 \end{bmatrix} \tag{9.43}$$

where the subscript $_0$ represents the origin's coordinates ϕ_0, λ_0.

The local secant system is a special case of the local vertical where the transformation includes an additional rotation of the X, Y plane about the λ axis through some angle α.

9.4.2.3.6.3 Geodetic-Plane. Transformations of geodetic coordinates to and from plane coordinates such as UTM and state plane are beyond the scope of this manual. For detailed descriptions, refer to

Department of the Army Technical Manual, TM5-241-8, *Universal Transverse Mercator Grid*, July 1958,

U.S. Department of Commerce, Coast and Geodetic Survey Special, Publication No. 251, *Conformal Projections in Geodesy and Cartography*, and

U.S. Department of Commerce, Coast and Geodetic Survey Special, Publication No. 193, *Manual of Plane-Coordinate Computation*.

9.4.2.4 CORRECTIONS TO OBJECT SPACE COORDINATES

As pointed out in the introduction (section 9.4.2) and in chapter II, the basic premise of the mathematical expressions (or transformations) of analytical photogrammetry is that the object point, the lens (represented as a point) and the resulting image in the camera all lie on a straight line. Deviations (image coordinate corrections) from this principal inside the camera were covered in section 9.4.2.2. This section discusses those perturbations that occur outside the camera which can be predicted and modeled mathematically with some certainty.

9.4.2.4.1 ATMOSPHERIC REFRACTION

Atmospheric refraction is the result of changes in density of the atmosphere which causes light to follow a curved path instead of a straight line. From figure 9-35, $\Delta \alpha$, the angle between a straight line and the tangent to the light path (shown as a single ray) at the lens is a function of altitude, terrain height, direction of the ray and atmospheric conditions along the path. The ultimate effect of refraction is to displace all images radially from the nadir point of the photograph.

In its simplest form, atmospheric refraction can be expressed by

$$\Delta d = K \tan \alpha$$

where Δd is angle of displacement, α is the angle the ray makes with the "true vertical" and K is a constant related to the atmospheric conditions. The constant K can be viewed as the amount of displacement attributable to a ray at 45° from the vertical.

Many articles have been published on the subject of atmospheric refraction, notably Brown (1957), Schmid (1959), Bertram (1966, 1969), Schut (1969), and Saastamoinen (1974). In most cases, the differences in treatment are mainly contained in the derivation and formulation of the constant K. In an unpublished article, Bertram (*see* 3rd edition, MANUAL OF PHOTOGRAMMETRY) derived the following expression for K using the 1959 ARDC model for the atmosphere.

$$10^6 R = \frac{2335}{H-h}\left[(1 - 0.02257h)^{5.256} - (1 - 0.02257H)^{5.256}\right] -277.0\,(1 - 0.02257H)^{4.256}. \qquad (9.47)$$

$$K = \left[\frac{2410\,H}{H^2 - 6H + 250} - \frac{2410\,h}{h^2 - 6h + 250}\left(\frac{h}{H}\right)\right]. \qquad (9.44)$$

In this expression, K is in microradians and H is the flying height in kilometres above sea level.

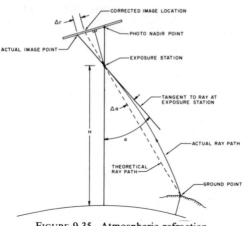

FIGURE 9-35. Atmospheric refraction.

The elevation of the point on the ground is h (in kilometres). In a later article, Bertram (1966, 1969) used the ARDC Model Atmosphere, 1959 to derive an expression using an integral involving the variation of altitude of the light ray with its velocity. The resulting equation is

$$\delta\theta = \frac{\tan \theta}{(Z_c - Z_g)c}\int_{Z_g}^{Z_c}(Z - Z_g)dv. \qquad (9.45)$$

Here, θ is the angle of the object from the vertical at the camera station and c is the velocity of light. For evaluation of the integral he uses the trapezoidal rule and derives a table of refraction (in microradians) for intervals of 1,000 metres of flying height from 500 to 20,500 metres.

Schut (1969) derives formulas for atmospheric refraction as a function of densities using Snell's law of refraction of light rays. His expression, identical with Bertram, is

$$\alpha = 0.000226\frac{\tan \theta}{Z_c - Z_g}\sum((Z - Z_g)d\rho). \qquad (9.46)$$

Here, Z_c is the camera height and Z_g is the elevation of the ground point. He lists a table of refraction at 50 metre intervals for flying heights of .50 to 90 kilometres for ground heights of 0 to 4 kilometres.

Saastamoinen (1973), carries the formulation one step further and introduces an expression for the variation of refractive index as a function of height. For flying heights up to 11 kilometres he gives

For flying heights to 9 kilometres he gives the more simple form

$$10^6 R = 13(H-h)[1 - 0.02(2H+h)] \qquad (9.48)$$

where both H (flying height) and h (ground point elevation) are in kilometres.

Table 9-1 lists photogrammetric refraction in microradians at 45° for all the above expressions along with Schmid (1959). For most applications, expression (9.48) appears to be the most convenient.

Several methods have been proposed for the correction of the effects of atmospheric refraction. The most popular is the one that assumes a vertical photograph and computes displacements of each image on the photograph from the principal point. That is, for any point the correction in radial distance is given by

$$\Delta r = K\left(r + \frac{r^3}{f^2}\right) \qquad (9.49)$$

where r is the radial distance of the measured point and f is the focal length. Obviously, this approach will work for near vertical photographs

TABLE 9-1. PHOTOGRAMMETRIC REFRACTION (IN μrad) AT 45°

H (km)	h = 0.0 (Km) (1)	(2)	(3)	(4_1)	(4_2)	h = 1 (km) (2)	(3)	(4_1)	(4_2)	h = 2 (km) (2)	(3)	(4_1)	(4_2)
.5	4.8	6.5	4.9	6.5	6.4								
1.0	14.5	12.6	9.8	12.7	12.5	0							
1.5		18.5	14.9	18.5	18.3	6.0	8.3	6.0	6.0				
2.0	24.2	24.1	19.9	24.1	23.9	11.7	15.0	11.7	11.7				
2.5		29.3	25.0	29.4	29.3	17.1	21.0	17.2	17.2	5.6	9.0	5.6	5.6
3.0	33.9	34.3	30.0	34.4	34.3	22.3	26.7	22.3	22.4	10.9	16.7	10.9	10.9
3.5		39.0	35.0	39.1	39.1	27.1	32.2	27.2	27.3	15.9	23.6	15.9	16.0
4.0	43.6	43.5	39.8	43.6	43.7	31.7	37.4	31.8	32.0	20.6	29.9	20.6	20.8
4.5		47.7	44.6	47.8	48.0	36.1	42.4	36.1	36.4	25.1	35.7	25.1	25.4
5.0	53.3	51.6	49.2	51.7	52.0	40.2	47.2	40.2	40.6	29.3	41.2	29.4	29.6
5.5		55.3	53.6	55.4	55.8	44.0	51.8	44.1	44.5	33.3	46.4	33.4	33.7
6.0	58.1	58.8	57.8	58.9	59.3	47.6	56.2	47.7	48.1	37.0	51.2	37.1	37.4
6.5		62.1	61.9	62.2	62.5	51.0	60.3	51.1	51.5	40.6	55.7	40.6	41.0
7.0	63.0	65.1	65.6	65.2	65.5	54.2	64.2	54.3	54.6	43.9	60.0	44.0	44.2
7.5		67.9	69.2	68.1	68.3	59.2	67.9	57.3	57.5	47.0	63.9	47.1	47.2
8.0	67.9	70.6	72.5	70.7	70.7	59.9	71.3	60.0	60.1	49.8	67.5	50.0	49.9
8.5		73.0	75.5	73.1	72.9	62.5	74.4	62.6	62.4	52.5	70.8	52.6	52.4
9.0	77.6	75.2	78.3	75.4	74.9	64.9	77.2	65.0	64.5	55.0	73.9	55.2	54.6
9.5		77.3	80.8	77.5		67.1	79.8	67.2		57.3	76.6	57.5	
10.0	77.6	79.2	83.1	79.4		69.1	82.1	69.2		59.5	79.1	59.6	

(1) and (3) from 3rd Ed. (Schmid and Bertram), (2) Schut, (4_1) and (4_2) Saastamoinen

only and need only be computed once during the data-correction phase (e.g.: correction for distortion, comparator error, differential shrinkage, etc.). Tables 9-2 and 9-3 give the correction Δr in micrometres for points on a photograph at radial distances of 10 to 160 millimetres taken at altitudes of 1 through 10 kilometres. All corrections are for points at sea level and table 9-3 is for a camera of 88 millimetres focal length and table 9-2 for 152 millimetres.

In precision applications, however, one must consider the "tilts" of the photograph and compute the correction during the iterative solution.

This can be accomplished quite simply by a change in the approximate Z (in a local system) coordinate.

From figure 9-36 the change in the Z ground coordinate which will produce the same effect at the plate as atmospheric refraction is given by

$$dZ = \frac{(Z_0 - Z_G)\tan\theta}{\sin^2\theta}\alpha \qquad (9.50)$$

where α is in radians and the subscripts $_0$ and $_G$ denote "camera station" and "ground." The angle θ is the difference in direction between

TABLE 9-2. IMAGE DISPLACEMENTS (IN MICROMETRES) FOR ATMOSPHERIC REFRACTION AT FLYING HEIGHTS (H) AND RADIAL DISTANCE (r). FOCAL LENGTH = 152 mm AND GROUND ELEVATION OF ZERO

r_{mm}	H (km) 1	2	3	4	5	6	7	8	9	10
10	.1	.2	.3	.4	.5	.6	.7	.7	.8	.8
20	.3	.5	.7	.9	1.0	1.2	1.3	1.4	1.5	1.6
30	.4	.8	1.1	1.4	1.6	1.8	2.0	2.2	2.3	2.5
40	.5	1.0	1.5	1.9	2.2	2.5	2.8	3.0	3.2	3.4
50	.7	1.3	1.9	2.4	2.9	3.3	3.6	3.9	4.2	4.4
60	.9	1.7	2.4	3.0	3.6	4.1	4.5	4.9	5.2	5.5
70	1.1	2.0	2.9	3.7	4.4	5.0	5.5	6.0	6.4	6.7
80	1.3	2.5	3.5	4.4	5.3	6.0	6.7	7.2	7.7	8.1
90	1.5	2.9	4.2	5.3	6.3	7.1	7.9	8.6	9.1	9.6
100	1.8	3.5	4.9	6.2	7.4	8.4	9.3	10.1	10.8	11.3
110	2.1	4.0	5.7	7.3	8.6	9.9	10.9	11.8	12.6	13.3
120	2.5	4.7	6.7	8.5	10.1	11.5	12.7	13.8	14.6	15.4
130	2.8	5.4	7.7	9.8	11.6	13.2	14.7	15.9	16.9	17.8
140	3.3	6.2	8.9	11.3	13.4	15.2	16.8	18.3	19.5	20.5
150	3.7	7.1	10.2	12.9	15.3	17.4	19.3	20.9	22.3	23.4
160	4.2	8.1	11.6	14.7	17.4	19.8	22.0	23.8	25.4	26.7

TABLE 9-3. IMAGE DISPLACEMENTS (IN MICROMETRES) FOR ATMOSPHERIC REFRACTION AT FLYING HEIGHTS (H) AND RADIAL DISTANCE (r). FOCAL LENGTH = 88 mm AND GROUND ELEVATION OF ZERO

r_{mm}	H (km)									
	1	2	3	4	5	6	7	8	9	10
10	.1	.2	.3	.4	.5	.6	.7	.7	.8	.8
20	.3	.5	.7	.9	1.1	1.2	1.4	1.5	1.6	1.7
30	.4	.8	1.1	1.5	1.7	2.0	2.2	2.4	2.5	2.7
40	.6	1.2	1.7	2.1	2.5	2.8	3.1	3.4	3.6	3.8
50	.8	1.6	2.3	2.9	3.4	3.9	4.3	4.7	5.0	5.2
60	1.1	2.1	3.0	3.8	4.5	5.2	5.7	6.2	6.6	7.0
70	1.4	2.8	3.9	5.0	5.9	6.7	7.4	8.1	8.6	9.1
80	1.8	3.5	5.0	6.4	7.5	8.6	9.5	10.3	11.0	11.6
90	2.3	4.4	6.3	8.0	9.5	10.8	12.0	13.0	13.8	14.6
100	2.9	5.5	7.8	10.0	11.8	13.5	14.9	16.2	17.2	18.1
110	3.6	6.8	9.7	12.3	14.5	16.6	18.4	19.9	21.2	22.3
120	4.3	8.3	11.8	14.9	17.7	20.2	22.3	24.2	25.8	27.2
130	5.2	10.0	14.2	18.0	21.3	24.3	26.9	29.2	31.1	32.8
140	6.2	11.9	17.0	21.5	25.5	29.1	32.2	34.9	37.2	39.2
150	7.4	14.1	20.1	25.5	30.2	34.4	38.1	41.4	44.1	46.4
160	8.7	16.6	23.6	30.0	35.5	40.5	44.8	48.6	51.8	54.6

true vertical and the point in question. The full correction then can be written as

$$dZ = \frac{\Delta X^2 - \Delta Y^2 - \Delta Z^2}{\Delta Z} R \qquad (9.51)$$

which is easily done in each iteration of the triangulation solution for every point.

9.4.2.4.2 EARTH CURVATURE

Most analytical triangulation systems are designed to compute the solution using a rectangular coordinate system for object space. When done in this manner, there is no need for the so-called *earth curvature* correction. Some systems, however, introduce the coordinates of the ground objects using some map projection and therefore must account for the curvature of the earth when elevation is involved. This is not an advisable approach but the subject is included here for reference when the user has no other alternative.

From figure 9-37 it is apparent that the correction that must be made is the one that accounts for the difference in elevation between the earth's curved surface and a plane tangent to the earth at the photograph's nadir. The effect is that the images are apparently displaced toward the nadir at increasing amounts with increasing aspect angles. Assuming the photograph to be perfectly vertical, the correction (in micrometres) can be computed from (using notation in the figure)

$$\Delta r = \frac{Hr^3}{2Rf^2} \qquad (9.52)$$

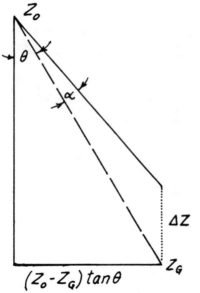

FIGURE 9-36. Correction for atmospheric refraction.

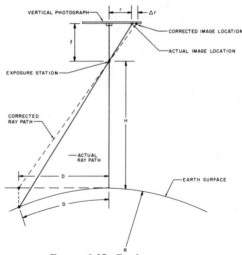

FIGURE 9-37. Earth curvature.

where r and f are in millimetres. The resulting correction is subtracted from the plate coordinate measurements. Table 9-4 demonstrates the magnitude (in micrometres) that the correction can reach for an ultra-wide-angle camera ($f = 88.5$ mm) at flying heights of 1 to 10 kilometres.

9.4.3 CHARACTERISTICS OF ANALYTICAL AEROTRIANGULATION ADJUSTMENTS

As pointed out earlier, the difference between various methods of analytical aerotriangulation are found mainly in the procedures and algorithms used in the adjustment. In general, the algorithms used depend either on collinearity or coplanarity which are covered adequately in chapter II and will not be discussed here. The adjustment procedure, on the other hand, varies with different users and most likely is chosen to reflect the particular requirements, configuration of cameras, computers and measuring equipment used by the organization. The adjustments also reflect the mathematical skills and ingenuity of the individuals responsible for developing the system. No system is static; that is, improvements and refinements in theory and programming are being made continuously. Consequently, a detailed description of each system is not warranted. Instead, the following discussion will cover the three basic methods in common use today; namely, sequential, independent models, and simultaneous (bundle) adjustments.

9.4.3.1 SEQUENTIAL ADJUSTMENT PROCEDURE

An early method of analytical aerotriangulation that evolved from the then current practices of aerotriangulation using analog devices, is the sequential adjustment or, as it is sometimes

called, the polynomial method. The procedure was developed by the National Research Council (NRC) of Canada (Schut, 1955–56) and the British Ordnance Survey (Thompson, 1959). As the name implies, the procedure is done in steps: first, strip aerotriangulation with respect to an arbitrary rectangular coordinate system (relative orientation); and second, transformation of the resulting strip coordinates to the ground control system (absolute orientation).

Before the strip aerotriangulation is begun, the measured photograph coordinates are reduced to an origin at the principal point, and radial corrections are applied for lens distortion, refraction, and usually earth curvature.

Strip aerotriangulation consists in assuming a convenient arbitrary orientation for the first photograph of the strip and computing in turn the orientation of each successive photograph with respect to the preceding one. This orientation is in two steps: the separate relative orientation of the photograph, and the scaling of the resulting model to the preceding one. The only criterion for relative orientation is that rays to corresponding points in two successive photographs must intersect. The model is scaled, after relative orientation, by means of points common to the preceding model (for the first model, by assuming a convenient arbitrary scale).

The relative orientation is not disturbed by errors in the identification of common points; and, if these points are chosen in fairly level terrain, the scaling is little affected by small differences in their identification. Furthermore, the result of the strip triangulation is independent of the choice of the first photograph and of the direction of the triangulation.

At the National Research Council, strip aerotriangulation was originally programmed (1953) for the minimum IBM 650 which contains a

TABLE 9-4. IMAGE DISPLACEMENTS (IN MICROMETRES) FOR FLYING HEIGHTS (H) AND RADIAL DISTANCE (r) CAUSED BY EARTH'S CURVATURE ($f = 88.5$ mm)

r_{mm}	H (km)									
	1	2	3	4	5	6	7	8	9	10
10					.1	.1	.1	.1	.1	.1
20	.1	.2	.2	.3	.4	.5	.6	.6	.7	.8
30	.3	.5	.8	1.1	1.4	1.6	1.9	2.2	2.4	2.7
40	.6	1.3	1.9	2.6	3.2	3.8	4.5	5.1	5.8	6.4
50	1.3	2.5	3.8	5.0	6.3	7.5	8.8	10.0	11.3	12.5
60	2.2	4.3	6.5	8.6	10.8	13.0	15.1	17.3	19.5	21.6
70	3.4	6.9	10.3	13.7	17.2	20.6	24.0	27.5	30.9	34.3
80	5.1	10.2	15.4	20.5	25.6	30.7	35.9	41.0	46.1	51.2
90	7.3	14.6	21.9	29.2	36.5	43.8	51.1	58.4	65.7	73.0
100	10.0	20.0	30.0	40.0	50.0	60.1	70.1	80.1	90.1	100.1
110	13.3	26.6	40.0	53.3	66.6	79.9	93.3	106.6	119.9	133.2
120	17.3	34.6	51.9	69.2	86.5	103.8	121.1	138.4	155.7	173.0
130	22.0	44.0	66.0	88.0	110.0	131.9	153.9	175.9	197.9	219.9
140	27.5	54.9	82.4	109.9	137.3	164.8	192.3	219.7	247.2	274.7
150	33.8	67.6	101.3	135.1	168.9	202.7	236.5	270.5	304.0	337.8
160	41.0	82.0	123.0	164.0	205.0	246.0	287.0	328.0	369.0	410.0

2,000-word drum. Later, the program was used with a simulator on the IBM 1620, Model II. The increasing use of analytical aerotriangulation, shown by the demand for copies of the program, dictated further revision and update of the program to facilitate its use in mass production. A FORTRAN version with a complete description of the system is given in Schut, 1973.

9.4.3.1.1 APPROXIMATE RELATIVE ORIENTATION

In both the NRC and the British Ordnance Survey methods, the aerotriangulation is based upon the coplanarity condition equation (for a detailed explanation *see* chapter II). Vectors in the image space are referred to as $\bar{x}_i$, the corresponding vectors in object space are $\bar{X}_i$, the orientation matrix is $\bar{A}_i$. The subscripts i denote the number of the photograph. The triangulation is performed in the X direction, and the x_i axes must be chosen roughly parallel to the strip direction.

The matrix $\bar{A}$ is composed of the conventional sequential rotations ω, ϕ, κ, as described in chapter II.[2] Then

$$\bar{X}_i = A_i \ x_i, \qquad (9.53)$$

and

$$\begin{vmatrix} B_x & B_y & B_z \\ X_i & Y_i & Z_i \\ X_{i+1} & Y_{i+1} & Z_{i+1} \end{vmatrix} = 0 \qquad (9.54)$$

is the basic coplanarity condition equation for the relative orientation of photograph $i + 1$ with respect to photograph i.

The base component B_x is given unit length, then B_y, B_z, and the matrix A_{i+1} must be determined. The base components B_y and B_z are computed as approximate values B_y° and B_z° plus corrections dB_y and dB_z. The values of B_y and B_z obtained in the preceding model, or zero, can be used as initial approximations B_y° and B_z°. The matrix A_{i+1} is computed by premultiplying an approximation A_{i+1}° by an orthogonal matrix R. Therefore, for relative orientation the corrections dB_y and dB_z, and the three independent parameters of R must be determined.

Equation 9.54 can now be replaced by

$$\begin{vmatrix} B_x & B_y^\circ + dB_y & B_z^\circ + dB_z \\ X_i & Y_i & Z_i \\ X_{i+1}^\circ + dX_{i+1} & Y_{i+1}^\circ + dY_{i+1} & Z_{i+1}^\circ + dZ_{i+1} \end{vmatrix} = 0. \qquad (9.55)$$

In this equation, X_{i+1}°, Y_{i+1}°, and Z_{i+1}° indicate approximations obtained with A_{i+1}°; and dX_{i+1}, dY_{i+1}, and dZ_{i+1} are corrections that are functions of the parameters of R.

The matrix A_i serves as the initial approximation A_{i+1}. Therefore, the orthogonal matrix R will be fairly close to the unit matrix. In the first approximation, R is equal to the sum of a unit matrix and a skew-symmetric matrix.

$$R_i = \begin{bmatrix} 1 & -\alpha_3 & \alpha_2 \\ \alpha_3 & 1 & -\alpha_1 \\ -\alpha_2 & \alpha_1 & 1 \end{bmatrix}. \qquad (9.56)$$

The elements of the orthogonal matrix R can be written in many different ways as functions of three parameters, as described by Schut (1958–59). In each case, differentiation of R with respect to the parameters shows that α_1, α_2, and α_3 may be identified either with the parameters of R or with simple functions of these parameters. Here they are identified with the sines of a primary rotation about the X axis, a secondary rotation about the Y axis, and a tertiary rotation about the Z axis. They are therefore equivalent to the sines of the conventional rotations ω, ϕ, κ.

The vector $\bar{X}_{i+1}$ is

$$\bar{X}_{i+1} = RA_{i+1}^\circ \bar{x}_i \qquad (9.57)$$
$$= R\bar{X}_{i+1}^\circ.$$

When the indicated multiplication is performed, the corrections to the initial approximation vector $\bar{X}_{i+1}^\circ$ are found as,

$$\begin{aligned} dX_{i+1} &= & + Z_{i+1}^\circ \alpha_2 - Y_{i+1}^\circ \alpha_3 \\ dY_{i+1} &= - Z_{i+1}^\circ \alpha_1 & + X_{i+1}^\circ \alpha_3 \\ dZ_{i+1} &= + Y_{i+1}^\circ \alpha_1 - X_{i+1}^\circ \alpha_2 \ . \end{aligned} \qquad (9.58)$$

Substituting these values in equation (9.55) gives the condition equation linearized with respect to the 5 parameters of relative orientation α_1, α_2, α_3, dB_y, and dB_z:

$$\begin{vmatrix} B_x & B_y^\circ & B_z^\circ \\ X_i & Y_i & Z_i \\ 0 & -Z_{i+1}^\circ & Y_{i+1}^\circ \end{vmatrix} \alpha_1 + \begin{vmatrix} B_x & B_y^\circ & B_z^\circ \\ X_i & Y_i & Z_i \\ Z_{i+1}^\circ & 0 & -X_{i+1}^\circ \end{vmatrix} \alpha_2 + \begin{vmatrix} B_x & B_y^\circ & B_z^\circ \\ X_i & Y_i & Z_i \\ -Y_{i+1}^\circ & X_{i+1}^\circ & 0 \end{vmatrix} \alpha_3$$

$$+ \begin{vmatrix} Z_i & X_i \\ Z_{i+1}^\circ & X_{i+1}^\circ \end{vmatrix} (B_y^\circ + dB_y) + \begin{vmatrix} X_i & Y_i \\ X_{i+1}^\circ & Y_{i+1}^\circ \end{vmatrix} (B_z^\circ + dB_z) + \begin{vmatrix} Y_i & Z_i \\ Y_{i+1}^\circ & Z_{i+1}^\circ \end{vmatrix} B_x = 0. \qquad (9.59)$$

This equation is linear with respect to the 5 elements of relative orientation dB_y, dB_z, α_1, α_2, and α_3. Each point whose coordinates x_i, y_i, x_{i+1}, and y_{i+1} have been measured yields one

---

[2] The NRC notation has been preserved in this description. For comparison with chapter II, A is equivalent to M^T.

such equation. The 5 orientation elements are computed from 5 such equations obtained from 5 suitably located points. This computation gives directly improved approximations for B_y and B_z. The cosines of the three rotations are computed from the sines; the orthogonal matrix $\mathbf{R}$ is computed, similarly to $\mathbf{A}$, from these sines and cosines; and an improved approximation for $\mathbf{A}_{i+1}$ is computed by premultiplying $\mathbf{A}_{i+1}^\circ$ by $\mathbf{R}$.

9.4.3.1.2 ADJUSTMENT OF RELATIVE ORIENTATION

When more than 5 points are available, improved values of B_y, B_z, and $\mathbf{A}_{i+1}$ are computed by a least-squares adjustment incorporating a maximum of 25 points.

The use of redundant equations requires weighting. For this purpose, the condition equation 9.54 must be differentiated with respect to the coordinates x_i, y_i, x_{i+1}, and y_{i+1}. This results in additional terms in equation 9.59.

Instead of 0, the right side of equation 9.59 becomes

$$
\begin{aligned}
&-
\begin{vmatrix}
B_x & B_y^\circ & B_z^\circ \\
\alpha_{11_i} & \alpha_{21_i} & \alpha_{31_i} \\
X_{i+1}^\circ & Y_{i+1}^\circ & Z_{i+1}^\circ
\end{vmatrix} dx_i
-
\begin{vmatrix}
B_x & B_y^\circ & B_z^\circ \\
\alpha_{12_i} & \alpha_{22_i} & \alpha_{32_i} \\
X_{i+1}^\circ & Y_{i+1}^\circ & Z_{i+1}^\circ
\end{vmatrix} dy_i \\[4mm]
&-
\begin{vmatrix}
B_x & B_y^\circ & B_z^\circ \\
X_i & Y_i & Z_i \\
\alpha_{11_{i+1}} & \alpha_{21_{i+1}} & \alpha_{31_{i+1}}
\end{vmatrix} dx_{i+1}
-
\begin{vmatrix}
B_x & B_y^\circ & B_z^\circ \\
X_i & Y_i & Z_i \\
\alpha_{12_{i+1}} & \alpha_{22_{i+1}} & \alpha_{32_{i+1}}
\end{vmatrix} dy_{i+1}
\end{aligned}
\tag{9.60}
$$

in which the α_{jk} are elements of matrix $\mathbf{A}$.

These terms are needed only for the computation of the weights and correlation of equations (9.59). Normal equations are formed, as described in chapter II, and solved for the five orientation elements.

If the observations for different points are not correlated, equations 9.59 are free from correlation. In this case, only the weights of the individual equations need be computed; as a result, the computation of normal equations is very simple. For the computer program, this case has been assumed. Furthermore, the four observations for a pair of corresponding image points are presumed to be of equal weight and free from correlation. Assuming, also, practically vertical photographs, it then follows from equations 9.54 that all equations 9.53 have equal weight.

If the solution of the normal equations gives values of α_1, α_2, or α_3 that are greater than a given test value (0.007, corresponding to about 25'), the adjustment is repeated. In practice, repetition occurs in less than 1 percent of the triangulated models.

9.4.3.1.3 SCALING OF MODELS AND COMPUTATION OF STRIP COORDINATES

After the relative orientation of photograph $i + 1$ with respect to photograph i, rays from corresponding image points through the respective projection centers should intersect. However, using the measured photograph coordinates, they will, in general, fail to intersect because they lie in different planes. Therefore, a point must be defined which will represent the point of intersection. A suitable point is one midway along a vector $\mathbf{D}$ between the vectors $\mathbf{X}_i$ and $\mathbf{X}_{i+1}$ in the region where the rays come closest together. The best choice for $\mathbf{D}$ is the cross product of $\mathbf{X}_i$ and $\mathbf{X}_{i+1}$. However, since the triangulation is performed in the X direction, it is possible to choose for $\mathbf{D}$ the unit vector in the direction of the Y axis. In this case, the scalars k_1 and k_2 become

$$
k_1 = \frac{Z_{i+1}B_x - X_{i+1}B_z}{Z_{i+1}X_i - X_{i+1}Z_i}
\tag{9.61}
$$

$$
k_2 = \frac{X_i B_z - Z_i B_x}{Z_{i+1}X_i - X_{i+1}Z_i}.
\tag{9.62}
$$

The coordinates of the required point are then

$$
\begin{aligned}
X &= X_{c_i} + k_1 X_i \\
Y &= 1/2\big[(Y_{c_{i+1}} - k_2 Y_{i+1}) + (Y_{c_i} + k_1 Y_i)\big] \\
Z &= Z_{c_i} + k_1 Z_i.
\end{aligned}
\tag{9.63}
$$

The residual Y parallax, is given by

$$
Y \text{ parallax} = (Y_{c_{i+1}} - k_2 Y_{i+1}) - (Y_{c_i} + k_1 Y_i).
\tag{9.64}
$$

Before these computations are performed for all measured points, equations 9.61, 9.62 and 9.63 are used to compute the Z coordinates of the points common to the preceding model.

These coordinates are compared with those in the preceding model and used for the computation of a scale factor. The three base components of the new model are multiplied by the scale factor and then added to the coordinates of projection center C_i to obtain those of projection center C_{i+1}.

9.4.3.1.4 ADJUSTMENT OF STRIP COORDINATES

Inasmuch as the triangulation is computed in an arbitrary coordinate system and at an arbitrary scale, it is necessary to adjust the computed coordinates to ground control. For strip transformation, linear conformal transformations are used if no corrections for strip deformation are needed. If such corrections must be applied, then polynomials of the second and higher degree in the strip coordinates are employed to conduct a sequence of transformations, each of which is conformal in two dimensions. In this way, the distortion of individual models is kept to a minimum.

9.4.3.2 INDEPENDENT MODELS METHOD

The basic unit of the independent models method is the relatively oriented photogrammetric model. The position of the model in space and its scale are arbitrary. The object points from each model are given in a three-dimensional Cartesian coordinate system of arbitrary origin. Because the scale of each model is arbitrary, no base change is needed and orientation can be performed using first or second order analog instruments. Likewise, stereo- or monoscopic comparators can be used as well. These facts allow for a universal application of the independent models method because the instrument requirements are met almost everywhere, even in small photogrammetric agencies.

In the block adjustment procedure, all models of the block are simultaneously transformed to the terrain, using all available control points and tie points (simultaneous absolute orientation). The perspective center of each photograph is used as an additional tie point with the special task of pitch control (*see* figure 9-38). Several methods are used in practice for the determination of the perspective center coordinates in analog devices. A general description of some of these methods follows.

1. If the position of the perspective center is not affected by the relative orientation (Wild A7, A8, A10 and other instruments) the coordinates can be determined once and used for the subsequent models, as long as the model coordinate system

is maintained (for instance, no freehand motion with the A8). For the perspective center determination itself one of the following two procedures can be recommended.

   a) The coordinates of 6 grid points are measured in one z level and the perspective center is determined by a resection in space using the known grid coordinates and the calibrated focal length. The left- and right-hand side perspective centers are determined separately.

   b) Using a grid plate on the left- and a right-hand side, a grid model is established by relative orientation and the spatial coordinates of 6 points of this grid model are measured. The theoretical coordinates of the 6 points and the two perspective centers of the grid model are known from the grid width and the calibrated focal lengths. The theoretical grid model is then transformed to the measured one by a spatial similarity transformation, which gives the coordinates of both perspective centers in the model coordinate system.

2. In some instruments (Zeiss Planimat and Planicart, Kern PG2 and PG3 and others) the perspective center coordinates vary with the relative orientation and must be determined separately for every model. Therefore the manufacturers have developed special procedures for the successive determination of the two perspective centers. With the aid of bubble levels (Zeiss) or autocollimation devices (Kern), the mechanical guide rod is set perpendicular to the horizontal x,y plane. Then z is varied until an annular mark of the guide rod (at a calibrated distance from the space point) coincides with an index, representing the projection cardan. Subsequent measurement gives the x and y coordinates of the perspective center directly. The z coordinate is derived from the recorded value by adding the calibrated distance between the annular mark and the space point.

3. When the measurement is performed by stereo- or mono-comparators and the model formation is done computationally the perspective centers can be computed together with the model points.

9.4.3.2.1 STANDARD PROCEDURE OF BLOCK ADJUSTMENT BY INDEPENDENT MODELS

9.4.3.2.1.1 The Mathematical Model. To allow for a simultaneous absolute orientation of all models in a block, a spatial similarity transformation is applied to every model. If a terrain point i is measured in model j (in an arbitrary coordinate system) the relationship between the model coordinates x_{ij} y_{ij} z_{ij} and the terrain coordinates x_i y_i z_i is given by:

$$\begin{bmatrix} x_i \\ y_i \\ z_i \end{bmatrix} = \lambda_j \mathbf{R}_j \begin{bmatrix} x_{ij} \\ y_{ij} \\ z_{ij} \end{bmatrix} + \begin{bmatrix} \Delta x_j \\ \Delta y_j \\ \Delta z_j \end{bmatrix} \quad (9.65)$$

where λ_j is a scale factor, $\mathbf{R}_j$ is a three dimensional orthogonal matrix and Δx_j, Δy_j, Δz_j are translations in the x, y, and z directions. Among

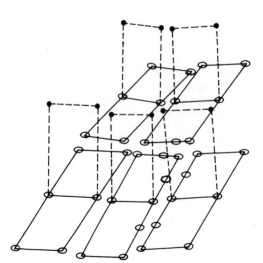

FIGURE 9-38. Graphic of independent models methods.

the different possible formulations of $\mathbf{R}$ the so-called "Rodrigues matrix" is recommended, because no trigonometric functions are required in its formation. This matrix is given by

$$v_{zc_i} = z_i - z_{c_i} \} \text{ height} \quad (9.70)$$

where v_{xc_i}, v_{yc_i}, v_{zc_i} are the residuals of the control point coordinates x_{c_j}, y_{c_j}, z_{c_j}, which are weighted in accordance with their accuracy. If a

$$\mathbf{R} = \frac{1}{1 + (a^2 + b^2 + c^2)/4} \begin{bmatrix} 1 + (a^2 - b^2 - c^2)/4 & -c + ab/2 & b + ac/2 \\ c + ab/2 & 1 + (-a^2 + b^2 - c^2)/4 & -a + bc/2 \\ -b + ac/2 & a + bc/2 & 1 + (-a^2 - b^2 + c^2)/4 \end{bmatrix}. \quad (9.66)$$

In cases where the quantities a, b, and c are small, they are identical with ω, ϕ and κ. Using equations (9.65) and (9.66) for each measured point in every model three observation equations can be formed as

$$\begin{bmatrix} v_{x_{ij}} \\ v_{y_{ij}} \\ v_{z_{ij}} \end{bmatrix} = \begin{bmatrix} x_i \\ y_i \\ z_i \end{bmatrix} - \lambda_j \mathbf{R}_j \begin{bmatrix} x_{ij} \\ y_{ij} \\ z_{ij} \end{bmatrix} - \begin{bmatrix} \Delta x_j \\ \Delta y_j \\ \Delta z_j \end{bmatrix} \quad (9.67)$$

where, v_x, v_y, v_z are the residuals of the transformed model coordinates. The linearization of the nonlinear observation equations (9.67) requires initial values λ_o and a_o, b_o, c_o for all models. If the models are roughly pretransformed with respect to the terrain coordinate system, initial values of $\lambda_o = 1$ and $a_o = 0, b_o = 0, c_o = 0$ can be assumed. The linearized observation equations are then written as:

control point is to be kept unchanged, an infinite weight (10^{10} or 10^{20}) is assigned to its coordinates. On the other hand, with a weight zero, a control point is treated as a check point without effecting the adjustment.

The solution of the system of observation equations (9.68), (9.69), and (9.70), which will be treated in 9.4.4.2.2, yields the $7 m$ parameters λ_j, a_j, b_j, c_j, Δx_j, Δy_j, Δz_j. Using these parameters, all models are transformed with the rigorous similarity transformations given by (9.65) and (9.66).

Because of the nonlinearity of the equations (9.67) the procedure must be iterated, but 2 or 3 iteration steps are sufficient. The unknown terrain coordinates are then computed after the last iteration step.

The 7 parameter treatment is contained in the computer programs called SPACE-M (Blais, 1973) and ALBANY (Erio, 1975). An efficient alternative, however, to the 7 parameter ap-

$$\begin{bmatrix} v_{x_{ij}} \\ v_{y_{ij}} \\ v_{z_{ij}} \end{bmatrix} = \begin{bmatrix} 1 & 0 & 0 \\ 0 & 1 & 0 \\ 0 & 0 & 1 \end{bmatrix} \begin{bmatrix} x_i \\ y_i \\ z_i \end{bmatrix} + \begin{bmatrix} -x_{ij} & 0 & -z_{ij} \\ -y_{ij} & z_{ij} & 0 \\ -z_{ij} & -y_{ij} & x_{ij} \end{bmatrix} \begin{bmatrix} y_{ij} & -1 & 0 & 0 \\ -x_{ij} & 0 & -1 & 0 \\ 0 & 0 & 0 & -1 \end{bmatrix} \begin{bmatrix} d\lambda_j \\ da_j \\ db_j \\ dc_j \\ \Delta x_j \\ \Delta y_j \\ \Delta z_j \end{bmatrix} - \begin{bmatrix} x_{ij} \\ y_{ij} \\ z_{ij} \end{bmatrix}. \quad (9.68)$$

Equations (9.68) are evaluated for all model points and the perspective centers. Usually all model coordinates are treated as uncorrelated observations, however, with different weights. Considering a block of m models and n terrain points, the block adjustment will contain $7 m$ unknown transformation parameters plus $3 n$ unknown terrain coordinates.

In general the available control is not error free and consequently, the control point coordinates are treated as observations. If terrain point i is a control point in planimetry (height), two (one) additional observation equations are written:

$$\begin{bmatrix} v_{xc_i} \\ v_{yc_i} \end{bmatrix} = \begin{bmatrix} 1 & 0 \\ 0 & 1 \end{bmatrix} \begin{bmatrix} x_i \\ y_i \end{bmatrix} - \begin{bmatrix} x_{c_i} \\ y_{c_i} \end{bmatrix} \} \text{ planimetry} \quad (9.69)$$

proach is the "planimetry-height-iteration" which is used in the computer program PAT-M43, developed at Stuttgart University (Ackermann, et al, 1973). Here the 7 transformation parameters are split into two groups of 4 and 3 parameters. In the planimetric iteration step by which the process starts, the parameters λ_j, c_j, Δx_j, Δy_j are determined from the observation equations of the x, y coordinates of the model points and control points. If a and b are neglected in (9.66) and λ and c are replaced by the new parameters,

$$\begin{aligned} d &= \lambda(1 - c^2/4)/(1 + c^2/4) \\ e &= \lambda c/(1 + c^2/4), \end{aligned} \quad (9.71)$$

the nonlinear observation equations (9.67) degenerate into the following linear system

$$\begin{bmatrix} v_{x_{ij}} \\ v_{y_{ij}} \end{bmatrix} = \begin{bmatrix} 1 & 0 & x_i \\ 0 & 1 & y_i \end{bmatrix} + \begin{bmatrix} -x_{ij} & y_{ij} & -1 & 0 \\ -y_{ij} & -x_{ij} & 0 & -1 \end{bmatrix} \begin{bmatrix} d_j \\ e_j \\ \Delta x_j \\ \Delta y_j \end{bmatrix} \tag{9.72}$$

Equations (9.72) are written for all model points but not for perspective centers, which are not used at this stage. Furthermore, observation equations of type (9.69) are written for all planimetric control points. The solution of this system gives the parameters d, e, Δx, Δy for all models. Using d and e, the original parameters λ and c can then be computed from

$$\lambda = (d^2 + e^2)^{1/2}$$
$$c = 2(\lambda - d)/e. \tag{9.73}$$

Finally, λ, c, Δx, Δy along with the rigorous spatial similarity transformations given by (9.65) and (9.66) are applied to all models. From this, transformed coordinates x, y, z are obtained for all model points, including the perspective centers.

In the successive height iteration steps, the parameters a_j, b_j, Δz_j are determined from the observation equations of the model heights z, the perspective center coordinates x, y, z and the control heights z.

The degenerate observation equation system is obtained from (9.67) by setting $\lambda = 1$ and neglecting c, Δx and Δy. This is justified by reason of the preceding planimetric iteration step. Due to the fact that parameters a and b are included together in the observation equations, they remain nonlinear and therefore require a linearization. With $a_o = 0$ and $b_o = 0$ we obtain a linearized system of observation equations which is a subsystem of (9.68), that is,

process is terminated following the second or third height iteration step, after which the final terrain coordinates are computed (see section 9.4.4.2.2).

With respect to accuracy and speed of convergence, the planimetric-height-iteration is equivalent to the 7 parameter approach. However, beyond this, the planimetry-height-iteration scheme has the following advantages.

1. The planimetric iteration step is linear and therefore requires no initial values λ_o and c_o (scale and azimuth). For that reason no procedures are necessary for obtaining pretransformed model coordinates.
2. In case a direct method is used for the solution of the reduced normal equation system (see section 9.4.4.2.2) the computer running time is reduced considerably (roughly at a factor 3) because two smaller systems of size $4m$ and $3m$ are solved occasionally instead of one large system of size $7m$.
3. Weighting of the observations is simpler, because the model coordinates x, y and z are used in two separate adjustments. From experience the following weight assumptions can be recommended for homogeneous data.
   Planimetric block adjustment
   $$w_x = w_y = 1 \quad \text{for model points,}$$
   Height block adjustment
   $$w_z = 1 \quad \text{for model points,}$$
   $$w_x = w_y = 0.25, w_z = 1 \text{ for perspective centers.}$$
4. The planimetric block adjustment has independent importance only when the x, y terrain co-

$$v_{z_{ij}} = z_i + \begin{bmatrix} -y_{ij} & x_{ij} & -1 \end{bmatrix} \begin{bmatrix} da_j \\ db_j \\ \Delta z_j \end{bmatrix} - z_{ij} \tag{9.74}$$

$$\begin{bmatrix} v_{x_{ij}} \\ v_{y_{ij}} \\ v_{z_{ij}} \end{bmatrix} = \begin{bmatrix} 1 & 0 & 0 \\ 0 & 1 & 0 \\ 0 & 0 & 1 \end{bmatrix} \begin{bmatrix} x_i \\ y_i \\ z_i \end{bmatrix} + \begin{bmatrix} 0 \\ z_{ij} \\ -y_{ij} \end{bmatrix} \begin{bmatrix} -z_{ij} & 0 \\ 0 & 0 \\ x_{ij} & -1 \end{bmatrix} \begin{bmatrix} da_j \\ db_j \\ \Delta z_j \end{bmatrix} - \begin{bmatrix} x_{ij} \\ y_{ij} \\ z_{ij} \end{bmatrix} \tag{9.75}$$

Equations (9.74) are written for all model points but not for perspective centers. For these, equations (9.75) are used. The height control points are introduced using observation equations of type (9.70).

After the system is solved, the solution parameters a, b, Δz are substituted back into (9.65) and (9.66) by which all models are rigorously transformed in x, y, and z.

Due to the mutual effect between planimetry and height and the nonlinearity of the problem, both the planimetric and the height block adjustment must be repeated alternately. Usually, the

ordinates are of interest (for instance in cadastre) and the independent models are preleveled computationally or in the instrument.

9.4.3.2.2 SYSTEM SOLUTION

The following procedures for the computation of the unknown transformation parameters of all models from the observation equations of model points and control points refer to one individual iteration step, containing 7 transformation parameters (spatial approach), 4 parameters (planimetric block adjustment) or 3 parameters (height block adjustment). Each leads to a system of observation equations of the following type

$$v = Ac + Bp - f \quad (9.76)$$

f = vector of observations

v = vector of residuals

c = vector of unknown terrain coordinates

p = vector of unknown transformation parameters

A,B = coefficient matrices.

System (9.76) is solved by least squares, minimizing $v^T W v$ (W = weight matrix of the observations). The resulting normal equations are

$$\begin{bmatrix} A^TWA & A^TWB \\ B^TWA & B^TWB \end{bmatrix} \begin{bmatrix} c \\ p \end{bmatrix} = \begin{bmatrix} A^TWf \\ B^TWf \end{bmatrix}. \quad (9.77)$$

Eliminating the unknown terrain coordinates from (9.77) results in the following reduced normal equation system

$$(B^TWB - B^TWA(A^TWA)^{-1}A^TWB) p = (B^TWf - B^TWA(A^TWA)^{-1}A^TWf). \quad (9.78)$$

When programmed for a computer, the formation of the observation equations (9.76) along with the normal equations (9.77) can be bypassed and the reduced normals (9.78) formed directly. Direct methods are mainly used today for the solution of the reduced system (9.78) even though sophisticated programming is required to obtain favorable computer running time. The most advanced procedure is recursive partitioning (see chapter II), which can be based on the method of Gauss or Cholesky. Because computing time is strongly dependent on the band width of the reduced normal equation matrix, a minimization of the band width is desirable. In most cases this can be done by automatic reordering of the models normal to the direction of flight.

The iteration process consists of a number of successive individual adjustments. Within each iteration step, reduced normal equations, according to (9.78) are directly formed and solved. With the aid of the transformation parameters the rigorously transformed model coordinates are then obtained, which are thereafter input to the succeeding iteration step.

The unknown terrain coordinates c are computed by substituting the transformation parameters p back into the normal equation system (9.77) associated with the last iteration step. The resulting terrain coordinates x_i y_i z_i are the weighted arithmetic mean of the finally transformed coordinates x_{ij} y_{ij} z_{ij} of point i in model j along with the control point coordinates x_{ci} y_{ci} z_{ci}, if the point in question is in addition a control point.

The residuals $v_{x_{ij}}$ $v_{x_{ij}}$ $v_{x_{ij}}$ are the differences between the terrain coordinates x_i y_i z_i and the finally transformed model coordinates x_{ij} y_{ij} z_{ij}. Finally, the variance factor σ_0^2 is computed from the residual vector v using

$$\sigma_0^2 = v^T Pv/r$$
$$r = \text{redundancy}. \quad (9.79)$$

9.4.3.2.3 APPLICATIONS

Independent model block adjustment has found worldwide applications, in particular, when using the time and cost saving form of planimetry-height iteration. It is estimated that several tens of thousands of models per year are adjusted using this method.

On practical projects the block adjustment must usually be repeated two or three times because of gross errors in the data. The additional computing time, however, can be reduced by starting the repetition runs using the transformed model coordinates of the preceding run, thus requiring only one new spatial iteration step.

In addition to the classical medium-scale applications, two important groups of block adjustment projects can be distinguished today. The first one covers engineering, cadastre and network densification projects where block adjustment by independent models is used for precise point determination. Here photo-scales of 1:4000 to 1:8000 are typical. In planimetry, dense perimeter control is used and absolute accuracies of 5 cm to 10 cm are attained. If an adequate height accuracy of 10 cm to 20 cm is required (not in cadastre) a relatively dense network of height control points must be provided.

The second group covers the small scale projects, where independent-model block adjustment is used to determine the minor control points needed for plotting individual models. Here, large blocks of 1000 or more models represent the most economic solution, because only a very small number of planimetric control points are required to meet the accuracy standards of 0.1 mm to 0.2 mm at map scale. To demonstrate, a general description of the Canadian planimetric test block COED is given (Gauthier et al, 1973).

Figure 9-39 shows the block, covering an area of 97,000 km² and consisting of 2193 models at a photo scale of 1:60,000 (in the plot only every 5th exposure station is marked). Twelve points were selected from the given control and used in the block adjustment, simulating an Aerodist network of grid width of approximately 130 km. The absolute accuracy was estimated as the RMS value $\mu_{x,y}$ of the x and y discrepancies at 185 check points and amounted to $\mu_{x,y} = 11.1$ m or, 0.22 mm at 1:50,000 map scale.

This accuracy is sufficient, in particular if one considers the limited accuracy of the check point coordinates which were estimated to be 5 m. The computing time for one run of the planimetric block adjustment of COED was 2838 central-processor seconds on a CDC 6600 computer, corresponding to 1.3 sec per model.

For small scale mapping, the z coordinate is usually desired at a higher accuracy than the x and y coordinates. To meet this accuracy requirement, an uneconomical number of height

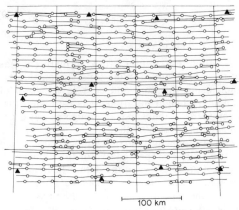

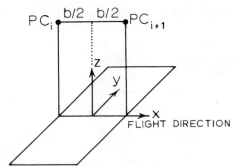

FIGURE 9-40. Model coordinate system for self calibration.

FIGURE 9-39. Planimetric test block.

control points are necessary. A similar result however, can be obtained with fewer height control points, when additionally APR (airborne profile recorder) and/or statoscope measurements are made and all data are processed in a simultaneous adjustment (*see* section 9.4.4.3.3).

9.4.3.2.4 EXTENSIONS OF THE STANDARD PROCEDURE

9.4.3.2.4.1 Self-calibrating Block Adjustment by Independent Models. The standard procedure of independent-model block adjustment presumes that the errors of the model coordinates are purely random. The actual model data, however, are usually superposed on systematic errors, which degrade the model accuracy and propagate in an unfavorable manner in the block.

The concept of self-calibration is based on a simultaneous compensation of these systematic errors with the aid of additional parameters in the block adjustment. In this way, only the random errors of the model coordinates remain and the final block accuracy is improved without additional measurements. The self-calibration concept, applicable to the 7 parameter approach as well as to the planimetry-height-iteration, is only briefly described here. A comprehensive treatment of the subject is given in Ebner and Schneider, 1974.

The formulation of equal corrections for different models requires the model coordinate system shown in figure 9-40. Implementing the concept of self-calibration, the observation equations of the model coordinates x, y, and z are extended by the effect of suitably selected additional parameters. Figure 9-41 shows the effect of 8 planimetric correction terms p_1 to p_8 and of 6 terms h_1 to h_6 for height correction. Together, with the 7 transformation parameters, they form a system of nearly orthogonal polynomials which allow for a compensation of all systematic deformations appearing in 8 model points (including the coordinates of the two perspective centers).

In cases of homogenous data, the additional parameters may be treated as block invariant terms. On the other hand, if different cameras, films, and/or measuring instruments are used in one project, the block will fall into several groups of models, for which different systematic deformations can be expected. Accordingly, separate sets of additional parameters should be attached to these groups. After the block adjustment, however, the values of the additional parameters must be checked by statistical significance tests. (For more details *see* Ebner and Schneider, 1974 and Ebner 1976.)

Using the terminology of collocation, the systematic deformations of the model coordinates can be interpreted as the statistical signal. Therefore, it is adequate to treat the additional parameters as properly weighted observations of value zero.

The extended observation equations lead to reduced normals, containing both the unknown transformation parameters and the unknown additional parameters. A favorable banded and bordered structure of the reduced normal equation matrix can be obtained with the additional parameters forming the border.

The efficiency of the concept of self-calibration has been demonstrated by practical results obtained with the well controlled OEEPE test block Oberschwaben (1:28,000 Zeiss RMKA photography, signalized tie and control points, Zeiss PSK measurements). From this extensive material a wide angle sub-block of 100 models (4 strips of 20 percent sidelap) was selected, covering an area of 20×62.5 km². In planimetry, perimeter control was used with a spacing of $i =$ 2, 4, 8 and 11 base lengths. As usual, in height, dense lines of control points were arranged across the strip direction. The bridging distance was chosen as $i = 4, 8, 12$ and 25 base lengths, with a reduced spacing of $i/2$ along the open block edges.

The block adjustments were based on planimetry-height-iteration and the additional parameters were written as block invariant terms. In figure 9-42 the systematic model deformations are reconstructed from the adjusted values

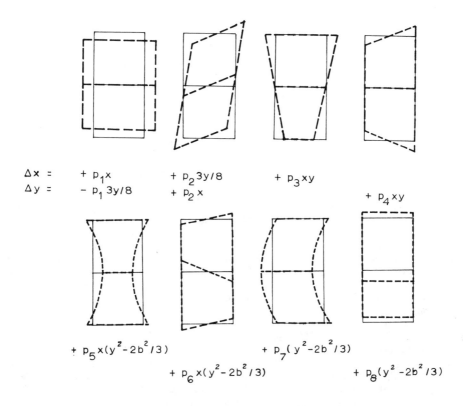

$$\Delta x = \quad + p_1 x \qquad + p_2 3y/8 \qquad + p_3 xy$$
$$\Delta y = \quad - p_1 3y/8 \qquad + p_2 x \qquad\qquad\qquad + p_4 xy$$

$$+ p_5 x(y^2 - 2b^2/3) \qquad\qquad + p_7(y^2 - 2b^2/3)$$
$$+ p_6 x(y^2 - 2b^2/3) \qquad\qquad + p_8(y^2 - 2b^2/3)$$

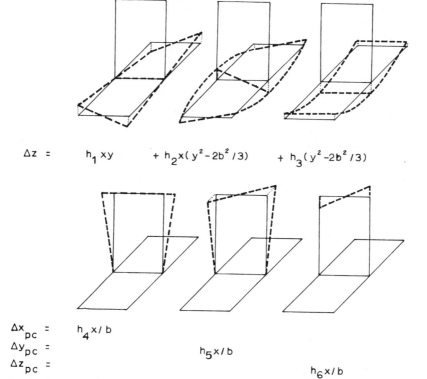

$$\Delta z = \quad h_1 xy \quad + h_2 x(y^2 - 2b^2/3) \quad + h_3(y^2 - 2b^2/3)$$

$$\Delta x_{pc} = \quad h_4 x/b$$
$$\Delta y_{pc} = \qquad\qquad\qquad h_5 x/b$$
$$\Delta z_{pc} = \qquad\qquad\qquad\qquad\qquad h_6 x/b$$

FIGURE 9-41. Effects of planimetric correction terms p_1 to p_8 and height terms h_1 to h_6.

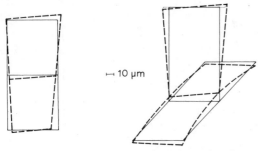

FIGURE 9-42. Systematic model deformation from additional parameters.

of the additional parameters. The maximum amounts are in the order of 10 μm at the photoscale.

The individual results are represented in tables 9-5 and 9-6. As table 9-5 shows, self-calibration leads to a considerable improvement of the planimetric accuracy. The RMS values μ_{xy} of the x, y discrepancies at the check points are improved by a factor of 1.6 to 2.9 which corresponds to 6.3 μm to 7.7 μm at the photo-scale (18 cm to 22 cm in the terrain). The measure σ_{op} represents the standard deviation of the model coordinates x and y for which the corresponding value of 4.3 μm is near the accuracy limit attainable in photogrammetry today.

As table 9-6 shows, the height accuracy shows less improvement. This is valid for σ_{oh} (standard deviation of the model heights) as well as for the absolute accuracy μ_z. The only exception appears with the extreme control distribution $i = 25$, where the relatively poor accuracy $\mu_z = 65.0$ μm improved by a factor of 2.4 and a very reasonable value $\mu_z = 26.7$ μm is obtained.

In practice, the accuracy improvement by self-calibration will vary from project to project, depending on the occasional proportion in size between the systematic and the random errors. In most cases, however, one can be assured that the inherent systematic deformations of the model coordinates are well compensated and that the remaining errors are mainly random.

9.4.3.2.4.2 *Implementation of Equal Elevation Constraints.* If arbitrary points on the shoreline of a lake are measured in one or several models, they can be used to improve model leveling, even if the absolute height of the water

level is unknown. There are several ways to use such measurements in the block adjustment. However, an approach which is applied in the computer program PAT-M43, (based on planimetry-height-iteration Ackerman *et al*, 1972) is as follows.

If a point is measured on the shoreline of lake *l*, the point number *i* is split into three components:

1. a code digit which indicates that the point is a shoreline point,
2. the lake number *l*,
3. a subnumber *k*, to distinguish the individual points along the same shoreline.

In the planimetric block adjustment the lakeshore points in model *j* have no particular function and are simply treated like all other points (*see* Ackerman, 1975). In the height block adjustment, on the other hand, the observation equations of the model heights are altered to

$$v_{z_{klj}} = z_l + \begin{bmatrix} -y_{klj} & x_{klj} & -1 \end{bmatrix} \begin{bmatrix} da_j \\ db_j \\ \Delta z_j \end{bmatrix} - z_{klj}. \tag{9.80}$$

Equations (9.80) are written for all measurements of lake shore points in all models. In the computation, equal elevation of all shoreline points of a lake is constrained by the formulation of one unique adjusted height z_l for all measurements z_{klj}. The equations (9.80) are weighted adequately and added to the hitherto existing observation equations (9.74), (9.75), and (9.70) of the height block adjustment.

At times, a lake will stretch over many models of the block. Because of the common height constraint of all shoreline points of this lake, all models concerned are directly interrelated, which will accordingly increase the band width of the reduced normal equation matrix. The computing time for solving this system of reduced normals, therefore, could increase beyond acceptable limits. In the PAT-M43 program this danger is avoided by subdividing a lake into a number of independent sublakes each covering only one model. In Canada, where an extraordinary number of lakes exist, this approach is applied regularly. Although the inherent efficiency of the equal-elevation constraints is in theory somewhat reduced by the sublake concept, very

TABLE 9-5. PLANIMETRIC ACCURACY WITH AND WITHOUT SELF-CALIBRATION.

Control Version	Control Points	Check Points	Without Selfcalibration		With Selfcalibration		Accuracy Improvement	
			σ_{op} [μm]	μ_{xy} [μm]	σ_{op} [μm]	μ_{xy} [μm]	σ_{op}	μ_{xy}
i = 2	32	226	6.8	9.9	4.4	6.3	1.5	1.6
i = 4	16	242	6.5	13.4	4.3	6.6	1.5	2.0
i = 8	8	250	6.2	20.0	4.3	7.1	1.4	2.8
(i = 11)	6	252	6.1	22.1	4.3	7.7	1.4	2.9

TABLE 9-6. HEIGHT ACCURACY WITH AND WITHOUT SELF-CALIBRATION.

Control Version	Control Points	Check Points	Without Selfcalibration		With Selfcalibration		Accuracy Improvement	
			σ_{oh} [μm]	μ_z [μm]	σ_{oh} [μm]	μ_z [μm]	σ_{oh}	μ_z
i = 4	47	178	8.4	14.7	7.6	14.1	1.1	1.0
i = 8	26	199	8.3	19.0	7.6	17.1	1.1	1.1
i = 12	19	206	8.3	22.1	7.6	18.9	1.1	1.2
i = 25	12	213	8.3	65.0	7.6	26.7	1.1	2.4

favorable results are obtained in practice. With the Reindeer Lake test block of 356 models, flown with 20 percent sidelap at a photo-scale of 1:24,000, 675 lake points were measured on 52 lakes. Using a minimum of only 15 vertical control points, an RMS value $\mu_z = 1.16$ m was obtained from the discrepancies at 117 height check points (Mosaad et al, 1975).

9.4.3.2.4.3 Combined Block Adjustment of Statoscope, APR and Independent Model Data. Statoscope measurements are related to the z coordinates of the perspective centers. These height data differ from ordinary vertical control by the fact that the isobaric surface, to which the statoscope reading refers, is not known. Consequently, separate additional parameters are needed for each strip to describe the height of the isobaric reference surface along the flight line. For instance, if strip s was flown with a statoscope, each perspective center i of this strip would require one additional observation equation of the following type,

$$v_{z_{is}} = z_i - (f_s + g_s t_{is}) - z_{is},\qquad (9.81)$$

where $v_{z_{is}}$ is the residual of the statoscope reading z_{is}. The term $(f_s + g_s t_{is})$ provides a constant shift and a tilt correction of the isobaric surface along the flight line s. The coefficient t_{is} represents the distance of the exposure station i from the (arbitrary) beginning of strip s. As this distance is not normally available, the elapsed time is used instead and constant speed of the aircraft is assumed.

APR (airborn profile recorder) data are related to the z coordinates of terrain points, which are obtained as the difference of statoscope readings and the vertical distance measurements between the airplane and the terrain. Usually the terrain profile is recorded continuously. In addition a 35 mm spotting camera, the axis of which is aligned to the measuring beam, takes photographs at short intervals. Thus a number of points of the APR terrain profile can be identified and transferred to the air survey photographs for measurement. APR profiles can be obtained simultaneously with the air survey photograph or from separate low altitude APR flights.

Again, the isobaric surface to which the APR recordings refer is not known and has to be

determined simultaneously with the block adjustment. For each terrain point i, measured by APR on profile p, one additional observation equation is written:

$$v_{z_{ip}} = z_i - (f_p + g_p t_{ip}) - z_{ip}\qquad (9.82)$$

where $v_{z_{ip}}$ is the residual of the APR height z_{ip}. The term $(f_p + g_p t_{ip})$ has the same meaning as in equation (9.81). Some of the APR heights can refer to vertical control (for instance water surfaces of known height). In this case further observation equations of type (9.70) are written. These height control points are not necessarily measured in photogrammetric models, because APR profiles can be directly tied to vertical control outside the photo-block.

The observation equations of type (9.81) and (9.82) are weighted according to the accuracy of the statoscope and APR measurements and added to the hitherto existing observation equations. It should be mentioned here that arbitrary APR cross-flights are allowed and the system automatically provides the interconnection of crossing profiles *via* the photogrammetric models.

The resulting reduced normals contain the unknown model transformation parameters and the unknown shift and tilt terms, belonging to the strips s, flown with statoscope and to the APR profiles, p. Again, a favorable banded and bordered structure of the reduced normal equation matrix can be obtained if these additional parameters are arranged at the end of the vector of unknowns.

The combined block adjustment of statoscope, APR and independent model data is realized in the computer program PAT-M43 (Ackermann et al, 1975). The additional data, however, have an effect on the height block adjustment only. From practical projects, treated by this program, accuracy figures for statoscope and APR are available.

In the OEEPE test block Oberschwaben (40 × 62.5 km², 1:28,000 photography) statoscope observations were also recorded so that the accuracy obtainable with statoscope supported block adjustment could be checked (Ackermann, 1975). A wide-angle and a super-wide-angle block were adjusted (20 percent sidelap, 200 models each). Statoscope readings (z_{is}) associated with individual perspective centers (i) were introduced according to (9.81). Only two cross lines of height control points were used in the most extreme case to bridge a block length of 62.5 km (1 = 25 base lengths). The RMS value of the residuals of the statoscope readings (z_{is}) amounted to 0.38 m in the wide-angle case and to 0.35 m with super-wide-angle. From the discrepancies at more than 400 checkpoints the absolute height accuracies could be estimated and the corresponding RMS values (μ_z) ranged from 0.84 m (wide-angle) to 0.58 m (super-wide-angle).

The efficiency of APR assisted aerial triangulation is demonstrated by the results obtained in the southwestern Ontario APR test block (Klein, 1975). In that test, five wide-angle strips having 20 percent sidelap were flown at a photo-scale of 1:33,000. The flights started at Lake Saint Clair and terminated at Lake Ontario. APR profiles were recorded simultaneously. Additional APR profiles were flown at a lower altitude of 2,000 m across strip direction. In both cases the APR system was of the radar type. The block consists of 380 models and covers an area of 25 × 250 km². Vertical control was restricted to two cross lines of six points each, at the beginning and at the end of the block. In the simultaneous adjustment the APR heights (z_{ip}) of the individual longitudinal and cross profiles (p) were introduced according to (9.82). In all, 170 check heights were available to determine the empirical mean accuracy (μ_z).

The following results were obtained with APR cross profiles, which were neither started nor closed on vertical control. (Their purpose was to control tilt and twist of the long strips). Using only two cross profiles at the beginning and end of the block (distance 250 km) μ_z = 2.45 m was obtained. One additional APR-cross profile, shortening the bridging distance to 125 km, led to μ_z = 1.66 m. Finally, the absolute accuracy was improved to μ_z = 1.46 m by using five cross profiles at a distance of 62 km.

The results demonstrated that APR-assisted block adjustment is capable of providing minor height control to an accuracy of 2 m even when no vertical control points are used inside the block. This accuracy is sufficient for mapping with contour intervals of at least 10 m or, in some case, even 5 m. A further improvement in economy can be realized by replacing the expensive simultaneous APR profiles of the longitudinal strips with statoscope readings.

9.4.3.3 SIMULTANEOUS (BUNDLE) ADJUSTMENT

In the two previous sections, the sequential method and the independent models method were discussed. Both methods are logical extensions and an outgrowth of aerotriangulation as practiced with analog instruments. A radical departure from standard practices came with the development of the simultaneous or bundle procedure in analytical photogrammetry which is to be discussed in this section. In the simultaneous method, the basic unit is the pair of coordinates x and y of an image on the photograph. Using these coordinates (along with some camera constants and the ground position of several of the images), ground coordinates of intermediate points and estimates of the camera's orientation are derived from a simultaneous adjustment. This method differs from the sequential adjustment and independent models in that the solution leads directly to the final coordinates in a single solution and does not treat the "absolute" and "relative" orientations separately. As a result, the solution and associated error propagation are more rigorous in that certain correlations are not ignored. The procedure is iterative and requires that initial approximations be assigned to the unknown elements of camera orientation and to the initially unknown coordinates of ground points. The computations are relatively simple and repetitive and as a result are well suited, when organized properly, for application to large- and medium-scale computers.

9.4.3.3.1 THE MATHEMATICAL MODEL

As pointed out earlier, a photogrammetric camera can be defined geometrically by introducing the concept of a lens (considered a point) as the center of projection along with an image point in the focal plane. The general photogrammetric problem can then be expressed mathematically as the central projection of a three-dimensional object space onto a two-dimensional image space (*see* Brown, 1957). A graphic depiction is shown in figure 9-43 where the image plane is shown as a diapositive, that is, falling between the lens and object space. In the figure, the center of projection has the coordinates X^c, Y^c, and Z^c and the camera records the image (X^I, Y^I, Z^I) of an object point with coordinates X, Y, Z all relative to an arbitrary Carte-

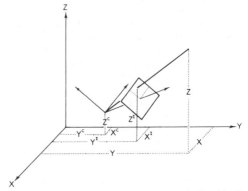

FIGURE 9-43. Graphic of projective relationship between image and object space.

sian coordinate system in object space. A second Cartesian coordinate system (*see* section 9.4.2) is used to define the image coordinates (x and y) in the focal plane of the camera. The problem, therefore, in the simultaneous adjustment, is to determine the geometric relationship that exists between the object-space coordinate system and the image-plane coordinate system. As shown in section 9.4.2 the plate coordinates are assumed to be translated from the fiducial system to the principal point by applying small known translations in the x and y directions.

From section 2.2.3.2.3 of chapter II, one finds that the projective relationship between the object space coordinates of a point and its corresponding image-plane coordinates at the focal plane can be expressed by

$$\begin{bmatrix} X - X^c \\ Y - Y^c \\ Z - Z^c \end{bmatrix} = \lambda \begin{bmatrix} m_{11} & m_{12} & m_{13} \\ m_{21} & m_{22} & m_{23} \\ m_{31} & m_{32} & m_{33} \end{bmatrix} \begin{bmatrix} x \\ y \\ c \end{bmatrix}$$

where the m_{ij} are direction cosines between the axes of the two systems and λ is a scalar. The various rotational systems that can be used to generate the m_{ij} are given in section 2.2.3.2 of chapter II and will not be repeated here. Rearranging and dividing by the third equation to eliminate the scalar results in two equations that express the relationship between image and object space as a function of three angles

$$x = \frac{c\left[m_{11}(X - X^c) + m_{12}(Y - Y^c) + m_{13}(Z - Z^c)\right]}{m_{31}(X - X^c) + m_{32}(Y - Y^c) + m_{33}(Z - Z^c)}$$

$$(9.83)$$

$$y = \frac{c\left[m_{21}(X - X^c) + m_{22}(Y - Y^c) + m_{23}(Z - Z^c)\right]}{m_{31}(X - X^c) + m_{32}(Y - Y^c) + m_{33}(Z - Z^c)}.$$

These, then, are the basic condition equations which form the basis of the simultaneous adjustment method.

9.4.3.3.2 PROBLEM SOLUTION

For each image point measured on a photograph, one pair of equations of the type (9.83) can be written. The typical photogrammetric triangulation problem is normally made up of a series of overlapping photographs, each having a certain number of measured coordinates (adequate for determinacy) from which a consistent solution is sought which best satisfies the basic set of condition equations. Since the number of equations are normally in excess of the number required for a unique solution, the method of least squares is used for the adjustment. Section 2.6 of chapter II developes the formulation of a simultaneous least squares adjustment including an algorithm for efficient computation of the large systems associated with blocks of photography.

9.4.3.3.2.1 Linear Form of the Collinearity Condition Equations. The differential coefficients in the observation equations as given by

(2.240) in section 2.6.1.1 of chapter II can be developed quite easily through the use of the skew-symmetric matrix as shown by Lucas (1963), Brown (1969), Pope (1970) and Elassal, *et al* (1970). The derivation is concise and avoids the algebraic "messiness" of a direct differention of expressions (9.83).

The collinearity condition equations (9.83) can be simply written as

$$x = c(\mathbf{m}/\mathbf{q})$$
$$y = c(\mathbf{n}/\mathbf{q})$$

where x and y are, as usual, the image coordinates of a point, c is the camera principal distance and $\mathbf{n}$, $\mathbf{m}$, $\mathbf{q}$ are functions of the rotation matrix ($\mathbf{M}$) and of the coordinates of the air-station and the object point, that is

$$\begin{bmatrix} \mathbf{m} \\ \mathbf{n} \\ \mathbf{q} \end{bmatrix} = \mathbf{M} \begin{bmatrix} X - X^c \\ Y - Y^c \\ Z - Z^c \end{bmatrix}. \qquad (9.84)$$

If, for convenience in notation, we denote

$$\mathbf{A} = \begin{bmatrix} \dfrac{\partial x}{\partial \mathbf{m}} & \dfrac{\partial x}{\partial \mathbf{n}} & \dfrac{\partial x}{\partial \mathbf{q}} \\ \dfrac{\partial y}{\partial \mathbf{m}} & \dfrac{\partial y}{\partial \mathbf{n}} & \dfrac{\partial y}{\partial \mathbf{q}} \end{bmatrix} = \dfrac{c}{\mathbf{q}} \begin{bmatrix} 1 & 0 & -\dfrac{\mathbf{m}}{\mathbf{q}} \\ 0 & 1 & -\dfrac{\mathbf{n}}{\mathbf{q}} \end{bmatrix}$$

then the differential form for dx, dy in terms of dX, dY, dZ, dX^c, dY^c, dZ^c and the differential changes in the parameters of $\mathbf{M}$ is given by

$$\begin{bmatrix} dx \\ dy \end{bmatrix} = \mathbf{A} \begin{bmatrix} d\mathbf{m} \\ d\mathbf{n} \\ d\mathbf{q} \end{bmatrix}$$

$$= \mathbf{A} \left\{ d\mathbf{M} \begin{bmatrix} X - X^c \\ Y - Y^c \\ Z - Z^c \end{bmatrix} + \mathbf{M} \begin{bmatrix} dX - dX^c \\ dY - dY^c \\ dZ - dZ^c \end{bmatrix} \right\} \qquad (9.85)$$

From equation (9.84),

$$\begin{bmatrix} X - X^c \\ Y - Y^c \\ Z - Z^c \end{bmatrix} = \mathbf{M}^{\mathsf{T}} \begin{bmatrix} \mathbf{m} \\ \mathbf{n} \\ \mathbf{q} \end{bmatrix}$$

so that (9.85) becomes

$$\begin{bmatrix} dx \\ dy \end{bmatrix} = \mathbf{A} d\mathbf{M}\mathbf{M}^{\mathsf{T}} \begin{bmatrix} \mathbf{m} \\ \mathbf{n} \\ \mathbf{q} \end{bmatrix} + \mathbf{A}\mathbf{M} \begin{bmatrix} dX - dX^c \\ dY - dY^c \\ dZ - dZ^c \end{bmatrix}. \qquad (9.86)$$

Since $\mathbf{M}$ is orthogonal, $\mathbf{M}\mathbf{M}^{\mathsf{T}} = \mathbf{I}$ and $-d\mathbf{M}\mathbf{M}^{\mathsf{T}} = \mathbf{M}d\mathbf{M}^{\mathsf{T}}$ which is skew-symmetric and can be expressed as

$$\begin{bmatrix} 0 & \delta_1 & \delta_2 \\ -\delta_1 & 0 & \delta_3 \\ -\delta_2 & -\delta_3 & 0 \end{bmatrix}. \qquad (9.87)$$

Substituting (9.87) into (9.86) and rearranging terms

$$\begin{bmatrix} dx \\ dy \end{bmatrix} = -\frac{c}{q} \begin{bmatrix} n & q-m^2/q & mn/q \\ -m & mn/q & q+n^2/q \end{bmatrix} \begin{bmatrix} \delta_1 \\ \delta_2 \\ \delta_3 \end{bmatrix} + \mathbf{AM} \begin{bmatrix} dX - dX^c \\ dY - dY^c \\ dZ - dZ^c \end{bmatrix}. \quad (9.88)$$

If the orientation angles are those demonstrated in section 2.2.3.2.3 of chapter II (ω, ϕ, κ), then equation (9.88) can be expressed in terms of differentials of those angles (*see* Pope 1970) since,

$$d\mathbf{MM}^T = \begin{bmatrix} 0 & (\sin\phi\,d\omega + d\kappa) & (-\sin\kappa\cos\kappa\,d\omega + \cos\kappa\,d\phi) \\ (-\sin\phi\,d\omega - d\kappa) & 0 & (-\cos\phi\cos\kappa\,d\omega - \sin\kappa\,d\phi) \\ (\sin\kappa\cos\kappa\,d\omega - \cos\kappa\,d\phi) & (\cos\phi\cos\kappa\,d\omega + \sin\kappa\,d\phi) & 0 \end{bmatrix}$$
$$(9.89)$$

and therefore

$$\begin{bmatrix} \delta_1 \\ \delta_2 \\ \delta_3 \end{bmatrix} = \begin{bmatrix} \sin\phi & 0 & 1 \\ -\sin\kappa\cos\phi & \cos\kappa & 0 \\ -\cos\phi\cos\kappa & -\sin\kappa & 0 \end{bmatrix} \begin{bmatrix} d\omega \\ d\phi \\ d\kappa \end{bmatrix}. \quad (9.90)$$

Equations (9.88) and (9.90) can then be used to compute the elements for the $\dot{\mathbf{B}}_{ij}$ and $\ddot{\mathbf{B}}_{ij}$ matrices given by equation 2.241 in chapter II.

9.4.3.3.3 SIMULTANEOUS ADJUSTMENT WITH ADDITIONAL PARAMETERS

The simplicity of the basic concept of the simultaneous adjustment (*e.g.*, the problem is developed from the two collinearity condition equations only), coupled with many ingenious concepts and algorithms for efficient computation and "bookkeeping" that have been developed through the years, provides a system that can be expanded to include solutions that would normally be prohibitive. The concept of the banded-bordered system of normal equations was presented at the ISP Commission III Symposium in Stuttgart, Germany in 1974 by Brown and further expounded upon by Brown (1975). Basically, Brown extends the concept of the banded reduced normal equation system of a block (*see* section 2.6.4 of chapter II), along with its solution using "recursive partitioning" (Gyer 1967, described in section 2.6.6 of chapter II) to include a border that accomodates parameters that are common to large subsets of the original observation equations. The elegance of the approach allows for the inclusion of these additional parameters with a great savings in computational time over that encountered with a "brute force" conventional adjustment. Brown suggests that the system can be applied to "self-calibration" of the cameras used in a block (Brown, *et al*, 1964) when using auxiliary sensors, in satellite applications of photogrammetry where orbital constraints are to be imposed, and in the form of "added parameters" in empirical error models which account for detectible systematic errors in the plate residuals from a block

adjustment. The error models can be general polynomials that regard the systematic error to be common to all photographs or to be present in a sub-set of all the photographs. In addition, the method allows for the general simultaneous adjustment to incorporate the concurrent adjustment of the geodetic network used as the basis of control for the block or for the inclusion of equal elevation constraints as may be encountered by the measurement of points on the shoreline of lakes of unknown elevation. All in all, the method presents an attractive extension of the basic simultaneous adjustment.

9.4.4 THEORETICAL ACCURACY OF ANALYTICAL TRIANGULATION

With the advent of analytical triangulation came the desire on the part of the photogrammetrist to study the error propagation associated with the adjustment in order to better determine a criterion for the accuracy of the system. An early study by Schmid (1961) was restricted to strips of photography because of computer and program limitations. In that study, however, Schmid found that for strips there is definitely an optimum number of models between control points (6 to 7), given a particular camera field of view and a constant length of the strip. That is, as the altitude is raised, fewer photographs are needed to cover the same area, but the scale is smaller and the accuracy decreases. Conversely, as one lowers the altitude, the scale increases for better accuracy, but the error propagation associated with a greater number of connections begins to degrade the results.

Blocks of photographs behave differently, however. Studies have shown (Ackermann, 1966) that, for a block, the error propagation to the center is minimal when the aerotriangulation is controlled with perimeter points. Additionally, the studies showed that the most favorable distribution in the perimeter was every two airbases. Using this as a basis (Ebner, Krack, Schubert, 1977), a theoretical study was conducted using block adjustments of idealized photography with 60% forward and both 60% and 20% side overlap. The terrain points formed a quadratic grid of $n \times n$ meshes where a mesh width was equal to the base length between exposures. Consequently each photograph covered 4 meshes or 9 points respectively, resulting in a total of $(n + 1)^2$ points per block. The investigation was made for horizontal control distributions as shown in figure 9-44 and for the 20% sidelap cases, with dense vertical control across the strip direction (*see* figure 9-45). The vertical-control distribution was prompted by the results of studies of Jerie (1968). In the 60% side lap cases, the dense vertical control was replaced by that regular pattern shown in figure 9-46.

As far as control requirements are concerned, there is no great difference between independent-model and simultaneous block adjustments. Therefore, investigations of both methods were based on the same control distributions. In addition, the control points were assumed to be error-free in the studies.

9.4.4.1 INDEPENDENT-MODEL RESULTS

The results (Ebner, 1972) of the investigation are based on wide-angle photography and the following accuracy assumptions for the independent-model coordinates x, y, z and the perspective center coordinates x_0, y_0, z_0 are made

$$\sigma_x : \sigma_y : \sigma_z : \sigma_{x_0} : \sigma_{y_0} : \sigma_{z_0} = 1:1:1.5:4:4:1.$$

All coordinates were treated as uncorrelated observations. The planimetric accuracy was found to be independent of the height control distribution and, analogously, the planimetric control has very little effect on the accuracy of the height. Therefore, the results for planimetry and height were treated separately. The mean accuracy (μ_{xy}) of the adjusted planimetric block coordinates is expressed in units of the standard deviation $\sigma_x = \sigma_y$ (called σ_{xy}) and the mean accuracy μ_z of the adjusted block heights is divided by the standard deviation σ_z. However, the accuracy measures μ_{xy} and μ_z as well as σ_{xy} and σ_z are related to the same scale, therefore, μ_{xy}/σ_{xy} and μ_z/σ_z directly represent the

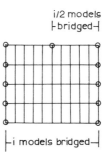

FIGURE 9-45. Vertical control distribution for blocks having 20% side overlap.

effect of the error propagation in the block adjustment.

Figure 9-47 shows the planimetric results for 20% side lap and figure 9-48, the corresponding ones for 60% side overlap. For each of the four control versions (P1, P2, P3 and P4), the value μ_{xy}/σ_{xy} is plotted against the block size, expressed by the number (n) of base lengths. In the case of 20% side overlap, the block consists of $n^2/2$ models ($n = 40:1640$ models).

If the number of control points is constant for increasing n (P2, P3, P4), then μ_{xy}/σ_{xy} can be approximated by a linear function of n. If the bridging distance along the block perimeter is kept constant (P1 : $i = 2$), the accuracy law is even more favorable and μ_{xy}/σ_{xy} can be approximated by a logarithmic function of n. This logarithmic accuracy law was confirmed by a separate investigation of blocks with up to 10,000 models (Ebner, 1970) and by a theoretical study of infinite block size by Meissl (1972).

In large blocks with few control points (cases P3 and P4), the open edges are the weakest places. This disadvantage can be reduced considerably by an extension of the block size by two base lengths on all sides of the block, with the control points then being situated inside the block (*see* Ackermann, 1966).

Because of the favorable error propagation in planimetry, large blocks of a thousand or more models are desirable from an economic point of

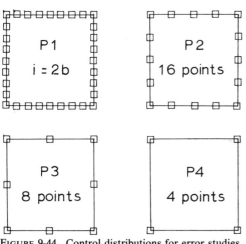

FIGURE 9-44. Control distributions for error studies.

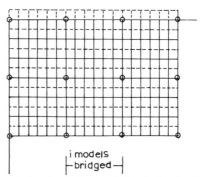

FIGURE 9-46. Vertical control distribution for blocks having 60% side overlap.

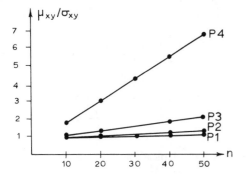

P1: $\mu_{xy} / \sigma_{xy} \approx 0.6 + 0.3 \log n$

P2: $\mu_{xy} / \sigma_{xy} \approx 0.8 + 0.01 n$

P3: $\mu_{xy} / \sigma_{xy} \approx 0.8 + 0.03 n$

P4: $\mu_{xy} / \sigma_{xy} \approx 0.5 + 0.13 n$

FIGURE 9-47. Independent models planimetric results for 20% side overlap.

view. When large areas have to be aerotriangulated with a given scale of photography and accuracy, the number of control points required can be reduced considerably when many small blocks are gathered into a few larger blocks. Another advantage of the large block appears when (in large scale applications) a more less dense perimeter control is available and the accuracy (μ_{xy}) must be optimized. This optimization is possible by selecting a very large photoscale which leads to many models and a small σ_{xy}. On the other hand μ_{xy}/σ_{xy} will remain relatively small due to the excellent accuracy law in the case of many perimeter control points and μ_{xy} will therefore become pleasingly small.

In height, the accuracy is directly dependent on the bridging distance (i) and not on the block size. The accuracy measure (μ_z/σ_z) can be well approximated by the following linear functions of i

$\mu_z/\sigma_z \approx 0.3 + 0.22 i$ for 20% side overlap, and
$\mu_z/\sigma_z \approx 0.25 i$ for 60% side overlap.

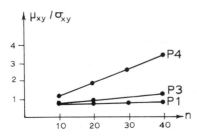

P1 : $\mu_{xy} / \sigma_{xy} \approx 0.5 + 0.2 \log n$

P3 : $\mu_{xy} / \sigma_{xy} \approx 0.5 + 0.02 n$

P4 : $\mu_{xy} / \sigma_{xy} \approx 0.3 + 0.08 n$

FIGURE 9-48. Independent models planimetric results for 60% side overlap.

The height accuracy properties are much poorer than the corresponding ones in planimetry because a considerable amount of height control points are necessary inside the block to arrive at an acceptable answer. An improvement is possible when using auxiliary data such as APR and statoscope.

A direct comparison of the accuracies obtained with 20% and 60% side overlap is possible in the case of planimetry. In height, where different control distributions were used, the comparison is based on the same number of control points. The result is that 60% side overlap is superior when the control distribution is poorer. In planimetry the ratio of accuracies is about 1.4 to 1.6 and in height 1.8 to 2.5.

9.4.4.2 SIMULTANEOUS ADJUSTMENT RESULTS

In the study of the simultaneous adjustment, the image coordinates were treated as uncorrelated observations having unit weight. Therefore, the standard error of unit weight for the adjustment (σ_o) is directly the accuracy of the image coordinates. If the accuracy indicators μ_{xy} and μ_z are reduced to the scale of the photograph, it is possible to express them in units of σ_o. Although planimetry and height effect each other, to some extent, a separation of the results in x, y and z is justified.

The results for planimetric accuracy are shown in figure 9-49 for 20% side overlap and in figure 9-50 for 60% side overlap. With the control distributions given by P2, P3 and P4, the accuracy measure (μ_{xy}/σ_o) is a linear function of n. This corresponds to the results obtained with independent models. In the case P1, the dependency on n is so small that the logarithmic accuracy law could not be determined significantly. Therefore, μ_{xy}/σ_o is approximated as a constant.

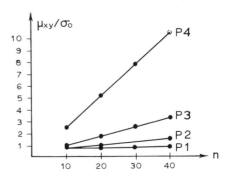

P1: $\mu_{xy} / \sigma_o \approx 0.9$

P2: $\mu_{xy} / \sigma_o \approx 0.5 + 0.025 n$

P3: $\mu_{xy} / \sigma_o \approx 0.3 + 0.08 n$

P4: $\mu_{xy} / \sigma_o \approx 0.27 n$

FIGURE 9-49. Simultaneous adjustment planimetric results for 20% side overlap.

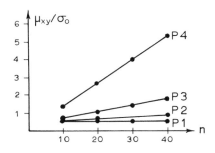

P1 : $\mu_{xy} / \sigma_0 \approx 0.6$

P2 : $\mu_{xy} / \sigma_0 \approx 0.4 + 0.015\,n$

P3 : $\mu_{xy} / \sigma_0 \approx 0.3 + 0.04\,n$

P4 : $\mu_{xy} / \sigma_0 \approx 0.14\,n$

FIGURE 9-50. Simultaneous adjustment planimetric results for 60% side overlap.

As with independent models, the height accuracy is practically independent of the size of the block and the accuracy measure (μ_z/σ_0) is a linear function of the bridging distance i, that is

$\mu_z/\sigma_o \approx 1.0 + 0.18\,i$ for 20% side overlap, and
$\mu_z/\sigma_o \approx 0.31\,i$ for 60% side overlap.

A comparison in accuracy between 20% and 60% side overlap again shows that 60% gains in superiority as the control distribution is decreased. In planimetry, the ratios in accuracy are about 1.5 to 2.0 and in height about 1.9 to 2.1 (relative to the same number of height control points).

The results obtained by the simultaneous adjustment are based on wide-angle photography. When the focal length (f_1) is replaced by focal length f_2 then the results (μ_{xy}/σ_o) remain the same and the height results (μ_z/σ_o) are changed by a factor f_2/f_1.

9.4.4.3 Concluding Remarks

The accuracy predictions for independent models are related to σ_{xy} and σ_z, respectively, while the simultaneous adjustment results are expressed in units of σ_o. To allow for a comparison between σ_{xy} and σ_o as well as between σ_z and σ_o the following ratios can be used

$$\sigma_{xy} \approx 1.5\,\sigma_o \text{ and } \sigma_z \approx 1.4\,(f/b)\sigma_o.$$

Additionally, we must remember that the two adjustment procedures are based on different stochastic assumptions. With the simultaneous adjustment, the only correlations neglected are between the image coordinates while with the independent models, the coordinates x, y, and z of the model points are also assumed to be uncorrelated in the study. However, in spite of these simplifications, both groups of accuracy models can be used successfully in planning. Experience is that they are realistic with 20% to 40%, provided that the systematic errors of the system have been minimized.

9.4.5 INSTRUMENTS FOR ANALYTICAL TRIANGULATION

The instrumentation used for analytical photogrammetry is designed to provide a set of x, y coordinates for all images of interest on each photograph. Measurements of these discrete images may be obtained from single stage monocomparators, or from two or more stage stereocomparators. Point transfer and marking devices are used to permanently identify corresponding images to be measured on a comparator by marking the diapositive.

9.4.5.1 Monocomparator Measurements

In this measuring philosophy, the point images to be measured are distinctively marked on photographic transparencies. There are three categories of such points.

1. Points marked on the object before the photographs are obtained—for example, paneled ground-control points, painted patterns on missiles, signalized points on objects for nontopographic photogrammetry.
2. Detail points which are readily identifiable on all photographs—for example, street intersections or building corners on aerial photographs, star images.
3. Artificial points marked in appropriate locations on the diapositives by point transfer and marking instruments. Such instruments are used, with stereo-observation, to mark corresponding image points on all photographs covering the area.

Because all these points are plainly marked or unmistakably identifiable on the diapositives, each photograph can be measured separately, and for this purpose monocomparators are employed.

Proponents of this approach believe that separating the identification and measuring operations increases accuracy. Also, points are permanently marked and cannot be lost in subsequent operations. On the other hand, when point marking devices are used to make holes in the emulsion, the original identification is the final one, and the opportunity to randomize the identification error by replicated measurements is lost.

Earlier monocomparators employed monocular viewing devices. However, modern monocomparators all use binoculars or large projection screens to present the imagery to the operator to ease eye fatigue and improve pointing precision.

9.4.5.2 Stereocomparator Measurements

It is widely accepted that the stereoscopic acuity of the human eyes exceeds the vernier (monocular) acuity by a factor of 2. Since the invention of the stereocomparator by Pulfrich in 1901, stereo-observation and measuring have been distinguishing characteristics of both aerial and terrestrial photogrammetry.

The stereocomparator technique permits the simultaneous identification and measurement of points on two photographs at a time. It is not limited to sharply defined points but can be used to select points anywhere on the photograph as long as there is sufficient texture to form a "hard" stereoscopic impression.

When points appear in the triple-overlap area or in the side-overlap area between adjacent strips of aerial photographs, it is generally advisable to mark them permanently on one of the photographs in order to minimize errors in identification. In some organizations, sketches or photographs are made of all points as an aid in assuring that the same point is measured on all photographs. Misidentification of points is a major source of error in analytical triangulation.

9.4.5.3 Point Transfer and Marking Instruments

Point transfer and marking instruments are designed for rapid stereoview setup and corresponding image point marking as a preparatory step to comparator measurement. These instruments are used to permanently mark the diapositives by making small holes in the emulsion with a drill or heated needle. The devices consist of two stages, that allow for rapid parallax removal in the operator's field of view, and high quality optical trains that permit stereoscopic viewing of the two corresponding image points. The modern point marking instruments allow for optical image rotation and zoom magnification. Thus, transfer of corresponding image points from different scale photography can be accommodated. This capability permits transfer of aerotriangulated control and other image points from high altitude photographs to larger scale compilation photography.

Table 9-7 summarizes the characteristics of the photogrammetric point transfer and marking instruments currently in production. All can accommodate film and glass plate diapositives. Some point markers allow for permanently marking an annular ring around images of interest or around the small marked hole to permit rapid location (*see* figures 9-51, 9-52, 9-53, and 9-54).

9.4.5.4 Comparator Point Marker Instrument

Kern has produced an instrument, the CPM1 (*see* figure 9-55), that serves as a monocomparator and a stereodigitizer, as well as a point transfer and marking instrument. This instrument was developed by adding linear encoders to the left stage plate of the PMG2 point transfer instrument and then interfacing a digitizing unit and a recording device. The CPM1 design is especially suited for fully analytical aerotriangulation where natural and artificial pass and tie points are located, marked, and simultaneously measured and recorded. Refer to PMG2 on Table 9-7

TABLE 9-7. CHARACTERISTICS OF PHOTOGRAMMETRIC POINT TRANSFER AND MARKING INSTRUMENTS

Manufacturer	Model No.	Picture Carrier Format	Allowable X, Y motion / Allowable ΔX, ΔY motion	Optical Rotation	Magnification Range
Kern	PGM 2	230 mm × 230 mm (Left picture carrier can be rotated ±15)	230 mm. common, Δx = 85 mm. Δy = 200 mm.	Individual thru 360°	Continuous 5×–25×
Wild	PUG 4	340 mm × 470 mm	Free hand plus ±10 mm. common, ±10 mm. right carrier	Individual thru 360°	Continuous 6×–24×
Zeiss (Oberkochen)	PM 1	425 mm × 480 mm	Free hand plus ±15 mm. common or, ±15 mm. individually left or right carrier	Individual thru 360° Common to 90° & 180°	Continuous 6.5×–26×
Jena	Transmark B	230 mm × 230 mm	168 mm × 225 mm common, ±10 mm individually left or right carrier	Individual thru 360°	10× (Optional 7×, 13×, 20×)

TABLE 9-7—*Continued*

Field of View Diameter	Measuring Mark Size @ Image Plane	Point Marking Method	Standard Mark Sizes	Finder Marking	Comments
38 mm. @ 5×	Luminous marks, individually variable size, continuous from 30–200 μm	Steel drills	40, 60, 100 μm	None	Vacuum hold-down for diapositives. Controls for drill speed and descent adjustments. Cold stage illumination by fibre optics
26 mm – 6.5 mm @ 6× 24×	Black marks 50 μm – 12 μm @ 6× 24×	Steel drills	40, 60, 100 180, 250 μm	6 mm scribe circle	Electric drive for drills and scriber. Optional non-reversing optics for negatives
28 mm – 7 mm @ 6.5× 26×	Concentric luminous/black marks 40, 60, 80 & 120 μm	Heated needle	40, 70, 100 200 μm	Annular mark	Automatic photo clamping. Two speed descent control of marking tools. Resolution of 120 lines/mm. @ 26×
25 mm @ 10×	Light green luminous marks 33, 50 & 66 μm	Laser rays	Continuous from 45–100 μm	Annular mark	Non-contacting laser rays vaporize the photo-emulsion and eliminate need for centering adjustments

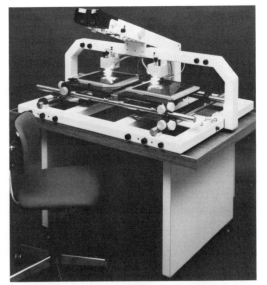

FIGURE 9-51. The Kern PMG2 Zoom Point Transfer Instrument (Courtesy of Kern & Co. Ltd.)

for the CPM1 point transfer and marking characteristics and to the CPM1 on table 9-8 for its monocomparator characteristics.

9.4.5.5 MONOCOMPARATORS

Monocomparators are designed to provide x, y image coordinates on single photographs. They are characterized by their suitability for rapid, simple and accurate measurement properties. All modern monocomparators are available with encoders for digital output for subsequent numerical adjustments and they can all accomodate diapositives and negatives on film or glass plates. Their primary application is to provide point measurements for analytical aerotriangulation; especially for measurement of clearly defined natural points, signalized points and artificially marked points.

Most employ glass linear scales that provide

FIGURE 9-52. The Wild PUG4 Point Transfer Device. (Courtesy of Wild Heerbrugg Ltd.)

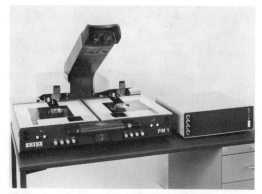

FIGURE 9-53. The Zeiss PM1 Point Transfer Instrument. (Courtesy of Carl Zeiss, Oberkochen.)

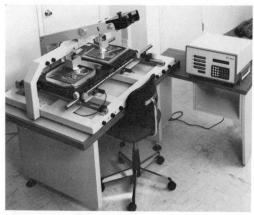

FIGURE 9-55. The Kern CPM1 Comparator Point Marker. (Courtesy of Kern & Co. Ltd.)

high accuracy while minimizing errors due to temperature. Some use precision lead screws, verniers and grid plates for the basic measurement system. Recent advances in dual frequency laser technology has been applied to some comparators for submicrometre accuracies.

Table 9-8 provides a comparative sampling of the primary characteristics of the photogrammetric monocomparators that are currently in production (*see* figures 9-56, 9-57, 9-58, 9-59, 9-60 and 9-61).

9.4.5.6 STEREOCOMPARATORS

The function of a stereocomparator is to measure simultaneously the coordinates of corresponding images on a stereopair of photographs.

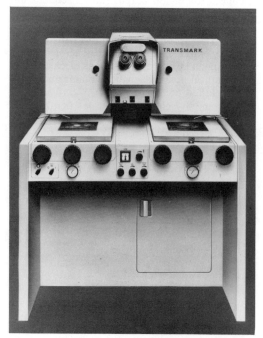

FIGURE 9-54. The Jena TRANSMARK B laser point transfer equipment (Courtesy of Zena Co.).

Separate measuring systems are provided for each photograph, and a binocular optical system allows for stereoscopic observation of the two photographs. The measuring systems, viewing systems, and readout systems employed in stereocomparators are similar to those employed in monocomparators.

For the modern techniques of analytical photogrammetry, it is advantageous to have completely separate measuring systems for the two photographs. However, for scanning the photographs without losing the stereoscopic impression, it is necessary to translate both photostages or their viewing optics simultaneously. Earlier stereocomparators were generally designed with a series of superimposed stages to accommodate both requirements of independent and common photostage translations. In recent designs, completely separate x, y stages are employed for each photograph. They may, however, be coupled together for simultaneous translation of both photographs.

The Jena Stecometer C stereocomparator (*see* figure 9-62) design allows for common x motion to an upper stage and a common y motion to a lower stage containing the collection optics. Differential parallax motions are permitted by separate photocarrier shifts, py direction on the left, px direction on the right, both mounted on the upper stage.

The basic design of the Jena Steko 1818 stereocomparator (*see* figure 9-63) is designed for low quantity, medium accuracy work, such as for education, training, and close range photogrammetry.

The Zeiss (Oberkochen) PSK2 stereocomparator (*see* figure 9-64) design differs from typical mono- and stereocomparator designs in that the diapositives are mounted vertically against two engraved grid glass photostages (grid line spacing of 5 mm). Coarse measurements on each stage are made to the nearest grid line with a precision measuring scale. Fine measurement is

TABLE 9-8. CHARACTERISTICS OF PHOTOGRAMMETRIC MONOCOMPARATORS

Manufacturer	Model	Measurement Range (mm.)	Measurement Encoding System	Least Count	Quoted Accuracy (RMS)
DBA	Series 700 Multilaterative	245 × 245	Precision linear scale with vernier encoder	1 μm	0.8 − 1.3 μm
Kern	MK 2	230 × 230	Linear incremental glass scale encoders	1 μm	1 μm
Kern	CPM 1	230 × 230	Linear incremental glass scale encoders	1 μm	1 μm
Keuffel & Esser	Mono Digital Comparator	229 × 458	Glass scales moved past linear optical encoders	1 μm	0.8 − 1.4 μm
OMI	TA1/P	230 × 230	Precision leadscrews with optical rotary encoders	1 μm	1.5 μm
Zeiss (Oberkochen)	PK 1	240 × 240	Ruled photostage reference lines moved past linear glass scales	1 μm (Optional 0.5 μm)	<1 μm
Jena	Ascorecord 3DP	300 × 300	Precision glass scales and spiral micrometres with optical translational and rotary encoders	0.1 μm	0.4 μm

TABLE 9-8—*Continued*

Viewing System	Magnification Range	Standard Measurement Marks	Max. Slew Speed	Comments
Monocular Microscope on pivot arm.	10× to 30× Zoom	Available	Freehand	Measurements are made along an arm from a fixed pivot point. Self calibrating as measurements are collected from four 90° placements of diapositives
Monocular	16×	Available	Freehand	
Binocular	5× to 25× Zoom	50 μm adjustable for differential zoom	Freehand	Second stage permits stereo-viewing Vacuum holddown for diapositives
Binocular	8.8× to 52.5× Zoom	8 operator selectable reticles	75 mm/sec	Stage moves on air bearings
Binocular	10×	25 μm opaque	20 mm/sec	Stage electrically driven via servo links
Binocular	12× or 20× (Optional 5×, 30×)	5 mark groups. Each can be changed online from luminous to black	Freehand 400 mm/sec	Concentric fine x, y adjustment with tilting plate stepped down 1:75
Binocular	18×, 29×, 36×, 45×, 58× and 90×	Available	Freehand	Maintains Abbe's comparator principle

achieved by displacement of a precision geared-down (30:1) spindle. The measuring cycle is controlled in accordance with an internal program, preselected by the operator at the control unit, which allows the following switch steps.

I Stereo
Stereoscopic viewing of the photograph pair by transmitted light, simultaneous and parallactic movement of the photographs.

II Binocular left
Binocular viewing of the left photograph by transmitted light, movement of the left photograph.

III Binocular right
Binocular viewing of the right photograph by transmitted light, movement of the right photograph.

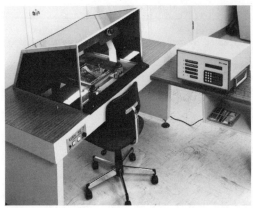

FIGURE 9-56. The Kern MK2 Monocomparator. (Courtesy of Kern & Co. Ltd.)

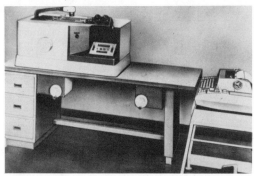

FIGURE 9-58. The O.M.I. TA1/P Monocomparator (Courtesy of Ottico Meccanica Italiana S.p.A.)

IV Measuring left
Binocular viewing of the left grid by reflected light, movement of the left measuring scale.

V Measuring right
Binocular viewing of the right grid by reflected light, movement of the right measuring scale.

VI Recording
Instrument coordinates. Indication of switch step.

Continued program control by means of functional key, interruption of program with erase key.

To assure continuity of measurements for aerotriangulation, OMI builds a three-photograph stereocomparator (*see* figure 9-65) which permits measurement of points in the triple overlap. The viewing system contains an optical switch that permits each eyepiece to be directed to any one of the three photographs. This allows for observation of the corresponding image in one stereopair and then immediate observation of the same image in the second stereopair. The routine for systematic measurement of consecutive photographs is illustrated in figure 9-66.

Characteristics of four photogrammetric stereocomparators that are currently in production are given in table 9-9. The future of photogram-

metric stereocomparators has evolved in the production of the numerous modern analytical stereoplotters that are discussed in chapter XIII.

9.4.5.7 COMPARATOR CALIBRATION

All photogrammetric mensuration instruments that have photostages and/or optics that move in or parallel to an x, y plane may be defined as comparators. This includes all monocomparators, stereocomparators and analytical plotters.

Each of these instruments relies on a precision transport system in or parallel to an x, y plane, usually along guides and ways. These comparators are mechanical and therefore are restricted in precision by the manufacturing process that produced them. These mechanical imperfections result in the creation of systematic linear and nonlinear measuring errors that, by proper analysis, may be modeled mathematically and removed by a calibration process. Properly administered, the calibration will also monitor the performance quality of the measuring machine and hence indicate when repair, readjustment, or maintenance is required. The linear systematic errors typically present in comparators are differential scalers (Sx, Sy) along the x and y axes

FIGURE 9-57. The K & E Mono Digital Comparator. (Courtesy of Keuffel & Esser Company)

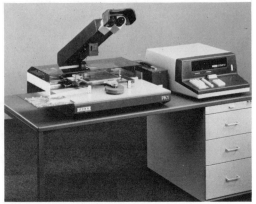

FIGURE 9-59. The Zeiss PK1—Precision Comparator. (Courtesy of Carl Zeiss, Oberkochen.)

FIGURE 9-60. The DBA Model 703 Multilaterative Comparator. (Courtesy of DBA Systems, Inc.)

FIGURE 9-62. The Jena Stecometer C precision stereo-comparator (Courtesy of Zena Co.).

and the lack-of-orthogonality angle (α) between the axes. The nonlinear systematic errors are usually present as long term weaves and curves along the x and y axes due to the combined imperfections in the ways and guides and in the measurement standard employed, *e.g.*, linear scale or leadscrew.

A complete comparator calibration method (Fritz, 1973) that uses precision grid plates can provide statistics that indicate; 1) operator pointing precision, 2) the inherent quality of measurements before calibration (RMS), 3) the comparator quality after removal of linear systematic errors, 4) the comparator quality after removal of nonlinear systematic errors, and 5) the quality of the given grid coordinates. A description of this method follows.

9.4.5.7.1 GRID PLATE

A square reseau or grid plate containing 25 symmetrically-spaced reseau or point images is required. The plate must be stable, microflat, optically clear and at least ¼ inch thick. Each of the 25 images should be identical in shape and easy to point on with a measuring mark. The side dimension of the grid plate should be approximately 200-215 mm for standard stage formats of 250 mm. This will enable the grid to be rotated 11.3° from the stage axes. Calibrated coordinates for each grid point are not required for

determination of the parameters that describe the linear systematic errors. However, calibrated grid coordinated are necessary to determine many of the nonlinear parameters.

9.4.5.7.2 MEASUREMENT

Redundant, repetitive measurements (pointings) on all 25 grid points should be performed such that the mean pointing on each of the grid points is of equal quality to the mean pointing on any other grid point (this allows for the introduction of equal weights for all grid points in subsequent least squares adjustments). The grid plate is then rotated 90° and redundant measurements on all 25 points are repeated. Similarly, the plate is measured in the 180° and 270° positions so that a complete set of measurements in four placements of the grid plate on the same photostage location is made. Additional four placement measurements of the grid plate are required for large format photostages.

9.4.5.7.3 DATA REDUCTION

All of the measurements are processed through a grand least squares adjustment that determines parameters for differential scalers (Sx, Sy), non-

FIGURE 9-61. The Jena Ascorecord 3 DP precision monocomparator (Courtesy of Zena Co.).

FIGURE 9-63. The Jena Steko 1818 stereocomparator (Courtesy of Zena Co.).

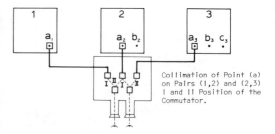

FIGURE 9-64. The Zeiss PSK2—Precision Stereocomparator. (Courtesy of Carl Zeiss, Oberkochen.)

orthogonality (α), rotations (θ_i), and translations (Tx_i, Ty_i). The adjustment assumes the grid plate coordinates to be known and adjusts the four measurements sets ($i = 1\text{-}4$) while allowing for each set one rotation (θ_i), and two translations (Tx_i, Ty_i) and, for all, a common nonorthogonality angle (α), scaler x (Sx), and scaler y (Sy). The observation equations for this adjustment are in the form of:

$$\begin{bmatrix} v_x \\ v_y \end{bmatrix} = \begin{bmatrix} a & b \\ c & d \end{bmatrix} \begin{bmatrix} X \\ Y \end{bmatrix} + \begin{bmatrix} e \\ f \end{bmatrix} - \begin{bmatrix} x \\ y \end{bmatrix}$$

or

$$v = AX + T - x$$

where:

$$X = \begin{bmatrix} X \\ Y \end{bmatrix} = \text{orthogonal (grid) system}$$

$$x = \begin{bmatrix} x \\ y \end{bmatrix} = \text{nonorthogonal (comparator) system}$$

$$v = v_x, v_y = \text{residual errors}$$

$$A = a, b, c, d, e, f = \text{linear parameters.}$$

It has been shown (Fritz, 1973) that an exact, linear, noniterative least squares solution for the

FIGURE 9-65. The O.M.I. TA3/P Stereocomparator. (Courtesy of Ottico Meccanica Italiana S.p.A.)

Collimation of Point (a) on Pairs (1,2) and (2,3) I and II Position of the Commutator.

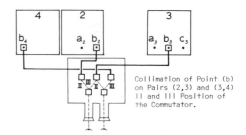

Collimation of Point (b) on Pairs (2,3) and (3,4) II and III Position of the Commutator.

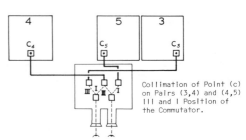

Collimation of Point (c) on Pairs (3,4) and (4,5) III and I Position of the Commutator.

FIGURE 9-66. Observation Sequence for Aerotriangulation on O.M.I. TA3/P.

photostage calibration, with residuals given in the measurement system, does exist. Furthermore, the physically identifiable comparator calibration parameters defined in the comparator coordinate system can be explicitly derived as:

$$Sx = \frac{(c^2 + d^2)^{1/2}}{(ad - bc)}$$

$$Sy = \frac{(a^2 + b^2)^{1/2}}{(ad - bc)}$$

$$\alpha = \tan^{-1}\left(\frac{-ac - bd}{ad - bc}\right).$$

These parameters define the systematic linear errors that should be removed from all stage measurements.

Most of the nonlinear systematic errors that may be present in the comparator can be determined from the same four sets of measurements only if the grid plate has been precisely calibrated. Since the grid plate was rotated approximately 11.3° on the stage, the projection of the 25 grid points onto each measurement axis provides an even spacing for data sampling (see figure 9-67). Normally, the following 4th degree polynomials are more than sufficient to describe

TABLE 9-9. CHARACTERISTICS OF PHOTOGRAMMETRIC STEREOCOMPARATORS

Manufacturer	Model	Number of Photostages	Measurement Range (mm)	Measurement Encoding System	Least Count	Magnification Range
OMI	TA3/P	3	240 × 240	Precision leadscrews with optical rotary encoders	1 μm	Slip-on eyepieces for 7× or 10× (1:4.5 zoom optional)
Zeiss (Oberkochen)	PSK 2	2	250 × 250	Precision scales and spindles aligned with precision grid	1 μm or 5 μm	Slip on eyepieces for 8× or 12× or 16×
Jena	Stecometer C	2	240 × 240	Precision leadscrews with rotary encoders	1 μm	Slip-on eyepieces for 6×, 12× or 18×
Jena	Steko 1818	2	186 × 186	Leadscrews and graduated drums	5 μm	8×

TABLE 9-9—Continued

Field of View Diameter	Optical Rotation	Standard Measurement Marks	Slew Speed	Comments
20 mm @ 7×	Doves for each photostage	35 μm diameter black point	18 mm/sec	Photostages are servo driven Stages are horizontal Polaroid camera on one optical train to record image measured is standard
15 mm @ 12×	Doves for each photostage	25 μm diameter black point	10 mm/handwheel revolution	Binocular viewing for individual stages Stages are vertical Ortho-pseudo commutation thru deflector disks Reversing prism for diapositive/ negative commutation
25 mm @ 12×	Doves for each photostage	20, 40 and 60 μm luminous points Choice of three colors on wheel for each of 2 ring, 3 points or 1 cross marks		x and y are servodriven Py movement on left stage Px movement on right stage Optional camera on left records image, measurement mark and point number RMS = 2 μm
16 mm	None but photostages rotate thru 360°	Black keyhole shape with pointed end plus 50 μm point		Often used for terrestrial photography Desktop unit and easy to transport RMS = 4.5 μm

the independent nonlinear errors along the stage axes.

$$lx + v_x = Ax^4 + By^4 + Cx^3 + Dy^3 + Ex^2 + Fy^2$$
$$+ Gx + Hy + J$$
$$ly + v_y = A'x^4 + B'y^4 + C'x^3 + D'y^3 + E'x^2$$
$$+ F'y^2 + G'x + H'y + J'$$

$$-l + v = Ax.$$

Where:

$l = lx, ly$ = residuals after linear calibration adjustment (previously denoted as v_x, v_y).

$v = v_x, v_y$ = residuals after the least squares nonlinear calibration adjustment.

$x = A - J, A' - J'$ = unknown linear coefficients of the polynomial.

$A = x, y$ = comparator measurements after correction for systematic linear errors.

9.4.5.7.4 CALIBRATION ANALYSIS

The standard error of a single observation of unit weight from each of the linear and nonlinear adjustments is the statistic that represents the accuracy of measurements from a comparator photostage. The precision of operator's measurements can be determined from a statistical analysis of his redundant pointings on each of the 25 gridpoints, e.g., the standard deviation of a single pointing observation and the standard deviation of a mean pointing observation.

For large format photostages an overlapping or adjacent grid placement technique is used to

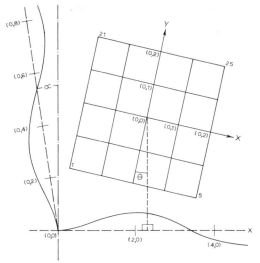

FIGURE 9-67. An exaggerated grid—comparator relationship.

cover the entire measurement area. In this case, the square grid plate must be measured in the four 90° rotation sets for each area of the stage. Simultaneous reduction of all placements will produce a single set of parameters that physically describes the systematic errors of the stage.

It must be emphasized that calibration of a large format stage should be made only with a square grid plate. The use of non-square grid plates can contribute significant grid coordinate biases that will invalidate the physical significance of the photostage calibration. Furthermore, the calibration of any format comparator is dependent on the collection of measurements from the four 90° placements of the grid plate on the same area of a comparator stage to minimize the influence of grid plate coordinate errors. An analysis of the adjustment residuals in the given grid coordinate system will provide a statistic that represents the accuracy of the given grid coordinates. In addition, a collective mean of the grid residuals from numerous comparator calibrations from a variety of photostages will provide the basis for a partial grid calibration.

9.4.5.7.5 APPLICATION OF COMPARATOR CALIBRATION PARAMETERS

The linear and nonlinear parameters derived from calibration of a comparator photostage should be the first coordinate refinement applied to all photostage measurements. Removal of linear comparator errors is performed first by the following equations:

$$X_a = Sx \cdot x + Sy \cdot y \cdot \sin \alpha$$
$$Y_a = Sy \cdot y \cdot \cos \alpha$$

Next, nonlinear comparator errors are removed from comparator measurements by the following equations:

$$x_a = x + Ax^4 + By^4 + Cx^3 + Dy^3 + Ex^2 + Fy^2 + Gx + Hy + J$$

$$y_a = y + A'x^4 + B'y^4 + C'x^3 + D'y^3 + E'x^2 + F'y^2 + G'x + H'y + J'$$

where

x_a, y_a = comparator measurements corrected for all systematic comparator errors.

x, y = comparator measurements corrected for all linear systematic comparator errors (Sx, Sy, α).

REFERENCES AND SELECTED BIBLIOGRAPHY

SECTIONS 9.2 AND 9.3

Bachmann, W. K., 1946, Théorie des erreurs et compensation des triangulations aériennes: Lausanne, Switzerland.

Baetsle, P. L., 1956, Compensation des blocs photogrammétriques en altimetrie par relaxation: Bulletin de la Société Belge de Photogrammetrie, no. 44.

Baussart, M., 1957, L'évolution de l'aérotriangulation a l'Institut Georgraphique National Francais: Photogrammetria, v. 14, no. 2.

Bjerhammar, A., 1951, Adjustment of aerotriangulation: Photogrammetria, v. 7, no. 4.

Blachut, T. J., 1956, Airborne control method of aerial triangulation: Photogrammetria, v. 12, no. 4.

——— 1957–58, Use of auxiliary data in aerial triangulation over long distances: Photogrammetria, v. 14, no. 1.

Blachut, T. J., and Lyytekainen, H. E., 1958, Airborne controlled aerial triangulation: National Research Council of Canada, report no. AP-PR13.

Bonneval, H., 1958, Rapport sur le problème de la compensation des blocs de bandes: Bulletin da la Société Belge de Photogrammetrie, no. 52.

——— 1960, Le Problème de la compensation des blocs de bandes: Photogrammetria, v. 16, no. 3.

Brandenberger, Dr. A. J., 1951, The practice of spatial aerial triangulation: commissioned by the Photogrammetric Institute of the Federal Institute of Technology, Zurich.

——— 1957–58, Some considerations about error propagation in strip triangulations; attainable accuracy: Photogrammetria, v. 14, no. 2.

——— 1959, Strip triangulations with independent geodetic controls; triangulation of strip quadrangles: Publications of The Institute of Geodesy, Photogrammetry and Cartography, no. 9, The Graduate School, Ohio State University.

——— 1960, Aerial triangulation with auxiliary data, General Report Sub-Commission III/4, International Society for Photogrammetry, London, 1960: Publications of The Institute of Geodesy, Photogrammetry and Cartography, no. 10, The Graduate School, Ohio State University.

Brandenberger, Dr. A. J., and Laurila, S., 1956, Aerial triangulation by least squares; third interim technical report: Mapping and Charting Research Laboratory, Ohio State University Research Foundation.

Brucklacher, W. A., 1959, Zur räumlichen Aerotriangulation von Bildstreifen, Beiträge zur numerischen Photogrammetrie: Verlag der Bayerischen Akademie der Wissenschaften, München.

Förstner, R., 1960, Probleme der Ausgleichung von Aerotriangulationen, der zufällige Fehler bei der

Aerotriangulation langer Flugstreifen: Mitteilung no. 36 des Institut für Angewandte Geodäsie, Frankfurt.

Gotthardt, E., 1944, Der Einfluss unregelmässiger Fehler auf die Lufttriangulation: *Zeitschrift für Vermessungswesen*, v. 73, no. 4.

von Gruber, O., 1935, Beitrag zur Theorie und Praxis von Aeropolygonierung und Aeronivellement: *Bildmessung und Luftbildwesen* nos. 3 and 4.

Hallert, Bertil, 1960, Photogrammetry, basic principles and general survey: McGraw-Hill Book Company, Inc., New York.

Helava, U. V., 1957, New principle for photogrammetric plotters: *Photogrammetria*, v. 14, no. 2.

—— 1959, Use of oblique photographs to control the bend of aerial triangulation: National Research Council of Canada, Report no. AP-PR18.

—— 1960, Analytical plotter using incremental computer: *Photogrammetric Engineering*, v. 26, no. 3.

Helmy, R., 1958, Die Aerotriangulation mit unabhängigen Bildpaaren: Dissertation, Swiss Federal Institute of Technology, Zurich.

Henry, T. J. G., Determination of topographic profiles by use of radar and pressure altimeters: Meterological Division, Department of Transport, Canada.

Jacobsen, C. E., 1951, High precision shoran test, phase 1, AF Technical Report no. 6611: U.S. Air Force, Wright Air Development Center, Wright-Patterson Air Force Base, Ohio.

Jerie, H. G., 1958, Block adjustment by means of analogue computer: *Photogrammetria*, v. 14, no. 4.

—— 1960, Block adjustment with analogue computer applying "composed sections;" evaluation of the accuracy to be expected: ITC Publications, no. A2, Delft, The Netherlands.

Karara, H. M., 1956, Fehlerfortpflanzung und Ausgleichung von Aerotriangulation-Streifen mit gemessenen Querstrecken: Dissertation, Swiss Federal Institute of Technology, Zurich.

—— 1959, About the character of errors in spatial aerotriangulation: *Photogrammetric Engineering*, v. 25, no. 3.

Laurila, Simo, 1976, Electronic surveying and navigation: J. Wiley, New York.

Mahoney, W. C., 1961, Proposal, development and testing of a system of analytical triangulation for medium scale digital computers: Dissertation, The Ohio State University.

Moore, R. E., 1960, Application of the Jerie block adjustment to small scale mapping: *The Canadian Surveyor*, v. 15, no. 6.

Nowicki, A. L., and Born, C. J., 1960, Improved stereotriangulation with electronic computers: *Photogrammetric Engineering*, v. 26, no. 4.

Roelofs, R., 1941, Theory of errors in aerial triangulation: *Photogrammetria*, v. 4, nos. 3 and 4.

—— 1952, Adjustment of aerial triangulation by the method of least squares: *Photogrammetria*, v. 8, no. 4.

—— 1953, Practical example of adjustment of aerial triangulation by the method of least squares: *Photogrammetria*, v. 10, no. 1.

Santoni, E., 1957, Aerial triangulation using the solar periscope: *Photogrammetria*, v. 14, no. 1.

Scher, M. B., 1955, Stereotemplet triangulation: *Photogrammetric Engineering*, v. 21, no. 5.

Schermerhorn, W., 1940, Introduction to the theory of errors of aerial triangulation in space: *Photogrammetria*, v. 3, no. 4.

Schwidefsky, K., 1959, An outline of photogrammetry: New York, Pitman.

Solaini, L., and Trombetti, C., 1959, Rapport sur les résultats des travaux d'enchainement et de compensation éxécutés pour la Commission A de l'OEEPE jusqu'au mois de Juin 1959 (texte et tableaux), no. 1071/OEEPE/A.

Trorey, L. G., 1947, Slotted templet error: *Photogrammetric Engineering*, vol. 13, no. 2, June 1947.

Waldhausl, P., 1959, Height adjustment by the ITC-JERIE method: *Photogrammetria*, v. 16, no. 1.

van der Weele, A. J., 1953−54, Adjustment of aerial triangulation: *Photogrammetria*, v. 10, no. 2.

—— 1956, Rational adjustment of blocks of aerial triangulation: *Photogrammetria*, v. 12, no. 4.

—— 1957, General problems in aerial triangulation: *Photogrammetria*, v. 14, no. 2.

Wiser, P., 1959, Considérations sur les erreurs de l'aérotriangulation: Bulletin de la Société Belge de Photogrammétrie, No. 58.

Zarzycki, J. M., 1952, Beitrag zur Fehlertheorie der räumlichen Aerotriangulation: Dissertation, Swiss Federal Institute of Technology, Zurich.

Zeller, M., 1952, Practical experience in determining a net of points by aerotriangulation of different parallel strips and their compensation: *Photogrammetria*, v. 8, no. 4.

SECTION 9.4

Ackermann, F., 1966, On the Theoretical Accuracy of Planimetric Block Triangulation. *Photogrammetria*, 21, 145−170.

—— 1975, Accuracy of Statoscope Data-Results from the OEEPE-Test Oberschwaben, Proceedings of the Stuttgart ISP Commission III Symposium. Deutsche Geodatische Kommission, Reihe B, Heft 214.

Ackermann, F., Ebner, H. and Klein, H., 1972, Combined Block Adjustment of APR Data and Independent Photogrammetric Models. *The Canadian Surveyor*, 26, 384−396.

—— 1973, Block Triangulation with Independent Models, *Photogrammetric Engineering*, 39, 967−981.

Bennett, Jean M., 1961, Method for Determining Comparator Screw Errors with Precision, *Journal of the Optical Society of America*, Vol. 51, No. 10, pp. 1133−1138.

Bertram, Sidney, 1966, Atmospheric Refraction, *Photogrammetric Engineering*, Vol XXXII, No. 1, pp. 76−84.

—— 1969, Atmospheric Refraction in Aerial Photogrammetry, Discussion Paper, *Photogrammetric Engineering*, Vol. XXV, No. 6, p. 560.

Blais, R., 1973, Program SPACE-M. Users Manual (Interim Version). Surveys and Mapping Branch. Dept. of Energy, Mines and Resources, Ottawa.

Brock, Robert H., 1966, Development of Calibration Technique for a Screw Type Comparator, (research proposal), Syracuse University Research Institute, Syracuse, N.Y., p. 12.

Brown, Duane, 1957, A Treatment of Analytical Photogrammetry with Emphasis on Ballistic Camera Applications. RCA Data Reduction Technical Report No. 39.

—— 1958a, A Solution to the General Problem of Multiple Station Analytical Stereotriangulation. RCA Data Reduction Technical Report No. 43.

—— 1958b, Photogrammetric Flare Triangulation. RCA Data Reduction Technical Report No. 46.

—— 1959, Results in Geodetic Photogrammetry I. RCA Data Processing Technical Report No. 54.

—— 1964, An Advanced Plate Reduction for Photogrammetric Cameras, AF Cambridge Research Laboratories Report No. 64-40.

—— 1966, Decentering Distortion of Lenses, *Photogrammetric Engineering*, Vol XXXII, No. 3, p. 444.

Brown, D. C., and Trotter, 1969, SAGA, A Computer Program for Short Arc Geodetic Adjustment of Satellite Observations, AF Cambridge Research Laboratories Report.

Brown, D. C., 1969, Advanced Methods for the Calibration of Metric Cameras, DBA Systems, Inc. Report.

—— 1974, Bundle Adjustment with Strip- and Block-Invariant Parameters, Proceedings, ISP Commission III Symposium, Stuttgart, Germany.

Byrd, W. O., 1951, Some Elementary Aspects of Computational Problems in Photogrammetry: Mapping and Charting Research Laboratory, Technical Paper No. 142, Ohio State University.

Case, James B., 1961, The Utilization of Constraints in Analytical Photogrammetry: *Photogrammetric Engineering*, Vol. 27, No. 5, 766–778.

Chitayat, A., 1960, Interference Comparators for Coordinate Measurement of Ballistic Plates: *Photographic Science and Engineering*, Vol 4, No. 5.

Church, Earl, 1945, Revised Geometry of the Aerial Photograph: Syracuse University, Bulletin No. 15.

—— 1948, Theory of Photogrammetry: Syracuse University, Bulletin No. 19.

Conrady, A. E., 1918, The Five Aberrations of Lens-Systems, *Monthly Notices of the Royal Astronomic Society*, Vol. 79, No. 1, pp. 60–66.

—— 1919, Decentered Lens-Systems, *Monthly Notices of the Royal Astronomical Society*, Vol. 79, No. 5, pp. 384–390.

DeMeter, E. R., 1962, Automatic Point Identification Marking and Measuring Instrument, *Photogrammetric Engineering*, Vol. 28, No. 1.

Dodge, H. F., Handwerker, D. S., and Eller, R. C., 1959, A Geometrical Foundation for Aerotriangulation, Progress Report 3, Topographic Division, U.S. Geological Survey.

Döhler, M., 1972, Standard Tests für Photogrammetrische Auswertegerate, Commission II, XII International Congress for Photogrammetry, Ottawa.

Doyle, Frederick J., 1964, The Historical Development of Analytical Photogrammetry: *Photogrammetric Engineering* Vol. 30, No. 2.

Ebner, H. 1972, Theoretical Accuracy Models for Block Triangulation, *Bildmessung und Luftbildwesen*, 40, 214–221.

—— 1970, Die Theoretische Lagegenauigkeit Ausgeglichener Blocke mit bis ze 10000 Unabhangigen Modellen, *Bildmessung und Luftbildwesen*, 38, 225–231.

—— 1976, Self Calibrating Block Adjustment, Invited Paper of Commission III, ISP Congress, Helsinki, Finland.

Ebner, H. and Schneider, W., 1974, Simultaneous Compensation of Systematic Errors with Block Adjustment by Independent Models, *BUL*, pp 198–203.

Ebner, H., 1977, Theoretical and Practical Accuracy of Photogrammetric Control Networks, Invited Paper, IAG Symposium, Sopron.

Erio, G. 1975, Three Dimensional Transformations of Independent Models, *Photogrammetric Engineering and Remote Sensing*, 41, 1117–1121.

Elassal, Atef A., Brewer, R. K., Gracie, G., and Crombie, M. A., 1970, Final Technical Report, MUSAT IV, DAAK02-67-C-0592, Prepared for USAETL, Fort Belvoir, Virginia.

Faulds, Archur H. and Brock, Robert H., 1964, Atmospheric Refraction and its Distortion of Aerial Photographs, *Photogrammetric Engineering*, Vol. XXX, No. 2.

Finsterwalder, S., 1899, Die Geometrische Grundlagen der Photogrammetrie, *Deutsche Math. Ver. Leipzig Jahresber.*, Vol. 6, No. 2.

—— 1900, Über die Konstruktion von Höhenkarten aus Ballonaufnahmen, *Kgl. Bayrische Akad. Wiss. München Sitzungsber., Math.-phys. Kl.*, Vol 30, No. 2.

—— 1903, Eine Grundaufgabe der Photogrammetrie und ihre Anwendung auf Ballonaufnahmen, *Kgl. Bayrische Akad. Wiss. München Abh., II. Kl.*, Vol. 22. Pt. 2.

—— 1932, Die Hauptaufgabe der Photogrammetrie, *Bayrische Akad. Wiss. München Sitzunsber., Math.-phys. Kl.*

Finsterwalder, S. and Scheufele, W., 1903, Das Rückwärtseinschneiden im Raum, *Kgl. Bayrische Akad. Wiss. München Sitzungsber., Math.-phys. Kl.*, Vol. 33, No. 4.

Fisher, T., 1921, Über die Berechnung des Räumlichen Rückwärtseinschnitts bei Aufnahmen aus Luftfahrzeugen, Jena, G. Fisher.

Fritz, Lawrence W., 1973, Complete Comparator Calibration, NOAA Technical Report, NOS 57, National Ocean Survey, NOAA, Rockville, Md., pp. 96.

Gauthier, J., O'Donnel, J. and Low, B., 1973, The Planimetric Adjustment of Very Large Blocks of Models: Its Application to Topographical Mapping in Canada, *The Canadian Surveyor*, 27, pp. 99–118.

von Gruber, O., 1942, Photogrammetry, Collected Lectures and Essays, American Photographic Publishing Co., Boston.

Gugel, Ralph A., 1963, Photogrammetric Data Reduction Analysis Calibration of Comparators for Ballistic Camera Data, Technical Report No. MTC-TDR-63-10, Air Force Missile Test Center, Air Force Systems Command, Patrick Air Force Base, Fla., pp. 116.

Gyer, Marice S., 1967, The Inversion of the Normal Equations of Analytical Aerotriangulation by the Method of Recursive Partitioning, RADC-TR-67-69, Rome Air Development Center, Rome, New York.

Hallert, B., 1961, Investigations of the Weights of Image Coordinates in Aerial Photographs. *Photogrammetric Engineering*, Vol. 27, No. 4.

—— 1962, Determination of the Geometrical Quality of Comparators for Image Coordinate Measurements, *GIMRADA Research Note No. 3*, U.S. Army Engineer Geodesy, Intelligence and Mapping Research and Development Agency, Fort Belvoir, Va., pp. 95.

—— 1963, Test Measurements in Comparators and Tolerances for Such Instruments. *Photogrammetric Engineering*, Vol. 29, No. 2.

Harris, W. D., Tewinkel, G. C. and Whitten, C. A.,

1962a, Analytic Aerotriangulation, U.S. Coast and Geodetic Survey Technical Bulletin No. 21.

—— 1962b, Analytical Aerotriangulation in the Coast and Geodetic Survey, *Photogrammetric Engineering*, Vol. 28, No. 1.

Herget, Paul, 1956, Computational Extension of Control Along a Photographed Strip, Mapping and Charting Research Laboratory Technical Paper No. 201, Ohio State University.

Herget, Paul and Mahoney, W. C., 1957, The Herget Method. Mapping and Charting Research Laboratory Technical Paper No. 209, Ohio State University.

Hugershoff, R., and Cranz, H., 1919, Grundlagen der Photogrammetrie aus Luftfahrzeugen. Stuttgart, K. Wittwer.

Jerie, H. G., 1968, Theoretical Height Accuracy of Strip and Block Triangulation With and Without Use of Auxiliary Data. *Photogrammetria*, 24, pp. 19–44.

Jeyapalan, K., 1972, Calibration of a Comparator. *Photogrammetric Engineering*, Vol. XXXVIII No. 5, pp. 472–478.

Kern & Co. Ltd., Generalized Program for the Calibration of Monocomparators and Coordinatographs, Aarau, Switzerland, pp. 23.

Klein, H., 1975, Results from the South-Western Ontario APR-Test Block, Proceeding of the Stuttgart ISP Commission III Symposium, Deutsche Geodätische Kommission, Reihe B, Heft 214.

Lambert, J. H., 1759, Freie Perspektive, Zürich.

Loewen, Erwin G., 1964, Coordinate Measurement, The Elusive Micron, *Photogrammetric Engineering*, Vol XXX, No. 6, pp. 962–966.

Lucas, James R., 1963, Differentiation of the Orientation Matrix by Matrix Multipliers, *Photogrammetric Engineering*, Vol. XXIX, p. 708.

Makarovic, B., 1972, Testing of Instruments. *ITC Textbook of Photogrammetry*, Vol. IV, 7 Part B, International Institute for Aerial Survey and Earth Sciences, Enschede, The Netherlands, pp. 164.

Schmid, Hellmut H., 1956–57, An Analytical Treatment of the Problem of Triangulation by Stereophotogrammetry, *Photogrammetria*, Vol. 12, Nos. 2 and 3.

—— 1959, A General Analytical Solution to the Problem of Photogrammetry, *Ballistic Research Laboratories Report No. 1065.*

—— 1961, The Propagation of Residual Errors in Rigorously Adjusted Aerial Triangulation Schemes, *Bolletino Di Geodesia e Scienze Affini*, XX, No. 4.

Schmid, Hellmut H., and Schmid, Erwin, 1965, A Generalized Least Squares Solution for Hydrid Measuring Systems, U.S. Coast and Geodetic Survey Technical Bulletin No. 24.

Schut, G. H., 1955–56, Analytical Aerial Triangulation and Comparison Between It and Instrumental Aerial Triangulation, *Photogrammetria*, Vol. 12, No. 4.

—— 1957–58, Analysis of Methods and Results in Analytical Aerial Triangulation, *Photogrammetria*, Vol. 14, No. 1.

—— 1958–59, Construction of Orthogonal Matrices and Their Application in Analytical Photogrammetry, *Photogrammetria*, Vol 15, No. 4.

—— 1959–60, Remarks on the Theory of Analytical Aerial Triangulation, *Photogrammetria*, Vol. 16, No. 2.

—— 1965, A Method of Block Adjustment for Heights with Results Obtained in the International Test, *Photogrammetria*, Vol. 20, pp. 35–51.

—— 1967, Polynominal Transformation of Strips Versus Linear Transformation of Models, A Theory and Experiments, *Photogrammetria*, Vol. 22, pp. 241–262.

—— 1969, Photogrammetric Refraction, *Photogrammetric Engineering*, Vol. XXXV, No. 1, pp. 79–86.

—— 1970, External Block Adjustment of Planimetry, *Photogrammetric Engineering*, Vol. XXXVI, No. 9, pp. 974–982.

—— 1973, Similarity Transformation and Least Squares, *Photogrammetric Engineering*, Vol XXXIX, No. 6, pp. 621–628.

—— 1974, On Correction Terms for Systematic Errors in Bundle Adjustment, *BUL*, Vol. 6, pp. 223–229.

Shewell, H. A. L., 1953, The Use of the Cambridge Stereocomparator for Air Triangulation. *The Photogrammetric Record*, Vol. 1, No. 2.

Smialowski, A. J. 1963, N.R.C. Monocomparator. *Canadian Surveyor*, Vol. 17, No. 2.

Thompson, E. H., 1959a, A Rational Algebraic Formulation of the Problem of Relative Orientation. *The Photogrammetric Record*, Vol. 3, No. 14.

Massachusetts Institute of Technology, 1962, Analytical Aerial Triangulation Error Analysis and Application of Compensating Equations for the General Block Triangulation and Adjustment Program. Civil Engineering Systems Laboratory Publication 153.

McNair, A. J., Dodge, H. R., and Rutledge, J. D., 1958, A Solution of the General Analytical Aerotriangulation Problem. Final Technical Report, ERDL Contract No. DA-44-009ENG2986, Cornell University.

Meissl, P., 1972, A Theoretical Random Error Propagation Law for Anblock Networks With Constrained Boundary, *Osterreichische Zeitschrift für Vermessungswesen*, pp. 61–65.

Merritt, Everett, 1958, Analytical Photogrammetry, New York, Pitman Publishing Co.

Mosaad Allam, M. and Chaly, C. K., 1975, Block Adjustment Using the New Version of PAT-M43-APR-Statoscope-Lake, Unpublished Manuscript, Dept. of Energy, Mines and Resources, Ottawa.

Poetzschke, Heinz G., 1967, A Comparator Calibration Method, *Ballistic Research Laboratories Report No. 1353*, U.S. Army Material Command, Aberdeen Proving Ground, Maryland, pp. 33.

Pope, Allen J., 1970, An Advantageous, Alternative Parameterization of Rotations for Analytical Photogrammetry, National Ocean Survey, Rockville, Md.

Proctor, D. W., 1962, The Adjustment of Aerial Triangulation by Electronic Digital Computers. *Photogrammetric Record*, Vol. 4, No. 19.

Pulfrich, C., 1919, Über Photogrammetrie aus Luftfahrzeugen, Jean, G. Fisher.

Robinson, G. S., 1963, The Reseau as a Means of Detecting Gross Lack of Flatness of Film At The Instant of Exposure, *Photogrammetric Record*, Vol. 4, No. 22.

Rosenfield, George H., 1961, The Application of Analytical Photogrammetry to Missile Trajectory Measurement, *Photogrammetric Engineering*, Vol. 27, No. 4.

—— 1963a, Photogrammetric Data Reduction Analysis, General Photogrammetric Adjustment, Technical Documentary Report, No. MTC-TDR-63-9.

———— 1963b, Calibration of a Precision Coordinate Comparator, *Photogrammetric Engineering*, Vol. 29, No. 1.

Saastamoinen, J., 1972, Refraction, *Photogrammetric Engineering*, Vol XXXVIII, No. 8, pp. 799–810.

———— 1974, Local Variation of Photogrammetric Refraction, *Photogrammetric Engineering*, Vol. XL, No. 3, pp. 295–301.

———— 1959b, A Method for the Construction of Orthogonal Matrices. *The Photogrammetric Record*, Vol. 3, No. 13.

Washer, Francis E., 1957, The Effect of Prism on the Location of the Principal Point, *Photogrammetric Engineering*, Vol XXIII, No. 3, pp. 520–532.

CHAPTER X

Stereoscopy

Author-Editor: GEORGE L. LAPRADE
Contributing Authors: DR. S. JAMES BRIGGS, RICHARD J. FARRELL, EARL S. LEONARDO

10.1 Introduction

STEREOSCOPY IS THE SCIENCE and art that deals with the use of images to produce a three-dimensional visual model with characteristics analogous to those of actual features viewed using true binocular vision. There are three primary applications of stereoscopy in photogrammetry and interpretation. One is an interpretation aid in recognizing the three-dimensional form of an object; another is in the estimation of slopes or relative heights; and the third is as an adjunct to contour plotting devices designed to accurately measure height differences. The basic principles of depth perception and stereoscopy required for an understanding of all three of these applications are discussed. The effect of stereoscope configuration and

image collection geometry on the perceived stereoscopic model are analyzed and descriptions are given of the simpler commercial stereoscopic devices, particularly those used for slope and relative height estimation.

Throughout this chapter it is assumed that the stereoscopic pair consists of two vertical photographs with negligible tilt or other distortions. Methods for eliminating or compensating for such distortions are covered in subsequent chapters which also describe in detail the more sophisticated stereoscopic plotting instruments used for accurate height measurements. Stereoscopic flight planning considerations are given in chapter VII.

10.2 Basic Principles of Stereoscopic Vision

10.2.1 CONCEPTS OF VISUAL DEPTH PERCEPTION

In three-dimensional space perception the observer uses a number of visual characteristics or cues to determine the depth and size of objects. These can conveniently be divided into monocular and binocular cues, depending on whether they are apparent with only one eye or require both.

10.2.1.1 BINOCULAR CUES

There are two binocular cues used in space perception. They are the convergence angles of the optical axes of the eyes and *retinal disparity*. In figure 10-1 consider the eyes fixed on an isolated point in space P, such that the rays from P pass along the optical axes of the eyes and focus at C, the center of the retinas.[1] The convergence muscles have rotated the eyes so that their optic axes are aligned with the rays from P, forming the convergence angle γ. The sensing of the

amount of muscular tension resulting from different convergence angles provides a cue to the absolute distance to the point P.

Retinal disparity can be explained using the concept of corresponding retinal points in the two eyes. In figure 10-2 the eyes are again fixed on the point P but another point A has been introduced into the field of vision. Rays from A focus on the retina at a in the left eye and at a' in the right eye. Consider the drawings of the two eyes to be rotated until their optic axes are aligned and then overlaid one on the other as shown.

The theory of retinal disparity, to a considerable extent confirmed by experimental evidence, states that if a and a' fall at the same place on the retina of the composite eye, then points A and P will be observed as being at the same distance from a line through the axes of the two eyes. If a and a' are not coincident, a retinal disparity exists and points A and P will be observed as being at different distances from the observer.

In figure 10-3 the eyes are fixating on point P of an object with height h. Rays from point P again focus on the center of the retina while the rays from A focus at a and a'. However, in this case a and a' do not fall at the same place on the retina

---

[1] For purposes of this discussion the optical characteristics of the eye can be considered analogous to those of a camera. Physiological details of the eye can be found in Graham (1965).

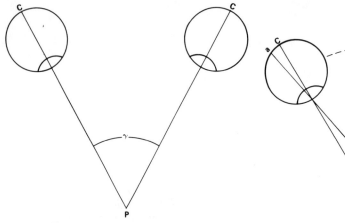

Figure 10-1. Convergence angle.

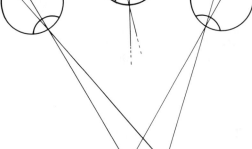

FIGURE 10-2. Corresponding retinal points.

of the composite eye; therefore, a retinal disparity exists. This phenomenon results in the perception of points P and A being at different distances from the observer. As shown in figure 10-3 the disparity angle to either eye is $\alpha/2$ and the total disparity angle for the composite eye is α.[2] This angle determines the magnitude of the retinal disparity and is equal to the difference between the convergence angles to points P and A.

If the disparity angle α is small enough, the observed object in figure 10-3 will be fused. However, if the disparity is too large when the eyes are fixed on the base, the top of the object will appear as two entities, or "split." If the eyes are fixed on the top of the object, the reverse effect occurs and the base is seen as two separate images when there is a large retinal disparity. This effect can be observed by holding a pencil about a foot in front of the eyes and fixed on a distant object. The pencil will appear to be doubled. The maximum disparity angle α, which can occur without producing double images is approximately one degree. At the other end of the scale, the minimum value of α, which will give a perceptible difference in depth, is approximately three seconds for a typical observer. By comparison, ordinary vernier acuity is about 12 seconds. While these angle values vary for different observers and viewing conditions, they indicate the sensitivity of retinal disparity and why it is so important in depth perception.

In the discussion given here the concept of retinal disparity has been somewhat idealized.

For example, in figure 10-2 it was implicitly assumed that zero disparity would be produced by any one of the points lying on a horizontal line through P. Such a zero disparity line is called an *horopter*. Careful experiments show that the perceived horopter is actually a line with a slight curvature. However, for the field of view and accuracy of depth determination achievable with stereoscopes, it is fully satisfactory to approximate the horopter with a straight line. A more comprehensive discussion of retinal disparity and the horopter can be found in Graham (1965).

10.2.1.2 MONOCULAR CUES

There are a variety of monocular cues, only one of which is subject to any significant control in stereoscopic viewing. This is the *focusing accommodation* of the individual eye. When fixed on a point, the muscles of the eye adjust focal length to provide a focused image on the retina. Points at different distances from the observer require different focusing accommodations. The sensing of the tension in the focusing muscles required for this accommodation can provide a cue to the distance to the object. The focusing

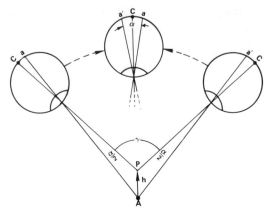

FIGURE 10-3. Retinal disparity concept.

---

[2] For convenience it will be assumed in the discussion which follows that objects are centered between the eyes, cameras and the stereoscope lenses. The same equations for vertical exaggeration and model magnification would be obtained using an asymetrical geometry, and it has been shown that the same apparent heights are produced with either configuration (LaPrade, 1973).

accommodation of just a single eye provides only very relative discriminations of depth distances, particularly at distances greater than a metre (Graham, 1965). However, its interaction with the retinal disparity and convergence accommodation when using both eyes can be a significant factor for stereoscopes.

In addition to focusing accommodation, the other principal monocular cues are as follows:

perspective—parallel lines such as railroad tracks appear to converge toward the horizon,

movement parallax—lateral head movement causes nearer features to appear to move a greater distance with respect to the horizon than do more distant features,

relative size of known objects,

overlap—a nearer object overlaps part of one at a greater distance,

highlights and shadows,

atmospheric obscuring of fine details as a function of distance.

While these monocular cues are sometimes of importance in stereoscopic viewing, they are essentially fixed for any given stereoscopic pair and are thus not subject to modification by adjustments of the stereoscope parameters.

10.2.2 STEREOSCOPIC ANALOGUE OF BINOCULAR VISION

In developing a stereoscopic analogue of true *binocular vision,* the obvious approach would be to provide a convergence accommodation, retinal disparity, and focusing accommodation compatible with each other and with a real viewing condition. In practice this objective is only partially achieved because of stereoscope design limitations.

The closest approximation to this ideal is that diagrammed in figure 10-4. In figure 10-4(A) the eye views a pyramid. Because of the angle α there is a retinal disparity which results in the perception of the height, h_T.

In figure 10-4(B) the eyes have been replaced by two cameras which produce photographs with the same base dimensions as the actual object. In the left photograph there is an elevation displacement of the top of the object so that it is recorded a distance $p_c/2$, to the right of the base. Similarly, the right-hand camera results in a displacement $p_c/2$, to the left.

In figure 10-4(C) images of the two photographs have been projected onto a viewing surface so that the base of the object for both exactly coincides. Of course, the images of the top are separated by the parallax distance p_c. To obtain a stereoscopic view of this combined image, it is necessary to provide a means whereby the left eye sees only the light rays from the left photograph and the right eye sees only the right photograph. A common way of doing this is with a *vectograph.* The two photographs are projected on top of each other using two light

sources polarized at right angles with respect to each other. *Polarizers* are then placed in front of each eye with the planes of polarization corresponding to those of the light sources. For example, if the left photograph is projected with horizontally polarized light and the right with vertically polarized light, then a horizontally polarized viewer at the left eye permits only light rays from the left photograph and blocks those from the right. Similarly, the right eye sees only the rays from the right photograph.

A similar method is to make a combined print, or *anaglyph,* of the two photos with the left one printed in one color and the right in a different color. When viewed through corresponding color filters, the left eye sees only the left photograph and the right eye the right one. A third technique is to use a *flicker* system to alternately project the images intended for the separate eyes while synchronously blocking the light to the opposite eye. In this way each of the eyes alternately sees only its appropriate image. However, with a relatively high flicker rate each eye perceives its image as being continuous.

Using any of these methods, the stereoscopic viewing geometry is as shown in figure 10-4(C). The base of the object has the same dimensions and is the same distance from the observer as in the actual binocular viewing case shown in figure 10-4(A). The tops of the object on the two photographs are represented by the horizontal arrowheads. Rays from these points and from the center of the base form the same angles at the eye and produce the same retinal disparity as in the true binocular vision case. Thus, the projected stereoscopic model height h, is the same as h_T, its true value.

Also note that the convergence angles and the convergence accommodation are the same in both cases. The only difference is that the focusing accommodation for the top of the object must be for the distance to the photograph D, in the stereoscopic case, while it is for the distance D-h_T, in the true binocular case.

10.2.3 VERTICAL EXAGGERATION AND MODEL MAGNIFICATION

Several transformations in the perceived stereoscopic model dimensions may occur as shown in figure 10-5. For the highly idealized analogue diagrammed in figure 10-4 the perceived height and base dimensions of a feature h and b would be equal to the true dimensions h_T and b_T as indicated by the vertical cylinders in (A) and (B) of figure 10-5. In practice this one-to-one correspondence is neither practical nor desirable for stereoscopic mapping applications.

In figure 10-4(C) assume that the scales of the two photographs are reduced by one-half. Then the dimensions of both the base and the parallax p_c will be one-half as great and perceived values

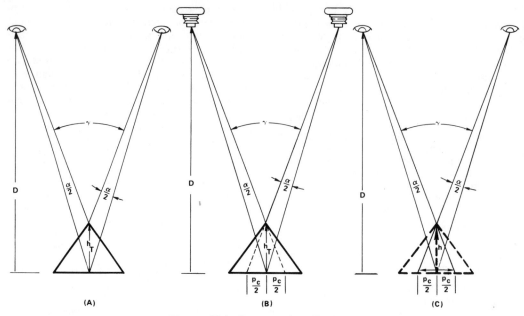

FIGURE 10-4. Stereoscopic analogue.

of the height and base will be reduced as indicated in figure 10-5(C).

Similarly, if the images are enlarged there will be a *magnification* of the dimensions of the perceived stereoscopic model as shown in figure 10-5(D). Note that for both reduction and magnification, the ratios of the vertical and horizontal dimensions of the perceived model are still equal to h_T/b_T, the value for the actual object.

In figure 10-4(B), if the cameras are moved farther apart, the recorded dimensions of the base will remain the same, but the value of the parallax p_c is increased. When these new images are viewed stereoscopically, the perceived height of the stereoscopic model is increased but the base remains the same. Thus, a *vertical exaggeration* of the model occurs such as that shown in figure 10-5(E). The vertical exaggeration q can be defined by the relationship

$$q = \frac{\text{perceived height-to-base ratio of stereoscopic}}{\text{height-to-base ratio of actual object}}.$$

$$(10.1)$$

Thus, for figure 10-5(E) the vertical exaggeration is 2.

Either magnification or vertical exaggeration will increase the apparent height of features in the stereoscopic model, permitting a more accurate determination of height differences. However, there is a significant distinction between the two. Model magnification maintains the true height-to-base ratio of objects but reduces the area that can be viewed at one time. Vertical exaggeration provides an increase in apparent height without changing the area viewed. Aerial photographs, of course, are at a relatively small scale and therefore produce a greatly reduced stereoscopic model. Partially for this reason a camera configuration is normally used which gives a vertical exaggeration between 2 and 3.

In nearly all cases where stereoscopes are used for estimation, it is the relative height with respect to some known reference which is of interest, because it is difficult to make accurate estimates of absolute heights. If the reference is a known horizontal distance, it is very important

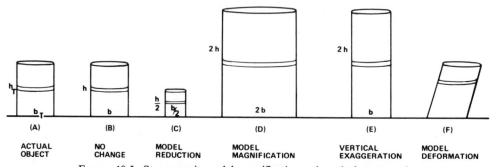

FIGURE 10-5. Stereoscopic model magnification and vertical exaggeration.

to know the vertical exaggeration. For example, with a vertical exaggeration of 2, it is necessary to take one-half of the observed height estimated with respect to some base reference. However, it is seldom necessary to know the model magnification because the apparent height and the base distance of known size to which it is being compared change proportionately.

With some stereoscope configurations model distortions or deformations may occur as shown in figure 10-5(F). These are principally of two types. One is a tilting of the feature and the other is a compression of some heights in the model with respect to others. Tilting occurs when a feature not midway between the camera positions is viewed midway between the stereoscope lenses or *vice versa*. It can normally be removed by moving the stereoscopic pair laterally so the viewing geometry is compatible with that of the photographs.

Relative compression of heights with respect to each other can lead to serious errors in slope or height estimations. In figure 10-5(F), the top of the tank has been compressed with respect to the bottom as shown by the displacement of the band around the center. In a model with this type of deformation, height differences at lower levels appear too large with respect to a given horizontal reference, while those at higher levels are compressed. Therefore, it is desirable to select stereoscope parameters which eliminate this effect.

10.2.4 PSEUDOSCOPIC ILLUSIONS

In stereoscopic viewing, or even with a single photograph, it is possible to obtain an apparent reversal of natural relief so that hills become valleys and rivers appear to flow along ridges. Such an effect is known as a *pseudoscopic illusion*. In stereoscopic viewing it occurs when the two images are presented to the left and right eye in the reverse order. With a single photograph, it occurs most frequently when the shadows fall away from the observer.

For the stereoscopic analogue discussed relative to figure 10-4(C), consider what happens when the left photographic image is presented to the right eye and *vice versa*. The stereoscopic viewing situation is now as shown in figure 10-6. Instead of the top of the feature being perceived at a height h above the surroundings, it now appears to be below the surface. This effect can be observed by viewing figure 10-7 with a pocket stereoscope. In one case the stereoscopic pair has been mounted to give the correct view, and in the other, the psuedoscopic view.

The existence of this illusion is very dependent on the type of features in the images. It is readily perceived when the pseudoscopically inverted relief resembles a feature which actually exists in nature. However, if the pseudoscopic model is completely unrealistic, for example, trees or buildings inverted below the surface,

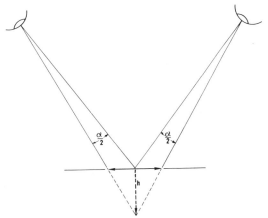

FIGURE 10-6. Pseudoscopic geometry.

then usually neither pseudoscopic nor true three-dimensional models are perceived; rather, the stereoscopic view appears the same as that of either photograph alone.

The pseudoscopic illusion in stereoscopic viewing is easily recognizable and seldom presents a practical problem. It can be avoided with conventional vertical photography by positioning the stereoscopic pair so the overlap areas of the photographs are toward the center and the areas not common to the two, are toward the outside.

A similar pseudoscopic illusion can often be observed with a single aerial photograph. To obtain the true relief effect most people prefer that the photograph be oriented with the shadows falling toward the observer. It has been proposed that this preference exists because in the normal environment the light source is above the eye level. In many instances, when the photograph is rotated 180 degrees and the shadows fall away from the observer, a pseudoscopic effect is observed, as shown in figure 10-8 which shows the same photograph in the two different orientations. The unusual nature of this pseudoscopic effect is that except in extreme cases many people (possibly up to 30 percent of the observers) will sense only the correct configuration while the remaining 70 percent will perceive the pseudoscopic effect. Individuals with sight in only one eye[3] develop a high degree of dependence on monocular cues such as highlights and shadows. Consequently, they are especially susceptible to this pseudoscopic illusion and have a particularly strong preference for photographs oriented with the shadows falling toward them.

As with the stereoscopic illusion, the shadow pseudoscopic effect is very dependent on the nature of the image. It most readily occurs when the pseudoscopically perceived relief closely

---

[3] Including the author-editor of this chapter.

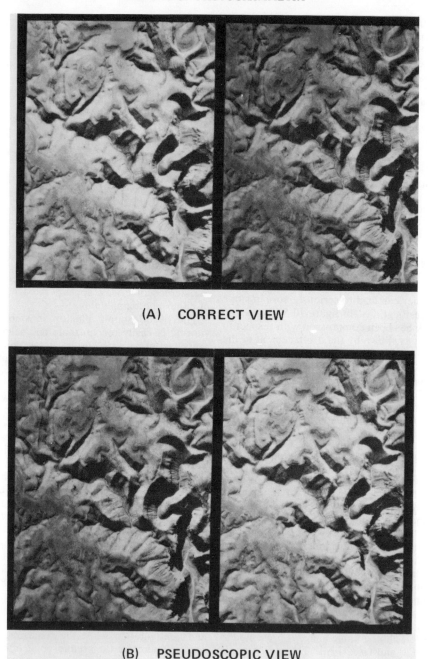

(A)   CORRECT VIEW

(B)   PSEUDOSCOPIC VIEW

FIGURE 10-7. Effect of photograph orientation on stereoscopic orientation.

corresponds to that of features which actually exist. When this is not true, there is a tendency for the observer's visual system to reject the pseudoscopic shadow image. In marginal cases the pseudoscopic and normal effect may often be observed alternately on the same portion of the image, or one effect may be evident on one part and the opposite effect on another part.

10.3 Stereoscopic Parameter Relationships

The vertical exaggeration, magnification and deformation of the model are determined by the stereoscope geometry, camera parameters, and flight configuration. These relationships have not been experimentally determined in all cases and they also vary to some extent for different individuals. However, they can be given with sufficient accuracy for most practical applications where it is desired to estimate terrain slopes or the heights of features.

(A)   CORRECT VIEW

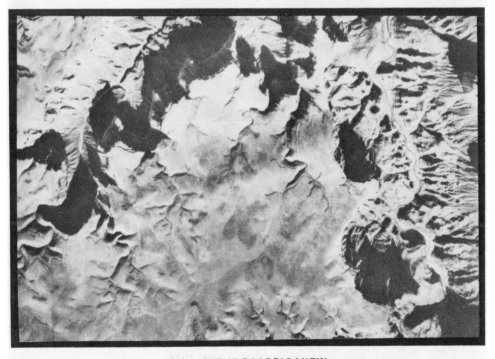

(B)   PSEUDOSCOPIC VIEW

FIGURE 10-8. Effect of shadow orientation.

10.3.1 SIMPLE LENS STEREOSCOPES

10.3.1.1 STEREOSCOPIC ANALOGUE

The stereoscopic analogue to binocular vision for a simple lens stereoscope is shown in figure 10-9. Through two lenses the observer views the stereoscopic photographic pair placed at a distance f_s below the lenses. The elevation displacement on either photograph is $p_c/2$, so the total *parallax* is p_c. The total base width on either photograph is b_c.

The rays from the photographs passing through the lens strike the eye. Again the parallax produces the angle α and a corresponding retinal disparity. If the rays are extended backward away from the observer, as shown by the dotted lines, they converge to form a projected figure of height h_p and base width b_p with the top a distance I from the lenses. It is natural to assume this projected figure would be the same as the visually perceived stereoscopic model. Unfortunately, the nearly exact analogue produced by the vectograph is not maintained with conventional lens stereoscopes.

In figure 10-9 the convergence angle γ and the retinal disparity are still compatible with values which could be obtained in true binocular vision. However, the focusing accommodation depends on the image distance produced by either stereoscope lens. This in turn is determined by the distance from the lens to the photograph f_s. A value could be selected for f_s so that the virtual image distance was the same as that to the projected stereoscopic model. The focusing and the convergence accommodations would then be for the same distance. There is some experimental evidence that this configuration provides improved discrimination for small height differences in the stereoscopic model (Farrell, Anderson, Kraft, and Boucek, 1975).

However, there is a common belief that over long periods of time there is less eye strain if the focusing accommodation is for infinity. Although this belief is suspect, stereoscope manufacturers nevertheless make f_s essentially equal to the focal length to satisfy this condition.

Figure 10-10 shows the typical viewing geometry for such focal plane lens stereoscopes with p_c, h_p, and I representing the same parameters as in figure 10-9. The stereoscopic pair is placed at the focal length f_s, below the lenses. Therefore, all the rays from any point on either photograph are parallel upon leaving the lens as shown. The focusing accommodation of the eyes is for infinity, but the convergence accommodation is set by γ. It is interesting to note that the convergence angle in this case is the same for all observers regardless of their eye spacing. Also, it is determined by the separations of the stereoscopic pair w, and of the lens centers E_L, and is not affected by the eye spacing.

To make the convergence accommodation

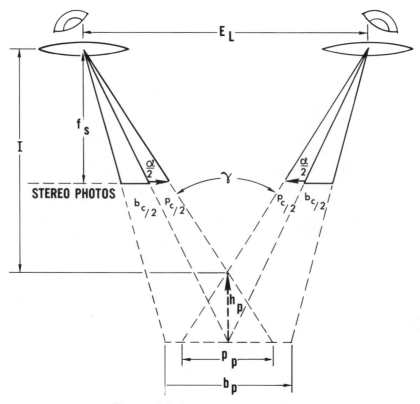

FIGURE 10-9. Lens stereoscope analogue.

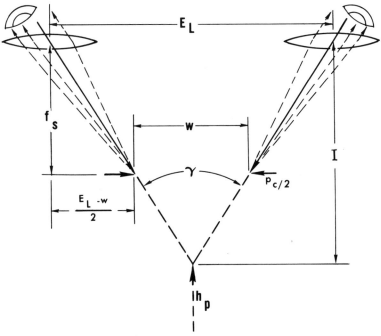

FIGURE 10-10. Focal plane stereoscope geometry.

correspond to the focusing accommodation in figure 10-10, it would be necessary to place each photograph directly beneath the center of its lens so that the axes of the eyes were parallel as they would be in viewing a distant object such as the moon. Even this does not conform to the true situation in actual binocular vision. For objects whose closest point is beyond about 400 metres, even the largest structures will not produce a detectable disparity angle α (Graham, 1965). Thus, for such objects true binocular vision produces no discernible retinal disparity with the focusing and convergence accommodations essentially for infinity. However, in the lens stereoscope the retinal disparity angle is determined by the parallax p_c and may be quite large.

For this reason it is not possible in a focal plane stereoscope to obtain a configuration where the focusing accommodation, retinal disparity, and convergence accommodation all correspond to binocular vision. Because of this discrepancy, an appreciable number of observers cannot fuse the two images into a stereoscopic model when the photographs are placed so the axes of the eyes are parallel, and nearly everyone experiences some eye strain under these conditions.

Since no true analog exists, it is necessary to determine the focal plane stereoscope configuration which provides the optimum results. To accomplish this the first requirement is to relate the stereoscopic pair parameters to the image collection geometry shown in figure 10-11. For simplicity, the film has been projected in front of the lens rather than behind it. The photographic parallax and base width are p_c and b_c. Their cor-

responding values on the object being photographed are p_T and b_T. B is the camera base and H the height above the peak of the object. The true height of the object is h_T, and f_c is the camera focal length. From similar triangles

$$\frac{p_c}{b_c} = \frac{p_T}{b_T} \tag{10.2}$$

and

$$\frac{p_T}{h_T} = \frac{B}{H}, \tag{10.3}$$

so

$$\frac{p_c}{b_c} = \frac{B}{H} \frac{h_T}{b_T}. \tag{10.4}$$

Because B/H is essentially constant for a given pair of stereoscopic photographs, the true *height-to-base ratio* is directly proportional to p_c/b_c.

To ensure there is no distortion of relative heights of the type discussed in connection with figure 10-5(F), it is necessary that the perceived height-to-base ratio h/b, be directly proportional to h_T/b_T. From equation (10.4) it is evident this will be true only if h/b is also directly proportional to p_c/b_c.

Experiments have shown that this condition is satisfied for only one particular value of I/E_L (*convergence angle*) in the stereoscope configuration of figure 10-9. Within the accuracy of the experimental data, this value of I/E_L was found to be ≈ 5 for all subjects tested except one with severe myopia (LaPrade; 1972, 1973). It is perhaps not coincidental that this distortion-free

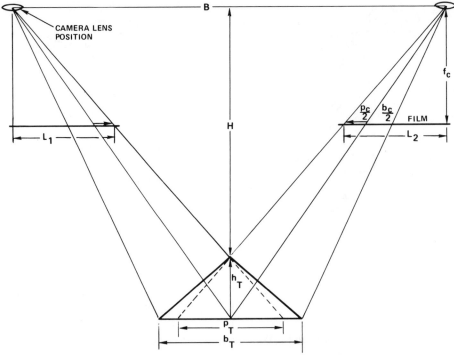

FIGURE 10-11. Image collection geometry.

value of I/E_L is approximately equal to the ratio of the distance for most distinct vision to the eye spacing. For a typical interpupillary distance of 2.5 inches a ratio of 5 corresponds to a viewing distance of 12.5 inches. This is also approximately the distance observers give when asked to estimate the distance to the stereoscopic model using focal plane stereoscopes (Miller; 1958, 1960).

A typical set of data are shown in figure 10-12. The perceived height-to-base ratios h/b are plotted against the ratio of the parallax-to-base width p_c/b_c. The same results were obtained for all focal plane stereoscopes regardless of the lens focal length. As discussed previously, h/b must be directly proportional to p_c/b_c if there are to be no model distortions. It is evident this is experimentally true only for the curve where $I/E_L = 5$. For the value of I/E_L greater than 5, the taller heights in the stereoscopic model are compressed with respect to the lower. For example, with a true height-to-base ratio such that p_c/b_c is 0.10, the perceived h/b is 0.75 as indicated when $I/E_l = 16$. If the true height were doubled, equation (10.4) shows that p_c/b_c would also be twice as great or equal to 0.20. However, the perceived height is not doubled because, as the graph illustrates, h/b only increases to 1.17 instead of 1.50 when p_c/b_c is equal to 0.20. This corresponds to more than a 20 percent compression of the greater height with respect to the lower. For the value of I/E_L less than 5, the lower heights are perceived as compressed with respect to the taller.

Another interesting conclusion drawn from the data is that the projected figure in figure 10-9 is the same as the perceived stereoscopic model only when $I/E_L \approx 5$. From similar triangles in figure 10-9,

$$\frac{h_p}{p_p} = \frac{I}{E_L}$$

and

$$\frac{p_p}{b_p} = \frac{p_c}{b_c}$$

so

$$\frac{h_p}{b_p} = \frac{I}{E_L} \frac{p_c}{b_c}. \qquad (10.5)$$

The perceived h/b will equal h_p/b_p only when h/b is also given by

$$\frac{h}{b} = \frac{I}{E_L} \frac{p_c}{b_c}. \qquad (10.6)$$

From the experimental data it is evident that equation (10.6) is true only when $I/E_L \approx 5$. With this condition satisfied not only are model deformations minimized but also the projected model in figure 10-9 gives a true representation of the perceived stereoscopic model.

Ideally, the optimum value of I/E_L should be established separately for each observer, but for the accuracy with which relative heights can be estimated a value of 5 will be satisfactory in most practical applications. Fortunately, when allowed to select their preferred spacing of the

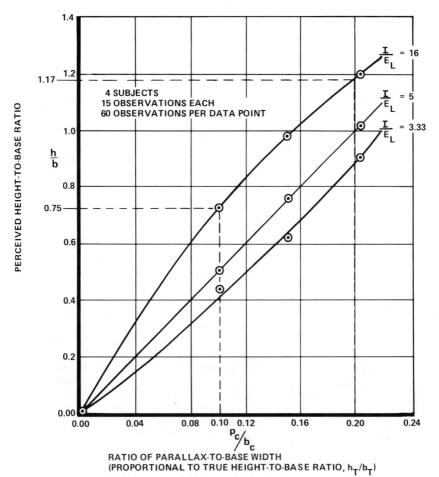

FIGURE 10-12. Experimental data (adapted from LaPrade, 1972).

stereoscopic pairs most observers choose a separation which provides a value of I/E_L somewhere near 5. However, there are appreciable
variations sometimes introducing distortion errors of as much as 30 percent. For accurate
height estimates it is recommended that the
stereoscopic pair separation be set as discussed
in section 10.3.3 so $I/E_L = 5$. Because this configuration provides the optimum viewing conditions, the value of the vertical exaggeration and
model magnification will be developed for this
case.

10.3.1.2 VERTICAL EXAGGERATION FOR LENS STEREOSCOPE

Under the above conditions, the height-to-
base ratio h/b of the perceived stereoscopic
model in figure 10-9 equals h_p/b_p, the values for
the projected figure. From equations (10.4) and
(10.6),

$$\frac{h}{b} = \frac{I}{E_L} \frac{h_T}{b_T} \frac{B}{H}. \qquad (10.7)$$

When B/H is not known, it can be determined
directly from the camera focal length and the

photographs. From figure 10-11, $B/H = (L_1 +
L_2)f_c$. The distance L_1 is measured from the nadir
point on the photograph to any selected point
lying along the ground track and L_2 is measured
from the nadir of the other photograph to the
same selected point.

Utilizing equation (10.1), the expression for
the *vertical exaggeration* is obtained from equation (10.7),

$$q = \frac{h/b}{h_T/b_T} = \frac{I}{E_L} \frac{B}{H}. \qquad (10.8)$$

For the case under consideration, $I/E_L \approx 5$, so
that

$$q \approx 5 \frac{B}{H}. \qquad (10.9)$$

This result is very significant. It indicates that
with the optimum stereoscope configuration the
vertical exaggeration of the stereoscopic model
is determined only by the camera base-to-height
ratio. It is not dependent on the stereoscope,
lenses, camera focal length, or scale of the
photographs. However, a careful distinction
must be made between vertical exaggeration and
reduction or magnification of the stereoscopic

model discussed relative to figure 10-5. As next will be shown, all these parameters are important in establishing the magnification of the stereoscopic model.

10.3.1.3 MODEL MAGNIFICATION FOR LENS STEREOSCOPE

From the equation (10.6),

$$\frac{h}{b} = \frac{I}{E_L}\frac{p_c}{b_c} . \qquad (10.6)$$

Thus, for the fixed value of I/E_L, the observed height-to-base ratio h/b depends only on p_c/b_c, the ratio of the parallax-to-base width for the stereoscopic pair. However, in figure 10-9 it is apparent that if p_c and b_c are increased proportionately, then both h_p and b_p will increase. Thus, the model is magnified even though the height-to-base ratio and the vertical exaggeration remain constant.

Figure 10-11 shows that changing the camera focal length f_c simply changes the photographic scale so that both p_c and b_c vary proportionately but their ratio remains constant. Therefore, increasing f_c produces a *magnification* of the stereoscopic model but does not affect the vertical exaggeration. The same relationship applies to a photographic enlargement made from the original camera negative. It is also apparent from figure 10-11 that the only way to change p_c/b_c and the vertical exaggeration is to change B/H. This conforms to the result obtained in equation (10.9).

The magnification produced by the stereoscope lenses also enlarges the stereoscopic model. In effect, the lens enables the eye to see objects distinctly at shorter distances than would otherwise be possible, thereby making them appear larger. In figure 10-13 (A), the eye is viewing the object with dimensions b_c at the distance

for most distinct vision d. The angle subtended by the object at the eye is θ_o. If the object is moved appreciably closer, the eye will no longer be able to focus a clear image on the retina. However, with the stereoscope lens in front of the eye as shown in figure 10-13 (B), the object can be moved closer to position f_s, and the eye can still focus. The subtended angle is increased to θ. This is the same angle an object of height b would subtend at the distance for most distinct vision. Thus, the lens has provided an effective enlargement of the object from b_c to b. A conventional definition of the lens magnification is the ratio b/b_c. From similar triangles the magnification M is equal to

$$M = \frac{b}{b_c} = \frac{d}{f_s} . \qquad (10.10)$$

The lens magnification increases the apparent photographic parallax p_c and base width b_c proportionately. Thus, there is an enlargement of the stereo model but again the ratio of p_c/b_c remains constant; therefore, the observed height-to-base ratio h/b and the vertical exaggeration are not affected by the lens magnification.

To illustrate, using a lens in figure 10-9 which permitted the photographs to be moved closer to the eye would cause the apparent size of both p_c and b_c to be increased producing an enlargement of the projected stereoscopic model.

As explained previously (section 10.2.3), it is seldom necessary to know the exact stereoscopic model magnification or reduction because it is only practical to estimate heights with respect to known references. However, if desired, an expression for model magnification k can be easily obtained. It is equal to the ratio of an apparent horizontal dimension in the stereoscopic model to its true value. Thus,

$$k = \frac{b}{b_T} . \qquad (10.11)$$

The scale of the photograph, S, is equal to

$$S = \frac{b_c}{b_T} , \qquad (10.12)$$

and from equation (10.10) for the stereoscope lens magnification

$$b = \frac{d}{f_s}\, b_c . \qquad (10.13)$$

Solving these three equations gives the model magnification to be,

$$k = \frac{Sd}{f_s} . \qquad (10.14)$$

Although the distance for most distinct vision d varies from approximately 9 to 14 inches for different subjects, it is fixed for a given observer. Increasing the camera focal length f_c or enlarging the negative photographically increases the scale S. It is evident from equation (10.14) that this will increase the model magnification in direct proportion. Using a stereoscopic

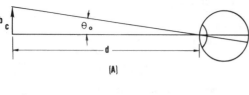

(A)

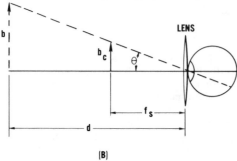

(B)

FIGURE 10-13. Magnification of lens stereoscope.

lens which decreases the distance to the image f_s, increases the magnification accordingly. These conclusions, of course, are the same as those presented previously in a more qualitative manner.

10.3.2 SIMPLE LENS MIRROR STEREOSCOPE

The simple lens mirror stereoscope is the exact equivalent of a regular lens stereoscope except that two mirrors (or prisms) are placed at a 45-degree angle in front of each lens to laterally offset the light path. This increases the physical separation of the stereoscopic pair so that larger

photographs can be used without their overlapping at the center.

Figure 10-14 shows the viewing geometry of the mirror stereoscope. The optical axes of the system are indicated by the dotted lines going vertically downward through the center of the lenses to the first mirror, horizontally to the second mirror, and then downward again. The stereoscopic pair is placed so that the total light path distance is essentially equal to the focal length f_s of the lenses. Thus, as with the regular lens stereoscope, the focusing accommodation is for infinity. In the focal plane of the lenses the distance between the two optic axes is w_o. In the focal plane of the lenses the distance between

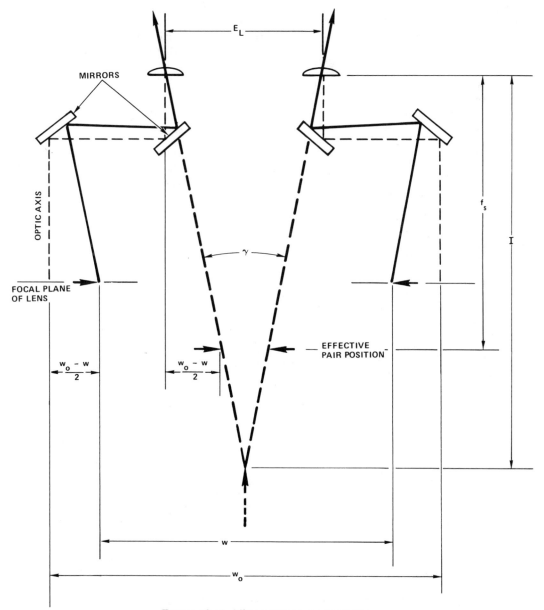

FIGURE 10-14. Mirror stereoscope geometry.

the two optic axes is w_o. The stereoscopic pair separation is w, normally somewhat less than w_o.

The observer, of course, cannot detect the light path offset and perceives the rays from the stereoscopic pair as if they came from the effective pair position indicated on the diagram. The projected image is at the distance I from the lenses and the effective convergence angle is γ.

Comparing this to figure 10-9 it is apparent the perceived cues are identical to those that would be produced by a regular lens stereoscope with the same values of f_s and I/E_L. Thus, the same distortions would be expected unless again I/E_L ≈ 5. In the same experimental studies this was found to be true; also, the perceived h/b of the model was the same for the mirror as for the lens stereoscope with the same I/E_L values. Therefore, equations (10.9) and (10.14) for vertical exaggeration and model magnification are also applicable to the mirror stereoscope.

The lens focal length f_s is usually greater for the mirror stereoscope than for the lens stereoscope, so there is a reduction in the model size, although the vertical exaggeration is unchanged. On the other hand, larger scale photographs of a given area can be used with the mirror than with the lens stereoscope without their overlapping at the center.

10.3.3 OPTIMUM STEREOSCOPIC PAIR SEPARATION

As discussed previously, most observers naturally select a stereoscopic pair separation which provides a value of I/E_L somewhere near 5. For casual stereoscopic viewing or rough approximations of height this procedure will normally be satisfactory. However, many experienced workers in the field become very proficient at making accurate estimates of slopes and relative heights. When using lens or mirror stereoscopes for these purposes, a pair separation should be determined which will make I/E_L ≈ 5 so that model distortions will be minimized.

From similar triangles in figure 10-10

$$\frac{I}{E_L} = \frac{f_s}{E_L - w} \approx 5 \, ,$$

so

$$w = E_L - \frac{f_s}{5} \, . \tag{10.15}$$

The value of f_s is determined by measuring the vertical distance from the center of the stereoscope lens to the stereoscopic pair. If the stereoscope lens spacing is variable, it should be adjusted to the interpupillarly distance so the eyes are aligned with the centers of the lenses. The distance between their centers is then E_L. If the lens spacing is fixed, then this fixed distance must, of course, be used for E_L. These values of f_s and E_L are then substituted in equation (10.15) to get the desired separation of the photographs.

For mirror stereoscopes the determination of

the separation of the photographs is slightly more involved. From similar triangles in figure 10-14

$$\frac{I}{E_L} = \frac{f_s}{w_o - w} \approx 5 \, ,$$

so

$$w = w_o - \frac{f_s}{5} \, . \tag{10.16}$$

The value of f_s is most easily determined by removing one lens and determining its focal length. This can be done with collimated light on an optical bench or by determining the distance from the lens to the point where it focuses the sun's rays. Some manufacturers give a value for the distance between the centers of the two outside mirrors which ideally would be equal to w_o. However, slight misalignments of the mirrors can appreciably alter the effective separation of the optic axes. A more precise method is to direct collimated light (or the sun) into the two lenses and note the two points where they focus in the stereoscopic pair plane. The distance between these two points is w_o. These values of f_s and w_o are then substituted in equation (10.16) to determine w.

From figures 10-9 and 10-14, the calculated pair separation should ideally be for the top of the feature being viewed to exactly conform to the conditions for which the equations for vertical exaggeration and model magnification were developed. However, the horizontal scale and the height of the model in these figures have been greatly exaggerated for clarity. In practice, the difference in convergence angle (I/E_L) for the top and bottom of a feature will have a negligible effect on model distortion. Therefore, any two common points at some average height can be chosen in setting the pair separation.

The above procedure is concerned only with determining the stereoscopic pair separations to provide an optimum value of I/E_L. Other considerations in arranging the photographs are discussed in section 10.7.

10.3.4 STEREOSCOPIC MICROSCOPES

Some conventional and mirror stereoscopes can be used either with or without microscopic eye pieces (see section 10.6). Also, some stereoscopes are made exclusively for microscopic viewing. Essentially, they provide the same analogue to binocular vision previously discussed but with additional model magnification caused by the microscopes. However, there are more severe limitations on the choice of the convergence angle γ and more uncertainty in determining the *focus accommodation*.

With lens stereoscopes the stereoscopic pair was placed in the focal plane so the rays striking the eye from any given point were always parallel. For the stereoscopic microscope, the observer adjusts the focus for his personal preference. Thus, the focusing accommodation is no longer necessarily for infinity and also may differ significantly for different observers.

In figure 10-9 the field of view for the single lens is relatively large. With the stereoscopic pair separated to provide an $I/E_L = 5$ ($\gamma = 11.4$ degrees), large areas of the photographs are visible. However, through a typical stereoscopic microscope with 8X magnification, the field of view may be only 0.25 inch. With a common point on the two photographs centered on the optic axes of the binocular microscope, exactly the same area is viewed in each eyepiece. Unfortunately, many stereoscopic microscopes are mounted with their optic axes parallel, so this configuration requires a convergence accommodation for infinity which has already been shown to be impractical. This objection can be overcome by moving the pair slightly together so the separation of the common point is less than that of the optic axes. The eye then looks along a line at a slight angle to the optic axis to provide a more optimum convergence angle.

A penalty is paid for this adjustment. When the photographs are moved off the optic axes, the areas visible in the two eye pieces are no longer exactly the same. Of course, only the portions of the areas which are common can be viewed stereoscopically. For the optimum convergence angle of 11.4 degrees ($I/E_L = 5$), the typical 8X microscope will reduce the common area coverage by as much as 50 percent. Thus, only approximately 0.125 inch of the photographs can be viewed at one time, requiring continuous shifting of the photographs, if any appreciable area is of interest.

In order to determine the pair separation required for a desired convergence angle, it is necessary to know the details of the microscope design. This information is normally not available to the working interpreter. In addition, the effect of the arbitrary focusing accommodation on model distortions is not known. For these reasons the optimum configuration for estimating relative heights cannot be specified for stereoscopic microscopes.

Sometimes the dual microscopes are mounted with their optic axes converging, with the degree of convergence proportional to the eye spacing of the observer. While this modification permits 100 percent common areas to be viewed with a convergence accommodation other than for infinity, it is even more difficult to determine the convergence accommodation produced by a given pair separation. Also, the unknown effects of the arbitrary focusing adjustment still exist.

If stereoscopic microscopes are to be used for slope and relative height estimation, perhaps the best compromise is to focus the eyepiece at the maximum distance from the fixed stereoscopic pair (closest to the eye) which still provides a distinct image. Next, select the pair separation which is most comfortable. These steps will result in a focusing accommodation for approximately infinity and a value of I/E_L somewhere near 5 so that the model will be relatively distortion free, and equation (10.9) for the vertical exaggeration can be used.

The previous discussion was concerned with methods for obtaining a distortion-free model for use in relative height estimations. Stereoscopic plotting devices require the operator to adjust an artificial reference and a terrain model height so they are equal (*see* section 10.4.1). In this case a configuration is desired which best permits the operator to discriminate small height differences and null them out. It is relatively unimportant if there are distortions in the model, providing they are the same for both the reference height and terrain model.

The effect of different focusing and convergence accommodations on stereoscopic depth difference discrimination has not been determined for a full range or combination of these variables. However, there is evidence the discrimination capability is optimum when the two accommodations are for the same distance (Farrell, Anderson, Kraft, and Boucek, 1975). This indicates it would be advisable to make the focusing accommodation and convergence angle compatible for stereoscopic plotting viewers wherever possible.

10.4 Stereoscopic Mensuration

In making direct measurements from stereoscopic pairs of photographs to determine actual heights, some application of the floating dot principle is normally used whether the device is a simple stereometer or the more sophisticated plotting machines. The basic principle will be described first, and then it will be shown how the *parallax* measurements are converted to actual heights.

10.4.1 FLOATING DOT PRINCIPLE

If two points or dots whose spacing can be varied are introduced into the plane of the

stereoscopic pair of photographs, they can be aligned with any two corresponding points on the pair. In figure 10-15 the parallax p_c produces a disparity angle α which is perceived by the observer as the height h. With the dots aligned with the top of the feature as shown, they provide the same retinal disparity with respect to the base. Thus, they are perceived as a single point at the same height above the base as the top of the feature. Increasing the separation of the dots by an amount equal to the parallax p_c causes them to have a corresponding retinal disparity with respect to the top, and the fused dot will now appear to be at the base level. If the dots were moved closer together than the tops of

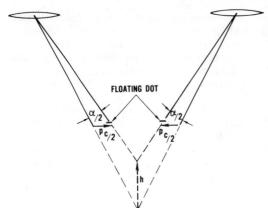

FIGURE 10-15. Floating dot principle.

the feature, there would be an additional disparity and the dot would appear to be completely above the feature or "floating."

The change in separation required to make the dot appear to move from the top to the bottom of a feature is exactly the same as the parallax. This change can be determined using a parallax bar (such as that shown in figure 10-23, section 10.6, being viewed with a mirror stereoscope). The dots are on the bottom of two pieces of glass which rest on the surfaces of the photograph and are connected by a calibrated bar with a micrometer adjustment to vary their spacing. The parallax corresponding to any two different heights is read directly.

An alternative approach used in many plotting devices is to hold the dot separation fixed and vary the photograph separation. A similar relative change in elevation of the dot with respect to the model occurs. In this case the parallax is the change in separation of the photographs required to produce the apparent change in position of the dot from one level of the model to the other.

The floating dot can also be used to obtain elevation contours. The separation of the dots and photographs is held constant but the photographs are free to move together. The operator always maintains the apparent position of the dot on the surface of the model. Because the dot position corresponds to a fixed height, this traces out an equal elevation contour. If the movement of the photographs is linked to a plotter, the contour can be automatically recorded. By changing the relative separation of the dots and photographs a new elevation is established, so any desired number of contour levels can be obtained.

In the sophisticated plotting devices, the dots are not actually in contact with the photographs but are optically made to appear in the same plane. With this configuration it is important to focus the dots or the photographs so the planes really do coincide. This can be accomplished by moving the head laterally to ensure there is no movement parallax between the dot and its background.

10.4.2 MENSURATION RELATIONSHIPS

From equations (10.2) and (10.3)

$$h_T = p_c \, \frac{b_T}{b_c} \, \frac{H}{B} \,, \qquad (10.17)$$

but b_c/b_T is the scale of the photographs S so

$$h_T = \frac{p_c}{S} \, \frac{H}{B} \,.$$

Thus, using the measured value of p_c, and knowing the scale and the ratio H/B, the true height of any feature can be calculated. If H/B is not known, it can be determined from the camera focal length and the stereo photographs as previously discussed in section 10.3.1.2. In all but the simpler plotting devices, this calculation is either made automatically or given in reference tables supplied with the equipment.

10.5 Color Stereoscopy Considerations

There are two visual phenomena which can create problems when making stereoscopic height measurements on color photography using floating dot mensuration equipment. Both of these are related to the facts that the eye has considerable *chromatic aberration* and that the optical geometry of the eye varies from individual to individual.

The first visual phenomenon is simply that for some individuals stereoscopic acuity and consequently the variations and repeatability of an individual's measurements may be different for different colors. In most cases this effect is probably too small to be of practical concern.

The second visual phenomenon can have a sufficiently large effect to be of practical con-

cern. It may also partially explain why studies comparing achromatic and color films are fairly consistent in showing larger vertical height errors with color. This phenomenon is called *chromostereopsis* and it is most simply described as seeing a set of different colored objects at different heights dependent on their color, even though no height difference, or parallax, is present. Some people see colors at the red end of the spectrum as being closer while others see blues as closer. Most people are not aware of this phenomenon, yet its magnitude under viewing conditions like those existing in floating dot mensuration equipment can be substantial. One study (Kraft, Booth, and Boucek, 1972) showed that subjects were much more successful in discriminating height differences in black and white, as opposed to color targets. Further, this study showed that there was only a very low correlation between stereoscopic acuity performance with black and white and that with color.

The magnitude, and even the direction (red near versus blue near) of chromostereopsis varies among individuals and as a function of changes in viewing conditions, such as luminance. In particular, chromostereopsis becomes very large when the light enters at opposite edges of the pupils of the two eyes. This would occur, for example, if the eyepiece separation in a small exit pupil display such as a binocular microscope was not correct for the operator. Separating the eyepieces too far causes red to appear closer, while too small a separation causes blue to appear closer.

In summary, the users of floating dot stereoscopic mensuration equipment with color imagery should be aware that an accurate operator with black and white may not do as well with color, and that operators should take particular care to set the equipment's eyepiece separation, focus, and brightness the same each time.

10.6 Basic Stereoscopic Equipment

10.6.1 GENERAL

The first recorded optical instrument incorporating the principles of stereoscopy was developed by *Robert Wheatstone* in 1838. His instrument consisted of two mirrors which reflected the images from a pair of stereoscopic pictures directly to the eyes. Figure 10-16 is a schematic of the Wheatstone stereoscope. This led, in 1857, to the development of the Helmholz four-mirror stereoscope design similar to that previously diagrammed in figure 10-14 but without the lenses. With minor design and engineering changes, it has endured to this day. A few years after Wheatstone developed the reflecting stereoscope, *Sir David Brewster,* in 1849, designed a lens stereoscope which consisted of two convex lenses whose axes are about 0.375 inch (9.52 mm) farther apart than the observer's interpupillary distance. His design is basically that discussed in respect to figure 10-9. These two basic ideas, both over 100 years old, represent the simplest optical instruments for stereoscopic viewing of two-dimensional images in use today.

Stereoscopic viewing devices described in this section fall into three general categories: lens, mirror, and microscopic.

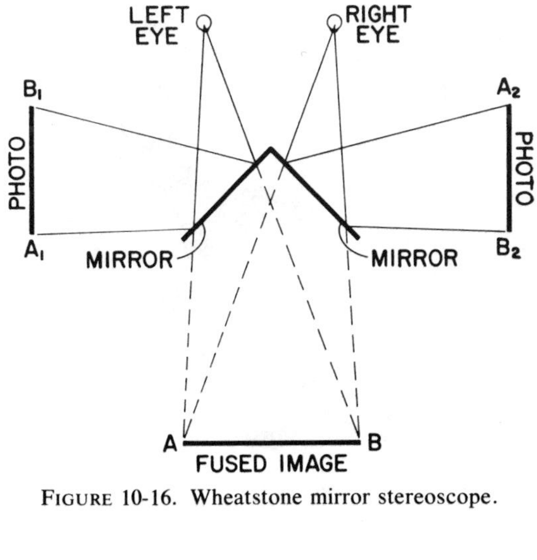

10.6.2 LENS STEREOSCOPES

Of the various kinds of stereoscopes in use, the simple lens stereoscope, basically as designed by Brewster, is still the most widely used because of its low cost, portability and simplicity of operation and maintenance. Its disadvantages are limited range of magnification, difficulty of annotating photographs during observation and limited capability of spreading photographs while observing (*i.e.,* distance of spread is usually less than interpupillary distance).

Most simple lens stereoscopes have essentially the same specifications: a fixed focus lens system with the stereoscopic pair in the focal plane and an adjustable interpupillary distance. Magnifications fall in the 2× to 4× range. Figure 10-17 is a typical 2× magnification folding stereoscope available throughout the industry. Variations of this basic design include a combination 2× to 4× model, figure 10-18. With this and similar styles, the observer snaps auxiliary lenses into place and lowers the legs to accommodate the new focal length. The model in figure 10-19 features individual ocular focusing accomplished by draw tubes which have a 0.5-inch (12.7 mm) range. Other variations include

FIGURE 10-16. Wheatstone mirror stereoscope.

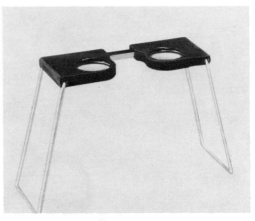

FIGURE 10-17. Abrams folding-type lens stereoscope, model CF-8.

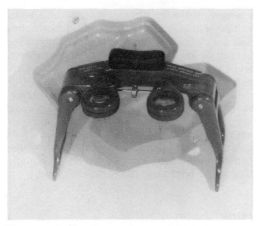

FIGURE 10-18. Abrams 2×-4× folding-type lens stereoscope, model CB-1.

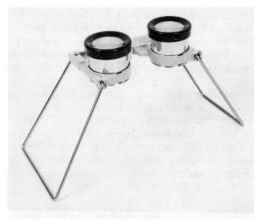

FIGURE 10-19. Individual focus lens stereoscope (Alan Gorden Enterprises "Eye Saver").

three-legged stands and, in figure 10-20, an attachment to fold the photographs to allow the observer to stereoscopically view any part of the overlap area. Still other variations mount the lenses on counterweighted bars to free the work surface of the limitations imposed by the legs.

One somewhat unique design is a combination lens-prism stereoscope, figure 10-21. The photo-graphs are mounted on two surfaces that fold like the pages of a book. The lens system is on an adjustable bar above the photographs. One photograph is viewed directly through the lens the other through a lens and prism. The entire overlap area can be scanned by sliding the lens system along the bar and rotating it from side to side.

FIGURE 10-20. Galileo outdoor stereoscope (model SGF1a).

FIGURE 10-21. Lens/single prism stereoscope (U.S. Geological Survey photograph).

10.6.3 MIRROR STEREOSCOPES

Mirror stereoscopes are designed to view larger areas of the stereoscopic pair, but because of the length of the optical path, at a smaller scale. Binoculars can be used on most mirror stereoscopes if a larger scale is needed. Figure 10-22 is a modern version of the original Helmholz design. Improvements and modifications to the original have been achieved using new materials and designs. The result has been improved operator comfort and utilization, and a clearer image. For example, figure 10-23 is a basic mirror stereoscope, with binoculars and parallax bar for elevation determination. The eyepieces are inclined for greater viewing comfort and the offset support permits free access to the work surface. An additional feature is the associated tracing stereometer for calculating differences in terrain elevations and determining

FIGURE 10-22. Wild mirror stereoscope (model ST4).

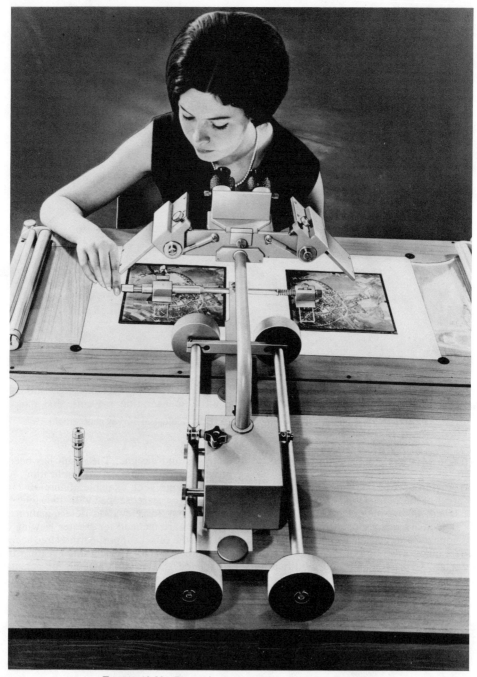

FIGURE 10-23. Zena mirror stereoscope with parallax bar.

heights of objects. Using conventional techniques, both contours and planimetric detail can be plotted.

Figure 10-24 is a small folding mirror stereoscope designed for field work. Its lens system provides a 2.3× magnification of a field of view of about 65 mm, and it folds to fit into a pocket.

Figure 10-25 is a scanning mirror stereoscope with inclined eyepieces. It affords two magnifications and, through a system of mirrors and

prisms, the ability to scan the entire overlap area of the photographs without moving either the stereoscope or the pictures. As shown, its design allows two stereoscopes to simultaneously view the same area for training or conference purposes.

Its disadvantages include a relative loss of illumination because of the number of glass surfaces and more maintenance, because of the number of mirrors and prisms.

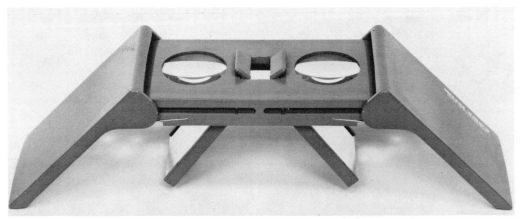

FIGURE 10-24. Wild pocket mirror stereoscope (model TSP-1).

Figure 10-26 is another device for dual viewing. This too has a complex lens-mirror-prism system. It was designed as a training device to enable instructor and student to view the same image simultaneously at 1.5× or 3× magnifications.

10.6.4 STEREOSCOPIC MICROSCOPES

This family of instruments provides many features: a wide range of magnification (up to 100× is commonly available), coupled zoom elements which provide continuously variable magnification, 360-degree image rotation, and a variety of reticles for mensuration and interpretation purposes. Figure 10-27 shows a portable zoom stereoscope for use with either a 9- or 5-inch film format. These stereoscopes work best with transparencies (film or glass; positive or negative), but can be used with opaque prints. When attached to light tables, their versatility increases with the capabilities of the table. For example, the table can incorporate film-flattening, vacuum systems, light adjustment control, mensuration systems, and motorized film drives for roll film.

Figure 10-28 shows a typical microscope-light

FIGURE 10-25. "Old Delft" scanning mirror stereoscope (model ODS5111).

FIGURE 10-26. Alan Gorden Enterprises Condor dual stereoscope (model T-22).

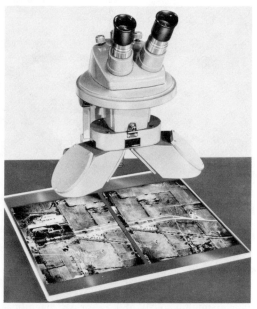

FIGURE 10-27. Zoom 95 stereoscope (Bausch & Lomb).

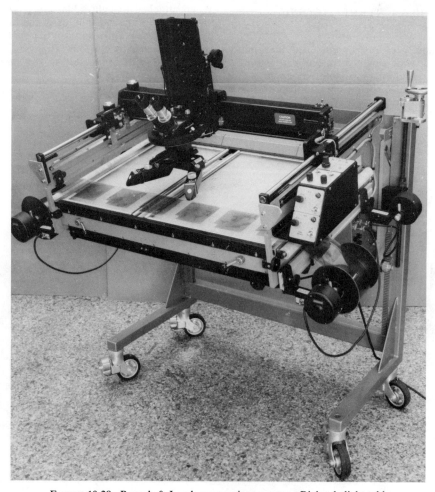

FIGURE 10-28. Bausch & Lomb zoom microscope on Richards light table.

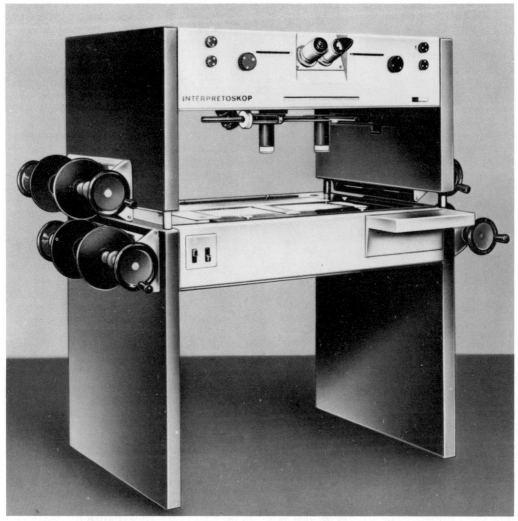

FIGURE 10-29. Zena Interpretoscope C.

table arrangement. With most of these micro- scopes, the rhomboid prisms can be removed to convert the microscope to a binocular instrument where both eyes view one photograph monoscopically.

A second instrument of this type is shown in figure 10-29. It features an enclosed optical system with the ability to compensate for scale variations up to 1:7.5 between photographs, and permits 360-degree image rotations. Thus, roll film from two different runs, with photographs at different scales can be viewed stereoscopically if the individual rolls do not provide complete stereoscopic coverage. It also has a built-in height measurement system and dual optics to allow simultaneous study by two observers. The distance between the optics and the table top is large enough to allow comfortable delineation or annotation on the photographs.

The various stereoscopes discussed above are typical of products available from many manufacturers. They were chosen as representative of a much wider selection of instruments. A complete listing of designs and options is beyond the scope of this MANUAL.

10.7 Stereoscopic Orientation Procedures

The various families of stereoscopes show evidence of good design principles and manufacturing practices. However, many of the benefits of these efforts are not utilized because too few people know how to use them to best advantage. As a result, one major problem is eye fatigue, usually caused by improperly oriented pairs of photographs. This section will attempt

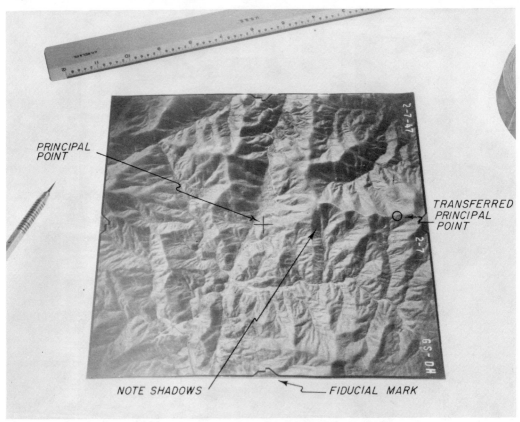

FIGURE 10-30. Locating principal point and transferred principal point.

to overcome these orientation deficiencies. It will tell how to align photographs so the center lines (principal point and transferred center) are in line and parallel, both to each other and to centerline of the stereoscope's lenses. Thus, the direction of parallax corresponds to that produced by the human eye and is the closest analogue to true binocular vision. The procedures discussed below will also avoid the pseudoscopic effect discussed in section 10.2.4.

The first step in proper stereoscopic observation is to locate the principal point of each photograph, then transfer it to the overlap area of the other photograph. The principal point can usually be located by intersection of two lines drawn from side or corner fiducial marks, as shown in figure 10-30. It can be transferred to the second photograph by observation of detail. Use a very fine pencil or sharp needle to locate the exact points. If desired, a fine line from principal to transferred point can be drawn on each photograph.

To avoid the pseudoscopic effect, position the photographs so the common overlap area is toward the center and the areas not common fall toward the outside. Then check to see where the shadows fall. If necessary, rotate the photographs 180 degrees, as a unit, so the shadows fall generally toward the observer.

The next step is to position the photograph on the drawing board or other viewing surface for comfortable viewing. Fasten photograph number 1 to the table so the line connecting the principal point and transferred center is parallel to the edge of the table. Fasten the photograph to the table. Place photograph number 2 approximately parallel to number 1. Lay a straightedge along the center line of photograph number 1. Adjust the position of alignment of number 2 until its center line also falls along the straightedge as shown in figure 10-31.

Holding all four points along the straightedge, adjust the distance between common points (figure 10-32) according to the value determined from the method described in section 10.3.3. The separation distance will vary from about two inches for lens stereoscopes to about 10 inches for mirror types.

Check alignment and distance, then fasten photograph number 2 to the table. Thus, the center lines of the stereoscopic pair will be in line and parallel with the proper separation.

Now place the stereoscope over the properly oriented and separated photographs. The axes of the eyepieces should be parallel to the center lines of the photographs. If the center lines have not been drawn on the photographs, lay a straightedge along the four principal points.

FIGURE 10-31. Alignment of principal points and transferred principal points.

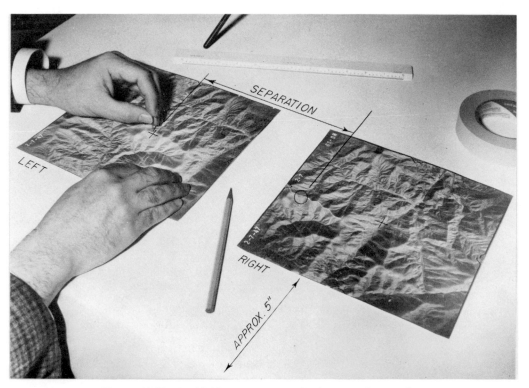

FIGURE 10-32. Establishing proper separation between photographs.

Then rotate the stereoscope, if necessary, so the line or straightedge viewed by one eye is superimposed on the other. When one continuous line is observed, proper orientation of all elements has been accomplished for viewing the three-dimensional model.

With experience, proper orientation of a stereoscopic model can usually be accomplished without performing all these steps. However, when high accuracy is required (as when using a parallax bar), the full procedure should be followed for better results and more comfortable viewing.

REFERENCES

Blakemore, G., 1970, The range and scope of binocular depth discrimination in man: *Journal of Physiology,* v. 211, p. 599–622.

Farrell, R. J., Anderson, C. D., Kraft, C. L., and Boucek, G. P. Jr., 1975, Effects of convergence and accommodation on stereopsis: *Document D180-19501-1,* The Boeing Co., Seattle, Wash.

Fry, G. A., 1942, Measurement of the threshold of stereopsis: *Optometric Weekly,* October 22.

Graham, C. H. (ed), 1965, *Vision and Visual Perception.* Wiley; New York.

Graham, C. H., 1951, Visual perception, chapter 23, in Stevens, S. S. (ed): *Handbook of Experimental Psychology;* Wiley; New York.

Julesz, B., 1971, *Foundations of Cyclopean Perception;* University of Chicago Press, Chicago, Ill.

Julesz, B., 1965, Texture and visual perception: *Scientific American,* v. 212, February 1965, p. 38–48.

Kraft, C. L., Booth, J. M., and Boucek, G. P., 1972, Achromatic and chromatic stereoscopic performance: a paper presented at 12th Congress of International Society of Photogrammetry, Montreal, Can., July 27, 1972, Aerospace Division, The Boeing Co., Seattle, Wash.

LaPrade, G. L., 1972, Stereoscopy—a more general theory: *Photogrammetric Engineering,* 38: 12, p. 1177–1187, Dec.

LaPrade, G. L., 1973, Stereoscopy—will data or dogma prevail?: *Photogrammetric Engineering,* p. 1271–1275.

Miller, Charles I., 1958, The stereoscopic space-image: *Photogrammetric Engineering,* 24: 5, p. 810–815.

Miller, Charles I., 1960, Vertical exaggeration in the stereo space-image and its use: *Photogrammetric Engineering,* 26: 5, p. 815–818.

Ogle, K. N., 1967, Some aspects of stereoscopic depth perception: *Journal of Optics Society of America,* v. 57, p. 1073–1081.

Ogle, K. N., 1964, *Researches in Binocular Vision.* Hafner; New York.

Thurrell, R. F. Jr., 1953, Vertical exaggeration in stereoscopic models: *Photogrammetric Engineering,* Sept.

Wright, W. D., 1954, Stereoscopic vision applied to photogrammetry: *Photogrammetric Record,* v. 1, p. 29–45.

CHAPTER XI

Double-Projection Direct-Viewing and Paper Print Instruments

Author-Editor: H. B. LOVING

Contributing Authors: S. J. FRIEDMAN, JOSEPH DANKO, JOSEPH THEIS

11.1 Introduction

THIS CHAPTER on double-projection direct-viewing plotting instruments and paper print type plotters is basically a condensation and up-dating of chapters twelve and thirteen of the Third Edition of the MANUAL OF PHOTOGRAMMETRY authored by B. Thomas Hopkins and John T. Pennington respectively. Further details on these plotters can be obtained by referring to the Third Edition of the MANUAL OF PHOTOGRAMMETRY.

The double-projection direct-viewing plotting instruments and the paper print type instruments are of interest to the modern photogrammetrists for a technical as well as an economical aspect. Technically, they represent a simple and direct solution to the problem of forming an analogic, three dimensional accurate stereomodel from two dimensional photographs exposed as stereo-pairs. In spite of the fact that accuracy attainable with these plotters is less than that from optical train type plotters, the advantages of simplicity and low cost have encouraged the improvement of this type of equipment to their present state of development.

11.2 Double-projection Direct-Viewing Instruments

11.2.1 HISTORY

The principle of double-projection direct-viewing stereoplotters was effectively born in 1898 when photographic slides of a three-dimensional model of a house were projected in stereopairs by *Theodore Scheimpflug* and the plans of the model house plotted from the projected images.

Although it is accepted that Scheimpflug and *Doležal* (1899) recognized that the principle of double projection was valid for aerial (balloon) photography, it was not until 1915 that *M. Gasser* applied Scheimpflug's ideas to aerial photographs. In this application, Gasser introduced the use of dichromatic anaglyphic (red-green) projection as well as the alternate blinking (flicker) system to determine the intersection of corresponding rays. The principle of dichromatic anaglyphic projection had already been demonstrated by *I. Ch. d'Almeida* in Paris as early as 1858.

Following Gasser's developments, *U. Nistri* in 1919 designed and constructed a double-projection direct-viewing stereoplotter called the "Photocartograph." It is of particular interest that in 1923 Nistri suggested that this type of equipment, by alternate orientation of successive projectors, could reduce the need for ground control. Here we have an indication of the advent of multiplex and its capability for aerotriangulation.

Another worker in this field, contemporary with Nistri, was *J. Predhumeau*, of Paris. Essentially, his work was similar to Gasser's and consisted of dichromatic anaglyphic projection of aerial photographs. Predhumeau is credited with engineering improvements on Gasser's basic ideas; however, there is no evidence that he developed any new concepts.

There is no evidence of any further significant progress in double-projection direct-viewing plotting equipment until the 1930's when the class of equipment generally designated as "multiplex" was introduced. The first multiplex is generally attributed to Zeiss, of Jena, in 1934. The Italian authorities claim, however, that U. Nistri, of Rome, conceived this same equipment simultaneously and there is valid documentary evidence to support that claim. In any event, the multiplex plotter was introduced to the photogrammetric profession about 1934.

Shortly thereafter, the multiplex of Zeiss was introduced in the United States. Its further refinement and development was encouraged by the U.S. Geological Survey and the U.S. Corps of Engineers. The Bausch and Lomb Optical Company, of Rochester, New York, manufactured it in the U.S.A., beginning about 1940, and produced an improved version based on

Geological Survey recommendations which became extremely popular with American photogrammetrists. A similar version was manufactured after World War II in England by the Williamson Company. Meanwhile continuous equivalent improvement and production was maintained in Italy by Ottico Meccanica Italiana, a Nistri-owned corporation. Zeiss, of West Germany, did not resume manufacture of this instrument after 1945, although Zeiss of East Germany, now a separate company, did continue to produce the prewar model.

A limitation of the multiplex, which was not acceptable to some photogrammetrists, was the necessity of using diapositives which were reductions (in the ratio of 4.5:1) of the original photographs. A solution to this problem was offered in the Photorestituteur designed by *M. R. Ferber*, of Paris. Ferber's development was a double-projection anaglyphic plotter which used full-sized diapositives but with only a small portion of the plate illuminated at any one time. This was accomplished in his design by a small but powerful projection lamp system swinging about the projection lens. A variable telecentric lens system was required in Photorestituteur to focus the projected image on the viewing platen. This was necessary because the focal length of the objective lens matched that of the aerial camera and therefore projected rays which focused only at infinity unless an auxiliary lens was used. *H. Kelsh* in the United States utilized the principle of the swinging light in the design of his plotter; however, the projection lens system is simplified by use of an objective lens of shorter focal length than that of the aerial camera, thereby permitting projection at a finite distance (30 inches). This simple solution of Kelsh's is a vital factor in the great popularity which his instrument has attained.

The great popularity of the Kelsh plotter in the United States after World War II stimulated the entire field of double-projection direct-viewing equipment. Under the leadership of *Russell K. Bean*, the United States Geological Survey developed the ER-55 projector utilizing an ellipsoidal reflector in its illumination system and diapositives approximately double the size of those for the multiplex, but about one-half the size of the aerial negative. At a nominal projection distance of 525 millimetres, the scale of the projected model is approximately 3.4 times the scale of the aerial negative. The commercial version of the ER-55 plotter is manufactured by Bausch and Lomb as the "Balplex."

Nistri, Williamson, and Kern all introduced stereoplotters of the double-projection direct viewing type utilizing full-sized diapositives prepared from the aerial negatives at approximately 1:1 scale. Each of these instruments had unique features advancing some improvement over earlier plotters.

11.2.2 PROJECTION SYSTEMS

A projection system is a basic requirement of the double-projection direct-viewing plotting instruments. In this projection system the images on the diapositives derived from the original aerial negatives are illuminated and projected through an objective lens system onto a viewing screen or platen, where the projected image is viewed by reflected light. This double-projection direct-viewing system permits the observer to see the reconstructed features of the terrain more nearly in their true relationship to the projectors than is the case with more complex stereoplotting instruments.

Projectors may be designed to project images from diapositives printed in the same size and format as the original aerial negative, or from diapositives that are printed as reductions of the aerial negative. Whatever the design, it is an essential requirement that the projected cone of rays conform exactly in angular relationship to the cone of rays that entered the aerial camera lens to form the original negative. It is a further requirement that the images formed by these rays be brought to a focus at a finite distance (the projection distance of the plotter). These geometric conditions are attained by precise calibration of the projector's principal distance and principal point and by careful attention to the requirements of interior orientation. The projector mounts must therefore be constructed so that each projector may be rotated about each of the three mutually perpendicular axes (x, y, and z), referred to internationally as *omega, phi*, and *kappa*, respectively. A translation motion along the x axis is also necessary as a means of achieving a desired scale in the stereoscopic model.

Once the stereoscopic model is formed by the process of relative orientation, the model as a unit must be properly related to the datum of the plotter, and the air base of the model (separation of the projectors' perspective centers along the x axis) must be adjusted to bring the model to a desired scale. Mutual rotational adjustments of the projectors may be used to tip and tilt the model into proper relationship with the datum plane. However, a preferred method is to utilize the tilting adjustments incorporated in the projector supporting frame to achieve this relationship. These adjustments constitute the procedures of absolute orientation.

11.2.3 ILLUMINATION SYSTEMS

The necessity of projecting a photographic image through a small aperture with light passing through a filter (dichromatic or polaroid) imposes severe demands upon the illumination system of double-projection direct-viewing plotters.

Two basic illumination systems are in common use today. One system illuminates and projects the entire photographic plate simultaneously. All instruments relying on this principle utilize reduced-size diapositives. This is necessary because no practical system is available for illuminating the entire surface of a 9- × 9-inch full-size diapositive without incurring problems of excessive heat and loss of light.

An alternative illumination system, which has proven very successful with plotters utilizing full-size diapositives, is the swinging light. In this arrangement a small narrow-angle projection-lamp system is rotated on a swinging arm about the emergent node of the lens, illuminating and projecting a relatively small area of the transparency. Thus the operator can view only a small area at any one time.

11.2.4 VIEWING SYSTEMS

In theory, the viewing system of double-projection direct-viewing plotters can consist of any device which will mutually exclude from each of the operator's eyes the view seen by the other eye. The most economical way of accomplishing this effect is by the anaglyphic system of projecting the two images of a stereoscopic pair in complementary colors (usually red and blue-green) and nearly in superposition, so that a stereoscopic image is obtained by viewing the projected images through spectacles having filters of the corresponding complementary colors. The system has two principal defects. First, it greatly attenuates or absorbs the available light. The efficiency of light transmission through the projector filters and then through the operator's spectacle filters is very low. A second defect is that it precludes the possibility of using color photography.

The Kelsh Instrument Division, Danko Arlington, Inc., Baltimore, Md. has designed a polarized light source for stereoscopic perception. It is called the Kelsh PPV Stereo Viewing System and was designed to overcome the early objections to the use of polarized light for this purpose.

As stated in the MANUAL OF PHOTOGRAMMETRY, Third Edition, a special surface was required for the tracing table platen to provide a polarized reflection. This surface by its nature, had to be grainy, and therefore the image was a diffuse one, giving poor resolution. The second objection was that the operator had to maintain a close degree of parallelism between his eye base and the base for the projector lenses. If the operator cocked his head just a few degrees, the linear polarization would be lost. Such a limitation on the operator's physical movements could cause him to be quite uncomfortable.

Both of the above objections have been overcome with the Kelsh PPV system. The aluminum viewing platen of the PPV is treated chemically to utilize the granular structure of the aluminum itself. The chemical treatment converts the platen surface to thousands of tiny, mirror-smooth aluminum mountains, in the order of 10 to 20 micrometres high. The mountain tops are rounded, and will reflect a considerable amount of light. A tiny DC motor, mounted under the platen, causes the platen to spin at about 900 RPM. The spinning action converts the grainy reflected image into a smooth one, because the reflections are being flashed to the operator's eye at thousands of times per second. Although a rate of only 15 mountains per second would suffice for the improvement in resolution, a speed of 900 RPM is the minimum speed at which any soil on the surface of the platen will blend into the reflected image. The operator does not then detect any perceptible rotation.

The second improvement was the use of circularly polarizing filters for the projection system instead of the linearly polarizing filters. The circular polarizer is a sandwich consisting of a piece of linear polarizer, bonded to a quarter-wave retardation sheet, which has been oriented to an angle of 45 degrees to the transmission direction of the polarizer. The retardation plate itself has two axes, slow and fast. Its S (slow) axis is placed to the right or left of the polarizer transmission axis. If to the right, the circularly polarized beam is a right-handed helix; to the left it is a left-handed helix. With a circular polarizer, a right handed beam cannot pass through a left-handed polarizer, no matter how the polarizer may be rotated.

When a circularly polarized beam reflects off the spinning platen, it reverses its direction of rotation. Therefore, with the use of spectacles properly equipped with left- and right-hand circularly polarizing filters, the operator may see a stereoscopic image. There is an effective extinction of the opposite image throughout an arc of plus or minus 25° of head movement by the operator. Beyond that, about 2% of the light is allowed to pass through. Therefore, the operator may comfortably view a stereomodel with improved resolution on the spinning platen, making much better use of the available light.

A third method which has been used with success is a mechanical arrangement of rotating or oscillating shutters. In this method, one eye and its associated projected image are simultaneously blocked while the other eye views its image without obstruction. Because of the ability of the retina to retain an image for a short interval, an alternate opening and closing of the shutter in approximately 1/15 of a second or faster permits the operator to view the stereoimages without being uncomfortably aware of the discontinuous character of the superposed images. This is analogous to the principle which permits motion pictures to be viewed without apparent flicker.

A fourth method, which actually is a simplification of and antedates the mechanical system described above, is the so called flicker system. This method does not provide true stereo perception, but permits three-dimensional mapping. The operator does not wear any glasses or devices in front of his eyes. Rather, the projectors are synchronized to alternately project their images at a very slow cycle of about 1/5 of a second. This permits the operator to be aware of any separation of corresponding images, because the images will appear to flicker or vibrate to a degree depending upon the separation due to parallax. Where two corresponding images are in coincidence, they will appear not to move, thus effectively marking the intersection point.

The mechanical shutter system has the advantages of very high efficiency and of using color photography. Earlier versions of this equipment have been expensive and quite cumbersome to use. However, a recently developed mechanical shutter system using stepping motors can be wholly mounted on the plotter.

11.2.5 MEASURING SYSTEMS

The projection systems and viewing systems of the double-projection direct-viewing plotting instruments permit the formation and observation of the stereo-model; however, observation of the stereo-model is relatively meaningless unless it is possible to make precise measurements within the model and to delineate mapworthy detail from the model onto the map manuscript. The measuring system of these plotters provides this important link.

Essentially, this system consists of a viewing screen or platen upon which the stereoimages are projected. A reference mark (usually an illuminated point of light, but sometimes a black circle, a cross, or an arrow) located in the center of the platen appears to float in the stereoscopic model and, therefore, is referred to as the floating mark. The viewing screen and associated floating mark are the means by which the coincidence of corresponding images is observed in the stereoscopic model. When the floating mark is brought into contact with any point of the apparent surface of the model as viewed on the platen, the horizontal and vertical position of such a point can be plotted and/or measured at the scale of the projected model. The viewing screen can be guided freely throughout the stereoscopic model by the compiler while maintaining the floating mark in contact with the surface of the model at the point of observation. A pencil lead or other drawing device located orthographically beneath the floating mark provides a means for plotting detail in a horizontal plane at the scale of the stereoscopic model; a precise lead screw, to which the platen is attached and which, through a mechanical linkage, actuates a digital counter, permits vertical displacements of

the platen to be read directly in feet or metres at the model's scale.

A common form of measuring system is the tracing table, which contains the viewing screen, lead screw, vertical displacement counter, and pencil carriage, and which slides directly on the reference plane of the plotting instrument while the stereoscopic model is observed on the platen. Direct delineation of detail from the stereomodel at model scale is frequently accomplished through the pencil lead attached to the tracing table. However, many double-projection direct-viewing plotters use a pantograph which is connected to the tracing table by means of a stud in the place of the lead in the pencil carriage. The usual practice is to reduce the model to some smaller scale as a matter of convenience; however, enlargement of the model's scale can be accomplished by some of the pantographs attached to these plotters.

Instead of a tracing table resting directly on the reference plane of the plotter, some of the double-projection direct-viewing plotters incorporate a platen mounted on cross-slides so that the ways of the cross slides form the reference plane. Details of the stereoscopic model are drawn on a table mounted on one side of these plotters through a mechanical linkage connected to the floating mark of the platen.

11.2.6 DISTORTION COMPENSATION

The effect of radial distortion introduced into the original aerial negative by the aerial camera objective lens can be corrected in one of three ways in the double-projection direct-viewing plotters.

One way is to eliminate the distortion in the preparation of the diapositive. When printing the glass diapositive from the original negative, it is feasible to incorporate distortion in the optics of the diapositive printing system equal, but opposite, to that caused by the aerial camera lens. This results in a diapositive free of distortion, and the corrected plate is then projected through a projection lens that is also distortion-free. In some cases the distortion correction is incorporated in the lens of the diapositive printer.

A second method is to design the projection lens with a radial distortion pattern which will be equal, but opposite, to the distortion in the uncorrected diapositive plate. If this distortion is properly computed, the projected image in its optimum plane of focus will be distortion-free.

The third method of distortion correction is accomplished by continuously varying the principal distance of the projection lens according to a definite pattern established by the radial distortion which is to be corrected. In this manner the bundle of rays emanating from each image on the diapositive is projected at an angle equivalent to the angle of the rays entering the camera lens when the image was formed on the aerial negative. The variation in principal distance is accomplished by a follower which traverses an aspheric cam surface as the model is scanned, and which, moves the objective lens along its optical axis or displaces the diapositive plate with respect to the lens.

11.2.7 DIAPOSITIVES

Double-projection, direct-viewing plotters, like other precise stereo-plotters, utilize positive transparencies printed on glass plates or film. These glass plates or film are customarily referred to as diapositives. Originally, the words "diapositive" and "dianegative" were used to describe a photographic image according to its position and format with respect to the aerial camera lens, that is, whether exterior or interior to the aerial camera. However, in common photogrammetric practice today the term "diapositive" is applied almost exclusively to mean a glass plate or film transparency with a positive image.

Diapositives are customarily printed by either of two distinct procedures. In a projection-printing system, light rays from a source which may be controlled in duration and intensity pass through the aerial negative, through a correction plate (if required), and through a high-resolution lens, and expose the negative images on the emulsion of the diapositive. The images from the negative are reversed in position when printed on the diapositive. Such diapositives are placed in the stereo-projectors with the emulsion down.

In the contact-printing system, the emulsion of the aerial negative is placed in contact with the emulsion of the diapositive during exposure, and the negative images are printed directly on the diapositive without reversal of position. Such diapositives are placed in the stereoprojectors with the emulsion up.

11.2.8 SUPPORTING FRAMES

The double-projection, direct-viewing plotter, regardless of type or manufacture, consists essentially of two or more optical projection devices which must be suspended in space; therefore, a supporting frame is essential. The design of the frame varies with the manufacturer; however, aside from supporting the projectors, the frame must be sufficiently rigid and stable to maintain the projectors in their orientation for extended periods of time without the slightest displacement. The frame may also, by virture of adjustable supporting screws, aid in the absolute orientation of the projection units. These are very severe requirements and considerable mechanical engineering ingenuity has been devoted to the design of the supporting frames.

11.2.9 SUPPORTING TABLE

The supporting table serves as a base for the supporting frame and also provides a support for the reference table (or cross-slide system) which is the reference plane of the plotter. In some plotters the supporting frame rests directly on the base frame, independent of the reference table. In other plotters the reference table rests on the base frame and, in turn, carries the supporting frame. In either case the supporting table must have sufficient strength and stability to carry the supporting frame and reference table and to suppress vibrations which might otherwise cause undesirable disorientation of the stereomodel. Adjustable feet are necessary so that the supporting table can be leveled.

The reference table serves as the datum to which the stereomodel is referred and upon which the map manuscript may be compiled; it is also the base on which the tracing table rests, and for these reasons the quality of the surface and its flatness are important. Reference tables have been manufactured of glass, aluminum, magnesium, granite, slate, and marble.

11.2.10 DESCRIPTION OF DOUBLE-PROJECTION, DIRECT-VIEWING PLOTTING INSTRUMENTS

11.2.10.1 MULTIPLEX

The name "multiplex" is applied to anaglyphic, double-projection stereo-plotters having the following general characteristics: (1) the stereomodel is projected from diapositives that are reduced from the aerial negative according to the ratio:

$$\frac{\text{focal length of camera}}{\text{principal distance of projectors}} ;$$

(2) the projection system illuminates the entire area of the diapositive; and (3) the stereomodel is measured and drawn by observation of a "floating mark" in a viewing screen of a tracing table on which the images are projected. It originated in Europe and was manufactured by Zeiss Aerotopograph prior to World War II. Early multiplex was designed for 8¼-inch normal-angle photographs. Although some of the normal-angle equipment may be still in use, equipment at this time is almost entirely wide-angle (90°).

11.2.10.1.1 BAUSCH AND LOMB MULTIPLEX

The only multiplex equipment manufactured in the United States is made by Bausch and Lomb, Inc., Rochester, New York. It is designed for 6-inch-focal-length wide-angle photographs. Two types of equipment are manufactured: one for the use of the military, and one for civil and commercial use. The main difference is in the principal distance of the projectors— 28.182 mm for military models and 30.00 mm for commercial models.

11.2.10.1.1.1 Projectors. The projector consists of two major parts: one, an illuminating unit, generally called the lamp house, containing the light bulb, a monochromatic filter, and the condensing lens elements; the other, the camera-body-supporting bracket assembly. The condensing lenses in the lamp house provide an even dispersion of light over the format area of the diapositive, and condense this light at the aperture of the projection lens in the camera body (figure 11-1). The supporting bracket assembly permits the entire projector to be moved linearly along the three major axes and rotated about any of these axes to recover the relative orientation of the aerial camera at the instant of exposure. The projector lens is nominally distortion-free. The principal distance of all the projectors of one type is the same, with variations in camera focal lengths reduced to the same value by ratio settings in the diapositive printer. The scale of the projected stereomodel is usually about 2 1/3 times the scale of the photographs, and is limited to near this magnification by the depth of focus of the projectors.

11.2.10.1.1.2 Tracing table. The tracing table is the instrument utilized for measuring differences of elevations and for drawing various map details from the stereomodel. It is also used during relative orientation to observe the displacement of corresponding images so that corrective action may be taken to remove *y* parallax. The images from the projectors are observed on the matte surface of a viewing platen with an illuminated reference mark in its center.

The platen can be raised or lowered for making vertical measurements, with the amount of movement indicated on a counter (or vernier). The platen travels vertically on parallel standards, and a pencil, located orthographically below the reference mark, can be raised or dropped for tracing map detail or contours.

11.2.10.1.1.3 Frames. Multiplex frames, including the bars, are available in several sizes: the single frame, which is about 6 feet long; the double frame, which consists of two single frames joined together; and the short frame (figure 11-2) which can accommodate only three projectors. The single frame can accommodate about nine projectors at nominal spacing, and the double frame twice as many. The standards at the ends of the bar have footscrews that allow the entire frame to be tilted from front to rear, or end to end. The bar has a track with a rack attached, which is engaged by a pinion gear on the projectors for movement in the *x* direction. The short frames do not have standards that rest on the plotting table, but the bar can be tilted for absolute orientation (on some models the table top can be tilted).

FIGURE 11-1. Diagrammatic sketch of multiplex model.

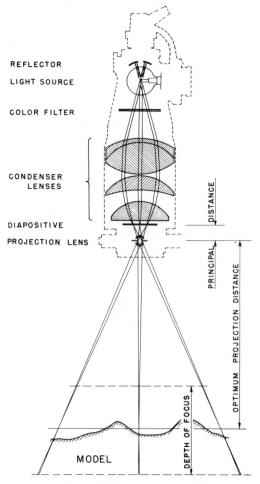

FIGURE 11-2. Multiplex projector optical system.

FIGURE 11-3. Bausch and Lomb Multiplex, short frame.

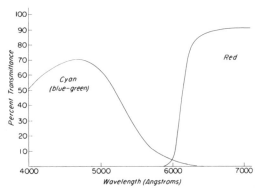

FIGURE 11-4. Transmission curves for red and cyan (blue-green) filters.

11.2.10.1.1.4 Plotting table. The "regular" plotting table for a single frame is about 7 feet long, and for the double frame two tables are placed end to end. Great care is necessary when this is done since the table surface serves as a reference plane for vertical measurements. It is therefore necessary to adjust the tables until both table tops lie in the same plane. Inasmuch as measurements are made to less than 0.1 millimetre, each table surface must be flat to a tolerance much less than this.

11.2.10.1.1.5 Filters and spectacles. For anaglyphic viewing, it is necessary that the colors be chosen to be as nearly complementary as possible. In Bausch and Lomb equipment, the colors are red and cyan (blue-green) (figure 11-4). A filter of one color fits in the condenser housing of one projector, and the other color in the other projector. The spectacles are made to match the filters, and therefore transmit light from only one projector to each eye.

11.2.10.1.1.6 Electrical system. The projector lamps are designed to operate at 20 volts; consequently, a transformer is furnished to allow the system to be connected to standard 115-volt outlets. Connected to the transformer is a control box with selector switches for any particular pair of projectors along the bar. It also has rheostats to permit variation of the light intensities of the projectors, to allow for disparities in density of the diapositives and lack of uniformity of light transmission by the projector.

11.2.10.1.1.7 Cooling system. Since the lamps in the projectors provide 100 watts at maximum output, the projectors become quite hot. Forced-air cooling is optional for the multiplex projectors of newer design (lamp bulb horizontal), but the projectors of older design (lamp bulb vertical with base up) should be cooled to avoid shorter bulb life and the risk of damaging the condenser lens. A blower system for cooling is preferred since this permits filtering the air before it enters the projector, thereby reducing dust.

11.2.10.1.2 NISTRI PHOTOMULTIPLEX

The Nistri Photomultiplex was manufactured by Ottico Meccanica Italiana, Rome, Italy. Although it is designed to accommodate 6-inch wide-angle photographs and is basically similar to other multiplex equipment, it is different in several mechanical aspects. The Model D III instrument is shown in figure 11-5. Unique features are discussed below.

11.2.10.1.2.1 Projectors. The Nistri projectors have seven, rather than the more common six, motions that may be used for relative and absolute orientation. The extra motion is rotation about the z axis, which is coincident with swing only when the projectors are vertical. The focal lengths of the lenses are such that the optimum projector magnification provides a stereomodel at a scale slightly less than 3 times the scale of the original photographs. The prin-

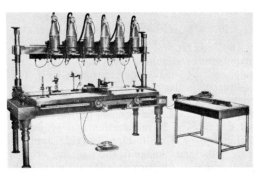

FIGURE 11-5. Nistri Photomultiplex Model-D III.

cipal distance is nominally 35.0 millimetres; the projectors are therefore a little larger than those manufactured in America.

11.2.10.1.2.2 Frame. The frame is approximately the same length as in American-built multiplex but, since the stereomodels are larger, the frame will accomodate only six or seven projectors. For the same reason, the frame must be high enough to hold the projectors a nominal 17 inches above the viewing platen. The frame may be tilted to level the model for absolute orientation.

11.2.10.1.2.3 Electro-coordinatometer. A unique feature of the Nistri Photomultiplex is the electrocoordinatometer and electrocoordinatograph. The viewing platen is carried on a cross-slide system, driven in x and y by handwheels and in z by a small electric motor which is controlled by the operator's foot. The x and y motions are transmitted through servo motors to the electrocoordinatograph, where the scale of the plot may be varied from that of the stereomodel. A flat plotting table and hand-guided tracing stand may be obtained in lieu of the electrocoordinatometer.

11.2.10.1.3 WILLIAMSON MULTIPLEX

In England, multiplex equipment is manufactured by the Williamson Manufacturing Company, Ltd. (figure 11-6). It is similar in principle and design to other multiplex, but has some unique and distinguishing features. It is designed for use with 6-inch wide-angle photography.

11.2.10.1.3.1 Projectors. The use of aspheric lenses in the condenser system reduces the overall height of the projectors. The light is located at the top of the projector and, as with other varieties, heat is dissipated by use of a blower system. A rotating shutter system is available for viewing in lieu of anaglyphic filters.

11.2.10.1.3.2 Tracing table. A distinctive feature of the Williamson multiplex equipment is the tracing table. Four "feet," or bosses, are in contact with the plotting surface rather than the usual three. A single column guides the platen as it is raised and lowered in the stereomodel, and elevation measurements are made by observation of the projected and enlarged image of a glass scale. Interchangeable scales are available to match various model scales.

11.2.10.1.4 SUPER-WIDE-ANGLE MULTIPLEX

The Zeiss (Jena) Company, of East Germany, manufactures a multiplex having a nominal angular coverage of 120 degrees (figure 11-7). This equipment is designed for use of super-wide-angle mapping photography, which has the same angular coverage. The nominal projector principal distance of 21 millimetres and projection distance of 270 millimetres provide approximately the same magnification of the original photographs as American-made multiplex. Light distribution is sufficiently uniform to provide a

FIGURE 11-6. Williamson multiplex.

usable stereomodel throughout. The projectors are marked to show whether they should be used with red or blue filters. Otherwise, the equipment is essentially equivalent to its narrower-angle counterparts.

11.2.10.2 ER-55 (BALPLEX) PLOTTER

The ER-55 plotter (figure 11-8) is manufactured by Bausch and Lomb, Inc., under the trade name "Balplex," and is based on a projector designed by Russell K. Bean, U.S. Geological Survey. The designation "ER-55" is derived from the use of ellipsoidal-reflector (ER) projectors with a principal distance of 55 millimetres. It is an anaglyphic projection instrument that illuminates the entire format area of the photograph, and is designed for use with 6-inch-focal-length wide-angle photographs. Although many parts are similar to, or even identical with, the multiplex manufactured by the same company, the ER-55 (Balplex) has some features that are quite unique.

A distinguishing feature of the ER-55 (Balplex) projector is its reflecting illumination system. The reflector is a concave front-surface mirror conforming to a prolate ellipsoid. Figure 11-9 is a schematic diagram of the arrangement of the light source (1); the reflecting mirror (2); the diapositive (3); and the projector lens (4). Light rays from a lamp filament positioned at one focus (1) of the ellipsoid pass through a

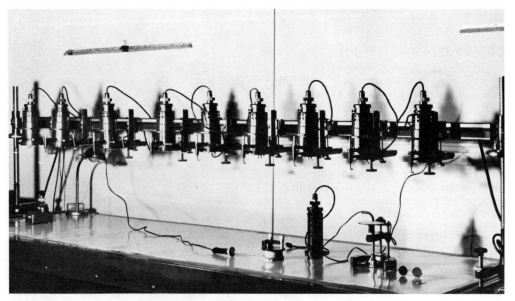

FIGURE 11-7. Zeiss (Jena) super-wide-angle multiplex.

curved red or blue filter and strike the reflecting surface (2). The reflected rays pass through the diapositive (3), converge at the aperture of the projector lens (4), and form an image on the platen of the tracing table. The surface extending from (5) to (6) represents that portion of the ellipsoid required for the reflection of the bundle of rays encompassing the entire diapositive area, with the center of the lens aperture as a perspective center for the reflected rays. The ellipsoidal-reflector illuminating system is highly efficient and free from the usual chromatic aberrations present in a condensing-lens system. All colors are reflected, including red, so that an efficient forced-air cooling system is mandatory to prevent damage to the projection lens, the reflector, the diapositive holder, or the diapositive.

The projection lens is of the hypergon type and is practically distortion-free. It has a nominal focal length of 49.8 mm and an effective lens aperture of $f/16$. At optimum projection distance, images are magnified approximately 3.4 times their size on the original photographic negative. In order to assure sharp imagery in the optimum projection plane when plotting oblique photography, the lens is mounted so that it can be canted to satisfy the Scheimpflug condition.

The projector has a variety of step-tilt settings to accommodate low-oblique or high-oblique photographs. The diapositives, which are prepared in a reduction printer that also provides correction for radial lens distortion, are 110 millimetres square, with the format approximately 82 by 82 millimetres, and are held in position by a plateholder. The diapositive is centered in the

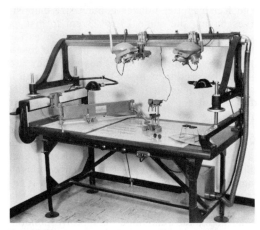

FIGURE 11-8. ER-55 (Balplex) plotter, convergent orientation.

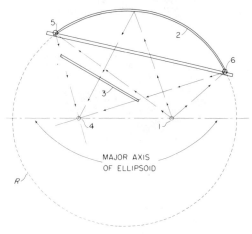

FIGURE 11-9. Schematic diagram of ER-55 (Balplex) projector.

plateholder by a special centering device (figure 11-10) that is a necessary auxiliary instrument.

11.2.10.2.1 760-Millimetre Balplex

The nominal focal length of the ER-55 (Balplex) projector lens is 49.8 millimetres which, in combination with a principal distance of 55 millimetres, fixes the optimum projection distance at 525 millimetres. This lens can be replaced with a lens of 51.2-millimetre focal length, which provides for an optimum projection distance of 760 millimetres, thereby permitting orientation of stereomodels (figure 11-11) at scales similar to those of the Kelsh plotter. At the increased projection distance, a larger plotting table is required.

11.2.10.2.2 Super-Wide-Angle Balplex

To accommodate photography having an angular coverage of 120 degrees, Balplex equipment having this capability has been developed (figure 11-12). The ellipsoidal reflector is increased in size to provide illumination throughout the larger angular coverage, and the nominal focal length of the lens (29.5 mm) is such that, at the optimum projection distance (360 mm), the scale of the projected images is about four times that of the original photographs. It is necessary to lower the frame to provide the 14-inch nominal projection distance for this configuration.

11.2.10.3 Kelsh Plotter

The Kelsh Plotter (figure 11-13) is a stereoscopic mapping instrument of the double projection type that was developed during the tenure of Harry T. Kelsh with the U.S. Department of Agriculture and the U.S. Geological Survey. The instrument is now made by the Kelsh Instrument Division of Danko Arlington, Inc. Although

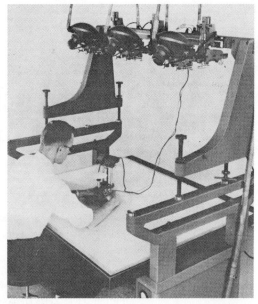

FIGURE 11-11. ER-55 (Balplex) plotter, vertical orientation.

similar in some respects to the Multiplex, the Kelsh Plotter has many characteristic differences. Basically, the current standard Kelsh Plotter is intended for use with 153mm focal length, wide angle photography. Although projector cones of other focal lengths may be obtained, it is possible to compile 88mm, 210mm, and 305mm photography on a standard (153mm) focal length Kelsh Plotter with the use of an affine solution, as long as the photography is reasonably level. With modern camera equipment mounted in newer aircraft, reasonably level photography (within about 3 degrees of differential tilt) is usually the rule rather than the exception.

The Kelsh Plotter was originally designed as a single model instrument. However, it is now available in 3, 4, and 5 projector units. The Kelsh Plotter utilizes contact size diapositives having the same 23 × 23 cm format as the aerial negative. Acceptable diapositives for use in the Kelsh Plotter may be made in an inexpensive contact printer, however they are generally produced in one to one projection printers, such as the LogE, which automatically dodge contact size diapositives of superior quality.

A principal and distinctive feature of the Kelsh Plotter is the means of illumination. The lamp-condenser assembly in modern Kelsh Plotters consists of a tungsten quartz halogen lamp with a finely wound filament which is focused by a condenser lens on the aperture of the projection lens. This lamp housing swings on a yoke about the rear node of its corresponding projector lens as directed by guide rods that are attached to the tracing table. The light emanating

FIGURE 11-10. ER-55 diapositive centering device.

FIGURE 11-12. Balplex super-wide-angle projectors on universal double-projection (UDP) frame.

from the lamp housing passes through a part of the diapositive, is condensed at the aperture of the projection lens and is directed toward the platen of the tracing table.

In the Kelsh Plotter system there are three principal methods for providing a separation of corresponding images. The oldest method is with the use of an appropriate red or blue filter to provide anaglyphic separation. The operator views the image on the tracing table with red and blue spectacles so that the left eye sees only the left image and the right eye only the right image. The resulting stereo model is seen by the operator as a somewhat grayish image.

The Stereo Image Alternator (SIA) is a second way of viewing a Kelsh stereo model. Stepper motors driving shutters under the projection lenses are synchronized with a rotating shutter viewer above the platen of the tracing table. The corresponding images are alternately flashed quite rapidly, giving the operator a sharp stereo model with increased illumination.

The third method that has been developed for viewing stereo on a Kelsh Plotter is the PPV system. Left and right circular polarizers are placed in the projector beams, and the respec-tive images are reflected from a specular platen surface that is spinning at about 900 RPM. The corresponding images are viewed on this spinning platen with right- and left polarizing spectacles. The operator views a crisp stereo model, also with greater illumination than is possible with the anaglyph system.

A standard Kelsh Plotter projector may be adjusted within a principal distance range of 150 to 156mm to match the aerial camera focal length. The optimum projection distance is 765mm, so that the image in the optimum plane of projection is magnified five times. Special cones and lenses are available for a 1070mm projection distance (for $7\times$ magnification). A 6mm extension ring is also available which will increase the principal distance range from 156 to 162mm. This makes possible the projection of a $4\times$ magnification model for 153mm as well as 88, 210, and 305mm photography, using an affine solution.

The tracing table is nearly identical to that of the Multiplex, except for the yoke by which the guide rods are attached to the platen, in a position which effectively directs the corresponding beams (projected images) to the platen area. A

FIGURE 11-13. Kelsh plotter, model KPP-3B.

new pencil-lift mechanism is used with the tracing table. It may be operated by either the operator's right or left thumb, when inserted in a fork-like lever.

The development of stable-base polyester-film diapositives was the motivation for the redesign of the Kelsh correction-cam system. Modern aerial cameras are essentially distortion-free, and there is little need for an aspheric ball cam to correct for the distortion of the taking camera. The new design was introduced to correct for the distortion caused by the lower glass plate of a film diapositive sandwich. The mechanism using the improved cam varies the principal distance of the projectors by small amounts, sufficient to compensate for the distortion at each differential portion of the diapositive as it is scanned. In addition to the distortion correction, when using the film diapositive sandwich, the principal distance at the optical axis of the projection lens must be shortened by ⅓ of the thickness of the lower glass plate of the sandwich. This will compensate for the displacement of the image, which also occurs with the introduction of the lower glass plate. The cam system can be deactivated when using distortion free photography emulsion down.

Projector motions on the basic 5030-B Kelsh Plotter are limited to one translational motion along the X axis (Bx) and rotational motion about the X axis, Y axis, and Z axis. The three to five projector KPP-B and K-5 Kelsh Plotters have additional translational motions along the Y and Z axes (By and Bz). A tip-tilt indicator equipped with precise levels and micrometer screws permits the transfer of X tilt and Y tilt to adjacent projectors which is required for successive model orientations in aerotriangulation. All modern Kelsh Plotters have the necessary range of Y-tilt to accommodate convergent low-oblique photography.

A conventional Kelsh Plotter frame is divided into three segments: the table frame, the supporting A-frame assembly, and the X-bar projector support. Four jack screws in the corners of the supporting frame provide the means for height adjustment and levelling. The X-bar is supported on three elevating screws which provide a means of tilting it, and consequently the model, in any direction. These screws are generally used to level the stereoscopic model to ground or analytical control. The 5030-B Kelsh Plotter does not have a Bx motion. Therefore, provision for tilting the X-bar is essential to permit levelling the model along the X axis.

All modern Kelsh Plotters are equipped with a polished granite plotting table which is supported and levelled on the table frame. Larger tables are required when additional projectors are used. On older Kelsh Plotters, a mechanical pantograph was used to reduce the model down to scales less than 1:1. However, modern Kelsh Plotters may be digitized with an X, Y, Z digitizer mounted on the table, or the Kelsh K-600 Digitizing System. Along with recording digital information from a stereo model, the digitizer may be used to operate an electronic pantograph next to the plotter, which can plot at scales from less than 1:1 to 12:5 to 1.

The current Kelsh Plotter system uses a solid state illumination power supply which op on standard 115 volt alternating current. stats in the tracing table control the illumination of the corresponding projectors.

11.2.10.4 ARMY MAP SERVICE MODEL-2 STEREOPLOTTER

The AMS M-2 stereoplotter (figure 11-14) is manufactured in accordance with specifications prepared by the Defense Mapping Agency. This plotter is found in either two- or three-projector models. In general, it may be said that the M-2 plotter is similar to the Kelsh plotter in that the diapositive plates are 9 × 9 inches, the swinging light sources are guided by space rods attached to the tracing table, and a pantograph is available for compilation.

The plotting table top of the M-2 plotter is a granite slab about 4 inches thick and large enough to accommodate two stereomodels of vertical photography at a scale slightly more than 5 times the scale of the pictures. The surface of this slab is ground flat by the same process used for surface plates and, therefore, its flatness is superior to other types of plotting surfaces. Its weight is about 1,600 pounds.

11.2.10.5 MILITARY HIGH-PRECISION PLOTTER

The Military plotter (figure 11-15) was developed for the Corps of Engineers by the Engineer Research and Development Laboratories to supplement the multiplex and allow greater exploitation of high-altitude photography. It was designed for field-troop and base-plant use. It is similar in most respects to the Kelsh plotter since it was found that, with certain modifications, that instrument would fulfill the military requirements. It is a two-projector instrument that has interchangeable truncated cones for use with 20-degree convergent photographs. Some of its features that were not available on the Kelsh plotters at the time of development include: the lightweight plotting table top, the level bubbles, and the bz motions.

11.2.10.6 THE NISTRI PHOTOCARTOGRAPH

The Nistri Photocartograph Model VI (figure 11-16), commonly referred to as the Nistri "Photomapper," is manufactured by Ottico Meccanica Italiana, of Rome, Italy. It is based on the anaglyphic principle and utilizes full-scale 9- × 9-inch diapositives. Radial lens distortions resulting from imperfections in the aerial camera lens are corrected by the projection lenses, which are designed with compensating distortions. In general performance and characteristics, the instrument is similar to the Kelsh plotter, although there are significantly different design details, such as the use of a single space rod for guiding the projection lamps, and projection cones with integrally mounted projection lenses. The electrocoordinatograph manufactured by

FIGURE 11-14. M-2 plotter.

Nistri can be mounted on the supporting table of this plotter to permit changing the scale between model and manuscript without recourse to a mechanical pantograph.

11.2.10.7 THE WILLIAMSON LSP PLOTTER

The Williamson LSP plotter is manufactured by the Williamson Manufacturing Company, Ltd., of London, England. This plotter is an anaglyphic-type projection plotter with an optimum projected model scale of 5 times the diapositive scale. Swinging lights for illumina-

tion are actuated by guide rods attached to the viewing platen. Diapositives are full-size 9 × 9 inches. Radial lens distortion is corrected in the diapositive printer. A unique "saddle" seat and cross-slide arrangement permits the operator to move directly into the projected model area for scanning and delineation of map detail. Tracing of details in the model is done by handwheels providing x and y motion. A separate z wheel provides height adjustments. Drawing of the model details is accomplished by a variable-ratio pantograph, which is attached to the tracing table.

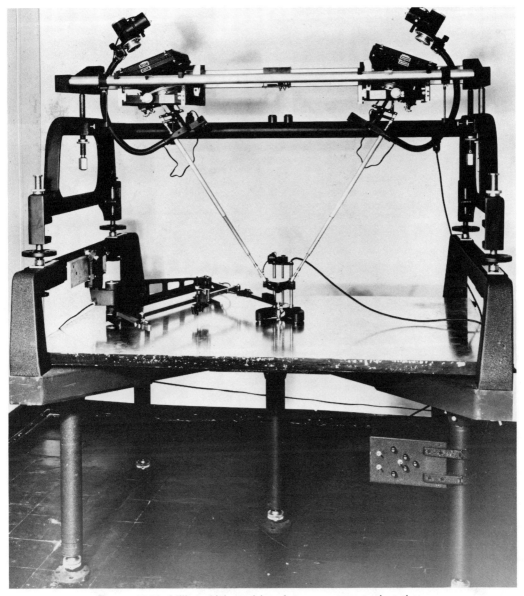

FIGURE 11-15. Military high-precision plotter, convergent orientation.

11.2.10.8 THE KERN PG1 PLOTTER

The Kern PG1 (figure 11-17) is manufactured by Kern & Company, Ltd., Aarau, Switzerland. This double-projection plotter is generally similar to the familiar Kelsh type in that it utilizes full-size diapositives and swinging lights for projection. Mercury-vapour lamps illuminate the projected area of the model so well that there is no need to install the plotter in a darkened room. This plotter does not utilize the familiar dichromatic anaglyphic filter system in projecting the stereomodel; rather, stereoscopic viewing is accomplished by means of a rotating filter which is mounted in front of the viewing platen. In a manner similar to the Williamson LSP plotter, the operator's seat can move into the field of

projection. A polar pantograph is utilized to draw the map detail, and to provide a means of changing scale between the model and the map.

11.2.10.9 THE GAMBLE STEREOPLOTTER

The Gamble stereoplotter (figure 11-18) was designed by Samual G. Gamble, Department of Mines and Technical Surveys, Canada, and manufactured by PSC Applied Research, Ltd., of Toronto, Canada. This instrument utilizes either multiplex or Balplex projectors to form the stereomodel in anaglyphic projection. A very close-spaced pattern of luminous dots is projected to form an elevation reference plane in the stereomodel. Raising and lowering of the projectors in unison moves the model datum with

FIGURE 11-16. Nistri Photocartograph, Model VI.

respect to the projected elevation reference plane. Contours and planimetric detail are traced freehand on a map sheet mounted directly upon the supporting table. The hand-held tracing pencil thus becomes the floating mark in the projected model.

11.2.10.10 THE TWINPLEX PLOTTER

The twinplex plotter (figure 11-19), while capable of projecting stereopairs of convergent or transverse low-oblique photographs for compilation, is designed specifically for strip aerotriangulation with twin low-oblique photographs. The Twinplex utilizes ER-55 (Balplex) pro-

jectors mounted in pairs in an arrangement that can be oriented to simulate the twin low-oblique camera installation in the aircraft. Stereopairs can be oriented in sequence for developing horizontal or vertical pass points in bridged strips. This instrument is a U.S. Geological Survey development and its use has been limited.

11.2.11 CALIBRATION AND ADJUSTMENT

Accurate adjustment and calibration of the various elements of double-projection, direct-viewing plotters are necessary to attain proper

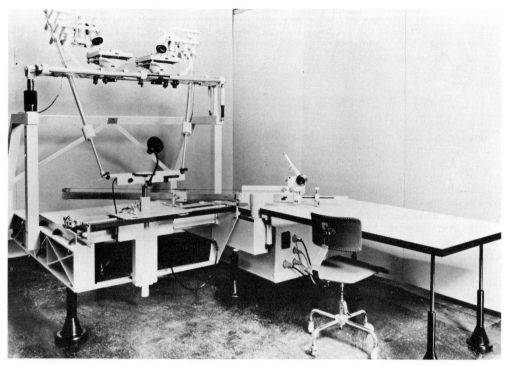

FIGURE 11-17. Kern PG-I plotter.

functioning of the instrument and maximum geometric fidelity of the resultant projected model. Special tools and the skill of trained technicians are usually required to calibrate

these plotters when adjustments are made that affect the instrument's geometry. There are tests, however, that may be performed by the operator, periodically or whenever the functioning of the instrument is in question, to determine which elements, if any, are improperly calibrated.

11.2.11.1 GENERAL

The methods of calibration and adjustment discussed herein vary in complexity and equipment required, but for the most part they can be performed with a minimum of equipment by a careful individual having some knowledge of the basic principles involved. It can be assumed that the manufacturer had good equipment and

FIGURE 11-18. Gamble Stereoplotter.

FIGURE 11-19. Twinplex plotter.

trained personnel; consequently, his calibration and adjustment probably are superior to those done by the user. This is particularly true of the projectors; therefore, caution should be exercised in altering any settings that are fixed at the factory.

11.2.11.2 OPERATIONAL CHECKS AND ADJUSTMENTS

There are a number of operational checks that are of a routine nature and should be made almost every day. Some of these, such as adjustments of the projector lamp and/or reflectors for optimum illumination, adjustment of guide rods to center the light from the projector lamp on the tracing table, and centering the bulb that illuminates the floating mark, have little appreciable effect on accuracy but do influence the efficiency of compilation. In addition, the telescoping action of guide rods should be checked at frequent intervals. On the other hand, the orthographic relationship of the floating mark and the drawing point of the tracing table is more likely to get out of adjustment than some other parts of the equipment, and may result in serious plotting errors.

11.2.11.3 CALIBRATION OF PROJECTORS

The primary consideration in any stereoplotter projector is the fidelity with which the perspective geometry of the photograph is recovered. This recovery is based on precise location of the principal point and exact setting of the projector principal distance. These are adjusted and calibrated at the factory with specialized equipment.

11.2.11.3.1 PRINCIPAL DISTANCE

There are several methods for checking the principal distance calibration of the projector, all of which are based on solution of similar triangles. All require a diapositive with a grid of known dimensions, or at least a known distance along a line through the principal point on the diapositive.

The simplest method, and probably the least accurate, involves only the use of a grid plate and the tracing table. The grid plate is carefully centered in the projector. The projector is then plumbed to the plotting surface by adjusting the tilt motions until a well-calibrated tracing table can be run through its entire z motion with the floating mark remaining in the center grid intersection, or until the projected grid pattern is precisely rectangular. The tracing-table platen is run to a low position near the lower limit of the projector depth-of-focus, and the positions of the ends of the x and y center grid lines are carefully plotted. The tracing table is raised an exact number of millimeters to near its upper range, and the process repeated. Figure 11-20 represents the configuration in one plane (xz or yz). The sum of distances d_1 plus d_2 is in the

same ratio to the corresponding grid distance ab on the plate as the h travel of the tracing table is to the projector distance:

$$\frac{\text{P.D.}}{h} = \frac{ab}{d_1 + d_2}, \text{ or P.D.} = \frac{h \times ab}{d_1 + d_2} \quad (11.1)$$

where;

ab = the distance on the grid plate and,
h = the vertical travel of the tracing table.

The average is taken of the values obtained from the two planes (xz and yz).

A method that can be used for the Bausch and Lomb multiplex requires, in addition to the grid diapositive, a means for accurately measuring the distance from the boss on the front of the projector which defines the tilt axis to the tracing-table platen, and knowledge of the location of the emergent lens node with respect to the tilt axes (some early projectors did not tilt about the emergent node). The projector tilts are adjusted so that the diapositive plane is made parallel to the table surface. With the projection distance near the optimum, selected grid intersections are plotted with the tracing table. Without altering the height of the platen, the distance from the top of the cylindrical boss on the front of the projector down to the platen, and the distances between the plotted points, are measured as accurately as possible. (On most projectors the boss has a conical hole in its center which can be used as a reference point, in which case the r term is dropped in the following equation.)

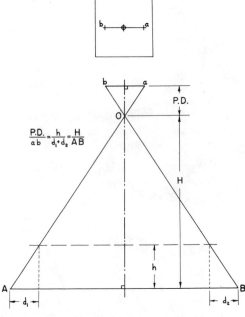

FIGURE 11-20. Diagram showing relationships, in one plane, for determination of projector principal distance.

The principal distance P.D. is computed by substitution in the expression:

$$P.D. = \frac{ab(H + e - r)}{AB} \qquad (11.2)$$

where;

ab = the x (or y) distance on the grid plate,
AB = the corresponding distance in the projection,
H = the distance from the top of the boss to the platen,
r = the radius of the cylindrical boss, and
e = the nominal distance from the tilt axis to the emergent node of the lens (on modern projectors the tilt axis and emergent node are coincident and the e term is therefore zero).

A mean is taken between the two determinations in the x and y directions. Some additional accuracy can be achieved if a special plate with diagonal lines is available, providing slightly longer distances and improved strength of figure.

Greater precision in the determination of principal distance requires greater accuracy of measurement of the projective geometry. To this end, a precise glass or metal scale graduated in the metric system with a least graduation of 0.10 mm can be supported horizontally in a reference plane at the optimum projection distance to measure the separation of the projected grid intersections with a high degree of precision. A rod of steel (or other stable material) with rounded ends, whose length is precisely known, can be used as a height gauge to accurately fix the distance along the principal ray between the external vertex of the lens and the reference plane upon which the grid is projected (figure 11-21). The length of the rod plus the distance from the lens vertex to the emergent node of the lens should be equal to the optimum projection distance of the plotter.

With the grid plate carefully centered and the projector axis perpendicular to the reference plane, the height of the external vertex of the projector lens above the reference plane is set with the steel rod. The ends of the rod should be covered with a thin material such as plastic tape of known thickness to protect the lens surface. The distance between selected grid intersections projected onto the reference plane is measured with the metric scale, and the principal distance is computed from the following relationship:

$$P.D. = \frac{ab(R + V)}{AB} \qquad (11.3)$$

where;

ab = the x (or y) distance on the grid plate,
AB = the corresponding distance measured in the reference plane,
R = the rod length plus the thickness of the protective tape at the ends, and
V = the distance from the external vertex of the lens to its emergent node.

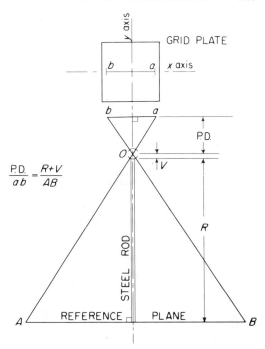

$$\frac{P.D.}{ab} = \frac{R+V}{AB}$$

FIGURE 11-21. Diagram showing use of precise steel rod for determination of projector principal distance.

For plotters of the Kelsh type the principal distance may be checked by direct measurement with a depth micrometer through a hole in the center of a special glass plate of known thickness positioned in the diapositive holder. The correction cam (if any) should be oriented to accommodate the axial ray of the projection and the principal distance adjustment (if any) set to the nominal value for the projector. A piece of soft tissue paper placed under the arms of the micrometer and another piece placed over the lens will protect the lens and will not affect the readings. The depth micrometer reading to the upper surface of the lens, plus the nominal distance of the entrance node of the lens from its upper surface, minus the thickness of the glass plate, equals the principal distance of the lens. (If the intended practice in the normal operation of the instrument is to project diapositives with the emulsion surface uppermost in the projector, two-thirds of the diapositive thickness must be added to the micrometer reading to obtain the projector's effective principal distance, and compensation must be made for the distortion caused by projection through the glass.)

The principal distance of projectors of the multiplex and Balplex types is adjusted by shifting the projector lens in its mount while the plane of the diapositive remains fixed. This should be done only in the laboratory, where appropriate tools and knowhow are available. On the other hand, the principal distance of projectors of the Kelsh type can be adjusted by moving the plane of the diapositive relative to the lens. This can be done by adjusting the screws on which the diapositive plateholder rests. However, this adjustment may result in disturbance of the condition of perpendicularity between the diapositive plane and the optical axis; it should always be followed by checking the latter element of calibration.

To achieve proper interior orientation in projectors arranged for the projection of vertical photographs, it is essential that the diapositive lie in a plane that is perpendicular to the optical axis of the projector lens. This condition is closely held in the assembly of multiplex and ER-55 (Balplex) projectors. However, in projectors of the Kelsh type the diapositive plateholders are supported on leveling screws which must be adjusted to meet this condition (these screws are also adjusted to achieve the nominal principal distance of the projector).

As the mechanical axis of the swing motion of Kelsh-type projectors is considered to be coincident with the optical axis of the projector lens, the bossed surface of the plateholder frame must be perpendicular to this mechanical axis. This condition must be achieved before the principal distance and the principal point of these projectors can be checked. The perpendicularity between the plane of the diapositive and the optical axis of the projector lens can be checked with a precise level (sensitive to 40 seconds) and a 0.25-inch diapositive plate selected for flatness and uniform thickness. With the level resting on the diapositive in its plateholder on the projector, the projector is carefully leveled about the photogrammetric x and y axes. The projector is then rotated 180 degrees about its swing axis and the level of the diapositive plate noted. If the glass plate has been maintained in a horizontal plane, the plane of the diapositive is considered to be perpendicular to the optical axis of the projector lens. Failure to meet this condition indicates need for adjustment of the leveling screws which support the plateholder. As adjustment of these screws affects the projector's principal distance, this element of calibration must be checked when the screws are disturbed. If, after adjustment, the plateholder fails to sit firmly on all four screws, the projector is no longer in proper calibration.

11.2.11.3.2 PRINCIPAL POINT

The principal point in a photogrammetric projector is defined as the foot of the perpendicular from the interior perspective center to the plane of the diapositive. This point is indicated in the projector by the intersection of the lines joining opposite pairs of the fiducial marks by which the diapositive is centered. This indicated position must be in agreement with the position of the projector's true principal point, which normally can be considered to lie on the projector's optical axis. (Canting the projector lens or tilting the plane of the diapositive to satisfy the Scheimpflug condition will alter this normal relationship of principal point and optical axis. Therefore, a check of principal-point calibration should be made with the projector arranged to project vertical photographs.) A position error of the principal point is often indicated by its effect on y parallax, and sometimes by its effect on elevations in the model. Like a principal-distance error, the effect of a given error increases in proportion to the relief in the model.

A more reliable check of the calibration of the principal point is independent of the coincidence of the mechanical swing axis with the optical axis. It is based on the theory that the principal-point ray of a precisely leveled projector must be plumb if the principal point is correctly located. The verticality of this ray can be readily determined by observing the projection of the center grid intersection of a carefully centered precise grid when projected in a leveled projector. The observation is made at two levels, near the upper and lower limits of usable imagery, and a displacement of either projected image with reference to the other indicates a principal-point error.

A still more reliable check, and one that is relatively simple, can be made by projecting a diapositive with a precise square grid onto a reference figure composed of three equally spaced points carefully plotted in a straight line on stable material and placed in the projection plane. The reference figure is intended to represent the projection of selected grid intersections, equally spaced about the center intersection, at nearly the maximum magnification for the projector. Therefore, the spacing of the plotted points should be computed to meet this condition. Figure 11-22 diagrams the relationship. (Bear in mind, however, that the displacement d can occur in any direction and not necessarily in the same direction as the base line of the reference

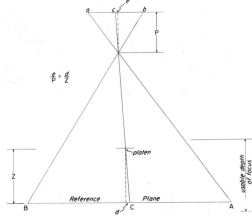

FIGURE 11-22. Schematic of the relationships for determination of principal-point error.

figure, as shown in the diagram.) With the diapositive centered in the projector, the tilt motions are adjusted until the projected grid is exactly square. Without disturbing the tilts, the reference figure is shifted in the plane of projection and the projection distance of the projector is adjusted until the three plotted points are centered on their respective projected grid images. A tracing table that has been carefully calibrated with respect to the verticality of its columns, and with extensions added to raise the platen to near the upper limit of usable, projected imagery, is then introduced. The floating mark is centered on the projected center grid intersection and this point carefully plotted; the tracing table is turned 180 degrees and the point plotted again to verify the calibration of the table. The mean of these two positions should fall on the center point of the reference figure. If not, an error of the principal point in the opposite direction from the plotted displacement is indicated. The error e of the principal point can be computed from the formula:

$$e = \frac{dP}{Z} \qquad (11.4)$$

where;

d = the mean of the displacements plotted in the plane of projection,
P = the principal distance of the projector, and
Z = the height of the tracing table platen above the lower plane of projection.

After correction of the error, the reference figure is rotated 90 degrees and the procedure repeated to check the accuracy of the principal-point correction.

Correction of a principal-point error is a laboratory procedure, which varies among the plotters of the double-projection type. Basically, however, it requires a shift of the index points to which the diapositive is referenced in interior orientation. In the various multiplex projectors it is accomplished by an adjustment of the stage plate on which the principal-point index is marked; or by an adjustment of the foot pads on which the centering microscope rests; or by an adjustment of the centering arm and the stop which controls its position within the projector. In Balplex projectors it is accomplished by adjusting the alining ring with its three balls on which the diapositive holder rests, and in projectors of the Kelsh type it is accomplished by adjusting the fiducial tabs to which the diapositive is referenced when centering the plate.

Correction of a principal-point error is always made in the opposite direction to the displacement observed in the plane of projection, and may be made by trial and error. However, a variety of measuring devices may be applied to control precise increments of correction to the index points which establish the principal point.

The Balplex projector is so designed that the aligning ring may be adjusted with the diapositive and its holder in place and while observing the movement of the projected center grid intersection.

There are many variations of the foregoing procedures, yielding various degrees of accuracy. One variation that offers considerable refinement, although it requires repeated model orientations to determine the amount and direction of the principal-point error, is based on observations of y parallax at two levels in a carefully oriented and leveled grid model. A step-by-step description of this method, applicable to the ER-55 (Balplex) projector, may be found in Anonymous 1961. The described procedure may be readily modified to suit other types of projectors.

11.2.11.4 ADJUSTMENT OF LIGHT SOURCE

For maximum model illumination, the image of the filament of the projector bulb must be focused on and centered in the aperture of the projector lens. This is a condition that must be met in all types of double-projection plotters. Adjustment of the light source for maximum illumination is important for efficient operation since these plotters suffer considerable light loss in their projection systems under the best of conditions.

For those projectors that employ condensing-lens systems to concentrate the illumination in the lens aperture, this condition is met when the filament is centered on the optical axis of the condensing system, at the appropriate distance from that lens. A clear diapositive or a clear glass plate of equal thickness must be in place in the projector during the adjustment operation. The relationship of the filament coil to the aperture of the projector lens can be conveniently observed by employing a short-focal-length auxiliary lens such as a pocket magnifier to focus the image of the filament and aperture on the table top. (For Kelsh and similar plotters this can be seen by direct observation of the image of the filament on the projector lens itself.) Adjustment screws in the lamp sockets of the various instruments provide for the necessary lateral motions to center the filament coil in the lens aperture.

For those instruments that employ a swinging light source, the cone of light emanating from the lamp housings must be centered as nearly as possible on the platen, for all positions of the tracing table within the working limits of the projected model. Provision for adjustment is made in the coupling of the guide rods and lamp yoke, or in the condenser housing itself.

The efficiency of illumination in the ER-55 (Balplex) projector is dependent not only on the centering of the lamp filament, but also on the positioning of the reflector so that one focus is centered in the projection lens aperture. An approximation of the condition can be achieved by trial adjustments while observing the distribution of light in the plane of projection, or by projecting the image of the lamp filament through the lens aperture as with other projection plotters. A more precise adjustment of the reflector can be made with auxiliary foci-finder equipment (figure 11-23). This consists of a metal crossframe with five light sources in two perpendicular rows. The light in the center is white and is common to both rows; the other two lights in each row are red and green, respectively. A viewing fixture (A) with a frosted glass viewing screen and center cross, adjustable along the lamp-house axis, is inserted in the projector lamp-house opening and is used to observe the reflected images of the lens aperture as illuminated by the lights on the cross-frame centered beneath the projector at a distance of 600 to 800 millimetres. The ellipsoidal reflector is adjusted laterally and vertically with respect to the projector body until the circles of red, green, and white light are made concentric and are situated close to the center cross on the viewing screen. The depth of the viewing screen will probably need adjustment to obtain concentricity.

Because of variations in manufacture, the refinements of reflector adjustments are limited, and complete concentricity of the circles with the center cross probably cannot be obtained. However, if any part of the group of concentric circles is touching the cross intersection, the reflector position should be within the range of adjustment of the lamp in its socket.

The greatest concentration of light is obtained in the aperture of the projector lens when the axis of the filament coil lies in a vertical plane which also contains the two focal points of the ellipsoid. The construction of the lamp socket is such that this orientation of the filament coil is assured when the lamp bulb is in place. The final centering of the lamp bulb is accomplished by the adjustment screws while observing the images of the lens aperture and filament coil projected onto the table top with the aid of a short-focal-length lens.

11.2.11.5 CALIBRATION FOR DISTORTION COMPENSATION

The system for compensation of the effects of radial lens distortion in the double-projection direct-viewing plotters must provide correction that closely matches the distortion introduced into the negative by the aerial camera lens. It is usually not practical to provide distortion compensation to match each individual camera; therefore, the usual practice is to design a compensation system to match a group of cameras that have nearly equal radial distortion characteristics. Modern low-distortion camera lenses have largely eliminated the need for distortion-compensation systems.

Figure 11-24 illustrates the pattern of deformation of a grid model in a Kelsh plotter resulting from the compensation of a nominal metrogon lens. If the stereoscopic grid test indicates that the desired compensation is not attained, laboratory procedures are necessary for measuring the amount of deformation introduced by the compensation system. This can be accomplished by precise comparator measurements of grid intersections on those diapositives which have been printed in a printer having an aspheric plate for distortion compensation; or by dial indicator measurements of the displacement of the projector lens resulting from the action of the ball cam in various angular zones from the projector's axial ray.

If the radial distortion characteristics of the aerial camera lens are known, the displacement of the projector lens along its axis required to compensate for this distortion can be computed by substitution in the following expression:

$$\Delta P = \frac{M}{M + 1} \times \frac{P\Delta d}{d} \qquad (11.5)$$

where;

$\Delta P =$ the desired lens motion (principal distance change),

FIGURE 11-23. Crosslights and foci-finder (A) for reflector calibration.

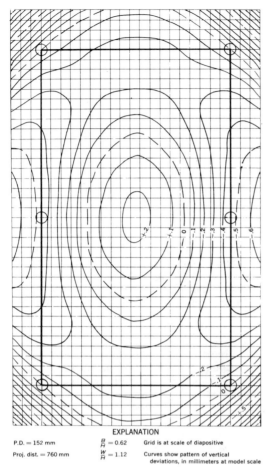

EXPLANATION

P.D. = 152 mm	$\frac{B}{H}$ = 0.62	Grid is at scale of diapositive
Proj. dist. = 760 mm	$\frac{W}{H}$ = 1.12	Curves show pattern of vertical deviations, in millimeters at model scale

FIGURE 11-24. Pattern of a grid model resulting from compensation of nominal metrogon distortion.

M = the magnification of the projection (ratio of projection distance to principal distance),

P = the principal distance,

d = the distance on the diapositive from the center to the point being considered, and

Δd = the radial distortion at that point.

11.2.11.6 EFFECT OF RELIEF ON CAM PERFORMANCE

In double-projection direct-viewing plotters, the projection distance varies with relief; therefore, the M-value (equation 11.5) for magnification is in reality a variable rather than a constant. The use of cams designed for a specific magnification results in slightly erroneous correction for distortion at other magnifications, and the amount of this error can be shown to vary with amount of relief departure from the nominal projection distance and the degree of compensation required at the point in question. The effect of the erroneous compensation depends on the position in the stereomodel, but

usually results in extraneous x and y parallax. The latter of these may cause erroneous relative orientation or may simply cause difficulty of observation in local areas, while the former may cause up to about 0.02 millimetre error in elevation reading when relief of 100 millimetres is present. Since errors of this magnitude can arise from distortion variations from camera to camera, calibration error of a given camera, error of cutting the cam surface, improper model scale, and other such sources, it is apparent that the approximation afforded by the cam correction is sufficiently accurate for practical purposes.

11.2.11.7 CALIBRATION OF THE PANTOGRAPH

Faithful mapping of the stereomodel onto the tracing sheet at a reduced scale is dependent on precise calibration of the variable-reduction pantograph. For this reason, periodic accuracy tests should be made on all pantographs at the various reduction ratios normally used in compilation.

Accurate reduction by a pantograph is dependent on two conditions: (1) the ratio of reduction from the original to the copy must be correct, and (2) the map must be a geometrically faithful to the original. To achieve these conditions, the arms of the pantograph must form a true parallelogram, they must be graduated with sufficient precision to permit accurate ratio settings, and they must move in horizontal planes parallel to the top of the plotting table for all operating positions of the tracing end of the pantograph. A corollary requirement is that all bearings must maintain a precise vertical orientation throughout all motion of the pantograph.

A variable-ratio pantograph can be calibrated in a variety of ways, and those described herein may be modified to some extent to suit the equipment available. Calibration based on fidelity of reproduction of a geometric figure requires a minimum of equipment and can be readily used on the plotting table; however, since the adjustments are made by trial and error, this can be time-consuming. A carefully constructed rectangle of known dimensions, approximating the size of the neat model, is reproduced with the pantograph at various reduction ratios, and the sides and diagonals of the resulting figures are measured and compared with the same elements of the original figure to determine the accuracy of reduction ratio and the geometric fidelity of reproduction. (The corners of a level model projected from grids of known dimensions can be used for this provided that the projection distance is carefully measured to determine the factor of magnification from grids to projected model.) Corrections to the pantograph verniers are made by trial and error until the reproduced figures are correct within tolerable limits.

There is a more efficient and direct method of calibrating, but it requires the use of a special jig. This procedure is based on the following relationships, which are necessary for faithful mapping.

1. The positions of the king post K, the drawing point B, and the tracing point T (figure 11-25), projected orthographically onto the plotting surface, must lie in a straight line in all operating positions, and at all ratio settings.
2. The ratio of KB to KT, as projected onto the plotting surface, must be correct at all ratio settings and in all operating positions.

Figure 11-26 shows a schematic diagram of a jig for the straightline calibration of a 720-millimetre pantograph (the jig will also accommodate an 840-millimetre pantograph). A straight baseline, KT, is scribed on the jig, with positions indicated for the tracing point of the pantograph in its extended position, and in its folded position. Crosses are also scribed on this baseline to represent the correct positions of the drawing point for these positions of the tracing point at several selected ratios. The scribed arc on the jig provides a means of precisely positioning point K of the jig on the extended axis of the king post. A microscope fastened to the cantilever arm is adjusted along the arm in conjunction with adjustments of the jig until the reference index of the microscope remains coincident with the scribed arc as the cantilever arm is rotated about the king post. In this position, the jig is oriented for calibration and, with the verniers set for an appropriate ratio, the tracing point (T) of the pantograph is set on the extended index point on the jig. The relationship of the drawing point to the appropriate cross on the baseline is observed through a microscope, and any deviation is corrected by adjusting the C-vernier. The tracing point is then set on the folded index point on the jig and the relationship of the drawing point to a cross on the baseline is observed as before. Any deviation is corrected by adjusting the B-vernier. These corrections should establish a true parallelogram, and any errors in reduction ratio can be corrected by repeating the procedure and applying equal adjustments to all three verniers to bring the drawing point into coincidence with the plotted index cross on the jig with the pantograph in its extended position. With the panto-

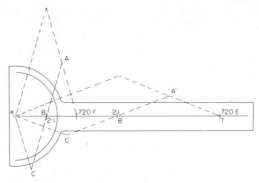

FIGURE 11-26. Schematic drawing of a pantograph calibration jig.

graph in its folded position, the drawing point as determined by the observing microscope should not deviate more than 0.005 inch from the plotted cross on the jig. Excessive deviation indicates some mechanical defect in the pantograph.

In addition to these methods of pantograph calibration, a mathematical analysis (*see* anonymous, 1960) may be applied to observed discrepancies to determine correct vernier settings. A more recent method determines refined vernier corrections from a least-square adjustment of the discrepancies in a plotted rectangle as compared to a master rectangle. The mathematical adjustment procedures will, in general, give a more precise calibration; however, a mechanical defect in the pantograph may be more difficult to isolate than when using the jig, and for this reason small discrepancies resulting from such a defect may be erroneously distributed to the vernier corrections.

11.2.12 STEPS IN STEREOSCOPIC MAP COMPILATION

The steps in compilation of a map by stereoscopic procedures are grouped into two or, in some instances, three major phases. The initial phase is orientation, or the process by which the model is brought into being and made to conform to a datum. The orientation phase can be divided into interior orientation and exterior orientation, with the latter further subdivided into relative orientation and absolute orientation. The second phase, stereotriangulation, is an extension of the orientation process to include several models when there is not sufficient control for the orientation of each model individually. When stereotriangulation is required, orientation becomes both an initial part of the stereotriangulation phase and, usually, an intermediate phase between stereotriangulation and compilation. Compilation is the final phase, which covers the delineation on the manuscript of the features seen in the model. When stereotriangulation is not required, the initial orientation is followed directly by compilation.

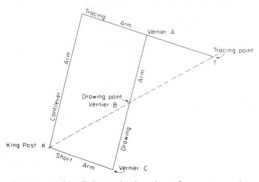

FIGURE 11-25. Schematic drawing of a pantograph.

11.2.12.1 INTERIOR ORIENTATION PROCEDURES

The purpose of interior orientation is to recover, insofar as possible, a projected cone of rays geometrically identical with the cone of rays that entered the camera lens to make the original exposure. Positioning of the projector so that this objective may be attained is referred to as interior orientation.

11.2.12.1.1 CONDITIONS AFFECTING INTERIOR ORIENTATION

Interior orientation imposes two requirements: (1) the perpendicular from the interior perspective center of the projection lens to the plane of the emulsion surface of the diapositive must be correct; and (2) the intersection of this perpendicular with the plane of the diapositive must fall at the principal point of the photograph, as defined by the intersection of opposite pairs of fiducial marks.

The procedure of interior orientation is dependent upon a number of interrelated characteristics of the camera, the stereoprojector, the negative, and the diapositive-printing operation. These factors are (a) the nominal focal length of the camera used, (b) the distortion of the camera lens, (c) the contour of the correction cam or diapositive-printer correction plate, (d) the nature of the diapositives used, (e) calibration of the principal distance and principal point of the projector, (f) the nature of the diapositive-printing operation, (g) shrinkage of the film base of the aerial negative, (h) registration of the negative fiducial marks with those of the printer, and (i) adjustment of the diapositive within the diapositive holder to center it on the principal point of the projector. For the most part, these factors can be controlled in the place where the diapositives are printed and by appropriate calibration of the projectors. In such cases, the centering of the diapositive in its holder is the only operation of interior orientation for which the operator is responsible.

11.2.12.1.2 STEPS IN INTERIOR ORIENTATION

Before the diapositive can be centered to establish the correct geometric relationship between the emulsion surface of the diapositive and the internal perspective center of the projector lens, a number of preliminary steps are essential, as described in the following sections.

11.2.12.1.2.1 Internal adjustments of the stereoprojector. In this operation, the stereocompiler should have a thorough knowledge of the conditions affecting interior orientation, as well as information on the extent to which these conditions have been controlled in the calibration of the projectors and in the printing of the diapositives. With this knowledge he can, if necessary, select and install appropriate distortion-correction cams to eliminate uncorrected radial distortion in the diapositives and adjust the principal distance of the plotter's projectors to agree with either the equivalent focal length of the camera or with the principal distance of the negative or diapositive as appropriate. It is essential that, for those plotters requiring selection and installation of appropriate correction cams (or other correction devices) and adjustment of principal distance, the operator follow exactly the instructions for these operations provided by the manufacturer of the plotting instrument.

If oblique photography is to be used, a further internal adjustment of the projector is required which is canting of the lens in accordance with the Scheimpflug principle. This adjustment is generally made at the same time as the setting of the projector attitudes, and reference is therefore made to it again in 11.1.6.2.2, below.

11.2.12.1.2.2 Relation of aerial photograph characteristics to projector attitudes. The characteristics of the aerial photographs to be projected, that is, whether vertical or twin low-oblique, determine the attitude of the projectors with respect to their supporting frame. Vertical photographs are exposed in a single camera in the aircraft with the optical axis of the camera vertical or near-vertical. Therefore, for projecting vertical photographs, the projectors are mounted with their optical axes in a similar orientation. Twin low-oblique photographs result from simultaneous exposures in twin cameras mounted in the aircraft with the optical axis of each camera tilted 20° from the vertical, as shown in figure 11-27. A 20° tilt from the vertical is most common, although other amounts of tilt are sometimes used. This twin-camera setup may be rotated so that the plane containing the camera's axes is parallel to the line of flight (convergent position) or perpendicular to the

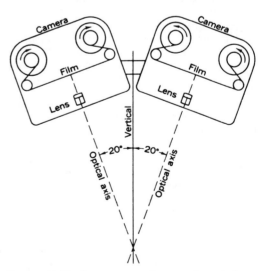

FIGURE 11-27. Arrangement of camera for twin low-oblique photography.

line of flight (transverse position). Therefore, in order to project twin low-oblique photographs in the correct spatial relationship, each projector must be adjusted so that its optical axis is tilted to the same extent as that of the corresponding camera in the mount in the aircraft.

Projection of vertical photographs is generally considered the normal operation on the double-projection, direct-viewing plotters; therefore, no major adjustments are necessary to adapt these plotters for such photography.

Projection of twin low-oblique photographs requires greater tilts relative to the x and y axes than are available in the x and y tilt motions of the projectors of double-projection plotters when in the normal near-vertical position. Accordingly, step-tilt adjustments are provided as part of the projector mounts in those instruments designed to accommodate twin low-oblique photographs. Of the double-projection, direct-viewing plotters, only the ER-55 (Balplex) can be adapted to project high-oblique photographs, both convergent and transverse low-oblique photographs, and vertical photographs. Step-tilt adjustments on the slide-and-yoke assembly allow the ER-55 (Balplex) projector to be tilted up to 60° relative to a horizontal plane through the x axis and 20° either clockwise or counterclockwise about the y axis. On the other hand, the multiplex and early-model Kelsh plotters are designed to project only vertical photographs. The later model Kelsh plotters, the AMS Model 2, the Nistri Photomapper, the Williamson LSP, and Kern PG1 incorporate in their design a simple release which allows the projector to tilt 20 degrees to accommodate convergent low-oblique photographs. However, the 20° transverse tilt required for transverse low-oblique photographs cannot be accommodated.

11.2.12.1.2.3 Centering the diapositive. The method of centering the diapositive in its holder is determined by the design of the diapositive holder and its relation to the projector.

Multiplex. The diapositive in the multiplex is supported upon four bosses (except in the Zeiss and U.S. Army types of projectors wherein the diapositive is supported on a glass plate), one under each corner, which duplicate those used to support the diapositive in the diapositive printer. In the later model of multiplex, these bosses are mounted in a frame (figure 11-28) which surrounds the diapositive and in which the diapositive can be securely fastened. Centering adjustment screws (1) slide this frame in its own plane (perpendicular to the lens axis) to align the principal point of the diapositive with that of the projector. It is important that the diapositive be in firm contact with all four bosses (or with the glass plate in the U.S. Army projector). This contact can be checked by tapping the diapositive lightly over each boss, or if the diapositive is supported by a glass plate, by looking for the interference patterns (Newton's rings), in the

FIGURE 11-28. Multiplex projector camera body.

light reflected from the two glass surfaces in contact. The presence of these rings is evidence of sufficient contact.

The procedure of centering the diapositive in the multiplex depends on which of the three previously described devises is used for indicating the principal point of the projector.

Centering arm. With the lamp house in place and the projector light turned on, the centering arm is released until it reaches its stop. The image of the black dot on the reticle of the centering arm and the image of the white cross or circle marking the principal point of the diapositive are abserved on the tracing table platen or the table top. These marks are brought into concentric alinement by means of the centering screws, which should be manipulated with care so as to make the final motion of the screws against their opposing springs.

Principal-point microscope. Centering the diapositive by means of the principal-point microscope is done with the lamp house removed from the projector. The microscope is placed in position on the projector with one foot (any one of the three) in the cone-shaped foot pad, one foot in the V-shaped pad, and the third foot on the flat pad. The principal-point mark of the diapositive is observed through the microscope against a light-colored background, and is brought into alignment with the cross-hairs of the microscope reticle by means of the centering screws. This orientation should be checked by placing the microscope successively in all three possible orientations.

Stage plate. With the lamp house in place, the projected images of the principal-point marks (one on the stage plate and the other on the diapositive) can be observed and brought into alinement by means of the centering screws.

ER-55 (Balplex). In the ER-55 (Balplex) projector, the diapositive plateholder (figure 11-29) is a frame for placing, holding, and locking the

FIGURE 11-29. ER-55 diapositive, diapositive plateholder, and holddown mask.

diapositive in its correct position in the projector. The orientation of the diapositive holder in the ER-55 (Balplex) projector is determined by the design of the diapositive holder. The positioning of the three balls on the projector aligning ring and the positioning of the corresponding V-slots on the diapositive plateholder are related so that they match in only one position. The correct position can best be recognized by the nature of the construction of the plateholder, since one of its sides is beveled or recessed to clear the lamp bulb. For this reason the diapositive holder is always placed in the projector with this side toward the lamp bulb. This fixed relationship determines the orientation in which the diapositive must be placed in the holder.

ER-55 (Balplex) diapositives are centered in their holders by means of a centering device (figure 11-30), which utilizes four magnifying lenses as an aid to precise centering. With the diapositive in place in the holder, the holder is placed on the centering device with the beveled (or recessed side) of the holder adjacent to the catch. The diapositive is shifted laterally by means of the centering screws, to bring its fiducial marks into coincidence with the fiducial marks on the glass plate of the centering device. The final motion of the centering screws should be made against the opposing springs.

Kelsh and similar plotters. The diapositive holders for the Kelsh plotter and the similar double-projection, direct-viewing plotters which use 9- by 9- inch format dispositives are frames which hold the diapositives in correct relationship to the projector lens. These diapositive holders have supporting surfaces which are precisely machined to a plane, and on which the diapositive rests. Also, the diapositive holders have accurately located fiducial marks, to which the diapositives are adjusted in the centering operation. Each diapositive holder is designed to be used with a particular projector of the plotter and in a fixed relationship thereto; therefore the holders are not interchangeable. The Scheimpflug condition is usually satisfied in these plotters by special diapositive holders with a built-in tilt, rather than by canting the lens. These diapositive holders are to be used only when projecting convergent low-oblique photographs. The fixed relationship of the diapositive holder with its projector governs the orientation in which the diapositive must be placed in its holder. It is well to mark the principal point of the diapositive by scratching a fine cross (with a sharp needlepoint) on the emulsion surface at the intersection of its fiducial lines. This operation is not an essential step in interior orientation; however, the identification of the principal point in the subsequent projected image is a convenience during compilation. Diapositives printed in optical projection-type printers will ordinarily have an identifying mark printed at the location of the principal point; therefore, the operation of scratching a cross at the principal-point location is necessary only for those diapositives printed by the contact process.

The diapositives are placed in the diapositive holders with the emulsion surface either up or down, depending on the photographic printing process by which the diapositives were printed. (If 0.06-inch plates are used, with the emulsion side up, a flat 0.13-inch clear glass cover plate should be placed over each of the diapositives.) With the plateholder frame placed on a light table capable of illuminating the fiducial marks of the diapositive through the fiducial marks of the diapositive holder, the diapositive is centered by shifting it laterally to bring its fiducial marks into coincidence with those of the plateholder. The holddown clamps of the plateholder frame are then tightened to secure

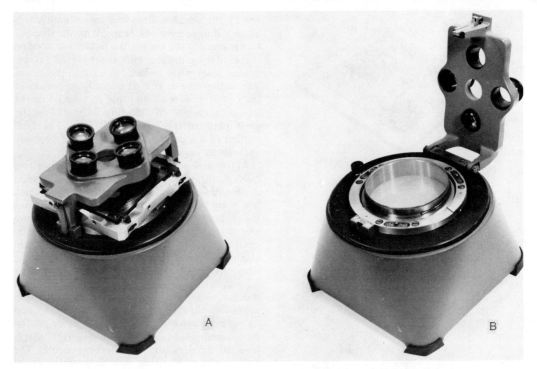

FIGURE 11-30. ER-55 diapositive centering device (*A*), closed for centering operation, (*B*) open.

the position of the glass plate. The plateholder frame is then placed on its corresponding projector cone.

11.2.12.1.3 EFFECTS OF INTERIOR ORIENTATION ERRORS

Figure 11-31 shows the deformations of a model caused by small errors of principal-point location. Principal-distance and principal-point errors have no effect upon a model that is a flat surface, but their effects increase in proportion to the relief in the model. When large tilts are present in addition to relief, these effects are compounded. For this reason, precise calibration is particularly important for projectors in which convergent or transverse low-oblique photographs are to be used, since the oblique orientation of the projectors has the effect of increasing the model deformations resulting from principal-distance and principal-point errors. Also, these deformations can become cumulative in control extensions by stereotriangulation; therefore, projectors used in this operation should be calibrated with special care. In general, only the vertical effects are important; the horizontal displacements are seldom large enough in a single model to be significant. In figure 11-31, the dashed lines indicate the normal shape of a model which, in this illustration, is assumed to be a block with rectangular faces. The solid lines depict the nature of each particular deformation, with magnitudes greatly exaggerated in comparison with those occurring in practice.

Both principal-distance error and principal-point error can result from improper calibration of either the aerial camera, the diapositive printer (if one is used), or the stereoprojector. Principal-point error can also result from a failure to aline each negative properly in the diapositive printer or from a failure on the part of the stereo-compiler to center each diapositive correctly in its projector.

Those errors having their source in the optical or mechanical performance of the aerial camera or diapositive printer are very likely to be duplicated in each projection. The errors of negligent operation are more likely to occur in one projection only.

11.2.12.1.3.1 Principal-distance error. The essential effect of principal-distance error is to alter the vertical scale of the model with respect to its horizontal scale. An error in the principal distance of one projector bears the same proportion to the total principal distance, as the resultant error in the vertical dimension of the edge of the block under the opposite projector bears to the total height of the block. Thus, when both principal distances are equally in error, the vertical dimension of the whole block is uniformly in error; if the principal-distance error is, say, 1 millimetre in a principal distance of 55 millimetres, the vertical error would amount to 1 foot in 55 feet, 10 feet in 550 feet, *etc.*

11.2.12.1.3.2 Principal-point error. When the principal points are displaced in such a manner that they have unequal x components, errors of vertical scale occur. In such an instance, the

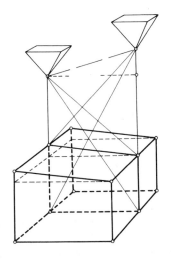

PRINCIPAL DISTANCE-ONE ONLY

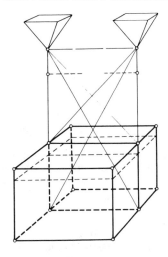

PRINCIPAL DISTANCE-BOTH EQUAL

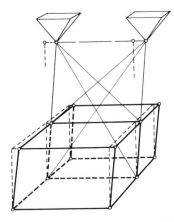

PRINCIPAL POINT-EQUAL IN X

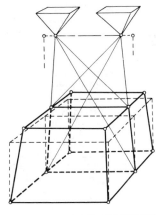

PRINCIPAL POINT-OPPOSITE IN X

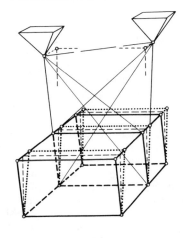

PRINCIPAL POINT-OPPOSITE IN Y

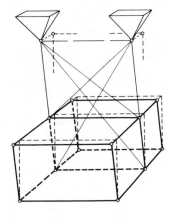

PRINCIPAL POINT EQUAL IN Y

FIGURE 11-31. Effects of interior orientation errors.

vertical scale is uniformly affected over the entire model. When both principal points are displaced equally in any direction, a condition is obtained wherein there is no scale error and no y-parallax error. The block simply leans slightly in the direction of the error. When the principal points are displaced so that they have unequal y components, y parallax will be evident in the model at all elevations above and below the plane of the points used for clearing y parallax in the relative orientation procedure.

11.2.12.2 RELATIVE ORIENTATION PROCEDURES

As stated earlier, relative orientation is one part of the exterior orientation phase; as such, it follows the interior orientation operation and generally precedes absolute orientation, although under certain conditions the adjustments of relative orientation and those of absolute orientation can be accomplished concurrently.

11.2.12.2.1 PURPOSE OF RELATIVE ORIENTATION

The purpose of relative orientation is to reconstruct the same perspective conditions between a pair of photographs that existed when the photographs were taken. As it is not generally known just what these conditions were, the procedure of relative orientation is founded on the fact that each point on the ground was the origin (and hence an intersection) of a pair of corresponding rays, one to each exposure station. Therefore, their reprojected counterparts in the double-projection direct-viewing plotting instruments must also be made to intersect, pair by pair. This is accomplished by a systematic procedure of applying rotational and translational movements to the projectors, at the same time observing the images on the platen of the tracing table until corresponding images are made to coincide over the entire model area. This procedure of relative orientation is familiarly referred to as clearing y parallax.

11.2.12.2.2 PERCEPTION OF Y PARALLAX

When corresponding images from a pair of projectors not in proper relative orientation are observed on the tracing-table platen, components of separation ordinarily are seen in both the x and y directions. (For this discussion of relative orientation, the model air base is considered to be parallel to the x direction unless otherwise stated.) The x component at any one point can be eliminated either by altering the air base (x motion of the projectors) or by changing the height of the platen. The separation then remaining at the point is strictly y parallax (often referred to among photogrammetrists simply as parallax), which can be eliminated only by some motion of the projectors other than x motion.

When relative orientation is first begun, y parallax is observed without the filter spectacles, and appears as a distinct separation between corresponding images. The color filters should be in place, but only as an aid in distinguishing one image from the other. Images affording good contrast and ready identification should be selected, such as cleared fields surrounded by woods, or prominent roads or fence patterns.

As relative orientation progresses and the y parallax diminishes to something less than a millimetre, it becomes preferable to view the parallax stereoscopically, using the filter spectacles. Viewed in this way, y parallax appears no longer as a separation of the two images, for they are fused into one three-dimensional model, but as a y-direction split in the floating mark, the apparent separation between the two dots being equal to the disagreement of the actual images on the platen. The smallest y parallax appears finally as a slight fuzziness of the dot, or perhaps as a slight fringe of colors upon its upper and lower edges. Moving the mark about a small area to obtain different backgrounds, an experienced observer can detect a y parallax of .01 millimetre or even less.

11.2.12.2.3 PROCEDURES FOR RELATIVE ORIENTATION

There are two basic procedures for relative orientation of stereoscopic models, the choice depending somewhat upon limitations that might be placed on certain projector movements. In addition, there are many variations of the basic procedures which may appeal to individual operators, or which have special application if unusual problems of relative orientation are encountered.

A specific procedure for relative orientation is mandatory only in the common orientation of successive projectors, as in the stereotriangulation of a strip of aerial photographs. In this case, succeeding projectors are brought into orientation with those of foregoing models by adjustments of the motions (translational and rotational) of the added projector only.

On the other hand, some of the double-projection instruments which utilize 9- by 9-inch diapositives are designed as single-model instruments. These instruments are not intended for stereotriangulation and, since orientation of a single model requires only the rotational motions about each of the three mutually perpendicular axes (x, y, and z) plus a translational motion along the x axis for adjustment of scale, the construction of such instruments can be simplified by providing only these motions. Therefore, relative orientation in such instruments is restricted to that method which employs the rotational motions of the projectors only (the swing-swing method or some variation thereof).

The time required to perform orientation (both relative and absolute) may often be decreased, if, at the start, the projectors are roughly adjusted to the best estimate of what their ultimate orientations will be. If nothing else is known, it may at least be assumed that the flight line is level, straight, and parallel with the supporting bar, and that each photograph is truly vertical. Accordingly, it is good practice to commence by leveling the supporting frame and each projector, not necessarily measuring these adjustments, but at least estimating them with care. The y slides of those plotters whose projectors have y motions should be set to a common value; similarly the z slides should be set to a common value. For conventional vertical photographs, the projectors should be spaced at an average of approximately 0.6 times the optimum projection distance of the projector, for convergent low-oblique photographs this spacing should be 1.2 times the optimum projection distance, and for super-wide-angle vertical photographs, it should be 1.0 times the optimum projection distance.

11.2.12.2.3.1 Relative orientation of vertical photography. The many alternative procedures for clearing y parallax are but slight variations of one fundamental procedure of relative orientation. The procedures given in this section are two of the standard variations.

The possible variations of procedure can more readily be appreciated if the six projector motions are examined, because the effects of these motions dictate the general procedure of relative orientation. (It should be remembered that some of the single-model instruments have only four motions, since they do not have translational motions along the y or z axes, that is, y motion and z motion.) Figure 11-32 shows how these motions would appear to distort the projected image of a grid. The plane of projection is assumed originally to be parallel to the diapositive grid plate, and the projected grid in this position is represented by dashed lines. The numbers locate the critical positions in a model at which y parallax is observed and cleared.

One-projector method. All the motions are applied to projector II.

(1) Clear y parallax at 2 with y motion.
(2) Clear y parallax at 1 with swing.
(3) Clear y parallax at 4 with z motion.
(4) Clear y parallax at 3 with y tilt. If the amount of tilt is large, it is oftentimes convenient, although not necessary, to restore the model approximately to its original projection distance by a suitable x motion.
(5) Overcorrect approximately 1½ times the y parallax at 6 with x tilt.
(6) Repeat the procedure until no y parallax remains at any of these five positions.
(7) Check position 5. The y parallax here indicates that some of the previous positions have not been completely cleared of y parallax.

If projector I had been used, the requirement obviously would have been to reverse positions 1, 3, and 5 with positions 2, 4, and 6, respectively. Positions 3 and 4 may be reversed with positions 5 and 6, respectively, in any procedure. This is a matter of choice.

In addition to these possible substitutions of positions, the order in which the motions are adjusted may also be varied to some extent. From the different combinations of these factors that are possible, the operator can perhaps choose an alternative procedure that he will prefer to the one given above. For example, an alternative that is preferred by some operators consists of adjusting the z motion in the third step to distribute y parallax equally between positions 4 and 6. The x tilt is next adjusted, overcorrecting the y parallax at either 4 or 6 approximately 4 times. If these adjustments are suitably made, positions 2, 4, and 6 can be cleared of y parallax by a readjustment of the y motion alone. Finally, y tilt is adjusted at position 3.

Swing-swing method. Insofar as relative orientation is concerned, it is evident that identical swings applied to both projectors have the same effect as y motion applied to one alone. Also, identical y tilts applied to both projectors have the same effect as z motion applied to one alone. If these substitutions are made in the foregoing procedure, a method of achieving relative orientation without the use of translational motions is obtained.

(1) Clear y parallax at 1 with swing of projector II.
(2) Clear y parallax at 2 with swing of projector I.
(3) Clear y parallax at 3 with y tilt of projector II. (Follow with x motion if desired. *See* step 4, one-projector method.)
(4) Clear y parallax at 4 with y tilt of projector I. (Follow with x motion if desired.)
(5) Overcorrect (1½ times) the y parallax at either 5 or 6 with x tilt of either projector.
(6) Repeat the procedure until all positions have been cleared of y parallax.

Rationalization method. Parallax removal by rationalization is suggested as an alternative procedure that in most instances is more efficient for accomplishing relative orientation. This method is based on certain of projector relationships resulting from the design of the plotter, or that can be controlled by appropriate projector settings, and on the nature of the variations of tilt and crab that generally occur in aerial photography. This method applies to the orientation of each model of a flight after the first model is oriented. It is based on the fact that an aircraft's transverse tilt (x tilt) can be expected to vary more between successive exposure stations than the angle of crab. In the interior orientation, each diapositive is centered in a holder whose relationship to its projector is fixed; therefore, the angle of swing between successive diapositives remains fixed until disturbed by the procedure of relative orientation. A necessary requirement for this method of relative orientation is that the air base of the model be coincident with the x axes of the two projectors, and that this condition be maintained for each successive model. In this way, y parallax introduced by the rotation of a projector about its y axis is symmetrical in relation to the air base. A further requirement in orienting successive models by this method is that the leading diapositive of each completed model be shifted to the trailing pro-

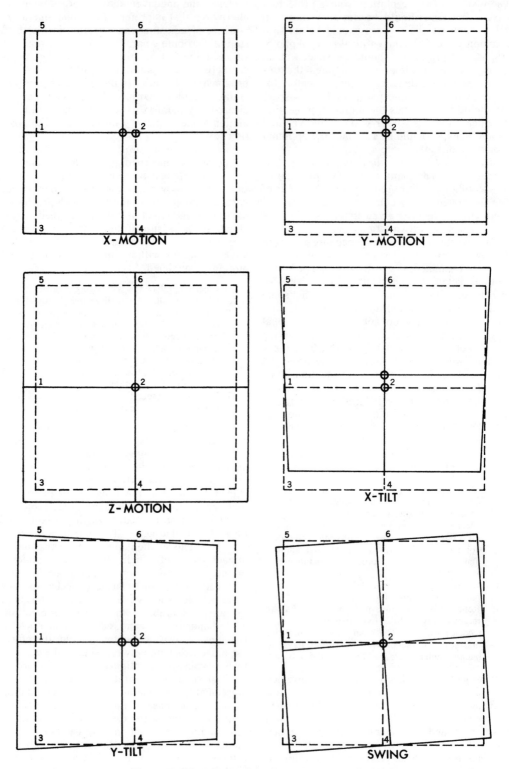

FIGURE 11-32. Effects of the six projector motions.

jector without disturbing the orientation of this projector. Relative orientation then proceeds as follows.

> Observe y parallax at the center of the model area (point 10, figure 11-33); clear approximately half of this parallax with an x tilt adjustment of one projector and the remaining half with the x tilt of the other projector.
>
> Clear y parallax at 1, using the swing motion of projector II.
>
> Clear y parallax at 2, using the swing motion of projector I.
>
> Observe conditions of y parallax at points 4 and 6, and proceed to (a) or (b).
>
> (a) If y parallax conditions at these two model corners indicate symmetry about the air base, clear y parallax at the point having the smaller y parallax with a y tilt adjustment of projector I.
>
> Overcorrect the remaining y parallax in the other corner to an amount equal to 1½ times its value, using the x-tilt motion of projector I.
>
> Clear y parallax introduced at point 2 with the swing motion of projector I.
>
> Observe y parallax conditions at points 4 and 6 and repeat this sequence until points 2, 4, and 6 are y parallax free.
>
> (b) If y parallax conditions at points 4 and 6 are asymmetrical in direction from the air base, overcorrect the y parallax at either corner an amount equal to 1½ times the sum of y parallax values at both corners using an x tilt motion of projector I.
>
> Clear y parallax introduced at 2 with the swing motion of projector I.

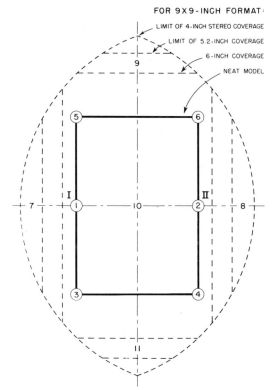

FOR 9X9-INCH FORMAT:
LIMIT OF 4-INCH STEREO COVERAGE
LIMIT OF 5.2-INCH COVERAGE
6-INCH COVERAGE
NEAT MODEL

FIGURE 11-33. Locations in the model at which to observe y parallax.

> Observe y parallax conditions at points 4 and 6, and repeat procedure (a) or (b), depending upon the y parallax conditions existing at these points, until points 2, 4, and 6 are free of y parallax.
>
> Clear y parallax at 1, using the swing motion of projector II.
>
> Clear y parallax at 3, with y tilt adjustment of projector II.
>
> Point 5 serves as a check position and should be free of y parallax if the y parallax at all the other observation points has been properly cleared. The entire procedure is repeated until all points are free of parallax.

11.2.12.2.3.2 Relative orientation of low-oblique photographs. Low-oblique photographs differ from vertical photographs in a number of ways that influence relative orientation procedures. In convergent photography, no single exposure is common to two successive models. Each convergent model is formed from a forward-looking exposure coupled with a backward-looking exposure from the succeeding exposure station, the stereoscopic overlap being 100 percent. For this reason, except in the Twinplex plotter, relative orientation of a series of successive models cannot be accomplished without disturbing the orientation of the foregoing models. On the other hand, in transverse photography single right-looking or left-looking exposures are common to two successive half models (right or left). Therefore, orientation of successive right or left half models (formed from a transverse pair) can be accomplished without disturbing the orientation of the foregoing models. However, orientation of the complete model (both right-looking and left-looking half models as a unit), either as single models or successive models, is possible only in the Twinplex plotter.

The inclination of the projectors' mechanical x tilt and swing axes with respect to the rectilinear axes, on which the procedures of relative orientation are based, increases the difficulty of relative orientation because components of x tilt are introduced as a projector is rotated about its swing axis and because components of swing are introduced as a projector is rotated about its x axis. Since the y axis of a projector remains perpendicular to the air base, no other rotational components are introduced as a projector is rotated about this axis. However, because of the increased air base and the angular relationship of corresponding rays in the convergent models, the y parallax observed as a result of a given degree of y tilt is approximately twice that observed in a vertical model for the same degree of y tilt.

11.2.12.2.3.3 Relative orientation of convergent photographs. A relative orientation procedure based on the swing-swing method is applicable for the orientation of single convergent models for compilation. As the translational motions are not disturbed in this method, the y and z slides of the projectors may be preset to maintain the air base parallel to the supporting bar, which tends to simplify the subsequent orientation of the models. The following procedure may be followed by referring to the numbered positions given in figure 11-34.

Clear y parallax at principal point 1 with swing of projector II.

Clear y parallax at principal point 2 with swing of projector I.

Clear y parallax at 3 with y tilt of projector II (recover approximate scale with x motion if desired).

Clear y parallax at 4 with y tilt of projector I (recover approximate scale with x motion if desired).

Clear 90 percent (approximately) of the parallax at 5 with the x tilt of either projector.

Repeat the procedure until all positions are free of y parallax (as the procedure is repeated, the swing adjustments should be refined by observing the y-parallax conditions along the line from 9 to 10 and the final swing adjustments should leave the model free of parallax along this line).

Finally, adjust the two y tilts.

Removal of y parallax by the rationalization method may be applied equally well to the orientation of convergent models but because of the inclination of the projectors' swing and x tilt axes, the sequence of projector adjustment varies. The procedure is described in detail as follows.

Observe y parallax at the center of the model area (point 8 of figure 11-34), and clear approximately one-half of this parallax with an x tilt adjustment of the right projector. Clear the remaining half with the x tilt of the other projector.

Clear x parallax at 1 with the swing motion of projector II.

Clear y parallax at 2 with the swing motion of projector I.

Observe y parallax at 4 and 6, and proceed with the following steps (1) and (2) as appropriate.

(1) If y parallax conditions at 4 and 6 indicate symmetry about the air base, clear y parallax

at the point having the smaller y parallax with a y tilt adjustment of projector I. Correct 90 percent of the parallax remaining in the other corner with the x tilt adjustment of projector I. Clear y parallax introduced at 2 with the swing motion of projector I. Clear y parallax at 1 with the swing motion of projector II. Observe again the y parallax at 4 and 6 and repeat this sequence until points 1, 2, 4, and 6 are free of y parallax.

(2) If the y parallax at 4 and 6 is asymmetric in direction from the air base, correct 90 percent of the sum with the x tilt adjustment of the left projector. Clear y parallax introduced at 2 with the swing motion of projector I. Clear y parallax at 1 with swing motion of projector II. Observe the y parallax conditions at points 4 and 6, and repeat procedure (1) or (2), dependent upon the y parallax conditions existing at these points, until points 1, 2, 4 and 6 are free of y parallax.

Clear y parallax at 3 with the y tilt adjustment of projector II.

Point 5 should be free of y parallax if the y parallax at the other observation points has been properly cleared. The entire procedure is repeated until all points are free of y parallax.

11.2.12.2.3.4 Relative orientation of transverse photographs.
The procedures for relative orientation of a model of transverse pairs are similar to those for relative orientation of vertical photographs. However, certain modifications are required as a result of the inclination of the y-tilt and swing axes of the projectors in the transverse attitude, and because of the oblique projection of the model. For orientation, models of transverse pairs are most conveniently viewed from a position at the back of the slate table. In this position the operator cannot bump the projectors with his head, yet can still reach them easily.

The numbered circles and the roman numerals in figure 11-35 represent the positions within the model area at which y parallax is observed and adjusted by the swing-swing method, as follows.

Set the y slides of both projectors equal. Also, set the z slides of both projectors equal.

Clear y parallax at principal point 1 with swing projector II.

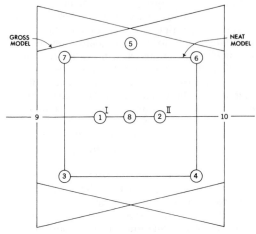

FIGURE 11-34. Locations in the convergent model at which to observe y parallax.

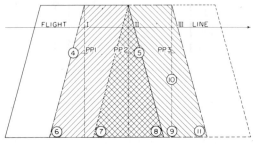

FIGURE 11-35. Locations in the model of transverse pairs at which to observe y parallax, Note: I, II, and III are approximate nadir positions for successive exposures.

Clear y parallax at principal point 2 with swing of projector I.

Clear y parallax at approximate nadir point I with y tilt of projector II.

Clear y parallax at approximate nadir point II with y tilt of projector I.

Overcorrect the y parallax at 7 by an amount approximately equal to the parallax observed, with x tilt of either projector.

Again clear y parallax at the principal points with the swing motions of opposite projectors. Refine the swing adjustments by observation of y parallax at points 4 and 5.

Refine the adjustments until the nadir points, principal points, and point 7 are free of y parallax.

Observe any residual y parallax at 6 and 8 in the extreme flare-out area, and clear with y tilt motions of opposite projectors.

Refine all adjustments as necessary to complete the removal of y parallax.

The y tilt solution of a model of transverse pairs is extremely critical, and further refinement may be required to level the model.

In most instances, only one pair of projectors is available on each bar for compilation. When this is the case, it is recommended that the diapositive be transferred from projector to projector, as necessary, to form successive models, and each succeeding model be oriented by the swing-swing method.

The orientation of successive models of transverse pairs without disturbing the orientation of the projector common to two successive models can be accomplished only by a method in which all orientation adjustments are applied to the added projector (the one-projector method). This procedure has application in stereotriangulation when successive exposures of the right or left wing of a strip of transverse photographs are brought to a common orientation. Another application is in the common orientation of two successive models to plot double-model stereotemplets.

The following procedure assumes that a third projector is to be oriented to the model formed by projectors I and II. All the adjustments are applied to projector III.

Clear y parallax at principal point 3 with y motion.

Clear y parallax at principal point 2 with swing.

Change the y motion setting to clear y parallax at nadir point III.

Clear y parallax at nadir point II with y tilt motion.

Clear y parallax at 9 with z motion.

Observe y parallax at 10 (approximately equidistant from nadir point III and point 9). Increase the observed parallax approximately 4 times with x tilt motion.

Clear y parallax at nadir point III with y motion.

Repeat procedure at points 9, 10 and nadir point III until the points concerned are free of y parallax.

Observe the condition of y parallax at the limit of stereoscopic vision beyond principal point 2, and clear any residual parallax with swing motion.

Clear any residual y parallax at point 7 with y tilt motion.

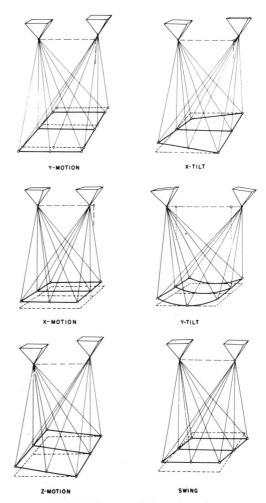

FIGURE 11-36. Effects of relative orientation errors.

Observe y parallax condition at point 11. Any residual parallax at this point indicates the original solution needs to be refined.

The new model may be brought to the same scale as the preceding model by comparing elevation readings of a common point in the vicinity of nadir point II and adjusting the air base of the new model (by means of the x motion of projector III) until the elevation readings of the common point are identical in both models. Additional points in the model are compared, on a line passing through the nadir (II) in the y direction, to evaluate the soundness of the bridge tie.

11.2.12.2.4 EFFECTS OF RELATIVE ORIENTATION ERRORS

The effects on the model datum of a small error in the setting of each motion of the stereoprojector are pictured in figure 11-36 (These effects, as indicated by the differences between the dashed and the solid outlines, are greatly exaggerated for illustrative purposes. Assuming a maximum residual y parallax of 0.1 millimetre, the magnitude of any effect is but a few tenths of a millimetre, excepting, of course, the effect of x motion, which involves no y parallax.) These errors affect elevations in the model more seriously than they do horizontal positions. The horizontal deformation resulting from a residual y parallax of 0.1 or 0.2 millimetre is not significant in a single model.

As indicated in figure 11-36, and assuming a base-height-width ratio of 2:3:3, errors of relative orientation affect the vertical datum of the model as follows.

Error from y motion.—The error has no effect upon the model datum.

Error from x motion.—The effect of this error is to change the scale of the model. The uniform raising or lowering of the model datum with respect to the tabletop, or instrument datum, is offset by reindexing the scale of the tracing table.

Error from z motion.—The principal effect of this error is to tilt the model datum about a y axis. As the datum remains flat, this effect is offset by the subsequent leveling of the model when absolute orientation is performed. The maximum tilt of the neat model produced by 0.1-millimetre y parallax in the neat model is 0.2 millimetre.

Error from x tilt.—The principal effect of this error is to produce a hyperboloidal warping of the model datum. The maximum curvature of the datum occurs along the diagonals of the model. In the x and in the y directions the datum remains uncurved; that is any xz or yz section is a straight line. The maximum warp produced by 0.1 millimetre y parallax in the neat model (deviation of 1 corner above or below a plane passing through the other three corners) is 0.8 millimetre.

Error from y tilt.—The principal effect of this error is to produce a cylindrical warping of the model datum. The maximum curvature of the datum occurs in the x direction, the datum remaining uncurved in the y direction. Convergent y tilt lowers the y direction centerline of the model with respect to the left and right edges, as shown in figure 11-36; divergent y tilt causes this centerline of the model to be raised. The maximum warp produced by 0.1 millimetre y parallax in the neat model (deviation of the center of the model above or below a plane passing through the four corners) is 0.1 millimetre.

Swing error.—The principal effect of this error is to tilt the model datum about an x axis. As the datum remains flat, this effect is offset by the subsequent leveling of the model when absolute orientation is performed. The maximum tilt produced by 0.1 millimetre y parallax in the neat model is 0.44 millimetre.

To summarize, it is seen that of the six errors only two, namely x tilt error and y tilt error, have a significant effect. Each of these warps the model datum in a characteristic manner, by which the error may be identified. It is also to be noted that no relative-orientation error produces any curvature of the model datum in the y direction.

Errors of relative orientation can result not only from failure of the stereocompiler to clear y parallax completely, but they may also result from uncorrected photographic distortion. As the y components of distortion at each of the four corner locations where y parallax is observed usually differ between the two photographs of the overlapping pair, erroneous projector settings will undoubtedly result from clearing y parallax completely at these positions. Fortunately, however, the effect of distortion in present day projector-printer-camera systems is not appreciable in a single model.

As relative orientation is but a means to an end, the ultimate objective of all the orientation processes being absolute orientation, it is permissible to use the flatness of the model datum instead of y parallax as a means of adjusting the x tilt and the y tilt settings. In fact, the ultimate settings of these two motions usually must depend on their effect on model flatness as revealed during the course of leveling. The x tilt motion, especially, creates a warp of the model datum more critical than its y parallax effect.

The improved resolution of the modern double-projection direct-viewing plotting instruments facilitates the perception of small residual y parallaxes. Consequently, errors of relative orientation are minimized, and their effects on the projected model datum are reduced. Conversely, the model datums in these plotters can be warped intentionally only a small amount before objectionable y parallax becomes apparent.

Errors in relative orientation are expected to produce similar deformation in all models, whether vertical, convergent, or transverse pairs. The base-height-width ratio is a determining factor in the degree of model deformation resulting from a given amount of residual y parallax.

The extent of the deformation of the datum of a convergent model for the same residual y parallax varies from the values given for vertical models, since the base-height-width ratio of convergent models is approximately 6:5:5. This base-height relationship aids perception of y parallax resulting from erroneous rotational settings of a projector about the y axis (y tilt) or about the swing axis. The x tilt solution in a convergent model may be weaker than the y tilt solution, since the x tilt and swing motions of the projector have a component in the rectilinear x, y, and z axes of the model. However, since the base-height ratio of the convergent model facilitates the precise reading of elevations, reliable elevations become an index of the angular accuracy of the projector adjustment, and may be used to refine the x tilt solution of the model.

The removal of x tilt parallax in models of transverse pairs is highly critical, since considerable model deformation can be expected to result if a residual x tilt remains in the model. Since the anticipated use of transverse low-oblique photographs is in reconnaissance mapping, it is unlikely that reliable elevations will be available as an index of model flatness. For this reason, extreme care must be used in refining the relative orientation of transverse pairs.

11.2.12.3 ABSOLUTE ORIENTATION PROCEDURES

The stereoscopic model formed by the completion of relative orientation has an undetermined relationship to the horizontal and vertical datums. The adjustments of the relatively oriented model, to make it conform in scale and in horizontal and vertical position with the datums of the map sheet, is called absolute orientation, and generally follows relative orientation.

11.2.12.3.1 PURPOSE OF ABSOLUTE ORIENTATION

As a final step, the model which has been created as a result of interior and relative orientation must be brought to a predetermined and desired scale and to a space position that is correctly related to the established horizontal and vertical datums if measurements or maps are to be readily obtained from it. With the exception of bringing the model to the desired scale, which is accomplished by a change in the air base of the model and therefore can be accomplished by a motion of one projector alone, the achievement of absolute orientation is accomplished by a rotation or translation of the entire model as a unit.

The model projected in a double-projection, direct-viewing plotting instrument is spatial in concept, embracing not only the visible surface of the terrain but also the space above and below this surface. In addition to the terrain surface, therefore, another surface can be conceived as having been reconstructed as part of the model, a surface representing the sea-level (geoid) of nature. This surface is called the model datum. When the interior and relative orientations of the photographs have been correctly recovered, the model datum is, for all practical purposes, a plane or flat surface over the extent of one model. When absolute orientation is also recovered, the model datum becomes parallel to the tabletop, or level. If residual photographic distortions or orientation errors cause this model datum to be deformed or warped, the datum cannot be leveled.

The vertical reference surface is provided by the top of the supporting surface or the plane of the cross-slide system which supports the viewing platen of the instrument. The slight difference between this surface, which is a plane, and the geoid in nature, which conforms to the curvature of the earth, is not significant within the area ordinarily covered by a single vertical model. However, in the case of vertical models of very-high-altitude photographs, models of super-wide-angle photographs, or low-oblique models of medium-altitude photographs, the effect of earth-curvature becomes significant and should be corrected. Empirical corrections to the height (V) of this curvature can be made to the vertical index of the tracing table, based on the expression

$$V = 1.97L^2 \qquad (11.6)$$

where V is in centimetres, and L is the length of the region in kilometres.

The horizontal datum is provided by the control points plotted on the base sheet, or by the pass points resulting from the stereotriangulation phase. The model may be related directly to the horizontal datum of the base sheet, placed orthographically beneath the stereomodel, in which case the tracing table slides directly upon it; or a pantograph may be interposed between the tracing table and the base sheet in order to reduce the model's scale to a smaller-scale on the base sheet. In this case, the base sheet is situated beneath the pantograph and not under the tracing table.

Absolute orientation comprises three operations: bringing the model to the proper size, or "scaling the model"; bringing the model datum parallel to the tabletop, called either "leveling" or "horizontalizing the model"; and positioning the model with references to the horizontal datum.

11.2.12.3.2 SETTING THE PANTOGRAPH

The use of a variable-ratio pantograph in stereocompilation to reduce, or sometimes enlarge, the model's scale to a different compilation scale requires that the pantograph setting be derived before starting the absolute orientation. This derivation requires a thorough knowledge of the relationship of the flight height of the photographs to the model's scale, and knowledge of the variations of model's scales permissible to achieve compatibility either with the available scales of the height-indicating devices of direct-reading tracing tables or with reference tables prepared for the conversion of tracing-table height readings from millimetres to metres or feet at selected scales.

For direct-reading tracing tables, the height-indicating devices, such as the Bausch and Lomb with selective gears, and the USGS wall-projecting height indicator, are adaptable to a choice of discrete common scales rather than to a continuous range of scales. For various scales of 1:6,000 or larger, the selective gear system of the Bausch and Lomb direct-reading tracing table provides height readings based on the ratio of one counter revolution to 10 feet. For a number of model scales smaller than 1:6,000, gear combinations are available to provide height readings based on the ratio of one counter revolution to 100 feet.

Film strips for the wall-projecting height indicator may be printed for any desired scale; however, the usual practice is to provide film strips at scales with denominators in multiples of 500.

The recommended procedure for determining the desired pantograph setting is as follows.

1. Compute the model's scale, using the formula:

$$\text{Model's scale} = \frac{d}{304.8H} \text{ or } \frac{d}{1000H}$$

where d is the optimum projection distance in millimetres, and H is the nominal flight height above ground in feet or metres respectively.
2. Select an appropriate height-indicating scale (determined by a combination of gears for the Bausch and Lomb tracing table or by a film strip for the wall-projecting height indicator) such that the scale of the height reading will be approximately equal to the computed model scale.
3. Substitute this scale for the model scale to recompute the projection distance d. (If the selected height-indicating scale is exactly equal to the computed model scale, this step is not required.)
4. Compare the computed projection distance with the optimum projection distance of the plotting instrument. If the computed projection distance varies by more than 25 millimetres from the optimum value, a new height-reading scale should be selected that will give a closer approximation to the computed model scale.
5. Compute the pantograph setting (P.S.) in millimetres from the relationship

$$\text{P.S.} = \frac{S_m X L}{S_c}$$

where S_m is the denominator of the representative fraction for model scale; S_c is the denominator of the representative fraction for compilation scale; and L is the length of the pantograph arms in millimetres (usually 720 millimetres or 840 millimetres).

11.2.12.3.3 SCALING THE MODEL APPROXIMATELY

The scale of a model is a direct function of its air-base dimension, and therefore can be varied by changing this dimension. To achieve this, one of the projectors is moved along a line representing the air base. (If the direction of the x motion is not along the air base, corrective increments of other motions, predominantly y motion and/or z motion must be included to maintain relative orientation undisturbed.) With convergent models or models of transverse pairs it is desirable to achieve and maintain an approximate scale of the model during the relative orientation procedure. Large adjustments of the air base can disturb the relative orientation solution to the extent that a repetition of the relative orientation procedure may be required.

The procedures for establishing an approximate scale and position are very similar for vertical, convergent, or transverse models projected in any of the double-projection, direct-viewing plotters.

Present practice is for map compilation in most of these plotters to be accomplished by using a pantograph; therefore the compilation area is offset laterally from the spatial model. The scale of the compilation is usually reduced from the scale of the projected model by the reduction ratio established with the pantograph. The position of plotted control points on the map sheet, as related to the position of the respective images in the model, is determined with the drawing pencil of the pantograph, since the tracing point of the pantograph is connected to the tracing table at a point perpendicularly beneath the floating mark. Congruence between the map sheet and the stereoscopic model can be achieved by positioning the floating mark of the tracing table in contact with the stereoscopic image of a plotted control point and shifting the map sheet until it is precisely positioned, as indicated by the pantograph's drawing pencil. If two widely separated control points or pass points are thus brought into agreement, the scale of the model can usually by considered adequate for this phase of the model orientation. The operator will, in general, find it advantageous to relate the map sheet to the model in this way rather than to shift the model into proper relationship with the map sheet by applying a combination of x and y motions to the projectors.

11.2.12.3.4 LEVELING THE MODEL

A vertical control point generally is made available in each of the four corners of the neat stereoscopic model. These points are the basis for relating the model to the vertical datum. Inasmuch as the stereomodel is a small-scale representation of the actual terrain, some means of relating the vertical control to height readings at the stereomodel scale is necessary. With the height-indicating devices of the direct-reading tracing tables, it is possible to read these elevations directly in feet or metres at most of the common scales. The older tracing tables have height indicators graduated in millimetres; for these the given elevations must be converted to their equivalents in millimetres at the model's scale.

Models projected in double-projection, direct-viewing plotters with projectors having all six orientation motions can be leveled by appropriate adjustments of these motions only. However, this procedure is rather laborious, and is required only on the infrequent occasions when the supporting frame cannot be tilted to level the model. Models projected in plotters with projectors having only the three rotational motions (plus an x motion) can be leveled only by tilting the model as a unit through appropriate adjustments of the projector supporting frame. Because leveling a model by tilting the frame does not disturb the relative orientation of the projectors, it is considered the simplest and preferred method. This method is equally applicable to vertical, convergent, or transverse photography. Excessive tilts of the supporting frame should be avoided, however, since such a condition can create instability in the stereomodel.

Inasmuch as the procedures for leveling the model by adjustments to the projectors only are rarely used, they are not given in detail here. The procedures for leveling the model by rotational adjustments of the projector's supporting frame are, however, applicable to all the double-projection, direct-viewing plotters, and can be considered the standard practice.

The first requirement for leveling the model is an index or reference elevation. It can be established by (a) adjusting the tracing-table platen to the height that places its floating mark in apparent contact with the spatial position of any one of the vertical control points, and (b) setting the vertical reading of the height counter to agree with the given elevation at that point. Vertical readings of the three remaining control points indicate, by the amount of their failure to agree with their given elevations, the lack of parallelism between the model datum and the reference plane. An examination of the discrepancies between the correct elevations of the control points and the vertical readings may indicate that the model datum fails to lie in a plane. Frequently, model deformation is due to the existence of minute residuals of y parallax that are beyond visual detection limits.

From an examination of the vertical discrepancies throughout the model, the slope of the model datum in the x and y directions, and the required corrections, can be determined. The leveling procedure is as follows.

Tilt the frame in the y direction as required, using the appropriate leveling screws of the projector supporting frame.

Tilt the frame in the x direction as required, using the leveling screws of the projector supporting frame that allow a tilt in the x direction.

The exact degree of tilt necessary to level the model can be determined by trial and error through a succession of model readings and tilt adjustments, until all vertical discrepancies are removed. A quicker method permits the tilt adjustment to be made with the precision of stereoscopic measurement while the model is under observation. In this method, the vertical corrections are preset on the height indicator of the tracing table, and the projector frame is tilted while observing the model, until the floating mark of the tracing table is in apparent contact with the surface of the model at the desired vertical control position. The amount of correction set on the height indicator of the tracing table will vary with various instruments, and will depend on the relationship of the model dimensions to the span between the leveling screws of the instrument.

An alternative method relates a turn of the leveling screw to the degree of tilt imparted to the model. With this knowledge, the model can be leveled easily by proportionate turns of the leveling screws.

11.2.12.3.5 FINAL SCALING AND POSITIONING OF MODEL

Each model pertinent to the compilation of a project area must be of the same scale, and bear a proper positional relationship to surrounding models. At least four horizontal-control points are required per model to provide common positions and distances in adjacent models.

The final scale is attained when the size of the model satisfies most accurately all given distances within the area it defines. To check the scale, the horizontal distances between the positions of image points in the model are compared with the corresponding dimensions defined by their plotted positions on the base sheet. The distances between diagonally opposite corner points in the model are the longest and are, therefore, the best dimensions for checking scale. Frequently, the available horizontal control points define distances of apparently inconsistent scales; the model should then be adjusted to the size that averages the discrepancy of the scales indicated by the two diagonal distances. As an approximate scale has been attained previously, only a limited amount of x motion is necessary to correct the air-base dimension of the model. The existing y parallax and level solutions should remain undisturbed.

The model, the base sheet, or both, are adjusted in their respective horizontal planes to obtain an orthogonal relationship between the model and the base sheet. The positions of horizontal-control image-points must plot in agreement, or as nearly so as possible, with their pre-established map positions.

The desired positional relationship in the flight direction can be achieved by adjusting the base sheet or by imparting identical x motions to each of the projectors. Positional correction in the y direction can be made by movement of the base sheet, or by adjustment of the position of the anchor point of the pantograph if one is in use. The base sheet must be rotated in the plane defined by the reference table to achieve a proper azimuthal relationship.

When all horizontal-control points within the model area are of the same order of reliability, positional discrepancies should be adjusted to reduce the value of residual errors of position to a minimum. When models are controlled by points differing in the order of accuracy with which their positions were established, the solution which favors points of higher reliability should be accepted. Consideration must be given to the position and alinement of certain features, such as railroads that have tangents of a length sufficient to extend beyond the area of a single model.

11.2.12.3.6 SEQUENCE OF OPERATIONS

It is practical, operationally, to do the relative and absolute orientations concurrently, since the absolute supplies a check on the adequacy of the relative and may indicate necessary refinements. The following routine is suggested as the most efficient, since it may save time by eliminating repetition of some operations.

(1) Achieve an approximate orientation by the swing-swing method.
(2) Identify two horizontal control points (or pass points) and adjust the size of the model so that two positions, well separated in the model, agree with the control. If a large adjustment to scale is necessary, and if the x motion is not parallel to the air base, additional parallax may be introduced. This can be removed by adjustment of the y and/or z motions.
(3) Identify vertical control and level the model against points in any three corners. If initial error is excessive, it may be necessary to use the adjusting motions on the projectors to bring the model datum within the normal range of the leveling screws of the supporting bar. Since relative orientation of the model is approximate, little time is consumed by this readjustment of the projector positions.
(4) When the model has been nearly leveled against three corner points, reset the model scale to the two horizontal points already used.
(5) Refine the y parallax elimination.
(6) Identify and read all elevations given. Improve orientation, if necessary, to bring all elevations into agreement with datum. Use leveling screws on supporting bar to make final leveling adjustments.
(7) Identify all horizontal control. Adjust the model's scale, if necessary, and the position of the manuscript, as required, to bring all points into best agreement.

The above routine is designed to avoid repetition of "fine" parallax removal and to reduce the necessary adjustments of the fine parallax removal to a minimum. The method is equally applicable to vertical, convergent, and transverse models.

11.2.12.3.7 WARPED MODELS

A model datum that is not flat obviously cannot be leveled. It was shown that only two errors of relative orientation affect the flatness of the model datum; y tilt error and x tilt error. Of these, x tilt error is the more difficult to control, and many models require further refinement of the relative x tilt before a final level can be achieved.

A model that cannot be made to agree exactly with the vertical control should not be compiled until all sources of error have been investigated and corrective measures taken. In general, one or more of the following may be found to cause the difficulty.

(1) Misidentification of a control point.
(2) An error in copying elevation figures.

(3) An image distortion of some sort.
(4) An imperfect clearing of y parallax.
(5) An error in the selection of scale-conversion gears or of the film strip for the direct-reading tracing table.
(6) An error in the control itself.

Possibilities (1), (2), and (5) should be eliminated by careful checks of the data and of the tracing table. If the errors in model readings indicate a warp of the model resulting from an error of relative orientation, the operator should attempt a further adjustment of the relative tilt settings of the projectors in an effort to flatten the datum. Dependent on the base height-width ratio of the photographs, it is seldom possible to curve a model datum more than 0.1 or 0.2 millimetre by y tilt, or to warp it more than 0.2 or 0.3 millimetre by x tilt without introducing objectionable y parallax. If greater errors than these exist, or if their pattern is irregular, the control elevations themselves should be suspected of error.

11.2.12.4 STEREOTRIANGULATION

Stereotriangulation affords a means of establishing supplementary horizontal and vertical control points when relatively few ground-survey control points are available. Stereotriangulation may be limited to a single strip of photographs or may encompass a number of adjoining parallel strips. In either case, the stereotriangulated strips are adjusted to a best fit to the existing control, and in the case of multiple strips these are, in turn, adjusted to each other. Stereotriangulation with double-projection, direct-viewing plotting instruments has generally been a bridging operation, in which successive models were brought to a common relative orientation and the entire strip adjusted to a uniform scale that fitted the existing horizontal control. Of the double-projection, direct-viewing plotting instruments, the multiplex and ER-55 (Balplex) are used the most extensively for bridging over long flights. Kelsh and similar plotters are usually intended for simultaneous orientation of not more than two stereomodels, and therefore other techniques are frequently used for extension of control. One effort to make this type of instrument suitable for bridging is represented by the "Sky Bar" (figure 11-37). This instrument (U.S. patent #2,744,442) which incorporates the use of five or more stereoprojectors, is used by the patentee, Robert A. Cummings, Jr., Pittsburgh, Pa. Other means of using plotters of this type for bridging, such as the utilization of levels, are also available.

Digital readout devices are now available that permit the recording of pass-point coordinates from stereomodels in these plotters, with subsequent block adjustment of the pass-point coordinates by an electronic computer. Detailed

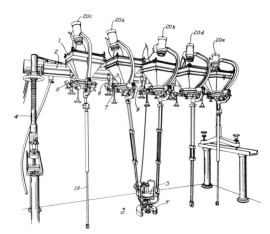

FIGURE 11-37. "Sky-bar" (Robert A. Cummings, Jr.).

procedures for various instrumental methods of aerotriangulation are given in chapter IX.

11.2.12.5 STEREOCOMPILATION

Once absolute orientation has been achieved, the compilation of information from the oriented models is the final procedure in stereoplotting. Although the plot may have any one of several forms, depending on the intended use of the compilation, the procedures for compiling topographic maps may be generally applied to specialized compilations. These procedures are very similar with all of the double-projection, direct-viewing plotters.

The stereocompilation of topographic maps is generally divided into two principal phases: compilation of planimetry, and compilation of contours. Both phases are usually (but not always) completed on a given model before the next model is begun. The compilation may be a direct orthographic tracing beneath the tracing table, or it may be drawn through a mechanical train such as a pantograph, which produces the line drawing at a different scale at a location removed from the position of direct orthographic projection of the model. Within each model, the planimetry is always compiled first, so that the contours can subsequently be shaped to conform with planimetric features. Since it is impossible to describe all possible variations in compilation procedures, some generalized methods are given, from which alternative procedures may be developed for specialized cases.

11.2.12.5.1 GENERAL COMPILATION PROCEDURES

The particular topographic details to be compiled depend on the type of map being prepared, and a compilation procedure which will provide the desired information in the most expeditious manner is always preferred.

The orientation of the model proceeds in the usual way (interior, relative, and absolute) ex-cept that, for compilation of planimetric detail only, an absolute level is not required. (Generally, a level solution within 1.0 millimetre across the model in both directions can be achieved quickly and affords sufficient accuracy.)

11.2.12.5.1.1 Compilation of planimetric features. Planimetric features are delineated by following the feature with the floating mark of the tracing table, adjusting the height of the platen so that the floating mark is always in contact with the model's apparent surface as the plotting proceeds.

An exception is the Gamble plotter in which the tracing table is eliminated and the manuscript, with a pattern of numerous fixed floating marks in the form of light dots projected thereon, becomes the viewing screen. Compilation of planimetric detail is accomplished by tracing each feature with a hand-held pencil while moving the model vertically up or down to keep the projected model surface coincident with the tracing point. (Since the manuscript itself is the viewing screen, nothing can be interposed between it and the projectors; therefore, a pantograph cannot be used with this plotter.) This operation requires careful coordination between model movements and tracing operations. However, since each line or symbol drawn on the manuscript serves as an additional floating mark, the checking of planimetry is easily accomplished by comparing map detail with model detail as the model is raised or lowered, as necessary, to keep the model surface coincident with the map feature.

Frequent reference to the contact prints with the aid of a stereoscope is helpful in interpreting desired information that may not be clearly visible in the stereomodel. In topographic mapping, the desired detail is frequently interpreted by observations in the field in advance of stereocompilation. This information is usually annotated on the contact prints, and must be incorporated in the map by the stereocompiler.

11.2.12.5.1.2 Compilation of contours. Contours are delineated by setting the tracing-table platen at the appropriate contour elevation, and tracing the contour with a lateral motion of the tracing table while the floating mark is maintained in constant visual contact with the apparent terrain surface in the model. Adjacent contours (above or below) are drawn with the tracing-table platen raised or lowered an amount representing the contour interval.

Contour compilation with the Gamble plotter is accomplished by tracing the contour directly on the manuscript with a hand-held pencil in reference to the projected pattern of floating marks, after the model has been adjusted to an appropriate height for the contour being compiled. This procedure of contour compilation facilitates the shaping of contours during compilation, so that little or no adjustment of contours is necessary during the finishing drafting.

11.2.12.5.1.3 Contouring difficult areas. Difficulty in contouring is often experienced in areas of relatively flat terrain, in densely wooded areas, in shadows, and in areas that offer either too much or too little image contrast. Generally, careful study of the contact prints, and a knowledge of topography, will enable the stereocompiler to overcome some of these difficulties.

11.2.12.5.1.4 Drawing the drainage. Regardless of the skill with which the floating mark is used in tracing the contour lines, some smoothing, respacing, or reshaping of the lines will be necessary if they are to express the characteristic features of the terrain with clarity and proper emphasis. As most land forms have developed to some extent through the effects of erosion, the representation of drainage is of great importance in obtaining the proper expression of these forms. Developing the proper expression in stereocompiled contours will always be facilitated if all the drainage, large and small, is lightly drawn prior to the final drafting of the contours.

11.2.12.5.2 USE OF PANTOGRAPH OR ELECTRONIC TABLES

The use of a variable-ratio pantograph or electronic table permits compilation of map manuscripts at scales different than model scales. The scale selected for compilation may be that desired for a particular field use, for map publication, or for other anticipated uses. In normal use, the pantograph is suspended from a king post attached to a support at one side of the plotting instrument (usually the left end-supporting frame) and connected to a stud located perpendicularly beneath the floating mark of the tracing table. In this position the compilation area is to the compiler's left as he sits at the front of the table, and is visible and accessible to the compiler.

Compilation of map detail with a pantograph or electronic table follows the same general procedures and sequence of operations as those applicable to direct tracing of detail by the tracing table. Observation and measurement of topographic features are performed at the scale of the model. Drafting or scribing operations are accomplished at a different and more efficient compilation scale.

11.2.12.5.3 FINISHING THE COMPILATION

Map-worthy detail drawn directly from the model by a tracing table or hand-held pencil, or through a pantograph, lacks the permanence and finished quality desirable either for a completed manuscript or as a medium in the process of reproduction as a published map. Therefore, some final drafting, scribing, or other finishing operation on the manuscript is necessary and desirable.

11.3 Paper-Print Plotters

11.3.1 INTRODUCTION

It is the purpose of this section to describe the principles and use of photogrammetric instruments designed for plotting details from paper prints. For many applications, this type of equipment is entirely adequate and, in fact is often preferred to instruments requiring glass transparencies. For the transfer of detail from photographs to a base manuscript, as for instance in map revision, the simple devices, based on the camera lucida principle or optical projectors, are often suitable. When the job requires height measurement or contouring, as in original map compilation, the relatively simple paper-print stereoscopic plotters can be used effectively in many instances.

No attempt will be made here to indicate specific appications for each instrument, since the choice of equipment must be based on a careful analysis of the particular project and the many interrelated parameters pertaining thereto. For the purpose of this discussion, these instruments will be classified in two broad categories: instruments for planimetric detail sketching, and stereoscopic plotting instruments.

11.3.2 INSTRUMENTS FOR PLANIMETRIC DETAIL SKETCHING

Instruments designed for sketching planimetric detail from photographs are of two basic types: those based on the camera lucida principle and those employing optical projection. These instruments are essentially tracing devices by means of which planimetric detail may be compiled on a control net or information revised on existing maps. They incorporate means for changing scale in the process of compilation and, in some types, provision is made for an approximate rectification for tilt in the photographs.

11.3.2.1 CAMERA LUCIDA TYPE

Instruments based on the camera lucida principle have been used as long as aerial photography has been available. In these instruments, the eye receives two superimposed images, one from the photograph and one from the base manuscript. The operator can adjust the instrument so that selected images on the photograph are made to coincide with their plotted positions on the manuscript in order that other features between the selected points may be traced in their relative positions.

11.3.2.1.1 VERTICAL SKETCHMASTER

The vertical sketchmaster manufactured by Aero Service Corp. (figure 11-38) makes use of a semitransparent mirror at the eyepiece and a first-surface mirror above the photograph. The observer views the manuscript through the

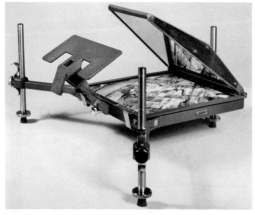

FIGURE 11-38. Vertical sketchmaster (Aero Service Corp.).

FIGURE 11-39. Aero-sketchmaster (Carl Zeiss, Oberkochen).

semitransparent mirror and the photograph by reflection from the semitransparent and first-surface mirrors. The instrument may be raised and lowered on its supports to adjust the photo-image points to correspond with points on the manuscript, and may be tilted by means of the foot screws, permitting the approximate removal of the effects of small tilts. Interchangeable meniscus lenses of various powers are provided for insertion beneath the semitransparent mirror when the instrument is used to transfer detail at a ratio other than 1:1. These lenses serve to bring the apparent manuscript plane into focus with the photo-plane, and to eliminate parallax or the apparent motion of the manuscript with respect to the photograph when the eye is moved.

11.3.2.1.2 AERO-SKETCHMASTER

The Aero-Sketchmaster (figure 11-39), manufactured by Zeiss Aerotopograph, makes use of a double prism with full reflecting and transmitting surfaces at the eyepiece. It is supported by a vertical column and base plate. The column is provided with a rack on which the clamp head can be displaced vertically by means of an adjusting knob and locked in place. The clamp head carries the photo-holder arm and the eyepiece. By adjusting the height of the eyepiece, the apparent size and scale of the map to be plotted can be varied. The photo-holder arm is provided with a ball-and-socket joint to permit tilting and rotation of the photo-holder in any direction. The desired motions are accomplished by a handle mounted at the left-hand side. A locking mechanism is provided. The eyepiece is mounted on a horizontal arbor angled in front and inserted in the clamp head. It is of the adjustable type and permits varying the distance with respect to the photo-holder and hence also the scale of projection.

In this instrument, the observer views an image of the aerial photograph superimposed on the map at a 45° angle of incidence at the eyepiece. A set of 14 interchangeable lenses is provided with the instrument.

The eyepiece may be displaced vertically with respect to the table surface and the photo-holder may be displaced in a direction parallel to the table surface, both movements having a range of from 4.7 to 13.4 inches. This means that the image-map scale ratio can be varied from 1:0.4 to 1:2.8. For approximate setting, the vertical column and the horizontal arm are marked with graduations. The angular field of view of the instrument is about 45°, permitting coverage of an area having a 10-inch diameter when the eyepiece is extended the full 13.4 inches.

11.3.2.1.3 STEREOSKETCH

The Stereosketch of Hilger & Watts, Ltd. (figure 11-40) is an instrument that provides the observer with superimposed views of a base map and a stereoscopic model. The stereoscopic feature allows an appreciation of relief that is advantageous in the identification and interpretation of detail and in certain other applications. The instrument can be used for plotting form lines and for hill shading (where the direction of illumination is assumed to be from the northwest and slopes are shaded accordingly). The Stereosketch accommodates 9- by 9-inch paper prints and can correct tilts up to ±5 degrees. It also provides means for varying the plotting scale.

The Stereosketch consists of a stereoscope mounted on a wooden stand incorporating two photograph tables and a drawing table. The binocular head of the stereoscope has two pentagonal prisms behind the eye lenses, to give a

FIGURE 11-40. Stereosketch (Hilger & Watts, Ltd.).

viewing angle of 35° from the horizontal. The first-surface mirrors beneath the prisms are semi-reflecting so that the base map on the surface of the drawing table is seen at the same time as the photographs. Slideways for the insertion of either a mask or a supplementary lens are located below the semireflector.

The drawing table is raised and lowered between vertical slides by a handwheel; the range of movement is 12 inches, which permits the ratio of photograph to drawing size to be varied from 1:0.45 to 1:1.25. The table can be tilted ±5° in both the direction of flight and laterally, by means of three adjusting screws located under the drawing table. It can also be pulled toward the operator as much as 7 inches to allow the drawing to be touched up and improved during plotting.

11.3.2.1.4 RECTOPLANIGRAPH

The Rectoplanigraph (figure 11-41), manufactured by the Fairchild Camera and Instrument Corp., is also based on the principle of the camera lucida, utilizing a double prism with full reflecting and semitransmitting surfaces. It can be used with photographs taken with cameras having any focal length from 6 inches to 15 inches. The circular easel, which holds a standard 9- by 9-inch print, allows rotational adjustments through 360 degrees. The instrument will accommodate oblique camera angles of 0 to 70 degrees, measured from the vertical. The eyepiece and easel may be raised or lowered by a slow-motion control and the easel distance from the eyepiece may be varied to match the map scale. The instrument also provides a lateral tilt range of ±10 degrees.

FIGURE 11-41. Rectoplanigraph (Fairchild Camera and Instrument Corp.).

11.3.2.2 OPTICAL-PROJECTION INSTRUMENTS

Optical-projection instruments provide projected images of photographic prints or other opaque material superimposed on a map or map manuscript. They are useful in transferring detail from near-vertical photographs or other source material. Five instruments of this type are shown in figures 11-42 through 11-46.

11.3.2.2.1 CAESAR-SALTZMAN REFLECTING PROJECTOR

The photograph or roll map to be traced is placed in this instrument at the front of the inclosed chamber. It can project photographs and sections of opaque maps up to 14 by 14 inches in size. Interchangeable felt-faced holders are provided for roll maps or photographs. Filtered air, delivered by an electric blower into the lamphouse and on the pressure plate, keeps the temperature at a safe operating level. The diagonal mirror for erecting the image is a siliconmonoxide-coated first-surface mirror producing no visible distortion or loss of light.

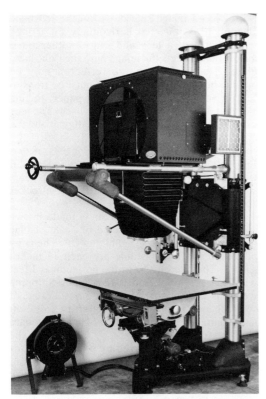

FIGURE 11-42. Caesar-Saltzman vertical reflecting projector (J. G. Saltzman, Inc.).

The lens is focused by either hand and has coarse and fine adjustment, with focusing counterbalanced. A high grade 12-inch, $f/4.5$ projection lens is standard for the instrument, providing a magnification range of from 0.25 times to 5.0 times. Greater reduction can be made by the use of lenses of shorter focal length which can be supplied. For reduction to 0.125, a 7½-inch lens with 13-inch cone is used. A 5½-inch lens with 13-inch cone will produce a reduction to 0.10. The entire lamphouse unit has a horizontal motion of 16 inches, with slow-motion control, to obtain coincidence at points on the tracing table. The vertical motion of the projector is actuated by an electric motor. Limit switches prevent overtravel. Projectors are counter-balanced with concealed weights aimed at smooth, vibrationless operation. The instrument is supplied with a metal table having a linoleum drawing surface and leveling jacks, or with a restitutional table which can be tilted to approximately 10° and rotated in any direction. It can also be obtained without any table. A rheostat or variable transformer controls the voltage on the projection lamps, allowing variation in brightness to suit individual needs. The overall height to the top of the column of the instrument is 9 feet 3 inches. The minimum required height to make a 5-time enlargement is 10 feet 4 inches.

Another Caesar-Saltzman vertical reflecting

projector is shown in figure 11-43. This instrument, with automatic focus, was designed as a portable unit for field use in a van. It can handle material as large as 10 inches square. Mounted in a gimbal arrangement, the projector may be leveled regardless of the position of the floor to which it is fastened. The moving elements have locking devices to prevent damage during transportation. The drawing table is a rigid, stainless-steel-covered panel 30 inches by 40 inches in size. A high grade 7½-inch $f/4.5$ projection lens is employed and automatic focus is maintained throughout the entire range of magnification—from 0.33 diameter to 3.5 diameters. Pointers mounted on a scale divide the range of magnification into increments of 0.25 diameter. The overall height of the instrument is 74 inches. The table top is 29 inches high.

Another Caesar-Saltzman model (not shown) is basically the same as the one just described but has a larger range of magnification—from 0.25 diameter to 4.0 diameters. Its overall height is 80 inches.

11.3.2.2.2 KARGL REFLECTING PROJECTOR (K & E PROD. NO. 720402)

The Kargl reflecting projector, figure 11-44, projects copy from an easel onto a glass tracing table. The easel is open at three sides for placement of large copy or roll material. The copy is illuminated by tungsten-iodine bulbs that remain in proper position at any range. The lens is a 6-inch, $f/4.5$ wide-angle type. Focusing is com-

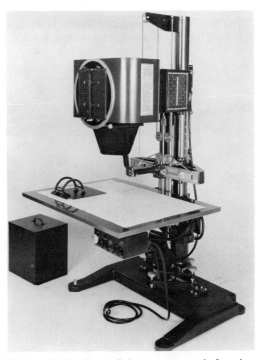

FIGURE 11-43. Caesar-Saltzman automatic-focusing reflecting projector (J. G. Saltzman, Inc.).

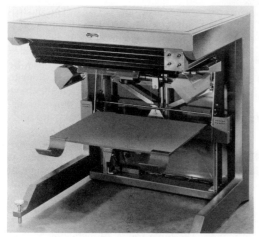

FIGURE 11-44. Reflecting projector (Kargl Instruments, Inc.).

pletely automatic through the entire scale range: reductions to one-fourth size and enlargements up to four times from copy of any type. All mechanisms, lights, and lenses are located underneath the tracing area to allow the operator full movement around the table without interference from shadows. The height of the table is 42 inches with a 27- by 38-inch tracing area. The instrument may be moved through standard door openings. An *x-y* tilt mechanism for slanting of the copy easel up to plus or minus 4° is an optional component of the instrument.

11.3.2.2.3 KAIL REFLECTING PROJECTOR

The Kail reflecting projector, figure 11-45, is a double-reflecting projector built into a table.

Copy placed image side down on the left-hand glass plate is reflected horizontally by a tilted mirror through a lens and onto another tilted mirror that reflects the image onto the right-hand plate. The material being reflected may be opaque but the tracing material must be translucent or semitransparent. Tracing paper, tracing cloth, and frosted plastics may be used. The lens used is a 12-inch, $f/4.5$ coated anastigmat and the mirrors are 12- by 12-inch first-surface chroluminum. The light source consists of two 150-watt reflector flood lamps cooled by an electric fan. Scale control is by worm gear, shaft, and cable, and focus control by sprocket and chain.

The Kail reflecting projector is made in two sizes. Model K-3 has a range of scale change from 3:1 reduction to 1:3 enlargement. The copy plate and the tracing table are each 14 inches square. The overall size of the projector is 3 feet wide, 6 feet long, and 3 feet high. Shipping weight is 250 pounds. Model K-5 has a range of scale change from 4½:1 reduction to 1:4½ enlargement. The copy plate is 14 inches square and the tracing plate is 14 by 36 inches. The right-hand mirror is moved by a control rod for larger scale changes. The focus is remotely controlled through a flexible shaft. The overall size is 3 feet wide, 8 feet long, and 3 feet high. Shipping weight is 310 pounds.

11.3.2.2.4 MODEL 55 MAP-O-GRAPH

The model 55 Map-O-Graph manufactured by Art-O-graph, Inc. (figure 11-46), projects an image of any 11- by 11-inch section of a photograph, map, or document directly onto any map or document placed on the drafting table. The

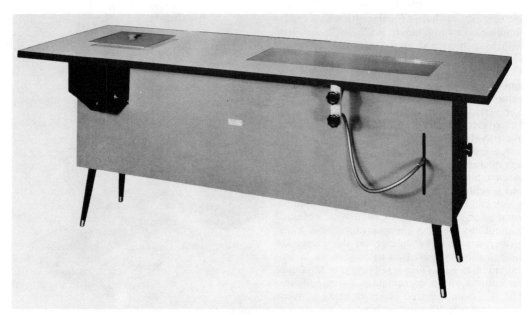

FIGURE 11-45. Reflecting projector (Philip B. Kail, Associates).

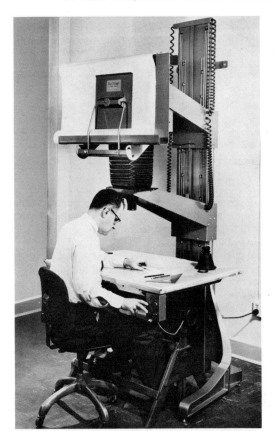

FIGURE 11-46. Model 55 Map-O-Graph (Art-O-Graph, Inc.).

portion projected can be set to the same scale as the original or to any different scale from 5-time reduction to 5-time enlargement. With an auxiliary lens, both can be increased to 7-time. Two-stage lighting provides a choice of either 450 or 750 watts. The lower wattage is used in maximum reduction or up to two-time enlargement. The 750-watt lighting is used for greater enlargement or when room light cannot be adequately controlled. Reflectors and a multi-facet diffusion system spread light over the entire copy area. Cooling is by natural convection. Copy is held flat against a $3/16$-inch-thick glass window by a pressure plate blanketed with sponge rubber. The bellows, of double-wall construction, extend to 42 inches. The projection lens is a coated $f/4.5$ Buhl, Tessar type, 4-element, of 9½-inch focal length. It is equipped with an iris diaphragm which stops down to $f/32$.

The lamp housing and lens carriage are driven through sprocket chains by two 115-volt reversible-gear motors with instant start-stop brakes. For manual operation, each motor can be disengaged with a simple lever. The remote control station has an "up" and "down" switch for each moving carriage with "maximum-up," "maximum-down," and "no-collide" safety

features. The transporting mechanism consists of a formed steel column, with the main housing and lens carriage components running on two tracks using a total of 18 grooved wheels with ball-race bearings. Counterbalancing is provided by metal weights suspended by airplane cable riding over ball-race bearing sheaves. Brackets for wall mounting are standard equipment; however, a stand with either leveling feet or brake casters is available as optional equipment. Dimensions of the wall mounted instrument are: width—18 inches, height—96 inches. Using the stand the dimensions are: width—42 inches, height—97 inches. The overall depth is 39 inches. The maximum operating height when the housing is fully extended to 103½ inches. The estimated net weight is 210 pounds. The stand weighs 36 pounds.

11.3.2.3 STEREOSCOPIC PLOTTING INSTRUMENTS

The instruments described in the following sections provide the capability for making stereoscopic measurements on aerial photographs. They range from simple devices limited in operation to the measurement of parallax differences to more complex instruments that provide photogrammetric mapping capabilities approaching or equaling those of some of the more elaborate stereoscopic instruments described elsewhere in this manual. The main features distinguishing the instruments described herein from other stereoscopic plotting instruments used in mapping operations are that these instruments are designed specifically for use with aerial photographic prints; and they are compact, lightweight, and portable. These instruments may be classified in two basic types: the stereometer type and the orthograpic projection type.

11.3.2.3.1 STEREOMETER TYPE

For certain photogrammetric operations wherein accuracy can be sacrificed in favor of low first cost, simplicity of operation, and mobility of equipment, the simple stereometer-type instrument can be used satisfactorily. Differences in elevation are determined with these devices by measuring the differences in absolute stereoscopic parallax in the stereoscopic pair and converting these to differences in elevation. Master parallax tables are available to aid in this conversion. Model deformations and image displacements due to tilt and relief are inherent and no provision is made for their correction.

11.3.2.3.1.1 Tracing stereometer. The Tracing Stereometer (figure 11-47), manufactured by Carl Zeiss, Oberkochen, is a simple device used in conjunction with a mirror stereoscope to determine elevation differences and to trace form lines and planimetry from vertical photographic steropairs by direct measurement of parallax differences. It consists of a rod shaped microm-

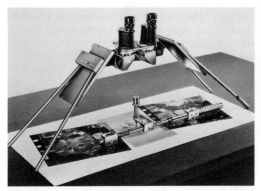

FIGURE 11-48. Stereopret (Carl Zeiss, Oberkochen).

FIGURE 11-47. Tracing Stereometer (Carl Zeiss, Oberkochen).

eter equipped with a glass measuring-mark plate at both ends. The measuring marks are etched circles with yellow dots at the centers. By means of the micrometer screw, the distance between the measuring marks can be varied through a range of 35 millimetres with displacements read directly to 0.05 millimetre and estimated to 0.01 millimetre. By an auxiliary adjustment, the range can be extended by 15 millimetres. One of the glass measuring-mark plates can also be displaced perpendicularly to the long axis of the micrometer to adjust for residual y parallax.

A tracing device, which consists of a lead holder that can be raised or lowered to bring the lead into contact with the plotting paper, is attached to the micrometer rod between the two measuring-mark plates.

When the measuring marks are placed on corresponding images of a stereopair and viewed through the stereoscope, they fuse with the photographs to form a floating mark. When the measuring marks are precisely on corresponding images, the floating mark will be in contact with the apparent ground surface at that point in the stereoscopic model. By making similar measurements of points throughout the model and recording the micrometer readings, differences in elevations can be computed. Form lines are traced by setting the micrometer for specified form-line intervals and moving the stereometer in the stereomodel area in such a way as to keep the floating mark in contact with the ground.

11.3.2.3.1.2 Stereopret. The Stereopret (figure 11-48), also manufactured by Carl Zeiss, Oberkochen, consists of an oblique viewing stereoscope, a stereomicrometer, a photocarriage, an instrument base plate, a plotting surface, and a pantograph. The stereoscope is of the parallel-mirror, reflecting type. It is attached to the stereoscope base plate by two tubular support members set at a 45° angle from horizontal. The stereoscope base plate is fastened by three "Tommy" screws through a supporting block to the aluminum instrument base plate. The optical viewing assembly rests on a bridge

supported by the tubular members and contains a pair of obliquely arranged viewing prisms which permit 1-to-1 scanning of most of the stereomodel without shifting the head. The optical viewing assembly is designed to accept 4-power (6-power on request) monoculars, correction prisms, and green filters for ease in viewing. The interocular distance may be adjusted by means of a small knurled knob located on the bridge. The stereoscope can also be used as a separate photointerpretation instrument.

The stereomicrometer is firmly attached to the stereoscope base plate by means of a support arm with means provided for raising or lowering the stereomicrometer relative to the photographs. The stereomicrometer is equipped with two measuring-mark glass plates and a micrometer adjustment which can vary the measuring-mark separation through a range of 35 millimetres. Horizontal parallaxes may be read directly to 0.05 millimetre on the micrometer.

The photo-carriage consists of a y carriage, an x carriage, and two plateholders. The y carriage is primary and consists of the entire photocarriage which travels on two parallel guides mounted on the instrument base plate. The x carriage travels on parallel guides within the photocarriage. To remove residual y parallaxes, the right-hand photo-holder can be displaced in y direction relative to the left-hand photoholder, accomplished by means of a knurled knob at the left of the photo-carriage which also serves as a means for moving the carriages for scanning. The photo-holders are stainless steel plates which will accommodate paper-print photographs up to 9 by 9 inches in size. Cross lines for centering photographs are engraved on the plates and glass cover plates are provided to hold the photographs in place.

The instrument base plate is made of aluminum and measures 16 by 26 inches. The plotting surface consists of a plate of Wasurit measuring 29 by 31½ inches and is attached to the right-hand edge of the instrument base plate.

If tracing is to be done at the scale of the left-hand photograph, a rigid tracing arm is attached at the right-hand end of the photo-carriage. The pantograph provides a scale range of 0.2 to 3.0 times the photograph scale.

For transport, the stereoscope and its accessories are packed in one carrying case and the other components are packed in a second case. The instrument in its carrying cases weighs approximately 163 pounds.

11.3.2.3.1.3 Stereocomparagraph. The Stereocomparagraph (figure 11-49), manufactured by Fairchild Camera and Instrument Corp., is essentially a mirror stereoscope mounted on a base with a parallax-measuring device or parallax bar, a drawing attachment, and a parallel-motion mechanism. The parallax-measuring device consists of two marks or dots on glass discs which are mounted on a base bar in such a manner that they may be moved laterally. The right-hand mark is attached to a micrometer screw, to allow changing and measuring the spacing between the marks. This micrometer is graduated from 0 to 25 millimetres, and may be read directly to 1/100 millimetre. It may also be clamped at any reading, as desired. The right-hand mark may also be adjusted through a small range in a direction perpendicular to the micrometer screw. The left-hand mark is held in place by a locating pin and may be adjusted in increments of 5 millimetres through a range of 25 millimetres. Nominal spacing between the marks is 150 millimetres.

The drawing attachment is a metal arm, one end of which is attached to the base of the instrument. There is a pencil chuck in the other end of the arm, with provision for raising the pencil lead from the map sheet. The compilation is essentially a tracing of the left-hand photograph of the stereoscopic pair and obviously contains the tilt and relief distortions of that photograph.

The instrument may be attached to any standard drafting machine or parallel-motion device, to provide parallel movement of the parallax bar at all times.

11.3.2.3.1.4 Stereo Line plotter. The Stereo-Line plotter (figure 11-50), developed by the Topographical Survey of Canada, enables a topographer, using elevations in the four corners of the stereomodel as control, to draw approximate contours rapidly on photographs. It does this by using a plane of parallel lines that intersect the model and serve as the measuring system. Instead of removing parallax from the model, the operator deforms the plane of lines to fit the deformed surface of the model.

The left side of the instrument is a plate with parallel scribed lines that can be moved in the x direction. Light passes through these parallel lines to the back of the left photograph. The left side then is the height-measuring part of the instrument and simulates a parallax bar. The observer viewing through a stereoscope sees a three-dimensional model of the terrain and a grid of parallel lines either above, through, or below the level of the terrain. All movements for relative and absolute orientation, or in effect a combination of the two, are carried out by moving only the right projector, which projects a set of parallel lines on the back of the right-hand photograph.

Since the tilted photographs are laid flat, there are height deformations in the stereomodel and it is necessary to deform the measuring surface—that is, the plane of the grid—to fit the deformed surface. By moving the plate under the left picture, the operator can then raise or lower this plane of lines and in effect get an intersection of this plane with the stereomodel. It is then a simple matter for the topographer to trace contours on the photographs that represent the intersections of this plane of lines and the stereomodel.

FIGURE 11-49. Stereocomparagraph (Fairchild Camera and Instrument Corp.).

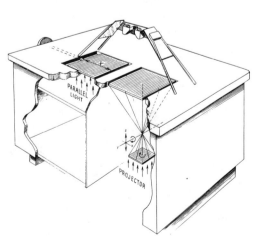

FIGURE 11-50. Stereo Line plotter (Topographical Survey of Canada).

11.3.2.3.2 ORTHOGRAPHIC PLOTTING INSTRUMENTS

At present, there are a number of paperprint plotters on the market, representing a variety of approaches to the solution of the problem of correcting for the image displacement due to tilt of aerial photography and simultaneously providing an orthographic projection at the required scale.

11.3.2.3.2.1 Radial Line Planimetric plotters. The Radial Line plotter of Hilger & Watts (figure 11-51) and the Radial Planimetric plotter of Kail (figure 11-52) use precisely the same principle for plotting planimetric detail from aerial photographs. The similarity in design of the two instruments can be seen in the photographs. Their design is based on the radial-line method, which makes use of the principle that, when aerial photographs are taken with the camera axis nearly vertical, the nadir and the principal point are nearly coincident, so that displacements resulting from differences in ground elevation or photoscale always occur radially (to a close approximation) from the principal point. Their intended use is for map revision and for plotting in connection with road construction, urban area development, and other applications in forestry, geology, archaeology, and teaching.

Each plotter consists, basically, of a precise first-surface mirror stereoscope mounted over two shiftable photo-tables, each containing a pair of radial arms which are connected to a parallel bar. The stereoscope is adjusted so that the line of sight is vertical onto the center of each photo-table. Over each of these photo-tables is an arm (cursor) with a thin line (radial index line) which radiates from and pivots at the center of the table. Each cursor is connected at the outer end to another arm (radial) underneath the photo-table which also pivots at the center of the table. The two radials are attached by means of sliding linkages to the parallel bar which can move in x and y directions, but which is constrained by parallel-motion linkage to remain parallel to the air base.

For operation, the photographs are placed on the photo-table with each principal point over the center of its respective table. A center pin is inserted through the cursor, the photograph, and

FIGURE 11-52. Radial Planimetric plotter (Philip B. Kail, Associates).

the photo-table. Each photograph is then rotated until the line joining the principal point of the photograph with the transferred principal point of the companion photograph coincides with the air base.

The parallel bar with which each radial arm is engaged carries the holder for the plotting pencil. The pencil can be moved in a vertical direction and is held in contact with the paper by the weight of its holder. In operation, this type of instrument is placed on a drawing board containing a mounted map sheet on which the control points are already marked.

When the photographs are viewed through the stereoscope, the radial index lines on the cursor are seen to intersect at a point on the ground. This point is plotted by the pencil. Movement of the pencil bar causes the cursors to swing about their pivots and the point of intersection of the radial index lines may be made to move over the area of overlap of the two photographs. Thus, in the plotting of detail, the bar is moved in such a way that the intersection traverses the linear features or the outlines of area features and the pencil reproduces on the map sheet a plan drawing corrected for height distortions and scale.

The drawing scale is changed by varying the distance between the two sliding linkages on the ends of the parallel bar, each plotter providing a means for coarse and fine adjustments. The scale range of this type of plotter is from approximately one-third to slightly over the full size of the photographs. This scale range may be exceeded for special applications; however, losses in coverage and accuracy may be expected. By using the scale adjustment and moving the plotter on the map sheet, a setting can be obtained in which, when the pencil is in coincidence with a point on the map sheet, the radial lines will intersect the corresponding image points on the photographs.

An accurate intersection cannot be obtained when the detail to be plotted is close to the air base because the radial index lines intersect very acutely. To overcome this difficulty provisions have been made for the tables to be displaced a fixed distance in the y direction on either side of

FIGURE 11-51. Radial Line plotter (Hilger & Watts, Ltd.).

the x axis. The radial arms then swing from false photo-centers and allow satisfactory intersections at these points. The error caused by this practice is negligible for the scales at which the plotters can be used.

11.3.2.3.2.2 K.E.K. stereoscopic plotter. The K.E.K. stereoscopic plotter (figure 11-53), manufactured by Philip B. Kail Associates, is portable and is capable of automatically correcting information from photographs so that planimetric and hypsographic features are drawn to scale in one operation. It consists of the following five basic elements: the stereoscope, two phototables, a floating mark, a drawing arm, and the frame.

The principle of the plotter entails the establishment of a horizontal datum plane with the floating mark. This plane can be raised or lowered by varying the distance between the two dots which make up the floating mark. The scale at which the plotter draws is also dependent on the distance between the two dots. With the datum plane established, the stereoscopic model is then brought into this plane by raising or lowering the photo-tables.

The stereoscope is mounted so that the principal line of sight from each eye falls vertically onto the center of each photo-table when the tables are level. When the tables are tilted to compensate for tilt in the photographs, the principal line of sight from each eye falls approximately onto the plumb point of each photograph. Therefore, when the dots fuse into the floating mark, and that mark is brought into coincidence with the apparent model, the intersection of two radial lines has been obtained and the solution of a radial-line plot is thereby accomplished optically.

The frame of the K.E.K. plotter is made up of three rigid aluminum castings forming the base and two sides. The sides are securely bolted to the base and are tied together with the sheet metal cowling, the stereoscope bars, and the rear tie rod. The base is provided with four knurled, stainless steel, adjustable leveling screws.

The stereoscope is mounted on stainless steel bars and is completely adjustable. The front-surface mirrors are large enough so that the entire stereoscopic model may be viewed. The wing mirror frames are adjustable and may be moved and locked along the stereoscopic bars as well as revolved in two directions. The eyepiece may also be moved and revolved about the stereoscope bars. It is equipped with three sets of interchangeable lenses of variable magnification and the eye mirrors are adjustable for both alignment and eye convergence. Adjustable head rests and aligning sights are also provided.

The drawing arm, composed of four aluminum castings which pivot on each other, is designed to keep the movements of the floating dots in a perfect plane. A precision dial gauge graduated in one-thousandths of an inch is furnished to detect deviations from a plane, and means are provided to make adjustments.

The photo-tables are moved as a unit along their vertical axes by means of a polished aluminum handwheel on the left-hand side of the plotter. Both photo-tables are moved simultaneously, along two vertical shafts, through the connection shaft at the rear of the plotter, and are balanced by adjustable, counterbalance springs on the shaft. The handwheel movement is provided with an adjustable brake for varying the resistance to movement, or it may be locked in position for drawing contours.

A vertical scale and its adjustments are mounted around the left vertical shaft. As the photo-tables are displaced vertically, the vertical scale moves with them past an indicator mark. The indicator mark and vertical scale are viewed through the left eyepiece, with the line of sight passing through the eyepiece lens to the eye mirror, thence through an etched hole in the left wing mirror, through a magnifying lens, and onto the scale. The scale itself is direct-reading in feet and is mounted on a duraluminum drum. It is provided with adjustments for mapping at all

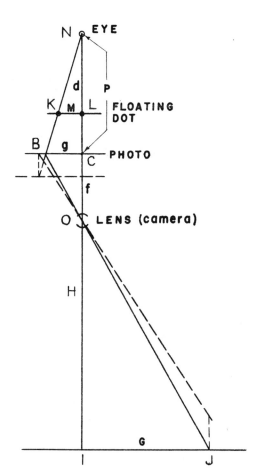

FIGURE 11-53. Schematic diagram of the K.E.K.

scales, for different camera focal lengths, and for the various settings of the pantograph. The settings of the pantograph give a drawing range of from 1.818 enlargement of the photo-scale to 0.55 reduction of the photo-scale. In combination with the settings of the plotter itself, the range is from 1.818 enlargement to 0.363 reduction. For proper illumination, the plotter is equipped with two 18-inch fluorescent tubes. The plotter is also equipped with flange rollers on the sides which enable it to be moved on a steel track mounted on the drafting table. The map sheet can then be moved around under the plotter for orientation.

11.3.2.3.2.3 The Stereotope. The Stereotope, manufactured by Zeiss-Aerotopograph, is a portable stereoplotter (figure 11-59) which utilizes a system of mechanical analogue computers called "rectiputers" for removal of tilt and relief displacements of near-vertical photographs, thereby providing the capability of compiling planimetric and hypsographic detail in one continuous operation. The rectiputers accomplish the rectification by corrective movements in only the xy plane.

The basic elements of the Stereotope are the stereoscope, the photo-carriage, the photo-holders, the parallel-motion linkage, the intermediate base plate, the tracing pad, and the pantograph.

The stereoscopic element of the plotter is the Zeiss oblique-viewing mirror stereoscope which is attached to the intermediate base plate. The optical system of the stereoscope consists of 2-, 4-, or 6-power monoculars, two pairs of rhomboidal mirrors, and one pair of prisms. The oculars are hinge-mounted and may be rotated out of position to allow 1-to-1 viewing of the model. The ocular field of view is 1.25 inches in diameter. All lenses are coated to reduce reflection and the mirrors are coated for abrasion resistance. Interpupillary range is controlled by an adjustment knob.

The stereomodel is scanned by moving the photo-carriage while the optical system remains stationary. The photo-carriage, which can be moved in the xy plane, is constrained to move parallel to the x axis by means of the parallel-motion linkage which is attached to the base-plate. The photo-carriage contains two photo-

holders and three internal rectiputers for correction of vertical and horizontal displacements, called rectiputers I, II, and III.

The measuring marks are engraved on transparent measuring-mark plates which are mounted on a holder called the measuring-mark bridge. The bridge is attached to the stereoscope base plate by a support arm that maintains the measuring marks at a fixed, but adjustable, distance above the photo-holders. The measuring marks are fixed at a distance of 210 millimetres apart.

The left-hand photo-holder is stationary relative to the photo-carriage, whereas the right-hand photo-holder is movable and controlled by an x parallax thimble, a y parallax thimble, and rectiputer I. Elevation differences are measured or set by the x parallax micrometer thimble which moves the right-hand photo-holder in the x direction. The x parallax micrometer thimble is located at the right-hand end of the photo-carriage. It provides a range of 30 millimetres and can be set or read directly to 0.05 millimetre. The y parallax thimble, which moves the right-hand photo-holder to remove residual y parallax, is located at the left-hand end of the photo-carriage. No angular or z motion of the photo-holders is possible or necessary.

Vertical deformation caused by tilt and differences in exposure altitudes is removed by the elevation computer termed rectiputer I. Once the computer is set for a specific solution, it provides a corrective x motion to the right-hand photo-holder as the model is scanned. The elevation computer is adjusted by means of four screws located at the front center of the photo-carriage. Each screw controls one corner of the model and is equipped with a graduated slide for relative settings. The computer adjustment is made by moving the floating mark to a point of known elevation at each corner of the model and raising or lowering the corner by means of the proper rectiputer I screw to bring the floating mark into coincidence with the earth's surface at the point.

Horizontal displacements caused by relief are removed by the perspective computer termed rectiputer II. The necessary input data to the perspective computer are: the photo-base measurement which is applied to the rectiputer II slide scale at the right front of the photo-carriage; the coordinate value under observation relative to the principal points; and the x parallax value (relative elevation) at a point under observation. The photo-base is measured directly on the photograph and the coordinate values are obtained continuously and fed automatically to the computer by movement of the photo-carriage. The x parallax value at any point is automatically provided to the perspective computer when the measuring mark is oriented vertically on the point and it also includes the vertical correction provided by rectiputer I. The

FIGURE 11-54. Stereotope (Carl Zeiss, Oberkochen).

perspective computer transforms the above values into a continuous horizontal adjustment of the pantograph linkage (terminal ends of rectiputers II and III).

Horizontal deformation caused by tilts in the left-hand photograph is corrected by the rectifying computer termed rectiputer III. This computer provides correction by means of a supplementary gear coupled to the perspective computer and is adjusted by means of a double slide scale located at the rear of the photocarriage. One scale rectifies for x tilt and the other for y tilt. Adjustments to be applied to the scales may be computed or done by trial and error and the computed correction is transmitted to the pantograph linkage.

The intermediate base plate, which is made of a light, nonferrous metal, measures 760 by 635 millimetres. It is 6 millimetres thick. Two steel plates, permanently mounted on the intermediate base plate and on which the rollers of the photo-carriage ride, are referred to as instrument base plates. The parallel-motion arm of the photo-carriage and the stereoscope base plate are attached to the intermediate base plate by three "Tommy" screws.

The tracing pad, made of a composition material, measures 740 by 800 by 4.5 millimetres and is attached to and situated at the right of the intermediate base plate. The pantograph provides a magnification ratio of 2.5 to 1 and a reduction of 1 to 5.

The Stereotope, which accommodates a wide range of photographic focal lengths, is also designed to accommodate diapositive plates.

11.3.2.3.2.4 The Stereoflex. The Stereoflex (fig. 11-55), manufactured by Société d'Optique et de Mécanique de Haute Précision, Paris, France, is a portable stereoplotter with which it is possible to form a parallax-free model from near-vertical aerial photographic stereopairs so that planimetric and topographic maps can be drawn to scale in one continuous operation. It consists of the following basic components: the frame, the base-carriages, the tilt cradles, the photograph holders, the altimetric attachment, the pantograph, and the plotting surface.

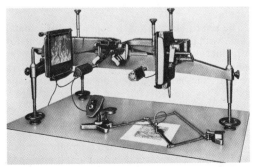

FIGURE 11-55. Stereoflex (Société d'Optique et de Mécanique de Haute Précision).

The Stereoflex will accommodate paper prints up to 9 by 9 inches and diapositive plates and photographic film up to 7.5 by 7.5 inches. The photographic prints are mounted vertically on photograph holders in parallel planes with the image sides facing each other. The photographs are viewed by means of two semi-aluminized mirrors each set between the photographic planes at opposing 45° angles. Reflected light from the aerial photographs travels horizontally to the mirrors, where it is reflected upward to the eye stations. Thus, the operator views the stereopair looking downward to the semi-aluminized mirrors by which means an apparent, horizontal spatial model is formed.

The frame of the Stereoflex consists of a central, cast iron beam supported on the plotting surface by three legs. Each leg is equipped with wheel-screws which are used for absolute orientation of the model. The base-carriages, each equipped with a semi-aluminized mirror, a tilt cradle, a photograph holder, and an illumination source, rest on the frame beam and are engaged to a lead screw. One half of the lead screw has left-hand threads and the other right-hand, thereby forming a bx motion for setting the instrument base. The bx motion knob is located at the right-hand end of the frame beam. The instrument base has a range between 56 and 110 millimetres and can be set to a least reading of 0.1 millimetre.

A forehead-rest and a pair of prisms for oblique viewing are provided on the frame between the two base-carriages. The prisms, however, cannot be used at base settings exceeding 90 millimetres.

Each base-carriage provides y-tilt which is a rotation of the cradle about an axis perpendicular to the base and through the theoretical eye station. The photograph holder and semi-aluminized mirror are attached to the cradle. The y tilt range for each cradle is 17 grads and may be set to a least reading of 0.1 millimetre. Knobs for y tilt motion are located at the top of the base-carriages.

The left-hand base-carriage provides a means for by motion of the photograph holder through a range of 40 millimetres.

Besides having a y tilt, the right-hand base-carriage has an x tilt (rotation about an axis parallel to the base and through the theoretical eye stations) with a range of 20 grads. It can be set to a least reading of 0.1 millimetre.

The Stereoflex is equipped with two sets of interchangeable photograph holders. Each holder of one set consists of a metallic plate with a centering mark. It accommodates paper-print photographs up to 9 by 9 inches. The other set accommodates diapositive plates, film, or paper prints up to 7½ by 7½ inches. Each holder is essentially a light-box with an opal-glass stage. Interchange of the photograph holders changes the eye-station-to-mirror-to-photograph distance, known as the instrument constant, and must be taken into account in computational preparations for plotting. The instrument constant for the 9- by 9-inch photograph holder is 365 millimetres and that for the 7½- by 7½-inch is 300 millimetres. An additional 5 millimetres is added to the instrument constant if the viewing prisms are used.

Each photograph holder is equipped to provide swing motion (rotation of the photograph about an axis through its principal point and through the theoretical eye station). The swing range of each photograph holder is 20 grads.

Vertical measurements of the spatial model are made by means of the altimetric attachment. It consists of a three-legged base, a vertical standard and lead screw, a black circular platen with an illuminated floating mark at its center, and an altimetric scale. The platen and the altimetric scale are attached to a carriage that can be displaced vertically through a range of 70 millimetres. As the operator views the stereopair, he also sees the illuminated floating mark through the semi-aluminized mirrors. Thus, once the model is oriented relatively and absolutely, the operator sees a spatial model and an illuminated floating mark which can be moved vertically and horizontally for measurement purposes. An adapter is located directly beneath the platen which accepts a tracing stylus centered on the floating mark or serves as a connection for the pantograph.

The altimetric scale provides for setting in the vertical scale of the model and for reading elevations in metres directly to 10 metres or in feet directly to 20 feet for scales from 1/13,000 to 1/30,000. It also provides for vertical readings to be made in millimetres or in inches directly to 1/32 inch.

Inasmuch as the viewing or the principal distance of the instrument is a constant, the focal length of the photography will differ from it. This results in an anamorphic model in which the vertical scale differs from the horizontal scale by a ratio factor of P/p where P is the instrument constant and p is the focal length of the photograph. The proper vertical scale must be computed using the ratio factor and the photographic scale and then set into the altimetric scale. The model may then be plotted to the photographic scale by direct use of the tracing stylus or it may be varied through a range of ratios from 1:8 to 3:1 by use of the pantograph.

11.3.2.3.2.5 The Cartographic Stereomicrometer. The Cartographic Stereomicrometer (figure 11-56), manufactured by Galileo-Santoni in Italy, is a portable-type plotter designed to form a parallax-free model from near-vertical aerial photographic stereopairs and to provide the capability for compiling planimetric and topographic maps to prescribed scales. It will accommodate photographs of format sizes up to 23 by 23 centimetres and focal lengths of 100 to 210

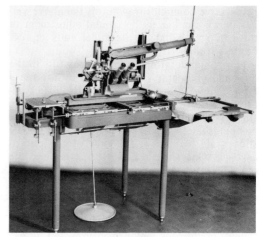

FIGURE 11-56. Cartographic Stereomicrometer (Officene Galileo S.p.A.).

millimetres. The Stereomicrometer utilizes a unique mechanical system for rectifying errors resulting from photographic tilts and relief displacements.

The primary components of the stereomicrometer are: base frame with an attached plotting table; photo-carriages; x, y, and z carriages; optical system; height corrector; and vertical (or polar) pantograph.

The base frame is constructed of cast metal and is supported by three adjustable legs. The plotting table is attached to the right end of the base frame and is supported by a cast metal arm extending from the base frame.

The photo-carriages, on which the photoholders are located, are mounted on the base frame. The left-hand photo-carriage can be translated in the x and y directions and also has provision for rotation of the photo-holder about its vertical axis (swing). The right-hand photocarriage is stationary; however, it does provide swing motion to the photo-holder. The motion knobs for swing and x motion of the left-hand photo-carriage are located at the top front of the base frame, easily accessible to the operator. For ease in removing residual y parallax while working a model, the y motion of the left-hand photo-carriage is accomplished through a foot treadle.

The photo-holders, each consisting of a metal frame with a glass stage plate, are removable for introduction of the photographs. Paper-print photographs are oriented on the photo-holders by four fiducials engraved on each glass stage plate. They are held flat and in place by plexiglass cover plates held firmly to each photoholder by four pressure clips. A light source for illuminating the paper prints is mounted on the x carriage whereas, for viewing glassplate diapositives or film negatives, the instrument is equipped with a light source beneath each photo-carriage.

The y carriage of this plotter is primary and it

travels on tracks mounted on each end of the base frame. Tracks on the *y* carriage support the *x* carriage and both the *z* carriage and the optical system are mounted on the *x* carriage. The *x* and *y* motions of the carriages are executed by means of a heavy metal ball about 2½ inches in diameter. The ball, which rests on a rubber-coated metal plate, is encircled by a metal ring which is attached to an arm from the *x* carriage. As the ball is rolled about, it provides a refined, highly controllable motion to the *x* and *y* carriages.

Three metal height-reading scales, graduated to be used with focal lengths of 4.5, 6, and 8 inches, are provided with the instrument. These special scales, which permit determination of elevations in millimetres, are inserted in a slot on the *z* carriage. A fourth scale, graduated in millimetres for use with any focal length within the range of the instrument, is also provided.

The sector scale consists of a rule, the lower end of which is pivoted and the other end (upper) slides along a slotted arc. The arc is graduated to set the slope angle of the rule in accordance with the photo-base-to-focal-length ratio. The *dx* element of the optical system is connected by a sliding linkage to the sector rule. When the *z* carriage is displaced vertically by means of a knob on the left side of that carriage, the *dx* element is displaced in the *x* direction through its sliding attachment with the sector rule.

The vertical pantograph consists of the main arm with a pivoting sleeve housing at each end, a connecting rod, two sleeves, and two sliding arms. The main arm is attached, at its center and left end, to two upright supports which are attached to the base frame. The main arm is free to rotate about its long axis. The connecting rod is connected to both sleeve housings to form a parallelogram. The left-hand sliding arm is connected to the *z* carriage by a ball socket and the right-hand sliding arm is similarly connected to the pencil carriage at the plotting table. The two pantograph supports are equipped with lead screws through which the main arm can be adjusted vertically at its center and left end. Adjustment at the left end is for setting focal length and that at the center is for setting the plotting scale.

The left pantograph sliding arm, being a space rod adjusted to the focal length of the photographs and attached to the *z* carriage, provides the means for correcting horizontal displacements due to relief. This is accomplished through the action of the sector scale, the *z* carriage, and the pantograph. With the slope angle of the sector rule set for the photo-base-to-focal-length ratio, a change in elevation (*dx*) results in a proportional change in focal length in accordance with the relationship.

$$\frac{bx}{f} = \frac{dx}{df}.$$ (11.7)

This corrective motion is transmitted to the pantograph through its connection with the *z* carriage.

For rectification of vertical errors due to photographic tilts, the Stereomicrometer is equipped with a height corrector which is attached to the left end of the base frame. The height corrector is a deformable, mechanical rectifying surface which in size approximates that of a stereomodel. The surface is made of a set of closely spaced cylindrical rods attached at each end to an articulated frame. By means of five adjustment screws, the surface can be deformed to duplicate model deformation caused by photographic tilts. A slider arm, mechanically linked to the *z* carriage, moves over the deformed surface, transmitting corrections to that carriage and consequently to the *dx* element of the optical system and to the pantograph. The instrument is also equipped with a means for reducing vertical errors caused by extreme relief.

Horizontal deformation caused by photographic tilts can be reduced by ϕ and w tilt-correction screws on the pantograph.

REFERENCES AND BIBLIOGRAPHY

Anonymous (1960) "Multiplex Plotter Procedures" *Topographic Instructions of the United States Geological Survey*. U.S. Geological Survey Center. Reston, Virginia. Chapter 3F4.

Anonymous (1960) "Kelsh Plotter Procedures" *Topographic Instructions of the United States Geological Survey*. U.S. Geological Survey Center. Reston, Virginia. Chapter 3F5.

Anonymous (1961) "ER-55 Plotter Procedures" *Topographic Instructions of the United States Geological Survey*. U.S. Geological Survey Center. Reston, Virginia. Chapter 3F6.

Ahrend, M. and Dreyer, G. (1968) "Der Doppelprojektor DPL, ein Stereokartiergerät der Ordnung IIb" *Bildm. u. Lufbildw.* 1968 (1): 16–22.

Birns, J. P. (1958) "A Comparison of the Kelsh and Balplex Plotters for Large-scale Mapping" *Photogram. Eng. 24 (1):* 74–76.

Coulthart, D. E. (1960) "The Army Map Service M-2 Stereoplotter" *Photogram. Eng. 26 (4):* 657–660.

Eden, J. (1959) "Considerations for the Design of a Projection Plotter" *Photogram. Eng. 25 (5):* 761–763.

Gamble, S. G. (1956) "A Combination Drawing and Tracing Table for Use with Multiplex" *Int. Archiv. f. Photogram.* 12 (4): G1.

Gruner, H. (1962) "History of the Multiplex" *Photogram. Eng. 28 (3):* 480–484.

Helmy, R. (1962) "Untersuchungen an einem Photokartographen Modell VI von Nistri" *Bildm. u. Luftbildw.* 1962 (4): 184–193; 1963 (1): 14–22.

Hopkins, B. Thomas (1957) "The Pantograph and Its Application to Stereocompilation" *Photogram. Eng. 23 (1):* 140–143.

Kelsh, H. T. (1949) "The Kelsh Plotter: Its Place in Photogrammetry" *Photogram. Eng. 15 (3):* 397–405.

Knauf, J. W. (1967) "The Stereo Image Alternator" *Photogram. Eng. 33 (10):* 1113–1116.

Knight, A. S. (1955) "An Analytical Method for the

Calibration of a Variable-ratio Pantograph" *Photogram. Eng. 21 (1):* 124–126.

Nistri, U. (1952) "Le Nouveau Photocartographe Nistri Mod. IV" *Proceedings VII International Congress of Photogrammetry, Washington.*

Odle, J. E. (1954) "A New Plotting Machine for Aerial Photographs" (Williamson Plotter). *Photogram. Record 1 (3):* 50–61.

Tewinkle, G. C. (1953) "Numerical Relative ORIENTATION" *Photogram. Eng. 29 (5):* 841–851 (1962) "Kelsh Plotter Notes" *Photogram. Eng. 28 (3):* 485–491.

Thompson, M. M. and Lewis, J. G. (1964) "Practical Improvements in Stereoplotting Instruments" *Photogram. Eng. 30 (5):* 802–809.

Wagner, N. B. (1954) "The Kelsh Plotter: A Five-year Review" *Photogram. Eng. 20 (4):* 669–671.

Anonymous *Handbook of Instruction for the Fairchild Stereocomparagraph.* Fairchild Camera and Instrument Corporation. Syosset, New York.

Brucklacher, Walter A. (1954) "New developments in the Field of Instrument Design at Zeiss-Aerotopograph from 1949 to 1954" *Photogram. Eng. 20 (4):* 650–000.

Cimerman, V. J. and Tomasegovic, Z. (1970) *Atlas of Photogrammetric Instruments.* viii + 216 pp., illustr. Elsevier. Amsterdam.

Fedoruk, G. D. (1963) "Problems in the Modernization of Processes and the Design of the STD-2 Stereometer" *Geodesy and Aerophot. 1963 (1):* 43–48.

Fischer, W. (1955) "Photogeologic Instruments used by the U.S. Geological Survey" *Photogram. Eng. 21 (1):* 35–000.

Jerie, H. G. (1954) "Stereogeräte 3. Ordnung in der Sowjetunion" *Photogrammetria 1954/55 (4):* 127–133.

Kail, Philip B. (1954). "The Double-reflecting Projector" *Photogram. Eng. 20 (4):* 700 (1949) "Improved K.E.K. Stereoscopic Plotter" *Photogram. Eng. 15 (3):* 405–000 (1949) "The Radial Planimetric Plotter" *Photogram. Eng. 15 (3):* 402–404.

Kusch, M. (1960) "Untersuchung und Vergleich von Stereometergeräten" *Vermessungsteknik 1960 (8):* 193–198.

Moore, R. E. (1964) "The Stereo Line Plotter—A New Third-order Instrument" *The Canadian Surveyor 18 (2):* 140–146.

Ray, Richard G. (1960) *Aerial Photographs in Geologic Interpretation and Mapping* U.S. Geological Survey Professional Paper 373. Superintendent of Documents. Washington, D.C. p. 230.

Schürer, K. and Kilpelä, E. "Über Radialkartiergeräte" *Bildm. und Luftbildw. 1965 (3):* 115–122.

CHAPTER XII

Plotting Machines with Mechanical or Optical Trains

Author-Editor: CARL J. BORN

Contributing Authors: BRANKO MAKAROVIC, EDWARD SPEAKMAN

12.1 General Principles

12.1.1 INTRODUCTION

CHAPTER XIV of the Third Edition of the MANUAL OF PHOTOGRAMMETRY, which was written by Prof. Dr. W. Schermerhorn and Dr. Branko Makarovic, provides a comprehensive coverage of the stereoplotting instruments with mechanical and optical trains. In this chapter on the same subject, much of the information that is applicable has been taken directly from the previous edition. The following significant changes, however, are to be noted: only those instruments which are currently manufactured or which are in popular use have been retained; new instruments which have been developed since the publication of the Third Edition have, of course, been added. For descriptions of the older models of stereoplotting instruments or for those which are obsolete, one should refer to the Third Edition.

Because of the current popularity of orthophotographs, many of the new stereoplotting instruments have attachments designed for their preparation. Only brief descriptions of these attachments have been included in this chapter. Chapter XV deals exclusively with the subject of orthophotography and its preparation, and, therefore, includes detailed descriptions of these auxiliary instruments.

It is desirable to distinguish between the instruments described in chapter XII and those described in chapters XI and XIII. Chapter XI includes two groups:

(a) those in which the bundle of rays of the two successive exposures are reconstructed by means of direct optical projection and from which the images are viewed directly;

(b) those instruments of limited precision which are called third-order instruments or paper print plotters. They are low in price, and give only approximate solutions to the geometrical relationships; however, their precision is quite satisfactory for special applications such as small-scale mapping.

Chapter XIII deals with the so-called analytical plotters; that is, those instruments realizing the relationship between the photograph and terrain coordinate systems not by reconstruction of the perspective bundles but by electronic computers in which the coordinate relationships are mathematically determined.

12.1.2 MAIN ELEMENTS OF ANALOG PLOTTING INSTRUMENTS

In the design of a complete plotting or restitution instrument, the following elements may be distinguished:

(a) a projection device;
(b) means of observation of images or of the model;
(c) devices for measuring coordinates or for plotting the model.

In the following sections we find that these three elements are sometimes represented by special devices which do not fit into a general classification; however, following is a general treatment of each of these subjects.

12.1.2.1 PROJECTION DEVICES

In all the instruments discussed in this chapter, the principle of reconstruction of the bundle of rays, *i.e.*, reproduction of the bundle of rays which form the images during exposure, is applied in one way or another. In figure 12-1, points O' and O'' are two exposure centers formed by the nodal points on the emulsion side of the camera lens. A_1 and A_2 are points in the terrain, of which a_1', a_2' and a_1'', a_2'' are the images in the first (left) and second (right) exposure. If we could bring two exactly similar cameras back into the air with the photographs in the same positions as they had during the exposure, lines through the points a_1' and O', a_1'' and O'' would intersect at A_1. The same applies to all corresponding points on two photographs. The locus of all intersection points forms a perfect geometric model of the terrain, on the same scale and in the same position. By shifting the left camera along the air base to position O_1' without changing the direction of the optical axis, the scale of the model will be reduced in the proportion $O_1' \, O'' : O' \, O'' = b : B$, where b is the reduced air base and B is the original air base.

It is obvious that we can place this pair of cameras anywhere in space. As soon as we bring

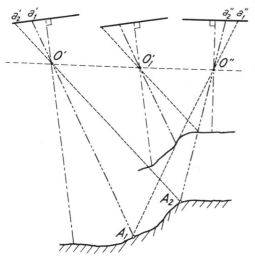

FIGURE 12-1. Scheme of the projection principle.

FIGURE 12-2. Centering device.

them into their correct relative positions, the model can be construed as the locus of the intersection points of corresponding projected lines. This model will have an orientation and a scale which are determined by the position and length of the base b and of the lateral inclination Ω of the model perpendicular to the vertical plane through the air base.

This principle of a restitution instrument can be realized by replacing the two assumed cameras at O' and O'' by:

(a) two projectors which, after introducing the correct inner and relative orientation with respect to the air base and to each other, form a model in space by direct optical projection (all the so-called double-projection instruments of chapter XI are based on this simple principle);

(b) two mechanical or partly optical, partly mechanical devices which artificially imitate the projection system of the two projectors of category (a).

In both cases, the model can be correct and free from deformation only if the bundles of rays produced by the projectors, either direct and optical or indirect and artificial, are identical with those of the cameras within the tolerances of each design.

To fulfill this requirement, the photograph must be centered in the projector and placed at the correct distance perpendicular to the real optical or hypothetical axis of the projector. This centering and introduction of the principal distance is the setting of the elements of inner orientation. The principal distance is normally constant for all photographs of the same mission. The centering of each film or plate, preferably in a special plateholder, is carried out on a centering device (fig. 12-2) using the four fiducial marks.

In the most general case, each projector has

six movements for the relative orientation, three rotations, and three translations. The rotations are around three axes which are normal to each other in their zero positions (see fig. 12-3). One of these is fixed and is called the primary axis. In some instruments, this is parallel to the x-axis (line of flight), while in others it is perpendicular to it. In the former case, the lateral tilt ω is the primary rotation, while in the latter it is the longitudinal tilt ϕ. The secondary axis rotates around the fixed primary axis, and the tertiary axis is almost always the axis of the projector around which the swing κ is imposed. Several instruments, however, do not have the full six freedoms of movement of each projector, but have the necessary five elements for relative orientation distributed over the two projectors. To obtain the correct absolute orientation, some instruments allow additional common inclinations Φ (in X-direction) and Ω (in Y-direction),

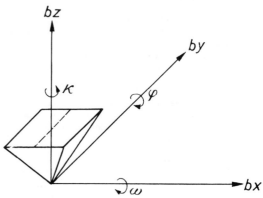

FIGURE 12-3. Orientation movements ($\phi, \omega, \kappa, bx, by, bz$).

while in others the same movements as for relative orientation are used for this purpose. Furthermore, there is always the capability of changing the base length to scale the model. Only in the so-called universal instruments are the six freedoms of movement needed for each projector.

12.1.2.2 MEASURING AND PLOTTING DEVICES

In each instrument, no matter how simple or how complicated it may be, there is a device for plotting and, in some cases, also for measuring coordinates of points in the model. This consists of a measuring mark, which is either a dot, a circle, or even a grid, connected with the system for scanning the model. As is shown in section 12.1.2.3, this mark is always observed stereoscopically and is called the floating mark. The scanning is often possible by freehand movement of the mark itself, using some transmission device which facilitates the movement. In many instruments the scanning is carried out with handwheels for traversing in x and y directions. In all instruments a third movement is necessary to give the floating mark the proper height for a given point in the model.

The result of this consideration is that in all instruments we must distinguish between two sets of movements:

(a) movements for orientation, consisting of three rotations of the projectors and three translations of the end points of the base line, of which the y-component can be replaced by a rotation and shift of the map sheet;
(b) movements for scanning the model, but which never disturb it. The scanning devices depend on the construction of the instrument, *e.g.*, the parallelogram of Zeiss as described in 12.1.4 below.

The study of the problems of orientation has taught us, however, that although we must distinguish between the two kinds of movements, no relative or absolute orientation is possible without the use of the floating mark and the scanning movements.

Different systems are used to measure the coordinates of points in a model:

(a) reading of precision scales with microscopes or loupes;
(b) transfer of the linear movements to rotations of spindles and readings on drums;
(c) automatic recording of the readings by mechanical and electronic means.

Not all instruments have the capability of reading the X- and Y-coordinates of any model point. However, for absolute orientation, there must always be the possibility of measuring and reading the heights of points in the model.

The double-projection instruments of chapter XI have, with a few exceptions, no other scanning device than a means of freehand movement of the mark, and have hardly ever a means of reading or registering X- and Y-coordinates.

12.1.2.3 OBSERVATION SYSTEMS

Almost all three-dimensional restitutions of photo-pairs are based on the stereoscopic observation of the model and of the floating mark, which is connected with a scanning system. There are three different methods of observation:

(a) direct observation of the projections;
(b) observation of the photographs with a telescope through the lens of the restitution camera;
(c) direct observation of the photograph with a binocular microscope.

These observation systems are so important in the design of the plotters that their differences, together with those of the projecting systems of section 12.1.2.1, contribute to the classification of instruments (*see* 12.1.6).

12.1.2.3.1 DIRECT OBSERVATION OF THE PROJECTION

Direct observation is possible only with instruments applying direct optical projection on a screen. This screen, with a floating mark in its center, is movable in the Z-direction (chapter XI). There are different methods in use for observation, including the well-known anaglyph system with red and green filters and spectacles, the polaroid method, the flicker-system with interrupted alternating illumination, and the latest—the Kern-Yzerman SDI-System.

12.1.2.3.2 OBSERVATION WITH TELESCOPES THROUGH THE PROJECTOR LENS

With the system of section 12.1.2.3.1 we can have only a restricted zone of reasonably sharp image. As a consequence, in models of mountainous terrain the image quality of the summits and in the valleys is less than in the plane of best definition, which lies somewhere in between. During the early 1920's when the first large plotting machines were designed, the possibilities for illumination of the diapositives were much more unfavorable to this system than at present. At that time, therefore, another solution was used to overcome this difficulty and to enable observation of a sharp image of all points in the negative for all positions in the model. If a projector lens is used with a focal and principal distance equal to that of the exposing camera, a bundle of parallel rays will pass out of the lens from each point on the plate. By observing a point with a telescope focused at infinity, we obtain a sharp image of this and of all other points of the negative on the focal plane of the telescope objective. A measuring mark is built into the telescope in that plane. If we now arrange the telescope in front of the lens of the restitution camera in such a way that it can swing around the center of the exit pupil of this lens, we can observe every point of the negative in focus. The line of sight of the telescope is always parallel to the parallel bundle coming from a photo-point (fig. 12-4). Using a telescope in front of each of the two restitution cameras, we cannot have a real intersection of the projecting rays, because the path of the rays is interrupted by the telescopes. An "artificial" intersection is obtained by the intersection of two steel rods which are rigidly connected with the telescopes in such a way that their axes coincide with the direction of the line of sight of the telescopes. The space rods represent the directions of the projecting lines in the model space and their intersection determines the geometric position of the corresponding model point.

The left and right telescopic images are presented to the left and right eye respectively. The two measuring marks appear as one floating dot in the stereoscopic image of the terrain. The X- and Y-movement of the mechanical intersection point is transferred to a pencil, which plots the planimetric position of the model points.

This type of instrument, often called the photogoniometer type, was used even before 1920, and is still in use today for the measurement of angles to points in the photograph with a theodolite or goniometer.

12.1.2.3.3 Observation of Directly Projected Images in Transmitted Light

In the instruments with double projection and direct viewing of the image on a screen (chapter XI), we have the difficulty not only of the definition of the image at different distances, which is avoided by the use of the telescope (section 12.1.2.3.2), but also, until the introduction of modern high intensity light sources, of the loss of light in the images reflected on the screen. As a result, another solution was introduced in the early twenties which overcomes the two difficulties simultaneously. In this solution, the images projected by the left and the right projectors are reflected by two separate mirrors and the light guided into the left and right parts of a

stereoscope. The floating mark is formed by dots in the center of the gimbal axes of the rotating mirrors. This system affords direct stereoscope observation of the photographs in transmitted light. The condition, however, is that each point of the photograph can be projected and focused sharply on the measuring mark, independent of its distance from the mark. For this purpose an automatic focusing system is built-in between the projector and the dot. The parallel bundles of light enter this system after they have passed through the objective lens of the projector. Although this automatic focusing system was formerly combined with a screen as in the British Barr & Stroud plotter and the French Gallus-Ferber instrument, it is used at present, in combination with mirrors, only in the stereoplanigraph of Bauersfeld-Zeiss (*see* 12.2).

12.1.2.3.4 Direct Observation of the Photographs

Another system of observation uses a stereoscope with built-in measuring marks set in such a position that the two lines of sight are perpendicular to the plane of the photograph. Relative displacement of stereoscope and photographs is necessary for scanning. One solution of this problem has a fixed stereoscope and movable plateholders (some of the Santoni instruments), while another has the plates fixed during scanning, but with movable lines of sight (Wild); the third solution is the combination (Kern PG 2 & PG 3).

In these instruments we have no optical reproduction of the projecting lines; instead, they are produced artificially by steel rods. These are called the plotters with mechanical projection. In principle, projection lines can also be formed by light rays emerging from a light source in the model point (Kuipersplotter). A combination of the two is also possible (Stereophot Baboz of SOM).

During scanning in instruments with mechanical projection, the upper joint of the space rod (the artificial photo-point) moves in the artificial photo-plane, which must be parallel to the photograph. This artificial plane is set at a distance equal to the principal distance of the exposure camera from the center of the gimbal axis, around which the space rod swings. In the solution now used in all Wild instruments, the artificial photo-point guides the line of sight during scanning, and remains in this line. In the solution used in some Santoni instruments at present, the space rod brings the plate with the artificial photo-plane into such a position that the actual photo-point falls on the fixed line of sight. The point of intersection of the two space rods, or another point bearing a fixed relation to this, again indicates the geometric position of a model point corresponding to the image point at which the floating mark is set in the stereoimage.

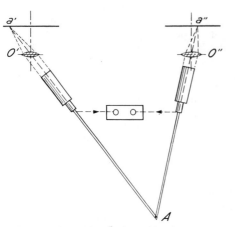

FIGURE 12-4. Observation with telescope.

12.1.2.3.5 PROBLEMS RELATED TO THE
METHOD OF OBSERVATION

The precision of the results, as well as the comfort of the operator, are influenced favorably by undisturbed stereoscopic vision. The requirements for this are:

(a) the scale of the two images offered to the eyes should be equal;
(b) epipolar rays should appear parallel to the eye base;
(c) the eye base and squint must be adapted to proper values for each individual operator; and
(d) there must be correct focusing and equal and good illumination.

12.1.2.3.5.1 Equal scale of the right and left image point. Although it is a well-known fact that slight deviations from these conditions can be overcome by our eyes, *e.g.*, 10 to 15 percent differences in scale, it is desirable to avoid the strain on the operator by fulfilling the required conditions as well as possible. In this respect the instruments with direct optical projection give no difficulty. With convergent photography, and even with arbitrary directions of the optical axes after the orientation of the model, the projections will show no difference in scale and no relative rotation. If the eye base is kept parallel to the air base (as is normal in the instruments described in chapter XI), the observations are made in epipolar planes, which is our natural viewing position.

Considering the other extreme, *i.e.*, direct perpendicular observation (section 12.1.2.3.4), these conditions are fulfilled only by correct vertical photographs. With convergent photographs, where epipolar rays are not parallel (*see* fig. 12-5) and the scales of the left and right images are unequal, Dove prisms are required for image rotation and pancratic ("zoom" or variable focus) systems for scale correction. It is obvious that these auxiliary tools must be guided automatically. Since the scale is deformed differently in different directions from a point, the pancratic system can become a complicated device for large angles of convergence (section 12.4). Instruments with direct observation of the photographs without these special correction devices can be used only for restitution of near-vertical photographs and oblique photographs with almost equal ω values.

In instruments where observation is through telescopes, pancratic systems are therefore necessary for all kinds of photography, even for purely vertical. This is because the object distance for the combination camera-lens/telescope-objective varies according to the direction of observation, whereas the image distance (focal distance of the telescope objective) is constant. In this type of observation, the distance D (fig. 12-6) will generally not be equal for corresponding image points. This causes a difference in scale which must be corrected with

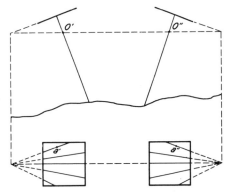

FIGURE 12-5. Epipolar rays with convergent photographs.

the pancratic system. Figure 12-6 shows schematically the formation of the image of a line element d (on the diapositive) on the plane B observed by the eyepieces. Figure 12-7 demonstrates the rotations α and i of the telescope relative to the camera. An easy geometrical derivation gives the following condition:

$$F_1 \cos \alpha_1 \cos i_1 = F_2 \cos \alpha_2 \cos i_2 \quad (12.1)$$

The setting of the pancratic system thus depends on both α and i. The influence of i,

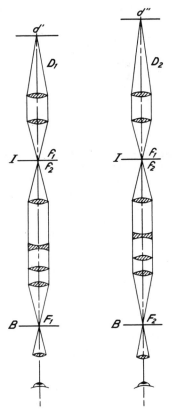

FIGURE 12-6. Pancratic system in instruments with telescopic observation.

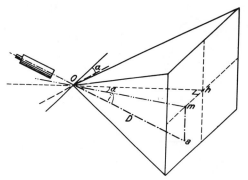

FIGURE 12-7. Rotation α and i of a telescope.

however, is rather small, since only the difference between i_1 and i_2 is important.

The influence of the greatest difference in i is about 5 percent and can be neglected. As a result, the setting of the pancratic system depends only on α. This is valid for all instruments of the photogoniometer type except the Thompson-Watts plotter. For convergent photography, a special arrangement for the same pancratic system would be necessary because of the difference in scale between the image points themselves, except in the isocenters. This arrangement, although possible, is usually not provided with this type of instrument, so that restitution of convergent photographs is in general not possible unless the convergence is so small that the difference in scale in corresponding points of the negatives is never greater than 15 percent.

12.1.2.3.5.2 Epipolar rays parallel to the eye base. Observation by telescopes has other characteristics with respect to image rotation and apparent shape of the observed terrain. In this method, as in that of observation of the transmitted projected images as described in section 12.1.2.3.3, apparent inclination of the image of the terrain increases with the angle between the optical axis and the direction of observation. Although this is not entirely correct, it appears that we are observing the model as if we were looking toward the corners of the floor of a room through a pinhole in the center of the ceiling, so that the floor rotates during scanning around the fixed pinhole. This impression does not exist in instruments with direct perpendicular observation and mechanical projection.

When observing a grid plate in some instruments, we see the lines parallel to the X-axis of the machine as parallel horizontal lines which are crossed at angles by the y-lines of the grid, the size of the angles decreasing toward the edges of the plate. In other instruments we find the y-lines parallel, and the x-lines making different angles with the X-direction.

This difference is due to the position of the primary and secondary axes of rotation of the telescope or the camera during the scanning of the model. The design of the instrument must be

such that the lines in the terrain parallel to the air base are observed parallel to the eye base. This condition can be fulfilled either by means of automatically rotating Dove prisms, or by the primary axis being parallel to the x-axis of the instrument.

12.1.3 METHODS OF LENS-DISTORTION COMPENSATION

Lens distortion is explained in chapter IV, which deals with photogrammetric optics. This distortion is a displacement of image points, and causes deformation of the bundle of rays at the image side in the camera. Our problem in this chapter is how the influence of this deformation can be compensated for, and how a correct undeformed model can be achieved in the restitution instrument. We will take into account only the most important part, the radial displacement of the image points (fig. 12-8) as expressed by:

$$+ \Delta r = r - c \tan \alpha = ha_2 - ha_1 \qquad (12.2).$$

This is the sign which is generally accepted, although in England and Sweden the opposite sign is used.

The methods of compensation are indicated here only in principle. Details of individual instruments are treated in the description of the instruments concerned.

12.1.3.1 THE PORRO-KOPPE SYSTEM

Independently of each other, the Italian, Porro, and much later, the German, Koppe, used the taking camera in measuring directions

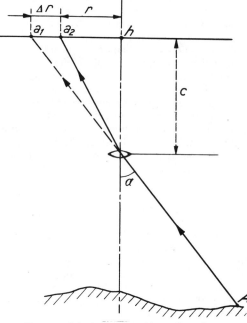

FIGURE 12-8. Influence of lens distortion.

to image points with a telescope (section 12.1.2.3.2). Porro designed his photogoniometer in 1865 in the early days of photography. It is obvious that this was the easiest method of eliminating the influence of the distortion of the camera lens on the observed directions. Since two cameras are needed in the restitution instrument, it is impossible to apply the original Porro-Koppe method using the original taking camera in the photogoniometer. Therefore, in these instruments the Porro-Koppe method now requires lenses with the same characteristics. The main point of this condition is that the distortion curves must be identical to those of the taking camera at the principal distance of that camera. This principle is applied in all instruments with telescopic observation and in those with double projectors and automatic focusing.

With modern wide-angle lenses, because of factors such as the difference in aperture between taking and restitution cameras, it is not practical to apply the pure Porro-Koppe principle with the same lenses in the two cameras. Thus, a kind of pseudo Porro-Koppe system is used for such photographs. The restitution camera is designed with a prescribed distortion and principal distance, and uses an optical system which bears no direct relation to that of the taking camera. In some instruments, a combination of lenses and aspheric compensation plates is used (Stereoplanigraph C8). It is obvious that arbitrary distortions can be introduced with this method, including the zero distortion necessary in the case described in section 12.1.3.2.

12.1.3.2 USE OF DISTORTION-FREE DIAPOSITIVES

Instead of being made by contact printing, diapositives may be made in projection printers. In these printers, compensation plates are introduced in such a way that the distortion of the printer lens plus that of the compensation plate just eliminates the distortion of the negative. As a result, no measures need be taken in the restitution instrument to compensate for distortion of diapositives. These printers are also suitable for standardizing principal distances or for reducing the size of the photographs, which was the original purpose of such printers.

With the use of modern aerial cameras that are nominally distortion-free (maximum radial distortion of less than ten micrometres), the need for and application of these reduction printers has been virtually eliminated. However, ratio printers for changing the size of the photograph are still required for use with instruments such as the Wild A9 autograph and B9 aviograph.

12.1.3.3 USE OF COMPENSATION PLATES

In plotting machines utilizing compensation plates, photographs are observed or projected through a glass plate having a thickness which varies in concentric circles around the principal point, but with one flat surface which is against the photograph (fig. 12-9). In its meridian section, which is represented in this figure, the plate functions like a wedge.

$$\sin \alpha = \eta \sin \alpha '$$
$$\eta \sin (\alpha - \alpha ') = \sin \beta \qquad (12.3)$$
$$\Delta r = d \tan (\alpha - \alpha').$$

The image displacement Δr depends on the refraction coefficient η, the thickness d of the plate at the point considered, and the inclination α of the perpendicular to the glass surface with respect to the perpendicular to the emulsion surface. Since the emulsion is in contact with the flat surface of the plate, the angle β is of importance only in the illumination system and not in the image displacement.

From the optical point of view, the compensation plates are less simple than they appear in this diagram, since in different planes through the vertical line of sight the angle between this line and the tangent to the glass surface in that plane is different. This causes phenomena in the images comparable to those occurring with non-axial bundles of rays in a normal lens. When a compensation plate is used in a projector or in a pseudo Porro-Koppe restitution camera, the direction of the bundle of rays is not constant, but is variable over the whole surface. This gives an additional complication which can result in a slight loss of quality of the images at the edges of the photographs, as in a wide-angle restitution camera where the entrance angle α approaches 50^g.

12.1.3.4 VARIATION OF THE PRINCIPAL DISTANCE

The change in the distortion can also be obtained by a variation of the principal distance over a distance Δc which is a function of the radial distance r from the principal point to the point observed and of the angle α (fig. 12-10).

$$\Delta c = \Delta r \cot \alpha. \qquad (12.4)$$

In some countries it is usual to express the

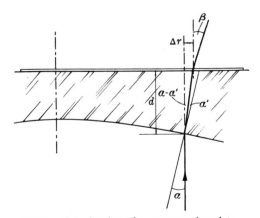

FIGURE 12-9. Section of a compensation plate.

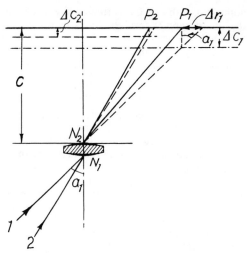

FIGURE 12-10. Compensation of distortion by continuous variation Δc of the principal distance c.

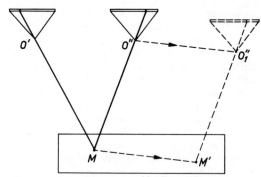

FIGURE 12-11. Displaced projector and split mark.

distortion of the camera lens by means of a diagram for Δc. To compensate for the distortion, a special Δc is required for each angle α. This can be realized in three different ways.

   (a) Movement of the perspective center. This requires displacement of the lens system of the projector along its optical axis (*see* Kelsh plotter, chapter XI).
   (b) Movement of the photograph in optical projection in the direction of the optical axis. This method is not applied in any existing instrument.
   (c) Displacement of the artificial photo point in the direction of the axis of the restitution camera. This system is introduced by Santoni in his mechanical projection instruments (section 12.4).

The required Δc values are obtained by means of a cam shaped in accordance with the distortion curve. The details of the design are given in the descriptions of the instruments. A similar solution is applied to the Poivilliers D.

12.1.4 THE PARALLELOGRAM OF ZEISS

We have seen in section 12.1.2.1 that the base line b which we introduce in a plotter determines the scale of the model. The question arises, "Are we free in the choice of this scale?" The construction of the instrument limits the choice and determines the largest and the smallest scale. Also, without special measures, the rather large projector bodies do not allow a base length smaller than their diameter plus a safety margin.

A solution which is applied to the construction of several instruments (fig. 12-11) is to shift one of the projectors, *e.g.*, O'', to the right over an arbitrary distance $O''O_1''$. As a result the intersection point (model point) M is divided into two parts (split model points). The projecting line

$O''M$ is shifted to $O_1''M'$ and the parallelogram $O''O_1''M'M$ determines the position of M'. The scanning is carried out by moving MM' as a whole, parallel to itself, in X, Y, and Z directions. The pencil connected to some point of the bridge MM' then carries out the same movements in X, Y as it did previously with a single intersection point. The displacement of O'' to O_1'' can have components in X, Y, and Z-directions. With a sufficiently large $O''O_1''$ the base line, and with it the scale of the model, can be chosen arbitrarily small.

In a few instruments (Santoni) this method is used also to avoid the difficulties of the construction of a mechanical intersection point. Then the line MM' is constant and is a few inches in the X-direction only. The X-component of the distance between the perspective centers of the two cameras is bx plus MM'.

One step further along the same line of thought leads to the possibility of keeping the restitution cameras in a fixed position. This is a great advantage with heavy cameras. The parallelogram of Zeiss was invented in 1909 by Pfeiffer and Bauersfeld, not, however, because of considerations of the weight of the projectors, but in order to obtain a fixed position of the perspective center in the von Orel-Zeiss Autograph. This arrangement will be explained not for its use in a plane as the inventors required, but for its application in space as it is used in many restitution machines. The base is introduced by varying the distance between M and M'. In figure 12-12, $O'O''$ is the base line with its three components bx, by, bz. The right projector O'' is placed at O_1''. $O'O_1''$ is constant and parallel to the x-axis of the instrument. Point M_2', determined by the parallelogram $O'O_1''M_2'M$ (parallelogram of Zeiss), now acts as the zero point of the actual base line. In order to shift $O''M$ parallel to itself to $O_1''M_2$, point M_2 is established by setting the three base components bx, by, bz, inwards from the zero point M_2'. Therefore the points M, M_2', M_2, with graduations for the base components, are built on the so-called base bridge. To scan the model, this bridge is shifted in space. It is obvious that with this construction

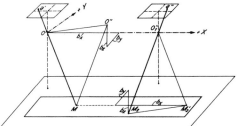

FIGURE 12-12. Parallelogram of Zeiss; base inside.

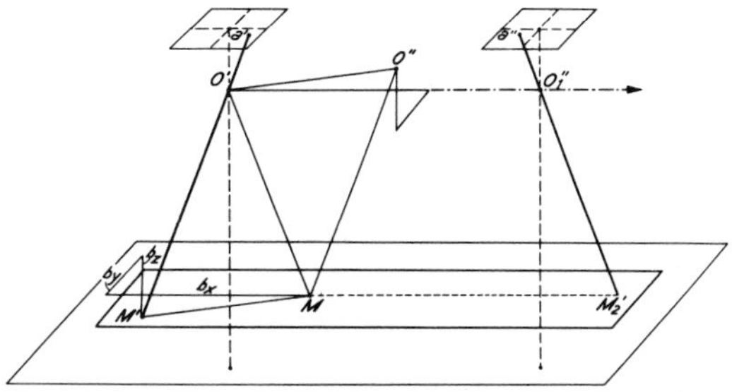

FIGURE 12-13. Parallelogram of Zeiss; base outside.

a varying base length, even base length zero, can be imposed. This construction has a third important advantage if the base components can also be introduced in the opposite direction from M_2'. Figure 12-13 shows this arrangement with plate 2 in the left projector and the base MM' now introduced on the base bridge "outside" from M. The space rods now take the positions $O'M'$ and $O_1''M_2'$, and the overlapping parts of the diapositives are inside. This arrangement allows continuous orientation of all pairs of photographs of a strip, always keeping the second photograph of the previous pair in its restitution camera without changing the elements of inner orientation and of ϕ and ω. Starting with base inside and plate 1 in the left projector, all odd numbers are set in the left projector and the even numbers are put in the right projector. All pairs of which the odd numbers are the lower of the pair are measured with base inside and all others with base outside.

In actual practice, we find that in most of these instruments we have the possibility of imposing base components bx, by, bz at the left as well as the right side of the machine, and generally the base bx is introduced automatically with half its value at the left and half at the right.

A base carriage with sufficient space for use with base inside and outside was first designed by Hugershoff in 1926 for aerial triangulation in his Aerocartograph. Since then, the same arrangement has been constructed in many other precision plotters.

12.1.5 AXIS OF ROTATION FOR ORIENTATION AND SCANNING A MODEL

12.1.5.1 GENERAL

In the previous sections, attention was given in several places to the movements necessary for orientation and those for scanning of the model. According to the method of projection, we must deal with the rotation of the projectors alone, or of projectors and space rods. Only instruments with direct optical projection have axes of rotation for the orientation of the projecting camera and not for scanning. In almost all systems, primary, secondary, and tertiary axes must be dis-

tinguished. The choice of axes has differing influences, as was mentioned in section 12.1.2.3.5.

In the next section we deal with the different arrangements of the axial systems in instruments using the Porro-Koppe principle, and their consequences in observation.

12.1.5.2 DISTRIBUTION OF ROTATIONS IN INSTRUMENTS USING THE PORRO METHOD WITH PHOTOGONIOMETERS

For the sake of simplicity, considering the optical elements for observation as telescopes, we distinguish three cases:

(a) The camera is fixed and the telescope is movable. This was the case in Porro's photogoniometer of 1865.
(b) The telescope is fixed and the camera is movable.
(c) The movements are shared between the camera and the telescope. In all three cases, however, the movements considered here are only those necessary for scanning the model. With a so-called fixed camera, there is also always a means for small rotations for orientation and even for imposing rather large angles of ϕ or ω on the cameras. After this setting, the measurement at points of the photograph, that is, the scanning can start. These movements are considered in this section. We must realize that we deal not only with the movements of telescopes and cameras, but, in the case of instruments of the photogoniometer type, also with the rotations of the space rods which represent the projecting ray in the image space in the camera. Only in case (a) are the movements of telescope and rod identical, because of the direct connection.

In figures 12-14 and 12-15, the horizontal axis through the perspective center is considered as the x-axis. Figure 12-14 shows case (a) as used in one of its few applications (Nistri Photostereograph).

Figure 12-15 shows case (c) schematically. In this design, both axes are fixed, and consequently there is no secondary axis (except the rotation of the camera for swing around the optical axis, which is used only for orientation). This system is accepted in the Poivilliers Stereotopograph and in the Thompson-Watts plotter. It is obvious that with this solution no direct fixed connection between a space rod and

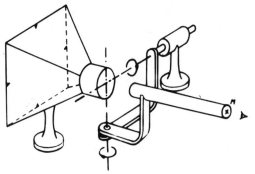

FIGURE 12-14. Telescope rotation with primary axis parallel to the x axis.

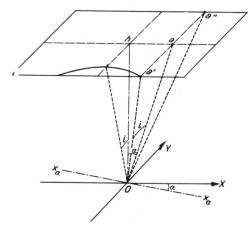

FIGURE 12-16. Influence of the position of the primary axis on the apparent images of a grid.

the telescope is possible as it is in case (a). The design necessary to overcome this difficulty depends on whether the camera rotates around a horizontal axis as in figure 12-15, or around a fixed vertical axis.

12.1.5.3 INFLUENCE OF THE POSITION OF THE SCANNING AXIS

Instruments applying the Porro system show a different apparent shape of a grid with different positions of the primary axis. We can demonstrate this in figure 12-16, which represents case (a) with the fixed camera. After having rotated the telescope and secondary axis through an angle α, around the primary y-axis to $x\alpha\,x\alpha$, if we rotate the telescope around $x\alpha\,x\alpha$, it will describe a plane which intersects the plane of the photograph in the line $a'a''$, which is parallel to the y-axis. All y-lines of a grid will be observed similarly, parallel to Y. Pointing now at a', which assumes rotation of the telescope through the angle around $x\alpha\,x\alpha$, the telescope is moved back to the main vertical plane. The line of sight describes a cone and the mark follows a hyperbola as the intersection of this cone with the plane of the image. The apparent x-directions at each point are the tangents to this hyperbola. Consequently, the x-lines of the grid are not seen perpendicular to the y-lines.

Similarly, we can prove that in a design with the primary axis parallel to the x-axis, those horizontal lines parallel to X (and to the flight line) will remain parallel in the observed images.

This difference affects the design of the instruments, since good stereoscopic fusion can be obtained only if the x-directions are observed parallel to the eye base in both photographs. To obtain this, a choice must be made between two solutions:

(a) construction of the instrument with the primary axis parallel to X (Zeiss Stereoplanigraph, Nistri Photostereograph $\beta2$, Thompson-Watts plotter); and

(b) use of Amici-Dove prisms which are rotated automatically in such a way that the x-lines remain parallel to the eye base (Poivilliers Stereotopograph type B).

12.1.6 CLASSIFICATION OF INSTRUMENTS

The old classification of instruments as first, second and third-order instruments is unacceptable and obsolete, since it connotes a diminishing degree of accuracy associated with each lower order; however, instruments which lacked universality were usually relegated to second-order, although they were equally accurate for the construction of a single stereo photogrammetric model. Makarovic avoided this mislabelling by using the terms universal plotters, precision plotters, topographic plotters, and third-order plotters. The term universal referred to plotters with such capabilities as base-in base-out (section 12.1.4) for the stereotriangulation of continuous strips, and the accomodation of oblique and convergent photography. As these capabilities have decreased in importance, so has the value of such classification, which

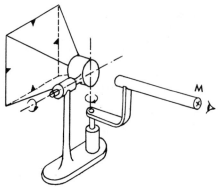

FIGURE 12-15. Movements shared between telescope and camera. Rotation of telescope around a fixed vertical axis.

gave special consideration to universality of plotters.

The proposal of Szongolies in 1968 at the Eleventh Congress of the International Society of Photogrammetry is perhaps a much more valid and meaningful classification system. It is based solely on the horizontal and vertical accuracy of the stereoscopic model formed from the projection of a pair of grid plates. Under this classification system, instruments fall into one of three categories.

(a) Precision plotters—mean horizontal error of $m_x = m_y \leqslant |\pm 15 \ \mu m|$ referred to the image plane and a mean vertical error of $m_h \leqslant |\pm 1/10,000|$ of the average projection distance (PD).
(b) Topographic plotters—mean horizontal error of $m_x = m_y \leqslant |\pm 50 \ \mu m|$ and a mean vertical error of $|\pm 1/10,000$ of $PD| \leqslant m_h \leqslant |\pm 2.5/10,000$ of $PD|$.
(c) Simple plotters—mean horizontal error of $m_x = m_y \leqslant |\pm 200 \ \mu m|$ and a mean vertical error of $m_h \leqslant |\pm 2/1,000$ of $PD|$.

On the other hand, such a classification system based only on accuracy is unsuitable for this text, and so a classification system based on design characteristics is used instead.

(a) Instruments achieving only an approximate realization of the geometrical relations.
(b) Double-optical-projection, direct-viewing instruments.
(c) Analytical plotters in which the relationships of coordinates of photo- to spatial-model are realized mathematically.
(d) Instruments with observation (through microscopes) of automatically focused, directly projected images.
(e) Instruments with partly optical and partly mechanical projection and observation through telescopes (photogoniometer type).
(f) Instruments with mechanical projection.
(g) Instruments with special solutions.

This chapter is concerned with the last four of these seven categories of instruments.

12.2 Instruments with Observation through Microscopes of Directly Projected Images

12.2.1 GENERAL

Automatic focusing, as discussed in section 12.1.2.3.3, was invented by Bauersfeld of Zeiss in 1921. It was one of the essential elements in his design of the Stereoplanigraph, for which Zeiss obtained a German patent in July, 1921. It was introduced to overcome the difficulties of the variable finite projection distance.

Sander mentions that the purpose of the auxiliary system of lenses is to present the mark to the diapositive in the projector as being located at infinity, and that for this purpose this lens system can be set at any place in the line between projector and mark. We will see that Bauersfeld chose a special position close to the projector lens as being most favorable.

12.2.2 PRINCIPLE OF AUXILIARY TELELENS SYSTEM AS USED IN THE STEREOPLANIGRAPH

A telelens system must have a variable focal distance. Therefore, it must consist of at least two lenses with a variable mutual distance e (fig. 12-17). This can be understood from the relation:

$$f_s = \frac{f_1 f_2}{f_1 + f_2 - e}. \qquad (12.5)$$

In the formula, f_1 and f_2 are the focal distances of the two lenses and f_s that of the system. This equation can be derived from the figure. A bundle of rays enters the positive lens, the angle of inclination at its focal point c is

$$\delta_1 = y_1/f_1.$$

The ray intersects the second lens at

$$y_2 = y_1 - e\delta_1.$$

A negative second lens gives, in general, a diverging effect

$$\delta_2 = -y_2/f_2.$$

The ray intersects the optical axis in the focal point F_s of the system at an angle $\alpha = \delta_1 - \delta_2$. Substituting the previous values for δ_1, δ_2, and y_2 into this equation for α, we get

$$\alpha = y_1 \frac{f_2 + f_1 - e}{f_1 f_2}. \qquad (12.6)$$

The focal distance

$$f_s = \frac{y_1}{\alpha} = \frac{f_1 f_2}{f_2 + f_1 - e}. \qquad (12.7)$$

Bauersfeld made $f_2 = -f_1$, giving

$$f_s = \frac{f^2}{e}. \qquad (12.8)$$

The auxiliary system acts as a positive lens with a variable focal length when e varies.

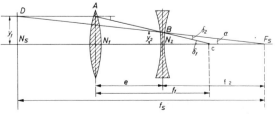

FIGURE 12-17. Auxiliary optical system of Bauersfeld.

The distance f_s from F_s to the left determines the back nodal plane of the system. We will now show that this plane has a fixed position with regard to the negative lens for all values of a.

From the similarity of triangles we find

$$y_2:y_1 = (f_1 - e):f_1$$
$$y_2:y_1 = (f_s - N_sN_2):f_s.$$

This gives together

$$N_sN_2 = f_s e/f_1. \qquad (12.9)$$

Substituting (12.7) into (12.9) gives

$$N_sN_2 = \frac{f_2 e}{f_1 + f_2 - e}. \qquad (12.10)$$

With $f_2 = -f_1 = -f$, equation 12.10 becomes

$$N_sN_2 = f = \text{constant.} \qquad (12.11)$$

This is an important quality due to the equal focal length of the two lenses.

In order to avoid the influence of lens distortion and other optical deviations, the auxiliary system is attached to the projector in such a manner that it is movable around the exit pupil of the projector lens. To avoid a wide aperture, and also to get the correct enlargement of the projection for all points, the optical axis of the auxiliary system is directed along the projecting ray, and its nodal point N_s coincides with the exit nodal point of the projector lens. This arrangement requires a fixed position for these two points. This is obtained by giving a fixed position to the negative lens and making e variable by shifting the positive lens along the optical axis of the auxiliary system. We show in later sections that this lens system is used in different ways. The most important is that for which it really was invented—the Zeiss Stereoplanigraph.

12.2.3 STEREOPLANIGRAPH OF ZEISS

12.2.3.1 History of the Development of the Stereoplanigraph

The construction of the first model C1 was finished in 1923, after a choice had been made from various solutions as described by Sander. Models C2 and C3 show improvements of essential elements of the construction, without changes of the basic design. These three models all have a horizontal orientation of the projecting cameras, which gives the entire setup a character related to terrestrial photogrammetry although it was definitely designed also for aerial photography. This latter aim was stimulated by the difficulties encountered in developing the Zeiss-von Orel Stereoautograph to a restitution instrument suitable for near-vertical photography.

The great change in the design of the Stereoplanigraph came with model C4 in 1930. This has a vertical orientation of the optical axes of the

cameras, with the primary axis in the direction of the flight line during scanning, and observation in epipolar planes. Just as the three models constructed before 1930 remained the same in principle, we can consider also all changes in models C5, C6, C7, and C8 as improvements of model C4, without real reorganization of either the principle on which it is based or of the basic features of its design. Recently, the manufacture of the stereoplanigraph has been discontinued. With the modern trend toward independent model triangulation, the universality of this plotter is a feature no longer in demand. However, the instrument is still in common use and therefore a brief description of the most recent model, C8, is included.

12.2.3.2 General Description of the Stereoplanigraph C8

The design of the instrument as a whole has not been changed since that of model C4 of 1930, with its vertical orientation of the axes of the cameras. The cameras are fixed on the camera bridge. For scanning of the model, this bridge describes a primary movement with the Z-column along the Y-rail, and along this column it describes a secondary movement in Z-direction. The movement in X-direction is carried out by the base carriage, with the floating marks on the cardan mirror (see fig. 12-18). The base components are imposed on the marks.

The most important element of the instrument since 1923 was, and still is, the auxiliary optical system. When comparing this design of an instrument for direct optical projection with a design applying pure mechanical projection, we must not forget that the Bauersfeld solution, although favorable to the quality of the observed images as compared with those of chapter XI, introduces requirements of adjustment to the mechanical parts of the auxiliary system. For

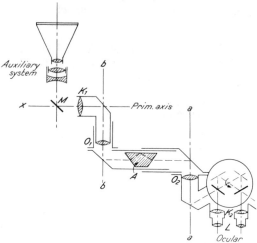

FIGURE 12-18. Plan of camera, mirror & scissor telescope of model C8.

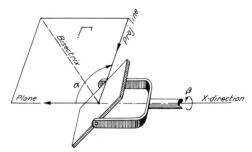

FIGURE 12-19. Rotation of the mirror.

discussion in the past about the basic difference between the Stereoplanigraph and those instruments with a mechanical projection has now lost its importance, since industry is capable of fulfilling the mechanical requirements with both solutions.

In the optical system of the C8, the bundle of rays emerging from any point in the photograph, after having passed through the auxiliary system, will always be reflected by the mirror into the X-direction of the instrument. In this way it can enter the stereoscope, which has its eyepieces in a fixed position.

instance, its gimbal axes and the movement of the center of the positive lens introduce requirements of adjustment of the same character as for instruments with mechanical projection. It can, for instance, be proved that the center of the positive lens must move along a straight line, which need not coincide with the optical axis, to avoid errors in the projection. Consequently, the

To achieve this it is necessary to guide the mirror in such a way that the surface remains perpendicular to the bisectrix of the projecting line and the x-direction through the rotation center of the mirror. The optical axis of the auxiliary system is guided in such a way that it always passes through the rotation center of the mirror, which coincides with the mark (*see* fig. 12-19).

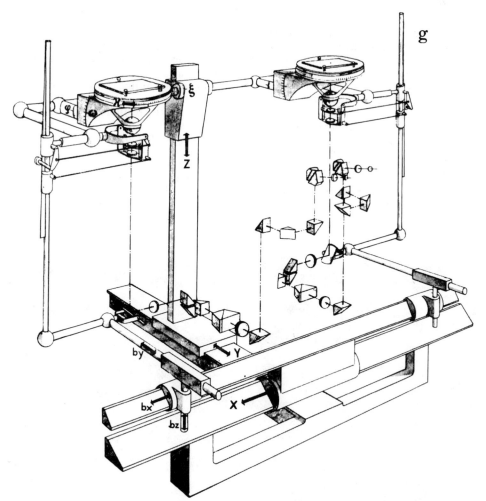

FIGURE 12-20. Exploded view of the optical system and base components of the Stereoplanigraph C8, showing orientation and scanning motions.

The movement in space of the projecting line can be seen as consisting of two components: a rotation α in the plane Γ, formed by the X-direction through the center of the mirror and the projecting line; and a second component, a rotation β with the plane Γ around the same X-direction. The bisecting line is always in the plane Γ, so that its movement is also a rotation with the plane around the X-direction and a rotation in this plane. The rotation of bisectrix and mirror together is derived from that of the projecting line. The movement of the projecting line is materialized in the instrument by the guiding rod g (*see* fig. 12-20) which is already mentioned as such for the auxiliary system. This rod has at its lower extremity a gimbal joint containing a fixed primary axis in X-direction and a rotating secondary axis perpendicular to it. The rotation around the primary axis is equal to the movement of the plane Γ. The device for mirror rotation contains a similar gimbal system, also with the primary axis in X-direction. The two systems are coupled in such a way that the rotations around the primary axis are equal and the secondary rotation is half the rotation α. With this arrangement, the mirror surface always remains perpendicular to the bisecting line, as required.

Because of the rotation of the mirror around the optical axis of the observing system, an image rotation over the same angle would be observed if this were not compensated by the rotation of an Amici prism. This rotation is automatically controlled by the primary rotation of the mirrors.

In the C8 there is a luminous mark introduced into the observing system by a separately fixed projection system *via* a half-silvered surface (*see* fig. 12-21). For this luminous mark, different sizes of the dots, or a circle with different colors, may be chosen. Circular marks are convenient

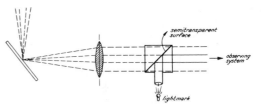

FIGURE 12-21. Projection of the luminous mark in the C8.

for pointing at presignalized points and dots are convenient for contouring.

We must take into account that there is an arrangement in the projecting cameras to make the principal distance equal to that of the taking camera multiplied by a factor representing the regular film shrinkage. This principal distance can differ a few millimetres from the focal distance of the lens of the projector. Because of this, the assumption that the rays between the projector lens and the auxiliary system are parallel will not be entirely correct. This makes it desirable that the nodal points coincide, and furthermore that there be freehand movement of the positive lens of the auxiliary system for sharp focusing on the photograph. This device is used only for focusing once for each model or strip.

An advantage of the Stereoplanigraph is that any type of photography, regardless of the kind of terrain, can be restituted so that observation is always sharp without strain on the eyes and without the need of any extra correction device. Before the super-wide-angle camera came on the market, it was possible to consider the Stereoplanigraph as the most universal restitution instrument, with its direct optical projection combined with the parallelogram of Zeiss. So far, however, the super-wide-angle photographs

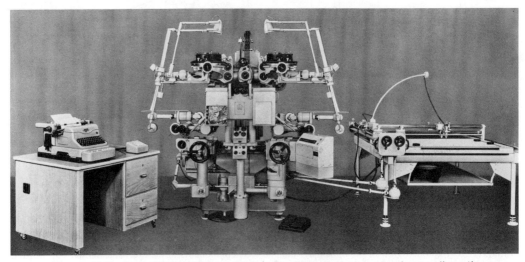

FIGURE 12-22. Latest Model C8 Stereoplanigraph w/Ecomat electromagnetic recording unit.

must be excluded, if no affine deformation can be accepted, owing to a combination of differences in height in the terrain and inclinations of the optical axes of the photographs.

The application of the Porro-Koppe principle requires a set of projecting cameras for each different kind of taking camera. Zeiss reduced this difficulty by the design of a universal plate carrier consisting of three separate parts: the plateholder, the central part and the lens cone. The central part, which has the device for small changes of the principal distance (up to ±7 millimetres), the dial, and movement for swing is constant for all different projectors. Only the lens cone and the plate holder must be changed. The plate holder also, if necessary, carries the compensating plate to equalize the distortion of the lens with that involved in the diapositive. The universal plate carrier was a successful effort to reduce the cost of projectors in case different types of photographs must be used. However, to accommodate the full range of focal lengths (80 millimetres to 610 millimetres), design data are available for the fabrication of twelve different projecting cameras.

Figure 12-22 shows the Stereoplanigraph with the electromagnetic recording device on the left and the coordinatograph on the right.

12.3   Instruments of the Photogoniometer Type

Instruments with partly optical and partly mechanical projection, and with observation through telescopes, are referred to as photogoniometer-type instruments.

12.3.1 HISTORY OF DEVELOPMENT

The further development of the first Zeiss automatic plotting instrument, the von Orel-Zeiss Stereoautograph developed about 1910, did not yield a solution suitable for restitution of aerial photographs with nearly vertical optical axes. As a result, with the increasing use of airplanes for aerial photography during the first World War, a method was developed for the use of restitution cameras applying the Porro-Koppe system and observation with a telescope. Figure 12-23 shows the simple way in which the mechanical intersection is realized. The cameras rotate around the fixed axes parallel to X. Assuming terrestrial photographs, the height is determined by transferring the inclination of the telescope to a rod L_1. This determination of height is carried out by transferring the distance from O_1 to the intersection point P_0 to the distance bridge Br_1 by means of a steel tape. Subsequently, Hugershoff designed another instrument in which the height was determined twice, using the inclination of the right-hand as well as of the left-hand projector.

In this instrument, with fixed position of the perspective centers O_1 and O_2 and without a Zeiss parallelogram, it is impossible to introduce a given map scale. This was improved in the model shown in figure 12-24, wherein the base could be introduced on the distance bridge. This Autocartograph of Hugershoff was used until 1926, when it was replaced by his Aerocartograph.

In the same period, in 1923, the French scientist G. Poivilliers designed his Stereotopograph Type A along similar lines. The first difference, however, is that the determination of height is carried out in the YZ-plane. The other difference is that Poivilliers never abandoned the basic principle of this instrument. He reshaped it in 1937, creating model B, which is still the most popular restitution instrument in France. It is described in section 12.3.2.

In both the Autocartograph of Hugershoff and the Stereotopograph of Poivilliers the intersection in space is realized on projections on vertical planes, which, in the instrument, are actually laid down in the XY-plane. Since 1926, however, all other designs have afforded intersection directly in space by means of steel space rods rotating around cardan axes. It cannot be denied that it is more difficult to produce a correct cardan system in which the two axes intersect at one point than to use single rotation axes in each of the two coordinate planes. Nevertheless, we find that modern techniques are fully capable of fulfilling the conditions of intersection of the space rods at a cardan-center.

It was probably the production by Zeiss of the Stereoplanigraph (section 12.2.3), an instrument that solved the problem of intersection in space, that stimulated Hegershoff to develop his instrument along the same lines. We see, furthermore, that the first photogrammetric restitution

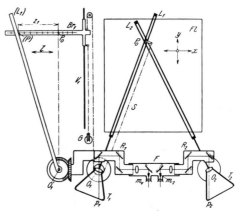

FIGURE 12-23. Scheme of the first design of the Hugershoff Autocartograph.

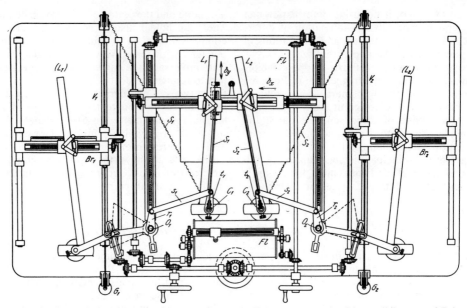

FIGURE 12-24. Plan of the final design of the Hugershoff Autocartograph with parallelogram of Zeiss.

instrument of Wild, the Autograph A2, belonging to the photogniometer type, accepted intersection in space instead of in the two coordinate planes.

Until recently, three instruments of this type were manufactured. They are the Poivilliers Stereotopograph, Nistri Photostereograph and the Thompson-Watts plotter. As previously noted, the universal instruments have declined in popularity. The capability for bridging continuous strips is no longer a requirement since the procedure has given way to independent model triangulation. This technique of setting single models and tying them together into a strip by sequentially transforming the coordinates of each model into the coordinate system of its predecessor is a numerical process that can be efficiently accomplished even on the smaller modern-day electronic computers. Thus an instrument with the base-in base-out capability (section 12.1.4) is not required; it is only necessary to precisely measure and record the coordinates of the model points and those of the projection centers. This can be carried out on a much simpler instrument than those previously mentioned. However, since these universal instruments are still in common use, a history of their development and a brief description of each will be presented.

12.3.2 POIVILLIERS STEREOTOPOGRAPH

12.3.2.1 HISTORY

The Stereotopograph was developed about 1923 by the French scientist G. Poivilliers, and was produced as model A by the Société d'Optique et de Méchanique de Haute Précision (S.O.M.) in Paris. In several respects, the Poivilliers design is an improvement on the Hugershoff Autocartograph. Both have the common properties that the mechanical representation of the projection is divided into two parts. The difference is that Hugershoff uses the projecting lines $O'A$ and $O''A$ (fig. 12-25), and Poivilliers the projections of these lines on the XZ and YZ planes.

Hugershoff abandoned this principle and, in 1926, introduced his Aerocartograph, which uses a spatial solution with space rods. Poivilliers and S.O.M., however, maintained the same principle after 1923, with the change of model A into model B in 1937.

12.3.2.2 GENERAL DESCRIPTION OF THE STEREOTOPOGRAPH TYPE B

As previously noted, the projection system of the Stereotopograph uses a double set of space rods; one for the projection of the homologous pencils of rays in the XZ plane and the other in the YZ plane. The relationship between the rods and the viewing telescopes is maintained by a special mechanical device. The position of the telescope can be considered as being obtained by a rotation a around a primary axis I in the Y-direction and a rotation i around a secondary axis II which is in the X-direction in the zero position and remains perpendicular to axis I during rotation about axis II. The corresponding position of the X-rod is determined by the same angle a; that of the Y-rod, however, by the angle i'. The relation between i and i' can easily be derived from figure 12-26

$$\tan i = \tan i' \cos \alpha. \qquad (12.12)$$

To facilitate the mechanical design, Poivilliers applied two additional rotations. Consider the z-axis of the projector to lie in the horizontal

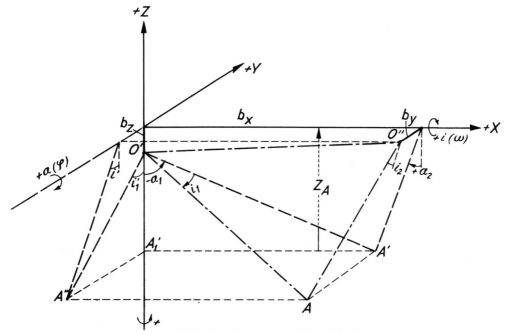

FIGURE 12-25. Base components; X- and Y-rods.

plane (fig. 12-27). The YZ-plane, meaning the Y-rod and the Y-axis, is brought into the horizontal plane by a 100^g rotation around the z-axis. The next step is an extra constant rotation of the projector telescope and telescope axis II around axis I. This is done in the opposite sense for the left and right projectors, which brings both plateholder sides of the projectors to the outside of the plotter. This is apparent for the right projector in figure 12-28. In this configuration, as shown in figure 12-27, for the left projection, the XZ- and YZ-planes are parallel horizontal planes. The projector and X-rod rotate through angle α around axis I, the telescope through angle i around axis II. The Y rod rotates through angle i' around axis III which is parallel to axis I. In the operation of the instrument, the spatial position of point P common to a pair of properly oriented photographs is determined by the orthogonal projections of the intersections of the pairs of space rods into the XZ- and YZ-planes respectively, yielding the position and the elevation of the point.

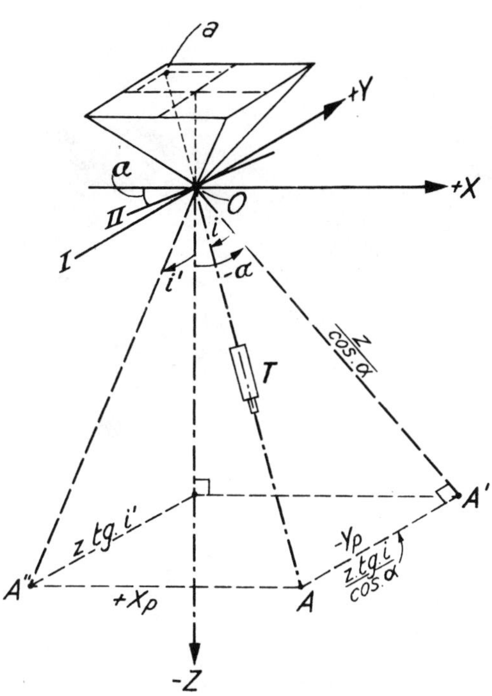

Since the Stereotopograph is based on the Porro-Koppe principle, which requires that the lenses of the plotting cameras be matched to that of the taking camera, the instrument is designed to accept interchangeable lens cones. Those, produced, accommodated the SOM lenses with focal lengths from 125 to 300 millimetres.

12.3.3 NISTRI PHOTOSTEREOGRAPH

12.3.3.1 HISTORY

Nistri, who with Santoni was one of the two Italian pioneers in the construction of restitution instruments, designed his Photocartograph in 1919 according to the principle of direct optical projection. Until 1930 he continued to improve this same instrument without trying to design one based on other principles. He never abandoned the principle of direct optical projection, of which his Photomapper (Fotocartografo, model VI) is his last creation.

FIGURE 12-26. Components of telescope rotation.

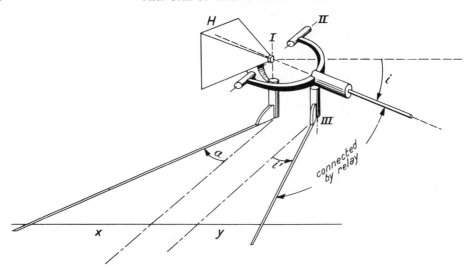

FIGURE 12-27. Perspective view of left projector, telescope, and X- and Y-rods.

In 1930, however, he wanted to complete his series of restitution instruments, consisting of the Fotocartografo and the Fotomultiplo, with an instrument which would have a wider range of possibilities.

At the Congress of the International Society for Photogrammetry in Paris in 1934, Nistri introduced the Photostereograph α as the result of his efforts between 1930 and 1934. In this design he accepted the use of:

    (1) the principle of Porro (after the Porro photogoniometer of 1865);

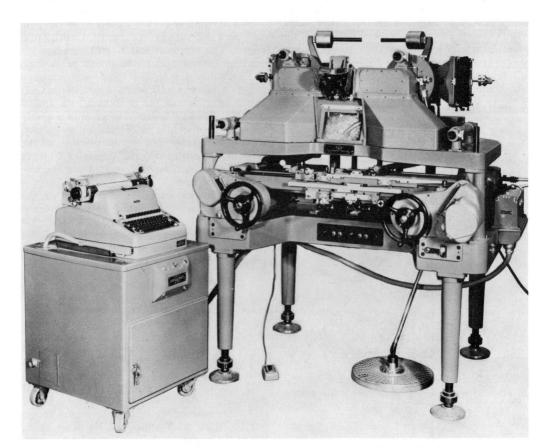

FIGURE 12-28. Poivilliers Stereotopographic B with automatic coordinate-registering device.

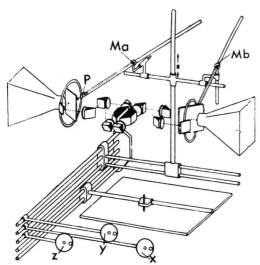

FIGURE 12-29. Optical system of Photostereograph Model α (1934).

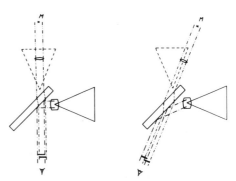

FIGURE 12-31. Relative displacement of exit pupils in model α due to the thickness of the mirror.

(2) a fixed position of the restitution cameras during scanning; and
(3) a modified Deville principle for the simultaneous observation of photograph and floating mark.

The third point is the most characteristic feature of the Nistri universal plotters. In this development, he replaced the telescope of the optical-mechanical type by a collimator from which a parallel beam of light emerges onto the camera lens. This gives an image of the mark M at point A on the diapositive. However, direct stereoscopic observation of the two diapositives would be possible only in a mirror stereoscope, without enlargement for viewing both diapositives as a whole. The variable distance (parallax) between corresponding points means a choice between this construction and that of a movable optical system before the camera. This latter system, however, was considered more difficult than the application of the Deville principle.

The principle as applied in its original form by Deville has the disadvantage that the mark and the photograph are observed at different distances, and consequently a simultaneous focusing of both is, in theory, impossible. Nistri,

however, has only parallel light coming from the lenses of both the camera and the collimator. Rotating the two beams together over 100^g, it is possible to observe both the photograph and the mark at infinity. This was the solution accepted by Nistri in his model alpha (fig. 12-29 and 12-30). He used a half silvered mirror, through which the beam from the floating mark passes, and the beam from P is reflected towards the eye.

The mirror is fixed to the camera at an angle of 50^g with its axis, thus giving a virtual image of the camera. The center of the gimbal system, about which the camera rotates for the orientation, coincides with the exit nodal point of camera lens, while the exit nodal point of the collimator lens coincides approximately with that of the virtual image of the camera lens.

Notwithstanding the originality of its design, the model alpha was not a success. The disadvantage of this special design was, first of all, the need for a large mirror in order to collect all beams emerging from the camera lens. The use of wide-angle photographs was hardly possible. The large mirrors could be kept flat only if their thicknesses were in proportion to their sizes. These heavy mirrors result in a displacement of the exit pupil of the beam of the mark relative to that of the image point (fig. 12-31).

It was not until several years after World War II that Nistri revised the design of his Photostereograph, presenting the beta model at the

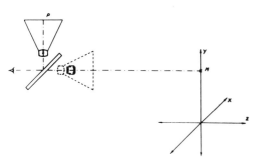

FIGURE 12-30. Explanation of system of figure 12-29.

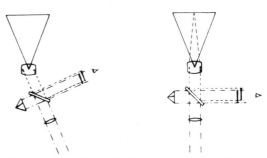

FIGURE 12-32. Principle of the projection in model β1.

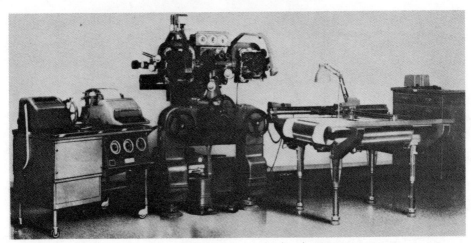

FIGURE 12-33. Photostereograph Nistri β2 (1956) with digital output.

International Congress on Photogrammetry in Washington, D.C., in 1952.

12.3.3.2 GENERAL DESCRIPTION OF THE PHOTOSTEREOGRAPH β

There were several great improvements in the application of the Deville principle to the projection system of the β model as compared with the α.

The most significant was that Nistri realized that the projecting collimator as well as the restitution camera could be considered as a Porro camera for the purpose of projecting the reference mark. This meant that a small, thin mirror could be placed in a fixed position on the collinator instead of on the camera. This avoided the need for a large mirror, and made the use of wide-angle and even of super-wide-angle photographs possible to the extent that other limitations such as stereoscopic observation and dimensions of the instrument were now decisive in this respect. Figure 12-32 shows the new arrangement in the model β1 of 1952.

The complete Photostereograph β2 is shown in figure 12-33. In contrast to many other instruments the xy-plane and the diapositives, in their zero positions, are vertical. The axes of rotation of the cameras are arranged in the same way as those of the collinators, i.e., the primary axis of each is parallel to the X-axis of model space.

One of the characteristic features of the Nistri constructions was the general use of electrical transmission of movements. Nistri was the first to apply this method to photogrammetric instruments (in his Fotomultiplo) in 1948. The β2 has hardly any mechanical transmission, but uses servomotors and electromagnetic transmissions. Because of this design feature, it is possible for the operator to scan and even contour by manipulating one handwheel which controls the direction of the movement of the floating marks, and a lever which governs the speed.

12.3.4 THOMPSON-WATTS PLOTTER

The Thompson-Watts plotter was manufactured by Hilger and Watts, London, from a design by Prof. E. H. Thompson of London University.

12.3.4.1 GENERAL DESCRIPTION OF THE THOMPSON-WATTS PLOTTER

In a description of this instrument published in 1954, two basic principles are mentioned. The first is the use of the air base as the primary axis of the instrument; this principle was first applied as long ago as 1926, when H. G. Fourcade incorporated it into the design of his

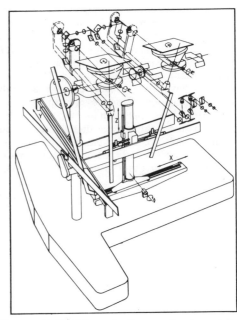

FIGURE 12-34. Schematic diagram showing rotations for relative orientation on the Thompson-Watts Plotter.

FIGURE 12-35. Thompson-Watts Plotter II.

Stereogoniometer. The second is the distinction made between the elements necessary for relative and absolute orientation, which together construct the model of the terrain, and scale and level the model for measuring and plotting. This distinction, which must be considered as the Fourcade principle proper, recognizes the two parts of the complete restitution operation as two separate manipulations.

The instrument has two cameras, which are mounted on horizontal beams consisting of two independent parts with a common support in the middle. The projectors have individual controls for the relative orientation of a pair of overlapping photographs. The viewing telescopes are mounted beneath the projectors, and their fixed axes are at right-angles to the axis of the beam and are designed to intersect the beam axis at points coinciding with the front nodes of the projector lenses. The photographs are scanned in the y-direction by rotating the beam, while in the x-direction by rotating the telescopes. Linked to each telescope in the x-direction is a lineal or space rod which converts the angular motion of the telescope into a linear measurement in the plane of the slide. Similarly, measurements in the Y-direction are effected by a mechanical linkage having a lineal attached to a slide on the beam and to a Y-carriage connected to the measuring screw. As a result, all measurements are taken in a fixed vertical epipolar plane with the projectors being rotated into their required position for observing and measuring any given point. The mechanical linkage can be best understood from the schematic in figure 12-34. A particularly useful feature of the instrument is that the absolute orientation either in leveling or

scaling will not disturb the relative orientation of the model. The overall appearance of the instrument is shown in figure 12-35. It is interesting to note that Thompson never designed his instrument with base-in base-out capability for stereotriangulation; and it was his opinion that this process should be carried out analytically using the comparator as the mensuration instrument.

12.4 Instruments with Mechanical Projection

12.4.1 HISTORY

In considering the development of this type of instrument, we meet the names of Santoni and Wild, and come to the conclusion that Santoni was the first to replace the lens as perspective center by the center of a system of gimbal axes. Starting with his "Autoriduttore" patented in 1920, he subsequently applied this system in all his photogrammetric restitution instruments, the projecting rays being represented by metallic space rods. The one exception is an instrument built for ballistic purposes in 1932.

In section 12.3.1 we saw that Wild produced an instrument of the photogoniometer type (Autograph A1 and later A2) in 1921. Not until 1935 did Wild accept the mechanical-projection principle in the A5 model. Since then Wild has also continued along the same line in all later restitution instruments.

The principle of mechanical projection has also been adopted by Kern with the introduction of the PG2 in 1960, and by Zeiss Oberkochen with the Planimat which was first presented at the 31st Photogrammetric Weeks 1967 in Karlsruhe. Similarly Zeiss Jena has introduced the Stereometrograph and Topocart, and SOP ELEM of France, the Presa 225 within the past ten years. The relatively recent growth in popularity of the mechanical projection principle can be attributed to several advantages of the system:

(1) the ease of accommodating a wide range of camera focal lengths;
(2) the ability to accommodate 23 × 23 centimetre and smaller formats for normal-, wide-, and super-wide-angle photographs (there are some exceptions to this generality among instruments);
(3) the simplicity of using aspheric plates on the projectors to correct for radial distortion and atmospheric refraction; and
(4) the improved resolution resulting from the orthogonal viewing of the photo-images.

Furthermore, with the obsolescence of oblique and convergent photography and the virtual exclusive use of vertical photography, the orthogonal viewing of the photo-images is ideal, and the panchratic system required to compensate for the apparent scale difference of the two

corresponding images as with convergent photography is no longer necessary.

12.4.2 GALILEO-SANTONI INSTRUMENTS WITH MECHANICAL PROJECTION

12.4.2.1 SANTONI STEREOCARTOGRAPH MODEL V

As previously noted, Santoni pursued the design of stereoplotters with mechanical projection. The present model was introduced at the 1964 International Congress on Photogrammetry in Lisbon. This instrument is radically different in design from its predecessor Model IV. The difference is immediately apparent from the horizontal mounting of its projection system, which was adopted to facilitate the adjustment and testing of the more sensitive units. The instrument like its predecessor is suitable for continuous strip triangulation, but will also accommodate 23 × 23 centimetre super-wide angle photographs.

12.4.2.1.1 MECHANICAL DESIGN

The parallelogram solution applied is shown in fig. 12-36. The two projectors are turned about the y-axes through 100^g, into a horizontal position (fig. 12-37), so that the operator faces the yz-planes. The optical units are movable for tracking, thus the photo-carriers are fixed. The tracking unit consists of two arms 5 and 6 (fig. 12-38), which can rotate about the parallel axes m and n. Axis m is fixed with respect to the projector, whereas axis n turns about m by means of arm 5. Arm 6 can turn about the secondary axis n, and at its end it is attached to the upper space rod by a ball joint 4. Assuming that the axes m and n are perpendicular to photo-plane, and that the assembly is stable, ball-joint 4 is forced to move in a plane (the mechanical photo-plane) which is parallel to the real photo-plane; it can be provided with aspherical correction cams for lens distortion. The photo-carrier, along with the tracking unit and the corresponding support, can be displaced along the nearly horizontal principal axis in order to set the principal distance. In addition, the position of the upper rod and the balance of

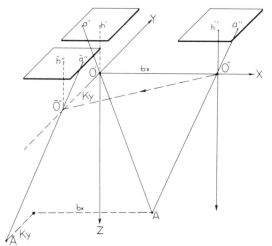

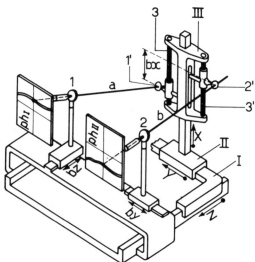

FIGURE 12-36. The parallelogram solution applied to the Santoni Stereocartograph Model V.

FIGURE 12-37. Schematic of the projectors as rotated into the horizontal position on the Santoni Stereocartograph Model V.

the space rod should then be readjusted. The principal distance can be changed continuously between $c = 86$ and $c = 220$ millimetres.

As an option, the projectors can be equipped with correction surfaces (figs. 12-39 and 12-40) which have the capability (after R. Zurlinden) to resolve arbitrary corrections in the x- and y-

components. For each component there is a surface which can be deformed mechanically, and on which a feeler slides, in the same way as the ball joint, 4, moves in front of the photo-graph. The deformations of the surface, with re-

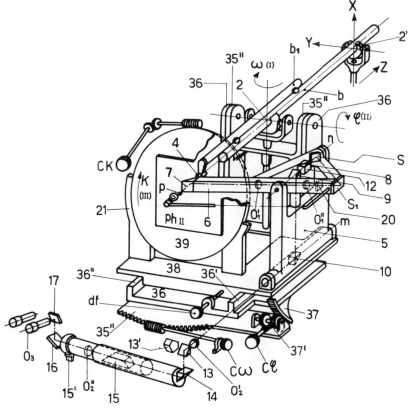

FIGURE 12-38. Schematic of fixed photo-carrier with movable space rod and optics for tracking on Santoni Stereocartograph Model V.

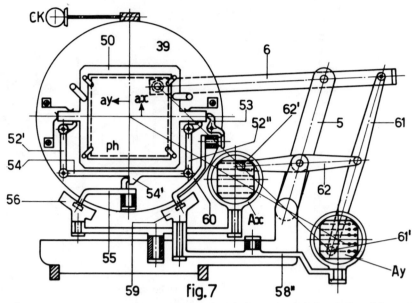

FIGURE 12-39. Schematic of plate holder and correction surfaces *Ax* and *Ay*, designed to compensate for non-calibratable distortions of the photography.

spect to a datum plane, are mechanically transferred, in a reduced ratio, to the photo-carriage as corrective shifts *x* and *y* or as corrective rotations to a plane parallel glass plate. This plate is located between the photograph and the obser-

vation system. In practice the errors are split into two categories: those which are precalibrated (S_x, S_y) and are constant for a whole strip or block, and those which can be detected from reseau photography (A_x, A_y) and are variable

FIGURE 12-40. Front view of Santoni Stereocartograph V showing the non-calibratable distortion correction devices located respectively in the lower left and right of the plate holders.

from photograph to photograph. The correction surfaces are 3× reduced with respect to the photographs, and the movements of the feelers are transmitted by means of a reduction parallelogram 5, 6, 61, 62. The two arms 5, 6, belong to the tracking unit. The feelers must be positioned on the correction surfaces on the straight line passing through the fixed axis m and the photo-point observed. When the measuring mark is set on the principal point, the feelers should be in the center of the correction surfaces. These are deformed by means of setting screws, thus permitting corrections for arbitrary, as well as symmetric, errors. The settings of S_x and S_y must be pre-determined, and can be introduced as displacements of the measuring mark from corresponding grid points. These displacements are then eliminated using the setting screws of the surfaces S_x and S_y. It is possible to set A_x and A_y only if special reseau photographs (fig. 12-41) are used. The photo-carrier is provided with a similar arrangement of crosses, and the discrepancies between the two sets can be eliminated using the setting screws of the correction surfaces A_x and A_y. In the instrument, the surfaces S and A are located one behind the other (fig. 12-42) and are pressed against the corresponding feelers. During tracking, the feelers displace the surfaces in the lateral direction. The displacements of the A-Surfaces are transferred, in reduced ratio (40:1), to the photo carriers, and those of the S-surfaces to the large plane-parallel glass plates 19 in front of the photographs (fig. 12-43).

The means for orientation are ω and ϕ setting and measuring devices for both projectors, κ-devices for the photo carriers, and devices for by and bz which cause displacements on a horizontal plane. The base, $+ bx$, is set on the vertical base bridge (fig. 12-37). The two base carriages I and II are displaced in opposite direc-

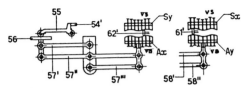

FIGURE 12-42. Deformation surfaces designed to compensate for camera lens distortion (S-surfaces) and measured film distortion (A-surfaces).

tions for this purpose. The situation is symmetric for the "base in" and "base out" settings. No common rotations are provided.

The space rods are divided into the main rod b, and the upper rod b_1 (fig. 12-38). The main rods are cylindrical, and are attached to the cardan joints which represent the projection centers; the rods slide through the cardan joints on the two bx carriages. The upper rods are semi-cylindrical, and are mounted on top of, and coaxial to, the main rods. They are attached to the tracking units by the ball-joints, 30 (mechanical photo-points), the centers of which lie on the axes of the corresponding main rods. These space rods are counter-balanced by a device which also actuates the antiflex unit, to prevent bending.

12.4.2.1.2 OBSERVATION SYSTEM

Small lamps moving with the tracking units provide illumination (fig. 12-38). After passing through the photograph (and the plane parallel glass plate of the correction device, see fig. 12-43) light is guided along the secondary arm (6) to the primary arm (5) of the tracking unit, and from there to the support of the projector. The length of the optical train remains constant during tracking. Prism (7) reflects the light along the arm where the two lenses O_1', O_1'' are mounted. These lenses form an image in the plane of the measuring mark (8).

Mirrors S_1 and S_2 reflect the mark along the rotation axis n, which is common to arms 6 and 5. The measuring mark (8) is a small black disk which can be replaced by a luminous dot projected into the train by a small projector (12) and merged with a semi-reflecting surface. All these units are fixed in the secondary arm (6) of the tracking unit.

Prism 9 is fixed in the primary arm (5); it reflects the light along this arm to prism 9 (fig. 12-38); from this prism the light is reflected along the axis n to prism 13 on the support of the projector. During tracking, prisms 7, 9, and 10 cause image rotations which cancel each other. The light is guided by the movable projector (for orientation) to the fixed eye piece by means of prisms 13 and 14 (replacing a cardan mirror), and the cardan mirror (16). A lens (O_2') in front of prism 13, forms parallel light, which passes through the Dove-prism (A) where image rotations caused by prisms 13, 14, and mirror 16 are

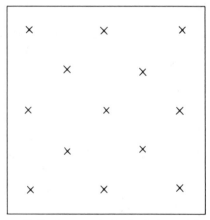

FIGURE 12-41. A reseau in the focal plane of the camera from which film deformation can be determined by comparing measured values of reseau intersections against calibrated values.

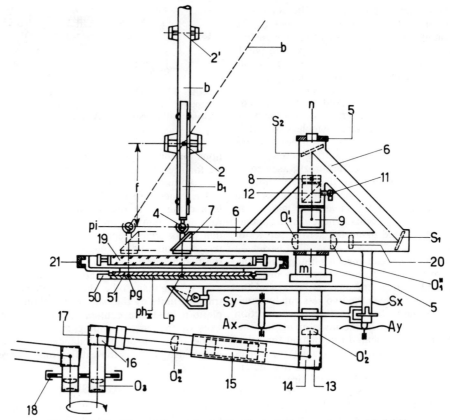

FIGURE 12-43. Observation system of the Santoni Stereocartograph Model V.

corrected. Lens O_2'' forms an image for observation in front of the eyepieces (O_3). The two eyepieces are attached to a revolving disk which acts as an optical switching unit. When turned through 200^g, the two optical trains will cross. The usual means for adjusting the eyepieces are provided. The magnification is $9\times$.

12.4.2.1.3 MEASURING AND PLOTTING SYSTEM

The carriages are supported by ball-bearings which move along thick chromium plated tubular rails. These carriage movements are controlled by spindles which incorporate a backlash-free system. The XY-movements can be made either by handwheels, coupled to a three-speed transmission, or by a pantograph which is electrically connected to the spindles. The pantograph allows rapid scanning of the model as well as ease of plotting. The Z-screw is normally moved by a foot wheel; however, the motion can be transferred to a third handwheel to provide higher sensitivity in stereoscopic measurements. The X- and Y-coordinates are displayed on graduated glass disks with least count of 0.01 millimetre. The Z-coordinate is displayed on a friction drum which allows for ease of resetting. Shaft encoders can be attached to the spindles for the automatic recording of

coordinates. The large, glass-top coordinatograph is connected to the plotter by means of telescopic rods whch permits considerable versatility in the placement of the drawing table (see fig. 12-44). A very large drawing table is also available, the stylus being driven by electric transmission through synchronous motors and steel tapes.

12.4.2.2 GALILEO SANTONI STEREOSIMPLEX IIId

The Stereosimplex IIId (see fig. 12-45) is a successor to the Model IIIa with some improvements. Models IIIb and IIIc were instruments designed for use with convergent photographs. The IIId is designed for use with wide and normal angle photographs with focal lengths ranging from 98 to 220 millimetres. Although the instrument cannot be used for continuous strip aerotriangulation with the base-in and base-out modes, it can be equipped with the "Santoni aeroplane," a device for sequentially transferring the rotational elements of orientation (ϕ,ω,κ) from one projector to the other, and thus triangulating continuous strips.

12.4.2.2.1 PROJECTION SYSTEM

The projection system (fig. 12-46) uses the Santoni parallelogram. It employs projectors

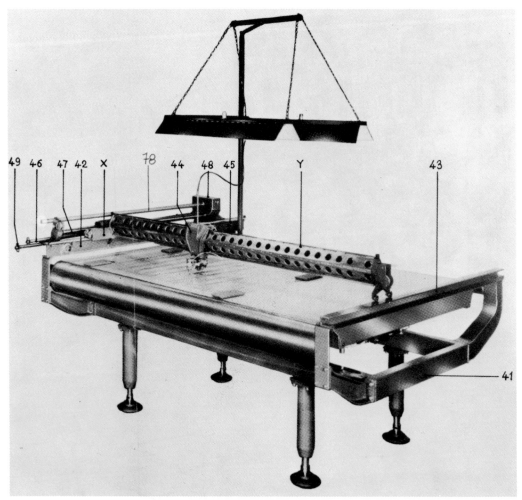

FIGURE 12-44. The large, glass-top coordinatograph provided with the Santoni Stereocartograph Model V.

with movable photo-carriers and fixed optical units. The projection centers are represented by the central cardan joints. The orthogonal cross-slides on which the photo-carrier is mounted are supported by the unit which carries part of the (fixed) observation system. The photo-carriers are, for mechanical reasons, symmetrically displaced in the X-direction with respect to the projection centers, and are attached to the upper ball-joints of the corresponding space rods. These ball-joints can be provided with aspherical cams to correct for lens distortion by continuously altering the principal distance. The user may make such cams himself, and a grinding machine is provided for this purpose. The functioning of a cam during tracking is illustrated in fig. 12-47. The photo-carriers are ordinary metal frames, with clamps, but without glass supporting plates. They must be set up on a special device for centering photographs. The principal distance can be changed continuously by engaging a crank for the two lead screws 39, 39′ (fig. 12-48). By turning the crank, the principal

distance can be changed, but only by full millimetres. The two screws (39, 39′) lie in the plane which passes through the projection center, so the α-screw has to be used after each change of c, in combination with a level, which is attached to the support of the tracking unit. After leveling with ϕ, the ϕ_0 reading must be adjusted to zero. In addition to the incremental device (millimetre steps), a device is provided for fine settings of the principal distance. This device (48) is attached to the photo-carriage. It can displace the upper ball-joint, g, perpendicularly to the photo-plane by small amounts (max. 25 millimetres), but when the principal distance is altered by a larger amount, tube n, supporting the upper rod i, must be extended or moved into the central part of the space rod.

Some of the means for orientation are attached to the projectors, and some to the split model points. All rotation axes pass through the projection centers. The rotations ω and ϕ have quite considerable ranges (ω: $+20^g$, -25^g; ϕ: $\pm 10^g$; $\pm 15^g$). The κ rotation ($\pm 10^g$) is performed

FIGURE 12-45. Galileo-Santoni Stereosimplex IIId.

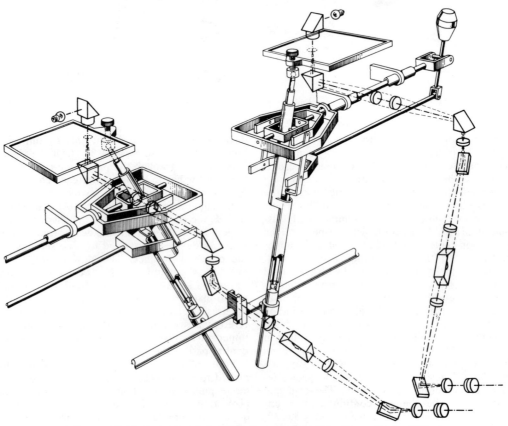

FIGURE 12-46. Projection and observation systems of the Stereosimplex IIId.

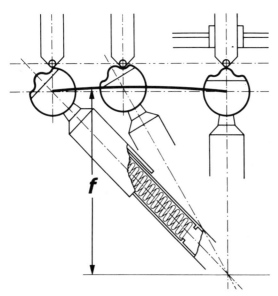

FIGURE 12-47. The spherical cam on the space rod of the Stereosimplex IIId. It compensates for lens distortion by varying the principal distances of the projector.

in a special way (fig. 12-49), by turning the whole support, together with the optical units and the cross-slide system, about the projection center V. Suppose the space rod is inclined (in the X-direction); the photocarriage is then forced to rotate about the upper joint, g. If the measuring mark is set initially on the photo-point P_1, it will be displaced after κ-rotation to the point, P_{1x}, while the photo-point set initially moves to position p. Hence, a y-parallax appears, similar to that which is observed when the photograph is

turned about the principal point h_x over κ. This arrangement is justified only when the Santoni aeroplane (previously mentioned), which can be placed on the cross-slide system for the transfer of κ rotations, is used.

Each projector can be shifted independently, by means of screws, in order to set the base components bx and by. The bz-component can be set to either of the split model points (lower joints) as a differential displacement of one Z-carriage with respect to the other. All the setting screws for orientation parameters are provided with drums (and scales) for measurement.

The space rods (fig. 12-46) are cylindrical, and made up of three parts. The middle section is attached to the central cardan joint. It carries a tube supporting the upper rod (moving inside the tube), and is extended into the eccentric main rod at the lower end. The two main rods are displaced eccentrically in the Y-direction (in opposite directions) in order to prevent them from touching in the model space. An "antiflex" device, to counter-balance and prevent bending, is attached to each main rod. This device consists of a parallelogram, with a counter-weight at its outer side, and the rod inside the main rod (inner rod) at the inner side. The upper arm of the parallelogram, which is coaxial with the ω-axis, carries a fork at each end. The extension of the inner rod is attached to the inner fork at the projection center, while the outer fork carries a lever to which the counter weight is attached. The lower arm of the parallelogram is attached to the lower end of this lever. This arm is connected to the inner rod below the fork. When the space rod is vertical, the antiflex device is in equilibrium, but when it is inclined, the counter-weight applies a lateral stress to the

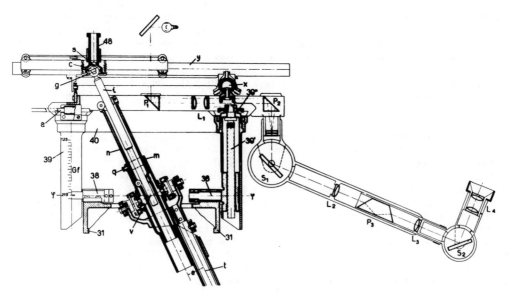

FIGURE 12-48. Detailed diagram of the observation system of Stereosimplex IIId.

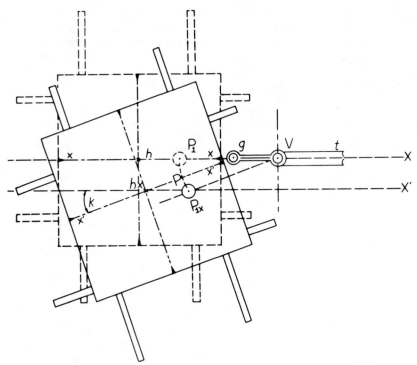

FIGURE 12-49. Schematic of the *kappa* rotation on Stereosimplex IIId.

inner rod *via* the lower arm. This stress is transferred to the main rod near the lower extension of the inner rod; thus, to some extent, it prevents bending and counter-balances the space rod at the same time.

12.4.2.2.2 OBSERVATION SYSTEM

During tracking, the observation system is fixed. Small lamps are used for illumination (fig. 12-46). The light from the photograph is reflected by a rectangular prism to a second prism, in front of the projector, which reflects the light downwards to the first cardan mirror. Two lenses, placed in front of the second prism, form an image in the plane of the measuring mark, which is located below the prism. These lenses are adjustable by means of a screw (for diapositives or negatives). All these optical units are attached to the tracking unit support. The light is guided from the movable projector (for orientation) to the fixed eyepiece, by means of two cardan mirrors which are interconnected by telescopic tubes. The half-angle rotations of the mirrors are controlled by gears (with the ratio of the diameters 2:1), and the primary rotations of the mirrors cause image rotations which must be corrected by use of the Dove prism. A lens in front of the Dove prism forms an image in front of the eyepiece for observation. The lower cardan mirror may be manually adjusted for squint

correction, and, in addition to the usual provisions for adjusting the eyepieces, extra facilities are provided for changing their location and the direction of viewing. The magnification is 8×.

12.4.2.2.3 MEASURING AND PLOTTING DEVICE

The movement of the plotting carriage is accomplished either through an arm attached to the X-component or through a reduction parallelogram pantograph with its pole attached to the base of the instrument. The latter yields a movement ratio of 1:8 or 1:12 between the operator's control and the pantograph, thus permitting the smooth and careful plotting of detail. The Z-carriages are driven by a hand wheel and two lead screws (*see* fig. 12-50). Linear scales are attached to the X- and Y-rails. Viewers and optical trains permit reading either scale by means of an eyepiece which is affixed near the binoculars. The rotation of a rhomboidal prism allows the operator to read individually the X- and Y-scales. The elevation drums are attached to the Z-hand wheel and interchangeable gears provide measurement over a wide range of scales.

The plotting table is located on the right side of the instrument. Plotting can be carried out at model scale by direct connection to the X-carriage or through a pantograph attachment for

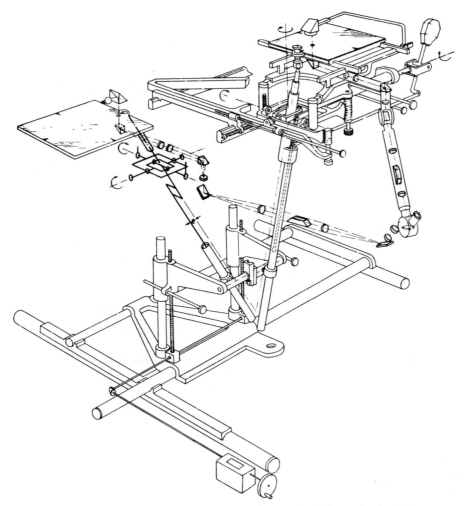

FIGURE 12-50. Schematic representation of the design of Stereosimplex IIId.

varying the ratio of model to plotting scale. The change in scale ratio is at fixed increments from a reduction of 4:1 to an enlargement of 1:3.3.

12.4.3 WILD INSTRUMENTS WITH MECHANICAL PROJECTION

As stated earlier, the A5 Autograph was the first Wild instrument featuring mechanical projection. The instrument, although capable of continuous strip triangulation, was also costly to use for compilation. It became readily apparent to Wild that a companion instrument of relatively simple design was required for compilation. The principle of mechanical projection was retained and the A6 Autograph was the outcome. With the advent of the high performance Aviator and Aviogon lenses, it became necessary to build instruments of higher precision both for aerotriangulation and for compilation to fully exploit the geometric and resolution char-

acteristics of those new cameras. The outcome in 1952 was the introduction of the A7 Autograph as the universal instrument, followed by the A8 autograph precision stereoplotter.

Similarly, along with the development of the super-wide-angle camera RC9 with the super-Aviogon lens around 1960, Wild introduced a new family of plotters. These were the Universal plotter, A9 Autograph and the B8 and B9 Aviographs. The main difference between the B8 and B9 is that the B8 accomodates photographs of the original size (23 × 23 centimetres), while in the B9, the plotting is carried out with the same reduced diapositives which were used for triangulation in the Autograph A9. The manufacture of the A7 and A9 Autographs by Wild has been practically discontinued. Again this must be regarded as the result of the modern trend toward independent model triangulation which can be executed for wide-angle photography on the A8 autograph or for super-wide-angle, as well as wide-angle photography on the A10 Autograph. This latter instrument is the most recent addition to the family of Wild plotters.

12.4.3.1 WILD AUTOGRAPH A7

12.4.3.1.1 GENERAL COMPARISON WITH THE WILD A5

The Wild A7 Autograph (*see* fig. 12-51) was displayed to the public in 1952. It replaced the A5 and incorporated many improvements, the aim of which was greater stability of the instrument particularly with respect to its adjustment. This was necessary because it was designed to accommodate photographs with a format of 23 × 23 centimetres instead of 18 × 18 centimetres. With this pupose in mind Wild had introduced the following changes:

(a) the distance between the floating mark and the observed photograph is fixed at 70 millimetres;
(b) the common camera rotations such as longitudinal and lateral tilt have been eliminated;
(c) the swinging guides of the observation system are supported by an accurately adjustable rail; and
(d) the space rods are counterbalanced by weights, and the cardan joints representing the perspective centers are heavy and rigid.

12.4.3.1.2 OBSERVATION SYSTEM

Figure 12-52 represents the optical path of the Autograph A7. Because of the need to raise or lower the plate carrier in order to change the principal distance, and because of the movement of the swinging girder containing the prism carriage with M, prism I, and the two lenses, M will be in the focal plane of these two lenses. This produces the required parallel light falling through prism II on the lens in front of prism III. The center of the bundle between prisms III and IV coincides with the (secondary) ϕ axis, which will be given small displacements by the vertical movement of IV due to an ω rotation as far as prism VI. The top and bottom surfaces of prism VII, both reflecting at an angle of 50^g, are separated by another surface reflecting to both sides at an angle of 50^g but in a direction perpendicular to the former. The rays coming from the right and left cameras are reflected downwards and upwards, and then continue in a vertical plane. Here they are transformed into parallel light in order to pass through the Dove prisms VIII, after which they pass through prisms IXa and X.

In X the light is reflected so that the two bundles are once more brought from a position in a vertical plane to a horizontal plane. Of greater importance, however, is the fact that prism IXa can be replaced by prism IX which has one more reflecting surface than IXa. This replacement may be made by moving a lever under the eye pieces which rotates a vertical axis. The effect is erection of a mirror-reversed inversion of the image. This is necessary to observe a correct image of diapositives as well as of negatives. The parallel bundles of light fall on the lens between XI and XII and are concentrated onto the objective plane of the eyepieces. The Dove prisms, and hence the bundles of light, can be rotated by means of two knobs on the main bridge above the eyepieces. These are used to correct the position of epipolar traces so as to give good stereoscopic vision.

Two further features must be mentioned which are especially useful for aerotriangulation. The

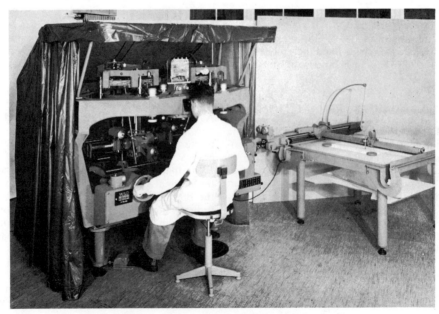

FIGURE 12-51. Front view of the Wild Autograph A7.

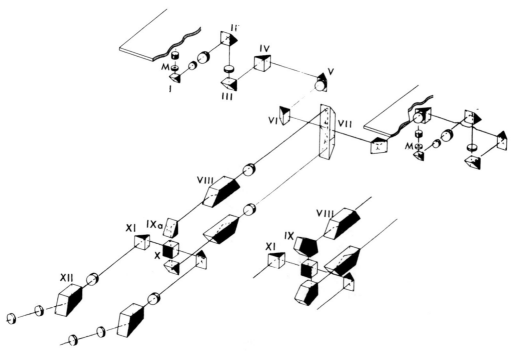

FIGURE 12-52. Diagram of the optical elements of the Autograph A7.

first is a lever under the main bridge close to the lever for the optical reversing device but working on a horizontal axis. This rotates prism X 100^g, interchanging the left and right images in the eyepieces. With this simple device, continuous triangulation of a strip is possible, maintaining normal stereoscopic observation. The second feature is the picture tumbling device, which makes it possible to convert y-parallaxes into x-parallaxes during the relative orientation. This tumbling device must be ordered specially.

12.4.3.1.3 FURTHER REMARKS ABOUT MECHANICAL DESIGN AND USE

To evaluate the design of the A7 instrument, we must realize that it was a consequence of the creation of the high performance Aviotar and Aviogon lenses. The aim was more precise aerotriangulation of the 15 centimetre Aviogon photography and a larger plotting scale than was possible in the A5. In section 12.4.3.1.1 a few of the differences which contributed to the required improvement were mentioned. It is also, however, the simplicity, the strong and rigid construction, and, above all, the high precision craftsmanship, which enable us to talk about micrometres when discussing such a heavy mass of metal.

Figure 12-53 shows the heavy construction of the primary axis which is common to both cameras and to the cardan joint II. Each space rod has three cardan joints. During scanning, the base carriage makes a primary Y, a secondary X,

and a tertiary Z movement along heavy cylindrical rails. The base component bx may be introduced by a crank (for symmetrical fast bx movements) and a knob (for fine movements). Knobs are available, with drums reading to 0.01 millimetre, for setting the secondary bz and the tertiary by base components. All knobs are within easy reach of the front of the machine.

The base carriage can be moved freely by hand (after releasing the X- and Y-spindles), by means of the corresponding handle. The four knobs for the ω and ϕ rotations are under the heavy frame in a position which can be easily reached by the operator. The corresponding scale drums are above the bridge (ω closest to the edge).

The drawing table, which has a working surface of either 100×100 centimetres or 140×100 centimetres, can be used separately as a coordinatograph for plotting grids or known points. It has two six-digit counters reading to 0.01 millimetre. The direction of rotation of X and Y can be reversed separately and the X- and Y-axes interchanged in the connection with the machine. Since the plotting of contours on a scribecoating is now becoming common, fluorescent illumination under the normal ground glass is to be recommended.

12.4.3.2 WILD A8 STEREOPLOTTING MACHINE

12.4.3.2.1 INTRODUCTION

The replacement of the A6 by the A8 was stimulated even more by the introduction of the

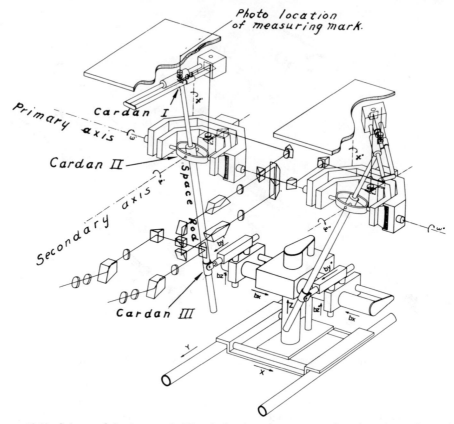

FIGURE 12-53. Scheme of the Autograph A7 optical train and movements for orientation and scanning.

high-performance lenses such as Aviotar and Aviogon than was the replacement of the A5 by A7. The misuse of the topographic plotter A6 for the production of large-scale maps at 1:5,000 was possible only as long as the scale of photography was not less than 1:10,000 to 1:12,000. After such lenses as the Aviogon came into use, a two-times enlargement from photograph to map did not exploit the possibilities of the new lenses sufficiently. A scale of photography of 1:20,000 to 1:25,000, which would be fully acceptable from the viewpoint of image quality, would require restitution in the Universal Precision Plotter A7 which was under construction during the same period, 1949-52. Would the expensive universal precision plotters be used for the many maps for engineering purposes at scales of 1:10,000, 1:5,000, and even 1:2,500, or would the Kelsh-type plotter, with its possibility of 5-times enlargement, which came on the market at that time, take over the greater part of this task? Undoubtedly this kind of reasoning led the Wild firm to design the A8 stereoplotter, which was presented together with the Autograph A7 at the Washington Congress of ISP in 1952.

12.4.3.2.2 GENERAL DESCRIPTION

Figure 12-54 illustrates the construction of the A8 diagrammatically. $O'O''P$ is the intersection triangle, in which the projection rays are replaced by space rods and the image points by the cardan centers L', L''. The actual microtelescopes and photographs have been moved outwards by means of the lazy tongs, as in the Autograph A6, so that the length of the base $O'O''$, in the instrument is independent of the size of the photographs. Since the central axes of the lazy tongs are firmly attached to the camera frames the foci of points L' and L'' can only describe planes, and the points P' and P'' follow them in exactly skew-symmetrical figures. The perpendicular distance from $O'O''$ to the planes described by points L',L'' is equal to the calibrated principal distance. Since the distances $O'L'$ and $O''L''$ vary with the movement of point P, each of the space rods slides in a bushing at points L' and L'', and another universally supported sleeve guides each rod respectively in the two projection centers O' and O''. If terrain point P is to be reconstructed, the point of intersection of the space rod axes is moved spatially in such a way that the points P' and P'' are simultaneously covered by the outer ends of the lazy tongs. These image points can be observed by the optical system, the line of sight of which passes through the outer ends of the lazy tongs.

The spatial movement of the point P is not freehand as in the Autograph A6, but is similar

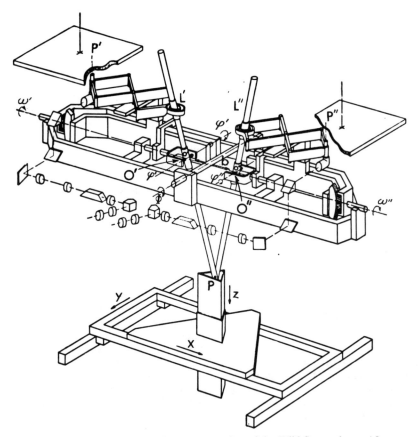

FIGURE 12-54. Diagram of the construction of the Wild Stereoplotter A8.

to the Wild A7, in which the XY-carriages are driven by handwheels through threaded spindles. This agreement, which is normal for almost all precision plotters, is one of the essential differences from topographic plotters. It allows the same enlargement ratio from photograph to map as in the universal precision plotters, and similar precision.

In figure 12-54 the axes for relative orientation are indicated. The two photographs can be rotated independently around the primary axis, lying in the base direction. The two secondary axes δ' and ϕ'' intersect the ω axis at points O' and O''. Finally each picture can be rotated around a κ axis passing through the principal point of each photograph, and perpendicular to the image plane.

The instrument has no by or bz base components. Consequently, only rotations can be used for relative orientation. For absolute orientation, there may be an inclination in Y-direction around the individual ω axes. In the X-direction, however, the longitudinal inclination of the projection system is imposed with a common movement. The main frame supporting the cameras is then rotated around the Φ axis located in the middle of the supporting arch, visible in figure 12-54. The handwheel for the rotation of the

model is conveniently located at the left front of the machine. The model may be scaled by changing the base length $O'O''$.

12.4.3.2.3 OPTICAL SYSTEM

The elements of the optical system of the A8 are shown schematically in figure 12-55. Through the eyepieces O_6, a part of the photograph, 40 millimetres in diameter, and magnified 6 times, is visible. In the plane of the photograph, the floating mark covers a circle of 60 micrometres in diameter.

The elements of the rotable telescope, from O_1 to O_2, are connected to the end of the lazy tongs. This part consists of the objective O_1, which focuses the plane of the photograph sharply in the plane of the floating mark M. The only adjustment of the optical elements which influences the geometrical precision of the instrument is the positioning of the floating mark relative to the photograph, as determined by the objective O_1. All other optical elements have no geometrical influence on the measuring accuracy. The prism P_1 and the objective O_2 are oriented by the auxiliary swinging girder in such a way that the pencil of rays always falls on prism P_2. Since the distance between prisms P_1 and P_2 is variable, the objective O_2 must have a focal distance equal

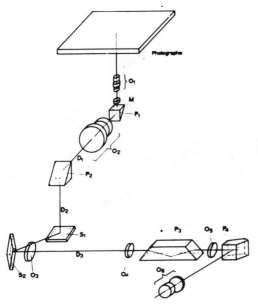

FIGURE 12-55. Optical system of the Wild A8.

These may be produced from black and white or color photography under normal room lighting conditions and with a format up to 500 by 280 millimetres.

The process is carried out by the optical scanning in the left-hand projector. Small elements of the photo are rectified by optical-electronic means and projected onto a film drum. The height information required for the differential rectification is taken from the stereomodel where the operator controls the floating mark with the right hand wheel while the model is being scanned. The scanning speed can be varied within the profile by a factor of 15, to accommodate the changes in steepness of terrain. By the proper selection of the slit mask (2 to 12 millimetres) according to the terrain, the operator can optimize the scanning time for the model.

The differential rectification of the photo-imagery is accomplished through the optical train which incorporates a Dove prism to rotate the image elements and zoom optics to correct for the changing scale. A gray wedge regulates the exposure as a function of the scanning speed and the zoom enlargement. An exposure meter incorporated in the optical train after the zoom optics allows the gray wedge to be adjusted to its initial setting. All of the preceding operations associated with the differential rectification are controlled by an analogue computer, which appears to the left of the Autograph A8 in figure 12-56.

to the distance to the floating mark M, so that a parallel bundle of light emerges from O_2. The prism P_2 directs the pencil of rays parallel to its axis of rotation onto mirror S_1. P_2 is located on the primary ω axis. The optical elements from the photograph to the prism P_2 are connected to the camera body, which has three degrees of freedom (ϕ, ω, and base). The purpose of mirrors S_1 and S_2 is to direct the pencil of rays from the movable camera body into the eyepiece box fixed to the supporting arch. To make this possible, mirrors S_1 and S_2 are controlled by a telescopic tube, attached by universal joints to the camera body and to the supporting arch.

From the mirror S_2, the pencil of rays passes through the objective O_3, which causes an intermediate image to be formed. Objective O_4 again establishes the parallelism of the rays, necessary before they can pass through the Dove prism P_3. This allows elimination of the image rotations which can occur in the case of big rotations of the cameras for relative orientation. Through prism P_4, the objective O_5 forms an image which is finally observed through the wide-angle eyepiece O_6.

The parellel path of the rays, established because of the Dove prism between O_4 and O_5, also makes it possible to adjust the separation of the eyepieces. The eyepieces have a centering mount which allows compensation for the vertical and horizontal components of any error due to squint in the observer's eyes.

12.4.3.2.4 WILD PPO-8 ORTHOPHOTO EQUIPMENT

The A8 Autograph may be equipped with the PPO-8 for the on-line generation of orthophotos.

12.4.3.3 WILD AUTOGRAPH A9

The Autograph A9 was designed by Wild as their first restitution instrument for super-wide-angle photographs. When the Super Aviogon came on the market, it was necessary also to present to the photogrammetric world an instrument suitable for the photographs taken with this camera. Otherwise, there would have been no other possibility than to apply the use of anamorphic bundles of rays, the solution accepted by Russian photogrammetrists.

Wild then tried to overcome the difficulties of designing a super-wide-angle restitution instrument applying the mechanical principle, by reducing the scale of the photographs twice, and also reducing the size and weight of all the construction elements of the instrument.

Thus, in its A9 Autograph, Wild created a high-precision instrument which can use reduced vertical wide-angle or super-wide-angle aerial photographs (fig. 12-57). It is designed principally for aerotriangulation and for mapping at scales from 1:25,000 to 1:100,000. The 11.5 × 11.5 centimetre dispositives for plotting are obtained from original negatives 23 × 23 centimetres by suitable reduction in the Wild A9 fixed ratio printer. The effects of residual distortion of the taking camera and of earth curvature and atmo-

FIGURE 12-56. Wild Autograph A8 with PPO-8 orthophoto equipment.

spheric refraction can be eliminated by printing through optical correction plates.

Although the average user may consider such reduction as a disadvantage, it must be admitted that this is not very important from the photographic point of view because of the use of fine-grain diapositives. Nevertheless, with a reduction 2:1, it can be assumed that the mechanical sources of errors in the projection system of the instrument will not be reduced in the same proportion. Consequently, the errors, expressed in micrometres in the original-size negatives, will be slightly larger than when original-size diapositives are used.

In 1960, Wild brought onto the market a topographic plotter, the Aviograph B9 (*see* 12.4.3.6), which also uses reduced-scale diapositives and is considerably cheaper than the Autograph A9. Thus, the logical equipment of an organization producing small-scale topographic maps from super-wide-angle photographs was one A9 for aerotriangulation and at least three or four Aviograph B9's for plotting from the same reduced

diapositives. This approach is no longer popular and so the manufacture of both the Autograph A9 and its companion plotter, the Aviograph B9, has been discontinued.

12.4.3.4 THE WILD AUTOGRAPH A10

12.4.3.4.1 INTRODUCTION

The Wild Autograph A10, figure 12-58 was developed by Wild Heerbrugg as a high-order precision photogrammetric instrument for aerotriangulation and large-scale stereoscopic plotting from photographs from all usual cameras, including super-wide-angle. The plotter will accomodate monochromatic and color photography.

The A10 was specifically designed to be used in conjunction with one or more Wild B8 Aviographs (section 12.4.3.5). The primary purposes of the A10 then would be to accomplish large scale plotting from super-wide-angle photography and to provide aerotriangulation (by the independent model method) for use by the B8S

FIGURE 12-57. Wild Autograph A9 equipped for aerotriangulation, with coordinate registration EK5 which has since been replaced by the EK 22.

Aviographs in the plotting of small and medium scales up to 1:2000.

The versatility of the A10 Autograph is such that stereoscopic plotting is possible not only from near-vertical and terrestrial photographs but also with photography from theodolites, stereometric cameras and microscopes.

12.4.3.4.2 GENERAL DESCRIPTION

The Wild A10 is designed to accept photography from cameras with angular fields of view up to 133^g, focal length from 86 to 308 millimetres, and formats up to 23 × 23 centimetres. The photography can be in the form of diapositives or negatives, glass plates or film base and either monochromatic or color with frame overlaps of 80% or less.

12.4.3.4.3 OPTICAL SYSTEM

The optical system is color corrected and provides an 8× image enlargement with orthogonal viewing across a 31 millimetre diameter field. Variable fluorescent strip lighting is used to give uniform illumination across the entire area of the diapositives or negatives. The measuring marks are .04 millimetre black circles built into the observation microscopes located directly under the photograph carriers. Parallel light rays are directed from the microscope carriage to a prism in the Y-carriage of the orthogonal XY-cross-

FIGURE 12-58. Wild Autograph A10.

slide system, then through two cardan mirrors to the eyepiece, as in the viewing system of the Wild B8S (section 12.4.3.5). Appropriate optical elements provide for image reversal (dia-negative) and rotation.

12.4.3.4.4 PROJECTION SYSTEM

The projection system employs the Wild par-allelogram in which the projection centers are separated by fixed distances. In this system, the conjugate light rays are simulated by two space rods which are suspended in the projection centers by means of cardan joints in such a manner that the space rods cross each other relative to the base setting.

The photograph carriers are in the negative position and displaced to the outside of the space rod projection centers, figure 12-59. Each space rod is connected at its upper cardan joint to a tie rod, which always moves in an epipolar plane and parallel to the picture plane at a fixed dis-tance from it. The other end of the tie rod is connected to the microscope. The principal dis-tance corresponds to the perpendicular distance between the projection center and the plane of motion of the tie rod. The ϕ- and ω-axes inter-sect in the projection centers and κ-motion is about the picture center.

There are no common tilts; bx is set on the base bridge, bz is set at the lower left cardan joint and by at the right lower cardan joint.

The Wild A10 has an integral sphere-lever system for the correction of earth curvature and refraction, figure 12-60. This system displaces the Z-spindle in the vertical direction by a cor-rective value based on the model scale set by the operator on a counter. The correction for lens distortion of the aerial camera is accomplished through aspheric plate photograph carriers when required.

Auxiliary equipment includes the Wild EK22 for coordinate registration, a choice of two plot-ting tables, and PRI profiloscope, which is an accessory designed to aid in rapidly locating, for measurement, points on cross-sections required for highway design.

12.4.3.4.5 MECHANICAL DESIGN AND USE

The cross-slide guidance system of the obser-vation microscope is illustrated in figure 12-59. The space rods are cylindrical steel rods at-tached at their centers of gravity to the central

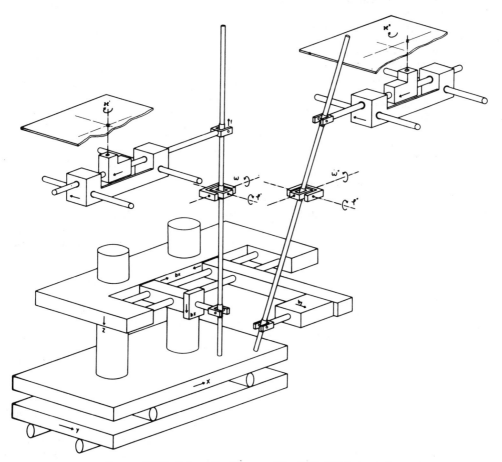

FIGURE 12-59. Schematic diagram of the Wild A10 Autograph.

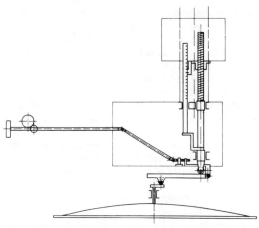

FIGURE 12-60. Earth curvature and atmospheric correction device on Wild A10.

cardan joints, thus eliminating the need for counter-balance weights. Sleeves located in the upper and lower cardan joints allow the space rods to slide through them during operation.

Two speed handwheels and a foot disk provide spatial drive and guidance of the measuring marks. Orientation is accomplished by the usual methods; the instrument dials permit the orientation elements and base components to be read respectively to .01^g or .01 millimetre. Release levers for the X- and Y-motions permit a freehand rapid movement during relative orientation. The Y- and Z-drives can be interchanged for plotting from terrestrial photographs.

The X- and Y-coordinates are displayed on six digit counters which can be set to any initial value with a least count of .01 mm. The digital Z-counter displays heights directly in feet or metres. The X- and Y-motions, controlled by handwheels, are transferred to the plotting table through a gearbox with interchangeable gears. An additional doubling gear is available, making a total gear ratio of 1.8 possible. The height counter is also provided with interchangeable gears.

The correction range for the earth curvature correction is from zero to a model scale of 1:100,000.

12.4.3.5 COORDINATE RECORDING FOR THE AUTOGRAPHS

Each of the Autographs as well as the Aviograph B8S may be equipped with the Wild EK 22 for the recording of coordinates. The model coordinates are transferred from the stereoplotting instruments to the EK 22 by means of incremental digitizers which are connected to the mechanical measuring system of the plotter. The measurements are shown as six-digit numbers in a visual display unit. The point number can consist of 14 digits, eight of which are set on a separate keyboard with a display unit. This keyboard

is on a separate stand and can be moved to suit the convenience of the operator. The remaining six digits are set on the control panel of the electronic unit.

A programming panel on the electronic unit permits the output sequence for point number, coordinates, blank spaces, tabular spaces, decimal points, and additional symbols.

The recording of coordinates can take place even while the measuring system of the plotter is in motion. This recording is activated either by a pushbutton or a foot switch. Thus it is possible to digitize the coordinates of features being plotted in pre-set intervals of distance or time. These variable intervals are 0.02 to 9.99 millimetres in increments of 0.01 millimetre, and 0.1 to 9.9 seconds variable in increments of 0.1 second.

12.4.3.6 THE WILD AVIOGRAPHS

12.4.3.6.1 INTRODUCTION

Wild produced the Aviographs B8 and B9 with the intention of creating new topographic plotters, after the production of the Autograph A6 was stopped in 1952. A second aim was to provide an instrument suitable for the restitution of super-wide-angle photographs, taken with the Wild camera RC9, having a principal distance of 88 millimetres and a photograph size of 23 × 23 centimetres.

A further characteristic desired was a correct geometrical solution as in the precision plotters, applying the same mechanical principle with observation perpendicular to the photograph. As these instruments would be used for small-scale topographic maps, medium-scale mapping for engineering projects, forestry, and so forth, they did not require the same precision, and should be less costly than the precision plotters, such as the Wild A8. Consequently, the designs could be as simple.

The main difference between the B8 and the B9 is that the B8 accommodates photographs of the original size, 23 × 23 centimetres, while in the B9 the plotting is carried out with the same reduced diapositives which were triangulated in the Autograph A9. The popularity of this latter concept has waned and so the manufacture of the B9 has been discontinued. This change in interest can be attributed to emphasis on independent model triangulation, which can be accomplished with super-wide-angle photography on the Autograph A10, or pure analytical triangulation. Either method can be used with the 23 × 23 centimetre plates and the plotting carried out on the Aviograph B8 at original photo-scale.

On the other hand, with the growing popularity of the B8, Wild has made modifications to the plotter which have not altered its basic design but have increased its versatility and accuracy. Specifically the ϕ setting screws have been relocated in front of the respective projectors for the convenience of the operator (fig. 12-64). The maximum projection distance has been increased to 350 millimetres and the height range to 138 millimetres. The components of the linear pantograph and the associated multi-position pole arm have been considerably strengthened. As a result the new instrument has been named the Wild B8S Aviograph Stereoplotter. It will be described in detail in the following paragraphs.

12.4.3.6.2 Projection System of the B8S

Figure 12-61 shows the basic principle of the design of the projection system. In diagram (b) of this figure we see the photographs in diapositive position. This arrangement is incorporated in the construction of the B8, as shown in diagram (c). The space rods, representing the corresponding rays, slide in cardan joints O' and O''. The mechanical photo-points are L' and L'', which are below the projection centers. As in the A8, the two plotting cameras are displaced outward on both sides. The swinging girder of the A7 and the lazy tongs of the A8 are, however, replaced by an orthogonal XY-cross-slide system (similar to the Autograph A10). Two tie rods of constant length a, which move on this crossrail system parallel to the photo-planes, are connected by cardan joints to the space rods in the mechanical photo-points L' and L''. The perpendicular distances from O' and O'' to the planes of L' and L'' represent the principal distances f. At their outer ends, the tie rods carry the viewing

microscopes with their measuring marks M' and M''.

Figure 12-62 shows the details of this projection system. The space rods are separated at their lower ends by a distance s, and mounted in a hinge, as shown in diagram (d) of figure 12-61. This is introduced for constructional reasons, and means that a Φ inclination of the base must be transferred to the distance s in the hinge. This transfer must take place around the primary axis of the hinge, which is parallel to Y in the vertical position of the projection plane.

The elements of orientation are the same as in the Wild A8: rotations of the plotting cameras are provided around three axes, of which ω is the primary axis, while, for absolute orientation, adjustments are available for base length, longitudinal tilt Φ, and lateral tilts ω' and ω''.

The instrument can also be used for 15 centimetre wide-angle photographs, but must be specially adapted for this purpose by interchanging parts of the plate-holder supports.

Correction plates can be used to compensate for the combination of lens distortion, earth curvature and atmospheric refraction. These plates must be uniquely ground according to the characteristics of the lens in the taking camera and the flight altitude at which the aerial photography is flown. This compensation is more important with super-wide-angle photography from the Super-Aviogon than with wide-angle photography from the Aviogon because of the lens characteristics of the former as well as the in-

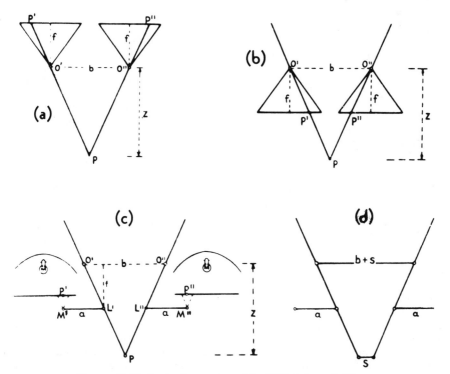

FIGURE 12-61. Projection system of the Wild Aviograph B8.

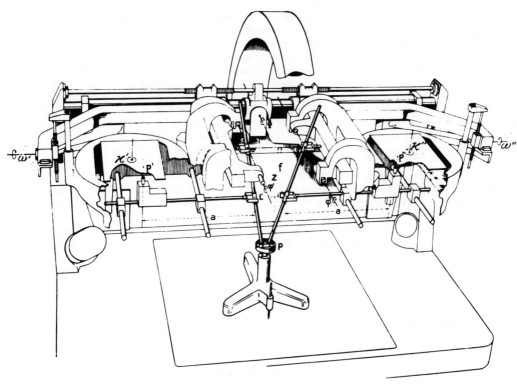

FIGURE 12-62. Schematic drawing of the Wild B8, without lamps, *etc.*

creased angle of coverage. A ratio printer may also be used to prepare diapositives in which the effects of lens distortion, earth curvature and atmospheric refraction have already been compensated.

12.4.3.6.3 OBSERVATIONAL SYSTEM OF THE B8S

The field of view of the eyepieces of the B8S has a diameter of 22 millimetres; the diameter of the measuring mark is 70 micrometres. The image presented to the eye has a magnification of 6×. Following the optical train (fig. 12-63), we start at the side of the photograph with the microscope carriage containing the two lenses, the measuring mark, *M,* and the pentagonal prism; the rays are transmitted through the rectangular prism which is moved in the *Y*-direction by the *Y*-carriage. The two mirrors are mounted in cardans and their function is to direct the bundle of parallel rays into the stationary part of the viewing optics and to keep it centered at all the camera positions as these vary with the setting of the orientation elements. The next lens collects the rays and passes them through the Abbé prism. The mirrors are steered by levers and a guide rod in such a way that the half-angle condition of the law of reflection is always maintained. The purpose of the Abbé prism is to convert for image rotation resulting from large tilt differences ($\Delta\omega$). This prism has the same

function as a Dove prism in the parallel rays in other plotting instruments.

12.4.3.6.4 PLOTTING AND MEASURING DEVICES OF THE B8S

The Wild B8S has the same orientation movements as the A8, but with freehand movement on the plotting table rather than handwheels. Plotting can be carried out directly at the model scale with the tracing stand shown in fig. 12-64, or a linear pantograph and plotting table can be attached for plotting at scales larger or smaller than model scale. This pantograph is of new design relative to the B8, and all components have been considerably strengthened, to increase the accuracy of the instrument. Non-slip and expansion-free punched steel tapes are now used for the pantograph extensions instead of the previous steel wire. The new model with the tape drive permits enlargement or reduction of model scale in 19 different ratios from 2:5 to 5:2. The raising and lowering of the plotting pencil is controlled by a foot pedal.

12.4.4 KERN STEREOPLOTTERS

Kern now manufacturers two plotters, the PG2 and the PG3. The prototype of the PG2 was first exhibited in 1960. It is a topographic plotter, designed for small- and medium-scale map production. Because of its rapid acceptance and popularity in the mapping community, Kern was

FIGURE 12-63. Optical system B8S (photograph courtesy of Wild).

FIGURE 12-64. Photograph of Wild B8S Aviograph with EK22 coordinate registration device on left.

encouraged to retain the geometrical principles of the plotter but develop a precision plotter for large scale mapping, which is known as the PG3. It made its first appearance at the International Congress on Photogrammetry at Lausanne, 1968.

12.4.4.1 KERN PG-2

12.4.4.1.1 GENERAL

The Kern PG-2 (*see* fig. 12-65) can be considered from the user's point of view as an instrument with an exact solution. It is designed for map production from near-vertical, wide- and super-wide-angle photographs having a principal distance of 80 to 155 millimetres. The largest suitable format for this instrument is 23×23 centimetres. Its main significant properties are:

(a) special mechanical solution to the projection system;
(b) the model is leveled by tilting the drawing table;
(c) polar pantograph with the parallelogram system; and
(d) relatively small dimensions.

12.4.4.1.2 PROJECTION SYSTEM

The projection is solved mechanically. In the model space, the lower ends of the normal space rods are attached to the base bridge, while the upper ends glide through the sleeves of the central gimbal joints which represent the perspective centers. The positions of the perspective centers are fixed, so that a Zeiss parallelogram is applied. The space rods (projecting rays) above the perspective centers (in the camera space) are projected onto two perpendicular vertical planes, XZ and YZ. These projections are realized mechanically with two rods (X and Y) for each camera. The rods are coupled to the central gimbal joints in such a way (fig. 12-66) that the above-mentioned projection onto two planes is accomplished. The camera tilts (ϕ,ω) are imposed, in their opposite sense, to the X- and Y-rods. The real photo-planes thus remain in a fixed (horizontal) position.

The number of orientation elements is restricted to a minimum. We may distinguish these according to the parts of the instruments on which they are imposed:

(1) the left-hand camera ($C_x{}'$, $C_y{}'$, κ', $\omega_x{}'$, and $\omega_y{}'$);
(2) the right-hand camera ($C_x{}''$, $C_y{}''$, κ'', $\phi_x{}''$, and $\phi_y{}''$);
(3) the base bridge (bx symmetrically, bz''); and
(4) the drawing table (Φ, Ω).

The two Y-rods are connected to the Y-carriages along which the observation system moves, and the two X-rods to the X-carriages for the plateholder movement. Consequently, the motions for scanning the photographs are made partly by the optical system and partly by the photographs themselves, so that a relatively simple optical system is possible.

Because of the fixed horizontal position of the X- and Y-carriages, no stress due to gravity is introduced into the projection system by the ϕ and ω tilts.

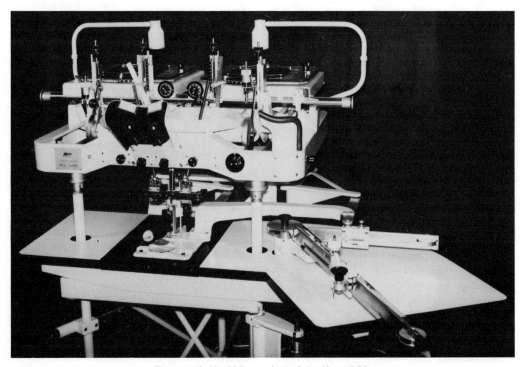

FIGURE 12-65. Oblique view of the Kern PG2.

FIGURE 12-66. Central gimbal joints in the Kern PG2 and their connections to the X and Y rods.

The geometrical relations on a tilted photograph (*e.g.*, for ϕ'') are shown in figures 12-67 and 12-68. The coordinates of an arbitrary photopoint B are:

$$x'' = c \tan (\alpha - \phi'') \qquad (12.13)$$
$$y'' = (c - \Delta c'') \tan\beta$$

where

$$\Delta c'' = (x'' + c \tan \phi''/2) \sin \phi''$$

or

$$\Delta c'' = c \sin\phi'' \tan (\phi''/2) + x'' \sin\phi''.$$

By substituting

$$c \sin\phi'' \tan (\phi''/2) = 2c \sin^2 (\phi''/2) = \Delta c_1''$$

and

$$x'' \sin\phi'' = \Delta c_2'',$$

we obtain

$$\Delta c'' = \Delta c_1'' + \Delta c_2''.$$

Thus, the variation of the principal distance $\Delta c''$ consists of a constant part for a given photograph ($\Delta c_1''$), and a part which varies according to the position of the point on the plate ($\Delta c_2''$). In practice, $\Delta c_1''$ should be considered only when tilts are larger than one grad.

Similar relations exist for the ω' tilt:

$$x' = (c - \Delta c') \tan \alpha \qquad (12.14)$$
$$y' = c \tan (\beta - \omega')$$

and

$$\Delta c' = \Delta c_1' + \Delta c_2' = 2c \sin^2 (\omega'/2) + y' \sin\omega'.$$

These relations must be solved correctly by the projection system. As mentioned above, the tilt of the photograph is replaced by the inclination in the opposite direction of the corresponding rod. If the space rod is inclined at an angle α, the X-rod is simultaneously inclined at $(\alpha - \phi'')$ and the above relation $x'' = c \tan (\alpha - \phi'')$ is fulfilled. The relation in y'' direction is more complicated because of the $\Delta c''$ correction. The value of the constant part $\Delta c_1''$ can be computed or taken from the appropriate nomograph (fig. 12-69) and introduced directly on the c_y'' dial, after the ϕ'' tilt is known. The variable part $\Delta c_2''$ is controlled automatically by the correction link K shown in figure 12-70. This link is tilted at an angle γ and slides, in the x-direction, along a feeler. The vertical displacements of the feeler

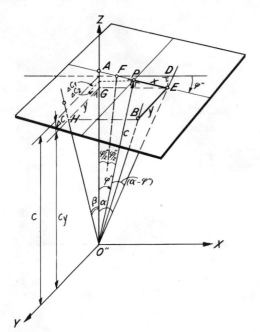

FIGURE 12-67. Geometrical relations on a tilted photograph.

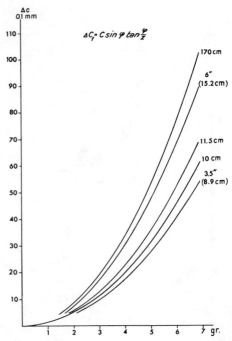

$$\Delta C_z = C \sin \varphi \ tan \frac{\varphi}{z}$$

FIGURE 12-69. Diagram for the Δc: corrections.

are proportional to the variations in x''. The pivot of the correction link coincides with the position of the feeler, for $x'' = O$. Consequently, the vertical shifts of the feeler are: $\Delta c_2'' = x'' \tan \gamma$. These are transferred to the slide by levers, for the principal distance c_y''. Theoretically, the required correction should be $\Delta c_2'' = x'' \sin\phi''$. Consequently, the condition is: $\sin\phi'' = \tan \gamma$, or $\gamma = $ arc tan $(\sin\phi'')$. The tilt γ is given by a screw actuating in the vertical direction. The screw is provided with a linear graduation for ϕ_2'' although, according to the above condition, it should be nonlinear.

The residual errors due to the accepted approximation are determined by the relation:

$$\partial = \sin\phi'' - \tan \gamma = \sin\phi'' - \phi'' \frac{\sin 5^g}{5^g}. \quad (12.15)$$

These can be represented graphically as shown in figure 12-71. For the values of $\phi'' = 0^g$, $+ 5^g$, or -5^g, the error is zero.

The maximum error in the range $\pm 6^g$ is about 25^{cc} (one fourth of a centigrad), or practically negligible. The relations between ω' and δ in the

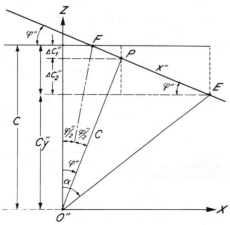

FIGURE 12-68. Projection onto the XZ-plane.

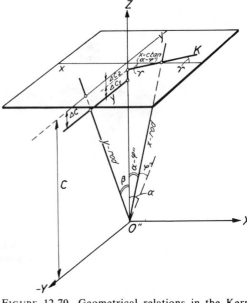

FIGURE 12-70. Geometrical relations in the Kern PG 2.

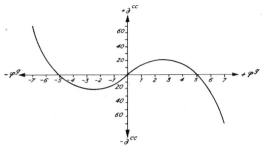

FIGURE 12-71. Residual errors in the Kern PG 2.

the first iteration. After the y parallaxes have been eliminated, the elements which influence only the x parallaxes have to be corrected (c'_x for $\Delta c = 2c \sin^2(\omega'/2)$; ϕ''_x and ω'_x are set on the same readings as ϕ''_y and ω'_y). These elements can also be used independently for the elimination of model deformations, without disturbing the relative orientation (introduction of y-parallaxes).

12.4.4.1.3 OBSERVATION SYSTEM

Because of the fixed horizontal positoin of the cross-rail system of the photographs and the shared movements in X- and Y-directions, a simple optical system can be used. This has the following significant properties:

(a) illuminated interchangeable floating marks (dot or diagonal cross) projected in the optical path;
(b) interchangeable magnification (2, 4, and 8 times) by a rotatable 2-lens element;
(c) the exit pupils of the eyepieces sufficiently far away to allow the operator to wear spectacles if he so desires; and
(d) Dove prisms are not required.

The lamp assemblies are attached to the Y-carriages (above the observation system) so that they are fixed with respect to the moving parts of the observation system.

The optical system is shown schematically in figure 12-72. The mirror 8 has a semisilvered surface on which the measuring mark is projected. The distance between the lenses 10 and element 11 is variable. Element 11 is the rotatable two-lens element for interchanging the mag-

y-direction of the left-hand camera are similar. Let us assume the correction link K' for the automatic corrections $\Delta c_2'$ is inclined at an angle δ. The condition $\tan \delta = \sin \omega'$ is solved in the same way as $\tan \gamma = \sin \phi''$ for the right-hand camera.

The shared orientation elements (c_x, c_y, ϕ''_x, ϕ''_y, ω'_x, ω'_y) are independent as far as their influence on the x- and y-parallaxes is concerned. For elimination of y parallaxes, only those elements which influence them, c_y, ϕ''_y, ω'_y, κ', κ'', and bz, can be used. In addition to the conventional empirical orientation procedure, the approximate value of ϕ''_y must be determined

$$\phi''_y = \frac{(\phi'')^3 + (\phi'')^5}{2}$$

for the constant part of the $\Delta c''_y$ correction ($\Delta c''_1 = 2c \sin^2(\phi''/2)$. This correction must be applied in

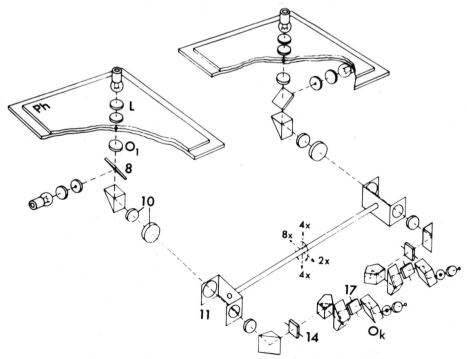

FIGURE 12-72. Plan of the observation system of the Kern PG2.

nification. Elements 14 and 17 are the plane-
parallel plates for the squint adjustment in horizontal and vertical directions (as visible in the
field of view of the eyepiece).

The illumination of the two measuring marks
(with yellow light) is adjustable simultaneously
with one rheostat, and the illumination of the
photographs is adjustable independently, by two
rheostats.

12.4.4.1.4 PLOTTING AND MEASURING DEVICE

The devices for measuring and plotting the
model set up by the projection system consist of
the following main elements:

(1) scanning carriage;
(2) drawing table;
(3) paralellogram system; and
(4) polar pantograph.

The scanning carriage glides with its guiding
plate on the gliding surface of the drawing table.
The movements are free-hand. The Z-columns
and the device for height measurement are both
attached to the guiding plate. This device consists of the driving disk for the Z-carriage, the
interchangeable-scale disk, and the driving
mechanism. The Z-carriage, with the base bridge
and its counterweight, moves along the Z-
column. The position of the counterweight with
respect to the Z-bridge is adjustable.

The drawing table has its normal drawing
surface at the right-hand side, and also has, in
front of the operator, the gliding surface for the
scanning carriage (fig. 12-73).

The longitudinal (Φ) and lateral (Ω) tilts for
leveling are imposed on the drawing table by appropriate screws which are attached to the legs
of the instrument. The Φ screw actuates rotations about the axis passing through the middle
of the gliding surface (Φ—Φ), and the Ω screw
actuates rotations about the eccentric axis
(Ω—Ω), displaced toward the operator, as shown
in figure 12-73.

The scanning carriage, the parallelogram system, and the polar pantograph are all tilted,
along with the drawing table, for Φ and Ω. The
influence of this tilt can be strongly reduced by
the use of a "friction brake." The area of the
gliding surface and the Z-range of the instrument
are limiting factors for variations of the model
scale, particularly with the super-wide-angle
photography.

The parallelogram system (aa' bb' cc' dd') in
fig. 12-73 has two functions:

(a) as a parallel guidance for the scanning carriage
    (because of the base bridge); and
(b) to transfer the movements of the scanning car-
    riage to the polar pantograph.

Because of the great separation between the
scanning carriage and the connection with the
polar pantograph, a stable parallelogram system
is needed to obtain the required, very precise,

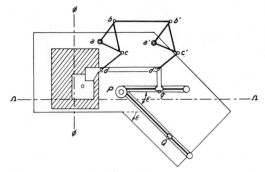

FIGURE 12-73. Kern PG 2 drawing table with parallelogram system and polar pantograph.

transmission of the movements. Plotting at
model scale may be accomplished with a simple
arm which is connected to the scanning carriage.
For plotting at small scales the instrument can be
equipped with a stable, one-arm, polar-type
pantograph (PG2-R) with tape and disk transmission. Two carriages move along the arm; the
outer carriage is connected to the parallel guide
(representing the model point), while the other is
connected to the tracing head. The two carriages
are interconnected by two steel tapes which
wrap around two disks connected to the polar
axis and also to the end of the arm. The lower
tape is driven by the movement of the parallel
guidance arm while the upper tape drives the
tracing carriage. The change in scale is determined by the difference in diameter of the two
disks on the polar axis and the initial distances of
the two carriages from the polar axis. For enlargement the polar axis is located "inside"
(near the instrument with respect to the carriages); for reduction the polar axis is located
"outside." The maximum scale variation with
this pantograph is 1:2 (enlargement) to 1.8:1 (reduction).

Large-scale plotting requires the use of the
PG2-L two-arm polar type pantograph with tape
and disk transmission. The arms are separated
by a fixed angle and the two carriages move
along the two arms respectively. With this configuration the polar axis is always "inside." The
enlargement range is from 1:1 to 1:3.33. The instrument for both pantographs is provided with
an electromagnetic remote control for the
pencil-lifter.

In March 1974 Kern introduced their new
semi-automatic stereo-plotting system, called
the PG2-AT. Basically it consisted of a small
coordinatograph attached to the base carriage.
The motion of the base carriage is transferred to
incremental encoders by means of precise rack
and pinion elements of the coordinatograph. The
Z-encoder is driven directly by the Z-disk. The
other major component of the system is the automatic plotting table. A computer serves as the
link between the PG-2 coordinatograph and the
automatic XY-plotter. With this arrangement the

X- and Y-axes can be individually scaled by any factor from 0.1 to 9.9. The following year Kern introduced the DC-2 Interacting Digitizing-Graphic System. By adding a Hewlett Packard Programmable Desk Calculator into the system, the capability has been expanded to automatically plot and orient twelve symbols as well as alpha-numeric characters. The system permits the automatic recording of discrete points (as required for aerotriangulation) or the digital recording of a continuous series of points representing profiles or linear map features.

12.4.4.1.5 KERN OP-2 ORTHOPROJECTOR

For the preparation of orthophotographs, Kern has designed the Kern OP-2. It consists of a projector, light source, and projection platform with Z-drive. The projector will accommodate photography with a focal length of 152 ± 2 millimetres, and projects at 1:1 mean image scale. A duplicate of either the left or right image of the pair can be used. The photo orientation requires two steps, *i.e.,* transfer of the camera inclination to the projector in the OP2, and coupling of the tracing stand of the PG-2 to the slit drive in the OP-2 in identical model points. The camera inclination is derived from the positions of the photo-center in relation to the nadir point. This is accomplished using the aerotriangulation accessories of the PG-2 and a small calculator. The model coordinate system of the PG-2 is then related to the nadir point of the image which is to be reprojected. With the aid of a digitizer, the starting point in the PG-2 can then be exactly related to the corresponding position of the slit mask.

The scanning motions of the OP2 are driven by the encoders in the model space and the electronic interface to the stepping motors which drive the slit and the projection platform. The exposure system is connected to the slit drive through a parallelogram system. The light source is a Xenon flashtube, and the intensity varies automatically with the part of the photo-format being exposed and the projection distance. The model is scanned manually in the PG-2 while the orthophoto is concurrently exposed on the OP-2. The schematic of the OP2 is shown in figure 12-74.

12.4.4.2 KERN PG-3

12.4.4.2.1 GENERAL

The Kern PG-3 is a precision plotter designed for the compilation of large maps, the generation of digital data for engineering and cadastral purposes, and for independent model aerotriangulation. The plotter will accommodate normal,

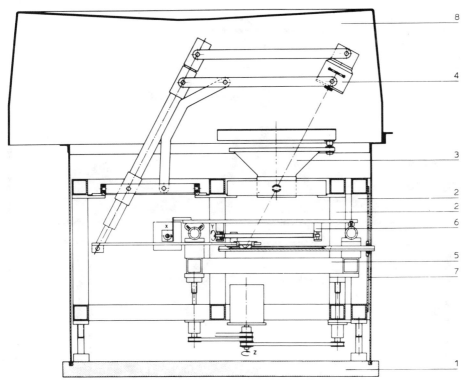

FIGURE 12-74. Sideview of the Kern OP2 orthophoto equipment

1. Base plate
2. Frame
3. Projector
4. Light source
5. Film platform with Z-drive
6. Slit mask guides with X- and Y-drive
7. Casing
8. Cover

wide, or super-wide angle photography with a 23 × 23 centimetre format or smaller and with focal lengths ranging from 84 to 310 millimetres. Although based on the same basic geometric principle as the PG-2, it has its own distinguishing characteristics:

(a) the projection system is inverted with the model space above the perspective centers and the mechanical projection extending upward;
(b) the photo-carriers are horizontal and coplanar;
(c) the conjugate rays are represented by rods in the model space and by pairs of straight-edge scales in the projection space;
(d) traversing the model is accomplished by movement of the photo-carriers in the x-direction and by movement of the prism-carriages in the y-direction;
(e) correction devices (mechanical computers) are used to compensate for the ϕ'' and ω' tilts;
(f) the orientation motions κ', κ'', and ω' are coupled for ease of parallax removal;
(g) common rotation Φ and Ω are provided for absolute orientation; and
(h) a cross-slide system driven by hand-wheels and a foot disk provides the means for measuring coordinates, while rapid scanning of the model is possible without loss of coordinate reference.

12.4.4.2.2 PROJECTION SYSTEM

The projection system uses the Zeiss parallelogram, but in an inverted position, where the projection centers are below the photo-carriers, but the split points of model space are projected upward (fig. 12-75).

The novel arrangement of the projection system can best be appreciated from the series of diagrams in figure 12-76. Starting with the photograph in negative and then positive position (diagrams A and B), the projection system is then inverted with model space projected upward (diagram C). In diagram D the space rods are separated by a fixed arbitrary distance and the bz is transferred from the projectors to the split-point base bridge. In diagram E it can be noted that the photo-carriers are retained in the horizontal position and the tilts in the photographs are transferred to the x- and y-rods in the projector space. Diagram F shows the relationship of the space rods, the x- and y-rods, and the coordinate reference system of the instrument. The bundle of rays in the photospace of each projector is represented by the x- and y-rods which are the projection of the rays in the XZ and YZ planes (fig. 12-77).

The x- and y-rods are connected to the photocarriage by a spring-loaded ball race. At distances away from the carriage equal to the principal distance are the respective rotating axes for the rods. Since the settings are made independently on each rod, connections for differential

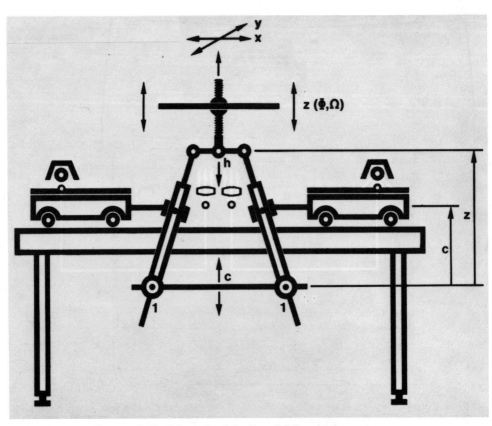

FIGURE 12-75. Schematic of the Kern PG-3 projection system.

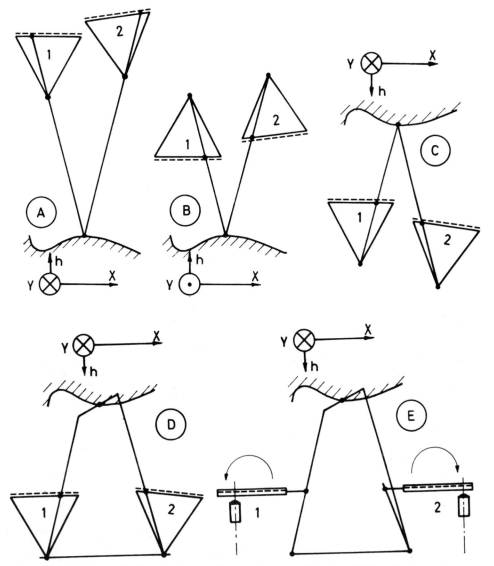

FIGURE 12-76. Schematic diagram showing the development of the Kern PG-3 projection system.

shrinkage can be taken into consideration by setting different values for the principal distance on each rod.

The rotational elements of relative orientation are divided between the two projectors. The κ rotation for both projectors is accomplished by rotating the photo-carriers about the principal points. On the other hand, ω is set only on the left-hand projector and ϕ on the right. Since the photo-carriers are not tilted, the connection devices to compensate for the relative tilts of the photographs are quite complicated; they are similar, but located at right angles to each other. The functional relationship between the photo-coordinates of an arbitrary point and the geometric principles of the plotter is given by the following equations

$$x'' = c_x \tan (\alpha'' - \phi''),$$
$$y'' = (c_y + \Delta c''_y) \tan \beta''$$

where

$$\Delta c''_y = (x'' - \Delta x'') \sin \phi''$$

and

$$\Delta x'' = c_x \tan (\phi''/2).$$

The mechanical realization of the relation for x'' in the projector, where the photo-plane is horizontal, is shown in fig. 12-78. The x''-rod is tilted with respect to the correct projection of the ray over $-\phi''$. This tilt is set by a screw, located at a distance, K, from the axis (fig. 12-79). Thus, the screw is displaced over $K \sin\phi''$ from its zero position.

For the mechanical realization of y'' resulting from the tilt ϕ'', the principal distance, c_y, set on

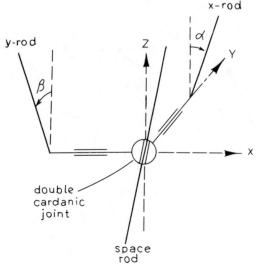

FIGURE 12-77. Diagram of the x- and y-rods with respect to the space rod. These rods form angles α and β with respect to the XZ- and YZ-planes. In a tilted photo these angles are not the same as the projections of the space rod in these planes.

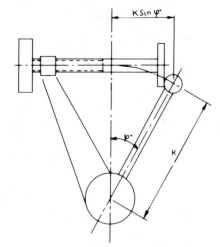

FIGURE 12-79. The mechanical setting for tilt, ϕ'', on the x-rod.

the y''-rod, must be altered by $\Delta c_y''$. A correction link (fig. 12-80) is attached to the photo-carriage for this purpose. The link can be inclined by an angle, γ, using a screw which is at a distance, K, from the link pivot. This screw is coupled with the X''-rod setting screw, and is therefore also displaced from its zero position by $K \sin\phi''$. Thus, the inclination of the correction link will be

$$\tan \gamma = K \sin(\phi'')/K = \sin(\phi''). \qquad (12.17)$$

A feeler is placed on top of the correction link, its vertical movements $\Delta c_y''$ being transferred

mechanically to the ball race on the y''-rod. When x'' and ϕ'' are zero, the feeler is located at the pivot of the correction link. However, in general it is separated from the pivot by a distance $(x'' - \Delta x'')$, where x'' is the displacement of the photo-carriage and the correction link from the zero position, and $\Delta x''$ is the differential shift of the correction link with respect to the photo-carriage (by means of a special mechanism). The mechanism for $\Delta x''$ is essentially a proportional link with a setting screw (for $c/10$) at a distance of 31.95 mm along one arm, the other arm being of variable length ($K \sin\phi''$). The screw, which sets the inclination of the correction link, has an upper and a lower extension. The upper end supports the correction link, while the lower end is located at the pivot of the proportional link ($K \sin\phi''$ equal to zero). The proportional link is inclined by the lower setting screw, where a tenth of the principal distance ($c/10$) is set. From similar triangles the following ratio is obtained

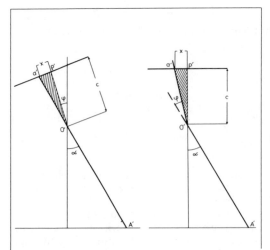

FIGURE 12-78. The right-hand diagram shows the x-rod forming an angle ϕ with the projection of the space-rod in the XZ-plane. This is due to the relative tilt ϕ between the photos.

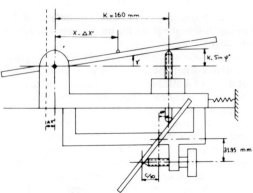

FIGURE 12-80. The mechanical correction to c_y for the mechanical realization of the relation to a photo coordinate y'' due to the tilt, ϕ''.

$$\frac{\Delta x''}{K \sin\phi''} = \frac{c/10}{31.95} \cdot \tag{12.18}$$

By substituting $K = 160$ mm and $\sin\phi'' = 2$ $\sin(\phi''/2)\cos(\phi''/2)$

$$\Delta x'' = c \frac{32}{31.95} \tan\left(\frac{\phi''}{2}\right) \cos^2\left(\frac{\phi''}{2}\right)$$

$$\approx c \tan\left(\frac{\phi''}{2}\right). \tag{12.19}$$

In this approximation, the maximum error is about 16 micrometres (for $\phi'' = 7^g$), which has a negligible effect on the $\Delta c_y''$ correction. The relations for the left-hand projector are similar. The ϕ' and ω'' tilts can be set separately for the x- and y-directions. Thus, model deformations can be corrected without affecting y-parallaxes.

The rotations κ', κ'' and ω' can be set by means of stepping motors, and they can be mutually coupled for speeding up the process of relative orientation: $\Delta\omega'$ and the two κ rotations ($\kappa' = \kappa''$) can be coupled, by means of proportional gears, in such a way that there will be no change in the y-parallax of a pre-selected point when changing ω'. The proportional factor $(Y^2 + Z^2)/Zbx$ must be numerically determined and set using the proportional gears.

The two split model points on the base bridge are set symmetrically to bx, but for $b\phi$ ($bz' = -bz''$) the whole base bridge is tilted. Electric motor drives are provided for the base components. The tracking device is set at the common tilts, ϕ and Ω. Appropriate measuring devices are attached to the setting screws for all orientation parameters. Cylindrical space rods are attached to the upper cardan joints.

12.4.4.2.3 Observation System

As in the PG-2, the horizontal arrangement of the photo-planes permits the use of a simple observation system. There is only one movement in the optical train, that of the prism-carriage in the y-direction. Image rotations do not occur. Therefore, Dove prisms or the equivalent are not needed. Fixed fluorescent tubes, emitting white light, are used for illumination. A prism with a semi-reflecting surface (below the photograph) combines the luminous measuring mark with the image without any intermediate lenses. A lens, attached to the prism carriages, forms parallel rays of light which enter the revolving two-lens unit. This unit permits stepped changes in the magnification: $2\times$, $4\times$, and $8\times$; or, with a different set of eyepieces: $2.5\times$, $5\times$, and $10\times$. The eyepieces are equipped with the usual adjustments, and the distance of the exit pupils from the lenses is sufficient to allow observation with spectacles (if desired).

12.4.4.2.4 Plotting and Measuring Device

Scanning in the PG-3 is normally done by means of X- and Y-handwheels and a foot disk

for Z as in the case of most precision plotters. The instrument is also equipped with a freehand drive for rapidly traversing the model. This mode can be used for relative orientation, and also for aerotriangulation, since the connection to the coordinate reference and readout is not interrupted by the declutching. The handwheels and foot disk activate the lead screws, which have a six-millimetre pitch and serve as the measuring spindles. The specially designed nuts with ball races can move freely about the spindles with a minimum of friction. The lead screws in turn activate the tracking device located at the top of the instrument and move through an X, Y, Z cross-slide system equipped with cylindrical rails. This cross-slide system is tilted through Φ and Ω for absolute orientation.

To keep the output of the model coordinates within six digits, 0.6 millimetres was chosen as the measuring unit, since it represents one-tenth of the pitch of the lead screws. This measuring unit is further subdivided into tenths and hundredths; however, for all practical considerations, the basic measuring unit should be used as though it were a millimetre. The instrument is equipped with X-, Y-, and Z-counters for visual readout. The least count in X and Y is 10 micrometres. The least count on the Z-counter depends upon the setting transmission lever system, which can be set for any model scale to read directly in metres, feet or decimal fractions thereof, if required. The spindles can be equipped with incremental shaft encodres enabling output on nixie tube, typewriter, punch card, or tape. The ERII coordinate registration device has been designed by Kern for this purpose.

Plotting is done on a drawing table with an effective area of 1200×1200 millimetres. The table is equipped with a magnetic pencil lifter. There is fluorescent back illumination of the plotting surface which provides light for scribing or plotting on transparent material. Transmission of motion from the instrument to the table is purely mechanical. The transmission ratios between the PG-3 and drawing table range from $2.4\times$ enlargement through $3\times$ reduction using exchangeable gears. These ranges can be doubled or halved by means of levers at the gear box. The PG-3 equipped with tracing table and ER1 is shown in figure 12-81.

12.4.5 CARL ZEISS (OBERKOCHEN) PLOTTERS WITH MECHANICAL PROJECTION

It can be assumed that the changes of modern technology compelled Zeiss to abandon the projection principle of the stereoplanigraph, *i.e.*, the observation through microscopes of directly projected images. This system required plotting cameras with principal distances that were nominally the same as the focal length of the taking camera; in other words, photography

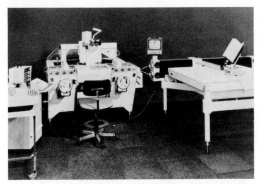

FIGURE 12-81. Kern PG-3 stereoplotter.

from 90 mm, 150 mm, and 210 mm focal-length cameras could not be accommodated by the same plotting cameras. The universality of the instrument to accommodate convergent and oblique photography was no longer an important consideration; however, the high resolution of modern photographic systems (the combination of lens and film) must be fully utilized by the plotter. The mechanical projection principle provides the best solution for present day requirements.

The first of this family of plotters, the Planimat, was introduced at the 31st Photogrammetric Week (1967) in Karlsruhe. It is the most versatile and precise of the three Zeiss plotters. It was also reasoned by Zeiss, from an economic standpoint, that the user should not have to invest in the purchase of plotters capable

of large scale map production when his requirements are to produce primarily medium or small scale maps. Thus, while adhering to the same basic design of a rigorous three dimensional mechanical solution, but with relaxed precision, two additional plotters were added to complete the series. We may classify these instruments as follows:

(a) planimat—a precision plotter of maximum versatility designed primarily for large scale mapping with 6× or more magnification from photograph to plotting scale;

(b) planicart—a lower-order precision plotter of lesser versatility designed for medium and medium-large scale mapping; and

(c) planitop—a topographic plotter designed for medium and small scale mapping.

These instruments will now be discussed in detail.

12.4.5.1 ZEISS PLANIMAT

12.4.5.1.1 GENERAL

The Planimat, figure 12-82, was developed by Carl Zeiss Oberkochen as a modern precision photogrammetric plotter which would provide a capability for small, medium, and large scale graphic plotting, digital data generation, and independent-model aerotriangulation.

The Planimat can accommodate camera focal lengths from 84 to 308 millimetres; negatives up to 23 × 23 centimetres; normal, wide and super-wide-angle photographs; near vertical and terrestrial photography; and diapositives or negatives on glass plates or film.

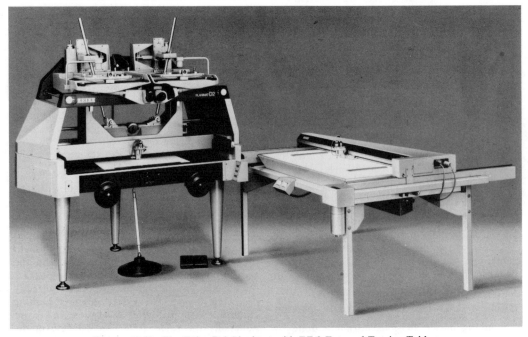

FIGURE 12-82. The Zeiss D 2 Planimat with EZ-3 External Tracing Table.

The projection system is distortion free with the standard photo-carrier plates; however, camera lens distortion can be compensated, if required, through aspheric correction plates. The photo-carrier plates are interchangeable. Grids of nine crosses, forming 90 millimetre squares, are provided on the photo-carrier plates to be used for determining the position of the perspective centers.

The instrument has both an internal and external tracing table. Auxiliary equipment is available for X-, Y-, Z-coordinate printing and electronic recording of discrete points, incremental, or automatic modes. Additional options provide the capabilities to drive the GZ-1 Ortho-projector through direct, storage, or automatic modes. Terrestrial photography plotting is accomplished with the aid of the terrestrial camera accessories.

12.4.5.1.2 PROJECTION SYSTEM

The design principle of the projection system used in the Planimat is shown in figure 12-83. A modified Zeiss parallelogram is used in which the rotation centers for ϕ and ω are fixed but the projection centers are not. Two mechanical guide rods L_A, L_B connect the perspective centers P_A, P_B, the image plane reference points D_A, D_B, and the space points G_A, G_B. Since the plotting cameras are in the positive position, the principal distance f is represented by the vertical distance from the upper cardan P_B to the central cardan D_B, and the projection distance Z is the vertical distance from the upper cardan P_B to the lower cardan G_B. The space points G_A, G_B are shifted in relation to each other for the introduction of base bx, space point G_A is displaced for introduction of bz, and space point G_B for by. The principal distance is set by displacing the perspective centers P_A, P_B vertically.

The photo-carriers (fig. 12-84) are eccentrically displaced in the Y-direction toward the operator. During scanning the photo-carriers move on orthogonal cross-slides while the observation system is fixed. Each photo-carrier can be rotated about the principle point for κ.

The primary axis of rotation, ϕ, and the secondary axis, ω, are eccentrically located with respect to the perspective center. This is best illustrated by figure 12-84 in which the perspective centers are at the back and above the plate holders while the centers of rotation of the plotting cameras are located beneath and toward the front of the plotting cameras. The fact that the cardan points of the cameras are located outside the perspective centers is a novel design. It has the advantages of yielding an especially simple optical system, a three point support of the plotting camera on the main, and it does not require a complicated mechanism to maintain the rotation of the plotting camera about the perspective center. Adjustments are accomplished by setting screws, all provided with dials, drums, and/or scales.

The space rods are attached to the lower cardan joints and slide through sleeves in the center and upper cardans.

12.4.5.1.3 OBSERVATION SYSTEM

With the exception of one mirror (fig. 12-84) which rotates with the plotting camera during ω movement, the viewing optical system is stationary. Illumination is provided by small lamps located above the photo-carriers. Light passes from the lamp through the photograph to a mirror located on the ϕ axis under the photo-carrier, then is reflected along the ϕ axis to a second mirror at the intersection of the ω and ϕ axes. This second mirror turns with the ω movement and reflects the light rays along the ω or primary axis to a series of lenses and prisms to the viewing eyepiece. Dove prisms are provided to compensate for image rotation. Luminous marks 40 micrometres in diameter are reflected into the light path as measuring marks. The observation

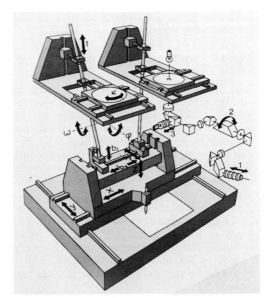

FIGURE 12-84. Principle of projection in Planimat.

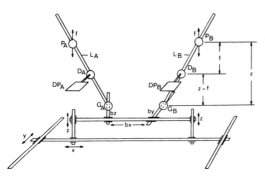

FIGURE 12-83. Schematic diagram of Planimat projection system.

system provides a 31-millimetre diameter field of view with normal viewing of 8× magnification.

12.4.5.1.4 PLOTTING AND MEASURING SYSTEM

The X- and Y-handwheels are equipped with gears for rapid and slow driving of the corresponding spindles. The guide spindles can be disengaged by lever for freehand motion during relative orientation. The spindle has steel wires inserted into the v-grooves so that the roller nuts which engage the carriages to the drive system run on hard and smooth surfaces.

Coordinates are measured with a mechanical X-, Y-, Z-coordinate counter. The least count for X and Y is .01 millimetre. The least count for Z is also .01 millimetre, although through the use of interchangeable gears, metres or feet in a variety of model scales may be read directly. Either mechanical or electronic recording devices can be used.

An earth-curvature correction attachment can be inserted into the transmission between the Y-counter and the Z-spindle. Since the magnitude of correction from the center of the model is only one-fourth as much in the X-direction as it is in the Y-direction, the attachment is designed for corrections based only on the Y-coordinate. Graphic plotting is accomplished on either the integral tracing table (part of the plotter) or external tracing tables. The 43 × 70 centimetre integral tracing table is attached to the frame beneath the tracking device. A tracing head is attached to the X-carriage for plotting at model scale. For plotting at other scales through the use of interchangeable gears, a 120 × 120 centimetre external tracing table connected to the Planimat by mechanical shafts or an 80 × 120 centimetre external tracing table connected through synchro-drives can be added.

Additional options provide the capability to acquire data which can be used to drive the GZ-1 Orthoprojector in direct, storage, or automatic modes. This is accomplished through the use of the Itek EC-5, which is an electronic image correlation attachment. This attachment will keep the floating mark in contact with the model surface while the stereoscopic profile is automatically scanned in the Y-direction and, depending on the selected swath width, the scanning mechanism with photomultiplier will automatically advance the required increment in the X-direction and proceed to scan the next swath.

12.4.5.2 ZEISS PLANICART

12.4.5.2.1 GENERAL

The Planicart (fig. 12-85) is a modern stereoplotter developed by Carl Zeiss, Oberkochen.

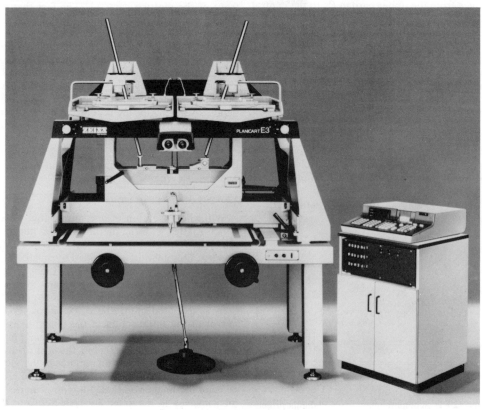

FIGURE 12-85. Zeiss Planicart C3.

The instrument is designed for graphic plotting at medium and medium-large scales, especially at 1:5,000 and 1:10,000.

The significant difference between the Planicart and the Planimat (section 12.4.5.1) is the limited versatility of the Planicart in comparison to the wide versatility of the Planimat. Specifically, the Planicart is a medium and large scale graphic plotter, whereas the Planimat covers the complete range of line map and orthophotographic plotting, and digital recording.

Many of the features of the Planimat are incorporated into the Planicart. Noteworthy are the use of single-arm guide rods, eccentricity of plotting-camera cardans, compound-slides in moving the photographs and terrain carriage, and a tracing table that is part of the plotter.

The versatility of the basic Planicart can be extended through the addition of auxiliary equipment such as an external tracing table, mini-computers, and coordinate-recorders.

Primary applications of the Planicart are the compilation of medium and medium-large scale maps, map revision and training. With auxiliary equipment, the applications can be extended to include large scale map compilation, digital measurement and recording, and on-line computations.

12.4.5.2.2 PROJECTION SYSTEM

The Planicart is similar in design and construction to the Planimat (fig. 12-86). The plotting cameras are supported at three points. Single-arm guide rods simulate the intersection of image-object bundles of light rays. As in the Planimat, the principle of eccentricity of camera cardans and perspective centers has been included in the design of the Planicart. The bz- and by-motions are applied at the lower left and right cardans respectively, and the bx-motion is applied between these same two cardans, *i.e.,* the split points in model space. A planar compound-slide system is used to displace the photographs, and a three dimensional

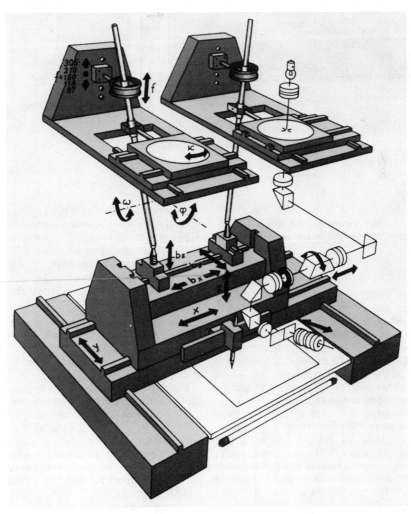

FIGURE 12-86. Principle of Projection in Planicart.

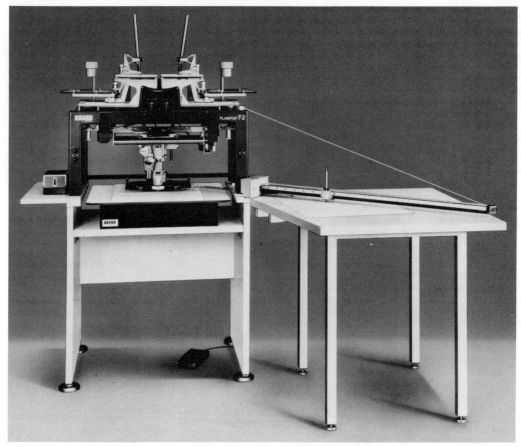

FIGURE 12-87. Zeiss Planitop F2.

compound-slide system is used for X-, Y-, Z-motions for model viewing. The normal κ, ϕ, and ω motions are available for model orientation.

12.4.5.2.3 OBSERVATION SYSTEM

The stereo model (fig. 12-86) is illuminated by a light which passes through the photograph in the plate holder through a lens to a mirror which reflects the light rays along the ϕ axis to another mirror which reflects the rays along the ω axis through a Dove prism to a series of lenses and mirrors on to the viewing eyepieces. The Dove prisms on either side are movable and can be used for correcting image rotation.

The Planicart can accept a range of focal lengths from 84 to 308 millimetres; a special accessory is required for plotting 308 millimetre photography. Photographs from normal, wide, and super-wide angle cameras in negative sizes up to 23 × 23 centimetres can be accommodated for plotting. The projection system with the standard photo-carriage plates is distortion free; however, the distortion of either camera lens can be corrected by aspheric plates if required. The optical system is stationary and provides normal viewing with a 6× magnification across a 31 millimetre diameter field of view. The measuring mark is a point 0.06 millimetre in diameter.

12.4.5.2.4 PLOTTING AND MEASURING SYSTEM

The Z-coordinate motion is accomplished through a foot pedal disk drive. Handwheels and spindles provide the X-, Y-motion. The X-, Y-spindles can be disengaged for freehand guiding during relative orientation.

Plotting by the tracing table is performed by a tool attached to the X-carriage. Since the tracing table is illuminated, the optional ZZ-4 tracing head can be attached as the plotting tool, and either pencil, ink, or scribing needles can be used for recording. In situations where the tracing table will not provide the necessary enlargement from model to map, an external tracing table is available which can be connected to the Planicart and, through change gears, provide additional enlargement ratios.

A Z-counter is mounted on the X-carriage and can be preset. Readout is either in feet or metres and (with the use of change gears) a wide range of model scales are available.

Other auxiliary equipment includes coarse/ fine gears for the X- and Y-handwheel drives, the ECOMAT-11 electronic-recording unit, and the DIREC-1 system for connecting the Planicart to desk computers. DIREC-1 can be used for absolute orientation and for the real-time transformation of model coordinates into ground coordinates.

12.4.5.3 ZEISS PLANITOP

12.4.5.3.1 GENERAL

The Planitop topographic plotter (fig. 12-87), developed by Zeiss (Oberkochen) is primarily designed for graphic plotting at medium and small scales. The Planitop is the third instrument in the Zeiss line of modern analog plotters which includes the Planimat and Planicart.

The primary differences between the three Zeiss plotters are the plotting scales: Planitop, medium and small scales; Planicart, medium and large scales; and Planimat, all scales.

One of the features of the Planitop is that the projection center is between the image cardan and the space point; while in the Planimat and Planicart, the projection center is above the image cardan and the space point. Other features included are eccentric displacement of camera cardan and projection center, a fixed viewing system, freehand guiding during plotting, and an accessory for earth curvature correction.

The basic applications of the Planitop are to small-scale graphic plotting, map revision, photointerpretation, and training. With additional accessories, the applications can include digitization, medium scale plotting and on-line computations. Triangulation by independent models is possible by determining the perspective center coordinates through space resection or with the aid of auxiliary devices.

12.4.5.3.2 PROJECTION SYSTEM

Double-arm guide rods (fig. 12-88) simulate the intersection of the image-object bundles of light rays. The three mechanical points on the guide rod are the image cardan, projection center cardan, and space point. The distance from image cardan to projection center is the focal length, and the distance from projection center to space point is the projection distance. The guide rods are supported at the lower point (space point). The base setting is performed at the space points with the aid of the Zeiss parallelogram. Displacement of the photographs is accomplished by a double-orthogonal slide system. Principle distances are changed by resetting the image cardans at the photo-carriage, followed by a fine adjustment.

12.4.5.3.3 OBSERVATION SYSTEM

The optical system is diagrammed in fig. 12-88. The stereo model is viewed through binoculars mounted on the front frame of the Planitop. Normal viewing is under 6× magnification, with a 30 millimetre diameter field of view. The luminous measuring mark is 0.06 millimetres in diameter.

The Planitop will allow plotting from wideangle and super-wide-angle photography (negatives or diapositives) up to 23 × 23 centimetres with standard fiducial marks. The focal length ranges from 84 to 90 millimetres and 150 to 156 millimetres. In special cases the instrument can be equipped to plot from normal-angle photographs at a 305 millimetre principal distance. The photo-holder plates are provided with a grid of nine crosses forming 90 millimetre squares for adjustment and testing. The plotting range will allow plotting of photography with 80 percent end-lap.

12.4.5.3.4 PLOTTING AND MEASURING SYSTEM

The Z-coordinate motion is driven by a motorized coarse adjustment with an additional fine adjustment screw. The Z-readout is by means of a glass scale and projector. Readout can be either in feet or metres. A variety of glass scales for different model scales are available. The X- and Y-motions are operated by freehand guiding of the tracing device.

The tracing device is mounted on the X-carriage and is equipped with an illuminator and provision for manual lowering of the tracing tool. Plotting can be done with pencil, pen or scribing needle.

The integral tracing table is the base for plotting. The plotting range is 240 × 320 millimetres. The Planimat can be set up on any solid table or on a special purpose table which is available with a luminous surface.

The enlargement range between photo and model for 23 × 23 centimetres wide-angle and super-wide-angle photography is from 0.5× to 1.5×. A secondary enlargement range of 0.75× to 4.0× is possible with the use of either the PP-2 or PP-3 polar pantographs.

Digitization of X-, Y-, Z-model coordinates can be accomplished with the addition of linear pulse generators connected to the ECOMAT-11 electronic recording unit or the DIREC-1 system for on-line calculations with mini-computers.

12.4.6 SOPELEM PRESA

12.4.6.1 GENERAL

The Presa is a precision instrument employing a mechanical projection system. The instrument was designed by J. Baboz, S.O.M. in France in 1964. The first model, Presa 224, was exhibited at the I.S.P. Congress in Lisbon. The later mod-

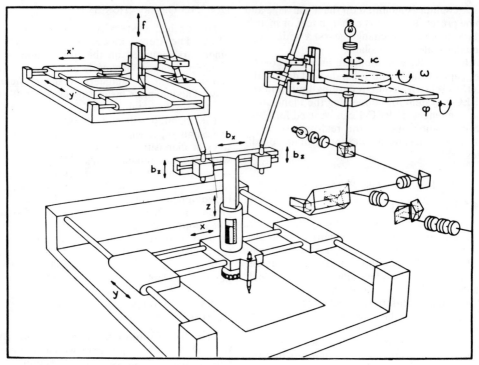

FIGURE 12-88. Mechanical and optical design of the Zeiss Planitop F2.

els, Presa 225, 225 RC, and 226 and 226 RC, are improved and extended versions developed by SOPELEM (Société d'Optique Précision, Electronique et Mécanique—after the merger between S.O.M. and O.P.L.). Models 224, 225, and 226 were developed for graphical outputs, whereas models 225 RC and 226 RC are adaptable also to digital outputs. Fig. 12-89 shows a general view of the Presa 225 RC.

The instrument uses a spatial projection in the model space and a planar projection inside the mechanical projectors. This reduces the

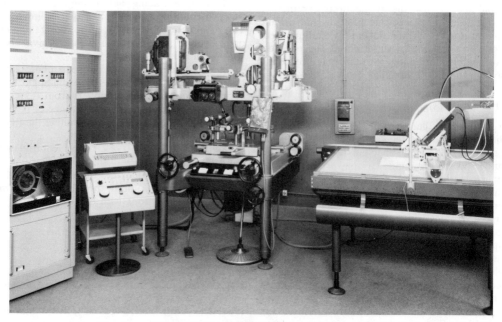

FIGURE 12-89. SOPELEM Presa 225 RC.

mechanical problem associated with extremely inclined rays in the case of super-wide-angle photographs. It also permits the application of a simple observation system.

The mechanical construction is arranged in two stages. The upper stage carries the projection system, whereas the lower stage supports the tracking device (X, Y, Z), drivers, and the measuring device (fig. 12-90).

Some typical features of the Presa are:

- the two photo-carriers are vertical and mounted asymmetrically;
- the projection centers are represented by five-axial cardan joints;
- tracking of photographs is carried out partly by moving the photo-carriages and partly by the observation optics; and
- the upper frame can be vertically displaced and tilted for absolute orientation.

The projectors are adapted for 23×23 centimetres and smaller photographs, taken by normal-, wide-, and super-wide-angle cameras with near vertical principal axes. Diapositives are negatives on glass plates or on film base.

The common outputs are large-scale line maps, digital data for cadastral and engineering applications, and aerotriangulation data for independent models.

12.4.6.2 PROJECTION SYSTEM

Some general features of the projection system have been mentioned above. The system employs a Zeiss parallelogram (fig. 12-90).

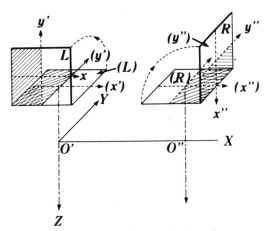

FIGURE 12-91. Arrangement of photo-planes.

The left-hand (LH) photo-plane is turned through 100^g in the vertical position parallel to the XZ-plane, and the right-hand (RH) photo-plane similarly in the vertical position parallel to the YZ-plane (fig. 12-91). The two photo-carriages are, in addition, located eccentrically with respect to the corresponding projection centers for mechanical reasons.

During tracking, one moves the photographs along (nearly) horizontal rails (x', y''), and the prism carriages of the observation optics along (nearly) vertical rails (x'', y'). These movements are actuated by the x- and y-rods (*i.e.* straight edged rulers), which represent the projections of conjugate rays on the XZ- and YZ-planes (fig. 12-92).

In the left-hand projector the y'-rod is turned 100^g from the real projection (y') in the same sense as the left-hand photo-plane (fig. 12-91). Similarly, the x''-rod is turned in the right-hand projector through 100^g from the real projection (x'') of the ray. Thus, in the left-hand projector

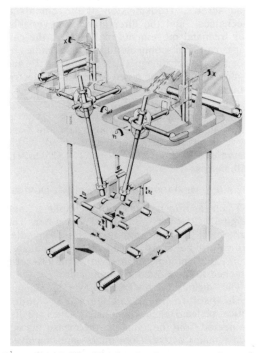

FIGURE 12-90. Mechanical construction of SOPELEM Presa 225 RC.

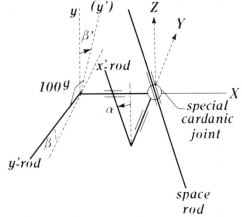

FIGURE 12-92. Planar solution inside the LH projector.

the x'-rod is inclined over α' with respect to the vertical, and the y'-rod over β' with respect to the horizontal.

The x'-rod is connected, via a spring-loaded ball race, with the photo-carriage at the principal distance c_x from the axis. Similarly, the y'-rod is connected with the prism carriage at the principal distance c_y. The arrangement in the right-hand projector is analogous but with interchanged x and y components.

For each projector the principal distance has to be set twice: c_x and c_y. It can be set for any value between 83 and 221 millimetres. The setting screws are provided with graduated drums. By introducing slightly different values for c_x and c_y, compensation can be made for regular film shrinkage.

Mechanical functioning can be visualized from fig. 12-90. While tracking in the Y-direction, one turns both cardan joints of the space rods. The left-hand joint displaces, via the y'-rod, the left-hand prism carriage in a (nearly) vertical direction. At the same time the right-hand cardan joint displaces, via rod y'', the right-hand photo-carriage in a (nearly) horizontal direction. While tracking in the X-direction, the left photo-carriage moves (nearly) horizontally and the right prism carriage (nearly) vertically. The photo-carriers are equipped with plane-parallel supporting glass plates, or, optionally, with aspherical correction plates for lens distortion.

The facilities for orientation are shared between the upper and lower stage of the instrument. Rotations are applied to the upper stage. The left projector can be tilted for ϕ and the right for ω (fig. 12-90). Hence, the corresponding photo-planes turn in their own planes. By means of an electric motor each photo-carrier can also be turned about its principal point for κ. The switches are on the instrument panel.

The upper stage can be tilted for absolute orientation (Φ, Ω) by using three vertical columns which also serve for coarse changes of the projection distance Z. The latter is required when altering the angular field of the photographs, e.g., from super-wide-angle to normal-angle. The leadscrews of the three columns are chain-driven, by a handwheel.

A special cross level has been provided for the transfer of rotation parameters from one projector to another.

The base components are set to the base bridge which is located on the lower stage of the instrument. The bx-component can be introduced symmetrically to both bx-carriages; by'' is set to the right and bz' to the left lower cardan joint (fig. 12-90). All setting screws are provided with graduated drums.

In model space, the projecting rays are represented by space rods attached to the lower cardan joints. These rods are counterbalanced by weights attached to the lower ends. The instrument is equipped with two pairs of space rods—one for normal- and wide-angle photographs, and another for super-wide-angle photographs.

12.4.6.2 Observation System

Because of the asymmetric arrangement of photo-carriages, the two optical trains are also asymmetric (fig. 12-90). However, each train is relatively simple since the rotations for relative orientation (i.e. ϕ_1, ω_2, κ_1, κ_2) cause the photographs to move in their own planes.

Fixed vertical fluorescent tubes serve for illumination. The prism-carriages are provided with conventional optical elements, including measuring marks which are black disks 40 micrometres in diameter. Parallel light is guided from the prism carriage downward to a rectangular prism, which is mounted on the projector support. This prism directs the light to the ϕ_1 (or ω_2 respectively) axis, where it is reflected through the axis into the upper frame. There the light is guided, by means of prisms, to the eyepiece. An optical transfer unit and a Dove prism, to compensate for image rotation caused by ϕ_1 (or ω_2), are built into the train.

Electric motor drives focus the first lens in each train on the emulsion of diapositives or negatives. The push-buttons for this are on the instrument panel. Units for image reversion, and zoom systems for magnification between $6\times$ and $12\times$, are also provided.

12.4.6.4 Driving and Tracking Devices

Tracking is controlled by means of X- and Y-handwheels, a Z foot-disk, and corresponding lead screws. Transmission of movements is mechanical, and the directions are reversible. For freehand movements in X and Y, the carriages can be disconnected from the spindles by means of electromagnetic relays. Fast X- and Y-movements can be given by means of electric motors which are controllable from a mobile panel.

The arrangement of the tracking device is shown in fig. 12-90. The carriages X, Y and Z can move along three orthogonal pairs of rails within the ranges of X—370, Y—360, Z—125 to 415 millimetres.

12.4.6.5 Measuring and Plotting Device

The measuring devices of the "graphical" and "digital" (RC) version of the Presa differ. The "graphical" version employs precision spindles (which also act as driving agents) and graduated drums (0.01 millimetre interval). The latter are attached to the X- and Y-handwheels.

The "digital" version RC, however, is equipped with absolute linear encoders, i.e., coded optical scales. Hence when the carriages are disconnected from the spindles, the measuring device remains active. Coordinates are displayed on nixie-tubes (fig. 12-89) with a least count of 5 micrometres.

A mini-computer converts instrumental heights Z into terrain heights H according to:

$$H = aZ + H_0$$

where a is a scale factor and H_0 is the height of the datum plane. The constants a and H must be entered in the computer. Automatic recording may be on paper tape, cards, or magnetic tape. It may be stationary, for discrete points, or continuous, for point streams.

For plotting, a large light table (120 × 120 centimetres) and a rectangular coordinatograph are provided. Transmissions are by means of electrical synchro-motors. Boxes with interchangeable gears are attached to the rear of the instrument. The X- and Y-movements are reversible and the axes can be interchanged. Linear scales and drums are provided for coordinate measurements. Optionally, counters for the common map scales can be attached to the coordinatograph.

12.4.6.6 ATTACHMENTS

Most of the customary attachments have been provided for the plotting device. These attachable devices are:

- a double tracing head for simultaneous plotting on two sheets near their edges;
- a televiewer, including a TV camera (directed to the drafting tool), a remotely controlled zoom system, a graticule (for coordinatograph positioning), and a TV monitor; and
- a magnifying glass, attachable to the tracing carriage, which permits the drafting tool to be observed from the operator's position.

These additional devices improve the operational characteristics and extend the versability of the instrument.

12.4.7 ZEISS JENA STEREOMETROGRAPH

12.4.7.1 GENERAL

The Stereometrograph (Zeiss Jena 1964) is an autonomous precision instrument designed primarily for large scale applications. Its model A was exhibited at the I.S.P. Congress in London in 1960. It was succeeded by models B to F, each one an amended version of its predecessor. Fig. 12-93 shows a general view of model F. Some typical features of the instrument are:

- the mechanical construction is heavy, stable, and large;
- the projection system and tracking device are positioned low, and are enclosed in a casing for protection; and
- transmission of motive power to the base carriages, tracking device (X, Y, Z) and the plotting coordinatograph is electrical.

Models D, E, and F have undergone several significant modifications and extensions as compared to their predecessors. These concern:

- capability of handling super-wide-angle and narrow-angle photographs;
- increased precision by the introduction of double leadscrews for the X- and Y-carriages;
- digital counters for coordinates;
- capacity to record automatically X, Y, bx, and by as an alternative to X, Y and Z;

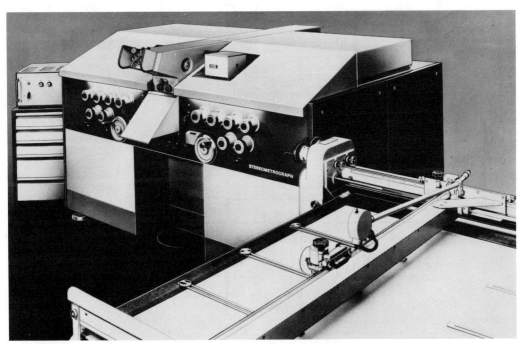

FIGURE 12-93. Zeiss Jena Stereometrograph F.

—extended ranges of model space; and
—several auxiliary devices and attachments.

The Stereometrograph F can handle 23 × 23 centimetre and smaller photographs, taken by narrow-, normal-, wide-, and super-wide-angle cameras with nearly vertical or horizontal principal axis. Diapositives or negatives on glass plates or film can be used.

The most common products are large-scale line maps and digital data for cadastral, engineering, and short range application, and aerotriangulation with independent models.

12.4.7.2 PROJECTION SYSTEM

Projection is done mechanically and in space (fig. 12-94). A Zeiss parallelogram is employed, though the base can only be set inside the parallelogram.

The central cardan joints, which represent the projection centers, are mounted in the fixed instrument frame. Each projector is provided with orthogonal cross slides, carrying the prism carriage of the observation optics. The slides are connected with the upper joint of the space rod for tracking the photograph. During tracking, the photograph, centered on a plane parallel supporting glass plate, is fixed. The supporting plate may be replaced by an aspherical correction plate for lens distortion and other concentric symmetrical distortions.

The principal distance is set by displacing the photocarrier, together with the cross slides, in the direction of the principal axis, i.e., along the three parallel columns which are coupled by means of a toothed ring. The corresponding setting screw and measuring unit are on the instrument panel. The transmission from the setting screw to the ring is mechanical. The principal distance c range is from 85 to 310 millimetres.

The orientation facilities are attached partly to the projectors and partly to the base-carriages (fig. 12-95). The rotations ϕ, ω, and κ are applied to the mechanical projectors about axes passing through the corresponding projection centers.

Thus, the cardan joints of a projector and of the corresponding space rod are concentric. The setting screws and measuring units for rotations are fixed on the instrument panel; transmissions to projectors are mechanical.

A change in ϕ, which is the primary rotation, causes a differential tilt of the projector with respect to the fixed instrument panel. Hence, all other settings, i.e., ω, κ, and c, with mechanical links from the panel, change accordingly. In order to prevent this, compensation gears are provided on the back of the panel. These gears couple the ϕ-screw to the ω-, κ-, and c-screws. No means are provided for the common rotations Φ and Ω.

The base components are set to the lower cardan joints on space rods. The bx-component is introduced by a relative displacement of the right X-carriage with respect to the left-hand one

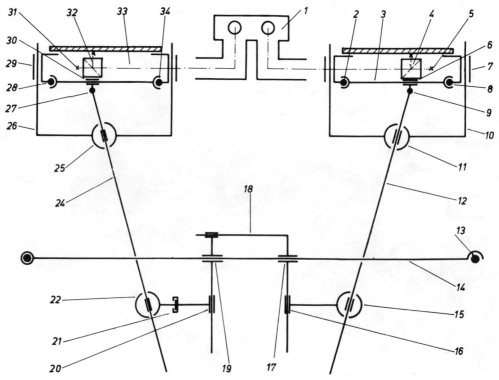

FIGURE 12-94. Projection system of the Stereometrograph.

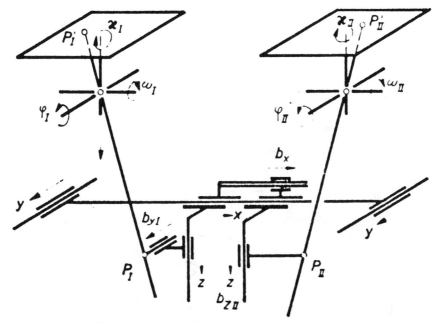

FIGURE 12-95. Arrangement of means of orientation.

(fig. 12-95). The by-component is set to the left cardan joint, whereas the bz-component is applied by shifting the right Z-carriage with respect to the left one. The setting and measuring devices for bx'', by' and bz'' are also attached to the instrument panel. However, transmission of power to the carriages is electrical.

The projecting rays are produced by one-piece cylindrical rods which are attached to the upper cardan joints (fig. 12-94).

12.4.7.3 OBSERVATION SYSTEM

For illumination, small movable lamps are employed (fig. 12-96). These are attached to the prism carriages which can move on orthogonal cross-slides.

Each prism carriage is provided with a prism (9) having a diagonal semi-reflecting surface. A luminous measuring mark (6, 7) is placed behind the prism at the same distance from the semi-reflecting surface as the photo-emulsion is above it. At the other end of the prism carriage is a telescope which produces a parallel light beam. The light is guided from the secondary x-carriage to the primary y-carriage and further, via prism 11, to the tiltable projector support. From the latter the light is guided via mirrors 14, 22, and 24, to the fixed instrument frame. These mirrors are controlled by telescopic tubes.

Between mirrors 14 and 22 an optical transfer unit forms an intermediate image and again a parallel light beam. The beam passes through Dove prism 25. The Dove prisms of the two optical trains can be operated individually or may be coupled. The unit, with interchangeable rectangular (26) and roof prisms (27), permits correct observation of diapositives and negatives. Lens 28 forms an image in front of the eyepieces for observation. The prism(s) 29 serves for setting the eyebase. The eyepieces (30, 31) are equipped for squint adjustment; they provide for 8× magnification.

12.4.7.4 DRIVING AND TRACKING DEVICES

Tracking in model space is generated by means of X- and Y-handwheels and a Z-foot disk. The revolutions are first magnified three times by gears and then transferred electrically to the leadscrews. The electrical signals, generated by synchro-transmitters, are first received by synchro receivers. The revolutions transferred are then triply reduced by gears and applied to the leadscrews. This enlargement and reduction of revolutions enables more accurate transmission of movement. For the same reason dual lead screws are provided for each coordinate axis, one on each side of the carriage.

The X- and Y-drives can be slow or fast, but freehand movements are not possible. The directions of movement (and of counting) are reversible. Moreover, the Y- and Z-drives can be interchanged. The handwheels (and measuring devices) for X, bx and Y, by are located symmetrically on the instrument panel.

Tracking is performed by means of the YXZ-cross-slide system (fig. 12-94). In order to minimize wear, narrow prismatic rails, made of hard steel, are inserted into stable cylindrical rails along the tracks where the wheels of the carriages run. The primary Y-carriage carries the X-rails on which two coupled X-carriages move. Each X-carriage carries, under the X-rails, a

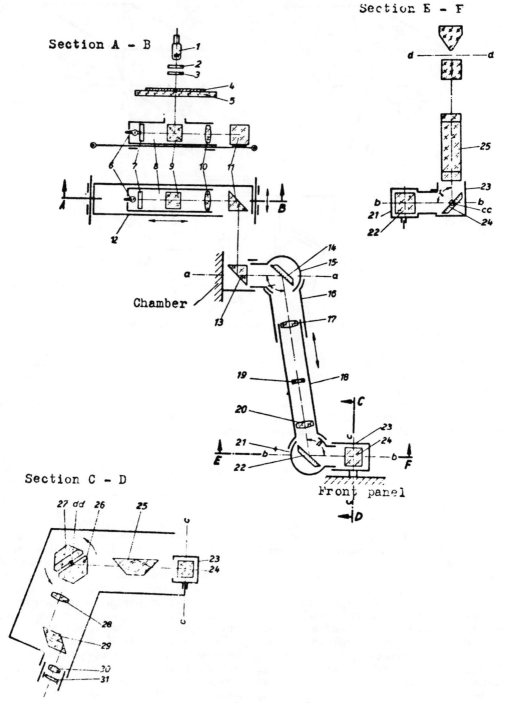

FIGURE 12-96. Observation System.

Z-column. When tracking in the Z-direction, the two Z-carriages move synchronously. The instrument casing, however, prevents the operator from observing instantaneously the position of the Z-carriages.

Below the eyepieces a location indicator is provided which enables the operator to observe the approximate location of the measuring mark on one photograph of the stereopair. The paper print concerned is placed on a light box divided by a square grid. The illuminated square indicates the approximate location of the measuring mark.

The ranges of the model space are: X: -180 to

+280 millimetres; Y: ±250 millimetres; Z: 130 to 350 millimetres (optional to 700 millimetres).

12.4.7.5 MEASURING AND PLOTTING DEVICES

The measuring device for the X- and Y-coordinates has two parts: precision spindles (which serve also as driving agents), and mechanical counters. The latter are connected to the handwheels. The last two decimals, i.e., tenths and hundredths of a millimetre, are displayed on drums. The counting directions are reversible.

The measuring device for heights is similar. However, it is provided with gears for changing scale, and with facilities for initial height settings.

Automatic recording is possible by means of the Coordimeter F which employs rotary incremental encoders. The device is applicable for point by point, and continuous operation. Recording may be on paper tape, cards, or magnetic tape. Coordimeter F permits the recording of two, three, or four quantities. If the Stereometrograph is used as a stereocomparator, the following quantities are recorded: X, Y, bx'', and by'. The Coordimeter F can also be connected to other Zeiss Jena instruments, e.g. the Stecometer or Ascorecord.

Two plotting tables (coordinatographs) are available: the large table (90 × 120 centimetres area), or the small table (80 × 80 centimetres). The motive power of the instrument is transmitted by electrical synchro-motors and can be connected to the XY-, XZ-, or YZ-axes of the instrument. Interchangeable gears are attached to the drawing table and extra gears are provided for doubling movements and for reversing direction.

The measuring device of the coordinatograph includes precision spindles and drums with masks for different scales which are used in combination with built-in gears. By attaching handwheels the coordinatograph can be used independently.

The table is equipped with the usual facilities, e.g., direct and transparent illumination, magnetic switch for drafting tool, setting microscope, etc.

12.4.7.6 ATTACHMENTS

Several attachments are available for the instrument and the plotting device. The attachments for the instrument are:

- an electrical corrector for cylindrical or spherical deformations of heights;
- a tilt computer (i.e. ω-gears) for plotting from oblique terrestrial photographs; and
- an electric motor drive, attachable either to the X- or Y-servo input of the instrument (or of the plotting coordinatograph).

The following attachments are available for the plotting table:

- a digitizer box which is essentially an interface unit for 4, 3, or 2 channels;
- a unit for marking points (on the map) with small circles;
- a tracing head for drawing dashed and dotted lines;
- double tracing carriages for plotting along the edges of two adjacent map sheets simultaneously;
- a mirror and magnifying glass for observing the drafting tool from operator's position; and
- a setting projector, similar to the Wild Profiloscope.

Most of these accessories are also available for other Zeiss Jena instruments.

12.4.8 ZEISS JENA TOPOCART B

12.4.8.1 GENERAL

The mathematical concept of the Topocart was formulated by F. Manek in 1958. The first model was shown at the I.S.P. Congress in Lausanne, in 1968.

A general view of the Topocart B, the successor of the model A, is shown in fig. 12-97.

The projection system is mechanical and implements a planar solution. The instrument is compact and enclosed in a casing. Some typical features of the Topocart B are:

- the photographs are horizontal and co-planar; thus a simple observation system can be used;
- the projection system is resolved into planes which are positioned horizontally; this eliminates the need for introducing counterweights;
- the mechanical parts are arranged into modules for easy transport and assembly; and
- various accessories and attachments permit of extended applicability of the instrument.

The photo-carriers are adaptable to formats ranging from 4 × 4 centimetres to 23 × 23 centimetres. The instrument can handle normal-, wide-, and super-wide-angle photographs with near-vertical or horizontal principal axis. Diapositives or negatives on glass plates or film base and paper prints can be used.

The main products are small- and medium-scale line maps, data and graphs for civil engineering, and various short-range applications and, by means of attachments, orthophotographs and drop-line charts.

12.4.8.2 PROJECTION SYSTEM

The conjugate rays of both bundles are projected onto the XZ- and YZ-planes. A Zeiss parallelogram is then applied to each plane. The projections of rays obtained are materialized by straight-edged rulers which are positioned in the instrument horizontally (fig. 12-98). Each projection center, in the XZ- and YZ-planes, is represented by a single vertical rotation axis because the two planes are arranged one above the other.

During tracking, the photo-carriers move while the observation system is fixed. Aspheric

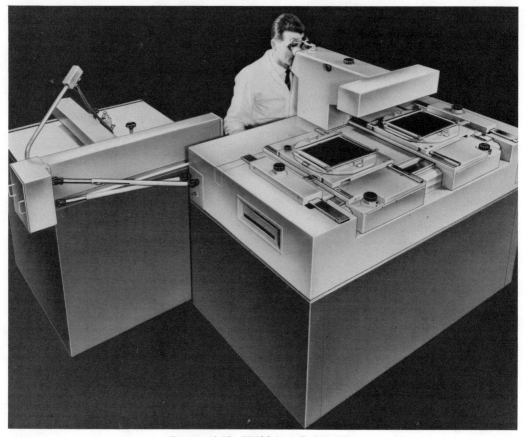

FIGURE 12-97. ZEISS Jena Topocart B.

correction plates for lens distortion are placed on top of the photographs, since the observation system is above. The principal distance, ranging from 50 to 215 millimetres, is set twice for each projector (c_x to the x-rod and c_y to the y-rod) to allow for compensation for regular film shrinkage.

Means of orientation are divided between the photo-carriers (κ', κ''), the mechanical tilt-computers (ϕ', ω' and ϕ'', ω''), and the X- and Y-carriages of the tracking device (bx'', by'' and bz_x'' and bz_y''). The swing rotations (κ' and κ'') are applied to the photo-carriers about the corresponding principal points. The base components

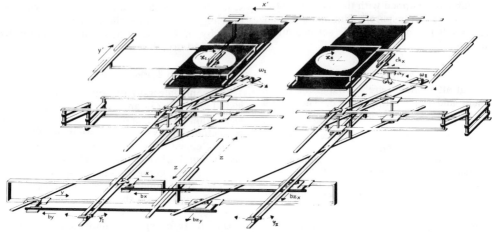

FIGURE 12-98. Schematic drawing of the mechanical projection system of the Topocart.

bx'' and bz''_x are set to the right-hand ball race on the X-carriage, whereas by'' and bz''_y are set to the right-hand ball race on the Y-carriage. Thus, the bz-component must be set twice, as it occurs in both planes (XZ and YZ).

Mechanical computers are provided for the ϕ- and ω-tilts. These computers allow for a collinearity transformation according to:

$$\begin{bmatrix} x \\ y \\ c \end{bmatrix} = \lambda\, A \begin{bmatrix} X \\ Y \\ Z \end{bmatrix} \qquad (12.20)$$

where the matrix A includes the terms of ϕ and ω. Both sets of coordinates refer to the projection center as origin.

For mechanical applications the above relationship can be transformed into (V.d. Hout, 1969):

$$x = c \tan(\alpha - \phi - \delta);$$
$$y = c \tan(\beta - \omega - \epsilon) \qquad (12.21)$$

where

$$\tan \alpha = \frac{X}{Z}; \ \tan \beta = \frac{Y}{Z};$$

$$\tan \epsilon = \frac{Y(Z \cos \phi + X \sin \phi) - YZ}{Z(Z \cos \phi + X \sin \phi) + Y^2};$$

$$\tan \delta = \frac{(X \cos \phi - Z \sin \phi)\left[(Z \cos \phi + X \sin \phi)\cos \omega + Y \sin \omega - (Z \cos \phi + X \sin \phi)\right]}{(Z \cos \phi + X \sin \phi)\left[(Z \cos \phi + X \sin \phi)\cos \omega + Y \sin \omega\right] + (X \cos \phi - Z \sin \phi)^2}.$$

Examining the equations for $\tan \delta$ and $\tan \epsilon$, it is apparent that the ϕ acts as the primary and ω as the secondary rotation. The geometrical principles of the mechanical tilt computers are shown in fig. 12-99.

The projections of a ray on the XZ- and YZ-planes are represented by the X- and Y-rods. These rods can turn around a common axis, which represents the projection center O. The upper part of each rod can be inclined differentially over $(\phi - \delta)$ and $(\omega - \epsilon)$ respectively. In addition, for mechanical reasons, the upper part of the Y-rod is bent through an angle of 100^g.

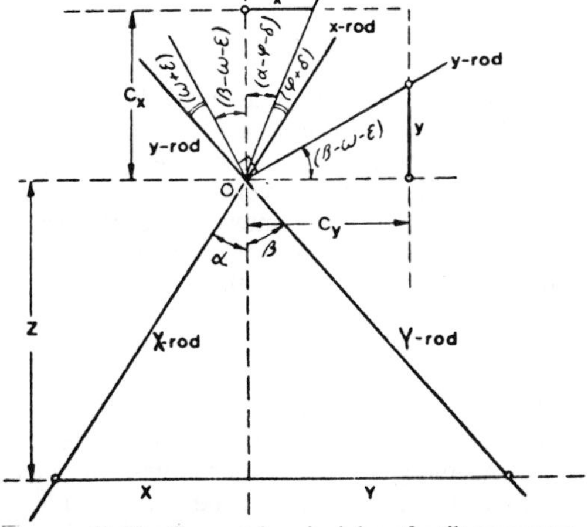

Hence the transformed x-coordinate is obtained at the principal distance c_x (in the z-direction) from 0, while the y-coordinate is similarly obtained at c_y. These two outputs of the tilt computer are mechanically transferred to the x- and y-carriages to move the photograph.

Tilts ϕ and ω are constant for a certain photograph. However, the small angles δ and ϵ are variable during tracking and should therefore be automatically controlled. The device is illustrated in fig. 12-100. The main parts of the tilt computers are the X-, ϕ- and δ- (i.e. x-) rods, the Y-, ω- (i.e., y-) and ϵ-rods, the four parallel rails I, II, III, IV, and the two cross rails ($1'$, $2'$) and $(3, 4)$.

The axes O and (O), together with the eight equal arms k, provide for parallel guidance of rails I to IV. When the X-rod turns, pivot 1 moves about 0 in a circle of radius k. Rail I is positioned at a distance $k \cos \alpha$ from 0. By setting ϕ, on the ϕ-rod, rail II is positioned at a distance of $k \cos (\alpha - \phi)$ from 0. This is accomplished by means of pivot 2. The ball race $1'$ represents the intersection of the X-rod and rail

I. A cross rail ($1'$, $2'$) crosses rail II at ball race $2'$. The upper arm of the ϵ-rod, which forms angle ϵ with the Y-rod, is connected to ball race $2'$. Its lower end controls, via pivot 3, the position of rail III. Hence, this rail is located at a distance $k \cos (\beta - \epsilon)$ from 0. The ω-tilt can now be set to the ω-rod (i.e. y-rod), which positions, by way of pivot $4'$, rail IV. Thus rail IV is located at $k \cos (\beta - \omega - \epsilon)$ from the projection center 0. The lower end of the ϕ-rod crosses rail III at ball race 3. A cross rail $(3, 4)$ connected to it, crosses rail IV at ball race 4. The lower end of the δ-rod (i.e. x-rod) is connected to this ball race. It forms an angle δ with the ϕ-rod, and thus an angle $(\alpha - \phi - \delta)$ with the Z-axis. The transformed x-coordinate is obtained on the upper extension of the δ-rod at the principal distance c_x from 0.

The y-rod is rigidly attached to the ω-rod at a right angle in 0. Thus it forms an angle $(\beta - \omega - \epsilon)$ with the perpendicular to the Z-axis. The transformed y-coordinate is thus found on the y-rod at principal distance c_y. The setting and measuring devices for ϕ-tilts are attached to the lower ends of the X-rods. To introduce an ω-tilt, the corresponding clamp should first be released, then either the photo-carriage should be slightly displaced in y-direction (if the Y-rod is fixed), or the Y-carriage slightly shifted, thus turning the Y-rod (if the photo-carriage is kept fixed). The

FIGURE 12-99. Geometric principles of a tilt computer.

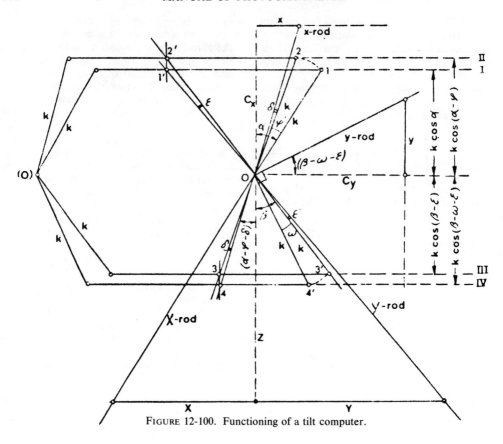

FIGURE 12-100. Functioning of a tilt computer.

appropriate measuring devices, *i.e.*, scales and verniers, are provided.

12.4.8.3 OBSERVATION SYSTEM

The observation system is positioned above the photographs. It is fixed and thus simple (fig. 12-101). Separate lamps are provided to illuminate transparencies or paper prints and the light is guided through several prisms to the eyepiece.

Prism *B* has a semi-transparent diagonal surface which combines a luminous measuring mark (60 micrometre diameter) with the image. The light beam of the measuring mark is guided from elements 14-17 to the prism 8 by a trapezoidal prism, 19. The optical distances from prism 8 to the photograph, 6, and to the measuring mark, 15, are equal. The luminous measuring mark can be replaced by a black disk. The eyepieces are provided with the usual adjustment facilities and their magnification is 6X with a field of view of 40 millimetres. When the Orthophoto B is connected, the right-hand optical train is branched using an extra semi-transparent surface.

12.4.8.4 DRIVING AND TRACKING DEVICES

Tracking is actuated by means of the *X*- and *Y*-handwheels and a *Z*-foot disc. Transmission to the *X*-, *Y*-, and *Z*-carriages is mechanical and the directions of movement are reversible.

Moreover, connections between the handwheels and foot disk, and the corresponding leadscrews are interchangeable. The *X*- and *Y*-movements can also be carried out freehand.

The *Z*-carriage is primary; the *X*-carriage is mounted on top of it, the *Y*-carriage below it thus the *X*- and *Y*-carriages are mutually independent. Since all carriages are mounted on horizontal prismatic rails, counter weights are not needed and the construction is very compact. The model space ranges are 240 mm in *X* and *Y* and 70-320 mm in *Z*.

12.4.8.5 MEASURING AND PLOTTING DEVICES

The device is comprised of measuring spindles (leadscrews) and mechanical digital counters for *X*- and *Y*-coordinates and for heights. The *X*- and *Y*-counters are coupled to appropriate handwheels with reversible directions and the height counter has interchangeable gears. Rotary encoders can be attached to the *X*-, *Y*- and *Z*-shafts, and linked to the Coordimeter for automatic coordinate recording.

For plotting, a separate table with a rectangular coordinatograph is provided. The standard table (60 × 60 centimetres net format) is equipped with mechanical transmissions and exchangeable gears but, larger tables with electrical transmissions can also be supplied (80 × 80 centimetres or 90 × 120 centimetres). The co-

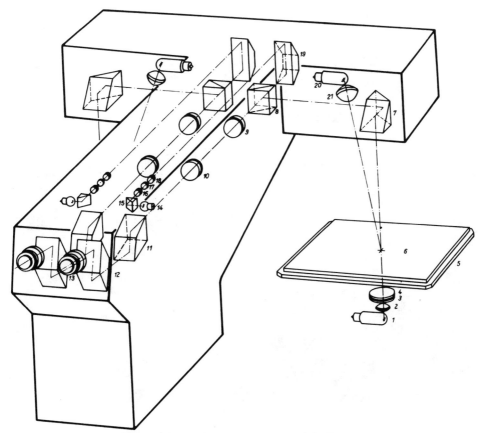

FIGURE 12-101. Observation system of the Topocart.

ordinatograph can be connected to any pair of the instrument axes: X, Y; X, Z; or Y, Z.

12.4.8.6 ORTHOPHOTO ATTACHMENT

The most important attachments are the Orthophoto B and the Orthograph (fig. 12-102).

These enable on-line production of orthophotographs and drop-line charts. Other attachments are essentially the same as those of the Stereometrograph and provide for a considerable broadening of the instrument's applicability.

12.5 Stereoinstruments with Anamorphic Bundles

12.5.1 INTRODUCTION

In the period 1956−60 in the USSR, plotters with anamorphic projection were introduced. They became known as "affine plotters." The reason for developing this new class of instruments was the increasing range of the angular field (27^g to 135^g), and thus of the principal distance ($c = 30$ to 1000 millimetres and more), of survey cameras.

An anamorphic bundle is obtained when the instrumental principal distance differs from that of the survey camera, assuming the photoformat is unchanged. Such bundles are said to be "affinely" distorted. If the photographs are vertical (*i.e.* the principal axes are vertical), the model formed by such bundles is accordingly distorted in height. Hence, the scales of height and planimetry differ.

In the period 1960−65, and later, a considerable amount of research was performed in the area of affine restitution. The main efforts were directed towards conceptual developments of the projection system (or parts thereof) along with theoretical and experimental investigations of performance. A distinction was therefore made between the geometrically exact solutions of the projection systems, and approximate solutions. The latter were introduced mostly in existing conventional instruments to adapt them for affine restitution (Makarović, 1966).

The important representatives of the class employing geometrically exact solutions are the Drobyshev Stereographs SD and Photocarto-

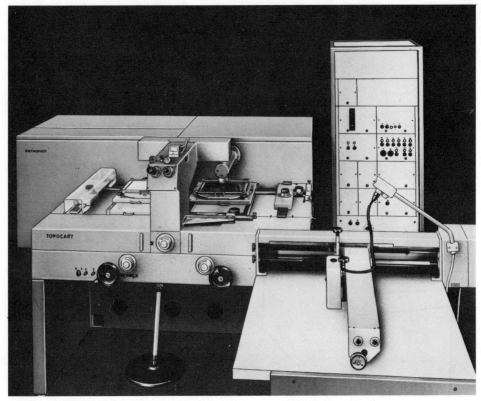

FIGURE 12-102. Topocart—Orthophoto.

graph, the Romanovsky Stereoprojectors SPR, the Zeiss Jena Stereotrigomat, and the Messerschmitt-Bolkov-Blohm-H.O.-1 Plotter. Since the Photocartograph is no longer produced and the M.B.B.-H.O.-1 Plotter belongs to the class of hybrid systems, further treatment here will be confined to the Stereographs SD, Stereoprojectors SPR, and the Stereotrigomat.

12.5.2 SD STEREOGRAPH

12.5.2.1 GENERAL

This instrument was designed in the USSR by Prof. F. V. Drobyshev (1957, 1957a). It appeared with some improvements in three successive models. The first two, SD I and SD II, are not suitable for strip triangulation, while the third, called the Universal Stereograph, is. The Stereograph was designed mainly for plotting medium- and large-scale maps from wide- or superwide-angle, near-vertical photographs. It uses an affine mechanical-projection system. The latest model has horizontal coplanar photocarriers with the observation system fixed and moving photographs for tracking. The plateholders can accept the standard Russian photograph format of 18 × 18 centimetres.

A peculiarity of the Stereograph is the special arrangement of the Zeiss parallelogram. The two projectors have a fixed separation in the Y-direction, while the X-separation is zero. Thus, one projector is behind the other as viewed by the operator. The instrument is compact, easily transportable, and simple to assemble. Figure 12-103 is a front view of the Stereograph SD II.

12.5.2.2 PROJECTION SYSTEM

The projection system of models I and II is adapted to the case of tilted photographs (Gerlach, 1962). Figure 12-104 shows this case in the projection on the XZ coordinate plane. The model $\bar{M}_o$ is deformed in the Z-direction only, from the conformal model M_o. In model $\bar{M}_o$, the planimetry is unchanged but the heights are deformed linearly and straight parallel lines and planes remain as such in the transformed model M_o. The direction of affinity coincides with the Z-direction; therefore the entire system, including the projectors and photographs, is deformed linearly in this direction. Consequently, the inner and outer orientations of the projectors differ from those of the camera, thus, conformal models may be considered as special cases with zero affinity. Figure 12-105 shows the vertical section along the line of maximum tilt of a photograph. Ph is the photo plane at exposure, tilted through v, with $\overline{Ph}$ (the corresponding plane at restitution tilted through $\bar{v}$. $\overline{Ph}$ is the

FIGURE 12-103. Front view of the Stereograph SD II.

affine transformation of Ph. O is the perspective center, and V the vanishing point. Points in the plane $\overline{Ph}$ are obtained as vertical displacements of the corresponding points in Ph.

The rays originally directed from Ph toward the conformal model M_o are directed, after transformation, from Ph toward the affine model $\overline{M}_o$. Only the rays within the horizontal plane VO, and the vertical ray passing through the nadir point n, remain unchanged. In figure 12-105, c is the principal distance of the photo plane Ph, and $\overline{c}$ of the plane $\overline{Ph}$. The distance $\overline{ab}$ is the affine transformation of ab. The corresponding orthogonal projection a_ob_o on the plane VO may be regarded as an affine transformation of ab for $\overline{c} = O$.

For geometrically exact solutions, it is convenient to use the desired ratio:

$$u = \frac{S_p}{S_h} \qquad (12.22)$$

where S_p and S_h are the scale denominators of the planimetric and height scales.

From figure 12-105 it follows that:

$$\tan \nu = \frac{aa_o}{Va_o} = \frac{bb_o}{Vb_o} = \frac{Z_b - Z_a}{a_ob_o}$$

and

$$\tan \overline{\nu} = \frac{\overline{a}a_o}{Va_o} = \frac{\overline{b}b_o}{Vb_o} = \frac{Z_{\overline{b}} - Z_{\overline{a}}}{a_ob_o}.$$

Substituting uZ_a and uZ_b for $Z_{\overline{a}}$ and $Z_{\overline{b}}$,

$$\tan \overline{\nu} = u \tan \nu. \qquad (12.23)$$

Further,

$$a_ob_o = ab \cos \nu = \overline{ab} \cos \overline{\nu}.$$

Magnification of the distance $\overline{ab}$ may be expressed by the ratio

$$\frac{\overline{ab}}{ab} = \frac{\cos \nu}{\cos \overline{\nu}}$$

or the affine deformation of the plane $\overline{Ph}$ by

$$\frac{\overline{ab} - ab}{ab} = \frac{\cos \nu}{\cos \overline{\nu}} - 1. \qquad (12.24)$$

The maximum deformation of $\overline{Ph}$ occurs in the direction of maximum tilt; there is no deformation perpendicular to it. Thus the plane $\overline{Ph}$ is

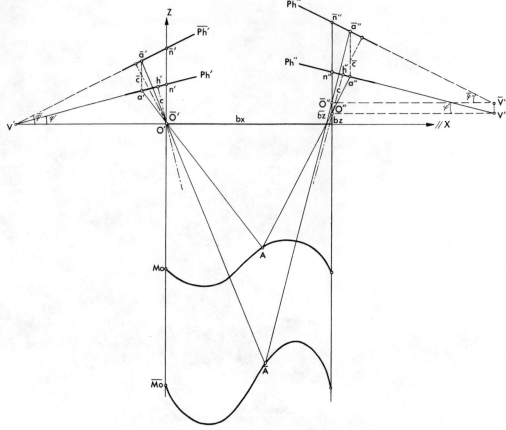

FIGURE 12-104. Affine projection of tilted photographs.

elongated along its slope. The inner orientation of $\overline{Ph}$ is defined by the principal distance $\bar{c}$ and the foot $\bar{g}$ of the perpendicular passing through O. Thus:

$$On = \frac{c}{\cos \nu}, \quad O\bar{n} = uOn = \frac{uc}{\cos \nu}$$

or, directly from figure 12-105

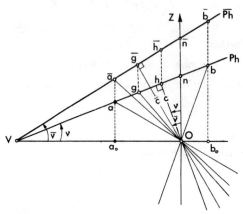

FIGURE 12-105. Vertical section along line of maximum tilt.

$$O\bar{n} = \bar{c}/\cos \bar{\nu}.$$

Combining these two equations:

$$\bar{c} = uc \frac{\cos \bar{\nu}}{\cos \nu}. \tag{12.25}$$

Comparing (12.22) and (12.25), it follows that

$$u = \frac{S_p}{S_h} = \frac{\bar{c} \cos \nu}{c \cos \nu}. \tag{12.26}$$

Thus the affinity factor u also depends on the tilt ν. From figure 12-105 some further important geometrical relations can be derived. Distance $nh = c \tan \nu$ and

$$\overline{nh} = c \tan \nu \frac{\cos \nu}{\cos \bar{\nu}}. \tag{12.27}$$

The foot-point $\bar{g}$ is defined by the distance $\overline{ng} = \bar{c} \tan \bar{\nu}$, and using (12.23) and (12.25)

$$\overline{ng} = u^2 c \tan \nu \frac{\cos \bar{\nu}}{\cos \nu}. \tag{12.28}$$

The corresponding distance in the photo-plane Ph is

$$ng = \overline{ng} \frac{\cos \bar{\nu}}{\cos \nu} = u^2 c \tan \nu \left(\frac{\cos \bar{\nu}}{\cos \nu} \right)^2. \tag{12.29}$$

Consideration of 12.27 and 12.28 shows that the required decentering of the point $\bar{h}$ is

$$\bar{g}\bar{h} = \bar{n}\bar{g} - \bar{n}\bar{h}$$

$$= c \tan v \left(u^2 \frac{\cos \bar{v}}{\cos v} - \frac{\cos \bar{v}}{\cos v} \right) \quad (12.30)$$

and the corresponding value in the plane Ph is

$$gh = ng - nh$$

$$= c \tan v \left[u^2 \left(\frac{\cos \bar{v}}{\cos v} \right)^2 - 1 \right]. \quad (12.31)$$

From the last two relations it follows that the inner and outer orientations are interdependent.

The geometrical relations may be simplified by introducing a fictitious horizontal plane Ph_o (fig. 12-106) passing through the isocenter i of the plane Ph; $\bar{c}_o$ is the corresponding fictitious principal distance of the projector. Thus the plane Ph_o represents a strictly vertical photograph.

For exact vertical affine deformation ($\bar{Z} = uZ$), the rays of the anamorphic bundle must pass through the points of the plane Ph and corresponding points of the plane Ph_o (e.g., the ray $\bar{A}$ $\bar{O}$ $\bar{a}$ a_o).

From figure 12-106, it follows that

$$\frac{\delta}{\Delta c} = \frac{d_o}{c}$$

$$\frac{\delta}{\Delta \bar{c}} = \frac{d_o}{c_o}$$

$$\Delta c = d_c \tan v$$

$$\Delta \bar{c} = d_c \tan \bar{v}.$$

Consequently

$$\frac{\Delta \bar{c}}{\Delta c} = \frac{\bar{c}_o}{c} = \frac{\tan \bar{v}}{\tan v}. \quad (12.32)$$

Equating expressions (12.26) and (12.32)

$$\frac{\bar{c}}{\cos \bar{v}} = \frac{\bar{c}_o}{\cos v}$$

or, according to figure 12-106, $\bar{O}g_o = \bar{O}n$.

Finally, the relationship between $\bar{c}_o$, $\bar{c}$, and c is obtained:

$$\bar{c}_o = \bar{c} \frac{\cos v}{\cos \bar{v}} = uc. \quad (12.33)$$

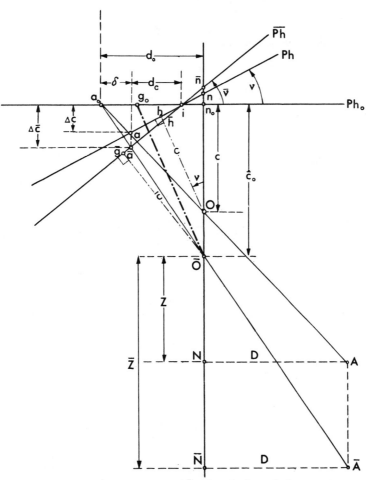

FIGURE 12-106. Geometry of fictitious horizontal plane.

Summarizing the above derivations, the following conditions must be fulfilled:

(a) the projector is tilted through v (or ϕ, ω);
(b) The correction plane $\bar{Ph}$ passes through the isocenter i, and is tilted through $\bar{v}$, where $\tan \bar{v} = (\bar{c}_o/c) \tan v$ (or $\tan \phi$, $\tan \bar{\omega}$); and
(c) The observation is perpendicular to the reference plane Ph_o.

The physical solution of the above conditions is shown schematically in figure 12-107. It represents the situation in the vertical section passing through both perspective centers O and $\bar{O}$. The tilts ϕ, ω and the principal distance c are applied to the projector ($c = Oh$). The observation system is stationary above point N_o, and the line of observation coincides with the line N_oOn. Ph represents the tilted photograph. The projector movements for scanning are restricted to the horizontal plane OO'. These are controlled by rod F, which is attached to the projector. This rod moves through a sleeve of arm G, which connects the feeler on the correction surface C with the central gimbal joint on the space rod. The connection is such that the central gimbal joint follows the same movements as the feeler on the correction surface C. Rod F follows the horizontal displacements $\bar{x}$ and $\bar{y}$ only. When the feeler is moved from N_o to A_o, the gimbal joint moves from $\bar{n}$ to $\bar{a}$ and the projector from O to O'. Point a' then appears under the observation system. Thus the geometrical conditions are fulfilled according to figure 12-106.

The correction plate C is tilted by $\bar{\phi}$ and $\bar{\omega}$. In the vertical position of space rod R, the feeler is at point N_o, which is vertically under the observation system Ob. The vertical separation between the fixed perspective center $\bar{O}$ (upper gimbal joint) and the isocenter i is the instrument principal distance $\bar{c}_o$. During tracking, point $\bar{a}$ describes the mechanical photo plane $\bar{Ph}$, which is parallel to the correction surface C. (In figure 12-106, triangles $\bar{O}$ i a and O i a are superimposed.) The lower end $\bar{A}$ of space rod R, which represents the mechanical model point, is connected to the base bridge. According to the geometrical assumptions, the correction plate C should rotate about isocenter i. As this point varies from photograph to photograph, its realization would introduce complications in mechanism and procedure. In reality C is rotated about a fixed point S (fig. 12-108), which is eccentric under the plate. When applying a tilt ϕ, ω, the plate C' deviates by $\bar{n}n'$ from its required position C, and consequently a correction $\delta \bar{c}_o = \bar{n}n'$ should be applied to the principal distance $\bar{c}_o$. This correction depends on the $\bar{\phi}$- and $\bar{\omega}$-tilt of the correction C':

$$\delta \bar{c}_s = d\bar{c}_\phi + \delta \bar{c}_\omega,$$

where

$$\delta c_\phi = \bar{n}n' = \bar{n}n_o + n_o n'$$

$$= c \tan \frac{\phi}{2} \tan \bar{\phi} + p \tan \frac{\bar{\phi}}{2} \tan \phi$$

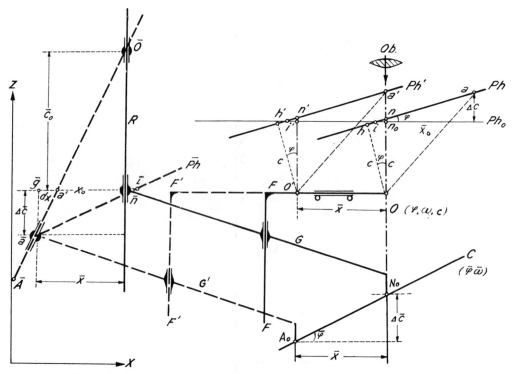

FIGURE 12-107. Plan of physical solution of affine restitution in SD Stereograph.

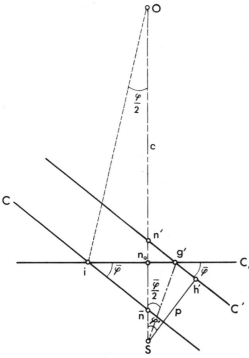

FIGURE 12-108. Rotation center of correction plate C.

and similarly,

$$\delta c_\phi = c \tan \frac{\omega}{2} \tan \bar{\omega} + p \tan \frac{\bar{\omega}}{2} \tan \bar{\omega}.$$

For small tilts;

$$\tan \bar{\phi} \approx \bar{\phi} \frac{\phi}{\phi} \approx \frac{c}{\bar{c}}, \ \tan \bar{\omega} \approx \bar{\omega}, \ \frac{\omega}{\bar{\omega}} \approx \frac{c}{\bar{c}}.$$

Thus:

$$\delta \bar{c}_\phi \approx \frac{\bar{\phi}^2}{2} \left(\frac{c^2}{\bar{c}} + p \right), \ \delta \bar{c}_\omega \approx \frac{\bar{\omega}^2}{2} \left(\frac{c^2}{\bar{c}} + p \right)$$

and finally:

$$\delta \bar{c}_s = \frac{1}{2} \left(\frac{c^2}{\bar{c}} + p \right) (\bar{\phi}^2 + \bar{\omega}^2). \tag{12.34}$$

The corresponding corrections can be obtained from tables or nomographs; the affinity factor u should be corrected accordingly. In practice only the difference of the corrections between the two projectors need be applied:

$$\Delta \delta \bar{c}_s = \delta \bar{c}'_s - \delta \bar{c}''_s$$
$$= (\delta \bar{c}'_\phi - \delta \bar{c}''_\phi) + (\delta \bar{c}'_\omega - \delta \bar{c}''_\omega). \tag{12.35}$$

These differences are applied to the right-hand mechanical photopoint. When $\Delta \delta \bar{c}_s < 0.06$ mm, it can be neglected.

Figure 12-109 is a diagram of the mechanical structure of the instrument. The projectors, 27 and 19, are one behind the other because of the way the Zeiss double parallelogram is used. Figure 12-110 shows the principle of this special arrangement. O' and O'' are the perspective centers and R' and R'' the corresponding rays to an arbitrary model point A. The ray R' remains in its original position, while the second ray R'' is shifted at first horizontally over Ky to the point $O''y$, then over the base component bx to the second point $(O''x)$. Finally, it is displaced by bz and by into the fixed point $O''x$. O' and

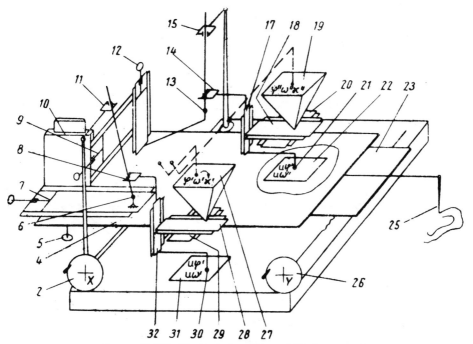

FIGURE 12-109. Mechanical structure of SD Stereograph.

$O''x$ are the fixed perspective centers in the instrument. The constant separation in x-direction is zero; therefore, the parallelogram on the XZ plane reduces to a line. However, there is a parallelogram in the YZ plane with a constant separation Ky, which must be somewhat greater than the diameter of the projectors.

The displaced rays are realized by space rods R' and $R''x$, points O' and $O''x$ by fixed gimbal joints, and the model points A and $A''x$ by the lower ends of the space rods. The base component bx is applied at the joint A of the first rod, whereas by and bz are applied at the joint $A''x$ of the second rod. This arrangement of the Zeiss double parallelogram permits a reduction in instrument size.

The projectors are moved for scanning (fig. 12-109) along the horizontal cross-slide systems 28, 29 and 18, 20. The correction plates are designated by 31 and 22, and the corresponding feelers by 30 and 21. The arms (G, fig. 12-107) connecting the feelers to the central gimbal

joints 8 and 14 of the space rods, are 17 and 32. Further, 11 and 15 represent the perspective centers ($\bar{O}'$ and $O''x$), and 6 and 13 are the mechanical model points ($\bar{A}$ and $\bar{A}''x$). The bx (7) motion is applied to the first model point 6, by (9) and bz (12) to the second (rear) model point 13. The base bridge is attached to the Z-carriage 10, which is movable along the corresponding column by actuating the foot disk 5. The Z-column is attached to the X-carriage 4, which moves along the X-rails on the primary Y-carriage 23. Planimetric movements are controlled by hand wheels 2 and 26. The drawing table is on the right-hand side of the instrument, and the pencil 25 is connected directly, in the model scale, or by means of a parallelogram pantograph, for a different plotting scale.

The plateholders are provided with separate supports for principal distances c from 55 to 210 millimetres, while the instrumental principal distance $\bar{c} = 130 \pm 3$ millimetres. Maximum tilts of the projectors are; $\phi_{max} = \omega_{max} = 2.7^g$. The

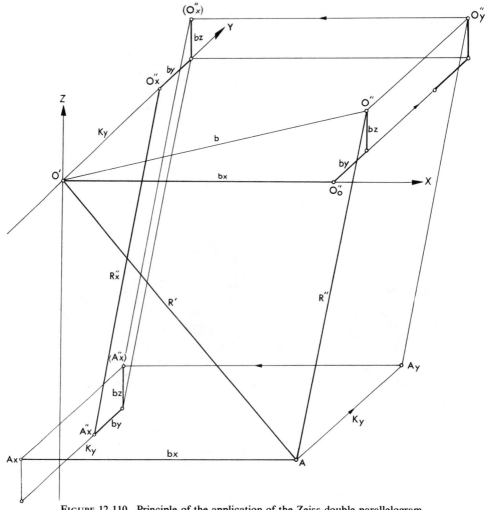

FIGURE 12-110. Principle of the application of the Zeiss double parallelogram.

affinity can be changed in discrete steps: 1.3, 1.9, and 2.4 times. Model scale is variable from 0.96 to 1.1 times the photo scale, while the pantograph ratios are within the limits of 0.5 to 2.2 times.

Between the X- and Y-handwheels (fig. 12-111), screws are situated with corresponding dials and transmission mechanisms for settings of tilts ϕ', ω', ϕ'', and ω'' to the correction surfaces C' and C''. Correction plates are made of glass and are supported on three points (fig. 12-112). Each correction plate C can rotate about its center support S; the $\bar{\phi}$ and $\bar{\omega}$ screws actuate the two side supports a and b, as shown in figure 12-112. Supports a', and b' and a'', b'' can be raised or lowered 8 millimetres. Distance $S'a' = S''a'' = 90$ millimetres, and $S'b' = S''b'' = 80$ millimetres. Consequently, the tilts are $\phi = h_\phi p/90$ and $\omega = h_\omega p/80$, where h_ϕ and h_ω are the height displacements of the corresponding supports. Thus, maximum tilts are about 7 grads. The dials are graduated for h_ϕ and h_ω.

When the tilts are less than 1.4^g, the projectors can be kept in their zero positions. However, the decentering elements must be applied in this case according to

$$d\bar{x} = c \tan \phi \approx \frac{h_\phi c^2}{90\bar{c}}$$

$$d\bar{y} = c \tan \omega \approx \frac{h_\phi c^2}{80\bar{c}} . \qquad (12.36)$$

The maximum projection error due to this approximation is less than 20 micrometres. Apart from this, an error (<80 micrometres) occurs in the principal distance c and causes a slight longitudinal tilt of the model. It is compensated for by leveling.

In the latest model, the Universal Stereograph, geometrical principles and the mechanical structure have been modified for horizontal (parallel) positions of photographs.

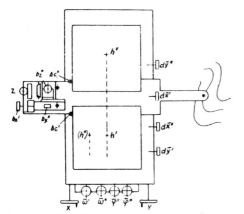

FIGURE 12-111. Tilt-adjustment screws for correction surfaces.

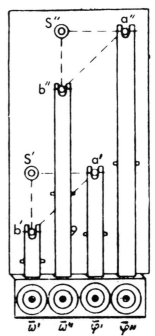

FIGURE 12-112. Correction-plate supports.

12.5.2.3 OBSERVATION SYSTEM

The special arrangement of the Zeiss double parallelogram causes asymmetry in the observation system. This is shown in figure 12-113 in (a) the horizontal, and (b) the vertical projection. Only the first four optical elements are symmetrical: the eyepiece, measuring mark (1), trapezoidal prism (2), and the lens (3). Lens (3) forms the image in the plane of the measuring mark consists of a rectangular prism (7″) above the photograph (8″) and a lens (4″) in front of (3). The focal length of (4″) is equal to the optical path from the photograph (8″); thus the light between (4″) and (3) is parallel.

In the left-hand system, the optical path is extended artificially by reflecting elements (5′), (6′), and (7′). The focal distance of lens (4′) is equal to the length of the optical path between it and the photo plane (8′); thus the light between (4′) and (3) is parallel also.

Diapositives are used with the emulsion toward the observation system; negatives, the other way about. If the photographs are tilted, the separation (8)-(4) is variable during scanning. Because of the long focal distances of lenses (4′) and (4″), the depth of focus is considerable; therefore, loss of image quality can be tolerated for small values of tilts ($\nu \leqslant 2^g$). With stabilized camera mounts for photographing, these tilts are completely negligible from the optical viewpoint.

With horizontal photographs, the decenterings $d\bar{x}$ and $d\bar{y}$ are applied directly to the observation system (fig. 12-113): $d\bar{x}'$ and $d\bar{x}''$ are applied to

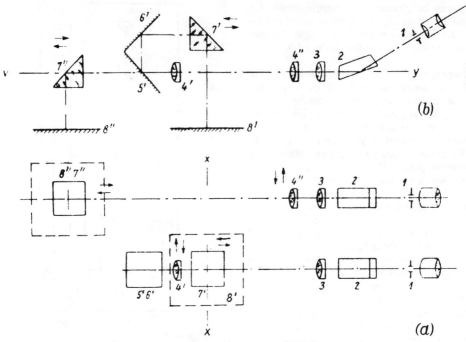

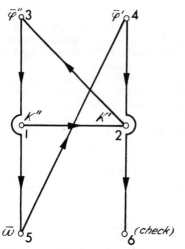

FIGURE 12-113. Observation system of SD Stereograph.

lenses (4′) and (4″), and $d\bar{y}'$, $d\bar{y}''$ to prisms (7′) and (7″). The corresponding setting screws are shown in figure 12-111.

The system has a 4.1 times magnification. The diameter of the field of view in the photo plane is 30 millimetres.

12.5.2.4 USE AND PRECISION

Relative orientation is carried out at the six conventional points, according to figure 12-114. For elimination of the y-parallaxes, the swing κ of the photographs and the tilts $\bar{\phi}$, $\bar{\omega}$ of the correction plates can be used. The tilts of the pro-

FIGURE 12-114. Relative orientation sequence.

jectors are derived from the tilts of corresponding correction plates ($\phi = \bar{\phi}/u$, $\omega = \bar{\omega}/u$). The tilts of the correction plates have a different effect on the y-parallaxes in the model from that in conventional instruments. Points (1) and (2) are not affected by the $\bar{\phi}$ and $\bar{\omega}$ rotations. Consequently, ω can be used for direct elimination of y-parallaxes without applying an over-correction. This simplifies orientation procedure. However, the corresponding $\Delta\delta c_s$ corrections must be applied due to the eccentric rotations of the correction plates. Furthermore, if the tilts ϕ, ω of the projectors are neglected, the decenterings dx and dy must also be taken into consideration.

It is convenient to begin relative orientation with the left-hand projector in zero position and to use the elements of the right-hand projector only. Thus no decenterings need be applied to the left-hand photograph before absolute orientation. The planimetric model scale is adapted to the assumed height scale according to:

$$S_p = uS_h = \frac{\bar{c}o}{c} S_h. \qquad (12.37)$$

The correction plates C are used for levelling together with the projectors (ϕ, ω or dx, dy) and the base component $b\bar{z}$.

The average time for a complete orientation is about 1 hour and 40 minutes, while plotting speed approximates that of the other one-man plotters. Planimetric accuracy, determined experimentally, is 0.2 millimetre on the map, while height accuracy is between 0.02 and 0.05

percent of Z, depending on the topography. Measurement on grid models (with zero tilts) have shown an accuracy of 0.008 percent of Z.

The optimum ratio between photograph and model scale, $S_{ph}:S_{mod} = 2:1$, although this is not economical for large scale mapping. In mountainous terrain, the accuracy can be improved if restitution is carried out in several height zones. If the photographs are tilted, the image quality, as observed by the operator, varies slightly owing to the lack of automatic focusing, causing an irregular accuracy in the restitution. This has been the main reason for modifying the SD for horizontal photographs.

12.5.2.5 SD-3 STEREOGRAPH

Model SD-3 is a table-top-universal instrument (Lovanov, 1971). The modifications concern, in addition to the horizontal arrangement of photocarriers, extension of the model space and possibility of handling photographs with maximum tilts of 3^g to 5^g, depending on the amount of affinity. The magnification of the observation system was increased to $4\times$ and $7\times$. The geometrical principles can be modified so as to allow a physical simplification of construction and higher performance. The following ratio is obtained (from fig. 12-115)

$$\frac{D}{Z} = \frac{ma}{mO} = \frac{d\cos v}{nO - nm} = \frac{d\cos v}{\dfrac{c}{\cos v} - d\sin v}$$

$$= \frac{d}{\dfrac{c}{\cos^2 v} - d\tan v}. \qquad (12.38)$$

The corresponding expression for anamorphic bundles is obtained by dividing relation (12.38) by the factor of affinity n:

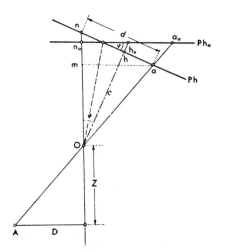

FIGURE 12-115. Modified geometric principles.

$$\frac{D}{nZ} = \frac{d}{\dfrac{nc}{\cos^2 v} - dn\tan v},$$

or

$$\frac{D}{nZ} = \frac{d}{\dfrac{\bar{c}}{\cos^2 v} - d\tan \bar{v}}. \qquad (12.38a)$$

For a strictly vertical photograph:

$$\frac{D}{nZ} = \frac{d}{\bar{c}_o}. \qquad (12.38b)$$

The realization of relation (12.38a) is shown in fig. 12-116. Ph_o is the horizontal reference plane and Ph the photo-plane in its original position, a is the image of an arbitrary point, a_o is its orthogonal projection, and a_1 its central projection on the reference plane Ph_o. Point a_2 is obtained by rotating plane ph into Ph_o about the isocenter i. The distance $na \approx n_o a_2 = d$.

O is the original perspective centre and $\bar{O}$ that relating to the anamorphic bundle of rays. The corresponding position of the correction plane is $\overline{Ph}$ and a_4 is the related "image" point obtained by the intersection of the ray $a_1\bar{O}$ with $\overline{Ph}$. Between the two planes there is the known relation:

$$\frac{\bar{c}_o}{c} = \frac{\Delta\bar{c}}{\Delta c} = \frac{\tan \bar{v}}{\tan v} = n. \qquad (12.39)$$

Point a_5 is obtained by the inverse orthogonal projection of the rotated point a_2, onto plane $\overline{Ph}$. The distance $qa_5 = n_o a_2 = d = d_i + n_o i$. According to relations 12.38a and 12.38b there follows the condition that the distance:

$$q\bar{O}_v = \frac{\bar{c}_o}{\cos^2 v} - d\tan v$$

$$= \left(\frac{\bar{c}_o}{\cos^2 v} - n_o i\ \tan \bar{v}\right) - d_i\tan v. \qquad (12.40)$$

By considering relation (12.38) and $n_o i = c_o \tan v/2$:

$$q\bar{O}v = \frac{\bar{c}_o}{\cos^2 v} - c_o\tan \bar{v}\tan\frac{v}{2} - d_i\tan \bar{v}$$

$$= \left(\frac{\bar{c}_o}{\cos^2 v} - \Delta\right) - d_i\tan \bar{v}. \qquad (12.41)$$

The term in brackets is constant for a given photograph, and represents the modified principal distance. Due to the eccentricity of the rotation center of the correction plate (with regard to i), the principal distance $\bar{c}_o$ must be corrected: $\bar{c} = \bar{c}_o + \delta\bar{c}_s$, where

$$\delta\bar{c}_s = \frac{v^2}{2}\left(\frac{c^2 + p}{c}\right).$$

$\bar{O}$ represents the new position of the perspective center and $\overline{Ph}'$ the corresponding correction plane. Point a_6 is obtained as the inverse orthogonal projection of point a_2 on plane $\overline{Ph}'$. Point a, on the ray Oa, is similarly obtained. The plane passing through a_3 is $\overline{Ph}'$. The tilts of

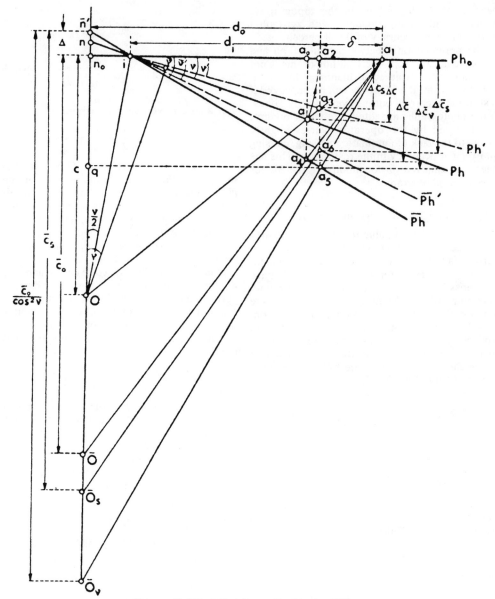

FIGURE 12-116. Principles realised in the USD.

planes $\overline{Ph}'$ and Ph' are slightly less than those of Ph and $\overline{Ph}$, but the differences are very small. From fig. 12-116 it follows that:

$$\frac{\delta}{\Delta c_s} = \frac{d_o}{c} \,, \quad \frac{\delta}{\Delta \overline{c}_s} = \frac{\delta}{\overline{c}_s - \Delta} \,,$$

where

$$\Delta = n_o \overline{n}'.$$

Consequently:

$$\frac{\Delta \overline{c}_s}{\Delta c_s} = \frac{d_o(\overline{c}_s - \Delta)}{c \delta} = n_s.$$

Further:

$$\Delta c_s = d_i \tan \nu', \quad \Delta \overline{c}_s = d_i \tan \overline{\nu}',$$

and:

$$n_s = \frac{\tan \overline{\nu}'}{\tan \nu'} \,.$$

From the above it follows that there is a solution for any ratio $(\overline{c}_i/c = n_i)$ if the correction plate is tilted over $\tan \overline{\nu}_i = n_i \tan \nu'$. The factor of affinity n_i must be equal for both projectors. However, the difference in $(\overline{c}_s - \Delta)$ and $\delta \overline{c}_s$ between the two projectors must be considered. If this difference is neglected, subsequent model deformations are compensated for by adequate $\overline{bz}$ translations. Equal variations of both modified principal distances cause a change in the affinity only.

In tilted photographs angular errors of the

radial directions from the nadir point are present. This causes projection errors and consequently distortions in geometry of the model. Adequate correction devices can be provided for the elimination of azimuthal errors. For this purpose Vasil'ev suggested inclining arm F (fig. 12-107) to make it parallel to the line connecting the perspective centre $\bar{O}$ to the isocenter i (fig. 12-116). The lines connecting all rotated points (a_2) in Ph_o with the corresponding points in the correction plane $\overline{Ph}$ (a_4) are parallel to $\overline{Oi}$. Consequently the $\Delta \bar{c}$ variations of the connection G cause corrective displacements $(a_o a_2)$ of the projector.

The orientation procedure is based on successive elimination of Y-parallaxes in the six conventional points, the swings of the photographs, the inclinations of the correction plates $\bar{\phi}$, $\bar{\omega}$, and consequently decentrations $d\bar{x}$, $d\bar{y}$ of the observation system.

12.5.2.6 EXPEDITION SD-3 STEREOGRAPH

In 1965 Drobyshev constructed a new version of the Stereograph—the "Expedition Stereograph ESD." This is basically a medium- and small-scale affine plotter suitable also for continuous restitution of strips. The projection system is based on the same principles as the SD-3. The basic difference lies in the dimensions and weight of the instrument. Photographs and instrument ranges are reduced twice—therefore the ESD is a compact and easily transportable instrument. The photo format is 9×12 centimetres, instrumental principal distance $\bar{c} = 70$ millimetres, and maximum photo-tilts can be $\pm 3^g$. The instrument uses a 5-time optical magnification; its dimensions are $60 \times 60 \times 60$ centimetres.

Recently the Stereograph SD-3 was adapted, by CNIIGAiK in Moscow, for large scale restitution. This successor of the SD-3 is called the Stereograph SC.

12.5.2.7 PHOTOSTEREOGRAPH FSD

In 1962 the prototype of another Drobyshev design was made (the CNIIGAiK-Photostereograph FSD). This instrument represents an adaption of the SD-2 to orthophoto-printing. The Photostereograph can be used as a plotter, for aerial triangulation, and for the production of orthophotographs. If desired, orthophotographs can be supplemented with contour lines. The FSD uses a constant instrumental principal distance of $\bar{c} = 135$ millimetres. The adaption for the orthophoto printer required a rearrangement of the XYZ-cross slide system. In the FSD, the Z-motion is primary, Y secondary, and the X-motion tertiary. Two X-carriages are movable on the Y-carriage and their mutual distance determines the bx-setting.

The orthophoto printer (fig. 12-117) is located vertically under the left photograph. Its projection axis coincides with the vertical line of sight of the observation system above the photo-graph. Both the observation system and the orthophoto printer are fixed during restitution. Since the photograph is located between them, two light sources are required for illumination: one under the photograph (for observation), and another above it (for projection). On the left carriage a light proof photo-cassette can be fixed. During tracking the cassette follows the XY- and Z-movements. A small slit is located on the photo-cassette vertically under the set photo-section; it can move only in Z-direction. The movements of the photographs are coupled with those of the XYZ-cross slides (carrying the cassette via the space rods of the projection system). The orthophoto printer includes a fixed lens located at its focal distance $f = \bar{c} = 135$ millimetres under the photograph, an automatic focusing system, an illumination system, a slit, and the photo-cassette.

Scale variations caused by the relief are adjusted automatically by changing $\bar{Z}$ during profiling. The rates $c/Z = \bar{c}/\bar{Z}$ are realized by the differential rectifier. An automatic focusing system forms sharp images in the projection plane and is controlled by a curved edge fixed to the base of the instrument and a transmission mechanism (as in the Stereoplanigraph).

A tracing stylus, which follows movements in Z-direction, is fixed under the position of the measuring mark on the right photograph. A map sheet (40×40 centimetres), provided with control points, can be fixed on the right carriage (X''), or more precisely on the bZ'' carriage. This facilitates absolute orientation of the model, and later—if desired—a plot of contour lines on a transparent sheet or directly on the orthophotograph. For orthophoto printing the Y-carriage is driven at a constant speed by an electrical motor. The operator can control the Z-carriage by turning the Z-handwheel or foot-disk. The widths of Y-strips, and thus the X-steps, are 2.5, 5, and 7 millimetres, depending on the type of terrain. The image quality of orthophotographs is satisfactory; contacts between adjacent strips are hardly noticeable.

For plotting of maps, a light, transportable electro-coordinatograph is employed. The instrument is enclosed in a tent which protects it against outside light. Some technical data for the instrument is as follows.

- Photo format 18×18 centimetres; photo tilts $\pm 3^g$.
- Instrument principal distance $c = 135$ millimetres; camera principal distance $c = 55$-200 millimetres.
- X, Y ranges ± 200 millimetres; $\bar{Z}$ range 80% $\bar{Z}$.
- Electro coordinatograph 300×400 millimetres.
- Height accuracy for mountainous terrain ($c = 70$ millimetres): 0.028% Z.
- Maximum locational error in orthophotographs: 0.4 millimetres.

12.5.2.8 STEREOGRAPH SD-3 WITH ORTHOPHOTO ATTACHMENT FPD-2

After developing the SD-3 a new orthophoto attachment was constructed. The attachment

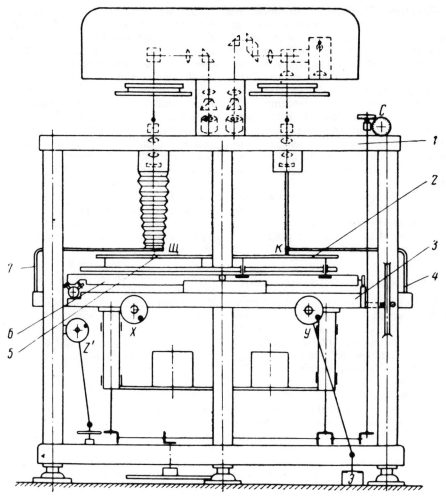

FIGURE 12-117. Orthophoto attachment.

FPD-2 is positioned on top of the plotting table of the SD-3. The right-hand optical train of the SD-3 is branched and guided, via mirrors and lenses, to the printer. The printer includes a zoom system, some prisms, a slit, and a plane film cassette. In operation the optical train is fixed. Traversing in the Y-direction and stepping in the X-direction are by means of electrical drives. Z-control is a separate hand wheel (after disconnecting the foot-disk). The control panel is in front of the orthophoto attachment, close to the operator and traversing speed may range between 2 and 2.4 mm/sec. The slit width can be 1.2 or 4 millimetres, and the printing time per model from 15 to 100 minutes.

12.5.3 ZEISS JENA STEREOTRIGOMAT

12.5.3.1 GENERAL

The first affine plotter designed outside the USSR was the Stereotrigomat (Zeiss Jena 1964), developed in East Germany in 1965. It represents one of the top examples of analogue resti-tution instruments. The Stereotrigomat is distinguished by its far-reaching versatility in regard to principal distances and output facilities, by the electrical transmissions between different parts of the instrument, and by its high mechanical performance. The extreme tilts of photographs are restricted to $\pm6^g$, but their principal distances may range from 35 to 600 millimetres. For principal distances between 35 and 90 millimetres an affine restitution must be applied. For the rest, congruent bundles are possible. The maximum format of photographs is 24×24 centimetres. The instrument is applicable as a precise plotter for aerial and terrestrial photography, for continuous triangulation of strips, independent model triangulation, and for the production of orthophotographs. It can also be used for high precision numerical restitution related to engineering and cadastral applications. The projection system is mechanical and an electrical transmission facilitates a high degree of freedom in the location of the system components. It also permits an application of simple geometrical principles which are not feasible in cases of mechanical transmission. These principles have been selected in such a way that the operation of the instrument is as simple as possible and can be operated by one person.

The main parts of the system are the measuring unit, the projection computers, the differential rectifier, the drawing table with coordinatograph, and the cabinet for electric components. This cabinet contains the electrical installation and is separate from other units of the system. The projection computers are also physically separated from the measuring unit and the computing system can, therefore, be extended arbitrarily or entirely replaced in the future. The instrument-system is divided into three large modules (fig. 12-118): the measuring unit (1), the computers (2), and the electric cabinet (3). The total weight is about 5600 kilograms.

12.5.3.2 PROJECTION SYSTEM

The geometric principles of the projection system are shown in fig. 12-119. An arbitrary photo point a is determined in the coordinate system XYZ by the photo-coordinates x and y, the principal distance $\bar{c}$, and the tilts ϕ and ω. The system x, y is assumed to be parallel to the XZ- or YZ-coordinate plane ($\kappa = 0$). The same point can be determined with coordinates x_o, y_o, and z_o which are in the system X, Y, Z. x_o, y_o, and z_o are the coordinates of photo-points in the XYZ system. The values of z_o, referring to all photo points, define the photo-plane tilted through ϕ

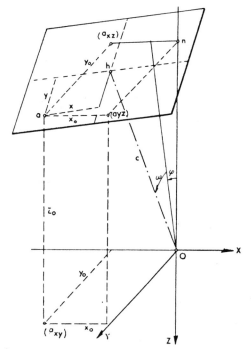

FIGURE 12-119. Photo-coordinates in the systems xyz and XYZ.

and ω. Assuming ϕ is the primary and ω the secondary rotation, the relationship between x, y, c and x_o, y_o, z_o can be derived from Fig. 12-120 and 12-121, representing orthogonal projection on the Yc plane and the XZ plane respectively:

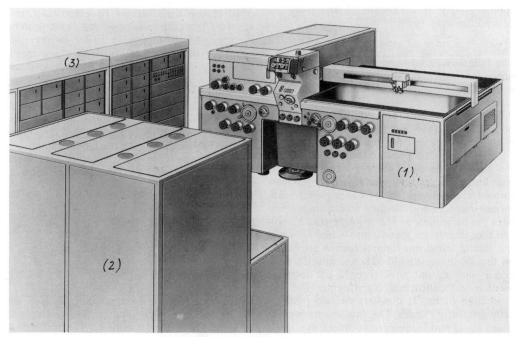

FIGURE 12-118. The Stereotrigomat.

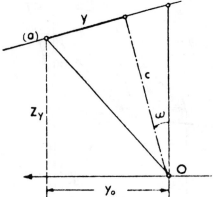

FIGURE 12-120. Projection on the Yc-plane.

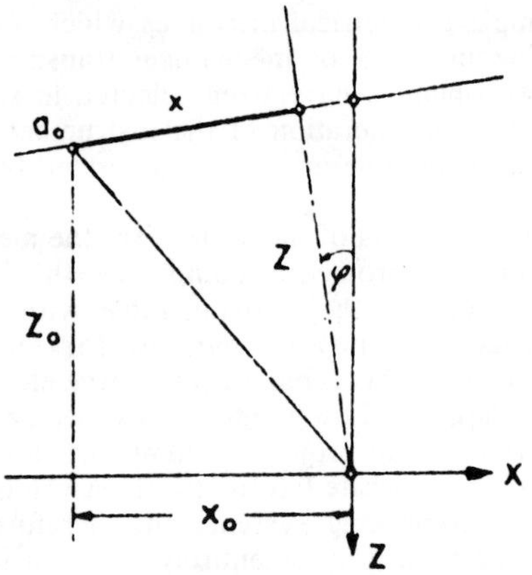

FIGURE 12-121. Projection on the XZ-plane.

$$y_o = y \cos \omega + c \sin \omega,$$

$$z_y = y \sin \omega + c \cos \omega, \qquad (12.42a)$$

$$x_o = x \cos \phi + z_y \sin \phi$$
$$= x \cos \phi + (c \cos \omega - y \sin \omega) \sin \phi,$$

and $\qquad (12.42b)$

$$z_o = x \sin \phi + z_y \cos \phi$$
$$= (c \cos \omega - y \sin \omega) \cos \phi - x \sin \phi.$$

The corresponding model coordinates are:

$$X = Z \frac{x_o}{z_o} = Z \frac{(c \cos \omega - y \sin \omega) \sin \phi + x \cos \phi}{(c \cos \omega - y \sin \omega) \cos \phi - x \sin \phi},$$

and $\qquad (12.43)$

$$Y = Z \frac{y_o}{z_o} = Z \frac{y \cos \omega + c \sin \omega}{(c \cos \omega - y \sin \omega) \cos \phi - x \sin \phi}.$$

An affinity in the vertical direction can be applied simply by multiplying Z and z_o by the desired factor n. Thus:

$$\bar{Z} = nZ \text{ and } \bar{z}_o = nz_o.$$

The fictitious photo-plane, representing the affinely deformed photograph, is thus tilted over $\bar{\phi}$ and $\bar{\omega}$. The horizontal components X and Y remain unchanged.

The projection is solved in two vertical planes. The input for the $X\bar{Z}$-plane are the values of x_o, $\bar{z}_o$ and for the $Y\bar{Z}$-plane are the values of y_o, $\bar{z}_o$. It is peculiar to this solution that the photographs need not be decentered. The transformation of x, y, c into x_o, y_o, z_o is performed successively by two mechanical conformal transformers. The first transformer simulates the projection on the yc-plane (fig. 12-120) and thus transforms y and c into y_o and z_y, while the second transformer simulates the projection on the XZ-plane (fig. 12-121) and transforms x and z_y into x_o and z_o. Figure 12-122 shows a mechanical conformal transformer for one projection plane. It consists of two pairs of orthogonal cross slides. The smaller, inner cross slides 11, 12 and 16, refer to coordinates y and c (or x and z_y respectively), since the larger (outer)

guides 2, 3 and 4 refer to the transformed coordinates y_o and z_y (or x_o and z_o). The former system (11, 12, and 16) can be rotated with respect to the system 2, 3, and 4 over ω (or ϕ respectively) by means of screw 19 about the axis 15, which represents the perspective center. Pivot 21 simulates the projected photo-point. The movements of 21 along rail 16 represent the changes in y (or x), and those along the parallel rails 11 and 12 the changes in c (or z_y). Similarly movements along 2 and 3 indicate changes in y_o (or x_o), and those along 4 in z_y (or z_o). The input for the first transformer (Yc) are the constants c and ω, and the variable photo-coordinate y. For the second transformer (XZ) the constant input is ϕ only; x is the variable photo-coordinate and z_y is the variable output of the first transformer. The photo-coordinates x and y are supplied to the transformers by electrical transmissions from the measuring unit, which can be positioned arbitrarily. In order to increase accuracy the coordinates x and y are multiplied by coefficients 1.6 (or 0.5 if $c > 300$ mm) in order to reduce the Z-range before they are supplied to the transformers. Obviously the principal distance c must be multiplied by the same coefficient. The outputs of the transformers (of each projector) $x'_o y'_o z'_o$ and $x''_o y''_o$, z''_o are transferred electrically to the model computers. These computers simulate the projections on the XZ- and YX-planes. They realize the following simple relations:

$$X = \frac{Z}{z_o} x'_o = \frac{\bar{Z}}{\bar{z}_o} x'_o, \ Y = \frac{Z}{z_o} y'_o = \frac{\bar{Z}}{\bar{z}_o} y'_o, \text{ and}$$

$$Z = \frac{\left(bx - x''_o \dfrac{bz}{z''_o}\right)}{\left(x'_o - x''_o \dfrac{z'_o}{z''_o}\right)} z'_o \qquad (12.44)$$

where $\overline{bz}$ is nbz. In these relations $\underline{X}, Y, \bar{Z}$ are the model coordinates and bx, by, bz the base components at the restitution. The orientation procedure is conventional.

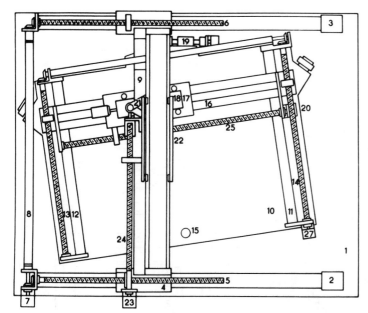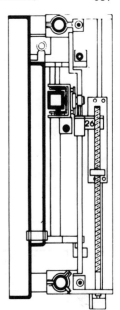

FIGURE 12-122. Mechanical transformer.

Since the projection is resolved into two coordinate planes ($X\bar{Z}$ and $Y\bar{Z}$), there are in total four projections of perspective centers lying in one vertical axis. Homologous rays, represented by their orthogonal projections onto the $X\bar{Z}$- and $Y\bar{Z}$-planes, are realized by four straight edged rulers which can rotate independently about a common vertical axis. They also have the $\bar{Z}$-carriage in common.

Figure 12-123 shows the location of the model computer (lower half) and of the two transformers (upper half). The second pair of transformers is located behind those shown in the figure. The functioning of a stage of the model computer, representing a single projection $X\bar{Z}$ (or $Y\bar{Z}$) of a ray, is shown in fig. 12-124. The projection of a photopoint (or its affine substitute) on the respective coordinate plane is realized by pivot $\bar{a}$, the projection of the perspective center by the axis z, and the projection of the corresponding model-point by pivot A. The movements of pivot $\bar{a}$ in its cross slides are controlled by the output of corresponding transformers (z_o, x_o or y_o respectively). For this purpose electrical transmissions with servomechanisms are employed. The values of z_o may be multiplied by an adequate factor of affinity ($\bar{z}_o = nz_o$ where $n = 1, 1.5, 2, 3, 4$) before they are supplied to the model computer. The movements x_o, $\bar{z}_o$ (or $y_o\bar{z}_o$) guide the rotations of the straight edged ruler about "2", and subsequently move pivot A on the cross slides which represent the $X\bar{Z}$ (or $Y\bar{Z}$) projection of the model. The distance ($\bar{Z}$) of the horizontal rail "1" (X or Y) depends in addition on x_o' (or x_o'' resp.) or on y_o' (and y_o''). The base components are imposed directly on pivot $\bar{A}$. Since the

projection of the perspective centers on the $X\bar{Z}$- and YZ-planes coincide physically, the corresponding base components bx (or by) are inserted as mutual distances between the respective pivots $\bar{A}$ in X (or Y) direction. These components are imposed as differential displacements of both X (or Y) carriages.

There is a symmetry between the "base in" and "base out" settings. Since the bz component occurs in XZ and YZ projections, it must be inserted to both corresponding pivots $\bar{A}$ as in Poevilliers B. Due to the electrical transmissions between setting screws and the corresponding slides, this is performed simultaneously with one setting screw.

The projection system is characterised by the following properties:

(1) system components (measuring unit, transformers, model computer and setting screws) can be arbitrarily located;
(2) simple geometrical principles can be realized;
(3) a solution in planes can easily be applied; and
(4) the system may be extended to other input output relations.

The mechanical construction of the projection system is based on two-dimensional cross slides and straight edged rulers rotating about a single axis. Attention was paid particularly to the mechanical performance of the measuring unit (the Abbe's comparator principle), to the carriages, guidances, spindles and nuts, and to the ball bearings. All cross slides and rulers are positioned horizontally in order to eliminate or reduce the disturbing effect of gravity. The secondary rail in each cross slide is driven by two synchronized spindles, one on each side.

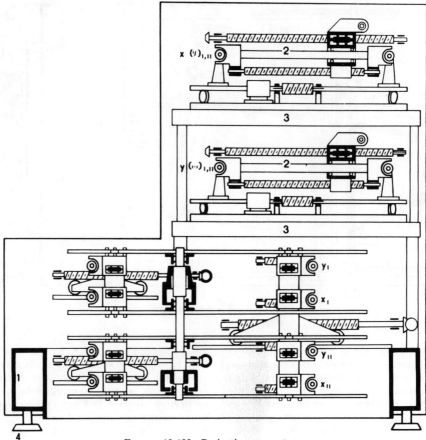

FIGURE 12-123. Projection computers.

The backlash and friction are reduced to a minimum.

Movements for tracking are transmitted electrically from the handwheels for X and Y, and the foot disk for $\bar{Z}$, to the cross slides carrying the pivots $\bar{A}$. Orientation and scale are introduced to the following parts:

(1) measuring unit, centration of photographs (δx, δy);
(2) mechanical transformers, c, ϕ, ω, and coefficient 1.6 (or 0.5); and
(3) model computers, factor of affinity n, base components bx, by, bz.

Movements in X, Y and $\bar{Z}$ are transformed via the model computer into x_o, y_o, and $\bar{z}_o$. These are electrically transmitted to the mechanical transformers for the conversion of x_o, y_o, z_o into x and y. There is a feedback between the transformers and the model computer. The x and y coordinates are transmitted electrically to the cross slides of the measuring unit, carrying the photographs. The electrical transmissions of movements are performed with a high accuracy, and are controlled by feedback circuits.

12.5.3.3 OBSERVATION SYSTEM

Since the photographs are mobile for tracking, a fixed optical system has been applied (fig. 12-125). It has the following three basic functions:

(1) binocular observation of stereoscopic images;
(2) differential rectification (in conjunction with the projection system); and
(3) photographic recording of the image (in the left-hand optical train).

The photograph, which is placed on the carrier, provided with a glass plate and fiducial marks, is illuminated through condenser lenses and a green filter. Its optical distance equals the focal distance of an optical transferring element.

A prism group containing two diagonal semi-transparent surfaces obtains a separate image for differential rectification (orthophotograph), and projects an illuminated measuring mark into the optical train. The respective projector consists of a lamp, colored filters, condenser lenses, and a diaphragm for different measuring marks. An Amicci prism is used for optical rotations of images. Two prisms can be shifted in opposite directions in order to interchange the left and

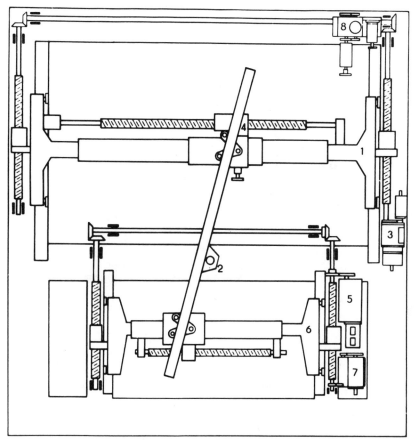

FIGURE 12-124. Model computer.

right optical trains and normal or mirror-reversed images are obtained via another prism. A mirror, which can be turned in or out, reflects the image for photographic registration with the small size camera. The number of the point measured and photographed can be set and projected on the edge of the record.

The brightness of the observed image can be adjusted by changing the mutual distance of plane-parallel plates. Between these plates there is a light-absorbing liquid and the absorption is proportional to the amount of the liquid between the plates. This arrangement facilitates a constant and maximum illumination of the separated images for differential rectification and photographic registration. The basic illumination can be adapted by rheostats. There are four sets of interchangeable eyepieces which provide optical magnifications of 6×, 9×, 12× and 18×. The description of the optical arrangement for the differential rectification is given in the following section.

12.5.3.4 Orthophoto Attachment

The orthophoto equipment represents an integral part of the Stereotrigomat. It provides a continuous correction for relief- and tilt-displacements in a photograph, and reduces the photograph to a uniform scale. That is, it produces a transformation of the central perspective into an orthogonal projection together with a desired change of scale. This task consists of two parts; the proper location and scaling of small elements of the image, and sharp optical imaging. The proper location is realized by spatial tracking of the orientated model. Planimetric tracking is performed automatically according to a meandering pattern of narrow strips in the Y-direction (fig. 12-126). The operator controls only the height of the floating mark so that it remains in contact with the surface of the terrain model. Planimetric movements in X and Y are synchronized with those of the film with respect to a fixed slit. The task of forming a sharp optical image for each successive element at a desired orthophoto scale is performed by the optical inversor.

The main parts of the orthophoto equipment are:

(1) the separated optical train;
(2) the optical inversor; and
(3) the film magazine.

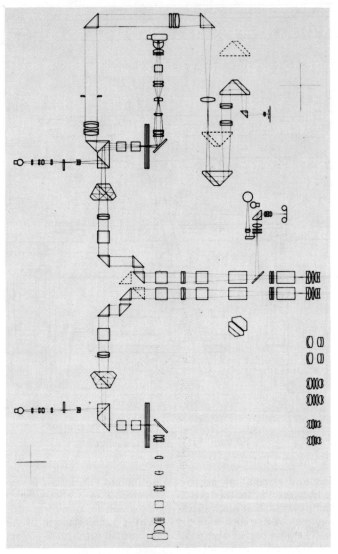

FIGURE 12-125. Observation system.

The separated branch of the left hand optical train transfers the image to the optical inversor (fig. 12-125). Two lenses represent an optical transferring element, which forms a real image with exactly 2× optical magnification. There is an automatically controlled diaphragm in the focal plane for adjusting the illumination intensity and its aperture is proportional to the orthophoto scale.

Input to the optical inversor is the twice enlarged image formed which may be regarded as the "object" for the inversor. The function of the inversor is to project a sharply focused and appropriately scaled image on the slit of the film magazine, and to control the aperture of the diaphragm for uniform illumination.

The adjustment of the scale should be in accordance with the momentary ratio $\bar{Z}: \bar{z}_o$. In addition, the standard focusing condition should be fulfilled:

$$\frac{1}{a} + \frac{1}{a'} = \frac{1}{f}, \tag{12.45}$$

where f is the focal distance of the lens, a is the object distance, and a' is the image distance. Taking into consideration the scale ratios between the orthophoto and the model, we obtain:

$$a = f + \frac{f \bar{z}_o}{Z s} \quad \text{and} \quad a' = f + \frac{f \bar{Z} s}{\bar{z}_o}. \tag{12.46}$$

The object distance a is computed and controlled by an electrical analogue computer which physically displaces the prism with respect to the lens. The corresponding image distance a' is controlled via a mechanical inversor accord-

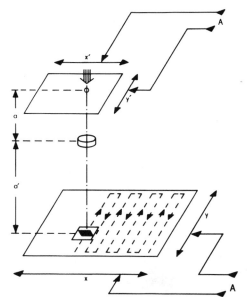

FIGURE 12-126. Differential rectifier scanning motions.

ing to Newton's lens equation. Variation of distance a' is performed physically by shifting the prism horizontally. The film magazine consists of a light-proof housing, provided with an adjustable slit, a cylindrical drum for the film, and the driving mechanism for its transport.

12.5.3.5 MEASURING AND PLOTTING DEVICE

The X and Y tracking is performed by means of hand wheels and the Z motion by a foot disk. For terrestrial photography the Y- and Z-axis can be interchanged and the scanning directions can be inverted. The extent of graduations on drums for X and Y may also be inverted by means of a mask. The reading accuracy (estimation) is 0.01 millimetres. There are fifteen different height scales possible, and a scale for approximate direct readings of the projection distance $\bar{Z}$ is also provided. Transmissions from the handwheels to the corresponding carriages are electrical. The drawing table is mounted on top of the housing on which the measuring unit and the differential rectifier are placed. The size of the drawing surface, which is semi-transparent and can be illuminated from below, is 90×120 centimetres. The output coordinates of the model computer are transmitted electrically to the electro-coordinatograph. Desired scale ratios may be introduced by means of gears and the directions of the coordinate axes are reversible. The coordinatograph can be connected with the XY-, XZ- or YZ-axes of the model and a second drawing table may be employed if desired. The drawing pencil can be replaced by a pricking needle, a setting microscope, or a pricking microscope and a device for the plotting of small circles can also be attached. A

recording device for coordinates can be connected to the instrument.

12.5.3.6 DROP-LINE RECORDING DEVICE

Simultaneously with the differential rectification, drop-line charts can also be produced using a special automatic attachment. The plotting stylus vibrates at three different amplitudes in order to indicate the adjacent contour intervals uniquely while profiling the model parallel to Y. Three thicknesses are used in order to avoid ambiguities regarding the direction of the slopes. The drop-lines can be engraved into a scribe coating with equipment consisting of an electronic control unit and a special plotting head. The electronic unit is located in the electrical cabinet, and the special head can replace the normal plotting head on the coordinatograph. Each lift and drop of the tracing stylus is the result of a signal from the controlling unit. The desired contour interval can be inserted by proper selection of an adaptor of which there are 35.

12.5.4 SPR STEREOPROJECTOR

12.5.4.1 GENERAL

The Stereoprojector, designed by Prof. G. V. Romanovskij (Buchholtz, 1961-62; Hoffmann, 1963; Romanovskij; Schoeler, 1957), was developed in three successive models; SPR-1, SPR-2, and SPR-3. In each successive model the performance and operational characteristics were improved. A special version of the SPR, the Stereophotoprojector, was adapted for orthophoto attachment.

The Stereoprojector is an affine plotter, specially developed for medium- and large-scale plotting from near vertical photographs (18×18 centimetres). Although the range of the instrument principal distance is only 150 to 300 millimetres, the SPR can handle photographs taken by cameras with principal distances from 35 to 350 millimetres.

12.5.4.2 PROJECTION SYSTEM

The projection system is spatial mechanical. The photo-carriers are horizontal and coplanar, and move in this plane for tracking and consequently, a simple stationary observation system is used. The photo carriers (fig. 12-128) are provided with x- and y-cross-slides for tracking while the perspective O' and O'' have a constant separation. They can be displaced simultaneously in the vertical direction for the setting of the principal distance c. The base bridge is attached to the Z-carriage and provides for settings of bx', by'', and bz''. The photo carriers are guided by the space rods L' and L'', which pass through the perspective centers O' and O'' and are attached at their lower ends to the base bridge. Two upper gimbal joints a' and a'', repre-

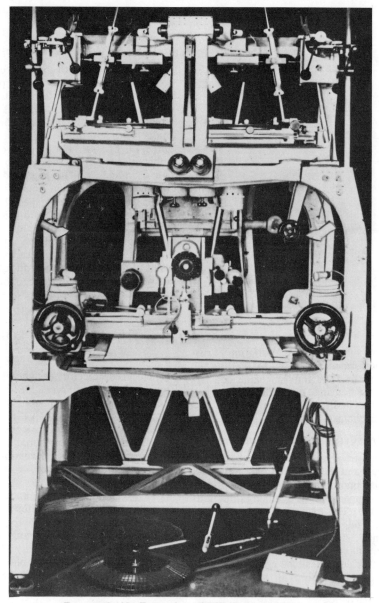

FIGURE 12-127. Front view of SPR Stereoprojector.

sent the mechanical photopoints and move during tracking in the fictitious horizontal photo-planes which are corrected for the tilt displacements δr. These corrections are realized by mechanical rectifiers situated above the corresponding photo-carriers.

The geometric principles of the mechanical rectifiers are as follows. In figure 12-129, Ph represents the original photograph and Ph_o its rectified equivalent. The figure shows the vertical section along the principal vertical where ν is the tilt of the plane Ph. Rotating the plane Ph through ν into Ph_o about the horizontal through the isocenter i, moves point a to (a). From similar triangles $a(a)a_o$ and Oia_o,

$$\frac{\delta r}{r_o} = \frac{a(a)}{Oi} = \frac{\delta c}{c}$$

and

$$\delta c = r \sin \nu. \qquad (12.47)$$

Thus,

$$\delta r = \frac{rr_o}{c} \sin \nu. \qquad (12.48)$$

A similar relation exists for an anamorphic bundle of rays. From triangles $\bar{O}ia_o$ and $\bar{a}(a)a_o$,

$$\delta r = \frac{\bar{r}r_o}{\bar{c}_o} \sin \bar{\nu}$$

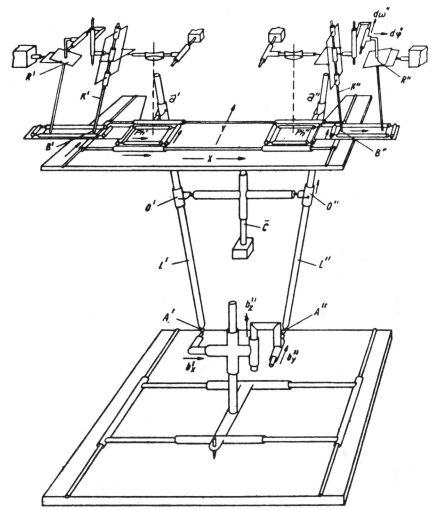

FIGURE 12-128. Mechanical diagram of the SPR Stereoprojector.

where

$$\bar{r} = r\,\frac{\cos \nu}{\cos \bar{\nu}}\,,$$

and consequently

$$\delta r = \frac{r r_o \cos \nu}{\bar{c}_o \cos \bar{\nu}} \sin \bar{\nu}.$$

Equating expressions (12.48) and (12.49) condition (12.38) is obtained by

$$\frac{\bar{c}_o}{c} = \frac{\tan \bar{\nu}}{\tan \nu} = u.$$

In both expressions (12.48) and (12.49) the value of r_o is unknown. From figure 12-129

$$\delta r = \delta_1 + \delta_2 \qquad (12.50)$$

where

$$\delta_1 = r(1 - \cos \nu) \qquad (12.51)$$

and

$$\delta_2 = r \cos \nu - r_o. \qquad (12.52)$$

Further,

$$\frac{\delta_2}{c} = \frac{(r_o - c \tan \nu/2)}{c}\,,$$

so

$$\delta_2 = \frac{\delta c}{c + \delta c}\left(r \cos \nu - c \tan \frac{\nu}{2}\right). \qquad (12.53)$$

Substituting expressions (12.50) and (12.52) into (12.49).

$$\delta r = r\left[1 - \cos \nu\left(1 - \frac{\delta c}{c + \delta c}\right)\right]$$

$$- \frac{c\,\delta c}{c + \delta c}\tan \frac{\nu}{2}\,.$$

Substituting further $\delta c = r \sin \nu$ (12.47) and the identity

$$\tan \frac{\nu}{2} = \frac{1 - \cos \nu}{\sin \nu}$$

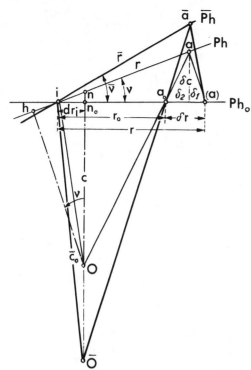

FIGURE 12-129. Geometry of the mechanical rectifiers.

and simplifying,

$$\delta r = \frac{r^2 \sin v}{c + r \sin v} = \frac{r \delta c}{c + \delta c}. \qquad (12.54)$$

The unknown distance r_o is obtained as the difference

$$r_o = r - \delta r = r \left(1 - \frac{\delta c}{c + \delta c} \right) \qquad (12.55)$$

$$= \frac{rc}{c + r \sin v}.$$

This expression can also be derived directly from figure 12-129. Relation (12.54) has to be realized automatically during tracking. Corrections δr are resolved into components in $\bar{x}$ and $\bar{y}$ directions (fig. 12-130). Thus

$$\delta c = r \sin v = \delta h = \bar{x} \sin \phi + \bar{y} \sin \omega \quad (12.56)$$

where $\bar{x}$ and $\bar{y}$ are the photocoordinates with the isocenter i as origin. Substituting (12.56) into (12.54) we obtain

$$\delta r = \frac{r(\bar{x} \sin \phi + \bar{y} \sin \omega)}{c + (\bar{x} \sin \phi + \bar{y} \sin \omega)} \qquad (12.57)$$

$$= \frac{(\bar{x}^2 + \bar{y}^2)^{1/2} (\bar{x} \sin \phi + \bar{y} \sin \omega)}{c + (\bar{x} \sin \phi + \bar{y} \sin \omega)}.$$

The relation between the tilt displacement δr and its components is

$$\delta r^2 = \delta \bar{x}^2 + \delta \bar{y}^2 = \frac{(\bar{x}^2 + \bar{y}^2)(\bar{x} \sin \phi + \bar{y} \sin \omega)^2}{(c + \bar{x} \sin \phi + \bar{y} \sin \omega)^2},$$

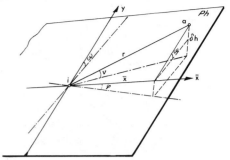

FIGURE 12-130. Components of δr correction.

and consequently

$$\delta \bar{x} = \bar{x} \frac{(\bar{x} \sin \phi + \bar{y} \sin \omega)}{(c + \bar{x} \sin \phi + \bar{y} \sin \omega)}$$

$$\delta \bar{y} = \bar{y} \frac{(\bar{x} \sin \phi + \bar{y} \sin \omega)}{(c + \bar{x} \sin \phi + \bar{y} \sin \omega)}. \qquad (12.58)$$

Tilt displacement corrections, δx and δy are applied by movements of the measuring mark with respect to the photograph. This is achieved optically by horizontally displacing the first lens L_1 of the observation system (fig. 12-131). These displacements are controlled automatically by the mechanical rectifiers. The light between lenses L_1 and L_2 is parallel; lens L_2 focuses the image in the plane of the measuring mark M. A displacement δr of lens L_1 causes a parallel shift of the line of sight by the same amount. Therefore the photograph, together with the upper gimbal joint, follows this shift so as to keep the measuring mark in optical coincidence with the image point.

Figure 12-131 represents schematically the mechanical rectifier in its function along the line of maximum tilt. Horizontal lever e, which is attached to lens L_1, is controlled by rod A_1, which rotates about cardan joint C_2. The vertical separation $C_1 C_2 = d_r$ depends on the tilt of the correction surface K and the eccentricity Δv of the feeler R. The tilt γ of plate K is equal to the inclination of rod A_2, to which it is attached. Both rotate as a unit about point C_4. The gimbal joints C_1 and C_4 are at the same level. Eccentricity Δv (or $\Delta \phi$, $\Delta \omega$) depends on the tilt v (or ϕ, ω) of the photograph at exposure. The tip of feeler R and the cardan joint C_2 are constrained to maintain the same level. Points C_1, C_4, C_5 and C_3 thus form a parallelogram. C_3 and C_5 are the lower gimbal joints of the rods A_1 and A_2.

The functioning is as follows. The photograph Ph is displaced through r (or $\bar{x}$, $\bar{y}$) by space rod L. Points C_3 and C_5, which are connected to the photo carrier, are shifted by the same amount. Consequently, rod A_2 with the correction plate K rotates about C_4, and the feeler R and the cardan joint C_2 are lowered or raised by d_r. Rod A_1 is inclined by $\gamma' = \gamma \pm d\gamma$ and point C is displaced δr (or $\delta \bar{x}$, $\delta \bar{y}$) to C_1. This translation is given by lever e to lens L_1.

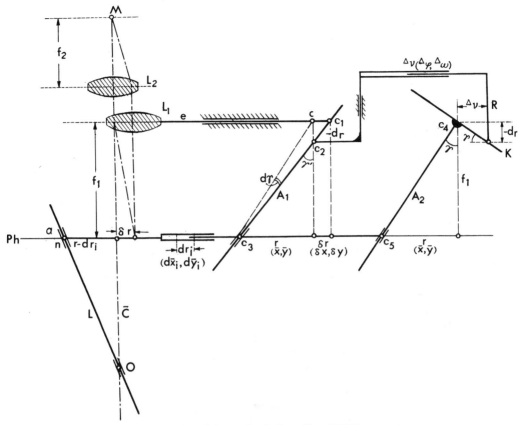

FIGURE 12-131. Plan of the mechanical rectifier, SPR Stereoprojector.

According to figure 12-129, the photograph should be centered with the nadir point at the fiducial center of the photocarrier. Therefore, the principal point must be decentered by $d\bar{x} = c \tan \phi$ and $d\bar{y} = c \tan \omega$. When space rod L is vertical, the measuring mark coincides optically with the image of the nadir point. Apart from this, the distance r (or $\bar{x}$ and $\bar{y}$), used as arguments for mechanical rectification, refer to the isocenter i as the origin. Therefore, points C_3 and C_5 of the mechanical rectifier have to be decentered by $d\bar{x} = c \tan \omega/2$ and $d\bar{y} = c \tan \phi/2$. From figure 12-131

$$\frac{-d_r}{f_1 - d_r} = \frac{\delta r}{r} \text{ and } \frac{-d_r}{\Delta\nu} = \frac{r}{f_1}.$$

Eliminating the vertical displacement d_r of the feeler R and C_2 we obtain

$$\delta r = r \frac{r\Delta\nu}{f_1\left(f_1 + \frac{r}{f_1}\Delta\nu\right)} = r^2 \frac{\Delta\nu}{f_1^2 + r\Delta\nu}. \quad (12.59)$$

Comparing (12.59) with (12.54) it follows that

$$\frac{\Delta\nu}{f_1^2 + r\Delta\nu} = \frac{\sin \nu}{c + r \sin \nu}$$

and consequently

$$= \frac{f_1^2}{c} \sin \nu. \quad (12.60)$$

Here, f_1 is the constant of the rectifier and equal to the focal distance of lens L_1. In practice, the total tilt ν eccentricity $\Delta\nu$ and d_r are resolved into components d_x and d_y in the $\bar{x}$ and $\bar{y}$ direction. According to figure 12-132

$$d_r = d_x + d_y = \frac{\bar{x}}{f_1}\Delta\phi + \frac{\bar{y}}{f_1}\Delta\omega$$

$$\Delta\nu^2 = \Delta\phi^2 + \Delta\omega^2 = \left(\frac{f_1^2}{c}\right)^2 (\sin^2 \phi + \sin^2 \omega)$$

and

$$\Delta\phi = \frac{f_1^2}{c} \sin \phi, \quad \Delta\omega = \frac{f_1^2}{c} \sin \omega. \quad (12.61)$$

Corrections for tilt displacements according to

$$\delta r = r \frac{d_r}{f_1 + d_r}$$

can also be resolved into components:

$$\delta\bar{x} = \bar{x} \frac{d_r}{f_1 + d_r} = \bar{x} \frac{(\bar{x}\Delta\phi + \bar{y}\Delta\omega)}{(f_1^2 + \bar{x}\Delta\phi + \bar{y}\Delta\omega)}$$

$$\delta\bar{y} = \bar{y} \frac{d_r}{f_1 + d_r} = \bar{y} \frac{(\bar{x}\Delta\phi + \bar{y}\Delta\omega)}{(f_1^2 + \bar{x}\Delta\phi + \bar{y}\Delta\omega)}.$$

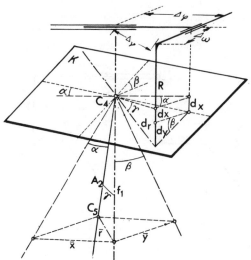

FIGURE 12-132. Components of tilt displacements.

By substituting relations (12.51) into the above, we obtain equations (12.58). Consequently, for further treatment the photographs may be considered strictly vertical. The apparently complicated rectification mechanism gives an impression of doubtful stability. In reality, the system is stable, because the variable input $(\bar{x}, \bar{y})$ are large motions, while the subsequent output consists of relatively small corrections $(\delta\bar{x}, \delta\bar{y})$.

A further consideration involves the base components bz. Since the mechanical photoplanes Ph' and Ph'' are rectified, an exact affinity exists in the vertical direction when both principal distances are changed simultaneously.

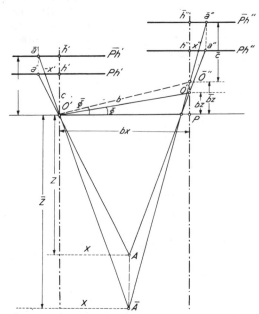

FIGURE 12-133. Affinely deformed bz component.

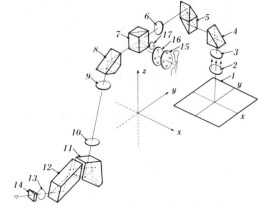

FIGURE 12-134. Observation system of SPR.

However, this is true only if $bz = 0$. If this is not so, the model will be deformed. This can be avoided by introducing the affinely deformed $\bar{bz}$ component. According to figure 12-133 $\bar{bz} = u \, bz$ and $\bar{Z} = u\,Z$, where u is the affinity factor. This factor must be selected in accordance with the existing gear transmissions of the height counter.

12.5.4.3 OBSERVATION SYSTEM

The projection system with movable photographs permits a stationary observation system; it is, in principle, a fixed binocular microscope. Dove prisms are not necessary. Between the photograph and the measuring mark are two lenses (fig. 12-134), the first one movable in its own plane to correct for tilt displacements. The measuring marks are interchangeable with three illuminated rings having diameters between 50 and 120 micrometres along with a T-shaped mark. The magnification of the system is 4 or 6 times and with the 4-times magnification, the diameter of the field of view on the photograph is 32 millimetres.

12.5.4.4 PLOTTING AND MEASURING DEVICE

For scanning motions in the model space, a three-axial cross-slide system is used (X, Y, Z). The carriages are controlled as usual by two handwheels and a foot disk. Planimetric coordinates can be read on dials, and the height counter is provided with ten gear transmissions. The drawing surface, 60×60 centimetres, is under the XYZ cross-slide system, and the tracing stylus attached to the X carriage. Thus, the map scale is equal to the model scale, which can vary between ½ to 2 times the photoscale. This restriction has obviously unpleasant consequences on economy in large-scale plotting. Therefore, for greater enlargements, a separate drawing table can be connected to the instrument.

12.5.4.5 STEREOPHOTOPROJECTOR

The Stereophotoprojector is a modified version of the SPR, adapted for orthophoto production. It is presumably the first instrument with an attached orthophoto printer. From one optical train the light is branched by means of a beam splitter, and guided to the slit which is placed in front of a photo-sensitive material. The projected image is continuously scaled and focused during traversing and printing is performed along parallel bands.

SELECTED REFERENCES

Ahrend, M., "The Planimat: a New Second-Order Photogrammetric Plotter;" *Bildmessung und Luftbildwesen*, No. 5; Karlsruhe, Germany, 1967.

Baboz, J., "Deux Conceptions Nouvelles d'Appareils de Restitution S.O.M.;" X I.S.P. Congress, Com. II; Lisbon, Portugal, 1964.

Buchholtz, A., "Zur stereoskopischen Auswertung von überweitwinkel Luftbildern;" *Phia*, Vol. 18, no. 1, pp. 15–32; USSR, 1961–62.

Drobyshev, F. V., "The SD Stereograph;" *Geod. i Aerofot*, no. 1; USSR, 1957.

———, "The new SD Stereograph;" *Geod. i Kartog*, no. 9; USSR, 1957a.

———, "Stereophotogrammetric instruments in the USSR;" *Phia*, Vol. 17, no. 2; USSR, 1960–61.

Gerlach, R., "Beitrag zur photogrammetrischer Auswertung von Bildpaaren mit affin 'überholtem Model';" *Vermussungstechnik*, Heft 3; German Democratic Republic, 1962.

Hofmann, O., "Der Stereometrograph, ein neues Zweibildkartiergerat aus Jena;" *Vermessungstechnik*, Heft 8; German Democratic Republic, 1960.

———, "Photogrammetrisher Geratebau in der Sowietunion;" *B.u.L.*, no. 4; German Democratic Republic, 1963.

———, "A new photogrammetric plotting system;" I.S.P. Congress Com. II, Ottawa, 1972.

v.d. Hout, C. M. A., "Point Transformation in Photogrammetry;" *Tijdschrift voor Kadaster en LandmeetKunde*, no. 1; The Netherlands, 1969.

Klauer, Jacob, "Practical Data on the Kern PG3;" Annual Convention of the American Society of Photogrammetry, Washington, D.C.; March, 1971.

Lobanov, A. N., "Aerofototopografia;" NEDRA, Moscow, 1971.

Makarovic, B., "Recent Instrument Development in Europe and Asia;" I.S.P. Congress Com. II, Ottawa, 1972.

———, *Stereoscopic Plotting Instruments of the Projection Type;* International Institute for Aerial Surveys and Earth Sciences, Enschede, The Netherlands; ©1972.

———, *Stereoscopic Instruments for Affine Restitution;* International Institute for Aerial Surveys and Earth Sciences, Delft, Enschede, The Netherlands; ©1966, 1973.

Romanovskij, G. V., A Stereoprojector for map compilation from vertical aerial photographs, Avt. sv., USSR, bl. 42 c 10/12, no. 136898, 25.03.61 (patent).

Schoeler, H., "Der Stereoprojektor SPR 2 von Romanovskij;" *Vermessungstechnik*, Heft 8; German Democratic Republic, 1957.

———, "Some Aspects of the Arguments for the Configuration of the Topocart-Orthophoto Instrument System;" Orthophoto Workshop III, San Antonio, Texas; June, 1975.

Schuebel, R., "E2 Planicart—a New Stereoplotter;" XII International Congress for Photogrammetry, Ottawa; 1972.

———, "The F2 Planitop Topographic Plotter;" *Bildmessung und Luftbildwesen*, Karlsruhe, Germany; April, 1974.

Schull, R., "The Wild A10 Autograph;" XI I.S.P. Congress; Lausanne, 1968.

S.O.P.L.E.M. Presa 225, Presa 225 RC descriptions; 1970.

S.O.P.L.E.M. Presa 226, Presa 226 RC, Doc. no. 20-13A; Paris, 1972.

Szangolies, K., "Stereotrigomat—a New Universal Instrument System for Photogrammetric Plotting;" Jena, German Democratic Republic, 1966.

———, "The Universality and Accuracy of Stereoplotters;" XI I.S.P. Congress Com. II: Lausanne, 1968.

———, "Topocart, Entwicklung und Mathematische Dartstellung;" *Kompendium Photogrammetric*, Band X; Jena, German Democratic Republic; 1974.

Zeiss, Jena, "Stereometrograph;" *Druckschriften*, No. N 14-360b-1; 1964.

———, "Topographische Auswertegerat Topocart B;" *Druckschriften*, No. N 14-368b-1.

Automation of the Photogrammetric Process

Author-Editor: S. JACK FRIEDMAN
Contributing Authors: J. B. CASE, U. V. HELAVA, G. KONECNY, H. M. ALLAM

13.1 Introduction

THE TITLE OF THIS CHAPTER, if taken literally, would require that its contents overlap into many other chapters of the MANUAL. Progress in the field of photogrammetry since the publication of the third edition (1966) has resulted in improvement and "automation" of all aspects of instrumentation and techniques so that the only chapters not affected would be those dedicated to basic mathematical theory and historical perspectives. Therefore the author has imposed limits on the subjects to be covered, giving due consideration to items discussed in other chapters and other items not covered elsewhere in the MANUAL. A further limitation will be that of discussing the photogrammetric process to the exclusion of the cartographic process. Industrial advances in recent years have been made in cartographic areas and the temptation to describe them is strong; however, this is an area beyond our scope.

We propose to review recent developments in a general manner as a sort of overview and follow with more details on those aspects which might be of greatest interest and further potential for photogrammetric specialists.

It has become customary in recent years for a paper to be presented to the quadrennial Congress of the International Society for Photogrammetry. Esten (1964), Nowicki (1968), Lorenz (1968), Bucci (1972), Makarovic (1972), Lorenz (1972), Thompson and Mikhail (1976), and Case (1980) all presented invited papers on aspects of automation of the photogrammetric process. Case's paper is the most recent and is a succinct but comprehensive review of progress to date. We quote freely from his paper.

This chapter will first consider the various photogrammetric functions and discuss how automation has affected them. These functions are: mensuration, rectification and orthorectification, elevation extraction, planimetric extraction and integrated systems. The chapter will then review and discuss in some detail the newest and, to our way of thinking, the most significant instrumental development in recent years, the analytical stereoplotter. Miscellaneous topics will follow and we will close with a discussion of problems, prospects and predictions.

13.2 Automation of Photogrammetric Functions

13.2.1 MENSURATION

Classically, mensuration has served as the initial step in aerotriangulation. Originally, mensuration and aerotriangulation were combined into a single operation in a stereoscopic plotting instrument (stereoplotter), the optical or optical/mechanical construction of the instrument itself providing the computation of the aerotriangulation. As analytical triangulation developed, it was found to be more efficient to do the mensuration separately. However, in recent years, with the development of analytical plotters and of stereoscopic plotting instruments with digitizers and with the refinement of independent-model aerotriangulation, many organizations have returned to making the measurements on stereoplotters.

Mensuration normally consists of several steps: (1) the identification of control points and/or the selection of pass points on a photograph, (2) the stereoscopic transfer of those points to the overlapping photographs upon which they appear, and (3) the actual measurement of photographic coordinates of the control points and/or pass points. In off-line mensuration, two different paths have been followed, each with some degree of automation. In the first, point selection and transfer are separated from mensuration, the former being performed on a stereoscopic point-transfer instrument such as the Wild PUG and the latter on a monocomparator. In the second, point selection, transfer, and mensuration are combined into a single instrument, the sterocomparator. Little has been done to automate the point-selection process, although some work has been done to equip point-transfer instruments with a coarse coordinate readout which could be used for pre-positioning on the monocomparator. The ultimate in automation of the monocomparator operation was reached with the Automatic Reseau Measuring Equipment or ARME (Roos, 1975). The ARME, when given calibration or approximate values of coordinates of grid intersections (reseaux) on a photograph, drives under computer control to the vicinity of the point and then automatically centers on and measures the point. Similarly, given a star catalog and following identification and measurement of at least two stars, the ARME will drive to the vicinity of, center on, and measure the remaining stellar images on a photograph. In the late 1960's attempts were made to automate the stereocomparator process, exemplified by the Automatic Point Transfer Instrument or APTI and the OMI/Bendix TA3/PA. However, these instruments, equipped with correlators, never achieved sufficient accuracy. More recent efforts have concentrated on providing, in the OMI/Bendix TA3/P1 and TA3/P11, assistance to the operator by performing relative and absolute orientation, pre-positioning, maintaining a y-parallax-free model, and allowing for editing of data (Seymore and Whitside, 1974; Helmering, 1977). Finally, many of these same functions have been incorporated into the analytical plotters for measuring. In particular, with the advent of stable base films and reseau equipped cameras, and because of the high accuracy of the stereocomparators and analytical plotters, it is no longer necessary to mark the points which are to be transferred and measured. The instrument is able to find the originally selected point solely from the previously measured coordinates of that point.

13.2.2 RECTIFICATION AND ORTHORECTIFICATION

Rectification is the process of projecting a photographic image from its plane to a horizontal plane in order to remove displacements due to tilt. In recent years the definition of rectification has somewhat broadened while, at the same time, rectification has become the most highly automated of photogrammetric operations. The major breakthrough was, of course, the development of differential rectification or orthorectification, in which the rectified photography is further corrected to remove displacements due to relief. And that process has been further refined in that transformations can be made to other planes than the horizontal plane, producing so-called three-dimensional or perspective views.

Rectification began as a straight optical projection with mechanical inversors to maintain the Scheimpflug condition. Some automation has been achieved, with rectifier settings controlled by a computer. Initial developments in orthorectification also employed optical projection. However, that projection was limited to a very narrow slit, with the slit being raised or lowered mechanically (i.e., manually by an operator) to make a profile across the surface of the terrain in the stereoscopic model. The next step towards full automation was the provision of automatic correlation to perform the profiling. And, instead of projecting it optically, the photograph is scanned and projected electronically. Such systems are exemplified chronologically by the Wild B-8 Stereomat, the UNAMACE, the Gestalt Photomapper GPM-2, and the Jena Topomat. These, of course, are on-line systems. However, in recent years there has been a trend towards off-line systems. Again, most of these systems use optical projection, most are computer controlled, and many are capable of employing profile data that are not necessarily obtained from the same photograph as that from which the orthophotograph is being produced. Because they are computer controlled, most of these instruments also are capable of producing stereomates, i.e., an orthophotograph into which x-parallax has been introduced such that, when the orthophotograph and its stereomate are viewed stereoscopically, the true relief is presented. Among these off-line systems are the Replacement-of-Photographic-Imagery Equipment or RPIE, the Wild Avioplan OR-1, and the Jena Orthophot-C. Descriptions of how such off-line systems operate have been given by Scarano and Jeric (1975) and Kraus et al. (1979)

The present direction being taken in rectification—a direction which is that of automation in all operations of photogrammetry—is that of digital image processing. This involves the use of already digitized imagery, such as that produced by the Landsat multispectral scanner (MSS) or the scanning and digitizing of imagery, i.e., photographic imagery. The imagery, then in digital form, can be transformed by any projection or perspective which can be modelled on the computer. Concepts for digital rectification have been given by Konecny (1979) and Keating and Boston (1979) whereas descriptions of actual digital rectification were reported upon by Murai (1978) and Tanaka and Suga (1979). The latter reference is of particular interest because it transforms a vertical perspective—a Landsat MSS image of Mt. Fugi—to a horizontal perspective as viewed from a point on the ground.

Finally, another concept of rectification, called "digital mono-plotting" by Makarovic (1974) and elaborated upon by Masry and McLaren (1979) and Besenicar (1978) is now being realized at the Defense Mapping Agency as the Extracted-Feature-Rectification-and-Processing-System (EFRAPS). In this concept, planimetric information is extracted from a single photograph and then digitized. The digitized information is transformed iteratively into ground coordinates. The advantages of this concept are that it utilizes inexpensive equipment (assuming that a computer is available) and that it allows for operations in parallel *i.e.*, planimetric information may be extracted before aerotriangulation is completed.

13.2.3 ELEVATION DETERMINATION

Elevations have traditionally been determined by an operator using an analog stereoscopic plotting instrument and manually maintaining a floating mark on the surface of the model while traversing the model along lines of equal elevation, *i.e.*, contours. The advent of computer-assisted instruments and analytical plotters has not materially improved this process (the operator must still manually follow the terrain with the floating mark); however, such equipment does offer several advantages over the conventional analog instruments, *i.e.*, they allow for very rapid set-up, they allow for the output of elevations (DTM), and, particularly in the case of the analytical plotters, they allow for the utilization of unconventional photography and coordinate systems. A thorough description of such systems was given by Dowman (1977a). The interest in analytical stereoplotters is evidenced by an entire issue of *Photogrammetric Engineerrin and Remote Sensing* (November 1978) devoted to those instruments; an issue of *The Canadian Surveyor* (June 1979) devoted to a single analytical plotter, the *Anaplot;* and the several symposia dedicated to analytical plotters.

The big jump in automation of the elevation extraction process came with the development of automatic correlation, described in the previous paragraph on rectification. Although automatic correlation was originally developed mainly to automate orthophoto production, it is now coming into its own as a cost-effective means of producing data for digital terrain-models (DTM). This is exemplified by the UNAMACE (Madison, 1975), the Gestalt GPM-2 (Allam, 1978), and the OMI-Bendix AS-11B-X (Scarano and Brumm, 1976). The UNAMACE and GPM-2 employ analog (electronic) correlation technology. However, the AS-11B-X utilizes on-line scanning along epipolar lines and digital correlation of those lines. Another digital on-line correlation concept, called RASTER, has been proposed by Hobrough (1978). However, the future of elevation determination, as with rectification, may well lie with off-line digital image processing as described by Panton (1978).

13.2.4 PLANIMETRY DELINEATION

The functional area of photogrammetry which has shown itself least amenable to automation is that of the delineation of planimetry. As with elevation determination, planimetry has been delineated manually, on a stereoplotter. And, again, the advent of computer-assisted analog stereoplotters and analytical plotters has not materially aided this process. However, the orthophotograph, in at least two disparate ways, has. Specifically, in one instance the orthophotograph has become a final product, thus bypassing most, if not all, need for planimetric delineation on a stereoplotter. Examples of this are the many orthophotographic maps produced by the U. S. Geological Survey (Southard, 1978). In the other instance, the orthophotograph serves as the basis for the delineation of planimetry. In this process, the operator would delineate, on an overlay of the orthophotograph, all of the visible planimetric detail. As an aid in interpretation, the operator may use a conventional stereoscope or zoom stereo microscope to view a steroscopic pair of photographs covering the same area as the orthophotograph, and he may then transfer the interpreted detail onto the overlay. Finally, the digital mono-plotting concept, mentioned in the section on rectification, provides another viable technique for the delineation of planimetry. In particular, the operator can view a stereoscopic pair of photographs under a stereoscope and simultaneously, on an overlay of one of the photographs of the stereopair, he can delineate the planimetric features. The planimetry on the overlay would then be digitized and transformed to the map.

The conventional topographic map is no longer the only product of the process. With the advent of digital data-bases, that process has been expanded to include the collection of data on land-use, the inventory of wetlands, the collection of data on features which can be used to simulate imagery from other sensors such as radar, terrain analyses to determine off-road routes, and so on. These have required the utilization of operators with interpretation skills beyond those required in the past, which is exactly what A. G. Williams was proposing when he said ". . . that a large part of the interest is going to come from participation in . . . Commission VII of the ISP on interpretation, which at one time was looked upon as being marginal to our profession." (Dowman, 1977b)

As in the other photogrammetric functions, the future of planimetric delineation appears to lie with digital image processing. (Bernstein and Ferneyhough, 1975; Andrews, 1977.) A great deal of effort is being expended in this area, especially with Landsat MSS imagery, as exemplified by the many articles now being published. The real problem in the automatic delineation of planimetry, as pointed out by Proctor, is "the software problem of persuading correlators not only to match pairs of images but to recognize roads and railways and buildings. . . ." (Dowman, 1977b). This problem is being attacked through a new discipline called "image understanding," a part of pattern recognition and artificial intelligence. Interestingly enough, most of the research in this area is being done by electrical engineers and computer scientists, not by photogrammetrists.

13.2.5 INTEGRATED SYSTEMS

The cost-effectiveness of automation is realized most dramatically in systems which combine one or more of the above photogrammetric functions with editing and data base management. Such systems typically are composed of several computer assisted stereoscopic plotting instruments or analytical plotters, a number of interactive displays for entering information and for editing, and a central computer to which the plotters and interactive terminals are linked and which provide for management of the data. A general description of such systems is given by Dowman (1977a) while examples of some systems are given below.

The most ambitious of such systems is the Integrated Photogrammetric Instrumentation Network or IPIN, developed by the Defense Mapping Agency Aerospace Center (DMAAC) in St. Louis for the collection of DTM data from photography. IPIN, which the characteristics of have been well documented by Elphingstone (1978a, 1978b) and others (Fornaro and Deimel, 1978; Kirwin, 1978; Bybee and Bedross, 1978; Elphingstone and Woodford, 1978), consists of the Pooled Analytical Stereoplotter System (PASS), the Automatic Compilation Equipment (ACE, the production version of the AS-11B-X epipolar stereoplotter), the IPIN Postprocessing System, and the IPIN Edit System. PASS consists of AS-11A and AS-11B-1 analytical plotters linked to central ModComp II/45 minicomputers, which act as the data base manager and input/output device for the analytical plotters. The ACE is employed as the high-speed generator of DTM, while the AS-11B-1's and AS-11A's collect DTM in those areas where the ACE correlators will not operate and they collect geomorphic data (ridges and streams) to be used for interpolation. The Postprocessor System manages the IPIN files and handles data flow among all the elements of the IPIN system, it converts ACE epipolar data into the geographic coordinate system, and it interpolates the data into a regular array. The Edit System includes an X-Y plotter and a number of interactive edit stations.

A similar system has been evolving at the Defense Mapping Agency Hydrographic/Topographic Center (DMAHTC) since the introduction of the UNAMACE some 15 years ago. At first the UNAMACE produced online, dropline contours and orthophotos. Then, some five years ago (Madison, 1975) the UNAMACE's were directed strictly to the production of digital terrain models (DTM), contours being generated off-line by computer from the DTM and orthophotographs being produced off-line by the RPIE (one RPIE can provide data for a number of UNAMACE's). The success of the UNAMACE has resulted in its now being improved with modern minicomputers and electronic circuitry. DMAHTC has instituted the same PASS as DMAAC: however, the various collection, processing, and edit systems are not yet directly linked as they are in IPIN. Finally, it should be mentioned that both DMAAC and DMAHTC employ orthophotographs (or maps) with overlays for the delineation of planimetry and Bausch & Lomb Zoom 240 stereoscopes for the interpretation and analysis of those data.

The U.S. Geological Survey (USGS) and the Surveys and Mapping Branch (Canada) have created integrated systems based on the Gestalt GPM 2 (Brunson and Olson, 1978; Allam, 1978; Zarzycki, 1978). As with the UNAMACE, the Surveys and Mapping Branch's GPM 2 was acquired for orthophotograph production; however, now the major product is the DTM, and the orthophotograph is secondary. Planimetric data are collected on computer-assisted analog stereoscopic plotting instruments (Wild B8s equipped with an M&S interactive-edit system), and drainage data are collected on those same plotters to aid in the interpolation of contours from the DTM. The system is linked to a computer to produce a digital topographic data base. At the USGS, data for DTM are collected with the GPM-2 and both planimetric data and data for supplementary DTM are collected on a number of computer-assisted analog stereoplotters (Kern PG2s and Wild B8s) as well as on analytical plotters (AS-11As and AS-11A-1s).

Another integrated system, the Wetlands Analytical Mapping System or WAMS (Autometric 1979), has been developed for the U.S. Fish and Wildlife Service. Used exclusively for the collection of planimetric data, it can also perform the mensuration and processing of analytical aerotriangulation. The WAMS consists of the Autometric APPS-IV analytical plotter, display terminals, plotters, and digitizers, all linked to a central computer which manages the WAMS geographic data base.

Finally, an integrated system typical of those in large commercial organizations has been developed by Hunting Surveys Limited (Leatherdale and Kier, 1979). The Hunting digital mapping system consists of Wild A8 stereoplotters linked on-line with a central computer. Both contours and planimetry are digitized with this system. However, editing is performed off-line on computer generated plots.

13.3 The Analytical Plotter

13.3.1 HISTORICAL PERSPECTIVE

The analytical plotter was invented by U. V. Helava while he was employed at the National Research Council of Canada. Helava announced his concept in a paper presented at an International Photogrammetric Conference on Aerial Triangulation held in Ottawa during August 28-31, 1957. Subsequently he was granted a patent by the U.S. Government (Patent Number 3,116,555, dated January 7, 1964).

A summary from Helava's landmark paper follows.

"All the existing photogrammetric plotters capable of solving the main photogrammetric problem are actually based on the same principle, which may be called the simulation principle. In this method real physical projection,—optical, mechanical or optical-mechanical—is used. A critical study shows that this principle cannot be considered as ideal since it results in inflexible and bulky instruments that require extremely high precision in manufacture. In addition, the economical factors, particularly the application of automation and the possibilities offered by electronics, should be considered. These thoughts were the starting point for studies which led to a new principle in which the physical projection is replaced by a mathematical one.

In the new principle the photographs or measuring marks are shifted in the xy plane by amounts corresponding to the displacements caused by central projection and by various other factors. These amounts are computed by a special computer and are executed by servo-mechanisms. Thus the plotter consists of two main parts—a viewing-measuring device and a computer. The viewing-measuring device resembles the stereocomparator except for the servo-mechanisms incorporated in the new instrument. The computer for the prototype instrument employs electronic analogies. This instrument, which is called the Analytical Plotter, is now being built by the National Research Council.

The new principle has several advantages over the simulation method and opens up new possibilities. For example, all known errors may be corrected, focal length and distortion characteristics of the camera used are easy to take into account, the size of the instrument is relatively small, etc. In addition, the principle lends itself well to automation of various operations. A semiautomatic relative orientation is planned to be incorporated in the prototype plotter."

Helava's innovative concept was met with somewhat indifferent interest by the photogrammetric community. This may be attributed to the state of development of control computers. In the late nineteen fifties, computers were very expensive and subject to frequent electronic failure which did not enhance their application to real time control, a necessary function for analytical plotter use. Furthermore, practising photogrammetrists were generally untrained in using computers and, therefore, understand-

ably suspicious of computer usage for such novel application. The time was right, however, and progress was not to be denied. A small group of people of diverse backgrounds, but with a common conviction that Helava's new instrument was desirable and could be realized, converged on the problem. Ottico Meccanica Italiana (OMI), led by its President, Umberto Nistri, sought out and obtained from the National Research Council an exclusive license to build the analytical stereoplotter. Nistri, a pioneer designer of photogrammetric instrument, recognized that OMI had no computer experience. Fortunately OMI had a close relationship with the Bendix Corporation. The Bendix Research Laboratories were assigned the task of designing and building the computer, electronic interface and computer programs to control the viewer and the coordinatograph which was to be designed and fabricated by OMI. The financial burden of realizing the prototype was greatly alleviated when an agency of the U.S. Air Force, the Rome Air Development Center provided a contract for the first analytical stereoplotter, designated the AP/1. Of course during the entire period of design and fabrication of the prototype system, NRC provided invaluable consultation. The AP/1 was delivered to RADC in 1961. The prototype was successful but additional innovations in design were considered desirable and two additional units, designated as AP/2's, were placed under contract. The AP/2 model more fully exploited the potential of the analytical plotter by providing computer control of zoom eyepieces and rotation of optical images. This feature allowed full exploitation of stereoscopic/models formed with panoramic or other non-metric photography. An excellent source describing these early days of the analytical plotter is the Proceeding of the Second International Photogrammetric Conference in Ottawa on the Analytical Plotter, April 1-4, 1963, jointly sponsored by the National Research Council and the Canadian Institute of Surveying.

The AP/2 model was decidedly successful. The AP/2 was precursor to a number of variations but always adhering to the same basic design, namely the principles laid down in Helava's basic concept. Table 13-1 summarizes the geneology of analytical plotters during the earlier period of time when the only models available were manufactured by the OMI-Bendix team (with the single exception of the UNAMACE manufactured by Bunker-Ramo Corporation).

The virtual monopoly of the OMI-Bendix team in the field of analytical plotters was terminated in 1976 when Bendix introduced their own version of an analytical plotter called the US1. About the same time OMI introduced a new model analytical plotter called the APC/4. Both

TABLE 13-1. EARLY GENEOLOGY OF ANALYTICAL STEREOPLOTTERS

Year	Model	Special Characteristics
1961	AP/1	First production unit.
1963	AP/2	Computer controlled optics, panoramic software.
1964	AP/C	For civil mapping applications, lower cost.
1965	AS11A	Improved version of AP/2.
1968	AS11B	AS11A with image correlation.
1968	UNAMACE	Non-optical, all electronic with image correlation.
1969	AS11A1	Improved AS11A with 9 × 18 inch format.
1969	AS11B1	AS11A1 with image correlation.

organizations relied upon commercially available mini-computers for control. The state of technology and lowering costs of the mini-computers made these feasible and desirable. Oddly enough, OMI and Bendix continued to collaborate on producing the AS11 series of analytical plotters for military customers while competing with each other in the civil applications.

The analytical plotter continued to be considered by the general photogrammetric community as a highly specialized, esoteric piece of equipment until the International Society of Photogrammetry Congress in Helsinki in 1976. At this congress seven manufacturers exhibited their versions of an analytical plotter. The majority of the world's manufacturers of stereoscopic plotting instruments were willing to commit themselves to the idea that analytical plotters were marketable. Table 13-2 lists the manufacturers, the model designation and operational status of these instruments.

It is interesting to note that of the seven instruments exhibited at the 1976 Congress very few were in production or being actively marketed. The others were prototypes. However at the Analytical Stereoplotter Workshop and Symposium in Reston, Virginia, in April 1980, sponsored by the American Society of Photogrammetry, eight manufacturers showed instruments, all of which were operational and all presumably in production. Subsequently in this chapter these instruments will be reviewed in more detail.

13.3.2 BASIC PRINCIPLES

The analytical plotter is a photogrammetric plotting system which solves mathematically the relationship between photographic image coordinates measured in the two dimensional photographic reference system and the ground coordinates of the object in the three dimensional "real" world. In contradistinction to the analytical plotter, the analog plotter attempts to establish this same relationship by utilizing mechanical and/or optical principles. This is usually accomplished by using space rods or optical projection or both. The analytical plotter, relying on pure mathematics rather than simulation through hardware, has some remarkable advantages.

The limitations of fine mechanics and optics for simulation are circumvented and a higher degree of accuracy is obtainable. For example, on the most precise analogue plotting instruments, measurement accuracies in the order of 20 micrometres at photographic scale are the best that one can normally expect. However the properly designed analytical plotter will permit accurate photographic scale measurements of 2 micrometres.

The conventional analogue plotting instrument is limited with respect to types of photography it can accept, that is, the capability to simulate the internal geometry of the camera. Focal length, film format size, lens distortion, film shrinkage are all important elements to be considered. A stereoplotter must account for all these elements if the resulting model is to be constructed with a high degree of accuracy. This can be very difficult and expensive and, in some cases, impractical when using analog instruments. However the analytical plotter accommodates for the effects of these elements in straightforward mathematical manner. The control computer of the analytical plotter calculates

TABLE 13-2. ANALYTICAL STEREOPLOTTERS EXHIBITED AT ISP, HELSINKI, 1976

Manufacturer	Model	Computer	Status
Bendix	U.S.1	DEC PDP 11/35	Operational
Galileo	D.S.	DEC PDP 11/05	Operational
Instronics	Anaplot	DEC PDP 11/45	Prototype
Keuffel & Esser	DSC	Data General Nova 3-12	Development
Matra	Traster 77	Telemechanique	Development
OMI	AP/C4	DEC PDP 11/03	Operational
OMI	AP/C3T	DEC PDP 11/35	Operational
Zeiss	Planicomp C-100	HP21MX	Operational

the necessary corrections required in real time and implements them in real time.

The analytical plotter is exceptionally well suited to digital exploitation of the model. Its basic configuration allows the operator to read, record and store coordinates in the local ground coordinate system or any other coordinate system that might be considered desirable. This is an extraordinary asset for digital terrain modelling and aerotriangulation. In the case of digital terrain modelling, the operator can instruct the computer to move the floating mark in a predetermined path in the model. The operator, by manual or automated (correlation techniques) methods, keeps the mark on the surface of the model and position coordinates can be recorded according to time, or distance travelled. In a strip triangulation, models can be joined without fear of misidentification of pass points. The pass points are identified unambiguously in each photograph by their coordinates and, therefore, recoverable without error.

Another unique advantage of the analytical plotter is the ability to store and recall data using the computer. Once a model has been acceptably formed it is a simple matter to store the constants and other data for the particular model. At a later time the model can be exactly reproduced on the same or a similar type of analytical plotter. This feature can be very useful in quality control of the compilation. It is also valuable for determining changes of topography or planimetry over a period of time.

The basic configuration of an analytical plotter is obvious and deceptively simple in concept. The system can be broken down into three main functional components; the stereoviewer, the control computer and the interface. A fourth component, the x-y plotter or coordinatograph, is normally considered as an option. Figure 13-1 is an elementary schematic of an analytical plotter.

The stereoviewer conventionally consists of a binocular viewing system, two stages for supporting the stereoscopic pair of photographs, control devices for moving the stages and/or the optical system and an illuminating system for the photographic imagery. The binocular system is of high quality, high resolution optics, with or without zoom lens capability. If the analytical stereoplotter system is intended for use with panoramic or other nonconventional photography, the zoom lenses will be independent of each other and controlled by the computer. Prisms (Dove) will also be included for rotating the images under computer control. A base-in, base-out device is conventionally provided to permit the left eye to view the right photograph and the right eye to view the left photograph. Figure 13-2 is a schematic of one type of optical system. Most models of analytical plotters utilize hand wheels and a foot wheel to control the apparent motion of the floating mark; however, optional control devices such as the joy stick or rotary ball are available as alternatives. The maximum format size is conventionally 254×254 mm; however, some analytical plotters are available with larger format size to accommodate nonconventional photography, such as panoramic, which is more efficiently exploited with a 254×254 mm format.

The control devices for moving the stages and/or optics are generally lead-screw and nut, tape drive or friction rollers. These positioning devices are normally actuated by d.c. servo motors receiving commands from the computer-stereoviewer interface. Actual location is determined by rotary or linear encoders capable of a resolution of the order of one or two micrometres. Command control devices, such as hand wheels, foot wheel or joystick, are usually coupled directly to rotary encoders. The encoders act through the interface to introduce data in the form of model coordinates to the computer which, in turn, releases stage positioning instructions to the servo motors.

The computer controlling the analytical plotter at the present stage of technology usually has a memory of 32,000 words of 16 bits. It is customary to find this computer provided with one or more cartridges or disks, a keyboard with CRT display and a printer. Additional memory, discs, tape drives, *etc*. can be added at the option of the buyer of a system. At the time of this writing the technology of computer design is advancing rapidly. It is reasonable to anticipate that the real-time functions of the analytical plotter control will be done by microelectronics. This will permit a number of plotters to use one computer. This computer will store, distribute and process data on an off-line basis while controlling other instruments such as X-Y plotters.

The programs provided with the analytical plotter are an essential part of the system. It is critical to the successful exploitation of the entire concept. The more developed and sophisticated the programs are, the greater will be the potential application of the analytical plotter to solving photogrammetric problems.

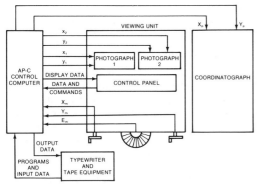

FIGURE 13-1. System diagram

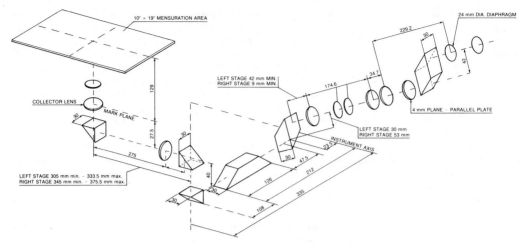

FIGURE 13-2. Optical system schematic

Programs for analytical plotters are usually categorized as operating systems and application programs. The operating systems provide for fundamental control of the components, that is, "housekeeping" or functions such as transmission of data, filing and recalling of data to and from memory, integrating and executing commands, basic mathematical routines, and myriad other details necessary to make a computer-controlled system function. Operating programs are usually provided by the manufacturer of the computer. However, because the analytical plotter is an integrated system, additional operating programs must be prepared by the system integrator who is normally the manufacturer of the analytical plotter.

The application programs deal with the photogrammetric use of the system. These programs are concerned with the setting up of the model and the exploitation of the properties of the model to produce data. Most analytical-plotter manufacturers categorize the application programs in a similar manner as follows.

13.3.2.1 INTERIOR ORIENTATION

This program provides for the establishment of the relationship between the photo-coordinates and stage-coordinates as well as the determination of the location of the principal point of the photograph. It is essential to include a photograph-to-stage transformation unless a photograph is always placed upon the stage in exactly the same position. Otherwise an important advantage of the analytical plotter is lost. This is the ability to reset a stereomodel or a single photograph to predetermined parameters. The principle point is determined by locating the fiducial marks of the photograph. Four or eight marks are normally utilized.

13.3.2.2 RELATIVE ORIENTATION

The relative-orientation program provides for semi-automatic establishment of the model. The program drives the photo-stages to theoretically desirable locations for removal of parallax. The operator must do the actual removal of parallax; the value of the parallax and the corresponding locations are stored by the computer. Upon elimination of parallax at six or more locations the computer calculates, by the method of least squares, a solution for the required exposure-station constants and a model is formed with residuals available as a measure of reliability of the solution. Different kinds of analytical plotters offer variations on the basic concept; however, fundamentally, they are based upon the classical equations of co-linearity for orientation.

13.3.2.3 ABSOLUTE ORIENTATION

This program rotates and translates the relatively oriented model to fit to ground survey data and establish the ground coordinate system in the model. The task is performed by placing the floating mark on control points identified in the model. The ground coordinates are known and stored in the computer. The model coordinates are recorded automatically. The accumulated data are utilized by the computer (assuming the data are adequate) to solve by the method of least squares. The new constants for each exposure station are determined and used. Reliability of the solution is indicated by the residuals. It is possible to combine relative and absolute orientation in one operation if desired.

13.3.2.4 CORRECTIONS AND ADJUSTMENTS

One of the remarkable advantages of the analytical plotter is the ease with which the model can be corrected for errors introduced by the photographic equipment or adjusted for external factors. The effect of any phenomenon contributing an error can be corrected by a suitable program if the effect can be expressed mathematically. For example, the following effects are commonly provided for on present analytical plotters; film shrinkage or affine displacement in the film, lens distortion in the camera, curvature of the earth's surface, atmospheric refraction and systematic errors inherent in the measuring unit.

13.3.2.5 SAVE AND RESTORE MODEL

The program allows the constants determining a model to be temporarily or permanently stored in the computer along with other data pertinent to the particular model. If a model is to be restored at a later time for any reason, it can be done so in seconds with an accuracy consistent with the original model. Further, the same model can be moved from one instrument to another with identical recapture provided the instruments are controlled by the same software.

13.3.2.6 VISIT POINTS

The visit points routine is very simple but valuable. It enables the operator to enter into the system photo-, model- or ground-point coordinates and the viewer will be automatically driven to the point desired. This capability allows an operator to re-observe an image point with absolute accuracy without error due to misidentification. It also allows viewing of the image of a ground point previously described only by coordinates.

13.3.2.7 TERRAIN MODEL DIGITIZING (DTM)

Programs for digitizing the model are finding expanding use with automation of photogrammetric and cartographic procedures. Without analytical photogrammetry, amassing the large amount of data required to reduce a model to a set of coordinates would be a tedious and expensive process. A DTM program for an analytical plotter provides a much improved procedure. The floating mark is driven in a preplanned X, Y pattern and the operator is required to either keep the mark on the ground continuously or at specific stopping points. The coordinates, so determined are transmitted to the computer for subsequent processing. The pattern can be varied according to the character of the terrain. Where civil engineering projects or physical inventories are concerned, the pattern can be altered and limited to the particular area of concern.

13.3.2.8 COMPUTATION OF AREA, VOLUME AND VECTOR

Programs can be provided which allow an operator to move from point to point while storing the ground coordinates of the observed points. Subsequently the vector determined by two points and the area and volume determined by three or more points can be instantly calculated.

13.3.2.9 AERIAL TRIANGULATION

Aerial triangulation programs perhaps vary the most among the programs provided by the manufacturers of analytical plotters. However, they all have the potential for greatly increasing the speed and accuracy of the procedure over the process with analog stereoplotters. Some programs are based upon the theory of continuous strips and others on independent-model theory. The rapid orientation procedures, unambiguous identification of pass points, earth curvature and atmospheric refraction correction, more accurate models and immediate recall of previous model in case of gross malfunction or error, all contribute to making aerial triangulation with the analytical plotter a vastly improved process.

The mathematical relationships which are basic to the application programs of an analytical plotter are the classical photogrammetric equations treated in chapter II. They are not discussed in detail in this chapter. Figure 13-3 demonstrates the basic relations between the model and photo-coordinate systems (for frame photography) conventionally utilized for programming. Model coordinates are designated in the lower part of the figure as X, Y and E. These coordinates must first be translated by distances representing the airbase components Bx, By and Bz so that the center of the new coordinate system is the perspective center of each photograph. The translated model coordinates are designated X_c, Y_c and Z_c. The coordinate system must then be rotated through the three angles kappa, omega and phi corresponding respectively to swing, roll and pitch. After the rotations have been performed, the three axes are termed X''', Y''' and Z'''. X''' and Y''' are in the plane of the photograph but are still in the scale of the model. The conversion to photo-scale is performed by multiplying each of the X''' and Y''' values by the focal length of the camera divided by the quantity Z'''.

Figure 13-3 also illustrates where the various corrections for distortion are made. The earth-curvature distortion and atmospheric-refraction distortion both occur in the portion of the figure that deals either with model coordinates or with coordinates that are in the scale of the model. For this reason, these distortions are compensated for by making corrections to the model coordinates. The other two sources of distortion, lens distortion and differential film shrinkage, are in the portion of the figure that represents photo-coordinates. Therefore, corrections for these distortions are made, in the computer, to the photo-coordinates.

Figure 13-4 is a flow chart for the real-time computation that is performed for one photograph. An identical computation, except for earth-curvature distortion, which is common to both photographs, must also be made for the second photograph. Starting at the lower left-hand portion of the figure, an earth-curvature correction is first calculated. It depends upon the X and Y model coordinates and the model scale. This calculation results in an elevation

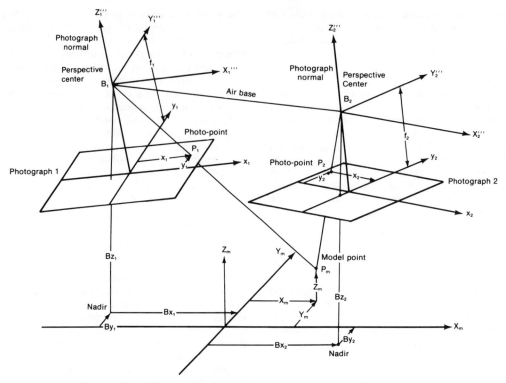

FIGURE 13-3. Diagram showing model and photograph coordinate systems

correction, which varies with the radial distance of the current model point from the center of the model. The earth-curvature correction is then added to an atmospheric-refraction correction, which is a function of the X, Y, and E model coordinates. The atmospheric-refraction correction depends on the radial distance of the current model point from the nadir point, *i.e.*, the location in the model that is directly beneath the camera. Also, to a secondary degree, the atmospheric-refraction correction depends upon the current elevation in the model.

The output from the atmospheric-refraction correction block is designated Z_m. It is the model

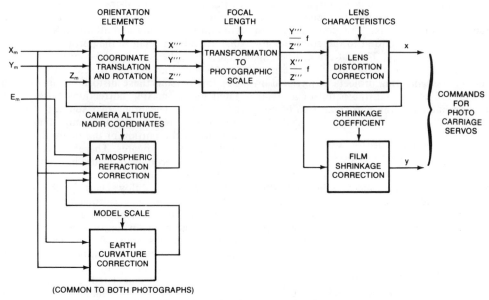

FIGURE 13-4. Flow chart of main computation for one photograph

elevation E_m, modified by the two elevation corrections. The three coordinate values X_m, Y_m, and Z_m are then processed in the coordinate-system translation and rotation block to produce X''', Y''', and Z''', which are the rotated and translated model coordinate system. The next block transforms the coordinates into photographic scale. Finally, a lens-distortion correction is applied to both of the resultant photo-coordinates. The corrected x-photo-coordinate is used directly as a command to a servo controlling a stage. The y-coordinate must have a differential film-shrinkage correction applied to it, after which that value is also used as a command for a servo motor.

13.3.3 CURRENT DESIGNS

Earlier in this chapter it was noted that at the Analytical Stereoplotter Workshop and Symposium in Reston, Virginia seven manufacturers demonstrated their versions of analytical stereoplotters. In somewhat more detail these instruments are described as follows.

13.3.3.1 THE ANALYTICAL PHOTOGRAMMETRIC PROCESSING SYSTEM IV (APPS-IV)

This instrument is manufactured by Autometric, Inc., Falls Church, Virginia. The APPS-IV (figure 13-5) is a medium accuracy (8-10 micrometres RMS) plotter. The stages will accommodate photographs up to 240 × 240 millimetres which is a standard size for most analytical stereoplotters. Transparencies or paper print viewing (optional) are provided for. Measurements are made using linear encoders with least count of 1 micrometre. The stages can be motor driven at rates of up to 0.02 inches per second and may be disengaged from the slow-motion drive and moved more rapidly without losing count. Slow motion drives are provided for both the overall motion, for Y translation of the left stage relative to the right and for X translation of right stage relative to the left. The operator controls the stages by rotating a ball for common movement of the stages and by using thumb wheels for differential motion of the stages.

The optical system is that of the Bausch and Lomb Stereo Zoom Transfer Scope. It has a range of magnification from 2.3× to 16.1×. Resolution at maximum magnification is 64 line pairs per millimetre. The reticle is an illuminated mark selectable by the operator in three sizes in the range of 25 to 100 micrometres.

The control computer is accessed through a standard RS-232C interface. The APPS-IV has a built-in microprocessor system (Intel 8085)

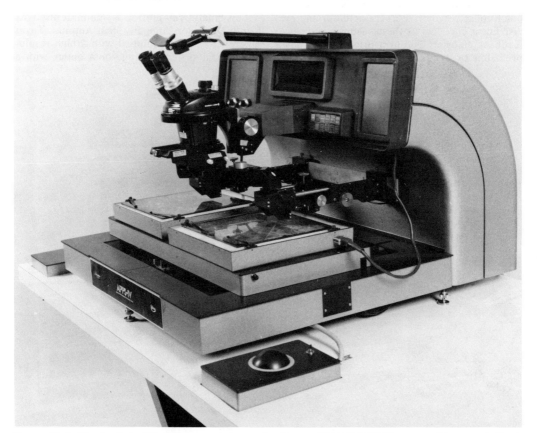

FIGURE 13-5. The Analytical Photogrammetric Processing System IV (APPS-IV) (Courtesy of Autometric, Inc.)

which allows one to operate independently of a host computer and provides a certain degree of independent operation.

13.3.3.2 THE US-2 ANALYTICAL STEREOPLOTTER

Helava Associates, Incorporated of Southfield, Michigan is the manufacturer of this system. The US-2 configuration (figure 13-6) normally consists of a stereoviewer, control interface, control computer with associated peripherals, plotting table, computer programs and CRT terminal. Operator control is exercised primarily through the CRT terminal, the hand wheels and/or free hand control, the foot wheel, various status switches and two foot-switches. Additional controls are provided for stage illumination, independent rotation and base-in/base-out viewing.

The stereoviewer consists of two precision X-Y photo-stages, a binocular viewing system, stage illuminators and major operator controls. It is suitable for use with black and white and color photography and accommodates film positives or glass plates up to 25 cm square. The X-Y stages are independently controlled and provide precise positioning with respect to stationary optical viewing axes. The measuring mechanism consists of a combination of low-friction ball bushings travelling on round shafts as guides, a pinch-roller servo-drive as the actuator and precise linear encoders as the measuring and servo feedback element. The measuring system has a precision of 1 micrometre with a maximum speed of 40 mm per second. Positional rms error in a stage is typically 0.0025 mm and is no greater than 0.004 mm per axis after removal of systematic errors by a computer-program.

The optical system provides a field of view of 24 mm with a fixed magnification of 8×. Other fixed magnifications can be provided as well as a 6×-27× zoom optical system at purchaser's option.

The control interface between viewer and computer incorporates a multiple microprocessor design which not only handles the servo motor interface and operator command inputs but also performs the projective transformations in real time.

A host computer as simple as a PDP-11/23 with dual floppy disks or as elaborate as a PDP-11/60 can be selected. Other minicomputers can be adopted. A typical control computer for the US-2 consists of a PDP-11/34 (manufactured by Digital Equipment Corporation) along with its associated equipment. It has 48K words of memory, 5.2 million bytes of interchangeable cartridge disk memory, a CRT terminal and a teleprinter. The CRT terminal allows the operator to enter and display data as well as to control the system.

13.3.3.3 DSC-3/80 ANALYTICAL STEREOCOMPILER

This analytical plotter is manufactured by Keuffel & Esser Company, San Antonio, Texas (figure 13-7). It is equipped with a high resolution optical system having zoom optics with a

FIGURE 13-6. US-2 Analytical Stereoplotter (Courtesy of Helava Associates, Inc.)

nominal range of magnification from 5× to 11× with low power eyepiece and 11× to over 55× with the higher power eyepiece. A 25 micrometre diameter floating mark is formed by a clear aperture on glass which is illuminated. The minimum viewing area at 11× magnification is 16 millimetres. Base-in/base-out is provided to view either stage through either eyepiece.

Control of the floating mark is accomplished with conventional handwheels for X and Y translation and a footwheel for control of displacement in the Z direction.

A unique feature of the DSC-3/80 is the use of the Erickson encoders for measuring locations on the photographs instead of the more conventional rotary or linear scale encoders. The Erickson encoder consists of a precisely calibrated grid coupled with a pick-up head that effectively doubles the actual line density of the grid. Since the grid and the stage are joined together with no relative motion between them, the measurement on the photo-stage is virtually independent of any mechanical drive or motion. This system permits micrometre accuracy measurement with minimum reliance on precision spindles, guides and drive trains.

The photo-carriages and the grids remain fixed and scanning of the photographic images is accomplished by mobile optics. Pick-up heads which travel with the mobile optics scan the grids and derive measurements in X and Y. The accuracy of each stage is plus or minus 2 micrometres standard deviation.

Operator communication with the computer is through an alpha-numeric keyboard. Instrument status is exhibited on a 256 character aphanumeric CRT. The control computer is a 700 NS cycle time C.P.U. with 64K byte bipolar memory. The C.P.U. has hardware multiply/divide, direct memory access interface, power minitor/auto restart with internal battery backup and a dual diskette storage unit with 630K × 8 byte capacity.

13.3.3.4 THE TRASTER ANALYTICAL STEREOPLOTTER

The Traster, figure 13-8, is manufactured by Matra Optique, Rueil-Malmaison, France. Among the current production of various types of analytical stereoplotters, the Traster is noticeably unique with respect to its optical viewing system. Whereas other designs rely upon optical train binocular viewing, the Traster utilizes rear projection with polarization to separate the two views and produce the stereo effect. The operator wears polarized glasses when observing the screen. The screen display is approximately 42 millimetres at photographic scale when magnifi-

FIGURE 13-7. DSC-3/80 Analytical Stereocompiler (Courtesy of Keuffel & Esser Co.)

FIGURE 13-8. Traster SST Stereocompilation System (Courtesy of Matra Optique)

cation is 10×. The floating mark is etched in the back surface of the center of the screen. As the photographs are moved the mark remains fixed and the projected images move. A CRT is provided which is located on the console to the left of the stereo projection screen. The CRT, with associated keyboard, is the communication link with the central processor. On the right side of the stereo-viewing screen is a CRT which is linked to a TV pickup to observe the X-Y plotting table.

The control computer of the Traster is normally the Eclipse S-130 or S-140, however, the manufacturer's literature indicates that other minicomputers are also available. The usual computer peripherals such as line printer, magnetic tape and disc storage are available.

Control for observing and measuring the stereomodel is accomplished by means of a trackball and drum. No handwheels or footwheel are offered. Linear optical encoders are utilized for measurement in the x and y directions and the overall error in the film plane is described as about two micrometres. Air bearings are used on the photo-carriage systems.

13.3.3.5 THE GALILEO D.S. DIGITAL STEREOCARTOGRAPH

The D.S. Digital Stereocartograph, figure 13-9, is manufactured by Officine Galileo S.p.A. of Florence, Italy. Some authorities do not categorize the Digital Stereocartograph as a fully analytical stereoplotter system due to its unique approach to the fundamental relationship between photo-coordinates and stereomodel or ground coordinates. For all other analytical stereoplotters described, the three dimensional coordinates are sent to the computer and the

photo-stages are positioned at the corresponding photo-coordinates by servo systems, usually based upon the equations of colinearity. The Digital Stereocartograph, on the other hand, requires that the photo-stages be positioned by the operator and the computer then derives the three dimensional coordinates in object space. This solution allows for a simpler electromechanical instrument but considerably complicates the real time operation such as in contouring and planimetric delineation.

The D.S. stereoviewer is the Galileo-Santoni stereocomparator fitted with shaft encoders to measure stage displacements and stepper motors to introduce Δx and Δy parallax to the right hand photo-carriage when required, either manually or under computer control. Handwheels and a footwheel provide control for moving the photo-carriages. Control for the plotting coordinatograph originates from the computer by means of stepping motors rather than the more commonly used servo systems of other manufacturers.

The D.S. utilizes the Digital Equipment Corporation PDP 11/05 as the control computer and presumably other models of the PDP 11 series can be substituted. Communication between operator and CPU is by means of dedicated control panel switches on the viewer and a line printer.

13.3.3.6 ANALYTICAL STEREOPLOTTER MODEL AP/C4

This analytical stereoplotter is manufactured by Ottico Meccanica Italiana S.p.A. (OMI) of Rome, Italy. The AP/C4, (figure 13-10) is a highly accurate analytical plotter designed to attain 2 micrometre rms error per axis. Precise measurement is attained by utilizing linear,

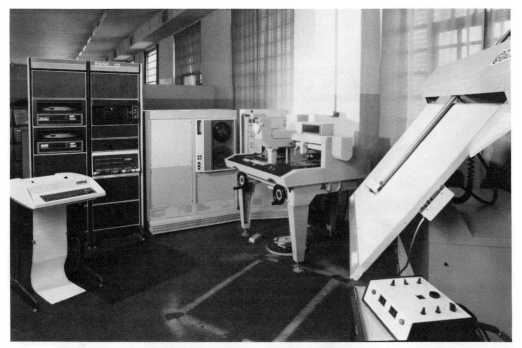

FIGURE 13-9. Galileo D.S. Digital Stereocartograph (Courtesy of Officine Galileo S.p.A.)

glass-scale optical encoders of the self-aligning type. The stages accommodate photographic-film transparencies and glass plates up to 250 mm square. The stages are moved by lead screws attached directly to D.C. servo motors.

The control elements of the stereoviewer are the conventional hand wheels and foot wheel plus a unique, free-hand pantograph device with variable-ratio displacement selector. Exclusion switches permit motion in only instrument X or Y. Additional control is exercised through push buttons on a control panel and a CRT.

The optical system provides standard magnification of $7\times$, $12\times$ and $20\times$, selectable by a control switch. Alternatively a zoom optical system with a range of 5 to $45\times$ is offered as an option. Binocular viewing of each stage in addition to conventional stereoscopic and pseudoscopic viewing is provided. Optics move along the X axis and the stages move along the Y axis. These mechanically independent motions preclude errors associated with stage-on-stage arrangements. The floating mark can be obtained either as an opaque disk of fixed diameter or an illuminated mark with adjustable diameter. The resolution of the viewing optics through all magnification ranges is not less than five line-pairs per unit of magnification.

The interface between viewer and control computer utilizes microprocessor technology to digitally control the servo motor systems. This provides positive control of the servos under varying conditions of ambient temperature and kinematic load on the drive mechanism.

The basic computer for controlling the AP/C4

is the PDP-11/03 with two floppy disks, 32K words of memory, a CRT and hard copy printout. The computer and peripherals are manufactured by the Digital Equipment Corporation. More sophisticated PDP-11 computers are directly substitutable for the 11/03, or other computers can be provided on special request.

13.3.3.7 THE AUTOPLOT

This instrument is manufactured by Systemhouse, Ltd., Ottawa, Canada. The Autoplot viewer (figure 13-11) consists of two independent, precisely positioned stages viewed by two high-resolution optical systems that are fixed relative to each other and to the instrument. The stages move both in X and Y while the optics remain stationary. The location of each stage is monitored by 1 micrometre resolution linear encoders. Each stage is moved along each axis by a friction drive which consists of precision bearings on a smooth, rotating shaft. The manufacturer claims this type of drive to be nearly maintenance-free, almost indestructable and capable of being set at any reasonable pitch at the factory.

The optical system is described as being of proven design with all rotational and magnification capabilities normally associated with a precise stereoviewing instrument.

The electronic interface between the operator and the system software is designed as state-of-the-art, microprocessor oriented, independent module relative to the overall system. The interface acts as the closed loop servo drive positional verifier as well as an input processing

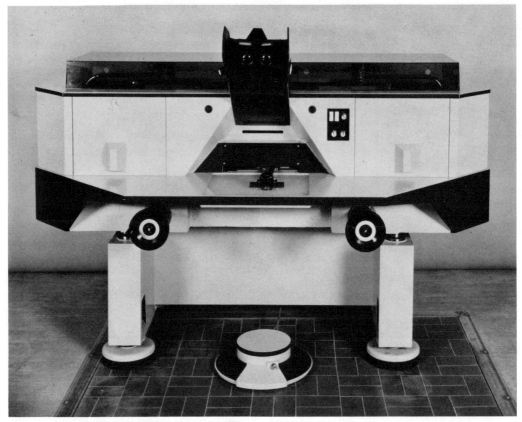

FIGURE 13-10. Analytical Stereoplotter Model AP/C4 (Courtesy of Ottico Meccanica Italiana)

and conditioning device from the operator controls to the system software modules. The controlling CPU of this system is the PDP-11/34 manufactured by Digital Equipment Corporation.

13.3.3.8 THE C100 PLANICOMP ANALYTICAL PLOTTER

The Zeiss C100 Planicomp System (figure 13-12) is manufactured by Zeiss, Oberkochen, Germany. The hardware of the system essentially consists of a measuring section, a control unit and peripherals. A basic opto-mechanical unit, the viewer, serves to view and measure the photographs. A photogrammetric control panel is located at the viewer front. The two units are monitored separately by an electronic controller which also contains an interface for the mini computer equipped with disk storage, which controls the entire system through suitable software.

The basic opto-mechanical unit contains two separate servo-driven systems of photo-carriages moving in X and Y and a fixed optical system for comfortable viewing. This is supplemented by a control panel with switches for all important settings and processes, coordinate display and drive elements, such as, hand wheels, pedal disk, joystick and speed control. The photo-carriages are designed as cross slides and driven by precision lead screws. Their travel is 240 mm $\times$ 240 mm, limited by a three-stage stop system. The photo-carriers are rigidly connected to the carriages and cannot be rotated. A nine point grid is provided for calibration. The viewing system consists of an illuminator, a measuring mark of 20 or 40 micrometres in diameter, which can be converted by a variable light source into a black or luminous floating mark to suit the picture content, Dove prisms, a prism control unit for change-over between orthoscopic and pseudoscopic stereo or binocular viewing of the left-hand or right-hand photography and slip-on eyepieces for 8$\times$ or 16$\times$ viewing magnification.

The electronic controller is the link between the basic opto-mechanical unit and the computer. It supplies power for the four photo-carriages, pulse counters, registers for carriage shift values, input registers, key illumination and interlocking of key arrays, data conversion, clock pulse generation as well as the control of computer input and output *via* interface.

The control computer for the system is the Hewlett Packard HP1000 mini computer. It includes 32K words and a disk storage unit of

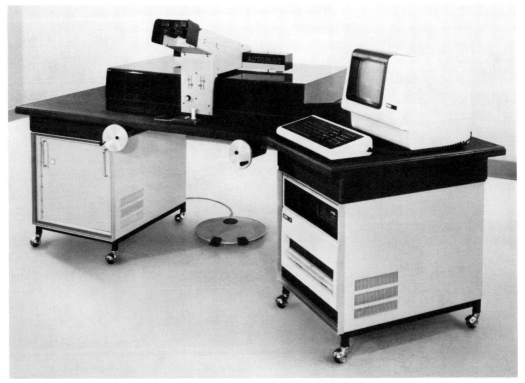

FIGURE 13-11. The Autoplot (Courtesy of Systemhouse, Ltd.)

4.9M bytes. A teleprinter or video terminal with on-line printer is used as computer console.

13.3.3.9 OTHER DESIGNS

There are other analytical stereoplotters in use today which because of highly specialized use are perhaps not well publicized, but represent large investments, produce tremendous amounts of photogrammetric data and are examples of the latest technology in this specialized field. Examples of such instruments are the AS11/A1, the TA3/P1 and TA3/P11 series of instruments. These instruments are utilized primarily by the U. S. Defense Mapping Agency and the U. S. Geological Survey.

The AS11/A1 stereoviewer is manufactured by Ottico Meccanica Italiana (OMI), Rome, Italy. The control computer is a ModComp II/25 manufactured by Modular Computer Corporation and the interface is manufactured by the O. M. I. Corporation of America. The system is designed to be as universal as possible with respect to input data. It is capable of setting stereomodels utilizing conventional frame photography or such non-conventional imagery as panoramic, strip or radar presentation. The stereoviewer will accept 240 × 480 mm format stereoscopic pairs. The viewing optics are computer controlled through servo systems so that each branch of the binocular optical train can attain independent magnification as well as in-

dependent Dove prism rotation. This capability allows for great versatility in stereoscopic observation. The independently controlled optics have a range of magnification from 7× to 63×. The measuring accuracy over the 240 × 480 area is 4 micrometre rms error. Rotary encoders and precision lead screws are the measuring elements. A more recent model, the AS11/ A1(NOS) has an interesting innovation. The floating mark is an illuminated dot which is computer controlled so as to stay constant in size regardless of magnification. This model also uses linear rather than rotary encoders. The AS11/A1(NOS) viewer (figure 13-13) is manufactured by OMI. It uses the PDP 11/45 and the interface was manufactured by the Bendix Research Laboratories.

The TA3/P series were originally designed as three stage stereocomparators. However, due to technological advances in computer design, the advantages of coupling these units to on-line computer control were realized. The TA3/P1 and TA3/P11 series of viewers are manufactured by OMI. The computers for control of these units were originally DEC PDP-15 units, however, they are being superseded by the PDP-11/60 units. Interfaces for these systems are manufactured by the Bendix Research Laboratories. The TA3/P1 and TA3/P11 series are very similar in their optical-mechanical characteristics to the AS11/A1, having similar stage sizes and

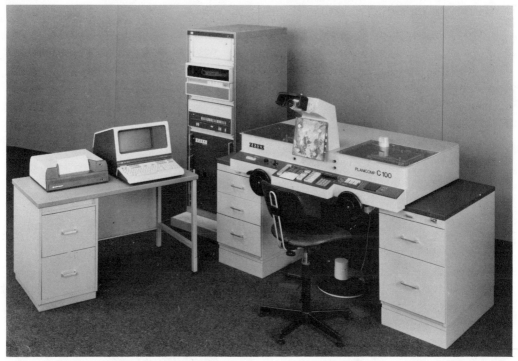

FIGURE 13-12. The C100 Planicomp analytical plotter (Courtesy of Zeiss, Oberkochen)

FIGURE 13-13. National Ocean Survey Analytical Plotter (NOSAP) (Courtesy of National Ocean Survey, NOAA)

optical performance. The P1's have rotary encoders and precise lead screws whereas the P11's have linear encoders. The P11's also have the computer-controlled, constant size floating mark.

It is interesting to note that a sizable group of three-stage (9.5″ × 9.5″) stereocomparators, the OMI TA3/P series which were used by DMA in the off-line mode, have recently been redesigned, coupled with ModComp II/25 computers and can now be classified as analytical plotters. This, of course, demonstrates the concept of the analytical plotter as a stereocomparator coupled with a digital computer.

13.3.4 GENERAL CONSIDERATIONS

In the descriptions of the various analytical plotters summarized in the previous paragraphs it is notable that very little space was devoted to programs provided by the individual manufacturers. The programs furnished with each system are, of course, extremely important to the proper functioning and utilization of the system concerned. However, even more than the pace of electronic hardware development, the development of programs for analytical plotters from all manufacturers is in an extremely dynamic and changing state. A listing of programs furnished by a particular manufacturer could be radically altered in a relatively short period of time as compared with equipment development. Of course there is a basic commonality to all programs now being supplied. These are the application programs which are fundamental to the photogrammetric process. As each manufacturer receives comments from his clients and observes what is successful amongst his competitors he will, no doubt, add to or alter his own programs. It is incumbent upon the purchaser of an analytical plotter to assess the current available programs and their applicability to his own special needs before committing to a particular plotter.

The trend of development of analytical plotters at this time appears to be toward very basic systems at a relatively lower cost. These basic systems are usually capable of being expanded or added to at a later date. For example, systems are offered without coordinatograph or X-Y plotter, without zoom magnification, with minimal computer size and accessories. This trend allows organizations with little money to try using analytical plotters and, after some experience, to decide better what additional equipment is desired.

The items common to all plotters now being made appear to be a stereoviewer that is compact in size, and two photo-carriages able to accommodate photographs 240 × 240 mm in size. Binocular viewing with fixed magnifications appear standard although some manufacturers offer zoom capability as a relatively low price option. An interesting exception to the binocular optics is the Traster viewer manufactured by Matra. The Traster has a rear-projection screen with stereoscopic viewing dependent upon polarizers to separate the two images. The advantages and disadvantages of this approach are yet to be fully evaluated.

Precise measurement of the stages' motions are usually dependent upon linear encoders. Some models rely upon rotary encoders and precise lead screws; however, the costs of manufacturing highly precise lead screws are making them less competitive in this application. The usual claimed precision of movement is 1 to 2 micrometres with a position accuracy of less than 5 micrometres. Motion of carriages is almost in all cases accomplished by D. C. servo motors controlled by the electronic interface. Power is transmitted by lead screws, friction linear drives or steel tape. Control of the system is similar on most units, depending basically upon hand wheels for X and Y motion and a foot wheel for Z control. Communication with the control computer depends upon a CRT terminal and in some models control switches as well. Several models include joystick or rotary ball as alternatives. OMI's AP/C4 offers, in addition, a free-hand pantograph device and the Bendix US-2 has available as an option an electronic cursor which provides free-hand motion.

Electronic interfaces between the stereoviewer and the control computer have perhaps the widest variety of design as there are numerous electronic designs available for adaptation, both digital or analog in technique. Manufacturers are reluctant to provide details which is understandable as they require extensive design engineering and are critical to proper functioning of the system. A trend is obvious however to utilizing microprocessor technology which provides a higher degree of reliability, reduction in components at lower cost. The use of microprocessors of course encourage placing real time transformation capabilities in the interface thereby relieving the control computer of considerable work. This trend allows the analytical stereoplotter to operate to a limited degree as a stand-alone system, independent of a control CPU. It also leads to multi-viewer control by a single CPU.

The computer that appears to be most popular with manufacturers of analytical plotters is the PDP-11 series manufactured by Digital Equipment Corporation. All versions of this computer from the 11/03 up to the 11/70 have at one time or another been applied to controlling analytical plotters. Zeiss makes use of the Hewlett Packard HP1000, and a very large number of ModComp II/25's have been used to control AS11/A's. All these computers are 16 bit, 32K word or more machines, have disk memory, and have Fortran compilers.

The analytical plotter is a largely unknown

device to the majority of photogrammetrists today. It has been in productive use since approximately 1960; however, due to high cost, it has been exploited primarily by military or research groups. It is only recently that a combination of increasing costs for skilled labor and drastic reduction in costs of electronic components has brought the costs of analog and ana-lytical plotters to a competitive status. This condition has intensified the interest in analytical plotters. The basic instrument of the not too distant future will be the analytical plotter. Microprocessor controlled units will work as independent stand-alone systems resorting to larger CPU's on an off-line basis for data storage processing and analog output.

13.4   Image Correlation

The automation of stereoscopic plotting instruments requires a means for the rapid and accurate sensing of parallax between stereoscopic images. This is generally achieved by the automatic scanning and matching (correlating) of conjugate imagery.

At present, different techniques may be used for the scanning of the stereoscopic pair. These include: electronic scanners (cathode-ray-tube devices, vidicon-type devices, or image-dissector tubes); solid-state scanners (self-scanned photodiode devices, charge-coupled devices, charge-injection devices or charge-coupled photodiode devices) and electro-optical scanners (lasers, light-emitting diodes or conventional lamps). The matching of the scanned imagery is done by electronic correlation, digital correlation or coherent optical correlation. In the following discussion, a review will be given of existing scanning systems and the methods employed for the correlation of conjugate imagery.

13.4.1 ELECTRONIC SCANNING USING CATHODE-RAY TUBES (CRT's)

CRT scanning is the most widely used system for scanning stereoscopic dispositives. The basic components of the system are a pair of flying-spot scanners, fixed optics (objective and field lenses), photomultipliers and raster shaping control modules (x and y deflection system).

The left and right diapositive carriers are positioned in the scanner by a computer so that the homologous areas are scanned simultaneously. The flying spot of light of the CRT (the raster) is imaged by the objective lens in the plane of the diapositive. The light transmitted through the diapositive is collected by a field lens and is converted to a signal by the photomultiplier. After corrections for scale and rotation have been made to the scanning pattern, the two areas are identical except for parallax due to height. To the correlator the parallax appears as a phase shift between the signals.

At present, several automated photogrammetric systems with CRT scanning are available. For example, the Gestalt Photo Mapper (GPM), the Universal Automatic Map Compilation Equipment (UNAMACE), the Topomat system and the Itek system. The area scanned varies from 0.25 × 0.50 mm in the UNAMACE to a 9 × 8 mm patch in the GPM.

13.4.2 CORRELATION OF SIGNALS

The processing of signals produced by electronic CRT scanning devices is normally achieved in an electronic correlator. Despite the similarity between the existing systems, they will be discussed separately.

In the GPM system, the signals from each scanner are filtered into six frequency channels, each of which responds to different rates of change in the magnitude of the signal. The process is similar to Fourier analysis in that the signal is analyzed into harmonic frequencies each of which is assigned an amplitude equal to the number of level changes detected by the channel of corresponding frequency.

Separate parallax measurements are made for each of the six channels by cross-correlating the digital coefficients generated by the analog-to-digital converter. Nearly 80,000 binary products are calculated for every scan of both patches. The signed magnitude of these products is proportional to the phase separation of similar patterns from the two diapositives and are summed to form 2444 12-bit numbers, each of which represents the parallax detected for homologous areas on the left and right diapositives.

In the UNAMACE, the signals from the CRT scanners are fed into a height-error sensor consisting of two delay units and two correlators. The displaced correlation functions are appropriately summed to provide net height-error signals. These signals are passed to a displacement-error measuring unit consisting of an integrator, a positive and negative threshold detector, a reversible counter and digital-to-analog (D/A) convertor. If the integrated error signal exceeds one of the threshold limits, the detector steps the reversible counter in a corresponding direction and resets the integrator to zero. The counter operates a D/A converter that produces a voltage to deflect the center of one of the diapositive scans in a direction to reduce the observed time differential. The operation continues until the height-error has been compensated.

In the Itek correlator, the signals are channeled to the correlator through video amplifiers, where they are split into several frequency bands to allow the low spatial frequencies to be correlated first, progressing to higher frequencies as the accuracy of alignment is improved. A channel selector is used to select the best channel depending on the degree of correlation. The output is analyzed and is passed to the integrators. The first-order signals from the integrator are applied to a scanning-pattern modulator to cause appropriate transformation to the scanning waveform. The modified scanning waveforms are added algebraically in the sum-and-difference circuit to provide equal but opposite transformation to the left and right rasters.

In the Topomat, the correlator processes the signals after signal shaping by three band-pass filters with different transmission bands. A delay line, whose delay time is adjusted to the cutoff frequencies, is associated with each band-pass filter. Using cross-correlation technique a discriminator characteristic is produced, which permits correlation between parallax polarity and voltage polarity. An integrator is used for the transition from digital information at the output of the correlator to analog control signals for the z motor.

13.4.3 SOLID-STATE SCANNERS

If an image is projected onto an array of light-sensitive elements, a system incorporating such an array can be constructed that is capable of detecting the location (x and y coordinates) and the intensity of light flux incident on the elements of the array. Grey values can already be distinguished in more than 16 levels.

Solid-state devices consisting of photodiode arrays in a linear or rectangular array have been used in remote sensing (Thompson 1973), and as sensors of satellite imaging systems (O'Conner 1972).

In automated photogrammetric instruments, such solid-state scanners are used in the Raster system (Hobrough 1978). The Raster system uses a linear array consisting of 1728 elements. The photo-images are enlarged slightly to give 10 micrometre diode spacing at photo-scale. The arrays are rotated by servo motors so as to lie substantially along epipolar lines on each photograph.

The basic components of the Raster are the video processors, transformation delay modules, loop processor and output processor. The correlation of the analog video signals from the self-scanned photodiode array is based on cross-correlation techniques (Hobrough 1978).

13.4.4 ELECTRO-OPTICAL SCANNERS

Image scanning is performed by mechanically deflecting a laser light across the left and right diapositives. The stage is mechanically moved with respect to the scanner to give a scan in the perpendicular direction. Mechanical deflection permits a very bright spot to be used, which provides a higher signal-to-noise ratio in the scanner output. The spot brightness is limited only by the detector or by heating of the film.

The photography is scanned and density measured on each photograph along epipolar paths across many profile lines parallel to the direction of carriage motion to obtain information on many parallel profiles or parallel lines of grid points.

The AS-11B-X system developed by Bendix uses this technique for scanning. Data on a total of 1280 points are collected along a path by scanning along successive epipolar lines across the path. The 1280 points are sampled along the epipolar lines at a spacing of 20 micrometres. The epipolar lines are spaced 50 micrometres apart.

The AS-11B-X uses a digital correlator consisting of analog-to-digital converters, a buffer memory, an address-modification circuit, a parallel processor and two minicomputers. The correlation is performed digitally by multiple-bit multiplication of corresponding image element densities. The products are added over desired image areas, producing data integration over these areas. The digital correlator computes the amount of shift or parallax between corresponding segments of imagery on the two photographs. The computation of elevation in an epipolar coordinate system is performed on-line. Further computations for the digital elevation model in the ground coordinate system is done off-line.

13.4.5 DIGITAL OFF-LINE CORRELATION

An algorithm for digital matching of image densities was described by Panton (1978). The algorithm includes an image matching procedure in which parallax components are determined by automatically correlating corresponding images. The basic idea behind the method is to set up a regularly spaced grid of points on one image and to find its conjugate point on the other image. The algorithm is implemented on a distributive network of parallel digital processors. Currently the primary output of the system is elevations.

13.4.6 COHERENT OPTICAL CORRELATOR

Two concepts were conceived and developed in the Bendix Research Laboratories. In particular, they are the image/image correlator and the image/matched-filter correlator. A comparison of the two correlators showed that the image/matched-filter is more suitable for image search problems and thus for relief determination (Krulikoski 1972).

The coherent optical image/matched-filter

processor is essentially an optical frequency-plane correlator in which one stereo-photograph is transformed into a matched filter for the other stereo-photograph, the filter having the form of a Fourier-transform hologram. Processor operation is based on the concept that the matching of conjugate imagery can be reduced to a process of signal detection. As the input photograph is scanned, the correlator detects conjugate image points, and the spatial position of the output light signals are direct measures of the differential coordinates.

From the photogrammetric point of view, the optical image/matched-filter processor is a special-purpose instrument that represents a departure from traditional stereoplotter design.

13.5 Automation of Orthophotograph Production

13.5.1 ELECTRONIC ORTHOPHOTOGRAPH PRINTING

Two methods are available for printing an orthophotograph electronically: the "slit" method and the "patch" method. In the slit method the film is traversed continuously by an exposing slit having a length approximately equal to the width of the correlation zone. The printing of the orthophotograph is a continuous process using the height correction generated by the correlator. The continuous printing by the slit method is suitable with instruments having a small correlation zone, such as the UNAMACE and Topomat.

In the patch method the film is exposed in an array of patches. The size of the patch is less than the width of the correlating zone and is normally dependent upon the variation of relief. In the patch method, all motion and computation ceases before each patch is printed. The GPM system employs the patch method.

13.5.2 DIGITAL ORTHOPHOTOGRAPH PRODUCTION

A program was described by Keating and Boston (1979) for the production of orthophotographs from digital elevation models (DEM's) and aerial negatives by using a microdensitometer as input/output device. From the known model coordinates and camera orientation parameters, the x and y image coordinates of either photograph in the stereopair are computed for each interpolated model point. The densitometer is then used to extract the film density at each image point corresponding to each model point. These film densities are rewritten by the densitometer in corrected orthographic positions onto orthophotographic film at any convenient scale.

13.5.3 DIGITALLY CONTROLLED OPTICAL ORTHOPHOTOGRAPH PRINTING

The OMI OP/C and the Wild Avioplan OR 1 systems are digitally controlled optical printing systems of similar nature. These systems operate in an on-line mode or off-line mode. Optical printing of the orthophotograph is made by exposing a film through a slit under control of the system's computer, which controls all the input and output operations, computes model-to-photo-transformation and drives the servo system in real-time. The system requires digital elevation profiles, camera parameters, and orientation data. These profiles may be obtained from DEM's produced by automated systems, analytical plotters or by profiling the model on analogue photogrammetric plotters. Profiles could be obtained by interpolating digitized contours from existing maps. The OR 1 could be interfaced to a photogrammetric system for the direct acquisition of profile data.

13.5.4 PRODUCTION OF STEREO-ORTHOPHOTOGRAPHY

The stereo-orthophotograph (stereomate) is similar to the orthophotograph except that it is printed with specified parallax restored. In automated systems employing CRT printing, such as the GPM, the rectified image has a specified parallax restored before printing. In systems using optical printing, such as the Wild OR 1 or the OMI OP/C, the profile data are preprocessed to compute a set of pseudo-profiles with the desired base-height ratio.

The stereomate and orthophotograph are normally generated from two successive photographs. The orthophotograph and stereomate produce a scaled model free from y parallaxes which could be used for direct map compilation.

13.6 Direct Digitization on Stereoplotting Instruments

Over the last few years there has been a gradual infiltration of digital technology into mapping for the direct digitization on stereoplotting instruments. The rationale for exploiting digital technology for mapping systems is related to perceived benefits derived by the use of digital technology as well as the increased capability anticipated within a digital base framework. Specifically a digital base involves the replacement of map manuscripts by digital files. By so doing the user has access to map data in a machine processable form which may be used for the generation of special types of maps and charts or for deriving other forms of terrain data (*e. g.* elevation profiles from digitized contours).

With regard to the method and extent of involvement of digital computers and peripheral units for direct digitization from stereoplotting instruments, several feasible systems can be drawn up. In principle they can be classified according to the number of stations (single or multiple), the capabilities for interactive display and editing, and the communication between the system components (uni-directional or bi-directional). The recent trend in the development of these systems is in the use of multiple stations with interactive display and editing terminals in a distributive computer network (Allam 1979).

The development of distributed systems has been a natural outgrowth of requirements to increase our computational abilities for faster and more efficient computing. A distributed system is usually characterized by a multiplicity of logical and physical elements which are interconnected to varying degrees *via* some communication mechanism. The essence of any distributed system is in the multiplicity of components, as well as specialized elements. Even though the individual components operate autonomously there is normally a unifying element, for example, the common data base. Individual system components are coupled in such a way as to provide some level of overall system integration.

SELECTED REFERENCES

Allam, M. M., 1978. The Role of the GPM-II/ Interactive Mapping System in the Digital Topographic Mapping Program, *Proceedings of the International Symposium, Commission IV, New Technology for Mapping*, Ottawa, Canada, 2-6 October 1978, pp. 5-23.

Allam, M. M., 1978. DTM Application in Topographic Mapping, *Photogrammetric Engineering and Remote Sensing*, Vol. 44, No. 12, December 1978.

Andrews, Harry C., 1977. An Educational Digital Image Processing Facility, *Photogrammetric Engineering and Remote Sensing*, Vol. 43, No. 9, pp. 1161-1168.

Autometric, Inc., 1979. *The Wetlands Analytical Mapping System*, Autometric, Inc., Falls Church, Virginia, July 1979, 7 p.

Baumann, Lee S. (Ed.), 1979. *Proceedings: Image Understanding Workshop*, Science Applications, Inc. Report Number SAI-80-974-WA, November 1979, 177 p.

Bernstein, Ralph, and Dallam G. Ferneyhough, Jr., 1975. Digital Image Processing, *Photogrammetric Engineering and Remote Sensing*, Vol. 41, No. 12, pp. 1465-1476.

Bertram, S., 1965. The Universal Automatic Map Compilation Equipment, *Photogrammetric Engineering*, Vol 31, No. 2, March 1965.

Besenicar, Jure, 1978. Digital Map Revision, *Proceedings of the International Symposium, Commission IV, New Technology for Mapping*, Ottawa, Canada, 2-6 October 1978, pp. 221-232.

Blachut, T. J., 1979. Basic Features, Design Considerations and Performance of the ANAPLOT, The Canadian Surveyor, Vol. 33, No. 2, pp. 107-124.

Brunson, Ernest B., and Randle W. Olsen, 1978. Data Digital Elevation Model Collection Systems, *Proceedings of the Digital Terrain Models (DTM) Symposium*, St. Louis, Missouri, 9-11 May 1978, pp. 72-99.

Bybee, Jesse E., and George M. Bedross, 1978. The IPIN Computer Network Control Software, *Proceedings of the Digital Terrain Models (DTM) Symposium*, St. Louis, Missouri, 9-11 May 1978, pp. 585-596.

Dowman, I. J., 1977a. Developments in On Line Techniques for Photogrammetry and Digital Mapping, *Photogrammetric Record*, Vol. 9, No. 49, pp. 41-54.

Dowman, I. J., 1977b. Discussion of "Developments in On Line Techniques for Photogrammetry and Digital Mapping," *Photogrammetric Record*, Vol. 9, No. 50, pp. 287-290.

Elphingston, Gerald M., 1978a. Integrated Photogrammetric Instrument Network, *Proceedings of the Digital Terrain Models (DTM) Symposium*, St. Louis, Missouri, 9-11 May 1978, pp. 568-575.

Elphingston, Gerald M., 1978b. Integrated Photogrammetric Instrument Network (IPIN), *Proceedings of the International Symposium, Commission IV, New Technology for Mapping*, Ottawa, Canada, 2-6 October 1978, pp. 102-110.

Elphingston, Gerald M., and Gene F. Woodford, 1978. Interactive Graphics Editing for IPIN, *Proceedings of the Digital Terrain Models (DTM) Symposium*, St. Louis, Missouri, 9-11 May 1978, pp. 597-609.

Fornaro, Robert J., and Lionel E. Deimel, Jr., 1978. The Edit Processor Component of IPIN, *Proceedings of the Digital Terrain Models (DTM) Symposium*, St. Louis, Missouri, 9-11 May 1978, pp. 481-492.

Hardy, J. W., Johnston, H. R., and Godfrey, J. M., 1968. An Electronic Correlator for the Planimat, *Presented Paper, ISP Congress*, Luzanne, Switzerland, July 1968.

Helmering, Raymond J., 1977. A General Sequential Alogrithm for Photogrammetric On-Line Processing, *Photogrammetric Engineering and Remote Sensing*, Vol. 43, No. 4, pp. 469-474.

Hobrough, G. L., and Hobrough, T. B., 1971. Image Correlator Speed Limits, *Photogrammetric Engineering*, Vol. 37, No. 10, October 1971.

Hobrough, G. L., 1968. Automation in Photogrammetric Instruments, *Photogrammetric Engineering*, Vol. 31, No. 4, July 1968.

Hobrough, G., 1978. Digital On-Line Correlation, *Bildmessung und Liftbildwesen*, Vol. 46, No. 3, pp.79-86.

Kirwin, Gary A., 1978. Pooled Analytical Stereoplotter System (PASS), *Proceedings of the Digital Terrain Models (DTM) Symposium*, St. Louis, Missouri, 9-11 May 1978, pp. 576-584.

Keating, Terrence J., and Dennis R. Boston, 1979. Digital Orthophoto Production Using Scanning Microdensitometers, *Photogrammetric Engineering and Remote Sensing*, Vol. 45, No. 6, pp. 735-740.

Kelley, R. E., McConnell, P. R. H., and Mildenberger, S. J., 1977. The Gestalt Photomapping System, *Photogrammetric Engineering and Remote Sensing*, Vol. 43, No. 11, November 1977.

Konecny, G., 1979. Methods and Possibilities for Digital Differential Rectification, *Photogrammetric En-*

gineering and Remote Sensing, Vol. 45, No. 6, pp. 727-734.

Kowalski, D. C., 1968. A comparison of Optical and Electronic Correlation Techniques, *Bendix Technical Journal*, Vol. 1, No. 2, Summer 1968.

Kraus, K., G. Otepka, J. Loitsch, and H. Haitzmann, 1979. Digitally Controlled Production of Orthophotos and Stereo-Orthophotos. *Photogrammetric Engineering and Remote Sensing*, Vol. 45, No. 10, pp. 1353-1362.

Krulikoski, S. J., and R. B. Forrest, 1972. Coherent Optical Terrain Relief Determination Using a Matched Filter, *Bendix Technical Journal*, Vol. 5, No. 1, Spring 1972.

Leatherdale, J. D., and K. M., Keir, 1979. Digital Methods of Map Production, *Photogrammetric Record*, Vol. 9, No. 54, pp. 757-778.

Madison, Harold W., 1975. Evolution of the UNAMACE at the Defense Mapping Agency Topographic Center, *Proceedings of the 41st Annual Meeting, American Society of Photogrammetry*, Washington, D. C., 9-14 March 1975.

Makarovic, B., 1974. From Digital Components to Integrated Systems in Photogrammetry, *Proceedings of the Symposium, The Role of Digital Components in Photogrammetric Instrumentation*, Torino, Italy, 2-4 October 1974.

Masry, S. E., and R. A. McLaren, 1979. Digital Map Revision, *Photogrammetric Engineering and Remote Sensing*, Vol. 45, No. 2, pp. 193-200.

Murai, Shunji, 1978. Digital Rectification of Oblique Photography in Coastal Mapping, *Proceedings of the International Symposium, Commission IV, New Technology for Mapping*, Ottawa, Canada, 2-6 October 1978, pp. 702-707.

Panton, Dale J., 1978. A Flexible Approach to Digital Stereo Mapping, *Photogrammetric Engineering and Remote Sensing*, Vol. 44, No. 12, pp. 1499-1512.

Parenti, G., 1976. Orthophoto Printing with the Analytical Plotter, *Photogrammetric Engineering*, Vol. 33 No. 4, April 1976.

Roos, Maurits, 1975. The Automatic Reseau Measuring Equipment (ARME), *Photogrammetric Engineering and Remote Sensing*, Vol.41, No. 9, pp. 1109-1115.

Scarano, Frank A., and Gerald A. Brumm, 1976. A Digital Elevation Data Collection System, *Photogrammetric Engineering and Remote Sensing*, Vol. 42, No. 4, pp. 489-496.

Scarano, Frank, and Anthony Jeric, 1975. Off-Line Orthophoto Printer, *Photogrammetric Engineering and Remote Sensing*, Vol. 41, No. 8, pp. 977-991.

Schoeler, Horst, 1977. Future Trends in Photogrammetric Instrumentation, *Proceedings of the Fall Technical Meeting, American Society of Photogrammetry*, Little Rock, Arkansas, 18-21 October 1977, pp. 241-253.

Seymour, Richard H., and Arliss E. Whiteside, 1974. A New On-Line Computer-Assisted Stereocomparator, presented at the *40th Annual Meeting, American Society of Photogrammetry*, St. Louis, Missouri, 10-15 March 1974, 14 p.

Southard, R. B., 1978. The USGS Orthophoto Program, A 1978 Update, *Proceedings of the International Symposium, Commission IV, New Technology for Mapping*, Ottawa, Canada, 2-6 October 1978, pp. 594-601.

Stewardson, P. B., 1976. The Wild Avioplan OR1 Orthophoto System, *Presented Paper, 13th ISP Congress, Comm. II*, Helsinki, Finland, 1976.

Szangolies, Klaus, 1979. *Rationalization of Map Production and Map Revision with Modern Automated and Digitized Photogrammetric Instruments and Technologies*, presented at the Second United Nations Regional Cartographic Conference for the Americas, Mexico City, 3-14 September 1979, 17 p.

Tanaka, Sotaro, and Yuzo Suga, 1979. Landscape Drawing from Landsat MSS Data, *Photogrammetric Engineering and Remote Sensing*, Vol. 45, No. 10, pp. 1345-1351.

Thompson, M. M., and E. M., Mikhail, 1976. Automation in Photogrammetry: Recent Developments and Applications (1972-1976), *Photogrammetria*, Vol. 32, pp. 111-145.

Williams, Owen W., 1980. Outlook on Future Mapping, Charting and Geodesy Systems, *Photogrammetric Engineering and Remote Sensing*, Vol. 46, No. 4, pp. 487-490.

Zarzycki, J. M., 1978. An Integrated Digital Mapping System. *The Canadian Surveyor*, Vol. 32, No. 4, pp. 443-452.

CHAPTER XIV

Rectification

Author-Editor: RANDLE W. OLSEN

Contributing Author: ROBERT E. ALTENHOFEN

14.1 Introduction

THIS CHAPTER discusses the theory of recti- fication and its application to photogrammetric practices. *Rectification* is the projection of a tilted photograph into one which is tilt-free and of a desired scale. It is mathematically equivalent to *projective transformation,* and much of the following discussion of rectification will be in terms of the equivalent projective transformation.

An initial section describes the general relationship of projective geometry (the theory of projective transformations) to rectification and photogrammetry. It also introduces the concept of the *cross-ratio,* which is at the basis of quantitative projective transformation. It is followed by a section on graphical methods of rectification, demonstrating the rudiments of the process and serving to introduce the analytical treatment. Analysis provides an algebraic, computa- tional basis for rectification and serves also as a basis for the discussion of the theory and design of rectifying equipment. A description of some of the currently used instruments is followed by an outline of operational procedures which cover both empirical and analytical orientation of rectifiers (orientation of the lens, easel and negative with respect to each other).

The rectification process using conventional equipment can be applied to the general problem of projective transformation. While full treatment of this subject is beyond the scope of this chapter, one example demonstrates the application of the multiple-stage technique to accomplish the affine transformation of a map requiring differential magnification. The multiple-stage theory is also applied to the rectification of high-oblique photographs.

14.2 Applications of Projective Geometry

Projective geometry is an important fundamental in photogrammetry. It is that branch of mathematics which deals with the geometric characteristics of projectively related figures, that is, figures related in such a way that straight lines drawn through corresponding points on the figures all intersect in one and the same point (*see* figure 14-1). An aerial photograph and the ground it images at the moment of exposure are projectively related. Principles of projective geometry are illustrated by projective figures arranged into their perspective position (14.2.2). This chapter applies these principles to develop the relationship of a single photograph to the flat map or to the undulating ground.

14.2.1 PROJECTIVE RELATIONSHIP BETWEEN PHOTOGRAPH AND MAP

Characteristics of projective geometry are derived from the relationship between a photograph and its two-dimensional projection onto the plane of a map. Figure 14-1 represents this relationship and serves as the basis for development of projective relationships.

Photo-plane quadrilateral *abcd* is projected through lens-point O into quadrilateral *ABCD.* The lens elevation h is measured along the plumb line *ON.* The photo principal distance f is along the photograph perpendicular Op, p being the principal point. The lines *ON* and *Op* determine the principal plane whose traces np and NP are principal lines in the photo- and map-planes respectively. The tilt t of the photo-plane is $<pOn$, or $<pSP$ between photo- and map-planes at their intersection, the axis of perspective. Isocenters i and I on the photo- and map-planes, respectively, are on the line through O parallel to the map plane in the principal plane, which bisects $<pOn$. A horizontal plane through O intersects the photo plane at the photo vanishing line or the true horizon. The projection of this vanishing line on the map plane is at infinity (i.e., the intersection of parallel planes). The photo vanishing point V is at the intersection of the photo vanishing line and the principal plane, and is equidistant from both the lens point O and the isocenter i. A plane parallel to the photo plane through O intersects the map plane at the map vanishing line (the

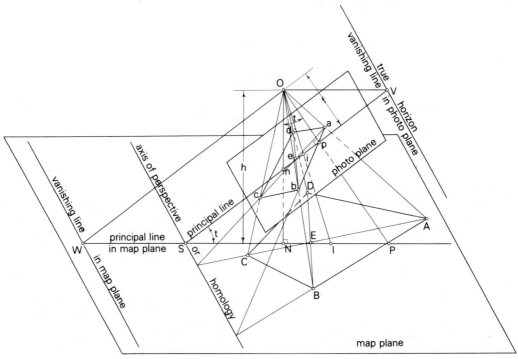

FIGURE 14-1. Projective relationship between photograph and map.

line projected from a line at infinity on the photo plane). The map vanishing point W is at the intersection of the map vanishing line and the principal plane, and is equidistant from both the lens point O and the isocenter I.

14.2.2 PLANE PERSPECTIVE GEOMETRY

Figures in plane perspective fulfill the conditions: (a) corresponding points are collinear with the center of perspective and (b) corresponding lines intersect on the axis of perspective.

Characteristics of plane perspective geometry are analyzed by rotation of the photo plane into the map plane about the axis of perspective (figure 14-1). Such a rotation places the corresponding quadrilaterals in the plane perspective (figure 14-2). Some unique relationships result which are important in future developments. The isocenters i and I coincide at the center of perspective. The parallelogram $OVSW$ of figure 14-1 is flattened into the principal line $WSIV$ thereby establishing the axis of perspective and the vanishing lines. Because opposite sides of the parallelogram are equal, $SV = IW$.

14.2.3 CROSS RATIO

The anharmonic- or cross-ratio of projective geometry is useful in its application to photogrammetry. The cross ratio of any four points p_1, p_2, p_3, p_4 along a line is, by definition,

$$\lambda \equiv (p_1 p_2 p_3 p_4) = \frac{p_1 p_3}{p_2 p_3} \bigg/ \frac{p_1 p_4}{p_2 p_4} . \quad (14.1)$$

The theory of the cross ratio states that the cross ratio of four distinct points on a line is invariant under projection. This theory is verified by showing the equality of cross ratios for four points on a line in the photo plane with the cross ratios for their corresponding points on the map plane. From figure 14-3, the cross ratio $(abcd)$ in the photo plane is

$$(abcd) = \frac{ac}{bc} : \frac{ad}{bd}$$

$$= \frac{ac}{bc} \cdot \frac{bd}{ad}$$

$$= \frac{\tfrac{1}{2} Op \, ac}{\tfrac{1}{2} Op \, bc} \cdot \frac{\tfrac{1}{2} Op \, bd}{\tfrac{1}{2} Op \, ad}$$

$$= \frac{\text{area } \Delta aOc}{\text{area } \Delta bOc} \cdot \frac{\text{area } \Delta bOd}{\text{area } \Delta aOd}$$

$$= \frac{\tfrac{1}{2} \, aO \cdot cO \cdot \sin < aOc}{\tfrac{1}{2} \, bO \cdot cO \cdot \sin < bOc} \cdot$$

$$\frac{\tfrac{1}{2} \, bO \cdot dO \cdot \sin < bOd}{\tfrac{1}{2} \, aO \cdot dO \cdot \sin < aOd}$$

$$= \frac{\sin < aOc}{\sin < bOc} \cdot \frac{\sin < bOd}{\sin < aOd} . \quad (14.2)$$

By a similar development, the cross ratio $(ABCD)$ in the map plane is

$$(ABCD) = \frac{AC}{BC} : \frac{AD}{BD}$$

$$= \frac{\sin < AOC}{\sin < BOC} \cdot \frac{\sin < BOD}{\sin < AOD} . \quad (14.3)$$

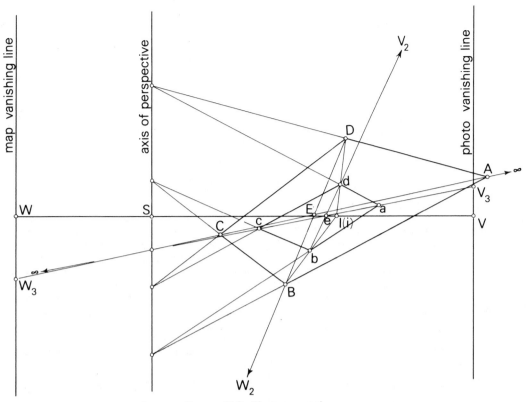

FIGURE 14-2. Plane perspective.

By identity of angles, it is easily shown that the cross ratios (*abcd*) and (*ABCD*) are equal. Thus the cross ratio is a function of only the angles between projected lines and is invariant for the same four points along any transversal across the projecting pencil of lines.

The Fundamental Theorem (Sommerville 1951) of projective geometry follows from the invariance of the cross-ratio under projection. It states that a unique projectivity is established by two projectively related three-point *ranges*, $O \cdot abc$ and $O \cdot ABC$ in figure 14-3. Assume the triplets *a*, *b*, *c* and *A*, *B*, *C* are known. If, in addition, a point *d* on the photo plane collinear with *a*, *b*, *c* is known, the cross ratio (*abcd*) can be determined. Since (*abcd*) = (*ABCD*), then

the corresponding position of *D* can be determined.

14.2.4 PROJECTIVE RELATIONSHIP BETWEEN PHOTOGRAPH AND GROUND

Principles of projective geometry are applicable to the exterior orientation of a photograph to the ground. In this case, the true projective relationship at the moment of exposure is accurately represented. The effect of relief prohibits

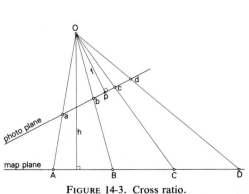

FIGURE 14-3. Cross ratio.

FIGURE 14-4. Projective relationship between a photograph and ground.

a projective relationship between the photograph and a plane figure such as a map. However, the principle of collinearity (section 2.9.2) of the camera lens (the center of perspective), a point on the photo, and its corresponding ground position is maintained. Figure 14-4 illustrates the collinearity condition and the effect of relief.

The center of perspective O represents the camera lens at an elevation H above a datum plane (*e.g.* the local horizontal plane). Points

a, b, c are images on the photo with corresponding points A, B, C on the ground. Z_A, Z_B, Z_C are corresponding elevations above the datum plane. Points A', B', C' are the orthographic positions of the ground points on the datum plane. A map may represent figures on this datum plane. Points a', b', c' are the projection of the datum positions on the photo plane and show the positions of a, b, c without relief displacement. *See* section 2.7 for a more complete analysis of the effect of relief.

14.3 Graphical Rectification

14.3.1 PAPER-STRIP METHOD

Figure 14-5 illustrates the paper-strip method of making a point-by-point graphical rectification based on the invariance of the cross ratio. $ABCD$ represents the map quadrilateral whose photo-image is $abcd$. It is required to locate the transformed position on the map of the point g on the photograph.

The pencil of rays $a \cdot bgcd$ having vertex a in the photograph is the projection of the map-plane pencil formed by the rays AB, AC, AD, and the unknown ray AG. Because one pencil is the projection of the other, they have equal cross ratios and are said to be equianharmonic. Therefore, the paper strip, laid as a transversal across the photo pencil and marked with intercepts b_1, g_1, c_1, d_1, is a graphical representation of the cross ratio. The strip is oriented over the map pencil with points b_1, c_1, d_1 falling respectively on rays AB, AC, AD. The ray Ag_1 fixes the direction of ray AG through the point sought. A similar procedure taking photo point b as the vertex will establish the direction of ray BG, which will intersect AG at the map position of G.

A modification of the paper-strip method that permits map detail to be revised from an oblique photograph is based on the projectivity of straight lines. In figure 14-6 A, B, C, D are the map positions of the photo points a, b, c, d.

Lines ab and cd are extended to intersect at j and lines ad and bc are extended to intersect at k. Corresponding map points J and K are the intersections of the corresponding line pairs AB, CD and AD, BC. Diagonals ac, bd and AC, BD intersect at corresponding points e and E in the photo and map planes. Thus, the corresponding lines je, ke, and JE, KE are determined in the photo and map planes. A continuation of this subdividing process will produce a network dense enough to permit the transfer of detail from photograph to map by reference to corresponding lines.

14.3.2 GRAPHICAL DETERMINATION OF TILT

If the map points form a parallelogram $ABCD$, as in figure 14-7, the orientation elements of the photograph can be determined graphically. In the plane of the photo, opposite sides ab and dc are extended to intersect at V_1; ad and bc intersect at V_2. Line V_1V_2 is the horizon or vanishing line, corresponding to the line at infinity in map plane. The principal line pV_o passes through the principal point p and is perpendicular to V_1V_2. If the principal plane is rotated about pV_o into the plane of the photo, the center of perspective O assumes a position on the perpendicular to pV_o at a distance $Op = f$, the principal distance. If the vanishing point V_o is

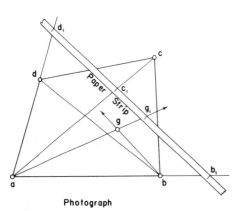

Photograph   Map

FIGURE 14-5. Paper-strip method.

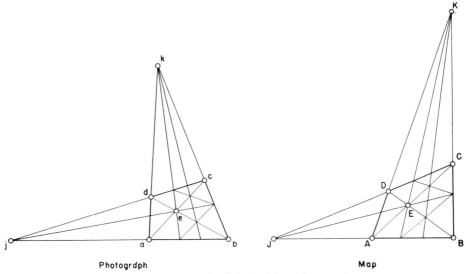

FIGURE 14-6. Transfer of detail from oblique photograph to map.

within construction distance, the nadir point n is fixed by drawing a line through O perpendicular to OV_o. From triangle pOV_o,

$$\tan t = \frac{f}{pV_o}. \qquad (14.4)$$

The distance from nadir to principal point is derived from triangle nOV_o,

$$\frac{np}{f} = \frac{f}{pV_o} \text{ or } np = \frac{f^2}{pV_o}. \qquad (14.5)$$

Two conditions for which this construction must be modified are diagrammed in figure 14-8: (1) the quadrilateral $ABCD$ on the map is not a parallelogram, and (2) the vanishing point V_o falls off the paper. In the map plane, draw parallelogram $ABCD'$. By the paper-strip method, locate point d' in the photo plane corresponding to D'. Vanishing points V_1 and V_2 will be the

intersection of line pairs $d'a$, cb and $d'c$, ab, respectively. For small tilts, V_1 and V_2 will fall beyond construction limits. The line $V_1'V_2'$, parallel to the horizon, is constructed by first drawing diagonal $d'b$. Draw through k, on $d'b$, a line parallel to ab to intersect $d'c$ extended at V_2'. Similarly, draw kV_1' parallel to cb. $V_1'V_2'$ is parallel to horizon V_1V_2.

The principal line nV_o' on the photo is drawn through p perpendicular to $V_1'V_2'$; it intersects the quadrilateral sides cd and ab at points e and g, respectively. Using the paper-strip method, locate P and either E or G in the map plane, to fix the principal line EPG. With O as center (figure 14-8) trace the pencil of rays Oe, Op, Og. Place this pencil on corresponding points E, P, G in the principal plane (considered as rotated about the principal line into the map plane). This resection fixes the center of perspective O' with

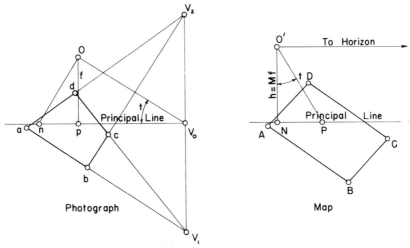

FIGURE 14-7. Graphical determination of tilt.

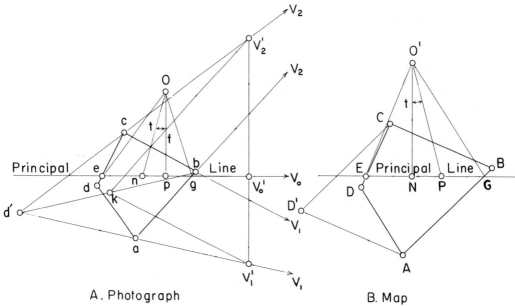

A. Photograph   B. Map

FIGURE 14-8. Modified graphical determination of tilt.

respect to the line EG. A perpendicular through O' to EG locates the map position of the nadir point N. From triangle $NO'P$,

$$\tan t = \frac{NP}{O'N} .$$ (14.6)

The distance $O'N$ represents the altitude h of the exposure station at the map scale. If M is the

magnification needed to bring the rectified photograph to map scale,

$$M = \frac{h}{f} .$$ (14.7)

The photo position of the nadir point n is located from p at the distance

$$np = f \tan t.$$ (14.8)

14.4 Analytical Rectification

14.4.1 ONE-DIMENSIONAL PROJECTIVE TRANSFORMATION

Principles of projective geometry are used to relate coordinate positions of points along a line to their corresponding projected positions along another line. From figure 14-2, three collinear points a, e, c on the photo plane are projected into points A, E, C on the map plane. One-dimensional coordinates are x and X on photo line ac and map line AC, respectively; positive direction is from a to c (and A to C). If the points a, e, c are numbered consecutively, and a fourth point is assumed, the equality of the photo and map cross ratios can be expressed in terms of the coordinates. From equation 14.1,

$$\lambda = \frac{x_3 - x_1}{x_3 - x_2} : \frac{x_4 - x_1}{x_4 - x_2} = \frac{X_3 - X_1}{X_3 - X_2} : \frac{X_4 - X_1}{X_4 - X_2},$$

or, if solved for the fourth map point,

$$X_4 = \frac{a_1'x_4 + d_1'}{a_4'x_4 + d_4'},$$ (14.9)

where coefficients a_1', a_4', d_1', d_4' are functions of the known coordinates of three points. The points must be distinct. Equation 14.9 can be

used for any unknown point-pair x, X and simplified by dividing numerator and denominator by d_4'. The result is the linear fractional transformation

$$X = \frac{a_1 x + d_1}{a_4 x + 1} .$$ (14.10)

Coordinates from two projective point-ranges of three points each (e.g. $o \cdot aec$ and $O \cdot AEC$) can be substituted by pairs into equation 14.10 and the three equations solved for a_1, a_4, d_1. By substitution of these coefficients into equation 14.10, map coordinates of any point along the line ac can be computed, given photo coordinates (and vice versa). In addition, the map horizon point W_3 (figure 14-2) is found by the transformation of an infinite photo-coordinate and the photograph vanishing point V_3 is found by the transformation of an infinite map-coordinate.

Projective transformations can be used for a practical method of tilt analysis with two-dimensional figures. In figure 14-2, another point on the vanishing line can be determined by transformations along corresponding diagonals bed and BED. the perpendicular from the photo principal point to the vanishing line establishes the direction of the principal line and the vanishing point V. From figure 14-1, the distance pV and a known focal length f determine the tilt of the photograph

$$t = \text{Tan}^{-1} (f/pV). \qquad (14.11)$$

14.4.2 TWO-DIMENSIONAL PROJECTIVE TRANSFORMATION

14.4.2.1 DERIVATION OF FRACTIONAL TRANSFORMATION

Positions of points on a plane can be related to their corresponding projected positions on another plane. The derivation of this relationship (von Gruber, 1930) combines projective geometry with two-dimensional systems of co-ordinates for photo- and map-planes. Modifications to the original derivation are made to provide right-handed coordinate systems and universal orientation parameters.

Figure 14-9 illustrates the geometry of the principal plane. Photo- and map-coordinate-axes x', y' and X', Y' respectively are oriented so that the origins are at the isocenters and the ordinates y' and Y' are along the principal lines toward the vanishing point (upward direction of tilt). The abscissas x' and X' are normal to the plane of the diagram. Point a and its projected position A is a point selected at random, not necessarily in the principal plane, with coordinates x_a', y_a' and X_A', Y_A' respectively. The basis for the development of projective relationships is the similarity of triangles aVO and OWA. In the principal plane,

$$\frac{WA}{WO} = \frac{VO}{Va}$$

or, by substitution

$$\frac{Y_A' + h'}{h'} = \frac{f'}{f' - y_a'} .$$

Thus, for points a, A,

$$Y_A' = \frac{h'y_a'}{f' - y_a'} ,$$

and for any point with ordinates y', Y',

$$Y' = \frac{h'y'}{f' - y'} . \qquad (14.12)$$

Normal to the principal plane,

$$\frac{X_a'}{x_a'} = \frac{OA}{aO} = \frac{WO}{Va} .$$

By a development similar to that of equation 14.12,

$$X' = \frac{h'x'}{f' - y'} . \qquad (14.13)$$

Figure 14-10 indicates the relationship of principal plane coordinate systems x', y' and X', Y' with arbitrary coordinate systems x, y and X, Y. Elements x_t', y_t', X_T, Y_T are translations between coordinate system origins and angles s and α are rotations swing and azimuth respectively (section 2.3). This relationship can be expressed analytically. For the photo coordinates,

$$\begin{aligned} x' &= x_t' - x \cos s + y \sin s \\ y' &= y_t' - x \sin s - y \cos s . \end{aligned} \qquad (14.14)$$

For the map coordinates,

$$\begin{aligned} X &= X_T + X'\cos \alpha + Y'\sin \alpha \\ Y &= Y_T - X'\sin \alpha + Y'\cos \alpha. \end{aligned} \qquad (14.15)$$

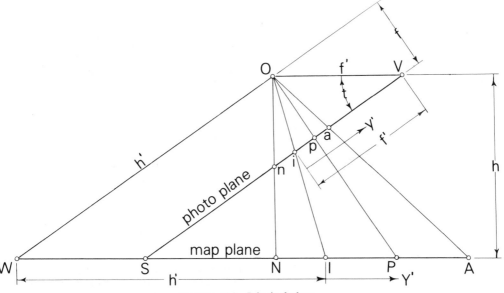

FIGURE 14-9. Principal plane.

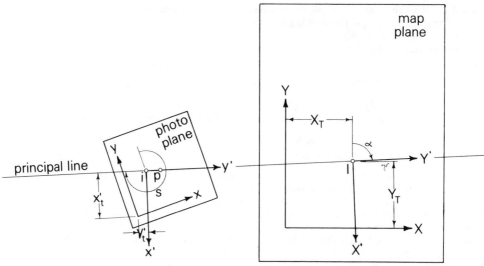

FIGURE 14-10. Photo and map coordinate systems.

Substitution of equations 14.12, 14.13, and 14.14 into equations 14.15 yields the linear fractional transformation

$$X = \frac{a_1 x + b_1 y + d_1}{a_4 x + b_4 y + 1}$$

$$Y = \frac{a_2 x + b_2 y + d_2}{a_4 x + b_4 y + 1},$$

(14.16)

where $a_1 = \dfrac{-h' \cos(s - \alpha) + X_T \sin s}{f' - y_t'}$

$a_2 = \dfrac{-h' \sin(s - \alpha) + Y_T \sin s}{f' - y_t'}$

$b_1 = \dfrac{h' \sin(s - \alpha) + X_T \cos s}{f' - y_t'}$

$b_2 = \dfrac{-h' \cos(s - \alpha) + Y_T \cos s}{f' - y_t'}$

(14.17)

$d_1 = \dfrac{h' (x_t' \cos \alpha + y_t' \sin \alpha)}{f' - y_t'} + X$

$d_2 = \dfrac{-h' (x_t' \sin \alpha - y_t' \cos \alpha)}{f' - y_t'} +$

$a_4 = \dfrac{\sin s}{f' - y_t'}$

$b_4 = \dfrac{\cos s}{f' - y_t'}$.

This projective transformation equation relates the coordinates of points on a photograph to their corresponding map coordinates.

14.4.2.2 COMPUTATION OF
TRANSFORMATION COEFFICIENTS

For each point, equation 14.16 in matrix notation is

$$A\,x = \ell,$$

(14.18)

where

$$A = \begin{bmatrix} -x & 0 & -y & 0 & -1 & 0 & xX & yX \\ 0 & -x & 0 & -y & 0 & -1 & xY & yY \end{bmatrix}$$

$$x = \begin{bmatrix} a_1 \\ a_2 \\ b_1 \\ b_2 \\ d_1 \\ d_2 \\ a_4 \\ b_4 \end{bmatrix} \qquad \ell = \begin{bmatrix} -X \\ -Y \end{bmatrix}.$$

There are eight unknowns in x, requiring four non-collinear points for a unique solution. Additional points can be used for an overdetermined solution. For J points, the least-squares solution for the unknowns is

$$x = \left[\sum_{j=1}^{J} (A^T A)_j \right]^{-1} \cdot \left[\sum_{j=1}^{J} (A^T \ell)_j \right].$$

(14.19)

These parameters may be used directly in equation 14.16 to transform any point from the photo coordinate system to its corresponding map coordinate.

14.4.2.3 INVERSE TRANSFORMATION

By algebraic manipulation, equation 14.16 can be written to solve for photo coordinates as a function of map coordinates. This inverse projective transformation is

$$x = \frac{(b_2 - b_4 d_2)X + (-b_1 + b_4 d_1)Y + (b_1 d_2 - b_2 d_1)}{(a_2 b_4 - a_4 b_2)X + (a_4 b_1 - a_1 b_4)Y + (a_1 b_2 - a_2 b_1)}$$

(14.20)

$$y = \frac{(-a_2 + a_4 d_2)X + (a_1 - a_4 d_1)Y + (a_2 d_1 - a_1 d_2)}{(a_2 b_4 - a_4 b_2)X + (a_4 b_1 - a_1 b_4)Y + (a_1 b_2 - a_2 b_1)}$$

This transformation requires the eight constants to be pre-determined from a forward solution such as equation 14.16. Its use is especially convenient for the determination of photo coordinates of grid and projection intersections for which map coordinates are known.

14.4.2.4 COMPUTATION OF ORIENTATION ELEMENTS

As shown by equations 14.17, the eight transformation parameters are functions of the eight orientation elements f', h', s, α, x_t', y_t', X_T, and Y_T. By algebraic manipulation of these equations, the orientation elements can be expressed as functions of the constants, as shown in table 14-1. The tilt t of the photograph can be computed if either the focal length or the photo coordinate of the principal point is known. From figure 14-9

$$\cos t = \frac{f' - y_p'}{f'} \tag{14.21}$$

$$\sin t = \frac{f}{f'} \,. \tag{14.22}$$

The flying height is

$$h = h'\sin t = h'\frac{f}{f'} \,. \tag{14.23}$$

14.4.3 THREE-DIMENSIONAL PROJECTIVE TRANSFORMATION

Coordinate positions of points on a photograph are projectively related to their corresponding positions on the ground, regardless of the amount of relief. This relationship is demonstrated with the orientation of a single photograph to the ground using a three-dimensional projective transformation. There are many methods of solving such a transformation. Space-resection, one of the more rigorous and systematic methods, determines the location

and attitude of the aerial camera. These constants can then be used to transform coordinates from photo-to-ground or ground-to-photo. Space-resection is a special application of the collinearity condition described in chapter 2.8.

14.4.3.1 DERIVATION OF EQUATIONS FOR SPACE-RESECTION

The derivation of the space-resection method is presented in the following paragraphs. From a development similar to that in 2.3.1.4.2, the projective relationship of photo coordinates in terms of the corresponding ground coordinates and camera constants is

$$x = \frac{1}{\lambda}\left[m_{11}(X - X^c) + m_{12}(Y - Y^c) + m_{13}(Z - Z^c)\right]$$

$$y = \frac{1}{\lambda}\left[m_{21}(X - X^c) + m_{22}(Y - Y^c) + m_{23}(Z - Z^c)\right]$$

$$-f = \frac{1}{\lambda}\left[m_{31}(X - X^c) + m_{32}(Y - Y^c) + m_{33}(Z - Z^c)\right]$$

$$\tag{14.24}$$

where x, y = photo coordinates of a point based on principal point as origin
f = camera focal length
X, Y, Z = ground coordinates of the point
X^c, Y^c, Z^c = ground coordinates of the camera
m_y = element of rotation matrix M containing direction cosines of the camera attitude
λ = scale factor at the point

$$= \frac{Z - Z^c}{m_{13}x + m_{23}y - m_{33}f} \,.$$

From equation 2.37, the relationships of the elements of the rotation matrix to omega-phi-kappa angles are

$$m_{11} = \cos \phi \cos \kappa$$
$$m_{12} = \cos \omega \sin \kappa + \sin \omega \sin \phi \cos \kappa$$
$$m_{13} = \sin \omega \sin \kappa - \cos \omega \sin \phi \cos \kappa$$

TABLE 14-1. ORIENTATION ELEMENTS OF PROJECTIVE TRANSFORMATION

$$\tan s = \frac{a_4}{b_4}$$

$$f' - y_t' = (a_4{}^2 + b_4{}^2)^{-1/2}$$

$$X_T = \frac{a_4(a_1 - b_2) + b_4(a_2 + b_1)}{a_4{}^2 + b_4{}^2}$$

$$Y_T = \frac{a_4(a_2 + b_1) - b_4(a_1 - b_2)}{a_4{}^2 + b_4{}^2}$$

$$\tan(s - \alpha) = \frac{a_4(a_4b_1 - b_4a_1) + b_4(a_4b_2 - b_4a_2)}{a_4(-a_4b_2 + b_4a_2) + b_4(a_4b_1 - b_4a_1)}$$

$$h' = (f' - y_t')^3.$$

$$\frac{a_4(a_4b_1 - b_4a_1) + b_4(a_4b_2 - b_4a_2)}{\sin(s - \alpha)}$$

$$x_t' = \frac{(f' - y_t')}{h'} \cdot \left[(d_1 - X_T)\cos \alpha - (d_2 - Y_T)\sin \alpha\right]$$

$$y_t' = \frac{(f' - y_t')}{h'} \cdot \left[(d_1 - X_T)\sin \alpha + (d_2 - Y_T)\cos \alpha\right]$$

$m_{21} = -\cos\phi\sin\kappa$
$m_{22} = \cos\omega\cos\kappa - \sin\omega\sin\phi\sin\kappa$
$m_{23} = \sin\omega\cos\kappa + \cos\omega\sin\phi\sin\kappa$
$m_{31} = \sin\phi$
$m_{32} = -\sin\omega\cos\phi$
$m_{33} = \cos\omega\cos\phi.$

Equation 14.24, in matrix notation, is

$$\begin{bmatrix} x \\ y \\ -f \end{bmatrix} = \frac{1}{\lambda}\,\mathbf{M}\,\mathbf{x}, \qquad (14.25)$$

where

$$\mathbf{M} = \begin{bmatrix} \mathbf{m}_1 \\ \mathbf{m}_2 \\ \mathbf{m}_3 \end{bmatrix} = \begin{bmatrix} m_{11} & m_{12} & m_{13} \\ m_{21} & m_{22} & m_{23} \\ m_{31} & m_{32} & m_{33} \end{bmatrix}$$

$$\mathbf{x} = \begin{bmatrix} X - X^c \\ Y - Y^c \\ Z - Z^c \end{bmatrix}$$

Rows 1 and 2 of equation 14.25 can be divided by row 3 to yield the collinear equations

$$\left. \begin{aligned} \frac{x}{-f} &= \frac{\mathbf{m}_1\,\mathbf{x}}{\mathbf{m}_3\,\mathbf{x}} \\ \frac{y}{-f} &= \frac{\mathbf{m}_2\,\mathbf{x}}{\mathbf{m}_3\,\mathbf{x}} \end{aligned} \right\} \qquad (14.26)$$

or, in determinant notation,

$$\begin{vmatrix} x & -f \\ \mathbf{m}_1\mathbf{x} & \mathbf{m}_3\mathbf{x} \end{vmatrix} = 0, \begin{vmatrix} y & -f \\ \mathbf{m}_2\mathbf{x} & \mathbf{m}_3\mathbf{x} \end{vmatrix} = 0 \cdot \quad (14.27)$$

New symbols F and G are introduced as functions of equations 14.27, resulting in

$$\begin{vmatrix} x & -f \\ \hat{\mathbf{m}}_1\hat{\mathbf{x}} & \hat{\mathbf{m}}_3\hat{\mathbf{x}} \end{vmatrix} = F = x\,\hat{\mathbf{m}}_3\hat{\mathbf{x}} + f\,\hat{\mathbf{m}}_1\hat{\mathbf{x}}$$

$$\begin{vmatrix} y & -f \\ \hat{\mathbf{m}}_2\hat{\mathbf{x}} & \hat{\mathbf{m}}_3\hat{\mathbf{x}} \end{vmatrix} = G = y\,\hat{\mathbf{m}}_3\hat{\mathbf{x}} + f\,\hat{\mathbf{m}}_2\hat{\mathbf{x}}. \qquad (14.28)$$

where the circumflex ($\hat{}$) indicates that approximate values are used. If all the quantities are properly evaluated, and if x and y are exact quantities, then F and G will be zero. These equations are non-linear and are difficult to solve in this form. To facilitate solution, a Taylor Series expansion of equations 14.28 provides the linear equations (see section 2.2.1)

$$-dF + \frac{\partial F}{\partial x}\,dx + \frac{\partial F}{\partial f}\,df + \frac{\partial F}{\partial X}\,dX + \frac{\partial F}{\partial Y}\,dY$$

$$+ \frac{\partial F}{\partial Z}\,dZ + \frac{\partial F}{\partial X^c}\,dX^c + \frac{\partial F}{\partial Y^c}\,dY^c + \frac{\partial F}{\partial Z^c}\,dZ^c$$

$$+ \frac{\partial F}{\partial \omega}\,d\omega + \frac{\partial F}{\partial \phi}\,d\phi + \frac{\partial F}{\partial \kappa}\,d\kappa = 0$$

and

$$-dG + \frac{\partial G}{\partial y}\,dy + \frac{\partial G}{\partial f}\,df + \frac{\partial G}{\partial X}\,dX + \ldots$$

$$+ \frac{\partial G}{\partial \kappa}\,d\kappa = 0.$$

Solution of these equations requires initial approximations for the six unknown constants of exterior orientation (X^c, Y^c, Z^c, ω, ϕ, κ).

Corrections (dX^c, dY^c, dZ^c, $d\omega$, $d\phi$, $d\kappa$) to the approximations are sought while minimizing a function of the errors (dx, dy) of the photo-coordinates. The terms dF and dG represent a function of the numerical errors of F and G due to the using the approximate values. The remaining differentials (df, dX, dY, dZ) are undesired corrections to fixed quantities and are set to zero. To simplify the notation, the preceding equations are rewritten to form observation equations for a point

$$v_x = l_1 + b_{11}\Delta\omega + b_{12}\Delta\phi + b_{13}\Delta\kappa - b_{14}\Delta X^c$$
$$\quad - b_{15}\Delta Y^c - b_{16}\Delta Z^c$$
$$\qquad\qquad\qquad\qquad\qquad\qquad\qquad (14.29)$$
$$v_y = l_2 + b_{21}\Delta\omega + b_{22}\Delta\phi + b_{23}\Delta\kappa - b_{24}\Delta X^c$$
$$\quad - b_{25}\Delta Y^c - b_{26}\Delta Z^c.$$

The symbol $\mathbf{v}$ is the customary notation for the residuals (i.e., those of the photo coordinates). The Δ-terms are the unknown corrections to the approximations. The coefficients b and the term ℓ can be computed by substituting the fixed and approximated quantities into appropriate formulas, shown in table 14-2. The remaining partial derivatives of the angular elements in the table are

$$\left. \begin{aligned} \frac{d\mathbf{M}}{d\omega} &= \begin{bmatrix} O & -m_{13} & m_{12} \\ O & -m_{23} & m_{22} \\ O & -m_{33} & m_{32} \end{bmatrix} \\[8pt] \frac{d\mathbf{M}}{d\phi} &= \\[4pt] \begin{bmatrix} -\sin\phi\cos\kappa & \sin\omega\cos\phi\cos\kappa & -\cos\omega\cos\phi\cos\kappa \\ \sin\phi\sin\kappa & -\sin\omega\cos\phi\sin\kappa & \cos\omega\cos\phi\sin\kappa \\ \cos\phi & \sin\omega\sin\phi & -\cos\omega\sin\phi \end{bmatrix} \\[8pt] \frac{d\mathbf{M}}{d\kappa} &= \begin{bmatrix} m_{21} & m_{22} & m_{23} \\ -m_{11} & -m_{12} & -m_{13} \\ O & O & O \end{bmatrix}. \end{aligned} \right\} \quad 14.30)$$

As an example for computation,

$$\frac{\partial \mathbf{m}_3\mathbf{x}}{\partial \omega} = \begin{bmatrix} O & -m_{33} & m_{32} \end{bmatrix} \begin{bmatrix} X - X^c \\ Y - Y^c \\ Z - Z^c \end{bmatrix}$$

$$= -m_{33}(Y - Y^c) + m_{32}(Z - Z^c).$$

14.4.3.2 SOLUTION OF EQUATIONS FOR SPACE-RESECTION

The observation equations 14.29 can be written in matrix notation as

$$\ell = \mathbf{B}\,\boldsymbol{\delta} - \mathbf{v} \qquad (14.31)$$

where

$$\mathbf{v} = \begin{bmatrix} v_x \\ v_y \end{bmatrix} \qquad \mathbf{B} = \begin{bmatrix} b_{11} & b_{12} & b_{13} & -b_{14} & -b_{15} & -b_{16} \\ b_{21} & b_{22} & b_{23} & -b_{24} & -b_{25} & -b_{26} \end{bmatrix}$$

$$\ell = \begin{bmatrix} -l_1 \\ -l_2 \end{bmatrix} \qquad \boldsymbol{\delta} = \begin{bmatrix} \Delta\omega \\ \Delta\phi \\ \Delta\kappa \\ \Delta X^c \\ \Delta Y^c \\ \Delta Z^c \end{bmatrix}$$

TABLE 14-2. COEFFICIENTS OF OBSERVATION EQUATIONS FOR RESECTION

$$l_1 = \frac{1}{m_3 x} \begin{vmatrix} x & -f \\ m_1 x & m_3 x \end{vmatrix} \qquad\qquad l_2 = \frac{1}{m_3 x} \begin{vmatrix} y & -f \\ m_2 x & m_3 x \end{vmatrix}$$

$$b_{11} = \frac{1}{m_3 x} \begin{vmatrix} x & -f \\ \dfrac{\partial(m_1 x)}{\partial\omega} & \dfrac{\partial(m_3 x)}{\partial\omega} \end{vmatrix} \qquad b_{21} = \frac{1}{m_3 x} \begin{vmatrix} y & -f \\ \dfrac{\partial(m_2 x)}{\partial\omega} & \dfrac{\partial(m_3 x)}{\partial\omega} \end{vmatrix}$$

$$b_{12} = \frac{1}{m_3 x} \begin{vmatrix} x & -f \\ \dfrac{\partial(m_1 x)}{\partial\phi} & \dfrac{\partial(m_3 x)}{\partial\phi} \end{vmatrix} \qquad b_{22} = \frac{1}{m_3 x} \begin{vmatrix} y & -f \\ \dfrac{\partial(m_2 x)}{\partial\phi} & \dfrac{\partial(m_3 x)}{\partial\phi} \end{vmatrix}$$

$$b_{13} = \frac{1}{m_3 x} \begin{vmatrix} x & -f \\ \dfrac{\partial(m_1 x)}{\partial\kappa} & \dfrac{\partial(m_3 x)}{\partial\kappa} \end{vmatrix} \qquad b_{23} = \frac{1}{m_3 x} \begin{vmatrix} y & -f \\ \dfrac{\partial(m_2 x)}{\partial\kappa} & \dfrac{\partial(m_3 x)}{\partial\kappa} \end{vmatrix}$$

$$b_{14} = \frac{1}{m_3 x} \begin{vmatrix} x & -f \\ m_{11} & m_{31} \end{vmatrix} \qquad\qquad b_{24} = \frac{1}{m_3 x} \begin{vmatrix} y & -f \\ m_{21} & m_{31} \end{vmatrix}$$

$$b_{15} = \frac{1}{m_3 x} \begin{vmatrix} x & -f \\ m_{12} & m_{32} \end{vmatrix} \qquad\qquad b_{25} = \frac{1}{m_3 x} \begin{vmatrix} y & -f \\ m_{22} & m_{32} \end{vmatrix}$$

$$b_{16} = \frac{1}{m_3 x} \begin{vmatrix} x & -f \\ m_{13} & m_{33} \end{vmatrix} \qquad\qquad b_{26} = \frac{1}{m_3 x} \begin{vmatrix} y & -f \\ m_{23} & m_{33} \end{vmatrix}$$

There are six unknowns in δ, requiring three non-collinear points for a unique solution. In order to reduce the effect of errors in the observations, it is recommended to utilize many more points. The principle of least-squares (chapter II) is used to minimize the sum of the squares of these residuals. For J points, the least-squares solution for the corrections to the approximated constants is

$$\delta = \left[\sum_{j=1}^{J} (\mathbf{B}^T\mathbf{B})_j \right]^{-1} \cdot \left[\sum_{j=1}^{J} (\mathbf{B}^T\ell)_j \right]. \qquad (14.32)$$

After solving for the corrections, the original approximations are corrected and the process is repeated for further corrections. Calculations are stopped when corrections become negligible, usually after three iterations.

After solution, the constants of exterior orientation (X^c, Y^c, Z^c, ω, ϕ, κ) are determined. Additional photo coordinates can be derived using these constants and known ground coordinates (X, Y, Z) using equation 14.24. Additional ground coordinates can be derived using these constants, known photo coordinates (x, y), and a known elevation (Z) using equation 2.27. Tilt and swing elements of camera orientation are related to the elements of the rotation matrix in equation 2.177a.

14.5 Optical-Mechanical Rectification

Rectification is the projective transformation of a tilted photograph into one which is tilt-free and of a desired scale. This may be done graphically, as described in 14.3; analytically, as described in 14.4; or optically and mechanically, as described in the present section, with a device called a rectifier.

14.5.1 THE SCHEIMPFLUG CONDITION

For proper rectification by optical-mechanical means, the rectifier must produce sharp imagery. This is attained by application of the lens law of optics

$$\frac{1}{A} + \frac{1}{B} = \frac{1}{F}, \qquad (14.33)$$

where F = focal length of lens
A = distance from object to incident node
B = distance from emergent node to image.

This law can also be expressed by Newton's form

$$X X' = F^2, \qquad (14.34)$$

where X and X' are distances to the object and image measured from the two foci, respectively.

The arrangement of object- and image-planes with respect to the lens plane is shown in Figure 14-11, as required by either form of the lens law. Point O represents a lens having focal points at D and D' in object and image spaces, respectively. The lens axis pierces the object plane at point e whose conjugate axial point E is at the distance determined from the lens law. The object-, image-, and lens-planes intersect along a line at S. This intersection of the three planes is called the Scheimpflug condition.

The Scheimpflug condition can be demonstrated graphically by systematic development of figure 14-11, given the object- and lens-

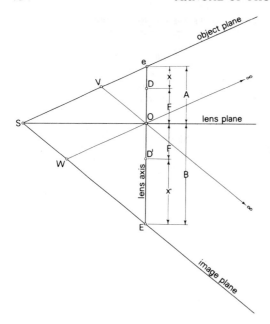

FIGURE 14-11. Scheimpflug condition.

cal conditions previously established. Strike an arc with V as center and the rectifier-lens focal length F as radius. Through the axis of perspective S, establish the plane of the rectifier-lens tangent to this arc. Strike another arc with V as center and VO as radius intersecting the lens plane at O', the rectifier lens point. Establish the easel plane of the rectifier by a line through S parallel to VO'. The resulting arrangement, with O' as the center of projection, will yield a transformation of an aerial negative identical to the map configuration that was projected into the tilted negative-plane by the camera lens at O.

It follows from the geometry of an aerial negative that the circle having V as center and passing through the camera lens O will intersect the negative-plane at the isocenter i. As this circle also contains the rectifier lens O', i is also the isocenter of the rectifier. Therefore, proper placement of the aerial negative in the rectifier requires that isocenters of negative and rectifier coincide.

planes and focal length. The object-plane vanishing point V is at focal length F from the lens plane to image at infinity. Furthermore, the image at infinity lies in a plane parallel to VO. The image-plane vanishing point W, also a distance F from the lens plane, is along a plane projected through the lens from an infinite object. Thus SV is parallel to WO and, by similar triangles, their lengths are equal. Because a) the image plane is parallel to VO and contains W, and b) $SV = WO$, the image plane must contain S, establishing intersection of the three planes at S.

14.5.2 GEOMETRY OF THE RECTIFIER

For proper rectification, both the Scheimpflug condition and the projective relationship between the aerial photograph and the ground (or its map) must be fulfilled in the rectifier. Figure 14-12 is a diagram of an aerial negative exposed at map-scale altitude h in a camera having a principal distance f. This diagram represents the principal plane section of an exposure having tilt t. Negative plane points p (principal point), i (isocenter), and n (nadir point) correspond respectively to map plane points P, I, and N. Lines through the camera lens point O parallel to the map- and negative-planes determine the vanishing line (horizon) trace V in the negative plane and the vanishing line trace W in the map plane.

The geometric arrangement of the rectifier necessary to project the negative-plane into the map plane may be superimposed on figure 14-12 by a construction based on projective and opti-

14.5.3 ORIENTATION ELEMENTS

The formulas for computing the orientation elements of rectification applicable to the non-tilting lens type of rectifier are derived from figure 14-12. In this type of rectifier, the lens is constrained to move in the direction of its fixed axis. The respective tilts of the negative- and easel-planes from the lens plane are α and β. The distances along the rectifier-lens axis from negative to lens and from lens to easel are A and B, respectively. Magnification M is the ratio of flying height to focal length (reciprocal of scale). From figure 14-12,

$$\sin \beta = \frac{F}{\overline{VO'}}$$

$$\sin \alpha = \frac{F}{\overline{VS}}$$

$$\overline{VO'} = \overline{VO} = f \csc t$$
$$\overline{VS} = \overline{WO} = h \csc t.$$

Therefore, by substitution,

$$\sin \beta = \frac{F}{f} \sin t \qquad (14.35)$$

$$\sin \alpha = \frac{F}{h} \sin t . \qquad (14.36)$$

From equation 14.35, it follows that the easel tilt β is independent of the magnification M. Dividing equation 14.35 by equation 14.36,

$$\frac{\sin \beta}{\sin \alpha} = \frac{h}{f} = M$$

$$(14.37)$$

or $\quad \sin \alpha = \dfrac{\sin \beta}{M} .$

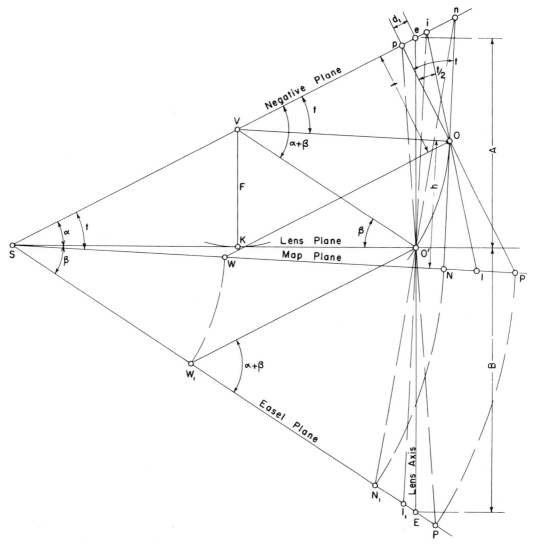

FIGURE 14-12. Geometry of a rectifier.

Therefore, for small angles of tilt, when $\sin \alpha$ is approximately equal to α,

$$\alpha = \frac{\beta}{M} . \qquad (14.38)$$

Equation 14.38 may be applied with negligible effect on the accuracy of rectification if neither α nor β exceed 15°. Applying a like simplification to equation 14.35, for values of $\beta < 15°$,

$$\beta = \frac{F}{f} t. \qquad (14.39)$$

The negative-to-lens distance $A = \overline{eO'}$ is derived by applying the law of sines to triangle eVO':

$$A = \overline{VO'} \cdot \frac{\sin eVO'}{\sin VeO'}$$

where
$$\overline{VO'} = F \csc \beta;$$
$$<eVO' = \alpha + \beta;$$
$$<VeO' = 90° - \alpha.$$

By substitution,

$$A = \frac{F \sin (\alpha + \beta)}{\cos\alpha \cdot \sin\beta} . \qquad (14.40)$$

The lens-to-easel distance $B = \overline{O'E}$ is derived by applying the law of sines to triangle $W_1O'E$:

$$B = \overline{W_1O'} \cdot \frac{\sin O'W_1E}{\sin W_1EO'} ,$$

where
$$\overline{W_1O'} = VS = F \csc \alpha;$$
$$<O'W_1E = \alpha + \beta;$$
$$<W_1EO' = 90° - \beta.$$

By substitution,

$$B = \frac{F \sin (\alpha + \beta)}{\sin\alpha \cos\beta} . \qquad (14.41)$$

The magnification M' along the rectifier lens axis is

$$M' = \frac{B}{A} \div \frac{\tan \beta}{\tan \alpha}. \qquad (14.42)$$

By substitution with equation 14.37,

$$M' = \frac{\cos \alpha}{\cos \beta} M. \qquad (14.43)$$

The separation of the rectifier-lens nodal points does not change the derivation of equations 14.40 and 14.41. If the design of the rectifier requires that the negative-to-easel distance be set, this separation must be added to the sum of A and B.

Because the rectifier-lens axis is fixed and pierces the negative-plane at e, the negative must be displaced to fulfill the requirement of coincidence between rectifier- and negative-isocenters. In figure 14-12 this displacement d_1 is considered negative if downward toward the vanishing point V, and positive if upward. In figure 14-12,

$$d_1 = \overline{ep} = \overline{Vp} - \overline{Ve}.$$

From triangle VpO,

$$\overline{Vp} = f \cot t.$$

From triangle VeO', by the law of sines,

$$\overline{Ve} = \overline{VO'} \frac{\sin VO'e}{\sin VeO'},$$

where
$$\overline{VO'} = \overline{VO} = f \csc t = F \csc \beta;$$
$$<VO'e = 90° - \beta;$$
$$<VeO' = 90° - \alpha.$$

By substitution,

$$d_1 = \frac{f}{\sin t}\left(\cos t - \frac{\cos \beta}{\cos \alpha}\right). \qquad (14.44)$$

This displacement can also be expressed as a function only of rectifier orientation elements. From figure 14-12

$$Ve = \sqrt{(A - F)^2 + (KO')^2},$$

where
$$A = F + \frac{F}{M},$$

$$\overline{KO'} = F \cot \beta.$$

thus
$$Ve = F\sqrt{\frac{1}{M^2} + \cot^2 \beta}.$$

From triangles VpO and VKO',

$$\overline{Vp} = \sqrt{(\overline{VO})^2 - f^2}$$

where
$$(\overline{VO})^2 = (\overline{VO'})^2 = F^2 + F^2 \cdot \cot^2 \beta.$$

By substitution,

$$d_1 = F\left[\sqrt{1 - \frac{f^2}{F^2} + \cot^2 \beta} - \sqrt{\frac{1}{M^2} + \cot^2 \beta}\right]. \qquad (14.45)$$

From development of equation 14.45 in a series, an approximation is

$$d_1 = -\frac{F}{2} \tan \beta \left(\frac{f^2}{F^2} + \frac{1}{M^2} - 1\right). \qquad (14.46)$$

Equations 14.35 to 14.46 impose certain limitations on the rectification process. Equations 14.35 and 14.36 show that the angles α and β are real angles if $F \leq f \csc t$ and $F \geq h \csc t$. Rectification would be impossible should the sine of either α or β prove to be imaginary (greater than 1). Figure 14-12 demonstrates this impossibility. If the circle described about V as center has a radius $F > \overline{VO}$, the line through S tangent to this circle could not intersect the circle through O; hence, there would be no real location for the rectifier lens O'. Also, if the circle about V has a radius $F > \overline{SV}$, the point S would fall within the circle and could not lie on a line tangent to it.

The mechanical and optical limitations of a rectifier generally limit α and β to 45°. If computation yields a value of these angles greater than 45°, it is necessary to use a rectifier lens of shorter focal length or to perform the rectification in several stages. The shorter the focal length of the rectifier lens, the more compact the apparatus and the smaller the tilt angles for a given range of magnification. Also, the tilt range of a rectifier is greater for enlargement than for reduction.

The computed value of d_1 may be greater than that recoverable in the rectifier. This may be due to the intolerable decentering of illumination, exceeding the mechanical range of the negative carrier displacement, or to displacement of some of the negative imagery beyond the field of projection.

Table 14-3 lists the orientation elements of rectification applicable to non-tilting lens rectifiers with focal lengths 150 mm and 180 mm for a camera of focal length 152.4 mm (6 inches). This tabulation verifies the statements concerning the limitations imposed on optical-mechanical rectification. Rectifier settings are more favorable for enlargement than for reduction.

14.5.4 MECHANICAL CHARACTERISTICS

14.5.4.1 NON-AUTOMATIC RECTIFIERS

Non-automatic rectifiers require the computation of elements of rectification, each of which must be set on its corresponding circle or scale. The simplest and least accurate type of non-automatic rectifier may be an enlarger modified to incorporate some or all of the degrees of freedom required by rectification theory. More accurate construction is based on optical-bench design, with provision for setting all elements precisely.

14.5.4.2 AUTOMATIC DETERMINATION OF OBJECT DISTANCE

Automatic rectifiers employ mechanisms to insure automatic fulfillment of the lens law of optics. These devices, called inversors, operate on the theory of inversion, a type of geometric

TABLE 14-3. ORIENTATION ELEMENTS OF NON-TILTING LENS RECTIFIERS

t	M'	α	β	A mm	B mm	d mm	α	β	A mm	B mm	d mm
			F = 150 mm; f = 152.4 mm					F = 180 mm; f = 152.4 mm			
0°	0.5	0° 0'	0° 0'	450.0	225.0	0.0	0° 0'	0° 0'	540.0	270.0	0.0
	1.0	0° 0'	0° 0'	300.0	300.0	0.0	0° 0'	0° 0'	360.0	360.0	0.0
	2.0	0° 0'	0° 0'	225.0	450.0	0.0	0° 0'	0° 0'	270.0	540.0	0.0
	3.0	0° 0'	0° 0'	200.0	600.0	0.0	0° 0'	0° 0'	240.0	720.0	0.0
	4.0	0° 0'	0° 0'	187.5	750.0	0.0	0° 0'	0° 0'	225.0	900.0	0.0
	5.0	0° 0'	0° 0'	180.0	900.0	0.0	0° 0'	0° 0'	216.0	1080.0	0.0
	6.0	0° 0'	0° 0'	175.0	1050.0	0.0	0° 0'	0° 0'	210.0	1260.0	0.0
5°	0.5	9°53'	4°55'	453.4	224.2	−26.4	11°53'	5°55'	545.9	268.5	−35.4
	1.0	4°55'	4°55'	300.0	300.0	−6.7	5°55'	5°55'	360.0	360.0	−6.7
	2.0	2°27'	4°55'	224.8	450.8	−1.8	2°57'	5°55'	269.6	541.4	0.3
	3.0	1°38'	4°55'	199.8	601.5	−0.9	1°58'	5°55'	239.7	722.6	1.6
	4.0	1°14'	4°55'	187.4	752.1	−0.6	1°28'	5°55'	224.8	903.6	2.1
	5.0	0°59'	4°55'	179.9	902.7	−0.5	1°11'	5°55'	215.8	1084.6	2.3
	6.0	0°49'	4°55'	174.9	1053.2	−0.4	0°59'	5°55'	209.8	1265.6	2.4
10°	0.5	19°59'	9°50'	464.5	221.5	−55.8	24°13'	11°50'	566.3	263.9	−77.6
	1.0	9°50'	9°50'	300.0	300.0	−13.3	11°50'	11°50'	360.0	360.0	−13.3
	2.0	4°54'	9°50'	224.2	453.4	−3.6	5°53'	11°50'	268.6	545.9	0.8
	3.0	3°16'	9°50'	199.3	606.0	−1.8	3°55'	11°50'	238.9	730.4	3.3
	4.0	2°27'	9°50'	187.0	758.4	−1.2	2°56'	11°50'	224.1	914.7	4.2
	5.0	1°58'	9°50'	179.6	910.8	−0.9	2°21'	11°50'	215.3	1098.8	4.6
	6.0	1°38'	9°50'	174.6	1063.1	−0.8	1°58'	11°50'	209.4	1282.8	4.8
15°	0.5	30°38'	14°46'	487.1	216.7	−93.0	37°41'	17°48'	613.2	254.8	−139.7
	1.0	14°46'	14°46'	300.0	300.0	−20.1	17°48'	17°48'	360.0	360.0	−20.1
	2.0	7°19'	14°46'	223.1	457.7	−5.3	8°48'	17°48'	266.7	553.7	1.5
	3.0	4°52'	14°46'	198.5	613.7	−2.7	5°51'	17°48'	237.4	744.2	5.2
	4.0	3°39'	14°46'	186.3	769.2	−1.8	4°23'	17°48'	223.0	934.0	6.5
	5.0	2°55'	14°46'	179.0	924.6	−1.4	3°30'	17°48'	214.3	1123.5	7.1
	6.0	2°26'	14°46'	174.2	1079.9	−1.2	2°55'	17°48'	208.6	1312.8	7.4

transformation. Inversors of this type determine the proper object distance given a magnification or image distance. Thus the operator can concentrate on geometric properties of the projected image which is automatically in focus throughout the full magnification range of the rectifier.

14.5.4.2.1 CAM INVERSOR

The cam inversor is a cam device which provides proper fulfillment of the lens law (equation 14.33). As shown in figure 14-13 its components are an immovable curved track and a lever arm L which rides along the track at R. The lever arm is attached to the carriage of the rectifier lens and pivots at P. For any given image distance or magnification, the lens carriage is positioned at a height B above the easel. As the lens carriage moves, the cam lever rotates as it rides along the track. This rotation is converted to linear motion by a linkage to the lever arm such as a cable attached to a wheel rotating with the lever (as illustrated), rack and pinion gears, or an elbow on the lever. The linear motion in turn controls the height of the negative carriage above the lens (object distance A).

14.5.4.2.2 PEAUCELLIER INVERSOR

The Peaucellier or scissors inversor (figure 14-14) provides another mechanical solution of

the Newtonian form of the lens law as in equation 14.34. This linkage consists of a rhombus $RJTL$, with sides of length U hinged at vertices T and R situated on the lens axis. A distance equal to the focal length F separates the easel and the negative from the vertices T and R. For the point

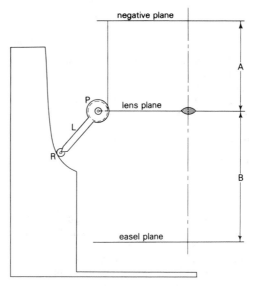

FIGURE 14-13. Cam Inversor

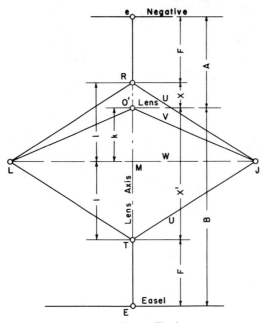

FIGURE 14-14. Peaucellier inversor.

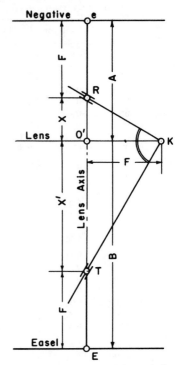

FIGURE 14-15. Pythagorean right-angle inversor.

e to image sharply at E, the lens O' must be located between R and T so that the distances X' and X satisfy equation 14.34. This location requires that the lens be carried by the apex of an isosceles triangle hinged to the rhombus vertices L and J. In this triangle $V = \overline{LO'} = \overline{O'J}$. The linkage will position the lens properly if its dimensions satisfy the relationship

$$U^2 - V^2 = F^2. \tag{14.47}$$

From right triangles in Figure 14-14

$$l^2 = U^2 - W^2 \text{ and } k^2 = V^2 - W^2.$$

Subtracting,

$$l^2 - k^2 = U^2 - V^2,$$

and factoring,

$$(l - k)(l + k) = XX' = F^2 = U^2 - V^2. \tag{14.48}$$

The Peaucellier inversor is actuated by keeping the point T fixed while driving the lens. Vertex R will assume a position such that $XX' = F^2$. When the lens occupies the midpoint position M, $X = X' = F$ and the negative to easel distance $eE = 4F$, neglecting nodal separation. This position represents unit magnification.

14.5.4.2.3 PYTHAGOREAN RIGHT-ANGLE INVERSOR

The Pythagorean right-angle inversor provides the simplest mechanical solution of the lens law. This device is diagrammed in figure 14-15. In the rectifier, the inversor is situated on one or both sides so that it remains clear of the projection field. A right-angle lever RKT is made to pivot around K, which is fixed in the plane of the lens O' at a distance from the lens axis equal to the focal length F. Sleeved pivots T and R are on

the lens axis. Connection T is fixed at the distance F above the easel, and R is constrained to move along the lens axis, carrying the negative above at the constant distance F. The inversor is made to act by moving the lens O' and with it the pivot K. Because sleeve T is fixed, this movement produces a rotation of the lever RKT about K which causes the distances X' and X to assume values satisfying equation 14.34. Therefore, the axial point e is imaged sharply at E.

14.5.4.3 AUTOMATIC DETERMINATION OF NEGATIVE TILT

Automatic rectifiers also employ inversors to insure automatic fulfillment of the Scheimpflug condition. Inversors of this type determine the tilt of the negative or object plane given the tilt of the easel or image plane and the position of the lens above the easel. Again, the operator can concentrate on geometric properties of the projected image which automatically satisfies optical and projective requirements. Therefore, the rectifier may be oriented by empirical means, with the practical advantage of eliminating the need to compute orientation elements.

14.5.4.3.1 CARPENTIER INVERSOR

The Carpentier inversor is a mechanical device which enforces the Scheimpflug condition. As shown in figure 14-16, tilted negative- and easel-planes intercept the lens plane along a common axis. The lens axis is fixed, and sleeved pivots D and J are constrained to move along guide rails LD and JK, which are located at the

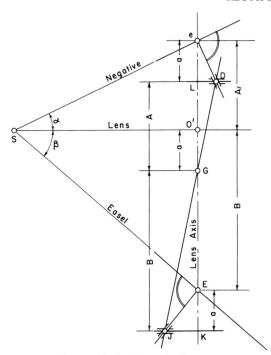

FIGURE 14-16. Carpentier Inversor.

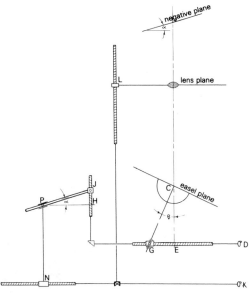

FIGURE 14-17. Simulator Inversor.

distance a below the negative and easel tilt axes e and E, respectively. Lineal space rod JGD pivots around G, which is fixed on the lens axis at the constant distance a below the lens. Tilts of the negative and easel are controlled by the perpendicular arms De and JE, which pass through the sleeved pivots D and J. If the easel is tilted, the points D and J move along their respective guide rails and are maintained collinear with G. It is easily proved by similar triangles that the distance $\overline{O'S}$ has the same value when computed from triangles $SO'e$ and $SO'E$. Therefore, negative-, lens-, and easel-planes intersect in the line through S. This collinearity exists regardless of the distance to S, which recedes to infinity as the planes approach parallelism.

14.5.4.3.2 SIMULATOR INVERSOR

The Simulator inversor is a mechanical device which simulates on a space rod the tilt required of the negative plane to satisfy the Scheimpflug condition. From figure 14-17, tilt crank D actuates easel tilt β by a rotating spindle driving a cross slide at G. Another spindle linked to the same crank drives a cross-slide at J such that $\overline{JH} = \rho\,\overline{GE}$, where ρ is a constant determined by the ratio of the respective pitches of the spindles. Magnification or image distance crank K controls the height of the lens above the easel through a rotating spindle and cross-slide at L. Another spindle linked to the same crank drives a perpendicular arm at N. Its pitch must be such that the ratio of distance moved at N to that at L is also the constant ρ. Perpendicular arm $\underline{PN}$ must be positioned such that $\overline{PH} = \rho\cdot M\cdot \overline{CE}$,

where M is the magnification. A space rod pivots around the cross-slide at J and slides through a sleeved pivot at P. It is tilted from horizontal by an angle α, where

$$\tan \alpha = \frac{\overline{JH}}{\overline{PH}} = \frac{\rho\cdot\overline{GE}}{\rho\cdot M\,\overline{CE}} = \frac{\tan\beta}{M}\,.$$

From identity with equation 14.37, where α is the tilt of the negative plane, the tilt of the space rod simulates the tilt required of the negative plane.

The tilt may be introduced into the negative carrier by direct mechanical linkage or by indirect linkage, such as servo-motors actuated by photo-electric cells sensitive to changes in the tilt of the space rod.

14.5.4.4 AUTOMATIC NEGATIVE PLANE DISPLACEMENT

Some automatic rectifiers have the capability of automatic fulfillment of negative plane displacement required for coincidence of rectifier and negative isocenters. Isocenter coincidence also insures coincidence of the rectifier and negative vanishing points (V on figure 14-12). Thus automatic negative plane displacement may also be called automatic vanishing point control. Since the required displacement is a somewhat complicated function of rectifier orientation elements (equation 14.46), it is difficult to provide automatic fulfillment with a mechanical inversor. However, through a small built-in electronic calculator, the variables f, F, M, and β can be introduced and transformed into the displacement as specified by equation 14.46. Variable resistors are used to transfer the computed distance to motors driving the displacement of the negative plane.

14.6 Rectification Instrumentation

14.6.1 BAUSCH & LOMB

Figure 14-18 illustrates the Bausch & Lomb autofocusing rectifier equipped to control dodging and exposure electronically. This design achieves compactness by keeping the plane of the negative in a horizontal position at a fixed distance above the base. Magnification is controlled by movement of lens and easel along the lens axis, which may assume any inclination from vertical to the maximum value of α recoverable in the instrument. The lens swings through the principal plane pendulumlike about the center e of the zero position of the negative-carrier, as diagrammed in figure 14-19. The easel tilts in one direction (the principal plane) and the non-tilting negative-plane swings about the lens axis such that the negative and rectifier principal planes coincide. Critical focus is maintained by the combination Peaucellier-Carpentier inversor (1, figure 14-18). It is evident from figure 14-19 that the complete assembly, including lens, inversor, and easel, swings with the lens axis. Geometrically, therefore, the Bausch & Lomb rectifier is based on a non-tilting or fixed-axis design.

FIGURE 14-18. Bausch & Lomb rectifier. (Photograph courtesy of Bausch & Lomb, Inc.)

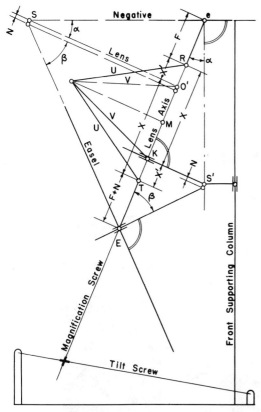

FIGURE 14-19. Diagram of Bausch & Lomb rectifier.

Manual controls are provided for negative-swing and x and y displacements. Power drive is provided for changing the magnification and tilting the lens axis. The Peaucellier component of the combination inversor diagrammed in figure 14-19 is actuated by a nut at K, which is displaced by the magnification screw. Lens O' is moved by a hollow screw (2, figure 14-18),

which partakes of the easel motion while sliding along a keyed shaft (3) geared to the magnification screw (4). This arrangement displaces the lens relative to the easel so that O' and K are maintained equidistant from M, the midpoint between negative and easel. The Carpentier component of the inversor establishes collinearity of planes of easel, lens, and negative by maintaining the perpendiculars $S'E$, $S'K$, and $S'e$ concurrent at S' (figure 14-19). The design of this inversor takes into account the nodal separation N. The five elements of orientation can be read from appropriate scales. The rectifying characteristics of the instrument are indicated in table 14-4.

14.6.2 H. DELL FOSTER RSS 200

The H. Dell Foster RSS 200, illustrated in figure 14-20, is an automatic non-tilting-lens type rectifier equipped with an on-line NOVA 1200 computer. The main components are a lens which moves vertically along its axis, an easel which tilts universally about x and y axes, and a negative-carrier which moves vertically and tilts about x and y axes. Specifications of the RSS 200 are listed in table 14-4.

Magnification and x and y camera tilt are entered into the computer manually from the console or automatically from magnetic tape. The lens is moved vertically above the easel by a ball nut attached to a leadscrew driven by a stepping motor. The tilt data are converted to x and y easel tilts by the computer using equation 14.39 with known camera and rectifier focal lengths. The easel is tilted by a perpendicular arm connected to a sliding ball and cylinder joint. The joint is translated along x and y axes by leadscrews driven by stepping motors. The computer instructs the stepping motors driving the lens and easel tilt leadscrews by comparing the desired positions with the actual positions

TABLE 14-4. RECTIFIER CHARACTERISTICS

Characteristics	Bausch & Lomb	H. Dell Foster RSS 200	K&E Kargl	Wild E-4	Zeiss SEG-5
Lens	Metrogon	Reprogon	Dagor	Reprogon	Topogon
Focal Length	5.5 in. (14 cm)	5.9 in. (15 cm)	6 in. (15 cm)	15 cm.	18 cm.
Lens Speed	f/6.3	f/5.6 to f/11	f/6.8 to f/32	f/8	f/6.8 to f/36
Maximum Format	9 × 9 in. (23 × 23 cm)	9.5 × 9.5 in. (24 × 24 cm)	9 × 9 in. (23 × 23 cm)	23 × 23 cm.	23 × 23 cm.
Easel Size	36 × 36 in. (91 × 91 cm)	42 × 50 in. (107 × 127 cm)	40 × 40 in. (102 × 102 cm)	106 × 106 cm.	100 × 100 cm.
Easel x-tilt range	0	±10°	0	±13.5°	±12.6°
Easel y-tilt range	−10° to +33°	±10°	0 to +22°	±13.5°	±12.6°
Swing range	360°	0°	360°	0°	0°
Magnification range	0.6 to 3.5	1.0 to 7.0	0.5 to 5.1	0.8 to 7.0	0.5 to 6.5
x-displacement range	± 90 mm.	0	0	±40 mm.	±45 mm.
y-displacement range	± 90 mm.	0	±2.1 in. (53 mm)	±40 mm.	±45 mm.

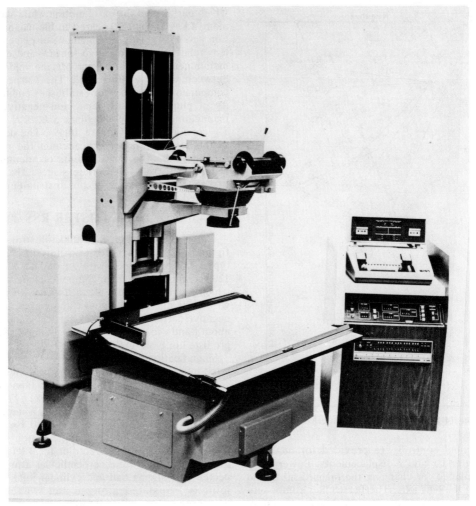

FIGURE 14-20. H. Dell Foster RSS 200 rectifier. (Photograph courtesy of H. Dell Foster Company)

indicated by encoders attached to the respective leadscrews.

The negative-carrier is automatically driven vertically above the lens by a stepping motor actuated by a cam inversor in accordance with the lens law. It is tilted about x and y axes mechanically with a modified Carpentier inversor, satisfying the Scheimpflug condition.

There is no provision in the RSS 200 for negative-plane displacement. For most tilted configurations, therefore, there is lack of coincidence of rectifier and negative isocenters. The resulting error in geometry of the projected image is negligible from a practical point of view. However, use of the RSS 200 for some special applications of rectification may not be possible.

14.6.3 K&E KARGL

The K&E Kargl rectifier, illustrated in figure 14-21, is a non-tilting lens type automatic rectifier which, like the Bausch & Lomb, has a fixed axis of tilt. The lens moves vertically along

its axis and the easel tilts about another fixed axis. The negative-carrier moves vertically along the lens axis, swings for alignment of negative and rectifier principal planes, and tilts about an axis parallel to the easel tilt-axis. The negative also displaces in its plane along the principal line. The specifications of the Kargl rectifier are shown in table 14-4.

The magnification setting utilizes a motor for coarse motion and a hand-wheel for fine motion. The values are set with a vernier on a precise scale. This setting effects proper positioning of both lens and negative carrier along the lens axis. A Peaucellier inversor (1) behind the rectifier provides mechanical coupling necessary for satisfaction of the lens law (section 14.33).

Easel tilt is set with a handwheel also on a precise scale. The handwheel rotates two lineals (2) which are linked perpendiculars to the easel and negative carrier. These lineals are the main components of a modified Carpentier inversor which provide proper satisfaction of the Scheimpflug condition. This modification links a

FIGURE 14-21. K&E Kargl rectifier.(Photograph courtesy of Keuffel & Esser Co.)

second Peaucellier (3) to the Carpentier to locate the intersection of the lineals on a perpendicular to the lens axis. As in the case of the Bausch & Lomb design (figure 14-19) this perpendicular must intersect the lens axis at a point K symmetrically equidistant with the lens O' from the midpoint M between negative and easel points e and E. The negative stage can be rotated through a full circle to provide proper swing or collinearity of the principal line with the direction of tilt. A graduated scale reading to the nearest degree of swing is provided.

The displacement of the principal point in the negative plane is done manually using a handwheel and scale. Its motion is limited to displacement along the principal line only.

14.6.4 WILD E-4

The Wild E-4 is an automatic non-tilting lens

type rectifier. Shown in figure 14-22, its main components are a lens which moves vertically along its fixed axis, an easel which tilts universally about x and y axes, and a negative-carrier which moves vertically above the lens, tilts about x and y axes, and displaces in its plane in x and y directions. Specifications of the E-4 are shown in table 14-4.

For introduction of magnification, a spindle is driven electronically or by a footdisk to move the lens carriage above the easel. To satisfy the lens law, the required object-distance is pro-

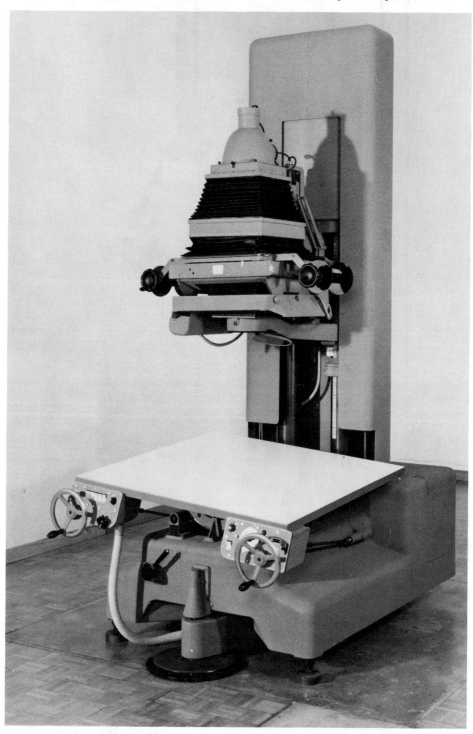

FIGURE 14-22. Wild E-4 rectifier. (Photograph courtesy of Wild Heerbrugg Ltd.)

vided by a cam inversor. The components of the inversor are a fixed, curved track and an elbow lever pivoted to the lens carriage. The lever rotates while riding along the inversor curve, causing its end to displace vertically with respect to the lens. An electronic circuit sensitive to this displacement actuates a servo-motor which drives a leadscrew effecting an equal vertical displacement to the negative carrier above the lens.

Easel tilt is introduced with spindles driving a sleeved cross-slide containing an arm perpendicular to the easel. The units of easel tilt are 1000 tan β, with negative values added to 1000. These units are proportional to the number of spindle rotations; thus the counters are driven by appropriate gearing from the spindles. The negative-plane is tilted in accordance with the Scheimpflug condition by a Simulator type of inversor. Figure 14-17 illustrates this inversor as used in the E-4. The tilt of the space rod PJ is transmitted to the negative-plane by servomotors actuated by photoelectric cells sensitive to changes in the tilt of the spacerod.

Displacement of the negative-plane is introduced manually in x and y directions with handwheels. The handwheels actuate servomotors which drive the displacement. The values of displacement must be determined analytically with equations 14.44 or 14.46. Nomograms are supplied with the E-4 to simplify this determination.

14.6.5 ZEISS SEG-5

The Zeiss SEG-5 rectifier, illustrated in figure 14-23, is an automatic rectifier with a fixed, vertical lens-axis. Easel, lens, and negative-carrier are arranged vertically one above the other. The lens and negative-carrier can be displaced verti-

cally and easel and negative-carrier can be tilted universally about x and y axes. The negative can be displaced on its carrier in x and y directions. Specifications of the SEG-5 are shown on table 14-4.

The mechanical characteristics of the SEG-5 are shown in figure 14-24. For setting the magnification, the lens (1) is positioned above the easel (2) by linkage of its carrier to a spindle rotated by a floor disk (3). A dial (4) is also connected to the spindle and is graduated into magnification units. The negative-carrier (5) is positioned above the lens automatically to satisfy the lens law using a cam inversor (6). The inversor curve is fixed to the rectifier column. The inversor lever attached to the lens carrier rides along the curve as the lens moves vertically. The rotation of the lever is fed to a spur rack over a geared segment. The rack actuates a microswitch controlling a motor which drives a chain linked to the negative-carrier.

The easel is tilted on the SEG-5 universally around the ball and socket mount (7) by means of the handwheels (8 and 9) which control its components of ω and ϕ. The values are read in tangential units similar to those of the Wild E-4 on dials (10) linked also the tilt handwheels. The negative-carrier is tilted in accordance with the Scheimpflug condition by a spatial Carpentier inversor (11).

The negative-plane displacement is controlled automatically so as to satisfy equation 14.46 by a small calculator (12). Focal length, easel tilt, and magnification are fed to this calculator which controls, by servomotors, the x and y displacements of the negative to insure coincidence of isocenters (and vanishing points) of the aerial negative and rectifier. This displacement may also be done manually (with the calculator disengaged) for special kinds of rectification.

14.7 Operational Procedures

The operation of a rectifier involves orientation by analytical or empirical methods followed by photographic exposure for adequate image quality. The analytical method provides a rectified image of the photograph by introducing pre-determined orientation elements. Requirements include a properly calibrated rectifier, known absolute orientation of the photograph to be rectified, and computation of the rectifier orientation elements. The empirical method requires a visual fit of points identifiable in the photograph image to a map or control template by systematic manipulation of the rectifier orientation elements. A combination of analytical and empirical methods is often used to optimize accuracy and production efficiency.

14.7.1 ANALYTICAL ORIENTATION OF THE RECTIFIER

14.7.1.1 CALIBRATION AND ADJUSTMENT

Analytical orientation of the rectifier requires periodic calibration and adjustment of the elements of orientation. The most important aspects of calibration include focusing the projected image, satisfying the Scheimpflug condition, and ensuring precision and accuracy of settings of easel tilt and magnification. Proper focus and magnification are achieved by the correct relationship of object and image distances (equation 14.33). This interdependence requires adjustment of focus initially such that resolution of the projected image is optimized over the

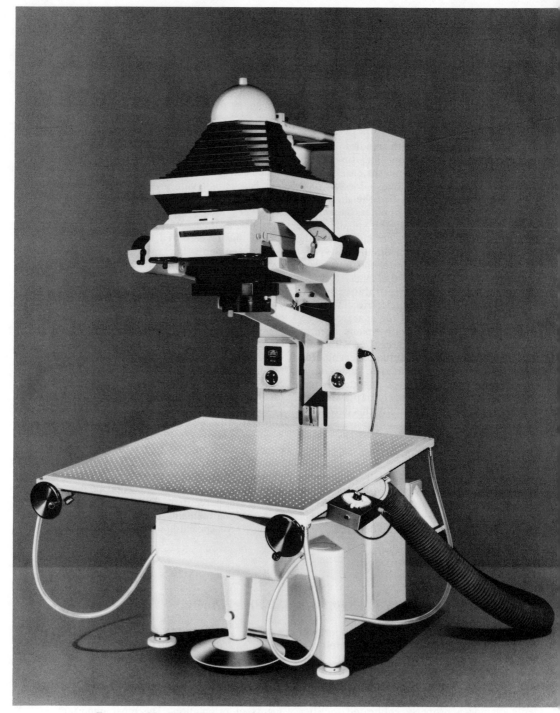

FIGURE 14-23. Zeiss SEG-5 rectifier. (Photograph courtesy of Carl Zeiss, Inc.)

magnification range to be used. For automatic rectifiers, fulfillment of the Scheimpflug condition is dependent on the accuracy of the inversor regulating the relationship of negative- and easel-tilts over the entire tilt range. Conventional inversors are adjusted at zero tilt by leveling the negative- and easel-planes and re-

indexing the tilt indicators to zero. The magnification element is calibrated using a grid plate and templates at various magnifications. On most automatic rectifiers, adjustments of the magnification indicator for index and rate of change can be made for an optimized fit.

Once the rectifier is calibrated mechanically,

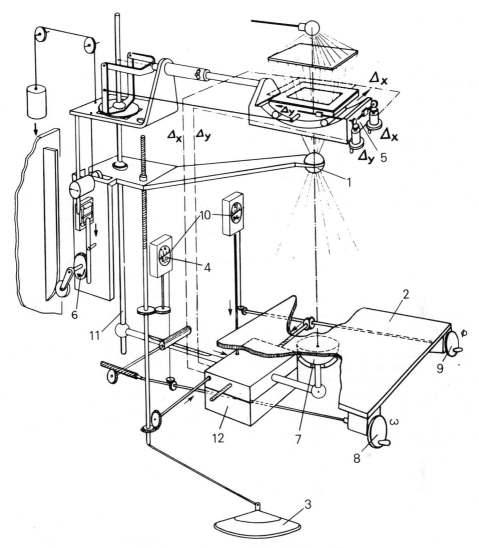

FIGURE 14-24. Diagram of Zeiss SEG-5 rectifier.

the limitations of accuracy and precision can be analyzed. Using projected grids at various magnifications and tilts, errors in accuracy can be determined by a comparison with templates of coordinate positions. Systematic errors consistent for a given magnification or tilt in repeated trials can be plotted to form error curves for the range of either element. These curves can then be used to correct computed orientation elements. Non-systematic errors resulting from precision limitations remain uncorrectable unless orientation is supplemented with empirical procedures. These limitations should be analyzed in determining the feasibility of analytical rectification for a given rectifier and for required accuracy of the rectified product.

14.7.1.2 COMPUTATION OF ORIENTATION ELEMENTS

Analytical rectification (section 14.4) requires initial computation of the rectifier elements of orientation. These elements are computed as functions of flying height, tilt, and swing of the photograph to be rectified. If not readily available, these data may be determinable from a projective transformation computation (section 14.4) or from the output of an aerotriangulation adjustment. The necessary orientation elements depend on the mechanical design and automatic capabilities of the rectifier. Automatic rectifiers, such as those described in section 14.6, are oriented with easel tilt β (equation 14.35), magnification M' (equation 14.43), negative-displacement d_1 (equation 14.46), and direction of tilt.

The direction of tilt orientation enforces coincidence of principal lines of photograph and rectifier. Analytical determination of this angle requires coordination of coordinate systems of photograph and rectifier, and a known swing s of the photograph. Conventionally, x and y axes of both the rectifier and the projection of the negative-plane stage-plate are to the right and back, respectively, on the easel. If the photograph coordinate system does not correspond to the negative-plane coordinate system, the swing must be corrected accordingly. Non-correspondence in intervals of 90 degrees occurs frequently in roll photography when the photograph coordinate system is established without knowledge of orientation on the roll itself. For rectifiers with a fixed axis of tilt, the direction of tilt is introduced by a rotation of the negative carrier numerically equal to the swing of the photograph being rectified. For rectifiers which tilt universally, the x and y components of easel tilt β are:

$$\beta_x = \beta \sin s$$
$$\beta_y = \beta \cos s. \qquad (14.49)$$

14.7.1.3 COORDINATION WITH AEROTRIANGULATION

Analytical rectification can be productively coordinated with an aerotriangulation adjustment of the same photography. Two applications which demonstrate this coordination are direct use of aerotriangulated camera orientation elements or use of a projective transformation to derive these camera elements using aerotriangulation coordinate data.

Camera orientation elements are generally included in the output data of analytical aerotriangulation adjustments using the simultaneous or bundle method (chapter 9.4.5). The elements which are used for rectification include the camera elevation, omega (ω), and phi (ϕ). For tilts less than 10 degrees, the tilt (t) is

$$t = \sqrt{\phi^2 + \omega^2}, \qquad (14.50)$$

where ϕ and ω are approximations of x and y components of tilt, respectively. In any case, from chapter 2.8.7.1.2

$$\cos t = \cos \omega \cos \phi. \qquad (14.51)$$

The elevation of the datum plane used for rectification must be subtracted from the camera elevation to derive the flying height used in analytical rectification.

If camera orientation elements are not available from the aerotriangulation, they are computed by projective transformation. The control data may be either aerotriangulated points, scaled map points, or ground control. The corresponding photo coordinates may be comparator data used as input to analytical aerotriangulation or scaled separately with a measuring device such as a coordinatograph or micro-rule. The projective transformation may be computed in two or three dimensions. Advantages of the two-dimensional method (equation 14.16) include computational simplicity and straightforward data input requiring no initial approximations. Use of the three-dimensional method

(equation 14.24) provides a more rigorous mathematical model for the determination of camera orientation. The computed camera elements, therefore, are generally more accurate than those computed indirectly as in table 14-1. The three-dimensional method also permits the use of control points in areas of large relief. Thus a photograph of a small area of a meandering valley with a known datum can be accurately rectified without necessarily having control points in the area. The two and three-dimensional methods require four and three control points, respectively, for unique solution. Additional points provide geometric coverage and redundancy desirable in a least-squares adjustment for both greater accuracy and error detection.

14.7.2 EMPIRICAL ORIENTATION OF THE RECTIFIER

14.7.2.1 PREPARATION OF MATERIALS

The rectifier is oriented empirically with the use of points or features common to the photograph being rectified and a map or control template at the desired scale. Systematic procedures for empirical orientation are simplified if these points are symmetrically distributed on the photograph and are easily identifiable in the projected image. Three non-collinear control points are required for a unique solution. In practice, at least four points, situated near the photograph corners, are used. The redundancy allows both detection of blunders and even distribution of small errors.

Points are marked on the photograph being rectified if distinct identification of natural features in the projected image is not possible. Commonly used methods of identification include circling of points with non-photographic blue ink, pin-holing the emulsion of the photograph, or drilling a small circle or dot on the emulsion with a point marking device. Artificial features such as aerotriangulation passpoints or fiducial marks are often used in empirical rectification. Corresponding control positions are plotted on a template using aerotriangulation coordinate data. Use of such points avoids the need to select and identify natural features.

The map or control template used in empirical rectification must clearly show its information without obscuring the projected imagery. Ideally, only line or point data which contributes to the orientation should be used. The map or template must be situated flat on the rectifier easel. A glossy, mylar-base material aids viewing of the photograph imagery by reflecting projected light.

14.7.2.2 EFFECTS OF ORIENTATION ELEMENTS

Each of the orientation elements of a rectifier produces a unique deformation of the projected imagery. The five degrees of freedom incorporated in the typical automatic rectifier enable the operator to establish the projective relationship between a pattern of points on the aerial negative and their corresponding positions on the map or control template. Figure 14-25 illustrates how the shape of a trapezoid ABCD is altered by manipulating the controls of a fixed lens-axis rectifier. It is assumed that ABCD is the image on the tilted easel of square abcd which is oriented in the negative plane with two sides parallel to the tilt axis of the rectifier easel. In accordance with projective theory, sides CB and DA pass through the intersection of the map vanishing line W_1 and the principal plane of the rectifier (figure 14-12). In figure 14-25, A'B'C'D' indicates the resultant easel figure after each change in an element of orientation. In diagrams B, C, D, and E, the sides AB and A'B' have been made to coincide by adjusting the magnification and translating the original configuration ABCD as necessary. This procedure reduces the effect of varying a particular element to a displacement of vertices C and D to the positions C' and D'.

As shown in figure 14-25A, an increase in magnification displaces vertices A, B, C, D radially outward from point E at which the lens axis pierces the easel.

For rectifiers with a fixed tilting axis, a swing of the negative about its principal point causes the vertices of easel figure ABCD to trace arcs of ellipses. The lengths of the arcs are approximately proportional to their distances from the lens. Therefore a point on the lower side of the easel is displaced farther than one on the upper side, as shown by the movement of C to C' in figure 14-25B. (A"B"C"D" represents the initial position of ABCD.) After sides AB and A'B' have been made coincident, quadrilateral A'B'C'D' shows that swing has deformed ABCD by lengthening BC, shortening AD, and changing the convergence of opposite sides.

An increase in easel tilt shortens AB and lengthens CD while the width of the trapezoid along the easel tilt-axis remains unchanged. After A'B' has been superimposed on AB by increasing the magnification (figure 14-25C), the resultant trapezoid A'B'C'D' shows an elongation in the direction of tilt together with an increase in the convergence of sides C'B' and D'A'. The change in convergence corresponds to the decrease in the distance from the easel tilt-axis to the map vanishing line caused by increase in the angle between the easel and negative-planes. For rectifiers with a fixed swing of the negative-carrier, the easel tilt deformation applies to both x and y directions of easel tilt.

A downward displacement of the negative parallel to the principal plane of the rectifier causes a corresponding downward movement of the easel image away from the vanishing line (figure 14-25D). On fixed tilt-axis rectifiers, this element is referred to as y-displacement. Lower vertices C and D are displaced more than A and B, which are situated above the easel tilt axis. After the coincidence setting of AB and A'B', the resultant trapezoid A'B'C'D' exhibits an elongation in the direction of tilt, but the convergence of sides B'C' and D'A' remains unaltered. An opposite displacement of the negative would shorten the trapezoid in the direction of tilt.

A displacement of the negative parallel to the axis of tilt (x-displacement on fixed tilt-axis rectifiers) causes shear deformation diagrammed in figure 14-25E. All displacements in the image are parallel to the easel tilt axis and increase in magnification with an increase in the distance from the easel vanishing line.

14.7.2.3 PROCEDURES

By employing the deformations diagrammed in figure 14-25, the rectifier operator can apply a systematic adjustment procedure to effect coincidence between points on the projected imagery and the map or control template. The specific orientation procedures depend on the distribution pattern of control points, rectifier design, and operator preference. Two procedures are outlined for empirical orientation of a photograph with a quadrilateral point pattern on automatic rectifiers of both fixed and universal tilt-axis design.

14.7.2.3.1 FIXED TILT-AXIS RECTIFIERS

Figure 14-26 illustrates the steps in adjusting the easel image A'B'C'D' of the aerial photograph quadrilateral abcd to the control template ABCD. The lens axis pierces the easel at E, which is the intersection of the easel tilt axis with the principal plane of the rectifier.

As shown in figure 14-26A, sides AB and A'B' are first brought into coincidence by adjusting the magnification while the angular elements of orientation are held at their zero positions. Normally, the remaining elements are adjusted in the sequence: easel tilt, swing, x-displacement, and y-displacement; after each adjustment, AB and A'B' are made to coincide by changing the magnification and shifting the template. The easel tilt is adjusted to make the length DC equal to D'C'. The swing is adjusted such that the points requiring the most outward displacement are directed toward the downward direction of tilt. A reversal of easel tilt may be necessary due to mechanical limitations or to operator convenience. This can be done if accompanied with a 180-degree rotation of the negative carrier. The easel tilt and swing adjustments do not need to follow the normal sequence. For example, the operator may note that vertex C' (figure 14-26A) must be displaced more than D'. Therefore, the second adjustment could be a clockwise swing to move C' to what becomes a lower position when the easel is subsequently tilted. The configuration resulting

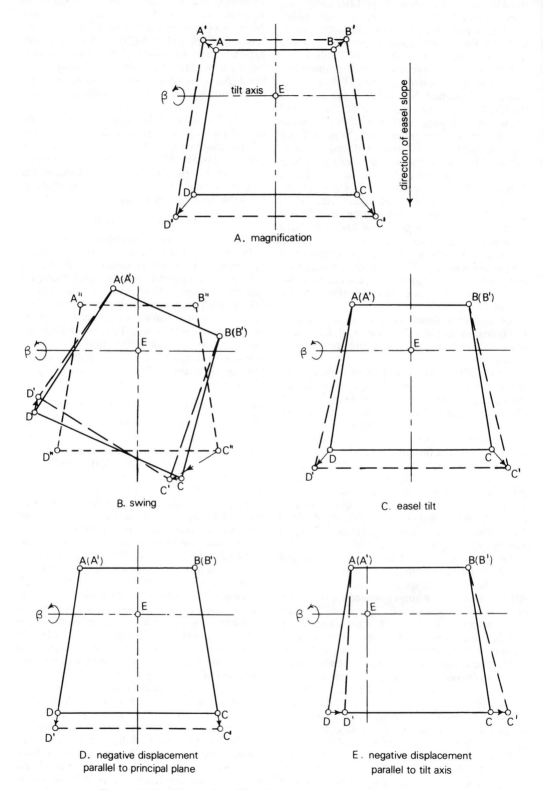

FIGURE 14-25. Effects of rectifier adjustments on shape of easel image.

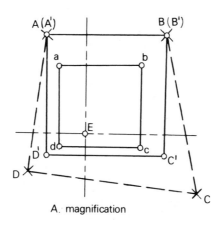

A. magnification

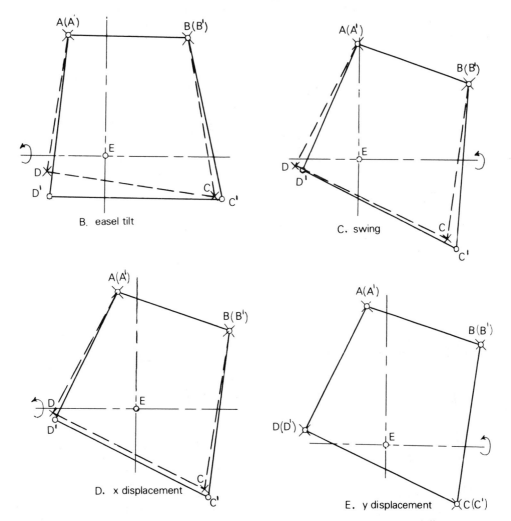

B. easel tilt

C. swing

D. x displacement

E. y displacement

FIGURE 14-26. Empirical orientation of automatic rectifier with fixed axis of tilt.

from the combined swing and easel tilt adjustments is shown in figure 14-26C. Shear deformation and elongation in the direction of tilt are removed by respective x and y displacements of the negative. They may also be interchanged. The x-displacement may be zero if ground points A, B, C, and D are approximately at the same elevation. By comparing the easel image and the template, a skillful operator can easily determine the most advantageous sequence of adjustments.

14.7.2.3.2 UNIVERSAL TILT-AXIS RECTIFIERS

The steps involved in the orientation of a rectifier with a universal tilt-axis are shown in figure 14-27. As shown in figure 14-27A, sides AB and $A'B'$ are first brought into coincidence by adjusting the magnification with the other elements initially at zero. While maintaining coincidence of AB and $A'B'$ with magnification adjustment, side DC is made to coincide in length to $D'C'$ by y-tilt easel adjustment (figure 14-27B). A length is increased by a downward direction of tilt. Length BC is made equal to

length $B'C'$ using x-tilt easel adjustment while maintaining coincidence of AD and $A'D'$ with magnification adjustment (figure 14-27C).

Empirical determination of negative displacement is difficult since x and y displacement directions do not normally correspond to the principal plane and axis of tilt directions. However, most rectifiers of this design incorporate automatic or graphical interpolation techniques for fulfilling the required displacement for conventional rectification. It is assumed, therefore, that the necessary negative displacements are incorporated during the empirical orientation process as magnification and tilt are introduced.

14.7.3 PHOTOGRAPHIC PROCEDURES

The rectification exposure employs photographic procedures similar to those of projection-type photographic enlargers. The method and time interval of exposure is dependent on the following variables:

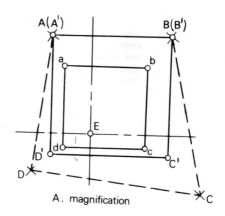

A. magnification

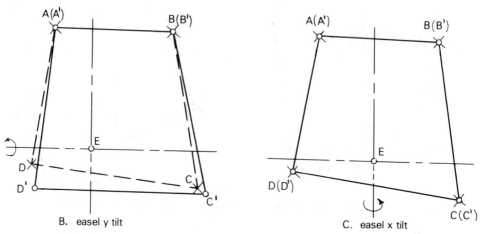

B. easel y tilt   C. easel x tilt

FIGURE 14-27. Empirical orientation of automatic rectifier with universal axis of tilt.

(a) tone and contrast of the negative being rectified,
(b) magnification,
(c) rectifier lens aperture,
(d) type of illumination,
(e) type of photographic emulsion being exposed,
(f) desired tone and contrast of the rectified imagery, and
(g) development procedures.

Uniformity of tone within a photograph or in relation to adjacent photographs may require manual or electronic dodging during exposure.

There are many types of photographic films and papers used in rectification. Negative or reversal films are commonly used when sub-sequent photographic generations are to be made from the rectified imagery. Scale-stable papers are used when dimensional stability is desired on an opaque base. Single weight glossy paper has been traditionally popular in mosaicking projects. This paper may be pre-soaked if the rectifier is equipped with a waterproof easel and if it is not objectionable to double or triple the exposure time. The wet-paper method eliminates differential paper expansion and subsequent adjustments of magnification. The mosaic is assembled from wet prints either taken directly from the rectifier or re-soaked after an intermediate drying period.

14.8 Multiple-Stage Rectification

Multiple-stage rectification is a series of projective transformations of a photograph with tilt exceeding the limitations of conventional rectification. This technique is most commonly applied using a standard rectifier with sufficient degrees of freedom to satisfy the requirements of orientation. Systematic operational procedures are based on projective geometry and the mechanical characteristics of the rectifier. The series of photographic steps involved in multiple-stage rectification usually diminishes the quality of the final print by increasing contrast and graininess and losing resolution. Offsetting this disadvantage is the improvement in exposure characteristics caused by minimizing the incident angles between light rays and easel. Single-stage high-oblique rectifiers may yield inferior prints on glossy surfaces because the angles of incidence exceed the maximum allowable for proper exposure of any paper not having a matte-surface.

14.8.1 ORIENTATION ELEMENTS

The elements of multiple-stage rectification can be determined by applying the principles of projective geometry. The nomenclature used in 14.5.3 is modified and expanded to include the following terms.

m = Magnification of a single stage.
n = Number of stages of the projective transformation, each producing a magnification m.
M = Magnification resulting from n stages ($M = m^n$).
q = Distance on the aerial negative from the isocenter to the true horizon; also called the perspective index of the exposure.
Q = Distance from the isocenter to the vanishing line in the negative plane of the rectifier.
d_i = Displacement of the negative isocenter from the rectifier lens axis.

The multiple-stage process is simplified by treating the final transformations as the product of a series of identical stages. This treatment has practical advantages because the rectifier set-tings, once determined, remain fixed for each succeeding step.

Figure 14-28 represents a rectifier fixed at a magnification $m < 1$. Repeating this projection yields a two-stage rectification with a magnification $M = m^2 < 1$. The perspective index of the aerial negative is $q = \overline{iV}$, the distance from the isocenter of the exposure to its true horizon, whose trace in the principal plane is V. From the geometry of the negative,

$$q = f \csc t.$$

The isocenter ray iO' makes equal angles with the negative and easel planes; therefore, triangle $iS'I_1$ is isosceles. The perspective index of the rectifier is $Q = \overline{iV'}$, the distance from the isocenter to the vanishing-line trace V'. As explained in 14.5.2, the isocenters of aerial negative and rectifier must coincide.

Equations for computing the elements of multiple-stage rectification are derived by applying the principles of projective geometry to figure 14-28. Points in the positive plane are symmetrically located about the lens point O' with respect to their corresponding points in the negative plane. Point ranges $iCVV'$ and $I_1CV_1\infty$ are equianharmonic (14.3.1). Therefore,

$$(iCVV') = (I_1CV_1\infty). \tag{14.52}$$

This equation expresses the fact that the vanishing point V' projects into the point at infinity. In a two-stage procedure, the first stage projects the perspective index q of the aerial negative into Q, the perspective index of the rectifier $Q = \overline{iV'} = \overline{I_1V_1}$. The first stage positive is then positioned in the unchanged rectifier so that I_1 coincides with the rectifier isocenter i and the first stage position of the horizon trace V_1 coincides with the vanishing point V' of the rectifier. The second stage projects the horizon trace into the point at infinity, in accordance with the requirement for correct rectification.

Expanding equation 14.52,

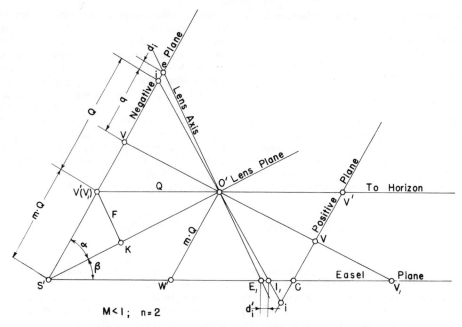

$M < 1; \quad n = 2$

FIGURE 14-28. Development of multiple-stage rectification equations ($M < 1$; $n = 2$).

$$\frac{iV}{CV} : \frac{iV'}{CV'} = \frac{I_1 V_1}{CV_1} : \frac{I_1 \infty}{C \infty}. \tag{14.53}$$

As $I_1\infty/C\infty = 1$, rearrangement of equation 14.53 gives

$$\frac{iV}{CV} = \frac{I_1 V_1}{CV_1} \cdot \frac{iV'}{CV'}. \tag{14.54}$$

From figure 14-28,

$iV = q;\ I_1V_1 - iV' = Q;$
$CV' = CV_1 = mQ;\ iC = (1 - m)Q.$

Substituting in (14.54)

$$\frac{q}{q - (1 - m)Q} = \frac{Q}{mQ} \cdot \frac{Q}{mQ}.$$

Solving,

$$Q = \frac{1 - m^2}{1 - m}q = \frac{1 - M}{1 - m}q = (1 + m)q \tag{14.55}$$

for $n = 2$. For three-stage rectification ($n = 3$) the formula becomes

$$Q = \frac{1 - m^3}{1 - m}q = \frac{1 - M}{1 - m}q = (1 + m + m^2)q.$$

For n stages,

$$Q = \frac{1 - M}{1 - m}q = (1 + m + m^2 + \cdots + m^{n-1})q.$$

When $m = M = 1$,

$$Q = nq. \tag{14.57}$$

That equation 14.56 also holds for enlargement may be proved by solving the following equations, which express the equivalence of anharmonic ratios as applied to the rectification

diagrammed in figure 14-29, where $m > 1$ and $n = 3$:

$$(CiVV') = (CIV_1\infty);\ (CiV_1V') = (CIV_2\infty).$$

V_1 and V_2 are, respectively, the first and second stage positions of the true-horizon trace V. The third-stage position of the true-horizon trace is at infinity.

The elements needed to set the nontilting-lens rectifier for the multiple-stage process may be computed from the value of Q. From figure 14-28,

$$\sin \beta = \frac{F}{Q}. \tag{14.58}$$

Also,

$$\sin \alpha = \frac{F}{mQ} = \frac{\sin \beta}{m}. \tag{14.59}$$

According to the established sign convention, d_i is plus or minus as the displacement of the isocenter i from the lens axis point e is away from or toward the rectifier vanishing-point trace V'.

Therefore,

$$d_i = \overline{ie} = \overline{V'i} - \overline{V'e}.$$

Substituting,

$$d_i = (1 - \frac{\cos \beta}{\cos \alpha})Q. \tag{14.60}$$

Because multiple-stage rectification usually is applied to a high-oblique exposure whose isocenter may be marked on the negative, the value of d_i suffices to fix the negative. The prin-

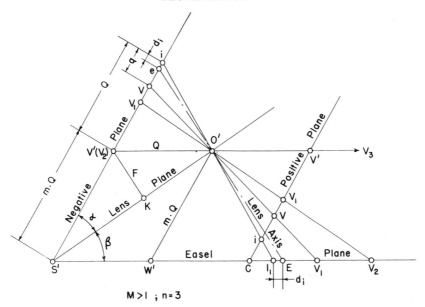

$$M > 1 \; ; \; n = 3$$

FIGURE 14-29. Multiple-stage rectification ($M > 1; n = 3$).

cipal-point displacement d_1 may, if needed, be computed from the formula

$$d_1 = d_i - f \tan \frac{t}{2}.$$

Magnification along the lens axis is computed as in the case of single-stage rectification from equation 14.43:

$$m' = \frac{\cos \alpha}{\cos \beta} m.$$

Applying the lens law,

$$B = (1 + m') F$$

$$A = (1 + \frac{1}{m'}) F.$$

Displacement of the negative may be checked by computing the projected length of d_i. The distance from the projected isocenter I_1 to the lens axis point E_1 in the easel plane is derived from Figure 14-28.

$$d_i' = \overline{W'E_1} - \overline{W'I_1}.$$

Substituting,

$$d_i' = m' Q (1 - \frac{\cos \beta}{\cos \alpha}) \qquad (14.62)$$

or

$$d_i' = m'd_i.$$

14.8.2 OPERATIONAL PROCEDURES

Operational procedures for multiple-stage rectification include determination of the tilt of the photograph, computation of orientation elements within the capabilities of the rectifier, and orientation in two or more stages using these elements.

An analysis of the photograph using analytical or graphical techniques provides a determination of the tilt necessary for further computations and for plotting of the principal line, isocenter, and isometric parallel. On high-oblique photography, the image of the apparent horizon is perpendicular to the principal line and parallel to the isometric parallel.

The following example shows the computation of the elements for two-stage rectification of a high-oblique exposure. Data: $f = 152.4$ mm; $F = 181.5$ mm; $t = 60°$; $M = 2$; $n = 2$. From these data:

$$q = f \csc t = 176.0 \text{ mm}; m = M^{1/n} = 1.4142;$$

$$Q = (1 + m)q = 424.9 \text{ mm}$$

$$\beta = \sin^{-1} \frac{F}{Q} = 25°18'; \alpha = \sin^{-1} \frac{\sin \beta}{m} = 17°35'$$

$$m' = \frac{\cos \alpha}{\cos \beta} m = 2.109$$

$$B = (1 + m')F = 564.3 \text{ mm}; A = \frac{B}{m'} = 267.6 \text{ mm}$$

$$d_i = Q (-\frac{\cos \beta}{\cos \alpha}) = +26.2 \text{ mm upward in negative plane.}$$

$$d_i' = m'd_i = +55.3 \text{ mm upward in easel plane.}$$

It is possible to do multiple-stage rectification by fixing the rectifier angles and varying the principal distance of the high-oblique negative to produce the required perspective index. Assume that the easel angle is fixed at $30°$ in the preceding example. Data: $\beta = 30°$, $n = 2$, $M = 2$, $m = \sqrt{2}$. Then, from equation 14.58,

$$Q = \frac{F}{\sin \beta} = 363.0 \text{ mm,}$$

and, from equation 14.55,

$$Q = (1 + m) q_r$$

or

$$q_r = \frac{Q}{1 + m} = 150.4 \text{ mm},$$

where q_r is the high-oblique perspective index required for a two-stage rectification with the easel angle fixed at 30°. As the perspective index of aerial negative is $q = f \csc t = 176.0$ mm, the required value is obtained by printing the positive in the ratio $q_r/q = 0.855$. The resulting positive has a principal distance $f_r = 0.855 f = 130.3$ mm; and its perspective index is such that the two-stage fixed-angle rectification will project its horizon into the line at infinity.

This fixed-angle procedure reveals the possibility of doing the rectification in two ways: (a) by considering the aerial negative fixed and adjusting the rectifier to give the required transformation, or (b) by considering the rectifier fixed and varying the scale of the negative to give it the perspective index required by the geometry of the instrument.

The two-stage rectification of a high-oblique is simplified when $M = m = 1$. Under these conditions, $n = 2$, $Q = 2q$, and $d_i = d_i' = 0$. Advantages of this procedure include (a) simplification of rectifier orientation, (b) adaptability to the ranges of available rectifiers, and (c) the opportunity of matching the rectified oblique to a contact print of its companion vertical negative. Any necessary scale changes are more conveniently made when the assembled mosaic is copied photographically. If an unusual project requires multiple-stage rectification at other than unit magnification, the photogrammetrist should compute the elements of orientation to determine the instrumental limitations that will govern the process. Generally, enlargement is limited by negative-carrier format or easel area, and reduction is limited by negative-plane tilt and displacement.

As the isocenter displacement d_i is zero, the positive should be positioned in the carrier with its isocenter on the lens axis; the projected image of the isocenter then coincides with the intersection of the easel tilt axis and the principal plane of the rectifier. Normal placement of the positive in the carrier will center the principal point on the lens axis. The isocenter is brought

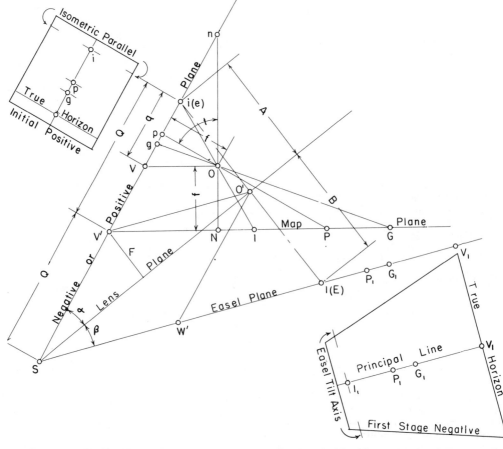

FIGURE 14-30. Rectifier set for two-stage, 1 to 1 rectification of high-oblique photographs.

to the lens axis by displacing the carrier downward (minus) a distance pi. If the maximum possible y-displacement downward is less than pi, the positive must be decentered in the carrier so that the correct isocenter location can be recovered by the y-displacement of the carrier. An approximate centering of the positive enables the operator to fix its displacement and swing by bringing the projected images of the isometric parallel and principal line into coincidence with easel tilt-axis and y-axis respectively.

Figure 14-30 represents this principal-plane section of a rectifier oriented for the two-stage, 1-to-1 rectification of high-oblique photographs. Plan views of the initial positive and the first-stage negative were drawn with the plane of the paper representing the carrier and easel planes respectively. In addition to the markings previously described, the initial positive of figure 14-30 shows a point g, which falls on a ray 10° beyond the principal-point ray. This ray, making

an angle of 70° with the plumbline, is considered a practical limit to the useful field of the high-oblique. The first stage position of g is G_1, and its second stage position is G, shown in the map plane. If the film positive is marked carefully with a needle, the displacement and swing of the carrier may be set with high accuracy by fulfilling the conditions diagrammed in figure 14-30—namely, coincidence between the projected image of the isocenter I_1 and the lens-axis-point E, and between the image of the isometric parallel and the easel tilt-axis. The image of the principal line as projected on the inclined easel is particularly sensitive to the swing adjustment because of the length of the ray $O'V_1$.

In the second stage of rectification, the first stage negative is placed in the carrier with its isocenter I_1 on the lens axis of the unchanged rectifier. According to theory, the first stage image of the true-horizon point V_1 will coincide with the vanishing point V' of the rectifier. This

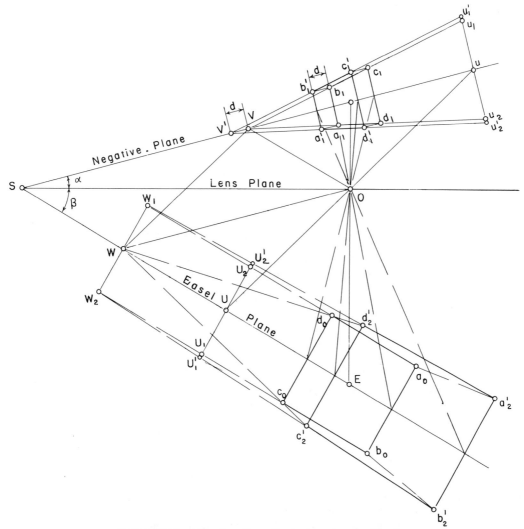

FIGURE 14-31. Affine transformation by two-stage rectification.

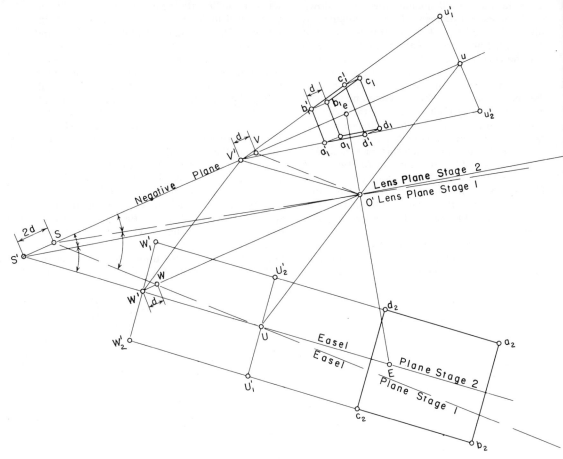

FIGURE 14-32. Final adjustment for affine transformation.

requirement is fulfilled because the perspective index q of the initial positive is projected into the distance I_1V_1 (figure 14-30), which is equal to the perspective index Q of the rectifier. Therefore, in the second stage the horizon point is projected into the point at infinity.

14.8.3 AFFINE TRANSFORMATION

The multiple-stage technique makes affine transformation possible. Such photographic alterations of map dimensions include differential magnification (transformation of one rectangle into another of different proportions) and shear (transformation of a rectangle into a parallelogram).

Figure 14-31 diagrams the use of the nontilting lens rectifier for two-stage differential magnification of a square $a_0b_0c_0d_0$. The geometric figures of the several stages are rotated into the plane of the paper to show their development. The first stage uses the rectifier as a camera which copies the square, and transforms it into the trapezoid $a_1b_1c_1d_1$ on the negative. The processed negative is replaced in the carrier, displaced downward a distance $d = \overline{a_1a'_1}$ and its trapezoid transformed into the elongated nearly rectangular figure $a_2'b_2'c_2'd_2'$. Minor adjustment of the rectifier will eliminate the slight nonparallelism to make $a_2'd_2'$ parallel to $b_2'c_2'$.

Figure 14-31 illustrates several properties of the projectively related negative and easel planes of the rectifier. A line through the lens point O parallel to the diagonal VW of parallelogram $OVSW$ intersects the easel and lens planes in corresponding points U and u whose magnification is one. Therefore, the side of the initial square $\overline{a_0b_0} = \overline{U_1U_2} = \overline{u_1u_2}$. Parallels to the principal line in the easel plane pass through the negative-plane vanishing point V. Therefore, sides c_1b_1 and d_1a_1 lie, respectively, on lines u_1V and u_2V. When trapezoid $a_1b_1c_1d_1$ is displaced downward its vertices generate parallels to the principal line in the negative-plane. Therefore, corresponding vertices in the easel plane are displaced along lines a_0a_2', b_0b_2', c_0c_2' and d_0d_2' passing through vanishing point W. Lines d_0a_0 and $d_2'a_2'$ in the easel plane corresponding to parallels d_1a_1 and $d_1'a_1'$ in the negative-plane pass through vanishing point W_1. Similarly

b_0c_0 and $b_2'c_2'$ pass through W_2. Displacement of the trapezoid $a_1b_1c_1d_1$ causes its sides $b_1'c_1'$ and $a_1'd_1'$ to intercept the length $u_1'u_2'$ on line u_1u_2. The corresponding and equal length in the easel plane is $U_1'U_2'$. U_1' and U_2' lie on lines $b_2'c_2'$ and $a_2'd_2'$ respectively.

Figure 14-32 diagrams the use of the rectifier when adjusted to eliminate the slight trapezoidal shape of $a_2'b_2'c_2'd_2'$. Displacement of $a_1b_1c_1d_1$ to $a_1'b_1'c_1'd_1'$ moves the vanishing points V and W to V' and W' respectively. $\overline{uV'} = \overline{V'S'}$. Parallelogram $OVSW$ is changed to $OV'S'W'$ to place these vanishing points in their proper positions. The unit points u and U remain fixed and the adjusted second-stage easel plane passes through U. The sides a_2d_2 and b_2c_2 are now parallel in the elongated rectangle $a_2b_2c_2d_2$ because they correspond to lines $a_1'd_1'$ and $b_1'c_1'$ which pass through the vanishing point V' in the negative plane. The rectangle width $\overline{a_2b_2} = \overline{U_2'U_1'} = \overline{u_2'u_1'}$, the length which undergoes unit magnification.

The two-stage affine transformation diagrammed in figures 14-31 and 14-32 produces approximately a 20 percent elongation in the initial square. In practice, the automatic rectifier is adjusted empirically to make the final figure fit a control templet. After the first-stage copying of the original map with the rectifier arbitrarily set to cause the negative image to fall within the format of the carrier, the second-stage projection is adjusted as diagrammed in figure 14-32 until the final quadrilateral fits the control templet. Actually the rectifier operator has complete freedom to alter the shape of the final-stage image.

REFERENCES

Cremona, L., *Elements of Projective Geometry,* Oxford University Press, London, 1913.

Finsterwalder, R., *Photogrammetrie,* Walter de Gruyter & Co., Berlin, 1952.

vonGruber, O., *Photogrammetry, Collected Lectures and Essays,* American Photographic Publishing Co., Boston, 1932.

Gruner, H. E., "A Two-stage Rectification System", *Photogrammetric Engineering,* Vol. 27, No. 4, 1961.

Hallert, B., *Photogrammetry, Basic Principles and General Survey,* McGraw-Hill, New York, 1960.

Hatton, J. L. S., *The Principles of Projective Geometry,* Cambridge University Press, London, 1913.

Keller, M. and Tewinkel, G. C., "Space Resection in Photogrammetry", *U.S. Coast and Geodetic Survey Technical Bulletin No. 32,* Washington, D.C., 1966.

Schmid, H., "An Analytical Treatment of the Orientation of a Photogrammetric Camera", *Ballistics Research Laboratories Report No. 880,* Aberdeen, MD, 1953.

Schwidefsky, K., *An Outline of Photogrammetry,* Pitman, New York, 1959.

Sommerville, D. M. Y., *Analytical Geometry of Three Dimensions,* Cambridge University Press, London, 1951.

Wild E4 Rectifier Instruction Manual, Wild Heerbrugg, 1965.

CHAPTER XV

Aerial Mosaics and Orthophotomaps

Author-Editor: ROY R. MULLEN
Contributing Authors: JOSEPH DANKO, ALDEN WARREN, MARIUS VAN WIJK, ERNEST BRUNSON

15.1 Introduction

THE INCREASED NEED for presenting a pictorial view of the earth's surface has led to the aerial mosaic as a means of showing a complete view of large areas. A single aerial photograph presents a picture of a portion of the earth's surface that is comprehensive and complete, and can generally be used and appreciated without need for training in photogrammetry or photographic interpretation. Because the single photograph is limited in area, groups of photographs are combined into mosaics to provide the aerial picture of larger areas.

A mosaic is an assembly of individual photographs fitted together systematically to form a composite view of the entire area covered by the photographs. The mosaic gives the appearance of a single photograph, producing a complete record of the area photographed. The usefulness and accuracy of the mosaic are governed by the methods used in its compilation, and, when properly constructed, the mosaic can meet the accuracy requirements of a good map.

Since a mosaic is a compilation of individual photographs, the compiler must know the fundamental characteristics of a single photograph if he wishes to construct mosaics which present a view of the earth's surface without excessive errors. Perfect matching of the images of adjoining photographs is virtually impossible because of variations in altitude of the airplane, large differences in elevation of the ground, divergence of the camera axis from the vertical at the time of exposure, and errors inherent in the photographic system of camera, lens, film print paper, and so on. By the use of proper techniques, however, these errors can be held to a minimum and not permitted to accumulate.

For greatest accuracy in the construction of a mosaic, all the individual photographs must be reduced to a common scale, must be rectified to remove the distorting effects of camera tilt, and must be assembled to accurate ground control at selected photoidentifiable points for control of azimuth, scale, and position. Such a mosaic is called a controlled mosaic. If only a general pictorial presentation is needed, without the accuracy of a controlled mosaic, an uncontrolled mosaic is sometimes useful. This may be assembled with little or no ground control by matching image detail in adjoining photographs, with tilt distortion only partially removed. A mosaic compiled to intermediate requirements is sometimes called semicontrolled.

Mosaics are a great value in all types of planning activities and have the advantage of showing as a single picture a wealth of detail of the entire area under study. The study of geologic features, flood control problems, irrigation projects, reservoir planning, and the investigation of natural resources can be simplified with the use of mosaics prepared from aerial photographs. For highway and railroad locations and for other right-of-way problems, such as for pipelines, transmission lines, and land acquisition, the mosaic facilitates selection of possible routes, often reducing or eliminating the need for extensive preliminary field surveys.

An aerial mosaic presents a complete and comprehensive view of the terrain, and, if the major and critical features of the area under study are carefully annotated, the clarity of presentation is greatly increased. A mosaic can be understood by users unfamiliar with the photogrammetry which produced it or with the symbolization used in other map forms. It is a rapid and economical method of producing map information. The aerial mosaic is not without its disadvantages, such as excessive detail, lack of topographic information, and errors due to topographic relief. The excessive detail can be minimized by proper selection of scale and by choice of contrast in printing the mosaic. Use of "stickup" lettering highlights the major features and minimizes the less important details. Topographic information can be added to the mosaics in the form of spot elevations of critical points and also in the form of contours. While it is true that in rough areas the complete effects of topographic relief are not usually removed (for economic reasons), nevertheless, with reasonable care and accuracy in preparation, a mosaic can be constructed to approach the National Map Accuracy Standards for horizontal position.

A mosaic of orthophotographs, in which tilt

and relief distortions have been removed through use of special instruments is called an orthophotomosaic. When properly assembled to accurate ground control, an orthophotomosaic is as accurate as a good planimetric map.

If map details are delineated on a photograph or aerial mosaic for emphasis and clarity, and/or contours are added, with names, boundaries, and/or other information, the photograph or mosaic becomes a photomap. Thus a photomap may have as much map information as is needed to serve its purpose, plus all the photographic detail of the original photographs.

The lines of demarcation between controlled, semicontrolled, and uncontrolled mosaics are not sharply defined. The differences are largely a matter of degree of control and rectification. The principal operations usually entering into the construction of a fully controlled mosaic are the following.

1. Procurement of aerial photography.
2. Procurement of ground control.
3. Preparation of map base, including plotting and map projections and the basic control.
4. Rectification of photographs.
5. Mosaic assembly.
6. Editing.
7. Reproduction.

These operations are described in detail in this chapter. For less well controlled mosaics, various details may be omitted insofar as they are not essential to the end product.

15.2 Planning

Many factors must be considered in a well-planned mosaic program. The intended use of the mosaic determines the accuracy required. Important factors include the type of terrain and its relief, the desired scale of the mosaic, the scale of the original photographs, the type and focal length of the camera, the method of copying or reproducing the finished mosaic and the amount of horizontal control required.

The preparation of a mosaic is frequently an adjunct to a topographic mapping program. In this case, the requirements of the mapping program usually govern and determine many of the above mentioned factors.

15.2.1 SCALE OF MOSAIC

In selecting the scale of the mosaic, the main consideration is usually the intended use of the mosaic, but size, shape, and cost may also influence the final choice. The following rather broad classification scheme illustrates general usage.

a. Small scale—1:20,000 and smaller—for use in geology, forestry, reclamation work, flood control, military, and other studies of large areas.
b. Medium scale—1:10,000 to 1:20,000—for use in city planning, preliminary location work for highways, railroads, transmission lines, pipelines.
c. Large scale—1:10,000 and larger—for special detailed investigations in urban highway work, railroad planning, housing and industrial building sites, bridge locations, preparation of tax maps and land-evaluation surveys, strip-mining plans, and other engineering studies.

15.2.2 AERIAL PHOTOGRAPHY

15.2.2.1 AERIAL CAMERA

Characteristics of lenses and cameras are described in chapters III and IV, but are mentioned here briefly because the selection of the camera and lens affects the flight height of the aircraft, the amount of image displacement due to ground elevation differences, and, to a large extent, the negative quality.

With a short-focal-length camera, the airplane can fly at a lower altitude, thus directly affecting the ease and cost of flying operations. Also, when small-scale photography is required, ceiling limitations of current aircraft dictate the use of a short-focal-length lens. Conversely, long-focal-length cameras produce photographs with reduced image displacement due to topographic relief, they produce large-scale photographs without flying too low, and, in general, they produce better quality negatives.

Mosaics have been compiled for photographs with practically every type of lens ranging in focal length from 88 mm to 122 cm, but generally 152 mm, 210 mm, and 304 mm photography is preferred. In rugged rural areas, particularly if the mosaic scale is small, the 152 mm lens is probably the best to use, while a 304 mm lens may best meet the requirements for a mosaic of an urban area. The 210 mm lens is effective for practically all types of mosaics.

For accurate mosaics, a precision camera is essential. When the mosaic project is part of a topographic mapping project, the mapping requirements govern the selection of the camera. In any case, the mosaic photography should be reasonably free of lens distortion and other camera errors.

15.2.2.2 SCALE

The scale at which the aerial negatives are to be exposed depends directly on the scale of the completed mosaic. The aerial negatives can be at the same scale as the final mosaic, but they are usually at a smaller scale. The enlargement of the aerial photographs to the final mosaic can be as great as 4 diameters, with the maximum enlargement limited by the quality of the aerial negatives, the techniques employed, and the equipment available for compilation of the mosaic. It is sometimes necessary to make a mosaic at greater than 4 times enlargement, but in most cases extreme enlargements are undesirable. If the negative quality is good and the laboratory techniques and equipment are adequate, the governing factor in the enlargement determination is usually the desired accuracy and quality of the mosaic.

The scale at which the mosaic is to be compiled must also be considered at this time. For speed and economy, the compilation scale should be the same as the photograph scale, but more accuracy can be obtained by compiling at a scale of 1¼ to 1½ times the scale of the photography. This also permits a smaller enlargement between mosaic compilation and final copy, thus assuring better photographic quality in the final copy. However, the problem of compensating for differential paper shrinkage becomes greater when the mosaic prints are enlargements; unless careful control is maintained over this factor, some of the advantages may be lost. In summary, best results are obtained with a photographic scale not less than ¼ of the final mosaic scale and with compilation at an intermediate scale.

15.2.2.3 OVERLAP

The amount of overlap required between photographs should be determined in the planning stage. Normal requirements are 60% forward overlap and 30% side overlap. Accuracy can be increased by increasing the amount of overlap, so that a smaller portion of each photograph is used, thereby reducing displacements due to relief. This also increases the number of photographs required and the cost of compiling the mosaic.

15.2.3 HORIZONTAL CONTROL

The prime purpose of controlling a mosaic is to assure that each feature is located in its proper horizontal position relative to all other features. The accuracy of the mosaic can be no better than the control used and the techniques employed in its compilation. The intended use of the mosaic is therefore the principal factor in determining the type, quality, and distribution of control needed.

Before planning new control, all existing control in the area should be thoroughly investigated. This control may exist as triangulation and traverse surveys, highway and railroad right-of-way plans, or reliable maps from which map positions can be accurately measured for points that can be positively identified on the photographs. Additional control surveys may then be planned as needed. Since a mosaic program is frequently combined with other mapping or engineering work, the additional ground surveys should be planned to serve all phases of the control development for the area.

The extent to which existing control must be supplemented by additional ground control is determined by the pattern of the existing control and how it can be used in extension of control by photogrammetric methods. Practically all mosaics depend on some form of establishing horizontal positions. A minimum of two horizontal control points is required to establish scale, azimuth, and position, but errors of scale and position increase with the distance from control. Therefore, for areas covered by a large number of photographs, it is necessary to use more than two points for basic control, the amount and distribution of which vary with the requirements of accuracy. Planning and execution of field control are discussed in chapter VIII.

There are many areas where excellent large-scale maps already exist. If maximum accuracy is not required, these may be a very good source of control, at least for semicontrolled mosaics.

Occasionally, a mosaic of an area is required but ground control cannot be obtained because of terrain character, location, time limitations, or cost. In such cases, a radial-line plot can be used to establish relative internal horizontal positions, even though adequate control is not available to establish correct scale or position of the mosaic (*see* chapter IX). The plot may be made by intersecting rays or slots at the approximate average scale of the photographs (or at any other desired approximate scale). In areas with large differences in elevation, the scale may vary somewhat in different parts of the assembly, but local relative position differences can be kept small.

15.3 Base Board

A base must be prepared for the control assembly. It can be masonite, plywood, aluminum, or other flat, rigid material and can later be used for assembly of the mosaic. Sometimes it is more convenient to use one base for the control plot and another for the mosaic assembly. Drawing paper can be mounted on the base board to facilitate the plotting of grid lines and control points.

The size of the base depends on the area to be covered and the scale of the control plot and compilation. It is usually easiest to compile at the same scale as the photographs, but, if the mosaic is to be reproduced at a larger scale, greater accuracy can be maintained by compiling at a scale intermediate between the two. If the area is too large to be handled on one base board at the selected scale it can be subdivided into two or more control-plot solutions, provided that each segment is surrounded by adequate ground control and enough overlap is allowed between adjoining segments to maintain continuity of control and detail matching.

15.3.1 MAP PROJECTION

15.3.1.1 CHOICE

Numerous types of map projections are available for controlling mosaics. The principles involved are essentially the same as those which govern the choice of a graticule for a map. Some of the more common map projections are discussed in chapter VIII. The choice is usually determined by the size, scale, and location of the area. In the United States, the polyconic projection, one of the State grid systems, or a local rectangular grid is usually selected. In any event, the selected graticule or grid system should be compatible with the system on which the control data are computed.

15.3.1.2 PLOTTING

The graticule and all the control points can be plotted using a rectangular coordinatograph. All projection or grid lines should be labeled at each end for easy identification.

15.4 Ratioing and Rectification

15.4.1 RATIO COMPUTATION

The control plot determines the correct location of each control point. The next step is to determine the enlargement or reduction necessary to make each print fit the control plot. To do this, a series of measurements is made on each photograph. The distance is measured on the control plot and also on the print. The distance on the plot divided by the corresponding distance on the print is the required enlargement factor for the point. The print is then ready for tilt analysis. If no rectifying camera is available and the mosaic is to be made from only ratioed prints, the average of the ratios is computed for each print and recorded on it.

15.4.2 TILT ANALYSIS

Various types of rectifiers are available. The final rectifier setting for each photograph is usually obtained by trial and error to obtain the best mean fit between the rectified print and the plotted positions of all the control points. If the terrain in a photograph is reasonably flat, the variation in enlargement ratio usually follows a simple pattern, which permits the approximate determination of the axis of tilt and determination of the tilt settings for the rectifier. As the elevation differences in the photograph become greater in proportion to the flight height, the pattern of the ratios becomes less symmetrical and the resulting tilt values are less accurate. It may then be better to determine tilt values based on critical points instead of the selected control points. Under these conditions, tilt determination is dependent more on personal experience and judgment than on a rigorous mathematical solution. In areas of extreme topographic relief, it may be difficult or impossible to adequately rectify a whole photograph in one ratioed print, and several prints may be needed, with different rectification settings, to take care of the large variations in scale.

To analyze a photograph for tilt, for the initial estimate of the rectifier settings, the enlargement ratios of the photograph are studied. A line is drawn through the center so that it passes through or near points of equal ratio with, if possible, higher ratios on one side and lower ratios on the other. This line is the axis about which the negative is to be tilted to effect the changes in the ratios during rectification, and the ratio on this tilt axis is the enlargement factor for the negative.

Numerical, mechanical, or graphical methods may be used for determining tilt. An example is the tilt slide rule illustrated in figure 15-1. To compute tilt, select a point on one side of the tilt axis, representing an average ratio value for that part of the photograph. Measure the distance from this point to the axis normally referred to as the photo-arm. The difference between the ratio value at this point and the one on the axis is called the ratio difference. Set these values together on the upper scales of the slide rule. Opposite the tilt axis ratio (on both center ratio scales) read the tilt settings for the negative carrier and for the easel. These settings apply to a fixed-lens rectifier. Similar computing devices are available for other types of rectifiers. If the tilt axis ratio is close to unity, the tilt values for the negatives and easel should be the same.

It should be emphasized that these tilt settings

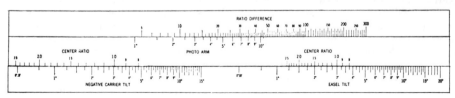

FIGURE 15-1. Tilt slide rule.

will produce rectified prints that fit the control at the point for which the settings were computed. If that point is representative of the average datum, the horizontal position and scale of the rectified print will be good. It is therefore important that the radial points be selected on images having elevations near the average elevation of the photograph. If the original radial-control points could not be selected to meet this requirement, additional points can be resected on the plot, their ratios computed, and these values used to determine tilt settings to produce satisfactory horizontal position. If it is not possible to compute a single tilt setting to satisfy the ratio values of all points, separate tilt settings may be necessary for each half of the photograph, using the same tilt axis and its ratio for both rectifications. The two rectified prints are then cut and matched along this axis when the mosaic is laid.

The tilt settings are recorded on the radial-control print, together with the azimuth of the tilt axis and the enlargement ratio at the axis. With these data and the ratios at all control points previously marked on the prints, the processing laboratory can provide the rectified prints needed for compilation of the mosaic.

15.4.3 RECTIFIED PRINTS

In processing the rectified prints, the tone and contrast should be carefully controlled so that adjacent prints in the mosaic will blend together to give the appearance of a single continuous photoimage. This tone matching calls for extra effort and skill on the part of the photographer but reduces the amount of touching up required later on the mosaic or the copy negative. Although this blending can be accomplished on the final copy negative, best results are obtained with mosaic prints well matched for tone and contrast. Automatic electronic-dodging rectifiers offer the best color, tone, and density control possible at the rectifying stage. If manual dodging is used, the development of the prints should be inspected very carefully because some of the differential in color and tone can be controlled in this stage.

The differential expansion of photographic paper presents a problem that can be partially overcome. In drying, the photoimage exposed on the photographic paper in the darkroom normally shrinks about 0.5 percent in the direction normal to the grain of the paper, but holds its size closely in the direction parallel to the grain. If the print is mounted with a water-content adhesive, the paper expands about 1 percent or more normal to the grain but expands only slightly parallel to the grain. To compensate for this, the computed enlargement can be reduced in proportion to the average net expansion that occurs between exposure and final mounting. This factor may vary with humidity in the laboratory, the type of photographic paper used, and the water content of the adhesive. Periodic tests should be made to determine the correct factor to be used. It is good practice to keep the direction of the paper grain constant in making the rectified prints. Usually this means running the grain in the general direction of the axis of tilt, so that the least change occurs along the axis.

If water-content adhesive is to be used, the wet-paper technique of exposure can help to obtain a photographic image having the correct scale after mounting. Before exposing, the photographic paper is immersed in water long enough to saturate it and is then carefully squeezed onto the easel. Sufficient light is necessary to permit a short exposure because the slightest shrinkage of the paper during exposure would result in a blurred image. Single-weight glossy paper is generally used for mosaic prints.

15.5 Mosaic Construction

15.5.1 LAYING THE MOSAIC

The mosaic is assembled on a smooth, hard, nonporous or semipermeable surface, such as masonite, plywood, chip board, or aluminum sheets. The size of the board is limited only by convenience and available space. In general, semicontrolled or uncontrolled mosaics are normally used as wall decoration and are usually large. Controlled mosaics are used for engineering purposes and as a result are usually divided into convenient sheet sizes. If the program includes topographic mapping, the mosaic sheets frequently conform to the size and scale of the map sheets. If the required mosaics is larger than the available size boards, and if space allows, several panels can be fastened together with suitable backing strips. Screw holes in the face of the boards should be countersunk, backfilled, sanded smooth, and covered with a coat of lacquer. Tape may be mounted over the panel seams and sanded smooth.

15.5.2 ASSEMBLING PHOTOS

15.5.2.1 TEARING

Starting in the center of the board, the first print is prepared for mounting by pricking all control points with a fine needle and circling them with a grease pencil. The print is trimmed just inside the border with a razor blade, and the edge is feathered to insure smoothness as layers of prints are mounted and to make the edges less perceptible to the eye and to the copy camera. The feathering is done as follows. Using a very light pressure and a very sharp blade, cut through the emulsion only but not into the paper. Gently fold back away from the emulsion the

FIGURE 15-2. Tearing the print.

part to be discarded. Hold the print in one hand (fig. 15-2) and, using the downward motion with the other, tear away the discard portion back from the cut line, leaving the edges of the retained portion thinner than the remainder of the print. Smooth out this feather edge with fine sandpaper. A good sanding tool may be made by rolling a magazine or other resilient material into a roll about 2 inches in diameter and covering it with sandpaper. The retained portion of the print may be held face down on a hardwood sanding block, about 12 by 12 by 4 inches with one edge rounded. Sand gently over the entire torn edge, being particularly careful not to remove too much paper near the edge; this might make it too thin and give a darkened appearance after mounting.

15.5.2.2 ADHESIVE

Gum arabic, glue, rubber cement, or paste are satisfactory adhesives. Gum arabic is commonly used and allows good working time before drying; it has the disadvantage that the print edges may become brittle and crack. This can be controlled somewhat by adding glycerine to the solution or by decreasing the amount of alum in the hardener when processing the prints. The gum arabic solution may be made by adding 6 pounds of purified gum acacia flakes to one gallon of water heated to 160°F, and stirring thoroughly until completely dissolved. Twelve drops of formaldehyde or oil of cloves, or 1/4 ounce of salicylic acid, may be added as a preservative, and 16 ounces of glycerine to protect the prints from excessive brittleness. This solution may be thinned slightly with water or used full strength with presoaked prints.

Rubber cement is satisfactory for single prints but not for large mosaics. It does not permit stretching the prints to fit control or to match detail, and its rapid setting makes speed essential in mounting. Glue is an excellent adhesive but presents a difficult problem of cleaning after the prints are mounted and also makes the print edges very brittle. Paste and water is an excellent adhesive as it allows easy control of the print and cleans up very easily.

15.5.2.3 MOUNTING

The adhesive is applied liberally to the back of the print or to the mosaic board, or to both, and the print is oriented to the control, using a pin point to fit the images to the plot. The print center point is positioned first and the pin left in place as a pivot while the outer control points are matched. As an alternative, if arm microscopes are available (figure 15-3), the microscope is first positioned over the control point on the base and the print is then slipped into position under the microscope with the control-point image centered under the microscope reticle. A squeegee is used to remove excess adhesive from under the print, and any adhesive remaining on the face of the print is removed with a damp cloth or sponge. The adhesive is allowed to set before the next print is placed.

Before applying adhesive to the next print, it is oriented to control in the same way and flipped over the previously laid print to permit visual selection of the proposed cut line. The cut line is usually about midway between print centers and is selected to give the best image match between the prints. Tone match and photographic detail are considered so that the cut line will not be apparent. As far as possible, areas of identical tone without specific detail should be selected, but excessive scalloping or straight-line cuts should be avoided except along roads or fence lines. If the cut parallels a road or fence, the feature is left in its entirety on the print being

laid. The cut line is marked with a grease pencil, and then the edge is cut, torn, and feathered as described in section 15.5.2.1.

After cutting, the match is checked by reorienting the print to the control, and any minor revisions that may be necessary in the cut line are made. The degree of fit between the control points and the plot indicates the amount of stretching that may be needed in the print. If the control fits well, the print must not be allowed to stretch; therefore, only a little glue should be used and the print quickly fastened into position. If the print is small, it may be wetted before the adhesive is applied. This causes it to expand in proportion to its saturation. Extra length may also be gained by working the print with the fingers or squeegee while fitting it into position. An infrared lamp may be used to speed up the setting and drying of the print to permit more rapid compilation or to prevent excessive shrinkage, especially when humidity is high; but too rapid drying may cause the print edges to raise slightly.

If the print is too large to fit the control, a new tilt analysis may be needed; especially if excessive relief is involved. If the tilt analysis calls for separate tilts for each half of the photograph,

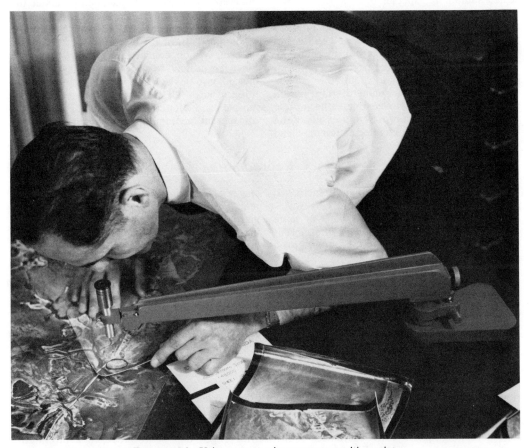

FIGURE 15-3. Using an arm microscope to position print.

one rectified print is laid to the control on the side of the tilt axis for which it was rectified, and the other rectified print is cut, generally along the common tilt axis, matched to the half already laid, and mounted to fit the control on its side of the axis.

After all the prints are laid, excess paste is removed with a damp cloth or sponge, the entire mosaic is checked for errors in image matching, edges are checked and repasted if necessary, and small ratioed patches are ordered to correct any errors in matching. These may be projected to size by orienting to an overlay on which pertinent detail has been transferred from the mosiac. Loose print edges which occur frequently, may require a stronger adhesive, such as casein glue, to make them adhere.

15.6 Blending

Before the mosaic is copied, a slight touchup or blending may be desirable. Since it is not always possible to obtain photographs with identical color and tone for mosaic work, an effort is made to obtain a blend of tones in the final mosaic even though the individual prints are not perfectly matched before assembly. It is highly desirable to have identical color and tone throughout, but this is very difficult because the original negatives may have been exposed through different haze conditions, or because of reflected light or hot spots.

Even with great care in the rectification process, some blending is frequently required on the final mosaic. This is accomplished by touching up with a photographic dye.

15.7 Cartographic Treatment

Mosaics are usually cartographically annotated, and descriptive information is added. These data may include ticks, grid designation, scale, authorship, special title, north arrows, and sheet layout (if several sheets are involved) and body information (names of various planimetric and cultural features and outlines of features to be especially noted on the mosaic). Township, County, State, or other political boundaries can be shown as required.

The amount and placement of information should be as required to complete the mosaic and render a clear, unconfused impression of the features shown. Too much information may lead to considerable confusion, while too little detracts from the usefulness of the mosaic. Border information is usually placed on white border material pasted on the mosaic board outside the desired image area. This information can be hand lettered (mechnaical lettering guides are recommended), or it can be printed "stick-up" lettering which can be floated into position. Since it is difficult to ink a clear grid line across a mosaic, grid ticks on the border are preferred to full grid lines. Designations on the body of the mosaic can be lettered, but printed stickup on a transparent medium is more satisfactory.

15.8 Reproduction

The usefulness of a mosaic may be increased greatly if many copies can be made. This is usually done photographically by means of large copy cameras. If final copy is desired at an enlargement over the compilation scale, three choices are available.

(a) Enlarge the copy negative and print in a contact printer.
(b) Photograph at one-to-one scale and print by enlargment.
(c) Partially enlarge at both the negative and printing stages.

15.8.1 SCALE AND SIZE OF COPIES

Scales and sizes of prints from copies of mosaic sections may vary widely to conform to different requirements and to the capability of the equipment available. Since the scale and size are determining factors in the usefulness of the final copy, these are usually fixed in the early stages of planning. At times, it is desirable to mount two or more sections of the mosaic on masonite or muslin to get a larger area for viewing, or the individual sheets may be mounted on muslin and bound in book form. The form of the final product depends, of course, on its intended use.

15.8.2 COPY CAMERA

There are many types of copy cameras. Those commonly used for mosaics can accommodate negatives at least 40 cm by 50 cm in size; many have a negative capacity of one metre square. The camera should be highly precise in both its mechanical movements and its optical quality. Common lens focal lengths range from 48 cm to 1 metre. With a narrow-angle lens and a properly angled light source, the shadows caused by the print edges can be reduced to a minimum. High-intensity lights reduce the required exposure time. Polarized lights and filters help reduce shadows and reflections.

15.8.3 COPYING PROCEDURE

Select a lens of as long a focal length as possible and place it in the camera. The easel plane and negative-holder plane must be parallel and mutually perpendicular to the lens axis. This can be checked by placing a grid on the easel and measuring it on a ground-glass screen placed in the negative plane. Any keystoning effect should be eliminated. With the camera checked out, place the mosaic on the easel. A makeline is usually marked on the mosaic board and is used to bring the negative plane to the proper scale position. Corner grid values may also be used for this purpose. The makeline is measured on the glass screen with a fine scale and an eye loupe. The lights are placed so that, when the mosaic is viewed from in front of the camera near the lens, glare and shadow are at a minimum. The mosaic is then viewed on the ground glass, and the lights further adjusted if necessary. The lens stop is set at the proper aperture for the light source and the film to be used, and the lens is covered with the lens cap. The ground-glass screen is swung aside, the film holder placed in its position, the lens cap removed, and the exposure made.

15.8.4 NEGATIVE BLENDING

After the copy negatives are developed, they are carefully inspected for evenness of tone, especially along all cut lines. Mosaics of low-contrast areas, such as desert regions, are especially difficult to tone match in the original and may require careful blending at this stage. This is the last opportunity to correct any tone-match failures in the earlier stages of the mosaic process. The blending is done on a light-table in a semidark area. Varying densities of dye are applied to the glossy side of the negative with a small brush to gradually darken the light areas to match the darker undyed areas. Care should be exercised to apply enough dye, but not too much. If too much dye is applied, it may be removed by washing the negative in water.

15.8.5 PRINTS

The blended negative is ready for producing the final prints, either by contact printing or projection. These prints may be single or double weight, glossy or matte finish. Sometimes low-shrink-base paper is used because of its dimensional stability. Screened film positives are useful for making prints by the diazo process.

15.9 Uncontrolled Mosaics

A mosaic requiring less accuracy than is demanded of a controlled mosaic can be compiled by less rigorous methods. For an uncontrolled mosaic, the photographs are brought to a common scale, and any existing control or map data may be used to maintain a reasonably accurate picture of the terrain. The effects of tilt and topographic relief tend to pull a mosaic in one direction or another when control or map data are not available. To counteract this tendency, an approximate azimuth method of compilation may be used. The photographs are laid out by strips in their proper order by matching image details. A straightedge is placed so as to pass as close as possible to all the principal points in the flight strip, and a straight line is drawn across the

assembled photographs. A similar line is drawn on the mosaic board. The line on the photographs is extended across each print and is transferred to any prints in the flight not included in the original laydown. The mosaic is assembled by orienting each print along the azimuth line on the mosaic board and matching detail between successive prints. To further reduce the effects of the various distortions, only the area of each print that is immediately adjacent to its principal point is used. In this area, the distortions are a minimum. Uncontrolled mosaics are annotated to show approximate scale and grid values and to give other information as may be required for the problem at hand.

15.10 Photomechanical Mosaicking

The aerial mosaic preparation described previously is being replaced by the method of mosaicking by the photomechanical process. The process is especially adaptable when orthophotographs are used in the preparation of the desired mosaics. The method insures high geometric accuracy and optimum image resolution, using stable base films. As in the previously described paper print method, images are scaled to control, adjoining print densities are matched and imagery detail is registered to ground con-

trol and adjacent imagery. Film images used in photomechanical mosaicking can be either positive or negative.

15.10.1 MOSAIC ASSEMBLY

Film images are prepared for assembly essentially the same way as paper prints; they are brought to scale and matched in contrast and density. The difference in assembly is that each film image is assembled and taped to a separate carrier sheet (figure 15-4). A carrier sheet is a sheet of clear stable-base material that carries the image. Two film images require two carrier sheets and two mask sheets. Two or more images can be assembled on the same carrier sheet, but, no two images can overlap on the same carrier sheet. Four carrier sheets and four mask sheets are the maximum needed to assemble a multiple-image mosaic.

To assemble two film images into a mosaic, the following materials and equipment are needed.

1. Light table.
2. Register studs.
3. A control base sheet, on clear stable material.
4. Four sheets of clear stable material.
5. Two film images at scale and processed to aim points.
6. A magnifying glass.
7. Tape, masking and red litho.
8. Opaque masking material, to keep exposing light from photographic materials.
9. A darkroom equipped with a tungsten light source, timer, vacuum frame, and diffusion sheet.
10. A processing room with temperature-control equipment.
11. Unexposed continuous-tone and litho film.
12. A punch register system.

All materials are prepunched to fit together on register studs—the control base sheet, the sheets of clear stable material, and the unexposed film. The first step is to place register studs in the punched holes of the control base sheet. Position

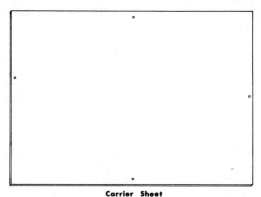

Carrier Sheet

FIGURE 15-4. A clean sheet of stable material, used for carrying the images.

the control sheet on the light-table, and tape the studs to the glass with masking tape (figure 15-5) to keep the studs in place when removing the punched material from the light-table.

Place the control base sheet on the studs and place a sheet of clear material over it. Label the bottom of the clear sheet *carrier sheet 1*, and at the top, mark it *top* (figure 15-5). Orient the main or first film image by preselected photoidentifiable points to corresponding points on the control base. Use the magnifying glass to be sure the points on the control base and the image detail match. Tape each corner of the image to the carrier sheet with masking tape, recheck the fit with the magnifying glass, and tape each side of the

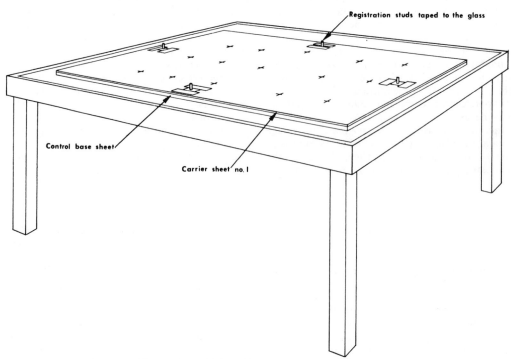

FIGURE 15-5. Light table with studs taped in place.

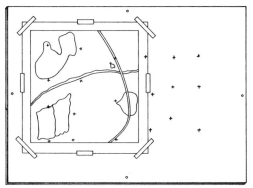

FIGURE 15-6. Image 1 taped to carrier sheet 1.

FIGURE 15-7. Image 2 registered to image 1 and to control base.

image to hold it securely in place (figure 15-6). Place a second sheet of clear material on the studs over carrier sheet 1 and label it *carrier sheet 2*. Orient the second image to the control points on the sheet and to detail in the imagery of the overlap area of the first image. Tape it to the carrier sheet after making sure of a good fit by inspection with the magnifying glass (figure 15-7). The images are now ready to be masked, the critical step in the process.

The key to joining the separate images into a continuous image is the diffusion, through a gap, made by constructing masks that overlap imagery from each exposure about 0.010 inch or less. Test exposures of different gap widths and a diffusion sheet are made in the vacuum frame to determine the correct width. The test can be performed with the images just assembled. Place another sheet of clear material on top, to become mask sheet 1. Look through the overlap of the two images and select an area in a straight line

from top to bottom that is matched in density and free, if possible, from water imagery. With a grease pencil, mark the area at the top and bottom of the clear sheet (figure 15-8). Position a strip of red litho tape in a straight line from the top mark to the bottom mark (figure 15-9). Completely mask the area of the clear sheet covering image 2, from the red litho tape to the side edge with opaque masking material (figure 15-10) to complete the mask for carrier sheet 1. It is used with the carrier sheet 1 for exposure in the vacuum frame.

The tests for the correct width of gap can be made on the mask for carrier sheet 2. Remove all sheets from the studs except mask sheet 1. Place a sheet of clear material on top. Then, using a shop microscope to measure widths, place short strips of red litho tape parallel to the litho tape on mask 1, with a gap of 0.010 inch between the edges of the two strips at the top, 0.006 in the

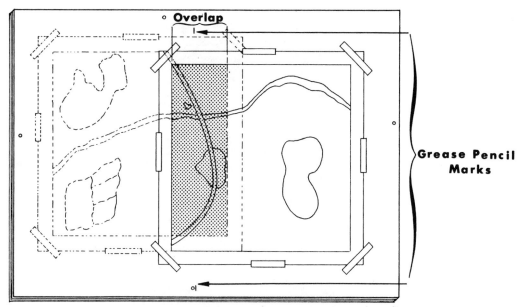

FIGURE 15-8. Marking the area for best possible join in the overlap area.

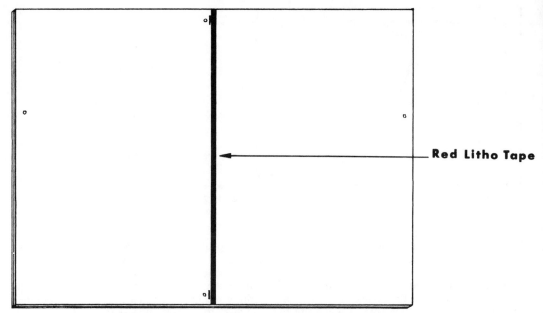

Red Litho Tape

FIGURE 15-9. Clear stable-base material to be used as the mask.

middle section, and 0.003 inch at the bottom (figure 15-11). Mask out the area of image from the litho tape strips to the side edge of the clear sheet.

In the vacuum frame in the darkroom, place a sheet of unexposed continuous-tone, pre-punched film with register studs inserted, place carrier sheet 1 with the emulsion of the image in contact with the emulsion of the unexposed film, and insert mask sheet 1 with the red-litho-tape side in contact with the back of the carrier sheet (figure 15-12). Close the frame and turn on the

vacuum, place the diffusion sheet between the frame and the light source, and expose for the predetermined time. Remove carrier sheet 1 and mask sheet 1, insert carrier sheet 2 and the test mask sheet, and expose. Process to 1.0 gamma. Examine the processed image on the light-table for the results of the three diffused gaps. A dark line indicates too wide a gap, a white or transparent line indicates too narrow a gap. Imagery appearing continuous without lines indicates the correct gap width for the system.

Return the two to the light-table with mask 1

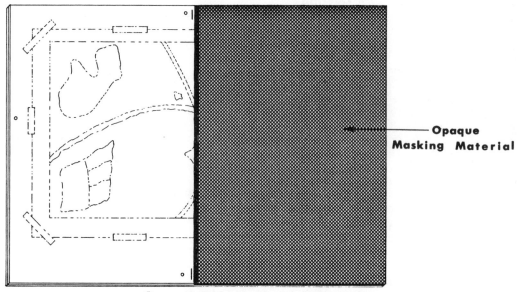

Opaque
Masking Material

FIGURE 15-10. Carrier sheet 1 and mask sheet 1.

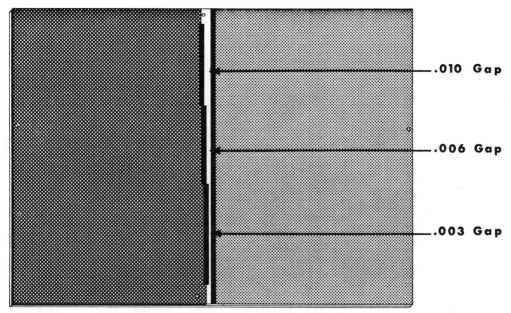

FIGURE 15-11. Mask for testing the diffused gaps.

on the bottom and the test mask over it. Remove the test strips of red litho tape and replace with a full-length strip positioned parallel to the strip on mask 1 with the gap between the edges of the two strips as determined by the test (figure 15-12). Mask the area covering image 1.

If indicated by the diffused gap test, a tone density difference between the two images can be corrected by adjusting the exposure timer of the carrier sheets. To make the final mosaicked image, expose the two carrier sheets with their corresponding mask, one at a time, with the emulsions of the images in contact with the emulsion of the continuous-tone film (figure 15-13). Place the side of the mask with red litho tape in contact with the back of its corresponding carrier sheet (figure 15-11). Adjust exposure if necessary to match density and process the

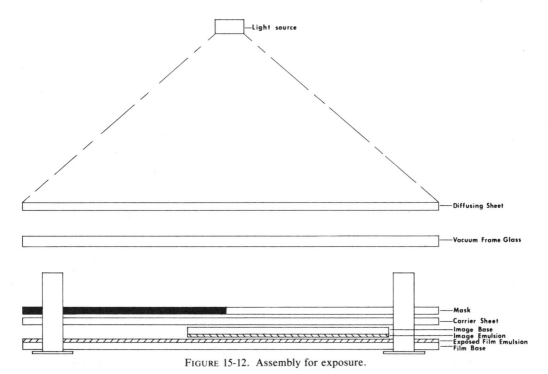

FIGURE 15-12. Assembly for exposure.

Mask #2 and Mask #1 creating the gap.

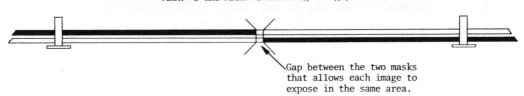

Gap between the two masks
that allows each image to
expose in the same area.

FIGURE 15-13. Exposing the two carrier sheets.

exposed film to 1.0 gamma. The result is a continuous-tone film image having the appearance of a single photograph. The mosaicked film can be halftone screened and combined with other line plates for lithographic reproduction.

15.10.2 MULTIPLE-IMAGE MOSAICS

Multiple images can also be mosaicked by the photomechanical technique, but the masking procedure is more difficult. To assemble a block of 4 images requires 4 carrier sheets (figure 15-14) and 4 mask sheets. A strip of 4 images requires only 2 carrier sheets and 2 mask sheets. A multiple mosaic of 9 images is shown in (figure 15-15).

Because no two adjoining images can be assembled on the same carrier sheet, detailed planning is needed to assemble images properly on carrier sheets.

The numbers 1 to 4 in the lower left corner of each image (figure 15-16) indicate the carrier sheet on which that image will be assembled. The letters in the upper left corner indicate the position of each image on the mosaic. Image E is the center or main image and is positioned on carrier sheet 1 (figure 15-17) to images B and D. Image C is fitted to images B and F, image G is fitted to images D and H, and image I is fitted to images F and H. All are assembled on carrier sheet 3 because any errors in the fit of the images

would be more apparent on the inner joins (images D, E, F, and E) than the outer joins (A to B and D, C to B and F, G to D and H, and I to F and H.) With the assembling of the images complete, the clear sheets to be used as masks are placed one at a time over the carrier sheets and marked where the best possible joins can be made. All the carrier sheets are removed except carrier sheet 1. A clear mask sheet is keyed to it, and red litho tape is used to outline the image that will be exposed (figure 15-21). Carrier sheet 1 is removed, mask sheet 2 is keyed to mask sheet 1, and a gap is formed between masks 1 and 2 with red litho tape where images B and E, and E and H join. The outlines of images B and H are completed with litho tape (figure 15-22). Mask 1 is removed, and mask sheet 3 is keyed to mask 2, gaps are formed between mask 3 and 2 where images A and B, C and B, G and H, and I and H join. The outlining of images A, C, G, and I is completed with litho tape (figure 15-23). Mask sheet 2 is removed, and the areas where no imagery is to be exposed are completely masked out. Mask sheet 4 is keyed to mask sheet 3, and gaps are formed where images D and A, D and G, F and C, and F and I join (figure 15-24). Mask sheet 3 is removed, and the areas where the imagery is to be exposed are masked out. Mask sheet 1 is returned, and mask 4 is keyed to it. Gaps are formed where images D and E, and F and E, join (figure 15-25), and the nonexposing

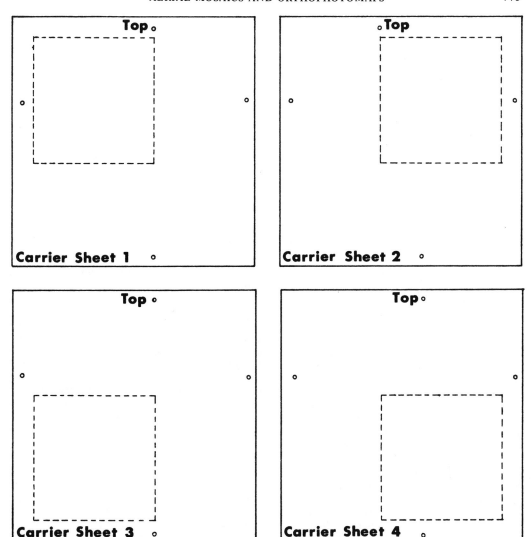

FIGURE 15-14. Four carrier sheets with images.

areas of the two are masked out. The procedure for compositing the carrier sheets with masks are the same as described in the two-image mosaic processes. The result will be a composite film with the appearance of a single continuous scene. This procedure is especially adaptable to preparing imagery for lithographic printing of mosaics.

15.11 Orthophotography

An orthophotograph is the equivalent of an orthographic photograph of the ground taken from above. It is made by removing the effects of tilt, relief, and many of the lens aberrations from standard perspective photographs of the terrain.

Orthonegatives are produced from geometrically rigid stereoscopic models by a scanning process which is controlled in such a way that contiguous scan widths are correlated thus minimizing scan lines. The requirement for relative orientation of the stereopair is identical to that for standard stereoscopic compilation. The absolute orientation relative to elevation within the model is not as critical as for standard stereoscopic compilation since a large error in model tilt can be tolerated before appreciable horizontal displacements result. Horizontal control is not required for the preparation of orthonegatives. The process of preparation of orthophotographs is usually referred to as differential rectification.

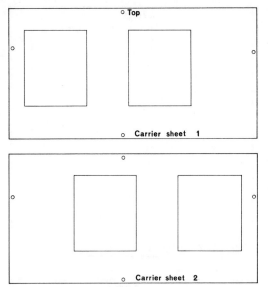

FIGURE 15-15. Two carrier sheets with four images.

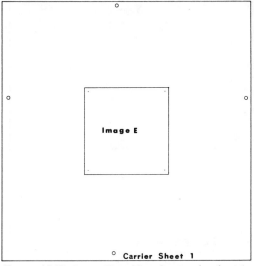

FIGURE 15-17. Image E positioned on carrier sheet 1.

15.11.1 PROJECT PLANNING

Factors to be considered in the preparation of orthophotographs are the aerial camera focal length, photographic scale, forward and side overlap, type of instrumentation on which the orthophotographs will be prepared, and the ultimate quality and use of the end product.

15.11.1.1 CAMERA FOCAL LENGTH

Wide angle cameras, having focal lengths of at least 152 mm are best suited for the production of orthophotographs. Errors which occur during the profiling process of producing the orthophotograph adversely affect the horizontal accuracy of the image when using superwide angle 88 mm focal length photography. Focal lengths longer than 152 mm produce even better horizontal results, however, some instrumentation for the production of orthophotographs can only effectively use the 152 mm focal length photography. Longer focal length cameras also reduce the amount of double imaging and hidden ground that adversely affect the image product.

15.11.1.2 PHOTOGRAPHY SCALE

The important factors to be considered in the planning for orthophotography projects are the ratio between the aerial photograph, the stereomodel, the orthonegative and the final map scale. It is necessary to plan the photo-scale so that a suitable enlargement or reduction ratio

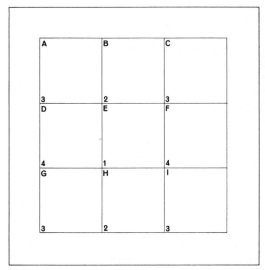

FIGURE 15-16. Multiple-image mosaic.

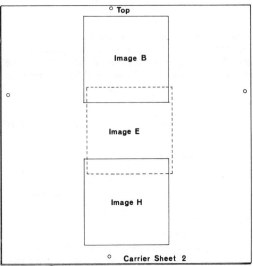

FIGURE 15-18. Images *B* and *H* positioned on carrier sheet 2.

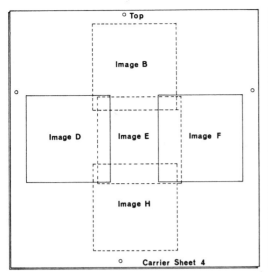

FIGURE 15-19. Images *D* and *F* positioned on carrier sheet 4.

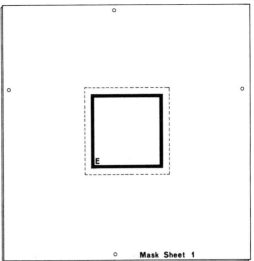

FIGURE 15-21. Carrier sheet 1 with image *E* is on the bottom and mask sheet 1 is on top. Red litho tape outlines the image that will be exposed in the vacuum frame.

to the final map- or photo-product results. Generally photographic enlargements greater than 4× should be avoided, however, better quality enlargers and higher resolution films may allow as much as 8× enlargements to be produced.

15.11.1.3 FLIGHT DESIGN

Factors to consider in flight design for orthophotographs are the flying height, the amount of ground relief and its variations in the project area and the orthophotograph instrumentation projection ranges. The shape of the resulting map sheet affects the chosen flight direction. Orthophotographs for mapping projections that

have long north-south dimensions are usually more economically flown in the north-south direction. With regard to forward overlap along the flight lines, standard overlaps, suitable for topographic mapping, are adequate for the production of orthophotographs. If terrain relief conditions vary considerably it may be advisable to procure aerial photography with forward overlap of up to 90% so that selection of overlapping stereopairs can be made to minimize the effects of areas of large relief differences falling in the corner areas of the scanned stereomodels.

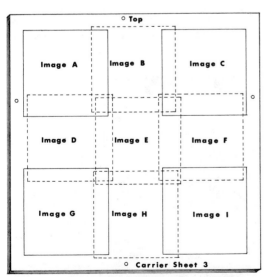

FIGURE 15-20. Images *A*, *C*, *G* and *I* positioned on carrier sheet 3.

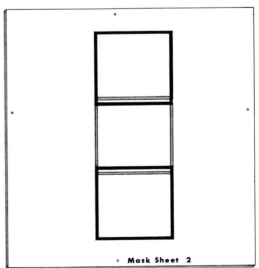

FIGURE 15-22. Mask sheet 1 is on the bottom and mask sheet 2 on top. The gaps to join *B* and *E* and *H* and *E* are made and the exposing images are outlined.

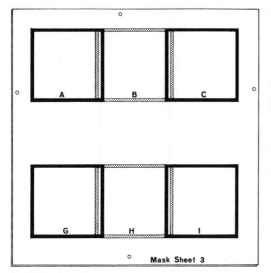

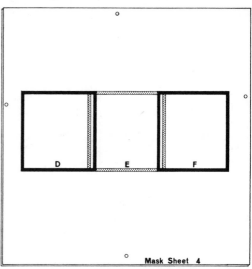

FIGURE 15-23. Mask sheet 2 is on the bottom and mask sheet 3 on top. The gaps to join A and B, C and B, G and H and I and H are made and the exposing images are outlined.

FIGURE 15-25. Mask sheet 1 is on the bottom and mask sheet 4 on top. Gaps are made where D and E and F and E join.

15.12 Instrumentation for Orthophotography

Instruments for producing orthophotography through the differential rectification process print the rectified image optically through a scanning slit or mask or they reproduce the image densities and patterns electronically on a cathode-ray tube (CRT). The CRT image is then transferred to a sensitized film which becomes the orthonegative. Instruments of the optical projection type are typified by the Kelsh K-320 Orthoscan, the Gigas Zeiss Orthoprojector GZ-1, the U.S. Geological Survey T-64 or-

thophotoscope, the Jena Orthophoto B, the Galileo-Santoni Orthophoto Simplex and the Wild OR-1. Instruments of the electronic scanning type are typified by the Gestalt Photomapper, and the UNAMACE.

15.12.1 GIGAS ZEISS ORTHOPROJECTOR

In the Gigas-Zeiss system, a conventional plotting instrument (for example, a C-8 Stereoplanigraph) is used in conjunction with a separate GZ1 Orthoprojector (figure 15-26). The plotting instrument drives the Orthoprojector either through direct mechanical or selsyn connection or through the readout of values deter-

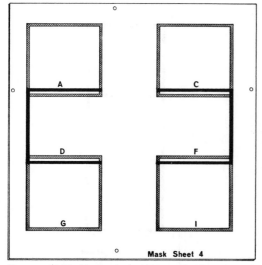

FIGURE 15-24. Mask sheet 3 is on the bottom and mask sheet 4 on top. Gaps are made where A and D, G and D, C and F and I and F join.

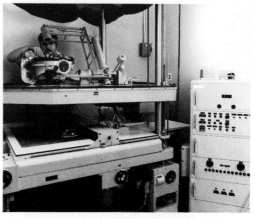

FIGURE 15-26. Gigas Zeiss orthoprojector.

mined in the plotting instrument and stored on tape for off-line application. The GZ1 projection system is made up of optical components of the C-8 Stereoplanigraph.

In operation, the exposure slit of the Ortho-projector travels in meander strips across the projection surface. As the slit travels, the scale of the projection image is continuously varied in accordance with the relief of the terrain by the z motion of the carriage which holds the projection system. Further functions, such as the control of the positive-lens carriage in the auxiliary system, the coordination between illumination and pupil of projection lens, as well as the position of pivots and principal points which determine optical independence follow known conditions.

15.12.2 K-320 ORTHOSCAN SYSTEM

The Kelsh K-320 ORTHOSCAN system (figure 15-27) is unique in that the projectors and the film plane are fixed and the exposing platen is moved in z to effect the scale changes. The image is tranferred from the exposing platen to the film bed by a flexible fiber-optic conductor.

A shutter is mounted on the platen and contains a slot that limits the size of the projected beam entering the fiber optic image conductor. This shutter is automatically actuated by two solenoids for an open or closed condition. Shutters with five different slot widths are available for the K-320. During the operation of the instrument, shutters of several different widths may be used during the printing of one or-thophoto-model.

The image is conveyed from the slot on the shutter to the film bed by means of the flexible fiber optic ribbon. This ribbon is 2 mm wide by 27 mm long, permitting a 24 mm maximum stepover in X with either a rectangular or parallelogram slot. The fiber optic ribbon itself is 910 mm long and is potted in epoxy for about 25 mm at both ends, at the platen surface and also where it contacts the flim. The film end of the ribbon is held very close to the emulsion surface of the film by means of a gimbaled air bearing.

In the K-320 system, the platen is stepped over the width of the aperture slot in the X direction, and the scanning is done by the operator in the Y direction.

The instrument can be operated on-line or off-line using a digital tape which is created during the scanning process. Off-line printing has the obvious advantage that the tape can be generated at a speed dictated by the complexity of the terrain and played back at a constant speed. In addition, the digital tape can be used to generate contours with computer programs that convert profile data into contours.

15.12.3 UNIVERSAL DIFFERENTIAL RECTIFIER, ORTHOPHOT B

The Jena Orthophot B is attached to the Topocart (figure 15-28). Its principal design features are that it uses orthogonal projection of the right hand photograph of the Topocart (figure 15-29) onto a cylindrical film carrier. The cylinder is moved back and forth in the y-direction by a motor and is rotated in steps for the x-direction. Since the cylinder is light-tight the orthophot B can be operated in a lighted work area. The instrument has a fixed optical system with two moveable inversor prisms for control of magnification. Movements in x and y take place in the photo-carriers of the Topocart and in the previously mentioned film drum of the Orthophot. The instrument has an enlargement range of from $0.7\times$ to $5.0\times$. Exposure control is through use of a gray wedge that is automatically controlled.

15.12.4 GESTALT PHOTOMAPPER II

The Gestalt Photomapper II (figure 15-30) is a computer controlled instrument that is primarily an analytical plotter equipped with two elec-

FIGURE 15-27. Kelsh K-230 ORTHOSCAN.

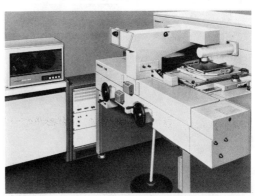

FIGURE 15-28. Jena Orthophot B attached to the Topocart.

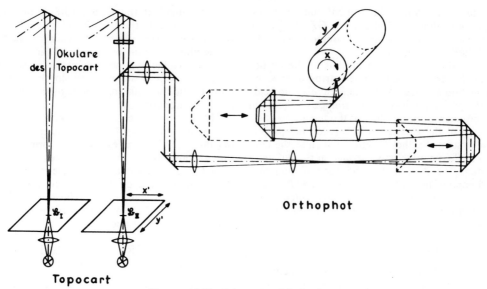

FIGURE 15-29. Schematic of Orthophot.

tronic printer output units in place of an output coordinatograph. The instrument uses automatic image correlation to measure spatial displacements between corresponding images in the left and right input stages and produces orthonegatives by systematic, patch by patch, differential rectification. The area being rectified is fixed at 8 × 9 mm, however, the recorded area is determined by a mechanical mask. The mask size is selected according to terrain characteristics. The GPM II consists of several modules; the control computer and its peripherals, the scanner units, the operators console, correlator and the output printers.

The control computer is a Nova 800 with

32,000 16-bit words of core-memory. The peripherals are two 9-track magnetic tape units, a CRT terminal, and a paper-tape reader. The computer is linked to all subsystems by a data cable for bidirectional control and data transfers.

The two scanner units are the coordinate measuring devices for the system. The scanning system consists of a cathode ray tube (CRT), an optical system, and a photomultiplier (PMT). The light for the CRT is transmitted through the diapositive and focused on the PMT. The PMT converts the light to an electric video signal that is sent to all other subsystems of the GPM-2. The video system and scanning rasters are controlled by electronic circuits that permit the scanned image to be linearly scaled, rotated, and translated.

The GPM-2 system is controlled at the operator's console which has a CRT terminal, a high-resolution television monitor, an x-y recorder, four handwheels, controls for adjusting correlator parameters, and a paper-tape reader. The CRT terminal is the main device for communicating with the computer. It is used to input data and display the status of all system modules. The high-resolution TV monitor displays the scanned images of the right and left diapositives. The images are displayed alternately to provide an accurate method of measuring parallaxes with respect to the measuring mark. The x-y recorder tracks the current scanner position on a paper print, necessary because the operator can only view a 9 × 8 mm area of the stereomodel.

The correlator measures the spatial displacement between corresponding images in the left and right video signals. The displacements represent x parallax and are proportional to ground heights. The correlator is a self-contained module with a memory for storing the measured parallax values. The parallax values are used as correction signals to the scanning systems, as data for generating a real-time contour display, and as data for the digital elevation model. Parallax values are calculated with each 9×8 mm patch on a 182 μm grid.

The two printing units consist of a transport system, CRT, optical system, and mechanical masks. The units can display alphanumeric text, contours, and orthophoto imagery. The printer and scanner transports are synchronized in their movement by the computer so that the differentially rectified patches are mosaicked on photosensitive emulsions. The GPM II system is capable of producing one orthonegative, a contour line-drop plot and a digital terrain tape of profile elevation data.

15.13 Stereo Orthophotos

Stereoscopic interpretation of the orthophotograph image is an essential requirement to make orthophotographs acceptable as an information source which would substitute for the original aerial photographs in mapping operations. Derivation of precise height information from the orthophotograph is also needed.

The geometry of orthophotographs, combined with their information content, offers the possibility of using the orthophotograph image as a basic information source in a wide range of mapping applications. This is particularly true for mapping of specialized terrain information in fields such as agriculture, forestry and geology, which in many cases has to be done by specialists who may not have access to photogrammetric plotting equipment or may lack the necessary photogrammetric expertise.

In order to interpret the terrain surface stereoscopically from an orthophotograph, a special photographic image has to be used which, together with the orthophotograph provides the x-parallaxes necessary for three-dimensional definition of the terrain surface. The stereoorthophotograph technique uses a stereomate for this purpose. The stereomate is a differential-

ly rectified photograph on which image shifts have been introduced in the x-direction proportional to the elevation differences in the terrain and parallel to the base of the original stereoscopic model.

The stereomate is produced following a similar differential rectification process as is used for the orthophotograph, the main difference being that the film on which the stereomate is recorded is continuously shifted with respect to the projected image in order to introduce the necessary x-parallaxes during the scanning procedure.

Generally, any orthophotograph generation device that scans in the x-direction can be converted to produce the stereomate. The GPM II system can use the height matrix generated by the automatic image correlation system to calculate the artificial x-parallaxes for production of a stereomate. The most advanced work on the production of stereo-orthophotographs, and their subsequent use has been done at the National Research Council of Canada where a complete system has been designed for the simultaneous production of the orthophotograph and its stereomate as well as a stereocompiler for the plotting of planimetry and contours.

15.14 Orthophotomaps

An orthophotomap is a topographic map on which the natural and cultural features of an area are depicted by color enhanced photographic images. Cartographic symbolization, contours and elevations, as well as lettering and boundary information are added as required by the user. On the orthophotomap the principal means of conveying the planimetric information is through the photographic imagery, therefore the orthophotomap contains much detail that is not found on a conventional line map. The following types of information generally are more fully depicted on orthophotomaps. The drainage patterns are more complete, vegetative cover is more accurate, cultivated areas are evident, field lines are shown and all features are shown in their true positions; there are no displacements due to cartographic symbolization.

Orthophotomaps, especially for medium scale mapping, are ideal for depicting those areas

where natural features are predominant, that is, where there is a low density of manmade features. However, they are also suitable for mapping in urban areas at scales of 1:10,000 and larger.

To lithographically reproduce the orthophotographic imagery without the use of screens requires the preparation of special image reproduction films. The continuous-tone imagery of the original orthophotomosaic is converted to minute random-dot patterns. This process of conversion is referred to as the image-tone process.

15.14.1 IMAGE REPRODUCTION FILMS

The image-tone process produces a positive film by exposing high-contrast litho film through a back-to-back (emulsions apart) sandwich of material consisting of a continuous-tone camera negative of the orthophotomosaic and a continuous-tone contact positive 70% as dense as the negative. The immediate result is a positive transparency on which the imagery consists of high-density random-dot patterns. The dot patterns are caused by a granularity or grain-clumping action during the preparatory steps and also the actual exposure. The product obtained from the exposure of the sandwich of materials is the image-tone positive film. From this positive, three final reproduction films are prepared for use in making the pressplates. These films are the image-tone negative, accent-tone negative, and surface-tone positive.

15.14.1.1 IMAGE-TONE NEGATIVE

The image-tone reproduction negative is a contact copy of the image-tone positive. This negative contains all of the photographic imagery shown on the positive. A pressplate made from it is used to print the basic map detail in selected colors.

15.14.1.2 ACCENT-TONE NEGATIVE

The accent-tone reproduction negative is a contact copy processed from the image-tone positive in such a way that the highlights and some middle-tone detail are eliminated and im-agery which accent the shadows and darker areas of the photographs left.

15.14.1.3 SURFACE-TONE POSITIVE

The surface-tone reproduction positive is a reverse-reading direct-positive contact copy of the image-tone positive and is used in printing the photographic highlights—the ground color or other surface characteristics of the area.

15.14.2 IMAGE-TONE PROCESSING

The steps in preparing the image-tone positive and the three reproduction films are shown graphically in (figure 15-31). For satisfactory results, extreme care must be taken to satisfy the critical requirements of each step.

15.14.2.1 CAMERA NEGATIVE

Production of the camera negative (step one, figure 15-31) requires a uniform density range and contrast balance. A typical suitable camera negative is obtained when the density-range readings from an imprinted gray scale on the camera negative are 1.21.

15.14.2.2 POSITIVE MASK

The positive mask is contact printed, emulsion to emulsion from the camera negative (step 2, figure 15-31). The density-range readings on the

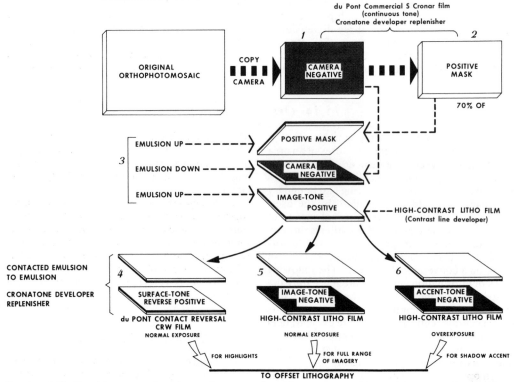

FIGURE 15-31. Graphic presentation of steps for preparation of image-tone positive and three reproduction films.

positive mask should be 0.86. The positive mask combined with the camera negative should produce a sum of densities of 0.35.

15.14.2.3 IMAGE-TONE POSITIVE

When the proper masks have been obtained the positive mask and camera negative are registered and taped with emulsions apart (step 3, figure 15-31). Exact register is important. This combination of positive and negative is exposed onto high-contrast litho film. The resulting product is a random-dot image-tone positive from which the previously mentioned surface-tone reproduction positive, the accent-tone reproduction negative and the image-tone reproduction negatives are produced. The combination of exposures of these reproduction films onto the lithographic plate completes the process of preparing image reproduction films for orthophotomap printing.

REFERENCES

Brumm, Maurice G.; Waters, James G. (1977) Photodensity Control System for Orthophoto Products, *Photogrammetric Engineering and Remote Sensing* Vol. 43, No. 9 pp 1177−1182.

Brunson, Ernest B. (1979) Operational Use of the Gestalt Photomapper II U.S. Geological Survey paper.

Clark, Clyde A.; Pumpelly, Jack W. (1968) Procedures for Processing Films for Orthophotomaps U.S. Geological Survey paper.

Danko, Joseph A. Jr. (1973) A New Concept in Orthophotography, *Photogrammetric Engineering* Vol. 39, No. 11 pp 1161−1170.

Hobbie, Dierk (1974) Orthophoto Project Planning, *Photogrammetric Engineering* Vol. 40, No. 8 pp 967−984.

Jeric, Anthony; Scarano, Frank (1975) Off-Line Orthophoto Printer, *Photogrammetric Engineering and Remote Sensing* Vol. 41, No. 8 pp 977−991.

Szangolies, Klaus (1971) Procedures and Accuracy of Differential Rectification with Topocart-Orthophot—Orograph, Collogue International Orthophotographie et Orthophotocartes, Paris pp 315−47.

van Wijk, Marius C. (1975) Stereo-Orthophotos Versus Conventional Orthophotos, Proceedings of the Orthophoto Symposium São Paulo, Brazil.

Warren, Alden (1977) Mosaicking By Photomechanical Method, Pan American Institute of Geography and History, Quito, Ecuador; 1977.

CHAPTER XVI

Non-Topographic Photogrammetry

Author-Editor: H. M. KARARA

Contributing Authors: M. CARBONNELL, W. FAIG, S. K. GHOSH, R. E. HERRON, V. KRATKY, E. M. MIKHAIL, F. H. MOFFITT, H. TAKASAKI, S. A. VERESS

16.1 Introduction

SINCE ITS INCEPTION over a century ago, photogrammetry has had its principal application to the production of topographic maps of the earths surface. As an efficient and convenient mapping tool, photogrammetry has developed into a highly reliable and precise measurement technique.

Since the early days of photogrammetry, photogrammetrists have tried applying their techniques outside the field of topographic mapping. Applications outside the field of topographic mapping have been referred to by some as "non-topographic." The development of non-topographic photogrammetry followed the general photogrammetric development in the mapping area. In recognition of the importance of non-topographic photogrammetry, the International Society for Photogrammetry has long devoted one of its seven technical commissions (Commission V) to this subject. Since its formation at The Hague ISP Congress in 1948, Commission V has had numerous titles including:

"Applications of Photogrammetry in Various Domains," "Special Applications of Photogrammetry," "Non-Cartographic Applications of Photogrammetry," "Non-Topographic Applications of Photogrammetry" and since 1972 "Non-Topographic Photogrammetry." This vacillation from one title to another reflects the continuing search for a satisfactory term to refer to the ever expanding fields of application of photogrammetry outside the field of topographic mapping.

For the lack of a better umbrella-term, this chapter is titled "*Non-Topographic Photogrammetry*" and covers the following topics: close-range photogrammetry, terrestrial photogrammetry, underwater photogrammetry, X-ray systems and applications, scanning electron microscope systems and applications, holographic systems and applications, moiré topographic systems and applications. A representative bibliography of the entire field of "special applications" of photogrammetry is included.

16.2 Close-Range Photogrammetry

16.2.1 INTRODUCTION

The term close-range photogrammetry is generally used when object-to-camera distances are not more than some 300 metres.

In any photogrammetric process, there are two major phases: a) acquiring data on the object to be measured by taking the necessary photographs, and b) converting the photographic data (perspective projection) into maps or spatial coordinates (orthographic projection). Thus, the total photogrammetric process can be subdivided into: data acquisition and data reduction.

The data acquisition system is concerned with procuring what may be termed the raw data or raw information. The raw data is realized in terms of the photograph. Hence the data acquisition system is concerned with obtaining necessary and suitable photography.

The data reduction system is used for converting the geometric information on the photo-

graphs into a final form suitable for the intended use. The final form may be analog, such as a map, or digital and recorded as printed or punched numbers.

When the output of the photogrammetric system is to be combined with other data and/or further modified, the term "data-processing system" is generally used for the system. An example would be the processing of the data to obtain the surface area or the volume of the object photographed.

16.2.2 PHOTOGRAPHIC DATA ACQUISITION SYSTEMS

Metric as well as non-metric[1] cameras are used for photography in close-range photogrammetry.

---

[1] Non-metric cameras are cameras not designed specifically for photogrammetric purposes. Amateur and professional cameras are included in this category.

16.2.2.1 METRIC CAMERAS

Close-range metric cameras include single cameras and stereometric cameras. Tables 16-1 and 16-2 list some of the pertinent characteristics of most of the currently available close-range metric cameras. These tables are based, in part, on information published by Carbonnell (1973).

16.2.2.1.1 SINGLE CAMERAS

Most modern photogrammetric single cameras consist of two main parts—an orientable camera support which can be mounted on a tripod and a tiltable metric chamber—which can be separated for transport. Most of such cameras operate with glass plates and some permit the use of roll and/or cut film. Table 16-1 lists most of the currently available single cameras.

Most of the single cameras listed in table 16-1 have variable principal distances to widen their focusing ranges. Focusing may be either in discrete steps or continuous. For example, through the use of adapter rings, the Wild P31 camera, which has otherwise a fixed focus of 25 metres, can be focused also to 7, 4, 2.5, 2.1, 1.8, 1.6 and 1.4 metres. Another way to vary the principal distance in a step-wise fashion is through the use of lenses which can be attached to the camera. For example, close-up lenses for camera-to-object distances of 0.5, 0.6, 0.75, 1, 1.5, and 2.5 metres can be attached to the lenses of Zeiss TMK-6 single camera.

With the exception of the Hasselblad MK70, single cameras must be supported on a tripod or some other suitable support. The Hasselblad MK70 can be held in the hand or supported on a tripod. In this camera the principal distance is continuously changeable, using the method found in focussable cameras designed for amateurs.

Not included in table 16-1 are metric single cameras designed, fabricated, and used for special purposes, such as those made by DBA Systems, Inc. of Melbourne, Florida. One of these cameras was designed for ultra-precise, close-range measurements of structures. It is equipped with a 19-inch focal length lens and accepts 190 × 215 mm × ¼-inch glass plates and produces a 7 × 7 inch picture. This camera is focused at infinity. Provisions for focusing at finite distances is made at the back of the camera and may be shimmed for focus at any distance from 25 feet to infinity. A pair of shorter focal length cameras has been built by DBA Systems Inc. for object distances less than 25 feet.

Another single camera system was built by DBA Systems Inc. specifically for work with space-environment simulators. The system includes three remotely-controlled, automatic plate-changing single cameras capable of carrying 16 plates 190 × 215 mm × ¼ inch or 32 plates of ⅛ inch thickness. Plates are loaded in a darkroom in individual cassettes which in turn are stacked in the removable supply section of the magazine. Once exposed, the plates are unloaded from the removable take-up section of the magazine. Lens systems are interchangeable so that lens systems of different focal lengths may be used as dictated by project requirements. Transfer of plates is completely positive allowing the single cameras to operate in any orientation. A remote controller operates the single cameras synchronously or independently.

On a few occasions, aerial cameras have been modified and used for close-range photogrammetric work. For example, Gracie (1971) used four K-24 aerial cameras for measurement of a spinning network (a model of a proposed large orbiting radio telescope). Each camera was refocused for an object-distance of seven metres by inserting a specially-machined metal spacer between the lens system and the body. Calibration constants were determined from measurements on photographs of the project.

Another example of custom-built close-range cameras is the Wild P72 (figure 16-1) which in-

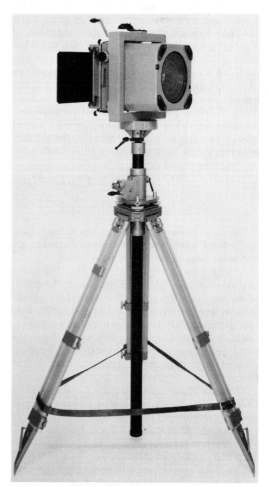

FIGURE 16-1. Wild P72 custom-built terrestrial camera (courtesy Wild-Heerbrugg, Inc.).

corporated a Wild RC5/RC8 cone mounted on a Wild C120 tripod. According to the customer's requirements the camera was focused at 13 m. An Aviogon lens was used, picture format was 18 × 18 cm, maximum aperture f/5.6, nominal principal distance 115 mm.

A brief discussion of five of the most recently introduced close-range single cameras is given below to convey some insight into the current trends in design of such cameras.

16.2.2.1.1.1 Hasselblad MK-70. This is a small and easy to use metric single camera fitted with a reseau (on a glass plate) located in the opening of the camera's back-plate immediately adjacent to the film plane. The reseau has 25 cross-shaped index marks. A 60 mm f/5.6 Biogon and a 100 mm f/3.5 Planar lens system is available for the MK70. Both lens systems have Synchro Compur shutters with speeds from 1 to 1/500 sec. and B (on command). While the Biogon has a focusing range from 0.9 m to ∞, the Planar lens system is permanently set at infinity. Upon request, the Planar lens system can be obtained prefocussed at any desired distance down to 2 metres. Two magazines are available for different 70 mm film loads with type II perforations. The magazine MK 70/70 is designed for standard-base film only, while magazine MK 70/200 can be used with standard- or thin-base film. Both magazines can be loaded with film length up to the maximum load and will shut off and obstruct exposure when film is out. Beside the standard quick coupling to a tripod, a special coupling is also available, which eliminates the side play and assures a good repeatability of camera position after removal and reinstallation of the camera.

16.2.2.1.1.2 Jenoptik UMK 10/1318 Universal Photogrammetric Camera System. This system is composed of four basic groups of functional units (metric chambers, magazine, mount, elec-tronic gear) which can be combined into four logical combinations. Two versions of the metric chambers are available; type F, equipped with a Lamegon 8/100 lens (f/8, f = 100 mm), and the other, type N, equipped with a Lamegon 8/100-N lens. The focusing range of both lenses is 1.4 m to ∞. The Lamegon 8/100 lens has less than 12 μm distortion for distances between infinity and 3.6 metres, while the Lamegon 8/100-N lens has less than 12 μm distortion for object distances between 1.4 and 4.2 m. The camera axis can be tilted between −30° and +90° with stops at 15° intervals. The magazine has two basic designs: one for 13 cm × 18 cm glass plates (version P) and the other (version F) for 190 mm roll film. Since both magazines can be optionally fitted with a Lamegon or a Lamegon-N lens, the system involves all together four types of cameras; two for film (10/1318FF and 10/1318NF) and two for glass plates (10/1318FP and 10/1318NP). For film flattening, a vacuum back, similar in design to those used in aerial cameras, is used and is connected to an external vacuum pump. A plate adapter frame can be attached instead of the film magazine, thus enabling 13 × 18 cm glass plates to be used in the film model. Single and double mounts are available. A special single mount for taking pictures vertically downwards is also available.

A combination of two single mounts on a support beam constituting the base results in a double camera system. A beam is commercially available with arrangements for base lengths of 320, 580, and 840 mm. The beam is mountable on a tripod of the type used for the Jenoptik SMK 5.5/0808 stereometric camera. A more versatile version of the double mount is used in the Jenoptic IMK 10/1318 stereometric camera.

16.2.2.1.1.3 Kelsh K-470 Terrestrial Camera. This camera (f = 90 mm, f/8, format 105 × 125 mm) contains a single 90 mm wide angle fixed-

FIGURE 16-2. Hasselblad MK-70 (courtesy Victor Hasselblad, Inc.).

TABLE 16-1. CHARACTERISTICS OF METRIC SINGLE-CAMERAS

Manufacturer	Model 1	Size* of Photographic Material (cm)	Nominal Focal Length (mm)	Total Depth of Field (m)	Range of Tilt of Camera Axis & Number of Intermediate Stops	Photographic Material	Comments
Galileo	Verostat	9 × 12 U	100		0→ ± 90° (2)	glass plates or cut film	variable principal distance (in steps)
Galileo	FTG-1b	10 × 15 H	155	10→∞	0→ ± 36° (continuous)	glass plates	variable principal distance (in steps)
Hasselblad	MK70 (Biogon lens)	6 × 6	60	0.9→∞	unlimited△	70 mm film	△ hand held or on tripod. variable principal distance (continuous mode) single frame exposure or sequence exposure
Hasselblad	MK70 (Planar lens)	6 × 6	100	15→∞▽	unlimited△	70mm film	▽ fixed focus at ∞ (upon request fixed focus at desired distances down to 2m). △ hand held or on tripod. motor driven; single frame exposure or sequence exposure.
Jenoptik Jena	UMK 10/1318 FP UMK 10/1318 NP	13 × 18 UH	99	1.4→∞	−30°→ + 90° (7)	glass plates	Lamegon 8/100 lens with distortion <12μm for object distances ∞→3.6m. Lamegon 8/100 N lens with distortion <12μm for object distances 4.2→1.4 m.
Jenoptik Jena	UMK 10/1318 FF UMK 10/1318 NF	13 × 18 UH	99	1.4→∞	−30°→ + 90° (7)	190mm roll film & glass plates (with adapter)	Lamegon 8/100 lens with distortion <12μm for object distances ∞→3.6m. Lamegon 8/100 N lens with distortion <12μm for object distances 4.2→1.4m.

Manufacturer	Model	Format		Range	Tilt	Material	Remarks
Jenoptik Jena	19/1318 Photo-theodolite	13 × 18 H	190	25→∞	none[8]	glass plates	δ lens can be shifted vertically (+30→ − 45mm) in snap-in steps of 5mm.
Kelsh	K-470	10.5 × 12.7 UH	90	2→∞	none	cut film, roll film, glass plates.	image format offset from the optical axis of the lens by 13mm.
Sokkisha	MK165	12 × 16.5 U	165	10→∞	0→ ± 30° (2)	glass plates	variable principal distance (in steps).
Wild	P32	6.5 × 9 UH	64	0.6→∞	on T1, T16 or T2: 0→ ± 40° (continuous) on GW1: 0→ ± 30° (continuous)	glass plates, cut film, roll film	variable principal distance (in steps—interchangeable spacers).
Wild	P31	10.2 × 12.7 UH (4″ × 5″)	100	6.6→∞ (f/22) 12.4→∞ (f/5.6)	0→ ± 30° (3) also +90°	glass plates & cut film	variable principal distance (in steps—interchangeable spacers)—wide-angle lens.
	,,	,,	45	1.5→∞ (f/22) 3.6→∞ (f/5.6)	,,	,,	Super-wide-angle lens.
	,,	,,	200	18-640 (f/22) 26-53 (f/5.6)		,,	Normal-angle lens. Standard focusing 35 m; adapter rings on request; minimum distance 8 m.
Zeiss (Oberkochen)	TMK-6	9 × 12 UH	60	5→∞	0→ ± 90° (2)	glass plates	6 close-up lenses are available for object-distances of 0.5m, 0.6m, 0.75m, 1m, 1.5m, and 2.5m.
Zeiss (Oberkochen)	TMK-12	9 × 12 UH	120	20→∞	0→ ± 90° (2)	glass	

* U/H: format Upright/Horizontal; UH: format Upright or Horizontal

TABLE 16-2. CHARACTERISTICS OF STEREOMETRIC CAMERAS

Manufacturer	Model	Size* of photographic Material (cm) for single camera	Nominal Focal Length (mm)	Base Length (cm)	Operational Range (m)	Range of Tilt of Optical Axes and Number of Intermediate Stops	Photographic Material	Comments
Galileo	Veroplast	13 × 18 H	150	56	1.6→∞	0→ ± 90° (continuous)	glass plates	variable principal distance (in steps)
Galileo	Veroplast	13 × 18 H	150	200	5→∞	0→ ± 90° (continuous)	glass plates	variable principal distance (in steps)
Galileo	Veroplast	9 × 12 U	100	120	2→∞	0→ ± 90° (continuous)	glass plates or cut film	variable principal distance (in steps)
Galileo	Technoster A	6.5 × 9 H	75	16→70	0.5→6	0→ ± 18° (continuous)	roll film	variable base length; convergence of individual cameras possible (0→ 13°); variable principal distance (in steps)
Galileo	Tecnoster B	23 × 23	150	30→70	2→5	$-45° → + 5°$ (continuous)	glass plates	variable base length
Jenoptik Jena	SMK-5.5/0808	8 × 8	56	40	1.5→10	0→ ± 90° (5)	glass plates	
Jenoptik Jena	SMK-5.5/0808	8 × 8	56	120	5→∞	0→ ± 90° (5)	glass plates	
Jenoptik Jena	IMK-10/1318	13 × 18 UH	99	35-160	1.4→∞	0→ - 45 (common & continuous)	glass plates or 190 mm film	variable base length; individual ϕ tilt (0→11°); common ω tilt (0→ - 45°)
Kelsh	K-460	10.5 × 12.7 U	90→120	23.7→92.0 (14.2→50.0 for table model)	0.36→∞	None	cut film, roll film, glass plates	variable principal distance (continuous); variable base length (continuous); 2 models
Nikon	TS-20	6.5 × 9 H	64	20	0.9→5	0→ ± 90° (2)	glass plates or cut film	
Nikon	TS-40	9 × 12 U	60	40	2.5→10	0→ ± 90° (2)	glass plates	

Nikon	TS-120	9 × 12 U and 6.5 × 9 U	60	120	5→50	0→±90° (2)	glass plates	designed primarily for bio-medical applications; variable principal distance (in steps)
Sokkisha	B-45	12 × 16.5 H	121	45	1→5	None	glass plates	
Sokkisha	SKB-40	6.5 × 9 H	67	40	2.5→10	0→±45° (continuous)	glass plates	
Sokkisha	SKB-120	6.5 × 9 H	67	120	5→∞	0→±45° (continuous)	glass plates	
Sokkisha	KSK-100	12 × 16.5 U	90	30→100△	1→∞	0→±15° (continuous)	glass plates	variable principal distance (in steps) △ base length settings: 30, 50 and 100 cm
Sokkisha	V-3	12 × 16.5 H	121	25→50▽	0.5→5	0→±27° (continuous)	glass plates	variable principal distance (in steps) ▽ base length settings: 25, 35 and 50 cm standard equipment
Wild	C 40	6.5 × 9 H	64	40	1.5→7	0→±90° (4)	glass plates	
Wild	2P32's with Base-Bar	6.5 × 9 UH	64	40,30, 20	0.9→9 0.6→2.5	horizontal only	glass plates, cut film, roll film	special
Wild	C 120	6.5 × 9 H	64	120	2.7→∞	0→±90° (4)	glass plates	
Zeiss (Oberkochen)	SMK-40	9 × 12 U	60	40	2.5→10	0→±90° (2)	glass plates	6 attachable close-up lenses are available for object distances of 0.5m, 0.6m, 0.75, 1m, 1.5m, and 2.5m
Zeiss (Oberkochen)	SMK-120	9 × 12 U	60	120	5→∞	0→±90° (2)	glass plates	

* U/H: format Upright/Horizontal; UH: format Upright or Horizontal

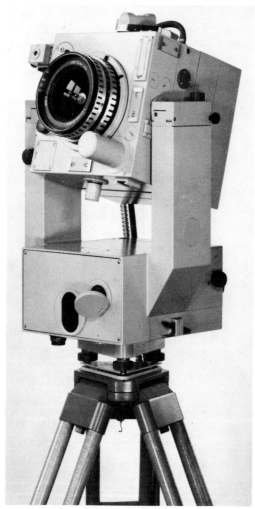

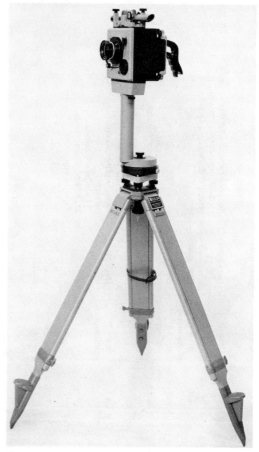

FIGURE 16-4. The Kelsh K-470 terrestrial camera (courtesy Kelsh Instrument Division/Danko Arlington, Inc.).

FIGURE 16-3. Jenoptik UMK 10/1318 FF universal measuring camera (film mode 1), with the standard single-camera mount (courtesy Jenoptic Jena, Ltd.).

focus lens with openings from f/8 to f/64. The focusing distance is from 2 metres at f/64 to infinity. Manual shutter speeds are 1/500 second to 1 second, with B (on command) setting. The K-470 uses elements of the K-460 system (section 16.2.2.1.2.4), along with a telescopic sight for convenience when establishing the photogrammetric network. The telescopic sight has 18× magnification and has a focusing range from 1.2 metres to infinity. The center of the frame of the 105 × 127 mm, stable-base, polyester-film used in the K-470 is offset from the optical axis of the lens by 13 mm. The photographer can mount the camera on its support in any one of four positions. For example, for photography of open pit mining, if the photographer is standing on the edge of a pit, he can position the camera so that the center of the film frame is higher than the optical center of the lens, more effectively capturing the imagery of the open pit mine below

the center of the lens. A quartz watch has been incorporated inside the body of the K-470. When the shutter is actuated, a battery-powered circuit automatically fires an internal flash in a light-sealed chamber within the body. The flash illuminates optical fibers which place four fiducial marks on the film at the instant of exposure. At the same time, the flash illuminates the face of the quartz watch, which is imaged by a tiny lens-system onto one corner of the film. The watch-face shows the day, month, year, hour, minute and second at the instant the film is exposed. With this information permanently recorded on the film, the exposure becomes a record for inventory purposes, as well as having an indication of the position of the exposure in the sequence. The watch itself is powered by a battery which needs to be replaced once per year. The magazine uses a trigger-operated pressure-pad mechanism and accepts single-film cassettes or roll-film cassettes. The Kelsh K-480 Stereoplotter has been designed to restitute imagery from the K-470 as well as the K-460.

16.2.2.1.1.4 Wild P31 Universal Terrestrial Camera. This camera (f = 100 mm, f/8, plate

FIGURE 16-5. Wild P31 universal terrestrial camera, with super-wide-angle lens cone (courtesy Wild-Heerbrugg, Inc.).

format 4 in. × 5 in.) was introduced in 1974. It is focused at a standard distance of 25 m. Sharp images can be obtained for object distances between infinity and 6.5 m (for an aperture of f/22 and circle of confusion of 0.05 mm). With additional, precise, adapter rings, which can be easily put on by the user himself, the P31 can now be focused on distances of 7 m, 2.5 m, 2.1 m, 1.8 m, 1.6 m, and 1.4 m. The radial distortion of the lens is less than ± 4 μm. The P31 operates with 4 in. × 5 in. × 3 mm glass plates and has an adapter for cut film.

In 1975, a significant improvement to the P31 was announced: an interchangeable, super-wide-angle lens system (Wild 4.5 SAgII) was developed for this camera. The new lens system has an extremely large depth of field, ranging from 1.5 m to ∞ for f/22 and a circle of confusion of 0.05 mm diameter (3.6 m to ∞ for f/5.6 and the same diameter of circle of confusion). The nominal principal distance of the lens system is 45 mm. Glass plate (cut film) size is 4 in. × 5 in.

(102 mm × 127 mm) and image size is 92 mm × 118 mm. The radial distortion is within ±4 μm, and the resolution at f/5.6 and infinite contrast is 70 lines per mm (area-weighted average).

At the ISP Helsinki Congress in 1976, a new normal-angle lens system for the Wild P31 was introduced. The focal length of this system is 20 cm, and the image is 10 × 13 cm. The system at the largest aperture of f/8, has a maximum distortion of 4 μm. The long camera resulting from the long focal length is lightened by the use of titanium. The single camera contains a glass plate in the image plane which carries a center cross and fiducial marks. In its standard version, the lens system is focused at an object distance of 35 metres. The user himself can exchange spacers in order to adapt the P31 for a shorter distance.

16.2.2.1.1.5 Wild P32 Terrestrial Camera. This is a small, light-weight single camera which

FIGURE 16-6. Wild P32 terrestrial camera, mounted on a Wild T16 scale reading theodolite (courtesy Wild-Heerbrugg, Inc.).

can operate with glass plates, cut-film (size 65 × 90 mm) or roll-film (standard size 120). It can be mounted on Wild T16 and T2 theodolites by means of special adapters. All theodolites adapted for the Wild D1-3 can also be used with the P32. The maximum aperture for the lens is f/8 and the nominal focal length is 64 mm. At maximum aperture and high contrast on Agfa Gaevert IP 15 film, the resolution is more than 100 lines/mm in the center of the picture and about 70 lines/mm in the corners. The principal point of the picture is offset by 10 mm relative to the principal parallel. The entire camera can be rotated through 360° around the optical axis, with fixed stops at 0°, 90°, 180° and 270°.

The P32 has a fixed standard focus at 25 m, which permits sharp photographs down to a minimum distance of 3.3 m (for f/22 and a circle of confusion of 0.05 mm diameter). Any change in focus between 25 m and 2.5 m can be performed at any Wild workshop. If a shorter focusing distance is required, the P32 must be modified in the factory in Heerbrugg. The focusing range then possible is 0.7 m to 1.40 m. Once this modification has been made, a change of focusing within the range 0.7 m to 1.40 m, a change to 2.5 m or more, and a change back to the range 0.7 m to 1.40 m can be made in any Wild workshop. A ground glass screen with shades, which is inserted into the camera instead of the film or plates, serves as a viewfinder.

A 40 cm bar is available for use in combination with two P32's, permitting stereoscopic horizontal photographs to be taken (the photogrammetric "normal case"). The bar has four adapters, making three base lengths possible: 40 cm, 30 cm, and 20 cm. The bar can be rotated around a vertical axis and locked with a lever. Synchronized release for both P32's is possible.

16.2.2.1.2 STEREOMETRIC CAMERAS

Table 16-2 lists some pertinent characteristics of most of the currently available stereometric cameras.

With a few exceptions, stereometric cameras operate with glass plates and consist of two single cameras rigidly mounted on a base of definite length, with the optical axes parallel to each other and perpendicular to the base (normal case of photogrammetry). Thus the minimum and maximum distances of the object and the maximum accuracy obtainable are set by the fixed geometry of the stereometric camera.

With the exception of the Jenoptik IMK-10/ 1318 and Galileo Tecnoster-A stereometric cameras, all currently available stereometric cameras are limited to the normal case of photogrammetry.

As is the case with most single cameras, most stereometric cameras have provisions for changing the principal distance to increase the range of focusing.

A small number of stereometric cameras will

be discussed in some detail to convey some insight in the modern trends of design of such cameras.

16.2.2.1.2.1 Zeiss SMK Stereometric Camera. Two versions of this stereometric camera are available: SMK-120 and SMK-40 with 120 cm and 40 cm base respectively. In both versions the lens system is the 60 mm wide-angle TOPOGON V, practically distortion free, with a fixed aperture of f/11. For the SMK-120, the depth of field is 5 m to ∞, and it is 2.5 to 10 m for the SMK-40. For both stereometric cameras, close-up lenses for camera-to-object distances of 0.5 m, 0.6 m, 0.75 m, 1 m, 1.5 m, and 2.5 m can be attached to the lenses. Both stereometric cameras operate with 9 × 12 cm glass plates (net negative area 8 × 10 cm). A special tripod is provided; it has extending legs, crank-operated vertical column and fixed control panel including a built-in battery. A wide-angle viewfinder is on the mid section of the tube holding the single cameras. In addition, mechanical framing marks are on the camera units. A bubble level in the control panel is used to level the stereometric camera. The picture is vertical (upright).

An adapter for architectural photography is available as an accessory. Using this adapter, photographs can be taken with the stereometric camera pointing vertically upwards, vertically downwards, inclined $\pm 30^g$, $\pm 70^g$, or with the base vertical.

Photographic transformation of photography taken at $\pm 30^g$ to the vertical plane, and of photography taken at $\pm 70^g$ to the horizontal plane is possible using the KEG-30 Transformation Printer, which is available as an accessory.

Other accessories include: an alignment mirror for aligning the base so that it is parallel to a facade; a telescope and an eccentric device for precisely extending the base line and obtaining

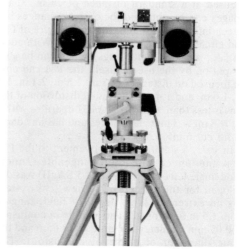

FIGURE 16-7. Zeiss SMK-40 stereometric camera (courtesy Carl Zeiss, Oberkochen).

FIGURE 16-8. Zeiss SMK-120 oriented for $+30^\circ$ oblique photography (courtesy Carl Zeiss, Oberkochen).

variable-base photography using only one stereometric camera.

16.2.2.1.2.2 Galileo-Santoni Tecnoster-A Stereometric Camera. This is composed of two single cameras, film size 6 × 9 cm and focal length 7.5 cm, mounted on a bar (fig. 16-9) with provisions to make the base length between 6.5 and 70 cm (in a special version of this stereometric camera the base length is variable from 1 to 2 metres). The camera is provided with a vacuum film-flattening device. The depth of field of the stereometric camera is between 0.5 and 6 m. The two single cameras can be tilted between $\pm 20^g$ and an angle of convergence up to 15^g between the optical axes can be introduced. The height of the base above ground can be varied between 0.5 and 2.3 m.

16.2.2.1.2.3 Jenoptik IMK 10/1318. This incorporates two UMK 10/318-N single cameras (section 16.2.2.1.1.2). The two single cameras are carried in fork-type mounts that are fastened to the base. The base length can be varied between 350 and 1600 mm. The mounts can be turned in two directions allowing, besides normal-case setting, a ϕ-tilt (convergence) of up to 12^g per single camera as well as a continuous ω-tip (inclination to horizontal) between 0^g and -50^g. The stereometric camera can be given a common inclination permitting optimum adaptation to the spatial position of the object at adjusted normal case or at a given convergence. This common convergence can be read on a scale, as is the case with the individual inclinations in ϕ and ω.

The base is connected to a vertical column and its height above the floor can be changed between 0.6 m and 2.10 m. A counterweight in the column permits vertical movement with ease. The column is held in a heavy foot and can be rotated within close limits. Wheels serve for coarse adjustment of direction.

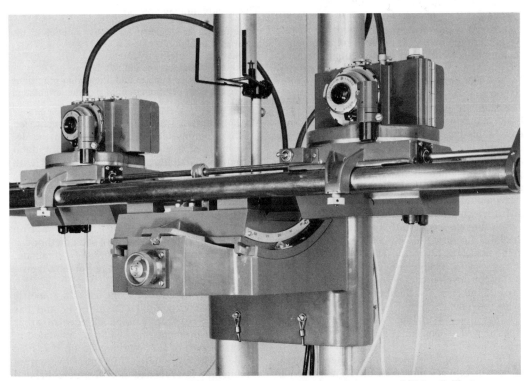

FIGURE 16-9. Galileo-Santoni Tecnoster-A stereometric camera (courtesy Officine Galileo).

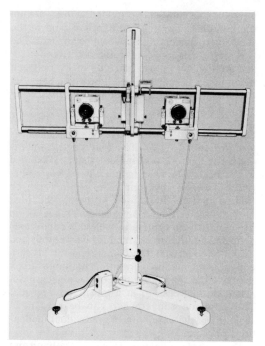

FIGURE 16-10. Jenoptik IMK 10/1318 (courtesy Jenoptic Jena, Ltd.).

The working ranges of the IMK 10/1318 correspond to those of the UMK single cameras incorporated in it; these have been discussed earlier. The IMK weighs 250 pounds and is 204 cm wide, 96 cm deep and 254 cm high.

16.2.2.1.2.4 The Kelsh K-460 Universal Stereometric Camera. This is available in two basic arrangements: (a) two single cameras mounted on a table and (b) two single cameras mounted on an X-bar supported on a tripod (fig. 16-11). The basic stereometric camera contains 90 mm wide-angle lenses and electronic shutters. The standard photo-size is 105 × 127 mm. The basic magazine has been designed to accomodate stable-base polyester negatives which are available in cut or in 105 mm roll form. Fiducials are illuminated by fiber-optic light-guides. The principal distance at the time of exposure is shown on the corner of the film. The basic stereometric camera will accept single film cassettes which are provided with it.

Roll-film cassettes, both upper and lower, are available for direct attachment to the magazine in the field. The lower cassette may be detached at any time to remove either a partial or full roll of film for development. A counter on the upper cassette indicates the number of available exposures remaining.

Other options include vacuum-plate magazines as well as magazines designed for use with glass plates.

The focusing distance of the stereometric camera is from 360 mm to infinity. The principal distance varies between 120 to 90 mm. X-bar base length is between 237 mm and 920 mm. Table base length is between 142 mm and 500 mm. Apertures are f/8 to f/32. The Kelsh K-480 Stereoplotter (section 16.2.3.1) is used to restitute imagery from the K-460 as well as the K-470.

16.2.2.1.2.5 NIKON TS-20 Stereo Terrestrial Photogrammetric Camera. This is a light-weight and compact stereometric camera of 20 cm base. Each camera has a Geo-Nikkor f/8-f/32 lens of 64 mm focal length, focused at 1.5 m. The depth of field is 0.9 m to 5 m. Distortion is less than 5 μm over the entire picture area. The camera operates with glass plates of 9 × 6.5 cm. The tripod has a tilting head with stops at 0, ±30° ±60° and ±90°.

16.2.2.2 NON-METRIC CAMERAS

Although various types of metric cameras are available, there is a considerable use for non-metric cameras that are immediately available. The use of non-metric cameras, as opposed to metric cameras, for photogrammetric purposes has the following advantages and disadvantages. The advantages are:

general availability,
flexibility in focusing range,
some are motor driven, allowing for quick succession of photographs,
they can be hand-held and thereby oriented in any direction, and
the price is considerably less than for metric cameras.

The disadvantages are:

the lenses are designed for high resolution at the expense of high distortion,
instability of interior orientation,
lack of fiducial marks, and the
absence of level bubbles and orientation provisions precludes the determination of exterior orientation before exposure.

Concentrated research and development, led by ISP Commission V, has resulted in the development of a number of methods for data-reduction particularly suitable for use with photographs from non-metric cameras. These methods are based on highly sophisticated analytical techniques which combine, in most cases, the calibration and evaluation phases. One of these methods will be outlined in section 16.2.3.2.1.2. Through the use of such methods, it is possible to reduce, or completely eliminate, the effects of the disadvantages of non-metric cameras listed above.

Since early in the 1970's, the use of non-metric cameras has expanded considerably and has made an impact on a large number of areas where measurements are required. As stated in the 1976 report of ISP Working Group V/2 (Faig 1976), "The non-metric/computer evaluation combination has reached its fullest potential and accuracies reaching the photogrammetric noise

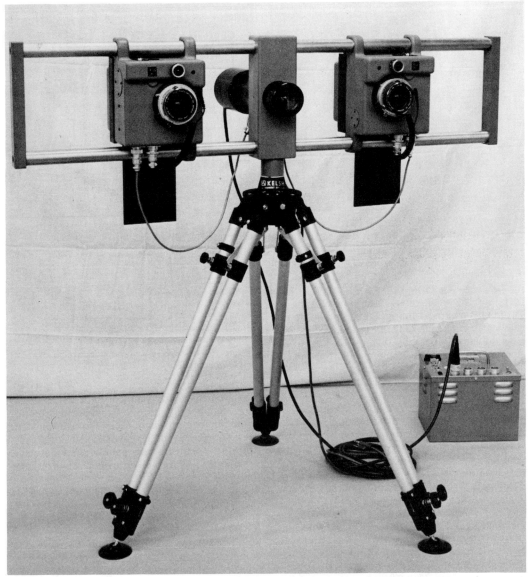

FIGURE 16-11. The Kelsh K-460 Universal stereometric camera (courtesy Kelsh Instrument Division/Danko Arlington, Inc.).

level have been achieved. It often depends on the individual project whether the low cost camera/expensive evaluation system or the metric approach is more suitable or financially advantageous, which leaves the decision to the user. Often project arrangements require versatility and light weight which can only be met by non-metric cameras, and with the progress that has been made in the evaluation phase this option now can be a high precision approach. The photogrammetric potentials of non-metric cameras are indeed very high.''

16.2.3 DATA REDUCTION SYSTEMS

Depending on the camera used and the desired final output, there are three alternatives for evaluation: analog, analytical and semi-analytical.

16.2.3.1 ANALOG APPROACH

The analog approach is recommended when the final product is to be in the form of a contour map[2]. This approach is suitable for data reduction from metric photographs but not recommended for precision evaluation of non-metric imagery in view of their rather large and sometimes irregular distortions (lens and film). It is important to note that, with a few notable ex-

---

[2] As mentioned in Section 16.2.3.2, the analytical approach can also be used for this purpose. The contours are then drawn by a computer-directed plotter.

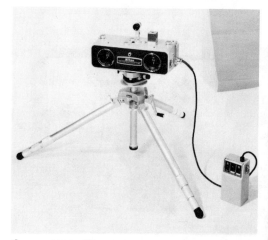

FIGURE 16-12. Nikon TS-20. Nikon TS-20 stereo terrestrial photogrammetric camera (courtesy Nikon, Inc.).

FIGURE 16-14. Zeiss terragraph plotter (courtesy Carl Zeiss, Oberkochen).

ceptions, most universal precision plotting instruments and most topographic plotters do not have a large enough range of principal-distances to evaluate photography taken with most of the metric and non-metric cameras. Plotting in such cases must therefore be done with an exaggerated principal distance (and an exaggerated "Z" scale).

Special plotting instruments, introduced for use with stereometric and metric single cameras, are available. Such plotters are, in general, designed for the normal case of photogrammetry and thus the term "normal case stereoplotters" is sometimes used to refer to this group of instruments. Table 16-3 lists some of the pertinent characteristics of "normal case" stereoplotters.

The Kelsh K-480 Stereoplotter (fig. 16-16) has been designed to restitute imagery from the Kelsh K-460 Universal Stereometric Camera as well as the K-470 Terrestrial Camera. The K-480 is equipped with projectors to accommodate the photographs from K-460 and K-470 single cam-

eras. Negatives from either single camera may be placed directly into the K-480 for restitution. If the photographs were taken through the glass pressure plate in the K-460 or K-470 single camera, the negatives are sandwiched between ground-glass plates of equivalent flatness and thickness in the plate-holders of the K-480, providing a correction for the projected imagery. If the vacuum platen is used on either single camera, each negative may also be placed in the K-480 between the two 3mm glass plates. Correction cams on the plotter may be moved to compensate for the thinner glass plate, and true imagery would be projected for the stereomodel.

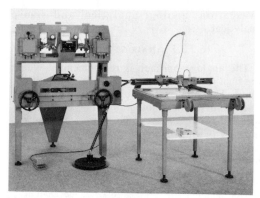

FIGURE 16-13. Wild A40 stereoplotter (courtesy Wild-Heerbrugg, Inc.).

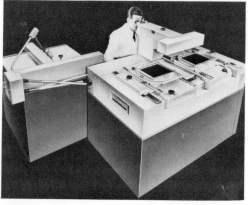

FIGURE 16-15. Jenoptik technocart plotter (courtesy Jenoptic Jena, Ltd.).

TABLE 16-3. CHARACTERISTICS OF "NORMAL CASE" STEREOPLOTTERS

Manufacturer	Model	Maximum Plate size (cm)	Range of Principal Distances (mm)	Range of Lateral Tilt	"Z" Range (mm)	Comments
Galileo	Verostat	9 × 12	PD=100*	±90°	60→400	*designed to plot from photographs taken by the Verostat stereometric camera
Jenoptik Jena	Technocart	23 × 23	50→215	±45°	35→350	
Nikon	TR-2	9 × 12	56→68	±60°	80→500	△designed for photography taken by stereometric cameras with horizontal principal axes
Sokkisha	ST-65	9 × 12	59→70	△	55→500	PD = principal distance
Wild	A40	9 × 12	54→100	±60°	40→(500-PD)	
Zeiss Oberkochen	Terragraph	9 × 12	52→67	±63°	60→610	

FIGURE 16-16. The Kelsh K-480 stereoplotter is designed to restitute imagery from the Kelsh K-460 universal stereometric camera and the K-470 terrestrial camera (courtesy Kelsh Instrument Division/Danko Arlington, Inc.).

The Hypergon lenses in the K-480 are designed to project imagery from a 6.5× to 10× magnification, from the normal focal lengths that are used in the single cameras. Projector illumination is provided by two 12-volt, 100-watt quartz halogen lamps. Also, the common-Z elevating screws on the K-480 are equipped with micrometer dials. Settings may be made on these dials to within 0.01 mm. The reference plane for the K-480 may be either a granite surface, or the "data tablet" for the Kelsh K-600 Digitizing System. In the latter case, the tracing table is fitted with a cursor around the pencil point, allowing the operator to move the table freely for contouring, or along a profiling bar for cross-sectioning. Stereo-viewing on the tracing table may be performed either with anaglyph filters, with the Kelsh M-20 Stereo Image Alternator System, or with the Kelsh PPV System.

Some stereoplotters, including Zeiss-Jena Stereometrograph F, Wild A7, Wild A9, Wild A10, Zeiss C-8 Stereoplanigraph, Zeiss Planimat D-2, Galileo Stereocartograph V, and Nistri Photostereograph Beta/2, and a few topographic plotters, including Galileo Stereosimplex IIC and Zeiss-Jena Topocart B are capable of plotting terrestrial as well as aerial photographs. Such plotters are suitable for close-range photogrammetry.

On the other hand, a few stereoplotters such as Kern PG3 and Sopelem Presa 225RC and most topographic plotters (such as Kern PG2, Nikon M5, Wild B8, Wild A8, Zeiss Planicart

E-2, Kelsh plotter and others) are capable of plotting from vertical or near vertical photographs only. Obviously, these plotters can also be used for plotting from horizontal and parallel or nearly horizontal and nearly parallel photographs, provided that the coordinate system is converted after the plotting is done. Also, topographic plotters capable of plotting from convergent aerial photograph, such as the Bausch and Lomb Balplex, have been used in close-range work.

For details about precision and topographic plotters, refer to chapter XII.

Analytical plotters (such as Nistri APC-3 and other models) are very suitable for close-range work. They combine in a most favorable way the advantages of the analog and analytical methods; on one hand retaining the capability of detailed mapping and on the other maintaining a high accuracy. For details about the analytical plotters, refer to chapter XIII.

Mention should be made of the Zeiss Stereocord G2, (fig. 16-17), a low-priced stereoplotter for simplified, computer-aided plotting. This instrument is suitable for those non-topographic applications which do not require the ultimate in precision. The Stereocord G2 is designed to make measurements (0.01 mm) from *one* photograph or *two* stereo photographs. A digitizer converts the position of a mark into digits and feeds these into an electronic calculator. The Stereocord G2 system has three basic components:

(1) a binocular viewing system and stage, which is linked to
(2) a digitizing unit (DIREC 1, the encoded x and y (one photograph) or x, y, and px (two stereo photographs) coordinates are displayed to 0.01 mm);
(3) a electronic calculator (typically Hewlett-Packard HP 9825). This calculator is connected to components 1 & 2 of the system. This calculator can be placed in communication systems, thus taking advantage of centralized-computer installations.

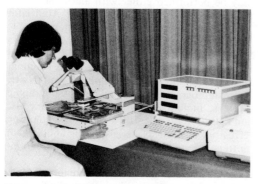

FIGURE 16-17. Zeiss stereocord G-2 (courtesy Carl Zeiss, Oberkochen).

The standard set of programmes which is supplied includes programmes for relative and absolute orientation, and for the determination of such quantities as distances in space, horizontal distances, vertical angles or slopes, areas, *etc.* Users may derive or design new programmes for their specific types of work.

16.2.3.2 ANALYTICAL APPROACH

An ever increasing use has been made of the analytical approach in which the measurements can be made either stereoscopically (using stereocomparators) or monocularly (using monocomparators).

There is essentially no difference in accuracy between results using data from stereocomparators or monocomparators in analytical photogrammetry (Karara & Marks, 1968). The difference between the two alternatives is primarily in the operational requirements. While the stereocomparator approach can be used for marked or unmarked stereophotography, the use of the monocomparator requires that all points to be measured be marked on the diapositives. There are three categories of such points: (a) points marked on the object before the photographs are taken, (b) "detail" points which are readily identifiable on all photographs, and (c) "artificial" points marked in appropriate locations on the diapositive by point transfer and marking instruments. Such instruments are used, observing stereoscopically, to mark conjugate points on all photographs.

Among the currently available stereocomparators are: Nistri TA3/P, Nistri TA2, Galileo M.S.1., Hilger-Watts 5B-50, Mann 2405 ST, Zeiss PSK-2, Zeiss-Jena Stecometer, Zeiss-Jena 1818, and Wild STK-1. Among the available monocomparators are: Galileo M.S.1, Zeiss-Jena Ascorecord, Zeiss PEK-21, Kern MK1, Kern MK2, Nistri TA1/P, Nistri P.C., Space Optics 102, Dell Foster RSS 600, Mann 880, and DBA plate comparator.

While all the above-mentioned stereo- and monocomparators are general purpose, the recently introduced Zeiss StR (Stereo Röntgen) and the Göttingen Reomat have been designed specifically for x-ray stereophotogrammetry and, as such, have important applications in numerous fields including biomedical sciences, clinical medicine, bioengineering, materials science, and metallurgy.

The Zeiss StR is a stereoplotter designed on the comparator principle and comes in three different models; StR-1, StR-2 and StR-3. While StR-1 has been intended only for stereoscopic observation of the x-ray radiographs, StR-2 is equipped with a simple measuring device, and StR-3 (fig. 16-82) is provided with a digital recording unit and a mini-computer. The Reomat is a stereocomparator developed at the University of Göttingen for quantitative evaluation of stereoradiographs.

The analytical method is very versatile and accurate and has important advantages over the analog method, such as increased accuracy of measurements and superior flexibility of procedure. In cases where the final output is to be continuous lines, a computer must be used to convert the point data into plotter commands. Thus the analog method might be preferrable in such cases.

The accuracy of different close-range photogrammetric procedures is often determined by experiments and checks. For standardized methods, one may use simple rules-of-thumb gained by experience for judging the accuracy in a specific case. For other methods, more complex means are applied to determine the accuracy of the results. In some methods a rigorous adjustment of the unknowns also includes the calculation of standard errors. In these cases, well-surveyed test fields are used, as in the experiments performed by Hottier (Hottier, 1972, 1974, 1976). For a single stereomodel and analytical data reduction, he found a considerable gain in accuracy as a result of measurement redundancy:

> when the number of repeated settings per image point is increased from one to three (gain of about 30% in accuracy);
> when the number of targets defining each object point is increased from one to three (gain of about 40%); and
> when the number of frames taken from each station is increased from one to three (gain of about 40% in accuracy).

Hottier found no correlation between the gains in accuracy, the geometry of the photographic set-up and the type of camera. He reported (Hottier, 1976) that the maximum gain in accuracy is about 50% and that this is obtainable only through mixing the parameters of measurement redundancy (setting per image point, number of targets defining an object point, and the number of frames per camera station). Hottier's extensive experiments revealed that there is practically no gain beyond 7 or 9 total comparator measurements. The optimum accuracy can be achieved in a number of combinations, for example by choosing one setting per image point, three targets per object point and three frames per camera station. If one chooses one setting per image point, one target per object point, and three frames per camera station, one would expect some 40% gain in accuracy over that of the 1-1-1 case (one comparator setting per image point, one target per object point, and one frame per camera station), according to Hottier's extensive experimentations.

In many special applications of photogrammetry a digital model of the object being mapped is often a more desirable output than the conventional contour map, on account of its flexibility, whether one is interested in measuring the surface geometry or in evaluating the deformations and movement of objects. A digital model provides the user with almost unlimited flexibility in the use of the electronic computer for design and analysis. Mathematical modeling of a surface and its behavior can be conveniently performed using a digital model. Such parameters as surface area and volume as well as deformation of surface points can all be computed directly from the digital model(s). Analytical methods have been developed to generate contours, cross-sections, profiles and perspective views directly from digital data.

The analytical approach can obviously be used in the rather simple case where the inner and outer orientation parameters of the cameras are known, but its advantages and versatility become much more pronounced in the most general case of photogrammetry where a simultaneous solution incorporates the inner and outer orientation parameters of the cameras (two or more), and the object-space coordinates of object points, all as unknowns. Significant advances have been made recently in perfecting the analytical method and in adapting this to solving problems in close-range photogrammetry. For example, through the use of advanced analytical methods, the photogrammetric reduction of photographs from non-metric cameras has become possible.

16.2.3.2.1 MATHEMATICAL FORMULATION IN CLOSE-RANGE PHOTOGRAMMETRY

16.2.3.2.1.1 For Photography from Metric Cameras. With little or no modifications, many of the computational methods which were developed over the years for aerial mapping (chapter II) and aerotriangulation (chapter IX) can be applied in close-range photogrammetric projects. Wong (1976) gave an overview of the mathematical formulation and digital analysis in close-range photogrammetry.

16.2.3.2.1.2 For Photography from Non-Metric Cameras. Because of the lack of fiducial marks in non-metric cameras, special techniques had to be devised. Among the approaches particularly suitable for non-metric photography is the *Direct Linear Transformation* (DLT) approach developed at the University of Illinois (Abdel-Aziz & Karara, 1971, 1974, Karara & Abdel-Aziz, 1974). The solution is based on the concept of direct transformation from comparator coordinates into object-space coordinates, thus by-passing the traditional intermediate step of transforming image coordinates from a comparator system to a photo-coordinate system. As such, the solution makes no use of fiducial marks. The method is based on the following pairs of equations:

$x + (x - x_0)(K_1 r^2 + K_2 r^4 + K_3 r^6 + ..) +$
$(r^2 + 2[x - x_0]^2) P_1 + 2(y - y_0)(x - x_0) P_2 =$

$$\frac{L_1 X + L_2 Y + L_3 Z + L_4}{L_9 X + L_{10} Y + L_{11} Z + 1} \qquad (16.1)$$

$y + (y - y_0)(K_1 r^2 + K_2 r^4 + K_3 r^6 + ..) +$
$2(x - x_0)(y - y_0) P_1 + (r^2 + 2[y - y_0]^2) P_2 =$

$$\frac{L_5 X + L_6 Y + L_7 Z + L_8}{L_9 X + L_{10} Y + L_{11} Z + 1}$$

where x and y are the measured comparator coordinates of an image point; x_0 and y_0 define the position of the principal point in the comparator coordinate system; $r^2 = (x^2 + y^2)$; K_1, K_2, K_3, P_1, P_2 are lens distortion coefficients; X, Y, Z are the object-space coordinates of the point imaged; and L_1 to L_{11} are eleven unknown constants.

When originally presented in 1971 (Abdel-Aziz and Karara, 1971) the basic equations of

terms (K_2, K_3, P_1, P_2). Eliminating these terms in equation (16.1), and substituting x' for $(x - x_0)$ and y' for $(y - y_0)$, equation (16.1) can be reduced to (assuming that $x_0 = y_0 = 0$):

$$x + x' K_1 r^2 = \frac{L_1 X + L_2 Y + L_3 Z + L_4}{L_9 X + L_{10} Y + L_{11} Z + 1}$$

$$\qquad (16.2)$$

$$y + y' K_1 r^2 + \frac{L_5 X + L_6 Y + L_7 Z + L_8}{L_9 X + L_{10} Y + L_{11} Z + 1}.$$

A complete documentation of the computer program for the DLT solution was written by Marzan and Karara (1975).

As explained by Wong (1976) (*see* chapter II) two different approaches can be used in solving equations (16.2). (a) Direct approach. The following equations can be derived directly from equation (16.2)

$$L_1 X + L_2 Y + L_3 Z + L_4 - xX L_9 - xY L_{10} - xZ L_{11} - x K_1 r^2 A - x = r_x$$

$$\qquad (16.3)$$

$$L_5 X + L_6 Y + L_7 Z + L_8 - yX L_9 - yY L_{10} - yZ L_{11} - y K_1 r^2 A - y = r_y$$

DLT did not involve any camera interior orientation constants, and represented a direct linear transformation between comparator coordinates and object-space coordinates. When the DLT mathematical model was later expanded to encompass interior orientation constants the title DLT was retained.

in which $A = L_9 X + L_{10} Y L_{11} Z + 1$, and r_x and r_y are the residual errors on the condition equations.

Letting $A K_1 = K_1'$, equation (16.3) can be reduced to the following matrix equation:

$$\begin{bmatrix} r_x \\ r_y \end{bmatrix} + \begin{bmatrix} -X & -Y & -Z & -1 & 0 & 0 & 0 & 0 & xX & xY & xZ & xr^2 \\ 0 & 0 & 0 & 0 & -X & -Y & -Z & -1 & yX & yY & yZ & yr^2 \end{bmatrix} \mathbf{L} + \begin{bmatrix} x \\ y \end{bmatrix} = 0 \qquad (16.4)$$

in which $\mathbf{L} = [L_1\, L_2\, L_3 \,..\, L_{11}\, K_1']^{\mathrm{T}}$.

Each object-point which has known object-space coordinates (X, Y, Z) gives rise to two such equations. A minimum of six control points are needed for a unique solution of the 12 unknowns in the L-matrix. It is a direct solution and no iteration is needed. (b) Iterative approach, Let V_x and V_y represent the small random errors in the measured (comparator) coordinates. The following equations can be derived from equation (16.2) by omitting the effects of V_x and V_y in the lens distortion constant K_1.

The number of constants of interior orientation taken into consideration in equations (16.1) depends on the degree of comprehensiveness desired. On the basis of experimental investigations, Karara and Abdel-Aziz (1974) concluded that, for all practical purposes, only the term K_1 needs to be taken into account in modeling lens distortions and film deformations. In the cases tested by them, no significant improvement in accuracy was gained by incorporating additional

$$V_x - \frac{X}{A} L_1 - \frac{Y}{A} L_2 - \frac{Z}{A} L_3 - \frac{1}{A} L_4 + \frac{xX}{A} L_9 + \frac{xY}{A} L_{10} + \frac{xZ}{A} L_{11} + xr^2 K_1 + \frac{x}{A} = 0$$

$$\qquad (16.5)$$

$$V_y - \frac{X}{A} L_5 - \frac{Y}{A} L_6 - \frac{Z}{A} L_7 + \frac{1}{A} L_8 + \frac{yX}{A} L_9 + \frac{yY}{A} L_{10} + \frac{yZ}{A} L_{11} + yr^2 K_1 + \frac{Y}{A} = 0$$

in which $A = L_9 X + L_{10} Y + L_{11} Z + 1$.

Since equation (16.5) is non-linear, it must be linearized and the subsequent least squares solution must be iterative.

In the DLT approach object-space control points should be well distributed. One must avoid having all object-space control points in one plane. As much deviation from the planar arrangements as can be allowed by the depth-of-field is highly recommended.

Having computed the coefficients from all the photographs, the object-space coordinates of any object point which is imaged in two or more photographs are then computed.

In the DLT method, the eleven constants (L_1, L_2, ..., L_{11}) are considered as independent. In 1977 Bopp and Krauss developed an exact solution to the DLT basic equations (referred to by them as the "11-parameter solution"), in which they took into account the dependency between these eleven constants and the nine independent constants necessary to represent the "perfect camera" (Bopp and Krauss, 1977). This was done by introducing two constraints, which are:

$$(L_1^2 + L_2^2 + L_3^3) - (L_5^2 + L_6^2 + L_7^2)$$
$$+ \left[(L_5 L_9 + L_6 L_{10} + L_7 L_{11})^2 \right.$$
$$\left. - (L_1 L_9 + L_2 L_{10} + L_3 L_{11})^2 \right]$$
$$(L_9^2 + L_{10}^2 + L_{11}^2)^{-1} = 0$$

and

$$(L_1 L_5 + L_2 L_6 + L_3 L_7)$$
$$- (L_9^2 + L_{10}^2 + L_{11}^2)^{-1} (L_1 L_9 + L_2 L_{10} + L_3 L_{11})$$
$$(L_5 L_9 + L_6 L_{10} + L_7 L_{11}) = 0.$$

(16.6)

The constraints are incorporated in the solution as additional observation equations with zero variances.

This method leads to a least squares adjustment with linear fractional observation equations and non-linear constraints. The constants are computed iteratively using three-dimensional, object-space control. With these constants, the object space coordinates of all image-points contained in at least two photographs can be determined in a second step. In contrast to the DLT, the "11-parameter solution" can also be applied to cases where the known interior orientation should be kept.

Another method particularly applicable for non-metric cameras is the self-calibration method developed at the University of New Brunswick (Faig, 1976).

16.2.3.2.2 ON-LINE ANALYTICAL PHOTOGRAMMETRY

On-line analytical photogrammetry is analytical photogrammetry in which the computer is connected directly to the comparator on which measurements are being made. (Use only of stereocomparators will be considered in this section.) The operator usually has control of the progress of computation not only through his use of the stereocomparator but also through a keyboard from which he can give instructions to the computer during the compilation. The theory for this kind of photogrammetry is identical to that used when measurement is completed before giving the data to the computer. However, because of the great variety of problems encountered in close-range photogrammetry, the more immediate control that the operator has over the computations in on-line analytical photogrammetry as compared with the "batch" method is sometimes advantageous.

16.2.3.2.2.1 Typical Features of Close-Range Photogrammetry. One of the main features of close-range photogrammetry is the absence of standard conditions for photography. For instance, the range of photographic scales is quite different in such applications as photography of human beings and the photography of architectural monuments. The object may be stationary, as in the photography of ships under construction, or moving so rapidly as to rapidly as to require special cameras, as for ballistics. The object may be very small and close, as in the photography of teeth, or vast and at the upper range of distance, as in the photography of open-pit mines.

This absence of standard conditions means, for analytical photogrammetry, that either the theory must be so complex that it can handle all situations, or it must be broken up into a number of separate theories each adapted to particular situations. For example, while the equations for central projection are suitable, with only slight modification, in aerial and terrestrial photogrammetry, they often require great modification to be applicable to problems in close-range photogrammetry. This is particularly true when photographs are taken with special lens systems such as fish-eye or anamorphic lenses.

A properly formulated program for on-line analytical photogrammetry has the capability of dealing efficiently with a variety of different problems. This may be an advantage when different problems must be solved rapidly or when the same kind of problem must be solved with a variety of different initial conditions.

16.2.3.2.2.2 Analytical Formulations. From the operational point of view there are three basic phases of an on-line analytical operation:

definition of the image geometry;
reconstruction of the photogrammetric model; and
detailed photogrammetric compilation of the model.

In defining the image geometry one actually chooses from existing models by determining the type of general conditions, and further specifies the characteristics of the image by its interior orientation and distortion parameters. Obviously, the image geometry essentially affects the process in the following operational phases. The reconstruction of a photogrammetric model results in a good description of relations between the images and the object by means of the parameters of exterior orientation. The reconstruction phase includes a one-step collection of measured data and usually an iterative solution of the parameters. The compilation phase represents an operation appearing practically continuous even though it is simulated by a fast repetitive cycle of digital computations. These proceed in a stream of densely spaced discrete data points defined by the operator's control of the floating mark in the observed optical model. In general, this computation is based on transformations between the model and image spaces with the use of all previously derived parameters of interior and exterior orientation. Included in the computations are corrections for image distortions.

In general, two different groups of parameters characterize the image-object relationship in on-line analytical systems:

—inner geometry $(\mathbf{x}', \mathbf{d})$,
—outer geometry $(\mathbf{X}, \mathbf{g})$.

Here, vector $\mathbf{x}' = (x', y')^{\mathrm{T}}$ represents the image coordinates, $\mathbf{d}$ is a vector of distortion parameters typical of the imaging system, $\mathbf{X} = (X, Y, Z)^{\mathrm{T}}$ is a vector of object coordinates, and $\mathbf{g}$ is a vector of parameters of exterior orientation. In a symbol form one can write for the imaging process

$$\mathbf{X} \xrightarrow{\mathbf{g},\mathbf{d}} \mathbf{x}' \qquad (16.7)$$

and for the reconstruction process

$$(\mathbf{x}',\mathbf{x}'',\mathbf{d}) \xleftarrow{\mathbf{g}} \mathbf{X}. \qquad (16.8)$$

After deriving the unknowns $\mathbf{g}$ with the use of suitable control points and with a previous knowledge of $\mathbf{d}$, one can start the routine intersection of individual model points $\mathbf{X}$ from stereo-observations $\mathbf{x}', \mathbf{x}''$

$$(\mathbf{x}',\mathbf{x}'') \xrightarrow{\mathbf{d},\mathbf{g}} \mathbf{X}. \qquad (16.9)$$

Image Geometry. The imaging process performs a suitable conversion of data from a three-dimensional object space into a two-dimensional image space. This conversion is always achieved by the use of physical means which represent a projecting system.

To unify various possibilities one can base the photogrammetric reconstruction on scaled projections using straight lines derived from corrected image coordinates. If applicable the corrections should also include the effect of the original curvilinear projection. This approach yields the basic equation of on-line photogrammetry representing the computer-controlled feedback link

$$\mathbf{x}' + \mathbf{c} = \frac{1}{\mu} \overline{\mathbf{P}}^{\mathrm{T}} \Delta \mathbf{X} \qquad (16.10)$$

where $\mathbf{c}$ is a vector of corrections for image distortions, μ is a scale factor which is variable in central projections and constant in parallel projections, and $\Delta \mathbf{X} = (\Delta X, \Delta Y, \Delta Z)^{\mathrm{T}}$ are object

coordinates reduced with respect to a suitable reference point. Projection matrix $\overline{\mathbf{P}}$ (3,2) represents the orientation of the projection bundle or beam in the object coordinate system, as analyzed by Kratky (1975, 1976).

Model Reconstruction. In off-line computations, the photogrammetric model can be reconstructed from corresponding sets of photo-coordinates $\mathbf{x}', \mathbf{x}''$ matched with control coordinates $\mathbf{X}$ as expressed in equation (16.8), using a suitable mathematical model for the expected relationship. In on-line analytical systems the same relationship is expressed in a slightly different form. The communication between photo- and object-coordinates is mediated by auxiliary model coordinates $\mathbf{x} = (x, y, z)^{\mathrm{T}}$

$$(\mathbf{x}',\mathbf{x}'',\mathbf{d}) \underset{\mathbf{g}_1 \; \downarrow \; \mathbf{g}_2}{\overset{\mathbf{g}}{\underset{\overleftarrow{\mathbf{x}}\to}{\longleftrightarrow}}} \mathbf{X}. \qquad (16.8a)$$

The coordinate system $(\mathbf{x})$ of the model becomes "master" for the remaining systems, extended by a graphical output $\tilde{\mathbf{x}}$.

Before the feedback $(\mathbf{x} \rightarrow \mathbf{x}', \mathbf{x}'')$ is established, photo-coordinates $\mathbf{x}', \mathbf{x}''$ are measured and the constants for $\mathbf{g}$ are computed. The computation could be arranged in several different ways. Here, we refer to the formulation given by Kratky (1975) and summarize its essential features.

In a single model the individual points are intersected from pairs of conjugated rays. A collinearity equation can be applied to each ray, in a general form $F(\mathbf{g}, \mathbf{l}) = 0$. Here, vector $\mathbf{l}$ represents ideal photo-coordinates. The real measurements $\mathbf{x}', \mathbf{x}''$ are not consistent with the equation F and yield residuals $\mathbf{u}$. Vector $\mathbf{g}$ is composed of photograph-related constants $\mathbf{g}_0$ representing interior and exterior orientation of both pictures, and by object-related constants $\mathbf{g}_r$ giving unknown coordinates for all points which are not control points. Elimination of $\mathbf{g}_r$ in the least-squares formulation leads to modified normal equations

$$\mathbf{B}_0^{\mathrm{T}}\mathbf{P}_0\mathbf{B}_0\mathbf{g}_0 + \mathbf{B}_0^{\mathrm{T}}\mathbf{P}_0\mathbf{u} = 0 \qquad (16.12)$$

where $\mathbf{B}_0$ is the matrix of coefficients related to the vector $\mathbf{g}_0$ and $\mathbf{P}_0$ a weight matrix representing weights introduced by the elimination of $\mathbf{g}_r$. The normal equations are formed sequentially from individual point contributions with the use of (4 × 4)-matrices

$$(\mathbf{P}_0)_j = \mathbf{P}_j\left[\mathbf{I} - \mathbf{B}_r(\mathbf{B}_r^{\mathrm{T}}\mathbf{B}_r)^{-1} \mathbf{B}_r^{\mathrm{T}}\right]_j. \qquad (16.13)$$

Here, $(\mathbf{B}_r)_j$ is a (4,3)-coefficient matrix associated with the coordinates of an object-point and $\mathbf{P}_j$ is a weight derived from the coordinate difference ΔZ_j between the object point and projection center according to

$$\mathbf{P}_j = 1/\Delta Z_j^2.$$

Any constraint $F(\mathbf{g}_0, \mathbf{c}) = 0$ among the parameters $\mathbf{g}_0$ is linearized to $\mathbf{B}_c \cdot \mathbf{g}_c - \mathbf{c} = 0$. Ultimately, the parameters $\mathbf{g}_0$ are computed from the accumulated normal equations

$$\mathbf{g}_0 = (\mathbf{B}_0^T \mathbf{P}_0 \mathbf{B}_0)^{-1} \mathbf{B}_0^T \mathbf{P}_0 \mathbf{u}. \qquad (16.14)$$

Since the formulation is based on linearized relations the solution must be iteratively updated. If required, the computation is completed by calculating the coordinates of individual intersected points p_j from updated residuals u_j

$$(\mathbf{g}_r)_j = - (\mathbf{B}_r)_j^{\dagger} \mathbf{u}_j. \qquad (16.15)$$

Derived parameters $\mathbf{g}_0$ can be used in the on-line mode with the aid of a proper decomposition $\mathbf{g}_0^T = (\mathbf{g}_1^T, \mathbf{g}_2^T)$ where $\mathbf{g}_1$ represents the return in the feedback link $(\mathbf{x} \rightarrow \mathbf{x}')$.

In general, the model coordinate system can be defined in an arbitrary manner with respect to the object, but it is advantageous to assume equal photo- and model-scales and introduce an auxiliary coordinate shift C. Thus, the working equations of an on-line analytical system which is physically control-driven *via* the model, can be given by

$$\begin{aligned} \mathbf{x}' &= \lambda \bar{\mathbf{P}}^T \, \Delta\mathbf{x} - \mathbf{c}, \\ \mathbf{X} &= \mathbf{C} + m\mathbf{x}, \qquad (16.16) \\ \tilde{\mathbf{x}} &= v\mathbf{x} \end{aligned}$$

where λ is a modified scale factor and v an arbitrary ratio to generate a graphical plot. The first formula in equation (16.16) represents transformations for both images. The exterior orientation is fully returned to the $(\mathbf{x} \rightarrow \mathbf{x}')$ link except for the scale factor m which is used in the $(\mathbf{x} \rightarrow \mathbf{X})$ computation. In operations, the floating mark is driven in the directions of the object coordinate system. Other possibilities of arranging the return are discussed in (Kratky, 1976).

Detailed Compilation. The first two operational phases in an on-line analytical process are concerned with the derivation of a valid analytical model for image geometry and, by using it, with the analytical reconstruction of the model for object geometry. In both phases, the on-line function is prepared and checked, but the computations required are performed off-line or in near-real time. Only after the models of both image and object geometries are derived, are they ready to be used in a process of detailed photogrammetric compilation which represents a typical real-time operation. The control of this operation is characterized by equation (16.8a) or in a modified form by

$$\mathbf{x} - \begin{cases} \mathbf{d}, \mathbf{g}_1 \rightarrow (\mathbf{x}', \mathbf{x}'') \\ \mathbf{g}_2 \rightarrow \mathbf{X} \\ v \rightarrow \tilde{\mathbf{x}} \end{cases}$$

With the computer performing all three computations in a high frequency, the operator retains dynamic control of the system through $\mathbf{x}$

and receives his feedback from stereo-observations of computer-positioned image details for $\mathbf{x}'$ and $\mathbf{x}''$.

16.2.3.2.2.3 Computer Program for General Orientation. Recently, the design of a new analytical plotter has been completed at the National Research Council of Canada. As a part of extensive photogrammetric computer programs supporting the functions of the ANAPLOT system, a program for the general orientation of a single model was developed, with particular emphasis on its use in close-range photogrammetry.

In accordance with the principles outlined above, the orientation problem solved by the program can be easily modified to fit the conditions of a given photogrammetric situation. This is achieved by a direct operator-computer interaction at the time of execution.

The program is capable of handling any photoscale and a variable number of unknown orientation parameters. The basic solution yields complete exterior orientation of two photographs with a total of 12 unknowns. It could be extended to provide an on-the-job calibration of interior orientation with three additional parameters common to both photographs or with three independent parameters for each of the photographs. The maximum number of unknowns is then 18. The number of parameters can be reduced by choosing a simplified orientation with certain elements preset to given values. An extreme case is represented by a relative orientation with five unknown parameters.

The computer is told which unknowns are to be included in the solution by a string of nine zeros or ones assigned in a standard way to individual elements of orientation. The string is used in the computations as a mask to compress or expand the vectors and matrices of the solution in an appropriate manner.

The problem can also include additional arbitrary conditions on any of the 18 unknowns. Since the constraints have to be entered in a simple way at execution time their formulation is restricted to linear forms. The operator enters the weight of the constraint, its numerical value and the coefficient vector associated with the constrained parameters. In its present form the program can accomodate up to seven constraints of two different types. The operator can either assign specific values to certain parameters and to their linear combinations or require a specific ratio of some computed values.

The reconstruction of a model is usually supported by a number of control points. In the on-line processing, measurements of points are made in an arbitrary sequence. The operator identifies each point by entering its code or number and specifies the corresponding control support with a suitable numeric code. Code "3" means that all three coordinates X, Y, and Z are known, code "2" denotes that X, and Y only are known, code "1" indicates that only the elevation Z is known and zero is the code that shows no control at all. Obviously, the combination of the control support and of the constraint formulation must provide a minimum information to avoid singular solutions.

The program is written in FORTRAN IV for the PDP 11/45 minicomputer controlling the ANAPLOT system, and is used in the near-real time mode after a conversational initialization of the conditions by the operator. The conversation proceeds from the keyboard of the computer terminal. The computer asks the operator to provide the required information in a series of steps as shown in figure 16-18 which is a facsimile of a computer printout.

In this example, the first instruction to the computer identifies the photographs as F-1 and F-2 and gives the scale factor 1 for the model computation, the principal distance as $f = 150$ mm, and a nine-digit string specifying the type of orientation. Here, zeros indicate which unknowns are to be retained in the solution while code one indicates which parameters are to be omitted from the solution. The string is coded for a standard sequence: X_c, Y_c, Z_c, κ, ϕ, ω, dx', dy', df, representing three groups of three values each, for the position of the projection center, for the rotation elements and for the elements of interior orientation, respectively. In the example, the full, twelve-parameter orientation is required, with no need to compute the elements of interior orientation. Next a single constraint is defined by determining the photogrammetric base, $b = 100.05$ mm, associated with a weight $P = 100$. The first ten values of the coefficient vector are given and the remaining eight zeros are truncated. The encoded constraint can be interpreted as $^2X_c - {}^1X_c = 100.05$. The following information specifies four of the measured

points, 11, 13, 31, 33, as supported by all three object coordinates whereas points 21 and 23 are used without any control support.

16.2.3.2.2.4 Practical Examples. Figures 16-19 through 16-26 give a series of examples illustrating different types of solutions obtained with the same general program from fictitious data. The data were computer-generated at 1:1 scale for two sets of regularly distributed groups of nine points at two different elevation levels. Point numbers 11 to 33 indicate the row-column position of points at the lower level while points 111 to 133 have corresponding positions at a 15 mm higher level.

Example 1 (Full Orientation with no Constraints—figure 16-19), is a straight forward solution with 12 unknowns. It was completed in four iterations. The indicated time figure includes the time used for printing intermediate results and should be reduced by three seconds per line of print in order to assess the real speed of solution. In this instance the time to obtain a solution is about 15 seconds. The resulting linear parameters are expressed in mm and the angular orientation is represented by rotation matrices. The resulting coordinates (X, Y, Z) and the parallax of the model (PY) are also in mm whereas all the photograph-related residuals are printed out in μm. Depending on the availability of control, the listing also shows discrepancies in control points. Since the computation is based on ideal data the resulting fit is perfect and all residuals equal zero.

In example 2 (full orientation with base constraint—figure 16-20), the solution is artificially constrained by enforcing a wrong length of the photogrammetric base. Since the weight used is 100 times larger than that given the measured photo-coordinates the constraint is very strong, as is obvious from $b = {}^2X_c - {}^1X_c = 200.026 - 99.976 = 100.05$ mm. As a result of the wrong length the orientation of the photographs is slightly disturbed and the listing shows residuals up to 9 μm.

In example 3 (full orientation with coordinate constraints—figure 16-21), both projection centers are falsely offset by 50 μm which again affects the quality of the reconstruction resulting in residuals up to 10 μm.

In example 4 (full orientation with relative constraint—figure 16-22) the constraint is very unnatural and is used only to demonstrate the potential and flexibility of on-line constraining. The absolutely enforced false condition $2\phi_1 + \phi_2 = 0$ can be checked by comparing the rotation elements m_{13} in both resulting matrices, *i.e.*

$$-2 \times 0.023993 + 0.047986 = 0.$$

Geometric relations are considerably disturbed and photo-residuals reach a maximum value of 161 μm.

Although artificial in this instance, relative constraints can be useful when one wants a

```
.R MODEL

SPECIFY MODEL - SCALE - F - MASK

 F-1 F-2 1, 150000,000000111

ADDITIONAL CONSTRAINTS?

YES

HOW MANY?

1

ENTER WEIGHT - VALUE - MATRIX

100, 100.05, -1,0,0, 0,0,0, 0,0,0, 1

ENTER POINT NUMBER - SPECIFY CONTROL SUPPORT

   11 3
   13 3
   21 0
   23 0
   31 3
   33 3
```

FIGURE 16-18. Information required in the computer program for general orientation.

```
ITERATION INCREMENTS

   XC        YC        ZC        KPPA      PHI       OME

-0.02528   0.04058   0.00151   0.00850  -0.01864   0.02790
 0.11565   0.04672  -0.00188  -0.04656   0.05122  -0.05952

-0.00183   0.00357  -0.00150   0.00145  -0.00140   0.00206
 0.00053  -0.00257   0.00186   0.00664  -0.00082  -0.00015

 0.00000  -0.00000  -0.00001   0.00004   0.00004   0.00003
-0.00000   0.00000   0.00002  -0.00006  -0.00040  -0.00034

 0.00000  -0.00000   0.00000  -0.00000   0.00000   0.00000
-0.00000  -0.00000  -0.00000  -0.00002   0.00000   0.00000


   PARAMETERS

              X          Y          Z        DX       DY       F
   F-1    100.000    199.999    250.000    0.000    0.000   150.000
   F-2    200.001    199.999    250.000    0.000    0.000   150.000


   0.999750   0.009997  -0.019997    0.997948  -0.039999   0.050004
  -0.009395   0.999506   0.029994    0.042978   0.997272  -0.060004
   0.020287  -0.029798   0.999350   -0.047468   0.062030   0.996945


   MODEL COORDINATES AND DISCREPANCIES.....   PHOTO DISCREPANCIES....

   PT      X          Y          Z        FY    VPY   VXL   VYL   VXR   VYR
   11   100.00     300.00     100.00     0.00    0     0     0     0     0
         0.00      -0.00       0.00
   13   200.00     300.00     100.00    -0.00    0     0     0     0     0
        -0.00      -0.00       0.00
   21   100.00     200.00     100.00    -0.00    0     0     0     0     0
        -0.00       0.00       0.00
   23   200.00     200.00     100.00    -0.00    0     0     0     0     0
                               0.00
   31   100.00     100.00     100.00    -0.00    0     0     0     0     0
         0.00       0.00       0.00
   33   200.00     100.00     100.00    -0.00    0     0     0     0     0
        -0.00       0.00       0.00


   REDUNDANCY =   9
   STANDARD UNIT ERROR =   0.4 MICRONS

   STANDARD ERRORS OF UNKNOWNS

   XC        YC        ZC        KPPA      PHI       OME
   0.0008    0.0008    0.0004    0.0000    0.0000    0.0000
   0.0009    0.0009    0.0004    0.0000    0.0000    0.0000
```

FIGURE 16-19. Example 1. Full orientation with no constraints.

symmetrical distribution of corresponding parameters in the left and right pictures by setting a condition $p_1 + p_2 = 0$.

In example 5 (full orientation and calibration—figure 16-23), all 18 unknowns are derived in this solution when the required string in the initial specification is reduced to a single zero. A minimum of two additional control points located above the level of remaining points guarantees an on-the-job calibration of both cameras. Since no wrong values were constrained in the solution, all residuals are equal to zero. A low degree of redundancy in control data causes higher uncertainty in the derived parameters than in some previous examples.

In example 6 (full orientation and calibration with constraint—figure 16-24), the solution from the previous example is degraded by assigning a wrong value of 148 mm to the principal distance of the left camera. Because principal distances

```
ITERATION INCREMENTS

    XC          YC          ZC          KPPA        PHI         OME

 -0.02781    0.03614     0.00094     0.00850    -0.02065     0.02518
  0.11581    0.05357    -0.00238    -0.04802     0.05160    -0.05523

  0.00055    0.00799    -0.00095     0.00151     0.00049     0.00477
  0.00055   -0.00940     0.00234     0.00812    -0.00100    -0.00437

 -0.00001    0.00001    -0.00004    -0.00001     0.00002     0.00004
 -0.00001   -0.00003    -0.00001    -0.00006    -0.00046    -0.00042

 -0.00000    0.00000    -0.00000    -0.00000    -0.00000    -0.00000
 -0.00000   -0.00000    -0.00000    -0.00003     0.00000     0.00000


    PARAMETERS

                X           Y           Z          DX          DY          F
    F-1      99.976     199.998     249.993      0.000       0.000     150.000
    F-2     200.026     199.998     249.992      0.000       0.000     150.000


   0.999747   0.010003  -0.020140      0.997941  -0.039986   0.050148
  -0.009396   0.999506   0.029993      0.042975   0.997272  -0.060010
   0.020430  -0.029796   0.999347     -0.047611   0.062041   0.996937


    MODEL COORDINATES AND DISCREPANCIES.....    PHOTO DISCREPANCIES....

    PT       X           Y           Z          PY    VPY   VXL   VYL   VXR   VYR
    11     100.00      300.00      100.00     -0.01    9     3     5     0    -4
            0.00        0.01        0.00
    13     200.00      300.00      100.00      0.01   -9    -1    -4    -3     4
           -0.00       -0.01        0.00
    21     100.00      200.00      100.00      0.00    0     0     0     0     0
    23     200.00      200.00      100.00     -0.00    0     0     0     0     0
    31     100.00      100.00      100.00      0.01   -8     3    -5     0     3
            0.00       -0.01        0.00
    33     200.00      100.00      100.00     -0.01    9     0     4    -3    -5
           -0.00        0.01        0.00


    REDUNDANCY =   7
    STANDARD UNIT ERROR =    5.7 MICRONS

    STANDARD ERRORS OF UNKNOWNS

    XC          YC          ZC          KPPA        PHI         OME
  0.0094      0.0151      0.0053      0.0000      0.0001      0.0001
  0.0094      0.0151      0.0053      0.0000      0.0001      0.0001
```

FIGURE 16-20. Example 2. Full orientation with base constraint.

are treated as negative values in the program, and their initial approximation is −150 mm the constraint must be formulated as $df_1 = 2.0$.

In example 7 (relative orientation—figure 16-25), only six angular parameters are allowed in the computation by choosing a mask "111000111." However, the relative orientation is defined by five elements and thus, one of the parameters must be eliminated by an additional single constraint. In this example, ω_1 was constrained to zero. Consequently, the errorfree model is appreciably titled, as is obvious from the resulting model coordinates. Because no control

support was available, the origin of the model coordinates is at the left projection center by default.

In example 8 (relative orientation—figure 16-26), the solution uses 12 unknowns and, obviously, seven of them must be controlled by suitable constraints. This is done by assigning coordinates to both projection centers and by specifying the correct value for ω_1. In contrast to the previous example, the resulting model is not tilted and is in addition error free. The printout (figure 16-26) contains additional information on the redundancy, on the standard error of unit

```
ITERATION INCREMENTS

    XC          YC          ZC          KPPA        PHI         OME

 -0.02678    0.03611     0.00124     0.00849    -0.01975     0.02516
  0.11650    0.05351    -0.00256    -0.04801     0.05222    -0.05526

 -0.00000    0.00801    -0.00110     0.00150     0.00001     0.00479
  0.00000   -0.00935     0.00247     0.00815    -0.00149    -0.00433

  0.00000    0.00000    -0.00004     0.00000     0.00003     0.00003
 -0.00000   -0.00003    -0.00002    -0.00009    -0.00045    -0.00042

 -0.00000    0.00000    -0.00000    -0.00000     0.00000     0.00000
 -0.00000   -0.00000    -0.00000    -0.00003     0.00001     0.00000


    PARAMETERS

                X           Y           Z         DX       DY        F
        F-1   100.050     199.996     250.016    0.000    0.000    150.000
        F-2   200.050     199.996     249.984    0.000    0.000    150.000


    0.999756   0.009985  -0.019711     0.997935  -0.039974   0.050287
   -0.009391   0.999506   0.029984     0.042972   0.997272  -0.060017
    0.020001  -0.029792   0.999356    -0.047751   0.062054   0.996930


    MODEL COORDINATES AND DISCREPANCIES.....    PHOTO DISCREPANCIES....

    PT       X           Y           Z        PY    VPY  VXL  VYL  VXR  VYR
    11    100.00      300.00      100.00     0.00    -2   -6  -10    1   -8
          -0.01       -0.00       -0.01
    13    200.00      300.00      100.00     0.00     0    2    8   -7    9
          -0.01       -0.00        0.01
    21    100.01      200.00      100.01    -0.00     0    0    0    0    0
    23    200.01      200.00       99.99     0.00     0    0    0    0    0
    31    100.00      100.00      100.00    -0.00     1   -7    9    2    8
          -0.01        0.00       -0.01
    33    200.00      100.00      100.00    -0.00     2    1   -8   -6  -10
          -0.01        0.00        0.01


    REDUNDANCY =   8
    STANDARD UNIT ERROR =    10.6 MICRONS

    STANDARD ERRORS OF UNKNOWNS

    XC          YC          ZC          KPPA        PHI         OME
  0.0011      0.0279      0.0082      0.0001      0.0000      0.0001
  0.0011      0.0280      0.0082      0.0001      0.0000      0.0001
```

FIGURE 16-21. Example 3. Full orientation with coordinate constraints.

weight and standard errors of the derived parameters.

16.2.3.2.2.5 Conclusions. The versatility of on-line systems is greater than that of any other photogrammetric system and the expansion of functions is mostly a matter of software development. This fact combined with the versatility of on-line analytical systems make them extremely attractive for close-range applications, especially in biomedicine and engineering.

16.2.3.3 SEMI-ANALYTICAL APPROACH

In this approach, a stereoplotter is used to form the stereomodel of the object. The scaling and absolute orientation of the model are then done analytically. The use of a stereoplotter rather than a comparator and the added flexibility resulting from the analytical phases makes this approach highly attractive.

16.2.4 AREAS OF APPLICATION OF CLOSE-RANGE PHOTOGRAMMETRY

The ever-expanding areas of application of close-range photogrammetry can be grouped into three major areas: architectural photogrammetry, biomedical and bioengineering

```
ITERATION INCREMENTS

   XC         YC         ZC        KPPA       PHI        OME

-0.03146   0.03621  -0.00035   0.00853  -0.02405   0.02522
 0.11196   0.05376  -0.00130  -0.04805   0.04810  -0.05513

-0.00017   0.00807  -0.00127   0.00162  -0.00013   0.00483
 0.00194  -0.00941   0.00213   0.00789   0.00026  -0.00440

 0.00017   0.00002   0.00002   0.00001   0.00019   0.00005
 0.00006  -0.00004  -0.00004   0.00002  -0.00038  -0.00038

 0.00000   0.00000   0.00000  -0.00001   0.00000   0.00000
 0.00000  -0.00000  -0.00000  -0.00002  -0.00000  -0.00000


   PARAMETERS

            X          Y          Z        DX       DY        F
   F-1    99.348   200.023   249.761    0.000    0.000   150.000
   F-2   199.669   200.023   250.118    0.000    0.000   150.000


   0.999661   0.010159  -0.023991     0.998040  -0.040166   0.047987
  -0.009435   0.999503   0.030090     0.043016   0.997277  -0.059907
   0.024285  -0.029854   0.999259    -0.045450   0.061854   0.997050


   MODEL COORDINATES AND DISCREPANCIES.....   PHOTO DISCREPANCIES....

   PT       X          Y          Z       PY    VPY  VXL  VYL  VXR  VYR
   11    100.00     300.00     100.00   -0.10   110   49  161  -23   51
          0.05       0.12       0.12
   13    200.00     300.00     100.00    0.02   -32  -53 -109   26  -76
          0.03      -0.04      -0.12
   21     99.95     200.01      99.88    0.00    -4    0   -2    0    2
   23    199.97     200.01     100.12   -0.00     4    0    2    0   -2
   31    100.00     100.00     100.00    0.11   -98   58 -154  -30  -56
          0.06      -0.13       0.13
   33    200.00     100.00     100.00   -0.03    21  -53  102   26   80
          0.02       0.05      -0.12


   REDUNDANCY =   7
   STANDARD UNIT ERROR = 122.7 MICRONS

   STANDARD ERRORS OF UNKNOWNS

   XC         YC         ZC        KPPA       PHI        OME
 0.1408     0.3224     0.0945     0.0006     0.0008     0.0014
 0.2566     0.3246     0.1214     0.0006     0.0015     0.0014
```

FIGURE 16-22. Example 4. Full orientation with relative constraint.

photogrammetry (biostereometrics) and industrial photogrammetry.

16.2.4.1 ARCHITECTURE

16.2.4.1.1 INTRODUCTION

It is noteworthy that the very first measurements ever made by photogrammetry (in the middle of the 19th century) had to do with monuments. It is also a fact that the term "photogrammetry" was introduced by an architect, Albrecht Meydenbauer, who made his first

photogrammetric surveys in 1867. For over a century, photogrammetric methods and equipment have continued to evolve. More recently, the field of architectural application of photogrammetry has undergone considerable expansion both in scope and diversity.

16.2.4.1.2 SURVEYS OF HISTORICAL MONUMENTS

Photogrammetric surveys of historic monuments can be grouped in three major categories: rapid and relatively simple surveys, accurate

ITERATION INCREMENTS

XC	YC	ZC	KPPA	PHI	OME	DX	DY	DF
-0.02409	0.03094	-0.00932	0.00899	-0.01954	0.02629	-0.00251	0.00745	0.01078
0.06219	0.06578	-0.04145	-0.04650	0.04956	-0.05601	0.04866	-0.01958	0.03948
-0.00284	0.01294	0.00274	0.00102	-0.00046	0.00375	0.00252	-0.00749	-0.00388
0.05347	-0.02112	0.03523	0.00657	0.00079	-0.00378	-0.05150	0.01869	-0.02953
-0.00002	-0.00002	-0.00007	-0.00001	0.00001	-0.00006	0.00001	0.00001	-0.00026
-0.00024	-0.00084	-0.00042	-0.00005	-0.00035	-0.00021	0.00263	0.00125	-0.00338
-0.00000	-0.00000	0.00000	0.00000	-0.00000	0.00000	-0.00001	0.00002	-0.00000
-0.00000	0.00001	-0.00000	-0.00002	0.00000	-0.00000	0.00023	-0.00032	0.00004
-0.00000	-0.00000	0.00000	0.00000	0.00000	-0.00000	0.00000	0.00000	0.00000
0.00000	-0.00000	0.00000	0.00000	0.00000	0.00000	-0.00002	-0.00001	0.00003

PARAMETERS

	X	Y	Z	DX	DY	F
C-1	99.999	200.000	249.996	0.002	-0.002	149.996
C-2	200.000	199.995	249.997	0.001	0.004	149.998

0.999750	0.009998	-0.019996	0.997948	-0.039999	0.050006
-0.009396	0.999506	0.029992	0.042979	0.997272	-0.060004
0.020286	-0.029797	0.999350	-0.047470	0.062030	0.996945

MODEL COORDINATES AND DISCREPANCIES..... PHOTO DISCREPANCIES....

PT	X	Y	Z	PY	VPY	VXL	VYL	VXR	VYR
11	100.00	300.00	100.00	0.00	0	0	0	0	0
	0.00	-0.00	0.00						
13	200.00	300.00	100.00	0.00	0	0	0	0	0
	-0.00	-0.00	0.00						
21	100.00	200.00	100.00	0.00	0	0	0	0	0
23	200.00	200.00	100.00	-0.00	0	0	0	0	0
31	100.00	100.00	100.00	0.00	0	0	0	0	0
	-0.00	-0.00	0.00						
33	200.00	100.00	100.00	0.00	0	0	0	0	0
	-0.00	0.00	0.00						
111	100.00	300.00	115.00	0.00	0	0	0	0	0
	0.00	-0.00	0.00						
131	100.00	100.00	115.00	0.00	0	0	0	0	0
	-0.00	-0.00	-0.00						

REDUNDANCY = 8
STANDARD UNIT ERROR = 0.5 MICRONS

STANDARD ERRORS OF UNKNOWNS

XC	YC	ZC	KPPA	PHI	OME	DX	DY	DF
0.0040	0.0039	0.0066	0.0000	0.0000	0.0000	0.0045	0.0040	0.0070
0.0062	0.0046	0.0068	0.0000	0.0000	0.0000	0.0063	0.0045	0.0068

FIGURE 16-23. Example 5. Full orientation and calibration.

and complete surveys, and very accurate surveys.

16.2.4.1.2.1 Rapid and Relatively Simple Surveys. These are used in preliminary studies for restoration and improvement, in inventory work (fig. 16-27), and in the study of the history of art. Stereometric cameras and other small-format photogrammetric cameras are used extensively, together with "normal case stereoplotters." Plotting is generally at a scale of 1:100.

To simplify the operations, inclined photography is taken at standard angles and slope calculators are used.

16.2.4.1.2.2 Accurate and Complete Surveys. These are used for systematic documentation of architectural heritage. Plotting scale is generally 1:50, while the details are mapped at 1:20 or 1:10. Large-format metric cameras with long focal lengths are preferred in this type of work in view of the accuracy requirements and the sizes

ITERATION INCREMENTS

XC	YC	ZC	KPPA	PHI	OME	DX	DY	DF
-0.02377	0.03353	-0.01127	0.00932	-0.01866	0.02730	-0.00202	0.00657	0.01333
0.06278	0.06164	-0.03450	-0.04677	0.04935	-0.05692	0.04868	-0.01534	0.03259
-0.00328	0.01051	-0.00108	0.00068	0.00000	0.00232	0.00349	-0.00720	-0.00000
0.05352	-0.01659	0.03484	0.00686	0.00103	-0.00282	-0.05120	0.01433	-0.02950
0.00000	0.00011	-0.00029	-0.00001	0.00003	-0.00004	0.00009	-0.00017	-0.00000
-0.00010	-0.00093	-0.00034	-0.00007	-0.00038	-0.00027	0.00229	0.00132	-0.00315
0.00000	-0.00001	-0.00001	0.00000	0.00000	0.00000	0.00000	0.00002	-0.00000
-0.00000	0.00002	-0.00000	-0.00002	0.00000	0.00000	0.00022	-0.00030	0.00002
0.00000	0.00000	0.00000	-0.00000	-0.00000	0.00000	-0.00000	-0.00000	-0.00000
0.00000	-0.00000	0.00000	0.00000	0.00000	0.00000	-0.00001	-0.00001	0.00003

PARAMETERS

	X	Y	Z	DX	DY	F
C-1	100.010	199.998	248.104	0.235	-0.117	148.000
C-2	200.004	199.997	250.001	-0.003	0.000	150.001

0.999777	0.009984	-0.018623	0.997948	-0.039995	0.050005	
-0.009431	0.999518	0.029583	0.042975	0.997272	-0.060008	
0.018909	-0.029400	0.999389	-0.047469	0.062034	0.996945	

MODEL COORDINATES AND DISCREPANCIES.....   PHOTO DISCREPANCIES....

PT	X	Y	Z	PY	VPY	VXL	VYL	VXR	VYR
11	100.00	300.00	100.00	0.08	-79	2	-79	0	0
	0.00	-0.12	0.00						
13	200.00	300.00	100.00	-0.00	1	0	2	0	0
	-0.00	0.00	0.00						
21	100.00	200.00	100.00	0.00	-1	0	0	0	0
23	200.00	200.00	100.00	-0.00	2	0	1	0	-1
31	100.00	100.00	100.00	-0.07	70	0	70	0	0
	-0.00	0.11	-0.00						
33	200.00	100.00	100.00	-0.00	1	0	2	0	0
	-0.00	0.00	-0.00						
111	100.00	300.00	115.00	-0.07	77	0	77	0	0
	0.00	0.10	0.00						
121	100.00	200.00	115.00	0.00	-1	0	0	0	1
	-0.00	-0.00	-0.00						
131	100.00	100.00	115.00	0.07	-73	-1	-72	0	0
	-0.00	-0.10	-0.00						

REDUNDANCY = 13
STANDARD UNIT ERROR =   41.7 MICRONS

STANDARD ERRORS OF UNKNOWNS

XC	YC	ZC	KPPA	PHI	OME	DX	DY	DF
0.3050	0.2914	0.0313	0.0002	0.0005	0.0004	0.3473	0.3233	0.0042
0.4873	0.3394	0.5664	0.0002	0.0005	0.0004	0.5015	0.3483	0.5662

FIGURE 16-24. Example 6. Full orientation and calibration with constraint.

of the buildings surveyed. The recently developed wide-angle cameras having focal lengths ranging between 100 mm to 150 mm are particularly suitable for this class of photogrammetric surveys.

Accurate surveys are used to document the technical history of the construction of the monument and its evolution as time passes, also to analyze its structural lines and to document its condition and its need for conservation and restoration. This is why one needs high accuracy and precision and as detailed a survey as possible (figure 16-28). The use of "first order" stereoplotters is, therefore, essential. Furthermore, normal case photography is often not possible due to the difficult conditions frequently

```
ITERATION INCREMENTS

  KPPA      PHI       OME

 0.01281  -0.03999  -0.00000
-0.03765   0.01084  -0.07982

-0.00241   0.02046   0.00000
-0.00605   0.03796  -0.01004

-0.00040  -0.00049   0.00000
 0.00357   0.00154   0.00014

-0.00000   0.00001   0.00000
 0.00016  -0.00032  -0.00016

 0.00000  -0.00000   0.00000
-0.00003  -0.00002  -0.00001


PARAMETERS

            X          Y          Z         DX        DY         F
   R-1    0.000      0.000      0.000     0.000     0.000    150.000
   R-2   78.507      0.000      0.000     0.000     0.000    150.000


  0.999750   0.009996  -0.020005      0.997948  -0.039996   0.049997
 -0.009998   0.999950  -0.000000      0.044380   0.994963  -0.089889
  0.020004   0.000200   0.999800     -0.046150   0.091923   0.994696


MODEL COORDINATES AND DISCREPANCIES.....   PHOTO DISCREPANCIES....

 PT      X          Y          Z        PY     VPY   VXL   VYL   VXR   VYR
 11    0.00      82.01    -115.35    -0.00      0     0     0     0     0
 13   78.51      82.01    -115.35    -0.00      0     0     0     0     0
 21    0.00       3.53    -117.71    -0.00      0     0     0     0     0
 23   78.51       3.53    -117.71    -0.00      0     0     0     0     0
 31    0.00     -74.94    -120.06    -0.00      0     0     0     0     0
 33   78.51     -74.94    -120.06    -0.00      0     0     0     0     0


REDUNDANCY =  1
STANDARD UNIT ERROR =   0.7 MICRONS

STANDARD ERRORS OF UNKNOWNS

  KPPA      PHI       OME
 0.0047    0.0019    0.0001
 0.0000    0.0000    0.0000
```

FIGURE 16-25. Example 7. Relative orientation.

encountered. In some countries, precision photogrammetric surveys have been made for "technical monuments" such as ancient bridges and viaducts of artistic value, as shown in figure 16-29.

A special case of accurate photogrammetric surveys is the survey of building exteriors (facades). Such surveys are carried out, particularly in central Europe, for the systematic documentation of harmonious architectural groups formed by series of houses in a street or on a square in ancient urban centers in towns and villages (figure 16-30). Because of space limitations, facade photography is often taken at an upward inclination (e.g. 30^g or 70^g) or from an elevatable platform on a special truck.

A second special case of accurate photogrammetric surveys is the partial detailed survey of particular part(s) of monuments. Such surveys are conducted in conjunction with restoration and consolidation projects. The highest possible accuracy is needed for these purposes. Depending on the needs, the final outputs of the survey can be in the form of plans, cross-sections, elevations, profiles (for arches), contours (for vaults and cupolas), and/or numerical

ITERATION INCREMENTS

XC	YC	ZC	KPPA	PHI	OME
3.33333	2.00000	2.33333	0.01692	-0.05110	0.03000
3.47662	2.00000	2.33333	-0.04779	0.01341	-0.07143
-0.00000	0.00000	0.00000	-0.00663	0.03141	0.00000
-0.00000	0.00000	-0.00000	0.00591	0.03622	0.01121
-0.00000	0.00000	0.00000	-0.00030	-0.00031	-0.00000
-0.00000	-0.00000	-0.00000	0.00186	0.00048	0.00031
-0.00000	0.00000	-0.00000	0.00001	-0.00000	-0.00000
0.00000	-0.00000	-0.00000	0.00003	-0.00011	-0.00009
0.00000	0.00000	0.00000	0.00000	0.00000	0.00000
-0.00000	-0.00000	-0.00000	-0.00001	-0.00000	-0.00000

PARAMETERS

	X	Y	Z	DX	DY	F
R-1	500.000	300.000	350.000	0.000	0.000	150.000
R-2	600.000	300.000	350.000	0.000	0.000	150.000

0.999750	0.009996	-0.020005		0.997948	-0.039997	0.049999
-0.009393	0.999506	0.030000		0.042976	0.997273	-0.060000
0.020295	-0.029805	0.999350		-0.047463	0.062026	0.996945

MODEL COORDINATES AND DISCREPANCIES..... PHOTO DISCREPANCIES....

PT	X	Y	Z	FY	VPY	VXL	VYL	VXR	VYR
11	500.00	400.00	200.00	-0.00	0	0	0	0	0
13	600.00	400.00	200.00	0.00	0	0	0	0	0
21	500.00	300.00	200.00	0.00	0	0	0	0	0
23	600.00	300.00	200.00	-0.00	0	0	0	0	0
31	500.00	200.00	200.00	-0.00	0	0	0	0	0
33	600.00	200.00	200.00	0.00	0	0	0	0	0

REDUNDANCY = 1
STANDARD UNIT ERROR = 0.2 MICRONS

STANDARD ERRORS OF UNKNOWNS

XC	YC	ZC	KPPA	PHI	OME
0.0000	0.0000	0.0000	0.0000	0.0000	0.0000
0.0000	0.0000	0.0000	0.0000	0.0000	0.0000

FIGURE 16-26. Example 8. Relative orientation.

data giving accurate dimensions between the main elements of the building or distances between these elements (figure 16-31).

The photogrammetric surveys conducted in the framework of UNESCO campaigns to salvage prestigious monuments such as Abou Simbel, Philae, Pétra, Borobudur (figure 16-32), are good examples of accurate and complete photogrammetric surveys.

16.2.4.1.2.3 Very Accurate Photogrammetric Surveys. These are needed for highly refined studies. Accuracy requirement is generally in the order of 1 mm and in some cases 0.1 mm.

The study of sculptures in monuments (figure 16-33) and the assessment of the evolution in the surface of defaced stones (in support of chemical and physical investigations into the "disease of the stone") require this very high accuracy. The principal difficulty in such cases is encountered in photography. Metric cameras permitting short object distances (*e.g.* by having variable principal distance or through the use of additional lenses) are of great help in this type of work.

16.2.4.1.2.4 Operational Procedures. Procedures for all of the above discussed types of photogrammetric surveys are well established

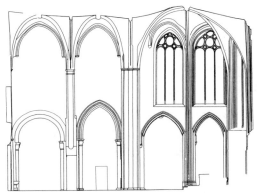

FIGURE 16-27. Example of photogrammetric surveys for general inventory works of monuments in France (courtesy Inventaire General, France).

FIGURE 16-28. Exterior elevation (facade) of the Church of Neresheim, Germany (courtesy Institut fur Baugeschichte und Bauaufnahme, Stuttgart).

and documented. Independent stereopairs of photographs are taken either horizontally, vertically or at some inclination using the camera(s) most suitable for the individual project. Base-to-distance ratio is kept rather small (1/5 to 1/15). External controls are kept as simple as possible (*e.g.* a number of distances and checks on the leveling bubbles of the camera). In case of complex objects, however, a network of reference points becomes necessary. Camera stations are normally located on the ground, on scaffoldings, on nearby buildings, on a hydraulic lift truck (figure 16-34), or even in helicopters which are sometimes used to take horizontal photographs of the upper portions of tall buildings.

The photographs are catalogued and stored in "photogrammetric archives" and are plotted only when the need arises. Plotting is mostly done using analog instruments.

In some photogrammetric surveys, photoplans are produced by rectifying and assembling a group of photographs. This technique is particularly suitable for plane surfaces of murals,

for mosaics (figure 16-35), for windows, and for facades, particularly when the streets are narrow. In this case, photography is systematically taken at a given inclination and rectified in simple instruments (*e.g.* Zeiss Oberkochen KEG-30 Transformer Printer). This approach is appealing both from the technical and economical points of view.

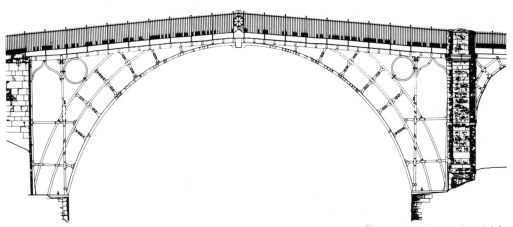

FIGURE 16-29. Photogrammetric survey of the "Iron Bridge," a British national monument in Coalbrookdale, Shropshire, England (1779) (Document of Department of the Environment, U.K.).

FIGURE 16-30. Survey of exterior of buildings (development of facades): Durstein, Austria (survey made by the Bundesdenkmalamt, Vienna).

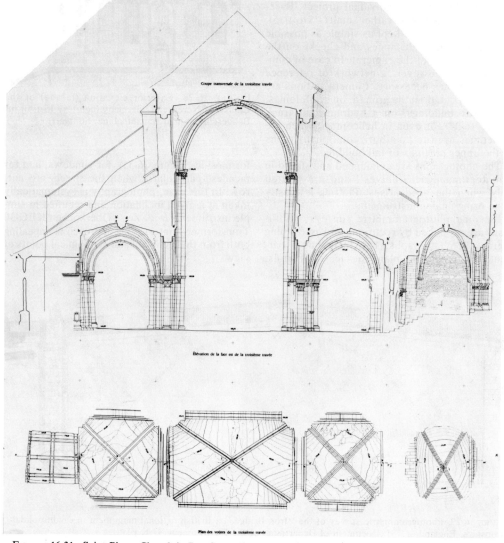

FIGURE 16-31. Saint-Pierre Church in Bur-Sur-Aube, France (courtesy IGN-direction de l'architecture).

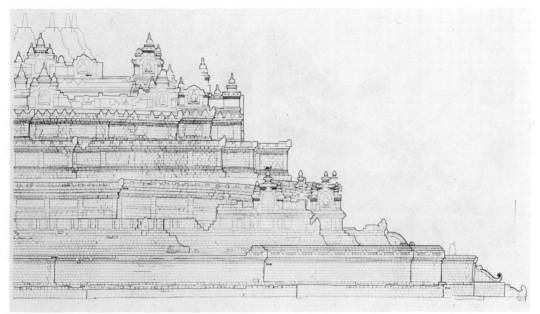

FIGURE 16-32. Temple borobudur in the island of Java, Indonesia (courtesy IGN-UNESCO).

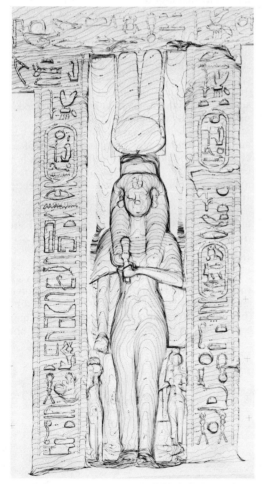

FIGURE 16-33. Temple of Abou Simbel, Egypt (courtesy IGN-UNESCO).

Orthophotography has recently been experimentally applied in Italy on photography of interiors of cupolas and in Germany and Poland for decorated surfaces with some relief. The application of orthophotography in architectural surveys remains, nevertheless, rather ticklish because of discontinuities in surfaces of buildings.

Some architectural surveys are made by analytical photogrammetry. In this approach, a certain number of points are accurately determined and then connected by architectural lines. Such a method is not applicable to complex and important monuments because it often involves too many assumptions on the course of the lines to be drawn between points thus giving a theoretical rather than a real representation. On the other hand, analytical methods can be advantageously applied in schematically treating groups of simple constructions, as has been done in the Scandinavian countries (figure 16-36).

The analytical approach is particularly suitable in studying the structure of monuments and in checking on their stability through the use of digital models. By forming digital models encompassing the monument's fundamental points and the skeleton of its structure, one can study the proportions, define the volumes, compare the forms, *etc*. By repeating these operations at intervals of time, one can follow and measure eventual deformations in the building and thus check on its stability. By targeting the points involved in the analytical and semi-analytical solutions, one significantly increases the precision of the observations and the accuracy of the solution.

Both the analog and the analytical approach lead to numerical data which is used to deter-

FIGURE 16-34. A hydraulic lift truck being used to position a Wild P31 camera in the spatial location desired (courtesy Wild-Heerbrugg, Inc.).

mine architectural forms. Using a computer, one can determine the curve or surface that best fits the group of points measured, according to the method of least squares. The principal works of this type were done for large cupolas of the Italian Renaissance (figures 16-37 and 16-38).

16.2.4.1.3 ARCHEOLOGICAL SURVEYS

As far as monuments are concerned, the same photogrammetric techniques used in architectural surveys are applicable to archeological surveys. For archeological excavations, however, the main characteristic is that the principal dimensions of the objects of interest are horizontal. Therefore, one often resorts to the use of either terrestrial photography taken with the camera axis inclined downward, or vertical aerial photography taken from low camera altitudes, with the camera suspended from a bipod or a tripod (about 15 to 20 metres high), a kite, or a captive balloon. American and Japanese archeological excavation missions often use captive ballons.

Photogrammetry is also used to study archeological finds in-situ, within the excavation camp, or in the museum. One can study fragments of sculptures, statutes of various dimensions, tomb monuments, coffins, jewelery, *etc.*

(figure 16-39). The main technical difficulty encountered in this type of work results from the lack of availability of metric cameras capable of being focused at object distances of 10-20 cm. One must often take photography at this range because of the smallness of the dimensions of the objects. In the absence of metric cameras suitable for this type of work, one often has to use non-metric cameras. Experience has shown that accuracy and fidelity of photogrammetric surveys are higher than what can be obtained by direct methods of mensuration.

Archeological photogrammetry has also been applied to underwater objects. Wrecks of ancient ships and ancient harbor installations have been surveyed. For the technical aspects of underwater photogrammetry, refer to section 16.4.

Large-scale aerial photography is the ideal vehicle for surveying extensive archeological sites (figure 16-40).

16.2.4.1.4 SURVEYS OF HISTORIC SITES

The restitution of terrestrial photography can sometimes provide interesting documentation on ancient architectural groups of urban buildings (historic centers) situated in the midst of more recent quarters or in isolated areas. Aerial pho-

FIGURE, 16-35. Mosaic in Gmayades Mosque, Damascus, Syria (courtesy IGN-UNESCO).

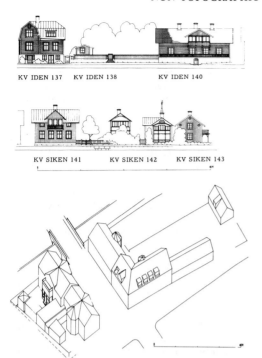

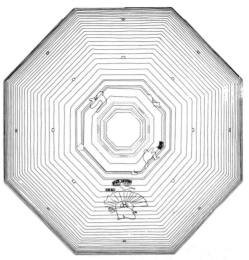

FIGURE 16-37. Photogrammetric survey of the cupola of Santa Maria del Fiori in Florence, Italy—plotting of the cupola. Contour interval: 1 metre (courtesy University of Florence and Officien Galileo).

FIGURE 16-36. Elevation drawing and axonometric view obtained by analytical procedures in a photogrammetric survey in Trosa, Sweden (courtesy Riksantikvarjeambete, Stockholm).

tography, however, is often more suitable for the study of historic centers. These ancient centers must be studied, preserved and, occasionally, developed and improved.

In addition to providing the plan, stereomodels offer other possibilities. One can obtain geometrically correct general elevations (figures 16-41 and 16-42), cross sections, perspective views, axonometric views of the object(s) photographed (*e.g.* historic centers). As these documents have accurate geometric properties, it is easy to add to the photogrammetric drawings proposed new buildings and to study their integration into the historic setting in order to consider the protection of the latter. Similar studies

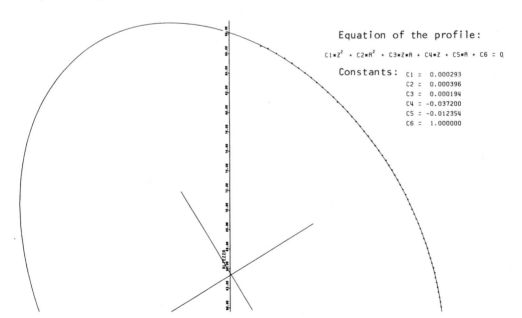

Equation of the profile:

$$C_1 \cdot Z^2 + C_2 \cdot R^2 + C_3 \cdot Z \cdot R + C_4 \cdot Z + C_5 \cdot R + C_6 = Q$$

Constants:
$C_1 = 0.000293$
$C_2 = 0.000396$
$C_3 = 0.000194$
$C_4 = -0.037200$
$C_5 = -0.012354$
$C_6 = 1.000000$

FIGURE 16-38. Photogrammetric survey of the cupola of Santa Maria del Fiori in Florence, Italy—profile of a rib, calculated on the basis of digital photogrammetric plotting of the points shown on the curve (courtesy University of Florence and Officine Galileo).

FIGURE 16-39. A bracelet and an ornament of a Gaullic Chariot (courtesy IGN).

can be undertaken for all kinds of new engineering works such as proposed roads, new bridges, viaducts, etc. Consequently the specialists have excellent data at their disposal for their study.

The introduction of a proposed new structure in the photogrammetric drawings is done on the basis of the coordinates of the fundamental points of the structure, computed in the coordinate system of the survey. To do this by analog instrument, one uses aerial photographs taken in flight lines oriented in the direction of, or perpendicular to, the chosen projection planes. During the plotting, the latter must be approximately parallel to the XZ or the YZ vertical planes of the instrument. The controls of the plotter are consequently modified by interchanging the X or Y with the Z outputs. Use is made of a differential which controls the displacement of the tracer on the tracing table.

The introduction of proposed structures on photogrammetric drawings can also be done through the use of digital tracing of elevations and perspectives, based on a set of data containing the coordinates of all important points of the proposed structures recorded by digital photogrammetric plotting. With the same digital data one can produce digital traces in plan, section, elevation and perspective from any viewpoints desired.

This approach is also useful to produce other kinds of documents which can be more easily interpreted by those who have to make decisions and need to judge the proposed transformation of the traditional aspects of the historic center. Such is the purpose of photomontage. Very often, however, these photomontages are not very accurate because they are produced by approximate methods; but photogrammetry can easily resolve this problem in an accurate fashion. Knowing the XYZ coordinates of the characteristic points of the proposed new development, together with those of the view point, the direction of the photography and the principal

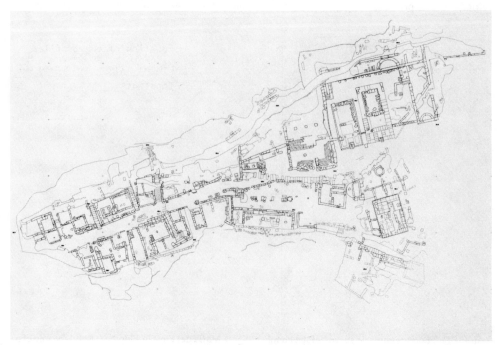

FIGURE 16-40. Archeological site of Glanum (Bouches du Rhone), France. Map made by aerial photogrammetry; scale of original plot 1:200; scale of aerial photography 1:3000 (courtesy IGN-Service des Fouilles).

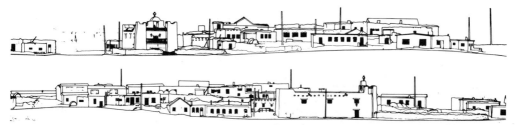

FIGURE 16-41. Indian pueblo in Zuni, New Mexico. Survey made by aerial photogrammetry (courtesy The Ohio State University).

distance of the photograph, it is easy to calculate the coordinates of the characteristic points on the photograph and to transfer them to a print or an enlargement. The proposed new structure is thus drawn on a photographic perspective of the historic center from the given viewpoint and one can see how it will look, if it is built (figure 16-43). This technique, known as "inverse photogrammetry" is in current use and has become an excellent means for analyzing problems connected with the protection of historic centers.

16.2.4.2 BIOSTEREOMETRICS (BIOMEDICAL AND BIOENGINEERING APPLICATIONS OF PHOTOGRAMMETRY)

16.2.4.2.1 INTRODUCTION

The study of biological form is one of the most engaging subjects in the history of human thought, which is hardly surprising considering the immense variety of living things. As new measurement techniques and experimental strategies have appeared, new fields of inquiry have been launched and more minds have become absorbed with the riddle of biological form. Discovery of the microscope and X-rays

prompted the development of microbiology and radiology, respectively. More recently, advances in electronics, photo-optics, computers and related technologies have helped to expand the frontiers of morphological research. Growing interest in the stereometric analysis of biological form typifies this trend.

Measurements of biological form and function were made from stereophotographs in the middle of the nineteenth century, shortly after the invention of photography. Why, you might ask, has it taken so long to establish a real place for photogrammetry in the biomedical world? Limited space does not permit a detailed discussion of this question; suffice it to say that the problem of bringing photogrammetrists and biomedical specialists together is a bit like trying to unite two tribes who speak different languages and are separated by uncharted territory. In this metaphorical setting, biostereometrics can provide the interpreter-guides needed to negotiate the no-mans land and make more durable connections than those which have occurred or are likely to occur by serendipity alone.

Over the years, contacts between photogrammetrists and biomedical specialists have

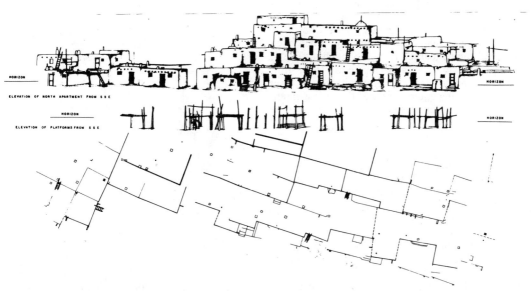

FIGURE 16-42. Indian pueblo in Taos, New Mexico. Survey made by terrestrial photogrammetry (courtesy The Ohio State University).

FIGURE 16-43. Introduction of proposed new buildings in the photograph of an urban area (Marseille, France) by inverse photogrammetry (courtesy IGN-Direction de l'Architecture).

been quite numerous (Herron, 1973) but most of the contacts involved trying to tie photogrammetry to a particular biomedical speciality. Unfortunately, these efforts generated surprisingly little sustained interaction. Recently, the more wide ranging approach of systematically relating stereometric analysis to biology and medicine in general has proved to be a more fruitful strategy. In this brief section, the rationale for and the scope of biostereometrics are outlined so that the important role which close-range photogrammetry has played, and continues to play, in this exciting development can be seen in a modern perspective.

16.2.4.2.2 THE RATIONALE

If biological structures were regular geometric shapes, there would be no great problem measuring them because simple lengths, breadths, and circumferences would be entirely adequate. But, as we all know, organisms have *irregular,* three-dimensional components and linear "atomistic" measures such as are produced by tapes and calipers cannot give an unambiguous, comprehensive, spatial quantification of a part or an organism as a whole. Of course, we must recognize that much has been and still can be learned from the judicious use of linear measurements, in spite of their limitations; and not until recently, with advances in data processing equipment, has the management of large quantities of stereometric data become practicable.

16.2.4.2.3 A DEFINITION

In a mathematical sense, the surface (internal or external) of any biological structure consists of an infinite number of points, all of which have a unique location in three dimensions, at any instant. By determining the three dimensional coordinates of enough of these points for the particular application (fig. 16-29), we can obtain a comprehensive, unambiguous set of measurements of a part or of the organism as a whole. Spatio-temporal changes due to movement or growth can be quantified in similar fashion, using serial measurements. Thus, bio-

stereometrics can be defined as the *spatial and spatio-temporal analysis of biological form and function, based on the principles of analytic geometry.*

16.2.4.2.4 SCOPE OF BIOSTEREOMETRICS

The scope of biostereometrics is outlined schematically in figure 16-30. The most important elements shown in the figure are briefly described below.

16.2.4.2.4.1 Stereometric Sensor. In the past, lack of convenient sensors retarded the growth of biostereometrics, but today there is a growing variety of stereometric sensors to choose from, figure 16-44, including: stereophotogrammetry, hologrammetry, moiré-fringe interfereometry, light-slit projection, stereometric X-rays, computerized axial tomography, and stereometric microscopy, among others. Each has its advantages and disadvantages and which technique is most appropriate depends on many considerations—object shape and size, environmental conditions, cost–effectiveness, available time, type of data required, research or clinical application and technical sophistication, to mention a few. Stereophotogrammetry is the most thoroughly studied and the most versatile stereometric sensing technique developed thus far. Among the advantages of this approach are: (1) it provides a permanent "holistic" record of the form which can be easily stored for later examination using the original or entirely new measurements; (2) the stereometric cameras are portable and amenable to use under varied environmental conditions; (3) the same equipment can be used to measure small parts as well as the whole organism and (4) stereophotogrammetry has a redundant capacity which can be used to determine what stereometric data are really

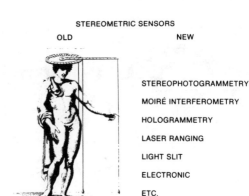

STEREOMETRIC SENSORS

OLD   NEW

STEREOPHOTOGRAMMETRY

MOIRÉ INTERFEROMETRY

HOLOGRAMMETRY

LASER RANGING

LIGHT SLIT

ELECTRONIC

ETC.

FIGURE 16-44. The coordinate data points are read off the object (such as a human body or body parts) using any suitable stereometric sensor. The first stereometric sensor was mechanical. The one shown on the left was used by Alberti in the fifteenth century. Today, there is a wide variety of stereometric sensors to choose from including those listed on the right.

needed for a particular research or clinical application. When the minimum data requirements have thus been defined, a simpler stereometric sensor can often be developed for the purpose at hand. As biostereometric research continues to grow, new sensing techniques can be expected to appear and others will be further refined. At present, there is little to suggest that any one approach will become a universal "method of choice". It seems more likely that a variety of techniques will be necessary to cope with the myriad subjects and conditions to be found within the realms of biology and medicine.

In every application, the stereometric sensor, the object being measured, and the data reduction procedure are integrally related. The importance of understanding this relationship was emphasized by Dr. Harold Morowitz, a Yale biophysicist, when he stated "Biological instrumentation cannot be separated from biological theory. The choice of appropriate instrumentation must be based upon a background of theory in the field of application and a broad knowledge of what is technologically feasible." This is a deceptively simple task, in that knowing what question to ask is perhaps the most crucial aspect of any area of research. Before he can tell the photogrammetrist, the photo-optical engineer or the mathematician what he would like to measure, the biologist or medical specialist must have a thorough understanding of what biostereometrics has to offer. Simply knowing that it is possible to make a contour map of the body or body part is rarely an adequate basis for definitive problem solving.

16.2.4.2.4.2 Computer Input and Output. Stereometric data can be stored in graphical (*e.g.,* contours, profiles, *etc.*) or in digital form. The digital form is more convenient for input to a computer, via punched cards, tape, or more directly using electronic, electro-optical, acoustic, or other signals. In this way, the versatile, analytical capabilities of the modern digital computer can be readily brought into play. With suitable programming, one can produce a wide variety of numerical outputs, including volumetric information, surface areas, slopes, rotations, angular deviations, and, of course, an infinite variety of statistical parameters. Once the details of body geometry have been entered in the computer, the computer can be made to generate a number of other potentially valuable outputs, using peripheral devices such as *X-Y* plotters, cathode ray tube (CRT) displays and numerically controlled (N/C) machine tools. Limited space does not permit a thorough treatment of these potentialities. A few examples are mentioned in figure 16-45.

16.2.4.2.5 CONCLUSION

Biologists and medical specialists are showing renewed interest in the stereometric analysis of

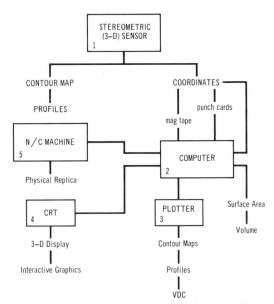

FIGURE 16-45. The scope of biostereometrics. This illustration is not meant to be exhaustive; it is merely intended to indicate the broad potentials of the subject.

biological form. Recent advances in computer technology and a growing range of stereometric sensing techniques have helped to expose the potential of biostereometrics. As a result, the use of photogrammetry is growing in such fields as: aerospace medicine, anthropometry, child growth and development, dentistry, marine biology, neurology, nutrition, orthodontics, orthopedics, pediatrics, physiology, prosthetics, radiology, and zoology, to mention a few.

The need for biostereometrics stems from the fact that linear tape and caliper measurements of inherently irregular three-dimensional biological structures are inadequate for many purposes. When stereometric data are used to fill this information gap the potentials for achieving more realistic models and making a more thorough analysis of biological form and function are far reaching. However, the best tools in the world confer no advantages unless they are used wisely. A petroleum geologist is expected to have the necessary training to select promising sites for oil exploration. Similarly, training in biostereometrics can be helpful in making decisions about "when" and "where" to use photogrammetry in biomedical research and clinical practice.

As the potentials of stereometric analysis in biology and medicine become more widely recognized, the role of biostereometrics in helping to unravel the complexities of organic form and function should continue to grow. Already, several photogrammetrists have chosen careers in biology and medicine and this number is expected to increase over the next few years.

16.2.4.2.6 A POSTSCRIPT

Recently, an important stage was reached with the very successful first international symposium on biostereometrics held in Washington, D.C., in September 1974. Over 50 papers were presented and the 160 participants came from twenty-two countries. There were several interesting and potentially valuable innovations in the hardware and software for stereometric analysis, but interested readers can find all the details in the proceedings (Herron and Karara, 1974) which was published by the American Society of Photogrammetry.

16.2.4.3 INDUSTRIAL PHOTOGRAMMETRY

Photogrammetry has been applied in numerous industrial fields and the potential for further expansion and growth is seemingly limitless. Meyer (1973) very aptly identified the fields encompassed by industrial photogrammetry as "application of photogrammetry in building construction, civil engineering, mining, vehicle and machine construction, metallurgy, ship-building and traffic, with their fundamentals and border subjects, including the phases of research, planning, production engineering, manufacture, testing, monitoring, repair and reconstruction. Objects measured by photogrammetric techniques may be solid, liquid or gaseous bodies or physical phenomena, whether stationery or moving, that allow of being photographed."

No attempt is made here to define the term "industrial photogrammetry" as such. It is felt that this term should be regarded as an open-ended rather than a definitive expression. This term, however, is useful for identity purposes and its systematic use in the literature should be helpful in furthering the application of photogrammetry in the various industrial fields. The experiences in the fields of architectural photogrammetry and biostereometrics clearly indicate the effectiveness of this strategy. The consistent use of the term "industrial photogrammetry" should be instrumental in drawing the attention of photogrammetrists and equipment manufacturers to this fertile field of application, and should be helpful in bringing the capabilities of photogrammetry to the attention of the various industries. This way, it is hoped that more and more industrial concerns would make full use of the economical and technical advantages of photogrammetry.

16.2.4.3.1 INDUSTRIAL REQUIREMENTS

While it is universally recognized that photogrammetric techniques have inherent potential and flexibility in non-contact and rapid spatial measurements, the fact remains that this technique will not gain general acceptance in industry unless it is cost effective and provides enough technical and economical advantages over other measuring techniques which are in

use or under consideration. Naturally, the factors that render industrial photogrammetric methods cost effective vary with the application. Each potential application requires careful review: the process must be defined, the accuracy must be investigated, direct cost must be analyzed and labor and time requirements must be considered. Only on the basis of estimates of this type can cost effectiveness of photogrammetric methods for a particular application be properly evaluated. Moreover, once a user has decided that photogrammetric techniques are indeed cost effective for his purposes, he must still weigh the merits of purchasing equipment and training operators against those of using a data reduction service.

In their article on the production of ship propellers, Knödler and Kupke (1974) state that the following factors combine to yield the economic benefit derived from the photogrammetric approach:

measurement time on the object is reduced by 90–95%;
saving in manpower;
reduced machine time for blade machining through optimization of the metal removal rate;
reduced material expenditure in the propeller casting manufacture through optimized molds;
a cut in recycling time for non-ferrous metals; and
a shorter production time for propeller manufacture.

Knödler and Kupke (1974) conclude that the photogrammetric technique outlined in their paper "is equally suited for other industries where work-pieces of a complex surface configuration are to be manufactured, which would be very time consuming to measure with conventional measuring tools."

A review of selected potential applications in metalworking industry by Bendix Corporation (1972) has confirmed that photogrammetric techniques can be both practical and economically feasible for industrial measurements and inspection tasks. The development of a systematic approach to implementing such applications is necessary to investigate the reduction of start-up costs, operating costs and equipment costs. Among the points brought up in the Bendix report (1972) was the necessity to develop large-format, focusable, metric cameras. Also, the need to simplify equipment setup and photogrammetric techniques was deemed essential for expediting data analysis.

16.2.4.3.2 EXAMPLES OF INDUSTRIAL APPLICATIONS

16.2.4.3.2.1 Automobile Construction. This is a fertile field which includes, for example, the measurement of pertinent dimensions of car body models to obtain production data, deformation measurements in test accidents, measurements of reaction to lateral wind forces, measurement of behavior after collision with road sign posts or barriers, measurement of tire deformation on the road, *etc*. Examples on these applications are given by: Ford Motor Company (1964), Kratky & van Wijk (1971), Santoni (1966), van Wijk & Pinkney (1972), Scholze, Töppler and Voss (1975), and others.

16.2.4.3.2.2 Machine Construction, Metalworking, Quality Control. This group encompasses measurements of complex machine parts, flow measurements in centrifugal pumps, measurement of vibrations, and measurements for quality control, *etc*. Examples on such applications are given by Wolfin (1966), Nauk & Voss (1972), Knödler & Kupke (1974), Higgins, Alice and McGivern (1972), Greggor (1973), Berling (1972), and others.

16.2.4.3.2.3 Mining Engineering. This includes process measurements in open-pit mining, tunnel cross sectioning and profiling, geological measurements, rock deformations, rock mechanics investigations, monitoring of production in mines, *etc*. The following references treat some of these subject areas: Allam (1975), Brandow *et al.* (1975), Chrzanowski & El-Masry (1969), Collins (1963), Gatu (1972), Grumpelt (1973), Kloss (1963), Linkwitz (1971), Ross-Brown *et al.* (1972, 1973), Torlegård and Dauphin (1975), Vlcek (1972), Woropajew (1973) and many others.

16.2.4.3.2.4 Objects in Motion. This includes tracking hand motions in industrial operations, studying glacier movements, assessing deformations of aerodynamic models in wind tunnels, tracking particle flow in bubble chambers, evaluation of displacements caused by explosions, *etc*. The following references treat some of these topics: Bednarski & Majde (1971), Bradner (1959), Brewer (1962), Brown *et al.* (1971), Flotron (1971), Garfield (1964), Gracie (1971), Maruyasu & Oshima (1968), Meyer (1964), Meyer & Will (1971), Ohman & van Wijk (1962), Reddy *et al.* (1969), Regensburger (1972), Waddell (1956), Zeller (1953).

16.2.4.3.2.5 Shipbuilding. This includes measurement of ship sections, measurement of propellers in its various stages of fabrication, wave measurements and position determination in model experiments, *etc*. References that deal with some of these topics include: Schmid (1973), Negut *et al.* (1972), Moffitt (1966), Moffitt (1968), Loomer & Wolf (1974), Faig (1972), Kenefick (1977).

16.2.4.3.2.6 Structures and Buildings. The use of photogrammetry to check on the construction and to measure the deformations of buildings, experimental structural models and large structures, such as reflector antenna and power dams, is covered by several papers, including: Argyris *et al.* (1972), Bernini *et al.* (1968), Borchers (1964), Brandenberger (1974), Brown (1962), Erlandson *et al.* (1974), Faig (1969), Foramitti (1966), Hallert (1954), Jaensch

FIGURE 16-46. View of a midship section taken with a Wild P31 by John F. Kenefick, Photogrammetric Consultant, Inc. Precise dimensioning of the midbody unit was one of several demonstrated applications of photogrammetry within shipbuilding, conducted in conjunction with the National Shipbuilding Research Program (Kenefick, 1977).

FIGURE 16-47. Photogrammetric determination of deformation of structures (Hottier, 1976).

FIGURE 16-48. Targets used in photogrammetric determination of deformation of structures (Hottier, 1976).

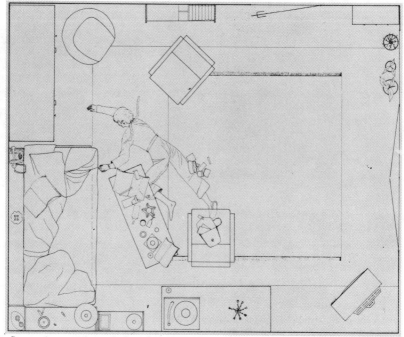

FIGURE 16-49. Stereophotographs and plot of a criminal act. Original scale of plot 1:10. By permission of the Cantonal Police of Zurich, Switzerland. Mock-up of a crime (courtesy Wild-Heerbrugg, Inc.).

(1970), Kenefick (1972), Konecny (1964), Marks (1963), Ockert (1959), Preuss (1972), Regenburger (1972), Shmutter & Etrog (1971), Wiedenhöft (1968), Yu (1959), and numerous others.

16.2.4.3.2.7 *Traffic Engineering*. This includes traffic accident investigations, profiling and cross-sectioning of roads and traffic tunnels, road surface roughness studies, ship path determination, *etc.* References in this group include: Berling (1970), Brandenberger (1974), Dach (1973), Fülscher (1963), Kobelin (1975), Kratky & van Wijk (1971), Sabey & Lupton (1967), among others.

16.2.4.3.2.8 *Other Industrial Applications.*

The bibliography includes a number of entries dealing with industrial applications not belonging to any of the above seven major groups. For example, Brinks *et al.* (1971) deal with predicting wholesale beef cuts, while Lydolph (1954) treats studies in animal husbandry. The velocity and direction of smoke trails are dealt with by Kiyoshi *et al.* (1966) and Pietschner (1964). The study of unsteady gas flow was the topic of Mann (1962), while the determination of the volume of huge pulpwood piles was treated by Young (1955). Gathering of evidence in criminal investigations is a process in which photogrammetry has been successfully applied in many countries (figure 16-49).

16.3 Terrestrial Photogrammetry

16.3.1 INTRODUCTION

While the term close-range photogrammetry is generally used in conjunction with object-to-camera distances of not more than some 300 metres, the term terrestrial photogrammetry is generally associated with object distances in excess of 300 metres. Another distinction between the two categories is the fact that in terrestrial photogrammetry cameras are focused at infinity while this is not the case in close-range photogrammetry as discussed in section 16.2.

16.3.2 DATA ACQUISITION

16.3.2.1 TERRESTRIAL CAMERAS

The photogrammetric single cameras discussed earlier in conjunction with close-range photogrammetry (section 16.2.2.1.1) are also used in terrestrial photogrammetry.

16.3.2.2 CONFIGURATION OF TERRESTRIAL PHOTOGRAPHY

16.3.2.2.1 THE "NORMAL CASE"

In this case (figure 16-50) the two camera axes (from camera stations O_1 and O_2) are parallel to each other and perpendicular to the base line B. The spatial coordinates of any point $P(X_p, Y_p, Z_p)$ can be expressed as:

$$X_p = \frac{Y_p}{f} x_1$$

$$Y_p = \frac{Bf}{p} \qquad (16.17)$$

$$Z_p = \frac{Y_p}{f} y$$

where p is the x-parallax, $p = x_1 - x_2$.

Through error propagation, one can determine σ_X, σ_Y and σ_Z, the standard errors in the spatial coordinates of any point. The error in the Y coordinate is primarily a function of the standard error of the parallax σ_p. Thus

$$|\sigma_Y| = \left| \frac{Bf}{p^2} \right| \cdot |\sigma_p| , \qquad (16.18)$$

from which the value of B can be computed as

$$B = \frac{(Y_{max})^2}{f} \left(\frac{\sigma_p}{\sigma_Y} \right). \qquad (16.19)$$

In equation (16.19), Y is taken as Y max because the required accuracy must be achieved at the maximum photographic distance. The length of base line is usually selected so that

$$\frac{Y_{max}}{20} < B < \frac{Y_{min}}{5}. \qquad (16.20)$$

16.3.2.2.2 TILTED PARALLEL PHOTOGRAPHY

In this case the camera axes are parallel and tilted relative to the base line as shown in figure 16-51. The geometry of this case can be related to the "normal case" through the following relation between B and b:

$$b = \frac{B}{f} (f \cos \phi - x_2 \sin \phi) .$$

Substituting in equations (16.17), one gets:

$$Y_p = \frac{b}{p} f = \frac{\frac{B}{f} (f \cos \phi - x_2 \sin \phi)}{p} f ,$$

$$X_p = \frac{\frac{B}{f} (f \cos \phi - x_2 \sin \phi)}{p} x_1 , \qquad (16.21)$$

$$Z_p = \frac{\frac{B}{f} (f \cos \phi - x_2 \sin \phi)}{p} y .$$

16.3.2.2.3 CONVERGENT PHOTOGRAPHY

If stereoscopic viewing is planned, the angle of covergence (ϕ in figure 16-36) should not exceed 7°. Figure 16-52 illustrates the "stereocon-

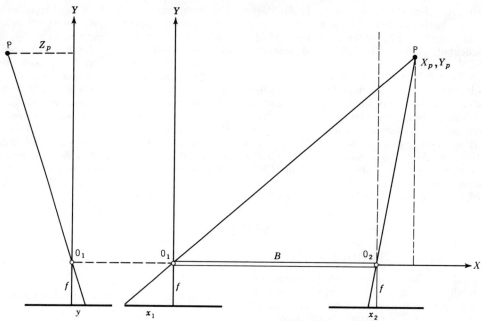

FIGURE 16-50. The "Normal Case" of terrestrial photogrammetry.

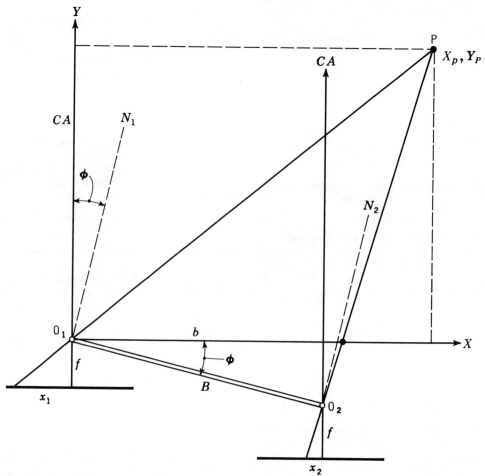

FIGURE 16-51. Tilted parallel terrestrial photography ($N_1 O_1$ and $N_2 O_2$ are normals to the baseline B).

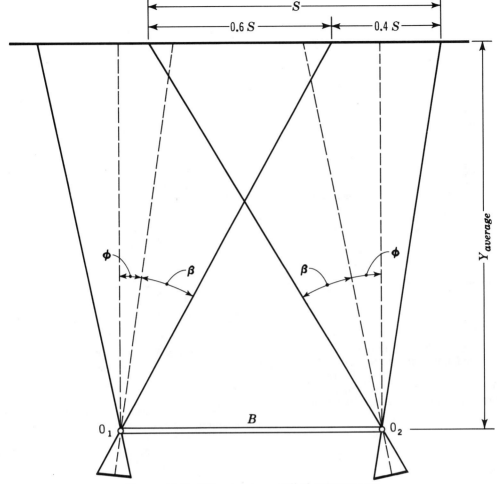

FIGURE 16-52. "Stereo-convergent" photogrammetry.

vergent" case. The overlap between the photographs is about 60% at the average photographic distance (Y_{av}); and the base length B can be computed as follows:

$$B = Y_{av} [1.4 \tan (\beta + \phi) - 0.6 \tan (\beta - \phi)] \quad (6.22)$$

where

β = 1/2 of the camera angle of view, and
ϕ = angle of convergence.

In general, however, convergent photography is obtained from two (or more) camera stations with about 100% overlap between photographs. Stereoscopic viewing is not possible with the geometry shown in figure 16-53; therefore, only analytical methods applied to individual points can be used.

Of the elements of exterior orientation, the rotational elements ϕ_1, ϕ_2, ω_1, ω_2 can be measured by the theodolite portion of the phototheodolite, or by a separate theodolite, and the elements κ_1 and κ_2 can be measured using a striding level. Another possibility for determining the elements of exterior orientation is to

computationally determine the orientation matrix of the photograph on the basis of the image-space and object-space coordinates of control points, as discussed in section 16.3.3.2.1.2.

The direct measurement of exterior orientation parameters introduces eccentricities (EC in figures 16-53 and 16-54) which need to be determined and taken into account. The camera of a phototheodolite rotates around an axis passing through its center of gravity to balance its weight. The eccentricity (EC) is the distance between the axis of rotation (R_1 or R_2) and the corresponding perspective center (O_1 or O_2).

The coordinates X_{R_i}, and Y_{R_i}, and Z_{R_i} of the axis of rotation can be determined by ground survey and the station coordinates can then be computed as:

$$\left.\begin{array}{l} X_{O_i} = X_{R_i} + EC \cos \psi_i \cos \omega_i \\ Y_{O_i} = Y_{R_i} + EC \sin \psi_i \cos \omega_i \\ Z_{O_i} = Z_{R_i} + EC \sin \omega_i \end{array}\right\} \quad (16.23)$$

where i refers to the particular camera station under consideration. Equations (16.23) are cor-

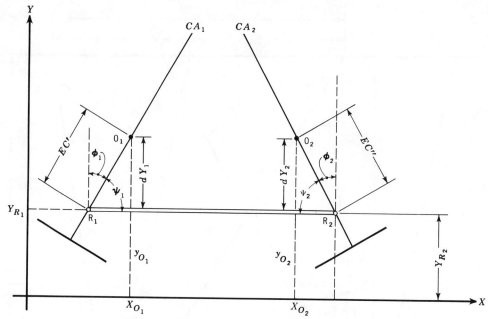

FIGURE 16-53. Convergent terrestrial photography.

rect only for those phototheodolites whose vertical and horizontal rotational axes intersect. This is not the case for all instruments. For a number of instruments, including the Wild P-30 phototheodolite, equations (16.23) modify to:

$$\left.\begin{array}{l} X_{O_i} = X_{R_i} + EC_h \cos \psi_i \cos \omega_i \\ Y_{O_i} = Y_{R_i} + EC_h \sin \psi_i \cos \omega_i \\ Z_{O_i} = Z_{R_i} + EC_r \sin \omega_i \end{array}\right\} \quad (16.24)$$

When high precision is required, for example in structural measurements, several photographs are taken from each camera station. The number of photographs necessary may be obtained from the following equation:

$$\sigma_0 = \frac{\sigma_i}{\sqrt{n}} \quad (16.25)$$

where σ_0 is the standard error of unit weight, σ_i represents the standard error of observation (for the photogrammetric instrument used σ_i is taken as the pointing error), and n is the number of photographs.

It can be shown using equation (16.25) that the economical number of photographs in the set of photographs taken from each camera station lies between 2 and 4. About 30% improvement in σ_0 is obtained by measuring on two photographs from each camera station and about 16% improvement is obtained by increasing the number of photographs to three. By increasing the number of photographs from 3 to 4 an improvement of about 7% in σ_0 is expected. Thus, multiple photographs from each camera station are justifiable. The improvements cited above for σ_0 are theoretical values which may vary in practice.

16.3.3 DATA REDUCTION

Data reduction in terrestrial photogrammetry can be done using analog or analytical methods.

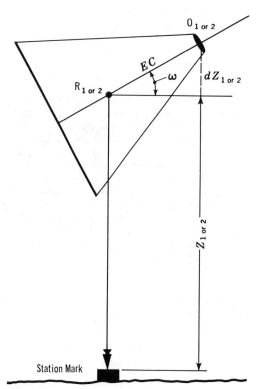

FIGURE 16-54. Eccentricity in phototheodolites.

16.3.3.1 THE ANALOG APPROACH

While some plotters, including the Wild A40 (figure 16-13) and Zeiss Terragraph (figure 16-14), are designed for plotting from terrestrial photographs only, other plotters are designed for plotting from aerial as well as terrestrial photography, such as the Galileo-Santoni IIC plotter, Wild A7 and A10 autographs and others.

The mechanical limitations of the analog plotter to be used in the reduction must be taken into account in designing the data acquisition set-up.

16.3.3.2 THE ANALYTICAL APPROACH

Terrestrial photographs (usually the original negatives) can be measured on mono- or stereo-comparators. The measured coordinates, in the comparator coordinate system, are then reduced to the center of the photographs, and corrected for lens distortion and atmospheric refraction. If film is used in the phase of data acquisition, then film shrinkage must also be included in the coordinate refinement.

There are two different analytical methods for determining the ground (object-space) coordinates of points of interest: approximate and rigorous. In the approximate method, the elements of exterior orientation, measured directly or indirectly, are regarded as *fixed* quantities on the basis of which the ground coordinates of object points are computed.

The rigorous method, on the other hand, involves no fixed quantities as such. All the quantities, those measured by ground surveys and those measured photogrammetrically, are all regarded as variable parameters in a simultaneous least squares solution. The rigorous method provides considerably better precision than the approximate method.

16.3.3.2.1 APPROXIMATE APPROACH

Data-reduction consists of (a) the determination of the locations of the perspective centers, (b) determination of the orientation matrix of each photograph, and (c) computation of the coordinates of points of interest.

16.3.3.2.1.1 Determination of Location of Camera Stations (perspective centers ie "frontal nodal points"). This is often done by resection. The geometry of space resection is shown in figure 16-55. The coordinates of the perspective center O are X, Y, Z; and the corresponding estimated approximate coordinates are X_0, Y_0, Z_0.

From figure 16-55 one can obtain

$$\cos (ij) = \cos (IJ) \qquad (16.26)$$

$$\cos (ij) = \frac{x_i x_j + y_i y_j + f^2}{S_i \, S_j}$$

where

$$S_i = \sqrt{x_i^2 + y_i^2 + f^2} \quad \text{and} \quad S_j = \sqrt{x_j^2 + y_j^2 + f^2}.$$

Cos (IJ) cannot be computed because X, Y, Z coordinates are unknown. The approximate value computed for cos (IJ) is

$$\cos (IJ)_0 = \frac{(X_0 - X_I)(X_0 - X_J) + (Y_0 - Y_I)(Y_0 - Y_J) + (Z_0 - Z_I)(Z_0 - Z_J)}{S_I \cdot S_J}$$

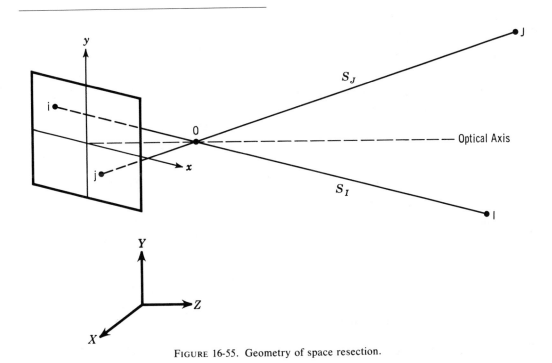

FIGURE 16-55. Geometry of space resection.

where

$$S_I = \sqrt{(X_0 - X_I)^2 + (Y_0 - Y_I)^2 + (Z_0 - Z_I)^2}$$
$$S_J = \sqrt{(X_0 - X_J)^2 + (Y_0 - Y_J)^2 + (Z_0 - Z_J)^2}.$$

Equation (16.26) can be rewritten as:

16.3.3.2.1.2 *Determination of the Orientation Matrix for each Photograph.* This depends on the way the elements of exterior orientation of the photograph were determined: direct or indirect.

$$\cos (IJ)_0 = \frac{\partial \cos (IJ)}{\partial X_0} \Delta X + \frac{\partial \cos (IJ)}{\partial Y_0} \Delta Y + \frac{\partial \cos (IJ)}{\partial Z_0} \Delta Z = \cos (ij).$$

Then the general observation equation may be shown as:

$$a_{ij}\Delta X + b_{ij}\Delta Y + C_{ij}\Delta Z - \cos (ij) - \cos (IJ)_0 = v_{ij}$$
$$\vdots \qquad \vdots \qquad \vdots \qquad \vdots \qquad \vdots \qquad \vdots$$

(A) *In case ω, ϕ and κ were measured in the field.* The orientation matrix is formed from the ϕ_i, ω_i and κ_i orientation angles.

$$\mathbf{M}_i = \begin{bmatrix} \cos \phi_i & 0 & \sin \phi_i \\ \sin \omega_i \sin \phi_i & \cos \omega_i & -\sin \omega_i \cos \phi_i \\ -\cos \omega_i \sin \phi_i & \sin \omega_i & \cos \omega_i \cos \phi_i \end{bmatrix} = \begin{bmatrix} m_{11} & m_{12} & m_{13} \\ m_{21} & m_{22} & m_{23} \\ m_{31} & m_{32} & m_{33} \end{bmatrix}$$

or

$$\mathbf{AX} - \mathbf{L} = \mathbf{V}, \qquad (16.27)$$

where the partial derivatives are:

$$a_{ij} = (X_0 - X_I)T_{JI} + (X_0 - X_J)T_{IJ}$$
$$b_{ij} = (Y_0 - Y_I)T_{JI} + (Y_0 - Y_J)T_{IJ}$$
$$c_{ij} = (Z_0 - Z_I)T_{JI} + (Z_0 - Z_J)T_{IJ}$$

and

$$T_{IJ} = \frac{\left(1 - \dfrac{S_I \cos (ij)}{S_J}\right)}{S_I S_J}.$$

If there are n points in the object space, the number of angle combinations is $(n(n - 1))/2$. Three points lead to a unique solution. For four points there are six angle combinations, thus six equations in (16.27).

Using the proper number of angle combinations, one computes the V matrix:

$$\mathbf{AX} - \mathbf{L} = \mathbf{V};$$

and letting

$$\mathbf{A}^T\mathbf{AX} - \mathbf{A}^T\mathbf{L} = 0,$$

one solves for

$$\Delta\mathbf{X} = (\mathbf{A}^T\mathbf{A})^{-1} \mathbf{A}^T\mathbf{L}.$$

The coordinates of the perspective center O are then

$$\left.\begin{array}{l} X' = X_0 + \Delta X \\ Y' = Y_0 + \Delta Y \\ Z' = Z_0 + \Delta Z \end{array}\right\} \qquad (16.28)$$

Start the second iteration with the new approximate values of X', Y' and Z'. The iteration should continue until

$$\Delta X = \Delta Y = \Delta Z = 0.$$

This orientation matrix takes into account that the κ angle is measured with the striding level thus equals zero.

(B) *In case the elements of exterior orientation are to be determined indirectly through computations.* The basic principle of computation in this step is the mathematical comparison of image-space vector $(\overrightarrow{S_a})$ between the perspective center of the lens and the image point, to the object-space vector $(\overrightarrow{S_A})$ between the perspective center and the object point (A).

Let's consider there are three given points: A, B, and C, for which the ground coordinates are X_A, Y_A, Z_A, X_B, Y_B, Z_B, and X_C, Y_C, Z_C. The photo-coordinates are x_a, y_a, x_b, y_b, x_c, y_c, and f. Further, let the camera station coordinates be X, Y, Z.

The vector of aO in the image space is defined as:

$$\overrightarrow{S_a} = \begin{bmatrix} \dfrac{-x_a}{\sqrt{x_a^2 + y_a^2 + f^2}} \\ \dfrac{-y_a}{\sqrt{x_a^2 + y_a^2 + f^2}} \\ \dfrac{f}{\sqrt{x_a^2 + y_a^2 + f^2}} \end{bmatrix} \begin{bmatrix} \cos xaO \\ \cos yaO \\ \cos zaO \end{bmatrix}$$

Similarly in the object space:

$$\overrightarrow{S_A} = \begin{bmatrix} \dfrac{X_0 - X_A}{OA} \\ \dfrac{Y_0 - Y_A}{OA} \\ \dfrac{Z_0 - Z_A}{OA} \end{bmatrix} \begin{bmatrix} \cos XAO \\ \cos YAO \\ \cos ZAO \end{bmatrix}$$

where

$$AO = \sqrt{(X_0 - X_A)^2 + (Y_0 - Y_A)^2 + (Z_0 - Z_A)^2}.$$

If **M** is formed for negatives, the orientation matrix, the relation between the two vectors, is

$$\mathbf{M}(\vec{S_A}) = (\vec{S_a}),$$

or in detail:

$$\begin{bmatrix} m_{11} & m_{12} & m_{13} \\ m_{21} & m_{22} & m_{23} \\ m_{31} & m_{32} & m_{33} \end{bmatrix} \begin{bmatrix} \cos XAO \\ \cos YAO \\ \cos ZAO \end{bmatrix} = \begin{bmatrix} \cos xaO \\ \cos yaO \\ \cos zaO \end{bmatrix}.$$

Because the **M** contains nine terms, a minimum of three points are required to solve for it. That is:

$$\mathbf{M}(\vec{S_A}\ \vec{S_B}\ \vec{S_C}) = \vec{S_a}\ \vec{S_b}\ \vec{S_c};$$

thus:

$$\mathbf{M} = (\vec{S_A}\ \vec{S_B}\ \vec{S_C})\ (\vec{S_a}\ \vec{S_b}\ \vec{S_c})^{-1}. \quad (16.29)$$

If more than three points are used in the solution, a least squares adjustment must be used. The observation equations in this case are

$$\cos XAO\, m_{11} + \cos YAO\, m_{12} + \cos ZAO\, m_{13} - \cos xaO = v_1$$
$$\cos XBO\, m_{11} + \cos YBO\, m_{12} + \cos ZBO\, m_{13} - \cos xbO = v_2$$

$$\vdots \qquad\qquad \vdots \qquad\qquad \vdots$$

$$Z_0 = \left[X_{0_2} - X_{0_1} - Z_{0_1} \frac{\cos XO_1P}{\cos ZO_1P} + Z_{0_2} \frac{\cos XO_2P}{\cos ZO_2P} \right] \left[\cfrac{1}{\left(\cfrac{\cos XO_2P}{\cos ZO_2P} - \cfrac{\cos XO_1P}{\cos ZO_2P} \right)} \right] \quad (16.31)$$

or in general form:

$$AX - L = V,$$

where

$$X = \begin{bmatrix} m_{11} \\ m_{12} \\ m_{13} \end{bmatrix}.$$

The rest of the matrices are easily identifiable.

The least-squares solution gives the orientation matrix as:

$$X = (A^{T}A)^{-1}\, A^{T}L.$$

16.3.3.2.1.3 Determination of Spatial Coordinates of Object Points. After determining the orientation of each photograph, one proceeds to determine the spatial object-space coordinates of observed object points. This determination is performed by intersection. The solution will be briefly described below.

Basically,

$$\left.\begin{array}{l} X_p = X_0 - \Delta X \\ Y_p = Y_0 - \Delta Y \\ Z_p = Z_0 - \Delta Z \end{array}\right\} \quad (16.30)$$

where X_p, Y_p, and Z_p are the point's coordinates in the object space, X_0, Y_0, Z_0 represent its approximate coordinates and ΔX, ΔY and ΔZ are the respective differences to be determined by the least-squares adjustments.

The given quantities are X_{0_1}, Y_{0_1}, Z_{0_1}, X_{0_2}, Y_{0_2}, Z_{0_2} coordinates of camera station O_1 and O_2 and the orientation matrices **M'** and **M''**.

The equation of the line from O_1 to point P is:

$$\frac{X_0 - X_{0_1}}{\cos XO_1P} = \frac{Y_0 - Y_{0_1}}{\cos YO_1P} = \frac{Z_{0_1} - Z_0}{\cos ZO_1P}.$$

Similarly for line O_2P:

$$\frac{X_0 - X_{0_2}}{\cos XO_2P} = \frac{Y_0 - Y_{0_2}}{\cos YO_2P} = \frac{Z_{0_2} - Z_0}{\cos ZO_2P}.$$

From these equations the approximate coordinates can be determined as:

$$X_0 = (Z_{0_1} - Z_0) \frac{\cos XO_1P}{\cos ZO_1P} + X_{0_1}$$

$$Y_0 = (Z_{0_1} - Z_0) \frac{\cos YO_1P}{\cos ZO_1P} + Y_{0_1}.$$

In equation (16.31) the direction cosines, as computed from the image coordinates and modified by the orientation matrix, are:

$$\cos XO_1P = m'_{11} \frac{x_1}{O_1P} + m'_{12} \frac{y_1}{O_1P} + m'_{13} \frac{f}{O_1P}$$

$$\vdots \qquad\qquad \vdots \qquad\qquad \vdots$$

and

$$\cos XO_2P = m''_{11} \frac{x_2}{O_2P} + m''_{12} \frac{y_2}{O_2P} + m''_{13} \frac{f}{O_2P}$$

$$\vdots \qquad\qquad \vdots \qquad\qquad \vdots$$

where

$$O_1P = \sqrt{x_1^2 + y_1^2 + f^2}$$

and

$$O_2P = \sqrt{x^2_2 + y^2_2 + f^2}.$$

It may be noted that there are a number of other ways to obtain approximate coordinates; some of these are particularly simple if the coordinate system is chosen so that one of its axes is parallel to the camera base line. The above solution is among the most general because it is independent of the orientation of the spatial coordinate system.

Having obtained the approximate coordinates, the space intersection can now be computed. Consider three camera stations for which the given quantities and the orientation matrices $\mathbf{M}'$, $\mathbf{M}''$ and $\mathbf{M}'''$ and the coordinates of the camera stations are $X_{0_1}, X_{0_2}, X_{0_3}, Y_{0_1}, Y_{0_2}, Y_{0_3}, Z_{0_1}, Z_{0_2}, Z_{0_3}$. Measured values are the point coordinates $x'_i y'_i f'_i, x''_i y''_i f''_i$ and $x'''_i y'''_i f'''_i$ for the first, second, and third negatives, respectively. The collinearity equations can be written as:

$$x' = f' \frac{\mathbf{M}'_1\mathbf{X}'}{\mathbf{M}'_3\mathbf{X}'}, \qquad y' = f' \frac{\mathbf{M}'_2\mathbf{X}'}{\mathbf{M}'_3\mathbf{X}'},$$

$$x'' = f'' \frac{\mathbf{M}''_1\mathbf{X}''}{\mathbf{M}''_3\mathbf{X}''}, \qquad y'' = f'' \frac{\mathbf{M}''_2\mathbf{X}''}{\mathbf{M}''_3\mathbf{X}''},$$

$$x''' = f''' \frac{\mathbf{M}'''_1\mathbf{X}'''}{\mathbf{M}'''_3\mathbf{X}'''}, \qquad y''' = f''' \frac{\mathbf{M}'''_2\mathbf{X}'''}{\mathbf{M}'''_3\mathbf{X}'''}.$$

(16.32)

The observation equations, in the general form, can be expressed as:

$$\mathbf{V} = \mathbf{AX} - \mathbf{L},$$

where

$$v_{x'} = \frac{\partial F'_1}{\partial X} \Delta X_i + \frac{\partial F'_1}{\partial Y} \Delta Y_i + \frac{\partial F'_1}{\partial Z} \Delta Z_i + F'_1 (X_0, Y_0, Z_0) - X',$$

$$v_{x''} = \frac{\partial F''_1}{\partial X} \Delta X_i + \frac{\partial F''_1}{\partial Y} \Delta Y_i + \frac{\partial F''_1}{\partial Z} \Delta Z_i + F''_1 (X_0, Y_0, Z_0) - X'',$$

$$v_{y'} = \frac{\partial F'_2}{\partial X} \Delta X_i + \frac{\partial F'_2}{\partial Y} \Delta Y_i + \frac{\partial F_2}{\partial Z} \Delta Z_i + F'_2 (X_0, Y_0, Z_0) - Y'.$$

The identities therefore are:

$$\mathbf{V} = \begin{bmatrix} v_{x'} \\ v_{x''} \\ v_{x'''} \\ v_{y'} \\ v_{y''} \\ v_{y'''} \end{bmatrix}, \quad \mathbf{X} = \begin{bmatrix} \Delta X_i \\ \Delta Y_i \\ \Delta Z_i \end{bmatrix},$$

$$\mathbf{A} = \begin{bmatrix} \dfrac{\partial F'_1}{\partial X} & \dfrac{\partial F'_1}{\partial Y} & \dfrac{\partial F'_1}{\partial Z} \\[2mm] \dfrac{\partial F''_1}{\partial X} & \dfrac{\partial F''_1}{\partial Y} & \dfrac{\partial F''_1}{\partial Z} \\[2mm] \dfrac{\partial F'''_1}{\partial X} & \dfrac{\partial F'''_1}{\partial Y} & \dfrac{\partial F'''_1}{\partial Z} \\[2mm] \dfrac{\partial F'_2}{\partial X} & \cdots & \cdots \\ \cdot & & \\ \cdot & & \\ \dfrac{\partial F'''_2}{\partial X} & \cdots & \cdots \end{bmatrix}$$

and

$$\mathbf{L} = \begin{bmatrix} x' - F'_1 (X_0, Y_0, Z_0) \\ \cdot \\ \cdot \\ \cdot \\ \cdot \\ \cdot \\ y''' - F'''_2 (X_0, Y_0, Z_0) \end{bmatrix}.$$

The normal equations, in the usual notations, are:

$$\mathbf{A}^T\mathbf{AX} - \mathbf{A}^T\mathbf{L} = 0$$

and

$$\mathbf{X} = (\mathbf{A}^T\mathbf{A})^{-1} \mathbf{A}^T\mathbf{L}.$$

The standard error of unit weight and the standard error of the most probable values of the coordinates are:

$$\sigma_0 = \sqrt{\frac{\mathbf{V}^T \mathbf{V}}{n - u}}$$

and

$$\sigma_x = \sigma_0 \sqrt{Q_{xx}}$$

respectively, where n is the number of observations, u is the number of unknowns and Q_{xx} is the diagonal of the inverse of the coefficient matrix of the normal equation.

16.3.3.2.2 THE RIGOROUS APPROACH

In this approach all the measured quantities (from the image coordinates to the elements of exterior orientation) are considered as unknowns and are determined simultaneously. Let us take an example and follow it through.

In a particular project, the base line and the orientation angles ϕ, ω at each camera station were measured. The general flow diagram of data reduction is given in table 16-4. The reduction of plate coordinates also involves their refinement as discussed in section 16.3.3.2. The computation of the camera-station coordinates is undertaken using equation (16.24). The orientation matrix for each photograph is formed using the measured values of the orientation angles ϕ and ω (figures 16-53 and 16-54). The orientation

angle κ is considered to be zero since the phototheodolite was leveled. The approximate coordinates X_0, Y_0, Z_0 of the perspective centers are computed using equation (16.31).

The combined observation and condition equations are then formed and solved, as discussed in

detail by Veress (1974) and as summarized below. The combined observations in condition form are:

$$\left. \begin{array}{l} x_i - f\dfrac{\mathbf{M}_1\mathbf{X}}{\mathbf{M}_3\mathbf{X}} = 0 \\[12pt] y_i = f\dfrac{\mathbf{M}_2\mathbf{X}}{\mathbf{M}_3\mathbf{X}} = 0 \end{array} \right\} \qquad (16.33)$$

The orientation matrix is regarded as:

$$\mathbf{M} = \begin{bmatrix} \mathbf{M}_1 \\ \mathbf{M}_2 \\ \mathbf{M}_3 \end{bmatrix},$$

and the $\mathbf{X}$ matrix is:

$$\mathbf{X} = \begin{bmatrix} X_0 - X_{0_i} \\ Y_0 - Y_{0_i} \\ Z_0 - Z_{0_i} \end{bmatrix}.$$

The X_0, Y_0 and Z_0 coordinates are the approximate coordinates of points to be determined. The X_{0_i}, Y_{0_i}, and Z_{0_i} are measured coordinates of the camera station "i".

Using the conventional symbols, the equation (16.33) can be written in detailed form as:

$$\left. \begin{array}{l} x_i = f\dfrac{m_{11}(X_0 - X_{0_i}) + m_{12}(Y_0 - Y_{0_i}) + m_{13}(Z_0 - Z_{0_i})}{m_{31}(X_0 - X_{0_i}) + m_{32}(Y_0 - Y_{0_i}) + m_{33}(Z_0 - Z_{0_i})} = w_{i_1} \\[14pt] y_i - f\dfrac{m_{21}(X_0 - X_{0_i}) + m_{22}(Y_0 - Y_{0_i}) + m_{23}(Z_0 - Z_{0_i})}{m_{31}(X_0 - X_{0_i}) + m_{32}(Y_0 - Y_{0_i}) + m_{33}(Z_0 - Z_{0_i})} = w_{i_2} \end{array} \right\} \qquad (16.34a)$$

Equation (16.34a) can be rewritten as

$$\left. \begin{array}{l} x_i - F_1(X_0, Y_0, Z_0) = w_{i_1} \\[6pt] y_i - F_2(X_0, Y_0, Z_0) = w_{i_2} \end{array} \right\} \qquad (16.34b)$$

Linearizing by Taylor's series, the following equations are obtained:

$$\left. \begin{array}{l} v_{x_i} - \dfrac{\partial F_1}{\partial \omega} v_\omega - \dfrac{\partial F_1}{\partial \phi} v_\phi - \dfrac{\partial F_1}{\partial \kappa} v_\kappa \\[12pt] \qquad + \dfrac{\partial F_1}{\partial X_{0_i}} v_X + \dfrac{\partial F_1}{\partial Y_{0_i}} v_Y + \dfrac{\partial F_1}{\partial Z_{0_i}} v_Z \\[12pt] \qquad\qquad - \dfrac{\partial F_1}{\partial X_0} \Delta X - \dfrac{\partial F_1}{\partial Y_0} \Delta Y - \dfrac{\partial F_1}{\partial Z_0} \Delta Z \\[12pt] \qquad\qquad\qquad + x_i - F_1(X_0, Y_0, Z_0) = 0 \\[18pt] v_{y_i} - \dfrac{\partial F_2}{\partial \omega} v_\omega - \dfrac{\partial F_2}{\partial \phi} v_\phi - \dfrac{\partial F_2}{\partial \kappa} v_\kappa \\[12pt] \qquad + \dfrac{\partial F_2}{\partial X_{0_i}} v_X + \dfrac{\partial F_2}{\partial Y_{0_i}} v_Y + \dfrac{\partial F_2}{\partial Z_{0_i}} v_Z \\[12pt] \qquad\qquad - \dfrac{\partial F_2}{\partial X_0} \Delta X - \dfrac{\partial F_2}{\partial Y_0} \Delta Y - \dfrac{\partial F_2}{\partial Z_0} \Delta Z \\[12pt] \qquad\qquad\qquad + y_i - F_2(X_0, Y_0, Z_0) = 0 \end{array} \right\} \qquad (16.35a)$$

TABLE 16-4. FLOW DIAGRAM OF DATA REDUCTION

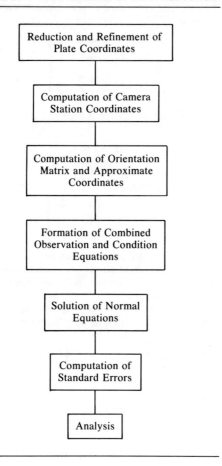

or in general matrix form:

$$BV + AX + W = 0. \qquad (16.35b)$$

In equation (34a) the v_{x_i} and v_{y_i} are the corrections rendered to the measured plate coordinates. The v_ω, v_ϕ and v_κ are the corrections to the measured orientation elements and the v_X, v_Y and v_Z are the corrections to the coordinates of camera station "i". The rest of the terms are readily identifiable.

The normal equations can be obtained from the combined observation and condition equations. The normal equation in matrix form is

$$BP^{-1}B^T K + AX + W = O,$$
$$A^T K \qquad\qquad = O,$$

or

$$\begin{bmatrix} BP^{-1}B^T & A \\ A^T & O \end{bmatrix} \begin{bmatrix} K \\ X \end{bmatrix} + \begin{bmatrix} W \\ O \end{bmatrix} = O.$$

In these equations the K is the correlate vector and P is the weight matrix. The P matrix can be selected according to the conventional least-squares procedures.

The solution of the normal equations is:

$$X = - [A^T M^{-1} A]^{-1} A^T M^{-1} W,$$

where

$$M = B P^{-1} B^T, \text{ and}$$
$$K = -M^{-1} (AX + W).$$

The standard error of unit weight is

$$\sigma_0 = \sqrt{\frac{V^T P V}{r - u}}.$$

where

$$V^T P V = -K^T W$$

and r and u are the number of conditions and parameters respectively. The standard errors for the observations and for the parameters are computed as follows:

$$\sigma_0 = \sigma_0 \sqrt{Q_{oo}},$$
$$\sigma_X = \sigma_0 \sqrt{Q_{XX}},$$

$$Q_{oo} = P^{-1} - P^{-1}B^T(M^{-1} - M^{-1} [A^T M^{-1}A]^{-1} A^T M^{-1}) BP^{-1},$$

and

$$Q_{XX} = [A^T M^{-1} A]^{-1}.$$

Having obtained the X vector, the most probable values of the coordinates can then be deduced by adding this vector to the approximate values of the coordinates. The results can be analyzed with respect to the geometry at each point. On the basis of the standard errors of coordinates, the error ellipsoid at each point may be computed and plotted.

16.3.4 GROUND CONTROL OPERATIONS

Ground control operations in terrestrial photogrammetry include the determination of elements of exterior orientation of the photographs, co-ordination and targeting of control points, as well as the targeting of object points. The matter of targeting is discussed below. All other ground control operations are undertaken in the regular manner (refer to chapter VIII for details).

16.3.4.1 TARGETS

Target design is very important in reduction of the pointing error; therefore the shape and size of targets should be carefully considered.

Among the various possible shapes for targets, circular and square targets are preferred by most users. In projects requiring high accuracy, circular targets have been extensively used. Erlandson and Veress (1975) used targets consisting of metal plates on which the circular target was placed photographically.

In choosing the size of the target, one should consider the geometry of the photograph, the photographic (object-to-camera) distance, and

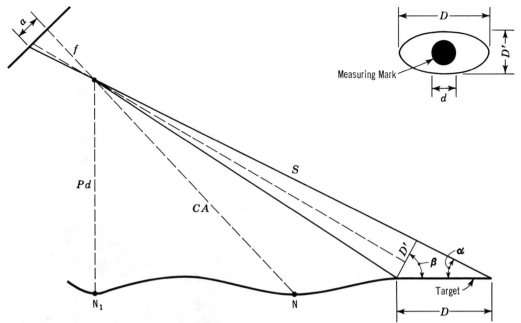

FIGURE 16-56. Targets in convergent terrestrial photography.

the diameter of the mark in the instrument to be used for measurement.

In the "normal case" of terrestrial photogrammetry the target size (diameter of the circular target, or side of the square target) can be determined as follows. The approximate distance can be obtained from an existing map, or estimated. Knowing the focal length of the camera, the scale of the photograph at the point to be targeted can be determined. Thus, the apparent size of the measuring mark at this point can be determined. The apparent diameter of the circular target must be at least 5/3 the apparent diameter of the measuring mark, thus providing an annulus which is about 1/3 of the diameter of the measuring mark.

In convergent photography, the determination of the target size is more involved. Figure 16-56 shows the geometry of the problem. In this respect, d is the diameter of the measuring mark. Due to the tilt of the photograph, a circular target with diameter D will be imaged as an ellipse with diameters D and D'. D' is a function of the angle β and angle α (figure 16-56).

The angle α can be obtained as follows:

$$\alpha = \tan^{-1} \frac{(Pd)}{B + (Pd)\frac{a}{f}}$$

where a is the half of the format size, f is the focal length of the camera, and B is the base length.

At the farthest point in the area photographed (point shown in figure 16-56) the relationship between D and D' can be written as:

$$D = D' \frac{\sin(\beta + \alpha)}{\sin \alpha} \qquad (16.36)$$

Since the minimum apparent diameter of the circular target cannot be smaller than the apparent diameter of the measuring mark of the photogrammetric instrument (i.e. the diameter of the measuring mark multiplied by the photo-scale at the point under consideration), D'_{min} can be determined as:

$$D'_{min} = d \frac{S}{f}$$

where

$$S = \frac{(Pd)}{\sin \alpha} .$$

Thus

$$D = \frac{d}{f}\left[\frac{\sin(\beta + \alpha)}{\sin^2 \alpha}\right](Pd), \qquad (16.37)$$

and

$$D'_{min} = \frac{d}{f \sin \alpha}(Pd).$$

Substituting this minimum value of D' in equation (16.36), the minimum size of the target can be obtained.

The upper limit for the target size is determined on the basis that the apparent size of the measuring mark should have an annulus of at most $(d/5)$ width between it and the circular target to maintain measurement accuracy. In other words:

$$D'_{max} = \frac{7d}{5 f \sin \alpha}(Pd).$$

Substituting this maximum value of D' in equation (16.36), the maximum size of the target can be obtained. The chosen size of the target diameter of the circular target or side of the square target) must then be between the maximum and minimum values just determined.

The targets whose size range has just been determined above are generally referred to as "measuring targets" since they are used mainly for measurements in the comparator. They cannot be seen on the plates, however, without 2-10 times magnification. For identification purposes, one normally uses an additional series of targets which can be of any design and whose images should be at least 1 mm in size on the photograph.

The sizes of the targets are to be determined individually for each project in order to obtain the required accuracy.

To overcome the perspective deformation difficulties associated with flat targets, particularly in case of convergent photography, spherical targets are sometimes used. Spherical targets can be manufactured from plastic material. The diameter of spherical targets can be computed as outlined above for flat targets in the normal case of terrestrial photogrammetry.

16.3.5 PRESENT-DAY APPLICATIONS

The classical application of terrestrial photogrammetry is cartographic, that is, in the production of topographic maps. Such maps are usually made for areas where the employment of aerial photogrammetry is not feasible. Consequently, terrestrial photogrammetry has played a complementary role in map production and did not become widely used.

In non-cartographic (or non-topographic) applications, both terrestrial and close-range photogrammetry have been widely used. Areas of application of close-range photogrammetry were discussed in section 16.2.4. Terrestrial photogrammetry has been extensively used in determining deformation and movement of large structures (*e.g.* Brandenberger and Erez (1972), Erlandson *et al.* (1973), Reynolds and Dearinger (1970), Richardson (1968), Roehm (1968), among others).

The accuracy obtainable in terrestrial photogrammetric projects depends on the equipment used and the degree of sophistication used in the measurements and computations. For example, it is possible to achieve accuracy in the order of 1/100,000 of the photographic distance (object-to-camera distance) if a long-focal-length camera and simultaneous adjustment are used. An example of long-focal-length cameras is the modi-

FIGURE 16-57. Modified Wild BC4 ballistic camera (courtesy U.S. Corps of Engineers, Seattle District).

fied Wild BC-4 Ballistic Camera (fig. 16-57) which was used by the U.S. Corps of Engineers for monitoring structural deformations of large structures (Erlandson and Veress, 1975). Naturally, less sophisticated equipment and adjustment procedures result in lesser accuracy. In general, the accuracy of structural monitoring ranges from 1/20,000 to 1/100,000 of the photography distance.

To achieve an accuracy in the order of 1/100,000 of the photographic distance in the object-space coordinates of targeted points, the following error levels (1σ) should not be exceeded (Erlandson and Veress, 1975):

Base-length error	$\leq \pm 0.01$ m
Angular error (ω, ϕ, κ)	$\leq \pm 5$ seconds of arc
Photogrammetric measurement error	$\leq \pm 6$ μm.

16.4 Underwater Photogrammetry

16.4.1 INTRODUCTION

Although highly developed for measuring and recording objects surrounded by air, photogrammetric procedures have played a secondary role in underwater surveys. This is mainly due to factors which influence the propagation of light within water, and—more critically—on water/air surfaces, thus disturbing central-perspective imaging, which is a fundamental concept of photo-grammetry.

Underwater photography has therefore often been employed qualitatively for photointerpre-tation and has gained widespread use for this purpose. The development of underwater photo-grammetry followed logically, aided by improved technology combined with analytical procedures.

At a time where world-wide attention focuses on the resources of the oceans, photogrammetry has become a valuable tool for mapping details of the ocean floor, supporting research in biol-ogy, hydrology and many other areas of interest. In this connection, primarily two types of pho-tography are encountered, one with both camera and object submerged, while the other uses air-borne cameras to record underwater objects. In both cases, similar optical difficulties have to be overcome.

16.4.2 DATA ACQUISITION

In order to understand certain features of underwater photography, it is necessary to con-sider some facts of optics.

16.4.2.1 UNDERWATER OPTICS

As light has to penetrate water before entering the camera lens, transmission behavior and re-fraction due to density changes are of great con-cern to underwater photographers.

16.4.2.1.1 OPTICAL CHARACTERISTICS OF WATER

The refractive index of water is approximately 1.33 and varies, particularly with salinity, tem-perature and pressure. As this index is required for analyzing the photography, it is essential to obtain data on these parameters. In addition, the refractive index is different for different wavelengths, leading to dispersion which pro-duces effects similar to chromatic aberration of lenses, and thereby reduces the resolution.

Attenuation, caused by a combination of ab-sorption and scattering, reduces the light inten-sity to a large extent. Depending on the compo-sition of the water, visibility varies from less than one metre in heavily polluted waters to 100 metres or more. As the amounts of absorption and scattering differ for different wavelengths, color filtering occurs, creating a blue-green ef-fect. Furthermore, contrasts are sharply reduced by scattering. It is therefore obvious, that underwater photography is necessarily close-range except when the camera is above the water.

16.4.2.1.2 GEOMETRIC OPTICS FOR LENSES IN IN WATER

When there are different media in object and image spaces, the positions of the cardinal points (focal-, principal- and nodal points) are different. With water in the object space, and air in the

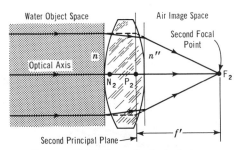

FIGURE 16-58. Second principal plane.

image space, both focal points and both principal points can be obtained by tracing appropriate rays as shown in figures 16-58 and 16-59. Par-axial rays from an infinitely distant axial object point converge to the second focal point F_2, while the first focal point F_1 is located on the optical axis such that it is imaged at infinity. For both traces, the principal planes are defined as passing through the intersection of incident and emergent rays. Their intersections with the opti-cal axis provide the principal points P_1 and P_2 of the lens. These two pairs of points define the focal lengths as:

$$f = \overline{P_1 F_1} \qquad f' = \overline{P_2 F_2}. \qquad (16.38)$$

The distances between principal and nodal points are

$$N_1 P_1 = f'(n - n'')/n \\ N_2 P_2 = f(n - n'')/n'' \qquad (16.39)$$

where n is the refractive index of the object space, and n'' is the one for the image space.

Together with the refractive index (n') of the lens, these constitute the thin lens equation:

$$n/f = n''/f' = (n' - n)/r_1 + (n'' - n)/r_2$$

where r_1 and r_2 are the radii of the first and sec-ond surfaces of the lens.

In lieu of a lens, often a window represents the optical component between the two media. Whether it is flat or dome-shaped, its surfaces are parallel so that any ray normal to them will pass through without being refracted. All other rays will be refracted such that they appear to be emanating from an apparent object located closer to the window than the real one (see fig-ures 16-60 and 16-61).

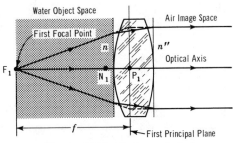

FIGURE 16-59. First principal plane.

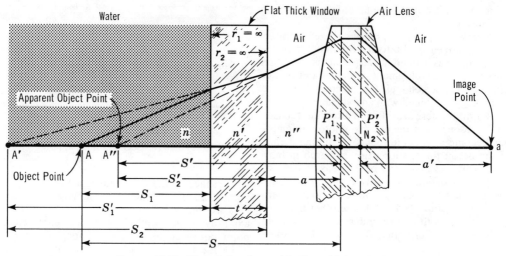

FIGURE 16-60. Flat window in combination with lens in air.

16.4.2.2 UNDERWATER CAMERAS

Virtually all underwater cameras on the market are non-metric cameras, with a choice ranging from simple and inexpensive to highly sophisticated and very costly models. The basic difference from other cameras is the watertight housing, including window, which has to withstand high pressures. Therefore, these cameras are limited to operation at a maximum depth. Some of the cameras have a corrective lens window (*e.g.* Ivanoff Corrector—Ivanoff *et al.* 1956)

which corrects for the refraction such that apparent and real locations of objects coincide. Such corrective windows make the use of wide-angle lenses possible, thereby reducing the number of photographs required to cover a certain area. Some underwater cameras have specially designed lenses, for which the calculations assume water in the object space. Distortion and other optical errors are much smaller than in the previous cases. Besides still cameras, rapid-fire movie and scanning versions of underwater cameras are also on the market.

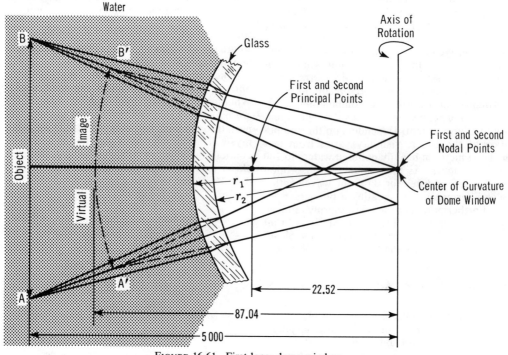

FIGURE 16-61. First lens: dome window.

Calibration, if done at all, consists mainly of on-the-job calibration, although an underwater-camera calibrator, comparable to a multi-collimator, has been developed by the U.S. Naval Photographic Center.

16.4.2.3 UNDERWATER LIGHTING

Visibility in water is limited by scattered light, which leads to loss of contrast. A diver with artificial light encounters a visibility problem similar to that of a person driving a car in dense fog. This effect can be reduced with a light shield, offsetting the light source, or by using crossed polarizers.

For a quantitative discussion of visibility, the concept of "attenuation length" is often used. One attenuation length is defined as the distance in which the radiant flux is reduced to $1/e$, or approximately 37% of its initial value. In water, one attenuation length averages 10 metres, but can be much lower.

When viewing and photographing within one attenuation length, acceptable contrast is obtained with the light-source close to the camera. With polarizers and light shielding, good images can be obtained at 3 to 4 attenuation lengths, even as far as 6, if the light is located near the object. Even with ideal arrangements, visibility cannot be extended beyond 15 attenuation lengths, because then any object disappears in its own forward scattered light.

High-energy, pulsed lamps—primarily stroboscopic lamps such as xenon- and continuous arc lamps (*e.g.* tungsten, halogen, mercury, metal-halide thallium, metal-halide dysprosium, high-pressure sodium) are most commonly used in underwater lighting.

16.4.2.4 FILM FOR UNDERWATER PHOTOGRAPHY

High-contrast black-and-white or color films with medium speed can be used for underwater photography. Special films with peak spectral sensitivities in the range of 480 to 550 nanometres have been developed. A two-color, reversal film of this type has been found to be advantageous for aerial, off-shore mapping because of better penetration ability.

16.4.2.5 UNDERWATER CONTROL

As absolute positioning of targets is usually difficult under water, a precoordinated control frame is often placed around the object, much like in conventional close-range photogrammetry (*see* figure 16-62).

This control frame can be used for the determination of both interior and exterior orientation of the camera. Especially for deep-sea missions it is often impossible to place any control underwater. In such cases, it is therefore necessary to determine the location and attitude of the camera as accurately as possible. For deep-sea photography this involves determining the position, depth, attitude and exposure interval. This can be achieved with a variety of instruments and navigatory systems. Even then, the absolute accuracy is much lower than the relative accuracy provided by photogrammetric means.

For airborne coastal mapping an inertial navigation system has provided good accuracy for the exterior orientation of the camera.

16.4.2.6 AUXILIARY EQUIPMENT

For underwater photography in shallow waters, the auxiliary equipment consists mainly of SCUBA (*S*elf *C*ontained *U*nderwater *B*reathing *A*pparatus) design gear and fins. For deep-sea mapping, manned or unmanned, submersible vehicles are used. These may range from simple to extremely complex units, equipped with a variety of sounding (echo sounder, side scan sonar) and navigatory devices of which photography is just a small component.

FIGURE 16-62. Stereoscopic test-square photographs of a vertical rock face at 20 m depth in the outer archipelago of the Swedish west coast. Visible are the control frame and various organisms (courtesy T. Lundalv).

16.4.3 DATA REDUCTION

16.4.3.1 ANALOG EVALUATION

16.4.3.1.1 EVALUATION OF UNDERWATER PHOTOGRAPHY

Analog evaluation of underwater photography is faced with significant disturbances of the central perspective caused mainly by lens distortion and the change in interior orientation parameters, especially the principal distance. This results in noticeable model deformations. Furthermore the image format is usually small and not very suitable for reduction in analog plotters.

An analog evaluation will therefore provide an approximate solution only, which clearly indicates that it is neither necessary nor economical to use precision plotters. If possible, distortion compensation should be used, and areas of large deformation avoided. A large amount of ground control can be used to determine the model deformation, which then provides the option of applying corrections during the plotting process or to the plotted map.

16.4.3.1.2 ANALOG EVALUATION OF TWO-MEDIA PHOTOGRAPHY

Two ways of evaluating aerial photographs of underwater objects have been suggested, one involves plotter modification while the other requires a reduction to the one-medium situation.

Plotter modification provides the best solution but is, however, limited to plotters designed on the optical principle. Although possible for plotters with two model points with a Zeiss-Parallelogram (by using optical transformations with plane-parallel plates in the optical train), this approach has gained acceptance in practical use only with instruments of the double-projector type having one model point. As indicated in figure 16-63, a correct solution is obtained when placing the measuring mark in water. This method has yielded good results.

The reduction to the one-medium situation is rather cumbersome and not very accurate. This is caused by the fact that even when properly oriented the respective rays do not intersect because of residual parallaxes. By defining a photogrammetric model as being free of horizontal parallaxes, a model can be obtained but is difficult to evaluate and is rather deformed. Plotting is possible if the vertical parallaxes are small. As the model needs to be corrected, the interactive use of an electronic calculator is required, even for approximate solutions. Strictly speaking, this is no longer an analog evaluation, and hybrid plotters appear to be better suited for this purpose.

16.4.3.2 ANALYTICAL EVALUATION

16.4.3.2.1 MATHEMATICAL FORMULATION FOR MULTI-MEDIA PHOTOGRAMMETRY

Analytical photogrammetry is based on either modelling an imaging ray from image point *via*

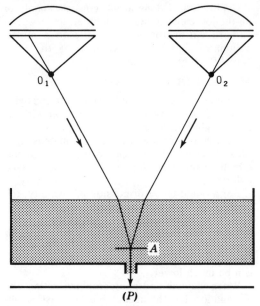

FIGURE 16-63. Principle of a double projector modified for the restitution of underwater objects.

projection center to object point (collinearity), or modelling two intersecting rays originating from the same object point but imaged on two images *via* two projection centers (coplanarity). Both approaches can be used for multi-media photography, of which two-media is a special case. The mathematical formulation can be done either by space resection or by model formation. The space-resection method was used in

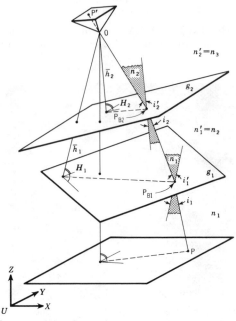

FIGURE 16-64. General case of underwater photographs.

Höhle's formulation (Höhle, 1971) which will be outlined below.

Referring to figure 16-64, the ray originating at the image point P' and passing through O hits the surface between media one and two at point P_{B1} and can be expressed as:

$$\frac{X - X_0}{\alpha_1} = \frac{Y - Y_0}{\beta_1} = \frac{Z - Z_0}{\gamma_1} = \rho \quad (16.40)$$

where X_0, Y_0, Z_0 are the coordinates of O, α_1, β_1, γ_1 the direction cosines of the ray, and ρ an auxiliary quantity.

where $\cos i_1 = \alpha_1 \lambda_1 + \beta_1 \mu_1 + \gamma_1 \nu_1$. The direction cosines of the normal to a surface $F(X, Y, Z) = 0$ are obtained using

$$\text{grad } F = \left(\frac{\partial F}{\partial X'} \quad \frac{\partial F}{\partial Y'} \quad \frac{\partial F}{\partial Z} \right) \quad (16.46)$$

and multiplication of the components of the normal by ϵ

$$\epsilon = \left[\left(\frac{\partial F}{\partial X} \right)^2 + \left(\frac{\partial F}{\partial Y} \right)^2 + \left(\frac{\partial F}{\partial Z} \right)^2 \right]^{-1/2}. \quad (16.47)$$

For the previously mentioned surfaces, this leads to:

Surface	λ	μ	ν
general plane	$\dfrac{a}{a^2 + b^2 + 1}$	$\dfrac{b}{a^2 + b^2 + 1}$	$\dfrac{1}{a^2 + b^2 + 1}$
horizontal plane	0	0	1
sphere	$\dfrac{X_B - a}{r}$	$\dfrac{Y_B - b}{r}$	$\dfrac{Z_B - c}{r}$

When the coordinates of P_{B1} are determined, the refracted ray can be expressed in a similar manner using Snells' Law of Refraction.

A regrouping of equations (16.40) provides the coordinates as functions of ρ which in turn depends on the border surface:

$$\begin{bmatrix} X_{B_1} \\ Y_{B_1} \\ Z_{B_1} \end{bmatrix} = \rho \begin{bmatrix} \alpha_1 \\ \beta_1 \\ \gamma_1 \end{bmatrix} + \begin{bmatrix} X_0 \\ Y_0 \\ Z_0 \end{bmatrix}. \quad (16.41)$$

For a plane: $aX + bY + Z + d = 0$, ρ becomes:

$$\rho = \frac{aX_0 + bY_0 + Z_0 + d}{a\alpha + b\beta + \gamma}. \quad (16.42)$$

For a horizontal plane,

$$\rho = \frac{Z_0 + d}{\gamma}. \quad (16.43)$$

For a sphere, $(X - a)^2 + (Y - b)^2 + (Z - c)^2 = r^2$, ρ assumes the following value

With the application of equation (16.40) to the object medium $M_o = M_3$ (in this case) two equations for the orientation unknowns can be formed:

$$\begin{bmatrix} X_p \\ Y_p \end{bmatrix} = \frac{Z_p - Z_{B2}}{\gamma_3} \begin{bmatrix} \alpha_3 \\ \beta_3 \end{bmatrix} + \begin{bmatrix} X_{B2} \\ Y_{B2} \end{bmatrix},$$

where X_p, Y_p, Z_p are the coordinates of an object-space point. Linearization and least squares adjustment are used to solve the problem which contains the unknowns of exterior (and possibly interior) orientation as well as surface parameters and indices of refraction.

For the model method, the surface parameters are determined together with the relative orientation. Strong correlations among the unknowns and changes in accuracy related to the depth-to-object distance ratios appear to restrict the use of this second approach.

$$\rho = - \left[\alpha(X_0 - a) + \beta(Y_0 - b) + \gamma(Z_0 - c) \right] \\ \pm \sqrt{\left[\alpha(X_0 - a) + \beta(Y_0 - b) + \gamma(Z_0 - c) \right]^2 - \left[(X_0 - a)^2 + (Y_0 - b)^2 + (Z_0 - c)^2 + r^2 \right]} \quad (16.44)$$

where the sign in front of the square root indicates the curvature in regard to the projection center (negative for convex surface).

The change in direction due to refraction is expressed as

$$\begin{bmatrix} \alpha_2 \\ \beta_2 \\ \gamma_2 \end{bmatrix} = \frac{n_1}{n_2} \begin{bmatrix} \alpha_1 \\ \beta_1 \\ \gamma_1 \end{bmatrix} - \left(\frac{n_1}{n_2} \cos i_1 - \sqrt{1 - \left(\frac{n_1}{n_2} \right)^2 + \left(\frac{n_1}{n_2} \cos i_1 \right)^2} \right) \begin{bmatrix} \gamma_1 \\ u_1 \\ \nu_1 \end{bmatrix} \quad (16.45)$$

16.4.3.2.2 TWO-MEDIA PHOTOGRAMMETRY
WITH THE ANALYTICAL PLOTTER

As mentioned before, the analog methods—except when using modified, double-projector instruments—provide approximate solutions only, even when applying corrections for the height deformations. It is, however, often desired to present underwater information in graphical form, e.g. with contour lines. The analytical plotters provide a most suitable means, because a rigorous analytical solution is used to carry out the graphical plotting. At the University of New Brunswick such a system was developed using the model approach (Masry & Konecny 1970). According to figure 16-65 the boundary between media M_1 and M_2 is assumed to be a plane π. Any object point above π is treated in the conventional manner. S_1F_1 and S_2F_2 are normal to the plane π passing through the photo-stations S_1 and S_2 respectively. A right-handed X, Y, Z coordinate system is chosen such that the x-axis passes through F_1 and F_2. P is an object point in medium M_2, O is a fictitious point located along S_1F_1 with the same

z-value as P, i_1 and i_2 are the angles of incidence and refraction of the imaging ray PP_1S_1 which—as previously stated—is usually not coplanar with PP_2S_2, and N is another normal to π, passing through P_1.

The law of refraction can be written as:

$$\sin i_1 \,/ \sin i_2 = n \qquad (16.48)$$

where n is the refractive index between M_1 and M_2.

Using triangles $S_1F_1P_1$ and P_1PD we get:

$$\sin i_1 = R \,/\, \sqrt{(h_1{}^2 + R^2)} \qquad (16.49)$$

and

$$\sin i_2 = (L - R) \,/\, \sqrt{Z^2 + (L - R)^2}. \quad (16.50)$$

L and R can be expressed in terms of the X and Y coordinates of P and p_1 as both are located either in π or a plane parallel to it. Therefore

$$\frac{R \, \sqrt{Z^2 + (L - R)^2}}{(L - R) \, \sqrt{h_1{}^2 + R^2}} = n$$

and

$$R^2 \left[(n^2 - 1)(L - R)^2 - Z^2 \right] + n^2 h_1{}^2 (L - R)^2 = 0. \qquad (16.51)$$

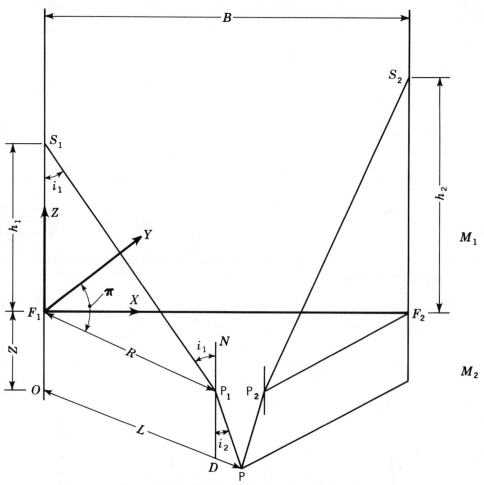

FIGURE 16-65. Geometry of two-media photography.

Using the known X, Y, Z coordinates of P, R can be determined from equation (16.51) by Newton's approximation. Of the four solutions, two are imaginary and one is larger than R and has no physical meaning, thus leaving just one practical result.

As a first approximation, $R \approx Lh_1 / (h_1 + z)$ is used to determine

$$X_{p_1} = X_p \cdot R/L ,$$
$$Y_{p_1} = Y_p \cdot R/L , \qquad (16.52)$$

and

$$Z_{p_1} = 0 .$$

Similarly the coordinates of p_2 are computed. Then, having determined p_1 and p_2, the image points are determined using the condition that the object point, the perspective center of the camera, and the image point are on a straight line. The photo-carriages are driven by the computer so that these image points are observed by the operator. These calculations are repeated continuously at a rate of about 30 times/second.

A matter which might need clarification is how the point P is determined before knowing the refraction. This matter becomes much clearer if we consider the conventional case of a point G on the ground. The operator drives in the three directions X, Y and Z in the model space to the vicinity of point G, and sees that the measuring mark is not on the point. He then moves the measuring mark closer to the point. The process can thus be thought of as an iterative process. The same concept holds for points in the water.

In the program developed at the University of New Brunswick, the model coordinates system was chosen with the XY-plane coincident with the boundary surface. This allows the program to determine from the sign of the z-coordinate the medium in which the model point lies. Model points above the water, such as point G, are dealt with in the conventional manner, a feature which may be of value for coastal mapping.

16.4.4 APPLICATIONS

There are numerous applications, many of which are not concerned with topographical mapping but rather with biology, archeology, hydrology and similar fields. As an illustration, two applications are described.

16.4.4.1 EXAMPLE FROM UNDERWATER ARCHEOLOGY

The wooden hull of a 4th century B.C. shipwreck excavated near Cyprus was photogrammetrically recorded with a stereocamera at a depth of 100 ft (*see* figures 16-66 and 16-67). The map was plotted with a Zeiss Stereotop using corrections obtained with the control frame. An accuracy derived from the control points of ±1.0 cm in planimetry and ±2.4 cm in height was obtained (Höhle 1971).

FIGURE 16-66. Underwater stereopair from a part of the ancient shipwreck (scale approximately 1 to 80, camera Calypso/Nikkor II with relative aperture *uw* Nikkor 28 mm, f/3.5 lens, film Kodak-Panatomic X DIN 17, exposure 1/60 second, relative aperture 1:5, 6, taken by J. Veltry, August 2, 1969, at about 2 P.M.).

16.4.4.2 EXAMPLE FROM COASTAL MAPPING

A test area in the Strait of Georgia, B.C., Canada, was photographed using a Wild RC8 Camera and Kodak 8442 color film with HF3 filter at a 1:10,000 photoscale. The camera orientation was determined with an inertial navigational platform. Using the analytical plotter, the coastal area was mapped up to 250 cm from the shore line in water up to 7 metres deep. The

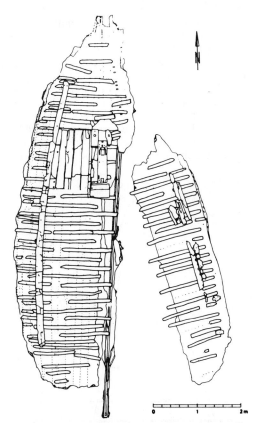

FIGURE 16-67. Planimetric map of a sunken trading ship dating from four centuries B.C. (scale of the original 1 to 10).

results were checked by comparing spot elevations against depth sounding for over 90 points, which yielded an accuracy of ±0.45 metres. This high accuracy, which was obtained despite the fact that the water surface was considered as plane without considering the ever present waves, shows the high potential of the analytical plotter for this type of application.

16.5 X-ray Systems and Applications

16.5.1 FORMATION OF X-RAY IMAGES

The basis for x-ray photogrammetry is the central projection of a bundle of x-rays from a focal spot in an x-ray generating tube through an object and onto a sensitized film. As such, it is no different from the basis of photogrammetry using cameras. In conventional photogrammetry, central projection is very closely approximated by applying systematic corrections to image points to take into account such influences as lens distortion, film deformation and atmospheric refraction. As now practised, x-ray photogrammetry assumes perfect central projection, and consequently no systematic corrections to image points are made.

Fig. 16-68 is a schematic of the relationship of the perspective center, the film, and the object in the making of an x-ray film or radiograph. The perspective center is assumed to be a point even though the focal spot has a finite size. Since the object lies between the focal spot and the film, the image or shadowgraph is *larger* than the object being x-rayed.

Fig. 16-69 is an enlarged view of the anode of the x-ray tube. The anode can be either stationary or rotating, which in the latter case prevents the anode from overheating under higher voltages than are used in stationary anode tubes. The size of the focal spot of the anode is fixed by the convergence of the electrons traveling from the focusing cup of the cathode and impinging on the target of the anode. The effective size of the focal spot is that area projected into a surface which lies parallel with the x-ray film, and is about 0.3 mm square in modern x-ray tubes. In fact, this projected area is larger or smaller than nominal, depending on whether one considers respectively the right or left side of the x-ray projection as diagrammed in figure 16-69.

Because x-rays tend to scatter when projected through the object and onto the film, they would ordinarily tend to diffuse the image, resulting in a loss of definition. If this diffusion presents a problem, a grid can be placed in front of the film holder or casette which contains lead vanes that are parallel with one another in one direction and which converge toward the nominal position of the focal spot of the x-ray tube. This is shown in figure 16-70. The grid is called a Bucky. The lead vanes absorb virtually all but the most direct primary rays and reduce diffusion which would otherwise exist in the image. The Bucky moves back and forth in its own plane during exposure in order to eliminate shadows of the vanes. The lineations appearing on some radiographs caused by stationary Buckys with very thin vanes are not particularly objectional from a photogrammetric standpoint.

Films for use in x-rays can be obtained either with an emulsion on one side only or on both sides of the film backing. Double emulsion film is faster and requires less exposure. However, it causes a diffusion of the image. X-ray films are relatively slow and require a fairly lengthy exposure to x-rays. This long exposure time is not objectionable in industrial applications and in medicine where the object is not a living thing. But for most medical applications, the object is alive, in which case the x-ray dosage must be held to a safe low level. In order to accommo-

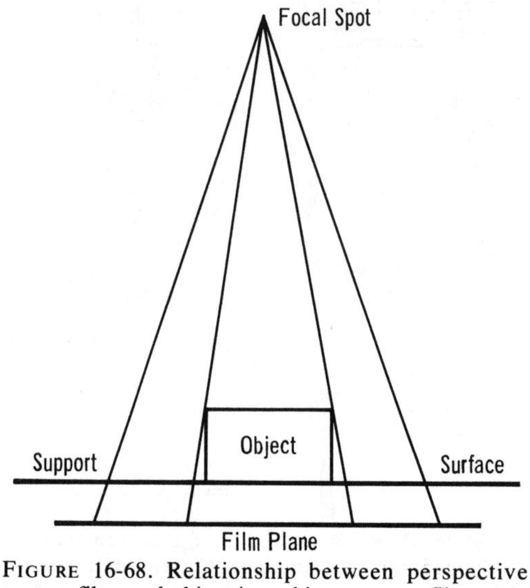

Focal Spot

Support   Object   Surface

Film Plane

FIGURE 16-68. Relationship between perspective center, film, and object in making an x-ray film.

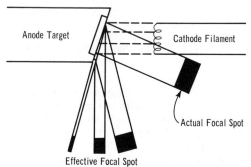

Anode Target   Cathode Filament

Actual Focal Spot

Effective Focal Spot

FIGURE 16-69. Anode of x-ray tube. The projected size of the focal spot varies from left to right as shown.

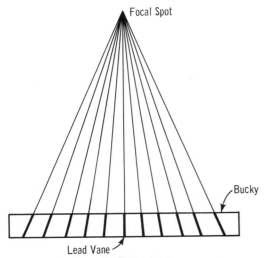

FIGURE 16-70. Grid lines composed of lead vanes held in the Bucky. The vanes converge toward the focal spot of the x-ray tube. Thus, the Bucky must be matched to the focal spot-film distance set off in the x-ray apparatus.

date the shorter exposures with slow film, the casette incorporates two intensifying screens in between which the film is held during exposure. A cross-section of the typical casette is shown in figure 16-71. The intensifying screen is coated with a lacquer containing fluorescent crystals which can be either coarse grained (fast screen) or fine grained (slow screen). The x-rays cause the screen to glow in proportion to the amount of x-ray exposure. This glow exposes the x-ray film in a much shorter period than do the x-rays themselves.

Image density in the radiograph is controlled by a combination of the amount of current in the cathode used to produce the electrons which strike the focal spot, and the duration or time of exposure. Image contrast is controlled by the voltage across the gap between the cathode and the focal spot. An increase in voltage causes an overall increase in density which has the effect of decreasing the image contrast. The proper combination of amperage-time-voltage for a given object is the concern of the x-ray technician. However, the photogrammetrist handling

radiographs must have a working knowledge of these factors in order to better advise the technician what changes are needed when experimenting to arrive at the optimum radiograph. The publication "The Fundamentals of Radiography" prepared by Eastman Kodak Company (1960) should be studied by anyone entering into any aspect of the field of radiography. It presents a very thorough treatment of this phase of the work.

16.5.2 IMAGE DEGRADATION IN X-RAY FILMS

The photogrammetrist working with radiographs is faced with conflicts which must be resolved or compromised in the process. The highest quality radiograph is obtained by reducing the physical size of the focal spot to say 0.1 mm; eliminating the intensifying screen; using a single emulsion film; and using a well-functioning Bucky. These conditions can be met in the case of inanimate objects, but not in the case of living specimens.

The practical limit to the effective size of the focal spot in medical work is the 0.3 mm figure given above. This allows a flow of enough electrons to generate an amount of x-rays sufficient

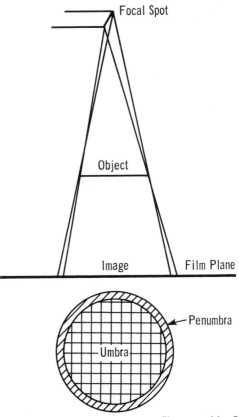

FIGURE 16-72. Penumbra on x-ray film caused by finite size of the focal spot.

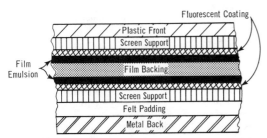

FIGURE 16-71. Cross-section of a casette containing intensifying screens.

to make an exposure in a reasonably short time. The finite size of the focal spot results in a penumbra throughout the x-ray film as diagrammed in figure 16-72. This effect tends to spread out edge gradients, and results in a soft image on the film. The effect of the penumbra has been studied in detail and reported by McNeil (1966).

A secondary effect of the finite size of the focal spot is the introduction of a pseudo distortion somewhat like a lens distortion. This distortion is caused by the subjective evaluation of the weighted center of an image point which is blurred because of the penumbra (Hallert 1960). The systematic nature of this distortion is caused by the variation in the effective size of the focal spot from left to right as indicated in figure 16-72. However, it is of such negligible magnitude as to have no effect on the measurement of radiographs.

The intensifying screens in the casettes used in medical applications cause a softening of the x-ray film image. This is more pronounced in the fast screen because of the coarse granularity of the screen. The alternative to the use of the intensifying screen is a longer exposure by a factor of as much as 25 times. This is permissible in projects where long exposure times are not objectionable, but not with living objects.

The double emulsion on x-ray film also has a softening effect on the film image. This effect is overcome by the use of a single emulsion at the expense of longer exposure time. Schernhorst (1969) reports the use of a single-emulsion film in a double-screen casette in which the emulsion faces the x-ray tube, and a sheet of black paper is inserted behind the back of the film in order to block off the glow from the rear intensifying screen.

In addition to the above mentioned sources of image degradation, there exist two possible sources of geometric degradation. The first is the possibility of film buckling if the casette is not in good condition. This would pose a problem only if the x-ray tube were aimed obliquely to the film plane at a sizeable angle rather than normal to the film as is the usual case. Under the usual circumstances, the x-ray coverage subtends a relatively narrow angle, and the film is held very firmly between the intensifying screens. Consequently, the possibility of film buckling can be ignored.

The second source of geometric degradation is shrinkage or expansion of the x-ray film between the time it was exposed and the time it is measured. All but the irregular and erratic distortions of this type can be evaluated and eliminated, if it is felt necessary, by locating appropriate controls in the object space.

16.5.3 STEREORADIOGRAPHS

The vast majority of x-ray films produced in medicine and industry are analyzed as single radiographs. Experienced radiologists can detect and interpret extremely subtle nuances in density, form, and size, in order to arrive at a diagnosis even without the use of any measuring device. On occasion, however, they resort to various kinds of scales, gages, and overlays in order to quantify lengths, areas, and angles. When a three-dimensional view is required for better interpretation, then two x-ray films must be exposed separately, one from each of two separate perspective centers. The two x-ray films are then viewed through a stereoscope in order to obtain the three-dimensional image. Since this is essentially a matter of interpretation of the object, the base-distance ratio must be kept small enough to allow the examiner to comprehend the third dimension without difficulty. The Kodak publication "The Fundamentals of Radiography" (1960) suggests the following values (table 16-5).

TABLE 16-5. STEREOSCOPIC TUBE SHIFTS FOR COMMON FOCUS-FILM DISTANCES UNDER PRACTICAL VIEWING CONDITIONS*

| Focus-Film Distances | For Eye-Image Distance of | | |
	25 inches Use Tube Shift of:	28 inches Use Tube Shift of:	30 inches Use Tube Shift of:
25 in.	2-9/16 in.	2-1/4 in.	2-1/16 in.
30 in.	3-3/16 in.	2-3/4 in.	2-9/16 in.
36 in.	3-7/8 in.	3-7/16 in.	3-1/8 in.
42 in.	4-5/8 in.	4-1/16 in.	3-3/4 in.
48 in.	5-3/8 in.	4-11/16 in.	4-5/16 in.
60 in.	6-13/16 in.	6 in.	5-1/2 in.
72 in.	8-5/16 in.	7-1/4 in.	6-11/16 in.
84 in.	9-3/4 in.	8-1/2 in.	7-7/8 in.
96 in.	11-3/16 in.	9-13/16 in.	9-1/16 in.

* (From *Fundamentals of Radiography*, published by Radiography Markets Division, Eastman Kodak Company)

These values result in a base-distance ratio of the order of 0.10 which is rather low. For the quantitative determination of the three dimensions by photogrammetry, the base-distance ratio should be as large as possible, limited only by the interpretability of points to be measured, angular coverage of the x-ray bundle, film size, and control location.

A stereoscopic pair of radiographs can be obtained in a number of ways, each way having its own advantages and disadvantages. In figure 16-73a, two x-ray tubes are fixed in relation to one another on a constant base. For practical reasons, the entire apparatus would be fixed, that is, both the casette holder and the x-ray tubes are a rigid unit. This has the advantage that once aligned and calibrated, the resulting photogrammetric reduction can be simplified because outer and inner orientation are known and fixed. This arrangement also permits the simultaneous firing of both x-ray tubes in order to obtain stereoscopic parallax of discrete points on a single radiograph. One disadvantage of the integral unit is its flexibility as regards to base-distance ratio. Cost and availability are the primary drawbacks to the two-tube system.

A more common arrangement for obtaining stereoradiographs is shown in figure 16-73b in which the x-ray tube is shifted parallel to the plane of the film. The distinct advantage of this method is its general availability. If the simplified geometry of the normal case (*see* section 16.5.8) is to be applied, either the shifting mechanism or the casette holder must be carefully aligned to ensure parallelism between the exposure base and the film plane. The base itself can be determined quite accurately by measuring the displacement of the tube housing with respect to some reference mark located to the side of the housing (Moffitt 1972). The chief disadvantage to this method is the general unwieldiness of the tube housing during the shift. Also, when x-raying a person, the commotion associated with the shifting operation has a disturbing effect on the person.

A technique which requires that the reference

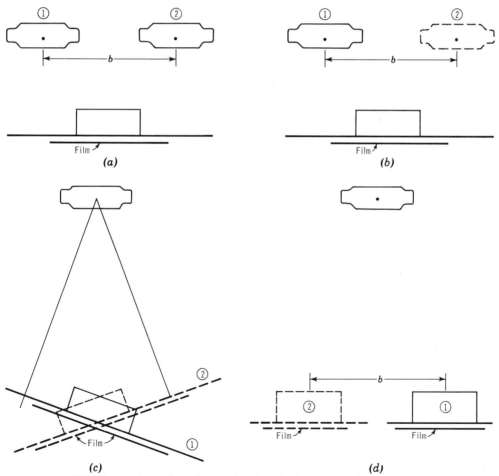

FIGURE 16-73. Geometric configuration employed to obtain stereoscopic pairs of radiographs.

plane and the casette holder rotate about an axis directly under the x-ray tube is shown in figure 16-73c. This results in the geometric configuration identical to that shown in figure 16-73a in which film-object position 1 is associated with the left tube position. This technique is described in detail by McNeil (1966). It is applicable for study of stationary objects, and in medicine for portions of the body which can be completely immobilized with respect to the film plane. However, rocking a living subject from one position to another will cause slight and differential shifts of both the internal and external parts, introducing false parallaxes which destroy the accuracy of the three-dimensional analysis.

The arrangement shown in figure 16-73d requires that the object being x-rayed be shifted under the x-ray tube in order to create the stereo base. This results in the geometric configuration identical to that shown in figure 16-73a in which film-object position 1 is associated with the left tube position. The advantage of this arrangement lies in the fact that the tube can be held fixed. However, this must be weighed against the necessity of having the appropriate mechanism for shifting the object-film array parallel with itself as a unit.

If a Bucky is used in any of the four techniques described above, the base must be parallel with the longitudinal direction of the vanes. Also, the Bucky must be chosen so that the convergence of the vanes as shown in figure 16-70 is compatible with the tube-film distance selected for the project.

A completely different arrangement of the x-ray tube-film relationship for obtaining three-dimensional information is shown in figure 16-74. This is an ideal arrangement under the condition that each point to be measured is discrete and unambiguous. There are two major drawbacks to this system: first, the radiographs cannot be viewed stereoscopically; and second, the positive identification of unmarked anatomical points or features on the two radiographs is virtually impossible. The singular advantage to this arrangement is the possibility of making two simultaneous exposures of a subject, thus eliminating the problem of subject movement between two exposures. In order to accomplish this, proper shielding between the two x-ray tubes and the opposing films must be provided so as to prevent fogging from secondary radiation, and there must be an ample source of electrical power. This arrangement is discussed in detail by Suh (1974).

16.5.4 INTERIOR ORIENTATION OF STEREORADIOGRAPHS

The interior orientation elements of an individual radiograph consist of the principal distance of figure 16-75 and the position O of the principal point. With a knowledge of the position

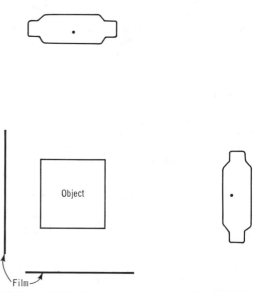

FIGURE 16-74. Arrangement of x-ray tubes and film planes for obtaining views at right angles to one another. This is analogous to 90° convergent photography.

of the film plane in the cassette, dimensions of the cassette and cassette holder, and the position of the focal spot inside the x-ray tube, it is possible to measure the principal distance directly with a fairly high degree of accuracy. If a reference surface, for instance the top of the table on which the object is placed, is interposed between the focal spot and the cassette holder, then the distance between this surface and the film plane must also be known.

Alternatively, the principal distance can be determined analytically if the proper control is located in the object space (Kratky 1975). The direct measurement of the principal distance *under a specific set-up* is more accurate than an analytical solution because of the limited depth of the control in the object space used for the latter determination. The advantage of the analytical solution lies in the possibility that the principal distance will change when a film cassette is changed or when the x-ray tube is shifted.

The principal point of the radiograph is defined as the foot of the perpendicular from the focal spot to the plane of the film. Its position can be determined in any number of ingenious ways. McNeil (1966) imbeds a pair of radiopaque reference axes into the reference platform as shown in figure 16-76. A right cylinder is then oriented on the platform such that the cylinder axis intersects the platform at the origin of the reference axis. The upper end of the cylinder contains five lead wafers with small conical pinholes, the central pinhole of which lies on the axis of the cylinder. A test exposure shows the amount and direction required to move the focal

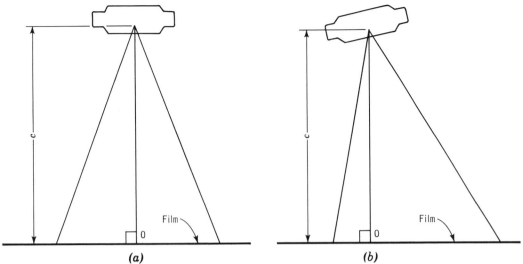

FIGURE 16-75. Elements of interior orientation of a radiograph. In (a), the x-ray tube is aimed normal to the film. In (b), the tube is aimed oblique to the film. In both cases, the principal distance c is measured from the focal spot along a normal to the film, and the principal point O is at the foot of this normal.

spot with respect to the reference platform in order to bring the principal point into coincidence with the origin of the reference axes.

A scheme used by Jonason and Hindmarsh (1975) is similar in principle to McNeil's. A control cage made of two plexiglass plates separated by steel rods sits on the table over the film cassette as shown in figure 16-77. The upper surface contains a lead ring and the lower surface contains a small radiopaque point directly beneath the ring. The x-ray tube is oriented over the control cage so that a small pinhole of light from the tube is projected vertically by means of a lens into the center of the lead ring. After the x-ray film is exposed, the shadows of the ring and the point in the lower plate should be concentric. If they are, the principal point has been determined. If they are not, the position of the principal point is calculated, based on the dimensions of the set-up.

If the plane of the film can be made strictly horizontal by means of level bubbles, then two different methods can be used to determine the position of the principal point. The first method is to plumb the position of the focal spot down onto the reference surface on which the object is to be placed and then mark this plumb point with a radiopaque mark. Plumbing can be done by an ordinary plumb bob or else by a plumbing rod containing a level bubble. This technique can prove quite tedious because of the general inaccessibility of the focal spot. However, it is possible to locate reference points on the exterior of the x-ray tube which will permit the plumbing operation to proceed without difficulty.

In the second method, wires with radiopaque weights, or else flexible chains (like keychains), are attached to the x-ray tube housing and allowed to hang into the object area. Since the film lies in a horizontal plane, the intersection of the shadows of the plumb lines will define the position of the principal point of the film. This is

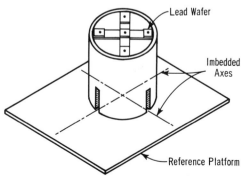

FIGURE 16-76. McNeil's apparatus for determining interior orientation elements.

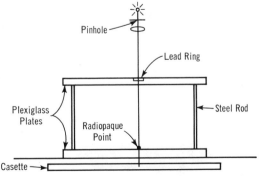

FIGURE 16-77. Jonason and Hindmarsh method for establishing principal point.

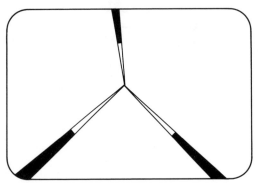

FIGURE 16-78. Convergence of plumb lines to locate principal point. The film plane is strictly horizontal.

shown in figure 16-78. The principal point can be marked directly on the film, or else its position can be computed analytically with respect to a set of radiograph axes by coordinate measurements of points on the plumb lines.

The use of control lines in the object space which are strictly normal to the film plane can be used in the manner described above using plumb lines. This permits the film plane to be oriented in any position including the vertical position. Moffitt (1972) and Baumrind (1975) employ lucite tubes, the upper and lower ends of which contain minute radiopaque marks. These tubes are shown in figure 16-79 in conjunction with a head holder used in dentistry and orthodontics. A line joining the two marks in each tube is normal to the plane of the film to a high degree of accuracy. The four pairs of marks registered on the film constitute four lines which intersect at the principal point. The measured coordinates of these marks are used first to obtain an approximate set of coordinates of the principal point by the intersection of two of the lines, followed by a

least squares adjustment to get the best value from all four lines.

In figure 16-80, let points 1-2, 3-4, 5-6, 7-8 represent four lines which should intersect in point O. If lines 1-2 and 3-4 are intersected to give approximate coordinates of the principal point, the results are:

$$x_0' = \frac{m_1 x_1 - m_3 x_3 + y_3 - y_1}{m_1 - m_3}$$

where

$$m_1 = \frac{y_2 - y_1}{x_2 - x_1} \text{ and } m_3 = \frac{y_4 - y_3}{x_4 - x_3}$$

and

$$y_0' = m_1(x_0' - x_1) + y_1 = m_3(x_0' - x_3) + y_3.$$

Then for each line, an observation equation is formed as follows. For line 1-2:

$$(y_2 - y_0')v_{x_1} + (x_0' - x_2)v_{y_1} + (y_0' - y_1)v_{x_2}$$
$$+ (x_1 - x_0')v_{y_2} + (y_1 - y_2)\Delta x_0 + (x_2 - x_1)\Delta y_0$$
$$+ \left[x_2 y_0' - x_1 y_0' + x_1 y_2 - x_2 y_1 - x_0' y_2 + x_0' y_1 \right] = 0.$$

The observation equations thus formed can be expressed as

$$\mathbf{Av} + \mathbf{B\Delta} + \mathbf{l} = 0.$$

The least squares solution for the corrections to the approximate coordinates of the principal point gives

$$\mathbf{\Delta} = -\left[\mathbf{B}^T (\mathbf{AA}^T)^{-1} \mathbf{B} \right]^{-1} \mathbf{B}^T (\mathbf{AA}^T)^{-1} \mathbf{l}.$$

In order to check any subsequent misalignment of the lucite tubes, the corrections to the coordinates of the end points can be obtained by

$$\mathbf{v} = -\mathbf{A}^{-1} (\mathbf{B\Delta} + \mathbf{l}).$$

If any of the residuals are inordinately large, the points associated with them must be checked for accuracy of alignment.

16.5.5 EXTERIOR ORIENTATION OF STEREORADIOGRAPHS

All of the arrangements of x-ray tube and film shown in figure 16-73 constitute the normal case

FIGURE 16-79. Moffitt's principal point locater. The front cylinders are 7 inches long.

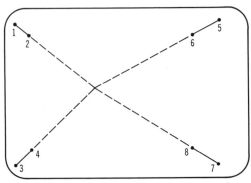

FIGURE 16-80. Four intersecting lines used to determine position of principal point when employing Moffitt's principal point locater. The best location is determined by a least squares solution.

of stereophotogrammetry, more or less, depending on the factors discussed in the following paragraphs. The ideal case of normal stereoradiogrammetry is realized by the apparently simple expedient of keeping the principal distance constant from one exposure to the other.

In figure 16-73a, if the two x-ray tubes are fixed in position so that their focal spots are the same distance from the film plane, the normal case is obtained. This implies that the cassettes used for the two exposures are matched as to thickness, spacing and positioning of the film, that the cassette holders or tracks are straight and true, and that the cassettes take the same orientation inside the tracks. The above conditions can be kept, with very little difficulty, well within the limits of accuracy required in radiography, considering the inaccuracies of measurement due to image degradation.

In figure 16-73b, the normal case is obtained if the movement of the x-ray tube is strictly parallel with the track of the cassette, and if the x-ray tube moves parallel with itself. In order to provide a more favorable base-distance ratio in the set up, the tube is sometimes rotated between the two exposures in order to provide x-ray coverage of the object. The x-ray tube may or may not rotate about an axis which passes through the focal spot. If it does, then the normal case is preserved. If it does not, then the eccentricity causes a slight change in the principal distance by an amount $\Delta c = e (1 - \cos \theta)$ as shown in figure 16-81. If the principal distance is large relative to the depth of the object, this eccentricity error can be approximately overcome by changing the scale of the radiograph in the reduction process. As the principal distance approaches the depth of the object, the shift in perspective center due to eccentricity must be taken into account either by a corresponding shift of the tube in the opposite direction, thus preserving the normal case, or else by computation.

If the tilting arrangement of figure 16-73c is employed, the axis of rotation passes through the position of the normal from the focal spot when the film plane is farthest from the focal spot, that is when the maximum principal distance is obtained. This would be when the film is midway between positions 1 and 2. In order to obtain the normal case, the tilt must then be exactly the same amount in both directions in order to make both principal distances the same.

The only requirement for obtaining the normal case with the arrangement shown in figure 16-73d is that the tracks or rails on which the object and film ride must be straight and coplanar.

By proper machining, assembling, and calibration, the normal case will hold, within the limits of accuracy of radiographic measurements. The only remaining elements of exterior orien-

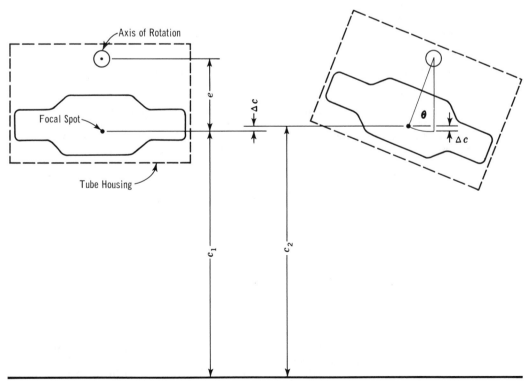

FIGURE 16-81. Eccentricity of focal spot causes a change in the principal distance if the tube is rotated to provide side coverage.

tation of a stereo pair are the length of the base and the direction of the base projected onto the film, that is, the direction of the radiographic x-axis. The base length can be measured directly as indicated in section 16.5.3 or else it can be computed from suitable control in the object space. The direction of the x-axis can be approximated by imbedding radiopaque control marks or wires in the reference plane or in the cassette which has been aligned with the direction of travel of either the x-ray tube or the object-film pair. This alignment is quite difficult and time-consuming and is suitable only when the stereo x-ray set-up is a rigid unit. Alternatively, the direction of the x-axis can be determined by observing the direction of parallactic displacement of a control point well elevated from the reference plane.

If some uncertainty exists in the conditions under which the normal case is fulfilled, both the interior and exterior orientation of a stereo-pair can be determined analytically if sufficient control is provided in the object space. This is discussed by Kratky (1975) and Lippert, et al (1975).

The interior and exterior orientation of the configuration shown in figure 16-74 can be determined by two single resection solutions based on the image positions of appropriate control located in the object space. One such solution is given by Suh (1974).

16.5.6 OBJECT SPACE CONTROL

The amount and distribution of object space control is dependent on many factors among which can be listed: (1) whether object points are to be determined only in one direction as heights or whether all three dimensions are necessary; (2) the total depth of the object to be measured; (3) the size of the film; (4) the convenience of locating the object in the control area; (5) accuracy required in final coordinates; and (6) whether exterior controls such as measured base and principal distances are considered adequate. The control configuration for the three-dimensional determination of object points is essentially the same whether or not the normal case is assumed.

If measurements are confined to the determination of object heights or depths with reference to the film plane, then a minimum of two points must be set. These are preferably imbedded in the reference plane, centrally in the object area, and as widely separated as possible in a direction normal to the base. Either one of the two points is then used as a reference for parallax measurement in a stereometer type instrument. The two points which have the same parallax, are used to align the photographs in the stereometer since the readings on both points must be the same. Additional points in the reference plane can be used to check any warpage of the refer-

ence plane due to any number of causes. One or more control points elevated above the reference plane can be used to check the value of the base.

For the three-dimensional determination of object points, at least three control points are required, two in the reference plane and one located above the reference plane as far as possible. If the principal point must be determined, an additional point elevated above the reference plane is required. These are the minimum conditions. In order to ensure more reliable results, an abundance of control is necessary. Thus, for the technique of locating the principal point shown in figure 16-79 and 16-80 a total of eight control points were used, although these need not be known in all three coordinates.

If an analytical double resection is employed (Kratky 1975) in order to determine all outer and inner orientation elements along with the solution of object coordinates, then a minimum of four well spaced control points, known in all three coordinates, are required. Here also, an abundance of control is desirable in order to test the reliability of the results.

Control takes the form of small radiopaque balls or wires secured to or embedded in some kind of a plexiglass frame or fixture which conveniently surrounds the object. The control points must be so positioned that their shadow or image does not fall off the x-ray film. Since each x-ray problem will be unique, there is no definite pattern to the placement of control. It should be three-dimensional, with the depth of control as great as or greater than the depth of the object being measured.

16.5.7 MEASUREMENT OF X-RAY FILMS

There are four general methods of measurement of x-ray films: 1) in the monocular mode using a comparator; 2) in the stereoscopic mode using a stereocomparator; 3) in the stereoscopic mode using a stereometer type instrument; and 4) in the stereoscopic mode using a stereoplotting instrument. In order to be able to measure a film with a monocomparator, the observer must be able to interpret those points to be measured with such accuracy as to render the measurements useful. For example, Baumrind (1971) studied the reliability of the identification of points in the human skull through statistical analysis and found a substantial variation between different points. Points to be measured on a monocomparator can be identified and marked stereoscopically for better interpretation as in a PUG point marker.

The monocomparator or stereocomparator used to measure the x-ray films must contain a measuring field large enough to accommodate x-ray films, otherwise the film must be reduced in size. The reduced image is frequently printed on glass in the form of a diapositive. One practical problem encountered in making diapositives is the limited latitude of the diapositive emulsion relative to the extremely wide latitude of x-ray films, resulting in contrasty images.

A technique used by some researchers employs the use of translucent or transparent overlays for marking points to be measured on the x-ray film. This preserves the film and permits the preparation and measurement of several overlays of the same film in order to arrive at a better statistical sample of point identification (Baumrind 1971).

The stereometer type instrument consisting essentially of a mirror stereoscope and parallax bar is used to advantage provided that the x-ray films can be oriented with sufficient accuracy to recover the normal case orientation. The Zeiss StR stereo x-ray comparator described by Seeger (1974) and shown in figure 16-82, is a stereometer with a stage large enough to accommodate film pairs up to 40 cm by 40 cm. The StR-3 model has provisions for measuring x and y coordinates to 0.1 mm as well as x-parallax to 0.01 mm. This permits a complete analytical solution in three dimensions from stereopairs taken in the normal configuration.

If the conventional mirror-stereoscope parallax bar is used in conjunction with a light table, the films must be reduced in size if they are much larger than 250 mm by 300 mm.

When a stereoscopic plotting instrument is employed to measure stereoradiographs, careful consideration must be given to the mechanical limitations of the plotter. The base-height ratio which can be accommodated in the projection type instruments such as the Kelsh, and in the optical train type instruments without the Zeiss parallogram such as the Wild A8 or B8, is mechanically limited by the base separation. This limitation sometimes requires that the x-ray tube be rotated as shown in figure 16-81 in order to arrive at a sufficient base-distance ratio to accommodate the plotting instrument.

A second limitation with the use of the stereoplotter is the principal distance of the plotter. If, for example, a Kelsh type instrument is used and if the focal spot-film distance is say 60 inches, a 10 times reduction is required to prepare diapositives in order to create a model with the same scale in all directions. If the reduction to diapositive scale does not match the principal distance, then an affine deformation will occur in the z-direction, the scale of which can be determined from control points or by computation (Schernhorst 1969). Since the stereopair will ordinarily be normal or nearly so, the deformation in the Z-direction will result in a constant Z-scale. The X and Y scales in the model will not be affected under the normal case.

16.5.8 REDUCTION OF X-RAY FILM MEASUREMENTS

The reduction of X-ray films in a stereoplotter is in all respects identical to the reduction of conventional aerial photographs except possibly for Z-scale determination, and so will not be discussed further. The majority of stereoradiographs are obtained in the normal configuration or nearly so within necessary accuracy limits, and discussion will concern this orientation.

In figure 16-83, let A represent an object-space, radiopaque, control point whose height above the film plane is known. This value, designated Z_A is the distance between the film plane and the support surface or some other reference surface plus the height of the point above the

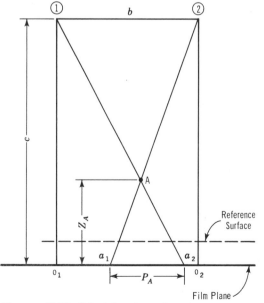

FIGURE 16-83. Principle of parallax determination from known base, principal distance, and height of control point.

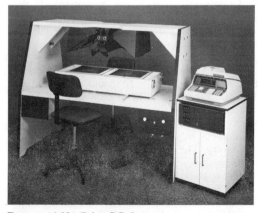

FIGURE 16-82. Zeiss StR-3 stereo x-ray comparator (courtesy Carl Zeiss, Oberkochen).

reference surface. The parallax of A, designated P_A is given by

$$P_A = \frac{bZ_A}{c - Z_A}$$

in which c is the principal distance and b is the base. Now by aligning the two x-ray films (or their reductions) in a stereometer type instrument, the difference in parallax between the control point and any other object point of interest can be measured. This permits an evaluation of the parallax p of the object point which is then used to determine the height of the point above the film plane by the relation

$$Z = \frac{pc}{b + p}.$$

If the principal points O_1 and O_2 of the left and right x-ray films have been identified, and if the direction of parallactic displacement has been established as the film x-axis, the film x and y coordinates of any object point as B in figure 16-84 can be measured on the left film. Then, by taking the left principal point as the origin of an object space three-dimensional coordinate sys-

tem, the X and Y coordinates of the point can be computed from

$$X = x\left(1 - \frac{Z}{c}\right)$$

$$Y = y\left(1 - \frac{Z}{c}\right). \,$$

The reduction of x-ray films when measured in the monocular mode and assuming the normal configuration is presented in detail by Moffitt (1972).

The reduction given by Kratky (1975) is made from measurements in a stereocomparator and in which the normal case is not assumed. In this method, all elements of interior and exterior orientation are determined analytically from redundant control. The solution of the object space coordinates is then performed using a general analytical approach.

16.5.9 MOVEMENT OF OBJECT BETWEEN EXPOSURES

X-rays in medicine are applied to living subjects for the most part. Consequently, one very serious difficulty faced by the x-ray photogrammetrist is the problem of the movement of the subject between the two exposures. The only practical way to make simultaneous exposures of living subjects onto two films is to adopt the configuration shown in figure 16-74. This arrangement requires elaborate shielding and also requires an enormous amount of electrical power for short periods. It is, of course, possible to make simultaneous exposures onto one film from two x-ray tubes oriented for the normal case (equal focus-film distances) provided that the object points are unambiguously marked beforehand with some kind of radiopaque marks (Hagberg 1961).

Fast-change cassette holders can reduce the problem of subject movement only slightly. Various methods of immobilizing the subject by means of mechanical restraints are also only moderately effective. Slight movements of the subject can take place in a fraction of a second; nevertheless, slight movements create false x and y parallaxes in the resulting film pair, destroying the accuracy of the system.

In order to overcome the effect of subject movement between the first and second exposure, the subject itself must contain an array of hard points of sufficient areal extent so as to permit a measurement of y-parallaxes similar to the analysis of a stereo model in aerial photogrammetry. The distance between at least two of these hard points must be known for scaling purposes. Based on the y-parallaxes caused by subject movement (*see* Kratky, 1975), the second film can be mathematically oriented by appropriate rotations ω, ϕ, and κ, and by shifts dy

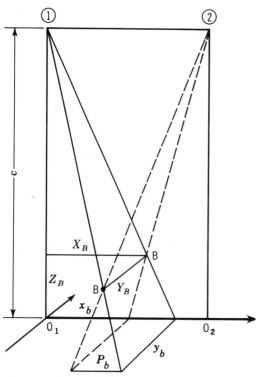

FIGURE 16-84. Relation between x-ray film coordinates (x, y, p) and object space coordinates (X, Y, Z). The origin of both systems is at the principal point O_1.

and dz in directions opposite to those of subject movement, from which corrected x-ray film coordinates of the second film are computed. This, of course, upsets the normal case and forces a general analytic solution. Unscaled object space coordinates are computed based on the original first film coordinates and the corrected second film coordinates. These must then be scaled up or down in order to correct for the dx shift of the subject between exposures. Correction for subject movement is discussed by Kratky (1975) and Moffitt (1972).

16.5.10 APPLICATIONS OF X-RAY PHOTOGRAMMETRY

Since there are so many different problems to which x-ray photogrammetry is applied, it would be redundant to list them here. Rather, the researcher is referred to the reference list (section 16.9.6). Two of the publications in particular appearing in the reference list (Herron 1972 and Singh 1970) contain quite extensive reference material on applications of x-ray photogrammetry.

16.6 Scanning Electron Microscope, Systems and Applications

The Scanning Electron Microscope (SEM) has opened new opportunities to various medical and non-medical applications. Single, two-dimensional SEM micrographs are generally satisfactory for many users. However, there is a steadily growing demand for three dimensional information. Many of these are microscopic versions of macroscopic problems commonly known to professional photogrammetrists.

The first commercial SEM was introduced in 1965 by the Cambridge Instrument Company.

Similar instruments by at least ten manufacturers in several countries are currently available. The SEM differs fundamentally from the Optical Light Microscope (OLM) and the Transmission Electron Microscope (TEM) in several ways (*see* table 16-6). The SEM, unlike the usual OLM and TEM, is able to image 3-D objects because it records not the electrons passing through the specimen but the secondary electrons that are released from the electrically conducting sample by the electron beam impinging

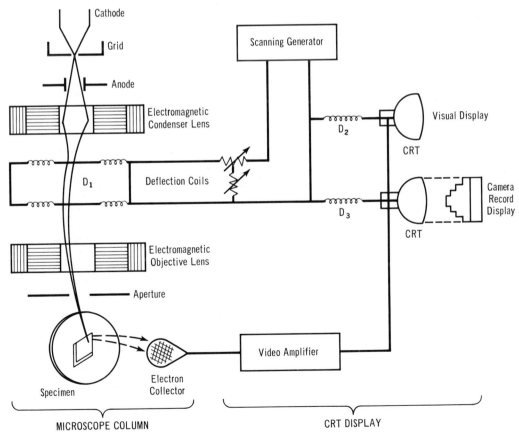

FIGURE 16-85. Schematic diagram of SEM.

on it. The sample can, therefore, be of any size and thickness that fits in the evacuated sample chamber of the SEM.

The imaging system of the SEM (*see* figure 16-85) uses electron radiation for localization and visible light radiation for information transfer. The resolving power is dictated by electron optical consideration and the information content is dictated by the interaction of visible light in matter. The electron radiation is used for two synchronous scanning beams, *viz.*, (a) a point source of radiation sweeping over the specimen in a well defined pattern of lines with a well defined velocity; and (b) a corresponding second point source of radiation over a fluorescent screen, which is eventually recorded with a camera (*e.g.* Polaroid Land Camera). A one-to-one correspondence in time and space exists between the specimen and image points. This, however, is achieved not by optical focusing but by time sequencing.

The primary factor causing variation in intensity in the SEM image are (*see* also Oately, 1972):

(1) factors which depend on the local angle of incidence of the electron probe,
(2) factors which cause variations in the fractions of secondary electrons collected by the detector, and
(3) factors which depend on the nature of the material.

The first two are similar to the angular relationship of the sun to the topographic features of the earth. The third factor is analogous to color in a visual image. The importance of these factors is

that something can be done about each of them to enhance the recorded image.

The stage plate containing the specimen (object) has four degrees of freedom, *viz.*:

(1) tilt, uniaxial, around X- (or Y-) axis, corresponding to ω (or ϕ) tilt in conventional photogrammetry;
(2) rotation around an axis parallel to the general direction of the Principal Electron Axis similar to κ rotation in conventional photogrammetry;
(3) X-translation; and
(4) Y-translation.

In the metric (quantitative) evaluation of SEM micrographs a major role is played by the resolution capability, which, with current technology, is greater than 10 nm. This may be the limiting factor for precision measurements, *e.g.*, at magnification 5000×, 10 nm corresponds to 0.05 mm on the micrograph. The measuring tool having a standard error of less than 0.05 mm at the photograph may then be considered adequate. Higher magnifications, giving comparatively unsharp image, will be of no help, while lower magnifications giving sharper images would demand precision measuring instrument, a stable photobase, *etc.*

The following photogrammetric characteristics of SEM imagery have been found from recent studies (*see* Ghosh, 1974, Maune, 1973, Nagaraja, 1974).

1. SEM system is better represented by a mathematical model for an effective central perspective projection, rather than for a parallel projection (figure 16-86), although not statisti-

TABLE 16-6. CHARACTERISTIC COMPARISON OF OLM, TEM AND SEM (AFTER BLACK, 1970)

Characteristics	OLM	TEM	SEM
Lens System	Glass or Quartz	Electromagnetic	Electromagnetic
Image Display	Eye retina, Film or Frosted glass	Fluorescent viewing plate or Film	CRT for viewing and photography
Energy	Radiation from UV through visible spectrum to IR	Thermonic emission of electrons accelerated by 50 to 1000 KV	Thermonic emission of electrons accelerated by 1 to 30 KV
Wave Lengths	200-700 nm	> 0.1 nm	> 0.1 nm
Resolution	100 nm	0.5 nm	10 nm
Depth of Field	Poor; 250 μm at 15× to 0.08 μm at 1200×	Good; 500 μm at 4000 × to 0.2 μm at 500,000× 200× − 300,000×	Excellent; 1000 μm at 100× to 10 μm at 10,000× 20× − 100,000×
Magnification	15× − 3000×	Stepped 0.1 − 1 μm thin	Continuous No thickness limit;
Sample Requirements	Thin sections only	section or only replica; 2-3 mm diameter	non-metals (better) metal coated (Au,Pd or Al); to 25mm dia.

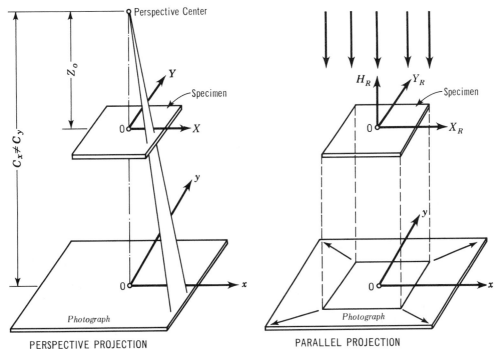

PERSPECTIVE PROJECTION   PARALLEL PROJECTION

FIGURE 16-86. Perspective and parallel projections in SEM.

cally significant at magnifications higher than 2000×.

2. Strict collinearity condition is disturbed by four types of systematic distortions, *viz.*, scale (differential), spiral, tangential, and radial (figure 16-87).

3. The difference between the perspective and parallel projections can be mathematically modelled such that the same computer program will work conveniently for either situation.

4. Scale distortion can be contained directly in the mathematical model for the projection.

5. For most applications, the tangential distortion can be effectively contained in the mathematical model for the spiral distortion.

6. The effects of radial and spiral distortions are best corrected by use of polynomials derived from theoretical considerations but experimentally checked.

7. The SEM can be stabilized long enough so as to calibrate the system and apply the results in subsequent 3-D information extraction of high precision.

Analogical 3-D continuous mapping can be performed on instruments in which near parallel projection is adaptable; *e.g.*, Zeiss Stereotope (*see* Ghosh, 1971); Nistri-Bendix AP/C (*see* Ghosh, 1971); Wild B9, modified, (*see* Wood, 1972). However, the highest limit of accuracy at discrete points is obtainable with multi-photo, computational, solutions (*see* Ghosh, 1974, Maune, 1973).

16.7 *Holographic Systems and Potential Applications*

16.7.1 INTRODUCTION

Coherent light can play a number of key roles in mapping systems of the future, particularly in close-range photogrammetry. Conventional photogrammetric systems use incoherent light. This incoherent light is polychromatic and has a random phase distribution. In coherent optical systems, the light source is usually a laser that produces monochromatic light that is temporally and spatially coherent. Coherent optical systems possess some important characteristics not found in incoherent systems. For example, since amplitude and phase of coherent light beams add vectorially, interferometric measuring systems and holographic display systems can be utilized. Further, coherent optical systems can utilize image processing by spatial filtering in the spatial frequency domain that is not available in incoherent optical systems.

Hologrammetry is an interdisciplinary field combining holographic techniques for purposes of interpretation, mensuration, mapping, and display. A hologram is a recording of the interference pattern of a reference beam and the object beam. An image is obtained when a coherent beam is used to illuminate the developed holographic plate.

There are various types of holograms. They differ primarily in the geometry of the optical

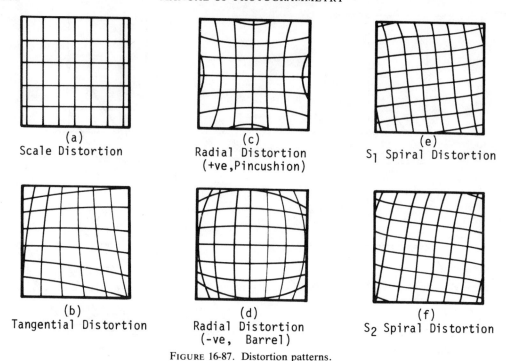

(a) Scale Distortion	(c) Radial Distortion (+ve,Pincushion)	(e) S_1 Spiral Distortion
(b) Tangential Distortion	(d) Radial Distortion (-ve, Barrel)	(f) S_2 Spiral Distortion

FIGURE 16-87. Distortion patterns.

arrangement used to record them. The schematic of making a *Fresnel* hologram is shown in figure 16-88a. When a duplicate of the reference beam (called the reconstruction beam) is used to illuminate the hologram a virtual image of the object is produced as shown in figure 16-88b. The geometry of this virtual image is identical to the object's geometry provided no appreciable emulsion shrinkage occurs during development.

In addition to the virtual image, three other images may be reconstructed from a Fresnel hologram. The real image, figure 16-89a, is obtained by using a reconstruction beam which is the complex conjugate of the reference beam. In case of a collimated reference beam, the complex conjugate is simply another collimated beam traveling in the opposite direction. The real image is pseudoscopically inverted, which makes it unattractive for use with three-dimensional objects. However, when used for planar objects, such as regular photographs, it has obvious potential since it can be projected on a screen and used for display.

Two other images may be generated from a Fresnel hologram by reflection of the reconstruction beam from the emulsion of the holographic plate as shown in figures 16-89b and 16-89c.

Another type is the so-called *focused image hologram*, the recording system of which is shown in figure 16-90a. The main difference is that a lens is placed between the object and plate such that the object's image falls near the hologram. The main advantage of focused-image holograms is that they may be reconstructed with a broadband source or diffuse white light as shown in figure 16-90b, as well as with a laser. Unfortunately, directly focused-image holograms of three-dimensional objects are not desirable for metric work because all object points focused off the hologram emulsion plane are distorted. However, the fact that white light is used for reconstruction makes it well suited for display. If the object is planar, such as a photograph, then focused-image holography has a good potential.

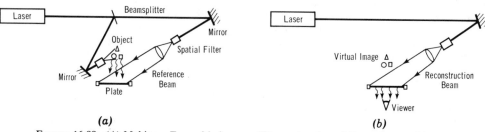

(a)   (b)

FIGURE 16-88. (A) Making a Fresnel hologram; (B) construction of Fresnel vertual image.

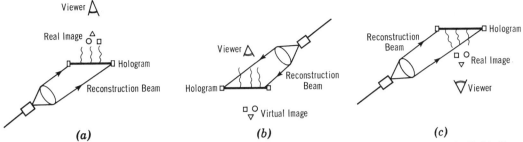

FIGURE 16-89. Fresnel hologram: (a) real image, (B) virtual-relief (reflection) image, and (C) real relief (reflection) image.

The different types of holographic images mentioned above have been considered by the photogrammetrist for purposes of mensuration and mapping as well as for display and interpretive purposes. Direct holograms of three-dimensional objects have been used as an alternative to the photogrammetric model for close-range applications since the image is usually the same size as the object. For larger objects, a hologram of a restituted photogrammetric model, called the holographic stereomodel, has been devised to extend the holographic capability to the wider applications.

16.7.2 DIRECT HOLOGRAMS OF THREE DIMENSIONAL OBJECTS

Fresnel holograms of three-dimensional objects offer an excellent alternative to the photogrammetric model for metric work.

Compared to a photogrammetric model, a Fresnel hologram has the following characteristics: (1) it affords the viewer many perspectives within the aperture fixed by the size of the hologram; (2) parallax exists in all directions, hence no restrictions on the placement of the observer's eyes; (3) problems of blind spots are less severe in holographic recording since multiple object beams may be used, as shown in figure 16-91; (4) unlike photography, particularly at close range, depth of field is not a critical quantity in holography; (5) a holographic image is usually the same size as the recorded object, while a photogrammetric model can be larger or smaller with proper design of the photography.

16.7.2.1 MEASUREMENT OF DIRECT FRESNEL HOLOGRAMS

A self-illuminated dot may be inserted in the space of the virtual image and its movements monitored by a 3-axis comparator (such as the Autograph A7, *see* figure 16-92). To determine the precision of pointing at a holographic virtual image, several researchers took repeated measurements on a well defined point. Table 16-7 gives the standard deviations in mm for a number of observers. The x, y plane is parallel to the hologram plate with x being parallel to the observer's eyebase. Auxiliary viewing magnifications ranged from 1.0 to 2.7 while the distance between observer and image was 10 cm to 20 cm.

Many observers noted increased pointing precision using small magnifications (less than $3\times$) with JPA reporting 0.008, 0.013, 0.070 mm, in x, y, z, respectively, when $20\times$ is used (Agnard, 1972).

While pointing precision depends essentially on the quality of the reconstructed image, accuracy of the image necessary for metric work depends on the geometric fidelity of reconstruction. If the reconstruction geometry is different from the recording geometry, the resulting image will be distorted. This difference in geometries arises from emulsion shrinkage and deviations

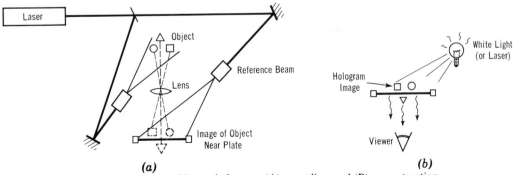

FIGURE 16-90. Focused-image hologram: (A) recording, and (B) reconstruction.

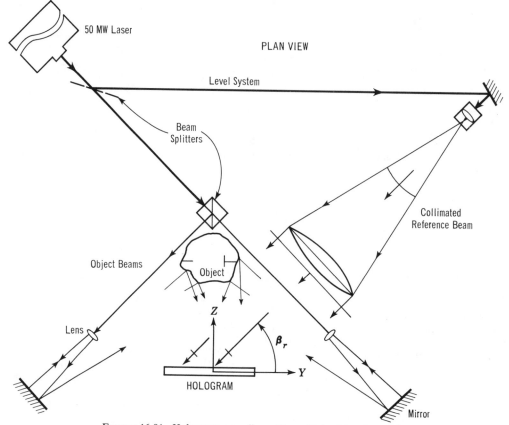

FIGURE 16-91. Hologram recording with multiple object beams.

between reconstruction and recording beams. Emulsion shrinkage can be effectively minimized (Kellie & Stevenson, 1973).

The other source of distortion would be completely eliminated if the same set up used for recording is also used for reconstruction and measurement. This is, however, a restriction which would make the use of holograms for mensuration and mapping limited to optical

FIGURE 16-92. Wild A7 modified for hologrammetric mensuration and mapping.

laboratories, thus curtailing their usefulness. Therefore, one must consider as more common the case of reconstructing the hologram using, in general, different optical components. The main task is to determine the acceptable tolerances within which the reconstruction geometry can approximate the recording geometry. If, as is conveniently done, the reference and reconstruction beams are collimated, only two angular elements remain. Careful experimentation showed that duplication of these two angles should be within 0.2° so that the resulting distortions are below the level of the pointing precision. In addition, simulation studies also showed that a reconstruction beam radius of 10^5 mm or larger would effectively mean a collimated beam with no significant distortions. If these tolerances are not met, Gifford's work (1974) showed that eye position would influence perceived image location. In fact, each eye would see a different image for the same original object point with possible resulting y-parallax.

While proper reconstruction geometry yields an image which correctly duplicates the object, its position and orientation may not be correct with respect to a reference system established in the object space. Therefore, for both targeting

TABLE 16-7. PRECISION OF POINTING ON A SINGLE POINT (DIRECT HOLOGRAM)

Observer	Standard Deviation in mm		
	x	y	z
CHG (Glaser & Mikhail, 1970)	0.018	0.035	0.084
	0.017	0.037	0.111
MKK (Kurtz et al, 1971)	0.013	0.023	0.030
	0.018	0.025	0.051
JPA (Agnard, 1970)	0.020	0.030	0.110
PVB (v. Berkefeldt, 1970)	0.022	0.012	0.074
DLG (Gifford & Mikhail, 1973)	0.018	0.040	0.078
RAG (Gifford & Mikhail, 1973)	0.023	0.026	0.071

and mapping a seven-parameter transformation is necessary to obtain data from the hologram which are referred to the desired system. This transformation may be performed either computationally or empirically. If the internal geometry of the hologram has been recovered sufficiently accurately (as explained above) the scale factor is unity and a six-parameter transformation is used. In an actual experiment the precision of measuring a hologram was found to be about the same as the precision of pointing.

An example of digital mapping from holograms involved the representation of the two surfaces of a bone-joint in a dog by a dense set of points. The research involved measuring the angular joint motion of the scapula and humerus of a German Shepherd as the animal walked a treadmill.

An external mechanical linkage with potentiometers was attached to the dog to accurately measure skeletal movements. Knowledge of how the two bones move, combined with a knowledge of the geometry of the humeral and scapual joint surfaces, can be used to determine the relative motion of these surfaces during walking.

While the angular motion can be directly measured, the determination of the internal surfaces of the bones at the joint made use of hologrammetric techniques. Holograms were made directly of the separated bone surfaces and the attached metal potentiometer mounting blocks which were used for control. Figure 16-93 shows the holographic virtual image of the scapula and mounting block. Of primary interest was an arthritic blister, for which coordinates were required relative to the mounting block. The coordinates of the mounting blocks were used in a seven-parameter transformation, to arrive at the control system.

Further efforts relate to the use of a computer to match the digitized representation of the surfaces of the scapula and humerus under varying conditions of angular orientation. Thus the points of contact can be determined as they were when the dog was walking. With a computer-controlled display the reconstructed surface of a specific area, such as that of an arthritic blister, can be studied.

16.7.2.2 MAPPING DIRECT FRESNEL HOLOGRAMS

Absolute orientation is performed instrumentally if one desires graphical maps from the holographic image. Once this is done, different projections of the object, such as plans, front views, and cross-sections, may be obtained by monitoring the movement of the illuminated dot when it is in apparent contact with the object's holographic image. In a like manner, contour lines can be plotted.

As an example, figure 16-94 shows a topographic map of a toothless dental casting.

The contouring of this object was especially interesting because it is a highly convoluted object. Note that the 6 mm and 8 mm contours pass beneath the others. The profile A-A (figure 16-94 shows this clearly.

Profiling was the most difficult task. When needed as a check, the profile was retraced starting from the opposite side. The dashed line represents the profile from the return plot.

As a final test one observer, who possesses only one eye, attempted to contour the same image. Figure 16-95 shows effective plotting except in regions of extremely steep slope. The

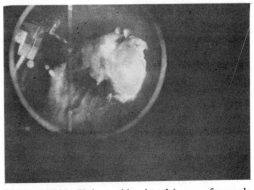

FIGURE 16-93. Holographic virtual image of scapula and mounting block.

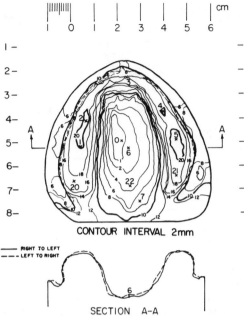

CONTOUR INTERVAL 2mm

―――― RIGHT TO LEFT
――――― LEFT TO RIGHT

SECTION A-A

FIGURE 16-94. Hologrammetric compiled topography of a toothless dental casting.

observer noted that there was never any question as to the shape of the object which he could discern from the image by moving his head about. He could not have used a stereomodel at all because this would require a subjective mental impression formed while simultaneously viewing two separate perspective photographs or projections. This test at least demonstrates the pictorial value of a true three-dimensional image.

16.7.3 HOLOGRAPHIC STEREOMODEL

A holographic stereomodel is a hologram of a restituted photogrammetric stereopair. It may be of the Fresnel or focused-image type.

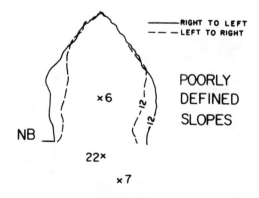

―――――RIGHT TO LEFT
――― ――LEFT TO RIGHT

POORLY
DEFINED
SLOPES

NB

12 mm CONTOUR

USING ONLY ONE EYE

FIGURE 16-95. Monocular contouring.

16.7.3.1 FRESNEL HOLOGRAPHIC STEREOMODEL

Figure 16-97 shows a schematic of recording a pair of overlapping transparencies. After the appropriate photogrammetric operation (either relative orientation or double rectification), the projected images will form a parallax-free stereomodel in space. A rear projection screen is used both for parallax removal and to eventually become the actual object to be recorded holographically. First, photograph 1 and reference beam 1 are used to expose the entire hologram plate. Then photograph 2 and reference beam 2 are used to expose *again* the same hologram. Thus, the hologram is actually a double-exposure hologram each of an equivalent photograph of a pair with parallel optical axes. The use of two reference beams is necessary for later image separation upon reconstruction. A half-wave retarder is used to rotate the polarization of one of the two reconstruction beams by 90°, and orthogonally oriented polarizing filters are used for viewing. These filters may be rotated to effect pseudoscopic viewing for purposes of improving pointing precision. Another scheme for viewing each photo-image by a separate eye is through selectively illuminating a different portion of the hologram for each eye.

16.7.3.2 FOCUSED-IMAGE HOLOGRAPHIC STEREOMODEL

The stereomodel is admirably suited for focused image holography because the object recorded is in a plane. The recording arrangement for a focused image holographic stereomodel is identical to the Fresnel type except for replacing the screen by the hologram plate itself. Therefore, the screen is first used in the procedure of parallax elimination, and once the model is restituted the screen is removed, the hologram plate placed in its position and recording made as in the Fresnel case. For reconstruction and viewing, the arrangement is essentially the same as for Fresnel case except that ordinary white light is used instead of a laser. Image separation for viewing can be done by polarization as before, or by proper arrangement such that at the selected viewing distance each image arrives at a different eye.

The focused image holographic stereomodel has the following advantages; (1) use of diffuse white light for reconstruction makes it a simple and less expensive operation; (2) removal of speckle associated with coherent illumination; (3) reduced reconstruction distortion since the object recorded is planar (the photograph); (4) removal of rear projection screen eliminates distortion due to its finite thickness. It, however, has some attendant disadvantages when compared to the Fresnel type, stemming from the fact that in focused-image holograms each object point generates only one image point on the hologram. Hence the hologram must be the same

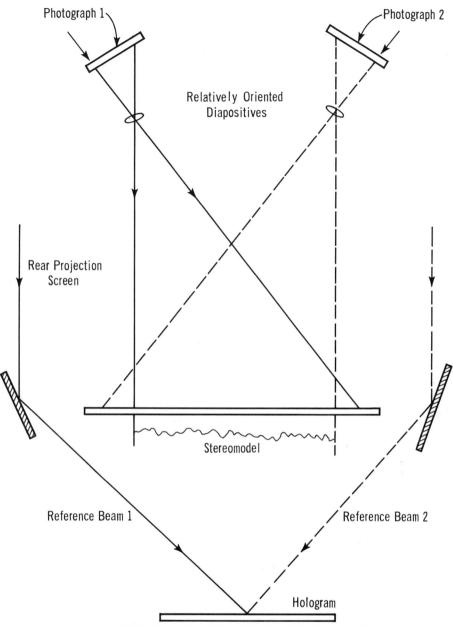

FIGURE 16-96. Recording of Fresnel holographic stereomodel.

size as the area of the photogrammetric model. Also, because of this one-to-one correspondence, any scratch on the hologram's emulsion results in destroying information from the stored model, which is not the case with the Fresnel type.

Because of the use of white light the focused image holographic stereomodel, and particularly the reflection type, is well suited for display and interpretive purposes. There are at present reflection focused-image holograms of terrain with high efficiency such that they can be viewed in regular ambient light. This is an attractive fea

ture for display and training purposes. On the other hand, Fresnel holographic stereomodels are better suited for mensuration and mapping since the use of a laser in reconstruction is no problem in the metrology laboratory. They are also usually more stable and exhibit less distortion than the focused-image kind.

For metric work from Fresnel holographic stereomodels, absolute errors in both the horizontal and vertical angles of each of the two reconstruction beams must be each less than $0.2°$ (the same as for the direct hologram). In addition, since there are essentially two holographic

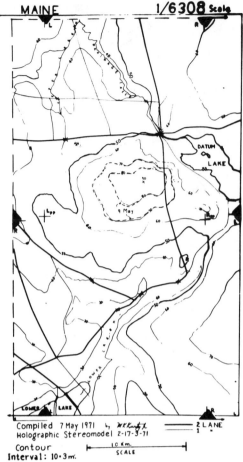

MAINE   **1/6308 Scale**

Compiled 7 May 1971 by *M.K.Kurtz*   2 LANE
Holographic Stereomodel 2-17-3-71   1

Contour
Interval: 10·3 m.   SCALE

FIGURE 16-97. Hologrammetrically compiled map.

early result of a topographic map made from a holographic stereomodel. Some contour lines (dashed lines in the figure) were plotted two days later to determine reproducibility. Taking the map of figure 16-97 as a demonstration of feasibility, there should be little doubt that accurate maps can be obtained from holographic stereomodels using an instrument which employs the two-dot system.

16.7.4 AUTOMATION IN HOLOGRAMMETRY

Coherent light can be used for correlating conjugate images. Balasubramanian (1974) is credited with introducing the technique of coherent optical heterodyning to this field. Coincidence detection of conjugate imagery may be effected through the use of the normalized cross-correlation coefficient (Balasubramanian, 1974). Such detection when performed automatically would lead to rapid plotting of contours or cross-sections.

16.7.5 CONCLUDING REMARKS

The different aspects of this section (16.7) can be considered as activities of a relatively recent interdisciplinary field called hologrammetry. It combines the techniques of photogrammetry and holography. Although consideration has been given to the application of photogrammetric methods and optical holography, the future may certainly bring other types of holography and procedures of coherent optics into this field.

Direct holograms of objects are useful recording alternatives to regular photographs. They may be used for objects measuring a few centimetres in each of three dimensions and suitable for 1:1 scale recording. Objects of complex shapes are better suited for holographic recording where vertical plane mirrors may be placed around the object in order to record otherwise invisible spots.

The holographic stereomodel (Fresnel or focused-image) stores and allows the retrieval of three-dimensional information; it is at an advanced stage compared to aerial photographs. This type of display offers distinct advantages for qualitative, interpretive, and training purposes, particularly if the focused-image process is used. For quantitative uses it also affords the user access to the three-dimensional-model which has heretofore been confined to the photogrammetric facility.

One distinct advantage of the Fresnel holographic stereomodel results from the information-storage capacity of a hologram. Not only can the equivalent of one pair of photographs be stored in the proper geometric relationship on one hologram, but also several stereomodels could be stored on the same hologram (multiplexing). One

images constructed, the relative error in the horizontal angles of the two reconstruction beams need to be less than 0.01° in order that y-parallax may be undetectable. Similarly, relative error in the vertical angles of only 0.01° or less can be tolerated. Finally, when one beam's collimation is in error, y-parallaxes are significant, but when compensating collimation errors occur, y-parallax becomes negligible. In the latter case, however, scale errors result in the model. In general, if $(1/R_c - 1/R_r) < 10^{-5}$, where R_c and R_r are radii of reconstruction and recording beams, respectively, scale errors will be negligible. Once these tolerances are met both mensuration and mapping are possible. Two half-dots in the plane of the image of the rear-projection screen must be used. In an experiment, measurements on a holographic stereomodel were fitted to the original object with the accuracies: $\sigma_x = 0.037$ mm, $\sigma_y = 0.037$ mm, $\sigma_z = 0.190$ mm which compare quite favorably to the results of strictly photogrammetric techniques which gave $\sigma_x = 0.030$, $\sigma_y = 0.030$, $\sigma_z = 0.115$ mm.

With respect to mapping, figure 16-97 is an

possible technique of storing models is to use only two small areas of the hologram plate for each model.

Another advantage of the holographic stereomodel is the expected simplification and ensuing economic savings in instrumentation for its mensuration and mapping. Although the equipment for making it would, by necessity, be complex and extensive, mensuration and mapping are by contrast much simplified. Systems for extracting digital or graphical information from holographic stereomodels require essentially an instrument of the type used with terrestrial normal-case photographs. This kind of instrument is considerably less complex and less expensive than regular plotters and requires operators with little photogrammetric training. Consequently, one can envision a central well-equipped facility for the production of holographic stereomodels, serving many smaller less expensive mensuration and mapping places.

The holographic stereomodel concept also offers unique advantages for subsequent, automated data, processing. Its configuration makes it adaptable to interferometric correlation techniques for automatic contouring or profiling, particularly when applying heterodyne detection.

16.8 Moiré Topography, Systems and Applications

16.8.1 INTRODUCTION

A powerful way of describing a three-dimensional shape is to draw contour lines. Formation of Newton's rings between an object and an optical flat is a direct means of observing and recording of contour lines, but the depth of the surface under test is limited to not more than several tens of the wavelength of light used. There has been no simple way of observing contour lines of an object with greater depth.

Tsuruta (1969) proposed an ingenious way of recording contour lines of an object with a diffusing surface. This method, however, produced contour lines only after several processes and requires high quality optics. The size and depth of the test object may be limited.

Drawing contour lines from a pair of stereo pictures with the aid of a stereoplotter is most commonly used to obtain contour lines of a large object but, like the Tsuruta's method, this is not a direct method either and requires expensive instruments.

Projection of two sets of fan-shaped beams at right-angles to each other produces *Multislit Lichtschnittverfahren* with 90° incidence angle visualize contour lines *in situ* but works well only on surfaces of simple curvature.

16.8.2 PRINCIPLE

Suppose a plane transmission grating with equally spaced lines s_0 apart is placed above an object to be tested (figure 16-98). The surface is illuminated by a point source S and observed through a small hole at E. The x axis is taken to lie parallel to the lines of the grating and the z axis is taken to be perpendicular to the grating surface (S and E lie on the Y-Z plane). The shadow of a small area of the grating around Q is projected around P. The shadow observed from E around R is the result of central projections applied twice on the grating around Q. Because the line spacing s_0 is very small compared with vertical distances S and E from the grating, the projections are practically parallel projections.

Therefore the shadow observed throught grating surface around R is a grating with different direction and spacing from the original but has the same phase relation to R as that of original grating to Q.

Assuming a sinusoidal grating, transmittance of the grating is

$$T_Q = \frac{1}{2}[1 + \cos 2\pi(\epsilon + y)/s_0] \qquad (16.53)$$

where s_0 is the line spacing of the grating and ϵ is initial phase of the grating.

The intensity of the projection of shadow around R is

$$I_s = \frac{1}{2}\{1 + \cos 2\pi[(\epsilon + y_q)/s_0 + \xi/s']\}I_0 \qquad (16.54)$$

where y_q is y coordinate of Q, ξ is a new coordinate with the origin at R, which is perpendicular to the projection of shadow of the grating lines on xy plane. I_0 is the intensity of the source. Using x and y, ξ is expressed as follows, assuming the angle of projected shadow relative to

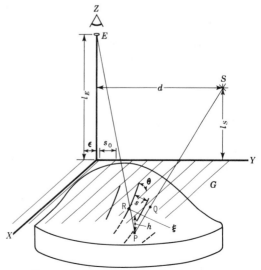

FIGURE 16-98. Formation of moiré pattern.

original lines as Θ, and the line spacing of projected shadow as s':

$$\xi = (y - y_R)\cos \Theta - (X - X_R)\sin \Theta. \quad (16.55)$$

Transmittance of the grating around R is expressed by T_R, which is represented by the same form as equation (16.53).

The moiré observed around R is obtained as a product of $I_S \times T_R$; thus

$$I_m = \frac{I_Q}{4} \Big\{ 1 + \cos 2\pi\big[(\epsilon + y_Q)/s_0 + (y - y_R)\cos \Theta/s' - (x - x_R)\sin \Theta/s'\big]$$

$$+ \cos 2\pi(\epsilon + y)/s_0$$

$$+ \frac{1}{2}\cos 2\pi\big[(y_Q - y_R)/s_0 + (y - y_R)(\cos \Theta/s' - 1/s_0 - (x - x_R)\sin \Theta/s'\big]$$

$$+ \frac{1}{2}\cos 2\pi\big[2\epsilon + y_Q + y_R)/s_0 + (y - y_R)(\cos \Theta/s' + 1/s_0) - (x - x_R)\sin \Theta/s'\big]\Big\}. \quad (16.56)$$

The first cosine term represents the projected shadow of the grating. This is not necessarily of high frequency, as will be mentioned later. In most cases, however, the frequency of this term is in about the same order as that of grating and can be separated from moiré. The second cosine term represents grating itself and the fourth cosine term the sum of grating and its shadow. These are also of high frequency.

The third cosine term represents moiré in the small area around R on the grating or around P on the surface. Lightness of moiré at P is obtained by making $x = x_R$ and $y = y_R$ where x_R and y_R are x and y coordinates of R:

$$I_p = \frac{1}{8}\,[1 + \cos 2\pi(y_Q - y_R)/S_0]I_0 \quad (16.57)$$

Note that the initial phase of the grating is dropped from the expression. This means that the moiré is stationary against parallel movement of the grating in its plane.

Elementary geometry gives

$$y_Q - y_R = [l_E d - (l_E - l_S)y_R]h/l_E(l_S + h) \quad (16.58)$$

where h is depth of P from the surface of grating, with h pointing downwards. By making $l_S = l_E = l$, equation (16.58) is simplified as

$$y_Q - y_R = hd/(l + h) \quad (16.59)$$

which is a function of h only for given d and l.

Thus the equal brightness line represents the equal depth line. The depth of the Nth bright line is obtained by making the argument of the cosine term of equation (16.57) equal to $2N$ as follows:

$$h_N = lN/(d/s_0 - N). \quad (16.60)$$

In most cases d is much larger than s_0 and the following relation may be used as a good approximation

$$h_N \approx l s_0 N/d. \quad (16.61)$$

When l is infinity, which means illuminating with collimated light and observing vertically

through a small hole on the focal point of a field lens, equation (16.60) is written as

$$h_N = s_0 N \cot \phi \quad (16.62)$$

where ϕ is incidence angle of illuminating light.

Hereafter, this technique will be called moiré topography. Figure 16-99 shows a contour line system of a 25-cent coin, on which a thin coat of white paint is applied. Collimated light and a field lens are used for illumination and observation. The grating used in a plane glass with Ronchi ruling of one to one ratio of width and space.

Use of a collimating lens and a field lens is not practical when the test object is large. Even in such a case equidepth lines are obtained by keeping $l_s = l_E$.

Figure 16-100 shows a contour-line system of a disk of cotton cloth revolving with flutter. A plane grating is placed in front of the disk with 5-cm clearance from the center of the cloth. One turn of a helical xenon flash-tube is used for illumination. Because the flashtube is not thin enough, some deterioration of visibility of contour lines with greater depth is observed. This

FIGURE 16-99. Contour line system of a 25-cent coin. Collimated light and a field lens are used for illumination and observation ($s = 0.125$ mm, illumination angle is 45°, and depth interval between successive fringes is 0.125 mm).

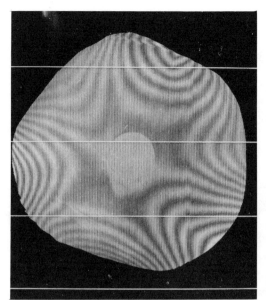

FIGURE 16-100. Contour line system of a cotton cloth disk of 36 cm diameter revolving with flutter. One turn of a helical xenon flash is used for illumination ($s = 1.0$ mm, $l_S = l_E = 100$ cm, $d = 20$ cm, $h = 5$ mm). The parts of the disk with fringe of greater visibility are close to the grating.

visibility change sometimes helps to judge which part of an object is closer to the grating.

An object which has a surface with large inclination introduces some difficulties. First, the width and spacing of the shadow of lines of grating increase on that part of the surface with a small angle to the illuminating light. This means that frequency of the first term of equation (16.56) is decreased and mixed up with the moiré. Also a moiré formed between higher harmonics of shadow and grating appears. This alias moiré does not appear in equation (16.56) because the equation is derived assuming a sinusoidal grating. These unwanted patterns change their shapes and positions with the movement of grating in its plane, whereas the equal-depth moiré is stationary. So by moving the grating in its plane during exposure, these unwanted patterns are washed away.

Figures 16-101 and 16-102 show a contour-line system of a mannequin taken with a stationary grating and with a moving grating.

By using two light sources arranged symmetrically about E, shadow-free illumination is possible without affecting the contour line system. Figure 16-103 is a contour line system of a mannequin taken by shadow-free illumination.

Moiré produced between two systems of contour line shows up the contour line system of equal depth difference of the two surfaces. The two contour line systems are expressed as

$$f_1(x,y) = p\Delta h$$
$$f_2(x,y) = q\Delta h \qquad (16.63)$$

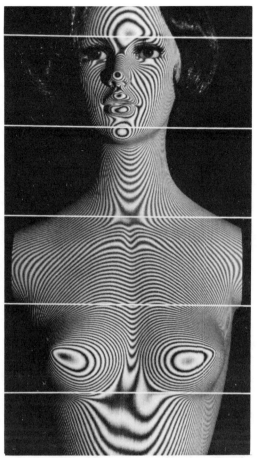

FIGURE 16-101. Contour line system of a life-size mannequin ($s = 1.0$ mm, $l_S = l_E = 200$ cm, $d = 100$ cm, $\Delta h = 2$ mm, and vertical distance of white lines is 100 mm). The grating is stationary during exposure. Note shadow of lines of grating on left cheek, and alias moiré on left and right cheeks.

where p and q are integral numbers and Δh is depth interval between successive contour lines. The moiré is expressed by a function which satisfies $q - p = K$, where K is another integral number. The moiré expressed as follows:

$$f(x, y) = [f_2(x, y) - f_1(x, y)] = K\Delta h \qquad (16.64)$$

which means that the moiré forms a set of curves of equal depth difference.

Figure 16-104 shows subtractively superimposed two contour line systems of a human back with raised and lowered right arm. Faint but observable moiré of the two contour line systems is observed on such part where original contour lines are close and the depth change is relatively small. The contour lines cross with large angle in the parts where deformation is large, and the moiré is hard to observe visually. The crossing points with same K's must be connected to give contour lines of equal depth difference.

Figure 16-105 shows a graphically drawn

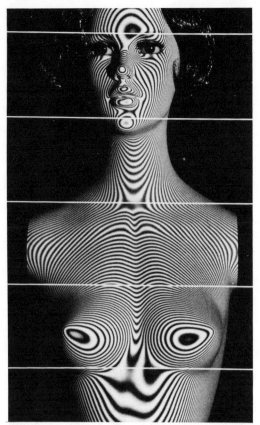

FIGURE 16-102. Contour line system of a life-size mannequin. Data are same as figure 16-101. The grating is maved parallel to itself in its plane during exposure. Note that shadows of lines of grating and alias moiré are washed away.

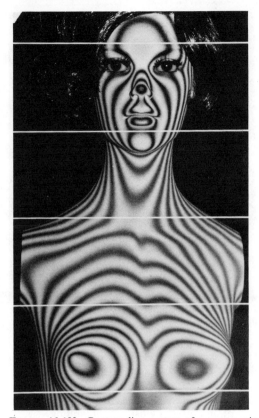

FIGURE 16-103. Contour line system of a mannequin taken by shadow-free illumination (s = 1.0 mm, $l_S = l_E$ = 200 cm, d = 50 cm, Δh = 4 mm) and the grating is moved during exposure.

contour line system of equal depth difference, which is made by using an enlarged transparency of the contour lines before and after deformation.

16.8.3 APPLICATION OF MOIRÉ TOPOGRAPHY TO LIVING BODY

In the course of development of this new technique, strong interests were shown from the medical field, because the moiré technique could offer a new means of measuring the shape of living body. There were, however, many problems to be solved.

In measurement of full size living body, the size and depth of the field required are considerably large, for instance, $1.8 \times 1.8 \times 0.9$ m³. To obtain sufficient illumination over such a field the light source must emit enough light flux and also be small enough to cast a distinct shadow of the grating to the required depth. A xenon short arc lamp of 2-4 KW would be ideal, but 500-W iodine lamps aligned along a straight line parallel to the lines of grating are chosen as a compromise between performance and economy.

The size of the pupil of the camera is another

important factor in increasing the range of good fringe contrast. It should be small enough to include within the depth of focus both the grating itself and its shadow cast on an object.

On the other hand, the shutter speed for a living body is preferably less than 1/8 sec. to avoid inevitable movement of the body. Combination of the limited aperture of the lens and a fast shutter speed results in an insufficient exposure. This is compensated for by adopting high speed film and hypersensitive development. Human skin is translucent, especially to red light and so the shadow of the grating is blurred by the diffusing of the light in the skin. This is the most unfavorable factor against obtaining high fringe contrast in case of a living body. The blurred shadow results in poor contrast of the contour moiré fringe.

If penetration of light is prevented by applying powder with good covering power on the skin, the fringe contrast is greatly improved. But in most cases, the application of powder on a subject is quite troublesome. So another technique to improve the fringe contrast on the natural skin is desired.

If the light penetrated into the skin is absorbed, the blur of the shadow due to diffusion will be less and will result in good fringe con-

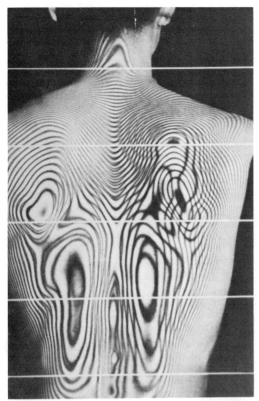

FIGURE 16-104. Subtractively superimposed two contour line systems with raised and lowered right arm.

FIGURE 16-105. Graphical drawn contour line system of equal depth difference.

trast. The bluish light is more readily absorbed in the skin than the reddish light. Therefore, the fringe contrast on natural skin is considerably improved under a bluish illumination.

In order to distinguish the light reflected at the skin surface from the light that has penetrated into the skin then and reflected back, the theory suggests the effectiveness of illuminating the object with appropriately polarized light while observing it through a matching analyzer. This is confirmed by experiment. But a combination of polarizer and analyzer that is effective on the human living body gives a drastic light loss which keeps the method from practical use.

A practical technique to improve the fringe contrast on a living body is to optimize the relation between width and clearance of shadow of the grating and to regulate the pitch of the grating and the obliquity of illumination. The combination of coarser pitch and more oblique illumination is preferred to finer pitch and less oblique illumination. Also, having the shadow width a little wider than the width of the illuminated line on the object is most favorable for improving the fringe contrast on the living skin.

After these factors were considered, an instrument large enough to measure a full size living body was constructed (Takasaki, 1974).

The grating is made by stretching nylon thread using two long screws as pitch guides. The tension on each thread is approximately 4 kg which summed up to 5 tons. A two-piece construction (the outer frame standing the tension and the inner frame holding the flatness) is adopted and proved to be very effective. Figure 16-106 shows the result. Camera distance is 160 cm, light source offset is 48 cm, pitch of the grating is 1.5 mm, and height difference between two successive fringes is approximately 5 mm. The light source is 3 × 0.5 KW iodine incandescent lamp, and the picture is taken using a 6 × 6 camera with f = 50 mm, F/11 lens. Exposure is 1/8 sec. for TRI × film rated ASA 1600.

The model exposes the natural skin on her arms and shoulders, wears white underwear on her body and thin colored hose on her legs. As is seen, the fringe contrast of contour moiré is good enough to trace on any part of the body.

A standard grid of 40 × 40 cm on the grating surface and a contour moiré on a V-shaped standard object are also pictured. The numbers on the standard object stand for the depth in cm of the corresponding marks from the grating surface. The shadow-free illumination is used to eliminate the alias moiré.

A middle-sized grating is seen beyond the model. This grating is a second moiré contouring system operated from the back of a subject at the

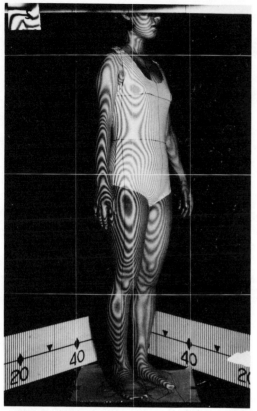

FIGURE 16-106. Contour moiré on a full size living body.

instant the picture from the front is taken. Thus the front and back relation of a living body can be determined.

16.8.4 CORRECTIONS OF THE CONTOUR

16.8.4.1 CORRECTIONS DUE TO THE ORDER OF A FRINGE

The contour moiré obtained by point source illumination is not equally spaced; the interval of depth from the grating increases with the order of the fringe. The contour is not the vertical projection of equal depth loci in contrast to the contour of a geographical map, but is the central projection about the point of the pupil of the camera to a plane of grating. Therefore a perspective correction must be applied.

The depth of the Nth contour moiré from the plane of the grating was given in equation (16.60) as

$$h_N = lN/(ds_0 - N).$$

The order of a fringe under consideration must first be determined. It is determined as follows.

A shadow of the central vertical line is seen on the bust and chest in figure 16-106. The hori-

zontal distance from a crossing point of the shadow and a contour moiré under consideration to the central vertical line is measured and translated to the plane of the grating. The situation is represented in figure 16-107, where the distance is shown as X_N. Using I, d, and X_N the depth h'_N of the contour moiré is obtained by simple geometrical relation as follows:

$$h'_N = l/(d/X_N - 1). \qquad (16.65)$$

The depth h'_N of the Nth fringe thus obtained may not be very accurate on account of the inaccuracy of determination of X_N. A series of fringe depths is tabulated by using a well calibrated instrument. A value close to the depth obtained from the shadow of a line can be picked from the table and the order of the fringe is determined. The order of the successive fringes is counted from this fringe. Thus the order of any fringe is determined, and the accurate depth h_N is obtained from the table.

16.8.4.2 PERSPECTIVE CORRECTION

Referring to figure 16-107, a point P on an object is seen at P' on the grating plane when it is seen through the lens L. Now what we measure from a picture is x'. To obtain a contour map of vertical projection, an actual coordinate x must be obtained from x', l, and h_p, which is the depth of the point from the grating. The same consideration is applied to the y coordinate, and the corrected coordinates are as follows:

$$x = x'(1 + h_p/l) \qquad (16.66)$$
$$y = y'(1 + h_p/l)$$

where the origin of the correction is the intersection point of the optical axis of the camera and the grating.

16.8.5 SUPPLEMENTAL TECHNIQUES FOR THE DEFINITION OF "HILLS" AND "VALLEYS"

One cannot judge from a single picture of contour moiré whether a part showing a concentric fringe is a hill or a valley. We could make this judgement if we had previous knowledge of the object. This sort of judgement cannot be expected with an unfamiliar object or in automatic processing of contours. There are several possible ways of making judgement that do not rely on previous knowledge.

(a) Using a second light source a very short distance from the camera, a coarse contour moiré is produced. This is added to the fine contour moiré produced by the first light source as shown in figure 16-108. If the coarse contour moiré fringe is broadened to cover the depth of interest, the hills will be bright centers and vice versa.

(b) The second method is to take a stereo picture pair of an object on which the contour moiré is

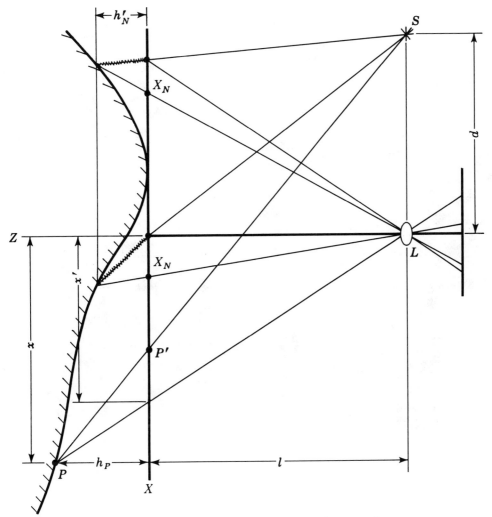

FIGURE 16-107. Order determination and perspective correction.

localized. A typical picture-pair is shown in figure 16-109. One can observe on the object not a meaningless random pattern, as is the case with a conventional stereo picture, but contour lines with stereoscopic view. There is a relative difference of position of contour lines of the two pictures due to parallax. There is a possibility of defining hills or valleys based on this parallax.

(c) The shadow of the reference thread that is running parallel to the thread of the grating (vertical in figure 16-106 and horizontal in figure 16-109) is an oblique section of the subject which enables one to make the judgement.

(d) If the camera is not situated on a plane including the light source and parallel to the grating, a set of moiré contours other than the equal depth contour is obtained. Two examples are presented in figure 16-110.

(e) An object like a living body cannot repeat exactly the same pose. So if an all-around measurement of a living body is required, at least

three contouring cameras must be used at the same moment from different angles. But this requires bulky instrumentation.

As a compromise, in the case of figure 16-106, two moiré contouring cameras, one placed in front and the other in back are shot simultaneously. The relation of the front and back contour pictures were determined automatically. But narrow regions at the sides cannot be contoured.

Another compromise is to take three-fourths of an all-around view of an object using one large grating and two wide angle cameras. The cameras are arranged in the front left and front right of an object situated just behind the grating. The optic axis of the camera is kept perpendicular to the grating. The grating is preferably the one with horizontal thread. If the two cameras are aligned in a horizontal line parallel to the

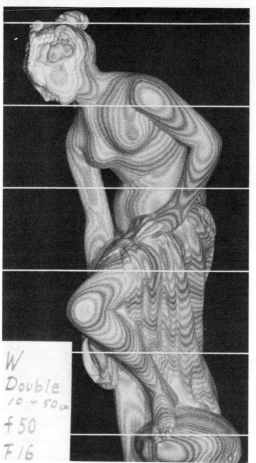

W
Double
10 ~ 50 cm
f 50
F 16

FIGURE 16-108. Picture of moiré contours obtained by using two light sources, one giving fine contour and the other course contour. The latter is added to the former. The result appears as if the fine contour is modulated by the coarse contour.

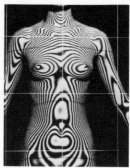

FIGURE 16-109. Stereo contour moiré pattern on a living body.

FIGURE 16-110. Two examples of moiré pattern that do not represent equal depth contour.

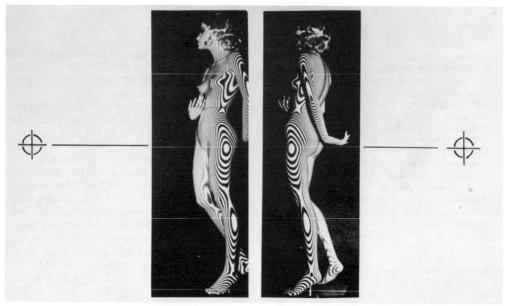

FIGURE 16-111. Example of taking a picture of three-fourths of an all-around view of the contour moiré by using two cameras and a grating. The two marks are origins for perspective correction.

thread of the grating and the light sources are aligned in another horizontal straight line with different height, the contour moiré on the object is the same to any one of the cameras, but with different perspective. Thus pictures of large perspective difference are obtained as shown in figure 16-111. The marks in the figure are origins for perspective correction for the two pictures, and three-fourths of an all-around view is measured using one grating and two cameras. (The pictures in figure 16-111 were taken separately using one camera.)

16.8.6 CONCLUSIONS

The accuracy of the Moiré technique has been proven to be within 0.2% as far as the depth is concerned. This high accuracy obtained by using rather crude instruments is the result of the simplicity of the theory and the fact that the grating itself interferes with its shadow to produce the contour moiré. The averaging effect of pitch errors caused by the motion of the grating also helps to improve the accuracy. The accuracy in the picture plane depends solely on the distortion of the camera lens.

Moiré topography has been successfully used in connection with manufacturing dummy bodies, and the method has proven to be of great potential.

16.9 BIBLIOGRAPHY

Abbreviations used:

ASCE: *American Society of Civil Engineers*
ACSM: *American Congress on Surveying and Mapping*
ASP: *American Society of Photogrammetry*
BSFP: *Bulletin de la Société Française de Photogrammétrie*
BSIFET: *Bulletino della Societá Italiana di Fotogrammetria e Topografia*
BuL: *Bildmessung und Luftbildwesen*
DGK: *Deutsche Geodätische Kommission*
ISP: *International Society for Photogrammetry*
PE: *Photogrammetric Engineering*
PE&RS: *Photogrammetric Engineering & Remote Sensing*
SPIE: *Society of Photo-Optical Instrumentation Engineers*
UI: *University of Illinois at Urbana-Champaign*

The following bibliography includes several entries not cited in the text. Entries are grouped in the following eight subdivisions:

16.9.1 CLOSE-RANGE AND TERRESTRIAL PHOTOGRAMMETRIC SYSTEMS
16.9.2 ARCHITECTURAL PHOTOGRAMMETRY
16.9.3 BIOSTEREOMETRICS
16.9.4 INDUSTRIAL PHOTOGRAMMETRY
16.9.5 UNDERWATER PHOTOGRAMMETRY
16.9.6 X-RAY PHOTOGRAMMETRY
16.9.7 SCANNING ELECTRON MICROSCOPY
16.9.8 HOLOGRAPHY AND MOIRÉ TOPOGRAPHY

9.1 CLOSE-RANGE AND TERRESTRIAL PHOTOGRAMMETRIC SYSTEMS

Abdel-Aziz, Y. I. and Karara, H. M., "Direct Linear Transformation from Comparator Coordinates into Object-Space Coordinates," *ASP Symposium on Close-Range Photogrammetry,* 1971.

Abdel-Aziz, Y. I., "Lens Distortion at Close Range," *PE, June 1973.*

Abdel-Aziz, Y. I., "On the Calibration of Close-Range Cameras," *ASP Proceedings,* March 1972.

Abdel-Aziz, Y. I., "Variation of Symmetrical Lens Distortion with Object Distance in Close-Range Photogrammetry," *ASP Proceedings, March 1973,* also *PE, June 1973.*

Abdel-Aziz, Y. I. and Karara, H. M., "Photogrammetric Potentials of Non-Metric Cameras," *Civil Engineering Studies, Photogrammetry Series No. 36, UI, 1974.*

Abdel-Aziz, Y. I., "Expected Accuracy and Convergent Photos," *PE, November 1974.*

Abdel-Aziz, Y. I., "Construction of Three-Dimensional Controls in Close-Range Photogrammetry," *ASP Proceedings,* March 1975.

Bopp, H. and Krauss, H., "A Simple and Rapidly Converging Orientation and Calibration Method for Non-Topographic Applications," *ASP Proceedings,* October 1977.

Brandenberger, A. J. and Erez, M. T., "Photogrammetric Determination of Displacements and Deformations in Large Engineering Structures," *The Canadian Surveyor, Vol. 26, No. 2, 1972.*

Brown, D. C., "Close-Range Camera Calibration," *PE, August 1971.*

Brown, D. C., "Calibration of Close-Range Cameras," *Invited Paper, 12th. ISP Congress (Comm. V), Ottawa, 1972.*

Carbonnell, M., "La Photogrammétrie Architecturale en 1972—Rapport du Comité International de Photogrammétrie Architecturale," *BSFP, No. 51, 1973.*

Döhler, M., "Nahbildmessung mit Nicht-Messkammern," *BuL, No. 1 & 2, 1971.*

Doyle, F. J., "Non Topographic Applications of Analytical Photogrammetry," *UI Photogrammetry Short Course, 1967.*

Ducloux, J., "Photogrammétrie à très courte distance. Etalonnage d' appareils de prise de vues," *BSFP, No. 48, 1972.*

Elassal, A. A., "Design of a General Data Reduction System for Analytical Photogrammetry," *Invited Paper, 12th ISP Congress (Comm. V)., Ottawa, 1972.*

Erlandson, J. P., Peterson, J. C., Veress, S. A., "The Modification and Use of the BC-4 Camera for Structural Deformations," *ASP Proceedings, Sept. 1974.*

Erlandson, J. P., Peterson, J. C., Veress, S. A., "Performance of Observations of Structural Deformations by Photogrammetric Methods," *Technical Reports, Parts I & II, U.S. Corps of Engineers, Seattle District, 1973.*

Erlandson, J. P. and Veress, S. A., "Contemporary Problems in Terrestrial Photogrammetry," *PE, Sept. 1974.*

Erlandson, J. P. and Veress, S. A., "Methodology and Standards for Structural Surveys," *ASP Symposium on Close-Range Photogrammetric Systems, 1975.*

Faig, W., "Single Camera Approaches in Close-Range Photogrammetry," *ASP Proceedings, March 1972.*

Faig, W., "Design, Construction, and Geodetic Coordination of a Close-Range Photogrammetric Test Field," *Civil Engineering Studies, Photogrammetry Series No. 32, UI, 1972.*

Faig, W. and Moniwa, H., "Convergent Photos for Close-Range," *PE, June 1973.*

Faig, W., "Precision Plotting of Non-Metric Photography," *ISP Symposium Biostereometrics' 74, 1974.*

Faig, W., "Photogrammetric Equipment Systems with Non-Metric Cameras," *ASP Symposium on Close-Range Photogrammetric Systems, 1975.*

Faig, W., "Calibration of Close-Range Photogrammetric Systems—Mathematical Formulation," *Invited Paper, Comm. V, 13th. ISP Congress, Helsinki, 1976. (PE&RS, Dec. 1975).*

Faig, W., "Photogrammetric Potentials of Non-Metric Cameras—Report of ISP Working Group V/2," *13th ISP Congress, Helsinki 1976. (PE&RS, Jan. 1976).*

Fry, J. M., "Methods of Graphical and Analytical Mensuration of Single Terrestiral Photographs," *Civil Engineering Studies, Photogrammetry Series No. 23, UI, 1969.*

Galileo-Santoni, "Industrial Photogrammetry," *Undated bulletin published by Leland Instruments, Ltd., London.*

Gracie, G., "Photogrammetric Measurement of a Spinning Network," *Proceedings of the ASP Symposium on Close-Range Photogrammetry, 1971.*

Hottier, P., "Contribution à l'Étude Expérimentale de la Precision de la Photogrammétrie Analytique à Courte Distance (7-12-m) dans le Cas du Couple," *Invited Paper, ISP 12th Congress (Comm. V), Ottawa, 1972.*

Hottier, P., "Nouvelle Contribution à l'Étude Expérimentale de la Photogrammétrie Analytique à Courte Distance (7-12 m) dans le Cas du Couple," *BSFP, No. 53, January 1974.*

Hottier, P., "Accuracy of Close-Range Analytical Restitutions: Practical Experiments and Prediction," *Invited Paper, Comm. V, ISP Congress, Helsinki 1976, (PE&RS, March 1976).*

Jacobi, O., "Kalibrieren gewöhnlicher Photoapparate und deren Verwendung als Messkammern," *BuL, June 1968.*

Jaksic, Z., "On-Line Computational Photogrammetry in Surveying and Mapping," *Proceedings of the XIVth FIG Congress, Commission 5, Washington, D.C., 1974.*

Jaksic, Z., "Analytical Instruments in Close-Range Photogrammetry," *ASP Symposium on Close-Range Photogrammetric Systems, 1975.*

Jenoptik Jena GmbH, "Aufnahmesysteme Universalmesskammer UMK 10/1318, Stereomesskammer SMK 5.5/0808," *Publication No. 14-404-1, 1973.*

Jenoptik Jena GmbH, "Industrial Photogrammetric Camera IMK 10/1318," *Publication No. 14-232a-2, 1972.*

Jenoptik Jena GmbH, "Phototheodolite Photheo 19/1318," *Brochure No. 14-3206-2, 1967.*

Karara, H. M., "Universal Stereometric System," *PE, November 1967.*

Karara, H. M., "On the Precision of Stereometric Cameras," *Invited paper, (Commission V), ISP 11th Congress, Lausanne, 1968.*

Karara, H. M., "New Trends in Close-Range Photogrammetry," *ASP Symposium on Close-Range Photogrammetry, 1971.*

Karara, H. M. and Marks, G. W., "Mono versus Stereo Analytical Photogrammetry," *Civil Engineering Studies, Photogrammetry Series No. 14, UI, 1968.*

Karara, H. M., "Simple Cameras for Close-Range Applications," *PE, May 1972.*

Karara, H. M. and Faig, W., "Interior Orientation in Close-Range Photogrammetry: An Analysis of Alternative Approaches," *Invited Paper, Commission V, ISP 12th Congress, Ottawa, 1972.*

Karara, H. M., "Recent Developments and Trends in Close-Range Photogrammetry," *Proceedings, First Pan American and Third National Congress of Photogrammetry, Photo-interpretation and Geodesy, Mexico City, 1974.*

Karara, H. M. and Abdel-Aziz, Y. I., "Accuracy Aspects of Non-Metric Imageries," *PE, September 1974.*

Karara, H. M., "Non-Topographic Photogrammetry, 1972-1976, *Report of ISP Comm. V, ISP 13th Congress, Helsinki, 1976, PE&RS, January 1976.*

Kenefick, J. F., "Ultra Precise Analytics," *PE, November 1971.*

Kenefick, J. F., Gyer, M. S. and Harp, B. F., "Analytical Self-Calibration," *PE, Nov. 1972.*

Kölbl, O., "Selbstkalibrierung von Aufnamekammern," *BuL, No. 1, 1972.*

Kölbl, O., "Metric or Non-Metric Camera," *Invited paper, Comm. V, 13th ISP Congress, Helsinki, 1976, (PE&RS, Jan. 1976).*

Kratky, V., "Digital Modeling of Limbs in Orthopaedics," *PE & RS, June 1975.*

Kratky, V., "Analytical Models from Parallel Beams of Rays," *Proceedings of the ASP/UI Symposium on Close-Range Photogrammetric Systems, 1975.*

Kratky, V., "Analytical On-Line Systems in Close-Range Photogrammetry," *Invited Paper, Comm. V, ISP Congress, Helsinki, 1976. (PE&RS, Jan. 1976).*

Kratky, V., "On-Line Analytics for Close Range Photogrammetry," *ASP Proceedings, Sept. 1976.*

Löfstrom, K. G. and Salmenpera, H., "Eine Analytische Methode mit Bundelausgleichung und ihr Einsatz in Sonderanwendungen der Photogrammetrie," *Invited paper, Comm. V, 12th. ISP Congress, Ottawa, 1972.*

Makarovic, B., "From Digital Components to Integrated Systems in Photogrammetry," *ITC Journal, No. 5, 1974.*

Malhotra, R. C. and Karara, H. M., "High Precision Stereometric Systems," *Civil Engineering Studies, Photogrammetry Series No. 28, UI, 1971.*

Malhotra, R. C. and Karara, H. M., "A Computational Procedure and Software for Establishing a Three-Dimensional Test Area for Close-Range Applications," *Proceedings, ASP Symposium on Close-Range Photogrammetric Systems, 1975.*

Marzan, G. T. and Karara, H. M., "A Computer Program for the Direct Linear Transformation Solution of the Colinearity Condition, and some Applications of it," *ASP Symposium on Close-Range Photogrammetric Systems, 1975.*

Marzan, G. T. and Karara, H. M., "Rational Design for Close-Range Photogrammetry," *Civil Engineering Studies, Photogrammetry Series No. 43, UI, 1976.*

Marzan, G. T., "Optimum Configuration of Data Acquisition in Close-Range Photogrammetry," *ASP Symposium on Close-Range Photogrammetric Systems, 1975.*

Masry, S. E. and Faig, W., "Utilization of the Analytical Plotter in Close-Range Applications," *ASP Symposium on Close-Range Photogrammetric Systems, 1975.*

Masry, S. E. and Faig, W., "The Analytical Plotter in Close-Range Applications," *PE&RS, January 1977.*

Merritt, E. L., "Procedure for Calibrating Telephoto Lenses," *ASP Symposium on Close-Range Photogrammetric Systems, 1975.*

Reynolds, J. D. and Dearinger, J. A., "Measuring Building Movement by Precise Survey," *Journal of Surveying and Mapping Division, ASCE, 1970.*

Richardson, J. T., "Measured Behaviour of Glen Canyon Dam," *Journal of Surveying and Mapping Division, ASCE, 1968.*

Robertson, K. D., "A Method for Reducing Refractive Index Errors in Length Measurement," *Technical Report, U.S. Army Engineer Topographic Laboratories.*

Roehm, L. H., "Deformation Measurement of Flaming Gorge Dam," *Journal of Surveying and Mapping Division, ASCE, 1968.*

Schwidefsky, K., "Precision Photogrammetry at Close-Ranges with Simple Cameras," *Photogrammetric Record, October 1970.*

Shernhorst, J. N., "Close-Range Instrumentation," *PE, April 1967.*

Schlienger, R., "Wild Cameras for Terrestrial and Close-Range Photogrammetry," *Presented Paper, Comm. V, ISP 12th. Congress, Ottawa, 1972.*

Schlienger, R., "Recent Improvements to the Terrestrial Cameras of Wild Heerbrugg," *ASP Symposium on Close-Range Photogrammetric Systems, 1975.*

Torlegård, A. K. I., "On the Determination of Interior Orientation of Close-up Cameras under Operational Conditions using Three-Dimensional Test Objects," *Dissertation, Stockholm, 1967.*

Torlegård, A. K. I., "State-of-the-Art of Close-Range Photogrammetry," *Invited Paper, Comm. V, ISP Congress, Helsinki, 1976. (PE&RS, Jan. 1976).*

Torrini, A. and Ferri, W., "Galileo-Santoni Equipment for Non-Topographic Applications of Photogrammetry," *Invited Paper, Comm. V, 10th ISP Congress, Lisbon, 1964.*

Trager, H., "The Zeiss Instrument System for Close-Range Photogrammetry," *ASP Symposium on Close-Range Photogrammetric Systems, 1975.*

Turpin, R. D., "Photogrammetric Techniques for Engineering Measurements," *PE, September, 1958.*

Turpin, R. D., "Testing of the Suitability of the Crown Graphic Camera for Photogrammetry," *PE, June 1960.*

Veress, S. A., "Adjustment by Least Squares," *ACSM Journal, 1974.*

Veress, S. A., and Tiwari, R. S., "Fixed-Frame Multiple-Camera System for Close-Range Photogrammetry," *PE&RS, September 1976.*

Veress, S. A. and Hatzopoulos, "A Plotting Instrument for Close-Range Photogrammetry," *PE&RS, March 1978.*

Voss, G., "Das Gerätesystem Industriephotogrammetrie aus Jena," *Vermessungstechnik, No. 7, 1971.*

Voss, G., "The Industrial Photogrammetry Instrument System of the Jena Optical Works," *Vermessungstechnik, No. 7, 1972.*

Voss, G., "The UMK 10/1318 Universal Photogrammetric Camera System," *Surveying News, No. 17, 1973.*

Voss, G., "The Stereometric Camera System SMK 55/0808 of the Jena Optical Work," *Jena Review, No. 3, 1974.*

Walker, J. W., Bowles, L. D., Yater, R. R., and Wilson, F. E., "A Unique Combined Geodetic and Photogrammetric High-Precision Survey," *ASP Symposium on Close-Range Photogrammetric Systems, 1975.*

van Wijk, M. C. and Ziemann, H., "The Use of Non-Metric Cameras in Monitoring High Speed Processes," *Invited Paper, Comm. V, ISP Congress, Helsinki, 1976. (PE&RS, Jan. 1976).*

Wolf, P. R. and Loomer, S. A., "Calibration of Non-Metric Cameras," *ASP Symposium on Close-Range Photogrammetric Systems, 1975.*

Wong, K. W., "Application of a Simultaneous Analytical Aerotriangulation Program (SAPGO-A) for Close-Range Problems," *ASP Symposium on Close-Range Photogrammetric Systems, 1975.*

Wong, K. W., "Digital Analysis Techniques," *ASP Symposium on Close Range Photogrammetric Systems, 1975.*

Wong, K. W., "Mathematical Formulation and Digital Analysis in Close-Range Photogrammetry," *Invited Paper, Comm. V, ISP Congress, Helsinki, 1976. (PE&RS, Nov. 1975).*

Ziemann, H., "Derivation of Spatial Coordinates From a 16 Movie," *ASP Proceedings, October 1972.*

9.2 ARCHITECTURAL PHOTOGRAMMETRY

Borchers, P. E., "Photogrammetric Recording of Cultural Resources," *National Park Service Publication No. 186, 1977.*

*Carbonnell, M., "L'histoire et la situation présente des applications de la photogrammétrie à l'architecture," *BSFP, No. 31, July 1968.*

*Carbonnell, M., "La photogrammétrie architecturale de 1968 à 1971, Rapport du Comité International de Photogrammétrie Architecturale," *BSFP, No. 45, January 1972.*

*Carbonnell, M., "La Photogrammétrie en 1972. Rapport du Comité International de Photogrammétrie Architecturale," *BSFP, No. 51, July 1973.*

*Carbonnell, M., "La Photogrammétrie architecturale en 1973 et 1974, Rapport du Comité International de Photogrammétrie Architecturale," *BSFP, No. 56, October 1974.*

*Carbonnell, M., "La Photogrammétrie Architecturale en 1975 et 1976, Rapport du Comité International de Photogrammétrie Architecturale," *BSFP, No. 67, July 1977.*

Comité International de Photogrammétrie Architecturale (CIPA), "Photogrammétrie des Monuments et des Sites," *A brochure published by ICOMOS, 1972.*

9.3 BIOSTEREOMETRICS

Currie, G. D., Leonard, C. D., Martonyi, "Photogrammetric Measurement of the Human Optic Cup," *PE&RS, June 1976.*

Herron, R. E., "Stereophotogrammetry in Biology and Medicine," *Invited Paper, Comm. V, 12th. ISP Congress, Ottawa, 1972.* (This paper contains an extensive bibliography on Biostereometrics of several hundred entries.)

Herron, R. E. "Biostereometric Measurement of Body Form" in *Yearbook of Physical Anthropology 1972,* American Association of Physical Anthropology, Vol. 16 1973, Washington, D.C.

Herron, R. E. and Karara, H. M. (Editors): *Biostereometrics '74,* Proceedings of the Symposium of Commission V, International Society for Photogrammetry, Sept. 10-13, 1974. American Society of Photogrammetry, Washington, D.C. 1974. (These proceedings contain fifty (50) papers on various aspects of Biostereometrics).

Herron, R. E., "Biostereometrics '74—A Report," *PE&RS, Jan. 1976.*

Renner, W. D., "The Photogrammetric Technique for Use in Radiation Therapy," *PE&RS, May 1977.*

Takamato, T., Schwartz, B., Marzan, G. T. "Op-

---

* These five publications of Mr. Carbonnell contain a total of 849 entries from throughout the World.

timum Conditions for Measurement of the Optic Disc by the Donaldson Retinal Camera," *ASP Proceedings, March 1978.*

Veress, S. A., Lippert, F. G., Takamato, T., "An Analytical Approach to X-Ray Photogrammetry," *PE&RS, Dec. 1977.*

9.4 INDUSTRIAL PHOTOGRAMMETRY

Allam, M. M., "Spatial Rock Face Deformations and Orientations of Geological Discontinuities in Open-Pit Mines Using Analytical Photogrammetry," *Proceedings, ASP Symposium on Close-Range Photogrammetric Systems, 1975.*

Argyris, J. H., Archer, W., Eberle, K. and Kirschstein, M., "Measuring of Three-Dimensional Deformations via Photogrammetric Methods and Electronic Data Processing," *Institut für Statik and Dynamic, Stuttgart, ISD Report No. 118, 1972.*

Atkinson, K. B., "A Review of Close-Range Photogrammetry," *Invited Paper, Comm. V, ISP Congress, Helsinki, 1976. (PE&RS, Jan. 1976).*

Bednarski, T. and Majde, A., "The Stereophotogrammetric Measurement of Displacements with Film Camera, in a Process of Explosive Forming," *BSFP, No. 42, 1971.*

Bendix Corporation (1972): *see Higgins & Alice (1972).*

Berling, D. W., "Use of Photogrammetry in Collision Investigation," *paper presented at the Collision Investigation Methodology Symposium, 1970. (K&E Publication S 51-542 e).*

Berling, D. W., "New Instruments and Methods for Measuring Static and Dynamic Processes in the Test Laboratory and Production Control," *paper presented at the 15th Survey Congress of the Institution of Surveyors, Australia, 1972.*

Bernini, F., Cunietti, M. and Galetto, R., "A Photogrammetric Method for Assessing the Displacements under Stress of Large Structure Models—Experimental Applications," *BSIFET, No. 3, 1968.*

Borchers, P. E., "The Photogrammetric Study of Structural Movements in Architecture," *PE, September 1964.*

Bopp, H., Krauss, H., Preuss, H. D., "Photogrammetric Control Survey of a Large Cooling Tower," *ASP Proceedings, October 1977.*

Bradner, H., "Precision Measurements of Bubble Chamber Film" *PE, June 1959.*

Brandenberger, A. J. and Erez, M. T., "Photogrammetric Determination of Displacements and Deformations in Large Engineering Structures," *The Canadian Surveyor, June 1972.*

Brandenberger, A. J., "Ship Collision Cases," *PE, January 1973.*

Brandenberger, A. J., "Deformation Measurements of Power Dams," *PE, September 1974.*

Brandow, V. D., Karara, H. M., Damberger, H. H., Krausse, H.-F., "Close-Range Photogrammetry for Mapping Geologic Structures in Mines," *Proceedings, ASP Symposium on Close-Range Photogrammetric Systems, 1975.*

Brandow, V. D., Karara, H. M., Damberger, H. H., Krausse, H.-F., "A Non-Metric Close-Range Photogrammetric System for Mapping Geologic Structures in Mines," *PE&RS, May 1976.*

Brewer, R. K., "Close-Range Photogrammetry—A Useful Tool in Motion Study," *PE, September 1962.*

Brinks, J. S., Clark, R. T., Kieffer, N. M. and Urick, J. J., "Predicting Wholesale Cuts of Beef from Linear Measurements obtained by Photogrammetry," *Journal of Animal Science, Vol. 13, 1971.*

Brock, R. H., Wasil, B. A. and Mandel, J. A., "Three

Station Analytical Photogrammetry for Stress Analysis of a Plate with Large Displacements," *Proceedings, ASP Symposium on Close-Range Photogrammetry, 1971.*

Brown, D. C., "Precise Calibration of Surfaces of Large Radio Reflectors by Means of Analytical Photogrammetry Triangulation," *Instrument Corporation of Florida, Research and Analysis Report No. 10, 1962.*

Brown, D. C., Kenefick, J. F. and Harp, B. F., "Photogrammetric Measurements of Explosive Bolts on the Canopy of the OAO Launch Vehicle," *DBA Systems, Inc., Photogrammetric Structural Measurements, No 31, 1971.*

Chrzanowski, A. and El Masry, S., "Tunnel Profiling Using a Polaroid Camera," *Canadian Mining and Metallurgy Bulletin, Vol. 62, 1969.*

Collins, T. L., "Inventories of Raw and Bulk Materials," *PE, July 1963.*

Cotovanu, E., "Messung von Brücken durch terrestrische Photogrammetrie," *Romanian Committee of Photogrammetry, Bulletin de Fotogrammetrie, Special Issue, 1972.*

Dach, H. J., "Terrestrial Photogrammetry Solves Special Tasks in Railway Engineering Surveys," *Jena Review, No. 6, 1973.*

Erlandson, J. P., Peterson, J. C., and Veress, S. A., "The Modification and Use of the BC-4 Camera for Measurements of Structural Deformations," *ASP Proceedings, March 1974.*

Erlandson, J. P. and Veress, S. A., "Monitoring Deformations of Structures," *PE&RS, Nov. 1975.*

Faig, W., "Vermessung dünner Seifenlamellen mit Hilfe der Nahbereichsphotogrammetrie, *DGK, Reihe C, Heft No. 144, 1969.*

Faig, W., "Shapes of Thin Soap Membranes," *PE, October 1971.*

Faig, W., "Photogrammetry and Hydraulic Surfaces," *Journal of the Surveying and Mapping Division, ASCE, No. SU2, 1972.*

Farmer, L. D. and Robe, R. Q., "Photogrammetric Determination of Iceberg Volumes," *PE&RS, February 1977.*

Flotron, A., "Photogrammetrische Messung von Gletscherbewegungen mit automatischer Kamera," *Schweizerische Fachblatt für Vermessung, Photogrammetrie, Kulturtechnik, No. 1, 1971.*

Foramitti, H., "La Photogrammétrie au Service de la Construction et de l'Architecture," *Deutsche Bauzeitung, No. 9 & 10, 1966.*

Ford Motor Company, "Photogrammetric Surface Measurement, Preliminary Test Program," *Report NCS-810-FR, 1964.*

Fulscher, P., "Ein neuer Profilmesswagen der SBB," *Schweiz. Zeitscrift für Vermessungswesen, Kulturtechnik und Photogrammetrie, May 1963.*

Garfield, J. F., "The Photogrammetry of the Tracks of Elementary Particles in Bubble Chambers," *PE, September 1964.*

Gatu, A., "Photogrammetric Measurement Accuracy of Wall Pillar Cracks in Rock Salt Minings," *Romanian Committee of Photogrammetry, Bulletin de Fotogrammétrie, Special Issue, 1972.*

Gracie, G., "Photogrammetric Measurement of a Spinning Network," *Proceedings of the ASP Symposium on Close-Range Photogrammetry, 1971.*

Gregor, K. N., "Analytical photogrammetry and numerical control in wind-tunnel model manufacture," *South African Journal of Photogrammetry, No. 1, 1973.*

Grumpelt, H., "Ein analytische Verfahren der Bündelverknüpfung in der Nahbildmessung, angewandt auf Deformationsmessungen an einer Tunnelröhre im Bergbaugebiet," *BuL, No. 6, 1973.*

Hallert, B., "Deformation Measurements by Photogrammetric Methods," *PE, December 1954.*

Higgins, P. T., Alice, M. B. and McGivern, R. F., "Industrial Applications of Photogrammetry," *The Bendix Technical Journal, Spring 1972.*

Hottier, P., "Analytical Photogrammetry with Homolog Image Curves," *Presented Paper, Commission V, 13th. ISP Congress, Helsinki, 1976.*

Jaensch, J. R., "Photogrammetric Checking of Structures," *Jena Review, No. 3, 1970.*

Karara, H. M., "Industrial Photogrammetry," *Proceedings, ASP Symposium on Close-Range Photogrammetric Systems, 1975.*

Kenefick, J. F., "Photogrammetric Measurement of Antenna Reflectors," *paper presented at the meeting of the International Union of Radio Science, 1972.*

Kenefick, J. F., "Applications of Photogrammetry in Shipbuilding," *PE&RS, September 1977.*

Kiyoshi, K., Nakamura, K., Yamakami, M. and Tenake, H., "Determination of both Wind Velocity and Direction through Synchronous Stereo-Photographs Taken of Smoke Trail," *Proceedings of the International Symposium of Photogrammetry (ISP Comm. V), Tokyo, 1966. (Special Volume No. 1, 1966, Journal of the Japan Society of Photogrammetry).*

Kloss, W., "Uber die Einfuhrung der Photogrammetrie in Bergbau," *Vermessungstechnik, No. 7, 1963.*

Knödler, G. and Kupke, H., "The Use of Industrial Photogrammetry in the Production of Ship's Screws," *Jena Review, No. 3, 1974.*

Kobelin, J., "Mapping Street Intersections for Traffic and Signal Design using a Close-Range Photogrammetric System," *Proceedings, ASP Symposium on Close-Range Photogrammetric Systems, 1975. (PE&RS, Aug. 1976).*

Konecny, G., "Application of the Wild C12 Stereometric Camera to Structural Engineering," *ASP Proceedings, March 1964.*

Kratky, V. and van Wijk, M. C., "Photogrammetry Used to Determine the Motion of a Vehicle Crashing into a Highway Barrier," *BSFP, No. 42, April 1971.*

Leydolph, W. K., "Stereophotogrammetry in Animal Husbandry," *PE, December 1954.*

Linkwitz, K. and Preuss, H. D., "Die photogrammetrische Vermessung der Modelle der olympischen Dächer München," *BuL, 1971, No. 4.*

Loomer, S. A. and Wolf, P. R., "Terrestrial Photogrammetric Measurements of Surface Water Velocities," *ASP Proceedings, September 1974.*

Mann, R. W., "Stereophotogrammetry Applied to Hydraulic Analogue Studies of Unsteady Gas Flow," *PE, September 1962.*

Marks, G. W., "Geometric Calibration of Antennae by Photogrammetry," *PE, July 1963.*

Maruyasu, T. and Oshima, T., "Short Range Photogrammetry of Objects in Motion," *Invited paper ISP Comm. V, Lausanne Congress, 1968.*

Meyer, R., "Eine Austrüstung für die stereophotogrammetrische Aufnahme bewegter Objekte," *Vermessungstechnik, No. 8, 1964.*

Meyer, R. and Will, G., "Photogrammetrische Bestimmung der Strömungsgeschwindigkeit im rotierenden Laufrad einer Kreiselpumpe," *BSFP, No. 42, 1971.*

Meyer, R., "The Present State in Industrial Photogrammetry," *Surveying News, No. 17, 1973.*

Moffitt, F. H., "Photogrammetric Definition of A Wave Survace," *Hydraulic Engineering Laboratory Bulletin HEL 12-3, University of California at Berkeley, 1966.*

Moffitt, F. H., "Wave Surface Configuration," *PE, February 1968.*

Nauk, W. and Voss, G., "Photogrammetry in Computer-Aided Design and Numerically-Controlled Production," *Surveying News, No. 25 1972.*

Negut, N., Radulescu, D. and Savulescu, C., "A Photogrammetric Process in Special Hydrotechnical Works," *Romanian Committee of Photogrammetry, Bulletin de Fotogrammetrie, Special Issue, 1972.*

Newton, I., "Close-Range Photogrammetry as an Aid to Measurement of Marine Structures," *Invited Paper Comm. V, ISP Congress, Helsinki, 1976. (PE&RS, Dec. 1975).*

Ockert, D. L., "A Photogrammetric Radio Telescope Calibration," *PE, June 1959.*

Ohman, L. H. and van Wijk, M. C., "A Photogrammetric Method of Determining the Mode Shapes of Vibrating Objects, and Results for a Thin Wing," *National Research Council of Canada, Report No. 7210, 1962.*

Oshima, T., "Recent Developments of Industrial Photogrammetry in Japan," *PE&RS, March 1976.*

Pietschner, J., "Die Ermittlung der Strömungsverhältnisse in bodennahen Luftschichten durch terrestischstereophotogrammetrische Vermessung von Rauchmarkierungen," *Vermessungstechnik, 1964, No. 4.*

Preuss, H. D., "Analytical Methods in Special Applications of Photogrammetry—Practical Experience from Model Measurements for the Roofs of the Olympic Sport Facilities," *Invited paper, ISP Comm. V, Ottawa Congress, 1972.*

Reddy, K. V. S., van Wijk, M. C., and Pel, D. C. T., "Stereophotogrammetry in Particle-Flow," *Canadian Journal of Chemical Engineering, February 1969.*

Regensburger, K., "Photogrammetric Measurements of Deformations and Motions in the Near Range," *Jena Review, No. 3, 1972.*

Ross-Brown, D. M., and Atkinson, K. B., "Terrestrial Photogrammetry in Open-Pits: 1-Description and Use of the Phototheodolite in Mine Surveying," *Inst. Mining & Metallurgy, London, Sec. A, Vol. 81, 1972.*

Ross-Brown, D. M., Wickens, E. H. and Markland, J. T., "Terrestrial Photogrammetry in Open-Pits: 2-An Aid to Geological Mapping," *Inst. Mining & Metallurgy, London, Sec. A, Vol. 82, 1973.*

Sabey, B. E. and Lupton, G. N., "Measurement of Road Surface Texture Using Photogrammetry," *Crowthorne Road Research Laboratory, Report LR57, 1967.*

Santoni, E., "Survey of Automobile-Body Models and Car Parts Photographed from Various Distance," *Proceedings of the Symposium of ISP Comm. V, Tokyo, 1966.*

Schmid, W., "Die Photogrammetrie im wasserbaulichen Versuchswesen," *Schweizer. Zeitschrift für Vermessung, Photogrammetrie, Kulturtechnik, No. 4, 1973.*

Scholze, F., Töppler, J., Voss, G., "Application of Photogrammetry in Motor Vehicle Construction," *Jena Review, No. 2, 1975.*

Shmutter, B. and Etrog, U., "Calibration of Storage Tanks," *PE, March 1971.*

Siefert, W., "Measuring the Trajectories and Velocities of Table-Tennis Balls Photogrammetrically," *Jena Review, No. 3, 1970.*

Torlegård, A. K. I. and Dauphin, E. L., "Deformation Measurement by Photogrammetry in Cut and Fill Mining," *Proceedings, ASP Symposium on Close-Range Photogrammetric Systems, 1975.*

Vlcek, J., "Instrument for Measurement of the Shaft Perpendicularity and Deformations," *Presented paper, ISP Comm. V, Ottawa Congress, 1972.*

Waddell, J. H., "Photogrammetry and the Photography of Motion," *PE, April 1956.*

Wiedenhoft, E., "Bauwerkskontrollmessungen durch Anwendung fotogrammetrischer Verfahren," *Signal und Schiene, No. 12, 1968.*

van Wijk, M. C. and Pinkney, H. F. L., "A Single Camera Method for the 6-Degree of Freedom Sprung Mass Response of Vehicles Redirected by Cable Barriers," *Paper presented at SPIE Seminar-in-Depth on Optical Instrumentation—A Problem Solving Tool in Automotive Safety Engineering and Bio-Mechanics', 1972.*

van Wijk, M. C. and Ziemann, H., "The Use of Non-Metric Cameras in Monitoring High Speed Processes," *Invited Paper, Comm. V, ISP Congress, Helsinki, 1976. (PE&RS, Jan. 1976).*

Veress, S. A. and Sun, L. L., "Photogrammetric Monitoring of a Gabion Wall," *PE&RS, Feb. 1978.*

Wolvin, J. H., "Quality Control Through Measurement by Photogrammetry," *Chicago Aerial Industries, Inc. TD-126, 1966.*

Wong, K. W. and Vonderohe, A. P., "Photogrammetric Measurement of Displacements Around Tunnels in Sandy Soils," *ASP Proceedings, March 1978.*

Woropajew, E., "Terrestrial Photogrammetry and Automatic Data Processing in Open-Cast Mine Surveying," *Jena Review, No. 6, 1973.*

Yu, W. W., "Photogrammetric Measurements in Structural Research," *PE, September 1959.*

Young, H. E., "Photogrammetric Determination of Huge Pulpwood Piles," *PE, December 1954 and March 1955.*

Zeller, M., "Stereophotogrammetry and Studies of Movements," *PE, September 1953.*

9.5 UNDERWATER PHOTOGRAMMETRY

Adekoya, O. L., "Application of Two Media Photogrammetry in Coastal Mapping," *M.Sc.E. Thesis, Dept. of Surveying Engineering, University of New Brunswick, 1973.*

Anderson, N. M., "Color and Infrared Photography as a Tool for Hydrographers," *Proceedings, 10th Annual Canadian Hydrographic Conference, 1971.*

Bassage, L. H., "Films for Underwater Photography," *PE, Oct. 1972.*

Deep Sea Photography. *The John Hopkins Press, Baltimore 1967.*

Harford, J. W., "Underwater Lighting Advancements," *PE, Oct. 1972.*

Höhle, J., "Zur Theorie und Praxis der Unterwasser Photogrammetrie," *Deutsche Geodätische Kommission, Reihe C, Heft 163, 1971. (a).*

Höhle, J., "Reconstruction of the Underwater Object," *PE, Sept. 1971 (b).*

Höhle, J., "Methoden und Instrumente der Mehrmedien-Photogrammetrie," *Invited Paper, Comm. II, 12th ISP Congress, Ottawa, 1972.*

Hopkins, R. E. and Edgerton, H. E., "Lenses for Underwater Photography," *Contribution No. 1196 from the Woods Hole Oceanographic Institution, 1961.*

IES Lighting Handbook, Section 26 "Underwater Lighting," *Illumination Engineering Society, 1972.*

Ivanoff, A., *et al.*, "Optical System for Distortionless Underwater Vision," *U.S. Patent 2, 730 014, 1956.*

Kreiling, W., "Einfache Auswertung von Zweimedien-Bildpaaren in Doppelprojektoren," *BuL, Nov. 1970.*

Masry, S. E., "Measurement of Water Depth by the Analytical Plotter," *International Hydrographic Review, 1975.*

Masry, S. E., Konecny, G., "New Programs for the Analytical Plotter," *PE, Dec. 1970.*

McNeil, G. T., "Underwater Photography," *PE, Nov. 1969.*

McNeil, G. T., "Underwater Photography," *PE, Oct. 1972.*

McNeil, G. T., "Optical Fundamentals of Underwater Photography," *Mitchell Camera Corp., 2nd Edition, 1972.*

McNeil, G. T., et al., "Underwater Photography," *Invited Paper, Comm. I., 12th ISP Congress, Ottawa, 1972.*

Meijer, J. G., "Formula for Conversion of Stereoscopically Observed Apparent Depth of Water to True Depths, Numerical Examples and Discussion," *PE, Nov. 1964.*

Okamoto, A., Höhle, J., "Allgemeines analytisches Orientierungsverfahren in der Zwei- und Mehrmedien-Photogrammetrie und seine Erprobung," *BuL, Feb. & May, 1972.*

Phillips, S., "Films for Underwater Photos," *PE, Sept. 1971.*

Pollio, J., "Application of Underwater Photogrammetry," *Marine Technology Society Journal, January 1969.*

Pollio, J., "Underwater Mapping with Photography and Sonar," *PE, Sept. 1971.*

Pollio, J., "Remote Underwater Systems on Towed Vehicles," *PE, Oct. 1972.*

Rebikoff, D., "Underwater Photogrammetry in 1975," *Proceedings, ASP Symposium on Close-Range Photogrammetric Systems, 1975.*

Rinner, K., "Problems of Two Medium-Photogrammetry," *PE, March 1969.*

Rixton, F. H., "Lighting for Underwater Photos," *PE, Sept. 1971.*

Rosencrantz, D. M., "Underwater Photography Systems," *PE, Sept. 1971.*

Seiffert, V. A., "Underwater Cameras, Lenses and Housings," *PE, Sept. 1971.*

Schmutter, B., Bonfiglioli, L., "Problems of Two Medium Photogrammetry," *Haifa. 1965.*

Twinkel, G. C., "Water Depths from Aerial Photographs," *PE, Nov. 1963.*

Torlegård, A. K. I., Lundlav, T. L., "Underwater Analytical System," *PE, March 1974.*

Van Wijk, M. C., Discussion Paper to "Water Depths from Aerial Photographs." *PE, July 1964.*

Welsh, J. J., "Bibliography on UW-Photography," *PE, Sept. 1971.*

Welsh, J. J., "Underwater Photography and Photogrammetry Bibliography," *PE, Oct. 1972.*

Welsh, J. J., "Underwater Photography and Photogrammetry Bibliography," *Eastman Kodak Publications, No. P-124, (1968) and P-124A, (1971).*

Williamson, J. R., "Analytical Reduction of Underwater Photography," *M.S. Thesis in Civil Engineering, University of Illinois, 1972.*

Zaar, K., "Zweimedienphotogrammetrie," *Öster-reichische Zeitschrift für Vermessungswesen*, 1948.

9.6 X-RAY PHOTOGRAMMETRY

Agnoletto, E., and Pagani, M., "X-Ray Photogrammetry with Special Reference to the Treatment of Uterine Cancer," *ISP Symposium Biostereometrics '74*, 1974.

Baker, P., Schraer, H., and Yalman, R., "The Accuracy of Human Bone Composition Determination from Roetgenograms," *PE, June 1959*.

Baumrind, S., "A System for Craniofacial Mapping through the Integration of Data from Stereo X-Ray Films and Stereo Photographs," *ASP Symposium on Close-Range Photogrammetric Systems*, 1975.

Baumrind, S., and Frantz, R., "The Reliability of Head Film Measurements: 1 Landmark Identification," *Am. Journal of Orthodontics, Vol. 60, 1971*.

Eastman Kodak Co., "*The Fundamentals of Radiography*," Rochester, N.Y., 1960.

Hagberg, S., "Roentgen Stereophotogrammetry in Studies of Liver Volume Variations in the Dog under the Effect of Hoemorrhagic Shock and Hypothermia," *Acta Chirurgica Scand. Suppl. Vol. 297, 1961*.

Hallert, B., "Determination of the Interior Orientation of Cameras for Non-Topographic Photogrammetry, Microscopes, X-Ray Instruments, and Television Images," *PE, December 1960*.

Hallert, B., "*X-Ray Photogrammetry*," Elsivier Publishing Co., New York, 1970.

Herron, R., "Biostereometric Measurement of Body Form," *Yearbook of Physical Anthropometry, Vol. 16, 1972*.

Jonason, C. and Hindmarch, J., "Stereo X-Ray Photogrammetry as a Tool in Studying Scoliosis," *ASP Symposium on Close-Range Photogrammetric Systems*, 1975.

Kratky, V., "Analytical X-Ray Photogrammetry in Scoliosis," *ASP Symposium on Close-Range Photogrammetric Systems*, 1975.

Lippert, F., Veress, S. A., Takamato, T., Spolek, G., "Experimental Studies on Patellar Motion Using X-Ray Photogrammetry," *ASP Symposium on Close-Range Photogrammetric Systems*, 1975.

McNeil, G. T., "X-Ray Stereo Photogrammetry," *PE, November 1966*.

Moffitt, F. H., "Stereo X-Ray Photogrammetry Applied to Orthodontic Measurements," *Invited Paper, Commission V, ISP 12th Congress, Ottawa, 1972*.

Moffitt, F. H., and Nasu, M., "Correction of Subject Movement in X-Ray Photogrammetry," *Unpublished Report, 1973*.

Schernhorst, J. N., "Medical Applications in Europe," *PE, May 1966*.

Schernhorst, J. N., "A Feasibility Study of Medical X-Ray Stereophotogrammetry with Second Order Plotters," *ASP Proceedings, September 1969*.

Seeger, E. and Arnu, M., "The StR1-3 Analytical System for Stereometric Evaluation of X-Ray Photographs," *ISP Symposium Biostereometrics '74*, 1974.

Singh, R. S., "Radiographic Measurements," *PE, November 1970*.

Singh, R. S., "Welding Defects from Stereoradiographs," *PE, December 1971*.

Suh, C. S., "The Fundamentals of Computer-Aided

X-Ray Analysis of the Spine," *Journ. Biomechanics, Vol. 7, 1974*.

Veress, S. A., Lippert, F. G., Takamato, T., "An Analytical Approach to X-Ray Photogrammetry," *PE&RS, Dec. 1977*.

Veress, S. A. and Lippert, F. G., "A Laboratory and Practical Application of X-Ray Photogrammetry," *ASP Proceedings, March 1978*.

9.7 SCANNING ELECTRON MICROSCOPY

Black, Temple J. "SEM, Scanning Electron Microscope," *Photographic Applications in Science, Technology and Medicine, March 1970*.

Boyde, A., Ross, H. F., and Bucknall, W. B., "Plotting Instruments for Use with Images Produced by Scanning Electron Microscopes," *Proceedings of the Biostereometrics' 74 Symposium, 1974*.

Ghosh, Sanjib K., "Volume Determination With An Electron Microscope," *PE, February 1971*.

Ghosh, Sanjib K., "Photogrammetric Calibration of A Scanning Electron Microscope," *Proceedings of the Biostereometric '74 Symposium, 1974*.

Howell, P. G. T. and Boyde, A., "Comparison of Various Methods for Reducing Measurements from Stereo-pair Scanning Electron Micrographs to Real 3-D Data," *Proceedings of SEM Symposium, Chicago, 1972*.

Klemperer, O., and Barnett, M. E., "Electron Optics," *3rd. Edition, Cambridge University Press, 1971*.

Maune, David F., "Photogrammetric Self-Calibration of a Scanning Electron Microscope," *Ph.D. dissertation, Ohio State University, 1973*.

Maune, D. F., "Photogrammetric Self-Calibration of Scanning Electron Microscopes," *PE&RS, September 1976*.

Nagaraja, Hebbur N., "Application Studies of Scanning Electron Microscope Photographs for Micromeasurements and Three Dimensional Mapping," *Ph.D. dissertation, Ohio State University, 1974*.

Oatley, C. W., "The Scanning Electron Microscope-Part 1-The Instrument," *Cambridge University Press, 1972*.

Oshima, T., Kimoto, S., and Suganuma, T., "Stereomicrography with a Scanning Electron Microscope," *PE, August 1970*.

Wood, R., "The Modification of a Topographic Plotter and its Application in the Three-Dimensional Plotting of Stereomicrographs," *Photogrammetric Record, October 1972*.

9.8 HOLOGRAPHY AND MOIRÉ TOPOGRAPHY

Agnard, J. P., "Hologrammétrie: Tolerence de l'Orientation Relative des Hologrammes pour Fines de Mesures," *M.S. Thesis, Laval University, Quebec, 1970*.

Agnard, J. P., "Hologrammétrie de Haute Precision," *Presented paper, Comm. V, 12th. ISP Congress, Ottawa, 1972*.

Balasubramanian, N., and Leighty, R. D. (Editors), "*Coherent Optics in Mapping*," *Proceedings, SPIE, Vol. 45, 1974*.

Balasubramanian, N., "Image Coincidence Detection Using Optical Correlation Technology," *Proceedings, SPIE Seminar, Vol. 45, 1974*.

Balasubramanian, N., "Coherent Optics in Photogrammetry," *tutorial notes, SPIE Seminar, 1974*.

Balasubramanian, N., "Comparison of Optical Contouring Methods," *Invited Paper, Comm. V, ISP Congress, Helsinki, 1976. (PE&RS, Jan. 1976).*

von Berkefeldt, P., "Versuche zur Herstellung, Rekonstrunktion und Anmessung von Hologrammen," *Diss., Hannover, 1970.*

Free, R. V., "Spinal Analysis Utilizing Moiré Topography," *Paper presented at the Biostereometrics '74 Symposium, 1974.*

Gates, J. W. C., "Three-Dimensional Location and Measurement by Coherent Optical Methods," *Invited Paper, Comm. V, ISP Congress, Helsinki, 1976. (PE&RS, Nov. 1975).*

Gifford, D. L. and Mikhail, E. M., "Some Close-Range Mensuration Techniques," *ASP Proceedings, October 1973.*

Gifford, D. L. and Mikhail, E. M., "Study of the Characteristics of the Holographic Stereomodel for Application in Mensuration and Mapping," *Final Technical Report, Part I, ETL-CR-73-14, 1973.*

Gifford, D. L., "Perceived Holographic Image Distortion," *Presented paper, SPIE Seminar, March 1974.*

Glaser, G. H. and Mikhail, E. M., "Study of Potential Applications of Holographic Techniques to Mapping," *Interim Technical Report, ETL-CR-70-8, 1970.*

Hildebrand, B. P. and Haines, K. A., "Multiple-wavelength and Multiple-source Holography Applied to Contour Generation," *Journ. of Opt. Soc. of America, Vol. 51, 1967.*

Kellie, T. F. and Stevenson, W. S., "Experimental Techniques in Real Time Holographic Interferometry," *Optical Engineering, February 1973.*

Kurtz, M. K., Balasubramanian, N., Mikhail, E. M., Stevenson, W. H., "Study of the Application of Holographic Techniques to Mapping," *Final Technical Report, ETL-CR-71-17, 1971.*

Leighty, R. D., "A Short Tutorial on Coherent Optics," *ASP Proceedings, March 1975.*

Leighty, R. D., "Potential Applications of Coherent Optics in Close-Range Photogrammetric Systems," *Proceedings of the ASP Symposium on Close-Range Photogrammetric Systems, 1975.*

Meadows, D. M., Johnson, W. O. and Allen, J. B., "Generation of Surface Contours by Moiré Patterns," *Applied Optics, No. 9, 1970.*

Mikhail, E. M. and Glaser, G. H., "Mensuration Aspects of Holograms," *PE, March 1971.*

Mikhail, E. M., "Hologrammetry: Concepts and Applications," *PE, December 1974.*

Mikhail, E. M., "Potential Automation in Hologrammetry," *ASP Proceedings, March 1975.*

Takasaki, H., "Moiré Topography," *Applied Optics, 1970.*

Takasaki, H., "Moiré Topography," *Proceedings of the ISP Biostereometrics '74 Symposium, 1974.*

Takasaki, H., "Simultaneous all-around Measurement of a Living Body by Moiré Topography," *Invited paper, Comm. V, 13th. ISP Congress, Helsinki, 1976. (PE&RS, Dec. 1974).*

Trustura, T. and Itoh, Y., "Interferometric Generation of Contour Lines in Objects," *Optical Communications, No. 1, 1969.*

CHAPTER XVII

Satellite Photogrammetry

Author-Editor: DONALD L. LIGHT
Contributing Authors: DUANE BROWN, A. P. COLVOCORESSES, FREDERICK J. DOYLE,
MERTON DAVIES, ATEF ELLASAL, JOHN L. JUNKINS, J. R. MANENT,
AUSTIN MCKENNEY, RONALD UNDREJKA, GEORGE WOOD

17.1 Introduction

The space age, born in the spirit of international competition of the late 50's, has matured, resulting in numerous useful contributions being made to Man's knowledge of the earth, the moon, the planets and the solar system. From the beginning of this space age, photogrammetrists have marveled at the tremendous possibilities offered by the view from orbital altitudes. This chapter explores those possibilities as offered by satellite photogrammetry and, more specifically, by analytical photogrammetry. Satellite photogrammetry, as distinguished from conventional aerial photogrammetry, consists of the theory and techniques of photogrammetry where the sensor is carried on a spacecraft and the sensor's output (usually in the form of images) is utilized for the determination of coordinates on the moon or planet being investigated.

The most significant satellite photogrammetry to date was done using imagery from the National Aeronautics and Space Administration's Apollo spacecrafts 15, 16, and 17 during the early 1970s—Doppler tracking produced an ephemeris, a stellar camera provided attitude, and an altimeter measured distance from the spacecraft to the lunar surface. The resulting data, combined with that obtained by the mapping camera, provided all the information necessary for satellite photogrammetry, not requiring surveyed ground control. Thus, satellites carrying similar kinds of equipment are uniquely suited to charting moons and planets. The mathematics in this chapter is intended to be general enough so that it will be transferable to future problems of satellite photogrammetry. The reader is referred to chapter II for a more general mathematical treatment.

The emphasis in this chapter, then, will be on the mathematics for determining 3-dimensional coordinates rather than on that for determining horizontal coordinates only. Landsat, for example, produces pictures that are useful primarily for 2-dimensional or planimetric mapping.

17.1.1 HISTORICAL DEVELOPMENT

Within the memory of many professional surveyors and cartographers, surveys for mapping and charting the earth were made by the surveyor lugging his theodolite, plane-table, and alidade across plains and up mountains. Likewise, for charting the moon and planets, astronomers throughout history have spent countless hours at their telescopes recording the surface features they could see. In the years following World War II the ground surveyor was largely replaced by the cartographic aerial camera and the photogrammetric plotting instrument. This combination of aircraft, camera, and plotting instrument greatly increased the rate of production, the accuracy, and the content of maps and charts.

The use of space satellites as sensor carrying vehicles is expected by many to provide an improvement in mapping and charting capability comparable to that which the airplane made over the ground surveyor. This new technology began with the launch of Sputnik 1 on October 4, 1957. Since then, space sensors have demonstrated an almost unique capability to chart areas whose environment is hostile to the ground surveyor, qualifying the moon and planets as ideal for applications. Thus, man's continuous search for answers to the great mysteries of the solar system has been given great impetus since the beginning of the space program in 1958.

Historic photographs obtained from the rockets launched at White Sands and elsewhere during the 1950's and from the U.S.'s early, manned spacecraft—Mercury, May 1961 to May 1963; and Gemini, March 1965 through November 1966—were examined by scientists throughout the world.

The earliest photographs of the moon from space were taken in October 1959 by the Soviet spacecraft LUNA 3. This spacecraft, which took off on 4 October 1959 on the second anniversary of the launching of SPUTNIK 1, took pictures of the far side of the moon, developed them automatically, and then transmitted them back to earth by radio. The quality of the pictures was judged to be much poorer than that of good telescopic photographs, but Soviet scientists were able to derive useful information from the noise and produce the first chart of the heretofore unknown back side of the moon. (Barabashov 1961)

In the United States the Ranger program, which began in 1961, was designed to obtain detailed photographs of the lunar surface from a spacecraft designed to subsequently crash onto the moon. The spacecraft carried two wide-angle and four narrow-angle television cameras. However, it was not until 28 July 1964 that Ranger 7 was successful in returning 4,316 pictures of high resolution. The last few pictures taken before the crash were described as clearly showing features less than one metre across compared to pictures taken by earth-based telescopic cameras that can only resolve features greater than 800 metres across. The Ranger 7 photographs also gave a view from a point about 800 times closer to the moon's Sea of Clouds where the spacecraft crashed. Ranger 8, launched on 17 February 1965, returned 7,137 pictures before striking the Sea of Tranquility. Ranger 9, launched 21 March 1965, returned 5,814 pictures before it crashed in the crater Alphonsus. Although the trajectory of the Ranger spacecraft

made photogrammetric mapping very difficult, a few photographic mosaics and maps for local areas of the moon were compiled using analytical methods. (Light 1966)

On 19 July 1965 the Soviet spacecraft Zond 3 passed the moon at 9,220 kilometres and took 25 pictures of the far side. These were of better quality than those taken by LUNA 3 in 1959. This flight was followed by that of LUNA 12 in October of 1966. LUNA 12 photographed the moon from an equatorial orbit. From the data obtained from this flight, the Soviets produced a new planimetric map at a scale of 1:5,000,000.

In February of 1966 the Soviet spacecraft LUNA 9 landed on the Sea of Storms. Four months later the United States Surveyor 1 landed near Flamsteed crater. Other lunar landings followed in rapid succession: LUNA 13 in December of 1966 and Surveyors 3, 5, 6, and 7 between 17 April 1967 and 7 January 1968. Though these landings contributed 70,000 photographs of the lunar surface, information about its bearing strength, and that manned landings were feasible, cartographic contributions were very limited. The Surveyor missions did provide the first full-color, close-up pictures of the moon's surface from which photo-mosaics of the landing sites and a topographic map (at scale 1:50 with 10-centimetre contours) were made. It has been reported that the Russian spacecraft LUNA 16, 20, and 24 returned sam-

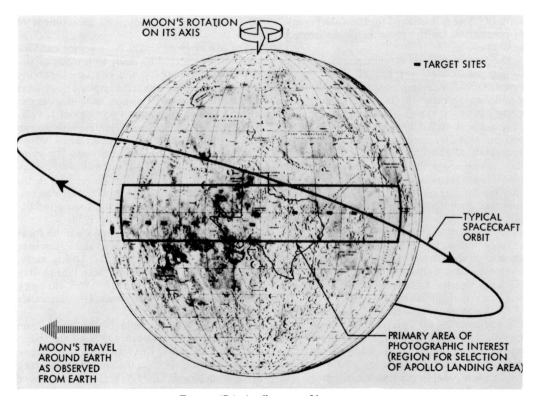

FIGURE 17-1. Apollo zone of interest.

ples of lunar soil back to earth. LUNA 17 and 21 left unmanned exploratory vehicles on the lunar surface.

Lunar Orbiter. The most successful of the pre-Apollo lunar satellites were the NASA Lunar Orbiters. The first of the five Lunar Orbiters was launched on 10 August 1966 and all five achieved nearly 100 percent of their objectives. Their primary objective was the return of high-resolution, small-area; and medium resolution, wide-area photography of proposed landing sites for Apollo spacecraft. These sites were distributed in a region between ±5° latitude across the visible face of the moon (figure 17-1). A second key objective was the photography of areas of significant scientific interest, including major craters like Copernicus and unusual formations. Figure 17-2 shows the regions photographed by each Lunar Orbiter. The missions were highly successful in satisfying both of the objectives. From the standpoint of the image quality recorded by the unique photographic system, the high-resolution (HR) camera was able to resolve 1 metre on the ground from an altitude of 46 kilometres while the medium resolution (MR) camera resolved 8 metres. Even the 8 metre resolution is phenomenal if we consider that telescopes on the earth can resolve no better than 800 metres. However the geometric quality of the pictures for photogrammetric applications left much to be desired. As yet we did not have photography of the high geometric fidelity needed for detailed mapping of the moon.

However, the unique photographic system utilized by the Lunar Orbiters merits elaboration since a considerable amount of triangulation and mapping of proposed landing sites for Apollo spacecraft was done using data from the Lunar Orbiter missions. Each medium-resolution photograph taken from the satellite was cut into 26 strips (framelets) in order to scan it for transmission back to earth by radio. A special film-camera system with two sets of optics was used (Kosofsky 1966; Kosofsky and El-baz 1970). The high resolution camera used a 610 mm, $f/5.6$ lens; while the medium-resolution camera used an 80-mm, $f/5.6$ lens. The film was 70-mm, SO-243 Eastman-Kodak aerial film with pre-printed reseau (with the exception of Mission I on which the reseau was not used). The reseau was used to compute corrections to the x,y-coordinates on the photographs. The 55 × 65 mm format of the film exposed a ground area of 31.6 × 37.4 km at 46 km altitude. The high-resolution system simultaneously exposed film with a format of 55 × 219.18 mm, giving a ground coverage of 4.15 × 16.6 km situated in the center of the region covered by the medium-resolution camera. The film was developed in the spacecraft using Eastman Kodak's Bimat System. One framelet of film of 2.54 × 57 mm was scanned at a time, the film being subsequently advanced 2.54 mm to scan the next individual framelet. A photomultiplier tube produced a signal proportional to the intensity of light for each picture-element, and a video signal was transmitted back to earth. Framelets were re-assembled on the ground on 35-mm film strips. A printer enlarged the framelets about 7.5 times, after which they were laid parallel to one another to make up a single exposure. A high-resolution photograph contained 96 framelets while a medium-resolution photograph contained 26 framelets.

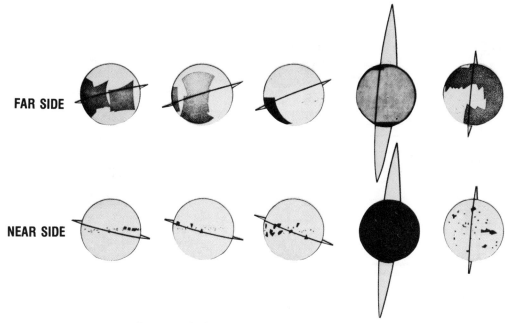

FIGURE 17-2. Lunar orbiter photographic coverage.

The framelet method of assembly turned the stereoscopic image obtained from a pair of photographs into an image having what was called the "venetian blind effect" (figure 17-3). Fortunately, each framelet contained up to 23 reseau marks and usually two to four fiducial marks. Transforming the measured coordinates into the coordinate system defined by the reseau and then transforming that system into the one defined by the fiducial marks, photogrammetrists were able to correct the imagery to within 20 to 30 micrometres. Figure 17-4 shows the Lunar Orbiter's imagery, transmission, and processing systems.

Lunar Orbiter Missions I, II, and III provided, for the most part, strips of photographs of candidate landing sites from a nominal height of 46 km and from orbits at an inclination of 12° from the equator. In one study involving four sites that were photographed on two or more missions, coordinates were determined independently on each mission. The purpose was to enable comparisons between the coordinates obtained for common points from the different missions in order to compare them. The root-mean-square deviation of individual common points from the average was as high as 1,500 metres. These errors were attributed to inconsistencies in the ephemerides for each mission as well as to the poor geometry involved. Using data from these missions, the proposed landing area for Surveyor III, one of a series of planned soft lunar landers, was mapped at scales of 1:2,000 and 1:5,000. (Schull and Schenk, 1968)

FIGURE 17-3. Photograph from lunar orbiter showing the "venetian-blind effect."

Subsequent Lunar Orbiter Missions IV and V were done from high altitude polar orbits while the perilunar altitude of Lunar Orbiter V was 100 km. Combined with the data obtained from the previous missions, nearly complete coverage of the lunar surface was obtained.

Analytical photogrammetry was done using the MUSAT program developed by A. Elassal. A version called GIANT is generally available. Exposure station positions obtained from the tracking-stations' data served as a control. Lunar triangulation programs that impose orbital constraints were also used. The craters Copernicus and Censorenius, as well as other craters and sites of scientific interest, were triangulated using the MUSAT program. Standard errors of some site coordinates ranged up to 1,500 metres.

The Lunar Orbiter Missions served as a great impetus to satellite photogrammetry. Using the

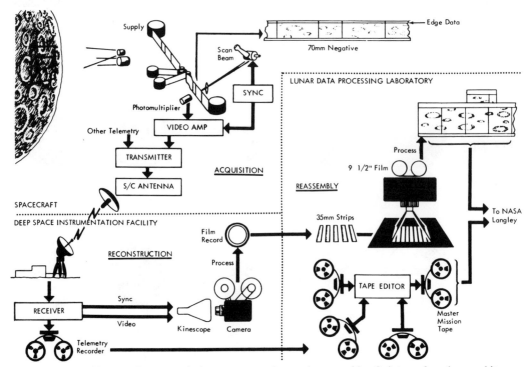

FIGURE 17-4. Photography, transmission, reconstruction, and reassembly of pictures from lunar orbiter.

data obtained from these missions, the Apollo moon landing sites were selected and experience was gained in satellite triangulation. This was also the first time a triangulation project had utilized a computer program with orbital constraints to establish control for mapping.

17.1.2 PLANETARY MAPPING

The Mariner series of spacecraft, whose objective was to obtain close-up pictures of the planets, utilized digital processing of imagery. Mariner 9, launched into orbit around Mars in May of 1971, provided an exceptional demonstration of the ability of orbiting spacecraft to provide data for planetwide reconnaissance mapping.

Processing of pictures from Mariner 9 was completely digital (Levinthal 1973). A variety of computer programs was developed to apply photometric and geometric corrections and to fit the pictures to selected grids. A control network with a planimetric accuracy of about 10 km was computed from the pictures (Davies 1973), mosaics were assembled, and airbrush renderings were produced to prepare a map series at scales of 1:25,000,000; 1:5,000,000; 1:1,000,000; and 1:250,000. (Batson 1973)

Gross topography was obtained by an unusual technique. The ultraviolet spectrometer on Mariner 9 recorded the intensity of solar ultraviolet radiation reflected from Mars. Since this radiation intensity is attenuated by the Martian atmosphere, the radiation-intensity profile, after appropriate corrections, is the inverse of the topographic profile if the top of the atmosphere is at constant elevation.

The U.S.S.R. was also launching spacecraft to Mars. Four spacecraft were put into orbit around the planet during July and August of 1973. Mars 5 was subsequently launched into an elliptical orbit with closest approach at about 1,700 km. Mars 5 carried two television cameras which could resolve 100 metres on the surface. The pictures obtained were quite similar to those obtained by the U.S.'s Mariner 9 and were used by Nepoklonov and others to compile a 1:5,000,000-scale map of portions of the planet.

The United States' Pioneer series of spacecraft was also designed to explore planetary and inter-planetary space. Pioneer 10, launched in March of 1972, was unique in that it was the first manmade object to leave the solar system. After 21 months of its voyage, which included passage through the asteroid belt, the spacecraft passed by Jupiter at a distance of 130,000 km. One of the 11 instruments on the spacecraft was a 2-channel, imaging photopolarimeter recording in the blue and red wavelengths (Gehrels 1974). About 80 images were obtained during the period of closest approach, in addition to the several hundred others obtained during two weeks when the planetary image was more than 40 resolution-elements in diameter.

Spacecraft Mariner 10 was launched in October of 1973 to fly by Venus and Mercury (figure 17-5). The television camera system was essentially the same as that carried on Mariner 9 to

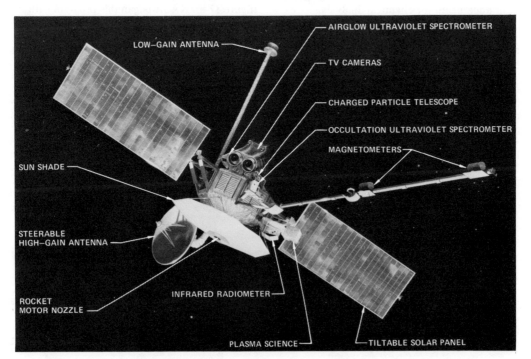

FIGURE 17-5. Spacecraft Mariner 10 (Photograph courtesy of Jet Propulsion Laboratories).

Mars, except that the focal length of the tele-photo lens was increased to 1,500 mm. As the spacecraft passed by Venus in February of 1974, about 3,400 pictures were taken, primarily in the ultraviolet band, in hope of penetrating the dense atmosphere (Murray 1974). Resolution varied from 130 km to 20 km. The pictures show enormous meteorological patterns. The space-craft then continued past Mercury at a distance of 1,000 km, where additional photographs were taken in late March of 1974. Mercury was found to have a crater-scarred surface with little evi-dence of volcanism.

Russian spacecraft in the Venera series have explored Venus since 1965, but few pictures have been published. Venera 2 passed Venus at 2,400 km, but communication failed. Venera 7 and 8 transmitted from the surface. Venera 9 landed on 22 October 1975 and Venera 10 landed on 25 October 1975. This spacecraft was the first to return pictures from the surface of Venus. Venera 11 landed on 21 December 1978 and Venera 12 on 25 December 1978. The spacecraft transmitted data for 95 minutes and 110 minutes, respectively and they were constructed to photograph the surface, but there is no informa-tion on whether they obtained pictures.

The Viking missions to Mars were very suc-cessful. Viking 1 landed 20 July 1976; it was closely followed by Viking 2 on 3 September 1976. As well as taking detailed photographs of Mars, the Viking orbiters checked the Martian cloud-patterns, the temperatures, and water vapor in the atmosphere. Both Viking landers pushed rocks aside and picked up soil samples from underneath for chemical analysis.

Photographic mapping of Mars' surface by the Viking landers continued in 1978. Analysis of the photography shows that Mars' *surface is cra-tered* much like the moon's. In figure 17-6, the large boulder to the left measures about 1 by 3 metres. Topographic mapping from images taken by the camera on Viking lander is reported by S. C. Wu (1976). Figure 17-7 shows the picture-taking process.

The Voyager missions, managed by the Jet Propulsion Laboratory, were another part of NASA's program of systematic, long-term ex-ploration of the planets and the solar system. A key mission began in 1977 with the launching of two very nearly identical spacecraft destined for Jupiter and then Saturn, with one intended for Uranus. Paving the way for these Voyager mis-sions were the successful initial investigations of the planet Jupiter by the Pioneer spacecrafts 10 and 11, in 1973 and 1974 respectively.

Voyager 2, launched on 20 August 1977 from Cape Canaveral, made its closest approach to Jupiter in July of 1979. Subsequently, Voyager 1 was launched on 5 September 1977. Its faster trajectory enabled it to make its closest ap-proach to Jupiter in March of 1979.

The two Voyager spacecraft each carried 11 scientific instruments including two television cameras. One camera had a narrow-angle 1,500 mm, f/8.5 optical system. The other had a wide-angle, 200 mm, f/3 optical system. These cameras were part of an imaging experiment and pro-duced high-resolution mosaics of the 4 major satellites of Jupiter. The Vidicon cameras had reseaux of 202 points to permit removal of geo-metric distortions. Pictures of stars were used to measure the focal length of the optical system and to calibrate the cameras geometrically. The control networks of the Galilean satellites have been computed photogrammetrically and mean radii measured.

17.1.3 SPACE PROGRAMS

17.1.3.1 THE APOLLO PROGRAM

The Apollo program was begun by NASA in 1960. It was to be the biggest and most complex of the programs for manned space-flight, with the landing of explorers on the moon (and their safe return to earth) as the primary objective (figure 17-8). NASA began this program with the testing of the mighty Saturn V rocket. This rock-et developed more than 3.4 million kg of thrust and was capable of sending more than 100,000 kg into orbit about the earth.

The first satellite of the Apollo program was launched on 11 October 1968. It made 163 revo-lutions. The principal source of photography was the Hasselblad 500 EL camera (chap. IV). This camera uses 70-mm roll film and has inter-changeable lens systems. Most photographs of the earth were made with systems of 80-mm focal length and with the camera mounted on a bracket in the hatch-window of the command module. Overlapping strips of nearly vertical photographs were made on SO-368 Ektachrome

FIGURE 17-6. Picture from Viking-1 lander—Martian landscape. (Picture courtesy of NASA).

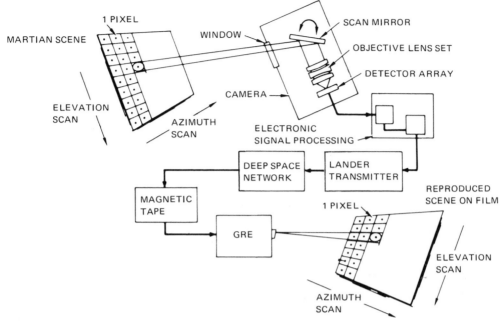

FIGURE 17-7. Imaging, transmission and reconstruction of pictures. (Picture courtesy of ITEK Corporation).

and on high-resolution, SO-242 color film. The camera was also used with a 250-mm lens system and held in the hand for photography of a few isolated scenes of special interest. The first cameras of quality good enough for cartographic work were not carried until Apollo missions 15, 16, and 17.

Next, on 21 December 1968 came Apollo 8, the first voyage to the vicinity of the moon. The eight-day flight brought the spacecraft within 80 km of the lunar surface. Apollo 9 came two months later, on 3 March 1969, with the first test, in space, of the moon-landing craft. The astronauts took color and black-and-white photographs of the earth.

Four Hasselblad cameras with 80-mm lens systems and 70-mm film were mounted on a metal frame fitting the hatch-window in the command module. Three of the cameras carried black-and-white film appropriately filtered to give green, red, and near-infrared pictures. The fourth camera carried false-color infrared film. Two pictures covered approximately the area of the standard 1:250,000-scale USGS map. These pictures were enlarged nearly 10 times and rectified using points identifiable on both the map and the photographs. The mosaic was then printed to serve as a base for the standard line map. A number of different color-schemes were tried and versions produced both with and without contours. Thirty years ago, there was little expectation that photographs taken from a height of 200 km and more could reveal anything new about our earth. Even in aerial photographs taken from about 20 km up, there was such a

deterioration in detail that it was thought impossible to produce satisfactory pictures from greater heights. As it turned out, atmospheric particles did not have nearly as great a detrimental effect on resolution as had been feared.

Apollo 10 followed on 18 May 1969 and went to within 20 km of the lunar surface. This mission was the final small step before the giant leap. The Hasselblad 500 EL with 80-mm and 250-mm lens systems were again used for photography.

Apollo 11 landed on the moon on 16 July 1969, and with the words, "That's a small step for a man, one giant leap for mankind", Neil A. Armstrong became the first man to step onto another celestial body. Again the Hasselblad 500 EL camera was used.

On the earlier Apollo missions, the cameras took many superb photographs, but the cameras were not designed for geodetic applications. After the historic landing of Apollo 11, it was realized that further scientific exploration would require more-specialized photography. So, for use on the last three missions, a totally new photographic system was assembled. One section of the service module in missions 15, 16, and 17 contained the panoramic camera built by Itek and the mapping camera system built by Fairchild. Included were the stellar camera and the laser altimeter. These instruments are shown in figure 17-9. The cameras are described in chapter IV, section 4.3.1.3. A detailed description of the panoramic camera is given in chapter IV, section 4.3.1.4. Figure 17-10 shows an example of the panoramic photographs taken from Apollo

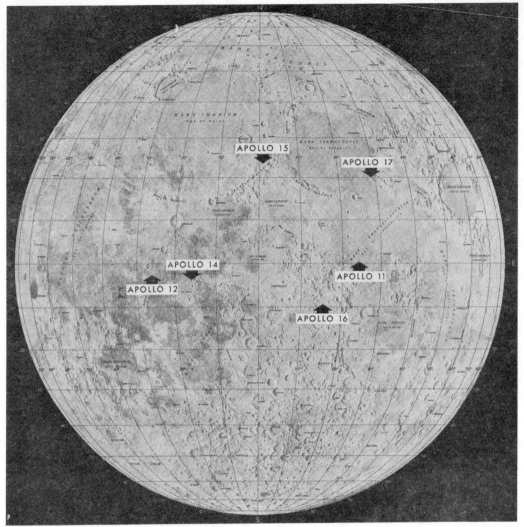

FIGURE 17-8. Apollo landing sites on the moon.

15 from 100 km above the surface of the moon. Figure 17-11 shows the mapping camera and laser altimeter. Figure 17-12 is a typical photograph taken by the mapping (frame) camera and

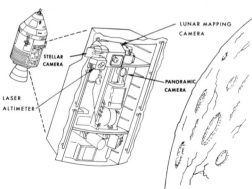

FIGURE 17-9. Scientific instrument module. (Courtesy of Fairchild Space and Defense Systems).

shows the landing site for Apollo 15. Hadley Rille is at the lower center of the picture. These pictures were used to establish control for the mapping program. On the three missions Apollo 15, 16, and 17, about 25% of the lunar surface was photographed. Figure 17-13 shows the photographic coverage achieved by the three missions and the relative sizes of the various maps published. The positions of the retroreflectors form a triangle which could be used to establish a lunar datum.

The Defense Mapping Agency, National Aeronautics and Space Agency, National Oceanographic and Atmospheric Administration, and the U.S. Geological Survey cooperated in determining the positions of 5,325 features (points) on the moon's surface. More than 51,000 measurements on 1,244 photographs were used in a block adjustment. The orientation given by the stellar photographs and the distance given by the altimeter were used as constraints;

FIGURE 17-10. Panoramic photograph taken from Apollo 15. (Courtesy of ITEK Corporation).

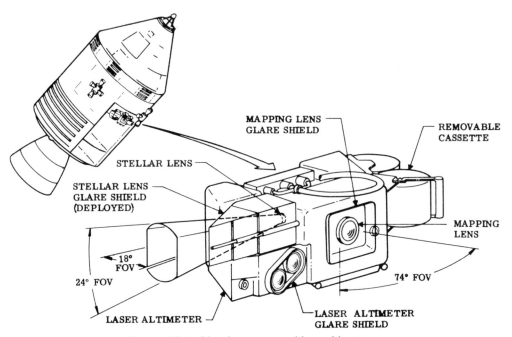

FIGURE 17-11. Mapping camera and laser altimeter.

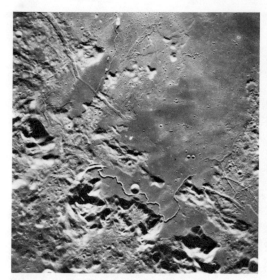

FIGURE 17-12. Photograph taken by frame camera on Apollo 15.

FIGURE 17-13. Apollo missions 15, 16, and 17—photographic coverage.

the spacecraft's locations were not used as constraints because the orbit proved to be unreliable. Covariance analysis indicated that 90 percent of the computed horizontal/vertical coordinates should be accurate to within 45 m (figure 17-14). *See* section 17.5 for details on the accuracy of the points. Some of the other results were:

topographic orthophotomaps at 1:250,000 scale with 50-m contours from the mapping photographs;
large scale maps at 1:50,000, 1:25,000, and 1:10,000 scales from the panoramic photographs;

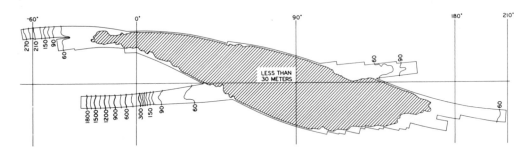

STANDARD DEVIATION OF HORIZONTAL POSITIONS

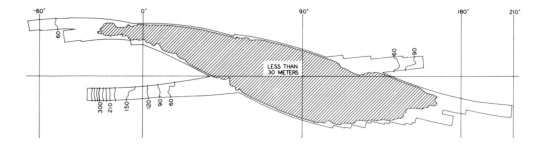

STANDARD DEVIATION OF ELEVATIONS
FIGURE 17-14. Accuracy of NOS/GS lunar coordinates.

topographic orthophotomaps of limited areas at scales as large as 1:10,000 with 5-m contours from the panoramic photographs; a mathematical reference surface (section 17.6).

17.1.3.2 THE ZOND PROGRAM

A series of spacecraft called Zonds were launched by the U.S.S.R. for planetary and lunar exploration, beginning in 1964. In April 1964, Zond 1 passed Venus at about 100,000 km. Zond 2, launched in November 1964, was sent to Mars but did not transmit data. During the period September 1968 to October 1970, Zond 5, 6, 7 and 8 returned photographic film to the earth from circumlunar orbits. The cameras had a focal length of 400 mm and a format 130 × 180 mm. Ground resolution was estimated to be 500 m, at best. From these photographs, the Russians compiled a strip mosaic at 1:2,000,000 scale and also made a few water-color maps at scale 1:1,000,000 for sites of particular interest.

17.1.3.3 SKYLAB PROGRAM

Skylab was the first space station put in orbit by the U.S.A. Launching dates for the four stations were:

1978 May 5 Skylab 1 (unmanned)
1978 May 25 Skylab 2 (manned 28 days)
1978 Jul 28 Skylab 3 (manned 59 days)
1978 Nov 16 Skylab 4 (manned 84 days).

The altitudes were, nominally, 435 km, the inclinations about 50°. Skylabs carried a variety of detecting and imaging instruments. Among the instruments were cameras for the optical and near-infrared regions and infrared spectrographs. A total of 25,407 pictures were returned. Table 17-1 gives some of the characteristics of the imaging instruments.

The scale of the photographs from the six multispectral cameras (S-190A) was 1:2,800,000. The S-190B camera was essentially the same as that carried on Apollos 13 and 14. Its 45.72 cm focal length produced photographs 115 × 115 mm at a scale of 1:950,000. Tests have shown that these photographs can be enlarged to about 1:100,000 without serious degradation of quality.

The multispectral scanner (S-192) covering 13

FIGURE 17-15. Skylab (Courtesy of NASA).

spectral bands from 400 to 1250 nm provided some cartographic results, but the cost of producing the results significantly reduced their value.

Skylabs were placed in repetitive orbits—*i.e.,* orbits such that the ground tracks, after a certain number of revolutions of the satellite, closely overlapped. Pictures of the same sites could therefore be taken again and again. But it also meant that there were wide spaces *between* ground tracks in which no photography existed. Consequently, no large areas could be mapped onto map-sheets of conventional size. On the

TABLE 17-1. CHARACTERISTICS OF IMAGING INSTRUMENTS IN SKYLAB

Experiment	Imaging Type	Focal Length	Ground Coverage	Spectral Range μm	1.6:1 Ground Resolution m
S-190A	Six Multispectral Cameras	15.24 cm	163 × 163 km	.4 to .9	38 to 79
S-190B	Recon-Frame Camera	45.72 cm	109 × 109 km	.4 to .88	17 to 30
S-192	Conical Multispectral Scanner	—	68.5 km Swath	.4 to 12.5	80 m*

* Pixel size.

other hand, the superb photography obtained with the film cameras has demonstrated the value of high-resolution photography in conjunction with Landsat imagery.

17.1.3.4 LANDSAT SATELLITES

While observations of the earth's surface were made from spacecraft Gemini, Apollo and Skylab, the satellite that began as Earth Resources Technology Satellite (ERTS-1, later changed to LANDSAT-1) has become the workhorse of NASA's Earth Observation Program. The success which LANDSAT has enjoyed is generally attributed to its long life and its repeated coverage of the same regions, rather than for its ability to give images of high resolution. Images restored directly from the telemetered data are at 1:1,000,000 scale; some users enlarge these up to four times for a planimetric photomap. Landsat-1 was launched on 23 July 1972 and died on 10 January 1978; Landsat-2 was launched on 22 January 1975 and died on 5 November 1979; Landsat 3 was launched on 5 March 1978. The ERTS program and satellites were renamed "Landsat" in 1975.

The Landsat program was experimental and was intended to establish the value of relatively coarse-resolution pictures (an element is 79 m square) for study of mineral and other resources. NASA funded several hundred investigations on the following subjects (among others): agricultural production, forestry, and mapping. Landsats 1 and 2 are in circular, nearly polar orbits about 920 km above the earth's surface, but they are no longer transmitting data. Landsat 3 has a period of 103 minutes and is still transmitting.

The instruments in Landsat which are of interest to photogrammetrists are the Return Beam Vidicon (RBV) and the Multi-spectral Scanner (MSS). Figures 17-16a, 16b, and 16c show Landsat and the two instruments, RBV and MSS.

Landsat-3 differs from Landsat-1 and Landsat-2 in the following respects.

The three RBV's of the earlier Landsats were replaced by a pair of single-band (0.505 to 0.750 micrometre) RBV's of twice the focal length. Together they cover twice the width, on the ground, as the MSS and with twice the resolution (picture elements 40 m square). The MSS is the same except that a thermal band of 10,400 to 12,600 nm is recorded with a resolution three times coarser (picture element 237 m square) than in the visible and near-infrared bands.

Mapping using images from Landsat is basically planimetric rather than topographic and thus is well adapted to use in a planimetric mapping system. However, Landsat has two definite limitations.
(1) Its low resolution precludes the recording of small cultural features.
(2) It cannot be applied to topographic mapping.

On the other hand, imagery from the first two Landsats has proven useful for line maps and

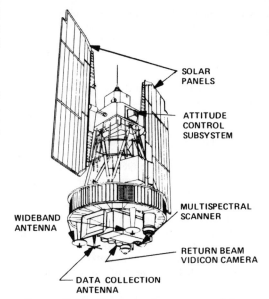

FIGURE 17-16a. Landsat and sensors complete.

charts at scales of 1:1,000,000 and 1:500,000. Imagery from Landsat-3 may allow useful reproduction at scales as large as 1:250,000 or perhaps even larger.

17.1.4 JUSTIFICATION FOR SATELLITE PHOTOGRAMMETRY

The usefulness of spacecraft for providing imagery for a closer look at the moon and the planets is unquestioned. There simply may be no other way that man can explore the solar system, and perhaps the galaxy, in the near future. Clearly, the technology is available to accomplish this. The opportunity exists to image and then map areas that are generally inaccessible, or that have an environment hostile to man. The moon and planets fall in the hostile-environment category. The challenge exists to utilize this great opportunity for the benefit of

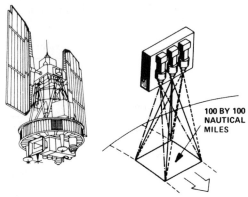

FIGURE 17-16b. Return beam vidicon subsystem.

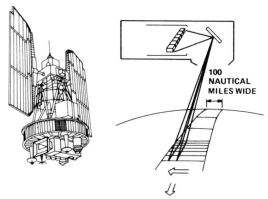

FIGURE 17-16c. Multispectral scanner subsystem.

mankind by continuing to reduce the cost and increase the utility of space systems.

Leatherdale (1978) points out an obvious economic advantage of space imagery for mapping. This is the reduction of the density of geodetic control points needed to achieve reasonable map accuracies. As an example of increased coverage and the need for less ground control points, Colvocoresses (1971) has shown that a spacecraft at 200 km altitude and carrying a 300-mm (12-inch) camera could photograph an area of 10,800 square km; only one stereoscopic model and 3 control points would be needed. Mapping using conventional aerial photography with a 6-inch camera for the same area would require 172 stereoscopic models and 516 control points. The 516 control points are certainly not a fixed number; in fact, they could probably be reduced significantly by analytical triangulation. However, the point is that photography from orbital altitudes covers more ground per photograph and therefore requires less ground control points. There are many more factors to be considered, but this is a basis for an economic comparison.

Photogrammetrists know that a stereoscopic model must be scaled, leveled and positioned before geometric information can be compiled on conventional stereoscopic plotters. Automated instruments require the position and orientation elements as inputs. Conventionally, this is done by reference to ground control points. In this regard, satellite-borne photographic systems have a significant advantage. The orbit is highly predictable from the laws of motion and from tracking data. If the precise time of each exposure is recorded, the position of the camera can be determined accurately. Furthermore, a photograph of the stellar field by a stellar camera can be taken at the same time as each photograph of the ground, and measurement of the stellar photographs will provide the orientation of the camera to a few seconds of arc. Having these data is equivalent to having ground control in every stereoscopic model. In

theory, these data can be completely substituted for ground control, which is not required except for checking the accuracy. Also, since the data are for each photograph, errors in the triangulation are controlled to a certain extent.

The economic justification depends on producing a high volume of imagery and making it readily available to the user as was imagery from Landsat.

Table 17-2 gives the salient characteristics of some satellites that have been giving or are expected to give photogrammetrically useful imagery.

17.1.5 CATEGORIES OF SATELLITE PHOTOGRAMMETRY

Imaging systems carried on satellites can be put into two basic categories, according to the requirements to be met by the application. These categories are (1) remote sensing (for interpretation) and (2) satellite photogrammetry. Remote sensing is characterized by the fact that the *nature* of the object imaged is of major importance, the *geometry* of minor importance. Satellite photogrammetry is characterized by the reverse order: geometry is of major, almost sole interest and the nature of the object is of minor interest. Satellite photogrammetry, in turn, can be divided into two sub-categories: planimetric (2-dimensional), in which only the horizontal geometry of the object is investigated, and topographic (3-dimensional), in which all three dimensions are investigated. Only the second of these subcategories will be considered in detail in this chapter. (A third sub-category, containing investigations of elevation or height only, is rarely met with and will not be considered.

17.1.5.1 TWO-DIMENSIONAL SATELLITE-PHOTOGRAMMETRY

Conventional photogrammetry involves mapping in two or three dimensions based on the analysis of one or more perspective images of the same scene recorded in two dimensions as photographs. The advent of imaging systems in space provided the opportunity to map the earth's surface directly in either two-dimensional (planimetric) or three-dimensional form. This simplification of the mapping problem is due to the great distance from which the earth can be viewed from space. If a sensor of limited overall field of view looks vertically at the earth from space, it produces a nearly orthographic projection and the resulting image can be recorded. This is typified by Landsat. Such a satellite records the earth's surface in approximately correct planimetric form. In so doing, it eliminates the capability to measure the third dimension of height, and thus has no direct value to three-dimensional (topographic) mapping. However, topography is generally static

TABLE 17-2. SATELLITES OF PHOTOGRAMMETRIC IMPORTANCE

Satellite	Sensor	Spectral Band	Altitude km	Nominal Scale	Ground Resolution Metres	Width of Cover km	World Land Coverage	Operational Period
LANDSAT 1 & 2	Multispectral Scanner	0.5 – 1.1 μm.4 bands	919	1:1,000,000	79	185	±82° N–S almost complete	1972–1978
	Return Beam Vidicon	.47 – .83 μm.3 bands					negligible	1975 →
SKYLAB	Multispectral Camera S190A	.4 – .7 μm	435	1:2,860,000	30*	150	negligible	1973
	Photographic Camera S190B	.4 – .88 μm		1:948,500	20*	108		
LANDSAT 3	Multispectral Scanner	.5 – 1.26 μm.5 bands	919	1:1,000,000	79	185	by request	1978 →
	Return Beam Vidicon	.5 – .75 μm broadband		1:500,000	40			
SEASAT	Synthetic Aperture Radar	Microwave L–band	794–808	1:500,000	25	100	72° N–S Selected Targets	1978 →
ESA[1] SPACELAB	Metric Camera RMKA 30/23	.4 – .8 μm	250	1:820,000	30*	190	Initially Negligible	1981 →
	Synthetic Aperture Radar	Microwave 6 band			30	9		
LANDSAT D	Thematic Mapper MSS	.45 – 12.5 μm	705		30		by request	1981 →
	Multispectral Scanner	.5 – 1.1 μm			79			
SPOT[2]	Multispectral Scanner	.5 – .9 μm	822	1:760,000	10	60	Global eventually	1984 →
Shuttle	Large Format Camera	.4 – .9 μm	227–417	1:912,000	≥10*	208 × 417	by request	1983
Shuttle	Shuttle Imaging Radar-A	L–Band	278	1:500,000	40 × 40	≥50	by request	Late 1981

[1] ESA is European Space Agency
[2] SPOT (System Probatoire d'Observation de la Terre) French System
* Pertains to film cameras where ground resolution pertains to one line pair in the image plane. Non-asteric values are pixel-size or ground-sample distance.

whereas the appearance of the earth is relatively dynamic. Thus, a satellite that can provide a nearly orthographic projection is of considerable importance because it greatly reduces the photogrammetric problems of planimetric mapmaking. The imaging system on a satellite such as Landsat records the energy scattered from the scene in a uniform manner whereas the response of an aerial photographic emulsion varies across the scene. Thus Landsat's sensors are effective radiometers. During the 1960's, a program was started to investigate the remote sensing of the earth's surface from space. Experiments were carried out on manned spacecraft. By 1968, a special-purpose satellite, ERTS-1 had been built. This satellite, re-named "Landsat" in 1975, is described in more detail in Anonymous (1975) and in Anonymous (1976) (and its revision).

17.1.5.1.1 Landsat

Of the four Landsat satellites, Landsat-1 and Landsat-2 (figure 17-16) were nearly identical, with the following characteristics.

Orbit
altitude	919 km
inclination	99 degrees
period of revolution	103 minutes
period for return to same point on earth	18 days
type	sun-synchronous
time at descending node	09:42 local solar time

Sensors (principal)
type	multispectral scanner (MSS)
spectral bands (micrometres)	
bands 4 and 5	0.5 − 0.6; 0.5 − 0.7;
bands 6 and 7	0.7 − 0.8; 0.8 − 1.1.
instantaneous field of view	73.4 m × 73.4 m (on the ground)
effective resolution (approx.)	86 m (on the ground);
smallest element displayed (pixel)	79 m
width of scan (on ground)	185 km
Transmission System	radio, S-band

Products:
before February 1979	images at 1: 3 369 000 and 1: 1 000 000 scales; magnetic tapes
after January 1979	images at 1: 1 000 000 scale; magnetic tapes

The satellites also carried three Return-beam vidicon cameras (RBV) covering three bands: 0.475−0.575 (band 1), 0.580−0.680 (band 2), and 0.5609−0.830 (band 3) micrometres, but they had very limited use on Landsat-1 and Landsat-2 as they had lower photometric accuracy than the multispectral scanners (figure 17-31) and only about the same resolution.

Landsat-3 had the same characteristics as the two previous satellites but covered a spectral band at 10.4−12.6 micrometres (but the MSS in this band did not perform satisfactorily) and had a pair of redesigned RBV cameras with the following characteristics.

type	RBV with image-plane scanner
spectral band	0.505 − 0.750 micrometre
number of lines per frame	4 125
number of points per line	5 375
field of view: instantaneous per frame	24 m (at ground) 98 km by 98 km
effective focal length	236 mm
geometric fidelity	20 to 40 m rms error within a frame.

Photographic copies of MSS and RBV images from Landsat-1, -2, and -3 are available at the EROS Data Center in Sioux Falls, South Dakota 57198. They and magnetic tapes are on sale without restrictions. They cover most of the land-areas of the world and also sizable portions of the shallow seas. Files exist in nearly every state of the U.S.A. where images can be viewed on microfilm or microfiche.

The success of Landsat is best measured by the number of stations involved, which by late 1979 consisted of 12 stations in operation or under construction.

Conventional electron-beam recorders have a more or less standard response to illumination so that, in theory, relative measurements of radiometric intensity can be made from the images. Unfortunately, this does not normally provide good contrast in any given band or scene; thus a great deal of information in the scene is lost when the images are converted to photographs.

When Landsat was first planned, mapping was not one of the primary applications provided for. Moreover, mappers with a primary concern for geometric precision were skeptical of the MSS as an imaging device and looked to the RBV's, which had been calibrated as described by McEwen (1972) as a more suitable source of data for mapping. However, since the RBV's were turned off because of a malfunction shortly after launching Landsat-1, investigators turned to the MSS images for their material. The MSS involves two basic limitations cartographically: (1) items too small to be usefully recorded as a 79-m square pixel (smallest resolved detail on the image) cannot be resolved or mapped, and (2) nearly orthogonal projections preclude determination of heights. In spite of these limitations, Landsat images have characteristics such that they do have important cartographic use. These characteristics include repeated coverage of the same scenes, good geometric accuracy, good radiometric coverage and accuracy (including part of the near infrared) and digital form of output. Colvocoresses (1975) provides a summary on application of Landsat to cartography, together with a comprehensive bibliography. Landsat images are conveniently indexed by path number and row. Starting in 1978, all images carried these indices, but unfortunately the system changed with LANDSAT-D, which had a different orbit from that of the previous Landsats.

17.1.5.1.1.1 *Geometric Aspects*. Chapman (1974) has shown that by using a modified version of the computer program developed by Bender (1970), for gridding satellite imagery, a Universal Transverse Mercator (UTM) grid can be fitted to the bulk (standard) processed multispectral scanner (MSS) image and the inexpensive cartographic product which results will meet or approach U.S. National Map Accuracy Standards for 1:250,000 or smaller-scale mapping depending on the method used. A summary of the MSS positional accuracy of well defined points has been presented by McEwen and by Colvocoresses (1975), as a result of tests applied to gridded maps produced by the U.S. Geological Survey with Landsat data. In an analysis based on an evaluation of both bulk and precise MSS image products, Colvocoresses (1975), summarized the expected errors after best fit to ground control.

Summary of Landsat MSS Positional Accuracy.

Mode and Form	Errors Range (rms)
Bulk processed through EBR by NASA	
• Best fit to UTM projection	150–350 m
• UTM grid fitted to single-band image	50–100 m
• UTM grid fitted to single-band mosaic of 2 to 4 images	100–150 m
• UTM grid fitted to multiband (colored) single image (litographed)	125–175 m
• UTM grid fitted to multiband (colored) mosaic of multiple images (lithographed)	200–300 m
Precisely fitted to UTM projection (NASA and IBM products)	50–100 m

Analytical triangulation experiments conducted with a strip of 15 Landsat images taken on a single orbit from Minnesota to Texas showed promise for extending control into unmapped areas and for preparing maps (McEwen and Asbeck 1974). By tying both ends of the strip to existing control and applying low-order adjustment equations, residuals of under 500 m rms were obtained. McEwen (1974) also described a block adjustment of 15 MSS images covering Florida which resulted in residuals of 105 m rms for control and tie points.

A method for identification of ground points in unmapped regions was developed by Evans (1974) in experiments using small (0.5 m) mirrors to reflect sunlight to the satellite and record the points so identified on the image.

A unique map-projection in which relative motion of earth, satellite, *etc.,* are involved was invented for Landsat by Colvocoresses (1973, 1974). The projection, called the Space Oblique Mercator, was developed mathematically by Snyder (1978). It is particularly suitable for use in an automated mapping system since Landsat images are projected by it with practically no distortion. From February 1975 to January 1979, all Landsat imagery processed by NASA on their electron-beam recorder was placed on this projection. Commencing in 1979, NASA used a variant, developed by M. Hotine, of the oblique Mercator map projection and fitted to Landsat orbits by Rowland. The geometric differences between the two map projections are very small (1:10,000) and of little concern to most users.

Return-beam Vidicon of Landsat-3. The geometric configuration of the RBV's on Landsat-3 is described by NASA (1976). The imaging surface of the vidicon contains 81 calibrated marks (crosses) of a reseau which permit the internal geometry to be accurately determined. From the launching of Landsat-3 (March 1978) to late January, NASA produced pictures from RBV images on an electron-beam recorder. Study of these pictures by U.S. Geological Survey indicates that the precision is on the order of 20–40 m rms. These pictures are those produced from RBV imagery and are not modified to take into account such external effects as obliquity of the camera or earth curvature. Beginning in January 1979, the RBV images were transformed to the oblique Mercator map projection by Hotine's method in order to bring them into congruence with images from the MSS. The few tests made on such pictures indicate that the geometric accuracies indicated by the pictures from the electron-beam recorder are retained but they are not fully congruent with MSS images.

17.1.5.1.1.2 Landsat-Image-Maps. By 1979, a variety of black-and-white and color image-maps had been compiled and printed by the U.S. Geological Survey. They vary in scale from 1:250,000 to 1:1,000,000 and in format from a single image to an image-map covering a whole quadrangle or state. Others have prepared small-scale image-maps of the United States, but these are without geographic references and are considered mosaics rather than maps. Warren (1977) describes techniques used in the production of a 1:500,000, gridded, color image-map of Florida.

Experimental image-maps of Antarctica at scales of from 1:250,000 to 1:1,000,000 have also been produced and have been useful in identifying and locating such features as glaciers, coastlines, islands and other geographic features. Significant movement and other changes have also been noted on successive images. MacDonald (1973) describes the use of these data in exploring and mapping Antarctica.

Map and Chart Revision. Experiments have

shown that Landsat images are useful for delineating areas in need of revision on large-scale and small-scale maps and for the actual revision of selected features such as bodies of water, vegetation, and bold cultural features on small-scale maps. Landsat images have been used successfully to revise the hydrographic features on a 1:500,000-scale, aeronautical chart. Canada has used Landsat images in revising and repositioning selected features on maps and charts as large as 1:50,000 scale (Fleming 1976).

Nautical Charting. Landsat images are of proven value to nautical charting through the water-penetration capabilities of radiation in MSS bands 4 and 5. Polycyn reported that under optimum conditions and with auxiliary data, shallow-sea bottoms can be charted to a depth of 22 m with accuracies within 10 percent of measured values (rms).

During 1976, the U.S.A. Defense Mapping Agency, which prepares nautical charts for civil as well as military use, applied Landsat images to revision of nautical charts. For the first area properly tested (Indian Ocean), "Notices to Mariners" and a revised chart based on Landsat data were issued. The revised chart (3rd edition of 61610, Chagos Archipelago, dated 28 August 1976) showed significant changes from the 2nd edition which was dated 21 February 1976. Landsat images were taken, during March 1976, in the high-gain mode (signals in spectral bands 4 and 5 were amplified by a factor of 3) and thus provided more information about the shallow-sea bottoms which return little light. The images showed that most of the underwater features in the Chagos, except for those near islands, were improperly positioned by as much as 16 km and an uncharted feature (Colvocoresses Reef) of considerable size was discovered. Hammack (1977) has described application of Landsat to the charting of the Chagos in considerable detail.

Thematic Mapping. Thematic maps in many forms have been derived from Landsat data. It has been demonstrated that open water, vegetation (infrared reflective), and snow and ice can be detected on Landsat images. This technique has been applied to preparation of an aeronautical chart at 1:500,000 scale.

17.1.5.1.2 Weather Satellites

Weather (meteorological) satellites are designed to monitor the atmosphere rather than the earth's surface but, in areas devoid of clouds, they do in fact record the earth's surface in considerable detail. Two types of orbits are involved (1) those that are near-polar and sun-synchronous, and (2) those that are near-equatorial and remain over the same general portion of the earth (geosynchronous). In both cases, sensors record sizable areas of the earth's surface at low resolution and relief displacement can be ignored. Thus, the earth can be directly mapped in two dimensions by transforming the images by a

map projection such as the space oblique Monitor, stereographic, or the simple perspective projection. The resulting maps are not highly accurate geometrically, but their principal purpose is to show cloud distribution and precision normally is not required.

The weather satellite that deserves special mention is that of the Defense Meteorological Satellite Program. These satellites in near-polar orbits produce images with relatively high resolution[1] and under good conditions of contrast such as in a snow scene, highways and towns are shown. The nighttime use of the satellite is quite spectacular as it records city lights, gas flares, and other sources of light. Imagery from the satellite is generally displayed on a cylindrical (Cassini) map-projection similar to that used for Landsat. Croft (1977) has described nighttime applications in considerable detail.

17.1.5.2 Three-Dimensional Satellite Photogrammetry

Satellite photogrammetry imaging-systems designed for photogrammetric geodesy can be characterized as stereoscopic imaging systems capable of producing data from which a photogrammetrist can determine 3-dimensional coordinates, and related topographic information.

The users' requirements must play a key role in the design of an earth, moon or planetary mapping system. Before attempting to specify a system which would collect data that would permit useful topographic mapping from space of the earth or of extraterrestrial bodies, it is necessary to examine some of the essential factors involved.

A topographic map contains three kinds of information (Doyle 1973).

1. Content—the cultural and natural features represented on the map.
2. Horizontal location—the reference graticule, grid, datum.
3. Elevation—spot heights, contour lines, profiles and elevations

A satellite photogrammetry imaging-system designed for mapping must be able to provide all three kinds of data.

Map content is determined, in part, by photographic resolution and scale, or more directly from resolved distance on the ground (resolution). It is difficult to establish a linear relation between map scale and resolution required, because some features, like roads and railroads, must be shown on a map regardless of its scale. Hence, these features are nearly independent of map scale; and therefore resolution required is not linearly related to scale. On the moon and planets, however, cultural features have, so far, not been detected so that a linear relation seems

---

[1] The high-resolution picture element is described as 0.33 nautical mile coverage on the earth.

more justified. Considering the photograph as a base (photomosaic or orthophotograph) for the map, a useful criterion can be produced.

The smallest feature which can be depicted on a map is assumed to have a least dimension of 0.25 mm (.010 inch). In order for an object to be photographically identifiable it must be imaged by about 5 resolution elements. It follows, then, that the resolution required for photography can be estimated as:

Ground Resolution = 0.2 × 0.25 mm × map scale number

or $$Rg = 5 \times 10^{-5}\, Sm$$

where

Rg = ground resolution required (metres) for the sensor
Sm = map scale number.

Example: for a 1:50,000 photograph, $Rg = 5 \times 10^{-5} \times 50,000 = 2.5$ m.

The merit of this equation may be judged by noting that it suggests the map should have a resolution of 20 l/mm. This is seldom achieved in either terrestrial or extraterrestrial mapping practice. The 20 l/mm is about twice the resolving power of the unaided human eye, so the need for 20 l/mm is subject to question. The human eye resolves approximately 10 l/mm (SPSE Handbook). Consequently an orthophoto, or image product made to the 20 l/mm criterion would contain two times more information than the unaided human eye can utilize. So in theory using Rg will provide imagery with a 2× safety factor, so that even 50% losses in resolution during normal photographic processes will generally assure an Rg in the final product of 10 l/mm. Using this 10 l/mm criterion, high-resolution imagery from space can be enlarged until its equivalent resolution is 10 l/mm. A useful equation is:

$$Ma = \frac{rp}{10}$$

where

Ma = allowable enlargement from photograph (on original image) to map scale
rp = resolution in l/mm.

The second kind of map information is horizontal location in an absolute coordinate system. For mapping the earth, this is generally provided by reference to ground control points. But such control does not exist on the other planets. However, orbital tracking data can provide spacecraft position to a high order of accuracy. When this is coupled with precise data on attitude, and time of exposure, an independent means of determining absolute horizontal location is available. This removes the need for ground control. Within a stereoscopic model, or a photogrammetric triangulation, the internal (relative) positional accuracy is approximately equal to the product of the scale of the photograph and the error in measuring, or identifying a point on the picture:

$$\sigma_p = S_p\, \sigma_m$$

where

σ_p = standard error in relative position of a ground point
σ_m = precision of measuring an image point
S_p = photo-scale number.

U.S. Map Accuracy Standards require that the standard error of horizontal positions should not exceed 0.3 mm at the map scale (Sm):

$$\sigma_p = 3 \times 10^{-4}\, Sm, \qquad \sigma_p \text{ is in metres.}$$

Example: for a 1:50,000 scale map
$$\sigma_p = 3 \times 10^{-4}\, Sp = 15 \text{ metres.}$$

The third kind of information on a topographic map is terrain relief depicted by contour lines, spot elevations and profiles, *etc.*

A statistical interpretation of U.S. Map Accuracy Standards for elevations requires that 90% (within 1.64 standard deviations) of the elevations must be accurate to one-half of a contour internal (CI):

$$1.64\, \sigma_h = 0.5 \text{ CI}$$

or

$$\sigma_h = 0.3 \text{ CI.}$$

Elevations are obtained from aerial photographs by means of image parallax which depends upon the sensor's geometric configuration, the image scale, and the measuring precision (σ_m):

$$\sigma_h = S_m \frac{H}{B} \sigma_m \sqrt{2}$$

H/B is the reciprocal of the base/height ratio.

The basic specifications particularly useful for planning extraterrestrial mapping at various scales may be derived by applying these previous formulas. The results are given in the following.

Map-Accuracy Requirements

Map-Scale Number	Map Ground Resolution	Sensor Output Ground Resolution	Std. Error in Position	Contour Interval	Std. Error in Elevation
S_m	MRg	Rg	σ_p	CI	σ_h
1,000,000	100 metres	50	300 metres	250 metres	75 metres
250,000	25	12.5	75	50	15
100,000	10	5	30	25	8
50,000	5	2.5	15	10	3
25,000	2.5	1.3	7.5	5	1.5

A fixed contour interval does not necessarily go with a given map scale, either. An interval fine enough to depict the character of the terrain should be chosen. The above values of CI are typical for the scale shown. Actual CI's vary with the roughness of the terrain. The values for Rg that should apply to the sensor output imagery will provide twice the resolution required for the photo-map product.

The Skylab S-190A, multispectral and S-190B, Earth Terrain Cameras, which both resolved about 80 line-pairs per millimetre, are utilized to indicate the utility of the estimating formulas given above.

dinates. One millisecond precision (.001 sec) was achieved with Apollo 15, 16, and 17.

3. Attitude—Orientation angles (ω, ϕ, κ) can be obtained from stellar sensors such as those on Apollo 15, 16, and 17 with accuracy on the order of 5″ to 10″.

4. Altimetric distance—Radar and laser distance measuring devices that are aimed down the optical axis of the mapping camera produce distance from the spacecraft to the ground with a precision below 1 m.

5. Calibration—Determination of the radial and decentering distortion should reduce distortion errors to less than 10 micrometres in the image plane of the sensor.

Cartographic Potential of Skylab Cameras

Item	S-190A	S-190B
Orbital Altitude	435 km	435 km
Focal Length	150 mm	460 mm
Photo Scale (S_p)	2,900,000	945,000
Ground Resolution (Rg)	36 m	12 m
Allowable Enlargement (Ma)	8	8
Max. Map Scale (S_m)	360,000	118,000
B/H Ratio	0.16	0.47 (25°Conv.)
Accuracy (relative) horizontal position σ_p	30 m	10 m
(σ_m = 10 μm) elevation σ_h	180 m	20 m
Contour Interval (CI)	550 m	60 m

The B/H ratios being much less than 1.0 significantly degrades the systems' capability to determine elevations from stereoscopic pairs that is, σ_h is large because H/B is greater than one.

An imaging system designed for photogrammetric surveys must be concerned with two basic categories. The first is the metric characteristics of the system and spacecraft which are as follows.

1. Coordinates (X^c, Y^c, Z^c) on the orbit as a function of time. If the Global Positioning System is used to determine the orbit, an accuracy of ± 10 m may be expected.

2. Time (t_i) of exposure correlated with the coor-

The second category is the geometrical configuration of the sensor. This takes into account the stereoscopic coverage of the surface, considering both forward and sidelap of the imagery. Also, the mode of operation whether it be vertical or convergent and its overall effect on precision in extracting relative height measurements (σ_h) from the stereoscopic pairs is important.

Figure 17-17 illustrates coverage for a moon mission designed for an orbital altitude of 80 km. Note that the 10% sidelap at the equator will continually increase toward the poles for a polar orbit. The stellar sensor is also shown. It derives attitude and relates it to the down-looking sensor. It is also informative to recognize that the totally cloud-free moon could be completely photographed in 28 days. Section 17.8 gives details on planning photography from space.

Within the context of the geometrical configuration, the mode of operation, Figure 17-18, is of vital importance. Three modes of vertical and convergent camera configuration are illustrated. The base/height (B/H) ratio and the convergence angle (α) are vital contributors toward reducing the error (σ_h) in determining relative heights from stereoscopic pairs. Figure 17-19 shows that σ_h will be much larger when using photographs N and $N + 1$; where using photographs $N + 1$ and $N + 7$ will greatly reduce σ_h as a function of a larger B/H ratio.

In figure 17-19,

$$\sigma_h = \frac{H}{B}\frac{H}{F}\sigma_m \sqrt{2}$$

or

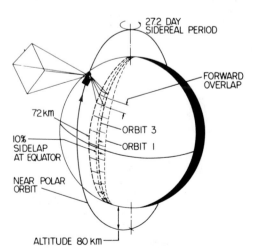

FIGURE 17-17. Moon mission coverage.

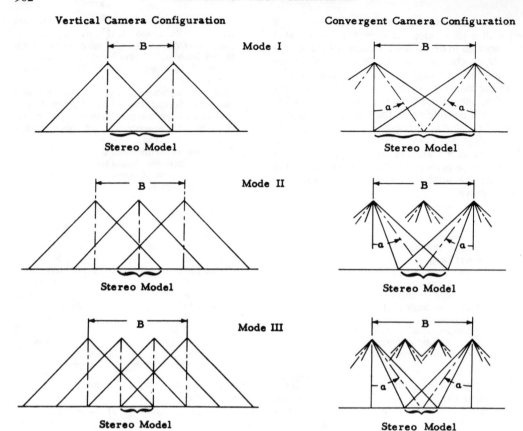

FIGURE 17-18. Mode of operation.

$$\sigma_h = Sp \frac{H}{B} \sigma_m \sqrt{2},$$

where

H = height of sensor above surface,
B = air-base,
f = focal length of sensor,
σ_m = precision of measurement of image, and
Sp = image scale.

A B/H ratio between 0.6 and 1.2 is a typical value depending on the requirements. Figure 17-20 shows a system designed for satellite photogrammetry. Its exterior orientation is provided by the satellite ephemeris, laser altimeter, and the stellar camera.

17.1.6 ELEMENTS OF SATELLITE PHOTOGRAMMETRY

In satellite photogrammetry, the perspective centers of the imaging systems may be assumed to be on the satellite's orbit. It follows, then, that the satellite must be observed to determine the imaging system's position in the orbit as a function of time. Examples of equipment used for observing satellites are the tracking stations of the Deep Space Network and the Spaceflight Tracking and Data Network (section 17.2). These stations are distributed around the globe. A

tracking station at Goldstone, California is shown in Figure 17-21.

These tracking stations determine direction, distance (range) and/or radial velocity (range-rate) of the object being tracked. These data are converted, by a process called orbit determination, to a set of six constants, called "orbital elements", that specify the orbit. The orbital elements most commonly used, the so-called "Eulerian elements", are defined in figure 17-22. More detail follows in section 17.2.7.1, but basically the elements

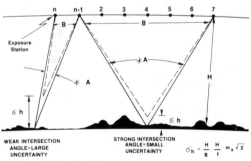

FIGURE 17-19. Relative photogrammetric accuracy vs angle of intersection.

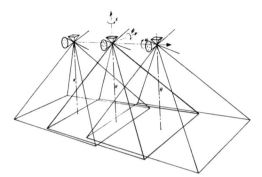

FIGURE 17-20. Exterior orientation of satellite photography.

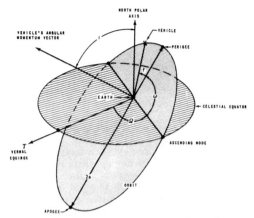

FIGURE 17-22. Eulerian elements of an orbit.

a and e determine the size and shape of the orbit, i, Ω and ω determine the orientation of the orbit, and T_0 is the instant at which the spacecraft was at perigee.

The satellite's coordinates are then usually the radius vector r, whose components are the radial distance from geocenter, and the true anomaly f. For satellite photogrammetry, the Eulerian elements are usually replaced by another set: the location and velocity of the satellite at a specific time. The satellite's coordinates are then its rectangular Cartesian coordinates. Section 17.2.7.4 gives a relationship between the Eulerian elements and the location/velocity elements (called the "state-vector"elements).

FIGURE 17-21. Tracking station, Goldstone, California. (Picture courtesy of Jet Propulsion Laboratory).

Thus the coordinates $(X^c, Y^c, Z^c)_i$ of the perspective centers of the camera (imaging system) along an orbital arc are not independent quantities, but instead are functions of the six orbital elements and time. A list of coordinates and corresponding times is called an *ephemeris*.

The other important quantities in satellite photogrammetry are the three angles of orientation (attitude) of the camera at the time t_i of exposure. Attitude is generally obtained by photographing a stellar field or by tracking a particular star or set of stars. The downward-looking (terrain) camera is calibrated to expose simultaneously with the stellar camera. As a result, the attitude of the stellar camera can be determined with respect to the stellar coordinates and transferred to the downward-looking camera.

In reducing the data from a stellar camera, one must first identify the stars on the photograph. Then the stars' right ascension and declination are calculated for the time of exposure. These are converted to direction cosines. The x, y-image-coordinates of the images of identified stars are then measured in a precise comparator, and the matrix of direction cosines of the image-rays are computed in the coordinate system of the stellar camera.

Then, assuming that the orientation matrix expressing the relationship between the stellar camera and the terrain camera is known by previous calibration, the relation between the two coordinate systems is then a simple 3-by-3 orthogonal rotation-matrix. Section 17.3.3 discusses this concept in detail. In future spacecraft, solid-state imaging-arrays may replace the stellar camera.

In summary, the tracking data are converted to an ephemeris of (X^c, Y^c, Z^c) coordinates as a function of time t_i. The stellar camera provides three orientation angles based on the star's positions. These are the six elements required to solve a photogrammetric problem. They may be used as first approximations in strip and block adjustments.

17.2 Orbit Determination

To do satellite photogrammetry, the location of the camera must be known for each instant that an image was formed. Because of the great height of the camera, it is usually not possible to locate the camera accurately by resection to points of known location on the ground. The camera's location must therefore be found by methods independent of the photography. Two such methods are practicable at present:

(a) measuring, at each instant of exposure and from points of known location, sufficient distances or directions to the satellite that the satellite's location is geometrically determinate;

(b) measuring, at suitable intervals of time, distances and/or directions and/or radial velocities of the satellite from a point or points of known location and using Newton's laws of motion to determine, from these measurements, the orbit of the satellite. The location of the camera is then at any instant determinable from the equation of the orbit.

The resection method used in aerial photogrammetry gives not only the location of the camera but also its orientation. The methods used in satellite photogrammetry do not give the orientation, and this must be found independently.

Method (a) has been used only infrequently up to the present time. It will probably be used much more extensively after the Global Positioning System of the U.S.A. Department of Defense becomes operational. Because of its minor importance, however, and because its theory is relatively simple, this method will not be further discussed in this section. Instead, we will concentrate on method (b). We provide here an overview of the principal kinds of equipment of present or potential usefulness for measuring the necessary distances, directions, and/or radial velocities and a summary of those parts of celestial mechanics which are important for satellite photogrammetry.

17.2.1 COORDINATE SYSTEMS

There are three types of coordinate systems commonly used in the computation of orbits: the astronomic right-ascension/declination, the astronomic Cartesian, and the Eulerian-element. The first of these is commonly used when giving observed coordinates, particularly in relation to the stars; the second is commonly used in computation, and the third in deriving the theory of motion of satellites.

Astronomers have handed down the tradition of conceiving of the distant stars as points fixed on the surface of an infinite sphere (the celestial sphere) (figure 17-23). The earth's equatorial plane (defined as the plane normal to the earth's axis of rotation) and the ecliptic plane (the plane defined by the earth's orbit about the sun) intersect the celestial sphere in great circles. The

direction of the line of intersection of the equatorial and ecliptic planes varies slowly (period of about 26,000 years) and plays an important role in the definition of non-rotating coordinate-systems. The ecliptic plane maintains, practically, a fixed orientation; observe that prescribing a specific reference time identifies a specific, fixed orientation of the slowly moving equatorial plane and therefore the vernal equinox[2] at that instant. This simple device of adopting the equator and equinox of a fixed date has been employed by astronomers to define the non-rotating coordinate systems to which the motion of all celestial objects can be referred (including the earth's motion). The fixed date and associated coordinate system are redefined every 50 years; the commonly used non-rotating system as of this writing is defined by the equator and equinox of 1950.0. The coordinate "right ascension" is measured from the vernal equinox in the plane of the equator, the coordinate "declination" is measured northward or southward from that plane (figure 17-23). The origin may be at the observer (topocentric) but, for satellite photogrammetry, is more usually taken at the barycenter of the primary.

The astronomic Cartesian coordinate system that is used for describing orbits about the earth has its origin at the earth's barycenter; the Z-axis is normal to the equatorial plane of 1950.0, the X-axis is directed toward the vernal equinox of 1950.0, and the Y-axis is normal to the X and Z axes to form a right-handed system. If the orbit is about some other body, that body's barycenter, rotational axis and equatorial plane are used instead. The plane of that body's orbit may be used instead of the plane of the ecliptic.

The Eulerian-element coordinate system (figure 17-22) is used when it is desired to specify the location of a satellite by only a single variable. The coordinate system is defined by six constants called the Eulerian elements of the orbit. It is described in section 17.2.6.

17.2.2 EQUIPMENT FOR MEASURING DIRECTIONS, DISTANCES, AND/OR RADIAL VELOCITIES

The equipment now in use or planned to be in use for measuring distances, directions, and/or radial velocities (range-rates) of spacecraft can be placed in four broad categories: cameras, lidars, radars, and frequency-measuring equipment.

---

[2] The vernal equinox(Y) is the point in the equatorial plane where the sun appears to an earth-centered observer to move from south to north up through the equatorial plane. This point is clearly along the instantaneous intersection of the equatorial and ecliptic planes and therefore precesses due to the precession of the equatorial plane.

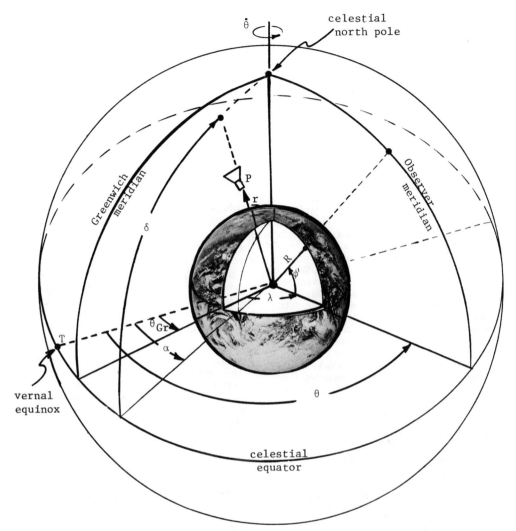

FIGURE 17-23. Celestial sphere.

(1) *Cameras* are historically the oldest type of equipment used. Their importance has considerably diminished in recent years, particularly for photography of imaging satellites, because unless the satellite carries an optical beacon, the satellite can be photographed only by reflected sunlight and from a station on the night side of the earth. This clearly is rather restrictive on the times and geometric conditions under which the satellite can be seen. When the often poor weather conditions are also taken into account, the camera's decline in importance is easy to understand.

(2) *Lidar,* also called "distance-measuring equipment" or "laser-ranging equipment," measures distance to a satellite by bouncing pulses of light off the reflecting surface of the satellite and converting the travel-time of the pulses to distance. The satellite (*e.g.,* GEOS-C or LAGEOS) is usually partly covered with an array of corner-cube reflectors which direct incident radiation back to its source. When atmospheric refraction is properly accounted for, lidars can measure distances to satellites with accuracies better than 1 metre—accuracies of a few decimetres have been reported.

(3) *Radar* is analogous, in its operation, to lidar. A sequence of pulses of radio waves is transmitted along a narrow (typically less than 2°) beam toward the spacecraft. The spacecraft returns the signal (either by reflection from the surface or from a corner-cube reflector, or by detecting the signal, amplifying it, and re-transmitting it) to the ground stations, where the travel-times of the pulses are measured and converted into distance. Many radars also can measure the direction in which the beam is pointed, allowing the direction of the satellite to be determined. While the accuracies of such measurements are considerably poorer than the distance-measuring accuracy of the radar, the measurements are often useful for checking the results determined from distance measurements. The accuracy of distance measurement by radar is somewhat less than that by lidar and is on the order of 1 to 2 metres. This may be lowered to 0.5 m under certain conditions.

A trend today is toward tracking satellites

from other satellites, rather than from the ground. As the most prominent example, NAVSTAR/GPS (U.S. Department of Defense) system now under development can be used in this way. It is to provide highly accurate, three-dimensional position, velocity, and time nearly instantaneously to suitably equipped users anywhere near the earth's surface, as well as to other spacecraft.

There are 18 satellites in the system, arranged in three groups of 6 satellites each, the satellites of a particular group being equally spaced along the same orbit. Each satellite carries an extremely stable atomic clock. The clock, a signal generator and a modulator generate, at two different carrier frequencies, a pseudo-random sequence of pulses which the satellite radiates continuously. These pulse-sequences are the signals which contain information regarding the satellite's location, satellite-clock error, and corrections for atmospheric refraction. Stations located in United States of America territory receive these signals and send the information to a master control station. This station processes the data and calculates future locations of the satellite, satellite-clock error, and atmospheric refraction. This information is then sent up to and stored in the satellite's memories for later broadcasting. The station also ensures that the satellites' clocks are synchronized to within a few microseconds. Users also receive the signals radiated by the satellites. If the user has a precise clock and timer, he can determine distances to the satellites and from these determine his location at the intersection of at least three spheres, each centered at a satellite. From 4 to 8 satellites will be within view of an observer simultaneously, allowing the user to determine his own location and the bias in his clock.

(4) *Frequency-measuring equipment* for measuring radial velocities (range-rates) to spacecraft comes in two varieties: one-way Doppler and two-way Doppler.

(a) One-Way Doppler: A sinusoidal radio signal is transmitted by the spacecraft at an approximately known frequency f_t. A ground station receives the signal with frequency f_r. The frequency shift $\Delta f = f_t - f_r$ is proportional to the radial velocity ($\dot{\rho}$) of the space craft with respect to the observer. The transmitter's frequency is subject to (sometimes poorly known) secular drifts and random variations in f_t, thus degrading the accuracy of this method. However, by incorporating methods for estimating the drift rate into the data reduction, this method has found wide application (*e.g.*, the Navy navigation satellite system).

(b) Two-Way Doppler: A sinusoidal radio signal is transmitted continuously by a ground station with frequency f_t. The spacecraft receives the signal at frequency f_{rs} and is equipped with a transponder which reproduces the received signal and re-transmits. The ground station receives the retransmitted signal with frequency f_r which is compared with the originally transmitted frequency f_t. The radial velocity then follows from the calculation

$$\dot{\rho} = (\frac{f_t - f_r}{2f_t}) \times \text{(speed of light)}$$
$$+ \text{(small refraction effects)}.$$

When refraction is properly accounted for, this method is capable of measuring (range-change)/(elapsed time) to within a few centimetres/second. Two-way doppler is the primary tracking scheme employed for interplanetary missions and is often used for tracking earth satellites.

Most equipment used for measuring direction, distance (range), and/or radial velocity (range-rate) of spacecraft is of the "tracking" variety—*i.e.*, is kept pointed toward the spacecraft while in operation. Such equipment, together with its auxiliary equipment, is known as a "tracking station". Tracking stations are frequently organized into "tracking networks", the stations in the network communicating with one or two centers (but usually not with each other) which act as message and computing centers. A large number of tracking networks are in operation throughout the world, most being maintained by various governmental agencies. The U.S.A. has the following networks, among others, in operation.

Spaceflight Tracking and Data Network (STDN). This network (managed by Goddard Spaceflight Center, NASA) is used for satellite tracking and communications. It tracks routinely virtually all manned and unmanned NASA spacecraft. Further information on this system and its use will be found in Gunther (1978).

Deep Space Network. This network is managed by the Jet Propulsion Laboratory. It is intended primarily for tracking of and communication with interplanetary spacecraft. The stations provide directions to the spacecraft but, more importantly, one- and two-way doppler determinations of range-rate. *See* Gunther (1978) for details.

Eastern Test Range and *Western Test Range.* These military radar networks are managed by the U.S. Air Force in cooperation with the U.S. Army, U.S. Navy, and NASA. The Eastern Test Range network is used primarily for tracking missiles launched from Kennedy Space Center whereas the Western Test Range network tracks missiles launched from Vandenberg Air Force Base in California. *See* Henriksen (1977) and Gunther (1978) for details on some of the stations in the network.

17.2.3 GEOMETRY OF OBSERVATIONS

17.2.3.1 RANGE[3] AND RANGE-RATE[3]

The triangle involved is shown in figure 17-24. From the geometry it is clear that the vector for the range is given by

$$\underline{\rho} = \underline{r} - \underline{R} \qquad (17.1)$$

---

[3] These terms, instead of "distance" and "radial velocity," were introduced by radar engineers and will be used here.

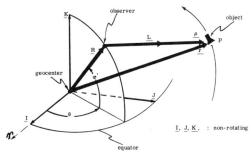

FIGURE 17-24. Fundamental observation triangle.

where

$\underline{r}$ = spacecraft position vector
$\underline{R}$ = observer position vector.

Equation (17.1) holds for any range-measuring instrument, regardless of its mount. Using methods of sections 17.2.4 and 17.2.5, the spacecraft's position $\underline{r}$ (t) and velocity $\underline{\dot{r}}$ (t) can be computed given initial conditions $\underline{r}$ (t_o), $\underline{\dot{r}}$ (t_o) or a set of "orbital elements". The position and motion of the observer $(\underline{R}(t)$ and $\underline{\dot{R}}(t))$ can be easily determined from the geometry of figure 17-24 as

$$\underline{R} = (R \cos \varphi' \cos \theta)\,\underline{I} + (R \cos \varphi' \sin \theta)\,\underline{J} + (R \sin \varphi')\,\underline{K} \qquad (17.2)$$

and

$$\underline{\dot{R}} = -R\dot{\theta}\,(\cos \varphi' \sin \theta)\,\underline{I} + R\,\dot{\theta}\,(\cos \varphi' \cos \theta)\,\underline{J} \qquad (17.3)$$

where

θ = right ascension of observer
$\quad = \theta_{Gr_0} + \dot{\theta}(t - t_o) + \lambda$
θ_{Gr_0} = right ascension of meridian of Greenwich
λ = longitude of observer's meridian
$\dot{\theta}$ = earth's speed of rotation $\approx 0.004\ 375\ 269$ rad/min
φ' = geocentric latitude of observor

We consider now the task of "simulating" (computing) the range (ρ) and range rate $(\dot{\rho})$, given spacecraft state $\underline{r}(t)$, $\underline{\dot{r}}(t)$, and observer station coordinates (R, φ', λ). If the components of spacecraft position $\underline{r}(t)$ and velocity $\underline{\dot{r}}$ are available in a geocentric, non-rotating frame, then the components of ρ (in the geocentric frame) can be computed from (17.1) as

$$\begin{aligned}\rho_x &= X - R \cos \varphi' \cos \theta \\ \rho_y &= Y - R \cos \varphi' \sin \theta \\ \rho_z &= Z - R \sin \varphi'. \end{aligned} \qquad (17.4)$$

The magnitude of ρ is then

$$\rho = [\rho_x{}^2 + \rho_y{}^2 + \rho_z{}^2]^{1/2} \qquad (17.5)$$

and the range-rate follows from differentiation of (17.5) as

$$\dot{\rho} = [\rho_x\dot{\rho}_x + \rho_y\dot{\rho}_y + \rho_z\dot{\rho}_z]/\rho \qquad (17.6)$$

and where $(\dot{\rho}_x, \dot{\rho}_y, \dot{\rho}_z)$ follow from differentiation of (17.1) or (17.4) as

$$\begin{aligned}\dot{\rho}_x &= \dot{X} + R\dot{\theta} \cos \varphi' \sin \theta \\ \dot{\rho}_y &= \dot{Y} - R\dot{\theta} \cos \varphi' \cos \theta \\ \dot{\rho}_z &= \dot{Z}. \end{aligned} \qquad (17.7)$$

Equations (17.4), (17.7), (17.5) and (17.6), evaluated sequentially (given $X, Y, Z, \dot{X}, \dot{Y}, \dot{Z}, t$, and observer's position coordinates R, φ', λ) to provide the mathematical model for the most common measurables, range (ρ) and range-rate $(\dot{\rho})$. For tracking equipment which measures angles and angular rates, it is necessary to consider the configuration of the instrument-mount and to relate these quantities to the spacecraft's and the observer's motion.

17.2.3.2 TRACKING-ANTENNA MOUNTS AND ASSOCIATED COORDINATE SYSTEMS

Tracking antennas generally have two degrees of rotational freedom. The three most common configurations are as follows.

Alt-azimuth Mount Referring to figure 17-25, the antenna rotates through the azimuth (AZ) about the local zenith and about a horizontal axis through the elevation angle (EL).

X-Y Mount Referring to Figure 17-26, the antenna rotates about a horizontal axis (directed either east or south, in the plane of the local horizon) through an angle θ_x and rotates through angle θ_y about a non-horizontal axis.

Polar Mount Referring to figure 17-27, the antenna rotates (about an axis parallel to the earth's axis of rotation) through the angle α_t and through an angle δ_t as shown.

The geometry associated with the fundamental observation triangle, for each of the above three mounts, is shown in figures 17-28 to 17-31. From these figures, the pairs of angles can be related to the satellite's and observer's coordinates as follows:

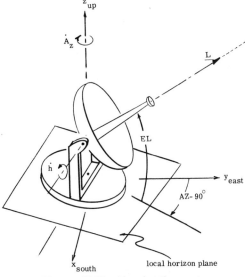

FIGURE 17-25. Alt-azimuth mount.

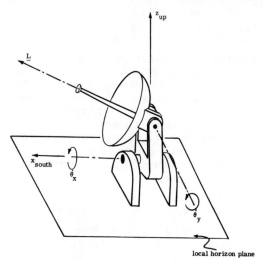

FIGURE 17-26. *X - Y* mount (+ *X* south).

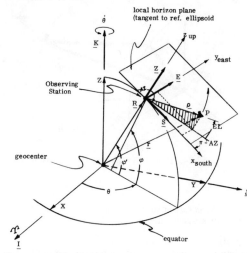

FIGURE 17-28. Orientation of the local horizon coordinate system relative to the geocentric, non-rotating coordinate system.

Alt-azimuth Mount

First observe from the figure 17-28 that the transformation between the non-rotating equatorial and the alt-azimuth coordinate systems is

$$[\underline{S}\ \underline{E}\ \underline{Z}]^T = [\mathbf{M}]\ [\underline{\mathbf{I}}\ \underline{\mathbf{J}}\ \underline{\mathbf{K}}]^T \qquad (17.8)$$

where the direction cosine matrix is

$$[\mathbf{M}] = \begin{bmatrix} s\varphi & 0 & -c\varphi \\ 0 & 1 & 0 \\ c\varphi & 0 & s\varphi \end{bmatrix} \begin{bmatrix} c\theta & s\theta & 0 \\ -s\theta & c\theta & 0 \\ 0 & 0 & 1 \end{bmatrix} = \begin{bmatrix} s\varphi c\theta & s\varphi s\theta & -c\varphi \\ -s\theta & c\theta & 0 \\ c\varphi c\theta & c\varphi s\theta & s\varphi \end{bmatrix}$$

where, for brevity, we employ the short-hand notation c = cosine, s = sine, and φ is the geodetic latitude. The horizontal components of the range vector can then be computed from the equatorial components (17.4) as

$$\begin{Bmatrix} \rho_{south} \\ \rho_{east} \\ \rho_{up} \end{Bmatrix} = [\mathbf{M}] \begin{Bmatrix} \rho_x \\ \rho_y \\ \rho_z \end{Bmatrix} . \qquad (17.10)$$

The azimuth (*AZ*) and elevation angles (*EL*) are then

$$AZ = \tan^{-1}\left[\frac{\rho_{east}}{-\rho_{south}}\right], 0 \le AZ \le 360° \quad (17.11)$$

and

$$EL = \sin^{-1}\left(\frac{\rho_{up}}{\rho}\right), -90° \le EL \le 90°. \quad (17.12)$$

X-Y Mount

The angles giving the antenna's orientation (figures 17-26 and 17-30) are

$$\theta_x = \tan^{-1}\left(\frac{-\rho_{east}}{\rho_{up}}\right), 0 \le \theta_x \le 180° \quad (17.13)$$

$$\theta_y = \sin^{-1}\left(\frac{\rho_z}{\rho}\right), -90° \le \theta_y \le 90°. \quad (17.14)$$

Polar Mount

The angles α_t (topocentric right ascension) and δ_t (topocentric declination) (*see* figures 17-27 and 17-31) are given by

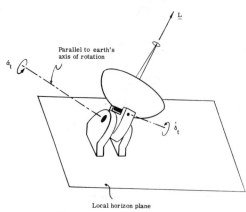

FIGURE 17-27. Polar mount.

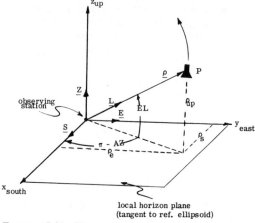

FIGURE 17-29. The range-vector geometry for instruments on an alt-azimuth mount.

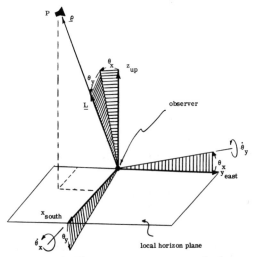

FIGURE 17-30. The range-vector geometry for instruments on X-Y mounts.

$$\alpha_t = \tan^{-1}\left(\frac{\rho_x}{\rho_y}\right), 0 \le \alpha_t \le 360° \qquad (17.15)$$

and

$$\delta_t = \sin^{-1}\left(\frac{\rho_z}{\rho}\right), -90° \le \delta_t \le 90°. \qquad (17.16)$$

17.2.4 KEPLER'S AND NEWTON'S LAWS OF PLANETARY MOTION

Using observations of planetary positions made by Tycho Brahe, Kepler derived the first essentially correct laws governing planetary motion; they were published in *Astronomia Nova* (1609). We paraphrase Kepler's laws (figure 17-32a and 32b) as follows.

Kepler's First Law (K1). Each planet moves along an elliptical path with the sun at one focus.
Kepler's Second Law (K2) Each planet moves in such a fashion that the radial line from the sun to the

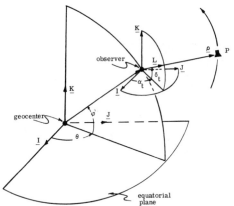

FIGURE 17-31. The range-vector geometry for instruments on polar mounts.

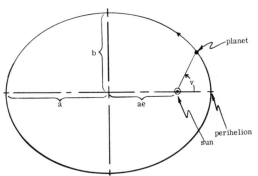

FIGURE 17-32a. Kepler's first law.

planet sweeps out area at a constant rate ("equal areas swept out in equal intervals of time").

If t_1, t_2, t_3, and t_4 are four instants at which a planet passed the labelled points of figure 17-32, and if $t_2 - t_1$ equals $t_4 - t_3$, then K2 requires that area A_1 equal area A_2.

Kepler's Third Law (K3). The squares of the periods (P_1^2 and P_2^2) of revolution about the Sun of any two planets have the same ratio as the cubes (a_1^3 and a_2^3) of their semi-major axes (a_1 and a_2).

In equation form, K3 is expressed simply as

$$(P_1 / P_2)^2 = (a_1 / a_2)^3. \qquad (17.17)$$

Newton formally verified that upon substituting his gravitational-force law into his second law of motion and solving the resulting differential equation, Kepler's laws could not only be verified, but generalized and extended. Newton's extensions of Kepler's three laws are as follows.

K1: Extended to include all conic sections (ellipses, parabolas, and hyperbolas) as possible paths.
K2: Established that "area being swept at a constant rate" is a necessary consequence of conservation of angular momentum.
K3: Showed that the third law is a special case (for one complete revolution) of a more fundamental relationship now known as "Kepler's Equation."

Newton's analysis also yielded a more general version of K3, namely

$$P^2 = 4 \pi^2 a^3 / G(m_1 + m_2). \qquad (17.18)$$

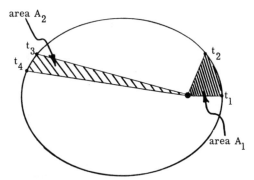

FIGURE 17-32b. Kepler's second law.

17.2.5 GEOMETRY OF THE ELLIPSE

Since most orbits of practical interest in orbital photogrammetry are elliptical, we direct our attention to elliptical orbits. The terminology and symbolism are shown in figure 17-33. In particular, notice the use of a reference circle of radius a, and the associated definition of eccentric anomaly, E. Table 17-3 provides formulas relating the more commonly encountered geometric quantities. Herrick (1972), should be consulted for detailed development of these results and more extensive lists of formulas.

tem $(\underline{\mathbf{I}}, \underline{\mathbf{J}}, \underline{\mathbf{K}})$ is given by the direction-cosine matrix $[\mathbf{C}]$ as

$$\begin{bmatrix} \mathbf{P} \\ \mathbf{Q} \\ \mathbf{W} \end{bmatrix} = [C(\Omega, i, \omega)] \begin{bmatrix} \mathbf{I} \\ \mathbf{J} \\ \mathbf{K} \end{bmatrix} \qquad (17.21)$$

and

$$\begin{bmatrix} x_\omega \\ y_\omega \\ z_\omega \end{bmatrix} = [C(\Omega, i, \omega)] \begin{bmatrix} X \\ Y \\ Z \end{bmatrix} \qquad (17.22)$$

where it follows from the geometry of figure 17-34b that

$$[C(\Omega, i, \omega)] = \begin{bmatrix} c\Omega c\omega - s\Omega cis\omega & s\Omega c\omega + c\Omega cis\omega & sis\omega \\ -c\Omega s\omega - s\Omega cic\omega & -s\Omega s\omega + c\Omega cic\omega & sic\omega \\ s\Omega si & -c\Omega si & ci \end{bmatrix} \qquad (17.34)$$

17.2.6 ORIENTATION OF THE ORBITAL PLANE

With reference to figure 17-34a, a coordinate system $(x_\omega, y_\omega, z_\omega)$ is defined with x_ω along unit vector $\underline{\mathbf{P}}$ toward perifocus (point of closest approach), z_ω normal to the orbital plane along $\underline{\mathbf{W}}$, and y_ω along $\underline{\mathbf{Q}} = \underline{\mathbf{W}} \times \underline{\mathbf{P}}$. The position vector $\underline{\mathbf{r}}$ can be written as

$$\underline{\mathbf{r}} = x_\omega \underline{\mathbf{P}} + y_\omega \underline{\mathbf{Q}} + (0)\underline{\mathbf{W}} \qquad (17.19)$$

or as

$$\underline{\mathbf{r}} = X\underline{\mathbf{I}} + Y\underline{\mathbf{J}} + Z\underline{\mathbf{K}}. \qquad (17.20)$$

The orientation of coordinate system $(\underline{\mathbf{P}}, \underline{\mathbf{Q}}, \underline{\mathbf{W}})$ with respect to the nonrotating coordinate sys-

The direction-cosine matrix is orthogonal, so that its inverse is simply its transpose

$$\begin{bmatrix} X \\ Y \\ Z \end{bmatrix} = [C^T(\Omega, i, \omega)] \begin{bmatrix} x_\omega \\ y_\omega \\ z_\omega \end{bmatrix}. \qquad (17.35)$$

It follows from (17.34) that, the orientation angles are given by

$$\Omega = \tan^{-1}(C_{31}/-C_{32}) = \tan^{-1}(W_x/-W_y)\ 0 \le \Omega < 360° \qquad (17.36)$$

$$i = \cos^{-1}(C_{33}) = \cos^{-1}(W_z)\quad 0 \le i < 180° \qquad (17.37)$$

$$\omega = \tan^{-1}(C_{13}/C_{23}) = \tan^{-1}(P_z/Q_z)\quad 0 \le \omega < 360°. \qquad (17.38)$$

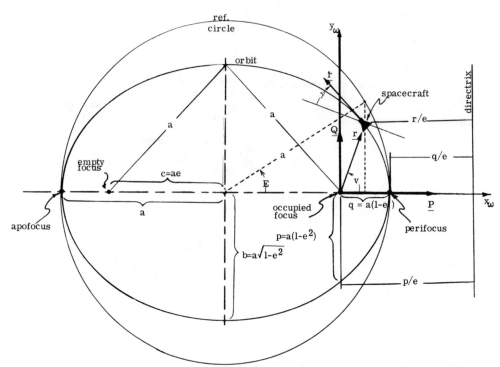

FIGURE 17-33. Detailed geometry of elliptic orbits.

TABLE 17-3. RELATIONS BETWEEN GEOMETRIC ELEMENTS OF AN ELLIPSE

$r = a(1 - e\cos E)$	(17.23)
$x_\omega = a(\cos E - e)$	(17.24)
$y_\omega = a\sqrt{1 - e^2}\ \sin E$	(17.25)
$\cos v = (\cos E - e)/(1 - e\cos E)$	(17.26)
$\sin v = \sqrt{1 - e^2}\ \sin E/(1 - e\cos E)$	(17.27)
$\tan v = \sqrt{1 - e^2}\ \sin E/(\cos E - e)$	(17.28)
$\cos E = (\cos v + e)/(1 + e\cos v)$	(17.29)
$\sin E = \sqrt{1 - e^2}\ \sin v/(1 + e\cos v)$	(17.30)
$\tan E = \sqrt{1 - e^2}\ \sin v/(\cos v + e)$	(17.31)
$\tan\gamma = e\sin E/\sqrt{1 - e^2}$	(17.32)
$b = a\sqrt{1 - e^2}$	(17.33)

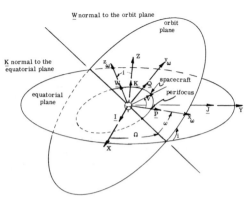

FIGURE 17-34b. Orientation of the orbit-plane system relative to the inertial coordinate system.

17.2.7 THE TWO-BODY PROBLEM

Newton placed celestial mechanics upon a firm analytical foundation, the cornerstone of which is his law of universal gravitation. It states that two point masses (m_1 and m_2) attract each other with a gravitational force whose magnitude is given by

$$f_{12} = \frac{G\,m_1\,m_2}{r^2} \qquad (17.39)$$

where r is the distance between m_1 and m_2. In vectorial form (figure 17-35)

$$\begin{aligned} \mathbf{f}_{12} &= -\mathbf{f}_{21} \\ &= G\,m_1\,m_2\,\mathbf{r}/r^3. \end{aligned} \qquad (17.40)$$

17.2.7.1 EQUATIONS OF MOTION

Newton's second law of motion requires that the vectorial sum of all forces acting on a particle equal its mass times its acceleration. This law applied to masses m_1 and m_2 yields the following equations of motion:

$$\begin{aligned} m_1\,\ddot{\mathbf{r}}_1 &= G\,m_1\,m_2\,\mathbf{r}/r^3 \\ m_2\,\ddot{\mathbf{r}}_2 &= -G\,m_1\,m_2\,\mathbf{r}/r^3. \end{aligned} \qquad (17.41)$$

Since the relative acceleration $\ddot{\mathbf{r}}$ is

$$\ddot{\mathbf{r}} = \ddot{\mathbf{r}}_2 - \ddot{\mathbf{r}}_1, \qquad (17.42)$$

substitution from equations (17.40) into (17.42) yields

$$\ddot{\mathbf{r}} = -\mu\mathbf{r}/r^3 \qquad (17.43)$$

where the gravitational-mass constant μ is $G\,(m_1 + m_2)$. To the extent that m_2 (e.g., the spacecraft's mass) is negligible compared to the mass m_1 (e.g., the earth's mass), μ is approximately $G\,m_1$.

Thus the objective is to determine a solution for the spacecraft's position of the form

$$\mathbf{r}\,(t) = \mathbf{r}\,(k_1, k_2, k_3, k_4, k_5, k_6; t - t_0).$$

The spacecraft's velocity $\dot{\mathbf{r}}\,(t)$ is of course found from this by differentiation. The six constants $\{k_i\}$ are frequently determined as the spacecraft's position and velocity in rectangular Cartesian coordinates at a specified time t_0 in the orbit — $(X_0, Y_0, Z_0, \dot{X}_0, \dot{Y}_0, \dot{Z}_0)$. It is also common, however, to use for these constants $\{k_i\}$ the Eulerian elements ($a, e, i, \Omega, \omega, T_0$), where a is the semi-major axis, e the eccentricity of the orbit, i the inclination, Ω the angle, in the equa-

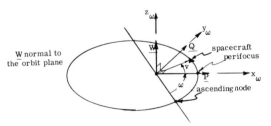

FIGURE 17-34a. The orbit-plane coordinate system.

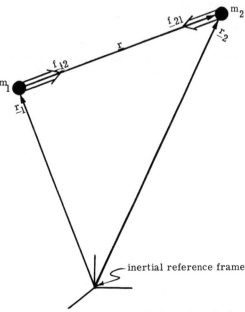

FIGURE 17-35. Newton's law of universal gravitation.

torial plane, from some fixed direction to the line of nodes, ω the angle from the line of nodes to perigee, and T_o the time of passage of the satellite through perigee.

The equation of motion (17.43) is the starting point for derivation of virtually every result in the two-body problem. The three ordinary differential equations implicit in (17.43) are non-linear and mathematically tractable, but solving them is not trivial. We will summarize the most fundamental integrals below, and then summarize algorithms which embody a number of dynamical formulas not derived here.

17.2.7.2 THE ENERGY (VIS-VIVA) INTEGRAL

Taking the dot-product of both sides in (17.43) with the velocity vector $\dot{\underline{r}}$ yields

$$\ddot{\underline{r}} \cdot \dot{\underline{r}} = -\frac{\mu}{r^3} \, \underline{r} \cdot \dot{\underline{r}} \qquad (17.44)$$

or

$$\ddot{\underline{r}} \cdot \dot{\underline{r}} = -\frac{\mu \dot{r}}{r^2} . \qquad (17.45)$$

Both sides of this equation are perfect differentials

since $\quad \dfrac{d}{dt}\left[\dfrac{\dot{\underline{r}} \cdot \dot{\underline{r}}}{2}\right] = \ddot{\underline{r}} \cdot \dot{\underline{r}}$ and $\dfrac{d}{dt}\left(\dfrac{\mu}{r}\right) = -\dfrac{\mu \dot{r}}{r^2}$.

Integration of (17.45) yields the energy equation

$$\frac{1}{2} V^2 - \frac{\mu}{r} = \text{constant} = k \qquad (17.46)$$
$$= \text{(total energy)/(unit mass)}.$$

Evaluation of this constant gives $k = \mu/2a$. The energy or *vis-viva* equation is often written as

$$V^2 = \mu\left(\frac{2}{r} - \frac{1}{a}\right). \qquad (17.47)$$

This equation plays a central role in approximating orbits. Observe that the total energy is a function of the semi-major axis only. One special case is noteworthy. To determine circular orbit speed, if $r = a = $ constant, then the energy equation (17.47) yields circular orbit speed as

$$V_c = \sqrt{\frac{\mu}{r}} . \qquad (17.48)$$

17.2.7.3 KEPLER'S EQUATION

Observe that the vector cross-product of the equation of motion (17.43) with $\underline{r}$ is

$$\underline{r} \times \ddot{\underline{r}} = -\mu \underline{r} \times \underline{r}/r^3 = 0, \qquad (17.49)$$

and since

$$d(\underline{r} \times \dot{\underline{r}})/dt = 2 \, \underline{r} \times \ddot{\underline{r}},$$

this integrates to give

$$\underline{r} \times \dot{\underline{r}} = h = \text{constant}$$
$$= \sqrt{\mu a \, (1 - e^2)} \, \underline{W} \quad \text{(Herrick, 1972).} \quad (17.50)$$

Now, in the orbit-plane coordinate-system

$$\underline{r} = x_\omega \underline{P} + y_\omega \underline{Q} \qquad (17.51)$$

$$\dot{\underline{r}} = \dot{x}_\omega \, \underline{P} + \dot{y}_\omega \, \underline{Q} \qquad (17.52)$$

so that (17.50) becomes

$$\underline{r} \times \dot{\underline{r}} = (x_\omega \dot{y}_\omega - \dot{x}_\omega y_\omega) \, \underline{W} = \sqrt{\mu a \, (1 - e^2)} \, \underline{W}$$

or

$$\sqrt{\mu a \, (1 - e^2)} = x_\omega \dot{y}_\omega - \dot{x}_\omega y_\omega. \qquad (17.53)$$

Then

$$x_\omega = a \, (\cos E - e) \qquad (17.54)$$
$$y_\omega = a \sqrt{1 - e^2} \sin E \qquad (17.55)$$

and therefore, in terms of eccentric anomaly E, equation (17.53) reduces to

$$\sqrt{\mu} \, a^{-3/2} \, dt = (1 - e \cos E) dE \qquad (17.56)$$

which, in turn integrates directly to

$$\sqrt{\mu} \, a^{-3/2} \, (t - t_0) = E - e \sin E - (E_0 - e \sin E_0). \qquad (17.57)$$

This result, known as *Kepler's Equation* was first derived by Newton and plays a central role in classical solutions for two-body position as a function of time. The classical form of Kepler's equation is

$$M = E - e \sin E \qquad (17.58)$$

where the classical definitions are

$$M = M_0 + n(t - t_0) = \text{mean anomaly} \qquad (17.59)$$
$$n = \sqrt{\mu} \, a^{-3/2} = \text{mean angular motion} \qquad (17.60)$$
$$t_0 = \text{an arbitrary reference epoch during the motion.}$$

Inversion of Kepler's equation for E has given rise to a large number of iterative root-solving methods, including the grandfather of modern differential correction processes, Newton's root-solving method. As applied to solution of Kepler's equation, this converges reliably in four or five iterations; we summarize this algorithm as follows.

First estimate E

$$E_c \cong M + e \sin M + \frac{e^2}{2} \sin 2 M + \ldots$$

Then calculate

$$M_c = E_c - \sin E_c$$
$$\frac{dM}{dE}\bigg|_c = 1 - e \cos E_c$$
$$\Delta E = (M - M_c)/\left(\frac{dM}{dE}\bigg|_c\right)$$
$$E_c = E_c + \Delta E$$

The loop is repeated until ΔE is less than a prescribed tolerance (typically 10^{-8}), then $E = E_c$ of the final iteration.

17.2.7.4 TRANSFORMATION FROM POSITION AND VELOCITY COORDINATES TO EULERIAN ELEMENTS

The following sequence of equations is provided as a useful algorithm. Many of the formulas have been encountered in the foregoing developments. Any new symbols may be taken as definitions in terms of the quantities on the *right-hand sides*. The Eulerian elements (a, e, i, Ω, ω, T) and related constants are computed from (X_0, Y_0, Z_0, $\dot{X}_0$, $\dot{Y}_0$, $\dot{Z}_0$, t_0) as follows.

$$r_0 = (X_0^2 - Y_0^2 + Z_0^2)^{1/2} = |\mathbf{r}_0|$$

$$V_0 = (\dot{X}_0^2 + \dot{Y}_0^2 + \dot{Z}_0^2)^{1/2} = |\dot{\mathbf{r}}_0|$$

$$1/a = (2/r_0 - V_0^2/\mu), \tag{17.47}$$

$$n = \sqrt{\mu}\, a^{-3/2} \tag{17.60}$$

$$D_0 \equiv \mathbf{r}_0 \cdot \dot{\mathbf{r}}_0 = X_0\dot{X}_0 + Y_0\dot{Y}_0 + Z_0\dot{Z}_0 \tag{17.61}$$

$$e \cos E_0 = 1 - r_0/a \tag{17.23}$$

$$e \sin E_0 = D_0/\sqrt{\mu\, a} \tag{17.62}$$

$$e = [(e \sin E_0)^2 + (e \cos E_0)^2]^{1/2} \tag{17.63}$$

$$E_0 = \tan^{-1}\left[\frac{e \sin E_0}{e \cos E_0}\right] \tag{17.64}$$

$$M_0 = E_0 - e \sin E_0 \tag{17.58}$$

$$T = t_0 - M_0/n \tag{17.65}$$

$$\dot{D}_0 = \mu(e \cos E_0)/r_0 \tag{17.66}$$

$$\mathbf{P} = \frac{1}{\mu e}(\dot{D}_0\mathbf{r}_0 - D_0\dot{\mathbf{r}}_0)$$

$$= C_{11}\mathbf{I} + C_{12}\mathbf{J} + C_{13}\mathbf{K} \tag{17.67}$$

$$p = a(1 - e^2)$$

$$H_0 = r_0 - p \tag{17.68}$$

$$\dot{H}_0 = D_0/r_0 \tag{17.69}$$

$$\mathbf{Q} = \frac{1}{e\sqrt{\mu\, p}}(\dot{H}_0\mathbf{r}_0 - H_0\dot{\mathbf{r}}_0) \tag{17.70}$$

$$= C_{21}\mathbf{I} + C_{22}\mathbf{J} + C_{23}\mathbf{K}$$

$$\mathbf{W} = \mathbf{P} \times \mathbf{Q} = C_{31}\mathbf{I} + C_{32}\mathbf{J} + C_{33}\mathbf{K} \tag{17.71}$$

$$i = \cos^{-1}(\mathbf{W} \cdot \mathbf{K}) = \cos^{-1}(C_{33}) \tag{17.72}$$

$$\Omega = \tan^{-1}\left(\frac{\mathbf{W} \cdot \mathbf{I}}{-\mathbf{W} \cdot \mathbf{J}}\right) = \tan^{-1}\left(\frac{C_{31}}{-C_{32}}\right) \tag{17.73}$$

$$\omega = \tan^{-1}\left(\frac{\mathbf{P} \cdot \mathbf{K}}{\mathbf{Q} \cdot \mathbf{K}}\right) = \tan^{-1}\left(\frac{C_{13}}{C_{23}}\right). \tag{17.74}$$

The use of this algorithm is restricted to computation of elliptical orbits and contains several singularities and associated numerical difficulties at zero eccentricity (e) and zero inclination (i). These singularities arise because the classical approach measures various angles from two lines which may be poorly defined (the major axis, and the line of nodes). These singularities may be avoided, using the method of section 17.2.7.6.

17.2.7.5 CLASSICAL SOLUTION OF THE TWO-BODY PROBLEM

Given initial position and velocity, we summarize here the classical procedure for calculating orbital position and velocity at any subsequent time. The solution here is a continuation of calculations based upon the foregoing formula summary of section 17.2.7.4, and assumes that the constants (a, e, n, M_0, p, [$\mathbf{C}$]) are available from those calculations. The solution for current position and velocity proceeds as follows:

$$M = M_0 + n(t - t_0) \tag{17.59}$$

$$M = E - e \sin E \text{ (solve for } E \text{ using} \tag{17.58}$$
$$\text{Newton's method as discussed}$$
$$\text{in section 17.2.7.3)}$$

$$x_\omega = a(\cos E - e) \tag{17.24}$$

$$y_\omega = a\sqrt{1 - e^2} \sin E \tag{17.25}$$

$$r = a(1 - e \cos E) \tag{17.23}$$

$$\dot{x}_\omega = -(\sqrt{\mu\, a} \sin E)/r \tag{17.75}$$

$$\dot{y}_\omega = (\sqrt{\mu\, p} \cos E)/r \tag{17.76}$$

$$\mathbf{r} = [\mathbf{C}]^{\mathrm{T}} \mathbf{x}\omega \tag{17.77}$$

and

$$\dot{\mathbf{r}} = [\mathbf{C}]^{\mathrm{T}} \dot{\mathbf{x}}\omega .$$

The above solution, since it is directly related to the calculations of section 17.2.7.4, does degrade near circular orbits (where $\mathbf{P}$ and $\mathbf{Q}$ are undefined when e goes to zero) and is, of course, valid only for closed orbits (since eccentric anomaly is used).

17.2.7.6 A NON-SINGULAR SOLUTION OF THE ELLIPTIC TWO-BODY PROBLEM

Herrick (1972) has developed several alternatives to the classical solution summarized above. Herrick's solutions accomplish the important practical objective of eliminating computational singularities. Restricting our attention here to the elliptic two-body problem, we summarize Herrick's "f and g solution" as follows

$$r_0 = (\mathbf{r}_0 \cdot \mathbf{r}_0)^{1/2} = (X_0^2 + Y_0^2 + Z_0^2)^{1/2}$$

$$V_0 = (\dot{\mathbf{r}}_0 \cdot \dot{\mathbf{r}}_0)^{1/2} = (\dot{X}_0^2 + \dot{Y}_0^2 + \dot{Z}_0^2)^{1/2}$$

$$D_0 = \mathbf{r}_0 \cdot \dot{\mathbf{r}}_0 = X_0\dot{X}_0 + Y_0\dot{Y}_0 + Z_0\dot{Z}_0$$

$$1/a = (2/r_0 - V_0^2/\mu). \tag{17.47}$$

Solve "Kepler's modified equation" in the form

$$\frac{\sqrt{\mu}\, (t - t_0)}{a^{3/2}} = \Phi - (1 - r_0/a) \sin \Phi \tag{17.78}$$

$$+ \frac{D_0}{\sqrt{\mu\, a}}(1 - \cos \Phi)$$

for Φ, using Newton's root-solving method (the starting estimate of $\Phi = \sqrt{\mu}\, a^{-3/2} (t - t_0)$ is adequate).

$$f = 1 - a(1 - \cos \Phi)/r_0 \tag{17.79}$$

$$g = (t - t_0) - a^{3/2} (\Phi - \sin \Phi)/\sqrt{\mu} \tag{17.80}$$

$$\mathbf{r} = f\mathbf{r}_0 + g\dot{\mathbf{r}}_0 \tag{17.81}$$

$$r = (X^2 + Y^2 + Z^2)^{1/2}$$

$$\dot{f} = \frac{-\sqrt{\mu\, a}}{\mathbf{r} \cdot \mathbf{r}_0} \sin \Phi \tag{17.82}$$

$$\dot{g} = 1 - \frac{a}{r}(1 - \cos \Phi) \tag{17.83}$$

$$\dot{\mathbf{r}} = \dot{f}\mathbf{r}_0 + \dot{g}\dot{\mathbf{r}}_0. \tag{17.84}$$

The essence of the above solution is that the reference direction for all angles is the initial position vector $\mathbf{r}_0$ instead of the perifocus direction $\mathbf{P}$. Now since $\mathbf{r}_0$ is well defined for all possible initial conditions, whereas $\mathbf{P}$ is not, the advantage is clear. It can be shown that $\Phi = E - E_0$.

For convenient reference, selected astrodynamic constants and data associated with characteristic two-body orbits about the earth, sun and moon are summarized in table 17-4.

17.2.8 INTEGRATION OF PERTURBED MOTION

Given the large number of existing integration procedures, perturbations, and their associated notational complexities, it is a rather challenging task to survey the subject adequately. The discussion is necessarily restricted to the most central concepts; references are cited that will provide detailed developments.

First of all, observe that setting up the equations for integration of a satellite orbit in the presence of non-two-body effects requires the following fundamental decisions:

a mathematical model of the perturbing effects must be defined;
a particular set of differential equations (of the many possible sets) must be selected which, upon integration, will give the orbit;
a specific numerical or analytical method (to carry out the integration) must be selected.

A wide variety of algorithms will satisfy the above general requirements. The decisions are coupled and affect in a complicated manner the efficiency and accuracy of the calculation. We consider only the most fundamental aspects of these decisions here.

17.2.8.1 NON-TWO-BODY ACCELERATIONS

Application of Newton's second law of motion to specific non-two-body force assumptions lead directly to equations of motion which assume the relation

$$\ddot{\underline{r}} = - \mu \frac{\underline{r}}{r^3} + \Delta \ddot{\underline{r}} \qquad (17.85)$$

where the perturbing accelerations ($\Delta \ddot{\underline{r}}$) are due to effects such as drag, higher harmonics in the earth's gravity field, gravitation of sun and moon, and radiation pressure.

17.2.8.2 DIRECT AND INTERMEDIARY METHODS

In direct methods, the total acceleration [equation (17.85), including an arbitrary set of perturbing effects] is integrated numerically to determine velocity and position as functions of time. This process is indicated formally as

$$\dot{\underline{r}}(t) = \int_{t_0}^{t} \ddot{\underline{r}}(\tau) \, d\tau + \dot{\underline{r}}(t_0)$$

$$\underline{r}(t) = \int_{t_0}^{t} \dot{\underline{r}}(\tau) \, d\tau + \underline{r}_0(t_0). \qquad (17.86)$$

A specific numerical method must be adopted to carry out the integrations (17.86), thus a direct method refers to "what" is integrated rather

than the numerical process employed in the integration process (there appears to be confusion on this point in the literature). In section 17.2.8.4, methods for carrying out these integrations are discussed. A direct method is the most straightforward and commonly used method, probably because the initial analytical work is minimal, and it applies to arbitrarily perturbed motion. On the other hand, no advantage is taken of the possible smallness of the perturbations and the fact that the gross nature of the motion (i.e., the two-body orbit) is analytically solvable.

In intermediary methods (of which Encke's is one) on the other hand, use is made from the outset of a nominal two-body orbit. For instance, let the *nominal* acceleration be denoted by the two-body acceleration

$$\ddot{\underline{r}}_n = \frac{-\mu}{r_n^3} \, \underline{r}_n. \qquad (17.87)$$

The *actual* acceleration is modeled in the form of equation (17.85) as

$$\ddot{\underline{r}} = \frac{-\mu}{r^3} \underline{r} + \Delta \ddot{\underline{r}} \qquad (17.88)$$

where $\Delta \ddot{\underline{r}}$ is short-hand for "all non-two-body terms."

Now, with reference to figure 17-36, suppose that the initial position and velocity at some instant t_0 are used to specify the nominal motion [solution of (17.88)]. The resulting nominal solution can be efficiently derived analytically from section 17.2.7.6. At any time t, after t_0, the actual motion $\underline{r}(t)$ will depart from the nominal motion $\underline{r}_n(t)$ by a vector $\delta \underline{r}(t)$ so that

$$\underline{r}(t) = \underline{r}_n(t) + \delta \underline{r}_n(t)$$

$$\dot{\underline{r}}(t) = \dot{\underline{r}}_n(t) + \delta \dot{\underline{r}}_n(t) \qquad (17.89)$$

$$\ddot{\underline{r}}(t) = \ddot{\underline{r}}_n(t) + \delta \ddot{\underline{r}}_n(t)$$

so the "departure-acceleration" of the actual from the nominal motion is

$$\delta \ddot{\underline{r}} = \ddot{\underline{r}} - \ddot{\underline{r}}_n. \qquad (17.90)$$

Substitution of the equations of motion yields the differential equation of "departure-motion"

$$\delta \ddot{\underline{r}} = \mu \left(\frac{\underline{r}_n}{r_n^3} - \frac{\underline{r}}{r^3} \right) + \Delta \ddot{\underline{r}}. \qquad (17.91)$$

Now this equation can, in principle, be integrated numerically to give departure-velocity and position and substituted into (17.89) to obtain the actual motion. However, the parenthetical term of (17.91) causes loss of significant figures since it is a small difference of two large terms. Herrick (1972) describes the classical re-arrangement of this equation to the final, numerically well-behaved form

$$\delta \ddot{\underline{r}} = \mu(\epsilon \underline{r} - \delta \underline{r}) + \Delta \ddot{\underline{r}} \qquad (17.92)$$

where

$$\epsilon = 3\, q - \frac{3 \cdot 5}{1 \cdot 2} q^2 + \frac{3 \cdot 5 \cdot 7}{1 \cdot 2 \cdot 3} q^3 \ldots \qquad (17.93)$$

TABLE 17-4. USEFUL ASTRONOMICAL CONSTANTS

Geocentric Orbits	Solar Orbits	Lunar Orbits
1 Earth Radius = $R_\oplus$ = 6 378.160 km = 3 963.205 st. mi.	1 astronomical unit (au) = 1.495 978 710 × 10⁸ km = 9.295 580 745 × 10⁷ st. mi.	1 Lunar Radius $R_{\mathrm{☽}}$ = 1 738.000 km = 1 079.943 st. mi.
$\mu_\oplus \equiv GM_\oplus$ = $3.986\ 012 \times 10^5\ \dfrac{\mathrm{km}^3}{\mathrm{sec}^2}$ = $1.536\ 216 \times 10^{-6}\ \dfrac{R_\oplus^3}{\mathrm{sec}^2}$	$\mu_\odot \equiv GM_\odot$ = $1.327\ 124\ 995 \times 10^{11}\ \dfrac{\mathrm{km}^3}{\mathrm{sec}^2}$ = $2.959\ 123\ 392 \times 10^{-4}\ \dfrac{\mathrm{au}^3}{\mathrm{day}^2}$	$\mu_{\mathrm{☽}} \equiv GM_{\mathrm{☽}}$ = $4.902\ 783 \times 10^3\ \dfrac{\mathrm{km}^3}{\mathrm{sec}^2}$ = $9.338\ 849 \times 10^{-7}\ \dfrac{R_{\mathrm{☽}}^3}{\mathrm{sec}}$
$\sqrt{\mu_\oplus} = \sqrt{GM_\oplus}$ = $6.313\ 487 \times 10^2\ \dfrac{\mathrm{km}^{3/2}}{\mathrm{sec}}$ = $1.239\ 442 \times 10^{-3}\ \dfrac{R_\oplus^{3/2}}{\mathrm{sec}}$	$\sqrt{\mu_\odot} = \sqrt{GM_\odot}$ = $3.642\ 972\ 685 \times 10^5\ \dfrac{\mathrm{km}^{3/2}}{\mathrm{sec}}$ = $1.720\ 210\ 276 \times 10^2\ \dfrac{\mathrm{au}^{3/2}}{\mathrm{day}}$	$\sqrt{\mu_{\mathrm{☽}}} = \sqrt{GM_{\mathrm{☽}}}$ = $70.019\ 88\ \dfrac{\mathrm{km}^{3/2}}{\mathrm{sec}}$ = $9.663\ 772 \times 10^{-4}\ \dfrac{R_{\mathrm{☽}}^{3/2}}{\mathrm{sec}}$
$V_{c\oplus} = \sqrt{\dfrac{\mu_\oplus}{R_\oplus}}$ = $7.905\ 359\ \dfrac{\mathrm{km}}{\mathrm{sec}}$ = $17\ 683.8\ \dfrac{\text{st. mi.}}{\mathrm{hr}}$	$V_{c\odot} = \sqrt{\dfrac{\mu_\odot}{\mathrm{lau}}}$ = $29.784\ 698\ 48\ \dfrac{\mathrm{km}}{\mathrm{sec}}$ = $66\ 626.472\ 98\ \dfrac{\text{st. mi.}}{\mathrm{hr}}$ = 1 "EMOS"	$V_{c\mathrm{☽}} = \sqrt{\dfrac{\mu_{\mathrm{☽}}}{R_{\mathrm{☽}}}}$ = $1.679\ 564\ \dfrac{\mathrm{km}}{\mathrm{sec}}$ = $3\ 757.077\ \dfrac{\text{st. mi.}}{\mathrm{hr}}$
$P_{c\oplus} = \dfrac{2\pi R_\oplus^{3/2}}{\sqrt{\mu_\oplus}}$ = 5 069.366 sec = 84.489 44 min = 1.408 157 hr	$P_{c\odot} = \dfrac{2\pi \mathrm{au}^{3/2}}{\sqrt{\mu_\odot}} \simeq 1\ \text{year}$ = 365.279 647 6 days	$P_{c\mathrm{☽}} = \dfrac{2\pi R_{\mathrm{☽}}^{3/2}}{\sqrt{\mu_{\mathrm{☽}}}} = 6\ 501.793\ \text{sec}$ = 108.363 2 min = 1.806 054 hr

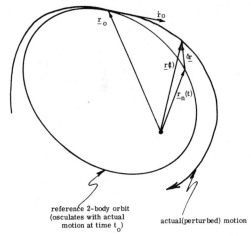

FIGURE 17-36. Integration of departure motion, intermediary method.

$$q = \frac{1}{2r_n^2}\left[\delta \underline{\mathbf{r}} \cdot (\underline{\mathbf{r}}_n + \underline{\mathbf{r}})\right] \tag{17.94}$$

which can be integrated formally as

$$\delta \dot{\underline{\mathbf{r}}}\,(t) = \int_{t_0}^{t} \delta \ddot{\underline{\mathbf{r}}}\,(\tau)\,d\tau$$

$$\tag{17.95}$$

$$\delta \underline{\mathbf{r}}\,(t) = \int_{t_0}^{t} \delta \dot{\underline{\mathbf{r}}}\,(\tau)\,d\tau$$

and, as in the direct method, these integrations can be performed by any convenient numerical device (see section 17.2.8.4). Since the actual motion osculates with the reference motion at t_0, then the additive constants of integration in (17.95) are zero.

Now, the fundamental advantage of the intermediary method over the direct is that, if the perturbations are small, then fewer significant figures are required in the integration of the small departure-acceleration (17.95) as compared to integration of the total acceleration (17.86) to maintain the same accuracy in $\underline{\mathbf{r}}\,(t)$ and $\dot{\underline{\mathbf{r}}}\,(t)$. Since requiring more significant figures (in numerical integration) usually results in smaller integration-time steps, one often finds that larger steps by Encke or similar integration lead to net savings in computer time. In the event that the integrated departures grow larger than some prescribed tolerance, they can be reduced to zero again by causing the intermediary orbit to osculate at the current position (this process of updating the reference motion is referred to as "rectification").

17.2.8.3 VARIATION OF PARAMETERS

The solution of the two body problem is often conveniently expressed in the form

$$S(t) \begin{cases} X = X(t,\,e_1,\,e_2,\,e_3,\,e_4,\,e_5,\,e_6) \\ Y = Y(t,\,e_1,\,e_2,\,e_3,\,e_4,\,e_5,\,e_6) \\ Z = Z(t,\,e_1,\,e_2,\,e_3,\,e_4,\,e_5,\,e_6) \\ \dot{X} = \dot{X}(t,\,e_1,\,e_2,\,e_3,\,e_4,\,e_5,\,e_6) \\ \dot{Y} = \dot{Y}(t,\,e_1,\,e_2,\,e_3,\,e_4,\,e_5,\,e_6) \\ \dot{Z} = \dot{Z}(t,\,e_1,\,e_2,\,e_3,\,e_4,\,e_5,\,e_6) \end{cases} \tag{17.96}$$

where $e_1, \ldots e_6$ is a set of orbital elements. Now, if perturbations are present, we can conceive of an osculating orbit defined by the actual position and velocity vectors at any instant. It is clear that the e_i's associated with any pair of neighboring times will not be exactly equal, since the osculating orbits will differ. Following LaGrange, we can conceive of the sequence of osculating orbits as evolving continuously in time and thus consider the e_i's to be functions of time. It is possible to derive rigorous differential equations (Brower and Clemence, 1961) which the time derivatives of the e_i's must satisfy and which have the form

$$\frac{de_i}{dt} = g_i\,(e_1,\,e_2,\,e_3,\,e_4,\,e_5,\,e_6,\,\Delta\ddot{\underline{\mathbf{r}}})\;\;i = 1,\,2,\,\ldots\,6$$

$$\tag{17.97}$$

which are valid for an arbitrary perturbation $(\Delta\underline{\mathbf{r}})$. Now equations (17.97), if integrated analytically or numerically, yield an osculating orbit with elements

$$e_i(t) = e_i(t_0) + \int_{t_0}^{t} g_i\,(e_1,\,e_2,\,e_3,\,e_4,\,e_5,\,e_6,\,\Delta\ddot{\underline{\mathbf{r}}})\,dt$$

$$\tag{17.98}$$

which, when substituted into (17.96) yield the actual position and velocity.

The motivation for variation of parameters lies in the fact that parameters of two-body-motion vary slowly in the presence of small perturbations; thus allowing approximations in carrying out the integrations (17.98). These integrations are often carried out on series in some small quantity associated with the perturbation. Such series when truncated often remain approximately valid over many revolutions. It is often overlooked, however, that while integrations of the variation-of-parameters type have received the most attention historically, the differential equations are attractive for numerical integration as well. The slow variation of the e_i's results in large integration steps.

17.2.8.4 NUMERICAL SOLUTION OF DIFFERENTIAL EQUATIONS

Integration procedures can be classified as either *single-step methods* or *multi-step methods*. A single-step method requires only knowledge of the state at time t_k to derive a state at subsequent time $t_{k+1} = t_k + h$. Foremost among single-step methods (in terms of frequency of use) are the Runge-Kutta methods. Multi-step methods, as the name implies, require a number of previous states to be saved along with differences and sums in a table; the number of previous states required defines the order of the method. Usually, a single-step method is used to generate the starting table, which can be extended automatically thereafter unless it is necessary to "restart" the table at a new step-size to maintain precision. Formost of the multi-step methods (for orbit integration) is the Gaussian "Σ^2" procedure (Herrick, 1972 and Merson, 1973). The Σ^2 procedure is employed in the majority of integrations by direct and intermediary methods (*see* section 17.2.8.2), exclusive of powered atmospheric-flight trajectory computations.

Single-step methods are simpler than multi-step methods, and are in fact computationally superior if the acceleration is highly non-linear (*e.g.*, as in powered, atmospheric flight). How-

ever, for the problem of nearly-circular orbits (in the absence of maneuvering) the Σ^2 multi-step method is between 25 and 50 times as efficient as Runge-Kutta, as measured by the typical number of evaluations required to maintain precision to eight 8-significant figures. Qualitatively, multi-step methods lose their advantage when the optimum step-size changes.

17.2.9 THE CORRECTION OF ESTIMATED ORBITS

Unless the orbit is particularly simple, the orbits of actual bodies are determined by comparing *measurements* of direction, distance, or range-rate with *predictions* based on theory and correcting the theory in such a way as to give better agreement between measurement and theory. The method of least squares is generally considered to provide results that satisfy the

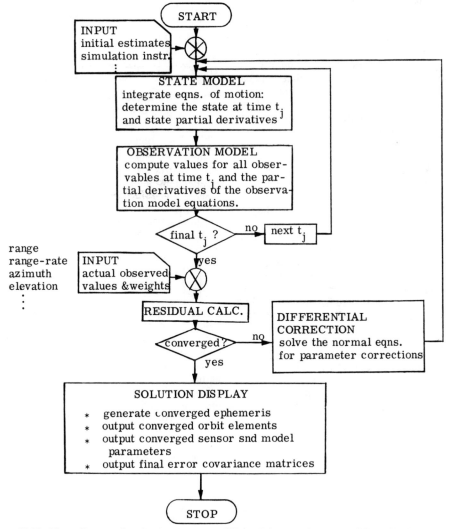

FIGURE 17-37. Flow diagram showing least-squares differential correction method for orbital determination.

concepts. The process of determining corrections to an orbit by the method of least squares is shown in figure 17-37. It is the same, except its specialization to orbit-determination, as the process described in chapter II and covered in most modern texts on statistical analysis.

In "short-arc" methods, observations are taken over a fraction of a revolution and the unknowns are generally only the orbital elements and, perhaps, the station coordinates. It is generally impossible to estimate forces (with the possible exception of drag) from observations over a single, short arc (fortunately, it is usually

unnecessary, since as the time spanned by the observations tend to zero, so do errors introduced by poorly modeled forces). "Long-arc" methods (methods involving observations taken over many revolutions), on the other hand, often allow for simultaneous determination of gravitational-field, station coordinates, and instrumental constants. Generally speaking, however, care must be exercised in choosing the force model parameters to be recovered. For example, equatorial orbits are insensitive to zonal harmonics; thus it is senseless to attempt estimation of zonal harmonics in this situation.

17.3 Theoretical Development of Satellite Photogrammetry

Highly significant advances have been made in the field of theoretical satellite photogrammetry during the past three decades. This is particularly evident in the area of simultaneous strip- and block-adjustments of aerial photography. Dr Hellmut Schmid (1951–1959) and D. C. Brown (1958) developed the mathematical foundation for a rigorous, least-squares, simultaneous adjustment of a large and general photogrammetric net. The mathematics here found immediate application in the area of space geodesy. This original development was expanded by Brown (1958b) to account for errors in locations of exposure stations. This extension was further refined by Brown to cover the case in which any of the elements of orientation and the coordinates of control points are considered as observed quantities with an error subject to adjustment. In the original formulation, these qualities were either completely known or completely unknown. Chapter IX discusses aerotriangulation and the variety of constraints that can be incorporated. The constraint which has greatest potential for extraterrestrial and satellite applications is the orbital constraint. The mathematical procedures for incorporating the orbital constraints into a photogrammetric adjustment were first formulated by Brown in 1961 in an unpublished, privately-circulated paper referred to by Case (1961).

17.3.1 DIFFERENCES BETWEEN AIRCRAFT AND SPACECRAFT MODES OF ANALYTICAL PHOTOGRAMMETRIC TRIANGULATION

The basic relationships of numerical photogrammetry are presented in chapter II as

$$x = x_p - c \frac{A(X - X^c) + B(Y - Y^c) + C(Z - Z^c)}{D(X - X^c) + E(Y - Y^c) + F(Z - Z^c)}$$

$$y = y_p - c \frac{A'(X - X^c) + B'(Y - Y^c) + C'(Z - Z^c)}{D(X - X^c) + E(Y - Y^c) + F(Z - Z^c)}.$$

The above equations are called the collinearity equations or projective equations where

x, y = image space coordinates of the object point,

$x_p y_p$ = image space coordinates of the principal point,

c = principal distance (focal length),

$\begin{bmatrix} A & B & C \\ A' & B' & C' \\ D & E & F \end{bmatrix}$ = rotation matrix giving the orientation of the image-space coordinate system in object-space,

X^c, Y^c, Z^c = object-space coordinates of the perspective center, and

X, Y, Z = object-space coordinates of the object points.

The nine elements $(A, B, \ldots, F)$ of the orientation matrix are unique functions of any three independent angles which render the matrix orthogonal. For this discussion, the three Eulerian angles, ϕ, ω, κ will be chosen.

Then for photogrammetry from aerial photography, the projective equations can be written in the form:

$$\begin{bmatrix} x \\ y \end{bmatrix} = f(x_p, y_p, c, \phi, \omega, \kappa, X^c, Y^c, Z^c, X, Y, Z).$$

Up to this point, the projective equations are independent of the motion of the vehicle, i.e., the exposure stations X^c, Y^c, Z^c could, in principle, be completely randomly distributed in space as could be expected from aircraft flight paths.

To develop the orbital version of these equations, it is desired to exploit the fact that the satellite proceeds along an orbit and all exposure stations lie on this orbit. Note that the ith exposure station along the kth segment of the orbit may be represented as:

$$\begin{bmatrix} X^c_{i_k} \\ Y^c_{i_k} \\ Z^c_{i_k} \end{bmatrix} = g(t_{i_k}, X_{o_k}, Y_{o_k}, Z_{o_k}, \dot{X}_{o_k}, \dot{Y}_{o_k}, \dot{Z}_{o_k},$$
other constants),

where $(X_{o_k}, Y_{o_k}, Z_{o_k}, \dot{X}_{o_k}, \dot{Y}_{o_k}, \dot{Z}_{o_k}, t_{i_k} - t_{o_k})$ is the state-vector (section 17.2) for the kth pass at an arbitrary epoch t_{o_k}; and t_{i_k} represents the time of exposure. If the values X^c, Y^c, Z^c in the projective equations are replaced by the values determined from $[X_{i_k}^c, \ Y_{i_k}^c, \ Z_{i_k}^c]^T$ above, the new equations functionally takes on the form:

$$\begin{bmatrix} x \\ y \end{bmatrix} = f[x_p, y_p, c, (\phi, \omega, \kappa,) (t, X_0, Y_0, Z_0, \dot{X}, \dot{Y}, \dot{Z}) X, Y, Z].$$

This, then, is the basic condition equation in functional form that forms the basis for satellite photogrammetric triangulation. Note the exposure station coordinates are replaced by the state vector and time.

Computer programs containing orbital condition equations in the adjustment of satellite photography have been developed by various people. See, for example, Davis and Riding (1970), D. Light (1970, 1972) and Doyle, Elassal and Lucas (1977).

17.3.2 LUNAR ORBITER ANALYTICAL TRIANGULATION AND THE APOLLO SYSTEMS

It has been previously pointed out that the Apollo 15, 16 and 17 imaging systems were the only ones that were specifically designed to collect photography that could be exploited in a rigorous photogrammetric sense. This section will use the Apollo imaging-system as an example in preparation for the rigorous theory of satellite photogrammetry that follows in section 17.4.

Figure 17-38 shows the general concept of the Apollo imaging system.

Let

x_p, y_p = image space coordinates of the camera's principal point

c = principal distance or focal length of the camera (often called f in the literature)

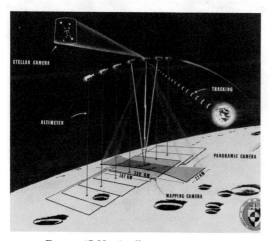

FIGURE 17-38. Apollo camera systems.

be known quantities, then, the functional form of the satellite photogrammetry equations can be stated as:

$$\begin{bmatrix} x \\ y \end{bmatrix} = F[(\phi, \omega, \kappa, t)_i,$$

| image coordinates | orientation angles & time |

$(X_0, Y_0, Z_0, \dot{X}_0, \dot{Z}_0, C_{nm}, S_{nm})$, $\qquad (X, Y, Z)_j]$. (17.99)

| position and velocity coordinates, gravitational constants | object-space coordinates of ground points |

Beginning at the left of the above equation and referring to figure 17-38, the data entering the above equations are as follows.

(x, y)—measured image coordinates from the mapping camera. Note the square format with dimensions 167 km square is the mapping camera's projected image on the terrain.

$(\phi, \omega, \kappa, t)_i$—three orientation angles and time of exposure at camera station i. These angles are determined by computing the stellar camera's orientation from the stellar photography and then computing the terrain camera's orientation from this. The orientation of the terrain camera with respect to the stellar camera is known from calibration. Section 17.4 gives equations relating the stellar camera to the terrain camera.

$(X_0, Y_0, Z_0, \dot{X}_0, \dot{Y}_0, \dot{Z}_0, C_{nm}, S_{nm})$—the position-and-velocity state-vector is computed by utilizing the doppler data from earth-based receivers. The state-vector represents the position and velocity of a satellite orbit at a given time. The C_{nm} and S_{nm} are constants defining the Moon's or planet's gravitational field.

(X, Y, Z)—object-space coordinates of points on the ground corresponding to points in image-space. These coordinates may be known.

D—distance from satellite to surface as measured by altimeter.

$$D = [(X^c - X^a)^2 + (Y^c - Y^a)^2 + (Z^c - Z^a)^2]^{1/2},$$

where

(X^c, Y^c, Z^c) = coordinates of spacecraft at time t.

(X^a, Y^a, Z^a) = coordinates of laser-illuminated point on the surface.

Notice that the doppler data provide positioning of the mapping camera system (MCS) on the orbit. The altimetric distance D is an excellent constraint on the satellite's height above the terrain (Light 1972). All of these data used simultaneously allow for a least-squares adjustment so that precise X, Y, Z coordinates can be determined.

The panoramic camera shown in figure 17-38 is independent of the MCS, and therefore depends on the X, Y, Z coordinates established by the MCS for its orientation to the lunar surface. Assume that at least 30 image-points are common to both the MCS and the panoramic images, as shown in figure 17-45. Orientation for the stereoscopic models from panoramic imagery depends on fitting the x,y-coordinates measured on panoramic images to the X,Y,Z ground-coordinates as established from the MCS. This combination improves the benefits from both systems. That is, the MCS is used to establish a network of accurate ground points, and the higher-resolution panoramic camera can be oriented to these points. Figure 17-39 illustrates 9 strips from 9 orbital segments assembled into a block of 27 photographs. Simultaneous adjustment of large blocks of photographs with their corresponding data allows the photogrammetrist to develop ground control networks for the moon or planets without the use of surveyed control points. Although ground control is not required to obtain a solution, it is important to realize that if it is available, it can be input as a known quantity (estimate with a small error), and is then a strong contributor to the solution. In fact, when sufficient ground-control is available, the attitude and ephemeris are not required.

17.3.3 DETERMINATION OF ORIENTATION ANGLES

A terrain camera in a satellite photogrammetry system must be associated with a precise means of determining attitude (the three angular orientation elements ϕ, ω, κ). Experience with the NASA Apollo 15, 16 and 17 missions has demonstrated that the use of a stellar camera to photograph the stars simultaneously with photography by the down-looking terrain camera is a good technique for finding the attitude. For such a system, precise knowledge of the angular relationship (lock-angle) existing between the stellar

camera and the terrain camera must be known from previous calibration.

Referring to figure 17-38, notice that the stellar camera is viewing the stars while the terrain camera is viewing the moon underneath. The lock-angle between the optical axes of the two cameras was approximately 96 degrees. This placed the stellar camera's field of view slightly above the moon's horizon so it would always view the sky. Figure 17-11 shows the fields of view of both cameras and the laser altimeter's location within the unit. The stellar camera and the terrain camera are rigidly coupled so that they retain their precisely calibrated relationship while in orbit.

17.3.3.1 THE STELLAR CAMERA

The stellar camera must be designed to photograph a portion of the star field and image at least three and preferably 20 to 30 stars per frame. The Apollo stellar camera imaged stars up to 6th magnitude. Since the angular coordinates of stars are precisely known, the attitude of the stellar camera at the instant of exposure can be determined to a very high degree of precision. (Usually to a few seconds of arc). Utilizing matrix multiplications as shown in 17.3.3.4, the orientation of the terrain camera can be derived and its orientation is, therefore, referred to the stars.

17.3.3.2 GEOMETRY OF THE STELLAR CAMERA

The Apollo 15, 16 and 17 stellar camera's essential geometrical characteristics are given below.

Focal Length: 76 mm, f/2.8
Image Format: 24 × 32 mm on 35 mm film

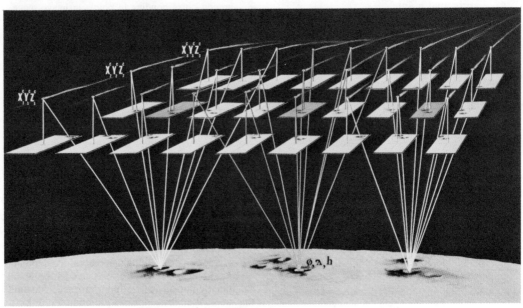

FIGURE 17-39. Block adjustment with altimetric constraints.

Field of View: 18° Vertical
                  24° Horizontal

Shutter Speed: 1.5 seconds
Eastman-Kodak film 3401

Film Capacity: 155 metres

The designer of a stellar camera must consider the following factors.

Pointing accuracy required.

Speed, film resolution, size of format.

Weight restrictions in the spacecraft.

Requirement to photograph at least a number of stars uniformly distributed over the photograph.

The following empirical formulas provide practical estimates for the angular precision of space-direction established by photogrammetry. These formulas can be used to evaluate stellar cameras.

$$\delta\phi = \frac{\sqrt{2}}{2n - p} \frac{m_x}{f} \qquad (17.100)$$

$$\delta\kappa = \frac{\sqrt{2}}{2n - p} \frac{m_x}{L} \qquad (17.101)$$

where

$\delta\phi$ = estimated angular error in pitch and roll (ϕ,ω) resulting from error in orientation to the stars,

$\delta\kappa$ = estimated angular error in yaw (κ),

n = number of stars used in computation. At least three are required. Three are arbitrarily selected for this formula,

p = number of elements of orientation carried as unknowns; ϕ, ω, κ equal 3,

m_x = standard deviation of the measured x,y-coordinates of stellar images,

f = principal distance of stellar camera, and

L = length of format or diagonal across frame.

Typical numbers from formulas (17-100) and (17-101) are

$$n = 3$$
$$p = 3$$
$$m_x = 0.003 \text{ mm}$$
$$f = 76 \text{ mm}$$
$$L = 32 \text{ mm.}$$

Then

$$\delta\theta = 0.000,018,5 \approx 4''$$
$$\delta\kappa = 0.000,044,1 \approx 9''.$$

The actual reduction of the Apollo 16 stellar photography yielded the following typical standard errors for the 3 angles of orientation, relating the stellar camera to the celestial 1950.0 system:

$$\sigma_\omega - 2''$$
$$\sigma_\phi - 2''$$
$$\sigma_\kappa - 12''.$$

Considering 6'' uncertainty in the lock angles connecting the stellar camera to the metric camera and the errors in the orientation of the metric camera to a local lunar datum, then the typical standard errors for the orientation of the mapping camera to the local lunar datum are as follows.

Standard Error	Celestial 1950.0	Local Lunar
σ_ω	4''	18''
σ_ϕ	13''	32''
ϕ_κ	3''	10''

These attitudes systematically differed by 3' to 5' from the values given in the NASA ephemeris. In order to be able to utilize the directional precision, inherent in metric cameras, the attitude of the camera must be established to a comparable precision. This is the function of the stellar camera.

The stars are assumed to be at an infinite distance with the earth as center of the celestial sphere. For the best stellar view, the stellar camera should be pointed towards the pole of the orbit. This minimizes the image smear caused by the pitch resulting from the forward motion of the vehicle. It also keeps the same stars in the field throughout the mission. On the Apollo missions, the stellar camera was aimed 96° up from the vertical as shown in figure 17-38. The error in rotation about the optical axis (κ) is the largest. When this error is transfered to the terrain camera, the error in κ becomes pitch (ϕ) in the terrain camera. This is unfortunate, since errors in pitch have a much larger effect on errors in the stereoscopic model than do errors in roll and yaw.

In order to overcome this large error in pitch, two stellar cameras can be used. Ideally the angle between their optical axes should be close to 90°. This may be difficult to achieve, but engineering should strive to balance the errors in the three angular values transferred to the terrain camera. This was not possible for the Apollo missions, but it should be a design consideration in other projects.

17.3.3.3 COMPUTATION OF RIGHT ASCENSION AND DECLINATION OF STARS

Let α be the right ascension (RA) and δ be declination (figure 17-22).

The positions of the photographed stars at the instant of exposure are required in the reduction. Since the positions of the stars are constantly changing due to proper motion, precession, nutation, and so on, the apparent places or the mean places of the stars are listed in catalogs only for certain epochs. The "Apparent Places of Fundamental Stars" at 10-day intervals in the year of observation and the "Vierter Fundamental katalog des Berliner Astronomischen Jahrbuchs (FK4)," are probably the most accurate catalogues of stars currently available.

If the 1535 stars of the preceeding catalogues are not adequate, the "Smithsonian Catalogue" containing positions of 258,997 stars may be employed. Since this catalogue lists only the mean places of the stars for the epoch 1950.0, the places must be transformed to mean places at the beginning of the year of interest, then reduced to apparent places at the instant of exposure, and, finally corrected to their topocentric places at the instant of exposure. Correction for stellar aberration must also be considered, and this is dependent on the celestial body being orbited and the characteristics of the orbit itself.

In addition to listing α, δ, the star catalogues usually provide magnitude and spectral class of each star. The primary references for star coordinates are Eichhorn (1974) and Nautical Almanac Offices (1961). In any case, the α, δ of 20 to 30 stars is needed for the attitude computation. Table 17-5 (Schreiter and Zeichel, 1958) gives the number of stars expected in a circular field. This table and formulas (17-100) and (17-101) can be used to assess the utility of a stellar camera.

17.3.3.4 RELATIONSHIP BETWEEN CAMERA ATTITUDE AND STELLAR DIRECTIONS

The mathematical model that establishes the relationship between the stellar camera's coordinate-system and the stellar coordinate system is expressed by the directional form of the equations

$$x_i = x_p + f \left[\frac{m_{11}U_i + m_{12}V_i + m_{13}W_i}{m_{31}U_i + m_{32}V_i + m_{33}W_i} \right]$$

$$y_i = y_p + f \left[\frac{m_{21}U_i + m_{22}V_i + m_{23}W_i}{m_{31}U_i + m_{32}V_i + m_{33}W_i} \right]$$

(17.102)

where

x_i, y_i = photographic coordinates of the ith star image
U_i, V_i, W_i = direction cosines of the ith star in the stellar coordinate system
x_p, y_p = photographic coordinates of the principal point
f = stellar-camera principal distance (focal length)
m_{ij} = (the i, j) component of a 3×3 orthogonal orientation matrix [M] which relates image space to the stellar reference system.

Photographic coordinates (x_i, y_i) are obtained from measurements on stellar images after correcting them for all the known systematic errors such as film deformation, comparator errors, *etc.* The stellar direction-cosines are computed from the expression

$$\begin{bmatrix} U_i \\ V_i \\ W_i \end{bmatrix} = \begin{bmatrix} \cos\alpha_i \cos\delta_i \\ \sin\alpha_i \cos\delta_i \\ \sin\delta_i \end{bmatrix}$$

(17.103)

where

α_i, δ_i are right ascension and declination of the ith star at the instant of exposure, corrected for stellar aberration.

The interior orientation parameters of the stellar camera (x_p, y_p, f) are obtained from previous calibration.

The [M]-matrix elements can be expressed in terms of three variables such as (ϕ, ω, κ). The stellar-camera attitude is fully determined in the stellar reference system once these three angles are computed. Due to the usually high degree of redundancy in a typical photograph, these quantities in [M] are computed by least squares.

The matrix [M] is the connecting link between the stellar camera's coordinate system and the stellar coordinates. Now by simple matrix multiplications, the orientation of the terrain camera with respect to the body it is orbiting can be computed. To illustrate the matrix manipulations, again the lunar Apollo case is utilized as an example. The stellar camera orientation in the celestial coordinate system is given by

$$(\mathbf{X}_s) = [\mathbf{M}_s] (\mathbf{U})$$
$$\phantom{(\mathbf{X}_s) = }{}_{3,1} \quad {}_{3,3} \quad {}_{3,1}$$

(17.104)

where

$(\mathbf{U})$ = direction cosines of stars $[U, V, W]^T$ in the celestial coordinate system,
$(\mathbf{X}_s)$ = direction cosines of stars in the stellar-camera coordinate-system, and
$[\mathbf{M}_s]$ = orientation matrix of stellar camera in celestial coordinate system.

Orientation of the terrain camera in the celestial coordinate system is given by

$$[\mathbf{M}_t] = [\mathbf{M}_{st}] [\mathbf{M}_s]$$
$$_{3,3} \quad {}_{3,3} \quad {}_{3,3}$$

(17.105)

TABLE 17-5. STARS IN A CIRCULAR FIELD AS A FUNCTION OF STELLAR MAGNITUDE

Diameter of Circular Field	Average Number of Stars Brighter Than Magnitude m						
	m = 7	m = 6.5	m = 6	m = 5.5	m = 5	m = 4.5	m = 4
4°	4	2.5	1.5	0.8	0.5	0.3	0.2
8°	7	10	5.9	3.4	2.0	1.1	0.6
12°	39	23	13	7.7	4.4	2.6	1.5
16°	70	41	24	14	7.9	4.6	2.6
20°	109	64	37	21	12	7.1	4.0
24°	156	92	53	30	18	10	5.8

where

[$\mathbf{M}_{st}$] = precalibrated relative orientation matrix between stellar camera and terrain camera, and

[$\mathbf{M}_t$] = terrain camera orientation matrix in celestial coordinate system.

The final objective is to obtain the orientation of the terrain camera in the coordinate system of the body being orbited. This is given by another simple matrix rotation

$$\left[\underset{3,3}{\mathbf{M}_{ct}}\right] = \left[\underset{3,3}{\mathbf{M}}\right]\left[\underset{3,3}{\mathbf{M}_t}\right] \qquad (17.106)$$

where

[$\mathbf{M}_{ct}$] = orientation of terrain camera in the coordinate systems of the body being orbited, and

[$\mathbf{M}$] = relative orientation matrix between the body centered and celestial α, δ coordinate system.

In the case of the moon, the moon's orientation [$\mathbf{M}$] with respect to the earth can be described in terms of three angles

$$[\mathbf{M}] = F(\gamma, i, \Omega'),$$

but these angles themselves contain quantities that must be calculated to determine the moon's position relative to the earth and the celestial sphere. These quantities are:

the moon's orientation in the celestial sphere,
the moon's rotation rate, and
the moon's physical librations.

There are several models that give values for these quantities. A recent model is by (Eckhardt, 1973). The subject of libration is the province of celestial mechanics and is not addressed here. The important point is that whether one is mapping the moon or a planet, it is essential to know the [$\mathbf{M}$] for that body and its relationship to the celestial sphere. In some cases it may be known very poorly or not at all.

Once [$\mathbf{M}_{ct}$] is determined, the matrix itself or the angles ϕ, ω, κ that it represents can serve as inputs in the triangulation process. That is, ϕ, ω, κ can become inputs to a large simultaneous adjustment of numerous photographs in a block computation.

17.4 Adjustment with Orbital Constraints

The problem of the rigorous photogrammetric adjustment of a strip or block of lunar orbiter photographs is addressed in detail by Davis and Riding (1970). Their development may be described as an extension of the bundle method originally developed in Brown (1958) and Brown, Davis and Johnson (1964) to use orbital constraints as suggested in Brown in 1960. A further development of the bundle adjustment with orbital constraints (for specific application to the Apollo photographic mission) is given in a final report to the U.S. Army by Hagg and Dodge. The presentation to follow is largely adapted from the foregoing references.

It is, of course, the applicability of orbital constraints that provides the essential distinction between a conventional strip of aerial photographs and a strip generated by a satellite-borne camera. Nothing prevents one from adjusting the latter in exactly the same manner as the former. In situations where ground control is abundant and well-distributed the specific use of orbital constraints may not prove to be particularly advantageous. However, in the case of lunar or other extra terrestrial photogrammetry, ground control is either nonexistent or of limited quantity and quality. In such circumstances the use of orbital constraints becomes essential to a favorable outcome. Three fundamentally different modes of utilizing orbital constraints exist:

A) rigidly enforcing coordinates of exposure stations as obtained from an external, independent orbital determination based on tracking observations;

B) treating coordinates of exposure stations from an orbital determination as subject to independent adjustment without regard to satisfying the pertinent differential equations of motion;

C) requiring that the adjusted coordinates of successive exposure stations on a given strip satisfy the pertinent differential equations of motion with state vector subject to appropriate, a priori statistical constraints.

Both A and B can be implemented within a conventional bundle adjustment. In the case of A one merely assigns insignificantly small variances to the given coordinates of the exposure stations whereas in the case of B one assigns more relaxed a priori variances in accord with the expected accuracies of the orbit. The adjusted coordinates of exposure stations resulting from B will no longer strictly lie on an orbital arc because each exposure station is allowed independent, but statistically constrained freedom to adjust. In practical application method A can be justified only when the independently determined orbit is of such high accuracy that its enforcement does not significantly degrade closures of photogrammetric triangulation. Method B is difficult to justify technically under any circumstances for it dispenses with valuable information implied in the equations of motion. Accordingly, the primary justification for B has to be based simply on expediency in permitting the use of a conventional bundle adjustment in lieu of a bundle adjustment that rigorously incorporates orbital constraints. Because of the particular circumstances attendant on extraterrestrial photogrammetry the only technically proper course for lunar and planetary applications is that indicated under C. Here, the given orbital elements (state vectors) that define a set of arcs generating overlapping photogrammetric strips are considered to be affected by significant error and hence are carried in the photogrammetric adjustment itself as observed quantities subject to appropriate statistical constraints specified by *a priori* covariance matrices. Inasmuch as it is the state vector for each arc that is adjusted, the coordinates of exposure stations following adjustment remain on orbital arcs, as indeed they should, rather than being allowed statistically controlled departure from prespecified arcs as in method B. Accordingly, method C may be regarded as the most appropriate and rigorous approach to the adjustment of photogrammetric strips and blocks generated by an orbiting camera. It is specifically this approach that is the topic of the present section.

17.4.1 OBSERVATIONAL EQUATIONS GENERATED BY THE kth STRIP

The individual strip associated with a given orbital arc constitutes the fundamental unit in a general bundle adjustment subject to orbital constraints. In the formulation of such an adjustment the symbology and system of subscripts and superscripts indicated in the following table will be adopted.

SYMBOL	INTERPRETATION
k	Refers always to k^{th} strip (orbital pass).
i_k	Refers always to i^{th} photograph in k^{th} strip.
j	Refers always to j^{th} lunar surface point.
$i_k j$	Refers always to j^{th} surface point on i^{th} photograph in k^{th} strip.
s	The total number of strips.
m_k	The total number of photographs in the kth strip.
m	The total number of photographs $m = \sum_{k=1}^{s} m_k$.
n	The total number of lunar surface points.
$(^0)$	Refers always to observed quantities.
$(^{00})$	Refers always to approximations or results involving approximations.

The photogrammetric observational equations appropriate to the jth point as measured on the ith photo of the kth strip may therefore be represented as

where

x, y	= image space coordinates of the projected object space point,
x_p, y_p	= image space coodinates of principal point,
c	= principal distance,
$\begin{bmatrix} A & B & C \\ A' & B' & C' \\ D & E & F \end{bmatrix} =$	orthogonal orientation matrix defining the rotational relationship between the image and object space coordinate systems,
X^c, Y^c, Z^c	= object space coordinates of the perspective center,
X, Y, Z	= object space coordinates of the object.

In the present development the elements of interior orientation will be regarded as precalibrated quantities not subject to adjustment. Later the implications of abandoning this assumption will be examined. The subscripts i_k on x_p, y_p and c admit the possibility that different cameras may have been involved in the generation of a given strip, although in practice this is not likely to be the case. Of greater likelihood is the possibility that not all strips comprising a given block are photographed by the same camera, as when strips from different missions are combined. This merely means that the appropriate values of x_p, y_p and c must be enforced for each strip, an eventuality admitted by the above formulation.

The linearized versions of the projective equations may be written as

$$\mathbf{v}_{i_k j} + \hat{\mathbf{B}}_{i_k j}^{(1)} \hat{\boldsymbol{\delta}}_{i_k}^{(1)} + \hat{\mathbf{B}}_{i_k j}^{(2)} \hat{\boldsymbol{\delta}}_{i_k}^{(2)} + \ddot{\mathbf{B}}_{i_k j} \ddot{\boldsymbol{\delta}}_j = \boldsymbol{\epsilon}_{i_k j}$$

(17.108)

in which

$$\mathbf{v}_{i_k j} = \begin{bmatrix} v_{x_{i_k j}} \\ v_{y_{i_k j}} \end{bmatrix}$$

(17.109)

are the residuals corresponding to the observed plate coordinates $x^0_{i_k j}$, $y^0_{i_k j}$ and

$$\boldsymbol{\epsilon}_{i_k j} = \begin{bmatrix} x^0_{i_k j} - x^{00}_{i_k j} \\ y^0_{i_k j} - y^{00}_{i_k j} \end{bmatrix}$$

(17.110)

is the vector of discrepancies between the observed plate coordinates $(x^0_{i_k j}, y^0_{i_k j})$ and the values $(x^{00}_{i_k j}, y^{00}_{i_k j})$ computed when the right hand sides of equations (17.107) are evaluated using approximations to the various parameters appearing in the projective equations. These pa-

$$x_{i_k j} = x_{p_{i_k}} - c_{i_k} \frac{A_{i_k}(X_j - X^c_{i_k}) + B_{i_k}(Y_j - Y^c_{i_k}) + C_{i_k}(Z_j - Z^c_{i_k})}{D_{i_k}(X_j - X^c_{i_k}) + E_{i_k}(Y_j - Y^c_{i_k}) + F_{i_k}(Z_j - Z^c_{i_k})},$$

$$y_{i_k j} = y_{p_{i_k}} - c_{i_k} \frac{A'_{i_k}(X_j - X^c_{i_k}) + B'_{i_k}(Y_j - Y^c_{i_k}) + C'_{i_k}(Z_j - Z^c_{i_k})}{D_{i_k}(X_j - X^c_{i_k}) + E_{i_k}(Y_j - Y^c_{i_k}) + F_{i_k}(Z_j - Z^c_{i_k})}.$$

(17.107)

rameters consist of the angular elements of orientation ϕ_{i_k}, ω_{i_k}, κ_{i_k} implicit in the elements of the orientation matrix $(A_{i_k}, B_{i_k}, \ldots, F_{i_k})$, the coordinates $X^c_{i_k}$, $Y^c_{i_k}$, $Z^c_{i_k}$ of the exposure station, and the object space coordinates X_j, Y_j, Z_j of the measured point. The remaining quantities in the linearized projective equations are defined by the expressions

$$\underset{(2,3)}{\mathbf{\dot{B}}^{(1)}_{i_kj}} = \frac{\partial(x_{i_kj}, y_{i_kj})}{\partial(\phi_{i_k}, \omega_{i_k}, \kappa_{i_k})}\bigg|_0 \quad , \quad \underset{(3,1)}{\mathbf{\delta}^{(1)}_{i_k}} = (\delta\phi_{i_k} \; \delta\omega_{i_k} \; \delta\kappa_{i_k})^T,$$

$$\underset{(2,3)}{\mathbf{\dot{B}}^{(2)}_{i_kj}} = \frac{\partial(x_{i_kj}, y_{i_kj})}{\partial(X^c_{i_k}, Y^c_{i_k}, Z^c_{i_k})}\bigg|_0 \quad , \quad \underset{(3,1)}{\mathbf{\delta}^{(2)}_{i_k}} = (\delta X^c_{i_k} \; \delta Y^c_{i_k} \; \delta Z^c_{i_k})^T,$$

(17.111)

$$\underset{(2,3)}{\mathbf{\ddot{B}}_{i_kj}} = \frac{\partial(x_{i_kj}, y_{i_kj})}{\partial(X_j, Y_j, Z_j)}\bigg|_0 \quad , \quad \underset{(3,1)}{\mathbf{\ddot{\delta}}_j} = (\delta X_j \; \delta Y_j \; \delta Z_j)^T,$$

in which the symbol $|_0$ is introduced to denote that the partial derivatives are evaluated at the approximations to the parameters (*i.e.*, at $\phi^{00}_{i_k}$, $\omega^{00}_{i_k}$, etc.) and the quantities $\delta\phi_{i_k}$, $\delta\omega_{i_k}$, etc., refer to the unknown corrections associated with the approximations used in the process of linearization.

The separation of the elements of exterior orientation into the two specific vectors $\mathbf{\delta}^{(1)}_{i_k}$ and $\mathbf{\delta}^{(2)}_{i_k}$ was done in anticipation of the introduction of orbital constraints, according to which the coordinates $X^c_{i_k}$, $Y^c_{i_k}$, $Z^c_{i_k}$ of the center of projection at a specified time of exposure t_{i_k} may be represented functionally by expressions of the form

$$\begin{bmatrix} X^c_{i_k} \\ Y^c_{i_k} \\ Z^c_{i_k} \end{bmatrix} = \begin{bmatrix} g_1(X_{o_k}, Y_{o_k}, Z_{o_k}, \dot{X}_{o_k}, \dot{Y}_{o_k}, \dot{Z}_{o_k}, t_{i_k} - t_{o_k}) \\ g_2(X_{o_k}, Y_{o_k}, Z_{o_k}, \dot{X}_{o_k}, \dot{Y}_{o_k}, \dot{Z}_{o_k}, t_{i_k} - t_{o_k}) \\ g_3(X_{o_k}, Y_{o_k}, Z_{o_k}, \dot{X}_{o_k}, \dot{Y}_{o_k}, \dot{Z}_{o_k}, t_{i_k} - t_{o_k}) \end{bmatrix}$$

(17.112)

in which X_{o_k}, Y_{o_k}, Z_{o_k}, $\dot{X}_{o_k}$, $\dot{Y}_{o_k}$, $\dot{Z}_{o_k}$ constitute the six elements of the state vector at an arbitrarily defined epoch. At this stage of the present development the elements of the state vector and the times of exposures will be regarded as unknowns having known initial approximations $X^{00}_{o_k}$, $Y^{00}_{o_k}$, $\ldots$, $\dot{Z}^{00}_{o_k}$, $t^{00}_{o_k}$. Further, it will be assumed that the initial approximations to the coordinates of exposure stations used in the above formation of the linearized projective equations (17-128) are generated from the use in the equations of motion of the approximate state vector and exposure times. Accordingly, this implies that

$$\underset{(3,1)}{\mathbf{\hat{\delta}}^{(2)}_{i_k}} = \begin{bmatrix} \delta X^c_{i_k} \\ \delta Y^c_{i_k} \\ \delta Z^c_{i_k} \end{bmatrix} = \underset{(3,1)}{\mathbf{\dot{B}}_{i_k}} \underset{(1,1)}{\mathbf{\delta}_{i_k}} + \underset{(3,6)}{\mathbf{\ddot{B}}_{i_k}} \underset{(6,1)}{\mathbf{\delta}_{i_k}}$$

(17.113)

in which

$$\underset{}{\mathbf{\dot{B}}_{i_k}} = \frac{\partial(X^c_{i_k}, Y^c_{i_k}, Z^c_{i_k})}{\partial(t_{i_k})}\bigg|_0 \quad , \quad \mathbf{\delta}_{i_k} = (\delta t_{i_k})$$

(17.114)

$$\underset{}{\mathbf{\ddot{B}}_{i_k}} = \frac{\partial(X^c_{i_k}, Y^c_{i_k}, Z^c_{i_k})}{\partial(X_{o_k}, Y_{o_k}, \ldots, Z_{o_k})}\bigg|_0 ,$$

$$\mathbf{\ddot{\delta}}_k = (\delta X_{o_k} \; \delta Y_{o_k} \ldots \delta \dot{Z}_{o_k})^T.$$

The actual process of generating the matrices specified in (17.114) would be accomplished as part of the particular reduction employed for numerical integration of the orbit.

The introduction of orbital constraints into the projective equations (17.107) corresponding to the kth strip is effected by replacing the vector $\mathbf{\delta}^{(2)}_{i_k}$ in (17.108) by the expression given by (17.113). This leads to revised projective equations of the form

$$\underset{(2,1)}{\mathbf{v}_{i_kj}} + \underset{(2,4)}{\mathbf{\dot{B}}_{i_kj}} \underset{(4,1)}{\mathbf{\delta}_{i_k}} + \underset{(2,6)}{\mathbf{\ddot{B}}_{i_kj}} \underset{(6,1)}{\mathbf{\delta}_k} + \underset{(2,3)}{\mathbf{\ddot{B}}_{i_kj}} \underset{(3,1)}{\mathbf{\ddot{\delta}}_j} = \underset{(2,1)}{\mathbf{\epsilon}_{i_kj}}$$

(17.115)

in which

$$\underset{(2,4)}{\mathbf{\dot{B}}_{i_kj}} = \begin{bmatrix} \underset{(2,3)}{\mathbf{\dot{B}}^{(1)}_{i_kj}} & \underset{(2,3)}{\mathbf{\dot{B}}^{(2)}_{i_kj}} & \underset{(2,3)}{\mathbf{\ddot{B}}_{i_k}} \end{bmatrix}, \quad \underset{(4,1)}{\mathbf{\delta}_{i_k}} = \begin{bmatrix} \underset{(3,1)}{\mathbf{\delta}^{(1)}_{i_k}} \\ \underset{(1,1)}{\mathbf{\delta}_{i_k}} \end{bmatrix}$$

(17.116)

$$\underset{(2,6)}{\mathbf{\ddot{B}}_{i_kj}} = \underset{(2,3)}{\mathbf{\dot{B}}^{(2)}_{i_kj}} \underset{(3,6)}{\mathbf{\ddot{B}}_{i_k}}.$$

It will be noted in (17.115) that the corrections to parameters corresponding to angular elements of orientation ϕ_{i_k}, ω_{i_k}, κ_{i_k} and the correction corresponding to the time t_{i_k} of exposure have been combined into the single vector $\mathbf{\delta}_{i_k}$. In addition, the vector of corrections to the coordinates of the exposure station $(\mathbf{\hat{\delta}}^{(2)}_{i_k})$ has been supplanted by the vector of corrections by the vector of corrections to the elements of the orbital state vector $(\mathbf{\delta}_k)$. In the conventional strip of m_k photographs a total of $3m_k$ parameters is generated corresponding to coordinates of exposure stations. On the other hand, through the introduction of orbital constraints, these $3m_k$ parameters are replaced by a fixed total of 6 parameters defining the specific orbital arc on which the exposure stations are constrained to lie. This serves to strengthen and to stabilize the photogrammetric adjustment while at the same time significantly reducing requirements for ground control.

17.4.2 EXTERNALLY GENERATED OBSERVATIONAL EQUATIONS

To this point in the development the only quantities classified as observations consist of the measured plate coordinates $x^0_{i_kj}$, $y^0_{i_kj}$ appearing in the projective equations. Other quantities appearing in the projective equations (elements of orientation, elements of orbital state vector and coordinates of ground points) have all been regarded as unknowns. In practice some of these quantities may also be subject to observation either directly or indirectly. For example, estimates of worthwhile accuracy may be available for the elements of orbital state vector. Here, the adjustment of the state vector would be properly governed by a 6×6 covariance matrix based on appropriate, *a priori* considerations. Similarly, it is likely that the times t_{i_k} of successive exposures would be measured to within fairly narrow limits by means of a satellite-borne timing system. In photographic systems such as were used on Apollo missions, independent estimates of angular elements of orientation (ϕ, ω, κ) may be available from auxiliary stellar cameras. In the present development specific consideration will be given to the possibility of independent observations of the following quantities:

angular elements of orientation (ϕ_{i_k}, ω_{i_k}, κ_{i_k}),
time of each exposure (t_{i_k}),
elements of state vector (X_{o_k}, Y_{o_k}, Z_{o_k}, $\dot{X}_{o_k}$, $\dot{Y}_{o_k}$, $\dot{Z}_{o_k}$), and
coordinates of ground control points (X_j, Y_j, Z_j).

Such observations give rise to additional observational equations and associated covariance matrices. It is convenient to combine for each photograph the observed angular elements of orientation and associated time of exposure into a common observational vector. Accordingly, the adjusted values of these quantities may be expressed by the equations

$$
\begin{aligned}
\varphi_{i_k} &= \varphi^0_{i_k} + v_{\varphi_{i_k}}, \\
\omega_{i_k} &= \omega^0_{i_k} + v_{\omega_{i_k}}, \\
\kappa_{i_k} &= \kappa^0_{i_k} + v_{\kappa_{i_k}}, \\
t_{i_k} &= t^0_{i_k} + v_{t_{i_k}}.
\end{aligned}
\tag{17.117}
$$

Here, the quantities on the left denote the adjusted values of the observations, the quantities with the superscripts $(^0)$ denote the observed values, and the v's denote the corresponding residuals. In the earlier process of linearization of the projective equations the adjusted values of the same quantities had been expressed in terms of initial approximations and corresponding corrections as

$$
\begin{aligned}
\varphi_{i_k} &= \varphi^{00}_{i_k} + \delta\varphi_{i_k} \\
\omega_{i_k} &= \omega^{00}_{i_k} + \delta\omega_{i_k} \\
\kappa_{i_k} &= \kappa^{00}_{i_k} + \delta\kappa_{i_k} \\
t_{i_k} &= t^{00}_{i_k} + \delta t_{i_k}.
\end{aligned}
\tag{17.118}
$$

Accordingly, the new observational equations (17.117) can be replaced by the equivalent set

$$
\begin{aligned}
\varphi^{00}_{i_k} + \delta\varphi_{i_k} &= \varphi^0_{i_k} + v_{\varphi_{i_k}}, \\
\omega^{00}_{i_k} + \delta\omega_{i_k} &= \omega^0_{i_k} + v_{\omega_{i_k}}, \\
\kappa^{00}_{i_k} + \delta\kappa_{i_k} &= \kappa^0_{i_k} + v_{\kappa_{i_k}}, \\
t^{00}_{i_k} + \delta t_{i_k} &= t^0_{i_k} + v_{t_{i_k}}.
\end{aligned}
\tag{17.119}
$$

and this, in turn, can be represented in matrix form as

$$
\dot{\mathbf{v}}_{i_k} - \dot{\boldsymbol{\delta}}_{i_k} = \dot{\boldsymbol{\epsilon}}_{i_k}
\tag{17.120}
$$
$$
{\scriptstyle(4,1)} \qquad {\scriptstyle(4,1)} \qquad {\scriptstyle(4,1)}
$$

where

$$
\dot{\mathbf{v}}_{i_k}_{(4,1)} = \begin{bmatrix} v_\varphi \\ v_\omega \\ v_\kappa \\ v_t \end{bmatrix}_{i_k}, \quad
\dot{\boldsymbol{\delta}}_{i_k}_{(4,1)} = \begin{bmatrix} \delta_\varphi \\ \delta_\omega \\ \delta_\kappa \\ \delta_t \end{bmatrix}_{i_k}, \quad
\dot{\boldsymbol{\epsilon}}_{i_k}_{(4,1)} = \begin{bmatrix} \varphi^{00} - \varphi^0 \\ \omega^{00} - \omega^0 \\ \kappa^{00} - \kappa^0 \\ t^{00} - t^0 \end{bmatrix}_{i_k}.
\tag{17.121}
$$

Under the assumption that errors in the observed angular elements of orientation may be mutually correlated but are independent of the error in the associated time, the covariance matrix associated with the residual vector $\dot{\mathbf{v}}_{i_k}$ assumes the form

$$
\boldsymbol{\Lambda}_{i_k}_{(4,4)} = \begin{bmatrix}
\sigma_{\varphi\varphi} & \sigma_{\varphi\omega} & \sigma_{\varphi\kappa} & 0 \\
\sigma_{\omega\varphi} & \sigma_{\omega\omega} & \sigma_{\omega\kappa} & 0 \\
\sigma_{\kappa\varphi} & \sigma_{\kappa\omega} & \sigma_{\kappa\kappa} & 0 \\
0 & 0 & 0 & \sigma_{tt}
\end{bmatrix}_{i_k}
\tag{17.122}
$$

The corresponding weight matrix may be designated as

$$
\dot{\mathbf{W}}_{i_k}_{(4,4)} = \dot{\boldsymbol{\Lambda}}^{-1}_{i_k}_{(4,4)}.
\tag{17.123}
$$

The observational equations generated by external measurements of elements of the orbital state vector can be developed in a similar manner. Thus the observational equations may be written as

$$
\begin{aligned}
X^{00}_{o_k} + \delta X_{o_k} &= X^0_{o_k} + v_{X_{o_k}}, \\
Y^{00}_{o_k} + \delta Y_{o_k} &= Y^0_{o_k} + v_{Y_{o_k}}, \\
Z^{00}_{o_k} + \delta Z_{o_k} &= Z^0_{o_k} + v_{Z_{o_k}}, \\
\dot{X}^{00}_{o_k} + \delta \dot{X}_{o_k} &= \dot{X}^0_{o_k} + v_{\dot{X}_{o_k}}, \\
\dot{Y}^{00}_{o_k} + \delta \dot{Y}_{o_k} &= \dot{Y}^0_{o_k} + v_{\dot{Y}_{o_k}}, \\
\dot{Z}^{00}_{o_k} + \delta \dot{Z}_{o_k} &= \dot{Z}^0_{o_k} + v_{\dot{Z}_{o_k}}.
\end{aligned}
\tag{17.124}
$$

With now obvious notation this system may be represented in matrix form as

$$
\dot{\mathbf{v}}_k - \dot{\boldsymbol{\delta}}_k = \dot{\boldsymbol{\epsilon}}_k
\tag{17.125}
$$
$$
{\scriptstyle(6,1)} \qquad {\scriptstyle(6,1)} \qquad {\scriptstyle(6,1)}
$$

The associated weight matrix is defined as

$$
\dot{\mathbf{W}}_k_{(6,6)} = \dot{\boldsymbol{\Lambda}}^{-1}_k_{(6,6)}
\tag{17.126}
$$

in which the covariance matrix $\dot{\boldsymbol{\Lambda}}_k$ of the *a priori* values of the orbital state vector would normally be generated in conjunction with the reduction of tracking observations.

The final system of external observations to be considered consist of coordinates of ground points. As above, these give rise to equations of the form

$$
\begin{aligned}
X^{00}_j + \delta X_j &= X^0_j + v_{X_j} \\
Y^{00}_j + \delta Y_j &= Y^0_j + v_{Y_j} \\
Z^{00}_j + \delta Z_j &= Z^0_j + v_{Z_j}.
\end{aligned}
\tag{17.127}
$$

which may be represented in matrix form as

$$\ddot{\mathbf{v}}_j - \ddot{\boldsymbol{\delta}}_j = \ddot{\boldsymbol{\epsilon}}_j \,. \qquad (17.128)$$
$$\underset{(3,1)}{} \quad \underset{(3,1)}{} \quad \underset{(3,1)}{}$$

The 3×3 covariance matrix associated with the jth point on the photographed surface is represented by the full matrix

$$\underset{(3,3)}{\ddot{\boldsymbol{\Lambda}}_j} = \begin{bmatrix} \sigma_{XX} & \sigma_{XY} & \sigma_{XZ} \\ \sigma_{XY} & \sigma_{YY} & \sigma_{YZ} \\ \sigma_{XZ} & \sigma_{YZ} & \sigma_{ZZ} \end{bmatrix}_j \,, \qquad (17.129)$$

and the corresponding weight matrix is given by

$$\underset{(3,3)}{\ddot{\mathbf{W}}_j} = \underset{(3,3)}{\ddot{\boldsymbol{\Lambda}}_j^{-1}} \,.$$

If it is assumed momentarily that the entire set of n surface points is subject to independent observation, the resulting set of observational equations will be

$$\underset{(3n,1)}{\ddot{\mathbf{v}}} - \underset{(3n,1)}{\ddot{\boldsymbol{\delta}}} = \underset{(3n,1)}{\ddot{\boldsymbol{\epsilon}}} \,, \qquad (17.131)$$

where

$$\underset{(3n,1)}{\ddot{\mathbf{v}}} = \begin{bmatrix} \ddot{\mathbf{v}}_1 \\ \ddot{\mathbf{v}}_2 \\ \cdot \\ \cdot \\ \ddot{\mathbf{v}}_n \end{bmatrix} , \quad \underset{(3n,1)}{\ddot{\boldsymbol{\delta}}} = \begin{bmatrix} \ddot{\boldsymbol{\delta}}_1 \\ \ddot{\boldsymbol{\delta}}_2 \\ \cdot \\ \cdot \\ \ddot{\boldsymbol{\delta}}_n \end{bmatrix} , \quad \underset{(3n,1)}{\ddot{\boldsymbol{\epsilon}}} = \begin{bmatrix} \ddot{\boldsymbol{\epsilon}}_1 \\ \ddot{\boldsymbol{\epsilon}}_2 \\ \cdot \\ \cdot \\ \ddot{\boldsymbol{\epsilon}}_n \end{bmatrix} .$$

$$(17.132)$$

Although correlation is admitted between the X, Y and Z coordinates of a given point, it is assumed that coordinates of different points are uncorrelated. Accordingly, the covariance and weight matrices associated with the vector $\ddot{\mathbf{v}}$ are block diagonal matrices of the form

$$\underset{(3n,3n)}{\ddot{\boldsymbol{\Lambda}}} = \text{diag} \left(\underset{(3,3)}{\ddot{\boldsymbol{\Lambda}}_1} , \underset{(3,3)}{\ddot{\boldsymbol{\Lambda}}_2} , \ldots , \underset{(3,3)}{\ddot{\boldsymbol{\Lambda}}_n} \right) , \qquad (17.133)$$

$$\underset{(3n,3n)}{\ddot{\mathbf{W}}} = \text{diag} \left(\underset{(3,3)}{\ddot{\mathbf{W}}_1} , \underset{(3,3)}{\ddot{\mathbf{W}}_2} , \ldots , \underset{(3,3)}{\ddot{\mathbf{W}}_n} \right) . \qquad (17.134)$$

It will be noted in (17.129) that errors in the X, Y and Z coordinates of a given point are allowed to be correlated. This admits utilization of the concept of the ellipsoidal control point which was originated in Brown, Davis and Johnson (1964). According to this concept, the adjusted rays corresponding to a given control point are constrained to intersect within the probabilistic domain of the error ellipsoid defined by the quadratic form

$$q = \begin{bmatrix} X - Y_j \ Y - Y_j \ Z - Z_j \end{bmatrix} \ddot{\boldsymbol{\Lambda}}_j^{-1}$$
$$\begin{bmatrix} X - X_j \ Y - Y_j \ Z - Z_j \end{bmatrix}^{\mathsf{T}} .$$

$$(17.135)$$

As pointed out in detail in Brown (1974), the size, shape and orientation of the error ellipsoid for a prescribed level of probability (as designated by the scalar q) is governed by the specified covariance matrix $\ddot{\boldsymbol{\Lambda}}_j$.

17.4.3 GENERAL NORMAL EQUATIONS FOR k^{th} STRIP

The linearized projective equations incorporating orbital constraints, namely (17.115), may be written

$$\mathbf{v}_{i_k j} + \mathbf{B}_{i_k j} \, \boldsymbol{\delta}_{i_k j} = \boldsymbol{\epsilon}_{i_k j} \qquad (17.136)$$

in which

$$\underset{(2,13)}{\mathbf{B}_{i_k j}} = \begin{bmatrix} \underset{(2,4)}{\dot{\mathbf{B}}_{i_k j}} & \underset{(2,6)}{\overset{\circ}{\mathbf{B}}_{i_k j}} & \underset{(2,3)}{\ddot{\mathbf{B}}_{i_k j}} \end{bmatrix}$$

$$\underset{(13,1)}{\boldsymbol{\delta}_{i_k j}} = \begin{bmatrix} \underset{(4,1)}{\dot{\boldsymbol{\delta}}_{i_k}^{\mathsf{T}}} & \underset{(6,1)}{\overset{\circ}{\boldsymbol{\delta}}_{k}^{\mathsf{T}}} & \underset{(3,1)}{\ddot{\boldsymbol{\delta}}_{j}^{\mathsf{T}}} \end{bmatrix}^{\mathsf{T}} . \qquad (17.137)$$

The covariance matrix of the plate measurements associated with the residual vector $\mathbf{v}_{i_k j}$ may be written as

$$\underset{(3,2)}{\boldsymbol{\Lambda}_{i_k j}} = \begin{bmatrix} \sigma_{xx} & \sigma_{xy} \\ \sigma_{xy} & \sigma_{yy} \end{bmatrix}_{i_k j} \,, \qquad (17.138)$$

and the corresponding weight matrix as

$$\underset{(2,2)}{\mathbf{W}_{i_k j}} = \boldsymbol{\Lambda}_{i_k j}^{-1}. \qquad (17.139)$$

To generate the normal equations for the strip one could first proceed in a strictly formal manner, disregarding considerations of determinacy, and attempt to form the normal equations generated solely by the single pair of observational equations implicit in (17.131). This would lead to the system

$$(\mathbf{B}_{i_k j}^{\mathsf{T}} \, \mathbf{W}_{i_k j} \, \mathbf{B}_{i_k j}) \, \boldsymbol{\delta}_{i_k j} = \mathbf{B}_{i_k j}^{\mathsf{T}} \, \mathbf{W}_{i_k j} \, \boldsymbol{\epsilon}_{i_k j} \qquad (17.140)$$
$$\underset{(13,2)}{} \underset{(2,2)}{} \underset{(2,13)}{} \quad \underset{(13,1)}{} \quad \underset{(13,2)}{} \underset{(2,2)}{} \underset{(2,1)}{}$$

which by virtue of the relations in (17.137) may be expanded to

$$\begin{bmatrix} \underset{(4,4)}{\dot{\mathbf{N}}_{i_k j}} & \underset{(4,6)}{\dot{\mathbf{N}}_{i_k j}} & \underset{(4,3)}{\dot{\mathbf{N}}_{i_k j}} \\ \underset{(6,4)}{\acute{\mathbf{N}}_{i_k j}} & \underset{(6,6)}{\acute{\mathbf{N}}_{i_k j}} & \underset{(6,3)}{\acute{\mathbf{N}}_{i_k j}} \\ \underset{(3,4)}{\overline{\mathbf{N}}_{i_k j}^{\mathsf{T}}} & \underset{(3,6)}{\acute{\mathbf{N}}_{i_k j}^{\mathsf{T}}} & \underset{(3,3)}{\ddot{\mathbf{N}}_{i_k j}} \end{bmatrix} \begin{bmatrix} \underset{(4,1)}{\dot{\boldsymbol{\delta}}_{i_k}} \\ \underset{(6,1)}{\overset{\circ}{\boldsymbol{\delta}}_{k}} \\ \underset{(3,1)}{\ddot{\boldsymbol{\delta}}_{j}} \end{bmatrix} = \begin{bmatrix} \underset{(4,1)}{\dot{\mathbf{c}}_{i_k j}} \\ \underset{(6,1)}{\overset{\circ}{\mathbf{c}}_{i_k j}} \\ \underset{(3,1)}{\ddot{\mathbf{c}}_{i_k j}} \end{bmatrix} \qquad (17.141)$$

in which

$$\underset{(4,4)}{\dot{\mathbf{N}}_{i_k j}} = \underset{(4,2)}{\dot{\mathbf{B}}_{i_k j}^{\mathsf{T}}} \underset{(2,2)}{\mathbf{W}_{i_k j}} \underset{(2,4)}{\dot{\mathbf{B}}_{i_k j}} \qquad \underset{(4,1)}{\dot{\mathbf{c}}_{i_k j}} = \underset{(4,2)}{\dot{\mathbf{B}}_{i_k j}^{\mathsf{T}}} \underset{(2,2)}{\mathbf{W}_{i_k j}} \underset{(2,1)}{\boldsymbol{\epsilon}_{i_k j}}$$

$$\underset{(6,6)}{\acute{\mathbf{N}}_{i_k j}} = \underset{(6,2)}{\acute{\mathbf{B}}_{i_k j}^{\mathsf{T}}} \underset{(2,2)}{\mathbf{W}_{i_k j}} \underset{(2,6)}{\acute{\mathbf{B}}_{i_k j}} \qquad \underset{(6,1)}{\overset{\circ}{\mathbf{c}}_{i_k j}} = \underset{(6,2)}{\acute{\mathbf{B}}_{i_k j}^{\mathsf{T}}} \underset{(2,2)}{\mathbf{W}_{i_k j}} \underset{(2,1)}{\boldsymbol{\epsilon}_{i_k j}}$$

$$\underset{(3,3)}{\ddot{\mathbf{N}}_{i_k j}} = \underset{(3,2)}{\ddot{\mathbf{B}}_{i_k j}^{\mathsf{T}}} \underset{(2,2)}{\mathbf{W}_{i_k j}} \underset{(2,3)}{\ddot{\mathbf{B}}_{i_k j}} \qquad \underset{(3,1)}{\ddot{\mathbf{c}}_{i_k j}} = \underset{(3,2)}{\ddot{\mathbf{B}}_{i_k j}^{\mathsf{T}}} \underset{(2,2)}{\mathbf{W}_{i_k j}} \underset{(2,1)}{\boldsymbol{\epsilon}_{i_k j}}$$

$$\underset{(4,6)}{\dot{\mathbf{N}}_{i_k j}} = \underset{(4,2)}{\dot{\mathbf{B}}_{i_k j}^{\mathsf{T}}} \underset{(2,2)}{\mathbf{W}_{i_k j}} \underset{(2,6)}{\acute{\mathbf{B}}_{i_k j}}$$

$$\underset{(4,3)}{\dot{\mathbf{N}}_{i_k j}} = \underset{(4,2)}{\dot{\mathbf{B}}_{i_k j}^{\mathsf{T}}} \underset{(2,2)}{\mathbf{W}_{i_k j}} \underset{(2,3)}{\ddot{\mathbf{B}}_{i_k j}}$$

$$\underset{(6,3)}{\acute{\mathbf{N}}_{i_k j}} = \underset{(6,2)}{\acute{\mathbf{B}}_{i_k j}^{\mathsf{T}}} \underset{(2,2)}{\mathbf{W}_{i_k j}} \underset{(2,3)}{\ddot{\mathbf{B}}_{i_k j}} \,.$$

$$(17.142)$$

The system represented by (17.141) is, of course, indeterminate inasmuch as it involves a total of

13 unknowns but was developed from only a single pair of observational equations. The reason for the derivation of this result lies in the fact that the nine matrices specified in (17.142) constitute the basic building blocks from which the general solution can be constructed. It is shown in Davis and Riding (1970) that the general system of normal equations for the kth strip is of the form (later in the discussion it will be indicated how this system can be formed directly from (17.141) through the exercise of a process called the method of *zero augmentation*).

$$\ddot{N}_{k_j} = \sum_{i=1_k}^{m_k} \ddot{N}_{i_{k}j} \qquad \ddot{c}_{k_j} = \sum_{i=1_k}^{m_k} \ddot{c}_{i_{k}j} \qquad (17.144)$$

It will be noted that all the quantities appearing in (17.144) are simply single or double sums of the corresponding quantities appearing in (17.142). The remaining quantities appearing in the general normal equation for the kth strip, namely those involving the various *a priori* weight matrices $\dot{W}_{i_k}$, $\hat{W}_k$ and $\ddot{W}_j$, are defined

	ANGULAR ELEMENTS OF ORIENTATION AND TIMES OF EXPOSURE			ORBITAL STATE VECTOR	COORDINATES OF SURFACE POINTS			
	Frame 1	Frame 2	Frame m_k	Strip k	Point 1	Point 2		Point n
	$\dot{N}_{1_k} + \dot{W}_{1_k}$	0	$\cdots$ 0	$\hat{N}_{1_k}$	$\bar{N}_{1_k 1}$	$\bar{N}_{1_k 2}$ $\cdots$		$\bar{N}_{1_k n}$
	0	$\dot{N}_{2_k} + \dot{W}_{2_k}$ $\cdots$	0	$\hat{N}_{2_k}$	$\bar{N}_{2_k 1}$	$\bar{N}_{2_k 2}$ $\cdots$		$\bar{N}_{2_k n}$
	0	0	$\cdots \dot{N}_{m_k} + \dot{W}_{m_k}$	$\hat{N}_{m_k}$	$\bar{N}_{m_k 1}$	$\bar{N}_{m_k 2}$ $\cdots$		$\bar{N}_{m_k n}$
	$\hat{N}_{1_k}^T$	$\hat{N}_{2_k}^T$ $\cdots$	$\hat{N}_{m_k}^T$	$\hat{N}_k + \hat{W}_k$	$\hat{N}_{k1}$	$\hat{N}_{k2}$ $\cdots$		$\hat{N}_{kn}$
	$\bar{N}_{1_k 1}^T$	$\bar{N}_{2_k 1}^T$ $\cdots$	$\bar{N}_{m_k 1}^T$	$\hat{N}_{k1}^T$	$\ddot{N}_{k1} + \ddot{W}_1$	0 $\cdots$		0
	$\bar{N}_{1_k 2}^T$	$\bar{N}_{2_k 2}^T$ $\cdots$	$\bar{N}_{m_k 2}^T$	$\hat{N}_{k2}^T$	0	$\ddot{N}_{k2} + \ddot{W}_2 \cdots$		0
	$\bar{N}_{1_k n}^T$	$\bar{N}_{2_k n}^T$ $\cdots$	$\bar{N}_{m_k n}^T$	$\hat{N}_{kn}^T$	0	0 $\cdots$		$\ddot{N}_{kn} + \ddot{N}_n$

$$\cdot \qquad (17.143)$$

$$
\begin{bmatrix}
\dot{\delta}_{1_k} \\
\dot{\delta}_{2_k} \\
\vdots \\
\dot{\delta}_{m_k} \\
\hline
\hat{\delta}_k \\
\hline
\ddot{\delta}_1 \\
\ddot{\delta}_2 \\
\vdots \\
\ddot{\delta}_n
\end{bmatrix}
=
\begin{bmatrix}
\dot{c}_{1_k} - \dot{W}_{1_k}\dot{\epsilon}_{1_k} \\
\dot{c}_{2_k} - \dot{W}_{2_k}\dot{\epsilon}_{2_k} \\
\vdots \\
\dot{c}_{m_k} - \dot{W}_{m_k}\dot{\epsilon}_{m_k} \\
\hline
\hat{c}_k - \hat{W}_k\hat{\epsilon}_k \\
\hline
\ddot{c}_{k1} - \ddot{W}_1\ddot{\epsilon}_1 \\
\ddot{c}_{k2} - \ddot{W}_2\ddot{\epsilon}_2 \\
\vdots \\
\ddot{c}_{kn} - \ddot{W}_n\ddot{\epsilon}_n
\end{bmatrix}
$$

The various elements of this system are defined directly in terms of the more basic elements defined in (17.142). Specifically, the $\bar{N}_{i_k j}$ are exactly as defined as in (17.142) and the remaining elements are defined as follows

$$\dot{N}_{i_k} = \sum_{j=1}^{n} \bar{N}_{i_k j} \qquad \hat{N}_{i_k} = \sum_{j=1}^{n} \bar{N}_{i_k j}$$

$$\dot{c}_{i_k} = \sum_{j=1}^{n} \dot{c}_{i_k j} \qquad \hat{N}_k = \sum_{i=1}^{m_k}\sum_{j=1}^{n} \bar{N}_{i_k j}$$

$$\hat{N}_{k_j} = \sum_{i=1}^{m_k} \hat{N}_{i_k j} \qquad \hat{c}_k = \sum_{i=1_k}^{m_k}\sum_{j=1}^{n} \dot{c}_{i_k j}$$

in conjunction with the external observational equations (17.120), (17.125) and (17.128), respectively. A few as-yet-unstated assumptions are involved in the derivation of the above system of normal equations. One assumption concerns the block diagonality of certain covariance or weight matrices. Specifically, if the weight matrix corresponding to the vector of observed plate coordinates for all measured points from the kth strip is denoted by $\hat{W}_k$, the weight matrix corresponding to the vector of angular elements of orientation and exposure times for all frames from the kth strip by $\dot{W}_k$, and the weight matrix corresponding to the vector of X, Y, Z coordinates for all surface points are denoted by $\ddot{\mathbf{W}}$, the following relations hold

$$\mathbf{W}_k \atop {(2m_kn,2m_kn)} = \text{diag} \; (\underset{(2,2)}{\mathbf{W}_{1_k1}} , \; \underset{(2,2)}{\mathbf{W}_{1_k2}} , \ldots ,$$

$$\underset{(2,2)}{\mathbf{W}_{1_kn}} , \ldots , \underset{(2,2)}{\mathbf{W}_{m_kn}}),$$

$$\dot{\mathbf{W}}_k \atop {(4m_k,4m_k)} = \text{diag} \; (\underset{(4,4)}{\dot{\mathbf{W}}_{1_k}} , \; \underset{(4,4)}{\dot{\mathbf{W}}_{2_k}} , \ldots , \underset{(4,4)}{\dot{\mathbf{W}}_{m_k}}),$$

$$\ddot{\mathbf{W}} \atop {(3n,3n)} = \text{diag} \; (\underset{(3,3)}{\ddot{\mathbf{W}}_1} , \; \underset{(3,3)}{\ddot{\mathbf{W}}_2} , \ldots , \underset{(3,3)}{\ddot{\mathbf{W}}_n}).$$

$$(17.145)$$

A second assumption is the temporary one that all measured points appear on all photographs. In practice, of course, this is not the case except possibly in special situations involving convergent photographs covering a common area. To undo this second assumption one merely needs to set the weight matrix $\mathbf{W}_{i_kj}$ equal to a null matrix when the jth point is not measured on the ith frame of the strip. This has two immediate logical consequences in the general normal equations (17.143). First, all $\mathbf{N}_{i_kj}$ corresponding to null $\mathbf{W}_{i_kj}$ reduce to null matrices (and hence do not have to be formed in the first place). Second, the various summations indicated in (17.144) have to be formed only for those particular points appearing on a given frame and/or for those particular frames containing a given point. Thus, for example, in (17.144) the $\mathbf{N}_{i_kj}$ are summed only over that subset of points that appear on frame i_k and the $\dot{\mathbf{N}}_{i_kj}$ are summed only over that subset of frames on which point j appears.

17.4.4 GENERAL NORMAL EQUATIONS FOR ORBITAL BLOCK

Now that the normal equations for an arbitrary orbital *strip* have been generated, the basic structural elements have been defined for the development of the normal equation of an arbitrary orbital *block* formed by overlapping strips. Corresponding to the partitioning indicated by the broken lines in (17.143), the normal equations for the kth strip can be represented more compactly as

(a) first, it is noted that the parametric vector corresponding to an orbital block of s strips is of the form

$$\boldsymbol{\delta} = (\dot{\boldsymbol{\delta}}_1^{\mathsf{T}} \dot{\boldsymbol{\delta}}_2^{\mathsf{T}} \ldots \dot{\boldsymbol{\delta}}_s^{\mathsf{T}} \overset{\circ}{\boldsymbol{\delta}}_1^{\mathsf{T}} \overset{\circ}{\boldsymbol{\delta}}_2^{\mathsf{T}} \ldots \overset{\circ}{\boldsymbol{\delta}}_s^{\mathsf{T}} \ddot{\boldsymbol{\delta}}_1^{\mathsf{T}} \ddot{\boldsymbol{\delta}}_2^{\mathsf{T}} \ldots \ddot{\boldsymbol{\delta}}_n^{\mathsf{T}});$$

(b) next, it is noted from (17.146) that the equations for the kth strip involve only a subset of the elements of $\boldsymbol{\delta}$;

(c) this situation can be remedied immediately by the formal expedient of appropriately expanding the coefficient matrix and constant column of (17.146) to operate on the full matrix $\boldsymbol{\delta}$ (this merely entails the insertion of zero matrices in positions corresponding to coefficients of those particular subvectors of $\boldsymbol{\delta}$ that do not appear in (17.146) and also of similarly augmenting the constant column with zeroes);

(d) suppose that the foregoing process of *zero augmentation* were performed for each of the s strips comprising the block (*i.e.*, for $k = 1$, $2, \ldots s$ in (17.146);

(e) then, except for one consideration, it could be asserted that all of the resulting augmented systems of normal equations are based on independent sets of observations;

(f) the consideration that compromises the independence of the sets of augmented normal equations is the existence in each set of the contribution of common *a priori* constraints on surface points (these give rise to the various matrices involving the $\ddot{\mathbf{W}}_j$);

(g) the foregoing difficulty can be removed by regarding the $\ddot{\mathbf{W}}_j$ in (17.146) as null matrices and writing a separate, suitably augmented system of normal equations to express the contribution of the *a priori* constraints on surface points;

(h) this latter system of normal equations, prior to zero augmentation to be conformable with the full matrix $\boldsymbol{\delta}$, would be simply

$$\begin{bmatrix} \ddot{\mathbf{W}}_1 & 0 & \cdots & 0 \\ 0 & \ddot{\mathbf{W}}_2 & \cdots & 0 \\ \cdot & & & \cdot \\ \cdot & & & \cdot \\ 0 & 0 & \cdots & \ddot{\mathbf{W}}_n \end{bmatrix} \begin{bmatrix} \ddot{\boldsymbol{\delta}}_1 \\ \ddot{\boldsymbol{\delta}}_2 \\ \cdot \\ \cdot \\ \ddot{\boldsymbol{\delta}}_n \end{bmatrix} = \begin{bmatrix} -\ddot{\mathbf{W}}_1 \ddot{\boldsymbol{\epsilon}}_1 \\ -\ddot{\mathbf{W}}_2 \ddot{\boldsymbol{\epsilon}}_2 \\ \cdot \\ \cdot \\ -\ddot{\mathbf{W}}_n \ddot{\boldsymbol{\epsilon}}_n \end{bmatrix} ;$$

(i) at this point the additive property of independent systems of normal equations may be invoked;

(j) according to this principle, systems of normal equations operating on a common parametric vector and formed from the adjustment of independent sets of observations may be added

$$\begin{bmatrix} \dot{\mathbf{N}}_k + \dot{\mathbf{W}}_k & \hat{\mathbf{N}}_k & \bar{\mathbf{N}}_{k1} & \bar{\mathbf{N}}_{k2} & \cdots & \bar{\mathbf{N}}_{kn} \\ \hat{\mathbf{N}}_k^{\mathsf{T}} & \overset{\circ}{\mathbf{N}}_k + \overset{\circ}{\mathbf{W}}_k & \hat{\mathbf{N}}_{k1} & \hat{\mathbf{N}}_{k2} & \cdots & \hat{\mathbf{N}}_{kn} \\ \bar{\mathbf{N}}_{k1}^{\mathsf{T}} & \hat{\mathbf{N}}_{k1}^{\mathsf{T}} & \ddot{\mathbf{N}}_{k1} + \ddot{\mathbf{W}}_1 & 0 & \cdots & 0 \\ \bar{\mathbf{N}}_{k2}^{\mathsf{T}} & \hat{\mathbf{N}}_{k2}^{\mathsf{T}} & 0 & \ddot{\mathbf{N}}_{k2} + \ddot{\mathbf{N}}_2 & \cdots & 0 \\ \cdot & \cdot & \cdot & & & \cdot \\ \bar{\mathbf{N}}_{kn}^{\mathsf{T}} & \hat{\mathbf{N}}_{kn}^{\mathsf{T}} & 0 & 0 & \cdots & \ddot{\mathbf{N}}_{kn} + \ddot{\mathbf{W}}_n \end{bmatrix} \begin{bmatrix} \dot{\boldsymbol{\delta}}_k \\ \overset{\circ}{\boldsymbol{\delta}}_k \\ \ddot{\boldsymbol{\delta}}_1 \\ \ddot{\boldsymbol{\delta}}_2 \\ \cdot \\ \ddot{\boldsymbol{\delta}}_n \end{bmatrix} = \begin{bmatrix} \dot{\mathbf{c}}_k - \dot{\mathbf{W}}_k \dot{\boldsymbol{\epsilon}}_k \\ \overset{\circ}{\mathbf{c}}_k - \overset{\circ}{\mathbf{W}}_k \overset{\circ}{\boldsymbol{\epsilon}}_k \\ \ddot{\mathbf{c}}_{k1} - \ddot{\mathbf{W}}_1 \ddot{\boldsymbol{\epsilon}}_1 \\ \ddot{\mathbf{c}}_{k2} - \ddot{\mathbf{W}}_2 \ddot{\boldsymbol{\epsilon}}_2 \\ \cdot \\ \ddot{\mathbf{c}}_{kn} - \ddot{\mathbf{W}}_n \ddot{\boldsymbol{\epsilon}}_n \end{bmatrix} . \qquad (17.146)$$

The formation of the normal equations for the orbital block comprised of s overlapping strips ($k = 1, 2, \ldots , s$) can be synthesized in the following manner:

directly together to produce the system of normal equations that would result from the simultaneous adjustment of all of the individual sets of observations;

(k) the direct application of this principle to the s independent systems of normal equations resulting from zero augmentation of (17.146) and to the additional independent system of normal equations (suitably augmented) indicated under (h) produces the general system of normal equations appropriate to the simultaneous adjustment of all observations from all strips;

(l) this system of normal equations turns out to be of the specific form

bital block differs substantially from those to be found in Davis and Riding (1970) and Haag and Hodge (1972). The final results, however, are equivalent. It is to be noted that precisely the same process, namely the direct addition of independently formed systems of normal equations rendered conformable by zero augmentation, could also have been used to generate the sys-

ANGULAR ELEMENTS OF ORIENTATION AND TIMES OF EXPOSURE			ELEMENTS OF ORBITAL STATE VECTORS			COORDINATES OF SURFACE POINTS		
Strip 1	Strip 2	Strip S	Strip 1	Strip 2	Strip S	Point 1	Point 2	Point n
$\bar{N}_1 + \dot{W}_1$	0 ···	0	$\bar{N}_1$	0 ···	0	$\bar{N}_{11}$	$\bar{N}_{12}$ ···	$\bar{N}_{1n}$
0	$\bar{N}_2 + \dot{W}_2$ ···	0	0	$\bar{N}_2$ ···	0	$\bar{N}_{21}$	$\bar{N}_{22}$ ···	$\bar{N}_{2n}$
0	0 ··	$\bar{N}_s + \dot{W}_s$	0	0 ···	$\bar{N}_s$	$\bar{N}_{s1}$	$\bar{N}_{s2}$ ···	$\bar{N}_{sn}$
$\bar{N}_1^{T}$	0 ···	0	$\mathring{N}_1 + \mathring{W}_1$	0 ···	0	$\hat{N}_{11}$	$\hat{N}_{12}$ ···	$\hat{N}_{1n}$
0	$\bar{N}_2^{T}$ ···	0	0	$\mathring{N}_2 + \mathring{W}_2$ ··	0	$\hat{N}_{21}$	$\hat{N}_{22}$ ···	$\hat{N}_{2n}$
0	0 ···	$\bar{N}_s^{T}$	0	0 ··	$\mathring{N}_s + \mathring{W}_s$	$\hat{N}_{s1}$	$\hat{N}_{s2}$ ···	$\hat{N}_{sn}$
$\bar{N}_{11}^{T}$	$\bar{N}_{21}^{T}$ ···	$\bar{N}_{s1}^{T}$	$\hat{N}_{11}^{T}$	$\hat{N}_{21}^{T}$ ···	$\hat{N}_{s1}^{T}$	$N_1 + \ddot{W}_1$	0 ···	0
$\bar{N}_{12}^{T}$	$\bar{N}_{22}^{T}$ ···	$\bar{N}_{s2}^{T}$	$\hat{N}_{12}^{T}$	$\hat{N}_{22}^{T}$ ···	$\hat{N}_{s2}^{T}$	0	$N_2 + \ddot{W}_2$ ··	0
$\bar{N}_{1n}^{T}$	$\bar{N}_{2n}^{T}$ ···	$\bar{N}_{sn}^{T}$	$\hat{N}_{1n}^{T}$	$\hat{N}_{2n}^{T}$ ···	$\hat{N}_{sn}^{T}$	0	0 ··	$N_n + \ddot{W}_n$

$$
\begin{bmatrix}
\bar{\delta}_1 \\ \bar{\delta}_2 \\ \vdots \\ \bar{\delta}_s \\ \hline
\hat{\delta}_1 \\ \hat{\delta}_2 \\ \vdots \\ \hat{\delta}_s \\ \hline
\ddot{\delta}_1 \\ \ddot{\delta}_2 \\ \vdots \\ \ddot{\delta}_s
\end{bmatrix}
=
\begin{bmatrix}
\bar{c}_1 - \bar{W}_1 \bar{\epsilon}_1 \\
\bar{c}_2 - \bar{W}_w \bar{\epsilon}_2 \\ \vdots \\
\bar{c}_s - \bar{W}_s \bar{\epsilon}_s \\ \hline
\hat{c}_1 - \hat{W}_1 \hat{\epsilon}_1 \\
\hat{c}_2 - \mathring{W}_2 \hat{\epsilon}_2 \\ \vdots \\
\hat{c}_s - \hat{W}_s \hat{\epsilon}_s \\ \hline
\ddot{c}_1 - \ddot{W}_1 \ddot{\epsilon}_1 \\
\ddot{c}_2 - \ddot{W}_2 \ddot{\epsilon}_2 \\ \vdots \\
\ddot{c}_n - \ddot{W}_n \ddot{\epsilon}_n
\end{bmatrix}
\tag{17.147}
$$

It will be noted that with but two exceptions all of the elements appearing in the above general system of normal equations for an orbital block also appear directly in the normal equations (17.146) appropriate to the strips forming the block. The exceptions are the elements $\ddot{N}_j$ and $\ddot{c}_j$ which consist of the sums

$$\ddot{N}_j = \sum_{k=1}^{s} \ddot{N}_{kj}, \quad \ddot{c}_j = \sum_{k=1}^{s} \ddot{c}_{kj}, \tag{17.148}$$

which, in turn, do involve elements appearing directly in (17.146).

The derivation just outlined for the development of the general normal equations for an or-

tem of normal equations for the kth strip expressed by (17.143). Here, the system defined by (17.141) would serve as the starting point for the process. With this understanding, then, the present treatment of the adjustment of orbital blocks may be viewed as being essentially self-contained.

It will be noted that the coefficient matrix of the general normal equations (17.147) is by no means solidly filled with non-zero elements. Quite the contrary, for it is seen that the $N + W$, the $\bar{N}$, the $\mathring{N} + \mathring{W}$ and the $N + \ddot{W}$ portions of the matrix are all block diagonal. In addition, the $\bar{N}$ and $\hat{N}$ portions of the matrix may be expected to be sparse in those situations wherein surface points appear in only a limited number of frames (as in blocks formed by frames having fairly regular forward and lateral overlap). As will be seen in the remaining sections, it is only through specific exploitation of the foregoing properties of the normal equations that it becomes possible to develop a practical reduction for blocks containing a large number of photographs.

17.4.5 REDUCED NORMAL EQUATIONS FOR AN ORBITAL BLOCK

As it stands, the general system of normal equations (17.147) is of order $4m + 6s + 3n$ where $m = m_1 + m_2 + \ldots + m_s$ denotes the total number of frames comprising the block. When m and n are large, a straightforward solution of (17.147) may be prohibitive. However, the block diagonality of some of the constituents of the covariance matrix can be exploited to effect an efficient reduction. The development of such a reduction begins with the representation of the general system in the following compact form corresponding to the partitioning indicated by the broken lines in (17.147)

$$\begin{bmatrix} \dot{N} + \dot{W} & \bar{N} & \ddot{N} \\ {\scriptstyle(4m,4m)} & {\scriptstyle(4m,6s)} & {\scriptstyle(4m,3n)} \\ \bar{N}^T & \hat{N} + \hat{W} & \tilde{N} \\ {\scriptstyle(6s,4m)} & {\scriptstyle(6s,6s)} & {\scriptstyle(6s,3n)} \\ \ddot{N}^T & \tilde{N}^T & \mathring{N} + \mathring{W} \\ {\scriptstyle(3n,4m)} & {\scriptstyle(3n,6s)} & {\scriptstyle(3n,3n)} \end{bmatrix} \begin{bmatrix} \dot{\delta} \\ {\scriptstyle(4n,1)} \\ \hat{\delta} \\ {\scriptstyle(6s,1)} \\ \mathring{\delta} \\ {\scriptstyle(3n,1)} \end{bmatrix} = \begin{bmatrix} \dot{c} - \dot{W}\dot{\epsilon} \\ {\scriptstyle(4m,1)} \\ \hat{c} - \hat{W}\hat{\epsilon} \\ {\scriptstyle(6s,1)} \\ \mathring{c} - \mathring{W}\mathring{\epsilon} \\ {\scriptstyle(3n,1)} \end{bmatrix}. \quad (17.149)$$

It will be noted that all of the submatrices comprising the coefficient matrix are block diagonal except for $\bar{N}$ and $\hat{N}$. The formal solution of the third matrix equation implicit in (17.149) for $\mathring{\delta}$ in terms of $\dot{\delta}$ and $\hat{\delta}$ is

$$\mathring{\delta} = (\mathring{N} + \mathring{W})^{-1} (\mathring{c} - \mathring{W}\mathring{\epsilon} - \ddot{N}^T\dot{\delta} - \tilde{N}^T\hat{\delta}). \quad (17.150)$$

The substitution of this result into the first two equations implicit in (17.149) yields the reduced system of normal equations

$$\begin{bmatrix} \dot{N} + \dot{W} - \ddot{N}(\mathring{N} + \mathring{W})^{-1}\ddot{N}^T & \bar{N} - \ddot{N}(\mathring{N} + \mathring{W})^{-1}\tilde{N}^T \\ \bar{N}^T - \tilde{N}(\mathring{N} + \mathring{W})^{-1}\ddot{N}^T & \hat{N} + \hat{W} - \tilde{N}^T(\mathring{N} + \mathring{W})^{-1}\tilde{N} \end{bmatrix} \begin{bmatrix} \dot{\delta} \\ \hat{\delta} \end{bmatrix}$$

$$= \begin{bmatrix} \dot{c} - \dot{W}\dot{\epsilon} - \ddot{N}(\mathring{N} + \mathring{W})^{-1}(\mathring{c} - \mathring{W}\mathring{\epsilon}) \\ \hat{c} - \hat{W}\hat{\epsilon} - \tilde{N}(\mathring{N} + \mathring{W})^{-1}(\mathring{c} - \mathring{W}\mathring{\epsilon}) \end{bmatrix} \quad (17.151)$$

The evaluation of this system is rendered feasible by the fact that the $3m \times 3m$ matrix $\mathring{N} + \mathring{W}$ is a block diagonal matrix of m 3×3 matrices. Hence, the required inversion of $\mathring{N} + \mathring{W}$ in (17.151) is easily accomplished. From this fact used in conjunction with the explicit structure of the other submatrices of (17.151) one can form the following practical algorithm for the forma-

tion of the reduced system of normal equations. First, one evaluates for point j and frame i_k the nine primary matrices in (17.142) for each frame in turn of the kth strip. These are then arranged to form the following matrices

$$\dot{N}_{kj} = \begin{bmatrix} \dot{N}_{1_kj} & 0 & \cdots & 0 \\ 0 & \dot{N}_{2_kj} & \cdots & 0 \\ \vdots & & & \vdots \\ 0 & 0 & \cdots & \dot{N}_{m_kj} \end{bmatrix}$$

$$\bar{N}_{kj} = \begin{bmatrix} \bar{N}_{1_kj} \\ \bar{N}_{2_kj} \\ \vdots \\ \bar{N}_{m_kj} \end{bmatrix}, \quad \dot{c}_{kj} = \begin{bmatrix} \dot{c}_{1_kj} \\ \dot{c}_{2_kj} \\ \vdots \\ \dot{c}_{m_kj} \end{bmatrix}$$

$$\hat{N}_{kj} = \begin{bmatrix} \hat{N}_{1_kj} \\ \hat{N}_{2_kj} \\ \vdots \\ \hat{N}_{m_kj} \end{bmatrix}, \quad \tilde{N}_{kj} = \begin{bmatrix} \tilde{N}_{1_kj} \\ \tilde{N}_{2_kj} \\ \vdots \\ \tilde{N}_{m_kj} \end{bmatrix}$$

$$\mathring{N}_{kj} = \mathring{N}_{1_kj} + \mathring{N}_{2_kj} + \ldots + \mathring{N}_{m_kj}$$
$$\mathring{c}_{kj} = \mathring{c}_{1_kj} + \mathring{c}_{2_kj} + \ldots + \mathring{c}_{m_kj}$$
$$\tilde{N}_{kj} = \tilde{N}_{1_kj} + \tilde{N}_{2_kj} + \ldots + \tilde{N}_{m_kj}$$
$$\tilde{c}_{kj} = \tilde{c}_{1_kj} + \tilde{c}_{2_kj} + \ldots + \tilde{c}_{m_kj}. \quad (17.152)$$

The above matrices thus generated for each of the s strips are employed, in turn, to form the following set of secondary matrices

$$\dot{N}_j = \begin{bmatrix} \dot{N}_{1j} & 0 & \cdots & 0 \\ 0 & \dot{N}_{2j} & \cdots & 0 \\ \vdots & & & \vdots \\ 0 & 0 & \cdots & \dot{N}_{sj} \end{bmatrix}, \quad \bar{N}_j = \begin{bmatrix} \bar{N}_{1j} \\ \bar{N}_{2j} \\ \vdots \\ \bar{N}_{sj} \end{bmatrix}, \quad \dot{c}_j = \begin{bmatrix} \dot{c}_{1j} \\ \dot{c}_{2j} \\ \vdots \\ \dot{c}_{sj} \end{bmatrix}$$

$$\hat{N}_j = \begin{bmatrix} \hat{N}_{1j} & 0 & \cdots & 0 \\ 0 & \hat{N}_{2j} & \cdots & 0 \\ \vdots & & & \vdots \\ 0 & 0 & \cdots & \hat{N}_{sj} \end{bmatrix}, \quad \tilde{N}_j = \begin{bmatrix} \tilde{N}_{1j} \\ \tilde{N}_{2j} \\ \vdots \\ \tilde{N}_{sj} \end{bmatrix},$$

$$\mathring{N}_j = \begin{bmatrix} \mathring{N}_{1j} & 0 & \cdots & 0 \\ 0 & \mathring{N}_{2j} & \cdots & 0 \\ \vdots & & & \vdots \\ 0 & 0 & \cdots & \mathring{N}_{sj} \end{bmatrix}, \quad \mathring{c}_j = \begin{bmatrix} \mathring{c}_{1j} \\ \mathring{c}_{2j} \\ \vdots \\ \mathring{c}_{sj} \end{bmatrix},$$

$$\mathring{N}_j = \mathring{N}_{1j} + \mathring{N}_{2j} + \ldots + \mathring{N}_{sj},$$
$$\mathring{c}_j = \mathring{c}_{1j} + \mathring{c}_{2j} + \ldots + \mathring{c}_{sj}. \quad (17.153)$$

In terms of these the following auxiliaries are evaluated

$$\dot{Q}_j = (\mathring{N}_j + \mathring{W}_j)^{-1} \ddot{N}_j^T$$
$$\hat{Q}_j = (\mathring{N}_j + \mathring{W}_j)^{-1} \tilde{N}_j^T$$
$$\dot{R}_j = \ddot{N}_j \dot{Q}_j$$
$$\bar{R}_j = \ddot{N}_j \hat{Q}_j = \dot{Q}_j^T \tilde{N}_j^T$$
$$\hat{R}_j = \tilde{N}_j \hat{Q}_j \quad (17.154)$$
$$\dot{S}_j = \dot{N}_j - \dot{R}_j$$

$$\bar{S}_j = \hat{N}_j - \bar{R}_j$$
$$\mathring{S}_j = \hat{N}_j - \mathring{R}_j$$
$$\bar{c}_j = \dot{c}_j - \dot{Q}_j^T (\ddot{c}_j - \ddot{W}_j \, \ddot{\epsilon}_j)$$
$$\mathring{c}_j = \mathring{c}_j - \mathring{Q}_j^T (\ddot{c}_j - \ddot{W}_j \, \ddot{\epsilon}_j).$$

As each of the above S's and c's is generated it is added to the sum of its predecessors and only the resulting sum is retained. When data for all of the n surface points have thus been processed, the following sums are produced:

$$\dot{S} = \dot{S}_1 + \dot{S}_2 + \ldots + \dot{S}_n$$
$$\bar{S} = \bar{S}_1 + \bar{S}_2 + \ldots + \bar{S}_n$$
$$\mathring{S} = \mathring{S}_1 + \mathring{S}_2 + \ldots + \mathring{S}_n \qquad (17.155)$$
$$\bar{c} = \bar{c}_1 + \bar{c}_2 + \ldots + \bar{c}_n$$
$$\mathring{c} = \mathring{c}_1 + \mathring{c}_2 + \ldots + \mathring{c}_n.$$

The reduced system of normal equations (17.151) can then be represented as

$$\begin{bmatrix} \dot{S} + \dot{W} & \bar{S} \\ \bar{S}^T & \mathring{S} + \mathring{W} \end{bmatrix} \begin{bmatrix} \dot{\delta} \\ \mathring{\delta} \end{bmatrix} = \begin{bmatrix} \bar{c} - \dot{W}\dot{\epsilon} \\ \mathring{c} - \mathring{W}\mathring{\epsilon} \end{bmatrix}. \qquad (17.156)$$

Once this system has been solved for $\dot{\delta}$ and $\mathring{\delta}$, each vector of corrections to coordinates of surface points can be evaluated from

$$\ddot{\delta}_j = (\hat{N}_j + \ddot{W}_j)^{-1} (\ddot{c}_j - \ddot{W}_j \, \ddot{\epsilon}_j) - \dot{Q}_j^T \dot{\delta} - \mathring{Q}_j^T \mathring{\delta}. \qquad (17.157)$$

The important consideration concerning the formulation of the reduced system of normal equations is that the processing of observations for each surface point can be performed sequentially and independently with the final system being generated by the cumulative addition of the various S_j and c_j submatrices. This means that the internal storage required of a computer depends primarily on the number of frames in the block and is essentially independent of the number of surface points being processed. Moreover, for a set number of frames, computing time for the formation and solution of the normal equations increases only in direct proportion to the number of surface points. As will be indicated presently, further computational efficiencies are to be gained when restrictions are placed on the number of frames in which a given surface point can appear.

17.4.6 ERROR PROPAGATION

The error propagation associated with the adjustment of an orbital block falls out largely as a by-product of the reduction. In particular, the covariance matrix of the combined vector $(\dot{\delta}^T \, \mathring{\delta}^T)$ is given directly by the inverse of the coefficient matrix of the reduced system of normal equations. Hence,

$$\begin{bmatrix} \text{cov}(\dot{\delta}, \dot{\delta}) & \text{cov}(\dot{\delta}, \mathring{\delta}) \\ \text{cov}(\mathring{\delta}, \dot{\delta}) & \text{cov}(\mathring{\delta}, \mathring{\delta}) \end{bmatrix} = \begin{bmatrix} \dot{S} + \dot{W} & \bar{S} \\ \bar{S}^T & \mathring{S} + \mathring{W} \end{bmatrix}^{-1}. \qquad (17.158)$$

The covariance matrix of the vector $\ddot{\delta}_j$ is then expressed in terms of the foregoing covariance matrix by

$$\text{cov}(\ddot{\delta}_j, \ddot{\delta}_j) = (\hat{N}_j + \ddot{W}_j)^{-1} + \qquad (17.159)$$
$$(Q^T \ddot{Q}^T) \begin{bmatrix} \dot{S} + \dot{W} & \bar{S} \\ \bar{S}^T & \mathring{S} + \mathring{W} \end{bmatrix}^{-1} \begin{bmatrix} \dot{Q}_j \\ \mathring{Q}_j \end{bmatrix}.$$

In this expression the term $(\hat{N}_j + \ddot{W}_j)^{-1}$ would alone represent the covariance matrix of the adjusted coordinates of the jth surface point in the event that the parameters in $\dot{\delta}$, $\mathring{\delta}$ were known perfectly. Accordingly, the second term on the right represents the contribution of errors in $\dot{\delta}$ and $\mathring{\delta}$ to the photogrammetric triangulation. In the ideal situation where sufficient control exists, the contribution of such errors can be suppressed to a relatively insignificant level. When this is not the case, the correlation between errors in different surface points can assume considerable significance. In such situations one may be concerned with the full covariance matrix of the entire vector of adjusted coordinates. The non-diagonal blocks of this matrix are given by

$$\text{cov}(\ddot{\delta}_i, \ddot{\delta}_j) = (\dot{Q}_i^T \, \mathring{Q}_i^T) \begin{bmatrix} \dot{S} + \dot{W} & \bar{S} \\ \bar{S}^T & \mathring{S} + \mathring{W} \end{bmatrix}^{-1} \begin{bmatrix} \dot{Q}_j \\ \mathring{Q}_j \end{bmatrix},$$
$$(17.160)$$
$$i = 1,2, \ldots, n; \quad j = 1,2, \ldots, n, \quad i \neq j.$$

From equations (17.159) and (17.160) one can evaluate the covariance matrix of the *differential* or *relative* positions of arbitrary pairs of points from

$$\text{cov}(\ddot{\delta}_i - \ddot{\delta}_j, \ddot{\delta}_i - \ddot{\delta}_j) = \text{cov}(\ddot{\delta}_i, \ddot{\delta}_i) + \text{cov}(\ddot{\delta}_j, \ddot{\delta}_j)$$
$$- 2\,\text{cov}(\ddot{\delta}_i, \ddot{\delta}_j). \qquad (17.161)$$

This result is of considerable importance to extra-terrestrial mapping, for here accuracies of points relative to one another are of dominant concern.

17.4.7 CONCLUDING CONSIDERATIONS

The reduced system of normal equations represented by (17.156) places no restrictions on the number of frames in which a given surface point can be imaged and measured. Indeed, the formulation admits the possibility that all surface points may be measured on all frames. Of greater practical interest is the situation in which the frames comprising an orbital block have a fairly regular degree of overlap within strips and between strips (as in a conventional aerial block). In such a situation, the restriction may be imposed that a given surface point can appear (and hence be measured) on no more than a predesignated number (m_0) of frames. A fairly generous value of 25 for m_0 would be sufficient to accommodate blocks having both forward and side overlaps of up to 80%. The limitation of m_0 to a relatively small number in comparison with m has important practical consequences in the foregoing development. Of especial importance is the consideration that the various matrices developed in (17.152) each actually involves only m_0 distinct, nonzero constituents. Accordingly, through the implementation of appropriate logical indexing, one can set up and operate on these matrices in a *compacted* form in which extraneous null matrices are never formed. A similar remark applies to the next stages of the development represented by equations (17.153) and (17.154). The final matrices in (17.154) resulting from this process, namely $\dot{S}_j$, $\tilde{S}_j$, $\overset{\circ}{S}_j$, $\bar{c}_j$ and $\tilde{c}_j$ would thus be generated in a compact form and would require appropriate logical augmentation with zero matrices prior to the accumulation indicated in (17.151). As long as m_0 itself is not excessively large, the formation of the reduced system of normal equations can be rendered efficient and practical through such compaction no matter how many frames may be involved in the block as a whole.

A second important consequence of the restriction that a surface point is measured only in a limited subset of m_0 frames stems from the consideration that this causes the matrices $\tilde{N}$, $\overline{N}$ and $\dot{N}$ in (17.149) to become sparse. But more especially, with suitable, regular ordering of the points, these matrices can be made to assume a regular, banded form. As a consequence, the coefficient matrix of the reduced system of normal equations (17.156) assumes a banded-bordered form. Indeed, in this matrix each of the submatrices $\dot{S} + \dot{W}$, $\tilde{S}$ and $\overset{\circ}{S} + \overset{\circ}{W}$ is individually banded. However, the dimensions of $\dot{S} + \dot{W}$ are likely to be far greater than those $\overset{\circ}{S} + \overset{\circ}{W}$. For example, in a block of 200 frames comprised of 10 strips of 20 frames each, the dimensions of $\dot{S} + \dot{W}$ would be 800 × 800 whereas that of $\overset{\circ}{S} + \overset{\circ}{W}$ would be 60 × 60. Accordingly, the banded character of $\overset{\circ}{S} + \overset{\circ}{W}$ is of less importance than that of $\dot{S} + \dot{W}$ and one loses little by treating the matrices $\tilde{S}^T$, $\overset{\circ}{S} + \overset{\circ}{W}$, and $\tilde{S}$ as if they were solidly filled matrices constituting the border of the banded matrix $\dot{S} + \dot{W}$. In this way it becomes possible to solve the system by means of the special algorithm termed *recursive partitioning*. This algorithm is developed in chapter II. It suffices here to mention that by virtue of the application of recursive partitioning, the adjustment of very large orbital blocks is rendered computationally feasible. It is noteworthy that the detailed structure of the banded portion of the matrix $\dot{S} + \dot{W}$ arising from a bundle adjustment with the exercise of orbital constraints corresponds precisely to that of the same block as

adjusted without the exercise of orbital constraints. The essential difference is that in the former case the band is constructed of elemental 4 × 4 submatrices, whereas in the latter case it is constructed of elemental 6 × 6 submatrices. Hence, the bandwidth of the former is only two thirds that of the latter.

The foregoing development of the bundle adjustment with orbital constraints is not intended to be all-encompassing. It does, however, provide the basic theoretical and practical framework for further elaborations and can clearly be expanded to embrace a variety of additional constraints, as, indeed, is done in Haag and Hodge (1972). Such constraints may arise from consideration of:

observations from a laser or radar altimeter;

recovery of elements of interior orientation x_p, y_p, c for each participating camera;

recovery of coefficients of radial and decentering distortion for each participating camera;

exercise of special geodetic constraints (distances, azimuths, height differentials) between ground points;

recovery of biases in precalibrated interlock angles between stellar and mapping cameras; and

recovery of coefficients of empirical error models formulated to account for residual systematic errors in measured plate coordinates.

By and large, the additional parameters associated with the introduction of such constraints can be relegated to an expanded border of the coefficient matrix of the normal equations. Hence, they need not disturb the essential framework of the basic reduction developed herein.

Although slanted toward extra-terrestrial applications, the foregoing development is directly adaptable to applications employing imagery from Skylab, Stereosat, Mapsat, Landsat, or the large format camera which is planned to fly in the Space Shuttle. Indeed, the potential, but-as-yet-untried application of the bundle adjustment with orbital constraints to imagery produced by the Return Beam Videcon (RBV) of Landsat could well provide the practical solution to the problem of small scale (1:250,000) mapping of large, undeveloped regions from a minimal network of ground control (*e.g.*, a few tens of control points for blocks of a few hundred frames).

17.5 *Geometric Reference System*

In general, a principal objective in satellite geodesy and photogrammetry is to provide a unified control network serving as a framework for compilation of maps and charts.

By analogy to terrestrial geodesy, it is appro-

priate to describe the size and shape of the moon or planet by means of a geometric reference surface. The establishment of an equipotential surface, equivalent to mean sea level on earth, is a separate problem.

The reference surface may be selected in several ways. Two of the most usual are (a) to specify only the general form (spherical, spheroidal, or general ellipsoidal) of the surface and let the surface's location and orientation be specified by the control network to which it is fitted, or (b) to specify not only the general form but also the orientation and/or location of the surface with respect to the network. The second alternative is generally selected for satellite photogrammetry. The reference surface is usually located by requiring that its center coincide with the center of mass of the body it represents. If, furthermore, the axes of the reference surface are required to be parallel to the axes of the coordinate system to which the control network belongs, then the reference surface Q has the equation

$$Q = \frac{X^2}{a^2} + \frac{Y^2}{b^2} + \frac{Z^2}{c^2} - 1 = 0 \qquad (17.162)$$

where

Q is a sphere if $a = b = c$
Q is an oblate spheroid if $a = b > c$
Q is a prolate spheroid if $a = b < c$
Q is a triaxial ellipsoid if $a = b = c$.

a, b, and c are then semi-axes of the ellipsoid.

Ideally, the reference-surface's equator should coincide with the body's equator and its minor axis should coincide with the body's axis of rotation. This will not necessarily result unless some constraints are imposed on the general solution. Also, it is desirable to reduce the lengths of the axes by a small factor to ensure that all elevations are positive.

As an example for Q as a reference ellipsoid when $a = b$, a technique for the mathematical determination of the reference surface follows. The general form of the equation for an ellipsoid is

$$\frac{(X_i - X_0)^2 + (Y_i - Y_0)^2}{a^2} + \frac{(Z_i - Z_0)^2}{b^2} = 1 \qquad (17.163)$$

where

$X_i\ Y_i\ Z_i$ = rectangular coordinates of surface points referred to the body's center of mass
$X_0\ Y_0\ Z_0$ = coordinates of the center of the reference ellipsoid.

Multiplying both sidses of equation (17.163) by $a^2 b^2$, we have the condition equation

$$(X_i - X_0)^2 b^2 + (Y_i - Y_0)^2 b^2 + (Z_i - Z_0)^2 a^2 - a^2 b^2 = F_i \qquad (17.164)$$

The objective is to find values for X_0, Y_0, Z_0, a, b to minimize $\sum_{i=1}^{n} F_i^2$. Since equation (17.164) is nonlinear, the solution must be obtained by successive approximation.

$$X_0^{k+1} = X_0^k + \Delta X_0, \text{ etc.}$$
$$a^{k+1} = a^k + \Delta a,\ b^{k+1} = b^k + \Delta b. \qquad (17.165)$$

Linearizing equation (17.164) gives the linearized observation equation

$$(F)^0 + \left(\frac{\partial F}{\partial X_0}\right)^0 \Delta X_0 + \left(\frac{\partial F}{\partial Y_0}\right)^0 \Delta Y_0 + \left(\frac{\partial F}{\partial Z_0}\right)^0 \Delta Z_0$$
$$+ \left(\frac{\partial F}{\partial a}\right)^0 \Delta a + \left(\frac{\partial F}{\partial b}\right)^0 \Delta b = 0. \qquad (17.166)$$

$(F)^0$ is the constant term derived from evaluating (17.164) with the latest approximation to a, b, etc. The partial derivatives in (17.166) are

$$\frac{\partial F}{\partial X_0} = -2b^2 (X_i - X_0)$$

$$\frac{\partial F}{\partial Y_0} = -2b^2 (Y_i - Y_0)$$

$$\frac{\partial F}{\partial Z_0} = -2a^2 (Z_i - Z_0)$$

$$\frac{\partial F}{\partial a} = 2a\left[(Z_i - Z_0)^2 - b^2\right]$$

$$\frac{\partial F}{\partial b} = 2b\left[(X_i - X_0)^2 + (Y_i - Y_0)^2 - a^2\right].$$

Equation (17.166) is linear in the five unknowns. Rewriting equation (17.166), in matrix notation, we have

$$\underset{(1,1)}{\mathbf{F^0}} + \underset{(1,5)}{\mathbf{B}}\ \underset{(5,1)}{\mathbf{\Delta}} = 0 \qquad (17.167)$$

where

$\mathbf{F^0}$ = the vector of constant term,
$\mathbf{B}$ = the matrix of partial derivatives, and
$\mathbf{\Delta}$ = vector of unknowns that define the reference ellipsoid.

Using the method of least squares, the normal equations are

$$[\mathbf{B^T B}]\mathbf{\Delta} = -\mathbf{B^T F^0}. \qquad (17.168)$$

Introducing a weight matrix $\mathbf{W}$, the normal equations are

$$[\mathbf{B^T W B}]\mathbf{\Delta} = -[\mathbf{B^T W F^0}] \qquad (17.169)$$

or more compactly,

$$[\mathbf{N}]\mathbf{\Delta} = (\mathbf{C}). \qquad (17.170)$$

In general, a weight matrix expressing the uncertainty in the five unknowns would not be available since no previous information is assumed. In this case, equation (17.168) is the normal equation.

Then the unknowns are given by

$$\mathbf{\Delta} = [\mathbf{N}]^{-1} (\mathbf{C}). \qquad (17.171)$$

Since the original equation is nonlinear, an iterative process is best. That is:

$$\mathbf{\Delta}^{k+1} = \mathbf{\Delta}^k + \mathbf{\Delta}. \qquad (17.172)$$

Finally, one must evaluate each $X_i\ Y_i\ Z_i$ in the equation to determine the quality of the fit.

Once the best-fitting reference figure has been determined, that is, X_0, Y_0, Z_0, a and b are known, then each $(X, Y, Z)_i$ surface point should have the (X_0, Y_0, Z_0) center of figure coordinates subtracted from it, yielding the rectangular coordinate referenced to the center of figure as follows:

$$X_i' = X_i - X_0,$$
$$Y_i' = Y_i - Y_0,$$
$$Z_i' = Z_i - Z_0.$$

It may be useful to rotate this set of coordinates about its rotational axis in order to establish a zero origin for datum purposes. Also, the rectangular $(X, Y, Z)'$ coordinates may be transformed to latitude longitude and height above the chosen reference ellipsoid using equations given in chapter VIII.

17.6 Coordinates Derived from Apollo Metric Camera Photographs

The metric camera system carried on Apollo lunar missions 15, 16, and 17 was the first photogrammetric system developed by the National Aeronautic and Space Administration (NASA). The system (figure 17-15) consisted of a 76-mm focal length mapping camera with 115 × 115 mm format, coupled with a stellar camera of 76-mm focal length and 18° × 24° angular field, and a laser altimeter with a 300 μ radian angular field and a least count of 1 metre. In theory, this system provided everything (focal length excepted) that a photogrammetrist could want. The position of each exposure station would be obtained from Earth-based tracking; the orientation of each photograph could be computed from the synchronized stellar exposure and the lock-angles determined by preflight calibration; and the scale of each stereoscopic model would be obtained directly from the altimetric data.

Operationally, the data obtained were adequate, but far less than optimum. Orbital ephemerides provided by NASA were found to have systematic deviations up to several kilometres from the photogrammetrically determined spacecraft positions. Attitude data for the mapping camera determined by reduction of the stellar photographs had a precision of about 5″ in roll and yaw and 15″ in pitch. The altimeter did not always function properly and ranges were available for less than half of the usable exposures. Most damaging was the limited lunar coverage. The Apollo program was reduced from 20 to 17 missions, anticipated, near-polar orbits were not occupied, and the photography acquired did not provide complete closure along the equator.

Two different, but related, solutions of the network resulted. The first triangulation solutions were performed by the Defense Mapping Agency (DMA, 1973 and Schimerman, 1976), and then a refinement was effected by a joint effort of between the National Ocean Survey and the U.S. Geological Survey (NOS/GS) (Doyle et al, 1977). The overall objective was an integrated simultaneous solution of data from all three missions.

In short, the learning process that took place among the photogrammetrists during the triangulation resulted in four significant differences in approach to the adjustment problem.

(a) DMA began by using orbital constraints provided by the tracking data to force a best fit between tracking and photogrammetry. The tracking data proved unreliable, so NOS/GS chose to abandon the tracking data for a purely photogrammetric solution.

(b) DMA transformed the camera's orientation from the inertial coordinate system into the selenocentric coordinate system of date using Koziel's theory for lunar librations. In the refinement, NOS/GS used a more recently developed theory by Eckhardt.

(c) DMA reduced the data from mission 15 to serve as a control network. Then photography from missions 16 and 17 were fit to this control network. NOS/GS, instead, performed a simultaneous adjustment of data from all three missions.

(d) The computer program used by DMA did not include a covariance propagation of errors as did the program used in the simultaneous adjustment.

Item (a) amounts to a fundamental difference in approach that was more obvious after evaluating the lack of ties between adjacent strips. Item (b) is explained by the fact that NOS/GS's solution was intended to be a refinement to take advantage of data editing, and more recent work on libration. Items (c) and (d) are both related to the same operational problem: the simultaneous solution for more than 23,000 unknowns with a complete covariance propagation requires a tremendous amount of "computer muscle," a program that can be tailored to the specific problem, the latest adjustment techniques, and a bit of luck.

17.6.1 RESULTS OF DMA REDUCTION

Starting with mission 15, DMA computed for each photographic sequence, a strip triangulation constrained by the stellar attitudes and the ephemerides for spacecraft positions. When photographs from adjacent sequences were compared, inconsistencies of up to several kilometres in tie point positions were discovered. Eventually it was found necessary to select a single sequence (revolution 44 of Apollo 15) to establish a control network, and other photographs were fitted to this (DMA 1973). After the completion of the network resulting from the vertical photographs, the controlled area was extended to the north and south by use of the side-oblique photographs from each mission.

After completion of the control network an interesting validation was performed by comparison with the positions of retroreflectors and transmitters computed from tracking data with the positions computed photogrammetrically.

(Using high-resolution panoramic photographs, the instruments were located on the mapping photographs).

The differences between photogrammetric and tracking data positions of surface instruments are as follows.

The reduction of so large a set of data presented a number of interesting problems (Doyle *et al* 1977) not least of which was interpretation of the results. The distribution of standard deviations in horizontal positions and elevations are shown in figure 17-24.

Instrument	Δ latitude	Δ longitude	Δ radius
Apollo 15 reflector	−206 m	−182 m	−169 m
Apollo 16 transmitter	−335 m	+163 m	+ 19 m
Apollo 17 transmitter	− 87 m	−277 m	−138 m
Lunachod II reflector	−793 m	−118 m	−174 m

17.6.2 RESULTS OF NOS/GS REDUCTION

The NOS/GS reduction started with a separate adjustment of data from each of the three Apollo missions. This provided:

(a) identification of measurement blunders,
(b) realistic estimates of measurement precision for each set of data, and
(c) initial estimates of exposure-station positions.

For the simultaneous solution of all three sets of data, the measured quantities were 51,138 image coordinates and 519 altimetric ranges. The adjusted quantities were three position components and three orientation components for each of 1244 photographs and three coordinates for each of 5234 ground points—a total of 23,436 unknowns. The coordinates of a single ground point centrally located in the area served as the only positional constraint. In order to provide consistency with the DMA solution, and to relate the solution to the center of mass of the moon, all positions were subsequently translated, but not rotated, to a best fit with the tracking ephemeris of revolution 44 of mission 15.

For 70 percent of the points the horizontal standard error (σ_H) is less than 30 metres, and for 74 percent the vertical standard error (σ_E) is less than 30 metres—a result that is quite respectable in comparison with errors in previous control. The area where these results were obtained coincides with the area of most dense photographic coverage and laser ranging. Near the western end of the strips all points were observed on only three photographs and laser-ranges were generally absent, with consequent rapid increase in both horizontal and vertical errors. Initially it seems strange that the horizontal errors increase much more rapidly than the vertical errors—contrary to the usual experience with cantilever extension. Experiments were conducted with simulated data to verify that, when photograph orientation is strongly constrained (as by stellar photographs) but scale transfer is dependent only on image points, horizontal errors do indeed increase more rapidly than vertical errors. Prevention of this was of course the fundamental reason for including the laser altimeter, and it is unfortunate that the instrument failed in critical areas of the photograph coverage.

17.7 Mariner Mars Missions

17.7.1 MISSION DESCRIPTION AND OBJECTIVES

Mars has been the subject of intensive study since the last century because of its earthlike characteristics and because of the possibility that the observed seasonal changes in the dark markings were due to vegetation. This continuing interest in the search for life on Mars is the driving force behind the Viking missions with their well-instrumented, soft-landing spacecraft. Although its diameter is about half that of the earth, Mars is indeed "earthlike" in many respects. It has an obliquity about the same as the earth and

a 24-hour 40-minute day. Its large bright polar caps can be seen forming during the winter months and fading with the coming of spring. The atmosphere of Mars is primarily carbon dioxide and has a surface pressure less than one hundredth that of the earth.

The exploration of Mars by spacecraft started with Mariner 4, which passed within 9845 km of Mars on July 15, 1965 after 228 days of flight. The payload consisted of a television system, a helium magnetometer, a solar-plasma probe, an ionization chamber, a trapped-radiation detector, a cosmic-ray telescope, and a cosmic-dust detector. All instruments recorded data; 21 television pictures were taken, stored on the tape recorder, and played back after the spacecraft became visible from earth again. The pictures merely sampled the surface, but revealed heavy cratering reminiscent of the lunar highlands.

The next missions to Mars were Mariner 6 and 7. Mariner 6 passed within 3510 km of Mars on July 31, 1969 and Mariner 7 passed within 3340 km on August 5, 1969. Each spacecraft carried a television system, an infrared spectrometer, an ultraviolet airglow spectrometer, and an infrared radiometer.

During the period from 48 hours to 28 hours before closest approach, Mariner 6 took 33 pictures of Mars, and another 17 pictures were taken from 22 hours to 7 hours before encounter. Twenty-five pictures were taken during the 18-minutes of closest approach. In a similar manner Mariner 7 acquired 93 pictures from far away and 33 pictures close by.

Mariner 9 was inserted into orbit about Mars on November 14, 1971. The orbit had an inclination of 64°.4 and a 11.98-hour period. The initial altitude at periapsis of 1387 km was later raised to 1650 km. The scientific instrumentation on Mariner 9 consisted of two television cameras, an ultraviolet spectrometer, an infrared spectrometer, and an infrared radiometer.

When Mariner 9 arrived at Mars, the planet was experiencing an extensive dust storm and it was not possible to see surface features in the television pictures. Early in January 1972, the storm had subsided sufficiently in the south polar region to proceed with the systematic mapping of the planet. In the course of the mission, Mars was to be mapped at high resolution (about one kilometre) for geologic studies and at low resolution (two to five kilometres) for photogrammetric computation of a geodetic control net. The geologic mapping sequences were quite successful; however, the geodetic control sequences contained gaps that resulted in difficulties in the development of the planet-wide control net.

On October 27, 1972, after Mariner 9 had been in orbit almost a year and had returned about 7000 pictures of Mars to earth, its telemetry was turned off because the stabilization gas supply was exhausted.

17.7.2 TELEVISION CAMERAS

All of the television cameras used the same vidicon tubes; however, the optics and electronics were changed so that the overall performance improved greatly on each successive mission. The advantage of this tube, developed by the General Electrodynamics Corporation, is that the image can be stored without degradation for many seconds, permitting slow readout without the need for buffer storage. Mariner 4's cameras had a 24-second readout time, and those on Mariner 6, 7 and 9 had a 42-second readout time.

Mariner 4 carried a single camera with an f/8, Cassegrain design, optical system with beryllium mirrors. The effective focal length was 305 mm, and the active format of the vidicon was a square 5.6 mm on a side. This area was scanned by 200 lines and each line was sampled 200 times, so that the pixel was a square, about 28 microns on a side. Six-bit encoding produced 64 (= 2^6) shades of gray.

Mariner 6 and 7 carried two cameras each, one with a focal length of 50 mm and the other with a focal length of 500 mm. The 50-mm lens was a standard Zeiss Planar f/2 designed for use on a 35-mm single lens reflex camera, remounted and stopped down to f/5.6. The 500-mm lens was specially designed and fabricated for this mission; it was an f/2.5 Schmidt-Cassegrain design. The 9.6 × 12.5 mm active format of the two identical vidicons was scanned by 700 lines and each line was sampled 945 times. The pixel was square, about 13 micrometres on a side, and 8-bit encoding gave 256 shades of gray.

The two cameras on Mariner 9 were modest redesigns of those carried on Mariner 6 and 7. The same lenses were used with the 50-mm one stopped down to f/4. The format of the pictures was 9.6 × 12.34 mm; it was scanned in 700 lines and each line was sampled 832 times. The pixel size was 13.7 × 14.8 micrometres; 9-bit encoding was used, recording 512 shades of gray.

The Mariner vidicon camera is a very poor photogrammetric instrument compared with the conventional metric film cameras flown in aircraft and the metric cameras flown in Apollo 15, 16, and 17, (Light 1972). The television format is very small compared with that of film and the resolution is somewhat lower than that of film. The electronic geometric distortions are large so that major corrections must be made to the picture measurements. The resulting standard error of image coordinates on Mariner frames is about 15 μm, compared with 10 μm for Apollo film. Because of the Apollo film format is 115 × 115 mm, image coordinates on the Apollo film can be determined with greater accuracy than on the Mariner vidicon pictures with more than one hundred times the area.

The reseau on the Mariner 6 and 7 cameras contained 56 marks approximately 3 pixels square. The reseau on the Mariner 9-A camera contained 111 marks, most of which were 3 pixels square; one was 5 pixels on a side to avoid any possible confusion in picture orientation.

The geometric distortions of the Mariner 9-A camera were measured by placing a grid in the collimator during preflight calibration (figure 17-40). Using the grid intersections and reseau locations it was possible to separate the electronic and optical distortions and to solve for the focal length of the optics and the coordinates of the principal point on the vidicon faceplate, (Snyder, 1971).

The pixel coordinates of the reseau points

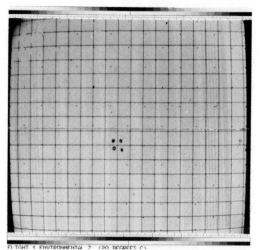

FLIGHT 1 ENVIRONMENTAL 7 (20 DEGREES C)

FIGURE 17-40. Grid for calibrating TV camera on Mariner 9-A.

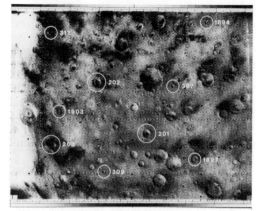

FIGURE 17-41. Mariner 9 television pixel grid.

vary from frame to frame because of the bending of the scanning beam by the remaining charge on the vidicon target. Because this shift can be as much as five pixels, depending on the picture exposure, a special program was written to locate the reseau points on each frame. Geometric corrections could then be made to the image coordinate measurements relative to the reseau. This procedure also automatically corrects for the vidicon's raster shift of a few pixels when it leaves the earth's magnetic field.

17.7.3 CONTROL NETWORK ON MARS

When television pictures began to be received from Mars, work started on both the control net and the map. There was, however, an immediate problem: there was no control for the map. Because of the rush to proceed with the map, preliminary analytical triangulation was done. (Berg 1970). In the meantime, work on the control system continued to incorporate new measurements, new computations of camera distortions, and a variety of improvements. By the time the control net was completed, the map had been finished for some time. This scheduling problem persisted throughout the planetary mapping program and accounts for the publication of a series of Mariner 9 control nets of Mars.

Measurements on the Mariner 9 television pictures were made by counting pixels on frames which had no geometric processing although a pixel grid had been superimposed to facilitate counting (figure 17-41). Using pixel coordinates of the reseau points, a special program (Kreznar 1973) was used to correct for the geometric distortions of the camera and to scale the pixel measurements to millimetre measurements in the focal plane with the origin at the principal point.

Figure 17-42 shows the geometric relationship between the picture-taking spacecraft and the planet Mars. With the camera focal length (f) and (ξ, η, ζ) the coordinates of a surface point in camera coordinates (in which the origin is at the camera and ζ is positive in the direction of the optical axis), the computed image coordinates (x_c, y_c) can be determined directly from the camera and planet geometry at the time the picture was taken through the following relations:

$$x_c = \frac{\xi}{\zeta} f,$$

$$y_c = \frac{\eta}{\zeta} f,$$

$$\begin{bmatrix} \xi \\ \eta \\ \zeta \end{bmatrix} = C\,M\,W \begin{bmatrix} R \cos \phi \cos (360°-\lambda) \\ R \cos \phi \sin (360°-\lambda) \\ R \sin \phi \end{bmatrix} - C \begin{bmatrix} S_x \\ S_y \\ S_z \end{bmatrix}.$$

(17.173)

The computed image coordinates correspond to the measured coordinates (x_0, y_0) of the same point on the picture. The symbols in (17.173) are defined as follows.

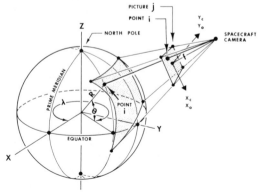

FIGURE 17-42. Geometric relationship—spacecraft camera to Mars.

R, ϕ, λ are the aerocentric radius, latitude, and longitude, respectively, of the surface point.

W is a 3 × 3 orthogonal matrix used to remove diurnal rotation, rotating an aerocentric vector from the *fixed* Mars-equatorial coordinate system to the *inertial* Mars-equatorial coordinate system. This is a simple rotation through the angle defined by Mars' central meridian (the fixed system) and Mars' vernal equinox (the inertial system). In both systems the third vector is in the direction of Mars' axis of rotation.

M is a 3 × 3 orthogonal matrix used to rotate an aerocentric vector from the inertial Mars-equatorial coordinate system to the 1950.0 earth-equatorial coordinate system. In this system the first horizontal vector is in the direction of earth's vernal equinox and the third horizontal vector is normal to the mean 1950.0 equator.

(S_x, S_y, S_z) are aerocentric 1950.0 earth-equatorial coordinates of the spacecraft.

C is a 3 × 3 orthogonal matrix used to rotate a camera-centered vector from the 1950.0 earth-equatorial coordinate system to the camera coordinate system. In this matrix the third horizontal vector is directed along the camera's optical axis and the first horizontal vector is in the direction of the central row of points in the reseau in 1950.0 earth-equatorial coordinates.

Both measured coordinates (x_0, y_0) and computed coordinates (x_c, y_c) exist for each control point identified on each picture. Usually, only approximate values of some of the parameters (such as latitude and longitude of the points) are known, and these values are improved using the method of least squares. In this computation, the residuals are the differences between the measured coordinates and the computed coordinates $(x_0 - x_c, y_0 - y_c)$ of each point on each picture.

The observation equations can be established in terms of differential corrections ΔP_k to specific parameters P_k as

$$x_{0,ij} - x_{c,ij} = \sum_{k=1}^{m} \frac{\partial x_{c,ij}}{\partial P_k} \Delta P_k,$$

$$y_{0,ij} - y_{c,ij} = \sum_{k=1}^{m} \frac{\partial y_{c,ij}}{\partial P_k} \Delta P_k,$$

(17.174)

where the subscripts i index the points on picture j (e.g., $(x_0, y_0)ij$ are image coordinates of the ith point on the jth picture). The differential corrections, ΔP_k, can be found by solving the normal equations formed from the observation equations (17.174). Improved values of the parameters P_k are then found by adding the corrections ΔP_k to the initial values of the parameters P_k. Computer programs have been written to improve the coordinates of the control points (R, ϕ, λ), the orientation angles in the camera orientation matrix C, and the planet's spin axis in the matrix M. Because it is important to have the option to solve for the spin axis, it is necessary to perform the computations in an inertial coordinate system (such as the 1950.0 earth-equatorial coordinate system) rather than the more conventional *fixed* Mars-equatorial coordinate system.

The Mariner 9 control net and cartographic products used a coordinate system for Mars which incorporated a new (Mariner 9) determination of the axis of rotation and a definition of the prime meridian as that meridian which passes through the center of a small crater called Airy-O at latitude $-5°19$ (de Vaucouleurs, Davies and Sturms 1973). This coordinate system for Mars was adopted by the International Astronomical Union in 1973 and was recommended for use in the *American Ephemeris and Nautical Almanac* and as the coordinate system for all new maps of Mars.

Recent computations of control nets use a single, large-block analytical triangulation in which corrections to the latitude and longitude of the control points and the three-camera orien-

TABLE 17-6. SUMMARY OF CONTROL NETS OF MARS

Control Net	Number of Points in Net	Number of Pictures in Net	Number of Observation Equations	Number of Normal Equations	Standard Error of Measurements (mm)
Mariner 6 and 7					
Nov. 1971	115	57	880	401	0.0135
Mariner 9					
Aug. 1972	323	122	1516	1012	0.0195
	156	113	1112	651	0.0174
	344	132	1638	1084	0.0150
	140	118	1018	634	0.0131
Nov. 1972	274	141	1652	971	0.0176
	335	140	1657	1090	0.0301
	363	140	1753	1146	0.0144
	231	142	1584	888	0.0157
	238	154	1514	938	0.0226
Apr. 1973	1340	613	8002	4519	0.0163
Jun. 1973	1645	660	9804	5270	0.0144
May 1974	2061	762	11678	6408	0.0155

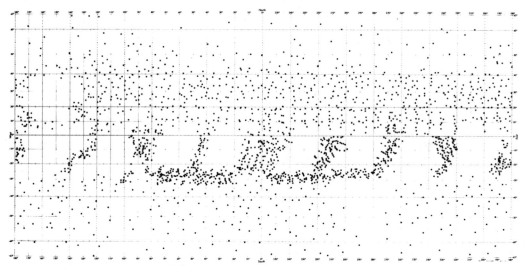

FIGURE 17-43. Distribution of control points on a Mars (Mercator) grid.

tation angles of the pictures are determined. The radii at the control points are derived from those measured by the Mariner 9 radio-occultation experiment (Kliore 1972). In the computation, each point contributes two unknowns (latitude and longitude) and each picture contributes three unknowns (camera orientation angles). The coordinates of the camera stations as determined by the JPL Navigation Team were used in the triangulation. Table 17-6 summarizes the results at various stages in the mapping and control program. Figure 17-43 shows the distribution of control points at the May 1974 stage.

It is always difficult to assess the accuracy of coordinates in a control net because unknown systematic errors are frequently larger than the computed standard errors. On Mars, this is almost certain to be the case; an error of $0°.05$ in the orientation of the axis of rotation would move the equator 3 km; the same shift would occur for an error in the inclination of the Mariner-9's orbit. The error of the axis might be of this magnitude; however, the error in the inclination is probably less. Other systematic errors could arise from navigational errors, changes in the camera's focal length, and electronic distortions due to launch stresses and the long operating lifetime. Although these errors are expected to be small, they might lead to errors of a few kilometres in the point coordinates; these errors are of the same magnitude as the random errors. Thus, as an estimate of the horizontal accuracy of the control net, most of the points have positional errors of less than 10 km and a few might run as high as 25 km.

17.8 Space Photographic Mission Planning

17.8.1 INTRODUCTION

In planning a photographic mission to be accomplished from a spacecraft, it is necessary to ascertain the photographic and photogrammetric goals, then to choose the appropriate camera types to be used, and then to set the flight parameters which make best use of the cameras' abilities. This section will discuss the formulas for three types of cameras most commonly used, *i.e.,* frame cameras, panoramic cameras and stellar cameras.

Frame cameras generally have wide-angle lenses (up to 90°) and have lens-distortions determined. They are fitted with between-the-lens shutters which expose the complete scene at a single instant. This eliminates the possibility of displacement of image-points because of motion of the camera. This type of camera is usually employed in cartographic missions.

Panoramic cameras normally have narrow-field-angle lenses, which are more suited to high resolution. They scan the scene across the line of flight. Since the images are placed on the film as a result of rotating the lens, there is a time delay proportional to the rate of rotation between the first and last images on the photograph. Any motions of the camera during this period will result in a distortion of the apparent relative position of objects in the photograph. The effects of these distortions can be controlled by adding calibration devices such as timing marks, scan and field-angle marks, *etc.,* and correcting the results. These cameras, because of their high resolution and wide field of view, are normally used for reconnaissance and can assist the cartographer by supplying the finer map-detail.

Stellar cameras are fixed-frame, metric cameras used to photograph stars in order to ascertain the attitude of the frame (or panoramic) camera at the instant of exposure. To accomplish this, the angle between the optical axis of the stellar camera and the frame camera must be determined to accuracies commensurate with mission objectives.

A spacecraft equipped with all three types of cameras would have photographic "footprints" as shown in figure 17-45. Figure 17-44 shows how frame-camera and panoramic-camera pictures would be combined.

More detailed descriptions of each type of camera will be found in chapter IV. The cameras described there are used in aircraft, but the essential features are identical to those used in spacecraft.

17.8.2 MAPPING PHOTOGRAPHY

17.8.2.1 Mission Considerations

The planning of a space mission intended to acquire photogrammetric and interpretive photography should be guided by certain basic ideas.

It is desirable that the orbit be nearly polar in order to be able to photograph every part of the planet.

It is desirable that the orbit be sun-synchronous in order that illumination conditions at each latitude remain unchanged during the mission.

It is desirable that the orbit be nearly circular in order to equalize ground resolution and photogrammetric accuracy over the entire planet.

As an example, the coverage plot for a 14-day photographic mission using the Large-format Camera (LFC) in a circular, 70-degree inclination orbit is pictured in figure 17-46. For this mission, the area of first priority is the United States of America, with Northern Africa having second priority. Because the U.S.A. has first priority, an October launch date was planned.

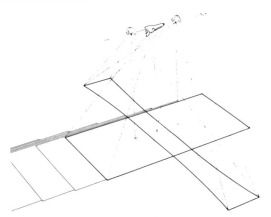

FIGURE 17-45. Photographic foot prints.

Statistics on weather governed this choice. The altitude was planned to be on the order of 250 kilometres. However, to achieve uniform lateral coverage, the LFC requires an altitude very close to 270 kilometres. At 250 kilometres, the ground track of the spacecraft would be repetitive and, therefore, a poor choice for a photographic mission.

17.8.2.2 Estimating Photographic Coverage

The ground coverage (A) per frame for either frame or panoramic photography can readily be estimated from the following equation:

$$A = 2h^2 \left(\frac{\tan \theta}{\cos^3 t} \right) \left[\frac{\tan \delta}{\cos \delta} + \ln \left(\frac{1 + \sin\delta}{\cos \delta} \right) \right]$$

where

h = altitude

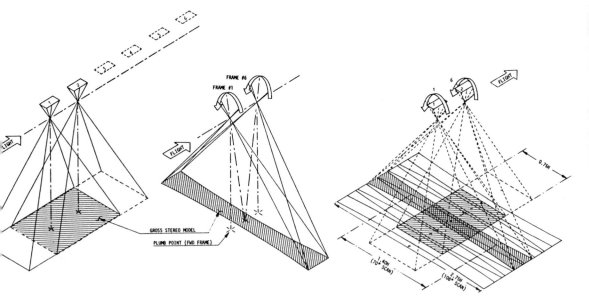

GE FORMAT CAMERA OVERLAP
AND FOOTPRINT GEOMETRY

PANORAMIC CAMERA CONVERGENT
STEREO FOOTPRINT GEOMETRY

COMBINED FOOTPRINT GEOMETRIES

FIGURE 17-44. Combination of imagery from frame and panoramic cameras.

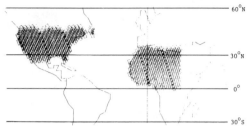

FIGURE 17-46. Coverage for 14-day mission.

t = effective ground tilt
θ = half field-angle in plane of tilt (intrack)
δ = half field-angle in plane perpendicular to plane of tilt (crosstrack)

This equation neglects planetary curvature but the error introduced by this assumption is less than 4% even in a ±60 degree-scan panoramic photograph (*see* footprint equations) and hence is negligible for virtually all practical cases.

For most systems the angular field of view of the lens is not sufficient to detect the curvature of the planet even from orbital altitudes. The one significant exception is a panoramic camera. In this case a suitable footprint equation including the effect of planetary curvature is obtained by assuming a cylindrical approximation to the planets surface. Referring to figures 17-47a and 17-47b, the footprint equations are as shown in figure 17-48. The error involved in neglecting curvature is given by:

$$\frac{\Delta X}{X} = \frac{h}{2R} \left(\frac{\sin \delta \cos \theta}{\cos \delta \cos \theta \cos t + \sin \theta \sin t} \right).$$

For a typical panoramic camera with $\theta = 0°$, $\delta = 60°$ and $t = 10°$ at an altitude of 150 nmi

$$\Delta X/X = 0.04.$$

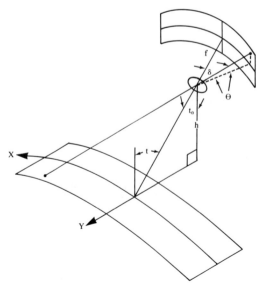

FIGURE 17-47a. Panoramic camera footprint.

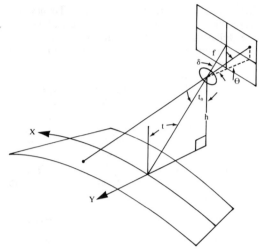

FIGURE 17-47b. Frame camera footprint.

A 4% error in a footprint calculation for a panoramic camera may be of interest but for a frame camera with a total field of view of 40° the resulting edge-of-footprint error, which would be approximately 0.8%, is negligible in most practical applications.

17.8.2.3 ESTIMATING EXPOSURE REQUIREMENTS

Exposure is defined as the quantity of light received by a photographic emulsion. It is the product of the rate at which light falls on the emulsion and the time during which the light is received. It is calculated by the equation

$$H = Et \qquad (17.175)$$

where

H = exposure (usually expressed in lux-seconds),
E = illuminance (usually in lux), and
t = time (usually in seconds).

The exposure required for high-altitude photography is dependent upon a number of factors.

1. The apparent brightness of the scene to be photographed (surface luminance and atmospheric effects).
2. The efficiency of the camera system to gather and transmit light.
3. The response of the recording medium (film) to the light reaching it.

Of the above three factors, the apparent scene brightness is the most difficult value to ascertain due to the many variables involved. The guidelines which follow will result in fairly accurate computations, but may have to be modified slightly if unusual conditions are encountered. These factors and their interrelationships are explored in the following.

17.8.2.3.1 SURFACE LUMINANCE

The amount of light reflected from a surface, *i.e.*, luminance, is dependent on three factors.

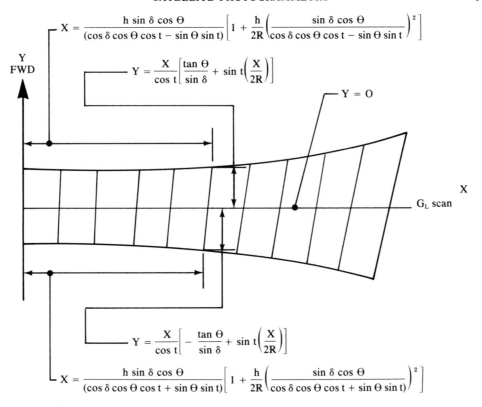

$$X = \frac{h \sin \delta \cos \Theta}{(\cos \delta \cos \Theta \cos t - \sin \Theta \sin t)} \left[1 + \frac{h}{2R} \left(\frac{\sin \delta \cos \Theta}{\cos \delta \cos \Theta \cos t - \sin \Theta \sin t} \right)^2 \right]$$

$$Y = \frac{X}{\cos t} \left[\frac{\tan \Theta}{\sin \delta} + \sin t \left(\frac{X}{2R} \right) \right]$$

$$Y = \frac{X}{\cos t} \left[- \frac{\tan \Theta}{\sin \delta} + \sin t \left(\frac{X}{2R} \right) \right]$$

$$X = \frac{h \sin \delta \cos \Theta}{(\cos \delta \cos \Theta \cos t + \sin \Theta \sin t)} \left[1 + \frac{h}{2R} \left(\frac{\sin \delta \cos \Theta}{\cos \delta \cos \Theta \cos t + \sin \Theta \sin t} \right)^2 \right]$$

NOTE: R = Earth Radius (3440 n.m.)
for δ = 0

$$Y = \begin{cases} \dfrac{h \sin \Theta}{\cos t \cos (t + \Theta)} \\[2ex] -\dfrac{h \sin \Theta}{\cos t \cos (t - \Theta)} \end{cases}$$

FIGURE 17-48. Footprint calculations considering planetary curvature.

1. The amount of light incident on the surface, *i.e.*, illuminace.
2. The spectral reflectance of the surface.
3. The reflection characteristics of the surface.

17.8.2.3.2 ILLUMINANCE

Illuminance, or illumination, is defined as the luminous flux incident on a surface per unit area. It is generally measured in terms of meter-candles (lux) or foot-candles, the latter being equal to 1 lumen per square foot.

The values of solar altitude versus geographic position, time of year and time of day are depicted in figures 17-49 through 17-56. The optimum solar altitude for high altitude earth-photography is 45°.

17.8.2.3.3 SPECTRAL REFLECTANCE

The spectral reflectance of the surface material is dependent primarily on its composition.

The spectral range of the most commonly used black and white panchromatic aerial films is from 360 to 720 nm. The blue cutoff is determined by the glass in the camera since optical glasses do not pass radiation below 360 nm. The red cutoff is determined by the spectral sensitivity of the aerial film. Many aerial films are not sensitive beyond 720 nm.

Over this spectral range, the distribution of radiant energy incident on the earth on a clear day closely approximates that of a blackbody temperature of 6,000°K. In figure 17-57, the blackbody curve for a 6,000°K radiator is plotted together with the relative solar distribution curve outside the atmosphere. Over the visible spectrum, the color temperature of the solar distribution is between 5,600 and 5,800°K. When the predominantly blue skylight from a clear sky (color temperature ranging from 10,000 to 60,000°K, depending on weather conditions and angle from zenith) is added to the sunlight, the color temperature increases to around 6,000°K.

These data become important for computing exposure for photography over narrow spectral

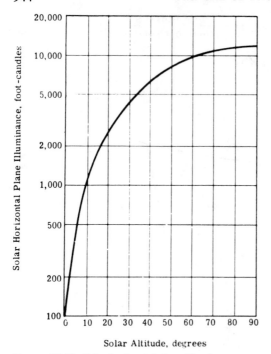

Solar Horizontal Plane Illuminance, foot-candles

Solar Altitude, degrees

FIGURE 17-49. Solar horizontal plane illuminance as a function of solar altitude in average clear weather.

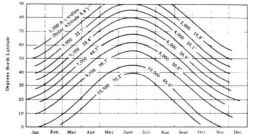

FIGURE 17-51. Constant solar horizontal plane illuminance as a function of north latitude and time of year (time: local apparent noon ± 1 hour).

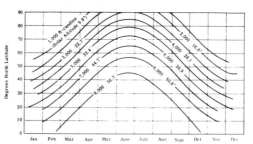

FIGURE 17-52. Constant solar horizontal plane illuminance as a function of north latitude and time of year (time: local apparent noon ± 2 hours).

ranges, *i.e.*, multi-band photography, but for most general exposure problems the average reflectance over the full spectral range is used.

It is not possible to give precise reflectance values for all specific objects, but representative values of the reflectances of a number of materials on the earth commonly seen from the air are known, and some of these values is given in table 17-7.

17.8.2.3.4 REFLECTION CHARACTERISTICS

In addition to the reflectance of the materials, the luminance range of the scene is determined by the topography and by the relative positions of the illuminant and camera. The two extremes of reflection are specular and diffuse, but the reflecting characteristics of most objects lie

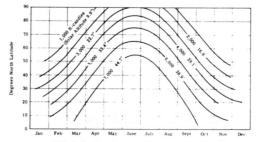

FIGURE 17-53. Constant solar horizontal plane illuminance as a function of north latitude and time of year (time: local apparent noon ± 3 hours).

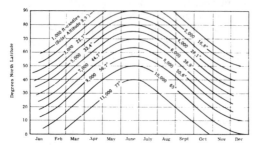

FIGURE 17-50. Constant solar horizontal plane illuminance as a function of north latitude and time of year (time: local apparent noon).

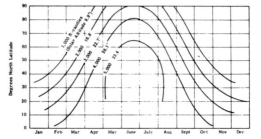

FIGURE 17-54. Constant solar horizontal plane illuminance as a function of north latitude and time of year (time: local apparent noon ± 4 hours).

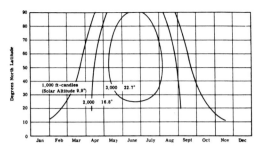

FIGURE 17-55. Constant solar horizontal plane illuminance as a function of north latitude and time of year
(time: local apparent noon ± 5 hours).

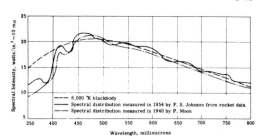

FIGURE 17-57. Comparison of solar spectral distribution outside the atmosphere with a 6,000° K blackbody.

somewhere between these limits. A perfectly
diffuse surface is described by Lambert's cosine
law, which states that the intensity of the radiation emanating in a given direction from any
small element of a diffuse surface is proportional
to the cosine of the angle between the normal to
the surface and the reflected light.

The earth is a specular-diffuse reflector; it is
primarily diffuse with a peak along the line of
specular reflection and a minor peak in the direction of the incident light. For angles less than
about 15 degrees to the surface normal, deviations from Lambert's law will not be significant.
Because of this, Lambert's law is assumed to
apply, and experience supports the validity of
this assumption.

Therefore, for most general exposure problems
the scene luminance can be predicted from the
scene illuminance charts figures 17-49 through
17-56 and the reflectance values of the scene
material content (table 17-7).

17.8.2.3.5 ATMOSPHERIC EFFECTS

As sunlight passes through the atmosphere,
strikes the earth, and is reflected back through
the atmosphere into space, a complicated process of scattering, attenuation, reflectance, and
absorption occurs. A simplified version of this is
shown in figure 17-58, an energy balance diagram
by Dylewski. The basic effects on ground objects viewed from above the atmosphere are a
reduction of reflected, image forming light and
an increase in the overall luminance of the objects due to scattering in the atmosphere. These
factors are treated as atmospheric transmission
and luminance.

17.8.2.3.6 ATMOSPHERIC TRANSMISSION

Data by Mazurowski and Walker show that
the transmission of the earth's atmosphere in the
normal direction, for clear weather conditions,
averages approximately 0.5 in the blue, 0.7 in
the green and 0.8 in the red regions of the spectrum. These values vary somewhat with wave
length and viewing angle but for all practical
purposes, a value of .65 is appropriate for high
altitude work.

17.8.2.3.7 ATMOSPHERIC LUMINANCE

For very high altitude photography the earth's
atmosphere has a luminance of its own which is
added to the ground scene illuminance when
viewed from above.

The value of this additional illuminance versus
solar altitude is depicted in figure 17-59.

17.8.2.3.8 APPARENT SCENE BRIGHTNESS

For high altitude photography the apparent
scene brightness is the scene luminance (computed by multiplying the scene illumination by
the scene reflectivity and the atmospheric transmission) plus the atmospheric illuminance.

$$\beta = ER_sX + I_A \qquad (17.176)$$

β = apparent scene luminance (foot lamberts)
E = illuminance (foot-candles) from figures 17-50
through 17-56
I_A = atmospheric luminance from figure 17-59
R_s = scene reflectivity (%) From table 17-7
X = atmospheric transmission from 0.5 to 0.8.

17.8.2.3.9 CAMERA CHARACTERISTICS

The camera features which must be known in
order to determine the proper exposure are:

the relative aperture,
the transmittance, and
the filter factor.

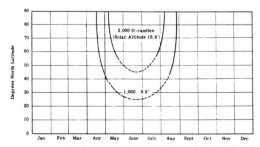

FIGURE 17-56. Constant solar horizontal plane illuminance as a function of north latitude and time of year
(time: local apparent noon ± 6 hours).

Most of these values are specified by the camera
manufacturer, but they may be estimated if the
definition of the terms is understood.

TABLE 17-7. TYPICAL VALUES OF TERRAIN REFLECTANCE

Material	Reflectance, percent	Material	Reflectance, percent
Black asphalt	2	Granite	18
Dark green woods	3	Yellowish sand	24
Water	5	Galvanized iron	27
Open grassland	6	Red paint	29
Streets and roofs	7	Steel	33
Slate gray paint	10	Concrete	36
Dark brown soil	11	Asbestos cement	42
Yellow brick	13	Aluminum	52
Dark concrete	17	Snow	80

17.8.2.3.10 RELATIVE APERTURE (F/NUMBER)

The f/number of a lens is defined as the ratio of the focal length to the diameter of the entrance pupil

$$f/D = f/\text{number} \qquad (17.177)$$
$$f = \text{focal length}$$
$$D = \text{diameter of the entrance pupil.}$$

example: a 12 inch focal length f/5.6 lens has an entrance pupil of 12/5.6 = 2.143 inches.

European manufacturers sometimes express the same property in the form of the so-called aperture ratio, writing it as a pure ratio, *i.e.*, 1:5.6. In either case, 5.6 is the f/number. The larger the relative aperture (the smaller the f/number), and the greater the level of illumination in the image plane for a given brightness in the scene.

17.8.2.3.11 TRANSMITTANCE

The transmittance of the lens is always less than 100%, since some light incident on the lens is reflected back toward the scene at the several glass-atmosphere interfaces. Transmittance is defined as the ratio of the light flux leaving a lens to the light flux entering the lens.

The lens manufacturer will specify the transmittance or the T/number of the lens.

The relationship between transmittance and T/number is expressed by the equation

$$\text{Transmittance } (T) = \frac{f/\text{no.}^2}{T/\text{no.}^2} \qquad (17.178)$$

17.8.2.3.12 FILTER FACTOR

When a filter is added to a camera, the exposure must be increased to compensate for the amount of light absorbed by the filter. The filter factor represents the increase in exposure required to produce the same density level on the film that would have been achieved without the filter. It is dependent on the spectral sensitivity of the film, the quantity of the illumination, and the filter characteristics. The values presented in table 17-8 were determined by comparing the characteristic curves for unfiltered and filtered

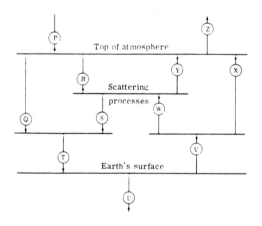

P   Total visible solar radiation that reaches the earth's atmosphere
Q   Portion of P that reaches the earth's surface without having been scattered, or sunlight
R   Scattered fraction of P
S   Scattered light that reaches the earth's surface, or skylight
T   Total light reaching the earth's surface
U   Fraction of T absorbed by the earth
V   Fraction of T reflected by the earth
W   Scattered fraction of V
X   Fraction of V that escapes into space without having been scattered
Y   Scattered light that escapes into space
Z   Total light from the earth that escapes into space

FIGURE 17-58. Energy balance diagram.

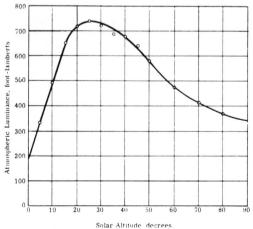

FIGURE 17-59. Atmospheric luminance *vs* solar altitude for clear weather conditions.

TABLE 17-8. FILTER FACTORS

Black & White Films	Wratten Filter				
	12	21	25	2E	89B
3414	2.0	2.2	2.5	1.6	—
3400	2.0	2.3	3.7	1.4	—
3401	1.8	2.3	3.8	1.3	—
3410	2.0	2.4	4.0	1.5	—
3411	2.0	2.4	4.0	1.5	—
2424	1.5	—	2.0	—	3.0
SO-289	1.5	—	2.5	—	4.0
* Color Films					
SO-242	Filters not recommended				
SO-127	Yellow filter layer coated over emulsion				
SO-397	No color-correction filters used				
3443	1.0	—	—	—	—

* The filter factor for color films SO-242, SO-127 and SO-397 is 1.0. H&D curves for color film 3443 include use of a Wratten-12 filter; therefore the filter factor for 3443 is also 1.0.

exposures at a density of 1.0 above base plus fog. The exposure difference is the filter factor. This technique determines the filter factor from the straight line portion of the H&D curve, the region of greatest interest for aerial photography. Because the filter factor is determined on the straight line portion of the curve, it is less subject to variations than if calculated by other techniques.

When evaluating filters for haze reduction in high altitude aerial photography, the filter factor is a necessary consideration. With many films, there is an increase in gamma (slope of the linear portion of the H&D curve) when the shorter end of the visual spectrum is filtered out, because gamma may be dependent on wavelength for a particular emulsion. However, the advantages of increased contrast with the filter may be offset by the resultant loss in photographic speed.

The filters listed cut out the short wavelength end of the spectrum to the following approximate wavelengths: Wratten 12, 500 nm; Wratten 21, 540 nm; and Wratten 25, 580 nm. As the speed of the film increases the filter factor (for minus blue filters) also increases. This result has been verified by previous spectral sensitivity determinations, where it was shown that as the speed of the film increases, the relative blue sensitivity also increases.

17.8.2.3.13 EMULSION SPEED

The efficiency of an emulsion in yielding density as a function of exposure is termed the photographic emulsion speed and is defined as the reciprocal of the exposure required to produce a given density.

The speed of an emulsion is dependent on several factors:

the chemical composition of the emulsion,
the size and frequency distribution of the silver halide grains, and
the spectral sensitivity of the emulsion.

Speed is also somewhat sensitive to the type of development and the spectral distribution of the illumination. A typical speed index for aerial films is the aerial film speed (AFS) defined by the relationship

$$AFS = 3/2H \qquad (17.179)$$

where H is the exposure required to produce a density on the H&D curve corresponding to a position where the density is 0.3 above base plus fog.

The H&D curve of an emulsion is important in order to ascertain the range of exposure the material will accommodate, i.e., the dynamic range of the film.

Figures 17-60 through 17-68 are H&D curves for the most commonly used aerial films, both

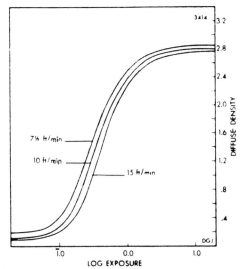

FIGURE 17-60. H&D curve for 3414 film, Versamat Model II processor and Versamat 641 developer at 95°F.

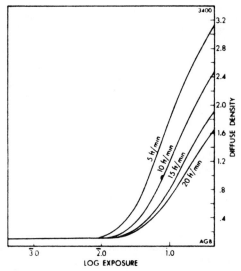

FIGURE 17-61. H&D curves for 3400 film, Versamat Model II processor and Versamat 641 developer at 85°F.

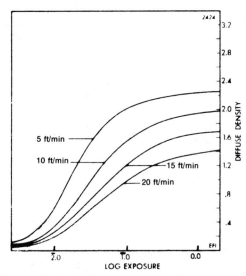

FIGURE 17-63. H&D curves for 2424 film, with Wratten filter no. 25, Versamat Model II processor and Versamat 641 developer at 85°F.

black and white and color. The curves for Infrared films (2424, SO-289, SO-127 and 3443) are derived empirically and therefore may require adjustment to ensure optimum exposure for specific circumstances.

17.8.2.3.14 EXPOSURE DETERMINATION

When a photographic film is exposed in a camera system, the simple relationship of equation (17.175) is expanded to encompass the transmission characteristics of the optical system. The expanded equation is

$$H = \frac{2.7t\beta\,T_1}{(f/no)^2 F} \qquad (17.180)$$

where

H = exposure (lux-seconds)
t = exposure time (seconds)
β = apparent scene luminance from equation (17.176)
T_1 = lens transmission
$f/no.$ = lens f/number
F = filter factor .

This relationship makes it possible to deter-

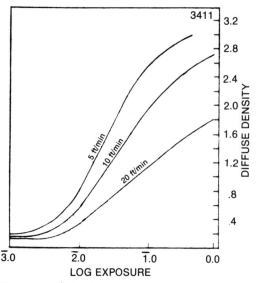

FIGURE 17-62. H&D curves for 3411 film, Versamat Model II processor and Versamat 885 developer at 85°F.

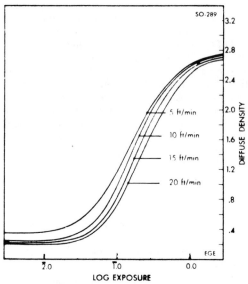

FIGURE 17-64. H&D curves for SO-289 film, Wratten filter no. 25, Versamat Model II processor and Versamat 885 Chemicals at 85°F.

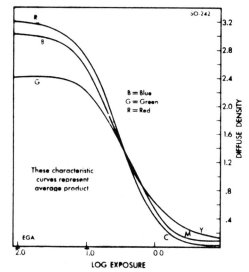

FIGURE 17-65. H&D curves for SO-242 film, Kodak process ME-4 (Modified).

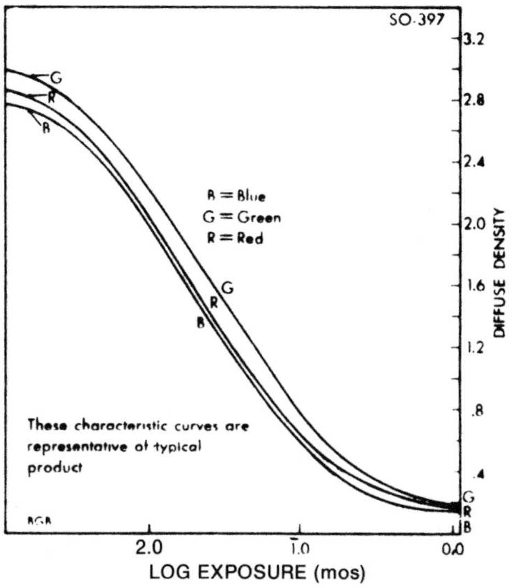

FIGURE 17-67. H&D curves for SO-397 film, Kodak EA 5 chemicals normal process (115F).

mine the exposure time required to produce a given density by considering the film H&D curve and the level of luminance. In order to do this the logarithm of H must be computed since most H&D curves are plotted as the log exposure versus density.

There is not complete agreement as to the best criteria for setting exposures, but a reasonable compromise is to set log H at a position where the mean apparent scene brightness will produce a density of 0.9 on the H&D curve. This criterion will generally permit the total scene illuminance range to be placed above base fog and below the maximum density of the material. If the expo-

sure is on or below base fog or if it has reached its maximum density, variations in density which create imagery will not be discernable. A nomograph which indicates how these factors interrelate is shown in figure 17-69.

A spacecraft exposure problem will acquaint the reader with the use of this nomograph.

A spacecraft achieves a sun-synchronous orbit at one hour past local noon on February 15th. It is desired to photograph a scene at 30° north latitude. The photographic equipment aboard is a camera which has an f/number of 5.6 fitted with a Wratten

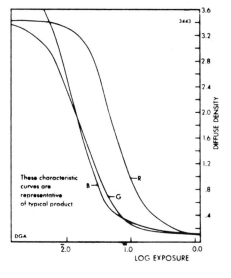

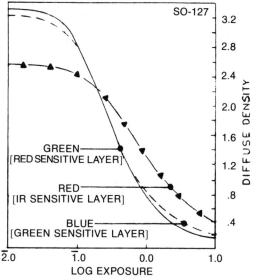

FIGURE 17-66. H&D curves for SO-127 film, Kodak process ME-4 (Modified).

B - Green Sensitive Layer
G - Red Sensitive Layer
R - IR Sensitive Layer

FIGURE 17-68. H&D curves for 3443 film with Wratten filter no. 12, EA-5 process.

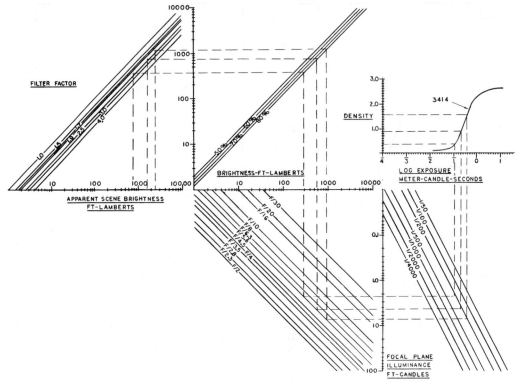

FIGURE 17-69. Exposure nomograph.

21 filter. The lens T/number is 6.26. The camera is loaded with Eastman Kodak 3414 black-and-white aerial film.

PROBLEM

Compute the shutter speed which will produce optimum photography.

SOLUTION

From figure 17-51 we find the solar altitude and solar illuminance = 44.7° and 7,000 foot-candles respectively.

From table 17-7 we estimate that the reflectivity of the scene will vary from 2% to about 40%. Since we wish to capture all the detail in the photograph we should set the mean exposure value determined at a density of 0.9 on the H&D curve for 3414 and check to ensure that the maximum and minimum exposure in the scene is on the straight line portion of the H&D curve.

Scene brightness will be from $7000 \times .02 = 140$ to $7000 \times .40 = 2800$ foot-lamberts. Multiplying these values by .65 (the average transmission of the atmosphere) and adding the atmospheric luminance from figure 17-59, 44.7° = 630 foot-lamberts, will determine the apparent scene brightness, *i.e.*, from 721 to 2450 foot-lamberts or a mean scene brightness of 1585 foot-lamberts.

Place this value on the scale labeled apparent scene brightness on the nomograph and draw a vertical line to the appropriate filter factor determined from table 17-8, *i.e.*, 2.2 for a Wratten 21 and 3414 film.

At the position of intersection of the filter factor move horizontally to the lens transmission of 80% as determined from the T/no. and equation (17.178).

At the position of intersection of the lens transmission, move vertically to the lens f/number, *i.e.*, 5.6.

At the position of f/number intersection move horizontally to the exposure time determination.

Now move to the H&D curve for 3414 and we note that the 0.9 density point is located at approximately log exposure of 1.3. Drop a vertical line from this point to intersect the mean apparent scene brightness line coming from the f/5.6 number intersection.

Note that the lines intersect at an exposure time of approximately 1/200 a second or 5 milliseconds. Since the exposure time must be used for the whole scene, repeat this process for the maximum and minimum apparent scene brightness values, *i.e.*, 721 and 2450 foot-lamberts. We see that the minimum density will be 0.35 and the maximum 1.6 which is well positioned on the straight portion of the H&D curve.

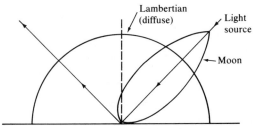

FIGURE 17-70. Reflection characteristics of the moon.

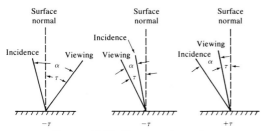

FIGURE 17-71. Photometric angles.

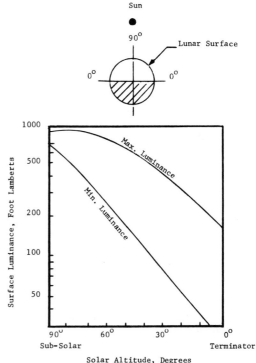

FIGURE 17-73. Solar altitude *vs* surface luminance (slope limits = ± 30°).

Exposure problems associated with lunar photography are similar to those for earth with the exception that there are no atmospheric effects and the lunar surface has a different reflective characteristic which must be accounted for. As opposed to a diffuse reflective characteristic on earth, the lunar surface is highly specular, figure 17-70. Virtually all the light is reflected back in the direction of incidence and there is a rapid reduction in intensity as the angle of reflectance departs from the angle of incidence.

The amount of light reflected back from the lunar surface (luminance) for all conditions of illuminance and viewing can be expressed by only two angles:

α – the phase angle, the angle between the incidence viewing vectors;

τ – the angle between the viewing vector and the normal to the surface slope when projected into the phase plane.

The angles and vectors along with the sign convention are illustrated in figure 17-71. Note that τ is positive only if the viewing vector is between the surface normal and incidence vector. The relationships between the amount of reflected light and the angles α and τ are described by a series of functions derived from data obtained by Fedorets. These functions are shown in figure 17-72. The photometric function, ϕ, determines the surface luminance by the relationship

$$\beta = \phi\rho E_s$$

where

β = luminance (foot-lamberts)
ϕ = Fedorets' photometric function

ρ = surface albedo
E_s = solar illuminance = 140,000 lux (constant).

The lunar surface albedo describes the percentage of light reflected, and is independent of surface geometry. It is determined primarily by the composition of the surface and can vary from 6 to 18 percent depending on the presence of such elements as iron and titanium. An albedo of 7 percent is representative of lunar maria and an albedo of 9 percent is characteristic for the total surface. The surface slope becomes important on the lunar surface as it directly effects the surface normal to which the photometric function is applied. The lunar surface rarely has slopes which exceed 30°. Figure 17-73 is a plot of surface luminance versus solar altitude which indicates the range one can expect when considering the variation of slopes on the surface and a camera with a field angle of about 40°. These data have been corroborated by Apollo Missions 15, 16 and 17. The optimum solar altitude for lunar photography is between 10° and 30°. These curves correspond very closely with actual measurements made from Apollo 15 photography.

Exposure determination uses the classic equation (17.180). β is the mean scene illuminance as determined from figure 17-73 and the exposure should be set on the H&D curve at a density equal to 0.9. Exposure on other planets will follow the

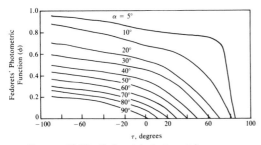

FIGURE 17-72. Fedorets' photometric curves.

same pattern after the reflectivity characteristics of the scene have been established.

17.8.2.4 ESTIMATING GROUND DISTANCE RESOLVED

The expected ground resolved distance (GRD) for either frame or panoramic photography can be estimated to a reasonable degree of accuracy with the use of figure 17-74, which is a plot of dynamic angular resolution (α) *vs* defocusing and image motion, and the equation for GRD:

$$GRD = h\alpha/\cos \delta$$

where

h = altitude
α = dynamic angular resolution (radians)
δ = angle between the line of sight and the local vertical.

The static angular resolution (α_0) used as an input to figure 17-74 is given by

$$\alpha_0 = 1/fR_0$$

where f = focal length and R_0 is the static system resolution (l/mm) of the camera.

As an example, assume a 300 mm f/5 lens system with a static resolution of 50 l/mm with a 3 millisecond exposure. Assuming a motion of 10 milliradians per second, a defocusing error of 60 micrometres, and a spectral band centered on 0.6 micrometres the input parameters for figure 17-74 are:

$$\alpha_0 = \frac{1}{(300)\,(50)} = 67 \ \mu\text{rad}$$

$$\frac{\mathring{\alpha}\,t_e\,D}{\lambda} = \frac{(10 \times 10^{-3})\,(3 \times 10^{-3})\,(60)}{(0.6 \times 10^{-3})} = 3.0$$

$$\frac{\Delta}{\lambda(f/\text{no.})^2} = \frac{(0.06)}{(0.6 \times 10^{-3})\,(5)^2} = 4.0 \ .$$

Entering figure 17-74 with these values gives

$$(\alpha^2 - \alpha_0^2)\left(\frac{D}{\lambda}\right)^2 \cong 15$$

which upon solving for α gives

$$\alpha = \left[15\left(\frac{.6 \times 10^{-3}}{60}\right)^2 + (67 \times 10^{-6})^2 \right]^{1/2} = 77 \ \mu\text{rad}.$$

Assuming an altitude of 300 km (300,000 m) and a vertical camera would give a ground-resolved distance of

$$GRD = 23 \text{ metres}.$$

If the static resolution alone had been used to predict GRD, neglecting the effects of motion and defocus introduced with figure 17-74, the predicted GRD would have been 20 metres. In this example the error introduced by neglecting the effects of motion and defocus would have been only 13%. However, under many conditions the error could be significantly larger.

17.8.2.5 BASE-TO-HEIGHT RATIO VERSUS OVERLAP

For height determination, it is essential to obtain stereo coverage. The accuracy of this determination is dependent upon the base-to-height ratio of the stereoscopic model *i.e.;*

$$\Delta H = (H/f)(H/B)\,\Delta P_x$$

where

ΔH = error of height determination
H = altitude of camera
f = focal length
B = distance between exposure stations
ΔP_x = measurement error.

It can be seen that as the B/H ratio increases, ΔH decreases or the greater the accuracy of height determination.

The relationship between the camera geome-

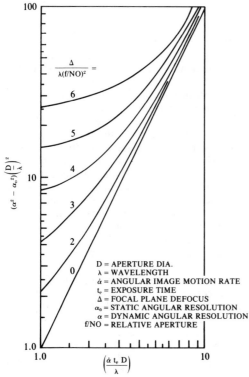

D = APERTURE DIA.
λ = WAVELENGTH
$\mathring{\alpha}$ = ANGULAR IMAGE MOTION RATE
t_e = EXPOSURE TIME
Δ = FOCAL PLANE DEFOCUS
α_0 = STATIC ANGULAR RESOLUTION
α = DYNAMIC ANGULAR RESOLUTION
f/NO = RELATIVE APERTURE

FIGURE 17-74. Estimated dynamic angular resolution *vs* image motion rate and defocus.

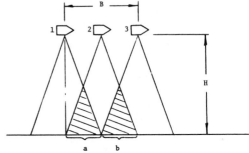

FIGURE 17-75. Stereoscopic coverage with 50% overlap (using adjacent pictures as stereoscopic pairs).

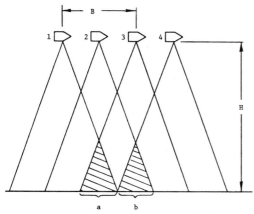

FIGURE 17-76. Stereoscopic coverage with increased B/H.

try and the B/H ratio for a vertical framing camera is given by

$$B/H = W_f (1 - OL)/f$$

where

B = distance between exposure stations
H = altitude of camera
W_f = film format width in the line of motion
f = focal length of the camera
OL = fractional overlap (*ie* 1.0 = 100%, 0.1 = 10% *etc.*)

Since stereoscopic coverage is a requirement, 50% overlap is the theoretical minimum for a vertical camera to achieve complete stereoscopic coverage (*see* figure 17-75).

In actual practice, 55 to 60 percent is recommended to allow for timing errors and pitch which could cause gaps in the coverage.

This approach, of course, limits the B/H ratio which may be obtained, but does require the smallest amount of film and the longest exposure cycle necessary to get the required coverage.

The B/H ratio can be increased by shortening the exposure cycle and utilizing photography from exposure stations 1 & 3 and 2 & 4, *etc.*, which results in reduced overlap and, therefore larger B/H ratios (*see* figure 17-76).

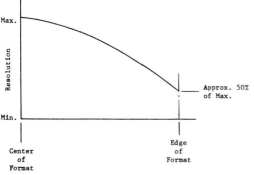

FIGURE 17-77. Typical resolution *vs* distance from center of photograph.

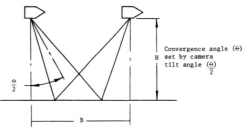

FIGURE 17-78. 100% overlap of stereoscopic pairs.

This approach can theoretically be extended to pairs 1 & 4, 1 & 5, *etc.*, up to the limit of the field angle of the camera. It is not recommended, however, that less than 30% of the frame be utilized since the resolution of the photography generally will be poorer at the extreme edges of the field. (*see* figure 17-77).

As stated before, more film will be needed if this approach is taken and one must ensure that the spacecraft can carry it. Also, the exposure cycle will have to be shortened. Therefore, one must trade the amount of allowable error in height against the camera's cycling ability and the spacecraft's capacity.

Unlike vertical photography, oblique photography with two cameras allows one to choose the stereoscopic angle so that there will be 100% overlap of a stereoscopic pair (*see* figure 17-78). This avoids having to use three photographs to construct a model of the area covered in a particular frame. To achieve 100% overlap it is necessary to select the pointing directions, which are generally directly opposite, so that the ground separation of the principal points of the forward and aft cameras is an integral multiple of the forward ground coverage of a single frame.

For two fixed cameras, one looking forward and the other backward, exposing simultaneously; the angle for 100% stereoscopic overlap is given by

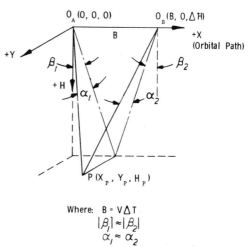

FIGURE 17-79. Spacial intersection model.

$$\theta/2 = \frac{1}{2} \sin^{-1}\left[\frac{NW_f(1-OL)}{f}\right].$$

For a nodding camera, alternately exposing forward and backward, the required angle is given by

$$\theta/2 = \frac{1}{2} \sin^{-1}\left[\frac{\left(N + \frac{1}{2}\right)W_f(1-OL)}{f}\right]$$

where f = focal length.

17.8.2.6 ACCURACY ESTIMATES

Figure 17-79 shows a simple and often used intersection model which is applicable to both fixed-frame and panoramic cameras and provides the equations:

$$X = H \tan \beta_1 \tag{17.181}$$

$$Y = \frac{H \tan \alpha_1}{\cos \beta_1} \tag{17.182}$$

$$H = \frac{V(\Delta T - \Delta t_1 + \Delta t_2) - \Delta H \tan \beta_2}{\tan \beta_1 + \tan \beta_2} \tag{17.183}$$

where:

V is orbital velocity,
ΔT is time on orbit between identical events at two sequential exposure stations (*eg.*, center of shutter opening in frame camera and start of scan in panoramic camera),
ΔH is change in orbital altitude between exposure stations,
β is defined in figure 17-79,
α is defined in figure 17-79,

Δt is time relative to start of scan within a panoramic exposure at which an image is recorded. Timing markers imaged on the film at regular intervals on the centerline of the exposure slit provides this. For frame cameras, Δt is zero.

Equations (17.181), (17.182) and (17.183), may be linearized by differentiation relative to the above defined variables and the following error equations constructed.

$$\sigma_x^2 = \tan^2\beta\,\sigma_H^2 + (H\sec^2\beta)^2\sigma_\beta^2 \tag{17.184}$$

$$\sigma_Y^2 = \frac{1}{2}\left[\left(\frac{\tan\alpha}{\cos\beta}\right)^2\sigma_H^2 + \left(\frac{H\sec^2\alpha}{\cos^2\beta}\right)^2\sigma_\alpha^2\right.$$
$$\left. + \left(\frac{H\tan\alpha\sin\beta}{\cos^2\beta}\right)^2\sigma_\beta^2\right] \tag{17.185}$$

$$\sigma_H^2 = \frac{1}{\tan^2\beta}\left[\left(\frac{\Delta T}{2}\right)^2\sigma_V^2 + \left(\frac{V}{2}\right)^2(\sigma_{\Delta T}^2 + 2\sigma_{\Delta t}^2)\right.$$
$$+ \left(\frac{\tan\beta}{2}\right)^2\sigma_{\Delta H}^2$$
$$+ \left(\frac{V\Delta T + \Delta H\tan\beta}{2\sin 2\beta}\right)^2\sigma_\beta^2\right]$$
$$+ \left(\frac{V\Delta T - \Delta H\tan\beta}{4\sin^2\beta}\right)^2\sigma_\beta^2 \tag{17.186}$$

where units are usually radians, metres, metres/sec and seconds.

To employ these equations for error estimation requires:

the calculation of coefficients through use of nominal values for the quantities, and
the estimation of individual sigmas of the quantities.

TABLE 17-9. FRAME/PANORAMIC CAMERA GEOMETRICAL COMPARISON

Item		Cartographic	Panoramic
1. Exterior Orientation			
Pitch	Exterior Orientation of Optical Axis	Fixed**	Time Variant***
Roll		Fixed**	Time Variant***
Yaw		Fixed**	Time Variant***
X	Exterior Position of Perspective Center	Fixed**	Time Variant
Y		Fixed**	Time Variant
Z		Fixed**	Time Variant
Earth Rotation		Fixed	Time Variant
2. Interior Orientation			
Focal Length		Fixed*	Fixed*
Longitudinal Field Angles		Fixed*	Fixed*
Lateral Field Angles		Fixed*	Time Variant
Lens Distortion (Longitudinal)		Fixed*	Fixed*
Lens Distortion (Lateral)		Fixed*	None
Recording Medium		Film	Film
Film Position		Fixed†	Time Variant†
Film Mensuration		Resolution Limited	Resolution Limited

  * Determined by calibration.
  ** Since exposure occurs in a few milliseconds, these items are considered to be unknown but fixed at exposure.
  *** Includes rotation of the platform and forward-motion-compensation.
  † Both are affected by film shrinkage.

The second is the more difficult. It requires a summation of errors from various sources that contribute to the total expected error in any quantity of equations (17.184), (17.185) and (17.186).

It is useful, first, to compare the geometrical properties of frame and panoramic cameras relative to time. Table 17-9 provides an overall comparison.

Each item in table 17-9 is affected by error. It is clear that the panoramic geometry is highly time variant and complicates the summation of errors. Because the panoramic geometry is more complex, the following is a check list of quantities that may be affected by error.

1. Location of spacecraft
   Time
   Orbit
2. Location of scene
   Atmospheric refraction
   Earth curvature
   Earth rotation
3. Position of camera
   a. Scale of geometry
      Velocity
      Time
   b. Attitude of support
      Pitch
      Roll
      Yaw

c. Attitude of nominal camera axis vs support
   Relative pitch
   Relative roll
   Relative yaw
d. Attitude of optical axis vs. nominal axis
   Convergence in pitch
   FMC in pitch
   Bearing runout in pitch
   Bearing runout in yaw
   Scan angle
4. Interior orientation
   a. Scale
      Focal length
   b. Optical axis location
      Principal point location
      Lens distortion
      Pointing errors
      Point location in time
5. Image-point location
   Point location along slit relative to principal point
   Point location along scan relative to scan indicators
   Point location in time
   (Note: lens distortion, film distortion, pointing errors and parallax clearance errors contribute to all of these items.)

When employing currently available technology, one can expect sigmas that are in the order of:

$$\sigma_V = 5 \times 10^{-3}\text{m/sec}, \qquad \sigma_{\Delta H} = 35 \text{ m},$$
$$\sigma_{\Delta T} = 2 \times 10^{-5}\text{second}, \qquad \sigma_\alpha = 5 \times 10^{-5} \text{ radians},$$
$$\sigma_{\Delta t} = 1 \times 10^{-5}\text{second}, \qquad \sigma_\beta = 5 \times 10^{-5} \text{ radians}.$$

17.9 Photography for Attitude Determination.

The geometric considerations involved have been covered in section 17.3.3. Here, the photometric considerations will be discussed.

Since attitude-determination accuracy will improve with the number of stars in the photograph and since the number of stars available decreases approximately exponentially with visual magnitude, it is desirable to be able to photograph as dim stars as possible. For this reason, a stellar camera will generally have a fast lens, such as an $f/2$, with a focal length selected to provide the required attitude-determination accuracy. The required exposure time for a stellar camera can be estimated from

$$t \approx \pi d^2 H/[\pi TD^2 I_m\xi - 4dHWf]$$

where

H = exposure in metre-candle-seconds,
f = focal length,
T = transmissivity of lens system,
D = lens aperture diameter
W = net angular velocity component perpendicular to the optical axis
I_m = intensity of illumination (lumens/m²) for m^{th} magnitude star at the entrance pupil of the lens. $(= 0.8 \times 10^{-6} \times 10^{0.4(1-m)}\text{lumens/m}^2)$
d = diameter of star image at the focal plane containing the fraction ξ of the total star image energy

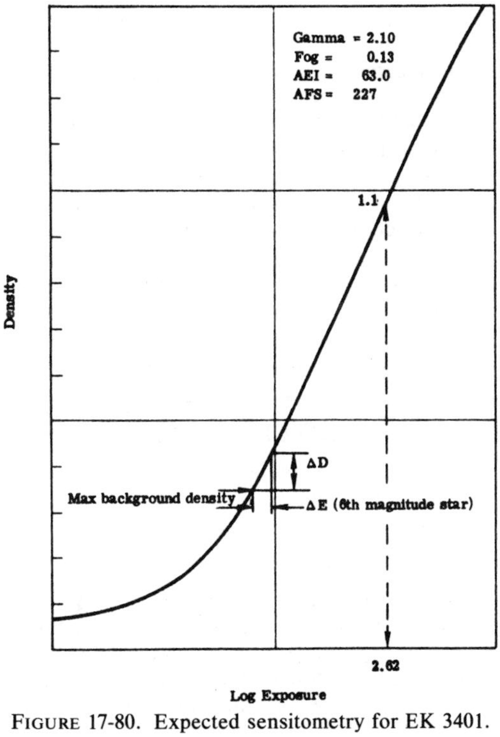

FIGURE 17-80. Expected sensitometry for EK 3401.

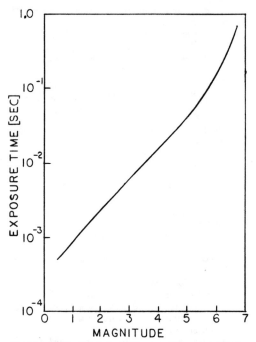

FIGURE 17-81. Exposure time *vs* stellar magnitude.

For a typical *f*/2 system that works over a reasonably broad spectral band (*eg* 0.4 to 0.75 micrometres) a spot 30 micrometres in diameter will contain 80% of the energy in the total image. However the value should be determined for the specific application at hand to obtain reasonably accurate estimates of exposure.

Experience has indicated that the best density for determination of image coordinates is approximately 30% above the background density. In addition, a stellar camera will have a reseau that must be pre-exposed on the film which is generally exposed to a density of approximately 0.3 above the base plus fog level (*see* figure 17-80). For 3401 type film, as an example, which has an AFS of 227 the required exposure to produce a density 0.3 above base plus fog is 6.6 × 10^{-3} metre-candle-seconds. For a stellar image density 30% higher (0.42) the incremental exposure required would be (assuming a γ of ~2)

$$\Delta H = 0.15 \times 6.6 \times 10^{-3}\text{mcs} \approx 10^{-3}\text{mcs}.$$

Assuming a lens transmission of 60% and a net angular motion rate for the image of approximately 1.3 milliradians/sec, the required exposure time as a function of stellar magnitude is as shown in figure 17-81. This graph is only good for the example cited, but a similar graph can readily be constructed for any specified system using the equations given above.

17.10 Future Satellite Remote Sensing Systems

Worldwide experience by many individuals and organizations with remote sensing data acquired from space—primarily Landsat-1, -2, and -3—has demonstrated the utility for land use classification, agricultural assessment, energy and resource exploration, and many other applications. There has been some reluctance to adopt these procedures in operational programs because of a lack of commitment to continuity of data collection, continuing changes in the systems, and some dissatisfaction with the types of data made available to users. Nevertheless there is sufficient conviction of the value of satellite remote sensing so that new systems will be developed in the decade of the 1980's.

17.10.1 DEVELOPMENTS IN THE UNITED STATES

Several major developments by NASA will have major impact on the future of remote sensing systems in the United States.

17.10.1.1 THE SPACE SHUTTLE

Far and away the largest NASA program is the development of the Space Shuttle—or Space Transportation System, as it will be called when it becomes operational. The major advantage of the Shuttle is that its principal component, the orbiter vehicle, can be returned to Earth and reused on successive missions. The orbiter (figure 17-82) is about the size of a DC-9 aircraft. The back of the vehicle opens up to a cargo bay which is 18.3 metres long, and 4.6 metres in diameter. On sortie missions of 2 to 12 day duration the cargo bay will be fitted with modular components developed by the European Space Agency in cooperation with NASA. These consist of crew transfer tunnel, with an access port for extra-vehicular activity (EVA), manned modules providing a shirt-sleeve environment for experiment operators, and instrument pallets for equipment to be operated in the open space environment. The other mode of operation of the Shuttle is to launch or recover independent satellites by means of a long articulated arm referred to as the Remote Manipulator System (RMS).

Initial launches of the Shuttle will be due east from Cape Canaveral permitting sensor coverage from latitudes 30° north to 30° south. The maximum inclination available from Cape Canaveral is 57° which will provide coverage of most of the world's populated areas. Polar orbits will not be available until after 1984 when the Western Test Range will become operational for Shuttle launches.

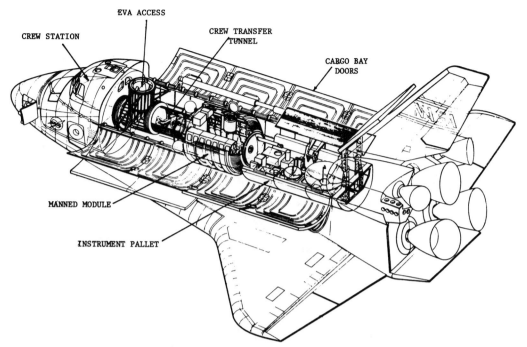

CREW STATION

EVA ACCESS

CREW TRANSFER
TUNNEL

CARGO BAY
DOORS

MANNED MODULE

INSTRUMENT PALLET

FIGURE 17-82. The Space Shuttle Orbiter is equipped with manned and unmanned modules for experiments.

17.10.1.2 THE MULTI-MISSION MODULAR SPACECRAFT

The second major development in the United States is the Multi-Mission Modular Spacecraft (MMS), which can be used for a wide variety of payloads. As shown in figure 17-83, it consists of a central core to which can be attached a power supply module, attitude control module, and command and data handling module. There is a propulsion module to provide in-orbit maneuvering, and an adapter between the spacecraft and its particular payload. The assembled spacecraft may be fitted with solar panels and various antennas for communications. The MMS is specifically designed for launch from the cargo bay of the Shuttle by means of the RMS. There is also a flight support system (figure 17-84) which mounts in the cargo bay so that the MMS may be retrieved by the RMS and the modules and payload be serviced on-board the Shuttle without the necessity of returning to earth. Wherever possible the MMS will be used as the bus for future independent spacecraft.

17.10.1.3 THE TRACKING AND DATA RELAY SATELLITE SYSTEM

The Tracking and Data Relay Satellite System (TDRSS) consists of two communication satellites in geosynchronous orbit at 41° and 171° west longitude and a single ground station located at White Sands, New Mexico. As shown in figure 17-85, TDRSS will provide 2 types of communication between low Earth-orbiting

spacecraft and the ground station. Multiple Access (MA) service at S-band will provide dedicated return link service at data rates up to 50 kb/sec for each of 20 user spacecraft simultaneously from each TDRS. Forward link MA service is time shared with a maximum data rate of 10 kb/sec and supports one user per TDRS at a time. User spacecraft requiring high data return rates will employ Single Access (SA) service on a priority scheduled basis to the articulated parabolic antennas which must be directed towards the user spacecraft. Forward link service at S-band (SSA) will provide data rates up to 300 kb/sec, and at ku-band (KSA) will provide data rates up to 25 Mb/sec. Return link service at S-band will provide 3.15 Mb/sec and at Ku-band will provide 300 Mb/sec. Each TDRS can support 2 SSA and 2 KSA data links simultaneously. Both MA and SA users can obtain range and range rate tracking data from the transmitted signals. Because of the location of the TDRS, there is a zone of exclusion over the Indian Ocean and extending up into the USSR from which low Earth orbiting spacecraft cannot communicate with TDRS. The size of the zone depends upon the altitude of the spacecraft being serviced.

17.10.1.4 THE THEMATIC MAPPER

A second generation optical mechanical multi-spectral scanner being developed by NASA is called the Thematic Mapper (TM). Its principal characteristics are given in table 17-10. As shown in figure 17-86, the instrument has an os-

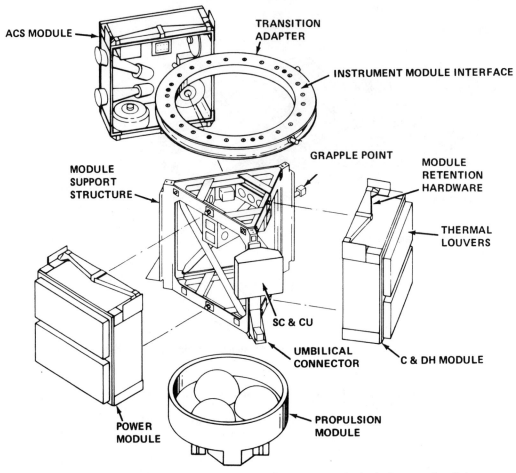

FIGURE 17-83. The Multi-mission Modular Spacecraft will be used as the basis for most free-flying remote sensing missions.

cillating mirror to scan the object space transverse to the direction of flight. A Cassegrain type telescope reflects the energy to visible and infrared detectors in the focal plane. Because the mirror scan is active in both directions an optical mechanical scan line corrector is required before the energy reaches the detectors.

17.10.1.5 THE GLOBAL POSITIONING SYSTEM

The Department of Defense is developing the Global Positioning System (GPS) which will eventually consist of an array of 18 separate satellites arranged in three different orbital planes. From any place on earth or in earth orbit, a minimum of four spacecraft will always be visible. By real time processing of the signals transmitted from the GPS it will be possible to establish the position of a vehicle on the ground or in space with a precision of a few metres in all three coordinates.

17.10.2 MAJOR POLICY DECISIONS

From 1977 to 1980, the U.S. Government conducted several related interagency studies which resulted in major policy decisions with regard to satellite remote sensing. These decisions are:

a. The Space Shuttle will replace all expendable launch vehicles.
b. All spacecraft data transmission will use the TDRS System and the existing NASA network of ground stations will be phased out.
c. Responsibility for civil operational Earth observation satellites is assigned to the National

TABLE 17-10. THEMATIC MAPPER PARAMETERS

Band number	Wavelength	IFOV
1	0.42–0.52 μm	30 m
2	0.52–0.60	30
3	0.63–0.69	30
4	0.76–0.90	30
5	1.55–1.75	30
6	10.4–12.5	120
7	2.10–2.35	30
Operating altitude	680–730 km	
Swath width	185 km	
Data transmission rate	85 Mb/sec	

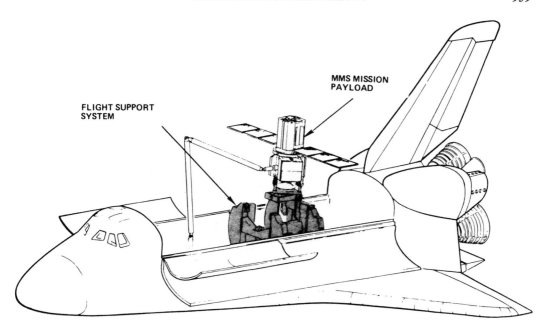

FIGURE 17-84. The Flight Support System mounts in the cargo bay of the Orbiter. The Multi Mission Modular Spacecraft and its payload can then be serviced on orbit, thus saving additional launch costs.

Oceanic and Atmospheric Administration (NOAA) which already operates the civil weather observation satellite systems.

c. The prices for satellite data to users will assure maximum recovery of the total system cost rather than simply the cost of data reproduction as at present.

e. Commercial ownership and operation of Earth observation systems is the eventual goal, and new systems should be planned for commercial viability.

17.10.3 LANDSAT-D

These technical developments and policy decisions come together in Landsat-D, the next

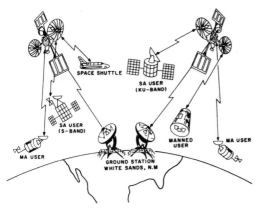

FIGURE 17-85. The two geosynchronous Tracking and Data Relay Satellites will provide forward and return data links to spacecraft operating in low Earth orbit.

spacecraft in NASA's experimental earth resources observation program. Landsat-D was originally scheduled for launch in September 1981, but will be delayed at least until June of 1982. The anticipated configuration of Landsat-D is shown in figure 17-87. The basic spacecraft is the Multi-Mission Modular Spacecraft (MMS) which supports an Instrument Module (IM) on which are mounted the Thematic Mapper (TM) and the Multi-Spectral Scanner (MSS) which is essentially the same instrument carried by Landsat 1, 2, and 3. There will be solar panels for power, several antennas for communication when the spacecraft is within range of a ground station, and a large articulated parabolic antenna for communication with the Tracking and Data Relay Satellite System (TDRSS). Spacecraft position will be established by communication with the Global Positioning System (GPS).

Landsat-D will be placed in a 705 km sun-synchronous orbit by the Delta 3920 expendable launch vehicle. This is necessary because near-polar orbits cannot be obtained with the Shuttle until Shuttle launch facilities are available at the Western Test Range, and this capability will not be available before 1984. Though the MMS is designed for retrieval by the Shuttle this is not presently contemplated for Landsat-D.

In operation the two sensors can acquire data simultaneously or independently—the MSS with an 80 m IFOV in 4 spectral bands, and the TM with a 30 m IFOV in 6 visible and near-infrared bands and 120 m IFOV in the thermal band. When the spacecraft is within range of a ground

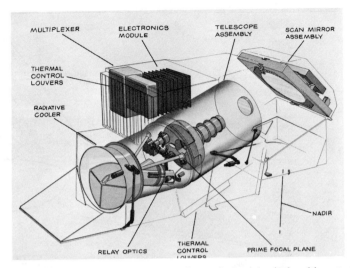

FIGURE 17-86. The Thematic Mapper is a second generation optical mechanical multispectral scanner using an oscillating scan mirror and reflective optics.

station equipped to receive it, the data can be transmitted directly. At all other times data will be relayed to TDRS and from there back to the receiving station at White Sands. It is worthy of note that presently (1980) no foreign ground station can receive the 85 Mb/sec data rate from the TM.

The Landsat-D project has encountered serious problems. The TM has had technical difficulties, cost overruns, schedule slippage and NASA is discovering major deficiencies in the ground processing system. The TDRSS has been delayed by the Shuttle. The currently operating Landsats 2 and 3 are beginning to show signs of failure. Consequently NASA is reevaluating the entire Landsat-D project. Most probably Landsat-D will be launched about June of 1982

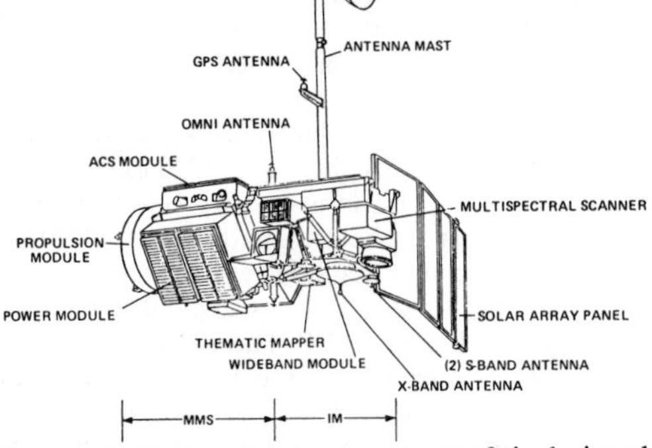

and will carry only the MSS. A backup spacecraft, Landsat D prime may be launched when the TM and its ground processing system are ready for operation, probably mid-1983. This spacecraft will carry both the TM and the MSS, and probably will have a wide band video tape recorder for use with the MSS. Tape recorders capable of handling the TM acquisition data rate are not currently feasible.

The Landsat series are considered "experimental" despite the fact that many user organizations make "operational" applications of the data. The National Oceanic and Atmospheric Administration (NOAA), which has been assigned responsibility for operational earth observation satellites, expects to assume control of the Landsat-D spacecraft and the MSS data sometime in 1983 when system operation has been successfully demonstrated by NASA. The TM will continue to be a NASA experimental system until full operational capability of the ground processing has been demonstrated, probably not before 1984 or 1985.

17.10.4 EXPERIMENTAL ELECTRO OPTICAL SENSORS

The development of experimental sensors and spacecraft continues to rest with NASA, and there are a number of projects which may influence the design of future satellite remote sensing activities.

17.10.4.1 THE OSTA-1 MISSION

The first several flights of the Space Shuttle are identified as Orbital Test Flight Missions. The second one of these, OFT-2, has been assigned to the NASA Office of Space and Terrestrial Applications (OSTA) to develop an earth

FIGURE 17-87. The Landsat D spacecraft is designed to carry the Thematic Mapper (TM) and Multi Spectral Scanner (MSS) and to return data via a high gain antenna which communicates with the Tracking and Data Relay Satellite System.

observation payload. The selected experiments are:

a. MAPS—Measurement of Air Pollution from Space. This is a thermal infrared gas filter correlator for measuring atmospheric carbon monoxide.
b. SMIRR—Shuttle Microwave Infra Red Radiometer. This is a 10 channel near infrared nonimaging radiometer with 100 m ground track.
c. OCE—Ocean Color Experiment. This is a 10 channel visible light digital imaging radiometer for estimating ocean bioproductivity.
d. NOSL—Nightime Observation of Storms and Lightning. This is a 16 mm movie camera in which the sound channel is used to trigger exposures of lightning flashes in severe storms.
e. AWSO—All Weather Surface Observations. This is an adaptation of the synthetic aperture L-band radar which was used on the Seasat mission. The antenna will be modified to provide a 47° look angle rather than 20.5° as on Seasat. The system will cover a ground swath of 50 km with a theoretical single look resolution of 5 × 40 m. This experiment is also known as Shuttle Imaging Radar-A (SIR-A).

The instruments will be mounted on the cargo bay pallet, and the OSTA-1 mission is presently scheduled for September 1981, at an altitude of 270 km.

17.10.4.2 SHUTTLE IMAGING RADAR (SIR-B)

NASA has also undertaken the development of a new radar antenna for use with the Shuttle. The antenna will consist of three panels. In its stored configuration it will be 2m × 3m; deployed it will be 2m × 9m with its long dimension transverse to the cargo bay axis. It can be mechanically scanned to provide look angles of 15°, 50°, or 70°, and will resolve 50 m at 15°, 25 m at 50°, and 20 m at 70°. It will operate at X-band or C-band and cover a 50 km swath. The principal research problem is how to handle the 150 Mb/sec data rate. A 10 minute data acquisition pass would acquire 10^{11} data bits. If transmitted directly to TDRSS this would require one high

data rate channel whenever the radar is operating. A conventional magnetic tape carries 10^9 bits, so 100 tapes would be required for on-board storage of a short pass. Some other mass storage system would be clearly necessary. The data could be recorded on film which would permit full mission coverage, but would result in degraded performance. The SIR-B development has not yet been assigned to a specific Shuttle mission, but it is hoped to have a total of 4 flights over a 2 year period.

17.10.4.3 LINEAR ARRAYS

One of the most promising sensors for Earth observation is the linear array of charge-coupled detectors (CCD). As shown in figure 17-88, an entire line from the ground scene is imaged instantaneously by an optical system to a line of detectors, rather than being scanned by a moving beam. The signals from the linear array of detectors must be recorded in the time the spacecraft takes to move to the next line of the ground scene. This is referred to as the "pushbroom" mode of data acquisition. Arrays of more than 2000 detectors are now available and they can be butt-joined to provide more than 10,000 detectors per line. The active element of each detector is on the order of 10 × 10 μm with detector spacing of about 13 μm. The ground resolution will depend on the ratio of sensor distance from the ground to optical system focal length. Several arrays with appropriate filters can be arranged in sequence to provide a multispectral system. Rectangular arrays are also manufactured, but are considered to have lower potential for earth observation sensors. NASA has undertaken studies of several sensor instruments based on linear array technology.

One concept called the Multispectral Resource Sampler, would have four arrays of 2000 detectors each in the focal plane of a 70 cm focal length telescope. Each array would have five separate spectral filters that would be selectable by command while in orbit. Spectral bands would be in the range 0.35 to 1.0 μm—the present sensitivity of linear array detectors. The system would have an IFOV of 15 m over a swath of 15 km and could be pointed ±40° in the across track direction and ±55° in the along track direction. Along track pointing permits stereo-coverage at variable base-height ratio, and across track pointing can provide more frequent repeat coverage of sample areas.

Another concept, not yet named, would use the MMS spacecraft in a 705 km Landsat D orbit to carry a 1 m focal length optical system and provide a 185 km swath in 7 spectral bands. Four bands in the visible (0.35 to 1.1 μm) would have 18,600 detectors each and provide 10 m IFOV. Two bands in the near infrared (1.5–2.5 μm) would have 9300 detectors each, and one band in

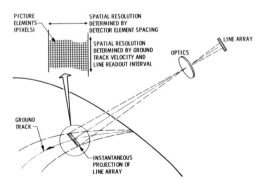

FIGURE 17-88. A linear array of detectors records an entire line of the ground scene instantaneously without moving parts. Forward motion of the spacecraft provides continuous coverage.

the thermal infrared $(10-12 \ \mu m)$ would have 2500 detectors. A two-position mirror could direct the field of view either to the nadir or to the rear in order to acquire stereo-coverage. There would also be a roll-pointing capability to provide more frequent coverage. A 50 percent on-board data compression technique would be used to reduce the data rate to 400–500 Mb/sec and the TDRSS would be used for data return.

NASA hopes to have a demonstration instrument available for a test flight, probably on the Shuttle, by 1986. One possibility under consideration is to abandon the TM entirely, and go directly to a system based on multispectral linear arrays.

17.10.5 THE NOAA OPERATIONAL EARTH OBSERVATION SYSTEM

As part of its planning for transition from experimental projects to operational programs, NOAA undertook a study of user data requirements, and described a provisional system to satisfy them. The principal characteristics of this system are listed in table 17-11. Two spacecraft would be in orbit in order to provide 8 or 9 day revisit capability, and a third spacecraft would be a spare either in orbit or on the ground. Variable resolution would be achieved by on-board clustering of the response from the linear array detectors. Spectral bands in the near infrared require new development since linear arrays are not currently sensitive to these wavelengths. It may be expected that these parameters will change as user requirements and costs of data are refined before the system is built. NOAA estimates that such a system could not be operational before 1989, and would cost around $150-250 million per year over a 10-year period.

17.10.6 OTHER PROPOSED U.S. SYSTEMS

Some users of satellite remote sensing data, concerned about the technical or economic viability of systems as complex as Landsat D, or with foreseen requirements for specialized types of data, have undertaken studies of other proposed space systems.

17.10.6.1 STEREOSAT

A group of commercial enterprises involved in mineral and energy exploration, called the Geosat Committee, placed a very high priority on high resolution stereo data which would be compatible with the multispectral data to be made available from Landsat D. In response to these requirements the Jet Propulsion Laboratory (JPL) conducted studies on a system called Stereosat. The principal characteristics of this system are listed in table 17-12 and the proposed spacecraft configuration is shown in figure 17-89.

In consonance with the policy decision that users will be expected to pay the total system costs, the Geosat Committee studied the feasibility of funding the development of Stereosat by contributions from its members. Though this remains a possibility, it awaits the resolution of questions regarding system management and exclusive rights to data for the subscribers. NOAA has not accepted Stereosat as part of its operational responsibility, and there is presently (1980) no NASA or other federal funding for its development.

17.10.6.2 MAPSAT

The U.S. Geological Survey is conducting a study (1980) of a system called Mapsat which has the objective of collecting both stereo and multispectral data for operational image mapping at 1:50,000 scale with 20 m contour intervals on a global basis. The principal characteristics of this system as presently defined are listed in table 17-13 and the proposed configuration of the spacecraft is shown in figure 17-90.

One of the noteworthy characteristics of Mapsat is the concept for deriving digital terrain elevation data from the stereo records. The very high attitude stabilization is specified so that the ground trace imaged by any one detector n on the forward looking array will also be imaged directly by detector n on the aft looking array as the spacecraft moves forward in its orbit. Therefore for any ground point, the two detectors n, the fore and aft positions of the lens as the ground point is observed, and the ground point itself, all lie in an epipolar plane, and the terrain

TABLE 17-11. NOAA PROPOSED OPERATIONAL SYSTEM

	IFOV (m)
Spacecraft —MMS, Shuttle launch and retrieve	
Orbit —705 km, same as Landsat D	
Sensors —linear arrays	
Spectral bands (μm)	
$0.45 - 0.52, 0.52 - 0.6, 0.76 - 0.9$	30 or 60
$0.63 - 0.69$	15 or 30
$1.55 - 1.75, 2.08 - 2.35$	30 or 60
Pointable $\pm$ 20 across track	
Data transmission via TDRSS	
Ground processing $-$ 185 $\times$ 185 km scenes	
40,000 low res + 20,000 hi res scenes/year	

TABLE 17-12. PROPOSED STEREOSAT PARAMETERS

Vehicle	—Multimission Modular Spacecraft
Launch	—Shuttle from Western Test Range
Orbit	—705 km, 98.2° inclination
Sensor system	

Sensor system
  3 telescopes – vertical, ±26.6 deg fore and aft
  f = 705 mm vertical, 775 mm fore and aft
  2096 element linear array for each telescope
    0.68 to 0.80 μm panchromatic response
Data handling
  1.85×10^6 pixels/sec/camera
  6 bit encoding – 64 grey levels
  2 tape recorders with 4×10^9 bit capacity
  $2\times$ on-board data compression
  16.6 Mb/sec transmission via TDRSS
Coverage
  61.4 km swath
  15 m IFOV for each telescope
  48 day repeat cycle

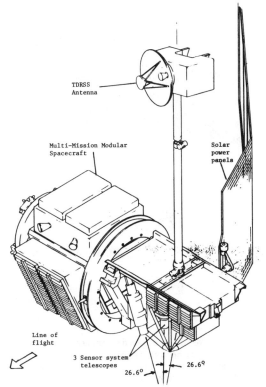

FIGURE 17-89. The proposed configuration of Stereosat would use the Multi-mission Modular Spacecraft to carry three telescopes with linear array sensors to provide convergent stereo data.

elevation along the ground trace of detector n can be found by time correlation of the output signals from the fore and aft detectors n. Correlation is thus reduced to a one-dimensional problem and ground data handling should be enormously simplified. Like Stereosat, Mapsat is not presently included in NOAA's plans for operational systems, and no hardware development funds have been allocated.

17.10.6.3 NATIONAL OCEANIC SATELLITE SYSTEM

The Department of Defense, the National Oceanic and Atmospheric Administration, and NASA have jointly funded a study of a National Oceanic Satellite System (NOSS) which will serve as a successor to the short-lived SEASAT and provide a limited operational demonstration of global sea surface observation based on re-

TABLE 17-13. PROPOSED MAPSAT PARAMETERS

Vehicle	—Multimission Modular Spacecraft 10^{-6} deg/sec attitude control
Launch	—Shuttle (?) from Western Test Range
Orbit	—919 km. 99.1° inclination same as Landsat 1, 2, and 3
Sensor system	

Sensor system
  3 refractive telescopes, approx. 1 m focal length vertical, ±23° fore and aft
  3 multispectral linear arrays, 17,500 elements/band
    0.47 to 0.57 m—blue green for water
    0.57 to 0.70 m—red for culture
    0.76 to 1.05 m—near IR for vegetation
    10 m elements can be clustered across track and along track for variable IFOV
Data handling
  96 Mb/sec maximum acquisition rate
  spectral bands commanded for each telescope.
  IFOV commanded for each spectral band
  6 bit encoding—64 grey levels
  wide band video tape recorders
  on board data compression.
  Spacecraft and sensor control via TDRSS
  data transmission to existing ground stations.
Coverage
  175 km swath
  10 to 60 m IFOV on command
  18 day repeat cycle

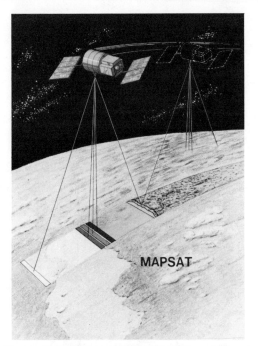

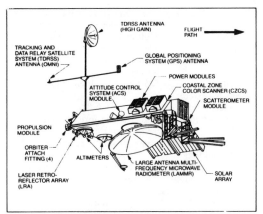

FIGURE 17-91. The proposed configuration of the National Oceanic Satellite System will carry the multifrequency microwave radiometer, the coastal zone color scanner, scatterometers, and altimeters to monitor global ocean waves, winds, temperature, and bioproductivity.

17.10.6.4 ICE AND CLIMATE EXPERIMENT

Notable by its absence on NOSS is a synthetic aperture radar (SAR) which provided some of the most useful data from the Seasat mission. It is particularly desirable for ice observation in the polar regions. The NASA Office of Space and Terrestrial Applications has formed a working group for an Ice and Climate Experiment (ICEX) which has proposed a satellite similar to NOSS but also carrying a SAR. The characteristics of the satellite are listed in table 17-15 and the pro-

FIGURE 17-90. The proposed configuration of Mapsat will have multispectral capability in both vertical and fore and aft arrays, but only one selected band will be used at any time to produce stereo data.

mote sensing from space. The principal characteristics of the proposed system are given in table 17-14 and a possible spacecraft configuration is shown in figure 17-91.

TABLE 17-14. PROPOSED NOSS PARAMETERS

Spacecraft —Multimission Modular Spacecraft (MMS)
Launch —Shuttle, Western Test Range, 1984-85
Orbit —700 km, near polar, sun synchronous
Lifetime —3 to 5 years
Sensors payload
  Large antenna multichannel microwave radiometer (LAMMR)
  4 m diameter 6.6 to 3.6 GHz
  Monitor temperature, wind speed, water vapor, ice

  Radar altimeter (ALT)
  1 m. diameter 13.5 GHz
  Monitor wave height and sea state

  Coastal zone color scanner (CZCS)
  9 channel visible and infrared scanner
  Measure ocean bioproductivity

  Radar scatterometer (SCAT)
  6 stick antennas 3 m. long at 14.6 GHz

Support payload
  Global Positioning System (GPS)
  Laser retroreflector (LRR)
  Surface data acquisition system (SDAS)

Data handling
  2 tape recorders of 4.5×10^8 bit capacity
  Transmission to TDRSS, 1.4 Mb/sec maximum

NOSS has been included in the FY 81 budget for system definition studies and hardware development may begin in FY 82.

TABLE 17-15. PROPOSED ICEX PARAMETERS

Spacecraft	—Multimission Modular Spacecraft (MMS)
Launch	—Shuttle, Western Test Range, 1986
Orbit	—700 km sun synchronous
	87° or 93° inclination

Sensor payload
    Large antenna multiband microwave radiometer (LAMMR) (same as NOSS)
    Radar and laser altimeters (IEAS)
    Radar scatterometer (SCAT) (same as NOSS)
    5 channel polar infrared mapping radiometer (PIMR)
    X band (9.6 GHz) synthetic aperture radar (SAR)
    100 m resolution over 360 km swath
    25 m resolution over 90 km swath
Support payload
    Global Positioning System (GPS)
Data handling
    Transmission to TDRSS
    Special ground facility at Goddard Space Flight Center

posed configuration of the spacecraft is illustrated in figure 17-92. Because of its similarity to NOSS, it is unlikely that ICEX will be funded as a separate mission. The Canadian Centre for Remote Sensing is developing a C-band radar for space applications and has proposed a joint mission for which CCRS would supply the radar and NASA would supply the spacecraft and launch. Alternatively CCRS may cooperate with the European Space Agency (ESA) by supplying the radar for a proposed ESA mission.

17.10.6.5 LARGE FORMAT CAMERA

One future space sensor system which has already (1980) been built is the Large Format Camera designed and constructed by Itek Optical Systems. The camera parameters are listed in table 17-16. Figure 17-93 shows schematic diagrams of the magazine assembly and lens cone assembly. The camera will be mounted in an environmentally controlled enclosure on a pallet which will be carried in the cargo bay of the Space Shuttle as shown in figure 17-94. The

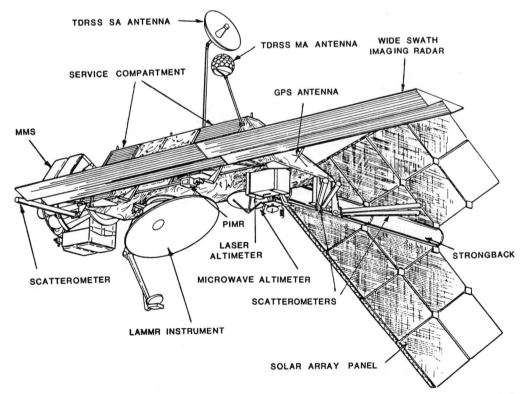

FIGURE 17-92. The proposed configuration of the spacecraft for the Ice and Climate Experiment (ICEX) includes a Synthetic Aperture Radar as the principal sensor.

TABLE 17-16. LARGE FORMAT CAMERA PARAMETERS

focal length	30.5 cm
aperture	f/6.0
film format	23 × 46 cm
12 fiducials for any 23 × 23 cm subframe	
5 cm spacing rear illuminated reseau	
lens resolution AWAR	92 1p/mm
radial distortion rms	8.7 μm
forward motion	
compensation	0 to 0.04 rad/sec
rotary shutter	0.004 to 0.032 sec
automatic exposure sensor	
overlap selection	10% to 80%
film capacity	4000 frames
2 position filter changer	

camera will be mounted with the long dimension of the format in the direction of flight and operated to produce 80 percent forward overlap between consecutive photographs. By selecting various combinations of frames it is possible to produce stereomodels with base-height ratios of 0.3 to 1.2 thus permitting topographic contour compilation in flat to rugged terrain. It is worthy of note that the center 23 × 23 cm portion of two consecutive frames will have a base-height ratio of 0.3 and can be accomodated by most conventional photogrammetric instruments in common use around the world.

From a nominal flight altitude of about 278 km (150 n.mi) the camera can produce a ground resolution of about 10 m with high resolution black/white film to 20 m with color infrared film. Ground positioning accuracy will be about 15 m and elevation accuracy from 7 to 28 m depending on choice of base-height ratios. This is adequate for compilation of maps at scale 1:50,000 with 20 to 80 m contour interval.

Unfortunately, the camera is now scheduled for only one flight in mid-1984 on a short lifetime mission at 28.5° inclination. This will not provide any extensive coverage of areas in the United States.

It has been proposed to mount the camera on the Multimission Modular Spacecraft, launch it with the Shuttle, leave it in orbit until film has been usefully expended, then collect exposed film and resupply with fresh film on a subsequent Shuttle rendezvous mission. This concept, shown in figure 17-95, has not been approved by NASA.

17.10.7 EUROPEAN PROGRAMS

Most European space programs are coordinated by the European Space Agency (ESA) which is a consortium supported by 13 western European countries. The contributions of France and the Federal Republic of Germany make up more than 60 percent of the budget, with Great Britain and Italy each contributing about an additional 10 percent. ESA has several programs which will affect the future of remote sensing by European nations.

17.10.7.1 THE ARIANE LAUNCH VEHICLE

The largest single project in ESA is the development of an unmanned launch vehicle called Ariane, and the corresponding launch facilities located at Kourou, French Guiana. The vehicle is a 3 stage rocket 3.8 m in diameter, 47.4 m high with a total lift-off weight of 200 metric *tonnes*. Ariane has a capability to place a payload of approximately 3600 kg into low Earth orbit at any inclination, or 800 kg to geostationary orbit. The first test launch in December 1979 was a complete success. The second launch in May 1980 failed because of an engine misfire. Two more test launches are planned in early 1981, and the first operational launch is expected in September or October 1981. Ariane will be the launch vehicle for all ESA earth observation spacecraft. Future developments of Ariane include increasing the payload to an eventual 4300 kg, the development of a "mini-Shuttle" (Hermes) for a crew of 2 to 5 persons, and an automated materials laboratory (Minos) for operation in earth orbit.

ESA has also considered several proposals for development of a standard spacecraft to be compatible with Ariane in the same way that the Multimission Modular Spacecraft (MMS) is designed to operate with the Shuttle. At the moment (1980) however, none of these have been accepted, and ESA missions will use the SPOT spacecraft being built by France (*see* section 17.10.7.5).

17.10.7.2 SPACELAB-1

As mentioned in section 17.10.1.1, ESA is building the modular components for the cargo bay of the NASA Shuttle. These components will be used in the joint ESA-NASA mission designated as Spacelab-1. This will be a Shuttle sortie mission of 1 week duration at an altitude of 250 km and an inclination of 57° to be launched in the Spring of 1983. Among many other experiments from both NASA and ESA, Spacelab-1 will carry two earth observing systems.

17.10.7.2.1 THE METRIC CAMERA EXPERIMENT

Mounted to an optical window in the manned module will be a standard Zeiss aerial camera modified for operation in space. The parameters for this experiment are given in table 17-17.

ESA distributed an announcement of opportunities and received more than 100 proposals for participation in the analysis of the photographs from various countries around the world. The photographs should be useful for map compilation at scales up to 1:50,000 and for many geoscience applications of photo-interpretation.

The Spacelab-1 camera is Phase A of an over-

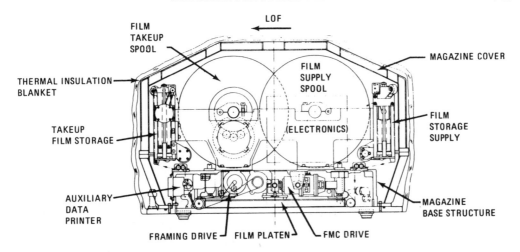

(a) Film magazine schematic diagram

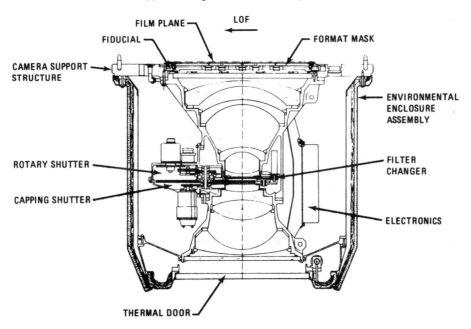

(b) Lens cone schematic diagram

FIGURE 17-93. Schematic diagram of the Large Format Camera built as a cargo bay payload for the Space Shuttle.

all ESA camera development program called ATLAS being conducted by the Federal Republic of Germany. Phase B, presently under discussion, will consider several options:

a. Add image motion compensation (IMC) to the RMK 30/23 camera and mount it in a pressurized container on the external Shuttle pallet
b. Increase the focal length to 60 cm, add IMC, increase the film magazine capacity, and mount on external pallet
c. Develop a new camera with 60 cm focal length, IMC, and 11.5 cm × 23 cm format for operation in the manned module.

Phase C of the ATLAS program will investigate the operation of these cameras on free-flying spacecraft launched and serviced by the Shuttle.

17.10.7.2.2 THE MICROWAVE REMOTE SENSING EXPERIMENT

The Microwave Remote Sensing Experiment (MRSE) for Spacelab-1 consists of a 2 m parabolic main reflector, a subreflector, and a feed horn mounted on a pedestal on the cargo bay pallet. The instrument, shown in figure 17-96, can operate in three different modes:

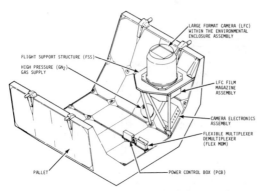

FIGURE 17-94. The Large Format Camera mounted on Shuttle pallet will have its own environmentally controlled enclosure.

TABLE 17-17. SPACELAB-1 CAMERA

Type	Zeiss RMK 30/23 frame camera
Focal length	30.5 cm
Photo format	23 × 23 cm
Film load	3 magazines 1650 frames
Image scale (250 km)	1:820,000
Coverage per frame	190 × 190 km
Ground resolution (AWAR)	20 m
Overlap	60 or 80 percent

a. Two frequency scatterometer to measure ocean wave spectrum with a 9 × 9 footprint.
b. Passive thermal radiometer to measure surface temperature with ±1 degree Kelvin sensitivity.
c. Synthetic aperture imaging radar with 9 km swath and 25 m resolution.

17.10.7.3 SPACELAB-13

The third Spacelab mission, called Spacelab 13 because it is the 13th Shuttle mission, currently planned for 1984-5, has been designated for a combined atmospheric science and Earth Observation payload. The final selection of instruments has not been made, but the probable earth observing instruments will include:

a. The metric camera from Spacelab-1, or one of the further camera developments from Phase B of the Atlas project.
b. The Microwave Remote Sensing Experiment (MRSE) from Spacelab-1.
c. A mechanical scanner, operating primarily in the mid-to thermal infrared wavebands.

d. A pushbroom linear array scanner operating in the visible and near infrared wavebands.
e. A large antenna synthetic aperture radar.

A proposed configuration for Spacelab 13 is shown in figure 17-97. Alternatives for the synthetic aperture radar include instruments currently being developed by the United States, Canada, and Federal Republic of Germany.

17.10.7.4 EUROPEAN REMOTE SENSING SATELLITES

ESA has committed to the development and launch of two European Remote Sensing Satellites: ERS-1 for launch late in 1985 or early 1986, and ERS-2 for launch about 2 years later. After extensive consultation with potential users, an advisory Remote Sensing Work Group has been formed, and contracts have been given to industry to establish requirements and propose spacecraft and sensor configurations. Two concepts have emerged: a Coastal Ocean Monitoring Satellite System (COMSS), and a Land Applications Satellite System (LASS).

Contemporary with these system studies, in-

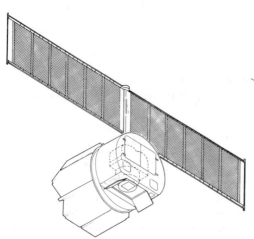

FIGURE 17-95. The Large Format Camera could be mounted on the Multimission Modular Spacecraft, launched and serviced by the Shuttle, to produce extensive coverage with photographs suitable for 1:50,000 scale mapping.

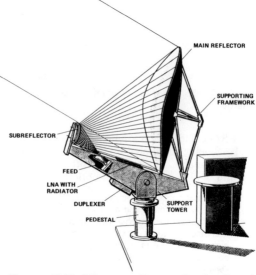

FIGURE 17-96. Microwave Remote Sensing Experiment for use on Spacelab 1.

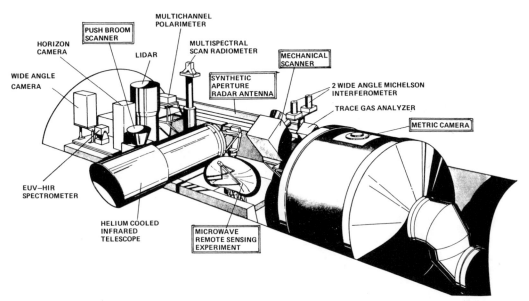

FIGURE 17-97. The proposed configuration for Spacelab 13 will include both atmospheric science and Earth observation instruments.

dustrial organizations have undertaken design or development of several candidate sensor systems. Among these are a Modular Opto-electrical Multispectral Scanner (MOMS), using linear arrays of detectors which will operate in the visible, near infrared, and eventually thermal infrared wavelengths, an Imaging Microwave Radiometer (IMR), a multichannel Ocean Color Monitor (OCM), and several versions of Synthetic Aperture Radar (SAR). The MOMS, and perhaps some of the other instruments, will be carried on test flights aboard the Shuttle, before being incorporated in the free-flying spacecraft.

By the end of 1980, ESA expects to come to a final decision on which system will be launched first and on the sensor complement for each spacecraft. System definition studies recommended SAR for both COMSS and LASS, but ESA doubts that an effective SAR can be delivered in time for the first satellite. It is probable that ERS-1 will be COMSS without SAR.

17.10.7.4.1 COASTAL OCEAN MONITORING SATELLITE SYSTEM

The objective of the Coastal Ocean Monitoring Satellite System (COMSS) is to provide fundamental data for oceanography, glaciology, and climatology. In addition, applications of direct economic value are expected to include: monitoring and forecasting sea state, detecting oil spills and other forms of ocean pollution, monitoring icebergs and maritime navigation, monitoring changes in coastline, and defining areas subject to flooding. The definition study for COMSS proposed the characteristics listed in table 17-18. The scatterometer and radar altimeter are recent additions to the proposed

payload, and they push the orbit to the higher altitude and near polar inclination. However they may well be included, since as noted in section 17.10.7.4, the SAR is not expected to be ready before 1987. Most probably COMSS will include the scatterometer and altimeter, but not the SAR.

17.10.7.4.2 LAND APPLICATIONS SATELLITE SYSTEM

The objectives of the Land Applications Satellite System (LASS) are to provide data for crop inventory and yield prediction, inventory and monitoring of forest productivity, land use classification and planning, water resources inventory and monitoring, and mineral and energy resources exploration. The definition study for LASS proposed the characteristics listed in table 17-19. There has been some expression of interest in adding one or more thermal channels to the infrared camera system. If this change is made the near infrared system may be changed to an optical mechanical scanner rather than a linear array camera.

ESA recognizes that without the SAR, LASS would be a near duplication of the capabilities of the USA Landsat-D. For this reason, LASS will probably be delayed until a suitable SAR is available. The suggested configuration of LASS is illustrated in figure 17-98.

17.10.7.5 SYSTÈME PROBATOIRE D'OBSERVATION DE LA TERRE

Independent of its other activities in support of ESA, the French Centre National d'Etudes Spatiales (CNES) is building a remote sensing satellite identified as Système Probatoire d'Ob-

TABLE 17-18. COASTAL OCEAN MONITORING SATELLITE SYSTEM

Launch vehicle	Ariane
Spacecraft	SPOT
Orbit	650-800 km altitude
	60°-90° inclination
	3 to 4 day repeat cycle
Sensors	
Synthetic aperture radar	SAR
resolution	30 × 30 m
swath width	100 to 200 km
look angle	20° to 40°
operational frequency	5.3 GHz
grey resolution	2.5 dB
polarization	HH or VV
Ocean Color Monitor	OCM
7 channels visible	0.415 to 0.750 μm
	800 m IFOV
1 channel near ir	1.02 μm
	2 km IFOV
1 channel thermal ir	11.5 μm
	800 m IFOV
	$\Delta t = 0.2°$ K
1 channel thermal ir	11.0 μm
	2 km IFOV
	$\Delta t = 0.05°$ K
Imaging Microwave Radiometer	IMR
frequency bands 6, 10.7, 15, 23, 31, 90 GHz	
swath	700 km
3dB footprint	90 × 130 km for 6 GHz to
	5 × 7 km for 90 GHz
polarization	H, V
Scatterometer	2 frequency C Band
Radar altimeter	1 m antenna
Data transmission	ESA Earthnet
	117 Mb/sec

servation de la Terre (Earth Observation Test System—SPOT). Sweden and Belgium are also contributing some space and ground station hardware. The characteristics of SPOT are listed in table 17-20.

The SPOT platform is designed for multimission use with clean separation between platform and sensor systems. It has been adopted by ESA as standard for use with the Ariane launch vehicle and will be employed on the COMSS (see section 17.10.7.4.1) and LASS (see section 17.10.7.4.2) missions. The SPOT spacecraft configuration is shown in figure 17-99.

Using the ±26° cross track pointing capability of the HRV, it is possible to locate the 60 km swath anywhere within ±400 km of the ground track. This makes it possible to look at a given area on several consecutive days rather than once per 26-day repeat cycle. It also makes it possible to acquire stereoscopic coverage by viewing an area from left and right orbital tracks.

The SPOT ground facility located in Toulouse, France, will provide spacecraft command and control, data reception and processing, product generation and dissemination. (The station is also being configured to receive X-band Thematic Mapper data from Landsat D.) The data

products to be produced are listed in table 17-21. CNES is considering use of a commercial operator to handle international distribution of SPOT products.

17.10.7.6 PROGRAMS IN THE USSR

Very little definitive information has been made generally available about the satellite remote sensing activities of the USSR. A major difficulty is the lack of a clear cut separation between military and civil programs. The Soviets have announced an electro optical earth observation system, and the MSUS and MSUM multispectral instruments on the Meteor weather satellite have demonstrated the ability to acquire data with a pixel dimension of 250 × 250 m. Cosmos 1076, launched February 1979, was identified as an ocean resources satellite, but no payload parameters were given.

The USSR relies primarily upon film return satellites for Earth observations. The manned spacecraft Salyut 4 was in orbit from December 1974 to February 1977. It carried the KATE-140 camera whose parameters are given in table 17-22. Salyut 4 was occupied by the crew of Soyuz 17 for 29 days, and by the crew of Soyuz 18 for 61 days. During these periods, photo-

TABLE 17-19. LAND APPLICATIONS SATELLITE SYSTEM

Launch vehicle	Ariane
Spacecraft	SPOT
Orbit	650 km altitude
	near-polar sun-synchronous
	14-day repeat cycle
Sensors	
Synthetic aperture radar	SAR
resolution	100×100 m
swath width	100 km
look angle	20°
operational frequency	5.3 GHz
grey resolution	1 dB
polarization	HH or VV
Visible camera system	
detectors	linear array
IFOV	30×30 m
spectral bands (4)	0.52 to 0.90 μm
	+ panchromatic
optics	f/5.4, 105 mm dia. dioptric
Near infrared camera system	
detectors	linear array or scanner
IFOV	30×30 m
spectral bands (2)	1.55 to 2.35 μm
optics	f/3 dioptric
Data transmission	ESA Earthnet
	178 Mb/sec

graphs (probably black/white) covering 22 M km² were acquired over the Soviet Union, Central Asia, the Kurile Islands, the Far East, and presumably other areas. Exposed film was brought back to earth in the Soyuz spacecraft. Map compilation at 1:100,000 scale has been accomplished. Earth photography was also an announced objective of other Soyuz missions and the same camera may have been used. Only a few frames of KATE-140 photographs have been released.

Soyuz 22, an eight day mission at 265 km altitude in September 1976, saw the introduction of the MKF-6 multispectral camera system built by Zeiss Jena in the German Democratic Republic. It consists of 6 matched cameras in a single mount, and is roughly comparable to the S-190A camera system used on the USA Skylab mission. More than 2000 multispectral photo-sets were acquired on the Soyuz 22 mission. The characteristics of the MKF-6 camera system are given in table 17-23. The most extensive use of this camera system has been aboard the manned space station Salyut 6 which at this time (Oc-

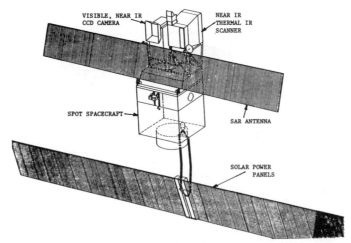

FIGURE 17-98. The Land Applications Satellite System being developed by the European Space Agency will carry two optical imaging systems and a Synthetic Aperture Radar.

TABLE 17-20. SYSTEME PROBATOIRE D'OBSERVATION DE LA TERRE (SPOT)

Launch	Ariane, March 1984
Spacecraft	SPOT
payload maximum	800 kg
solar power	1200 W
lifetime	2 yrs.
pointing accuracy 3σ	0.15 deg.
attitude rates 3σ	3×10^{-4} deg/sec.
attitude stabilization	3 axis reaction wheels
command and data handling	S band, 2 Kb/sec
Orbit	
altitude	822 km
inclination	sun synchronous 98.7 deg
repeat cycle	26 days
equator crossing descending	1030 hrs
Payload 2 identical HRV	(haute resolution visible)
detectors	3000 element linear arrays
swath width	60 km
pointing ± 26° cross track in 0.6 deg increments	

operating mode	multispectral	panchromatic
bands	0.5−0.6, 0.6−0.7, 0.8−0.9	0.5 to 0.9 μm
IFOV	20 m	10 m
sensitivity		0.5 percent
grey levels		8 bit = 256
optical system		
focal length		1082 mm
aperture		f/3.5

Telemetry	
on board storage	250 scenes 60 × 60 km
data rate	24 Mb/sec/HRV
frequency	X band 8.025 − 8.400 GHz
ground station	Toulouse, France

tober 1980) has been in orbit for more than 2½ years and has been serviced by several Soyuz missions to transfer crews, and by several unmanned Progress vehicles to resupply expendables. Salyut 6 is in an orbit at 260 km altitude and 65° inclination which permits photographic coverage of nearly all the inhabited areas of the globe. MKF-6 photographs from Salyut 6 have a ground resolution of about 15 m and each frame covers 114 km along track by 168 km across track. The USSR has acquired extensive coverage under bilateral agreements with other countries. A few prints of MKF-6 color composites have been released to the international scientific community for evaluation.

Zeiss Jena has made the PKA Automatic Precision Printer to make density compensated high resolution contact copies in quantity from MKF-6 original film. They have also built the MSP-4 Multispectral Projector to prepare en-

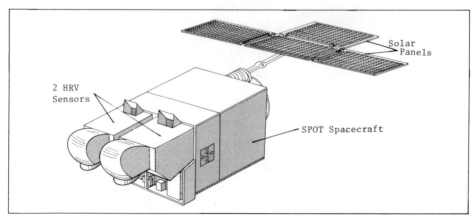

FIGURE 17-99. The French SPOT spacecraft will use a multimission vehicle, and be equipped with solar panels and two Haute Resolution Visible (HRV) sensor instruments.

TABLE 17-21. SPOT DATA PRODUCTS

Quick look products
  Uncorrected images on 70 mm film
  Uncorrected image data on High Density Digital Tape (HDDT)
  Auxiliary data on Computer Compatible Tape (CCT)

Level 1 products
  Radiometrically corrected for sensor variability
  Geometrically corrected for earth curvature and rotation,
    spacecraft altitude and velocity variations, off-nadir
    angle of view
  Available on 241 mm film, HDDT, and CCT

Level 2 products
  Geometrically corrected and geographically located by analog or digital correlation to identified
    ground control points
  Available on 241 mm film and CCT

Level 3 products
  Orthophotographs by use of digital elevation model produced from SPOT stereo data or from other
    sources
  Available on 241 mm film or CCT
  Film enlargements
  Color composites

larged color composites from up to four of the spectral separation negatives. Special color presentations can also be prepared by combining all six bands. For example, bands 2, 3, and 5 can be combined as positives, while bands 1, 4, and 6 are combined as negatives. The two composites are then combined to produce a new color presentation. The calibrated 9 point focal plane reseau marks are used to register the separate images rather than depending upon image detail.

The very high image resolution of the MKF-6 photographs exceeds the performance of conventional photogrammetric plotting instruments. For this reason 2.4× enlargements are made with an equivalent focal length of 300 mm and an image format of 132 × 195 mm which is compatible with the regular line of Zeiss Jena and USSR photogrammetric instruments. Aerotriangulation, topographic mapping, and map revision at scale 1:100,000 are performed on a routine basis.

The MKF-6 camera system is apparently used also in the unmanned photographic reconnaissance satellites. Ten such missions with an announced earth resources objective took place between May and October 1979. There is every indication that MKF-6 photography will continue to be acquired for the next several years.

TABLE 17-22. KATE-140 CAMERA ON SALYUT 4

Orbit:	350 km altitude, 51.6° inclination
focal length	144 mm
film format	18 × 18 cm
image scale	1:2,500,000
coverage per frame	450 × 450 km
ground resolution	35-40 m

17.10.7.7 THE NETHERLANDS PROGRAM

The Netherlands Agency for Aerospace Programs (NIRV) has initiated a study of a remote sensing satellite which would carry a Dutch built multispectral linear array sensor in a near equatorial orbit. The project is being planned in cooperation with Indonesia where the ground data reception station will probably be located. The sensor parameters will be specifically selected for the weather conditions and vegetation types in equatorial areas.

17.10.8 OTHER NATIONAL SPACE PROGRAMS

Other national space remote sensing programs have been undertaken by Brazil, Canada, India, and Japan.

17.10.8.1 BRAZIL

The Space Research Institute (INPE) in Brazil has published its intention to develop a national remote sensing satellite which might be launched in 1985 or 1986. The sensors would include a multi-spectral linear array and a high resolution panchromatic camera, but no details have been released.

17.10.8.2 CANADA

The Canadian Centre for Remote Sensing (CCRS) in cooperation with the Atmospheric Environment Service (AES) of the Energy Mines and Resources have conducted a study program called SURSAT (Surveillance Satellite). The primary purpose was to determine the applicability of the data from the NASA Seasat to the monitoring of ice conditions in the polar seas. These studies concluded that the essential

TABLE 17-23. THE USSR MKF-6 CAMERA SYSTEM

Lens: 125 mm focal length, f/4
Lens distortion: ±3 μm RMS
Resolution: 160 lp/mm on T-18 film
Format: 55 mm along × 81 mm across track
Dielectric interference glass filters

band peak	0.48	0.54	0.60	0.66	0.72	0.80 μm
half peak width	0.04	0.04	0.04	0.04	0.04	0.10 μm

Calibrated 9 point focal plane reseau
10 step density wedge in each frame
Automatic exposure sensor
Shutter speed 7 to 56 msec
Shutter synchronization <0.2 exposure time
Cycle rate 4 to 20 sec/frame
Overlap selection 20, 60, or 80 percent
Parallelism of lens axes <10 arc sec
Rocking mount IMC 16.8 to 39.8 m rad/sec
Film magazines 2500 frames/camera (thin base TA-18)

TABLE 17-24. PARAMETERS OF INDIAN
"BHASKARA" SPACECRAFT

Launch:	7 June 1979 by USSR from Kapustin Yar
Orbit:	512-557 km altitude, 50.7° inclination
Sensors	

(1) TV cameras (2) for land observation
  spectral bands: 0.54 to 0.66 μm
                   0.75 to 0.85 μm
  resolution: 1 km
(2) two frequency radiometer for ocean survey
  19.350 GHz with 125 km swath
  22.235 GHz with 200 km swath

TABLE 17-25. JAPANESE EARTH
OBSERVATION SATELLITES

Satellite	Launch	Purpose
MOS-1	1984	Marine bioproductivity
MOS-2	1985	Marine science
LOS-1	1987	Agriculture, forestry, land use
MOS-3	1989	All weather marine observation
LOS-2	1991	High resolution land observation
MAGRAS	1992	Magnetic, gravity satellite

TABLE 17-26. MARINE OBSERVATION SATELLITE MOS-1

Launch:	N series vehicle from Tanegashima, 1984
Orbit:	altitude 909 km
	inclination 99.1 degrees
	equator crossing time 1030 hours
	cycle time 17 days

Sensors:
(1) MESSR—Multispectral Electronic Self-Scanning Radiometer
  purpose: sea surface color
  linear array detectors
  4 spectral bands 0.51 to 1.10 μm
  50 metre IFOV
  100 km swath width
(2) VTIR—Visible and Thermal Infrared Radiometer
  purpose: sea surface temperature
  object space mechanical scanner
  visible band
    0.5−0.7 μm, 900 metre IFOV
  thermal bands
    6.0−7.0, 10.5−11.5, 11.5−12.5 μm, 2.6 km IFOV
      sensitivity 0.5° K
  500 km swath
(3) MSR—Microwave Scanning Radiometer
  purpose: atmospheric water content
  object space mechanical scanner

bands:	frequency	IFOV	sensitivity
	23.8 GHz	47 km	1.5° K
	31.4 GHz	38 km	1.5° K

  320 km swath
Total data rate: 8.8 Mb/sec

sensor is a synthetic aperture radar (SAR) capable of producing 25 to 30 m resolution for a swath width of approximately 100 km. Beginning in 1980, CCRS has undertaken a two-year study program to develop the operational parameters for a C-band radar which would meet these requirements. The study will include the mission and orbit parameters and the ground processing capabilities. The program might entail two or three satellites in orbit simultaneously to provide the necessary coverage for adequate monitoring of the arctic seas and coastlines.

At the moment, Canada does not envision its own launch program, but is negotiating with NASA and ESA on the terms of a cooperative agreement to fly the radar. It would be a likely candidate for ICEX (*see* section 17.10.6.4) or for COMSS (*see* section 17.10.7.4.1).

17.10.8.3 INDIA

The Indian Space Research Organization (ISRO) developed an Earth observation satellite "Bhaskara" which was launched in June 1979. The characteristics of this spacecraft are listed in

table 17-24. The two TV cameras failed shortly after launch, and no data from the radiometer has been released. ISRO is developing a second Bhaskara with essentially the same instruments which is planned for launch in 1981.

ISRO has also announced its intention to develop a second generation Satellite for Earth Observation (SEO) with increased sensor capabilities. This will be launched into a sun-synchronous orbit in 1985. ISRO is negotiating with both NASA and the USSR for launch service for this spacecraft.

17.10.8.4 JAPAN

In 1978 the Science and Technology Agency in Japan initiated a study to define Marine Observation Satellites (MOS), Land Observation Satellites (LOS), and geophysical satellites. The proposed program is listed in table 17-25.

An overall plan for instrumentation of these satellites has been developed, but detailed specifications have been published for only the first three. The characteristics of the proposed MOS-1 are given in table 17-26.

On MOS-2, the MESSR sensor will be eliminated and replaced by a radar altimeter and microwave scatterometer to determine sea-surface wave height, currents, and wind conditions.

The characteristics of the proposed LOS-1 are given in table 17-27.

On MOS-3 it is planned to carry a synthetic aperture radar as well as improved versions of

TABLE 17-27. LAND OBSERVATION
SATELLITE LOS-1

Launch:	N series vehicle from Tanegashima, 1987
Orbit:	altitude 700 km
	inclination sun synchronous
	cycle time 18 days

Sensors:

(1) Stereo cameras (2)
purpose: stereo observation
linear array detectors
panchromatic response 0.6−1.1 μm
25 metre IFOV
6 bit grey level encoding
50 km swath, pointable left and right

(2) VNIR—Visible and Near Infrared Radiometer
purpose: land cover classification
linear array detectors
4 spectral bands 0.45 to 1.10 μm
25 metre IFOV
8 bit grey level encoding
200 km swath

(3) VIR—Visible and Infrared Radiometer
purpose: vegetation and geology
object space mechanical scanner
spectral bands

0.55−0.70 μm	2.08−2.35 μm
0.80−1.10 μm	3.55−3.90 μm
1.55−1.75 μm	10.50−12.50 μm

IFOV 50 m except thermal 150 m
8 bit grey level encoding
200 km swath

(4) Infrared sounder
purpose: atmospheric correction
object space mechanical scanner
8 channels 12 to 19 μm
25 km IFOV
750 km swath

TABLE 17-28. MINING AND ENERGY RESOURCES
EXPLORATION SATELLITE MERES

Orbit: 560 km sun synchronous
Sensors:

(1) Visible and Near Infrared Radiometer
purpose: geological surface conditions
linear array detectors pointable by rotating
mirror to provide stereo
spectral bands

0.51−0.59 μm	0.80−0.95 μm
0.64−0.72 μm	0.90−1.10 μm
0.72−0.80 μm	

30 m IFOV
swath not defined

(2) Near and Thermal Infrared Radiometer
purpose: ground thermal and rock classification
object space mechanical scanner

spectral bands	IFOV
1.3− 1.9 μm	50 m
1.9− 2.5 μm	50 m
10.5−11.5 μm	130 m
11.5−12.6 μm	130 m

swath not defined

(3) Synthetic Aperture Radar
purpose: geologic structure, moisture and
rock classification
L-band (1.3 GHz frequency)
Resolution 25 metre
Swath not defined

the MOS-2 sensors. For LOS-2 a laser sensor is planned for measuring atmospheric constituents, and the resolution of the land surface observation sensors will be increased to a 15 metre IFOV.

In September 1979, the Ministry of International Trade and Industry (MITI) submitted a separate program plan to the Space Activities Commission. A study was undertaken by the Agency of Industrial Science and Technology to develop a Mining and Energy Resources Exploration Satellite (MERES). The characteristics of this proposed satellite, insofar as they have been defined, are given in table 17-28.

While intensive definition studies have been funded for both LOS and MERES proposals, it is not probable that both spacecraft will actually be built. The National Space Development Agency (NASDA) will evaluate the results of the studies and decide upon the final configuration, which may well combine characteristics from both proposals.

17.10.9 INTERNATIONAL COOPERATION

It is evident that there is a great deal of near duplication in the various proposals for future earth observation satellites. Minor differences in orbit parameters, number and limits of spectral bands, resolution, and swath widths result from the perceptions of research scientists in the various agencies. There are differing views on how difficult it will be to implement an effective synthetic aperture radar for operation in space. Finally there are different priorities placed upon use of space sensors for agriculture, geologic exploration, and marine sciences. But it is inordinately expensive to build and operate space systems, and equally expensive to process, disseminate, and apply the data which can be acquired.

Scientists and managers from all countries involved in space activities realize that satellites are inherently international in operation, and have begun to talk to one another about making their systems complementary rather than competitive. It would probably be advantageous to all concerned if many of the proposed systems discussed in this section were consolidated into a very few spacecraft operating under international agreements.

REFERENCES

Barabashov, Mikhailov and Lipsky, Atlas of the Moon's Far Side. Interscience. New York. (1961).

Batson, R. M. "Cartographic Products from the Mariner 9 Mission" *J. Geophys. Res. 7(20)*: 4424–4435 (1973).

Bender, L. V. An Algorithm for Gridding Satellite Photographs. Report of the Department of Geodetic Science No. 135. Ohio State University. Columbus, Ohio. (1970).

Berg, R. A. *Reduction of MARS 1969 (MM69) Television Photography*. Aeronautical Chart and Information Center, St. Louis. (1970).

Brouwer, D. and Clemence, B. M. Methods of Celestial Mechanics. Academic Press. New York (1961).

Brown, D. C. Photogrammetric Flare Triangulation—A New Geodetic Tool. RCA Data-reduction Technical Report No. 46, Patrick Air Force Base, Florida. National Technical Information Center. Springfield, Virginia (1958)b.

Brown, D. C. A Solution to the General Problem of Multiple-Station Analytical Stereotriangulation. RCA-MTP Data-reduction Technical Report No. 43. Patrick Air Force Base, Florida (AFMTC TR58-8). National Technical Information Center. Springfield, Virginia (1958).

Brown, D. C. "Evolution, Application and Potential of the Bundle Method of Photogrammetric Triangulation." *Proceedings of International Society of Photogrammetry, Commission III, Symposium in Stuttgart 1974*. International Society of Photogrammetry.

Brown, D. C.; Davis; and Johnson, The Practical and Rigorous Adjustment of Large Photogrammetric Nets. RADC TRD-64-092, Rome Air Development Center, Griffith AFB. National Technical Information Center. Springfield, Virginia. (1964).

Case, James B. "The Utilization of Constraints in Analytical Photogrammetry" *Photogram. Eng. 27(5)*: 766–788 (1961).

Chapman, W. H. "Gridding ERTS Images" *Proc. ACSM Fall Convention 1974* American Congress on Surveying and Mapping. Falls Church, Virginia. pages 15–19. (1974).

Colvocoresses, A. P. "EROS Cartography Program" Cento Seminar on Remote Sensing. Central Treaty Organization. Ankara, Turkey. (1971).

Colvocoresses, A. P. "Unique Characteristics of ERTS" *Symposium on Significant Results Obtained from ERTS*. NASA SP-327. Volume I. pages 1523–1525. Government Printing Office. Washington, D.C. (1973).

Colvocoresses, A. P. "Space Oblique Mercator, A New Map Projection of the Earth" *Photogram. Eng. 40(8)*: 921–926 (1974).

Colvocoresses, A. P. "Evaluation of the Cartographic Applications of ERTS-1 Imagery" *The Am. Cartographer. 2(1)*: 5–18 (1975).

Colvocoresses, A. P. "Overall Evaluation of Skylab (EREP) Images for Cartographic Application" *Surv. & Mapping 36(4)*: 351–360 (1976).

Craft, T. A. Nocturnal Images of the Earth from Space. Final Report, SRI Project 5993, to U.S. Geological Survey. National Technical Information Center, Springfield, Virginia. (1977).

Davies, M. E. "Mariner 9 Primary Control Net of Mars" *Photogram. Eng. 39(12)*: 1297–1302 (1973).

Davies, M. E. and Arthur, D. W. G. "Martian Surface Coordinates" *J. Geophys. Res 72(20)*: 4355–4394 (1973).

Davies, M. E. and Berg, R. A. "A Preliminary Control Net of Mars" *J. Geophys. Res. 76(2)*: 373 (1971).

Davis, Riding The Rigorous and Simultaneous Adjustment of Lunar Orbiter Photography Considering Orbital Constraints. RADC-TR-70-274. Rome Air Development Center. National Technical Information Center. Springfield, Virginia. (1970).

De Vaucouleurs, G.; Davies, M. E.; and Sturms, F. M.

Jr. "The Mariner 9 Aerographic Coordinate System" *J. Geophys. Res.* 72(20): 4395–4404 (1973).

Doyle, F. J. "Photographic Systems for Apollo" *Photogram. Eng.* 36(10): 1039–1044 (1970).

Doyle, F. J. "Can Satellite Photography Contribute to Topographic Mapping" *USGS J. Res.* 1(3): 000 (1973).

Doyle, F. J.; Elassal, Atef A.; and Lucas, J. R. Selenocentric Geodetic Reference System. NOAA Techn. Report NOS 70 NGS 5. vi + 53 pp. National Technical Information Center. Washington, D.C. (1977).

Eckhardt, D. H. "Physical Librations Due to the Third and Fourth Degree Harmonics of the Lunar Gravity Potential" *The Moon 6:* 127–134 (1973).

Eichhorn, H. Astronomy of Star Positions. Frederick Ungar. New York. (1974).

Evans, W. E. "Marking ERTS Images with a Small Mirror Reflector" *Photogram. Eng.* 40(6): 665–672 (1978).

Fleming, E. A. *Positioning Off-shore Features with the Aid of Landsat Imagery* Dept. of Energy, Mines and Resources. Ottawa, Canada.

Gehrels, T. et al. "Imaging Photopolarimeter Experiment on Pioneer 10" *Science* 183(4): 318–320 (1974).

Gunther, T. NASA Directory of Station Locations (4th edition). NASA Directory, Operational Orbit Support Branch, Code 572, Goddard Space Flight Center. Greenbelt, Maryland. (1978).

Hammack, J. C. "Landsat Goes to Sea" *Photogram. Eng. and Remote Sensing* 43(6): 683–691 (1977).

Henriksen, S. W. (ed.) The National Geodetic Satellite Program (2 vols.). U.S. Government Printing Office. Washington, D.C. (1977).

Herrick, S. Astrodynamics. (2 vols.) Van Nostrand Reinhold. London. (1971–1972).

Junkins, J. L. An Introduction to Optimal Estimation of Dynamical Systems. Noordhoff International Publishing. Leyden. (1978).

Kaula, W. Theory of Satellite Geodesy. Blaisdell. Waltham, Mass. (1966).

Kliore, A. J. et al. "The Atmosphere of Mars from Mariner 9 Radio Occultation Measurements" *Icarus* 17(2): 484–516 (1972).

Kosofsky, L. J. "Topography from Lunar Orbiter Photos" *Photogram. Eng.* 32: 277–285 (1966).

Kosofsky, L. J. and El-Baz, Farouk The Moon as Viewed by Lunar Orbiter NASA SP-200. Government Printing Office. Washington, D.C. (1970).

Kreznar, J. "User and Programmer Guide to the MM'71 Geometric Calibration and Decalibration Programs. JPL Report 900-575. Jet Propulsion Laboratory. Pasadena, California. (1973).

Leatherdale, J. D. "The Practical Contribution of Space Imagery to Topographic Mapping" *1978 Symposium of ISP Commission IV.* Ottawa, Canada (1978)

Levinthal, E. C., et al. "Mariner 9 Image Processing and Products" *Icarus* 18(1): 75–101 (1973).

Light, D. L. "Ranger Mapping by Analytics" *Photogram. Eng.* 32(5): 792–800 (1966).

Light, D. L. "Extraterrestrial Photogrammetry at TOPOCOM" *Photogram. Eng.* 36(3): 260–272 (1970).

Light, D. L. "Altimeter Observations as Orbital Constraints" *Photogram. Eng.* 39(4): 339–346 (1972).

Light, D. L. "Photo Geodesy from Apollo" *Photogram. Eng.* 39(6): 574–578 (1972).

Macdonald, W. R. "New Space Technology Advances Knowledge of the Remote Polar Regions" *NASA Proceedings of Third ERTS Symposium* (1973).

Masursky, H. et al. "Television Experiment for Mariner Mars 1971" *Icarus* 12(1): 10–45 (1970).

McEwen, R. B. *RBV Coordinates for ERTS-1:* National Technical Information Service. Springfield, Virginia (1972).

McEwen, R. B. and Asbeck, T. A. "Analytical Triangulation with ERTS" Proc. Am. Soc. Photogram. Washington, D.C. 1975. American Society of Photogrammetry. Falls Church, Virginia. pp. 490–503. (1975).

Merson, R. H. *Numerical Integration of the Differential Equations of Celestial Mechanics.* Royal Aircraft Establishment, Farnborough, England (1973).

Milne, W. E. Numerical Solution of Differential Equations. Dover Publ. New York. (1970).

Murray, B. C. et al. "Venus Atmospheric Motion and Structure from Mariner 10 Pictures" *Science* 183 (4131): 1302–1314 (1974).

NASA/Goddard Space Flight Center *Landsat Data User's Handbook.* Goddard Spaceflight Center, Greenbelt, Maryland (1976).

Nautical Almanac Offices Explanatory Supplement to the Astronomical Ephemeris and the American Ephemeris and Nautical Almanac. Her Majesty's Stationery Office. London. (1961).

Schmid, H. H. A General Analytical Solution to the Problem of Photogrammetry. BRL Report No. 1065, Aberdeen Proving Grounds. National Technical Information Center. Springfield, Virginia (1959).

Schreiter, J. B. and Zechiel, L. N. Determination of Aircraft Orientation by Airborne Stellar Photography. Part I. MCRL Techn. Paper No. (684)-1-283. Ohio State University Research Foundation. Columbus, Ohio. (1958).

Shull, C. W. and Schenk, L. A. "Mapping Surveyor-III's Landing Area" Proc. ASP Convention Washington, D.C. 1968. American Society of Photogrammetry. Falls Church, Virginia, pp. 205–214 (1968).

Snyder, J. P. "The Space Oblique Mercator Projection" *Photogram. Eng. and Remote Sensing* 44(5): 585–596 (1978).

Snyder, L. M. Mariner 9 TV Subsystem Calibration Report. JPL Report 610-101. Jet Propulsion Laboratory. Pasadena, Calif. (1971).

Stephens, J. M. "A New Advance in Planetary Mapping" *Photogram. Eng.* 39(4): 389–394 (1973).

Warren, A. "Photomosaicing by Photomechanical Method". PAIGH XIII Pan-American Consultation on Cartography. Quito, Ecuador. (1977).

Wu, S. C. "Stereo Mapping with the Viking Lander Camera Imagery". *Archives I.S.P. XIII Congress.* (1976).

CHAPTER XVIII

Training, Education and Professional Status

Author-Editors: ROBERT D. BAKER & ROBERT D. TURPIN
Contributing Authors: L. R. EVANS, JAMES P. LATHAM

18.1 Introduction

THE PURPOSE of this chapter is to provide an overview of the three phases of professional development—training, education, and professional recognition.

Training includes those activities through which a person is prepared to operate different types of equipment and to do the associated calculations and/or drafting. Training may be formalized, as is the case with schools that have technician training programs, or it may be "on-the-job," as is true for most stereo plotter operations.

Education, as used here, includes all aspects of formal schooling which provides the individual with an understanding of the many aspects of photogrammetry. Education may include, but would not be limited to, Bachelor's, Master's, or Ph.D. degrees which in some measure are related to photogrammetry. Many degree programs provide one or more elective courses in photogrammetry, but only a few include an "in-depth" study of photogrammetry. In addition to the formalized college and university courses, there has been a large number of "continuing education" programs developed both by universities and by organizations, including the American Society of Photogrammetry.

Professional status is the position photogrammetry enjoys with relation to other professions such as engineering and surveying. Efforts by the American Society of Photogrammetry in the 1970's have resulted in a Certification Program and a wider recognition of the photogrammetric profession by state licensing agencies for engineering and surveying.

The material which follows has been provided by the contributing authors and edited to be included here as an amplification of the above introduction.

18.2 Training

Photogrammetric training may be classified according to the orientation of the programs. Therefore training may be formalized, as for example in colleges (including many junior colleges) where the trainee attends regularly scheduled classes, or as for example in the military where personnel (both military and civilian) attend a regularly scheduled instruction for a fixed period of time. Another procedure is usually designated as on-the-job training, in which regular employers of a company or organization assist new employees to become competent in one or more aspects of photogrammetry, as for example in training stereo-plotter operators.

18.2.1 FORMAL TRAINING

One of the best organized specific training programs related to photogrammetry has been carried out for many years by the military in what is now the Defense Mapping School, Fort Belvoir, Va.

The forward to the Defense Mapping School Catalog, dated July 1975, reads in part as follows:

The Defense Mapping School, a component of the Defense Mapping Agency, is responsible for instruction in military cartography and related subjects. It provides direct support of the Army, Navy, Air Force, Marine Corps, and Allied Nations, by providing instruction and advisory services in the functional areas of mapping, charting and geodesy (MC&G). The Defense Mapping School conducts resident instruction in the following fields: Surveying, Cartography, Graphic Arts, Topographic Sciences, and related subjects.

Some courses offered by this school are listed below:

Photogrammetric Compilation 8 Weeks 1 Day

Provides Army enlisted (and selected civilian) students with a working knowledge of the principles and techniques of photogrammetric compilation, using multiplex and military high precision stereoplotters; provides a general knowledge of operation of the reduction printer and of stereo-triangulation. Includes orientation in analytical photogrammetric point positioning and orthophoto production for cartographic updating.

Basic Photogrammetric-Cartographic
    Techniques   8 Weeks 5 Days

Provides USAF enlisted (and selected civilian) students with a working knowledge of the computations and construction of projections and grids; plotting control data; selecting and marking points on aerial photography; computing tip and tilt, interpret, selects and delineates planimetric and reflectivity detail; compiling, revising and updating planimetric, hypsometric and reflectivity manuscripts on Air Target charts from cartographic source and Aerial Photography; performing color separation of planimetric, hypsometric and reflectivity data through the process of negative engraving; performs functions involved in the production of flight information publications; assembles aerial photo mosaics. Includes an orientation in the operation of the equipment used in the Analytical Photogrammetric Positioning System.

Advanced Photogrammetric-Cartographic
    Techniques   8 Weeks

Provides USAF career NCO (and selected civilian) students with education and training in the sciences related to the photogrammetric-cartographic career field including physical geography, datums, cartographic control, intermediate level photogrammetry, structural and industrial analysis, radar reflectivity, reconnaissance camera systems and the support functions of the photographic and lithographic areas. Include orientation in analytical photogrammetric point positioning for cartographic update.

Special Products Compilation   2 Weeks

Provides enlisted (and selected civilian) students with a working knowledge of the cartographic applications of remote sensor and multiband imagery to include; use, measurement and interpretation of radarscope, infrared, side looking airborne radar and high altitude small scale imagery; and the preparation and use of associated photo interpretation keys as applied to the compilation of special cartographic products in support of current weapons systems.

Analytical Photogrammetric Positioning
    System (APPS)   5 Days

Provides students with the photogrammetric concept, limited theory, and a working skill in using the APPS for point positioning from aerial photographs. Prepares selected students to present subsequent training on the APPS. Specifically provides specialized training in rapid and accurate determination of UTM or geographic coordinate positions with emphasis on training of other potential APPS users. Provides examples of appropriate APPS applications.

Although each of the courses listed above are from military training programs, they are good illustrations of (1) titles, (2) time, and (3) descriptions of formalized training programs.

18.2.2 ON-THE-JOB-TRAINING

18.2.2.1 GENERAL

Many organizations engaged in photogrammetric work have inaugurated programs for training new employees in this field. This may well apply regardless of the employee's prior education, training, or practical experience in photogrammetry.

These programs are not capable of being standardized because of the wide variation in the nature of work undertaken by the different organizations, in the equipment used, and in the background of the trainees. The training program of any one company may even change from year to year depending upon its current projects and the use of newer equipment. Programs vary from rather informal, brief introductory lectures and demonstrations of perhaps two days to a week's duration, to the more formal lecture-type courses which to some extent resemble college work. Some of these programs may be in the nature of employement participation in a cooperative education plan.

Because of this wide variation, it does not seem appropriate to outline any detailed program or course of study, but rather to give some suggestions and possible aids which may be adapted by an organization to its own peculiar situation. Obviously, these programs will depend upon the size of the company or organization, background of the trainees, equipment and materials used, and the kind of photogrammetric projects performed.

18.2.2.2 OBJECTIVES

The objectives of on-the-job training are usually more definite and clear-cut than is the case in a general education program. The trainee is generally being fitted for a specific job, at least as a starting assignment, although this may not be true in large organizations where openings may exist in several categories. In the latter case, one purpose of the training might be to discover the kind of work for which the trainee is best fitted, and consequently such training would be somewhat broader in scope. Even here, however, the program would be directed almost exclusively toward the methods and instruments used by the particular company.

18.2.2.3 TRAINEE REQUIREMENTS

Although some prerequisite attainments may be set, the nature of the training program is often determined by the prior education and experience of the employees themselves. Probably the minimum requirement would be a high school diploma, but more often the prerequisite might be set at a college degree. This would usually be in civil engineering, although a degree in geology, geography, or forestry might be satisfactory, especially if some work in elementary photogrammetry is included in these curricula. The college background preferred would undoubtedly be influenced to some extent by the nature of the work in which the employer is principally engaged. A strong background in the fundamentals of surveying and mapping is preferred by some.

It is held by many companies that on-the-job training is advisable even for employees who come with a good college background in surveying, mapping, and photogrammetry. While these employees may have a sound knowledge of the basic theory, they need supplementary training in the practical applications, especially in connection with the methods and equipment used by the employer. This, of course, is in keeping with the training programs for college graduates sponsored by many engineering and industrial organizations.

18.2.2.4 ORGANIZATION

Training programs are commonly organized as a function of the personnel department, although this may vary with the company; in fact, they could be placed in the department which can best spare the personnel and space for handling it.

Regardless of where the main responsibility for organization is placed, cooperation must be obtained among all departments so that the trainee receives instruction in all phases of the company's work. Contributions to the program may be necessary from all major departments in the way of space for classrooms, lectures, and equipment. Some equipment may be entirely satisfactory for training purposes long after it has become outmoded for regular production work.

Although various members of the company may be utilized for the instructional staff, the over-all organization of the program must necessarily rest in one person who coordinates the sequence of topics, arranges for the proper instruction at the proper time, sees that the appropriate room and equipment are available, and in general takes care of the mechanical details.

18.2.2.5 INSTRUCTION

Instruction is usually given by various members of the company's own staff, probably one person from each department. This may be the department head himself (for short courses) or someone whom he may assign to this task. Occasionally, a professional teacher, perhaps a college faculty member, is brought in to teach. An instructor should be chosen with care and consideration for his teaching abilities, and for his interest in the program and in new employees. He should, of course, be a person who has had considerable experience in photogrammetry himself.

It is conceivable that one person might serve as instructor for the entire program especially in small companies with limited personnel available for this assignment. This person would probably have complete charge of the program, although he would undoubtedly have the privilege of calling on others in the company for help in special instances or when some special equipment is under discussion.

Fairly detailed outlines of the work to be covered should be prepared, and the amount of time set for each item. Also, lesson plans for each meeting of the training class are helpful, so that the objectives and procedures are defined at the beginning of each class period or exercise. The time may be somewhat more flexible in such a program than in formal education, where the length and number of class meetings are rigidly fixed by the college or school calendar. Although a target date for completion should be set for an on-the-job program, the actual time taken may be shortened or increased as the needs of the individual trainees warrant.

The level at which instruction should begin naturally depends upon the background of the starting trainees. Some organizations assume no previous knowledge of photogrammetry at all and may even start with basic work in surveying and mapping before approaching photogrammetry. If the trainee has already had this basic training, the instruction could start at a more advanced level. However, some rapid review of elementary material is almost always essential. The upper limit of instruction would be the point at which the trainee is ready to assume his assigned tasks with a minimum of supervision.

The length of program ranges, as a consequence of the many variables involved, from just a few days to as much as two years. Undoubtedly, the former would require much more supervision of the trainee at his assigned job than would the latter, which probably in fact absorbs some of the supervision as a part of the program itself.

18.2.2.6 TEXTS AND MATERIALS

Texts and materials also vary with the type of instruction given on any particular program. If basic theory is to be covered, one of the standard college texts might be used, as well as the MANUAL OF PHOTOGRAMMETRY. For the application of theory to the needs of the company, special instruction manuals would probably be prepared by the company's own staff to supplement lectures and demonstrations and also to explain or outline whatever work is given in the nature of a workshop or laboratory. The amount of this type of material would depend upon the number of trainees, frequency of the program, and the instructors themselves. The larger the class, the more reliance would probably be placed upon written instructions, and this would also apply where the course of training is repeated at frequent intervals. The time necessary to prepare the material may not be justified for a course given, for example, once in five years, but would be if it is repeated every six months or so.

18.2.2.7 EQUIPMENT

The equipment used in training is largely determined by that which is used in the company's own work program. However, additional units may be necessary, especially of the smaller pieces such as stereoscopes, and possibly of some of the larger instruments which are widely used. These units would then be available exclusively for the training program, thereby avoiding conflict with regular production schedules. In some cases, a type of equipment which has been replaced by more modern units for production work may very well be retained for training purposes, having greater value for this purpose than as trade-in toward the new equipment.

18.2.2.8 TESTING

Some programs may require periodic tests similar to those given in formal school courses.

Written examinations would apply mostly to the theoretical parts of the training, to assure that the basic principles are mastered before proceeding with advanced work. Furthermore, the passing of these tests might be a condition of continued employment. They could also serve to determine the trainee's aptitudes, including the possibility of reassigning him to a different department of the firm for which he seems better suited.

Testing of the applicants can be accomplished by evaluating practice in the methods concerned. Certain levels of achievement can be established at various intervals in the training which must be met before advancing to the next level. In a short program designed mainly for training in one specific kind of work, the testing may be simply the demonstration of satisfactory results in work on the particular instrument, after which the trainee is placed on production work. The demonstration of satisfactory work would be initially concerned with quality, and speed would not be an important consideration for some time.

18.3 Education

18.3.1 INTRODUCTION

The dividing line between training and education is really a hazy area rather than a line. Indeed there is much overlap between training and education.

Also, there is no clearly defined line between remote sensing or photo interpretation and photogrammetry in the classical sense. Therefore, much of the following discussion will relate to education in both remote sensing and photogrammetry.

Few other fields of study are interrelated with such a diversified group of academic disciplines, and departments or research institutions; and few other fields have contributed so actively to the instrumented discovery of both scientific realities and data for applied environmental management both on earth and in space.

The extensive collateral training activities carried on by government agencies and private enterprises actively involved in photogrammetry and remote sensing may be suggested by some of the data in the surveys reported by the American Society of Photogrammetry, but no attempt has been made to survey these very significant contributions to the advancement of photogrammetric skills. As one survey demonstrates, the value of the short and "mini" courses, symposia, and summer institutes—often supported by the National Science Foundation—has been outstanding in expediting the transfer of knowledge and information about new technology and the potential of its application. These programs, which often have been scheduled in association with

the time and place of national meetings by various disciplines, have been particularly successful in dispersing awareness of remote sensing capabilities. Some professional associations have established their own committees or commissions on remote sensing to encourage communication, and some, such as the Association of American Geographers' Committee on Remote Sensing, have established periodical publications, which vary from newsletter to pamphlet size, and emphasize educational activities and methods.

Those interested in pursuing the historical perspective will note the great expansion of education related to the use of imagery. The advance of detecting and recording systems, the improvement in reproduction methods, and the advance of interpretation techniques and measurement methods have stimulated the growth in courses, programs, and career majors. However, growth in education has been so rapid that many significant dimensions are not yet well developed. The increasing importance of certification, as recognized by the Society, seems to suggest a greater attention to such tasks as defining the core requirements for a professional photogrammetrist and remote sensing practitioner. This also requires an improved structure of continuing education to maintain and update professional skills in a dynamic field. Many teachers of remote sensing are delighted with the recent publication of the MANUAL OF REMOTE SENSING, but still feel the need for a text that organizes learning experiences for students entering the field. Specialized texts relating to particular disciplines are developing, but because of the cost of color printing and the limited markets served, their effectiveness is often inhibited. Teaching aids which can introduce students to automatic measurement and interpretation systems are needed, because only a few students have access to the high-capability electronics of elaborate laboratories. Simplified interpretation systems which can relate to computer terminal access now found in almost all educational centers would advance educational development and could also encourage smaller agencies, such as county planning groups, to integrate imagery data in their information systems.

18.3.2 FORMAL EDUCATION

18.3.2.1 INTRODUCTION

Remote sensing and photogrammetry are an integral part of many programs at colleges and universities in the United States and Canada. In 1975, there were at least 470 courses in the United States and 64 in Canada that stressed remote sensing/photogrammetry. Thirty-eight universities in the United States and six in Canada had or planned to initiate majors, minors, or areas of specialization in remote sensing/ photogrammetry. Many of the courses include field trips and several are offered in the evening. At least 63 books have been adopted as textbooks and reference books. The American Society for Engineering Education reported at least 110 basic engineering research projects underway at 31 member institutions in 1974 and 1975.

18.3.2.2 RECENT SURVEYS

Remote sensing/photogrammetry is a critical tool in present-day research. It can reduce field reconnaissance time in a remote area of Alaska from a few months to a few days; tell a HUD planner where to site for greater soil and rock competence; watch wheat in the United States, USSR, and the Peoples Republic of China; reveal the crop pattern of pre-historic Sinagua farming in Arizona; and make mapping possible for the search for life on Mars.

The status of remote sensing and photogrammetry in college and university curricula has been the subject of several works in the past few years. Stanton (1971) discussed photogrammetry and photogrammetry-related programs at 125 universities in the United States. In 1972, Eitel published a list of 80 remote sensing courses, and Bidwell (1975) and Morain (1975) researched colleges and universities for courses and programs in this field. The data presented here were obtained between May 1975 and February 1976, mainly from responses of faculty members of colleges and universities to a questionnaire on courses and programs in remote sensing/ photogrammetry sent to more than 1200 institutions in the United States and Canada. In the

United States today, there are at least 187 courses in remote sensing[1], 74 in photointerpretation[2], 23 in photogeology, 8 in astrogeology, 96 in photogrammetry, 18 in image processing, and 59 in other related subjects in the United States. Of these, 23 are programmed for evening classes and 113 include trips. Courses are taught in 178 institutions in 24 academic areas.

Canadian programs in 1975 included at least 27 remote sensing courses, 10 photointerpretation courses, 25 photogrammetry courses, and two other related courses, taught at 13 institutions in seven academic areas. At least one of them is offered in the evening and 13 include trips.

About 75 percent of the remote sensing/ photogrammetry courses in the United States are taught by departments of geography (22 percent), geology (19 percent), civil engineering (20 percent), and forestry (13 percent) (Table 1). Another four percent of the courses are in the curriculum of civil engineering departments (if the Ohio State University's Department of Geodetic Science can be included in the category of civil engineering). Approximately one-third of all remote sensing courses are taught by geography departments, and 72 percent of all photogrammetry courses are taught by civil engineering departments (Ohio State included). The majority of the image processing courses are taught in electrical engineering departments, which reflects the infancy of computer applications to remote sensing. In the future, image processing courses are expected to increase in number and to be introduced to applications-oriented departments, i.e., geology, forestry, and geography.

The number of students enrolled in remote sensing/photogrammetry courses could not be accurately determined from the information received in response to the questionnaire. Many responses were received that did not include enrollment figures, and many of these were from large universities having fairly comprehensive programs in remote sensing and photogrammetry. A tally of the responses shows that at least 4000, and perhaps as many as 6000, students are enrolled annually in a remote sensing/ photogrammetry course, but there is no way to determine from this information how many of them enroll in two or more such courses.

Trips taken as an integral part of at least 113 courses include visits to local aerial survey firms or government agencies. Field trips provide the students with practice in ground-truth verification, and occasionally in actual data acquisition from remote-sensing platforms.

---

[1] Used in the general sense of the term, the courses cover the electro-magnetic spectrum from the ultraviolet through the radio range and include nonphotographic sensor systems.

[2] Courses mainly concerned with aerial photography and camera systems.

TABLE 18-1. RELATIONS OF TYPES OF COURSES TO DEPARTMENTS IN WHICH THEY ARE OFFERED (U.S.).

Type Course	DEPARTMENT									
	Geography	Geology	Forestry	Civil Engineering	Geodetic Science	Electrical Engineering	General Engineering	Other	Total	Percent
RS	72	42	22	13	1	5	4	28	187	40
RSr	1	1		2				5	9	2
PI	17	7	25	19			1	3	74	16
PIr	2	1		2	2				5	1
PG	1	1	13	54	15		10	2	96	20
PGr				5	1		1		7	1
MPI	9	7	2					1	19	4
PGe		23							23	5
AG		8							8	2
SD						10		4	14	3
IP	1	1		1		8	1	6	18	4
OP						4		6	10	2
Total	103	91	62	96	19	27	17	55	470	
Percent	22	19	13	20	4	6	4	12		

RS—Remote Sensing
RSr—Remote Sensing related
PI—Photo-Interpretation
PIr—Photo-Interpretation related
PG—Photogrammetry
PGr—Photogrammetry related
MPI—Map & Photo-Interpretation
PGe—Photogeology
AG—Astrogeology
SD—Systems Design
IP—Image Processing
OP—Optics

TABLE 18-2. THE OHIO STATE UNIVERSITY DEPARTMENT OF GEODETIC SCIENCE MASTER'S DEGREE PROGRAM (PHOTOGRAMMETRY) BY THESIS

Qtr	Geodetic Science Course No.	Name of Course	Credit Hours
Autumn	508	Fundamentals of Geodetic Surveying	5
	505	Photogrammetry and Photointerpretation	4
	645	Applied Math Methods, G.S.I.	3
	650	Adjustment Computations I	3
Winter	646	Applied Math Methods, G.S.II	3
	660	Geometric Photogrammetry	3
	664	Geodetic Astronomy	3
		Electives	
Spring	613	Introduction to Advanced Geodesy	5
	636	Mathematical Cartography	4
	778	Analogy Photogrammetry	5
Summer	998	Thesis	
		Field Courses	
		Math Requirement	10*
Autumn	637	Introduction to Advanced Cartography	5
	779	Computational Photogrammetry	4

* Math. At least 10 hours of graduate level mathematics, which may include that taken in Geodetic Science 645 and 646, are required. This requirement may be fulfilled any quarter.

Thirty-eight universities in the United States and six in Canada have or plan to initiate majors, minors, or areas of specialization in remote sensing, photogrammetry, or astrogeology (planetary geology). Several universities that offer more than one course in remote sensing/photogrammetry do not provide degree programs in these areas. Colorado State University, for instance, has at least 13 undergraduate and graduate courses in remote sensing and photogrammetry but does not offer a degree in either field, whereas the University of Miami, with only one remote sensing course, offers a minor and Ph.D. in remote sensing.

An example of the curriculum required for a Master's degree in photogrammetry is included as Table 18-2, taken from the pamphlet Curriculum Information, the Department of Geodetic Science, The Ohio State University.

There are no known programs in the United States or Canada similar to the South Australian Institute of Technology's graduate diploma in remote sensing, a two-year part-time graduate program that teaches remote sensing to professionals within the framework of their discipline. The program contains six courses: remote sensing I & II, applied interpretation I & II, and field assessment A & B. The first year of the program is concerned with the physical, environmental, and human factors of remote sensing data acquisition and interpretation, and the interpretation of visual imagery. The second year covers non-photographic remote sensing techniques

and the analysis of digital data. A good place for programs of this kind is the junior college, an excellent facility for training remote sensing and photogrammetric technicians.

Two American universities have developed innovative teaching techniques. Colorado State University videotapes its photogrammetry classes. The tapes are used by nonresident students at 21 cooperating institutions and seven county libraries in Colorado and Wyoming. And Oregon State University's School of Forestry has developed a self-instruction approach to aerial photo-interpretation instruction. This course is "self-paced and is built around the unit mastery concept." The student must obtain a "B" (80 percent) in each unit, and may retake an exam twice. In addition to the unit exams, two midterms, a final, a photo-mission report, and a landform map report are included in the grading scheme. The faculty at the University believes that this approach produces:

(1) An increased mastery and longer retention of material over the lecture-lab approach;
(2) A higher percentage of A's, B's, and C's, and fewer D's and F's;
(3) More highly motivated students and greater student satisfaction, and;
(4) More material covered in the same amount of time.

Classroom lectures of the various institutions are reinforced and supplemented by the use of readings in at least 64 textbooks. The most widely used remote sensing/photointerpretation text is T. Eugene Avery's *Interpretation of Aerial Photographs* (1968). When included in the category of remote sensing, it is used in 39 percent of the courses. As a text on photointerpretation, it is used for 50 percent of the courses. The next most used text on photointerpretation is U.S. Geological Survey Professional Paper No. 373 by Richard G. Ray (1960), Aerial photographs in geologic interpretation and mapping. The photogrammetry text most widely used is Paul Wolf's *Elements of Photogrammetry* (1974).

Many instructors find no single text satisfactory for all their needs, and consequently employ two required texts. Several instructors utilize only readings in various journals, such as *Photogrammetric Engineering and Remote Sensing*, symposia proceedings, and textbooks.

The MANUAL OF REMOTE SENSING (American Society of Photogrammetry, 1975) has been used at several institutions. The cost, size (two volumes), and complexity of this work will probably preclude its becoming the leading remote-sensing text in the United States, but it will continue to be used extensively as a reference book and for additional reading assignments by virtue of its excellent technical papers by leading researchers. Several remote-sensing texts are being prepared for publication.

Visual aids are available in formats that provide the instructor with selected 35-mm slides of satellite, aircraft, ground, and microscope data of various areas from several sensors and involving many scientific problems. Facilities where slides can be obtained without permission of the author or the U.S. Geological Survey include: Pilot Rock Inc., Arcata, California; the EROS Data Center USGS, Sioux Falls, South Dakota; the Technology Applications Center, University of New Mexico, Albuquerque, New Mexico; John Wiley & Sons (slides by Norman Gillmeister and Barry Siegal); McGraw-Hill (slides by John S. Shelton); and Purdue University, Laboratory for the Applications of Remote Sensing, West Lafayette, Indiana.

A wide range of remote sensing and photogrammetry equipment, from pocket stereoscopes to analytical stereoplotters, is available to students at institutions in the United States and Canada, and a few schools utilize their own aircraft to acquire specialized data. In addition to internal resources, several institutions maintain a close working relation with Federal, State, and commercial agencies. Only one formal internship was found in the survey, an arrangement of South Dakota State University with the U.S. Geological Survey EROS Data Facility, Sioux Falls, South Dakota.

The American Society for Engineering Education (ASEE) annually publishes a summary and analysis of engineering research and graduate study activities of the 195 ASEE member institutions in its journal, *Engineering Education*. The list does not represent all engineering research projects, since all institutions are not members, and all of those surveyed do not subdivide their projects into specific disciplines such as remote sensing and photogrammetry. Many of the subdivisions have peripheral applications to remote sensing. Readers interested in specialized areas are referred to the journal of the American Society for Engineering Education and to the various engineering departments.

The ASEE indicated that remote sensing/ photogrammetry engineering research projects were underway at at least 27 institutions in the 1973-74 school year (Engineering Education, 1974) and 31 institutions in the 1974-1975 school year (Engineering Education, 1975). There were more than 122 research projects in the 1973-1974 time period, and 110 in the 1974-75 time period.

18.3.3 CONTINUING EDUCATION: SYMPOSIA, CONFERENCES, WORKSHOPS, SHORT COURSES, AND TECHNICAL SESSIONS

18.3.3.1 INTRODUCTION

The American Society of Photogrammetry, in itself, has had a firm hand in improving the state-of-the-art in photogrammetry and remote sensing. The large number of concurrent technical sessions held at the annual and semi-annual ASP-ACSM meetings with pre-printed proceedings are ample demonstration. The Society has been a sponsor or co-sponsor of numerous meetings, symposia, and training sessions during the last several years. Proceedings of such sessions dot the personal bookshelves of photogrammetric professionals in this country. They offer excellent source material for generally keeping abreast of the field or for specifically penetrating a specialized subject. In addition, ASP members form the nucleus of a group of professionals who are striving to spread the word about the field in their separate professions and disciplines.

18.3.3.2 FORMS OF CONTINUING EDUCATION

Specialized training in photogrammetry and remote sensing takes many forms, from the formalized work for which special credit is given, to the presentation of new techniques and methods before technical sessions of learned and professional societies. Since there are such varied ways to reach the goal of adding to the practitioner's storehouse of knowledge, only a general discussion will occur in this section.

The American Society of Photogrammetry and other technical societies have taken the

---

NOTE: Research projects of an applications nature are not included in this paper; only basic engineering research projects are listed.

forefront in expanding the knowledge of the field of photo measurements and image interpretation. The layman who might see a reference to a technical symposium in photogrammetry or remote sensing is amazed by the variety of subjects. During 1978 the pages of *Photogrammetric Engineering and Remote Sensing* mentioned the following topics for specialized symposia and conferences:

Implementation of a Modern Multipurpose Land Data System

Remote Sensing (Canada)

Interaction of Marine Geodesy and Ocean Dynamics

Applications of Human Biostreometrics (NATO)

Computer Mapping Software and Data Bases

Digital Terrain Models

Measurement, Mapping and Management in the Gulf Coastal Zone

Coastal Mapping

Geological Applications of Satellite Remote Sensing (Canada)

Application of Remote Sensing Data to Wildlife Management

Large Area Crop Inventory Experiments

First Landsat Conference (Australasian)

Thermosense I: Conference on thermal infrared sensing technology in Energy Conservation Programs

The following short courses or workshops were announced in the journal:

Aerial Photography/Aerial Photo Interpretation

Advanced Topics in the Analysis of Remote Sensing Data

Remote Sensing Course (Alberta)

Infrared Technology/Advanced Infrared Technology

Applications of Remote Sensing Data

Optical Science and Engineering Short Course

Color Aerial Photography in the Plant Sciences

Introduction to Renewable Resource Inventory Methods

Numerous other symposia, conferences, short courses, and workshops were offered through direct mailing exclusive of being announced in *Photogrammetric Engineering and Remote Sensing*. Some of these were:

Vegetation Remote Sensing Workshops

Design of Optical Systems

Remote Sensing Applications in Resource Management

Symposium on Remote Sensing for Vegetation Damage Assessment

Symposium and Workshops on Photogrammetry

Digital Image Processing of Earth Observation Sensor Data Short Course

Remote Sensing and Digital Information Extraction Short Course

Auto-Carto IV: Cartography and Computing Applications in Health and Environment

Remote Sensing Technology & Applications Short Course

Machine Processing of Remotely Sensed Data Symposium

Photogrammetry Correspondence Courses

Learning to Use Our Environment Conference

The formats are varied. Some of the offerings are a continuing series; others are a one-shot effort. Some of the offerings are for advanced workers, some are for beginners, and some accept all levels of expertise.

Most symposia and conferences offer a variety of papers and poster sessions. The typical program offers a mixture of invited and contributed papers. Usually, several key speakers are invited to set the tone for the meeting, and the other positions are filled by the most promising presentations based on pre-submitted abstracts and paper outlines. Quite often, the subject matter for symposia and conferences are chosen in advance, with session titles named in the "call for papers."

Workshops are intended to allow the participant to "roll up his sleeves" and become involved in learning several accepted techniques for applying photogrammetry and remote sensing in his discipline. Short courses are even more helpful to the participant, who is led through carefully planned exercises and can check results with the "textbook" solution and discuss it with the instructor.

A phenomenon of recent times has developed in professional societies (other than photogrammetric) whose members employ photogrammetry and remote sensing in their work. These people organize special sessions at the regular meetings of these societies which offer a program of interest to practitioners and students of photogrammetry.

18.3.3.3 ARRANGEMENTS

With so many continuing education efforts taking place, there are many "old pros" in the continuing education system. For them, getting the germ of an idea for a special meeting, seeking out the proper sponsoring organizations, naming a program committee, choosing the speakers and lecturers, selecting the proper date and location, preparing all arrangements for the affair, and finally producing the proceedings, has become a routine operation. For others, with an excellent idea for a meeting theme, but with little experience in getting such a job done, a few notes or a check list on arranging such a meeting or short course seems in order.

(a) Setting the exact date: This should depend on what will be a convenient time for the prospective participants; what will be a good time for obtaining the necessary facilities; and what will be convenient for the prospective lecturers or staff.

(b) Choosing subjects for the lectures: these should depend upon the nature and academic backgrounds of the individuals to whom the course is primarily directed.

(c) Selecting the staff for a short course: the staff may be chosen to some extent based on the subjects selected. Early selection is especially important if some of the staff members are to be recruited from another campus or from industry.

(d) Arranging for hotel accommodations or dormitory space and meals if these are to be provided by an institution: hotel space in busy cities may need to be booked over a year in advance.

(e) Printing and mailing advance notices and programs: this will require preparation of a suitable mailing list to reach the people for whom the course is intended. Considerable advance notice must be furnished if a notice is to be printed in *Photogrammetric Engineering and Remote Sensing.*

(f) Arranging meeting rooms or classroom space with due regard for any visual-aid material and equipment that may be required by the speakers: Be sure to have a room which can be *darkened* if projection equipment is needed.

(g) Arranging for equipment, workshop materials, or any other items that may be needed.

(h) Arranging for coffee-break refreshments, and a banquet if one is to be held in connection with the program.

(i) Arranging for individual photographs for speakers, or group photographs of participants and staff.

(j) Financial arrangements: Will the course be self-supporting from participants? Can it be partially or wholly subsidized? What are the expenses of the speakers? What other expenses must be met?

(k) Printing of final programs, proceedings and other materials.

18.3.3.4 Short Course Topics

The following are some suggestions for topics that might be appropriate for a short course in photogrammetry. Articles in this Manual should serve to stimulate ideas for suitable topics.

History of photogrammetry
Fundamentals of photogrammetry
Fundamentals of photointerpretation
Aerial cameras and flight planning
Economics of photogrammetry
Stereoplotting equipment
Aerotriangulation
Analytical photogrammetry

A recent course in aerial photointerpretation for natural resource land managers had the following topics:

Stereoscopy
Preparing photos for viewing
Principles of photo interpretation
Color oblique stereo aerial photos
Maps and aerial photo orientation
Scale measurement
Height measurement on photos
Interpretation of photo detail
Landform analysis
Vegetation interpretation
Transfer of photo detail
Area measurement
Color and color infrared aerial photography
Slope measurement and road layout

Multistage sampling
Remote sensing systems
Satellite systems

A course in machine processing of remotely sensed data covered the following broad areas:

Digital representation and understanding of remotely sensed scenes
Utilization of digitally processed earth resource data
Extraction of information primarily from digital remotely sensed earth resource data

Examples of special topical areas in the course were (1) new results in classifier training strategies, and (2) precision rectification of multi-spectral scanner data.

These examples are merely presented to demonstrate the wide range of topics which can be covered in short courses in photogrammetry, image interpretation, and remote sensing.

18.3.3.5 Evening Extension Courses

Evening extension courses in photogrammetry are designed particularly for those who are engaged in the practice of surveying or mapping and who want a knowledge of the basic principles of photogrammetry for application to their surveying and engineering problems. (Similar illustrations could be made for evening extension courses in image interpretation and remote sensing.) Essentially these courses are similar to college courses in photogrammetry, cover about the same material in the same length of time, and are usually taught by a regular college faculty member. Because they take place in the evening, however, the sessions usually are held once a week for a 3-hour period.

Another difference between extension courses and regular college courses is in the usual seriousness and maturity of the extension course students, who are studying for a definite purpose, and are probably realizing immediate benefits in their daily work. Because of the similarity in content of the extension course to the college course for undergraduate students, it is not considered necessary to present here a detailed outline of the subjects to be covered. The following list of suggested topics may be modified or augmented according to the needs of the community:

Uses and interpretation of photographs
Aerial cameras and flight planning
Flight height and scale of photographs
Relief displacements
Tilt
Radial-line plotting
Planimetric mapping
Stereoscopy and parallax
Mosaics
Stereoplotting instruments
Ground control

18.4 *Professional Status*

Photogrammetry as an increasingly highly scientific and technical profession has grown dramatically during the past decade. From the early days of taking pictures and measuring them, through large scale topographic mapping, to use of sophisticated methods in the space program, the professional applications of photogrammetry continue on the move to greater heights. The possibilities for future applications and techniques remain virtually unlimited and are a fertile area for work by young scientifically trained individuals who wish to bring their initiative to bear on helping solve the space, land use, and environmental problems of mankind.

The professional status of photogrammetry and surveying and mapping within civil engineering was first and meaningfully addressed by a special *ad hoc* committee of the American Society of Civil Engineers. In February of 1959, the ASCE Board of Direction issued a policy statement, part of which follows:

"The American Society of Civil Engineers, on the basis of thorough studies carried out by a Task Committee on the Status of Surveying and Mapping, declares that the following four major categories in the field of activity commonly designated as surveying and mapping are a part of the Civil Engineering profession:

    I. Land Surveying
   II. Engineering Surveying
  III. Geodetic Surveying
   IV. Cartographic Surveying"

The ASCE Board of Direction also authorized the publication of the Committee report as a supplement to the ASCE Manual and Report on Engineering Practice No. 45, "Consulting Engineering—A Guide for the Engagement of Engineering Services." This supplement is further identified as "Supplement 45 A, Professional Practice of Surveying and Mapping Within Civil Engineering." Photogrammetry, the principal scientific and engineering technique used in achieving surveying and mapping results, was thereby apparently assured a recognized professional status within Civil Engineering. The American Society of Photogrammetry and the American Congress on Surveying and Mapping have both taken the same position as highly respected members of the scientific and engineering community.

Subsequent to this action by the American Society of Civil Engineers, the practice of photogrammetry proceeded on an increasing professional basis for several years. However, there then began to be increasing instances of a return to the competitive bidding, non-professional approach to offering and requesting photogrammetric service. Much confusion about this entire matter has resulted from the recently rendered Supreme Court decision on the National Society of Professional Engineers case with the United States. Department of Justice on competitive bidding provisions in the NSPE Code of Ethics. As a result of this Supreme Court decision, all technical societies have removed from their codes any mention of prohibiting or disciplining its members if they participate in competitive bidding. This *does not mean*, however, that competitive bidding is the required method of requesting or furnishing services, though unfortunately it is being interpreted in that way very widely today. Accordingly, a major effort remains ahead of re-education about the reasons for a full return, at every level, to a professional, non-competitive bidding approach to requesting and furnishing service, to best insure the expected results that all clients for photogrammetric service deserve to receive.

In the early 1970's, the Board of Direction of the American Society of Photogrammetry established its Professional Activities Committee. Traditionally this committee has been comprised of members, most of whom are Registered Engineers and/or Surveyors, and who also are representative of the various functional areas of the overall American Society of Photogrammetry membership. This includes vocational involvement from such sectors as education, private practice, government service and industrial-manufacturing and suppliers. The goals of the Professional Activities Committee were numerous, and several years work followed leading to achievement of an American Society of Photogrammetry Code of Ethics, and the American Society of Photogrammetry Classification Chart for Professional Aspects of Photogrammetry.

18.4.1 CODE OF ETHICS OF THE AMERICAN SOCIETY OF PHOTOGRAMMETRY

Honesty, justice, and courtesy form a moral philosophy which, associated with mutual interest among men, should be the principles on which ethics are founded.

Each person who is engaged in the use, development, and improvement of photogrammetry should accept those principles as a set of dynamic guides for his conduct and his way of life rather than merely for passive observance. It is his inherent obligation to apply himself in his profession with all diligence, and in so doing to be guided by this Code of Ethics.

Accordingly, each person in the photogrammetric profession shall have full regard for achieving excellence in the practice of his profession and for the essentiality of maintaining the highest standards of ethical conduct in his responsibilities and in his work for his employer, his clients, his associates, and society at large, and shall. . . .

1. Be guided in all of his professional activities by the highest standards, and be a faithful trustee or agent in all matters for each client or employer.
2. At all times function in such a manner as will bring credit and dignity to the professions in photogrammetry.
3. Not compete unfairly with anyone who is engaged in the photogrammetric profession by:
    a. advertising in a self-laudatory manner,
    b. monetarily exploiting his or another's employment position,
    c. publicly criticizing other persons working in or having an interest in photogrammetry, and
    d. exercising undue influence or pressure, or soliciting favors through offering monetary inducements.
4. Work to strengthen the profession of photogrammetry by:
    a. personal effort directed toward improving his personal skills and knowledge,
    b. interchange of information and experience with other persons interested in and using photogrammetry, with other professions, and with students and the public,
    c. seeking to provide opportunities for professional development and advancement of persons working under his supervision, and
    d. promoting the principle of appropriate compensation for work done by persons in his employ.
5. Undertake only those assignments in the use of photogrammetry for which he is qualified by education, training, and experience, and employ or advise the employment of experts and specialists whenever and wherever his clients or his employer's interests will be best served thereby.
6. Give appropriate credit to other persons and/or firms for their professional contributions.
7. Recognize the proprietary interests and rights of others.

18.4.2 PROFESSIONAL ASPECTS OF PHOTOGRAMMETRY

The American Society of Photogrammetry accepts without change the definitions used in the ASCE report for professional level and technician level work, as follows:

Professional Level: Work that involves the exercise of professional judgment, frequently based on knowledge acquired through higher learning, generally non-routine in character. The term implies one who can plan, perform, and/or direct all such operations in the category; this person is responsible for work performed by those under him.

Technician Level: Work that is primarily routine, of a technical nature, often demanding a high degree of skill, done under the direction of a professional person who is responsible for its outcome. Such work is preprofessional when performed by a professional trainee who, having completed courses of specialized intellectual instruction and study, is seeking to attain professional status.

Journal of the Surveying and Mapping Division ASCE, ASCE Proceedings dated Sept., 1950 (NO. 2166).

In the classification chart that follows, work listed under the heading of Technician Level also includes work that is preprofessional as defined above.

In the classification chart, several professional-level activities are listed, such as geographer, geologist, forester, and archaeologist, in which it is intended to connote that photogrammetry is used in this particular activity in a professional manner by a professional man. (The occupations mentioned are examples only and the list is not to be considered as comprehensive.) A practitioner in one of these disciplines may acquire professional competence in photogrammetry, and only when he possesses this competence is his use of photogrammetry to be construed as professional.

CLASSIFICATION CHART FOR PHOTOGRAMMETRY

I.   EDUCATION IN PHOTOGRAMMETRY
    A. Administration of instruction in Photogrammetry
    B. Undergraduate teaching
    C. Graduate teaching
    D. Technical writing
    Professional Level: Dean, department chairman, professor, technical writer.
    Technician Level: Teaching Assistant

II.   RESEARCH AND DEVELOPMENT
    A. Materials: Photographic emulsions and bases, chemicals, and drawing and reproduction materials.
    B. Instruments and Equipment: Lenses, cameras, sensors, platforms, rectifiers, enlargers, printers, measuring and plotting instruments, automation hardward and calibration devices.
    C. Systems: Mapping systems, photographic and other image interpretation systems. (Development of an integrated series of functions and techniques to produce a given result using photogrammetric principles.)
    D. Investigations and Research: Operations research, concept determinations, cost effectiveness studies, techniques studies and investigations.
    Professional Level: Research chemist, research physicist, research engineer, technical writer, cartographer, mathematician, electro-optical systems design engineer.
    Technician Level: Laboratory or shop assistant, test technician.

III. MANUFACTURE
    A. Materials: Photographic emulsions and bases, chemicals, and drawing and reproduction materials.
    B. Instruments and Equipment: Lenses, cameras, camera and other sensor platforms; rectifiers, enlargers, printers; sensor systems; viewing, measuring and plotting instruments; automation hardware and calibration devices.
    Professional Level: Manufacturing Engineer, quality control engineer, electro-optical

systems engineer, physicist, chemical engineer.

Technician Level: Shop technician, draftsman.

IV.   PHOTOGRAPHY (Includes aerial, terrestrial, underwater and space photography)

A. Technical Planning: Flight or exposure station parameters; photography specifications; camera calibration.

B. Procurement and Inspection: Technical negotiations; technical administration of contracts; inspection and acceptance.

C. Photographic Mission: Operation of camera-bearing vehicles; maintenance and operation of cameras; flight or course navigation.

D. Photographic Processing: Development, inspection and reflight requirements.

Professional Level: Planning engineer, aerospace engineer, photographic scientist, photographic engineer.

Technician Level: Draftsman, inspector, photographer, laboratory technician.

V.   REMOTE SENSING AND INTERPRETATION

A. Instrumentation selection and operational planning: Instrument carrying vehicles; space platforms; radar and thermal infrared sensors; scintillometers, radiometers, magnetometers; multi and special sensor combinations; viewing and scanning equipment; image enhancement and image data processing systems; operation and maintenance of sensor systems; preparation of imagery for end use.

B. Interpretation for general purposes and mapping: Conventional mensuration and interpretation of photographic and other imagery, pattern recognition; reporting and documenting results.

C. Interpretation for specific purposes and disciplines: geology, forestry, agriculture, land use, archaeology, water resources, meterology, mineral and aggregate resources, urban planning, industrial development, transportation facilities, volcanic and earthquake surveys and investigations, environmental and pollution surveys; reporting and documenting results.

D. Military Intelligence

Professional Level: Cartographer, electro-optical systems engineer, geologist, forester, archaeologist, hydrologist, planner, engineer, agronomist, soil scientist, materials engineer, resources scientist and engineer, earth scientist, environmentalist, analyst, etc.

Technician Level: Interpretation technician, laboratory technician, image analyst, draftsman.

VI.   CADASTRAL SURVEYS

A. Property and boundary surveys

B. Subdivision surveys and plats

C. Public land surveys

D. Surveys for plans and plats: Architectural (building sites); tax maps.

Professional Level: Land surveyor, cadastral engineer, geodetic engineer, engineer of surveys.

Technician Level: Stereoscopic instrument or plotter operator, operator of other photogrammetric equipment, laboratory technician, computations technician, draftsman, field survey assistant (instrument, tape, rod).

VII.   ENGINEERING SURVEYS FOR LOCATION DESIGN AND CONSTRUCTION

A. Location and design data surveys: Control, basic and supplemental (horizontal and vertical); culture and topography; profile and cross sections; measurement and digitization of topography and other vital detail.

B. Construction Surveys: Location surveys and staking on the ground of the designed facilities and/or structures; quantity and measurement surveys; "as built" surveys; utility surveys; surface mines.

Professional Level: Engineer, survey engineer, land surveyor.

Technician Level: Stereoscopic instrument or plotter operator, operator of other photogrammetric equipment, computations technician, draftsman, field survey assistant (instrument, tape, rod).

VIII.   TOPOGRAPHIC AND PLANIMETRIC MAPPING

A. Project Planning

B. Control, basic and supplemental (horizontal and vertical): Analog procedures, analytical methods, instrumental methods; field measurement methods.

C. Map compilation: Orientation of plotting instruments, delineation of planimetry and contours; measurement of spot elevations, profile and cross sections, and other terrain data; identification and annotation of principal topographic features and cultural details.

D. Field edit and completion surveys.

Professional Level: Planning engineer, topographic engineer, production engineer, mathematician, cartographer.

Technician Level: Stereoscopic instrument or plotter operator, operator of other photogrammetric equipment, laboratory technician, computations technician, draftsman, field survey assistant (instrument, tape, rod).

IX.   SPACE SURVEYS

A. Geodetic surveys: Figure of the earth and control extension from satellite triangulation, from ballistic camera photography of earth satellites, and from earth-satellite-borne synoptic photography; documentation of results.

B. Lunar surveys: Surveying and mapping of lunar surface using data derived from lunar photography and probes; documentation of data.

C. Planetary surveys: Surveying and map-

Activity for which either photogrammetric procedures, field survey procedures or a combination of the two may be applicable.

Operation of certain highly sophisticated equipment may in some instances be considered as professional-level activity.

ping of other planets using data derived from space photography and space probes; reporting and/or documenting data.

Professional Level: Geodetic engineer, geodesist, space scientist, topographic engineer, mathematician, cartographer.

Technician Level: Stereoscopic instrument or plotter operator, or operator of other photogrammetric equipment, laboratory technician, computations technician, draftsman.

X. SPECIAL APPLICATIONS

A. Photographic maps: photographic maps with contours, orthophotographic maps, photographic mosaics.

B. Operations and maintenance: Surveillance; elimination of hazards; condition and inventory surveys; quantitative measurements and evaluation.

C. Close Range Photogrammetry: Architecture; biomedical applications including conventional and x-ray stereophotogrammetry, photogrammetric solution of biological problems; hydrography, structural engineering; oceanography, geography; police work including crime detection, traffic and accident surveys.

D. Lasers and holography.

Professional Level: Cartographer, physicist, physician, dentist, archaeologist, geographer, oceanographer, architect, engineer, hydrographer.

Technician Level: Stereoscopic instrument or plotter operator, operator of other photogrammetric equipment, laboratory technician, draftsman.

Subsequent to these accomplishments, the Professional Activities Committee then focused its attention more directly on the matter of professional registration for practicing photogrammetrists. Registration as professionals in engineering, surveying, or photogrammetry of course must involve the legal and administrative control and licensing of the professional by the 50 respective states. It was determined early that resolution of this question by all the 50 states could require a lengthy process of negotiation and effort, since the rules and regulations, as well as practices, vary so widely from state to state. The Professional Activities Committee, as an interim effort to further the stature and professional status of photogrammetry, formulated, and the American Society of Photogrammetry Board of Direction approved, the institution of the American Society of Photogrammetry Voluntary Certification Program, outlined below.

18.4.3 ASP VOLUNTARY CERTIFICATION PROGRAM

18.4.3.1 GENERAL INFORMATION

The Articles of Incorporation of the American Society of Photogrammetry state that it will exert its influence toward the betterment of standards and ethics.

To this end, the Society's Professional Activities Committee, after several years of careful study, developed a program for the certification of photogrammetrists. This program was approved by the ASP Board of Direction at its meeting in Washington, D.C. on March 13, 1975.

In simplest terms, certification is official recognition by one's colleagues and peers that he has truly demonstrated professional integrity and competence. As such, the ASP Voluntary Certification Program complements, but is not a substitute for, registration which is a legal act on the part of the several states to protect the life, health, and property of their people.

The program as approved is entirely voluntary. It applies equally to persons associated with the several subdivisions of photogrammetry, which, by Society definition, include aerial photography, photogrammetric surveys, remote sensing, and photographic interpretation. In accordance with the Society's Code of Ethics, persons certified must, however, refuse to undertake any work within or related to the field of photogrammetry that is outside their range of competence.

Information on how to apply for certification, basic requirements, educational credits, and administrative review procedures are outlined on the pages that follow.

18.4.3.2 PURPOSE AND OBJECTIVES

A growing number of scientific and technical disciplines depend on photogrammetry for reliable measurements and information. It is in the interest of those who provide photogrammetric services, as well as of the users of those services, that such information and data be accurate and dependable.

The ASP Certification Program has as its purpose the establishment and maintenance of high standards of ethical conduct and professional performance among photogrammetrists.

The primary objectives of the program are:

1. to identify and recognize those persons who, after careful and just appraisal by their peers, are considered to have met the requirements established by the Society for certification;
2. to provide a basis for weighing the validity of allegations and complaints that involve practicing photogrammetrists, and for taking appropriate action in connection therewith; and
3. to encourage persons as yet not fully qualified to work toward certification as a goal of professional achievement.

18.4.3.3 ELIGIBILITY

The ASP Certification Program is open to all persons. It is not necessary that an applicant be a member of the American Society of Photogrammetry.

The program is applicable to persons associated with one or more of the functional areas of photogrammetry. Work at the professional and technician levels in each of these areas is outlined in the "Classification Chart for Photogrammetry."

18.4.3.4 BASIC REQUIREMENTS

In view of the wide variety of skills and disciplines studied and practiced by photogrammetrists, the ASP Certification Program is based primarily on evidence of demonstrated capability. Within prescribed limits it permits experience to be substituted for education and vice-versa. Basic requirements for certification are:

1. nine years of experience in photogrammetry, five years of which were in a position of professional responsibility demonstrating professional knowledge and competence;
2. the names of four persons who are in or who have held responsible positions in public or private practice, who have first-hand knowledge of the applicant's professional and personal qualifications;
3. declaration of compliance with the CODE OF ETHICS of the American Society of Photogrammetry; and
4. successful completion of an oral or written examination, if and when required by the Evaluation Committee.

Eminent members, who have been previously recognized for outstanding achievement in or major contributions to the field of photogrammetry, may be nominated by the President of ASP and awarded certification if approved by the Society's Board of Direction.

By July 1976, the first group of applicants to be "*certified* photogrammetrists," 50 in number, were reviewed and approved by the Evaluation Committee. By the end of 1978, the number of certified photogrammetrists had increased to 160. At present, about 800 individuals have inquired about and shown interest in the "Certification Program."

Following the general decline of recognition of the professional nature of photogrammetric activity in the late 1960's and early 1970's, the Boards of Direction of the American Society of Photogrammetry, American Congress on Surveying and Mapping, American Society of Civil Engineers, American Consulting Engineers Council, and National Society of Professional Engineers appointed an *ad hoc* committee to study the entire problem and prepare a report and policy statement concerning the professional nature of photogrammetry. Since the issue of competitive bidding bears so heavily on any discussion of professionalism, this effort by the special *ad hoc* committee was delayed, awaiting again the recent Supreme Court decision on the National Society of Professional Engineers case.

Now, hopefully, this effort will proceed, and the efforts of the American Society of Photogrammetry Professional Activities Committee can also proceed to the determination of the position of the photogrammetrist in the practice of the profession in the 50 states. Overcoming recent setbacks in the recognition of professionalism in photogrammetry now appears to involve a lengthy and formidable struggle for the future. This will have to involve a much larger group effort than that of a small Professional Activities Committee of the American Society of Photogrammetry. It will require effort by members of all the technical societies mentioned above, as well as increased effort in the university level education programs in science and engineering. In addition, a major effort will be required to clearly inform the members of the registration boards of the various states of the important place of top level professional photogrammetric practice in the future protection of the public health, safety and welfare.

18.4.4 PROFESSIONAL OPPORTUNITIES

Opportunities for employment in the field of photogrammetry, both in government and in private practice, have increased greatly over the last decade. The Federal government is probably the largest employer of photogrammetrists in the United States. Considerable numbers of such personnel are employed by the Defense Mapping Agency, the U.S. Geology Survey of the Department of the Interior, the National Geodetic Survey of the Department of Commerce, the Navy Oceanographic Office, several agencies within the Department of Agriculture, the Central Intelligence Agency, the Bureau of Land Management of the Department of the Interior, the National Aeronautics and Space Administration, and many other agencies, particularly in the Departments of the Army, the Navy and the Air Force. It is also significant to note that an increasing number of photogrammetrists are in demand within agencies involved with the space program. Great reliance is placed upon very sophisticated photogrammetric systems in the area of missile and spacecraft tracking and alignment, and thus an entire new field has been opened to photogrammetrists. The services of photogrammetrists are also being increasingly called upon in the area of determining precise geodetic locations on the surface of the earth through observations of manmade earth satellites.

Some of the most interesting and challenging positions in the field of photogrammetry lie in private practice of the profession. There is a large and vigorous practice throughout the United States, and professional and subprofessional personnel are in constant demand. Of particularly high need are experienced technicians capable of production work in field survey, stereo-compilation, drafting-scribing, and photographic laboratory operations. Privately practicing companies range in size from small, employing as few as six or eight individuals, to large, employing several hundred employees. Many architectural, engineering, and planning consultants have also established photogrammetric and remote sensing departments. These firms are located in all parts of the country, but the heavier concentrations are in the major population centers. The types of projects undertaken by the typical firm are varied. It is often said that no two projects are similar; and each new project involves differences in planning and execution. This variety of work presents a continual challenge, and the interest level of those requiring and furnishing the service remains high. Representative projects involve preliminary and design mapping for highway and railroad location, cross-section work for earth quantity determination, plant site mapping, pipeline and transmission line location, dam and reservoir work, stockpile inventories, municipal mapping, sewer and drainage work, tax mapping, airfield design, and many other projects in which the horizontal and vertical positions of points on the surface of the earth must be inventoried and documented with great precision. Much work is also being done in the field of non-topographic and close-range photogrammetry in areas such as X-ray photogrammetry and other medical applications; and traffic studies, architectural, and industrial photogrammetry in which objects are precisely measured. In many instances photogrammetric techniques have been successfully used to precisely reconstruct partially destroyed objects, such as landmarks and historical buildings. Computational, automated, and remotely controlled photogrammetric techniques are coming of age, and the new field of interactive graphics will have an important place in future applications of photogrammetry.

Other employment opportunities in photogrammetry lie in state governments and in the field of education. College level education in photogrammetry and survey engineering must be expanded as the demand continues for completely trained professional entry-level individuals. Members of the academic community must also make their awareness of the importance of trained professionals clear to the various State Registration Boards, as many of these currently do not recognize photogrammetry as a professional part of either surveying or engineering, but rather consider it a sub-professional service that is used from time to time by surveyors, engineers, and other professionals.

Research in photogrammetry continues in certain colleges and universities, and much valuable work is resulting from this effort.

In summary, it can be stated that photogrammetry in the United States continues to become increasingly important as a factor in the future well-being of the public. Since its beginnings in this country a few short decades ago, the profession has made tremendous advances in both equipment and technique, and it is now a major asset in solving many engineering, scientific and technical problems. Its future is indeed bright and practically unlimited.

REFERENCES

American Society for Engineering Education, "Engineering College Research and Graduate Study Survey," *Engineering Education*, Vol. 65, No. 6, March 1975.

American Society for Engineering Education, "Engineering College Research and Graduate Study Survey," *Engineering Education*, Vol. 66, No. 6, March 1976.

American Society of Photogrammetry, *Manual of Photogrammetry* 3rd edition, Chapter 23.

American Society of Photogrammetry, *Voluntary Certification Program*.

Bidwell, Timothy G., "College and University Sources of Remote Sensing Information," *Photogrammetric Engineering and Remote Sensing*, Vol. 41, No. 10, October 1975, pp. 1273–1284.

Defense Mapping School, *Catalog and Course Descriptions*, Defense Mapping Agency, Department of Defense, July 1975.

Eitel, Dean F., "Remote Sensing Education in the U.S.A.," *Photogrammetric Engineering*, Vol. 38, No. 9, September 1972, pp. 900–906.

Latham, James P., "Perspective on Education in Photogrammetry and Remote Sensing," *Photogrammetric Engineering and Remote Sensing*, Vol. 43, No. 3, March 1977, pp. 257–258.

Morain, S. A., *Geographic Education in Remote Sensing at the University Level*, Special Report for the American Association of Geographers Committee on Remote Sensing, March 1975.

Nealey, L. David, "Remote Sensing/Photogrammetry Education in the United States and Canada," *Photogrammetric Engineering and Remote Sensing*, Vol. 43, No.3, March 1977, pp. 259–284.

Reeves, Robert G., "Education and Training in Remote Sensing," *Photogrammetric Engineering*, Vol. 40, No. 6, June 1974, pp. 691–696.

Stanton, B. T., "Education in Photogrammetry," *Photogrammetric Engineering*, Vol. 37, No. 3, March 1971, pp. 293–303.

The Department of Geodetic Science, The Ohio State University Curriculum Information, April 1975.

CHAPTER XIX

Definitions of Terms and Symbols Used in Photogrammetry

Author-Editor: PAUL R. WOLF
Contribution Authors: STEVEN D. JOHNSON, TERRENCE J. KEATING, WALLACE E. KERR,
DAVID F. MEZERA, LEONARD T. PIMENTEL, FREDERICK A. SIEKER

19.1 Introduction

THE following represents an update and expansion of the definitions of terms and symbols as published in the Third Edition of the MANUAL OF PHOTOGRAMMETRY prepared by the Committee on Nomenclature of the American Society of Photogrammetry.

Corrections and additions to the list of terms are based on the suggestions received from photogrammetrists world-wide, and on studies by the committee of those changes in usage which seemed warranted by progress in photogrammetry. This list includes approximately 1500 photogrammetry terms. No attempt has been made to include terms which are primarily remote sensing, since these terms are defined in the glossary of the American Society of Photogrammetry's MANUAL OF REMOTE SENSING.

In the field of photogrammetry, research and development are pursued continually. New techniques and procedures introduce new terms and result in different connotations for old terms. Thus, the task of the Nomenclature Committee is a continuing one which requires the assistance of all photogrammetrists. Their corrections, comments, and suggestions are invited.

As with the 3rd Edition of the MANUAL OF PHOTOGRAMMETRY, boldface and italic print are used frequently. Boldface italic type is used for terms being defined, or to cross-reference terms. Italic type is used to emphasize significant terms in a definition, or in lieu of quotation marks.

19.2 Acknowledgements

The compilation of the definitions given in section 19-3 represents an activity of ASP's Nomenclature Committee which has spanned nearly four years. In addition to the input of the members of the Nomenclature Committee, many other photogrammetrists, world-wide, have willingly provided assistance in this effort.

In addition to chapter 24 of the Third Edition of the MANUAL OF PHOTOGRAMMETRY, the following sources have also been consulted to compile the definitions given in section 19.3.

1. American Society of Civil Engineers, 1972, Definitions of Surveying, Mapping and Related Terms: Manual No. 34: New York.

2. Department of Defense Glossary of Mapping, Charting, and Geodetic Terms, Defense Mapping Agency Topographic Center, 1973.
3. Research Studies Institute, Air University 1956, The United States Air Force Dictionary, Woodford A. Heflin, Editor: Air University Press.
4. H.O. Pub 220, Navigation Dictionary, U.S. Navy Hydrographic Office, 1963.
5. Pictorial Cyclopedia of Photography, First American Edition.
6. Basic Photography, A Primer for Professionals, 2nd Edition by M. J. Langford.
7. Photo Lab Index, Morgan and Morgan, Publishers.
8. Photogrammetric Engineering (Index), Vol. XL No. 12, December 1974.

19.3 Definitions of Common Terms Used in Photogrammetry

aberration (optics)—Failure of an optical system to bring all light rays received from a point object to a single image point or to a prescribed geometric position. ***astigmatism***—An aberration affecting the sharpness of images for objects off the axis in which the rays passing through different meridians of the lens come to a focus in different planes. Thus, an extra-axial point object is imaged as two mutually perpendicular short lines located at different distances from the lens. *lateral chromatic aberration*—An aberration which affects the sharpness of images off the axis because different colors undergo different magnifications. *longitudinal chromatic aberration*—An aberration which affects the sharpness of all parts of an image because different colors come to a focus at different distances from the lens. *spherical aberration*—An aberration caused by rays through various zones of a lens coming to focus at different places along the axis. This results in a point object being imaged as a blur circle (*see circle of confusion*). *coma*— An aberration affecting the sharpness of images off the axis in which rays from a point object off the axis passing through a given circular zone of the lens come to a focus in a circle rather than a point, and the circles formed by rays through different zones are of different sizes and are located at different distances from the axis. Therefore, the image of a point object is comet-shaped. *curvature of field*—An aberration affecting the longitudinal position of images off the axis in such a manner that objects in a plane perpendicular to the axis are imaged in a curved or dish-shaped surface. *distortion*—An aberration affecting the position of images off the axis in which objects at different angular distances from the axis undergo different magnifications. Frequently referred to as *lens distortion* (*see distortion curve.*)

absolute altitude—*see flight altitude.*

absolute error—Absolute deviation, the value taken without regard to sign, from the corresponding true values.

absolute humidity—*see humidity.*

absolute orientation—*see orientation.*

absolute parallax—*see absolute stereoscopic parallax* under *parallax.*

absolute stereoscopic parallax—*see parallax.*

accidental errors—*see error.*

accommodation—The faculty of the human eye to adjust itself to give sharp images for different object distances. (*stereoscopy*)—The ability of the eyes to bring two images into superimposition for stereoscopic viewing.

accumulative error—*see constant error* under *error.*

accuracy—The degree of conformity with a standard, or the degree of perfection attained in a measurement. Accuracy relates to the quality of a result, and is distinguished from *precision* which relates to the quality of the operation by which the result is obtained.

accurate contour—*see contour line.*

acetate—A nonflammable plastic sheeting used as a base for photographic films or as a drafting base for overlays where critical registration is not required.

achromatic lens—A lens that has been partly corrected for *chromatic aberration*. Such a

lens is customarily made to bring green and red light rays to approximately the same point focus. Also called *achromat.*

actinic light—A part of the spectrum that causes chemical changes to take place in light sensitive photographic emulsions. The light that creates images on light sensitive material. The blue or violet portion of the spectrum is considered the most actinic band of light. The actinic value also depends on the sensitiveness of the emulsions.

active system—A system which transmits an electromagnetic signal. A system with the capability to transmit, repeat, or re-transmit electromagnetic information. Contrasted with *passive system.*

active tracking system—A satellite tracking system which operates by transmission of signals to and receipt of responses from the satellite.

acuity, visual—A measure of the human eye's ability to separate details in viewing an object. The reciprocal of the minimum angular separation, in minutes of arc, of two lines of detail which can be seen separately.

acutance—An objective measure of the ability of a photographic system to show a sharp edge between contiguous areas of low and high illuminance.

adaptation (ophthalmology)—The faculty of the human eye to adjust its sensitivity to varying intensities of illumination.

additive color process—A method for creating essentially all colors through the addition of light of the 3 additive color primaries (blue, green, and red) in various proportions through the use of 3 separate projectors. In this type of process each primary filter absorbs the other 2 primary colors and transmits only about one-third of the luminous energy of the source. It also precludes the possibility of mixing colors with a single light source because the addition of a second primary color results in total absorption of the only light transmitted by the first color.

additive color viewer—Projector for positive transparencies secured in multiband photography, with each image superimposed on the other and illuminated with a different colored light.

adjusted angle—*see angle.*

adjusted elevation—*see elevation.*

adjusted position—*see position.*

adjustment—The determination and application of corrections to observations, for the purpose of reducing *errors* or removing internal inconsistencies in derived results. The term may refer either to mathematical procedures or to corrections applied to instruments used in making observations.

adjustment of observations—The determination and application of corrections corresponding to *errors* affecting the observations, making the observations consistent among themselves, and coordinating and correlating the derived data.

aerial camera—*see camera.*

aerial exposure index (A.E.I.)—The reciprocal of twice the *exposure,* expressed in metre candle seconds, at the point on the toe of the *characteristic curve* where the slope equals 0.6 *gamma* when recommended processing conditions are used.

aerial film—Specially designed roll-film supplied in many lengths and widths, with various *emulsion* types for use in aerial cameras.

aerial mosaic—*see mosaic.*

aerial photogrammetry—*see photogrammetry.*

aerial photography—Any *photograph* taken from the air. Sometimes called *aerial photo* or *air photograph.*

aerial photographic interpretation—*see photointerpretation.*

aerial photography—The art, science, or process of taking *aerial photographs (see photography).*

aerial reconnaissance—The collection of information by visual, electronic, or photographic means from the air.

aerial survey—*see survey.*

aeroleveling—As applied to model orientation during phototriangulation, barometric height measurements of the camera air stations which have been recorded during the photographic mission are used to represent the *bz* values during the orientation of the successive models on the stereoplotting instrument. Only differences in flight height are required and these are provided by the *statoscope.*

aerometeorograph—An instrument that records the pressure and temperature of the air, the amount of moisture in the air, and the rate of motion of the wind.

aeronautical chart (or *air navigation chart*)—A chart especially designed for air navigation use, on which—in addition to essential topography—are shown obstructions and aids to navigation and other pertinent information. They are sometimes referred to as *airway maps,* and are published at scales ranging from 1:1,000,000 to 1:250,000 and larger.

aeronautical data—Aids to air navigation, such as isogonic lines, compass rosettes, hour angles, airports, beacons and direction finders, major radio stations, airports, emergency landing fields, *etc.* Customarily applied to *aeronautical charts* and occasionally to other type *maps.*

aerotriangulation—*see phototriangulation.*

affine—A geometrical condition in which the scale along one axis or reference plane is different from the scale along the other axis or plane. *affine deformation*—a deformation in

which the scale along one axis or reference plane is different from the scale along the other axis or plane. *affine transformation*—a *transformation* in which straight lines remain straight and parallel lines parallel. Angles, however, may undergo changes and differential scale changes may be introduced.

air base (photogrammetry)—The line joining two *air stations,* or the length of this line; also, the distance (at the scale of the *stereoscopic model*) between adjacent perspective centers as reconstructed in the plotting instrument. *photobase*—the length of the air base as represented on a photograph. The distance between the principal points of two adjacent prints of a series of vertical aerial photographs. It is usually measured on one print after transferring the principal point of the other print (*see camera station*).

air navigation chart—*see aeronautical chart.*

air photo interpretation—*see photointerpretation.*

airborne control (ABC) system—A survey system for horizontal and vertical *control surveys* involving electromagnetic distance measurements and horizontal and vertical angle measurements from two or more known positions to a helicopter hovering over the unknown position. The elevation of the unknown position is determined by the use of a special plumbline cable.

airborne electronic survey control—*Control surveys* accomplished by electronic means from an airborne vehicle or platform, such as by hiran and shoran.

air speed—The speed of an aircraft, along its longitudinal axis, relative to the surrounding atmosphere.

air station—*see camera station.*

albedo—The ratio of radiant energy reflected to that received by a surface, usually expressed as a percentage. The term generally refers to energy within a specific frequency range, as the visible spectrum. Its most frequent application is to the light reflected by a celestial body.

almanac—A periodical publication of astronomical coordinates useful to a navigator. It contains less information than an ephemeris and values are generally given to less precision.

altimeter—(1) An instrument that indicates directly the height above a reference surface (*see barometer*). (2) air navigation: An aneroid barometer used for determining the altitude of an airplane above a specified datum. (3) surveying: An aneroid barometer specially constructed and calibrated for the purpose of determining differences of elevation within the ranges usually encountered in ground surveys. (4) radar: An instrument, called a *radar altimeter* used for determining an aircraft's flight height above terrain by the measurement of time intervals between the emission and return of electromagnetic pulses (*see* **APR**). *statoscope*—A sensitive barometer used in aerial photography for measuring small differences in altitude between successive air stations. *recording statoscope*—a statoscope equipped with a recording camera whose shutter is synchronized with that of the aerial camera and the image of the statoscope is recorded on each individual frame.

altimetry—The art and science of measuring altitudes and interpreting the results.

altitude (aerial photography)—Vertical distance above the datum, usually mean sea level, of an object or point in space above the earth's surface (*see* **elevation** and **angular altitude**).

altitude-contour ratio—*see* **C-factor**.

amici prism—A prism that deviates the rays of light through 90° and, because of its shape, inverts the image. An amici prism is a type of *roof prism*

anaglyph—A stereogram in which the two views are printed or projected superimposed in complementary colors, usually red and green. By viewing through filter spectacles of corresponding complementary colors, a stereoscopic image is formed.

analog instruments—devices that represent numerical quantities by means of physical variables; for example, by translation; by rotation, as in a mechanical gear system; and by voltage or current as in analog networks that use resistance to represent mechanical losses, capacitors and inductors to store energy and simulate the action of springs, etc. Analog is contrasted with *digital*. *Stereoscopic plotters* are examples of photogrammetric analog instruments (*see* **computer**).

analytical aerotriangulation—*see* **analytical phototriangulation**.

analytical nadir-point triangulation—*see* **radial triangulation**.

analytical orientation—Those computational steps required to determine tilt, direction of principal line, flight height, preparation of control templets at *rectification* scale, angular elements, and linear elements in preparing aerial photographs for rectification. Developed data are converted to values to be set on circles and scales of rectifier or transforming printer.

analytical photogrammetry—Photogrammetry in which solutions are obtained by mathematical methods.

analytical phototriangulation—A phototriangulation procedure in which the spatial solution is obtained by computational routines. When performed with aerial photographs, the procedure is referred to as *analytical aerotriangulation*.

analytical radial triangulation—*see* **radial triangulation**.

anastigmatic lens—A lens that has been corrected for *astigmatism* and *curvature of field* (*see* **aberration**).

aneroid barometer—*see* **barometer**.

angle—The inclination to each other of two intersecting lines, measured by the arc of a circle intercepted between the two lines forming the angle, the center of the circle being the point of intersection.

angle—The inclination to each other of two intersecting lines, measured by the arc of a circle intercepted between the two lines forming the angle, the center of the circle being the point of intersection. *adjusted angle*—an adjusted value of an angle which may be derived either from an observed angle or from a deduced angle.

angle of convergence—*see* **angular parallax** under **parallax**.

angle of coverage—*see* **lens**.

angle of depression—The complement of **tilt**. The angle in a vertical plane between the horizontal and a descending line. Also called *depression angle; minus angle; descending vertical angle*.

angle of incidence (optics)—As measured from the normal, the angle at which a ray of light strikes a surface.

angle of inclination—An angle of *elevation* or angle of *depression*.

angle of reflection (optics)—As measured from the normal, the angle at which a reflected ray of light leaves a surface.

angle of refraction—*see* **refraction**.

angle of tilt—*see* **tilt**.

angle of view—1. When the format is square—the angle between two rays passing through the perspective center (rear nodal point) to two opposite sides of the format. 2. When the image format is rectangular—it is necessary to define the sides of the format to which the angle refers. 3. Photogrammetrically, it is twice the angle whose tangent is one-half the length of the diagonal of the format divided by the calibrated focal length. Also called *covering power; field of view*.

angle of yaw—The angle between a line in the direction of flight and a plane through the longitudinal and vertical axes of an aircraft. It is considered positive if the nose is displaced to the right. Also called *yaw angle*.

Angstrom unit (*A*)—A unit of measure for the wavelengths of light, equal to one ten-millionth of a millimetre; for example, the visible spectrum extends from about 4,000A to 7,000A (400 to 700 nm, 0.4 to 0.7 μ, or 0.0004 to 0.0007 mm).

angular calibration constants—In a *multiple-lens camera*, or *multiple-camera assembly*, the values of angular orientation of the lens axes of the several lens-camera units to a common reference line. For example, in a *trimetrogon camera,* the angular relationships of the wing-camera axes with respect to the axis of the central (vertical) camera.

angular altitude—A measure in degrees of a given object above the horizon, taken from a given or assumed point of observation, and expressed by the angle between the horizontal and the observer's line of sight. Sometimes shortened to *altitude* (*see* **coaltitude**).

angular coverage—The angle made by the camera lens with the borders of the field of a vertical photo.

angular distance—1. The angular difference between two directions, numerically equal to the angle between two lines extending in the given directions. 2. The arc of the great circle joining two points, expressed in angular units. 3. Distance between two points, expressed in angular units of a specified frequency. It is equal to the number of waves between the points multiplied by 2π if expressed in radians, or multiplied by 360° if expressed in degrees.

angular distortion—*see* **distortion curve.**

angular magnification—*see* **magnification.**

angular parallax—*see* **parallax.**

angulator—An instrument for converting angles measured on an oblique plane to their corresponding projections on a horizontal plane.

annex point—A point used to assist in the relative orientation of vertical and oblique photographs, selected in the overlap area between the vertical and its corresponding oblique about midway between the pass points. Alternate sets of photographs only will contain annex points.

annotated photograph—A photograph on which planimetric, hypsographic, geologic, cultural, hydrographic, or vegetation information has been added to identify, classify, outline, clarify, or describe features that would not otherwise be apparent in examination of an unmarked photograph. The term generally does not apply to photographs marked only with geodetic control or pass points. (Contrast with **cartographic annotation**).

anti-curl backing—An unsensitized layer of gelation coated to the base side of the *film* to counter the "pull" of the *emulsion* side of the film during the drying process.

antihalation coating (photography)—*see* **halation.**

aperture—*see* **relative aperture** and **aperture stop.**

aperture stop (optics)—The physical element (such as a *stop, diaphragm,* or lens periphery) of an optical system which limits the size of the pencil of rays traversing the system. The adjustment of the size of the aperture stop of a given system regulates the brightness of the image without necessarily affecting the size of the area covered. *field stop*—The physical element (such as a stop, diaphragm, or lens periphery) of an optical system which limits the field of view covered by the system. *entrance pupil*—The image of the aperture stop formed by all the lens elements on the object side of the aperture stop. *exit pupil*—The image of the aperture stop formed by all the lens elements on the image side of the aperture stop. *entrance window*—The image of the field stop formed by all the lens elements on the object side of the field stop. *exit window*—The image of the field stop formed by all the lens elements on the image side of the field stop.

apochromatic lens—A lens that has been corrected for **chromatic aberration** for three colors.

apogee—*see* **apsides.**

apparent horizon—*see* **horizon.**

apparent position—An astronomical term applied to the observable position of a star, planet, or the sun. The position on the celestial sphere at which a heavenly body (or a space vehicle) would be seen from the center of the earth at a particular time. Compare with **astrometric position.** Also called *apparent place.*

approximate contour—*see* **contour** and **contour line.**

APR—An abbreviation for "Airborne Profile Recorder"; also called "*Terrain Profile Recorder*" (*TPR*). An electronic instrument that emits a pulsed-type radar signal from an aircraft to measure vertical distances between the aircraft and the earth's surface.

apsides—The points in the orbit of one celestial body about another at which the distance between the two bodies is greatest or smallest. For a body moving round the earth, *apogee* is the point farthest from the earth and *perigee* the nearest point. Referring to the sun, the two corresponding points are *aphelion* and *perihelion.* For a star revolving about the centre of the galaxy we have *apogalacticum* and *perigalacticum,* and for a component of a double star system we have *apastron* and *periastron.* The straight line connecting the apsides is called the *line of apsides,* and if the orbit is elliptical half the line of apsides is identical with the semi-major axis of the orbit and is thus one of the orbital elements.

architectural photogrammetry—Encompasses the application of photogrammetry in documenting and preserving information on the sizes and shapes of historic buildings and monuments (*see* **close range photogrammetry**).

arc measurement—A survey method used to determine the size of the earth. A long arc is measured on the earth's surface and the angle which subtends this measured arc is determined. By assumptions and mathematical formula the size and shape of the earth can then be determined.

arc navigation—A navigation system in which the position of an airplane or ship is maintained along an arc measured from a control station by means of electronic distance-measuring equipment, such as *shoran* (*see hiran* and *loran*).

arc triangulation—*see triangulation.*

area weighted average resolution (*AWAR*)—see *resolution.*

ascending node—That point at which a planet, planetoid, or comet crosses the ecliptic from south to north, or a satellite crosses the equator of its primary from south to north. Opposite of *descending node.*

aspherical lens—A lens in which one or more surfaces depart from a true spherical shape. For instance, *condenser* lenses are sometimes ground with a parabolic surface, thus making possible a practical elimination of spherical aberration near the outer zones. With such a condenser, it is possible to concentrate the light within a very small aperture. One type of *multiplex* projector is equipped with a condenser system of this nature.

assumed ground elevation—*see elevation.*

assumed plane coordinates—*see coordinates.*

astigmatism—*see aberration.*

astrometric position—The position of a heavenly body (or space vehicle) on the celestial sphere corrected for aberration but not for planetary aberration. Astrometric positions are used in photographic observation where the position of the observed body can be measured in reference to the positions of comparison stars in the field of the photograph. Also called *astrographic position.* Compare with *apparent position.*

astronomic control—*see control.*

astronomic station—A point on the earth whose position has been determined by observations on celestial bodies.

astronomical azimuth—*see azimuth angle.*

astronomical latitude—The angle between the plumb line and the plane of celestial equator. Also defined as the angle between the plane of the horizon and the axis of rotation of the earth. Astronomical latitude applies only to positions on the earth and is reckoned from the astronomical equator (0°), north and south through 90°. Astronomical latitude is the latitude which results directly from observations of celestial bodies, uncorrected for *deflection of the vertical.*

astronomical longitude—The angle between the plane of the celestial meridian and the plane of an initial meridian, arbitrarily chosen. Astronomical longitude is the longitude which results directly from observations on celestial bodies, uncorrected for *deflection of the vertical.*

astronomical refraction—The apparent displacement of an object that results from light rays from a source outside the atmosphere being bent in passing through the atmosphere. This results in all objects appearing to be higher above the horizon than they actually are. The magnitude of this displacement is greater when the object is near the horizon and decreases to a minimum assumed to be zero when the object is at the zenith. Sometimes shortened to refraction. Also called *astronomical refraction error; celestial refraction.*

astronomical triangle—The triangle on the *celestial sphere* formed by arcs of great circles connecting the celestial pole, the zenith, and a celestial body. The angles of the astronomical triangle are: at the pole, the *hour angle;* at the celestial body, the *parallactic angle;* at the zenith, the *azimuth angle.* The sides are: pole to zenith, the *co-latitude;* zenith to celestial body, the *zenith distance;* and celestial body to pole, the *polar distance.* Also called *PZS triangle* (*see polar coordinates* and *spherical coordinates* under *coordinates*).

asymmetrical—Not symmetrical.

atmospheric drag—A major perturbation of close artificial satellite orbits caused by the resistance of the atmosphere. The secular effects are decreasing eccentricity, major axis, and period. Sometimes shortened to drag.

atmospheric refraction—The refraction of light passing through the earth's atmosphere. Atmospheric refraction includes both astronomical refraction and terrestrial refraction.

atran—Acronym for "Automatic Terrain Recognition And Navigation." A navigational system which depends upon the correlation of terrain images appearing on a radar cathode-ray tube with previously prepared maps or simulated radar images of the terrain.

attenuation—The reduction in the strength of magnitude of a signal or action.

attitude (photogrammetry)—The angular orientation of a camera, or of the photograph taken with that camera, with respect to some external reference system. Usually expressed as *omega* (*x*-tilt), *phi* (*y*-tilt) and *kappa* (*z*-rotation); tilt, swing, and azimuth; or roll, pitch, and yaw.

autocollimator—*see collimator.*

autofocus rectifier—A precise, vertical photoenlarger which permits the correction of distortion in an aerial negative caused by tilt. The instrument's operations are motor driven and are interconnected by mechanical linkages to insure automatically maintained sharp focus.

autograph—A first-order stereoplotting instrument. Negative or glass positives (diapositives) may be used.

autopositive film and *paper*—A material which gives a positive copy from a positive transparency (or a negative from a negative) by direct processing. Also called *direct copy* or *direct positive*.

autoscreen film—A photographic film embodying a halftone screen which automatically produces a halftone negative from continuous-tone copy.

average deviation—In statistics, the average or arithmetic mean of the deviations, taken without regard to sign, from some fixed value, usually the arithmetic mean of the data. Also called *mean deviation*.

average error—see error.

axis—see optical axis; camera axis; fiducial axes.

axis of homology—(1) projective geometry: The intersection of two projectively related planes. (2) photogrammetry: The intersection of the plane of the photograph with the horizontal plane of the map or the plane of reference of the ground. Corresponding lines in the photograph and map planes intersect on the axis of homology. Also called the *axis of perspective*.

axis of lens—see optical axis.

axis of perspective—see axis of homology.

axis of tilt—see under *principal plane.*

azimuth—(1) surveying: the horizontal direction of a line measured clockwise from a reference plane, usually the *meridian*. Contrast with *bearing*. (2) Photogrammetry: azimuth of the principal plane. The clockwise angle from north (or south) to the principal plane of a tilted photograph.

azimuth angle—(astronomy) The angle 180° or less between the plane of the celestial meridian and the vertical plane containing the observed object, reckoned from the direction of the elevated pole. In astronomic work, the azimuth angle is the spherical angle at the zenith in the astronomical triangle which is composed of the pole, the zenith, and the star. In geodetic work, it is the horizontal angle between the celestial pole and the observed terrestrial object. Also called *astronomical azimuth*. (surveying) An angle in triangulation or in a traverse through which the computation of azimuth is carried. In a simple traverse, every angle may be an azimuth angle. Sometimes, in a traverse, to avoid carrying azimuths over very short lines, supplementary observations are made over comparatively long lines, the angles between which form azimuth angles. In triangulation, certain angles, because of their size and position in the figure, are selected for use as azimuth angles, and enter into the formation of the azimuth condition equation (azimuth equation).

back azimuth—If the *azimuth* of point *B* from point *A* is given, the back azimuth is the azimuth of point *A* from point *B*. Because of the *convergence of the meridians*, the forward and backward azimuths of a line do not differ by exactly 180°, except where *A* and *B* have the same geodetic longitude.

back focal length or *back focal distance—see focal length.*

backing (photographic)—*see antihalation coating.*

backup—An image printed on the reverse side of a map sheet already printed on one side. Also the printing of such images.

band—A set of adjacent wavelengths in the electro-magnetic spectrum with a common characteristic, such as the visible band.

barometer—An instrument for measuring the pressure of the atmosphere. *aneroid barometer*—A hollow corrugated metal box from which the air has been partially exhausted and whose walls are so thin that it changes form when the air pressure changes. Most aneroid barometers have two scales, one graduated to correspond to the height of a mercury column and the other to feet or meters of altitude. *mercury barometer*—Basically, a vertical glass tube containing mercury, with the upper end closed to form a vacuum above the mercury and the lower end resting in a reservoir of mercury. The column of mercury is sustained by the pressure of air on the reservoir. The height of the mercury column is read on a scale (*see altimeter*).

barometric elevation—An elevation determined with a barometer (*see elevation*).

barrel distortion—A type of geometric distortion found in scanning imagery (*see scanner*) in which elements crossing the flight direction are distorted by a combination of scanner-mirror rotation.

basal coplane—see coplanar.

basal orientation—The establishment of the position of both ends of an air base with respect to a ground system of coordinates. In all, six elements are required. These are essentially the three-dimensional coordinates of each end of the base. In practice, however, it is also convenient to express these elements in one of two alternative ways: (1) The ground rectangular coordinates of one end of the base and the difference between these and the ground rectangular coordinates of the other end of the base. (2) The ground rectangular coordinates of one end of the base, the length of the base, and two elements of direction (such as *base direction* and *base tilt*). *base direction*—The direction of the vertical plane containing the air base, which might be expressed as a *bearing* or an *azimuth*. *base tilt*—The inclination of the air base with respect to the horizontal.

basal plane—see epipolar plane under *epipoles.*

base-altitude ratio—see base-height ratio.

base color—The first color printed of a polychrome map to which succeeding colors are registered.

base direction—see basal orientation:

base-height ratio—The ratio (B:H) of the air base and the flight height of a stereoscopic pair of photographs. Also referred to as *base-altitude ratio; K-factor*.

base line—1. (survey) A surveyed line established with more than usual care, to which surveys are referred for coordination and correlation. 2. (photogrammetry) *see* *air base; photobase*. 3. (navigation) The line between two radio transmitting stations operating in conjunction for the determination of a line of position, as the two stations of a *loran* system.

base manuscript—*see* *compilation manuscript*.

base map—*see map*.

base, photo—*see photobase* under *airbase*.

base sheet—A sheet of dimensionally stable material upon which the *map projection* and *ground control* are plotted, and upon which *stereotriangulation* or *stereocompilation* is performed.

base tilt—*see basal orientation*.

basic control—*see control*.

bathymetric chart—A *topographic map* of the floor of the ocean (*see chart*).

bathymetric contour—*see depth curve*.

beam of light—*see ray of light*.

beam splitter—An optical device, such as a semi-reflecting mirror or a prism arranged so as to transmit different spectral bands along separate axes to various films, detectors, or other analyzing/recording devices.

bearing—*Direction* of a line measured as the acute angle from a reference *meridian;* usually expressed in the form "N 30° W" or "S 87° E." Contrast with *azimuth*.

bellows—A folding cloth or leather tube, generally square or pyramidal in shape, providing flexible, light-tight enclosure between camera lens and film.

benchmark (BM)—A marked *vertical control* point which has been located on a relatively permanent material object, natural or artificial, and whose *elevation* above or below an adopted *datum* has been established. It is usually monumented to include bench mark name or number, its elevation, and the name of the responsible agency.

between-the-lens shutter—*see shutter*.

binocular vision—*see stereoscopy*.

biostereometrics—The spatial and spatio-temporal analysis of biological form and function based on principles of analytic geometry. (Also referred to as *Biomedical* and *Bioengineering photogrammetry*.)

block adjustment—The adjustment of *strip coordinates* or *photograph coordinates* for two or more contiguous strips of photographs (*see strip adjustment*).

block of photographs—Two or more contiguous strips of photographs.

blueline—A nonreproducible blue image or outline usually printed photographically on paper or plastic sheeting, and used as a guide for drafting, stripping, or layout. Sometimes called blind image.

blunder—*see error*.

bridging—Synonym for *stereotriangulation*.

brightness scale (photography)—The ratio of the brightness or *luminance* of highlights to the deepest shadow in the actual terrain (as measured from the camera station) for the field of view under consideration.

Bx curve—Similar to Bz curve, except errors in the x-direction on horizontal control points are plotted as ordinates *versus* x-coordinates of these points (*see Bz Curve*).

By curve—Similar to Bz curve, except errors in the y-direction on horizontal control points are plotted as ordinates *versus* x-coordinates of these points (*see Bz Curve*).

Bz curve (photogrammetry)—A graphical representation of the vertical errors in a stereo-triangulated strip. In a Bz curve, the x-coordinates of the vertical control points, referred to the initial nadir point as origin, are plotted as abscissas, and the differences between the known elevations of the control points and their elevations as read in the stereotriangulated strip are plotted as ordinates; a smooth curve drawn through the plotted points is the Bz curve. The elevation read on any pass point in the strip is adjusted by the amount of the ordinate of the Bz curve for an abscissa corresponding to the x-coordinate of the point.

C-factor—An empirical value which expresses the vertical measuring capability of a given stereoscopic system; generally defined as the ratio of the *flight height* to the smallest *contour interval* accurately plottable. The C-factor is not a fixed constant, but varies over a considerable range, according to the elements and conditions of the photogrammetric system. In planning for aerial photography, the C-factor is used to determine the flight height required for a specified contour interval, camera, and instrument system. Also called *altitude-contour ratio*.

cadastral map—*see map*.

cadastral survey—A survey relating to land boundaries and subdivisions, made to create units suitable for transfer or to define the limitations of title. Derived from *cadastre* (meaning register of the real property of a political subdivision with details of area, ownership, and value), the term is now used to designate the surveys of the public lands of the United States, including retracement surveys for the identification, and resurveys for the restoration, of property lines; it may also be applied properly to corresponding surveys outside the public lands, although such surveys usually are termed *land surveys* or *property surveys* through preference.

*calibrated focal length—see **focal length**.*

calibration—The act or process of determining certain specific measurements in a camera or other instrument or device by comparison with a standard, for use in correcting or compensating errors of for purposes of record. *field calibration*—A term generally applied where only a combination of field and office computer techniques are available to check instrument accuracy. Adjustments, other than normal operator adjustments, cannot be made during field calibration. *camera calibration* (photogrammetry)—The determination of the *calibrated focal length,* the location of the *principal point* with respect to the fiducial marks, the *point of symmetry,* the resolution of the lens, the degree of flatness of the focal plane, and the lens distortion effective in the focal plane of the camera and referred to the particular calibrated focal length. In a multiple-lens camera, the calibration also includes the determination of the angles between the component perspective units. The setting of the fiducial marks and the positioning of the lens are ordinarily considered as adjustments, although they are sometimes performed during the calibration process. Unless a camera is specifically referred to, distortion and other optical characteristics of a lens are determined in a focal plane located at the *equivalent focal length* and the process is termed *lens calibration.* In *close-range photogrammetry,* calibration may be performed directly or indirectly on the camera, the individual photograph (particularly in case of non-metric cameras) or on the total system. Camera calibration may be done in one of three forms: laboratory, on-the-job, and self-calibration. Laboratory calibration of a camera is performed separately from the photography phase and is undertaken with goniometers or test areas of various sophistication. On-the-job calibration of a camera or a photograph utilizes object photography and object-space control points. Self-calibration of a camera or a photograph utilizes object photography and well-defined object points. Calibration of a system may be accomplished by including parameters such as affinity and non-perpendicularity of comparator axes (*see **angular calibration constants** and **collimate***).

calibration constants (photogrammetry)—The results obtained by calibration, which give the calibrated focal length of the lens-camera unit and the relationship of the principal point to the fiducial marks of a camera.

calibration correction—The value to be added to or subtracted from the reading of an instrument to obtain the correct reading.

*calibration error—see **instrumental error** under **error.***

calibration plate—A glass negative exposed with its emulsion side in the same position as is occupied by the service emulsion (on film or glass) at the time of exposure. This plate provides a record of the distance between the fiducial marks of the camera; it is sometimes called a *master glass negative* or *flash plate*.

calibration table—A list of calibration corrections or calibrated values.

calibration templet (photogrammetry)—A templet of glass, plastic, or metal made in accordance with the calibration constants to show the relationship of the principal point of a camera to the fiducial marks; used for the rapid and accurate marking of principal points on a series of photographs. Also, for a multiple-lens camera, a templet prepared from the calibration data and used in assembling the individual photographs into one composite photograph.

camera—A lightproof chamber or box in which the image of an exterior object is projected upon a sensitized plate or film, through an opening usually equipped with a lens or lenses, shutter, and variable aperture. *aerial camera*—A camera specially designed for use in aircraft. The prefix *aerial* is not essential where the context clearly indicates an aerial camera rather than a terrestrial camera. *ballistic camera*—A precision terrestrial camera, usually employing glass plates, used at night to photograph such objects as rockets or satellites against a star background. If the camera is mounted so that it tracks the stars (as with a sidereal mount) or the object, it may be called a *tracking camera. continuous-strip camera*—A camera in which a continuous strip exposure is made by rolling the film continuously past a narrow slit opening at a speed proportional to the ground speed of the aircraft. *copy camera* or *process camera*—A precision camera used in the laboratory for copying purposes. Also called process camera. *frame camera*—A camera in which an entire frame or format is exposed simultaneously through a lens that is fixed relative to the focal distance. *horizon camera* (aerial photography)—A camera used in conjunction with an aerial surveying camera in vertical photography to photograph the horizon simultaneously with the vertical photographs. The horizon photographs indicate the *tilts* of the vertical photographs. *mapping camera* or *surveying camera*—A camera specially designed for the production of photographs to be used in surveying. The prefixes *mapping* and *surveying* indicate that a camera is equipped with means for maintaining and indicating the interior orientation for the photographs with sufficient accuracy for surveying purposes. A mapping camera may be either an aerial mapping camera or a terrestrial camera. *multiband camera*—A camera that exposes different areas of one film, or more than one film, through one lens and a beam splitter, or two or more lenses equipped with different filters, to provide two or more photographs of the same scene in different special bands. *multiple-lens camera*—A camera with two or more lenses, with the axes of the lenses systematically arranged at fixed angles in order to cover a wide field of simultaneous exposures in all chambers. *precision camera*—An indefinite term sometimes applied to any camera used for photogrammetric purposes. May be construed as meaning a metric camera. *metric camera*—A camera whose *interior orientation* is known, stable, and reproducible. *multiple-camera assembly*—An assembly of two or more cameras mounted so as to maintain a fixed angle between their respective optical axes. *convergent camera*—An assembly of two cameras which take simultaneous photographs and are mounted so as to maintain a fixed angle between their optical axes. The effect is to increase the angular coverage in one direction, usually along the longitudinal axis of the aircraft. *fan camera*—An assembly of three or more cameras systematically disposed at fixed angles relative to each other so as to provide wide lateral coverage with overlapping images. *panoramic camera*—A camera which takes a partial or complete panorama of the terrain. Some designs utilize a lens which revolves about a vertical axis; in other designs, the camera itself is revolved by clockwork to obtain a panoramic field of view. *photogrammetric camera*—A general term applicable to cameras used in any of the several branches of photogrammetry. *stellar camera*—A camera for photographing the stars. If it is used for photographing the sun, it may be called a *solar camera*. If the stellar camera has been rigidly mounted and calibrated with respect to one or more mapping cameras in an airborne vehicle, the absolute attitude of a photograph taken with the mapping camera can be computed from the attitude of the stellar-camera photograph taken at that same instant. *stereometric camera*—A combination of two cameras mounted with parallel optical axes on a short, rigid base; used in terrestrial photogrammetry for taking photographs in stereoscopic pairs. *terrestrial camera*—A camera designed for use on the ground. (*see* **phototheodolite**). *trimetrogon camera*—An assembly of three cameras equipped with wide angle Metrogon lenses, in which one of the cameras is vertical and the other two are 60-degree obliques. *zenith camera*—A special camera so designed that its optical axis may be pointed accurately toward the zenith. It is used for the determination of astronomic positions by photographing the position of the stars.

camera axis—A line perpendicular to the focal plane of the camera and passing through the *interior perspective center* or *emergent nodal point* of the lens system.

camera base—*see* **air base.**

camera calibration—*see* **calibration.**

camera, horizon—*see* **horizon camera** under **camera.**

camera lucida—A monocular instrument using a half-silvered mirror, or the optical equivalent, to permit superimposition of a virtual image of an object upon a plane. The *sketchmaster* is such an instrument used for superimposing the image of a photograph upon a map or map manuscript. In the process, the image may be rectified.

camera, multiband—*see* **multiband camera** under **camera.**

camera obscura—*see* **camera lucida.**

camera, panoramic—*see* **panoramic camera** under **camera.**

camera station (photogrammetry)—The point in space occupied by the camera lens at the moment of exposure; also called *air station* or *exposure station*.

Canadian grid—*see* **perspective grid.**

candela (formerly *candle*)—The international unit of luminous *intensity;* the *luminance* of a blackbody radiator at the temperature of solidification of molten platinum is 60 candelas per sq. cm. The candela corresponds to one *lumen per steradian*.

candle power—Luminous intensity expressed in terms of the *candela*.

cantilever extension—Phototriangulation from a controlled area to an area of no control. Also, the connection by *relative orientation* and *scaling* of a series of photographs in a strip to obtain strip coordinates.

cardan link—A universal joint. An optical cardan link is a device for universal scanning about a point.

cardinal points—1. Any of the four principal astronomical directions on the surface of the earth: north, east, south, west. 2. (optics) Those points of a lens used as reference for determining object and image distances. They include *principal planes and points, nodal points,* and *focal points*.

carpentier inversor—*see* **inversors.**

carrying contour—*see* **contour line.**

cartesian coordinates—*see* **coordinates.**

cartographic annotation—The delineation of additional data, new features, or deletion of destroyed or dismantled features on a *mosaic* to portray current details. Cartographic annotations may include elevation values for airfields, cities, and large bodies of water; new construction and destroyed or dismantled roads, railroads, bridges, dams, target installations, and cultural features of landmark significance. (Compare with **annotated photograph**).

cartographic compilation—*see* **compilation.**

cartographic film—Film with a dimensionally stable base, used for map negatives and/or positives. Usually referred to by trade name.

cartographic license—The freedom to adjust, add, or omit map features within allowable limits to attain the best cartographic expression. License must not be construed as permitting the cartographer to deviate from specifications.

cartographic photography—see mapping photograph.

cartography—The art and science of expressing graphically, by maps and charts, the known physical features of the earth, or of another celestial body, often includes the work of man and his varied activities.

cartometric scaling—The accurate measurement of geographic or grid coordinates on a map or chart by means of a scale. This method may be used for plotting the positions of points, or determining the location of points.

catadioptric system (optics)—An optical system containing both refractive and reflective elements.

catoptric system (optics)—An optical system in which all elements are reflective (mirrors).

celestial coordinates—Any set of coordinates used to define a point on the *celestial sphere.*

celestial navigation—A means of navigation by which a geographical location is determined by reference to celestial bodies.

celestial sphere—An imaginary sphere of infinite radius, described about an assumed center, and upon which imagined positions of celestial bodies are projected along radii passing through the bodies. For observations on bodies within the limits of the solar system, the assumed center is the center of the earth. For bodies where the parallax is negligible, the assumed center may be the point of observation (*see polar coordinates* and *spherical coordinates* under *coordinates*).

center of projection—see perspective center.

center to center method—see mosaicking.

characteristic curve (photography)—A curve showing the relationship between exposure and resulting density in a photographic image, usually plotted as the density (D) against the logarithm of the exposure (log E) in candlemeter-seconds. It is also called the *H and D curve,* the *sensitometric curve,* and the *D log E curve. density*—A measure of the degree of blackening of an exposed film, plate, or paper after development, or of the direct image (in the case of a print-out material). It is defined strictly as the logarithm of the *optical opacity,* where the opacity is the ratio of the incident to the transmitted (or reflected) light. It varies with the use of scattered or specular light. *gamma*—The tangent of the angle which the straight-line portion of the characteristic curve makes with the log-exposure axis. It indicates

the slope of the straight-line portion of the curve and is a measure of the extent of development and the contrast of the photographic material. *speed* (film, plate, or paper)—The response or sensitivity of the material to light, often expressed numerically according to one of several systems (*e.g.,* H and D, DIN, Scheiner, and ASA exposure index) (*see relative aperture*). *gradient*—The slope of the characteristic curve at any point. *gradient speed*—The speed of a photographic material determined on the basis of the exposure corresponding to a particular gradient of the characteristic curve. (*See contrast, exposure,* and *brightness scale.*)

chart (mapping)—A special-purpose map, generally designed for navigation or other particular purposes, in which essential map information is combined with various other data critical to the intended use (*see map* and *aeronautical chart*).

check profile—A profile plotted from a field survey and used to check a profile prepared from a topographic map. The comparison of the two profiles serves as a check on the accuracy of the contours on the topographic map.

chopping (star or satellite trails)—Interrupting the photographic image of a star or satellite trail by a shutter or other device to provide precise timing and orientation data for geodetic observations of aerospace vehicles against a stellar background (see *ballistic camera* under *camera*).

chronograph—An instrument for producing a graphical record of time as shown by a clock or other device. In use, a chronograph produces a double record: the first is made by the associated clock and forms a continuous time scale with significant marks indicating periodic beats of the timekeeper; the second is made by some external agency, human or mechanical, and records the occurrence of an event or of a series of events.

chronometer—A portable timekeeper with compensated balance, capable of showing time with extreme precision and accuracy.

chromatic aberration—see lateral and *longitudinal chromatic aberration* under *aberration.*

chromostereopsis—The phenomenon in which a set of different colored objects are perceived at different heights dependent on their color, even though no height difference, or parallax, is actually present. (Contrast with *stereoscopy*).

cine-theodolite—A photographic tracking instrument which records on each film frame the target and the azimuth and elevation angles of the optical axis of the instrument.

circle of confusion (optics)—The circular image of a distant point object as formed in a focal plane by a lens. A distant point object (*e.g.,* a star) is imaged in a focal plane of a lens as a circle of finite size, because of such conditions as (1) the focal plane's not being placed at the point of sharpest focus, (2) the effect of certain aberrations, (3) diffraction at the lens, (4) grain in a photographic emulsion, and/or (5) poor workmanship in the manufacture of the lens.

Clarke spheroid (ellipsoid) of 1866—A reference ellipsoid having the following approximate dimensions: semimajor axis—6,378,206.4 metres; semiminor axis—6,356,583.8 metres; and the flattening or ellipticity—1/294.97.

close-range photogrammetry—A branch of photogrammetry wherein object-to-camera distances are not more than 300 metres. The field of Close-Range Photogrammetry encompasses the following three major areas of application: *Architectural Photogrammetry, Biostereometrics* (Biomedical and Bioengineering Photogrammetry) and *Industrial Photogrammetry*.

clinometer—A simple instrument used for measuring the degree of slope, in percentage or in angular measure.

closure or *closing error*—*see error.*

coaltitude—The complement of altitude, or ninety degrees minus the altitude. The term has significance only when used in connection with altitude measured from the celestial horizon, when it is synonymous with *zenith distance. (see angular altitude* and *astronomical triangle*).

coated lens—A lens whose air-glass surfaces have been coated with a transparent film of such thickness and index of refraction as to minimize the light loss due to reflection. The reflection loss of uncoated lenses amounts to about 4 percent per air-glass surface.

coherent optics—That branch of optics utilizing coherent radiation, produced from sources such as *lasers.* In photogrammetry coherent radiation is used to produce *holograms* which are used to obtain three-dimensional images of objects.

colatitude—The complement of the latitude, or 90° minus the latitude. Colatitude forms one side, zenith to pole, of the *astronomical triangle.* It is the side opposite the celestial body.

collimate—(1) physics and astronomy: To render parallel to a certain line or direction; to render parallel, as rays of light; to adjust the line of sight or lens axis of an optical instrument so that it is in its proper position relative to the other parts of the instrument. (2) photogrammetry: To adjust the fiducial marks of a camera so that they define the principal point (*see calibration*).

collimating marks—Marks on the stage of a reduction printer or projection equipment, to which a negative or diapositive is oriented (see *fiducial marks*).

collimation adjustment—*see collimate.*

collimation axis—In an optical instrument, the line through the rear nodal point of the objective lens that is precisely parallel with the center line of the instrument (*see line of collimation*).

collimation error—The angle by which the line of sight of an optical instrument differs from its *collimation axis.* Also called *error of collimination.*

collimation plane—The plane described by the *collimation axis* of a telescope of a transit when rotated around its horizontal axis.

collimator—An optical device for artificially creating a target at infinite distance (a beam of parallel rays of light); used in testing and adjusting certain optical instruments. It usually consists of a converging lens and a target (a system or arrangement of crosshairs) placed at the principal focus of the lens.

autocollimator—A collimator provided with a means of illuminating its crosshairs so that, when a reflecting plane is placed normal to the emergent light beam, the reflected image of the crosshairs appears to be coincident with the crosshairs themselves. This device is used in calibrating optical and mechanical instruments.

color composite—A composite in which the component images are shown in different colors (*see composite photograph*).

color photography—Photography in which either the direct-positive or the negative-positive color process is used.

color plate—A general term for the pressplate from which any given color is printed. Normally, the term is modified to reflect a special color or type of plate, such as brown plate or contour plate (*see combination plate*).

color separation—1. The process of preparing a separate drawing, engraving or negative for each color required in the production of a lithographed map or chart. 2. A photographic process or electronic scanning procedure using color filters to separate multicolored copy into separate images of each of the three primary colors. *Color-separation drawing*—One of a set of drawings which contains similar or related features, such as drainage or culture. There are as many drawings as there are colors to be shown on lithographed copy.

color-proof process—A photomechanical printing process which makes possible the combining of negative separations by successive exposures to produce a composite color proof on a vinylite sheet. The method is usually referred to by the manufacturer's trade name of the materials used.

coma—*see aberration.*

combination plate—Halftone and line work on one plate. Also, two or more subjects combined on the same plate (*see color plate*).

comparator—An optical instrument, usually

precise, for measuring rectangular or polar coordinates of points on any plane surface, such as a photographic plate. *monocomparator*—A precision instrument, consisting of a measuring system, a viewing system, and a readout system designed for the measurement of image coordinates on a single photograph. *stereocomparator* (photogrammetry)—A stereoscopic instrument for measuring parallax; usually includes a means of measuring photograph coordinates of image points.

compass—An instrument for indicating a horizontal reference direction relative to the earth (*see* **magnetic north**).

compensating lens (Photogrammetry)—A lens introduced into an optical system to correct for *radial distortion*.

compensation plate (Photogrammetry)—A glass plate having a surface ground to a predetermined shape, for insertion in the optical system of a diapositive printer or plotting instrument, to compensate for *radial distortion* introduced by the camera lens.

compilation—(1) cartography: The production of a new or revised map or chart, or portion thereof, from existing maps, aerial photographs, surveys, new data, and other sources (*see* **delineation**). (2) photogrammetry: The production of a map or chart, or portion thereof, from aerial photographs and geodetic control data, by means of photogrammetric instruments, also called *stereocompilation*.

compilation manuscript—The original drawing, or groups of drawings, of a map or chart as compiled or constructed from various data on which cartographic and related detail is delineated in colors on a stable-base medium. A compilation manuscript may consist of a single drawing called a *base manuscript*, or because of congestion, several overlays may be prepared showing vegetation, relief, names, and other information. Since the latter is usually the case, the base together with its appropriate overlays are collectively termed the compilation manuscript. The general term "manuscript" is not recommended without adequate qualification.

compilation scale—The scale at which a map or chart is delineated on the original manuscript. This scale may be larger than reproduction scale.

complementary colors (optics)—Two colors are complementary if, when added together (as by projection), they produce neutral-hue light.

composite map—A map which portrays information of two or more general types. Usually a compiled map, bringing together on one map, for purposes of comparison, data which were originally portrayed on separate maps.

composite photograph (aerial photography)—A photograph made by assembling the separate photographs, made by the several lenses of a multiple-lens camera in simultaneous exposure, into the equivalent of a photograph taken with a single wide-angle lens (*see* **multiple-lens camera** under **camera**).

computer, analog computer—A physical device or system which behaves in a manner analogous to some system under study, simulating the processes and producing results which are measured in terms of physical quantities. Often, an electrical system. *digital computer*—A machine for carrying out mathematical processes by operations based on counting, as distinct from an analog computer.

condenser (optics)—A lens or lens system designed to concentrate the illumination from a light source on a limited area. A reflector of ellipsoidal shape having the light source concentrated at one of its foci is optically equivalent to a condenser-lens system (*see* **aspherical lens**).

condition equation—An equation which expresses exactly certain relationships that must exist among related quantities, which are not independent of one another, exist a priori, and are separated from relationships demanded by observation. *Condition*—A term used in setting up equations for computation and adjustment of triangulation, trilateration, or traverse and analytical photogrammetry.

conjugate distance—The corresponding distances of object and image from the *nodal points* of the lens. The conjugate distances O and I and the focal length f of the lens are related by the equation

$$\frac{1}{f} = \frac{1}{I} + \frac{1}{O} \ .$$

This relation may also be expressed in Newton's form as

$$XX' = f^2$$

where $X = I - f$ and $X' = O - f$. The total distance from object to image equals the sum of the two conjugate distances plus or minus (depending on lens design) a small distance called the *nodal-point separation*.

conjugate image points—A term formerly used to denote the images of a single object point on two (or more) overlapping photographs. Use of this term is not recommended as it may lead to confusion with accepted usage in optics, wherein *conjugate points* (object and image) apply to one lens or lens system and are physically related according to the equations given under the definition of *conjugate distance*. The preferred term is *corresponding image points* (*see* **corresponding images**).

conjugate image rays—The preferred term is *corresponding image rays* (*see* discussion under *conjugate image points*).

conjugate principal point—*see corresponding principal point.*

contact glass—*see focal-plane plate* under *focal plane.*

contact plate—*see focal-plane plate* under *focal plane.*

contact print—*see print.*

contact printing frame—In photography and platemaking, a device for holding the negative and the sensitive material in contact during exposure. The light source may or may not be a separate element. If the frame contains a vacuum pump to exhaust all air within the frame to insure perfect contact between the negative and the sensitive material, it is known as a contact vacuum printing frame.

contact screen—1. A halftone screen made on a film base and used in direct contact with the film to obtain a halftone image from a continuous-tone original. 2. A pattern image on a film base used in contact with the film or plate to obtain a pattern image from an open window negative.

contesimal system—The division of the circle into 400 grads. Thus a grad equals .9 of a degree. Grads are divided into 100 minutes, and each minute into 100 seconds; written as 50^g $17'$ $98.8''$ or 50^g $1798''8$ or $50,^{17988}$.

continuous-strip camera—*see camera.*

continuous-strip photography—Photography of a strip of terrain in which the image remains unbroken throughout its length along the line of flight (*see continuous-strip camera* under *camera*).

continuous tone—An image which has not been screened and contains unbroken, gradient tones from black to white, and may be either in negative or positive form. Aerial photographs are examples of continuous-tone prints. Contrasted with *halftone; line copy.*

continuous tone gray scale—A scale of tones from white to black or from transparent to opaque, each tone of which blends imperceptibly into the next without visible texture or dot formation. Also called *continuous wedge.* Contrasted with *step wedge.*

contour (surveying and cartography)—An imaginary line on a land surface connecting points of equal elevation; also, the line representing this feature on a map or chart (properly called a *contour line*). *depression contour*—A closed contour inside of which the ground is at a lower elevation than outside (*see form lines*). *contour interval*—The difference in elevation between adjacent *contours.*

contour line—A line on a map or chart connecting points of equal elevation. *accurate contour*—A contour line which is accurate within one-half of the basic *contour interval;* also called *normal contour. approximate contour*—A contour line substituted for a normal contour whenever its accuracy is questionable.

carrying contour—A single contour line representing two or more contours, used to show vertical or near-vertical topographic features, such as steep slopes and cliffs (*see check profile*).

contour map—*see map.*

contour value—A numerical value placed upon a contour line to denote its *elevation* relative to a given datum, usually mean sea level.

contrast (photography)—The actual difference in density between the highlights and the shadows on a negative or positive. Contrast is not concerned with the magnitude of density, but only with the difference in densities. Also, the rating of a photographic material corresponding to the relative density difference which it exhibits (*see density* under *characteristic curve*).

control (mapping)—A system of points with established positions or elevations, or both, which are used as fixed references in positioning and correlating map features. Control is generally classified in four orders (with first order denoting highest quality) according to the precision of the methods and instruments used in establishing it, and the accuracy of the resultant positions and elevations. *basic control*—Horizontal and vertical control of third—or higher—order accuracy, determined in the field and permanently marked or monumented, that is required to control subordinate surveys. *horizontal control*—Control with horizontal positions only. The positions may be referred to the geographic parallels and meridians or to other lines of reference, such as plane coordinate axes. *vertical control*—Control with elevations only; usually referred to mean sea level. *astronomic control*—Control determined from astronomic observations. *geodetic control*—Control which takes into account the size and shape of the earth; implies a reference spheroid representing the *geoid* and horizontal- and vertical-control *datums. ground control*—Control established by ground surveys, as distinguished from control established by photogrammetric methods. The term usually implies *geodetic control* or *basic control. supplemental control*—Points established by subordinate surveys, to relate aerial photographs used in mapping with the system of geodetic control. The points must be positively *photoidentified,* that is, the points on the ground must be positively correlated with their images on the photographs.

control—photogrammetry: Control established by photogrammetric methods as distinguished from control established by ground methods. Also called *minor control.*

controlled mosaic—*see mosaic.*

control point—(photogrammetry)—Any *control station* (in a horizontal- and/or vertical-control system) that is identified on a photograph and used to aid in fixing the attitude and/or position of a photograph or group of photographs. More specific terms are *supplemental control point, photocontrol point, picture control point,* and *ground control point.*

control station—An object or mark on the ground of known, or accurately determined horizontal position (*horizontal control station* with *x*- and *y*-grid coordinates or latitude and longitude) or elevation (*vertical control station* or *benchmark*), or both, in a network of ground control. Control stations constitute the framework by which map details are fixed in their correct position, azimuth, elevation, and scale with respect to the earth's surface. Also called *control point; ground control point.*

control strip (aerial photography)—A strip of aerial photographs taken to aid in planning and accomplishing later aerial photography, or to serve as control in assembling other strips. Also called *control flight; tie flight; tie strip* (*see* **cross-flight photography**).

control survey—A survey which provides positions, horizontal and/or vertical, of points to which supplementary surveys are adjusted. The fundamental control survey of the United States provides the geographic positions and plane coordinates of triangulation and traverse stations and the elevations of bench marks which are used as the base for subordinate surveys.

convergence—*see* **angular parallax** under **parallax.**

convergence of meridians—The angular drawing together of the geographic meridians in passing from the equator to the poles. At the equator, all meridians are mutually parallel; passing from the equator, they converge until they meet at the poles, intersecting at angles that are equal to their differences of longitude. The term convergence of meridians is used to designate also the relative difference of direction of meridians at specific points on the meridians. Thus, for a geodetic line, the azimuth at one end differs from the azimuth at the other end by 180 degrees plus or minus the amount of the convergence of the meridians at the end points.

convergent photography—*see* **camera.**

convergent photography—Aerial photography using a *convergent-camera* installation.

convergent position—A split camera installation so positioned that the plane containing the camera axis is parallel to the line of flight (*see* **convergent camera** under **camera**).

converging lens—*see* **positive lens.**

coordinate axes—*see* **cartesian coordinates** under **coordinates.**

coordinate transformation—A mathematical or graphic process of obtaining a modified set of coordinates by scaling, rotating, and translating.

coordinates—Linear or angular quantities which designate the position that a point occupies in a given reference frame or system. Also used as a general term to designate the particular kind of reference frame or system, such as *plane rectangular coordinates* or *spherical coordinates.* **cartesian coordinates**—Values representing the location of a point in a plane in relation to two perpendicular intersecting straight lines, called coordinate axes. The point is located by measuring its distance from each axis along a parallel to the other axis. This system is extended to represent the location of points in three-dimensional space by referencing to three mutually perpendicular coordinate axes which intersect at a common point of origin. ***plane rectangular coordinates*** (also called simply *plane coordinates*)—A system of coordinates in a horizontal plane used to describe the positions of points with respect to an arbitrary origin by means of two distances perpendicular to each other. The two reference lines at right angles to each other and passing through the origin are called the *coordinate axes.* The *y*-axis may be in the direction of astronomic (true) north, magnetic north, or an assumed (arbitrarily assigned) north and is also called the *easting line.* The distances from the origin and parallel to the north-south axis are called the *ordinates,* the *y coordinates,* the *total latitudes* or the *northings.* The distances from the origin and parallel to the true (or arbitrarily assigned) east-west axis (or *x*-axis) are called the *abscissas,* the *x coordinates,* the *total departures* or *eastings.* A plane rectangular coordinate system is used in areas of such limited extent that the errors introduced by substituting a plane for the curved surface of the earth are within the required limits of accuracy. The north and east directions are usually taken as positive, and the south and west directions are usually taken as negative. To avoid the use of negative coordinates, the origin of the system may be established at a point to the southwest of the area, or the coordinates of the origin may be assigned large positive values instead of zero; these large positive coordinate values are sometimes called *false northings* and *false eastings.* The merit of a rectangular coordinate system is that positions of points, distances, and directions on it can be computed by the use of plane trigonometry. Plane rectangular coordinates may or may not be adjusted to a map projection. ***assumed plane coordinates***—A local plane coordinate system set up at the convenience of the surveyor. The reference axes are usually assumed so that all coordinates are in the first quadrant. The *y*-axis may be in the direction of astronomic (true) north, magnetic north, or an assumed north. A plane rectangular coordinate system based on, and mathematically adjusted to, a map projection so that geographic positions in terms of latitude and longitude can be readily transformed into plane coordinates, and the computations relating to them made by the ordinary methods of plane surveying. *state plane coordinate systems*—A series of grid coordinate systems prepared by the U.S. Coast and Geodetic Survey for the entire United States, with a separate system for each state. Each state system consists of one or more zones. The grid coordinates for each zone are based on, and mathematically adjusted to, a *map projection*. The *Lambert conformal conic projection* with two standard parallels is used for zones of predominant east-west extent and limited north-south extent. The *transverse Mercator projection* is used for zones of predominant north-south extent and limited east-west extent. *polar coordinates*—A system of coordinates used to define the position of a point in space with respect to an arbitrarily chosen origin by means of two directions, and one distance (i.e., the vectorial angles and radius vector). The primary axis of direction is the *polar axis*. *polar coordinates*—A system of coordinates used to define the position of a point in space with respect to an arbitrarily chosen origin by means of two directions and one distance (*i.e.*, the vectorial angles and radius vector). Any plane containing the polar axis may be called a meridional plane, and the plane perpendicular to the polar axis containing the origin is called the *equatorial plane* or *equator*. As any point must lie on a *meridional plane*, one coordinate of a point in this system is the angle formed by the intersection of its meridional plane with the reference meridional plane. This is called the *polar angle* or *polar bearing*. The second coordinate of a point is the angle in its meridional plane subtended at the origin between the line to the point and the polar axis, also, the arc of the great circle between the point and the pole. This angle is called the *polar distance*, and its complement, the angle between the line to the point and the equator, is the *declination*. The third coordinate is the distance between the origin and the point. *plane polar coordinates*—A system of polar coordinates in which the points all lie in one plane. In the terminology of analytical geometry, the distance from the origin to the point is the *radius vector* and the polar distance is the *vectorial angle*. *spherical coordinates*—A system of *polar coordinates* in which the origin is the center of a sphere and the points all lie on the surface of the sphere. The polar axis of such a system cuts the sphere at its two poles. In photogrammetry, spherical coordinates are useful in determining the relative orientation of perspective rays or axes and make it possible to state and

solve, in simple forms, many related problems. For example, as used in determining the exterior orientation of a single photograph, the origin is the camera station and the polar axis is the vertical. The polar bearing is the horizontal bearing (azimuth) of the principal plane, and the polar distance is the tilt. In the determination of the relative orientation between pairs of photographs by the method originated by Fourcade, the polar axis of the coordinate system is the air base, and the origin is one of the camera stations. A meridional plane in this case is called a *basal* or *epipolar plane*, and the reference meridional plane may be arbitrarily chosen but is usually the vertical (see also *celestial sphere*). *geographic coordinates*—A system of *spherical coordinates* for defining the positions of points on the earth. The *declinations* and *polar bearings* in this system are the *geographic latitudes* and *longitudes* respectively. *astronomical coordinates* (earth)—*astronomical latitude*—The angle between the plumb line and the plane of celestial equator. Also defined as the angle between the plane of the horizon and the axis of rotation of the earth. Astronomical latitude applies only to positions on the earth and is reckoned from the astronomical equator (0°), north and south through 90°. Astronomical latitude is the latitude which results directly from observations of celestial bodies, uncorrected for *deflection of the vertical*. *astronomical longitude*—The angle between the plane of the celestial meridian and the plane of an initial meridian, arbitrarily chosen. Astronomical longitude in the longitude which results directly from observations on celestial bodies, uncorrected for *deflection of the vertical*. *geodetic coordinates*—The quantities of *geodetic latitude*, and *longitude*, which define the position of a point on the surface of the earth with respect to the reference spheroid. *Geocentric Coordinates*—(terrestrial) Coordinates that define the position of a point with respect to the center of earth. Geocentric coordinates can be either *Cartesian* (x, y, z) or *spherical* (geocentric latitude and longitude, and radial distance). Also called *coordinate system; geocentric position*. *photograph coordinates* (photogrammetry)—A system of coordinates (either *rectangular* or *polar*) to define the positions of points on a photograph. If a two-dimensional system is used, the origin is usually the principal point, but it may be the nadir point, isocenter, one of the fiducial marks, or (on a high-oblique photograph) the intersection of the horizontal and principal line. The coordinate axes are usually either the fiducial axes, or the principal line and a photograph parallel. If a three-dimensional system is used, the origin is either the principal point or the perspective center. *space coordinates* (photogrammetry)—May refer to any general three-dimensional coordinate system used to define the position of a point in the object space, as distinguished from the image of the point on a photograph. *model coordinates* (photogrammetry)—The *space coordinates* of any point imaged on an overlapping pair of photographs, which define its position with reference to the air base. They correspond, in respect to the position of origin and direction of axes, to a system of spherical coordinates in which an air base is the polar axis. Consequently, one such system (as suggested by Fourcade) can be defined as follows: *origin:* The left-hand air station. *X axis:* The line of the air base to the right. *Z axis:* The line perpendicular to the *X* axis, in the basal plane containing the principal point of the left-hand photograph. (The ground is considered as being in the negative direction.) *Y* axis: The line perpendicular to the *X* and *Z* axes. The positive direction is toward the top side of the strip when viewed as running from left to right. *strip coordinates*—The coordinates of any point in a strip, whether on the ground or actually an air station, referred to the origin and axes of the coordinate system of the first overlap.

coordinatograph—An instrument used to plot in terms of plane coordinates. It may be an integral part of a stereoscopic plotting instrument whereby the planimetric motions (*x* and *y*) of the floating mark are plotted directly.

coplanar—Lying in the same plane, *basal coplane* (photogrammetry)—The condition of exposure of a pair of photographs in which the two photographs lie in a common plane parallel to the air base. If the air base is horizontal, the photographs are said to be exposed in *horizontal coplane.*

copy camera—*see camera.*

corresponding principal point—The principal point of one aerial photograph where it appears on the adjacent overlapping area of a stereo pair of photos (assuming more than 50% endlap). The use of the term conjugate principal point is not recommended (*see* discussion under *conjugate image points).*

correlation—1. (general) The statistical interdependence between two quantities (*e.g.*, in geodesy, gravity anomalies are correlated with other gravity anomalies, with elevation, with elevation differences, with geology, and so forth). 2. (surveying) The removal of discrepancies that exist among survey data so that all parts are interrelated without apparent error. The terms coordination and correlation are usually applied to the harmonizing of surveys of adjacent areas or of different surveys over the same area. Two or more such surveys are coordinated when they are computed on

the same datum; they are correlated when they are adjusted together.

correspondence (stereoscopy)—The condition that exists when corresponding images on a pair of photographs lie in the same epipolar plane; the absence of *y* parallax (*see* **y** *parallax* under *parallax).*

corresponding images—A point or line in one system of points or lines homologous to a point or line in another system. *corresponding image points*—(the use of the term conjugate points is not recommended; see discussion under *conjugate image points)* are the images of the same object point on two or more photographs (*see* *homologous images).*

corresponding image rays—Rays connecting each of a set of corresponding image points with its particular perspective center.

course (air navigation)—The direction in which a pilot attempts to fly an aircraft; the line drawn on a chart or map as the intended *track.* The direction of a course is always measured in degrees from the true meridian, and the true course is always meant unless it is otherwise qualified (*e.g.*, as a magnetic or compass course) (*see* *track* and *flight line).*

covariance—The arithmetic mean or expected value of the product of the deviations from their respective mean values of corresponding values of two variables.

coverage—The ground area represented on aerial photographs, photomaps, mosaics, maps, and other graphics.

crab (aerial photography)—The condition caused by failure to orient a camera with respect to the track of the airplane. In vertical photography, crab is indicated by the edges of the photographs not being parallel to the airbase lines (*see* *yaw).*

critical angle (optics)—The angle beyond which total internal reflection of a ray takes place when passing from a medium of higher refractive index to a medium of lower index. The angle is expressed by the equation $\sin A = N'/N$, in which A is the critical angle, N is the higher index of refraction, and N' is the lower index of refraction (*see* *Snell's law of refraction).*

crop—To trim or cut off parts of a photograph in order to eliminate superfluous portions and thus improve balance or composition. Usually accomplished by masking the image area during printing.

cross-flight photography—Single photographic strips having stereoscopic overlap between exposures and having a flight direction at right angles to that of coexistent area-coverage photography. When applied to shoran, the term implies that each of the cross-flight exposures is accompanied by recorded shoran distances (*see* *control strip).*

crosshairs—A set of wires or etched lines

placed on a *reticle* held in the focal plane of a telescope. They are used as index marks for pointings of the telescope such as in a transit or level when pointings and readings must be made on a rod.

cross section—A profile of the ground taken transverse to a centerline. It consists of recording elevations and corresponding distances, both left and right, from the centerline. Cross sections may be measured by field methods using an engineers' level and tape, but they are often more rapidly and conveniently taken using a stereoplotter equipped with a digitized coordinatograph (*see profile*).

cross tilt—An error introduced into stereotriangulation due to the inability to recover the exact camera stations for successive pairs. This condition is generally due to variations in equipment, materials, or to imperfect relative orientation.

culture (mapping)—Features of the terrain that have been constructed by man. Included are such items as roads, buildings, and canals; boundary lines; and, in a broad sense, all names and legends on a map (*see details*).

curvature of earth—The offset from the tangent to the curve, as a result of the curvature of the earth.

curvature of field—*see aberration.*

cylindrical lens—A lens in which the surfaces are segments of cylinders.

dark slide—A thin plate (usually metal or fiber, rigid or flexible) which, after insertion in a camera magazine, renders it light-tight. The employment of dark slides makes it possible to interchange camera magazines in daylight.

datum: horizontal-control datum—The position on the spheroid of reference assigned to the horizontal control of an area and defined by (1) the position (latitude and longitude) of one selected station in the area, and (2) the azimuth from the selected station to an adjoining station. The horizontal-control datum may extend over a continent or be limited to a small area. A datum for a small area is usually called a *local datum* and is given a proper name. The horizontal-control datum for the North American continent is known as the *North American Datum of 1927,* the initial station for which is MEADES RANCH, in Kansas, with the azimuth to station WALDO. All geodetic positions on the *North American Datum of 1927* depend on the position of MEADES RANCH and the azimuth to WALDO. vertical-control datum—Any *level surface* (usually *mean sea level*) taken as a surface of reference from which to reckon elevations; also called the **datum level** (*see benchmark*). Although a level surface is not a plane, the vertical-control datum is frequently referred to as the datum plane (*see geoid*).

horizontal plane—A plane perpendicular to the direction of gravity; any plane tangent to the geoid or parallel to such a plane (*see ground plane*). mean datum plane—A hypothetical reference plane at the average elevation of an area.

datum level—*see geoid,* and **vertical-control datum** under **datum.**

datum plane—*see datum.*

declination (geometry)—*see polar* and **spherical coordinates** under **coordinates.** (*see magnetic declination.*)

definition (photography)—Degree of clarity and sharpness of an image.

deflection of the vertical—The angular difference, at any place, between the upward direction of a plumb line (the vertical) and the perpendicular (the normal) to the *reference spheroid.* This difference seldom exceeds 30 seconds. Often expressed in two components, meridian and prime vertical. Also called *deflection of the plumb line; station error.*

delineation (cartography)—The visual selection and distinguishing of mapworthy features on various possible source materials by outlining the features on the source material, or on a map manuscript (as when operating a stereoscopic plotting instrument); also, a preliminary step in compilation. *photodelineation*—The delineation of features on a photograph (*see compilation*).

densitometer—Instrument for measuring the density of a photo image.

density—*see characteristic curve.*

density exposure curve—*see characteristic curve.*

depression angle—The complement of the angle of *tilt* (*see angle of depression*).

depression contour—*see contour.*

depth curve—A line on a map or chart connecting points of equal depth below the hydrographic datum. Also called *isobath* or *bathymetric contour.*

depth of field—The distance between the points nearest and farthest from the camera which are imaged with acceptable sharpness.

depth of focus—The distance that the focal plane can be moved forward or backward from the point of exact focus, and still given an image of acceptable sharpness. Also called *focal range.*

descending node—The point at which a planet, planetoid, or comet crosses the ecliptic from north to south, or a satellite crosses the equator of its primary from north to south. Opposite of **ascending node.** Also called *southbound node.*

details (mapping)—The small items or particulars of information (shown on a map by lines, symbols, and lettering) which, when considered as a whole, furnish the comprehensive representation of the physical and cultural features of the earth's surface. The greater the omission of details, the more generalized the map.

detection—An awareness that a pattern or object exists on the photo or remote sensing imagery. The first step in the process of *photo-interpretation*.

develop and *development* (photography)—To subject to the action of chemical agents for the purpose of bringing to view the invisible or *latent image* produced by the action of light on a sensitized surface; also, to produce or render visible in this way.

developer (photography)—The solution used to make visible the *latent image* in an exposed emulsion. In black and white photography the process is one in which the silver halide grains which were exposed to light are reduced to metallic silver (*see* *develop*).

diagonal check—Measurements made across the opposite corners of the basic frame of a *map projection* to insure the accuracy of its construction, or to establish and/or check the scale of *reproduction*.

diaphragm—The physical element of an optical system which regulates the quantity of light traversing the system. The quantity of light determines the brightness of the image without affecting the size of the image (*see* *aperture stop* and *relative aperture*).

diapositive (photogrammetry)—A positive photograph on a transparent medium, usually polyester or glass. The term is generally used to refer to a transparency used in a plotting instrument, a projector, or a comparator.

diapositive printer—A photographic device for producing diapositives from aerial negatives.

differential distortion—The resultant dimensional changes in length and width in any medium (*see* *differential shrinkage*).

differential leveling—*see* *leveling*.

differential shrinkage (mapping)—The difference in unit contraction along the grain structure of the material as compared to the unit contraction across the grain structure; frequently refers to photographic film and papers and to map materials in general (*see* *differential distortion*).

diffraction (optics)—The bending of light rays around the edges of opaque objects. Due to diffraction, a point of light seen or projected through a circular aperture will always be imaged as a bright center surrounded by light rings of gradually diminishing intensity. Such a pattern is called a *diffraction disk, Airy disk,* or *centric*.

diffuse reflection—The type of reflection obtained from a relatively rough surface (such as a *matte photographic print*), in which the reflected rays are scattered in all directions.

digital—Pertaining to data in the form of digits.

digital computer—*see* *computer*.

digitize—To use numeric characters to express,

or represent data, *e.g.,* to obtain from an *analog* representation of a physical quantity, a digital representation of the same quantity.

digit, binary—In binary digit, either of the characters, 0 or 1.

digit, octal—A digit that is a member of the set of eight digits: 0, 1, 2, 3, 4, 5, 6, 7, used in a numerical notation system that has a radix of eight.

dihedral angle—The angle between two intersecting planes.

dimensional stability—Ability to maintain size; resistance to dimensional changes caused by changes in moisture content and temperature (*see* *differential distortion, differential shrinkage* and *film distortion*).

diopter—A unit of measurement of the *power* of a lens, especially a spectacle-type lens. The power in diopters equals the reciprocal of the focal length in meters; thus, a lens whose focal length is 20 cm has a power of 5 diopters.

dip angle—The vertical angle, at the air station, between the *true* and the *apparent horizon,* which is due to flight height, earth curvature, and refraction.

direction—The position of one point relative to another without reference to the distance between them. Direction may be either three-dimensional or two-dimensional, the horizontal being the usual plane of the latter. Direction is usually indicated in terms of its angular distance from a reference direction (*see* also *azimuth* and *bearing*).

direction instrument theodolite—A theodolite in which the graduated horizontal circle remains fixed during a series of observations, the telescope being pointed on a number of signals or objects in succession, and the direction of each read on the circle, usually by means of micrometer microscopes. Direction instrument theodolites are used almost exclusively in first- and second-order triangulation. Also called direction theodolite (*see* *transit*).

direction of flight—The heading of an aircraft, or the direction in which it is flying.

direction method of observation—A method of observing angular relationships wherein the graduated circle is held in a fixed position, and the directions of the various signals are observed around the horizon. Thus, directions are pointings whereby angles are found by the differences in directions. Also called direction method of measuring horizontal angles.

direct linear transformation (DLT)—An analytical procedure through which the *comparator coordinates* of image points are transformed directly into *object-space coordinates* without the need for fiducial marks or initial approximations for the inner and outer orientation parameters of the photograph.

direct positive—A positive image obtained directly without the use of a negative.

direction of tilt—The direction (azimuth) of the

principal point of a photograph. Also, the direction of the *principal line* on a photograph.

***direct radial plot**—see **radial triangulation**.*

***direct radial triangulation**—see **radial triangulation**.*

discrepancy—A difference between results of duplicate or comparable measures of a quantity. The difference in computed values of a quantity obtained by different processes using data from the same survey.

dispersion—The separation of light into its component colors by its passage through a *diffraction grating* or by *refraction* such as that provided by a prism.

displacement—Any shift in the position of an image on a photograph which does not alter the perspective characteristics of the photograph (*i.e.*, shift due to tilt of the photograph, scale change in the photograph, and relief of the objects photographed). Contrast with *distortion*. ***relief displacement**—* Displacement of images radially *inward* or *outward* with respect to the *photograph nadir*, according as the ground objects are, respectively, *below* or *above* the elevation of the *ground nadir*. ***tilt displacement**—* Displacement of images, on a tilted photograph, radially *outward* or *inward* with respect to the *isocenter*, according as the images are, respectively, on the *low* or *high* side of the *isometric parallel* (the *low* side is the one tilted closer to the earth, or object plane). ***refraction displacement**—*Displacement of images radially *outward* from the *photograph nadir* because of atmospheric refraction. It is assumed that the refraction is symmetrical about the nadir direction.

***distance, principal**—see **focal length**.*

distortion—Any shift in the position of an image on a photograph which alters the perspective characteristics of the photograph. Causes of image distortion include lens aberration, differential shrinkage of film or paper, and motion of the film or camera. Contrast with ***displacement***.

distortion compensation—That correction applied to offset the effect of distortion.

distortion curve—A curve representing the linear distortion characteristics of a lens. It is plotted with image radial distances from the lens axis as abscissas and image radial displacements as ordinates. The distortion curve is based on the existence of *angular distortion*, which is defined as the difference between the object-space and image-space values of the angle between a ray and the lens axis. If the image-space value of this angle is greater than the object-space value, the angular distortion is positive; if less, it is negative. Geometrically, the radial distance of a ray trace in the focal plane is computed from the formula $r = f \tan \alpha$, in which f is the focal length and α is the object-space value of the ray angle. The actual radial distance will differ from the geometric value because of angular distortion. This difference, which may be considered a radial displacement of the ray trace, is termed *linear distortion*. It is considered positive when outward, and negative when inward. By varying the value of f, it is possible to make this difference equal to zero at any given radial distance. The value of f used in lens calibration for the purpose of determining linear distortion is called the *equivalent focal length*. For a wide-angle lens, it is computed from the above formula by measuring, in the plane of best average definition, the radial distance to the image at about 10° from the lens axis.

diurnal—Having a period of, occurring in, or related to a day.

***diverging lens**—see negative lens.*

***d log e**—see **characteristic curve**.*

dodging—A process used while enlarging photographs by projection. Light which passes through certain parts of the negative is held back, and prevented from striking the sensitized paper. Manual dodging is done by holding a piece of opaque material between the enlarger lens and the easel. Electronic dodging is produced by feedback of signal voltage through the negative or positive to be printed to minimize density variations of produced materials.

Doppler effect—The alteration in frequency of a wave radiation caused by a relative motion between the observer and the source of radiation, —acoustic Doppler effect applies to the propagation of source waves, —optical Doppler effect depends on the relative velocity of the light source and the observer, —thermal Doppler effect causes a widening of the spectral lines.

Doppler navigation—A system which measures ground speed and drift by means of electronically generated signals emitted from aircraft and reflected from the terrain. The system depends on the difference in frequencies between emitted and reflected signals caused by aircraft movement.

Doppler satellite positioning—The determination of geodetic positions through the measurement of Doppler shifts of radio signals emitted from passing satellites.

dot grid—Film positive with regularly spaced dots used in determining areas.

double weight paper—Heavy duty photographic paper. More durable and dimensionally stable than single-weight paper (*see **dimensional stability***).

double burn—The intentional exposure of two or more line and/or halftone negatives in succession and register on the same sensitized surface. Not to be confused with double exposure, which is usually unintentional. Also called *double shooting*.

double-model stereotemplet—*see* **templet**.

double-projection direct-viewing stereoplotter—A class of stereoplotters employing the principle of projecting the images of two correctly oriented overlapping aerial photographs onto a reference datum so the resultant images may be viewed directly without additional optical system support. Also called *double-optical projection stereoplotter*.

dove prism—A prism which reverts the image but does not deviate or displace the beam. A given angular rotation of the prism about its longitudinal axis causes the image to rotate through twice the angle. Also called a *rotating prism*.

drift (air navigation)—The lateral shift or displacement of an aircraft from its *course*, due to the action of wind or other causes.

duplicate negative—A negative made from an original negative or from a positive. The duplicate negative may be a true reproduction of the original or a reproduction possessing greater or less contrast. With direct positive materials, chemical reversal process, and duplicating film, it is not always necessary to make a positive to obtain a duplicate negative.

easting—*see* **plane rectangular coordinates** under **coordinates**.

earth-fixed coordinate system—Any coordinate system in which the axes are stationary with respect to the earth.

edge fog—*see* **fog**.

editing (cartography)—The process of checking a map or chart in its various stages of preparation to insure accuracy, completeness, and correct preparation from and interpretation of the sources used, and to assure legible and precise reproduction. Edits are usually referred to by a particular production phase, such as compilation edit, scribing edit, *etc*.

efficiency of a shutter—The relationship between the total time a shutter remains open (counting from half-open to half-closed position) and the time required for the shutter to reach the half-open and the half-closed positions.

effective area of aerial photograph—That central part of the photograph delimited by the bisectors of overlaps with adjacent photographs. On a vertical photograph, all images within the effective area have less displacement than their corresponding images on adjacent photographs.

effective focal length (EFL)—*see* **principal distance** under **focal length**.

electromagnetic radiation (EMR)—Energy propagated through space or through material media in the form of an advancing disturbance in electric and magnetic fields existing in space or in the media. The term *radiation*, alone, is used commonly for this type of energy, although it actually has a broader meaning. Also called *electromagnetic energy* (*see* **spectrum** and **radiation**).

electronic distance-measuring equipment (EDM)—Devices that measure the phase difference between transmitted and returned (*i.e.*, reflected or retransmitted) electromagnetic waves, of known frequency and speed, or the round-trip transit time of a pulsed signal, from which distance is computed. A wide range of such equipment is available for surveying and navigational use (*e.g.*, Geodimeter, Tellurometer, and *hiran*).

electromagnetic spectrum—*see* **spectrum**.

elevation—Vertical distance from the datum (usually mean sea level) to a point or object on the earth's surface. Sometimes used synonymously with altitude, which in modern usage refers particularly to the distance of points or objects above the earth's surface. *field elevation*—an elevation taken from the field computation of a line of levels. *adjusted elevation*—The elevation resulting from the application of an adjustment correction to an orthometric elevation. Also, the elevation resulting from the application of both an orthometric correction and an adjustment correction to a preliminary elevation. *assumed ground elevation*—The elevation assumed to prevail in the local area covered by a particular photograph or group of photographs. Used especially to denote the elevation assumed to prevail in the vicinity of a critical point, such as a peak or other feature having abrupt local relief.

electron microscope—A microscope of extremely high power that uses beams of electrons instead of rays of light; the magnified image being formed on a screen or recorded on a photographic plate.

electronic surveying—Any survey utilizing electronic equipment.

elevation meter—A mechanical or electromechanical device on wheels that measures slope and distance, and automatically and continuously integrates their product into difference of elevation.

ellipsoid—A surface whose plane sections (cross sections) are all ellipses or circles, or the solid enclosed by such a surface (*see* **geoid**).

emergent nodal point—*see* **nodal point**.

empirical orientation (rectification)—*see* **transformation**.

emulsion (photography)—A suspension of a light-sensitive silver salt (especially silver chloride or silver bromide) in a colloidal medium (usually gelatin), which is used for coating photographic films, plates, and papers.

emulsion speed—A property of photographic emulsions which determines how long they must be exposed to a given light source to secure equal density when developed. This speed may be given in USASI, DIN, Weston, Scheiner, or AEL scales.

emulsion-to-base—A contact exposure in which the emulsion of the copying film is on the side of the film opposite to that in contact with the sheet being copied (*see* **print**).

emulsion-to-emulsion—A contact exposure in which the emulsion of the copying film is in contact with the emulsion of the sheet being copied (*see* **print**).

endlap—*see* **overlap.**

enlargement—The production of a negative, diapositive or print at a *scale* larger than the original.

enlarging—The process of making a print or negative larger than the original by projection printing (*see* **transformation**).

enlarging lens—*see* **process lens.**

entrance pupil—*see* **aperture stop.**

entrance window—*see* **aperture stop.**

epipolar plane—*see* **epipoles.**

epipolar ray—*see* **epipoles.**

epipoles—In the perspective setup of two photographs (two perspective projections), the points on the planes of the photographs where they are cut by the *air base* (extended line joining the two perspective centers). In the case of a pair of truly vertical photographs, the epipoles are infinitely distant from the principal points. *epipolar plane*—Any plane which contains the epipoles; therefore, any plane containing the air base. Also called *basal plane*. *epipolar ray*—The line on the plane of a photograph joining the epipole and the image of an object. Also expressed as the trace of an epipolar plane on a photograph.

equation, normal—*see* **normal equation.**

equator (geometry)—In a system of polar or spherical coordinates, the great circle of a sphere which is perpendicular to the polar axis.

equivalent focal length—*see* **focal length.**

equivalent vertical photograph—A theoretically, truly-vertical photograph taken at the same camera station with a camera whose focal length is equal to that of a camera taking a corresponding tilted photograph.

error (statistics)—The difference between an observed or computed value of a quantity and the ideal or true value of that quantity. Errors are defined by types or by causes. *absolute error*—Absolute deviation (the value taken without regard to sign) from the corresponding true value. *accidental (irregular* or *random) errors*—Those whose occurrence depends on the law of chance only; they are unpredictable in regard to both magnitude and algebraic sign. In formulating adjustments for these errors, it is assumed that the relation between the magnitude and the frequency of the individual errors are regulated by some law, usually the *normal frequency function* (or *Gauss law*). Such errors tend to be *compensating* in their effect. The theory of errors and the method of least squares are based on the behavior of accidental errors. *systematic error*—An error which, for known changes in field conditions, undergoes proportional changes in magnitude and which, for unchanging conditions, remains unchanged, both in sign and magnitude. A systematic error always follows some definite mathematical and physical law, and a correction may be determined and applied. Also called *regular error*. *constant error*—The result of the observational conditions remaining unchanged, so that there is no change in either the sign or the magnitude of the error. A number of readings under these conditions produce an *accumulative error*, equal to the number of readings multiplied by the error in one reading. *gross error* (or *blunder*)—The result of carelessness or a mistake; *may* be detected through repetition of the measurements. *personal error*—The result of the inability of an observer to perceive or measure dimensional values exactly. Personal errors may be either random or systematic in behavior. *instrumental error*—Generally systematic in nature; the result of an imperfect condition of an instrument. However, such an error may be accidental or random in nature and result from the failure of an instrument to give the same indication when subjected to the same input signal. *index error*—An instrumental error caused by the displacement of the zero or index mark or vernier; constant in behavior. *closure* or *closing error*—The amount by which a quantity obtained by a series of related measurements differs from the true or fixed value of the same quantity. Hence, it is a cumulative measure of the various individual errors and blunders in the measurements. *average error*—The arithmetic mean, taken without regard to sign, of all the errors of a set. *mean error*—In American usage, synonymous with *average error;* in European usage, especially German, synonymous with *root mean square error*. *root mean square error (RMSE)*—The expression for the accuracy of a single observation; defined as the square root of the quantity; the sum of the squares of the errors divided by the number of errors. The square of this term is the *mean square error*. *standard error* or *standard deviation*—The root mean square value based on errors corrected by subtracting the average error (assumed to be a constant) from all the errors of a set. It is generally denoted by the Greek letter *sigma* (σ) and is a measure of precision (or accuracy) of a single observation. Its square is termed *variance*.

probable error—An error (or deviation from the mean) of such magnitude that the likelihood of its being exceeded in a set of observations is equal to the likelihood of its not being exceeded; its value is that of the *standard error* multiplied by 0.6745. The use of *standard error* is sometimes preferred in statistical studies. *residual* (or *residual error*)—The difference between any value of a quantity in a series of observations, corrected for known systematic errors, and the value of the quantity obtained from the combination or adjustment of that series. Also called *errors; residuals*. The latter term is generally used in referring to actual values in a specific computation. When a measurement is referenced to some arbitrary or estimated value, the difference generally is denoted as *deviation* instead of *error*. An error thus constitutes a deviation from a particular reference.

error, propagated—An error that occurs in one operation and spreads through later operations.

eulerian angles—A system of three angles which uniquely define, with reference to one coordinate system (*e.g.*, earth axis), the orientation of a second coordinate system (*e.g.*, body axes). Any orientation of the second system is obtainable from that of the first by rotation through each of the three angles in turn, the sequence of which is important.

exaggeration, vertical—see vertical exaggeration.

exchange agreement—An approved agreement between two or more organizations to furnish each other specified mapping, charting, and geodetic data as published, or on a request basis.

exit pupil—see aperture stop.

exit window—see aperture stop.

exposure—The total quantity of light received per unit area on a sensitized plate or film; may be expressed as the product of the light intensity and the exposure time, in units of (for example) metre-candle-seconds or watts per square metre. The act of exposing a light-sensitive material to a light source (*see characteristic curve*).

exposure factors—A large number of factors which must be taken into consideration in arriving at correct exposure time. Following is a list of some of the important factors: (1) relative aperture of lens; (2) type of film used; (3) reflecting power of the subject; (4) season of the year; (5) time of day; (6) color of the light; (7) geographical location; (8) flight altitude; (9) atmospheric condition.

exposure interval—The time interval between the taking of successive photographs (*see intervalometer*).

exposure meter—An instrument used to measure the intensity of light and thus determine correct exposure.

exposure station—see camera station.

exposure time—The time during which a light-sensitive material is subjected to the action of light.

extension of control—see phototriangulation.

exterior orientation—see orientation.

exterior perspective center—see perspective center.

eye base—Synonymous with *interpupillary distance* and *interocular distance*.

eyepiece—In an optical device, the lens group which is nearest the eye and with which the image formed by the preceding elements is viewed.

fan camera—see camera.

featheredging—see mosaic.

fiber optics—A device for relaying an image by means of a large number of transparent fibers (filaments) by multiple total internal reflection. The fibers are most commonly glass and less often a highly transparent plastic. Each fiber carries only one element of the image, so that the image is a mosaic in which the cell size is the fiber cross section rather than a continuous picture.

f number—see relative aperture.

fiducial axes (photogrammetry)—The lines joining opposite *fiducial marks* on a photograph (*see photograph coordinates* under *coordinates*).

fiducial marks (photogrammetry)—Index marks, usually four, which are rigidly connected with the camera lens through the camera body and which form images on the negative and usually define the *principal point* of the photograph. Also marks, usually four in number, in any instrument which define the axes whose intersection fixes the principal point of a photograph and fulfills the requirements of interior orientation.

field calibration—see calibration.

field check—The operation of checking a map *compilation manuscript* on the ground. Compare with *field classification* (*see check profile*).

field classification—Field inspection and identification of features which a map compiler is unable to delineate; identification and delineation of political boundary lines, place names, road classifications, buildings hidden by trees, and so forth. Field classification may be included as part of the control survey effort and normally is completed prior to the actual stereocompilation phase.

field completion—A combination of field inspections or surveys, either before or after compilation, to classify and complete the map content, correct erroneous data, and add information such as names, civil boundaries, and similar classification data. Its purpose is to fill in or confirm that portion of a map manuscript prepared by *stereocompilation*.

field contouring—Contouring a topographic map by field methods accomplished by planetable surveys on a prepared base. Generally, this operation is applied to terrain unsuitable for contouring by photogrammetric methods.

field correction copy—A map or tracing prepared in the field, delineating corrections for subsequent reproduction of a map (*see* also *field check* and *field completion*).

field elevation—*see* *elevation*.

field inspection (photogrammetry)—The process of comparing aerial photographs with conditions as they exist on the ground, and of obtaining information to supplement or clarify that not readily discernible on the photographs themselves (*see* *field classification* and *field completion*).

field stop—*see* *aperture stop*.

film—A plastic base which is coated with a light sensitive emulsion for use in a camera or printing frame.

film base—(photography)—A thin, flexible, transparent sheet of stable plastic material.

film distortion—The dimensional changes which occur in photographic film with changes in humidity or temperature, or from aging, handling, or other causes (*see* *differential shrinkage*).

film speed—That property of film which determines how much exposure must be allowed for a given light source in order to secure a negative of correct *density* and *contrast*. Speed is measured on the H. & D., Scheiner, DIN, Weston, USAIS, or AEI, scales.

filter—Any transparent material which by absorption, selectively modifies the light transmitted through an optical system.

filter factor—A number indicating the exposure increase necessary when using a filter, as compared to the exposure necessary under the same conditions without the filter.

filter ratio—The ratio between the filter factors of two or more filters with same film.

fix—(1) photography: To render a developed photographic image permanent by removing the unaffected light-sensitive material. (2) surveying: To establish the position of a point of observation by a surveying procedure, usually *resection*. Also, the point thus established.

fixed-ratio pantograph—see *pantograph*.

flash plate—see *calibration plate*.

flat—see *optical flat*.

flicker method—The alternate projection of corresponding photographic images onto a tracing-table platen or projection screen, or into the optical train of a photogrammetric instrument.

flight altitude—The vertical distance above a given datum, usually mean sea level, of an aircraft in flight or during a specified portion of a flight. In aerial photography, when the datum is mean ground level of the area being photo-

graphed, this distance is called *flight height, flying height,* or sometimes *absolute altitude*.

flight height—*see* *flight altitude*.

flight line—A line drawn on a map or chart to represent the track of an aircraft (*see* *course*).

flight map—The map on which are indicated the desired lines of flight and/or the positions of exposure stations previous to the taking of air photographs, or the map on which are plotted, after photography, selected air stations and the tracks between them.

flight plan—A plan, prepared before aerial photography is executed, to provide photographs suitable for subsequent photogrammetric processes or operations.

flight strip—A succession of overlapping aerial photographs taken along a single course.

floating dot—*see* *floating mark*.

floating mark (photogrammetry)—A mark seen as occupying a position in the three-dimensional space formed by the stereoscopic fusion of a pair of photographs and used as a reference mark in examining or measuring the stereoscopic model. The mark may be formed (1) by one real mark lying in the projected object space; (2) by two real marks lying in the projected or virtually projected object spaces of the two photographs; (3) by two real marks lying in the planes of the photographs themselves; (4) by two virtual marks lying in the image planes of the binocular viewing apparatus. *index mark* (photogrammetry)—A real mark (such as a cross or dot) lying in the plane or the object space of a photograph and used singly as a reference mark in certain types of monocular instruments, or as one of a pair to form a floating mark (as in certain types of stereoscopes).

focal length—The distance measured along the optical axis from the rear nodal point the lens to the plane of critical focus of a very distant object (*see* *back focal length* and *front focal length*). *nominal focal length*—An approximate value of the focal length, rounded off to some standard figure, used for the classification of lenses, mirrors, or cameras. *equivalent focal length*—The distance measured along the lens axis from the rear nodal point to the plane of best average definition over the entire field used in an aerial camera. In general usage, the term also applies to the distance from the rear nodal point to the plane of best axial definition; in photogrammetry, however, this meaning is rarely used and will not be understood unless the term is accompanied by a qualifying phrase. *back focal distance* or *back focal length*—The distance measured along the lens axis from the rear vertex of the lens to the plane of best average definition. This value is used in setting the lens in an aerial camera. *front focal length* or *front focal distance*—The distance measured from the vertex of the front surface of lens to the front focal point. *calibrated focal length*—An adjusted value of the *equivalent focal length* computed to distribute the effect of lens distortion over the entire field used in an aerial camera. This value also is stated as the distance along the lens axis from the interior perspective center to the image plane, with the interior center of perspective selected to distribute the effect of lens distortion over the entire field. The calibrated focal length is used in the determination of the setting of diapositives in plotting instruments and in photogrammetric computations based on linear measurements on the negative (such as those made with a precision comparator). *principal distance*—The perpedicular distance from the internal perspective center to the plane of a particular finished negative or print. The distance is equal to the calibrated focal length corrected for both the enlargement or reduction ratio and the film or paper shrinkage or expansion. The same perspective angles to points on the finished negative or print as existed in the taking camera at the moment of exposure are maintained at the internal perspective center. This is a geometrical property of each particular finished negative or print.

focal plane (photography)—The plane (perpendicular to the axis of the lens) in which images of points in the object field of the lens are focused. *focal-plane plate*—A glass plate set in the camera so that the surface away from the lens coincides with the focal plane. Its purpose is to position the emulsion of the film in the focal plane when the film is physically pressed into contact with the glass plate. Also known as *contact glass* or *contact plate*.

focal-plane plate—*see focal plane.*

focal-plane shutter—*see shutter.*

focal spot—The point of an X-ray tube through which X-rays emanate.

focus—The point toward which rays of light converge to form an image after passing through a lens. Also defined as the condition of sharpest imagery.

fog—A darkening of negatives or prints by a deposit of silver which does not form a part of the image. Fog tends to increase density and decrease contrast. It may be caused by exposure to unwanted light, exposure to air during development, forced development, or impure chemicals. *edge fog*—fog on film caused by leakage of light between the flanges of the spool on which it is wound.

foot-candle—A unit of illumination equivalent to one *lumen* of incident light per square foot.

foot-lambert—A unit of photometric luminance equivalent to $1/\pi$ candle per square foot. It is also the luminance of a perfectly plane diffuse surface radiating one lumen per square foot.

form lines—Lines drawn to represent the shape of terrain; unlike contour lines, they are drawn without regard to a true vertical datum or regular vertical interval.

forward lap—*see overlap.*

frame—Any individual member of a continuous sequence of photographs.

frame camera—*see camera.*

front focal length—*see focal length.*

front nodal point—*see nodal point.*

front surface mirror—An optical mirror on which the reflecting surface is applied to the front surface of the mirror instead of to the back; *i.e.*, to the first surface of incidence.

function, transfer—A mathematical expression that relates the input signals or variables to the output signals or variables, usually showing the operations performed on the variables by the process or the control element.

fusion—*see stereoscopic fusion* under *stereoscopy.*

gamma—*see characteristic curve.*

gap (aerial photography)—Any space where aerial photographs fail to meet minimum coverage requirements. This may be a space not covered by any photograph or a space where the minimum specified overlap was not obtained.

gelatin (photographic)—A natural or artificial protein or protein like substance used as a binding medium for the silver halide crystals in the common type of photographic emulsion.

geocentric—Relative to the earth as a center; measured from the center of the earth.

geocentric coordinates—*see coordinates.*

geocentric latitude—The angle at the center of the earth between the plane of the celestial equator and a line to a point on the surface of the earth. Geocentric latitude is used as an auxiliary latitude in some computations in astronomy, geodesy, and cartography, in which connection it is defined as the angle formed with the major axis of the ellipse (in a meridional section of the spheroid) by the radius vector from the center of the ellipse to the given points. In astronomic work, geocentric latitude is also called *reduced latitude*, a term that is sometimes applied to parametric latitude in geodesy and cartography. The geocentric and isometric latitudes are approximately equal.

geodesic line—A line of shortest distance between any two points on any mathematically defined surface. On a spheroid, a geodesic line is a line of double curvature, and usually lies between the two normal section lines which the two points determine. If the two terminal points are in nearly the same latitude, the geodesic line may cross one of the normal section lines. It should be noted that, except along the equator and along the meridians, the geodesic line is not a plane curve and cannot be sighted over directly. However, for conventional triangulation the lengths and directions of geodesic lines differ inappreciably from corresponding pairs of normal section lines. Also called *geodesic; geodetic line*.

geodesy—The science which treats of the determination of the size and figure of the earth (*geoid*) by such direct measurements as triangulation, precise traverse, trilateration, leveling, gravimetric observations, satellite triangulation and doppler methods. The applied science of geodesy is called geodetic surveying (*i.e.*, surveying which takes account of the figure and size of the earth.

geodetic control—*see* **control**.

geodetic coordinates—*see* **coordinates**.

geodetic datum—*see* **datum**.

geodetic latitude—is the angle which the normal at a point on the reference spheroid makes with the plane of the geodetic equator. Geodetic latitudes are reckoned from the equator, but in the horizontal control survey of the United States they are computed from the latitude of station Meades Ranch as prescribed in the *North American Datum of 1927*.

geodetic longitude—is the angle between the plane of the geodetic meridian and the plane of an initial meridian, arbitrarily chosen. A geodetic longitude can be measured by the angle at the pole of rotation of the reference spheroid between the local and initial meridians, or by the arc of the geodetic equator intercepted by those meridians. In the United States, geodetic longitudes are numbered from the meridian of Greenwich, but are computed from the meridian of station Meades Ranch as prescribed in the *North American Datum of 1927*. A geodetic longitude differs from the corresponding astronomical longitude by the amount of the prime vertical component of the local deflection of the vertical divided by the cosine of the latitude.

geodetic position—A position of a point on the surface of the earth expressed in terms of *geodetic latitude* and *geodetic longitude*. A geodetic position implies an adopted geodetic *datum*.

geodetic surveying—*see* **geodesy**.

geographic coordinates—*see* **coordinates**.

geographic latitude—A general term, applying alike to *astronomic latitudes* and *geodetic latitudes*.

geographic longitude—A general term, applying alike to *astronomical* and to *geodetic longitudes*.

geoid—The figure of the earth considered as a sea-level surface extended continuously through the continents. The actual geoid is an equipotential surface to which, at every point, the plumbline (direction in which gravity acts) is perpendicular. It is the geoid which is obtained from observed *deflections of the vertical* and is the surface of reference for astronomical observations and for geodetic leveling. *reference spheroid* or *ellipsoid*—A

spheroid determined by revolving an ellipse about its shorter (polar) axis and used as a base for geodetic surveys of a large section of the earth (such as the *Clarke Spheroid of 1866*, which is used for geodetic surveys in the United States). The spheroid of reference is a theoretical figure whose dimensions closely approach the dimensions of the *geoid;* the exact dimensions are determined by various considerations of the section of the earth's surface concerned. *level surface*—A surface which, at every point, is perpendicular to the direction of gravity; the *geoid,* or, in general, any surface parallel to it. If changes in elevation due to tides, winds, *etc.* are neglected, the surface of the sea is a level surface. A level surface is not a plane surface, but it is sometimes so regarded in surveys of limited areas (*see* **vertical control datum** under **datum**).

goniometer—An instrument for measuring angles. *photogoniometer*—An instrument for measuring angles from the true *perspective center* to points on a photograph.

gradient—*see* **characteristic curve**.

gradient speed—*see* **characteristic curve**.

grain (photography)—One of the discrete silver particles resulting from the development of an exposed light-sensitive material. *granularity*—The graininess of a developed photographic image, evident particularly on enlargement, that is due either to agglomerations of developed grains or to an overlapping pattern of grains.

graticule—1. A network of lines representing the earth's parallels of latitude and meridians of longitude (*see* **map projection**). 2. A scale at the focal plane of an optical instrument to aid in the measurement of objects (*see* **reticle**).

gravimeter (*gravity meter*)—An instrument designed to measure relative differences in the acceleration due to gravity at different locations.

gray scale—*see* **step wedge**.

great circle—A circle on the surface of a sphere, the plane of which passes through the center of the sphere. Also called *orthodrome*.

Greenwich civil time (*GCT*) or *Greenwich mean time* (*GMT*)—Mean solar time of the meridian of Greenwich, adopted as the prime basis of standard time throughout the world. Also called *universal time*.

Greenwich siderial time (*GST*)—Local sideral time at the Greenwich meridian. The arc of the celestial equator, or the angle at the celestial pole, between the upper branch of the Greenwich celestial meridian and the hour circle of the *vernal equinox*, measured westward from the upper branch of the Greenwich celestial meridian through 24 hours; Greenwich *hour angle* of the vernal equinox, expressed in time units.

grid—A uniform system of rectilinear lines

superimposed on aerial photographs, mosaics, maps, charts, and other representations of the earth's surface; used in defining the coordinate positions of points.

grid coordinates—see coordinates.

grid method (photogrammetry)—A method of plotting detail from oblique photographs by superimposing a perspective of a map grid on a photograph and transferring the detail by eye, that is, by using the corresponding lines of the map grid and its perspective as placement guides (*see perspective grid*).

grid plate—A glass plate on which is etched an accurately ruled grid. Sometimes used as a *focal-plane plate* to provide a means of calibrating film distortion; used also for calibration of plotting instruments. Sometimes called a *reseau.*

ground camera—The preferred term is *terrestrial camera* (*see camera*).

ground control—see control.

ground control point—see control point and *control station.*

ground distance—The great-circle distance between two ground positions, as contrasted with slant range, the straight-line distance between two points. Also called *ground range.*

ground photogrammetry—see terrestrial photogrammetry.

ground photograph—The preferred term is *terrestrial photograph.*

ground plane (photogrammetry)—The horizontal plane passing through the ground nadir of a camera station.

ground plumb point—The preferred term is *ground nadir* (*see* under *nadir*).

ground resolution—see resolution.

ground speed (air navigation)—The rate of motion of an aircraft along its *track* with relation to the ground; the resultant of the *heading* and *air speed* of an aircraft and the direction and velocity of the wind. (*see air speed*).

ground survey—see survey.

ground trace—The preferred term is *ground parallel* (*see* under *principal plane*).

ground truth—Data and observations on the earth's surface normally to quantify simultaneously recorded remote sensing imagery.

gyrocompass—A compass which functions by virtue of the couples generated in a rotor when the latter's axis is displaced from parallelism with that of the earth. A gyrocompass is independent of magnetism and will automatically align itself in the celestial meridian. However, it requires a steady source of motive power and is subject to dynamic error under certain conditions. Certain aircraft compasses also use gyroscopes to gain stability, while relying basically on the magnetic meridian; these are to be distinguished from the true gyrocompass.

gyroscopic stabilization—Equilibrium in the attitude and/or course of a ship or airborne vehicle maintained by the use of gyroscopes. Also, the maintenance (by the use of gyroscopes) of a camera in a desired attitude within an airborne vehicle.

H and D curve—see characteristic curve.

halation (photography)—A spreading of a photographic image beyond its proper boundaries, due especially to reflection from the side of the film or plate support opposite to that on which the emulsion is coated. Particularly noticeable in photographs of bright objects against a darker background. *antihalation coating* (photography)—A light absorbing coating applied to the back side of the support of a film or plate (or between the emulsion and the support) to suppress halation.

halftone—Any photomechanical printing surface or the impression therefrom in which detail and tone values are represented by a series of evenly spaced dots of varying size and shape, varying in direct proportion to the intensity of the tones they represent. Contrast with *continuous tone.*

halftone screen—A grating of opaque lines on glass, crossing at right angles, producing transparent apertures between intersections. Used in a process camera to break up a solid or *continuous tone* image into a pattern of small dots. Also called *crossline glass screen* (*see contact screen*).

hand-templet—see templet.

hand-templet method—see templet and *radial triangulation.*

hand-templet plot—see radial triangulation.

hand-templet triangulation—see templet and *radial triangulation.*

haze—A lack of transparency of the atmosphere, caused by the presence of foreign matter, such as dust or smoke.

heading (air navigation)—Azimuth of the longitudinal axis of an aircraft.

height displacement—see relief displacement under *displacement.*

height finder—A stereoscopic range finder so constructed as to indicate vertical heights rather than slant range (*see stereometer*).

high-oblique photograph—see oblique photograph.

hiran—An electronic distance-measuring system similar to *shoran,* but with improved accuracy, for measuring distances from an airborne station to each of two ground stations. The term *hiran* is a contraction of "high-precision sho*ran*" (*see shiran*).

hologram—A negative produced by exposing a high-resolution photographic plate without camera or lens, near a subject illuminated by mononochromatic, coherent radiation as from a laser: when placed in a beam of coherent light, a true three-dimensional image of the subject is formed.

homologous images—The images of a single

object point that appears on each of the two or more overlapping photographs having different perspective centers (*see* **corresponding images**).

homologous photographs—are two or more overlapping photographs having different camera stations.

horizon—In general, the apparent or visible line of demarcation between land/sea and sky, as seen from any specific position. Also called the *apparent horizon, local horizon, topocentric horizon,* or *visible horizon.* **true horizon**—A horizontal plane passing through a point of vision or a perspective center. The *apparent* or *visible* horizon approximates the *true* horizon only when the point of vision is very close to sea level.

horizon camera—*see* **camera**.

horizon, dip of—*see* **dip angle**.

horizon photograph (aerial photography)—A photograph of the horizon taken simultaneously with another photograph for the purpose of obtaining an indication of the orientation of the other photograph at the instant of exposure (*see* **horizon camera** under **camera**).

horizon trace—*see* **principal plane**.

horizontal control—*see* **control**.

horizontal-control datum—*see* **datum**.

horizontal control station or *horizontal control point*—*see* **control station**.

horizontal coplane—*see* **coplanar**.

horizontal pass point—*see* **pass point**.

horizontal parallax—*see* **absolute stereoscopic parallax** under **parallax**.

horizontal photograph—A photograph taken with the axis of the camera horizontal.

horizontal plane—*see* **datum**.

hour angle—Angular distance west of a celestial meridian or hour circle; the arc of the celestial equator, or the angle at the celestial pole, between the upper branch of a celestial meridian or hour circle and the hour circle of a celestial body or the vernal equinox, measured westward through 360°.

humidity—Degree of wetness, especially of the atmosphere. *relative humidity*—Ratio of water vapor present, at a given temperature, to the greatest amount possible at that temperature. *absolute humidity*—The weight of water vapor contained in a given volume of air, in grains per cubic foot or grams per cubic metre. *specific humidity*—The weight of water vapor per unit weight of the moist air.

hydrographic map—*see* **map**.

hypsograph—An instrument of the slide-rule type used to compute elevations from vertical angles and horizontal distances.

hypsography—*see* **topography**.

hypsometry—The determination of elevations above sea level. Generally applied to the determination of elevations through the measurement of air pressure by observing the boiling point of a liquid.

hyperstereoscopy—Stereoscopic viewing in which the scale (usually vertical) along the line of sight is exaggerated in comparison with the scale perpendicular to the line of sight. Also called *appearance ratio; exaggerated stereo; relief stretching; stereoscopic exaggeration.*

image—The permanent record of the likeness of any natural or manmade features, objects and activities. Images can be acquired directly on photographic materials using cameras, or indirectly if non-imaging types of sensors have been used in data collection.

image, latent—*see* **latent image**.

image motion—The smearing or blurring of imagery on an aerial photograph because of the relative movement of the camera with respect to the ground during exposure.

image-motion compensator—A device installed with certain aerial cameras to compensate for the forward motion of an aircraft while photographing ground objects. True image-motion compensation must be introduced after the camera is oriented to the flight track of the aircraft and the camera is fully stabilized.

image motion factors—Those factors wherein the image motion varies directly with the aircraft ground speed and lens focal length and inversely with the altitude.

image plane—The plane in the camera in which the plate or film is held. It is not exactly the primary focal plane of the lens, but is a plane placed so as to secure the best balance of sharp focus on all parts of the plate or film. Also called *photograph plane.*

image point—Image on a photograph corresponding to a definite object point.

imagery—The visual representation of energy recorded by cameras and other remote sensing instruments.

independent model triangulation—*see* **semi-analytic aerotriangulation**.

incident nodal point—*see* **nodal point**.

index correction—1. A correction applied to the reading from any graduated measuring device to compensate for a *constant error* such as would be caused by misplacement of the scale; the reverse of the *index error.* 2. (leveling) That correction which must be applied to an observed difference of elevation to eliminate the error introduced into the observations when the zero of the graduations on one or both leveling rods does not coincide exactly with the actual bottom surface of the rod.

index error—The *instrumental error* which is constant and attributable to displacement of a vernier or some analogous effect.

index map—(1) A map of smaller scale on which are depicted the locations (with accompanying designations) of specific data, such as larger-scale topographic quadrangles or geodetic control. (2) photography: A map showing the location and numbers of flight strips and photographs. *photoindex*—A *mosaic* (not an *index map*) made by assembling individual photographs, with accompanying designations, into their proper relative positions and copying the assembly photographically at a reduced scale.

index mark—*see floating mark*.

index mosaic—*see photoindex* under *index map*.

index of refraction—*see Snell's law of refraction*.

indirect photography—Photography in which the camera records an *image* cast upon a screen or similar display surface by electronic (television, radar, *etc*.) or other means.

industrial photogrammetry—Encompasses the application of photogrammetry in building construction, civil engineering, mining, vehicle and machine construction, metallurgy, ship-building and traffic, with their fundamental and border subjects including phases of research, planning, production engineering, manufacturing, testing, monitoring, repair and reconstruction (*see* **close-range photogrammetry**).

inertial surveying—The determination of geodetic positions through the use of gyroscopically oriented accelerometers which measure the three-dimensional components of changes in position as they are moved from point to point.

infrachromatic—A term used to describe *emulsions* sensitive to *infrared* radiations as used for infra-red photography.

infrared (photography)—Pertaining to or designating the portion of the electromagnetic *spectrum* with wavelengths just beyond the red end of the visible spectrum, such as radiation emitted by a hot body. Invisible to the eye, infrared rays are detected by their thermal and photographic effects. Their wavelengths are longer than those of visible light and shorter than those of radio waves.

infrared film—Film containing an *emulsion* especially sensitive to infrared and blue light.

instrument phototriangulation—*see stereotriangulation*.

interference fringes—*see test glass*.

interior orientation—*see orientation*.

interior perspective center—*see perspective center*.

interocular distance—Synonymous with *interpupillary distance* and *eye base*.

interpretation of photographs—*see photointerpretation*.

interpreter—A *computer* program that translates and executes each source language statement before translating and executing the next one.

interpupillary distance—The distance between the pupils of the eyes of an individual. Also called *eye base* and *interocular distance*.

intersection—The procedure of determining the position of an object point by intersecting lines of direction obtained photogrammetrically. The lines of direction may be obtained analogically by stereoplotter restitution or by graphic or mathematical means.

intervalometer—A timing device for automatically operating the shutter of a camera at selected intervals (*see* **exposure interval**).

inversors (photography)—Mechanical devices used to maintain correct *conjugate distances* and collinearity of negative, lens, and easel planes in autofocusing optical instruments, such as copy cameras and rectifiers. *Carpentier inversor*—One of the inversors which corrects for the Scheimpflug condition in a rectifier if the negative, lens, or easel planes are tilted and not parallel.

inverted image—An image that appears upside down in relation to the subject.

inverted stereo—*see pseudoscopic image*.

iris diaphragm—A continuously variable circular aperture in a lens which makes it possible to control the amount of light passing through the lens. Also called a *stop* (*see* **aperture stop**).

isocenter—(1) The unique point common to the plane of a photograph, its *principal plane,* and the plane of an *equivalent vertical photograph* taken from the same camera station and having an equal principal distance. (2) The point of intersection on a photograph of the *principal line* and the *isometric parallel*. (3) The point on a photograph intersected by the bisector of the angle between the plumbline and the photograph perpendicular. The isocenter is significant because it is the center of radiation for displacements of images due to tilt.

isocenter plot—*see radial triangulation*.

isocenter triangulation—*see radial triangulation*.

isoline—A line representing the intersection of the plane of a vertical photograph with the plane of an overlapping oblique photograph. If the vertical photograph were tilt-free, the isoline would be the *isometric parallel* of the oblique photograph.

isometric parallel—*see photograph parallel* under *principal plane*.

isoradial—*see radial*.

lambert—A photometric unit of surface brightness or *luminance* which is defined as the brightness of a perfectly diffuse plane surface radiating *one lumen per square centimetre; also* equivalent to *1/π candela per square centimetre.*

laser terrain profile recorder—An electronic instrument that emits a continuous wave laser beam from an aircraft to measure vertical distances between the aircraft and the earth's surface.

laser—An acronym for "Light Amplification by Stimulated Emission of Radiation." A device producing coherent-energy beams in the spectrum of light-or-near-light frequencies.

latent image (photography)—An invisible image produced by the physical or chemical effect of light upon matter (usually silver halide or halides), which can be rendered visible by the subsequent chemical process of photographic *development.*

lateral chromatic aberration. (Lens)—*see aberration.*

lateral magnification—see magnification.

lateral-oblique photograph—see oblique photograph.

lateral refraction—The horizontal component of the refraction of light through the atmosphere.

lateral tilt—see roll.

latitude—The range of photographic *exposure* which will result in a satisfactory negative. It is measured by the ratio of the maximum exposure which will yield a satisfactory negative to the minimum exposure. The latitude of a film is greatest when used to photograph a subject with low *contrast*. The term is also applied to printing papers (*see characteristic curve*).

laydown—see mosaic.

least count—(Micrometer or vernier) The finest reading that can be made directly (without estimation) from a vernier or micrometer.

least squares—A method of adjusting observations in which the sum of the squares of all the *deviations* or *residuals* derived in fitting the observations to a mathematical model is made a minimum.

legend—A description, explanation, table of symbols, and other information, printed on a *map* or *chart* to provide a better understanding and interpretation of it. The title of a map or chart formerly was considered part of the legend, but this usage is obsolete.

lens (optics)—A piece, or combination of pieces of glass or other transparnt material shaped to form an image by means of *refraction*. *angle of coverage*—The apex angle of the cone of rays passing through the front nodal point of a lens. Lenses generally are classified according to their angles of coverage, as follows: *narrow-angle*—less than 60°, *normal-angle*—60° to 75°, *wide-angle*—75° to 100°,

super-wide-angle or *ultra-wide-angle*—greater than 100° (*see aperture stop*). *positive lens*—A lens that converges a beam of parallel light rays to a point focus. Also called *converging lens. negative lens*—A lens diverging a beam of parallel light rays, with no real focus being obtained. Also called *diverging lens.*

lens calibration—see camera calibration under *calibration.*

lens component—see lens element.

lens distortion—see distortion under *aberration.*

lens element—One lens of a complex lens system. In a photographic lens, the terms *front element* and *rear element* are often used.

lens paper—A fine soft tissue paper used to clean or polish lenses.

lens speed—see relative aperture.

lens stereoscope—see stereoscope.

level surface—see geoid.

leveling (photogrammetry)—In *absolute orientation*, the operation of bringing the model datum parallel to a reference plane, usually the tabletop of the stereoplotting instrument. Also called *horizontalizing the model; leveling the model* (*see absolute orientation* under *orientation*).

light ray—see ray of light.

linear distortion—see distortion curve.

linear magnification—see magnification.

linear parallax—see absolute stereoscopic parallax under *parallax.*

line of collimation (optics)—The line through the second nodal point of the objective lens of a telescope and the center of the reticle. Also called *line of sight; sight line; pointing line; aiming line of the instrument* (*see collimation axis* and *collimation error*).

line of constant scale—Any line on a photograph which is parallel to the *true horizon* or to the *isometric parallel*. Also called *line of equal scale.*

list—The preferred term is *x tilt* (*see tilt*, and *omega* (ω) under *roll*).

local horizon—see horizon.

locating back (aerial photography)—A plane surface in an aerial camera parallel to but out of the focal plane by an amount equal to the thickness of the film. The film is held against the locating back by vacuum or by air pressure so that the emulsion surface lies in the focal plane at the instant of exposure. Locating backs are usually of metal; they are perforated or slotted to allow for the building up of a differential pressure or for the removal of air in the formation of a vacuum. A locating back which uses a vacuum is known as a *vacuum back*, and one which uses pressure is known as a *pressure back.*

longitudinal chromatic aberration—see aberration.

longitude—A linear or angular distance measured east or west from a reference meridian (usually Greenwich) on a sphere or spheroid.

longitudinal chromatic aberration—*see aberration*.

longitudinal tilt—*see tilt*, and *phi* (ϕ) under *pitch*.

loran—A method of applying pulse techniques to navigation; an acronym of the phrase "*lo*ng-*ra*nge *n*avigation." A pulsed transmitter (known as the master station) triggers one or more other pulsed transmitters (known as slave stations) which may be as far as several hundred miles away. A mobile receiver is provided to measure the difference in time of arrival of the coded signals or pulses from the master and slave stations. If a receiver is moved in a manner to keep a constant time difference, it follows a hyperbolic path. A number of such paths may be drawn on a chart for each set or pair of the high-power, permanent, land-based stations used. Several sets of stations make it possible to prepare a chart containing many intersecting families of such hyperbolas. By noting where these hyperbolas intersect on the charts, the navigator can obtain a position fix (*see* **hiran** and **shoran**).

louver shutter—*see shutter*.

low-oblique photograph—*see oblique photograph*.

loxodromic curve—Synonymous with *rhumbline*.

lumen—The unit of luminous flux, equal to the flux through a solid angle (steradian) from a uniform point source of one candela, or to the flux on a unit surface all points of which are at unit distance from a uniform point source of one candela.

luminance—In photometry, a measure of the intrinsic luminous intensity emitted by a source in a given direction; the illuminance produced by light from the source upon a unit surface area oriented normal to the line of sight at any distance from the source, divided by the solid angle subtended by the source at the receiving surface. Also called *brightness* (luminance is preferred) (*see* **lambert**). It is assumed that the medium between source and receiver is perfectly transparent; therefore luminance is independent of extinction between the source and receiver. The source may or may not be self-luminous. Luminance is a measure only of light; the comparable term for *electromagnetic radiation* in general is radiance.

lux—A unit of illumination equivalent to one *lumen* of incident light per square metre.

magazine (photography)—A container for rolled film or photographic plates, attached to the camera body and usually equipped with automatic mechanisms that advance and position the photographic material for exposure.

magnetic declination—The angle between *true* (geographic) north and *magnetic* north (direction of the compass needle). The magnetic declination varies for different places and changes continuously with respect to time.

magnetic north—The uncorrected direction indicated by the north seeking end of a compass needle. Also called *compass north* (*see* **magnetic declination**).

magnification (optics)—The ratio of the size of an image to the sice of the object. *linear magnification*—The ratio of a linear quantity in the image to a corresponding linear quantity in the object. It may be *lateral magnification* or *longitudinal magnification*. *lateral magnification*—The ratio of a length in the image, perpendicular to the lens axis, to a corresponding length in the object. *longitudinal magnifiction*—The ratio of a length in the image, parallel to the axis, to a corresponding length in the object. *angular magnification*—The ratio of the angle subtended at the eye by the image formed by an optical device, to the angle subtended at the eye by the object itself without the optical device. This is convenient where a distance in the object cannot be measured for expressing a *linear magnification*, as in using a telescope.

manuscript map—*see map*.

map—A representation (usually on a flat medium) of all or a portion of the earth or other celestial body, showing the relative size and position of features to some given scale or projection; also, a representation of all or part of the celestial sphere. A map may emphasize, generalize, or omit the representation of certain features to satisfy specific requirements. Maps are frequently categorized and referred to according to the type of information which they are designed primarily to convey, to distinguish them from maps of other types. *topographic map*—A map which represents the horizontal and vertical positions of the features represented; distinguished from a *planimetric map* by the addition of relief in measurable form. A topographic map show mountains, valleys, and plains; and, in the case of hydrographic charts, symbols and numbers to show depths in bodies of water. *contour map*—A topographic map which portrays relief by means of *contour lines*. *planimetric map*—A map which presents only the horizontal positions for the features represented; distinguished from a topographic map by the omission of relief in measurable form. *base map*—A map showing certain fundamental information, used as a base upon which additional data of specialized nature are compiled. Also, a map containing all the information from which maps showing specialized information can be prepared; a source map. *cadastral map*—A map showing the boundaries of subdivisions of land, usually with the bearings and lengths thereof and the areas of individual tracts, for purposes of describing and recording ownership. A cadastral map may also show culture, drainage and other features relating to the value and use of land. *hydrographic map*—A map showing a portion of the waters of the earth, including shorelines, the topography along the shores and of the submerged portions, and as much of the topography of the surrounding country as is necessary for the purpose intended (*see nautical chart*.) *manuscript map*—The original drawing of a map as compiled or constructed from various data (such as ground surveys or photographs), (*see compilation manuscript*). *special-purpose map*—Any map designed primarily to meet specific requirements. Usually the map information portrayed on a special-purpose map is emphasized by omitting or subordinating nonessential or less important information. A word or phrase is usually employed to describe the type of information which a special-purpose map is designed to present—for example, *route, tax,* or *index* map (*see composite map*).

map (verb)—To prepare a map or engage in a mapping operation.

map compilation—*see compilation.*

map nadir—*see nadir.*

map parallel—The intersection of the plane of a photograph with the plane of the map (*see axis of homology*). *ground parallel*—The intersection of the plane of the photograph with the plane of reference of the ground (*see axis of homology*).

map projections—A systematic drawing of lines on a plane surface to represent the parallels of latitude and the meridians of longitude of the earth or a section of the earth.

mapping camera—*see camera.*

mapping photography—Aerial photography obtained by precisely calibrated *mapping cameras* and conforming to mapping specifications, as distinguished from aerial photography for other purposes. Also called *aerial cartographic photography, cartographic photography; charting photography; survey photography.*

map substitute—A hasty reproduction of wide-coverage aerial photographs, photomaps, or mosaics, or of provisional maps, or any document used in place of a map, when the precise requirements of a map cannot be met.

mark, floating—*see floating mark.*

matching—The act by which detail or information on the edge, or overlap area, of a *map* or *chart* is compared, adjusted, and corrected to agree with the existing overlapping map or chart (*see compilation* and *revision*).

matrix—A rectangular array of numbers or functions called *elements,* arranged in rows and columns and used to facilitate the study or solution of simultaneous linear equations.

matrix algebra—The science of the treatment of the fundamental mathematical operations involving matrices.

matte print—Print made on photographic paper with a dull finish; more suitable for pencil or ink annotations than a gossy print but less suitable for interpretation than a semi-matte print.

mean datum plane—*see vertical-control datum* under *datum.*

mean sea level—(*MSL*) The average height of the surface of the sea for all stages of the tide, usually determined by averaging height readings observed hourly over a minimum period of 19 years (*see vertical-control datum* under *datum*).

mechanical templet—*see spider templet* under *templet.*

mechanical-templet plot—*see radial triangulation.*

mechanical-templet triangulation—*see radial triangulation.*

mensuration—The act, process, or art of measuring.

mecury barometer—*see barometer.*

meridian—A north-south reference line, particularly a great circle through the geographical poles of the earth, from which longitudes and azimuths are determined; or a plane, normal to the geoid or spheroid, defining such a line.

meridional plane—Any plane containing the *polar axis* of the earth (*see meridian* and *polar bearing*).

metal templet—*see spider templet* under *templet.*

metre—(abbr. m) The basic unit of length of the metric system, defined as 1,650,763.73 wavelengths in *vacuo* of the unperturbed transition $2_{p10}-5d$ in krypton μ. Effective 1 July 1959 in the U.S. customary system of measures, 1 yard = 0.9144 metre, exactly, or 1 metre = 1.094 yards = 39.37 inches. The standard inch is exactly 25.4 millimetre.

metre-candle-second—A unit of exposure in sensitometry; one second of exposure at a distance of one metre from a light source of one candle-power.

metric camera—*see camera.*

metric photograph—A photograph taken by a *metric camera.*

metric photography—The recording of events by means of photographs, either singly or sequentially, together with appropriate coordinates, to form the basis for accurate measurements.

microdensitometer—A special form of densitometer for reading densities in very small areas; used for studying astronomical images, spectroscopic records, and for measuring image edge gradients and graininess in films.

micrometer—An auxiliary device to provide measurement of very small angles or dimensions by an instrument.

micrometre—A unit of length equal to 1 millionth of a metre.

micron—An abbreviated term equivalent to *micrometre*.

mil—A unit of length equal to 1/1,000 of an inch.

millimetre—A unit of length; 1/1,000 of a metre; mm. Standard abbreviation for millimetre.

millimicrometre—A unit of length in the metric system; the thousandth part of a micrometre, 10 Angstrom units (mμ). Standard abbreviation for Millimicron.

minor control—*see photogrammetric control* under *control.*

minor-control plot—*see radial triangulation.*

minus color—A *complementary color.* For example, minus blue is *complementary* to blue.

mirror stereoscope—*see stereoscope.*

mistake—*see blunder* under *error.*

model coordinates—The space coordinates of any plot in a stereoscopic model which define its position with reference to the air base or to the instrument axes.

model datum—That surface in a stereoscopic model conceived as having been reconstructed as part of the model representing the sea-level datum of nature.

model, mathematical—A mathematical representation, usually of a process, device, or concept, which permits mathematical manipulation of variables as a means of determining how the process, device, or concept would behave in various situations, such as under the application of a specific stimulation.

model scale—The relationship which exists between a distance measured in a stereoscopic model and the corresponding object distance.

modulation transfer function (*MTF*)—An optical analogue to general systems theory whereby the trial resolution of the components of an optical system can be measured in terms of brightness as a linear function.

moiré effect—A pattern which is produced when two or more screens of similar pattern are placed one over the other, but slightly out of register.

monochromatic—Containing light or wavelength of one color.

monocomparator—*see comparator.*

monocular vision—Vision related to or adapted to the use of one eye.

mosaic (photogrammetry)—An assembly of aerial photographs whose edges usually have been torn, or cut, and matched to form a continuous photographic representation of a portion of the earth's surface. Often called *aerial mosaic. featheredging*—The thinning of overlapping edges of photographs before assembling into a *mosaic* in order to make match lines less noticeable. When overlapping edges feathered, shadows and sharp changes in contrast are reduced or eliminated. Also called *feathering. laydown*—Often used to designate a mosaic temporarily assembled from uncropped prints. *controlled mosaic*—A mosaic oriented and scaled to horizontal ground control; usually assembled from rectified photographs. *semi-controlled mosaic*—A mosaic composed of corrected or uncorrected prints laid to a common basis of orientation other than ground control. *uncontrolled mosaic*—A mosaic composed of uncorrected prints, the detail of which has been matched from print to print without ground control or other orientation. *strip mosaic*—A mosaic consisting of one strip of photographs or images taken on a single flight. *orthophotomosaic*—An assembly of orthophotographs forming a uniform scale mosaic.

mosaicking—The assembling of photographs or other images, the edges of which are cut and matched to form a continuous photographic representation of a portion of the earth's surface. *center to center method*—A method of assembling a strip mosaic from aerial photographs with a more than 50% overlap by matching a point near the center with corresponding points in the overlap area of adjacent pictures.

most probable value—That value of a quantity which is mathematically determined from a series of observations and is more nearly free from the effects of blunders and errors than any other value that might be derived from the same series of observations.

multiband camera—*see camera.*

multiband photography, multiband color photograph or *multispectral photograph*—Photography using a camera or other device that gives simultaneous imagery in each of several portions of the spectrum (*e.g.,* blue, 400-500 millimicrometres; green 500-600, red 600-700, infrared 700+).

multiple-camera assembly—*see camera.*

multiple-lens camera—*see camera.*

multiple-lens photograph—A photograph made with a multiple-lens camera (*see multiple-lens camera* under *camera*).

multiplex—A name applied to anaglyphic double-projection *stereoplotters* with the following characteristics: (1) the stereomodel is projected from diapositive reduced from an aerial negative according to a fixed ratio; (2) the projection system illuminates the entire diapositive form area; and (3) the stereomodel is measured and drawn by observation of a floating mark.

multiplex triangulation—*see stereotriangulation.*

multispectral photography—*see multiband photography.*

nadir—The point on the *celestial sphere* directly beneath the observer and directly opposite to the *zenith*. *photograph nadir* (photogrammetry)—The point at which a vertical line through the perspective center of the camera lens pierces the plane of the photograph. Also referred to as the *nadir point*. *ground nadir*—The point on the ground vertically beneath the perspective center of the camera lens. *map nadir*—The map position for the *ground nadir*.

nadir-point plot—*see radial triangulation*.

nadir-point slotted-template plot—*see radial triangulation*.

nadir-point triangulation—*see radial triangulation*.

nadir radical—*see radial*.

narrow-angle lens—*see angle of convergence* under *lens*.

nautical chart—A map especially designed for the mariner, on which are shown navigable waters and the adjacent or included land, if any, and on which are indicated depths of water, marine obstructions, aids to navigation, and other pertinent information. Also called *hydrographic chart*.

near infrared—The preferred term for the shorter wavelengths in the infrared region extending from about 0.7 micrometres (visible red), to around 2 or 3 micrometres. The longer wavelength end grades into the middle infrared. The term really emphasizes the radiation reflected from plant materials, which peaks around 0.85 micrometres. It is also called solar infrared, as it is only available for use during the daylight hours.

neat model—the portion of the gross overlap of a pair of photographs that is actually utilized in photogrammetric procedures. Generally, the neat model approximates a rectangle whose width equals the air base and whose length equals the width between flights.

negative—A photographic image on film, plate, or paper, in which the subject tones to which the emulsion is sensitive are reversed or complementary.

negative altitude—Angular distance below the horizon.

negative lens—*see lens*.

neutral density filter—A gray filter used to reduce exposure when a lens cannot be stopped down sufficiently.

newton ring (Photography)—Concentric bands of colored light sometimes seen around the areas where two transparent surfaces are not quite in contact. The rings are the result of interference and occur when the separation between the surfaces are of the same order as the wave length of light. These concentric bands appear as dark and light rings on photographic film and paper.

nodal point (optics)—One of two points on the optical axis of a lens (or a system of lenses) such that, when all object distances are measured from one point and all image distances

are measured from the other, they satisfy the simple lens relation (conjugate-foci formula): $1/f = 1/I + 1/O$. A ray emergent from the second point is parallel to the ray incident at the first. The first nodal point is referred to also as the *front nodal point* or *incident nodal point*, and the second point as the *rear nodal point* or *emergent nodal point*. Also called simply *node*, as *front node* (*see conjugate distance*). *nodal plane*—A plane perpendicular to the optical axis at a nodal point. *principal points*—When the initial and final media have different indexes of refraction, another set of points is introduced, known as *principal points*. These points possess the following property: When a small object is placed at right angles to the camera axis at one of these points, the size of the image formed at the other point equals the size of the object. When the two media have the same index of refraction, the principal points and the nodal points coincide.

node—*see nodal point*.

nominal focal length—*see focal length*.

non-metric camera—A camera whose *interior orientation* is completely or partially unknown and frequently unstable.

non-metric photograph—A photograph taken by a *non-metric camera*.

non-shrink paper—Photographic paper treated to be dimensionally stable under a wide variety of humidity and temperature conditions. Most expensive of all photographic paper (*see dimensional stability*).

nontilting lens rectifier—A class of rectifier wherein the lens is constrained to move in the direction of its fixed axis.

non-topographic photogrammetry—Photogrammetry applied outside the realm of topographic mapping.

normal-angle lens—*see angle of coverage* under *lens*.

normal contour—*see accurate contour* under *contour line*.

normal equation—One of a set of simultaneous equations derived from *observation, condition,* or *correlate equations,* and expressing a condition for a *least-squares* adjustment. In a least-squares adjustment, values obtained from the solution of normal equations (either directly or through the correlate equations) are applied to the observation or condition equations to obtain the desired corrections.

normal-angle lens—*see angle of coverage* under *lens*.

North American Datum of 1927—*see datum*.

objective lens—In telescopes and microscopes, the optical component which receives light from the object and forms the first or primary image. In a camera, the image formed by the objective lens is the final image. In a telescope or microscope used visually, the image formed by the objective lens is magnified by the eyepiece.

oblique photograph—A photograph taken with the camera axis intentionally directed between the horizontal and the vertical. *high-oblique photograph*—An oblique photograph in which the apparent horizon is included within the field of view. *low-oblique photograph*—An oblique photograph in which the apparent horizon is not included within the field of view. *lateral-oblique photograph*—An oblique aerial photograph taken with the camera axis as nearly as possible normal to the flight line. Also called *lateral oblique*.

oblique plotting instrument—An instrument (usually monocular) for plotting from oblique photographs. An oblique sketchmaster is such an instrument.

observation equation—A condition equation which connects interrelated unknowns by means of an observed function, or a condition equation connecting the function observed and the unknown quantity whose value is sought.

occupy (surveying)—To observe with a surveying instrument at a *station*.

opacity—*see characteristic curve*.

operator, machine—The person who actually manipulates the computer controls. Places data media into the input devices, removes output, mounts reels of tape, pushes initiate buttons, and performs other similar duties.

optical axis—In a lens element, the straight line which passes through the centers of curvature of the lens surfaces. Also called *principal axis*. In an optical system, the line formed by the coinciding principal axes of the series of optical elements.

optical center—The point within a simple thin lens at which the light rays are assumed to cross.

optical flat—A surface, usually of glass, ground and polished plane within a fraction of a wavelength of light. An optical element or glass blank with an optical flat is used to test the flatness of other surfaces. *parallel plate*—An optical disk with optically flat, parallel surfaces; used especially in optical micrometres.

optical rectification—The process of projecting the image of a tilted aerial photograph onto a horizontal reference plane to eliminate the image displacements caused by tilt of the aerial camera at the time of exposure (*see transformation*).

optical system—All the parts of a compound lens and accessory optical parts which are designed to contribute to the formation of an image on a photographic emulsion, or of a visual image, or of an image on a projection screen.

optical-projection instruments—A class of instruments which provide projected images of photographic prints or other opaque material superimposed on a map or map manuscript. Often used for transferring detail from near-vertical photographs or other source material (*see camera length*).

orientation (photogrammetry): *absolute orientation*—The scaling, leveling, and orientation to ground control (in a photogrammetric instrument) of a relatively oriented stereoscopic model or group of models. *exterior orientation*—The determining (analytically or in a photogrammetric instrument) of the position of the camera station and the attitude of the taking camera at the instant of exposure. In stereoscopic instrument practice, exterior orientation is divided into two parts, *relative* and *absolute orientation*. Also called *outer orientation* (*see resection*). *interior orientation*—The determining (analytically or in a photogrammetric instrument) of the interior perspective of the photograph as it was at the instant of exposure. Elements of interior orientation are the calibrated focal length, location of the calibrated principal point, and the calibrated lens distortion. Also called *inner orientation*. *relative orientation*—The determining (analytically or in a photogrammetric instrument) of the position and attitude of one of a pair of overlapping photographs with respect to the other photograph.

origin (surveying)—The reference position from which angles or distances are reckoned (*see coordinates*).

orthochromatic (photography)—(1) Of, or pertaining to, or producing tone values (of light or shade) in a photograph, corresponding to the tones of nature. (2) Designating an emulsion sensitive to blue and green light, but not to red.

orthogonal—At right angles; rectangularly; meeting, crossing, or lying at right angles.

orthographic projection—*see projection*.

orthophotograph—A photograph having the properties of an *orthographic projection*. It is derived from a conventional perspective photograph by simple or differential rectification so that image displacements caused by camera tilt and relief of terrain are removed.

orthophotographic map—A map produced by assembling orthophotographs at a specified uniform scale in a map format.

orthophotomap—An orthophotographic map with contours and color-enhanced cartographic treatment, presented in a standard quadrangle format and related to standard reference systems.

orthophotomosaic—*see mosaic*.

orthophotoscope—A photomechanical device, used in conjunction with a double-projection anaglyphic instrument, for producing orthophotographs.

orthostereoscopy—A condition wherein the

horizontal and vertical distances in a stereoscopic model appear to be at the same scale.

outer orientation—*see* **exterior orientation** under **orientation**.

overlap (photography)—The amount by which one photograph covers the same area as covered by another, customarily expressed as a percentage. The overlap between aerial or space photographs in the same flight is called the *end lap,* and the overlap between photographs in adjacent parallel flights is called the *side lap.*

overlapping pair (photogrammetry)—Two photographs taken at *different* exposure stations in such a manner that a portion of one photograph shows the same terrain as shown on a portion of the other photograph. This term covers the general case and does not imply that the photographs were taken for stereoscopic examination (*see* **stereoscopic pair** under **stereoscopy**).

overlay (mapping)—A record on a transparent medium to be superimposed on another record; for example, maps showing original land grants (or patents) prepared as tracing cloth overlays so that they can be correlated with the maps showing present ownership. Also, any of the several overlays that may be prepared in compiling a manuscript map; usually described by name—for example, *lettering overlay.*

orthophotoquad—A monocolor orthophotographic map presented in a standard quadrangle format and related to standard reference systems. It has no contour lines and little or no cartographic treatment.

panchromatic (photography)—Sensitive to light of all colors, as a film or plate emulsion.

pancratic system—A variable-power optical system. Also called *zoom system.*

panel—An element of a target used for control-station identification on aerial photography. Panels are made of cloth, plastics, plywood, or masonite, and are positioned in a symmetrical pattern centered on the station (*see* **target**).

panel base—(cartography) The completed assembly of pieces of film positive onto a grid or projection which is used as a base for *compilation.* Also called *film mosaic; panel.*

paneling—The placement of panels on a control station to facilitate station identification on aerial photography.

panoramic camera—*see* **camera**.

panoramic distortion—The displacement of ground points from their expected perspective positions, caused by the cylindrical shape of the negative film surface and the scanning action of the lens in a *panoramic camera* system.

panoramic photograph—Photography obtained from a *panoramic camera.*

pantograph—An instrument for copying maps, drawings, or other graphics at a predetermined scale. Pantographs capable of adjustment for several scales are known as *fixed-ratio pantographs.*

paper-strip method—(rectification) A graphical method of making a point-by-point rectification based on the invariance of the cross ratio. A modification of this technique permits map detail to be revised from an oblique aerial photograph based on the projectivity of straight lines (*see* **transformation**).

parallactic angle—1. (astronomy) The angle between a body's hour circle and its vertical circle. Also called *position angle.* 2. (photogrammetry) The angle subtended by the eye base of the observer at the object viewed. Also called *angle of convergence; angular parallax.*

parallactic grid (photogrammetry)—A uniform pattern of rectangular lines drawn or engraved on some transparent material, usually glass, and placed either over the photographs of a stereoscopic pair or in the optical system of a stereoscope, in order to provide a continuous floating-mark system.

parallax—The apparent displacement of the position of a body, with respect to a reference point or system, caused by a shift in the point of observation. **absolute stereoscopic parallax**—Considering a pair of aerial photographs of equal principal distance, the absolute stereoscopic parallax of a point is the algebraic difference of the distances of the two images from their respective photograph nadirs, measured in a horizontal plane and parallel to the air base. Generally shortened to *parallax;* also called *absolute parallax, horizontal parallax, linear parallax, stereoscopic parallax,* and *X parallax;* parallax also is used to denote such measurements, as above, in the plane of a photograph and in the direction of flight. **parallax difference**—The difference in the absolute stereoscopic parallaxes of two points imaged on a pair of photographs. Customarily used in the determination of the difference in elevations of objects. **y parallax** (photogrammetry)—The difference between the perpendicular distances of the two images of a point from the vertical plane containing the air base. The existence of *y* parallax is an indication of tilt in either or both photographs and/or a difference in flight height and interferes with stereoscopic examination of the pair. Also called *want of correspondence* and *vertical parallax,* though the latter term is not preferred. **angular parallax**—The angle subtended by the eye base of the observer at the object viewed. Also called *parallactic angle* or *angle of convergence.*

parallax bar—*see* **stereometer**.

parallax difference—see **parallax**.

paraxial ray—A ray whose path lies very near the axis of a lens and which intersects the lens surface at a point very close to its vertex and at nearly normal incidence.

parallax wedge—A simplified *stereometer* for measuring object heights on stereoscopic pairs of photographs. It consists of two slightly converging rows of dots or graduated lines printed on a transparent templet which can be stereoscopically fused into a single row or line for making parallax measurements.

parallax, x—*see absolute stereoscopic parallax* under *parallax*.

parallax, y—*see y parallax* under *parallax*.

pass point—A point whose horizontal and/or vertical position is determined from photographs by photogrammetric methods and which is intended for use (as in the manner of a *supplemental control point*) in the orientation of other photographs.

passive system—A system which records energy emitted or reflected but which does not produce or transmit energy of its own. Contrast with *active system*.

pattern—In a photo image, the regularity and characteristic placement of tones or textures. The arrangement of objects or areas in a systematic fashion (*see photointerpretation*).

pattern recognition—The identification of patterns forms, or configurations by automatic means.

pencil of light—*see ray of light*.

perigee—*see apsides*.

personal error—*see error*.

personal equation—The time interval between the sensory perception of a phenomenon and the motor reaction thereto. A personal equation may be either positive or negative, as an observer may anticipate the occurrence of an event, or wait until he actually sees it occur before making a record. This is a *systematic error*, treated as the constant type (*see personal error* under *error*).

perspective axis—*see axis of homology*.

perspective center—The point of origin or termination of bundles of perspective rays. The two such points usually associated with a survey photograph are the *interior perspective center* and the *exterior perspective center*. In a perfect lens-camera system, perspective rays from the interior perspective center to the photographic images enclose the same angles as do the corresponding rays from the exterior perspective center to the objects photographed. In a lens having distortion, this is true only for a particular zone of the photograph. In a perfectly adjusted lens-camera system, the exterior and interior perspective centers correspond, respectively, to the front and rear *nodal points* of the camera lens.

perspective grid (photogrammetry)—A network of lines, drawn or superimposed on a photograph, to represent the perspective of a systematic network of lines on the ground or datum plane (*see grid method*).

perspective plane—Any plane containing the perspective center. The intersection of a perspective plane and the ground will always appear as a straight line on an aerial photograph.

perspective projection—*see projection*.

perspective ray—A line joining a *perspective center* and a point object.

photoalidade—A photogrammetric instrument having a telescopic alidade, a plateholder, and a hinged ruling arm mounted on a tripod frame. It is used for plotting lines of direction and measuring vertical angles to selected features appearing on oblique and terrestrial photographs.

photoangulator—*see transformation*.

photobase—*see air base*.

photocontrol index map—Any selected *base map* or *photoindex* on which ground control and photo identified ground points (pass points and photogrammetric points) are depicted and identified.

photocontrol point—*see control point*.

photogoniometer—*see goniometer*.

photogrammetric camera—*see camera*

photogrammetric control—*see control, photogrammetric*.

photogrammetric survey—*see survey*.

photogrammetry—The art, science and technology of obtaining reliable information about physical objects and the environment, through processes of recording, measuring, and interpreting images and patterns of electromagnetic radiant energy and other phenomena. (For specific branches of photogrammetry, see under the proper name, as: *analytical photogrammetry, close-range photogrammetry, etc.*)

photograph—A general term for a positive or negative picture made with a camera on sensitized material, or prints from such a camera original. (For specific types of photographs, see under the proper name, as: *aerial photograph, multiple-lens photograph, etc.*)

photograph axes—The preferred term is *fiducial axes*.

photograph center—The center of a photograph as indicated by the images of the *fiducial mark or marks* of the camera. In a perfectly adjusted camera, the photograph center and the *principal point* are identical.

photograph coordinates—*see coordinates*.

photograph, horizon—*see horizon photograph*.

photograph meridian—*see principal plane*.

photograph nadir—*see nadir*.

photograph parallel—*see principal plane*.

photograph perpendicular—The perpendicular from the *interior perspective center* to the plane of the photograph (*see principal distance* under *focal length*).

photograph plumb point—The preferred term is *photograph nadir* (*see* under *nadir*).

photograph pyramid—A pyramid whose base is a triangle formed by three point images on a photograph, and whose apex is the *perspective center* of the photograph.

photographic interpretation—*see photointerpretation.*

photography—The art, science, and process of producing images on sensitized material through the action of light. The term *photography* is sometimes incorrectly used in place of *photographs;* however, the distinction between the *process* and the *product* is a valuable one and should be observed.

photography, tricamera—Photography consisting of the simultaneous exposure of three cameras systematically arranged at fixed angles to each other in such a way that overlap is provided between adjacent photographs. Generally, the cameras are arranged so that a center vertical photograph and two high-oblique photographs are obtained. This assembly is often referred to as a *trimetrogon camera* assembly because of the wide use of Metrogon lenses in early tricamera photography.

photoindex—(photointerpretation) *see index map.*

photointerpretation or *photographic interpretation*—The detection, identification, description, and assessment of significance of objects and patterns imaged on a photograph. *air photo interpretation* or *aerial photographic interpretation*—Photointerpretation applied to images on aerial photographs or to other aerial remote sensing imagery.

photomap—A photomosaic of a specified land area, which also contains marginal information, descriptive data, and a reference grid and/or projection (*see mosaic*).

photomosaic—*see mosaic.*

photometry—The study of the measurement of the intensity of light. At one time photometry referred only to the measurement of luminous intensity, intensity of light in the wavelengths to which the eye is sensitive. This restriction has proved difficult to maintain in practice.

photosensitive—A term used to describe substances whose chemical composition is altered by exposure to light (*see emulsion*).

photo-revised map—A topographic or planimetric map which has been revised by photo-planimetric methods.

phototheodolite—A ground-survey instrument combining a theodolite and a surveying camera, in which the relationship between the camera axis and the *line of collimation* of the theodolite can be measured.

phototopography—The science of surveying in which the detail is plotted entirely from photographs taken at suitable ground stations (*see terrestrial photogrammetry*).

phototriangulation—The process for the extension of horizontal and/or vertical control whereby the measurements of angles and/or distances on overlapping photographs are related into a spatial solution using the perspective principles of the photographs. Generally, this process involves using aerial photographs and is called *aerotriangulation* or *aerial triangulation* (*see* **analytical phototriangulation, radial triangulation,** and **stereotriangulation**).

pictomap—A color reproduction of a standard *photomosaic* on which the photographic imagery has been converted into interpretable colors and symbols by means of tonal masking techniques.

picture control point—*see control point.*

picture plane—A plane upon which can be projected a system of lines or rays from an object to form an image or picture. In perspective drawing, the system of rays is understood to converge to a single point. In photogrammetry, the photograph is the picture plane.

pitch—(1) air navigation: A rotation of an aircraft about the horizontal axis normal to its longitudinal axis so as to cause a nose-up or nose-down attitude. (2) photogrammetry: A rotation of the camera, or of the photograph-coordinate system, about either the photograph y axis or the exterior Y axis; *tip* or *longitudinal tilt*. In some photogrammetric instruments and in analytical applications, the symbol *phi* (ϕ) may be used.

plan-position-indicator (PPI) radar—A radar system employing a rotating antenna to scan all or part of a complete circle, in which blips produced by signals from reflecting objects are shown in plan position, thus forming a maplike display.

plane (photogrammetry)—*see principal plane.*

plane coordinates—*see coordinates.*

plane polar coordinates—*see coordinates.*

plane, principal—*see principal plane.*

plane rectangular coordinates—*see coordinates.*

planimeter—An instrument used to measure the area of any figure by passing a tracer around its boundaries and recording the area encompassed.

planimetric map—*see map.*

plat—A cadastral map (*see* **cadastral map** under *map*).

platform (remote sensing)—The objects, structure, vehicle, or base upon which a remote sensor is mounted.

plumb line—*see vertical line.*

plumb point—*see* the preferred term **nadir point** under **nadir.**

pocket stereoscope—*see* **lens stereoscope** under **stereoscope.**

pointing—1. (mensuration) Placing the reticle or index mark of a precision measuring instrument, such as a comparator, within the symmetrical center or center of gravity of a point being measured to determine its position relative to the position of other points in some system of coordinates. 2. (stereocompilation) A general term applied to the movement of the tracing table of a stereoplotting instrument to specific control and/or picture points on the datum during orientation of a stereomodel (*see line of collimation*).

pointing accuracy—The exactness, in surveying or photogrammetry, with which the line of sight or floating mark can be directed toward a target or image point.

pointing errors—Errors which reflect the accuracy with which the floating mark of a stereoplotting system can be located on a sharp model point. These errors generally follow a more or less random distribution but show a systematic trend with progressive working time on the instrument due to eye fatigue and its effect on stereoscopic perception.

point marker—A device used for identifying points on diapositives by either marking a small hole in the emulsion or marking a small ring around the detail point itself.

point of symmetry—The point in the focal plane of a camera about which all lens distortions are symmetrical. If the lens were perfectly mounted, the point of symmetry would coincide with the *principal point*.

point-transfer device—A stereoscopic instrument used to mark corresponding image points on overlapping photographs.

point, vanishing—see vanishing point under *principal plane.*

polar axis—see polar and *spherical coordinates* under *coordinates.*

polar bearing—see polar coordinates under *coordinates; meridional plane.*

polar coordinates—see coordinates.

polar distance—see polar coordinates under *coordinates.*

polarization (optics)—The act or process of modifying light in such a way that the vibrations are restricted to a single plane. According to the wave theory, ordinary (unpolarized) light vibrates in all planes perpendicular to the direction of propagation. On passing through or contacting a polarizing medium (such as *Polaroid*) ordinary light becomes plane-polarized, that is, its vibrations are limited to a single plane.

polarizing filter—A filter which passes light waves vibrating in one polarization direction only. Used over camera lenses to cut down or remove, (plane) rays of any or all other polarization direction(s) when they may constitute objectionable reflections from glass, water, or other highly reflecting surfaces.

Polaroid—A manufactured, plastic polarizing filter; on passing through it, ordinary light becomes plane-polarized.

Porro-Koppe principle—The principle applied in some photogrammetric instruments to eliminate the effect of *camera-lens distortion*. The photographic positive or negative is observed through a lens or optical system identical in distortion characteristics to the camera objective which made the original exposure. In effect, this method of observation is a reverse use of the camera, with the focal plane becoming the object which is imaged at infinity by parallel bundles of rays emerging from the lens. The chief ray of each bundle assumes its correct direction, and the cone of rays is identical to that whose vertex was the incident node of the camera lens at the instant of exposure. The parallel bundles may be observed by means of a telescopic system focused at infinity and made rotatable about the incident node of the lens. This method of eliminating lens distortion is utilized in photogrammetric instruments of both the monoscopic type, such as the photogoniometer, and the stereoscopic type used for map plotting.

Porro prism—A prism that deviates the axis 180° and inverts the image in the plane in which the reflection takes place. It may be described as two right angle prisms cemented together.

position—The location of a point with respect to a reference system, such as a *geodetic datum* (*see geodetic position*). The coordinates which define such a location. The place occupied by a point on the surface of the earth. Often construed as *horizontal position* when *elevations* are considered separately. *adjusted position*—An adjusted value of the coordinate position of a point.

positive lens—see lens.

power of a lens—see diopter and *magnification.*

precision—A quality associated with the refinement of instruments and measurements, indicated by the degree of uniformity or identity of repeated measurements. In a somewhat narrower sense, the term refers to the *spread* of the observations, or some measure of it, whether or not the mean value around which the spread is measured approximates the *true* value. Contrast with *accuracy.*

precision camera—see camera.

pressure back—see locating back.

pressure plate (photography—A flat plate (usually of metal but frequently of glass or other substance) which, by means of mechanical force, presses film into contact with the focal-plane plate of a camera.

principal axis—see optical axis.

principal distance—see focal length.

principal-distance error—In a stereoplotting system, an instrument error resulting from improper calibration of the aerial camera, diapositive printer, or projector. The error is of little importance in a flat surface model but the effects are increased in proportion to the relief in the model.

principal line—*see principal plane* (photogrammetry).

principal meridian—*see principal plane* (photogrammetry).

principal parallel—*see principal plane* (photogrammetry).

principal plane (optics)—A plane through a *principal point* and perpendicular to the optical axis.

principal plane (photogrammetry)—The vertical plane through the internal perspective center containing the photograph perpendicular of a tilted photograph. In the case of a truly vertical photograph, the principal plane and the other planes and lines discussed below lose their significance. *principal line*—The trace of the principal plane upon a photograph (*e.g.*, the line through the principal point and the nadir point). *horizon trace*—An imaginary line, in the plane of a photograph, which represents the image of the true horizon; it corresponds to the intersection of the plane of a photograph and the horizontal plane containing the *internal perspective center* or *rear nodal point* of the lens. *point, vanishing*—The point in the plane of the photograph at which a system of parallel lines in the object space converge. Since any system of parallel lines in the object space will meet at infinity, the image of the meeting point will be formed by the ray through the perspective center parallel to the system. The vanishing points of all systems of parallel lines parallel to one plane will lie on a straight line on the photograph called a "vanishing line." The vanishing line for all systems of horizontal parallel lines in the object space is the horizon trace. *photograph meridian*—The image on a photograph of any horizontal line in the object space which is parallel to the principal plane. Since all such lines meet at infinity, the image of the meeting point is at the intersection of the principal line and the horizon trace and all photograph meridians pass through that point. The *principal line*, sometimes called the *principal meridian*, is the only photograph meridian perpendicular to the photograph parallels, or lines of constant scale. *photograph parallel*—The image on a photograph of any horizontal line in the object space which is perpendicular to the principal plane. All photograph parallels are perpendicular to the principal line. The photograph parallel passing through the principal point is the *principal parallel*, and that passing through the isocenter is the *isometric parallel*. Thus, the isometric parallel is the intersecting line between the plane of a photograph and a horizontal plane having an equal perpendicular distance from the same perspective center.

axis of tilt—A line through the perspective center perpendicular to the principal plane. (The term is usually restricted to this definition.) The axis of tilt could be any of several lines in space (*e.g.*, the *isometric parallel* or the *ground line*), but the present definition is the only one which permits the concept of tilting a photograph without upsetting the positional elements of exterior orientation. *map parallel*—The intersection of the plane of a photograph with the plane of the map (*see axis of homology*). *ground parallel*—The intersection of the plane of the photograph with the plane of reference of the ground (*see axis of homology*).

principal point (optics)—*see nodal point.*

principal point (photogrammetry)—The foot of the perpendicular from the interior perspective center to the plane of a photograph.

principal point assumption—*see radial triangulation.*

principal-point error—A personal error in which the principal points in a stereoplotting system are displaced in such a manner that they have unequal *x*-components with a resultant error in vertical scale. Such errors are usually introduced into the system by either improper orientation of the diapositive plate in the printer, in the projector, or both.

principal-point radial—*see radial.*

principal-point triangulation—*see radial triangulation.*

print (photography)—A copy made from a transparency by photographic means. *contact print*—A print made with a transparency in contact with a sensitized surface. *ratio print*—A print in which the scale has been changed from that of the transparency by projection printing (*see positive*).

printer, line—A device that prints all characters of a line as a unit.

process camera—*see camera.*

process lens—A type of lens used in large *copy cameras* (*e.g.*, those used in map reproduction); usually of low aperture (*f*/10, approximately), narrow angle, long focal length, symmetrical construction, and limited range of magnification (0.5× to 2×).

profile—Elevations of the terrain along some definite line such as a project centerline. Positions along the line are given in terms of stationing, or distances from a starting point. Elevations are measured at a sufficient number of points to enable defining the configuration of the ground surface. This includes high points, low points, and points of slope changes.

projection—In geometry, the extension of lines or planes to intersect a given surface; the transfer of a point from any surface to a corresponding position on another surface, by graphical or analytical methods. *perspective projection*—The projection of points by straight lines drawn through them from some given point to an intersection with the plane of projection. Unless otherwise indicated, the point of projection is understood to be within a finite distance of the plane of projection. For example, a photograph is formed by a perspective projection of light rays from the rear node of the lens (the point of projection) to the emulsion (the plane of projection). As applied to the geometry of a photograph, the term *perspective projection* is preferable to the term *conic projection*. *orthographic projection*—A perspective projection of points by straight lines from a point of projection at an infinite distance from the plane of the drawing. It is regularly used in mechanical drawing and, when so used, the two vertical planes are revolved about their respective lines of intersection with the horizontal plane so as to show all three views on the plane of the paper.

pseudoscopic image—One in which the normal impression of relief is reversed (*see* **stereoscopy**).

quadrangle—A rectangular, or nearly so, area covered by a map or plat, usually bounded by given meridians of longitude and parallels of latitude. Sometimes shortened to quad; also called *quadrangle map*.

radar—(1) The principle of locating targets or objects by the measurement of reflections of radio-frequency energy from them. (2) A term applied to devices and systems which make use of this principle. Acronym for "RAdio Detection And Ranging."

radar altimeter—*see* under *altimeter*.

radargrammetry—The science of obtaining reliable measurements by means of *radar*.

radar photography—A combination of the photographic process and radar techniques. Electrical impulses are sent out in predetermined directions and the reflected or returned rays are utilized to present images on cathode-ray tubes. Photography is then taken of the information displayed on the tubes.

radar shadow—A condition in which radar signals do not reach a region because of an intervening obstruction.

radial (photogrammetry)—A line or direction from the radial center to any point on a photograph. The radial center is assumed to be the principal point, unless otherwise designated (*e.g., nadir radial*). *nadir radial*—A radial from the nadir point, or *isoradial* from the isocenter. *radial center*—The selected point on a photograph from which radials (directions) to various image points are drawn or measured (*i.e.*, the origin of radials). The radial center is either the principal point, the nadir point, the isocenter, or a substitute center. *radial plot*—*see* **radial triangulation**.

radial distortion—Linear displacement of image points radially to or from the center of the image field, caused by the fact that objects at different angular distances from the lens axis undergo different magnifications (*see* **distortion** under **aberration**).

radial line—A line drawn radially from the center point of a vertical photo.

radial line plot—*see* **radial triangulation**.

radial plotter—A device whereby two overlapping photographs are viewed stereoscopically, and the *planimetric* details in their common area can then be transferred to a map or base sheet through a mechanical linkage utilizing the radial line principle. Also called *radial-line plotter*.

radial triangulation—The aerotriangulation procedure, either graphical or analytical, in which directions from the radial center, or approximate radial center, of each overlapping photograph are used for horizontal-control extension by the successive intersection and resection of these direction lines. A radial triangulation also is correctly called a *radial plot* or a *minor-control plot*. If made by analytical methods, it is called an *analytical radial triangulation*. A radial triangulation is assumed to be graphical unless prefixed by the word *analytical*. A graphical radial triangulation is usually laid out directly onto ground control plotted on a map, map projection, or map grid; but it may be first laid out independently of such control and later adjusted to it as a unit. In the latter case, the scale and azimuth of the radial-triangulation unit are not known until it is adjusted to the ground control. The radial center for near-vertical photographs may be the *principal point*, the *nadir point*, or the *isocenter*. A radial triangulation is assumed to be made with principal points as radial centers unless the definitive term designates otherwise (as, for example, *nadir-point triangulation* or *nadir-point plot*, and *isocenter triangulation* or *isocenter plot*. The adjective *radial* is not necessary in these four terms). The adjective *analytical* is required to designate that the triangulation is by analytical and not graphical methods (*e.g., analytical nadir-point triangulation*). A *graphical radial triangulation* may be made by several methods, as follows: **slotted-templet triangulation** or **slotted-templet plot**—A graphical radial triangulation using slotted templets. **spider-templet triangulation, spider-templet plot, mechanical-templet triangulation**, or **mechanical-templet plot**—A graphical radial triangulation using spider (mechanical) templets. **hand-templet triangulation** or **hand-templet plot**—A graphical radial triangulation using any form of hand templets. In the preceding eight terms, it is assumed that the radial center is the principal point, unless the term includes the words *nadir point* or *isocenter* (*e.g., nadir-point slotted-templet plot*) or unless the context states that a radial center other than the principal point was used. (For definitions of various templets, *see* **templet**.) ***direct radial triangulation*** or ***direct radial plot***—A graphical radial triangulation made by tracing the directions from successive radial centers directly onto a transparent plotting sheet rather than by laying the triangulation by the templet method. ***strip radial triangulation*** or ***strip radial plot***—A direct radial triangulation in which the photographs are plotted in flight strips without reference to ground control and the strips are later adjusted together to the ground control. ***principal-point assumption***—The assumption that, on near-vertical photographs, radial directions are correct if measured from the *principal point*.

radiance—The accepted term for radiant flux in power units (*e.g.* watts) and not for flux density per solid angle (*e.g.* watts/cm² sr) as often found in recent publications.

radiation—The emission and propagation of energy through space or through a material medium in the form of waves; *e.g.,* the emission and propagation of electromagnetic waves, or of sound and elastic waves. The process of emitting radiant energy.

radiograph—An image produced from X-rays (*see* **x-rays** and **x-ray photogrammetry**).

radiometer—Instrument for measuring radiant energy.

random error—*see* **error**.

rate of climb—The rate of ascent from the earth's surface; the vertical component of the velocity of the center of gravity of an aircraft, usually expressed in feet per minute.

ratiometer—An instrument used to help solve the mathematical relationship of a photograph to a mosaic. It determines scale ratios from which, through mathematical formulas, a rectified print can be made on a properly calibrated rectifying printer.

ratio print—*see* **print**.

ray of light—The geometrical concept of a single element of light propagated in a straight line and of infinitesimal cross-section; used in analytically tracing the path of light through an optical system. ***pencil of light***—A bundle of rays originating at, or directed to, a single point. ***beam of light***—A group of pencils of light, as those originating at the many points of an illuminated surface. A beam of parallel light rays is a special case in which each pencil is of such small cross section that it may be regarded as a ray.

ray tracing (optics)—A trigonometric calcula-

tion of the path of a light ray through an optical system.

real time—Generally associated with data transmission at the time of occurrence with no delay.

rear nodal point—*see* **nodal point**.

reconnaissance—A general examination or survey of the main features, or certain specific features, of a region, usually as a preliminary to a more detailed survey.

reconnaissance survey—A preliminary survey, usually executed rapidly and at relatively low cost. The information obtained is recorded, to some extent, in the form of a reconnaissance map or sketch.

recording statoscope—*see* **altimeter**.

rectification—*see* **transformation**.

rectified print—A photograph in which tilt displacement has been minimized from the original negative, and which has been brought to a desired scale (*see* **transformation**).

rectoblique plotter—*see* **transformation**.

reduction—The production of a negative, diapositive or print at a smaller scale than the original (*see* **ratio print** under **print**).

reduction printer—*see* **diapositive printer**.

reference datum—A general term applied to any datum, plane, or surface used as a reference or base from which other quantities can be measured.

reference line—Any line which can serve as a reference or base for the measurement of other quantities. Also called *datum line*.

reference spheroid—*see* **geoid**.

reflectance—The ratio of the radiant energy reflected by a body to that incident upon it. The suffix (ance) implies a property of that particular specimen surface.

reflecting prism—A prism that deviates a light beam by internal reflection. (Practically all prisms used in photogrammetric instruments are of this type.)

reflecting projector—An instrument which is used to project the image of photographs, maps, or other graphics onto a copying table. The scale of the projected image can be varied by raising or lowering the projector or, in some models, the copy board. These latter models also allow the tilting of the copy board in *x*- and *y*-directions in order to compensate for tip and tilt distortion in aerial photographs.

refracting prism—A prism that deviates a beam of light by refraction. The angular deviation is a function of the wavelength of light; therefore, if the beam is composed of white light, the prism will spread the beam into a spectrum. Refracting prisms can be used in optical instruments only for small deviations (*see* **wedge**).

refraction—The bending of light rays in passing from one transparent medium into another which has a different index of refraction. The *angle of refraction* is the angle which the refracted ray makes with the normal to the surface separating the two media (*see Snell's law of refraction*).

refraction displacement—*see displacement.*

relative accuracy—(general) An evaluation of the random errors in determining the positional orientation (*e.g.*, distance, azimuth) of one point or feature with respect to another.

relative aperture—For a photographic or telescopic lens, the ratio of the *equivalent focal length* to the diameter of the *entrance pupil*. Expressed as *f*/4.5 or *f:4.5*; also called *f number* or *speed of lens*.

relative humidity—*see humidity.*

relative orientation—*see orientation.*

relative position—The location of a point or feature with respect to other points or features, either fixed or moving.

relative tilt—The tilt of a photograph with reference to an arbitrary plane, not necessarily a horizontal plane, such as that of the preceding or subsequent photograph in a strip. Also defined as the angle between the photograph perpendicular and a reference direction, such as the photograph perpendicular of the preceding or subsequent photograph in a strip.

relief—The elevations or the inequalities, collectively, of a land surface; represented on graphics by contours, hypsometric tints, shading, spot elevations, hachures, *etc.*

relief displacement—*see displacement.*

relief model—A three-dimensional representation of a portion of the earth's surface at a reduced scale. Normally some vertical exaggeration is used to accentuate relief features.

remote sensing—(1) In the broadest sense, the measurement or acquisition of information of some property of an object or phenomenon, by a recording device that is not in physical or intimate contact with the object or phenomenon under study; *e.g.*, the utilization at a distance (as from aircraft, spacecraft, or ship) of any device and its attendant display for gathering information pertinent to the environment, such as measurements of force fields, electromagnetic radiation, or acoustic energy. The technique employs such devices as the camera, lasers, and radio frequency receivers, radar systems, sonar, seismographs, gravimeters, magnetometers, and scintillation counters. (2) The practice of data collection in the wavelengths from ultraviolet to radio regions. This restricted sense is the practical outgrowth from airborne photography. Sense (1) is preferred and thus includes regions of the EM spectrum as well as techniques traditionally considered as belonging to conventional geophysics. Also called *rapid reconnaissance*. French: *teledetection;* German: *Fernerkun-*

dung; Portuguese: *sensoriamento remoto;* Russian: *Distantsionnaya;* Spanish: *percepcion remota.*

remote sensing imagery—The photographic and electronic image secured from platforms such as aircraft and satellites. The common types include panchromatic, infrared black-and-white, color and infrared color photographs; thermal infrared, radar and microwave imagery.

reproduction (mapping)—The processes involved in printing copies from an original drawing. The principal processes are photography, lithography (or engraving), and printing. Also, a printed copy of an original drawing, made by any of the processes of reproduction.

reseau—*see grid plate.*

resection—(1) The graphical or analytical determination of a position as the intersection of at least three lines of known direction to corresponding points of known position. (2) photogrammetry: The determination of the position and attitude of a camera, or the photograph taken with that camera, with respect to the exterior coordinate system.

residual or *residual error*—*see error.*

residual parallax—Small amounts of *y*-parallax which may remain in a model after relative orientation is accomplished.

resolution—The minimum distance between two adjacent features, or the minimum size of a feature, which can be detected by remote sensing. For photography, this distance is usually expressed in lines per millimeter recorded on a particular film under specified conditions; as displayed by radar, in lines per millimeter. If expressed in size of objects or distances on the ground, the distance is termed *ground resolution. area weighted average resolution (AWAR)*—A single average value for the resolution over the picture format for any given focal plane.

resolving power—An expression of lens definition, usually stated as the maximum number of lines per millimetre that can be resolved (*i.e.*, seen as separate lines) in the image. The resolving power of the lens varies with the contrast of the *test chart* and normally varies also with the orientation and position of the chart within the field.

resolving power target—A test chart used for the evaluation of photographic, optical, and electro-optical systems. The design usually consists of ruled lines, squares, or circles varying in size according to a specified geometric progression (*see test chart*).

restitution—The determination of the true (map) position of objects or points; the image of which appears distorted or displaced on aerial photographs. Restitution corrects for distortion resulting from both tilt and relief displacement. Restitution in photogrammetry is commonly achieved by *analytical methods* or through the use of *stereoscopic plotting instruments*.

reticle—A mark, such as a cross or system of lines, lying in the image plane of a viewing apparatus and used singly as a reference mark in certain types of monocular instruments or as one of a pair to form a floating mark, as in certain types of stereoscopic instruments (*see* **parallactic grid** and **floating mark**).

revert (optics)—To interchange the right and left sides of an image without altering the relative positions of the top and bottom, as occurs in certain prisms and mirrors.

rhomboidal prism—A prism that displaces the axis of the beam of light only laterally.

rhumb line—A line (curved) on the surface of the earth, crossing all meridians at a constant angle. Also called a *loxodromic curve*. On a *Mercator projection,* the *rhumb line* is represented by a straight line.

right-angle prism—A prism that turns a beam of light through a right angle. It inverts (turns upside down) or reverts (turns right for left) according to the orientation of the prism.

right ascension—The angular distance measured eastward on the equator from the *vernal equinox* to the hour circle through the celestial body, from 0 to 24 hours.

roll—(1) air navigation: A rotation of an aircraft about its longitudinal axis so as to cause a wing-up or wing-down attitude. (2) photogrammetry: A rotation of a camera or a photograph-coordinate system about either the photograph x axis or the exterior X axis. In some photogrammetric instruments and in analytical applications, the symbol *omega* (ω) may be used.

roof prism—A type of prism in which the image is reverted by a roof—that is, two surfaces inclined at 90° to each other.

rotating prism—*see* **dove prism**.

safelight—A lamp, for use in the darkroom, which supplies light of a color which will not affect the photographic material within a reasonable time. Different photographic materials require different safelight filters.

satellite triangulation—The determination of the angular relationships between two or more stations by the simultaneous observation of an earth satellite from these stations. Observational techniques have included the use of precise *ballistic cameras*.

scale—The ratio of a distance on a photograph or map to its corresponding distance on the ground. The scale of a photograph varies from point to point due to displacements caused by tilt and relief, but is usually taken as f/H where

f is the principal distance of the camera and H is the height of the camera above mean ground elevation. Scale may be expressed as a ratio, 1:24,000; a representative fraction, 1/24,000; or an equivalence, 1 in. = 2,000 ft.

scanner—(1) Any device that scans, and by this means produces an image. (2) A radar set incorporating a rotatable antenna, or radiator element, motor drives, mounting, *etc.* for directing a searching radar beam through space and imparting target information to an indicator.

Scheimpflug condition—The requirement that object, lens, and image planes be collinear for sharp focus in any direct-projection system. Sharp focus is achieved in a rectifier when the Scheimpflug condition is fulfilled and when the negative-to-lens and lens-to-easel distances satisfy the conjugate-distance formula (*see* **conjugate distance**). These conditions may be fulfilled automatically by the use of inversors (*see* **inversor**).

scribing—The process of preparing a negative which can be reproduced by contact exposure. Portions of a photographically opaque coating are removed from a transparent base by scraping with specially designed tools.

secor—A phase comparison electronic long range distance measuring system used to determine positions and orbits of satellites or flight vehicles that contain the necessary transponders. This term is an acronym for "SEquential COllation of Range."

semi-analytic aerotriangulation—Aerotriangulation in which a stereoplotting instrument is used to read x, y and z model coordinates of each stereopair in a strip or block; each model in its own unique coordinate system. This is followed by numerical formation of strips and blocks and adjustments to ground control utilizing a digital computer.

semi-controlled mosaic—*see* **mosaic**.

semimajor axis—One-half the longest diameter of an ellipse. Also called *mean distance*.

semiminor axis—One-half the shortest diameter of an ellipse.

semi-micrograph—Imagery produced by a scanning electron microscope.

sensitometer—An instrument which exposes a photographic film in a known manner so that its light-sensitive properties may be measured (*see* **characteristic curve**).

sensor—An instrument used to detect and/or record electromagnetic energy.

shiran—An electronic distance-measuring system for measuring distances with geodetic accuracy from an airborne station to each of four ground stations. This term is an acronym for "S-band HIgh precision short-RAnge electronic Navigation." (*see* **hiran**).

shoran—An electronic measuring system for indicating distance from an airborne station to each of two ground stations. The term is an acronym for the phrase "SHOrt RAnge Navigation." (*see* **hiran** and **loran**). *shoran straight-line indicator*—A shoran device for assisting the pilot to fly a straight flight line. *shoran control*—The control of aerial photographs by registration of the distance of the exposure station from two ground stations. *shoran line crossing*—A method of determining distance between two points by flying across the joining line.

shutter (photography)—The mechanism of a camera which controls the length of time the emulsion is exposed. *focal-plane shutter*—A shutter located near the focal plane; usually consisting of a curtain with a slot which is pulled across the focal plane to make the exposure. *between-the-lens shutter*—A shutter located between the lens elements of a camera; usually consisting of thin metal leaves which open and close or revolve to make the exposure. *louver shutter*—A shutter consisting of a number of thin metal strips or louvers which operate like a venetian blind to make the exposure; usually located just in front of or just behind the lens.

side lap—*see* **overlap**.

sidereal time—Time based upon the rotation of the earth relative to the *vernal equinox*.

side-looking airborne radar (*SLAR*)—A radar system using a stabilized antenna oriented at right angles to the aircraft's flight path. The acronym *slar* is derived from "Side Looking Airborne Radar."

sketchmaster—*see* **camera lucida**.

slotted templet—*see* **templet**.

slotted-templet method—*see* **templet** and *radial triangulation*.

slotted-templet plot—*see* **radial triangulation**.

slotted-templet triangulation—*see* **radial triangulation**.

Snell's law of refraction—This law states that, for a ray of light passing from one medium to another, the sine of the angle of incidence divided by the sine of the angle of refraction equals a constant. This constant is called the *index of refraction* when the medium containing the incident ray is air or a vacuum. The index of refraction can also be defined as the ratio of the velocity of light in one medium to that in another medium. The indexes of glass range from 1.46 to 1.80 (*see* **refraction** and *critical angle*).

software—A set of *computer* programs, procedures, and possibly associated documentation concerned with the operation of a data processing system.

solarization (photography)—A reversal of the gradation sequence in the (usually very dense) image obtained on the normal development of films, plates, and papers after a very intense or long-continued exposure. A still greater exposure appears to restore the original sequence of gradation.

sortie—An operational flight by one aircraft; also, photography obtained on a flight.

space coordinates—*see* **coordinates**.

special-purpose map—*see* **map**.

specific humidity—*see* **humidity**.

spectrum—An array of waves ordered in accordance with the magnitude of wave length. The electromagnetic spectrum extends from the shortest cosmic rays, through *gamma* rays, X-rays, ultraviolet radiation, visible radiation, infrared radiation, and including microwave and all other wavelengths of radio energy.

specular reflection (optics or microwave theory)—The type of reflection characteristic of a highly polished plane surface from which all rays are reflected at an angle equal to the angle of incidence.

speed (photography)—*see* **characteristic curve**.

speed of lens—*see* **relative aperture**.

spherical aberration—*see* **aberration**.

spherical coordinates—*see* **coordinates**.

spherical lens—A lens in which all surfaces are segments of spheres. Most photographic lenses belong in this class (*see* **aspherical lens**).

spheroid of reference—*see* **reference spheroid** or **ellipsoid** under **geoid**.

spider templet—*see* **templet**.

spider-templet plot—*see* **radial triangulation**.

spider-templet triangulation—*see* **radial triangulation**.

spirit level—A closed glass tube (vial) of circular cross section, its center line also forming a circular arc, its interior surface being ground to precise form; it is filled with ether or liquid of low viscosity with enough free space being left for the formation of a bubble of air and gas at the topmost point of the circular arc.

spirit leveling—*see* **leveling**.

split camera—An assembly of two cameras disposed at a fixed overlapping angle relative to each other. Also called *split-vertical camera*.

split photography—Aerial photography taken using a split camera installation. Also called *convergent photography; split-vertical photography*.

spot elevation—A point on a map or chart whose height above a specified reference datum is noted, usually by a dot or a small sawbuck and elevation value. Elevations are shown, wherever practicable, for road forks and intersections, grade crossings, summits of hills, mountains, and mountain passes, water surfaces of lakes and ponds, stream forks, bottom elevations in depressions, and large flat areas. Also called *spot height*.

standard—An exact value (a physical entity or

an abstract concept) established and defined by authority, custom, or common consent to serve as a reference, model, or rule in measuring quantities or qualities, establishing practices or procedures, or evaluating results. A fixed quantity or quality.

standard deviation—see error.

State plane coordinate systems—see coordinates.

station (surveying)—(1) A point whose position has been (or is to be) determined. A station may be a marked station (*i.e.*, a point more or less permanently marked for recovery) or an unmarked station, one which is not recoverable. (2) A length of 100 ft, measured along a given line, which may be straight or curved. (3) Any point on a straight (or curved) line, whose position is indicated by its total distance from a starting point, or zero point (*e.g.*, station 4 + 47.2, meaning 447.2 ft from zero) (*see camera station*).

statoscope—see altimeter.

stellar camera—see camera.

step wedge—A strip of film or a glass plate whose transparency diminishes in graduated steps from one end to the other; often used to determine the density of a photograph. Also called *gray scale* or *step tablet*. Contrast with *continuous tone gray scale*.

stereo—1. Contracted or short form of stereoscopic. 2. The orientation of photographs when properly positioned for stereoscopic viewing. Photographs so oriented are said to be "in stereo."

stereocomparagraph—A relatively simple and mobile stereoscopic instrument used for the preparation of topographic maps from photography. Differences in elevation are determined by measuring parallax difference on a stereoscopic pair.

stereocomparator (photogrammetry)—*see comparator.*

stereocompilation—see compilation.

stereogram—see stereoscopy.

stereometer—A measuring device containing a micrometer movement by means of which the separation of two *index marks* can be changed to measure parallax difference on a stereoscopic pair of photographs. Also called *parallax bar*.

stereometric camera—see camera.

stereomodel—The three-dimensional model formed by the intersecting homologous rays of an overlapping pair of photographs.

stereoplanigraph—A precise stereoscopic plotting instrument, especially valuable for extension of control, and capable of handling most types of stereoscopic photography, including terrestrial.

stereoplotter—see stereoscopic plotting instrument.

stereoscope—A binocular optical instrument for assisting the observer to view two properly oriented photographs or diagrams to obtain the mental impression of a three-dimensional model. *lens stereoscope*—An instrument consisting of two semi-convex simple lenses mounted in a frame a few inches above a pair of overlapping aerial photographs. Common magnifications are 2, 4, or 6 times. Only a small section of a stereopair of photographs may be viewed at any one time. *mirror stereoscope*—An instrument for viewing a stereopair of photographs that uses mirrors in addition to simple lenses, so that a relatively large area of each pair of photographs is viewed. This avoids frequent moving of photographs and flipping up the sides of photographs required when using a lens stereoscope since all the stereo-overlap can be seen.

stereoscopic base—see photobase under *air base.*

stereoscopic coverage—Aerial photographs taken with sufficient overlap to permit complete stereoscopic observation.

stereoscopic exaggeration—see hyperstereoscopy.

stereoscopic fusion—see stereoscopy.

stereoscopic image—see stereoscopy.

stereoscopic model—see stereomodel.

stereoscopic pair—see stereoscopy.

stereoscopic parallax—see parallax.

stereoscopic perception—Ability to perceive a three dimensional stereoscopic image.

stereoscopic plotting instrument—or stereoscopic plotter—An instrument for plotting a map or obtaining spatial solutions by observation of stereoscopic models formed by stereopairs of photographs.

stereoscopic vision—see stereoscopy.

stereoscopy—The science and art that deals with the use of *binocular vision* for observation of a pair of overlapping photographs or other perspective views, and with the methods by which such viewing is produced. *stereoscopic pair* (photogrammetry)—Two photographs of the same area taken from different camera stations so as to afford stereoscopic vision; frequently called a *stereopair*. *stereogram*—A stereoscopic pair of photographs or drawings correctly oriented and mounted or projected for stereoscopic viewing. *binocular vision*—Simultaneous vision with both eyes. *stereoscopic vision*—The particular application of binocular vision which enables the observer to obtain the impression of depth, usually be means of two different perspectives of an object (as two photographs taken from different camera stations) (Contrast with *chromostereopsis*). *stereoscope*—An optical instrument for helping an observer to view photographs, or diagrams, to obtain the mental impression of a three-dimensional model. *stereoscopic fusion*—The mental process which combines two perspective views to give an impression of a three-dimensional model. *stereoscopic image*—The mental impression of a three-dimensional model which results from viewing two overlapping perspective views. Also called *stereoscopic model* or *stereomodel*.

stereotemplet—see **templet.**

stereotemplet triangulation—see **templet.**

stereotriangulation—A triangulation procedure that uses a stereoscopic plotting instrument to obtain the successive orientations of the stereoscopic pairs of photographs into a continuous strip. The spatial solution for the extension of horizontal and/or ventical control using these strip (or flight) coordinates may be made by either graphical or computational procedures. Often called *bridging*.

stereotriplet—Three photographs, the center photo having a common field of view with the two adjacent photos, to permit complete stereoscopic viewing of the center photograph.

stop—see **aperture stop.**

strength of figure—(triangulation) The comparative precision of computed lengths in a triangulation net as determined by the size of the angles, the number of conditions to be satisfied, and the distribution of base lines and points of fixed position. Strength of figure in triangulation is not based on an absolute scale but rather is an expression of relative strength. Also applicable to the individual geometric figures within a given net.

striae (optics)—Threadlike filaments within a piece of glass caused by improper mixing of the molten glass during manufacture. These filaments are composed of glass having a slightly different index of refraction from that of the surrounding glass. The extreme fineness of *striae* often makes their detection difficult.

strip—see **flight strip.**

strip adjustment—Similar to a **block adjustment,** but limited to a single strip of photographs.

strip mosaic—see **mosaic.**

strip coordinates—see **coordinates.**

strip radial plot—see **radial triangulation.**

strip radial triangulation—see **radial triangulation.**

substitute center—A point which, because of its ease of identification on overlapping photographs, is used instead of the principal point as a *radial center*.

sun-angle—Elevation of the sun above the apparent horizon, usually measured in degrees. It is a critical parameter in aerial photography since sun-angle affects shadows.

super-wide-angle lens—see **angle of coverage** under **lens.**

supplemental control—see **control.**

survey—The act or operation of making measurements for determining the relative positions of points on, above, or beneath the earth's surface; also, the results of such operations; also, an organization for making surveys. *photogrammetric survey*—A method of surveying that uses either ground photographs or aerial photographs. *aerial survey*—A survey using aerial photographs as part of the surveying operation; also, the taking of aerial photographs for surveying purposes. *ground survey*—A survey made by ground methods, as distinguished from an *aerial survey*. A ground survey may or may not include the use of photographs.

surveying—Surveying may be defined as the science and art of determining relative positions of points above, on, or beneath the surface of the earth, or establishing such points. In a more general sense, however, surveying can be regarded as that discipline which encompasses all methods for gathering and processing information about the physical earth and the environment. Conventional ground systems are most frequently used, but aerial and satellite surveying methods are also common.

surveying camera—see **camera.**

swing—A rotation of a photograph in its own plane around the photograph perpendicular from some reference direction (such as the direction of flight). May be designated by the symbol *kappa* (κ). Also, the angle at the *principal point* of a photograph which is measured clockwise from the positive *y* axis to the *principal line* at the *nadir point* (see **yaw**).

symbol—A diagram, design, letter, or abbreviation, placed on maps and charts, which (by convention, usage, or reference to a legend) is understood to stand for or represent a specific characteristic or object.

systematic error—see **error.**

tangential distortion—Linear displacement of image points in a direction normal to radial lines from the center of the field (see **distortion** under **aberration**).

target—The distinctive marking or instrumentation of a ground point to aid in its identification on a photograph. In photogrammetry, *target* designates a material marking so arranged and placed on the ground as to form a distinctive pattern over a geodetic or other control-point marker, on a property corner or line, or at the position of an identifying point above an underground facility or feature. A target is also the image pattern on aerial photographs of the actual mark placed on the ground prior to photography (see **test chart** and **panel**).

telemeter (surveying)—An instrument for determining the distance from one point (position) to another. Some such instruments employ a telescope and measure the angle subtended by a short base of known length. An *electronic telemeter* measures the phase difference or transit time between a transmitted electromagnetic impulse of known frequency and speed and its return.

telemetry—The science of distance measurement by use of a telemeter. Also, the transmission of recorded data by means of radio, telephone, or telegraph.

telephoto lens (optics)—A lens comprising a *positive front element* and a *negative rear element;* the focal length of the combination is greater than the distance from the front lens surface to the focal plane. This construction is used to make relatively compact long-focus cameras.

templet (photogrammetry)—A graphical representation of a photograph; a templet records the directions, or *radials,* taken from the photograph. *hand templet*—A templet made by tracing the radials from a photograph onto a transparent medium, as on sheet plastic; hand templets are laid out and adjusted by hand to form the *radial triangulation. slotted templet*—A templet on which the radials are represented as slots cut in a sheet of cardboard, metal, or other material. *spider templet*—A mechanical templet fabricated by attaching slotted steel arms, representing radials, to a center hub. The spider templet can be disassembled and the parts used again. *stereotemplet*—A composite slotted templet adjustable in scale and representative of the horizontal plot of a stereoscopic model. *double-model stereotemplet*—A templet representative of the horizontal plot of two adjacent stereoscopic models that have been adjusted to a common, though random scale. *stereotemplet triangulation*—Aerotriangulation by means of an assembly of stereotemplets which allows horizontal positions to be obtained with a stereoscopic plotting instrument not designed for *bridging.* The method permits scale solutions by area and is not restricted to solutions along flight strips.

templet method—*see* general description under *radial triangulation; see* *templet.*

terrain—An area of ground considered as to its extent and topography.

terrestrial camera—*see* *camera.*

terrestrial photogrammetry—Photogrammetry utilizing terrestrial photographs.

terrestrial photograph—A photograph taken by a camera located on the ground. Sometimes called a *ground photograph,* although this is not a preferred term.

test chart—A chart for testing the performance of optical systems. The design usually consists of ruled lines, squares, or circles of various

sizes so arranged that the quality of a lens can be determined by examining the image of the chart (*see* **target, resolving power** and **resolving power target**).

test glass—An optical element used for checking the curvature of lens surfaces during final polishing. The test glass has a curvature equal to and opposite that of the desired lens. When the two surfaces are placed in contact and viewed in monochromatic light, *interference fringes* are formed. The fringe pattern (also called *Newton's rings*) is really a contour map of the air film between the two glasses, the contour interval being one-half a wavelength of light (about 2500A or 0.00001 in.).

thick lens—A term used in geometrical optics to indicate that the thickness of a lens is considered and that all distances are being measured from the *nodal points* instead of the lens center.

thin lens—A term used in geometrical optics to indicate that the thickness of a lens is ignored and that all distances are measured from the lens center; used for approximate computations.

three-point resection—Three-point resection *radial triangulation* is a method of computing the coordinates of the principal points of overlapping aerial photographs by resecting on three horizontal control points appearing in the overlap area.

tie—A survey connection from a point of known position to a point whose position is desired. A tie is made to determine the position of a supplementary point whose position is desired for mapping or reference purposes, or to close a survey on a previously determined point. To "tie-in" is to make such a connection.

tie point—Image points identified on photographs in the side lap area between two or more adjacent strips of photography. They serve to tie the individual sets of photographs into a single flight unit and to tie adjacent flights into a common network.

tilt—The angle at the perspective center between the photograph perpendicular and the plumbline (or other exterior reference direction); also, the dihedral angle between the plane of the photograph and the horizontal plane. The direction of *tilt* is expressed by *swing* (when referred to the axes of the photograph) or *azimuth* (when referred to the exterior coordinate system). In aerial photography, tilt may be separated into its component angles, referred to the fiducial axes, with the x axis being the one more nearly in the direction of flight. In aerial-camera orientation, a positive x tilt results from the left wing of the aircraft being lowered, displacing the nadir point in the positive y direction. Similarly, a positive y tilt results from the nose of the aircraft being lowered, displacing the nadir point in the positive *x* direction (*see* **roll, pitch, yaw**, and *relative tilt*).

tilt displacement—*see* **displacement**.

tip—Another term for *y* tilt (*see* **tilt**).

t-number, t-stop—A system of marking lens apertures in accordance with their actual light transmission, rather than by their geometrical dimensions as in the *f*-stop system.

toe—The portion of the *characteristic curve* below the straight-line section of the curve. It represents the area of minimum useful exposure.

tone—For imagery, each distinguishable shade variation from black to white.

tolerance—The allowable variation from a standard or from specified conditions.

topoangulator—An instrument used to measure vertical angles in the principal plane of an oblique photograph.

topographic feature—*see* **topography**.

topographic map—*see* **map**.

topography—Features of the surface of the earth considered collectively as to form. A single feature (such as a mountain or valley) is called a *topographic feature*. Topography is subdivided into *hypsography* (relief features), *hydrography* (water and drainage features), and *culture* (manmade features).

track (air navigation)—The actual path of an aircraft over the surface of the earth. The azimuth of this path generally is referred to the true meridian (*see* **course**).

transfer function—*see* **function, transfer**.

transformation—The process of projecting a photograph (mathematically, graphically, or photographically) from its plane onto another plane by translation, rotation, and/or scale change (*see* **coordinate transformation**). When a photograph is transformed to a horizontal plane, so as to remove displacement due to tilt, the process is termed *rectification;* however, relief displacement cannot be removed by this process. *empirical orientation*—The composited rectified adjustments of magnification, swing, easel tilt, *y*-displacement, and *x*-displacement used to correctly recreate the exact conditions in the projected image that existed in the negative at the instant of exposure. *transforming printer*—A projection printer, designed especially for use in transforming photographs according to fixed parameters. A rectified virtual image can be produced by a monocular instrument, such as a *sketchmaster* or *camera lucida*. Rectification of individual rays on tilted or oblique photograph may be made by analytical or graphical methods (*see* **paper-strip-method**) or by mechanical devices, such as the *rectoblique plotter* and the *photoangulator*.

transformed print—A photographic print made by projection in a transforming printer (*see* **multiple-lens camera** under **camera**, and **transforming printer** under **transformation**).

transforming printer—*see* **transformation**.

transit—A surveying device which consists of a telescope and graduated circles and is used primarily to measure horizontal and vertical angles.

translation—Movement in a straight line without rotation. The systematic movement of projector assemblies in line-of-flight directions in a *stereoplotting instrument*.

transmission (optics)—The ratio of transmitted light to the incident light. If 100 units of light fall upon a translucent material and 10 of them succeed in passing through, then it can be said that the material has 1/10 or 10 percent transmission.

transparency—A photographic print on a clear base, especially adaptable for viewing by transmitted light. Also, the light-transmitting capability of a material.

traverse—A method of surveying in which the lengths and directions of lines connecting a series of stations are measured. A traverse may be closed or open, according as it does or does not end on a known position or return to the starting point. Traverses may be of many kinds, such as *stadia, compass* or *transit traverse*.

triangulation—A method of surveying in which the stations are points on the ground at the vertices of a chain or network of triangles whose angles are observed instrumentally and whose sides are derived by computation from selected triangle sides called base lines, whose lengths are obtained from direct measurement on the ground. *arc triangulation*—A system of limited width designed to progress in a single general direction. Arc triangulation is executed for the purpose of connecting independent and widely separated surveys, coordinating, and correlating local surveys along the arc, furnishing data for the determination of a geodetic datum, providing a network of control points for a country-wide survey, *etc*.

triangulation station—A point on the earth whose position is determined by triangulation. Sometimes shortened to *trig point*.

tricamera photography—*see* **photography, tricamera**.

trigonometric leveling—The determination of differences of elevations from observed vertical angles combined with lengths of lines. A type of indirect leveling.

trilateration—A method of determining horizontal ground positions by measuring the sides of triangles in lieu of angles.

trimetrogon—*see* **trimetrogon camera** under **camera**; also **photography, tricamera**.

trimming and mounting diagram (photography)—A sketch which indicates how the print of a transformed multiple-lens photograph should be corrected to obtain, in effect, a photograph made by a single lens. The information is given in the form of distances referred to the fiducial marks on the photograph, and it is the result of a calibration for the particular camera used.

true horizon—see horizon.

ultra-wide-angle lens—see angle of coverage under *lens.*

uncontrolled mosaic—see mosaic.

universal sketchmaster—A type of sketchmaster in which vertical or oblique photographs may be utilized (*see sketchmaster*).

vacuum back—see locating back.

vanishing line—The straight line on a photograph upon which lie all the *vanishing points* of all systems of parallel lines parallel to one plane. The vanishing line for all systems of horizontal parallel lines in the object space is the horizon trace.

vanishing point—see vanishing point under *principal plane.*

variance—see standard error under *error.*

variate—In contradistinction to a *variable,* a variate is a quantity that may take on any of the values of a specified set with a specified relative frequency or probability often known as a *random variable.* It is to be regarded as defined not merely by a set of permissible values like an ordinary mathematical variable, but by an associated frequency (probability) function expressing how often those values appear in a given situation.

variation of coordinate method—A method of adjusting measurements in which the coordinates of geodetic points are varied so as to best fit the observations and retain mathematical homogeneity.

vectograph—A stereoscopic photograph composed of two superimposed images that polarize light in planes 90° apart. When these images are viewed through *Polaroid* spectacles with the polarization axes at right angles, an impression of depth is obtained.

vernier—A device which enables readings of a scale to be made to smaller units than the smallest division of the scale.

vernier acuity—A measure of the ability to perceive when one segment of a straight line has been displaced laterally with respect to the rest of the line. It is equal to the reciprocal of the angle subtended at the eye by a lateral displacement that can barely be discriminated by the observer.

vertical angle—A angle measured in a vertical plane which is referenced to horizontal. Vertical angles above horizontal are called *eleva-*

tion angles; those below horizontal are called *depression angles.*

vertical control—see control.

vertical-control datum—see datum.

vertical control point or *vertical control station—see control point, control station; bench mark.*

vertical coordinates—The vertical distance of a point above or below a reference datum. Points may be plus or minus according to whether the point is above or below the datum.

vertical datum—see datum.

vertical deformation—In relative orientation, the cumulative model warpage affecting the vertical datum from x-tilt error and y-tilt error.

vertical exaggeration—An increase or decrease in the vertical dimension of the perceived *stereomodel* when compared to its horizontal dimension ratio of the actual object.

vertical line or *plumb line*—A line which coincides with the direction of gravity.

vertical parallax—see y parallax under *parallax.*

vertical pass point—see pass point.

vertical photograph (aerial photography)—An aerial photograph made with the camera axis vertical (or as nearly vertical as practicable) in an aircraft.

viewfinder—An auxiliary device which shows the field of view of a camera.

vignetting (photography)—A gradual reduction in density of parts of a photographic image due to the stopping of some of the rays entering the lens. Thus, a lens mounting may interfere with the extreme oblique rays. An antivignetting filter is one that gradually decreases in density from the center toward the edges; it is used with many wide-angle-lenses to produce a photograph of uniform density by cutting down the overexposure of the center of the photograph.

visible spectrum—The part of the electromagnetic spectrum designated as "light": 4,000-7,000 angstroms (400-700 millimicrons 0.4-0.7 microns, 0.0004 to 0.0007 millimetres, 40-70 nanometres)

want of correspondence—see parallax.

wedge (optics)—A refracting prism of very small deviation, such as those used in the eyepieces of some stereoscopes (*see step wedge*).

weight (statistics)—The measure of relative reliability (or worth) of a quantity as compared with other values of the same quantity or with comparable quantities.

weighted mean—A value obtained by multiplying each of a series of values by its assigned weight and dividing the sum of those products by the sum of the weights.

wide-angle lens—see angle of coverage under *lens.*

wing photograph—A photograph taken by one

of the side or wing lenses of a multiple-lens camera (*see* **multiple-lens camera** under **camera**).

x axis—*see* **fiducial axes** and **tilt**.

x-motion—In a stereoplotting instrument, that linear adjustment approximately parallel to a line connecting two projector stations; the path of this adjustment is, in effect, coincident with the flight line between the two relevant exposure stations.

x parallax—*see* **parallax**.

x tilt—*see* **tilt**.

x-rays—That radiation within the electromagnetic spectrum having very short wavelengths (between 0.1 and 100 angstroms) and which is capable of penetrating solids.

x-ray photogrammetry—That branch of photogrammetry that deals with analysis and measurements from radiographs derived from X-rays.

y axis—*see* **fiducial axes** and **tilt**.

y-motion—In a stereoplotting instrument, that linear adjustment approximately perpendicular to a line connecting two projectors.

y parallax—*see* **parallax**.

y tilt—*see* **tilt**.

yaw—(1) air navigation: The rotation of an aircraft about its vertical axis so as to cause the aircraft's longitudinal axis to deviate from the flight line. Sometimes called *crab*. (2) photogrammetry: The rotation of a camera or a photograph coordinate system about either the photograph z axis or the exterior Z axis. In some photogrammetric instruments and in analytical applications, the symbol *kappa* (κ) may be used.

zenith camera—*see* **camera**.

zenith distance—*see* **coaltitude** and **astronomical triangle**.

z-motion—Movement of a stereoplotting projector in a vertical direction.

zoom system—*see* **pancratic system**.

19.4 *Symbols for Photogrammetry Terms*

Many of the common terms listed in section 19-3 frequently occur in photogrammetric formulations and computations, making it convenient to use symbols to represent them. If the equivalence of a term and a symbol is universally recognized, the symbol may also be substituted for the term in text (for the sake of brevity) without loss of clarity. The symbol usage recommended in this section is intended to enhance communication between photogrammetrists.

19.4.1 SYMBOL COMPONENTS

A new symbol developed to represent a photogrammetric term may consist of several different components. The classification of symbol components can be explained using the following general expression:

$$\text{SYMBOL} = \text{LETTER(S)} \, {}^{\overset{\text{overscore}}{\text{Superscript}}}_{\underset{\text{underscore.}}{\text{Subscript}}}$$

Each component in the above expression can take many variations, combinations of which create a resultant symbol. It is recommended that only the following variants be used when combining components to create symbols in photogrammetry. All symbols should be represented in italics.

LETTER(S)

    origin: English or Greek
    cases: Upper or lower case
    typeset: Boldface italics or normal italics

Overscore

    dots: ˙, ¨
    bar: −
    carets: ˆ

Underscore

    bar: −
    letters: English or Greek, upper or lower case

Superscripts

    letters: English or Greek, upper or lower case
    numerals: Roman or Arabic
    characters: Prime ('), double prime (")

Subscripts
>   letters: English or Greek, upper or lower case
>   numerals: Roman or Arabic

19.4.2 CATEGORIES OF SYMBOL COMPONENTS

Each of the symbol component variations given above could be used to represent innumerable terms. In the interests of communication ease, the following uses for symbol component variations in photogrammetry are recommended.

SYMBOL COMPONENT	RECOMMENDED USAGE
English Letter	
– upper case	Points, distances, lines, coordinates, *etc.,* in the object space
– lower case	Points, distances, lines, coordinates, *etc.,* in the image space
– upper case, bold	Matrices
– lower case, bold	Vectors
Greek Letter	
– upper case	Rotation angles in absolute space
– lower case	Dimensionless quantities or angular measurement
Letter (multiple)	for example dh - difference in height, or Δx - change in x coordinates
Dot Overscores	Derivatives
Bar Overscores	Average value, length between endpoints, least upper limit
Bar Underscore	Greatest lower limit
Letter Underscore	Row & Column contained in a matrix
Letter Superscript	n^{th} power, i^{th} value
Numeral Superscript	Powers, degrees of freedom, order of iteration, inverse functions (-1)
Character Superscript	Order of differentiation, photo designation
Letter Subscripts	Matrix row/column designations, i^{th} point, photo designation
Numerical Subscripts	Position in a sequence, set, or matrix

19.4.3 GREEK ALPHABET

As a convenient reference when developing new symbols or interpreting existing ones, the letters of the Greek alphabet are listed below.

Greek Letter (upper and lower case)	Greek Name	English Equivalent
A α	*Alpha*	a
B β	*Beta*	b
Γ γ	*Gamma*	q
Δ δ	*Delta*	d
E ϵ	*Epsilon*	e
Z ζ	*Zeta*	z
H η	*Eta*	$\bar{e}$
Θ ϑ	*Theta*	th
I ι	*Iota*	i
K κ	*Kappa*	k
Λ λ	*Lambda*	l
M μ	*Mu*	m
N ν	*Nu*	n
Ξ ξ	*Xi*	x
O o	*Omicron*	o
Π π	*Pi*	p
P P	*Rho*	r, rh
Σ σ	*Sigma*	s
T τ	*Tau*	t
Y υ	*Upsilon*	y, u
Φ ϕ	*Phi*	ph
X χ	*Chi*	ch
Ψ ψ	*Psi*	ps
Ω ω	*Omega*	$\bar{o}$

19.4.4 LIST OF PHOTOGRAMMETRY SYMBOLS

Many different symbols have been developed to represent photogrammetric terms. For the sake of uniformity, usage of the following common symbols is recommended.

a	Semimajor axis of the earth ellipsoid.
A	Matrix of the coefficients of parameters in a least squares solution.

abc	Image space plane containing points a, b, and c.
ABC	Object space plane containing points A, B, and C.
a, b, c, etc.	Sometimes used instead of numbers to identify image points corresponding to the ground points A, B, C, etc.
A, B, C, etc.	Sometimes used instead of numbers to identify successive ground points or camera stations.
$\mathbf{A}^T$	Transposed coefficient matrix.
$\mathbf{A}^{-1}$	Inverse coefficient matrix.
b	Base length at the scale of the stereomodel; Coefficient of a parameter (*i.e.*, an element of B); semiminor axis of the earth ellipsoid.
B	Base length (*i.e.*, the physical distance between camera stations).
$\dot{\mathbf{B}}$	Partial derivatives within observation equation matrix from exposure station parameters (partitioned matrix).
$\ddot{\mathbf{B}}$	Partial derivatives within observation equation matrix from object point parameters (partitioned matrix).
b'	Image distance between principal point and corresponding (conjugate) principal point on the right hand photograph.
b_x, b_y, b_z	Components of the base length at the scale of the stereomodel
B_x, B_y, B_z	Components of the base length in an exterior coordinate system.
c	As a superscript or subscript, used to identify the coordinates of a camera station (*e.g.*, X^c, Y^c, Z^c, or the letters L or O also may sometimes be used; Focal length, often called camera constant.
C	Grid convergence; Degrees centigrade.
d	Displacement of a photographic image due to any cause, Diaphragm.
D	Determinant; Density.
dh, dp	Difference in height or elevation, difference in parallax; d is used here to imply difference, not differential.
e	Eccentricity of the earth ellipsoid.
E	Endlap, (overlap) between successive exposures along the flight line; exposure; easting in ground or grid coordinates.
f	Flattening of the earth ellipsoid; focal distance in optics; Alternative for (c).
F	Degrees Farenheit.
h	Height or elevation of a ground station or object above sea level datum (unless specified otherwise).
H	Height or elevation of camera stations above sea level datum (unless specified otherwise); alternative for (Z°).
$h'h'$	Photograph, horizon.
i	Isocenter on a photograph; As a subscript, identifies the coordinates of the ith camera station (*e.g.*, X_i^c, Y_i^c, Z_i^c); image distance in optics.
$\mathbf{i}$	Unit vector in direction of x-axis.
I	Apparent ground position of the isocenter.
$\mathbf{I}$	Unit or identity matrix.
j	As a subscript, identifies the coordinates of the jth ground point (*e.g.*, X_j, Y_j, Z_j).
$\mathbf{j}$	Unit vector in direction of y-axis.
$\mathbf{J}$	Jacobian matrix.
$\mathbf{k}$	Unit vector in direction of z-axis.
K	Degrees Kelvin.
L	Lens point or camera station; may be used as a superscript or subscript to identify the coordinates of a camera station (*e.g.*, X_L, Y_L, Z_L).
L_1	Left image exposure station.
L_2	Right image exposure station.
m	Scale number; alternative for standard error or standard deviation, (*see* also σ); metre.
M	Magnification; A factor which denotes the ratio of a dimension on a photographic copy to the corresponding dimension on the negative.
$\mathbf{M}$	A 3-by-3 orthogonal matrix of direction cosines, usually specifying the angular orientation between the photograph coordinate system and an exterior coordinate system.
m_{11}, m_{12}, etc.	Elements of $\mathbf{M}$.
m_b	Photograph scale number.
m_k	Map scale number.
m_m	Model scale number.
n	Nadir point on a photograph; Index of refraction; number of measurements or observations.
N	Ground nadir point; Northing in grid coordinates; Radius of curvature for the prime vertical of the earth.
$\mathbf{N}$	Matrix of the coefficients of a set of normal equations.
$\dot{\mathbf{N}}$	Coefficient matrix of image exterior orientation elements (partitioned matrix).

$\ddot{N}$ Coefficient matrix of passpoint object space coordinates (partitioned matrix).

o Sometimes used to designate a photograph principal point or zero positions; Object distance in optics.

O Origin point of perspective, lens point, camera station; Origin of a system of space coordinates relative to a photograph.

p Photograph principal point.

P Ground principal point.

$\mathbf{P}$ Alternative for weight matrix (inverse of covariance).

P_x x-parallax corresponding to elevation difference.

P_y y-parallax.

P_z Vertical parallax in terrestrial stereomodel.

$\mathbf{Q}$ Covariance matrix; Matrix of cofactors or weight coefficients.

r Radial distance from any specified photograph center to an image; Redundancy, or degrees of freedom.

s Swing angle.

S Sidelap between adjacent flight strips; scale (either fractional/ratio/decimal representation).

t Tilt of the camera axis from the vertical; Time.

T As a superscript, used to indicate the transpose of a matrix (*e.g.*, M^T); Transmission.

t_x x-tilt, lateral tilt; Angular component of the resultant tilt, measured about the x-axis.

t_y y-tilt, longitudinal tilt; Angular component of the resultant tilt, measured about the y-axis.

u Number of unknowns.

v Speed; The coefficient of a residual, (V).

λ, μ, ν Direction cosines (*lambda, mu, nu*).

μm Micrometre or micron, 10^{-6} metre.

ρ (*rho*)   Conversion factor from radian to angular measure (*e.g.*, $\rho'' = 206{,}264{,}8''$).

σ (*sigma*)   Standard deviation or standard error.

σ^2 Variance.

$\dot{\sigma}$ Least squares correction vectors to exposure station parameters (partitioned matrix).

$\ddot{\sigma}$ Least squares correction vectors to object point coordinates (partitioned matrix).

ϕ (*phi*)   Pitch, longitudinal tilt, tilt about the y or Y axis; Latitude.

Φ (*PHI*)   Model rotation about Y-axis.

ω (*omega*)   Roll, lateral tilt, tilt about the x or X axis; In missile photogrammetry, elevation angle; Rotational velocity of the earth.

Ω (*OMEGA*)   Model rotation about X-axis.

Index